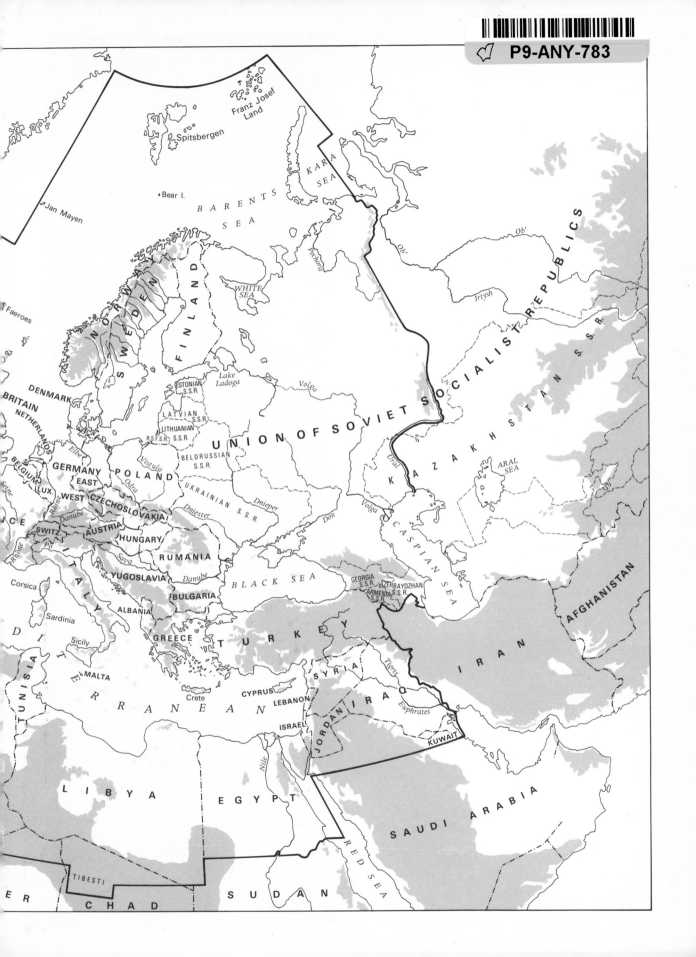

P9-ANY-783

Franz Josef Land

Spitsbergen

Jan Mayen • Bear I.

KARA SEA

BARENTS SEA

Faeroes

NORWAY

SWEDEN

FINLAND

WHITE SEA

Ob'

Irtysh

DENMARK

BRITAIN

NETHERLANDS

GERMANY EAST WEST

BELGIUM LUX.

POLAND

ESTONIAN S.S.R.

Lake Ladoga

Volga

LATVIAN S.S.R.

R.S.F.S.R.

LITHUANIAN S.S.R.

BELORUSSIAN S.S.R.

UKRAINIAN S.S.R.

UNION OF SOVIET SOCIALIST REPUBLICS

KAZAKHSTAN S.S.R.

Ob'

CZECHOSLOVAKIA

Elbe

Odra

Vistula

Dnieper

Ural

ARAL SEA

SWITZ.

AUSTRIA

HUNGARY

Danube

Dniester

Don

Volga

Rhône

ITALY

YUGOSLAVIA

RUMANIA

BULGARIA

Sava

Danube

BLACK SEA

CASPIAN SEA

IRAN

AFGHANISTAN

Corsica

ALBANIA

Sardinia

GREECE

TURKEY

GEORGIA S.S.R.

AZERBAYDZHAN S.S.R.

ARMENIA S.S.R.

Sicily

MALTA

CYPRUS

SYRIA

Tigris

Crete

LEBANON

IRAQ

Euphrates

TUNISIA

ISRAEL

JORDAN

KUWAIT

MEDITERRANEAN

Nile

LIBYA

EGYPT

SAUDI ARABIA

RED SEA

TIBESTI

CHAD

SUDAN

Handbook of the
**Birds of Europe
the Middle East and
North Africa**

**The Birds of the
Western Palearctic**

Volume V

Handbook of the
Birds of Europe
the Middle East and
North Africa

The Birds of the Western Palearctic

Volume V · Tyrant Flycatchers to Thrushes

Stanley Cramp *Chief Editor*
Duncan J Brooks   Euan Dunn   Robert Gillmor
Joan Hall-Craggs   P A D Hollom   E M Nicholson
M A Ogilvie   C S Roselaar   P J Sellar
K E L Simmons   K H Voous   D I M Wallace
M G Wilson

OXFORD   NEW YORK

OXFORD UNIVERSITY PRESS · 1988

DEDICATED TO THE MEMORY OF

H F WITHERBY

(1873-1943)

EDITOR OF *THE HANDBOOK OF BRITISH BIRDS*
(1938-41)

Oxford University Press, Walton Street, Oxford OX2 6DP
Oxford New York Toronto
Delhi Bombay Calcutta Madras Karachi
Petaling Jaya Singapore Hong Kong Tokyo
Nairobi Dar es Salaam Cape Town
Melbourne Auckland
and associated companies in
Beirut Berlin Ibadan Nicosia

Oxford is a trade mark of Oxford University Press

Published in the United States
by Oxford University Press, New York

British Library Cataloguing in Publication Data
Handbook of the birds of Europe, the Middle
East and North Africa: the birds of the
Western Palearctic.
Vol. 5: Tyrant flycatchers to thrushes
1. Birds—Europe 2. Birds—
Mediterranean Region
I. Cramp, Stanley
598.294   QL690.A1
ISBN 0-19-857508-4

Library of Congress Cataloging in Publication Data
(Revised for volume V)
Handbook of the birds of Europe, the Middle East and
North Africa.
Includes bibliographical references and indexes.
Contents: —v. 2. Hawks to bustards.—v. 3.
Waders to gulls.—v. 5. Tyrant flycatchers to thrushes.
1. Birds—Europe—Collected works. 2. Birds—
Mediterranean Region—Collected works. I. Cramp, Stanley.
QL690.A1H25   598.29'182'2   79-42914
ISBN 0-19-857505-X (v. 2)

Typeset by the Oxford Text System
Printed in Hong Kong

CONTENTS

Contents

Contents

INTRODUCTION

This volume begins the final part of this work—the passerines (Volumes V–VII). In Volume I a full explanatory Introduction was given and this was modified or extended in the Introductions to Volumes II–IV, but it has been decided to give here, at the start of the passerines, a full Introduction comprising all material, including new matter, relevant to the group.

The Handbook of British Birds (Witherby *et al.* 1938–41) represented a pioneer effort to review and present clearly the entire up-to-date knowledge of birds of Britain and Ireland. Its scholarly standard and concise presentation made it a firm basis for countless and diverse advances in ornithological research. Due in no small part to the stimulus it provided, ornithological knowledge has shown remarkable advances since it appeared. There has been a clear need for some years for a work which would incorporate the mass of new knowledge, now scattered and often difficult of access in the journals of many countries, to provide a work of reference both for the professional scientist and the ever-growing body of amateurs whose range of interest continues to widen. Such a work should also focus attention on the still considerable gaps in our information, even of common species, and on the correlations between different aspects of our knowledge, so providing a stimulus for further studies.

However, *The Handbook of British Birds* reflected the tendency, prevalent as late as the middle of this century, for serious interest in birds to be concentrated largely within the national boundaries of a country, leaving the rest to be dealt with peripherally. Parallel with this, there was a widespread reluctance to regard the study of birds as an essential and major part of biological science. Since then ornithology has formed a main growing point in studies of speciation and evolution, population dynamics, migration and dispersal, navigation, and ethology, while the impact of *The Handbook* impressed on ornithologists the need for, and the feasibility of, a more comprehensive approach. Thus *Ptitsy Sovietskogo Soyuza* (Birds of the Soviet Union) (Dementiev and Gladkov 1951–4), while constructed on similar general lines, covered a major part of a zoogeographical region, and this important Soviet initiative was followed by Palmer in the *Handbook of North American Birds* (1962, 1976) and more recently by the *Handbuch der Vögel Mitteleuropas* (Bauer and Glutz 1966–9, Glutz *et al.* 1971–7, Glutz and Bauer 1980–5, and in progress), dealing primarily with a compact group of central European countries. Still more recently, there has been published the first volume of a valuable new handbook of the birds of the USSR, *Ptitsy SSSR* (Ilyichev and Flint 1982). In the light of such progress, the mere revision of *The Handbook of British Birds*,

focused on a small group of islands off western Europe, could not be justified. It became clear that any new approach must deal with the zoogeographical province of which Britain and Ireland form a part—the western Palearctic. Our original intention was to define this so as to fill the obvious gap between the Soviet and North American works, but after hearing the comments of many ornithologists (including those of the USSR, who were anxious for us to extend the boundaries as far east as possible) we decided to include the whole of European USSR.

The Palearctic, like all faunal regions, is not susceptible to precise definition. Charles Vaurie, the author of the most recent authoritative ornithological work on this region, *The Birds of the Palearctic Fauna* (1959, 1965), adopted a more liberal interpretation than that of his great predecessor Ernst Hartert in *Die Vögel der palaärktischen Fauna* (1903–22). We have been somewhat more restrictive than Vaurie in delimiting the western and southern boundaries of the Palearctic. In the west, we have excluded Greenland (eastern Greenland was included by Vaurie in 1959 and the whole of it in 1965), but have followed him in covering all the eastern Atlantic islands south to the Cape Verde Islands, to which we have added the Banc d'Arguin group (but not the adjoining mainland) where the extensive recent researches by Naurois (1969*a*) have clearly demonstrated the Palearctic character of the avifauna. The important habitats of the eastern North Atlantic and Arctic Oceans are also included. In the south, we have, like Vaurie, covered the Sahara south to the northern borders of the Sahel region, but, while including the mountain massif of Tibesti (northern Chad), we have excluded those of Aïr (Niger) and Ennedi (Chad), where in our view the Afrotropical element predominates; further east, in Egypt and Saudi Arabia, our boundary runs to the north of his. If the overall boundary of the whole Palearctic region eludes any logically unquestionable demarcation, then the determination of the eastern limits of its western part is clearly largely arbitrary. For the reasons given earlier, we have included all European USSR, using the limits internationally agreed for the *Flora Europaea* (Tutin *et al.* 1964–80) (i.e. east to the Ural mountains and Ural river) with the addition of the remaining regions of the USSR between the Caspian Sea and the Black Sea. Further south, our area extends east to Turkey, Iraq, Kuwait, and northernmost Saudi Arabia, but we have excluded the whole of Iran.

There have been considerable differences of opinion and practice regarding the sequence of families, genera, and species (see, e.g., Lack 1966). In view of the existing confusion in this respect, K H Voous prepared his *List of Recent Holarctic Bird Species* (1977), aiming to promote a badly-needed uniformity and short-term stability without being inflexible or interfering with taxonomic research. We have so far followed his list precisely; if fresh research provides us with overwhelming reasons to make changes, such changes will be fully explained.

The main body of the work consists of the species accounts, but there are also general summaries covering the main taxa (taxonomic categories) above the level of the genus—order, family, and, if necessary, subfamily—compiled on a world basis with special reference to west Palearctic forms. The summary for each order is brief, usually consisting of (1) a characterization of the type of birds concerned together with indication of size, (2) a list of the constituent families, and (3) a comment on its relationship, if any, to other orders; where there is just one family in an order, the treatment is transferred to that for the family. The summary for each family usually briefly covers (1) the type and size of birds, (2) the number of species, genera, and, if necessary, subfamilies), (3) representation in the west Palearctic, and (4) certain, mainly external, morphological characters (for more detailed taxonomic diagnoses, see such works as Witherby *et al.* 1938-41). Where there is a division into subfamilies, some more-specific information usually given under the family summary is transferred to the subfamily summary. The order, family, and subfamily summaries have been prepared by Dr K E L Simmons and C S Roselaar.

The scientific names of the species treated are in accordance with the *International Code of Zoological Nomenclature* (3rd edition, 1985) and with subsequent rulings of the International Commission on Zoological Nomenclature. The Code and rulings cover only the formal side of nomenclature and leave room for independent taxonomic judgement on the boundaries of genera and species, which may affect the actual names that are used. So, in the heading of some species a synonym of the species name is given; this is an acceptable alternative dependent upon different generic or specific allocation of the species. Many species show geographical variation to such an extent that more than one race (subspecies) is recognized; such species have been termed *polytypic*. Species in which no races are recognized are said to be *monotypic*. Races are groups of similar populations, belonging to a single species, inhabiting a geographical subdivision of the species' range, and recognizably differing from other populations of the same species (Mayr 1969). We have tried to avoid naming stages in a cline, or populations differing from others in averages only but overlapping extensively in ranges of measurements or colours. Our concept of races is a wide one.

For polytypic species, a list is given of the breeding ranges of races occurring in the western Palearctic; races which do not breed in our area but are either migrants through it or of only accidental occurrence are indicated.

A separate sentence mentions *extralimital* races, giving names of any which are particularly important for our purposes. In a polytypic species, one of the races (the *nominate* race) has the same name as the species itself: e.g. under Blue Tit *Parus caeruleus*, the race *P. caeruleus caeruleus* is referred to as 'nominate *caeruleus*'. It is not necessarily the most 'typical', geographically central, or most widespread race, but merely the one first named.

Vernacular names in the various languages have been suggested by country correspondents, or have been taken from national checklists, from *A Field Guide to the Birds of Britain and Europe* (Peterson *et al.* 1983), or from *Nomina Avium Europaearum* (Jørgensen 1958).

The geographical scope of the work and limitations of space have forced us to be concise and this has been welcomed by most readers. In particular, we have tried to avoid references to previous literature which are not essential both by referring wherever possible to key papers which contain bibliographies enabling the interested reader to pursue the topic further, and especially (e.g. on aspects of distribution or populations in individual countries of the western Palearctic), by drawing attention to authoritative works where fuller details can be obtained. Pre-eminent among the latter is the *Handbuch der Vögel Mitteleuropas* already referred to, a magnificent and successful project which, when complete, will contain not only full surveys of all the species occurring within a wide area of central Europe, but also detailed and scholarly summaries of all aspects of their biology.

The species accounts are divided into the following sections, each the special responsibility of the editor named:

Field Characters   D I M Wallace
Habitat   E M Nicholson
Distribution and *Population*   S Cramp
Movements   D J Brooks
Food   D J Brooks
Social Pattern and Behaviour   Dr E K Dunn and M G Wilson
Voice   Dr E K Dunn, M G Wilson, Mrs J Hall-Craggs, and P J Sellar
Breeding   Dr M A Ogilvie
Plumages, *Bare Parts*, *Moults*, *Measurements*, *Weights*, *Structure*, and *Geographical Variation*   C S Roselaar

D J Brooks is responsible for the editing of the entire volume.

The scope of each section, together with the terms and conventions used and a discussion of special problems, is outlined in the following pages. There are, at the time of writing, 609 species (of which 266 are passerines) breeding in the western Palearctic (including introduced species with viable feral populations), and for these the species accounts include all the above sections. For regular non-breeding migrants to the western Palearctic (11 species), the sections on Population, Social Pattern and Behaviour, and Breeding are omitted—also the section on Food for accidental visitors which have occurred in the area since 1900 (152 species, and increasing annually).

All the above categories (breeding species, regular migrants, and accidentals since 1900) are illustrated by paintings showing every plumage which is identifiable in the field, including breeding and non-breeding of both sexes, and juvenile and other immature plumages. Flight patterns are illustrated where appropriate. The paintings in Volume V are the work of Norman Arlott, Hilary Burn, Dr P J K Burton, Alan Harris, Viggo Ree, the late Laurel Tucker, and D I M Wallace; their initials appear at the end of the caption for each plate. James Alder, John P Busby, K H E Franklin, Martin J Hallam, Ian Lewington, Norman McCanch, Darren Rees, and Chris Rose have prepared the line drawings for the Social Pattern and Behaviour sections and are credited at the end of each. Robert Gillmor is the editor with general responsibility for artwork. R J Connor and A C Parker have most generously provided photographs of eggs.

Finally, there are a number of species which do not merit full treatment or any illustration. They fall into two main categories: (1) those for which valid records exist but which have been extinct or otherwise not recorded in the western Palearctic since 1900 or have occurred only on ships at sea (chiefly Nearctic species in the eastern Atlantic); (2) those with only a dubious western Palearctic status, such as specimens of doubtful origin, disputed identifications, and escapes from captivity which have not yet established viable feral populations. Species in both categories are each treated in a single short paragraph concerned with distribution and migration only; those in the second category are distinguished by being in small type. Fossil species are not covered in this work.

We have not attempted to make independent judgements on the validity of records of vagrants, but have followed national checklists and the advice of our country correspondents (see p. 39), except in a few instances where authoritative criticisms have been published.

FIELD CHARACTERS

This section summarizes information on size, shape, and plumage as they appear in the field, differentiating where relevant between the sexes and between breeding, non-breeding and immature plumages, and making comparisons as necessary with similar species or others liable to be confused; it also includes brief notes on general habits and behaviour where these aid field identification. The presentation is in four paragraphs.

The first paragraph opens with ranges of total length and wing-span, both in centimetres. The measurement of total length is the conventional one of the bird on its back measured from tip of bill to tip of tail; it is thus not likely to be in a natural position, but the method has the great virtue of being widely used and understood. As explained in Volume I, wing-spans are rarely measured on live or freshly dead birds and the estimates in earlier volumes have been found to contain exaggerations of $c.$ 10% in a wide range of non-passerines (Noakes 1984; Holden 1985). The attempt to define size ranges and structural differences will be continued, however, and using a selection of actual wing-spans presented by dead birds (supplied by P Holden) every effort will be made to make size and wing-span estimates comparable within families and genera—though in the case of smaller passerines, the perception of wing-span differences is, in any case, not easy. The remainder of the first paragraph includes comparisons of size and structure with similar species (with particular attention given to the structure when plumage differences are limited), summaries of general character, main plumage patterns and colours of adults (both ♂ and ♀ if dissimilar) and immature (if this the most likely age-class to be seen), and brief comment on flight action and voice if important. The paragraph finishes with an indication of the extent and separability of sexual, seasonal, age, and racial differences.

The second paragraph gives detailed field descriptions, under separate headings of race, sex, breeding and non-breeding plumage, and as many immature stages as necessary; where distinct plumage morphs exist these are also described. Characters which are common to male and female, or adult and immature, are not usually repeated after the first mention. Information on the colours of bare parts is normally given at the end of the paragraph except where there are marked differences between the various sex- or age-classes. One of the aims of this paragraph is to encourage observers to distinguish sexual, seasonal, and age differences in plumages as far as field conditions allow. Too often a specific identification is considered adequate and there is little attempt to distinguish more than the most obvious plumages, but important biological data on migration patterns, age of first breeding, and social behaviour are obtainable with fuller sexing and ageing.

The third paragraph opens with a judgement of the ease of identification of the subject species, with notes on the sexual and age characters given in the preceding paragraph, particularly where ♀♀ and immatures invite more confusion with other species than do the adult ♂♂ (as is frequently the case in passerines). Comparisons with confusion species may be made immediately after such notes or, to avoid repetition, placed within the full descriptions in the preceding paragraph. The third paragraph continues with notes on flight, gait, and other actions that contribute to the bird's general character. It should be noted that published material on such potentially important features is often sparse and there is much scope for greater concentration on the topic. This paragraph closes with an indication of the bird's tameness, gregariousness, and, if useful in field identification, habitat preferences.

The final paragraph gives descriptions of song and common calls (though these are more fully discussed under Voice) and an indication of any potential confusion with utterances of other species.

References are not normally given in this section, except where a disputed character is mentioned or where there have been special studies which are still considered to be reliable and up-to-date analyses. In order to avoid repetition, directions to the sections on Voice, Plumages, and (importantly) Geographical Variation are freely given, particularly where attempts at racial identification should be tempered with a knowledge of the relevant systematics. For a full understanding of the intricacies of plumage succession, particularly its timing and the different feather tracts involved, readers should also study the Moults section carefully.

HABITAT

Observers who have trained themselves to note in the field precise, uniformly defined features of avian form and plumage have not yet generally accustomed themselves to paying similarly critical attention to describing the components and characteristics of habitat. Casual observation has long since shown the probability of encountering particular bird species within certain recognizable habitats. Most of these have loose popular names, often confused and ambiguous, which the majority of observers have been content to use. Similar problems have arisen throughout the environmental sciences. A scientific approach to habitat is slowly emerging as precise standard terms gain currency in the description of, for example, vegetation, local climates, hydrobiology, and land-forms. It is evident that ornithologists cannot continue to use terminology and concepts in the description of environment which botanists or other specialists concerned would find unacceptably loose and misleading. On the other hand, field ornithologists cannot be expected to master all these subjects, or to acquire more than a superficial knowledge of the ecological principles and techniques by which they are harmonized for use in the biosphere. A compromise approach is therefore required, which will assist progressively to eliminate unscientific terminology and attitudes and to substitute more exactly defined and systematically organized working methods, understandable to those specializing in the other relevant life and earth sciences.

Recent advances in these sciences have largely been reviewed and digested through the International Biological Programme. Its results would now enable a radically new format and terminology to be adopted here which would be widely accepted among interested branches of the scientific community, but would be strange and difficult for ornithologists.

After careful consideration, such an approach has not been adopted for three main reasons. First, it would fail to take account of the extent to which field observers have empirically, however loosely and unscientifically, evolved their own practical and reliable way of making instant judgements on the suitability of whatever terrain may be before them to one or other species of bird. Given this largely subconscious capacity for taking a bird's-eye view, birdwatchers might justly feel that they were being asked to abandon quick, proved, and convenient methods for slower, untested, and awkward ones.

Secondly, a great volume of imprecise data gathered over many decades on this basis would prove unusable in a more precise system, because it would be impossible to reconstruct with any accuracy the old data in the fresh terms.

Thirdly, the replacement of anecdotal and traditional by scientifically tested and standardized terms presupposes that a complete and agreed basis for such terms is available from the environmental sciences. That, unfortunately, is not yet the case. The task is complicated not only by international and other differences of approach but by the rapidity with which important recognizable habitats are being drastically modified, especially by changing land use and development, and by the spread of blended or degraded variants which are difficult to diagnose and define.

So, in a work of this character, the best course is an intermediate one, conserving so far as possible familiar terms but assigning to them reasonably precise meanings consistent with modern scientific usage, and treating the habitat of each species in a regular order of presentation and on a more comprehensive basis. Underlying the species treatments therefore is a systematic framework, derived from the World Checksheet Survey of the Conservation Terrestrial Section of the International Biological Programme, and a Glossary of terms newly designed to bring the description of habitats into line with international scientific practice so far as is consistent with meeting ornithological requirements and with a commonsense degree of simplicity. As the framework provided by the International Biological Programme does not cover unvegetated aquatic habitats or human artefacts it has been used here in the fully comprehensive expanded form developed in the Geogram system (see Nicholson 1973). It is of course impossible to relate such an approach fully to the many overlapping systems which have been proposed within particular branches of study or countries, but as many as possible of these have been taken into account. Those familiar with other methods, for example in the treatment of vegetation, should find no insuperable obstacle in achieving some degree of reconciliation.

A further parallel is offered by the effective systematization and adoption of precise methods and terms in making sight identifications. Descriptions of habitat have hitherto tended to dwell on such conspicuous points as obvious land-forms (including water) and the height, spacing, and general character of vegetation, to the neglect of past and present human influences unless crudely obvious, and of ecological factors. Use of airspace and underwater zones has also been widely neglected. Even in careful and detailed studies, results have not usually dealt fully with the varying attraction to different species of birds of specific habitat situations, or the precise linkages and distinctions in requirements which govern the habitat selection of different species of birds at different stages of their life-cycle. Not much has yet been done to build up a picture of total use, season by season, of a total range of habitats.

To define where each species should not be found is no less essential than to forecast where it should. No other class of living organisms can match birds in extent of occupancy of the entire range of habitats on earth, or in serving to monitor the capacity of these for supporting life. The study of bird habitat is therefore not merely an aid to finding and recognition of each species, but provides valuable indicators for the state of the biosphere.

In many cases attempts at full standard presentation are frustrated by gaps in material, for example in relation to altitudinal range and to moulting habits and displacements. Over the west Palearctic it often happens that regional variations can make a precise statement partially misleading unless it is lengthily qualified or explained, and discretion has had to be used in omitting details of trivial or local significance, in order not to swamp or confuse the presentation of what is normal.

On the other hand, habitat descriptions of many species are less complete and more superficial than they should be for lack of studies in depth. This is well illustrated by the studies made by Fjeldså (1973a, b) of the Slavonian Grebe *Podiceps auritus*. In order to test correlations between subspecific and other genetic distinctions and habitat selection he made thorough investigations of a number of isolated populations in northern Norway and Iceland for comparison with Swedish and Finnish stocks. He visited 966 sites in brackish bays, slow-flowing rivers, pools, and lakes, besides using several thousand literature sources. In the lakes, primary factors were morphometry (such as extent, depth, and altitude, and the profile under and above water); topography, including islands; geology; vegetation (marginal, emergent, floating, and submerged); water quality (eutrophic, mixotrophic, dystrophic, and oligotrophic, and hard or soft); and biological productivity. Distance from the sea or from other lakes, marshes, and small water bodies was also relevant. Analysis of emergent vegetation revealed the high importance for nest siting of distance from shore and from open water, of vegetation type, of density and cover, of anchorage, and of water depth. In all cases records were made of water colour, sight depth, plankton turbidity, and bottom sediment. The percentage of water area carrying emergent vegetation was a further relevant factor. The significance

of depth of water from surface to bed might be less than the depth from surface to top of submerged vegetation. The age of birds occupying territories of different types also proved significant, marginal sites tending to be left to the youngest. Account had to be taken of correlations with neighbourhood of other birds and specific plants. With all this wealth of detail the investigation was still incomplete. It ignored precise reactions to human presence or influence, and frequency and height of flight movements, for example. Such a case serves to emphasize how inadequate is our knowledge of factors determining habitat for almost all species of birds, and to show what opportunities exist for comparable studies in depth in countless other instances.

In the absence of such up-to-date comprehensive studies, reliance has to be placed largely on the literature, most of which consists of sketchy or fragmentary contributions, few of which are sufficiently precise and well-informed about habitat factors to enable specific conclusions to be drawn. Many are not, or are only partly, primary sources. It is necessary to cross-check and condense a very large number of references, few of which stand independently as sole or principal sources for any particular fact. This creates difficulties in deciding which out of countless sources consulted to select for citation. There is no problem for the small number of contributions tackling in depth the habitat aspects of a single species or group, or where some significant fact comes directly from a particular observer, especially in a source not much quoted and liable to be overlooked.

From the nature of the Habitat section, however, it is more frequently a case of producing valid generalizations from many different records, which cannot all be cited and no one of which stands out as of major significance or authenticity. Attempts to distil such generalizations from the literature have been previously made at different times and with differing degrees of success by many authorities, whose judgement and perception has often made our task easier. Some, such as the *Handbuch der Vögel Mitteleuropas* (Bauer and Glutz 1966-9, Glutz *et al.* 1971-7, Glutz and Bauer 1980-5, and in progress), have included copious lists of references, and those who wish to pursue habitat records in such detail would do well to consult them.

While ornithological sources must be responsible for defining habitat requirements of each bird species, the occurrence and distribution of these habitats often mainly depends on plant ecology and the distribution of vegetation. The absence of any comprehensive and authoritative treatment of these aspects for the west Palearctic proved a considerable handicap for earlier volumes of this work, but this has now been partly made good by the publication of *A Guide to the Vegetation of Britain and Europe* (1985) by O Polunin and M Walters which describes (with maps and photographs) arctic, boreal, Atlantic, central European, Mediterranean, steppe-border, alpine, freshwater wetland, and coastal plant communities. It also includes a glossary of terms, over 40 of which overlap with our own Glossary which follows, and we have taken the opportunity to check and where necessary revise these against it in order to promote standardization of usage. Although essentially botanical, the *Guide* is written in language which should be readily understandable by ornithologists, and may confidently be recommended as a most valuable aid to fuller understanding of bird habitats, containing as it also does introductory explanations of ecological concepts and terms and of the soils and climates of Europe, and a history of the development of its vegetation. For example, while Mediterranean dwarf shrub or garigue can be dealt with here only as a single type, its diversity is treated in Polunin and Walters by descriptions, with outlines of distributions and composition, of no less than twelve sub-types. With such an aid, much more precise studies of bird habitats should become possible over wide areas.

In the absence of any such complementary work for the arid zones of the west Palearctic, reference should be made to the *Ecology and Flora of Qatar* (1981) by K H Batanouny. Although Qatar itself is outside the west Palearctic, much of its flora is characteristic of the arid zone generally, and the descriptions of landforms, soils, climate, water, surface geology, and vegetation can prove of the utmost assistance for habitats over much of North Africa as well as south-west Asia. The numerous accurately titled coloured illustrations of desert vegetation and landforms are especially helpful.

We can see habitat not merely as the kind of terrain on which a given species of bird may be expected to be found but as a theatre in which each competes to identify, locate, and, where necessary, possess the most appropriate set of natural resources for its subsistence and reproduction. These resources comprise not only access to suitable diet, both in bulk and in trace elements, but perches and song-posts, places to run, wade, swim, or dive, airspace for mobility, foraging, or display, dusting and loafing places, and, on migration, spots where rapid refuelling can be done in safety at the right season. The list is not exhaustive, but it indicates how many cues or signals in the environment must be precisely read by a bird choosing its habitat, and consequentially by the field observer wishing to read and understand its choices. Many kinds of change—some gradual, some rapid—have influenced the evolutionary fit of each species to its current habitat, and consequently its success in competing with others. At the same time, pronounced and sometimes rapid changes demand corresponding adaptations, sometimes climatically, sometimes through effects of the spread or loss of vegetation, and ever more frequently through human impacts. Given sufficiently precise knowledge of their habitats, birds can prove valuable indicators of trends and changes which have significance also for human land uses. While the present Habitat accounts

will no doubt look primitive and superficial in the light of advances within coming decades it is hoped that they may stimulate not only more detailed but more comprehensive lines of study, linking them more plainly to other subjects of research and application which have hitherto been regarded as unrelated. A concise and masterly review of habitat selection in birds has been provided by Hildén (1965).

With certain exceptions in the interests of convenience, the Habitat section for every species follows a common pattern. Normally it begins with the biogeographical setting of the distribution of the species in terms of its latitudinal and altitudinal range and climatic limits and preferences, and the broad terrestrial and aquatic groupings in which it occurs. Then follow more details (including land-form and vegetation) of the habitats used for breeding, foraging, roosting (though more detail on this topic is usually given under Social Pattern and Behaviour), assembling to moult, stopping on migration, wintering, and other purposes; also of the zones used in or under water or in the air, and of relations with man.

Care has been taken to minimize duplication with other sections of this work, but in certain cases some overlap is inherent in the plan. For example, where conservation problems stem from or must be considered in relation to habitat, they are treated in this section. Where, however, they are more or less self-contained (e.g. shooting pressure or mortality from toxic chemicals) they are handled only under Population. Although specific recommendations have not normally been included, efforts have been made to bring out opportunities and needs for further observation and research, which are plentiful and important.

The following Glossary of terms aims to relate the wording in descriptions to ecological and geographic usage.

GLOSSARY

In attempting to attain greater precision and completeness in describing the habitat preference of each bird species, a major obstacle is presented by the vague and ambiguous terms in common use, even in scientific records, for many of the elements in the environment. That environment is composed of a multiplicity of components, each of which can be scientifically described, not necessarily in some new jargon but by giving coherent, clearly defined meaning to many words in everyday use. This approach has been adopted over several decades by leading plant ecologists, whose definitions for such words as bog and fen are largely followed here. It is hoped that in this way readers will be enabled to check more accurately the characteristics and limits of observed habitats and to acquire a more discriminating attitude to the semantics of the various terms employed, thus gradually bringing about a greater measure of clarification and standardization in their use. Owing to the great variations of these elements in the field, and the different, often conflicting practice in the use of terms, complete standardization must be recognized as unattainable. *As far as possible, usage of terms in the Habitat accounts follows the definitions given here, but divergences are sometimes inevitable, for example in instances which involve direct quotation from the literature.*

The function of this glossary is to define principal terms, leaving their refinements and subdivisions to be treated in the text as they arise. For those less familiar with the English language, the glossary should prove more enlightening than reference to normal dictionaries, which inevitably suffer from lack of a precision adequate for the study of bird habitats.

The guiding principle has been to conform in all important respects to the best recent authorities in each field, especially ecologists, geographers, and hydrologists, but not to hesitate to define more fully and to set quantitative criteria wherever advisable to facilitate accurate recording and comparison of ornithological data. Special care has been taken to minimize ambiguity and unconscious overlapping of terms. For convenient reference, words defined in this glossary are printed in capitals wherever else they occur in it.

ALKALI FLAT. A seasonally dried-up area of barren hard mud covered with alkali (pH above 7·2) which, owing to lack of rainfall has not been washed away. Occurs particularly in STEPPES. (Equivalent to *solonets* in Russian.)

ALLUVIUM. Fine material, mainly soils, brought down in suspension by a river and deposited on its floodplain or DELTA, creating levels which are often markedly more fertile than their surroundings. *See also* FLOODLAND.

AQUATIC VEGETATION. Grouped into four classes: (1) *marginal* for emergent plants forming a more or less narrow ZONE along the edge of standing or flowing open water, (2) *emergent* for plants other than marginal which are rooted on the bottom and project above the surface, (3) *floating* for plants forming mats, streamers or other patterns on the water surface, and (4) *bottom* for plants of which no part reaches or comes near to the surface.

ARABLE LAND. Land under or available for cultivation, excluding permanent GRASSLAND or pasture.

ARCTIC-ALPINE. The climatic ZONE, above the TREELINE, situated at increasing altitude towards lower latitudes, which carries many dwarf woody plants forming a HEATH cover below the snowline. The term TUNDRA is sometimes misleadingly extended to this ZONE. In exceptional areas, plants characteristic of this ZONE descend locally much lower, even to sea-level in upper mid-latitudes.

ARCTIC and SUBARCTIC. The boundary between these in the west Palearctic is taken as passing through southern Iceland, extreme northern Norway, and the entry to the White Sea, thence ranging down to as low as 65°N where it crosses the Urals. The southern boundary of the subarctic (*see* BOREAL) runs from 60°N 30°W in the Atlantic to the Norway coast just above the Arctic Circle, thence down near Bergen, Oslo, north of Stockholm, through Helsinki, and near Leningrad to the Urals (see map in Stonehouse 1971).

ARID. A climate in which rainfall is insufficient regularly to grow crops, but not necessarily to inhibit sparse natural vegetation.

AVALANCHE. The sudden fall down steep mountainsides of a large tonnage of material, leaving a cleared corridor which may persist for many years.

BADLANDS. Steep naked eroded ridges, rarely of hard rock, usually in ARID climate and typically *c.* 20–50 m high, often flanking open plains or valleys.

BARCHAN. A type of naked, barren crescentic desert sand-dune, with ends pointing downwind, up to 30–40 m high and 150–350 m broad, windward slope being gradual and leeward slope

steep. Upper layers are almost dry, and systems are mobile and unstable, sometimes coalescing into broad shapeless dune-fields. *See also* ERG, DESERT, DUNE.

BARRAGE. A major dam across a RIVER, usually in steep TERRAIN but often applied to resulting man-made LAKE, frequently more than 10 m deep and 1 km² in extent, and often much larger. *See also* RESERVOIR.

BARRANCA (or BARRANCO). A steeply sloping, water-worn valley on a mountainside, initially often shallow but becoming a deep narrow RAVINE although rarely of large scale and often soon terminating at SEA, especially in Atlantic SUBTROPICAL islands.

BIOMASS. The total weight of living organisms in a given area, thus reflecting its relative biological productivity. Sometimes includes also dead organic matter.

BOG. An area of wet, acid, poorly drained spongy peat, too soft to bear much weight, composed chiefly of decaying moss or other vegetation, receiving water only from precipitation. (Equivalent to *moss* in northern and central Britain.) Botanists distinguish between *valley bogs* (in local impermeable depressions), *raised bogs* (growing with a convex surface), and *blanket bogs* (covering varied TERRAIN, often UPLAND and with more *Molinia* grass and less *Sphagnum* moss). *See also* MIRE.

BOREAL. Not a precise geographical term. Used principally in relation to a large group of plants and animals whose distribution pattern can be matched with northerly climatic factors, mainly within the SUBARCTIC. It overlaps the ARCTIC and extends into the coolest parts of the north TEMPERATE deciduous forest.

BOTTOM AQUATIC VEGETATION. *See* AQUATIC VEGETATION.

BOTTOMLANDS. The low-level, usually fertile or wooded, and sometimes seasonally flooded lands on margins of large lowland RIVERS.

BRACKISH. *See* LAKE.

BRECK. A type of undulating or flattish lowland, often sandy and marginal for traditional cultivation, accordingly relapsing to HEATH or a kind of WOODLAND SAVANNA between intermittent and long-separated periods of farm use.

BRUSH (not used in Britain). THICKET or layer of small TREES or shrubs, often in or by a WOOD or FOREST.

BUSH. A more or less dense shrub, usually 1–3 m tall, with many woody branches and seasonally ample foliage providing good cover; sometimes a small group of shrubs forming a single mass. (Not used here in the sense, probably borrowed from the Dutch *bosch*, of wild forest or scrub country.)

CALDERA. An extra wide crater, typically several kilometres across, usually circular and of volcanic origin and often but not always with steep walls.

CANAL. Artificial channel of fresh or saline water, from 5 m to 50 m or occasionally much wider, sometimes on one level, otherwise on different levels linked by locks or lifts, with no consistent flow. Disused canals tend to become blocked with AQUATIC VEGETATION and their banks become overgrown to constitute an entirely different HABITAT.

CANOPY. The more or less continuous network of branches and foliage forming the top layer of FOREST, any gaps in which are normally narrower than the crowns of the separate TREES.

CANYON. *See* RAVINE.

CARR. Preferred use implies more or less mature tree cover (typically including alder *Alnus* and willow *Salix*), either fringing or often succeeding a FEN or BOG after drainage or other local change. Eventually forms WOODLAND on wet, usually flooded site. Also applicable to wet alluvial WOODS (cf. German *Bruch*).

CHAPARRAL. *See* MAQUIS.

CHOTT. *See* SABKHA.

CIRQUE. A natural rocky amphitheatre, typically in mountains at head of a stream or glacier, often with precipitous rock faces. (Equivalent to Scots *corrie* and Welsh *cwm*.)

CLEARING. The counterpart of a GLADE, but produced by tree-cutting or other human intervention and tending to be larger, up to 10 ha or more.

CLOSED VEGETATION. *See* SPARSE VEGETATION.

CONTINENTAL CLIMATE. A climate with limited rainfall and large daily and annual ranges of temperature, characteristic (except in humid TROPICAL areas) of a continental interior, or other region protected from or unaffected by maritime influences.

CONTINENTAL SHELF. The underwater marine plateau, usually taken as extending from the coast to *c.* 180 m (*c.* 100 fathoms) deep, but figures from 110 m upwards have been suggested as alternatives. Beyond this the *continental slope* falls more or less steeply towards the OCEAN bed. *See also* OFFSHORE, PELAGIC.

CONTINENTAL SLOPE. *See* CONTINENTAL SHELF.

COPPICE. Cutting woody stems on a rotational basis, often in association with a dispersed silviculture of taller, trees, is coppicing and produces coppice with standards, usually in small WOODLANDS. Accordingly a counterpart of COPSE, into which a coppice may become converted if it ceases to be managed as above.

COPSE. A small stand of trees, generally not larger than 5 ha, usually on level or gently sloping ground or in a lowland valley, not obviously planted, and mainly deciduous with substantial undergrowth.

CRAG. An abrupt, precipitous rock face, not necessarily high, characteristically backed by a gentle slope behind it. Commonly applied to inland precipices forming a local and limited counterpart to a SEA cliff.

CREEPER. *See* VINE.

DELTA. A complex RIVER or STREAM mouth where much silt is deposited, splitting the flow into numerous channels separated by emerging banks and (sometimes) REEDBEDS, and often giving rise to separate LAGOONS. *See also* ALLUVIUM.

DESERT. An infertile mainly unvegetated extensive tract of largely flat or gently sloping land in low or mid latitudes, characterized in west Palearctic by high day temperatures, abundant sunshine, and very low and irregular rainfall, inhibiting occurrence of perennial RIVERS and normally also of WETLANDS. Desert surfaces are mainly rock and gravel, only about a quarter of the world total being sandy. Soils are at an arrested mineral stage of development, without decomposers or other essentials for organic processes even if rapid evaporation did not take up any moisture arriving, thus restricting any vegetation to sparse perennials, locally and seasonally complemented after rains by a short-lived cover of annuals. Rainfall in cooler seasons is more effective for plant growth than in greater heat, but when heavy and accompanied by winds of over 20 km per hr rain can bring drastic erosion, sculpting distinct features. In some regions, however, periods of slight rainfall can support transient vegetation cover or plants capable of tapping groundwater *c.* 50 cm or more beneath surface.

DIKE (or DYKE). Artificial watercourse formed by excavation, commonly for drainage or water supply, at least 2 m wide and often regulated by a sluice or spillway. If less than 2 m wide it is a *ditch* or *trench*; if used for navigation or irrigation and over 5 m wide it is a CANAL. (Not used here in the derived sense of a bank of earth produced by excavation, although the bank may be regarded as part of the dike.)

DITCH. *See* DIKE.

DUNE. A hill of blown sand, often part of an extensive system, either on the coast or in a DESERT (*see* BARCHAN). Dunes are often up to *c.* 10 m high (exceptionally to 100 m), depending on volume and regularity of supply of fresh sand. They are initially mobile, and evolve through incipient, juvenile, mature, senile, and degraded forms sometimes ending under FOREST. Dunes sometimes have moist or wet depressions called *slacks*.

DUST. Fine dry pulverized material formed by sunlight and warmth on bare soils, and attractive to various birds for dusting. In DESERT, large dust storms of very hot, excessively dry, electrically charged particles can rise to *c.* 3000 m and travel inter-continentally, in contrast to true sandstorms which rarely rise above *c.* 30 m and travel only short distances.

DYKE. *See* DIKE.

DYSTROPHIC. *See* LAKE.

ECOSYSTEM. *See* HABITAT.

EMERGENT AQUATIC VEGETATION. *See* AQUATIC VEGETATION.

ENVIRONMENT. *See* HABITAT.

ERG. A pure sandy DESERT, especially in north-west Africa, often with prominent elevations above level base.

ESTUARY. Formed where a RIVER entering the SEA broadens into a tidal inlet. Thus distinct from a DELTA.

EUTROPHIC. *See* LAKE.

FALLOW. Farmland being rested from cropping for one or more seasons.

FEN. A waterlogged area of peatland, not normally acid, receiving its water supply largely from outside surface sources. *See also* BOG, MIRE.

FJELL. High ground, typically in BOREAL Scandinavia, rising above the TREELINE but often fringed by SCRUB, especially of birch *Betula*. (Equivalent roughly to *fell* in English. Norwegian variant *fjeld* is also applied to an elevated rocky plateau.)

FLOATING AQUATIC VEGETATION. *See* AQUATIC VEGETATION.

FLOODLAND. Land so sited in relation to normal run-off of water that it becomes inundated, usually only to shallow depth, at regular or frequent intervals, but never long enough wholly to destroy terrestrial vegetation cover (including farm crops where appropriate) and terrestrial animals.

FOREST. Continuous area of natural or semi-natural tall TREE cover, extending over not less than 10 km² and often much more, with mainly or wholly closed CANOPY at more than 5 m and typically above 20 m, depending on climate for regular moisture on fairly large scale. (Not to be confused with ancient usage for large area, not necessarily wooded, reserved for royal hunting under game laws.) *See also* FOREST TUNDRA, GALLERY FOREST.

FOREST TUNDRA (or WOODED TUNDRA). The ZONE, several or sometimes many kilometres wide, of stunted FOREST mixed with lichens and GRASSLAND and, in favoured valleys, with stands of well-grown TREES, beyond the limit of continuous TAIGA, but by definition incompatible with true TUNDRA, which begins strictly beyond the TREELINE, and not merely beyond the FOREST edge.

GALLERY FOREST. A fringing strip of TREES along both banks of a RIVER or STREAM traversing open country, usually in or near TROPICAL regions. *See also* RIVERAIN.

GARIGUE. Open, low-growing calcareous SCRUB on lowland, often rocky slopes or cliffs, in MEDITERRANEAN CLIMATE, and due to human interference with earlier FOREST. Includes more herbage, grasses, spiny and aromatic herbs, and bare patches and fewer tall woody plants than MAQUIS, but may be transitional towards that type or towards colonization by WOODLAND TREES such as Aleppo pine *Pinus halepensis*.

GEYSER. *See* SPRING.

GLADE. Small, open, but sheltered and usually elongated space within a FOREST or other WOODLAND, often with a relatively benign microclimate, and either natural or resembling a natural opening. Vegetated with grass, herbage or BUSHES. *See also* CLEARING, MUSKEG.

GORGE. *See* RAVINE.

GRASSLAND. Broad general term for areas of low CLOSED VEGETATION, predominantly of grasses and excluding woody plants, aquatics, mosses, or lichens. *See also* STEPPE.

GRAVEL-PIT. An area, commonly 5–50 m deeper than surrounding land surface, exploited for extraction of a loose, detrital sediment containing a large proportion of pebbly material 2–50 mm in diameter and therefore much coarser than sand. A worked pit may remain unvegetated or may become overgrown or inundated with water or both.

GROVE. More-or-less closed stand of mature TREES, natural or planted, with little or no undergrowth, usually in level, open landscape and of maximum extent 1 ha.

GULLY. A shallow, unstable, or miniature linear depression. *See also* BARRANCA, RAVINE.

HABITAT. A precise complex of features and conditions within which a given animal or plant species is adapted to survive. Thus differs from *environment*, which embraces the whole range of available natural resources and features from which specific habitats are chosen, and from *ecosystem*, embracing the entire reciprocally related complex of animals and plants whose habitats overlap or adjoin.

HALOPHYTE. A plant adapted to surviving on land impregnated with salt.

HAMADA. (Meaning in Arabic 'the unfruitful'.) A desert area in the Middle East covered by stones and gravel, remaining when finer particles have been blown away by the wind. One form is gently sloping while the other is a flat slightly elevated plateau, sometimes of exposed bedrock. Lichens grow freely on the stones, and in places sparse shrublets occur, locally also *Acacia* and *Zygophyllum*. (Equivalent to *reg* in north-west Africa.)

HEATH. A dry, normally lowland area of usually flat sandy or gravelly and porous soil, fairly free from peat and carrying evergreen dwarf shrubs such as heather *Calluna* or *Erica*, normally under 0·5 m high; sometimes interspersed with scattered pines *Pinus*, birches *Betula*, or other TREES. Often traceable to earlier deforestation or abandoned cultivation. Corresponding UPLAND areas dominated by *Calluna* with much peat are *heather moors*. *See also* BRECK.

HEATHER MOOR. *See* HEATH.

HEDGE (or HEDGEROW). Artificial, narrow, dense, often straight linear barrier of woody vegetation, normally kept 1–3 m high, especially for separating farm fields; variant type interspersed with shade TREES.

HILL. A more or less steeply rising natural elevation above lowlands, usually applied to eminences not more than *c.* 600 m high, above which, especially when more rugged and abrupt, they become MOUNTAINS. A large complex of hilly country ranks as *uplands*. *See also* FJELL, PLATEAU.

HOT SPRING. *See* SPRING.

ICE. The various types have been classified by Dunbar (1955) and Armstrong and Roberts (1956, 1958). The most relevant are as follows. (1) *Pack-ice*: more or less hummocky ice, always formed at sea and unattached, varying from open to fully consolidated, but often divided by channels termed *leads*. (2) *Fast ice*: sea-ice attached directly or indirectly to the shore, from which it may extend far out, up to *c.* 400 km. (3) *Icesheet*: layer of ice covering large area of land, continuously moving along defined channel from higher to lower ground, and usually eventually to sea. (4) *Iceberg*: mass of ice broken away from a glacier, at least 5 m above sea surface, and thus distinct from much thinner but often more extended (5) *icefloes* or (6) *icefields*, always of marine origin.

INLET. Used as a general term to cover any bay, gulf, fjord, outer ESTUARY, or smaller indentation in a coastline of SEA or large LAKE which produces a more sheltered or richer HABITAT than neighbouring unindented coast.

INSHORE. This would ideally be defined in terms of a complex of factors such as depth, temperature gradient, and littoral influences, but it is more practical to adopt a standard distance relevant to the observation of movements of land-based birds. This may be arbitrarily set at 5 km from low-water mark, rather than at 'maximum of 4 or 5 miles out to sea' in Witherby *et al.* (1941). Where there are islands or ISLETS well in sight of the shore, all intervening water and the ZONE up to 5 km beyond them should be included, as should neighbouring SEAS less than 6 m deep.

ISLET. A very small island, typically under 10 ha.

KARST. An area of weathered limestone marked by abrupt ridges, caverns, and underground water flows causing solution. The bare pavement surface is broken by deep clefts and interspersed with GRASSLAND or woody vegetation, and locally by cultivation.

KRUMMHOLZ. Twisted, gnarled, and distorted woody vegetation, especially coniferous, produced within marginal ZONE for tree growth, normally on MOUNTAINS or coast. *See also* TREELINE.

LAGG. The fringe of a MIRE, originally simply a drainage channel, but by extension including the area of its influence. Often distinguished from the rest of the mire by different water quality and the growth of moisture-loving TREES and shrubs, usually at SCRUB height.

LAGOON. An enclosed shallow coastal LAKE, POOL, or landlocked marine inlet of either fresh or saline water contained by sand, shingle, or earth banks, or coral REEFS; strictly of natural origin, but often extended to similar waterbodies partly or whollyman-made.

LAKE. A natural body of standing fresh water arbitrarily defined as exceeding 1 km² in surface area; subject to wind and wave action, but to currents only near entry and exit of running water, and where wind action is significant. Chemically, lakes are characterized as *oligotrophic* or unproductive (including *dystrophic* for particularly shallow waters with winter pH below 6 which are peat-stained), *mesotrophic* or moderate to high in productivity at a winter pH of *c.* 7, and *eutrophic* or high in productivity both for algae and macrophytes and often discoloured by the former. *Marl lakes*, clear, calcareous, and of high alkalinity, are rather rare. *Brackish lakes* are intermediate between fresh and SEA water. Ornithologically, area of open water, depth, and abundance of MARGINAL AQUATIC VEGETATION tend to be more significant than chemical variations. *See also* POOL, SALT LAKE.

LEAD. *See* ICE.

LITTORAL. The intertidal ZONE of a tidal SEA or OCEAN; sometimes applied to non-tidal waterbodies to include the underwater limit to which the more abundant attached plants can grow, sometimes to a depth of 30 m or more, corresponding to light penetration. May thus overlap INSHORE ZONE at sub-surface levels.

MACHAIR. A belt of usually vegetated coastal DUNES formed of fertile shelly sand, especially in the Outer Hebrides (Scotland).

MANGROVE. Although applied by some to TREES or shrubs of a few genera, used ecologically (often under variant *mangrove swamp*) to mean any population of TREES adapted to live along muddy LITTORALS in TROPICAL regions, where they can resist strong wave action and exhibit a dense habit of growth, seldom reaching any great height (normally well below 20 m).

MAQUIS. Type of Mediterranean SCRUB forming a dense community of prickly or spiny plants 1–3 m tall, and of others such as *Pistacia lentiscus*. Differs from GARIGUE in being typical of non-calcareous soils. While distinction is valuable for HABITAT description, countless variations and blends occur in practice, and boundaries with other habitats attract a wider range of bird species, for example where patches of cultivation occur. Both types are characterized by a large element of broad-leaved SCLEROPHYLLOUS plants. (Equivalent to *chaparral* of western USA and *matorral* of Spain, but latter also includes GARIGUE.)

MARGINAL AQUATIC VEGETATION. *See* AQUATIC VEGETATION.

MARL. *See* LAKE.

MARSH. WETLAND area in which standing water does not usually cover the surface, apart from scattered POOLS or fringes of LAKES, RIVERS, and other waterbodies (thus differing from a SWAMP). Usually excludes peatland (see MIRE) and is associated with mineral soil, coastal levels, and depressions in RIVER basins, usually at low or moderate altitude. May be of any size, but often small.

MATORRAL. *See* MAQUIS.

MEADOW. More or less compact and often enclosed stand of usually permanent GRASSLAND and mixed herbage, normally free of TREES or BUSHES, not currently subjected to grazing, and in traditional farming practice usually cut for hay. Applied also to FLOATING AQUATIC VEGETATION and to high Alpine GRASSLANDS which may be seasonally grazed, and are therefore strictly pastures. *See also* WATER-MEADOW.

MEDITERRANEAN CLIMATE. Also known as western margin warm TEMPERATE climate. Characterized by hot summers with high sunshine and almost complete drought, as well as mild winters, usually with no month below 6°C mean, but with substantial rainfall creating overall sub-humid state.

MERSE. *See* SALT-MARSH.

MESOTROPHIC. *See* LAKE.

MIRE. An arbitrary redefinition by modern botanists of a loose country word, now used to cover all peatlands classified according to whether their water supply (1) comes from outside, bringing in adequate nutrients, or (2) is derived from precipitation on them, and is accordingly poor in nutrients, with low pH value; also transition types. In HABITAT terms the first group are FENS or *fen-mires*, which may be sloping or flat; if flat they vary according to whether they are flooded permanently or for long or brief periods, giving rise to sedge, SCRUB, alder *Alnus*, or other vegetation (*see also* LAGG), while some in the TUNDRA ZONE are permanently frozen at subsoil level (*see* PERMAFROST). The second group includes blanket, raised, and flat BOGS, coastal or UPLAND, continental or OCEANIC.

MOOR. A general term applied to acid peaty, usually UPLAND tracts which are unenclosed, typically covered with heather *Calluna*, moor grass *Molinia*, *Sphagnum* moss, etc., and not used primarily as pasture. *See also* HEATH. (*Moor* in German is equivalent to MIRE as here defined.)

MOUND. Any artificial pile or rounded bank of earth or other material not exceeding 1 ha and rising more than 1 m (including large pit heaps, spoil heaps, refuse heaps, and archaeological or landscaping eminences) and similar natural eminences.

MOUNTAIN. A massive elevation of hard and often bare rock rising more or less steeply, even precipitously to at least 600 m and often much more. Culminates in a sharp peak, an extended ridge, or a more-or-less level PLATEAU; frequently forms part of a major mountain chain, range, or massif attaining high altitudes over a wide or extended area. Associated

vegetation exhibits marked zonation, from broad-leaved TREES at base through conifer FOREST to stunted conifer growth at TREELINE in middle latitudes above which alpine MEADOWS and HEATH may ascend to a snow-line: beyond this, permanent snow normally precludes use as a HABITAT. At higher latitudes, and higher altitudes at mid-latitudes, glaciers may occur frequently, while in lower latitudes, especially in DESERTS, rock surfaces become too hot and dry for vegetation except in shady GULLIES. *See also* HILL.

MUDFLAT. An often-extensive level area of fine clayey particles usually containing much water and having sticky and slippery properties, wholly or mainly unvegetated, sometimes dissected by a network of small channels. Exposed either regularly by tides or seasonally or periodically by ceasing to be submerged with water.

MUSKEG. A clear, low-lying area, usually within a coniferous FOREST, which often was formerly open water but has become filled with peat moss, over which fringing TREES are gradually advancing. Typically fairly narrow and small, and muskeg areas often appear from the air as a mosaic of GLADE-like sheltered spaces. *See also* TAIGA.

OASIS. An area ranging from small to very large, surrounded by DESERT but rendered fertile by presence of fresh water stored underground naturally. Resulting rich vegetation, usually including groves of PALMS and often sheltering human settlements, is an attraction to trans-desert migrants and to colonization by certain non-desert species.

OCEAN. The main worldwide body of interconnected salt waters, with a normal salinity range of 3·3 to 3·7%.

OCEANIC. Used in two senses: (1) relating to OCEAN, as distinct from SEA, and (2) relating to a climate or region significantly influenced by the neighbourhood of the OCEAN. *See also* SUBOCEANIC WATERS, PELAGIC.

OFFSHORE. The ZONE extending from the limits of INSHORE waters to the edge of the CONTINENTAL SLOPE.

OLIGOTROPHIC. *See* LAKE.

OPEN VEGETATION. *See* SPARSE VEGETATION.

ORCHARD. An enclosure planted with fruit TREES which are, or will become when fully grown, at least 2–3 m tall, excluding stands of planted shrubs such as coffee or tea, which are without TREE cover. Here understood to include widely spaced stands (e.g. GROVES of olive *Olea* in certain Mediterranean countries) which are similarly cultivated but may be undersown with cereal or other ground crops, and which might not be called orchard in common usage.

ORNAMENTAL WATERS OR PONDS. Purposely created for amenity in PARKS and other landscapes, these rarely extend over 1 km², are often loosely called LAKES.

OUED. *See* WADI.

PALM. TREE-like monocotyledonous plant of family Palmaceae.

PARK (or PARKLAND). Originally any enclosed, small or medium-sized tract of land reserved for hunting, but has acquired a confusing range of meanings. For HABITAT and landscape, most relevant meaning is as TEMPERATE, often artificial counterpart of SAVANNA, blending scattered TREES and BUSHES singly or in clumps or SPINNEYS with predominantly GRASSLAND cover. Frequently forms small managed unit in humid climates, especially on undulating landforms. Many areas managed as national parks or city parks contain no such HABITATS.

PELAGIC. The ZONE of the deep OCEAN beyond the CONTINENTAL SLOPE.

PERMAFROST. Permanently frozen subsoil, and hence the ZONE over which that condition persists. *See also* TUNDRA.

pH. A measure of acidity and alkalinity in terms of the concentration (on an inverse logarithmic scale) of hydrogen ions in a solution, affording a serviceable means of assessing a soil's limitations for supporting various kinds of vegetation. For water, reliability, especially in summer, is much more limited. A pH value of 7·2 indicates neutrality; higher values (up to 14) indicate alkalinity, and lower values acidity.

PLANTATION. As FOREST, WOOD, SPINNEY, and COPSE are defined as consisting of natural, semi-natural, or spontaneous TREE cover, it follows that all TREE stands planted by man during the past 400 years (other than ORCHARDS and minor distinctive forms such as roadside avenues and narrow shelterbelts alongside fields and gardens) should be classed as plantations.

PLATEAU. A more or less extensive tract of generally level land at much higher elevation than a plain, and thus prone to become dissected by STREAMS carrying run-off.

POND. A small body of fresh standing water, having less than 10 ha of open water, formed artificially by hollowing out or impoundment. (By analogy or confusion, sometimes wrongly used for similar POOLS of natural origin, while ORNAMENTAL WATERS are often wrongly called LAKES.)

POOL. Natural body of standing fresh water with less than 1 km² of open water (see also LAKE, TARN). Also used for a small body of still water in a watercourse, often seasonally cut off (see also RIVER).

QUARRY. Site of open-air extraction usually of deposits of hard rock so placed that they can be excavated in quantity out of slope or cliff by working from a lower level, generally by blasting or cutting. Quarries do not often exceed 1–2 ha.

RAVINÈ (or GORGE). A narrow valley, deep in proportion to its width, usually with rocky precipitous sides but sometimes between abrupt grassy slopes, often carrying a torrent below. A miniature RAVINE is a GULLY; a very large one is a CANYON. See also BARRANCA.

REEDBED. A level area of sluggish drainage, usually with shallow standing water during much of the year, carrying dense tall uniform stands of emergent aquatic plants such as reed *Phragmites*, reedmace *Typha*, reed meadowgrass *Glyceria*, and *Papyrus*. Applicable not only to extensive tracts but to fragments down to a fraction of a hectare, and to sites where small areas of open water are enclosed by such vegetation.

REEF. Typically a narrow or very narrow ridge of rock or coral, just above or below the surface of the water, often forming a natural breakwater enclosing a relatively calm LAGOON of protected water. Also applied to shorter and less linear rock structures nearly or just emerging above the water surface.

REG. See HAMADA.

RESERVOIR. An artificial body of open fresh water, normally over 10 ha in area, formed by excavation or impoundment in a depression or by enlargement of a natural pool or lake. See also BARRAGE.

RIPARIAN. See RIVERAIN.

RIVER. A major flowing watercourse, more than 5 m and up to 100 m or more broad between banks, having either a single channel or multiple channels which may be braided into a network of interlaced STREAMS between wide banks of sand, earth, or gravel. Seasonal flow fluctuates, even ceasing entirely for periods more or less extended and regular, leaving occasional standing POOLS. Where the banks are low there may be ancillary backwaters, oxbows, and small WETLANDS, in a RIVERAIN ZONE. See also STREAM, WADI.

RIVERAIN. Relates to the entire strip adjoining and influenced by a RIVER—in contrast to *riparian*, which is limited to its immediate banks. See also GALLERY FOREST.

SABKHA (or SEBKHA). An Arabic term denoting inland or coastal saline flats with fine silt and calcareous sands and high water table. The surface is sometimes covered with brine or with HALOPHYTIC vegetation, such as *Arthrocnemum* or *Juncus*, but is often entirely unvegetated. Sometimes vulnerable to dune encroachment. (Equivalent to *chott* in north-west Africa, and to *qa* in Jordan.)

SAHEL. ZONE of thorn SAVANNA south of the Sahara in western and central Africa.

SALINA. Saltpan, salt POND, or LAGOON, or other seasonally or generally flooded area of high enough salinity for economic exploitation, usually applied in Spain to those in which the water area is dissected by artificial banks.

SALT LAKE. A natural body of standing water, usually without an outflow, containing over 0·03% dissolved solids (in extreme cases up to 10%). High evaporation and fluctuating inflow are common features. Some of these LAKES may be more specifically defined as soda LAKES or alkali LAKES. Properly the latter term refers to those containing alkali of pH 7·3 and higher, but shallow, muddy, seasonal LAKES drying out to barren hard mud with a saline surface are often called ALKALI FLATS or sometimes 'white alkali'. The term 'salt lake' is best reserved for those which at least seasonally attain 1 km² or more.

SALT-MARSH. Low-lying coastal or ESTUARINE land regularly or intermittently overflowed by spring tides and covered with HALOPHYTIC vegetation, being distinct from the everyday intertidal ZONE occupied by MUDFLATS and seaweeds. Apparently no clear difference between salt-marsh and *salting*. (Equivalent to Scottish *merse*.)

SALTING. See SALT-MARSH.

SAVANNA. A controversial term best used only in a broad sense for landscapes influenced by alternating dry and rainy seasons in lower and mid-latitudes. Generally these are mainly level or undulating, with a closed cover of grass and scattered well-spaced TREES and BUSHES, singly or in small fairly open groups permitting distant views in all directions. They tend to be semi-ARID, with no or infrequent wetland elements, and should be at least semi-natural and in sizeable tracts. Man-made landscapes of similar appearance are PARKLANDS. See also WOODLAND SAVANNA.

SCLEROPHYLLOUS. Applies to plants with hard or stiff leaves and similar organs which reduce the loss of water by transpiration in dry climates.

SCREE (or TALUS). An unstable but persistent mass of loose stones or boulders lying on or below steep mountainsides or slopes, having become detached from above. Provides countless crevices of various sizes.

SCRUB. Characterized by low TREES, often below 5 m, whose growth is checked or much retarded by adverse natural or human influences, or which are unsuited to create tall FOREST, at least in the particular climatic, soil, or other conditions. Scrub is also often found for long periods after fire or clear-felling and, for example, along margins of BOGS and ARCTIC-ALPINE HEATH. BUSHES are often interspersed among scrub. MAQUIS and THICKET are types of scrub.

SEA. One of the smaller divisions of an OCEAN, adjoining a coastline and sometimes nearly landlocked. Very large bodies of salt water not communicating with the OCEAN, such as the Caspian, are also in common usage loosely called seas. A sea may have a much higher or lower salinity than that of the OCEAN, ranging in the Red Sea up to 4·1%.

SKERRY (usually grouped as SKERRIES). Cluster or archipelago of rocky ISLETS over which the SEA breaks in stormy weather.

SLACK. See DUNE.

SPARSE VEGETATION. Plants separated by more than their diameter. Distinguished from *open vegetation*, where the plants are on average separated by less than their diameter (but not overlapping), and *closed vegetation*, where they are in contact or overlap. These terms relate to whichever layer of vegetation shows the most complete coverage.

SPINNEY. A small, often elongated, stand of TREES with undergrowth, sometimes a relict of former FOREST, on a steep slope or in a valley head, not normally larger than 1 ha and typically of deciduous species.

SPIT. A narrow, low ridge or bank of any type of sediment, attached to the land at one end and terminating in open water at the other.

SPRING. A site where enough water issues through a natural opening to make an appreciable but not necessarily continuous current. If the temperature is more than 10°C above the yearly mean of the surrounding air, it is a *hot spring*; if the water is projected into the air, a hot spring becomes a *geyser*.

STEPPE. A broad band of lowland covered with dry GRASSLAND and XEROPHYTIC shrubs (e.g. *Artemisia*), with many scattered seasonal POOLS of fresh or saline water, and also ALKALI FLATS. It extends over continental mid-latitudes of the Palearctic east of the Danube, and is thus warm in summer but cold in winter and at night (*see* CONTINENTAL CLIMATE). Sandwiched between FOREST and DESERT ZONES, steppe is deeply encroached upon, creating many hybrid HABITATS. Its great carrying capacity, far exceeding that of DESERT, has made it the home base for many species of birds, though its instability and frequent inhospitality tend to compel them to be migratory or nomadic, while its treelessness and lack of rocks have encouraged underground reproduction and shelter, often in burrows made by marmots *Marmota*, susliks *Citellus*, and other rodents. In places, steppe is interspersed with, or grades into, DESERT. Some species have adapted to wide new HABITATS created by FOREST clearance and cultivation. The use of the term 'steppe' for other regions is not favoured by leading geographers, with the possible exception of the narrow transitional ZONE between the Mediterranean FORESTS and the Sahara.

STREAM. A watercourse whose bed seasonally or permanently contains appreciable water, mostly or wholly less than 5 m wide, having a sloping profile sufficient to produce flow but not to create persistent turbulence or rapids. *See also* RIVER, TORRENT.

SUBARCTIC. *See* ARCTIC.

SUBOCEANIC WATERS. Taken to include any beyond the CONTINENTAL SLOPE which are within 500 km of the nearest land, and therefore subject to significant terrestrial influences in terms of bird distribution and of pollution. *See also* PELAGIC.

SUBTROPICAL. *See* TROPICAL.

SWAMP. Differs from a MARSH in that its saturated surface is permanently, or at least throughout summer, below water level; nevertheless richly vegetated, typically with reed *Phragmites*, *Papyrus*, or other dense-growing tall aquatic plants, and frequently with moisture-loving trees. Applicable in such conditions to MIRES as well as to non-peatland HABITATS. May be very large. *See also* MANGROVE.

TAIGA. Dense, predominantly coniferous FOREST, little disturbed or dissected and often containing moist areas such as MUSKEGS. It occupies the broad cold dry BOREAL belt south of the TUNDRA in upper mid latitudes; frequently buffered from it by the intermediate, more open and fragmented transition ZONE, sometimes quite broad, which is often contradictorily referred to as FOREST TUNDRA.

TALUS. *See* SCREE.

TARN. A natural POOL, often at some altitude, normally small and at most not much more than 1 ha in area, with low pH value and no inflow stream of significant volume or length, often on tableland or undulating ground without developed drainage pattern, or alternatively in a glacial mountain CIRQUE.

TEMPERATE. The mid-latitude climatic ZONE, including *cool temperate* and *warm temperate*. In the west Palearctic a humid cool temperate climate with a short cold season (fewer than 6 months with mean temperature below 6°C) is confined to Britain, Ireland, and a band from northern Spain through France and the Low Countries to Scandinavia below about 62°N. A similar climate with a longer winter cold season covers most of the remaining land between 40°N and 60°N except the Mediterranean coast and part of the USSR between the Black Sea and southern Urals.

TERRAIN. A tract of land regarded in terms of its physical form, surface, and cover as relevant to its use, access, or attractiveness. Thus less specific than HABITAT but more so than ENVIRONMENT, with emphasis on topographical rather than ecological features.

THERMAL. A rising current of warmer air, in avian context providing lift for soaring birds.

THICKET. A dense stand of low trees, usually under 5 m high, forming close SCRUB in which deciduous and thorny species are often blended with shrubs, bracken *Pteridium*, and other plants tolerant of shade and moisture. Accordingly a TEMPERATE counterpart of MAQUIS, and a successional stage.

TORRENT. A perennial watercourse, in mountainous or UPLAND regions, with rocky bed and banks and characterized by a steep gradient, sudden fluctuations in flow, and usually many emerging rocks and gravel banks. Breadth variable, mainly less than 10 m. Rapids or waterfalls may be frequent.

TREE. A large non-climbing woody plant, usually not less than 5 m tall, with a complex underground root system and one or more hard bark-clad stems carrying a network of branches and leaves. PALMS are not true trees: they should be referred to separately, and their groupings should be termed 'clusters' or (if appropriate) PLANTATIONS. Bamboos are woody grasses, and are also excluded, even when of FOREST height.

TREELINE. (Equivalent to *timberline* in North America.) The limit of tolerable exposure to cold, wind, and other adverse climatic elements, either at higher altitudes or at higher latitudes, beyond which TREES are unable to grow, and TUNDRA or ARCTIC-ALPINE communities take over. Near the treeline there may be a ZONE of KRUMMHOLZ.

TRENCH. *See* DIKE.

TROPICAL. Properly, the region between the Tropics of Cancer ($23\frac{1}{2}$°N) and Capricorn ($23\frac{1}{2}$°S), but 'the tropics' often in the literature include adjacent areas of similar climate roughly between 30°N and 30°S, otherwise referred to here as *subtropical*, and characteristically showing tropical conditions for only a part of the year.

TUNDRA. Loosely associated, like the ARCTIC region, with the 10°C isotherm of the warmest month, north of which TREE growth ceases to be possible. The ZONE of treeless tundra within the west Palearctic is mainly between the Kara and White Seas north of the Arctic Circle, with small outliers along the Urals and the coasts of Murmansk and northern Norway, as well as on islands from Novaya Zemlya to Iceland. This tundra is mainly lowland and near coasts of the Arctic Ocean; it contains a wide variety of vegetation types, all low-growing, and is typically conditioned by PERMAFROST. *See also* FOREST TUNDRA.

UPLANDS. *See* HILL.

UPWELLING. Displacement of warmer surface water, usually by offshore winds, causing cooler water rich in nutrients to rise to the surface in well-marked ZONES, often generating much increased biological productivity and thus attracting a substantial population of marine birds.

VINE (or CREEPER). A climbing plant, which may be woody (i.e.

a liana) and ascend to substantial height on host TREES, sometimes stretching between them.

WADI (or OUED). A watercourse, sometimes flowing only once a year or less, in ARID countries. The term is customarily extended to include the immediate valley which it occupies. When flow does occur through intense rainfall, it is liable to be sudden, copious, and destructive, often briefly becoming a braided RIVER, and leaving surface POOLS and sumps of groundwater which support substantial vegetation, including TREES.

WASTELAND. Not a valid ecological term but widely used elsewhere for vacant and derelict urban and industrial land.

WATER-MEADOW. Riverside or streamside MEADOW, naturally or artificially inundated for part of year.

WETLAND. A generic term used to cover all non-marine aquatic HABITATS, large or small, whether permanently under standing or flowing water, or intermittently inundated, or brought into a moist state for a significant period, either by natural or artificial means. The Ramsar Convention on Wetlands of International Importance (1971) extends it to cover marine waters to depths of 6 m at low tide, though such waters are here referred to as INSHORE and are not included in 'wetland'.

WOOD. An area of natural or spontaneous TREE cover, smaller than a FOREST but larger than a COPSE, and not necessarily having a closed CANOPY.

WOODLAND. Used as a broad general term to cover all types and scales of TREE cover where density of stands exceeds those falling within PARKLAND and SAVANNA. This corresponds to the French 'terrains boisés' and in international professional usage to the clumsy 'forested land'; it does not correspond to North American practice which characterizes woodland as an open stand of TREES, usually under 15 m tall, with an intervening good growth of grasses or shrubs; this is classed here as WOODLAND SAVANNA.

WOODLAND SAVANNA. Intermediate between SAVANNA and FOREST, consisting of GRASSLAND or shrubs interspersed with frequent usually fairly low TREES, singly or in groups, too widely spaced to form anything like a continuous CANOPY or to attract strictly FOREST birds.

XEROPHYTE. A plant structurally adapted to an ARID or semi-ARID climate or HABITAT, or to places where access to moisture is at least seasonally difficult.

ZONE. An elongated belt or region distinguished by some common characteristic indicated by its name, which differentiates it from other adjoining zones, e.g. STEPPE zone.

DISTRIBUTION

The basic information on the distribution and populations of species in the west Palearctic has been supplied by the correspondents for each country, whose names are listed in the Acknowledgements and their initials given in the relevant texts after any data supplied by them. It must again be stressed that the maps represent the state of our current knowledge, which is still limited for some areas even in the west Palearctic; in this volume, information is particularly limited for those species living in desert areas.

On the maps, RED is used to indicate breeding distribution and GREY the areas where the species is regularly present in normal winters. Each map should be used in conjunction with the relevant Movements section.

We are particularly indebted to G G Buzzard for much

information on the distribution of species in some east European countries, and, for world distribution, to S C Madge and Dr D A Scott for details from Afghanistan and Iran respectively, and to G Bundy, M D Gallagher, M C Jennings, and Mrs F E Warr for data for various parts of Arabia.

Finally, it should be stressed that the Accidental paragraph under Distribution does not cover accidental occurrences outside the west Palearctic region.

POPULATION

Despite the growing interest in census work, detailed accounts and even estimates for breeding numbers are available for only a limited number of species in relatively few countries. National and regional estimates are given where available, or local counts where information is scanty, together with any data which illustrate trends. Apart from the references to published work, many of the country correspondents have helped; as in Distribution, this is indicated by the use of their initials. The section concludes with details of any post-fledging survival and mortality statistics (breeding success is covered in the Breeding section), and the age of the oldest known wild ringed bird.

MOVEMENTS

This section deals essentially with migration patterns and seasonality of movement, and largely ignores the still-speculative subject of orientation; it is thus concerned with 'where' and 'when' rather than 'how'. For discussions of orientation theory, see Griffin (1965), Matthews (1968), Evans (1966b, 1968), and Baker (1982, 1984).

Each species account begins with a brief statement of whether the species is *migratory* (all or most individuals make regular seasonal movements between breeding and wintering ranges), *partially migratory* (populations contain substantial migratory and non-migratory elements), *dispersive* (makes more or less random movements, lacking direction bias other than topographical limitations), *resident* (basically non-migratory though some individuals may move long distances), or *sedentary* (individuals not normally moving more than 50 km).

In specialized habitats, there may also be *nomadism* (irregular movements by desert and semi-desert species in response to rainfall or drought) or *vertical displacement* (altitudinal movements by mountain breeders which descend to foothills or valleys in winter). In the Palearctic, migration seasons are governed by temperature and day length and their effects on food supply, but in the Afrotropics, from where several species straggle north into the west Palearctic, migration seasons are less well defined and are for some species governed to a large extent by seasonality of rainfall; such species are *rains migrants*. In the interval between the end of a breeding season and the onset of autumn migration, birds may make random movements, sometimes over long distances, the short duration distinguishing them from normal dispersal as defined above; these movements are termed *post-fledging dispersal* in the case of juveniles, and *post-breeding dispersal* in the case of adults. True migrations involve *emigration* (departures) and *immigration* (arrivals), and in Europe these movements tend to be basically from north to south in autumn and the reverse in spring (with longitudinal displacements also); some species show evidence of *loop migration*, which is a spring return by a different route from that used in autumn. Though each migratory population tends to have a *standard direction* (the average direction of movement between breeding and wintering areas), there may be deflections along *leading lines* (visible and roughly linear topographical features followed on occasion by migrants), even causing *retromigration* (movements temporarily in an inappropriate direction for the season). The latter should not be confused with *reversed migration*, which is general and sustained movement in a seasonally inappropriate direction. Migrants may also be deflected by *drift* (displacement from normal route by cross-winds), leading to *re-determined migration* (movements of drifted migrants back towards the normal route or quarters); related terms are *one-direction navigation* (flights on a constant heading without compensation for drift or artificial displacement), and *goal-orientation* (flights that do compensate for displacement, however caused, so that birds arrive in their normal quarters).

Special types of movement believed to apply only to limited groups of species are: *abmigration*—a spring migration by a bird that had wintered in its natal area, best known among ducks (Anatinae) where due to early pairing; *aberrant migration*—a winter visitor migrating in spring to a breeding area other than that from which it came the previous autumn, also best demonstrated by ducks; and *eruptions*—irregular (not annual) departures from the breeding range, sometimes *en masse* (called *irruptions* in the areas to which the birds go).

Generally, for each species, only those populations which spend some part of their annual cycle in the west Palearctic are dealt with in detail; populations which are the origins of stragglers to our area are treated less fully, while extralimital populations are mentioned only briefly or ignored—although all breeding and wintering areas are shown on the distribution maps.

The subject of the timing of movements is only briefly covered. In such a huge area there are, inevitably, major variations and it would take much more space than is available to give comprehensive coverage. Only broad outlines are given, and readers with particular interest in this aspect should consult national avifaunas. Most species have an ANNUAL CYCLE DIAGRAM (normally at the head of the Breeding section) which summarizes the timing of movements of birds from a specific region of the west Palearctic, always the same region as that named under Breeding.

Brief information on specifically migratory behaviour (e.g. whether movements are diurnal or nocturnal) is given if available, but other aspects (e.g. degree of flocking, territoriality on migration and in winter, year-to-year fidelity to breeding and wintering areas) are normally covered under Social Pattern and Behaviour.

FOOD

For many species, useful information on foods and feeding behaviour is sparse and scattered, and may be difficult to evaluate. There are, for instance, numerous observations of feeding behaviour, though they are rarely related statistically to changes in diet, availability of foods, locality or season. Anecdotal observations, particularly of unusual or unidentified foods, are widespread, but their importance can seldom be assessed. Quantitative data based on analyses of the contents of the digestive tract are obviously more reliable as guides to what the normal diet may be. However, the methods of assessing the contents vary considerably, and have been well reviewed by Hartley (1948) and Bartonek (1968). The importance of each food category can be expressed in a number of ways: as a percentage of the total number of food items analysed, as a percentage of the total volume, as a percentage of the total weight (wet or dry), or as a percentage of the total number of individuals in which the food category occurred; some (particularly Soviet) workers use more devious analyses combining two or more of these basic methods. Each method has inherent faults and virtues, which vary with species and composition of diet. Where the results of studies are quoted in the text, the method of analysis used is given wherever possible; if it is not given this will normally be because the information was omitted from the published source. Much work has been based mainly on the contents of the gizzard, though this may reflect to a large extent the digestibility of different food items (Dillon 1959; Bartonek and Hickey 1969; Swanson and Bartonek 1970; Bengtson 1971): there is commonly an over-emphasis of the importance of items which take longer to digest, e.g. seeds. The contents of the oesophagus are thought to provide a more accurate picture of the diet (Sugden 1973; Bengtson 1975). The data may also be influenced by the time which elapses between death and removal of the contents of the digestive tract, owing to post-mortem digestion and bacterial action (Koersveld 1951; Dillery 1965). Studies based on an analysis of birds' regurgitated pellets are also prone to bias due to differential digestion of different classes of food—even skulls of small mammals can be completely digested, with susceptibility to this varying between prey species (e.g. Short and Drew 1962, Raczyński and Ruprecht 1974, Clark 1975, Hardy 1977, Lowe 1980).

Information on the variations in diet with changes in season, sex, and age are available for few species and is often based on an inadequate and/or biased samples. In many studies, practical rather than statistical requirements dictate sample size. It is equally rare for quantitative correlation to be shown between changes in diet and changes in locality, or consideration given to the main ecological factors which may influence the availability and abundance of preferred foods. Few attempts have been made to evaluate daily intake, or the nutritional and calorific values of food items. Ideally, the relative importance of different foods should be expressed in terms of metabolizable energy contributed, but these figures are rarely available. Only a few comparative studies have been made on species living in the same area, particularly in relation to availability of foods and possible competition between species, especially closely related species (e.g. Olney 1965, Goszczyński 1981).

The main emphasis has been given to data from the west Palearctic, and extralimital data are given only when of special interest or when west Palearctic information is sparse. A brief statement of the main foods taken is followed by a description of feeding methods, which where possible are related to the type of food being taken and to any significant correlation with anatomical and physiological adaptations. For most species, a list of major food items recorded is then given with, when available, their size range. This is followed by summaries and (where possible) a comparative assessment of quantitative studies. If available, an estimate of daily food intake is given. The final paragraph deals with the food of the young.

It has not been practicable to standardize fully the common and scientific names of food items, and in most cases the scientific names are as given in the original publication concerned. However, the following are useful for reference.

Plants
TUTIN, T G *et al.* (1964–80) *Flora Europaea*. Cambridge.
KOMAROV, V L *et al.* (1934–64) *Flora USSR*. Leningrad and Moscow.
CZEREPANOV, S K (1973) *Flora USSR: Supplements and Corrections*. Moscow.

Vertebrates
ELLERMAN, J R and MORRISON-SCOTT, T C S (1966) *Checklist of Palearctic and Indian Mammals 1758–1946*. London.
MERTENS, R and WERMUTH, H (1966) *Die Amphibien und Reptilien Europas*. Frankfurt-am-Main.
WHEELER, A (1969) *Fishes of the British Isles and Northwest Europe*. London.
BAILEY, R M *et al.* (1960) A list of common and scientific names for fishes from the United States and Canada. *Spec. Publ. Am. Fish. Soc.* **2**.

Invertebrates
ILLIES, J (ed) (1967) *Limnofauna Europaea*. Stuttgart.
NORDSIECK, F (1968) *Die europäischen Meeres-*

Gehäuseschnecken (Prosobranchia) vom Eismeer bis Kapverden und Mittelmeer. Stuttgart.
NORDSIECK, F (1968) *Die europäischen Meeres-(Bivalvia), vom Eismeer bis Kapverden, Mittelmeer und Schwarzes Meer*. Stuttgart.

SOCIAL PATTERN AND BEHAVIOUR

This section is divided into two parts: (1) social pattern (aspects of society) and (2) social behaviour (mutual relationships, including interaction and communication, between two or more individuals), the term *social* being used here in a restricted sense and not as the equivalent to words such as 'gregarious' or 'sociable'. As before, the accounts have been prepared when possible in consultation with ornithologists having specialist knowledge of the species concerned (in many cases, they have also provided information on vocalizations which have been used in this section and that on Voice). The treatment of the Passeriformes, in this and the two following volumes, is essentially the same as that in the previous four dealing with the non-passerines and outlined in the Introductions to Volume I (1977) and Volume III (1983). The aim in part 1 is to characterize the basic, evolutionarily stable social system (or organization) and dispersion pattern, principally of adults but also of young birds and family groups. For general reviews of this topic, see for example Lack (1954a, 1968), Crook (1965, 1970), Selander (1965), Orians (1969, 1971), Kunkel (1974), Wilson (1975a), Emlen and Oring (1977), and Krebs and Davies (1981). The term *dispersion* is used in the special sense of the 'state of spatial separation' between individuals or larger assemblages of birds within their habitat; it should not be confused with 'dispersal' (see Movements). Dispersion is of three types: (1) *random*, (2) *aggregated* (individuals, pairs, or family groups less dispersed than they would be if distributed at random), and (3) *over-dispersed* (more dispersed and more evenly distributed than if at random); see Salt and Hollick (1946); and Hinde (1956). Although treatment is concerned primarily with the intraspecific social structure, any significant relationships between different species are noted (flocking associations, interspecific territoriality). The actual descriptions of the social behaviour patterns involved, however, are given in part 2, the aim there being to cover, in the situations in which they occur, the major aspects of the interactions between conspecific individuals (flock members, mates, parents and offspring, siblings, etc.). Though much initial analysis has been undertaken during the selection of information and its allocation to given categories within the text, the treatment in both parts 1 and 2 is largely factual, with the avoidance of detailed interpretative comments (e.g. on topics such as motivation). Being particularly interested, however, in the influence of ecological factors on the evolution of social pattern and behaviour, we hope to provide source material for the better understanding of the behavioural ecology of each species, bearing in mind that various, apparently disparate elements of a bird's biology—involving (e.g.) habitat selection, population dynamics, movements, feeding, and breeding as well as its social pattern and social behaviour—are often inter-related. Such linked adaptations evolved through the process of natural selection acting on the individual enable it to survive and reproduce in the face of the exigencies and vicissitudes of the environment.

1 (Social pattern). This paragraph starts with an outline of the kind of social pattern (association and spatial dispersion) typical *outside the breeding season*, plus any variation due to environmental factors, such as weather and food supply, and any differences in social pattern between birds on migration and in their winter quarters. For species that are gregarious outside the breeding season, information is given on flock type, size, composition (age and sex structure), and regular association with other species (e.g. in mixed flocks). For species that are territorial outside the breeding season, the information includes type and size of territory, etc. (see further, below). Treatment then usually proceeds under the following standard sub-headings: BONDS, BREEDING DISPERSION, and ROOSTING.

BONDS. This sub-section starts with a classification of the *mating system* characteristic of most members of the species, the term being used to cover both the type of heterosexual association (including pair-bond, if any) and the respective participation of the associating birds in parental care. Information is also given when available on sex-ratios and the age, for each sex, of first pairing or breeding. Basically, mating systems are of two types: (1) monogamous; (2) polygamous, i.e. either polygynous or polyandrous (see below). (The term 'polygamy' is also sometimes used to indicate the state of having three or more mates, i.e. as an extension of bigamy, but this definition is far too restrictive for a term much better employed as a generic name of the states of polygyny and polyandry.) As the passerines all have altricial and nidicolous young requiring complex parental care, monogamy is far more common than polygamy in this order while a *rapid multi-clutch breeding strategy* (in which a female lays two or more clutches for simultaneous incubation by herself and her mate(s)—see Introduction to Volume III) is not known to occur in any species (monogamous or polygamous). In a *monogamous mating system*, one male and one female have a breeding association on a seasonal or long-term (even life-long) basis, usually with joint care of the young; such a system typically involves a strong pair-bond and leads to the production of a single brood or of two or more broods in sequence. In a *polygynous mating system*, one male has a breeding association with two or more females either simultaneously or successively during one breeding season. In the case of successive (or serial, or sequential) polygyny, the male may assist one or more of his mates—

especially if the system is based on his ownership of a large breeding territory (*resource-defence polygyny*). In some polygynous passerines, a male and a group of females associate together (*harem* or *mate-defence polygyny*); in others, the males gather at traditional display areas and are visited there by a number of females solely for the purpose of copulation (*lek* or *male-dominance polygyny*). In the last type of polygyny, there is no real bond between the associating males and females (i.e. the relationship is essentially promiscuous); in the others, however, the bond between the polygamous male and each female may be more definite and is sometimes strong. (It should be noted here that mating systems and pair-bonds are usually classified only from the point of view of the sex, typically the male, that initiates or plays the major role in the sequence of events leading to breeding; in the case of harem polygyny for instance, though the male is polygamous, each of his females maintains a monogamous pair-bond with him.) In a *polyandrous mating system*, one female has a breeding association with two or more males either simultaneously or successively during one breeding season. Uncommon in birds generally, polyandry is poorly documented in passerines and may not, in fact, be nearly as rare as once believed—though probably still existing mainly as the opportunistic (facultative) mating strategy of some individuals in species that are normally monogamous. Similarly, birds of species that are normally monogamous may engage in multiple relationships with members of the opposite sex other than the mate, as either an alternative or opportunistic strategy, such associations involving (e.g.) promiscuous matings or true bigamy. Bigamy perhaps arises most often from local variations in the sex-ratio or as a result of variation in mate quality, but promiscuous matings are probably a fairly widespread phenomenon among passerines, as suggested by the prevalence of mate-guarding. In the case of passerine species which form definite pair-bonds (the majority), the type of social organization within the pair is outlined if known—e.g. whether the two birds establish a rank-order (with one sex taking the dominant role in heterosexual situations and the other a subservient one). Finally, a summary of the roles of the sexes in parental care is given, together with information on post-fledging care and any subsequent associations, and on any social systems within the family group affecting parent-young or inter-sibling relationships. The latter include *brood-division*, i.e. the formation of two family sub-groups with each parent feeding only certain young (Simmons 1974). Though reliably documented for relatively few species, brood-division is probably not uncommon in passerines; for recent reviews, see (e.g.) Moreno (1984*a*), Harper (1985*b*), and McLaughlin and Montgomerie (1985). The phenomenon of *communal* or *co-operative breeding* is also covered under Bonds, with details of the number of 'helpers', their status and kinship to the breeders, and their role in the care of the young, etc. A few species with such a breeding system were treated in Volume III (certain *Charadrius* plovers) and Volume IV (Pied Kingfisher *Ceryle rudis* and bee-eaters *Merops*), though such cases were not fully typical. More common in passerines, true communal breeding has been reported from over 30 families or subfamilies—though mostly in tropical regions and Australia, being rare in Europe and North America (Brown 1978). The accentors (Prunellidae) provide an example in the present volume.

BREEDING DISPERSION. This sub-section starts with a statement on the main type of dispersion system characteristic of the species during the breeding season, i.e. whether *solitary* (over-dispersed) or *colonial* (gregarious), whether *territorial* or *non-territorial*—any variation in pattern due to (e.g.) variations in local features of the habitat (including food supply) also being indicated. It should be noted that both solitary- and colonial-nesting species may be territorial, these categories not being mutually exclusive as is sometimes supposed. Though details of sites and nests will be found in the Breeding section, information is given here on nest-spacing and breeding densities: in colonial nesters, the dispersion of the nests (loose or dense) within each colony and also the dispersion of different colonies in relation to one another; in solitary nesters, the distances between neighbouring nests and between home-ranges or territories. A *home-range* is an undefended area, typically ill-defined and often overlapping with the home-ranges of conspecific birds, occupied by an individual or larger social unit for part or all of the annual cycle; it may contain a territory within it, usually a small one (see types 3–8, below). A *territory* is a defended area, typically well-defined and discrete, occupied by a pair, family, or individual for all or part of the breeding cycle, or all year round in some species, and maintained—typically to a much more elaborate extent than is a home-range—by social conventions. (N.B. The distinction between territories and home-ranges is not always made in the literature, or possible to make in cases when the area is isolated: if the area frequented is not certainly defended, it is best termed a home-range, not a territory; if most or all of the area is defended, it is better termed a territory not a home-range.) Both territories and home-ranges tend to be aggregated by restriction of suitable habitat. Where the territories of non-colonial species are clumped, and especially where they do not occupy all the available habitat locally, the collection (mosaic) of territories is termed a *neighbourhood group*, though in some studies this usage is restricted to special cases, e.g. to individuals sharing the same song dialect (Village Indigobirds *Vidua chalybeata*: Payne and Payne 1977). In territorial species (solitary or colonial), the type of territory held is indicated together with the various key activities known to be conducted, more or less exclusively, within its boundaries (e.g. pair-formation, courtship, copulation, nesting, self-feeding, feeding of

young); also whether separate territories are held at the same or different stages in the breeding cycle (e.g. feeding territories, pairing territories, brood territories). The following are among the various types of territory held by passerines when breeding (the young being reared there, at least to fledging, unless otherwise indicated). (1) *All-purpose territory* ('large' or 'Type-A' territory): a sizeable area, usually well-defined, used for feeding, pairing, courtship, mating, nesting, and the rearing of the young; typical of many solitary nesting passerines, and especially of communal breeders. (2) *Breeding-only territory* ('Type-B' territory): a fairly large area, usually well-defined, used for all or most reproductive activities but not for feeding to any extent (i.e. the adults obtain food, both for themselves and for the young, wholly or mainly elsewhere); again, not uncommon in passerines that nest solitarily. (3) *Nest-area territory*: a small area round the nest not used regularly for feeding; found in both solitary and colonial species. (4) *Nest-site territory*: a small area confined to the nest itself; most typical of colonial species. (5) *Pairing territory*: a small area used for pairing, i.e. obtaining a mate; often an 'individual-territory' (see below) until pair-formation occurs, when it may be abandoned unless used also for mating. (6) *Mating territory*: a small area used for mating (i.e. copulating)—may also be used for pairing. (7) *Brood-territory*: an area of variable size used only for the rearing of the young, i.e. after fledging. (8) *Feeding territory*: an area of variable size, separate from any breeding territory, used mainly or solely for feeding. (N.B. The term 'breeding territory' is also used as a generalized term for types 1–4 when precise information is not available, 'individual-territory' for an area occupied by a single bird (e.g. for self-advertisement prior to pairing or for feeding—see 5 and 8 above), and 'permanent territory' for one held throughout the year (i.e. for breeding and wintering); individual-territories may also be held in winter, usually for feeding, the term 'winter territory' is often used in the literature to cover all types of territory held outside the breeding season.) Where possible, the actual size of the territory is given—with data in square metres (m^2), hectares (1 ha = 10 000 m^2), or square kilometres (1 km^2 = 100 ha); also, the approximate size of home-ranges.

ROOSTING. Social aspects of *roosting* (sleeping) and *loafing* (resting at times other than the main period of sleeping) are the primary subjects here (loafing, however, mainly for gregarious species). Roosting/loafing patterns during and outside the breeding season are treated separately and classified as (1) non-gregarious (solitary or in pairs or single families) or communal (in gatherings of single birds, pairs, families, single flocks, multiple flocks, etc.) and (2) nocturnal or diurnal. For breeding birds, any differences between members of the pair are noted, also the place of roosting in relation to the nest, territory, colony, etc.; for wintering birds, whether there is evidence of the roost being used as a site in a dispersal system (e.g. in relation to finding food). Information is given on typical sites, classified as sheltered (i.e. from the weather), protected (e.g. by being sited in inaccessible places), concealed, traditional, temporary, etc.; also on times of arrival and departure and on any other relevant aspects (including, for convenience, resting and sleeping postures, and social behaviour at the roost). Finally, some basic information is given on comfort behaviour, if available.

2 (Social behaviour). This paragraph usually starts with mention of behaviour or other characteristics of general significance in the biology of the species, including any features of its plumage pattern or feather structures that play a wide role in display. In some cases, comments will be made on any features of the species' behaviour that facilitate or make difficult the study of its behaviour and any that pose problems of interpretation. Note is made too of how wild or tame the birds are, approachable or otherwise, in relation to their contact with man. Also given here are descriptions of the basic forms of anti-predator avoidance involved in individual survival, both active (fleeing, etc.) and passive (concealment, etc.); this includes information on interactions with predators and other intruding animals (including man) other than that covered later under Parental Anti-predator Strategies. In the treatment of mobbing behaviour, here and elsewhere, special attention will be given to the mobbing of owls (Strigiformes) as this is a particularly striking feature of the behaviour of many passerines. Treatment then usually proceeds under the following standard headings: FLOCK BEHAVIOUR, SONG-DISPLAY, ANTAGONISTIC BEHAVIOUR, HETEROSEXUAL BEHAVIOUR, RELATIONS WITHIN FAMILY GROUP, ANTI-PREDATOR RESPONSES OF YOUNG, and PARENTAL ANTI-PREDATOR STRATEGIES. Particular attention is given to *displays* (species-characteristic behaviour patterns functioning in communication as social signals), both visual and vocal—though the actual descriptions of calls and other auditory signals are given separately in the Voice section. Visual displays consist of postures and movements derived largely from more basic, everyday, non-signal behaviour (e.g. feeding actions, locomotory intention-movements, thermoregulatory plumage postures, comfort behaviour, etc.) and have typically been modified (ritualized) to serve their new signal function, so much so at times that their origins may be obscured; in some cases, however, the displays still so closely resemble their precursors that they can be identified as signals only by their regular and formal occurrence in certain social situations or by comparison with the homologous but more highly ritualized displays of closely related species. Displays may be wholly visual or also have vocal components, or they may be more or less wholly vocal (the latter sometimes being distinguished as *vocalizations* in the literature). We have aimed for clarity by giving distinctive names (non-interpretative ones as far as

possible) to all the most important displays of each species, using capital initial letters (note, however, that small initial letters are used for categories—thus, the same name may be used in some species as the name of a particular display, e.g. Song-display, and in others as the name of a category of displays, e.g. song-displays). That the same terminology has also been used for the similar and homologous displays of less closely related taxa does not necessarily imply homology (though it should be remembered that we are dealing with but a single order of birds in these last three volumes). As before, our approach is basically ethological (concerned with the objective, biological study of behaviour), but we have mostly avoided the use of the more technical language of that and related disciplines.

FLOCK BEHAVIOUR. Here we are concerned with behaviour promoting individual survival shown by birds as members of a group, largely outside the breeding season but covering flocking at the very beginning and end of it and flocking by non-breeders and (e.g.) feeding birds during it. Note that flock sizes (etc.) will have been treated in part 1 of Social Pattern and Behaviour and that details of any heterosexual behaviour shown in the flock, including pair-formation, will be under Heterosexual Behaviour; descriptions of aggressive encounters, threat displays, etc., may be given here (especially if most characteristic of flocking birds) or under Antagonistic Behaviour. Among topics covered are: the extent of flocking (feeding, loafing, etc.), flock density in various situations, disputes between members of the same flock and the reasons for it, any other interactions related to the social organization of the flock (including information on individual-distance, hierarchies, etc.—see below), the extent and audibility of contact-calling (both on the ground and in flight), pre-flight and post-flight signals, flock reactions to predators (including mobbing), flight manoeuvres (in response to predators or not), and relationships with other species in mixed flocks. The term *individual-distance* is used for that small, exclusive area of space that each bird of certain species maintains around itself when associating with others (e.g. in a flock or roost). This area has no permanent topographical restriction (as has a territory) but moves with the bird; in some cases, spacing may be maintained round a pair or family group. Such spacing is achieved by the showing of overt aggression towards other individuals or by avoidance, often mutual. In a minority of passerines, the opposite condition also occurs—individuals associating in close body contact at times (*clumping*). Flocking birds may also be organized on the basis of *hierarchy*, i.e. by the establishment of a social rank-order (so-called 'peck-order'), initially through combat and then through the mutual acceptance of status on the basis of individual recognition; such a system, however, is usually possible only in permanent flocks of small or relatively small size. (N.B. As the concept of hierarchy has been much abused in studies of animal behaviour, especially those based on captive birds in conditions of artificial crowding, its application to natural populations requires caution.)

SONG-DISPLAY. As vocal display in the form of singing plays such an important role in the biology of passerines, and is such a distinctive feature of the social behaviour of the males in so many species, we have decided to emphasize this by creating a new standard heading. The behaviour involved has been termed *song-display* whether or not the singing is accompanied by posturing. Both *territorial-song* (advertising song given, e.g., to attract the attention of potential rivals and/or mates) and *courtship-song* (song directed specifically to the individual being courted) may be treated, also details of when, how, and where the bird sings and the relationship of song to the breeding cycle, territory, etc.—though the actual descriptions of the sounds themselves will be given under Voice. In many species, song-display consists entirely of the vocal component, though the use of special song-posts, and the structuring of the singing into song-bouts, for example, are also part of the ritualization of song as display; in others, song has important postural components. Among the most dramatic performances, aerial *song-flights* may be mentioned as especially characteristic of many passerines. Singing by the female and any antiphonal singing between mates (or other forms of duetting) are also covered here, but only if the behaviour is homologous with song-display by the male of the species. Further aspects of song-display behaviour are picked up under later headings, as necessary.

ANTAGONISTIC BEHAVIOUR. This deals with the whole complex of aggressive (hostile) and escape responses, and their intermediates, that occur during interactions between 'rivals'—i.e. conspecific birds (usually but not invariably of the same sex) that have no mutual bonds—involved in (e.g.) disputes over food, mates, or the establishment and maintenance of territory, social hierarchies, etc. The term *antagonistic behaviour* is here applied to such behaviour; it was coined to replace the ethological term 'agonistic behaviour', a motivational one of wider application indicating the presence of attack/escape tendencies ('drives') in the interactions, not only of rivals but also of birds of the opposite sex that have formed a bond or are in the process of establishing one (see separately under Heterosexual Behaviour, below). The term *fighting* is sometimes used in the literature to cover all overtly aggressive aspects of antagonistic behaviour, but is employed here in its more usual sense to indicate actual physical hostile contact, i.e. mutual fighting ('combat'). As well as (1) overt attack and fighting (including all significant components from hostile approach to full combat) and (2) overt escape behaviour (fleeing, withdrawal), the forms of antagonistic behaviour include (3) displays of various types—e.g. *threat display* (signalling the likelihood of attack in order to induce withdrawal by the opponent), *advertising display* (drawing the attention of rivals to the displaying bird, often an occupant of a territory), *appeasement display* (signalling subservience in order to forestall or inhibit attack), and *defensive-threat display* (signalling subservience but also the possibility of attack on provocation). Some of these displays (e.g. appeasing, advertising) appear, in the same or closely similar form, in both antagonistic and heterosexual situations; in such cases, the main description of the behaviour is usually given here. The same applies to those special threat-cum-advertisement displays (other than song-display) that have the dual function of repelling rivals and attracting mates; termed 'Imponiergehaben' in German, such behaviour has no established name in English—though *self-assertive display* is perhaps the most suitable. The topics of antagonistic behaviour are usually treated in the following sequence (with or without the use of informal, numbered headings): (1) general features—the nature of aggression in the species in question, situations in which advertisement and other displays occur (establishment and maintenance of territory, obtaining a mate, etc.), roles of the sexes, seasonal chronology, etc.; (2) threat; (3) fighting. Any important interspecific aspects are also noted—especially social contact with closely related species.

HETEROSEXUAL BEHAVIOUR. This deals with those many interactions between birds of the opposite sex that are involved in the formation and maintenance of pair-bonds, copulation, etc. The term *heterosexual behaviour* covers a much wider field than what is usually implied by 'sexual behaviour' (mating and other 'friendly' interactions) and includes those more disruptive elements arising from hostility and fear that often occur (e.g.) during pair-formation and the establishment of any dominance relationship between the sexes, especially early in the breeding season. Though there can be considerable overlap between categories, the behaviour includes (e.g.) courtship displays and other pair-bonding behaviour between members of a pair (chasing, courtship-feeding, allopreening, etc.), mating displays, appeasement displays, and nest-showing displays. The term *courtship* is used broadly to cover that species-characteristic behaviour involved in both the establishment and maintenance of the pair-bond and which facilitates successful *mating* (copulation), directly or indirectly (e.g. by stimulating the physiological processes leading to ovulation). Courtship may take different forms before pair-formation and after it (during the 'engagement period', especially if this is prolonged); it may be *mutual*—with the sexes performing similar or different roles, simultaneously or reciprocally (sometimes in the form of elaborate *ceremonies*)—or, more typically in passerines (even in those species that are alike in plumage characters), *unilateral* (usually male performing to female). If performed by a number of birds that have gathered expressly to display, courtship is further characterized as *communal*; such lek behaviour, however, is rare among passerines (especially outside the tropics), most species engaging in courtship as single pairs. *Chasing* (in which the male pursues the female in flight, often persistently and rapidly) is a particularly characteristic form of passerine heterosexual behaviour, though its exact nature (sexual or aggressive) is not always clear; male passerines will also accompany the female persistently at other times, probably mainly in order to prevent her copulating with another male (*mate-guarding*). In *courtship-feeding*, which may continue after egg-laying when only the female incubates, the male gives food to his mate (*food-presentation*); the exchange may be preceded and accompanied by special calls and displays, especially on the part of the female—who often begs in the manner of a young bird (*food-solicitation*)—but in some species the behaviour is now so highly ritualized that food is seldom or never actually passed. Though courtship-feeding, especially in those species that practise it in a simple (unritualized) manner, may often be directly functional (i.e. provide the female with extra food during the critical periods of egg-formation and incubation), it would appear to have a wider significance, as part of pair-bonding, in all species in which it occurs. The term *allopreening* is now given to that behaviour which used often to be called 'mutual preening'; during such preening, both members of the pair preen one another at the same time or take it in turns, or the preening is confined to one sex. The behaviour may be ritualized and involve special soliciting postures; in some species, it is now so highly ritualized that no contact between bill and feathers is made. Like courtship-feeding, allopreening is clearly bond-maintaining as well as directly functional at times (i.e. helping in feather care). The same applies to mating behaviour and any displays associated with it; in some species, for example, mating occurs well in advance of ovulation and egg-laying as part of pair-courtship. As well as distinct pre-copulatory displays (including soliciting by the female), some species show distinctive post-copulatory behaviour (in the form of display or, e.g., fleeing). The *appeasement-displays* shown by one sex to the other may be of the same type shown during antagonistic encounters (and will have already have been described under the appropriate heading) or may be specific to heterosexual situations; included in this category is any conciliatory behaviour shown when pair-members meet after separation—in the form of *greeting-displays* or more elaborate *meeting-ceremonies*, which may contain at least some elements of the soliciting behaviour shown in other types of heterosexual behaviour (see above). *Advertising-display* (see also Antagonistic Behaviour, above) involves behaviour, including song-display (treated in detail separately), whereby a lone bird (often a territory-holder) attracts members of the opposite sex to it. Some species of passerine, especially hole-nesters, also practise another type of advertisement in the form of *nest-* (or *site-*) *showing display*, thereby attracting the mate or a potential mate

to a likely site or showing it the position of the nest. The topics of heterosexual behaviour are usually treated in the following sequence for each species (with or without the use of informal sub-sections with numbered headings): (1) general features (including sex-recognition, the timing and progress of pair-formation in relation to migration, winter flocking, etc.; (2) pair-bonding behaviour (courtship, etc.), (3) courtship-feeding (for convenience, treated separately from other pair-bonding behaviour); (4) mating; (5) behaviour at the nest (including nest-site selection, nest-showing display, behaviour at nest-relief, etc.). In some species, nest-site selection is so elaborate that the topic is treated either separately on its own or, if associated closely with pair-bonding behaviour (e.g. when the female chooses a mate by selecting the site he is advertising), under that heading.

RELATIONS WITHIN FAMILY GROUP. This deals with communication and other social interactions between parents and their young (brooding, feeding, begging, etc.) and between siblings (disputes over food, etc.). Information is given on nest-sanitation (removal of faecal sacs, etc.) and its development during the nestling period. Behavioural aspects of the development of the young are also described, including changes in begging responses, the fledging process, and the development of self-maintenance.

ANTI-PREDATOR RESPONSES OF YOUNG. This deals with the reactive measures taken by dependent young birds (both nestlings and fledglings) in the presence of predators or other intruders (including man) that threaten their safety—either in direct reaction to danger or indirectly in response to parental warning calls, etc. Such behaviour is either passive (concealment, etc.) or active (avoidance, threat, etc.). Special attention is given to behaviour in the nest, including premature fledging.

PARENTAL ANTI-PREDATOR STRATEGIES. This deals with the measures taken by breeding birds to protect their eggs and young against predators, as opposed to those involved in individual survival—though the two often interact. As with other groups, the *parental anti-predator strategies* of the passerines consist of both *passive* and *active* measures, our classification being adapted from that devised for the waders (Charadrii) in Volume III (see also Simmons 1985c). Passive measures are of two main types: (1) precautionary and (2) reactive. Passive *precautionary measures* include a wide variety of adaptations (related, e.g., to the siting, concealment, and protection of the nest) and behaviour (involving the cautious routines of parent birds in its vicinity, incubation by one sex, etc.) which reduce the likelihood of the nest, eggs, and young even being located by a predator; however, because of limitations of space, these are not treated in detail—though some relevant information can be obtained from the Breeding section. Passive *reactive measures* include the responses shown directly to the predator, but mainly before it has detected the existence or location of the nest, eggs, or young, e.g.: the giving of warning and other calls (alerting the mate or young) and the performance of other alarm signals; adoption of concealing postures on the nest; furtive departure from the nest; concealment of eggs or young before departure; and inconspicuous or concealing behaviour by the adults in the vicinity of the nest. Active measures are also of two main types: (1) distraction behaviour and (2) expelling attack. The aim of *distraction behaviour* is to deflect the predator or divert its attention from the eggs or young by means of (a) stratagems, (b) distraction-threat display, (c) demonstration, and (d) diverting attack. Some authors have used the term 'distraction-display' for what is here called 'distraction behaviour' but we use 'distraction-display' as a generic name for distraction behaviour of two types (see also Simmons 1985c): distraction-lure display (a stratagem) and distraction-threat display; these two types are often confused in the literature, and do intergrade in some cases, but the functional distinction is clear—the first aims to induce the predator to approach the performing bird, the second to induce it to withdraw. Similarly, attack takes two main forms: diverting attack (a form of distraction behaviour) and expelling attack; the distinction between the two is not absolute, however, the form and intensity of the attack much depending— as in the case of other reactions—on the kind of predator and its activity, the degree of danger to the eggs or young, and the 'cost' to the parent in terms of parental investment and of its own survival (see also below). *Stratagems* are 'ruses' or 'tricks' intended to deceive the predator. Often accompanied by calls, they include distraction-flight but, in passerines, mainly take the form of *distraction-lure display* in which the parent bird presents stimuli to the predator that elicit and direct its prey-seizing responses. Such lure-display is of two main types: (1) the *disablement type* ('injury-feigning', etc.), in which the performing bird appears to be incapacitated in some way, and (2) the *small-mammal type*, in which it looks and behaves like a small rodent ('rodent-run', etc.). Except in certain ground-nesting species, such behaviour is relatively uncommon or poorly developed in passerines, though it has been recorded even in some species that have concealed nests well off the ground, the sitting bird performing amongst the foliage or dropping to the ground and performing there (Skutch 1976, which see for other stratagems practised by certain passerines). *Distraction-threat display* may also be shown to certain (mainly small) predators by breeding passerines, both at the nest (in the form of *nest-protecting display*) and away from it, as the bird confronts the predator in order to alarm it; though most such displays have important visual components, some hole-nesting species for example produce frightening sounds out of sight from within the cavity. *Demonstration* consists of loud and persistent calling from perch or ground or in the air, the demonstrating bird sometimes also making itself conspicuous

while doing so and approaching the predator in a hostile manner, even in the form of a *mock attack-flight* (i.e. without actually striking it). Some species perform *song-flight*, of the same type used during intraspecific advertisement, when predators approach the nest or young; this may be a further form of aerial demonstration, or possibly act as a form of distraction-flight. *Diverting attack* (the last type of distraction behaviour considered here) involves inhibited ('mock') attack on the predator serving to deflect it—occurring (e.g.) at the climax of aerial demonstration or from nest-protecting display (when the threatening bird may flutter up into the face of the predator). *Expelling attack*, on the other hand, is uninhibited and aims to drive the predator right away by force; as with diverting attack, the behaviour may take the form of aerial *dive-attack* (so-called 'dive-bombing'). The motivational, evolutionary, social, and environmental factors that determine the optimum strategy, sequence of strategies, or changes in tactics shown at any stage of the cycle or in any situation are highly variable, so much so that parental anti-predator behaviour (and especially distraction-lure display) has earned itself a reputation for unpredictability, even capriciousness. However, in the majority of passerines, the most intense reactions appear to occur when the young (a) leave or are close to leaving the nest (naturally or prematurely) or (b) are discovered or seized by a predator (see also above). The parental anti-predator strategies of each species are treated in the text in the following sequence, usually with the aid of informal, numbered headings: (1) passive measures; (2) active measures: against birds; (3) active measures: against man; (4) active measures: against other animals. Any difference between the sexes in the type or intensity of the behaviour shown is noted.

Where possible in part 2, text-figures are included for each species, showing characteristic postures and displays. The sources for these figures are given at the end of the section.   KELS

VOICE

Exceptionally complex and difficult problems are met in attempting to develop a scientific method of describing and interpreting the vocal utterances of birds. Primarily, bird vocalizations are conditioned by anatomical variations in the syrinx, the functional mechanism of which is still imperfectly understood. Certain species have evolved an elaborate communication system of distinct calls, each having a specific signal function within an observable situation. Others use voice much less frequently, and often on a more primitive level. The task of description and interpretation is complicated not only by the wide scope for variation offered by the parameters of sound, pitch, loudness, duration, and tone quality but also by differences linked to sex, age, emotional state, individuality, and region. Voice accordingly must be considered in the light of Social Pattern and Behaviour.

Songs are vocal displays. The human eye, when well trained, is capable of resolving and fixing much of the visual information afforded by birds and their activities, except the fastest. The human ear is much more limited and fallible. It is also much harder to train in promptly resolving and accurately recalling bird sounds. Birds themselves can resolve bird sounds at least ten times faster than man. Moreover, at the highest frequencies used by birds, the hearing of most observers, although reaching similar levels, becomes undependable. Even within lower frequency bands, there are often inexplicable gaps or weaknesses, commonly unrecognized by the observer and not necessarily constant. Moreover, apart from musicians and linguists within their special fields, probably far fewer people have well-trained ears than well-trained eyes.

While the risks of distortion of a visual image due to weather, topography, distance, or poor light are fairly familiar, corresponding interferences with sound reception are less well known. An example is the Doppler effect which, when the sound source and the hearer are moving towards or away from each other, leads to an apparent rise or fall in the pitch of the sound. Bird utterances exist only within a moment of time, and for practical purposes only in so far as they are, at that moment, received, acted upon, memorized, or recorded by an interested listener. Few items of ornithological information are so likely to be lost or subject to error. Further confusion results from the lack of progress so far in assigning to descriptive words in ordinary use scientifically clear and distinct meanings. From the standpoint of semantics the entire range of words suitable for describing bird utterances is in all languages more or less strongly influenced by onomatopoeic echoes often carrying a misleading message.

Attempts to express sound in letters and syllables, supported by accents or other indications of emphasis, pitch, and tempo, may lead to a bewildering range of alternative renderings. Such is the fallibility of the human ear that it persists in 'hearing' birds voicing consonants which objective methods do not confirm, although it is perhaps too late to rename *Phylloscopus collybita* the 'iffaff' or, in German, the 'ilpalp'. Where an utterance happens to approximate to a human voice or a familiar instrument, as in the traditional 'cuckoo', 'coucou', or 'kuckuck', passable results are obtainable by use of musical notations. Proposals for standard methods of bird voice recording in the field, such as those of Voigt (1894, and numerous later editions, including 1933 and 1961) and Saunders (1951) (see p. 30), have unfortunately not so far led to sufficient agreement to become established, though they have at least helped to clarify the problems of accurate description. Any competent descriptions accompanied by syllabic renderings can tell us something useful about the pattern, pitch, and tone of a call.

In studying bird utterances, the hopes of many have recently become centred on sound recordings. These have great advantages. They can be stored and replayed at will. They can be analysed at convenient speeds with sophisticated equipment enabling their components and characteristics to be more clearly perceived, and they can be converted into visible sonagrams and oscillograms (see below). Much progress has recently been made in recording on tape a wide range of utterances of many species of birds in various countries. The recordists deserve praise for their dedication and skill, and no less for co-operating so amicably and helpfully in an international network for exchange and dissemination of results. The immediate effect, however, in view of the youthful state of bio-acoustic studies, is to throw up a bewildering array of new queries and unresolved problems, which can only be treated tentatively in the present state of knowledge.

Such problems, which directly impinge on the preparation of the Voice section, are aggravated by external complications. The rapid growth of bio-acoustic studies, of which research into bird vocalizations forms part, gives rise to a need for co-ordinating techniques, standards, and terminology, and for establishing new concepts and methods, going beyond the concerns of ornithology. Such needs have been taken into account only indirectly in the present work, but they will doubtless become more important in future.

REQUIREMENTS FOR ADEQUATE TREATMENT

A more serious immediate challenge is the emerging awareness in behavioural studies of the significance of vocal signals and communications, which have hitherto tended to be assigned no more than an ancillary role in interpreting many performances in visual terms. In field identification also, the potential usefulness of recognition through utterances has often been underrated. Treatment of voice as a self-contained branch of study is accordingly inadmissible. Every utterance has to be evaluated in relation to its role in behaviour, communication, and location, to its relevance and reliability in specific identification and individual recognition, to its significance for geographically isolated populations and for taxonomy, and to the anatomically specialized organs which govern it, as well as to its linkage with sex, age, physiological state, and surrounding circumstances.

Many obstacles hinder the integration of bird voice analysis with such other branches, having their own incompatible requirements, practices, and vocabulary. The task of producing a comprehensive, fully digested and standardized scientific treatment of bird voice, as well as non-vocal signals, can be broken down into stages, at each of which serious difficulties arise. They are:

1. Understanding the actual production of sounds as determined by the anatomy and physiology of the bird.

2. Variations in reception of the resulting utterance due to the position of the bird in relation to the observer or recorder, and to topography, weather, and background noise.

3. Variations in reception due to differing capabilities, age, education, and training of observers.

4. Limitations inherent in different types of recording equipment and ancillary devices such as parabolic reflectors and microphones, and in capacity to discriminate between noise and signal, which may distort or confuse material available for analysis.

5. Limitations in media and equipment for playback of recorded sound.

6. Limitations of conversion of sound into a visual image by, for example, a sonagraph, oscillograph, or use of computer techniques.

7. Use of techniques and discretion in editing a sonagram to distinguish signal from background noise.

8. Scrutiny of available descriptions, discs, and tapes and preparation of a critical list of hitherto recognized utterances.

9. Critical comparison of data from description with data from tapes and other recordings, and with relevant behavioural studies.

10. Preparation of a digested comprehensive account, with sonagraphic illustrations, after analysis and reconciliation of inconsistencies.

11. Summation of general voice characteristics of the species or group, and role of voice in its life-history, indicating points or areas needing further study.

ORGANIZATION OF THE VOICE SECTION

For each of these stages data are partly lacking and techniques are capable of improvement. To bring together all the varied types of expertise required, and to co-ordinate and process their inputs, has proved a formidable task. Some of the difficulties and points for decision should be briefly outlined. Not only is the number of observers interested in bird voice unduly small in relation to those active in other fields, but the use of their results is compromised by the continuing lack of any generally accepted and readily understood method of comparative verbal description of bird utterances, partly owing to widespread ignorance of the use of phonetics.

For many species, available data are limited to a few descriptions, based perhaps on a single individual and even on a single utterance. Comprehensive and critical studies exist as yet only for a small minority. Most authors have felt compelled by this situation to confine themselves to piecemeal assembly of odd scraps of information, too often merely copied, with variations, in one work after another. While subject to similar constraints, we have had the advantage of access to a large selection of documented utterances recorded on tape. By analysing these in comparison with a newly ordered abstract of conventional verbal descriptions and renderings in syl-labic form, a synthesis has for the first time been achieved within the west Palearctic between the experience of field observers and the evidence of recorded sound, itself almost entirely from the field.

Although no rigid formal structure has been maintained in the text, it usually begins with a brief general outline of the voice of the species, any special features, its relation to allied forms, its significance in the life-pattern, and the adequacy or otherwise of data available. The various calls are then listed, wherever appropriate separately for male, female, and young. In this and subsequent volumes, special attention is paid, both in text and sonagrams, to the highly developed nature of passerine song. According to structure and function, song may be subdivided into (e.g.) primary (territorial, advertising, or full) song, courtship-song, subsong, and winter song. Note is taken of whether just the male, or both sexes, sing. Individuals typically have a repertoire of phrases distinctive in structure and in rate and character of delivery; phrases comprise units, in turn composed of sub-units and syllables (see Glossary, p. 32). In song, as in other calls, the extent of variation within and between individuals and races is treated briefly, as data permit. Descriptions of song and calls include, objectively, the duration, frequency, amplitude, and 'timbre' or tonal quality (which depends upon distribution and relative strength of harmonics, presence or absence of transients, modulation rate in frequency and amplitude, etc.) and subjectively, time, pitch, loudness and range of audibility, and tone quality which includes character of utterance and circumstances conditioning the utterance as heard. Account is also taken of influences on and limits of the production mechanism, and of the social background and signal value of the sound pattern, detailed treatment of which occurs in Social Pattern and Behaviour, notably under the sub-heading Song-display.

Even where one or more of these aspects may be inapplicable, they need to be taken into account in evaluating the vocabulary of the species. Uniform utterances on the one hand, and variable utterances on the other need to be clearly distinguished. Attention must also be given to individual, local, and regional variations in utterances, or to the absence of particular utterances in one or other sex, or at certain seasons or ages, or in certain parts of the geographical range. Fully reliable generalized accounts of the language of a species have in most cases yet to be produced. Analysis of data shows that certain species and groups, and certain areas, have been badly neglected, and that even where primary data about voice appear fairly complete there may be shortcomings in standards of description and analysis or recording.

The attainment of 'word-consistency' in descriptions has presented severe difficulties. In face of the ambiguities, confusions, and vaguenesses often enshrined in the literature, we have found it acceptable neither to continue on the same lines, nor to attempt at a single leap to adopt a new set of rigorously scientific descriptive terms unfamiliar to readers and even to workers in this field. As in the Habitat section, however, we look forward to a time when all descriptive terminology, however loosely the words may be employed in normal usage, will be screened against clearly formulated scientific definitions, together affording a comprehensive coverage of ornithological requirements for accurate communication.

For the present we have had to be content to begin by eliminating clearly misleading or unserviceable expressions in order to build up some approximation to an acceptable vocabulary. In order to help readers, especially those for whom English is not their first language, we have tried wherever possible to include a number of synonyms or partial equivalents, which may afford fuller clues to the nature of each utterance. We have also, wherever possible, included complementary physical indications, such as duration, intervals of time and pitch and numbers and rates of repetitions, and comparisons with other sounds, as well as sonagrams, to assist in what can at best only be a groping towards communication. Often even this has been frustrated by uncertainty over the intention of the writer of an original description, and whether his chosen words accurately conveyed his intention. Where translation from one language to another has been involved the difficulties are compounded.

The greatest recent advance in the study of bird voice has been the widespread success in obtaining tape recordings, both in the wild and in captivity, for comparisons with verbal descriptions which they may for some purposes replace. The reconciliation of the two has been accorded the highest priority in preparing the present work. In planning it, much thought was given to the inclusion of reproducible sound material, for instance on disc or tape, but the disadvantages outweighed the advantages. We have therefore relied upon visual processing in the form of sonagrams.

VOICE ILLUSTRATIONS

Two forms of diagrammatic illustration are used, both derived from precise measurements made by electronic instrumentation: (1) sound spectrograms (sometimes called audio-spectrograms), usually abbreviated to 'sonagrams', and (2) melograms (used in Volumes I and II only; for full explanations see Volume I and Hjorth 1970).

Sonagrams are produced by the Kay Sona-Graph, designed and made by the Kay Elemetrics Corp., New Jersey, USA; from Volume II onwards, the model used is the 6061B with Sona-Counter 6079A: see Hall-Craggs (1979). They are basically graphs showing the distribution of sound energy in three parameters: (a) frequency (pitch), vertical scale, (b) time (duration), horizontal scale, (c) amplitude (loudness), depth of shading from pale grey to dense black. Frequency is measured in kilocycles, now usually referred to as kilohertz (kHz), and it is important

to note that the frequency scale is linear, i.e. it proceeds by evenly spaced divisions from 80 Hz to 8 kHz (except in the lowest division, 80 Hz to 1 kHz, where, with the wide-band filter in operation, there is a slight downward shift resulting in a narrower band). Since the frequency ratio of the octave is 2:1, the vertical scale expands by a corresponding extent for each ascending octave: thus, 1–2 kHz, 2–4 kHz, and 4–8 kHz each represent one octave.

Since many—probably most—avian vocalizations fall within the range 1–8 kHz, the coarseness of the analysis within the first 1 kHz is of little consequence. The best frequency resolution of the Sona-Graph is 45 Hz; this, too, is coarse in the lower reaches of the frequency spectrum where the semitone, the smallest musical interval *easily* perceived by all humans, may measure only 2 Hz or less; but in the region of 1 kHz the semitone measures about 60 Hz, and so a fair degree of accuracy may be obtained in the upper octaves. Either of two band-width filters may be used, 45 Hz or 300 Hz, the former giving better frequency resolution and the latter better temporal resolution (Figs 1a–b, 2a–b). Time resolution is excellent at 0·0015 s. This is important in the analysis of bird sounds since it is now accepted that the temporal resolution of birds is much finer than that of man, probably ten times better, though higher figures have been suggested. Thus the avian ear may distinguish 200 discrete sounds per s while, to the human ear, sounds fuse at about 20 per s.

The maximum duration of a single sonagram is about 2·5 s. However, for the purpose of illustrating sections of continuous song exceeding 2·5 s it is a simple matter to make several overlapping sonagrams and thereafter match and join them.

Flexibility of the analytical process may be obtained by varying the playback speed of recordings fed to the Sona-Graph (Figs 3a–c). Half-speed playback results in a clearer picture of temporal detail (the duration of a complete sonagram being reduced to 1·25 s) and may also be used to halve the frequency of sounds exceeding 8 kHz, the maximum frequency shown on the usual sonagram. Contrary to some expectations, such high-frequency sounds occur rather rarely in bird songs and calls. Doubling the playback speed of a recording obscures some of the temporal detail and extends the horizontal scale to 5 s but is useful in obtaining a clearer picture of low-frequency sounds.

Some of the delicate shading showing intensity of sound may be lost in reproduction but, when change of intensity is important (for purposes of description or recognition of vocalizations), a running amplitude display can be added to a sonagram. This correlates exactly with the time axis but the vertical scale of the subsidiary graph is in decibels (Fig 4).

Sonagrams show the acoustic spectrum of each sound so that much information about the nature of the sound and 'timbre'—or tonal quality—of a tone may be deduced.

Fig 1. Pure tones from a signal generator at 1 kHz and 2 kHz: sound spectrographic analysis, (a) with wide-band filter (300 Hz), (b) with narrow-band filter (45 Hz).

Fig 2. Human whistle: (a) wide-band filter, (b) narrow-band filter, (c) in musical notation.

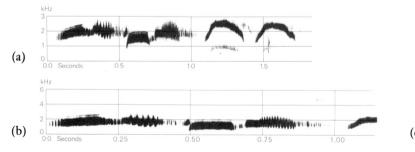
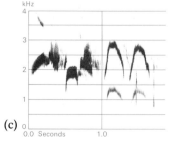

Fig 3. Wide-band sonagrams of song-phrase of Blackbird *Turdus merula*: (a) normal playback speed, (b) half-speed playback of first part, (c) double-speed playback.

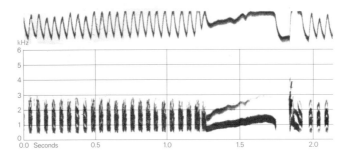

Fig 4. Wide-band sonagram of 'churring' of Leach's Storm-petrel *Oceanodroma leucorhoa*. The upper curve shows relative loudness.

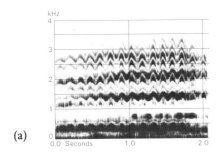

(a)

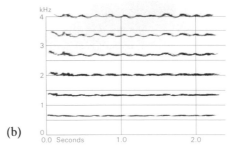

(b)

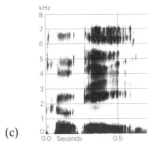

(c)

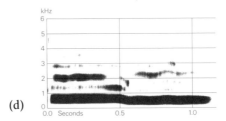

(d)

Fig 5. Spectral analysis of tones and speech. (a) Wide-band sonagram of human singing vowel sound 'ay' on note B, 246 Hz, a semi-tone below middle C. (b) Narrow-band sonagram of violin tone at E (659 Hz). Both (a) and (b) have strong harmonics and similar rates of vibrato—about 6 Hz. (c) The call 'cooee' spoken quietly by human ♀; the dense regions are formants which reflect individual characteristics of the vocal tract; the initial, discrete, thin vertical line is produced by the consonant 'c'. (d) 'Cooee' sung by the same voice on E (659 Hz), gliding down a third to C sharp (554 Hz); harmonics replace the formants and the consonant is now closer to the succeeding vowel. Form (d) is used for hailing at a distance while the form (c) is adequate for close quarters. Similar functional attributes appear in many bird vocalizations (see Hall-Craggs 1978).

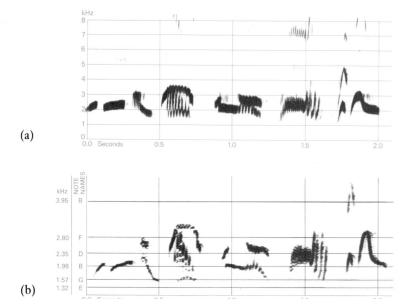

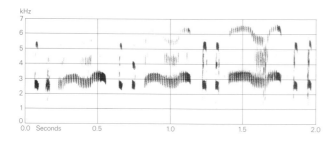

Fig 6. Features of bird song. (a) In this phrase of a Blackbird *Turdus merula*, notes 3, 4, and 6 show frequency modulation (vibrato) at rates of 100 Hz, 50 Hz, and 80 Hz respectively. Notes 4 and 5 illustrate the simultaneous but independent use of 2 sound sources (see 'Diad' in Glossary). Notes 6-8 reveal change of emphasis between 2nd and 3rd harmonics, note 7 having particularly strong 2nd harmonic. (b) A narrow-band scale magnification of (a) demonstrating lack of harmonic relationship between constituent notes of each diad; note 7 is a useful example of the precise 2:1 ratio between a fundamental and its 2nd harmonic; note that the time resolution of the narrow-band filter is too coarse to resolve the rapid frequency modulation revealed in (a).

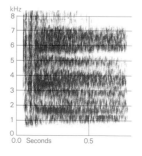

Fig 7. Amplitude modulation in 'kititikeeri' calls of Arctic Tern *Sterna paradisaea*. The rate is high at 80-90 Hz and remains almost stable throughout the pitch undulation in each long note.

Fig 8. Threatening hiss of Blue Tit *Parus caeruleus*. This sound appears to have all the characteristics of white noise excepting some low frequency components.

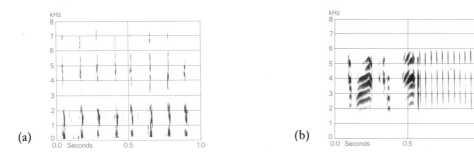

Fig 9. (a) Click-like sounds made by keys of electric typewriter. (b) Rapidly repeated clicks which constitute the rattle in the song of the Swallow *Hirundo rustica*.

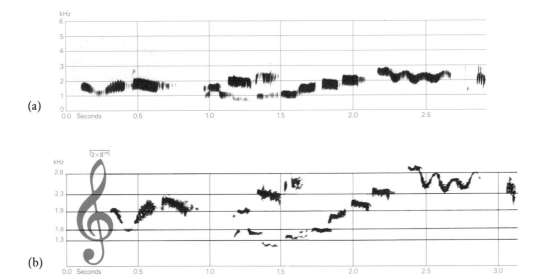

Fig 10. Song-phrase of Blackbird *Turdus merula*: (a) as wide-band linear-scale sonagram; (b) analysed with narrow-band log-scale and scale magnifier to effect a simple musical notation. Frequency lines in (b), inserted by using the digital counter, are equivalent to the 5 lines of a treble stave transposed up 2 octaves. Scale magnification limits *c.* 1·0–3·0 kHz (for further explanation, see Hall-Craggs 1979).

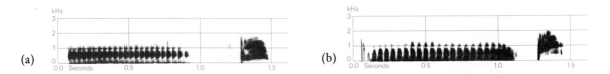

Fig 11. (a) Sonagram of Inciting-call of Ruddy Shelduck *Tadorna ferruginea*. (b) Human imitation of same call with initial consonant 'k' (initiating consonants in bird vocalizations are often imagined, probably because man cannot start vowel sounds as quickly as birds).

'White noise', containing virtually all frequencies, spreads over the entire spectrum (Fig 8), while the thin, pure tones from a signal generator appear as narrow perfectly horizontal lines (Fig 1). Tones produced by the voice and musical instruments are usually rich in harmonics (Fig 5). The average ear does not analyse harmonics into discrete entities as it analyses (or can usually be trained to analyse) the several simultaneously sounding notes in a chord. Instead of hearing harmonics as they appear on a sonagram we are aware of a particular tonal quality which differs from one instrument to another according to the relative strength of the harmonics. Formants (see Glossary, p. 31, and Fig 5c) are also heard as particular qualities of voices or instruments and, as such, are valuable aids to individual recognition. Tonal quality is also much affected by starting transients—the brief initiating sounds which precede, and may follow, the steady state of a tone. Other factors which affect the quality of a tone and which may be read from a sonagram are frequency modulation and amplitude modulation. The former is akin to the vibrato of the string player or singer (Fig 5) but the modulation rate is generally much faster in bird sounds; it appears on the sonagram as rapid and regular frequency alternation of fairly uniform range (see Fig 6a, also Fig 3). Amplitude modulation is a waxing and waning of the amplitude of a sound (Fig 7); it may be coupled to frequency modulation.

Noises, i.e. those sounds lacking a sufficient concentration of energy within narrow margins of a principal frequency to give tones of identifiable pitch, appear as broad band patches of energy. The frequency range is often wide and may encompass the available spectrum, as in the hiss of the Blue Tit *Parus caeruleus* (Fig 8) or the noise made by the key depression of an electric typewriter (Fig 9a) and similar clicks which make up the rattles in the song of the Swallow *Hirundo rustica* (Fig 9b).

The figures will help in the mental realization of sonagrams. With practice, it becomes possible to read and interpret the diagrams and gain a fair impression of the sounds portrayed. In many cases, they provide at a glance much information which could only be conveyed, if at all, by lengthy and complex verbal description. It is best to begin by looking at temporal factors, establishing the time pattern in the mind, and then to note the most obvious pitch movements—ascending, descending, steady, or variable. Distinguishing between familiar tones and noises presents no problems (Figs 1, 2, 5, 8, 9) but it is often difficult to imagine the quality of the sound; for example, whether the tone is 'reedy' or 'flute-like', or the noise is growling, grunting, screeching, clucking, etc. In such instances, much can be learned from studying the diagrams of sounds of well-known species and noting similarities and differences between these and unfamiliar sounds.

By using a Sona-Counter 6079A with a scale magnifier, frequency counter, and log scale (as in Fig 10b), many sonagrams (e.g. Fig 10a) can be made to resemble simple musical notation in the pitch domain without forfeiting the strict objectivity of the analytical process (Hall-Craggs 1979).

PHONETIC RENDERINGS

As an aid to realizing the sound quality of sonagrams, phonetic renderings are given in the text. Phonetic, or onomatopoeic, representation is beset by pitfalls on account of its subjective nature; no major studies have been made of the phonetic content of bird sounds (except a few with reference to the imitative abilities of talking birds) and it is readily apparent that even within a group of ornithologists experienced in recording and describing avian sounds there is considerable disagreement in this sphere, especially in connection with the supposed consonantal content of sounds. The rapid initiation of many bird sounds gives a strong impression of the presence of a consonant but it is often impossible to decide which consonant. Hence where consonants are given they should be treated as no more than suggestive of the kind of sound that may be heard (Fig 11).

It is generally much easier to reach agreement about vowel sounds, probably on account of their longer duration but here, too, there are problems which may often be attributed to individual variation of voice within a species; thus one member of a flock may be heard to pronounce 'a' as in cat and another 'e' as in pet. Study of the recordings suggests that this may, at times, be attributable to sexual dimorphism of voice, but little work has been done in this field and it could provide a fruitful project for research. It must also be emphasized that in using material from a variety of sources and languages, full consistency is unobtainable. Except in a few cases where it is especially helpful, no attempt is made to translate foreign renderings into English, and the reader is referred to the pronunciation of the language in question for satisfactory interpretation. This approach reflects the belief that in a work of this scope it is useful for the reader to have a variety of renderings at his disposal.

Many requirements for further study emerge from what has been said above. As the study of bird voice lags so far behind some other facets of ornithological research, a few lines must be devoted to explaining its requirements. These include:

1. Systematic appreciation of the extent of the field and of the descriptive and experimental techniques for exploring it, including all necessary standardization of terms and methods.
2. Further development of existing laboratory facilities, archives, and networks of specialist communication, with a view to expanding resources and filling in gaps in structure, including closer relations with other

branches of bio-acoustics and improved international provision for co-operation and publication.

3. Increased facilities for training and opportunities for research in this area of bio-acoustics.

4. Compilation of an authoritative newly worked compendium of the vocabulary of as many species as possible on an international basis, to lay a foundation for future work, on lines sketched in the present Voice section.

5. Initiation of a series of co-operative studies designed to bring together research on voice, behaviour, and communications theory, and to demonstrate the potential of a modern scientific effort in bio-acoustics.

By way of illustration we include a few possible topics for specific studies which have come to our attention in preparing this work:

1. Intraspecific variation of vocalization: (a) regional, (b) group, (c) individual.

2. Relation between habitat type and vocalization (e.g. it is claimed that species calling from a forest floor use lower frequencies than those in the canopy and that inhabitants of coniferous woods produce quieter sounds than those of deciduous woodland).

3. Relation between territory size and vocalization type.

4. The study of racial characteristics of voice as an aid in taxonomy.

5. An investigation of female song and the possibilities of more duetting in species of temperate regions.

6. Diurnal variation in song.

7. A study of song outside the breeding season.

8. Development of song in the individual.

9. Interspecific imitation.

10. The extent of imitation within different environments.

BIBLIOGRAPHY

A unique collection of sonagrams and voice descriptions of over 400 European bird species is to be found in:

BERGMANN, H-H and HELB, H-W (1982) *Stimmen der Vögel Europas.* Munich.

Other general literature on avian bio-acoustics:

ARMSTRONG, E A (1973) *A study of bird song.* New York.

BERGMANN, H-H and HELB, H-W (1980) Vogelstimmen schwarz auf weiss. Series of articles in *Welt der Tiere* from issue 5 of volume 7.

BONDESEN, P-G (1977) *North American bird songs—a world of music.* Klampenborg.

BUSNEL, R-G (ed) (1963) *Acoustic behaviour of animals.* Amsterdam.

CATCHPOLE, C K (1979) *Vocal communication in birds.* London.

GAILLY, P (1984) Communication acoustique et chants des oiseaux. *Cah. Ethol. appl.* 4(1), 73–120.

GREENEWALT, C H (1968) *Bird song: acoustics and physiology.* Washington.

HALL-CRAGGS, J (1978) In Merson, J (ed) *Investigating music.* Australian Broadcasting Commission, 8–20 (with tape).

HALL-CRAGGS, J (1979) Sound spectrographic analysis: suggestions for facilitating auditory imagery. *Condor* 81, 185–92.

HARTSHORNE, C (1973) *Born to sing: an interpretation and world survey of bird song.* Bloomington.

HINDE, R A (ed) (1969) *Bird vocalizations.* Cambridge.

HJORTH, I (1970) A comment on graphic displays of bird sounds and analyses with a new device, the Melograph Mona. *J. theoret. Biol.* 26, 1–10.

JELLIS, R (1977) *Bird sounds and their meaning.* London (with disc).

KROODSMA, D E and MILLER, E H (eds) (1982) *Acoustic communication in birds.* New York.

SAUNDERS, A A (1951) *A guide to bird songs.* New York.

TEMBROCK, G (1977) *Tierstimmenforschung.* Wittenberg Lutherstadt.

THIELCKE, G (1970) *Vogelstimmen.* Berlin.

THIELCKE, G A (1976) *Bird sounds.* Ann Arbor.

THORPE, W H (1961) *Bird-song.* Cambridge.

VOIGT, A (1894) *Exkursionsbuch zum Studium der Vogelstimmen.* Leipzig; 12th edition (revised by E Bezzel) published in Heidelberg in 1961.

Sound recordings:

PALMÉR, S and BOSWALL, J (1969–80) *A field guide to the bird songs of Britain and Europe.* Sveriges Radio, Stockholm. 15 discs with songs and calls of 585 species. The most comprehensive work.

ROCHÉ, J-C (1964–70) *Guide sonore des oiseaux d'Europe.* Institut Echo, Aubenas-les-Alpes. 1: western Europe, 27 discs, 256 species. 2: southern Europe, 13 discs, 82 species. 3: northern Europe, 11 discs, 129 species.

See also important list of discs by C CHAPPUIS annotated in *Alauda* 1974–85, 42–53.

Other regional compilations and selections:

KIRBY, J (1973–5) *Wild life sound tracks.* Middlesborough. 9 cassettes.

KOSMOPHON (1958–73) *Stimmen einheimischer Vögel.* Stuttgart. 18 discs and several additions. Mostly not arranged in systematic order.

LEWIS, V C (1980) *British bird vocabulary (an aural index).* Published by V C Lewis, Rosehill House, Lyonshall, Herefordshire, England. 12 cassettes with descriptive text in 6 albums.

NORTH, M and SIMMS, E (1969) *Witherby's sound-guide to British birds.* London. 2 discs, calls and songs of 194 species; accompanying text in book form.

ROCHÉ, J-C (1968) *A sound guide to the birds of north-west Africa.* Institut Echo, Aubenas-les-Alpes. 5 discs, 43 species.

SCHUBERT, M (1968–84) *Stimmen der Vögel Mitteleuropas.* Eterna. 6 discs, species grouped by habitat.

SCHUBERT, M (1975–84) *Stimmen der Vögel Südosteuropas.* Eterna. 2 discs.

SCHUBERT, M (1982) *Stimmen der Vögel Zentralasiens.*

Eterna. 2 discs, 59 species from Irkutsk (USSR) and Mongolia.

SVENSSON, L (1984) *Soviet birds*. Stockholm. 1 cassette and booklet, 34 passerines from Siberia and Kirgiziya.

VEPRINTSEV, B (1960-74) *The voices of birds in nature*. Moscow. 6 discs, 115 species.

VEPRINTSEV, B (1982) *Birds of the Soviet Union: a sound guide*. Moscow. First 3 in a projected series of 25 discs intended to accompany the new Soviet handbook edited by Ilyichev and Flint (*Ptitsy SSSR*, 1982).

WAHLSTRÖM, S (1962-71) *Danske Fuglestemmer*. 18 discs, accompanying text in Danish, book.

GLOSSARY

ACCELERANDO. A gradual quickening.

AMPLITUDE. The increase or decrease of air pressure at a given point during a sound. The maximum value of the displacement in an oscillatory motion (waveform). *See* INTENSITY, LOUDNESS, VOLUME.

ANTIPHONAL SINGING. Precisely timed alternating singing or calling normally restricted to the performance of male and female for the formation and maintenance of the pair-bond. Thus a type of DUETTING.

CADENCE. A short, easily recognized mode of bringing to a close a PHRASE in a SONG, e.g. the terminal flourish of Chaffinch *Fringilla coelebs* song and the characteristic descending pitch interval at the end of Yellowhammer *Emberiza citrinella* song.

CODA. The terminal section of a PHRASE: longer than a CADENCE, rarely stereotyped, and useful in the analysis and description of multi-phrase REPERTOIRES.

COMMUNAL SINGING. Tendency of social groups or adjacent territory-holders to sing together in response to the same stimulus or as a result of social facilitation.

COUNTER SINGING. *See* SONG-DUEL.

CRESCENDO. A gradual increase in LOUDNESS.

CYCLE. When a waveform repeats itself each complete repetition is called a cycle. Thus, a string or other sound source vibrating at 440 cycles per s (Hertz) produces tuning A. *See* HERTZ.

DECIBEL (dB). A measure comparing the INTENSITY or LOUDNESS of two or more sounds. The decibel scale is logarithmic. It ranges from the threshold of hearing, reaching an intensity that is painful to the human ear at about 130 dB.

DIAD. Two synchronous, harmonically unrelated TONES resulting from the independent action of two sound sources in the SYRINX.

DIALECT. Produced when there is a tendency among local populations of a species to approximate their SONGS to a common pattern as a result of mutual IMITATION.

DIMINUENDO. A gradual decrease in LOUDNESS.

DOPPLER EFFECT. The apparent change in the PITCH of a sound when the sound source and the listener are moving towards or away from each other. When the distance between the two is decreasing the apparent pitch is raised and vice versa.

DUETTING. Normally occurs when both members of a pair sing simultaneously or antiphonally as part of the courtship display, to maintain the pair-bond or to maintain contact. *See* ANTIPHONAL SINGING, POLYPHONIC SINGING, UNISON SINGING.

FIGURE. A short succession of sounds producing a single impression. *See* MOTIF.

FORMANT. Distinctive component of a sound, occurring at FREQUENCY regions reinforced by the resonance characteristics of the vocal tract.

FREQUENCY. The number of CYCLES occurring per unit of time.

The unit of time is normally 1 s, and frequency is then measured in HERTZ. *See* PITCH.

FUNDAMENTAL FREQUENCY. The lowest component FREQUENCY of a periodic wave; the basic factor that determines the PITCH of a sound.

GLISSANDO. An extremely rapid succession of TONES either ascending or descending in PITCH. *See* PORTAMENTO.

HARMONIC. A component of a periodic wave having a FREQUENCY which is a simple multiple of the FUNDAMENTAL, e.g. the component whose frequency is twice that of the fundamental is called the second harmonic. *See* OVERTONE, TONAL QUALITY.

HERTZ (Hz). Synonym for CYCLES per s. 1 kHz = 1000 Hz.

IMITATION. Vocal copying of any sound pattern. *See* MIMICRY.

INSTRUMENTAL SOUND (MECHANICAL SOUND). Non-vocal sound used for communication, e.g. drumming, bill-snapping, wing-clapping, etc.

INTENSITY. A measure of the magnitude of the sound stimulus, usually expressed in DECIBELS. *See* AMPLITUDE, LOUDNESS, VOLUME.

KILOHERTZ (kHz). 1000 CYCLES per s (= 1000 Hz).

LEGATO. Sounds performed in a smooth or connected manner. The opposite is STACCATO.

LOCATORY SOUND. One containing cues by means of which the position of the source can be determined.

LOUDNESS. The relative magnitude of two or more sounds as perceived by the listener. *See* VOLUME.

MECHANICAL SOUND. *See* INSTRUMENTAL SOUND.

MELODY. Change of PITCH in time.

MELOGRAM. A graph showing synchronous curves of principal FREQUENCIES and their relative AMPLITUDES in real time.

MELOGRAPH. An instrument for measuring and displaying changes in principal FREQUENCIES and their relative AMPLITUDES in time.

MIMICRY. Vocal copying of a sound pattern, especially of other species, avian or otherwise. *See* IMITATION.

MNEMONICS. Words or sentences designed to aid the recall of sound patterns.

MODULATION. Any periodic alternation of FREQUENCY, or AMPLITUDE of a sound wave. *See* VIBRATO.

MOTIF. The French equivalent of FIGURE; often used in bird vocalization literature.

NOISE. Non-periodic sound waves.

NOISE, WHITE. A sound with an approximately equal amount of power at every component FREQUENCY over the audible range.

NOTE. A single TONE of definite PITCH.

OCTAVE. The interval between two FREQUENCIES having the ratio 2:1. The second HARMONIC is one octave above the FUNDAMENTAL.

OVERTONE. A PARTIAL having a FREQUENCY higher than that of the basic frequency. Often used as a synonym for HARMONIC.

PARTIAL. A component of a complex TONE.

PHASE. State of change or development; hence in an oscillatory motion, the position and character of a wave at any given moment.

PHRASE. Short musical passage forming part of a longer passage. In bird SONG, a number of song-phrases (or songs) comprise a song REPERTOIRE.

PITCH. The pitch of a NOTE is equal to the FREQUENCY of the source VIBRATION. Pitch refers to the auditory sensation, frequency to the physical measurement.

POLYPHONIC SINGING. A type of DUETTING in which the contributions from the 2 (or more) birds synchronize or overlap, each part having its own temporal and PITCH patterning.

PORTAMENTO. A continuous gliding from one PITCH to another; the whistle of the Starling *Sturnus vulgaris* is often thus described. *See* GLISSANDO.

PRIMARY SONG. Sometimes used to refer to the normal loud

specific SONG. Also called advertising, territorial, or full song.

RALLENTANDO. A gradual slowing.

RATTLE. An atonal (i.e. noisy) TREMOLO.

REACTION TIME, AUDITORY. The briefest period of time between stimulus and response.

REPERTOIRE. The total vocal and INSTRUMENTAL output of any individual. Often applied only to SONG but should, strictly, include all vocalizations and instrumental sounds used in communication.

RHYTHM. In musical usage the term is often restricted to temporal patterning. However, a time pattern—with or without PITCH patterning—may or may not be performed rhythmically. Rhythm embraces those minute gradations of time and INTENSITY which give to abstract sound patterns a forward impetus to points of climax and rest. Its expression does not require a regularly recurring strong beat and many bird SONGS are profoundly rhythmic.

SEGMENT. A succession of uniform UNITS within a PHRASE.

SEMITONE. The smallest musical interval in common use in Western music. There are 12 semitones in an OCTAVE. Bird SONGS often incorporate smaller intervals than the semitone but it is not known whether birds can distinguish much smaller PITCH intervals than man is able to.

SONAGRAM (abbreviation of Sound Spectrogram). A precise visual representation of a sound pattern (see p. 24). Distribution of sound energy is shown as a function of time. LOUDNESS (INTENSITY) is indicated by the depth of shading. Much of the information on a sonagram is not immediately apparent to the ear, e.g. HARMONICS are not heard as discrete sounds occurring simultaneously but as a single TONE with a TONAL QUALITY which is largely dependent upon the relative strength and distribution of the harmonics.

SONG. A relatively complex pattern of sounds in time which may be repeated exactly and is, consequently, recognizable not only at the specific level but often at the group and individual levels. Song has both tonal and temporal form. Advertising (territorial) song is usually under the control of the sex hormones and its utterance is therefore largely confined to the breeding season.

SONG-DUEL (COUNTER SINGING). Alternation of SONG which may occur when two or more territory-holders are responding to one another.

SOUND SPECTROGRAPHIC ANALYSIS. The sound spectrograph is an instrument for making spectral analyses (SONAGRAMS) of single sounds or series of sounds. Sounds are recorded magnetically on the edge of a 30-cm metal turntable which is then revolved. At each revolution the signal is scanned by a 45-Hz or 300-Hz filter, beginning with the low FREQUENCIES and working upwards. The output is recorded on dry facsimile paper attached to a drum which revolves synchronously with the turntable. A recording stylus shifts gradually up the frequency scale in step with the scanning oscillator, thereby recording the frequency components at any given instant. AMPLITUDE variations are shown by fluctuations of INTENSITY at the output of the filter, the darker regions on the paper indicating the louder sounds and vice versa. Precise measurement of the relative amplitudes of the components of a sound may be obtained by means of a sectioner. This gives a graph of relative amplitude against frequency at any required instant (c. 0·04 s).

SOUND SPECTRUM (ACOUSTIC STRUCTURE). An illustration of the distribution of energy among the component FREQUENCIES and INTENSITIES, and PHASES of a sound at any given moment in time.

STACCATO. See LEGATO.

SUBSONG. A useful general term to denote forms of quiet SONG. More specifically it refers to a quieter, extended warbling in which fragments of the full (advertising) song may be heard, often with imitations of other species. A bird uttering subsong is usually perched low and in an inconspicuous place; the bill may be closed while singing. Subsong is not confined to out-of-season utterances and may readily be heard during the period of full song.

SUB-UNIT. A discrete section of a complex UNIT. The term is useful for describing minor divisions that sometimes occur in repeated complex units. See SYLLABLE.

SYLLABLE. This has been used to describe various successive sections of vocalizations. If, in verbal use, the term describes a section of a word, then in avian vocalizations, by analogy, it may be used to denote a non-discrete section of a discrete UNIT or SUB-UNIT as shown on a SONAGRAM.

SYRINX. The avian organ of voice. Unlike the human larynx it is situated at the lower end of the trachea at the junction of the two bronchi. Although variable in structure and complexity between groups of birds, it is generally a bony and cartilaginous chamber containing membranes which are activated by the passage of air from the lungs.

TEMPORAL RESOLUTION. The minimum time lapse between successive sounds which are still perceptible as discrete.

TIMBRE. See TONAL QUALITY.

TONAL QUALITY (or TIMBRE). That aspect of a sound which distinguishes it from other sounds of the same PITCH. Physically, tonal quality is determined by the presence or absence of PARTIALS and their relative INTENSITY and distribution, by TRANSIENTS, and by the rate and range of FREQUENCY and/or AMPLITUDE MODULATION.

TONE. In general use, a musical sound (i.e. with a definite PITCH) as distinct from a NOISE. In musical theory the term *whole tone* means a PITCH interval of one sixth of an OCTAVE or 2 SEMITONES. A *pure tone* is a sound of sinusoidal wave form, produced by artificial means (the simplest being a tuning fork), not by the voice.

TRANSCRIPTION. The representation of avian vocalizations in symbolic form, e.g. words, notation, onomatopoeia, etc.

TRANSIENT. The brief sounds which precede and follow the steady state of a TONE. The FREQUENCY of the transient vibrations is the resonant frequency of the system; the frequency of the steady state vibrations is the frequency of the driving force.

TRANSPOSITION. The exact repetition of a sound pattern at a higher or lower PITCH while the pitch and time relationships within the pattern remain constant.

TREMOLO. Rapid reiteration of one NOTE. A very rapid tremolo may have the appearance of AMPLITUDE MODULATION.

TRILL. A rapid and regular alternation of two NOTES, usually with a small PITCH interval.

UNISON. Sounding at the same PITCH and in the same pattern.

UNISON SINGING. Synchronous singing by 2 or more birds in identical time and PITCH patterns.

UNIT. A discrete section of a FIGURE. May be simple or complex: a simple unit comprises a single continuous sound sometimes divisible into SYLLABLES; a complex unit comprises 2 or more discrete SUB-UNITS and is distinguished by exact repetition of the entire complex.

VIBRATO. Rapid and continuous periodic rise and fall in PITCH. A form of FREQUENCY MODULATION.

VOLUME. The subjective experience of INTENSITY. See LOUDNESS.

WHISTLE. A term often used to describe those TONES which seem to approach most nearly the extreme simplicity of PURE TONES.

BREEDING

The BREEDING SEASON is taken as that period within which the species lays and incubates its eggs and rears its young to the flying stage. Some authorities would also include pair-formation and nest-building but these activities cannot be pinned down as accurately in time as the laying of an egg and may take place over a prolonged period that would be misleading to define as the breeding season.

The ANNUAL CYCLE DIAGRAMS show the normal season for the occurrence of eggs (E) and unfledged young (Y) with margins for early eggs and late broods. The diagram is in each case for a specified geographical area which is usually a large one taking in a considerable proportion of the species' range. Good data on variations in breeding season in different parts of the range are rarely available, although a south to north, or perhaps south-west to north-east, cline undoubtedly exists for many species. For some species the information on season is insufficient to compile a meaningful diagram; exceptionally, all that is known is a handful of dates when eggs or young have been recorded.

The timing of the breeding season is clearly affected by such factors as climate, latitude, food availability, and the physiological condition of the bird. The timing of the onset of breeding is also affected by the proximate factors of spring temperatures and rainfall, and the dependent conditions of vegetation growth and existing food supply. While this general assumption is probably true for the majority of species, comments have been restricted to those species for which actual studies on this aspect have been carried out.

NEST: its site, construction, and building. The description of the nest-site is restricted to its precise location, the broader aspects being covered in Habitat. The emphasis has been placed on the normal situation, omitting lists of exceptional, often aberrant, sites. Also included here is the re-use of nests in subsequent years.

The materials used in the construction of the nest are given fairly precisely in those relatively few cases where they have been examined in detail. For the majority of species, however, the information is rather generalized. As with nest-site, the exceptions have not been given undue prominence. Average dimensions of nests are usually quite easy to take but published accounts deal with mostly small samples, if these are stated at all. Detailed observations are still required on the building of nests of most species; in particular information is lacking on the roles of the sexes, the time taken, and the time of day.

EGGS. These are described briefly in terms of shape, texture, and colour. The first and last of these should be read in conjunction with the colour plates of eggs (reproduced actual size). The principal terms used in describing shape are oval (or ovate), elliptical, pointed (or pyriform), and rounded (or sub-elliptical), which should be self-explanatory. These may be qualified by adjectives such as blunt, broad, short, or long.

Measurements and weights of eggs are given to the nearest 0·1 millimetre and gram respectively. Range and sample size are included wherever possible, and a sample size of 100 has been regarded as acceptable though by no means always achieved. The most important single source of information on egg sizes is the work of Schönwetter (1967, 1979, and in preparation), which also includes information on eggshell weights, and geographical variation.

CLUTCH. The ideal information under this heading is a large sample of clutches (100 as a reasonable minimum) set out giving the percentage distribution of the different clutch sizes, together with the mean. This has been achieved only for a minority of species, however, and for the remainder it has been possible to give only the limits of normal clutch sizes, plus extremes. Sufficiently large samples from different parts of the range of a species to allow regional comparisons are given where available, as are examples of seasonal variation in clutch size.

The extremes of clutch size must be treated with great caution. At the lower end of the scale there is always a strong possibility that one or more eggs may have disappeared before the observer recorded the size, while at the upper end there is the complication of more than one female laying in the same nest. Both these occurrences can, of course, alter the true percentage distribution as well as the mean, yet neither can be completely allowed for. For a few species, for example among the ducks, where laying by two or more females in the same nest is quite regularly recorded, information is given separately for these *dump* nests.

The number of broods normally reared is given together with any known regional variation. The occurrence of a replacement for a lost clutch is mentioned, though only as a probability in the many instances where it is likely but has not been proved. Most single-brooded species seem capable of laying a replacement clutch when their first is lost, particularly when this happens early in the incubation period. In a small number of species the replacement clutch is known to be smaller than the first, but for many more information is lacking. The normal interval between laying eggs is stated where known.

INCUBATION. The incubation period is given as accurately as has been recorded, though this is usually a range of days. The period refers always to that taken for one egg to hatch from the time incubation starts, rather than for the full clutch, though some authors do not make clear which they are giving. The division of incubation between the parents is stated where known, as is the stage of laying at which incubation begins, and its subsequent effect on the hatching pattern. These items tend to follow family, or at least generic, lines and consequently where it has not been reported it can usually be inferred. Similarly, disposal of eggshells after hatching is in general a family characteristic.

YOUNG. The first terms under this heading refer to the state of maturity of the young at hatching, as follows. (1) *Precocial*: well developed at hatching—covered in down, eyes open, and capable of coordinated movements, including a measure of terrestrial or aquatic locomotion; may feed themselves, or be fed by their parents. (2) *Altricial*: poorly developed at hatching—little or no down, eyes closed, capable of little movement; always fed by parents. (3) *Semi-altricial*: much as (2) but with a better covering of down, eyes open, or closed for only a short time after hatching, and rather more active.

These three terms are coupled with three more. (4) *Nidifugous*: leaves nest and its vicinity soon after hatching. (5) *Nidicolous*: stays in or near the nest for some time after hatching. (6) *Semi-nidifugous*: stays in or near the nest for some time after hatching although physically capable of leaving altogether.

The parts played by the parents in feeding and caring for the young are dealt with next. Where the parents regurgitate food for the young this is divided into *complete* (food deposited on the nest or ground) and *incomplete* (food retained in parent's throat or mouth, and obtained by the young there).

FLEDGING TO MATURITY. The fledging period is given as precisely as possible. It is not recorded for a surprising number of species, and authorities differ about others. Equally, the dependence of the young on the parents after fledging is largely unknown, being particularly difficult to establish in those species which embark on a migration together but arrive at their destination separately. The age of first breeding is stated where known (more detail on this, if available, is given under Bonds in the Social Pattern and Behaviour section).

BREEDING SUCCESS. This can be given only for a minority of species. Even a study lasting several years may not be sufficient to give a reliable figure in the case of long-lived species.

PLUMAGES AND RELATED TOPICS

PLUMAGES. This section is primarily intended for use with the bird in the hand, although it should also be of value when studying close-up photographs or detailed field notes. Usually one race is described, the characters of others being given in the final section on Geographical Variation. The description of a plumage is presented in the sequence: head, neck, upperside and underside of body, tail, and wings (for names of plumage areas, see figures on pp. 36-7; see also *Br. Birds* 1981, **74**, 239-42)

and *Dutch Birding* 1985, **7**, 37-48). The plumages are discussed in the following sequence: adult breeding, adult non-breeding, nestling, juvenile, and immatures. The descriptions refer to fully developed plumages, but it should be kept in mind that elements of several plumages may be found in the same bird, e.g. during active moult or when moult is arrested.

The *breeding plumage* is defined as the plumage worn during part or all of the nesting season, but sometimes acquired long before. In many species, the breeding plumage regularly alternates with a *non-breeding plumage*, acquired during the post-breeding moult (see below); in others, the same plumage is worn during the whole year and simply termed *adult*.

The first full plumage following the nestling stage is called *juvenile* in all species, even in those, such as in many warblers (Sylviidae), in which it is identical with the adult. The juvenile plumage may be followed by one or more recognizable *immature* plumages. If the plumage following the juvenile is identical with the adult, there are no named immature plumages, and the plumage is called 1st adult, but this does not imply that the bird breeds in this plumage for sexual maturity may be deferred. The precise sequence of plumages will be clear from the descriptions.

A new terminology for plumages and moults was proposed by Humphrey and Parkes (1959) in which the breeding plumage is termed 'alternate' and the non-breeding 'basic'. Stresemann (1963) criticized these terms because it is not always clear whether the non-breeding plumage is the basic one—but nevertheless, this system of names is widely used in North America. To us, however, the terms given above seem more straightforward and easier to understand; they link the various plumages with the phases in the life-cycle for which they have been evolved. For an alternative system, see *Br. Birds* 1985, **78**, 419-27.

BARE PARTS. These are described for the same race as the plumages in the sequence: iris, bill, bare skin on head (if present), leg, and foot. Bare-part colours cannot be studied satisfactorily in museum skins for they often fade after death; we have therefore relied on notes on specimen labels, descriptions of live and fresh dead birds, and colour photographs.

MOULTS. The following moults are recognized: adult post-breeding, adult pre-breeding, post-juvenile, and a variable number of moults of immature plumages. The main moult of the annual cycle is almost invariably the *post-breeding moult*. This is usually *complete*, involving not only the body plumage, but also the flight- and tail-feathers. The *pre-breeding moult* is normally *partial*, the flight-feathers, tail, and other parts of the plumage being retained. The replacement of nestling down by juvenile plumage could be termed *pre-juvenile moult*; it is not described in the Moults section, but information (if available) is given in the description of the nestling under

Plumages. The *post-juvenile moult* is the first moult in which contour feathers are replaced. It is variable in timing and extent, being complete or partial (sometimes involving only a few feathers), ranging in timing from early summer of the first calendar year to summer of the second.

The description of a moult contains information about timing, sequence of primary replacement, and, for partial moults, about the parts of the plumage which are replaced. Timing of moult in adults in compared, where appropriate, with timing of breeding and migration in the ANNUAL CYCLE DIAGRAM, where P = primary, B = body. Sequence of primary replacement may be *descendant* (from the carpal joint outwards), rarely *ascendant* (from the outermost primary inwards), *irregular*, or *simultaneous*. The progress in time of flight- and tail-feather moult is usually recorded with a system in which each feather is scored 0 to 5 according to its age and development: 0 is an unmoulted old feather, 5 a fully-grown new one (see, e.g., Ginn and Melville 1983). Thus, for primaries, if one wing is scored, the moult score will lie between 0 (all feathers old) and 50 (all feathers new)—or 45 if vestigial p10 is not accounted for; if both wings are scored, 0-100 (not often done, as moult is usually virtually symmetrical). For tail, scores are 0-60 in species with 12 tail-feathers and if both halves of tail are scored. In this work, except where stated, scoring involves all 10 primaries of one wing and both halves of tail.

MEASUREMENTS. In most species, data are given for length of wing, tail, bill, and tarsus. All measurements are given in millimetres (the unit of measurement is omitted from the text). The wing is measured by pressing it against a rule and stretching it fully; *wing length* is the distance from carpal joint to the tip of the longest primary. The tail is measured with dividers, *tail length* being the distance from the point where the central tail-feathers emerge from the skin to the tip of the longest feather. Bill and tarsus are measured with vernier calipers. *Bill length* is usually given as the chord of the culmen from the nasal-frontal hinge in the skull to the tip—bill (S); often also from the distal corner of the nostril to the tip—bill (N). *Exposed culmen* is the chord of the culmen from the frontal feathering to the bill-tip, but this is often difficult to measure in Passeriformes and so is little-used in Volumes V–VII (throughout Volumes I–IV, 'bill' was generally used for this measurement). *Tarsus length* is measured from the middle point of the joint between tibia and tarsus behind the leg to the middle point of the joint between tarsus and middle toe in front of the leg. Occasionally, length of *middle toe* is given; it is measured from the joint at the base of the middle toe (in front of the leg) to the tip of the middle claw. In the tables of measurements, length of the middle toe with claw appears simply as 'toe'.

All measurements are taken from study skins except where stated. For tail, bill, and tarsus, measurements of live or freshly dead birds are virtually identical, but ex-tensive data collected on many species measured freshly dead and again after skinning (ZMA) show that lengths of wing and middle toe decrease *c.* 2% after skinning (see, e.g., Engelmoer *et al.* 1983).

For each measurement, the following parameters are given: mean (m), standard deviation (s), number of specimens measured (n), lowest value of series (l), and highest value of series (h). These are presented in a standard form: m (s; n) l-h. If any of these values is unknown, the sequence is shortened to m (n) l-h, to m (l-h), or to l-h (n). Means over 200 are given to the nearest whole number, means under 200 are given to one decimal place. Range is given in whole numbers, except when the values are under 20. The statistical significance of the difference between two means is tested with a *t*-test (see, e.g., Sokal and Rohlf 1969). Where values differ markedly from those given elsewhere, this is discussed.

At the head of each table, the sources of the material are specified, using the following abbreviations:
BMNH, British Museum (Natural History), Tring
BTO, British Trust for Ornithology, Tring
IRSNB, Institut Royal des Sciences Naturelles Belges, Brussels
MNHN, Muséum National d'Histoire Naturelle, Paris
RMNH, Rijksmuseum van Natuurlijke Historie, Leiden
ZFMK, Zoologisches Forschungsinstitut und Museum Alexander Koenig, Bonn
ZMA, Zoölogisch Museum (Instituut voor Taxonomische Zoölogie), Amsterdam
ZMK, Universitetets Zoologiske Museum, København
ZMM, Zoological Museum, Moscow
ZMO, Zoologisk Museum, Universitetet i Oslo

WEIGHTS. These are taken from notes on specimen labels, from published or unpublished data on birds captured for ringing, and from other literature sources. They are presented in the same way as measurements. All weights are given in grams (the unit of measurement is omitted from the text). If information exists on biological relevance of seasonal or daily weight variations, this is briefly summarized.

STRUCTURE. The following points are covered: shape of wing; number of primaries; wing formula (expressed as distances from tips of primaries to tip of wing); shape of tail, bill, and leg; proportions of toes; other structural peculiarities. Primaries are numbered from the carpal joint outwards (*descendantly*), as has become common practice. Individual primaries are indicated by p and a number, p1 being the innermost, p10 the outermost functional primary in most species. The secondaries are numbered from the carpal joint inwards, abbreviated as s1, s2, etc. The elongated innermost secondaries (usually s7–s9 in Passeriformes) differ in function and moult from the others; they are commonly called tertials. The tail-feathers are numbered from the central pair outwards: t1, t2 etc.

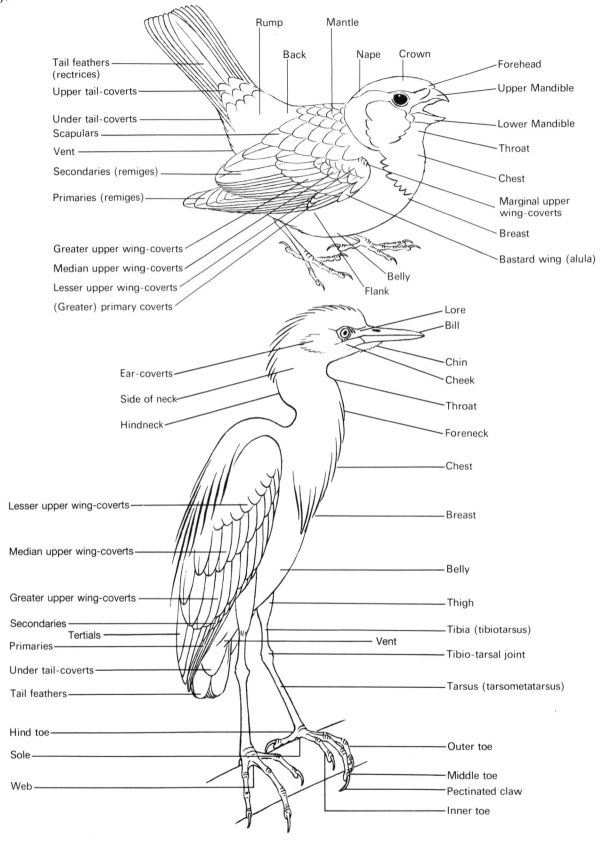

Rump

Mantle

Back

Nape Crown

Tail feathers (rectrices)

Upper tail-coverts

Forehead

Upper Mandible

Under tail-coverts

Lower Mandible

Scapulars

Throat

Vent

Chest

Secondaries (remiges)

Marginal upper wing-coverts

Primaries (remiges)

Breast

Greater upper wing-coverts

Bastard wing (alula)

Median upper wing-coverts

Belly

Lesser upper wing-coverts

Flank

(Greater) primary coverts

Ear-coverts

Lore

Bill

Side of neck

Chin

Cheek

Hindneck

Throat

Foreneck

Chest

Lesser upper wing-coverts

Breast

Median upper wing-coverts

Belly

Thigh

Greater upper wing-coverts

Tibia (tibiotarsus)

Secondaries

Tertials

Vent

Tibio-tarsal joint

Primaries

Under tail-coverts

Tarsus (tarsometatarsus)

Tail feathers

Hind toe

Outer toe

Sole

Middle toe

Pectinated claw

Web

Inner toe

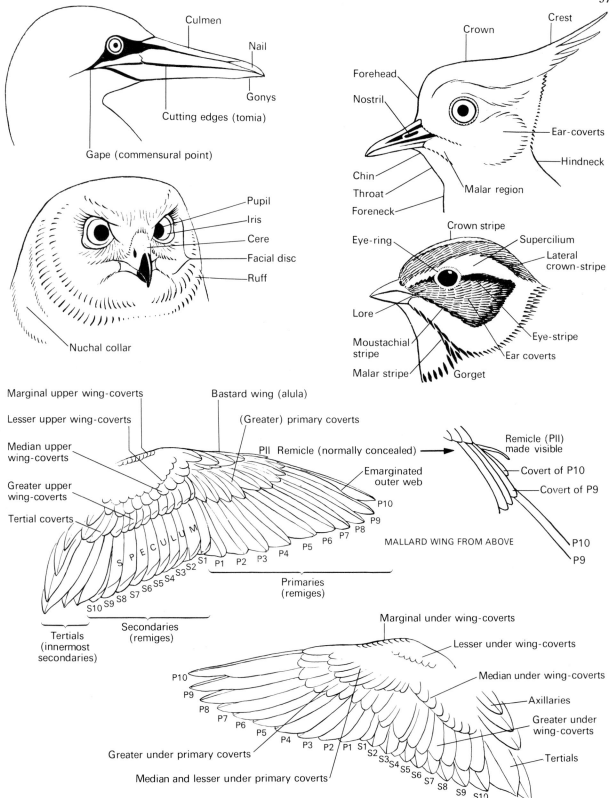

Culmen

Nail

Gonys

Cutting edges (tomia)

Gape (commensural point)

Crown

Crest

Forehead

Nostril

Ear-coverts

Hindneck

Chin

Throat

Malar region

Foreneck

Pupil

Iris

Cere

Facial disc

Ruff

Nuchal collar

Crown stripe

Eye-ring

Supercilium

Lateral crown-stripe

Lore

Moustachial stripe

Malar stripe

Eye-stripe

Ear coverts

Gorget

Marginal upper wing-coverts

Lesser upper wing-coverts

Median upper wing-coverts

Greater upper wing-coverts

Tertial coverts

Bastard wing (alula)

(Greater) primary coverts

PII Remicle (normally concealed)

Emarginated outer web

Remicle (PII) made visible

Covert of P10

Covert of P9

MALLARD WING FROM ABOVE

P10

P9

P10

P9

P8

P7

P6

P5

P4

P3

P2

P1

S1

S2

S3

S4

S5

S6

S7

S8

S9

S10

SPECULUM

Primaries (remiges)

Secondaries (remiges)

Tertials (innermost secondaries)

Marginal under wing-coverts

Lesser under wing-coverts

Median under wing-coverts

Axillaries

Greater under wing-coverts

Tertials

Greater under primary coverts

Median and lesser under primary coverts

MALLARD WING FROM BELOW

GEOGRAPHICAL VARIATION. The general nature of the geographical variation is indicated (even where no formal races are recognized) and differences between the recognized races are summarized. Reasons are given for rejection of races accepted by other authors.

GLOSSARY

Many of the terms used in the Plumages and associated sections are shown in the diagrams on pp. 36–7; others are given below.

ADULT. In practice, defined as birds in plumage no longer changing with age.

BAR. Feature of colour pattern oriented transversely on feather.

BARBULES. Tiny branches of barbs of feather.

BODY FEATHERS. All CONTOUR FEATHERS with exception of tail-feathers and feathers of wing.

BOOTED TARSUS. Covered with single horny sheath.

BRISTLE. Stiff, hair-like feather usually with a few barbs at base of shaft.

BULLA. Tympanic element of the SYRINX when it has a bubble-like appearance (plural: bullae).

CARPAL JOINT. Wrist joint.

CLINE. Series of populations of a species showing gradation in one or more characters from one end of its range to the other (hence *clinal variation*).

CONSPECIFIC. Belonging to the same species.

CONTOUR FEATHERS. All feathers with well defined webs belonging to the outer cover of the body, including flight-feathers and tail-feathers (opposite: *down feathers*).

DIASTATAXIS. Arrangement of feathers in wing in which 5th greater upper covert has no corresponding secondary (*see also* EUTAXIS).

DIMORPHISM. The existence of two differing MORPHS.

DISTAL. Pertaining to part of feather, leg, wing, etc., which is farthest from body (*see also* PROXIMAL).

DOWN FEATHERS. *See* CONTOUR FEATHERS.

EUTAXIS. Arrangement of feathers in wing in which secondary corresponding to 5th greater upper covert is present (*see also* DIASTATAXIS).

FENESTRA. Membranaceous 'window' in tracheal BULLA (plural: fenestrae).

FILOPLUMES. Fine, hair-like feathers with small tuft of barbs at tip (occasionally at other places along shaft).

FORM. Neutral term indicating an individual variant or taxonomic unit (e.g. species, race).

GONYS. Ventral ridge of lower mandible from tip to forking of RAMI.

LAMELLAE. Small plates or scales.

LAPPET. Wattle at corner of mouth.

MORPH. One of two or more well-defined forms belonging to the same population.

PAPILLA. Small, conical eminence.

PLUME. Type of ornamental feather.

POWDER-DOWN. Soft and friable DOWN producing fine dust particles used in plumage care.

PROXIMAL. Pertaining to part of feather, leg, wing, etc., which is closest to body (*see also* DISTAL).

PTERYLOSIS. The way in which feathers are arranged on the skin, usually in feather tracts wearing CONTOUR FEATHERS (pterylae), and featherless spaces (apteria) wearing at most only DOWN.

RAMI, MANDIBULAR. Halves of lower jaw.

REMICLE. Vestigial outermost primary.

RETICULATE TARSUS. Tarsus with numerous small scales, randomly arranged.

SCUTELLATE TARSUS. Tarsus with one or more distinct rows of relatively large scales.

SHAFT-STREAK. Narrow streak of contrasting colour along shaft of feather.

SIGNIFICANT. Shown by statistical test as unlikely to be due to chance (said of difference between means of two or more samples).

STANDARD DEVIATION (SD). Statistical parameter estimating the scatter around the mean in a sample of data. Theoretically, 99% of sample lies within $2 \cdot 58 \times SD$ from the mean, 95% within $1 \cdot 96 \times SD$.

STREAK. Feature of colour pattern oriented longitudinally on feather.

SYRINX. 'Voice box' of birds, lying at the lower end of the TRACHEA.

TRACHEA. Windpipe.

VARIATION. Differences in any character between animals of the same species. The following broad types of variation may be recognized: individual (between individuals belonging to the same population, same sex, and same age and studied at the same season); seasonal; sexual; age; geographical.

WING FORMULA. Configuration of tips of primaries relative to each other, expressed as distances from tip of each primary to tip of longest.

ACKNOWLEDGEMENTS

The preparation of this volume has involved again a major commitment of both money and time. The heavy financial burden, which has been aggravated by inflation in recent years, could not have been contemplated without the generous help and understanding shown by the Delegates of the Oxford University Press throughout the work. We are indebted also to the Commission of the European Economic Communities, the Ernest Kleinwort Charitable Trust, and the Royal Society for the Protection of Birds.

Time has been given generously by many ornithologists throughout the world. First, by members of the Editorial Board, and we are grateful to the British Trust for Ornithology, the Edward Grey Institute of Field Ornithology (University of Oxford), the University of Amsterdam, and the Wildfowl Trust for facilitating their labours. Secondly, valuable assistance was given to the editors by experts who prepared various accounts: Movements—Dr L A Batten, the late R A Cawthorne, P Clement, J H Elgood, Dr D G C Harper, J Hellmich, R Hudson, M G Kelsey, T Lloyd-Evans, C J Mead, and Miss D F Vincent; Food—Dr P J Edwards, A G Gosler, I L Gray, A Gretton, J Lunn, the late B D S Smith, Dr T A Stevens, Miss C A Thomas, and Miss D F Vincent; Social Pattern and Behaviour and Voice—P J Conder, Dr L Cornwallis, Dr P G H Evans, Dr P J Greig-Smith, J Hellmich, Mrs J Hall-Craggs, Dr D W Snow, and Miss D F Vincent; Breeding—D R Collins; Plumages—Dr D W Snow and Dr J Wattel; their initials are given at the end of the relevant accounts. As in all previous volumes, Miss R E Wootton has typed the entire text and compiled the References, foreign names, and indexes.

A key role was again played in the Voice section by P J Sellar who located and assembled the best available recordings of each species' repertoire; his comments and annotated indexes were invaluable. Mrs J Hall-Craggs has continued to make detailed aural and spectrographic analyses of these tapes and of further recordings held at the Sub-Department of Animal Behaviour, University of Cambridge; from the many sonagrams produced, the figures for publication were chosen. The generosity of the British Broadcasting Corporation in providing copies of their entire collection of natural history discs, and the enduring cooperation of the Sub-Department of Animal Behaviour in giving her access to these and to all necessary instrumentation and facilities, were of inestimable value. The British Library of Wildlife Sounds and its curator, R Kettle, and collaborators in many countries most generously provided tape-recordings; details are given in the captions to the relevant sonagrams. Where recordings are available as gramophone records or cassettes, references are given as follows:

Cornell Laboratory of Ornithology (1977) *Beautiful bird songs of the world*. Cornell.

Melodiya (1964) *The voices of birds in nature* 4. Disc 14867/68. Moscow.

Roché, J-C (1966) *Guide sonore des oiseaux d'Europe* 2; (1970) 3. Institut Echo, Aubenas-les-Alpes, France.

Roché, J-C (1968) *A sound guide to the birds of north-west Africa*. Institut Echo, Aubenas-les-Alpes, France.

Schubert, M (1982) *Stimmen der Vögel Zentralasiens*. Eterna.

Svensson, L (1984) *Soviet birds*. Stockholm.

Sveriges Radio (1972-80) *A field guide to the bird songs of Britain and Europe* by S Palmér and J Boswall 1-15 (discs).

Swedish Radio Company (1981) *A field guide to the bird songs of Britain and Europe* by S Palmér and J Boswall 1-16 (cassettes).

For recordings which have not been published commercially, assistance in contacting the original recordists may often be obtained from the Curator, British Library of Wildlife Sounds, National Sound Archive, 29 Exhibition Road, London SW7 2AS.

A vital part was played by the correspondents who, as stated earlier, provided much of the basic data on status, distribution, and populations for species occurring in their country:

ALBANIA   Dr E Nowak
ALGERIA   E D H Johnson
AUSTRIA   Dr H Schifter, P Prokop
AZORES   Dr G Le Grand
BELGIUM   Dr P Devillers
BRITAIN   Dr J T R Sharrock, Dr P C Lack
BULGARIA   T Michev, B Ivanov, Dr T Petrov, Dr L Profirov, Dr S Simeonov, Z Spiridonov, I Vatev, Dr S Dontchev, J L Roberts
CANARY ISLANDS   K W Emmerson
CHAD   C Erard
CYPRUS   P Flint, P F Stewart
CZECHOSLOVAKIA   Dr K Hudec
DENMARK   T Dybbro
EGYPT   P L Meininger, W C Mullié, S M Goodman
FAEROES   Dr D Bloch, B Olsen
FINLAND   Dr O Hildén
FRANCE   R Cruon, P Nicolau-Guillaumet
GERMANY (EAST) DDR   Mrs S Schnabel
GERMANY (WEST) BRD   A Hill
GREECE   W Bauer, H J Böhr, G Müller, G I Handrinos
HUNGARY   Dr L Horváth, L Haraszthy
ICELAND   Dr A Petersen

IRAQ   H Y Siman
IRELAND   C D Hutchinson
ISRAEL   Prof H Mendelssohn, U Paz
ITALY   P Brichetti, B Massa
JORDAN   P A D Hollom, D I M Wallace
KUWAIT   P R Haynes, Prof C Pilcher
LEBANON   Lt-Col A M Macfarlane
LIBYA   G Bundy
LUXEMBOURG   Dr P Devillers
MADEIRA   A Zino, G Maul
MALI   Dr J M Thiollay, B Lamarche
MALTA   J Sultana, C Gauci
MAURITANIA   R A Williams, J Trotignon, B Lamarche
MOROCCO   Dr M Thévenot, J D R Vernon
NETHERLANDS   C S Roselaar
NIGER   Dr J M Thiollay
NORWAY   V Ree
POLAND   Dr A Dyrcz, Dr L Tomiałojć
PORTUGAL   R Rufino
RUMANIA   Prof W Klemm, G G Buzzard, Dr V Ciochia
SAUDI ARABIA   M C Jennings, G Bundy
SPAIN   A Noval
SVALBARD   I Byrkjedal
SWEDEN   L Risberg
SWITZERLAND   R Winkler
SYRIA   Lt-Col A M Macfarlane
TUNISIA   M Smart
TURKEY   M A S Beaman, R F Porter
USSR   Prof V E Flint, D V M Galushin, H Veromann, G G Buzzard
YUGOSLAVIA   V F Vasić

We also wish to thank J Zaagman for preparing the Annual Cycle diagrams and all those who made available photographs, sketches, and published material on which the drawings illustrating the Social Pattern and Behaviour section were based; their names are given at the end of the relevant accounts.

We are grateful to R Frumkin, W Grabiński, Dr S Harding, Dr H Källander, Miss A Martínez Fernández, and Miss I Silva Bohórquez for valuable help with translations.

Finally, we are greatly indebted to the following, who assisted in many ways too diverse to specify in detail though credits are given in the text where appropriate: J Alder, G Alenius, A M Allport, M Amphlett, S Andersson, Arabian American Oil Company, A Aragüés, O Arheimer, Dr C Askenmo, D R Aspinwall, T Bacon, M A S Beaman, I Baker, K L Béres, Dr H-H Bergmann, Dr C J Bibby, R G Bijlsma, G Bishton, T Bjerke, T I Bjønnes, I Boldt, K Borg, I de Boroviczény, J H R Boswall, Dr W R P Bourne, C G R Bowden, H J Boyd, British Trust for Ornithology, R K Brooke, G K Brown, G Bundy, I Byrkjedal, A Carlson, Prof E Cătuneanu, E A Chapman, H Christiansen, P A Clancey, P Clement, Dr N J Collar, D R Collins, P R Colston, C J Colthrup, R J Connor, Dr L Cornwallis, E Cowley, H W E Croockewit, T Csörgo, M Dallmann, N N Danilov, Dr C E Davies, Dr N B Davies, P G Davis, W R J Dean, G Dick, A Dicksen, S Dicksen, M Dornbusch, R F Durman, Dr A Dyrcz, M B R Eagles, R A Earlé, S Efteland, J Ekberg, Dr T Elfström, R E Emmett, Prof Y Espmark, Dr A A Estafiev, M I Evans, P J Ewins, Fair Isle Bird Observatory, G D Field, Dr J C Finlayson, Flamborough Ornithological Group, Prof V E Flint, Prof V E Fomin, R Frumkin, C Fuller, R J Fuller, H Galbraith, I C J Galbraith, Major M D Gallagher, Dr I M Ganya, Dr P J Garson, C Gauci, Dr E I Gavrilov, Z Głowaciński, S M Goodman, P J Grant, A Green, Dr R E Green, A Gretton, Dr L G Grimes, Dr R Günther, G H Gush, A Håland, J Halcide, K Hall, P Hams, D Hansford, O Hasson, P J Hayman, P V Hayman, C J Hazevoet, J B Heigham, Dr H-W Helb, J Hellmich, I R Hepburn, C M Herrera, J Herrmann, Dr D A Hill, Dr U Hirsch, O Hogstad, Prof G Högstedt, F Holgado, P Hope Jones, M B Horan, C Hoth, Prof V D Ilyichev, Dr S Ince, Dr E V Ivanter, E Jaakkola, O J Jansen, Dr C Jarry, E D H Johnson, F J S Jones, O H Jørgensen, Dr E de Juana, J A Kålås, Dr H Källander, Dr J Karlsson, M Kasparek, B King, J L Klein, Dr A Knox, Dr A J Knystautas, T Köhl, Dr C König, F J Koning, Dr V A Korovin, P Koskimies, the late Dr Yu V Kostin, Dr A F Kovshar', K Kraüter, Dr P Kurlavičius, Dr E N Kurochkin, K Kyllingstad, P A Lassey, R Lenwink, Y Leshem, F LeSueur, V C Lewis, B Little, K Lockwood, Dr H Löhrl, F de Lope Rebollo, K-H Loske, Dr V M Loskot, Dr G L Lövei, W Lübcke, C-F Lundevall, C Lynch, J A McGeoch, Prof A W G Manning, S Marchant, Prof P Marler, R P Martins, C J Mead, Dr G F Mees, K Metz, R Möckel, L Molnow, Dr A P Møller, Dr J Moreno, J H Morgan, K Mukai, Abbé R de Naurois, Dr I A Neufeldt, Dr B Nicolai, G Olioso, V Olsson, T Oxelsen, J Palfery, D C Palmer, B Pambour, Dr E N Panov, D Parker, J Paul, D J Pearson, J Pearson, Dr J-C Pedroli, C Persson, B Petersen, B N Phillips, M Pickers, R F Porter, F Post, R Prodon, P Profus, M Pryl, R J Prytherch, Prof F S Purroy, R Raby, Dr D Radu, Dr R J Raines, L Raner, Dr Yu S Ravkin, V Ree, Dr G Rheinwald, N J Riddiford, G Rinnhofer, S Rjukan, D Robel, C R Robson, P Sackl, V A D Sales, S Scebba, Dr E R Scherner, Dr L Schifferli, R K Schmidt, Dr M Schubert, Dr D C Seel, L L Semago, G Shaw, H Shirihai, Prof C G Sibley, H Siman, Dr K E L Simmons, M Sivacusa, I Skoog, B E Slade, T Slagsvold, Prof P J B Slater, R Sluys, E Smith, P E Smith, Dr B K Snow, Dr J Sorjonen, A Stagg, Miss J Stannard, P Steele, J Sultana, L Svensson, N van Swelm, Dr J Talik, Dr V S Talposh, P B Taylor, W Tilgner, Dr L Tomiałojć, P Triplet, E L Turner, Dr A Tye, Dr S J Tyler, Dr R Van den Elzen, F J Walker, Dr D Wallschläger, Dr T Wesołowski, Dr K Wheatley, U Widemo, C G Wiklund, C Williams, Dr B Wood, M V Yakovleva, Dr Y Yom-Tov, Prof R G Zhordaniya, Dr V I Zinoviev, Dr V A Zubakin.

CITATION

The editors recommend that for references to this volume in scientific publications the following citation should be used: Cramp, S (ed.) (1988) *The Birds of the Western Palearctic*, Vol. V.

Order PASSERIFORMES

The perching birds, known colloquially as passerines. Largest order of class Aves, comprising well over half (*c.* 5300) of the total of living species. Diverse group of tiny to fairly large landbirds of many adaptive types, mainly arboreal but also terrestrial and aerial. Well characterized by possession of a syrinx (resonating chamber at lower end of trachea consisting of bony rings supplied with complex system of muscles and vibrating membranes) and of perching feet suitable for gripping slender branches (equipped with set of 4 toes joined at same level, hind toe of which is often stronger than others and non-reversible). Cosmopolitan, except Antarctica and some oceanic islands; represented in all terrestrial habitats throughout the world, from high Arctic to low Antarctic (South Georgia) and from almost waterless desert to tropical rain-forest. Some species are long-distance migrants.

No universal agreement on number of passerine families, total varying from *c.* 40 (Sibley 1985; Sibley and Ahlquist 1985*b*) to 105 (Wolters 1975-82), though most authors recognize *c.* 60 (e.g. Storer 1971, Morony *et al.* 1975). These formerly classified in 3-6 suborders arranged in 2 groups: suboscines, comprising 14-16 families in 2-5 suborders (see further, below); and oscines or true song-birds, comprising remaining families and four-fifths of the species (*c.* 4200) in a single suborder—Passeres (or Oscines). Now, however, with some re-allocation of families, suboscines too considered to form just a single suborder—Deutero-Oscines (Voous 1985) or Oligomyodi (Sibley 1985; Sibley and Ahlquist 1985*b*) with 12-13 families. Passeres represented in west Palearctic by 33 families, but Deutero-Oscines extralimital except for single species recorded as a straggler.

Subdivision of order traditionally based on morphology and anatomy of leg and foot and on position and structure of syrinx (including number of syringeal muscles). As currently classified, suboscines include the following families: (1) Eurylaimidae (broadbills), (2) Philepittidae (asities, etc.), (3) Furnariidae (ovenbirds), (4) Dendrocolaptidae (woodcreepers), (5) Formicariidae (antbirds), (6) Rhinocryptidae (tapaculos), (7) Cotingidae (cotingas), (8) Pipridae (manakins), (9) Tyrannidae (tyrant flycatchers), (10) Oxyruncidae (sharpbills), (11) Phytotomidae (plantcutters), and (12) Pittidae (pittas) (Voous 1985); to these may be added (13) Acanthisittidae (New Zealand wrens) (Sibley and Ahlquist 1985*b*). According to Stresemann (1927-34) and others (e.g. Ames 1971), suboscines likely to be of polyphyletic origin and a number of authors (e.g. Wetmore 1960) have proposed an arrangement of 3 suborders: Eurylaimi (family 1, above), Tyranni (families 2-13), and Menurae, latter comprising Menuridae (lyrebirds) and Atrichornithidae (scrub-birds). Storer (1971) later arranged them in 5 suborders: Eurylaimi (family 1), Furnarii (families 3-6), Tyranni (families 7-11), Menurae, and an unnamed suborder (families 2, 12, and 13). Subsequent biochemical and morphological studies, however, established that Menurae do not belong with rest of this assemblage (Sibley 1974); indeed, that group now placed in Passeres on evidence of DNA-DNA hybridization data (Sibley and Ahlquist 1985*b*). In addition, Feduccia (1974, 1975, 1977) showed that, again excepting Menurae, suboscines share an important common derived character—a non-reptilian condition of the structure of the bony stapes (columella) of the ear found otherwise amongst birds only in kingfishers (Alcedinidae). Consequently, Voous (1977) treated the suborders Eurylaimi, Furnarii, and Tyranni, plus Pittidae, as a monophyletic assemblage (his earlier 'Deutero-Oscines'), leading eventually to today's simpler arrangement in a single suborder (see above).

Following Voous (1977, 1985), oscine families represented in west Palearctic are: (1) Alaudidae (larks), (2) Hirundinidae (swallows), (3) Motacillidae (pipits, wagtails), (4) Pycnonotidae (bulbuls), (5) Bombycillidae (waxwings, Grey Hypocolius *Hypocolius ampelinus*), (6) Cinclidae (dippers), (7) Troglodytidae (wrens), (8) Mimidae (mockingbirds), (9) Prunellidae (accentors), (10) Turdidae (chats, thrushes), (11) Sylviidae (Old World warblers, kinglets), (12) Muscicapidae (Old World flycatchers), (13) Timaliidae (babblers), (14) Aegithalidae (long-tailed tits), (15) Paridae (typical tits), (16) Sittidae (nuthatches), (17) Tichodromadidae (Wallcreeper *Tichodroma muraria*), (18) Certhiidae (typical treecreepers), (19) Remizidae (penduline tits), (20) Nectariniidae (sunbirds), (21) Oriolidae (Old World orioles), (22) Laniidae (typical shrikes), (23) Corvidae (jays, magpies, crows, etc.), (24) Sturnidae (starlings), (25) Passeridae (Old World sparrows, snow finches), (26) Ploceidae (weavers), (27) Estrildidae (waxbills, etc.), (28) Vireonidae (Vireos), (29) Fringillidae (finches), (30) Parulidae (American warblers), (31) Thraupidae (tanagers), (32) Emberizidae (New World sparrows, buntings, etc.), and (33) Icteridae (New World orioles, etc.). Of these, Mimidae, Vireonidae, Parulidae, Thraupidae, and Icteridae are essentially Nearctic and only accidental in west Palearctic; Estrildidae are Afrotropical, Asian, and Australasian but introduced into west Palearctic.

Passeres usually thought to be of monophyletic origin, but relationships within order highly complex. Some families (e.g. Alaudidae, Hirundinidae) well-characterized and have all members easily allocated, but demarcation between many others poorly defined, accounting for great variation in number of families recognized, e.g. from 36 (Mayr and Amadon 1951) to 91 (Wolters 1975-82). By general consent, however, these fall into 3 broad adaptive groups (Mayr and Greenway 1956; Storer 1971; Voous

1977, 1985): (1) Old World insect-eaters and relatives, including families 3–20 above; (2) Corvidae and related families, including Ptilonorhynchidae (bowerbirds), Paradisaeidae (birds of paradise), and families 21–7; and (3) assemblage of 9-primaried insect- and seed-eaters, including families 28–33. Marked variation in sequence of families between various checklists, field-guides, and handbooks due mainly to some authors considering group 2 the most advanced—sequence then 1–3–2 (so-called Basle sequence: see Mayr and Greenway 1956); most, however, now consider group 3 the most recently developed and diversified group and list oscine families in sequence 1–2–3 (as above), following Storer (1971). Difficulties remain, nevertheless, in fitting some families of uncertain affinities into these 3 groups, notably the Alaudidae and Hirundinidae (which are usually listed first though are not necessarily primitive), while other families have been shifted from one group to another, causing further complications. For survey of problems of sequence of families and of sequence of species within passerine families, see Voous (1977) whose classification is followed here.

Traditional views of oscine classification have now been radically changed by DNA-DNA hybridization data which will necessitate sweeping future changes if accepted—as seems likely, for this method is not based on morphological characters and thus avoids pitfalls of resemblances due to convergent evolution in unrelated groups (see Sibley and Ahlquist 1985a, and literature cited there). This work indicates that Passeres fall into 2 broad assemblages ('parvorders'), 'Corvi' and 'Muscicapae', each comprising 3 main groups (super-families). In 'Corvi' are found: (1) 'Menuroidea'—e.g. Climacteridae (Australasian treecreepers), Menuridae, and Ptilonorhynchidae; (2) 'Meliphagoidea'—Maluridae (fairy-wrens, emu-wrens), Meliphagidae (honey-eaters), and Acanthizidae; and (3) 'Corvoidea'—'Corvidae' (drongos, crows, birds of paradise, monarch flycatchers, etc.), 'Laniidae' (typical shrikes), 'Vireonidae' (vireos), and a number of other taxa. In 'Muscicapae' are found: (1) 'Turdoidea'—Bombycillidae, Cinclidae, 'Turdidae' (thrushes, flycatchers, chats), and 'Sturnidae' (starlings, mockingbirds); (2) 'Sylvioidea'—'Sittidae' (nuthatches, wallcreepers), 'Troglodytidae' (typical treecreepers, wrens, etc), 'Paridae' (penduline tits, typical tits), Aegithalidae, Hirundinidae, 'Regulidae' (kinglets), Pycnonotidae, 'Cisticolidae' (African warblers), Zosteropidae (white-eyes), and 'Sylviidae' (Old World warblers and allies, including babblers); and (3) 'Fringilloidea'—Alaudidae, 'Nectariniidae' (sugarbirds, flowerpeckers, sunbirds, etc.), 'Ploceidae' (Old World sparrows, waxbills, wagtails, pipits, accentors, weavers), and 'Fringillidae' (typical finches, buntings, American warblers, New World orioles, tanagers, etc.). Some of these relationships, as they affect west Palearctic taxa, will be discussed further in the family summaries.

All passerines share a number of skeletal and myological characters (Stresemann 1927–34), including aegithognathous palate, absence of basipterygoid processes, structure of sternum, arrangement of flexor muscles of toes, and 14 cervical vertebrae (15 in Eurylaimidae); differ from remainder of class Aves (colloquially known as non-passerines) in shape of spermatozoa. 9–10 functional primaries, with an additional reduced outermost one (i.e. 10–11 altogether) in some species. Generally 9 secondaries, inner 3 of which form tertials, but 13 in Menuridae and 10–11 in some Corvidae and allies (Stephan 1970); eutaxic. Generally 12 tail-feathers, but 10 or 14 in some and up to 16 in Menuridae. Intestinal caeca present but rudimentary. Oil-gland present; unfeathered. Powderdown feathers found in Artamidae (wood-swallows). Suboscines characterized by rather simple structure of syringeal muscles (basically mesomyodian or anisomyodian) with generally fewer than 4 pairs of intrinsic muscles; syrinx of oscines complex (basically acromyodian or diacromyodian) with 5–8 pairs of intrinsic muscles, matching higher vocal capacities. Perching foot (see above) with all toes entirely free in most oscines but some of front toes usually fused in suboscines (Eurylaimidae characterized by entirely fused middle and inner toes, partly fused middle and outer toes, and by tendon joining flexor muscle of hind toe to those of front toes). Type of tarsal scutellation shows much variation within order and this and shape of tarsus have been used as taxonomic characters (see, e.g., Ridgway 1901–11). Aftershaft much reduced or absent. Generally a single (post-breeding) moult in course of annual cycle but some variation (for details, see accounts for individual families). Moult of primaries generally descendant, but ascendant in Spotted Flycatcher *Muscicapa striata* and complex in dippers *Cinclus* and some Old World warblers (e.g. *Cettia*, *Locustella*). Moult of tail generally centrifugal, but many exceptions. Young atricial, nidicolous, and food-dependent; blind at hatching with sparse down (dense in Menuridae) or body virtually naked, especially in hole-nesting species; beg for food by gaping, typically exposing brightly coloured pattern of mouth, often with contrasting pale or dark spots—in non-passerines, bright gape present only in hoopoes (Upupidae), mousebirds (Coliiformes), and cuckoos (Cuculiformes).

Anting behaviour recorded from many groups, ants being applied in bill to plumage, usually underside of wing-tip (direct or active anting, ant-application), and/or allowed access to plumage (indirect or passive anting, ant-exposure); this habit unique to Passeriformes (e.g. Simmons 1966, 1985a). Other behavioural characters include head-scratching by indirect (or 'overwing') method in majority of families, oiling of head by head-scratching, and dissipation of heat by gaping and panting (not by gular-fluttering, so far as known). Bathing widespread, mainly by use of stand-in method in shallow water, but some groups use in-out method (jumping into and out of water repeatedly) or are flight- or plunge-bathers, while others bathe only or mainly in rain or amongst wet foliage (see, e.g., Simmons 1985*b*). Dusting confined to only a few groups, but sunning—both for gaining heat (sun-basking) and other purposes (sun-exposure)—widespread, usually by means of lateral and spread-eagle postures (details in Simmons 1985*d*, 1986). Sipping is most usual drinking method, with dip-and-tilt action, but a few groups drink by sucking while others drink only from flight (Maclean 1985). Foot used in feeding by some species or larger groups, either for scratching up earth or ground debris (with or without flicking action of bill) or for holding food items—gripped between toes (like hand) or clamped to perch.

Piciformes (woodpeckers, etc.) usually considered to be closest relatives of Passeriformes. Cracraft (1981), however, indicated that Coliiformes and Coraciiformes (kingfishers, bee-eaters, etc.) may be closer—but all three of these near-passerine orders included by him within the same major 'division' of birds as the Passeriformes.

Family TYRANNIDAE tyrant flycatchers

Small to medium-sized suboscine passerines (suborder Deutero-Oscines); mainly arboreal, often feeding on insects caught by sallying out from perch, though some species terrestrial. About 360 species in numerous genera. Occur over whole of New World but main diversity in Neotropics. Northern forms migratory. A single representative (Acadian Flycatcher *Empidonax virescens*) in west Palearctic, recorded only as a straggler.

Arboreal species quite similar in form to Old World flycatchers (Muscicapidae), with similar feeding habits, but differ in partly fused front toes, circular tarsal scutellation (scutes interrupted on inside of tarsus), fewer syringeal muscles (see Passeriformes), and several other anatomical characters. Sexes usually of similar size. Aerial feeders have bill usually flattened and mouth wide (giving triangular outline to bill as seen from above); upper mandible with distinct hook at tip; strong bristles at base of bill. Wing variable in size and shape but long and pointed in flycatching species. Tail of medium length in most species but short in some and long in others; usually square-tipped or slightly forked. Legs and feet usually small and weak. Head-scratching by indirect method. Plunge-bathers, diving into water from perch. Anting recorded in at least one species.

Plumage generally dull-coloured (brown, grey, or olive) but some species have contrasting patch on crown (usually red, orange, or yellow) and some crested. Sexes usually similar; no seasonal variation.

Closest relatives within suboscines are Pipridae (manakins), Cotingidae (cotingas), Oxyruncidae (sharpbills), and Phytotomidae (plantcutters), as revealed by anatomy, egg-white proteins, and DNA data (Sibley 1970; Sibley and Ahlquist 1985*b*). For relationships within Tyrannidae, see Traylor (1977).

Empidonax virescens Acadian Flycatcher

PLATE 37
[between pages 544 and 545]

Du. Groene Elftiran   Fr. Moucherolle de l'Acadie   Ge. Buchentyrann
Ru. Американская восточная мухоловка   Sp. Papamoscas verde

Platyrhynchos virescens Vieillot, 1818

Monotypic

Field characters. 11·5 cm; wing-span 21–24 cm. 20% smaller than Spotted Flycatcher *Muscicapa striata*; slightly larger (especially about head) than Red-breasted Flycatcher *Ficedula parva*. Rather small, high-crowned bird with flycatching habits. Upperparts essentially olive-green with striking double pale wing-bar. Underparts dusky-olive to yellow-white. Flight like *Ficedula* flycatcher; wags tail but does not hold it cocked. Voice important in distinction from congeners. Sexes similar; no seasonal variation. Juvenile separable.

ADULT BREEDING. Head down to below eye, nape, back, and rump olive-green, uniform except for narrow pale yellow eye-ring. Wings dark dusky-olive, strikingly relieved by broad pale yellow tips to median and greater coverts which create much more obvious double wing-bar than on *M. striata* (on some birds, faint 3rd bar on longest lesser coverts). Pale yellow fringes to tertials and secondaries show as obvious lines or panel on folded wing. Tail dusky olive, lacking obvious marks (thus like *Muscicapa*, not *Ficedula*). Chin and throat dull white, becoming faintly yellow-olive on breast and upper flanks, and yellow then dull white again on rear flanks, belly, and vent. Underwing bright yellow on axillaries, paler and whiter on coverts. ADULT NON-BREEDING. On some, underparts heavily suffused dull yellow, this colour restricting or covering white throat and vent; these birds not easily distinguishable from Yellow-bellied Flycatcher *E. flaviventris*. JUVENILE. Brownish-olive above with buff feather-fringes and wing-bars. Underparts whitish with olive wash on breast. At all ages, bill brown-horn above, yellow-flesh below; legs dark grey.

Above description will distinguish *E. virescens* from west Palearctic flycatchers but quite insufficient to separate it from 4 congeners which are equally or even more likely to cross North Atlantic. Judged by breeding distributions in eastern North America, likelihood of occurrence of these 5 is ordered (1) *E. flaviventris*, (2) Alder Flycatcher *E. alnorum*, (3) Least Flycatcher *E. minimus*, (4) Willow Flycatcher *E. traillii*, (5) *E. virescens*, and all must be considered once generic identification made. *E. virescens* is largest, with longest bill and wings (folded primaries extend 1·7–2 cm beyond tertials; in other 4 species, extension less than 1·3 cm), and has upperparts less brown or grey than *E. alnorum*, *E. traillii*, and *E. minimus*, and less yellow than *E. flaviventris*. *E. virescens* less yellow above (except in autumn) than *E. flaviventris*, wing-bars and wing-panel rather duller, and upper mandible paler. See Phillips *et al.* (1966) and Phillips and Lanyon (1970). Flight and behaviour recall *Ficedula* flycatcher; wags tail but without lifting it above wing line, unlike full cocking action of *Muscicapa* and *Ficedula*.

Only call likely from vagrant an emphatic 'weece' (*E. flaviventris* gives a distinctive loud sneezy 'chew', *E. alnorum* a loud piping 'peep', *E. minimus* a sharp 'whit', and *E. traillii* a liquid 'wit').

Habitat. Breeds in Nearctic in lower middle latitudes, in moist or swampy lowland woods, especially deciduous floodplain forests, usually staying below canopy, and particularly favouring beech trees. Prefers deep shade of fairly mature stands with ample open areas for flycatching. Not normally outside woodlands in breeding season. Sometimes in wooded uplands, and often along streams or riversides, where nests may be sited above water. (Peterson 1947, 1960; Pough 1949; Godfrey 1966; Johnsgard 1979; Robbins *et al.* 1983.) In winter in Venezuela, occurs at 900–1200 m in forest edge, clearings, second growth, swampy woods, and haciendas, hunting among low branches of densely foliaged low trees (Schauensee and Phelps 1978). Flies only in lowest airspace, except possibly on migration. Contact with man slight, but locally affected by loss of habitat.

Distribution. Breeds mainly from south-east North Dakota, southern Michigan, extreme southern Ontario, southern New York, and south-west Connecticut south to central Texas, the Gulf coast, and central Florida. Winters from Nicaragua to Ecuador and Venezuela.

Accidental. Iceland: specimen, perhaps ♂, 4 November 1967 (AP).

Movements. Migratory.

Occurs on southward migration in south-east USA, casually west to western South Dakota and Nebraska, regularly on Caribbean coast of Mexico, and in Belize, Guatemala, and Honduras (American Ornithologists' Union 1983); rare (occurring in October) in Bahamas and western Cuba (Bond 1961). Winters in eastern Nicaragua and on Caribbean (most frequent) and Pacific (scarce) slopes of Costa Rica and Panama, also in Colombia, Ecuador, and western Venezuela (American Ornithologists' Union 1983; Rappole *et al.* 1983). Spring migration apparently follows similar route, with eastern-most Caribbean records again in western Cuba and Bahamas where rare (Bond 1961).

Post-breeding dispersal and autumn migration in August and early September may coincide with hurricane season in eastern USA, migrants lingering in deep south of USA into early October. Spring migration commences April, reaching Florida (probable records) late April to early May (Phillips *et al.* 1966), breeding areas in USA by May, and northern regions as late as early June. There are, however, only a few accidental records well to north of breeding range (e.g. Todd 1957, Godfrey 1966), and this, combined with comparatively southern North American breeding distribution, might indicate transatlantic vagrancy is unlikely. Difficulty of identification, even in the hand, complicates evaluation of records. TL–E

Voice. See Field Characters.

Plumages. ADULT BREEDING. Entire upperparts uniform olive-green, centres of feathers of crown sometimes slightly darker olive. Sides of head to just below eye, sides of neck, and upper flanks like upperparts, but lores and ear-coverts sometimes slightly brighter green and eye surrounded by a rather distinct ring of pale yellow feathers; feather-bases on lores slightly paler, yellowish, sometimes showing as a pale patch, not extending to eye-ring. Chin and upper throat white with faint pale grey mottling, gradually merging into olive-green on lower cheeks and chest; olive-green of chest sometimes interrupted by pale yellow in middle. Central belly white, merging into pale yellow on sides and on lower flanks, vent, and under tail-coverts; sometimes entirely pale yellow. Tail-feathers dusky grey, outer webs tinged olive-green. Flight-feathers dark grey, secondaries with broad pale green-yellow fringes along outer webs (narrow or absent on tips of outer secondaries); outer webs of tertials with broader green-yellow fringes, soon bleaching to white; outer web of p10 with narrow yellowish fringe at base. Basal ⅓ of secondaries without pale fringe, and basal parts of secondaries and tertials darker than remainder of flight-feathers, showing as dull black bar. Upper wing-coverts including primary coverts dull black, greater and median coverts with broad pale yellow tips, showing as 2 very distinct wing-bars; lesser coverts with broad green tips, sometimes showing as less distinct bars, tips of longer lesser coverts occasionally yellow, forming a fairly distinct 3rd bar. Marginal under wing-coverts and axillaries bright pale yellow, remainder of underwing whitish-yellow. *In worn plumage*, upperparts slightly duller and greyer green; underparts distinctly paler, yellow-white or white; olive-green patches on sides of chest more pronounced; some dusky grey of feather-bases sometimes showing on belly; wing-bars and fringes of secondaries and tertials bleached to off-white and partially abraded. ADULT NON-BREEDING. Similar to adult breeding, but usually a more prominent yellow wash on flanks and lower belly (less white) and wing-bars sometimes richer yellow-buff. JUVENILE. Rather like adult, but upperparts duller greenish- or brownish-olive, feather-tips with narrow dull ochre-buff fringes, sometimes joining to form faint pale half-collar on rear and sides of neck; buff fringes on rump slightly wider, appearing slightly paler than remainder of upperparts. Feathers of underparts softer than in adult, white or (sometimes) faintly suffused pale yellow; chest-band greyish-olive, less olive-green than adult. Tail, tertials, and wing as adult breeding, but tips of greater and median upper wing-coverts broadly margined deep buff or clear chestnut-buff, wing-bars buffier than in most adults. FIRST ADULT NON-BREEDING. Like adult non-breeding, but juvenile wing, tail, and occasionally part of body feathers retained until arrival in winter quarters: bars on juvenile wing rich buff, slightly bleached when compared with fresh juvenile, but often buffier than yellow-white or yellow-buff of adult non-breeding. Underparts usually pale greenish-yellow with white chin and olive-grey chestband (as in adult non-breeding), but occasionally throat suffused pale yellow and remainder of underparts entirely lemon-yellow (Mengel 1952).

Bare parts. ADULT. Iris brown. Upper mandible deep or dark horn-brown, lower mandible entirely lilac-white or flesh-colour; mouth flesh-colour or pale dull yellow. Leg and foot grey. JUVENILE Similar to adult, but mouth sometimes brighter yellow. (Ridgway 1901–11; Phillips *et al.* 1966.)

Moults. ADULT POST-BREEDING. Complete, primaries descendant. Starts late July and early August, head first; by early September, usually all moult completed or 1–2 outer primaries still growing, but occasionally moult hardly started (Mengel 1952; Mumford 1964; Phillips *et al.* 1966). ADULT PRE-BREEDING. Partial, in winter quarters. No information on timing and extent (Mengel 1952). POST-JUVENILE. Complete, primaries descendant. Timing rather variable, depending on hatching date; starts shortly after fledging as soon as tail full-grown. Mantle, scapulars, chest, and flanks moulted first, throat, belly, and crown last; wing and tail in winter quarters. Among those examined, specimens with head and body in 1st adult non-

breeding but with wing and tail still juvenile occur from early August; some, however, still fully juvenile mid-August, and others show mixture of juvenile and 1st adult non-breeding on body up to mid-September—birds in such a plumage occur on migration. No information on further progress of post-juvenile moult in winter quarters and no data on extent and timing of pre-breeding moult (C S Roselaar; BMNH).

Measurements. ADULT. Eastern USA, April–September; skins (BMNH, RMNH). Bill (S) to skull, bill (N) to distal corner of nostril. Exposed culmen on average *c.* 5 less than bill (S).

WING	♂ 76·2 (1·77; 21)	73–80	♀ 71·6 (1·00; 10)	68–75
TAIL	58·5 (2·12; 16)	56–62	55·0 (1·21; 10)	53–57
BILL (S)	16·7 (0·51; 16)	15·5–17·7	16·2 (0·40; 9)	15·4–16·7
BILL (N)	9·4 (0·35; 16)	8·8–9·9	9·1 (0·32; 9)	8·5–9·4
TARSUS	15·6 (0·53; 16)	14·8–16·6	14·8 (0·66; 9)	14·2–16·0

Sex differences significant, except for bill.

JUVENILE. Wing on average *c.* 0·4 shorter than adult, tail *c.* 0·1; tarsus and bill similar to adult from shortly after fledging.

Weights. Belize, March–April: ♂♂ 10·0, 12·2, 16·5; ♀ 12·9 (Russell 1964). Kentucky (USA): May–June, ♂♂ 12·3, 12·7, ♀ 15·5; September ♂♂ 12·3, 12·9, 14·1; ♀♀ 12·2, 13·0 (Mengel 1965). Breeding, USA: ♂ 13·2 (11) 11·9–13·9, ♀ 12·7 (19) 11·1–13·8 (Walkinshaw 1966).

Structure. Wing rather long, tip slightly rounded. 10 primaries: p8 longest, p9 0–1 shorter, p10 7–9, p7 1–3, p6 6–11, p5 10–16, p1 17–24 (BMNH); p6 rarely only 5·5 shorter than p8 (Phillips *et al.* 1966). Outer web of p7–p9 and inner of p8–p10 emarginated. Tail rather long, tip square; 12 feathers, rather narrow and with pointed tips. Bill markedly flattened, distinctly triangular as seen from above, 7–8 mm wide at base, sides slightly convex; culmen with blunt ridge, ending in fine nail at tip; lower mandible flattened, with blunt ridge at tip of gonys. Nostrils small, rounded, largely covered by tuft of short bristly feathers. Base of upper mandible with *c.* 5 bristles of up to 10 mm long projecting obliquely forward above gape. Leg and foot weak and slender, tarsus partly scutellated; middle and outer toe basally fused. Middle toe with claw 12·3 (10) 11·6–13·3 mm; outer toe with claw *c.* 77% of middle with claw; inner *c.* 73%, hind *c.* 85%.

Recognition. *E. virescens* belongs to a genus whose members are notoriously difficult to identify. See Field Characters. CSR

Family ALAUDIDAE larks

Rather small oscine passerines (suborder Passeres); terrestrial, feeding on insects, seeds, and plants. About 80 species in 15 genera: (1) *Mirafra* (bush-larks), *c.* 30 species—mainly Afrotropics, with outliers in Madagascar, southern Asia, and Australia; (2) *Certhilauda* (long-billed larks), 3 species—southern Africa; (3) *Eremopterix* (finch-larks), 7 species—Africa and south-west Asia; (4) *Eremalauda* (Dunn's Lark *E. dunni*), monotypic—North Africa and Arabia; (5) *Ammomanes* (desert larks), 4–5 species—Africa and Asia; (6) *Alaemon* (hoopoe-larks), 2

species—northern Africa and Middle East; (7) *Chersophilus* (Dupont's Lark *C. duponti*), monotypic—Spain and North Africa; (8) *Pseudalaemon* (Short-tailed Lark *P. fremantlii*), monotypic—East Africa; (9) *Rhamphocoris* (Thick-billed Lark *R. clotbey*), monotypic—North Africa, Middle East, and Arabia; (10) *Melanocorypha* (calandra larks), 6 species—southern Europe, North Africa, and Asia; (11) *Calandrella* (short-toed larks and sandlarks), 12–13 species—southern Europe, Africa, and Asia; (12) *Galerida* (crested larks), 5–6 species—Europe, Africa, and Asia; (13) *Lullula* (Woodlark *L. arborea*), monotypic—Europe and Middle East; (14) *Alauda* (skylarks), 3–4 species—Europe, North Africa, and Asia; (15) *Eremophila* (horned larks), 2 species—Holarctic, Himalayas, and northern South America. Occur mainly in open, arid or semi-arid habitats of Old World, hardly penetrating into Australia and oceanic islands; only one species (Shore Lark *Eremophila alpestris*) in North America and northern South America, but Skylark *Alauda arvensis* introduced into North America and locally elsewhere. Many forms migratory. Family represented by 21 species in west Palearctic, all but one breeding.

Differ from all other Passeres in structure of tarsus and syrinx: back of tarsus rounded (latiplantar), covered with small scutes (not sharply edged and smooth); syrinx without ossified pessulus (small knob at point of fusion of bronchiae) and with only 5 pairs of muscles. Body rather robust; neck short. Sexes often differ in size, with ♂ larger. Bill shape highly variable, even within species (as in *Melanocorypha*) or between sexes (as in Razo Lark *Alauda razae*); adapted to special feeding method and/or diet (e.g. long and curved for digging or short and stout for crushing hard seeds). Wing fairly long and often pointed; wing-area greatest in migrant forms (Meinertzhagen 1951a). Flight typically strong and undulating, with periodic closure of wings, but rather fluttering and wavering over short distances; steeply climbing or circling song-flights are characteristic. Tails short or of medium length. Legs short to fairly long; usual gait a walk or run, less frequently a hop.

Hind claw straight and often long, especially in species living on soft soil with short close vegetation, but short in those on bare and hard ground; toes and claws relatively shortest in expert runners, such as Hoopoe Lark *Alaemon alaudipes*. Head-scratching by indirect method. Bathing in standing water does not occur, but rain-bathing and dusting typical. Sunning recorded but not anting.

Plumage usually cryptic, unmarked or streaked; often adapted to colour of local soil, species occurring in large range of habitats tending to have several races differing in colour, particularly when non-migratory (e.g. Vaurie 1951a). Some species show conspicuous marks on wing or outer tail-feathers, especially in flight. Sexes usually similar. Single complete annual moult typical (adult post-breeding); juveniles somewhat unusual among non-tropical oscines in having a complete rather than a partial post-juvenile moult immediately after fledging. Nestlings have rather scanty down, confined to upperparts; often spotted. Mouth with some contrasting dark spots. Young usually leave ground nest before able to fly.

Relationships to other Passeres, as indicated by traditional anatomical studies, obscure. Hence, now usually put at beginning of sequence, though previously sometimes placed near 9-primaried oscines because bill similar to that of buntings (Emberizinae) and finches (Fringillidae) and p10 is reduced in adult. Affinities with these and other members of 9-primaried group supported by egg-white protein data (Sibley 1970) and more recently by that of DNA-DNA hybridization (see Passeriformes).

Eremopterix nigriceps Black-crowned Finch Lark

PLATES 1 and 11
[facing page 160 and between pages 184 and 185]

Du. Zwartkruinvinkleeuwerik   Fr. Alouette-moineau   Ge. Weissstirnlerche
Ru. Чернобрюхий вьюрковый жаворонок   Sp. Alondra cabecinegra   Sw. Svartkronad Finklärka

Pyrrhalauda nigriceps Gould, 1841

Polytypic. Nominate *nigriceps* (Gould, 1841), Cape Verde Islands; *albifrons* (Sundevall, 1850), southern Sahara east to Nile valley; *melanauchen* (Cabanis, 1851), Red Sea coast and eastern Africa, east to north-west India.

Field characters. 10–11 cm; wing-span 20–22 cm. 20–25% smaller than any other west Palearctic lark; hardly larger but broader-winged than Serin *Serinus serinus*. Tiny, rather dumpy lark, with stubby, conical bill, short, broad wings, rather short legs, and (on ground) compact form suggesting small finch (Fringillidae). ♂ more boldly patterned than any other west Palearctic lark, with totally black underparts and sandy upperparts; grey-white head decorated with black. ♀ strikingly different from ♂, with sandy-brown upperparts and no obvious characters except for dull dark median-covert bar and black under wing-coverts. Sexes dissimilar; no seasonal variation. Juvenile

difficult to separate from ♀. 3 races in west Palearctic; Red Sea and Arabian race, *melanauchen*, described here (see also Plumages and Geographical Variation).

ADULT MALE. Forehead white, cheeks, sides of neck, and collar on nape grey-white, contrasting markedly with black crown, band through eye to bill, and lower border of cheek. Rest of upperparts warm dun, with indistinct soft, dusky streaks on mantle and wing-coverts, and paler grey-buff edges to tertials and wing-coverts; darker-centred median coverts form dull bar across wing. Tail dark, with sandy-brown central feathers and black outer feathers unmarked except for dirty white webs on outermost pair; undertail black. All upperparts except tail may bleach to stone colour. Shoulder outlined with white but underparts from chin to vent solidly black. In flight, wide black under wing-coverts show well on broad wings. ADULT FEMALE. Pale tawny above, buff-white below; indistinct brown streaks and mottling on upperparts, lower throat, and chest, showing only at close range; most obvious feature at rest is darker median-covert bar, as in ♂. Tail and under wing-coverts similar to ♂, but underwing with less black. JUVENILE. Resembles adult ♀ but feathers distinctly pale-edged on mantle and wing-coverts and faintly barred grey on crown. At all ages, bill horn; legs pale grey-horn.

Unmistakable in west Palearctic, with small size, black under wing-coverts, and mainly black tail diagnostic (south of Sahara, confusable with 3 congeners). Flight action noticeably light, with rapid flutter on take-off (when almost invisible pale ♀♀ and juveniles suddenly become visible alongside ♂♂), easy bursts of fast wing-beats, and then undulating descent. Song-flight consists of fluttering ascent, undulating flight, and stepped descent with wings slightly raised. Escape-flight variable: usually short if disturbance slight but long and panicky if flock frightened. Walks, and runs quickly, crouching close to ground on stopping; will also stand upright, recalling alert sparrow *Passer*. Highly gregarious.

Song twittering, even chirping in quality, recalling smaller pipits *Anthus*. Commonest call a quiet 'jip'; chorus of flock sounds thin and distant.

Habitat. A tropical species based on saharo-sahelian savannas, extending into west Palearctic only along lowland coastal belts with moisture regime favouring vigorous seasonal growth of herbage, especially such grasses as *Aristida*, *Lasiurus*, and *Panicum*, and succulents such as *Aizoon*; also occasional *Acacia* trees. Associated soils tend to be fine and sandy, often in depressions or beds of wadis, or on flats subject to occasional brief flooding. Saline, stony, and rocky terrain, and broken or strongly sloping ground are avoided. On islands of Dahlac archipelago in Red Sea, common on beaches near low scrub (Clapham 1964). Near coast of southern Morocco, attracted to tussocky vegetation (Valverde 1957), but on Red Sea coast just within tropics favours patches of prostrate ephemeral herbage (including *Arnebia*) subject to camel grazing (E M Nicholson and P A D Hollom). In eastern Saudi Arabia, breeds in open, lightly vegetated sandy desert (G Bundy), but afterwards frequents margins of desert ponds formed by sewage and oasis runoff water (J H Morgan and J Palfery). On Taif plateau (Saudi Arabia) found sparsely from sea-level to *c.* 1800 m (Meinertzhagen 1949c). Needs little water, and on Somali coast even drinks seawater (Archer and Godman 1961a). In Indian subcontinent, occurs plentifully on arid sandy wastes and near canal-fed desert cultivation (Ali and Ripley 1972). On Cape Verde Islands, occurs on dry

plains, also on stubble and an aircraft runway (Bannerman and Bannerman 1968). High-density concentration in breeding territories, indicated by tight circles of song-flights in lower or lower-middle airspace, necessitates higher level of biological productivity than true desert can afford. As a tropical savanna–desert ground bird, has to accept some disturbance by grazing animals, especially camels, but little affected by human activities at very low population levels.

Distribution. ISRAEL. May breed occasionally (HM). IRAQ. Only one breeding site known (Allouse 1953). KUWAIT. Breeds (PRA). CHAD. Tibesti: possibly breeds (Siman 1965).

Accidental. Israel, Algeria.

Population. CAPE VERDE ISLANDS. Fairly common (Bannerman and Bannerman 1968).

Movements. Resident and partial short-distance migrant.

Present all year in Sahel breeding zone, though in central Chad at least some birds move further north, to breed, in wet season August–October (Newby 1980). Nomadic tendency indicated by extensions to breeding range when conditions suitable, e.g. recent spread into Sénégal following drought conditions there (Morel and Ndao 1978). Etchécopar and Hüe (1967) considered this a Sahel species in which some birds move north into eastern North Africa (Libyan Desert and Egypt) in winter, which is rainfall season there; inclusion of Libya unsubstantiated however (Bundy 1976). Extent of movement obscured by uncertainty over northern limits of breeding range, which (as in Sahel) probably vary from year to year according to rainfall. Breeding range certainly extends into southern Morocco, and Valverde (1957) thought these birds moved south for winter.

Summer visitor to interior deserts of Arabia where it does not occur in same areas each year and seems to be extremely nomadic (Jennings 1980, 1981a). In eastern Saudi Arabia, apparently 2 distinct populations. Birds south of *c.* 25°N stay all year around breeding localities and breed earlier (March) than those to north, where, apart from occasional January–March coastal records, the species is absent late September to late April, sometimes even from June. Large flocks occurring near coast of eastern Saudi Arabia in May are perhaps migrants crossing Persian Gulf. (G Bundy.) Incomplete information on wintering areas of migrants, but range certainly includes Oman where substantial passages occur July–November and March–June (Gallagher and Woodcock 1980; Walker 1981a, b). Also locally migratory in Pakistan and north-west India (Ali and Ripley 1972), and some Arabian migrants perhaps cross Gulf of Oman. RH

Food. Seeds and insects. Feeds either by picking up food from ground or by taking it directly from vegetation—usually by reaching up from ground, but also

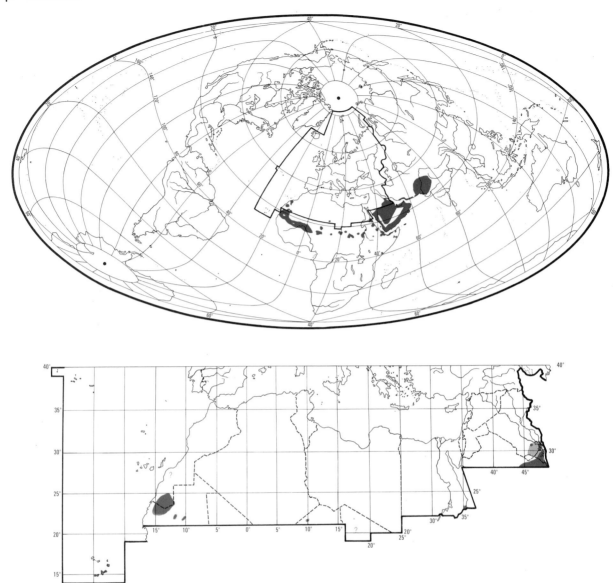

by fluttering up to grass heads, presumably to snatch seeds or, possibly, insects. May run erratically and occasionally chases flying insects, showing considerable speed and agility, and one bird seen using this method persistently. (Gallagher and Woodcock 1980; J H Morgan and J Palfery.) In Somalia will enter shops to pick up grain among customers, and apparently drinks brackish and salt water from coastal lagoons and sea (Archer and Godman 1961a).

On Cape Verde Islands, 4 birds collected by Naurois (1969b) at unspecified time of year contained small seeds (many), spiders (Araneae), grasshoppers (Orthoptera), 2 small beetles (Coleoptera), assassin bugs (Reduviidae), 2 caterpillars (Lepidoptera), and other insect remains; also grit. For comparison with Bar-tailed Desert Lark *Ammomanes cincturus* in same area, see that species (p.

61). In southern Morocco, 6 adults and 4 juveniles contained only seeds, including those of grass *Panicum* (Valverde 1957). In southern Sahara, Bates (1933–4) found shot birds invariably contained seeds, generally grass seeds. In Ennedi (Chad), feeds especially on seeds of *Panicum turgidum* (Gillet 1960). In Niger, July or August, stomachs contained insects (Villiers 1950). In Eritrea (Ethiopia), feeds readily on small locust-hoppers (Orthoptera) (Smith 1955b). At Jiddah (Saudi Arabia), April, ♂ fed ♀ with seeds of *Zygophyllum simplex* (Trott 1947). In India and Pakistan, takes grass and weed seeds, and insects (Ali and Ripley 1972).

♀♀ on Cape Verde seen carrying insect larvae, presumably for young (Bourne 1955), and identifiable items brought to nestlings in eastern Saudi Arabia were moths (Lepidoptera), caterpillars, and other insect larvae.

Young, in and out of nest, eastern Saudi Arabia, were fed sunrise to sunset. At one nest, 261 feeding visits in one day: 29·5% of visits 05.00–08.00 hrs, 37% 16.00–18.00 hrs; only 3% 11.00–13.00 hrs. Food delivery to young (in and out of nest) tends to be in bouts lasting 9–18(–25) min. Adults sometimes visit nest apparently without food but spend long time with bill, almost closed, deep in nestlings' mouths; perhaps delivering water. (J H Morgan and J Palfery.) DJB

Social pattern and behaviour. Unless otherwise acknowledged, account based on detailed notes supplied by J H Morgan and J Palfery from extralimital study at Dhahran (eastern Saudi Arabia).

1. Gregarious outside breeding season; flocks relatively small and loose—up to 60 birds (Ripley and Bond 1966; Erard and Etchécopar 1970; Gallagher and Woodcock 1980; Jennings 1980; Walker 1981a, b). In May, eastern Saudi Arabia, assemblies of up to 4500 or more birds, probably migrants (G Bundy). Flock size perhaps somewhat dependent on temperature as said to occur generally alone or in pairs in Somalia, but in flocks of 20–30 in cold weather (Archer and Godman 1961a). One flock (no details), western Sahara, contained numerous independent young and some adults (Valverde 1957). BONDS. No information on mating system. Once established, territory defended by both members of pair. Incubation and brooding mostly by ♀, less so by ♂. Both sexes share feeding of young; for further information on parental roles in care of young, see part 2. BREEDING DISPERSION. Solitary and territorial, but, at least in some areas, forms neighbourhood groups (Alexander 1898; Strickland and Gallagher 1969; P A D Hollom). No information on size of territory, though it extends beyond immediate nest-area and apparently overlaps with neighbours'; boundaries not strictly defended. Breeding ♀♀ especially tend to trespass on neighbouring territories, while unestablished birds not uncommonly enter territories to feed. After hatching, boundaries even more flexible, birds then defending only immediate nest-area and sometimes collecting food (for young) well outside their own 'territory'. In 2 years, nearest nests 60 m and 70 m apart. In area of *c.* 75 ha, of which just over 30 ha densely occupied, *c.* 10–11 pairs and 5 unpaired (but territorial) ♂♂; in the following year, 12 nests in same area. At Abu Ramad (Egypt), singing ♂♂ often less than 100 m apart

(P A D Hollom and E M Nicholson). Site-fidelity seems to vary with region. In central Arabia, birds extremely nomadic, breeding in different areas every year (Jennings 1980; J H Morgan and J Palfery). In eastern Saudi Arabia, appear to breed in same areas year after year, but no information on site-fidelity of individuals. ROOSTING. Nocturnal but no details. In middle of day, regularly seeks shade of vegetation or rocks (J H Morgan and J Palfery), e.g. at Khartoum (Sudan), February–May, parties of 5–6 crouched beneath bushes (Witherby 1901). In Saudi Arabia, July, *c.* 20 adults and juveniles loafed in early afternoon; sought shade of rocks, then flopped on bellies in sand and, with wings spread and fluttering, made hollows (J H Morgan and J Palfery). On Cape Verde Islands, dust-bathes freely (Bourne 1955). For roosting of young, see Relations within Family Group (below).

2. Often allows fairly close approach, and in Somalia enters shops to feed among customers (Archer and Godman 1961a; see also Alexander 1898, Witherby 1901, Ripley and Bond 1966). Crouches if threatened, often creeping under large stone, and may also give Warning-call (Alexander 1898: see 3 in Voice); by crouching thus, small flocks can escape notice of harrier *Circus* passing nearby (Meinertzhagen 1954). When excited, e.g. during confrontation, raises crest (see Heterosexual Behaviour, below). FLOCK BEHAVIOUR. No details. Flocks usually silent (Bourne 1955). SONG-DISPLAY. ♂ sings (see 1 in Voice) from start of breeding season, usually in Song-flight, sometimes from ground (Erard and Etchécopar 1970; Jennings 1981b) or bush (J H Morgan and J Palfery) or even tree-top (Alexander 1898). Song-flight also rarely performed by ♀ (see below). ♂ reported to perform Song-flight above sitting mate (Ripley 1951; Archer and Godman 1961a), but this not necessarily so, and one ♂ sang regularly at opposite end of territory from nest-site. Typically, ♂ takes off and, with legs dangling, ascends fairly steeply but linearly with rapidly beating wings, usually singing as he climbs. At peak of ascent, usually *c.* 6–10(–20) m, flight becomes undulating: rapid fluttering on outstretched wings interspersed with slight drops with wings briefly and partially folded. Bird typically describes roughly circular path of relatively small diameter (Fig A) but sometimes follows straight line from one part of territory to another. Duration of Song-flight varies, but usually not more than 1 min. In 3 flights (different birds), 11 song phrases in 50 s, 7 phrases in 35 s, 3 phrases in 12 s. At end of flight, bird initiates descent sometimes by a dipping action of body and outspread wings, or by briefly closing wings. Descends in stepped series

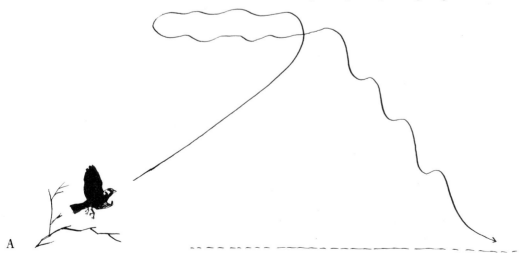

A

of swooping glides, wings held stiffly out and slightly above horizontal; at end of each glide, bird rises slightly before gliding further. On nearing ground, lands immediately or continues in level flight at *c.* 1 m for some distance before landing. Sometimes begins and ends Song-flight on a branch, and one such bird favoured bushes *c.* 50 cm high on top of mound in territory, returning there after each of several successive Song-flights repeated at intervals of a few minutes. Another ♂ sang intermittently for *c.* 2 min from similar perch, interspersed with 'tchip' calls (see 1 in Voice). 2 ♀♀, both incubating, recorded performing Song-flights near their nests. Singing most frequent early in morning (beginning at first light) and during 60-90 min before sunset. Mostly February-July in Middle East, varying with region (Bates 1936-7; Gallagher and Rogers 1978; Jennings 1980; Walker 1981*b*). In eastern Saudi Arabia, earliest 9 March, commonest late April through May, declining late May, but not infrequent in June, and noted up until 22 October. Common in October in North Yemen (D J Brooks). Display in September reported for Oman and United Arab Emirates (Stanford 1969). Occurrence of display in July, Bahrain, when young on the wing, interpreted as indicating 2nd broods (Ripley 1951); however, no evidence of 2nd broods at Dhahran. ANTAGONISTIC BEHAVIOUR. Both sexes defend nest-area, sometimes also wider part of territory. In boundary dispute between pair with young and newly formed pair, quite a lot of 'skirmishing' occurred. Another ♀ interrupted feeding her young to intercept and chase off intruding ♂♂, and mate sometimes did likewise. One bird flew at and chased trespassing Crested Lark *Galerida cristata* and Hoopoe Lark *Alaemon alaudipes*. Once, when House Sparrow *Passer domesticus* trespassed, ♂ flew at it and both rose squabbling into the air before landing and facing each other; ♂ raised and fluttered wings over his back (exposing black under wing-coverts) and *P. domesticus* shortly flew off. Also sometimes evicts Bar-tailed Desert Lark *Ammomanes cincturus* from preferred daytime loafing sites. HETEROSEXUAL BEHAVIOUR. (1) General. Timing varies markedly between regions and years. At Jiddah (western Saudi Arabia), birds appear to pair at beginning of December prior to breeding in February (Trott 1947). In eastern Saudi Arabia, where breeding apparently later, birds arrived (in 2 successive years) in late April and took up territories immediately. Nest-building started at most within 2 days of arrival, probably within 24 hrs. (2) Pair-bonding behaviour. (a) Aerial display. Chasing-display common in apparently established pairs: pair suddenly take off and ♂ chases ♀ in rapid flight, path twisting from side to side but staying fairly level *c.* 1 m above ground. Sometimes, during chase, pair rise up vertically *c.* 1 m, facing each other; once, ♂ overtook ♀ at end of chase before landing. On landing, Charging-display (see below) often occurs. Once, after Chasing-display, ♂ performed 'mini' Song-flights, rising *c.* 5 m before dropping above ♀ on ground. In another pair (with 3 eggs), ♂ on 2 occasions dive-bombed ♀ on ground. Unaccompanied ♀♀ passing over ♂'s territory frequently chased by him. ♂♂ sometimes also leave their territories and fly over territories of neighbouring ♂♂ to reach ♀♀. (b) Ground display. Most frequently occurring behaviour

is Charging-display. ♂ stands *c.* 1 m or less from ♀, angles head down, raises crest-feathers, holds wings slightly out from body and sometimes half-spreads them (Fig B), and runs rapidly and directly at ♀. On reaching her, ♂ often calls excitedly (probably 2 in Voice) and immediately turns away and jumps (less commonly runs) back to his starting point. Occasionally, instead of thus retreating, ♂ stands over ♀, mantling her with spread wings. Sometimes ♂ flies in, lands beside mate, and charges her, all in one movement. When charged, ♀ mostly stays put, sometimes crouching, but occasionally makes short dash at approaching ♂, causing him to retreat. ♂ may perform Charging-display up to 6 times in rapid succession. Charging-display most frequent at beginning of breeding season and often at end of Chasing-display. Performed most frequently by site-prospecting pairs; persists in established pairs apparently until hatching, although observed once in September. For evidently same or closely related display, Cape Verde, see Bourne (1955). On 2 occasions, ♂♂ performed Tail-cocking display: wings drooped and quivering slightly, and tail cocked so as to present black under tail-coverts to ♀. In one case, ♂ performed display as he walked away from ♀ (Fig C).

C

Significance of display not clear but both incidents occurred around laying time. ♂ may sidle up to ♀, possibly to unpaired ♀♀ as preliminary to courtship, though status of participants usually not known. Sidling ♂ approaches ♀ indirectly and often in stiff low Horizontal-posture: e.g. ♂, with crown raised, approached ♀ sideways, presenting first his right side, then his left; when in front of ♀, ♂ flattened crown and prostrated himself with head low and wings slightly away from body (Fig D); ♀ then preened and ♂ soon apparently lost interest and

D

left to feed. On another occasion, 2 ♂♂ landed *c.* 3 m from 2 feeding ♀♀; one ♂ ran toward ♀♀, appeared to stiffen and carry himself lower and closer to ground, and teetered forwards. Approaching one ♀, he presented his left side and attempted a circling advance with wings held slightly out. No visible response from ♀, and when both ♀♀ flew off, displaying ♂ pursued them and landed nearby, following again when one ♀ flew further. In another case, ♂, in Horizontal-posture, made sideways approach, then ran parallel to ♀ for *c.* ½ m. Sometimes, Horizontal-posture adopted by ♂ towards established mate: e.g. once, when ♀ feeding, ♂ shadowed her 2-3 m away in Horizontal-posture with bill forwards and crest raised above line of back; when ♀ flew off, ♂ followed and landed by her again. (3) Courtship-feeding. Only one report, and frequency not known. Once, during breeding season, ♂ picked up seeds

B

of *Zygophyllum simplex* and offered them to 3 ♀♀, one of which took them from his bill (Trott 1947). (4) Mating. No information. (5) Behaviour at nest. ♀ appears to select nest-site, accompanied by ♂. Once, over 2 days, pairs moved restlessly and excitedly around territory, inspecting bushes, perhaps for potential nest-sites. After a few minutes in one spot, one of the pair (usually ♂) flew *c.* 10 m then landed; ♀ followed and hovered over bushes. Twice, after performing Charging-display, ♂ ran into a bush as if trying to entice ♀ to follow. ♀ also entered bushes without encouragement from ♂. ♀ builds nest, attended by ♂. This also reported on Cape Verde (Bourne 1955). ♀ probably selects ready-made hollow, though she subsequently shuffles about in it with partly open wings. Once, sight of ♀ carrying twig apparently caused ♂ to become greatly excited, resulting in vigorous Chasing-display. Nest-building seems to begin just before sunset, thereafter mostly in early morning. ♀ once added nest-material after incubation had begun. After nest built, both ♂ and ♀ fly directly to and from it, sometimes rising vertically off it. ♂ recorded calling ♀ off nest prior to relieving her and vice versa. RELATIONS WITHIN FAMILY GROUP. From hatching, young brooded by both parents, more so by ♀. At one nest, between dawn and dusk, pair brooded for 288 min (*c.* 25% of daylight), of which ♀ contributed 61·5%, ♂ 38·5%. Brooding most continuous when nest exposed to strong sun. Parents regularly remove faecal sacs during early nestling period. Both sexes feed young. At one nest, ♀ appeared to resist approach of ♂ to feed young, chivvying him with pecks to bill and nape, and at times appearing to climb on his back; both birds had crests raised. Well-grown young beg with upstretched necks, open mouths, and flapping wings, but do not call in nest. Towards end of period in nest, young sometimes go outside it to be fed. Usually leave nest at 8 days; once after 1st nestling left, 6-day-old siblings also attempted to clamber out but were not fed outside nest and soon re-entered it. On first vacating nest, one brood of 2 moved *c.* 20 m to shade, and apparently both parents fed them there; young ran *c.* 30 cm towards parents approaching with food. The following morning, brood split up, ♀ leading one chick away and thereafter caring for and feeding it, while ♂ attended the other. As young develop, parents lead them further afield, perhaps beyond nest-territory. After dispersing properly from nest, offspring never return; roost under or within small bushes, sometimes led there by parent at sunset. Young mostly remain in shade of bush or tussock and call continuously (see Voice). May be fed there or may walk after parent, which may summon young by calling. After feeding excursion, parent leads young back to shade. Young begin to peck at vegetation 3–4 days after leaving nest and run strongly *c.* 1 day later. Young probably dependent on parents for some time after fledging (Trott 1947; Gallagher and Rogers 1978; Jennings 1980; J H Morgan and J Palfery). ANTI-PREDATOR RESPONSES OF YOUNG. After leaving nest, young mostly take refuge in bush or tussock. Once, when 2 chicks left nest temporarily, they lay flat and motionless, side by side, heads outstretched on sand, and bills closed; eyes half-closed or closed, effecting excellent camouflage. From leaving nest until fledging, this flattening response elicited by parent's Warning-call (see 3 in Voice). Older young run or fly on close approach. PARENTAL ANTI-PREDATOR STRATEGIES. (1) Passive measures. For Warning-call of adult, see above. Rarely heard during incubation. When well-grown unfledged young are closely approached by man, parent summons them with Warning-call to lead them from danger. (2) Active measures: against birds. Shrike *Lanius* which perched near nest was immediately mobbed by resident pair. (3) Active measures: against man. Both aerial and ground stratagems used. Aerial

demonstration apparently a lure stratagem: once, after sunset, when young roosting, ♂ accompanied intruder moving through territory, gave Warning-calls, and was apparently trying to lead intruder away. Ground display probably not uncommon; involves distraction-lure displays of perhaps 2 kinds, rodent-run and disablement type, separately or in combination. In one sequence, incubating ♂, suddenly flushed, flew off *c.* 10 m, lay prostrate on ground, and moved along, wings spread, dragging in sand and fluttering weakly. When closely approached again, flew a further 10 m, crouched, and ran like mouse ahead of intruder, twice trailing left wing. On every subsequent approach, flew *c.* 10 m further off and finally, when quite far from nest, landed, stood upright, and flew off normally. Aerially demonstrating bird (see above) once landed to perform scurrying mouse-like run on ground. In 4 cases of distraction-display at Dhahran, only one bird of pair performed. In northern Oman, April, ♀ with food performed distraction-display (not described) while ♂ fluttered around intruder's feet (Walker 1981*a*). (4) Active measures: against other animals. When horned vipers *Cerastes cornutus* passed near offspring, pair became very excited and alarmed, openly hopping in front of predators, calling loudly and continuously. When small lizard (Lacertidae) approached crouching chick, ♀ chased after it. Dogs passing near nest elicited alarm-calls.

(Figs by J P Busby: after drawings by J Palfery.)   EKD

Voice. Freely used in breeding season, especially by ♂, but birds mostly silent at other times (e.g. Bourne 1955). Account based on extralimital material supplied by J H Morgan and J Palfery for Dhahran (eastern Saudi Arabia).

CALLS OF ADULTS. (1) Song of ♂. A regularly repeated single phrase (P A D Hollom), typically given by ♂, rarely by ♀ (see Social Pattern and Behaviour). Final unit usually prolonged and descending; overall effect sweet and plaintive. Phrase variable in structure and pitch, but in North Yemen, at least, variation apparently largely geographical: individual birds always repeated same phrase and neighbours used similar one, but different phrase used at each of many sites visited (D J Brooks). Phrase usually comprises 2–3(–5) units. In phrase of 2 units, 1st typically rises in pitch, while 2nd is drawn out and falls, e.g. 'whit-teeoo' or 'psit pechew'. In our recordings, phrase repeated at intervals of *c.* 1 s; renderings (by J Hall-Craggs) include 'woo chee-a-wee peeu' (Fig I); 'wichAwee' with no terminal 'peu'; 'wich-a-wee ee(u)' (Fig II); 'wa-ter-di-ti neeu' (Fig III). In Saudi Arabia, more complicated songs include 'phuteet teu teeu', 'te-WEE te-WEEyup', and 'zree tulitit weeh tisweeeeee' (J H Morgan and J Palfery); for rendering of song similar to 2nd of these, see King (1978). In our recordings, most complicated song roughly 'pituee-ueewee', interspersed with 'pituwee' (E K Dunn). For other renderings of 3- or 4-unit songs, see Bates (1936-7), Bourne (1955), and Erard and Etchécopar (1970). Low conversational warbling, heard once when ♂ and ♀ together at nest, perhaps also song (J H Morgan and J Palfery). Song serves to demarcate territory and attract mate. For diurnal and seasonal occurrence, and modes of delivery, see Social Pattern and Behaviour. (2) Contact-alarm calls. Oft-repeated 'jip', 'tchip', 'jeep', or 'tchup',

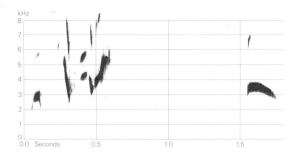

I   P A D Hollom   Egypt   March 1984

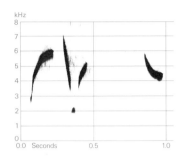

II   C Chappuis   Niger   October 1971

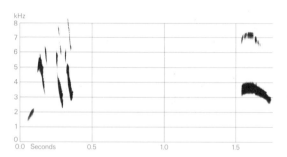

III   P A D Hollom   Egypt   March 1984

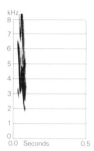

IV   P A D Hollom   Egypt
March 1984

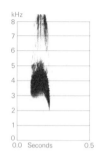

V   C Chappuis   Niger
October 1971

given by both sexes, in the air or on ground. Serves various functions, notably pair-contact, to signal alarm prior to hatching of young, or agitation, excitement, etc., e.g. when ♂ sees strange ♂ intruding on his territory or when ♀ flies over it (J H Morgan and J Palfery); given also outside breeding season (D J Brooks). During chase involving ♂ and 2 ♀♀, 'chet', almost like sparrow *Passer*; also 'chep' (Fig IV) by both sexes from ground, apparently indicating alarm at presence of intruder (P A D Hollom); 'chipi' of ♀♀ (Valverde 1957) presumably same or variant. (3) Warning-call. Similar to call 2, but a more reedy and lower pitched 'dzeep' or 'jreep', or 'zree'. In recording (Fig V), 'zreeu' with abrupt, nasal, buzzing quality (J Hall-Craggs, P J Sellar). Usually given in flight, often in combination with call 2, principally as warning to young in or out of nest, but occasionally during incubation and then presumably directed at mate. Call causes young out of nest to crouch (J H Morgan and J Palfery). Very loud 'pii' or 'cri' like sparrow *Passer* (Valverde 1957) is probably this call. (4) Other calls. (a) A trill of rather poor quality, rather like Skylark *Alauda arvensis* or Crested Lark *Galerida cristata*; given by ♀ as she flew into a feeding area, and again after landing (J H Morgan and J Palfery). Soft chuckling 'flight-note' reminiscent of *A. arvensis* (Bourne 1955) perhaps the same. (b) A very low, fruity 'chuc tuc tuc' given by ♂ as he flew off nest (J H Morgan and J Palfery). (c) A very quiet, subdued 'chirrup' when ♀ arrived at nest where ♂ brooding; not known which of pair gave call (J H Morgan and J Palfery). (d) A deep hoarse 'prieu prieu' by ♀ (Valverde 1957); no further details.

CALLS OF YOUNG. Apparently no calls of any kind given in nest. After leaving nest, food-call of young typically a persistent 'cheeop cheeop cheeop', audible from c. 25 m in still conditions. Evidently serves to help parent locate chick. (J H Morgan and J Palfery.)   EKD

Breeding. SEASON. Cape Verde Islands: eggs laid September to January or February (Bannerman and Bannerman 1968). Southern Morocco: breeding begins late April (Valverde 1957). Eastern Saudi Arabia: north of c. 25° N (where migratory), eggs laid by end of April; south of this (where resident), breeds March (G Bundy). North Yemen and Oman: chicks found October (D J Brooks, G Bundy). SITE. On ground in shelter of tussock or stone. Nest: scrape, saucer-like depression (hoof-print can be used), or quite deep cup-shaped hole (apparently dug specially) with top of nest flush with ground; lined with grass, hair, or feathers on base of fine twigs; may have rim of pebbles surrounded by twigs, with centre of nest bare so that eggs laid directly on sand (D J Brooks, J H Morgan and J Palfery). Building: by ♀ (Bourne 1955), though ♂ usually accompanies her; usually completed in 24 hrs (J H Morgan and J Palfery). EGGS. See Plate 79. Sub-elliptical, smooth and fairly glossy; greyish or dirty white, evenly speckled and blotched light brown, with underlying purplish-grey spots. 2 eggs, Egypt, 17·6 × 13·5 mm and 17·5 × 13·4 mm; calculated weight 1·67 g (Schönwetter 1979); 2 eggs, Sudan, 18·5 × 13·5 mm and 18·5 × 13·0 mm (Hüe and Etchécopar 1970). Eastern Saudi Arabia: 18·4 × 14·2 mm (17·9–18·8 × 13·8–14·7), n = 7 (J H Morgan and J Palfery). Clutch: 2, rarely 3, in Cape Verde Islands and North Africa (Bannerman and Bannerman 1968; Hüe and Etchécopar 1970); of 18 clutches, Saudi Arabia, 8 of 2 eggs and 10 of 3 (J H Morgan and J Palfery). Only one brood in eastern Saudi Arabia north of *c.* 25° N (G Bundy); see also Social Pattern and Behaviour. No further information from west Palearctic; all remaining details from Saudi Arabia (J H Morgan and J Palfery). 1st egg usually laid 2 days after nest completed, remainder at 24-hr intervals in early morning. INCUBATION. 11(-12) days ($n = 5$). By both sexes but mainly ♀. Probably begins with 1st egg. At one nest, 04.45-11.10 hrs, eggs incubated 69·6% of time; ♀ did 78·4% of this, ♂ 21·6%. YOUNG. Parents probably split brood between them on leaving nest (see Social Pattern and Behaviour). FLEDGING TO MATURITY. Fledging period *c.* 12-14 days. Young usually leave nest at 8 days; dependent on parent for some time after fledging. BREEDING SUCCESS. Of 10 clutches of 3 eggs, 6 hatched 3 young but in only one did 3 chicks survive to leave nest, and then 3rd chick found dead same day, suggesting maximum rearing ability of 2 chicks, particularly with each parent caring for and feeding single chick after leaving nest.

Plumages (nominate *nigriceps*). ADULT MALE. Forehead including feather-tufts over nostrils white, but easily discoloured by dirty soil. Lores, cheeks near gape, streak through and just above eye, and crown (with rounded rear end) black. Rear of cheeks, ear-coverts, and hindneck white, hindneck slightly tinged pale grey-brown when plumage fresh. Mantle, scapulars, and back pale grey-brown tinged cinnamon-pink (paler grey without cinnamon when worn and bleached), fringes of feathers slightly paler grey-brown (almost white when worn), narrow shaft-streaks blackish. Rump pale grey-brown merging into cream-buff or isabelline-white of upper tail-coverts. Underparts from chin to under tail-coverts black, except for white patch on sides of breast, white thighs, and variable amount of white on lower flanks; black of throat extends laterally up into broad black band on side of neck. Central pair of tail-feathers (t1) pale grey-brown, like upperparts; remainder of tail black, outer web of t6 usually broadly fringed isabelline-white, other tail-feathers narrowly fringed pink or cream along tips when fresh. Flight-feathers dark grey or dark brownish-grey, narrowly but sharply fringed isabelline-white when fresh; tertials pale grey-brown with broad paler grey fringes. Upper wing-coverts similar to mantle and scapulars. Under wing-coverts and axillaries black. See also Moults for alleged existence of non-breeding plumage. ADULT FEMALE. Upperparts entirely rufous-cinnamon, tinged vinaceous when plumage fresh, cinnamon-buff when worn (not grey-brown as in adult ♂); shafts on crown and scapulars greyish-black, hindneck and upper tail-coverts often slightly paler isabelline-buff (especially when worn). Lores, patch round eye, and sides of neck pink-buff, cheeks and ear-coverts deeper cinnamon-buff. Sides of breast and band across chest cinnamon-buff, remainder of underparts pale cream-buff or (when worn) off-white; sometimes a few tiny black or grey specks on chin and narrow grey streaks on upper chest. Tail as in adult ♂, but t1 cinnamon like upperparts, not grey; t6 black with poorly defined isabelline outer web, t5(-t4) often with isabelline fringe along outer web. Flight-feathers, tertials, and greater upper wing-coverts either like adult ♂ (thus rather greyish) or with rufous-cinnamon

fringes and then more similar to mantle and scapulars; lesser and median upper wing-coverts like mantle and scapulars. Axillaries and under wing-coverts black, deepest along leading edge of wing, but rather variable elsewhere—sometimes greyish-black or strongly suffused buff along borders. NESTLING. No information. JUVENILE. Like adult ♀, but crown, mantle, scapulars, and upper wing-coverts have contrasting pale buff or warm buff feather-fringes, each fringe subterminally bordered by narrow dull grey spot or arc; tertials cinnamon with broad off-white fringe (whole tertials soon bleached to off-white); fine and indistinct dusky spots on rump and upper tail-coverts; chest rather heavily marked with poorly defined dull black spots. Flight-feathers narrowly and evenly fringed pink-cinnamon. FIRST ADULT. Similar to adult, but at least some immatures retain all or part of juvenile flight-feathers and greater upper wing-coverts; in ♂♂, such birds have head and body like adult ♀, but feathers of throat, chest, and breast (not crown) all or partially black. Subsequent plumages as adult.

Bare parts. ADULT. Iris brown or dark brown. Bill pale horn-brown or bluish-white, culmen usually slightly darker horn-brown. Leg and foot pale flesh, greenish-white, or whitish-horn. (BMNH, ZMA.) NESTLING. No information. JUVENILE. Iris dark ochre-brown or brown. Bill greenish-grey or pale greyish-horn. Leg and foot sandy-grey. (BMNH.)

Moults. Based on nominate *nigriceps* from Cape Verde Islands, *albifrons* from Sudan, and *melanauchen* from Arabia, Kuwait, and Iraq. ADULT POST-BREEDING. Complete; primaries descendant. Plumage worn by April-May in all races. Most *albifrons* from Sudan start flight-feather moult late April to mid-May; primary moult score 10-22 reached 2nd week of May, when head and body still heavily worn except for some new scapulars, tertials, upper wing-coverts, or feathers of crown and mantle; *melanauchen* from Arabia reaches this stage mid-May to mid-June. In all races, plumage new mid-August or September, but occasionally some outer primaries or tail-feathers still growing. Stated by Vaurie (1951a) to have non-breeding plumage, in which ♂ similar to adult breeding ♀ except for variable amount of black on underparts; allegedly replaced by full breeding during partial pre-breeding moult. Among specimens examined, however, none showed indications of moulting into or out of such a plumage, all in fresh full breeding by September-October (but only a few June-August birds seen). POST-JUVENILE. Not fully elucidated; stated to be partial (e.g. by Ticehurst 1923), but probably complete in some early-fledged juveniles. At least some 1st-year birds retain juvenile flight-feathers, greater upper wing-coverts, and all or part of tail; among these, ♂♂ have plumage similar to adult ♀ (showing cinnamon face and crown) but with partly black underparts; among those examined, this plumage developed January-March when 2-3 months old (BMNH). All immatures similar to adult from following September, hence this plumage replaced by full breeding during May-August when retained juvenile flight-feathers perhaps also moulted.

Measurements. ADULT. Nominate *nigriceps*. Boa Vista and São Tiago (Cape Verde Islands), October-May; skins (BMNH, RMNH). Bill (S) to skull, bill (N) to distal corner of nostril; exposed culmen on average 2·6 mm less than bill (S).

	♂		♀	
WING	77·6 (1·64; 19)	74-80	74·7 (1·25; 15)	73-77
TAIL	43·6 (1·93; 14)	41-48	42·2 (1·36; 15)	40-45
BILL (S)	12·9 (0·50; 13)	12·4-13·6	13·4 (0·64; 11)	11·9-14·3
BILL (N)	8·2 (0·31; 14)	7·8-8·6	8·3 (0·44; 12)	7·6-9·0
TARSUS	16·5 (0·57; 13)	15·6-17·4	16·6 (0·45; 12)	15·7-17·3

Sex differences significant for wing.

E. n. albifrons. Northern and central Sudan, October–May; skins (BMNH, ZMA).

WING   ♂ 80·3 (1·15; 15)   79–82   ♀ 76·6 (2·24; 9)   74–80

E. n. melanauchen. Ethiopia and Arabia plus 1 from Iraq, February–September; skins (BMNH, RMNH, ZMA).

WING   ♂ 81·4 (1·63; 28)   78–84   ♀ 77·9 (2·44; 9)   74–81

JUVENILE. Wing and tail as adult.

Weights. *E. n. albifrons.* Aïr (Niger) and Ennedi (Chad), January–April: 3 ♂♂ 14–16, 1 ♀ 12 (Niethammer 1955*b*).

Structure. Wing rather short and broad, tip rounded. 10 primaries: p8 longest, p9 0–2 shorter, p10 38–46, p7 0(–1), p6 1–5, p5 6–10, p1 18–20; p10 reduced, 2(8)0–4 longer than upper greater primary coverts in both adult and juvenile. Outer web of p6–p8 and inner of (p7–)p8–p9 emarginated. Tail rather short, tip shallowly forked; 12 feathers, t1 2–5 shorter than t5–t6. Bill finch-like, short, conical, heavy at base; culmen strongly decurved; cutting edges slightly decurved, sometimes with slight notch in middle; gonys straight or slightly convex. Nostril small, rounded; covered by tuft of short bristly feathers. Tarsus rather short, toes short and rather thick. Middle toe with claw 11·8(13) 10·3–12·9; outer toe with claw *c.* 70% of middle with claw, inner *c.* 70%, hind *c.* 80%. Claws short, decurved, rather slender and sharp, hind claw 5·4 (4) 4·8–6·2 mm.

Geographical variation. Slight in size, more pronounced in colour. Nominate *nigriceps* from Cape Verde Islands smallest, Saharan race *albifrons* has slightly longer wing and tail, Ethiopian and Arabian populations of *melanauchen* larger still,

Indian *melanauchen* largest and sometimes separated as *affinis* (Blyth, 1867) on account of this. Bill and tarsus of *albifrons* and *melanauchen* similar to nominate *nigriceps*, not larger, e.g. bill (S) of Arabian ♀♀ of *melanauchen* 12·8 (0·87; 16) 11·3–14·2. ♂ *albifrons* like ♂ nominate *nigriceps* in colour, but upperparts more uniform sandy-grey with less dark grey-brown of feather-bases showing, appearing less spotted dull grey; white patch on forehead on average slightly larger, black on lores narrower. ♀ *albifrons* shows less contrast between darker rufous feather-centres and paler rufous fringes on mantle and scapulars than ♀ nominate *nigriceps*; dark shafts of crown and scapulars less pronounced, upperparts appearing slightly darker and deeper rufous, but difference not marked and not all birds separable. ♂ *melanauchen* differs from both nominate *nigriceps* and *albifrons* in having more extensively black crown and black upper mantle, latter largely separated from black of hindcrown by white band across hindneck; white patch on forehead smaller than in *albifrons*, about similar to nominate *nigriceps*; remainder of upperparts grey, similar to nominate *nigriceps* but slightly paler and less spotted, less sandy-cinnamon than in *albifrons*. ♀ *melanauchen* not as rufous above as ♀♀ of other races, feathers grey with buff fringes, upperparts appearing greyish-buff, not rufous-cinnamon, upper wing-coverts mainly grey; band across chest grey (pale pink-cinnamon in *albifrons*, rufous-cinnamon in nominate *nigriceps*).

Forms superspecies with Ashy-crowned Finch Lark *E. grisea* from Indian subcontinent, Chestnut-headed Finch Lark *E. signata* from eastern Ethiopia, Somalia, and northern Kenya, Fischer's Finch Lark *E. leucopareia* from East Africa, and Grey-backed Finch Lark *E. verticalis* from southern Africa (Hall and Moreau 1970).   CSR

Eremalauda dunni Dunn's Lark

PLATES 3 and 11
[between pages 160 and 161, and 184 and 185]

DU. Dunns Leeuwerik   FR. Ammomane de Dunn   GE. Einödlerche
RU. Чернохвостый пустынный жаворонок   SP. Alondra de las dunas   SW. Streckad ökenlärka

Calendula dunni Shelley, 1904

Polytypic. Nominate *dunni* (Shelley, 1904), Sahara; *eremodites* (Meinertzhagen, 1923), Arabia.

Field characters. 14–15 cm; wing-span 25–30 cm. 15% smaller than Skylark *Alauda arvensis* and 10% smaller than Desert Lark *Ammomanes deserti*; in spite of overlapping measurements, noticeably bulkier than Bar-tailed Desert Lark *A. cincturus*. Rather small but bulky lark, with strikingly heavy bill contributing to large-headed appearance, broad, well-rounded wings, and (at times) long-looking legs. Plumage essentially pale sandy with moustached and bold-eyed face, and tail pattern recalling short-toed larks *Calandrella*. Streaking of upperparts and chest variably distinct but important to separation from *Ammomanes* larks. Wing-beats noticeably flapping. Call distinctive. Sexes similar; no seasonal variation. Juvenile separable at close range. 2 races in west Palearctic; Arabian race, *eremodites*, described here (see also Geographical Variation).

ADULT. Upperparts sandy to pink-isabelline, with rufous tone on folded wing (particularly on tertial-centres, greater coverts, and bases of flight-feathers) and grey tone on nape; fairly dense, regular dark brown streaks on crown, and more diffuse, paler streaks from nape to lower back, but no obvious contrasts on wings. Appearance of head dominated by heavy, conical bill (as broad at base as it is deep and almost half as long as crown) and complex facial pattern. Latter consists of (a) dark brown to black lines forming distinct moustachial and (particularly) malar stripes and surround to cheeks, (b) almost white eye-surround and rear supercilium, and (c) pale sandy rear cheeks; pattern creates unique bold-eyed, at times staring, yet somehow depressed expression. Tail strikingly patterned with pale red-brown central feathers contrasting boldly with black outer panels which have only indistinct buff outer margins. Underparts cream to satin-white, noticeably paler on some birds than others and unmarked except for buff wash and streaks on chest; chest-marks may be dark and particularly discrete on sides of chest (hence clearly visible in the field) or pale and diffuse (then difficult to see even at close range). Bill dull but warm white, with brown tip and culmen. Legs usually glowing pink or sandy. JUVENILE. Not studied in the field (see Plumages).

All too easy to overlook among populations of *A. deserti* and *A. cincturus*, with distribution in west Palearctic incompletely known. Previously thought to be indistinguishable from *A. cincturus* but now known to possess several invariable, diagnostic characters, of which massive bill, head-marks, and tail pattern are the most visible. In addition, structure differs from both *A. deserti* and *A. cincturus* in larger head, distinctly broad and round wings, and broader tail. Folded wings of *E. dunni* have short wing-point (less than $\frac{1}{4}$ tertial length), rather as in Lesser Short-toed Lark *C. rufescens* and quite unlike long wing-points ($\frac{1}{2}$ tertial length) of the *Ammomanes* larks. Strikingly tame, allowing close approach and often content to run or creep away from observer, using rocks and shrubs as cover. Flight has rhythm typical of family but wing-shape yields characteristically flapping or even floppy action, recalling *Mirafra* larks of Afrotropics. In flight, tail also looks rather short and thus flight silhouette noticeably bulky. Escape-flight low and short. Gait varies from a creeping shuffle in cover to a fairly upright walk or run on flat ground.

Song recalls *A. cincturus* but more developed and longer sustained, with short, sad melodious phrases contrasting with long, scratchy warbles. Calls include thin, liquid 'prrp' and, in alarm, loud, ringing 'chee-oop'.

Habitat. In very warm lower latitudes, tropical and subtropical, on flat arid lowlands, not rocky, stony, or of broken ground, and apparently overlapping little into full desert, but data inadequate and critical habitat factors not yet determined. In eastern Saudi Arabia breeds in small lightly vegetated sandy wadis and shallow depressions within undulating gravelly desert, apparently avoiding sand desert. Winters sparsely on firm stony desert, apparently living on plant seeds left over from spring growth of grass; also on old middens where nomad bedouin have corralled sheep (J Palfery). Near Jiddah (western Saudi Arabia), January, always on flat desert, never near rocks or stones (Meinertzhagen 1949c). Often feeds among slight grass with occasional low bushes (Meinertzhagen 1951). At Azraq (Jordan), findings in 3 localities indicated preference for rough sandy ground with widely scattered but thick and sizeable shrublets in wadi spreads of limestone desert, overlapping with Lesser Short-toed Lark *Calandrella rufescens*, Crested Lark *Galerida cristata*, and Hoopoe Lark *Alaemon alaudipes*, in contrast to Bar-tailed Desert Lark *Ammomanes cincturus*

which is less tolerant of vegetation (Hemsley and George 1966). On southern edge of Sahara, inhabits dry grassy plains (Cave and Macdonald 1955; Etchécopar and Hüe 1967). In Mauritania, found breeding on steppe with close pattern of grass *Aristida plumosa* giving generally green appearance, on loose soil in shallow depression near well; not really a desert species (Naurois 1974).

Absence from steppe, savanna, and mountainous areas, and adaptation to desert fringe only partly explain survival in competition with other Alaudidae apparently sharing same or closely similar habitat. Understanding of habitat distinctions and survival factors must await fuller data.

Distribution. Probably overlooked due to problems of identification, also possibly a nomadic breeder (Round and Walsh 1983; Wallace 1983a).

JORDAN. Bred Azraq 1965-6 (Wallace 1983a; S Cramp). SAUDI ARABIA. May breed near Haql, Gulf of Aqaba (A J Stagg). MAURITANIA. Bred Zemmur 1970 (Naurois 1974). One seen near Chegga 1930 (Heim de Balsac and Mayaud 1962). EGYPT. Possibly breeds occasionally in north-east Sinai, where 4-5 (including 2 singing) 20 March 1978 and one March 1981 (SMG, PLM, WCM).

Accidental. Lebanon (Harrison 1962). Israel: 1 (Round and Walsh 1981). Kuwait: 4, 17 March 1987 (CP). Records for Ahaggar massif (Algeria), 1953-4 (Niethammer and Laenen 1954), refer to Black-crowned Finch Lark *Eremopterix nigriceps* (Niethammer 1963a).

Population. No information on populations or trends.

Movements. Resident; also some dispersal of uncertain, but probably small, extent. As in other desert Alaudidae, likely that main component of movement is nomadism during drought conditions. Though present all year in southern Sahara and Sahel, local fluctuations in numbers occur, e.g. Lamarche (1981) for Mali. Casual August-February records further north, but 1 in north-east Mauritania, February 1930, and proven breeding in north-west Mauritania (see Distribution) are sole records for North Africa. Short-distance movements occur in Arabia also (Gallagher and Woodcock 1980; Jennings 1980)—at least once, related to abnormal rainfall (Walker 1981b). Apparently nomadic, and perhaps only wintering, in eastern Saudi Arabia (G Bundy).   RH

Food. Seeds and insects. Picks food from surface of ground, and also digs (for seeds according to Jennings 1981b) with bill: uses hammering or rapid sideways flicking movements, throwing up spurts of sand and creating a pit. Also examines tussocks and undersides of shrubs, sometimes jumping up to snatch at items. Turns over and pecks at goat (etc.) droppings with bill. Once seen feeding by sewage stream and flicking over pieces of water-weed with bill to pick up food items. Typically

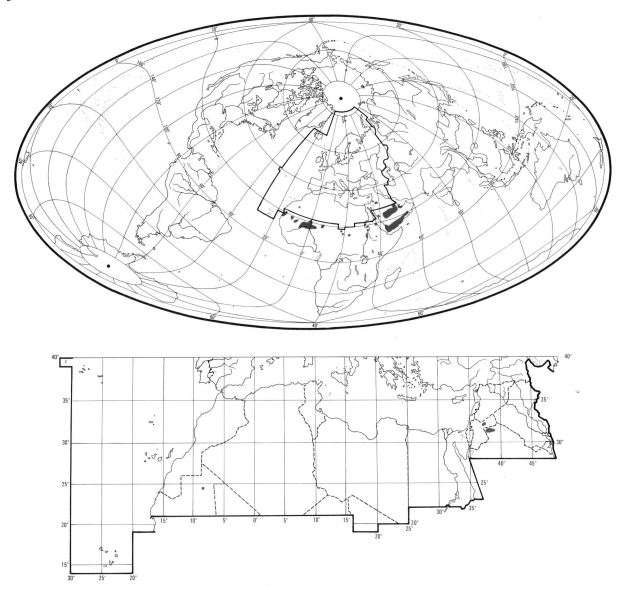

active while feeding, alternately running and pausing, though also adopts rather shuffling gait and squatting posture, with legs well bent. (G K Brown, J Palfery.)

In central Mali, August–November, 6 birds contained seeds, largely of grass *Panicum turgidum*; 1 also contained insects (Bates 1933–4). The few stomachs examined by Meinertzhagen (1954) contained only small seeds.

Young at one nest, eastern Saudi Arabia, were brought green caterpillars by both parents (G K Brown).   DJB

Social pattern and behaviour. Little known.

1. Gregarious outside breeding season, occurring in groups of 2–20 (though more than 3 perhaps unusual), often with other Alaudidae (Bates 1936–7; Jennings 1980, 1981*b*; Lamarche 1981; Round and Walsh 1981; G K Brown, J Palfery). BONDS. No information. BREEDING DISPERSION. Territorial. In northern Mauritania, 4–5 nests found in 25 ha (Naurois 1974). Since

birds probably nomadic in search of suitable habitat (Jennings 1980), site-fidelity is unlikely. ROOSTING. No information on nocturnal roosting. By day, usually keeps to shade, close under bushes and low shrubs (Round and Walsh 1981). In Saudi Arabia, often seeks shade under overhanging rock, at foot of bushes; in 2 cases inside holes of lizards (Lacertidae), returning repeatedly to same holes; birds may also resort habitually to same bushes, leading to formation of several hollows in sand beneath them. Birds sometimes move restlessly from one hollow to another and, once settled, doze with bills and chins resting on ground, wings half spread. In light breeze, bird once seen sprawled on top of low plant, adopting a very relaxed posture with half-spread wings, and was easily approachable. (J Palfery.) For shade-seeking by young, see Relations within Family Group.

2. Typically confiding, allowing close approach, e.g. to within 5 m (Meinertzhagen 1954; Jennings 1980; Round and Walsh 1981). FLOCK BEHAVIOUR. No information. SONG-DISPLAY. Apart

from Song-flight, ♂ also sings from ground and from perch (G K Brown, J Palfery). Song-flight performed from start of breeding season, in central Arabia beginning March–April (Jennings 1980). Song-flight variously described. Following account compiled from observations by G K Brown. Bird typically rises into wind to 30 m, sometimes 50 m or more. While singing, stays more or less stationary, but swinging from side to side, with slow, lazy, almost owl-like wing-beats, effecting rather floppy appearance. In one Song-flight, closed wings every 2–3 beats, in another with every wing-beat. At end of one Song-flight, plunged to ground and chased conspecific bird; at end of another, made gradual floating descent on broad outspread wings. Song-flight also said to be slowly rising, followed by rapid return to ground (Jennings 1981b). During display, usually hovers *c.* 6–10 m above ground, fluttering wings fairly rapidly and closing them now and again without appearing to lose height; one Song-flight lasted *c.* 2 min (J Palfery), and in another, bird stayed at 50 m above ground for 4–5 min (G K Brown). At Azraq, bird performed 'loose and floppy' Song-flight: made brief ascent to *c.* 1·5 m and flew fairly directly, legs dangling, over line of small shrubs; thought to be singing as it took off (D I M Wallace). At end of Song-flight one bird landed on top of small bush and continued to sing from there. Another sang from ground as it fed (J Palfery). Singing from ground also reported by Round and Walsh (1981) and Wallace (1983a); see also 1 in Voice. Once, bird gave apparent Subsong as it crept about under shrubs (Wallace 1983a: see 1 in Voice). Song-flights performed mainly early in morning, once at least 45 min before dawn when scarcely light but moon full (G K Brown). ANTAGONISTIC BEHAVIOUR. Nothing known, other than Song-display. HETEROSEXUAL BEHAVIOUR. For onset of pairing, associated with performance of Song-flights, see above. No further information. RELATIONS WITHIN FAMILY GROUP. Report of chicks 'toddling about' but unable to fly, and thought to be this species (Bates 1936–7), suggests that young may leave nest relatively early, as in other Alaudidae. By day, fledged young seek shelter under rocks, bushes, etc. (J Palfery). ANTI-PREDATOR RESPONSES OF YOUNG. In 2 cases, fledged young disturbed from daytime shade preferred to run rather than fly from human intruder. One bird which had been sheltering under a rock thus ran around looking for another suitable refuge, eventually hiding under a large flat slab. (J Palfery.) PARENTAL ANTI-PREDATOR STRATEGIES. Following account for eastern Saudi Arabia by G K Brown. (1) Passive measures. One pair much more wary at nest than Bar-tailed Desert Lark *Ammomanes cincturus*. (2) Active measures: against man. On approach to 3 m, bird disturbed from nest (with young) flew *c.* 20 m, landed, then performed distraction-lure display of rodent-run type: a crouched shuffling run. On another occasion at same nest, approach to 5 m caused bird to fly off low over ground, then perform rodent-run (as above), with wings dragging to feign injury. Before returning to nest, uttered a few short song-phrases on ground.   EKD

Voice. Freely used during breeding season. No information for other times. Other than song, several closely similar calls reported, especially by J Palfery and G K Brown; repertoire therefore not fully understood, and following scheme tentative.

CALLS OF ADULTS. (1) Song of ♂. A scratchy warbling, interspersed with a melodious but melancholy whistling of short phrases; a rather sad warble, confusable with Bar-tailed Desert Lark *Ammomanes cincturus*, but more developed, with intermediate phrases linking bell-like notes (Round and Walsh 1981). Variable, including a subdued warbling sound, double whistles reminiscent of Black-crowned Finch Lark *Eremopterix nigriceps*, and an occasional churring wheeze (J Palfery). In Song-flight, a series of fairly short sweet rambling phrases; one phrase began 'pee-pee-pee-peoo', became more elaborate, and ended with a long 'chee'; another phrase 'chee-chee-chee-cheeree-cheereer-cheeree-cheeee', final unit long-drawn. One bird returning to nest uttered a few short 'cheek-cheek-a-cheek' phrases. Overall, quality of short rather simple phrases not unlike Skylark *Alauda arvensis* (but not so loud or vehement) or Crested Lark *Galerida cristata* (G K Brown). Song also described as a soft repeated 'dree-dree-up' (Walker 1981b); a creaky twittering (Gallagher and Rogers 1980); like a squeaking gate (Jennings 1981b). Delivered in Song-flight, from perch, or from ground (see Social Pattern and Behaviour). Song from ground a series of quick rambling notes, both sweet and harsh, each phrase lasting 2–4 s; also a shorter phrase ending with a long 'cheee' (G K Brown). In recording (Fig I), song from ground a repeated short

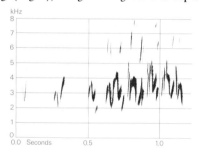

I C Chappuis Niger November 1971

warbling phrase, something like 'twit wit wit-weare-rear-re-ree' (J Hall-Craggs). Probably similar song from ground described as a scratchy little phrase, interspersed with 'peeuw' like song-unit of Hoopoe Lark *Alaemon alaudipes* (J Palfery; see also call 4c, below). Song of bird creeping about under shrubs described as more like subsong of a scratchy-voiced warbler (Sylviinae) than a typical lark (Alaudidae); rendered a sort of 'screedle-screedle-scri-rit-rit-screedle-screedle...' *ad infinitum* (Wallace 1983a). (2) Contact-calls. (a) A single or repeated 'ziup' or 'chiup chiup' given in flight (Round and Walsh 1981). Also rendered 'cheeoop' or 'teeoop' (G K Brown); a descending 'shreeup' or 'treeup', reminiscent of Skylark *Alauda arvensis* (J Palfery); 'tree-chup' or 'two-weep' (Walker 1981b); 'tchi-oui', like White Wagtail *Motacilla alba* but much softer (Lunais 1983). (b) Slightly shrill, chirpy 'chrruit chrruit', again reminiscent of *A. arvensis* and given in flight (J Palfery), perhaps also a contact-call. (c) Various other calls, perhaps similar or related: in flight, a rather hoarse 'chleep', or 'cheelip'; on ground, 'cheep', 'chleek', and 'cheet-cheet' (G K Brown). (3) Alarm-calls. (a) Evidently a more emphatic variant of call 2. In full alarm, a loud ringing 'chee-oop' (Round and Walsh 1981; G K Brown); 'tu-wep' (Walker 1981b). (b) On flushing, bird called 'chup-chup-chee-oo' (Round and Walsh 1981). Very similar call expressing alarm and rendered 'chip-chip-twee', of which last unit almost a whistle, described by G Bundy. Bird flushed from daytime loafing site flew off nervously, calling 'chipchipchipchipchip chip chip', rather like Blackbird *Turdus merula*; on another occasion, probably same call 'jipjipjip' (J Palfery). (c) A sweet descending 'steeuw' or 'seeuw' given by 2 birds on ground as they were approached, probably an alarm-call; high-pitched 'seeuw seeuw sweet' by another bird just before landing, and 'seeuw seeuw seeuw-tk-tk' given by adult accompanying fledgling (J Palfery); on ground, a soft 'teu', like Bullfinch *Pyrrhula pyrrhula* (G K Brown). (d) Call roughly rendered 'trrp-trrp', not unlike Contact-alarm call of Short-toed Lark *Calandrella brachydactyla* but less dry (G Bundy); in flight, quite hard 'chirrp' when disturbed, also 'cheerp-a-cheerp' (G K Brown); presumably similar to thin liquid 'prrp', unlike dry sharp 'prrrt' of Lesser Short-toed Lark *C. rufescens* (Round and Walsh 1981). Calls 3c and 3d may evidently be combined; thus, birds in flight called 'seeu-prt', of which 1st unit high-pitched and sweet, the 2nd lower-pitched, shorter, and fruity in quality; on other occasions rendered 'seeu-trrt' and 'seeuw-cht seeuw-cht' (J Palfery). (e) When anxious, 'cheelee' (G K Brown). (4) Other calls. (a) Repeated high-pitched 'pee' or 'spee' directed by adult at fledgling (J Palfery), perhaps a warning-call. (b) Rather 'electronic' 'zzrrp' given along with 'seeu-prt' call by 3 birds in flight (J Palfery). (c) One bird imitated mournful piping note of *A. alaudipes* (G K Brown).

CALLS OF YOUNG. No information for nestlings. On approach of human intruder, fledged young gave repeated, thin, high-pitched, rather soft call, variously rendered: 'pee pee'; 'peeuw peeuw peeuw' with plaintive quality, trailing off on 2nd syllable; 'swee swee'; also give a low soft 'hwee', once in response to call 4a of adult, to which similar but thinner, weaker, and shorter (J Palfery). EKD

Breeding. SEASON. Mauritania: breeding once recorded in January, following autumn rains (Naurois 1974). SITE. On ground in shelter of tussock. Nest: scrape lined with fresh vegetation. Building: no information. EGGS. Similar in shape and form to those of *Mirafra* larks; white, with blackish and lavender spots and blotches (Naurois 1974). Clutch: one of 2 eggs, another of 3 (R de Naurois). No further information.

Plumages (*E. d. eremodites*). ADULT. Entire upperparts including upper wing-coverts bright pink-cinnamon; fringes of feathers of crown, hindneck, mantle, scapulars, back, and rump slightly tinged grey when plumage fresh; feather-centres of forehead and crown dark brown ending in black or dark grey points towards tips; feather-centres of mantle, back, and scapulars slightly deeper cinnamon, giving indistinct streaky effect. Sides of head buff (palest near eye), ear-coverts and sides of neck dark pink-cinnamon, some indistinct grey mottling in front of and below eye and on ear-coverts. Sides of throat often with distinct, though broken, black malar stripe. Chin and throat pink-white. Chest and upper flanks pale pink-buff, sides of breast darker pink-cinnamon, all marked to variable extent with short and narrow faint dark brown streaks or dots. Remainder of underparts pale pink-cream. Central pair of tail-feathers (t1) bright pink-cinnamon like upperparts, t2 black partially washed cinnamon, t3–t6 black; tips of t2–t6 with narrow and poorly defined cinnamon fringes, t4–t5 with narrow cinnamon outer edge, outer web of t6 pink-cinnamon, grading to isabelline along outer edge. Flight-feathers bright pink-cinnamon, tips of p4–p9 black for 6–10 mm, but black rather faint and poorly defined on p4 and p8–p9; all flight-feathers narrowly fringed pale cinnamon along tips. Under wing-coverts and axillaries pale pink-buff, almost similar to pale cinnamon-pink undersurface of flight-feathers. *In worn plumage*, crown more distinctly streaked black; mantle and scapulars more distinctly streaked red-brown; chin and throat whiter; belly paler, isabelline-white; pale fringes along tips of tail-feathers and primaries lost by abrasion. NESTLING. No information. JUVENILE. No information on *eremodites*. Juvenile of nominate *dunni* similar to adult, but each feather of crown, mantle, and scapulars with white spot on tip; hindneck and back to upper tail-coverts with off-white narrow feather-fringes; upper wing-coverts (including greater primary coverts) with rather broad and poorly defined cream-white fringes; feathers of underparts narrower and looser than adult, whiter, chest paler pink-cream; tail-tips more extensively suffused cinnamon, outer webs of t5–t6 with less sharply defined isabelline colour. Flight-feathers uniform rufous-cinnamon, without black on primary-tips and without pale fringes, only outer web of p10 pale isabelline and tips of inner secondaries and tertials sometimes narrowly white. In worn plumage, white spots and fringes on upperparts lost through abrasion, except for bleached white ones on greater wing-coverts and greater primary coverts. FIRST ADULT. Like adult, indistinguishable once relatively long and rounded juvenile p10 moulted.

Bare parts. ADULT. Iris dark brown or pale sepia. Bill off-white, tinged yellow, grey, or pink and with brown tinge on culmen and tip. Leg and foot pale flesh-colour, pink-white, or sandy-white. (Hartert and Steinbacher 1932–8; BMNH.) NESTLING, JUVENILE. No information.

Moults. ADULT POST-BREEDING. Complete; primaries descendant. *E. d. eremodites* from Arabia slightly worn in January, moderately worn in March, heavily worn May–June, with face and crown often quite bare; one from 4 June had some growing feathers on forehead, crown, and neck, but flight-feather moult not started; 2 from 30 June had primary moult scores of 5 and 9 and showed more extensive moult on head and upperparts; specimens from August–September had plumage new, one from 5 August still growing outer primaries (score 39) (BMNH). Nominate *dunni* from Sudan apparently moults earlier, as plumage already heavily worn by February–March and all new in August (BMNH); Mauritanian birds, breeding January (Naurois 1974), probably have similar moult schedule. POST-JUVENILE. Complete, primaries descendant. Timing approximately as in adult, but information limited. Some nominate *dunni* from Sudan in slightly worn juvenile in February, starting moult to adult plumage with some feathers of mantle and scapulars; others still fairly fresh in May. In both races, moult completed and plumage indistinguishable from adult by August.

Measurements. *E. d. eremodites.* ADULT. Saudi Arabia and Oman, all year; skins (BMNH). Bill (S) to skull, bill (N) to distal corner of nostril. Exposed culmen on average 3·2 mm less than bill (S).

WING	♂ 97·1 (1·69; 16)	95–101	♀ 87·9 (1·89; 9)	85–92
TAIL	54·9 (1·93; 15)	52–58	49·7 (2·17; 9)	47–54
BILL (S)	16·3 (0·51; 15)	15·6–17·2	14·9 (0·80; 8)	13·8–16·3
BILL (N)	10·7 (0·44; 14)	9·8–11·2	9·6 (0·46; 8)	9·3–10·4
TARSUS	22·1 (0·85; 15)	20·5–23·4	21·8 (0·76; 9)	20·3–22·6

Sex differences significant, except for tarsus.

Nominate *dunni*. ADULT. Mali and Sudan, all year; skins (BMNH).

WING	♂ 83·7 (1·59; 13)	82–87	♀ 78·6 (1·89; 4)	76–81
BILL (S)	14·8 (0·74; 12)	13·8–15·8	14·2 (0·34; 4)	13·8–14·6

Sex differences significant for wing.

JUVENILE. Wing on average 2·5 mm shorter than adult, bill not full-grown until completion of post-juvenile moult; tail and tarsus similar to adult from shortly after fledging.

Weights. No information.

Structure. Wing short and broad, tip rounded. 10 primaries: p8 longest, p9 1–4 shorter, p7 0–1, p6 1–2, p5 3–6, p4 8–17, p1 16–26. Adult p10 reduced, 44–60 shorter than wing-tip, 0–4 shorter than longest upper primary covert; juvenile p10 longer and with more rounded tip. Outer webs of p5–p8 and inner webs of p6–p9 emarginated. Tertials long, almost reaching wing-tip (short in desert larks *Ammomanes*). Tail short, tip square; 12 feathers, t1 1–3 shorter than t4–t5. Bill heavy, rather long (about half head length), straight for basal $\frac{1}{3}$, culmen and cutting edges decurved over terminal $\frac{2}{3}$ toward blunt tip; depth of bill at base *c.* 6–7 mm, width *c.* 6 mm. Nostril covered by dense tuft of short feathers, ending in some fine bristles. Tarsus rather long and slender, toes short. Middle toe with claw 15·0 (6) 13·6–16·2; outer toe with claw *c.* 76% of middle with claw, inner *c.* 77%, hind *c.* 94%. Hind claw short and curved, 7·4 (6) 6·6–8·0.

Geographical variation. Rather marked, both in colour and size. Nominate *dunni* from Sahara distinctly smaller than Arabian race *eremodites* (see Measurements): tail on average *c.* 4 mm shorter, tarsus *c.* 2 mm; bill distinctly shorter and stubbier. Thus, together with plumage characters, this race resembles large ♀ finch lark *Eremopterix* even more than does *eremodites*. Streaks on crown red-brown instead of black-brown, hindneck streaked rufous (not grey as in *eremodites*), mantle and scapulars with broader deep rufous centres and without pink-grey fringes, appearing distinctly deeper rufous than *eremodites* when plumage fresh and more heavily streaked when worn, not unlike bleached birds of rufous race *carolinae* of Thekla Lark *Galerida theklae*. Sides of head and neck paler without grey specks, black malar stripe absent; underparts uniform pale cream-white apart from pale cinnamon breast-sides (slightly streaked with darker rufous, not brown-black as in *eremodites*) and pale pink-cinnamon central chest and flanks. Dark tips to outer primaries much less marked than in *eremodites*, p5–p7 only slightly dusky on tips. CSR

Ammomanes cincturus Bar-tailed Desert Lark

PLATES 3 and 11
[between pages 160 and 161, and 184 and 185]

Du. Rosse Woestijnleeuwerik   Fr. Ammomane élégante   Ge. Sandlerche
Ru. Чернополосный пустынный жаворонок   Sp. Terrera colinegra   Sw. Sandökenlärka

Melanocorypha cinctura Gould, 1841

Polytypic. Nominate *cincturus* (Gould, 1841), Cape Verde Islands; *arenicolor* (Sundevall, 1850), Sahara, Arabia, and Middle East. Extralimital: *zarudnyi* Hartert, 1902, Iran and Pakistan.

Field characters. 15 cm; wing-span 25–29 cm. 5–10% smaller and distinctly slimmer than Desert Lark *A. deserti*; noticeably smaller-headed and less bulky than Dunn's Lark *Eremalauda dunni*. Rather small, neatly made lark, with character rather like bunting *Emberiza*, especially in its distinctive small bill and head shape. Plumage essentially pale grey-buff, more uniform than in other desert-dwelling larks, but with obvious orange glow on upper- and underwing and on tail-base. Primaries and tail-feathers variably tipped black, with marks on tail usually strong enough to recall wheatear *Oenanthe*. Sexes similar; no seasonal variation. Juvenile separable at close range. 2 races in west Palearctic; North African and Arabian race, *arenicolor*, described here (see also Geographical Variation).

ADULT. Upperparts buff, with grey tone overall and orange-pink tones on folded wing; virtually unpatterned except for faintly paler supercilium, tips to wing-coverts, and fringes to tertials, and almost black primary-tips. Tail-base as folded wing but with deep dark brown to almost black terminal marks on central feathers and short similarly coloured marks on outer feathers creating quite distinct inverted T. Underparts pale buff-white with buff tones strongest on slightly mottled chest and palest on chin, throat, and vent. In flight, wings show strong orange-pink glow (particularly below) and dark trailing edge to primaries. Bill short, stubby; sandy-pink. Legs sandy-horn to pale pink-brown. JUVENILE. Not studied in the field (see Plumages).

Although congeneric with *A. deserti*, general character

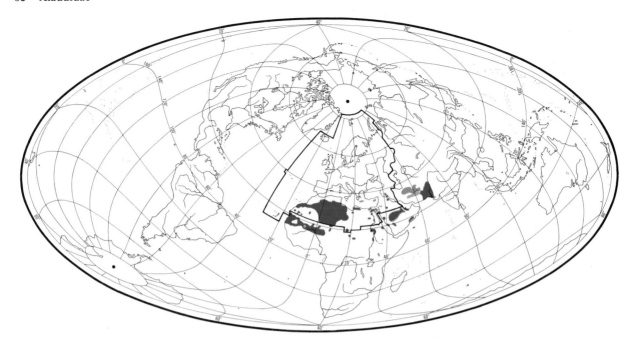

and behaviour distinctive. *A. cincturus* often recalls a bunting *Emberiza* more than does *A. deserti*, being generally more active with freer gait and faster flight. Thus while at first sight confusable with *A. deserti*, *E. dunni*, and juvenile Temminck's Horned Lark *Eremophila bilopha*, *A. cincturus* shows the following distinctive characters: (1) relatively small, bunting-like bill; (2) rounded, bunting-like head; (3) long, dark wing-point ($\frac{1}{2}$ length of tertials); (4) almost unpatterned plumage except for (5) dark transverse, terminal marks on outer wing and rather short tail. Of these, points 1, 2, and 5 unique to *A. cincturus*. Usually rather wild, not allowing close approach and escaping with fast, long-legged run or lengthy flight. Flight involves flickering, even jerky, bursts of wing-beats giving noticeable speed. Song-flight either a series of steep undulations or a quite high and circular sweep, with action more flapping than in normal flight. Escape-flight low, fast, and not short. Gait most free of desert-dwelling larks, with length of legs usually obvious and easy walk and nimble run faster than both *A. deserti* and *E. dunni*. Stance frequently upright, emphasizing spindly legs and recalling *Emberiza* and even *Oenanthe*.

Song a pure, almost ethereal series of simple phrases, comprising quiet fluty notes and louder, characteristic, mournful 'see-oo-lee' like creaking door. Commonest calls a thin, high 'peeyu' and a chirrupy 'chweet'.

Habitat. In subtropical and tropical lower latitudes, in contrast to Desert Lark *A. deserti* on wide flat or gently sloping bare stony or sandy deserts, rarely with more than sparse vegetation, and without access to water or relief from intense heat. In such terms ranks as most essentially desert-dwelling of west Palearctic Alaudidae, and penetrates furthest into interior of Sahara, although highest density found near Atlantic coast (Valverde 1957). In north-west Africa also frequents low dunes and hollows between them, or sandy wadis amidst mountains (Heim de Balsac and Mayaud 1962). At Azraq (Jordan), although associated with open stony limestone hammada, tends to prefer parts with sandy patches and at least some shrublets (Hemsley and George 1966). On Cape Verde Islands, occurs on reddish soil covered with small pebbles and rocks on a tableland near the ocean (Bannerman and Bannerman 1968). In some parts of range, replaces Crested Lark *Galerida cristata* as common roadside bird, especially at lower elevations, but in Baluchistan (Pakistan) abundant up to *c.* 1700 m or more after breeding season (E M Nicholson). In Arabia, a bird of flat absolute desert, seldom among bushes or even rocks (Meinertzhagen 1954). In Indian subcontinent, in open stony scrub-and-bush plains and plateaux, and ploughed fields and fallow, perching freely on overhead wires, but in Baluchistan on more rocky and barren wastes (Ali and Ripley 1972). Wide-ranging, mobile, and even nomadic, often shifting ground, flying in lowest airspace. Erratic distribution attributed to varying state of vegetation (Valverde 1957). In view of low biological productivity of habitat and fluctuating climatic conditions, is probably nomadic within extensive areas, often at thin overall density except in temporarily favourable circumstances. Precise criteria for habitat choice accordingly remain obscure, as do reasons for relative success in maintaining populations on terrain apparently so unfavourable to survival. Few terrestrial species have so little close contact with man in most parts of extensive geographic range.

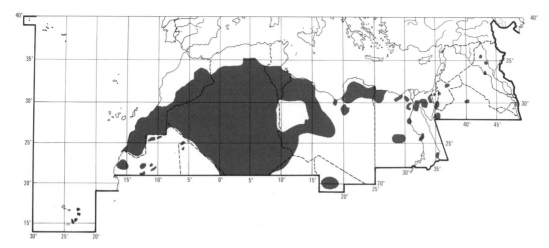

Distribution. Inadequately known, partly because of identification problems (Wallace 1983a).

IRAQ. Only one record of breeding (Allouse 1953). Accidental. Italy, Syria, Kuwait, Malta.

Population. No information on trends.

MOROCCO. Common in south (Valverde 1957). LIBYA. Tripolitania: widespread and probably commoner than Desert Lark *A. deserti*. Cyrenaica: common. (Bundy 1976.) EGYPT. Uncommon and local (SMG, PLM, WGM). CAPE VERDE ISLANDS. By no means plentiful; local (Bannerman and Bannerman 1968). JORDAN. One of commonest Alaudidae at Azraq, especially on hamada (Wallace 1983a).

Movements. Resident, also dispersive to an uncertain extent. In North Africa there are complex short-distance movements after breeding (Etchécopar and Hüe 1967). These mainly concern early-breeding birds of Saharan zone (e.g. recorded Ahaggar only December–March), which make post-breeding dispersal, February–May, to spend dry summer months in less arid regions; such movement more marked in drought years (Heim de Balsac and Mayaud 1962). Nomadic behaviour also noted in Saudi Arabia (Jennings 1980, 1981a) and southern Oman (Gallagher and Woodcock 1980). Scarce and irregular visitor to Gulf territories of Kuwait, Qatar, Dubai, and northern Oman, mainly March–July (Bundy and Warr 1980; Gallagher and Woodcock 1980; Walker 1981a).

RH

Food. Seeds and other plant material, and insects. Will feed by digging into ground, but less often than Dunn's Lark *Eremalauda dunni* (at least when accompanying that species) and picks from surface more; also seen to examine plants (G K Brown). Feeding flock will move rapidly forward, alternately running and pausing (J Palfery), and birds watched by Bannerman and Bannerman (1968), apparently feeding thus, took *c.* 5–6 quick steps at a time,

occasionally pecking at something or opening wings and making a little jump, perhaps after an insect; also seen picking up seeds. Will dash about after passing flies (Bourne 1955).

Food said by Meinertzhagen (1954) to comprise small seeds and also small insects. In southern Morocco, April and July, 6 adults contained seeds (including those of *Aizoon*) and insects (Valverde 1957). Stomachs of 2 birds from Algeria, March, were filled with seeds (Roche 1958). On Cape Verde Islands, October, 4 birds collected by Naurois (1969b) contained small seeds and much other plant material, as well as insects including grasshoppers (Orthoptera) and a beetle (Coleoptera); also grit. Thus believed to take fewer insects than Black-crowned Finch Lark *Eremopterix nigriceps* in same area, though during several hours' watch on Cape Verde in September, Bourne (1955) saw them take insects only, in contrast to accompanying *E. nigriceps*.

Will apparently give insects to young (Gallagher and Rogers 1980).

DJB

Social pattern and behaviour. Little known.

1. Gregarious outside breeding season, though less so than some other Alaudidae, generally occurring in small flocks: e.g. in Arabia, often 2–3 together, though flocks of 35 recorded (Meinertzhagen 1954; Jennings 1980, 1981b; J Palfery); in Algeria, flocks of 4–7, but mostly singly or in pairs (Koenig 1895); in Libya, flocks of 'about a dozen' (Bulman 1942). Once, in western Sahara, concentration of 50 (Valverde 1957). Often associates with other Alaudidae, e.g. in central and eastern Saudi Arabia especially with Dunn's Lark *Eremalauda dunni* (Jennings 1980; G K Brown); in Morocco, November–February, in small flocks with Temminck's Horned Lark *Eremophila bilopha*, Lesser Short-toed Lark *Calandrella rufescens*, and *Galerida* larks (Blondel 1962a); in Algeria and Libya, July, with *E. bilopha* (Koenig 1895; Whitaker 1902). BONDS. On several occasions, winter, Morocco, birds always in pairs, even when in small parties (Smith 1962–4), suggesting pair-bond maintained outside breeding season. No further information. BREEDING DISPERSION. Territorial, probably in neighbourhood groups. At Azraq (Jordan), transects yielded *c.* 2 birds per km on stony hammada and *c.* 1 per km in wadis; some territories

only 20-30 m from Desert Lark *A. deserti*, such neighbours sharing airspace (Wallace 1983*a*). In Libya, April, birds performed Song-flights *c.* 100 m apart (G Bundy). On Cape Verde Islands, *c.* 10 nests with eggs or young in 2 km² (Naurois 1969*b*). ROOSTING. No information on nocturnal roosting. Following account of daytime loafing by J Palfery. In Saudi Arabia, birds typically shelter from sun under slabs of rock in middle of day. Up to 5 birds together under one broken concrete culvert. One bird squeezed itself into gap *c.* 8 cm high between concrete and ground, and flopped on sand with half-spread wings. Birds also make loafing hollows in sand, kicking out sand vigorously with feet, then prostrate themselves in hollows with head outstretched, wings half-spread, and tail flat.

2. Rather shy (Jennings 1981*b*; Wallace 1983*a*), but not invariably so, e.g. at nest rather shy (Koenig 1985) or quite confiding (Hartert 1915); readily approachable when with young out of nest, Tunisia (M G Wilson). Confiding manner also reported out of breeding season, Morocco (Bannerman and Bannerman 1953). On Cape Verde, allowed approach only to 20-30 m (Bannerman and Bannerman 1968). In Morocco, March, flushed bird rose and flew in straight, downward sloping path, with undulating, steeply dipping action, giving 'see-ou' call (see 2a in Voice) in rhythm with dips; soon landed again (P A D Hollom). Similar incident, thought perhaps to be territorial display-flight (J Palfery), but possibly simple alarm-reaction, reported in Saudi Arabia: bird flew *c.* 50 m in regularly undulating flight *c.* 3 m above ground, alternately rising with a few wing-beats and, with 'tzoi' call (see 2d in Voice), gliding down, rising again, etc. FLOCK BEHAVIOUR. Flying flocks constantly twist and turn in perfect unison, exposing reddish wing colour not visible when perched (Bulman 1942). SONG-DISPLAY. From start of breeding season, ♂ performs Song-flight similar to *A. deserti* and Black-crowned Finch Lark *Eremopterix nigriceps*. Flight-path deeply undulating and markedly meandering but roughly circular, *c.* 25-40 m above ground (G Bundy); slow, and undulating like yo-yo, at 6-10 m, bird holding head raised and chest puffed out (G K Brown); altitude varies and path not uncommonly a straight line (P A D Hollom). Wing action described as butterfly-like (Walker 1981*b*); a series of undulating strokes alternating with forced downward glide (Bourne 1955). Undulations steep but relatively brief: e.g. in Jordan, 2 strongly undulating Song-flights at up to *c.* 10 m, in roughly straight line, bird dropping at intervals of *c.* 1 s to accompaniment of 'wee' notes (P A D Hollom: see 1 in Voice). In Song-flight, Iran, bird ascended from bush and began singing ('wee' notes) at *c.* 3 m above ground; gained height despite dropping steeply with every note; descended again soon after reaching *c.* 12-13 m above ground (P A D Hollom). For further details of song in Song-flight, see 1 in Voice. Song-flight may last up to 3 min and ends with steep descent to ground (G Bundy). Also sings from a low perch (Gallagher and Woodcock 1980), often a bush (P A D Hollom). Once, sang from ground: ♂ of pair, foraging near nest, flew to nest-area where he sang (see 1 in Voice) from a low bank; ♀ followed him carrying nest-material in bill (J Palfery). In central Arabia, commonly sings March-April (Jennings 1980); in Jordan, April until at least 9 May (D I M Wallace); in Oman, reported late January (Walker 1981*b*). ANTAGONISTIC BEHAVIOUR. ♂♂ chase each other off territorial boundaries (D I M Wallace). Birds sometimes chase one another out of favourite daytime loafing sites (see Roosting, above); chasing bird may utter harsh sounds (J Palfery: see 2e in Voice). HETEROSEXUAL BEHAVIOUR. No details. RELATIONS WITHIN FAMILY GROUP. No information. ANTI-PREDATOR RESPONSES OF YOUNG. No information. PA-RENTAL ANTI-PREDATOR STRATEGIES. Tame at the nest in eastern Saudi Arabia, flushing at *c.* 3 m and flying only *c.* 10 m away (G K Brown). When human intruder approached nest containing single, almost fledged young, both parents ran quickly towards nest, ♂ uttering a continuous plaintive 'wheet' (Alexander 1898: see 2c in Voice). EKD

Voice. Freely used, especially by ♂, in breeding season. Renderings by different authors, especially of song, not always easy to reconcile, and variability indicated in the following may sometimes be more apparent than real.

CALLS OF ADULTS. (1) Song of ♂. A relatively short phrase of clipped trilling sounds, or alternatively pure single notes, repeated during Song-flight. Bird may switch from trilling song to single notes (see below), but perhaps a tendency, as in Short-toed Lark *Calandrella brachydactyla*, for trilling sounds to feature more in initial ascent phase of Song-flight (P A D Hollom). In recording by J-C Roché, begins with continuously repeated phrase *c.* 3·5 s long, commonest sequences roughly 'turr-ee tre-le tree-tree-you' (Fig I) and 'tre-le tre-le tree-tree-you'; in Fig I, 2nd unit ('tre-le') higher-pitched than 1st, and this usually so (G Bundy). 3rd unit ('tree-tree-you') rather invariable and regularly occurring, but other units more variable: e.g. 'tre-le' latterly gives way to a more sustained tremolo 'trrrrrrrrrreeeee' (J Hall-Craggs: Fig II), and 3 units—in varying order—may separate each 'tree-tree-you'. Apparently 'tree-tree-you' given at peak of each ascent within undulating Song-flight, the shorter units during the intervening drops. About half-way through recording, last sustained trill gives way to series of 'zoo-ee' and 'zoo-it' sounds, then to 8 high-pitched 'seeeeeee' notes (Fig III) at intervals of *c.* 1 s. In Song-flights, Iran and Jordan, April-May, only these sounds heard, rendered quiet high 'wee', and uttered with cessation of wing-beats causing steep drop in height; also given for shorter time on ground (P A D Hollom). Hoarse, rising 'dzeeep' from displaying bird, Saudi Arabia (G K Brown), is apparently related. Song from ground near nest, a simple squeaky 'see-wee' (J Palfery), evidently similar. Also rendered a high-pitched 'tseep', apparently delivered at top of each ascent (G Bundy). Recording by E D H Johnson (Algeria, February) similar to phrase in Fig I, but comprises 4 syllables, thus 'whee-tree-tree-you' (E K Dunn). Other renderings of song include: at Azraq (Jordan), a mournful 'see-oo-lee', like creaking door (D I M Wallace and I J Ferguson-Lees); in Saudi Arabia, 'sutzee' (J Palfery) or 'peeyu-pee' and 'see-tiou', in which rather trilling '-tiou' given on upward part of flight undulation (G K Brown); in Iran, 20 'cher-holf wee' phrases in *c.* 30 s—'cher-holf' brief, quiet and low-pitched, 'wee' clear, high, and stronger (P A D Hollom); see also Bourne (1955) and Walker (1981*b*). Subdued warbling Subsong reported during steep descent to ground at the end of one Song-flight (G Bundy). (2) Contact-alarm calls. (a) Various renderings of (presumably) the same call. Thin, high 'peeyu' from birds in

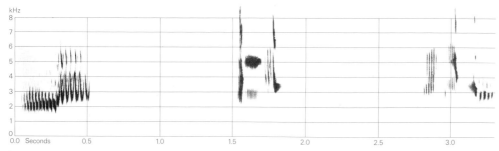

I Roché (1968) Algeria February 1967

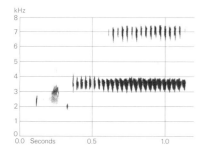

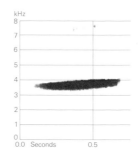

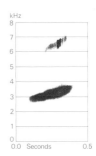

II Roché (1968) Algeria February 1967   III Roché (1968) Algeria February 1967   IV E D H Johnson Algeria February 1969

feeding flock; sweet, rather pure, descending 'pheeyou', almost a whistle and querulous in timbre, from bird which had just alighted on top of regular daytime loafing site; repeated 'heeoo' from birds disturbed from midday loafing sites (J Palfery); 'teu' or 'peu', also 'teeoup' (G K Brown); 'see-ou', in which 'see-' higher-pitched than 'ou', given during dips in undulating flight by bird flushed by human intruder (P A D Hollom); 'hu-ee' from pairs in flight (Smith 1962–4). (b) As well as call 2a, birds in feeding flocks use a rather chirrupy 'chweet' or 'chwit', also 'kweep' and a hoarse 'shweep' (G K Brown); dry, rattling, chirrup, 'rreep', and, in flight, a similar but higher-pitched, short, reedy 'jreep' (J Palfery). (c) A thin high note rising in pitch, often uttered on ground when bird was approached (G Bundy). A continuously repeated plaintive 'wheet', given by ♂ near nest threatened by intruder (Alexander 1898). Mournful 'divee-ee' and 'dweet' of birds with young, running about by intruder's feet (M G Wilson), evidently the same call, as also a soft 'piou' rather like Bullfinch *Pyrrhula pyrrhula* from anxious bird near nest (G K Brown). In recording (Fig IV), call rendered 'huit' (J Hall-Craggs), not unlike 'seeeeeee' (Fig III) but starting off lower-pitched and given at intervals of 2–3 s (E K Dunn). This call, rendered 'coo-ee', thought to express alarm (Smith 1962–4). (d) A harsh, loud 'tchoi-tchou', rather like sparrow *Passer* (Bourne 1955), described as alarm-call; 'tzoi' sound made by bird in low undulating flight (J Palfery) probably this call. (e) A low-pitched, not very loud, gravelly 'zrrt zrrt' by bird chasing another; grating 'tchairz', also by pursuing bird, harsher than 'zrrt' sound (J Palfery), probably a variant. Such calls presumably denote excitement, anger, etc., of

pursuer. Harsh 'wij' given on ground (Smith 1962–4) perhaps the same or a related call. (3) Bird landing near nest and running towards it gave rather sharp 'cheeip' (G K Brown).

CALLS OF YOUNG. No information.   EKD

Breeding. SEASON. Cape Verde Islands; eggs laid September–April, probably affected by occurrence of rains (Bannerman and Bannerman 1968). North Africa: February–April(–May) (Heim de Balsac and Mayaud 1962; Hüe and Etchécopar 1970). Eastern Saudi Arabia: from March (J Palfery). SITE. On ground usually in shelter of tussock or small stone. Nest: shallow depression lined with vegetation, often with rim of small stones all round or on exposed side. Building: no information. EGGS. See Plate 79. Sub-elliptical, smooth and glossy; white, lightly spotted with black, grey, and purple. 21·5 × 15·3 mm (19·3–23·7 × 13·2–16·9), n = 60; calculated weight 2·59 g (Schönwetter 1979). Clutch: 2–4, once 5. Of 27 clutches, Tunisia and Algeria: 2 eggs, 8; 3, 16; 4, 3; mean 2·8 (Heim de Balsac and Mayaud 1962). No further information.

Plumages (*A. c. arenicolor*). ADULT. Forehead, crown, hindneck, sides of neck, mantle, and scapulars pink-cinnamon with variable amount of pale grey tinge on tips of feathers; feather-centres of crown darker grey-pink or rufous-pink with dark grey shafts. Tertials, back, rump, and upper tail-coverts uniform bright rufous-cinnamon. Lores, supercilium, and rather broad ring around eye pale pink-buff or cream-white, often with indistinct grey patches just before and behind eye. Cheeks cream-buff, merging into pink-cinnamon or pale brown on ear-coverts; ear-coverts and lower cheeks at sides of throat often indistinctly spotted grey. Chin and throat pale cream-buff or off-white. Chest, sides of breast, flanks, and thighs buff or pink-buff, darkest on sides of breast; sides of breast usually with rather indistinct grey shaft-streaks or spots, occasionally extending as faint specks to centre of chest. Remainder of underparts cream-white or white. Tail pink-cinnamon or rufous-cinnamon; slightly paler, isabelline or cream, towards outer web of t5–t6; tips of tail-feathers with contrasting black or brown-black spot *c.* 10 mm long, fringed cinnamon when fresh; black on outer webs of tail-tips slightly reduced outwards; outer web of t6 uniform isabelline or cream without black. Flight-feathers rufous-cinnamon, p5–p9 with ill-defined black or brown-black tip 5–10 mm long, bordered by narrow cream-pink fringe; p1–p4 with broader but less distinct pale fringe, subterminally washed dusky grey; reduced p10 grey with cinnamon wash on centre and narrow cinnamon outer edge. Upper wing-coverts pink-cinnamon like mantle and scapulars; median and greater coverts and longest feathers of bastard wing with grey or brown-grey centres; inner webs of greater primary coverts grey except on tips. Under wing-coverts and axillaries pink-cream to pink-cinnamon, paler than pale rufous-cinnamon undersurface of flight-feathers. *In worn plumage*, upperparts less strongly suffused grey, appearing deeper rufous-cinnamon; dark feather-centres of crown and longer upper wing-coverts more distinct; outer webs and tips of tertials bleached to pinkish-isabelline; pale fringes of tips of tail-feathers and primaries worn off, black tips less deep and less contrasting, sometimes hardly discernible, especially on t1 and primaries; underparts off-white with rather indistinct cream-buff band across chest. Some individual variation: some birds show hardly any grey on upperparts even in fresh plumage, others rather sandy-yellow above in worn plumage. NESTLING. No information. JUVENILE. Closely similar to adult. Feathers of body shorter and looser than in adult, often paler and more cream-coloured, less greyish pink-cinnamon, especially on fringes of upperpart feathers; black on tips of tail and flight-feathers less sharply defined and less extensive, sometimes completely absent on primaries; no grey on outer web of p10; p10 slightly longer and with broader tip than in adult, 6–9 mm longer than greater upper primary coverts (1 mm shorter to 5 mm longer in adult). FIRST ADULT. Like adult, but at least some birds distinguishable by remnants of juvenile feathering, especially some flight-feathers.

Bare parts. ADULT, JUVENILE. Iris dark brown or yellow-brown. Upper mandible pink-flesh, sandy-grey, or sandy-yellow; lower mandible yellow-pink, salmon-pink, sandy-yellow, or pinkish-grey with grey tip. Leg and foot sandy-grey, sandy-yellow, isabelline-white, or pale greyish-horn with pink tinge; claws slate or pink-horn. (Stanford 1954; BMNH.) NESTLING. No information.

Moults. ADULT POST-BREEDING. Complete; primaries descendant. Only a few specimens in moult examined. In both *arenicolor* and nominate *cincturus*, most adults in fresh plumage October–January, rather worn to heavily worn March–May; from late February or March onwards, some birds start replacement of face, mantle, chest, or sides of breast; one from May had inner 2 primaries shed, 4 others had not yet started, but were heavily worn. Main moult period probably June–September. In particular, some nominate *cincturus* (with variable breeding season) deviate from this pattern, being heavily worn in December or in body moult in January. POST-JUVENILE. Probably usually complete with primaries descendant, but not always so; moult apparently arrested during adverse conditions

or when new breeding season starts after sudden rains. In nominate *cincturus*, fresh juveniles examined date from November to January, and some in body moult February–March. 4 specimens from April–May had apparently arrested moult: one had all feathering new except for juvenile p6–p10, other 3 had body all or largely new, flight-feathers, tail (in one, except t1–t3), tertials, and greater upper wing-coverts old. Birds with all plumage new April–May probably immatures with moult completed. 2 specimens of *arenicolor* in 1st adult plumage, November, had moult arrested: one with all plumage new except flight-feathers, the other with p7–p10 new but p1–p6 still juvenile (showing moult pattern similar to some immature waders Charadrii). One from March had primary moult arrested: p1–p4 new in left wing, p1–p3 and p7 new in right.

Measurements. ADULT. Nominate *cincturus*. Cape Verde Islands (Boa Vista, São Tiago, Maio, Sal), October–May; skins (BMNH, RMNH). Bill (S) to skull, bill (N) to distal corner of nostril; exposed culmen on average 3 mm less than bill (S).

	♂		♀	
WING	93·6 (1·80; 14)	91–97	87·8 (2·42; 10)	84–91
TAIL	52·8 (2·57; 11)	49–57	49·3 (1·78; 10)	48–52
BILL (S)	14·3 (0·42; 11)	13·7–14·9	13·5 (0·69; 9)	12·2–14·5
BILL (N)	8·6 (0·43; 11)	7·9–9·2	8·3 (0·52; 9)	7·5–9·1
TARSUS	21·6 (0·64; 11)	20·8–22·8	21·0 (0·41; 10)	20·5–21·7

Sex differences significant, except for bill (N).

A. c. arenicolor. Southern Morocco, Algeria, and Tunisia, October–May; skins (BMNH, RMNH, ZMA).

	♂		♀	
WING	95·4 (2·45; 16)	91–100	89·6 (2·50; 19)	86–93
TAIL	53·5 (2·76; 9)	48–56	48·3 (2·53; 11)	45–51
BILL (S)	14·1 (0·46; 9)	13·4–14·8	12·7 (0·98; 11)	11·6–13·8
BILL (N)	8·1 (0·35; 9)	7·7–8·6	7·9 (0·60; 11)	7·2–8·6
TARSUS	21·9 (0·70; 9)	21·1–23·3	21·9 (0·70; 9)	19·7–22·6

Sex differences significant for wing, tail, and bill (S).

JUVENILE. Wing on average 4·5 mm shorter than adult, tail similar to adult.

Weights. Algeria: ♂ 21–23·5 (10). Aïr (Niger), January–February: ♂♂ 14, 17; ♀ 16. Ennedi (Chad), April: ♀ 18. (Niethammer 1955b.)

Structure. Wing rather long and broad, tip rounded. 10 primaries: p8 longest, p9 2–5 shorter, p10 46–56, p7 0–1, p6 2–9, p5 9–17, p1 23–27 in ♂, 20–26 in ♀; adult p10 1 shorter to 5 longer than greater upper primary coverts, on average 1·7 longer (*n* = 12), juvenile p10 7 (3) 6–9 longer. Outer webs of (p5–)p6–p8 and inner of (p7–)p8–p9 emarginated. Longest tertials reach to tip of p4(–p5) in closed wing. Tail rather long, tip shallowly forked; 12 feathers, t1 3–8 shorter than t5–t6. Bill short and rather slender, straight except for decurved culmen, tip sharp; nostrils covered by tuft of short soft feathers, some tipped with fine bristles; similar bristles along gape and on lower cheeks. Tarsus rather short and slender, toes short and rather thick. Middle toe with claw 12·9 (13) 11·4–14·5; outer toe with claw *c.* 82% of middle with claw; inner *c.* 74%, hind *c.* 96%. Front claws decurved, short; hind claw slightly decurved, rather short, 6·2(6)5·3–6·8.

Geographical variation. Slight, in colour only. Nominate *cincturus* from Cape Verde Islands deeper rufous-cinnamon on upperparts, flight-feathers, and tail than *arenicolor* from Sahara, Arabia, and Middle East; in fresh plumage, crown more brownish-cinnamon, mantle dark greyish-cinnamon (dull greyish with darker streaks in worn plumage); sides of neck and all chest distinctly streaked grey on vinous-cinnamon ground (almost uniform pink-cinnamon in *arenicolor*), throat almost completely mottled grey on cream-buff ground (uniform cream-white in *arenicolor*), remainder of underparts deeper pink-cream. Black spots on tips of tail-feathers larger and deeper black, up to 15-20 mm long, bordered by narrower pale fringes; black extends partly on to outer web of t6 (length of black on inner web of t6 *c.* 20 mm in nominate *cincturus*, on outer web 5-10; 6-12 on inner web in *arenicolor*, none on outer web). Black on tips of longer primaries slightly more extensive, deeper, and more sharply defined. Populations of *arenicolor* from eastern Chad and Sudan average slightly paler and smaller than typical populations further north and are sometimes separated as *pallens* Le Roi, 1912 (e.g. Grant and Mackworth-Praed 1958), but similarly pale birds occur occasionally elsewhere (Niethammer

1955*b*). *A. c. zarudnyi* from Iran and Pakistan distinctly darker than both *arenicolor* and nominate *cincturus*; upperparts dusky brown-grey, underparts buff with grey spots on grey-brown chest, tail with broader black tips, flight-feathers pale grey-brown, less bright rufous; wing on average 5 mm longer than in *arenicolor* (Hartert 1903-10).

Forms superspecies with Rufous-tailed Desert Lark *A. phoenicurus* from India, which is larger, darker rufous-brown (also below), has more extensive black on tail-tip, more black on outer webs of flight-feathers, and heavy bill; sometimes considered conspecific (Hartert 1903-10; Hartert and Steinbacher 1932-8; Ali and Ripley 1972), as differences between *arenicolor* and *A. phoenicurus* in part bridged by *zarudnyi*; treated separately here on account of different wing pattern, wing formula, and relatively heavy bill.　　　CSR

Ammomanes deserti Desert Lark

PLATES 4 and 11
[between pages 160 and 161, and 184 and 185]

Du. Woestijnleeuwerik　　FR. Ammomane du désert　　GE. Steinlerche
Ru. Пустынный жаворонок　　SP. Terrera sahariana　　Sw. Stenökenlärka

Alauda deserti Lichtenstein, 1823

Polytypic. *A. d. payni* Hartert, 1924, south-east and southern Morocco; *algeriensis* Sharpe, 1890, Algerian Sahara north of 31°N, southern Tunisia, and western Tripolitania (Libya); *mya* Hartert, 1912, central Algerian Sahara between *c.* 27°N and 30°N; *whitakeri* Hartert, 1911, western Algerian Sahara through Ahaggar region and Tassili N'Ajjer (southern Algeria) to Fezzan and central Tripolitania (Libya) and perhaps Tibesti (Chad); nominate *deserti* (Lichtenstein, 1823), Nile valley from south of Wadi Halfa (Sudan) to Qena (Egypt), and coasts and mountains bordering Red Sea (in Egypt, Sinai, southern Jordan, north-west Saudi Arabia, and perhaps coasts of Sudan, Eritrea, and rest of Saudi Arabia); *isabellinus* (Temminck, 1823), central and eastern Egypt (in Nile valley south of Sohag and on Red Sea coast interdigitating with nominate *deserti*) and northern Sinai, north to Dead Sea region and east through northern Saudi Arabia and southern Iraq to Tigris and *c.* 48°E; *annae* Meinertzhagen, 1923, black lava deserts of Azraq area (Jordan) and probably neighbouring Syria; *coxi* Meinertzhagen, 1923, Syria and northern Iraq east to Samarra and Al Fallujah; *cheesmani* Meinertzhagen, 1923, Iraq east of Tigris and Iran west of Zagros mountains, south to Bandar e Bushehr. Extralimital: *geyri* Hartert, 1924, Mauritania and Aïr (Niger); *kollmannspergeri* Niethammer, 1955, Ennedi (Chad); *erythrochrous* Reichenow, 1904, Dongola area (northern Sudan); *azizi* Ticehurst and Cheesman, 1924, Al Hufuf area (north-east Saudi Arabia); *darica* Koelz, 1951, southern Zagros mountains (Iran); *iranica* Zarudny, 1911, Iran east from Tehran and Baluchistan (except extreme north-east), southern Afghanistan, and western Pakistan; *parvirostris* Hartert, 1890, western Turkmeniya (USSR); *orientalis* Zarudny and Loudon, 1904, north-east Iran, northern Afghanistan, and USSR east from eastern Turkmeniya; *phoenicuroides* (Blyth, 1853), south-east Afghanistan, north-west India, and Pakistan east of *iranica*; 3 further races in Arabia and 2-3 in Ethiopia and Somalia.

Field characters. 16-17 cm; wing-span 27-30 cm. 15% shorter but no less bulky than Skylark *Alauda arvensis*; 10% larger than Bar-tailed Desert Lark *A. cincturus* and Dunn's Lark *Eremalauda dunni*. Medium-sized, robust lark with usually long, evenly pointed bill, rather large head, full body, and quite long wings, but relatively short tail. Plumage dull and little patterned; subject to wide tonal variation with habitat, being sandy on sand, grey on rocks, and almost black on dark basalt. Plumage usually shows pale fore-supercilium and eye-ring, crown-streaks, dark mottling on throat and chest, and dark tail-panel. Voice distinctive. Sexes similar; no seasonal variation. Juvenile separable at close range. 9 races in west Palearctic; extreme forms easily distinguishable but only Red Sea area race, nominate *deserti*, described here (see also Geographical Variation).

ADULT. Upperparts dull brown-isabelline, with no obvious warm tones and patterned only by rather faint head-marks, paler tips and fringes to wing-coverts and tertials, and dark centres to central tail-feathers. In some (probably bleached) birds, back may appear mottled; in all, plumage looks coarse-textured. In flight, pattern of spread tail less distinctive than that of either *A. cincturus* or *E. dunni*, but some birds show confusing dull black triangle on rufous ground suggestive of *A. cincturus*. When closed or incompletely spread, tail usually appears rufous-buff on central feathers and on outer edges, with dark grey-brown panels in between. Wing pattern differs from *A. cincturus* in lacking distinctly darker tips to primaries and in having faint wing-bar, caused by paler tips to median wing-coverts. Underparts pale buff, with spotted chest obvious on some but not on others and

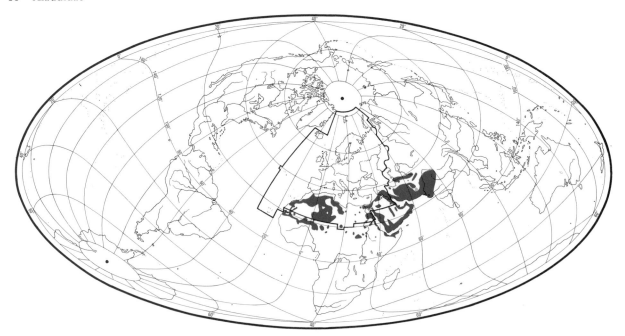

contrasting little with paler, mottled throat. Facial marks also variable: on some, fore-supercilium and eye-ring almost white and malar and moustachial stripes dark brown, creating fairly distinct pattern on front of face, but others have both paler and dark marks less intense and not contrasting, creating only faint pattern. Bill always evenly pointed but size variable: always longer and less stubby than in *A. cincturus* and never as deep or as blunt-ended as in *E. dunni*; dark horn above but with yellow on cutting edges and base of lower mandible. Legs yellow-horn, close in tone to *A. cincturus*, less pink than *E. dunni*. JUVENILE. Lacks dull black triangle on tail. Not fully studied in the field (see Plumages).

Extremely variable plumage tone, amorphous pattern, and variation in bill length often provoke mis-identification. Important therefore to remember that (1) it is essentially a lark of broken, rocky, and, above all, usually sloping habitats, not overlapping those of *A. cincturus* and *E. dunni*, and (2) its general character is less reminiscent of other desert-dwelling larks, being more like *Alauda arvensis* (or even a large pipit *Anthus* or small thrush *Turdus*). Has long wing-point (like *A. cincturus* but not *E. dunni*) and proportionately longer tail than either *A. cincturus* or *E. dunni*. Of plumage marks, pale fore-supercilium, breast-mottling, and relatively well-marked median wing-coverts show most constantly. Remarkably tame, allowing very close approach and escaping only by short move or flight. Flight action recalls *Alauda arvensis* but wing-beats more constant, so that progress lacks uneven quality of that species but is slower than *A. cincturus*. Song-flight not high—circular and long-sustained, with undulations in time with song. Escape-flight quite high, bird usually moving to higher ground. Usually shuffles or walks unhurriedly, but occasionally runs. Stance as *A. arvensis*, usually less upright than *A. cincturus* but more so than *E. dunni*.

Song melodious, with soulful quality, and, unlike *A. cincturus* and *E. dunni*, far-carrying. Call a full, slightly hollow monosyllable, 'cuu' or 'puu', slowly repeated.

Habitat. Despite misleading name, not found in flat open desert, being closely attached to low rock-faces, flanking escarpments, boulder-strewn or stony slopes, and scree; habitats without some nearly vertical element are generally avoided. Apparently preferred terrain in southern Morocco is covered with large boulders, almost bare and near a hill (Valverde 1957). Appears wherever large stones cover the soil; not on sand, sand-dunes, or beaches, or in palm oases (Heim de Balsac and Mayaud 1962). At least in parts of range finds rock faces surrounding springs especially attractive, e.g. in Baluchistan; also readily attaches itself to such structures as an isolated police post or even a desert monastery such as St Anthony's in Wadi Araba, Egypt (E M Nicholson). In Anti-Atlas (Morocco), occupies open wooded slopes with Chaffinch *Fringilla coelebs* and Coal Tit *Parus ater* (Heim de Balsac and Mayaud 1962), but this atypical. Sedentary and usually reluctant to move far either on foot or in flight. Will, however, range alongside tracks and roads and over bare ground especially after breeding season. In Tibesti (Chad), ranges up to 3000 m (Heim de Balsac and Mayaud 1962); in Saudi Arabia and Baluchistan, up to c. 2000 m (Meinertzhagen 1949c; E M Nicholson); in Afghanistan, numerous up to 1400 m and higher, in deserted stony foothills, on stony slopes, and on a dry area close to tamarisk *Tamarix* scrub by riverside (Paludan 1959). In USSR, birds from higher elevations descend for winter to adjoining plains (Dementiev and Gladkov

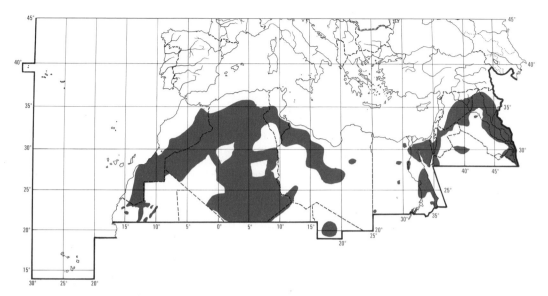

1954*a*). In south Caspian Sea area, sometimes forages amongst washed up seaweed (Isakov and Vorobiev 1940). In Indian subcontinent favours stony hill slopes, dry narrow valleys, rocky ground along base of foothills, and fallow land in desert canal cultivation, resorting to water sources where possible (Ali and Ripley 1972).

Distribution. TURKEY. Recently found in south-east; probably a resident breeder in very small numbers (MB, RFP).

Accidental. Lebanon, Cyprus.

Population. No information on population trends.

JORDAN. Azraq: commonest lark on basalt (Wallace 1983*a*). EGYPT. Locally common (SGM, PLM, WCM). MOROCCO. Common in south in some areas (Valverde 1957).

Movements. Resident. Described as sedentary in Palearctic Africa, Middle East, and Indian subcontinent (Heim de Balsac and Mayaud 1962; Etchécopar and Hüe 1967; Ali and Ripley 1972; Jennings 1980). This lack of movement further evident from high degree of subspeciation and adaptation of plumage colour to local soil or rock type (see Geographical Variation). Dispersal, where this occurs, is very local. In Saudi Arabia, large numbers can gather at water during summer (Jennings 1980), and in eastern Arabia the species is a casual July–March visitor to the coastal plain and other areas where it does not breed (Bundy and Warr 1980; G Bundy). Local movement occurs in Azraq area of Jordan, birds leaving Shaumari reserve for winter (Clarke 1980). In northernmost populations (Soviet Central Asia), only altitudinal movements recorded, with upland birds descending to plains for winter (Dementiev and Gladkov 1954*a*). RH

Food. Seeds and insects. Seeds taken from ground and by hammering at goat droppings. Insects taken from ground and among stones (Simmons 1952*a*, 1970), also from small bushes, by reaching up from ground below or by clambering about in them (D J Brooks, G K Brown, J Palfery). Normal gait a quick walk, and will turn head sideways to stare down like thrush *Turdus*. Will fly sharply up from ground as high as *c*. 6 m to take flying insects. (Simmons 1952*a*, 1970.) Seen to swoop on 5-cm-long grasshopper (Acrididae) from some distance, and break it up by shaking vigorously before eating. Readily breaks open buds of *Convolvulus trabutianus* to eat parasites inside (Pasteur 1956), and recorded picking bits of meat from bone of gazelle (Gazellinae) and eating moths (Lepidoptera) discarded by Geyr von Schweppenburg (1918). Will drink water but, except as dew, presumably often lives without it (e.g. Ripley 1951, Simmons 1952*a*, Meinertzhagen 1954, Pasteur 1956, Sage 1960*a*, Ali and Ripley 1972).

In central Morocco, October–November, stomachs contained soft insects and soft green food (Meinertzhagen 1940). In Algeria, February–April, 8 ♂♂ and 8 ♀♀ contained numerous seeds (Roche 1958). In Tunisia, adult ♂, February, contained small seeds and sand; another, May, held grasshoppers (Orthoptera) and sand (Blanchet 1951). In central Sahara, diet includes small beetles (Coleoptera), seeds of *Astragalus* and *Artemisia*, and young shoots of grasses (Gramineae) (Meinertzhagen 1934). In western Niger, May, 5 birds contained mostly small seeds; also a few ants (Formicidae) and 1 beetle (Bates 1933–4). Also in Niger, August, one adult contained insects (Villiers 1950). In central Saudi Arabia, takes grain dropped on roads; stomachs contained largely seeds, also ants, Orthoptera, and Hemiptera (Bates 1936–7). In south Caspian area (USSR), winter, stomachs contained small seeds of *Suaeda* and *Salsola*, and in one case grains of wheat (Isakov and Vorobiev 1940). In southern Turkmeniya, summer, all of 11 stomachs contained invertebrates (mainly Orthoptera and Odonata) and 9 contained seeds (Polygonaceae, Chenopodiaceae, Guttiferae, Cistaceae, Lythraceae, Boraginaceae) (Kekilova 1969). Of 29 further stomachs from southern Turkmeniya, spring and summer, 97% contained seeds and 86% contained invertebrates: Lepidoptera larvae in 41%, Orthoptera 31%, beetles (largely Curculionidae) 24%; also bugs (Hemiptera), termites (Isoptera), flies (Diptera), ants, and Arachnida (Medvedev and Esilevskaya 1973). In Zagros mountains (Iran), 4 birds contained seeds and grubs (Paludan 1938). In India and Pakistan, takes seeds (e.g. of *Setaria verticillata* and *Panicum ramosum*) and insects (Ali and Ripley 1972).

At one nest in Morocco, adults brought mainly grasshoppers to feed young (Sage and Meadows 1965). In Algeria, one chick contained seeds (Roche 1958).

DJB

Social pattern and behaviour. Quite well studied outside breeding season (Simmons 1952a), less well during it. Account includes material on extralimital *phoenicuroides* (Ali and Ripley 1972).

1. Only mildly gregarious outside breeding season, never forming large dense flocks. Often encountered alone or in twos and threes, at most in loose parties of 5-8(-20) birds (e.g Meinertzhagen 1930, Paludan 1938, Simmons 1952a, Ali and Ripley 1972, Jennings 1980, 1981b); after young fledge, birds occur in family parties of 3-4 (Ali and Ripley 1972) or 4-8 (Koenig 1895). According to Meinertzhagen (1951a), *Ammomanes* larks defend territories throughout the year, perhaps explaining relatively high incidence of single birds and pairs outside breeding season. In Somalia, even 2 birds seldom seen together except in breeding season (Archer and Godman 1961a). For observations on winter flock, see part 2. Sometimes associates with other species, e.g. in Algeria, February, with House Buntings *Emberiza striolata* and Trumpeter Finches *Bucanetes githagineus* (Gaston 1970, which see for proportions in mixed assemblages—also for dispersion, in birds per km, of *A. deserti* in winter). BONDS. No evidence for other than monogamous mating system (K E L Simmons). For ♂-♂ 'friendship pair', see Niethammer (1954). Both members of pair share in nest-building and feeding young (Pasteur 1956). Incubation by ♀ (Orr 1970). BREEDING DISPERSION. Solitary and territorial, usually well-dispersed (Simmons 1970; Walker 1981b). Territory size of *Ammomanes* larks said to vary enormously; may be c. 0.4 ha, or up to 40 ha where food scarce (Meinertzhagen 1951a); no hard evidence presented, however, for these estimates, or that such areas defended. At Azraq (Jordan) territories often appeared to be large (Nelson 1973). During breeding season, Azraq, c. 3-6 birds per km (of walked transect), according to habitat; on ground, separated from Bar-tailed Desert Lark *A. cincturus* by as little as 20-30 m, and both species may use same air-space (Wallace 1983a). In northern Sahara (Morocco), March-May, 5 sample counts yielded average of c. 25 pairs per km²; transects, April-June, yielded c. 1-3 birds per km (Blondel 1962a). No information on site-fidelity, but does not appear to use same nest twice, within or between seasons (Pasteur 1956, which see for evidence of 2nd clutch of one pair). ROOSTING. Always on ground, and usually quite in the open (Meinertzhagen 1951a). In Jebel el Sauda (Libya), active mainly in morning and afternoon, resting under shade of boulders in heat of the day (Whitaker 1902). Shelters from rain under stones, in crevices, etc. (Koenig 1924). Typically late to roost (Pasteur 1956). Reported singing at night, Krasnovodsk, USSR (Portenko 1954), and calling at 04.00 hrs (when still dark) in Saudi Arabia (J Palfery). Drinking probably most frequent in morning, but also evening (e.g. Meinertzhagen 1934). In Algeria, birds drank most frequently 07.00 hrs (1 hr after sunrise) and 11.00 hrs (Gaston 1970). In India, drank at desert rain-puddle regularly at c. 09.00 hrs; not in flocks but in relays of 2-3 birds at a time (Ali and Ripley 1972). In Iraq, flocks of up to 15 came regularly to drink at pools, often near, but never closely associated with, drinking parties of Crested Larks *Galerida cristata* (Sage 1960a).

2. Often tame (Whitaker 1905; Meinertzhagen 1934; Simmons 1952a): on Ahaggar plateau (Algeria) allows approach to within a few feet, then flushing only a short distance before landing again (Meinertzhagen 1940); very shy in north-east Sudan, however (Rothschild and Wollaston 1902). When disturbed, typically runs away in crouched posture (Koenig 1924) and ascends a small stone or other prominence (Simmons 1970), and, when threatened, said to squat (Zedlitz 1909; Archer and Godman 1961a); according to Meinertzhagen (1954), however, never squats, preferring to run or fly. Rapid wing-flicking of bird running on ground (Neophytou 1974) perhaps an alarm reaction. In Egypt, the following possible escape reactions noted: bird constantly flew in, out, and around a pit in ground in deliberate moth-like flight with somewhat abrupt changes in direction; on another occasion, bird flew around erratically in small space, turning, rising slightly, veering off again at sharp angle, and settling quickly nearby (Simmons 1952a). FLOCK BEHAVIOUR. For flocking to drink, see Roosting (above). One flock in Egypt studied October-March by Simmons (1952a). Group contained c. 12 members, typically loosely dispersed, and seldom in close flock formation; flock ranged over 1 square mile (259 ha), but usually occurred in a smaller area (c. 50 ha). In flock, birds maintained individual-distance, seldom closer than 30-60 cm, and more usually c. 2 m. Gave Contact-calls (see 2a in Voice) while feeding on ground. Flight Contact-call (see 2b in Voice) of one bird stimulated other birds to give same call, to gather together, and to fly up. Upflight also sometimes triggered by 2 birds initiating a pursuit-flight (see Antagonistic Behaviour, below). Flock probably dispersed late February to March. SONG-DISPLAY. ♂ has brief, undulating Song-flight. Consists either of horizontal flight between 2 eminences, or of steep ascent from and return to ground, with or without horizontal flight in between; descent may be vertical; horizontal phase has undulations c. 2 m deep due to alternation of bouts (c. 0·8 s) of deep floppy wing-beats with equal periods of closed wings. Maximum height usually 5-10(-40) m (Guichard 1955; Pasteur 1956; P A D Hollom, J Palfery). Also described as involving flight in wide circles (E N Panov), and in another description bird took off from hill and flew slightly downhill along shallow valley (Gallagher and Rogers 1970; M D Gallagher). May sing throughout Song-flight (P A D Hollom) or only after reaching maximum height (Guichard 1955). For synchronization of song with Song-flight, see 1 in Voice. Also sings during a feeble, quick flight (Simmons 1970), perhaps seasonal precursor of proper Song-flight. Not uncommonly sings when perched, especially on vantage points (Wallace 1983a; P A D Hollom: see 1 in Voice), and rival ♂♂ will sing at each other from (e.g.) tops of tents in February (Simmons 1952a). Said to sing from ground more than *A. cincturus* (Meinertzhagen 1954). In Oman and United Arab Emirates, Song-flight performed February-April (Stanford 1969); in

Bahrain, March–May (Rogers and Gallagher 1973). In Krasnovodsk, sings especially April–May, waning from mid-July (Portenko 1954). In eastern Saudi Arabia, Song-flights made throughout day, most frequently around dawn (J Palfery). ANTAGONISTIC BEHAVIOUR. No information for breeding season, though disputes not uncommon at other times; the following account from study in Egypt by Simmons (1952). If individual-distance infringed (e.g. when intruder forced birds to bunch), departure of one bird typically elicited a pursuit-flight from its nearest neighbour: low zigzag aerial chase over varying distance, terminated by one bird (often the leading one) landing. Pursuit-flights stimulated Excitement-calls (see 3 in Voice) in nearby birds. Chasing also occurred on ground, 2 birds running with lowered heads. Sometimes, when 2 birds met and neither gave ground, crouching and then physical contact resulted— e.g. vertical breast-to-breast upflight, or grappling fight; at other times, merely moved out of each other's way. As winter progressed, close approach less likely to lead to pursuit. In Zagros mountains (Iran), pursuit-flights also frequent among parties of 3–5 birds during March, but not evident by beginning of April when birds in pairs (Paludan 1938). The following incident in Bahrain, February, perhaps antagonistic: 2 birds on ground faced each other, one (♂) stationary and forebody lowered, the other moving forward whilst pivoting body from side to side (M D Gallagher). HETEROSEXUAL BEHAVIOUR. (1) General. In Oman, courtship begins early January (Walker 1981*b*), but no details. (2) Pair-bonding behaviour. 2 birds seen in probable 'courtship chase', one apparently singing (G K Brown). Once, when 2 birds engaged in antagonistic pursuit-flight, another nearby crouched with tail fanned along ground and wings drooped (Simmons 1952*a*); after singing, ♂ seen to circle ♀ with wings held slightly away from body (Sudhaus 1970); this reminiscent of ground-display in other Alaudidae. ♂ approaching ♀ on ground may mock-feed (E N Panov), suggesting tentative advance. (3) Courtship-feeding. No information. (4) Mating. No information. (5) Behaviour at nest. Both members of pair help to build nest (Pasteur 1956). RELATIONS WITHIN FAMILY GROUP. Virtually nothing known. Both members of pair feed young (Pasteur 1956; Sage and Meadows 1965). Young probably leave nest before able to fly, as in other Alaudidae: one ⅔-grown bird found sheltering under bush where being fed by parents (G K Brown). After young fledge, parents continue to feed them (G K Brown) and family parties of 3–4 birds occur (Ali and Ripley 1972). ANTI-PREDATOR RESPONSES OF YOUNG. A ⅔-grown bird out of nest hopped clumsily away when handled (G K Brown). PARENTAL ANTI-PREDATOR STRATEGIES. (1) Passive measures. Once eggs hatched, ♀ may sit very tightly, e.g. ♀ remained covering brood of 2, her wings extended, despite being examined at close quarters (Wallis and Pearson 1912*b*). One ♀, flushed off nest, was very wary, and would not return until intruder withdrew (Bolster

1922). For Anxiety-call of parent when human intruder near chick, see 4 in Voice. (2) Active measures. When human intruder near nest, India, ♀ *phoenicuroides* allowed close approach, but ran freely and wandered about in erratic fashion, giving no clue to location of nest (Whistler 1922).   EKD

Voice. Variation between races reported (see call 1), but no study to substantiate suggested differences or to establish complete repertoire.

CALLS OF ADULTS. (1) Song of ♂. Melodious, rather sad, and (unlike Bar-tailed Desert Lark *A. cincturus* and Dunn's Lark *Eremalauda dunni*) far-carrying (D I M Wallace). A repeated phrase, typically of 2–3 syllables: e.g. 'chur-rer-ee' (P A D Hollom), 'trreeooee' (J Hall-Craggs: Fig I), 'too-too-wee' (Vere Benson 1970); also 'ther chew choorup' (J Palfery). Given during Song-flight. Often also on ground, when typically incomplete (see below) and often leads directly into Song-flight (P A D Hollom): e.g. bird on ground began singing 'trrrreeee trrrreeee', with *c.* 1 s between each unit (J Hall-Craggs: Fig II)—likened to squeaky pulley wheel (P J Sellar); then, with considerable intensification of song, made strongly undulating ascent at fairly steep angle to *c.* 5–6 m; 'trrrreeee' units changed to 'chip chip chip...', presumably during descent to ground (P A D Hollom). In Fig II, no evident variation in 'trrrreeee' units, but in other descriptions song typically more variable, with distinct 2nd part of phrase starting with closure of wings in each undulation of Song-flight (E N Panov): e.g. in Oman, each undulation accompanied by 'chew' on ascent and 'chup-chup choo-oo-ee' on descent (Gallagher and Rogers 1978; M D Gallagher). In description of presumed Song-flights, Morocco, each undulation accompanied by a rolling 'trrri trrru', duration *c.* 1·6 s; most of phrase given on rising part of undulation, but may also accompany the drop, or overlap both (Pasteur 1956). In Tunisia, similarly rendered 'twerr-reep' (M G Wilson). More complicated renderings suggest that variation, both within and between songs, may be substantial. Thus, in Saudi Arabia, song a loud series of varied, short, sweet whistled phrases of 1–4 notes each, e.g. 'cha-wee cheew chew-eee chee-a-whew' (King 1978). For other variations, see Simmons (1952*a*), Moore and Boswell (1956), and Walker (1981*b*). In Bahrain, individual calls of *insularis* varied, and distinct from those of *taimuri* in Oman (Gallagher

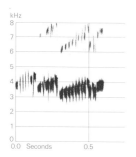

I C Chappuis Morocco April 1966

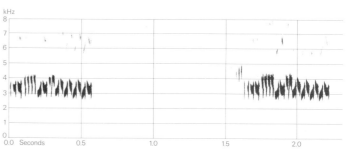

II P A D Hollom Morocco March 1978

and Rogers 1978). Song of *iranica* in Zagros (Iran) apparently different from that at Fezzan, Morocco (Erard and Etchécopar 1970, which see for renderings of both). Song on ground evidently contains elements of, or is closely related to, full song: e.g. a throbbing, trilled repeated 'treee-trooo' or 'reetr-rooo' (P J Sellar: Fig III)

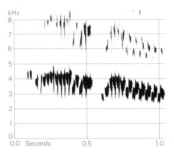

III P A D Hollom Morocco March 1978

given by bird as it perched on stones between running around and giving call 2 (P A D Hollom); 5 disyllabic units, each *c.* 1 s long, given in *c.* 12 s (E K Dunn). (2) Contact-calls. (a) On ground, a soft quiet undertone, 'chu'; sometimes a slurred 'chee-lu', when feeding (Simmons 1952*a*); also rendered 'heeoo' or 'hee-ew' (J Palfery). Commonly a soft piping repeated 'teup' or 'cheup' (Fig IV) which does not carry far (P A D Hollom); also

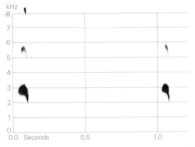

IV P A D Hollom Morocco March 1978

rendered a low measured 'tyup tyup tyup' (E N Panov), 'wheeb' (P J Sellar), or soft 'wheep' (J Palfery). For similar renderings, see Gallagher and Rogers (1978). (b) In flight, rapidly repeated 'chu' or 'wu'; also 'chee-wit' and numerous variants, including 'chWEE-u chWEE-u chWEE-u', 'chu-lit chee-wu chee-wu', 'chu-lu-weet chu-lit' (Simmons 1952*a*). Far-carrying 'hew hew kitew kitew kitew' probably a similar sequence (J Palfery). This probably the querulous piping note mentioned by Tice-hurst *et al.* (1921–2). For other renderings, evidently of same or similar calls, see Gallagher and Rogers (1978). (3) Excitement-calls. Rapid twittering sounds, closely related to, and hard to distinguish from, call 2 and variations; include 'chee-u', 'chee-u chee-u chee-wit', and 'chul-wit-chutle', repeated 4–5 times, and given when intensity of interactive behaviour in flock increases. Similar fast liquid twittering sounds given during pursuit-flights (Simmons 1952*a*). Other descriptions, pro-bably of such calls, include 'sweehirip-a sweshirrip',

producing a babble of sound when several birds present; also 'tischiripp-tsichiripp-tschirip' from birds in pairs (M D Gallagher). Low 'chirra-ra', which sometimes follows song (Vere Benson 1970), probably the same or a related call. (4) Anxiety-call. Soft 'piou' and 'teeoup' when young out of nest disturbed by human intruder (G K Brown).

CALLS OF YOUNG. Recently fledged bird called 'cheerp' (G K Brown).   EKD

Breeding. SEASON. North Africa: start of season varies from January–February in south and near west coast to March–April in north (Heim de Balsac and Mayaud 1962; Hüe and Etchécopar 1970). Eastern Saudi Arabia: recently fledged young seen in April, and young seen in nest in mid-June (D J Brooks, G K Brown). SITE. On ground, usually in shelter of tussock or stone. Nest: shallow scrape lined with available vegetation, with rim or ramp of stones, all round or on exposed side. Building: by both sexes (Pasteur 1956). EGGS. See Plate 79. Sub-elliptical, smooth and glossy; greenish-white to pink, finely spotted dark or reddish-brown, and some purplish-grey, spots sometimes concentrated at broad end. 22·0 × 16·0 (20·6–23·5 × 15·0–17·5), *n* = 25; calculated weight 2·91 g (Schönwetter 1979, eggs from Algeria and Tunisia). Clutch: 1–5, but 1–4 in desert areas of North Africa, and 3–5 further north. Of 81 clutches, Algeria and Tunisia: 3 eggs, 27; 4, 48; 5, 6; mean 3·86. Of 12 clutches from desert areas: 1 egg, 1; 2, 5; 3, 5; 4, 1; mean 2·5 (Heim de Balsac and Mayaud 1962). Sometimes 2 broods (Portenko 1954; Pasteur 1956). INCUBATION. Period un-known. At one nest, Israel, by ♀ only, with typical daily pattern as follows: ♀ incubated overnight but left nest immediately after sunrise (*c.* 05.00 hrs) and was absent until 07.00 hrs, then periodically absent *c.* 07.00–16.00 hrs, leaving eggs uncovered (Orr 1970). No further information.

Plumages (*A. d. algeriensis*). ADULT. In fresh plumage (about July–February), upperparts deep pink-cinnamon with slight pale vinous-grey cast on crown, mantle, and scapulars; rump brighter, deep vinous-pink; upper tail-coverts and tertials purer rufous-cinnamon. Sides of head cinnamon-buff or greyish-buff with rather indistinct cream-pink supercilium (hardly extending behind eye), pale buff ring round eye, and dull grey line along lores (not reaching eye) bordered by pale buff near base of upper mandible; some faint dull grey spots from lower mandible backwards over cheeks, some faint grey streaks on ear-coverts. Chin and throat cream or isabelline-white, pale colour extending upwards slightly to behind ear-coverts; sides of neck cinnamon-buff, merging into pink-cinnamon on chest, which has rather indistinct grey spots. Remainder of underparts pink-cinnamon (less deep than upperparts, not vinous), paler pink-buff or cream on central belly and vent. Central pair of tail-feathers (t1) dull black with broad rufous-cinnamon fringe all round and with rufous-cinnamon basal half of inner web; t2–t4 dark grey with narrower rufous-cinnamon fringes at tip and along distal part of outer web, base of outer web fully rufous-cinnamon; t5–t6 dark grey with narrow rufous-cinnamon fringe at tip and wholly rufous-cinnamon outer web. In spread tail, dull black or grey forms dark triangle (pointing towards base of t1) on rufous-cinnamon ground: 3–4 cm of t1 black, *c.* 1–1·5 cm of inner web of t6. Undersurface of tail largely glossy dark grey, except for rufous outer web of t6. Flight-feathers dark grey, outer webs narrowly fringed rufous-cinnamon near tip, broadly towards base, inner webs with rufous wedge on base; secondaries faintly edged rufous. P9 narrowly edged rufous on outer web only, p10 with narrow rufous fringe on outer web near tip and sometimes faint rufous edge along outer web. Lesser upper wing-coverts vinous pink-cinnamon, like upperparts; median and greater upper wing-coverts, bastard wing, and upper primary coverts purer rufous-cinnamon, like tertials, but tinged grey or dull grey-brown on feather-centres. Under wing-coverts and axillaries pink-cinnamon, like under-surface of flight-feathers (but tips of secondaries and inner primaries grey, primaries becoming more extensively grey towards outermost). *In slightly worn plumage* (about September–April), upperparts still rather deep pink-cinnamon, but slightly duller, sandy vinous-grey; centres of tertials, median and greater upper wing-coverts, greater upper primary coverts, and longest feather of bastard wing darkened to dark olive-brown or dark grey-brown, terminal fringes bleached to isabelline and partly abraded; dark portions of tail and flight-feathers changed to dark olive-brown, terminal fringes grey-buff or buff and largely abraded; short supercilium and pale eye-ring paler isabelline, more contrasting; grey line on lores and grey spots on cheeks and chest often more distinct; underparts less deep pink-cinnamon, more pink-buff or cream-buff. *In heavily worn plumage* (about March–August), upperparts dull cinnamon-brown, pink remaining only on rump; tail, tertials, and flight-feathers grey-brown, with rufous restricted to base of outer webs (except for whole outer web of t6); supercilium and dark line on lores indistinct; underparts isabelline-buff, often with much mottled grey of feather-bases visible, grey spots on cheeks and chest largely worn off. NESTLING. Down long and dense; white, greyish-white, or buff-white. JUVENILE. Upperparts buff-brown (rather like worn adult), feathers partly fringed pale buff on tips; crown, mantle, and scapulars tinged vinous-grey subterminally; rump rufous-buff, not vinous-pink as adult. Pale buff supercilium even shorter than adult, just reaching above front of eye; isabelline eye-ring broken by buff at rear. Chest and sides of breast pink-buff, remainder of underparts isabelline-buff to cream-white; grey spots on cheeks and chest very faint or absent. Tail rufous-cinnamon, like adult, but t1 without grey, t1–t5 with grey restricted to inner web, and t6 cinnamon-buff with grey streak on inner web; spread tail without blackish triangle, undersurface rufous, not grey. Flight-feathers as adult, but tips evenly fringed rufous-buff, also on inner webs and on secondaries; p10 with broad but ill-defined cinnamon or buff fringe along outer web and tip. Upper wing-coverts (including primary coverts) grey or vinous-grey with broad and regular cinnamon to pale buff fringes. Soon influenced by bleaching and abrasion; upperparts and upper wing-coverts markedly paler sandy-pink-buff, rufous of rump, tail, and flight-feather bases bleached to pink-buff; pale fringes of tips of tail and flight-feathers worn off (except of p10); underparts isabelline-white with pale buff chest-band. FIRST ADULT. Like adult; indistinguishable once juvenile p10 replaced. Unlike many other Alaudidae, adult p10 broad and rather long (as in juvenile), but juvenile p10 has broadly rounded tip (bluntly pointed in adult) and broad but ill-defined pale fringe along outer web and tip (in adult, faint sharp pale edge along outer web, slightly wider near tip but not extending beyond shaft).

Bare parts. ADULT. Iris hazel, dull light brown, or brown. Upper mandible dark horn-brown, brown, yellow-brown, yellow-green, or olive-green with darker brown or grey culmen and tip; lower mandible pale horn-yellow, whitish-horn, or grey-flesh. Leg and foot pale yellow-brown, pale flesh-brown, yellow-white, or greyish flesh-colour. NESTLING. Mouth orange-yellow; dark spot on tip of upper and lower mandible, 1 spot on each side of base of tongue. JUVENILE. Iris grey-brown or brown. Bill brown, middle and basal portion of lower mandible dirty yellow or flesh-brown. Leg and foot dirty brown, grey, or flesh-colour. (Hartert 1910; Ripley 1951; BMNH, ZMA.)

Moults. ADULT POST-BREEDING. Complete; primaries descendant. Starts with scattered feathers of head, mantle, scapulars, and median upper wing-coverts, followed by p1. In Iran, Afghanistan, and USSR starts late June to early August; p1 shed approximately in second half of July, body and wing-coverts virtually new at primary moult score *c.* 30, and moult completed with p9-p10 late August to early October (Vaurie 1951*a*; Dementiev and Gladkov 1954*a*; Paludan 1959; BMNH). In Ahaggar (southern Algeria), starts late May and early June (ZMA). No data for other populations, but plumage heavily worn in *payni*, *algeriensis*, and *isabellinus* in June; these races (as well as *annae* and nominate *deserti*) have plumage completely new in October, and hence moult during same period as in populations cited above. POST-JUVENILE. Complete; primaries descendant. Starts shortly after fledging; timing thus highly variable and apparently independent of race. Starts from late May with some feathers of forehead, crown, mantle, scapulars, sides of chest, and p1 (e.g. some *geyri*, *isabellinus*, *cheesmani*), but others still fully juvenile up to early September (Vaurie 1951*a*); immatures nearing completion of moult examined from late July to early October (BMNH, ZMA).

Measurements. ADULT. *A. d. algeriensis.* Northern Sahara of Algeria and southern Tunisia, all year; skins (RMNH, ZMA). Bill (S) to skull, bill (N) to distal corner of nostril; exposed culmen on average *c.* 3·5 less than bill (S).

WING	♂ 100·8 (1·81; 26)	98–104	♀ 94·1 (1·60; 13)	92–97
TAIL	63·5 (1·85; 26)	60–67	58·2 (2·50; 13)	54–62
BILL (S)	17·5 (0·57; 26)	16·4–18·3	16·2 (0·53; 15)	15·6–17·1
BILL (N)	10·7 (0·62; 25)	9·9–12·1	9·6 (0·31; 15)	9·2–10·3
TARSUS	22·1 (0·69; 25)	20·9–23·3	21·6 (0·90; 15)	20·3–22·9

A. d. payni. Southern Morocco, all year; skins (BMNH).

WING	♂ 103·7 (1·91; 13)	101–108	♀ 98·3 (2·67; 7)	95–102

A. d. mya. Central Algerian Sahara, 28–30°N: wing, ♂ 107–111, ♀ 97–101; bill (S) ♂ 19·1–20·9, ♀ *c.* 18–19·5 (Hartert 1921-2; BMNH, ZMA).

A. d. whitakeri. Central Tripolitania (Libya), Ahaggar (southern Algeria), and western Algeria; tail, bill, and tarsus Ahaggar only (Hartert 1921-2; Hartert and Steinbacher 1932-8; BMNH, ZMA).

WING	♂ 106·5 (2·47; 19)	103–112	♀ 98·5 (2·37; 17)	95–103
TAIL	68·6 (1·67; 9)	66–71	64·2 (1·81; 8)	62–66
BILL (S)	18·4 (0·96; 10)	17·3–19·9	17·1 (0·31; 7)	16·7–17·6
BILL (N)	12·2 (0·43; 7)	11·6–12·8	10·7 (0·74; 7)	9·7–11·9
TARSUS	23·9 (0·59; 10)	23·1–24·6	23·2 (1·04; 9)	21·9–24·4

Bill (S) in central Tripolitania 18–20·5 (Hartert 1921-2). Single unsexed bird, Tibesti (Chad): wing 103, exposed culmen 13·5 (Berlioz 1950).

A. d. geyri. Aïr (Niger), wing ♂ 99–105 (4), ♀♀ 89, 96 (Niethammer 1955*b*).

Nominate *deserti*. Southern Nile valley in Egypt, southern Sinai, and Aqaba (Jordan): wing, ♂ 98-103, ♀ 92-98; bill (S), ♂ 16·5-17·5, ♀ 14·5-16 (Hartert 1921-2; BMNH, RMNH, ZMA).

A. d. isabellinus. North-west Egypt, southern Jordan, Dead Sea region, northern Arabia, and southern Iraq (Vaurie 1951*a*; BMNH, RMNH, ZMA).

WING	♂ 98·5 (2·15; 19)	94-103	♀ 94·6 (2·19; 19)	92-99
BILL (S)	16·8 (1·03; 12)	15·5-18·5	14·8 (1·31; 12)	13·0-17·5

A. d. annae. Lava deserts of Azraq (Jordan): wing, ♂ 106·0 (1·84; 5) 104-108, ♀ 100·8 (1·89; 4) 98-102; bill (S), ♂ 14·5-16·5, ♀ 14·5-16 (Hartert and Steinbacher 1932-8; BMNH).

A. d. coxi. Syria and north-west Iraq: wing, ♂ 104·9 (2·62; 7) 102-108, ♀ 97·5 (3) 95-100; bill (S), ♂ 16-18, ♀ 16·5 (Hartert and Steinbacher 1932-8; BMNH).

A. d. cheesmani. Eastern Iraq and Iran south-west of Zagros mountains (Paludan 1938; Vaurie 1951*a*; BMNH).

WING	♂ 104·8 (2·45; 13)	100-108	♀ 97·1 (1·75; 12)	95-100
BILL (S)	17·0 (0·58; 10)	16·0-18·0	16·0 (0·74; 9)	14·5-17·0

A. d. darica. Yazd and Fars, eastern Zagros mountains, Iran (Vaurie 1951*a*).

WING	♂ 107·7 (3·61; 15)	101-113	♀ 100·1 (4·40; 9)	93-106
BILL (S)	17·5 (0·66; 16)	16·0-18·0	16·3 (0·78; 11)	14·5-17·5

A. d. iranica. Eastern and south-east Iran (west to Tehran) and southern Afghanistan (Vaurie 1951*a*).

WING	♂ 106·4 (4·22; 39)	98-112	♀ 100·3 (2·95; 29)	96-106
BILL	(S) 17·6 (0·68; 43)	15·5-18·5	16·6 (0·72; 37)	15-18·2

Sex differences significant for all measurements in all populations of which standard deviation known.

JUVENILE. Wing on average 2·5 mm shorter than adult, tail 2 mm; tarsus full-grown from shortly after fledging, bill full-grown once post-juvenile moult completed.

Weights. *A. d. payni*. Figuig (Morocco), April: probable ♂♂ 25, 26, 26, 28; probable ♀ 25·5 (BTO).
A. d. algeriensis. Southern slopes of Atlas Saharien (Algeria): ♂ 27-28 (5), ♀ 24-25 (3) (Niethammer 1955*b*). Birds intermediate between *algeriensis* and *mya*, central Algeria: ♂ 27-30 (4), ♀ 23-26 (3) (Niethammer 1955*b*).
A. d. whitakeri. Southern Algeria: ♂ 25-26·5 (7), ♀ 20-25 (6); another sample, ♂ 24-29·5 (25), ♀ 22-26 (16) (Niethammer 1955*b*).
A. d. geyri. Aïr (Niger) ♂ 22-25 (4), ♀♀ 21, 21 (Niethammer 1955*b*).
A. d. cheesmani. South-west Iran, mid-March to early April: ♂ 22·9 (1·36 ; 4) 21-25 (Paludan 1938).
A. d. iranica. Western and south-west Afghanistan: March, ♀ 24; early July, adult ♂♂ 22, 27, adult ♀ 25, juvenile ♂♂ 23, 24, 25 (Paludan 1959).

Structure. Wing rather long, broad at base, tip rounded. 10 primaries. In adult, p7 longest; in *algeriensis*, *isabellinus*, and other short-winged races, p6 and p8 0-1(-2) shorter, p9 6-10 shorter, p10 47-54 (♂) or 43-49 (♀), p5 1·5-4, p4 6-11, p1 20-25 (♂) or 15-21 (♀); in *whitakeri*, *iranica*, and other long-winged races, p10 50-55 (♂) or 48-52 (♀), p1 22-26 (♂) or 20-24 (♀). In juvenile (all races), p7 and p8 longest, p9 3-8 shorter, p10 36-47 (♂) or 35-44 (♀), p6 1-3, p5 2·5-5, p4 9-12, p1 18-23 (♂) or 17-21 (♀); hence, juvenile p9 and especially p10 relatively longer than adult, p4-p6 relatively shorter. Adult p10 not as strongly reduced as in many other Alaudidae, rather broad and with blunt tip, 8·3 (40) 4-11 longer than tip of longest greater upper primary covert in short-winged races (exceptionally up to 14), 12 (5) 8-15 longer in long-winged races; in juveniles, p10 13·2 (5) 11-15 longer in short-winged races, 15·5 (15) 12-18 in long-winged races. Outer web of p4-p8 and inner of p5-p9 emarginated. Tail rather short, tip square; 12 feathers. Tertials short, longest equal in length to secondaries, tip reaching p1-p3 in closed wing. Bill rather long; basal half and cutting edges straight, terminal half of culmen decurved, gonys curved slightly upwards; bill heavier and rather deep at base in races *mya*, *darica*, and *iranica*, short but heavy in *parvirostris*. Short feathers with bristle-like tips covering nostrils; short bristles in front of and below eye and on chin. Tarsus rather short and thick, toes short and heavy. Middle toe with claw 15·7 (22) 14-18 mm; outer and inner toe with claw both *c*. 71% of middle with claw, hind *c*. 91%. Front claws short, rather heavy, and decurved; hind claw rather short, slightly decurved; hind claw 6·9 (6) 5-8.

Geographical variation. Marked and complex. 17 races have been described for west Palearctic alone, with 10 more just outside its limits which may straggle into our area or influence populations living just within our borders. Not all these now considered worthy of racial status, however. Colour often directly related to colour of local soil and intensity of sunlight. Birds from sandy habitats mostly buff-coloured, those of stony ground various shades of grey, rufous, or brown. In some areas, pale and dark birds live side-by-side (notably in Ahaggar in southern Algeria, in Nile valley, and along shores of Gulf of Aqaba), though no real dimorphism, as intermediates usually occur; in adjacent areas, pale or dark birds may occur alone, similar to one of nearby variants. Bleaching and abrasion have marked effect and produce further complications. Basic division as advocated here based on size, followed by length and shape of bill. Long-winged populations (*mya*, *whitakeri*) inhabit central Sahara from Mauritania to central Libya, characterized also by long and rather heavy bill. Similarly long-winged birds occur from Zagros mountains in Iran and from Turkmeniya eastwards (*darica*, *iranica*, *parvirostris*, *orientalis*), but bill of these shorter than in those of central Sahara, though deep at base. Geographical division between long- and short-winged birds not sharp, however, ranges of long-winged populations being surrounded by birds of intermediate size, into which they gradually merge (though some are separable by colour): encircling central Sahara, *payni* (south-east Morocco), 'intermedia' (Algerian Sahara near El Golea), *geyri* and *kollmannspergeri* (southern Sahara in Aïr and Ennedi mountains, respectively); around Iranian plateau, *coxi* (Syria and northern Iraq), *cheesmani* (eastern Iraq and neighbouring south-west Iran), and *phoenicuroides* (south-east Afghanistan, Pakistan east of *c*. 67°E, and north-west India). Remaining populations of Mauritania, northern Algeria, Tunisia, Egypt, Sudan, Middle East, and Arabia are small in size. As marked sexual size-dimorphism occurs, ♂♂ of small-sized races can be as large as ♀♀ of large-sized ones.

A. d. algeriensis from northern Algeria (between *c*. 31°N and Atlas Saharien), southern Tunisia, and western Tripolitania (Libya) rather pale and sandy, upperparts pink-cinnamon with vinous-grey cast. *A. d. payni* of southern Morocco east to Figuig darker greyish-cinnamon above, also with vinous tinge when fresh, underparts darker pink-buff. *A. d. mya* from central Algerian Sahara between *c*. 27°N and 30°N closely similar to *algeriensis* (apart from large size and heavy bill) but upperparts slightly more sandy, slightly less vinous-grey; birds similar in colour to *mya* but intermediate in size between *mya* and *algeriensis* occur in El Golea area (named *intermedia* Heim de Balsac, 1925). Typical *whitakeri* from central Tripolitania (Libya) larger than preceding races and with long bill, upperparts distinctly darker grey-brown, only slightly tinged vinous when fresh; rump and upper tail-coverts contrastingly rufous-cinnamon; underparts buff-brown with prominent dark grey-brown streaks; tail dark brown except for rufous outer edges of t5–t6; equally large and dark birds inhabit Ahaggar mountains in southern Algeria, sometimes separated as '*bensoni*' Meinertzhagen, 1933, but difference slight and separation from *whitakeri* impracticable. Situation in Ahaggar complicated, as pale birds also occur there, as well as in neighbouring plains of southern Algeria (south of *c*. 25°N); pale birds sometimes stated to occur on lower levels in Ahaggar or restricted to ♀ only (e.g. by Hartert and Steinbacher 1932–8), and these sometimes separated as '*janeti*' Meinertzhagen, 1933. According to Niethammer (1955*b*) and specimens examined, geographical overlap in Ahaggar is complete, however, and *whitakeri* hence considered to be polymorphic. Pale Ahaggar birds darker than *mya* and with relatively shorter and thicker bill, upperparts sandy-brown, less vinous-sandy, underparts deeper cinnamon-buff and with more heavily streaked throat; juveniles distinctly paler buff-brown to sandy-buff above. Similarly, large dark grey birds occur in western Algeria, intergrading with *payni*; dark grey birds also occur in Mauritania (named *monodi* Dekeyser and Villiers, 1950) and in Aïr area in northern Niger, and these sometimes combined into single race with Ahaggar birds (e.g. by Vaurie 1959), but considered separate here (under name *geyri*), as size apparently smaller (Niethammer 1955*b*; Vaurie 1955*c*). Status of Tibesti birds uncertain, as only a single unsexed specimen known: separated as *mirei* Berlioz, 1950; colour dark grey-brown; may belong either to equally dark *whitakeri* or to nominate *deserti*. Nominate *deserti* dark grey-brown or brownish-grey above, like *whitakeri*, but much smaller; occurs locally in Nile valley between Wadi Halfa (northern Sudan) and Qena (central Egypt); also recorded from Red Sea coasts of Quseir (Egypt), southern Sinai (where named *katharinae* Zedlitz, 1912), Aqaba (Jordan), and Tabuk and Ha'il, northern Saudi Arabia (Meinertzhagen 1930; BMNH, RMNH); birds from Red Sea coast of Sudan, Eritrea, and Saudi Arabia (named *samharensis* Shelley, 1902, and *hijazensis* Bates, 1935) appear inseparable; situation again complicated, as much paler grey-brown or sandy birds also occur within geographical range of nominate *deserti*, though apparently not really overlapping, but interdigitating. These pale birds (*isabellinus*) are rather like *mya*, but small like nominate *deserti* and without vinous cast of *mya* and *algeriensis*; upperparts pale

buff-brown with slight pale grey or pale olive-grey tinge, underparts buff or cream-buff, less pinkish than *mya* or *algeriensis*, pink-cinnamon rump contrasting less than in nominate *deserti* or *whitakeri*; tail rufous with black distal triangle like *mya* and *algeriensis*, not almost wholly dark as in nominate *deserti*. *A. d. isabellinus* occurs in northern Egypt, northern Sinai, and Dakhla oasis, south in Nile valley to Sohag and locally further south to Wadi Halfa; birds from Nile valley between Sohag and Bir Abad (*c*. 25°N to *c*. 26°30′N) are grey like *algeriensis* and sometimes separated as *borosi* Horvath, 1958, but included in *isabellinus* here. Further east, *isabellinus* extends from Negev and Dead Sea region (where slightly greyer and more heavily streaked—named *fraterculus* Tristram, 1864) through Jordan, northern Saudi Arabia, and southern Iraq east to the Tigris; in north-east Saudi Arabia, gradually paler towards pale cream race *azizi* in Al Hufuf area. *A. d. annae* from black lava deserts north of Azraq (Jordan) is darkest race; upperparts dark slate-grey or dark sooty-grey except for dull rufous longer upper tail-coverts and some grey or buff fringes along upper wing-coverts; chin to breast and flanks dark grey, streaked white on chin and buff on throat and upper chest; central lower belly, vent, and under tail-coverts dark cinnamon-pink, streaked dusky on tail-coverts; tail dark grey-brown with rufous fringe along outer web of t6, flight-feathers almost fully black-brown; slightly larger than *isabellinus*. *A. d. coxi* inhabits deserts from Syria and northern Iraq between Jebel ed Druz, Homs, and Raqqa east to Samarra and Al Fallujah; colour as in *isabellinus*, but size larger. East from Tigris, *coxi* replaced by *cheesmani*, which is about similar in size to *coxi*, but with darker upperparts and darker pink-brown underparts; extends east to foothills of Zagros mountains in south-west Iran and south to Bandar e Bushehr. Further east, *darica* and *iranica* occur; both large and with heavy but rather short bill, dark grey above (*darica* slightly paler) with contrasting dull rufous rump, greyish-buff-brown below with extensive grey spots and streaks on chin, throat, and chest. Extralimital *parvirostris* from western Turkmeniya similar to *darica*, but bill small; *orientalis* from central Asia pale sandy like *isabellinus*, but size near that of *iranica*; *phoenicuroides* from north-west corner of Indian subcontinent is rather large and dark brown. Extralimital southern races include *kollmannspergeri* from Ennedi (northern Chad) and *erythrochrous* from Dongola area (Sudan), both rufous-cinnamon above in varying degree, a pale grey race on island of Bahrain, and several dark races in Arabia and north-east Africa.

Forms superspecies with Gray's Lark *A. grayi* from Namib desert (Hall and Moreau 1970). CSR

Alaemon alaudipes Hoopoe Lark

Du. Witbandleeuwerik   Fr. Sirli du désert   Ge. Wüstenläuferlerche
Ru. Пестрокрылый пустынный жаворонок   Sp. Sirli desértico   Sw. Ökenlöplärka

Upupa alaudipes Desfontaines, 1789

Polytypic. Nominate *alaudipes* (Desfontaines, 1789), North Africa from Mauritania to Sinai and Sudan, merging into *doriae* in Middle East and Arabia; *boavistae* Hartert, 1917, Cape Verde Islands; *doriae* (Salvadori, 1868), Iran east to north-west India. Extralimital: *desertorum* (Stanley, 1814), coasts of Red Sea from Port Sudan and Jiddah south to north-west Somalia and Aden.

Field characters. 18–20 cm, of which bill up to 4 cm; wing-span 33–41 cm. Body size close to Skylark *Alauda arvensis* but bill, wings, and legs much longer. ♂ up to 20% larger than ♀. Quite large, attenuated lark, with character on ground first recalling large pipit *Anthus* but with long, decurved bill often held above horizontal. Plumage essentially sandy-grey above and stone-white below, with (on ground) little relief except for face-marks, dark spots on breast, and transverse dull black and white bands on folded wing. Appearance in flight dramatically different, with black and white bands on both wing surfaces (recalling Hoopoe *Upupa epops*) and black, white-edged tail. Sexes similar; no seasonal variation. Juvenile separable. 3 races in west Palearctic; North African race, nominate *alaudipes*, described here (see also Geographical Variation).

ADULT MALE. Upperparts pale sandy, with grey tone in some lights (and some birds greyer overall—see Plumages); streaked on crown but virtually uniform on hindneck, back, and lesser wing-coverts, becoming paler on rump and upper tail-coverts. White lores, eye-ring, and supercilium, narrow black eye-stripe, narrow black moustache, and black border to lower cheeks create open face pattern. Across folded wing, 3 dull black bands run almost horizontally and contain 2 white bands near shoulder and 1 under tertials, but pattern often interrupted by feather disturbance particularly of innermost greater coverts. Tertials dark brown in centre, sand-white on edges; do not fully cloak primaries, so that folded wing ends in noticeably black point. Rather long tail shows narrow white edge, deep black panels, and grey-brown centre. Underparts stone-white, with chest streaked and blotched black. From moment of take-off, appearance of bird changes dramatically, as striking black and white bands over both surfaces of extended wing become fully visible; bold contrast also occurs between pale rump and black panels on uppertail, and between pale vent and black undertail. From below, wing strikingly white, with transverse band usually showing at most as black patch on primaries. ADULT FEMALE. Up to 30% shorter-billed and 15% shorter-winged than ♂, with weaker head- and chest-marks. On some, chest-marks may be dull and diffuse, and only black eye-stripe sharply defined. JUVENILE. Bill less long and decurved; plumage

differs distinctly from adult in being spotted and warmer-toned above, and in lacking ♂'s heavy breast pattern. At all ages, bill dark grey; legs pale grey, grey-white in juvenile.

Adult unmistakable, though at distance in haze confusion with large pipit *Anthus* and even courser *Cursorius* possible. Juvenile unmistakable in flight but on ground could suggest rufous morph of Dupont's Lark *Chersophilus duponti* (but that species always spotted on throat and chest, with typical lark wing pattern). Flies reluctantly except when singing or displaying, appearing to explode off ground and making usually short, low, somewhat undulating and half-circular escape-flight. Flight action loose and free but with illusion of weak flutter caused by wing pattern. Song-flight comprises short vertical climb and descent. Walks and runs; often runs for several hundred metres. Apparently confident of its cryptic plumage, stands erect in alarm with bill and head held up often recalling gaunt, attenuated Mistle Thrush *Turdus viscivorus*. Restricted to arid steppe and shrubby areas in desert.

Voice distinctive, with remarkable fluted, piped, and whistled song which can have marked ventriloquial quality. Calls with a thin tremulous whistle.

Habitat. From arid tropics and subtropics to fringe of Mediterranean zone, mainly in flat or gently undulating lowlands, but including plateaux up to 2000 m in Baluchistan, Pakistan (E M Nicholson), and to *c.* 1000 m in Afghanistan (Paludan 1959). Attracted to coastlines, even down to seashore; in north-west Africa, highest density recorded in sublittoral zone (Valverde 1957). Requires ready access to sand or soft soil for probing, and thus favours wadis with sandy beds, small or even large dunes, abandoned cultivation, vehicle tracks, roadsides, sandy steppe with *Stipa* grass and saltmarsh, with either dense or sparse vegetation, especially where plants pile up heaps of wind-blown sand. Tolerates some blend of gravel with sand, but avoids hammada and other stony or rocky desert, broken ground, and frequent tall trees or shrubs. Among best equipped passerines for rapid terrestrial movement over mixed terrain, and covers a fairly extensive home range, often including mounds, ridges, orgentle knolls affording convenient look-out

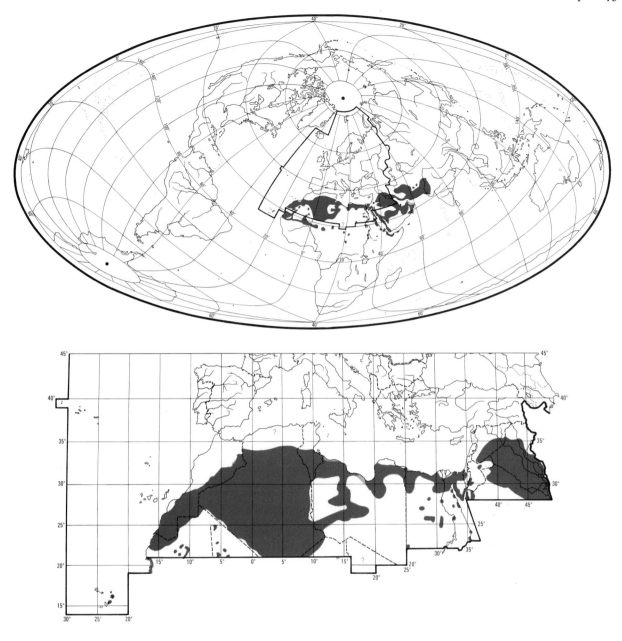

posts. At Azraq (Jordan), almost entirely confined to flat sandy wadis with extensive and sometimes fairly dense growth of low to medium-sized shrublets (Hemsley and George 1966). Common in areas of flat powdered coral on islands of Dahlak archipelago in southern Red Sea (Clapham 1964). In eastern Saudi Arabia, occurs throughout desert areas including gravel plains and sand desert; absent only from extensive bare salt-flats and larger stretches of high dunes (G Bundy). Subject, however, to wide differences in density between habitat types (Valverde 1957; Heim de Balsac and Mayaud 1962). Flies freely, mostly in lower airspace. Tolerant of human presence and even neighbourhood of buildings. Rarely forced to shift habitat by adverse changes.

Distribution. Few range changes reported.

CHAD. Possibly breeds Tibesti (Simon 1965). MAURITANIA. Banc d'Arguin: may breed but no proof (Naurois 1969a). ALGERIA. May breed near Lake Boughzoul (François 1975).

Accidental. Italy, Greece, Lebanon, Malta.

Population. IRAQ. Common (Allouse 1953). MOROCCO. Common in south in some areas (Valverde 1957). ALGERIA.

Most common between 28°N and 35°N (EDHJ). EGYPT. Locally common (SGM, PLM, WCM).

Movements. A true desert bird, resident, and with movement mainly apparent from extralimital occurrences.

Described as very sedentary in North Africa (Heim de Balsac and Mayaud 1962), while Etchécopar and Hüe (1967) made no mention of dispersal. However, there have been several November–February records from coastal plains of western Morocco and from northern Algeria (Heim de Balsac and Mayaud 1962; Smith 1965a), and small parties not infrequent in winter on coastal dunes of Tripoli, Libya (Bundy 1976). Also, at least 26 records from Malta, most August–December though singles in April, May, and July (Sultana and Gauci 1982). Capacity for dispersal in response to changing climatic conditions shown by recent extension of breeding range to Sénégal (Morel and Ndao 1978) and possibly northern Algeria (François 1975).

Similarly resident in Middle East breeding areas (Bundy and Warr 1980; Clarke 1980; Jennings 1981a). Undefined seasonal movements reported from Oman (Gallagher and Woodcock 1980); seen only occasionally in northern Oman, in September–October and March (Walker 1981a).   RH

Food. Largely insects and, where available, snails; also a little plant material. Much animal food obtained by digging with bill. Walks slowly along and, without appearing to pause and listen, suddenly stops to dig fiercely, sometimes for 3–4 min (Meinertzhagen 1951a); will dig into 5 cm of sand, often quite hard and rock-like (Harrison 1970b). Prey thus taken are normally insect larvae, but has been seen to dig out and kill a small gecko *Stenodactylus khobarensis* (Gallagher and Rogers 1978). Meinertzhagen (1951a) never saw an excavation fail. Picks items from ground or plants (Ali and Ripley 1972) and while scrambling on low scrub (Gallagher and Woodcock 1980). Recorded picking maggots from a carcass (Jennings 1981b) and catching a butterfly *Vanessa cardui* (Smith 1965a), a hawk moth (Sphingidae) (Sage and Meadows 1965), and a large dragonfly (Odonata) (G K Brown). Deals with snails (Gastropoda) by flying with them to height of 6–23 m and dropping them on to a stone; if this does not break shell, will beat them against stone directly (Hegazi 1981). One bird on unvegetated coastal sabkha ran about continuously for 4 min, making frequent changes of direction and pausing briefly from time to time; presumably foraging, though not seen to bend down or otherwise take food (D J Brooks). Once seen hovering over water, picking item from surface, and taking it back to dry land to eat (G K Brown). Typically a solitary feeder, but on Cape Verde Islands birds said to gather at a favourite spot towards sunset to feed (Bannerman and Bannerman 1968), and in Oman loose parties frequent rubbish tips (Walker 1981b).

On Cape Verde Islands, diet said by Alexander (1898) to be chiefly locusts (Orthoptera); one February stomach contained insects and spiders (Araneae) (Naurois 1969b). In southern Morocco, April, 3 adults contained larvae of beetles (Tenebrionidae), and a July adult contained ants (Formicidae) and other insects (Valverde 1957). In Algeria, stomachs (no dates) contained small beetles (Coleoptera), sand-flies, and hard seeds (Tristram 1859–60); 3 ♂♂ and 1 10-day-old bird, March, contained chopped up plant matter and grasshoppers (Orthoptera) (Roche 1958). Bird from Tunisia contained ants, bits of grass, and grains of sand (Bannerman 1927). Also in north-west Africa, takes termites *Psammotermes* (Geyr von Schweppenburg 1918; Blondel 1962a). In Chad, takes large numbers of grasshoppers, young locusts, and (especially) ants (Newby 1980). In Egypt, said to take mainly the mantis *Eremophila* (Koenig 1924), and bird from Palestine, December, contained insects (Hardy 1946). In desert near El Hammam (Egypt), where snails *Eremina* abundant, these comprised major part of diet: from February to late June, c. 6–11 birds removed 1090 snails (6·5 kg of flesh) from area of 2·4 ha (Hegazi 1981). Of 8 stomachs from central Saudi Arabia, all contained larvae of ant-lions (Myrmeleontidae) and 7 also seeds (Bates 1936–7). In India and Pakistan, takes tiny beetles and other insects, also seeds of (e.g.) *Suaeda* (Ali and Ripley 1972).

Apart from stomach contents of bird noted above, no information on food of young.   DJB

Social pattern and behaviour. Account includes material on extralimital *doriae* from Iran (Erard and Etchécopar 1970; Ali and Ripley 1972).

1. Not markedly gregarious outside breeding season, at most usually congregating into small parties (Meinertzhagen 1930; Moore and Boswell 1956); at Azraq (Jordan), said not to be gregarious at any time (Nelson 1973). In Israel and Arabian Gulf region, usually solitary, sometimes 2–3 together (Jennings 1981b; Inbar 1982). In Algeria, spring, mostly in pairs; also quite often in groups of 3–6 (Koenig 1895). After breeding, occurs in small family parties, in western Sahara of 2–4(–6) birds (Valverde 1957); in Seistan (Iran), family parties comprised pair with 2 young (Erard and Etchécopar 1970). In December, Sudan, at least 20 in fairly close company, but mostly in pairs (Butler 1905). In Dhofar (Oman), flocks at rubbish tips September–February, maximum c. 30 birds in loose party (Walker 1981b). BONDS. No information on mating system. Apparently only ♀ incubates (Alexander 1898; Ali and Ripley 1972), but at one nest both sexes said to have incubated (Currie 1965; see also subsection 5 in Heterosexual Behaviour, below). Fledged young remain with parents for 1 month or more (Inbar 1982: see above), presence of both parents indicating that ♂, as in other Alaudidae, helps to feed young. BREEDING DISPERSION. Solitary and territorial. Territory typically large (Nelson 1973; Wallace 1983a). In Israel, c. 500–600 m² in Negev desert, up to several km² in Sinai (Inbar 1982). Said sometimes to exceed c. 250 ha (Meinertzhagen 1954), 'up to several square miles' (Meinertzhagen 1951a), though density data (below) suggest that territory must often be much smaller than either of these. In eastern Saudi Arabia, territories apparently 100–150 m

across (G K Brown). In northern Sahara, late March to early June, *c.* 1·25 birds per km² (Blondel 1962*a*). In central Arabia, up to 4–5 pairs per km² in optimum conditions (Jennings 1980). At Azraq, walked transect yielded *c.* 2 birds per km in wadis (Wallace 1983*a*); at Azraq, territories 'extremely large', overlapping with those of Temminck's Horned Lark *Eremophila bilopha*, and sometimes with Desert Lark *Ammomanes deserti* and other species (Nelson 1973). ROOSTING. Nocturnal; on ground. Once among rocks by sea-shore, once under tuft of grass (Meinertzhagen 1951*a*). In Iraq, August, birds crouched by day in shade afforded by troughs in gravel (Moore and Boswell 1956). In eastern Saudi Arabia, will use bill and feet to dig scrape under bush and crouch there, wings half spread (J Palfery); birds near sea often stand at water's edge during hot periods (G K Brown). Sand-bathing reported during period of Song-display (G K Brown). In Iran, observed most often in early morning and evening when temperatures lower (Erard and Etchécopar 1970), and this presumed to reflect activity rhythm. In Arabia, found up to 110 km from water and evidently capable of going for long spells without drinking; thought likely, however, to take dew condensed on vegetation (Ticehurst and Cheesman 1925). Infrequency of drinking also reported by Brosset (1961).

2. Typically tame and confiding, especially when paired, allowing approach to within a few metres. Often reluctant to fly, preferring to run ahead of intruder, at most flying low for short distance (Bannerman 1927; Meinertzhagen 1930, 1954; Gallagher and Woodcock 1980). Never squats, according to Meinertzhagen (1954), but said to do so by Kittenberger (1960), and one bird in Oman crouched in sand and allowed approach to within 3 m before moving (Walker 1981*b*). Characteristic style of running from danger frequently described: runs fast and long, like a courser *Cursorius* (Bannerman 1927; Stanford 1954), easily sprinting *c.* 200 m or more (Nelson 1973), up to 300–400 m (Meinertzhagen 1954). Runs ahead of pursuer with head bent low (Alexander 1898) or bill uplifted (Gallagher and Woodcock 1980: Fig A). Between runs, typically halts, takes 1–2 more steps, and stops, bobbing body up and down (Bannerman 1927) and raising its head as if to listen (Alexander

A

1898). Human intruder on territory may also elicit wing-spreading (Inbar 1982: see also Antagonistic Behaviour, below). Fond of perching on a bush or sand-knoll when alarmed (Stanford 1954). For Alarm-call see 4 in Voice. FLOCK BEHAVIOUR. No information. SONG-DISPLAY. Numerous descriptions of ♂'s Song-flight, from which it appears that considerable variation occurs in nature of ascent and descent, and height reached. For part of song preliminary to Song-flight (Ali and Ripley 1972), said to summon mate (Alexander 1898), see 1 in Voice—also for details of synchronization of song with elements of Song-flight display. Following account appears to be representative of many Song-flights: from top of bush or sand-knoll, ♂ springs up, singing, on fluttering wings and with tail spread, rising more or less vertically for *c.* 1–4(–10) m, flips over, even somersaulting (thus displaying striking plumage pattern), then, holding wings close to body, nose-dives back to same perch; sings on descent and opens wings at last moment before landing (Fig B). Song-flight performed repeatedly at brief intervals (Lynes 1930; Rowntree 1943; Stanford 1954; Schüz 1957*b*; Archer and Godman 1961*a*; Ali and Ripley 1972; M G Wilson). Sometimes flies close to ground with quivering wings, and, once perched, quivers slightly spread wings and tail (Inbar 1982). In returning to original perch, may land right beside mate (Alexander 1898; Rowntree 1943). Bird landing on ground usually runs (Gallagher and Woodcock 1980). Once, displayed over same spot for almost 1 hr (Tristram 1859–60). One ♂ in Algeria displayed over area of *c.* 500 m diameter (Lynes 1930). Although ascent to *c.* 25 m above ground apparently common, height may vary markedly; bird said to fling itself into the air with closed wings, then turn over for descent (Stanford 1954), presumably reaching no great height.

B

In Saudi Arabia, typically ascends to 1-2 m (King 1978b). In much higher Song-flights, pattern of ascent and descent varies from that described above: e.g. may ascend spirally, sometimes to over 60 m, then execute slow spiral 'sail-down' on outstretched fluttering wings; song continues until final vertical plummet to ground with wings closed, checking only at last moment (Meinertzhagen 1951a, 1954). Spiral 'moth-like' ascent also reported by Nelson (1973), and final plummet noted from c. 30 m by Harrison (1970b). Song characteristically heard at dawn (Butler 1905). In many regions, Song-display extends from March to May (Stanford 1954; Gallagher and Rogers 1978; Jennings 1980; Walker 1981b) but may begin in January (Rowntree 1943; Meinertzhagen 1954). In Oman, also heard September (Stanford 1969). ♀ occasionally sings (Currie 1965) but context and function unknown. ANTAGONISTIC BEHAVIOUR. Territory is said to be guarded jealously by both sexes (Meinertzhagen 1951a, 1954), or by ♂ only (Harrison 1970b). Threatens conspecifics and other birds intruding on territory by wing-spreading (Inbar 1982: see above). Once, 2 birds crouched horizontally, facing each other with tails lowered and fanned, plumage ruffled, giving Threat-calls (see 3 in Voice); soon after, fought in the air (G K Brown). At mixed loafing site, Saudi Arabia, dominant over Bar-tailed Desert Lark *Ammomanes cincturus* and Dunn's Lark *Eremalauda dunni*, chasing both out of scrapes (J Palfery). No further information. HETEROSEXUAL BEHAVIOUR. (1) General. In Arabia, birds pair in January (Meinertzhagen 1954; see also Song-display, above). (2) Pair-bonding behaviour. Little known. Song-display (see above) evidently directed strongly at ♀, and may start and finish alongside her. Ground-display (no details) occurs around the nest during egg-laying, sand often being churned up into regular 'dancing channels' (Meinertzhagen 1954). In Israel, ♂ and ♀ fly up to c. 20-25 m and ♂ then chases ♀ for several hundred metres; both birds 'spread and fold' wings while giving whistling sounds (Inbar 1982), perhaps call 2; aerial chase often followed, after both land, by Song-flight of ♂ (Inbar 1982). In Libya, pursuit-flights (presumably of heterosexual nature) common in early morning, one bird chasing another, twisting and turning with great agility (Bulman 1942). (3) Courtship-feeding. No information. (4) Mating. Only one observation, of pair feeding and displaying on ground: ♂ sang regularly and ♀ twice solicited him from 20-50 m away by crouching, partly spreading and fluttering wings rapidly, so making bold wing pattern conspicuous. After c. 1-2 min, copulation followed (Marchant 1963b). (5) Behaviour at nest. Role of sexes in nest-building not known. Bird approaches nest furtively, typically evading notice of observer (Inbar 1982). According to Archer and Godman (1961a), however, typically descends almost vertically on to nest. ♀ leaves nest several times per day to feed. While ♀ incubating, ♂ guards nearby, sometimes performing Song-flights over nest-area (Inbar 1982). RELATIONS WITHIN FAMILY GROUP. Little known. Both parents feed young (Inbar 1982). In one nest with 2 young, the droppings of the young lined the nest (Ticehurst *et al.* 1921-2). Young can leave nest before fledging (Ali and Ripley 1972; Inbar 1982); wander about in immediate vicinity of nest for several days until able to fly well, and stay with parents 1 month or more (Inbar 1982). In Egypt, young bird over 1 month old was still accompanied by a parent; suspected that parent inserted bill right down into mouth of this young to feed it (Hachisuka 1924). ANTI-PREDATOR RESPONSES OF YOUNG. One brood of well-grown young 'froze' in nest on approach of human intruder (J Palfery). While still flightless, young scatter if alarmed and hide in vegetation or natural hollows (Inbar 1982); capable of swift running (Ali and Ripley 1972) and thus flee danger (S

Marchant). PARENTAL ANTI-PREDATOR STRATEGIES. (1) Passive measures. May be quite confiding at nest, as at other times. At one nest, bird approached to within 3m of human intruder, running away when he came nearer (Ticehurst 1926). If disturbed, incubating ♀ may flush and run away (Ali and Ripley 1972). Suspicious parents leave nest and wait until safe to return (Inbar 1982). (2) Active measures: against man. When nest disturbed, parents sometimes fly towards intruder or even stoop at close quarters (Meinertzhagen 1954). Although distraction-display of disablement type never seen by Meinertzhagen (1954), this reported by Ali and Ripley (1972), ♀ running ahead of observer with wings dragging. While ♀ sits, ♂ may use 'antics' to 'decoy' intruder away from nest (Alexander 1898); no details, but perhaps no more than the running just ahead of intruder, and periodic halting, described in introduction to part 2 (above). When intruder caused incubating ♀ to leave nest, ♂ performed apparent Song-flight nearby (Bolster 1922), perhaps as a distraction-display. (3) Active measures: against other animals. In Saudi Arabia, when snake *Malpolon* near nest containing 3 well-grown young, parent dived repeatedly at snake, flying up to c. 1·2 m above ground to avoid strikes, continuing thus until snake took cover (J Palfery).

(Figs by J P Busby: A from drawing in Chabot 1932; B from drawings in Lynes 1930 and Valverde 1957.)   EKD

Voice. Freely used in breeding season, when song of ♂ is characteristic feature of desert. Account includes material on *doriae* (Ali and Ripley 1972), but differences, if any, from nominate *alaudipes* not known.

CALLS OF ADULTS. (1) Song of ♂. Variously described as ringing and far-carrying (Alexander 1898), musical, melodious, plaintive, piping, or fluting (e.g. Whitaker 1905, Meinertzhagen 1930, 1954); quality frequently likened to Nightingale *Luscinia megarhynchos*. Song typically of 3 parts: a series of fluting sounds, then a trill, and finally another series of fluting sounds; these delivered during Song-flight, but extent to which song given while airbone evidently varies. According to Stanford (1954) and King (1978b), first 2 parts delivered when perched, 3rd part during ascent and descent of Song-flight; trill also thought to be given only when perched (J Palfery). This pattern probably typical of Song-flights to no great height: 1-2 m (King 1978b; J Palfery); account by Stanford (1954) also indicates a low Song-flight. In other descriptions (e.g. Alexander 1898, Zedlitz 1909), 1st series of fluting sounds given on ascent, trill as bird flips over at peak of ascent, and final part on descent; though no firm evidence, this pattern presumably associated with higher Song-flights (see Social Pattern and Behaviour). No details of Song-flight accompany our recording (Fig I), which begins with 3 low fluting sounds—a plaintive and measured 'whee whee whee' (E K Dunn); this followed by trilling 'trrr-eeeeee' in which 'trrr-' about same pitch as preceding 'whee' notes but extended 'eeeee' sound markedly higher; there follow 2 'whee' notes on descending scale, immediately followed by series of c. 17 'whee' notes, the first 4 very short, of uniform pitch, and rapidly repeated, becoming longer-drawn and

I   C Chappuis/Sveriges Radio (1972–80)   Morocco   April 1966

I   *cont.*

I   *cont.*

II   P A D Hollom   Egypt   March 1984

more slowly repeated towards end; pitch of last few notes also rises. Duration of whole song in Fig I *c.* 12 s and, after *c.* 5 s of silence, identical song repeated. In same recording, bird, now presumably perched, then delivers 23 phrases of 'whee' notes and latterly some 'wheetlip' sounds, of varying pitch, lasting *c.* ¾ min before song repeated again. In *doriae* probably this mixture of notes while perched rendered 'too-too-te-te', whistled rather plaintively, deliberately, and softly (Ali and Ripley 1972); probably this or similar phrase, given 3–4 times, said to summon ♀ mate prior to performance of Song-flight (Alexander 1898). Song rendered thus by King (1978*b*): 1st part a series of *c.* 8–10 loud clear whistled notes of uniform pitch, running into 2nd part—a loud musical trill; 3rd part *c.* 6 loud clear whistled notes of uniform pitch; whole song rendered 'whuu too-too-too too-too too-too-too titititititi teew teew teew teew teew teew'. During plummet, a long 'kreeee' (G K Brown). For other renderings of all or part of song, see Mountfort (1965), Ali and Ripley (1972), Gallagher and Woodcock (1980), and Walker (1981*b*). At a distance, song may sound mechanical, like something tapping very regularly against a telegraph wire; also hard to determine direction of source (Rowntree 1943). Timbre quite human (i.e. whistle) or instrumental (Meinertzhagen 1930). ♀ also said to sing occasionally (Currie 1965), but no details. (2) Contact-call. A buzzing sound, difficult to render, but roughly 'zeee'

(E K Dunn: Fig II), resembling buzzing component of song of Trumpeter Finch *Rhodopechys githaginea* (P A D Hollom). Also rendered a buzzing or wheezing note (M G Wilson) or a soft jangling 'jinzing' reminiscent of Siskin *Carduelis spinus* (Moore and Boswell 1956). (3) Threat- and Excitement-calls. A harsh 'shweee' given by 2 birds confronting each other. Same or related harsh calls: 'chweek' from 1 of 4 feeding birds, likewise once in flight; rather hoarse trilling 'shreerp' when 2 birds came into close contact at edge of lagoon; 'schirrr' in pursuit-flight; bird perched on bush gave long harsh 'skeerrk', rather like Jay *Garrulus glandarius*, before singing (G K Brown). (4) Alarm-call. A 'too' (Ali and Ripley 1972). No further information. (5) Other calls. Common call described as a 'thin tremulous whistle' (Gallagher and Woodcock 1980), perhaps call 2.

CALLS OF YOUNG. No information.   EKD

Breeding. SEASON. Cape Verde Islands: from October, perhaps to February or March (Bannerman and Bannerman 1968). North Africa: begins from end of February in southern desert areas, to May in north (Heim de Balsac and Mayaud 1962; Hüe and Etchécopar 1970). Iraq: laying from mid- to end of May (Ticehurst 1926; Marchant 1963*b*). Eastern Saudi Arabia: eggs seen only in March, and young still in nest in mid-April (J Palfery). SITE. In or on top of tussock or low shrub, 30–60 cm above ground; also, perhaps less often, on ground in shelter of rock or bush. Nest: when above ground, constructed of twigs woven into bush, and lined with softer material; on ground, in shallow scrape lined with twigs and with soft inner lining of wool and plant down; one nest, Libya, had soft lining coated with smoothed sand (Norris 1964); average internal diameter 7·5 cm, depth 4·0 cm (Ali and Ripley 1972). In Israel, edge of nest usually paved with row of small, flat stones forming possibly protective wall (Inbar 1982). Building: no information on role of sexes. EGGS. See Plate 79. Sub-elliptical, smooth and glossy; white to pale buff, variably spotted and blotched reddish-brown, with underlying lavender and grey markings, particularly at broad end. Nominate *alaudipes*: 24·3 × 17·1 mm (22·3–27·9 × 16·0–19·0), $n = 25$; calculated weight 3·65 g; *doriae* very similar in size (Schönwetter 1979). Clutch: 2 on Cape Verde Islands (Bannerman and Bannerman 1968). Of 17 clutches in North Africa: 2 eggs, 7; 3, 5; 4, 5; mean 2·88; possibly 2 broods (Heim de Balsac and Mayaud 1962). In Israel, 2–3(–4) eggs; apparently 1 brood per year (Inbar 1982). INCUBATION. About 2 weeks. By ♀. (Alexander 1898; Inbar 1982.) In one case where both ♂ and ♀ apparently incubated, spells on nest up to 40 min (Currie 1965). Incubation begins with penultimate or last egg (Inbar 1982). FLEDGING TO MATURITY. Young stay in nest 12–13 days, fledging several days later; stay with parents apparently for 1 month or more (Inbar 1982).

Plumages (nominate *alaudipes*). 2 colour morphs occur, connected by intermediates; rufous morph common in North Africa, where grey morph rare; in Syria and Jordan, grey morph perhaps more common (2 of 10 examined). ADULT MALE. RUFOUS MORPH. Forehead, crown, mantle, scapulars, tertials, and back cinnamon-buff, brightest (pink-cinnamon) on tips of feathers and on tertials, duller and browner towards feather-bases; brown feather-shafts on crown sometimes partly visible. Hindneck and sides of neck mouse-grey with slight buff suffusion, often contrasting somewhat with buff crown and mantle. Rump and upper tail-coverts pale grey with off-white feather-fringes, partly suffused buff when plumage fresh. Distinct pale buff or cream-white supercilium, extending from nostrils to above rear of ear-coverts, bordered below by blackish line through eye (often not reaching base of bill on lores) and sometimes bordered above by dusky feathers on sides of crown above eye. Pale cream streak reaching backward from base of upper mandible, widening into cream-buff patch on ear-coverts; rear of ear-coverts often slightly darker buff or partly mottled dusky; narrow and indistinct blackish streak from gape ends in distinct blackish spot on lower cheek below ear-coverts. Chin and throat pale cream-pink or white, often with some fine black spots on sides of throat. Chest, sides of breast, and flanks pale buff or pink-cinnamon, chest with distinct rounded or triangular spots 1–4 mm across (size strongly variable: in some, mere specks and chest largely pale; in others, spots large and almost joining, showing as distinct chest-band); remainder of underparts pale cream or white. Central pair of tail-feathers (t1) cinnamon-buff like upperparts; shaft black, a streak of variable width on feather-centre grey or olive-brown; t2–t5 black or brown-black with narrow cinnamon-buff or cream-white outer edge and (when feather new) tip; t6 black with broader white outer edge and tip, outer web often largely white. Primaries largely black; p1–p6 with white on bases extending 1·5–2 cm beyond tips of greater upper primary coverts, p7–p8 with white basal inner web and some white or pale grey on basal outer edge; *c.* 1·5 cm of tip of p1 white, tips of p2–p3 with gradually less white, on p4 restricted to broad fringe along tip only. Tips of p5–p10 and outer webs of p9–p10 narrowly

edged buff or off-white. Secondaries white with black band *c.* 2 cm wide (*c.* 1·5–2 cm from secondary-tips); broad outer fringes along middle and terminal portions of secondaries buff. Much individual variation in extent of black band across secondaries: black usually extends over both webs, but occasionally inner webs largely or fully white or width of black band much reduced on middle secondaries. Greater upper primary coverts and longest feather of bastard wing greyish-black with cream-white or pale buff edge along outer web and tip; greater and median upper wing-coverts cinnamon-buff like scapulars and tertials, olive-brown or black-brown central bases often visible, greater with broad cream-pink or white tips; lesser upper wing-coverts pale grey with some buff suffusion when fresh, usually markedly greyer than neighbouring median wing-coverts and scapulars. Under wing-coverts and axillaries cream-buff or white. *In worn plumage*, crown, mantle, scapulars, and t1 duller greyish-buff, hindneck purer grey, shorter lateral upper tail-coverts off-white, white supercilium and black marks on head and chest more pronounced, underparts purer white (in some birds, stained rufous by soil, however), dark bases of greater and median upper wing-coverts bleached and less contrasting, pale edges along tip of tail-feathers and longer primaries abraded. GREY MORPH. Upperparts uniform grey (similar in tinge to hindneck and lesser upper wing-coverts of rufous morph), only tertials and greater and median upper wing-coverts suffused cinnamon on tips, and ground-colour of underparts cream-white or white. ADULT FEMALE. RUFOUS MORPH. Like adult ♂, but hindneck, upper tail-coverts, and lesser upper wing-coverts less pure grey, more extensively washed buff, not or hardly contrasting with remainder of upperparts; ground-colour of chest on average slightly deeper buff; black band across secondaries narrower, 1–2 cm. GREY MORPH. Apparently like adult ♂. NESTLING. Almost entirely covered with long whitish or pale buff-white down (Norris 1964; Ali and Ripley 1972). JUVENILE. RUFOUS MORPH. Upperparts and upper wing-coverts rufous-cinnamon or deep pink-cinnamon (deeper rufous and less pink than adult, virtually without grey on hindneck and lesser upper wing-coverts), crown, mantle, scapulars, back, rump, and lesser and median upper wing-coverts with rather broad buff or isabelline tips, subterminally bordered by dark grey crescent (latter occasionally faint or partly lacking). Supercilium less distinctly defined than in adult, black streaks on sides of head reduced, usually only just visible on lores and above buff ear-coverts, but lower ear-coverts often more distinctly spotted grey. Underparts dirty white, except for pale buff chest, sides of breast, and upper flanks, chest with tiny grey specks, sometimes hardly visible. Tail as in adult, but feathers pointed at tips, t2–t5 more broadly fringed buff along tip, t6 with fully pale buff or white outer web and 4–8 mm of tip of inner web buff. Flight-feathers as in adult, but black bar across secondaries always fully developed, though sometimes less distinctly defined; ground-colour of outer 2–3(–4) primaries grey-buff rather than black; p5–p9 and all greater upper primary coverts with 1·5–2 mm wide buff fringes along tip; p10 relatively longer (see Structure) and with more rounded tip, narrow buff outer edge extending to tip of inner web. Sexes similar (see Measurements). GREY MORPH. Like rufous morph, but feathers of upperparts pale buff or grey with off-white fringes and dark grey subterminal crescents. FIRST ADULT. Variable, depending on moult; some birds (especially ♀♀) moult completely and are probably indistinguishable from adult. At least some birds retain juvenile outer primaries (including characteristic p10), outer greater upper primary coverts, and part of tail; others retain all juvenile flight-feathers, greater coverts, and tail.

Bare parts. ADULT. Iris brown. Eyelids yellow. Upper mandible plumbeous-grey, dark horn, or dark green-grey, lower mandible similar but with paler grey, pink-grey, or olive-grey base. Mouth black. Leg and foot grey-white or china-white, claws greenish-brown. NESTLING. Mouth yellowish-orange. JUVENILE. Iris brown. Bill paler than adult, pale brown-grey, flesh-brown, or greenish-grey, darker towards tip. Mouth yellow. Leg and foot dirty white, claws pale horn. Not known at what age bill and mouth darken to adult colour; some birds with plumage like adult but with bill and mouth like juvenile are perhaps 1st adult (as, e.g., a July bird from Saudi Arabia: Ripley 1951), but bill and mouth of full adult in non-breeding season may also be pale instead of dark. (Ticehurst 1923; Ripley 1951; Norris 1964; Ali and Ripley 1972; BMNH, ZMA.)

Moults. ADULT POST-BREEDING. Complete; primaries descendant. In nominate *alaudipes* and in *doriae* from Middle East, starts with p1 late June to early August, completed with p10 late August to early October; some feathering on crown or sides of body occasionally moulted from late May; body still largely old at primary moult score 10 (usually July), body and central tail-feathers new at moult score 30-35 (usually August), when outer tail-feathers, many wing-coverts, and part of belly still growing. No *boavistae* of Cape Verde Islands in moult examined; all October birds had plumage new, all May birds heavily worn. POST-JUVENILE. Complete or partial; primaries descendant. Moult of head, body, tertials, and lesser and median upper wing-coverts complete or virtually so, starting after hatching, late May to September; mantle, scapulars, head, and chest first. Number of flight-feathers replaced depends on hatching date: early-hatched birds start with p1 from late May, reaching primary moult score 30 by late July to early September and completing all moult (including secondaries, greater coverts, and tail) from late August onwards. Juveniles fledged late July and August suspend flight-feathers moult in late autumn, those fledged late August and September retain all old flight-feathers during winter, as well as variable number of greater upper wing-coverts and tail-feathers. Of 9 recognizable 1st-winter birds examined, 4 had all primaries old, 1 had 3 inner new, 2 had 4, 1 had 5, and 1 had 7 inner ones new; specimens with all primaries new indistinguishable from adult.

Measurements. ADULT. Nominate *alaudipes*. Algeria and Tunisia, all year; skins (BMNH, RMNH, ZMA). Bill (S) to skull, bill (N) to distal corner of nostril; exposed culmen on average 5·0 mm less than bill (S).

WING	♂ 128 (3·19; 20)	123-134	♀ 115 (2·84; 13)	111-120
TAIL	84·4 (3·88; 21)	79-92	76·4 (2·62; 14)	72-80
BILL (S)	32·4 (1·47; 26)	29·5-34·6	29·0 (0·92; 14)	27·4-30·3
BILL (N)	22·0 (1·17; 26)	20·1-24·1	19·7 (1·01; 14)	18·5-21·4
TARSUS	34·3 (1·14; 28)	32·0-36·2	31·2 (0·92; 15)	29·8-32·7

Sex differences significant.

Syria and Jordan, all year; skins (BMNH, ZMA).

WING	♂ 134 (2·96; 7)	130-139	♀ 124 (— ; 2)	123-124
BILL (S)	32·6 (1·54; 7)	30·5-34·3	31·8 (— ; 2)	31·0-32·5
BILL (N)	22·9 (0·71; 7)	21·6-23·6	21·3 (— ; 2)	20·8-21·8

A. a. boavistae. Boa Vista and Maio (Cape Verde Islands), October-May; skins (BMNH).

WING	♂ 128 (2·11; 15)	124-132	♀ 115 (2·84; 9)	111-119
BILL (S)	31·5 (1·61; 14)	28·0-33·4	28·1 (1·47; 9)	26·1-30·6
BILL (N)	20·8 (1·09; 14)	18·9-22·4	18·6 (1·01; 7)	17·1-19·7

A. a. doriae. Iraq and Kuwait, all year; skins (BMNH).

WING	♂ 135 (2·94; 10)	131-139	♀ 125 (1·77; 7)	122-127
BILL (S)	34·7 (2·22; 10)	31·5-36·9	30·7 (1·20; 5)	29·5-32·6
BILL (N)	23·8 (2·02; 10)	21·3-25·7	21·4 (1·36; 5)	20·0-23·5

Iran and Afghanistan; skins (Vaurie 1951a; Paludan 1959; BMNH).

WING	♂ 141 (2·08; 27)	138-146	♀ 131 (5·64; 9)	124-138
BILL (S)	36·4 (1·78; 20)	33-40	33·2 (1·78; 7)	30-36

JUVENILE. Wing on average 4·6 shorter than adult, tail 1·2 shorter; bill and tarsus similar to adult when post-juvenile body moult about completed. Wing of 1st winter birds with retained juvenile outer primaries, Algeria and Tunisia (RMNH, ZMA):

WING	♂ 124 (1·81; 8)	120-127	♀ 110 (— ; 2)	109-111

Weights. Nominate *alaudipes*. Sahara. ♂ 39-47 (10), ♀ 30-39 (6) (Niethammer 1955b). Saudi Arabia, July: ♂ 46·5 (Ripley 1951).

A. a. doriae. Afghanistan, late March and early April: ♂ 48·7 (1·63; 6) 47-51; ♀♀ 46, 47 (Paludan 1959).

Structure. Wing long and broad, tip rounded. 10 primaries: p8 longest; in adult, p9 5-9 shorter, p10 64-70 (♂) or 54-60 (♀), p7 0-1, p6 1-2, p5 4-8, p4 19-24, p1 34-39 (♂) or 26-33 (♀), in juvenile, p9 4-9 shorter, p10 58-63 (♂) or 53-57 (♀), p7 0-1, p6 0-4, p5 3-8, p1 32-37 (♂) or 27-31 (♀); adult p10 small, rather narrow and with pointed tip, 6·3 (36)1-9 longer than greater upper primary coverts (exceptionally up to 12 longer); juvenile p10 slightly longer and with more broadly rounded tip, 8·7 (10) 4-11 longer (usually 9-11, but 4-5 in 2 birds). p5-p8 and inner of p6-p9 emarginated. Longest tertials reach to tip of p4-p5 in closed wing. Tail rather long, tip square; 12 feathers. Bill long, slightly longer than head; basal half almost straight, distal half with culmen and cutting edges distinctly decurved, gonys slightly decurved; tips of both mandibles acutely pointed. Nostrils rather small, rounded, not covered by feathers (unlike many other Alaudidae). Some fine bristles along gape, most pronounced at base of upper mandible. Tarsus long, toes and claws short and rather thick, hind claw slightly decurved. Middle toe with claw 17·1 (23) 15·7-18·9; outer toe with claw c. 79% of middle with claw, inner c. 75%, hind c. 96%. Hind claw 7·3 (10) 6·2-8·6.

Geographical variation. Rather slight, both in colour and size. Birds of Cape Verde Islands and North Africa rather small, size clinally larger through Middle East towards Iran (see Measurements). Extralimital race *desertorum* from Red Sea coasts about as small as Cape Verde *boavistae* and North African nominate *alaudipes*; *doriae* from Pakistan and north-west India smaller than birds from Iran, wing ♂ 126-137, ♀ 116-119 (Ali and Ripley 1972). Rufous morph of *boavistae* has broader and deeper rufous pink-cinnamon fringes along feathers of upperparts than nominate *alaudipes* of North Africa, not sandy-pink; grey morph darker and purer grey than buffish-grey North African birds, but tertials and t1 rather deep rufous; spots on chest generally large, but some nominate *alaudipes* similar; bill of *boavistae* on average slightly more slender than nominate *alaudipes*, especially tip. Colour of birds from Syria and Jordan like those of North Africa, so these included here in nominate *alaudipes*, though size slightly larger; birds from Iraq have narrower pink-cinnamon fringes on upperparts, with more grey of feather-bases visible; both colour and size near typical *doriae* from Iran which is about as grey as grey morph of *boavistae*. Upperparts of Red Sea race *desertorum* also grey, but slightly paler, purer, and less brown than *doriae* and *boavistae*; spots on chest variable in size, as in other races.

CSR

Chersophilus duponti Dupont's Lark

Du. Duponts Leeuwerik   Fr. Sirli de Dupont   Ge. Dupont-Lerche
Ru. Тонкоклювый пустынный жаворонок   Sp. Alondra de Dupont   Sw. Smalnäbbad lärka

Alauda Duponti Vieillot, 1820

Polytypic. Nominate *duponti* (Vieillot, 1820), Iberia, Morocco, northern Algeria, and north-west Tunisia; *margaritae* (Koenig, 1888), southern slopes of Atlas mountains from about Biskra (Algeria) eastward, south-east Tunisia, northern Libya, and north-west Egypt.

Field characters. 18 cm, of which bill 2 cm; wing-span 26-31 cm. Apart from bill, 15% shorter than Skylark *Alauda arvensis* but noticeably more robust than short-toed larks *Calandrella*; 20% smaller than Hoopoe Lark *Alaemon alaudipes*. Small to medium-sized uncrested lark, with quite long, decurved bill, rather long neck, bulky body, and spindly legs; folded wing-points of rather short, broad wings, almost cloaked by long tertials, do not extend down tail as in most larks. Plumage pattern much as *A. arvensis* but noticeably scaled or evenly fringed on scapulars and folded wing; copious streaks on head interrupted by pale crown-stripe, supercilium, eye-ring, and half-collar below cheeks. Ground-colour of plumage varies from dark brown to chestnut. Inveterate runner. Voice distinctive. Sexes similar; no seasonal variation. Juvenile separable. 2 races in west Palearctic, distinguishable in the field. Adult. (1) Spanish and north-west African race, nominate *duponti*. Upperparts basically brown, copiously and extensively streaked black-brown and patterned with (a) cream crown-stripe (not always visible), (b) long, buff-white supercilium and bold eye-ring, (c) pale grey surround to brown-streaked cheeks, forming obvious half-collar, (d) regular scaling of mantle feathers and scapulars, with narrow but complete buff margins to feathers overlapping as in small juvenile *Calidris* wader, (e) well-marked median coverts, with pale buff margins contrasting with black-brown centres and forming obvious feature, (f) distinctly pale-margined greater-coverts and tertials, tertials almost cloaking primary-tips, and (g) white edges to tail. Underparts essentially white, with indistinct moustachial stripe but black-brown spots on throat becoming full, well-spaced streaks on chest and flanks. Underwing grey. In flight, only white edges to tail show well; wings noticeably uniform and dark above, grey below. (2) Central and eastern North African race, *margaritae*. Plumage pattern as nominate *duponti* but colours distinctly paler, with feather-centres cinnamon or chestnut and margins cream, even white when bleached. Bird appears less streaked, more mottled than nominate *duponti*, with paler face and more diffuse chest-marks. In both races, wear and changes in feather overlap can cause appearance of pale lines on back. Birds in northern Algeria show intermediate appearance (see Geographical Variation). Juvenile. Like adult but less heavily streaked

above and below, upperparts with broader cinnamon scaling, feathers tipped white. At all ages, bill horn-colour, lower mandible paler; at least $\frac{2}{3}$ length of head, quite markedly decurved on both mandibles. Legs long, often showing tibia, but spindly, flesh-brown to whitish.

Unmistakable when seen well, but *margaritae* subject to risk of confusion with juvenile *A. alaudipes* when size, wing pattern, and plain chest of *A. alaudipes* obscured. When bill not visible, build may provoke confusion with crested larks *Galerida* but *C. duponti* lacks crest. Flight fast, with wing shape producing more flapping action than *Alauda*; normal flight and song-flight both end in sudden rapid perpendicular plunge, with flutter 2-3 m above ground. Prefers to escape by persistent running among ground cover. Climbs up low plants. Inconspicuous except when singing and calling in first and last hours of daylight.

Song resembles twittering of Linnet *Carduelis cannabina*. Commonest calls a whistling 'hoo-hee', 2nd syllable nasal and rising in pitch, and 'dweeje' like harsh spring call of Greenfinch *C. chloris*.

Habitat. Ecologically as well as geographically restricted; confined to dry and usually warm Mediterranean lowlands and plateaux. In Spain, on northern fringe of range, occurs both near coasts at heights of 50-120 m and far inland at 340-1200 m, at temperatures ranging from −10° to 40° C and with very low rainfall. Always in open country, breeding on flat areas or on slopes not exceeding 25% gradient, with vegetation cover not more than 30-50 cm tall, including especially feather-grass *Stipa*, mugwort *Artemisia*, and in one case *Salicornia*, growing in more or less dense clumps separated by open corridors suitable for rapid running. Such sites, of *c.* 40-60 ha, form islands in midst of large expanses of cultivated land, which has recently been encroaching severely on them. Outside breeding season, such areas abandoned in favour of cereal fields (especially barley *Hordeum* and oats *Avena*), where birds assemble in mixed flocks with Skylark *Alauda arvensis* and Calandra Lark *Melanocorypha calandra*. Breeding sites may be shared with Thekla Lark *Galerida theklae*, Short-toed Lark *Calandrella brachydactyla*, and Lesser Short-toed Lark *C. rufescens*. (Suarez *et al.* 1982; Aragüés and Herranz 1983; A Aragüés.) In Morocco, breeds on *Artemisia* steppe and

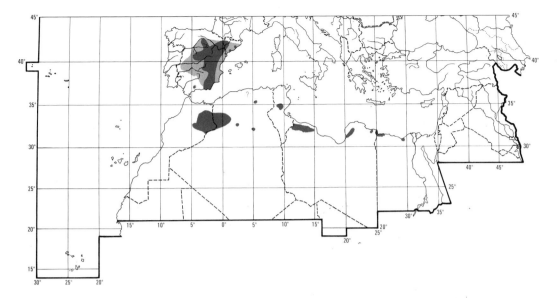

in feather-grass above 1000 m, where ground-frost occurs at times of earliest nesting (Smith 1965a). Also occurs in arid scrub country where sufficient concealment available at times on sand, stony desert, or semi-desert (Etchécopar and Hüe 1967; Harrison 1982). Mainly a ground bird.

Distribution. Inadequately known due to secretive habits.

SPAIN. Present known distribution mapped, but likely to be more extensive (Suárez *et al.* 1982; Aragüés and Herranz 1983; AN). PORTUGAL. Has bred Lisbon (Irby 1875); no records since (RR).

Accidental. France, Italy, Malta.

Population. SPAIN. Relatively abundant (Suárez *et al.* 1982); seems to have declined in Aragón since *c.* 1960 (AN). EGYPT. Few recent records (PLM, WCM). LIBYA. Tripolitania: thinly distributed; Cyrenaica: scarce and local (Bundy 1976). ALGERIA. Nowhere very common (EDHJ). TUNISIA. Not common (MS).

Movements. Resident. Some dispersal occurs, but very little known of this and it may be irregular. Has been found occasionally (1 dated record, for March) in Algerian Sahara, south of breeding range and outside normal habitat (Heim de Balsac and Mayaud 1962; Etchécopar and Hüe 1967). Further, vagrants known from Mediterranean basin, from Malta and Italy westwards, mainly September–December; these not necessarily all of North African origin, however, since also breeds in eastern Spain. In particular, autumn records on Spanish side of Straits of Gibraltar are probably Spanish-bred birds (normally resident). RH

Food. Mainly insects and small seeds according to Brock (1970). 2 birds from eastern Spain contained a Lepidoptera larva and at least 12 adult beetles (Carabidae, Tenebrionidae, Scarabaeidae, Silphidae, Curculionidae); 2 others, September, contained 90 seeds (mainly *Asphodelus*), at least 6 beetles, ants (Formicidae), and grasshoppers (Orthoptera) (A Aragüés). Feeds by using bill like woodpecker (Picidae) to dig into ground or into sandy bases of plant tufts; also by splitting open balls of horse dung (Smith 1965a; A Aragüés). No further information. DJB

Social pattern and behaviour. Poorly known. Account includes material supplied by A Aragüés from study in Ebro valley (Spain).

1. Not very sociable, and usually occurs singly or in pairs (e.g. Whitaker 1905 for Tunisia), though small parties sometimes recorded (Gaston 1970). See also comments by Meinertzhagen (1951a), who lumped *Chersophilus* with *Certhilauda*. In Ebro valley, outside breeding season, birds leave breeding areas by day and disperse into cereal crops where closely associate with large flocks of Skylark *Alauda arvensis* and Calandra Lark *Melanocorypha calandra* (A Aragüés). BONDS. No information on mating system. In Tunisia, early May, family party apparently comprised ♀ and 2 fledged young (Koenig 1893). BREEDING DISPERSION. According to Meinertzhagen (1951a), '*Certhilauda*' larks (see above) occupy large territory which is defended by both sexes, but no information available specifically for *Chersophilus duponti*. In Iberia, line transect counts (mostly in breeding season but some at other times) showed densities of 8·9 birds per 10 ha in *Gypsophilla hispanica* to 20·5 per 10 ha in mixed *Linum* and *Genista pumila* (see Suárez *et al.* 1982). In Ebro valley, densities generally lower and variation between 8 study areas considerable. In breeding season (mainly March–May), 1·7-6·6 encounters recorded per 10 ha. Birds mostly in open areas of *c.* 50-200 ha, fewer in smaller plots (30-40 ha) bordered by cereals. Up to 6 other species of Alaudidae share same sparsely vegetated areas (A Aragüés). ROOSTING. In Ebro

valley, outside breeding season, birds return from feeding grounds to breeding areas in evening (A Aragüés); presumably roost there. No information on roosting habits. A very early riser with peak activity early and also late in the day (Aragüés and Herranz 1983). In Ebro valley, will sing all day, though mostly in morning and at dusk. Also sings in early part of night and before dawn—e.g. in breeding season at 22.30 hrs and at 04.30 hrs (A Aragüés). In Cyrenaica (Libya), late April, incessant song given only up to c. 08.00 hrs; none at other times (Hartert 1923). In Morocco, February–March, birds called frequently during moonlit nights, and again in early morning; in early February, vocal activity much reduced from c. 2 hrs after sunrise (Smith 1962–4, 1965a). Not seen at water in eastern Morocco and probably never drinks (Brosset 1961).

2. Typically rather shy and retiring (Mountfort 1954), more so than other Alaudidae (Hartert 1915); shyness less marked in breeding season (A Aragüés). Can be virtually impossible to observe on foot, but takes no notice of vehicle (Smith 1965a). In Algeria, birds allowed approach to c. 7–8 m and flew only then, landing c. 50 m further on (Le Fur 1975). Reluctant to fly when approached by man, tending rather to make for cover. Often runs with remarkable speed (may use tracks between vegetation: Aragüés and Herranz 1983), rather like rodent. May stop and look back at intruder or bob up in a very slim and erect posture from behind stone or plant (Bannerman 1927; Mountfort 1954, 1958). One bird called on flushing (Bannerman 1927: see 3 in Voice). Normally runs into thickest available cover on landing (Whitaker 1905) and extremely loath to fly a 2nd time (Brock 1970). FLOCK BEHAVIOUR. Nothing recorded. SONG-DISPLAY. Song of ♂ (see 1 in Voice) given in flight, on ground, or from stone or low plant (Zedlitz 1909; A Aragüés). In strong wind, given from ground whilst hidden in vegetation (A Aragüés). Song delivered during ascent of Song-flight in which bird may rise to c. 100–150 m, thus sometimes higher than *A. arvensis* and lost to view. May remain aloft for c. 30 min or even nearly 60 min. Descent sudden, vertical, and very rapid, with closed wings which are opened 2–3 m from ground to come in at angle of c. 50–70° (Hartert 1915, 1923; A Aragüés); see also Stanford (1954). According to Meinertzhagen (1930), Song-flights more frequent than in Hoopoe Lark *Alaemon alaudipes*, but resemble that species; see descriptions under that species (p. 78) from Meinertzhagen (1951a, 1954). However, slow parachute descent never recorded in Ebro valley (A Aragüés). Song-flight may be repeated after short interval (Hartert 1923). Early-morning song often aerial (Erlanger 1899; Whitaker 1905). Bird singing on ground has head tilted up, with bill at c. 45–70° (A Aragüés: Fig A). According to Zedlitz

(1909), may run about on ground, stop and give burst of song, then run on, etc.; perhaps refers to behaviour of disturbed bird. In Ebro valley, birds sing vigorously and more or less continuously in breeding season, more from ground after young leave nest, and overall reduction noticeable during rearing period (A Aragüés). In southern Cyrenaica, incessant song in late April was possible prelude to 2nd brood (Hartert 1923). ANTAGONISTIC BEHAVIOUR. No information. HETEROSEXUAL BEHAVIOUR. In Ebro valley, pair-formation and courtship activity start in early spring: e.g. in 3 years, 20, 23, and 25 March (A Aragüés). Only other information refers to apparent pair seen near half-built nest in Morocco, early February, only 1 bird actually building (Smith 1962–4). RELATIONS WITHIN FAMILY GROUP. No data on early stages. Observations suggest young leave nest early, as in other Alaudidae (A Aragüés). In Tunisia, early May, ♀ recorded accompanying 2 fledged young and apparently still feeding them (Koenig 1893). ANTI-PREDATOR RESPONSES OF YOUNG. Fledged juvenile crouched motionless close to ♀ when sibling shot (Koenig 1893). PARENTAL ANTI-PREDATOR STRATEGIES. One pair slow to return to nest containing complete clutch of 3 fresh eggs after disturbance (Hartert 1915).

(Fig by J P Busby: from photograph in Aragüés and Herranz 1983.)　　　　　　　　　　　　　　　　　　　　MGW

Voice. Completely distinctive (A Aragüés).

CALLS OF ADULTS. (1) Song of ♂. Recordings indicate song not like other Alaudidae but, in timbre and overall quality, closer to twittering of Linnet *Carduelis cannabina* (P J Sellar). Equal to Hoopoe Lark *Alaemon alaudipes* in quality (Meinertzhagen 1930) and more varied (Zedlitz 1909). Nominate *duponti* of Hauts Plateaux (Algeria) and *margaritae* of southern Tunisia use phrases 'tsii dida diii', with final unit rather like call of Bar-tailed Desert Lark *Ammomanes cincturus*, and 'tsii didla didla diii', with 'tsii' high-pitched, fine, and not very far-carrying. In southern Cyrenaica (Libya), *margaritae* gave a clear 'dii-drii', 'dii-dii-drii', or 'dür-drii', often followed by a trilling 'drrrrrr' (Hartert 1915, 1923). Recordings by C Chappuis and J-C Roché in Morocco reveal medley of sounds probably covering all the above renderings, so that further corroboration perhaps required to support claim in Hartert (1923) that marked local variation exists. Songs in 2 recordings remarkably similar in timbre (highly variable in both cases), also in patterns of ascent and descent of pitch intervals, and, to lesser extent, in time patterning (J Hall-Craggs). Evident in recordings are light, rather fluty tremolos or very fast liquid bubbling sounds and an oft-repeated, very distinctive 'whee-ur-wheeee' or longer 'hoo hee-ur hoo-eeee' in which final (sub-)unit has characteristic nasal or twangy timbre and rising pitch; also, slow 'wzeep', 'pizzeep', or 'huwip' not unlike Reed Bunting *Emberiza schoeniclus*, rather ventriloquial 'dwur-dee' somewhat reminiscent of song of Bullfinch *Pyrrhula pyrrhula* and followed by faint popping sound, and occasionally a rasping or buzzing trill (M G Wilson). Fig I shows 2 pure-sounding whistled units— 'whew heee'—followed by 'der-zwee-zweep' and 'vzee-

A

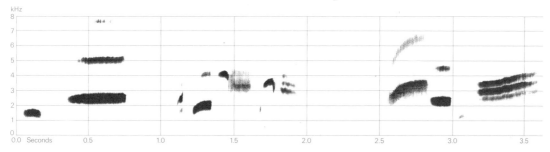

ur-d'wheeee'; last motif also given alone (J Hall-Craggs). Description of song-phrase as a loud fluting 'totiii-totiii totiii' (A Aragüés) thus perhaps an over-simplification or rendering of only certain components. Song before dawn usually begins with a few whistled calls (J-C Roché). Rather nasal song given from ground included call 2b (Smith 1965a), and further descriptions in the literature suggest that most if not all other calls may occur in song. For further descriptions of song, see Stanford (1954) and Jonsson (1982). (2) Contact-calls. (a) Commonest call, given in flight and from ground, a very human, fluting 2-syllable whistle, 2nd syllable rising sharply in pitch and rather creaky and nasal: 'hoo hee', 'coo-chic', or 'pu-chee'; sometimes 'hoo hoo hee' or just 'hoo' (Smith 1962–4, 1965a; Bundy and Morgan 1969). Recording (Fig II) suggests 'HOO te-HEEEE' and similar to 1st

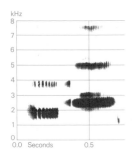

II   Roché (1968)   Morocco
March 1967

motif in Fig I, but 2nd (louder) sub-unit of 2nd note clearly shows independent use of 2 internal tympanic membranes (J Hall-Craggs, M G Wilson). (b) A 'dweeje', 'sjwee-er', 'wheeje', or twangy 'sj-wee' like Greenfinch *Carduelis chloris*, often following disyllabic or monosyllabic variant of call 2a (Smith 1962–4, 1965a). Drawn-out 1st unit in 'twee tee-wit-war' (Whitaker 1905) probably the same, though whole call rather like a song-phrase (see call 1). (3) Alarm-calls. Soft, rather quiet 'tsii' when running (Hartert 1923); shrill 'tsii' (Meinertzhagen 1930) perhaps the same or a related call. Flying bird, probably when disturbed, may give call similar to *Galerida* larks (Smith 1962–4). Whistling call, given by flushed bird (Bannerman 1927) presumably also

belongs here. (4) ♂ coming to nest once gave a churring 'terrrrrrr' (Hartert 1915).

CALLS OF YOUNG. No information.   MGW

Breeding. SEASON. Iberia: one nest of 3 eggs, 13 April (A Aragüés). Northern Algeria and Tunisia (nominate *duponti*): eggs laid 1st week March to early June (Heim de Balsac and Mayaud 1962). Libya (*margaritae*): nest with 3 eggs found March, southern Cyrenaica; in late April, birds must have had young and were presumably about to nest a 2nd time (Hartert 1923). SITE. On ground, set into tussock, hidden under bush, or sheltered by stone. Nest: scrape lined with available vegetation, including rootlets, fibres, small twigs, and hair, sometimes litle or no softer lining; no rampart of stones seen at 2 nests in Algeria (Hartert 1915). External diameter 9·4–9·6 cm, internal diameter 6·9–7·1 cm, height 4·1 cm, depth of cup 3·2 cm, weight 18·8 g (A Aragüés). Building: no information. EGGS. See Plate 79. Sub-elliptical, smooth and glossy; white to pink, densely spotted reddish-brown, with purplish-grey undermarkings. Nominate *duponti*: 23·6 × 17·2 mm (22·4–24·6 × 16·6–18·0), $n=14$; calculated weight 3·59 g. *C. d. margaritae*: 23·4 × 17·2 mm (22·6—24·2 × 16·6–18·0), $n=10$; calculated weight 3·57 g (Schönwetter 1979). Clutch: 3–4 (2–5). Of 75 clutches, Algeria and Tunisia: 3 eggs, 57; 4, 18; mean 2·76 (Heim de Balsac and Mayaud 1962). 2nd brood and/or replacement clutch possible (Heim de Balsac and Mayaud 1962). YOUNG. One young 10-11 days old found outside nest, suggesting precocious departure from nest, as in other Alaudidae (A Aragüés). No further information.

Plumages (nominate *duponti*). ADULT. Feather-centres of forehead, crown, mantle, scapulars, and back have broad full black or blackish-grey streak, bordered at side by some rufous-cinnamon or olive-brown; sides of feathers of central crown and upper mantle paler, pink-cinnamon to off-white, those on central crown joining to form pale median crown-streak; fresh feathers narrowly and evenly fringed cream-pink white, forming pale scaly pattern on upperparts. Rump rufous-brown with cream feather-fringes and black subterminal crescent; upper tail-coverts rather bright rufous-cinnamon. Distinct though narrow supercilium, lores, and ring round eye pink-white or white, faintly mottled grey on lores and just above front of eye; supercilium extends back across hindneck, where narrowly streaked black. Ear-coverts dark brown, faintly streaked rufous-buff. Cheeks, sides of neck, and chin isabelline or white closely streaked black on sides of neck, spotted black in 2 mottled stripes, one from gape to lower ear-coverts (sometimes indistinct) and one along side of chin. Throat, chest, and sides of breast isabelline or cream-white, finely spotted dark grey on throat, marked on chest with bold rounded black spots and on sides of breast with regular black streaks, which are bordered rufous-cinnamon. Flanks and upper breast isabelline with rufous-brown shaft-streaks (latter partly dusky grey towards feather-tips); thighs grey-brown; remainder of underparts isabelline or off-white, sometimes stained grey or rufous by soil. Central pair of tail-feathers (t1) rufous-brown with darker centre and pink-cinnamon or off-white fringe, shafts white; t2 dark brown; t3–t5 black, t5 with sharp but rather narrow cream or white fringe along outer web; t6 white washed pale tawny and with grey basal wedge on inner web. Flight-feathers dark grey-brown, darkest and almost greyish-black towards outer primaries; tips with narrow and ill-defined cinnamon-grey or isabelline fringe, outer webs with more sharply defined tawny or pink-brown fringes; inner webs basally bordered pink-cinnamon; broad sharp fringe along outer web of p9 and along minute p10 cream-white. Tertials black-brown grading to rufous-brown on central tips; broad and even fringes contrastingly cream or white, shafts contrastingly whitish-horn. Lesser upper wing-coverts rufous-brown or olive-brown with broad pink-grey fringes; median coverts rufous-brown with black subterminal crescents and broad pale cinnamon or off-white fringes along tips (together forming narrow wing-bar); greater coverts and bastard wing dark brown with broad rufous-cinnamon fringes along outer webs and tips. Greater upper primary coverts dull black with cinnamon fringes. Under wing-coverts and axillaries grey, lesser and median coverts broadly fringed isabelline. *In worn plumage*, pale scaling formed by feather-fringes of crown, mantle, scapulars, and tertials lost through abrasion, forehead and crown appearing black with limited rufous on feather-bases and with more distinct median crown-streak; lores, supercilium, cheeks, sides of neck, chin, and throat bleached to white; supercilium usually distinct, but white extension across hindneck sometimes disappears due to abrasion; dark spots and stripes on sides of neck, cheeks, and throat sometimes less distinct, partly abraded; rufous borders of black spots on chest and sides of breast bleached to buff, spots forming more conspicuous streaks; belly to under tail-coverts purer white (if not soiled); pale fringes of tips of median upper wing-coverts and primaries largely abraded. NESTLING. No information. JUVENILE. Upperparts rather like fresh adult, but scaly feather-fringes broader, up to 2 mm wide, cinnamon at sides, white at tip; black feather-centres reduced to subterminal mark, merging into grey-brown or rufous-brown bases; pale median crown-streak sometimes indistinct or absent. Spots on cheeks and chest paler than adult, greyish, small, often faint on cheeks. Feathers of belly and vent short and soft, cream-white. Tail-feathers as adult, but even narrower and more sharply pointed, evenly fringed buff or off-white, submarginally marked black. Flight-feathers as adult, but narrowly and evenly fringed pale buff or white, partly bordered black or dull grey subterminally; p10 longer and broader than in adult (see Structure); upper wing-coverts and tertials evenly fringed buff or white, like mantle and scapulars. FIRST ADULT. Like adult; indistinguishable once juvenile p9–p10 replaced.

Bare parts. ADULT. Iris light brown or brown. Upper mandible dark brown or dark greyish-horn, cutting edges of upper mandible and whole lower mandible pale flesh-colour to pale horn-brown. Leg and foot brownish-flesh, flesh-brown, or pale flesh-colour. (Hartert 1910; Berg 1984; BMNH). NESTLING. No information. JUVENILE (*C. d. margaritae*). Iris yellow or brown. Upper mandible straw-brown, lower pale brown or flesh-grey. Leg and foot flesh-grey or almost white with grey tinge; claws almost pink. (Stanford 1954; BMNH.)

Moults. ADULT POST-BREEDING. No information, none in moult examined. 8 birds from May and early June heavily worn, but not yet moulting; 10 from mid- and late October had plumage completely new (BMNH, RMNH, ZMA). POST-JUVENILE. 2 recently fledged juveniles from early May, Cyrenaica (Libya), had just started moult with growth of some scapulars (BMNH). No juvenile feathers remaining on birds examined from mid-October and later.

Measurements. ADULT, FIRST ADULT. Nominate *duponti*. Southern Spain, Balearic Islands (locality probably wrong: Hartert 1910), northern Algeria, central Tunisia, October–June; skins (BMNH, RMNH, ZMA). Bill (S) to skull, bill (N) to distal corner of nostril; exposed culmen on average 3·2 less than bill (S).

WING	♂	103·3 (1·71; 15)	99–106	♀ 91·7 (2·36; 10)	88–95
TAIL		62·5 (3·90; 8)	58–66	54·6 (2·15; 6)	52–58
BILL (S)		23·9 (0·91; 14)	22·5–25·1	21·6 (1·29; 9)	19·3–23·2
BILL (N)		16·2 (1·16; 14)	14·3–17·8	14·5 (0·87; 9)	13·4–15·8
TARSUS		24·3 (0·65; 8)	23·1–25·2	23·0 (0·72; 6)	22·2–23·6

Sex differences significant.

Unsexed live birds, Spain: wing 99·4 (3·79; 34) 88–104, tail 65·8 (3·30; 29) 56–70, bill (S) 23·3 (1·24; 30) 20–25; of these, 24 birds with wing 99–104·5 probably ♂, 6 with 88–93·5 probably ♀, 4 with 97–99 uncertain; none with wings 94–96 (A Aragüés).

C. d. margaritae. Southern Tunisia, Cyrenaica (Libya), and north-west Egypt, December–May; skins (BMNH, RMNH, ZMA).

WING	♂	102·2 (2·18; 29)	96–107	♀ 91·4 (1·27; 8)	89–93
TAIL		56·9 (1·06; 8)	55–58	50·2 (— ; 3)	49–52
BILL (S)		26·6 (1·08; 29)	24·5–29·2	23·7 (1·18; 7)	22·1–25·5
BILL (N)		18·6 (0·83; 19)	16·9–20·0	15·6 (1·32; 5)	14·2–17·2
TARSUS		25·3 (0·90; 18)	23·9–26·9	23·6 (0·68; 4)	22·8–24·4

Sex differences significant. Bill significantly longer than nominate *duponti*, tail significantly shorter.

Weights. Unsexed birds, Spain (see Measurements for probable sex composition): 39·4 (3·22; 33) 32–47 (A Aragüés). See also Aragüés and Herranz (1983).

Structure. Wing rather short, broad, tip rounded. 10 primaries: p7 and p8 longest, p6 and p9 each 0–3·5 shorter (mainly 0·5–1·5), p10 59–73 shorter in ♂, 54–58 in ♀, p5 3–7 shorter, p4 13–17 (♂) or 8–12 (♀), p3 18–23 (♂) or 13–17 (♀), p1 22–28 (♂) or 17–22 (♀). Adult p10 reduced, rather narrow, tapering to pointed tip, 7·3 (10) 4–11 shorter than longest upper primary covert; juvenile p10 with broad and rounded tip, about equal in length to longest covert. Outer web of p5–p8 and inner of p6–p9 emarginated. Longest tertials reach to about tip of p4 in closed wing. Tail short, tip square; 12 feathers, narrow and rather pointed towards tips. Bill long, about equal to head length, slender; basal half straight, distal half distinctly decurved, ending in sharp point. Nostril covered by tuft of short feathers. Many short soft bristle-like feather-tips in tufts covering nostrils, on lores, along base of upper mandible, and on chin. Feathers of hindcrown slightly elongated, but not narrow, forming short and broad crest as in Woodlark *Lullula arborea*. Tarsus and toes rather short and thick, not as slender as in (e.g.) Skylark *Alauda arvensis*; claws short and broad, but hind claw long and almost straight. Middle toe with claw 17·0 (10) 16–19 mm; outer toe with claw *c.* 74% of middle with claw, inner *c.* 78%, hind *c.* 102%. Length of hind claw 10·2 (52) 8–12·5 in nominate *duponti*, exceptionally 15 (BMNH; A Aragüés); 7·6 (10) 6–9 in *margaritae*.

Geographical variation. Marked. *C. d. margaritae* from southern Atlas mountains to north-west Egypt very distinctly rufous, not black and rufous like nominate *duponti* from Iberia; tail and hind claw distinctly shorter, bill distinctly longer (see Measurements). Differences in colour in part bridged by populations from Morocco to north-west Tunisia, which are locally rather variable. In typical *margaritae*, centres of all feathers of upperparts (including upper wing-coverts, tertials, and t1) deep rufous-cinnamon, fringes pale pink-cinnamon, cream, or off-white; throat to chest pale pink-buff to off-white, rather narrowly streaked rufous, not boldly streaked black; plumage rather similar to sandy-rufous races *superflua* and *carolinae* of Thekla Lark *Galerida theklae* and even more so to Dunn's Lark *Eremalauda dunni* (these species differ markedly in structure, however). Some *margaritae* have centres of feathers of crown, mantle, chest, and sometimes scapulars partly tinged with olive-brown (occasionally even black-brown: Stanford 1954), tending somewhat towards nominate *duponti*, though appearance still mainly rufous. Most birds from northern Algeria less deep black on upperparts than typical nominate *duponti*; feather-centres of upperparts and chest narrower, olive-black or olive-brown; sides of feathers of upperparts paler pink-cinnamon. Even as far east as central Tunisia, however, some breeders are as dark as Spanish birds, and hence birds of northern Algeria included in nominate *duponti* rather than being considered intermediate.   CSR

Rhamphocoris clotbey Thick-billed Lark

PLATES 2 and 12
[facing page 160 and between pages 184 and 185]

DU. Diksnavelleeuwerik   FR. Alouette de Clotbey   GE. Knackerlerche
RU. Толстоклювый жаворонок   SP. Alondra de Pico Grueso   SW. Tjocknäbbad lärka

Melanocorypha clot-bey Bonaparte, 1850

Monotypic

Field characters. 17 cm; wing-span 36–40 cm. Close in size to Crested Lark *Galerida cristata* but with much heavier bill and head and 20% longer wings; at least 10% shorter than Calandra Lark *Melanocorypha calandra* but with much deeper bill. Medium-sized to large lark with remarkably deep, heavy bill and broad, head, upright stance and long legs, and relatively long-winged and short-tailed flight silhouette. Adult has uniform pink grey-brown upperparts, black-splashed face, bold black spotting from upper chest to vent, and broad white trailing edge to wings. Appears strikingly dark head-on, with pale bill, white throat, white eye-ring, and white vent obvious in contrast. Sexes dissimilar; no seasonal variation. Juvenile separable.

ADULT MALE. Crown, nape, hindneck, mantle, and rump pink-isabelline with grey tones, becoming pale, almost buff-white in upper tail-coverts. Front and sides of head strongly marked, with blackish mottles above bill, black line on upper lores, white fore-supercilium joining prominent white eye-ring, and white line running towards bill, black cheeks and lateral throat-patches punctuated by white spot at base of lower mandible, and off-white patch dividing rear cheeks; makes bird appear both dark- and deep-headed, particularly head on. Lateral throat-patches join in band under white chin and upper throat; black marks extend downwards in form of heavy black spots and splashes to cover chest, belly, and flanks, coalescing and appearing dense, even uniform, in centre of body and remaining most discrete on buff-washed flanks. Lower part of body pattern contrasts markedly with white rear belly and ventral area. Wings rather more rufous on coverts than back and boldly patterned with (1) dark brown rufous-fringed tertials, (2) apparently bleached tips to greater coverts, (3) black secondaries with white tips, and (4) black primaries with noticeable white terminal margins. In flight, upperwing strongly patterned with dull pink coverts and pale central wing-bar formed by pale tips, black flight-feathers, and broad white trailing edge to secondaries and inner primaries; underwing black with broad white trailing edge. Uppertail isabelline, with more rufous centre, almost black sub-terminal marks on all but central feathers (forming dull subterminal band), and dull white outer feathers; below, basically dull white but with almost black subterminal band enhanced by sharper contrast. Bill finch-like in form; though not as massive as that of Hawfinch *Coccothraustes coccothraustes*, much deeper and broader than those of *Melanocorypha* larks; bright yellow-horn, with dark grey tip. Legs long, with tibia fully exposed in most attitudes; brown-grey. ADULT FEMALE. Usually smaller than ♂, with black areas less intense and underbody less heavily marked. JUVENILE. Wing pattern as adult but otherwise much more uniform, with only faint face and underbody markings. May suggest large Bar-tailed Desert Lark *Ammomanes cincturus*.

Adult unmistakable; given clear sight of bill, juvenile also. Confusion with *Melanocorypha* larks possible at distance, particularly when only dark underwing catches eye, but unlikely to persist when bill and plumage details show. Flight powerful, with stronger action than in Skylark *Alauda arvensis* and recalling *Melanocorypha* larks, as does flight silhouette with plumage pattern increasing long and pointed appearance of wings. Escape-flight short and low, ending in abrupt stop, bird often then becoming invisible, as plumage pattern markedly cryptic against most backgrounds. Song-flight apparently involves slow descent, but poorly described so far. Walks and hops when feeding, but will run from danger. Stance noticeably upright, bird reaching up to tug at plants and standing noticeably erect in early stages of alarm. Often in small flocks.

Song quiet and jingling. Calls include a low 'coo-ee' and 'sree'; 'oo-ep' in flight. Some birds remarkably silent, even in parties.

Habitat. In Mediterranean and arid subtropical zones bordering or overlapping edges of desert, especially on open stony hammada where some sparse vegetation may be present. Accordingly ranked as a typical desert lark, but also found on wadi beds with grass and other plants or on stony-clay terrain with relatively rich vegetation, including succulents such as *Aizoon*, on clay beds with hillocks, and on enclosed areas between villages (Valverde 1957). Inhabits deserts with solid stony soils, preferring rocky plateaux or rocky slopes and slightly more vegetated terrain. Avoids sand-dunes (Heim de Balsac and Mayaud

1962). At Azraq (Jordan), conspicuous on black rocks of basalt lava desert, but also found sharing scrubby sandy and stony wadi bed with Temminck's Horned Lark *Eremophila bilopha*, Hoopoe Lark *Alaemon alaudipes*, Lesser Short-toed Lark *Calandrella rufescens*, and Crested Lark *Galerida cristata*. Largely a ground bird, running and hopping; flies in lower airspace (Nelson 1973). Tends to occur more frequently on edge of wadis, and to make trips to water sources (Hemsley and George 1966). In Arabia, inhabits rocky slopes and grassy wadis but mainly a bird of absolute desert (Meinertzhagen 1954).

Distribution. Range uncertain, especially in east.

EGYPT. Possibly rare breeder in north, but no proof (SMG, PLM, WCM).

SYRIA, JORDAN. Bred Syrian desert 1930–1 but exact site not known (Kumerloeve 1969a).

Population. No information on numbers or trends.

Movements. Basically resident, though subject to no-madic dispersal outside breeding season. Few precise data, but indication that some birds move towards coastal plains for winter.

Erratic movements occur in north-west Africa in winter, including one record of thousands south of Col de Jerada, north-east Morocco, in late February; such movements irregular and in very variable numbers, and perhaps in response to desert conditions (Heim de Balsac and Mayaud 1962). Nomadic tendency in Morocco reflected in apparent absence from Sahara in winter (November onwards) when it occurs instead, sometimes in large flocks, on plains of western and northern Morocco and on Hauts Plateaux; some birds back south of Atlas by February (Smith 1965a). Comparable irregular occurrences elsewhere in North Africa east to Suez (Etchécopar and Hüe 1967).

Probably local resident in northern desert of Saudi Arabia, and occurs as uncommon winter visitor south to a line Medina–Riyadh–Harad (Jennings 1981a); in Eastern Province, regular only at Harad, October–March (G Bundy). Small numbers occur in Kuwait January–April (Bundy and Warr 1980). Not recorded southern or south-east Arabia. RH

Food. Seeds, green plant material, and insects. Sometimes hops when feeding (Meinertzhagen 1954). Reaches up to plant heads to pull them down to ground (Ferguson-Lees 1970a), and uses bill as clippers to cut green shoots (Heim de Balsac 1924). Will dig for food (Ferguson-Lees 1970a). Hard seeds apparently swallowed whole, not crushed, and weak bite given when in the hand suggests bill muscles not as strong as large bill would suggest. In central Morocco, birds seen gathering to feed on abundant caterpillars of painted lady *Vanessa cardui*, cramming them into their mouths. (Meinertzhagen 1940.) When

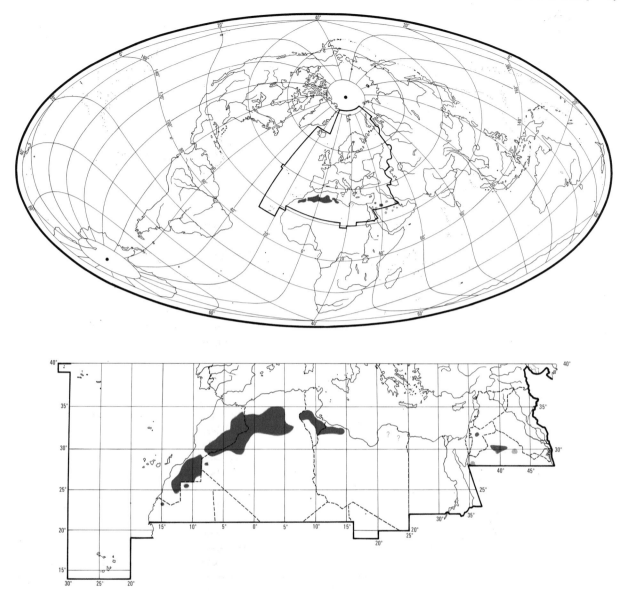

seeds eaten, considerable grit taken to aid digestion, but when insects eaten stomachs contain no grit (Meinertzhagen 1951a).

In central Sahara and southern Algeria, stomachs and observation showed diet to be green plant material (mostly), insects, and seeds (Heim de Balsac 1924, 1926). Other stomachs from Algeria, October, contained seeds; in one, a small lizard, in another some green plant material (Rothschild and Hartert 1923). 2 birds from Tunisia contained minute seeds; one also a locust (Orthoptera) and the other a good-sized beetle (Coleoptera) (Whitaker 1905). 2 others from Tunisia, September, and 3 from southern Morocco, June, contained seeds and insects including ants (Formicidae) (Bédé 1918; Valverde 1957). At Azraq (Jordan) in April, food included small green

fruits of spurge *Euphorbia kahirensis* and material from plantain *Plantago oveta* (Mountfort 1965; Wallace 1983a).

Records of food brought to young, in and out of nest, north-west Africa, are all of large and small locusts and other Orthoptera (Wallis and Pearson 1912a; Rothschild and Hartert 1914; Heim de Balsac (1924a).   DJB

Social pattern and behaviour. Little studied.

1. Dispersion outside breeding season appears to vary, perhaps largely with food availability. Usually occurs in small scattered parties, not in bigger assemblages (Whitaker 1905; Meinertzhagen 1951a; Nelson 1973). In Morocco, singly or in pairs in autumn (Meinertzhagen 1951a); in winter, however, regularly in parties of 6-12 and often in large flocks (Smith 1962-4, 1965a). Once, in mid-February, Morocco, numerous flocks of 4 to *c.* 50 birds each, totalling tens of thousands of

birds over 2000 km² (Brosset 1961). Indications are, therefore, that even in optimal conditions, birds are typically dispersed in relatively small groups. In Morocco, once associated with Corn Buntings *Miliaria calandra* and Goldfinches *Carduelis carduelis* (Meinertzhagen 1940). BONDS. No firm information on mating system but birds said to retain pair-bond when in small parties outside breeding season (Meinertzhagen 1940). ♀ incubates (Aharoni 1931), but not known if ♂ also participates. Both parents feed young (Wallis and Pearson 1912a; Heim de Balsac 1924). Family bonds maintained after fledging (Jourdain *et al.* 1915) but duration not known. BREEDING DISPERSION. Solitary and territorial. In southern Algeria and central Sahara, breeding pairs scattered (Heim de Balsac 1926). In hammada terrain, *c.* 1 pair per 10 km (Heim de Balsac 1924). No further information. ROOSTING. Nocturnal and terrestrial, as in other Alaudidae, but no details. Birds seek shade at hottest time of day; incubating birds protected by orientating nests towards rising sun, thus shaded by midday (Heim de Balsac 1924). Drinks regularly when water freely available, e.g. birds came daily to one watering-place in Morocco (Brosset 1961), likewise at Tindorf (Sénégal) where birds noticeably drank only one drop of water at a time (Heim de Balsac and Heim de Balsac 1954). Evidently capable, however, of not drinking for lengthy period (e.g. Meinertzhagen 1951a), and water derived from insect food probably important (Valverde 1957).

2. Usually readily approachable, but more wary when intruder near nest (Whitaker 1905; Heim de Balsac 1924; see Parental Anti-predator Strategies, below). Apparently prefers to run rather than fly from danger, and does not fly far before landing again (Whitaker 1905). When alarmed often runs and mounts a stone or other vantage point, adopting a distinctively erect stance, with fore-body high, head raised or drawn back, and tail bobbing (Heim de Balsac 1924; Meinertzhagen 1940, 1951a, 1954). When approached, one bird ran from bush to bush, keeping watchful eye on intruder (Bédé 1918). Said never to squat when alarmed (Meinertzhagen 1951a), but one bird, which had evidently seen intruder from a distance, flattened itself on ground, remaining motionless, with head turned towards intruder, and flew only when approached within 1m (Whitaker 1905). In possible escape-flight, one bird flew away very fast, swerving and twisting in jerky flight, giving 'co-ep' call (Smith 1962-4: see 2 in Voice). For Alarm-call, perhaps confined to encroachment on nest, see Parental Anti-predator Strategies (below) and 3 in Voice. Probably this the call given by wounded bird fluttering along ground trying to evade human captor (Bates 1936-7). When handled, wounded bird bit captor's hand (Koenig 1895). FLOCK BEHAVIOUR. For calls given in flocks see especially 4 in Voice. No further information. SONG-DISPLAY. Song-flight of ♂ seldom witnessed and no detailed description available. ♂ rises from ground, reaches 'a certain' height, and sings (see 1 in Voice) before descending in gliding flight, as in Calandra Lark *Melanocorypha calandra* (Heim de Balsac 1924). ♂ flies high in the air, calls (see 1 in Voice) and throws itself right and left in zigzag flight (Rothschild and Hartert 1914); according to Heim de Balsac (1924) this quite aptly describes gliding descent of Song-flight. However, said by J-C Roché to descend 'parachute' fashion, like pipit *Anthus*. Singing on ground, as bird walks or runs, evidently not uncommon (see 1 in Voice). ANTAGONISTIC BEHAVIOUR. Apart from Song-display, nothing known. HETEROSEXUAL BEHAVIOUR. In mid-March, Morocco, where several pairs occurred, ♂♂ chased ♀♀ around (Smith 1962-4); this presumably heterosexual pursuit typical of other Alaudidae. No further information. RELATIONS WITHIN FAMILY GROUP. Little known. Young presumably leave nest before fledging, as in other

Alaudidae; this implied by 4 half-grown young found dispersed at varying distances from empty nest. After fledging, both parents feed young for unknown period. (Wallis and Pearson 1912a; Jourdain *et al.* 1915.) ANTI-PREDATOR RESPONSES OF YOUNG. No information. PARENTAL ANTI-PREDATOR STRATEGIES. (1) Passive measures. ♀ sits tightly on hard-set eggs, but flushes more readily at other times, running from nest (Rothschild and Hartert 1914), and then markedly wary (see below). (2) Active measures. When human intruder near nest, running of agitated adults has element of distraction-lure display: adults carrying food refused to go near nest for hours on end, and tried to lead intruders away from nest, uttering Alarm-calls with increasing intensity; calling attracted conspecific birds in the vicinity, up to *c.* 15 gathering around intruders near nest (Heim de Balsac 1924). Evidently similar response, without however attracting other birds, described by Rothschild and Hartert (1914) when intruder near nest or young. ♂ perhaps near nest, Algeria, called continuously and strutted nervously about, both ♂ and ♀ finally flying off *c.* 300 m or more (Lynes 1930). In Syria, bird flew low over ground towards nest in 'lame flight' and, when *c.* 2m from nest, rose straight up in the air and dived down to sit quietly by nest (Aharoni 1931).   EKD

Voice. Rather silent during breeding season (Rothschild and Hartert 1911), but repertoire relatively wide (Smith 1965a). Though several calls described, context and function not fully known, and following scheme therefore tentative.

CALLS OF ADULTS. (1) Song of ♂. During Song-flight, ♂ gives sounds described as not especially harmonious, and reminiscent of the distorted notes of Calandra Lark *Melanocorypha calandra* (Heim de Balsac 1924). In recording (Fig I), birds in Song-flight give 'a frenzied little jingle of sweet notes' (P J Sellar); also described as tinkling and warbling, sweet and rather quiet (J Hall-Craggs). In descending zigzag phase of Song-flight (see Social Pattern and Behaviour), ♂ gives a repeated 'djup djup...' units much shorter than in call 3 (Rothschild and Hartert 1914); this possibly call 2. Several descriptions of singing on ground: in recording by C Chappuis, song comprises varied phrases of liquid notes, with quality of Corn Bunting *Miliaria calandra* (P J Sellar), each phrase separated by several seconds. Quiet little song, with some notes like song of Linnet *Carduelis cannabina*, given by bird as it walked around (P A D Hollom); conversational 'woot-e-toot' (P A D Hollom) perhaps also an isolated phrase of song. In recording (Fig II), apparently of bird on ground, 'tk tl-eeeeeee-t-p-rrit' (J Hall-Craggs), interposed twice between series of call 2 and series of call 3; this perhaps similar to preceding. Short warbling notes given by ♂ running along ground (Rothschild and Hartert 1911, 1914), possibly also song. (2) Contact-call. Commonly associated with flight, and rendered variously: a sharp 'prit', not unlike *Miliaria calandra* (Bundy and Morgan 1969); a musical chipping 'wick wick' (Meinertzhagen 1940); 'co-ep' (Smith 1965a); a faint 'kooip' or 'quip', somewhat like Quail *Coturnix coturnix* (P A D Hollom); same liquid quality evident in recording (Fig III) of perched bird, rendered an abrupt

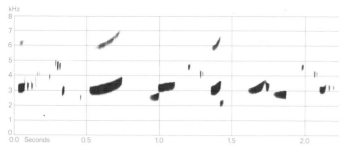

I Roché (1968) Morocco March 1966

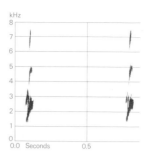

III Roché (1968) Morocco March 1966

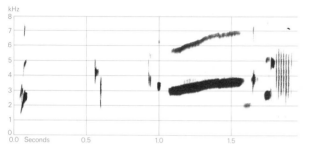

II Roché (1968) Morocco March 1966

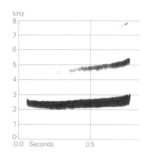

IV Roché (1968) Morocco March 1966

rising 'wuk' (P J Sellar), and thought by recordist to express anxiety as intruder approached nest. (3) Alarm-call. A long-drawn plaintive whistling 'tsee' or 'tsi tsi', not unlike Bar-tailed Desert Lark *Ammomanes cincturus* but stronger and louder; given repeatedly and with varying intensity by birds running on the ground when nest threatened (Rothschild and Hartert 1911, 1914). All other renderings express flute-like quality: e.g. a fluting 'pee' (Lynes 1930); a slight musical whistle, like opening syllable of call of Curlew *Numenius arquata* (Meinertzhagen 1940). In recording (Fig IV), a regularly repeated long low melancholy whistle: 'tweee', mostly clear with slightly rising pitch, but also sometimes nasal and then slightly descending (E K Dunn). This call probably the 'drawn-out little whistling note', ascending in pitch, described by Bundy and Morgan (1969); apparently also the call, similar to call of Hoopoe Lark *Alaemon alaudipes*, given by wounded bird fluttering along ground when intruder nearby (Bates 1936–7). 2 calls reported, with no details of context, by Smith (1962–4) also clearly the same or related: a low far-carrying 'coo-ee', softer than call 2—rather like plover (Charadriidae), and very similar to *A. cincturus*; also 'wheet wheet wheet', with variations. (4) Other calls. Repertoire described by Smith (1962–4, 1965a) includes: a hard indescribable 'splitting' noise, given on ground; quiet 'sree' on ground and in flight. Perhaps the same as 'shrreeep' given on landing, Jordan (P A D Hollom), and similar sounds occur in recording by J-C Roché. Function not known.

CALLS OF YOUNG. No information. EKD

Breeding. SEASON. North-west Africa: late March to late May, exceptionally February or even January (Heim de Balsac and Mayaud 1962). Jordan: nest found late April (Wallace 1983a). SITE. On ground, usually under bush or stone; less often in the open. Nest: shallow depression filled with lining of vegetation, often with rim of stones on open side. Building: no information. EGGS. See Plate 79. Sub-elliptical, smooth and glossy; creamy-white to pink, finely speckled with red-brown or chestnut, with underlying greyish mottling; when fresh, characteristic pink or bright pink tint. 25·5 × 18·3 mm (23·6–29·0 × 16·9–19·0), $n = 40$; calculated weight 4·36 g (Schönwetter 1979). Clutch: 3–5 (2–6). Of 53 clutches, north-west Africa: 2 eggs, 3; 3, 27; 4, 17; 5, 5; 6, 1; mean 3·51 (Heim de Balsac and Mayaud 1962). No further information.

Plumages. ADULT MALE. Forehead, crown, hindneck, mantle, scapulars, back to upper tail-coverts, and lesser upper wing-coverts pink-cinnamon with pale vinaceous-grey tinge on tips of feathers and narrow dark grey shaft-streaks on crown, close in appearance to upperparts of Desert Lark *Ammomanes deserti*. Small white spot on side of forehead near base of feather-tuft over nostril. Narrow black streak from base of upper mandible (below white forehead-spot) extending over eye to sides of crown, bordered below by white lores and irregular white ring around eye (widest below eye). Narrow black line from corner of mouth over upper cheeks widens into black-and-buff streaked ear-coverts and extends down into broad black band over lower cheeks to upper throat (black meeting or nearly meeting in centre), with another black band (partly mottled cream or white when plumage fresh) down over rear of lower cheeks from ear-coverts to throat, isolating white spot on centre of lower cheeks, white spot near base of lower mandible, and white chin. Sides of neck pale grey with buff tinge and black spots.

Lower throat, chest, breast, and belly pale cream or white, sides of breast and upper flanks pinkish-grey, each feather with black subterminal blob with narrow white terminal fringe; blobs rather small on lower throat and on sides of chest, breast, and belly, appearing pale with some dark spots; blobs large on middle of chest, breast, and belly, merging when narrow pale feather-fringes become worn. Lower flanks pink-cinnamon; vent, thighs, and under tail-coverts cream or white. Central pair of tail-feathers (t1) pink-cinnamon with ill-defined buffish-grey or olive-brown terminal half; t2 rufous-cinnamon with broad black tip (2-2·5 cm on inner web, somewhat less on outer web), tip broadly fringed cinnamon and cream-white; t3-t4 gradually paler cinnamon at base and with less black at tip; t5-t6 largely white including narrow terminal fringe, black tip of inner web of t5 c. 1·5-2 cm long, extending to outer web, that of t6 1-1·5 cm long and not reaching outer web. Outer primaries (p5-p10) dark grey-brown (darkest on bases and middle portions of inner webs) with broad cream or buff fringe along terminal inner border and tip of p5-p7 and along outer web of p9, narrower cream border along tips of p8-p10 and along outer webs of p6-p8. P1-p4 and secondaries brownish-black with broad white tips, c. 1·5-2 cm wide on secondaries, narrowing to c. 0·5 cm on outer web of p4, often slightly tinged pink-buff when plumage fresh. Median and greater upper wing-coverts and tertials rufous-cinnamon with olive-black or olive-brown centres (sometimes largely hidden) and broad paler pink-cinnamon fringes along tips, latter soon bleaching to sandy or pinkish-isabelline; tertial coverts mainly rufous-cinnamon without black or brown. Greater upper primary coverts and feathers of bastard wing olive-black with rather broad pink-cinnamon fringes along distal halves. Under wing-coverts and axillaries black, small coverts along leading edge of wing broadly fringed cream-white. *In worn plumage*, upperparts deeper rufous-cinnamon without grey cast; black on sides of head, throat, and on mid-chest to mid-belly less mottled white, head-markings more prominent, white patch below ear-coverts sometimes almost completely lost through abrasion, sides of breast and belly more heavily mottled black; pale fringes to tips of tail and outer primaries abraded; secondaries more contrastingly black and white. Some individual variation in extent of black face marks and black of underparts, perhaps related to age; some birds (perhaps 1st adults) have black bands on sides of head narrower and more heavily mottled than others, black band across throat widely interrupted in middle (even in worn plumage), and black feather-centres on chest, sides of breast, and sides of belly rather narrow and pointed in shape, not broad and rounded as in full adults. ADULT FEMALE. Bands on sides of head duller black and more heavily mottled white than in adult ♂, not or hardly extending towards rear of ear-coverts, cheeks, and to throat; feather-centres on chest, breast, and belly dark grey rather than black, narrow and pointed, not tending to join on mid-breast and mid-belly, grey centres on mid-breast 2-3 mm wide (c. 5 mm in adult ♂); ground-colour of chest pink-buff rather than whitish-pink. Under wing-coverts and axillaries black-brown or grey-black rather than deep black. In heavily worn plumage, black on throat and mid-breast to mid-belly may form coalescent patches, and most heavily marked ♀♀ (supposedly adult) inseparable from least-marked ♂♂ (supposedly 1st adults), but see Measurements. NESTLING. No information. JUVENILE. Upperparts, including t1, upper wing-coverts, primary coverts, and bastard wing uniform cinnamon-pink, except for more or less grey centres of longer tertials, greater coverts, greater primary coverts, and bastard wing (unlike adult, no dark shaft-streaks on crown, no vinaceous-grey tinge on mantle and scapulars, and no blackish

centres on tertials, median and greater coverts, greater primary coverts, and bastard wing). Sides of head and neck uniform cinnamon-pink, except for faintly paler eye-ring, pink-white spot below eye, dusky grey mottling on lores, and faint grey spots on cheeks. Chest, sides of breast, and flanks cinnamon-pink, faintly spotted grey on sides of breast, remainder of underparts pink-cream, whitest on vent (unlike adult, no distinct black marks on sides of head and underparts). Tail as adult, but dark tail-tips grey rather than black, smaller, and less distinctly defined. Flight-feathers as adult, but inner webs and tips of primaries very broadly fringed cinnamon-pink (least so on outermost); outer webs of p5-p8 rather broadly fringed cinnamon-pink, all outer webs of p9-p10 cinnamon-pink. Under wing-coverts and axillaries dark grey with broad cinnamon-pink fringes. FIRST ADULT. Indistinguishable from adult once juvenile p9-p10 replaced (but see Adult Male).

Bare parts. Iris brown. Bill yellowish-horn with blackish tip in ♂, purer yellow in ♀ (Hartert 1910); dull pink (Goodman and Watson 1983); when nesting, sometimes milky-white (Hartert and Steinbacher 1932-8). Leg and foot dirty bluish-grey (Hartert 1910) or pale straw (Goodman and Watson 1983). NESTLING. No information. JUVENILE. Like adult, but bill dull pink with slightly greyish-horn culmen and cutting edges. (BMNH, ZMA).

Moults. ADULT POST-BREEDING. Complete, primaries descendant. Plumage fresh or rather fresh in specimens from October-February, slightly worn March-April, heavily worn May-June; one adult ♀ from early June, central Algeria, not yet moulting; adult ♂ from 22 June, western Algeria, had many new feathers growing and primary moult score 9. Adults or 1st adults from late August, central Algeria, had completely new plumage with scattered feathers of body still growing. POST-JUVENILE. Complete, primaries descendant. Birds in full fresh juvenile plumage examined from late March to late June, Algeria and Tunisia; one from 11 July, western Algeria, in worn plumage, p1 shed (primary moult score 1), new t1 growing; one from 28 August, central Algeria, in worn juvenile with many 1st adult feathers growing on forehead, mantle, chest, and lesser coverts, and primary moult score 6.

Measurements. ADULT. South-east Morocco, Algeria, and Tunisia, all year; skins (BMNH, RMNH, ZMA). Bill (S) to skull, bill (N) to distal corner of nostril; exposed culmen on average 2·5 less than bill (S).

WING	♂ 129·6 (2·15; 26)	125-134	♀ 122·3 (1·78; 12)	119-125
TAIL	59·4 (2·06; 10)	57-64	56·0 (1·66; 12)	53-58
BILL (S)	20·0 (1·25; 10)	17·8-22·0	19·3 (0·86; 15)	18·2-20·9
BILL (N)	14·2 (0·48; 10)	13·6-15·2	13·7 (0·69; 13)	12·2-14·6
TARSUS	23·3 (0·79; 10)	22·0-24·8	22·9 (0·87; 15)	21·4-24·3

Sex differences significant for wing and tail.

Only a few from Egypt and Middle East examined; these similar to Maghreb sample above, e.g. wing ♂ 129·7 (3) 128-131.
JUVENILE. Wing and tail on average both c. 3 shorter than adult.

Weights. Algerian Sahara, late November and early December: ♂♂ 52, 55; ♀♀ 45, 45 (ZFMK).

Structure. Wing rather long, broad at base, tip rather pointed. 10 primaries: p8 longest, p9 1-3 shorter, p10 79-85 (♂) or 62-75 (♀), p7 1-3, p6 8-13, p5 18-27, p4 26-38, p1 48-53 (♂) or

42-47 (♀). P10 reduced in adult, pointed and narrow, 9 (8) 4-12 shorter than longest upper primary covert; longer and with rounded tip in juvenile, 1 shorter to 2 longer than primary coverts, on average equal to longest covert ($n=4$). Outer webs of p6-p8 and inner of p7-p9 emarginated. Tail rather short, shallowly forked; 12 feathers, t1 4-8 shorter than t5-t6. Bill rather short, very massive, *c.* 13 mm deep at base, terminal half rather compressed laterally; culmen strongly decurved, cutting edges slightly decurved with marked notch in middle of upper mandible (absent in juvenile); basal half of lower mandible straight, gonys curved upwards to blunt tip, middle of cutting edge with blunt tooth fitting in notch of upper mandible. Some individual variation in bill shape: basal half of bill sometimes straight and only terminal half curved to tip, bill appearing very massive, in others culmen decurved from base onwards and shape more semicircular. Nostrils small and rounded, covered by small tuft of feathers. Some small bristles project obliquely down from lateral base of upper mandible. Tarsus and toes short, rather slender; middle toe with claw 17·4 (10) 16·0-18·6, inner and outer toe with claw both *c.* 75% of middle with claw, hind *c.* 96%. Claws rather short and blunt, slightly decurved; hind claw 7·8 (9) 6·7-8·5.   CSR

Melanocorypha calandra Calandra Lark

PLATES 5 and 12
[between pages 160 and 161, and 184 and 185]

Du. Kalanderleeuwerik   Fr. Alouette calandra   Ge. Kalanderlerche
Ru. Степной жаворонок   Sp. Calandria común   Sw. Kalanderlärka

Alauda Calandra Linnaeus, 1766

Polytypic. Nominate *calandra* (Linnaeus, 1766), southern Europe and North Africa, east to Ural steppes, Transcaucasia, eastern Turkey, north-west Iran, and southern shore of Caspian sea; *hebraica* Meinertzhagen, 1920, Levant from Gaza strip, Israel, and western Jordan north to western Syria and neighbouring parts of southern Turkey; *psammochroa* Hartert, 1904, northern Iraq, Iran (except north-west), northern Afghanistan, and southern USSR from Turkmeniya to eastern Kazakhstan.

Field characters. 18-19 cm; wing-span 34-42cm. About 10% longer and 25% bulkier than Skylark *Alauda arvensis*, with proportionately broader wings, shorter tail, and much heavier bill; about 10% larger than Bimaculated Lark *M. bimaculata*. Large, robust, heavy-billed lark, with large wing area but no crest. Upperparts essentially brown and well streaked; underparts off-white, little streaked but with large black patches on sides of upper breast. Face dominated by heavy, rather conical bill and quite prominent supercilium and eye-ring. White trailing edge to wing; underwing dark; tail has white outer feathers. Flight lacks hesitancy of *A. arvensis*. Sexes similar; no seasonal variation. Juvenile separable. At least 2 races in west Palearctic but differences indistinct; only nominate *calandra* described here.

ADULT. Crown, rear face, hindneck, back, and wings buff- to grey-brown with black-brown feather-centres forming streaks on head and back, line of spots across median coverts, and panels on greater coverts and inner secondaries. None of latter patterns particularly striking and markings most obvious on face, where buff-white rear supercilium and narrow eye-ring combine to create 'spectacle', and on sides of upper breast, where black patches show as either almost continuous crescent or series of up to 3 joined blotches and extend downwards into diffuse dark grey-brown streaks on sides of breast. Some ♀♀ have less obvious breast patches. Rest of underparts cream to ochre-white, strongest-toned along flanks, and virtually unmarked though faint moustachial and malar stripes may show. Tail dark brown with all feathers fringed buff and outermost white. Folded wing-point obvious, at least half length of tertials. In flight, all characters so far described become difficult to see except for obvious white sides and corners to tail, but wings show bold white trailing edge formed by tips to secondaries and inner primaries. From below, dark underwing striking, appearing almost black in shadow; darkness increases apparent size of wing area. When worn, bird becomes paler, with more grey or olive tones to plumage and slightly stronger markings on face and wing. Bill pale buff-horn, with dark culmen and tip and red tinge at base. Legs pale red-brown. JUVENILE. Plumage pattern essentially as adult but feather-fringes and feather centres contrast more, making bird appear more speckled above and on chest. Supercilium, throat, and neck-sides noticeably paler than adult; breast patches less solid.

Distinctive character, with weight of bill exceeding all other larks except Thick-billed Lark *Rhamphocoris clotbey*, and body bulk and leg length always obvious. Of the 4 *Melanocorypha* in west Palearctic, *M. calandra* the most widespread but subject to confusion with smaller, yet closely similar *M. bimaculata*, particularly in Near East where they may intermingle on passage. Hence neither *M. calandra* nor *M. bimaculata* instantly identifiable; see *M. bimaculata* for detailed discussion. Flight free, with long and broad-based wings affording easy, even powerful flight action which lacks hesitancy of *A. arvensis* but has typical lark undulations and sweeps. Escape-flight usually low and fast, in half circle, ending in plummet to low cover. Song-flight variable; can recall that of *A. arvensis*

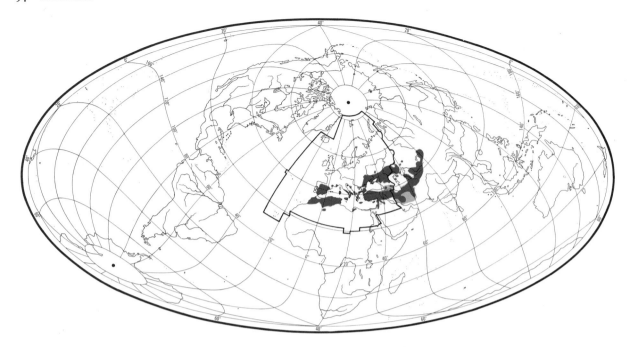

with bird ascending to great height and hovering there, but can also be low and circular, with more flapping action. Gait powerful; walks and runs more than hops and perches freely on shrubs. Will hide in cover by squatting but stance generally fairly upright. Feeding actions include digging.

Voice loud and distinctive. Song recalls *A. arvensis* but louder, less scratchy, and more melodious, with frequent interjection of 'kleetra' note which is commonest call and has jangling quality of Corn Bunting *Miliaria calandra*.

Habitat. In lower, middle, and marginally upper middle latitudes, subtropical, Mediterranean, steppe, and temperate, on open lowland plains and upland plateaux. Avoids rocky, gravelly, saline, and other infertile or degraded soils, and semi-deserts, but tolerates low and uncertain rainfall and regular summer heat up to 32°C (Voous 1960). Shows little attachment to water and normally keeps clear of wetlands and forests as well as bare sand and rock, especially steep faces and broken terrain. Essentially a steppe bird, at opposite end of spectrum from desert-living *Ammomanes* larks. Occurs primarily on grasslands, ranging from virgin steppe to cultivated crops, areas of profuse mixed herbage, and even water-meadows. Sometimes occurs among shrubs, bushes, or even well scattered low trees, but prefers openness, avoiding proximity of shelter-belts, although in USSR sometimes found close to forest edges. There occupies open steppe with well-developed grass cover, cornfields, fallow land, and some steppe with sparse *Artemisia* (Dementiev and Gladkov 1954a). In Andalucía (Spain), found commonly throughout plains and cultivated

valleys and on margins and islands of Guadalquivir river (Mountfort 1954). In Spain and North Africa, occurs almost exclusively in cultivated areas, including meadows (Bannerman 1953). In Algeria and Tunisia inhabits cultivated plains and plateaux up to 1000 m and higher, avoiding bare steppe and desert (Heim de Balsac and Mayaud 1962). In Anatolia (Turkey), inhabits open cultivated country, flocking in winter to feed on newly sown cornfields (Wadley 1951). In Arabia, a bird of cultivation and grassland, avoiding high elevations; roosts on ground but often sits on telegraph wires or small bushes; shifts in winter to more desert terrain (Meinertzhagen 1954). Lives at high density in territories maintained by persistent song-flights rising from lower to upper airspace.

Distribution. Range decrease in France.

FRANCE. Now absent in former breeding areas, e.g. Gard, Camargue (RC). Bred rarely Corsica 19th century (Thibault 1983). ITALY. Formerly bred in coastal areas of Friule, Venecicia Guîlic, and Veneto (Arrigoni degli Oddi 1902) and in Pachana valley (PB, MS). USSR. Has spread north in Voronezh region (Wilson 1976).

Accidental. Britain, Belgium, Netherlands, West Germany, Denmark, Norway, Sweden, Finland, East Germany, Poland, Czechoslovakia, Austria, Switzerland, Malta, Canary Islands, Madeira.

Population. Decreased France, Italy, Greece, Israel, and probably Spain.

FRANCE. Under 1000 pairs, decreased due to hunting (Yeatman 1976). SPAIN. Apparently decreasing (AN). ITALY. Sardinia: decreased since 1965 (Schenk 1980).

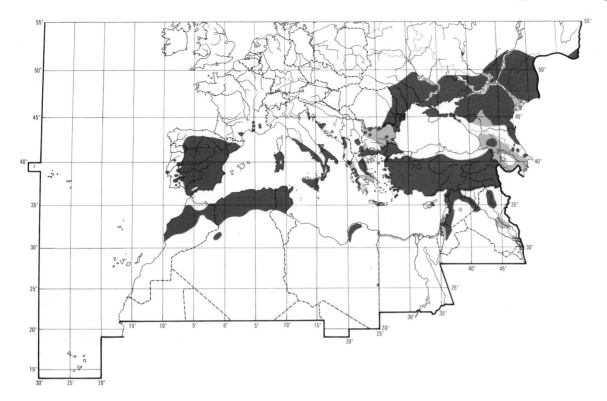

GREECE. Some decline (GIH). USSR. Very common, abundant in some places (Dementiev and Gladkov 1954*a*); has declined in Kazakhstan between Volga and Ural rivers (Shishkin 1976). CYPRUS. Common (Flint and Stewart 1983). ISRAEL. Recent decline due to habitat changes (UP).

Movements. Resident in southern Europe, Near East, and North Africa, though migratory to partially migratory in USSR.

Though resident over much of west Palearctic range, birds not necessarily sedentary. Forms large flocks in autumn and winter, often mixing with Corn Bunting *Miliaria calandra*, and these wander to unknown extent (e.g. Voous 1960, Heim de Balsac and Mayaud 1962, Bernis 1971). Reported to be summer visitor, mid-March to October, to plains of Sardinia (Mocci Demartis 1973). A rare and irregular passage migrant to Malta, March to early April and mid-September to early November (Sultana and Gauci 1982); such birds probably too far west to be USSR emigrants. Flocks seen in October at Laghouat, south of Atlas Saharien in Algeria, may have come from some distance away (Heim de Balsac and Mayaud 1962). However, sole ringing recovery is a movement of 220 km south-east on Adriatic coast of Italy, ringed April, recovered August (Zink 1975).

In USSR, normally only very small numbers winter in northern (especially north-eastern) parts of range (e.g. in Transvolga, Astrakhan, and northern Kazakhstan), and

southern race *psammochroa* is mainly migratory on central Asian steppes, some birds emigrating even as far south as Tadzhikistan. Not necessarily resident anywhere in USSR, for birds in southern breeding areas perhaps move out to be replaced by migrants from further north (Dementiev and Gladkov 1954*a*). Winter range of migrants poorly defined, in absence of ringing data, but extends from southern USSR through Iran, Iraq, and Middle East to Sinai (Egypt); specimens of both nominate *calandra* and *psammochroa* have been taken in Sinai (Vaurie 1959; Etchécopar and Hüe 1967). Claimed sight records from northern Arabia questioned by Jennings (1981*a*), since Bimaculated Lark *M. bimaculata* occurs there, but 2 November–December records from eastern Saudi Arabia at *c*. 26°N (G K Brown).

Main passages in USSR appear to be in October and March, though in autumn cold-weather exodus continues into November (e.g. mass departures then with cold weather in Sea of Azov region); spring return movement often obscured by nomadic movements of locally-wintering flocks (Dementiev and Gladkov 1954*a*). Visible migration in Turkey occurs October–November and March, on small scale only; very large mid-winter numbers there doubtless include high proportion of resident local birds (Vittery *et al.* 1972; Beaman *et al.* 1975). RH

Food. In summer, largely insects; in winter, seeds and grass shoots. Runs during feeding, taking items from ground. Digs for pupae with bill and can use it to crack frozen snow crust. Occasionally flies up to inspect tops of bushes from the air. Will drink brackish water. (Dementiev and Gladkov 1954a; Meinertzhagen 1954.)

The following invertebrates recorded in diet, some quite large: damsel flies, etc. (Odonata), grasshoppers (Acrididae), bugs (Hemiptera, e.g. Pentatomidae, Cicadidae), termites (Isoptera), Lepidoptera larvae, flies (Diptera: e.g. Syrphidae), Hymenoptera (ichneumons Ichneumonoidea, ants Formicidae, wasps Vespidae, bees Apoidea), adult and larval beetles (Coleoptera: Carabidae, Scarabaeidae, Elateridae, Tenebrionidae, Bruchidae, Chrysomelidae, Curculionidae), spiders (Araneae), woodlice (Isopoda), molluscs. Plant food includes hemp *Cannabis* and seeds or fruits of docks (Polygonaceae), goosefoot *Chenopodium*, *Amaranthus*, Rosaceae, Leguminosae, *Hypericum*, centaury *Centaurium*, Boraginaceae, Compositae, grass *Echinochloa*, and cereals; also grass shoots, including wheat. (Dementiev and Gladkov 1954a; Pek and Fedyanina 1961; Samorodov 1968; Kekilova 1969; Anon 1970; Averin and Ganya 1970; Kostin 1983.)

In summer, USSR, takes largely insects, plant food (seeds, grain, grass shoots) comprising no more than 10% of diet; takes more Orthoptera than other Alaudidae. In autumn, as insects disappear, switches to plant material. (Dementiev and Gladkov 1954a; see also Kekilova 1969.) Thus, in Moldavia during May, 3 stomachs contained mainly insects, including beetles (5 Curculionidae, 1 Tenebrionidae, 1 *Anisoplia*), 3 Orthoptera, and 3 ants; 5 October–February birds held only plant food—seeds of wild plants, wheat and barley grain, and shoots of winter wheat (Averin and Ganya 1970). Similarly, insects become more important in spring: in Crimea, January–February, 66 stomachs contained 99·1% seeds (including 62·4% *Echinochloa* and 13·9% rice) with animal material (beetles) in only 2 stomachs; 2 from March contained equal proportions of seeds (rice and sorghum) and ants; in June–July, 3 contained 24% plant material (15% rice), 34% beetles, 25% ants, and 16% molluscs (Kostin 1983). On steppes north of Caspian Sea, diet included shieldbugs (Pentatomidae) and beetles (*Otiorrhynchus*, *Taphoxenus*) (Shishkin 1980). See also Pek and Fedyanina (1961). During December–January in Kalmytskaya ASSR (north-west of Caspian Sea), took more cereal grain than other Alaudidae in same area: of 132 birds, none contained animal food; 79·5% contained cultivated plant seeds (largely wheat, barley, and millet) and 34·8% contained seeds or fruits of wild plants—largely *Rumex* (in 18·2%), Compositae (9·1%), *Chenopodium* (6·8%), and *Polygonum* (4·5%); also grass shoots (in 3·8%); all stomachs contained grit, mostly sand—average 69 pieces per bird (Samorodov 1968). At area of grazing near Kherson, flock once consumed all grass shoots available in snow-free patches and under thin snow cover (Dementiev and Gladkov 1954a). In Morocco, autumn, birds contained hard seeds and green shoots; one bird contained a grasshopper

(Meinertzhagen 1951a). In Tunisia, January, one bird contained wheat, barley, and small pebbles; another, April, contained small seeds, barley, and grit (Blanchet 1951). Bird from Jenin (Palestine), May, contained beetles and vegetable remains (Hardy 1946). Birds in winter, southern Israel, were full of small hard seeds (Meinertzhagen 1951a).

In April 1975 in semi-desert of north Caspian region, birds fed on huge hatch of caterpillars (Pyralidae): each ate average 615 caterpillars (12·3 g dry weight) per day, in total (with density of 160 birds per km²) removing 6·3% of available caterpillars per day; (presumably total energy consumption of free-living birds 321 kJ per bird per day (Shishkin 1980).

No information on food of young. DJB

Social pattern and behaviour. No detailed studies.

1. Often markedly gregarious outside breeding season: e.g. flock of *c.* 2500, central plateau of Turkey, mid-December (Porter *et al.* 1969); see also Macfarlane (1978) for Syria and Koenig (1888, 1893) and Zedlitz (1909) for Tunisia. In Kalmytskaya ASSR, late December to mid-January, average flock size 51·8 (4818 birds seen over 28 days) (Samorodov 1968). On Cyprus, flocks of up to *c.* 1000 typical of post-breeding period (June to October or November), usually smaller later; flocks of up to 50, late September to early October, probably migrants (Flint and Stewart 1983). In USSR, small roving flocks typically occur from July, with up to *c.* 200 in August and many thousands (of eastern *psammochroa*) in peak passage periods and in winter (Blaszyk and Steinbacher 1954; Dementiev and Gladkov 1954a; Dolgushin *et al.* 1970); in north-west Kazakhstan, late summer, 280 birds in 1 km² (Shishkin 1976). Near Garmusch (Iraq), mid-April, flock of *c.* 1000 recorded resting on migration (Weigold 1912), but in Jebel ed Druz (Syria), migrant flocks normally contained 10–20 or up to 200 (Meinertzhagen 1951a). Large flocks also occur for watering—e.g. in Jordan, early October (Meinertzhagen 1925); see also Brosset (1961) and Dolgushin *et al.* (1970). Not uncommonly associates with other Alaudidae (e.g. Dolgushin *et al.* 1970); also with Corn Bunting *Miliaria calandra* (e.g. Meinertzhagen 1951a for Morocco). BONDS. Little information. Apart from suggestion by Took (1972) of possible polyandry (see Breeding Dispersion and part 2), other accounts apparently indicate monogamous mating system though no details available. Vagrant (possibly ♂), Reusstal (Switzerland), early May, recorded associating with ♀ Skylark *Alauda arvensis* (Christen and Jenny 1983). ♂ reported to be very devoted to ♀ (Dolgushin *et al.* 1970). No information on role of sexes in care of young, nor on length of care. BREEDING DISPERSION. Solitary and territorial, though neighbourhood groups may be formed in areas of high density: e.g. in Ukraine, in lush secondary steppe, also not infrequently in fields of maize *Zea* and sunflowers *Helianthus* (Blaszyk and Steinbacher 1954). 6–8 ♂♂ reported displaying in 'small area' (Charlemagne 1912; see also part 2); on Cyprus, possibly 'loosely colonial', even polyandrous (Took 1972), but no hard evidence. In USSR, 1–2 pairs per ha of suitable habitat and nests *c.* 50–100 m apart (Dementiev and Gladkov 1954a); in area between Ural and Emba rivers, 0·1–4·8 pairs per 1-km transect, higher figure referring to grass–*Artemisia* associations (Poslavski 1963). Other transect counts in north Caspian region showed 12·0–44·0 pairs along 10 km, higher value in steppes (Poslavski 1974). In area between Volga and Ural rivers, 40

birds in 1 km² (Shishkin 1976). In Afghanistan, ♂ reported to have defended territory of *c.* 3¼ ha in crops (Meinertzhagen 1951*a*). ROOSTING. Solitary and, outside breeding season, probably gregarious, though no details; see Bimaculated Lark *M. bimaculata* (p. 105). Always on ground, with body pressed flat, according to Dolgushin *et al.* (1970). Occasionally in Morocco (Meinertzhagen 1951*a*) and frequently in Arabia (Meinertzhagen 1954), loafs on telegraph wires and bushes. Sometimes sings well before dawn (J-C Roché). On Cyprus, song given from ground in evening until nearly dark (Took 1972; see also part 2). In late afternoon or evening comes regularly to watering places to drink—e.g. in Jordan, early October (Meinertzhagen 1925, 1951*a*). Habit also regular in USSR, summer and autumn: in steppes near Lake El'ton (eastern Volgograd region) birds visit freshwater pond and saltwater stream, most often either around midday, or before 10.00 hrs and at 15.00–17.00 hrs (Dementiev and Gladkov 1954*a*; Dolgushin *et al.* 1970). In eastern Morocco, flocks visit watering place twice a day (Brosset 1961). Digs pits for dust-bathing (Dolgushin *et al.* 1970).

2. Shy and not easily approached in open country but will squat where cover available (Meinertzhagen 1954). According to Bannerman (1953), often crouches first when alarmed rather than flying off. 4 vagrants, Netherlands, October, not shy and once allowed approach to *c.* 5 m; after flushing, flew fast, low, and in zigzags, before dropping abruptly to ground (Verkerk 1961). Vagrant in England shy and nervous, and often flew up vigorously to considerable height, circled, then plunged down to glide before landing; often crouched in hollows; alarm apparently indicated by upright stance with neck extended— sometimes flew with neck extended (Ash 1962). Dark neck-patches prominently displayed in antagonistic and heterosexual interactions; presumably have signal function (Jonsson 1982, which see for illustration). FLOCK BEHAVIOUR. Comes to water in dense flocks, circling many times before landing to drink hastily (Meinertzhagen 1951*a*). Migrating flocks 'disorderly', though once *c.* 200 Alaudidae (60% *M. calandra*) more tight-packed like Starlings *Sturnus vulgaris* (Bub 1955). Large flocks typically very noisy (e.g. Dementiev and Gladkov 1954*a*: see 2b in Voice). Near Dushanbe (Tadzhikistan, USSR), winter, constant hubbub from birds on ground; flocks easily alarmed (Ivanov 1969). SONG-DISPLAY. ♂ will sing (see 1 in Voice) on ground or from bush, etc. (see below), but more often initially when flying in regular circles at height of *c.* 10 m, then ascending and continuing to sing (Dementiev and Gladkov 1954*a*); ascends in spirals according to Zedlitz (1909). Song-flight sometimes very high—higher than *A. arvensis* (Zedlitz 1909; Meinertzhagen 1951*a*). May also climb steeply to *c.* 50 m, then circle in various directions (Dolgushin *et al.* 1970), like Woodlark *Lullula arborea* (Alexander 1927). During

circling flight wings fully extended and slow, deep wing-beats (loose-winged flapping: M G Wilson) recall wader (Charadrii); dark undersurfaces prominent (Bergmann and Helb 1982: Fig A). Wing-action may appear to emphasize down-beat of wing (P A D Hollom). Description by Stanford (1954) refers to bird rotating or fluttering wildly in one place; alternate beating of wings also recorded. Performance may last many minutes (Bergmann and Helb 1982); *c.* 25–30 min recorded (Charlemagne 1912; Alexander 1927). Descent may be vertical and silent for last few hundred feet (Mountfort 1954); final gentle glide and horizontal flight before landing reported by Dolgushin *et al.* (1970). For sounds given on ground after Song-flight, see 4a in Voice. Also recorded descending to a certain height, then beating wings jerkily below body like displaying Redshank *Tringa totanus* (Koenig 1888) or Common Sandpiper *Actitis hypoleucos* (Stanford 1954). According to Reiser and Führer (1896), bird then sings with especial vigour, while excited flights from bush to bush may refer to performance in presence of ♀. In low circling form of Song-flight, deep and even wing-beats periodically interrupted by glides with wings above or below body (Dolgushin *et al.* 1970; see also Wadley 1951 and Heterosexual Behaviour, below). According to Zedlitz (1909), this variant commoner than high Song-flight in Tunisia and bird gives rather fragmented song and frequent loud calls (see 2 in Voice). Singing birds apparently stimulate one another to perform Song-display (Bergmann and Helb 1982). In Ukraine, Song-flight commonly performed in small groups and flushed birds would sing near individual already in Song-flight (Blaszyk and Steinbacher 1954). Up to 10 noted singing together in southern USSR (Ivanov 1969); see also Breeding Dispersion (above). Vigorous and protracted singing from ground frequent (Mountfort 1954), especially in morning (Schubert and Schubert 1982); on Cyprus, for up to *c.* 40 min in evening (Took 1972). Bergmann and Helb (1982) mentioned only occasional singing by territorial ♂ from rock or bush near nest. ♂ on ground may have head feathers ruffled (to form crest), wings slightly drooped, and tail raised. Throat swollen and dark neck-patches thus conspicuous while singing (Dolgushin *et al.* 1970: Fig B).

B

In Bulgaria, late May, one ♂ adopted upright posture with wings slightly drooped but almost completely closed; tail raised at *c.* 50° to ground, *c.* 100° to back; during song (quieter than that given in flight), neck repeatedly extended and shortened, with brief pauses at top and bottom of each movement; no ♀ seen nearby (Mauersberger 1964). In Rome province (central Italy), Song-flights mainly mid-March to mid-July, some also in October–November (Alexander 1927). In Morocco (Thévenot *et al.* 1982) and central Turkey (Wadley 1951), mainly April-May. In Tunisia, from February, though breeding not before mid-April (Koenig 1888); high Song-flight performed late afternoon and at sunrise in mid-March (Zedlitz 1909). In

A

USSR, will sing on migration (Dolgushin *et al.* 1970), but mostly early March to late June (or mid-July) in fine weather, also in autumn (Dementiev and Gladkov 1954a); for Ukraine see also Blaszyk and Steinbacher (1954) and for southern USSR see Ivanov (1969). In Afghan Turkestan, Song-flights recorded on 19 May, after young fledged (Meinertzhagen 1938). Song-flights also performed by 2 of 4 autumn vagrants in Netherlands (Verkerk 1961). ANTAGONISTIC BEHAVIOUR. ♂ occupies territory on arrival and sings vigorously; often, several ♂♂ sing fairly close together (see above) and brief chases may ensue if their flight-paths cross (Dolgushin *et al.* 1970). On Cyprus, no rivalry seen between 2–3 ♂♂ performing Song-flight in close proximity, though some chasing occurred after 6–7 ♂♂ took off to sing; these birds afterwards returned to same area to continue feeding together (Took 1972). According to Meinertzhagen (1951a, 1954), territory closely guarded in breeding season, ♂ spending much time in vigorously repulsing attempted intrusions; in another study, most antagonism said to arise when paired ♂ drives rival ♂♂ away from his mate (Koenig 1888). HETEROSEXUAL BEHAVIOUR (1) General. In Pamiro-Alay (southern USSR), pair formation (at latest) from early April (Ivanov 1969); in eastern part of range, in Kazakhstan, from early March (Dolgushin *et al.* 1970). In Tunisia, full courtship-display recorded from end of March (Zedlitz 1909); see also Song-display (above). No antagonism recorded between pairs during courtship (Zedlitz 1909). (2) Pair-bonding behaviour. Involves aerial chases and courtship display (Dolgushin *et al.* 1970). Few other details. ♀ said to be usually present at spot where ♂ descends after Song-flight (Charlemagne 1912). On Cyprus, ♂ recorded following presumed ♀ as she descended, then rising up to sing above her; first ♂'s place then taken by another ♂ who eventually landed to feed by presumed ♀. Once, mid-March, when 6–7 ♂♂ performed Song-flight (see above), 2–3 ♀♀ were present in field below them (Took 1972). ♀ said often to take off some time after ♂'s low circling Song-flight (see above) and to fly in elegant gliding curves, ♂ following with tail fanned and looking bigger than usual, also giving frequent rattling and other calls (Zedlitz 1909). On ground, ♂ may run about with light dancing steps near ♀ or circle her (while singing) in posture assumed for Song-display (see above); body sometimes swayed, and raised tail may also be fanned; will also take off to perform low circling and gliding Song-flight (Dolgushin *et al.* 1970; Sudhaus 1970). While ♀ incubating, ♂ reported to spend most time singing (in air or from ground) (Reiser and Führer 1896). No further information but see Bonds (above). RELATIONS WITHIN FAMILY GROUP. Virtually nothing known. Young fed in nest for *c.* 10 days and leave before able to fly—in USSR, from second half of May to late July (Dementiev and Gladkov 1954a; Dolgushin *et al.* 1970). On Cyprus, late May, one such young bird was less than 2 m from nest at which parent continued nest-sanitation though it contained only 2 addled eggs (Took 1972). ANTI-PREDATOR RESPONSES OF YOUNG. No information. PARENTAL ANTI-PREDATOR STRATEGIES. (1) Passive measures. Generally cautious on breeding grounds (Reiser and Führer 1896) though not shy at nest according to Ticehurst *et al.* (1921–2). ♀ sits very tightly (Dolgushin *et al.* 1970). (2) Active measures. When nest threatened, ♀ performs distraction-lure display of disablement type, stumbling, knocking against ground, dragging one wing, etc. (Dolgushin *et al.* 1970). Flushed ♀♀ also gave an (undescribed) alarm-call and one feigned injury, but no nests found (Charlemagne 1912).

(Figs by J P Busby: A from drawing in Nelson 1973; B from photograph in Dolgushin *et al.* 1970.)   MGW

Voice. No detailed study of repertoire but ♂'s song (given in spring, also to some extent in autumn) fairly well known. No appreciable variation between songs of birds from Spain and Greece (J-C Roché), and rattling components (see below) apparently similar in Turkey and Iran (J Hall-Craggs). Other calls—frequently given (e.g.) by winter flocks—often harsh and with trilling, twittering, or dry jingling quality. For extra sonagrams, see Bergmann and Helb (1982).

CALLS OF ADULTS. (1) Song of ♂. Not unlike Skylark *Alauda arvensis* but louder, more complex, and richer (e.g. Dementiev and Gladkov 1954a, Mountfort 1958); more fluting but lacking jubilant quality of *A. arvensis*, and grating sounds constantly interpolated (Hartert 1923). Thus more guttural and probably with longer units than in *A. arvensis* (Schubert and Schubert 1982). Normal song a continuous flow of short phrases or single notes (often imitative: see below) and characteristic rapid chirping or vibrating sounds—e.g. 'pitjur-ir-ir-ir-ir' or 'klitra-a'a'a'a' (see call 2); latter sounds are like song of Corn Bunting *Miliaria calandra* (Jonsson 1982); see also Schubert and Schubert (1982) who mentioned noisy, high-pitched rattling or jingling elements reminiscent of song of Starling *Sturnus vulgaris* or *M. calandra*. Fig I shows 3 very fast rattles increasing in loudness; similar sounds in recording from Iran by P A D Hollom are more liquid due to slower sub-unit rate, but similarity perhaps suggests rattles are species-specific song-elements or calls (J Hall-Craggs). High-pitched and rapid trilling sounds (Bergmann and Helb 1982) may at times resemble Canary *Serinus canaria* in purity of tone (Mountfort 1958). Rich 'chrrr' and clear, whistling 'klirí' frequently given in song (Dementiev and Gladkov 1954a); probably the same as this is 'sillEE' (M G Wilson) or clearly trisyllabic 'si-too-eee' (J Hall-Craggs) shown in Fig II; see also call 2. In recording (Fig III), part of song roughly 'weedl weedl weedl weedl seesee didly-dee-de-doo', suggesting Dunnock *Prunella modularis* and possibly mimicry. Sounds illustrated in Fig IV strongly reminiscent of song of Pied Wagtail *Motacilla alba*. Also evident in recording are longer and flute-like sounds, e.g. 'see' and/or 'pee' preceded by a ripple (Fig V), quiet buzzes or rasping snarls, a low-pitched musical gurgle, and rather quiet 'tweep' sounds like finch (Fringillidae) or sparrow *Passer* (J Hall-Craggs, M G Wilson). Apart from claim by Dolgushin *et al.* (1970) that song has no great imitative content (in contrast to Bimaculated Lark *M. bimaculata*), and suggestion by Hartert (1923) that mimicry characteristic of only some individuals, other authors consider *M. calandra* an excellent mimic, with according to Alexander (1927), song composed largely of mimicry. Frequently and convincingly mimicked sounds (e.g. 'wid' of Swallow *Hirundo rustica* and 'kroarr' of Carrion Crow *Corvus corone*) may be delivered in series and with much repetition (Bergmann and Helb 1982). Species imitated in southern Spain and Greece include

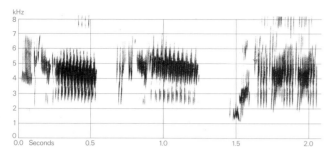

I   E D H Johnson   Turkey   March 1972

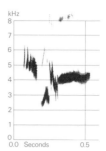

II   E D H Johnson   Turkey   March 1972

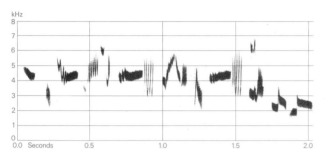

III   E D H Johnson   Turkey   March 1972

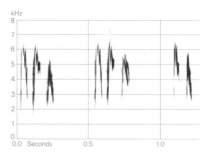

IV   E D H Johnson   Turkey   March 1972

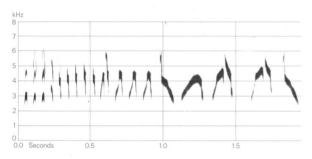

V   E D H Johnson   Turkey   March 1972

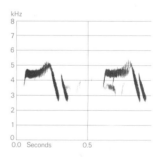

VI   P A D Hollom   Iran   May 1977

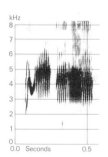

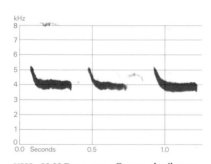

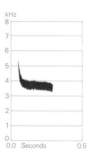

VII   H-H Bergmann
Greece   April 1979

VIII   H-H Bergmann   Greece   April 1979

IX   V C Lewis   England
May 1962

Collared Pratincole *Glareola pratincola*, Short-toed Lark *Calandrella brachydactyla*, *H. rustica*, Goldfinch *Carduelis carduelis*, Linnet *C. cannabina*, and *Miliaria calandra* (J-C Roché); in recording by C Chappuis (Greece), additionally Green Sandpiper *Tringa ochropus* and Green Woodpecker *Picus viridis* (M G Wilson). Recording (Fig VIII) also reveals apparent mimicry of long 'pew' notes (with remarkable purity and steadiness of pitch) from song of Wood Warbler *Phylloscopus sibilatrix*; Fig IX shows 1 unit from *P. sibilatrix* song for comparison (J Hall-Craggs). For further details of bird species imitated, see Charlemagne (1912), Alexander (1927), and Dementiev and Gladkov (1954a); also Reiser and Führer (1896), Weigold (1912), and Passburg (1966). (2) Contact- and flight-calls. Shrill and liquid sounds given on ground, also at take-off for Song-flight and on landing after it (J-C Roché). (a) Recording (Fig VII) reveals a quite long and buzzing 'ZWEEArrr' leading into song (J Hall-Craggs). This, or similar sound, frequently mentioned: loud, rather nasal 'kleetra' (Mountfort 1954, 1958); buzzing 'zreea' (M G Wilson); 'kltra', louder than *A. arvensis* (Christen and Jenny 1983); distinctive 'klytra' (Koenig 1888); 'klŷrt' as contact- and courtship-call (Zedlitz 1909); rather chirping or twittering 'kliera', sharper than *A. arvensis* (Verkerk 1961). Evidently occurs in song, especially in low circling Song-flight, according to Zedlitz (1909); see also call 1. Fig VI shows 2 apparently closely related calls ('KLEEtra'), with a rasping snarl below 2nd; more like 'ke-le-tra' in recording by E D H Johnson (J Hall-Craggs). However, Schubert and Schubert (1982) considered 'klitra', with many variations and imitations, to be characteristic of *C. brachydactyla* rather than of *M. calandra*. (b) Flight-calls all more or less sharp and trilling and may precede song: recording by H-H Bergmann, Greece, suggests 'pstIri' lasting just over 0·3 s and 'si-si-tRRRRR' (J Hall-Craggs); for alternative renderings and sonagrams, see Bergmann and Helb (1982). Often coarse and raucous (e.g. 'kchyrryk') and with jingling tone rather like *A. arvensis* (Jonsson 1982). Single vagrant, southern England, gave succession of loud 'prrrruuuup' sounds, impossible to describe adequately, while circling; also rather long 'trrrreeeep' during downward plunge (Ash 1962). In Turkey, winter, flocks gave continual 'chirrup chirrup' sounds as they rose and resettled (Wadley 1951). In Iran, once a very distinctive, ringing 'üib' (Schüz 1959); perhaps same as 'kliri' given in song (see above). (3) An (undescribed) alarm-call given by ♀ on flushing (Charlemagne 1912). (4) Other calls. (a) Warbling sounds given on ground after full aerial song (Alexander 1927); possibly refers to elements described under calls 1 and/or 2.

CALLS OF YOUNG. No information.　　　　MGW

Breeding. SEASON. Spain: laying begins early April (Bannerman 1953). Algeria: laying from early April to early June (Heim de Balsac and Mayaud 1962). Greece: laying from first half of April to early June (Bannerman 1953; Kinzelbach 1969). Cyprus: small young in nest early April (Took 1972). Southern USSR: 1st eggs late March in southern Ukraine, early April in Aral-Caspian area, mid-April in north-west Kazakstan (Dementiev and Gladkov 1954a; Shishkin 1976). SITE. On ground, under tussock. Nest: shallow depression, lined with grass stems and leaves, with inner lining of softer vegetation. Building: by both sexes (Bub and Herroelen 1981). EGGS. See Plate 79. Sub-elliptical to oval, smooth and slightly glossy; whitish, sometimes greenish or yellowish, heavily spotted and sometimes blotched dark brown or red-brown, and pale purple; blotching sometimes concentrated at broad end. 24·5 × 18·0 mm (22·0-27·2 × 16·2-19·5), $n = 150$; calculated weight 4·09 g (Schönwetter 1979). Clutch: 4-5 (3-6). Of 88 clutches, Algeria: 3 eggs, 3; 4, 70; 5, 12; 6, 3; mean 4·2 (Heim de Balsac and Mayaud 1962). 2 broods; replacements also laid (Makatsch 1976). INCUBATION. 16 days (Dolgushin *et al.* 1970). By ♀ only according to Dementiev and Gladkov (1954a); see also Reiser and Führer (1896). However, presence of brood-patches in ♂ suggests he may take some part (Dolgushin *et al.* 1970). YOUNG. Altricial and nidicolous. Cared for and fed by both parents (Bub and Herroelen 1981). FLEDGING TO MATURITY. Fledging period unkown but young fed in nest for *c.* 10 days (Dementiev and Gladkov 1954a). BREEDING SUCCESS. In Volgograd region (USSR), heavy losses of clutches to predators and bad weather; even in relatively favourable conditions, *c.* 50% of eggs lost (Golovanova 1967).

Plumages (nominate *calandra*). ADULT. Forehead, crown, mantle, inner scapulars, and back grey-brown with slight olive-buff tinge and broad dull black shaft-streaks; hindneck paler, buff-brown, and with narrow dark grey streaks; outer scapulars like inner, but slightly brighter buff or rufous-brown; all feathers with narrow pale buff or off-white terminal arcs when fresh (about September–October). Rump and upper tail-coverts olive-brown or buff-brown, longer coverts with blackish centres, feathers fringed rufous-buff to off-white along tips. Lores, feather-tufts above nostrils, and indistinct narrow ring round eye mottled grey, off-white, and buff; short supercilium from just above eye to over ear-coverts buff, sometimes finely speckled grey, rather indistinct; often a trace of black along upper ear-coverts at border of supercilium. Narrow and mottled dark grey or olive-brown stripes run from gape below eye and from base of lower mandible along sides of chin and over cheek, both ending in uniform olive-brown or buff-brown ear-coverts; areas between streaks and sides of neck uniform buff or cream-buff or with some indistinct grey specks; stripe along sides of chin and across cheeks occasionally absent or indicated by a few grey spots only. Chin and upper throat cream or pale buff. Side of chest with large and conspicuous black crescent-mark, rather variable in width and shape; marks almost meet on lower throat, though usually separated by *c.* 5-10 mm in centre (lower throat sometimes spotted black and both crescents then appear to meet narrowly). Chest and sides of breast buff, each feather with rounded or triangular black spot on tip *c.* 1-2 mm wide; spots usually larger, greyer, and less well-defined towards sides of breast and lower chest; sides of breast often tinged rufous-cinnamon; central lower chest sometimes with indistinct uniform grey patch. Flanks and thighs olive-brown, sometimes tinged rufous; remainder of underparts white or cream-white, breast and longer under tail-coverts often partly tinged grey-buff or pale buff. Central pair of tail-feathers (t1) olive-brown with black centre and rufous fringe; t2-t4 olive-black to black with 2-6 mm of tip white (widest on inner web), and narrow cream-pink or off-white edges; t5 black with 5-11 mm of tip white and with broad cream-pink or off-white fringe along outer web (sometimes, all outer web white); t6 cream-pink or white with dusky grey wedge on base of inner web; undersurface of tail dusky grey except for white t6 and indistinct white tips of t2-

t5. Flight-feathers greyish-black, secondaries with off-white tip 4-12 mm wide, p1-p3 with somewhat narrower tip, replaced by rufous-buff fringe on tip of p4-p7(-p8). Outer edge of secondaries and of outer primaries (except emarginated parts) narrowly fringed pink-buff; remainder of feathers faintly edged grey-buff, outer web of p9 rather broadly fringed pale buff or off-white. Greater upper primary coverts and bastard wing greyish-black with rather narrow pink-buff fringes along tips and outer webs. Lesser upper wing-coverts olive-brown with pale buff or off-white tips; median and greater upper wing-coverts black or olive-black, broadly fringed rufous-cinnamon along tips and more narrowly at sides (tips showing as 2 pale bands across wing when plumage fresh). Tertials like greater coverts but tinged olive-brown subterminally. Under wing-coverts, axillaries, and undersurface of flight-feathers dusky grey (appearing black in poor light), except for white tips of secondaries and innermost primaries and for pale buff fringes along marginal coverts. Bleaching and wear have marked effect: pale terminal fringes of upperpart feathers disappear within 1-2 months of moult, and warm rufous or buff colours of outer scapulars, tertials, upper wing-coverts, primary-tips, t1, sides of breast, and flanks bleach soon afterwards (last on parts of upper wing-coverts, where sometimes retained until March-April). By about mid-winter, upperparts are dark grey-brown or dark olive-grey with blackish feather-centres and paler grey feather-sides, with purer grey hindneck and only faint traces of buff on fringes of scapulars, upper tail-coverts, and t1; short supercilium off-white, lores and cheeks mottled grey and off-white, ear-coverts grey-brown, chin and (usually) lower sides of neck white; black patch on side of chest more distinct; chest pale grey (darkest on centre), only slightly buff towards sides of breast. Flanks and thighs olive-grey, breast to under tail-coverts off-white; tail dull black, pale fringes partly abraded, t6, outer edge of t5, and tips of t5 and (narrowly) t4(-t3) white; tips and fringes of flight-feathers bleached to white. In worn breeding plumage, upperparts streaked black and dull grey, supercilium and stripes on cheeks hardly visible, pale tips of tertials, tail, upper wing-coverts, and primaries virtually lost through abrasion; spots on chest often indistinct, partly abraded. Sexes indistinguishable (but see Measurements). NESTLING. No information. JUVENILE. Forehead and crown with narrow white feather-fringes on centre of crown, buff elsewhere. Rear and sides of head like adult, but each feather with small black subterminal spot or narrow arc and narrow buff or white terminal fringe (thus appearing spotted); distinct supercilium, chin, throat, and lower sides of neck cream-white or white; lores and feathers at base of lower mandible uniform buff. Mantle, shorter scapulars, and back to upper tail-coverts black with greyish feather-bases; feather-tips narrowly fringed white, sides buff. Longer scapulars and tertials olive-brown with black submarginal U-mark, fringed white along tip, buff along sides. Chest and sides of breast buff with distinct brown-black spot 2-3 mm across near tip of each feather, narrowly bordered buff or white at feather-tip; spots tend to join to form larger patch at upperside of chest, but no distinct solid black crescent as in adult. T1 like tertials, t2-t4 similar to primaries, t5 dull black with pale buff outer web and tip, t6 buff-white. Flanks greyish-buff, feathers with some dusky subterminal suffusion; breast to under tail-coverts cream-white or white. Primaries, feathers of bastard wing, and greater upper primary coverts olive-black or brown-black, evenly and sharply fringed buff along sides and tips (on tips, soon fading to white); p10 relatively long compared with adult (see Structure), tip broadly rounded, outer web pale buff, tip fringed pale buff. Secondaries and tertials dark olive-brown, rather narrowly bordered pale

buff at sides, more broadly fringed pale buff or off-white on tip; tertials with black-brown submarginal U-mark. White tips of secondaries narrower and less well-defined than in adult, 2-6 mm wide on outer web, somewhat broader on inner, hardly visible from below. FIRST ADULT. Like adult; indistinguishable once last juvenile feathers (usually p9-p10 and middle secondaries) replaced. Birds with outer web of t5 fully white or cream and with pale tips of secondaries ill-defined perhaps in 1st adult plumage, but no known-age specimens examined, nor known whether all 1st adults show such characters.

Bare parts. ADULT. Iris hazel, grey-brown, light brown, or brown. Upper mandible and tip of lower mandible dark horn-grey or dark horn-brown, cutting edges and base of lower mandible pale yellow-horn or pale brown-yellow; mouth pale orange or yellow, tongue pink. Leg and foot flesh-pink, light pink-straw, fleshy-horn, yellow-horn, light brown, or brown; soles occasionally orange-yellow. (Hartert 1903-10; Ash 1962; Slings 1982; BMNH, RMNH.) NESTLING, JUVENILE. No information.

Moults. Based on 15 moulting specimens (BMNH, RMNH, ZMA) and on Vaurie (1951a) and Bub and Herroelen (1981). ADULT POST-BREEDING. Complete; primaries descendant. Starts late June to late July with p1 and scattered feathers on head or body; body and wing-coverts virtually all new at primary moult score 30-40; moult completed with p9-p10 and middle secondaries late August to mid-October. POST-JUVENILE. Complete; primaries descendant. Only a few in moult examined. Birds in fresh juvenile plumage examined June-August (RMNH, ZMA) and some occur September (Dementiev and Gladkov 1954a). Probably starts soon after fledging, as September birds can have moult almost complete (Vaurie 1951a; RMNH, ZMA). Assuming moult starts shortly after fledging and speed of moult as in adult post-breeding, late-fledging birds may not finish until November.

Measurements. ADULT. Nominate *calandra*. Southern Europe and north-west Africa, March-August; skins (Stresemann 1920; BMNH, RMNH, ZMA). Bill (S) to skull, bill (N) to distal corner of nostril; exposed culmen on average 3·7 less than bill (S).

WING	♂	132·4 (3·43; 63)	126-141	♀ 118·7 (1·84; 31)	115-122
TAIL		63·8 (2·51; 36)	59-68	56·5 (2·76; 25)	52-62
BILL (S)		20·8 (1·30; 37)	17·7-23·0	18·1 (1·08; 24)	15·9-20·2
BILL (N)		13·8 (0·81; 37)	11·9-15·3	11·9 (0·63; 24)	10·6-13·5
TARSUS		28·6 (1·10; 39)	26·8-30·2	26·9 (1·00; 26)	25·5-28·3

Sex differences significant.

Wing. Nominate *calandra*: (1) Iberia (RMNH, ZMA). (2) Italy (RMNH, ZMA), (3) Balkans and Greece (Stresemann 1920; Makatsch 1950; RMNH, ZMA), (4) Tunisia (BMNH, RMNH, ZMA), (5) Hauts Plateaux of Algeria (RMNH, ZMA), (6) Volga steppes and Caucasus area (Vaurie 1951a; RMNH, ZMA), (7) West and central Asia Minor (Weigold 1914; Kumerloeve 1961, 1968; RMNH, ZMA), (8) Van Gölü and extreme south-east Turkey (Kumerloeve 1969a) and Azerbaijan, north-west Iran (Vaurie 1951a). *M. c. hebraica*: (9) Birecik, Amik Gölü, and Ceylânpinar (southern Turkey), and Levant south to Gaza (Kumerloeve 1963, 1969b, 1970a, b; BMNH). *M. c. psammochroa*: (10) Iraq and western Iran (BMNH), (11) Iran (Kermanshah and Tehran to Fars) and north-west Afghanistan (Vaurie 1951a), (12) USSR (Bub and Herroelen 1981).

(1)	♂ 129·4 (2·25; 7)	126–134	♀ 118·4 (1·72; 7)	116–121
(2)	131·6 (2·22; 7)	128–134	118·4 (1·68; 7)	116–121
(3)	131·7 (3·08; 25)	126–139	119·0 (2·13; 12)	115–122
(4)	133·6 (3·20; 15)	128–139	118·8 (2·10; 4)	117–122
(5)	136·5 (4·20; 6)	132–141	—	
(6)	135·4 (3·14; 7)	130–140	119·0 (1·41; 4)	117–120
(7)	130·5 (3·00; 19)	125–136	119·0 (4·12; 9)	115–128
(8)	134·5 (— ; 13)	130–138	118·2 (— ; 2)	115–121
(9)	133·4 (2·84; 34)	129–141	120·2 (3·29; 11)	115–126
(10)	134·6 (2·98; 7)	130–138	118·5 (— ; 3)	112–126
(11)	134·1 (— ; 50)	128–141	120·2 (— ; 37)	114–126
(12)	130·9 (— ; 70)	122–136	116·0 (— ; 19)	112–121

JUVENILE. Wing and tail on average both *c.* 5 shorter than adult; bill full-grown once post-juvenile moult completed, tarsus shortly after fledging.

Weights. Nominate *calandra*. Greece: January, ♀♀ 65, 66; April, ♂ 64 (Makatsch 1950). Central Turkey: April, ♂♂ 62, 65; May, ♂ 64 (Kumerloeve 1968). South-east Turkey: June, ♂ 64·4 (5·65; 8) 57–73, ♀ 57 (Kumerloeve 1969a). Northern Iran: March, ♂ 58 (Schüz 1959); July, ♂♂ 65·5, 54·4, ♀ 44·2 (Paludan 1940). USSR: ♂ 59·8 (4) 54–63, ♀ 61 (Dementiev and Gladkov 1954a).

M. c. hebraica. Birecik (southern Turkey): May, ♂♂ 60, 64, 64 (Kumerloeve 1970a). Jabbul (northern Syria): June, ♂♂ 66, 72 (Kumerloeve 1969b). Ceylânpinar (Turkey-Syria border 4 ♂♂ 59–63(–75), ♀ 54 (Kumerloeve 1970b).

M. c. psammochroa. Turkmeniya (USSR), winter: ♂ 60–69 (6), ♀ 52–56 (7) (Rustamov 1958). Kirgiziya (USSR): ♂ 54–71 (18), ♀ 50–61 (5) (Yanushevich *et al.* 1960). Near Tehran (Iran), late May–June: ♂♂ 57·7, 55·9 (Paludan 1940).

Nominate *calandra* or *psammochroa*. Kazakhstan (USSR): February, ♂ 62; March, ♂ 60–70 (4), ♀ 57·5; May, ♂♂ 62·5, 65, ♀♀ 53, 65; June, ♂ 55·7 (Dolgushin *et al.* 1970).

Structure. Wing rather long, base broad, tip fairly pointed. 10 primaries: p8 longest, p9 0–3 shorter, p10 86–97 (adult ♂), 76–83 (adult ♀), or 74–78 (juvenile), p7 1–4, p6 8–15, p5 19–26, p4 26–36, p1 44–52 (♂) or 39–45 (♀). P10 strongly reduced in adult, narrow and pointed, 14·8 (12) 12–18 shorter than longest upper primary covert; in juvenile, slightly longer, broader, and with more rounded tip, 6·5 (5) 3–11 shorter than longest primary covert. Outer web of p6–p8 and inner web of (p6–)p7–p9 emarginated. Longest tertials reach to tip of p3–p4 in closed wing. Tail rather short, tip almost square; 12 feathers, t1 1–5 shorter than t4–t6. Bill heavy, deep at base, usually rather gradually tapering to sharp tip, except for decurved culmen, but occasionally length about similar to depth at base, culmen markedly decurved, and tip blunt; cutting edge of upper mandible sometimes with slight notch; lower mandible very deep at base. Some short and rather soft bristles along base of upper mandible, near gape, on chin, and in feather-tufts

covering nostrils. Tarsus and toes rather short, strong; claws rather slender, slightly decurved. Middle toe with claw 22·5 (28) 19–25; outer toe with claw *c.* 70% of middle with claw, inner *c.* 75%, hind *c.* 109%; length of hind claw 14·2 (42) 11–16, range exceptionally 10–19.

Geographical variation. Slight, both in colour and size. Slight clinal increase in size towards east, irrespective of races recognized, but birds of Algerian Hauts Plateaux large and *psammochroa* of USSR apparently small (see Measurements). Variation in colour difficult to assess, as geographical trends relatively minor compared with enormous colour changes caused by bleaching and abrasion; only birds at same stage of wear comparable. Nominate *calandra* from southern Europe is darkest race, feather-centres of upperparts black, sides dull grey-brown, brighter olive-brown or rufous-brown only when newly moulted; chest greyish, distinctly spotted black; sides of breast and flanks dull olive-grey, hardly rufous. Populations from Algeria, Tunisia, Libya, and Volga-Ural steppes similar to south European nominate *calandra*, but dark feather-centres of upperparts slightly narrower and fringes slightly paler grey, fringes with pink tinge when fresh; hindneck paler grey and less streaked; very similar to European nominate *calandra*, however, and thus included in that race. Birds from Van Gölü, extreme south-east Turkey, and neighbouring Iranian Azerbaijan slightly paler still, more narrowly streaked dark grey above, tending towards south-eastern race *psammochroa*, but feather-fringes of upperparts grey, not pinkish or sandy-yellow; included here in nominate *calandra*, but perhaps separable as *hollomi* Kumerloeve, 1969. Breeders from western and central Asia Minor rather variable even within one area: some similar to European nominate *calandra*, others like North African birds, still others similar to Van Gölü birds. *M. c. psammochroa* occurs east from Iraq, Zagros mountains, Khorasan (Iran), and Turkmeniya (USSR); feather-centres on upperparts narrow and rather pale grey-brown, feather-fringes broad, pink-cinnamon when fresh, sandy-grey or sandy-yellow when worn; chest and flanks pale grey-buff, spots on cheeks and chest indistinct and small, virtually absent in worn plumage. *M. c. hebraica* from Levant south to Gaza has rather broad and blackish shaft-streaks on upperparts and rather heavy spots on cheeks and chest, as in nominate *calandra*; feather-fringes on upperparts extensively pink-cinnamon, however, pale grey only when plumage heavily worn, deep pink-cinnamon or pale rufous-cinnamon when plumage newly moulted; these fresh birds formerly separated as *gaza* Meinertzhagen, 1919, e.g. by Vaurie (1951a), but *hebraica*, as other races, shows marked change from rufous to greyish within a short period after moult. Further east along Syrian-Turkish border, birds tend towards *psammochroa*, and have been separated as *dathei* Kumerloeve, 1970; here included in *hebraica*, however, as not separable from typical *hebraica* from Birecik (Turkey), Aleppo (Syria), and Lebanon. CSR

Melanocorypha bimaculata Bimaculated Lark

PLATES 5 and 12
[between pages 160 and 161, and 184 and 185]

Du. Bergkalanderleeuwerik   Fr. Alouette monticole   Ge. Bergkalanderlerche
Ru. Двупятнистый жаворонок   Sp. Calandria bimaculada   Sw. Asiatisk kalanderlärka

Alauda bimaculata Ménétries, 1832.

Polytypic. Nominate *bimaculata* (Ménétries, 1832), north-east Turkey, southern Transcaucasia, and mountains of north-west, northern, and western Iran, east to Elburz mountains and Kerman; perhaps this race in southern Urals; *rufescens* C L Brehm, 1855, central, southern, and south-east Turkey, Syria, Lebanon, and Iraq; *torquata* Blyth, 1847, from Khorasan (north-east Iran) and Turkmeniya (USSR) eastward.

Field characters. 16-17 cm; wing-span 33-41 cm. About 10% smaller and relatively longer-winged than Calandra Lark *M. calandra*; slightly shorter but somewhat bulkier than Skylark *Alauda arvensis*. Medium-sized to large lark, of closely similar appearance to *M. calandra* but differing in whiter supercilium, narrower but usually longer black crescents on upper breast, and absence of obvious white trailing edge to wings. Tail shows white terminal band. *M. b. rufescens* distinctly redder-brown than any form of *M. calandra*. Sexes similar; no seasonal variation. Juvenile separable. 2 races breeding in west Palearctic, clearly differing in basic plumage tone.

ADULT. (1) Race of north-east Turkey and northern Iran, nominate *bimaculata*. Distinctly paler, sandier, and greyer above than sympatric *M. calandra* and differing as follows: (a) supercilium and eye-ring whiter, creating more obvious 'spectacle' and contrasting more with dark brown eye-stripe and surround to cheek; (b) streaks on back more contrasting, emphasizing pale nape and rump; (c) black crescents on sides of upper breast narrower and extend towards centre, sometimes meeting; (d) narrow cream tips to secondaries and inner primaries, not forming obvious pale trailing edge; (e) edges to outer tail-feathers buff (not white); (f) all tail-feathers except central and outermost pairs tipped white, usually forming quite distinct terminal band. In flight, last 3 characters create obvious differences from *M. calandra* though white-tipped tail not always easy to see on worn birds. Underwing dark grey-brown, lacking pale trailing edge but with almost translucent flight-feathers contrasting with dark coverts in one autumnm vagrant to England (Flumm 1977). (2) Race of central Turkey to Iraq, *rufescens*. Distinctly darker and more rufous-brown above than nominate *bimaculata* and any dark form of *M. calandra*, appearing very dark in the field. Bill more pointed than *M. calandra*; horn, with lower mandible yellower. Legs flesh-brown. JUVENILE. Differs from adult as in *M. calandra*.

Wing and tail pattern allow ready separation from *M. calandra* in close flight view but care always needed in observations of bird on ground or flying at distance. General character differs most in more pointed bill, shorter tail, and lesser bulk, but such differences difficult to judge on solitary vagrant. Flight, gait, and habits said to be similar to *M. calandra*. Escape-flight certainly similar but proportionately longer wings noticeable on take-off.

Song resembles *M. calandra*. In flight, 'prrp' or 'chirp' like *A. arvensis* but more mellow.

Habitat. In lower-middle warm continental latitudes, in higher and rougher parts of steppe zone, replacing Calandra Lark *M. calandra* up to at least 1650 m in Afghanistan (Paludan 1959), 2000 m in Armeniya, USSR (Dementiev and Gladkov 1954a), and 2150 m in north-east Iran (P A D Hollom). Ascends mountains to limit of crops, and occurs on cultivated plateaux at high altitudes; also on dry heath, shrubland, stony tracts, and almost bare steppes, replacing *M. calandra* on gravelly areas and in uplands. In Anatolia (Turkey), found most frequently in spring in semi-desert from which *M. calandra* absent, e.g. in thorny desert north-west of Tuz Gölü; elsewhere in Turkey, however, the 2 species occur together (Wadley 1951).

During migration uses lake shores, and fields of rice and stubble (Dementiev and Gladkov 1954a). Winters in India in barren semi-desert, sparse cultivation, harvested and fallow fields, margins of jheels, and on dry tidal mudflats (Ali and Ripley 1972).

In Armeniya, frequently feeds on fields where harvest completed but never attacks standing crops (Adamyan 1963a), though this does happen in Sudan winter quarters (Beshir 1978).

Distribution. IRAQ. Breeding only Sammara-Takist area: further investigation required (Allouse 1953). KUWAIT. Bred 1978 (PRH).

Accidental. Britain, Sweden, Finland, Italy.

Population. No information on population trends.

TURKEY. Common in east (Kumerloeve 1969a). USSR. Breeds in large numbers in suitable habitats (Dementiev and Gladkov 1954a).

Movements. Migratory. Some winter on southern edge of breeding range—e.g. Muganskaya steppe in Azerbaydzhan and around Dushanbe in Tadzhikistan (Dementiev and Gladkov 1954a)—but majority migrate south

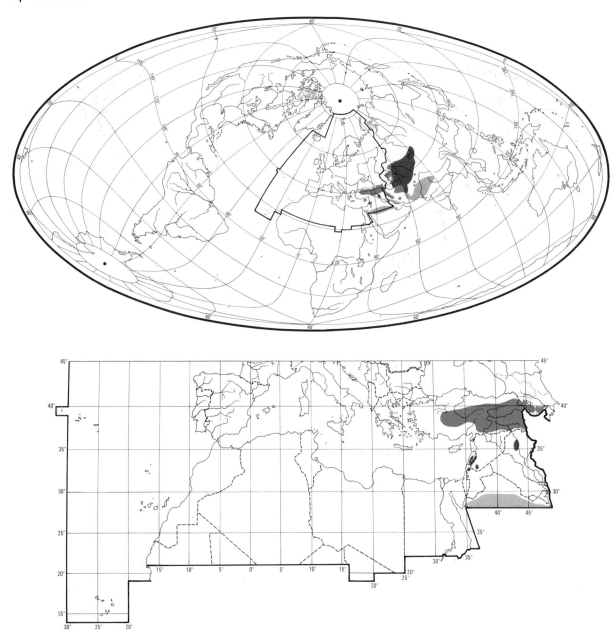

to Pakistan and north-west India, Middle East (including Iran), and north-east Africa (Vaurie 1959; Ali and Ripley 1972). No ringing data, but evidence from skins indicates that population east of Caspian (*torquata*) winters in India, Pakistan, and Iran, while western populations (nominate *bimaculata* and *rufescens*) overlap in Middle East and north-east Africa (Cave and Macdonald 1955; Ali and Ripley 1972; Moreau 1972). Extent of wintering in Levant uncertain; probably mainly a passage migrant there, as on Cyprus (P Flint, P F Stewart).

Timing of passages poorly known but, as in Calandra Lark *M. calandra*, main periods appear to be October–November and March–April. Extreme dates in Turkey,

where a summer visitor, are 1 April and 21 November (Vittery *et al.* 1972; Beaman *et al.* 1975). 'Immense flocks' reported at Wadi Gaza in southern Palestine from mid-August to late October, but none after mid-December (Meinertzhagen 1920*b*); presumably these included local breeders as well as passage migrants. This movement continues into north-east Africa; occurs only on passage in Egypt (P L Meininger, W C Mullié, S M Goodman), but many winter in Sudan where the species is an agricultural pest (Beshir 1978); some also reach northern Ethiopia, especially Eritrea where flocks of up to 30 reported (Smith 1957; Urban and Brown 1971). Regular winter visitor to northern Arabia, and occurs uncommonly or irregularly south to Jiddah on Red Sea and to Oman on Persian Gulf, mainly November–March (Bundy and Warr 1980; Gallagher and Woodcock 1980; Jennings 1981a). RH

Food. Insects and seeds. When taking cultivated *Sorghum*, feeds either by nipping out grain from low seed heads (especially when seed still soft), by perching on large seed heads (more often as seed ripens), or by collecting fallen grain (Beshir 1978). In Sudan, recorded doing considerable damage by digging up freshly sown grain (Schmutterer 1969).

The following items have been recorded in diet in Palearctic. Invertebrates: grasshoppers, etc. (Orthoptera: Tettigoniidae, Acrididae), termites (Isoptera), bugs (Hemiptera), Lepidoptera (Lasiocampidae, Noctuidae, Geometridae, including larvae and pupae), flies (Diptera), Hymenoptera including ants (Formicidae), beetles (Coleoptera: Carabidae, Silphidae, Alleculidae, Scarabaeidae, Tenebrionidae, Cerambycidae, Chrysomelidae, Curculionidae), Arachnida. Plant material: seeds of *Amaranthus*, *Malcolmia*, Leguminosae, *Hypericum*, bedstraw *Galium*, *Heliotropium*, grasses (Gramineae, including cereal grain); plant leaves recorded once. (Meinertzhagen 1951a; Adamyan 1963a; Kekilova 1969; Medvedev and Esilevskaya 1973.)

In Armeniya (USSR), diet during breeding season apparently mainly insects; before and after this period, mainly weed seeds, only rarely cereal grain. 40 adults obtained at various times contained insects and weed seeds. (Adamyan 1963a.) In southern Turkmeniya, spring and summer, all of 21 stomachs contained invertebrates (mainly Lepidoptera, Coleoptera, and locusts; fewer Curculionidae taken than by other Alaudidae in the area) and 11 also contained seeds (Medvedev and Esilevskaya 1973); in autumn, all of 12 stomachs contained seeds (mostly *Amaranthus*, with barley *Hordeum* in 6) and 8 also contained invertebrates (mostly termites and ants) (Kekilova 1969). One *rufescens*, spring, contained seeds, much green food, and a few remains of grasshoppers (Orthoptera). Bird from Kashmir held only insect eggs and pupae. (Meinertzhagen 1951a.)

Nestlings studied in Armeniya ate only insects: 130 items in collar samples comprised 77% (by number) Coleoptera (largely *Blitopertha*, *Omophlus*, and *Galeruca*), 18% Lepidoptera, and 5% grasshoppers (Acrididae). Food brought to nest every 3–15 min; 5–6 insects per visit. (Adamyan 1963a.) DJB

Social pattern and behaviour. Most complete study by Adamyan (1963a) in Armeniya (USSR). Several aspects still not well known.

1. Like other *Melanocorypha* larks, often in large flocks outside breeding season, e.g. up to *c.* 500 for watering (Meinertzhagen 1951a). In Butana semi-desert (Sudan), feeding flocks of up to 200 occur, rarely exceeding 400–500 (Beshir 1978); see also Macleay (1960). One gathering of *c.* 3000 in central Turkey, late September (Beaman 1978). Near Erevan (Armenia), ♂♂ arrive slightly before ♀♀ and remain in flocks of *c.* 20–50 for first few days, especially if cold (Adamyan 1963a). In Turkey, spring, usually in pairs or singly (Wadley 1951). In Armenia, small roving flocks, often in company of Lesser Short-toed Lark *Calandrella rufescens*, typically recorded near breeding grounds after breeding; flocks of up to 300 occur by late August, prior to departure (Adamyan 1963a). In Kazakhstan (USSR), flocks of juveniles noted from July and more general flocking of adults and independent young takes place from August (Dolgushin *et al.* 1970); flocks of *c.* 1000 occur on passage (Dementiev and Gladkov 1954a). Will associate with Calandra Lark *M. calandra* (e.g. Wadley 1951). BONDS. Mating system presumed monogamous though no detailed information. Incubation by ♀ alone but care of young (feeding and nest-sanitation) by both sexes (Adamyan 1963a); see also Relations within Family Group (below). BREEDING DISPERSION. Solitary and territorial. In Kazakhstan, mostly in isolated pairs, though neighbourhood groups (some quite dense) occur where habitat suitable (Dolgushin *et al.* 1970). ROOSTING. Nocturnal and, outside breeding season, communal. In Butana, winter, birds fly *c.* 8 km from feeding grounds (sorghum fields) to open desert, roosting in pits *c.* 4 cm deep, dug with feet. One roost covered *c.* 3·2 × 1·6 km. Birds also roost during day from *c.* 10.00–11.00 hrs, and again *c.* 13.00–14.00 hrs. Pits tend to be sited under plants and birds also defecate there (Beshir 1978). In Jebel ed Druz (Syria), scattered flocks recorded roosting on cut cropland (Meinertzhagen 1951a). ♀ roosts on nest during incubation (Adamyan 1963a). In early spring, will sing (see 1 in Voice) until *c.* 23.00 hrs (Johansen 1952), sometimes also at night, especially if moonlit (Dementiev and Gladkov 1954a; Dolgushin *et al.* 1970). Often lives in waterless areas and possibly not an obligatory drinker (Dolgushin *et al.* 1970; Beshir 1978); elsewhere, large flocks visit watering places, especially in late summer and autumn (Dolgushin *et al.* 1970); in western Zeravshan mountains (Uzbekistan/Tadzhikistan, USSR), mainly 05.30–07.00 hrs (Ivanov 1969). Will bathe in dust or loose earth (Dolgushin *et al.* 1970); for sand-bathing by captive bird, see Werner (1975).

2. Little information on degree of shyness towards man. Vagrants in Britain not particularly shy (Jones 1965; Flumm 1977); perched singing bird in Iran permitted approach to 22 m (P A D Hollom). FLOCK BEHAVIOUR. Flocks formed for watering are tight-packed on ground (Meinertzhagen 1951a). Feeding flocks in Sudan not well integrated, birds continually joining and leaving (Beshir 1978). For calls given by small flock when approached by man, see 3 in Voice. SONG-DISPLAY. According to Meinertzhagen (1938), Song-flight exactly as in *M. calandra*, while Ramsay (1914) referred to a low 'hovering'. Variants possibly occur, as in *M. calandra*. Normally sings while ascending and has tail closed (P A D Hollom); then, with quivering wings, slowly describes circles *c.* 150–200 m across, and glides smoothly down to perch on some eminence (Adamyan 1963a). In slow

A

Song-flight, performed at greater altitude and for longer than *M. calandra*, resembles bat (Chiroptera) (Dementiev and Gladkov 1954a); tail widely fanned in still or moderately windy conditions, wing-action like slow breast-stroke swimming (P A D Hollom). Will also sing from low perch (P A D Hollom) or from ground—in an upright posture (Fig A) or (also while walking or running about) in posture (Fig B) typically

B

assumed by displaying *Melanocorypha* larks (Vere Benson 1970; see *M. calandra*, p. 97); song from ground sometimes accompanied by wing movements (P A D Hollom). In Armeniya, Song-flight performed soon after arrival of ♂ flocks, particularly if weather fine, from late March or early April, though increased vocal activity especially associated with pair-formation from mid-April (Adamyan 1963a). In southern USSR, noted from last 10 days of March (Ivanov 1969). In Afghanistan, ♂ recorded performing Song-flight while ♀ carrying food for young (Meinertzhagen 1938). ANTAGONISTIC BEHAVIOUR. Little information. Apart from Song-flight (see above), ♂-♂ chases (accompanied by loud song) are another typical feature of early breeding season (Adamyan 1963a; Dolgushin *et al.* 1970). Vagrant to Shetland (Scotland), June, showed aggression toward 2 Skylarks *Alauda arvensis* (Whitehouse 1978), but no details. HETEROSEXUAL BEHAVIOUR. (1) General. In Armeniya, pair-formation from mid-April. Most heterosexual activity takes place in sunny weather (sharp increase particularly following rain) and in morning, little around midday; renewal in evening, though none after dusk (Adamyan 1963a). (2) Pair-bonding behaviour. ♀ usually nearby when ♂ performs Song-flight. ♀ will approach ♂ on ground but, when ♂ rushes towards her, ♀ quickly flies *c.* 50–100 m and, on landing, begins to circle in a rather crouched posture. ♂ follows, imitating her every movement. After a few circles, ♂ usually manages to seize ♀'s tail or wing, this then leading to copulation (Adamyan 1963a). (3) Courtship-feeding. No information. (4) Mating. See subsection 2 (above) for pre-copulatory sequence. No further details. (5) Nest-site selection. Presumably by ♀ who does all nest-building (in Armeniya, from late April or early May: Adamyan 1963a); see Breeding. (6) Behaviour at nest. Incubating ♀ flies off to feed every *c.* 20–45 min for breaks of *c.* 10–20 min. While ♀ incubating, ♂ usually sings from rock near nest or circles in vicinity (Adamyan 1963a). RELATIONS WITHIN FAMILY GROUP. Following account from Adamyan (1963a) which see for fuller details of physical development of young. Eggs pipped at *c.* 11–12 days, hatch at *c.* 12–13 days. Both parents start to feed young *c.* 3–4 hrs after hatching. Eyes of nestlings about half-open at *c.* 1 week. At *c.* 9 days, young move about easily and support themselves on their toes; at this stage, leave nest and hide under bushes nearby. Able to flutter for *c.* 10 m and eyes are fully open. Still fed regularly by parents and even when able to feed independently (at *c.* 25–30 days), remain with ♀. Nearly full-grown and able to fly well at *c.* 1 month. ANTI-PREDATOR RESPONSES OF YOUNG. At 1 week, may crouch in nest when

threatened, or even attempt to escape (Adamyan 1963a). PARENTAL ANTI-PREDATOR STRATEGIES. (1) Passive measures. Some ♀♀ sit very tightly. At approach of man, ♂ gives an alarm-call (not described) and ♀ normally moves several metres away from nest before flying up (Adamyan 1963a). (2) Active measures. ♀ not infrequently performs distraction-lure display of disablement type (Adamyan 1963a), e.g. feigning broken wing (Erard and Etchécopar 1970).

(Figs by J P Busby: A from photograph in Dolgushin *et al.* 1970; B from drawing in Vere Benson 1970.) MGW

Voice. Full study of repertoire required, including more detailed comparisons with Calandra Lark *M. calandra*.

CALLS OF ADULTS. (1) Song of ♂. Substantially similar to *M. calandra* (see also below). Recording suggests jingling or jangling trills or buzzes and rattles dancing up and down in pitch (Fig I; compare *M. calandra*, Fig I, p. 99), followed by a rather musical phrase, almost a rippling motif (Fig II); also a softer, fluting 'tiu pu pu pu' (Fig III) and occasional series of rather quiet '(s)ip', 'chip', or 'cheep' sounds like domestic chicks and in irregular temporal pattern, as well as 'dweee' not unlike Greenfinch *Carduelis chloris*. In recording by C Chappuis, pattern of phrasing (and some sounds) sometimes suggest song Thrush *Turdus philomelos*. Other features: sounds strongly reminiscent of Skylark *Alauda arvensis*, cat-like purrs and descending 'trip' sounds (Fig IV), sweet, rather thin whistling melody (Fig V), attractive deep warbling, and peculiar nasal 'chir' or like the word 'you' spoken with a metallic timbre and menacing tone (J Hall-Craggs, M G Wilson: Fig VI). Loud and not unpleasant repetition of short, warbling strains with few clear notes and with pauses after every few phrases (Ramsay 1914). Less beautiful and rich than *M. calandra* according to Dolgushin *et al.* (1970), while Johansen (1952) found it louder and more varied; shorter, less sustained, with fewer harsh grating sounds than *M. calandra* (Erard and Etchécopar 1970). Often recalls Short-toed Lark *Calandrella brachydactyla* in churring twitter or rattle (Vere Benson 1970). Song delivered from ground, Cyprus, early April, a hard, clicking, and continuous 'tchup tchup tchup tchup' (F J Walker); in Turkey in May, ♂ apparently on breeding grounds gave much more animated and varied song from ground (P A D Hollom). 6–7 bouts of song given initially with pauses of *c.* 10–20 s, then of *c.* 2–10 min (Adamyan 1963a). Possible Subsong by vagrant in Britain a chirpy trill more rattling than *A. arvensis*; little rise and fall in pitch. Abrupt, monotonous, subdued, and rather squeezed-out; each phrase disjointed. Contained 'prrp', 'cheewit', 'che-wit-che', and 'chirp' sounds (Jones 1965). Song may contain mimicry of calls and songs of other species in area: e.g. in Armeniya (USSR), Bee-eater *Merops apiaster*, *C. brachydactyla*, Isabelline Wheatear *Oenanthe isabellina*, and Crimson-winged Finch *Rhodopechys sanguinea*; also calls of snowcocks *Tetraogallus* and Pale Rock Sparrow *Petronia brachydactyla* (Adamyan 1963a). (2) Contact- and flight-calls. A 'tchup turrup' like

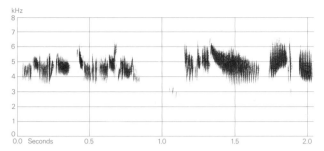

I   B N Veprintsev and V V Leonovich   USSR   April 1974

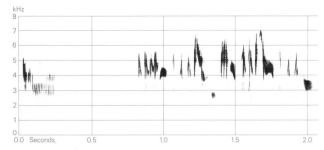

II   B N Veprintsev and V V Leonovich   USSR   April 1974

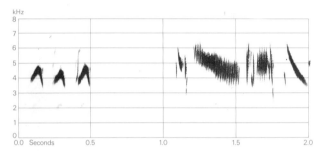

III   B N Veprintsev and V V Leonovich   USSR   April 1974

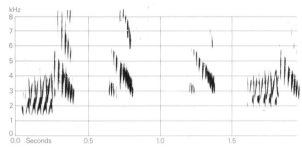

IV   C Chappuis   Iran   May 1977

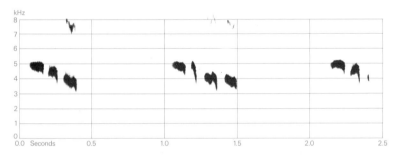

V   C Chappuis   Iran   May 1977

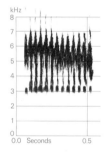

VII   B N Veprintsev and V V Leonovich USSR   April 1974

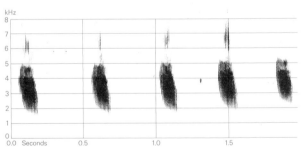

VI   C Chappuis   Iran   May 1977

cheewit-chewit' (Jones 1965). (3) Alarm-calls. On Cyprus in April, small flock feeding on ground gave soft 'twe-arp twe-arp twe-arp'; on closer approach, 4 'chup' sounds and 'chulp chulp cher lit' (F J Walker). (4) Other calls. (a) Recording (Fig VII) suggests a quite loud, high-pitched, metallic buzzing trill—like running nail over teeth of metal comb (M G Wilson); strongly recalls churr of *Acrocephalus* warbler (P J Sellar). Possibly same as flight-call described by Beshir (1978), but context not known.

CALLS OF YOUNG. No information.　　　　MGW

A. arvensis (Walker 1981a, b); deep, not chirpy 'knoop knoop' (Wadley 1951) presumably the same. Melodious, repeated, rising and falling 'ter lu lup chup' given in flight (F J Walker). A loud, cheery trill (Beshir 1978) perhaps the same. Gives 'crryit' and 'tjurrup' sounds often recalling *A. arvensis*; also 'pchi-pcherp' and 'tyipp' (Jonsson 1982). Short 'prrp' or 'chirp' by British vagrant, like *A. arvensis*, but more mellow and subdued; less often 'prrp-

Breeding. Most information from Adamyan (1963a) for Armeniya (USSR). SEASON. Armeniya (USSR): laying begins first few days of May; latest fresh clutches mid-July. SITE. On ground in shelter of tussock or low bush, usually shaded from sun at hottest time of day. Nest: shallow scrape, lined with grasses and rootlets, walls of coarser material than floor; also clad in bits of old dung, rag, paper, etc. Average internal diameter 7·6 cm (7·1-7·8), external diameter 12·2 cm (11·0–13·5), depth of cup 5·8 (5·0–6·7), total height 7·1 cm (5·5–8·3), thickness of wall 1·7 cm (1·1–2·3); sample size not given. Building: by ♀ only; first clears away stones and dead vegetation, then digs scrape 10–15 cm across and up to *c.* 7 cm deep over 1–2 days; lining added in further 4–5 days. EGGS. See Plate 79. Sub-elliptical to oval, smooth and glossy; white, greyish, or brownish, occasionally with olive tinge, heavily or lightly spotted and speckled, sometimes in patches, with light or dark brown and some pale purple. 24·2 × 18·0 mm (22·8–24·5 × 16·8–18·9), *n* = 37; calculated weight 4·04 g (Schönwetter 1979). 25·8 × 17·8 mm (24·8–27·5 × 17·0–18·5), *n* = 20; weight 4·09 g (3·8–4·6), *n* = 20 (Adamyan 1963*a*). Clutch: 3–5, rarely 6. Of 26 clutches, Armeniya: 3 eggs, 6; 4, 15; 5, 5; mean 3·96. Possibly 2 broods (Dementiev and Gladkov 1954*a*). INCUBATION. 12–13 days. By ♀ only, beginning with penultimate egg. YOUNG. Altricial and nidicolous. Cared for and fed by both parents. FLEDGING TO MATURITY. Leave nest at *c.* 9 days, before able to fly. BREEDING SUCCESS. Of 37 eggs laid, Armeniya, 23 disappeared, mainly due to predators, and 1 clutch of 4 destroyed in heavy rain; 9 young hatched.

Plumages (*M. b. rufescens*). ADULT. Closely similar to Calandra Lark *M. calandra*, differing mainly in more distinct marks on side of head with more pronounced supercilium, mainly dark outer tail-feathers, and only narrowly white-tipped secondaries and inner primaries. Dull black shaft-streaks on upperparts rather narrow and sharp; broad lateral fringes grey, distinctly tinged tawny, pale rufous-cinnamon, or greyish-cinnamon when fresh, paler pinkish-grey or sandy-grey when worn; shaft-streaks narrower and fringes paler than *M. c. calandra*, but resembling races *hebraica* and *psammochroa* of *M. calandra* (which, however, often show almost unstreaked hindneck); rump almost uniform pale olive-grey, often contrasting slightly with tawny or pink-cinnamon upper tail-coverts and fringes of t1 (some *hebraica* and *psammochroa* similar, however). Distinct white supercilium, rather narrow in front of eye towards nostrils (where contrastingly bordered by narrow blackish lines above and below), wider behind eye (where bordered below by blackish streak along upper ear-coverts). Narrow white eye-ring. White patch from upper gape to below eye; bordered above by blackish line on lores, below by dusky line from gape to behind eye. Ear-coverts cinnamon-buff to sandy-grey, faintly spotted darker grey, dark spots sometimes joined to dark band along rear of ear-coverts and towards sides of chin. White patch near lower mandible, often bordered below by mottled dusky line along sides of chin. *M. calandra* also has pale supercilium, but only behind eye and not distinctly bordered by black lines; dark lines on sides of head and bar along rear of ear-coverts occasionally distinct in *M. calandra*, but mainly in nominate *calandra* and in *hebraica*, not in paler *psammochroa*. Black crescent on upper side of chest slightly narrower than in *M. calandra*, but extending further towards centre, narrowly meeting in front in about half of specimens examined (meeting in only 10% of *M. calandra*). Chest and sides of breast pale buff or off-white, sides of breast rather indistinctly spotted dark brown, chest virtually uniform apart from grey wash to centre (*M. calandra psammochroa* closely similar, however). T2–t6 olive-black or black, outer web of t6 dark olive-brown with narrow pink-cinnamon or off-white outer fringe; feather-tips cream or white for 5–9 mm on inner webs,

2–4 mm on outer webs (*c.* 1 mm only on tip of outer web of t6) (*M. calandra* has similar pale tips on t4 and inner web of t5, but tips of t2–t3 narrower, and outer web of t5 and all t6 largely white). Flight-feathers as in *M. calandra*, but greyer, less blackish and contrasting less with upper wing-coverts; outer edges and tips narrowly fringed tawny or greyish-cinnamon, fringes not exceeding 1 mm in width (unlike *M. calandra*, secondaries and inner primaries not broadly tipped white). Undersurface of tail, under wing-coverts, and axillaries slightly paler than *M. calandra*, dark grey-brown rather than greyish-black. Bleaching and abrasion have same effect as in *M. calandra*, dark streaks on upperparts (including hindneck) and patches on chest becoming more distinct; unlike *M. calandra*, white supercilium, dark lines on lores and from gape backwards, and dark surround to ear-coverts become more pronounced. White tips of tail and secondaries disappear through abrasion in heavily worn plumage; tip of tail then similar to *M. calandra*, but t6 dark, not white (beware *M. calandra* missing t6); *M. calandra* always retains white secondary tips, even when heavily worn. NESTLING. Down sparse, straw-yellow (Adamyan 1963*a*, which see for development). JUVENILE. Closely similar to juvenile *M. calandra* and often hard to distinguish. Both species may show pronounced supercilium and both often do not show dark patch or crescent on side of chest; spots on centre of chest of *M. bimaculata* smaller than in most *M. calandra*, but *M. calandra psammochroa* often has spots equally small. Main distinguishing features are pattern of outer tail-feathers and secondaries, as in adult, though differences less pronounced: all tail-feathers of *M. bimaculata* dark grey-brown or olive-brown, rather narrowly and evenly fringed pale buff or white, only outer web of t6 largely white (in *M. calandra*, outer web of t5 and t6 buff or white); secondaries similar to tail-feathers, hence tips only narrowly and sharply fringed white (in *M. calandra*, white fringes on secondary-tips often broader, but sometimes equally narrow; not sharply defined, however, merging gradually into grey-brown of feather-centres). Juvenile p10 almost equal in length to longer upper primary coverts, rounded at tip, fringed cream on outer web (in adult, tiny, narrowly pointed, fully grey-brown). FIRST ADULT. Similar to adult; indistinguishable once last juvenile feathers (usually p9–p10) replaced.

Bare parts. ADULT. Iris light brown, brown, or dark brown. Upper mandible horn-brown to horn-black, darkest on culmen and tip; cutting edges of upper mandible and all lower pale horn or greenish-horn, changing to yellow or yellow-horn at base and gape. Leg and foot bright pink-flesh, flesh, yellow-flesh, brownish-flesh, or yellow-brown, joints more or less dusky; claws dusky. (Sharpe 1890; Jones 1965; Flumm 1977; BMNH.) NESTLING. Bare skin yellowish-pink. Bill pale brown. Leg, foot, and claw pale yellow. (Adamyan 1963*a*.) JUVENILE. No information.

Moults. ADULT POST-BREEDING. In USSR, moult starts with loss of p1 from 17 July; moult of secondaries and greater upper wing-coverts starts when p6 shed, late July; tail starts with shedding of p4–p5; all moult usually completed by second half of August; in Iran, moult mainly late June and early July (Adamyan 1963*a*). In Iran and Afghanistan, all specimens 8 July to 30 August were in moult; one from 9 July had started with crown, remainder of plumage (including wing) old; body moult virtually complete 6–30 August, but wing and tail only about ¾ grown; 2 from 11 September completed or virtually so, 4 others still with some wing- and tail-feathers growing; all moult completed October (Vaurie 1951*a*).

Measurements. ADULT. *C. b. rufescens.* Central southern Turkey (mainly Gaziantep) and Levant south to Jebel ed Druz (southern Syria), April–June; skins (BMNH). Bill (S) to skull, bill (N) to distal corner of nostril; exposed culmen on average 4·2 less than bill (S). Wing includes some data from Vaurie (1951*a*) and Kumerloeve (1961, 1970*b*), tail, bill, and tarsus include birds from central Anatolia, Turkey (BMNH, ZMA).

	♂		♀	
WING	123·5 (2·41; 19)	119–129	114·4 (1·94; 10)	111–117
TAIL	53·2 (2·34; 17)	49–57	49·6 (1·95; 11)	47–53
BILL (S)	21·3 (0·90; 19)	20·4–23·2	20·1 (0·85; 18)	19·2–21·4
BILL (N)	14·2 (0·76; 17)	13·1–15·7	13·2 (0·58; 11)	12·4–14·1
TARSUS	26·7 (0·55; 17)	25·9–27·6	26·6 (0·67; 11)	25·4–27·6

Sex differences significant, except for tarsus.

Wing of some other populations; breeding season. *C. b. rufescens*: (1) central Anatolia (Ankara to Taurus and Beyşehir Gölü) (Kumerloeve 1970*a*; BMNH, ZMA), (2) Van Gölü area (south-east Turkey) (Kumerloeve 1968, 1969*a*). Nominate *bimaculata*: (3) north-east Turkey (Kumerloeve 1968), Armenia, USSR (Nicht 1961), north-west Iran, and Zagros mountains, Iran (BMNH). *C. b. torquata*: (4) Turkmeniya and Tadzhikistan (USSR), and Afghanistan (Vaurie 1951*a*; Paludan 1959; RMNH, ZMA); (5) Khorasan (eastern Iran), Pakistan, and north-west India (Vaurie 1951*a*).

	♂		♀	
(1)	122·9 (1·67; 5)	121–126	113·4 (1·49; 4)	111–115
(2)	121·5 (1·86; 15)	118–124	114·8 (— ; 3)	113–116
(3)	124·6 (2·18; 16)	119–128	113·1 (2·55; 9)	110–118
(4)	123·4 (1·94; 9)	120–126	113·6 (2·02; 7)	110–116
(5)	121·0 (— ; 20)	118–125	114·1 (— ; 15)	110–119

Wing of ♂ from Lorestan (Zagros mountains, Iran): 125·1 (10) 122–130 (Vaurie 1951*a*).

JUVENILE. Wing and tail on average both *c.* 4 shorter than adult.

Weights. All races combined: (1) February–April, (2) May, (3) June, (4) July–October (Paludan 1940; Kumerloeve 1961, 1968, 1969*a*, 1970*a*, *b*; Nicht 1961; Dolgushin *et al.* 1970).

	♂		♀	
(1)	56·8 (4·98; 5)	52–62		
(2)	54·1 (2·97; 13)	48–59	54·1 (4·33; 7)	48–62
(3)	53·1 (2·72; 12)	49–57	52·0 (5·39; 5)	47–59
(4)	52·5 (2·72; 5)	49–56	50·3 (— ; 3)	49–53

Structure. Wing relatively longer than in *M. calandra*, *c.* 2·3 times tail length rather than *c.* 2·0. P8 longest, p9 0–2 shorter, p10 80–87 (♂) or 73–78 (♀), p7 1–5, p6 10–17, p5 20–27 (♂) or 17–22 (♀), p4 27–33 (♂) or 23–29 (♀), p1 38–48 (♂) or 32–39 (♀). Adult p10 13·3 (7) 12–15 shorter than longest upper primary covert. Outer web of p7–p8 and inner of (p7–)p8–p9 emarginated. Tail short, tip square. Bill as in *M. calandra*, but relatively longer, less deep at base, more acutely pointed. Middle toe with claw 22·0 (7) 21–23 mm. Hind claw 13·1 (6) 11–15. Remainder of structure of wing, bill, leg, and toe as in *M. calandra*.

Geographical variation. Slight in colour, virtually none in size. *M. b. rufescens* from central southern Turkey, northern Syria, Iraq, and Lebanon is most rufous race: feathers of crown, hindneck, mantle, scapulars, and back with black centres, olive-grey sides, and rather deep pink-cinnmamon or tawny fringes; ear-coverts, chest, and fringes of flight-feathers tinged tawny-cinnamon; supercilium white or (occasionally) pale cinnamon; general colour close to *M. calandra hebraica*. Birds from Jebel ed Druz (southern Syria) average slightly paler than those farther north, those from central Anatolia, southern Taurus, and Van Gölü area (Turkey) are slightly darker and tend towards nominate *bimaculata* (Kumerloeve 1968, 1969*a*, 1970*a*; BMNH, ZMA). Nominate *bimaculata* from north-east Turkey, Transcaucasia, and northern and western Iran through Elburz and Zagros mountains has sides of feathers of upperparts mainly pale grey, only slightly pink along fringes, not as cinnamon as *rufescens*; ear-coverts and chest tinged grey-buff, chest faintly spotted cinnamon-brown, fringes of flight-feathers creamy or white, less tawny; supercilium often more pronounced than in *rufescens*; in general, colour close to *M. calandra psammochroa*. *M. b. torquata* breeds from north-east Iran and Turkmeniya (USSR) eastwards, but in winter occasionally occurs in west Palearctic (specimen from Jordan in BMNH); upperparts paler and greyer than nominate *bimaculata*, streaks on upperparts less dark and heavy; rump purer grey, less olive or brown; ground-colour of upperparts paler sandy-grey in worn plumage, not as dull grey as in nominate *bimaculata*, but otherwise similar.

Closely related to *M. calandra*, but overlap in geographical range too extensive to allow classification as superspecies. CSR

Melanocorypha leucoptera White-winged Lark

PLATES 6 and 12
[between pages 160 and 161, and 184 and 185]

Du. Witvleugelleeuwerik   Fr. Alouette leucoptère   Ge. Weissflügellerche
Ru. Белокрылый жаворонок   Sp. Calandria aliblanca   Sw. Vitvingad lärka

Alauda leucoptera Pallas, 1811. Synonym: *Melanocorypha sibirica*.

Monotypic

Field characters. 18 cm; wing-span 33–37 cm. Close in size to Calandra Lark *M. calandra* but with more attenuated silhouette, due to relatively longer, even slightly forked tail; 15% longer and much bulkier than Snow Bunting *Plectrophenax nivalis*. Upperparts chestnut on crown and wing-coverts; underparts essentially white with streaked chest and flanks. Wings black and chestnut with thick white trailing edge. In flight, shape and action recall Skylark *A. arvensis* as much as *M. calandra*. Sexes closely similar; some seasonal variation. Juvenile separable.

ADULT MALE. Hindneck, back, rump, and scapulars pale brown-grey, with dark brown feather-centres creating streaks and some pinker fringes most obvious on sides of rump. Against these plumage areas, the following characters stand out: (1) chestnut crown, upper cheeks, and lesser and primary coverts; (2) cream oval surround to eye; (3) black-brown-centred, chestnut- or buff-fringed median and greater coverts, with median creating quite obvious feature; (4) white panel of bunched secondaries, contained by almost black, buff-fringed tertials and black-brown, buff-tipped primaries (last creating long wing-point, up to $\frac{1}{2}$ length of tertials); (5) white outer web to long outermost primary. Underparts white, with chestnut tinge to sides of breast and flanks; overlaid with variable dark grey-brown marks most obvious in scattered necklace on sides of and (less obviously) centre of upper breast and in streaks along flanks. In spring, abrasion makes chestnut areas more uniform, mantle darker, and marks of underparts less obvious. In flight, upperwing shows bold tri-coloured pattern of chestnut coverts, black central panel formed by secondary-bases and most of primaries, and thick white trailing edge reaching to inner primaries; leading edge of outermost primary also white. Underwing white. Tail has mainly black-brown centre contrasting with white edges and rump. ADULT FEMALE. Essentially as ♂ but all contrasts subdued, with markedly less chestnut crown and paler chestnut wing-coverts. Upper breast more uniformly and heavily streaked. JUVENILE. Shows pale tips to upperpart feathers typical of juvenile larks, but otherwise resembles dull ♀, with crown and wing-coverts brown, not chestnut. At all ages, bill short and conical; brown-horn. Legs rather short; yellow-brown.

Distinctive lark, but not free of confusion. 2 traps exist for unwary observer: (1) similarity of plumage pattern worn by juvenile and 1st-winter ♂ *P. nivalis* and (2) close similarity of both character and plumage of Mongolian Lark *M. mongolica*. Different form and voice of *P. nivalis* soon obvious, but *M. mongolica* presents more serious problem, through either escape or potential vagrancy (has plumage colours even more intense than those of *M. leucoptera*, with chestnut tones stronger on head and rump but adult has diagnostic bold black patches on sides of upper breast; see photographs in Piechocki *et al.* 1982). Flight has typical lark action and, together with longer tail than in other *Melanocorypha*, recalls *Alauda arvensis*. Escape-flight low and usually short, bird plummeting into cover and running on before looking back. When landing on bare ground (or disturbed there), adopts upright alarm posture before running away to hide in cover. Song-flight can be of soaring type, or involve circling. Gait as *M. calandra*; perches on bushes and tussocks. In spite of bold plumage, often inconspicuous on ground.

Song continuous, high-pitched, and repetitive. Calls little known: in one vagrant resembled *A. arvensis*.

Habitat. In continental warm arid middle latitudes, lowland and to some extent upland, avoiding rocky and mountainous, forested, and wetland areas, and those humanly settled or much disturbed, although sometimes overlapping cultivation, especially on migration or in winter. Tolerates saline soils, but also lives on sand, clay, gravel, and sometimes on dark earth. Prefers vegetation neither tall nor dense, such as *Festuca–Artemisia* associations with patches of bare clay, but also occurs in feather-grass *Stipa* and herbaceous steppe and may inhabit shores of fresh water and salt lakes in hills. For breeding density transects in various habitats, see Social Pattern and Behaviour; see also Dolgushin *et al.* (1970) and Koshelev (1980). Visits water, including saline water, where opportunity offers. In winter occurs in forest-steppe (Dementiev and Gladkov 1954a). More dependent than Black Lark *M. yeltoniensis* on *Stipa* steppe, preferring its tufts among short grass of meadow-like character. Avoids damp grasslands, but in summer frequents neighbourhood of salt-marshes; returns late in spring and avoids snow as well as highly saline ground. Sometimes rises high to sing, but mostly remains below 15 m (Bannerman 1953).

Distribution and population. Little information on population trends.

USSR. Invariably less numerous than other Alaudidae; by no means ubiquitous and where it does occur only a few pairs (Dementiev and Gladkov 1954a); declined between Volga and Ural rivers, Kazakhstan (Shishkin 1976). Range decreased in last 100 years with ploughing up of steppes and creation of shelter-belts (GGB).

Accidental. Britain, Belgium, West Germany, Finland, Poland, Czechoslovakia, Austria, Switzerland, Italy, Yugoslavia, Greece, Bulgaria, Malta.

Movements. Short-distance migrant, mainly within USSR. Following based on Grote (1936), Dementiev and Gladkov (1954a), and Vaurie (1959).

Most of breeding range vacated in winter, exceptions being southern and western extremities where, however, winter birds may have come from further north or east. No evidence that any are resident (but no ringing data). Winter range extends south and (especially) south-west: to Turkestan, Transcaspia, northern Iran, Caucasus, Black Sea region of USSR, Ukraine, and Rumania. Exceptionally, has wandered even further west in winter, and has reached North Sea countries.

Birds flock in midsummer, when breeding completed, and some movement begins then. Exodus protracted, and some birds still present in southern Urals in December; however, main passage occurs mid-August to October. Return movement proceeding in March; southern areas reoccupied by mid-April, though in northern parts main arrivals in late April or even early May. As in other migrant Alaudidae, moves on broad front, with flock succeeding flock when passage at peak.   RH

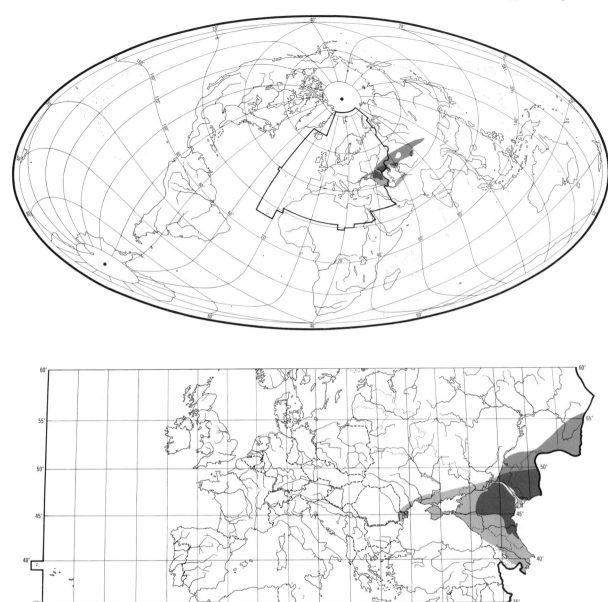

Food. Insects and seeds in summer, seeds in winter. No information on feeding methods. Does not avoid saline water (Dementiev and Gladkov 1954a).

The following recorded in diet. Invertebrates: grasshoppers, etc. (Orthoptera: Tettigoniidae, Acrididae), bugs (Hemiptera: Pentatomidae, Miridae), Lepidoptera (including larvae of Pyralidae and Noctuidae), Hymenoptera (sawflies Tenthredinidae, Ichneumonoidea, ants Formicidae), beetles (Coleoptera: Carabidae, Hydrophilidae, Scarabaeidae, Buprestidae, Melyridae, Anthicidae, Tenebrionidae, Nitidulidae, Cerambycidae, Chrysomelidae, Curculionidae), spiders (Araneae). Plant material mainly seeds or fruits: of docks and knotweeds (Polygonaceae), goosefoots (Chenopodiaceae), sea-

lavender *Limonium*, Compositae, sedges (Cyperaceae), and grasses (Gramineae) including cereal grain; also vegetative parts. (Dementiev and Gladkov 1954a; Ryabov and Mosalova 1967a; Samorodov 1968; Shishkin 1980.)

In northern Kazakhstan (USSR), Ryabov and Mosalova (1967a) compared diet in 1938 (presumably in summer) with diet in summer 1960–2, i.e. before and after cultivation of steppes. Before cultivation, animal food predominant: present in 100% of 152 stomachs, plant material in 29·6%. After cultivation, only slightly more animal than plant food in stomachs (by wet weight and volume): animal in 85·4% of 48 stomachs, plant in 81·2%. Animal food analysed from before cultivation comprised 555 items, after cultivation 156 items: beetles (largely Curculionidae) 49·7% and 23·1% respectively of total animal items (by number), Orthoptera 35·3% and 11·5%, Lepidoptera 9·5% and 0, spiders 2·5% and 0, Hemiptera 2·2% and 12·8%, Hymenoptera (largely ants) 0·7% and 49·4% (in 1960-2 data, 3·2% not accounted for). Plant food before cultivation comprised seeds of wild plants (in 26·3% of stomachs) and vegetative parts (in 3·3%); after cultivation, wheat and millet in 56·2% of stomachs, seeds of wild plants (largely *Polygonum*) in 25·0%, and vegetative parts in 6·3%. In Kalmytskaya ASSR, December-January, 31 stomachs contained 5 beetles and 477 seeds and fruits: plant material included 49·1% (by number) Polygonaceae, 41·9% *Chenopodium*, 5·2% cereal grain, and 3·8% *Artemisia* (Samorodov 1968). Birds from Crimea, winter, contained only seeds (Meinertzhagen 1951a). See also Kistyakivski (1926), Dementiev and Gladkov (1954a), and Ryabov and Samorodov (1969).

In April 1975 in semi-desert of north Caspian region, birds fed on huge hatch of caterpillars (Pyralidae): each ate average 458 caterpillars (9·2 g dry weight) per day, in total (with density of 25 birds per km^2) removing 0·7% of available caterpillars per day; (presumably total) energy consumption of free-living birds 263 kJ per bird per day (Shishkin 1980).

No information on food of young.   DJB

Social pattern and behaviour.

1. Normally gregarious outside breeding season, flocks (often large) are formed for migration, local movements, and winter feeding—e.g. in Kherson steppes (Ukraine, USSR) from November (Dementiev and Gladkov 1954a)—also for roosting (Meinertzhagen 1951a). In Kalmytskaya ASSR, winter, flock of c. 4000 recorded resting in steppes (Samorodov 1967); average feeding flock size there 201 (Samorodov 1968). Late-February passage flocks, Kazakhstan (USSR), contained 30-100 birds (n = 40 flocks) (Dolgushin et al. 1970), and in area between Sea of Azov and Volga river, late October and early November, up to 100-150 (Bub 1955). Small flocks of juveniles (up to c. 100 recorded: Grote 1936) formed from mid-July (Dementiev and Gladkov 1954a; Dolgushin et al. 1970). In Barabinskaya steppe (Novosibirsk, USSR), less numerous than in Kazakhstan and no juvenile flocks or pre-departure gatherings recorded (Koshelev 1980). In autumn, frequently associates with other Alaudidae (Dolgushin et al. 1970), notably Black Lark *M. yeltoniensis* (Portenko 1954). BONDS. No information on mating system, but in Barabinskaya steppe, mid-June, transect-counts apparently showed sex-ratio skewed toward ♂♂ (Koshelev 1980; see also part 2 and *M. yeltoniensis*, p. 118). Young tended by both parents and family unit remains intact until young able to fly (Dolgushin et al. 1970). BREEDING DISPERSION. Solitary and territorial. Nesting territory said by Grote (1936) to be small, so that almost appropriate to speak of 'extensive breeding colonies'. In steppes of northern Kazakhstan, late 1930s, most numerous bird together with Skylark *Alauda arvensis*: in favoured habitat, 12-14 birds in 1 km^2. By late 1950s, c. 5 times more common in fallow and ploughed-up areas than in virgin steppe—122 birds along car transect of 100 km; where virgin land abutted on fallow arable, 355 birds per 100 km (Ryabov and Mosalova 1967a). In area between Volga and Ural rivers, 20 birds in 1 km^2 (Shishkin 1976). For transect

counts in Volgograd region, early April to early August, see Golovanova (1967). In Barabinskaya steppe, shows preference for *Stipa-Festuca* steppe, *Artemisia*, and solonchak (loose-textured saline soil) near lakes, so that dispersion rather fragmented. Transects (from motorcycle, belts c. 50-100 m wide), mid-June, showed 4 pairs along 1 km, 3 pairs in 2 ha of solonchak and sparse, low *Artemisia* and *Festuca*, 3 pairs in 8 ha of dry, fallow ridge with similar vegetation, 3 pairs in 6 ha of *Artemisia-Festuca* steppe. Pairs c. 100-150 m apart, and c. 100-200 m from nests of *A. arvensis* (Koshelev 1980, which see for detailed comparisons with *A. arvensis*). In Zaysan depression (Kazakhstan), in very good years, 2-3 pairs along 1 km (Dolgushin et al. 1970), and in Ural-Emba watershed area, 0·1-3·8 pairs with meadow-steppe as favoured habitat (Poslavski 1963). Other transect counts in north Caspian region produced 8-12 pairs along 10 km, higher values relating to *Artemisia-grass-Anabasis* habitat (Poslavski 1974). ROOSTING. Nocturnal and, outside breeding season, communal. At Balaclava (Crimea, USSR), in large, very dispersed flock on short grass (Meinertzhagen 1951a). In Kustanay (northern Kazakhstan), not infrequently roosts in lofts and barns like *A. arvensis*, especially in bad weather (Ryabov and Samorodov 1969). In Kazakhstan, birds said to have special 'shade-nests' for diurnal roosting in hot weather (Ryabov 1949); not recorded in Barabinskaya steppe (Koshelev 1980). Regularly visits watering places to drink: at Lake El'ton (Volgograd, USSR), mainly 11.00-12.00 hrs, then 15.00-16.00 hrs, sometimes 13.00-14.00 hrs (Dementiev and Gladkov 1954a). Dust-bathes on roads, etc. (Dolgushin et al. 1970).

2. Warier than Calandra Lark *M. calandra* (Dementiev and Gladkov 1954a). Vagrant ♀, southern England, generally shy, spending much time crouched in cover; having landed in bare area, would run to nearest cover and crouch there until flushed again (Sage and Jenkins 1956). Such behaviour typical of *Melanocorypha* larks (Meinertzhagen 1951a). FLOCK BEHAVIOUR. No information. SONG-DISPLAY. Song-flight usually performed at height of c. 10-20 m and not very long (but see below); sometimes sings while flying even lower over ground (Grote 1936; Dementiev and Gladkov 1954a: see 1 in Voice). In Barabinskaya steppe, Song-flight at c. 30-40 m less common than Song-display on ground (Koshelev 1980; see below). Ascends for Song-flight with fairly rapid and shallow wing-beats (Esilevskaya 1967); wings then beaten slowly and deeply so that tips meet above and below body (Dolgushin et al. 1970); in fairly long flight, slow and deep wing-beats alternate with rapid and shallow ones; will also hover briefly (Esilevskaya 1967, 1968). English vagrant ♀ (see above) recorded gliding down like *A. arvensis*: wings rigid, turned slightly down at tips, tail just above level of back (Sage and Jenkins 1956), but not known if this is, or resembles, typical descent from Song-flight. Often sings from some eminence, e.g. in western Kazakhstan, commonly on mounds created by susliks *Citellus*, or from bush (Dolgushin et al. 1970). According to Johansen (1952), invariably lands on bush and continues singing after descent from Song-flight. In Barabinskaya steppe, early June, ♂ recorded running about with wings drooped (i.e. half-opened) and tail cocked, jumping on to hummock and singing (Koshelev 1980); see other *Melanocorypha* larks for descriptions of similar display postures. Will sing on migration (Grote 1936); on breeding grounds, mainly from April through to end of July (Dementiev and Gladkov 1954a). ANTAGONISTIC BEHAVIOUR. Virtually nothing known. First to arrive on Barabinskaya steppe, mid- to late May, are lone ♂♂, followed by flocks of 3-7 ♂♂ and pairs. ♂♂ normally occupy territories from end of May and pairs on territories from c. 5-10 June. (Koshelev

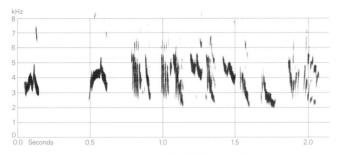

I   B N Veprintsev and V V Leonovich   USSR   April 1974

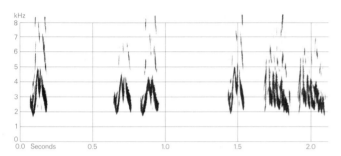

II   B N Veprintsev and V V Leonovich   USSR   April 1974

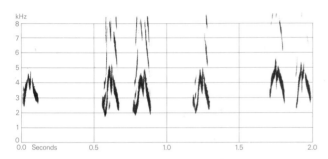

III   B N Veprintsev and V V Leonovich   USSR   April 1974

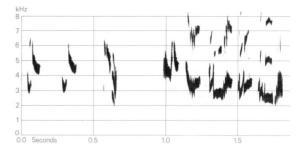

IV   B N Veprintsev and V V Leonovich   USSR   April 1974

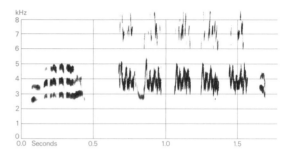

V   B N Veprintsev and V V Leonovich   USSR   April 1974

1980.) HETEROSEXUAL BEHAVIOUR. (1) Pair-bonding behaviour. In Barabinskaya steppe, pair-formation takes place on migration or soon after arrival. Many ♂♂ remain unpaired, at least in some years, and leave breeding grounds in mid-June (Koshelev 1980). On landing from Song-flight, will adopt posture as described above, with head feathers also slightly ruffled to give effect of crest, and run about ♀ thus (Dolgushin *et al.* 1970). (2) Courtship-feeding. No information. (3) Mating. No information. (4) Nest-site selection. No information. (5) Behaviour at nest. ♂ usually *c.* 5 m away while ♀ incubating (Koshelev 1980). RELATIONS WITHIN FAMILY GROUP. Young leave nest before flight-feathers fully developed. Tended by both parents during nestling period and afterwards until able to fly (Dolgushin *et al.* 1970). ANTI-PREDATOR RESPONSES OF YOUNG. If frequently disturbed, may leave nest at 6–7 days old; run *c.* 1–2 m, then crouch, relying on very effective crypsis (Shishkin 1982). Older young out of nest also crouched when approached by man (Koshelev 1980). PARENTAL ANTI-PREDATOR STRATEGIES. (1) Passive measures. Some ♀♀ very tight sitters, leaving nest only at last moment. One flew up silently, landed *c.* 5 m from nest, and then started to run off (Dementiev and Gladkov 1954a; Koshelev 1980). (2) Active measures. Readily performs distraction-lure display of disablement type, sometimes

both parents together. One ♂ tending juvenile ran about, hid in grass, and called (not described) upon landing at certain distance (Koshelev 1980).   MGW

Voice.

CALLS OF ADULTS. (1) Song of ♂. Of fine quality and similar to Skylark *Alauda arvensis* (Grote 1936) or even more melodious (Kozlova 1975), though considered by Dementiev and Gladkov (1954a) and Portenko (1954) to be simpler, louder, harsher, and less sonorous. Recording (Fig I) reveals 2 quite discrete 'tweep' units, then a cascade like water running over stones, rocks, etc. (J Hall-Craggs), or suggesting a continuously high-pitched song Thrush *Turdus philomelos* (P J Sellar). Overall, less flowing than *A. arvensis*, with fairly regular repetition of sound in a 1–2 or 1–2–3 pattern: e.g. 'tweep tweep-tweep tweep cheer-cheer' (J Hall-Craggs, M G Wilson: Fig II). Includes rich and loud cheeping and chirruping like sparrow *Passer* or, at times, Chaffinch *Fringilla coelebs*: in recording of simpler song, irregular series of 'cheep' sounds with slightly varying pitch suggest singing ♂ House Sparrow *P. domesticus* (J Hall-Craggs, M G Wilson: Fig III). Sweeter, fluting sounds recall Woodlark *Lullula arborea*, though do not attain full lilting beauty of that species (M G Wilson); recording suggests 'sip sip tyu', then rather musical cadence not unlike *L. arborea* but shorter and not descending as much (J Hall-Craggs: Fig IV). Recording suggests a mewing sound (1st unit, level of which has been boosted, in Fig V) much weaker than rest of song and probably a refined variant of call 2; this followed by fast, buzzing 'zree' sounds and a twittering (last 2 units of Fig V) like Swallow *Hirundo rustica*—possibly mimicry (J Hall-Craggs, M G Wilson). Song contains mimicry of other birds—e.g. Teal *Anas crecca*, Red-footed Falcon *Falco vespertinus*, Quail *Coturnix coturnix*, various waders (Charadrii), Black Tern *Chlidonias niger*, swallows (Hirundinidae), and thrushes (Turdidae); also suslik *Citellus* (Dementiev and Gladkov 1954a; Dolgushin *et al.* 1970). (2) Contact- and flight-calls. (a) Slightly tinny, clear 'wed' or 'wäd' given at intervals (Schüz 1959). Drawn-out, quiet and rather hoarse squeal, rather like angry cat (Dementiev and Gladkov 1954a), not typical of Alaudidae (Dolgushin *et al.* 1970). (b) Calls of vagrant ♀ in England resembled *A. arvensis*, but distinguishable (Sage and Jenkins 1956).

CALLS OF YOUNG. No information. MGW

Breeding. SEASON. Southern USSR: laying starts late April, continuing to mid-July; most eggs laid in May (Dementiev and Gladkov 1954a; Golovanova 1967; Dolgushin *et al.* 1970; Koshelev 1980). SITE. On ground, usually well-concealed and sheltered by tussock or other plant. Nest: shallow scrape, lined with carelessly woven basket of vegetation; average external diameter 8-12 cm, internal diameter 6-8 cm, depth of cup 5-11 cm, thickness of wall 1-2 cm (Dementiev and Gladkov 1954a; Koshelev 1980). Building: by ♀ only (Shishkin 1982). EGGS. See Plate 79. Sub-elliptical, smooth and glossy; very variable—whitish, pale green, or pale yellow, variably speckled and spotted olive, grey-brown, and grey, usually with ring of spots at broad end. 22·8 × 16·2 mm (21·0-24·5 × 15·3-17·2), n = 120 (Schönwetter 1979). Weight 3·23 g (2·6-3·8), n = 18 (Koshelev 1980). Clutch: 5-6 (4-7) (Dolgushin *et al.* 1970; Koshelev 1980). Also variously reported as 3-5 (Grote 1936), 6-8 (Zarudnyi 1888), and 7 (Golovanova 1967) . One brood (Grote 1936; Makatsch 1976), or 2 (Zarudnyi 1888); 2 seems likely in view of prolonged laying season. INCUBATION. 12 days (Dementiev and Gladkov 1954a; Dolgushin *et al.* 1970); 13 days (Makatsch 1976). By ♀ only (Dementiev and Gladkov 1954a), though ♂ may do some (Dolgushin *et al.* 1970). YOUNG. Altricial and nidicolous. Cared for and fed by both parents (Dolgushin *et al.* 1970). FLEDGING TO MATURITY.Fledging period unknown but young leave nest still unable to fly and continue to be cared for by parents (Dolgushin *et al.* 1970).

Plumages. ADULT MALE. Forehead, crown, hindneck, and ear-coverts deep rufous-cinnamon, feathers narrowly tipped buff when plumage just moulted; hindcrown, hindneck, and ear-coverts with dark grey shaft-streaks varying in extent. Small bristles over nostril and along upper mandible black; short feathers over nostril white or cream. Broad supercilium and ring around eye white or cream, feathers often mottled grey on lores, in front of and below eye, and on rear of supercilium above ear-coverts; often only part of supercilium directly above and behind eye pure white. Sides of neck and all upperparts pale grey-buff with dark brown shaft-streaks; sides of feathers at border of shaft-streaks often brighter buff or pink-buff; upper tail-coverts deep rufous-cinnamon with narrow black

shaft-streaks. Cheeks, sides of chin, chest, and sides of breast white or pale cream, each feather with small triangular dark grey spot at tip; spots sometimes join to form mottled stripe below white eye-ring, on lower cheek, or on upper sides of chest; spots on sides of breast often bordered rufous, sides of breast sometimes largely rufous, and grey spots on sides of chin and chest sometimes wholly or partly rufous. Flanks white or cream with narrow blackish shaft-streaks bordered rufous; thighs grey-buff; remainder of underparts pale cream or white. Tail largely black or brown-black, central pair of feathers (t1) broadly fringed rufous-cinnamon; t2-t4 narrowly edged white on outer web and buff on tip, t5 with largely white outer web, t6 fully white. Primaries brown-black or grey-black, terminal third of inner web of p1-p3 white; broad fringes along tips of p1-p7 and narrower ones along outer edges white or buff; narrow fringe along outer edge of p8 and along tip of p8-p9 cinnamon-buff or cream, outer web of p9 largely white, minute p10 white. Secondaries basally dull black, terminal halves contrastingly white; tertials olive-brown with blackish base, broad off-white fringe along tip, and broad rufous fringe along base of outer web. Lesser and median upper wing-coverts and greater upper primary coverts deep rufous-cinnamon; upper wing-coverts with white tips when plumage fresh, primary coverts and longer lesser and median coverts with dark grey or blackish bases (mainly on inner webs), often largely hidden. Greater upper wing-coverts buff-brown or olive-brown, more or less tinged rufous, fringed off-white. Under wing-coverts and axillaries white, small coverts along leading edge of wing tinged rufous. Individual variation rather marked: some birds (perhaps 1st adults) are duller, with shaft-streaks on upperparts blacker, broader, and more triangular than described above; crown, ear-coverts, upper tail-coverts, and upper wing-coverts less uniform deep rufous-cinnamon; cheeks, sides of breast, and chest with less rufous tinge and more distinctly spotted black, t6 white with dusky spot on inner web near tip. Differences caused by wear not as strong as those of individual variation. Feather-fringes of upperparts all tinged pink in newly moulted plumage. Worn plumage of brightest rufous birds less pale grey buff on mantle, scapulars, and rump, more distinctly streaked; grey spots on cheeks, sides of chin and breast, and chest sometimes largely lost through abrasion, sides of breast and chest appearing off-white with indistinct rufous mottling. Duller birds are heavily streaked black on upperparts when worn, rufous of forehead and crown often largely abraded, but upper tail-coverts and upper wing-coverts still largely rufous (more heavily mottled dusky than in bright ♂♂); cheeks, ear-coverts, and sides of breast rather heavily streaked black. ADULT FEMALE. Like adult ♂, but feather-fringes of forehead, crown, hindneck, ear-coverts, and upper tail-coverts pale pink-grey or grey buff, and feather-centres dull black or dark grey, like remainder of upperparts, without deep rufous-cinnamon; rufous of t1 and upper wing-coverts usually less deep than in adult ♂; cheeks, sides of chin, chest, and sides of breast more distinctly streaked black than adult ♂, less finely spotted; ground-colour of chest, sides of breast, and flanks white, only limited amount of rufous present on sides of breast; inner web of white t6 with large grey wedge along border. Some individual variation (perhaps age-related) in width of dark streaks on upperparts and in amount of rufous on greater upper primary coverts and lesser and median upper wing-coverts; brightest rufous ♀♀ may show more rufous on forewing than dull ♂, but latter usually shows more rufous on forehead, sides of crown, upper tail-coverts, and sides of breast; some virtually indistinguishable, however, especially in worn plumage. NESTLING. Covered with yellowish down-feathers 1·5-2 cm long, closely resembling grasses of environment (Johansen 1952). JUVENILE. Feathers of upperparts dark brown with yellow-buff bases and buff-white or cream fringes along tips. Sides of head and neck down to sides of chin white or cream-buff, finely speckled dark brown on cheeks and ear-coverts. Chest, sides of breast, and flanks yellow-buff or cream, spotted dark brown or grey-brown. Remainder of underparts off-white. Tail brown with buff mottling, each feather with even buff fringe submarginally bordered by dark brown line; t6 uniform buff-white, except for dark grey wedge on inner web. Primaries and tertials like central tail-feathers; inner webs of p1–p3 cream; secondaries with broad white or buff-white tips, showing large pale wing-panel, as in adult. Lesser upper wing-coverts and primary coverts brown with even cream-buff or cream-white fringes; median and greater upper wing-coverts with dark submarginal crescent like tail, but tips cream or off-white. FIRST ADULT. Indistinguishable from adult once last juvenile feathers (usually p9–p10) replaced. Variation in adults described above perhaps age-related, but no known-age birds examined.

Bare parts. ADULT. Iris brown or dark brown. Bill horn-brown, dark horn, or blackish, base of lower mandible yellow-horn or yellow-cream. Leg and foot yellow-brown, pale brown, or brownish-flesh with darker toes. (Hartert 1910; Dementiev and Gladkov 1954*a*; BMNH.) NESTLING. 2 black spots at base of tongue, 1 on tip (Neufeldt 1970). JUVENILE. No information.

Moults. Little-known; a few in moult examined but these undated. ADULT POST-BREEDING. Complete; primaries descendant. In August, flight-feathers and tail still old, but new or growing feathers on head, wing-coverts, and underparts (Dementiev and Gladkov 1954*a*); moult completed September-October (Bub and Herroelen 1981). Generally no moult in birds collected between November and early June, but some mid-winter birds had new feathers growing on forehead, crown, ear-coverts, and chest, perhaps indicating a partial pre-breeding moult in at least some birds; undated birds had body feathers and wing-coverts new or growing, middle primaries, outer secondaries, outer tail-feathers, and tertials growing, outer 2–3 primaries and inner secondaries old. POST-JUVENILE. Complete; primaries descendant. No information on timing; all moult completed at same time as adult post-breeding (Bub and Herroelen 1981).

Measurements. ADULT. Mainly Volga steppes, Caucasus, and Rumania, November–June; skins (RMNH, ZMA). Bill (S) to skull, bill (N) to distal corner of nostril; exposed culmen on average 3·4 shorter than bill (S).

WING	♂ 122·4 (1·90; 20)	119–127	♀ 113·0 (1·80; 15)	111–117
TAIL	63·7 (2·53; 20)	60–69	59·2 (2·30; 15)	56–63
BILL (S)	15·6 (0·66; 20)	14·7–17·1	15·1 (0·60; 15)	14·2–15·9
BILL (N)	9·7 (0·66; 20)	8·7–10·9	9·1 (0·47; 15)	8·7–10·0
TARSUS	24·6 (0·70; 18)	23·2–25·6	23·6 (0·63; 14)	22·6–24·6

Sex differences significant.

JUVENILE. Wing on average *c.* 7 shorter than adult, tail slightly shorter; bill and tarsus probably full-grown once post-juvenile moult completed.

Weights. USSR: adult, ♂ 46·1 (3) 44–48, ♀ 43·5 (3) 39–46 (Dementiev and Gladkov 1954*a*); adult, ♂ 46·5 (18) 40–52, ♀ 41·6 (6) 36–48 (Bub and Herroelen 1981). Kazakhstan (USSR): March, ♂ 49; April, ♂ 46–49 (6), ♀♀ 45, 48; May, ♂ 43–46 (3), ♀ 36·5; June, ♂ 40–47 (5), ♀ 38; July, ♀ 43; August, ♂ 48; September, ♂♂ 44, 52·5, ♀ 38·8 (Dolgushin *et al.* 1970).

Structure. Wing long, broad at base, tip rather pointed, 10 primaries: p8 longest, p9 0–2 shorter, p10 81–90 (♂) or 74–80 (♀), p7 4–7, p6 13–18, p5 20–27, p1 47–55 (♂) or 42–47 (♀). Adult p10 minute, narrow and pointed, 11–16 shorter than longest upper primary covert; juvenile p10 slightly longer and with broader and rounder tip, length equal to longest primary covert or slightly shorter. Outer web of p7–p8 and inner of p8–p9 emarginated. Longest tertials reach to about tip of p4 in closed wing. Tail rather long, tip slightly forked; 12 feathers, t1 3–8 shorter than t4–t6. Bill strong but short, conical; smaller than other west Palearctic *Melanocorypha*; culmen decurved, cutting edges slightly decurved; gonys curved upwards. Many fine bristles along base of upper mandible. Tarsus and toes rather short and slender. Middle toe with claw 18·3 (22) 17–19, outer toe with claw *c.* 71% of middle with claw, inner *c.* 76%, hind *c.* 118%. Claws rather slender, straight or slightly decurved; length of hind claw 12·1 (28) 9–14, exceptionally 7–16.

Geographical variation. None.

Forms superspecies with Mongolian Lark *M. mongolica* from southern Transbaykalia, Mongolia, and neighbouring parts of China, which differs mainly in larger size and large black patches at sides of breast. CSR

Melanocorypha yeltoniensis Black Lark

PLATES 6 and 12
[between pages 160 and 161, and 184 and 185]

DU. Zwarte Leeuwerik FR. Alouette nègre GE. Mohrenlerche
RU. Чёрный жаворонок SP. Calandria negra SW. Svart lärka

Alauda yeltoniensis Forster, 1768

Monotypic

Field characters. 19–20 cm; wing-span 34–41 cm. Largest of genus in west Palearctic, exceeding Calandra Lark *M. calandra* in average wing and tail length. Large, robust lark, of similar form to *M. calandra* except for slightly less deep-based bill. ♂ essentially black, except for yellow-horn bill and wide off-white or buff tips to head- and body-feathers in autumn and winter. ♀ much less distinctive, lacking uniformity of ♂ but extensively black-brown above and on flanks, with pale rump. Juvenile even less distinctive but displays adult's diagnostic dark under wing-coverts. Sexes dissimilar; marked seasonal variation in ♂. Juvenile separable.

ADULT MALE. All feathers jet-black, looking dusky when worn and well fringed and tipped buff to grey-white when fresh. Width of pale margins immediately following moult sufficient to give bird close-packed hoary or sandy scales on crown, sides of chest, and flanks, and almost to hide black on lower back and rump. Uniformity of plumage increases with wear, and few tips visible by breeding season. Underwing and undertail dark, with wing-coverts fully black. Bill short and decurved above; yellow, with black tip when breeding, yellow-horn at other times. Legs black, unlike any other lark. ADULT FEMALE. Head and upperparts basically dark brown, but even when worn never having uniformity of ♂; when fresh, fully fringed and tipped pale buff, with noticeably paler head, sides of back, and rump. Wings and tail black-brown, with pale buff or white margins obvious on edges of tail and tertials and again rarely completely lost. Underparts buff-white, with white patches below cheeks, black splashes below them, and variable black or dark brown spots and streaks on rest of body; pattern of marks open on chest but dense, even blotched, on and between fore-flanks. All dark areas increase with wear but never join to create uniform darkness of ♂. Pattern of underwing and undertail as ♂ but duller. Bare parts as non-breeding ♂. JUVENILE. Resembles ♀ but cheeks paler, back has more black, rufous-buff, and white speckling, and flanks are less heavily streaked and have whiter ground-colour.

♂ almost unmistakable at any season but beware possibility of melanistic congener. ♀ and juvenile less distinctive and subject to some risk of confusion with Thick-billed Lark *Rhamphocoris clotbey* (smaller in body but with larger bill, and with uniform pinkish crown and back), juvenile White-winged Lark *M. leucoptera* (distinguished by thick white trailing edge to wing), and Bimaculated Lark *M. bimaculata*. Identification of *M. yeltoniensis* best made on underwing pattern. All other large larks with dark underwing have either obvious white trailing edges, or coverts uniform with undersurface of flight-feathers; in *M. yeltoniensis*, under wing-coverts distinctly darker than dusky undersurfaces of flight-feathers which show no obvious pale trailing edge. Tail of *M. yeltoniensis* also almost uniformly dark, lacking (even in ♀ and juvenile) obvious white edges of many larks and white tips of *M. bimaculata*. Flight, gait, and habits much as *M. calandra* but usually sings at lower height, as in *M. leucoptera*.

Song rich and varied with much mimicry. Calls little known.

Habitat. In middle continental latitudes, concentrated in steppe zone of warm dry summers and snowy winters, on broad open plains and rolling country, where wormwood *Artemisia* (especially) or feather-grass *Stipa* are dominant, with plenty of short grass, often forming a mosaic pattern of various associations, including fescue *Festuca* and orache *Atriplex*, sometimes on saline or alkaline soils, and on clay, usually in neighbourhood of water, including freshwater or saline lakes. Avoids mountains and in most cases even foothills, sandy or bare steppe, and deep rich black earth; also usually avoids cultivation, subsisting on wild plants. In winter, spends most time in areas where snow thinnest, either through wind or artificial action; thus will forage on roads, follow herds of horses, or even straggle into human settlements. Prefers singing from low mounds to making song-flights, which are confined to lower airspace. Ecologically conservative and unadaptable. (Bannerman 1953; Voous 1960; Krivitski 1962; Dolgushin *et al.* 1970.)

Distribution and population. Little information on range changes or population trends.

USSR. Generally less numerous than Skylark *Alauda arvensis* but abundant in some localities (Dementiev and Gladkov 1954a); has declined in Kazakhstan between Volga and Ural rivers, Kazakhstan (Shishkin 1976).

Accidental. Belgium, West Germany, Austria, Italy, Greece, Rumania, Lebanon, Malta.

Movements. Dispersive and perhaps nomadic, mainly within USSR. Also evidence of age and sex variation in degree of dispersal.

Throughout breeding range, present all winter in roaming flocks (which may include birds from more distant breeding areas). Birds wander beyond breeding range in winter (sometimes in large numbers), extent probably varying from year to year according to severity of weather (Dementiev and Gladkov 1954a). Fairly regular in winter even north of breeding range (occasionally as far as Moscow), though most such dispersal is south to south-west: as far as Turkestan (vagrant to Mongolia), northern Iran, Black Sea region, and Ukraine. Adult ♂♂ predominate in winter in breeding areas, almost exclusively so in colder seasons, which suggests that ♀♀ and young birds tend to disperse furthest (Grote 1936; see also Social Pattern and Behaviour).

For increase in flock sizes prior to spring return movement, see Social Pattern and Behaviour. In Karatau foothills (southern Kazakhstan), where birds occur only in years of exceptional emigration, migrant flocks up to 500 birds wide passed every 20–30 min during peak passage in mid-March (Dolgushin *et al.* 1970). In Transvolga region, where some present all year, conspicuous passage in March (especially second half) and more diffuse autumn movement in September–October (Dementiev and Gladkov 1954a). Cold-weather movements continue later than this: e.g. in Kurgal'dzhin reserve (northern Kazakhstan), flocks of 1000 or more make local movements typically associated with severe blizzards towards end of year (Krivitski 1962).   RH

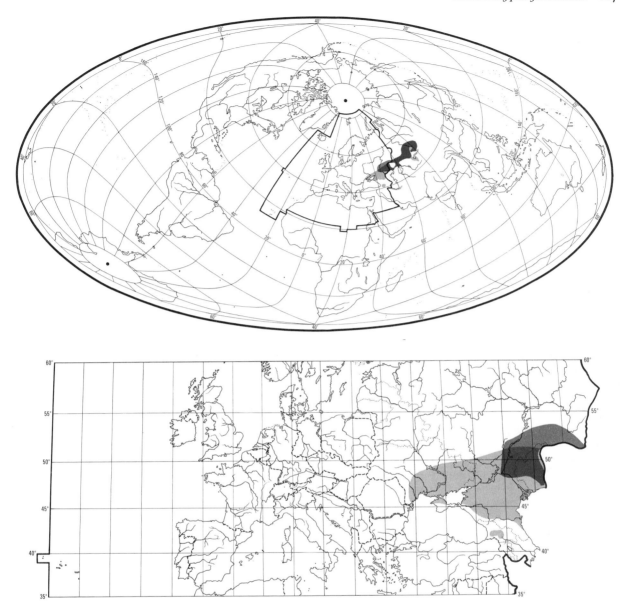

Food. Insects and seeds. Digs through snow to reach food. In loose snow, makes distinctive channels 15–20 cm long, probably by moving snow with head and breast. If necessary, digs down up to 8 cm, then makes side tunnels of *c*. 10–12 cm to reach seeds. Turns over snow using rapid hammering action with bill (Krivitski 1962; Dolgushin *et al.* 1970) and said to be able to crack frozen surface of ground (Dementiev and Gladkov 1954*a*). Breaks off top of grass *Stipa* stems with bill (apparently by reaching up from surface of deep snow) and also eats from seed heads already on ground (Krivitski 1962). In late winter, especially, gathers by roads or around animal herds to feed on disturbed ground (see Habitat). Will drink saline water (Dementiev and Gladkov 1954*a*).

Invertebrates recorded in diet comprise grasshoppers, etc. (Orthoptera: Tettigoniidae, Acrididae), bugs (Hemiptera: Pentatomidae, Lygaeidae, Cicadidae, Cercopidae, *Tarisa*), butterflies and moths (Lepidoptera: Pieridae, Geometridae, Noctuidae), flies (Diptera: Tipulidae, Culicidae, Tabanidae), Hymenoptera (sawflies Tenthredinidae, Ichneumonidae, Braconidae, Chalcidoidea, ants Formicidae, digger wasps Pompilidae, bees Apoidea), adult and larval beetles (Coleoptera: Cicindelidae, Carabidae, Haliplidae, Dytiscidae, Histeridae, Staphylinidae, Scarabaeidae, Buprestidae, Elateridae, Phalacridae, Pythidae, Meloidae, Tenebrionidae, Nitidulidae, Cerambycidae, Bruchidae, Chrysomelidae, Curculionidae), spiders (Araneae), millipedes *Julus*, Isopoda

(Crustacea). Plant food includes seeds and fruits of docks (Polygonaceae), goosefoots (Chenopodiaceae), *Amaranthus*, Cruciferae, Leguminosae, Boraginaceae, Labiatae, and grasses (Gramineae); cultivated grain includes buckwheat *Fagopyrum*, sunflower *Helianthus*, oats *Avena*, wheat *Triticum*, barley *Hordeum*, and millet *Panicum*; also leaves and stalks. (Dementiev and Gladkov 1954a; Pek and Fedyanina 1961; Ryabov 1967; Samorodov 1968.) Stomachs usually contain grit: in 97% of 153, May–July, average 56 pieces in each; in 100% of 43, December–February, average 241 pieces (Ryabov 1967).

During May–July on Kustanay steppes (northern Kazakhstan) took both plant material (in 89% of 136 stomachs) and invertebrates (in 83%). Plant food comprised 60% by weight: largely seeds or fruits of wild plants especially *Polygonum* and *Rumex* (in 41% of stomachs, or 81% by number of total 3797 seeds and fruits); also of wheat, barley, and millet (37%, 13%—taken mainly on roads), grasses (3%, 2%), Boraginaceae and Labiatae (1%, 1%), and Chenopodiaceae (1%, 0·2%); plant leaves and stalks recorded in 13% of stomachs. Most important invertebrates were those most abundant in the area; comprised Coleoptera (in 65% of stomachs, or 60% by number of total 350 invertebrates—mainly Curculionidae, Tenebrionidae, Scarabaeidae), Hymenoptera (in 24%, 27% of total invertebrates—mainly ants), Diptera (6%, 3%), adult and larval Lepidoptera (5%, 5%), Orthoptera (3%, 3%), Hemiptera (1%, 1%), and Arachnida and Isopoda (1%, 1%). In spring, before young leave nest, adults took mainly insects (recorded in 90% of stomachs, plant material in 30–40%); ate especially Curculionidae, Scarabaeidae, Chrysomelidae, and plant leaves and stalks. No differences between diets of ♂♂ and ♀♀, and adults and juveniles took roughly the same proportions of animal and plant food, though juveniles took more cereal grain as they tended to frequent roads more than adults; also took more Curculionidae, but fewer plant leaves and stalks, Hymenoptera, Scarabaeidae, Chrysomelidae, and Lepidoptera. In same area during late 1930s, breeding-season diet was much less varied: Orthoptera much more numerous and formed main food, and little cereal grain available. (Ryabov 1967). At Lake El'ton (eastern Volgograd region), eats mainly beetles (Dementiev and Gladkov 1954a).

Winter food largely or exclusively plant material. On Kustanay steppes, December–February, 43 stomachs contained only plant fruits and seeds: largely of *Polygonum* and *Rumex* (in 72% of stomachs, or 70% by number of the 2492 items); also Chenopodiaceae (72%, 23%), cereal grain (33%, 3%), *Amaranthus* and Cruciferae (5%, 3%), Boraginaceae and Labatiae (5%, 1%), and grasses (2%, 0·1%) (Ryabov 1967). In Kalmytskaya ASSR, December–January, 6 of 22 stomachs contained grass shoots; remaining 738 items comprised seeds or fruits of *Chenopodium* (57% by number), Polygonaceae (39%, almost wholly *Rumex*), and *Salsola* (3%), and 1% Coleoptera

(Samorodov 1968). In Kurgal'dzhin reserve (Kazakhstan), where cereal grain otherwise important in diet, can be more difficult to obtain in winter and birds turn to *Stipa* seeds. Despite severe conditions of long winters, shot birds all well-fed. (Krivitski 1962.)

No information on food of young.   DJB

Social pattern and behaviour. Most aspects still inadequately studied. Fullest summary in Dolgushin *et al.* (1970). Important additional data collected by Moiseev (1980) in northern and central Kazakhstan in summer, and on Buzachi river (eastern Caspian) and lower Chu river (southern Kazakhstan) in winter.

1. Gregarious outside breeding season, often markedly so. In Kustanay steppes (Kazakhstan), wintering flocks normally of 10–20 to several hundred (Ryabov and Samorodov 1969), but flocks of 1000 or more (presumably ♂♂—see below) occur in local movements (see Movements). In Kalmytskaya ASSR, December–January, average flock size 59 (Samorodov 1968); in mid-January, size of flocks 28–600 (Samorodov 1967); see also below. Wholly single-sex flocks the rule; only exceptionally odd ♀♀ in ♂ flock. ♀ flocks generally smaller (Dementiev and Gladkov 1954a; Ryabov and Samorodov 1969; Dolgushin *et al.* 1970). In Kustanay, during 3 winters (1961–2, 1963–4, 1964–5), ♂:♀ ratio 5:1, in hard winter (1965–6), 1:5; in northern Kustanay, in late January of 1966, 1:7, further south 1:2 (Samorodov and Ryabov 1969); in central and northern Kazakhstan, generally 8–9:1, according to Moiseev (1980). In Kalmytskaya ASSR, *c.* 1:3 (Samorodov 1967). In Kurgal'dzhino reserve (northern Kazakhstan), flocks of up to 100 ♂♂ present by roads and in steppes from early winter. In first half of winter, ♀♀ represented only by a few flocks of 5–6 birds each; total numbers of ♀♀ increase markedly from mid-February and flocks of 50–100 appear in March (Krivitski 1962). In late May (peak breeding season), flocks of up to 7 ♂♂ and ♀♀ occur at favourable feeding grounds—e.g. on lakeside where flies (Diptera) abundant (E N Panov). At Lake Tengiz (Kazakhstan), flocks of unpaired ♂♂ occur May–June, apparently attracted by abundance of washed-up invertebrates (Moiseev 1980). In Kushum area (north-west Kazakhstan), young birds recorded flocking (also with other Alaudidae) from mid-June; this occurs widely from early July, adults also beginning to flock at this time. Often forms mixed flocks with White-winged Lark *M. leucoptera* and other Alaudidae, including, for winter foraging, Shore Lark *Eremophila alpestris* (Dolgushin *et al.* 1970). BONDS. Said to be generally in pairs on breeding grounds but far more ♂♂ present than can be explained by relatively unobtrusive nature of ♀♀ (Dementiev and Gladkov 1954a; Dolgushin *et al.* 1970); mating system possibly thus not wholly monogamous. One brood of 4 were all ♂♂, and of 56 juveniles trapped in August, 38 were ♂♂ (Moiseev 1980). Both sexes share in feeding of young (Dolgushin *et al.* 1970). BREEDING DISPERSION. Solitary and strictly territorial (E N Panov). Dispersion very uneven, both locally and over large area; compact neighbourhood groups occur and similar habitat nearby may hold none or only a few scattered pairs (Dolgushin *et al.* 1970). Preliminary dispersion of 'territories' apparently linked with presence of bushes, mounds, etc., suitable for display (see part 2), and singing ♂♂ then 30–300 m apart with no clear boundaries between them where ♀♀ not present (see also Antagonistic Behaviour, below). By start of nest-building, territories concentrated on lower ground near water where pairs *c.* 300–700 m apart (Moiseev 1980). Dolgushin *et al.* (1970) similarly recorded displaying ♂♂ *c.* 75–100 m apart in suitable habitat, but considered nesting territories probably bigger; this suggests some displaying ♂♂ fail to breed. However, at Lake

B

El'ton (eastern Volgograd) nests *c.* 100 m apart (Dementiev and Gladkov 1954*a*). Adults forage for young both within and well outside territory; in one case, ♂ feeding young out of nest flew *c.* 2 km away across river (Moiseev 1980). In area of Ural-Emba watershed, 0·1-4·6 pairs per 1-km transect, higher value on *Artemisia-Anabasis* associations (Poslavski 1963); for further transect counts in north Caspian region, see Poslavski (1974). Territories may overlap with those of *M. leucoptera* and Skylark *Alauda arvensis* (Moiseev 1980). ROOSTING. Nocturnal and, outside breeding season, usually communal. Roosting groups may be fairly dense, birds crouching with heads into wind, *c.* 10-15 cm apart (Dolgushin *et al.* 1970). In Kurgal'dzhino, winter, birds tend to roost where they have fed during the day, preferring areas with thin snow-layer, especially roads. Dig holes *c.* 10-13 cm across and *c.* 5-10 cm deep, always attempting to reach ground. Will also use holes previously dug under grass clumps in searching for seeds (see Food). Birds roost in groups, as indicated by many holes being sited close together (Krivitski 1962; also Dolgushin *et al.* 1970). In bad weather (blizzards, etc.), shelter in hay or straw stacks or use dense grass cover in hollows. In midday heat, seek shade of bushes and lie there with bill open and wings spread. In summer, visit water regularly to drink, normally in morning and evening. Marked fondness for dust-bathing (Dolgushin *et al.* 1970).

2. Some birds surprisingly confiding in breeding season, allowing very close approach. ♂♂ recorded coming in to towns in severe winter weather and not shy then (Grote 1919, 1936); in Kurgal'dzhino, ♀♀ (almost never ♂♂) recorded feeding in villages (Krivitski 1962). Where cover available, *Melanocorypha* larks sometimes squat until almost trodden on. In river mouth, Crimea (USSR), *M. yeltoniensis* recorded squatting on black seaweed (Meinertzhagen 1951*a*). FLOCK BEHAVIOUR. In small late-May flocks (see part 1) birds defend individual-distance of *c.* 50 cm (E N Panov). SONG-DISPLAY. Peculiar and striking owing to variety of postures adopted, types of movement, and beauty of flight accompanied by rather melodious song (Dolgushin *et al.* 1970: see 1 in Voice). ♂ sings from ground or some eminence—hummock, clump of earth or snow, pile of straw—with wings drooped and tail cocked (Fig A) (Grote 1936;

A

Dementiev and Gladkov 1954*a*; Krivitski 1962). At higher intensity, gradually changes to posture shown in Fig B: wings extended more to side, then forwards and flapped (E N Panov). Description of bird singing occasionally while running about, feathers ruffled and wings open (Grote 1936), probably refers to same high-intensity phase; may have been performed in presence of ♀ (see Portenko 1954 and below). Song perhaps given more often from ground than in flight, but, at least in interaction with ♀, ♂ regularly interrupts bouts of courtship (see Heterosexual Behaviour, below) to perform Song-flight: ascends while singing, beating wings evenly and deeply so that they meet over back in soft wing-clap; whole effect like displaying pigeon (Columbidae). At height of *c.* 20-30 m, ♂ briefly stalls and initially glides down with wings raised or held below body, then plummets almost to

ground, brakes, and glides horizontally for a few metres before landing gently, feet pushed forwards; descent variable (Dolgushin *et al.* 1970). Said also to ascend (while singing) at gentle angle with much flapping of not-fully-open wings (not clear whether ascent significantly different from that described above), then to perform low, circling Song-flight recalling displaying wader (Charadrii) or owl (Strigidae) (Grote 1936; Portenko 1954); sharp upward wing-beats alternate with gliding on horizontally extended wings (E N Panov). Will sing also in winter, from October (Portenko 1954; Moiseev 1980), and in early spring while on passage, either from ground or in low Song-flight (Dolgushin *et al.* 1970). In Kurgal'dzhino, song delivered on clear, sunny days in early February (some display even in blizzards), when ♀♀ certainly not always present; in March, also by ♂♂ separating from main flock (Krivitski 1962); or still within it (Grote 1936). In Transvolgan steppes near Saratov, song (and courtship) from about 20 March to mid-July (Dementiev and Gladkov 1954*a*). ANTAGONISTIC BEHAVIOUR. (1) General. In winters with little snow, ♂♂ display (no details, but see below) and defend territories in December (Moiseev 1980). Birds move on to breeding grounds in early spring, ♂♂ taking up territories and performing Song-display (see above) when thawed patches begin to appear (Dolgushin *et al.* 1970). Some ♂♂ are defending territories by late March (Moiseev 1980). (2) Threat and fighting. Before territory boundaries established, one ♂ will chase another up to *c.* 0·5 km, then (not returning) land and continue to sing on nearest bush (Moiseev 1980). During boundary conflicts, adopts same postures as described above for song on ground (E N Panov). No more-detailed reports of intraspecific threat or fights but sometimes aggressive towards other species: will chase *M. leucoptera* and *A. arvensis* out of territory but often ignores them if they return (Moiseev 1980). One pair also chased Calandra Lark *M. calandra* and ♂ attacked 2 Black-headed Gulls *Larus ridibundus* (Dolgushin *et al.* 1970); also recorded attacking *Calandrella* larks, Wheatear *Oenanthe oenanthe*, and even susliks *Citellus* (E N Panov). HETEROSEXUAL BEHAVIOUR. (1) General. In Kurgal'dzhino, March, when larger flocks of ♀♀ appear (see part 1), ♂♂ hardly ever recorded near them (Krivitski 1962). Most birds on breeding grounds by mid-April; some pair-formation apparently takes place on migration, but also as late as May. In area between Volga and Ural rivers, pairs form from late March in south to early April in north; similar on lower Irgiz (Dolgushin *et al.* 1970; also Dementiev and Gladkov 1954*a*; Moiseev 1980). (2) Pair-bonding behaviour. Several ♂♂ may attempt to court one ♀ (Dementiev and Gladkov 1954*a*; see Bonds, above). ♂ flying near ♀ when patrolling territory tends to hover over her with unusually rapid wing-beats, calling (Moiseev 1980: see 2c in Voice). In display to ♀ on ground, ♂ bows so that breast almost touches ground; tail half spread and raised vertically, wings closed but drooped to touch ground, feathers on neck and breast ruffled, head raised; may run or jump thus around ♀ and also call (Dolgushin *et al.* 1970: see 2a in Voice). Postures and movements generally as in ♂ singing on ground (E

N Panov; see above). From time to time during performance, takes off in Song-flight (see above) (Dolgushin *et al.* 1970). (3) Courtship-feeding. Not known to occur. (4) Mating. No information. (5) Nest-site selection. Not known which sex chooses site, but scrape dug by ♀. One ♀ fed with ♂ when not building (Moiseev 1980). (6) Behaviour at nest. While ♀ incubating, ♂ spends most time on look-out *c.* 50–100 m from nest. At one site, ♀ tended to walk *c.* 15–20 m on leaving nest, then to call (see 2c in Voice) with neck extended while looking about; immediately joined by ♂ with whom she then fed, though also flew off alone to drink. On return, landed *c.* 2–3 m from nest than ran in (Moiseev 1980). RELATIONS WITHIN FAMILY GROUP. See Bonds (above). The following based on observation at one nest in northern Kazakhstan. ♂ tended to start feeding young before sunrise, announcing arrival with call 2c which caused brooding ♀ to leave. ♀ would return immediately after ♂ left, this pattern continuing until mid-morning. Both fed young later in day when less brooding required. When both arrived at nest together, ♀ would feed young first, ♂ singing nearby without releasing food. ♂ frequently displayed when near ♀. Adults swallowed chicks' faecal sacs (young restless if any faeces left in nest) and carried away their pellets. Young left nest at 9 days old, over several hours; dispersed in grass, parents locating them by call. When begging out of nest, stretched up and gave food-call. (Moiseev 1980.) ANTI-PREDATOR RESPONSES OF YOUNG. No information. PARENTAL ANTI-PREDATOR STRATEGIES. (1) Passive measures. ♀ normally a tight-sitter, allowing very close approach (Dolgushin *et al.* 1970). When hide set up near one nest, ♀ would not return for *c.* 1 hr. ♂ apparently led her back: hopped in front of her, tail raised and wings drooped, then moved (while Mock-feeding) towards nest. ♀ approaching nest to feed young crouched then ran off when Pallid Harrier *Circus macrourus* flew over (Moiseev 1980). (2) Active measures: against man. Sometimes performs distraction-lure display of disablement type (Esilevskaya 1967). (3) Active measures: against other animals. When suslik *c.* 5 m from nest, ♂ hovered over it, landed nearby, and ran off (Moiseev 1980).

(Figs by J P Busby: from drawings by E N Panov.) MGW

Voice. Poorly known.

CALLS OF ADULTS. (1) Song of ♂. Said by Dementiev and Gladkov (1954a) to be less attractive than Calandra Lark *M. calandra*, though rich in high-pitched trills. Recording reveals, as in White-winged Lark *M. leucoptera*, an astounding vocal virtuosity, richness of repertoire, and capacity for mimicry (J Hall-Craggs). Contains rich purring trills and twitters, also a mew like Buzzard *Buteo buteo* (M G Wilson) or short 'cui' of Curlew *Numenius arquata* (J Hall-Craggs). Fig I shows sounds resembling Goldfinch *Carduelis carduelis*, Dunnock *Prunella modularis*, Song Thrush *Turdus philomelos*, and Wren *Troglodytes troglodytes*; last 2 units are perfect copy of sweet fluting song notes of Blackbird *Turdus merula*. Recordings also contain sound like scolding of tit *Parus* and a rippling, ascending scale; also 2 segments and refined terminal flourish from song of Chaffinch *Fringilla coelebs* (Fig II). Rather thin, drawn-out whistles (Fig III) and faint snarls or growls (Fig IV) strongly recall Starling *Sturnus vulgaris*. Overall, song has much in common with Skylark *Alauda arvensis*: resemblance closer in more structured segment shown in Fig V (strong 2nd and 3rd har-

monics probably due to use of reflector by recordist); Fig VI suggests a generally more complicated, less structured, and higher-pitched song than *A. arvensis* (J Hall-Craggs, M G Wilson). One caged bird sang loudly and continuously for *c.* 15 min or longer (Meinertzhagen 1951a). (2) Other calls. (a) Short trills given during display on ground (Dolgushin *et al.* 1970). (b) Recording suggests a fizzing trill rather like call of Corn Bunting *Miliaria calandra*, sometimes closer to call of wagtail *Motacilla*: 'tslreep' (P J Sellar, M G Wilson: Fig VII); possibly the same as call 2a, but context not known. (c) Either sex may give short, sometimes loud, call, apparently primarily as contact with mate (Moiseev 1980). (d) Sonagram from recording by E N Panov indicates existence of distinct flight-call which may be given between songs; not described, but steep ascent and descent of pitch (from over 3 kHz to over 5 kHz to under 3 kHz in *c.* 0·25 s) apparently typical.

CALLS OF YOUNG. Food-call of bird out of nest a slightly hoarse cheep, easily distinguished from that of nestling (Moiseev 1980). MGW

Breeding. SEASON. Transvolga (USSR): first eggs laid late March. Orenburg region (USSR): first eggs laid late April (Dementiev and Gladkov 1954a). Kazakhstan (USSR): first eggs late April, latest clutch beginning of August (Dolgushin *et al.* 1970; Moiseev 1980). SITE. On ground in shelter of tussock. Nest: shallow depression lined with grass stems and other vegetation; external diameter 12–15 cm, internal diameter 7–10 cm, depth of cup 3·5–9 cm (Bostanzhoglo 1911; Dolgushin *et al.* 1970). Of 48 nests examined, Kazakhstan, 33 were surrounded by pieces of animal dung, some of them brought in by ♀ (Moiseev 1980). Building: by ♀, first digging scrape with bill (Moiseev 1980). EGGS. See Plate 79. Sub-elliptical, smooth and glossy. In Orenburg region, pale blue with light brown, often olive, mottling and blotching, co-alescing at broad end (Zarudnyi 1888); on Caspian-Aral steppes, light olive-green with light brown markings, partly evenly distributed, partly forming band at broad end (Bostanzhoglo 1911). 25·5 × 18·1 mm (23·0–27·8 × 17·2–19·0), *n* = 26; calculated weight 4·34 g (Schönwetter 1979). Weight of 8 eggs, 3·3–4·2 g (Dolgushin *et al.* 1970). Clutch: 4–7; 8 reported by Zarudnyi (1888). Of 15 1st clutches, Kazakhstan: 3 eggs, 2; 4, 4; 5, 6; 6, 2; 7, 1; mean 4·7 (Moiseev 1980). INCUBATION. 15–16 days; by ♀ only (Dementiev and Gladkov 1954a). No further information.

Plumages. ADULT MALE. Entirely black but in fresh plumage with broad yellow-buff or off-white feather-fringes on head, neck, mantle, scapulars, back, rump, chest, and lower flanks, almost hiding black of feather-centres and -bases; remainder of body and upper wing-coverts with narrower pale fringes, appearing largely black. Tertials black with some variable dusky grey or buff-brown subterminal spots or marks, narrow pale yellow-buff outer fringes, and wider pale yellow fringe along tips; flight-feathers, tail, and greater upper wing-coverts with

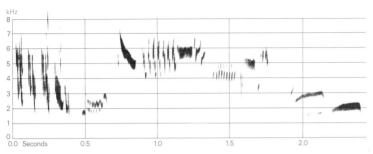

I   B N Veprintsev   USSR   March 1981

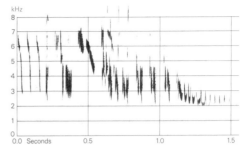

II   B N Veprintsev   USSR   March 1981

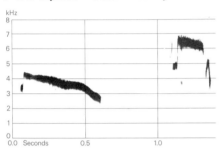

III   B N Veprintsev   USSR   March 1981

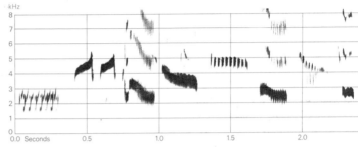

V   B N Veprintsev   USSR   March 1981

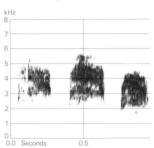

IV   B N Veprintsev   USSR   March 1981

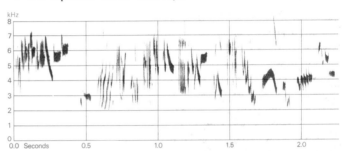

VI   B N Veprintsev   USSR   March 1981

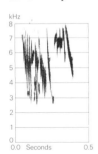

VII   B N Veprintsev   USSR   March 1981

narrow and ill-defined grey or white fringes along tips, t1 also narrowly fringed white at sides. Under wing-coverts and axillaries black. Wear has marked influence, all pale fringes disappearing and whole bird becoming black. However, much individual variation in width of fringes and in speed of abrasion: some birds fully black by December, others still broadly fringed in late spring. In general, fringes of tertials, flight-feathers, wing-coverts, tail, and belly abrade rapidly, these parts appearing fully black from November; fringes of ear-coverts, lower mantle, back, rump, and lower flanks retained longest, some birds notably pale-rumped though otherwise fully black. ADULT

FEMALE. Mainly brown, not black as ♂. In fresh plumage, feathers of forehead, crown, hindneck, and sides of neck with black or black-brown triangular centres and broad yellow-white or grey-white fringes, appearing largely hoary-white with blackish spots; streak over eye, ear-coverts, and tufts of feathers covering nostrils uniform yellow-white. Mantle, scapulars, and back to upper tail-coverts hoary-grey with some black-brown of feather-centres visible; in particular rump largely pale; feather-centres of lower mantle and scapulars often partly grey-brown, bordered by dark arc inside pale fringe. Lores, cheeks, and ear-coverts faintly streaked buff-brown and buff-yellow. Central chin and upper throat uniform cream-white; lateral chin, lower throat, chest, and sides of breast pale buff or cream-white with large black-brown spot on feather-centres, latter sometimes joining to form blackish patch on upper sides of chest. Remainder of underparts cream or off-white, flanks, thighs, and under tail-coverts with black-brown streaks. Tail dark olive-brown, tips of feathers sometimes partly mottled grey-brown or buff-brown; central pair (t1) broadly fringed buff to off-white, t2–t5 narrowly fringed on outer web and tip only; outer web of t6 largely white. Flight-feathers and greater upper primary coverts dark grey-brown; distal part of outer webs and tips of secondaries fringed white; primaries and primary coverts faintly fringed white along tips, outer web of (p7–)p8–p9 broadly fringed white.

Lesser and median upper wing-coverts black-brown with broad white fringes; greater coverts and tertials similar, but black-brown shading to olive-brown towards outer edges and tips, narrowly bordered by black line submarginal from pale fringe. Under wing-coverts and axillaries dark grey-brown, lesser coverts fringed off-white. In fresh plumage (about August–October), pale fringes of head, upperparts, chest, tail, and flight-feathers tinged pink-cinnamon, especially at border of dark feather-centre. In worn plumage, pale feather-fringes largely lost through abrasion, forehead, crown, mantle, and scapulars with extensive blackish centres and restricted grey buff or grey-white fringes; sides of chest and breast almost uniform black; chin, throat, central chest, and flanks off-white, heavily spotted black, with some black of feather-bases often visible on remainder of underparts; flight-feathers, tail, and tertials almost completely black-brown except for off-white outer fringes of p8-p9 and t6; pale fringes on rump and often on secondaries and on lesser and median upper wing-coverts retained longest. In fresh plumage, head, scapulars, and rump distinctly paler and less streaked than Calandra Lark *M. calandra* and Skylark *Alauda arvensis*, chest more broadly spotted black. In worn plumage, all head and body darker than these species, especially underparts blacker; no pale supercilium. In all plumages, no broad white tips on secondaries and tail-feathers (unlike *M. calandra*), only outer web of t6 white (unlike *A. arvensis*), and lesser and median upper wing-coverts black-brown with contrasting off-white fringes (unlike any other Palearctic Alaudidae). NESTLING. Down long, yellowish-buff (Harrison 1975). JUVENILE. Ground-colour of upperparts dull black, tinged grey on hindneck and upper mantle and olive on feather-centres of lower mantle and scapulars; each feather has short pale yellow-buff or white fringe on tip *c.* 1 mm wide, crown appearing spotted, mantle to rump showing short bars. Supercilium rather broad and long, pale cream; remainder of sides of head pale cream-buff with coarse black mottling on front and upper part of ear-coverts. Entire underparts pale cream-buff or off-white with rounded dull black spots on chest (*c.* 1 mm across) and sides of breast (*c.* 3 mm); flanks with dull black or olive-black spots. Tail black with broad pale buff-white tips and outer edges on t5-t6 (extending to inner web on t6). Lesser upper wing-coverts as mantle and scapulars; median and greater dull black with cream-white fringes *c.* 2-3 mm wide along tips. Flight-feathers as adult (black in ♂, dark grey-brown in ♀), but pale outer edges and tips wider, especially on p8-p10; p10 broad with rounded tip, 0-6 mm shorter than longest upper primary covert (10-17 mm shorter in adult). FIRST ADULT. Apparently like adult, and not separable once juvenile p10 shed. Perhaps differs in some minor details (e.g. pattern of tips of tertials, tail, or flight-feathers), but no known-age individuals examined. See also Moults.

Bare parts. ADULT. Iris brown or dark brown. Bill pale yellow with grey-brown, dark horn-brown, or blackish tip; culmen sometimes blackish. Leg and foot grey-black or black. (Hartert 1910; Dementiev and Gladkov 1954a; Toschi 1961; BMNH.) NESTLING, JUVENILE. No information.

Moults. ADULT POST-BREEDING. Complete, primaries descen-

dant. Starts late June, at peak in July, completed in August; head and neck sometimes almost bare due to intense moult (Dementiev and Gladkov 1954a). POST-JUVENILE. Complete or nearly so; primaries descendant. Starts presumably soon after fledging, late May to late August; completed from July onwards. Single ♀ from 1 September in last stage of moult of flight-feathers, tertials, and tail (BMNH). Moult occasionally incomplete, probably in late-fledged young: at least one winter ♂ had some juvenile greater upper wing-coverts, inner secondaries (s5-s7), and part of tail retained (RMNH).

Measurements. ADULT. Mainly Volga steppes (south European USSR), all year; skins (BMNH, RMNH, ZMA). Bill (S) to skull, bill (N) to distal corner of nostril; exposed culmen on average 3·6 shorter than bill (S).

WING	♂ 136·1 (2·43; 40)	132–142	♀ 128·6 (1·79; 19)	117–125
TAIL	70·6 (3·72; 12)	65–74	60·5 (1·79; 19)	58–63
BILL (S)	19·2 (1·35; 12)	16·8–20·8	17·5 (1·23; 18)	15·1–19·1
BILL (N)	13·3 (0·47; 12)	12·7–14·1	11·4 (0·34; 18)	10·8–12·2
TARSUS	24·7 (0·77; 12)	23·8–26·2	24·0 (0·72; 18)	22·1–24·9

Sex differences significant.

Wing, USSR: ♂ 134·1 (294) 124–144, ♀ 116·7 (87) 111–127 (Bub and Herroelen 1981).

JUVENILE. Wing on average *c.* 6 shorter than adult, tail *c.* 2; bill and tarsus probably full-grown once post-juvenile moult completed.

Weights. USSR: ♂ 63·7 (44) 56–76, ♀ 56·4 (20) 51–68 (Bub and Herroelen 1981). Kazakhstan (USSR): January, ♂ 60–75 (6), ♀ 51–61 (5); February, ♀ 57; March, ♂ 62–76 (6), ♀ 68; April, ♂ 57–67 (7), ♀ 54–57 (6); May, ♂ 58–68 (4), ♀♀ 53, 56; June, ♂ 56–65 (6); July, ♂♂ 60, 65; August, ♂ 60–72 (5), ♀♀ 55, 55; September, ♂ 56–64 (6); November, ♂♂ 62, 65; December, ♀ 50–59 (3) (Dolgushin *et al.* 1970).

Structure. Wing long, broad at base, tip rather pointed. 10 primaries: p8 longest, p9 0–3 shorter, p10 91–99 (♂) or 78–87 (♀), p7 1–3, p6 9–13 (♂) or 7–10 (♀), p5 20–25 (♂) or 17–20 (♀), p1 47–52 (♂) or 37–46 (♀). Adult p10 minute, narrow and pointed, hard to find beneath greater primary coverts, 14–17 (♂) or 10–14 (♀) shorter than longest primary covert; juvenile p10 broader and longer, tip narrowly rounded, equal in length to longest greater upper primary coverts or slightly shorter. Outer web of p6-p8 and inner of p7-p9 emarginated. Longest tertials reach to about p4 in closed wing. Tail rather short, tip slightly forked; 12 feathers, t1 4–10 shorter than t5-t6. Bill very strong (though often less deep at base than in *M. calandra*), short; culmen distinctly decurved, cutting edges slightly so. Many fine bristles among feathering covering nostrils and along base of upper mandible. Tarsus and toes short and rather slender; claws strong, straight or slightly decurved. Middle toe with claw 20·6 (14) 18–24; outer toe with claw *c.* 67% of middle with claw; inner *c.* 76%; hind with claw about equal to middle with claw, but toe shorter and claw longer—hind claw 12·6 (17) 11–15, middle claw 8·4 (8) 6–11.   CSR

Calandrella brachydactyla **Short-toed Lark**

Du. Kortteenleeuwerik	Fr. Alouette calandrelle	Ge. Kurzzehenlerche
Ru. Малый жаворонок	Sp. Terrera común	Sw. Korttålärka

Alauda brachydactila Leisler, 1814

Polytypic. Nominate *brachydactyla* (Leisler, 1814), Europe north to Yugoslavia and southern Rumania, and apparently local on Mediterranean coast of north-west Africa; *hungarica* Horváth, 1956, Hungary; *rubiginosa* Fromholz, 1913, North Africa and Malta; *hermonensis* Tristram, 1864, Levant from Sinai north to Syria (except north-west) and just into southern Turkey along border of central and eastern Syria; *woltersi* Kumerloeve, 1969, north-west Syria and neighbouring southern Turkey (Amik Gölü, Birecik, Gaziantep); *artemisiana* Banjkovski, 1913, Asia Minor, Transcaucasia, and north-west Iran, in Zagros mountains south to Lorestan and Fars; *longipennis* (Eversmann, 1848), plains north of Caucasus and perhaps Ukraine, east through steppes of lower Volga and Ural rivers and from Turkmeniya east to northern Mongolia and north-east China. Extralimital: *dukhunensis* (Sykes, 1832), Tibet and central China.

Field characters. 13–14 cm; wing-span 25–30 cm. About 30% smaller than Skylark *Alauda arvensis*, with shorter, more finch-like bill, no crest, and rather more compact form. Small, rather flat- and square-headed lark, usually of pale, cryptic coloration and (when adult) lacking streaks on chest. Within west Palearctic, colour variable with western birds essentially warm sandy-buff above and eastern ones pale grey-ochre. Upperparts have typical lark pattern; underparts usually little-marked except for buff breast and sometimes-prominent small dark patch at shoulder. Tertials almost overlap tips of primaries, unlike Lesser Short-toed Lark *C. rufescens*. Flight lighter than *A. arvensis*. Voice distinctive: calls dry (not rattling like *C. rufescens*). Sexes similar; no seasonal variation. Juvenile separable. 7 races in west Palearctic, 2 described here (see also Geographical Variation).

ADULT. (1) South European race, nominate *brachydactyla*. The 2nd most rufous race, with rufous most obvious on crown, giving capped appearance exaggerated by blocked end to head (though no real crest). Rest of upperparts basically sandy-buff, broadly streaked dull black-brown and most obviously relieved on face by buff-white supercilia (not quite joining over bill) and browner cheeks, and on wings by bold transverse line of median coverts with dark brown centres and pale buff margins. Nape and rump paler than rest, lacking obvious streaks. In flight, shows pale rump, and black tail faintly divided by brown and boldly edged white. Underparts white, with sandy-buff wash over chest and along flanks, and normally marked only with small brown-black mark near shoulder which often expands when neck stretched but can also be hidden under buff fringes in fresh autumn plumage. On a few (probably immature) spring vagrants and breeding birds, chest may show more streaks in scattered gorget and lack obvious discrete patch (see also Plumages). (2) North Asian race, *longipennis*. 2nd most grey race, lacking cap and with heavier, even more finch-like bill. Upperparts sandy brown-grey, with relatively heavier streaking but no differences in other plumage marks. JUVENILE. Upperparts distinctly more spotted than adult, with subterminal black and terminal buff-white to white marks on feathers producing sharply contrasting speckles at close range. Underparts whiter, with irregular gorget of dark brown spots and streaks across upper chest, usually but not always coalescing into dark patch by shoulder and only rarely extending to fore-flanks. FIRST WINTER. Although plumage normally completely moulted by September, some apparently retain characteristic juvenile pattern of chest-marks and thus make distinction from *C. rufescens* difficult. At all ages, bill horn, with yellow to buff base to lower mandible; legs brown-flesh.

Commonest member of genus in west Palearctic and thus a key species in lark identification, but unobtrusive and sometimes difficult to observe closely. However, troublesome overlaps exist in size and appearance with *C. rufescens* (see that species) and with 6 other species of juvenile or adult lark: Dunn's Lark *Eremalauda dunni* (with similar size and long tertials but with relatively more massive bill and no chest-marks); ♀ and juvenile Black-crowned Finch Lark *Eremopterix nigriceps* (20% smaller, with relatively deeper bill, only faintly marked plumage, and black under wing-coverts); Desert Lark *Ammomanes deserti* (15% larger, with longer, more pointed bill, no obvious streaking, no chest-marks, and dull tail pattern); Bar-tailed Desert Lark *A. cincturus* (same size but with bunting-like, not finch-like, character, no obvious streaking, no chest-marks, and dark terminal tail-band); juvenile Temminck's Horned Lark *Eremophila bilopha* (10% larger, with no obvious streaking, no chest-marks, and longer tail); *Alauda arvensis*, particularly of *dulcivox* group (25% larger, with short crest, and heavy streaks above and over all chest); Woodlark *Lullula arborea* (15% larger, with short crest, heavy streaks above and over chest, and bold supercilium). Also liable to be confused with Hume's Short-toed Lark *C. acutirostris* of Asian highlands which could conceivably wander to west Palearctic (slightly larger, with longer and more slender bill; plumage greyer above and duller and browner below, lacking white edges to tail), and Red-capped Lark *C.*

cinerea of Arabia and Afrotropics with which sometimes treated as conspecific (see Geographical Variation). Given this plethora of pitfalls, important to recognize that within *Calandrella* and other genera of Alaudidae, voice and song-flight (see below) more helpful in diagnosis than variable plumage and shape. Flight light and easy, with rapid take-off, fast undulating progress, and sudden descent without terminal flutter of *Alauda*. Flying flocks keep tight formation, like small finches *Carduelis*. Song-flight begins with fluttering ascent, and continues into high but small circles; usually ends with plummeting descent. Uses shuffling walk, faster, leggier run, and active hopping. Does not perch on plants; generally unobtrusive on ground.

Song a regular, quickening, high-pitched phrase of 7 notes, uttered with persistence; falling, plaintive terminal cadence characteristic. Commonest calls: short, dry 'tchirrup' or 'chichirrp', less rippling than similar note of *A. arvensis*; tone of disyllabic note recalls House Martin *Delichon urbica*, House Sparrow *Passer domesticus*, and Pied Wagtail *Motacilla alba*, and lacks sharper, rattled timbre of *C. rufescens*; also plaintive 'seeou' in alarm.

Habitat. Ecologically intermediate between Alaudidae of desert or semi-desert and those adapted to more vegetation cover, breeding in middle and lower middle latitudes in steppe, Mediterranean, and fringing temperate zones. Basically a steppe bird, favouring dry open plains and uplands, terraces, slopes, and undulating foothills, which can be of sand or clay, sometimes stony or gravelly. In Mediterranean Europe, where few other Alaudidae compete for varied habitat opportunities, demonstrates versatility and evolutionary potential by reaching high densities in several contrasting lowland types, especially near coast. On island of Mallorca, Spain (where Thekla Lark *Galerida theklae* is the only other breeding lark), extends from low bushy *garrigue* with mosaic of small patches of bare soil to neglected farmland, weedy fallows, stubble, harvest fields, and cereal crops growing up to *c.* 70 cm but with some openings, and usually flanked by dirt farm-tracks accessible for dusting and social behaviour. May occupy areas with soil-moving and foundation-laying in progress. Will watch and sing from tops of dry-stone walls, and sometimes power-lines and supporting poles, but only exceptionally tops of shrubs and apparently not yet using buildings or trees. Flies freely, mostly in lowest airspace, but in song-flights up to 100 m or more (E M Nicholson). In marismas of Guadalquivir (mainland Spain), breeds on low flat grassy islands, sharing neighbouring lower *Salicornia* patches with Lesser Short-toed Lark *C. rufescens*. Both there and in Camargue (France) breeds on dried mud with sparse plant cover and high ground temperature, while on torrid stony plains of the Crau, adjoining Camargue, it is by far the most numerous breeding species, everywhere at high density (Hoffmann 1958; Valverde 1958). In late autumn in Camargue, spreads to stubbles and fallows. Tolerates saline areas, even drinking saline water; will nest near water, even on seashore. On Biscay coast of France breeds on young mobile sand-dunes among marram grass *Ammophila*, with *Artemisia* and other pioneer plants, and elsewhere on strip of heathland behind new dunes and in front of forest edge (E M Nicholson). In Tunisia, as on Mallorca, overlaps marginally with *G. theklae* where grassland abuts on *Tamarix* and olive plantations (E M Nicholson). In inland Turkey occurs on open cultivated land, even occasionally round farm buildings (Wadley 1951).

On Hortobágy plain (Hungary), breeds on a few small dry dusty sites scattered *c.* 10-30 km apart on vast grey alkali flats of degraded soil and scattered pools, covered with dominant *Artemisia* and sub-dominant *Festuca* over areas of a few hundred square metres. Avoids more extensive tracts of apparently suitable habitat and, as elsewhere, bunches in closely grouped territories at regular localities (Horváth 1956). In southern Ukraine (USSR), overlaps with Skylark *Alauda arvensis* and Calandra Lark *Melanocorypha calandra* in dense low grass, but in USSR normally keeps to open steppe with sparse vegetation, avoiding virgin lands with good grass cover. In Armeniya, occurs in rocky desert steppes and semi-deserts with *Artemisia*. Despite association with arid lands usually avoids full desert, as well as tall dense herbage, wetlands, and close stands of trees; on migration will cross mountainous rocky and broken terrain, and will even settle freely on tarmac highway (P A D Hollom and E M Nicholson). Migrates in lowest airspace in loose parties or flocks, landing apparently rather to rest than to feed. Generally unconcerned about human presence unless at very close quarters, and little affected over most of range by human activities. Winters on steppe-like semi-desert plains of Africa south of Sahara.

Distribution. Range decreased in France; some northward expansion in USSR.

FRANCE. In 20th century disappeared from centre and much reduced on Atlantic coast; new colony of 10-15 pairs recently between Paris and Orléans (RC). AUSTRIA. Bred 1966 (HS). EGYPT. Has bred Wadi Natrun (Meinertzhagen 1930); no evidence of breeding in north since (SMG, PLM, WCM). IRAQ. Distribution inadequately known because of confusion with Lesser Short-toed Lark *C. rufescens* (Moore and Boswell 1956; Marchant and Macnab 1962; Marchant 1963b). USSR. Voronezh region: has expanded northwards (Wilson 1976).

Accidental. Iceland, Ireland, Channel Islands, Belgium, Netherlands, West Germany, Denmark, Norway, Sweden, Finland, East Germany, Poland, Austria, Switzerland, Chad, Canary Islands.

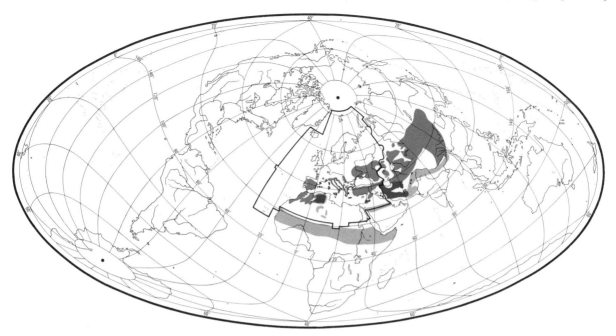

Population. Decreased France; no information on trends from elsewhere.

FRANCE. 1000-10 000 pairs (Yeatman 1976). Bretagne: decreased, near extinction (RC). Corsica: scarce (Thibault 1983). SPAIN. Numerous (AN). LIBYA. Rather scarce (Bundy 1976). JORDAN. Azraq: locally relatively dense (Wallace 1983a). CYPRUS. Common (Flint and Stewart 1983). MALTA. Very common (Sultana and Gauci 1982). USSR. Plentiful in Transvolga, Askaniya steppe area, and coastal Black Sea reserve, rare in Azov and Kharkov (Dementiev and Gladkov 1954a); declined between Volga and Ural rivers, Kazakhstan (Shishkin 1976).

Movements. Migratory in Palearctic, except perhaps locally in southern parts of range where may be only partially migratory.

European and Middle East races (nominate *brachydactyla*, *hungarica*, *artemisiana*, *woltersi*, *hermonensis*) mainly if not wholly migratory, though nominate *brachydactyla* has been reported as resident in Transcaucasia and Muganskaya steppe, southern USSR (Dementiev and Gladkov 1954a). Passage occurs on broad front across Mediterranean, Sahara, and Middle East to winter quarters in Africa south to Sahel and Red Sea, mainly within arid zone 14-17°N. Not all African migrants can be subspecifically identified (Vaurie 1959), and apparently no confirmation yet for occurrence of *artemisiana* (of Caucasus/Caspian/Iran), though this should occur in north-east Africa at least (Moreau 1972). Nominate *brachydactyla* (from southern Europe) and *hermonensis* (from Middle East) certainly overlap in winter in north-east Africa (Vaurie 1959; Moreau 1972). Only *brachydactyla* listed for Sénégal (Morel and Roux 1973), though

the mainly migratory North African population *rubiginosa* presumably occurs in West Africa in winter. Central Asian race *longipennis* winters from Turkmeniya (USSR) to India (south to 16°N); also twice collected (November, December) in Kenya (Britton 1980), but its status in Red Sea region remains unknown (Moreau 1972). Small numbers (races unknown) winter in Arabia, e.g. winter average of *c.* 40 birds at Thumrait, Oman (Walker 1981b); on Masirah island (Oman), winters only after rainfall (Griffiths and Rogers 1975).

After flocking during July, autumn movement begins mid-August, at peak late August and early September in Europe and Turkey though continuing into October. See Telleria (1981) for Straits of Gibraltar data and comments on later peak movements claimed by other authors. Smith (1965a) reported big passage on Atlantic coast Morocco, mid-September to late October, which cannot be allied with earlier migration timing at Straits of Gibraltar and hence may refer to migratory North African population. At height of passage, flocks pass in waves; *c.* 10 000 over Tuz Gölü (central Turkey) on 5 September 1968 in continuous south-westerly movement, when largest numbers passed 07.00-11.00 hrs though passage continued into late afternoon (Vittery *et al.* 1972).

Spring passage is early. Begins late January in southern Morocco (Heim de Balsac and Mayaud 1962; Zink 1975), but these probably returning North African *rubiginosa*, since passage of northern and eastern breeders through Malta and Cyprus (where spring migration more conspicuous than autumn) does not begin until early March (Sultana and Gauci 1982; Flint and Stewart 1983). However, passage reported early January to early April at Elat, Israel (Safriel 1968). Even in USSR, vanguard

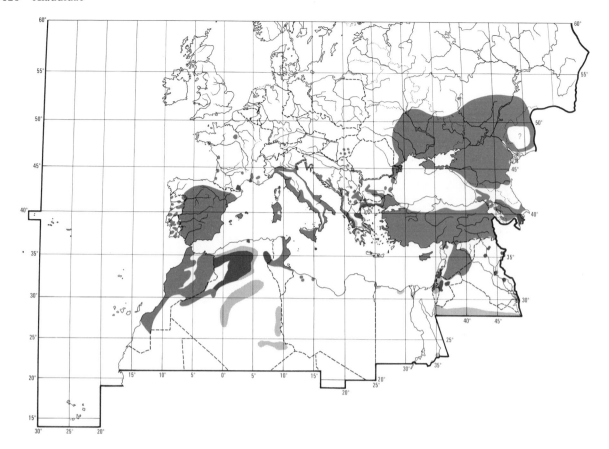

returns in March, with immigration continuing into April
(Dementiev and Gladkov 1954*a*). Spring passage over
Coto Doñana (southern Spain) is from west to south-west
direction, whereas other passerines arrive from south; this
indicates that *C. brachydactyla* makes longer sea-crossings
involved in reaching Iberia to west of Straits of Gibraltar
(Henty 1975).   RH

Food. Insects and seeds in summer, mostly seeds only at
other times. Takes food from ground or low plants (Endes
1970), and often digs with bill (Meinertzhagen 1951*a*).
Once seen to fly up *c.* 2 m to take flying insect (Endes
1970), and recorded hovering over flowers to take insects
from them (Olioso 1974). May squat when feeding, tarsi
on ground (Gallagher and Woodcock 1980). Said to be
able to live for months without water but visits it regularly
where available; drinks brackish water (Dementiev and
Gladkov 1954*a*; Endes 1970).

The following recorded in diet. Invertebrates: dam-
selflies, etc. (Odonata), grasshoppers, etc. (Orthoptera),
termites (Isoptera), bugs (Hemiptera: Miridae, Penta-
tomidae, Cicadidae, Aphididae), adult, pupal, and larval
Lepidoptera (Geometridae, Noctuidae, Aegeriidae), flies
(Diptera: Bibionidae), Hymenoptera (Tenthredinidae,
Ichneumonoidea, Chalcidae, Scoliidae, ants Formicidae,
Sphecidae, bees Apoidea, Psammcharidae), adult and

larval beetles (Coleoptera: Carabidae, Histeridae,
Scarabaeidae, Buprestidae, Elateridae, Dasytidae, Tene-
brionidae, Coccinellidae, Cerambycidae, Chrysomelidae,
Bruchidae, Curculionidae), spiders (Araneae), woodlice
(Isopoda), snails (Gastropoda). Plant material mainly
seeds or fruits: of docks and knotweeds (Polygonaceae),
goosefoots (Chenopodiaceae), *Amaranthus*, chickweeds
Stellaria, poppies (Papaveraceae), Cruciferae, buttercups
(Ranunculaceae), fumitory *Fumaria*, Rosaceae, Leguminosae, loosestrifes (Lythraceae), *Hypericum*, sea-lavender
Limonium, broomrapes (Orobanchaceae), Boraginaceae,
Labiatae, Compositae, lilies (Liliaceae), sedges (Cyperaceae), and grasses (Gramineae) including cereal grain;
also plant leaves and stems. (Dementiev and Gladkov
1954*a*; Pek and Fedyanina 1961; Ryabov and Mosalova
1967*b*; Kekilova 1969; Endes 1970.)

In spring at Askaniya-Nova (Ukraine, USSR), stomachs filled with weed seeds (*Amaranthus albus*, *Polygonum
aviculare*, *P. novoscanium*) and shoots of grass *Poa bulbosa*.
In summer, insects much more important and may
comprise most of diet. Thus, at Askaniya-Nova birds
contained ants *Myrmica* and Orthoptera (*Calliptamus*,
Metrioptera) (Dementiev and Gladkov 1954*a*). In southern Turkmeniya, however, few seeds taken in spring,
food then comprising mostly Orthoptera, termites, bugs,
and ants; seeds much more important in summer (Kekilova

1969). Birds from steppe by Sea of Azov (Ukraine), June, contained beetles (*Pedinus, Pachybrachis, Cassida, Chaetocnema, Phyllotreta, Spermophagus, Baris, Dolychosoma*), ants (*Cataglyphus, Tetramorium, Messor, Lasius*), bugs (*Nisius, Brachycarenus*), ichneumons, and pupae of moths; also grain. In El'ton steppe (Volgograd region), birds contained Orthoptera (*Metrioptera, Celes*), beetles (*Harpalus, Sphenoptera, Cylindromorphus, Cerambycidae, Ischironota, Othiorrhynchus, Procus, Polydrosus, Eusomus, Tylacites, Psalidium, Baris, Aphodius*), bugs (*Cotysus*, Miridae), flies, Hymenoptera (ants, ichneumons, *Scolia*), butterflies, and caterpillars (Dementiev and Gladkov 1954a). In Crimea, June–July, 43 stomachs contained 60·9% invertebrates (including 19·3% beetles, 13·4% ants, and molluscs), 35·6% seeds of herbs (including 27·8% *Amaranthus*), and 3·5% grain (wheat, rice, sorghum) (Kostin 1983). In Kustanay steppes (northern Kazakhstan), May–June, adults (50 stomachs) took animal and plant food in about equal proportions (by weight and volume): 274 invertebrates comprised 50·4% (by number) ants, 32·5% beetles (largely Curculionidae), 10·6% bugs (largely Cicadidae), 2·6% other Hymenoptera, 1·8% Orthoptera, 1·1% Lepidoptera, and 1·1% spiders; plant material largely *Polygonum* seeds and grain, but 7·5% of stomachs contained leaves and stems. Fledged juveniles took more plant material than adults, but proportions of different food categories otherwise similar. (Ryabov and Mosalova 1967b.) Plant food eaten in summer includes also (presumably seeds of) grasses *Panicum* and *Setaria* (Dementiev and Gladkov 1954a). Bird from Moldavia contained wheat and Chenopodiaceae (Averin and Ganya 1970). See also Pek and Fedyanina (1961), Lakhanov (1966), and Medvedev and Esilevskaya (1973). On Canary Islands, April, bird seen to catch small grasshopper (Shirt 1983). In Algeria, February–March, 4 stomachs were crammed with seeds (Roche 1958). In central Saudi Arabia takes millet and other grain (Bates 1936–7), and birds from Hijaz (north-west Saudi Arabia) contained mainly small seeds and a few grasshoppers and beetles (Meinertzhagen 1954). In India and Pakistan, eats grass and weed seeds (e.g. *Eleusine, Eragrostis, Trianthema*), beetles (e.g. Scarabaeidae, Curculionidae), ants, caterpillars (Geometridae), and small snails (Ali and Ripley 1972).

Basal metabolism (presumably of captive birds) in north Caspian region (USSR): late February, 40·7 kJ per bird per day (25–55, *n*=16) ; May, 42·0 (35·5–51, *n*=4); August, 29·0 (21·5–32·5, *n*=4) (Shishkin 1980).

Young given only animal food (Endes 1970). Fed dawn to dusk, average 10–13 times per hr (2–10 min between feeds), though rate may drop to 6 times per hr around midday (Dementiev and Gladkov 1954a). For diet of fledged young, see above.   DJB

Social pattern and behaviour. Major studies in Italy (Dathe 1952) and in Hortobágy (Hungary) (Endes *et al.* 1967; Endes 1969–70, 1970).

1. Highly gregarious outside breeding season, typically occurring in flocks of a few birds or up to several thousand (e.g. Niethammer and Laenen 1954; Endes 1970; Flint and Stewart 1983), occasionally many more, e.g. in Mali, flocks of up to 700, December, 10 000 plus in March–April (Lamarche 1981). At Attika (Greece), flocks usually of 10–30, rarely 50–60, once 400 (Steinfatt 1954). In spring, migrant flocks vary from a few to several hundred birds (e.g. Guichard 1960, Endes 1970). Soon after arrival on breeding grounds, birds tend to be in parties of 2–5, usually several ♂♂ and 1 ♀ (Endes *et al.* 1967; Endes 1970). At end of season, family parties gather into flocks. At Hortobágy, entire breeding population of a given nesting area, plus fledged offspring, rove together unaccompanied by other species (Endes *et al.* 1967). In Malta, juveniles often congregate into large flocks, some of which start leaving mid-August (Sultana and Gauci 1970, 1982). Wintering flocks frequently associate with other Alaudidae, notably Lesser Short-toed Lark *C. rufescens* (Bannerman and Bannerman 1953), Crested Lark *Galerida cristata* (Bodenham 1944), Calandra Lark *Melanocorypha calandra*, and Skylark *Alauda arvensis* (Endes 1970); on Balearic Islands, with Corn Bunting *Miliaria calandra* (Munn 1931). BONDS. Mating system probably monogamous, but no hard evidence. ♀ alone incubates and broods young, but both sexes feed young (Endes *et al.* 1967; Endes 1970). Family bonds maintained for a while after fledging; duration not known, but 1st broods independent quite quickly, 2nd broods less so (Endes *et al.* 1967). Age of first breeding 1 year (Endes 1969–70). BREEDING DISPERSION. Solitary and territorial. Territories typically clustered, forming neighbourhood groups of 10–20 pairs; usually 10–20(–30) km between groups. Territory relatively small, 40–50 m in diameter; in 1 case, 2 occupied nests 15 m apart (Guichard 1960; Rucner 1960; Endes *et al.* 1967; Endes 1970). Neighbouring territories sometimes overlap despite vigorous defence (Endes 1970). Density, Hortobágy, 2–3 pairs per ha (Endes 1970), 4–5 pairs per ha (Endes *et al.* 1967); density does not change between 1st and 2nd broods (Endes 1970). In Rhône (France), territory appeared to be very small, with 10–12 singing ♂♂ in 6 ha (Olioso 1974). In Rimini (Italy), *c.* 8 pairs in 198 ha (Dathe 1952). In steppes (USSR), up to 2–3 pairs per ha (Dementiev and Gladkov 1954a). In north-west Turkmeniya, up to 8–10 birds per km of transect in mid-April, and 15–20 late April, though not all of latter known to be breeding (Rustamov 1954). Territory said to be chosen by ♀, but defended by ♂; serves for courtship, nesting, and rearing young to fledging. Resident pair forages in immediate vicinity of nest but also outside territory. (Endes *et al.* 1967; Endes 1970.) Never uses 1st nest for 2nd clutch; nest of 2nd clutch sometimes quite near 1st, but may be up to several hundred metres away (Endes 1970). In Cyprus, 'song territories' overlap with those of *M. calandra* which feeds alongside *C. brachydactyla*; *G. cristata* also share territories but do not feed alongside (Took 1972). In Coto Doñana (Spain) may form neighbourhood groups with *C. rufescens* (Mountfort 1958). ROOSTING. Nocturnal, on ground. In breeding season, ♂ scoops out shallow, smooth-sided, unlined hollow in ground, at variable distance (40–60 m recorded) from nest. If pair remain near 1st nest when building new one for 2nd clutch, ♂ may retain same roost-hollow, otherwise makes new one. Hollow vacated by day, except often in persistent bad weather. After young fledge, nest not used by ♀ or young for roosting. In July–August, when young dispersing, number of roosting-hollows in breeding area increases, indicating that

♀♀ then also make them (fledged young not known to make hollows). (Endes 1970.) Large roosting flocks—presumably of migrant or wintering birds—sometimes make hundreds of roosting-hollows in relatively small area (see Bannerman 1953). For loafing, etc., of post-breeding flocks, see Flock Behaviour (below). In Malta, not uncommonly sings on moonlit nights (Sultana and Gauci 1982); singing around midnight also reported by Reiser and Führer (1896), but probably not very common, as not heard by Dathe (1952) or Endes (1970). For singing from ground before dawn, see Song-display, below. In southern Ustyurt (Kazakhstan, USSR), singing begins 1-2 hrs before dawn, early May. Birds less active 10.00-15.00 hrs when sun hotter, tending to fly little and rest in shade of bushes; resurgence of activity at 16.00 hrs until almost dark (Rustamov 1954). For development of roosting and comfort behaviour in young, see Relations within Family Group (below).

2. Quite approachable, but also wary, often stopping to straighten up and stretch neck to scan for possible danger (Guichard 1960); often mounts small eminence for surveillance (Dathe 1952). Squats on landing, effecting excellent camouflage, and maintains posture close to ground while feeding, etc. (Wadley 1951; Mountfort 1954). Squatted when Red-footed Falcon *Falco vespertinus* flew overhead (Nadler 1974: Fig A).

A

Raises crest when excited (Dathe 1952). For other responses when disturbed, see Flock Behaviour. For Alarm- and Warning-calls, see 2b and 3 in Voice. FLOCK BEHAVIOUR. In April-May, Coto Doñana, flocks compact and highly coordinated, rising as one (Mountfort 1954); constantly on the move as they feed, flying low with undulating flight; small party may squat before flying off together (Ferguson-Lees 1970b). Wariness varies with time of year. On arrival on breeding grounds, flocks easily approachable, and come within 8-10 m of motionless intruder; after breeding, loath to fly. When driven, run swiftly, stopping and starting in sudden jerky manner and watching intruder intently. Once flushed, describe wide undulating arcs, then dive suddenly to ground. Wary and restless shortly before migration, often flying up for no apparent reason. (Endes *et al.* 1967; Endes 1970.) In winter, Chad, passage flocks wild and flighty, readily flushing to re-settle singly or in small groups; become larger or more compact in early March, and then behave more in unison (Holmes 1974). SONG-DISPLAY. Well studied. Sings (see 1 in Voice) mostly in flight but also not uncommonly on ground (see below). Elaborate Song-flight of ♂ performed over nest-territory. Following account compiled from detailed descriptions by Dathe (1952), Endes *et al.* (1967), Endes (1970), and Bergmann and Helb (1982); see also Voice for (e.g.) further relationship of song with phases of Song-flight. Bird ascends steeply with rapidly beating wings to *c.* 8-15 m; flies, frequently in spiral path, drifting from side to side (see below), and giving introductory part of song (see 1 in Voice); thereafter, gives main part of song (see below), this continuing up to 30-50 m, whereupon bird extends wings, motionless, and, on final note of song-phrase, closes wings and descends, or may open wings to effect slower gliding descent. Before reaching ground, beats

wings again a few times to achieve a much lesser and usually silent ascent, then drops down again and initiates a new major ascent, repeating sequence of song as for 1st ascent, and so on. Song-flight thus a series of deep undulations regularly interspersed with shallow ones. According to Bergmann and Helb (1982), bird silent on initial descent with wings closed, starts singing on opening wings to glide, and continues to sing on re-ascent until peak reached. Bird often described as rising and falling vertically on the spot (Guichard 1960)—as on a thread (Hüe 1952; Bagnall-Oakeley 1955), or like a yo-yo (Vere Benson 1970): e.g. in Song-flight in Israel, *c.* 14 undulations per min, bird gliding down during song, and recovering height during short gaps between song-phrases (P A D Hollom). However, pattern variable, and often (perhaps usually) more complex than simple vertical oscillation. Singing bird flies into wind, effecting balance between forward movement and backward drift; moreover, in calm conditions, bird may describe irregular horizontal circles (Bergmann and Helb 1982). In this main part of song, at peak of ascent, bird may follow irregular path of arcs and spirals (Fig B, showing plan view), all the

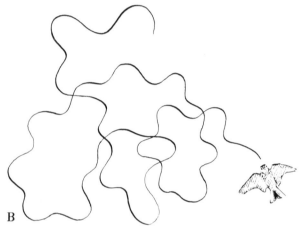

B

while meandering 30-50 cm from left to right, undulating up and down, with associated song pattern (Dathe 1952; Endes *et al.* 1967; Endes 1970). Final descent to ground, usually a silent headlong dive, arrested *c.* ½ m above ground by spreading wings and performing slightly undulating flight like pipit *Anthus* for 10-20 m before landing. More rarely, descends obliquely or makes stepped descent, dropping by (e.g.) 4 stages of 10 m, and singing at each step (Dathe 1952; Endes *et al.* 1967; Endes 1970, in which see Fig 8; Nelson 1973). For incidence of these variants of descent, associated with intruder, see Parental Anti-predator Strategies (below). Occasionally, bird descends singing all the while, without any final plunge, or instead of landing may spiral up to start again (Endes *et al.* 1967; Endes 1970). Compared with *A. arvensis*, does not ascend as high, and flight-path much more erratic, notably in horizontal plane (Endes 1970); more undulating than *C. rufescens* (Mountfort 1958). Duration of Song-flight usually 3-5(-15) min (Dathe 1952) According to Dathe (1952), time of day and stage of breeding season have little influence on duration of Song-flight. However, increase in length during season reported by Endes *et al.* (1967) and Endes (1970), as follows: during pair-formation *c.* 1-2(-3) min; during nest-building, 5-6(-8) min; during incubation and nestling stage, 6-8(-26) min. During first few days after arrival on nesting grounds, no Song-flights performed but singing on ground not uncommon, especially before dawn, and at dusk; bird uses any available vantage point, e.g. clump of earth, telegraph wire (Dathe 1952; Steinbacher 1952; Endes *et al.* 1967; Endes 1970). Singing ♂ sometimes walks towards incubating ♀ (Endes 1970). On completion of Song-flight, ♂ also sometimes delivers final phrase of song on landing (Dathe 1952). Song (see 1 in Voice) given throughout the day, mostly in morning (Dathe 1952, which see for seasonal variation), less at midday (Endes 1970); See also Roosting. Song given throughout breeding season until after young leave nest (Nagy 1935). In North Africa, Song-flights reported from January (Endes *et al.* 1967; Endes 1970). In Attika, song first heard early April, *c.* 10 days after arrival in nesting area (Steinfatt 1954). Song declined (for details see 1 in Voice) early May among early breeders, but resurgence occurred among pairs whose 1st nests destroyed (Dathe 1952); song wanes markedly from mid-July as flocks develop (Dathe 1952; Endes *et al.* 1967; Endes 1970). In Malta, ♂♂ occasionally sing late August to early September (Gibb 1946; Sultana and Gauci 1982). Song once reported during spring passage (Witherby *et al.* 1938a). ANTAGONISTIC BEHAVIOUR. Territory-owner highly aggressive towards conspecifics encroaching on territory (Nadler 1974). Especially at start of breeding season, ♂♂ thus often chase one another and then fight (Guichard 1960). Chasing described as 'reciprocal', involving only ♂♂ (Dathe 1952); ♀ only rarely helps to drive off intruder. Dominant ♂ may chase rival for appreciable distance (Endes 1970). *A. arvensis* also chased in short but persistent pursuit-flight by singing ♂ (Christen 1983), though birds other than conspecifics usually ignored (Endes 1970). Fight usually begins on ground, territory-owner dropping down from Song-flight to attack. Fight often 1–2 m above ground, sometimes an ascending tussle up to considerable height; does not always involve physical contact. Fighting most frequent early in season, recurring in birds which lose 1st clutch; ♂ markedly aggressive during incubation and nestling period, continuing to defend territory until young fledge (Endes 1970). In some contexts, Song-display explicitly antagonistic. Thus, in repeated observations of Song-flights by 2 ♂♂ near common territorial boundary, birds kept singing while staying only 2–3 m apart for 5 min; when one turned in flight, other followed suit, shadowing it; if one attacked, other tended to take evasive action. On another occasion, rivals flew on parallel paths, sometimes only 80 cm apart, before shortly returning to respective territories (Dathe 1952). If conspecific intrudes into territory during owner's Song-flight, this usually elicits a stepped descent (see above), with song at each step (Endes 1970). HETEROSEXUAL BEHAVIOUR. (1) General. Pair-formation evidently begins immediately on arrival at breeding grounds, pairs forming from and within small flocks (see introduction to part 1). For apparent courtship behaviour between 2 ♂♂ outside breeding season in Algeria, see Niethammer (1954). (2) Pair-bonding behaviour. Song-flight of ♂ sometimes has evident relation to ♀, since ♀ not infrequently accompanies singing ♂ in flight. In such cases, ♂'s flight-path lacks horizontal curves and he delivers a particularly rapid succession of phrases, as if stimulated by proximity of ♀. Before pair-bonds established, several ♂♂ sometimes chase single ♀. In group of several ♀♀ and 1 ♂, ♂ finally focused attention on 1 ♀, during which there was constant interference from other ♂♂. As pair-bond develops, ground-display gains prominence; ♂ persistently follows ♀ who flies or walks away; at times they fly or glide towards each other just above ground. ♂ struts around with tail raised, raising and lowering crest from time to time, hops around ♀, sometimes bouncing up and down on the spot. At high intensity, runs in tight circles around ♀, spreading tail and holding wings slightly out from body and quivering them. Sometimes makes nodding or bowing movements towards ♀ or may crouch. ♀

apparently indifferent to ♂'s advances and sometimes runs aggressively at him, though stopping short of physical contact. From time to time, ♂ rises and hovers 1–2 m over ♀, sometimes circling 25–30 m above her while singing. At first ♀ may be somewhat hostile to descending ♂, pecking at him. (Endes *et al.* 1967; Endes 1970.) (3) Courtship-feeding. Little information, and none available for period before laying. One ♂ apparently brought food for ♀ sitting on 4 eggs (Nadler 1974), and brooding bird also fed thus (Olioso 1974). Courtship-feeding at nest perhaps not, however, regular or widespread, as Endes (1970) never saw ♂ feed ♀ there; instead, ♀ left nest periodically to self-feed. (4) Mating. Usually occurs after prolonged bout of ground courtship (Endes *et al.* 1967) in which both birds may play soliciting roles. In typical sequence, ♂ performs pre-copulatory display: hops around ♀, nodding and then gently pecking at her head; after 2–3 min, ♀ crouches, shivering wings, and copulation follows, lasting a few seconds (Endes 1970). Copulation reported during nest-building (Endes 1970) and after laying: 4 days after clutch complete, disturbed ♀ left nest and landed 1 m from ♂, and *c.* 10 m from nest; ♂ flew on to her back and rapid copulation followed; ♀ then crouched again, shivering wings, but when ♂ was about to mount her again, she resisted, pecking him; ♂ settled next to her and ♀ ran back to nest (Dathe 1952). Marked reduction in ground-display and copulation during bad weather (Endes 1970). (5) Behaviour at nest. ♀ chooses nest-site, collects almost all nest-material, builds nest, and incubates; ♂ keeps look-out (Dathe 1952; Guichard 1960; Endes 1970), only occasionally bringing nest-material (Endes 1970). Pair always land some way from nest and approach on foot. At one nest, ♂ performed Song-flight while ♀ collected nest-material, and apparently summoned him from time to time with Contact-calls (see 2a in Voice); ♂ would then land and walk round ♀. ♀ continues to build after laying (Dathe 1952). RELATIONS WITHIN FAMILY GROUP. Following account mainly after Endes *et al.* (1967) and Endes (1970). Only ♀ broods, also does most feeding of young. Young gape a few hours after hatching and give quiet calls (see Voice). ♀ broods young for first 2 hrs, then begins feeding them. Bird bringing food usually gives Contact-calls 15–20 m from nest and young beg loudly. When sated, young crouch motionless in nest. Eyes begin to open at 4 days, and young increasingly active from 5 days. According to Guichard (1960), young do not leave nest until they can fly, and this perhaps true where disturbance minimal. Most leave nest at 8–10 days (Endes *et al.* 1967; Endes 1970; Sultana and Gauci 1970). Usually leave at time of afternoon feed, leaping from nest on return of adult. If runt left in nest, parents continue to feed it. Young begin to fly at 11–12 days. In captive study, young started self-preening at 10 days. At 14 days, slept with head buried in mantle feathers, sand-bathed at 25 days. By 34 days, much more mobile, but came together for roosting. Young of 1st brood often stay together after fledging and tend to be tamer than those of 2nd brood which have longer contact with parents (Endes *et al.* 1967; Endes 1969–70). PARENTAL ANTI-PREDATOR STRATEGIES. (1) Passive measures. When intruder enters territory, birds usually utter Alarm- and Warning-calls and snatches of song in the air or from some vantage point on ground (Sultana and Gauci 1982). Almost impossible to surprise sitting ♀ who typically leaves nest when intruder some way off, trotting discreetly through stones, vegetation, etc. (Guichard 1960). Sits tighter on eggs or young in bad weather, flushing only on close approach (Munn 1931; Endes 1970). ♂ reported on nest by Shnitnikov (1949) said to have stayed put until closely threatened by dog, then neither flew nor ran off but held ground *c.* 1 m from nest. (2) Active measures: against man. If intruder stays by nest for some time, pair begin to circle intruder, sometimes running rapidly along ground or making short flights; little or no vocal demonstration (Guichard 1960). ♂ the bolder, always first to return to nest area; ♂ or ♀ may hover just above nest or fly over intruder in wide arc. When young hatch, ♂ changes behaviour: during incubation, he interrupts Song-flight to plunge to ground, but during nestling period he initiates Song-flight under threat, as in *A. arvensis* (Nadler 1974). When intruder in territory, ♂ sings particularly loudly and then, falling silent and with wings closed, makes steep descent from Song-flight to land near but never at nest (Dathe 1952; Endes 1970). As intruder approaches nest, sitting ♀ walks or runs up to several hundred metres, maintaining some distance from intruder who follows (Endes 1970); this effectively a distraction-lure stratagem. Both ♂ and ♀ participate to lead intruder away from young out of nest (Endes *et al.* 1967). If ♀ closely approached before flushing from eggs or (especially) young, she may perform mobile distraction-lure display of disablement type: appears to fall over after moving a few metres from nest, dragging herself along on erratic path and feigning broken wing. When she reaches a certain distance, flies up and lands quite far away, beginning to walk around slowly and cautiously peck from time to time (Endes 1970); this perhaps mock-feeding. (3) Active measures: against other animals. Apparently as for intrusion by man: e.g. ♀ runs from nest, effecting distraction-lure stratagem; ♂ initiates Song-flight when quadruped intrudes on nest-territory (Endes *et al.* 1967; Endes 1970). In Malta, adult feigned injury to lure snake *Coluber* away from nest (Sultana and Gauci 1970).

(Figs by J P Busby: A after photograph in Nadler 1974; B after diagram in Endes 1970.) EKD

Voice. Used throughout the year, but especially in breeding season when song of ♂ dominant feature. No information on differences, if any, between races.

CALLS OF ADULTS. (1) Song of ♂. Given in Song-flight or from ground. A clear musical sound (Sultana and Gauci 1970); less melodious and varied than Lesser Short-toed Lark *C. rufescens*, and, in pitch, not unlike twitter of Swallow *Hirundo rustica* (Mountfort 1958). Shrill, jingling, and melodious (Endes 1970). Resembles Black Redstart *Phoenicurus ochruros* and Whinchat *Saxicola rubetra*, and—in being divided into phrases—differs from Skylark *Alauda arvensis* or Woodlark *Lullula arborea*; *contra* several sources, Contact-call (call 2) not interpolated, despite some resemblance (Dathe 1952). For criticism of some other (notably fluting) interpretations, see Schubert and Schubert (1982). Comprises short phrases similar to *C. rufescens* but less rapidly trilling. In ascent phase of Song-flight, often begins with single accelerating 'dip-dip...', not infrequently interspersed with mimicked calls, e.g. of Kestrel *Falco tinnunculus* (thus 'dip dip ki ki') or Crested Lark *Galerida cristata* (Bergmann and Helb 1982, which see for sonagrams; see also Dathe 1952 and Steinfatt 1954 for mimicry of *G. cristata*). Other species mimicked include Pied Wagtail *Motacilla alba*, House Sparrow *Passer domesticus*, *L. arborea* (Dathe 1952), and Collared Pratincole *Glareola pratincola* (Endes 1970). For apparent mimicry of song of Chaffinch *Fringilla coelebs*, see below. Many notes

are variants on 'klitra', also given by Calandra Lark *Melanocorypha calandra* (Schubert and Schubert 1982). Introductory part of song also rendered 'di di'; in course of song, fuller variant, 'wiehe wiehe', indicates that Song-flight soon to end (Dathe 1952); introductory units also rendered 'zik zik' or 'tschirp tschröp', most reminiscent of Tree Sparrow *Passer montanus* (Endes 1970). Rarely, introductory part of song omitted; one bird gave 6 introductory notes in succession, while another gave such notes only, omitting main song altogether; such irregularities more prevalent towards end of song-period (Dathe 1952; see Social Pattern and Behaviour for seasonal duration). After ascending to a certain height (*c.* 15 m: Endes 1970), introductory notes give way to main song: series of 10–20(–60) phrases (Dathe 1952), each phrase *c.* 8–10 units, and phrases repeated persistently at short intervals (Mountfort 1954). In our recording, song a short (*c.* 1·7–2·2 s) distinct phrase repeated with significant variation at intervals of *c.* 2·5–3 s; phrase (Fig I) starts with 3 short 'tik' units, followed by a rapid musical series of tinkling warbling sounds. 10th phrase (Fig II) given in recording closely resembles song of *F.*

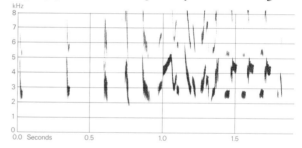

I P A D Hollom Israel April 1980

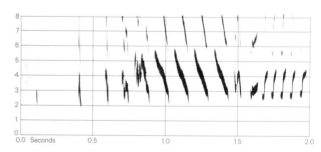

II P A D Hollom Israel April 1980

coelebs, without terminal flourish of that species (J Hall-Craggs). In Iran, a rattling 'stek-tek tair-air-air te-o', the 1st 2 syllables dry, the rest more tuneful, descending at end. In southern Turkey, song more varied, continuous and tuneful at beginning of Song-flight, becoming repetitive with regular breaks once maximum height reached (P A D Hollom). Rendered 'tsee-tsit-si-WEE tsi-WICHoo' (Witherby *et al.* 1938a); also 'tschöp-zöp-zirile-ürilézirüé', first 2 units brief, remainder sounding like long-drawn trill (Nagy 1935). Average duration of phrase 1·3 s (0·8–1·7) (Bergmann and Helb 1982), interval between phrases 1-1·5 s (Schubert and Schubert 1982), $3\frac{1}{2}$-$4\frac{1}{4}$ s (Dathe 1952). Average *c.* 12 phrases per min (Bergmann and Helb 1982). One ♂ sang 48 phrases in $1\frac{3}{4}$ min, maximum (different ♂) 68 phrases in 5 min 10 s; songs usually 3-5(-15) min long, (Dathe 1952, which see for variation among ♂♂). On final unit of phrases, coinciding with peak of ascent, bird commences descent from Song-flight. For further synchronization of song with Song-flight, see Social Pattern and Behaviour. Ground-song quieter, with tendency towards several pauses (Endes 1970; for context, etc., see Social Pattern and Behaviour). Presumably this, or a variant, is the descending musical trill, *c.* 2-3 s long, rendered roughly 'tee-tee-ti-ti-te-te-too-too-too-too' and delivered from ground on Malta, early May to end of June; said to be quite different from 'normal' song (Gibb 1946). (2) Contact-alarm calls. (a) Contact-calls. Commonest calls, frequently given in flight, at all times of year. A soft 'chup' or a chirrup like sparrow *Passer* (e.g. Vere Benson 1970). Rendered 'tchi-tchirrup' (Mountfort 1958); a short 'djij' or 't-r-t', slightly similar to but much shorter than 't-rrr-t' of *C. rufescens* (Smith 1962-4). Jingling 'girrtirtir', reminiscent of *A. arvensis* and Serin *Serinus serinus* but harder and louder than latter (Dathe 1952); 'tschirr' or multisyllabic 'trtr...' (Bergmann and Helb 1982, which see for sonagrams). In recording by P A D Hollom (Morocco) of small flock flushed by recordist, a variety of such calls, e.g. 'dji', 'ti', 't-t-t-t', and 'tur' (J Hall-Craggs), suggesting Sand Martin *Riparia riparia* (P J Sellar). (b) Excitement and alarm expressed by slight variants of contact-calls, e.g. 'girrtititt', 'tschrrt' (by 2 fighting ♂♂), 'quät quät' (Dathe 1952); soft low 'kirk kirk kirk' (Wadley 1951). (3) Warning-call. A clear, piercing 'psië' during disturbance (Bergmann and Helb 1982, which see for sonagram); a similar 'psie' often given after ascent into Song-flight following disturbance (Bergmann and Helb 1982). In recording (Fig III), 'psiee' thus inserted

III P A D Hollom Israel April 1980

between phrases of song depicted in Figs I-II (J Hall-Craggs); slightly plaintive 'seeou' during Song-flight (P A D Hollom) probably the same. Distinctive,plaintive, whistling 'see-eer', seldom heard on first arrival on breeding grounds, Malta, but very frequently late April to early August, and commoner than song in July (Gibb 1946). Shrill 'zri-zrii' or 'zii' given by bird returning to nest with food for young probably this call (Endes 1970, which see for variants, including possible mimicry). (4) Other calls. In song, fluting 'züi-zö', 'wüi-zü', 'tuli',

'tuüli', 'huüit' (Endes 1970); these and numerous other sounds given in Song-flight may be mimicked calls. When flushed, 'teu' (P A D Hollom).

CALLS OF YOUNG. During period in nest, food-call a quiet 'zi' or 'si' sound; at time of leaving,also rendered 'zii' and 'siü'. At 10 days, a harder 'tziüp tiüp'. By 28 days, food-call similar to Contact-call of adult— 'tschiritschip'. (Endes 1969-70; Endes 1970, which see for similar variants.)   EKD

Breeding. SEASON. North Africa: laying from beginning of April to early June. Spain: laying from early May to July. South-eastern Europe, including Cyprus: laying mid-April to June. Hungary: first eggs early May. USSR. Laying April-July. (Bannerman 1953; Dementiev and Gladkov 1954a; Heim de Balsac and Mayaud 1962; Endes *et al.* 1967; Took 1972). SITE. On ground, usually in shelter of tuft of vegetation; sometimes in the open. 1st nests, Hungary, mid-May, placed on south side of tussock or at random; 2nd nests, mid-June, on north side of tussock (Endes *et al.* 1972; L Horváth). Nest: shallow depression, lined grass leaves and stems, rootlets, etc. (including seaweed), with inner lining of softer vegetation, feathers, thistle down, wool, etc.; mean external diameter 8·7 cm, internal diameter 6·2 cm (5·5-8·0), *n* = 30, inner depth 4·6 cm (3·0-6·0), *n* = 30 (Endes *et al.* 1967; Endes 1970). Frequently surrounded by small lumps of soil or stone. Building: despite some earlier reports, ♂ plays minor role in nest duties, ♀ choosing site and building nest (Dathe 1952; Guichard 1960); according to Endes (1970), ♂ sometimes brings material. ♀ starts building 10-14 days after arrival on nesting grounds, taking 7-10 days to complete nest; builds from dawn to dusk, first fashioning scrape or adopting hoof-print, wagon-track, etc.; collects lining material from outside territory and brings it on foot, less commonly flying. (Endes *et al.* 1967; Endes 1970). Not uncommonly continues building during incubation (Dathe 1952; Nadler 1974) and even in nestling period (Nadler 1974). Many nests abandoned before completion, especially late in season (Endes *et al.* 1967; Endes 1970) though ♀ may lay in unfinished nest (Nadler 1974). EGGS. See Plate 79. Sub-elliptical, smooth and glossy. Whitish, creamy-white, or sometimes greenish or greyish, variably marked in 2 types: (1) heavily but evenly mottled pale brown and lavender grey; (2) spotted and blotched darker brown, with some pale purplish-grey, marks larger and denser at broad end, forming zone or cap. Nominate *brachydactyla*: 19·8 × 14·8 mm (17·0-22·8 × 13·4-15·6), *n* = 130; calculated weight 2·24 g. *C. b. rubiginosa* not consistently different (Schönwetter 1979). Average weight of 30 fresh eggs, 2·21 g; 1st egg laid in clutch smallest, last largest (Endes *et al.* 1967). Clutch: 3-5(-6); reported by Bannerman (1953) that 3-egg clutches more usual in western populations, 4-egg in eastern, but evidence slight. Of 69 clutches, Tunisia and Algeria: 3 eggs, 40; 4, 23; 5, 6; average 3·51 (Heim de

Balsac and Mayaud 1962). Reported as 4–5 in USSR (Dementiev and Gladkov 1954a). In Hungary, usually 5 eggs, rarely 4 (Endes *et al.* 1967). 2 broods; in Hungary, 2nd similar in size to 1st (Endes *et al.* 1967). Replacement clutch laid. Eggs laid daily. INCUBATION. 13 days (Endes *et al.* 1967). By ♀ only. Eggs laid early in morning, ♀ very rarely returning to nest during day. Incubation begins with last egg. Brood hatches over 24 hrs. (Endes 1970; Makatsch 1976). YOUNG. Altricial and nidicolous. Cared for and fed by both parents, but ♀ takes the larger role. FLEDGING TO MATURITY. Fledging period 12–13 days; young leave nest at 9–10 days. 1st-brood young become independent very shortly after fledging, but 2nd-brood young remain with parents for some weeks (Endes *et al.* 1967). Age of first breeding 1 year (Makatsch 1976). BREEDING SUCCESS. No data, but losses known due to predators (50% of nests on Malta predated by western whip snake *Coluber viridiflavus*: Sultana and Gauci 1982), while in Hortobágy (Hungary), many nests succumbed to grazing sheep (Endes 1970).

Plumages (nominate *brachydactyla*). ADULT. Forehead, crown, mantle, scapulars, and back with rather broad greyish-black or black streaks on feather-centres and broad pink-buff, buff, or cinnamon-buff fringes along sides of feathers, fringes merging into deeper rufous towards blackish centres; rufous deepest and black streaks sharp and narrow on crown; forehead and forecrown occasionally hardly streaked, uniform grey-brown or rufous-brown. Rump and upper tail-coverts cinnamon-buff or tawny-buff, longer tail-coverts with narrow black shaft-streaks. Small feather-tufts covering nostrils off-white (except for part bordering base of culmen, which is brown); broad patch on lores pale cream-buff to off-white, extending into broad cream or white supercilium and into short pale buff streak below eye. Line behind eye over upper ear-coverts black-brown, cheeks and remainder of ear-coverts pale buff-brown or olive-brown, sometimes with indistinct dark grey line from gape backwards or with some grey or olive-grey spots on rear of cheeks and ear-coverts. Chin and throat uniform pale cream or white, extending up to pale cream-buff or off-white sides of neck. Chest and sides of breast buff, feather-centres often slightly darker rufous-buff on chest and brown on sides of breast; upper side of breast with prominent brown-black or dark olive-brown crescent-mark; occasionally, feather-centres of upper chest and sides of breast with narrow but contrasting black spots, forming narrow mottled gorget (see also Lesser Short-toed Lark *C. rufescens*, p. 142). Flanks and thighs uniform pink-buff or pale buff; belly, vent, and under tail-coverts cream-white or white. Central pair of tail-feathers (t1) dull black with broad pink-cinnamon or rufous-cinnamon fringe (soon wearing off); t2–t5 greyish-black or black, t5 with broad pink-buff or off-white fringe along outer web; t6 pink-buff or off-white with dark grey base and broad dark grey streak along border of inner web; undersurface of tail (except much of t6) blackish. Flight-feathers dark grey or dark brown, tips and outer webs indistinctly and narrowly fringed grey-buff or pale grey (soon wearing off on primaries); outer web of p9 more broadly and sharply fringed buff or off-white. Tertials dark olive-brown or dark grey-brown, broadly fringed pink-cinnamon, fringes soon bleaching to pale buff-grey or grey and partly lost through abrasion. Lesser upper wing-coverts uniform

buff or greyish-buff; median coverts with prominent dull black centre and broad cinnamon-buff fringe along outer web and tip; greater coverts like median but centres slightly less contrasting olive-black, and fringe along tip wide; fringes of median and greater coverts rapidly affected by bleaching and abrasion, forming prominent bars across wing when fresh, but narrow and broken pale buff or grey-buff bars when worn. Greater upper primary coverts and most feathers of bastard wing greyish-black with rather broad buff fringe along outer web and narrower one along tip; longest feathers of bastard wing fully greyish-black. Under wing-coverts and axillaries white with slight grey-buff tinge. Some individual variation in width of dark streaks and in ground-colour of upperparts: some hardly streaked on forehead and crown, others evenly streaked; some with crown distinctly brighter rufous-cinnamon than remainder of upperparts, others with ground-colour of crown and upperparts more uniform greyish-buff. Abrasion and bleaching have marked influence. Slightly worn birds (November–May) described above distinctly and broadly streaked black on upperparts. In fresh plumage (September–November) feather-fringes on upperparts broad, rufous-cinnamon to sandy-buff, dark feather-centres largely hidden, in particular on crown, rump, and lesser upper wing-coverts; supercilium indistinct, pale pink-buff; broad chest-band buff (deepest towards sides of breast), black crescent often indistinct, partly hidden under buff feather-fringes; tail narrowly tipped cinnamon-buff. In heavily worn plumage, May–July, pale feather-fringes on upperparts strongly abraded, bleached to dull grey, only rump still uniform greyish-buff; lower throat and chest greyish-buff; often some grey of feather-bases visible on underparts; abraded flight-feathers dark grey-brown, secondaries indistinctly edged grey at tip, p9 broadly fringed off-white. NESTLING. Down rather long but sparse, on upperparts and flanks only; yellow-white or pale buff (Witherby *et al.* 1938a; Endes 1970). For development, see Endes (1970). JUVENILE. Rather like adult, but all feathers of upperparts, including upper wing-coverts, dark brown or earth-brown with even cream-buff fringes, merging into contrastingly white fringe along rounded tip; centres of some longer feathers tinged olive or buff. Sides of head as in adult, but ear-coverts partly speckled white, and white or cream supercilium broader and more contrasting; underparts white, chest and sides of breast buff or pale buff, marked with rounded dark grey or brown spots, latter sometimes merging to form rather indistinct black patch on upper sides of breast; t1–t5, flight-feathers, and greater upper primary coverts evenly fringed buff or white; p10 longer than in adult and with broadly rounded tip, evenly fringed pale buff on outer web and tip (in adult, narrow and sharply pointed, faintly fringed off-white on outer web only—see also Structure). FIRST ADULT. Like adult; indistinguishable once last juvenile feathers (usually p9–p10) replaced.

Bare parts. ADULT. Iris brown or dark brown. Upper mandible horn-brown, darkest on culmen and tip; cutting edges and lower mandible greenish-yellow, yellow-horn, or pale horn. Leg and foot pale brownish-flesh, flesh-red with brown tinge, flesh-brown, or light brown. (Hartert 1910; Witherby *et al.* 1938a; Endes 1970; BMNH.) NESTLING. Bill horn-brown; mouth deep yellow with orange palate; 2 large black spots on each side of base of tongue, one near tip of tongue, one on tip of upper mandible (Witherby *et al.* 1938a; Endes 1970). Bare skin of eyelids, crown, and upperparts bluish-black, gape-flanges yellow-white, neck purple-red, underparts orange-red; leg and foot yellowish-pink, claws white; colour of mouth changes to yellow-pink at age of $1-1\frac{1}{2}$ months (Endes 1970). JUVENILE. Iris dark brown. Bill horn-brown with pale yellow gape-flanges. Leg and foot yellow-pink to pale brown-flesh. (Endes 1970.)

Moults. Based on Vaurie (1951*a*), Dementiev and Gladkov (1954*a*), and specimens (BMNH, ZMA). ADULT POST-BREEDING. Complete; primaries descendant. Starts between late June and late July, sometimes early August; body and tail virtually fully replaced at primary moult score 35-40; moult completed with growth of p9-p10 between mid-August and mid-September. Some birds in last stage of moult as early as mid-July or as late as early October (also, one early November), but these probably immatures with characteristic juvenile p10 already lost. ADULT PRE-BREEDING. In *dukhunensis*, partial, involving body and upper wing-coverts January to mid-March (Vaurie 1951*a*). Usually no moult in other races, but a few *rubiginosa* from Tunisia had crown and part of mantle and sides of breast contrastingly new in April (ZMA). POST-JUVENILE. Complete; primaries descendant. Starts at age of *c.* $1\frac{1}{2}$ months with p1 and scattered feathers of mantle and sides of breast; birds just starting examined from early June to July, but later-fledged juveniles probably start from August or early September; no data on time of completion, but see adult post-breeding.

Measurements. ADULT. All data April–August; skins. Nominate *brachydactyla*. Wing and bill to skull of (1) Iberia and southern France (RMNH, ZMA), (2) Italy, Yugoslavia, and Greece (Stresemann 1920; Makatsch 1950; BMNH, ZMA); other measurements combined. Bill (N) to distal corner of nostril; exposed culmen on average 2·9 less than bill to skull.

	♂		♀	
WING (1)	93·4 (1·72; 19)	91-96	89·8 (2·08; 5)	88-93
(2)	96·2 (2·44; 37)	92-102	90·1 (2·14; 17)	86-94
TAIL	56·9 (1·61; 11)	54-59	53·9 (2·33; 9)	50-57
BILL (1)	14·3 (1·05; 5)	13·2-15·4	13·7 (0·54; 5)	13·1-14·6
(2)	14·3 (0·56; 6)	13·5-15·2	14·2 (0·57; 7)	13·4-15·1
BILL (N)	9·1 (0·47; 11)	8·5-9·8	8·7 (0·69; 9)	7·9-9·6
TARSUS	21·1 (0·57; 7)	20·4-21·9	20·2 (0·74; 7)	19·3-21·2

Sex differences significant, except for bill.

Wing and bill to skull of (1) *rubiginosa*, north-west Africa and Malta; (2) *hermonensis*, Israel, Jordan, Lebanon, and Syria; other measurements combined (BMNH, RMNH, ZMA).

	♂		♀	
WING (1)	94·1 (1·67; 14)	92-97	89·0 (1·89; 7)	86-91
(2)	91·8 (0·89; 6)	91-93	89·3 (1·45; 6)	86-90
TAIL	55·4 (1·51; 12)	53-58	53·2 (2·25; 8)	51-56
BILL (1)	14·1 (0·54; 8)	13·4-14·9	13·8 (0·39; 6)	13·4-14·3
(2)	13·7 (0·31; 6)	13·1-13·9	13·9 (—; 3)	13·1-14·4
BILL (N)	8·8 (0·49; 12)	7·7-9·4	8·8 (0·43; 8)	8·3-9·5
TARSUS	21·1 (0·71; 11)	20·0-22·0	20·0 (0·64; 7)	19·3-20·9

Sex differences significant, except for bill.

C. b. woltersi. Amik Gölü and Gaziantep, central-south Turkey (Kumerloeve 1963, 1969*c*; BMNH, ZFMK). Bill (S) to skull.

	♂		♀	
WING	95·5 (1·86; 10)	93-99	89·6 (1·39; 6)	87-92
BILL (S)	14·3 (0·51; 6)	13·5-15·1	14·2 (0·94; 4)	13·4-15·3

C. b. artemisiana. (1) Eastern Turkey, Caucasus, and Transcaucasia (Jordans and Steinbacher 1948; Nicht 1961; Kumerloeve 1969*a*; RMNH, ZFMK); (2) western and central Asia Minor (Vaurie 1951*a*; Rokitansky and Schifter 1971; BMNH, ZMA). Bill to skull.

	♂		♀	
WING (1)	95·4 (1·44; 7)	94-98	90·4 (2·88; 7)	85-94
(2)	97·8 (3·02; 8)	93-101	93·0 (—; 3)	92-95
BILL (2)	14·6 (0·48; 9)	13·6-15·2	14·6 (0·63; 4)	14·0-15·5

C. b. hungarica. Hungary (Endes 1972).

	♂		♀	
WING	92·8 (—; 9)	90-97	89·8 (—; 7)	86-94

C. b. longipennis. Wing and bill to skull of (1) Crimea and Volga steppes, (2) Turkmeniya, Uzbekistan, and Kazakhstan (BMNH, RMNH, ZFMK, ZMA). Other measurements combined.

	♂		♀	
WING (1)	95·1 (1·67; 5)	93-97	92·2 (3·67; 5)	87-96
(2)	97·1 (1·77; 7)	95-101	91·2 (2·96; 7)	87-95
TAIL	55·6 (2·15; 7)	53-58	55·8 (2·51; 5)	53-60
BILL (1)	13·1 (0·43; 4)	12·7-13·5	12·8 (—; 3)	12·4-13·4
(2)	13·4 (0·23; 7)	13·0-13·7	13·4 (0·44; 7)	12·5-13·9
BILL (N)	8·2 (0·60; 6)	7·7-9·1	8·2 (0·48; 5)	7·7-8·9
TARSUS	21·0 (0·46; 6)	20·3-21·4	20·2 (0·68; 5)	19·2-21·1

Sex differences significant for wing (2).

C. b. dukhunensis. Indian subcontinent (Vaurie 1951*a*).

	♂		♀	
WING	99·9 (—; 79)	93-104	95·3 (—; 24)	92-100

JUVENILE. Wing on average *c.* 3 shorter than adult.

Weights. Nominate *brachydactyla*. Bulgaria, Greece, and European Turkey, late April to June: ♂ 21·8 (1·53; 9) 20-25, ♀ 22·7 (3) 21-26 (Niethammer 1943; Makatsch 1950; Rokitansky and Schifter 1971; Bub *et al.* 1982). Morocco, on spring migration (probably partly *rubiginosa*): 20·7 (161) 16-24 (Ash 1969).

C. b. hermonensis. Ceylanpinar (southern Turkey), May: 23-25 (12) (Kumerloeve 1970*b*). Ahaggar (Algeria) and Aïr (Niger), winter: ♂ 21-26 (9), ♀ 20-21 (5) (Niethammer 1963*b*).

C. b. woltersi. Amik Gölü (southern Turkey), May: ♂ 24·5 (0·58; 4) 24-25, ♀♀ 26, 27 (Kumerloeve 1969*c*).

C. b. artemisiana. Asia Minor (USSR), and Kerman (Iran), late April to July: ♂ 22·2 (1·42; 17) 20-25, ♀ 23·6 (2·67; 5) 20-27 (Nicht 1961; Kumerloeve 1967, 1969*a*; Rokitansky and Schifter 1971; Desfayes and Praz 1978).

C. b. longipennis. Kazakhstan (USSR): March-April, ♂♂ 25·2, 28·5, ♀ 25-30 (7); May-June, ♂ 22-27 (7), ♀ 23 (3) 22-24; August-September, ♂ 23·8 (3) 22-26, ♀ 24·2 (Dolgushin *et al.* 1970). Afghanistan: March and early April, ♂ 22·3 (23) 19-25, ♀ 19·9 (9) 19-22; May, ♀ 19; October, ♀ 20 (Paludan 1959). Northern Iran, April and early May: ♂ 23·0 (3) 20-25, ♀ 18·5 (Schüz 1959).

C. b. dukhunensis. India, winter: 20 (14) 18-23 (Ali and Ripley 1972). Mongolia, May-July: ♂ 24·3 (1·89; 7) 22-28, ♀♀ 20, 21 (Piechocki and Bolod 1972).

Race unknown. Belgium, October: ♂♂ 23·0, 27·8, ♀♀ 17·1, 20·0; unsexed 21·0 (Herroelen 1979).

For weight development of juvenile, see Endes (1970).

Structure. Wing rather short, broad at base, rather pointed at tip. 10 primaries: p8-p9 longest or either one 0·5 shorter than other; p10 62-67 (♂) or 58-65 (♀) shorter than longest, p7 1-2, p6 8-11, p5 16-20 (♂) or 14-19 (♀), p4 20-25 (♂) or 17-24 (♀), p1 30-34 (♂) or 27-33 (♀). Adult p10 narrow and pointed, strongly reduced. 8·8 (14) 7-11 shorter than longest upper primary covert; juvenile p10 broader and with rounded tip, 4·3 (4) 1-6 shorter than longest covert; juvenile p7 4 shorter than wing-tip. Outer web of p7-p8 and inner of p8-p9 emarginated. Longest tertials usually reach to between tips of p6 and p7 in closed wing, occasionally equal to p7 or between p7 and p8; when worn, tertials often equal to p6 or halfway between p5 and p6. Tail rather short, tip slightly forked; 12 feathers, t1 3-5 shorter than t5-t6. Bill rather short and slender, straight; cutting edges virtually straight apart from slightly decurved tip of upper mandible, culmen gradually decurved towards sharp tip; lower mandible slender, gonys slightly curved upwards. Some geographical variation in bill shape: bill of *artemisiana* relatively longer and more decurved, tip laterally compressed; in *hungarica*, *longipennis*, and *dukhunensis*, relatively shorter and more conical than in nominate *brachydactyla*, *rubiginosa*, *hermonensis*, and *woltersi*. Nostrils covered by short tufts of feathers; short bristle-like feather-tips show on lores, along base of upper mandible and on sides of chin. Feathers of crown broad and rather short, not elongated, crown not crested, but feathers can be raised. Tarsus rather short, slender; toes short and slender. Middle toe with claw 14·9 (21) 13-16 mm, outer toe with claw *c.* 67% of middle with claw, inner *c.* 72%, hind *c.* 96%. Front claws short, slightly decurved; hind claw rather short, straight or slightly decurved; length of hind claw 7·8 (34) 7-9 mm, exceptionally 3-10 mm.

Geographical variation. Slight and predominantly clinal; boundaries between most races arbitrarily drawn. Variation especially difficult to assess because of marked individual variation and marked influence of abrasion and bleaching; racial identification often only possible when series of specimens in equal stage of plumage abrasion directly compared. Birds in freshly moulted plumage are often closely similar, independent of race: e.g. upperparts of fresh *rubiginosa* only slightly brighter pink-cinnamon than cinnamon-buff of fresh nominate *brachydactyla*, and difference in width of streaking not noticeable, and even 'grey' eastern races fairly bright pink-buff when fresh. In the following discussion, only birds in fairly fresh spring plumage compared, but even here plumage may rapidly change colour: e.g. ground-colour of upperparts of *hungarica* dark ochre on arrival in breeding area, changing to grey within 2-3 weeks (Endes 1970). In general, upperparts and ear-coverts become greyer and less rufous towards east, and supercilium and underparts become whiter, less pale buff or cream, but in central and eastern Asia trend reversed, with upperparts becoming browner again. Nominate *brachydactyla* from southern Europe east to western Yugoslavia generally heavily streaked black or black-brown on upperparts. In 85% of birds (*n*=62) ground-colour of crown rufous-cinnamon, contrasting with pale rufous-brown, buff, grey-buff, or grey remainder of upperparts, in remaining 15%, crown like upperparts (Vaurie 1951*a*). Bill rather long, wing short but increasing in length towards east. In birds from south-east Yugoslavia, Greece, Bulgaria, and southern Rumania, ground-colour often slightly greyer than in western birds, more greyish-buff, streaks on average slightly narrower and less deep black, and crown often similar in colour to remainder of upperparts; sometimes separated as *moreatica* (Mühle, 1844), e.g. in Stresemann (1920), but best included in nominate *brachydactyla*. Birds from eastern Rumania and probably Ukraine are similar to those of Balkans, but bill shorter and more conical, similar to birds of Volga-Ural steppes, though plumage not as pale as these. *C. b. rubiginosa* from north-west Africa (except locally on Mediterranean coast, where nominate *brachydactyla* occurs: Hartert 1921-2; Hartert and Steinbacher 1932-8) distinctly more extensively rufous on upperparts than nominate *brachydactyla*; crown virtually always rufous, remainder of upperparts pink-cinnamon or sandy with narrower

and paler dark olive-brown streaks. *C. b. hermonensis* from Sinai north to Lebanon, Syria (north-west to Membij), and Ceylanpinar, southern Turkey (Kumerloeve 1970*b*), rather similar to *rubiginosa* in showing relatively narrow and olive-brown streaks, but crown in 78% of birds (*n*=18) similar to upperparts, not bright rufous (Vaurie 1951*a*), and upperparts slightly more buff or grey-buff, less pink-cinnamon. North-west Syria (west from El Bab) and neighbouring southern Turkey (Amik Gölü, Gaziantep, Birecik) (Kumerloeve 1969*c*) inhabited by *woltersi*, which is similar to *hermonensis* in size, bill structure, and streaking of upperparts, but which shows much paler ground-colour; crown and upperparts greyish, not buffish, paler and less heavily streaked than *artemisiana*; intergrades with *artemisiana* in Silifke-Adana area and with *hermonensis* in Urfa, southern Turkey (Kumerloeve 1969*c*). *C. b. artemisiana* from Asia Minor (in south to Maras, Diyarbakir, and Siirt), Transcaucasia, and north-west Iran (in Zagros, south to Lorestan and Fars: Vaurie 1951*a*) rather heavily streaked dark brown on upperparts, distinctly more heavily than *rubiginosa*, *hermonensis*, and *woltersi*, but not as heavy and black as south-west European nominate *brachydactyla*; ground-colour of upperparts grey-buff or pale grey, on crown often slightly rufous; closely similar to nominate *brachydactyla* from Balkans, but differing from this race and all others by long and laterally somewhat compressed bill-tip. *C. b. hungarica* from Hungary also rather heavily streaked dark brown on buff or grey ground, close to *artemisiana* and Balkan population of nominate *brachydactyla*, but bill shorter and more slender than these and underparts (including underwing) greyer (Horváth 1956; Endes 1970, 1972); apparently closely similar to Volga-Ural steppe population of *longipennis*, which also has small bill, but which has longer wing, paler underparts, and somewhat narrower streaks on upperparts. Volga-Ural steppe populations of *longipennis* sometimes included in *artemisiana* (e.g. by Vaurie 1951*a*), but streaks of upperparts narrower, ground-colour of crown, hindneck, and rump more sandy-grey, and bill small; somewhat duller and more sandy-grey above than *hermonensis*, not as pale grey as *woltersi*. Birds similar to those of Volga-Ural steppes occur in Afghanistan, Turkmeniya, Uzbekistan, and most of Kazakhstan (USSR); those of Caucasus and eastern Iran are intermediate between *artemisiana* and *longipennis*, but nearer to *artemisiana*. East from eastern Kazakhstan, populations become more heavily streaked black on pale brown ground, and these may deserve recognition, but united with paler birds from remainder of Kazakhstan and Volga-Ural steppe into *longipennis* for sake of nomenclatorial stability, as darker birds are typical *longipennis* and separation would leave pale birds unnamed. Tibetan *dukhunensis*, which winters in India subcontinent and may reach west Palearctic, has rather small bill like *longipennis*, and rather dark upperparts like Chinese and Mongolian populations of latter, but wing distinctly longer.

Forms superspecies with Red-capped Lark *C. cinerea* of Arabia and Afrotropics, sometimes treated as conspecific (Hall and Moreau 1970).

Recognition. See Lesser Short-toed Lark *C. rufescens* (p. 144).

CSR

Calandrella rufescens Lesser Short-toed Lark

PLATES 7 and 13
[facing pages 161 and 185]

Du. Kleine Kortteenleeuwerik   Fr. Alouette pispolette   Ge. Stummellerche
Ru. Серый жаворонок   Sp. Terrera marismeña   Sw. Dvärglärka

Alauda Rufescens Vieillot, 1820. Synonym: *Calandrella pispoletta*.

Polytypic. Nominate *rufescens* (Vieillot, 1820), Tenerife (Canary Islands); *polatzeki* Hartert, 1904, Gran Canaria, Fuerteventura, and Lanzarote (eastern Canary islands); *apetzii* (A E Brehm, 1857), Spain; *minor* (Cabanis, 1851), North Africa (except Nile delta) and Middle East from northern Sinai north to southern Turkey (along Syrian border), east to western and central Iraq; *nicolli* Hartert, 1909, Nile delta; *pseudobaetica* Stegmann, 1932, eastern Turkey, north-west Iran (including Mazandaran), and southern and eastern Transcaucasia; *heinei* (Homeyer, 1873), north-east Rumania, USSR from Ukraine, northern Caucasus, and Turkmeniya east to north-west Altai mountains and Kirgiziya, and northern Afghanistan; *aharonii* Hartert, 1910, central Turkey; *persica* (Sharpe, 1890), east and south-east Iraq (south from Mosul), Zagros mountains, central and eastern Iran, and southern Afghanistan; *leucophaea* Severtzov, 1872, salt deserts from southern Kazakhstan and Turkmeniya east to Lake Balkhash area and Fergana basin. Extralimital: 4-6 races from south-east Altai mountains and Sinkiang (China) to Manchuria, of which *cheleensis* (Swinhoe, 1871) is easternmost.

Field characters. 13-14 cm; wing-span 24-32 cm. Size more variable than Short-toed Lark *C. brachydactyla*, but most races are of similar bulk to that species and all show longer wing-point (since tertials fall noticeably short of primary-tips) and even shorter bill. Small lark with very similar character to *C. brachydactyla* but differing from adult of that species in more heavily streaked and browner upperparts and chest. Lacks discrete dark marks at shoulder; pale supercilia fully join over bill, forming pale forehead. Voice distinctive: calls rattling (not dry like *C. brachydactyla*). Sexes similar; no seasonal variation. Juvenile separable. 10 races in west Palearctic, 2 described here (see also Geographical Variation).

ADULT. (1) Spanish race, *apetzii*. Compared to *C. brachydactyla brachydactyla* of southern Europe, ground-colour of plumage more rufous-brown than sandy-buff, and darkened further by strong black-brown feather-centres on upperparts. Heavier streaks most obvious on crown and mantle; conversely, bar across median coverts less distinct, due to wider pale tips. Chest copiously and regularly streaked in quite deep gorget which extends on to flanks (and sometimes as far as lateral tail-coverts); unlike *C. brachydactyla*, never shows completely discrete dark patch at shoulder or unspotted throat and neck-sides. Pale forehead-band formed by joining of fore-supercilia is obvious, not broken as in *C. brachydactyla*. Chest-streaks noticeably darker and usually longer than those of even juvenile *C. brachydactyla*. (2) South Russian race, *heinei*. Plumage greyer and less heavily streaked than *apetzii* and north Asian *C. brachydactyla longipennis*. Compared to *apetzii*, chest less heavily streaked but marks still more copious than any *C. brachydactyla*. Important to recognize that structure of wing-point totally diagnostic and, at close range, is as easy to see as chest pattern: tertials fall at least 15 mm short of longest primary, at least 3 primary-tips showing as slim extension of folded wing (in *C. brachydactyla*, tips of tertials almost cloak those of

primaries and thus hide wing-point). JUVENILE. All races. Shows pale spotting (or dark and white speckles) on upperparts. Indistinguishable from adult by September. At all ages, short, deep-based bill horn, with darker tip and paler base; legs bright ochre.

Eminently confusable with juvenile congeners but less likely than *C. brachydactyla* to be mistaken for other small or medium-sized larks, since plumage pattern far less variable and closely matches only that of much larger Skylark *Alauda arvensis*. Best separated from *C. brachydactyla* by voice (see below). Normal flight and gait as *C. brachydactyla* but relatively rounder wing sometimes visible in flight silhouette. Song-flight lower and in wide circles; lacks obvious undulations but involves changes in speed.

Song quite unlike *C. brachydactyla*: can at times be almost hysterical and is always a jaunty, jerky jangle of continuous phrases, with only brief pauses of interjections of calls and rare plaintive notes. Commonest call less dry, more protracted, and sharper than *C. brachydactyla*: loud, rattled or rippled 'prrt' or 'prrirrick', often 'chirrit'; clipped double 'r' and terminal 't' or 'ck' much more audible than in *C. brachydactyla*. Calls of single bird and escaping flock can recall Lapland Bunting *Calcarius lapponicus*.

Habitat. Breeds in middle latitudes, in continental steppe, Mediterranean, and semi-desert zones, over-lapping widely with Short-toed Lark *C. brachydactyla*, and sometimes sharing breeding territories, as on *Salicornia* patches in marismas of Coto Doñana (Spain), although not on grassier lowlands used exclusively by *C. brachydactyla* (Valverde 1958). Another area of overlap, around Azraq (Jordan), finds *C. brachydactyla* in small minority, while *C. rufescens* flourishes so much on sandy or silty ground with low to medium shrub cover as to be ranked the commonest of all local passerines (Hemsley

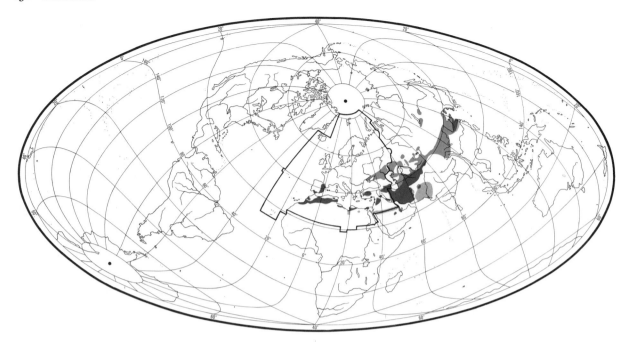

and George 1966). Seems to subsist on barer, poorer, drier, more saline, or more clayey or gravelly sites than *C. brachydactyla*, thus reducing real degree of competition even where they exist side by side, filling subtly complementary roles. Although normally lowland, exception provided by Transcaucasian race *pseudobaetica* which ascends to alpine meadows at 3000 m, inhabiting also rocky steppes and clay-covered lowlands and foothills. Greater hardihood also reflected in readiness to overwinter on breeding grounds in face of severe conditions (Dementiev and Gladkov 1954a). In north-west Africa prefers uncultivated steppe and stony plains, and needs near access to water sources (Heim de Balsac and Mayaud 1962). In central Turkey, noted only on flat sandy steppe, held together by short tufty grass, while *C. brachydactyla* widespread in open country, both cultivated and 'desert' (Wadley 1951). Wintering birds in India occur in open sandy semi-desert country, but are considered indistinguishable in habits from *C. brachydactyla* (Ali and Ripley 1972). Summed up, however, as preferring more clayey, drier, and hotter terrain than *C. brachydactyla*, more salinity, and stony wastes rather than grassy steppes (Voous 1960). While this tendency fairly apparent, habitat distinctions are far from clear-cut.

Distribution. Little information on range changes.
PORTUGAL. Not found breeding during atlas work (AR). 2 singing Algarve 1987 (G P Catley). BULGARIA. May have bred in north; no proof (Königstedt and Robel 1973). IRAQ. Range incompletely known due to confusion with Short-toed Lark *C. brachydactyla* (Moore and Boswell 1956; Marchant and Macnab 1962; Marchant 1963b). LIBYA. Cyrenaica: southern limits unclear

(Bundy 1976). CANARY ISLANDS. Gran Canaria: range decreased due to habitat changes (Bannerman 1963).

Accidental. Ireland, West Germany, Finland, Italy, Bulgaria, Rumania, Malta.

Population. Decreased Egypt; little other information on population numbers or trends.
USSR. Common (Dementiev and Gladkov 1954a). JORDAN. Azraq: most common lark (Wallace 1983a). EGYPT. Common, declined (SMG, PLM, WLM). LIBYA. Common (Bundy 1976). CANARY ISLANDS. Common, little apparent change (Bannerman 1963); more local on Gran Canaria and Tenerife (KWE).

Movements. Resident to dispersive (nomadic) in western parts of range, dispersive to migratory in centre and east.
WESTERN POPULATIONS (Iberia, North Africa, Canary Islands). Sedentary in Spain (Bernis 1971). Any wandering there at individual rather than population level, e.g. a few records of individuals in Tarifa area (Cadiz) in August–September (Bruhn and Jeffrey 1958; Tellería 1981). A scarce and irregular passage bird, September–April, on Gibraltar (Cortés *et al.* 1980). In North Africa, large movements reported Tiznit (Morocco) in November, and at Ouarzazate (Morocco) and Colomb-Béchar (northwest Algeria) in March; also report from Berguent (Morocco) as being more numerous there in winter than in breeding season (for summaries, see Heim de Balsac and Mayaud 1962, Smith 1965a). However, such reports perhaps refer to relatively local (perhaps nomadic) movements by post-breeding flocks of North African population. Scant evidence for winter presence further south; several records, of parties up to 15, on Banc

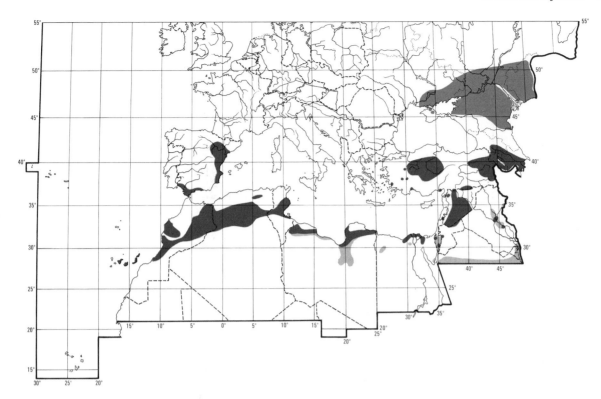

d'Arguin (Mauritania) in September and especially November (Knight 1975*a*), and several birds near Jos (northern Nigeria) in December 1961 (Elgood *et al.* 1966). North African race, *minor*, occurs as vagrant on Malta, where 7 spring and 4 autumn records of the species (Sultana and Gauci 1982). Endemic Canary Islands races (nominate *rufescens*, *polatzeki*) are resident there (Bannerman 1963).

CENTRAL AND EASTERN POPULATIONS (Turkey, Middle East, and southern USSR, east to Manchuria). Mainly migratory over much of USSR breeding range; winters only in small numbers in Eruslan steppes, and around Chkalov winters only in milder years, but resident and nomadic in Soviet Central Asia. Autumn departures from Transvolga begin August; spring passage noted Syr-Dar'ya in late February and early March and around 20 March at mouth of Ural river (Dementiev and Gladkov 1954*a*). Apparently a summer migrant to breeding areas of eastern Turkey, and winter numbers much reduced (apparently absent in severe seasons) on central plateau; majority of birds present early April to late September. More widespread in Turkey in winter, and especially during passage periods when quite common on south coast (Beaman *et al.* 1975). Scarce and irregular passage migrant on Cyprus, mid-November to April (most frequent February–April) and Turkey the likely area of origin though some may originate in USSR (Flint and Stewart 1983). In Iraq and Iran, winter immigrants attributed to *heinei* (breeding south-east and south-central USSR) reported

plentiful, mid-October to late February (Allouse 1953; Vaurie 1959), and this race has occurred rarely as far south-west as Egypt (Meinertzhagen 1930). Resident and winter visitor in Saudi Arabia, though evidently nomadic in arid centre where it breeds only in years when conditions suitable; more numerous and widespread outside breeding season, especially in passage periods, September–November and February–March (Bundy and Warr 1980; Gallagher and Woodcock 1980; Jennings 1980). Birds occurring on Persian Gulf are attributable to race *persica*, breeding eastern Iraq to southern Afghanistan (Meinertzhagen 1954; C S Roselaar). From eastern parts of range, migrants reach southwards into northern parts of Indian subcontinent and China.　　　RH

Food. Largely insects in summer, more seeds in spring and autumn, and presumably largely seeds in winter. Said by Gallagher and Woodcock (1980) to feed like Short-toed Lark *C. brachydactyla* (see p. 126). Often flies up from ground to take flying insects (Ticehurst 1926).

The following recorded in diet. Invertebrates: damsel flies, etc. (Odonata), grasshoppers, etc. (Orthoptera), termites (Isoptera), bugs (Hemiptera: Scutelleridae, Lygaeidae, Cicadidae), larval Lepidoptera (Pyralidae, Noctuidae, Geometridae), flies (Diptera, including Tabanidae), Hymenoptera including ants (Formicidae), adult and larval beetles (Coleoptera: Cicindelidae, Carabidae, Histeridae, Geotrupidae, Buprestidae, Tenebrionidae, Chrysomelidae, Curculionidae), spiders

(Araneae: Thomisidae, Lycosidae, Gnaphosidae). Plant material mainly seeds or fruits: of goosefoots (Chenopodiaceae), *Amaranthus*, Cruciferae, fumitory *Fumaria*, Rosaceae, *Hypericum*, mallows (Malvaceae), Zygophyllaceae, Lythraceae, Boraginaceae, and grasses (Gramineae) including cereal grain; also shoots. (Kekilova 1969; Shishkin 1980.)

In semi-desert steppes of north Caspian region (USSR), 54 stomachs analysed April–August: in mid-April, caterpillars 81·5% of food (by dry weight), declining through to late summer; from late June onwards, Orthoptera comprised 55·5–67·5%; other main food adult and larval beetles (mainly Carabidae and Curculionidae), 8–59%, most in late May; seeds, bugs, and ants taken throughout the period, each usually less than 10% of total dry weight, but seeds especially were much more important by number, comprising c. 50% of items in mid-May and August samples; this pattern not constant, however— in previous year, caterpillars taken only in mid-May, Orthoptera became important earlier (from late May), and more ants taken (c. 80% by number in mid-April) (Shishkin 1980). In southern Turkmeniya, spring, 7 stomachs contained mainly invertebrates, especially Odonata, Orthoptera, and termites; 9 in autumn contained mainly seeds, especially *Amaranthus*, grass, and *Heliotropium* (Kekilova 1969). In Crimea, stomachs contained seeds, ants, and small beetles (Kostin 1983). On Canary Islands, said to eat mainly seeds (especially *Chenopodium*) and insects (Bannerman 1963). The only definite winter data are from south Caspian region: stomachs contained mainly seeds of *Salicornia* (Isakov and Vorobiev 1940).

In April 1975 in semi-desert of north Caspian region, birds fed on huge hatch of caterpillars (Pyralidae): each ate average 339 caterpillars (6·8 g dry weight) per day, in total (with density of 27 birds per km²) removing 0·6% of available caterpillars per day; (presumably total) energy consumption of free-living birds 195 kJ per bird per day. Basal metabolism (presumably of captive birds), February–July, c. 31 kJ per bird per day (21·5–49, $n = 52$). (Shishkin 1980.)

No information on food of young.   DJB

Social pattern and behaviour. Most aspects less well studied than Short-toed Lark *C. brachydactyla*.

1. Gregarious outside breeding season, sometimes markedly so. In USSR, generally more gregarious than *C. brachydactyla* (Portenko 1954). In Morocco and USSR, flocks noted throughout breeding season, though most flocking starts after completion of nesting and moult (Lynes 1925b; Portenko 1954; Dolgushin *et al.* 1970). Large winter flocks and aggregations may number thousands—e.g. in Iraq (Ticehurst *et al.* 1921–2); in Morocco, c. 2000 *minor* recorded moving and feeding as loose flock (Meinertzhagen 1951a). Such numbers may be associated with abundant food (e.g. rice in Saudi Arabia: Ticehurst and Cheesman 1925), but also occur at watering places (Dementiev and Gladkov 1954a; Dolgushin *et al.* 1970). For further reports of late autumn and winter flocks (mostly hundreds), see (e.g.) Bourne (1959), Nelson (1973), Rogers and Gallagher (1973), and Kozlova

(1975). In Iraq, some break-up of flocks takes place from early March (Marchant and Macnab 1962). In central Turkey, occasionally in small flocks in March and early April, always singly or in pairs later (Wadley 1951). Not infrequently associates with other Alaudidae outside breeding season: e.g. in Iraq with *C. brachydactyla* (Marchant and Macnab 1962); in Libya, with Skylark *Alauda arvensis* (Bulman 1942); in Monts des Ksours (northern Algeria), most often with Bar-tailed Desert Lark *Ammomanes cincturus* and Temminck's Horned Lark *Eremophila bilopha* (Blondel 1962). On eastern Canary Islands, April, also with Trumpeter Finch *Bucanetes githagineus*, Berthelot's Pipit *Anthus berthelotii*, and Linnet *Carduelis cannabina* (Shirt 1983). BONDS. Mating system probably monogamous (e.g. Dolgushin *et al.* 1970), but no details. Care of young by both sexes (Dementiev and Gladkov 1954a); see also Relations within Family Group (below). BREEDING DISPERSION. Solitary and territorial, though neighbourhood groups also occur—in Iran (P A D Hollom), in central Arabia (Jennings 1980), in Kazakhstan (Dolgushin *et al.* 1970), and, with *C. brachydactyla*, in Spain (Mountfort 1958). Dispersion perhaps often dictated by particular habitat requirements: preference for solonets (alkali soil) type mentioned by (e.g.) Shnitnikov (1949) and Shishkin (1976). In Syrian desert, where food relatively scarce, 'territory' c. 80 ha (Meinertzhagen 1951a). In Kazakhstan, up to 5 singing ♂♂ in small area and nests may be c. 100 m apart (Dolgushin *et al.* 1970). Pair forage 10–200 m or more from nest (Dementiev and Gladkov 1954a). By far the most abundant bird on eastern Canary Islands; counts exceeded 40 birds per transect-hour in several localities on central plains of Fuerteventura (Shirt 1983). At Azraq (Jordan), commonest of Alaudidae on overgrown edges of pools (c. 9 birds per km); by far the commonest within wadis (c. 7 birds per km). Locally very dense within such habitats, with 12·5–17·5 pairs per ha (Wallace 1983a, which see for further details of dispersion by habitat). In area between Volga and Ural rivers (north-west Kazakhstan), where formerly rare but now commonest of 8 species of Alaudidae, 60 birds in 1 km² in breeding season (Shishkin 1976). Transect counts also in north Caspian region showed 26–63 pairs along 10 km, higher value relating to *Artemisia*—*Anabasis* habitat (see Poslavski 1974). Apart from in Spain (see above), association with *C. brachydactyla* on breeding grounds reported from Tunisia (Heim de Balsac 1924a; Bannerman 1927), Iraq (Marchant and Macnab 1962), and (also with Bimaculated Lark *Melanocorypha bimaculata*) from Kazakhstan (Dolgushin *et al.* 1970). ROOSTING. Nocturnal and (outside breeding season at least) apparently communal, though few details. In Sinkiang (China), February, birds roosted nocturnally in pits on mounds thrown up by rodents (Kozlova 1975). In southern Caspian region (Iran), migrants apparently rested with *A. arvensis* on short, damp grass in lea of dune (Schüz 1959). Song sometimes given at night (Shishkin 1976). No *Calandrella* larks seen at water by Meinertzhagen (1951a), though in Syrian desert *minor* seen drinking dew from grass in early morning. Can go without water for months but will drink readily (in central Asia, summer and winter: Kozlova 1975) if water (including brackish or salt) is available. At Lake El'ton (eastern Volgograd region, USSR), 11 June, *C. rufescens* and *C. brachydactyla* both arrived at freshwater pool from c. 07.00 hrs, numbers peaking at 12.00–13.00 hrs; at Soroch'ya balka, 14 June, peaks at 09.00–11.00 and 13.00–17.00 hrs; at Smoroda river, 1 July, large numbers arrived in a steady flow 08.00–17.00 hrs, with maximum 15.00–16.00 hrs (Dementiev and Gladkov 1954a).

2. Like *C. brachydactyla*, normally squats immediately after landing (Mountfort 1954). In Tunisia, apparently less shy and wary than *C. brachydactyla*; often allowing close approach before

taking flight, especially in spring when paired or pairing (Whitaker 1905; Zedlitz 1909); large winter flocks allowed Erlanger (1899) to walk into their midst; however, rather wild if approached on foot, according to Bannerman (1927). In Iraq, not confiding at any season; birds also tend to turn backs towards observer (Marchant and Macnab 1962). Generally shy also in Semirech'e, USSR (Shnitnikov 1949). If pursued, will first run off, concealing itself behind vegetation (Dolgushin *et al.* 1970), or may call (probably calls 2–3a) while running and when watching alertly from (e.g.) large stone (H-W Helb). Flock of 30 in Ireland, January, frequently crouched behind tussocks and often stretched up necks for surveillance when being stalked. Allowed approach to *c.* 3 m or less and when disturbed (frequently), flew only short distance (Anon 1960). Relative wariness in *Calandrella* larks depends very much on cover; tend to be wild in open country. In flat desert, birds constantly on the move and hardly attempt to hide; will squat in grassland (Meinertzhagen 1951a). FLOCK BEHAVIOUR. In southern Caspian region, large flocks typically tight-knit, birds mixing with *A. arvensis* but tending to keep slightly apart while feeding. Often do not flush as quickly as *A. arvensis*, but join with them in flight later. Migrant flocks tend to be more compact than *A. arvensis*. One flock of *c.* 250 flew about constantly, then broke up into 2–3 units for onward migration (Schüz 1959). Feeding flocks often noisy (e.g. Ticehurst *et al.* 1921–2, Smith 1965a: see 2 in Voice). See also Meinertzhagen (1951) for general comments on *Calandrella* larks. SONG-DISPLAY. Sings mostly in flight but also from low perch or ground (Whitaker 1905; Stresemann *et al.* 1937; Volsøe 1951; Dolgushin *et al.* 1970). Following description of Song-flight based, where not indicated otherwise, on material supplied by H-H Bergmann and H-W Helb from observations on Canary Islands (see also Bergmann and Helb 1982). During steep and rather jerky ascent—like *A. arvensis*, with abrupt beating of wings and spreading of tail (Koenig 1888; Dementiev and Gladkov 1954a)—usually gives series of rhythmless sounds (see 1–3a in Voice) followed by continuous and varied song with many imitations. Several authors suggest that *C. rufescens* usually sings lower down than *C. brachydactyla*; however, records of birds singing at (e.g.) 3–4 m (Hüe 1952; Anon 1960) probably not typical. For claims of Song-flights higher than *C. brachydactyla*, see (e.g.) Koenig (1888), Johansen (1952), and Portenko (1954); see also below. On Canary Islands, Song-flight performed at 20–30 m and similarly structured to that of *C. brachydactyla* (see that species, p. 128), bird alternating periodically between ascent and gliding descent phases, and song-phrase fitting in with this pattern. Undulations in Song-flight said to be less marked than in *C. brachydactyla*, though changes of pace occur (Dennis and Wallace 1975; also Hüe 1952, Mountfort 1954); yo-yo effect (see *C. brachydactyla*) nevertheless present according to Etchécopar and Hüe (1957), though this denied by Ferguson-Lees (1970). Wind strength and direction modify frequency and form of Song-flight. Bird normally heads into light wind and flutters in undulations while gaining height, then glides down (while singing) with shallow, shivering wing-beats; repeats this several times while moving forward, then descends; in strong wind, probably hovers without undulating. In calm conditions may make undulating flight in irregular circles *c.* 30 m across; appears to move with effort, as if pushing itself round (M G Wilson). In southern Spain, bird sang in close and rising spirals, alternately clockwise and anticlockwise, at *c.* 100 m (Mountfort 1958). Other reports refer to singing while climbing in wide circles (Dennis and Wallace 1975), and to slow circling with slow wing-beats recalling Song-flight of Serin *Serinus serinus* (Jonsson 1982). Song-flight ends in silent, vertical plummet to ground, or in downward glide with tail raised while singing. Also recorded singing and descending while beating wings slowly and deeply, tail widely spread. Song-flight may last several minutes. On Tenerife (Canary Islands), birds gave 75 phrases in 6·5 min and 38 phrases in 3 min. Once described singing on ground while feeding, but perhaps influenced by presence of observer. Performance of Song-flight infectious amongst neighbours. (Bergmann and Helb 1982; H-H Bergmann, H-W Helb.) On Tenerife, paired ♂♂ sang 12–24 March (Volsøe 1951). In Tunisia, song noted from early February (Koenig 1888); in Iraq, from early March (Marchant and Macnab 1962). In western Kara-Kum (USSR), singing at peak intensity during first half of April (Rustamov 1954). In Morocco, song noted also in mid-November and early January (Smith 1962–4). ANTAGONISTIC BEHAVIOUR. Few details. Often, several ♂♂ sing fairly close together (Dolgushin *et al.* 1970). Birds react vigorously to neighbours by singing from perch and giving high-pitched calls—presumably 3a in Voice (H-H Bergmann). In north-west Kazakhstan, intra- and interspecific (with other Alaudidae) territorial fights take place mid-March to end of April when some birds paired, others in small flocks (Shishkin 1976). ♂ engaged in courtship will drive off and pursue rival ♂, but interaction not especially violent (Koenig 1888). HETEROSEXUAL BEHAVIOUR. (1) General. In Sous (Morocco) where probably resident, birds paired and performing courtship up to late May (Lynes 1925b). In Tunisia, pairs recorded from February (Koenig 1888). On Lanzarote, all birds paired and holding territory 30 March (H-H Bergmann). In Kazakhstan, pair-formation and breeding start soon after arrival on breeding grounds (Dolgushin *et al.* 1970). (2) Pair-bonding behaviour. When ♂ performing Song-flight, ♀ usually on ground and may call quietly (see 3e in Voice) or feed. ♂ landing near ♀ usually moves round her with light dancing steps as part of courtship (Koenig 1888). (3) Nest-site selection. No details. While ♀ constructs nest, ♂ accompanies her silently or may feed nearby (Dolgushin *et al.* 1970). RELATIONS WITHIN FAMILY GROUP. At Lake El'ton, ♀ brooded young overnight and left nest at 04.10 hrs; carried faeces of young away and dropped them *c.* 200 m from nest. First brought food 10 min later. ♂ flew around, sang, and moved about on ground before first bringing food some 3 hrs after ♀. Adults foraged in separate areas. No feeding 13.45–15.30 hrs, but ♀ shaded young by standing over them with wings spread. At sunset, ♀ again settled to brood young while ♂ continued to sing. Young normally leave nest on 9th day (Dementiev and Gladkov 1954a). In Tunisia, one brood left on morning of 8th day, apparently not prematurely (Jarry 1969). Young remain near nest initially, fed by parents (Polatzek 1908; Lynes 1925b). ANTI-PREDATOR RESPONSES OF YOUNG. Crouch when disturbed, relying on highly effective crypsis (Shishkin 1982). PARENTAL ANTI-PREDATOR STRATEGIES. (1) Passive measures. Sitting bird left nest 'in good time' at approach of man (Shnitnikov 1949). (2) Active measures: against man. Breeding ♂ will sweep towards observer and hover in the air very close to him, also give alarm-call (Stanford 1954: see 3a in Voice). The following are perhaps distraction-displays. (a) In Dobrogea (Bulgaria), early June, 2 birds (not definitely breeding) ran ahead of observers giving Contact-alarm calls (see 2 in Voice); 1 performed Song-flight (Königstedt and Robel 1973). (b) On Canary Islands, ♂ descended rapidly from Song-flight and landed *c.* 10 m from observer; beat widely-spread wings demonstratively and evenly 4 times while facing observer; repeated performance several times after running short distance (H-W Helb). (3) Active measures: against other animals. Once performed distraction-lure display against dog, which chased it for several hundred metres (H-H Bergmann).

MGW

Voice. Birds in Libya generally less vocal than Short-toed Lark *C. brachydactyla* (Stanford 1954), in USSR more so (Portenko 1954). 2 vagrants in Ireland, May, mainly silent apart from brief song by one (Anon 1960). For extra sonagrams, see Bergmann and Helb (1982).

CALLS OF ADULTS. (1) Song of ♂. Of fine quality, but volume, duration, and richness vary. Especially rich in pitch variation (Koenig 1888). Similar to *C. brachydactyla* but longer, more continuous, richer, more varied, and more melodious (e.g. Dementiev and Gladkov 1954a, Mountfort

1958, Ferguson-Lees 1970). During ascent for Song-flight, bird usually gives series of rhythmless 'wihd', 'wierr', and 'trt-wihd' sounds—probably mostly variants of calls 2–3a (H-H Bergmann, H-W Helb). Short, hoarse, and disyllabic sounds described by Dementiev and Gladkov (1954a) are presumably the same, as are those illustrated in Fig I: 'srieh', though penultimate unit more like 'srrrrh' owing to frequency modulation (J Hall-Craggs). Such sounds are usually followed immediately by continuous song which contains much high-quality

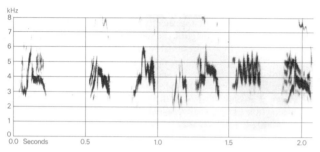

I   P A D Hollom   Iran   April 1972

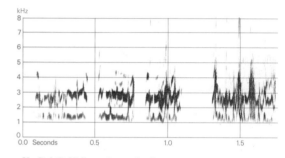

II   P A D Hollom   Iran   April 1972

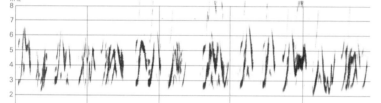

III   P A D Hollom   Iran   April 1972

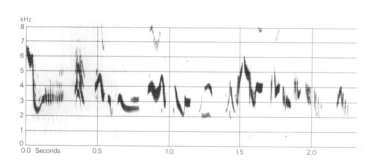

IV   P A D Hollom   Iran   April 1972

V   P A D Hollom   Iran   April 1972

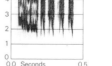

VI   P A D Hollom   Iran   April 1972

imitation (see below) and only then by rapidly delivered, dry, trilling, and mostly short (*c.* 1·5 s or up to *c.* 3 s or more) phrases which often begin with accelerating single 'tjü' sounds or (at least in *polatzeki* of Lanzarote, Canary Islands) 'dip-dip', followed by very rapid chattering with timbre of call 2—like rubbing pebbles together (Bergmann and Helb 1982; H-H Bergmann, H-W Helb). Phrase beginning with 'dip' sounds in recording by J-C Roché (Spain) may be imitation of *C. brachydactyla*. In Saudi Arabia, mid-April, song contained some sweet sounds, but mostly churred or grating whistles: e.g. 'tchik tchik tchik tweek-tweek-tweek-tweek-tweek-cha-wee-chip'; 'tchik' thin and high pitched, 'tweek' lower pitched and of hard, musical quality, while high, clear whistling 'cha-wee-chip' recalled Wood Sandpiper *Tringa glareola*. Another variant rendered 'tik tik tik tidik-tidik tch-tch-tchew-tch-a-you'—hard, somewhat grating, becoming a rapid jumble and descending in pitch (King 1978*b*). In recording of bird singing initially on ground then in flight, persistent, shrill 'shreep' sounds rather like sparrow *Passer* or nasal, squeaky chuckles (Fig II) change without break to louder and more piercing sounds (Fig III), some suggesting 'pseeeet' of wagtail *Motacilla*, then to more continuous medley of chattering chirrups and sweet warbling (Fig IV). Various rattling sounds (see also call 2) clearly important in repertoire; this confirmed also in recording by J-C Roché (J Hall-Craggs, M G Wilson). In Song-flight, longer or 'double' phrases sometimes given at change-over from ascent to undulating flight. Bird usually gives same phrase type about 3 times before changing to another type or to harsh or purer sounding calls, some protracted, always rhythmless. (H-H Bergmann, H-W Helb.) Song has jaunty, jerky, at times almost hysterical jangling quality; momentary pauses may herald change in style and tone to throaty rattles alternating with call 2 or occasional plaintive sounds (Dennis and Wallace 1975), and clicking sounds may occur in song (Anon 1960). At end of Song-flight, bird may give trills like Corn Bunting *Miliaria calandra*, abbreviated phrases, or various calls in rhythmless series, including trills with frequency modulation, and a drawn-out 'tried' ('ie' pronounced 'ee' in English) (H-H Bergmann, H-W Helb). Mimicry includes songs and calls of other passerines—including *C. brachydactyla* and Crested Lark *Galerida cristata* (Jonsson 1982). Song recorded in Spain shows perfect pitch and timbre reproduction of *C. brachydactyla* but different rhythm as pauses omitted; overall, *c.* 50% *C. brachydactyla* and *c.* 50% *Galerida* larks (J-C Roché); higher-pitched rattles in same recording may be copy of alarm-calls of House Martin *Delichon urbica* (J Hall-Craggs). Also mimics waders (Charadrii) (Ferguson-Lees 1970) and human sounds—e.g. whistling (Bergmann and Helb 1982). On eastern Canary Islands, birds convincingly mimicked included Kestrel *Falco tinnunculus* (also in Spain: recording by J-C Roché), Stone Curlew *Burhinus oedicnemus*, Whimbrel *Numenius phaeopus*, Black-bellied Sandgrouse *Pterocles*

orientalis, swifts *Apus*, Berthelot's Pipit *Anthus berthelotii*, Spectacled Warbler *Sylvia conspicillata*, and Linnet *Carduelis cannabina*. Captive birds may become remarkable mimics. (Etchécopar and Hüe 1957.) Song from ground typically has longer pauses (*c.* 2 s) between phrases; such song usually like *M. calandra* and may be broken off after introductory calls or shortly after start of rapid chattering (H-H Bergmann, H-W Helb). In recording (Fig V), 2 motifs—'weet weet weeohooo' and 'tree tui weeyuh'—perhaps copy of *Galerida* lark (M G Wilson), though also suggest Song Thrush *Turdus philomelos* (P J Sellar, J Hall-Craggs). (2) Contact-alarm call. Given frequently; in flight, including before and during song (Mountfort 1954; Anon 1960; Bergmann and Helb 1982) and on ground (Hüe 1952) when alarm element perhaps present (Königstedt and Robel 1973). Well known and usually distinctive, but occasionally resembles *C. brachydactyla* (Dennis and Wallace 1975)—e.g. 'diürrrüp' (Jonsson 1982). Sounds normally sharp, quite loud, dry, and of rattling or rippling type; some variants more drawn-out, others short, harder, and explosive; sometimes given in series. Renderings include 'prrrt', 'tchrrr(y)', 'tiirr', 'prr-prt', 'tch-chrrr', 'chirrik', 'prrirrick', 'prrit', 'chrit', and 'drrrit' (Schüz 1959;' Dennis and Wallace 1975; Gallagher and Woodcock 1980; Bergmann and Helb 1982; Jonsson 1982). Recording (Fig VI) suggests 'chit-it-it-it-it'; discrete parts uttered with extreme rapidity to give effect of rattle (J Hall-Craggs). At times an almost voiceless 'trrt' or (with sharper onset) 'tzrrt' or 'tschrrrt' (Schüz 1959). Doubled units or couplets sometimes given—e.g. 'chirrick-chirrit' and flushed birds may suggest Lapland Bunting *Calcarius lapponicus* (Dennis and Wallace 1975). A 'rick-kick-kick' not unlike Turnstone *Arenaria interpres* but sharper, possibly due to several birds calling at once (Anon 1960). Unlike *C. brachydactyla*, constant twittering from flock typical (Smith 1965*a*); a mixture of Contact-alarm calls and other sounds (Schüz 1959; see below). (3) Other calls, some used in alarm. (a) Sharp, repeated 'zisie' (trisyllabic), 'zied' (Bergmann and Helb 1982). In Tunisia, a mournful 'tee-ou' from 2 birds apparently anxious (M G Wilson). Sometimes combined with trilling sounds—'tr-zied' (Bergmann and Helb 1982); disturbed bird may give call 2 and purer sounds (song fragments) at take-off (H-W Helb); apart from call 2, actively migrating birds in Middle East gave a softer 'tseeü' very like 'migration call' of Skylark *Alauda arvensis* (Bourne 1959); in south Caspian region (Iran), sonorous, ringing 'üt üt üt' given by flocks and interspersed with call 2 (Schüz 1959). (b) Squeaky 'chik' (Wadley 1951); shrill 'whit' like call given by Stonechat *Saxicola torquata* when disturbed (Stanford 1954). (c) Shrill 'shreee' like Yellow Wagtail *Motacilla flava iberiae* (Smith 1962-4). (d) Imitated call of *F. tinnunculus* (see above) thought by Polatzek (1908) to have alarm function. (e) Quiet twittering sounds given by ♀ on ground (Koenig 1888).

CALLS OF YOUNG. No information.   MGW

Breeding. SEASON. Canary Islands: eggs laid from 2nd week of March; eggs found in May (Bannerman 1963). Algeria and Tunisia: eggs laid from 2nd week of April to early June (Heim de Balsac and Mayaud 1962). Spain: eggs laid from April (Makatsch 1976). Southern USSR: eggs laid April–June (Dementiev and Gladkov 1954a). Eastern Saudi Arabia: recently-fledged young occur as late as early August (G Bundy). SITE. On ground in shelter of tussock. Nest: shallow scrape, lined with vegetation. Building: by ♀ only (Dolgushin et al. 1970). EGGS. See Plate 79. Sub-elliptical, smooth and glossy; very variable, usually whitish, yellowish, or buff, more or less spotted and blotched dark brown. C. r. polatzeki: 19·7 × 14·3 mm (17·6–22·5 × 13·0–15·5), n = 80; calculated weight 2·08 g; rufescens not significantly different. C. r. heinei: 19·1 × 14·7 mm (17·0–21·3 × 13·0–15·7), n = 80; calculated weight 2·13 g (Schönwetter 1979). Clutch: (2–)3–5. Of 11 clutches, Gran Canaria: 2 eggs, 2; 3, 7; 4, 2; mean 3·0 (Bannerman 1963). Of 84 clutches, Algeria and Tunisia: 2 eggs, 4%; 3, 42%; 4, 38%; 5, 15%; average 3·6 (Heim de Balsac and Mayaud 1962). Probably 2 broods (Bub and Herroelen 1981). Eggs laid daily. INCUBATION. By ♀ only (Polatzek 1908; Bub and Herroelen 1981), or with ♂ taking some part (Dolgushin et al. 1970). YOUNG. Cared for and fed by both parents (Bub and Herroelen 1981). FLEDGING TO MATURITY. Young leave nest at 9 days (Dementiev and Gladkov 1954a). No further information.

Plumages (C. r. minor). ADULT. Upperparts pale rufous-cinnamon or pink-cinnamon when fresh to greyish sandy-buff or pink-buff when worn; forehead and crown narrowly but distinctly streaked black, hindneck and sides of neck more faintly dark grey; black or black-brown streaks or triangles on feather-centres of mantle, scapulars, back, and upper tail-coverts, broader and less well-defined than streaks on crown, showing as dark mottling rather than regular streaking (except for outer scapulars and upper tail-coverts), especially in worn plumage; rump either uniform pink-cinnamon to greyish sandy-buff, or marked with narrow and faint dark grey shaft-streaks. Tuft of small feathers above nostril off-white, mixed with longer blackish bristles; broad pink-buff or pale buff patch in front of eye, extending into narrow pale buff ring round eye and narrow pale buff supercilium; lores, eye-ring, and supercilium cream-white or white in worn plumage, contrasting more with buff crown and ear coverts than in fresh plumage. Narrow dark grey line from gape backwards below eye to ear-coverts; another mottled dark grey line along sides of chin across rear of cheeks to lower ear-coverts (sometimes indistinct); ear-coverts indistinctly streaked grey and cinnamon-buff or pale buff, often with dark grey border along supercilium and at rear. Patch at base of lower mandible cream-buff or white, mottled grey towards rear. Chin and throat white or slightly tinged pink-buff, often with some faint grey specks, usually extending into pale streak behind ear-coverts. Chest, sides of breast, and flanks cinnamon-buff, cream-buff, or pale buff, depending on wear of plumage, ground-colour often cream-white or white when abraded. Chest and sides of breast with distinct black or greyish-black streaks 1–1·5 mm wide, forming distinctly streaked chest-band, usually well-defined from white of throat, but merging into streaked sides of neck; streaks often rufous-cinnamon or ochre-cinnamon

on lower chest and lower sides of breast, narrow and olive-brown or grey on flanks; streaked parts of underbody not as sharply divided from uniform cream or white belly, vent, and under tail-coverts as from throat. Central pair of tail-feathers (t1) grey-brown or olive-brown with broad rufous-cinnamon, pink-cinnamon, or buff fringes and horn-black shafts; t2 dark grey-brown; t3–t5 black-brown or black, all narrowly and faintly edged buff, especially on outer edge and tip, outer web of t5 white or cream except for narrow dark border along shaft; t6 white, partly tinged buff when fresh, inner border with broad dark grey streak. Flight-feathers and greater upper primary coverts greyish-black, narrowly fringed pink-cinnamon or pale buff; fringes poorly defined on inner webs and on tips of secondaries, more sharply on outer webs and on tips of primary coverts and primaries (on primaries, soon lost through abrasion); tertials olive-brown with broad cinnamon or buff fringes (soon bleaching and partly abraded). Lesser upper wing-coverts cinnamon-buff or sandy-buff, feather-centres often slightly darker rufous-olive or brown; median coverts olive-black (fresh) or dark olive-grey (worn); greater coverts dark olive-brown or olive-grey; all coverts broadly fringed pale rufous-cinnamon, pink-cinnamon, or cinnamon-buff, especially on tips, forming 2 cinnamon bars across wing; in worn plumage, dark centres of coverts less contrasting, fringes bleached and partly abraded, sometimes largely lost. Under wing-coverts and axillaries pale grey, lesser coverts fringed pale buff, others partly tinged pink-buff. In fresh plumage, upperparts and upperwing pink-cinnamon, only dark streaks on forehead and crown and dark centres of median and great coverts distinct; pale supercilium and dark lines on sides of head not marked, blackish streaks on chest and sides of breast rather poorly defined. Dark streaks on upperparts, sides of neck and breast, and on chest distinct when plumage worn, supercilium distinct, but dark lines on cheeks and spots on cheeks and throat often less so; ground-colour of upperparts and ear-coverts more sandy-buff, of sides of head and underparts virtually white. NESTLING. Down on upperparts only; long and dense, pale yellow-brown, pale buff, or sandy (Hartert and Steinbacher 1932–8; Harrison 1975). JUVENILE. Rather like adult, but upperparts, upperwing, t1–t5, and flight-feathers with even and contrasting pink-buff, cream, or white feather-fringes, on longer feathers submarginally bordered black; chest marked with rather small and indistinct grey spots; p10 longer and more rounded at tip than in adult (see Structure), distinctly fringed buff along tip (unlike adult). FIRST ADULT. Like adult; indistinguishable once juvenile p10 replaced.

Bare parts. ADULT. Iris grey-brown or brown. Bill horn-grey, often darker on culmen (horn-brown or black-brown); base of lower mandible yellowish. Leg and foot brownish-flesh, flesh-brown, or yellow-brown (Hartert 1910; Ali and Ripley 1972; BMNH). NESTLING. According to Hartert and Steinbacher (1932–8), mouth flesh with yellow edges along bill and a distinct black spot on both upper and lower mandible, leg pale. According to Harrison (1975), mouth orange with 1 spot on either side of base of tongue and 1 on tip. JUVENILE. Like adult, but bill paler horn, tinged flesh-colour (BMNH).

Moults. Based on data from Vaurie (1951a), Dementiev and Gladkov (1954a), Piechocki and Bolod (1972), Bub et al. (1982) and specimens (BMNH). ADULT POST-BREEDING. Complete; primaries descendant. Starts mid-June to late July with inner primaries and scattered feathers of head and body; in heavy moult July–August; all moult completed late August to mid-September, sometimes October. POST-JUVENILE. Complete; primaries descendant. Starts soon after fledging, but this varies due to prolonged breeding period; thus starts late May to early August. In Aral Sea region, starts from mid-June, some in heavy moult late June and July, but others start late July. Moult probably completed between late July and early October.

Measurements. ADULT. Wing and bill to skull of (1) nominate *rufescens*, Tenerife, and (2) *polatzeki*, Fuerteventura and Lanzarote (eastern Canary Islands); all year, skins (BMNH, RMNH, ZMA). Other measurements combined. Bill (N) to distal corner of nostril; exposed culmen on average 2·2 less than bill to skull.

WING (1)	♂ 90·1 (1·68; 12)	87–93	♀ 84·5 (—; 3)	82–86
(2)	88·1 (1·77; 15)	85–91	83·3 (1·76; 16)	80–86
TAIL	51·7 (1·91; 12)	48–54	47·6 (1·25; 7)	46–50
BILL (1)	12·5 (0·36; 9)	11·9–13·2	12·0 (—; 3)	11·6–12·4
(2)	12·1 (0·55; 16)	10·9–12·7	12·1 (0·67; 13)	11·2–13·1
BILL (N)	7·7 (0·41; 7)	7·3–8·3	7·5 (0·34; 6)	7·0–7·9
TARSUS	19·4 (0·83; 13)	17·9–20·4	19·3 (0·38; 7)	18·8–19·8

Sex differences significant for wing and tail.

C. r. apetzii. Spain, all year; skins (BMNH). Bill (S) to skull.

WING	♂ 89·0 (1·69; 14)	87–92	♀ 83·7 (1·74; 10)	81–87
BILL (S)	12·3 (0·73; 11)	11·1–12·8	11·8 (0·61; 10)	10·8–12·3

Sex differences significant for wing.

C. r. minor. Wing, (1) Algeria and Tunisia, December–June; (2) Sinai, Syria, and central-south Turkey, summer; skins (RMNH, ZFMK, ZMA). Other measurements Algeria and Tunisia only. Bill (S) to skull.

WING (1)	♂ 91·1 (2·05; 18)	87–93	♀ 84·9 (1·93; 13)	82–89
(2)	92·1 (1·34; 7)	90–94	87·2 (—; 3)	87–88
TAIL	53·2 (1·67; 16)	50–56	49·2 (2·49; 12)	46–53
BILL (S)	12·3 (0·62; 14)	11·2–13·2	12·0 (1·02; 12)	10·1–13·0
BILL (N)	7·8 (0·55; 15)	6·9–8·5	7·7 (0·46; 12)	6·9–8·6
TARSUS	20·4 (0·58; 15)	19·5–21·3	19·7 (0·75; 11)	18·4–20·7

Sex differences significant, except for bill.

C. r. nicolli. Nile delta (Egypt), all year; skins (BMNH, RMNH).

WING	♂ 88·0 (1·63; 11)	85–90	♀ 83·3 (—; 3)	82–84
BILL (S)	12·4 (0·63; 9)	11·7–13·4	12·4 (—; 3)	11·7–13·2

Sex differences significant for wing.

C. r. pseudobaetica. Eastern Turkey and Azerbaijan (Iran), summer; skins (Vaurie 1951a; Kumerloeve 1968, 1969a; ZFMK).

WING	♂ 97·7 (2·22; 23)	94–101	♀ 91·2 (2·22; 12)	88–94
BILL (S)	12·3 (0·77; 4)	11·5–13·3	11·9 (0·36; 9)	11·5–12·6

Sex differences significant for wing.

C. r. heinei. Volga–Ural steppes (south European USSR), March–May; skins (BMNH, RMNH, ZMA).

WING	♂ 98·1 (2·13; 15)	95–102	♀ 94·1 (3·10; 9)	89–98
TAIL	58·5 (1·51; 13)	56–61	56·0 (1·79; 9)	53–59
BILL (S)	12.6 (0·59; 13)	11·3–13·3	12·6 (0·49; 8)	11·5–13·0
BILL (N)	7·9 (0·42; 13)	7·2–8·4	7·8 (0·34; 8)	7·2–8·3
TARSUS	20·7 (1·22; 13)	18·9–22·6	20·3 (0·94; 8)	18·9–21·3

Sex differences significant for wing and tail.

C. r. aharonii. Central Turkey, summer; skins (Hartert and Steinbacher 1932–8; Kumerloeve 1963; BMNH, ZFMK).

WING	♂ 99·8 (2·84; 11)	96–104	♀ 95·0 (—; 2)	93–97
BILL (S)	13·7 (—; 3)	12·9–14·3	12·8 (—; 1)	—

Sex differences significant for wing.

C. r. persica. Iraq and Kuwait, winter; skins (BMNH).

WING	♂ 103·0 (2·26; 12)	99–106	♀ 98·6 (3·05; 7)	96–103
TAIL	63·0 (2·50; 12)	59–66	58·5 (2·84; 7)	55–62

BILL (S)	13·8 (0·62; 12)	12·8–14·7	13·0 (0·65; 7)	12·0–13·5

Sex differences significant.

JUVENILE. Wing on average *c.* 10 shorter than adult, but sample very small.

Weights. ADULT. *C. r. minor*. Morocco, March: 19, 19·1 (Kersten *et al.* 1983). Syria, June: 20, 20, 21 (Kumerloeve 1969b). Southern Turkey, May: 20–25 (7) (Kumerloeve 1970b).

C. r. pseudobaetica. Van and Agri region (eastern Turkey), late May and June: ♂ 24·4 (1·38; 19) 22–27, ♀ 23·2 (0·96; 4) 22–24 (Kumerloeve 1968, 1969a).

C. r. heinei. Northern Iran: 1st half March, ♂♂ 23, 24, 26; early April, ♂ 25, ♀ 21 (Schüz 1959). Kazakhstan (USSR): March–April, ♂♂ 25·2, 28·5, ♀ 25–30 (6); May–June, ♂ 22–27 (7), ♀ 23 (3) 22–24; August–September, ♂ 23·6 (3) 22–26, ♀ 24·2 (Dolgushin *et al.* 1970).

C. r. aharonii. Central Turkey, June: ♂ 24·2 (0·98; 6) 23–26 (Kumerloeve 1963).

C. r. persica. South-east Afghanistan, May: ♂ 22·0 (1·41; 4) 20–23, laying ♀ 27 (Paludan 1959).

C. r. leucophaea. Kazakhstan: March–April, ♂ 19–24 (5), ♀ 21–25 (5); May–June, ♂ 22·0 (3) 20–23, ♀♀ 22·7, 24; October, ♀ 25 (Dolgushin *et al.* 1970).

NESTLING. For growth curve, see Jarry (1969).

Structure. Wing rather short, broad at base, tip rather rounded. 10 primaries: p8 longest (rarely, up to 1 shorter than p9), p9 0–4 shorter, p10 60–71 (♂) or 53–61 (♀), p7 0–2·5, p6 3–8, p5 12–18 (♂) or 9–16 (♀), p4 17–23 (♂) or 14–21 (♀), p1 25–34 (♂) or 22–30 (♀). Adult p10 reduced, narrow and pointed, 9·2 (20) 7–12 shorter than longest upper primary covert; juvenile p10 longer and with rounded tip, almost equal in length to longest primary covert. Outer web of p6–p8 and inner of (p7–)p8–p9 emarginated. Tertials short, longest usually reaching tip of p4–p5 in closed wing when fresh, often only (p2–)p3–p4 when worn; rarely, tip of fresh longest tertial between p5 and p6 or almost reaching p6. Tail rather short, tip slightly forked; 12 feathers, t1 2–5 shorter than t5–t6. Bill very short, thick at base; culmen distinctly decurved (see also Recognition). Tarsus and toes short, slender; claws short, slender, slightly decurved, hind claw rather short. Length of middle toe with claw 14·1 (0·76; 25) 13–15·5; inner toe with claw *c.* 73% of middle with claw, outer *c.* 70%, hind *c.* 103%. Hind claw 7·58 (1·75; 27) 6–10, exceptionally 2–12.

Geographical variation. Marked, complex. Involves colour (mainly ground-colour and width of shaft-streaks of upperparts and chest, and amount of white in tail), size (as expressed in length of wing and tail) and shape of bill (heavy and conical or small and fine). Nominate *rufescens*, *apetzii*, *nicolli*, and *pseudobaetica* are darkest races, with upperparts with broad and heavy black streaks, chest also heavily marked. Ground-colour of upperparts of nominate *rufescens* rufous-brown or cinnamon-brown, on chest rufous-cinnamon. Iberian *apetzii* has even more extensive black marks than nominate *rufescens*, below extending to throat and even chin, but ground-colour paler and more strongly contrasting, pale grey-brown or cream-buff on upperparts, cream or off-white on chest. Egyptian *nicolli* has streaks slightly narrower than nominate *rufescens*, but distinctly broader than neighbouring race *minor*; ground-colour of upperparts greyish cinnamon-brown or olive-grey with cinnamon tinge, of chest pale buff. *C. r. pseudobaetica* rather similar to *apetzii*, but black chest spots smaller and tail with more white; ground-colour of upperparts greyish-brown, of chest cream-buff or white. Races with rather narrow black or black-brown streaks on upperparts and chest are *polatzeki* from eastern Canary islands, *minor* from North Africa and Middle East, and *heinei* from southern USSR. Streaks on upperparts on average finer in *minor* than in other 2, but those on chest in all 3 equally narrow and distinct. Ground-colour of upperparts of *polatzeki* pink-cinnamon or pale sandy-rufous, of *minor* slightly paler sandy-pink or cinnamon with isabelline feather-fringes, but *minor* appears paler than *polatzeki* through narrower dark feather-centres, rump in particular more uniform; *heinei* also close to *minor*, but ground-colour of upperparts pale olive-grey, only slightly tinged cinnamon when fresh, less sandy-cinnamon than *minor*, rump in particular sandy-grey instead of cinnamon; streaks of *heinei* slightly heavier and more distinct, especially when plumage worn; *heinei* also rather similar to *nicolli*, but dark streaks distinctly narrower. Remaining races of west Palearctic very pale with narrow streaks on upperparts and almost obsolete ones on chest. *C. r. aharonii* (synonym: *niethammeri* Kumerloeve, 1963) from central Turkey has streaks on upperparts black, rather narrow and sharp; rather indistinct when plumage fresh, distinct when worn; ground-colour of upperparts pale ash-grey or stone-grey, occasionally faintly pinkish; ground-colour of underparts white, slightly tinged greyish-cream on chest when fresh; chest-streaks narrow but distinct. Iranian *persica* mainly differs from *aharonii* in pale cinnamon-pink, pale rufous-sandy, or yellowish-sandy ground-colour of upperparts, not mainly grey; dark streaks on upperparts and chest as in *aharonii* or even narrower; t6 white except for broad grey wedge on base of inner web; birds from south-east Afghanistan, attributed to *persica*, have small bill and less white in tail; these perhaps a separate race (Vaurie 1951a; Paludan 1959). Transcaspian *leucophaea* has upperparts pale ash-grey with very narrow and indistinct grey shaft-streaks, in autumn especially appearing uniform pale grey; supercilium broad, white; spots on chest sparse and tiny; t6 largely white. Differs from *heinei* from same area not only in paler colour, but also in different wing shape: wing-tip formed by p7–p8(–p9) (not p8–p9), with p6 2–4 mm shorter than wing-tip (in *heinei*, 5 or over); also, tail less deeply forked. Extralimital races grade from pale sandy with faint streaks in that from Sinkiang towards brown with heavy streaks in easternmost.

Probably better split into 2 species, as breeding ranges of *heinei* and *leucophaea* overlap widely in Kazakhstan, Uzbekistan, Turkmeniya, and perhaps north of Caucasus, apparently without interbreeding; hence most Russian authors now separate *leucophaea* with 4–6 other races from central and eastern Asia as *C. cheleensis* (e.g. Stepanyan 1968, 1983). Some races may also overlap in Middle East, notably rather small, dark, and fine-billed *minor* with large, pale, and heavy-billed *aharonii* (both recorded in breeding season in same locality in Syria: Hartert 1921–2) and equally large, pale, and heavy-billed *persica* in Iraq; no details yet available on whether these large and small races merge into each other or whether they overlap without interbreeding. If treated as separate species, not known to which member of *rufescens–cheleensis* species-pair each should be linked: though *leucophaea* (considered to belong to 'C. cheleensis') is pale like *aharonii* and *persica*, bill is notably fine and wing shape different; wing length and shape of *aharonii* and *persica* agree with *heinei*,

but *heinei* darker and with finer bill. Position of *pseudobaetica* unclear, being a small-billed race occurring north-east Iraq, Iranian Azerbaijan, and neighbouring eastern Turkey and Trans-caucasia, between ranges of heavy-billed races *aharonii* and *persica*; wing length of *pseudobaetica* variously reported to be small, like *apetzii* and *minor* (Stegmann 1932), intermediate (Dementiev and Gladkov 1954a), or large like *heinei* and *aharonii* (Vaurie 1951a; Kumerloeve 1968, 1969a); probably most closely related to *heinei*, as stated to intergrade into that race in eastern Caucasus region (Hartert and Steinbacher 1932–8; Stegmann 1932).

Forms superspecies with Rufous Short-toed Lark *C. somalica* from east and north-east Africa and with Sand Lark *C. raytal* from Indian subcontinent (Hall and Moreau 1970).

Recognition. Closely similar to Short-toed Lark *C. brachydactyla*. In both, ground-colour and amount of streaking strongly variable due to individual and geographical variation and strong influence of abrasion and bleaching. Feather-tuft covering nostril fully white in *C. rufescens*, forming unbroken continuation of supercilium along forehead; in *C. brachydactyla*, inner part of feather-tuft usually dark, separating supercilia at base of culmen. Sides of neck, sides of breast, and chest closely streaked in *C. rufescens*; in adult *C. brachydactyla*, sides of neck, chest, and flanks uniform, but sides of breast with distinct dark crescent-mark; in *C. rufescens*, however, streaks sometimes merge to form ill-defined dark spot on sides of breast, however, especially in darkest races (see Geographical Variation), while in palest races streaking of sides of neck and chest is narrow and almost obsolete; on the other hand, *C. brachydactyla* frequently shows narrow streaks on sides of neck and on upper chest, those on upper chest sometimes forming complete band across; throat of *C. brachydactyla* never spotted. Best distinguishing characters as follows. (1) Relative length of tertials: short in *C. rufescens*, long in *C. brachydactyla* (beware of specimens in abraded plumage, however—see Structure of both species). (2) Emargination of outer webs of outer primaries: p6–p8 emarginated in *C. rufescens*, though emargination of p6 sometimes indistinct; in *C. brachydactyla* p7–p8 emarginated, but p6 sometimes faintly also. (3) Shape and length of bill: in *C. rufescens*, bill shape varies with race, but bill always short, deep at base, appearing stubbier than in *C. brachydactyla*, and even strong and conical in some pale eastern races; in *C. brachydactyla*, bill usually rather long, appearing relatively slender and acutely pointed. Bill length is valid character for most populations outside USSR (also for Caucasus birds), except for bills of *C. r. persica* and *C. r. aharonii*, which are near *C. brachydactyla* in length, but much deeper at base; bill to skull in *C. rufescens* 10·1–13·2, to nostril 6·9–8·6; bill to skull in *C. brachydactyla* 13·2–15·6, to nostril 8·1–10·0. Difference in bill length in birds of plains of southern USSR much less marked: bill to skull in *C. r. heinei* 11·3–13·3, to nostril 7·2–8·4; bill to skull in *C. b. longipennis* 11·5–14·4, to nostril 7·7–9·1; bill of *C. r. heinei* deeper at base, however, almost triangular, of *C. b. longipennis* fine and slender. Juveniles of the 2 species hard to separate, differing mainly in emargination of primaries and in bill shape. CSR

Galerida cristata Crested Lark

Du. Kuifleeuwerik   Fr. Cochevis huppé   Ge. Haubenlerche
Ru. Хохлатый жаворонок   Sp. Cogujada común   Sw. Tofslärka

Alauda cristata Linnaeus, 1758

Polytypic. EUROPEAN GROUP. Nominate *cristata* (Linnaeus, 1758), central Europe south to north-west Spain, Pyrénées, north-east Italy, northern Yugoslavia, and north-west Hungary, in USSR in Baltic republics, Belorussiya, and northern Ukraine east to Kiev; *pallida* C L Brehm, 1858, Iberia (except north-west Spain); *kleinschmidti* Erlanger, 1899, north-west Morocco south to Rabat and Azrou, east to Er Rif; *neumanni* Hilgert, 1907, Toscana to Rome area (Italy); *apuliae* Jordans, 1935, south-east and southern Italy and Sicily; *meridionalis* C L Brehm, 1841, south-west and southern Yugoslavia, Albania, Bulgaria, mainland Greece, Ionian islands, Cyclades, Crete, and westernmost Asia Minor; *cypriaca* Bianchi, 1907, Karpathos, Rhodos, and Cyprus; *tenuirostris* C L Brehm, 1858, north-east Yugoslavia, eastern Hungary, Rumania, and southern European USSR, east to lower Ural river, south to Crimea and plains north of Caucasus; *caucasica* Taczanowski, 1887, Caucasus, west and central Transcaucasia, north-east Turkey, in Anadolu and Taurus mountains west to at least Samsun and Pozanti, and Aegean islands from Samothraki to Samos. NORTH-WEST AFRICAN GROUP. *G. c. riggenbachi* Hartert, 1902, western Morocco from Casablanca to at least Sous valley; *carthaginis* Kleinschmidt and Hilgert, 1905, coastal north-west Africa, north from Tellian Atlas, west to Oujda (north-east Morocco), in Tunisia south to about Sousse; *randonii* Loche, 1858, Hauts Plateaux of Algeria north of Atlas Saharien, west to upper Moulouya valley in eastern Morocco and (perhaps this race) further west along northern slopes of High Atlas to about Marrakech; *macrorhyncha* Tristram, 1859, Atlas Saharien and northern Algerian Sahara from Laghouat and Ghardaïa west to Figuig and Béchar; also Atar area in Mauritania; *arenicola* Tristram, 1859, north-east Algerian Sahara (east from about Biskra and Ouargla), southern Tunisia (south from Gafsa and Sfax), and Tripolitania (Libya); *helenae* Lavauden, 1926, Illizi (north of Tassili N'Ajjer, south-east Algeria), and (probably this race) south-west Libya; *festae* Hartert 1922, Cyrenaica (Benghazi to Tobruk). SUDAN-ARABIAN GROUP. *G. c. nigricans* C L Brehm, 1855, Nile delta, north of delta barrage, east to Dumyat; *maculata* C L Brehm, 1858, Nile valley from Cairo to Aswan, El Faiyum, and borders of Nile delta from Alexandria to Suez Canal; *halfae* Nicoll, 1921, Nile valley, south of Aswan, to Wadi Halfa (northern Sudan); *brachyura* Tristram, 1864, Cyrenaica (north-east Libya, south of *festae*), coastal north-west Egypt east to Alexandria and Wadi el Natrun, Egyptian Red Sea coast, Sinai, and in Israel, Jordan, and northern Saudi Arabia from Lod, Jerusalem, and Jordan valley east to southern and central Iraq; *cinnamomina* Hartert, 1904, coastal Levant from Mount Carmel and Haifa to Beirut; *zion* Meinertzhagen, 1920, Judaean highlands north from Jerusalem, area round Lake Tiberias, Lebanon (east of *cinnamomina*), Syria, and southern border of Turkey between Mersin and Ceylanpinar. ASIATIC GROUP. *G. c. subtaurica* (Kollibay, 1912), central Turkey, west to Kutahya and Elmali, and east to Van Gölü, but not in Anadolu and Taurus mountains (*caucasica*), extreme west (*meridionalis*), and south (*zion*); also in Armeniya, southern Azerbaydzhan (eastern Transcaucasia), and north-west Iran, south in Zagros mountains to about Lorestan and in south Caspian districts east to westernmost Turkmeniya (USSR), north to Kara-Bogaz Gol; *magna* Hume, 1871, Kazakhstan eastwards from lower Ural river and Mangyshlak peninsula round Aral Sea and Lake Balkhash to Sinkiang (China).

Extralimital: 1-2 races of north-west African group in northern Niger and Chad (perhaps including Tibesti highlands); 2-6 in Sahel group (Sénégal and Mauritania to western Sudan), 5-8 races in Sudan-Arabian group, including *altirostris* C L Brehm, 1855, Nile valley from Dongola to Berber (northern Sudan) and *somaliensis* Reichenow, 1907, northern Somalia and Lake Turkana (northern Kenya); 5-8 races in Asiatic group, including *iwanowi* Loudon and Zarudny, 1903, Iran and Turkmeniya (east of *subtaurica*) east to southern Tadzhikistan, Afghanistan, and north-west Pakistan, which perhaps reaches Iraq, Kuwait, and Saudi Arabia in winter.

Field characters. 17 cm; wing-span 29-38 cm. Slightly shorter overall than Skylark *Alauda arvensis* but distinctly bulkier about head and body, with rather long, strong bill and rather short, broad tail; same size as Thekla Lark *Galerida theklae*. Medium-sized lark, with long spiky crest on rear crown, portly character on ground stemming most from usually deep belly and rather short tail, and with rather compact silhouette in flight due to broad wings and short tail. Plumage pattern and colours recall dull *A. arvensis* but differ distinctly in stronger facial marks, heavy moustaches, more open chest-streaks on paler ground, more uniform upperparts and buff outer tail-feathers. Underwing of European races glows orange-buff. Flight undulating, with obvious flaps. Sexes similar; little seasonal variation. Juvenile separable. 24 races in west Palearctic; clines obscured by variable correlation of plumage and soil colours and by wear; racial identification in the field largely impractical; see Geographical Variation. Only nominate *cristata* discussed here.

ADULT. Ground-colour of both upper- and underparts

essentially pale buff, so bird appears more uniformly coloured than *A. arvensis* and other similarly sized larks. Upperparts also toned grey to brown, with black-brown streaks obvious on feathers of crown and crest-spike, quite well marked on mantle, but much less so on hindneck and lower back and rump. Underparts darkest on chest and flanks, palest on belly. At distance, noticeable features are dark crest, black-brown streaks on chest, dark brown centres to tertials, and black-brown, buff-edged outer panels to tail. At close range, general appearance less uniform, and form and plumage exhibit the following distinctive characters: (1) rather long, strong bill with noticeably decurved culmen—dark horn above, pale yellow-flesh below; (2) long, narrow-feathered, dark crest, projecting 2 cm above rear crown when erect (at least twice as long as that of *A. arvensis*); (3) obvious facial pattern, with buff-white supercilium, eye-ring, and upper cheeks contrasting with buff- or grey-brown loral line, rear eye-stripe, and surround to lower cheeks, and with darker, black-brown moustachial and malar stripes; (4) unstreaked chin and central throat (finely streaked in *A. arvensis*); (5) streaked chest, with rather narrow or diffused straight black-brown striations radiating evenly from lower throat and broadening only on sides near upper flanks (pattern neither as dense nor as regular as in *A. arvensis*); (6) thinly and dully streaked nape and collar below cheeks; (7) rather dull margins to wing-coverts, inner secondaries, and tertials, not forming obvious wing-bars; (8) strong, pale flesh-horn legs and feet. In flight, most characters blur but appearance remains distinctive due to compact silhouette (rounded at times) and contrast between relative uniformity of wing (lacking white trailing edge of *A. arvensis*) and marked pattern of tail (noticeably buff on outer edges, with black panels on each side of dark brown central feathers). On retreating bird, tail pattern usually the only upperpart character to catch the eye, but when bird flies overhead or turns, almost glowing orange-buff under wing-coverts can be even more striking. Lower underparts very dull in some birds, as if dirty. In summer, abrasion of feather-fringes causes black-brown feather-centres of upperparts to become more obvious (but they may also become dull, so that worn birds appear almost plain-backed); underparts become paler. Shorter-crested ♀ may be separable in comparison with ♂. JUVENILE. Crest distinctly shorter and less spiky. Ground-colour of upperparts rather browner than adult, but underparts paler. Upperparts less streaked than adult but heavily spotted white on crown, mantle, and scapulars. Chest spotted rather than streaked and marks less dense; rest of underparts distinctly whiter.

In Europe (north of Iberia) and throughout Middle East, the only representative of its genus and unmistakable to experienced observer, though perhaps liable to confusion (on ground) with dull *A. arvensis* showing more crest than usual; confusion quickly dispelled when differences in wing and tail pattern visible. In Iberia, southern France, and North Africa (east to Libya), often occurs close to Thekla Lark *G. theklae* and all too easy to confuse with it, since both specific and racial characters subject to overlap and even reversal. For distinctions, see *G. theklae* (p. 163). Flight over short distances noticeably flapping and floating, with broad wings and short tail obvious in low glides and heavy flutter on landing. Action essentially as *A. arvensis* but bird appears noticeably bulkier when wings closed and does not show waving or weak wing-beats of that species, with result that undulations more marked over longer distances or on migration, like Woodlark *Lullula arborea*. Escape-flight variable: sometimes low and short but often at fair height (up to 50 m), and long (up to 300 m); may flush rapidly, with energetic wing-beats, but lacks panic manoeuvres of *A. arvensis*. Song-flight used less than in *A. arvensis* but ♂ will ascend to heights of up to 150 m and sing for 3 min or more while circling; flight-action far less rapid than in *A. arvensis*, with individual wing-beats distinguishable. Gait an even-paced walk or swift run; shares with *G. theklae* a characteristic shuffle when feeding, this action being exaggerated by deep, often loose belly feathering. Perches freely on rocks, low bushes, etc., but rarely higher up. Relative shortness reflected in less horizontal, more half-upright posture than *A. arvensis*. Does not sit close, frequently rising well ahead of observer, and not prone to flocking, even on migration when only small parties occur.

Song essentially sweet and plaintive with phrases of 4-6 notes continually repeated and interspersed with a few trills; lacks vehemence of *A. arvensis* and loud musical variety of *G. theklae*, being less loud and arresting than either; usually given in shorter bursts than *G. theklae* but can be sustained for 10 min or more. Calls include plaintive and slurred whistles, 'klee-tree-weeoo'; in alarm, sharp 'klee-vee'.

Habitat. Widely spread across and beyond continental west Palearctic, from fringe of boreal zone through temperate, steppe, Mediterranean, arid semi-desert, and desert zones, including oases. Prefers lowland plains and levels, although ascending in Atlas mountains of north-west Africa to 1260 m (Pätzold 1971) and breeding in Switzerland towards 700 m (Glutz von Blotzheim 1962); extralimitally in USSR along wide valleys and up roads to *c.* 1500 m or even 2000 m, especially near human habitations or artefacts (Dementiev and Gladkov 1954a). Has failed to colonize marine islands, even where large, and rich in apparently suitable habitat (Meinertzhagen 1951a). Avoids mountainous, broken, forested, wetland, muddy, and most coastal terrain, and is more of a steppe bird in Africa than in Asia, where coldness of winters would conflict with preference for sedentary existence (Pätzold 1971). In USSR, however, tolerates up to 144 days annually of snow cover (Dementiev and Gladkov

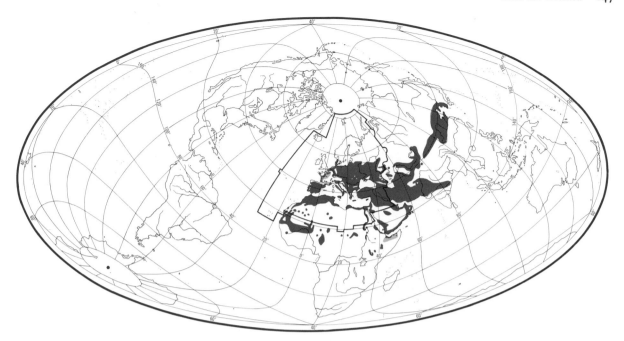

1954*a*), although avoiding severe frost, and high winds or precipitation. Basic habitat is open, dry, often warm and dusty, flat or gently sloping, with very low or sparse vegetation not covering more than *c*. 50% of territory, any trees or shrubs present being widely spaced and not hemming it in. Attracted by various man-modified areas simulating semi-desert features, such as railway yards, parade grounds, airfields, harbour surrounds, gravel pits, refuse dumps, and urban and industrial wastelands. Many indications suggest gradual transition, over recent centuries up to present, from earlier attachment to steppe-semi-desert zones towards areas modified by cultivation (or its abandonment), settlement, deforestation, road and rail construction, or extractive and other uses creating relatively open land. Transition more pronounced towards west of range, while in east more apt to be driven by winter snows into human settlements, filling role of House Sparrow *Passer domesticus*. For roosting on buildings, see Social Pattern and Behaviour. As a ground bird, is ruled by local type of soil and vegetation cover, apparently reacting against excess moisture and incompatible soil colour or texture, as well as steep or broken slopes (especially if rocky), loose sand, coarse gravels, wet mud, and close or tall herbage. Pale, firm, sandy, clay, or loamy soils present least obstacle to movement. Avoids modern closely-grown cornfields, but will nest in those traditional fields in Spain with short stalks spaced at wide intervals. Similarly occupies pastures only where grass is sparse, and also asparagus fields and vineyards (Abs 1963). In North Africa, noted as tied to terrain where it can probe for food with bill (Heim de Balsac and Mayaud 1962). Some African races are adapted to desert conditions (Valverde 1957). At Azraq (Jordan), breeds on silt-dunes and in rough wadi beds in limestone desert (Hemsley and George 1966). In India, favours open sandy or salt-encrusted semi-desert plains (often near cultivation), grassy sandy riverain tracts, dry tidal mudflats, grassy stony hill-slopes with rock outcrops in dry and moist-deciduous situations; open sparse shrub jungle, forest clearings, and treeless downs (Ali and Ripley 1972). In Iberia, except on its core habitat of (e.g.) level open farmland, saltpans, and edges of marismas, may compete with Thekla lark *G. theklae* (Wallace 1965) though even where found in close proximity, actual overlap appears exceptional, and perhaps coexists more commonly with Skylark *Alauda arvensis*, which seems dominant over *G. cristata*, especially on moister ground with more vegetation cover (Meinertzhagen 1951*a*); see also Social Pattern and Behaviour. Despite apparently sucessful accommodation to various modern land uses, can fluctuate inexplicably in numbers and distribution, e.g. in Switzerland (Glutz von Blotzheim 1962); sometimes colonization may have outrun sustainable limits and become vulnerable to reverse. Locally (e.g. in Belgium), diminution attributed to reduced horse population (Lippens and Wille 1972). Apart from frequent song-flight and local excursions (usually below *c*. 100 m) is very much a ground-bird, and is less ready than *G. theklae* to perch on trees, shrubs, fences, or walls (E M Nicholson).

Distribution. No longer breeds in Norway or parts of northern USSR.

FRANCE. Range decreased (Yeatman 1976). NETHERLANDS. Colonized 16th in century (CSR). NORWAY. Last known breeding 1972 (Fremming 1984). FINLAND. Bred 1870-2 (Merikallio 1958). AUSTRIA. Some decline in west

(HS, PP). USSR. Baltic: some spread north; spread into Lithuania first half 19th century, then to Latvia and Estonia (HV); disappeared from Estonia and Leningrad region in last 30 years (Mal'chevski and Pukinski 1983; HV).

Accidental. Britain, Finland, Chad, Malta, Canary Islands.

Population. Decreased France, Netherlands (colonized 16th century), Denmark, Sweden, Poland, Czechoslovakia, Austria, Switzerland, and Rumania.

FRANCE. 1000–10 000 pairs (Yeatman 1976); decreased Somme due to mechanical agriculture and use of pesticides (Triplet 1981). BELGIUM. About 400 pairs (Lippens and Wille 1972). NETHERLANDS. 300–500 pairs 1976–7 (Teixeira 1979); increased 1830–1900, then decreased, now stable or slight decline (CSR). WEST GERMANY. About 15 000 pairs (Rheinwald 1982). DENMARK. Decreasing (TD); for population changes northern Jutland 1968–76, see Møller (1978a), some decrease. SWEDEN. Spread from *c*. 1850 reaching to 60°N; strong decrease before 1940; 30–40 pairs in 1960s, 20–25 pairs 1977, *c*. 10 pairs 1979, 5 pairs 1981, 4–5 pairs 1984 (LR, P Alström). POLAND. Decreasing (AD). CZECHOSLOVAKIA. Strong decline 1950–70, but stable from 1970 (KH). AUSTRIA. Common in mid-1950s, at least in east, now

very scarce (HS, PP). SWITZERLAND. Marked decrease in 20th century; 1–2 pairs (RW). RUMANIA. Decreased (VC).

Movements. Largely migratory in north of USSR breeding range. Mainly resident elsewhere; some dispersal occurs, but scale uncertain. Apparently sedentary in North Africa, Middle East, and northern India, where birds show much subspeciation and adaptation of plumage colour to that of local soils.

Mainly resident in Europe, even as far north as Sweden (Rendahl 1964), though number of stragglers to Baltic coasts, Norway, Helgoland, West German and Dutch Frisian Islands, and Britain show that some birds move (Abs 1963). Labitte (1957b) suggested that adults are resident, with 1st year birds prone to wander in autumn and winter, sometimes well away from natal area. Limited ringing data, but single Swedish nestlings found Belgium (October) and Lot-et-Garonne in France (winter), latter 1500 km SSW; West German (Westfalen) bird found Belgium (October); Belgian nestling found Charente-Maritime in France (October); Swedish and Belgian-ringed birds all recovered in 1st year (Labitte 1957b; Rendahl 1964; Zink 1975). However, of 25 other recoveries of German-ringed birds (Abs 1964), 23 moved only 0–12 km and singles to 18 km and 34 km, and the few Spanish-ringed recoveries were all within 10 km of ringing site (Bernis 1971). However, very small numbers of *Galerida* which cross Straits of Gibraltar in autumn are probably mainly this species (see Thekla Lark *G. theklae*, p. 165), though origins unknown. Many records from Gibraltar itself (Cortés *et al.* 1980), and total 17 diurnal migrants crossed Straits from Tarifa (Cadiz) between 16 September and 27 October 1977 (Tellería 1981). In Sardinia (where it does not breed), some numbers occur in autumn and winter, specimens being main European race, nominate *cristata* (Steinbacher 1956, Vaurie 1959). Only a vagrant (2 autumn records) to Malta, however (Sultana and Gauci 1982).

Seasonal movements more marked in European USSR, though probably for most part only short-distance displacements (Dementiev and Gladkov 1954*a*). Wholly migratory in northernmost breeding areas, e.g. Alatyr in Chuvash ASSR (55°N), and partially so further south, e.g. Poltava in Ukraine (49°30′N) where small numbers overwinter and then only in milder seasons. Yet will occur regularly in winter even north of breeding range, e.g. Kursk region (51°45′N). Even in European USSR, possible that local birds move out, to be replaced by migrants from further north. In valley of Terek river (Dagestan), numbers increase noticeably from September. South Caspian race *subtaurica* is resident, but local movements occur and birds have been collected in December in Sumbar valley (Turkmeniya) alongside *iwanowi* which is resident there (Dementiev and Gladkov 1954*a*). Central Asian race *magna* (breeding Kazakhstan to Afghanistan and Baluchistan) is certainly migratory in north, and this race reaches Sind (Pakistan) and Punjab (northern India) in winter (Ali and Ripley 1972).

In central Saudi Arabia, where resident, numbers increased in winter by arrivals of a more boldly streaked form (Jennings 1980). At Awamir in northern Oman, 4 September, flock of 5 dark and heavily-streaked birds flew off south-east when flushed (Walker 1981*a*). Origins of these birds not known. RH

Food. Plant material (seeds, also leaves) and invertebrates (especially beetles); fewer invertebrates in winter. Most food taken from on or below ground surface. Digs with blows of bill to left and right, making funnel-shaped hole *c.* 2 cm across and *c.* 2 cm deep; towards the end, both hacks and pushes bill into ground. Especially likely to dig where dung lying (Abs 1963; Pätzold 1971) and will peck grain out of it (Averin and Ganya 1970; Ferguson-Lees 1970*d*). Bird feeding on larvae of ant-lions (Myrmeleontidae) in Nigeria moved systematically from one depression in the sand to another, stabbed with bill, then swallowed (Fry 1966*a*). Often extracts seeds from seed heads on ground and sometimes (like Thekla Lark *G. theklae* with stones) turns over leaves (Sudhaus 1966*b*). From low plants (up to 10–12 cm high) will take seeds directly, also tips of shoots (etc.) and invertebrates (Abs 1963; Sudhaus 1966*b*; Averin and Ganya 1970). Will take

insects by aerial-pursuit (Abs 1963; Krüger 1970), and 3 times seen taking dragonflies *Anormogomphus* and stripping wings off before eating body (Sage 1960*a*). Softens large insects by beating them, and they often lose their wings in the process (Abs 1963). Most seeds and fruit swallowed whole but will de-husk cereal grain and grass seed by knocking it against ground (Abs 1963; Sudhaus 1966*b*). In captivity, ejects pellets of seed husks (Abs 1963). Will drink salt water (Rustamov 1958) and recorded apparently eating snow (Sudhaus 1966*b*).

The following recorded in diet in west Palearctic. Invertebrates: adult dragonflies (Odonata), stoneflies (Plecoptera), grasshoppers, etc. (Orthoptera), bugs (Hemiptera), larval Lepidoptera (e.g. Noctuidae), flies (Diptera: Muscidae), Hymenoptera (ants Formicidae, bees Apoidea), beetles (Coleoptera: Carabidae, Staphylinidae, Scarabaeidae, Elateridae, Cantharidae, Tenebrionidae, Coccinellidae, Cerambycidae, Chrysomelidae, Curculionidae), spiders (Araneae), snails (Mollusca), earthworms (Oligochaeta, mostly small). Plant material mainly seeds or fruits: of willows *Salix*, limes *Tilia*, hemp *Cannabis*, knotweeds (Polygonaceae), goosefoots (Chenopodiaceae), Portulacaceae, *Amaranthus*, pinks (Caryophyllaceae), Cruciferae, poppies *Papaver*, bramble *Rubus*, storksbill *Erodium*, flax *Linaria*, plantain *Plantago*, Labiatae, speedwells *Veronica*, scabious (Dipsacaceae), Compositae (including sunflower *Helianthus*), and grasses (Gramineae) including cereal grain; also leaves. Will eat bread, banana, chocolate (etc.), and recorded pecking at Brussels sprouts (Kovačević and Danon 1952; Abs 1963; Sudhaus 1966*b*; Averin and Ganya 1970; Pätzold 1971.)

Compilation of casual observations in Schleswig-Holstein (West Germany) indicated seeds and bread taken all year, invertebrates (including earthworms) mostly March–November (Sudhaus 1966*b*). In Moldavia (USSR), takes mainly invertebrates from early April to June, mainly plant food at other times: invertebrates mainly beetles (especially Curculionidae, Elateridae, Tenebrionidae, Carabidae), with Hemiptera, Orthoptera, ants, spiders, and caterpillars also important; seeds include mignonettes (Resedaceae), *Echinochloa*, goosefoots, cereals, and sunflower (Averin and Ganya 1970). In Crimea, March–July, 35 stomachs contained 54·9% invertebrates (mainly ants, beetles, and molluscs) and 45·1% seeds; in October–February, 7 contained 15·6% invertebrates (beetles and molluscs) and 84·4% seeds (Kostin 1983). In Kalmytskaya ASSR (north-west of Caspian Sea), December–January, 16 stomachs contained almost wholly seeds and fruits (99·8%, by number, of 1177 items), mostly *Chenopodium* (70·5%) and *Amaranthus* (21·2%); also 0·2% beetles, and 1 bird contained green plant material (Samorodov 1968). No clear seasonal pattern of variation in data from Turkmeniya: in spring, invertebrates present in 86·6% of 134 stomachs, plant material in 75·4%; summer, 73·6% and 93·6% respectively (*n* = 110); autumn, 42·3% and 90·8% (*n* = 142); winter,

45·8% and 62·7% (n = 59) (Bel'skaya 1974; see also Medvedev and Esilevskaya 1973). For Yugoslavia, see Kovačević and Danon (1952). In west-central Spain, March–May, 19 stomachs all contained invertebrates but only 1 contained these alone: mainly beetles (Curculionidae in 58% of birds, Carabidae 28%), ants (58%), caterpillars (52%), and spiders (28%); seeds in 95%, mainly of grasses (in 47%) and Labiatae (37%); green plant material in 84% (Abs 1963). Stomachs from Palestine contained the following (sample sizes unknown): May, insects (including Orthoptera) and seeds; October, insects; winter, seeds (Hardy 1946). In Zagros mountains (Iran), March–May, 5 of 8 stomachs contained invertebrates including grubs (Paludan 1938). Near Casablanca (Morocco), December, 5 stomachs contained mainly ants, also grain; though presumably feeding in different habitat, birds had thus eaten more insects and fewer seeds than Thekla Larks G. theklae from same area; see that species, p. 166 (Niethammer 1937). In southern Morocco, May, 5 birds contained seeds and grain (Valverde 1957). In Tunisia, 5 stomachs, February–March and June, contained seeds (including grain) and, in 1, grass (Blanchet 1951). Near Cairo (Egypt) in autumn, 10 birds contained mostly seeds (largely of grasses, cereals, and *Chenopodium*); also ants, spiders, and insect larvae (Abs 1963). In Niger, August or September, one bird contained insects (Villiers 1950), and another, January, contained grass seeds (Niethammer 1937). In Somalia, eats seeds and succulent shoots of *Suaeda*, and in south-west Arabia often seen on seashore pecking at seaweed (Meinertzhagen 1951a), though perhaps taking associated invertebrates.

Young are fed mainly on insects, especially caterpillars and small Orthoptera (Ferguson-Lees 1970d). In Nakhichevanskaya ASSR (Transcaucasia, USSR), mainly Orthoptera (E N Panov). In West Germany, 2 fledged young seen to be given bread, another given seeds (Sudhaus 1966b).   DJB

Social pattern and behaviour. Well known. For summaries and most important earlier references, see Abs (1963) and Pätzold (1971). More recent study in Bayreuth (Bayern, West Germany) by Gubitz (1983). Following account includes material supplied by A P Møller from study of small population (10 pairs) in northern Jutland (Denmark).

1. Not markedly gregarious. Outside breeding season, sometimes occurs singly, though generally in pairs or small parties of 3–4 or up to 10–15 (Abs 1963; Kozlova 1975). In northern Jutland, average 'flock' size, September–April, 1·2–2·5; maximum during study 7 and at any time 25. Roosts held average 3·4 birds (2–5), n = 5 (A P Møller). Larger gatherings generally loose-knit (e.g. Dolgushin et al. 1970) and may be attracted by favourable food supply, sometimes in bad weather: e.g. in East Germany, January, at least 30 on rubbish tip c. 200 × 300 m (Kummer 1983); in Turkey, parties of up to 20 in severe conditions (Porter et al. 1969). Flocks also occur for drinking: e.g. up to 25, eastern Iraq (Sage 1960a); c. 60–70 at small pool c. 3 m across in Turkmeniya, USSR (Rustamov 1958). Migrant parties also usually small (e.g. Bub 1955), though 18 recorded in Schleswig-Holstein, West Germany (Pätzold 1971), and

largest gatherings—up to 100 in Netherlands (Abs 1963) and allegedly up to 150 in Bayern (Ries 1908)—possibly referred to migrants. Such numbers exceptional (Gubitz 1983). No flocks reported from Egypt by Meinertzhagen (1930), but more recently flocks of 30–40 recorded in cultivated areas of Nile Valley, mid- to late October (Vuilleumier 1979). In Jutland, family parties of adults and dependent young occur from early summer to early autumn (A P Møller). In Kazakhstan (USSR), small nomadic parties particularly evident from late summer to early autumn (Dolgushin et al. 1970). In Dhofar (Oman), flocks occur from October, maximum 40 in December; disperse late February (Walker 1981b). Similarly, in Macedonia, parties break up with first improvement in weather (Chasen 1921). Sometimes associates with other Alaudidae: e.g. in Spain, often with Thekla Lark G. theklae (Mountfort 1954; Niethammer 1954); in Tunisia and Israel, with Short-toed Lark Calandrella brachydactyla (Bannerman 1927; Bodenham 1944); in Cyrenaica (Libya), with Lesser Short-toed Lark C. rufescens (Hartert 1923). Recorded associating with Turnstones Arenaria interpres in coastal towns of Somalia (Archer and Godman 1961a). Shows no special tendency to associate with other birds in flight (Pätzold 1971). BONDS. Monogamous mating system (A P Møller); see, however, Lynes (1912) whose suggestion that this perhaps not always so was based on frequent occurrence of 3 birds together in breeding season and record of 8 eggs in a nest. In Turkmeniya (USSR), ♂♂ apparently predominate all year: ♂:♀ ratio in winter 4·3:1, in spring 2·7:1, in summer 2·1:1, in autumn 2·2:1; fledged young 1·9:1. Some members of population evidently do not breed (see Bel'skaya 1974). Duration of pair-bond not established by ringing but possibly long-term (Niethammer 1937; Sudhaus 1965), and in some cases apparently maintained throughout winter (Chasen 1921; Portenko 1954), probably involving older birds (Sudhaus 1965; Dolgushin et al. 1970); pairs can be obvious in winter flocks (Schuster 1944). In northern Jutland, most pairs remain within territory all year (A P Møller). See also Heterosexual Behaviour (below). Pair-fidelity possibly linked with site-fidelity (Labitte 1957b). For hybridization with G. theklae, see that species (p. 166); G. cristata has also hybridized with Skylark Alauda arvensis in captivity (Abs 1963). In Algerian Sahara, winter, 2 ♂♂ recorded associating closely (Niethammer 1954); see also Heterosexual Behaviour (below) and G. theklae. In Pakistan, apparently shared nest (containing full clutch of each species) and incubation with Indian Sand Lark Calandrella raytal (Newton 1936). Incubation by ♀ alone (e.g. Hartley 1946b; Müller and Gass 1973). Young fed and nest-sanitation performed by both sexes (Löhrl 1944; Hartley 1946b; Hinsche 1960); ♀ probably does most early feeding, ♂'s contribution becoming significant only after brooding stops at c. 5–7 days (Pätzold 1971; A P Møller); some older reports claimed ♂ fed young only after they had left nest (see Abs 1963). Young fed for long period (albeit irregularly) after fledging, until beginning of moult—by ♂ alone if ♀ starts 2nd brood (Abs 1963). Family party may remain intact for up to c. 1 month after young leave nest. Age of first breeding 1 year (Pätzold 1971; A P Møller). BREEDING DISPERSION. Typically solitary and territorial. In Jutland, average size of 21 territories 0·58 ha (0·39–1·09), centres 281 ± SD83 m apart (A P Møller). In Somme (northern France), territory c. 4·6 ha; feeding area of c. 1 ha shared with neighbouring pair (Triplet 1981). In Bayreuth, 2 territories c. 15 ha and c. 30 ha (Gubitz 1983). Near Dresden (East Germany), territory c. 2 ha; in Spain and Rheinland (West Germany), where density low, c. 4 ha (Abs 1963). In Israel, territories of c. 2·8–4·0 ha recorded (Meinertzhagen 1951). In USSR, near human habitations, nests sometimes c. 100 m apart or less (Dementiev and Gladkov 1954a). In Nile delta (Egypt), where birds abundant, nests not uncommonly *c.* 50 m apart (Meinertzhagen 1951). On Cyprus, late March, singing ♂♂ also *c.* 50 m apart, though density probably higher later (Bennett 1982). Nests for 2nd and subsequent broods or replacement clutches recorded 5–80 m from that previously used (Gengler 1903; Labitte 1957b; Müller and Gass 1973). Territory used for courtship and feeding, though some feeding also done elsewhere (Abs 1963; A P Møller). Defence of territory probably in most cases by ♂ alone (A P Møller), but in northern France where territory apparently defended also in winter, both birds of pair involved (Kerautret 1979). Density in Oberfranken (Bayern, West Germany), 10–15 pairs in 4300 km^2 (Gubitz 1982). In northern Jutland, 210 pairs in 7458 km^2; in study area (1·68 km^2), 5·4–7·1 pairs per km^2 in 2 years (Møller 1978a, b; A P Møller). In Schleswig-Holstein, 450±100 pairs in 15 680 km^2 (Sudhaus 1966a; Pätzold 1971). In Hoyerswerda (East Germany), highest density *c.* 3 pairs per ha (Krüger 1967). Transects in Kara-Kum desert (Turkmeniya, USSR) revealed (e.g.) 3–4 pairs along 4–5 km (Rustamov 1954); for further census results, see Rustamov (1958). In Turkmeniya, 2 1-year-olds recorded nesting *c.* 200 m from natal site (Bel'skaya 1974). Often nests close to *G. theklae* (Mountfort 1954; Niethammer 1955a; see also Antagonistic Behaviour, below), or shares territory with *C. brachydactyla* and *M. calandra* (Took 1972; see also Meinertzhagen 1938). Territory may overlap or be dovetailed with *A. arvensis* (Witsack 1968; Gubitz 1983). ROOSTING. Major study by Krüger (1970) in Hoyerswerda (East Germany). Nocturnal, and solitary or in pairs (Krüger 1970), though usually communal outside breeding season (e.g. Gengler 1903). Normally on ground though flat roofs of 4–5-storey buildings now used regularly in Hoyerswerda and Leipzig (East Germany); new building completed one summer was occupied the following autumn despite lights (Krüger 1967, 1970). Sites on ground often near paths or tracks (Krüger 1970) and include cropland (Meinertzhagen 1951), scattered vegetation (Kozlova 1975), plantation of poplars *Populus*, short grass at edge of field, sand-pit (Krüger 1967), and stones of hill edge (Hardy 1946). In summer, usually in the open with no cover, but in late autumn and winter, especially in bad weather, sites normally snow-free and sheltered: e.g. by wall, under bicycle shed, under tufts, bushes or stones, etc. (Abs 1963; Krüger 1970; also Meinertzhagen 1951); will also use open fields with little snow cover (A P Møller). Roosting in snow said to be normal in frosty weather (Dementiev and Gladkov 1954a). May roost on flat surface or use already existing hollows, but pits normally made in sand or loose soil; pits elongated or rounded and shallow. May first clear away leaves, then normally scratches with feet and rotates body; nestles down with feathers ruffled and wings slightly spread and may draw in leaves close to body. Birds did not scratch or rotate where depressions already present in sand-pit, but one did so, inappropriately, on flat roof (Abs 1963; Krüger 1967, 1970). Young birds apparently roost in natural hollows after leaving nest (Dementiev and Gladkov 1954a); old report referred to ♂ covering them with dead leaves and grasses (see Schmied 1969, also Relations within Family Group, below). Captive young started to enlarge depression by scratching with feet at *c.* 12 days old (Pätzold 1971). Birds also defecate in pits and accumulation of droppings indicates pits probably used regularly. 3 captive birds in large aviary used 33 different pits over 1 year. In period 2–24 October, each probably used 2–3 pits; 1 pit used for maximum 6 nights (Abs 1963). Roost-sites often used for several days in succession but frequent changes typical and probably due to disturbance, bad weather, stage of

breeding cycle, etc. In communal roosts, pits often only a few cm apart. 3–6 birds recorded in small area of 1 to several m^2 and other small groups close by (Krüger 1970; also Witsack 1969); in Bayern, once 10–15 huddled close together (Gengler 1903). Normally 2–5 m apart (A P Møller). In summer, birds roost in nesting territory; paired birds not yet breeding, 0·5–2 m apart. On flat roof, ♂ (or possibly ♀ no longer brooding) probably roosts close to nest, but site clearly imposes restrictions on adults and young out of nest but not yet able to fly. Fledged young usually roost close to one or both parents, at least initially. Birds usually in vicinity of roost for some time (maximum *c.* 45 min) before moving to it. In winter, normally feed then; may sand-bathe (especially in summer) or sit motionless; little or no self-preening. Sleep with bill tucked into ruffled back feathers. Typically goes to roost after sunset, and leaves early. In Hoyerswerda, average arrival time (after sunset) at roost (*n* = 137 observations) in clear to cloudy conditions: in breeding season (March–July) 20 min; during moult (early August and late September) 16 min; during October Song-display period 24 min; in winter (November to early March) 20 min. Up to 10 min earlier in overcast conditions. Temperature has little influence but level of satiety probably plays a part. Fledged young probably go to roost first, then ♀♀, finally ♂♂. From late January or early March (depending on weather) to end of July, and from late September to early November, awakening marked by song or repeated loud calls. In main breeding season (and often from late January or early February), first Song-flight performed when still dark. Time of waking more dependent on season (see above) than is time of going to roost. In clear to cloudy conditions, start of calling in different seasons as defined above: 72, 54, 54, and 48 min before sunrise, earliest 80, latest 33 min before. In overcast conditions: 62, 34, 48, and 38 min before, earliest 77, latest 23 min before. (Krüger 1967, 1970.) In Bonn (West Germany), start of activity as indicated by first calls: August, 18 min before sunrise; September, 34 min; December, 43 min (Abs 1963). In Turkmeniya, winter, birds active 09.00–10.00 hrs onwards. At peak of breeding activity, sing from *c.* 2 hrs before dawn to *c.* 14.00–15.00 hrs, then in evening from 17.00 hrs to dusk (Rustamov 1954, 1958). ♂'s song given before end of activity in breeding season ceases about mid-July, but some resume in September–October (Krüger 1970). In Bayreuth, on 14 June, maximum morning song-activity on ground 02.40 hrs (sunrise at 04.05 hrs); on 17 July, evening maximum 21.30 hrs (sunset at 20.31 hrs). On 12 April (sunrise at 05.34 hrs) Song-flight first performed at 05.45 hrs, and on 18 May (sunset at 20.10 hrs) last at 19.30 hrs (Gubitz 1983). Birds also rest during day; head normally lowered and depressed crest barely visible. In Denmark, peaks for this in morning during April–January, and in evening throughout year. Lack of resting February–March possibly due to food stress. Non-incubating ♂ frequently rests on small mound or other elevation (also used as song-post) in territory (A P Møller). At hottest time of day, birds seek shade under bushes, behind poles, etc. (Ripley 1951; Moore and Boswell 1956; Archer and Godman 1961; Kozlova 1975). In Dhofar, birds used small pits under grass tussocks or rocks (Walker 1981b). Shows marked fondness for dust-bathing on or by roads (Chasen 1921; Bannerman 1927; Meinertzhagen 1930), also in sand on seashore (Kozlova 1975). Captive young sand-bathed from *c.* 13 days old: lay panting, bill half-open, wings spread and trembling (Pätzold 1971). Snow-bathing recorded once (A P Møller). See also Arndt (1981).

2. Sometimes allows close approach. In Africa, most confiding and fearless of all Alaudidae (Archer and Godman 1961a). In

Macedonia, birds tame in breeding season, running about in grass *c.* 4 m from intruder (Chasen 1921). Will come as close as 2 m to take bread from man (Lenggenhager 1954). Tameness may be more pronounced in winter (Nicoll 1914; Witsack 1968) or not (Gubitz 1983). In Jutland, flies from pedestrian at average $4·4 \pm SD2·8$ m ($n=17$), cyclist at 1·8 m ($n=4$), dog at 2·9 m ($n=3$), and from cat at 2·5 m ($n=2$) (A P Møller). Not always this tame though (e.g. Whitaker 1905, Niethammer 1955a). When threatened, may at first crouch slightly, with crest depressed and wings held away from body, and call (Barrett *et al.* 1948; Gubitz 1983: see 4 in Voice); mock-feeding movements also recorded (E N Panov). More vocal than *G. theklae* when approached by man (Niethammer 1955a). If closer approach made, often simply trots away, body more erect and crest raised (Barrett *et al.* 1948; Gubitz 1983). If pursued, may run and hide in cover (Abs 1963), but will also fly off several hundred metres at height of *c.* 50 m (Niethammer 1955a; E M Nicholson). Birds may move around a predator, give call 4, and perch conspicuously (A P Møller). *Galerida* larks seldom squat, preferring to watch intruder from upright position; will, however, squat when threatened by falcon *Falco*, though try to escape from harrier *Circus* (Meinertzhagen 1951). Loud, infectious Flight-calls (see 4 in Voice) given during flight to roost, though this less marked outside breeding season. Before flying to roost and on arrival there, birds restless with much running about and calling (see 2 in Voice). May appear to reject first-chosen site but often return to it (Krüger 1967, 1970). FLOCK BEHAVIOUR. Birds in flocks (perhaps pairs) often only *c.* 20-30 cm apart (A P Møller), though *c.* 30 cm said by Gerber (1949) to be in any case the normally defended individual-distance. Amount of calling in small feeding parties proportional to dispersion: generally quiet calls given when a few feeding together, but excited calling by one bird may set off long, loud chorus (Barrett *et al.* 1948). SONG-DISPLAY. ♂'s song (see 1 in Voice) given from ground, perch, or in flight; proportions vary (Gubitz 1983; also Meinertzhagen 1938, Barrett *et al.* 1948, Harrison 1966). Weather may also exert influence. In Bayreuth, 75% of Song-flights ($n=60$) performed in sunny conditions; 58% of total number of songs given from perch—quite often in poor weather (Gubitz 1983). On one sunny May day, 32% of song given in flight (Pätzold 1971). In Denmark, 80% of morning song is given in flight, 78% of evening song. Proportion highest at noon and during May (75%) and July (85%). Song from perch frequent in early morning and commoner February–April and October–November (A P Møller). Quiet song from ground typical of spring before breeding, also of autumn when feeding (Abs 1963); may be protracted, ceasing only momentarily while pecking (Barrett *et al.* 1948). Loud, full song often given from ground or perch, however, and may be prelude to Song-flight (Barrett *et al.* 1948; Abs 1963). Singing bird recorded squatting on tarsi, crest fully raised, bill pointed almost vertically up, throat vibrating (Chasen 1921). For further details, see below and 1 in Voice. In Song-flight, ♂ usually takes off from elevated point (A P Møller), ascends at angle into wind and starts loud song only at *c.* 30-70 m (Barrett *et al.* 1948; Abs 1963; Pätzold 1971; A P Møller). May continue ascent to *c.* 100-200 m (Gubitz 1983); see also Hartert (1913) and Harrison (1966). Where singing starts at *c.* 15-20 m, ensuing Song-flight lower and shorter (Gubitz 1983). Height more or less constant throughout bout of song (Barrett *et al.* 1948). Singing bird uses slow fluttering wing-beats, frequently hovering in wind, then continues silently on undulating path, sings again, etc. (Barrett *et al.* 1948; Abs 1963). Overall, flight less hovering than *A. arvensis* and flight-path involves more circles, arcs, and undulations with

some straight sections (Pätzold 1971); an 'aimless wandering' (Chasen 1921). Generally describes wide circles over territory (Abs 1963), song occupying about half of each circuit (E M Nicholson). May cover *c.* 300-400 m or up to 600 m during Song-flight, flying in relatively small circles at territory limits (Pätzold 1971; Gubitz 1983). Unlike *A. arvensis*, descent angled, and rapid spiralling used mainly in final phase for braking (Gubitz 1983). Barrett *et al.* (1948) recorded steep, silent descent with wings half-closed and immediate resumption of song on ground; see also Abs (1963). In one study in East Germany, longest Song-flight (in May), 4·3 min, average 1·4 min (Pätzold 1971). In Bayreuth, equivalent figures for May 10 min and 4·3 min ($n=14$). Overall average (not including exceptional record of 37 min) 3·4 min, shortest 40 s (Gubitz 1983). In another study, average 3 min, but once 25 min by unpaired ♂ (Abs 1963, 1970). In Denmark, 30-875 s, average 200 s ($n=42$); longer in early spring and autumn; 1·4-2·5 Song-flights per hr, average interval 32 min, $n=19$ (A P Møller). In European range, start of song-period influenced by weather (Abs 1963) but normally (January-)February-July: e.g. in Denmark, 25 February-19 July, peak May-July (A P Møller); in Bayreuth, peak April-May (Gubitz 1983). See also Alexander (1927) for central Italy, Barrett *et al.* (1948), and Krüger (1970). Song-flights mainly in breeding season—e.g. in Bayreuth, 10 February-18 July (Gubitz 1983) but occur also on fine autumn days (Abs 1963). In autumn and winter, song normally from ground or perch. In Denmark, autumn period 18 September-15 November (A P Møller); similar in East Germany (Krüger 1970) and slightly shorter (to end of October) in central Italy (Alexander 1927). In Cyprus, song noted from early February, and until late August or September (Flint and Stewart 1983); in Israel, mid-March to mid-June, also in autumn to late September (Hardy 1946); in Oman and United Arab Emirates, February-May (Stanford 1969). In Turkmeniya (USSR), recorded from late January; increasing from mid-March when most birds paired; decrease before 2nd clutch, then increase again for peak of 2nd cycle (Rustamov 1954; Bel'skaya 1974). ANTAGONISTIC BEHAVIOUR. (1) General. *Galerida* larks typically active in territorial defence, with much chasing of intruders (Meinertzhagen 1951). In some areas, low density and territory size (see Breeding Dispersion, above) of *G. cristata* ensure that fights rare (Abs 1963; Gubitz 1983). In Egypt, autumn, apparently aggressive chases reported (Simmons 1952); in northern France, winter, chases probably in defence of territory (Kerautret 1979). In Denmark, aggression rare in winter flocks (A Møller), though apparently quite marked in exceptionally large gathering (Ries 1908). Peak antagonism early in breeding season coinciding with start of song (Gengler 1903; Barrett *et al.* 1948; A P Møller). In USSR, 2 ♂♂ recorded fighting over ♀ in January and February (Rustamov 1958; Kostin 1983), also at Dhahran (Saudi Arabia) in early May when each ♂ attempted to display (see below) at ♀ (J Palfery). (2) Threat and fighting. Neighbouring ♂♂ may advertise possession of territory by giving call 2 or apparently challenge with presumed call 3 (Barrett *et al.* 1948). To rebuff intruder, ♂ may first run, then fly, towards it and pursue it in the air until it has left area (A P Møller). Twittering-calls (see 3 in Voice) given in aerial chases (Abs 1963); 3-4 birds may be involved—presumably pair chasing single intruder or rival pair (Kerautret 1979; Gubitz 1983). Bird recorded threatening another by running at it sideways, head up, crest vertical, and tail raised but not fanned (Hartley 1946b). In similar incident, ♂ gave full, loud song on ground with wings held away from body, drooped towards tip, and very rapidly vibrated; crest erect and tail raised at *c.* 45°. Ran towards 2nd bird and shuffled back and forth in front of it for *c.* 30 s, apparently showing off pale flanks. When 2nd bird moved off, ♂ flew to roof and sang loudly, back and rump feathers ruffled. In crouched Threat-posture commonly assumed during squabbles over food, wings half-spread and head thrust forward. Threatened bird frequently depresses crest (Barrett *et al.* 1948). In Nigeria, Song-duels and associated aggression noted February–March. 2 birds may chase in area *c.* 30 m across, then face up at *c.* 3 m or more, one or both singing vigorously. Bird adopting Threat-posture chases off more erect rival (Mundy and Cook 1972). Rivals may advance simultaneously with bills open and give Rattle-call (Abs 1963; see 5 in Voice). May feint at close quarters, jump up with slight collision, then chase. Some fights prolonged, birds moving far from starting point (Chasen 1921). Fights vigorous (but never damaging), with much calling and bill-snapping; rarely, a few feathers fly (Gengler 1903). Reaction to playback of *G. theklae* calls suggested territories defended against that species (Abs 1963). In Sudan, also attempt to rebuff intrusions by Hoopoe Lark *Alaemon alaudipes* and finch larks *Eremopterix* (Meinertzhagen 1951). In Denmark, recorded chasing off *A. arvensis* (A P Møller); no interspecific antagonism with *A. arvensis* in Bayreuth (Gubitz 1983). Speed and greater agility of *A. arvensis* probably give it dominance in most situations (Meinertzhagen 1951). *G. cristata* dominant over sparrows *Passer*—House Sparrow *P. domesticus* a frequent food-competitor (Sudhaus 1966*b*; Müller and Gass 1973; Gubitz 1983); also, in winter, over Great Tit *Parus major* and Greenfinch *Carduelis chloris*, but not Blackbird *Turdus merula* (A P Møller). Once successfully attacked Black Redstart *Phoenicurus ochruros* (Gubitz 1983). HETEROSEXUAL BEHAVIOUR. (1) General. Courtship as described below does not definitely constitute pair-formation and, in birds known to be paired, perhaps only strengthens bond and, like song and pursuit, probably serves as sexual stimulus and to synchronize breeding (Sudhaus 1965). In Denmark, almost all courtship in morning, 25 February–19 June, mostly February–March; recorded only between mates (A P Møller). In Nakhichevanskaya ASSR (south of Caucasus, USSR), from late March to late April (E N Panov). Courtship also recorded in warm weather in late September (Abs 1963); in Bahrain, recorded from December (Gallagher and Rogers 1978), and in East Germany, in mid-January (Kummer 1983). In Coswig (East Germany), performed while ♂ and ♀ feeding young (Hinsche 1960); may precede 2nd brood (Sudhaus 1965). (2) Pair-bonding behaviour. ♂ pursues ♀ in low zigzag flight and gives loud Twittering-calls (Gengler 1903; Abs 1963; see 3 in Voice). Many descriptions of courtship on ground, most detailed by Sudhaus (1965). ♂ holds crest fully erect with head held slightly forward, breast swollen, body horizontal, tail fanned and raised, and body feathers ruffled, more so at higher intensity (Figs A–B). In this Tail-up posture, normally moves towards ♀ in tripping or mincing gait, appearing to glide over ground (E N

Panov), and dances thus around or in front of her; according to Barrett *et al.* (1948), ♂'s 'light buoyant hops' (bouncing, both feet together, as if on a pogo stick: J Palfery) take him in either direction, though also reported to circle ♀, almost always clockwise (Gengler 1903). ♂ gives Courtship-song (see 1b in Voice) with bill open. Different degrees of intensity apparent in ♂'s Tripping-dance: at lower intensity, closed tail raised at *c.* 45° and wings held close to body; at high intensity, wings widely and horizontally spread—more so on side nearer ♀ (Barrett *et al.* 1948) or away from her (A P Møller)—and fluttered or vibrated in small-amplitude movement (Fig C). ♀

C

usually crouches flat and silent, with crest depressed, or occasionally raised vertically as in ♂ (E N Panov), wings sometimes partly opened (Gengler 1903; Moore and Boswell 1956). If ♀ moves away, ♂ will chase her (Sage 1964*a*), and may give call 2 (Gengler 1903). If ♀ quiet and remains still, ♂ usually turns to face away from her in high-intensity Tail-up posture (Fig D): tail raised at *c.* 70–80°, wings drooped, and

D

vent feathers ruffled to form cushion and draw attention to bright white cloacal region. ♂ may turn head to one side and look back at ♀; often freezes in Tail-up posture. ♂'s bout of display may last up to 10 min (Gengler 1903) and ♀ normally waits for ♂ to finish before getting up, then usually bill-wiping and sand-bathing. Courtship usually ends abruptly, ♂ running off in normal posture to feed and soon joined by ♀. ♂ once began to sing afterwards. (Sudhaus 1965.) Variants reported by other authors: ♀ may walk about, apparently feeding, while ♂ displays (Gengler 1903); ♂ may run rapidly with occasional wing-flapping, fly up briefly, and bow while circling ♀ (Dolgushin *et al.* 1970). ♂'s head may be bent back to emphasize streaked throat (Moore and Boswell 1956); see also Hartley (1946*b*). In bout of winter courtship lasting 2–3 min, one or both birds gave soft contact-calls (Kummer 1983: see 2 in Voice). In interaction observed by Sage (1964*a*), ♂ always faced ♀, wings drooped to ground and partly open to show orange-buff undersides; ♂ circled ♀ in simple run. At Dhahran, ♂ similarly moved wings so as to flash undersurfaces at ♀ (J Palfery). See also Stresemann (1957) for captive bird displaying to Thick-billed Lark *Rhamphocoris clotbey*. In 'hide-and-seek' of early courtship, ♀ recorded flying from ♂

A

B

and crouching behind stones as if on nest; ♂ in an erect posture with head high, wings slightly raised and held away from body, moved about giving call 2 and apparently searching for ♀; when ♀ (still crouched) walked into view, ♂ suddenly alert, but then Mock-fed and flew away, still calling (Barrett *et al.* 1948); see also subsection 5. (3) Courtship-feeding. ♀ not fed by ♂ on nest (Gengler 1903; Abs 1963). (4) Mating. May follow pursuit-flight (see above). ♀ drops abruptly to ground and lies flat, tail slightly to half spread, opened wings trembling vigorously. ♂ may stand crossways to ♀ and peck repeatedly at her neck, or, with crest erect and tail closed and slightly depressed, may sing loudly and make big hops back and forth several times before mounting; climbs rather than flies on to ♀, wing-trembling less than before (E N Panov: Fig E). Presses

E

tail under ♀'s for rapid copulation. During copulation, ♂ has crest erect and gives Copulation-call (see 7 in Voice). Copulation repeated, ♀ wing-shivering between acts (Gengler 1903; Sudhaus 1965). See also Sage (1964*a*) who recorded birds self-preening after copulation then flying off. ♀ once gave call 2 as she flew off, while ♂ sang loudly and for long time afterwards (E N Panov: Fig F). ♀ may frustrate ♂'s mounting

F

attempts by running forward (Hartley 1946*b*). In Jutland, most copulation takes place in morning during May (A P Møller). (5) Nest-site selection. The following behaviour does not definitely constitute nest-site selection. ♂ may walk into low scrub, calling vigorously, then stop and slowly crouch with tail spread and depressed. ♂ repeatedly bows towards ♀ just outside cover. ♀ may crouch, and pick up grass, and drop it. ♂ may emerge, or ♀ may follow him into cover where bowing repeated. (Barrett *et al.* 1948.) While ♀ builds nest, ♂ usually nearby and may accompany her (Pätzold 1971) or perform Song-flight (Bennett 1977). (6) Behaviour at nest. ♀ said to spend most of day on nest once 1st egg laid (Haun 1930). May add material

during laying, incubation, or even after hatching (Löhrl 1944; Marchant 1963; Pätzold 1971). Incubation stints of up to *c.* 1 hr in morning; during frequent breaks (usually less than 10 min, though up to 20 min), ♀ self-preens or feeds, often accompanied by ♂, and pair may move up to *c.* 400 m away (Abs 1963; Pätzold 1971). While ♀ on nest, ♂ often sings from one of several favoured perches (*c.* 20–50 m from nest) also used as look-outs (Haun 1930). On return, ♀ may land *c.* 1–1·5 m away and call (Schmied 1969: see 9d in Voice). Normally lands some distance (1–15 m) from nest and looks around before running or walking in directly, crouched with crest depressed; makes good use of cover (Gengler 1903; Lenggenhager 1954; Labitte 1957*b*; Pätzold 1971). May use different route each time (Hinsche 1960). Leaves in same manner (Gengler 1903) or flies up directly (Hinsche 1960). RELATIONS WITHIN FAMILY GROUP. Young brooded by ♀ for *c.* 5–7 days (A P Møller). Until young 2 days old, brooding ♀ leaves nest every *c.* 20–30 min to collect food; at 3–4 days old, broods for *c.* 15–20 min at a time. In cool weather, young fall into death-like torpor if not brooded, but recover if warmed, even after several hours (Pätzold 1971). Adult approaching nest with food gives Feeding-call (see 8 in Voice) to which young respond from 1st day by begging: neck extended, bill open, giving quiet food-calls (Pätzold 1971; Bel'skaya 1973). Adult tends to feed nestling which gapes widest and leans furthest forward; may result in death of some brood-members by *c.* 6–8 days, corpses then normally carried away by ♀ (Bel'skaya 1973); no inter-sibling aggression recorded by A P Møller. Faeces of young initially swallowed by ♀, but from *c.* 3–5 days carried away (Abs 1963; A P Møller); pellets of nestlings treated similarly (Abs 1963). Eyes of young opened to slit by 4th day, fully by 6th day when nestlings turn purposefully towards adult and give clear food-calls (further details, also of physical development, in Pätzold 1971 and Bel'skaya 1973). Leave nest at *c.* 8–11 days old (Hartley 1946*b*; Pätzold 1971); return to it during next *c.* 3 days (Dementiev and Gladkov 1954*a*) or never (Hartley 1946*b*). ♀ may attempt to brood roosting young (Dementiev and Gladkov 1954*a*). According to Löhrl (1944), young run skilfully only a few hours after leaving nest; other authors reported an initially clumsy hopping or leaping, then running after *c.* 3 days (e.g. Abs 1963, Schmied 1969, Kozlova 1975). Probably leap anyway when excited or to overcome obstacles (Pätzold 1971). Abs (1963) and Schmied (1969) reported fledging at *c.* 15–16 days; however, probably capable of sustained flight only after *c.* 20 days, and fed until this time, perhaps longer (Pätzold 1971; A P Møller). Adults frequently give call 2 when accompanying young out of nest; following disturbance, young call and family re-unites (A P Møller). Captive young separated at *c.* 9 days old gave contact-calls and ran together for warmth; kept close together after 10 days and often performed mutual allopreening (Pätzold 1971). ANTI-PREDATOR RESPONSES OF YOUNG. Captive young crouch flat in nest (Teschemaker 1912*b*). If surprised out of nest, crouch with head pressed flat in grass (Gengler 1903); tend to flee and try to hide in cover when pursued (Abs 1963). PARENTAL ANTI-PREDATOR STRATEGIES. (1) Passive measures. Sudden disturbance early in cycle, or even when with small young, may cause desertion (Pätzold 1971). Brooding ♀ watched from hide drew grasses across to form screen (Pätzold 1971). Sentinel ♂ gives call 2 (with pauses) when man or other predator near nest. ♂ then runs off (Haun 1930; A P Møller). Once clutch complete, ♀ may allow approach to *c.* 1 m before leaving (Löhrl 1944). According to Haun (1930), ♀ usually flushes at 10–15 m, much earlier if ♂ not present. (2) Active measures: against man. On sighting predator, both adults often perch conspicuously and fly about giving call 2 (A P Møller). ♀ recorded fluttering about near threatened nest (Bates 1936–7). From start of incubation, ♀ performs distraction-lure display of disablement type (Löhrl 1944). In Iraq, left nest containing slightly incubated eggs, ran like rodent, fluttered wings, and called (S Marchant). Recorded blundering through grass late in incubation. Adult with young *c.* 8–9 days old grovelled on ground a few metres away, head low, tail depressed and back feathers ruffled, and called (not described); also fluttered through grass and covered *c.* 50 m on ground in rapid scuffling with wings widely spread (Hartley 1946*b*). See also Haun (1930) for possible low-intensity distraction-flight by ♂. (3) Active measures: against other animals. Pair recorded attacking weasel *Mustela nivalis* (Triplet 1981). Will hop in front of snakes, calling loudly and continuously (J H Morgan and J Palfery).

(Figs by J P Busby: A–B and E–F from drawings by E N Panov; C from drawing in Sudhaus 1965; D from drawings in Abs 1963, Sudhaus 1965, and by E N Panov.)   MGW

Voice. Rich repertoire; more diverse than Thekla Lark *G. theklae* (Abs 1963); see that species (p. 167) and below for further comparisons. More vocal than Skylark *Alauda arvensis* and Woodlark *Lullula arborea*; calls 1–2 given more or less throughout the year (see Social Pattern and Behaviour and below for details); other calls also frequent (Gengler 1903; Pätzold 1971; A P Møller). Descriptions suggest considerable variation within some calls and situation rendered more complex by existence of intermediate calls (2–3, 2–4, 2–3–4, etc.) resulting from superimposition of characters; see Bergmann and Weiss (1974) for full discussion, including sonagrams. In south-west USSR, song and especially calls of *tenuirostris* perhaps thinner and more piping than nominate *cristata* (Berndt 1944), but *arenicola* of Tunisia showed no significant differences (Zedlitz 1909). No other firm evidence of geographical variation, but see below. Near Erlangen (West Germany), birds closely mimicked shepherd's whistles (Tretzel 1965*a*, which see for sonagrams of this phenomenon and other vocalizations). For additional sonagrams, see also Abs (1963) and Bergmann and Helb (1982).

CALLS OF ADULTS. (1) Song of ♂. (*a*) Loud song given in flight or from perch or ground comprises long, soft fluting sounds, short whistles, pulsating elements, loud double notes and tremolos, and odd twittering sounds. Characteristically includes many imitations of other bird species (Barrett *et al.* 1948; Abs 1963); see below. Intermediate between *A. arvensis* and *L. arborea*; lacks vehemence of *A. arvensis* and not so continuous, though more coherent, than *L. arborea*. Phrases sometimes typical of *A. arvensis*, although shorter (Mountfort 1954) and song equal to it in richness of melodies when in full flow. More rounded and fluting than *A. arvensis* so that timbre closer to *L. arborea* as are short trills (Barrett *et al.* 1948; Pätzold 1971). In Tunisia, main phrase 'utz-utz-utz-zwee-du-du' (E M Nicholson). In Libyan desert, song given from ground in spring was quite unlike *A.*

arvensis and comprised musical, rather mournful, mainly 3-unit phrases: 'wee-too-wee too-chee-too twee-too-chee twee chee' (Bulman 1942); may refer to loud song (see below) or perhaps only call 2 given as a series. In recording, bird starts with rather loud 4–5-unit phrases separated by pauses of 3–4 s. Resembles Mistle Thrush *Turdus viscivorus* up 1 octave (P J Sellar). At least some phrases are probably call 2 (see below). Shorter pause (0·86 s) precedes change to much quieter continuous or rambling song (Figs I–II) more like *A. arvensis*. This

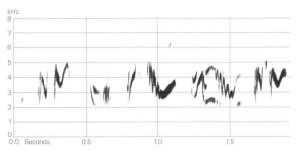

I P Szöke Hungary

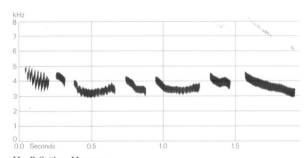

II P Szöke Hungary

variant contains chirruping, at times like song of Pied Wagtail *Motacilla alba*, various imitations (see below), rasping or nasal buzzes like *A. arvensis*, whistle like shepherd, and some rather petulant mewing sounds (J Hall-Craggs, M G Wilson). For comparisons with *G. theklae*, see that species. In loud song, units 0·1–0·3 s long; frequency 1·5–5 kHz. In Song-flight, series of linked phrases lasts 4–12 s; *c.* 12 series (6–20: Pätzold 1971) given per flight, 3–5 s apart. From perch, series longer at *c.* 22 s but pauses similar (Abs 1963). Species mimicked in Bayreuth (Bayern, West Germany) were all local passerines: House Sparrow *Passer domesticus*, Black Redstart *Phoenicurus ochruros*, and Blackbird *Turdus merula*; less commonly, Greenfinch *Carduelis chloris*, Great Tit *Parus major*, and Starling *Sturnus vulgaris* (Gubitz 1983); in Denmark, *S. vulgaris* and *A. arvensis* (see below), also Budgerigar *Melopsittacus undulatus* (A P Møller); in Middle East, *Tringa* waders, Rufous Bush Robin *Cercotrichas galactotes*, and Red-wattled Plover *Hoplopterus indicus* (Simmons 1951*a*; Moore and Boswell 1956). Fig III shows imitation of call of Greenshank *Tringa nebularia* and Fig IV shows imitation of Alarm-calls

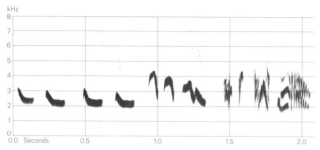

III P Szöke Hungary

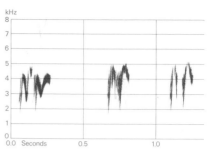

IV P Szöke Hungary

of Swallow *Hirundo rustica* (J Hall-Craggs, M G Wilson). In East Germany and Poland, birds imitated calls (and song in some cases) of various passerines, also Partridge *Perdix perdix* (Barrett *et al.* 1948). For further examples of mimicked bird species, see Schüz (1959) for northern Iran and Walker (1981*a*) for northern Oman. Captive birds trained to mimic human speech (see Abs 1963). (b) Courtship-song. Quiet, rambling and chattering, not unlike twittering song of *H. rustica* (Abs 1963). Strangled, more or less squeaky and rich in imitations, at times very like *T. merula* (Sudhaus 1965). Rambling, soft Subsong (given with bill closed) may contain calls 2 and 4, trills like Linnet *Carduelis cannabina*, grating or sweet 'chirruee', and various imitations (Barrett *et al.* 1948). (c) Quiet song sometimes given on ground while feeding. Can be difficult to hear at 6–15 m (Pätzold 1971); at *c.* 4 m, bird assumed to be singing, as bill moved, but nothing audible (Abs 1963). ♀ noted singing in captivity (Bub and Herroelen 1981). (2) Contact-alarm calls. Commonly given by both sexes from ground, perch, or in flight, particularly in early morning and throughout the year, mainly February–April and September–December (A P Møller), or March–May and August–October (Barrett *et al.* 1948). Characteristically pure, musical, liquid, and lilting; rather thin and ethereal. Renderings include 'peeleevee', 'di-dji-djii', 'twee-tee-too' (Mountfort 1954, 1958; Bergmann and Helb 1982; D J Brooks, E N Panov). Recording (Fig V) suggests 'dear dreer deer', with terminal 'r' indicating sudden drop in pitch (J Hall-Craggs). In Tunisia, firm and plaintive like *G. theklae*, but quieter, less musical and throatier, although not harsh; 'eee-ee-dweer' and 'ee-gur-fee-ur' when disturbed at nest (E M Nicholson). Further descriptions in (e.g.)

Niethammer (1955*a*), Wallace (1965), and Bergmann and Weiss (1974). Normally 3–4 units (exceptionally 2–5), though 2 calls often given in rapid succession (Abs 1963); 1–80 calls per sequence, averaging 6·1–7·4 in December–March, only 2·0–3·1 May–November (A P Møller). Duration of units (0·1–0·4 s) generally increases within call which is usually *c.* 1 s long. Pitch normally falls slightly within each unit and from 1 unit to next (Abs 1963; Bergmann and Weiss 1974). Average frequency 4 kHz (3–5 kHz). Highly variable. Final unit usually long with vibrato structure: e.g. 'wee wee wee werrrr' apparently as part of discontinuous song; end of final unit might be heard as a rolled 'r', though 'r' also used to suggest pitch descent (J Hall-Craggs: Fig VI), but call sometimes ends with 2 short units with sharp rise and then fall in pitch (Abs 1963); 2nd of these variants may possibly be intermediate with another call (see Bergmann and Weiss 1974). Shrill and loud when excited, soft and sweet when few birds feeding together; gradually softens as bird relaxes (Barrett *et al.* 1948). Rather sibilant 4-unit call in definite rhythm described by Barrett *et al.* (1948); less full and clear than typical call 2, an almost hissing trill. See Abs (1963) and Pätzold (1971) for full discussion, including possibility of geographical variation; further illustrations also in Bergmann and Weiss (1974). (3) Twittering-call. Probably a higher-intensity development of call 2; apparently an abbreviated, more urgent alarm. Individual sounds are loud, strangled whistles, like chirruping of sparrows *Passer* (Abs 1963). Shrill twittering 'zi ti zlie trü' (Bergmann and Helb 1982). Short and harsh, trilling 'prrrr' given by ♂ chasing ♀; more gentle variant, not unlike *A. arvensis*, given towards intruders (A P Møller). Frequency 4–5 kHz, thus higher pitched

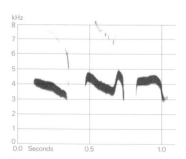

V P A D Hollom Iran April 1972

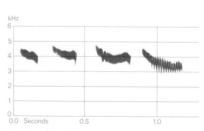

VI P Szöke Hungary

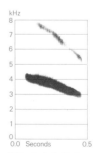

VII P A D Hollom Iran April 1972

than call 2. Noise components intermixed with tonal frequencies. Typically shrill due to marked, rapid, and irregular pitch fluctuations in each unit (Abs 1963; Bergmann and Weiss 1974). Call up to 2 s long (Pätzold 1971). Loud, forced 'chirrup' sound, shrieked at high intensity (Barrett *et al.* 1948); see also Alexander (1927) for possibly same call; this presumably one unit of Twittering-call (Abs 1963), but see call 9e. (4) Flight-call. Gentle fluting 'djui' (Niethammer 1937; Abs 1963); 'too-hea' (E M Nicholson); short and nasal 'schwüid' (Bergmann and Helb 1982). Soft onset and low average frequency of 2–3 kHz (Abs 1963). Relatively quiet and unobtrusive. Harshness at close quarters due to characteristic presence of 4 or more formants (not simple harmonics); lowest weakest at onset but becoming much louder towards end. Unlike other calls, also very often a single unit, though may be given several times at higher intensity (when frequency modulation also typical: see Fig 3b in Bergmann and Weiss 1974). Initial fall in pitch as shown in Fig 61 of Abs (1963) apparently not typical, though pitch usually rises at end (Bergmann and Weiss 1974). Given by both sexes shortly before or after take-off (Abs 1963; A P Møller). Slow, soft, pure-toned whistling 'hooeee' or short 'huyp' given by single bird or party near predator, particularly in August; up to 24 calls per sequence, average 6·4, $n = 30$ (Barrett *et al.* 1948; A P Møller). (5) Rattle-call. A rattling sound given in antagonistic context. No sonagrams, nor more detailed description available, but see *G. theklae* (Abs 1963). (6) Distress-call. Shrill, loud screech given when handled or captured by predator (Abs 1963; A P Møller); indistinguishable from call of *A. arvensis* in similar situation (Pätzold 1971). (7) Copulation-call of ♂. Quiet and thin piping or cheeping (Gengler 1903; A P Møller). (8) Feeding-call. Short, very quiet whistling sound given before feeding young (Pätzold 1971). (9) Other calls. (a) Shrill, whistling 'seeoo' given in antagonistic context (Barrett *et al.* 1948). (b) In recording, bird gives a long series of simple descending 'heeooo' sounds (J Hall-Craggs: Fig VII); clearly similar to (or perhaps the same as) call 9a, and compare also last unit in Fig II. (c) In Tunisia, 'tirra-loo' when flushed (E M Nicholson). (d) Short trilling call given by ♀ returning to nest (Schmied 1969). (e) Strikingly harsh call of 2 or more units given when alarmed or excited before take-off; noted frequently only in Cyprus (Bergmann 1983, which see for sonagram).

CALLS OF YOUNG. Low piping or cheeping sounds given in rhythm of breathing, also as food-call, from 1st day, though slightly more distinct only from 4–5 days. From *c.* 6 days, food-calls clear and some disyllabic with 2nd syllable slightly lower pitched; quiet cheeping also given after feeds (Pätzold 1971; A P Møller). At 7 days, captive young called 'zidü ziziedü'; when gaping, 'zieh'; after feed, 'zieh zieh zit'. A 'zirrr' given at 7–8 days when handled probably expressed fear. Lower-frequency

sound appended to higher ('zidü') only during last stages of nestling period (Abs 1963). 4 different calls given at 11 days old (Pätzold 1971). Contact-calls from young out of nest resemble early food-calls (A P Møller). Young out of nest being fed called 'zeeeee-zee' (Anon 1946a). Calls given by fledged young transitional between nestling and adult calls (Pätzold 1971), and include at least 2 whistling calls (Bergmann and Weiss 1974): shrill 'siäiep' 0·3 s long and 3·0–5·5 kHz (Abs 1963) in which pitch first falls, then rises, then falls again (Bergmann and Weiss 1974); shrill, soft, through-the-teeth whistling 'seeee-eep' (Barrett *et al.* 1948) presumably the same; whistling 'huit', with pitch first rising, then falling steeply (Bergmann and Weiss 1974).   MGW

Breeding. SEASON. France and West Germany: March–July. Spain: April–June. North Africa: March–June (Abs 1963). Southern USSR: from mid-April (Dementiev and Gladkov 1954a). Turkmeniya (USSR): from early March; main laying period April–June (Bel'skaya 1974). Eastern Saudi Arabia: laying begins mid-March, possibly earlier; fledged young occur from mid-April (G Bundy). SITE. On ground in the open, or in shelter of low shrub or tussock, also under low bank. Nest: shallow depression with untidy lining of grass or other vegetation; can be domed, giving shelter from sun, dome incorporating lowest branches of shrub (Clancey 1944). Average dimensions (no sample given): external diameter 135 mm (110–166); internal diameter 63 mm (44–78); depth of cup 79 (64–94); walls average 36 mm thick. New nest built for each clutch, including replacements, when nest much less solid, sometimes just walls and little or no bottom (Bel'skaya 1974). Building: by ♀, taking 1–4 days (Labitte 1957b; Abs 1963; Bub and Herroelen 1981). EGGS. See Plate 79. Sub-elliptical, smooth and glossy; off-white to grey-white, finely spotted and speckled buff-brown and grey, markings sometimes gathered into zone or cap at broad end. Nominate *cristata*: 22·9 × 16·6 (19·0–24·7 × 15·0–18·3), $n = 200$; calculated weight 3·24 g. *C. c. magna*: 22·4 × 17·1 mm (20·2–24·2 × 15·3–18·4), $n = 167$; calculated weight 3·38 g. *G. c. iwanowi*: 22·2 × 16·0 mm (20·0–24·4 × 15·0–17·3), $n = 94$; fresh weight 3·4 g (3·1–3·9), $n = 13$ (Bel'skaya 1974). Samples of other races mainly small but no apparent differences from nominate *cristata* in *meridionalis*, *cypriaca*, *riggenbachi*, or *carthaginis*. *G. c. nigricans*: 21·7 × 16·2 mm (20·2–24·2 × 15·3–18·4), $n = 22$; calculated weight 2·94 (Schönwetter 1979). *G. c. arenicola*: 21·9 × 17·0 mm (20·0–23·4 × 16·0–17·8), $n = 16$, calculated weight 3·27 (Schönwetter 1979), and 23·8 × 16·9 mm (22·1–24·9 × 16·4–17·8), $n = 21$ (Makatsch 1976). Clutch: 3–5(–7). Of 115 clutches, North Africa: 3 eggs, 20; 4, 81; 5, 14; mean 3·9 (Heim de Balsac and Mayaud 1962). Of 24 clutches, Turkmeniya: 3 eggs, 1: 4, 9; 5, 10; 6, 3; 7, 1; mean 4·75 (Bel'skaya 1974). 2–3 broods. Replacements laid. Eggs laid daily, before 07.30 hrs (Hartley 1946b). INCUBATION. 11–13 days. By ♀ only, though ♂ may stand covering eggs while ♀ away from nest (Hartley 1946*b*). Begins with last egg, or 3rd egg (Bel'skaya 1974); hatching takes place over 2 days (Hartley 1946*b*). In period of hot weather, Palestine, ♀ left eggs uncovered for *c.* 52% and 32% of observation periods of 73 and 95 min (Hartley 1946*b*). YOUNG. Altricial and nidicolous. Cared for and fed by both parents (Hartley 1946*b*; Hinsche 1960). FLEDGING TO MATURITY. Young usually leave nest at 9 days, either returning to nest over next 3 days (Dementiev and Gladkov 1954*a*) or abandoning it altogether (Hartley 1946*b*). Fledge at 15–16 days (Abs 1963; Schmied 1969); 16–18 days (Dementiev and Gladkov 1954*a*), but do not fly well until 20 days and fed until this time, perhaps longer (Pätzold 1971; A P Møller). Exact age of independence not recorded (but see Social Pattern and Behaviour). BREEDING SUCCESS. Of 36 eggs laid, Palestine, 18 hatched and 18 young were reared to fledging (Hartley 1946*b*). Of 48 eggs laid, Turkmeniya (USSR), 85·4% hatched (2·1% infertile, 12·5% perished) and 70·8% of those (62·5% of eggs laid) reared to fledging; 12·5% of nestlings died in bad weather, 9·8% were eaten by ants; in years of low rodent numbers, many eggs and young taken by predators (Bel'skaya 1974).

Plumages (nominate *cristata*). As in other Alaudidae, strongly subject to bleaching, especially in first month after moult; change in colour during winter rather slight, but more rapid again in breeding season, when abrasion is a particularly important factor. When freshly moulted in September, much more rufous on upperparts than 1 month later, almost resembling another race (Stresemann 1920); account below and text of Geographical Variation do not refer to these freshly moulted birds. ADULT. Entire upperparts olive-brown, sides of feathers slightly brighter buff (especially on crown, hindneck, and outer scapulars), feather-tips slightly more olive-grey (especially when worn); rather narrow and ill-defined dull black streaks on forehead, crown, hindneck, scapulars, back and upper tail-coverts, broader ones on mantle—these streaks hardly visible in fresh plumage (except for crown, hindneck, and mantle), but much more distinct in worn plumage. Upper tail-coverts slightly tinged rufous, especially on bases of longest coverts. Small tuft of elongated feathers on hindcrown contrastingly black, forming distinct pointed crest. Supercilium buff; rather wide above lores and behind eye, narrow and indistinct above eye. Incomplete feather-ring below eye mottled pale buff and grey, hardly contrasting with more heavily mottled lores, but bordered behind by narrow dull black line extending backwards along upper ear-coverts. Ear-coverts mottled and streaked olive-brown or black-brown and buff, brown often most intense at rear, bordered behind by almost uniform pale grey-buff or cream-buff upper sides of neck. Cheeks buff or cream with mottled dull black lines backwards from gape and along sides of chin. Central chin and throat off-white, grading to buff-white on lower throat and towards sides of neck. Lower sides of neck olive-brown with ill-defined dull black streaks. Chest and sides of breast dull pink-cinnamon or buff-brown, each feather with rather broad and ill-defined dark olive-brown or dull black triangular mark on tip; dark marks form ill-defined spotting rather than distinct streaking. Flanks and under tail-coverts pale buff-brown or dull greyish-buff, narrowly and indistinctly streaked olive-brown on flanks and longest coverts; remainder of underparts cream-buff or pale buff. Central pair of tail-feathers (t1) dark olive-brown with slightly blacker centres, faint rufous tinge on lateral bases, and narrow buff fringes (soon worn off); t2–t5 black, t5 with narrow dull rufous-buff fringe along outer web; t6 dull rufous-buff, tinged grey on inner web, with darker grey wedge along inner border, and with dark shaft. Flight-feathers and greater upper primary coverts dark olive-brown with pale cinnamon wedge on base of inner webs; narrow and ill-defined pale buff fringes along tips, broad and ill-defined fringes along outer webs of secondaries, and narrow but sharp rufous-buff to pale cinnamon fringes along outer webs of primaries (except emarginated parts) and along outer webs and tips of greater primary coverts. Lesser upper wing-coverts greyish-olive-brown (like mantle and scapulars), faintly tipped pink-buff or pale buff; median and greater upper wing-coverts dark olive-brown or brown with broad and contrasting cinnamon or buff fringe along tip; tertials dark olive-brown (darkest on centres) with rather broad cinnamon fringes at sides and tips; pale fringes of upper wing-coverts and in particular tertials soon bleach to pale buff or off-white, strongly subject to abrasion. Axillaries and under wing-coverts pale buff-brown with slight grey tinge; marginal coverts spotted dull black. During winter, rather bright olive-brown of upperparts, tertials, and t1 becomes more greyish, especially on abraded feather-tips; dark streaks more prominent; pale ground colour of supercilium, cheeks, chin, throat, and belly paler buff-white. *In heavily worn plumage*, June–August, upperparts dark olive-grey, distinctly streaked dull black on crown, mantle, scapulars, and back, with only limited amount of olive-brown visible on some feather-bases; ground-colour of sides of head and underparts dull white, dull black streaks on chest and sides of breast distinct, often joining to form larger dark patches on upper sides of chest; t6 pale grey-brown with pale buff outer web. Plumage easily discoloured by soil: birds of humid areas may have dirty grey tinge all over, especially during breeding season; birds in more arid climates tinged rufous, brown, grey, or mealy-white, depending on local soil. NESTLING. Down long and dense on upperparts, upperwing, and flanks, sparse on underparts; whitish straw-colour (Witherby *et al.* 1938*a*; Abs 1963). Down paler and denser than Skylark *Alauda arvensis*, bare skin of upperparts darker (Pätzold 1983). For development of feathers and growth, see Heinroth and Heinroth (1924–6) and Bel'skaya (1973). JUVENILE. Ground-colour of upperparts as adult, but dark feather-centres reduced or absent, except in crest; smaller feathers of forehead, hindneck, and rump with narrow white terminal fringes, longer feathers of crown, mantle, scapulars, and upper wing-coverts with broad pink-buff fringe at sides and contrastingly white spot or fringe on tip, subterminally bordered by black-brown spot or (on median and greater wing-coverts) W-mark. Sides of head as adult, but ear-coverts and cheeks faintly speckled white. Underparts paler than adult; feathers of chin, throat, and belly soft and loose, pale buff; chest and sides of breast buff with small dark brown spots; flanks and under tail-coverts buff, hardly streaked. Tertials and t1 as adult, but evenly fringed pale buff (soon bleaching and abrading); outer web and tip of t5 more extensively buff. Flight-feathers as adult, but pale fringes along outer webs broader, paler, and extending round tip, also on p10 (unlike adult), submarginally bordered by black line. FIRST ADULT. Like adult; indistinguishable when last juvenile feathers (usually p9–p10) replaced.

Bare parts. ADULT. Iris brown or dark brown. Bill horn-brown or dark brown on culmen and tip, cutting edges and base of lower mandible yellow-brown, green-horn, pale horn-grey, brownish-flesh, or pale flesh; mouth yellow-flesh. Leg and foot pale brown, brownish-flesh, or flesh-coloured, claws pale brown or horn-brown. Some variation with race, dark races having darker bare parts, pale races paler: e.g. blackish *nigricans* (Nile delta) has dark brown iris and horn-brown bill and leg, rufous *ruficolor* (Morocco) has red-brown iris, grey-brown bill, and pale brown leg, pale *magna* (central Asia) has hazel-brown iris, grey-white, green-horn, or brown-flesh bill, and very pale brown or brown-flesh leg, foot, and claws. NESTLING. Mouth drab orange-yellow with black spot inside tip of upper mandible, large bifurcated spot on tip of tongue and inside tip of lower mandible, and (according to Heinroth and Heinroth 1924-6, but no other sources) black spot on each side of lower mandible. Pattern of spots rather variable, those on tongue sometimes absent (Heinroth and Heinroth 1924-6). Gape-flanges yellow-white, dull ivory-white, or pale yellow; bare skin purple-blue, leg and foot pink-flesh with bluish scutes, claws whitish. JUVENILE. Like adult; for nominate *cristata*, black-brown bill and red-brown leg also recorded. (Hartert 1910; Meinertzhagen 1930; Witherby *et al.* 1938a; Abs 1963; Ali and Ripley 1972; BMNH, ZMA.)

Moults. Based mainly on data from USSR, Iran, and Afghanistan (Vaurie 1951a; Dementiev and Gladkov 1954a; Paludan 1959), with additional information on other races from Abs (1963), Bub and Herroelen (1981), and specimens in BMNH and ZMA. ADULT POST-BREEDING. Complete; primaries descendant. In European and Asiatic groups, moult starts with scattered feathers on head, mantle, scapulars, or sides of breast mid-June to mid-August, followed by p1 late July to late August. By mid-August to late September, body and tail largely new but outer primaries and some secondaries still growing; all moult completed late August to mid-October. In North Africa and Middle East, some birds moult at same time as those further north, but others start with p1 from mid-June and complete moult late July or later. In Mauritania, moult occurs April-June, in Chad March-August, in northern Sudan March-April. POST-JUVENILE. Complete; primaries descendant. Moult starts from age of 4-6 weeks (Heinroth and Heinroth 1924-6; Abs 1963) with p1 and scattered feathers of mantle and breast. Moulting juveniles in North Africa recorded from late May onwards, in central Europe from early June. Immatures in advanced moult (but still distinguishable from adult by retained juvenile p10 and scattered feathers of body) recorded mid-July to mid-September. Some birds still fully juvenile by late August and first days of September, and these may complete moult as late as October-November. In Mauritania, Nile valley, and Niger, moult starts about March, in Chad and Sudan from about December.

Measurements. ADULT. EUROPEAN GROUP. Nominate *cristata*. Netherlands, all year; skins (RMNH, ZMA). Bill (S) to skull, bill (N) to distal corner of nostril; exposed culmen on average 3·1 mm shorter than bill (S).

	♂			♀		
WING	107·8	(1·65; 31)	105-111	101·3	(2·97; 14)	97-106
TAIL	62·4	(1·87; 31)	59-66	58·3	(2·64; 13)	55-62
BILL (S)	19·9	(0·78; 30)	18·6-21·1	19·0	(0·85; 13)	18·1-20·1
BILL (N)	13·1	(0·61; 30)	12·2-14·1	12·6	(0·58; 14)	11·6-13·4
TARSUS	25·3	(0·65; 28)	24·3-26·5	25·0	(0·59; 14)	24·3-26·1

Sex differences significant, except tarsus.

Wing and bill (N) of ♂. (1) Nominate *cristata*, central Europe. (2) *G. c. pallida*, Iberia. (3) *G. c. kleinschmidti*, north-west Morocco. *G. c. meridionalis*: (4) western Yugoslavia, Bulgaria, and northern Greece, (5) Crete. *G. c. caucasica*: (6) Caucasus, (7) eastern Taurus mountains, Turkey. (Abs 1963.)

	WING			BILL (N)		
(1)	107·8	(2·05; 29)	104-111	13·9	(1·00; 25)	13-15
(2)	106·8	(2·72; 88)	102-114	14·3	(0·75; 88)	13-16
(3)	105·6	(2·17; 7)	102-108	14·0	(0·66; 7)	13-15
(4)	107·1	(1·53; 28)	102-110	14·2	(0·53; 28)	13-15
(5)	106·2	(2·05; 25)	102-111	13·3	(0·55; 25)	12·5-14·5
(6)	108·6	(2·12; 13)	103-113	13·9	(0·90; 13)	12·5-15
(7)	105·6	(1·95; 14)	102-108	14·2	(0·71; 14)	13-15

Wing and bill (S) of ♂. *G. c. meridionalis*: (1) mainland Greece, (2) Cyclades, (3) Crete, (4) western and southern Asia Minor (east to Mersin). *G. c. cypriaca*: (5) Karpathos and Rhodos, (6) Cyprus. (7) *G. c. caucasica*: islands in northern Aegean Sea (Samothraki to Samos). (Watson 1962.)

	WING			BILL (S)		
(1)	104·9	(17)	99-109	20·1	(17)	19-21
(2)	102·4	(26)	97-106	19·9	(26)	18·5-21·5
(3)	102·1	(8)	99-105	20·3	(8)	19·7-21
(4)	103·8	(21)	99-109	19·8	(21)	18·5-20·5
(5)	103·0	(5)	99-105	20·5	(5)	20-21
(6)	102·1	(7)	99-104	19·3	(7)	18·5-20·7
(7)	103·9	(14)	101-106	20·2	(14)	19-21

Some other samples. *G. c. meridionalis*: (1) southern Yugoslavia (Stresemann 1920), (2) near Izmir, Turkey (Weigold 1914). *G. c. tenuirostris*: (3) Volga-Ural steppes, European USSR (Vaurie 1951a; RMNH, ZMA). *G. c. caucasica*: (4) Taurus mountains, Turkey (Jordans and Steinbacher 1948; Kumerloeve 1961, 1968).

	♂			♀		
(1) WING	108·0	(2·34; 39)	101-115	100·6	(2·15; 44)	96-105
(2) WING	106·4	(2·77; 8)	101-110	98·4	(2·37; 7)	95-102
(3) WING	109·2	(2·39; 5)	105-111	101·8	(2·02; 5)	100-105
BILL (S)	20·1	(0·63; 5)	19·5- 21	19·4	(0·58; 5)	19·0-20·4
(4) WING	105·3	(— ; 16)	102-109	99·4	(— ; 10)	94-106

G. c. neumanni. Wing ♂ 104·2 (6) (Abs 1963). Wing ♂ 105·8 (3) 105-108, ♀ 99·3 (3) 98-100 (RMNH, ZMA).

G. c. apuliae. Wing ♂ 109·2 (6) (Abs 1963). Wing ♂ 106·5 (3) 104-110, ♀♀ 100, 106 (RMNH, ZMA).

G. c. tenuirostris. Crimea, 14 birds: wing 92-106, bill (S) 17-20·4, bill (N) 11·8-15·2 (Hartert and Steinbacher 1932-8).

NORTH-WEST AFRICAN GROUP. *G. c. arenicola*. North-east Algerian Sahara and southern Tunisia, all year; skins (RMNH, ZMA).

	♂			♀		
WING	109·8	(1·75; 11)	107-112	103·7	(1·56; 8)	101-106
TAIL	62·2	(1·42; 10)	60-64	58·4	(1·86; 8)	56-61
BILL (S)	22·5	(0·81; 11)	21·3-24·0	21·4	(1·04; 8)	20·4-22·7
BILL (N)	15·4	(0·60; 11)	14·5-16·4	14·8	(0·58; 8)	14·0-15·7
TARSUS	25·9	(0·55; 11)	25·1-26·9	25·8	(0·72; 8)	24·7-26·6

Sex differences significant, except for tarsus.

Wing and bill (N) of ♂: (1) *carthaginis*, coastal northern Algeria and Tunisia; (2) *randonii*, Hauts Plateaux of Algeria; (3) *macrorhyncha*, north-west Algerian Sahara and bordering Atlas Saharien; (4) *arenicola*, north-east Algerian Sahara and Tunisia south from Gafsa; (5) *festae*, Cyrenaica (Abs 1963).

	WING			BILL (N)		
(1)	109·7	(2·81; 12)	105-113	15·7	(0·90; 12)	13·2- 16
(2)	115·7	(1·64; 6)	113-117	17·3	(1·64; 6)	16-19·5
(3)	109·3	(2·97; 10)	105-114	16·3	(0·85; 10)	15-18
(4)	107·8	(2·56; 33)	104-112	16·0	(0·76; 31)	14-17·5
(5)	107·1	(1·98; 15)	105-109	14·7	(0·58; 15)	13·5-15·5

Wing and bill (S) of ♂: (1) *riggenbachi*, (2) *carthaginis*, (3) *randonii*, (4) *macrorhyncha*, (5) *arenicola* (Vaurie 1959).

	WING			BILL (S)		
(1)	109	(5)	104-114	22	(5)	21-22

(2)	110	(5) 108-111	21	(5) 21-22
(3)	117	(5) 115-119	24	(5) 23-25
(4)	111	(5) 107-114	24·5	(5) 23-26
(5)	110	(5) 106-112	22	(5) 21-23

G. c. riggenbachi. Western Morocco, average of 5 ♂♂: wing 107·4, bill (N) 14·6 (Abs 1963). Morocco: wing ♂ 107-113, ♀ 98-102; bill (S) ♂ (19) 20-22, ♀ 16-18 (Hartert 1910).

G. c. carthaginis. Northern Tunisia: wing ♂ 108·3 (3) 107-110, ♀ 104·1 (4) 102-107; bill (S) ♂ 21·6 (3) 21-22, ♀ 22·2 (4) 22-23; bill (N) ♂ 14·9 (3) 14·4-15·3, ♀ 15·2 (4) 14·9-15·6; tail 57-65, tarsus 25-28 (RMNH, ZMA).

G. c. randonii. Hauts Plateaux (Algeria) and Missour (Morocco): wing ♂ 115·8 (0·94; 9) 114-117, ♀ 109·8 (3) 107-112; bill (S) ♀ 22·6, bill (N) ♀ 15·4 (Hartert and Steinbacher 1932-8; RMNH, ZMA). Marrakech (intermediate with *riggenbachi*), average of 4 ♂♂: wing 112·5, bill (N) 16·8 (Abs 1963).

G. c. macrorhyncha. North-west Algerian Sahara: wing ♂ 110-118 (16), ♀ 102-106(-110) (8) (Hartert 1921-2).

Tail of *randonii* and *macrorhyncha* 61-69, tarsus 26-29 (RMNH, ZMA).

G. c. helenae. Illizi (south-east Algeria), 3 unsexed birds: wing 104-107, bill (S) 18-19 (Hartert and Steinbacher 1932-8). This race or related *jordansi*, Ennedi (north-east Chad): wing ♂♂ 103, 106, ♀♀ 95, 97 (Niethammer 1955b); bill (N) ♂♂ 14·3, 16 (Abs 1963).

G. c. festae. Cyrenaica: wing ♂ 105-109 (15), ♀ 98-102 (11); bill (S) both sexes (20-)21-22(-22·5) (Hartert and Steinbacher 1932-8).

SUDAN-ARABIAN GROUP. *G. c. brachyura.* Jordan valley (Vaurie 1951a).

WING	♂ 105·9	(10)	101-109	♀ 101	(5) 99-104
TAIL	57·5	(10)	54-60	56·0	(5) 55-58
BILL (S)	20·0	(10)	17-21·5	20·4	(5) 19-21

Wing and bill (N) of ♂: (1) *nigricans*, Nile delta; (2) *maculata*, Nile valley (Cairo to Aswan) and El Faiyum; (3) *halfae*, Wadi Halfa area (northern Sudan); (4) *brachyura*, coastal north-west Egypt, Sinai, and Dead Sea region (Abs 1963).

	WING		BILL (N)	
(1)	103·4 (2·49; 12)	99-106	13·9 (1·18; 12)	12·5-17
(2)	105·1 (2·45; 24)	102-110	14·0 (0·83; 24)	13-15·5
(3)	102·7 (2·18; 10)	100-107	13·3 (0·70; 10)	12-14
(4)	106·3 (1·80; 10)	104-110	14·5 (0·44; 10)	14-15

G. c. nigricans. Range of 22 birds: wing ♂ 104-107, ♀ 97-100 (Hartert 1910).

G. c. maculata. El Fayium: wing, ♂ 100-106 (41), ♀ 92-100 (19) (Hartert 1921-2).

G. c. brachyura. Sexes combined: northern Saudi Arabia and south-west Iraq (n=5), wing 97-114, bill (S) 19-21; Transjordania (n=12), wing 99-115, bill (S) 18-22; Dead Sea region (n=7), wing 99-114, bill (S) 17-20; Sinai (n=14), wing 103-111, bill (S) 16-21; Wadi el Natrun (n=17), wing 101-108, bill (S) 20-22; coastal north-west Egypt (n=22), wing 103-114, bill (S) 20-22 (Meinertzhagen 1930).

G. c. cinnamomina. North-west Israel and south-west Lebanon: wing ♂ 100-109, ♀ 97-99 (Hartert 1921-2).

G. c. zion. North-east Israel, northern and eastern Lebanon, and Syria (n=44): wing ♂ 98-109, ♀ 96-104 (Hartert 1921-2). Amik Gölü and Ceylanpinar (southern Turkey near Syrian border): ♂ 106·5 (11) 104-109, ♀♀ 100, 100 (Kumerloeve 1961, 1963, 1970b).

ASIATIC GROUP. *G. c. subtaurica.* Eastern Iraq and north-west Iran (Kermanshah and Azerbaijan) (Vaurie 1951a).

WING	♂ 110·6	(20)	107-117	♀ 105·4 (10)	100-115
TAIL	62·4	(20)	58-66	60·2 (10)	56-70

BILL (S) 21·2 (20) 20-22·5 20·2 (10) 19-21·5

♂♂ from Ankara and Konya (Anatolian plateau, Turkey): wing 110·4 (14) 107-113, bill (S) 20·6 (14) 18·5-22 (Watson 1962). Anatolian plateau from Elmali and Burdur north to Bolu, east to Eregli and Tuz Gölü: wing ♂ 110·2 (29) 104-116, ♀ 103·9 (16) 99-108 (Jordans and Steinbacher 1948; Vaurie 1951a; Kumerloeve 1961; Rokitansky and Schifter 1971; RMNH). Area round Van Gölü (south-east Turkey): wing ♂ 108·8 (1·78; 13) 106-112, ♀♀ 100, 101 (Kumerloeve 1968, 1969a). Elburz mountains (northern Iran): wing ♂♂ 107, 108, ♀ 101·0 (3) 98-103 (Paludan 1940; Schüz 1959).

G. c. magna. Sinkiang (China): wing ♂ 116·6 (9) 114-121, ♀♀ 106, 108; bill (S) ♂ 22·9 (9) 22-24, ♀♀ 21·3, 21·4 (Vaurie 1951a; ZMA).

G. c. iwanowi. Northern Iran and southern USSR from southern Turkmeniya to eastern Kazakstan: wing ♂ 113·9 (23) 110-121, ♀ 107·4 (10) 101-112; bill (S) ♂ 22·3 (23) 21-24, ♀ 21·7 (10) 20·5-23 (Vaurie 1951a).

JUVENILE. Wing c. 3 shorter than adult, tail c. 5 shorter; tarsus full-grown from shortly after fledging, bill when post-juvenile moult completed.

Weights. ADULT. Nominate *cristata*. East Germany: ♂ 45·2 (3·35; 17) 40-52, ♀ 44·1 (1·84; 11) 40-47; lean and wounded ♂♂ 33, 37·5; laying ♀ 55 (Bub and Herroelen 1981). North-east France, winter: ♂ 40-48 (5), ♀ 42-46(3) (Bub and Herroelen 1981). France: ♂ 40·8 (1·89; 5) 38-43, ♀ 40·7 (3·77; 9) 37-48 (Mayaud 1931). Central Europe: ♂ 45·9 (2·70; 12) 42-50 (Abs 1963).

G. c. pallida. Mainly from breeding season, Spain: ♂ 41·4 (3·00; 20) 36-47 (Abs 1963).

G. c. meridionalis. Greece and European Turkey, May: ♂♂ 40·2 (3) 37·5-42 (Niethammer 1943; Rokitansky and Schifter 1971). Southern Yugoslavia, Bulgaria, and northern Greece: ♂ 41·5 (2·50; 6) 39-45 (Abs 1963). Mainland Greece, ♂ 47·3 (17) 41-53; Cyclades, ♂ 40·2 (26) 36-45; Crete, ♂ 40·6 (8) 37-44; western and southern Asia Minor (east to Mersin), ♂ 41·0 (21) 36-46 (Watson 1962).

G. c. cypriaca. Rhodos and Karpathos: ♂ 38·9 (3) 36-42 (Watson 1962). Cyprus: probable ♀, October, 36 (Hallchurch 1981).

G. c. caucasica. Eastern Taurus mountains (Turkey): ♂ 35·5 (2·36; 14) 32-40 (Abs 1963); May, ♂ 42; November, ♂ 37 (Kumerloeve 1968). Islands in northern Aegean Sea: ♂ 41·4 (14) 39-46 (Watson 1962).

G. c. randonii. Hauts Plateaux (Algeria): ♂ 46-54 (Niethammer 1955b); ♂ 52 (Abs 1963).

G. c. macrorhyncha. North-west Algerian Sahara: ♂ 37-42 (16) (Niethammer 1955b); ♂ 47·6 (3·14; 8) 44-54 (Abs 1963). Figuig (south-east Morocco), March-April: probable ♂ 35, probable ♀ 37 (BTO).

G. c. helenae or closely related *jordansi*. North-west Niger (near Algerian border), January: ♂ 39·7 (3) 37-42 (Niethammer 1955b; Abs 1963). Ennedi (northern Chad), April: ♀ 38 (Niethammer 1955b).

G. c. brachyura. Azraq (Jordan), April-May: probable ♂♂ 36, 36; probable ♀♀ 36·0 (3) 35-39·5 (BTO).

G. c. zion. Southern Turkey near Syrian border, May: ♂ 38; range of 3 ♂♂ and 1 ♀, 39-47 (Kumerloeve 1963, 1970b).

G. c. subtaurica. Ankara and Konya, Anatolian plateau (Turkey): ♂ 46·4 (14) 40-52 (Watson 1962). Western Anatolia, July: ♂♂ 47·2, 36·2; ♀♀ 38·5, 36·2 (Rokitansky and Schifter 1971). Van Gölü (south-east Turkey), late May to early July: ♂ 42·4 (3·42; 14) 38-48, ♀♀ 39, 44 (Kumerloeve 1968, 1969a).

PLATE 1. *Eremopterix nigriceps* Black-crowned Finch Lark (p. 46). *E. n. albifrons*: **1** ad ♂ fresh (autumn), **2** ad ♂ worn (spring), **3** ad ♀ fresh (autumn), **4** ad ♀ worn (spring), **5** juv. Nominate *nigriceps*: **6** ad ♂ fresh (autumn), **7** ad ♀ fresh (autumn). *E. n. melanauchen*: **8** ad ♂ fresh (autumn), **9** ad ♀ fresh (autumn). (DIMW)

PLATE 2. *Alaemon alaudipes* Hoopoe Lark (p. 74). Nominate *alaudipes*: **1** ad rufous morph fresh (autumn), **2** ad rufous morph worn (spring), **3** juv rufous morph. *A. a. doriae*: **4** ad fresh (autumn). *A. a. boavistae*: **5** ad rufous morph fresh (autumn). *Rhamphocoris clotbey* Thick-billed Lark (p. 87): **6** ad ♂ fresh (autumn), **7** ad ♂ worn (spring), **8** ad ♀ partly worn, **9** juv. (DIMW)

PLATE 3. *Eremalauda dunni* Dunn's Lark (p. 54). *E. d. eremodites*: **1** ad fresh (autumn), **2** ad worn (spring), **3** juv. Nominate *dunni*: **4** ad fresh (autumn), **5** ad worn (spring).　　*Ammomanes cincturus* Bar-tailed Desert Lark (p. 59). *A. c. arenicolor*: **6** ad fresh (autumn), **7** ad worn (spring), **8** juv. Nominate *cincturus*: **9** ad fresh (autumn). (DIMW)

PLATE 4. *Ammomanes deserti* Desert Lark (p. 65). *A. d. payni*: **1** ad fresh (autumn), **2** ad worn (spring), **3** juv. *A. d. algeriensis*: **4** ad fresh (autumn), **5** ad worn (spring). Nominate *deserti*: **6** ad fresh (autumn). *A. d. isabellinus*: **7** ad fresh (autumn), **8** ad worn (spring). *A. d. annae*: **9** ad fresh (autumn). (DIMW)

Zagros mountains (western Iran), March–May: ♂ 38·0 (2·44; 6) 35–41, ♀♀ 35, 36·6 (Paludan 1938). Elburz mountains (northern Iran): March, ♂ 42, ♀♀ 37, 40 (Schüz 1959); July, ♂ 36·2, ♀ 35·5 (Paludan 1940).

G. c. iwanowi. Afghanistan: early March to mid-April, ♂ 39·0 (18) 31–43, ♀ 41·9 (9) 37–51; late April to July, ♂ 38·2 (2·45; 13)34–43, ♀ 34·8 (3·65; 8) 32–41; September–October, ♂ 43·7 (3) 43–44, ♀ 37·0 (2·12; 5) 34–39 (Paludan 1959).

JUVENILE. 72% of average adult weight reached on leaving nest (at 9–10 days); similar to adult from *c.* 3 weeks (Abs 1963).

Structure. Wing rather short, broad at base, tip rounded. 10 primaries. In adult, p7-p8 longest or either one 0–1 shorter than other, p9 2–4 (1–5) shorter than longest, p10 58–68 (♂♂ of small and intermediate sized races), 64–74 (♂♂ of *randonii*, *magna*, and other large races), 55–63 (♀♀ of smaller races), or 61–66 (♀♀ of larger races); p6 0–2 shorter than wing-tip, p5 4–9, p4 16–20 (smaller ♂♂), 18–23 (larger ♂♂), 14–17 (smaller ♀♀), or 15–19 (larger ♀♀), p1 24–29 (smaller ♂♂), 28–34 (larger ♂♂), 21–26 (smaller ♀♀), or 26–30 (larger ♀♀). In juvenile, p8 longest, p9 1–3 shorter, p10 48–64, p7 0–3, p6 1–5, p5 6–12, p4 15–19, p1 24–35 (n = 3). Adult p10 reduced, narrow, with pointed tip; 4·9 (72) 3–8 shorter than longest upper primary covert in *pallida*, *meridionalis*, and nominate *cristata*, 3·2 (31) 0–6 shorter in North African races, 4·3 (29) 2–8 in Asiatic races (exceptionally, 1 longer). Juvenile p10 broader, longer, and with rounded tip; 2 shorter to 6 longer than longest covert (average 1·5, n = 6). See also Structure and Recognition of Thekla Lark *G. theklae* (pp. 172–3). Outer web of p5-p8 and inner of p6-p9 emarginated. Longest tertials reach to p5-p6 in adult, to about p4 in juvenile. Tail rather short, tip square; 12 feathers. Bill rather long, relatively slender; about half length of head in most races, but almost as long as head in some North African ones; culmen distinctly decurved towards pointed tip, cutting edges slightly decurved, lower mandible straight, but in North African and Asiatic groups whole bill slightly decurved. Nostril covered by small tuft of feathers, ending in fine bristles; some more fine bristles along base of upper mandible and on chin. Feathers of central hindcrown elongated, forming spike-like crest; longest feathers 19·5 (15) 16–23 in adult ♂, 19·1 (10) 16–24 in adult ♀, 11–20 in juvenile; rather broad and with rounded tip when plumage fresh, narrow and pointed when worn. Tarsus and toes rather short and slender. Middle toe with claw 19·6 (54) 18–22 (all races); outer toe with claw *c.* 68% of middle with claw, inner *c.* 69%, hind *c.* 107%. Front claws rather short and slender, slightly decurved; hind claw rather long, straight or slightly decurved 11·5 (70) 9–14 (all races), range exceptionally 5·1–17·4.

Geographical variation. Marked and complex, involving mainly ground-colour and intensity of streaking, less so size

PLATE 7 (*facing*).
Calandrella brachydactyla Short-toed Lark (p. 123). Nominate *brachydactyla*: **1** ad fresh (autumn), **2** ad worn (spring), **3** juv. *C. b. rubiginosa*: **4** ad fresh (autumn). *C. b. longipennis*: **5** 1st autumn. *C. b. artemisiana*: **6** ad fresh (autumn), **7** ad worn (spring). *Calandrella rufescens* Lesser Short-toed Lark (p. 135). *C. r. minor*: **8** ad fresh (autumn), **9** ad worn (spring), **10** juv. *C. r. nicolli*: **11** ad fresh (autumn). *C. r. apetzii*: **12** ad fresh (autumn). *C. r. heinei*: **13** ad fresh (autumn), **14** ad worn (spring). *C. r. aharonii*: **15** ad fresh (autumn). *C. r. persica*: **16** ad fresh (autumn). Nominate *rufescens*: **17** ad fresh (autumn). *C. r. polatzeki*: **18** ad fresh (autumn). (DIMW)

and bill-shape. In resident populations, colour variation mainly related to aridity or humidity of environment and amount of sunshine, but also to local variations in colour of soil; in migrant populations (mainly those of USSR), colour rather constant over large areas despite distinct local variations in soil colour (Vaurie 1951a). Following account based mainly on Vaurie (1951a) for Iran to central Asia, Dementiev and Gladkov (1954a) for USSR, Watson (1962) for north-east Mediterranean, Abs (1963) for Europe and Africa, and Sasvári-Schäfer (1966) for Balkans, with additional information from Hartert (1921–2), Hartert and Steinbacher (1932–8), Kumerloeve (1961, 1963, 1968, 1969a, 1970b), large collection of specimens in BMNH, and smaller ones in RMNH and ZMA. Differences between races slight, often visible only in series of skins. Generally, no regard given here to very local so-called 'aberrant soil variations' (populations with aberrant colour and with very restricted range occurring on aberrantly coloured soil within range of widespread race), e.g. ash-grey birds of eastern Calabria, Italy (named *heraelaciniae* Foschi, 1978) within range of olive-grey *apuliae*; rufous-brown birds from Drama (northern Greece) within range of greyish-brown *meridionalis*; pale sandy-grey birds from Milos within light grey-brown *meridionalis* of remainder of Cyclades; rather local rufous *cinnamomina* of coastal Levant perhaps also better considered an aberrant soil variant of more widespread pale grey-cinnamon *zion*, as size similar. 5 main groups of races recognized here (not necessarily similar to groupings by previous authors), each with certain size characteristics, but some races transitional between groups; all races of Sahel group (Sénégal and southern Mauritania to Sudan) extralimital.

EUROPEAN GROUP (northern Morocco and Europe east to Ural river, Caspian Sea, and Asia Minor, except for Anatolian plateau and Turkey south-east of Taurus mountains). Wing of intermediate length, bill rather short: average wing of ♂ 104–110, average bill (to skull) of ♂ 19·5–21·0. Nominate *cristata* rather dark grey-brown or olive-brown on upperparts, streaks on upperparts and chest broad and black but poorly defined. *G. c. pallida* from Iberia averages paler on upperparts than nominate *cristata*, dark streaks paler and narrower, not markedly contrasting, ground-colour of upperparts rather pale olive-brown with variable buff tinge; chest with rather heavy but ill-defined streaks on cream-buff ground-colour. Pronounced local variation: birds from higher parts of central Spain generally larger and paler than those of Portugal and coastal Spain; those of southern Spain particularly small and red-brown, tending towards next race (see Abs 1963 for details). *G. c. kleinschmidti* from north-west Morocco belongs to European group, not to north-west African group, though intergrading with races *riggenbachi* and *carthaginis* of that group; closely similar to nominate *cristata* (and included in that race by Vaurie 1959), but wing shorter, upperparts darker greyish-olive-brown with more contrasting black streaks, chest rather heavily and contrastingly streaked black on pale cinnamon ground. *G. c. neumanni* restricted to western Italy in Toscana and Rome area; wing shorter than nominate *cristata*, upperparts rufous-brown (less olivaceous-grey) with more distinct streaks, chest more sharply marked on deeper buff ground. *G. c. apuliae* from south-east and southern Italy and Sicily close to nominate *cristata*, but upperparts on average paler and greyer, dark olive-grey with broad but ill-defined streaks; chest pink-buff with rather sharp and heavy streaks. *G. c. meridionalis* from Dalmatia and Danube river south to Cyclades, Crete, and westernmost Asia Minor darker than nominate *cristata*, but with similar rather heavy but ill-defined streaks on upperparts and chest. Birds from Dalmatia, Bulgaria, and northern Greece typically tawny-olive or brown with slight rufous tinge on upperparts. Towards south, ground-colour gradually paler and size slightly smaller: birds from central Greece and western Asia Minor greyish-brown above, light buff below; those of Crete sandy grey-brown above, cream-buff below. Distribution interrupted in North Aegean Sea, where darker *caucasica* breeds from Samothraki to Samos; *meridionalis* gradually merges into *zion* east from about Elmali on coast south of Taurus mountains, but apparently no intergradation with larger and paler *subtaurica* from Anatolian plateau. *G. c. cypriaca* from Karpathos, Rhodos, and Cyprus is paler and more buffish above than *meridionalis*; rather like *pallida* but with heavier streaks on upperparts, ground-colour of upperparts pale sandy-grey; bill of Cyprus birds short and thin; chest cream-buff with rather sharp and heavy streaks. *G. c. tenuirostris* from eastern Hungary and Rumania east to Ural river slightly paler and colder grey than nominate *cristata*, dark streaks on upperparts and chest somewhat better defined; ground-colour of upperparts dark brownish-grey with buff tinge, less warm olive-brown; wing on average slightly longer than nominate *cristata*, bill longer and often thinner. *G. c. caucasica* similar in colour to *tenuirostris* and sometimes considered inseparable, but bill usually shorter and heavier; size gradually smaller from Caucasus westwards through western Transcaucasia and Taurus mountains (Turkey); in southern Taurus, occurs west to Solak (Jordans and Steinbacher 1948), and reappears on islands of North Aegean Sea, where small like neighbouring *meridionalis*, but with darker brownish-grey upperparts and with heavier and more distinct black streaks above and on chest (Watson 1962).

NORTH-WEST AFRICAN GROUP (western Morocco to Libya, south to northern parts of Mauritania, Niger, and Chad). Wing rather short (in southern Sahara) to long (on Algerian Hauts Plateaux), bill longer than other races of similar size. Bill relatively shortest in *riggenbachi* from western Morocco and in *festae* from Cyrenaica, which are connecting links with short-billed European and Sudan-Arabian groups respectively. *G. c. riggenbachi* rich cinnamon-brown on upperparts, cinnamon-buff on underparts; sharp and heavy black streaks on crown, mantle, and chest; not as dull brown as *kleinschmidti* from northern Morocco, with which it intergrades between Rabat and Casablanca; extends east to foothills of Atlas; in Atlas proper replaced by much larger birds (e.g. above Marrakech) similar to or intergrading with *randonii*. *G. c. carthaginis* from plains along northern coast between Oujda (north-east Morocco) and Sousse (Tunisia) rather warm colour on upperparts, as in *riggenbachi*, but tinge more buff-brown, less cinnamon, and rump more olive, dark streaks narrower and less sharp; underparts paler, buff, with narrower black streaks; wing and bill slightly longer than in typical *riggenbachi*. Uncommon *randonii* from Hauts Plateaux of Algeria west to upper Moulouya valley in eastern Morocco similar to *carthaginis*, but much larger and streaks on upperparts heavier and sharper; upperparts as *riggenbachi*, but ground-colour pinkish-cinnamon-buff with more contrasting streaks, underparts paler buff and with narrower streaks on chest. *G. c. macrorhyncha* has bill almost as long as *randonii*, and ground-colour of upperparts pinkish-cinnamon-buff as in that race, but streaks much paler olive-brown or pale rufous-brown, entire upperparts appearing more uniform sandy-brown; underparts cream-buff, chest-streaks narrow and grey (contrastingly black in *randonii*); occurs in Sahara of north-west Algeria and in Atlas Saharien, east to Laghouat and Ghardaïa west to Béchar, but also in Atar in Mauritania and perhaps inhabits intervening areas south along Anti-Atlas. *G. c. arenicola* from north-east Algerian Sahara (east from Ouargla and Biskra), southern Tunisia (south from Gafsa and Sfax), and north-west Libya closely similar to *macrorhyncha*, showing similarly pale cinnamon upperparts and narrowly and faintly streaked chest, but smaller,

in particular bill shorter. *G. c. festae* from Cyrenaica (Libya) is cinnamon-rufous on upperparts, marked with rather narrow and sharp dark streaks, cinnamon-buff on underparts, marked with rather heavy black streaks on chest; slightly darker and better marked than *arenicola*, paler cinnamon and with less heavy streaks than *riggenbachi* and *carthaginis*. Races inhabiting southern Sahara little known: *helenae* occurs Illizi (Fort Polignac, south-east Algeria) and birds in neighbouring south-west Libya are probably this race; upperparts almost uniform red-brown, underparts rufous-buff with rather large dark brown spots on chest; *jordansi* from northern Niger (north-west of Agadez) closely similar to *helenae* on upperparts, but chest faintly and narrowly spotted only; birds similar to *helenae* or *jordansi* occur in Ennedi (northern Chad) and perhaps also Tibesti.

SUDAN-ARABIAN GROUP. Wing short or rather short, bill short, both similar in size to European group or slightly smaller. Variation marked but very gradual, and boundaries between most races difficult to establish; in Nile valley, gradually paler towards south and from valley floor towards bordering deserts. *G. c. nigricans* in Nile delta darkest race of the species; upperparts dark olive-grey with heavy and contrasting brown-black streaks, underparts pale cinnamon with broad sharp black streaks on chest, extending to breast and flanks. *G. c. maculata* from borders of delta, El Faiyum, and Nile valley between Cairo and Aswan rather dark olive-grey above, as in *nigricans*, but streaks narrower and browner, still strongly contrasting; chest-spots slightly smaller and less extensive; upperparts paler, more buff, and streaks narrower towards south, merging into small race *halfae* south of Aswan, which is slightly paler and greyer than *maculata* and which is in turn replaced by paler sandy-brown *altirostris* in Nile valley between Dongola and Berber (northern Sudan). *G. c. brachyura* from inland Cyrenaica (Stanford 1954), north-west Egypt (east to Alexandria), Sinai, Negev, and Dead Sea region east through northern Arabia to southern and western Iraq (north-east to Baghdad area) sometimes united into single race with *altirostris* (e.g. by Vaurie 1959), but upperparts pale olive-grey with buff tinge, greyer and less brown than *altirostris*, less yellowish-sandy than *arenicola*; dark streaks on upperparts and chest narrow, rather distinct in northern Egypt and Dead Sea region, faint in interior of Saudi Arabia and Kuwait. *G. c. cinnamomina* from north-west Israel and south-west Lebanon (Mount Carmel to Beirut) rather deep sandy-cinnamon on upperparts, narrowly streaked brown, chest with heavier and blacker streaks than *brachyura*. *G. c. zion* from Jerusalem and Jebel ed Druz north through eastern Lebanon and Syria to Ceylanpinar (southern Turkey near Syrian border) has similar markings, but ground-colour of upperparts paler, pink-cinnamon or pale cinnamon-grey; black chest-spots narrow but sharp, streaks on upperparts vary from rather broad and brown-black in Jerusalem area to faint and grey in eastern Syria, where *zion* perhaps intergrades with *subtaurica* of northern Iraq; probably intergrades with greyish-brown *meridionalis* in southern Turkey between Elmali and Amik Gölü.

ASIATIC GROUP (Anatolian plateau and eastern Iraq, east through Kazakhstan and Iran to central and eastern Asia). Wing long and bill decurved, as in north-west Africa, but bill relatively shorter; populations of Indian subcontinent and east Asia much smaller; however, some with rather long bill. *G. c. subtaurica* inhabits Asia Minor, but not in north-east and in Anadolu and Taurus mountains, where *caucasica* occurs, nor in extreme west and south, where replaced by *meridionalis* and *zion*; further east, occurs southern Transcaucasia, north-west Iran, and southern shores of Caspian Sea from Muganskaya steppe (Azerbaydzhan) to Kara Bogaz Gol (Turkmeniya), intergrading with more eastern race *iwanowi* in southern Zagros mountains (Iran) and

Kopet Dag (north-east Iran and Turkmeniya); in south, distributed to Mosul area (Iraq) and south-east Iraq. Upperparts of *subtaurica* pale olive-grey, pale sandy-grey, or isabelline-grey, rather narrowly but distinctly streaked brown, underparts cream with rather heavy black spots on chest; greyer and more uniform above than *zion*, less sandy than *brachyura*, paler and less well-marked than *caucasica*, and larger than any of these. *G. c. iwanowi* from central Iran and central Turkmeniya east to north-west Pakistan and southern Tadzhikistan is warmer buff on upperparts than *subtaurica*, dark streaks narrow and sharp, chest cream-buff with narrow and sharp dark streaks; some populations more sandy-buff ('*vamberyi*' Härms, 1907), others greyer above and whiter below (typical *iwanowi*). North of *iwanowi*, range of *magna* extends from shores of north-east Caspian Sea to south-east Kazakhstan and western China; upperparts soft pale buff, hardly streaked; underparts cream-buff with narrow and restricted shaft-streaks. For extralimital races, see Abs (1963) for Africa, Meinertzhagen (1954) for Arabia, and Vaurie (1951a) for Asia.

Sometimes included in genus *Alauda*, but see Voous (1977).

CSR

Galerida theklae Thekla Lark

PLATES 8 and 13
[facing pages 184 and 185]

Du. Thekla Leeuwerik   Fr. Cochevis de Thékla   Ge. Theklalerche
Ru. Короткоклювый хохлатый жаворонок   Sp. Cogujada montesina   Sw. Lagerlärka

Galerita Theklae C L Brehm, 1858

Polytypic. Nominate *theklae* C L Brehm, 1858, southern France and Iberia (including Balearic islands); *erlangeri* Hartert, 1904, northern Morocco from Tanger south to Oulmès and Azrou, east to Algerian border; *ruficolor* Whitaker, 1898, Morocco south of *erlangeri*, south to High Atlas, and coastal strip of northern Algeria and northern Tunisia, in latter south to *c.* 36°N; *aguirrei* Cabrera, 1922, Morocco south from Anti-Atlas to northern Mauritania; *superflua* Hartert, 1897, Hauts Plateaux and Atlas Saharien in Algeria and neighbouring eastern Morocco west to middle Moulouya river, east to central Tunisia (Sousse to Gafsa and Gabès); *carolinae* Erlanger, 1897, north Algerian Sahara, west to Figuig (Morocco), east through Chott Melrhir area and southern Tunisia to northern Libya and north-west Egypt. Extralimital: 6 races in north-east Africa.

Field characters. 17 cm; wing-span 28–32 cm. Same size as Crested Lark *G. cristata* but with taller, more fan-like crest, rather slighter form, longer legs, and (in Europe) shorter bill. Medium-sized lark, closely similar to *G. cristata* and differing constantly only in structure, heavier marks on chest, on and around cheeks, and on hindneck and upper mantle, and (in Europe) grey underwing. Usually restricted to inclines, rocky areas, and interfaces of scrub or heath with grassy plains or cereal plots. Sexes similar; little seasonal variation. Juvenile separable. 6 races in west Palearctic; cline obvious, with Iberian form easily distinguished from all but north Moroccan form (see Geographical Variation); only nominate *theklae* described here.

Adult. Ground colour of plumage not so uniform as *G. cristata*; upperparts appear both greyer and darker (darkness due to more prominent black-brown feather-centres) and underparts show darker chest and (by contrast) paler belly. At distance, noticeable features are dark crest, obvious spots on chest, and dark-streaked sides of lower neck and upper mantle. At close range, form and plumage exhibit the following characters which allow distinction from *G. cristata cristata*: (1) 10% shorter bill, with almost straight culmen conveying tapering and spiky, not long and decurved form; (2) fuller crest rising from centre of crown, not just at rear; (3) more distinct black-brown centres to feathers of crown, nape, hind- and side-neck and upper back; (4) darker cheeks, with streaks visible on at least lower ear-coverts; (5) darker, indistinctly streaked throat; (6) more conspicuous and noticeably splashed pattern of bold drop-shaped spots on chest (recalling that of Song Thrush *Turdus philomelos*), extending towards upper flanks and along lower edge of streaked collar under cheeks and terminating sharply across chest (recalling pattern of Skylark *Alauda arvensis* more than loose distribution of streaks on *G. cristata*); (7) rather more contrast between margins and centres of wing-coverts and tertials; (8) rump contrastingly rufous. In flight, most characters blur and bird then indistinguishable from dark individuals of *G. cristata*, except when brown-grey under wing-coverts and axillaries show; in strong illumination, underwing may even appear silvery (quite unlike *G. cristata*). With wear, bird becomes darker as buff-brown margins to many feathers become narrower, but dull uniformity of *G. cristata* apparently does not result. Juvenile. Easily distinguished from adult by shorter crest and white-spotted upperparts but easily confused with similarly aged *G. cristata*, which also shows spots on chest (unlike adult). At all ages, bare-part colours similar to *G. cristata*.

In areas also inhabited by *G. cristata* (applies to all parts of range in west Palearctic), *G. theklae* is always a bird to be worked at hard rather than merely identified. As already made clear, differences between *G. theklae* and *G. cristata* neither striking nor easy to see and much care needed to allow for geographical variation in plumage within the 2 species. Since adult's chest pattern is a much used character, important to realise that forms of *G. cristata* with heavy chest-streaks are restricted to Turkey and highlands of Israel, Jordan, and Syria where *G. theklae* does not occur; conversely, forms of *G. theklae* with less heavy streaks occur widely in east, central, and southern Morocco and in coastal districts of Algeria, and these, with their more rufous or paler, sandier plumage tones, invite serious confusion with sympatric races of *G. cristata*. Identification in North Africa, particularly in region of Tanger (where darkest forms of both species occur within sight of each other), must therefore rely heavily on differences in habitat, structure, behaviour, and voice. Normal flight and song-flight similar to *G. cristata* but that associated with song usually high, circling or hanging but sometimes lower and more fluttering. Escape-flight shorter, with bird tending to break back behind observer (perhaps due to more pronounced territoriality). Gait and feeding actions much as *G. cristata* but perhaps a little nimbler on ground, while most postures definitely less hunched in appearance (due to less deep belly). Perches on plants more than *G. cristata* and is less easy to flush from ground cover. Not gregarious. Generally prefers rocky and grassy slopes or interfaces between open ground and adjacent scrub or heath; thus occurs patchily within range of (often within sight of) *G. cristata*.

Song and many calls similar to *G. cristata*, but commonest call (often included in song) is distinctive: a soft, fluted whistle of 2–4 notes, lower pitched than *G. cristata*, with long final syllable rising then falling—'doo-dee-doo-deeee'.

Habitat. In lower middle latitudes, mainly Mediterranean, but overlapping warm temperate and subtropical, with widely separated extralimital population in north-east African tropics (Voous 1960). Overlap and competition with Crested Lark *G. cristata* thus limited to restricted parts of range, where problems of field identification have sometimes aggravated difficulties in distinguishing habitat differences (Wallace 1965). Characteristic of *G. theklae* is requirement for mixture of contrasting elements within compact territory. Thus occupies (e.g.) ample dry bare open soil, side by side with rocks and boulders or dense shrub cover, even with trees; roadsides alongside walled cornfields or quiet heaths; or dry stream beds with clumps of oleander *Nerium oleander* or flanked by bushes on steep slopes. Such edge-effect mosaics often ecologically impermanent. Where they have succeeded clearance of earlier forest cover, they imply historical expansion of range over land previously unsuitable, as in south-east France where birds live on shrub-heath of kermes oak *Quercus coccifera* with rosemary *Rosmarinus*, *Cistus*, and other second growth plants. In Spain, also on open shrub-heath, often used for grazing by goats and sheep which are in process

of destroying appropriate plant cover. Modern vineyards, olive-groves, and cornfields are habitats representing more advanced stage in same process (Abs 1963). In Tunisia, occurs within mature olive plantations where ground sufficiently clear and edge not too distant; in Mallorca, will enter fringe of Aleppo pine *Pinus halepensis* woods (E M Nicholson). Readily accessible firm bare soil and some short herbage usually essential, but unlike *G. cristata*, will thread through head-high herbage if necessary. On heathland, requires shrub or tall herbage cover to be amply broken up by patches of bare or stony ground, rock slabs, or stands of sparse dry plants. On arable land, prefers stubble or fallow to ploughed fields. Abandoned farmland creates especially favourable habitat, as for Woodlark *Lullula arborea*, which is closest ecological counterpart among Alaudidae. Less attached to level sites than *G. cristata*, wherever moderate slopes afford above requirements; although liking for rocky and hilly terrain sometimes over-emphasized, near southern edge of Palearctic range in north-west Africa extends up hills and high plateaux to at least 1200 m compared with c. 800 m for *G. cristata* there (Bannerman 1963). Also in North Africa, settles in steppes of *Stipa*, on plantings of cactus and *Euphorbia cactoides*, and in vegetated hills. In Ethiopia, occurs on steppe-type lands on high plateaux above 2300 m (Abs 1963), and at 1550–1700 m on steppe of low *Acacia* bushes scattered over stony ground and in dry wadi beds with more rocky terrain (Erard and Jarry 1973); recorded on arid steppes at 2550 m and in mountains at 3200 m (Erard and Naurois 1973). In Kenya, occurs on lava covered by grass steppe (Abs 1963). In Andalucía (Spain), occurs on arid scrub-dotted hillsides up to c. 1200 m (Mountfort 1954). Often occurs down to sea-level on coast, living on sand-dunes or even foreshore within c. 50 m of waves, as on Mallorca (E M Nicholson). Appears to overlap more often with Short-toed Lark *Calandrella brachydactyla* than with Crested Lark in Tunisia, although not in Coto Doñana, Spain (E M Nicholson). Perches freely on bushes, singing on them as alternative to song-flight (Mountfort 1958). Flies mainly in lower airspace and less widely than *G. cristata*. Concentrates within compact areas and avoids constantly disturbed spots, as well as neighbourhood of busy settlements. Thus shows more complex and marginal habitat preferences than *G. cristata*, which are evidently not imposed on it by competition with that species; this apparent in Mallorca where absence of *G. cristata* does not lead to spread beyond habitats characteristic of *G. theklae* elsewhere.

Distribution. EGYPT. Common near Solum 1920 (Meinertzhagen 1930). Not seen there since 1928; area not visited recently (SGM, PLM, WCM).

Population. FRANCE. Under 100 pairs (Yeatman 1976). SPAIN. Locally very common (AN). LIBYA. Locally common (Bundy 1976).

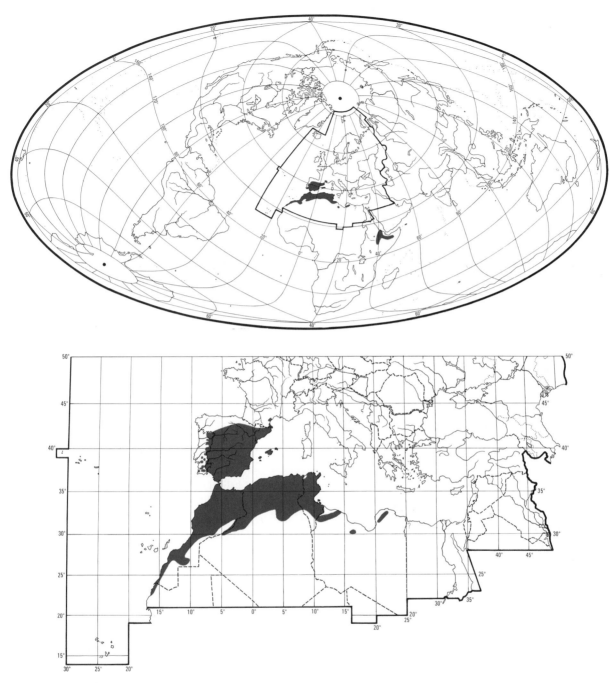

Movements. Resident. Reported as very sedentary in North Africa and Iberia (Heim de Balsac and Mayaud 1962; Bernis 1971; Juana Aranzana 1980), but no ringing data on possible local dispersals. Birds breeding on Sierra Carbonera (Andalucía) believed to descend to lower ground for winter, and collected then at Algeciras (Stenhouse 1921). Hence possible that this species is involved in small-scale autumn movements of unidentified *Galerida* at Straits of Gibraltar (Tellería 1981). However, *G. theklae* only once identified on Gibraltar (September 1971), though many Crested Larks *G. cristata* recorded there (Cortés *et al.* 1980). RH

Food. Insects and seeds. Most food taken from ground, often by searching under stones: pushes bill under stone then flips it over with quick sideways jerk. Unlike Crested Lark *G. cristata*, not known to dig for food, though this perhaps a consequence of more stony habitat on which usually seen. Will leap up to *c.* 1 m to take insects in flight or from vegetation. (Pasteur 1958; Abs 1963.)

The following recorded in diet in west Palearctic. Invertebrates: grasshoppers, etc. (Orthoptera), earwigs (Dermaptera), bugs (Hemiptera, including Cicadidae), lacewings, etc. (Neuroptera), larval Lepidoptera, adult and larval flies (Diptera: Tipulidae, Stratiomyidae), Hymenoptera (including gall wasps Cynipidae, ants Formicidae), adult and larval beetles (Coleoptera: Carabidae, Histeridae, Scarabaeidae, Buprestidae, Tenebrionidae, Chrysomelidae, Curculionidae), spiders (Araneae), millipedes (Myriapoda), and small snails (Mollusca). Plant material mainly seeds: of knotweed *Polygonum*, goosefoot *Chenopodium*, Portulacaceae, spurrey *Spergula*, catchfly *Lychnis*, Leguminosae, storksbill *Erodium*, flax *Linaria*, centaury *Centaurea*, Boraginaceae, Labiatae, and grasses (Gramineae) including cereal grain; also plant shoots. (Blanchet 1951; Abs 1963.)

In Spain, of 29 stomachs, all contained invertebrates: mostly beetles (including Curculionidae in 45%, Scarabaeidae in 41%, Tenebrionidae in 21%), ants (in 41%), caterpillars (in 28%), Hemiptera (in 24%), grasshoppers (in 10%), and spiders (in 10%); 93% contained seeds (including grain in 31%, *Lavandula* in 28%, *Spergula* in 14%) and 31% green plant material (Gil 1927, 1944). Similarly, of 17 stomachs from west-central Spain, April-May, 95% contained invertebrates: mostly beetles (including Curculionidae in 58%, Scarabaeidae larvae in 23%), ants (in 47%), spiders (in 47%), caterpillars (in 23%), grasshoppers (in 17%), and flies (in 17%); all contained seeds (including Labiatae in 47%, *Spergula* in 41%, *Linaria* in 41%) and 95% contained green plant material (Abs 1963). Near Casablanca (Morocco), December, 9 stomachs contained mainly seeds of *Echium* and Portulacaceae and some of grass, also ants, Curculionidae, insect eggs, and fly pupae; though presumably feeding in different habitat, birds had thus eaten fewer insects than *G. cristata* from the same area; see that species, p. 150 (Niethammer 1937). In Tunisia, November-June, 18 stomachs of adults all contained small seeds; 3 (November-February) contained green grass, 1 a grain of wheat, 2 (February-March) a small snail, and 1 (March) small beetles; 2 juveniles, June, contained black seeds and a small snail (Blanchet 1951). In Tripolitania (Libya), April, 14 stomachs contained many grasshoppers and 1 contained a grain of barley (Moltoni 1935).

Food of young includes larger insects and fat larvae (Ferguson-Lees 1970e). Young at one nest in central Morocco, late March, were brought very small food items, mostly unidentifiable; once a beetle, once Myriapoda. During 08.30-09.30 hrs, *c.* 10 feeding visits (unusually, by only 1 parent) to brood of 2 (Pasteur 1958). See above for food of juveniles.   DJB

Social pattern and behaviour. Much less well known than Crested Lark *G. cristata*; see Abs (1963) for comparative study.

1. Little information on dispersion outside breeding season, but according to Meinertzhagen (1951a) *Galerida* larks seldom form flocks. *G. theklae* usually in pairs, according to Bannerman

and Bannerman (1983). Party of *c.* 20 recorded in Morocco, early November, and flocking tendency more pronounced than in *G. cristata* both there and in Eritrea (Smith 1962-4). In north-west, also in some places in East Africa, recorded associating with *G. cristata* (Abs 1963); during winter in Monts des Ksours (northern Algeria), also in flocks with Lesser Short-toed Lark *Calandrella rufescens* and Temminck's Horned Lark *Eremophila bilopha* (Blondel 1962b). BONDS. No indication that mating system other than monogamous. Mixed pair (♂ *G. theklae*, ♀ *G. cristata*) produced fertile eggs in captivity (Abs 1963), and in Spain ♀ *G. theklae* and ♂ *G. cristata* recorded associating, possibly as pair (Niethammer 1955a). In Algerian Sahara, winter, ♂ *G. theklae* formed 'friendship-pair' with ♂ *G. cristata* (Niethammer 1954). Young fed by both parents (Abs 1963; Ferguson-Lees 1970e), family remaining together for unknown period after young leave nest (Munn 1931). BREEDING DISPERSION. Solitary and territorial, though in Roussillon (southern France), neighbourhood groups of 2-4 pairs apparently typical and only 1 solitary pair reported (Mayaud 1931; Guichard 1963). According to Guichard (1963) such groups fairly well spaced, though territory size may be 'restricted' (Mayaud 1931). Small home-range normally occupied in typical habitat; in Murcia (Spain), late summer, family fed in area of *c.* 1 ha (E M Nicholson). With population density generally low, territory of *c.* 4 ha the rule, as in *G. cristata* (Abs 1963), though said by Meinertzhagen (1951a) to have larger territory than *G. cristata*. On Coto del Rey (Andalucía, Spain), April, *c.* 3-4 pairs on *c.* 1 ha (E M Nicholson). In degenerate woodland, Morocco, 4·7 pairs per km[2] (Thévenot 1982). In Monts des Ksours, average of 18 per km[2] for 12 visits to one study area (March-June), and 10 per km[2] from 3 visits (March-June) in larger study area of different habitat. On fixed motor vehicle transect 22 km long and *c.* 50 m wide, 1 April-2 June, average 32 birds over 9 counts; on 14-km transect, 24 March-25 June, average 12 birds over 15 counts (9 birds in first 6 km); on 24-km transect, 1 April-30 May, average 50 birds over 8 counts. (Blondel 1962b, which see for discussion of habitat differences.) May nest close to *G. cristata*, e.g. 1 nest *c.* 50 m from courting pair of that species (Niethammer 1955a), territories recorded overlapping with those of Short-toed Lark *C. brachydactyla* and Skylark *Alauda arvensis* (Abs 1963). ROOSTING. Little information, but all *Galerida* larks roost on ground and, in North Africa, *G. theklae* recorded doing so among rocks (Meinertzhagen 1951a). Like *G. cristata*, captive birds scratch out shallow pits in loose sand for nocturnal roosting (Abs 1963). Generally perches more often on bushes and trees than *G. cristata* (Abs 1963), spending long time immobile but alert (Guichard 1963). In south-east Morocco, birds came to water holes to drink but able to do without water (Heim de Balsac and Heim de Balsac 1954). Recorded dust-bathing (E M Nicholson).

2. Not shy, frequently allowing approach to *c.* 2-3 m and showing no alarm; see (e.g.) Mayaud (1931) for southern France, Mountfort (1954) for Spain, Whitaker (1905) and Blanchet (1951) for Tunisia, and Bannerman and Bannerman (1983) for various parts of range. Such close approach (to *c.* 2 m) permitted in open terrain (Hüe 1952). Often stated that *G. theklae* tamer and more confiding than *G. cristata*, but this perhaps not generally valid. In Somalia, both species tame and confiding (Archer and Godman 1961a). On Mallorca (Spain), *G. theklae* varies between very tame and quite shy; when disturbed, may try to work round behind observer rather than flying off (E M Nicholson); see also *G. cristata*. According to Guichard (1963), *G. theklae* more closely resembles Woodlark *Lullula arborea* than *G. cristata* in several aspects of behaviour.

FLOCK BEHAVIOUR. No information. SONG-DISPLAY. Song (see 1 in Voice) delivered from ground, rock, top of bush, or in flight. Most song given in Song-flight in Algeria (Rothschild and Hartert 1911) and southern France (Guichard 1963), but frequently uses perches 30 cm to *c.* 3 m high in Spain (Niethammer 1955a; Mountfort 1958), southern Morocco (Valverde 1957), Tunisia (Whitaker 1905), and north-east Africa (Zedlitz 1911). Will take off from bushes for Song-flights (Niethammer 1955a; Mountfort 1958). Occasionally sings while hanging in the wind at *c.* 10-12 m (Mountfort 1958), but may also ascend until lost to view (Rothschild and Hartert 1911; see also Blanchet 1923); in southern France, also quite high, though not as high as *A. arvensis* (Guichard 1963). Song-flight resembles that of *G. cristata* (Abs 1963): bird describes fairly big circles, flying with weak, slow, fluttering wing-beats, or moves horizontally back and forth, tail and wings fully spread, at times quite high up (Niethammer 1955a; Guichard 1963; Ferguson-Lees 1970e; Bergmann and Helb 1982). According to Aplin (1896), song not continuous while circling; bird sings chiefly while hanging in the air with beating wings, then flies on before singing again, etc. As in *G. cristata*, average duration of Song-flight 3 min; longest 7 min (Abs 1963); 3-5 min noted by Niethammer (1955a). Birds of race *carolinae* reported frequently to have stayed aloft for *c.* 20-30 min (Rothschild and Hartert 1911). In Spain, bird invariably ceased singing high up, then plunged vertically to ground or perch (Niethammer 1955a); sudden vertical descent also noted by Whitaker (1905). For quiet song and Courtship-song delivered on ground, see below and Voice. No information on song periods in Europe. In northern Algeria, song given from early February (Blondel 1962b), though none noted in autumn (Rothschild and Hartert 1923). In southern Morocco, much song in early April, continuing mid-month when many pairs feeding young, but reduced by early May when moult in progress (Valverde 1957). On Asmara plateau (Eritrea), birds singing vigorously in February and March (Zedlitz 1911). ANTAGONISTIC BEHAVIOUR. (1) General. Song-flight probably important for territory demarcation. In Salamanca area (Spain), 3 birds still feeding together late March and early April—territories presumably not yet delineated. As in *G. cristata*, low density and small territory size result in few territorial fights. Playback experiments suggested that interspecific territoriality occurs with *G. cristata* (Abs 1963). (2) Threat and fighting. ♂ on look-out in territory often gives long series of Contact-alarm calls (Abs 1963: see 2a in Voice), though this not definitely or exclusively a form of territory advertisement. In southern France where neighbourhood groups occur (see Breeding Dispersion, above), ♂-♂ chases frequent but no fights seen (Guichard 1963). ♂ once flew purposefully from centre of territory to limit to drive off rival (Abs 1963). In Somalia, recorded chasing away Desert Lark *Ammomanes deserti* (Meinertzhagen 1951a). HETEROSEXUAL BEHAVIOUR. (1) General. On Coto del Rey, almost all birds paired in April (E M Nicholson). (2) Pair-bonding behaviour. As in *G. cristata*, ♂ gives Twittering-calls while pursuing ♀ low over ground (Abs 1963: see 3 in Voice). ♂ also displays near ♀ on ground: raises tail and gives Courtship-song (Bergmann and Helb 1982: see 1b in Voice). Performance clearly similar to *G. cristata*, but differs in that ♂ *G. theklae* makes curtseying or bobbing movement as in *A. arvensis* (Abs 1963); see that species (p. 198) for discussion of Bowing (etc.). No further information. RELATIONS WITHIN FAMILY GROUP. Probably much as in *G. cristata*. ♂ may sing while carrying food to nest (Abs 1963). In south-west Morocco, adult feeding 2 young in nest (which also contained 1 fresh egg) normally approached it in stages,

last being longest; briefly hovered at *c.* 5-6 m (probably not as an anti-predator stratagem) before dropping directly to nest or nearby (Pasteur 1958). Like other Alaudidae, young leave nest at *c.* 9 days old, *c.* 1 week before able to fly (Ferguson-Lees 1970e). Members of family keep fairly close together on ground and in flight (E M Nicholson). ANTI-PREDATOR RESPONSES OF YOUNG. If handled, will go rigid (Abs 1963). When attempt made to catch young *c.* 8-9 days old, birds ran and hopped alternately in escape bid. Hopping probably used generally when excited (Abs 1963). PARENTAL ANTI-PREDATOR STRATEGIES. (1) Passive measures. ♂ will often rise up and perform fluttering Song-flight over man intruding on territory (Abs 1963). In southern Morocco, bird leaving nest with young tended to move away a certain distance before flying up. See above for staged approach. If bird sighted observer while hovering, it would fly off but remain in the air and not come closer than 4-5 m if observer fairly near nest. Frequent alarm-calls given (see 2a in Voice), these increasing in intensity when nest acutely threatened (Pasteur 1958). Variant Contact-alarm calls (see 2a in Voice) given at end of long sequence possibly an expression of ♂'s increased agitation when Magpie *Pica pica* flew over (Abs 1963). (2) Active measures. When highly agitated as observer left after *c.* 1 hr, both birds of pair flew above him, at times descending to less than 3 m, moved off, returned, perched intermittently, etc.; constant alarm-calls given (Pasteur 1958); see also Dorst and Pasteur (1954). Distraction-lure display of disablement type also recorded: bird runs away, wings drooped and spread (Whitaker 1905); may follow simple aerial demonstration as described above (Dorst and Pasteur 1954). MGW

Voice. Repertoire more restricted than in Crested Lark *G. cristata* (Abs 1963); for best distinction from that species, see call 2. As in *G. cristata*, intermediates between basic calls occur; these result from superimposition of characters and, in *G. theklae*, involve calls 2-3, 2-4, and 3-4 (Bergmann and Weiss 1974, which see for sonagrams of these and some simple calls). For most detailed study, based on comparison with *G. cristata* and including sonagraphic analysis, see Abs (1963). For further sonagrams, see Bergmann and Helb (1982).

CALLS OF ADULTS. (1) Song of ♂. (a) In loud song, long fluting sounds alternate with short whistles, tremolos, and double notes; numerous good imitations of other bird species interpolated. Inclusion of call 2, in which fluting final unit especially drawn-out at 0·6 s, may provide major aid to identification. Song comprises series of linked phrases forming chains of varying length or a persistent outflow of melodious and complex chattering and whistling sounds with imitations. Quieter at beginning of series, and call 4 may also be given then (Abs 1963; Bergmann and Helb 1982). According to Guichard (1963), continuous outpouring of melodious but slightly sharp sounds like *Sylvia* warbler given by some birds is possibly an embellished and modified form (with imitations) of the short phrases with brief pauses given without much variation by others. Several authors make comparison with *G. cristata*, but Hüe (1952) warned that song of the 2 species of same basic type and that differences subtle and difficult to describe—and difficult to appreciate in

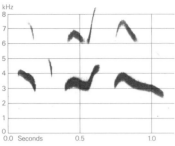

I   C Chappuis/Sveriges Radio (1972–80)
Morocco   April 1966

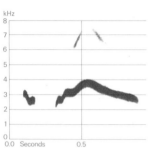

II   C Chappuis/Sveriges Radio (1972–80)
Morocco   April 1966

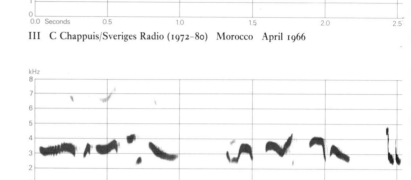

III   C Chappuis/Sveriges Radio (1972–80)   Morocco   April 1966

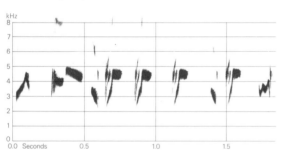

V   Roché (1966)   France   April 1960

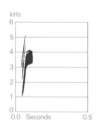

VI   Single call of Chaffinch *Fringilla coelebs*
(Marler and Hamilton 1966)

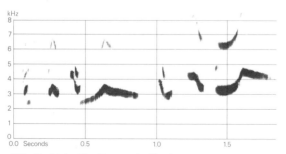

VII   Roché (1966)   France   April 1960

IV   E D H Johnson   Algeria   March 1973

practice according to Jonsson (1982). Variations in descriptions perhaps point to regional differences but this not proven. Song said by Jonsson (1982) to be more intense than in *G. cristata*, with rapid warbling more like finch (Fringillidae). In southern France, characteristic, strikingly musical song comprised short, melodious phrases interrupted by pauses and said to be quite unlike *G. cristata* (Niethammer 1955a). In Spain, song-phrases briefer, harder or more metallic, with scratchy or brittle quality; overall less rich in tone and variety than *G. cristata* and with fewer fluting or bell-like sounds (Wallace 1965; also Mountfort 1958); in some populations, greater variation derives from mimicry of other Alaudidae (Ferguson-Lees 1970e). In Portugal, song-phrase 'ootsi-kwar-weet-weet', soft, gentle, and slow, not unlike Wood-lark *Lullula arborea*, but thinner and less fluting. In Algeria (*carolinae*), aerial song far-carrying, with beautiful melancholy fluting quality (Rothschild and Hartert 1911; Blanchet 1923). In Tunisia, gentle, undemonstrative, and unstressed with some pulsating sounds like Skylark *Alauda arvensis* but less shrill and vibrant (E M Nicholson). Also in Tunisia, from perch, a soft and plaintive 'tweet a tweet a twee' given 2–3 times and ending with 'twee twee' (Whitaker 1905); perhaps only variants of calls 2 and 4. Recordings from North Africa by C Chappuis and E D H Johnson and from France by J-C Roché reveal most attractive and extraordinarily musical songs with many thin but loud and drawn-out whistles, some like those used by human to call dog; also rattling and sweeter fluting reminiscent of *Sylvia* warblers, fragments recalling song of Song Thrush *Turdus philomelos*, very quiet 'chook' sounds like Blackbird *T.*

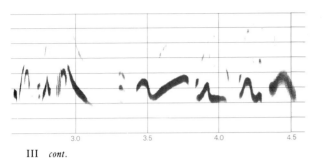

III *cont.*

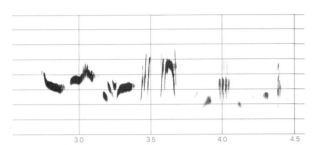

IV *cont.*

merula, some elements like *A. arvensis* and others like twittering of finch (Fringillidae), an occasional hard, pebble-rubbing buzz, and some plaintive mews. In recording by C Chappuis, bird starts in leisurely fashion, giving several highly tonal motifs with much portamento: 'sweee swee-eee sweeeoo' (Fig I), at times with rather squeaky quality like rubber toy, followed by wild and abandoned whistle 'chew wheeeeeoo' (Fig II), in which high pitch of 2nd unit more obvious to the ear than fading and portamento tail. Such motifs probably (variants of) calls 2 and/or 4 (see below for further discussion). Shift to more warbling song begins with 2 super-portamento tones, then changes to a warble and ends with a tonal motif and a unit showing pronounced vibrato (Fig III); warbling preceded by 2 tonal motifs shown in Fig IV (J Hall-Craggs, M G Wilson). In recording by C Chappuis, bird gives 5 bursts of song in 43 s, with pause of 3–5 s between these; in recording by E D H Johnson, bird gives 2 (1st 20 s, 2nd 25 s), with pause of 2 s between. Brief pauses (less than 0·5 s) occur during long bouts of song. Considerable variation in rate of delivery of notes, motifs, and phrases (J Hall-Craggs). For further description and comparison, see (e.g.) Aplin (1896), Hartert (1910), and Hüe (1952). On Mallorca (Spain), late April, remarkable bout of mimetic song included Woodchat Shrike *Lanius senator* alternating with Sardinian Warbler *Sylvia melanocephala*, Great Tit *Parus major*, Rock Thrush *Monticola saxatilis*, Chaffinch *Fringilla coelebs*, and (less successfully) Marmora's Warbler *S. sarda* (Jordans 1928). Also mimics calls of swifts *Apus* (Bergmann and Helb 1982). In recording (Fig V), bird mimics 'pink' sounds of *F. coelebs* (J Hall-Craggs); for

actual call of *F. coelebs*, see Fig VI. (b) Courtship-song indistinguishable to the ear from that of *G. cristata* (Abs 1963). Like typical song (above), rich in imitations but slightly strangled and relatively quiet (Bergmann and Helb 1982). (c) Quiet song once given on ground while feeding (Abs 1963). (2) Contact-alarm calls. (a) Given by both sexes. Soft, melodiously fluting whistle, of 2–4 notes with average frequency 2–3 kHz and thus lower pitched than equivalent in *G. cristata*. Most importantly, last note is longest of series (0·4 s, others 0·13–0·2 s; see also call 1) and first rises in pitch, then falls; marked emphasis on final note due to its length and frequency pattern (Abs 1963; Bergmann and Helb 1982). Renderings include 'dju dü düIë' and 'dju djuIe' (Bergmann and Helb 1982); 'dädüdie' and a variant 'dädüdje' in which pitch rises through call (2–4 kHz); these 2 variants sometimes given by ♂ in long, alternating series from look-out in territory; sonagrams indicate both variants comprise 3 separate units rather than syllables as suggested by these renderings (see Abs 1963); also 'uituit-ú' with stressed and drawn-out final note given mainly during chick-rearing (Pasteur 1958). Not so loud, more muffled, and less shrill than *G. cristata* (Niethammer 1955a; Wallace 1965). Reference to higher-intensity, more disorderly and vehement, higher-pitched alarm calls by Pasteur (1958) and Guichard (1963) is possible further indication of variation. In recording by J-C Roché, Contact-alarm calls given in song as a phrase of 6 discrete notes with, sometimes, other brief ancillary notes interpolated. Phrase clearly divided into 2 3-note sub-phrases (motifs), 1st showing overall descending, 2nd overall ascending pitch, but every note of portamento type, gliding up or down or both (compare *G. cristata* Fig VI, p. 156), and this to some extent disguises overall pitch pattern in sonagram. 1st motif 'tu te t'weeer', 2nd 'tiu te(r) weeee' (Fig VII). Last note of a motif usually of smooth ascending and descending portamento type, but occasionally shows frequency modulation (J Hall-Craggs, M G Wilson). Other renderings of calls: in Portugal, plaintive and liquid 'wee-te-tee'; on Mallorca, 'hwee hwee hwee' or 'weet-wee-mee', not unlike feeble Wood Sandpiper *Tringa glareola*, also shrill 'shree-urr-wee' or quieter, low, slurred and plaintively piping 'ti-turr-wee'; in Tunisia, clear, musical 'ee-reev-eee' (E M Nicholson). (b) Low, croaking 'quirk' said by Aplin (1896) to be given by alarmed birds; in Coto Doñana (Spain), monosyllabic 'brit' (Wallace 1965) possibly the same or a related call, but context not known. (3) Twittering-call. Indistinguishable from that of *G. cristata* and sonagrams very similar; as in *G. cristata*, loud twittering sounds given by ♂ in pursuit of ♀ (Abs 1963). A 'da tlide tri tilü' (Bergmann and Helb 1982). (4) Flight-calls. Before taking off or in flight, 'dschüi' given singly or as a series (Bergmann and Helb 1982). Fig 5e in Bergmann and Weiss (1974) indicates prescence of up to 7 formants in a call (see *G. cristata*, p. 157). Rendered 'dwoid dwoid'; indistinguishable to the ear from *G. cristata*, but sonagram showed only 1 rising glissando—thus similar to 2nd half of 1 unit in equivalent call of *G. cristata* (Abs 1963). In Tunisia, soft and agreeable 'twowp'; also, apparently as warning or on take-off, deliberate and emphatic 'eeer-eeer' or 'dwirr-dee' (E M Nicholson). (5) Rattle-call. See below.

CALLS OF YOUNG. Based on study of captive birds by Abs (1963). On 2nd day, only a few 'zip' sounds; at 3 days 'zip zep zip-zip-zip'; at 5 days 'zip-zip-zrii'; at 7 days 'schrii zeck zip-zip-zip' in quite long series; at 8 days 'zeck' when gaping and 'trü' after feed; at 9 days 'diü'. At 3 weeks, 1 gave a 'ziäip' similar to *G. cristata* but lower pitched (2·0–3·6 kHz) and shorter (0·2 s); in hard onset, reminiscent of nestling call. At 7 weeks old 'dji-djie-djiä' (2–4 kHz), probably developed from nestling call at 5 days; last unit longest (0·2 s) with widest frequency range, and stressed; at 3 months, final unit (0·38 s) even more stressed and expanded with rising glissando—this the forerunner of adult call 2. Food-call at 7 weeks old, a short (0·13 s), sharp, whistling 'huit' with energy concentrated in 2nd harmonic and pitch rising steeply in glissando (1·7–3·5 kHz); longer (0·2 s) at 3 months but frequency range narrower—perhaps forerunner of adult call 4. Snarling 'drrä' given at 7 weeks possibly forerunner of an adult Rattle-call (not yet described from *G. theklae* but known to be in repertoire of *G. cristata*; see p. 157). Juvenile song developed as follows: at 3 weeks old (following feed), birds gave gentle fluting or whistling 'didüdidüdüdidü'; at 3 months, also short whistles and rattling sounds. After moult at 6 months only pulsating 'trr trr trr' added, while long whistling sounds had ceased; forerunner of adult call 2 also interwoven. Though isolated from song and calls of their own species, captive juveniles were able to hear songs of other species, and though one soon gave excellent imitation of *F. coelebs*, none gave typical adult song of *G. theklae* nor adult calls; essential elements of these presumably therefore acquired only by learning.   MGW

Breeding. SEASON. Spain and Portugal: eggs laid February–June (Wallace 1965). North Africa: laying period in central, northern, and eastern Algeria and Tunisia early April to early June; in western Morocco, mid-February to late May (Heim de Balsac and Mayaud 1962). SITE. On ground in the open or in shelter of tussock. Nest: very similar to Crested Lark *G. cristata*, though generally smaller (Abs 1963). Building: no information. EGGS. See Plate 79. Sub-elliptical, smooth and glossy; very similar to *G. cristata*, perhaps indistinguishable. Nominate *theklae*: in Spain, 22·8 × 16·5 mm (21·6–25·0 × 15·4–17·7), *n* = 35, calculated weight 3·21 g; on Balearic Islands, 22·8 × 17·3 mm (21·5–24·5 × 16·5–18·1), *n* = 24, calculated weight 3·53 g. Other races no different. (Schönwetter 1979.) Clutch: 3–4 (2–7). Of 21 clutches, central northern Algeria: 4 eggs, 13; 5, 5; 6, 2; 7, 1; mean 4·6. Of 65 clutches, eastern Algeria and Tunisia: 3 eggs, 15; 4, 42;

5, 7; 6, 1; mean 3·9 (Heim de Balsac and Mayaud 1962). 2 broods (Bub and Herroelen 1981). INCUBATION. No information. FLEDGING TO MATURITY. Fledging period *c.* 15 days, young leaving nest at 9 days and spending next 6 days in surrounding cover (Abs 1963). No further information.

Plumages (nominate *theklae*). ADULT. Forehead, crown, hindneck, and mantle dull grey-brown with slight cinnamon-buff tinge, each feather with broad dull black central streak; streaks sharpest and deep black on crown, less sharply defined elsewhere; scapulars, back, and upper rump similar to mantle, but ground-colour usually dull olive-grey or pure grey, less brown and cinnamon. Lower rump and upper tail-coverts uniform rufous-cinnamon, contrasting strongly with grey-brown back; longest upper tail-coverts usually with ill-defined olive-brown central streak. Rather narrow and indistinct supercilium pale cinnamon-buff to pale cream-buff, extending from tuft of feathers above nostrils over eye to above ear-coverts. Narrow pale buff ring of feathers round eye. Streak from side of upper mandible backwards below eye mottled dull black and buff, widening behind eye into buff-brown ear-coverts which are spotted or streaked dull black. Patch on lower cheek near base of lower mandible buff to off-white, bordered below and behind by rather broad but mottled dull black line. Chin and throat cream-white, faintly spotted grey on chin, more distinctly on throat. Sides of neck cream-buff, coarsely spotted and mottled grey, merging into cream-buff chest and sides of breast, which show large and conspicuous dull black spots, rather triangular in shape and together joining to form distinct streaks. Remainder of underparts cream-buff, shading to cream-white on vent; flanks narrowly and sharply streaked dull black. Central pair of tail-feathers (t1) with ill-defined dull black centre and broad olive-grey fringe along tip; fringe gradually wider and shading to rufous-cinnamon towards base. T2 dark olive-brown; t3–t5 black, t5 with broad pink-cinnamon fringe along terminal half of outer web and tip; t6 pink-cinnamon or rufous-cinnamon with grey wedge on base of inner web, cinnamon of inner web sometimes faintly tinged grey, occasionally part of outer web also. Flight-feathers and greater upper primary coverts dark olive-brown, darker and duller olive-black toward outer primaries, slightly more rufous-brown near bases of outer webs of secondaries; narrow fringes along outer webs pink-cinnamon, ill-defined and restricted to terminal parts of feathers on secondaries and primary coverts, sharper and slightly paler on primaries, particularly wide on p9, but virtually absent from emarginated parts of outer primaries. Flight-feathers with large pink-cinnamon wedge at base of inner web. Tertials dark olive-brown or sepia-black, narrowly fringed pink-cinnamon (soon bleaching to white). Lesser upper wing-coverts dull olive-grey or grey with blackish centres, like scapulars and back; median and greater coverts olive-brown with ill-defined black streak near tip and broad contrasting pink-cinnamon fringe (soon bleaching to pale buff or isabelline, and soon lost through abrasion on tips and outer webs of greater coverts). Bastard wing dull black with narrow pink-cinnamon fringe along outer webs. Under wing-coverts and axillaries pale grey with pink tinge, greyer than pinkish undersurface of flight-feather bases; coverts along leading edge of wing dull black with pale buff fringes. Birds often liable to staining by rusty, sandy, brown, or grey soil. *In worn plumage*, black feather-centres of crown and crest more prominent; mantle, scapulars, and back duller brownish-grey, more heavily streaked black, contrasting more distinctly with rufous rump and upper tail-coverts; supercilium paler, more contrasting; black streaks below eye and over lower cheek to ear-coverts more distinct; ground-colour of underparts paler cream to off-white (if not contaminated by soil); black streaks on chest more prominent, but dark grey spots on throat and lower sides of neck often largely abraded, lower throat showing off-white bar across; pale fringes of tertials and median and greater upper wing-coverts lost through wear, tips and outer webs often bleached to grey. Differs from Iberian race *pallida* of Crested Lark *G. cristata* in darker and browner ground-colour of upperparts; prominent black streaks on forehead and crown (not on a few elongated feathers of hindcrown only); deeper rufous and more contrasting rump; heavily streaked and mottled ear-coverts, sides of neck, and hindneck; more distinct black streaks on sides of head; grey-mottled throat; narrower but sharper and blacker chest-spots; narrow but distinct black streaks on flanks; darker tertials and upper wing-coverts, with pale fringes appearing more distinct; slightly brighter rufous outer tail-feathers; greyer axillaries and under wing-coverts; duller pink basal undersurface of flight-feathers; see also Structure and Recognition. Most of these distinguishing characters valid only in fresh or moderately worn plumage and refer only to nominate *theklae*. NESTLING. Down long and dense, on upperparts and flanks only; yellow-white or dusty-grey (Hartert 1910; Abs 1963). JUVENILE. Ground-colour of upperparts, tertials, and upper wing-coverts as in adult or slightly more buff, but all feathers with narrow and even pink-cinnamon fringe on sides and broader white fringe on tip; no dark shaft-streaks, but shorter feathers with subterminal black-brown or olive-brown spot of W-mark, longer feathers with narrow black submarginal U-mark; elongated feathers of central hindcrown black with broad white tips. Supercilium as in adult or slightly broader and whiter. Tertials broadly fringed pink-cinnamon, fringes soon bleaching to white. Underparts off-white, chest and flanks tinged yellow-buff, chest with numerous small rounded spots. Tail as in adult, but t1 largely rufous-buff, often with slight grey suffusion sub-terminally and with broad but indistinct pink-buff fringes; t5 with more rufous on tip than adult. Flight-feathers and greater upper primary-coverts as in adult, but evenly fringed pink-cinnamon or buff along outer web and tip, often bordered by narrow dark lines submarginally. P10 longer than in adult, tip rounded; buffish-grey with ill-defined buff outer fringe and tip (in adult, more sharply pointed, inner web darker grey, outer web more sharply bordered buff; no buff along tip). Virtually indistinguishable from juvenile *G. cristata*, but chest spots slightly larger, sharper, and more rounded, and t1 deeper rufous, less olive-buff; bill shorter and with relatively heavier base (see Structure). FIRST ADULT. Like adult; indistinguishable once juvenile p10 replaced.

Bare parts. ADULT, JUVENILE. Similar to *G. cristata*: iris hazel to dark brown, bill dark horn-brown with paler cutting edges and pale horn, flesh-horn, or pale flesh base of lower mandible (in *carolinae* sometimes almost white); leg and foot flesh-brown, flesh-tan, pale yellow-flesh, or pale flesh (BMNH). NESTLING. Largely similar to *G. cristata*, mouth orange-yellow or yellow with black spots, but spots arranged slightly differently: rather small spots inside tips of both mandibles (in *G. cristata*, spot inside tip of lower mandible large and bifurcated), single spot on each side of base of tongue (none in *G. cristata*), and 3 tiny spots and thin black line near tip of tongue (in *G. cristata*, large bifurcated spots on tip of tongue) (Abs 1963).

Moults. ADULT POST-BREEDING. Complete; primaries descendant. Only a few in moult examined. Specimens of nominate

theklae, *ruficolor*, and *superflua* from June and early July have plumage heavily worn but moult not started (BMNH, ZMA); *ruficolor* starts July, *aguirrei* in June (Abs 1963). Nominate *theklae* with moult almost completed examined late September; moult fully completed in *aguirrei* and *superflua* from mid-September. No traces of moult in October specimens (all races). POST-JUVENILE. Complete; primaries descendant. In captive juveniles, starts at age of 11 weeks (in *G. cristata*, at 4–6 weeks) (Abs 1963), but probably starts at about same time as *G. cristata* in the wild, as juveniles of both species are equally scarce in collections (i.e. plumages probably equally short-lived). Recently-fledged juveniles (not yet moulting) present mid-May to mid-July (Stanford 1954; BMNH, ZMA). Specimens of *superflua* in fresh juvenile plumage, but with some inner primaries shed, examined from mid-July (BMNH, ZMA).

Measurements. ADULT. Nominate *theklae*. Central and southern Spain, all year; skins (BMNH, RMNH, ZFMK, ZMA). Bill (S) to forehead, bill (N) to distal corner of nostril; exposed culmen on average *c.* 3·0 less than bill (S).

	♂		♀	
WING	104·7 (2·05; 22)	102–108	99·1 (3·09; 24)	92–104
TAIL	57·4 (2·54; 15)	54–62	54·6 (1·96; 15)	51–58
BILL (S)	18·0 (0·97; 19)	16·5–19·6	17·2 (0·76; 19)	15·6–18·8
BILL (N)	11·3 (0·77; 15)	10·1–12·4	10·9 (0·60; 15)	10·1–12·1
TARSUS	25·3 (0·78; 15)	23·9–26·7	25·2 (0·83; 14)	23·7–27·1

Sex differences significant for wing, tail, and bill (S). Wing, southern France: ♂ 101–103 (3), ♀ 94–98 (4). Bill (N), Mallorca (Spain): ♂ 11·0 (0·39; 15) 10·1–11·6, ♀ 11·1 (0·54; 8) 10·1–11·8. (Mayaud 1931.)

G. t. erlangeri. Tanger, Oulmès, and Azrou (northern Morocco), December–February; skins (BMNH).

	♂		♀	
WING	103·8 (2·11; 10)	101–107	97·2 (1·77; 10)	94–100

G. t. ruficolor. Central Morocco and coastal strip of northern Algeria and northern Tunisia, all year; skins (BMNH, RMNH, ZFMK, ZMA).

	♂		♀	
WING	105·8 (1·93; 25)	102–110	99·2 (2·33; 11)	95–102
TAIL	60·7 (2·19; 10)	58–64	53·6 (3·07; 8)	48–58
BILL (S)	18·1 (0·63; 15)	16·6–19·2	17·4 (0·79; 8)	16·3–18·5
BILL (N)	11·3 (0·42; 10)	10·3–11·9	11·1 (0·62; 8)	10·3–11·9
TARSUS	25·7 (0·69; 10)	24·3–26·9	25·2 (1·01; 8)	23·9–26·1

Sex differences significant for wing and tail.

G. t. superflua. Hauts Plateaux and Atlas Saharien (Algeria), all year; skins (RMNH, ZMA).

	♂		♀	
WING	106·6 (1·18; 11)	105–109	100·7 (1·33; 13)	99–103
TAIL	63·0 (2·54; 12)	60–68	58·6 (3·22; 10)	55–64
BILL (S)	17·8 (0·64; 12)	17·1–18·9	17·5 (0·56; 10)	16·8–18·5
BILL (N)	11·1 (0·54; 12)	10·5–12·1	10·9 (0·55; 10)	10·1–11·8
TARSUS	25·8 (0·64; 12)	24·8–26·7	25·6 (0·77; 10)	24·6–26·9

Sex differences significant for wing and tail.

G. t. carolinae. El Golea and Chott Melrhir (Algeria), southern Tunisia, Cyrenaica (Libya), and north-west Egypt, January–May; skins (BMNH, RMNH, ZMA).

	♂		♀	
WING	101·6 (2·49; 11)	99–105	95·9 (1·62; 8)	94–98

Sex differences significant for wing and tail.

Wing and bill (N) of ♂. Nominate *theklae*: (1) Salamanca (Spain), (2) southern Portugal, (3) Balearic Islands. (4) *G. t. erlangeri*. *G. t. ruficolor*: (5) central Morocco, (6) northern Algeria and Tunisia. (7) *G. t. superflua*. (8) *G. t. aguirrei*. *G. t. carolinae*: (9) central Algeria and southern Tunisia, (10) Cyrenaica and north-west Egypt. (Abs 1963). Bill (N) apparently measured slightly differently from samples above, as average *c.* 0·5 mm less.

	WING		BILL (N)	
(1)	105·6 (2·00; 16)	100–106	11·6 (0·80; 16)	10·5–13
(2)	101·8 (2·40; 9)	97–107	12·2 (— ; 9)	11·5–13
(3)	99·9 (1·94; 15)	96–104	11·1 (0·43; 15)	10·5–12
(4)	100·4 (2·33; 5)	97–103	12·0 (— ; 5)	11–13
(5)	101·4 (1·91; 13)	99–105	11·3 (0·87; 13)	11–12
(6)	103·7 (2·44; 16)	99–107	12·1 (0·93; 15)	11–14
(7)	104·7 (2·58; 36)	99–111	12·0 (0·65; 35)	11–13·5
(8)	104·8 (— ; 8)	99–109	11·5 (0·54; 8)	10·5–12
(9)	101·7 (2·91; 20)	97–106	12·1 (0·63; 20)	11–13
(10)	100·1 (2·46; 12)	95–103	11·4 (0·48; 12)	11–13

JUVENILE. Wing on average *c*. 2 mm shorter than adult, tail *c*. 4 mm. Bill and tarsus full-grown at start of post-juvenile moult.

Weights. Nominate *theklae*. Southern France: ♂ 34·0 (3) 31–36, ♀ 37·2 (3·41; 4) 35–42 (Mayaud 1931). Spain, unsexed: 36·8 (23) 34–41 (Bub and Herroelen 1981). Salamanca (Spain), breeding season: ♂ 34·7 (1·88; 7) 32–37 (Abs 1963). Balearic Islands (Spain), unsexed: 30·8 (10) 25–35 (Mester 1971).

G. t. ruficolor. Central Morocco, winter ♂ 36·2 (2·59; 5) 32–39; northern Algeria, winter, ♂ 40 (Abs 1963).

G. t. superflua. Hauts Plateaux and Atlas Saharien (Algeria), winter: ♂ 39·6 (2·34; 21) 33–43 (Abs 1963).

G. t. carolinae. Figuig (Morocco), April: probable ♂♂, 36, 41 (BTO). Algerian Sahara, winter: ♂ 38·9 (3) 38–39·5 (Abs 1963).

Structure. Wing short, broad at base, tip rounded. 10 primaries: p7–p8 longest or either one 0–0·5 shorter than other, p9 1·5–4·5 shorter than longest; in adult, p10 58–66 (♂) or 50–59 (♀), p6 0–2, p5 2–5, p4 14–18·5 (♂) or 12–15 (♀), p3 19–23 (♂) or 16–19 (♀), p1 23–27 (♂) or 20–22 (♀) (all races; *n* = 35); in juvenile, p10 49–57 shorter, p6 1–2, p5 5–8, p4 15–18, p3 19–23, p1 23–27 (both sexes; *n* = 4). Adult p10 reduced (though less so than in *G. cristata*), narrow and with pointed tip; in all races, p10 3 shorter to 4 longer than tip of longest greater upper primary covert, on average 0·3 shorter in ♂ (*n* = 26), 1·0 longer in ♀ (*n* = 16); juvenile p10 slightly longer, broader, and with rounded tip, 4·2 (4) 2–6 longer than longest covert. Outer web of p5–p8 and inner of (p6–)p7–p9 emarginated. Longest tertials reach to tip of about p5 when fresh, to about p4 when worn. Tail rather short, tip square; 12 feathers. Compared to *G. cristata*, bill distinctly shorter (*c*. ½ head length), relatively deeper at base, slightly less slender, and less decurved. Bill width at 10 mm from tip 30–40% of bill length (to nostril) in adult (15–30% in *G. cristata*); in juvenile, 36–62%, mainly *c*. 45% (in juvenile *G. cristata*, 24–47%, mainly *c*. 34%) (Abs 1963). Feathers of central crown and hindcrown elongated, forming a full fan-like crest up to 2 cm long (in *G. cristata*, feathers of hindcrown only are elongated, forming spike-like crest). Tarsus and toes relatively slightly longer and more slender than in *G. cristata*. Middle toe with claw 18·7 (21) 18–20; outer toe with claw *c*. 72% of middle with claw; inner *c*. 75%, hind *c*. 112%. Front claws short, rather slender, decurved; hind claw rather long, slightly decurved, 11·1 (34) 9–13, exceptionally 5–14. For other details of structure and differences from *G. cristata*, see Abs (1963).

Geographical variation. Marked in colour, rather slight in size; mainly clinal and some populations rather variable, but variation not as complex as in *G. cristata*. Nominate *theklae* in higher parts of central Spain long-winged, populations of low coastal parts and Balearic Islands smaller. In north-west Africa, *erlangeri* of northern Morocco smallest; size gradually increases

towards south and east, but note marked difference between long-winged *superflua* of Atlas area of Algeria and Tunisia and short-winged *carolinae* in Sahara immediately south of it. Extralimital populations from north-east Africa small with relatively short tail (Abs 1963). In all populations, plumage much affected by bleaching and wear, and often coloured by soil; following account based on moderately worn and bleached specimens (about November–April) without soil contamination. Nominate *theklae* from Roussillon (France) and Iberia rather dark rufous-brown or grey-brown above, heavily streaked black; underparts pale buff or cream-yellow with prominent black spots on chest. Birds from Balearic Islands average browner on sides of neck, and have fine bill and short wing; sometimes separated as *polatzeki* Hartert, 1912 (e.g. by Abs 1963), but overlap large and some Iberian populations inseparable. *G. t. erlangeri* from northern Morocco is darkest race, ground-colour of upperparts browner than in nominate *theklae*, streaks on upperparts and chest broader and deeper black; ground-colour of underparts often paler cream, however, less buff. Middle and High Atlas as well as coastal strip of Morocco between Rabat and Essaouira inhabited by *ruficolor*; this race shows rather broad black-brown feather-centres on upperparts, as in nominate *theklae*, but fringes of feathers are contrastingly pale greyish cinnamon-brown, upperparts appearing heavily streaked rufous and black-brown; black spots on chest slightly smaller than in nominate *theklae*, but ground-colour buff or cream as in that race; populations from coastal strip of northern Tunisia and Algeria sometimes separated as *harterti* Erlanger, 1899 (e.g. by Hartert and Steinbacher 1932–8), upperparts on average slightly more cinnamon-rufous, black-brown streaks of upperparts and chest on average slightly narrower, but many indistinguishable when large series compared. *G. t. superflua* occurs from middle Moulouya river (Morocco) and south slopes of Tellian Atlas in Algeria and central Tunisia south to Atlas Saharien, Gafsa, and Gabès (Tunisia); feather-centres of upperparts narrow, dark olive-brown, fringes olive-grey with variable pink-cinnamon tinge; distinctly less heavily streaked than *ruficolor* and contrast with uniform rufous-cinnamon rump less marked; underparts isabelline or cream-white with narrow but distinct dark grey or dull black spots on chest, rather like north Algerian populations of *ruficolor*. Strong individual variation, some birds paler cinnamon-brown above, others more sandy olive-grey, but such birds generally occur side-by-side, though proportion of sandy olive-grey birds larger in eastern part of range, from about Bou Saâda (Algeria) to Sousse (Tunisia); these latter populations sometimes separated as *hilgerti* Rothschild and Hartert, 1912. Small-sized *carolinae* occurs immediately south of *superflua*, extending through northern Sahara from Figuig (Morocco) to north-west Egypt; differs from *superflua* by rufous-cinnamon feather-centres on upperparts, fringed by pale pink-cinnamon, but feathers of crown and mantle sometimes narrowly streaked olive-grey or fringes partly tinged pale olive-grey; in general, upperparts paler and more uniform rufous than in *superflua*, especially in bleached populations of sandy north-east Algerian Sahara, which are sometimes separated as *deichleri* Erlanger, 1899. Underparts of *carolinae* pale cream to off-white, like *superflua*, but spots on chest narrower, less extensive, often all or partly rufous instead of dark grey; birds from Cyrenaica (Libya) sometimes greyer above than typical *carolinae* and have been separated as *cyrenaica* Whitaker, 1902, but most birds indistinguishable from *carolinae*. *G. t. aguirrei* (synonym: *theresae* Meinertzhagen, 1939) inhabits Morocco south from Anti-Atlas; rather like *ruficolor*, with which it merges clinally, but dark feather-centres of upperparts and chest narrower, upperparts more extensively rufous; southernmost populations closely similar to *carolinae*, but differ in slightly more prominent dark streaking and in longer wing and shorter bill. Extralimital races generally differ in more saturated colours; see Colston (1982).

Forms superspecies with Malabar Crested Lark *G. malabarica* from India (Abs 1963); this species sometimes united with *G. theklae* into single species under name *G. malabarica* (Scopoli, 1786) (e.g. by Colston 1982), but differences between *G. malabarica* and *G. theklae* larger than among races of *G. theklae*, and *G. malabarica* has some characters in common with *G. cristata*; hence considered a separate species here, following Vaurie (1959) and Abs (1963), though some of the differences between *G. malabarica* and Mediterranean races of *G. theklae* are bridged by north-east African races of *G. theklae*.

Recognition. Hard to separate from *G. cristata*, even in the hand; some specimens cannot be identified and some of these are perhaps hybrids. Differences in colour (e.g. heavier streaks on hindneck, heavier spots on chest, greyish underwing, contrastingly uniform rufous rump and upper tail-coverts, more rufous outer tail-feathers, etc.—see Plumages) only valid for Iberian nominate *theklae* when compared with Iberian race

pallida of *G. cristata*; these distinctions in part also valid for populations of coastal north-west Africa, but not for more rufous and less heavily marked inland populations. Differences in crest shape difficult to assess in the hand, and shape also changes through wear. Best distinctions of *G. theklae* are: short bill with relatively wider and deeper base (see Structure), relatively longer p10, and more rounded wing. Bill to skull in *G. theklae* 16·3–18·9, in *G. cristata* from Iberia and North Africa 19·0–25·8; bill to distal corner of nostril in *G. theklae* 10·1–12·1 (10·5–13 according to Abs 1963, rarely up to 14), in *G. cristata* from Iberia and North Africa 12·6–17·5 (13–19·5 according to Abs 1963). Adult p10 on average 0·2 mm longer than longest greater upper primary covert in *G. theklae* (ranging from 3 shorter to 4 longer); p10 5·0 (13) 3–8 shorter in *G. cristata* from Iberia, but only 3·2 (31) 0–6 shorter in north-west Africa. In juvenile *G. theklae* p10 4·2 (4) 2–6 longer; in *G. cristata* from north-west Africa 1·3 (3) 0–3 longer, but note very small samples. P5 3·6 (31) 2–5 shorter than wing-tip in adult *G. theklae*, exceptionally 6; p5 5·9 (38) 5–9 shorter in adult *G. cristata*, rarely 4; p5 5–7 shorter in juvenile *G. theklae*, 9–12 in *G. cristata*. CSR

Lullula arborea **Woodlark**

Du. Boomleeuwerik   Fr. Alouette lulu   Ge. Heidelerche
Ru. Лесной жаворонок   Sp. Totovía   Sw. Trädlärka

Alauda arborea Linnaeus, 1758

Polytypic. Nominate *arborea* (Linnaeus, 1758), Europe south to Portugal, northern Spain, northern Italy, northern Yugoslavia, north-west Rumania, and Ukraine (USSR); *pallida* Zarudny, 1902, north-west Africa, southern Europe, south of nominate *arborea*, Crimea, Caucasus, Turkey, and Levant east to Iran and Turkmeniya (USSR).

Field characters. 15 cm; wing-span 27–30 cm. 20% shorter than Skylark *Alauda arvensis* and more delicately built; 10% larger and distinctly bulkier than *Calandrella* larks. Smallest of medium-sized larks with erectile crest, short square tail, broad rounded wings, and fine bill, all combining into subtly but distinctly different form from *A. arvensis*. Plumage basically buff, well streaked above and on chest, with bold white supercilium which reaches nape and conspicuous black and white marks on carpal feathers. Tail tipped (not edged) white. Flight buoyant but strikingly hesitant, even recalling small bat. Sexes similar; no seasonal variation. Juvenile separable. 2 races in west Palearctic, separable in the field at extremes.

ADULT. (1) North European race, nominate *arborea*. Upperparts basically warm buff, though strikingly pale, almost white, on hindneck and more olive-toned on rump; patterned essentially as in all streaked larks but with black-brown feather-centres more noticeable, due to increased contrast with fringes overall, and paler, whiter ground on hindneck, surround to cheek, and chest. Most obvious characters are: (a) bold supercilium, narrow and pale buff in front of eye but broad and near-white behind it, reaching nape (and thus visible from behind); (b)

variable but usually marked dark brown surround to cheeks which emphasizes supercilium and contains pale cream-buff central ear-coverts; (c) broad necklace of sharp black streaks continuous from upper breast to hindneck; (d) sharply etched breast pattern which however lacks warm buff ground-colour of *A. arvensis* and therefore lacks such sharp contrast with flanks and belly as in that species; (e) small but noticeable white/black-brown/white mark formed by primary coverts; (f) quite short wing-point, extending only ½ of tertial length; (g) dull tail with no white edges and with terminal white spots not usually visible. Lower underparts buff-white, with distinctly yellow tinge overall when fresh and brown wash on flanks. In flight, most characters described above become invisible; supercilium and (more often) primary coverts may show, but attention best given to obvious lack of white trailing edge to wing (tips of secondaries merely buff) and lack of white outer tail-feathers, such absences instantly excluding even runt or juvenile *A. arvensis*. White tips to tail-feathers not easy to see, being restricted to inner 3 of outer 4 pairs and subject to wear; tail does however appear rather uniformly dark on both surfaces. Underwing grey-white on coverts, grey on axillaries and flight-feathers. Bill exceptionally fine for a lark; dark horn, with base to lower mandible pink when breeding, duller at other times. Legs pink or even pale red-brown. JUVENILE. Plumage marks less distinct due to copious white spotting of upperparts; face pattern less contrasting. Primary-covert patch present. (2) South European, African, and Middle East race, *pallida*. Paler than nominate *arborea*, with greyer ground-colour to upperparts and white ground-colour to underparts.

May be confused with unusually marked adult and (particularly) short-tailed juvenile *A. arvensis*. Look first for *L. arborea*'s diagnostic carpal marks. Flight distinctive: bird exhibits blunt-winged and short-tailed silhouette, with markedly undulating, hesitant action, producing jerky and wavering progress; wing-beats come in even looser bursts than those of *A. arvensis*. Escape-flight low, short, and fluttering, slower than *A. arvensis*. Song-flight consists of fluttering, spiral ascent, then less constantly fluttering wide circles, then a descent, occasionally with final plummet; flight lower but covering wider area than *A. arvensis*. Gait as *A. arvensis*, but lighter-stepping, even when shuffling with body close to ground. Stance appears less horizontal but this almost wholly an illusion due to short tail. Perches freely on trees and bushes, even walking along branches.

Voice sweet and musical, with liquid, yet ringing tone. Song contains diagnostic fluty 'lu-lu-lu' phrase. Flight-call 'lit-loo-eet'.

Habitat. Breeds in milder upper and lower middle latitudes of west Palearctic, scattered over the more benign situations between July isotherms of *c.* 17°C and 31°C (Voous 1960). Avoids severe cold and windy wet conditions and also hot arid areas. Mainly a temperate and Mediterranean species, but overlaps boreal and steppe zones. Distribution in Britain critically examined by Yapp (1962) who found it confined to that part of the country having either 1500 hrs or more of sunshine per year, or average January minimum temperature above 5°C; it was absent from certain areas where sunshine met the supposed requirement but average January minimum temperature fell below 1°C; distribution thus rather similar to that of Cirl Bunting *Emberiza cirlus*. On such climatic constraints are superimposed topographical requirements for mainly lowland (but often sloping) well-drained sites on sand, gravel, or chalk, with short grass for feeding, longer grass or heather for nesting, and scattered trees, bushes, or other suitable fixed song-posts as alternative to delivery of song in flight. Avoids intensive agriculture, but often favours neglected or abandoned farmland. For proportion of different habitat types in territories, see Harrison and Forster (1959). Widespread association with areas close-cropped by large populations of rabbits *Oryctolagus cuniculus* affected since *c.* 1950 by impact on them of myxomatosis, leading to growth of coarser and taller vegetation. Afforestation of such mar-

ginal lands and growing recreational disturbance have been further adverse influences (Sharrock 1976). Impact of hard winters and inclement breeding seasons has also contributed to mainly downward trend in numbers and growing discontinuity of occupied areas.

Habitat requirements, outlined above for Britain, differ somewhat for other western, eastern, and southern regions of west Palearctic. In west, most characteristic are heathlands, low moors, rundown farmlands with ragwort *Senecio* or similar weeds, scrubby hillsides, thinly timbered parkland, outskirts of woods, and felled woodland (Witherby *et al.* 1938a). Burnt-over areas, open woods of birch *Betula* or oak *Quercus*, golf courses, and sand-dunes with trees or shrubs also favoured (Harrison 1982). In central Europe, optimal habitat is dry warm open pine-heath, often with heather *Calluna* and with 2–8 year old trees, fitting between habitats of Tree Pipit *Anthus trivialis* and Tawny Pipit *A. campestris*, with inclination towards *A. trivialis*. Prefers presence of hills and hollows to flat areas. Likes abandoned fields and orchards near woods and vineyards providing ground conditions are right. Will nest near fringe of cities (Pätzold 1971). In Rzepin forest (western Poland), optimum habitat is 2-year-old pine *Pinus* plantations of *c.* 2–3 ha adjacent to older forest on at least 1 side (Mackowicz 1970). In Switzerland, prefers loamy and humus-rich as well as sandy soils, moraines, fallow fields, and vineyards. Needs less open and smaller territory than Skylark *Alauda arvensis* and less rich vegetation. Breeds at times even above treeline, at 1600 m or higher. (Glutz von Blotzheim 1962.) Southern race, *pallida*, frequents maize fields, often near water, and occupies open oak-woods to edge of dunes. Ascends to 2280 m in Pyrénées (Pätzold 1971). In North Africa, occurs on stony hillsides with forests of cork oak *Quercus suber* and holm oak *Q. ilex* (Bannerman 1953), and maquis or open woods at 600–1800 m, but in High Atlas up to 3000 m; common on plateaux but avoids *Stipa* steppes (Heim de Balsac and Mayaud 1962). In USSR, more characteristic of open and fringe areas of forests (e.g. dry pinewoods with some birch and aspen *Populus tremula*), mixed or broad-leaved forest clearings, and regrowth after fire, windthrow, or felling; sometimes also sandy ridges in floodland with willows *Salix*, or treeless slopes. In south, however, will occupy steppe-like plains with some grass cover, thickets of juniper *Juniperus* in high rugged rocky localities with few trees, and semi-desert mountain slopes. Ascends to 2000 m in Armeniya, but in winter descends to orchards in valleys. (Dementiev and Gladkov 1954a.)

Past changes or impacts, either humanly or naturally caused, play important role in evolution of habitat, which is essentially transitory and unstable unless fortuitously sustained by management. Demand for commanding perches and for feeding areas, neither bare nor densely overgrown, leads to choice of beginnings or aftermath of woodland growth. Although still basically a ground-

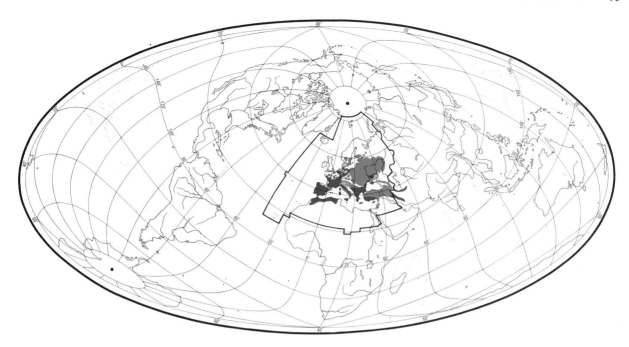

dweller, needing clear space for movement, is dependent on trees or such substitutes as posts, walls, fences, and overhead wires, while retaining habits of flight through both lower and upper airspace, especially for singing. Changing land use and to some extent human disturbance accordingly play a more than locally important role in affecting survival, expansion, and stability of population.

Distribution. Ranges decreased in Britain, Ireland, France, Belgium, Netherlands, and Sweden (see also Population).

BRITAIN, IRELAND. Marked fluctuations. In early 19th century, breeding north-east England and sparsely in Ireland; disappeared from both areas in mid-19th century (except Ireland, where nested *c.* 1905 and 1954). Expanded 1920–51, then range decreased, breeding ceasing in several English counties; possible reasons hard winters and habitat and climatic changes (Parslow 1973, which see for maps). FRANCE. Marked decrease in range (Yeatman 1976). TURKEY. Thinly distributed (Vittery *et al.* 1971).

Accidental. Iceland, Faeroes, Kuwait, Libya, Malta.

Population. Decreased over much of north-west and central Europe: Britain, Ireland, France, Belgium, Netherlands, West Germany, Denmark, Sweden, Czechoslovakia, and Switzerland.

BRITAIN. Population levels have followed range changes (see Distribution): decrease first half 19th century, increase *c.* 1920 to early 1950s; marked decrease since to 200–450 pairs 1968–72, 160–180 pairs 1975, 400–430 pairs 1981, and 210–230 pairs 1983 (Sharrock 1976; Sitters 1986). FRANCE. 10 000–100 000 pairs; decreased

(Yeatman 1976). BELGIUM. About 400 pairs (Lippens and Wille 1972). Marked decrease in less than 70 years, upper Belgium (Ledant and Jacob 1980); decreased Turnhoutse Kempen (Meeus 1982). NETHERLANDS. 800–900 pairs 1976–7 (Teixeira 1979); rather common 1900, strong decline since 1950, but slight recent increase (CSR) to 2500–3000 pairs 1980–4 (R G Bijlsma, R Lansink, and F Post), though collapsed again 1985 after severe winter (CSR). WEST GERMANY. Decreased (Yeatman 1971). Local counts showed decline in 8 states, due to habitat destruction (Bauer and Thielcke 1982). DENMARK. Decreasing (TD). NORWAY. 150–200 pairs 1983 (O J Hanssen). Decreased since mid-1950s; slow increase since 1970 (Hanssen 1984). SWEDEN. 3000 pairs (Ulfstrand and Högstedt 1976). Marked decline in 1950s and 1960s, shown by visible migration counts; slight increase in south in late 1970s (LR). CZECHOSLOVAKIA. Marked decline after 1960 (KH). AUSTRIA. Sparse and local (HS, PP). SWITZERLAND. Marked decline (Géroudet 1983; Biber 1984). GREECE. No changes noted (GIH).

Movements. Migratory in northern half of breeding range, partially so in central Europe and southern USSR, but mainly resident in maritime climate of western Europe and in Mediterranean basin (including North Africa). Winters within southern half of breeding range, or a little beyond it into Egypt and northern Middle East. Less conspicuous as a migrant than many other Alaudidae, since seldom forms large and compact flocks for diurnal passage.

Northern race (nominate *arborea*) mainly migratory in Fenno-Scandia, USSR (where some winter north of Black Sea, and thence westward across northern Europe to

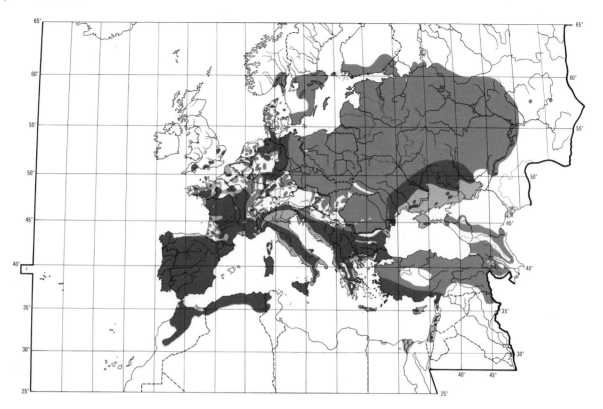

West Germany); partially migratory in Low Countries and central Europe. Long-distance ringing recoveries (more than 500 km) with October–March finding dates include the following (see Spaepen and Cauteren 1961, Rendahl 1964, Bernis 1971, Paevski 1971, Zink 1975): Sweden to Belgium, France (3), and Italy; Finland to Netherlands and France (Charente Maritime); Soviet Baltic states to Netherlands (2), Belgium (2), France (16), and Italy (1); Poland to Italy (2); Germany (West and East) to Netherlands, France (3), and Italy (2); Belgium to France (Manche, Basses Pyrénées) and Spain (Guipuzcoa); Switzerland to Spain (Castellón); Czechoslovakia to Spain (Cordoba); Hungary to Italy (3) and Cyprus. Apart from the last, all showed SSW–WSW movement and were found near western seaboard (Netherlands to Biscay), under Atlantic climate, or in southern Spain, southern France, and Italy under Mediterranean regime. Some of foregoing birds (notably those ringed Baltic states, Germany, Switzerland, and Belgium) were ringed while on passage and so likely to have originated from further north or north-east. No ringing evidence for British breeders, but population apparently partially migratory, with most breeding sites (especially Breckland) deserted from September or October. Sparse records from coasts (mostly southern and eastern England but north to Shetland) occur mostly September to mid-November, December to mid-January, and March to early May; most (especially in north and east) are presumably from Continent, but some are probably emigrating British birds. British breeding sites re-occupied late January to mid-March. (Sitters 1986.)

Southern race (*pallida*) migratory to partially migratory in southern USSR and perhaps Asia Minor, but mainly resident or dispersive elsewhere. Summer visitor to USSR east of Black Sea, except in Turkmeniya where resident and apparently in Armeniya where there are seasonal altitudinal movements (Dementiev and Gladkov 1954a). More widespread in Turkey outside breeding season (though eastern and northern areas vacated in winter), and small-scale diurnal passage reported Black Sea and south coast (Beaman *et al.* 1975), but no proof that Turkish-bred birds emigrate, and migrants may be of Russian origin. Irregular October–February reports from northern Persian Gulf (Bundy and Warr 1980) likely to be *pallida*, as certainly were some winter specimens from Egypt (Meinertzhagen 1930). Scarce migrant, mainly autumn, through Malta, 4 skins being *pallida* (Bannerman and Vella-Gaffiero 1976; Sultana and Gauci 1982); this small movement likely linked to scarce and irregular occurrences in coastal zone of Tripoli, Libya (Bundy 1976). Such birds perhaps originate in southern Europe rather than USSR, judged from situation further west. There, small-scale movement also occurs across Straits of Gibraltar in autumn (mid-October to early November): 38 birds counted over Tarifa (Cadiz) in 1976, 133 in 1977 (Tellería 1981). These more likely to be Spanish *pallida* than north European *arborea* which has never been identified in north-west Africa (Heim de Balsac and

Mayaud 1962). Hence circumstantial evidence that, while south European *pallida* are mainly resident, small numbers cross Mediterranean in autumn. North African *pallida* become more widespread in autumn and winter, but scale of dispersal unclarified.   RH

Food. In breeding season, largely medium-sized insects and spiders; at other times, apparently mostly seeds. Direct observation, and analysis of food brought to young, indicate food collected from ground surface, uppermost layer of soil, and from low parts of plants. Brooding ♀ seen trying to catch fly flying by, but not otherwise recorded taking prey in flight or by perching anywhere other than on ground. Feeding bird bustles about swiftly, pecking among plants and in earth. Rubs caterpillars repeatedly against ground. (Mackowicz 1970.) Most seeds swallowed complete with husks; only coarser seeds with awns de-husked by beating against ground (Pätzold 1971). Beetles to be fed to young have front part of body torn off (Prokofieva 1972b). Adult gathering food for young makes pile of insects on ground as described for Skylark *Alauda arvensis* (Ashford 1915: see p. 192). When breeding, forages only within 100 m of nest (Steinfatt 1939a; Harrison and Forster 1959; Mackowicz 1970).

The following recorded in diet in west Palearctic. Invertebrates: dragonflies (Odonata), grasshoppers, etc. (Orthoptera), cockroaches *Blattella*, bugs (Hemiptera: Pyrrhocoridae, Aphidoidea), thrips (Thysanoptera), larvae of ant-lions *Myrmeleon*, adult, pupal, and larval Lepidoptera (Pyralidae, Noctuidae, Geometridae), adult, pupal, and larval flies (Diptera: Tabanidae, Empididae, Asilidae, Tachinidae), Hymenoptera (sawflies Diprionidae, ants Formicidae), adult and larval beetles (Coleoptera: Cicindelidae, Carabidae, Staphylinidae, Scarabaeidae, Byrrhidae, Elateridae, Chrysomelidae, Curculionidae, Scolytidae), spiders (Araneae) and their cocoons, harvestmen (Opiliones), centipedes (Chilopoda), millipedes (Diplopoda), small snails (Gastropoda), and an earthworm (Oligochaeta). Plant material mainly seeds: of Scots pine *Pinus sylvestris*, knotweeds (Polygonaceae), hemp *Cannabis*, *Amaranthus*, pinks (Caryophyllaceae), poppies *Papaver*, Leguminosae, cranesbills *Geranium*, Boraginaceae, Compositae, and grasses (Gramineae); also leaves and buds and, in winter, catkins of birch *Betula* and hazel *Corylus*. (Baer 1909; Steinfatt 1939a; Kovačević and Danon 1952; Tutman 1962; Averin and Ganya 1970; Mackowicz 1970; Pätzold 1971; Prokofieva 1972b.)

No detailed studies of adult diet. Small-scale stomach analyses and other data as follows. Poland: Mackowicz (1970). Germany: Baer (1909), Niethammer (1937), Pätzold (1971). Yugoslavia: Kovačević and Danon (1952), Tutman (1962). USSR: Ptushenko and Inozemtsev (1968), Averin and Ganya (1970), Kostin (1983). Spain: Gil (1927). Generally held to eat much less plant material than *A. arvensis*, though winter diet perhaps largely

seeds: of the 5 birds examined from late autumn and winter only 1 (Spain, 2 November) contained any invertebrates (Gil 1927; Tutman 1962; Kostin 1983). At other times, apparently little plant material, and 12 faecal samples from spring and summer at Thetford (England) showed Curculionidae to be the favourite food item (C G R Bowden and R E Green). At Rzepin (western Poland), plant food is pine seeds only, except for early spring (after birds' arrival and before insects appear) and in autumn when also takes weed seeds (Mackowicz 1970).

Food of young wholly or largely medium-sized invertebrates. In Leningrad region (western USSR), 70 food samples from young comprised 91 items including 59% (by number) grasshoppers *Omocestus* (mostly nymphs), 19% spiders (mainly *Pardosa paludicola* with body length 6-8 mm) and their cocoons, 14% Lepidoptera (adults, pupae, and larvae), and 4% small beetles (Prokofieva 1972b). According to Steinfatt (1939a), food of young in descending order of importance: caterpillars (probably including sawfly larvae), beetles, flies, small butterflies, spiders, dragonflies, and snails. At Thetford, faeces from 5 broods contained the following in descending order of abundance: caterpillars, spiders, Curculionidae, Diptera, grasshoppers, sawfly larvae, adult beetles *Byrrhus*, millipedes, and a small snail (C G R Bowden and R E Green). At Rzepin, where diet studied by collar-samples, 349 items included 21·2% (by number) spiders, 18·3% seeds of *Pinus sylvestris* (mostly germinating), 16·9% Lepidoptera (adults, pupae, and larvae, largely Noctuidae), 15·5% beetles (apparently largely adults, and largely *Byrrhus pilula*), 10·3% harvestmen, 7·7% larvae of pine sawfly *Diprion pini*, and 5·2% Diptera (adults, pupae, and larvae, largely larvae of Tachinidae). Chicks in pine plantations given considerably more spiders and caterpillars than those in forest clearings but fewer germinating seeds, pine sawfly larvae, and harvestmen. When seeds consumed, pebbles 4-9 mm across also fed to young. Over the 11 days in nest, average brood brought average 3660 invertebrates weighing *c.* 170 g. At 11 days old, average wet weight of single feeding to young 0·16 g ($n=12$), average 3·4 food items (1-16) per feeding ($n=76$ feedings). (Mackowicz 1970.) Each adult brings food every 12-20 min; at 6 days old, average 7 visits per hr; at 7 days, 113 feeds per day (Pätzold 1971).   DJB

Social pattern and behaviour. Most aspects well known. Major studies by Krampitz (1952) in Frankfurter Stadtwald (West Germany), Koffán (1960) at Budaörs (Budapest, Hungary), and Mackowicz (1970) in Rzepin forest (western Poland). For summary, see Pätzold (1971).

1. Overall, less gregarious than some other Alaudidae, notably Skylark *Alauda arvensis*. Loose-knit parties of *c.* 15-20 (usually less) occur outside breeding season (Meinertzhagen 1951a). In Rzepin, birds in flocks on arrival from winter quarters and parties of 10-12 re-formed after pairing if weather bad (Mackowicz 1970). In Frankfurter Stadtwald, single birds or small parties on breeding grounds for up to 10 days before

first song (Krampitz 1952). With aggregation of family parties in autumn, flocks of up to 50 occur, though some pairs apparently remain on territory until late autumn, not joining flocks (Mackowicz 1970); see also Koffán (1960). In Lüneburger Heide (West Germany), largest flock in spring 32 (in early March); flocks bigger than family parties recorded from late August and largest gathering (c. 150) in October (Schumann 1959). In Voronezh region (USSR), c. 30-40 recorded together as early as 30 June (Wilson 1976). In England, small feeding flocks also noted in breeding season (Nethersole-Thompson 1932). For further reports of, and comments on, late-summer and autumn flocks, see (e.g.) Moore and Boswell (1956) for Iraq, Labitte (1958) for Eure-et-Loir (France), and Steinfatt (1939a) for Rominter Heide (Poland/USSR). Sometimes associates with other birds on migration, e.g. *A. arvensis* or finches (Fringillidae) (Wyss 1947; Mackowicz 1970), though such associations loose and accidental (Steinfatt 1939a). BONDS. Nothing in major studies to indicate mating system other than monogamous, though one case of bigamy reported (without details) by Witherby *et al.* (1938a). In some cases, pair-bond maintained for more than 1 breeding season; probably due primarily to site-fidelity and not known definitely whether birds remain together throughout winter and on migration. Pair-bond unlikely to be life-long (Nethersole-Thompson 1932). Site-fidelity and attachment to territory evidently a common feature, particularly of ♂♂. Although ♀ in one case showed territorial fidelity for at least 2 years and took different mate in 2nd year, unpaired ♀ normally pairs up with ♂ who has returned to his territory of the previous year. Thus, marked ♂ bred for at least 4 consecutive years in same territory but paired with different ♀ each year. Another marked ♂ returned to natal area to breed for first time then showed territorial fidelity for 6 years, but similarly changed mate each year (Koffán 1960; Pätzold 1971). Pair-bond evidently strong once established, birds rarely apart (e.g. Labitte 1958, Harrison and Forster 1959). ♀ normally does all incubation and brooding (Mackowicz 1970), though exceptionally ♂ broods (Ashford 1922). Young fed by both sexes (more by ♀), by ♂ alone if ♀ has started incubation of 2nd clutch (Koffán 1960; Mackowicz 1970). Complete lack of feeding by ♂ (Krampitz 1952; Labitte 1958) probably due to disturbance (Pätzold 1971). Family may remain together more or less until autumn departure (Labitte 1958; Mackowicz 1970). BREEDING DISPERSION. Solitary and territorial. Several pairs often nest fairly close together while apparently suitable areas alongside not occupied. At Budaörs, same territory and same small nesting area used by different pairs over years (Koffán 1960). In Dorset (southern England), groups of 3-4 pairs c. 1·6 km apart; some solitary pairs in between (Ashford 1922); see also (e.g.) Nethersole-Thompson (1932) and Harrison and Forster (1959). In Rzepin, plantation of up to 3 ha usually occupied by only 1 ♂; larger (3-5 ha) plantations (normally elongated rectangle) may hold 2 ♂♂ with diagonal boundary (Mackowicz 1970). In Cornwall (southern England), 2 nests only c. 20 m apart, though sites separated by fence (Meinertzhagen 1951a). Territories require to have suitable trees (etc.) used by ♂ as song-post and look-out. Tops of young conifers often bent from constant use (Haun 1930; Krampitz 1952; Leuzinger 1955). Territory in Surrey contained 4 such look-outs; that most frequently used afforded good view over almost whole territory (Harrison and Forster 1959). In Rzepin, 20 000-30 000 m², 3 times that of *A. arvensis* (c. 7000-11 000 m²) in same area (Mackowicz 1970). In Surrey, before nesting, birds ranged over maximum c. 10·5 ha, but nesting territory in 2 cases much smaller at c. 4-5 ha. Nests recorded c. 50 m and c. 200 m from ♂'s song-posts (Harrison and Forster 1959). In

Frankfurter Stadtwald, birds use relatively small 'song territory' and larger neutral feeding grounds (Krampitz 1952). In Rzepin, first areas occupied after arrival are old territories or adjacent younger plantations, then areas probably not suitable in previous year, and finally, barely suitable and often small areas. Nest usually sited near open area (c. 50% less than 8 m away) and on edge of territory, though proximity of trees (of suitable height) is probably determining factor in some cases. If more mature plantation used, nest normally located outside it (Mackowicz 1970). At Budaörs, where density low, territory indistinct and nests situated c. 100-200 m away from area favoured by ♂ (also for song) but also used by ♀ for feeding (Koffán 1960). Similar report in Mackowicz (1970), though nests less far (up to c. 27 m) 'outside territory'; exceptionally, also sited thus following destruction of 1st nest. Territory normally defended by ♂ and used for copulation, courtship, and nesting, and presumably for most rearing of young (see Relations within Family Group, below); parents forage for young mostly within c. 50-100 m of nest (Mackowicz 1970; also Steinfatt 1939a). Particularly if both pair-members use same territory in consecutive years, nests may be very close (e.g. 1·5 m) to one previously used, or actually in a scrape (see part 2) made in preceding year; further away (8-650 m) if ♀ new (see also subsection 5 in Heterosexual Behaviour, below). Within a season, new nest normally built for each brood or replacement clutch; may be c. 40-500 m away from that previously used. Exceptionally, same nest used for 3 successive broods (Koffán 1960; Mackowicz 1970; Pätzold 1971; also Labitte 1958). In the Heller (Leipzig, East Germany) formerly 2 pairs in 10 ha (20 pairs per km²); in more recent years, optimal habitat there (c. 1·2 km²) held 6 pairs (5 pairs per km²). At Budaörs, in c. 4 km² of suitable habitat (orchards, vineyards, fields) average density 2·7 pairs per km², maximum 6 pairs per km² (Pätzold 1971). In Mecklenburg (East Germany), density generally low. In one wooded area of 115·5 km², 25 singing ♂♂; in another pinewood of 165 km², 26-30 ♂♂. Density over larger areas significantly influenced by proportion of pine plantations 1-5 years old: thus 3·0 pairs per km² where 9·4% pine plantations, 1·7 pairs per km² where only 2·5% (Klafs and Stübs 1977). For Brandenburg (East Germany), see Rutschke (1983). In varied habitat in Königs Wusterhausen (East Germany), at least 14 singing ♂♂ in c. 49 km², i.e. 0·29 pairs per km², local maximum 3 pairs in 1 km²; pairs unevenly dispersed (Pätzold 1971). In Belgian uplands, highest density c. 7·7 pairs per km² of sandy heath with pine *Pinus* and birch *Betula* (Ledant and Jacob 1980). In southern England during late 1960s and early 1970s, occupied areas probably held c. 0·1-0·25 pairs per km² (Sharrock 1976); for earlier data, see Raines (1945) and Christian (1965). In south of Leningrad region (USSR), singing ♂♂ c. 200-400 m apart (Prokofieva 1972b). In Morocco, 2·4 pairs per km² in degenerate woodland, 3·5 pairs per km² in forest with trees over 7 m (Thévenot 1982). In Israel, once 2 pairs in c. 2 ha (Meinertzhagen 1951a). At Budaörs, recorded nesting 4·5 m from *A. arvensis* and c. 25 m from Crested Lark *Galerida cristata* (Koffán 1960). ROOSTING. Nocturnal and, outside breeding season, apparently communal, though few details. In Estonia (USSR), winter flock roosted in stubble (Meinertzhagen 1951a). In Rzepin, birds squeeze in between small pine twigs or grass; site usually covered from behind. Normally face west; several sites used over a season— changes probably due to weather (Mackowicz 1970). Longer grass, heather *Calluna*, and earth clods used for shelter in bad weather (Krampitz 1952; Harrison and Forster 1959). Use of pits for roosting confirmed only in captive birds but suspected also in wild birds; pits deep so that bird's back at or below ground level (Pätzold 1971). Breeding pair roost close together in territory; site (near later nest) marked by accumulated faeces (Mackowicz 1970; also Meinertzhagen 1951a). ♀ roosts on nest before clutch complete and continues until young c. 8 days old. Recorded roosting c. 2 m from nest after cessation of nocturnal brooding (Mackowicz 1970). Sleeping bird normally has bill tucked into scapulars. Posture rarely adopted by incubating ♀ during day; head usually just drawn in for naps of c. 3 s to 1½ min. On 12 April, in Rzepin, pair active from c. 05.30 hrs; first 4 hrs spent mostly feeding (Mackowicz 1970). In central Morocco, birds continued feeding in heat of day, but stayed in shade (Meinertzhagen 1940). No water-bathing recorded, but bathing in loose sand frequent (Mackowicz 1970). Occasionally drinks from puddles (Mackowicz 1970).

2. Not shy where free from persecution and usually flies up only at close approach (Naumann 1900; Labitte 1958). Fighting birds, also singing ♂ and nearby ♀, may be almost oblivious of observer (Krampitz 1952; Mackowicz 1970). When disturbed, sometimes shuffles along, stretching and rapidly retracting neck. ♂ sighting danger gives alarm-call (see 3b in Voice); ♀ then also calls (Mackowicz 1970). Excitement may be expressed in raising of crest and bobbing movements accompanying calls (Harrison 1970a; Pätzold 1971). ♂ tends to take off first, ♀ following. On landing, ♀ usually goes straight into cover, ♂ to exposed perch (Krampitz 1952). If ♀ flushed, ♂ tends to interrupt Song-flight and drop down to join her. Birds normally stay in territory, not moving far (Labitte 1958). Usually quiet near nest-scrape; no alarm-calls given when closer than 20-40 m (Mackowicz 1970). Will squat where cover available if approached. In presence of raptor, will squat in grass; in open ploughland, may at first freeze, then slowly squat. Recorded flying up into tree when Sparrowhawk *Accipiter nisus* passed (Meinertzhagen 1951a); said by Meinertzhagen (1940) to be possibly not typical, but a normal reaction when disturbed according to Harrison (1970a). Difficult to flush from roost or shelter (Krampitz 1952). FLOCK BEHAVIOUR. Large autumn flock less tightly packed than Starling *Sturnus vulgaris*, more like Fieldfare *Turdus pilaris* (Schumann 1959). Birds fairly close together for winter roosting (Meinertzhagen 1951a). Feeding flocks generally quiet; if alarmed, bird gives call 5 (see Voice), others usually join in, this then signalling departure (Mackowicz 1970). SONG-DISPLAY. ♂ sings (see 1 in Voice) in flight or from tree, overhead wire, mound, or ground (Nethersole-Thompson 1932; Dementiev and Gladkov 1954a). Song-flight important for self-advertisement and territory proclamation; performed mainly by single ♂♂ looking for mate, also in period soon after pairing (Mackowicz 1970). ♂ makes angled ascent, normally from tree-top; spirals up, then circles at fairly constant height— usual maximum c. 100 m if unpaired, c. 50 m if paired. Paired ♂♂ generally sing less intensely, less continuously, and in briefer Song-flights. If bird takes off from ground, song starts c. 10-20 m up (or at least when above trees). Sings all the while, moving in irregular loops and spirals, dipping and rising in sweeping curves. Flight often rather weak and hesitant, fluttering and bat-like; less commonly, hovers over one spot (Nethersole-Thompson 1932; Witherby et al. 1938a; Krampitz 1952; Mackowicz 1970; Pätzold 1971). Song, to some extent coinciding with rhythm of rapid wing-beats, given while ascending slightly; ensuing silent descent also slight and made with closed wings (E N Panov). In early stages (see below), ♂ may simply cross territory and give brief song-phrases. Later (especially in fine, sunny weather), circles round territory may be 30-80 m across (Mackowicz 1970); often moves far beyond confines of nesting area according to Wadewitz (1957), over much wider area than *A. arvensis* (Witherby et al. 1938a).

Descent may be gradual and spiralling with continuous song, or (less frequently), song broken off at 50-30 m, bird then plummeting with closed wings to same or neighbouring perch or ground, wings opened just before landing (Dallas 1928; Witherby et al. 1938a; Dementiev and Gladkov 1954a; Pätzold 1971). According to Mackowicz (1970), faster drop only from tree-top height. Rare variant of Song-flight involves almost vertical ascent and descent to starting point (Witherby et al. 1938a). Song-flight usually lasts c. 2 min (Mackowicz 1970); average in May 1·6 min (Pätzold 1971). Unpaired ♂♂ sometimes stay aloft for much longer: 70 and 94 min recorded (Mackowicz 1970); c. 30 min at night (Davenport 1922). 48% of song given in flight on a sunny May day (Pätzold 1971). Singing bird may stimulate others to perform Song-flight (Krampitz 1952), so that up to c. 12 noted at once (Nethersole-Thompson 1932). Brief Song-flight on arc-like flight-path occasionally performed by ♀ (Krampitz 1952). ♂'s Song-flight often preceded or followed by song from perch or ground; one performed Song-flight for c. 30-35 min, during brief pauses singing with same intensity on tree (Krampitz 1952). Song from look-out given predominantly from start of incubation; also in autumn when normally brief, though up to 34 min recorded (Mackowicz 1970). Vigorous song from perch towards end of nestling phase (after temporary break during brooding) possibly prelude to 2nd brood (Krampitz 1952); see also Koch (1936). Song from ground given by unpaired and paired ♂♂ apparently to mark occupancy of territory; no longer given a few days after start of incubation. Usually lasts 1-2(-14) min with intervals between bursts of up to 1 min. May replace Song-flight in poor weather (Mackowicz 1970). For song on ground by ♀, see Relations within Family Group (below). Maximum song activity in early morning: peak for unpaired ♂♂ 07.00 hrs in April and 04.00 hrs in June; paired ♂♂ c. 1 hr later and tend to cease at c. 10.00 hrs. Unpaired ♂♂ often sing before sunset (c. 18.00-19.00 hrs), especially May-June (Mackowicz 1970). Autumn song also mainly in morning (Schumann 1959). Nocturnal song not infrequent, including in flight; warm, windless, and moonlit nights favoured; occasionally sings throughout night (Davenport 1922; Labitte 1934; Dementiev and Gladkov 1954a; Wadewitz 1957). Start of singing strongly influenced by weather and, especially, day-time temperature (Mackowicz 1970; Pätzold 1971). In Britain, main song-period March to mid-June; irregular but fairly frequent song given late January and February, and mid-September to mid-October; exceptional, or only Subsong, late October and November (Witherby et al. 1938a). In Eure-et-Loir (France), normally from early February; autumn and (exceptional) winter song possibly given by young birds (Labitte 1958); see also Voice. In Frankfurter Stadtwald, Song-flights performed by unpaired ♂♂ from about mid-February, high Song-flights, late March and April, probably only from late passage migrants (Krampitz 1952). In Rzepin, song gradually increases from March with rise in temperature to reach peak in 2nd half of June (Mackowicz 1970). In Morocco, song apparently general from 1st week of April (Lynes 1933). In Iraq, fragmented song given from ground 18 February (Moore and Boswell 1956). Will sing while actively migrating (Hasse 1965). ANTAGONISTIC BEHAVIOUR. (1) General. In Rzepin, first territories normally occupied c. 10 days after arrival. Most defence of territory (fights, chases) falls during copulation period and early incubation but then gradually wanes, and ceases with hatching, e.g. intruding ♂ may be allowed to sing directly above. Fights may, however, take place at any season including in autumn flocks (Mackowicz 1970). In Surrey, territory generally occupied well in advance of breeding. In one case, most intrusions from contiguous territory took place in corner not overlooked from defending ♂'s main look-out. When pair left territory after loss of young during fledging, different ♂ occupied it within 24 hrs (Harrison and Forster 1959). Territorial antagonism may be pronounced (Almásy 1896) or almost non-existent (Meinertzhagen 1951a; Krampitz 1952). (2) Threat and fighting. In low-intensity advertisement of occupied territory, ♂ may stand on elevation, with feathers ruffled, and indulge in nervous, ritualized self-preening (Mackowicz 1970). In Transcaucasia (USSR), 2 birds (probably ♂♂) recorded singing in flight along boundary and occasionally fighting briefly; feathers on back of head ruffled (E N Panov). In Rzepin, neighbouring ♂♂ often sing without regard to each other, and boundary disputes rare. However, advent of strange ♂ within territory usually means resident ♂ will cease singing and attempt to drive intruder to territory limit. Particularly fierce fight (c. 5 hrs with breaks) ensued when 2nd ♂ arrived during bout of courtship between territorial pair; ♀ did not participate (Mackowicz 1970). Non-participation by ♀ typical; ♂ usually returns to her afterwards and may 'caress' her (no details), then fly up to sing (Nethersole-Thompson 1932). In aerial chase (often protracted), birds usually fly low; pursuer may have dangling legs, and gives rather vehement song (variant of that given on ground) alternating with call 6 and Bill-snapping. On ground, birds advance rapidly in Threat-posture: plumage ruffled, head held low and thrust forward, tail spread; wings quivered (rapid up-and-down movement of wing-tips) and bowing movements made (Mackowicz 1970); small downward pecking movements (Harrison 1970a) perhaps the same. In antagonistic encounter lasting c. 4 min, birds faced one another in Threat-posture (Fig A). Slow advance by 1 bird accompanied by slight impulsive fanning and closing of tail in both. From time to time, one or other would turn round, then again face opponent bill-to-bill. Both made several attempted attacks, and gave loud calls (see 3b in Voice) and occasional short song-phrases throughout conflict (E N Panov). Combatants also recorded high wing-lifting (2–3 times), nodding, ground-pecking, and running about; also sang in short bursts (May 1947). Another dispute between 2 ♂♂ involved breast-to-breast buffeting just above ground, chasing c. 10 m above trees (birds once making physical contact), and fighting on ground in which one bird called (see 8c in Voice) on being pecked. No damaging fights recorded (Mackowicz 1970). In experimental study to determine reactions of territorial ♂♂ to playback of normal and modified song-phrases, reactions varied from conspicuous turning towards sound-source, through song and calls, to swooping attack and landing by sound-source (Tretzel 1965b). Much violent competition and chasing occurred where breeding in same field as *A. arvensis* (Meinertzhagen 1951a). One ♂ recorded attacking Great Grey Shrike *Lanius excubitor* in territory (Krampitz 1952), and another swooped at *S. vulgaris* which was finally driven off by ♂ and ♀ in Threat-posture. Yellowhammer *Emberiza citrinella* also threatened and attacked (Mackowicz 1970). For experimental work to test responses to various stuffed birds (e.g. Cuckoo *Cuculus canorus*) at nest, see Smith and Hosking (1955) and Pätzold (1971). HETEROSEXUAL BEHAVIOUR. (1) General. In Rzepin, c. 50% of ♂♂ paired in about 3rd week after arrival in mid-March, ♀♀ normally appearing (or at least becoming obvious) in territories towards

end of March (Mackowicz 1970); similarly in Frankfurter Stadtwald, where ♀♀ apparently arrive on breeding grounds at same time as ♂♂, but not confirmed that pair-formation takes place on or before migration (Krampitz 1952); see also Pätzold (1971) who noted that pair-formation does not take place territory; scrapes 8 m apart, 38 m from roost, but site eventually used was c. 40 m away from these and c. 10 m from roost. Another pair made 2 scrapes and nest built in one 4 days later (Mackowicz 1970; also Krampitz 1952). In Surrey, following copulation, ♀ immediately started to make scrape on bare ground, gyrating with feathers ruffled; nest eventually built c. 440 m away (Hoffman 1951). At Budaörs, each pair usually makes 2–3(–6) or more scrapes c. 2 m or up to 100–800 m apart (Koffán 1960). In Rzepin, 2–4 scrapes usual; often in twos (made by ♂ and ♀) and then c. 5–10 m apart (Mackowicz 1970). Actively scraping bird has breast pushed to ground and tail pointed straight up. ♂ and ♀ recorded working in same scrape and carrying material. Even after nest-site chosen, birds apparently still visit other scrapes and will remove snail shells placed in them (Koffán 1960). (6) Behaviour at nest. ♀ may sit on nest irregularly before laying; often restless, leaving nest frequently (Mackowicz 1970) and spending most time away feeding (Krampitz 1952). ♂ may call ♀ off to feed. Once, attacked her vigorously; fight involved head-pecking, upward leaps, and buffeting. Afterwards, ♂ flew off, ♀ shook herself. Aggression possibly elicited by ♀'s crouching to take grit (Mackowicz 1970). When ♀ incubating, ♂ mostly on look-out where may sing quietly; only occasionally performs Song-flights. Some quiet reciprocal calling occurs. ♀ leaves in response to ♂'s calls, or perhaps when she sees him approaching silently. ♀ leaves c. 12 times per day to feed, etc. (Mackowicz 1970, which see for incubation shifts). Breaks thus about every hour; rarely exceed 20 min (Nethersole-Thompson 1932), though up to c. 2 hrs recorded early in incubation (Pätzold 1971). ♂ usually escorts ♀ while she is not incubating, also returning to nest with her (Nethersole-Thompson 1932; Trimingham 1956). At one Swiss site, ♀ always approached nest from same direction (Leuzinger 1955). In Rzepin, ♀ usually landed c. 6 m or less from nest on return (Mackowicz 1970). At Budaörs, where nest exceptionally c. 200 m from 'territory', ♂ apparently only fed with ♀, rarely escorted her back to nest, etc. (Koffán 1960). Incubating ♀ continues to shape nest intermittently until hatching. Also preens, sleeps (but alert when ♂ not on look-out, and will drink dew or rain drops without leaving, and may rush out to take (e.g.) passing ants (Formicidae) (Mackowicz 1970). RELATIONS WITHIN FAMILY GROUP. Duration of day-time brooding varies with number of young: with 4–5 young, ceases more or less on 5th day, 3 young on 6th, 2 young on 8th day (Mackowicz 1970); brooded at any stage if weather bad (Leuzinger 1955). Nocturnal brooding may continue for up to 7 days (Pätzold 1971) or for 8–11 days (Mackowicz 1970); only 3–4 days according to Krampitz (1952). Brooding stints brief (a few minutes). ♀ also shelters young from rain and hail, and spreads wings and ruffles feathers to protect them from sun. At 6 days, young pant and gape in heat. Adults also remove ectoparasites from heads of young and nest. Feeding begins in first few hours after 1st young hatched. ♀ invariably starts, ♂ (apparently stimulated by sight of ♀ doing so) following c. 20 min to 6½ hrs later. Before leaving nest, ♀ normally thrusts

A

head forward and up, looking about alertly (Mackowicz 1970). ♂ and ♀ together more or less constantly once brooding ceases (Koch 1936). Often (80% of cases) both parents come to nest simultaneously (Mackowicz 1970), announcing arrival with call 4 (Steinfatt 1939a; P A D Hollom). Cautious approach to nest typical, birds landing c. 5-12 m away and making good use of cover while running to nest (Ashford 1922; Leuzinger 1955). Almost crawl if sense danger. May use regular route (Pätzold 1971), or make wide detour (Leuzinger 1955). Usually move c. 1-10(-15) m from nest before flying up (Barber-Starkey 1909; Steinfatt 1939a; Labitte 1958; Koffán 1960). Recorded flying (sometimes pair together) to tree c. 30 m from nest, then moving in on ground over c. 15-20 m (Leuzinger 1955). In early stages, ♀ generally approaches first and settles to brood young after feeding them; rises immediately, however, if ♂ approaches. ♀ may then sing c. 0·5-1·5 m from nest, (mock-)peck at ground and (mock-)preen. ♂ once sang when ♀ reached nest ahead of him (Mackowicz 1970) and sometimes gives snatches of song at departure (Steinfatt 1939a). In another locality, ♂ tended to raise crest more than ♀ and to call more vigorously and loudly immediately after arrival. In Britain, courtship display (see below) also recorded in autumn (Witherby et al. 1938a). In 3 cases, c. 6 days between pair-formation and laying (Koffán 1960). (2) Pair-bonding behaviour. In first encounter, ♂ may cease singing and dive excitedly at ♀; crest raised, but tail not spread. Chases ♀ away, giving loud but strangled song, and sometimes Bill-snapping. Afterwards, ♂ may make abrupt wing- and tail-jerking movements for some time (Krampitz 1952; Mackowicz 1970). Heterosexual chases apparently common during courtship and may even involve established pairs (Krampitz 1952). ♂ may also sand-bathe between bouts of song; possibly displacement activity resulting from courtship excitement. During pairing, neighbours may answer each other's contact-calls (see 3a in Voice), leading to 'conversation' between all members of a population; this rare later (Krampitz 1952). Once ♀ accepted, ♂ makes frequent (and lower—see Song-display, above) Song-flights, invariably landing by her (Mackowicz 1970). Following probably early courtship (21 March): ♂ and ♀ called (see 3a-b in Voice); ♂ moved on to elevation and repeatedly quivered wing-tips, bowed, and sang; moved trippingly forward, ♀ then calling excitedly and advancing towards ♂; when birds c. 20-30 cm apart, ♂ turned, moved away with slightly ruffled breast and tail slightly raised, and ♀ followed; ♂ finally squatted low, tail widely spread; birds disturbed after c. 11 min. ♂ typically quivers wings before and during courtship, bows low and assumes Tail-up posture (see *G. cristata* and *A. arvensis*); plumage ruffled, head drawn back and rear thrust out; also turns in this posture. ♂ follows ♀ or solicits copulation (see below) (Mackowicz 1970). Brief description in Moore and Boswell (1956) referred to ♂ drooping half-spread wings before passive ♀. In another account, both birds 'ducked and bobbed' excitedly, crests erect, tails opened and closed. ♀ may call (see 8b in Voice) and ♂ may give snatches of song or calls, or may flutter over ♀ (Witherby et al. 1938a). After pairing, ♀ may interpolate contact-calls into ♂'s song or perform Song-flight herself. ♀ also once recorded standing erect on mound, breast feathers ruffled. Birds may flutter up together in apparent mock-battle; tails spread (Krampitz 1952); ♂ sings and ♀ may also give a few phrases (Pätzold 1971). Between pair-formation and beginning of incubation (in Rzepin, for c. 3-4 weeks from end of March) ♂ often follows ♀ closely, sometimes for hours at a time. Birds normally quiet but soft contact-calls sometimes given. ♂ frequently pauses on elevation, preens, or sings; body fairly erect, tail closed and pointed vertically down; wings and tail trembled (Krampitz

1952; Mackowicz 1970). This following of ♀ by ♂ considered by Mackowicz (1970) to be important for territory demarcation, feeding, and nest-site selection, but link with nest-site selection not proven according to Krampitz (1952), and behaviour continues after laying (Pätzold 1971). Most likely due simply to ♂'s need to guard ♀ from insemination by intruding ♂. See also subsection 5 (below). (3) Courtship-feeding. According to Niethammer (1937), Dementiev and Gladkov (1954a), and Portenko (1954), ♂ feeds ♀ on nest. Not recorded in Rzepin, though in one case, ♂ came to nest with food at start of incubation as if to feed young; ♀ left (Mackowicz 1970). (4) Mating. Few details. In Rzepin, ♀ approached displaying ♂ then crouched and fluttered wings, but copulation did not take place due to arrival of another ♂ (Mackowicz 1970). In Surrey, one copulation followed bout of feeding and display (no details) by ♂ (Hoffman 1951). In one case, unsuccessful mounting attempts by ♂ led to reversed copulation (Witherby et al. 1938a). At Budaörs, ♂ frequently attempted copulation during apparent nest-site selection, but his advances rejected by ♀ (Koffán 1960). (5) Nest-site selection. During pair's period of close association, will apparently examine depressions, pushing into grass clumps, picking up material, etc., before site chosen. Occasionally show aggression with excited calls, raising of crests, and repeated feather shaking. Selection probably gradual; difficult to determine how and when choice made. In one case, ♀ apparently solicited copulation, then both moved away and started making scrapes; ♂'s site eventually chosen (Koffán 1960). In Rzepin, ♀ thought to select scrape to be used. One pair started making scrapes from c. 10 days after occupation of (Wyss 1947). Up to 4th day, ♀ spends much time brooding young so that ♂ does more foraging; ♀'s share greater overall, however. Later on, ♂ more frequently approaches nest first. From 7th day, parents often feed young together, standing side by side (Mackowicz 1970). For feeding, adult stands over nest and calls quietly. Initially, young beg by raising head and gaping silently; when able to see, head-sway also (eyes open to slit by 5th, wide by 6th-7th day). ♀ may stimulate wider gaping by touching inside of nestling's mouth and nodding. Food placed into nestling's mouth. Food-calls then reach peak, subside, and cease after adult departs (Mackowicz 1970). Later, food-calls given on hearing parents c. 10-15 m from nest (Steinfatt 1939a). Young silent if adult merely awaits defecation. Faeces of young initially swallowed; later, usually carried away and dropped (Mackowicz 1970) at c. 4-5 m before flying up; exceptionally fed to young (Pätzold 1971). Young tend to move closer together in nest after first few feeds; by 3rd day, turn towards middle of nest, and from 5th day keep heads directed to where parents normally approach. Remain in nest 10-13 days (Mackowicz 1970) or 12-14 days (Koch 1936); leave earlier (8th day) if disturbed (Mackowicz 1970). Once, moved to another depression, staying there 1-2 days as if in nest; parents at first made mock-feeding movements at abandoned nest (Pätzold 1971). From 9th day, young raise crest when excited (Mackowicz 1970); may move c. 1 m outside nest to meet incoming parent. Beg by gaping wide (yellow gape and black tongue spots prominent) and shivering wings. Wing-stretch before exit and after return to nest (Steinfatt 1939a). Young eventually walk away from nest, normally in direction from which parents usually return with food; any remaining runt still fed (Mackowicz 1970). In one case, left nest over 1 day (Krampitz 1952). May develop and use runs through grass, and occasionally squat together in hollows (Harrison 1970a) though typically scatter rapidly over small area staying mobile and dispersed (Mackowicz 1970). Feeding continues, and parent may walk along and feed young almost without stopping (Harrison 1970a). ♀ may start new nest when young out of nest but still dependent (Koffán 1960). Young fly moderately well from 16th day (Koch 1936). May stay in general area for at least 1 month (Labitte 1958), but do not feed in same territory if another brood started (Mackowicz 1970; also Nethersole-Thompson 1932). 2nd-brood family remains together longer. In one case, only ♀ still fed full-grown young; ♂ often separated off to sing from trees or in flight (Krampitz 1952). Family may eventually move to fresh feeding grounds and unite with others in autumn flock (Mackowicz 1970). For more details, including physical development of young, see Mackowicz (1970) and Pätzold (1971). ANTI-PREDATOR RESPONSES OF YOUNG. Will run away from nest on 9th day (Mackowicz 1970). Young removed from nest at 10 days old crouched with head down when placed on ground. Moved into cover as observer left. If approached from front will move backwards into cover (Pätzold 1971). Young out of nest usually crouch silently if threatened or on hearing parental alarm. May allow close approach after adults have flushed, then scatter and freeze again in cover, leaving only when parents' calls cease. May use rather jerky hopping motion and flap wings in escape bid; give loud fright-calls or are silent, similarly when handled. If one of brood attacked by Red-backed Shrike *Lanius collurio* or Jay *Garrulus glandarius*, others scatter (Krampitz 1952; Meinertzhagen 1951a; Pätzold 1971). PARENTAL ANTI-PREDATOR STRATEGIES. (1) Passive measures. Some ♀♀ will desert during nest-building if disturbed (Koffán 1960). ♂ may sing at approach of man; otherwise gives alarm-calls from look-out. ♀ then also leaves or, if ♂ has departed earlier, crouches low on nest (Trimingham 1956; Mackowicz 1970) and may call (Haun 1930). ♂ once joined ♀ immediately she left, birds then attempting to get behind observer; ♀ gave quiet calls (see 3a in Voice) and birds perched conspicuously, sometimes close together (Krampitz 1952). 3rd bird may join in vocal demonstration (Nethersole-Thompson 1932). See also Schmitt and Stadler (1915) and Haun (1930). ♀ generally sits tighter than *G. cristata* or *A. arvensis* (Pätzold 1971), and more likely to do so towards end of incubation and just after hatching, then permitting very close approach (less than 1 m) and even allowing herself to be touched (Koffán 1960; Christian 1965; Mackowicz 1970). Flushes more readily if approached from behind (Mackowicz 1970). One ♀ sat tight also at night, tolerating use of flash with observer at *c.* 0·5 m (Pätzold 1971). Sometimes slow to return after disturbance (Nethersole-Thompson 1932). In one case, ♀ remained away while observer in hide; ♂ sang constantly nearby and finally came to nest but did not incubate (Wadewitz 1957). (2) Active measures: against birds. Tends to freeze in presence of raptors. ♀ did so also near nest when *C. canorus* nearby. ♂ may give alarm-calls in response to those of certain other species in area (Mackowicz 1970), or, when *A. nisus* overhead, fragment of song (Wyss 1947). Loud and persistent alarm-calls given spontaneously by ♂ on sighting Black Kite *Milvus migrans*, Buzzard *Buteo buteo*, *L. excubitor*, and *G. glandarius*, though Kestrel *Falco tinnunculus* regularly hunting in area ignored. ♀ avoided nest when raptor present, but recorded crouching and sand-bathing when *M. migrans* near (Krampitz 1952). When young out of nest attacked by *L. collurio*, ♀'s vigorous defence led to fight which ended only with intervention of observer (Mackowicz 1970). (3) Active measures: against man. On flushing from nest, ♀ likely to perform distraction-lure display of disablement type; this especially likely late in incubation (but even from start: Labitte 1958), and generally not used after brooding ceases (Krampitz 1952). Bird may fly low over ground (Krampitz 1952) or flutter wildly but briefly over crops (Leuzinger 1955). Either directly

from nest or after low flight, ♀ may shuffle or drag herself slowly along in crouched posture with head close to ground, tail widely fanned and depressed, and half-open wings fluttered rapidly; or may run rapidly, wings trailing loosely, and swish through grass. Wing and tail markings prominent in posture assumed by agitated ♀ (see Fig B). May stop briefly every few

B

metres, breast on ground, wings quivering, and look back at observer, or move 15-20 m without pause then disappear into cover and finally fly up to join ♂ (Witherby 1936; Krampitz 1952; Trimingham 1956; Mackowicz 1970). If observer stays by nest, ♀ may land first at *c.* 6-8 m, then fly on for repeat; gives up after *c.* 10-15 s if lure not effective (Mackowicz 1970). Birds may then return, ♂ giving loud alarm-calls; ♀ recorded flying low round intruder and suddenly fanning tail, also landing with food and sand-bathing (Krampitz 1952). Parents may try to entice young away from nest subjected to serious disturbance (Pätzold 1971). Both adults gave alarm-calls when family party away from nest was closely approached (Trimingham 1956); in this situation, ♂ usually flushes first, then ♀ (Krampitz 1952).

(Figs by J P Busby: A from drawing by E N Panov; B from drawing in Krampitz 1952.)   MGW

Voice. Most calls characteristically melodious (Wadewitz 1957b). ♂ generally more vocal, and more likely to give variants, than ♀ (Krampitz 1952; Pätzold 1971). For musical notation and analysis, see Schmitt and Stadler (1915), also Voigt (1933). For extra sonagrams, and experimental attempt to determine characteristic and response-releasing components of song, see Tretzel (1965b); see also Bergmann and Helb (1982). For Bill-snapping, see Social Pattern and Behaviour.

CALLS OF ADULTS. (1) Song of ♂. Unlikely to be confused with other west Palearctic Alaudidae. Series of remarkably rich and full-sounding phrases separated by pauses. Clear, liquid, mellow, pleasantly fluting sounds given in descending pattern—pitch (2-4 kHz) normally falls gradually through each phrase (Tretzel 1965b); effect extraordinarily mellifluous, melodious, rather melancholy, and soulful (Nethersole-Thompson 1932; Voigt 1933; Witherby *et al* 1938a; Wadewitz 1957b; Bergmann and Helb 1982). According to Mackowicz (1970) whose study in western Poland involved no sonagraphic analysis, 5 different phrase-types may be distinguished based on sound of particular units and rhythm of phrase: slow, low-pitched disyllabic (see below) units—'di lee', 'de lee', or 'd'lee'; slightly faster, higher-pitched 'dee yah', 'diyah', or 'dyah', resembling part of song of Tree Pipit *Anthus trivialis*; low-pitched 'loo'; higher-pitched 'dli'; 'li' sounds given in fast tremolo. Birds give mostly phrases with 'dli' units (40-45%), and 'loo' type also common.

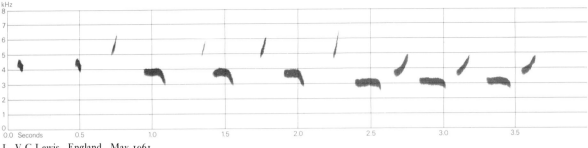

I V C Lewis England May 1961

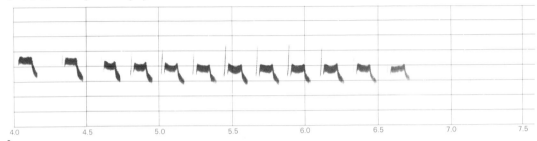

I *cont.*

Contra (e.g.) Heinroth and Heinroth (1924-6), many different combinations of 5 phrases produce very varied song. No set rule governing sequence of phrases (see Mackowicz 1970). Phrase often begins with a few modified introductory units and may contain mixture of unit-types (Tretzel 1965*b*). Recording of bird singing from perch analyzed in detail by J Hall-Craggs. Although recording is of only 1 bird, repertoire revealed to be far more complex than indicated in Polish study (see above); support for this in Tretzel (1965*b*). Song comprises 13 phrases (1st probably incomplete) with total of 22 different segments (runs of uniform units) and 24 unit-types; only 3 cases of repetition of unit-types over whole bout of 13 phrases. If syllable used to mean discrete and non-discrete sub-units, units mostly 3-syllable (13 out of 24), but disyllabic (7), monosyllabic (3), and (possibly) 4-syllable units occur; however, some syllables not audible unless playback speed reduced. Phrases mostly of 1 (long) segment (8 out of 13); also 1 of 2 segments, 2 of 3 segments, and 1 of 4 segments (see Fig I). Monosyllabic units of 'ti', 'te', or lower-pitched 'tu' type. Disyllabic units include 'oo-li' (as a trill), also slow 'uip' (developing through accelerando 'yip' to fast 'y-i-i-i-i...') and, with falling pitch, 'dio'. Some 3-syllable units rise in pitch—e.g. 'tlooeee' (sharp onset, then steep fall before a portamento rise); others show fall (varyingly steep)—'k-teeoo', '(t)sioo', 'de leeoo', also 'siyah' descending in same segment to 'sioo'; some units have sharp end transient—'deelet' and 'te leit'. Fig I thus rendered 'ti ti k-teeoo k-teeoo k-teeoo k foo-ee foo-ee foo-ee (t)sioo (t)sioo (t)sioo (t)sioo (t)sioo (t)sioo (t)sioo (t)sioo (t)sioo (t)sioo (t)sioo (t)sioo'. Phrases tend to start quietly followed by crescendo or by crescendo, diminuendo, and accelerando; pitch usually descends gradually from start to finish. Single-segment phrases in particular start quietly; this followed by crescendo, and always by accelerando. Multi-segment phrases do not show same overall gradual change in intensity, delivery rate of units, and pitch change, but variety seen rather in transition from one segment type to another: see (e.g.) in Fig I, abrupt change from segment 3 to 4 and subtle transition via pivot note (last unit of one segment same as, or very similar to 1st unit of next) from segment 2 to 3. Occasionally, transition takes form of sudden drop in pitch. Also in 12th phrase, 2 introductory units followed by unusually long pause; 2 units repeated but preceded by different unit which then reappears as introduction to segment of 6 trisyllabic units and apparently functions as lead into something stronger or of greater importance. In Fig II (best treated as 1 segment), 4 'tsee' units followed by 14 'twee' units in good example of drift. (J Hall-Craggs.) Each ♂ has many different phrase-types in repertoire (Bergmann and Helb 1982). Usually 5-8 phrases per bout, more than 10 rare according to Schmitt and Stadler (1915). Phrases comprise 7-15 units, each of 2-3(-4) sub-units (see above); in typical unit, 1-2 short, usually higher-pitched sub-units precede longer, normally lower-pitched sub-unit (Tretzel 1965*b*). Phrase lasts 2-3 s (Tretzel 1965*b*), 3-5(-8) s (Witherby *et al.* 1938*a*), and average in Polish study 9.7 s, with average 6.9 disyllabic units per phrase ($n = 13$) and 11.0 'dli' units ($n = 4$) (Mackowicz 1970). Omitting probably incomplete phrase in recording, mean phrase duration 4.49 s, median 4.18 s ($n = 12$) (J Hall-Craggs). In Song-flight, average 8.7 phrases per min (Mackowicz 1970). Pauses between phrases *c.* 2-3 s, sometimes even shorter (Bergmann and Helb 1982); see also Witherby *et al.* (1938*a*). Song from perch may be given with longer pauses; average 4.4 phrases per min (Mackowicz 1970); in recording of perched bird, however, pauses mainly 0.2-1.0 s, in 2

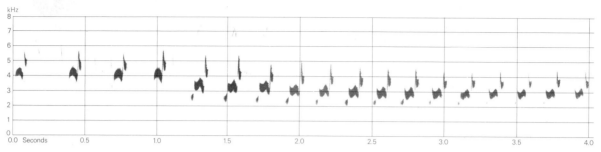

II   V C Lewis   England   May 1961

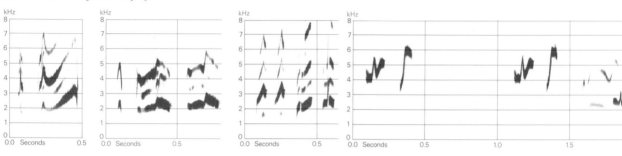

III   V C Lewis
England
April 1961

IV   A J Williams   England
June 1977

V   P A D Hollom
England   April 1980

VI   P A D Hollom   England   April 1980

cases much longer at 2·5 s and 3·5 s (J Hall-Craggs). Song given on ground varied but usually quiet and can be difficult to locate (Krampitz 1952); normally brief and with pauses of up to 1 min (Mackowicz 1970). Rather quiet variant given during incubation comprises fragments of song mixed with call 3 and long pauses (Mackowicz 1970). For other contexts, see Social Pattern and Behaviour. Mimicry discussed in detail only by Krampitz (1952); possibly given only in song from perch. Includes song-phrases of Swallow *Hirundo rustica*, Great Tit *Parus major*, Skylark *Alauda arvensis*, and Chiffchaff *Phylloscopus collybita*, also whole song of *A. trivialis*. (2) Song of ♀. In Song-flight, a combination of call 3 and twittering sounds (Krampitz 1952). On ground, each rapid phrase comprises call 3 followed by a trill; given mostly during nestling phase (Mackowicz 1970). ♀♀ begin to sing in 1st autumn (Heinroth and Heinroth 1924-6). (3) Contact-alarm call. Harder or softer in tone depending on mood (Pätzold 1971). Best-known and most frequently uttered call. Given throughout the year in a variety of situations (perched or in flight): when feeding, resting, on nest, or tending young; also during courtship and to express anxiety (V C Lewis); frequently given by migrants (Sultana and Gauci 1982). 2 (1-3) syllables. Shows individual variation. Renderings include 'd'lui', 't'tluiii', 'dit-tlui', 'tew leet' (Mackowicz 1970); similar to last is 'plew-lip' given by adult approaching nest to feed young (P J Sellar: last unit of Fig VI). Shorter variants: 'dli', 'tli', and 'lui'; very quiet when given at or near nest— not audible beyond c. 5 m (Mackowicz 1970); 'tüje', 'liüt', and 'üt' given mainly by ♀ when disturbed and as warning to young (Krampitz 1952). For further renderings

and situations, see (e.g.) Voigt (1933), Krampitz (1952), Wadewitz (1957b), Koffán (1960), and Bergmann and Helb (1982). In some variants, usually given on ground, mainly in territory, and especially during period of territorial antagonism, 2nd syllable (or 2nd and 3rd in 3-syllable calls) carries stress, particularly evident as excitement increases: 'tch'tui', 'didluiit', or 't'luiiluit'; final 'uii' loud and far-carrying (Mackowicz 1970). Recording (Fig III) suggests 'tiTUEEi(t)' in which final syllable faint and barely audible; similar unit occurs in song (J Hall-Craggs); 'dipchuEE(t)' (M G Wilson). When disturbed, melodious, multisyllabic calls often given as a series: 't-dlülerid-tlilerid', etc. (Bergmann and Helb (1982). Such calls evident in recording: e.g. 'ke kooee kooii' (Fig IV)—compare pitch pattern with Fig III (J Hall-Craggs). See also Nethersole-Thompson (1932) and Steinfatt (1939a) for trilling calls given by disturbed birds. Broken song-phrases also typically given by excited ♂♂ (Tretzel 1965b). For further descriptions and variants, many associated with alarm and excitement, see Krampitz (1952), Wadewitz (1957b), and Pätzold (1971). (4) Rather

PLATE 8 (*facing*).
Galerida cristata Crested Lark (p. 145). Nominate *cristata*: **1** ad fresh (autumn), **2** ad worn (spring), **3** juv. *G. c. subtaurica*: **4** ad fresh (autumn). *G. c. riggenbachi*: **5** ad fresh (autumn). *G. c. macrorhyncha*: **6** ad fresh (autumn). *G. c. arenicola*: **7** ad fresh (autumn). *G. c. festae*: **8** ad fresh (autumn). *G. c. nigricans*: **9** ad fresh (autumn). *G. c. maculata*: **10** ad fresh (autumn). *G. c. brachyura*: **11** ad fresh (autumn).
Galerida theklae Thekla Lark (p. 163). Nominate *theklae*: **12** ad fresh (autumn), **13** ad worn (spring), **14** juv. *G. t. ruficolor*: **15** ad fresh (autumn). *G. t. erlangeri*: **16** ad fresh (autumn). *G. t. carolinae*: **17** ad fresh (autumn), **18** ad worn (spring). (DIMW)

PLATE 11. *Eremopterix nigriceps* Black-crowned Finch Lark (p. 46): **1–2** ad ♂, **3–4** ad ♀. *Eremalauda dunni* Dunn's Lark (p. 54): **5–6** ad. *Ammomanes cincturus* Bar-tailed Desert Lark (p. 59): **7–8** ad. *Ammomanes deserti* Desert Lark (p. 65). *A. d. annae:* **9** ad. *A. d. algeriensis:* **10–11** ad. *Alaemon alaudipes* Hoopoe Lark (p. 74): **12–13** ad. *Chersophilus duponti* Dupont's Lark (p. 82). Nominate *duponti:* **14–15** ad. *C. d. margaritae:* **16** ad. (DIMW)

PLATE 12. *Rhamphocoris clotbey* Thick-billed Lark (p. 87): **1–2** ad ♂, **3–4** ad ♀. *Melanocorypha calandra* Calandra Lark (p. 93): **5–6** ad. *Melanocorypha bimaculata* Bimaculated Lark (p. 103): **7–8** ad. *Melanocorypha leucoptera* White-winged Lark (p. 109): **9–10** ad. *Melanocorypha yeltoniensis* Black Lark (p. 115): **11** ad ♂ fresh (autumn), **12–13** ad ♂ worn (spring), **14** ad ♀ worn (spring), **15** 1st-autumn ♂. (DIMW)

quiet falsetto 'priheet', 2nd syllable higher pitched. Often alternates with call 3 (Mackowicz 1970). A 'quit-quet' or 'kwe-kwe' and 'kwe-kwe-quoy-quoy' (or '-quer'), sometimes a little shriller; rather unlike typical *L. arborea* call (P A D Hollom); recording (Fig V) suggests 'kwi kwi kwu kwu', sometimes only 2(-1) upper or lower notes uttered alone, or order reversed (J Hall-Craggs); quiet, very like Sanderling *Calidris alba* at nest (P J Sellar). Common. Frequently given by both parents approaching nest to feed young, also by ♂ moving towards nest where ♀ incubating. Given additionally some distance from nest, also outside breeding season, and reciprocally between pair-members, when change to call 3 indicates relaxation of excitement (Mackowicz 1970). (5) Trilling 'peerrr' given by disturbed migrants; completely different from 'alarm-call' (higher-intensity call 3) given after pairing (Mackowicz 1970); 'bübübübü' or 'bibibibi' as contact between flock members in autumn (Krampitz 1952) possibly the same or a related call. (6) Rather coarse grating sounds, not unlike young Starling *Sturnus vulgaris*. Given (infrequently) during antagonistic interactions and may be combined with Bill-snapping (Mackowicz 1970). (7) At take-off, a subdued 'düg-düg...' (Bergmann and Helb 1982). (8) Other calls. (a) In courtship-display, ♀ may give single low call or quiet warble (Witherby *et al.* 1938a); possibly variants of calls 2 and 3. (b) Brief squeak once given by pecked bird during fight (Mackowicz 1970).

CALLS OF YOUNG. Initially quiet single food-calls barely audible *c.* 1 m from nest (Mackowicz 1970). Still quiet 'chip' and 'chep' sounds given by nestlings 5 days old (P A D Hollom); 'tee tee tee' eventually developing into persistent chorus (Mackowicz 1970). Nestlings expecting approach of adults give a 'siss-siss-siss'; food-calls of 2–3 syllables given from 10th day (Pätzold 1971). Shortly before leaving nest, young give fuller and quite loud 'twe-chep' or 'ee(t)-chip' (with incisive, positive 2nd syllable) when parents absent (P A D Hollom, P J Sellar: 1st 2 (discrete) sub-units of Fig VI); similar 'tseep tseeseep' and 'whee tsit sits' serve as contact-calls from *c.* 9th–10th day, also given intermittently from cover

when out of nest (Mackowicz 1970); see also Steinfatt (1939a). Sibilant, wheezy, rapidly repeated and insistent 'tseet', 'tseep', or 'seee' (Fig VII) apparently indicate

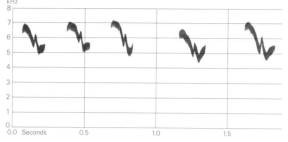

VII P A D Hollom England April 1980

food being passed (J Hall-Craggs, P A D Hollom); 'tsee tsee tsee tsee' like young *S. vulgaris* or woodpeckers (Picidae) (Mackowicz 1970); food-calls given by older young (also out of nest) still quiet and rather high-pitched according to Krampitz (1952) and Wadewitz (1957b). From 12th day, young occasionally give a coarse, quiet 'geet'; probably forerunner of adult call 3 (Mackowicz 1970). Young out of nest accompanied by adults gave a distinct 'pewdik-peedik', higher pitched than adult call 3; also when flushed, 'sisisisi' (Trimingham 1956). Fear expressed (when handled, etc.) by quiet but shrill 'geerreet', like adult call 6 (Mackowicz 1970); loud, anxious cheeping (Pätzold 1971) or fluting (Wyss 1947); 'fiet' sounds like pipit *Anthus* (Krampitz 1952). Song develops fully in 1st autumn (Pätzold 1971); 2 juveniles performing Song-flights at *c.* 10 m gave muted sounds with nasal timbre (see Schmitt and Stadler 1915). MGW

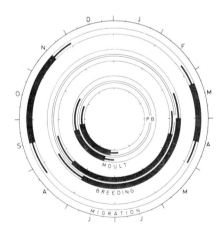

Breeding. SEASON. Britain and western Europe: see diagram; in Britain, exceptionally, eggs occur early March (Witherby *et al.* 1938a). USSR: laying starts up to 1 month later (Dementiev and Gladkov 1954a). North Africa: main laying period end of March to end of May (Heim de Balsac and Mayaud 1962). SITE. On ground, in shelter of scrub, bracken, or grass, sometimes at base of tree stump or sapling; rarely, on bare ground in the open. Majority of nests face between north-west and

PLATE 13 (*facing*).
Calandrella brachydactyla Short-toed Lark (p. 123). Nominate *brachydactyla*: **1-2** ad. *C. b. artemisiana*: **3** ad.
Calandrella rufescens minor Lesser Short-toed Lark (p. 135): **4-5** ad.
Galerida cristata Crested Lark (p. 145). Nominate *cristata*: **6-8** ad. *G. c. altirostris*: **9** ad.
Galerida theklae Thekla Lark (p. 163). Nominate *theklae*: **10-12** ad. *G. t. carolinae*: **13** ad.
Lullula arborea Woodlark (p. 173): **14-15** ad.
Alauda arvensis Skylark (p. 188). Nominate *arvensis*: **16-17** ad. *A. a. dulcivox*: **18-19** 1st ad non-breeding (1st winter).
Alauda razae Razo Lark (p. 207): **20** ad.
Eremophila alpestris Shore Lark (p. 210). *E. a. flava*: **21** ad ♂, **22** ad ♀. *E. a. bicornis*: **23** ad ♂, **24** ad ♀.
Eremophila bilopha Temminck's Horned Lark (p. 225): **25-26** ad. (DIMW)

south-east to avoid direct sun (Koffán 1960). Over 60% of nests distanced from nearest tall trees by factor of 1–2 times tree height (Mackowicz 1970). Solitary. Nest: deep depression in ground, lined with vegetation. Depression of variable shape, though generally circular (Koffán 1960), with diameter at mouth 110 mm (750-120), depth 55 mm (34-73), n = 11 (Mackowicz 1970). Base layer (c. 25% of total weight) of thicker stems, leaves, pine-needles, and some moss; main layer (70%) of grass leaves and stems; inner lining (5%) of finer grasses and some hair (Mackowicz 1970); average width 66 mm, depth 59 mm, with walls up to 99 mm thick at sides and 22 mm at base (Koffán 1960). Natural hollows not used (Mackowicz 1970). Usually 2–4(–6) scrapes made by each pair, in twos c. 10 m apart, one by each bird (Mackowicz 1970). Building: hollows built by both birds of pair, but actual nest solely by ♀; when excavating hollow, bird moves round in circles, moulding and digging with feet and pecking out loose material with bill (Mackowicz 1970, which see for detailed description and timings of nest-building). Hollows generally take 1–2 days to excavate, with nest-building up to 5 days, carried out in morning only (Mackowicz 1970). EGGS. See Plate 79. Sub-elliptical, smooth and fairly glossy; whitish, sometimes tinged olive, with fine spots and sometimes larger blotches of brown or grey-brown, often formed into band round broad end. 21·6 × 16·3 (19·7-23·5 × 15·1-17·4), n = 100, Britain (Witherby et al. 1938a); 20·7 × 15·5 (19·5-22·0 × 14·9-16·5), n = 46, central Europe (Rey 1912); eggs of pallida not significantly different from those of nominate arborea (see Schönwetter 1979). Weight of fresh eggs 3·4 g (2·9-4·0), n = 21 (Mackowicz 1970). Eggs laid at 24-hr intervals, in morning, from 1–3 days after nest-completion (Mackowicz 1970). Clutch: 3–5 (2–6). Of 27 clutches, Poland: 2 eggs, 2; 3, 7; 4, 9; 5, 8; 6, 1; mean 3·97 (Mackowicz 1970). 2 broods, occasionally 3. 2nd nest initiated c. 11·5 days after fledging of 1st brood (Koffán 1960). 2nd or later clutches may be larger than 1st (Mackowicz 1970). INCUBATION. 12–15 days. By ♀ only, beginning with last egg, though penultimate egg may be covered for much of the time; hatching near-synchronous. See Mackowicz (1970) for detailed account of incubation behaviour. YOUNG. Altricial and nidicolous. Cared for and fed by both parents. FLEDGING TO MATURITY. Fledging period 10–13 days, though flight poor for first few days. May leave nest as early as 8 days if disturbed. Families remain together into autumn, 1st-brood young staying in general vicinity of breeding area while parents rear 2nd brood, then combining with them into larger group (Mackowicz 1970). BREEDING SUCCESS. No data.

Plumages (nominate arborea). ADULT. Forehead and crown black streaked cinnamon-brown, bordered below by a broad pale buff or cream-white supercilium extending from small feather-tufts above nostrils to centre of nape; supercilium sometimes faintly speckled grey above lores, narrowly streaked black on nape, where hidden under elongated crown-feathers when these not raised. Hindneck and upper mantle streaked greyish-cinnamon and black; lower mantle, scapulars, and back streaked black and cinnamon-brown. Rump and upper tail-coverts uniform pale olive-brown or pale rufous-brown. Lores and stripe below eye mottled grey, cream, and cinnamon, widening towards cinnamon-brown ear-coverts; black stripe over upper ear-coverts behind eye, rear of ear-coverts sometimes spotted black. Sides of neck greyish-buff with short black streaks, almost uniform buff behind ear-coverts. Cheeks and sides of chin buff with rather narrow black shaft-streaks, forming mottled moustachial and malar stripe. Central chin and upper throat uniform pale buff or cream, or very faintly speckled grey. Lower throat, chest, and sides of breast buff, closely streaked black, shaft-streaks widening into triangular spots on feather-tips, spots often bordered rufous laterally. Flanks buff with narrow and ill-defined dark grey streaks, broadly bordered cinnamon-brown; remainder of underparts pale cream-yellow or off-white. Central pair of tail-feathers (t1) olive-brown with blackish centre; t2-t5 black or olive-black with 3-6 mm of tip pale and rather ill-defined grey-buff, pale olive, or partly ochre-brown; t6 dark olive-brown with slightly paler centre and faint and narrow cream outer edge and tip. Flight-feathers brown-black with sharp but very narrow pink-cinnamon outer edges (broader near emarginations of outer primaries), rather poorly defined narrow grey-buff tips, and ill-defined pale grey bases of inner webs. Tertials dull black or olive-black with rather narrow cinnamon-brown fringes along sides, and broad ill-defined cinnamon-brown triangle on tip. Lesser upper wing-coverts olive-brown, like rump, but outermost black with broad and contrasting cream-white tips; median coverts olive-black with broad cinnamon-brown fringes; greater coverts olive-brown merging into olive-black near tip and with rather narrow cinnamon-brown fringe at sides and broad triangular cinnamon-brown patch at tip (like tertials). Upper primary coverts contrastingly marked, together with black-and-white outer lesser coverts forming boldly patterned forewing: lesser and median primary coverts cream-white or white (sometimes hidden under black cinnamon-tipped bastard wing), greater primary coverts black with contrasting pale cinnamon tips. Axillaries grey; under wing-coverts pale grey to cream-white, lesser along leading edge of wing black with pale buff tips. Some individual variation in width of black streaks on upperparts (those on crown, inner scapulars, and back always widest) and in depth and extent of rufous on ear-coverts, sides of breast, and chest, but more marked variation due to bleaching and wear: by mid-winter, cinnamon-brown streaks of upperparts bleached to buff-brown or greyish-cinnamon, cinnamon-brown fringes of tertials and upperwing-coverts to grey-buff or pale buff, and cinnamon-brown of tips of tertials and t1 largely abraded; pale tips of tail-feathers, pale fringes of flight-feathers, and bold cinnamon or cream patches on outer lesser coverts and upper primary coverts bleached to off-white; streaks on cheeks, chest, and sides of breast more sharply defined (those on sides of chest sometimes joining to form small uniform black patches). In spring and early summer, pale streaks on upperparts bleached to buff or off-white, sometimes largely lost through abrasion, forehead and crown in particular becoming largely black; virtually all cinnamon of tips and fringes of tertials and greater upper wing-coverts abraded; supercilium off-white, but in heavily worn plumage often speckled dusky, like sides of neck and cheeks and then not as distinct as in remainder of year; chest sharply streaked, but ground-colour faded to pale buff or off-white; flanks hardly streaked; remainder of underparts off-white, often tinged dusky due to grey feather-bases showing through or to contamination by soil; in some cases, even off-white tips of tail-feathers and greater upper primary coverts virtually worn off. NESTLING. Down dense and long on upperparts and head, shorter and more scanty below; down white with light yellow-brown tinge on 1st day; grey on head and grey with buff tips elsewhere on 3rd day (Witherby *et al.* 1938*a*; Mackowicz 1970). For further development, see Mackowicz (1970). JUVENILE. Rather like adult, but black feather-centres on forehead, crown, mantle, scapulars, and back broader and rounded distally; sharp cream or off-white fringes at tip, irregular and ill-defined buff-brown at sides; crown largely black. Rump and upper tail-coverts with faint dark subterminal crescent and pale buff terminal fringe; supercilium broad and off-white, but not extending to feather-tufts above nostril nor to central nape (unlike adult), and broken by black just above front of eye. Spots on chest and sides of breast smaller and more rounded, less deep black; ground-colour paler, cream-buff or off-white, not partly rufous. White or pale buff on tail-tips more extensive, especially on outer feathers, extending into sharp point along shaft; t6 with large white or buff wedge on inner web and largely off-white outer web (except for base), sometimes mottled dusky near tip. Flight-feathers with rather narrow but even pale buff or off-white fringes, tertials with broader pale fringes and narrow dark submarginal crescents (pale fringes soon wear off, especially on tertials and longer primaries). All upper wing-coverts and greater upper primary coverts olive-brown with broad and even pale buff or off-white fringes along rounded tips, submarginally bordered dull black; forewing not as contrastingly patterned as adult. P10 longer and with more broadly rounded tip than adult, tip fringed off-white (unlike adult); 1–7 mm longer than longest greater upper primary covert (narrow and pointed in adult, 0–7 shorter than longest covert). FIRST ADULT. Like adult; indistinguishable when last juvenile feathers (usually p9–p10) replaced.

Bare parts. ADULT. Iris brown or dark brown. Upper mandible dark brown or dark grey-brown, lower horn-brown with pink-brown or flesh-coloured base (base brighter pink during nesting). Leg and foot brownish-flesh, dull flesh-yellow, or pale brown, brightest during breeding (RMNH, ZMA.) NESTLING. Mouth deep yellow or sulphur-yellow with single black spot on tip of tongue, 2 tiny black spots on basal side of tongue and 1 spot inside tip of each mandible; gape-flanges yellowish-white (Witherby *et al.* 1938*a*; Mackowicz 1970). At *c.* 2 weeks, spots inside mandibles faint, spots on tongue still distinct. At hatching, bare skin yellowish, except for blue-grey crown; bill pale brown; leg pale yellow. Bare skin yellow-flesh on 2nd day, darkening to dark grey on head and back and to dark violet on rear body by 3rd-6th day; on leaving nest (*c.* 12th day), bill pale grey with darker tip, gape-flanges small, cream-yellow. (Mackowicz 1970.)

Moults. ADULT POST-BREEDING. Complete; primaries descendant. Starts late June to late July (on average slightly earlier in southern populations, but much individual variation; in some as late as mid-August); p1 shed first, soon followed by wing-coverts, head, body, and tail; tail-feathers often grow more or less simultaneously; all plumage largely new at primary moult score *c.* 35; moult completed with re-growth of p9–p10 early September to early October, occasionally late October; nasal bristles grow last, sometimes up to early November. POST-JUVENILE. Complete, primaries descendant. Starts with p1 at age of 4-6 weeks (Bub and Herroelen 1981), soon followed by feathers of crown, mantle, and sides of breast. As recently-fledged juveniles occur early May to late August, date of start of moult highly variable. Among specimens examined, moult

started early June to mid-August, completing mid-August to mid-October.

Measurements. ADULT. Nominate *arborea*. Netherlands, all year; skins (RMNH, ZMA). Bill (S) to skull, bill (N) to distal corner of nostril; exposed culmen on average 2·9 mm less than bill (S).

WING	♂ 97·2 (1·94; 46)	94–101	♀ 93·4 (1·73; 37)	91–97
TAIL	51·5 (1·53; 24)	49–54	48·5 (2·20; 17)	45–52
BILL (S)	14·7 (0·72; 27)	13·6–16·0	14·4 (0·30; 17)	14·0–14·9
BILL (N)	8·8 (0·60; 26)	8·0–9·8	8·8 (0·46; 17)	8·0–9·7
TARSUS	21·5 (0·68; 27)	20·6–23·1	21·2 (0·50; 17)	20·2–22·0

Sex differences significant for wing and tail. No difference between breeding, migrant, and wintering birds, all combined above.

Wing of birds from Sweden: ♂ 93-100 (31), ♀ 87-97 (14) (Svensson 1984*a*); ♂♂ 99·4 (5) 98-101, ♀ 95·6 (7) 92-100 (Vaurie 1951*a*). Wing in USSR: ♂ 96·3 (743) 92-102, ♀ 91·0 (621) 87-95 (Bub and Herroelen 1981).

L. a. pallida. Spain and Portugal, March–June; skins (RMNH, ZMA).

WING	♂ 95·6 (1·85; 10)	94–98	♀ 91·1 (2·16; 5)	88–94
TAIL	50·4 (1·65; 10)	48–53	46·3 (1·72; 5)	45–48
BILL (S)	14·4 (0·57; 10)	13·6–14·9	14·7 (0·38; 5)	14·2–15·1
BILL (N)	8·9 (0·35; 10)	8·3–9·4	8·5 (0·21; 5)	8·2–8·7
TARSUS	21·8 (0·68; 10)	21·1–22·8	21·7 (1·11; 5)	19·9–22·6

Sex differences significant for wing and tail.

Birds from: (1) Italy, Greece, and Turkey (RMNH, ZMA); (2) Makedonija (Yugoslavia), range excluding ♂ of 93 (Stresemann 1920); (3) Iran (Vaurie 1951*a*).

WING (1)	♂ 98·0 (1·85; 16)	95–102	♀ 93·3 (1·72; 13)	90–96
(2)	97·4 (1·60; 50)	95–100	95·3 (2·30; 11)	90–97
(3)	99·7 (—; 12)	97–103	95·3 (—; 18)	91–100

Wing of birds from Crete: ♂ 90-96 (18), ♀ 87-92 (8) (Niethammer 1942).

JUVENILE. Wing and tail both on average *c.* 6 shorter than adult; bill and tarsus similar to adult once post-juvenile moult completed.

Weights. ADULT. Nominate *arborea*. Netherlands and Belgium, all year: ♂ 28·0 (9) 23-35, ♀ 32·1 (5) 30-35; exhausted frost victims, ♂♂, 20·7, 23 (Verheyen 1957; RMNH, ZMA). USSR: ♂ 27·4 (513) 23-28, ♀ 26·3 (448) 24-32 (Bub and Herroelen 1981). Belgium, October-November: 25-39 (188), of which only 15 were 34 or over (Bub and Herroelen 1981). Switzerland: 26·6 (2·68; 40) 21·0-30·5, once 36·5 (Bub and Herroelen 1981).

L. a. pallida. Greece and Turkey, May-July: ♂ 24·1 (5) 20-29, ♀♀ 22, 27·5 (Niethammer 1943; Rokitansky and Schifter 1971). Northern Greece, Turkey, and northern Iran, December-February: ♂♂ 31, 32, 33; ♀♀ 29, 30, 30, 30 (Makatsch 1950; Schüz 1959; Kumerloeve 1970*a*). Spain: 26·1 (19) 23-29 (Bub and Herroelen 1981).

NESTLING. At hatching, *c.* 2·8; rapid increase until 9th-11th day, when *c.* 21·3; slower increase later on (Mackowicz 1970).

Structure. Wing rather short, broad at base, tip rounded. 10 primaries: p7 longest, p8 0-1 shorter, p9 1-4, p10 53-62, p6 1-3, p5 5-10, p4 15-20, p3 19-25, p1 24-31. Juvenile p10 relatively longer than in adult, 45-51 shorter than longest primary (p7-p8), 3·2 (8) 1-7 longer than longest upper primary covert—in adult, 3·4 (40) 0-7 shorter than longest covert. Outer web of (p5-)p6-p8 and inner of (p6-)p7-p9 emarginated. Tips of longest tertials fall between tips of p5 and p6 in closed wing. Tail short, tip square; 12 feathers. Bill sharp and slender, short; both culmen and cutting edges straight except for slightly decurved tip; lower mandible straight and slender. Nostril covered by small feather-tuft; some long and rather soft bristles along base of upper mandible and on chin. Feathers of crown elongated, but not narrow, up to *c.* 15 mm long, forming short broad crest. Tarsus and toe short and slender; front claws slender, short, decurved; hind claw long and slender, slightly decurved. Middle toe with claw 18·0 (13) 17–19 mm; outer toe with claw *c.* 69% of middle with claw, inner *c.* 75%, hind *c.* 117%. Length of hind claw 12·6 (22) 10–15, exceptionally 7–17.

Geographical variation. Slight, predominantly clinal, in colour only (variation in size negligible: see Measurements). Typical *pallida* from Iran as well as populations from Turkey, Levant, Greece, southern Italy, and north-west Africa have fringes of crown, scapulars, and back greyish pink-cinnamon or grey-buff in fresh plumage, not cinnamon-brown as in nominate *arborea*; rump and upper tail-coverts and ground-colour of hindneck and upper mantle distinctly grey or olive-grey, less olive- or rufous-brown; ground-colour of sides of breast and chest pale cream or off-white, similar to remainder of underparts, not buff and rufous as in nominate *arborea*; black streaks on chest sometimes narrower. Birds from southern Portugal, central Spain, Corsica, Sardinia, central Italy, central Rumania, and possibly also Crimea and Caucasus are intermediate to varying degree between *pallida* and nominate *arborea*; in general, nearest *pallida*, and included in that race, but some specimens indistinguishable from nominate *arborea*, especially in worn plumage. Subspecific names have been applied to some of those intermediate populations, such as *familiaris* Parrot, 1910, Corsica (e.g. by Hartert 1921–2) and *flavescens* Ehmcke, 1903, Rumania, but neither recognized here. Close to *Alauda*, but retained in monotypic genus *Lullula* mainly on behavioural and traditional grounds (K H Voous).   CSR

Alauda arvensis Skylark

PLATES 10 and 13
[between pages 184 and 185]

Du. Veldleeuwerik   Fr. Alouette des champs   Ge. Feldlerche
Ru. Полевой жаворонок   Sp. Alondra común   Sw. Sånglärka

Alauda arvensis Linnaeus, 1758

Polytypic. Nominate *arvensis* Linnaeus, 1758, Azores and Europe from Wales, England, and Norway east to Ural mountains, south to central France, Alps, north-west Yugoslavia, north-west Hungary, Czechoslovakia, and in European USSR to *c.* 50°N; *scotica* Tschusi, 1903, Ireland, north-west England, and Scotland, north to Faeroes; *guillelmi* Witherby, 1921, northern Portugal and north-west Spain; *sierrae* Weigold, 1923, central and southern Portugal (south from Sierra da Estrêla) and southern Spain; *harterti* Whitaker, 1904, north-west Africa; *cantarella* Bonaparte, 1850, north-east Spain (south to Sierra de Gredos, Sierra de Guadarrama, and Aragón, in Pyrénées west to at least Pamplona), southern France, Italy (including Mediterranean islands), central and southern Yugoslavia, central Hungary, and Greece, east through south European USSR to *c.* 42°E and south to northern slopes of Caucasus; *armenicus* Bogdanov, 1879, Transcaucasia, eastern Turkey (not known which race breeds in central and western Turkey), and south-west and northern Iran in Zagros and Elburz mountains; *dulcivox* Hume, 1873, steppes of lower Volga and Ural mountains east to Yenisey, south to northern Kazakhstan, and in mountains of central Asia from western Turkmeniya and north-east Iran east to Tien Shan and Tarbagatay. Extralimital: *kiborti* Zaleski, 1917, central Siberia east of *dulcivox*, from Yenisey, to Lena and Zeya rivers, south to Altai, northern Mongolia, Manchuria, and Korea; *intermedia* Swinhoe, 1863, south-east Siberia north to lower Amur river; *pekinensis* Swinhoe, 1863, north-east Siberia; 1–4 further races central and eastern Asia.

Field characters. 18–19 cm; wing-span 30–36 cm. 30% larger than short-toed larks *Calandrella*; 20% larger than Woodlark *Lullula arborea*; as long but less bulky than Crested Lark *Galerida cristata*; shorter and less bulky than *Melanocorypha* larks. Commonest and most widespread lark of west Palearctic, encapsulating form and character of family for most observers; this most exemplified by strong bill, fairly stout and long body, quite long legs, wings, and tail, streaked brown plumage, and terrestrial habits. Chief marks of *A. arvensis* are short crest, open-faced appearance (due to pale lores and supercilium), fully streaked and quite sharply demarcated chest, white trailing edge to wings, and white tail-sides. Behaviour includes characteristic hover before landing and familiar high-level song-flight in breeding season. Sexes similar; no seasonal variation. Juvenile separable. 8 races in west Palearctic, but only 2 described here (see also Geographical Variation).

ADULT. (1) North European race, nominate *arvensis*. Upperparts buff-brown (sometimes yellower or greyer), extensively streaked with black-brown. Underparts buff-white (sometimes yellower), boldly streaked with black-brown on chest. At distance on ground, bird looks uniformly brown and streaked except for usually obvious pale belly. At close range, general appearance breaks up into distinctive combination of the following subtle characters: (a) well-streaked crown, ending in short crest, clearly visible when erect; (b) buff lores and cream supercilium forming soft line over cheeks and extending round them to join lightly streaked pale throat; (c) paler, finely streaked nape; (d) bold, regular, vertical streaking and spotting on chest, extending down upper flanks; (e) quite well-marked buff edges to black-brown median coverts and dark brown greater coverts, forming distinct upper and (usually) lower wing-bar; (f) long, warm buff-edged and black-brown-centred tertials, providing long cloak to flight-feathers; (g) white-edged tail, largely black with brown centre. In flight, subtler characters blur but upperparts show dull white trailing edge to wing (fading out on middle primaries) and (more obviously) white-edged tail; below, shows dark chest and buff-grey under wing-coverts which contrast with paler belly and vent. (2) West Asian and Siberian race, *dulcivox*. Distinctly paler than nominate *arvensis*. Upperparts more sandy-grey, emphasizing narrow dark centres of larger feathers, especially on median coverts. Underparts noticeably whiter, lacking yellowish or buff tinge of nominate *arvensis*; this difference in colour most evident September–December. JUVENILE. All races. Differs distinctly from adult not only in shorter tail (in fledglings) but also in remarkably spotted and speckled appearance of upperparts, cleaner throat, and less dark, more drop-shaped streaks on chest. At all ages, bill horn, darker on upper mandible; legs yellow-brown.

Often the only lark in many habitats throughout west Palearctic, and thus a key species in lark identification. However, can suggest many other smaller or larger larks and can also be confused with ground-living Lapland Bunting *Calcarius lapponicus* and Snow Bunting *Plectrophenax nivalis*, and even with Corn Bunting *Miliaria calandra*. Most specific characters are short crest (less obvious than on *Lullula arborea* and, particularly, *Galerida*), rather sharp lower border to chest-marks (sharper than on any other west Palearctic lark), and prominent, long, white outer tail-feathers (shared as single most obvious mark only by much smaller *Calandrella*, slimmer Shore Lark *Eremophila alpestris*, and much bigger-winged and bigger-headed Calandra Lark *Melanocorypha calandra*). For distinctions from Small Skylark *A. gulgula*, see that species (p. 205). Flight over short distances more weaving and fluttering than most larks, with rather laboured take-off and characteristic loose hover before landing; more purposeful over long distances and on migration but with characteristically erratic rhythm, involving bursts of loose, rather floppy or noticeably flapped wing-beats alternated with short glides or almost complete wing-closures. Actions combine into loose, slightly rising and falling and above all hesitant progress displayed by no other common passerine of open country and diagnostic even at long range. Flight silhouette alternately cruciform and elliptical (as wings are closed), with heavy-chested look balanced by tail length. Escape-flight usually short and wavering, but migrants often shyer, tearing off as if in panic. Song-flight much more confident: bird rises with rapid wing-beats, and after prolonged hovering falls by 'parachute' (of both wings and tail) or circling sweep, to end with sudden drop to ground or just above it; astonishing outpouring of song lasts through ascent and hover and may recommence at lower level during descent. Gait essentially a noticeably even-paced walk, but becomes a shuffle when legs are flexed and body lowered—even a mouse-like creep when feeding in short cover or high winds; can run freely and will hop over obstacles. Perches freely on rocks, low artefacts, and bushy plants, but only rarely on wires and trees. Somewhat irascible when in flocks, disputing food sources with fellows and other passerines, even as large as Fieldfare *Turdus pilaris*. Crouches to escape detection, so often providing sudden explosion at observer's feet.

Song has great spirit and vehemence, strongly appealing to human ear: a loud, far-carrying, high-pitched but pleasantly modulated warble, interrupted by scratchier phrases, call-notes, and more plaintive sounds; frequently sustained for minutes. Calls variable and far less distinct: commonest a liquid, rippling, disyllabic, 'chirrup', but also varied into huskier 'skilloch' or 'skirrick'; less commonly, dry, less protracted 'skip' or 'chip', slurred 'prrp', and whistled 'skee-oo'. Autumn birds resembling *dulcivox* heard to give distinctive short, clipped 'chirp' (D I M Wallace). Beware overlap of all these calls with those of other larks.

Habitat. Breeds in upper and lower middle latitudes across and beyond west Palearctic, spreading from continental to oceanic climates and from temperate into boreal zone. Markedly less Mediterranean than Woodlark *Lullula arborea* and Crested Lark *Galerida cristata*, which need more warmth and are more patchily distributed owing to more specific habitat requirements. Inhabits open surfaces of firm, level or unobstructed soils, neither arid nor muddy, although often moist, and preferably well clothed with grasses (including cereals) or low green herbage. Presumed to have spread from natural steppe grasslands with deforestation and expansion of crops and pastures encouraging massive habitat changes, especially through 19th century.

As the most widely distributed bird species in Britain and Ireland, occurs freely wherever any kind of farming practised, and also on managed open spaces such as golf links, playing fields, airfields, and ballast pits, also sand-dunes, salt-marshes, and even upland rough pastures and moorland up to arctic-alpine zone at 1000 m or more. Upland densities, however, tend to be much lower than those on lowland, except where those are limited by small fields, narrow valleys, disturbance, or building. For densities in particular habitats, see Social Pattern and Behaviour. Recent agricultural changes have had both favourable and unfavourable effects; in Britain, prompt regulation of use of harmful pesticides has averted significant deterioration in status (Sharrock 1976). More detailed habitat investigation in Britain has shown that in lowlands the species is characteristic of raised mires

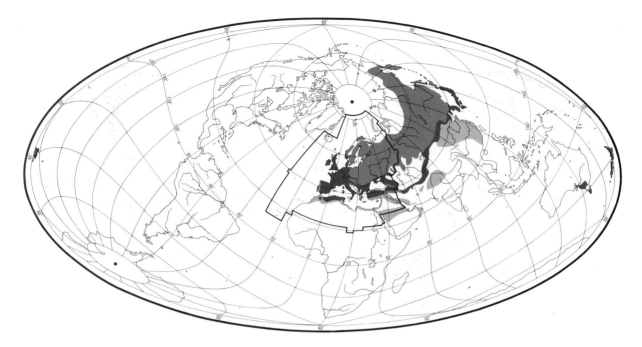

and raised bogs, but only of grazed stage of fen vegetation; in uplands, occupies whole spectrum from paramaritime moorland and bog through blanket bog and gently and steeply sloping heather or grass moorland to montane heaths, and is generally one of the 3 commonest bird species occurring on most sites. Also ubiquitous on chalk downland but in contrast to *L. arborea* occurs twice as numerously on thicker grass as on short turf, and predominates where bracken *Pteridium aquilinum* present. (Fuller 1982.) Vulnerability as a ground-nester reduced by relative privacy of intensively managed open habitats, and by tolerance of denser and taller plant cover than is acceptable to most *Alaudidae*. Avoids neighbourhood of even isolated trees, tall hedges, bushes, rock-faces, boulders, and gravel patches. Infrequently on extensive sand or mud surfaces. Perches rather little on overhead wires, fences, or posts, although more often on walls, banks, and mounds. Advantage of being able to feed, nest, and pursue most other activities within same habitat is enhanced by unusual capabilities in sustained song-flight (Nicholson 1951). In USSR, typical species of steppe and forest-steppe zone, which it enters along river valleys and in wake of expanding agriculture. Common in dry meadows, extensive forest clearings, grassy forest margins, open heaths, pastureland, and steppes in which cereals and various grasses grow; also in actively worked parts of peat deposits, and in semi-desert region; heard singing over centres of large cities such as Kharkov. In Armeniya and elsewhere, nests in fields and mountain meadows up to 2650–2750 m, sometimes in alpine meadows and occasionally in sparse juniper forests. Extralimitally in Soviet central Asia even extends to *Artemisia* steppes, and occurs with Shore Lark

Eremophila alpestris in alpine meadows at 3000 m; also in meadow patches amid morainal debris and occasionally amid arctic birch thickets; in Kamchatka on maritime tundra. (Dementiev and Gladkov 1954*a*.) In Swiss Alps, also breeds up to 2000 m or more, in meadows above treeline (Glutz von Blotzheim 1962). Fondness for dusting, often by minor roads or tracks, and lack of interest in water presumably reflect steppe and semi-desert origins. In autumn, resorts commonly to stubbles and fallows, but continues to avoid disturbed and built-up areas except under stress of hard weather, and does not enter gardens, orchards, or small enclosures of any sort.

Distribution. No changes reported.

TURKEY. Very thinly distributed (Vittery *et al.* 1971).

Accidental. Bear Island, Iceland, Kuwait, Azores, Madeira.

Population. Decreased France, Denmark, Switzerland, and Rumania, and possibly Ireland and Norway; increased northern Finland; little or no change in Britain, Sweden, and Czechoslovakia.

FAEROES. 5–15 pairs (Bloch and Sørensen 1984). BRITAIN, IRELAND. 2–4 million pairs. No evidence of marked widespread change, though some local reductions; limited fluctuations 1965–74 (Parslow 1973; Sharrock 1976). IRELAND. Possible general decrease suggested (Ruttledge 1966). FRANCE. Over 1 million pairs; some decrease (Yeatman 1976). BELGIUM. About 350 000 pairs (Lippens and Wille 1972). NETHERLANDS. 600 000–750 000 pairs (Teixeira 1979). WEST GERMANY. 2–4 million pairs (Rheinwald 1982). DENMARK. Decreased (Klug-Andersen 1984; Vickholm and Väisänen 1984). NORWAY.

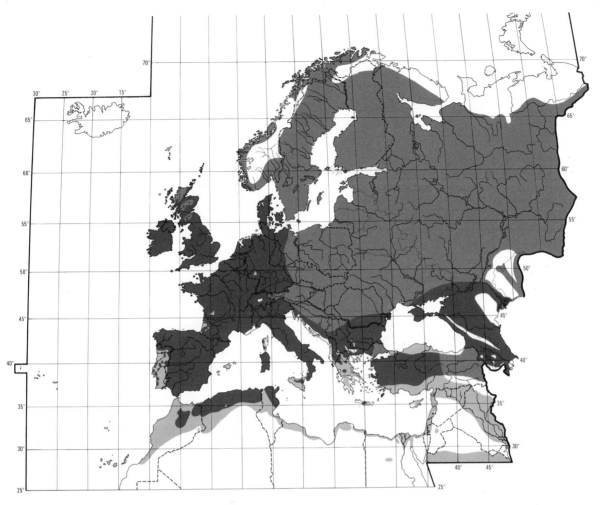

Possibly decreased (VR). SWEDEN. About 800 000 pairs (Ulfstrand and Högstedt 1976). No changes known (LR). Migrants decreased, Falsterbo, since mid-1970s (Roos 1983). FINLAND. About 200 000 pairs (Merikallio 1958); increased in north 1941–77 (Väisänen 1983); little change (Vickholm and Väisänen 1984). CZECHOSLOVAKIA. No changes reported (KH). SWITZERLAND. Decreased (RW). AUSTRIA. No indications of drastic change (HS, PP). GREECE. Slight decline (GIH). RUMANIA. Decreased (VC). USSR. Declined between Volga and Ural rivers, Kazakhstan (Shishkin 1976).

Mortality. England: average annual adult mortality 33·5%; average mortality of young during 1st year after independence 38% (Delius 1965). Oldest ringed bird 8 years 5 months (Rydzewski 1978).

Movements. Shows gradation from being wholly migratory in north and east of breeding range to making no more than local movements in south. Europe populations (except those of southern fringe) studied through large-scale ringing; results summarized by Spaepen and Cauteren (1962, 1968), on which much of the following based.

NORTH EUROPEAN RACE, nominate *arvensis*. Northern and central Europe largely vacated in winter, when recoveries grouped in western Europe from Britain, Ireland, and Low Countries to Iberia, and in southern France (especially Rhône valley) and northern Italy; however, distribution of recoveries influenced to unquantified extent by local variation in hunting pressure. Autumn migration essentially west to south-west across Europe, and birds reaching western seaboard include many from northern and central European USSR east to about 50°E. Other birds presumed from this population winter (apparently in moderate numbers only) around Black Sea and Transcaucasia (Dementiev and Gladkov 1954*a*), but no recoveries there. Majority of birds from West and East Germany, Baltic region, and northern half of European USSR join Fenno-Scandian birds in migrating along west European seaboard; hence USSR birds must migrate initially due west to join this main stream. These populations have produced many fewer winter recoveries in south-east France and Italy than in western France and Iberia. Birds from southern half of European USSR, Hungary, and Czechoslovakia also show south-westerly trend in autumn, but this takes them into Italy and south-east France (one Czechoslovakian bird recovered in Spain). Extent to which main continental race, nominate *arvensis*, reaches north-west Africa in winter remains unclear. Regular autumn migration across Straits of Gibraltar, and large numbers wintering in Maghreb (Heim de Balsac and Mayaud 1962; Tellería 1981) may involve nominate *arvensis*, but perhaps Iberian *sierrae* and *cantarella* also or instead—though nominate *arvensis* has been collected in Tunisia (Heim de Balsac and Mayaud 1962; Etchécopar and Hüe 1967).

In contrast to continental situation, where mainly migratory (though degree inadequately quantified for Low Countries and France), British and Irish birds are mainly resident. Upland and northern Scottish breeding areas often vacated in winter, but such birds do not necessarily move long distances; recoveries of British-bred birds mainly within 5 km, though a small number up to 130 km (Hardman 1974); also one recovery (possibly atypical) of a Scottish bird in Portugal in 1st autumn (Spencer and Hudson 1981). British winter population augmented by continental immigrants, but few foreign-ringed birds recovered; hence suspected that large numbers of autumn migrants which pass through Britain and Ireland do so quickly *en route* to France and Iberia (Hardman 1974).

European migration on broad front, though denser along certain zones, and mountain chains cause passage concentrations (e.g. Alps and Pyrénées); guiding effect of coast lines occurs but less important than for other passerines, since this species readily crosses open sea, e.g. Bay of Biscay (Spaepen and Cauteren 1968). Large literature on regional studies of directions of visible passage not reviewed here, since such results influenced by local factors. For Danish studies of weather influencing migration, see Rabøl (1974) and Rabøl and Noer (1973). ♂♂ tend to migrate later than ♀♀ in autumn, and return earlier in spring, though no complete segregation (Spaepen and Cauteren 1968). In West Germany, where population largely migratory, ♂♂ predominate amongst those remaining behind (Niethammer 1970; Senk *et al.* 1972; Vauk 1972*b*). Autumn passage begins September, peaking in northern Europe in first half of October; lasts into first half of November in southern Europe. Large-scale cold-weather movements can occur at any time during winter. Return passage begins January in south, becoming heavy in February and early March. Arrivals in Scandinavia mid-February to late March, in USSR in March and early April. For further details, see Spaepen and Cauteren (1968).

OTHER RACES. No detailed studies. In west Palearctic, southern races largely resident, with local flock movements in winter: *harterti* (north-west Africa), *guillelmi* (north-west Iberia), *sierrae* (most of southern Iberia), and *cantarella* (north-east Spain to northern Caucasus). However, some *cantarella* at least, perhaps from more easterly breeding areas (e.g. Balkans, south European USSR), are migratory, reaching North Africa, Cyprus, Iraq, and Persian Gulf (Vaurie 1959), and eastern Turkey and north-west Iran (*armenicus*) are vacated in winter. Siberian populations certainly migratory: westernmost race, *dulcivox* (breeding western Siberia), reaches Iran, Afghanistan, Pakistan, and north-west India, and the more easterly races overlap in winter in Far East (China, Korea, Japan). Though the name *intermedia* is sometimes used for migrants or stragglers to west Palearctic, that race (breeding far-eastern USSR and north-east China) is extremely unlikely to occur.   RH

Food. Plant and animal material taken at all times of year, but insects especially important in summer, cereal grain and weed seeds in autumn, leaves and weed seeds in winter, and cereal grain in spring. Walks over ground taking items from soil surface or pecking at leaves, flowers, or seed heads. Seems to locate all food visually and to peck at it directly, though sometimes digs with bill (not feet) in loose soil for newly sown, partly exposed grain, and uproots cereal seedlings up to *c*. 4 cm tall to peck off attached grain (shoot often breaks during uprooting, and bird then leaves it). When eating leaves of wheat, pecks off pieces *c*. 6 mm long. When searching for seeds, seems to make many unsuccessful pecks, apparently not just turning over debris to uncover food or taking very small items; proportion of unsuccessful pecks closely correlated with proportion of empty seed husks, suggesting birds have difficulty distinguishing between them and seeds. (Green 1978*b*; R E Green.) Unlike finches (Fringillidae), does not remove seed husks before swallowing, except for rough hairs, etc., of some species. Exceptionally (but especially when feeding young), takes flying insects in aerial-pursuit. (Pätzold 1983.) Adult gathering food for young recorded making pile of insects on ground until 'bundle large enough to carry away comfortably' (Turner 1915). Large items may be beaten against ground and only soft parts fed to young (Prokofieva 1980). At one nest in Moscow region (USSR), all food obtained within 50 m (Ptushenko and Inozemtsev 1968). Wild birds not recorded drinking by Delius (1969), but in Kazakhstan (USSR) do so regularly if water reasonably close; able to go without, however (Dolgushin *et al.* 1970); on southern Caspian (USSR), flocks not infrequently drink sea water, November–December (Isakov and Vorobiev 1940).

Diet in west Palearctic includes the following. Invertebrates: springtails (Collembola), mayflies (Ephemeroptera), damsel flies (Odonata: Lestidae), stoneflies (Plecoptera), grasshoppers, etc. (Orthoptera), earwigs (Dermaptera), bugs (Hemiptera: Miridae, Cercopidae, Aphidoidea), adult and larval Lepidoptera (Pyralidae, Noctuidae, Geometridae), caddis flies (Trichoptera), larval and pupal flies (Diptera: Tipulidae, Bibionidae, Stratiomyidae, Rhagionidae, Therevidae, Syrphidae,

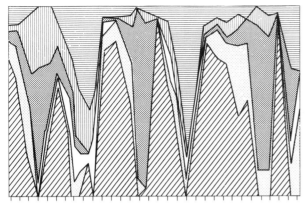

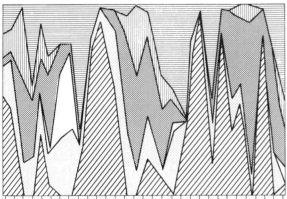

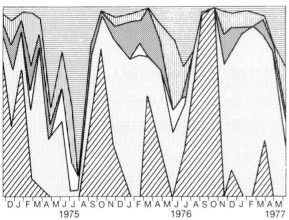

DJFMAMJJASONDJFMAMJJASONDJFMAM
1975   1976   1977

Invertebrates	Cereal grain	Monocotyledonous leaves
Grass flowers and seeds	Dicotyledonous weed seeds	Dicotyledonous leaves

Fig 1 Seasonal variation in the diet of Skylarks *Alauda arvensis* at 3 sites on arable land in East Anglia (England). Vertical extent of the enclosed areas indicates proportions by dry weight. (Green 1978*b*.)

Lonchaeidae, Tachinidae, Muscidae), adult, pupal, and larval Hymenoptera (sawflies Tenthredinidae, Ichneumonoidea, ants Formicidae), adult and larval beetles (Coleoptera: Carabidae, Silphidae, Staphylinidae, Ge-

Table A   Main foods of Skylarks *Alauda arvensis* on different feeding sites in East Anglia, England (Green 1978*b*).

Feeding site	Main foods
Stubble	Grain
Cereal sowing	Exposed and germinated grain
Ploughed land	Weed seeds from soil surface
Winter wheat fields (November–April)	Wheat leaves
Leys, harvested beet and vegetable fields (April–August)	Weed seeds from seedlings
Growing crops of all types (April–August)	Insects, grass flowers, leaves, weed seeds

otrupidae, Scarabaeidae, Byrrhidae, Elateridae, Cantharidae, Tenebrionidae, Chrysomelidae, Curculionidae), spiders (Araneae), harvestmen (Opiliones), millipedes (Diplopoda), centipedes (Chilopoda), slugs and snails (Pulmonata), eggs of earthworms (Oligochaeta). Plant material mainly seeds and leaves: of nettles *Urtica*, docks and knotweeds (Polygonaceae), goosefoots (Chenopodiaceae), *Amaranthus*, pinks (Caryophyllaceae), Cruciferae, buttercups (Ranunculaceae), fumitory *Fumaria*, poppies (Papaveraceae), Leguminosae, violets (Violaceae), pimpernel *Anagallis*, Boraginaceae, Labiatae, nightshades (Solanaceae), speedwells *Veronica*, Compositae, grasses (Gramineae) including cereal grain; also fallen fruit. (Hammond 1912; Kovačević and Danon 1952; Ptushenko and Inozemtsev 1968; Samorodov 1968; Green 1978*b*, 1980; Pätzold 1983.) Adults recorded eating bread, cheese, and fried black pudding (Carter 1961; Low 1961).

For diet on arable land in East Anglia (England), studied from faeces, see Table A and Fig 1. Cereal grain main food in spring and autumn, suggesting birds preferred it, resorting to leaves and weed seeds when it was scarce (though leaves and weed seeds still taken even when grain abundant); leaves, especially of cereals, predominant in diet in mid-winter at 2 sites but not at the other where weed seeds important (especially *Polygonum* and *Galeopsis*); invertebrates taken in large numbers in summer, especially adult beetles (Elateridae, Curculionidae, Carabidae, Chrysomelidae), though Fig 1 underestimates their importance as many are also fed to young; grass seeds and flowers taken for short period in summer. Tended to prefer larger seeds (Table B), not

Table B   Sizes of seeds taken by Skylarks *Alauda arvensis* from ploughed land at 2 sites in East Anglia (England) compared with sizes of seeds available in the soil (Green 1978*b*).

Seed dry wt (mg)	Site 1 (% by wt)		Site 2 (% by wt)	
	Soil	Diet	Soil	Diet
0·5–1·0	78	10	6	1
1·7–5·1	22	90	62	51
6·8	0	0	32	48
No of fields sampled	10		5	

taking those of less than 0·3 mg dry weight at all (e.g. *Veronica*, *Urtica*, *Matricaria*) although present in large numbers; preference apparently not due to relative conspicuousness. (Green 1978*b*.) Can cause considerable damage to sugar beet by feeding on leaves of seedlings (see Green 1980; also Edgar and Isaacson 1974, Hardman 1974). In Crimea (south-west USSR), January–February, 16 stomachs contained only seeds (including 36·1% *Amaranthus*, 22·3% Polygonaceae, 11·9% Labiatae, 11·7% Chenopodiaceae); in March–July, 19 contained 37·5% plant material (mainly weed seeds) and 62·5% invertebrates (mainly ants and their eggs, and beetles) (Kostin 1983). In Moscow region, June–July, 40 adults contained mainly insects (these mostly beetles, especially Carabidae and Curculionidae) and also seeds (Ptushenko and Inozemtsev 1968). Further data on adult diet as follows. Britain: Hammond (1912), Florence (1914), Collinge (1924–7). France: Lebeurier and Rapine (1935*a*). Hungary: Dandle (1957*b*), Sterbetz (1971). Yugoslavia: Kovačević and Danon (1952). Turkey: Wadley (1951). USSR: Samorodov (1968), Averin and Ganya (1970), Prokofieva (1980). Important extralimital study in Kazakhstan (USSR) by Rjabow (1968); for other data from outside west Palearctic, see Isakov and Vorobiev (1940), Pek and Fedyanina (1961), Won *et al.* (1968*b*), Kekilova (1969), and Moeed (1975).

Feeding rates on arable fields, East Anglia, highly variable except when grazing winter wheat: on cereal sowings and stubbles, 0·05 items per s (0·02–0·06); on ploughed fields, 0·06 items per s (0·01–0·10); on winter wheat, 0·87 items per s (0·76–0·92); average energy intake rates respectively 30·5, 3·5, and 3·3 J per s; feeding rates on ploughed fields increased with seed density up to *c.* 5–10 seeds per m² (*c.* 0·9 items per s), but little increase thereafter (Green 1978*b*). In April 1975 in semi-desert of north Caspian region, birds fed on huge hatch of caterpillars (Pyralidae): each ate average 424 caterpillars (8·5 g dry weight) per day, in total (with density of 347 birds per km²) removing 9·4% of available caterpillars per day; (presumably total) energy consumption of free-living birds 243 kJ per bird per day (Shishkin 1980).

Young fed almost entirely on insects for 1st week of life (R E Green). At 2 nests in Leningrad region (western USSR), only 5·7% of items brought to young were plant material, mostly oats (presumably grain); remainder mainly Diptera (these mostly Muscidae, Tipulidae, and Tachinidae) and beetles (mostly larval Silphidae); smaller young given slightly fewer hard items (beetles) than large young, though diets otherwise similar (Prokofieva 1980). Young recorded being fed on bread and (probably) cheese (Carter 1961; Low 1961). See also Ptushenko and Inozemtsev (1968), Bösenberg (1969), and Sikora (1980).

DJB

Social pattern and behaviour. Major study at Ravenglass (Cumbria, north-west England) by Delius (1963). For summary, see Pätzold (1983). Account includes material supplied by R E

Green from study in Cambridgeshire and Suffolk (eastern England).

1. Often in flocks outside breeding season. In eastern England, dispersion showed 2 distinct patterns: cohesive flocks (probably immigrants) present on arable farmland, December–March. Up to 1000 recorded, but average flock size 23 (10–61, *n* = 14); maximum 50% in flocks, mid-winter. Others solitary; known local breeders joined flocks only in cold weather (R E Green); see also Flock Behaviour (below). Similar pattern in Basse-Bretagne (France) where first true flocks occur late October with arrival of migrants; few flocks in February, and these small and probably of local birds (Lebeurier and Rapine 1935*a*). Wintering birds in West Germany apparently mostly ♂♂ (Niethammer 1970; Senk *et al.* 1972; Vauk 1972*b*). At Ravenglass, ♂♂ arrive in small single-sex parties (10–20); comprise birds which take up territories locally as well as passage birds. Bad weather causes re-flocking after territories occupied (Delius 1963; Weber 1970*a*). After breeding, small parties build up to *c.* 20–40 from July into August (Delius 1963; Pätzold 1983). Independent young associate in flocks. In south-east Ireland, autumn, birds migrate in flocks of 2–50, occasionally singly (Goodbody 1950); in USSR, gatherings of several hundred typical of autumn migration period, but migrant flocks smaller (Dementiev and Gladkov 1954*a*), e.g. 50–100 in Kazakhstan (Dolgushin *et al.* 1970). See also Bub (1955) and Kummerlöwe and Niethammer (1934). High counts in late winter and early spring include flocks of *c.* 1000 at Sea of Galilee (Israel) in February (Meinertzhagen 1951*a*), exceptional flock of *c.* 4000 near Ararat (eastern Turkey), mid-March (Beaman *et al.* 1975), and flocks of thousands on spring passage in Kazakhstan (Dolgushin *et al.* 1970). True association with other species in mixed flocks not recorded by Meinertzhagen (1951*a*) and phenomenon certainly rare in Danish migrants (Rabøl and Noer 1973). However, in Iraq, often recorded associating with other Alaudidae, Spanish Sparrow *Passer hispaniolensis*, and Water Pipit *Anthus spinoletta*, and sometimes with Linnets *Carduelis cannabina* and Corn Bunting *Miliaria calandra* (Moore and Boswell 1956); on Crete, with *M. calandra* (Stresemann 1956*a*). Perhaps in all cases chance associations at favourable feeding grounds. BONDS. Monogamous mating system. At Ravenglass, 93 monogamous pairs recorded, 2 cases of bigamy: in one case, paired ♂ took over widowed ♀ and her (neighbouring) territory; soon neglected 1st ♀ so that no 2nd brood produced (Delius 1963). According to Meinertzhagen (1951*a*), non-migratory populations—e.g. in western Ireland and Outer Hebrides (Scotland)—maintain pair-bond throughout winter. ♀♀ tend to arrive *c.* 10–15 days after ♂♂ (Pätzold 1983) or at Ravenglass, up to 1 month later; pair-members there also often in different flocks after breeding. Mate-fidelity strong through breeding season: 43 out of 44 pairs stayed together for at least 3 months; 1 ♀ deserted ♂ after being trapped on eggs (Delius 1963). Similarly, in eastern England, 4 pairs making total of 6 repeat nests after predation nevertheless stayed together (R E Green). At Ravenglass, 56% of 101 birds surviving over consecutive seasons returned to territory of previous year. Birds choosing new territory (44%) moved on average 130 m; ♀♀ perhaps more likely to change. 10 out of 18 survivors returned to same territory over 3 years, 3 out of 12 survivors over 4 years, and 1 out of 3 survivors over 5 years. Site-tenacity clearly promotes mate-fidelity, but marked correlation also found between breeding success of pair in given years and likelihood of its remaining a pair in following year. 16 out of 30 pairs maintained pair-bond (and territory) for 2nd season; of other 14, in 6 cases both partners changed territory, in 5 the ♀♀, in 3 the ♂♂. Data from later in Ravenglass study showed that of 36 pairs, members

of which survived for over 2 years, 17 remained faithful, 1 over 3, another over 4 seasons; in all cases mate-fidelity linked with site-tenacity. In 16 cases in which ♂♂ changed territory, only 3 of their ♀ ex-partners returned to territory of previous year and paired with new ♂, one such ♀ still showed interest in mate of previous year (in adjoining territory). Other 13 ♀♀ changed territory but none moved to new territory of former mate; even in 2 cases where ♀♀ (newly arrived and unpaired) visited former mate (still unpaired) in new territory, birds did not pair up. Of 25 cases in which ♂♂ had died, 14 of the ♀ ex-partners returned to old territory and took new mate, other 11 settled elsewhere. 1-year-olds also show fidelity to natal area. New settlers (possibly this age-group) showed no preference in mate choice between other new settlers and older birds. (Delius 1963, 1965.) Nestlings fed and nest-sanitation performed mainly by ♀ (Delius 1963); ♂ does 30-50% of feeding (Pätzold 1983), in closely related Japanese Skylark *A. japonica* 36% (Haneda and Obuchi 1967). Young out of nest tended by both parents until *c*. 30 days old. Split-brood care recorded. ♂ does more (or all) feeding if ♀ starts another brood (Labitte 1957; Delius 1963; R E Green). Age of first breeding was 1 year in 6 marked birds (Delius 1965). BREEDING DISPERSION. Solitary and territorial. In favourable habitat, territorial boundaries contiguous. In Dresden area (East Germany), 4 nests in 1·2 ha of lucerne averaged 70 m apart. Highest densities in legumes where territories *c*. (60-)80 m across; in cereals, *c*. (80-)100 m (Pätzold 1983). On Ushant (France), territory of barely 0·25 ha recorded; even less where food abundant and density unusually high, or may be up to *c*. 12 ha (Meinertzhagen 1951a). At Ravenglass, average territory size *c*. 0·5 ha (0·25-0·8), $n = 41$; nearby (probably mainly unsuitable habitat), 3 territories averaged 1·5 ha, and solitary bird in Lake District (north-west England) sang over *c*. 20 ha (Delius 1963). Territory defended by both sexes (more by ♂) and used for pair-formation (although ♂ and ♀ recorded associating outside it just prior to laying: R E Green), copulation (probably), nesting, rearing of young, and (normally) for feeding. At Ravenglass, territory normally area over which ♂ sings and where birds feed; nearby, also in case of solitary ♂ (see above), feeding territory much less than 1·5 ha (Delius 1963). In eastern England, 95% of song given within 50-80 m of nest ($n = 9$ ♂♂), but some ♂♂ ranged up to 300 m, ♀♀ up to 350 m, to feed; *c* 70% of feeding (by 11 marked birds) outside ♂'s main singing area, and pattern maintained throughout breeding cycle including for feeding of young (R E Green). Territory limits often marked by topographical features, being clearer where features prominent and where density high and territories small. Boundaries generally constant, some remarkably so over years; slight changes may occur when mates of previous year return and re-pair (see Bonds). Territory appears to regulate population levels, birds not acquiring territory being excluded from breeding (Delius 1963; also Pätzold 1983). At Peine (Niedersachsen, West Germany), pair apparently requires open area of minimum 5-10 ha (3-4 times normal territory size) if surrounded by woods or buildings. Territories on average 160 m from such features, distance proportional to their area and height; if either feature exceeds 500 ha, at least 220 m away (Oelke 1968a). In farmland of Aare plain (Switzerland), density higher in large fields practically devoid of trees and bushes, but not clear if lower densities in other areas due solely to presence of copses, woods, hedges, etc. (Christen 1984), and territories elsewhere located in pine *Pinus* plantations where trees *c*. 1-2 m tall (Mauersberger 1979). Numerous data available on densities in various habitats. See Berger and Gössling-Bednarek (1973) for calculation of minimum number of pairs required for reliable determination of average density. In Britain, overall density *c*. 500-1000 pairs per occupied 10-km square (5-10 pairs per km²); highest on coastal dunes, saltmarsh, and chalk downland, where average 75 pairs per km². On farmland, average 18·4 pairs per km². (Sharrock 1976.) On arable farmland, eastern England, over 3 years, 22-49 ♂♂ per km², positively correlated with proportion of cereals (R E Green). Elsewhere, densities on farmland apparently greatly exceed this: e.g. in Mecklenburg (East Germany), 20-165(-200) pairs per km² of arable farmland (Klafs and Stübs 1977); in USSR, 'fields' hold 100-200 pairs per km², i.e. singing ♂♂ 50-100 m apart (Dementiev and Gladkov 1954a). Similar range of densities in West Germany: e.g. barely 10 pairs per km² on *Calluna* heath (Pätzold 1983) to 154 pairs per km² on cultivated lowland marsh (Busche 1982). For further data from West Germany, see Berg-Schlosser (1975); for Alsace (France), see Kempf (1982); for Switzerland, Schifferli *et al.* (1982); for Mecklenburg, Klafs and Stübs (1977); for Brandenburg (East Germany), Rutschke (1983). Densities reported in other habitats as follows: in pine plantations up to 5 years old, Mecklenburg, 4-26 pairs per km² (Klafs and Stübs 1977), but new pine plantations in Brandenburg supported 100-120 pairs per km² (Rutschke 1983); in wasteland and sewage farms, Brandenburg, respectively 2-14 and 1·2-5·5 pairs per km² (Rutschke 1983). In steppe, Moravia (Czechoslovakia), 300 birds per km² (Lelek and Havlín 1957); in grass *Stipa* sand steppe of northern Kazakhstan (USSR), birds preferred border between ploughed and unploughed steppe, with maximum 25 pairs per km² (Kozhevnikova 1962; Pätzold 1983). Near Black Sea (northern Bulgaria), territories recorded overlapping with those of Calandra Lark *Melanocorypha calandra* and Short-toed Lark *Calandrella brachydactyla* (Pätzold 1983). ROOSTING. Nocturnal and always on ground. Normally solitary but winter flocks roost communally. Generally avoids dense cover, preferring shallow depressions, natural or otherwise (R E Green). In Kazakhstan, sometimes roosts in pits made earlier for dusting (Dolgushin *et al.* 1970). In Dreux (France), winter, nocturnal and diurnal roosting mainly in old stubble (Labitte 1937). Near Dresden during week of frost in January (minimum −17°C), *c*. 200 birds used field strip *c*. 300 × 6 m sheltered by trees, making hollows *c*. 15 cm deep and across; moved on once temperature rose to 0°C (Pätzold 1983). ♀ normally roosts on nest from night after laying of penultimate egg. May doze on nest during day but normal sleeping posture (head tucked under wing) not adopted then (Delius 1963, 1969). In early May, incubating ♀ slept from 20·00-06·30 hrs (Delius 1963). In Kazakhstan, seeks shade at hottest time of day (Dolgushin *et al.* 1970). In eastern England, January-February, when food scarce, comfort behaviour occupies less than 4% of daylight hours; in October-December (grain available on stubbles, etc.), birds rest for *c*. 40-80% of time, but mostly active in 2 hrs after dawn and before dusk (R E Green). Other data on activity rhythms relate to song. Typically sings very early and retires later than most passerines (Hardy 1964); will also sing on moonlit nights (Dementiev and Gladkov 1954a). Dusting regular. Other comfort behaviour recorded includes sunning and rain-bathing; see Delius (1969) for details.

2. Rather wary for a passerine (R E Green). On Mellum island (West Germany), resident birds apparently shyer than winter visitors; permit approach to *c*. 10-15 m, others to *c*. 6-8 m (Rittinghaus 1948). At Ravenglass, birds accustomed to observer also flew at *c*. 10-15 m. Sometimes fly up suddenly when man *c*. 2-3 m away (Delius 1963). Birds scavenging at campsites remarkably tame, allowing observer to pass at *c*. 1 m (Kooiker 1978). Normally squats if approached; in ploughland, will freeze in an upright posture (Meinertzhagen 1951a). ♀♀

more likely to crouch than ♂♂ (Delius 1963). Early in breeding season, ♂ tends to sing when flushed, then circle where ♀ on ground. If ♀ flies off low and calls (not described), ♂ usually follows (Lebeurier and Rapine 1935a). When falcon *Falco* overhead, normally crouches or freezes, keeping an eye on danger; may walk slowly into cover then crouch or freeze in take-off posture. Sometimes gives call 8 quietly. Flying birds may drop abruptly and enter cover. Attacked by Merlin *F. columbarius*, bird made off giving call 9. Birds performing Song-flight will fly even higher without interrupting song to escape from Hobby *F. subbuteo*, and continue to sing while evading stoops of *F. columbarius*. *F. columbarius* eating *A. arvensis* mobbed (Delius 1963; Stolt 1971; Pätzold 1983). Kestrel *F. tinnunculus* normally elicits little reaction, though singing bird may fly higher, and desultory dive-attacks recorded. Will chase Cuckoo *Cuculus canorus* and Red-backed Shrike *Lanius collurio* (Delius 1963; R E Green). Trapped birds occasionally give call 9. When handled, lie quiet and do not peck; may make sudden escape attempts (Delius 1963) or not (Madge 1967). FLOCK BEHAVIOUR. In Denmark, migrant flocks show pattern of break-up (more frequent) and fusion. On encounter, flocks moving in different directions often shift and/or interchange individuals. Flocks of over 10 tend to break up after flying a few hundred metres; normally loose-knit, birds *c.* 5-15 m apart. Large flocks dense and restless, flying low and calling (see 7 in Voice). Take off in dense mass but soon split up, especially before emigration over sea (Rabøl and Noer 1973). Flock approaching sea at *c.* 60-80 m made sudden whirling descent, then flew on fast and low over waves (Cretté de Palluel 1903-4). Flocks forming after moult give only calls 5 and 7a; shyer than in breeding season (Pätzold 1983). See also Lebeurier and Rapine (1935a) for wildness of migrant flocks. In eastern England, solitary birds (see part 1) feed independently and show no antagonism. Concentrate at good food source but move independently, giving call 7a on flying off (R E Green). Cohesive (immigrant) flocks feed and roost fairly close together; fly as compact flock and give call 7b (R E Green). Birds in winter flocks often close together when feeding (R E Green); defend individual-distance of *c.* 30 cm using forward run and call 4, but no real fights recorded. May sing quietly while feeding (Delius 1963). Chases lasting *c.* 5-10 s frequent during autumn song-period; if the 2 birds involved belong to moving flock, tend to keep up with it (Clark 1948). In South Uist (Scotland), flock drove off Snow Buntings *Plectrophenax nivalis*; chased them far, then returned to feed (Meinertzhagen 1951a). SONG-DISPLAY. Song of ♂ (see 1 in Voice) given from ground, open perch (e.g. post, tree) or in flight. For Song-flight, usually takes off silently and ascends at *c.* 20-70°, always into wind. After 1-2 s, may give series of Contact-calls (see 5 in Voice). From *c.* 10-20 m, steep spiralling ascent accompanied by song, or starts singing immediately after take-off (Pätzold 1983). Vigorous song during ascent typical of period with pair-formation and territorial disputes (Lange 1951). Ascends, with tail spread and characteristically fluttering wings, to *c.* (20-)50-100 m (Radig 1914; Delius 1963; Pätzold 1983). In *dulcivox*, 45% of Song-flights over 50 m (Dementiev and Gladkov 1954a). In *A. japonica*, height said to be dependent on ground temperature: 80-100 m at 24-28°C, 30-70 m at 16°C (Suzuki *et al.* 1952). Next (normally longest) phase not preceded by any noticeable change in wing action or song; involves hovering (10-12 wing-beats per s), often alternating with slow horizontal circling over territory or, not infrequently, beyond (Dementiev and Gladkov 1954a; Delius 1963; Pätzold 1983). According to Radig (1914), wind tends to induce more hovering, though individual differences probably also involved. Normally

returns to same spot after circling. In strong wind, periodically closes wings as typically in forward flight, this breaking up song (Delius 1963). Descends in slow spirals, wings fully spread and motionless, and continues to sing while gliding down to land, or (more often) suddenly stops singing at *c.* 10-20 m, then plummets with nearly closed wings, braking just before landing by spreading wings and tail (Dementiev and Gladkov 1954a; Delius 1963; Pätzold 1983); see also Gröbbels (1909). After long Song-flight, descent typically silent (Lange 1951). According to Meinertzhagen (1951a), usually returns to starting point. More Song-flights performed when ♀ absent. 1 ♂ showed peak before pair-formation, 2 others during incubation (Delius 1963); see also Clark (1947). Brief Song-flights (less than 30 s) frequent (Lange 1951); average duration 2-2·5 min (Radig 1914; Rollin 1931; Delius 1963); according to Radig (1914), only few sing more than 5 min, 10 min rare, 20 min exceptional— though 57 min (Brown 1986a) and 68 min (Rollin 1943a) recorded. At Ravenglass, variation through season: average 3·4 min (n=11) during territory occupation and pair-formation, 2·4 (n=84) during incubation of 1st clutch, 1·3 (n=34) during feeding of 1st brood, 2·1 (n=26) during incubation of 2nd clutch, 2·0 (n=9) during feeding of 2nd brood (Delius 1963). Another study showed increase through season (Lange 1951); see also Rollin (1931) who found high average of (poorer quality) song in autumn. Very long and continuous song usually from one individual (Lange 1951). Length reduced where much rivalry in high-density population (Radig 1914; Rollin 1943a; Clark 1947), though Lange (1951) found little influence. After weather has inhibited Song-flights, rapid increase typical, then return to normal rate (Delius 1963). On one sunny May day, 93% of all song given in flight (Pätzold 1983). ♀ will sing (see 2 in Voice) in the air and on ground; at Ravenglass, *c.* 15 records related to period of territory occupation and pair-formation or sexual readiness (Delius 1963). ♂ singing on ground often just sits quietly, feathers ruffled. Will sing quietly from ground or perch as continuation of aerial song, also while feeding. Sometimes sings loudly in more upright posture on elevation: bill opened wide to reveal bright mouth, throat feathers ruffled to show white bases, crest raised (Delius 1963). At Ravenglass, February-April, most aerial song in morning. Starts 03.30 hrs, April-May, and frequency about constant throughout day (Delius 1963). In central European USSR, June, song given from 02.30 or 04.00 to 11.00 hrs; usually little or none up to 16.00 hrs, then through to dusk (Dementiev and Gladkov 1954a). In southern Sweden, sings from *c.* 04.20 hrs, mid-April; starts progressively earlier until most before 03.00 hrs, May-July (earliest 01.00 hrs, late May and mid-July). Ends *c.* 18.00-19.00 hrs in April, 20.00-21.00 hrs in May, 20.00-21.30 hrs in June and 19.30-21.20 hrs in July (J Karlsson). See also Lange (1951) for Denmark. Song delivered on ground mainly in morning and evening (Delius 1963; Pätzold 1983), though amount may increase markedly when bad weather inhibits Song-flights (Rollin 1943a), and in *A. japonica* more frequent than aerial song throughout season (Haneda and Obuchi 1967). In Dresden area, on July afternoons, birds often sang for *c.* 4-5 min from tops of trees 1-2 m tall (Pätzold 1983). In Japan, 14% of daylight hours spent singing in breeding season (Haneda and Obuchi 1967). For further detailed data on song output in April-May and July, north-east England, see Rollin (1943b, 1956). In Britain, main song-period late January to (early) July, also October; irregular but fairly frequent late September, November, late December and early January; exceptional or only Subsong late July to early August and first half of December (Witherby *et al.* 1938a); 222-47 song days per year (Hardy 1964). Break or marked reduction in aerial and ground song during (July-)August-early September associated with moult (Delius 1963; R E Green); in Denmark, most birds stop from about 13 July or 4 August, then sing again September–November (Lange 1951). Aerial song given from late September generally weaker than in breeding season; song from ground possibly begins earlier, and also frequent in early February (Delius 1963). Age of birds singing on ground in September unknown (R E Green). For France, see Lebeurier and Rapine (1935a) and Labitte (1957). On Mellum, full and loud song given in November, also (when sunny and 0°C), December–January (Rittinghaus 1948). In USSR, migrants will sing loudly in high Song-flights (Dementiev and Gladkov 1954a) or (in autumn) make only brief ascents (Bub 1955). ANTAGONISTIC BEHAVIOUR. (1) General. At Ravenglass, newly-arrived ♂♂ feed together peaceably, but local birds suddenly turn aggressive after only a few minutes. Fights early in day initially give way to re-flocking later on. Pattern (though rendered more complex by pair-formation) similar in later-arriving ♀♀. Territorial aggression increases from arrival up to peak in early April, then shows steady decline (little in bad weather) until virtually absent in July; adults trying to feed own offspring in another territory usually attacked, however. Some (already moulted) ♂♂ temporarily revive aggression in mid-August (Delius 1963), and frequent chases and supplanting attacks occur October–November (R E Green); see also Flock Behaviour (above). Normally no trespass into neighbouring territory even when owner absent, and personal acquaintance generally inhibits aggression between neighbours. New arrivals trying to occupy vacated territories elicit unusually vigorous attacks by neighbours, and others from further away will cross intervening territories to join in attempted repulsion; newcomers also highly aggressive, often withstanding attacks (Delius 1963). In eastern England, feeding birds tolerated within a territory; only strange ♂♂ trying to sing or chase resident ♀ evicted (R E Green). Territory limits ignored when birds unite to chase predator. Bird on own side of boundary more likely to attack, other to flee, but tendencies in balance on boundary and roles can change. 'Confident' intruders (possibly with mate, nest, or young nearby) commonly intimidate owners. ♂♂ generally more aggressive; birds usually attack others of their own sex. ♂ frequently intervenes to help own mate repel intruder. If ♂ displays at another ♀, his mate will attack her vigorously; occasionally attacks also strange ♂♂. Antagonistic behaviour patterns highly variable (e.g. in degree of completeness) and frequently merge. (Delius 1963.) (2) Threat and fighting. In Advertising-display, ♂ stands erect on elevation, breast feathers ruffled (appear paler), crest raised, tail fanned, and carpal joints slightly raised; silent or gives call 3. Similar posture accompanied by loud song (see above) also occurs (Delius 1963). Rivals recorded in Advertising-display c. 4–5 m apart for over 6 min; intruder finally fled and was chased 30–50 m (Pätzold 1983). Bowing and (less often) Hopping (see subsection 2 in Heterosexual Behaviour, below) also performed by aggressive birds of both sexes. Territory-owner may fly in short arc towards intruder and land; usually gives call 3 in flight. Close to intruder, changes to Threat-flight: series of short upward glides with wings widely spread and slightly raised, interrupted by brief fluttering; rarely, gives call 3. Pursuit-flight (with call) often follows: e.g. frenzied and accelerating with sharp turns, birds c. 7–10 m up and c. 1 m apart; pursuer (sometimes both) may perform Song-flight afterwards (Clark 1948). On ground, rivals adopt Threat-Posture sometimes less than 1 m apart (Delius 1963), even touching bills (Pätzold 1983): body low and thrust forward, feathers slightly ruffled, crest more or less raised, tail fanned jerkily and usually depressed, sometimes

raised; wings half-opened and raised over back. Call 4 (less often 3) given with bill wide open. Rival threatened frontally or (when ready to break off) side-on; will also threaten overflying bird. Rivals stationary or run forward (see Flock Behaviour, above), wings outspread and waving. Various incomplete forms common (see Delius 1963). Aerial fight (Fig A) normally

A

develops from threat on ground: usually flutter up and use bill, claws, and (probably) wings as weapons. Fights not damaging, though a few feathers may fly. Only rarely tumble to ground interlocked, but may flutter up and fight repeatedly (Delius 1963; Pätzold 1983; see also Selous 1901). 2–3 ♂♂ sometimes fight over 1 ♀ (Dementiev and Gladkov 1954a). 2 ♂♂ recorded flying at c. 30 m, then ascending to c. 50 m, never more than 1 m apart, both singing when not actually attacking; one finally ascended to c. 150 m in Song-flight while other sang directly below it at c.10–30 m (Radermacher 1977); see also Radig (1914) for apparent Song-duel. Bird ready to flee may adopt an erect posture, fixing gaze on rival; sleeking of feathers indicates escape imminent. May then jump, run, or fly off, or turn away and peck at ground or tear out grass and toss it aside. Bird making running escape stops frequently to crouch in hollows and watch rival (Delius 1963). For aggression with *L. arborea*, see that species (p. 180). At Ravenglass, Meadow Pipit *Anthus pratensis* frequently attacked where territories overlap, but also often left alone as almost always Wheatear *Oenanthe oenanthe* (Delius 1963). Rock Pipit *A. spinoletta* chased 'playfully' on island where no rival *A. arvensis* (Rollin 1956). HETEROSEXUAL BEHAVIOUR. (1) General. In Kazakhstan, some newly-arrived flocks apparently contain already paired birds (Dolgushin *et al.* 1970). At Ravenglass, ♂♂ have taken up territories when ♀♀ arrive; most pairs formed by c. 10 days later. ♂♂ temporarily off their territories show no interest in ♀♀. Older ♀♀ returning to previous year's territory and chased off by new resident ♀, also any ♀ newcomers, move singly around territories until paired. ♂♂ without territories and some ♀♀ (together c. 10% of population and probably mainly 1-year-olds) remain unpaired, move about irregularly and openly over quite wide area. 2 such ♂♂ took over territories with widows and 1 ♀ replaced a lost bird. ♂ deserted by mate remained on territory for 2 months; failed to attract another ♀ but paired successfully in following year. Pair-members apparently recognize each other at c. 30 m (Delius 1963, 1965). Closest association between ♂ and ♀ a few days before laying, otherwise mainly solitary in spring and summer (R E Green). ♀ approaching territorial ♂ before pair-formation not usually treated as intruder. Once ♂ paired, strange ♀♀ normally attacked, only rarely elicit display; most behaviour described below thus between mates. Pair-formation may be seen as ♀ settling in ♂'s territory. Bond probably firmly established after birds together several hours (Delius 1963). (2) Pair-bonding behaviour. In first approach, ♀ may land a few metres away, then move, anxiously alert, into cover. When ♀ begins to feed, or before, ♂ approaches and immediately begins courtship-display. Most ♀♀ move on at this stage; if ♀ stays, she keeps close to ♂ who displays frequently. If ♀ moves off, ♂ chases her and attempts (occasionally with success) to get her back into territory. Sometimes, ♂ turns back at boundary. If chase moves outside territory, other ♂♂ join in. ♂ tends to give call 5b if ♀ leaves. Courtship on ground usually follows fragmented song by ♂. Proximity of ♀ most commonly elicits 2 closely linked behaviour patterns: with head up and feathers sleeked, ♂ Hops (Fig B) several times 1–2 cm high front-on to ♀, on the spot

B

or towards her; intention-movement to Hop is brief upward head-jerk. Frequently changes from Hopping to Bowing (very close or up to several metres from ♀): ♂ may move or turn away, head bowed forward and tail raised; tail-raising often occurs in isolation (Delius 1963). Makes stiff pecking movements at ground (not touching it) every few seconds (Stresemann 1956b; Pätzold 1983). Further reports of tail-raising by 2 birds, or only 1 while other slowly raised wings every 5–8 s, not definitely heterosexual (Pätzold 1983), but see Clark (1948). Bowing sometimes accompanied by shivering of wing-tips (small-amplitude movements also shown in isolation); wing closer to ♀ may be held out and flapped (Delius 1963). ♂ also preens and feeds so that whole process may last several minutes. At higher intensity, ♂ has ruffled feathers, those on throat and neck form 'ruff', crest raised (more if ♀ near), tail drooped, and trembled wing-tips at times brush ground; Bows, 'dances', and Hops right and left in semicircles. ♀ may then approach on zigzag course, but ♂ apparently oblivious until birds touch. May then whirl around each other, flutter up c. 2–3 m, and fly off c. 50–200 m. After landing, performance repeated or ♂ adopts high-intensity Tail-up posture: wings drooped and trembled, slightly spread tail raised vertically and bent slightly forwards; cloacal region with bright white feathers directed at ♀. ♂ may remain thus, motionless, for up to 2 min, singing quietly. ♀ recorded Bowing without tail-raising (Stresemann 1956b; Pätzold 1983); see also Selous (1901). Wing closer to ♀ held slightly away from body as ♂ runs round her, bill wide open to give loud song. When Tail-up posture shown to strange ♀♀, usually combined with aggression. ♂ also recorded running at ♀ then pecking for food under her bill. By flying towards partner, either may initiate aerial chase of maximum 1 min in which ♂ sings quietly in fast flight close behind ♀. After landing, only ♂ performs apparently ritualized cooling: sleeks feathers, holds wings away from body, and pants (Delius 1963). While ♂ performs Song-flight, ♀ normally moves about inconspicuously on ground (Dementiev and Gladkov 1954a). Rarely, antagonism between mates (neither sex dominant) arises if displaying ♂ is too near ♀, if ♀ solicits and ♂ is unwilling, or if both after same food: threat mostly brief and incomplete, involving call 4. ♂ also recorded driving ♀ back from strange territory after her eviction from there (Delius 1963). (3) Courtship-feeding. Apparently does not occur. ♂ not recorded feeding ♀ on nest (Delius 1963). (4) Mating. ♀ runs towards ♂ in posture similar to ♂'s high-intensity Tail-up posture, but

body horizontal and low, feathers (especially on back) slightly ruffled; often sings quietly. ♂ Hops and stands next to ♀ (now still), vibrates wing on her back (Fig C), then mounts, beating

C

wings rapidly for balance. Pushes tail under ♀'s as she moves it sideways for cloacal contact. Afterwards, ♂ jumps off and moves quickly away; ♀ retains posture briefly then resumes previous activity. ♂ apparently disregards soliciting ♀ for several days. Most copulation takes place in morning, closely linked with nest-building, i.e. over 3–4 days before laying. 1 successful copulation evidently enough for fertilization of all eggs. Occasionally, ♂ attempts copulation when ♀ sand-bathing or self-preening (Delius 1963). (5) Nest-site selection. Apparently by ♀. ♀ recorded moving about (more flying than usual) over periods of up to ½ hr, nestling down and shuffling about to deepen available hollows, mainly in denser vegetation. Up to 20 hollows visited, and material later carried to 4–5 (Delius 1963). Scrapes sometimes made where no natural depression (Pätzold 1983). ♂ usually nearby but not seen to participate (Nethersole-Thompson and Nethersole-Thompson 1943). Birds (presumably ♂♂) recorded making 'forms' (not used as nests) after Song-flight (Suffern 1951). During later part of nest-building, ♂ may accompany ♀ closely, flutter over nest, perch nearby, and sing (Delius 1963), but degree of attachment varies (Pätzold 1983). (6) Behaviour at nest. Some building may continue during laying throughout which ♀ visits nest briefly for up to 5 min. Most laying 05.00–08.00 hrs. Bird approaching nest for incubation (from building: Lebeurier and Rapine 1935a), normally lands several metres away and, for departure, walks some distance before flying up. On nest, ♀ may give call 5a (mostly at start of shift), sometimes in response to ♂'s call or on sighting him. ♀ alert or about to leave has neck forward, feathers sleeked but crest raised. More details in Delius (1963) and Pätzold (1983). ♀ off nest for feeding sometimes accompanied by ♂ who afterwards performs Song-flight (Lebeurier and Rapine 1935a). RELATIONS WITHIN FAMILY GROUP. Clutch normally hatches within 8 hrs. Except in rain and cold, young not normally brooded after 5th day; stints shorter than for incubation, with little fluctuation during day (Delius 1965). ♀ recorded shading young from hot sun (Willford 1925). Feeding by ♀ begins only a few minutes after 1st young hatched, ♂ following at latest 2 hrs afterwards (Pätzold 1983). Near nest, especially when young still small and blind, adult gives Feeding-call (see 6 in Voice); adult at nest leaves immediately at approach of mate. ♂ (very rarely ♀) recorded coming to nest with food early in incubation and giving Feeding-call. Blind nestlings gape in response to acoustic stimuli: neck extended with slight swaying, black and yellow mouths prominent. Eyes open on 4th day when young also give food-call. On 5th day, turn towards various objects and gape, eventually only to parent (Delius 1963). Young fed (usually 1 per visit) bill-to-bill, seeds possibly regurgitated for older nestlings (Delius 1963; R E Green). Faecal sacs initially swallowed by adult, later carried away and dropped in flight

(Pätzold 1983) or after landing; hideaway used for some time by young out of nest also cleaned thus. Nestling recorded gaping and calling possibly to threaten sibling (Delius 1963). For development of comfort behaviour, see Delius (1969). At Ravenglass, young spent average 8·5 days (7–11, $n = 32$) in nest (Delius 1963). Come out to meet food-carrying parent on 6th day but always return; stay out longer on 7th but still roost in nest, similarly on 8th when hop and run further from nest and give different food- and contact-call (see Voice). Leave spontaneously, all within 4 hrs (Delius 1963); sometimes lured away by parents (Groebbels 1940). Adults (without food) recorded frequently hovering over young *c.* 11 days old (Pätzold 1983). Feeding rate declines between 12 days old and fledging (at *c.* 15 days) when young also start self-feeding; recorded self-feeding constantly at 19 days (Delius 1963) and fly well at 20 days (Dementiev and Gladkov 1954*a*). Tend to hop for some distance after parent who repeatedly feeds them single items, then retreat into cover (where they spend most time) or continue self-feeding; beg by gaping and wing-shivering. ♀ building 2nd nest may alternate this with feeding 1st brood. Adults do not feed strange young, but able to locate own offspring even if these silent. Straying juvenile once attacked, but young tend to remain in parental territory. Sand-bathe regularly (Delius 1963; Pätzold 1983). Young suddenly attacked (mainly by ♂ at 28 days, final separation coming at 30 (Delius 1963) or 32 days (Pätzold 1983). In another study, adults and young associated for only *c.* 5–6 days after fledging (R E Green). ANTI-PREDATOR RESPONSES OF YOUNG. From *c.* 4 days old, crouch on hearing call 8 from adult (mainly ♀), also when nest suddenly in shadow or (later) when something unfamiliar approaches, or other bird species signal alarm (Delius 1963). If removed from nest at 7 days and placed on ground, will slowly crawl away (Pätzold 1983). Independent juveniles shy and inconspicuous (Delius 1963). PARENTAL ANTI-PREDATOR STRATEGIES. (1) Passive measures. If disturbed, may desert during nest-building (Lebeurier and Rapine 1935*a*), also later if disturbance sudden (Pätzold 1983, which see for habituation to man). If nest in cover over 3 cm high, incubating bird allows very close approach; crouches then flies up directly from nest; in sparse cover, also early in incubation, flies when man still distant (Delius 1963; Pätzold 1983; R E Green); for cautious approach and departure, see subsection 6 in Heterosexual Behaviour (above). Alarmed birds recorded landing *c.* 20 m from nest, running low, making detours, and occasionally standing erect with crest raised (Pätzold 1983), also apparently mock-brooding and mock-pecking (Delius 1963). When with young, adults give call 8 frequently; otherwise, the few such calls given are mainly by stationary ♂ who moves tail slightly up and down. Crows *Corvus* near nest or young elicit crouching or call 8 (Delius 1963). (2) Active measures: against birds. Distraction-lure display performed towards waders (Charadrii) approaching nest—e.g. Ruff *Philomachus pugnax* which is also subjected to flying attack, bird veering away just in front of it (Schildmacher 1966). (3) Active measures: against man. ♂ tends to ascend for Song-flight (Delius 1963), circling fairly low (Dolgushin *et al.* 1970); may land *c.* 100 m from nest, sing from perch, swallow food, wait, or repeat Song-flight (Wyss 1947; Pätzold 1983). If cut off from nest, ♀ often flies up and hovers near it (Delius 1963) or *c.* 30 m away constantly looking back (Pätzold 1983). Distraction-lure display of disablement type performed by ♀ throughout incubation, also after hatching (Delius 1963), mainly late in incubation and just after hatching (Dolgushin *et al.* 1970); occurs especially if bird suddenly disturbed early in stint (Delius 1963). Departure with very rapid, shallow wing-beats is perhaps low-intensity variant (R

E Green). Otherwise, bird flies off short distance, then runs away fast, with body low, or flutters off, tail spread, one wing raised, and other dragging; allows man to *c.* 1–2 m, then flutters *c.* 5–6 m, feigns injury while landing, pauses, and again allows close approach, etc., until intruder lured up to *c.* 20 m from nest (Delius 1963; Pätzold 1983). Young allegedly carried away from disturbed nests in feet (Hesse 1917) or bill (Pätzold 1983). (4) Active measures: against other animals. Weasels and stoats (Mustelidae) sometimes ignored, but often followed in mixed flock (with *A. pratensis* and *O. oenanthe*): in one case, birds fluttered low over them, landed close, and flew up again, but gave call 8 only infrequently and quietly. Flutters at sheep and rabbits *Oryctolagus* only when near nest, rabbit then even pecked (Delius 1963).

(Figs by J P Busby: A from photograph in Pätzold 1983; B–C from drawings in Delius 1963.)   MGW

Voice. ♂ generally more vocal than ♀. Calls 1, 5b, and (less so) 8 given almost exclusively by ♂, all others except 2 by both sexes (Delius 1963). For examination of possible relationship between respiratory and syringeal movements and song production, see Brackenbury (1978) and Csicsáky (1978). For extra sonagrams, see also Bergmann and Helb (1982) and Pätzold (1983).

CALLS OF ADULTS. (1) Song of ♂. (a) In flight. Loud and melodious, but with fairly restricted range of sounds, although the many different combinations and long, apparently unbroken flow (see Brackenbury 1978 and Csicsáky 1978) give it attractive quality (Dementiev and Gladkov 1954*a*). Typically a long-sustained burst at frequency 2–5 kHz. Few if any other bird species of west Palearctic sing as persistently in flight (Pätzold 1983). Loud, clear, shrill warbling, pleasingly modulated (Witherby *et al.* 1938*a*). Usually gives 1–2 'trii' or 'trli' sounds (see also call 5b) soon after take-off, then continuous song containing frequent series of varyingly modulated 'trli' or 'dji' whistles in variety of pitch patterns (Voigt 1933, which see for modified musical notation). Frequent repetition typical (Bergmann and Helb 1982): e.g. of a high-pitched 'tee-ee tee-ee', then lower-pitched 'tyu-yu tyu-yu' (Dementiev and Gladkov 1954*a*) which may be call 5b; see also Radig (1914) who additionally noted a short 'wit-wit'. Evident in recording by M Palmér are trills and tremolos with much variation in length, speed, pitch, and timbre, and a 'whee-ou'; also drawn-out, rather nasal sounds, fragments reminiscent of Song Thrush *Turdus philomelos*, and a 'chup chup chup' not unlike Nightingale *Luscinia megarhynchos*, then a wild whining or wailing sound (M G Wilson); latter probably the long fluting sounds woven in during descent (Delius 1963). In recording (Fig I), fragment from sustained, lilting song shows undulating pitch pattern; rapid trills and tremolos interspersed with relatively long tones (units a, b, c). In different part of same recording (Fig II), pitch pattern of apparently more urgent song lacks broad ascents and descents of Fig I; buzzing character due to rapid frequency modulation visible in units a, b, c, d; section shown in Fig I is more musical (J Hall-Craggs).

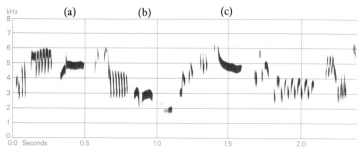

I   W H Thorpe   England   July 1966

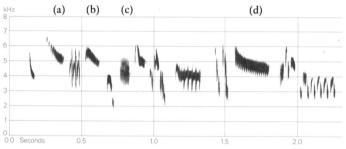

II   W H Thorpe   England   July 1966

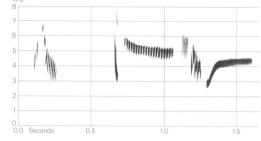

III   V C Lewis   England   June 1967

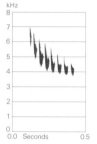

IV   V C Lewis   England   June 1967

Call 7a frequently interpolated (Pätzold 1983), especially towards end of song (Bergmann and Helb 1982), though Lange (1951) found song to be normally louder and more rippling (due to inclusion of call 7a) during ascent, slower and rather melancholy during descent. Marked individual variation (Delius 1963). Will mimic other birds and animals, e.g. songs and/or calls of other Alaudidae, waders (notably Redshank *Tringa totanus*), Marsh Warbler *Acrocephalus palustris*, and Linnet *Carduelis cannabina* (Radig 1914; Lange 1951; Bergmann and Helb 1982; Pätzold 1983); see details for *dulcivox* in Dementiev and Gladkov (1954a). (b) On ground. Basically similar to song in flight but extremely variable; usually shorter, quieter, more warbling and melodious and with more pauses (Lebeurier and Rapine 1935a; Delius 1963; R E Green). In recording (Fig III), bird (sex unknown) approaching nest to feed young gives call 6 (last unit of Fig III), but also apparent song with attractive purring or tremolos, some high-pitched, and a soft churr like House Sparrow *Passer domesticus* (M G Wilson) or Whitethroat *Sylvia communis* (P J Sellar). Weak, hesitant song with pauses of 3 s or more between bursts typical of autumn period (Lange 1951). Subdued, fragmented variant of flight-song given in chases (Delius 1963). (2) Song of ♀. Monotonous and fragmented, quieter than that of ♂ (Delius 1963). (3) Rapidly repeated 'tschrr tschrr' given by ♂ in Advertising-display; harder and harsher than call 5a (Delius 1963). Sharp 'zerrerer' (Radig 1914) and vibrato 'zerr' (Pätzold 1983) probably the same. (4) Quiet, hoarse 'gjj gjj' given in threat (Delius 1963); according to Radig (1914), a sharply vibrato variant of call 5b. In fights, gives drawn-out and harsh 'wäd' or 'dsä', also 'srrrt' (Bergmann and Helb 1982);

last possibly same as call 3 or related. (5) Contact-calls. (a) Rapidly repeated 'schrr' often like call 3, but usually softer, more melodious. If given by one pair-member, other normally responds similarly (Delius 1963). The following presumably all alternative renderings: soft, full 'giergiergier' given at nest (Radig 1914); rounded, almost fluting 'gürr gürr' said to be like *C. cannabina* and given (e.g.) by ♂ near nest where ♀ incubating (Pätzold 1983); soft, melodious, trilling 'trrü' (Bergmann and Helb 1982). (b) Plaintive, drawn-out 'juuu...' given at intervals, mostly by ♂ deserted by (prospective) mate, also rarely by incubating ♀ (Delius 1963). Following possibly same or related calls: clear, metallic whistling 'tree-ree-ree' or 'tree-eek' (Dementiev and Gladkov 1954a); sharp 'trieh', 'tried', or 'trlie' ('ie' pronounced 'ee' in English) as contact soon after arrival (Radig 1914; Pätzold 1983). (6) Feeding-call. Repeated soft 'tju tju' (Delius 1963). Similar to call of Woodlark *Lullula arborea* given in similar situation (Pätzold 1983). Recording suggests 'teweee' with rising pitch (final unit of Fig III), apparently combined with fragmented song (see call 1b); also a remarkably beautiful rippling descent (J Hall-Craggs: Fig IV). (7) Flight-calls. (a) Liquid, rippling 'chїrrup' (Witherby *et al.* 1938a); 'tschirit' (Delius 1963); hard, smacking 'tschrl' or 'tschr', also when flushed 'trri-tu' or 'tschritju' (Bergmann and Helb 1982). Given throughout the year, also when disturbed (R E Green). Richer 'chuddle-ee' from eastern birds in Iraq unlike typical call of nominate *arvensis* (Moore and Boswell 1956). See also Field Characters. Recording of bird presumably about to fly reveals variety of sounds: short 'ker-tik' followed by 'ti-ki-te-rierrrrr' in which first 3 sub-units short and hard and final sub-unit ends in rolled 'r' (J Hall-Craggs: Fig

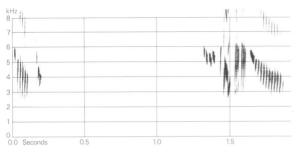

V   V C Lewis   England   June 1967

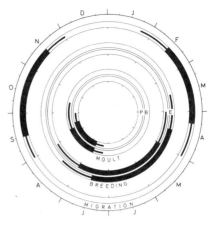

V), buzzing 'tit-terr-reep', shorter, sawing 'wherree' or 'wherru', 'tcherr', and short, dry, rattling 'trt', 'prrt', or 'prirrip' (M G Wilson). (b) Thin 'seep' rather like Redwing *T. iliacus* and characteristic of birds in cohesive winter flocks (R E Green); drawn-out 'migration call' (Rittinghaus 1948) probably the same, as are sharp, plaintive 'tjii' or 'zijh', at distance a more muted, disyllabic 'zie' with melancholy fall in pitch, given when disturbed (Bergmann and Helb 1982), and thin whistling 'seeerrrr' or 'seeirrrr' given by perched bird and interspersed with fragmented song in recording by V C Lewis. (8) A 'tlütütü tlütütü' typically given by disturbed bird during rearing of young (Delius 1963); loud fluting 'tluutoi' from captive ♂ (Pätzold 1983). (9) Shrill sound given by frightened bird (Delius 1963).

CALLS OF YOUNG. None heard from very small nestlings. When gaping at *c*. 4 days old, a hoarse 'ch ch ch' (Delius 1963) or 'chj chj' (Pätzold 1983); very like adult call 4 and possibly forerunner to it. From 8 days old, food-call 'iuiu iuiu', perhaps later developing into adult call 5b (Delius 1963). Birds removed from nest at this age gave quiet piping sound (Pätzold 1983); not necessarily an expression of fear as frightened juveniles give call indistinguishable from adult call 9. Well-developed song given on ground at *c*. 10 weeks old (Delius 1963); if removed from nest early, never learn typical song but may become great mimics (Pätzold 1983).   MGW

Breeding. SEASON. Britain and north-west Europe: see diagram. Little variation over rest of range, except for delay in onset of breeding related to increasing latitude or altitude (Dementiev and Gladkov 1954a; Heim de Balsac and Mayaud 1962; Makatsch 1976). Laying date related to temperature and to age of bird (Delius 1965). SITE. On ground in the open or among short vegetation such as grass or growing crops. Solitary. Nest: shallow depression lined with grass leaves and stems, with inner lining of finer material; rarely with rampart of small stones (Congreve 1950). Building: Probably by ♀ only, though ♂ may help excavating depression. EGGS. See Plate 80. Sub-elliptical, smooth and fairly glossy; grey-white, often tinged greenish, thickly spotted brown or olive. Nominate *arvensis*: 23·4 × 16·8 mm (20·5–26·5 × 15·3–18·5), $n = 300$; calculated weight 3·35 g. Small samples of other

races (*sierrae*, *harterti*, *cantarella*, *dulcivox*) suggest no significant variation from above (Makatsch 1976; Schönwetter 1979). Clutch: 3–5(–7). Of 89 clutches, England: 3 eggs, 33; 4, 51; 5, 5; mean 3·69 (Delius 1965). Significant difference between this mean and those from France (3·75, $n = 85$: Labitte 1957), Germany (3·87, $n = 55$: Haun 1931), and Estonia, USSR (3·97, $n = 36$: Randla 1963), suggesting increasing size from west to east. In England, clutch size increases during season with mean 3·3 for 1st clutch ($n = 48$), 3·8 for 2nd ($n = 28$), and 4·0 for 3rd ($n = 19$), but no significant difference between clutches laid by 1st-year birds (3·5, $n = 22$) and older birds (3·55, $n = 40$) (Delius 1965). Up to 4 broods per season, fewer in northern latitudes. Mean 2·7 clutches per pair (1–4, $n = 34$) in England (Delius 1965). Replacements laid after loss of eggs or young; period between loss and replacement related to timing of loss, with mean 4 days if during egg-laying period, 6 days if during incubation, and 5–6 days if during nestling stage (Delius 1965). INCUBATION. 11 days. By ♀ only; for details of incubation stints, see Delius (1963) and Pätzold (1983). Begins with last egg; hatching synchronous. YOUNG. Altricial and nidicolous. Cared for and fed by both parents. FLEDGING TO MATURITY. Fledging period *c*. 18–20 days, but young usually leave nest at 8–10 days. Become independent at *c*. 25 days (R E Green). Age of first breeding 1 year (see Social Pattern and Behaviour). BREEDING SUCCESS. Of 319 eggs laid, England, 67% hatched and 46% reared to leaving nest; significant difference in success related to clutch size, with 64% hatching and 32% fledging from 93 clutches of 3 eggs, and 72% hatching and 50% fledging from 180 clutches of 4 eggs; main losses during incubation and rearing to predators (Delius 1965).

Plumages (nominate *arvensis*). ADULT. Rather variable, some birds having ground-colour of upperparts and chest rufous, others greyer above and more buff below, and some variation in width and contrast of dark streaks on upperparts and chest. Abrasion also has marked influence, pale feather-fringes soon wearing off and olive-brown colour fading to grey. Many intermediate birds occur, however, and thus no real morphs separable, though this sometimes advocated. In fresh plumage of dark rufous birds, feathers of forehead, crown, mantle, inner scapulars, and back with broad black centres, merging into olive-brown at sides; feathers narrowly fringed pink-rufous or buff at sides, narrowly white or pale buff at rounded tips. Outer scapulars similar, but feather-sides more extensively olive-brown and buff, hindneck buff or pale cinnamon with narrow and indistinct dull black streaks. Feathers of rump and upper tail-coverts olive-brown with black centres, broad pale olive-grey sides, and narrow buff or pale grey fringes along tips. Tertials dark olive-brown, grading to black near base, broadly fringed pink-cinnamon at sides and white along tip. In paler rufous birds, black centres of upperpart feathers are hardly wider and hardly more contrasting than those on hindneck of dark rufous birds, except for crown; sides of mantle feathers and scapulars extensively tinged rufous-buff or rufous-cinnamon, narrow fringes along tips buff rather than white; rump and upper tail-coverts olive-grey with buff tinge, hardly streaked dark except for longest coverts; tertials rufous-brown or olive-brown, less blackish. In dark greyish variants, black feather-centres of upperparts are broad, as in dark rufous birds, but feather-sides are pink-cinnamon or sandy-buff and tips narrowly white, sides soon bleaching to grey, especially on hindneck, scapulars, rump, and tertials. Pale grey birds are similar, but feather centres only narrowly black, upperparts appearing pale olive-grey with slight buff tinge and limited amount of black blotching on crown and mantle. Tufts covering nostrils and broad patch on lores pale buff to off-white (tufts black just at sides of culmen), extending into narrow eye-ring and narrow supercilium widening slightly above ear-coverts; colour of eye-ring and supercilium variable, rufous-buff to buff in more rufous variants (hardly contrasting), pale cream-buff to off-white in grey variants. Cheeks and ear-coverts narrowly and indistinctly streaked black, rufous-olive, and buff in rufous birds, olive-brown and pale buff in paler birds. In rufous birds, ear-coverts bordered behind by black-spotted buff or pale cinnamon band down side of neck; in paler birds, band cream-buff to off-white, hardly spotted, contrasting more with ear-coverts and sides of chest. Upper chin uniform cream-buff to off-white, lower chin, throat, and lower cheeks similar but finely speckled grey or black. Ground-colour of chest buff or cream-buff, merging into rufous-cinnamon or cinnamon-buff on breast and sides of body in more rufous birds, tawny or pink-buff in paler birds. Chest, sides of breast, and upper flanks marked with distinct black streaks 1–1·5 mm wide, but streaks occasionally duller, brown-black or grey-black, less sharply defined; streaks frequently join at upper sides of chest to form black bar-like patch. Lower flanks cinnamon-buff, olive-buff, or buffish-olive-grey, narrowly and indistinctly streaked darker olive-brown or olive-grey. Remainder of under-parts white, usually with yellow-buff, pale buff, or cream tinge on under tail-coverts; longer under tail-coverts sometimes with dark central streak. Central pair of tail-feathers (t1) black, broadly fringed pale olive-brown or pale buff-brown at sides, narrowly bordered white along tip; t2–t5 black, t2 with olive or buff tinge along outer web and tip (often faintly on t3–t4 also), t5 with outer web white (except base); t6 white with broad black wedge towards base of inner border. Flight-feathers greyish-black, secondaries and inner primaries with narrow and ill-defined pale grey or grey-buff tips; outer webs of secondaries with broad but ill-defined rufous-cinnamon or pink-buff fringes, those of primaries with narrow and sharp pale cinnamon or pale buff fringes (except on emarginated parts). Greater upper primary-coverts and tiny pointed p10 greyish-black with rather narrow and sharp pale cinnamon or buff fringe along outer web and tip. Lesser upper wing-coverts buff-brown with olive or grey tinge and pale buff fringes; median and greater coverts darker, brown-black or olive-black, tips broadly fringed cinnamon, terminally grading to off-white. Under wing-coverts and axillaries pale grey, tinged buff and with ill-defined paler tips. *In slightly worn plumage* (about November–December), narrow white fringes along feather-tips of crown, mantle, scapulars, back, and tail, and broader white fringes of tertials disappear. When more heavily worn (about mid-winter to April), olive-brown tinges of feather-tips on upperparts bleach to grey (especially on tertials and t1) and black streaks become sharper and more contrasting, in particular on cheeks, throat, and chest. In breeding season, upperparts duller, heavily streaked and mottled black, crown and mantle in particular largely black; hindneck tawny-buff (in rufous variants) to pale buff or cream-white (in pale variants) with sharp black streaks; rump dusky olive-grey, often broadly streaked, hardly tinged buff; band behind ear-coverts off-white, virtually without speckles, lower cheeks and throat with small and indistinct speckles or virtually uniform; ground-colour of chest pale buff, slightly deeper rufous towards sides of breast, chest streaks narrower and sharper. Plumage sometimes tinged grey or rufous by soil, especially on underparts. NESTLING. Down long, fairly dense, but underparts virtually naked; straw-yellow or buff-yellow. For development with age, see Pätzold (1983). JUVENILE. Upperparts and upper wing-coverts a mixture of black or black-brown and tawny-buff or yellow-buff, but amount of each strongly variable: some birds largely black, with buff restricted to hindneck, basal feather-sides, and indistinct subterminal patches on longer scapulars; others much more extensively buff, feather-centres brown-buff with black U-mark subterminally; still others about equally mottled buff and black; in all variants, feathers have contrasting white or cream fringes along tips (rather similar to pale fringes of fresh adult). Supercilium wider and paler than adult; ear-coverts often finely speckled white; in paler birds, often conspicuous white band from side of throat along rear of ear-coverts (more distinct than band of pale adults). Underparts as adult, but chin and throat without grey specks; ground-colour of chest, sides of breast, and flanks paler, tawny-buff or yellow-buff; streaks on chest more broken into spots, each spot broader, less sharply defined, and more rounded at tip than in adult, narrowly fringed white; in some pale birds, chest-spots small, pale buff-brown, hardly visible except on sides of breast, where bordered black near tip. Tail rather variable, but always with broader and more sharply defined buff fringes than adult, all round on t1, on outer web and tip only on t2–t4; feather-centres of t1–t5 sometimes black, but more often t1 with buff-brown notches at sides or buff-brown with indistinct black bars, t2–t5 dark grey-brown, sometimes with faint buff marbling; tail-feathers narrower and more pointed than in adult. Outer webs and tips of flight-feathers and greater upper primary coverts with narrower and more sharply defined buff fringes than adult, subterminally blacker than adult; fringes along tips of outer primaries even and sharp (unlike adult). P10 distinctly longer, broader, and with more rounded tip than in adult, 6 mm shorter to 1 mm longer than longest upper primary covert (in adult, 10–17 mm shorter, narrow and pointed, hidden under reduced outermost greater upper primary covert). Upper wing-coverts (including lesser) darker than in adult, fringes broad, sharp, and contrastingly pale buff to off-white, centres dark grey or olive, ending black subterminally, and often with buff spot. Sometimes very different from adult, and sometimes hard to distinguish, but always with marbled t1, rounded chest-spots, broad and sharp pale fringes on outer upper wing-coverts, pale even fringes along tips of outer primaries, and long p10. FIRST ADULT. Like adult; indistinguishable when last juvenile feathers (usually p9-p10) shed.

Bare parts. ADULT, FIRST ADULT. Iris hazel-brown, cinnamon-brown, brown, or dark brown. Upper mandible dark horn-brown or horn-black, cutting edges and lower mandible pale horn-brown or yellow-brown. Leg and foot yellow-brown, pale brown, pale red-brown, or flesh-brown, joints sometimes blackish. NESTLING. Bare skin of upperparts dark violet, of underparts orange-flesh. Mouth yellow; dark spot on tip of tongue, and another (sometimes absent) on each side of base; one spot inside tip of each mandible; gape-flanges white or pale yellow. Leg and foot greyish-flesh. JUVENILE. Iris brown; bill, leg, foot, and claws flesh-colour. (Hartert 1910; Heinroth and Heinroth 1924-6; Witherby *et al.* 1938a; Dementiev and Gladkov 1954a; Pätzold 1983; BMNH, ZMA.)

Moults. ADULT POST-BREEDING. Complete; primaries descendant. Starts with p1 and scattered feathers of body between mid-July and mid-August, occasionally early July or late August; moult completed after *c.* 50-58 days (Davies 1981b; Ginn and Melville 1983), mid- or late September. At primary moult score *c.* 35, body and tail largely new, p6-p8 growing, p9-p10 shed. POST-JUVENILE. Complete; primaries descendant. Starts 2-3 weeks (Pätzold 1983) or 4 weeks (Heinroth and Heinroth 1924-6) after fledging, hence timing strongly variable. P1 first, shed between last days of May and early September. Moult completed with p10, approximately early August to mid-October at age of *c.* 3 months (Pätzold 1983); some small feathers occasionally still growing November. Up to score 15-20, plumage largely juvenile, except for some scapulars, median coverts, and tertials; body moult rapid at score 20-35, not separable from adult after score *c.* 35 (when juvenile p10 shed), except when a few other juvenile feathers still present. (C S Roselaar, RMNH, ZMA.)

Measurements. ADULT, FIRST ADULT. Nominate *arvensis*. Netherlands: wing (1) from breeding season, April-August, (2) September-February; other measurements from all year round, including also Scandinavian breeders; skins (BMNH, RMNH, ZMA). Bill (S) to skull, bill (N) to distal corner of nostril; exposed culmen on average 3.2 mm less than bill (S).

	♂		♀	
WING (1)	113.5 (2.38; 55)	108-119	104.8 (2.55; 52)	100-110
(2)	115.5 (2.73; 226)	109-123	105.1 (2.64; 193)	99-112
TAIL	66.2 (2.35; 37)	62-70	61.5 (1.87; 19)	59-64
BILL (S)	15.6 (0.80; 38)	14.1-17.2	14.8 (0.95; 20)	13.6-16.1
BILL (N)	9.9 (0.63; 29)	9.1-11.0	9.2 (0.61; 18)	8.5-10.0
TARSUS	24.9 (0.95; 37)	23.5-26.7	24.0 (0.84; 21)	22.6-25.5

Sex differences significant.

Freshly dead lighthouse victims, Netherlands, October-November: 116.0 (3.15; 104) 108-123, adult ♀ 106.5 (2.86; 58) 102-112, 1st adult ♀ 105.3 (3.04; 134) 98-114 (♀♀ aged by inspection of gonads; difference significant) (ZMA). Estimated average of live birds, Gibraltar Point (Lincolnshire, England): wing, ♂ 115.3 (2.3; *c.* 130), ♀ 105.4 (2.7; *c.* 130) (Davies 1981b).

A. a. scotica. Scotland, skins: wing, ♂ 112 (2.42; 46) 107-116, ♀ 102.3 (12) 98-106 (Niethammer 1971). Estimated average of live birds, North Ronaldsay (Orkney): wing, ♂ 114.3 (1.5; *c.* 100), ♀ 103.8 (2.8; *c.* 200) (Davies 1981b).

Wing of breeding birds (unless stated otherwise). Sources: (V) Vaurie (1951a), (K) Kumerloeve (1968, 1969a), (M) skins in BMNH, RMNH, ZMA. Nominate *arvensis*: (1) Poland, Fenno-Scandia, and north-west USSR; (2) England. *A. a.*

scotica: (3) Faeroes and Scotland. *A. a. cantarella*: (4) southern France and north-east Spain; (5) Italy; (6) Hungary, Yugoslavia, and Rumania. *A. a. armenicus*: (7) eastern Turkey, Armeniya (USSR), and Elburz mountains (northern Iran). *A. a. dulcivox*: (8) Ural mountains; (9) eastern Kazakhstan and south-west Siberia; (10) Tien Shan mountains. *A. a. sierrae*: (11) central and southern Portugal and southern Spain. *A. a. guillelmi*: (12) northern Portugal and north-west Spain. *A. a. harterti*: (13) Morocco; (14) north-west Africa. Race unknown, migrants and winter birds: (15) Iraq; (16) north-west Iran and Zagros mountains south-east to Kerman (Iran).

	♂		♀		
(1)	114.5 (1.80; 17)	110-117	105.8 (— ; 3)	104-108	(M)
(2)	112.9 (2.13; 5)	111-116	105.7 (2.93; 8)	102-108	(M)
(3)	113.6 (3.01; 11)	110-119	104.6 (2.06; 4)	102-106	(M)
(4)	112.0 (2.35; 18)	109-117	102.7 (2.54; 8)	99-106	(M)
(5)	114.3 (— ; 10)	112-119	104.5 (— ; 2)	104-105	(V)
(6)	113.2 (2.64; 19)	108-117	103.8 (1.94; 6)	101-107	(M)
(7)	120.4 (2.82; 16)	116-126	109.5 (1.07; 8)	108-111	(K,V)
(8)	117.0 (— ; 7)	106-119	104.7 (— ; 3)	102-106	(V)
(9)	116.1 (— ; 16)	112-121	109.2 (4.86; 6)	103-116	(V)
(10)	116.7 (— ; 17)	114-120	107.2 (— ; 9)	105-109	(V)
(11)	112.0 (1.97; 23)	108-115	104.6 (1.29; 5)	103-107	(M)
(12)	109.5 (2.32; 12)	106-114	102.0 (3.67; 4)	97-105	(M)
(13)	112.4 (1.02; 7)	111-114	102.0 (— ; 2)	101-103	(M)
(14)	112.7 (— ; 11)	106-	103.0 (2.94; 4)	100-106	(V)
(15)	119.7 (— ; 5)	114-124	107.2 (— ; 3)	106-108	(V)
(16)	117.2 (— ; 31)	113-123	107.3 (— ; 27)	102-117	(V)

Bill (S), mainly of breeding birds, origins as above.

	♂		♀		
(4),(6)	15.6 (0.57; 16)	14.8-16.6	14.6 (— ; 3)	14.2-15.3	(M)
(5)	15.7 (— ; 10)	15-16	15.2 (— ; 2)	14.5-16	(V)
(8)	16.2 (— ; 7)	15.5-17	15.6 (— ; 3)	14.5-16.2	(V)
(9)	16.9 (— ; 16)	15.2-18.2	16.4 (0.85; 6)	15-17.2	(V)
(10)	16.9 (— ; 19)	15-18	16.1 (— ; 9)	15.5-16.5	(V)
(14)	16.9 (— ; 11)	16-17.5	16.0 (0.33; 4)	15.8-16.5	(V)
(15)	17.0 (— ; 5)	16.5-17.5	15.3 (— ; 3)	15-16	(V)
(16)	15.9 (— ; 31)	14.5-17.5	15.2 (— ; 27)	14.2-16.5	(V)

A. a. cantarella, northern Caucasus: wing, ♂ 116.1 (4) 115-117, ♀ 107.5 (3) 106-109; bill (S), ♂ 15.8 (4) 15.1-16.3, ♀ 14.5 (3) 13.5-15.6. *A. a. dulcivox*, lower Volga: wing, ♂ 116.5 (4) 115-117; bill (S), ♂ 15.7 (4) 15.2-16.2. *A. a. scotica*, Scotland: bill (S), ♂ 16.1 (7) 15.5-17.1. (RMNH, ZMA.)

JUVENILE. Wing on average 9.0 shorter than adult and 1st adult, tail 6.8 shorter; tarsus full-grown at about fledging, bill when post-juvenile moult almost completed.

Weights. ADULT, FIRST ADULT. Nominate *arvensis*. Southern England: (1) January-February, killed by frost (Ash 1964). Netherlands: (2) December-January, killed by frost; (3) May-August, killed by traffic (ZMA). Killed at lighthouses: (4) September, (5) October, (6) November, (7) December (in part killed during mass-migrations at onset of frost, some apparently near exhaustion), (8) January, (9) February, (10) March (ZMA). (11) Killed at lightship in south-west Baltic, March; (12) East Germany, early April (Bub and Herroelen 1981).

	♂		♀	
(1)	28.7 (— ; 3)	28-29	24.4 (2.00; 7)	17-30
(2)	30.3 (2.52; 7)	27-34	29.0 (3.93; 7)	23-32
(3)	38.2 (3.68; 7)	34-43	34.6 (— ; 3)	32-36
(4)	38.0 (2.95; 6)	34-41	35.3 (2.36; 5)	33-39
(5)	42.7 (4.70; 102)	32-51	37.2 (3.63; 286)	29-47
(6)	44.1 (4.70; 90)	34-55	37.6 (4.20; 68)	28-47
(7)	37.5 (7.64; 24)	24-50	31.7 (6.57; 21)	22-42
(8)	40.8 (3.90; 10)	33-45		
(9)	44.9 (3.56; 37)	39-54	35.8 (6.02; 16)	26-43
(10)	42.7 (— ; 3)	41-45	34.3 (5.94; 28)	24-43

| (11) | 38·5 (6·37; 39) | 33-45 | 34·5 (3·39; 52) | 29-42 |
| (12) | 37·9 (— ; 23) | 30-46 | 34·3 (— ; 35) | 27-40 |

Sex differences (5)-(11) significant.

For unsexed birds from Lincolnshire, see Davies (1981*b*); monthly variations in unsexed samples often caused in part by variable proportion of sexes, e.g. note high number of ♀♀ in Netherlands sample in October and March and high number of ♂♂ in January and February. No statistical difference between 1st adults and full adults (Davies 1981*b*; ZMA).

A. a. scotica. New Zealand, January: ♂ 40·8 (9) 39-46, ♀ 38·9 (5) 34-51 (Niethammer 1971).

A. a. cantarella. Northern Greece, November-April: average ♂ 43·6 (14), ♀ 40·7 (6) (Makatsch 1950). Syria, January: ♂ 44 (Kumerloeve 1969*b*).

A. a. sierrae. Southern Spain, May: ♂♂ 33, 38 (Niethammer 1957).

A. a. dulcivox. Mongolia, late April to early August: ♂ 39·6 (4·39; 5) 35-46, ♀ 41 (Piechocki and Bolod 1972). Kazakhstan (USSR): February-March, ♂ 39-48 (7), ♀ 38; April-May, ♂ 33-48 (21), ♀ 30-39 (6); June-August, ♂ 30-42 (52), ♀ 28-39 (13); September-October, ♂ 34-46 (9), ♀♀ 32·5, 33·8 (Dolgushin *et al.* 1970).

A. a. armenicus. Eastern Turkey, Armeniya (USSR), and Elburz mountains (Iran), May-July: ♂ *c*. 44 (14) 38-49, ♀ *c*. 42 (9) 38-46 (Paludan 1940; Nicht 1961; Kumerloeve 1968, 1969*a*).

A. a. armenicus or *dulcivox*. Turkey and northern Iran, winter: ♂ 40·7 (3·89; 10) 35-46, ♀ 36·2 (4·09; 5) 32-43 (Schüz 1959; Kumerloeve 1968, 1970*a*).

NESTLING, JUVENILE. For development, see Pätzold (1983).

Structure. Wing rather long and pointed, broad at base. 10 primaries. In adult, p8 and p9 longest or either one 0-2(-4) shorter than other; p7 0·5-2 shorter than wing-tip, p6 6-11 (♂) or 4-8 (♀), p5 18-23 (♂) or 12-19 (♀), p1 35-42 (♂) or 30-36 (♀). P10 strongly reduced, narrow and pointed; in ♂, 75-87 shorter than wing-tip, 14·9 (8) 14-17 shorter than longest upper primary covert; in ♀, 64-75 shorter than wing-tip, 12·0 (8) 10-14 shorter than covert. Juvenile wing slightly more rounded at tip than adult, p10 relatively longer, p3-p6 slightly shorter; juvenile p10 6 shorter to 1 longer than longest primary covert, on average 2·5 shorter (*n*=10). Outer web of p6-p8 and inner of p7-p9 emarginated. Longest tertials reach to about tip of p5 in closed wing of adult, to about p4 in juvenile. Tail rather long, shallowly forked; 12 feathers, t1 5-9 shorter than t4-t5; t6 often slightly shorter than t4-t5; juvenile tail-feathers narrow and pointed, t1 shorter than in adult. Bill short, rather slender, straight; culmen gradually sloping towards slightly decurved tip, cutting edges straight or faintly decurved. Nostril covered by tuft of short feathers ending in fine bristles; similar short and fine bristles along base of upper mandible and on chin. Feathers of crown slightly elongated, broad and with rounded tips, rather crest-like when raised, but not ending in narrow spike as in Crested Lark *Galerida cristata*. Tarsus and toes rather short and slender. Middle toe with claw 20·2 (14) 18-23 mm; outer toe with claw *c*. 69% of middle with claw; inner *c*. 77%, hind *c*. 124%. Front claws slender and sharp, decurved. Hind claw very long and sharp, straight or slightly decurved; in *scotica* and nominate *arvensis*, ♂ 15·6 (25) 13-19, ♀ 15·0 (15) 12-16; in southern races, ♂ 15·0 (27) 11-18, ♀ 12·6 (7) 11-14.

Geographical variation. Slight, involving size and colour. Size increases slightly towards east, with races of western Siberia, Transcaspia, and Transcaucasia largest; clinally smaller again towards Pacific, but north-east Siberian populations large. In western Europe, cline of decreasing wing length runs towards south. Differences in colour slight and obscured by strong individual variation; most races separable only by comparing series of birds at same stage of plumage wear. Nominate *arvensis* in western part of range usually olive-brown above with broad black streaks, while paler and more greyish birds predominate towards east; variation strong everywhere, however. *A. a. scotica* from Ireland, north-west England, Scotland, and Faeroes similar to dark and brown variants of nominate *arvensis*, but streaks on crown and mantle on average slightly wider and deeper black, ground-colour of upperparts deeper rufous-brown (especially on hindneck), rump and upper tail-coverts rufous, less grey; chest deeper rufous, less buff; apparently no grey variants. *A. a. cantarella* from north-east Spain, southern France, Italy, and Balkan countries east to lower Don and Greater Caucasus closely similar to nominate *arvensis*, but both dark and pale variants paler than corresponding variants of nominate *arvensis*; ground-colour of upperparts pink-cinnamon or sandy-grey rather than olive-brown or olive-grey, dark feather-streaks of upperparts slightly narrower, contrasting more with paler ground-colour; ground-colour of chest cream-buff, less deep buff, dark streaks slightly narrower and more contrasting. Birds from north-west Spain, southern France, and Italy are on average more yellow-buff and less contrasting on upperparts than those of central Yugoslavia eastwards; latter sometimes separated as *lunata* Brehm, 1842 (Horváth *et al.* 1964) or perhaps as *coelipeta* Pallas, 1827 (Matvejev and Vasić 1973), but difference slight. In Spain, *cantarella* breeds south to Aragón, Sierra de Guadarrama, and Sierra de Gredos west to Pamplona area, intergrading with *guillelmi* in Salamanca and probably Cantabrian mountains and probably with *sierrae* south of river Tagus. Birds from northern slopes of Caucasus average paler and have slightly narrower streaks than *cantarella*, tending towards *dulcivox*, but still closest to *cantarella*; birds of Transcaucasia and neighbouring eastern Turkey are similar to *cantarella*, but distinctly larger, warranting separation as *armenicus*. Situation in central Asia Minor, Zagros mountains, and northern Iran not yet fully elucidated: winter birds from these areas are smaller than *armenicus*, similar to *cantarella* in size and colour, but not known whether these are local breeders or migrants from elsewhere. *A. a. dulcivox* occurs from steppes along lower Volga and Ural mountains to northern Kazakhstan, foothills of Altai mountains, and Yenisey river; streaks on upperparts and chest narrower and even more contrasting than in *cantarella*, ground-colour of upperparts pink-buff when fresh, sandy-grey to pale grey when worn; ground-colour of chest cream or off-white. Occurs also in mountains of southern USSR from Bol'shoy Balkhan and Kopet Dag (Turkmeniya) to Pamir and Tien Shan and in neighbouring northern Iran and Afghanistan, but here on average slightly darker and more broadly streaked than typical *dulcivox*, tending towards *cantarella* or *armenicus*; these mountain populations sometimes separated as *schach* Ehmcke, 1904, *almasyi* Keve, 1943, or *dementjevi* Korelov, 1953 (see, e.g., Horváth *et al.* 1964). *A. a. sierrae* from central and southern Portugal (south from Serra da Estrêla) and southern Spain is darker and browner above than *cantarella*, but not as bright as nominate *arvensis*; black streaks on upperparts broader than in *cantarella*, feather-fringes duller olive-grey; ground-colour of chest rather pale cream-buff, paler than *cantarella*. *A. a. guillelmi* from northern Portugal (Aveiro to Braga) is distinctly darker on upperparts than *sierrae*, black feather-centres more extensive, fringes rufous-brown; ground-colour of chest distinctly deeper rufous-buff. Some birds are paler, fringes of upperparts more olive, chest paler buff, tending towards *sierrae*, though not as pale; population of Galicia (north-west Spain) consists only of dark birds. *A. a. harterti* from north-west Africa buff-brown with rather narrow streaks above; paler than *sierrae* and nominate *arvensis*, less olive than *cantarella*; differs from similar-sized *sierrae*, *guillelmi*, and *cantarella* by longer and thinner bill. Situation in north-west Africa not fully elucidated, as some birds from Hauts Plateaux (Algeria) and central Tunisia are large and pale with normal bill, similar to *dulcivox* and perhaps migrants of that race, though collected late April and May. For extralimital races, see Vaurie (1951*a*, 1959).

Forms species-group with Small Skylark *A. gulgula* and (if considered a separate species, as advocated by Stepanyan 1983) with Japanese Skylark *A. japonica*.   CSR

Alauda gulgula Small Skylark

PLATE 10
[between pages 184 and 185]

Du. Kleine Veldleeuwerik   Fr. Alouette gulgule   Ge. Kleine Feldlerche
Ru. Малый полевой жаворонок   Sp. Alondra india   Sw. Mindre sånglärka

Alauda gulgula Franklin, 1831

Polytypic. *A. g. inconspicua* Severtzov, 1873, south-west Siberia, Transcaspia east to Zaysan basin, and from central Iran east to plains of Pakistan and northern India; perhaps this race vagrant to Israel. Extralimital: *lhamarum* Meinertzhagen and Meinertzhagen, 1926, Pamir mountains and western Himalayas from Gilgit and Kashmir to northern Punjab; nominate *gulgula* Franklin, 1831, Sri Lanka, south, central, and south-east India, and Burma; 8 further races from south-west India, central Himalayas, and Tibet east to China, Taiwan, south-east Asia, and Philippines.

Field characters. 15·5–16·5 cm; wing-span 26–30 cm. About 15% smaller than Skylark *A. arvensis*, with structure recalling Woodlark *Lullula arborea* and differing distinctly from *A. arvensis* in longer bill, shorter, broader-based, and rounder wings (lacking obvious extension of primaries when folded), and shorter and rather narrower tail; legs proportionately longer than *A. arvensis*. Plumage closely similar to *A. arvensis* but shows rusty rear cheeks and wing-panel, more narrowly streaked breast-band on buffier underparts, and sandy (not white) trailing edge to wings in flight. Flight slow and flapping. Call distinctive. Sexes similar; litle seasonal variation. Juvenile separable. West Palearctic records cautiously assigned to western and most migratory race *inconspicua* (but see below).

Race unknown (description based on autumn and winter birds from Israel: Shirihai 1986). ADULT. Not dissimilar to pale west Asian and Siberian *A. arvensis dulcivox*, but at close range shows: (1) dark olive rather than black-brown streaks; (2) rusty ear-coverts, striking off-white rear supercilium, and weak marks on lower cheeks and throat (head pattern thus recalling *L. arborea*); (3) less developed crest; (4) distinctive rusty fringes to secondaries and primaries, forming warm-coloured panel when tightly folded and producing sandy or rusty (not white) trailing edge to wing; (5) rusty rump; (6) tail with dull, sandy outer feathers, and brown (not black-brown-black) centre (unlike both *A. arvensis* and *L. arborea*); (7) generally buff or sandy underparts, with (8) neater breast-band usually more narrowly streaked. Note that plumage differences from *A. arvensis dulcivox* in flight-feathers, rump, and tail are not evident in typical skins of adult *inconspicua* (see Plumages) and such characters were previously thought to be shown only by less migratory or sedentary eastern races. However, birds recorded in Saudi Arabia, just outside west Palearctic (Brown and Palfery 1986), were more like typical *inconspicua* (e.g. no rusty fringes to flight-feathers, but still differing from *A. arvensis* in structure and voice). JUVENILE. Differs from adult as in *A. arvensis*. No information on field distinction of juveniles of the 2 species. At all ages, bare parts coloured as *A. arvensis*, but bill rather longer and more pointed and legs proportionately longer.

Little studied, but recent Israeli observations have shown it to be more distinctive than previously thought. Thus post-juvenile bird in late autumn and winter has noticeably smaller, more compact, and more neatly streaked appearance than robust, long-winged, long-tailed, and boldly streaked *A. arvensis dulcivox*, with differences in head pattern, wing colours, and underpart tones obvious at close range and even suggesting *L. arborea*. In flight, beats of short, rounded wings produce slower, more flapping action, while they and somewhat shorter tail contribute to more compact silhouette, again suggesting *L. arborea* as much as *A. arvensis*. Gait and behaviour much as *A. arvensis* but Israeli birds easier to approach.

Song similar to *A. arvensis* but including buzzing notes. 2 common and distinctive flight-calls: (1) harsh, buzzing 'pzeebz', final part with twanging quality (Brown and Palfery 1986); 'baz baz' or 'baz-terrr', reminiscent of

Richard's Pipit *Anthus novaeseelandiae* (Shirihai 1986); (2) soft 'pyup', similar to Ortolan Bunting *Emberiza hortulana* (Brown and Palfery 1986).

Habitat. Breeds in east Palearctic and Oriental regions where it largely replaces Skylark *A. arvensis* in equivalent habitats, showing similar adaptation from moist natural grasslands to field cultivation at fairly wide range of altitudes, and similar tendency to avoid closed, wooded, or broken terrain. In USSR, breeds in steppe zone in low-lying grassy fields and meadows, often in valleys or on floodlands or in damp sedge, but also on dry steppe with tamarisks *Tamarix* and small steppe reeds. Has colonized both cultivated and abandoned farmlands and stubbles, and also rice-fields, and ascends mountains to 2000 m or even to 3600 m in Pamirs. (Dementiev and Gladkov 1954*a*.) In Afghanistan found between 440 and *c*. 2700 m, often sharing cultivated fields and grass-covered areas along rivers with Crested Lark *Galerida cristata* (Paludan 1959). In Baluchistan (Pakistan), strongly localized on irrigated terraced wheat fields at *c*. 2300 m until after breeding season (E M Nicholson). In Kashmir, breeds from *c*. 1500 m to at least 3000 m, in meadows and especially on grasslands dotted with beds of irises or *Trollius*, or with coarse herbage, as well as fallows or rough ploughed land with patches of stones; rice-fields, like forest, are avoided (Bates and Lowther 1952). In subtropical and tropical India, ranges from coastal plain in south up to 4300 m in Ladakh, seasonal altitudinal movements being regular from higher breeding areas, involving shift from upland pastures and grasslands or crops of lucerne and barley to damp grass by water, stretches of rank grass bordering salt-pans and tidal mudflats, crops of cereals and vetches, dry rice stubbles, or even to urban playing-fields (Ali and Ripley 1972).

Even song-perches are generally confined to clods or stones rather than taller plants or artefacts, but much song delivered in middle airspace, up to 150-200 m (Dementiev and Gladkov 1954*a*). Despite somewhat greater altitudinal range than *A. arvensis* and similar spectrum of adaptability, has not achieved a comparably dominant status within its distribution.

Distribution. Breeds from Iran, Transcaspia, and Soviet Turkestan to Afghanistan, Baluchistan, Himalayas, Tibet, and China south to India, Sri Lanka, Indo-China, and Philippines. Northern populations migratory, others more or less resident.

Accidental. Israel: 16 (10 trapped) at Elat, 28 September 1984-5 April 1985, up to 8 wintering; 10 at Elat from 1 October 1985 (Shirihai 1986). USSR: occurs occasionally in Caucasus according to Dementiev (1934) and Vaurie (1959), but not referred to by Dementiev and Gladkov (1954*a*). Kuwait: one, November 1986 (CP). Egypt: bird said to be of race *inconspicua* collected Wadi Natrun, October 1914 (Meinertzhagen 1930); specimen

examined by S M Goodman and found to be Skylark *A. arvensis*.

Movements. Resident or migratory. Northern populations of north-western race, *inconspicua* (which breeds just outside south-east boundary of west Palearctic), and Pamirs population of *lhamarum* are present in USSR only during breeding season. Depart mainly during early October though some remain until mid-December; return February to mid-March (Dementiev and Gladkov 1954*a*; Ivanov 1969). In USSR, *inconspicua* replaced in winter by Skylark *A. arvensis dulcivox* (Abdusalyamov 1973). Wintering range of these north-western populations and migratory status of birds breeding Iran and Afghanistan not clear: e.g. *inconspicua* said by Ali and Ripley (1972) to be resident in north-west India and Pakistan. *A. g. inopinata* of central China and south-east Tibet winters in Nepal, northern Burma, and plains of northern India (Vaurie 1959). Other populations essentially resident or subject to altitudinal and other local migratory and nomadic movements (Vaurie 1959; Ali and Ripley 1972).

As well as recent records from Israel and Kuwait (see Distribution), 2 wintered on Bahrain November-January 1978-9 and 5 at Dhahran (eastern Saudi Arabia) October-March 1984-5, both just outside west Palearctic (Brown and Palfery 1986). Observation of ringed bird in Israel (presumably one of birds ringed at same site in previous season) suggests some year-to-year fidelity to wintering areas (Shirihai 1986).   JHE, DJB

Voice. See Field Characters.

Plumages (*A. g. inconspicua*). ADULT. Closely similar to Skylark *A. arvensis dulcivox*. Ground-colour of upperparts pink-cinnamon to sandy-grey with rather narrow dark shaft-streaks, as in *A. a. dulcivox*, but shaft-streaks of mantle and scapulars dark olive-brown, not blackish-brown; sides of head and neck, chest, and flanks more extensively tinged buff than *A. a. dulcivox*; pale buff supercilium long and rather broad, often rather contrasting; narrow off-white eye-ring and pale band behind ear-coverts less contrasting; dark streaks on lower cheeks faint or absent; chest and sides of breast with narrow but distinct shaft-streaks, forming more regularly streaked and better defined chest-band; throat uniform pale buff. Outer webs of t5-t6 and tips of secondaries and inner primaries tinged sandy-pink, less white than in *A. arvensis*; thus, tail pattern less contrasting and pale trailing edge to wing less conspicuous. See also Structure. JUVENILE. Differs from adult in same characters as juvenile *A. arvensis*.

Bare parts. ADULT. Iris hazel-brown or brown. Upper mandible horn-brown, brownish-grey, or dark grey, lower pinkish-flesh or greyish-yellow with darker horn-grey tip; mouth pink or yellowish-flesh. Leg, foot, and claws flesh-pink, yellowish-flesh, or brownish-flesh. (Ali and Ripley 1972; BMNH, H Shirihai.) JUVENILE. Similar to juvenile *A. arvensis* (BMNH).

Moults. Timing and sequence as in *A. arvensis dulcivox* (Vaurie 1951*a*). In USSR, adult in post-breeding moult late July to late August; moulting juveniles recorded 29 June (Dementiev and Gladkov 1954*a*). In Afghanistan, moult completed (or nearly so) by late September and early October; juveniles starting mid- to late July (Paludan 1959). 3 birds from Afghanistan and western Pakistan, second half of August, had moult almost complete (BMNH).

Measurements. *A. g. inconspicua.* ADULT. Transcaspia (USSR), Afghanistan, and western Pakistan, March–October; skins (BMNH, ZMA). Bill (S) to skull, bill (N) to distal corner of nostril.

WING	♂ 98·3 (2·56; 16)	94–102	♀ 91·5 (3·39; 12)	85–96
TAIL	54·1 (2·54; 10)	50–58	51·7 (2·02; 12)	48–56
BILL (S)	16·4 (0·52; 16)	15·6–17·3	16·0 (0·66; 12)	15·3–16·9
BILL (N)	10·7 (0·40; 10)	10·1–11·3	10·4 (0·47; 12)	9·7–11·1
TARSUS	24·9 (0·57; 10)	24·2–26·0	24·5 (0·84; 12)	23·2–25·8

Sex differences significant for wing.

JUVENILE. Afghanistan: wing, ♂ 91·8 (2·68; 5) 89–96 (Paludan 1959).

Weights. *A. g. inconspicua.* Afghanistan: late May to July, ♂ 26·6 (13) 25–30, ♀ 25·5 (1·29; 4) 24–27; late September and early October, ♂ 26·0 (3) 25–27, ♀ 27·3 (3) 26–29 (Paludan 1959). Kazakhstan (USSR): ♂ 24–28 (9), ♀ 29·2 (Dolgushin *et al.* 1970).

Structure. Closely similar to *A. arvensis*, but wing shorter and with more rounded tip, tail shorter, and bill longer and more attenuated; measurements of leg and foot as in *A. arvensis*, hence relatively heavier. 10 primaries: (p6–)p7–p9 longest or 1–2 of them 0–1 mm shorter than other; p6 0–5 shorter than wing-tip (on average 1·3; in *A. arvensis*, 5–9·5, on average 7: H Shirihai), p5 6–10 shorter, p4 15–18, p1 23–27; p10 reduced in adult, 58–66 shorter than wing-tip, 9·5–12 shorter than longest upper primary covert (on average 10·7 shorter; in *A.*

arvensis, 13–18 shorter: H Shirihai). Differs from *A. arvensis* in wing-tip index (distance between wing-tip and tip of p1, divided by wing length): 0·22–0·26 in *A. gulgula*, 0·32–0·36 in *A. arvensis*. In fresh adult plumage, tip of longest tertial falls 1–8 mm (average 3·2) short of wing-tip when wing closed (H Shirihai); in worn plumage, spring, longest tertial often strongly abraded and up to 25 mm shorter than wing-tip. Bill length (to skull) divided by wing length 0·175 (0·154–0·201) in *A. gulgula*, 0·142 (0·134–0·156) in *A. arvensis* (Vaurie 1951*a*).

Geographical variation. Marked; most races distinctly darker and often smaller than *inconspicua*; see Vaurie (1951*a*). Race described in plumages is *inconspicua*, breeding nearest to west Palearctic (e.g. in Turkmeniya and central Iran) and thus the most likely race to straggle to our area—and birds apparently of this race have occurred nearby in Saudi Arabia (Brown and Palfery 1986). However, vagrants in Israel (Shirihai 1986; see Field Characters) appear more heavily marked black on upperparts than typical *inconspicua* and rusty tinge on ear-coverts and on outer fringes of most flight-feathers more pronounced than in *inconspicua* (and most *A. arvensis*); rather similar to *lhamarum* of Pamirs and western Himalayas in these respects, but perhaps an undescribed race (C S Roselaar).

Sometimes included in *A. arvensis*, e.g. by Meinertzhagen (1951*a*), but the 2 forms differ markedly in structure and overlap widely in Transcaspia and central Asia apparently without interbreeding. However, difference in structure bridged by Japanese Skylark *A. japonica* from Japan, which is variously considered a race of *A. arvensis*, a race of *A. gulgula*, or (as here) a separate species; last course also taken by Stepanyan (1983), who considered the 3 species to form species-group.

CSR

Alauda razae Razo Lark

PLATES 10 and 13
[between pages 184 and 185]

DU. Razo Leeuwerik   FR. Alouette de Razo   GE. Razalerche
RU. Жаворонок острова Разо   SP. Terrera de Cabo Verde   SW. Razolärka

Spizocorys razae Alexander, 1898

Monotypic

Field characters. 12–13 cm; wing-span 22–26 cm. Less than ¾ size of Skylark *Alauda arvensis*, with 30–40% shorter and rounder wings (primaries entirely hidden beneath tertials of closed wing), and shorter tail, but with rather heavier and longer bill and rather long sturdy legs; thus appears curiously 'front-heavy'. Rather small, compact, short-crested *Alauda* lark restricted to Razo island in Cape Verdes. Plumage mealy-grey, streaked dull black above; pale cream below, with dark-streaked buff chest. Short tail noticeably dark, with prominent white outer edge, but short rounded wings of adult lack obvious pale trailing edges of *A. arvensis*. Sexes similar; no seasonal variation. Juvenile separable.

ADULT. Upperparts dull grey, usually with warmer brown fringes to mantle and scapulars, also dull black streaks on crown, back, and scapulars, becoming shorter and finer on hindcrown and crest, rump, and above tail.

Face patterned as *A. arvensis*, with pale cream or dull white lores and indistinct supercilium and grey-streaked, buff lower cheeks. Wings brown-grey, with blackish centres to tertials and median coverts and pale buff or off-white edges to larger coverts; inner primaries and secondaries narrowly fringed white but no obvious trailing edge (unlike *A. arvensis*); underwing cream-buff. Tail dull black, contrasting with pale rump and showing prominent white outer feathers. Chin and throat white. Chest and flanks cream or buff with sharp black streaks, streaks widest on sides of breast and fore-flanks; rest of underparts pale cream-buff. When plumage worn, upperpart streaks appear heavier and belly becomes almost white. Bill dark horn, with almost white base to lower mandible. Legs brown-flesh. JUVENILE. Resembles adult more than in *A. arvensis* but with rather buffer ground-colour and more black-spotted appearance to upperparts, better-defined pale fringes to wing-coverts and flight-feathers, and much less distinctly marked chest.

Little described lark, but its isolation on single island of Atlantic archipelago with only 3 other larks (on other islands) allows easy identification. Would probably suggest a cross between *A. arvensis* and Lesser Short-toed Lark *Calandrella rufescens*. Flight and gait not described but structural differences from *A. arvensis* suggest more flapping wing action but similar walk.

Voice similar to *A. arvensis* but song much less varied.

Habitat. Entire species concentrated in less than 100 ha on small windswept waterless island of Razo (Cape Verde Islands). Most live on a flat area of decomposing lava and tufa carrying meagre growth of herbs and low scrub, but a few spread to wide ravines with fine sandy bottoms, or feed on stretch of black rocks close to the ocean, involving an altitude range of *c.* 150 m. For plant species found on the island, see Bannerman and Bannerman (1968).

Distribution. CAPE VERDE ISLANDS. Restricted to the small island of Razo (Bannerman and Bannerman 1968); reason for limited distribution not clear, perhaps climatic (Naurois 1969b); may once have been more widespread (Bourne 1955).

Population. CAPE VERDE ISLANDS. Razo: declined; under 50 pairs 1965, under 40 pairs 1968 (Naurois 1969b), *c.* 20 pairs 1979-82 (Collar and Stuart 1985), over 150 birds (*World Birdwatch* 1985, 7 (2), 4).

Movements. Sedentary. No records from anywhere other than Razo. RH

Food. Based on material gathered by N J Collar. Food thought by Alexander (1898a) to be grass seeds, and 2 stomachs collected by Naurois (1969b) contained ants (Formicidae), beetles (Coleoptera), seeds, small germinating plants, and other vegetable matter, as well as grit. However, Meinertzhagen (1951a) and Hall (1963) considered large bill adapted to digging, e.g. for insect larvae; Meinertzhagen's (1951a) claimed observation of this is open to doubt (N J Collar), but has now been recorded digging, using bill to prise pebbles free of soil, presumably to expose food items (R P Martins and M A S Beaman). Bill length difference of 20·7% exists between sexes, and feeding ecologies thus likely to differ (Burton 1971). DJB

Social pattern and behaviour. Very little information. Account based on material compiled by N J Collar.

1. Occurs mainly in groups of (1-)5-10 birds. Large gathering of adults and immatures recorded in early January (Naurois 1969b; R de Naurois); perhaps most of the population. Recorded also in flocks (including for feeding) late April when birds in moult and breeding almost over (Alexander 1898a). In flock of 15-20, mid-June 1981, several, perhaps most, immatures (Collar

and Stuart 1985). BONDS. Virtually no information, but nothing to suggest mating system other than monogamous. Both sexes said to incubate (Alexander 1898b). BREEDING DISPERSION. Only information is on density. Formerly 40-50 pairs mostly in *c.* 1-2 km² (central plateau and some adjacent areas) of the 7 km² island; now probably *c.* 20 pairs in *c.* 1 km² (Naurois 1969b; Schleich and Wuttke 1983; N J Collar, R de Naurois). ROOSTING. No details. In early October, birds called most persistently from ground shortly before dusk (Alexander 1898b; see 3 in Voice).

2. Easily approached; shows no fear of man (Alexander 1898a; Bourne 1955; Naurois 1969b). Recent observations indicate perhaps now slightly less approachable (Collar and Stuart 1985). When ♀ in flock pursued by man, ♂, with crest erect, would immediately appear and run close by ♀ (Alexander 1898a). FLOCK BEHAVIOUR. In January, juveniles associated more closely than adults and called (no details) constantly (R de Naurois). SONG-DISPLAY. Song of ♂ (see 1 in Voice) given in flight and when 'stationary' (not known whether this refers to song on ground or perch). Song-flight may take place after bout of courtship on ground (see below). ♂ sings while ascending with gentle wing-beats; ascent vertical, unlike spirals of Skylark *A. arvensis* (Alexander 1898b); continues to sing while hovering at *c.* 10 m (Harreveld 1985). Will sing 'in chorus' in early morning. Song-flights recorded in early October (Alexander 1898b) and in early March (Harreveld 1985), but see Breeding for erratic season. HETEROSEXUAL BEHAVIOUR. ♂ and ♀ generally keep fairly close together while feeding, etc. (R de Naurois). In courtship (clearly similar to other Alaudidae), ♂ has wings drooped to brush ground, approaches and circles ♀, then performs Song-flight (Alexander 1898b). No further information. MGW

Voice. CALLS OF ADULTS. (1) Song of ♂. Said by Alexander (1898b) to consist of constant repetition of call 2; rather monotonous but somewhat less so when several birds singing simultaneously. Said by R de Naurois to be not unpleasant and similar to Skylark *A. arvensis*. In recording by C J Hazevoet, song from ground or perch comprised jingles like *A. arvensis c.* 1 s long separated by pauses of *c.* 2 s; in short Song-flight, became more continuous but still rather repetitive and much less varied than *A. arvensis* (P J Sellar). (2) Call given at take-off similar to Flight-call of Skylark *A. arvensis* (Alexander 1898a). (3) Ventriloquial call given constantly by birds on ground (Alexander 1898b).

CALLS OF YOUNG. No information. MGW

Breeding. SEASON. Probably erratic, governed by rains; well-grown young seen in nest in April; birds courting and eggs found in October of same year (Alexander 1898a, b). SITE. On ground among short grass, under boulder or creeping plants (Alexander 1898a, b; Naurois 1969b). Nest: frail structure of grass, placed in small depression (Alexander 1898a, b). Building: no information on role of sexes. EGGS. Not fully described but reported as similar to Woodlark *Lullula arborea* in colour and dimensions (Alexander 1898b). Clutch: one of 3 reported (Alexander 1898b). INCUBATION. 13 days; by both sexes (Alexander 1898b); confirmation required. No further information.

Plumages. ADULT. Entire upperparts dull grey with dull black feather-centres on forehead, crown, mantle, scapulars, and back, and narrower black shaft-streaks on hindcrown, rump, and upper tail-coverts. Sides of head dark grey-brown, finely speckled dusky grey, especially on rear of ear-coverts. Patch on lores and rather narrow and faint supercilium cream-white or white. Lower cheeks and sides of neck pale buff with some fine dusky grey specks. Chin and upper throat white; chest, breast, and flanks cream to buff with narrow and sharp black shaft-streaks (*c.* 0·5 mm wide on mid-chest, 1–1·5 mm on sides of breast and flanks). Remainder of underparts pale cream-buff. Tail dull black, all feathers with narrow and indistinct grey fringes; outer web of t5 white, t6 largely white except for broad dusky wedge on basal and middle portion of inner web. Flight-feathers blackish-grey with white edges along outer webs (except for emarginated parts of outer primaries); fringes along tips of secondaries and inner primaries white. Tertials dull black with rather narrow and indistinct pale grey-brown or buff fringes. Upper wing-coverts dark brown-grey with pale buff or off-white fringes along tips. Under wing-coverts and axillaries pale cream-buff. When freshly moulted, feathers of crown and mantle with narrow white fringes along tip, longer scapulars and tertials with narrow and regular white fringes; chest and breast purer pink-buff. *In worn plumage*, upperparts duller, appearing more heavily streaked dull black or brownish-black; underparts paler, belly almost pure white. NESTLING. No information. JUVENILE. Like adult, but feathers of forehead, crown, hindneck, mantle, and scapulars short, buff merging into black towards rounded tips, tips narrowly fringed white; back to upper tail-coverts and upper wing-coverts narrowly fringed white, usually bordered by some dusky black suffusion subterminally; chest and sides of breast only faintly marked with some dull black spots. Tail as adult, but t1 pointed, dark grey with slight buff suffusion in middle, merging into dull black at sides, edges broadly pale buff or white; flight-feathers as in adult, but white fringes slightly broader and more sharply defined.

Bare parts. ADULT. Iris dark hazel. Bill blue-grey with dusky tip or dark horn-colour with whitish base to lower mandible. Leg dull pink-flesh, grey-flesh, or brown-flesh, foot brown-flesh or blue-grey; claws blackish-horn. (Alexander 1898a; C J Hazevoet.) NESTLING, JUVENILE. No information.

Moults. ADULT POST-BREEDING. Complete; primaries descendant. According to limited number of specimens examined, probably late April to September. Plumage in October fresh or slightly worn; 4 birds, late April and May, had variable number of feathers of forehead, crown, hindneck, and mantle replaced, one also some scapulars and part of breast-sides, remainder of plumage heavily worn. Flight-feather moult just started in ♀ from late April, but, though heavily worn, not in 3 ♂♂ from late April and May. (BMNH.) Heavily worn birds seen mid-March (R P Martins and M A S Beaman). POST-JUVENILE. Complete, primaries descendant. Timing not exactly known, but undoubtedly starts shortly after fledging (late October to May: Bannerman and Bannerman 1968). One still fully juvenile in May, another had body in fresh 1st adult by May, but flight-feathers, tertials, and tail still juvenile (BMNH).

Measurements. ADULT. Razo (Cape Verde Islands), October, late April, and May; skins (BMNH). Bill (S) to skull, bill (N) to distal corner of nostril; exposed culmen on average 3·0 less than bill (S).

WING	♂ 85·4 (1·34; 13)	83–89	♀ 78·6 (1·24; 9)	76–80
TAIL	48·1 (2·65; 10)	45–53	43·8 (0·84; 10)	42–45
BILL (S)	18·0 (0·55; 10)	17·1–18·7	14·8 (0·54; 10)	13·8–15·6
BILL (N)	12·5 (0·42; 10)	11·9–13·1	9·9 (0·28; 10)	9·5–10·3
TARSUS	21·9 (0·43; 9)	21·3–22·4	20·2 (0·51; 10)	19·7–21·2

Sex differences significant. For marked size dimorphism, see also Burton (1971).

JUVENILE. Wing and tail within range of adults in 2 examined.

Weights. No information.

Structure. Closely similar to other *Alauda*, except for more rounded wing and longer bill (Hall 1963), and neither separation of monotypic genus *Razocorys* nor inclusion in *Calandrella* warranted. 10 primaries: p7 and p8 longest, p9 1–2 shorter, p10 51–56, p6 0·5–2, p5 4–7, p4 8–14, p1 18–25; p10 tiny and pointed in adult, hidden by greater primary coverts; slightly longer and with more rounded tip in juvenile. Outer web of p6–p8 and inner of p7–p9 emarginated. Tips of longest tertials reach to tip of p4. Tail rather short, 12 feathers; t3–t4 longest, t1 3–5 shorter, t6 2–4. Bill as in Skylark *A. arvensis*, but longer, *c.* $\frac{2}{3}$ head length; culmen and cutting edges slightly and gradually decurved towards sharply pointed tip; base deeper than in *A. arvensis*, 5·5–6 mm in ♂, 5–5·5 in ♀; bill of ♂ distinctly heavier than *A. arvensis*, of ♀ less so. Feathers of crown slightly elongated, forming short crest as in *A. arvensis*; longest crown feather *c.* 10 mm. Leg and foot similar to *A. arvensis*; hind claw only slightly decurved (as in that species), though relatively shorter; middle toe with claw 17·9 (6) 16·5–19·3; outer toe with claw *c.* 71% of middle with claw, inner *c.* 75%, hind *c.* 110%; hind claw 10·0 (6) 8·5–11·6. For other details of structure and relationships, see Hall (1963) and Harrison (1966). CSR

Eremophila alpestris Shore Lark

Du. Strandleeuwerik   Fr. Alouette hausse-col   Ge. Ohrenlerche
Ru. Рогатый жаворонок   Sp. Alondra cornuda   Sw. Berglärka   N. Am. Horned Lark

Alauda alpestris Linnaeus, 1758

Polytypic. *E. a. flava* (Gmelin, 1789), arctic Eurasia, in Asia east to Anadyrland, south to Stanovoy mountains and northern end of Lake Baykal; *brandti* (Dresser, 1874), steppes of lower Volga river and northern Transcaspia, east through plains of Kazakhstan to northern Mongolia and western Manchuria; Altai, Tarbagatay, and eastern Tien Shan mountains; *atlas* (Whitaker, 1898), Morocco; *balcanica* (Reichenow, 1895), Balkan countries and Greece; *penicillata* (Gould, 1838), Asia Minor, Caucasus, Transcaucasia, north-east Iraq, and western and northern Iran east to about Gorgan area; *bicornis* (C L Brehm, 1842), Lebanon; nominate *alpestris* (Linnaeus, 1758), northern Quebec and northern Labrador, south to James Bay, south-east Quebec, and Newfoundland; straggler to Britain. Extralimital: *albigula* (Bonaparte, 1850), mountains of eastern Iran and southern Transcaspia, east through Afghanistan to western Tien Shan, Pamirs, and Gilgit; *c.* 6 further races in central Asia and *c.* 25 in the Americas.

Field characters. 14–17 cm; wing-span 30–35 cm. About 10% smaller and slighter than Skylark *Alauda arvensis*; 15% bigger than Temminck's Horned Lark *E. bilopha*. Medium-sized lark with form like *Alauda* except for shorter bill, proportionately longer tail and (in adult ♂) thin 'horns' above eyes. Adult essentially pink-brown above and white below, with face and upper breast strikingly decorated black and yellow. Juvenile shows similar ground-colours, heavily spotted except on pale yellow face and white belly. Sexes similar; obvious seasonal variation. Juvenile separable. 7 races in west Palearctic; north Eurasian race, *flava*, described here (see also Geographical Variation).

ADULT MALE. Ground-colour of face and throat pale yellow, marked with black on front and centre of crown, lores, and fore-cheeks, conspicuous black gorget across upper breast, and pink-brown rear crown, rear cheek-surround, and nape. Tufts of black feathers on crown sides form narrow erectile 'horns'. Rest of upperparts and flanks pink-brown; marked with lengthy mottlings rather than streaks; paler edges to wing-feathers noticeable only on tertials. Tail rather long; pink-brown in centre, with black panel and white outer edge on either side. Underparts below gorget and flanks white, tinged dusky-pink on sides of chest and faintly spotted dusky on chest. In spring, abrasion enhances contrasts of face which becomes highly decorated with full yellow forehead and supercilium, intense black mask and pink hindcrown and nape. In flight, shows no obvious wing pattern but contrast of long white vent and broad black centre to undertail often catches eye. ADULT FEMALE. Resembles ♂; at close range, distinguished by duller facial colours and more obviously streaked upperparts. JUVENILE. Much more reminiscent of other larks (especially *Alauda*) than adult, showing characteristic, boldly spotted upperparts, with black-brown and buff ground-colours. Underparts also spotted but show yellow wash on face and clusters of dusky marks in place of adult's gorget and flanks. Assumes adult plumage in August. At all ages, bill grey-horn, with black tip; legs black.

North of Mediterranean, adult unmistakable, but in Levant and across North Africa, liable to confusion with *E. bilopha* (see that species, p. 225). Almost complete geographical separation of arctic communities of *E. alpestris* present few chances to confuse juvenile with young *A. arvensis*, but close or sympatric breeding of *E. alpestris* with Woodlark *Lullula arborea* and *A. arvensis* in alpine habitats from south-east Europe east to Turkey and Iraq could lead to more frequent risk of mistake. Short bill and yellowish face of *E. alpestris* are, however, diagnostic. No evidence of westward vagrancy of eastern races, but nominate *alpestris* of Nearctic, with larger bill and more rufous lesser wing-coverts, has crossed North Atlantic to Britain. Flight action light, recalling (with length of tail) large pipit *Anthus* as much as other medium-sized larks. Wing-beats noticeably rapid, with almost complete closures, causing shooting accelerations and first cruciform, then slim silhouette, as bird or flock flickers over ground. Contrast of white body and black undertail obvious overhead. Escape-flight usually fairly short but occasionally high and long. Song-flight uncommon; begins with high fluttering ascent like *A. arvensis* but then gives way to wavy rises and falls; never long sustained. Gait includes both shuffling and high-stepping walk, crouching and more-normal run, and (particularly on shingle) hopping. Feeding bird constantly on the move, with quiet and unobtrusive progress hiding considerable speed. Perches freely on raised ground, rocks, and buildings but not on plants.

Voice varied, with both song and calls lacking volume of most larks and having characteristic sibilant, piping, even rippling quality. Song more warbling than skirling, with frequently repeated, drawn-out note recalling that of Corn Bunting *Miliaria calandra*. Calls include 'tseep', like long note of Meadow Pipit *Anthus pratensis* (but less anxious in tone and more incisive), and rippling 'tsee-sirrp', often uttered by members of escaping or excited flock.

Habitat. Breeds in west Palearctic in high latitudes or

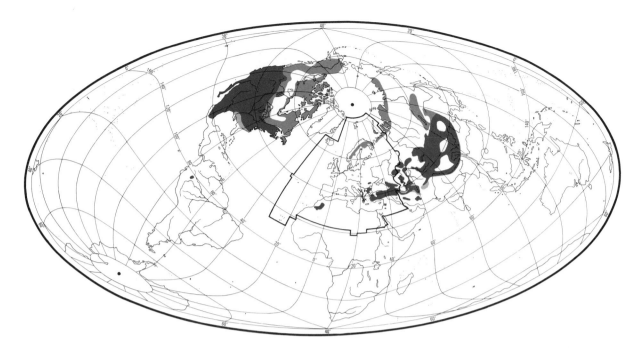

at high altitudes, in regions widely separated by forested, cultivated, wetland, and other unsuitable terrain. Mainly in subarctic or arctic lowland tundra, or in barren steppes and arctic-alpine zones of middle and lower-middle latitudes, much of intermediate area being occupied by Skylark *Alauda arvensis*. No such geographical separation exists in Nearctic, where absence of competing Alaudidae enables other habitats (pre-empted by them in Palearctic) to be colonized, including ploughed fields, airports, short-grass prairies, sparse sage-brush flats, planted grasslands, cultivated fields, and deserts (Godfrey 1966; Johnsgard 1970). Everywhere, however, avoids dense or tall vegetation and steep or broken terrain. In far northern west Palearctic favours dry stony patches in lichen tundra, nesting mostly on level ground amid thinnest herbage, or along dry rock-strewn shores of fjords, even by small fishing villages. Elsewhere, breeds on more or less bare high plateaux and upland ranges, in Morocco at 2000–3500 m (Heim de Balsac and Mayaud 1962) and in Himalayas up to 5300 m, touching snowline, where it occurs on barren stony sometimes windswept steppe country with scanty grass tufts and herbaceous plants such as *Artemisia*, and *Caragana* bushes; occasionally also on alpine meadows (Ali and Ripley 1972). Where found in geographical proximity to Temminck's Horned Lark *E. bilopha*, maintains total habitat separation as a montane rather than semi-desert species (Heim de Balsac and Mayaud 1962). In winter, occurs on coastal dunes, salt-marshes, and beaches, and some arable lands in lowlands which are free of snow, or have plants emerging above snow cover (Voous 1960). Does not usually occur on same terrain as other Alaudidae. Generally less aerial

than other Alaudidae, making less use of airspace over territory in song-flight.

Distribution. Spread in Sweden and Rumania.

BRITAIN. Summered Scotland 1972–3, possibly breeding 1973 (Watson 1973*a*); bred 1977 (Batten *et al.* 1979). NORWAY. Breeding Rauma 1978 (VR). SWEDEN. Believed to have spread into Lapland in 19th century, continuing well into 20th century (LR). ALBANIA. May breed; no proof (EN). RUMANIA. Expanding range (VC).

Accidental. Spitsbergen, Iceland, Faeroes, Ireland, Czechoslovakia, Austria, Malta, Spain.

Population. Decrease in Finland.

Sweden. 10 000 pairs (Ulfstrand and Högstedt 1976). Numbers of migrants at Falsterbo have decreased since mid-1970s (Roos 1984*a*). FINLAND. About 10 000 pairs (Merikallio 1958). Marked continuing decline in passage birds suggests decrease in breeding population (OH). GREECE. No changes noted (GIH).

Oldest ringed bird 7 years 1 month (Clapp *et al.* 1983).

Movements. Northern races migratory, southern ones resident or make altitudinal movements.

NORTHERN RACES. Eurasian arctic and subarctic race *flava* wholly migratory, wintering coasts of southern North Sea and western Baltic (including southern Sweden), sparingly inland in north-central Europe, and in large numbers across southern USSR (e.g. reaching Sea of Azov, northern Caucasus, Kazakhstan steppes, Altai, Amur), Mongolia, and northern China (Dementiev and Gladkov 1954*a*; Zink 1975). Passage details poorly

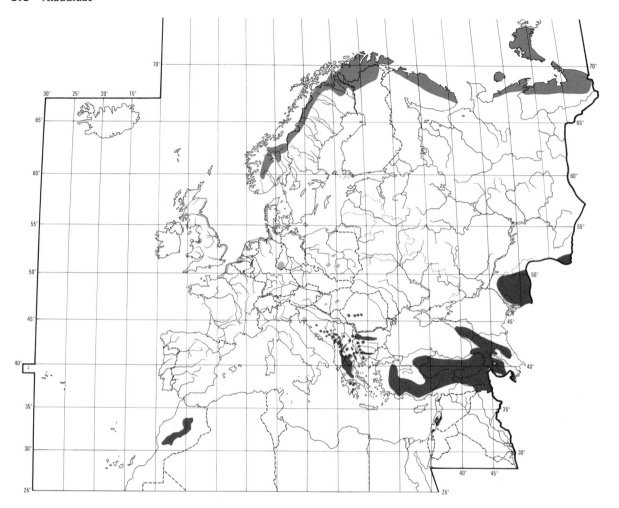

known in virtual absence of ringing recoveries, while European population relatively small and peripheral. Regular wintering in eastern Britain dates from about 1870, and comparable 19th century increase occurred on German side of North Sea; not clear whether this was due to changed breeding range or density in Scandinavia, or altered migration pattern (Witherby *et al.* 1938a; Bannerman 1953). Visible passage records indicate that movement to and from southern North Sea wintering area is across Sweden, Denmark, and West Germany rather than through Norway, hence scarcity in Scotland. Only 3 long-distance recoveries published: Swedish Lapland (November) to Kalmar, southern Sweden (December) (Rendahl 1964); bird hatched Swedish Lapland found in Huelva, Spain (October) (Saez-Royuela 1952; Bernis 1971), though validity of this recovery has been questioned (Zink 1975); West Germany (January) to Komi, USSR (May) (Zink 1975).

Autumn departures from Finnmark (Norway) in late September and first half October, but earlier from USSR breeding grounds: e.g. begins exodus in August on Yenisey and all gone by end of September. Passage through Baltic area and arrivals in North Sea mid-October to mid-November. Onward movement can occur in severe weather, becoming more numerous in southern wintering areas and stragglers penetrating beyond normal range. Return movement begins in March, with breeding areas reoccupied from late April in Finnmark to second half May in Novaya Zemlya and Yamal peninsula, often while ground still snow-covered. ♂♂ predominate amongst first waves of arrivals, ♀♀ amongst later ones (Bannerman 1953).

West-central Asian race *brandti* is partial migrant. Flocks regularly occur south of breeding range in winter (e.g. in Turkmeniya, Mongolia, northern China), but in some breeding areas a few birds are present all year in regions where most are migratory (e.g. around Krasnoyarsk). Migrants in winter quarters mid-November to March. In Soviet Central Asia (e.g. Alma-Ata, Balkhash steppes, Altai) basically resident, though descending from mountains to winter in valleys (Dementiev and Gladkov 1954a).

SOUTHERN RACES. All populations of Balkans, Morocco, Asia Minor, Middle East, and south-central Asia basically resident, as to be expected from marked degree of subspeciation. These breed in mountainous regions, and many birds descend in autumn to foothills, plains, and cultivations, therefore becoming more widespread during winter (Dementiev and Gladkov 1954a; Heim de Balsac and Mayaud 1962; Beaman *et al.* 1975). However, no detailed studies on extent of individual dispersal.

NORTH AMERICAN RACES. These show same trends as in Palearctic. Northern (i.e. Canadian and Alaskan) races are migratory, wintering from southern Canada into USA. Upland and desert races of western North America (including Mexico) are variously resident, or show altitudinal movement which can involve some individuals dispersing considerable distances into lowland regions. See American Ornithologists' Union (1957) and Godfrey (1966). RH

Food. In summer, insects and some seeds; in winter, seeds. Takes food from ground, walking about (Bent 1942) or running in short spurts (Ali and Ripley 1972). Takes seeds from plant stems, pulling vigorously if necessary; 2–3(–4) birds may work one plant, stripping seeds from bottom to top. To reach tall seed heads will (1) flutter up and seize stem with bill, pulling it down to ground, (2) step on base, bending it over, or (3) fly up to take seeds. Uses quick munching bill movements before swallowing seeds. Takes flies (Diptera) on ground with quick side-to-side and forward bill movements. (Hinsche 1958; King and Ladhams 1970.) To reach food under snow, scratches with feet, tossing it aside with bill (Feindt *et al.* 1956). However, in northern Kazakhstan (USSR), digs in snow much less than Black Lark *Melanocorypha yeltoniensis*, and less deeply, taking food mainly from snow surface (Krivitski 1962) and associating with *M. yeltoniensis* to take seeds left exposed after that species has been digging (Dolgushin *et al.* 1970). In North America, especially during snow cover, will pick at horse droppings, feed on compost heaps, and enter farmyards for grain and food scraps; recorded digging up larva of moth *Apantesis* (Bent 1942) and 'doubling up' cutworm-like larva before taking it to young (Goodpasture 1950).

The following recorded in diet in Palearctic. Invertebrates: springtails (Collembola), grasshoppers, etc. (Orthoptera), earwigs (Dermaptera), termites (Isoptera), bugs (Hemiptera), adult and larval Lepidoptera, caddis flies (Trichoptera), flies (Diptera: Tipulidae, Culicidae, Chironomidae, Tabanidae, Asilidae), Hymenoptera (sawflies Tenthredinidae, Ichneumonoidea, ants Formicidae, bees Apoidea), adult and larval beetles (Coleoptera: Carabidae, Staphylinidae, Scarabaeidae, Tenebrionidae, Curculionidae), spiders (Araneae), small Mollusca, small Crustacea, worms (Oligochaeta: Enchytraeidae, earthworms Lumbricidae). Plant material mainly seeds or fruits: of water starwort *Callitriche*, waterweed *Elodea*, docks and knotweeds (Polygonaceae), goosefoots (Chenopodiaceae),

alder *Alnus*, pinks (Caryophyllaceae), Cruciferae, barberry *Berberis*, fumitory *Fumaria*, Rosaceae, chick-pea *Cicer*, *Hypericum*, crowberry *Empetrum*, plantain *Plantago*, Boraginaceae, Compositae, *Gagea*, sedges (Cyperaceae), and grasses (Gramineae) including cereal grain; also buds of willow *Salix*, shoots of sedges and grasses, and lichens. (Seguin-Jard 1922; Witherby *et al.* 1938a; Dementiev and Gladkov 1954a; Pek and Fedyanina 1961; Kekilova 1969; King and Ladhams 1970; Medvedev and Esilevskaya 1973; Rinnhofer 1974; Bachmann 1983; Danilov *et al.* 1984.)

In Lebanon, April–May, seen feeding on flies and other insects in snowfields (Meinertzhagen 1935). Data from Yamal peninsula (north-central USSR) indicate that on breeding grounds proportion of invertebrates in diet increases through spring to peak in mid-summer and then falls again. Thus, in spring 11 stomachs contained seeds and shoots of sedges and grasses, buds of willow, seeds of Compositae, bramble *Rubus arcticus*, and alder, and crowberries *Empetrum*; invertebrates comprised Staphylinidae, Chironomidae larvae, Trichoptera, and Enchytraeidae. In mid-July, 2 stomachs contained only insects: over 40 Chironomidae, several Staphylinidae, 1 Noctuidae larva, 1 Trichoptera, 1 fly, and 1 Tipulidae. In August, seeds again predominant: of 9 stomachs, 5 contained only seeds; insects comprised 3 Tipulidae, 1 Carabidae, 1 fly, and 1 Chironomidae. In late September, 1 bird contained 2 Carabidae and seeds of sedge. (Danilov *et al.* 1984.) Similar pattern in western Tien Shan mountains (south-central USSR): 2 stomachs, April, contained only seeds—of *Gagea* and knotweed *Polygonum*; stomach from May contained seeds of *Polygonum* and grasses and many small beetles and spiders; July juvenile contained small beetles, earwigs, 2 small caterpillars, and a very few seeds; adult and 2 juveniles from late August contained beetles, larval Diptera, Hymenoptera, and spiders, but adult and 1 juvenile had eaten mainly seeds including *Polygonum*, *Gagea*, *Cicer*, goosefoot *Chenopodium*, Potentilla, and grasses (Lobachev and Kapitonov 1968). In Kirgiziya (south-central USSR), presumably in summer, 62 stomachs contained mainly insects: many locusts and grasshoppers (Orthoptera), adult and larval beetles (mainly Curculionidae, also Carabidae and Scarabaeidea), ants, Asilidae, and adult and larval Lepidoptera; also a mollusc, vegetative parts of plants, and seeds of wheat, barley, oats, Compositae, Polygonaceae, and barberry *Berberis* (Pek and Fedyanina 1961). In foothills of Kopet Dag (southern Turkmeniya) during spring and summer, 82% of 34 stomachs contained seeds; 79% contained invertebrates, mainly Lepidoptera (in 50%), Coleoptera (in 24%, mainly Curculionidae), locusts (in 15%), and termites (in 15%—significant part of diet in early June but rarely recorded in spring) (Medvedev and Esilevskaya 1973). Also in southern Turkmeniya, presumably in summer, 4 stomachs contained seeds of Boraginaceae, grass, and insects; 3 others contained 5 ants, 1 fly pupa, and other insects, also seeds comprising 90% *Heliotropium* and a few *Hypericum*, *Arnebia*, *Fumaria*, and wheat (Kekilova 1969). For El'ton steppe (eastern Volgograd region), see Dementiev and Gladkov (1954*a*). In Labrador (Canada), July, 3 birds contained 27% (by volume) adult and larval Lepidoptera, 7% ants, 6% other Hymenoptera (including ichneumons), 4% Diptera, 3% spiders, 3% bugs (including leaf hoppers and aphids), and 2% molluscs (mostly small *Mytilus edulis*); the 48% plant material included fruit of bilberry *Vaccinium uliginosum* (32%) and seeds of sedges (Cottam and Hanson 1938).

Bird wintering inland at Cheddar reservoir (south-west England), October–April, ate seeds of chickweed *Stellaria media*, mouse-ear *Cerastium holosteoides*, pearlwort *Sagina apetala*, crucifers (*Capsella bursa-pastoris*, *Cardamine hirsuta*), groundsel *Senecio vulgaris*, *Callitriche palustris*, and *Elodea canadensis*; in late winter also took many 'ground-loving' Diptera (King and Ladhams 1970). In Somme (France), eats seeds of sea kale *Crambe maritima* (F Sueur); stomach of bird from Vendée, November, contained only lichens (Seguin-Jard 1922). In southern and central East Germany, birds on waste land in December took seeds of *Chenopodium* and crucifer *Sisymbrium*; large numbers attracted to waste grain at farms, and also took seeds of plantain *Plantago major* and dock *Rumex* (Hinsche 1958; Rinnhofer 1972, 1974). One January day (−17°C) in Austria, bird took millet *Panicum* at feeding station on house terrace (Haar 1975). In Moldavia (USSR), November, one bird contained oats *Avena* and (presumably seeds of) Chenopodiaceae and other weeds (Averin and Ganya 1970). In Transvolga region and Ukraine, wintering birds contained seeds of Chenopodiaceae, knotgrass, sedge, millet, and wheat (Dementiev and Gladkov 1954*a*). In Kalmytskaya ASSR (north-west of Caspian Sea), December–January, 5 birds contained 370 seeds and fruits: 98·1% Chenopodiaceae, 1·4% *Polygonum*, and 0·5% *Onosma* (Boraginaceae) (Samorodov 1968).

On dry shrub-steppe in south-east Washington (USA), of 179 birds, those from February–October contained 1·1–67·5% seeds, remainder invertebrates (monthly averages by dry weight); in November–January, 66·3–100%. Average (monthly) length of food items thus greater during February–October: varied from $2 \cdot 07 \pm SD0 \cdot 17$ mm ($n = 16$) in January to $7 \cdot 06 \pm SD1 \cdot 07$ mm ($n = 22$) in May; larger birds took significantly larger items than smaller birds. (Rotenberry 1980; see also for seasonal comparison with diet of other passerines in the area.) For further (mostly lowland) studies in USA, see (e.g.) McAtee (1905) and review by Martin *et al.* (1951).

In Morocco, ♀ recorded carrying stonefly to young (Lynes 1920). On Yamal peninsula, diet of nestlings contained more earthworms and fewer spiders than for other passerines in the area: 352 items (all invertebrates) included 50·4% Tipulidae, 25·0% Tenthredinidae larvae,

7·1% earthworms, 5·7% flies, 4·7% Chironomidae, 4·5% caterpillars, and 4·5% beetles (Danilov *et al.* 1984). In western Tien Shan, faeces of young showed food almost entirely animal, mainly small insects; large mosquitoes important for 2nd broods (Lobachev and Kapitonov 1968). On prairies of Canada, nestlings studied by Maher (1979) ate largely caterpillars (Lepidoptera) and grasshoppers, but during May and June seeds comprised 6% and 10% (by number) respectively of the 613 and 314 items fed to them; items mostly 5–16 mm long. In Vitosha mountains (Bulgaria), 4 young *c*. 7 days old fed 12 times in 95 min; at *c*. 8 days, fed every 18 min (7–23) (Pätzold 1979). In Colorado (USA), food brought every 2–3 min to nest with 3rd brood young 5–6 days old (Kelso 1931).   DJB

Social pattern and behaviour. No major studies in west Palearctic. Following account draws heavily on North American data, notably studies of *praticola* by Pickwell (1931) and Beason and Franks (1974); see also Sutton and Parmelee (1955) and summary in Bent (1942).

1. Usually gregarious outside breeding season. Flocks migrating along coast of Poland mostly 10–20(−50) (Górski 1982), and winter gatherings in various parts of central Europe, also (at high altitude) in Haut-Atlas (Morocco), of this order or, less commonly, up to 200 or more (Leege-Juist 1906; Hinsche 1958; Hammerschmidt 1966; Klawitter and Lenz 1967; Lenz 1969; Juana and Santos 1981). In southern East Germany, waste grain attracted large numbers (Rinnhofer 1974). Post-breeding season flocks and autumn gatherings (5–200) reported from high altitudes of Elburz Mountains (northern Iran) (Norton 1958) and Turkey (Gaston 1968; Beaman *et al.* 1975). In USSR, very large flocks often occur on passage and in winter (Dementiev and Gladkov 1954*a*): e.g. flocks of thousands of *flava* on spring migration (Dolgushin *et al.* 1970); see also Danilov *et al.* (1984) for Yamal peninsula. Winter gatherings in North America sometimes huge: e.g. *c*. 15 000 *lamprochroma*, California, in exceptionally harsh conditions (McLean 1936). On Juist (West Germany), few adult ♂♂ noted in wintering flocks (Leege-Juist 1906), and birds usually arrive on breeding grounds in single-sex flocks, ♂♂ first (e.g. Pickwell 1931), though not always (see Heterosexual Behaviour, below). Reflocking typically occurs after dispersal to breeding grounds if weather turns bad (Pickwell 1931; Piechocki *et al.* 1982). Flocking pattern through year illustrated by study in Karzhantau (western Tien Shan, USSR). Feeding flocks of 10–20 sometimes forced down by heavy snow in mid-April, but birds occur mostly as pairs or singles by mid-May. Flocks of independent 1st-brood young (3–8) occur from early June or July; swell to 20–30 (adults and young) by mid-August and to winter norm of 50–80 by late September (Lobachev and Kapitonov 1968). In central Ohio (USA), nominate *alpestris* characteristically gregarious throughout its stay (October–March) but rarely mixes with locally breeding *praticola* (Walker and Trautman 1936). For further details of dispersion patterns in North America, see (e.g.) Terrill (1917), Taylor (1925), and Pickwell (1931). In Kalmytskaya ASSR, winter, recorded resting with large flock of White-winged Lark *Melanocorypha leucoptera* (Samorodov 1967). Skylark *Alauda arvensis*, also buntings (Emberizidae), notably Snow Bunting *Plectrophenax nivalis* and Lapland Bunting *Calcarius lapponicus*, as well as various finches (Fringillidae), often mentioned as feeding

associates; in most cases, however, not a true association (Leege-Juist 1906; Feindt *et al.* 1956; Hinsche 1958; Rinnhofer 1974) although this does occur; in West Germany, late December, *c.* 30 *E. alpestris* fully integrated with *c.* 1000 Twite *Carduelis flavirostris* and *c.* 10 *P. nivalis* (Gloe 1982). BONDS. Nothing to suggest mating system other than monogamous. Little information on pair-bond. At Hildesheim (West Germany), ♂ and ♀ always close together, mid- to late February (Feindt *et al.* 1956) suggesting early pair-formation or prolongation of bond outside breeding season. However, ♂♂ of migratory populations tend to arrive on breeding grounds before ♀♀ (see above) and pairing for life considered unlikely (Sutton 1932). Bond apparently strong at least during breeding season; in one study, change of mate occurred only when partner lost (Beason and Franks 1974). Incubation and brooding by ♀ (Pickwell 1931; Sutton 1932; Verbeek 1967); reports of ♂ doing some incubation (Lynes 1920; Dolgushin *et al.* 1970) perhaps require confirmation. Both adults feed young and perform nest-sanitation (Verbeek 1967), ♂'s share of feeding usually about equal to ♀'s (Pickwell 1931). Brood perhaps sometimes split between parents after leaving nest (Criddle 1920). ♂ may, however, do most or all feeding of 1st brood out of nest (e.g. Pätzold 1979), presumably in some cases when ♀ re-lays and rears 2nd brood (see Spjøtvoll 1970). BREEDING DISPERSION. Solitary and territorial. Report of 10 pairs close together at 1400-1550 m, north-east of Kursunlu, Turkey (Schubert 1979), suggests neighbourhood groups may occur. In Karzhantau, closest nests *c.* 15 m apart; birds collected food *c.* 30-150 m from nest, indicating considerable overlap in territories (Lobachev and Kapitonov 1968) or small nesting territory separated from neutral feeding grounds. In Pamirs (USSR), pairs *c.* 70-100 m apart in suitable habitat, 1-3 km apart where conditions less favourable (Potapov 1966). Other reports on territory size and inter-nest distances are from North American studies. On Bylot Island (Canada), centres of 4 territories *c.* 400 m apart (Drury 1961). In alpine tundra of Wyoming (USA), territory *c.* 1·5 ha if nest taken as centre; average distance between nests in continuous habitat 142 m (Verbeek 1967). Observations on *praticola* in Illinois and New York (USA) indicated ♂ sings over area considerably larger than territory on ground—boundaries not vertical for any great height. Song-posts often up to *c.* 150 m from nest. Territories measured by song-posts 0·4-5 ha, smaller where density higher (Pickwell 1931); or, also in Illinois, average 1·6 (0·6-3·1) ha (Beason and Franks 1974). Territory established and defended by ♂. Young reared within it, but adults obtain some food elsewhere (Pickwell 1931; Beason and Franks 1974). Neutral feeding areas occur (Pickwell 1931), and apparently separated nesting and feeding territories recorded, probably influenced by cultivation (Goodpasture 1950). Where 4 nests built in a season, 3rd was 41·5 m from 1st, 4th *c.* 335 m from 1st and *c.* 65 m from 3rd (Goodpasture 1950); in Manitoba (Canada), 2nd-brood nest 'a few feet' from 1st (Criddle 1920). Data on densities limited. In Rondane-Dovrefjell area (south central Norway), at least 15 pairs within area *c.* 20 km across at 1200-1450 m above sea-level (Spjøtvoll 1970). On Hardangervidda (southern Norway), 0·5 pairs per km² (I Byrkjedal and J A Kålås). Various kinds of open tundra, Yamal peninsula (USSR), hold 2-20 pairs per km². Density generally rather low in damper habitats (some variation depending on census method), much higher (24-32 pairs per km²) at edge of shore zone (Danilov *et al.* 1984). In Karzhantau, at *c.* 2200-2800 m, usually at least 10 birds along 1 km, often more; density *c.* 400-700 pairs per km² (Lobachev and Kapitonov 1968); see also Dolgushin *et al.* (1970). On Bylot Island, 4 pairs in 1 square mile (*c.* 1·5 pairs

per km²) (Drury 1961). In central Illinois, densities generally higher away from interstate and county highways despite uniformity of habitat (see Clark and Karr 1979). ROOSTING. Nocturnal and probably mainly solitary (but see below); always on ground. Rail tracks used in Britain and North America (Shannon 1974). Normally roosts in shelter of grass clump or earth clod on day-time feeding grounds (Pickwell 1931); short grass preferred according to Kelso (1931). Study of *actia* and desert-dwelling *ammophila* indicated roost-holes dug (captive birds used bill) on cold nights, frequently at site offering shelter from wind. Fresh holes (*c.* 8 × 5 × 4 cm) probably excavated every night in soft soil. 1 site had 25 holes within circle of 10 m diameter; another had 5 in a depression 20 cm wide (Trost 1972). At Hildesheim, mid- to late February (temperature down to -20°C), apparent pair flew *c.* 1·5 km from feeding site to field with deep snow but protected from east wind. At first apparently sheltered in animal tracks but for roosting made pits in snow; burrowed down in snow until invisible, rotating whole body and helping with wings (Feindt *et al.* 1956). Birds also recorded burrowing into haystacks during blizzard (Krivitski 1962). ♂ roosts within a few metres of nest when ♀ incubating or brooding young (Criddle 1920; Lobachev and Kapitonov 1968; Dolgushin *et al.* 1970). Seeks shade of bushes, etc., in intense summer heat. Crouches facing into wind in strong wind and dust-storm (Behle 1942). Birds observed emerging from roost beside or beneath grass clumps first stood on pebbles for self-preening and shaking, etc., then started to feed. In warm winter sun, will doze while standing (one-footed) on stone or hunched up on turf (Sutton 1927). In Karzhantau, in hot weather of late July and August, birds vocal from dawn (05.00-05.30 hrs). Fly in loose-knit flocks to eastern slopes to feed; this followed by rest period *c.* 11.00-15.00 hrs and renewed activity through to sunset (Lobachev and Kapitonov 1968). In hot desert, recorded perching on bushes, presumably to benefit from slightly lower temperature compared with ground. No drinking recorded in June (insect diet probably provides sufficient moisture) but in Mojave desert after breeding season, drinks daily just before dawn: e.g. on 10 August 06.30-07.30 hrs (Trost 1972, which see for further experimental work on adaptations to heat). In Karzhantau, from mid- or late July, drinks at stream, usually 15.00-18.00 hrs. Water bathing also recorded (Lobachev and Kapitonov 1968; also Rustamov 1958).

2. Moderately confiding at times. When newly arrived in Britain, often very approachable, though may become wild by January (Shannon 1974). Always alert, even when apparently feeding busily (Behle 1942; Feindt *et al.* 1956). In winter, may allow approach to *c.* 20-40 m. In cold and snow, large flocks tend to flush at *c.* 12-15 m, smaller flocks at *c.* 8 m. Often do not fly far. Less shy than *A. arvensis* where both occur together (Feindt *et al.* 1956; Hinsche 1958; Stahlbaum 1978). Feeding birds said to keep their backs towards observer (Dubois 1936). May first give call 3a and crouch when alarmed, then flee (Beason and Franks 1974) or, at closer approach, run for short distance and adopt an alert posture—legs and neck fully extended, horns raised—often on slight elevation (Leege-Juist 1906; Dubois 1936; Feindt *et al.* 1956). Various reports of greater tameness include birds coming into camps in search of food (Whistler 1925a) and, in hard weather, taking scattered seed in towns (e.g. Bent 1942). Will seek refuge in rocks or under bushes to escape from raptors (Shannon 1974). FLOCK BEHAVIOUR. Flocks generally widely spread for feeding, resting, and roosting, more compact in flight (Pickwell 1931; Lobachev and Kapitonov 1968). At Zerbst (East Germany) in January, birds mostly kept close together while flying and when moving across ploughland to feed. All rested at same time, standing on

clods, and some self-preened. Warmer weather induced more (undisturbed) flying activity, with splitting up of flock, chases by 2-3 birds, and rudimentary Song-flights. Call 4 given constantly in flight. Birds kept apart from other species (Hinsche 1958); see also Leege-Juist (1906). Song may also be given from ground (Walker and Trautman 1936), especially if warm (Sutton 1927). In Yamal, old ♂♂ dominant over younger ♂♂ and (lowest in hierarchy) ♀♀. *P. nivalis* also subordinate. Birds generally keep 10-15 cm apart. For details of threat postures used to defend individual-distance and/or feeding territory, see Antagonistic Behaviour (below). Bird having fed for some time quite easily displaced by new arrival (Danilov *et al.* 1984). See also Feindt *et al.* (1956). Where birds fed artificially near house, one defended territory of *c.* 1 m² for *c.* 1 hr (Bent 1942). In Kurgal'dzhino (northern Kazakhstan, USSR), Black Lark *M. yeltoniensis* obvious food competitor and not tolerated by flocks of *flava* and *brandti* (Krivitski 1962). In Illinois, early April, flock of *c.* 10-15 nominate *alpestris* allegedly had sentinels posted while others fed (Pickwell 1931). In Orenburg area (USSR), one of several flocks flew up and circled; then joined by others, birds spiralling higher in dense mass until lost to view (Dolgushin *et al.* 1970); presumably starting next stage of migration. In Mojave desert, flocks (of 100 or more) usually spend only *c.* 5 min at watering place (Trost 1972). Mixed flock of *E. alpestris* and *C. lapponicus* recorded mobbing large raptor (Bent 1942). SONG-DISPLAY. Song of ♂ (see 1 in Voice) given from ground or slight elevation (rarely from tree or building), and in flight (Pickwell 1931; Kelso 1931; Verbeek 1967). Song from ground probably serves mainly to demarcate territory (Pickwell 1931; Beason and Franks 1974); Song-flights probably mainly heterosexual (Beason and Franks 1974). Relative proportions of ground and aerial song apparently vary geographically. Song-flights rare or absent in Bulgaria (Pätzold 1979), Moroccan Atlas (Heim de Balsac 1948), Iran (P A D Hollom), Kazakhstan (Dolgushin *et al.* 1970), Pamirs (Potapov 1966), and Kashmir (Osmaston 1927). On Bylot Island (*hoyti*), short bursts given on ground apparently as preliminary to Song-flight. Just before take-off, bird stands erect on regularly-used grass tuft or rock, horns raised and feathers sleeked (Drury 1961). Ascent usually silent and fairly steep (*c.* 60°) into strong wind, in wide spirals if wind light (Drury 1961; Beason and Franks 1974). Climbs to *c.* 80-250 m, alternating bouts of rapid fluttering wing-beats with brief periods of coasting on closed wings, so that ascent undulating (Dubois 1936; Drury 1961; Beason and Franks 1974). 25 Song-flights by *praticola* averaged *c.* 155 m (90-270), height varying individually or with weather (Pickwell 1931); 50-70 m recorded in *hoyti*, lower (10-20 m) in strong wind (Drury 1961). At peak of ascent, bird glides with wings and tail widely spread, giving phrased song (see 1a in Voice). Continuous song (see 1b in Voice) given while beating wings slowly and deeply; may intermittently close wings and tail (E N Panov). According to Pickwell (1931), continuous song typical of early phase of Song-flight, though also given at other times. While singing, usually heads into wind and may remain almost stationary. Between bursts of song gives 10-20 wing-beats to ascend and repeat gliding phase. If blown backwards, may stop singing to fly forward close to original position; bird moving forward while singing will pause then circle back to previous spot. Altitude usually increases slightly after first few bursts of song, then decreases through rest of flight. Sometimes sings constantly over centre of territory or (in stiller conditions) circles or flies slowly back and forth across it. At conclusion, closes wings and drops head-first, more or less vertically or at steep angle, with audible whizzing sound; close to ground (*c.* 7-10 m up:

Pickwell 1931), opens wings to fly horizontally for some distance. Sometimes lands on starting point. (Sutton 1927; Pickwell 1931; Dubois 1936; Drury 1961; Verbeek 1967; Beason and Franks 1974; P A D Hollom, E N Panov.) For variants— e.g. song given from start of ascent—and summary of earlier descriptions, see Pickwell (1931). Low and brief Song-flights (sometimes with song during ascent) perhaps normal in parts of USSR (see Dementiev and Gladkov 1954*a*, Dolgushin *et al.* 1970). Average length of 32 Song-flights 2·34 min (1-5) (Pickwell 1931); 2·46 min (25 s to 8 min, *n* = 17) (Beason and Franks 1974). In another study, up to 11 min recorded, and all probably over 5 min (Verbeek 1967). One ♂ nominate *alpestris* gave 24 bursts of song in 1·5 min, another gave 32 bursts during flight of 3 min (Bent 1942). ♂ singing from ground normally has horns raised (see Fig A). Bouts of song

A

from ground by *praticola* 1·5 s-45 min (Beason and Franks 1974); see also Pätzold (1979). Song-flights generally little affected by weather, but calm, mild, and overcast conditions perhaps favoured (Pickwell 1931; Sutton and Parmelee 1955). Song from ground reduced in wind and cloud, but temperature of -25°C had no effect (Beason and Franks 1974). In Wyoming, tends to sing earlier and later than other species (Verbeek 1967). On Bylot Island, June, evening song-period 20.00-22.00 hrs (Drury 1961); see also Dubois (1936) for Montana (USA). In Illinois, May, song given from ground 1·6-1·9 hrs before sunrise, with peak *c.* 15 min later. Little activity around sunrise when birds probably feeding; minor peaks *c.* 2 hrs after sunrise and *c.* 1-2 hrs before sunset (Beason and Franks 1974). Earlier study in Illinois indicated song from ground regular throughout day, Song-flights mostly around midday and near sunset. In mid-June, first song at *c.* 04.00 hrs, last *c.* 20.10 hrs. For further details, including break-down of day's activity in mid-June by one ♂, see Pickwell (1931). Little information available on song-periods. Birds sing occasionally in winter (e.g. Rinnhofer 1974). On Hardangervidda, starts almost immediately after arrival (I Byrkjedal). Similarly in northern USSR (e.g. Timanskaya tundra and Gydanskiy peninsula), from early or late May. In Krasnoyarsk (where resident), song given from February (Dementiev and Gladkov 1954*a*). Other data from North America. On Southampton Island (Canada), ♂♂ start only when ♀♀ arrive in early June (Sutton 1932). On Baffin Island (Canada), none noted after mid-July (Sutton and Parmelee 1955). Main period for *praticola* mid-January to early July in Illinois and mid-February to late June in New York. Most Song-flights in May, from nest-building through to incubation. Occasionally sings in autumn (Pickwell 1931). Most song from ground March-April. Phrased song mostly January-February; more continuous variant from mid-March (Beason and Franks 1974). ANTAGONISTIC BEHAVIOUR. (1) General. ♂♂ usually occupy territories soon after arrival: e.g. at Vadsö (northern Norway), flock dispersed within a week (Bannerman 1953). In winters with much snow, birds still in flocks and silent 1-2 weeks before taking up territories (I Byrkjedal). Territory defended throughout breeding cycle, from its establishment until young leave nest (Verbeek 1967). Initially, fights and chases (often involving several ♂♂: e.g. Verbeek 1967) fairly general; later mainly near contiguous boundaries (Pickwell 1931). In Karzhantau, much activity, with chases, etc., in first half of May (Lobachev and Kapitonov 1968). In Illinois, most territorial fights February–March, chases frequent in April. When ♂ disappeared and was replaced, fights took place over *c*. 1 week to establish limits roughly coinciding with those of old territory (Beason and Franks 1974). In Mongolia, territorial ♂♂ drove off small flocks of migrants (Piechocki *et al.* 1982); ♂ *praticola* (♀ brooding) ignored flocks of *c*. 10–15 nominate *alpestris* 6–7 m away (Pickwell 1931). Earlier reports indicated that Song-flight in a given area performed by only 1 ♂ at a time (Pickwell 1931; Drury 1961), but 2 ♂♂ recorded in simultaneous Song-flight *c*. 100 m apart (Beason and Franks 1974). In Illinois, territorial fights involved only ♂ against ♂; if both birds of pair trespassed, only offending ♂ chased. When 2 ♂♂ flew over a trespassing ♀ during chase, she would follow them back to her own territory (Beason and Franks 1974). (2) Threat and fighting. Territorial ♂ gives call 3a on sighting rival/intruder and runs towards it, stopping every 30–40 cm to call (Beason and Franks 1974). On boundary, may just stand erect, or occasionally call and strut about with horns raised, tail spread, and wings drooped. Approaches may lead to vigorous ground-pecking by both birds (Pickwell 1931). When closer—e.g. 30–45 cm apart (Cottrille 1950)—birds more crouched, with heads lowered, horns raised, back hunched, and (sometimes) wings drooped. Remain thus in Threat-posture or walk about rather mincingly, making rather jerky up-and-down (perhaps ground-pecking referred to above) or side-to-side head movements. Excited twittering sometimes develops into perfunctory or full song (Meinertzhagen 1938; Sutton and Parmelee 1955; Danilov *et al.* 1984). Fights on ground rare according to (e.g.) Pickwell (1931) and Sutton and Parmelee (1955), but frequently recorded in another study: 2 ♂♂ stood close together and suddenly extended one wing to strike; after each blow, both retreated *c*. 30 cm then approached for repeat performance (Verbeek 1967; see also Dubois 1935). Birds often run at each other, wings raised above back (see drawings in Danilov *et al.* 1984 showing threat in feeding flock), or fly together, then ascend to *c*. 5–20 m, pecking and clawing all the way, dashing against each other, tumbling over, and perhaps using wings as weapons (Pickwell 1931; Sutton and Parmelee 1955). Fights not usually damaging but can be very persistent (Dubois 1936). In one case, bird in moribund state (possibly exhausted) still pecked by rival (Cottrille 1950); in another, bird stood over rival aiming pickaxe blows at its head (Whistler 1925a). Typical aerial fights often alternate with high-speed chasing (Pickwell 1931; Verbeek 1967); zigzag pursuit over *c*. 40 m once followed dive-attack on boundary (Drury 1961). During chase, pursuer may sing (Potapov 1966) or give call 4 (E N Panov). Having chased intruder to boundary, territory-owner typically circles back and sings from song-post (Beason and Franks 1974) or in flight (Drury 1961). Also recorded raising and lowering horns repeatedly for some time (Lobachev and Kapitonov 1968). Short (*c*. 1 m) chases on boundary performed in 'tit-for-tat' pattern; on landing, both may mock-feed (see Fig B), and one or both may sing from ground after bout of chasing (Pickwell 1931; Beason and Franks 1974; E N Panov). ♀ once recorded chasing strange ♂ away from young out of nest but fled at his 2nd approach (Pickwell 1931). One ♀ attacked Starling *Sturnus vulgaris* near nest, and a ♂ drove away Vesper Sparrow *Poecetes gramineus* (Lovell 1944). In Vitosha mountains (Bulgaria), ♂ contested song-perch with ♂ Water Pipit *Anthus s. spinoletta* and ♂ Rock Thrush *Monticola saxatilis* (Pätzold

B

1979). HETEROSEXUAL BEHAVIOUR. (1) General. In some populations, ♀♀ arrive at same time as first ♂♂ and pairs perhaps already formed then (Pickwell 1931); see also Bonds (above) for close association of ♂ and ♀ in winter. For timing of pair-formation and courtship in various parts (and altitudes) of USSR, see Dementiev and Gladkov (1954a) and Dolgushin *et al.* (1970). Courtship (with Song-flights and courtship-display and -feeding, also sexual chases) begins after territories are established and occurs periodically through breeding season (Beason and Franks 1974); recorded when feeding young out of nest (Bannerman 1953). ♂ attends ♀ closely during building and laying, but ♀ said to be 'curiously indifferent' to ♂ for most part (Pickwell 1931). ♂ losing mate performed many Song-flights for 2 weeks afterwards (Beason and Franks 1974). (2) Pair-bonding behaviour. ♂ recorded following ♀ constantly for up to *c*. 2 hrs while feeding (Sutton 1927); probably mate-guarding. Chases frequent. Usually start on ground, some in flight (Beason and Franks 1974). ♂ may sing while pursuing ♀ (Potapov 1966). Apparently tries to seize ♀'s head or tail-feathers; if he succeeds in flight, both birds drop to ground. Chase not followed by copulation (Beason and Franks 1974). Close to ♀ on ground, ♂ adopts Courtship-posture: body more horizontal than usual, wings drooped, tail spread and raised, horns also raised. May make short runs, not necessarily towards ♀ who, however, tends to fly off if determined rush made at her; ♂ may then sing from stone. Otherwise, ♂ struts for up to *c*. 1 min or rotates before ♀, calls (see 3c in Voice), quivers wings, and spreads black chest-patch (once while squatting *c*. 30 cm from ♀: Sutton and Parmelee 1955). ♀ feeds, apparently ignoring ♂. Interaction not followed by copulation (Pickwell 1931; Meinertzhagen 1938; Beason and Franks 1974; Piechocki *et al.* 1982). ♂ also recorded pecking gently at ♀'s head and, after birds had moved apart, stealing furtively round ♀ whilst wing-shivering. Same posture (Fig C) also adopted by ♂ in

C

interaction (no details) with another pair, when call 3a also given (E N Panov). (3) Courtship-feeding. ♀ recorded begging (no details) from ♂ carrying food. ♂ mounted ♀, dismounted, walked about briefly, and then repeated performance; ♀ then took food (Beason and Franks 1974). ♂ also reported occasionally to feed ♀ on nest (Sutton 1932; Sutton and Parmelee 1955; Rustamov 1958). (4) Mating. ♀, presumably soliciting, once seen to 'flutter and crouch' before ♂ like ♂ House Sparrow *Passer domesticus* (Pickwell 1931). When young already fledged, ♀ (no previous display by ♂) held body horizontal, wings slightly drooped, and moved tail from side to side. ♂ ran up and copulated. When ♂ performed courtship-display shortly afterwards, ♀ moved off (Beason and Franks 1974). In East

Finnmark (Norway), copulation recorded when birds still feeding 1st-brood young just out of nest (Blair 1936). (5) Nest-site selection. Evidently by ♀ (Sutton 1927; Pickwell 1931). ♀ recorded making short flights (less than 1 m), ♂ walking after her. ♀ approached grass tufts, apparently inspecting them; also flew back and forth between 2 places (Beason and Franks 1974). The following perhaps connected with nest-site selection when performed in presence of ♀: ♂ sings with horns raised, simultaneously making mock-feeding movements to left and right (Fig D), then tramples for long time at base of small

D

bushes (E N Panov). Nest-scrape excavated by ♀ who exhibits general restlessness during nest-building; takes 4 days for 1st nest, 2 days for replacement nest. ♂ usually nearby at this time but does not assist (Sutton 1927; Pickwell 1931). (6) Behaviour at nest. Eggs normally laid in early morning (e.g. Verbeek 1967, Lobachev and Kapitonov 1968). ♀ leaves nest to feed, etc., taking breaks of 5-10 min (Lovell 1944; Beason and Franks 1974), but see also subsection 3 (above). ♂ usually far from nest while ♀ incubating (Lobachev and Kapitonov 1968), often in separate feeding territory (Goodpasture 1950). ♂ recorded bringing flakes of peat (see Breeding) to nest where ♀ incubating (Bannerman 1953), also once came to nest before hatching and made thrusting movements at eggs (Lovell 1944). RELATIONS WITHIN FAMILY GROUP. Young brooded by ♀ for first few days (Pickwell 1931); average stint 11 min (6-17). Brooding and shading from sun not recorded after 8 days by Verbeek (1967), or brooding may stop after c. 6 days, ♀ then roosting near nest (Lobachev and Kapitonov 1968; Dolgushin et al. 1970). ♀ also recorded sheltering young from rain (Dubois 1936). ♀ brooding young c. 10 days old in colder conditions became completely snowed in. Young fall into torpor if chilled but revive when warmed. Pair feeding 4 young c. 7 days old collected food within c. 200 m of nest. Normally came to nest together; landed habitually on rock c. 2 m high, waiting there 1-2 min before dropping (♀ usually first) to ground, to pause again before running (with detours) c. 40 m to nest. (Pätzold 1979.) See also Mousley (1916) for use of regular perches. In another study, birds tended to land 3-5 m from nest, ♀ approaching silently, ♂ giving call 3a. ♂ tended to fly up directly from nest, ♀ to walk away some distance first. When brooding, ♀ would step off, allow ♂ to feed young, then resume brooding (Verbeek 1967). Eyes of young open at c. 3-4 days (Verbeek 1967; Lobachev and Kapitonov 1968). Turn and gape in response to human sounds and movements initially but discriminate by 5-6 days old (Kelso 1931; Pickwell 1931; Verbeek 1967). Faint call noted at c. 3 days old (Potapov 1966, which see for further details of physical development). Faecal sacs eaten or carried away c. 15-35 m (Pickwell 1931), e.g. after about every 2nd feed (Pätzold 1979). Dead young also removed (Verbeek 1967). In south-central Norway, young left nest at c. 12-13 days old (Spjøtvoll 1970). Average in most North American studies c. 10 days (9-14). Shorter periods (to 7 days) perhaps typical of later broods or due to disturbance (Kelso 1931; Pickwell 1931;

Verbeek 1967); see also Lobachev and Kapitonov (1968) and Danilov et al. (1984) for USSR. At c. 8 days old, may move c. 30-40 cm from nest when parent c. 1 m away, but return afterwards (Verbeek 1967). Normally follow food-carrying parent (Pickwell 1931). At first only hop (Pickwell 1931), though Verbeek (1967) reported them walking readily on leaving. 2 young recorded huddling together c. 3 m from nest after leaving at c. 9-10 days old (Kelso 1931). Sometimes stay nearby for up to 4 days (Lobachev and Kapitonov 1968; Beason and Franks 1974), though recorded up to 150 m away (still being fed) at c. 11 days (Kelso 1931). Able to fly c. 40-50 m at height of c. 6 m by 17-19 days old (Pätzold 1979); 75-100 m at c. 15 days according to Pickwell (1931) and Verbeek (1967). Fed for at least 6 days out of nest (Sutton and Parmelee 1955). In one case, both parents still with self-feeding juvenile of this age and all roosted c. 1 m apart and c. 10 m from nest c. 9 days later. When juvenile c. 36 days old, ♂ interrupted courtship to chase it far out of territory. Eviction not final however, as juvenile still in territory when pair had 2nd brood in nest (Kelso 1931). 2 young 12 days old intruded into territory of pair with 3 young and were subsequently fostered (Beason 1984). ANTI-PREDATOR RESPONSES OF YOUNG. Remain silent and immobile or crouch in nest, normally closing eyes (Kennedy 1913; Verbeek 1967), e.g. when observer c. 7 m away (Kelso 1931). Crouching develops after eyes open (Drury 1961), or only at c. 7-9 days, handled birds reacting thus when placed on ground, while younger nestlings tended to wriggle (Pickwell 1931). Brood of 4 quiet when handled but 1 called in response to ♂'s alarm-call (see 5 in Voice) when being replaced (Sutton and Parmelee 1955). Bird removed from nest at c. 9 days and placed on ground ran c. 2-3 m through grass before crouching (Dubois 1936). Birds handled when out of nest may call and struggle to escape (Pickwell 1931) or regurgitate food (Taylor 1925). Normally crouch, allowing approach to c. 3-7 m, even when fledged (Drury 1961; Plucinski 1973). Fledged young typically make mock-feeding movements when alarmed (E N Panov); face intruder on landing and bow (as if looking at feet) every c. 4 s (Verbeek 1967). PARENTAL ANTI-PREDATOR STRATEGIES. (1) Passive measures. Tends to desert if disturbed during nest-building (Pickwell 1931); wary then, also during laying, flying at c. 20-25 m from man (Lobachev and Kapitonov 1968). Shyness at nest varies. From start of incubation (less commonly after hatching) often leaves when intruder c. 25-100 m away: flies low with steady wing-beats and without marked undulation (Pickwell 1931), sometimes moving well away to separate feeding territory (Goodpasture 1950). ♀ may sit tight even at very close approach, almost allowing herself to be touched (Sutton and Parmelee 1955); reluctant to leave when cold or late in day (Pickwell 1931). ♀ may leave with exaggerated slowness (Terrill 1917), ruffle and sleek feathers, or walk with head up and give an alarm-call occasionally (Sutton and Parmelee 1955), or flush with explosive flutter of wings (Sutton 1927), also when young out of nest (Drury 1961); sometimes both birds simply hide (Dubois 1936). Generally more likely to call after hatching (Pickwell 1931) and to sit tight (Lobachev and Kapitonov 1968). Common reaction (before and after hatching) when disturbance long is to stay close (even when observer near nest), then to feign indifference, primarily by mock-feeding—sometimes pair together (e.g. Terrill 1917, Pickwell 1931, Dubois 1936, Drury 1961); perhaps in some cases actually collects food (Goodpasture 1950). ♀ also recorded tearing at plant-down when prevented from returning to nest with eggs; eventually settled to incubate when man c. 2 m away; when flushed again, returned and hopped over nest (Sutton 1932). One bird ascended to c. 100-135 m and circled silently (Pickwell 1931). ♂ often sings (sometimes in Song-flight), especially during incubation, when man approaches or when driven about in territory, though may just call (Criddle 1920; Pickwell 1931; Drury 1961). ♀ recorded moving quickly towards nest when ♂ singing, but exhibited 'studied aimlessness' when ♂ silent (Sutton and Parmelee 1955); one ♂ sang every time ♀ settled on eggs (Sutton 1932). ♂ often tamer when intruder approaches closely, though noisy departure from nest by ♀ (see above) may cause ♂ to stop singing (Sutton 1927). After hatching, ♂'s agitation expressed mainly in calls, though may feed very close, apparently unconcerned, or stay away until safe (Pickwell 1931). Pair with young out of nest alternated rapid movement in grass with standing alert on stone (Watson 1973*a*). (2) Active measures: against birds. Recorded successfully driving off domestic chickens *Gallus* by flying at them and landing on their backs (Brooks 1908; Criddle 1920). When American Kestrel *Falco sparverius* nearby, birds ceased feeding for over 30 min and called constantly (Pickwell 1931). (3) Active measures: against man. May perform distraction-lure display of disablement type. In *praticola*, considerable variation in behaviour, but no link found with particular stage of cycle (Pickwell 1931). Distraction-display performed by *hoyti* only when tending unfledged or weakly flying young (Drury 1961). ♀ recorded fluttering and rapidly running from nest, horns raised (Pickwell 1931); dragging one wing when near young out of nest (Dubois 1936); running away in crouched posture, wings quarter to half spread, flank feathers ruffled, then doubling back (Watson 1973*a*), also taking off after display, then circling and landing *c*.10 m away, occasionally calling (Sutton and Parmelee 1955). Bird giving call 3a and mock-feeding (see above) walked rapidly away when followed, then flew (Beason and Franks 1974); possibly also a lure. ♂ said also to perform distraction-display (e.g. Rustamov 1958), but no details. ♂ will flutter close by when young handled (Kennedy 1913; Mousley 1916), and in exceptional incident ♂ gave alarm-calls and attacked observer handling young: thrashed about excitedly, touching man's arms and legs with wings, also pecking; ♀ also approached, but did not attack (Sutton and Parmelee 1955); see also Terrill (1917). Ringed nestlings sometimes carried away by adult in flight (Berger 1953), or possibly dragged from nest (Lovell 1944). ♂ (carrying food) with young out of nest repeatedly approached ♀ and fluttered wings, then moved aside; departed when fledged juvenile landed near ♀ who only gave alarm-calls all the while (Sutton and Parmelee 1955). (4) Active measures: against other animals. Either parent will make dive-attacks on ground squirrels *Spermophilus* (Dubois 1936).

(Figs by J P Busby: from drawings by E N Panov.) MGW

Voice. Most vocal during nest-site selection and early nesting, quieter during incubation and much more so after young leave nest (Terrill 1917). Unlike other Alaudidae, calls exhibit almost exclusively a pure-toned 'i' (pronounced 'ee' in English) (Bergmann and Helb 1982); see also Feindt *et al.* (1956). Further study required to establish number of calls in repertoire and situation in which they are given (Bergmann and Helb 1982, which see for extra sonagrams); see also Beason and Franks (1974). Most information for extralimital races (North America). Data not sufficient to make detailed comments on geographical variation, nor to assess its extent.

CALLS OF ADULTS. (1) Song of ♂. (a) Overall more common variant ('intermittent song': Pickwell 1931) given mainly in flight comprises phrases lasting 1·0-2(-2·5) s, with fairly high pitch and rapid delivery. Variable and subdued twittering; initial 3 (2-4) units deliberate or slightly hesitant, then accelerating into a final jumble of trills or liquid chittered sounds. Falls then rises in pitch (Terrill 1917; Pickwell 1931; Beason and Franks 1974; Bergmann and Helb 1982; J Hall-Craggs). Variously rendered as follows: 'pit-pit pit-wit pittle wittle little little leeee' in *praticola* (Pickwell 1931); thin, unmusical 'tsip tsip tsee-didi' in nominate *alpestris*, jingling and metallic sounds like distant sleigh bells, also squeaks like old gate (Bent 1942); weak 'chit-chi-chiddle chee-la' (P A D Hollom). In *brandti* of USSR, 'tsee-tsee-(tsee) tee-lee-lee', quiet and tinkling and like bunting *Emberiza* (Dementiev and Gladkov 1954*a*); description of song in *albigula* similar—melodious and high-pitched trill—but long-drawn 'eeeee' said to precede accelerated phase (Lobachev and Kapitonov 1968). In recording (Fig I),

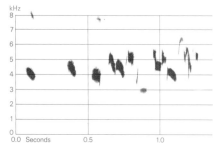

I C Chappuis Turkey May 1977

simply structured but musical and jingling phrase delivered rapidly after slow start. Not unlike Reed Bunting *E. schoeniclus*: simple, descending 'cheee' or 'chew' note followed by 'stip stip steedle leee', final unit sometimes more rolled or trilled. Apparent rudimentary song (Fig II)—perhaps the emergence of phrased song from Sub-

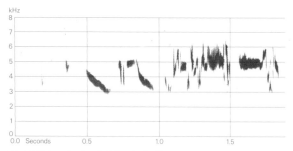

II J Kirby England March-April 1970

song—has similarly slow start, but introductory sounds more like quiet 'tik'; units achieve more tonal character and show rapid frequency modulation. Phrases generally sound much alike, and sonagrams reveal tendency to ascend and descend more or less in same order, although differences also apparent (J Hall-Craggs, M G Wilson). Similarly short, shrill, and rather feeble phrased song described for *elwesi* of Kashmir and Tibet; that of *longirostris* quite loud and sweet (Osmaston 1927; Ali 1946). In rudimentary Song-flights performed by *flava* in East Germany, mid-March, 'ziep' sounds (Hinsche 1958) presumably same as introductory units described above. Average by *praticola* 11·9 phrases per min in remarkably regular repetition (Pickwell 1931); similarly in *bicornis* of Lebanon, phrases of *c.* 2·5 s alternated with pauses of *c.* 3 s (P A D Hollom). Up to 12 units per phrase (Terrill 1917); rapid repetition almost like continuous outburst at times (Sutton 1927). (b) Apparently random sounds, and same jumbled trill as in 1st variant, lasting from a few seconds to over 1 min. 2 variants often interspersed (Beason and Franks 1974). Sweet and liquid when given on ground early in season (Terrill 1917). In study by Pickwell (1931), vigorous, evenly discursive 'pit-wit wee-pit pit-wee wee-pit' ('recitative') given very often as a brief prelude (a few seconds) to 1a in flight, also (sometimes quieter and incomplete) on ground. In *balcanica*, not very melodious and rather quiet bursts lasting 5–7 s reminiscent of Dunnock *Prunella modularis* (Pätzold 1979). Recording (Fig III) of more continuous variant is rather like Skylark

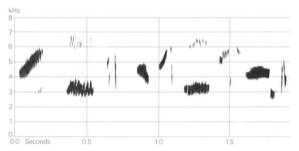

III Roché (1970) Finland June 1968

Alauda arvensis, though delivery less ebullient and near-repetition more marked. Sounds less musical than in 1a, slightly buzzy owing to much relatively slow frequency and amplitude modulation. Delivery rate slower than in 1a, but rather jerky at times, with an irregular tempo. Contains also a thin, high-pitched 'whee-teeou', possibly call 3a (J Hall-Craggs, M G Wilson). Imitation of other birds mentioned (without details) only by Dementiev and Gladkov (1954a). For further descriptions of song variants and possible indications of racial differences in North America, see Terrill (1917), Sutton (1927, 1932), Dubois (1936), Walker and Trautman (1936), and Drury (1961). (2) ♀ occasionally gives short and weak snatch of song in response to that of ♂ (Sutton 1927). (3) Contact-alarm calls. (a) Long-drawn, rather melancholy 'zieh', 'tieh', or 'zi-eh'; sometimes tending more toward 'ü' and suggesting *P. modularis* or Bullfinch *Pyrrhula pyrrhula*. Given singly or as a series as contact-call on ground or in flight (Leege 1910; Feindt *et al.* 1956). Recording (Fig IV) suggests high-pitched, long-drawn 'seee' or 'seer' not unlike Redwing *Turdus*

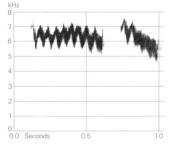

IV J Kirby England October 1969

iliacus (J Hall-Craggs, P J Sellar); some sounds more trilling—'pseerrrr' (M G Wilson). Pure-sounding 'psiit' or 'wiit' (Bergmann and Helb 1982), 'p-seet' (Pickwell 1931), 'tseeeep' (Terrill 1917), 'chlit' or 'sleep' (P A D Hollom), 'zuweet' or 'zur-reet' (Pickwell 1931) given mostly when disturbed. A 'tchle-see' in pair-contact (P A D Hollom), and in recording of feeding flock 'SEEsi' (P J Sellar) or 'TSEEdit' (M G Wilson: last 2 units of Fig V) presumably same as 'p-seet-it' of Pickwell (1931).

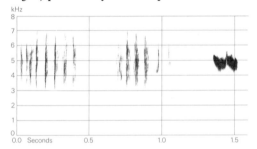

V J Kirby England March–April 1970

Disyllabic renderings such as 'didü' (Kroymann 1967), 'hidü' (Rinnhofer 1965), 'pee-u' (Terrill 1917), plaintive, rather quiet 'sirou' (P A D Hollom), or 'chipew' (Taylor 1925) probably also belong here. Commonest call 'su-weet' or 'tsee-ree' (Shannon 1974) given loudly as challenge by ♂ and more abruptly but softer than this in alarm; softer 'weet' or 'su-weet' given as contact between adult and young, also in flight, and softest variant by extremely alarmed or distressed bird (Beason and Franks 1974). Soft 'tjreeh' given by ♂ approaching nest to feed young (Verbeek 1967); quiet tinkling sounds from bird feeding young out of nest (Goodpasture 1950) probably closely related if not same call, as is 'tsveen...tsveen' given during interaction between ♂ and pair (E N Panov). (b) Excited birds just prior to or at take-off may give more rippling call (Watson 1973a): 'zirr', 'tirr', 'tirrlitt' suggesting Serin *Serinus serinus* (Feindt *et al.* 1956); 'tsr' given in flight (Bergmann and Helb 1982) perhaps the same, as are quiet, dry rattling sounds shown as 1st 2 units in Fig V. (c) Apparently distinct 'check' or 'cheek' said to be given by ♂ as contact-call to ♀ (Pickwell 1931); 'chittering' sounds given by ♂ during courtship (Beason and Franks 1974) perhaps the same. (4) Calls given mainly (or exclusively) in flight. High-pitched, mournful 'zeet-eet-it', 'ee-zee-dip' or 'zeet-it-a-weet' when flushed (Pickwell 1931; Sutton and Parmelee 1955). Gentle 'tsee tseetsee tsee ...' like tit *Parus* given in chase (E N Panov). Other renderings: 'siit-di-dit' or 'psiid sit sidit' (Bergmann and Helb 1982); 'zilili' (Stahlbaum 1978). Recording (Fig VI)

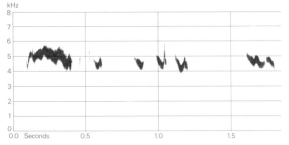

VI J Kirby England March–April 1970

suggests 'TSEE ti tititi titi' (J Hall-Craggs). See also (e.g.) Terrill (1917) and Bent (1942); trisyllabic calls mentioned by Rinnhofer (1965) and Kroymann (1967) probably also belong here. (5) Harder calls apparently indicating alarm. (a) An 'ee-chek' like Wheatear *Oenanthe oenanthe* (Lobachev and Kapitonov 1968); 'vzeet-vzet' (E N Panov); sharp 'tí-sick' (Bent 1942); anxious chirps from both sexes (Sutton and Parmelee 1955); see also Pätzold (1979). (b) A 'jrrrreeeet' like wader (Charadrii) when accompanying young out of nest (Terrill 1917). (6) Other calls. (a) In recording by J Kirby, some more rasping or buzzing 'tzip' sounds from feeding birds (M G Wilson).

CALLS OF YOUNG. Food-call of small nestlings very quiet even when adult at nest: e.g. squeaky 'ihh' at *c.* 2–3 days old (Kelso 1931; Pickwell 1931). Clear chirp or 'whit' given when handled only *c.* 45 min after hatching (Dubois 1936). Single loud 'peep' (Verbeek 1967) or 'weet' (Beason and Franks 1974) like adult call 3a (Pickwell 1931) serves as contact with parent after leaving nest: 'rrree' and 'teep...teep' (E N Panov)—age not stated. At *c.* 16 days, thin and high-pitched 'chirp' as food-call and when disturbed (rather like adult call 3a), then 'tsit-tseet' like adult at *c.* 36 days (Kelso 1931). According to Sutton and Parmelee (1955), food-call of well-grown young (flying strongly) still a barely audible 'sweet'.   MGW

Breeding. Little information from west Palearctic, but several studies in North America. SEASON. Northern Scandinavia: for double-brooded birds, laying of 1st clutch usually starts in last week of May, 2nd clutch at end of June; single clutches usually laid in 2nd–3rd week of June; season up to 3 weeks later in arctic USSR (Bannerman 1953; Dementiev and Gladkov 1954a). SITE. On ground in the open; often in short vegetation, or sheltered by small tussock; occasionally on slight hummock; nest-entrance usually protected from prevailing wind or direct sun (Verbeek 1967). Solitary. Nest:

depression in ground, usually excavated but sometimes natural, average diameter 86 mm (75–95), *n* = 8; lined with small twigs, rootlets, grass stems, and leaves, with finer lining of grass and *Salix* down; average nest diameter 64·3 mm (58–75) and depth 41·9 mm (37–50), *n* = 8 (Beason and Franks 1974). Small stones and other items placed round nest, sometimes forming path (Blair 1936; Bannerman 1953). Building: by ♀ only, excavating depression with bill, and sometimes feet, over 1–2 days, adding lining 2–4 days later (Beason and Franks 1974). In Tien Shan (USSR), 2 records of pairs seen gathering thistle down as nest-lining (Lobachev and Kapitonov 1968). EGGS. See Plate 79. Sub-elliptical, smooth and glossy; variable in colour but most greenish-white, heavily spotted yellowish-brown, often with brown hair-streaks, also may have larger spots or dark zones. 22·7 × 16·2 mm (20·5–26·0 × 14·7–18·0), *n* = 100 (Witherby *et al.* 1938a). Calculated weight 3·15 g (Makatsch 1976). Clutch: 2–4(–5). Of 14 clutches, Wyoming (USA): 2 eggs, 1; 3, 10; 4, 3; mean 3·14 (Verbeek 1967). 1–2 broods, usually only 1 in higher latitudes. Replacements laid after egg or nestling loss, generally within 2 days; in one case of double-brooding, 2nd clutch started 1 week after 1st brood fledged (Beason and Franks 1974). Laying interval 24 hrs. INCUBATION. 10–11 days. By ♀ only. Begins with last egg or occasionally penultimate; hatching synchronous or nearly so. YOUNG. Altricial and nidicolous. Cared for and fed by both parents. FLEDGING TO MATURITY. Young leave nest at 9–12 days but do not fly until 16–18 days. Age of independence not known. BREEDING SUCCESS. Of 82 eggs in 24 nests, Illinois (USA), 79% hatched and 60% of hatched young fledged; overall success 48% (Pickwell 1931). Of 47 eggs in 15 nests, Wyoming, 87% hatched and 61% of hatched young fledged; overall success 53% (Verbeek 1967). Of 26 eggs in 8 nests, Illinois, 50% hatched and 46% of hatched young fledged; overall success 23% (Beason and Franks 1974).

Plumages (*E. a. flava*). ADULT MALE. Forehead yellow, extending laterally into broad yellow supercilium; feathers narrowly tipped olive when plumage fresh, supercilium and especially forehead appearing slightly mottled. Broad black band across forecrown, partly hidden by pale olive-brown feather-tips in fresh plumage; black extends laterally into elongated 'horns' above rear of supercilium. Remainder of crown, hindneck, upper sides of neck, and upper mantle dark vinous-pink, but vinous on crown largely hidden by feather-tips which are yellowish olive-grey with faint dusky shaft-streaks and dusky terminal rim; hindneck and upper mantle with narrower grey feather-tips, vinous more exposed, sometimes with sharp narrow dark brown shaft-streaks. Lower mantle, scapulars, and back olive-brown with greyer feather-sides and rather poorly defined dark brown or brown-black shaft-streaks; upper rump similar to back, but with vinous tinge; lower rump and upper tail-coverts dark vinous-pink, longer coverts with brown tip and black-brown shaft-streak. Feather-tufts covering nostrils black, extending into rather narrow black stripe on lores and to broad black patch on rear of cheeks; black finely speckled yellow when plumage fresh. Ear-coverts yellow (usually paler than forehead and supercilium), sometimes faintly streaked olive; separated from yellow of supercilium and from pale yellow lower sides of neck by broad olive band running back from eye (where speckled yellow) and across rear of ear-coverts. Chin and upper throat yellow or pale yellow, extending below cheeks to pale yellow lower sides of neck; lower throat and chest black, feather-tips narrowly fringed white in fresh plumage. Sides of breast dark vinous-pink with grey suffusion; flanks pale vinous-brown with dark shaft-streaks. Remainder of underparts white; feather-tips on breast with grey-black triangular spots varying in depth and size, sometimes showing as distinct blackish spots, more often as grey mottling or suffusion, and virtually lost through wear in abraded plumage. Central pair of tail-feathers dark vinous-pink with browner tip and broad black shaft-streak; t2-t5 black with narrow white outer edge along tip of t5; t6 greyish-black with largely white outer web. Flight-feathers largely dark grey, blacker towards outer primaries; basal parts of outer webs of secondaries fringed vinous-brown, terminal parts and tips white; outer webs and tips of primaries narrowly edged white (except tips of outer primaries), widest along outer web of p9. Tertials dark brown, merging into olive-brown and vinous-brown on outer webs and tips; tips narrowly fringed white in fresh plumage. Lesser upper wing-coverts deep vinous-pink, slightly paler on tips; median upper wing-coverts with vinous-pink or vinous-brown outer web and shaft, dark olive-brown or black inner web, and broad and poorly defined white fringe along tip; greater upper wing-coverts olive-brown or black-brown with broad and poorly defined pale olive-brown or vinous-brown fringe along outer web and poorly defined white tip. Greater upper primary coverts greyish-black with brown outer webs; bastard wing dark olive-brown. Under wing-coverts and axillaries white; longer coverts tinged grey, shorter coverts along leading edge of wing partly dark brown. Wear and bleaching have marked effect: in fresh plumage (up to about December), black of face and chest partly hidden by narrow yellow or white feather-fringes and crown and hindneck distinctly tinged green-yellow; in worn plumage (from about April), crown to upper mantle purer and deeper vinous-pink, lower mantle, scapulars, and back more heavily streaked, black of face and chest sharply demarcated, breast white with slight traces of grey spots; yellow of head may bleach to yellowish-white by June-July (exceptionally, yellow parts of head fully white, even in fresh plumage). ADULT FEMALE. Rather similar to adult ♂ and best distinguished by narrow black triangle on feather-centres of central crown (black band on forecrown thus less sharply divided from central crown), pale brown or yellow-brown hindneck and upper mantle, only slightly tinged vinous-pink (in ♂, extensively deep vinous-pink), and darker brown lower mantle, scapulars, and back, with heavier black-brown streaks, appearing less uniform greyish olive-brown. See also Measurements. Black of forecrown, lores, cheeks, throat, and chest usually more restricted on average, more extensively hidden under broader pale feather-tips; as in adult ♂, fringes wear off, and though fully black facial pattern usually not attained until breeding season, some acquire it by February. Black patch on cheeks often slightly narrower, less broadly rounded. Lesser upper wing-coverts often browner than in adult ♂, less deep vinous-pink. NESTLING. Down confined to dense and long tufts on upperparts; cream-white or straw-yellow. For plumage development, see Drury (1961), Verbeek (1967), and Beason and Franks (1973). JUVENILE. Markedly different from adults. Upperparts entirely black-brown with variable amount of buff of feather-bases visible, each feather with contrasting triangular or square buff to off-white spot on tip. Sides of head mottled

buff and black-brown; lores, ring round eye, and sometimes stripe behind eye paler, buff with indistinct grey mottling, but no distinct supercilium; lower cheeks often blacker, forming ill-defined dark patch. Lower throat, chest, and flanks dull greyish-buff, marked with large and poorly defined blackish spots on chest and flanks; remainder of underparts dirty buff-white. T1 buff-brown with broad buff fringe and irregular blackish submarginal mark, t2-t6 dark buffish-grey with sharply defined buff fringe along outer web and tip (widest on t6). Flight-feathers and tertials dark grey with buff-brown suffusion; tertials and secondaries with broad buff fringes all around (except bases), submarginally bordered by dark grey or black; primaries with buff fringe along outer web and tip, partly bordered dark grey subterminally. Upper wing-coverts black-brown or greyish-black with buff to off-white tips. Under wing-coverts and axillaries buff-white. FIRST ADULT. Similar to adult, often indistinguishable once last juvenile feathers (usually p9-p10) shed. Fringes of innermost primaries rufous, especially on inner web, not as greyish-white as adult (but too few known-age birds examined to be certain if character always valid). Black on face and chest more broadly hidden by feather-fringes than in adult ♂ (especially in 1st adult ♀; 1st adult ♂ more closely similar to adult ♀). Sexing best done by crown-spots (present in ♀ only), amount of vinous on hindneck and upper mantle (virtually none in ♀, more in ♂, but adult ♀ sometimes as much as 1st adult ♂), by amount of streaking on upperparts, and (most reliably) by size (see Measurements).

Bare parts. ADULT. Iris brown. Bill horn-black, bluish-black, or dark horn-grey; tip black, base of lower mandible plumbeous-grey; bill of ♀ usually paler than ♂, entire bill plumbeous-grey or blue-black, except for black tip and pale flesh, cream, horn-colour, or bluish-white base of lower mandible. Mouth plumbeous-grey. Leg, foot, and claws black, soles dirty white, cream, or horn-brown; in ♀, leg and foot often browner (sepia, umber-brown, or red-brown), claws dark horn-brown, soles pale horn or yellow. NESTLING. Bare skin brown. Mouth and gape-flanges yellow, tongue with 1 black spot on tip and 3 at base. JUVENILE. Iris medium brown or brown. Bill horn-yellow, tip black in ♂; mouth yellow. Leg and foot yellowish-flesh on 25th day, changing to grey-violet and plumbeous by 40th day; soles yellow. Bill gradually darkens from tip towards base, yellow base of lower mandible retained longest. No information on when yellow of bill-base and mouth replaced by plumbeous of adult. (Whistler 1925a; Spjøtvoll 1970; Ali and Ripley 1972; Harrison 1975; Pätzold 1981; BMNH, RMNH, ZMA.)

Moults. ADULT POST-BREEDING. Complete; primaries descendant. In most races, including southern *balcanica*, *penicillata*, *atlas*, and *bicornis*, usually starts with p1 between mid-July and mid-August, but some feathers of body or wing-coverts occasionally moulted in June; all moult completed mid-September to early October (Vaurie 1951a; Dementiev and Gladkov 1954a; Paludan 1940, 1959; BMNH, ZFMK, ZMA). Captive ♀ *balcanica* started 22 June with p1; all moult completed late September (Pätzold 1981). Lowland populations of *brandti* in part start earlier than arctic and montane races, p1 shed about late June to mid-July (occasionally late May), all moult completed late August to late September (Dementiev and Gladkov 1954a; Piechocki and Bolod 1972; BMNH, RMNH). POST-JUVENILE. Complete; primaries descendant. Starts within a few weeks after fledging, p1 first, followed by scattered feathers of body and wing-coverts from primary moult score c. 20; in heavy moult of head, body, tail, and wing-coverts at score 30-40. Timing highly variable, depending on hatching date. Moult mainly starts July to mid-August, completed late August to mid-October, but (as in adult post-breeding) some *brandti* start in late May, and some *penicillata* and *albigula* still fully juvenile late August to mid-September. (Dementiev and Gladkov 1954*a*; BMNH, RMNH, ZMA.) Assuming same speed of moult as adult, immatures nearing completion are to be found from about late July to mid-November. In captive ♀ *balcanica*, moult started at 38 days old with shedding of some scapulars; p1 shed at 40 days, gradually followed by other primaries; p9 shed at 96 days, fully grown at 118 days; secondaries shed at 43-102 days; tail (sequence: t1 to t6) shed at 65-89 days (all new at 112 days); last body feathers regrown at 118 days (mid-October) (Pätzold 1981).

Measurements. ADULT, FIRST ADULT. *E. a. flava*. Fenno-Scandia and north-west USSR (breeding) and western and central Europe (winter); skins (RMNH, ZMA). Bill (S) to skull, bill (N) to distal corner of nostril; exposed culmen on average *c.* 3·6 less than bill (S).

WING   ♂ 111·8 (2·25; 47) 108-116   ♀ 103·0 (2·02; 33) 100-107
TAIL   68·4 (2·73; 47) 64-74   60·7 (3·09; 33) 56-67
BILL (S)   14·7 (0·53; 40) 13·6-15·5   13·8 (0·63; 32) 12·6-14·8
BILL (N)   8·8 (0·41; 40) 8·0-9·5   8·1 (0·37; 30) 7·5-8·8
TARSUS   22·6 (0·66; 38) 21·4-23·6   21·8 (0·69; 32) 20·6-22·8

Live birds, East Germany (Bub and Herroelen 1981).

WING   ♂ 113·1 (2·68; 40) 107-119   ♀ 105·2 (1·95; 45) 101-109

Samples of breeders from northern Eurasia: (1) Fenno-Scandia to Yamal peninsula and lower Ob'; (2) lower Ob' to lower Lena; (3) Yakutia, Wrangel Island, and north-east Siberia from river Lena to Anadyr (Portenko 1973, probably measured by different method).

WING (1) ♂ 108·8 (33) 102-116   ♀ 101·1 (9) 98-105
(2)   110·2 (28) 105-115   103·5 (9) 101-106
(3)   113·6 (19) 111-117   102·7 (10) 99-107

E. a. brandti. Wing of (1) lowlands from lower Volga to eastern Kazakhstan (BMNH, RMNH, ZMA), (2) eastern Tien Shan mountains (BMNH, RMNH, ZMA). (3) Mongolia (Piechocki and Bolod 1972); other measurements are (1) and (2) combined.

WING (1) ♂ 113·8 (1·95; 14) 110-117 ♀ 106·9 (2·82; 9) 103-111
(2)   118·2 (2·31; 19) 114-123   109·4 (2·58; 10) 105-113
(3)   110·7 (2·78; 21) 107-117   101·0 (2·59; 12) 95-105
TAIL   76·0 (2·60; 24) 72-82   66·4 (2·60; 11) 63-70
BILL (S) 15·7 (0·55; 26) 14·9-16·8 15·2 (0·69; 11) 14·3-16·2
BILL (N) 9·6 (0·41; 23) 8·9-10·4 9·2 (0·52; 11) 8·4-9·8
TARSUS 22·8 (0·78; 23) 21·2-24·3 21·9 (0·51; 12) 21·3-22·6

E. a. atlas. Morocco (BMNH).

WING   ♂ 113·6 (2·67; 14) 109-118 ♀ 105·7 (1·83; 11) 102-108

E. a. balcanica. (1) Yugoslavia and Bulgaria (Stresemann 1920; Bub and Herroelen 1981); (2) Greece (Niethammer 1943).

WING (1) ♂ 116·6 (2·55; 20) 112-120 ♀ 109·4 (2·41; 5) 106-112
(2)   113·2 (1·92; 5) 111-116

E. a. penicillata. (1) Western and central Asia Minor (BMNH, ZFMK); (2) eastern and north-east Turkey (ZFMK); (3) Caucasus (BMNH, RMNH, ZFMK, ZMA); (4) Azerbaijan (north-west Iran); (5) Zagros mountains (south-west Iran) (Vaurie 1951*a*).

WING (1) ♂ 118·4 (2·73; 8) 114-121 ♀ 107·9 (1·72; 6) 106-111
(2)   120·2 (2·77; 9) 117-124   110·5 (—; 2) 110-111
(3)   121·5 (2·65; 13) 117-126   109·5 (2·00; 5) 107-112
(4)   122·3 (—; 5) 117-126   112·0 (—; 8) 109-114
(5)   119·2 (—; 13) 115-124   109·6 (—; 14) 104-112
TAIL (3) 78·3 (3·59; 8) 73-84   66·2 (3·63; 5) 63-72

BILL (S) (1,2) 17·8 (0·71; 10) 16·9-18·8 16·2 (0·39; 5) 15·8-16·7
(3)   17·3 (0·37; 7) 16·9-17·8 16·2 (0·41; 4) 15·7-16·7
(4)   18·0 (—; 15) 17-19 17·4 (—; 8) 16·2-18·5
(5)   18·4 (—; 13) 17-20 17·1 (—; 12) 16·5-17·5
TARSUS (1,2) 24·6 (0·57; 10) 24·0-25·5 23·3 (0·87; 5) 22·9-24·2
(3)   24·1 (0·56; 7) 23·6-25·2 23·9 (0·70; 4) 23·0-24·6

E. a. bicornis. Lebanon: (1) BMNH; (2) Vaurie (1951*a*).

WING (1) ♂ 115·8 (2·26; 23) 112-120 ♀ 107·6 (1·60; 10) 105-110
BILL (S) (2) 17·6 (—; 12) 16-19 16·2 (—; 8) 15-18

E. a. albigula. (1) Gilan (north-east Iran) (Stresemann 1928); (2) Afghanistan (Paludan 1959); (3) central and eastern Iran, Turkmeniya, and Afghanistan (Vaurie 1951*a*).

WING (1) ♂ 116·7 (2·21; 10) 113-120 ♀ 107·3 (2·31; 3) 106-110
(2)   120·3 (—; 10) 116-124   110·7 (—; 10) 104-116
(3)   118·3 (—; 29) 114-124   110·9 (—; 20) 105-120
BILL (S) (3) 17·4 (—; 57) 16-19 16·9 (—; 32) 16-18·2

Nominate *alpestris*. Eastern North America (Ridgway 1901-11).

WING   ♂ 111·5 (—; 15) 108-113 ♀ 103·8 (—; 15) 101-109

In all races, sex differences significant in samples for which standard deviation known.

JUVENILE. Wing and tail both shorter than in adult, but no exact information.

Weights. *E. a. flava.* (1) Helgoland (West Germany), migrants (Krohn 1915; Weigold 1926); (2) East Germany, winter (Bub and Herroelen 1981); (3) USSR (Dementiev and Gladkov 1954*a*); (4) Yamal peninsula, summer (Danilov *et al.* 1984).

(1)   ♂ 39·0 (—; 10) 32-46 ♀ 32·7 (—; 5) 26-37
(2)   39·9 (2·12; 16) 36-45   36·9 (3·12; 21) 30-44
(3)   37·3 (—; 38) 31-43   36·5 (—; 19) 27-43
(4)   39·9 (—; 36) 34-48   36·8 (—; 15) 31-41

Lapland, July: ♂ 34·5 (3) 34·2-34·8, ♀ 32·3. Southern Norway, February: ♂♂ 45, 45·5. (Bub and Herroelen 1981.) Exhausted birds, Netherlands, winter: ♂ 26·0, ♀ 22·8 (ZMA).

E. a. brandti. Mongolia: (1) late April to June; (2) late July to mid-August (Piechocki and Bolod 1972). (3) USSR (Dementiev and Gladkov 1954*a*).

(1)   ♂ 31·6 (3·18; 16) 27-40 ♀ 29·6 (1·56; 12) 27-32
(2)   34·7 (2·34; 6) 32-38   —
(3)   32·7 (—; 5) 30-34   34·3 (—; 3) 31-36

E. a. balcanica. Greece, June: ♂ 40·5 (4) 40-41 (Niethammer 1943).

E. a. penicillata. Eastern Turkey: June, ♂ 41·8 (2·04; 6) 39-45, ♀ 40; November, ♂ 46 (Kumerloeve 1968, 1969*a*). Iran, April-July: ♂♂ 37·2, 37·8, 39·2; ♀♀ 32·4, 34·1 (Paludan 1938, 1940).

E. a. bicornis. Mt Hermon (Lebanon/Syria): ♂♂ 38, 39 (ZFMK).

E. a. albigula. Iran, May-June: ♂ 39 (3) 38-40 (Desfayes and Praz 1978). Afghanistan: (1) June and early July; (2) late September and early October (Paludan 1959).

(1)   ♂ 35·7 (10) 30-39 ♀ 34·6 (5) 30-40
(2)   38·9 (13) 29-42   36·4 (9) 34-42

For North American races, see Behle (1943) and Montagna (1943).

JUVENILE. For development of *balcanica*, see Pätzold (1981); for growth curves of North American races, see Drury (1961), Verbeek (1967), and Beason and Franks (1973).

Structure. Wing rather long, broad at base, tip rather rounded. 10 primaries: in *flava*, p8 longest, p9 (0-)1-3 shorter, p10 74-78 (♂) or 66-71 (♀), p7 1-2, p6 7-12, p5 17-23, p4 24-31, p1 39-44 (♂) or 34-41 (♀); in *penicillata*, *balcanica*, *albigula*, *bicornis*, and mountain populations of *brandti*, p8 longest, p9 1-3 shorter, p10 80-88 (♂) or 71-77 (♀), p7 0-2, p6 4-7, p5 13-18, p4 20-27, p1 35-42 (♂) or 31-39 (♀). Juvenile wing has more rounded tip than adult; p10 slightly longer and broader, less pointed, 18 mm long or $\frac{2}{3}$ of length of primary coverts (in adult, 9-11 mm long: Stresemann 1920). Outer web of p6-p8 and inner of (p6-)7-p9 emarginated. Longest tertials reach to about p4 in closed wing. Tail rather long, tip square or shallowly forked; 12 feathers, t1 2-7 shorter than t3-t4. Bill short, rather thick at base; culmen distinctly and cutting edges slightly decurved; bill relatively shorter, stouter, and deeper at base than in Skylark *Alauda arvensis*. Feathers at sides of crown elongated, tips pointed, forming 'horns'; longest in ♂, especially in *penicillata* and relatives; virtually absent in juveniles. Tarsus rather short, slender. Toes short and rather thick; middle toe with claw 17·6 (16) 17-19 in *flava*, up to 2 mm longer in larger races *penicillata*, *albigula*, etc; outer toe with claw *c.* 74% of middle toe with claw, inner *c.* 78%, hind *c.* 108%. Claws strong, almost straight; front claws rather short; hind claw long, 10·7 (23) 8-13 in *flava*, 11·8 (5) 9-15 in nominate *alpestris*, 14·2 (13) 10-19 in *penicillata* and *albigula*.

Geographical variation. Marked and complex. 2 main groups discernible: (1) *alpestris* group (northern Eurasia and the Americas) in which black of cheeks separated from black chest by white or yellow streak from throat up to sides of neck; (2) *penicillata* group (central Asia west to Balkans) in which broad black cheek-patch broadly connected with extensively black chest, isolating small pale throat-patch (though black connection at lower sides of neck sometimes narrow and indistinct in ♀). Intermediates between the groups occur in central and eastern Tien Shan mountains (USSR). Marked variation occurs within both groups, mainly involving colour of pale areas of face and throat (yellow or white), general colour and streaking of upperparts, colour of lesser upper wing-coverts, and size; in *penicillata* group, also variations in amount of white on forehead and in bill shape. Sexual dimorphism not equally marked in all populations; differences thought to reflect phylogeny (see Kozlova 1981). *Alpestris* group comprises *flava*, *brandti*, *atlas*, and (as a straggler from eastern Canada) nominate *alpestris*—also 1 extralimital race in Tsaidam basin (west-central China) and many American races (see, e.g. Oberholser 1902), not considered further here. *Penicillata* group comprises *balcanica*, *penicillata*, *bicornis*, and (just beyond west Palearctic border) *albigula*—also 6 races in central Asia, not further considered here. *E. a. flava* from northern Europe rather small, but size increases clinally towards eastern Siberia (see Measurements), though difference not large enough to warrant separation of eastern race *euroa* Thayer and Bangs, 1914; pale face-marks and throat distinctly bright yellow, crown tinged yellowish-olive (especially in ♀ and fresh-plumaged ♂), mantle and scapulars dark olive-grey with rather faint streaking (♂) or olive-brown with distinct streaks (♀). *E. a. brandti* from plains of southern USSR east to Mongolia and western Manchuria (China) similar in size to *flava*, except for populations of central and eastern Tien Shan, which are large and intermediate in plumage between *brandti* and *albigula*; pale face-marks and throat cream-white or white; crown, hindneck, and upper mantle pale vinous-pink; lower mantle scapulars, and tertials with narrow and indistinct olive-brown feather-centres and broad light grey fringes (in particular, autumn ♂ appears uniform pale grey

above apart from vinous-pink hindneck and upper mantle); narrow strip of black feathers, 1-2(-3) mm wide, along base of upper mandible; black of chest-band reaches higher up sides of neck, white stripe between chest and cheeks narrower; lesser upper wing-coverts pale vinous-pink; upperparts of ♀ buffish-grey with pink tinge on hindneck and more distinct brown streaks on crown, lower mantle, and scapulars. *E. a. atlas* from north-west Africa is close to *flava*, showing similar yellow face-marks and throat and with distinct greenish-yellow tinge on crown and hindneck in fresh plumage; mantle and scapulars of freshly-moulted ♂ dull grey with faint pink-buff tinge, heavily streaked grey-brown and dark brown when worn (contrasting with vinous-cinnamon crown and hindneck), ♀ always rather heavily streaked, but less brown than *flava*; lesser upper wing-coverts sandy-brown (not vinous); black horns longer than in *flava*. Nominate *alpestris* from eastern Canada hardly separable from *flava*, but lesser upper wing-coverts darker vinous-red, less pinkish. *E. a. penicillata* from Caucasus south to northern and south-west Iran, north-east Iraq, and eastern Turkey differs markedly from *flava* in large size and in extensive black on sides of head and neck (though black forecrown of ♀ often indistinct); upperparts rather similar to *brandti*, ♂ light grey with vinous-pink hindneck and upper mantle and with limited streaking on mantle and scapulars (vinous-pink slightly greyer and paler than in *brandti*); ♀ extensively buff-grey with more distinct dull black streaks; pale areas on face and throat pale yellow when fresh (distinctly paler than *flava*, but not white as in *brandti*), bleaching to yellowish-white or fully white in spring; black horns on sides of crown long and distinct; lesser upper wing-coverts light grey with slight pink tinge (pale vinous-pink in *brandti*). *E. a. albigula* from Gorgan area (north-east Iran) and southern Transcaspia (USSR) east to western Tien Shan, Pamirs, and north-west Himalayas is close to *penicillata* (and, apart from black sides of lower neck, also to *brandti*), but face-marks and throat white; upperparts of ♂ sandy-grey, not as pure grey as *brandti* and *penicillata*, hardly streaked (except when worn); upperparts of ♀ sandy-buff with more distinct streaks, hardly separable from *brandti* (but black on cheeks of *albigula* usually broader, often just meeting sides of chest, sometimes broadly meeting), or *penicillata* (but *albigula* less pure grey with narrower streaks and whiter chin). *E. a. balcanica* from Balkans and Greece similar to *penicillata* but pale face marks and chest slightly deeper yellow in fresh plumage (though not as deep as *flava*); upperparts and fringe of t1 slightly purer grey. As in *penicillata*, ♀ similar to ♂ but black patch on forecrown often indistinct, broken into spots, not sharply divided from black-streaked central crown; black of cheeks and chest duller, stripe on cheek narrower and partly mottled; crown, hindneck, mantle, and scapulars sandy-buff, heavily streaked dull black; crown and hindneck tinged yellow when fresh, not contrasting uniform pale vinous-pink as in ♂; in heavily worn plumage, both sexes heavily streaked on upperparts (but hindneck of ♂ still vinous) and pale areas of face and throat bleached to white, virtually indistinguishable from *penicillata* in worn plumage. *E. a. bicornis* of Lebanon area completely uniform vinous-buff on upperparts in fresh adult ♂ (no contrast between vinous-pink of hindneck and grey of mantle and scapulars, unlike *balcanica* and *penicillata*), upper tail-coverts and t1 pink-cinnamon or buff-cinnamon; pale areas of face and throat pale yellow, soon bleaching to white (intermediate in colour between *penicillata* and *albigula*); black band along base of upper mandible *c.* 2 mm wide; upperparts of ♀ sandy-buff with pink tinge, faintly streaked on mantle and scapulars (less streaked than *penicillata*); bill generally more slender than in *penicillata* and *albigula*.

Populations from Asia Minor problematical: those of southern Taurus mountains sometimes included in *bicornis* (Stresemann 1928; Hartert and Steinbacher 1932–8; Vaurie 1959; Kumerloeve 1961), or these and others considered identical to *penicillata* (Vaurie 1951*a*). Specimens examined (Ankara area and Taurus) belong to neither race: independent of wear, all ♂♂ from Caucasus streaked on mantle and scapulars, those from Asia Minor virtually unstreaked; general appearance in Asia Minor uniform greyish pink-buff, in Caucasus light buffish-grey with slightly darker feather-centres; ♀♀ from both Caucasus and Asia Minor are streaked above, but ground-colour in Caucasus greyer (and crown more yellow in fresh plumage),

in Turkey paler buffish-grey. *E. a. bicornis* similar to birds of Asia Minor in showing unstreaked upperparts (unless heavily worn), but colour of crown and hindneck similar to mantle, scapulars, and tertials (no contrast between vinous-pink hindneck and remainder of upperparts, unlike birds of Asia Minor and Caucasus); upper tail-coverts and t1 cinnamon (not pink-grey), pale areas of face and throat yellow (not white); upperparts of ♀ sandy-buff (not buffish-grey). Pending further research on birds in equally worn plumage, birds from Asia Minor included here in *penicillata*.

For relationship with Temminck's Horned Lark *E. bilopha*, see that species (p. 230). CSR

Eremophila bilopha Temminck's Horned Lark

PLATES 13 and 14
[facing pages 185 and 232]

Du. Temmincks Strandleeuwerik Fr. Alouette bilophe Ge. Hornlerche
Ru. Малый рогатый жаворонок Sp. Alondra cornuda de Temminck Sw. Ökenberglerche

Alauda bilopha Temminck, 1823

Monotypic

Field characters. 13–14 cm; wing-span 26–31 cm. 20% smaller and shorter-tailed than Shore Lark *E. alpestris*; close in size to Bar-tailed Desert Lark *Ammomanes cincturus*, but less upstanding. Rather small, delicate lark, sharing 'horns' and basic plumage pattern of *E. alpestris*, but has pale areas of face white, upperparts sandy-pink, and underparts buff-white. Juvenile lacks all facial pattern and shows faint pale spots above. Gait and flight even lighter than *E. alpestris*. Sexes similar; little seasonal variation. Juvenile separable.

ADULT MALE. Ground-colour of head and throat white, with black band on crown above eye, black base to upper mandible, black lores and mask-like patch under eye, narrow but conspicuous black band across chest, and pink-isabelline rear crown and nape. Head pattern thus even more contrasting than *E. alpestris*. Tiny black 'horns' on sides of crown, as in *E. alpestris*. Rest of upperparts pink-isabelline, appearing unmarked except at close range when slight mottling on back and wing-coverts and paler fringes to tertials may show; thus generally cleaner and brighter above than *E. alpestris*. Underparts essentially white, with faint sandy-pink flush on breast and along flanks. Tail pattern even more contrasting than in *E. alpestris*, with pale pink-isabelline of rump extending down centre, and dark underside contrasting with vent. In worn plumage, upperparts slightly more rufous, underparts often discoloured by soil. ADULT FEMALE. Closely resembles ♂ but black areas of head and chest often more restricted and duller and horns shorter.

JUVENILE. Lacking head pattern and having white-spotted upperparts, markedly dissimilar from adult and more reminiscent of other larks, especially *A. cincturus* and ♀ and immature Black-crowned Finch Lark *Eremopterix nigriceps*. Best distinguished by generic character, behaviour, and longitudinal 3-toned tail pattern of adult.

In Levant and across North Africa, may overlap with wandering *E. alpestris* but most birds of both species usually well divided by altitude and habitat. *E. bilopha* essentially a smaller, paler bird adapted to level steppe or desert regions, not undertaking any marked movements. Adults in any case unmistakable, with slightness, white face, and plumage uniformity all obvious to experienced observer. Juvenile much paler than *E. alpestris* and best distinguished from similarly coloured desert larks by tail pattern. Flight and gait recall *E. alpestris* but even more free. Escape-flight usually short and low. Song-flight not infrequent: involves rather weak, fluttering climb to medium height followed by steep descent.

Voice similar to *E. alpestris* but quieter.

Habitat. In lower middle latitudes in arid, warm, usually level lowlands, bare or sparsely vegetated, extending from Mediterranean to oceanic climate in Morocco. There common on plains covered with *Anabasis* (Chenopodiaceae), on stony patches in *Stipa* grass, in thin wispy grass on plateaux and on flat wormwood *Artemisia* steppe. In north-west Africa, only below 1000 m, whereas Shore Lark *E. alpestris* lives above 2000 m and is completely

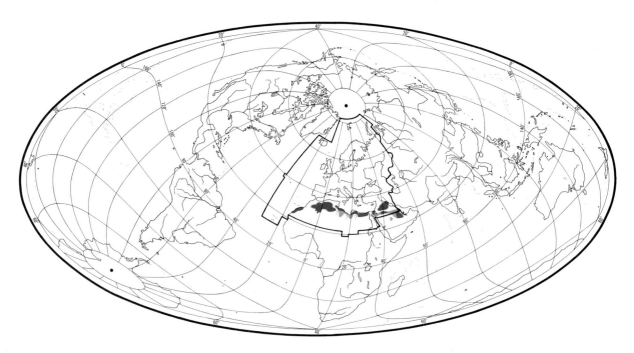

separated in habitat. Although occurring on stony plains and deserts, prefers stony plateaux and steppes with solid soil; in sandy areas stays on patches where soil compact, avoiding pure sand (Heim de Balsac and Mayaud 1962). Does not penetrate much into desert, although able to survive on completely bare plains where other birds infrequent. Especially associated, however, with slopes carrying low cover of succulent *Aizoon* (Valverde 1957). In Arabia, bird of absolute desert, often on pebble-strewn patches or bare sand; occasionally in thin grass desert (Meinertzhagen 1954). At Azraq (Jordan), prefers gravel hammada or bouldery tracts (Nelson 1973). Also at Azraq accepts only areas where shrublet cover low and thinly scattered (Hemsley and George 1966). Although not a typical desert bird, avoiding extensive or loose sand and being closely related to arctic-alpine *E. alpestris*, is associated rather with desert than steppe group of Alaudidae, and appears independent of supply of surface water. Apart from seasonal song-flights, largely a ground bird, rarely using raised perches and flying in low airspace, only as necessary. Has minimal contact with man.

Distribution. No range changes reported.

LIBYA. Distribution uncertain in south (Bundy 1976). Accidental. Malta.

Population. No information on population trends.

MOROCCO. Common in some areas in south (Valverde 1957). EGYPT. Locally not uncommon (SMG, PLM, WCM). JORDAN. Azraq: 3rd commonest lark (Wallace 1983a). IRAQ. Common in western deserts (Meinertzhagen 1924a).

Movements. Resident. Some dispersal occurs; this probably at individual rather than population level in North Africa, but may be a more regular feature in northern Arabia.

Balsac and Mayaud (1962) cited a few autumn-winter records to north and south of breeding range in Algeria, and a February occurrence at Port Etienne in Mauritania; occasionally recorded at Nouakchott in Mauritania, October and December (Gee 1984). Also, 4 records from Malta, October–March (Sultana and Gauci 1982). A scarce resident in north-central Saudi Arabia, but small flocks visit the region in winter, departing by March (Jennings 1980). In north-west and western Saudi Arabia, a winter visitor in large flocks to Tabuk ($28\frac{1}{2}°$N) and has occurred also at Nugrah ($25\frac{1}{2}°$N) and Jiddah ($21\frac{1}{2}°$N), including flock of 25 at the last (Jennings 1981a). However, no evidence yet of movement in Persian Gulf states, and not recorded in Bahrain, Qatar, or United Arab Emirates (Bundy and Warr 1980).   RH

Food. Seeds and occasional beetles (Coleoptera); ignores grasshoppers (Orthoptera) (Meinertzhagen 1954). 4 adults from southern Morocco, April–July, contained seeds, including those of *Aizoon* (Valverde 1957). One bird at Azraq (Jordan) seen feeding on fruits of *Ephedra transitoria* (Clarke 1980). Said also to eat plant shoots (Anon 1970b). Feeding flocks in Morocco keep tightly together, and birds seen turning over stones many times their own weight (Smith 1965). Birds shot in early morning, Iraq, had throat moist and crop full of water; had presumably been drinking dew or eating moisture-secreting plants (Meinertzhagen 1924a).   DJB

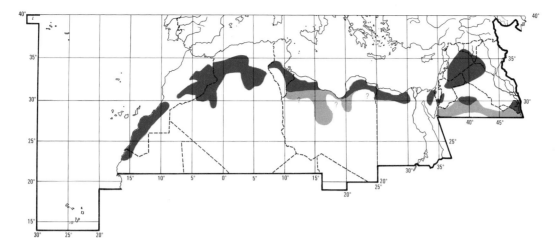

Social pattern and behaviour. Poorly known, but most aspects probably similar to Shore Lark *E. alpestris*.

1. Generally in small flocks outside breeding season: e.g. in Morocco, early February, average flock size 4 (1–7) (Smith 1962–4). Reports from Saudi Arabia mostly similar: see Meinertzhagen (1954) and Jennings (1980, 1981*b*). In Algerian Sahara, flocks loose knit and widely distributed; also occurs not uncommonly in twos (Niethammer 1954). In north-east Egypt, January, scattered in pairs (Meinertzhagen 1921). Larger flocks recorded in winter: in Saudi Arabia (see Movements), at Monts des Ksours in northern Algeria (Blondel 1962*a*), and (over 100) in east-central Morocco in November (Smith 1965*a*). In southern Morocco, June, flocks of 15–20 or more recorded, but small flocks the rule (Valverde 1957). Large summer flocks reported from Libya, frequently associating then with Bar-tailed Desert Lark *Ammomanes cincturus* (Whitaker 1902, 1905); recorded associating with other Alaudidae also in Algeria (Blondel 1962*a*). BONDS. No detailed information on mating system. Pair-bond apparently maintained throughout summer (Brosset 1961). In Algerian Sahara, late November, 2 ♂♂ recorded associating closely (Niethammer 1954); see also Stanford (1954) and part 2. No information on role of sexes in incubation and brooding, but both evidently feed nestlings (Shannon 1974) and accompany (presumably also feed) them for some time after they leave nest (Heim de Balsac 1924). In Jordan, pair fed well-grown young Lesser Short-toed Lark *Calandrella rufescens* (Shannon 1974). BREEDING DISPERSION. Probably solitary and said by Jennings (1981*b*) to be highly territorial. No information on territory size or its uses. In southern Morocco, dispersion apparently dependent on distribution of *Aizoon* (important food plant). Vehicle transects covering belt 30–50 m either side of road showed 9, 11, and 28 birds per km in pure *Aizoon*, 2 and 4 per km in adjoining region with *Salsola* and 'teza', and 0·6 per km in adjacent barren plains; area with some *Salsola* and grasses, but with clumps of *Aizoon*, held about 1·5 birds per km over *c.* 70 km (Valverde 1957). In steppe south of Laghouat (Algeria), early April, over 50 along 16 km (Haas 1969). At Azraq (Jordan), walked transects produced *c.* 4 birds per km in wadi spreads and *c.* 3 birds per km on hammada (Wallace 1983*a*). ROOSTING. No information.

2. Usually tame, e.g. allowing approach to *c.* 10 m (Bannerman 1927). See also (e.g.) Hartert (1915), Heim de Balsac (1924), and Smith (1962–4). According to Chabot (1932), flies only if pushed to limit and often runs about rather mouse-like, close to observer; however, such behaviour possibly typical of birds with eggs or young. Other reports refer to tameness in still conditions and wildness in wind (Meinertzhagen 1954) and to great shyness of winter flocks (Erlanger 1899). According to Whitaker (1905), calls while making typical curving ascent and (slightly differently) when descending (see 5 in Voice). FLOCK BEHAVIOUR. Feeding flocks often tightly packed and swift-running (Smith 1962–4, 1965*a*). SONG-DISPLAY. Song of ♂ (see 1 in Voice) given from ground and in the air. According to Shannon (1974), aerial component more marked than in *E. alpestris*; at Azraq, however, brief song given mainly from ground (Nelson 1973). Song-flight similar to *E. alpestris* (Anon 1970*b*): ascent rather feeble and never to any great height (Meinertzhagen 1954; Etchécopar and Hüe 1967). In Jordan, bird ascended to *c.* 25 m and gave short phrase with wings outspread and stationary; fluttered in between. Song-flight ended in plummet dive (P A D Hollom; *contra* Meinertzhagen 1954 and Etchécopar and Hüe 1967). In southern Morocco, early June, birds ascended only to *c.* 2–4 m; dropped back almost vertically to starting point or slightly ahead, against wind (Valverde 1957). ANTAGONISTIC BEHAVIOUR. No information on intraspecific antagonism. Pair sharing feeding of young *C. rufescens* (see Bonds, above) attempted to drive off parents (Shannon 1974). HETEROSEXUAL BEHAVIOUR. (1) Pair-bonding behaviour. Clearly similar to *E. alpestris*. ♂ adopts Courtship-posture with head up, 'horns' raised, feathers ruffled, tail spread, and wings drooped. Dances with rapid, light steps in front of ♀ (Koenig 1895; Meinertzhagen 1935). In Algerian Sahara, late November, one of 2 ♂♂ sang quietly at the other, horns raised; performed Courtship-display. Birds moved rapidly and with light steps around each other. Both were shot, and 1 had gonads more developed than other (Niethammer 1954). (2) Behaviour at nest. Incubating bird very agitated in dust-storm; repeatedly left nest or moved about on eggs and ruffled feathers (Shannon 1974). No further information. RELATIONS WITHIN FAMILY GROUP. At exposed nest in Jordan, one adult remained constantly to shade young from sun. Both adults recorded at nest together (Shannon 1974), but not known if this regular. In central Arabia, young recorded leaving nest at *c.* 16–17 days old (Jennings 1980); possibly not typical (compare *E. alpestris*, p. 218). ANTI-PREDATOR RESPONSES OF YOUNG. No information. PARENTAL ANTI-PREDATOR STRATEGIES. Pair recorded running round observer, picking up and dropping feathers, but no nest found (Stanford 1954). Incubating bird flushed only when man

very near nest, also returned when nest being examined (Koenig 1895). Bird shading young on nest (see above) also allowed very close approach (Shannon 1974). MGW

Voice. Overall, similar to Shore Lark *E. alpestris* (Smith 1965a; J-C Roché), though quieter (D I M Wallace). Further study required to test validity of claim by (e.g.) C Chappuis that voice indistinguishable in all respects from *E. alpestris*. The following scheme provisional.

CALLS OF ADULTS. (1) Song of ♂. Disconnected bursts of soft, melodious twittering or quiet, very fine warbling (Koenig 1895; Hartert 1915; Meinertzhagen 1935). Bright and pleasant sounding (Whitaker 1905), though rather thin, at least when given in flight (P A D Hollom). Less vigorous than *E. alpestris* (Shannon 1974). Rendering 'chiri-bí chiri-bí' (Valverde 1957) presumably an attempt to describe only part of a phrase. In recording by H Roché, song starts with series of calls (probably call 2 or 3). Mostly a somewhat monotonous repetition (with some variation) of short phrases: 'dee dee-eeee', 'chep seee-eee', or 'chep-ep seeee'. Thin, drawn-out whistles, at times quite musical, alternate with richer, more twittering or chirruping sounds. Warbling at times not unlike mournful Robin *Erithacus rubecula*, but thin whistles distinct (M G Wilson). Dry rattles closely resemble those given by *E. alpestris* (see Fig V of that species, p. 220; J Hall-Craggs). Recording (Figs I-II) reveals melodically very varied song; phrases less than 2 s up to more than 4·5 s. Delivery rate comparable to that of song variant 1b in *E. alpestris* (J Hall-Craggs). Rather slow, measured (almost stuttered) delivery of shorter sounds and long whistles (some slightly rougher or more

nasal). Much variation in duration of units, some also in timbre: 'chep-chep-chew-eeee', etc. Call 2 perhaps also woven into song (M G Wilson). (2) Quiet 'see(y)oo', metallic 'chee-u' or 'chee-yoo' with variations (Smith 1962-4, 1965a); sometimes apparently more trisyllabic 'sweee-teee-ooo' while running about on ground (M G Wilson). Possibly the same or related sounds in song include musical 'SEEEEu teeU see' (Fig III) and 'TIU see' followed by a warble (J Hall-Craggs: Fig IV). Soft nasal 'tzew' when flushed (P A D Hollom) also presumably the same, as is perhaps the short whistle, more drawn-out and slightly lower-pitched than Bar-tailed Desert Lark *Ammomanes cincturus*, mentioned by Hartert (1915). (3) Undistinguished, quite loud 'tsip' or 'sweeeep' like wagtail *Motacilla* (Nelson 1973; M G Wilson). In recording, such sounds given quietly by pairs moving about on ground, once a slightly louder 'shree-ep' (M G Wilson: Fig V). A 'chizz' given by flock members as well as call 2 (Smith 1962-4) probably the same. (4) Oft-repeated, weak 'tri' given in flight (Meinertzhagen 1930); 'tiri-tiri-tiri-tiri' as contact-call by ♀ (Koenig 1895) perhaps related. (5) Other calls. Short, sharp, and more prolonged sounds given in flight (Whitaker 1905) unlikely to be distinct.

CALLS OF YOUNG. No information. MGW

Breeding. SEASON. Algeria: eggs laid April-May. Western Morocco: eggs laid mid-February to April (Heim de Balsac and Mayaud 1962). Jordan: eggs and young found late April and early May; fledged young found end of April and early May (Wallace 1983a). Eastern Saudi Arabia: nest seen March (R Raby). SITE. On

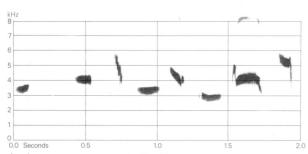

I C Chappuis Morocco April 1966

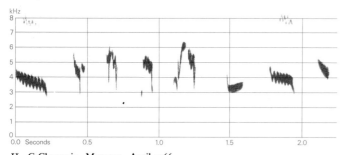

II C Chappuis Morocco April 1966

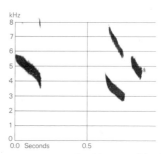

III Roché (1968) Morocco March 1966

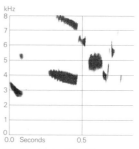

IV Roché (1968) Morocco March 1966

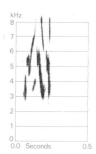

V P A D Hollom Morocco March 1978

ground in the open, or in shelter of tussock. Nest: shallow depression lined with grass, twigs, and rootlets, with inner lining of soft grass-heads; mud lining with rag and wool also recorded (Jennings 1980); usually a rampart of small stones. One nest, Algeria: circumference 37 cm, external diameter 11 cm, internal diameter 7 cm, height 4 cm, depth of cup 2·5 cm (Koenig 1895). Building: no information on role of sexes. EGGS. See Plate 80. Sub-elliptical, smooth and glossy; virtually indistinguishable from Shore Lark *E. alpestris*. 22·0 × 15·3 (19·7-24·7 × 14·8-16·4), $n=40$; calculated weight 2·65 g (Schönwetter 1979). Clutch: 2-4. Of 17 clutches, North Africa: 2 eggs, 4; 3, 11; 4, 2; mean 2·88 (Heim de Balsac and Mayaud 1962). YOUNG. One parent sheltered young from hot sun (Shannon 1974). FLEDGING TO MATURITY. Recorded leaving nest at *c.* 16-17 days (see Social Pattern and Behaviour). No further information.

Plumages. ADULT MALE. Feathers at base of upper mandible and broad band on lores and through and below eye black, ending in large rounded blob on lower ear-coverts and hindcheeks; black on both sides of upper mandible sometimes narrowly meets on top. Forehead and broad supercilium white (sometimes faintly tinged pink or cream in fresh plumage), extending over ear-coverts to sides of neck. Rear of forehead and front of crown black (feathers narrowly and faintly tipped pink-cream when fresh), extending into black elongated 'horns' at sides of crown. Remainder of crown, hindneck, lower sides of neck, mantle, scapulars, and back to upper tail-coverts bright rufous-cinnamon, feathers broadly fringed pink-grey or pale vinous-grey; vinous-pink cast very distinct in fresh plumage (more so than in, e.g. Bar-tailed Desert Lark *Ammomanes cincturus* which otherwise has rather similarly coloured upperparts), slightly less so when worn. Upper wing-coverts, tertials, and central pair of tail-feathers (t1) bright rufous-cinnamon, feathers narrowly tipped cream or white when fresh, fringes bleached to off-white when worn; wing-coverts and t1 with vinous-pink tinge, tertials often distinctly rufous, more so than adjacent vinous-pink scapulars and back; longest feather of bastard wing, basal and middle portion of inner web of longest tertials, and centre of terminal part of t1 dark grey or greyish-black. Feathers at base of lower mandible, chin, and throat white; chest black, extending to sides of upper breast, separated from black lower cheek by white band. Sides of breast and flanks pink-cinnamon, flanks partly streaked white; remainder of underparts white, sometimes tinged grey or rufous by soil. T2-t4 black with partial and narrow pale buff or off-white outer edge, t5 black with slightly broader white outer edge, t6 black (dark grey towards tip) with white outer web and narrow white tip. Flight-feathers dark grey, slightly darker towards tips, basal and middle portions of outer webs of secondaries extensively washed pink-cinnamon, primaries with less extensive cinnamon wash near base of outer webs; distal ⅓ of secondaries and of p1-p4 fringed white, outer edge and tip of p5-p8 washed cinnamon, outer web of p9-p10 largely white. Under wing-coverts and axillaries white, small coverts along leading edge of wing pink-cinnamon. In worn plumage, upperparts slightly more rufous, less vinous-pink, contrasting less with rufous tertials; tail and flight-feathers browner, less black; white band between black cheeks and sides of chest narrower, black sometimes almost continuous; underparts dirty white, often coloured by soil, some black of feather-bases occasionally

visible. ADULT FEMALE. Closely similar to ♂ and often indistinguishable (unlike Shore Lark *E. alpestris*); black band across rear of forehead and forecrown narrower on average, often mottled pink-cinnamon, especially on forecrown; 'horns' on average shorter, sometimes not fully black, but partly cream-buff. Black band from side of upper mandible to rear of cheek and black band across chest on average narrower and less sharply defined, sometimes tinged brown and partly mottled grey-brown and white (perhaps especially in 1st adult). NESTLING. No information. JUVENILE. Upperparts, including upper wing-coverts, tertials, and t1 sandy-cinnamon, feathers with indistinctly defined whitish tips, appearing more yellowish than adult, less distinctly vinous-pink, and without black face and chest marks of adult. Forehead, narrow supercilium, and narrow eye-ring isabelline or cream-white, indistinctly defined from sandy-cinnamon crown and ear-coverts and from cream-buff cheeks and sides of neck. Lower cheeks faintly spotted dusky grey. Chest and sides of breast sandy-cinnamon or cream-buff, some dark grey spots on sides of chest hardly visible; chin, throat, and underparts down from breast and flanks pale cream-yellow or off-white. Tail as adult, but dark grey-brown rather than black, pale and narrow outer edges less distinctly defined, outer web of t6 more extensively isabelline on tip. Outer webs of flight-feathers pink-cinnamon, inner webs washed grey, much less black than in adult; broad fringes at tips of primaries buff or isabelline, subterminally bordered by some dusky grey mottling; p9 grey-buff with cream-white outer web and tip (in adult, greyish-black with white outer web which is reduced to narrow white outer edge near tip). P10 somewhat broader and with more rounded tip than adult, but reduced and difficult to find at all ages. FIRST ADULT. Like adult; indistinguishable when last juvenile feathers (usually p9-p10) lost. Some variation in colour of fringes of secondaries: birds with fully white fringe along distal end of secondaries and inner primaries perhaps adult, those with fringe partly cinnamon and white of tip less sharply defined from dark grey centre perhaps 1st adult (wing of latter also relatively more worn), as in *E. alpestris*, but no known-age specimens examined.

Bare parts. ADULT. Iris hazel, brown, or dark brown. Bill dark grey-horn, blue-horn, or greyish-black, base of lower mandible paler, greenish-grey, blue-grey, or blue-flesh. Mouth slate-blue. Leg and foot dull grey, flesh-grey, or dark purplish-brown; claws dark horn. NESTLING. No information. JUVENILE. Iris hazel or brown. Bill pale pink-horn or yellow-horn with darker terminal cutting edges and tip; gape and mouth yellow. Leg and foot horn-grey, joints slate-grey, soles yellow-horn, claws dusky horn. (Stanford 1954; BMNH, ZMA.)

Moults. ADULT POST-BREEDING. Only a few in moult examined, all from Algeria and Morocco. 15 birds from April fairly worn, 6 from May heavily worn, not yet moulting. 3 birds from 5 June had shed 1-2 inner primaries (moult scores 1-3), head and body worn except for some growing feathers on upper back; one from 26 June with some scattered growing feathers on body and many new median upper wing-coverts and tertials, primary moult score 16, head and tail very worn; one from 18 July had about half plumage of head, body, and upperwing new or growing, remainder (including tail) heavily worn, moult score 17. None in more advanced moult seen; moult probably completed late August to mid-September, as 2 birds from late September and 17 from October had all plumage new. In Mauritania, 1 adult had started moult on 10 June (Stresemann 1926). POST-JUVENILE. Complete; primaries descendant. Only a very few examined. 4 from late April and May of year of hatching had not yet started flight-feather moult; head and body juvenile, except for some growing feathers on mantle, scapulars, and sides of breast. One from August had head, body, and tail new, but outer 5 primaries juvenile, moult score 21. In Mauritania, 2 had started moult on 8 and 10 April; 1 nearing completion 13 April (Stresemann 1926). None of those examined late September and October retained any juvenile feathers.

Measurements. ADULT. (1) Algeria and Tunisia, (2) Libya and Egypt, (3) Lebanon and Syria, (4) Jordan, Iraq, and Arabia, (5) all areas combined, all year; skins (BMNH, RMNH, ZMA). Bill (S) to skull, bill (N) to distal corner of nostril; exposed culmen on average 2·7 less than bill (S).

WING(1)	♂ 100·2 (2·38; 21)	97–104	♀ 92·4 (2·36; 9)	88–95
(2)	98·1 (1·54; 7)	96–100	90·9 (1·52; 5)	89–93
(3)	101·5 (2·85; 10)	98–106	93·2 (1·92; 5)	91–96
(4)	100·3 (2·25; 12)	97–104	93·8 (1·44; 5)	92–96
(5)	100·2 (2·50; 50)	96–106	92·5 (2·09; 24)	89–96
TAIL	65·6 (2·14; 18)	62–69	60·0 (2·50; 9)	57–64
BILL (S)	15·1 (0·58; 17)	14·2–16·1	13·9 (0·82; 9)	12·7–14·9
BILL (N)	9·7 (0·42; 10)	9·1–10·2	9·2 (—; 3)	8·7–9·7
TARSUS	21·2 (0·69; 16)	20·0–22·1	20·3 (1·02; 10)	18·6–21·6

Sex differences significant, except bill (N).

JUVENILE. Wing and tail on average c. 5 shorter than adult; bill and tarsus similar when post-juvenile moult completed.

Weights. Algerian Sahara, November: ♂♂ 38, 39 (ZFMK).

Structure. Wing rather short and broad, tip rounded. 10 primaries: p8 longest, p7 and p9 both 0–1 shorter, p10 64–72 shorter, p6 4–6, p5 12–16, p4 20–24, p1 29–36. P10 tiny in adult, narrow and pointed, hidden under primary coverts; 10·8 (9) 9–13 shorter than longest greater upper primary covert; in juvenile, slightly broader and with more rounded tip, 54–59 shorter than longest primary (p8), 1–4 shorter than primary coverts. Length of 'horns' on sides of crown 16–24 in ♂, 10–19 in ♀. Bill rather slender and sharp, longer than *E. alpestris flava*, more slender

than *E. alpestris penicillata*. Middle toe with claw 14·3 (10) 13·4–15·9; outer toe with claw c. 76% of middle with claw; inner c. 78%, hind c. 95%; length of hind claw 7·0 (12) 5·5–8·2.

Geographical variation. Slight, involving size only. Birds from Mauritania on average slightly smaller than those from Algeria and Tunisia, wing of Mauritanian ♂ 94·7 (3) 93–96, ♀ 90·3 (3) 88–92, and these birds sometimes separated as *elegans* Stresemann, 1926. Sample small, however, and birds from Libya and Egypt also rather small in size (see Measurements); hence *elegans* not recognized.

Sometimes considered a race of *E. alpestris*; though strongly different from some southern races of *E. alpestris* (*atlas* from Morocco and *penicillata* from Caucasus region), differences bridged by *E. a. bicornis* of Lebanon mountains. *E. a. bicornis* has upperparts closely similar to *E. bilopha* (vinous-rufous, not greyish as in other races of *E. alpestris*), but differs as follows: upper wing-coverts and tertials olive-brown (rufous in *E. bilopha*); face marks and chin pale yellow, bleaching to white when worn (not as yellow as *E. a. penicillata*, not as white as *E. bilopha*); black patch on lower cheeks connected with black of chest (not separated by white band as in *E. bilopha*, but several other races of *E. alpestris* also have separated patches); small pale throat patch; much longer wing (105–120, only 2 ♀♀ below 107; in *E. bilopha* 88–106, only 2 ♂♂ from Syria over 104). In addition, *E. bilopha* differs from all races of *E. alpestris* in ecology, largely uniform juvenile plumage, and absence of pronounced sexual dimorphism (for instance, ♀ of *E. a. bicornis* differs from ♂ by narrower and more mottled black head markings, black of cheeks often separated from that of chest, and upperparts distinctly marked with narrow black-brown shaft-streaks); hence, considered separate species, in accordance with Stresemann (1926) and Vaurie (1959). However, some hybridization may occur, as one ♀ near Damascus (Syria), March, had size and most colour characters similar to ♀ *E. a. bicornis*, but sharply black head pattern and unstreaked upperparts like ♀ *E. bilopha* (BMNH). CSR

Family HIRUNDINIDAE swallows

Highly specialized group of small oscine passerines (suborder Passeres); flight-feeders on flying and wind-borne invertebrates, spending much time on the wing. 70–80 species in 16–19 genera, of which the following occur in Old World: (1) *Pseudochelidon* (river martins), 2 species of very limited distribution—Zaïre and Thailand; (2) *Riparia* (sand martins), 4 species—one species (Sand Martin *R. riparia*) widely in Holarctic, others African and Asian; (3) *Cheramoeca* (White-backed Swallow *C. leucosterna*), monotypic—Australia; (4) *Pseudhirundo* (Grey-rumped Swallow *P. griseopyga*), monotypic—Afrotropics; (5) *Phedina* (Malagasy martins), 2 species—Malagasy region, Afrotropics; (6) *Psalidoprocne* (rough-winged swallows), 9–13 species—Afrotropics; (7) *Ptyonoprogne* (crag martins), 3 species—southern Europe, Africa, and Asia; (8) *Hirundo* (typical swallows), c. 30 species (including those previously placed in *Cecropis*

and *Petrochelidon*)—Old and New Worlds but with most in Africa and rest of Old World; (9) *Delichon* (house martins), 3 species—Europe, Africa, and Asia. *Pseudochelidon* rather aberrant and sometimes placed in separate subfamily or even family; further subdivision of rest of family mainly based on type of nest, which varies from simple cup in hole to complicated, retort-shaped construction of mud (Mayr and Bond 1943; Brooke 1972, 1974; Voous 1977). Cosmopolitan but absent from polar regions and most oceanic islands. Many highly migratory. Family represented by 7 species in west Palearctic, all breeding.

Differ from other Passeres in distinctive syrinx with more or less complete bronchial rings (in contrast with half rings bearing a membrane across inner face found in other oscines). Body slender, neck short. Sexes similar in size. Bill very short and flattened and mouth wide; a few weak rictal bristles present in some species. Wing very long and pointed; 10 primaries but p10 extremely reduced. Flight light and graceful, with high manoeuvrability. Tail often forked, outer feathers greatly elongated in some species; 12 feathers. Legs and toes short and slender, but claws strong; front toes more or less united at base. Tarsus sharply ridged at rear (acutiplantar). Gait a shuffling, mincing walk or brief run with aid of wings at times. Head-scratching by indirect method. Bathe by dipping in flight in series of brief splashes without landing on surface. Drink from flight and, less often, by sipping from puddles (etc.). Sunning common, sometimes with one wing lifted. Dusting reported but confirmation needed. Anting not recorded.

Plumage compact. Usually dark (black, blue, or brown), often with metallic sheen and contrasting underparts (white, chestnut, or greyish) and/or rump (white, buff, or rufous); underparts sometimes with chest-band, tail sometimes with contrasting spots or patches. Tarsus and toes sometimes partly or fully feathered. Sexes usually alike or nearly so. Typically a single complete moult in non-breeding season (all ages), but some species have partial 2nd moult which is inconspicuous as no change in appearance is involved (except in House Martin *D. urbica*). Nestlings have rather scanty 1st down soon after hatching, but denser 2nd down develops after *c.* 1 week.

Relationships to other Passeres often considered uncertain (e.g. Sibley 1970), and hence family often now usually placed at beginning of sequence. DNA data, however, suggest affinities lie with Old World warblers (Sylviidae) and allies (see Sibley and Ahlquist 1982; also Passeriformes, p. 42).

Riparia paludicola Brown-throated Sand Martin

PLATES 15 and 18
[between pages 232 and 233]

Du. Vale Oeverzwaluw   Fr. Hirondelle de la Mauritanie   Ge. Braunkehluferschwalbe
Ru. Малая ласточка   Sp. Avión zapador africano   Sw. Brunstrupig backsvala

Hirundo paludicola Vieillot, 1817

Polytypic. *R. p. mauritanica* (Meade-Waldo, 1901), north-west Morocco. Extralimital: *minor* (Cabanis, 1850), from Eritrea and Sudan west to Sénégal, north along Nile to Wadi Halfa; nominate *paludicola* (Vieillot, 1817), southern Africa; *chinensis* (J E Gray, 1830), southern Asia; 3-4 further races in Afrotropics. Not known to what race stragglers to Middle East belong; probably *minor*, but *chinensis* not impossible.

Field characters. 12 cm. wing-span 26-27 cm. Fractionally smaller than Sand Martin *R. riparia*. Smallest hirundine of west Palearctic with character and behaviour of *R. riparia*, but rather more compact form. Both upperparts and underparts duller than *R. riparia*, contrasting less; no chest-band. Sexes similar; no seasonal variation. Juvenile separable at close range. 2 races in west Palearctic; north-west Moroccan race, *mauritanica*, described here (see also Geographical Variation).

ADULT. Upperparts basically dull mouse-brown, lacking sandy wash of *R. riparia*. Underparts from chin to flanks and fore-belly pale mouse-brown fading to dull white on rear belly and vent; rather variable (at all ages), however, some birds appearing dark-chested and a few lacking white vent. Underwing dusky brown, contrasting with rear underbody. Bill and legs black. JUVENILE. Resembles adult but, at close range, inner wing and rump feathers show dark buff margins and tips. Bill black, legs flesh-brown.

Instantly separable from *R. riparia* when underparts visible, but liable to confusion with *Ptyonoprogne* martins when small size and uniformly dark tail not apparent. Crag Martin *P. rupestris* is 20% larger (with greater bulk obvious in head, body girth, and wing and tail width) and shows contrasting dusky black under wing-coverts and bold white spots on spread tail; other differences include pale speckled throat emphasizing head-cap, and pale buff fore-body but darker (buff) vent. African Rock Martin *P. fuligula* is only 5-10% larger (with slightly greater bulk obvious only in head and body) and shows distinctly greyer (less brown) plumage and, like *P. rupestris*, white spots on spread tail; other differences include uniform buff-white throat and fore-body and darker underwing. Flight less fast and agile than *R. riparia*; action more fluttering and stiffer-winged, at times recalling small bat (A M Allport); lacks obvious momentum of *P. rupestris* and dash of *P. fuligula*. Gait restricted, as in *R. riparia*. Behaviour as *R. riparia* but less sociable.

Song twittering. Call disyllabic, recalling *R. riparia* but clearer toned—'sree-sree'.

Habitat. Replaces Sand Martin *R. riparia* in low latitudes near southern limits of west Palearctic. Consequently tolerates much warmer climates, but exposed to oceanic influences in Moroccan breeding area which reaches up to Mediterranean climatic zone. No significant differences in habitat from *R. riparia*.

Distribution and population. Restricted to Morocco. No information on numbers or trends. Extension of breeding range near Rabat (*Br. Birds* 1986, **79**, 289).

Accidental. Israel, Saudi Arabia.

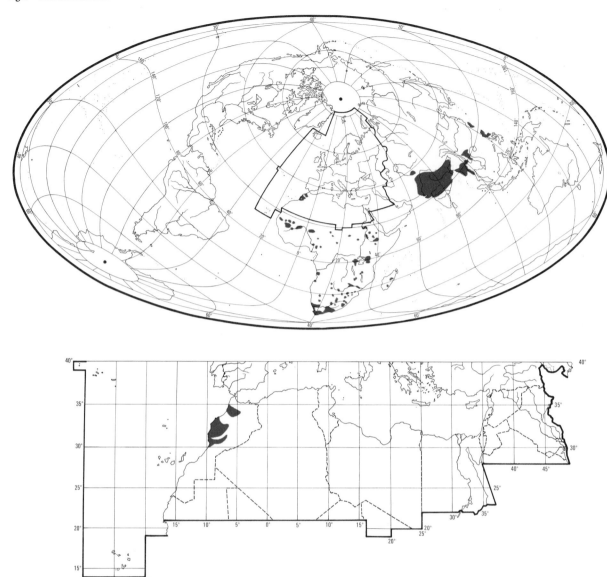

Movements. Moroccan population sedentary, breeding during winter (Etchécopar and Hüe 1967). In East and South Africa, at least partially migratory though such movements not properly understood (Mackworth-Praed and Grant 1963; Britton 1980); in Sudan, apparently moves south after breeding in north of country(Cave and Macdonald 1955).

Resident in Pakistan, Bangladesh, and India, though subject there to local movements, and (as in Morocco) breeds in winter months (Ali and Ripley 1972). Isolated population in south-central USSR migratory, with autumn exodus in progress in August and colonies reoccupied by beginning of June (Dementiev and Gladkov 1954b). Vagrants of unknown origin have occurred within west Palearctic in Israel and at Tabuk (north-west Saudi Arabia) in February (Jennings 1981a), and outside it at Abqaiq (eastern Saudi Arabia) in May (G K Brown) and in Oman September–October (Gallagher and Woodcock 1980).   RH

Food. No information from west Palearctic. Elsewhere, diet consists of small flying insects. Takes prey in flight,

PLATE 15. *Riparia paludicola mauritanica* Brown-throated Sand Martin (p. 231): **1** ad, **2** juv. *Riparia riparia* Sand Martin (p. 235). Nominate *riparia*: **3** ad, **4** juv. *R. r. diluta*: **5** ad. *R. r. shelleyi*: **6** ad. *Delichon urbica* House Martin (p. 287): **7** ad breeding, **8** ad non-breeding, **9** juv. (LT)

PLATE 16. *Ptyonoprogne fuligula* African Rock Martin (p. 248). *P. f. obsoleta*: **1** ad, **2** juv. *P. f. spatzi*: **3** ad. *P. f. perpallida*: **4** ad. *Ptyonoprogne rupestris* Crag Martin (p. 254): **5** ad fresh (autumn), **6** ad worn (spring), **7** juv. (LT)

PLATE 17. *Hirundo rustica* Swallow (p. 262). Nominate *rustica*: **1** ad ♂ fresh (autumn), **2** ad ♂ worn (spring), **3** ad ♀ fresh (autumn), **4** juv. *H. r. savignii*: **5** ad fresh (autumn). *H. r. transitiva*: **6** ad. *Hirundo daurica* Red-rumped Swallow (p. 278). *H. d. rufula*: **7** ad fresh (autumn), **8** ad worn (spring), **9** juv. Nominate *daurica*: **10** ad fresh (autumn). (LT)

PLATE 18. *Riparia paludicola mauritanica* Brown-throated Sand Martin (p. 231): **1-2** ad. *Riparia riparia riparia* Sand Martin (p. 235): **3-4** ad. *Ptyonoprogne fuligula obsoleta* African Rock Martin (p. 248): **5-6** ad. *Ptyonoprogne rupestris* Crag Martin (p. 254): **7-8** ad. *Hirundo rustica rustica* Swallow (p. 262): **9-10** ad. *Hirundo daurica rufula* Red-rumped Swallow (p. 278): **11-12** ad. *Delichon urbica* House Martin (p. 287): **13-14** ad breeding. (LT)

PLATE 21. *Anthus spinoletta* Rock Pipit and Water Pipit (p. 393). *A. s. petrosus* (Rock Pipit): **1** ad breeding fresh, **2** ad breeding worn, **3** ad non-breeding fresh (autumn), **4** juv. *A. s. kleinschmidti* (Rock Pipit): **5** ad breeding fresh. *A. s. littoralis* (Rock Pipit): **6** ad breeding fresh, **7** ad non-breeding fresh (autumn). *A. s. rubescens* (Water Pipit): **8** ad breeding fresh. (DIMW)

PLATE 22. *Anthus spinoletta* Water Pipit (p. 393). Nominate *spinoletta*: **1** ad breeding fresh, **2** ad breeding worn, **3** ad non-breeding fresh (autumn), **4** juv. *A. s. coutellii*: **5** ad breeding fresh, **6** ad non-breeding worn (winter). *A. s. rubescens*: **7** ad non-breeding fresh (autumn). (DIMW)

PLATE 23. *Anthus hodgsoni yunnanensis* Olive-backed Pipit (p. 337): **1** ad worn (spring), **2** 1st ad non-breeding fresh (1st autumn) with strong head pattern, **3** 1st ad non-breeding fresh (1st autumn) with weak head pattern, **4** juv.   *Anthus trivialis* Tree Pipit (p. 344): **5** ad fresh (autumn) brown type, **6** ad fresh (autumn) olive type, **7** ad worn (summer), **8** ad with strong head pattern, **9** juv. (DIMW)

PLATE 24. *Anthus berthelotii* Berthelot's Pipit (p. 327). Nominate *berthelotii*: **1** ad fresh (autumn), **2** ad worn (spring), **3** juv.   *Anthus pratensis* Meadow Pipit (p. 365). Nominate *pratensis*: **4** ad fresh (autumn and spring, olive type), **5** ad fresh (autumn and spring, brown type), **6** ad worn (winter and summer, olive type), **7** juv. *A. p. whistleri*: **8** ad fresh (autumn and spring). (DIMW)

PLATE 25. *Anthus gustavi* Pechora Pipit (p. 359): **1** ad breeding, **2** ad non-breeding fresh (autumn), **3** juv. *Anthus cervinus* Red-throated Pipit (p. 380): **4** ad ♂ breeding (dark type), **5** ad ♂ breeding (typical), **6** ad ♀ breeding, **7** ad non-breeding, **8** 1st ad ♂ breeding (1st summer), **9** juv. (DIMW)

PLATE 26. *Motacilla citreola* Citrine Wagtail (p. 433). Nominate *citreola*: **1** ad ♂ breeding, **2** ad ♀ breeding, **3** ad ♂ non-breeding, **4** 1st ad ♂ and ♀ non-breeding (1st winter). *M. c. werae*: **5** ad ♂ breeding. *Motacilla cinerea* Grey Wagtail (p. 442). Nominate *cinerea*: **6** ad ♂ breeding, **7** ad ♀ breeding, **8** ad ♂ non-breeding, **9** juv. *M. c. canariensis*: **10** ad ♂ breeding. *M. c. schmitzi*: **11** ad ♂ breeding. (NA)

often over water; often breaks surface in dipping to catch low-flying insects (Mackworth-Praed and Grant 1963; Clancey 1964). In southern Asia, regularly visits grass fires (Bilby 1957). At Simla (India), takes flies, mosquitoes, and other winged insects (Jones 1947-8). Several stomachs from Zaïre, June, contained small beetles and hard parts of similar insects (Chapin 1953). In South Africa, bird recorded eating grasshopper, retrieving it from water surface when dropped (Taylor 1942). DJB

Social pattern and behaviour. Little known about *mauritanica*, and following account based largely on extralimital *chinensis* and nominate *paludicola*.

1. Highly gregarious at all times of the year. Only information on flock size for southern Africa where often in large assemblages, exceeding 50 birds (Brooke 1971). Frequently associates with other Hirundinidae: e.g. in late January, Malawi, c. 300 feeding with more than 3000 Sand Martins *R. riparia* (Long 1959); in November, Malawi, 100 sitting on reeds with more than 2000 *R. riparia* (Hanmer 1977). BONDS. No information on mating system. Both members of pair help to excavate nest-tunnel (Oo-u-kijo 1936). 2 ♀♀ sometimes lay in same nest (Dementiev and Gladkov 1954b). Young dependent on parents for some time after fledging (Oo-u-kijo 1936). BREEDING DISPERSION. Solitary or in small colonies (Heim de Balsac and Mayaud 1962). In South Africa, usually colonial but solitary nesting not uncommon (Clancey 1964). In west Palearctic, maximum size of colonies apparently less than elsewhere: e.g. at Massa (Morocco), groups of 3-6 pairs dispersed along c. 2-3 km of river (U Hirsch). In southern Africa, usually c. 6-12 pairs (Vincent 1946-9) but up to 500 birds reported (Mackworth-Praed and Grant 1963; McLachlan and Liversidge 1970). One colony, Cape Province, contained 137 nests (W R J Dean). In USSR, nests in large colonies, sometimes with *R. riparia* (Portenko 1954). Within colonies, nests close together (McLachlan and Liversidge 1970; Etchécopar and Hüe 1983). No data on distance between nests, but for photograph of colony in Taiwan showing more than 30 nest-entrances, see Oo-u-kijo (1936). ROOSTING. No information for breeding season. In southern Africa, large flocks roost in reedbeds outside breeding season while 'small populations' roost in old breeding holes (McLachlan and Liversidge 1970). In South

Africa, one of the latest birds to roost, typically flying about over water until almost nightfall (Taylor 1942).

2. FLOCK BEHAVIOUR. No information. ANTAGONISTIC BEHAVIOUR. During mating sequence (see below), intruding conspecific alighted and settled between pair engaged in pre-copulatory display; one of the disturbed birds, probably ♂, adopted a crouched posture, as in his initial soliciting approach to ♀, and slowly approached intruder which then flew off. Crouched posture thus perhaps signals threat in this context (Broekhuysen and Stanford 1954). HETEROSEXUAL BEHAVIOUR. (1) Pair-bonding behaviour. In Morocco, nest-tunnels occupied November-January (U Hirsch). No further information. (2) Mating. Occurs apparently in the air (see below) or on ground (Meinertzhagen 1940), but relative frequencies not known. Detailed description of one mating on ground by Broekhuysen and Stanford (1954); sexes of participants not definitely known, but inferred from behaviour. 2 birds, sitting close, began to circle each other, and one (A, presumably ♂) adopted a crouched posture towards other (B, presumably ♀), and eventually started to nibble B's breast and cloacal feathers. At first, B did not seem to respond but then appeared to crouch down slightly whereupon A mounted B, climbing on from the side. Once mounted, A opened its wings, and spread and lowered its tail; after some time, dismounted. Procedure repeated several times and copulation evidently attempted or successful. Once, pair tried unsuccessfully to copulate in flight, fell into lake, rose again immediately, and flew to dry land where they copulated successfully on ground (Taylor 1942). (3) Behaviour at nest. After choosing appropriate site, ♂ and ♀ (pair-members) take turns at perching and digging, periodically fly around (as if checking site is safe), go off to feed for a while, then return to recommence digging. Birds shuffle sand backwards as they withdraw from nest-tunnel, pushing it out as they emerge. One pair differed markedly in contributions, one digging much more enthusiastically than the other. At beginning of excavation, departures from nest-site relatively brief, but get longer as tunnel lengthens; site abandoned completely if any obstruction encountered. (Oo-u-kijo 1936.) No further information. RELATIONS WITHIN FAMILY GROUP. After fledging, young remain with parents for unknown period (Oo-u-kijo 1936). No further information. ANTI-PREDATOR RESPONSES OF YOUNG, PARENTAL ANTI-PREDATOR STRATEGIES. No information. EKD

Voice. Little known, and available material largely extralimital. No detailed information on variation between races.

CALLS OF ADULTS. (1) Song. No description for west Palearctic *mauritanica*. In *chinensis*, a soft twittering (Severinghaus and Blackshaw 1976). In nominate *paludicola*, a twittering warble in breeding season (Mackworth-Praed and Grant 1963). (2) Contact-call. In nominate *paludicola*, a thin little 'svee-svee', given apparently throughout the year (Mackworth-Praed and Grant 1963); disyllabic 'sree-sree' reminiscent of Sand Martin *R. riparia* but clearer in tone (D I M Wallace). (3) Other calls. (a) In *mauritanica*, a chittering 'skirrr' from feeding flock, also from 2 birds perched amicably near each other (P A D Hollom); 'rit' of *chinensis* (Etchécopar and Hüe 1983) perhaps the same. (b) In Morocco, late March, birds jockeying for position on perch bickered with weak harsh sounds like distant Jay *Garrulus glandarius*, with some variation in pitch; calls more emphatic with upsurges in squabbling (P A D Hollom), suggesting expression of annoyance or threat.

CALLS OF YOUNG. No information. EKD

Breeding. SEASON. Morocco: laying begins November–December and lasts to February, exceptionally April (Heim de Balsac and Mayaud 1962; U Hirsch). SITE. In excavated tunnel in river or gorge bank, or side of road bank or pit; typically 0·8–3 m above ground (Oo-u-kijo 1936). Nest: in chamber at end of tunnel 30–80 cm long, often rising slightly; chamber 12·7–14·0 cm high (Oo-u-kijo 1936); nest cup of feathers and grass. Building: by both sexes; one tunnel completed in c. 20 days (Oo-u-kijo 1936). EGGS. See Plate 80. Sub-elliptical, smooth and glossy; white. *R. p. mauritanica*: 17·3 × 12·3 mm (16·4–18·5 × 12·0–13·5), $n = 14$; calculated weight 1·49 g (Schönwetter 1979). Clutch: 3–4. No information on number of broods. INCUBATION. About 12 days. By both sexes. FLEDGING TO MATURITY. Fledging period c. 20 days. No further information.

Plumages (*R. p. mauritanica*). ADULT. Upperparts, including crown down to eye, uniform fuscous-brown, about similar in colour to upperparts of Sand Martin *R. riparia riparia*; rump and upper tail coverts slightly paler, dark grey-brown; upper wing-coverts slightly darker, brownish-black, fringes along tips of median and greater coverts slightly paler grey-brown. Cheeks, ear-coverts, sides of neck, breast, and flanks uniform grey-brown, slightly paler and greyer than upperparts; chin and throat dull grey, feather-tips slightly mottled grey-brown, not contrasting with cheeks and breast. Grey-brown of breast and flanks gradually merges into white lower belly, vent, and under tail-coverts. Primaries and tail dull black; secondaries greyish-black with faint narrow white edges along outer webs and tips; tertials fuscous-brown (like upperparts), faintly edged off-white. Under wing-coverts and axillaries grey-brown, like breast and flanks, but slightly glossy, appearing paler in some lights; small coverts along leading edge of wing indistinctly fringed pale grey. In worn plumage, upperparts slightly paler and greyer, less saturated fuscous-brown; breast paler, washed pale grey, sometimes scarcely darker than belly. NESTLING. Down long but sparse, on head and back only (Oo-u-Kijo 1936); white (Mackworth-Praed and Grant 1973). JUVENILE. Like adult, but ground-colour of upperparts slightly paler, greyish, and with narrow grey-buff or greyish-cinnamon feather-fringes; fringes rather narrow and ill-defined on crown, mantle, and scapulars, but wider and deeper rufous-buff or sandy-buff on forehead, back, rump, upper tail-coverts, upper wing-coverts, and tertials, sometimes showing some contrast between buff-washed rump and greyish-black tail. Outer webs and tips of tail-feathers narrowly edged buff, inner webs with paler and wider edges; narrow buff fringes along outer webs and tips of flight-feathers; feather-tips on underparts variably fringed rufous, showing as buff wash. Lesser under wing-coverts rufous or buff. In worn plumage, rufous or sandy-buff fringes and edges bleached to off-white or pale grey, narrower ones usually worn off; pale fringes (frayed) remain only on tertials, greater upper primary coverts, and upper wing-coverts, with more faint ones along tail- and flight-feathers.

Bare parts. ADULT. Iris dark brown. Bill, leg, and foot slate-black or black. (BMNH.) NESTLING. No information.

JUVENILE. Iris dark brown. Bill black; mouth chrome-yellow. Leg and foot reddish-brown or brownish-flesh. (BMNH.)

Moults. ADULT POST-BREEDING. Complete; primaries descendant. In *minor* from Sudan (nesting October–January), starts late March to late April. About half-way through moult (primary moult score 16–30) in late April and mid-May, when forehead, crown, mantle, and chest largely new, as well as variable amount of remainder of body; tail old, except for new or growing t1–t2. All moult completed August–September. In *mauritanica*, starts immediately after nesting, March–May; completed June–July (Stresemann 1927–34; Heim de Balsac and Mayaud 1962; BMNH, ZFMK). All wing, tail, and body new in November (perhaps after a partial pre-breeding moult), in one t1 still growing; all worn in March (BMNH). In USSR, some specimens of *chinensis* had started moult of crown and nape by 10 July, while others replacing wing-coverts and some feathers of upperparts then (Dementiev and Gladkov 1954b). POST-JUVENILE. Complete; primaries descendant. In *minor* (fledging November–January), plumage worn by April–May; apparently starts moult a few months later than adult. In *mauritanica* (fledging January–March), juvenile plumage fresh or slightly worn in second half of May, and one from 30 May had just started moult with mantle and scapulars (ZFMK); body slightly worn and flight-feathers heavily abraded in November when prospecting nest-sites with older birds and gonads slightly enlarged; flight-feather moult apparently delayed until c. 1 year old. (BMNH, ZMA.) In *chinensis* from India (fledging February–May), moult starts with forehead May–June (Vaurie 1951c).

Measurements. ADULT, JUVENILE. *R. p. mauritanica*. Morocco, November–June; skins (BMNH, ZFMK, ZMA). Bill (S) to skull, bill (N) to distal corner of nostril; exposed culmen on average c. 3·8 less than bill (S). Tail measured to tip of t6; fork is tip of t1 to tip of t6.

	♂		♀	
WING	100·7 (2·34; 7)	97–104	100·2 (1·68; 7)	98–103
TAIL	41·5 (2·14; 7)	38–45	42·1 (1·46; 7)	40–44
FORK	4·4 (1·48; 7)	3–7	3·5 (1·08; 4)	2–5
BILL (S)	9·4 (0·78; 7)	8·7–10·3	9·6 (0·58; 7)	9·0–10·5
BILL (N)	4·8 (— ; 3)	4·7–4·9	4·7 (0·28; 6)	4·4–5·1
TARSUS	10·9 (0·49; 7)	10·4–11·8	10·7 (0·56; 6)	9·9–11·5

Sex differences not significant. Juvenile combined with adult in table above, though juvenile wing and tail on average c. 2·6 shorter than adult and fork 1·4 shorter. Morocco: wing, 102 (10) 100–105 (Vaurie 1959); wing, ♂ 103–109, ♀ 101–108 (Hartert 1910).

R. p. minor. Nigeria and Chad to Sudan and Ethiopia, all year; skins (BMNH, RMNH).

	♂		♀	
WING	96·3 (2·48; 13)	92–100	96·5 (2·67; 12)	92–101

Sudan: wing, 97 (10) 95–103 (Vaurie 1959); wing, 95–102·5 (12) (Hartert 1910).

R. p. chinensis. (1) India, all year (Vaurie 1951c); (2) Taiwan, February–September; skins (BMNH, RMNH).

	♂		♀	
WING(1)	93·1 (1·67; 22)	90–96	93·1 (1·72; 17)	89–96
WING(2)	94·1 (2·33; 12)	90–99	93·5 (2·14; 8)	90–96

Middle Amu-Darya (USSR): wing, ♂ 91·8 (6) 89·5–94, ♀ 89·1 (10) 84·5–92 (Dementiev and Gladkov 1954b). Pakistan and north-west India: wing, 90–96; tail 37–45 (Ali and Ripley 1972).

Nominate *paludicola*. South Africa: wing, 106 (15) 101–110, tail 49—59 (McLachlan and Liversidge 1970); wing up to 113 (RMNH).

Weights. No information on *mauritanica*. In slightly larger nominate *paludicola*, South Africa: adult, (August-)September, 13·3 (0·73; 26) 12·8–14·5; February 11·5 (0·91; 38) 9·2–13·1; juvenile, (August-)September, 13·3 (0·86; 7) 12·9–14·6 (R Earlé). In smaller *chinensis*, Taiwan, February–April: nesting adult ♂ 8·3 (1·33; 6) 7–11, ♀ 7·3 (3) 7–8; juvenile ♂ 7·9 (1·53; 6) 7–11, ♀ 7·8 (3) 7–8 (RMNH).

Structure. Wing long and pointed, narrow at base. 10 primaries: p9 longest, p8 0·5–3 mm shorter, p7 5–9, p6 12–17, p5 17–24, p1 41–48. P10 reduced, narrow and pointed, hidden under slightly longer outer greater upper primary coverts; in adult, p10 73–75 shorter than p9, 10–11 shorter than longest upper primary covert; in juvenile, p10 66–70 shorter than p9, 8–9 shorter than longest primary covert (sample small, however; some overlap would probably occur in larger sample). Flight-feathers not emarginated. Tertials very short. Tail rather short, tip slightly forked, 12 feathers. Bill as in *R. riparia*, but smaller, slightly less flattened, culmen less sharply ridged. Tarsus and toes entirely bare, weak; toes rather long and slender. Middle toe with claw 12·6 (4) 11–14; outer and inner toe with claw both *c.* 65% of middle with claw, hind *c.* 70%. Claws sharp, decurved, length changing with stage of nesting activity (short when digging).

Geographical variation. Rather slight. Birds from highlands of north-east and eastern Africa as well as nominate *paludicola* from southern Africa larger, *minor* from Sahel zone, *chinensis* from southern Asia, and Madagascar race smaller. Birds from more humid areas darker, less extensively white on belly, those of arid areas paler. *R. p. minor* from Sahel zone differs from *mauritanica* of Morocco in smaller size and slightly paler and more greyish-brown upperparts; *R. p. chinensis* from southern Asia small; mouse-grey or pale grey-brown on upperparts and chin to breast.

Forms superspecies with *R. riparia* and with Congo Sand Martin *R. congica* from middle and lower Congo and Ubangi rivers (Hall and Moreau 1970). CSR

Riparia riparia Sand Martin

PLATES 15 and 18
[between pages 232 and 233]

Du. Oeverzwaluw   Fr. Hirondelle de rivage   Ge. Uferschwalbe
Ru. Береговушка   Sp. Avión zapador   Sw. Backsvala   N. Am. Bank Swallow

Hirundo riparia Linnaeus, 1758

Polytypic. Nominate *riparia* (Linnaeus, 1758), north-west Africa and Europe, east through Siberia to Kolyma river and Kamchatka, south to northern Kazakhstan, northern Altai mountains, and northern parts of Baykal region; also Turkey and Levant east to western and northern Iran, and North America; *diluta* (Sharpe and Wyatt, 1893), lower Ural river, Transcaspia, and eastern Iran, east to north-west India, Burma, China, and Mongolia; *shelleyi* (Sharpe, 1885), Nile valley in Egypt and northern Sudan. Extralimital: *ijimae* (Lönnberg, 1908), south-east Siberia to Sakhalin, Kuril Islands, and Hokkaido.

Field characters. 12 cm; wing-span 26·5–29 cm. 20–25% smaller than Swallow *Hirundo rustica*, with distinctly slighter structure and little-forked tail. Rather small, slightly built hirundine, smallest in Europe and 2nd smallest of west Palearctic. Dark brown above, white below; only obvious characters dark brown chest-band, but dark dusky brown underwing and undertail also noticeable. Flight typical of hirundine but rather weaker than *H. rustica*. Sexes similar; no seasonal variation. Juvenile separable at close range. 3 races in west Palearctic, separable in the field.

ADULT. (1) North Eurasian and North American race, nominate *riparia*. Upperparts and flight- and tail-feathers dark earth-brown, with grey tone (when fresh) and sandy wash (when worn) occasionally visible on forehead, rump, wing-coverts, and tertials. Dark brown of upperparts extends over quite deep cheeks and into quite broad and complete chest-band. Rest of underbody white, except for brown smudges on flanks. Underwing dusky brown, with (when fresh) white tips on coverts visible only at close range. Plumage pattern remarkably cryptic against many backgrounds; bird flickers in and out of sight. (2) Egyptian race, *shelleyi*. Noticeably smaller than nominate *riparia* with wing up to 15% narrower. Paler above; chest-band narrower but quite distinct. (3) South Eurasian race, *diluta*. Close in size to nominate *riparia*. Distinctly paler above, with grey tone obvious; chest-band paler, narrower, and less well-defined than nominate *riparia* and *shelleyi*. Chin and throat washed buff, even speckled brown. JUVENILE. All races. Similar to adult but dark plumage areas less uniform. In flight at close range, narrow buff to dull white fringes make rump and inner wing-coverts appear paler than on adult. Perched at close range, buff-tinged (even speckled) throat and buff fringes on centre of chest-band provide ready distinction from nearby adult. At all times, bill and feet black-brown.

Unmistakable in clear, close view. From above, may suggest Brown-throated Sand Martin *R. paludicola* and *Ptyonoprogne* martins but none of these shows dark chest-band above white breast and belly. From below, plumage pattern might recall Alpine Swift *Apus melba* but that bird relatively huge and powerful, with proportionately longer wing and quite different flight action. Flight fast and agile but without grace of *H. rustica* or steadiness of House Martin *Delichon urbica*; involves less gliding, more fluttering, fewer changes in height, and

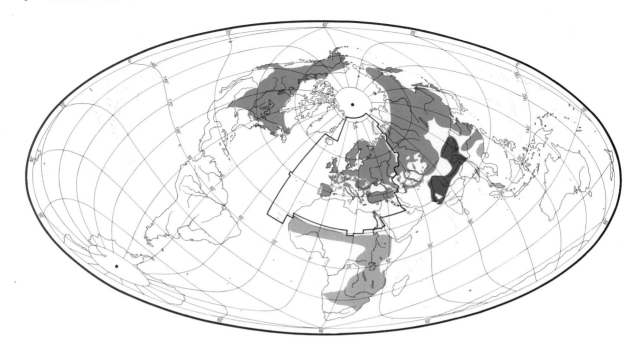

more sudden changes in direction. Capable of slow progress when hunting torpid insects over water, unlike *H. rustica*. At roosts, distant wheeling and twisting flocks resemble hosts of midges. Perches and clings, but gait restricted to shuffling walk.

Voice quieter than most hirundines. Song a harsh twittering, little more than a repetition of commonest call—a grating 'tschr'.

Habitat. Mainly aerial, usually in low airspace. Breeds from Mediterranean through steppe, temperate, and boreal to upper subarctic zones, in continental and also oceanic climates, with summer temperatures ranging down almost to 10°C, and varying regimes of wind and rainfall (Voous 1960). Rarely on ground, spending much of day in flight, resting on overhead wires and other suitable perches; roosts gregariously, often in reedbeds. As an aerial feeder, largely independent of nature of underlying terrain, provided it supports adequate flying insects. Avoids densely wooded or built-up areas, mountains and broken terrain, deserts, and narrow valleys, preferring neighbourhood of sandy, loamy, or other workable banks, excavations, cliffs, and earth-mounds, suitable for tunnelling to form nest-chambers. Such sites often occur naturally along rivers and streams, or by lakes and on sea coasts, or artificially where sand or other materials are extracted in conditions leaving some faces undisturbed. Occasionally uses holes in quay walls, support walls of vineyards or highways, and drain-pipes, where conveniently situated to permit exploitation of favourable food source. Will when necessary fly to forage at a distance from nest-hole. Generally in lowlands, but in Switzerland some colonies up to 600 m or even slightly

over 900 m (Glutz von Blotzheim 1962), and extralimitally in Himalayas to *c.* 4500 m (Ali and Ripley 1972). In USSR, readily lives in villages or even cities, but this is uncommon in western Europe (Dementiev and Gladkov 1954*b*). Requirements lead to patchy distribution.

On migration and in winter, much less circumscribed, but still vulnerable to human impacts on habitat, particularly where, as in Sahel zone, over-grazing, over-cropping, and other unwise land uses lead to desertification of important winter quarters.

Distribution. No marked changes recorded.

ALGERIA. Bred 19th century and perhaps 1978 (EDHJ). MOROCCO. Bred 1969 (Ruthke 1971). TUNISIA. Said to have bred 19th century (Heim de Balsac and Mayaud 1962); no recent evidence.

Accidental. Faeroes, Madeira, Cape Verde Islands.

Population. Decreased Britain, Netherlands, Poland, Switzerland, and Rumania, and locally in West Germany, Sweden, and East Germany.

BRITAIN. No evidence of marked widespread change (Parslow 1973). About 250 000 pairs but nearly 1 million pairs before crash 1968–9, probably due to drought in Sahel (Sharrock 1976; Cowley 1979). In 1984, under 10% of mid-1960s population due to mortality in Sahel (Mead 1984*b*). FRANCE. 10 000–100 000 pairs (Yeatman 1976). 4700 pairs along Loire from Bonny to Blois, same as before 1960 (*Bull. Orléanais* 1982, 1, 59). BELGIUM. About 15 000 pairs (Lippens and Wille 1972). In south, *c.* 400 pairs 1972–3 (Billen and Tricot 1977). In Lorraine, number of pairs increased by nearly 45% in 1983 compared with 1972–3 but decreased by 44% in 1984

(Pierre (1985). NETHERLANDS. Strong decline, mainly due to habitat changes. 25 000 pairs 1963-4, 5000-8000 pairs 1976-7 (Teixeira 1979). Still declining (Wammes *et al.* 1983; CSR). WEST GERMANY. Decline some areas; see Bauer and Thielcke (1982) for local counts. Significant increase in Peine (Niedersachsen) 1951-83 (Streichert 1984). DENMARK. Marked decrease 1978-80 (Nøhr *et al.* 1983). NORWAY. About 1100 pairs (Johansen 1984). SWEDEN. About 300 000 pairs (Ulfstrand and Högstedt 1976), declined by $33\frac{1}{3}\%$ in one province of central Sweden 1960-75 (Frycklund 1980). FINLAND. About 300 000 pairs (Merikallio 1958). EAST GERMANY. Marked decline in Oberlausitz in last 25 years (SS). POLAND. Decrease in west (Tomiałojć 1976). CZECHOSLOVAKIA. No decline reported (KH). SWITZERLAND. 4600 pairs in

1980; declined 37% since 1960 (Sieber 1982). RUMANIA. Decreased (VC).

Mortality. Britain: average annual adult mortality 65%; average annual 1st-year mortality 77% (Mead 1979b); in Kent, average annual adult mortality 60%, average annual 1st-year mortality 80% (Harwood and Harrison 1977); in Nottinghamshire, average annual adult mortality 81% 1968-73 (a period of decline), 58% 1973-8, 71·6% 1968-83, and average annual 1st-year mortality 92% 1968-73, 77% 1973-8 (Cowley 1979; E Cowley). Sweden: average annual adult mortality 54·4% for ♂♂, 60·9% for ♀♀; average annual 1st-year mortality 58·2% for ♂♂, 60·8% for ♀♀ (C Persson). Oldest ringed bird 9 years (Peterson and Mueller 1979).

Movements. Migratory; most of breeding range vacated in winter. Main west Palearctic populations plus all those of Siberia (nominate *riparia*) winter in African Sahel zone (from Sénégal eastwards) and in East Africa south to Moçambique. Central and east Asian populations (*diluta*, *ijimae*) winter in India and south-east Asia, though *diluta* (breeding east and south from north Caspian region) has been collected also in Palestine and Egypt. Localized race *shelleyi* (breeding Egypt and northern Sudan) is resident in part, though in winter ranges further south into Sudan and recorded also on Red Sea coast of Eritrea (Vaurie 1959). Nearctic populations winter in South America south to Argentina.

USSR populations move through Nile Valley and across Arabia, and extend much further south in Africa in winter than do western populations. Over 10 000 reported from Zambia in one mixed roost (Tree 1966). These populations breed much later, and large numbers of birds remain on wintering ground into late May (Pearson 1971); passage through eastern Mediterranean peaks in late April or early May, compared with early April in west. Occasional out-of-season records in Mediterranean do not amount to regular wintering.

Extensive ringing, particularly in Britain and Ireland (Mead and Harrison 1979a), has revealed much detail of movements within western Europe. During late summer and autumn, early-brood juveniles undertake local movements, which, by beginning of August, become oriented southwards; those leaving Britain choose short sea crossing, departing between Kent and Dorset. Passage of British and Irish birds continues SSW to Biscay coast of France and skirts north-west end of Pyrénées. Some birds then move down Ebro valley to Mediterranean coast of Spain, and others appear to move overland to Coto Doñana area. Birds cross Mediterranean into Morocco, reaching Sahel region by October or early November. Adults seem to travel faster than young birds. There is good evidence (Mead and Harrison 1979a) that 'traditional' routes are used by birds originating from a given breeding area and that they may keep together on passage. Birds travel by day and roost communally at night, generally in reedbeds. Similar migration pattern documented for populations of West Germany (Bub and Klings 1968) and southern Scandinavia (Persson 1973).

In wintering areas, birds congregate where food available, which is often (but not always) associated with water (Moreau 1972). Appear to be nomadic, and ringing evidence from British birds suggests that many move east through Sahel region during winter to make their return northwards via Niger inundation zone (Mali) or further east. Huge numbers seen passing north at Lake Chad, with daily totals estimated at over a million birds (Ash *et al.* 1967), though this was before drought-related population crash during 1970s. Elgood (1982) referred to 100 000 per day passing north at Lake Chad in March, with autumn passage there only slightly less spectacular.

Spring arrivals of birds at northern edge of Sahara indicate severe strain of desert crossing, with very low weights and sometimes considerable mortality (Ash 1969). British birds have been found in spring east to Tunisia, Malta, Italy, and eastern France (in contrast with Iberia-Morocco route used in autumn). Passage of experienced birds is about 3 weeks before 1st-years, and ♂♂ precede ♀♀. In a normal year, first arrivals in southern Britain occur before end of March and heavy mortality sometimes caused by snow in mid- to late April. In such cirumstances, birds may stay in burrows at a colony in torpid state and, if weather does not improve, die there (Mead and Harrison 1979b).

Both adults and 1st-years show significant degree of area- and colony-fidelity, but with plasticity expected of a species breeding in unstable sites which may become less suitable or disappear completely between seasons. In Britain, median distance of settlement in subsequent summers is 6 km for birds ringed as juveniles and 3 km for adults; 87% of juveniles and 93% of adults return to within 10 km of where ringed (Mead 1979). Also a sexual difference: in a Nottinghamshire study, recaptures in subsequent years of birds ringed as juveniles at same colony were 66·7% ♂ and 33·3% ♀ (Cowley 1979). Some birds change colony between 1st and 2nd broods in same summer, though frequency uncertain (Cowley 1983). A West German population study (Loske 1983b) found comparable age-related differences in colony-fidelity, but no clear evidence of the sexual variation found in Britain.

CJM

Food. Small airborne invertebrates. Feeds almost exclusively in flight, but at one site in England birds regularly hunt on ground for maggots and pupae (Calliphoridae) dropped by anglers, running about vigorously and searching among vegetation and in cracks in stonework (Clegg 1984); also recorded feeding on ground by Schulze (1971). Once recorded taking small Diptera from surface of sea by hovering (King 1967). Said also to have been recorded taking insect larvae from water (Dementiev and Gladkov 1954b); something similar observed by Löhrl (1969) and larva of water beetle (Dytiscidae) recorded in stomach of bird in Ukraine, USSR (Myasoedova 1965). Average feeding height $15·0 \pm 15·0$ m, $n=328$ (Waugh 1979). While actually foraging, proportion of time spent flapping $86·5 \pm SD9·8\%$, $n=91$ (Bryant and Turner 1982). Recorded following tractor ploughing field covered with tall weeds and taking insects disturbed (Schmidt 1967b). Near Stirling (Scotland), birds fed up to 1 km from nest, but mostly up to 0·25 km, average 0·19 km, $n=91$ (Bryant and Turner 1982); at Peine (West Germany), up to 1·3-1·5 km (Oelke 1968b); near Albany (New York, USA), mostly within c. 0·8 km (Stoner and Stoner 1941); in Lapland, preferred hunting area of birds from one colony was in river delta 6 km away (Svensson 1969).

The following recorded in diet in west Palearctic. Mayflies (Ephemeroptera), dragonflies (Odonata), stoneflies (Plecoptera: Leuctridae), grasshoppers (Acrididae), Psocoptera, bugs (Hemiptera: Heteroptera, Cicadellidae, Psyllidae, Aphidoidea), Neuroptera, moths (Lepidoptera: Noctuidae), caddis flies (Trichoptera: Phryganeidae), flies (Diptera: Psychodidae, Tipulidae, Trichoceridae, Culicidae, Anisopodidae, Simuliidae, Mycetophilidae, Chironomidae, Bibionidae, Scatopsidae, Stratiomyidae, Tabanidae, Asilidae, Empididae, Dolichopodidae, Lonchopteridae, Phoridae, Trypetidae, Syrphidae, Sepsidae, Opomyzidae, Agromyzidae, Chloropidae, Tachinidae, Calliphoridae, Muscidae, Hippoboscidae), Hymenoptera (sawflies Tenthredinidae, Ichneumonoidea, Cynipoidea, Chalcidoidea, Proctotrupoidea, Mutillidae, ants Formicidae), adult and (once) larval beetles (Coleoptera: Carabidae, Dytiscidae, Hydrophilidae, Staphylinidae, Scarabaeidae, Heteroceridae, Mordellidae, Anthicidae, Cerambycidae, Chrysomelidae, Curculionidae, Scolytidae), spiders (Araneae) (Csiki 1904; Witherby *et al.* 1938*b*; Pavlova 1962; Myasoedova 1965; Waugh 1979; Clegg 1984). No records of plant material as food in west Palearctic, but such records (e.g. of seeds) in stomachs in North America said to be due to accidental ingestion (Bent 1942).

No west Palearctic studies concerned solely with food eaten by adults, but for data on items in throat-pouches (thus destined for young) and stomachs of adults and/or fledged juveniles, see Csiki (1904) for Hungary (6 birds, items largely beetles) and Myasoedova (1965) for Ukraine (71 birds, largely Diptera and beetles); see also Ptushenko and Inozemtsev (1968) and, for Kirgiziya, Pek and Fedyanina (1961). All other detailed information on diet comes from studies of food of young. In winter at Lac de l'Upemba (Zaïre), food said to be principally very small flies (Verheyen 1952).

Food of young in New York State (USA) said to be similar to adults' but to contain more soft-bodied insects: thus no stoneflies, fewer beetles, and more flies and Hemiptera (Stoner 1936). In central Scotland, 2623 items from faeces of young comprised 69·3% (by number) Diptera (including 33·2% Schizophora), 12·2% aphids, 10·7% beetles (average length 3·05 mm), 5·0% Hymenoptera, and 2·5% other insects; for prey size relative to other Hirundinidae and to that available, see Swallow *Hirundo rustica* (p. 267); adults brought fewer large social Hymenoptera than available to them, but otherwise selected prey groups according to their availability, with no apparent bias towards groups particularly associated with water (Waugh 1979). Latter finding not repeated in study in Nottinghamshire (central England) in which food contained massive preponderance of aquatic insects (E Cowley). In Ryazan' region (USSR), collar-samples from Oka valley contained 579 items of which 48·2% leaf hoppers (Cicadellidae), 37·3% Diptera, 5·5% aphids, 2·6% ants, 3·8% other Hymenoptera, 1·2% beetles, and

1·4% other insects; in Pra valley, 2488 items comprised 49·9% Diptera, 9·8% leaf hoppers, 9·6% aphids, 9·4% other Hemiptera, 7·9% beetles, 5·9% spiders, 5·1% Hymenoptera (few ants), 2·2% mayflies, and 0·2% other insects; no great difference in food brought to young of different ages (Pavlova 1962). For North American studies, see Beal (1918), Stoner (1936), and Beyer (1938). See Tables A–B in *H. rustica* (p. 267) for data on size of food boluses collected for young. In Ryazan', boluses *c.* 100–200 mg (presumably wet weight) irrespective of age of young (Pavlova 1962). In Nottinghamshire, each contained from 30 (mayflies) to 100 (aphids) items (E Cowley). Unlike in *H. rustica*, bolus weight linked more strongly to distance of food source than to weather: positively correlated with distance of food source, time between nest-visits, time spent actually foraging (see below), flight speed when foraging, and daily rainfall; negatively correlated with maximum daily temperature; not correlated with prey weight or food abundance (Bryant and Turner 1982, which see for further optimal foraging data). At colony in Denmark, young fed average 4·2 times per hr (0·8–8) in rainy weather, 9·0 (4·5–18) when dry (sample sizes not given); mostly 1–5 min between feeds, rarely more than 25 min; no correlation with age or number of nestlings (Asbirk 1976). Similar rates in Wołk (1964), though these rather lower than in many other studies, and highest rates recorded by Moreau and Moreau (1939*a*): in fine weather, 10.00–18.00 hrs, 4 nests with large young received 12–49 (mostly 25–43) feeds per hr. Rate fairly regular through the day but highest late morning and in afternoon (Stoner and Stoner 1941). In central Scotland, $5·20 \pm SD2·59$ min ($n = 88$) between visits, of which $4·75 \pm SD2·69$ min ($n = 91$) spent actually foraging (Bryant and Turner 1982). In North America, 115 feeding visits recorded on one day (Bent 1942), and in Ryazan' never more than 200 per day (Pavlova 1962). Typical brood thus brought *c.* 7000 insects (total dry weight *c.* 7 g) per day.   DJB

Social pattern and behaviour. Based partly on North American material, including major studies by Stoner (1936), Petersen (1955), and Hoogland and Sherman (1976). For important recent studies in west Palearctic, see Asbirk (1976), Sieber (1980), and Kuhnen (1985).

1. Often highly gregarious, at all times of year. Flocks of up to several hundreds, even thousands, common during migration periods and in winter quarters, sometimes associated with other Hirundinidae. For roosting assemblies of up to *c.* 2 million, see Roosting (below). At Azraq (Jordan), flocks of up to 1000, once tens of thousands, on stopovers during spring passage (Wallace 1982). In Malawi, late January, *c.* 3000 feeding with *c.* 300 Brown-throated Sand Martins *R. paludicola* (Long 1959). In southern Turkey, of 683 birds passing on autumn migration, average flock size was 4·3 (Sutherland and Brooks 1981). In Ohio (USA), one post-breeding flock of *c.* 250000 birds (Campbell 1932). In Michigan (USA), 2 post-breeding flocks each *c.* 2000 birds, all apparently juveniles (Crockett and Nickell 1955). Ringing studies suggest that some groups of birds remain together all year (Mead and Harrison 1979*b*).

BONDS. Monogamous mating system of seasonal duration probably the rule, but ♂♂ commonly attempt to copulate with ♀♀ other than mates; no information on success of such attempts (Beecher and Beecher 1979); see also Heterosexual Behaviour (below). In study in Nottinghamshire (England), at least some birds (probably 4 out of 8 marked ♀♀, and 2 ♂♂) changed mate for 2nd broods; not known whether others did or not. Change of mate for 2nd brood more likely after failure of 1st, but unlikely in parts of colony where breeding highly synchronized. (Cowley 1983; E Cowley.) Both sexes share in brooding and feeding of young (e.g. Hoogland and Sherman 1976). After fledging, young dependent on parents for c. 1 week (Asbirk 1976). Age of first breeding 1 year (Stoner 1936; Mead and Harrison 1979a). BREEDING DISPERSION. Typically colonial, occasionally solitary. Colonies divided into sub-colonies (Hoogland and Sherman 1976; Mead and Harrison 1979a; Sieber 1980), often making colony difficult to demarcate (Rogers and Gault 1968): e.g. 'colonies' of several thousand pairs extend for tens of km along Don river, Voronezh region, USSR (Barabash-Nikiforov and Semago 1963); discrete colonies here evidently smaller than implied, though one exceeding 2000 pairs reported from Voronezh (Wilson 1976). Throughout range, most colonies less than 50 pairs, but 100 pairs not unusual. In Britain, average 37·6 pairs ($n=57$), perhaps larger in north than south (Morgan 1979). Average size of 126 colonies, southern Belgium, 31 pairs (Billen and Tricot 1977); in northern West Germany, 25 pairs, $n=10$ (Kuhnen 1978); in central-southern Sweden, c. 46 pairs, $n=136$ (calculated from Andersson 1982). In Denmark, of 134 colonies, those near coast averaged 19 'holes', compared with 42 holes inland (Møller 1979). Of 64 colonies, Westfalen (West Germany), 56·3% of less than 50 pairs, 21·9% of 51–100, 21·9% of more than 100; largest 450 pairs (Loske 1983a). Of 123 colonies, Switzerland, 50% of less than 25 pairs, 25% of 26–50, 22% 51–100, 3% more than 100 (Sieber 1982). For similar data, Finland, see Lind (1964); for Poland, Józefik (1962); for Hungary, Marián (1968); for Rumania, Kohl et al. (1975). For colony sizes in Michigan (USA), see Hoogland and Sherman (1976). Breeding dispersion as follows. Along Mureş river (Rumania), c. 2300 pairs on 650 km (Kohl et al. 1975); along Tisza river (Hungary), c. 26000 pairs on 370 km (Marián 1968); along San river (Poland), c. 4600 pairs on 442 km (Józefik 1962). In Westfalen, shortest distance between colonies 2·3 km, greatest 15·9 km (Loske 1983b). In 2221 km² of Nordrhein-Westfalen (West Germany), 0·75 pairs per km² (Kuhnen 1975). In Peine (West Germany), 700–1050 pairs per km² (Oelke 1968b). In south-west Scania (Sweden), 2·54 pairs per km² (C Persson). Nest-densities as follows. In one Michigan colony, 5·5 nests per m², $n=45$ nests (calculated from Johnson 1958b). Derivation of distance between nests dependent on definition of colony or sub-colony, on certainty that holes are occupied, and on stage of season. On average, c. 50% of holes occupied (Kohl et al. 1975), c. ⅔ (C Persson); at one colony, Poland, c. 27·6% of holes occupied for raising of 1st broods (Wołk 1964). According to Mead and Harrison (1979a), 1-year-olds often occupy peripheral sites and more often have to excavate new nest-sites, but Sieber (1980) found that late arrivals tended to nest below or among existing nesters, so that, on average, such nests closer to neighbours than earlier on. In 1st week of breeding season, average 26–45 cm between nests; in 5th week 17–20 cm; average over 3 years 27–30 cm (Sieber 1980). In Michigan colony, minimum spacing 10 cm, average 18·5 cm, $n=72$ (Petersen 1955). In colony on River Usk (Wales), 6·3 cm–1·5 m (Rogers and Gault 1968). Small nest-area territory defended around mouth of tunnel (Asbirk 1976), but

size difficult to establish as aggression wanes with increasing distance (Sieber 1980); serves as focus for pair-formation, also for protection from interference by conspecifics (see Antagonistic Behaviour, below). In study of 2nd broods, some birds (4 ♀♀, 2 ♂♂) retained original nest-site, others (3 ♂♂, 14 ♀♀) changed site (Cowley 1983). In another study, 2nd broods usually raised in same nest as 1st, ♂♂ showing stronger fidelity to site than ♀♀ (Sieber 1980). Change of nest-site by ♀ usually follows nesting failure (E Cowley). In southern Sweden, birds often changed colonies between broods in order to join a breeding group with which they were synchronized (C Persson). When nest disturbed, ♀♀ more likely than ♂♂ to change colony within season (Persson 1978). Sometimes new holes dug for 2nd broods but always in previously occupied part of colony. ♀ that lost mate and well-grown young re-dug same hole, also visited and dug elsewhere, but finally bred with new mate in 1st hole (Sieber 1980). Adults markedly faithful to colony, even sub-colony, between years, more so than 1st-year birds to natal colony (Leys and Wilde 1970; Scherrer and Deschaintre 1970; Mead 1979a; Petersen and Mueller 1979). ♂♂ more faithful than ♀♀ (Mead 1979a). Numerous species reported nesting in disused tunnels within occupied colony of R. riparia: of 16 such species, Britain, Tree Sparrow Passer montanus, House Sparrow P. domesticus, and Starling Sturnus vulgaris common associates (Mead and Pepler 1975, which see for review); Bee-eaters Merops apiaster reported in France (Salvan 1961; Lévêque 1964), Syria (Meinertzhagen 1935), and Rumania (Kohl et al. 1975); Rock Sparrow Petronia petronia in steppes of USSR (Grote 1932). ROOSTING. In breeding season, roosts in nest; outside breeding season communally in large numbers. From start of nest-building, usually both members of pair roost in nest (Asbirk 1976; Sieber 1980), but away from nest-hole after young c. 12 days old (Petersen 1955). At colonies subject to heavy predation, ♂♂ commonly roost outside nest (C Persson). Small cluster once observed at start of breeding season: 6–8 birds huddled together in nest-tunnel, and thought able to withstand low temperatures by becoming torpid (Mead 1970b). Goes to roost later than D. urbica, making more use of twilight (Church 1958, which see for June–September roosting times). Departure from colony in morning markedly later in cold weather (C J Mead). For sleeping posture of young in nest, see Relations within Family Group (below). Young roost in nest-tunnel for c. 1 week after fledging (Sieber 1980); according to Stoner (1942), only c. 15–20% of dependent young return to their own nest-site for roosting, instead often associating with other juveniles in any vacant tunnel in vicinity. Young fledging from colony in Nottinghamshire, September, roosted in colony for c. 10 days after fledging (Cowley 1983). Independent young excluded by parents from natal nest-site, but may use others for daytime loafing (Asbirk 1976). After fledging, young and also adults regularly loaf together on wires, etc. (Petersen 1955), typically with much jostling over perch sites (Fig A). Comfort behaviour often a flock activity: e.g. in early May, Somerset (England), c. 300 sunbathed in lee of wall; birds extended one or both wings, some lying on sides and raising wing to expose underparts to sun, while others flew to shallow water, bathed, and preened, before returning to flock (King 1970). Bathe by flying low over water, then rising to c. 0·5 m before dipping one or more times (Lind 1960). In flight, birds preen by turning heads to reach backwards (Sparks 1961). Fledged young at colony (see Relations within Family Group, below), and post-breeding flocks of adults and juveniles, regularly land on ground to sunbathe, dust-bathe, and preen (Thom 1947; Petersen 1955; Barlow et al. 1963; Nicholson 1981). Dust-bathing once performed on top of wall (Clegg

A

1984). In May, flock seen preening and shuffling bodies in wet sand (Carr 1968). Also once bathed in dew-covered grass (Oliver 1979). Outside breeding season, typically roosts communally in reedbeds, osiers, and other swamp vegetation (Witherby *et al.* 1938*b*; Verheyen 1956*a*; Aspinwall 1975). At one reedbed roost, birds typically perched halfway up emergent stems at junction of leaf and stem; sometimes 2 birds on same reed (Labitte 1937). Post-breeding migrants also use unoccupied holes in colonies (Mead 1962). In August, southern England, birds recorded roosting in hollows on shore, some apparently formed with own bodies (Cawkell 1950). In September, West Germany, large numbers apparently roosted densely on dried mud bed (Alkemeier 1985). At Azraq, hundreds of spring migrants roosted on bare sand with bee-eaters (Meropidae) (Nelson 1973). For mixed winter roosts with *H. rustica*, see that species (p. 269). In bad weather, early September, Estonia (USSR), 2 groups of *c.* 120 and 100 roosted in dense cluster like 'swarm of bees', some on top of others, on inner wall of dovecote; cluster silent and passive, even when touched or illuminated (Keyserlingk 1937). After dispersal from breeding grounds, main roosts occupied continuously July–September (Mead and Harrison 1979*a*). Pattern is of relatively isolated, but often very large, regional roosts, or, for given region, series of smaller roosts (Mead 1962). At Chichester (England), peak of *c.* 50 000 birds; at roost in Cambridgeshire (England), 1968, estimated 2 million birds, this exceptional. At Chichester, mostly juveniles, proportion of adults in different years 7·1–17·0%. Individual marked adults recorded using given roost for same period (typically *c.* 10–14 days) in successive years (Mead and Harrison 1979*a*), thus moving leisurely from one roost to another along migration route (Mead 1962). Juveniles use roosts in near vicinity of natal colony, but also more catholic in choice than adults, being more likely to use roosts distant from natal colony, and staying for shorter periods than adults (Mead and Harrison 1979*a*). Following account of Chichester reedbed roost (*c.* 80 × 20 m) based on Mead (1962). In mid-September, birds started approaching roosts *c.* ½ hr before sunset. Large flocks, apparently feeding, formed over reedbed, then, after 10–15 min, began flying low over water around roost. After a few such 'passes', some birds landed on tops of reeds, calling, and balancing with outstretched wings, sometimes only to take off again as another wave passed over. Gradually more birds landed. Remainder of flock flew more compactly, and either entered roosts after 1–2 low passes, or else plummeted in after spectacular towering movement. Initially, birds in roost bustled noisily, only settling in silence when almost dark. None left until well after dawn when majority 'erupted' almost simultaneously from roost, while a few stayed until 1 hr after dawn (Mead 1962). Pre-roosting manoeuvres also include spiral descents (Witherby *et al.* 1938*b*).

2. Whenever Collared Dove *Streptopelia decaocto* flew over roost in Humberside (England), birds rose in tight flock and chased it, calling loudly, for up to 800 m before returning to roost. Same response elicited by Kestrel *Falco tinnunculus* (Catley 1982). For other alarm reactions, see Parental Antipredator Strategies (below). FLOCK BEHAVIOUR. In postbreeding flocks, probably mostly juveniles, wide variety of activities in addition to comfort behaviour reported, notably attempted copulation, and incipient excavation, nest-building, and brooding (Thom 1947; Petersen 1955; see also Heterosexual Behaviour, below). Within one large flock, apparently of juveniles, birds congregated in small groups and individuals took turns to 'brood' small stone; also repeatedly picked up small feathers, pieces of grass, etc., and chased one another in the air (Crockett and Nickell 1955). For similar behaviour, see Thom (1947). In autumn, settled flocks also exhibit periodic 'panic' upflights, all birds taking off and circling around before landing again (Alder and James 1951). In apparent migratory exodus, flock of several thousand left Chichester roost before dawn, and, instead of flying around low down for several minutes as usual when resident at roost, flew directly south, climbing until lost from sight (Mead and Harrison 1979*a*). In Zaïre, November, arrival of a new flock of migrants caused evident excitement among those already assembled; mutual attraction of the 2 groups led to marked aerial evolutions, new arrivals often swooping down and skimming water (Verheyen 1952, which see for other flock manoeuvres). For other flock activities, see Heterosexual Behaviour (below). ANTAGONISTIC BEHAVIOUR. In early stages of nest-site excavation, residents of both sexes vigorously defend site, both vocally and physically, against intruding conspecifics (Petersen 1955; Asbirk 1976; Sieber 1980), though Hickling (1959) and E Cowley reported little antagonism. For attacks on *H. rustica* trying to enter nest-hole, see O'Connor (1961). When nest-tunnel longer, resident sits in entrance facing outwards, sites higher in bank being the more effectively defended (Petersen 1955). ♂ sings (see 1b in Voice) throughout excavation phase. If another bird (of either sex) approaches, ♂ performs Advertising-display (Fig B, rear): sings, ruffles head and throat plumage, and vibrates closed wings. Once ♀ takes up residence, she challenges intruders, especially ♀♀. Resident of either sex confronts intruders with Threat-display (Fig C, left): faces rival, ruffles

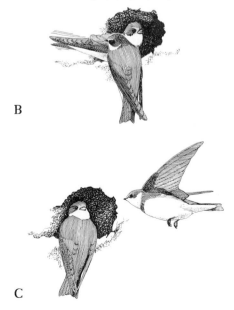

B

C

D

neck plumage, gives Excitement-call (see 3 in Voice), and, at higher intensity, gapes, pecks, and sometimes fights fiercely. In Fig D, ♀ (left), still aggressive after expelling rivals, gapes at her mate who responds with Advertising-display. (Kuhnen 1985.) Resident may push intruder forcefully with bill or, hovering, grasp nape and pull intruder away, often leading to fight in which both peck, and fall locked together by feet, sometimes continuing struggle for several seconds on ground; pursuit-flights also occur (Beyer 1938; Petersen 1955; Sieber 1980). Intruder nearly always (98% of 88 cases) retreats if threatened or attacked, less often (62%, $n=93$) if resident only calls (Sieber 1980). One bird defended nest-site strongly after disappearance of mate, but once paired anew was apparently more tolerant of nearby birds (Petersen 1955). Residents attack and expel independent young (their own or from other broods) landing in nest-site (Petersen 1955; Hoogland and Sherman 1976). Frequent displays among birds in post-breeding flock, interpreted as threat by Thom (1947), are probably pre-copulatory display (C J Mead: see Heterosexual Behaviour, below—also for antagonism during Mate-guarding chases, and for 'feather-fighting'). HETEROSEXUAL BEHAVIOUR. (1) General. Birds arrive at colony-sites in flocks of usually unpaired ♂♂ and ♀♀ (Kuhnen 1985). Began regularly visiting colony in Denmark c. 2 weeks after arrival in area (Asbirk 1976). 1st-year birds arrive at colony c. 2–3 weeks after adults (Mead and Harrison 1979a). Pairing linked to nest-site selection; completed before end of excavation. In Britain, song (see 1 in Voice) given late April to early July, occasionally up to early September (Alexander and Alexander 1908). (2) Nest-site selection. ♂ takes initiative in selecting site. On arrival at colony, over 90% of first landings are at old holes (Sieber 1980, which see for factors affecting nest-site selection). Majority of these landings by ♂♂ (E Cowley). Early-arriving ♂♂ may use hole from previous year, later arrivals selecting any suitable place where they can find a foothold. Typically, individuals or small loose groups make wide circling flights away from colony area and return to move along vertical face of colony in hovering flight, pausing at defended sites, or landing to cling to bank. No unpaired bird digs more than a shallow hole at this stage (Petersen 1955). Birds hovering in front of holes and landing are ♀♀ looking for resident ♂♂, as well as site-prospecting ♂♂ (Asbirk 1976; Kuhnen 1985). Silent birds (normally ♀♀) are welcomed or tolerated by ♂♂ claiming sites (Kuhnen 1985). Site-inspection highly animated, accompanied by much calling which climaxes as birds land and start running into holes and excavating; often 3–4 or more birds visit given hole, entering it one after the other (for behaviour at holes, see Antagonistic Behaviour, above). Calling ceases abruptly when all birds wheel away; after a few minutes, return to repeat performance, successive visits following at ever-shorter intervals and with mounting intensity for an hour or more, before dispersal. Initially, site-prospecting occurs 1–2 hrs after dawn, and again in evening, but at almost any time in ensuing days. (Hickling 1959.) In one colony where a few ♀♀ had 2nd broods, some changed site and mate to do so (♀ thought to choose nest-site

rather than resident ♂ as such: Kuhnen 1985; E Cowley). From week prior to their 1st broods starting to appear at nest-entrance, these ♀♀ prospected at other sites in adjacent parts of colony; began typically by approaching then veering away from holes, later sometimes entering occupied holes. Prospecting behaviour attracted other birds to display and settle in neighbouring unused holes. (Cowley 1983.) (3) Pair-bonding behaviour. ♂ little concerned with attracting a mate until his nest-tunnel c. 30 cm long. Then attempts to lure ♀ into hole: faces outwards, performing Advertising-display (Kuhnen 1985); waits there until ♀ approaches, then flies out singing (see 1b in Voice), with rather fast wing-beats (E Cowley), described as distinctive 'butterfly-flight' (Kuhnen 1985); ♂ makes repeated Circling-flights; returns each time to hole, typically calls (see 2 in Voice) and lands (Kuhnen 1985), or hovers opposite entrance (Nest-showing). Sometimes flies to another part of colony, knocks ♀ away from nest-hole, and tries to coax her back to his (E Cowley). If ♀ in front of ♂, he often overtakes her and lands first at hole, apparently to entice her inside. In Fig B, ♂ (at rear) perched at nest-site after Circling-flight directs Advertising-display at ♀ attracted. Distinctive song (see 1a in Voice) by both birds signals growing pair-bond. (Kuhnen 1985.) Pair-formation said to have occurred when pair enter and leave nest-site together; before then, birds fly in and out independently (Beecher and Beecher 1979). For other indications of pair-bond development, see Kuhnen (1985). (4) Mate-guarding. For 7–8 days after pair-formation (i.e. from start of mating phase 3–5 days before 1st egg laid, and ending when 4th laid), ♂ stays within 1 m of mate on every flight from nest-hole, up to 100 times per day. ♂ often sings (see 1c in Voice). Flights may last over $\frac{1}{2}$ hr, and unescorted flights by ♀ rare. Pair typically attract 1–5 other chasing ♂♂ (hereafter called interlopers) who attempt to follow ♀ as closely as does her mate, probably aiming to achieve promiscuous matings; some such copulation attempts seen (see also subsection 5, below). Interlopers may be paired or unpaired. ♀ often doubles back to challenge interlopers, sometimes bumping them or engaging in vigorous fights (see Antagonistic Behaviour, above). If pair emerge from nest-site into group of potential interlopers, ♂ of pair may overtake mate and attempt to steer her back into nest; may bump her once or twice, knock her to ground, or even engage her in fight, then head her back. Interlopers chase ♀♀ likely to be, or appearing to be, in their fertile period—such ♀♀ chased on 64% of unescorted flights. (Beecher and Beecher 1979; see also Tooby 1947, Petersen 1955, Hickling 1959.) Mate-guarding ♂♂ much less aggressive during nest-building than later on when ♀ receptive (Kuhnen 1985). (5) Mating. Occurs from 3–5 days before laying of 1st egg until start of incubation (Kuhnen 1985). Thought perhaps to occur mostly in nest-chamber (Tooby 1947; Petersen 1955; see also King 1955), but no evidence, and in study by Asbirk (1976) in which nest-chambers constantly observed, no copulation seen. Most accounts refer to mating on ground, but also on wires, and sometimes—perhaps unsuccessfully—in the air. In occurrence on ground, 2 birds crouched low and motionless c. 1 m apart, with bills and heads touching ground. Presumed ♂ then arose and, with wings 'waving wildly' and body swaying from side to side, walked toward ♀, mounted her, and copulated, wings now raised rigidly. Afterwards, ♀ shook herself vigorously and both finally flew off (Watson 1946). On 3 occasions, ♂ made hovering approach to land on back of ♀ perched on ground; she quivered one or both wings, with head turned sideways and slightly up, while ♂ balanced by continuously flapping wings (King 1955). For similar copulation attempt between House Martin *Delichon urbica* (♂

role) and *R. riparia*, see King (1958). No mention of calls in these accounts, but ♂ gives variant of song (see 1b in Voice) whilst mounting (Sieber 1980). In flock of 60-80 birds, May, several copulations or attempted copulations occurred on ground; once, after copulating, 2 birds sitting side by side touched bills (Carr 1968). Attempted copulation on ground common in post-breeding flocks (e.g. Alder and James 1951, King 1955), perhaps mostly of juveniles: e.g. in large flock (ages unknown), July, birds hovered over others, landed, and approached with wings spread and lowered; birds circling each other on ground with spread wings and tails (Thom 1947) were probably also indulging in pre-copulatory behaviour. Attempted copulation by ♂♂ induced on ground in experiments with wings-spread decoys, demonstrating promiscuous tendencies of ♂♂ (Petersen 1955; Hoogland and Sherman 1976); copulation attempts occurred only in period from nest-site excavation until incubation (Hoogland and Sherman 1976). In mating on wires, pair alighted beside each other; ♂, singing with increasing excitement, sidled towards ♀, sometimes shivering his wings; ♀ twice took off, closely followed by ♂, returned to same spot, and, on 2nd occasion, ♂ landed on her back. Once, leading bird (sex unknown) of chasing pair flew with wings held in V, and attempted copulation followed on wire fence. In chases, ♂ sometimes takes lead briefly before landing; thought to solicit mating. (Tooby 1947.) After ♀ landed, ♂ sometimes approached her in slow run before, once, mounting briefly (Anon 1952). Once, ♂, giving harsh 'jid-did-did' sounds, apparently attempted copulation in the air (Tooby 1947), but possibly a confrontation as these calls have aggressive quality (see 3 in Voice). (6) Behaviour at nest. ♂ does not complete excavating tunnel until paired (Kuhnen 1985). 1st egg laid in unlined nest, but lining of feathers added during egg-laying and incubation (Stoner 1936), sometimes also after hatching (Beyer 1938; Petersen 1955); feathers brought by ♂ when young hatching (E Cowley). Fights often start when birds carrying feathers approached colony, up to 100 birds surrounding feather-bearer; in mêlée, feather often lost to another bird, and once changed owners 7 times (Hoogland and Sherman 1976). Feathers thought by Beyer (1938) to be dropped playfully, but their interception (feather-fighting) interpreted as competition for a limited resource by Hoogland and Sherman (1976, which see for relationship between frequency of feather-fighting and colony size). During incubation, nest-relief almost instantaneous (Moreau and Moreau 1939a). Usually, relieving bird lands in hole-entrance and runs into tunnel uttering distinctive sound (see 2 in Voice); when half-way along tunnel, sitting bird rises, runs past mate, and leaves hole, and relieving bird immediately sits on clutch (Asbirk 1976). RELATIONS WITHIN FAMILY GROUP. Young brooded almost constantly on day of hatching, parents alternating at nest. On successive days, brooding diminishes steadily (Petersen 1955; Pavlova 1962; Asbirk 1976), almost stopping in daytime by 7th day (Beyer 1938). Until 10 days, young may be brooded by either or both parents at night: of 36 nests, ♀ in 23, ♂ in 6, both in 7. Parents rarely spend night in nest after young 12 days old. During first days after hatching, both parents swallow faecal sacs, but, from c. 4 days, start carrying sacs away (Petersen 1955). At 9 days, young leave nest to defecate in tunnel (Beyer 1938; Pavlova 1962). By 14 days, run almost to tunnel entrance and turn round to defecate, but do not eject faeces outside entrance. Parents remove faeces less promptly when young older but continue removal at irregular intervals until young depart. Both parents feed young (Petersen 1955). However, c. 1 week before fledging, some ♀♀ leave mates to raise young while they pair with new ♂♂ to raise 2nd broods; while pairing anew, such ♀♀ occasionally

visit or feed unfledged young of 1st brood (Cowley 1983). Arriving parent gives Feeding-call (see 7 in Voice), young giving similar call from 5 days whenever tunnel darkened by bird entering (Beyer 1938). Change in light stimulates gaping (Fulk 1967). When gorged, chick's head lolls in nest (Cowley 1977). If small young (up to 4 days) apparently satiated and unresponsive, ♂ parent may lightly trample them, twittering (Beyer 1938). At 9 days, young mobile and vociferous, rushing towards arriving parent; at 12 days, wait near tunnel entrance to be fed, sometimes jockeying for position (Petersen 1955). Parents thought to recognize own young by voice at c. 15-17 days (Hoogland and Sherman 1976). When parent approaches, each chick gives incessant food-calls until fed or adult leaves. Recipient of food usually retreats deeper into tunnel (Petersen 1955). Well-grown young exercise by supporting their bodies on wings and tail pressed on to nest (E Cowley); also peck at and beg from one another; may sleep with head tucked into shoulder (Cowley 1977). Young sometimes topple out of nest-hole, and in one instance were fed by adults on ground until they fledged, then returned to nest-hole (Pauler 1972). At point of fledging, parents circle around in the air outside (Asbirk 1976), sometimes calling (Petersen 1955: see 2 in Voice). Fledged young fed in the air or, for first few days, at nest-entrance (Stoner 1936; Petersen 1955); period spent by 1st brood at entrance perhaps prolonged if ♀ parent has started 2nd (overlapping) brood (Cowley 1983). Post-fledging dependence c. 1 week (Petersen 1955; Sieber 1980), during which time fledglings also try to catch insects for themselves (Asbirk 1976). Fledged young spend much of day loafing, sunning, and dust-bathing; also scratch at bank and play with nest-material in incipient nest-building behaviour (Stoner 1936; Petersen 1955). Commonly play with feathers, dropping them in flight and retrieving them (C J Mead). Also take part in aerial chases, including 3-4 or more birds, but less persistent than mate-guarding chases of adults. May hover in front of holes (not necessarily their own), calling and singing (Tooby 1947). Fledged young swim quite well, supporting and propelling themselves mainly by wings (Stoner 1936). Young leave colony shortly after fledging (Emlen and Demong 1975), then spend a period exploring area around colony (Mead and Harrison 1979a). ANTI-PREDATOR RESPONSES OF YOUNG. On approach of human intruder, young (raised in captivity) crouched down in nest (Cowley 1977); likewise if illuminated in nest (Asbirk 1976). From 8-15 days, have marked tendency to crawl under ledges or other objects. When removed a little way from nest-tunnel, almost invariably move towards it, often shuffling backwards. Occasionally, well-grown young removed from nest will struggle violently and utter Distress-calls (see 6 in Voice), summoning adults who give Alarm-calls (Stoner 1936: see 5 in Voice). On hearing Alarm-calls and Warning-calls (see 4 in Voice) of adults, young retreat backwards into nest-tunnel (Windsor and Emlen 1975). PARENTAL ANTI-PREDATOR STRATEGIES. (1) Passive measures. While adults not especially shy during incubation and first few days after hatching, they soon become more wary and reluctant to enter nest-holes when human intruder nearby (Stoner 1926). (2) Active measures: against birds. Response varies with raptor, and thus threat it poses. When Hobby *Falco subbuteo* hunting, take avoiding action, swooping down to level out near ground; if in vicinity, but not hunting, *F. subbuteo* escorted away from colony by swirling flock of perhaps 20-30 birds, mostly above and behind it (C J Mead). Presence of passing hawk (Accipitridae) or crow (Corvidae) elicits Warning-calls (Tooby 1947) and often mobbing (Mead and Pepler 1975). At Ithaca (USA), approach of American Kestrel *Falco sparverius* first stimulated Warning-calls which caused other adults to fly out from colony in loose flock uttering Alarm-calls, these often continuing during subsequent mobbing attack; when *F. sparverius* in flight, mobbing birds flew very close, though alarm subsided when it perched. Both Alarm- and Warning-calls shown experimentally to induce upflight (Windsor and Emlen 1975). Mobbing response usually initiated by 1-2 adults whose nests are immediately threatened. Many or most adults (but not recently-fledged juveniles) at colony join mobbing flock and regularly challenge predators threatening nests distant from their own. If predator gets close to nest-holes, birds begin flying in horizontal ring several layers of birds deep. Individuals fly within ring, hover in front of predator for a few seconds, then resume circling; typically no physical contact with predator (Hoogland and Sherman 1976, which see for relationship of mobbing response to colony size). (3) Active measures: against man. When man appears suddenly near nest, birds give Warning-calls (Tooby 1947). Rarely mobs man, except in most alarming circumstances, e.g. birds may swirl around mist-net from which captured birds calling (C J Mead). (4) Active measures: against other animals. Swirling flock once escorted fox *Vulpes* away from colony (C J Mead). Both avoiding and mobbing responses, as above, reported for weasel *Mustela nivalis*, whose appearance in colony stimulated Alarm-calls and formation of flock which skimmed over face of colony and dived towards its base; weasel continued killing young in nests despite harassment (Robert 1979); similarly with stoat *M. erminea* (Baudoin 1980). Calls for ground predators same as for aerial ones (E Cowley). Black rat snakes *Elaphe obsoleta*, which had been preying on adults and young in nests, mobbed whenever they emerged from nest-holes (Blem 1979). However, no mobbing or other response seen when *E. obsoleta* made similar inroads into another colony (Plummer 1977).

(Figs by M J Hallam: A from photograph by F Pölking in *Orn. Mitt.* 1983, 35, 117; B-D from drawings in Kuhnen 1985.)   EKD

Voice. Freely used throughout the year. Generally hoarser, less liquid, and less musical than House Martin *Delichon urbica* (P J Sellar). Account includes North American material; differences in voice, if any, not known.

CALLS OF ADULTS. (1) Song. A harsh twittering, little more than a sequence of call 2 (Witherby *et al.* 1938b; Bergmann and Helb 1982, which see for sonagram). Song usually given near nest-site by ♂ as self-advertisement in flight or from nest-hole (Sieber 1980; Bergmann and Helb 1982), during mate-guarding chases (Petersen 1955), and when mounting ♀ (Sieber 1980); by both sexes in exchanges during pair-formation, and to threaten conspecifics (Petersen 1955). Quality of song varies with function and context; 3 variants distinguished by Petersen (1955). (a) During pair-formation, a soft subdued pleasing twittering by both members of pair sitting side by side or facing each other in nest-entrance. (b) In threat, a loud coarse twittering, broken into long irregular phrases, and continuing for 1 min or more. More generally used by site-advertising ♂, perched or in Circling-flight, to attract ♀♀ and rebuff rivals: described as a rapid series of soft 'dsch' sounds ('territorial song': Kuhnen 1985). Described by Sieber (1980) as several syllables of equal pitch getting progressively shorter during sequence. (c) In mate-guarding chases, not as loud and harsh as in

threat, nor as soft and murmuring as in pairing. In recording of aerial chase, song an urgent, almost continuous, harsh bubbling chatter (P J Sellar); at one point (Fig I), 'chik-ik chik-ik cheik cherk cherk cherk cherk',

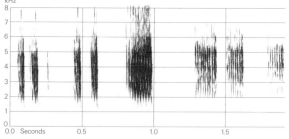

I   P J Sellar   Sweden   May 1973

with crescendo to 5th unit, then diminuendo to last, this resembling song-type 1b (J Hall-Craggs); later in same recording, song an almost continuous chattering 'ch-cher ch-cher cher chi-chi-chi-chi-chi-ch-chi-chi-i-i-i-i' (Fig II), possibly song-type 1c (J Hall-Craggs). Recording by

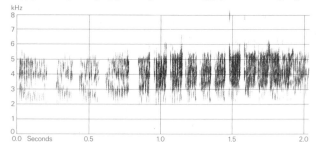

II   P J Sellar   Sweden   May 1973

P A D Hollom of song from 2 birds flying together includes several rattle sounds, though not certain that these given by singing bird; rattles less uniform in loudness and less regular than those given by singing *D. urbica* (J Hall-Craggs). Song given throughout breeding season, also reported during spring migration (Alexander 1917). (2) Contact-call. Commonest call, given perched or in flight: a harsh grating 'tschrd' or 'tschr', not as trilling as *D. urbica* (Bergmann and Helb 1982, which see for sonagram). In recordings, a brief single rasp (Fig III), also a disyllabic 'brrabit' (P J Sellar). Strident 'dsch' or 'dsch-dsch' given by ♂ before landing at site after Circling-flight (Kuhnen 1985), and low-pitched 'buzz' of ♂ flying near nest, possibly to stimulate fledging of young (Petersen 1955), are probably this call. 'Humming' sound, given by bird landing in nest-entrance and running up tunnel to relieve sitting mate (Asbirk 1976) is perhaps a subdued version. (3) Excitement-call. Series of 'dschäd' sounds, given in confrontations with conspecific birds, e.g. when contesting perch-sites in pre-migratory flocks (Bergmann and Helb 1982, which see for sonagram). This is probably the 'trrp' or 'schrrp' sounds given in rapid succession by ♀ towards intruding conspecifics during nest-building (Kuhnen 1985); probably also the

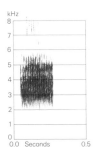

III S Palmér/Sveriges Radio (1972–80) Sweden June 1959

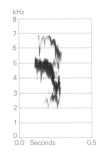

IV V C Lewis England May 1978

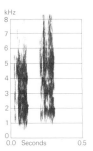

V R Margoschis England August 1968

call given by both sexes and described by Sieber (1980) as noisy with several units of equal length, higher than call 1; probably also the harsh excited 'jid-did-did', louder than call 2 or song, given as ♂ apparently attempted to copulate in the air (Tooby 1947). (4) Warning-call. 'Tsee-er' or 'tsee-ip', similar to *D. urbica* but typically much more creaky, lower-pitched, and less shrill (Tooby 1947); also rendered a sharp 'zier', falling in pitch (Bergmann and Helb 1982, which see for sonagram). In Fig IV, sounds like 'TREEip'. Said to be given typically 3 times, by 1st bird to detect danger (Windsor and Emlen 1975). Elicited by sudden appearance of anything alarming (C J Mead; see also Social Pattern and Behaviour). (5) Alarm-call. 'Ret' (Witherby *et al.* 1938*b*); lower-pitched than Warning-call, and given singly by birds alerted to danger by call 4 (Windsor and Emlen 1975). (6) Distress-call. A loud shrill cry, given a few times but never in a long series, by some birds when handled, and combined with violent struggles to escape (Stoner 1936). (7) Feeding-call. A series of peculiarly sweet fine notes, much higher in pitch than usual, given by adult entering nest-hole to feed small young, softly to older young unresponsive to proffered food (Beyer 1938).

CALLS OF YOUNG. Soft low food-calls given from day of hatching, becoming louder with age (Asbirk 1976). In recordings, evidently small young give irregular 'pop' sounds, presumably bill-snapping (J Hall-Craggs); then, when food-bearing parent arrives, an explosive squealing, giving way to squeaking sounds (P J Sellar). In recording of older young, food-calls a clamour of much hoarser and deeper sounds, rather like damp chamois leather rubbed hard and abruptly against glass (E K Dunn). When parent approaches nest entrance, series of loud sharp chirps given incessantly by older young until fed, or until parent leaves (Petersen 1955). At *c.* 15–17 days, food-calls show individual differences, by which parents recognize own young (Hoogland and Sherman 1976; Bergmann and Helb 1982). For a few days prior to fledging, young often appear at nest-entrance, twittering and warbling much like adults, but lower and harsher; within a few more days, however, no pronounced difference in timbre. Some well-grown nestlings give Distress-call, like adult, when handled (Stoner 1936). Contact-call of juvenile 'bra-bit' (P J Sellar: Fig V), softer than adult (Tooby 1947). EKD

Breeding. SEASON. Britain and western Europe: see diagram. Little variation across range, though up to 2 weeks earlier in south (Makatsch 1976). SITE. Hole in river bank, sand quarry, or sea-cliff, occasionally in drain pipe or other artificial hole. Of 104 nests in southern England, 40% in river banks, 44% in quarries, remainder miscellaneous; of 58 nests in northern England, 52% in river banks and 33% in quarries; mean height of nest-hole above ground 1·8 m in river banks, 3·9 m in quarries, 3·7 m in coastal sites (Morgan 1979). Colonial. Nest: excavated hole, mean length 65 cm (35–119, but most 46–90), *n* = 29 (Hickling 1959), ending in chamber 4–6 cm in diameter. Nest-cup made of feathers, grass, leaves, etc. Building: by both birds, taking up to 14 days; mean 4·4 days, *n* = 96 (Sieber 1980). Once burrow *c.* 8 cm deep, proceeds at *c.* 13 cm per day (Petersen 1955); overall rate 8–10 cm per day (Stoner 1936). Nest built in further 1–3 days, started immediately tunnel complete (Asbirk 1976; Sieber 1980). Begun by ♂, continued by both, completed by ♀ (Kuhnen 1985). For detailed description of digging and associated behaviour, see Hickling (1955), Asbirk (1976), and Sieber (1980). EGGS. See Plate 80. Sub-elliptical, smooth and fairly glossy; white. Nominate *riparia*: 17·5 × 12·6 mm (15·2–20·0 × 10·7–13·7), *n* = 250; calculated weight 1·43 g. *R. r. shelleyi*: 16·9 × 12·2 mm (15·2–19·0 × 11·7–13·0), *n* = 25; calculated weight 1·37 g (Schönwetter 1979). Clutch: 4–6 (2–7). Of 56 clutches, Britain: 2 eggs, 2%, 3, 5%; 4, 27%; 5, 45%; 6, 21%; mean 4·78 (Morgan 1979). 2 broods, except in north and east of range where only 1;

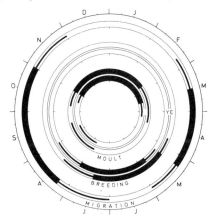

2nd clutch normally laid in refurbished 1st-brood nest, though tunnel sometimes lengthened a little (Sieber 1980). 1st egg laid 5 days (0–8) after nest completed (Asbirk 1976). Period between fledging of 1st brood and start of 2nd clutch 7·5 days (Morgan 1979). INCUBATION. 14–15 days. By both parents, starting with last or penultimate egg; hatching near-synchronous. Parents incubate in roughly equal shares during day; at night, of 32 nests, ♀ alone found on 21, ♂ alone on 2, and both on 9 (Petersen 1955). YOUNG. Altricial and nidicolous. Cared for and fed by both parents. FLEDGING TO MATURITY. Fledging period 22·3 ± SD2·1 days, $n = 30$ (Turner and Bryant 1979). Age of independence c. 30 days. Age of first breeding 1 year. BREEDING SUCCESS. Of 187 eggs laid in 39 clutches in artificial burrows, Denmark, 88% hatched and 78% fledged, or 69% overall success (Asbirk 1976). Breeding success correlated with length of burrow and with synchrony of broods in sub-colonies; of 312 eggs laid by 64 pairs in burrows less than 70 cm long, 50·9% produced fledged young; of 250 eggs laid by 48 pairs in burrows over 70 cm in length, 73·2% produced fledged young (Sieber 1980). In south-west Scania (Sweden), over 11 years, 10–25% or more of nests destroyed, half attributed to sand-falls, rest to predation by badgers *Meles meles*; at one colony, 73·5% of 408 burrows destroyed by badgers (C Persson).

Plumages (nominate *riparia*). ADULT. Entire upperparts including upper wing-coverts dark earth-brown or dark grey-brown; slightly greyer and feather-tips faintly edged pale grey in fresh plumage, duller and more uniform olive-brown when worn; pale fringes of fresh plumage widest on forehead, rump, upper tail-coverts, and median and greater upper wing-coverts, usually still showing on arrival at breeding grounds, when forehead may appear pale grey, buff, or pale grey-brown; colour all uniform June–July, but some scattered new white-tipped feathers may appear again from August onwards. Lores and feathering near gape pale grey-brown or pale grey, often slightly tinged buff when fresh; dull black spot in front of eye. Upper cheeks and ear-coverts earth-brown or grey-brown, colour similar to crown or slightly paler, faintly speckled paler grey in fresh plumage. Chin and throat silky-white, usually with faint pale buff tinge when fresh (in particular on chin), extending to sides of neck and behind ear-coverts as narrow white band, gradually merging into grey-brown of hindneck; chin and lower cheek occasionally speckled grey-brown. Chest with distinct and contrasting grey-brown band, widening towards sides of breast; in fresh plumage, grey-brown feather-tips of chest-band fringed pale grey or pale grey-brown, especially on central chest, and band then often less distinct in middle. Flanks dark grey-brown; remainder of underparts white, feathers of central breast with dark grey-brown spots and white fringes, showing as mottled patch of variable extent. Tail, flight-feathers, and greater upper primary coverts black-brown with slight olive tinge and faint green gloss; tail-feathers and secondaries faintly edged white along outer webs and tips when fresh, outer web of p9 narrowly edged white; outermost primaries and tips of others blacker, less tinged olive. Under wing-coverts and axillaries dark grey-brown, lesser coverts fringed pale grey along tips; undersurface of flight-feathers slightly paler and greyer than under wing-coverts,

shafts of primaries white below, dark horn-brown above. NESTLING. At hatching, down rather long, consisting of sparse tufts on crown and back; pale grey or grey-white. By 10th day, appears spiny due to growth of closed feather sheaths with almost full coat of dense, short grey-brown down between (Heinroth and Heinroth 1924–6); see also Bent (1942). JUVENILE. Rather like adult, but feathers of upperparts (including upper wing-coverts and tertials) fringed pink-cinnamon, rufous-cinnamon, or (sometimes) cream-white along tips; fringes ill-defined, widest on forehead, back to upper tail-coverts, longer upper wing-coverts, and tertials, narrower on crown, mantle, and scapulars. Fringes strongly subject to bleaching and abrasion: bleached to off-white at age of 1–2 months (slightly later on rump and upper tail-coverts), and largely worn off by 2–4 months (but traces still visible on tertials, greater upper wing-coverts, and back to upper tail-coverts). Sides of head and underparts as in adult, but lores, lower cheeks, sides of neck, and chin to chest strongly washed cinnamon or rufous-buff, remainder of underparts less strongly tinged pink-buff or cream; dark chest-band often hidden under extensive rufous wash on centre. Tail- and flight-feathers as adult, but narrow edges along tail-feathers and along tips of inner secondaries slightly wider than in fresh adult; tail- and flight-feathers show narrow pale edges when those of adult are usually worn off. Some individual variation in extent and colour of pale fringes and rufous wash: fringes and wash usually extensive and rufous, but in some birds fringes largely absent on upperparts (except for tertials, greater upper wing-coverts, rump, and upper tail-coverts), more closely resembling adult, or fringes are pale instead of rufous, even in fresh plumage; rufous wash below occasionally almost absent, even in fresh plumage, restricted to some feather-fringes on chin, cheeks, sides of neck, and central chest. In worn plumage, similar to adult, but traces of off-white fringes remain along longer upper wing-coverts, tertials, rump, secondaries, and inner primaries, fringes of inner greater coverts and tertial coverts still tinged buff; partly moulted adult at same time of year often shows white fringes along new inner primaries and feathers of rump, but not along all secondaries, greater coverts, and tertials, and these are not partly tinged buff.

Bare parts. ADULT. Iris dark brown. Bill black or brown-black. Leg and foot black-brown or dark brown. NESTLING. Bare skin flesh-pink at hatching; mouth yellow, unspotted; bill flanges pale yellow. JUVENILE. Iris dark brown. Bill horn-brown, flanges along gape yellow in first months after fledging. Leg and foot flesh-brown or horn-brown at fledging, soles paler; gradually darken after fledging, soles last, but no information on timing. (Heinroth and Heinroth 1924–6; Witherby *et al.* 1938*b*; BMNH, ZMA.)

Moults. ADULT POST-BREEDING. Complete; primaries descendant. Starts on breeding grounds with part of mantle, scapulars, tail-coverts, tail, and occasionally tertials, mainly from August-September; rarely a few inner primaries replaced in northern part of breeding range, more often in south. In colonies in Oldenburg (West Germany), 3·8% of adults had started body moult in June, 2·2% tail, 0·2% tertials ($n = 600$); in July, 3·6% body, 10% tail, 0·6% tertials ($n = 500$); in August, 29% body, 14% tail, 2% tertials, 0·4% primaries ($n = 450$); birds in a roost, 21 August, had up to 6 tail-feathers new and 8 out of 11 showed some body moult; more ♀♀ than ♂♂ had started (Bub and Herroelen 1981). In England, 1·9% of 3465 birds had 1–2(–5) primaries new in autumn (Mead 1975, 1980). In Sweden, 1 of c. 10 birds had new or moulting p1–p2 early

September (Persson 1979). Of 445 adult migrants in western Tien Shan (USSR), mid-August to late September, 33% had 1-3(-6) primaries replaced (most often p1-p2, which tend to be shed together); through the period, no increase in proportion of birds which had started moult; moult suspended in 73% of birds; of those examined, 33% of *diluta* had started ($n=424$), 24% of nominate *riparia* ($n=21$) (Gavrilov 1971). Suspended moult also recorded on autumn arrival at Lake Chad (Dowsett 1971) and during autumn in Spain, Greece, Cyprus, and Saudi Arabia (Mead 1980). In Manchuria, all of 18 *ijimae* in primary moult in second half of August, with up to 5 primaries replaced (Piechocki 1958). In central Africa, moult starts with longer upper wing-coverts and tertials in October, soon followed by innermost unmoulted primary; tail moulted from t1 to t6, starting with t1 when primary moult about half-way; body starts mainly when t1(-t2) full-grown; all moult completed February or later (Bub and Herroelen 1981). Secondaries start (with s1 first) when $c. \frac{2}{3}$ of primaries completed (Bub and Herroelen 1981) or with p4 (Mead 1980). In Zambia, moult of primaries starts approximately mid-October to early December, completed p9 about mid-February to early April; estimated moult duration 135 days (Ginn and Melville 1983). In Kenya and Uganda, moult starts late October or November, flight-feathers completed mid-March to mid-April, body $c.$ 2 weeks later (Ginn and Melville 1983; D J Pearson). In birds examined from Surinam, moult starts with p1, some wing-coverts, and tertials mid-October to mid-November; t1 follows from about primary moult score 10-20, head and body mainly from score 30-40; all completed about early February to mid-March (RMNH). Moult of *shelleyi* from Egypt apparently earlier, birds with moult completed or nearly so occurring from mid-December (BMNH). POST-JUVENILE. Complete; primaries descendant. Starts later than adult; much variation in timing, in relation to variable hatching period. In roost at Oldenburg, 21 August, 48% of 42 birds showed some moult of body feathers (Bub and Herroelen 1981); in England, none of $c.$ 25 000 in primary moult in autumn (Mead 1975, 1980). Some records of birds starting primary moult from Spain and Greece (Ginn and Melville 1983), but moult mainly in winter quarters. Moult on average 4-6 weeks later than adult, completing April or later (Bub and Herroelen 1981); $c.$ 6 weeks later in Surinam (RMNH), but samples small and probably much variation. Duration of primary moult estimated $c.$ 141 days in West Africa, 120-150 days in East Africa (Mead 1980; Ginn and Melville 1983).

Measurements. Nominate *riparia*. West and central Europe (mainly Netherlands), April-September; skins (RMNH, ZMA). Tail to t6; fork is tip of t1 to tip of t6. Bill (S) to skull, adult only; exposed culmen of adult on average 4·1 mm shorter than bill (S), bill to nostril 5·1 shorter; juvenile bill (S) in August 0·6 shorter than adult.

WING AD	♂ 106·7 (2·15; 25)	103-111	♀ 106·5 (1·64; 16)	103-110	
	104·4 (2·31; 12)	102-109	103·0 (2·20; 13)	99-107	
TAIL AD	51·0 (1·80; 24)	48-54	51·0 (1·61; 17)	48-54	
JUV	47·4 (1·82; 10)	45-50	47·0 (2·48; 8)	44-50	
FORK AD	9·1 (1·49; 24)	7-13	9·7 (1·48; 17)	7-13	
JUV	7·9 (1·70; 11)	5-10	7·6 (0·94; 9)	6-9	
BILL (S)	9·9 (0·49; 10)	9·2-10·5	9·9 (0·86; 15)	8·9-11·2	
TARSUS	10·3 (0·43; 13)	9·6-10·8	10·6 (0·55; 18)	9·8-11·5	

Sex differences not significant. Juvenile wing, tail, and fork significantly shorter than adult; bill full-grown from shortly after fledging, tarsus on 11th day (Turner and Bryant 1979).

Wing of live adults: (1) Finland, (2) Oldenburg (West Germany),

(3) northern France, (4) Hungary (Vie 1975; Bub and Herroelen 1981).

(1)	♂ 107·3 (2·56; 183)	102-115	♀ 108·0 (2·77; 111)	102-117
(2)	107·1 (2·45; 154)	101-115	107·3 (2·49; 206)	100-115
(3)	106·2 (2·80; 390)		106·1 (2·68; 343)	
(4)	106·7 (2·76; 118)		107·6 (3·15; 174)	

Wing of 12 611 spring migrants, Tunisia, 99-117, exceptionally 96-120 (Bub and Herroelen 1981). North American adults: wing 102·9 (2·86; 14) 99-107, tail 48·3 (1·39; 11) 45-50 (RMNH, ZMA).

R. r. diluta. Wing. USSR: ♂ 101·6 (18) 97-105, ♀ 100·1 (20) 96-108 (Dementiev and Gladkov 1954*b*); ♂ 102 (10) 96-108 (Vaurie 1959). India: (99-)102-108 (Ali and Ripley 1972); 88-98 (Hartert 1921-2; Ticehurst 1922*b*). Spring migrants, Afghanistan: 104·6 (5) 102-108 (Paludan 1959).

R. r. shelleyi. Egypt and northern Sudan, December-May; skins (BMNH).

WING AD	♂ 94·6 (1·43; 7)	92-96	♀ 93·3 (3·53; 6)	88-98
TAIL AD	42·9 (2·59; 7)	39-47	42·6 (2·56; 6)	39-45
FORK AD	6·6 (0·93; 7)	6-8	5·9 (1·43; 6)	4-8

Sex differences not significant.

R. r. ijimae. Wing, 97-111 (Hartert and Steinbacher 1932-8; Ali and Ripley 1972; Piechocki and Bolod 1972); Manchuria, adult 105·2 (18) 100-109, juvenile 105·0 (19) 100-109 (Piechocki 1958).

Weights. Nominate *riparia*. Britain, range of over 4000 birds: 10·2-19·1, mainly 11·4-14·5 (Pepler 1966). Münsterland (West Germany), July-August: adult ♂, 12·7 (1·10; 19) 11·4-16·5, adult ♀ 13·6 (1·0; 32) 11·2-15·6, juvenile 12·2 (0·78; 76) 10·5-13·7 (Berger and Kipp 1966). Netherlands, July-August: juvenile ♂ 13·2 (0·84; 7) 11·8-14·3, juvenile ♀ 13·4 (0·87; 9) 11·5-14·5 (ZMA). Finland: ♂ 11·5-17 (183), ♀ 11-19·5 (177); average before 20 June, ♂ 15·0 (48), ♀ 15·6 (39); average after 20 June, ♂ 14·1 (53), ♀ 14·6 (65) (Bub and Herroelen 1981). Yamal peninsula (USSR): 16·1 (8) 14-19 (Danilov *et al.* 1984).

On spring migration, Portland (southern England): 12·8 (11) (Ash 1969). Malta: 13·2 (1·2; 50) 11·0-16·0 (J Sultana and C Gauci). Morocco: 11·3 (255) 8·9-17·3; average in morning 11·1 (122), in afternoon, 11·5 (128); after a cold spell, 3 dead birds averaged 8·9 (Ash 1969). Asia Minor: 13·5 (33) 10-20 (Vauk 1973*a*). Lake Chad (Nigeria), late March to mid-April: 13·2 (2·18; 32) 10·8-20·4 (Fry *et al.* 1970).

On autumn migration. Malta: 14·2 (1·5; 50) 11·5-17·5 (J Sultana and C Gauci). Northern Iran: 14·4 (4) 13·6-16·0 (P J K Burton, BTO). In winter quarters, exhausted ♀, shortly after arrival in South Africa, November: 9·4 (Skead and Skead 1970). Zambia, February: 11·4 (0·96; 43) 9-13 (K-H Loske). Uganda: November-February 12·8 (0·6; 107) 10·5-14·4, March 13·6 (0·9; 50) 11·4-15·7, April 14·4 (1·2; 385) 12·5-19·8, May 17·9 (1·3; 4) 16·5-19·5 (Pearson 1971).

R. r. diluta. Kirgiziya (USSR): ♂ 11·5-16·5 (8), ♀ 13·0 (1) (Yanushevich *et al.* 1960). USSR: ♂♂ 11·1, 12·9; ♀ 12·4 (6) 10·1-14·8 (Dementiev and Gladkov 1954*b*). India: 11·5 (18) 10-14 (Ali and Ripley 1972). Afghanistan, spring migrants: 14·4 (5) 14-15 (Paludan 1959).

R. r. ijimae. Manchuria (China), second half of August: adult 14·6 (19) 13-16, juvenile 14·5 (19) 12-16 (Piechocki 1958).

NESTLING. For growth, see Petersen (1955) and Turner and Bryant (1979).

Structure. Wing long and pointed, narrow at base. 10 primaries: p9 usually longest, p8 1-4 shorter, p7 7-12, p6 14-21, p5 21-29, p1 50-57; rarely, p8 longest and p9 up to 1 shorter. P10 reduced, narrow and pointed, hidden below greater primary coverts; in adult, 70-77 shorter than p9, 7-10 shorter than longest upper primary covert; in juvenile, 65-74 shorter than p9, 8-10 shorter than longest primary covert. Flight-feathers not emarginated. Tertials shorter than secondaries. Tail rather short, forked; t1 7-13 shorter than t6 in adult, 5-10 in juvenile. Bill broad, short, flattened, triangular when seen from below; culmen with blunt ridge, ending in slightly decurved tip; gape wide. Nostril partly covered by membrane. Tarsus short and slender, bare except for small feather-tuft on lower rear end; toes rather long and slender. Length of middle toe with claw 13·3 (25) 12-16; outer and inner toe with claw both *c.* 65% of middle, hind *c.* 70%. Claws fine and sharp; length varies with digging activity.

Geographical variation. Slight, mainly clinal; involves size (as expressed in wing length), depth of ground-colour of upperparts and chest-band, and width and contrast of chest-band. Size probably clinally smaller towards south. Nominate *riparia* from northern and central Europe east to Siberia largest, birds from southern Europe, north-west Africa, Turkey, and Levant perhaps slightly smaller (but only a few certain breeders examined). *R. r. diluta* and *ijimae* from central Asia (Ural river

and Transcaspia to Kuril Islands and Hokkaido) rather small: average wing of various populations *c.* 101-104, range 96-111. *R. r. shelleyi* from Egypt and apparently Indian populations of *diluta* smallest, average wing *c.* 93, range 88-98; Indian birds perhaps separable as *indica* Ticehurst, 1916; see, e.g., Whistler (1916) and Ticehurst (1922b). Nominate *riparia* from North America about as large as *ijimae* (see Measurements). Nominate *riparia* rather dark, and chest-band, though variable in width, generally rather dark and clear-cut. *R. r. diluta* distinctly paler on upperparts than nominate *riparia*, pale grey-brown or dark mouse-grey, chest-band paler, narrower, and less clear-cut. *R. r. shelleyi* has slightly paler upperparts and chest-band than nominate *riparia*, but not as pale as *diluta*; chest-band narrower than in nominate *riparia*, but clear-cut (unlike *diluta*). *R. r. ijimae* from south-eastern part of species' range is slightly darker on upperparts and chest than nominate *riparia*, blacker, less brown; chest-band distinct and broad; pale edges to flight-feathers and rump feathers more conspicuous. Races with contiguous boundaries have broad intergradation zones.

Forms superspecies with Congo Sand Martin *R. congica* and Brown-throated Sand Martin *R. paludicola* (Hall and Moreau 1970). CSR

Ptyonoprogne fuligula African Rock Martin (includes Pale Crag Martin)

PLATES 16 and 18
[between pages 232 and 233]

Du. Vale Rotszwaluw FR. Hirondelle isabelline GE. Steinschwalbe
RU. Африканская скалистая ласточка SP. Avión roquero africano SW. Blek klippsvala

Hirundo fuligula Lichtenstein, 1842

Polytypic. *P. f. obsoleta* (Cabanis, 1850), Egypt, Sinai, Dead Sea region, Arabia (except south-west and north-east), and south-west Iran; *perpallida* (Vaurie, 1951), north-east Arabia and southern Iraq; *presaharica* (Vaurie, 1953), foothills of Atlas Saharien in Algeria and Morocco south to Timimoun and El Golea, and (perhaps this race) Mauritania; *spatzi* (Geyr von Schweppenburg, 1916), Ahaggar and Tassili N'Ajjer (southern Algeria), Fezzan (southern Libya), Tibesti and Ennedi (Chad), and Mali. Extralimital: *pallida* Hume, 1872, central and eastern Iran, Afghanistan, and Pakistan; *buchanani* (Hartert, 1921), Aïr (Niger); *arabica* (Reichenow, 1905), Gebel Elba, Red Sea province of Sudan, south-west Arabia, northern Somalia, and Socotra, grading into *obsoleta* in western Saudi Arabia; *pusilla* (Zedlitz, 1908), from Mali to western and central Sudan, Eritrea, and central Ethiopia, merging into *obsoleta* in Nile valley of northern Sudan; *rufigula* (Fischer and Reichenow, 1884) (synonym: *fusciventris* Vincent, 1933), southern Sudan and East Africa, south to Zimbabwe and northern Moçambique; *bansoensis* (Bannerman, 1923), west-central Africa; nominate *fuligula* (Lichtenstein, 1842), eastern Cape Province (South Africa); 2 further races in central and southern Africa.

Field characters. 12·5 cm; wing-span 27·5-30 cm. Only marginally larger but more robust than Sand Martin *Riparia riparia*; *c.* 15% smaller than Crag Martin *P. rupestris*. Similar to *P. rupestris* in structure and flight, and in rather uniform plumage with white tail-spots, but paler and greyer with white (unstreaked) throat and less contrasting dark under wing-coverts. Sexes similar; no seasonal variation. Juvenile separable at close range. 4 races in west Palearctic, differing most in tone of upperparts and fore-underparts; darkest and palest west Palearctic races described here (see also Geographical Variation).

ADULT. (1) Race of Saharan massifs, *spatzi*. Darkest west Palearctic race. Upperparts, vent, and undertail dusky brown. Like *P. rupestris*, lateral under tail-coverts marked with dull white spots, and inner webs of all but central tail-feathers spotted almost white below tips (though tail-spots rather less obvious than in *P. rupestris* and visible from above only when tail fully spread). Unlike *P. rupestris*, chin and throat unstreaked, dull white, at close range contrasting with black-brown lores and brown head-cap. Underwing mainly pale buff-grey, with coverts less dark than blackish feathers of *P. rupestris* though still appearing dusky in wing-pit and behind

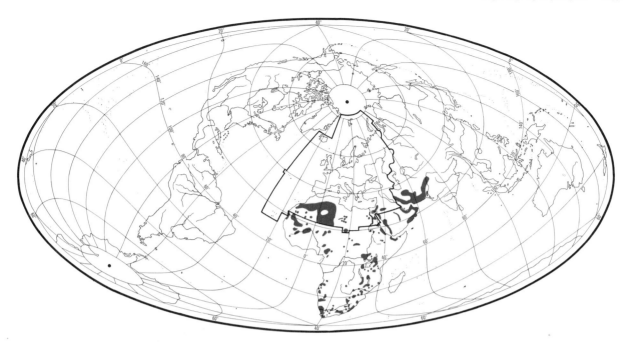

leading edge. (2) South Iraq and east Arabian race, *perpallida*. Palest of all races but not most uniform, with distinctly greyer upperparts than *spatzi* and *P. rupestris* (showing almost chalky 'saddle' between browner wings) and much whiter underparts except for dusky undertail and more contrasting dark under wing-coverts than *spatzi*. JUVENILE. All races. Paler above than adult, with buff to off-white feather-fringes visible at close range; buffier below than adult, appearing even more uniform. At all ages, bill black; legs brown to black (flesh-toned in juvenile).

Darker races invite confusion with *P. rupestris* but with experience, smaller size, less dark underwing, and pale throat and forebody allow ready distinction. Paler races do not immediately suggest *P. rupestris*, and can be mistaken for *R. riparia* (when lack of chest-band not obvious) and Brown-throated Sand Martin *R. paludicola* (which lacks chest-band but has darker throat and forebody). Separation from *Riparia* best based on vent and tail patterns, bulk, and flight action. Like *P. rupestris*, flight involves much steady gliding; not fluttering like *Riparia*. Behaviour typical of family but essentially a desert species, and locally more gregarious than *P. rupestris*.

A quiet bird, though song and calls apparently resemble *P. rupestris*.

Habitat. In warm dry lower latitudes, from fringe of Mediterranean through subtropical to tropical, forming counterpart of Crag Martin *P. rupestris* in tropical and arid mountainous and rocky areas. In North Africa frequents gorges and ravines in desert areas, but in Egypt found near monuments (e.g. Abu Simbel) and in certain desert towns (e.g. Aswan). In Sudan, nests on face of crags (Cave and Macdonald 1955; Etchécopar and Hüe 1967). In Sierra Leone, resident on crags, and on rock faces at lower elevations; frequents summit slopes and cliffs but avoids open plateaux (G D Field). In eastern Saudi Arabia, present all year at breeding sites (rocky jebels and escarpments in desert), but in winter also occurs at pools well away from these (G Bundy); in Hejaz and Asir apparently prefers lower and drier hills than *P. rupestris* breeding there (Jennings 1981). In Morocco, hundreds spend winter over swamps and lake at Rincon (Smith 1965a). Otherwise not known to differ significantly in habitat from *P. rupestris*.

Distribution. Desert distribution often inadequately known.

IRAQ. In gorges in north—not sure whether this species or Crag Martin *P. rupestris* (Moore and Boswell 1956); *P. rupestris* identified there by Chapman and McGeoch (1956) and McGeoch (1963).

Population. Little known on numbers or trends.

EGYPT. Locally common (PLM, WCM). ISRAEL. In last 10 years, breeding on houses in the Aravah and Negev (UP).

Oldest ringed bird 6 years 9 months (Dejonghe and Czajkowski 1983).

Movements. Mainly resident, but also some seasonal movement of unclarified extent.

North African (Saharan) populations often resident, though some movement is indicated by local changes in numbers, particularly October–January (Heim de Balsac

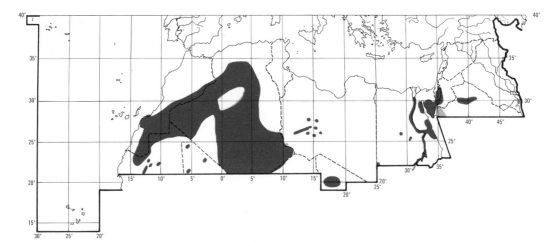

and Mayaud 1962; Etchécopar and Hüe 1967); breeding sites reoccupied from February. Not known how far such birds move, though *presaharica* (breeding in western Morocco and northern Nigeria) has been collected in Ahaggar (southern Algeria) in winter (Vaurie 1959).

In Saudi Arabia, largest flocks occur December–January, when birds move around to exploit food concentrations; in interior, these midwinter numbers larger than those present September–November (Jennings 1980). In Oman, where common resident, numbers augmented in autumn and winter, with passages in August–November and February–March (Gallagher and Woodcock 1980; Walker 1981*a,b*); origins of such birds unknown, but possibly include Iranian ones. In Pakistan and northern India resident with local movements (Ali and Ripley 1972). RH

Food. Insects, taken in flight. Like Crag Martin *P. rupestris*, flight involves much steady gliding (e.g. Meinertzhagen 1930, Archer and Godman 1961*b*). Prey typically taken in aerial-pursuit, though birds watched at Jiddah (Saudi Arabia) fed by 'hovering' in airstream just above crest of ridge (Bates 1936–7). Apparently hunts usually at no great height, e.g. within 10 m of ground at Amani, Tanzania (Moreau 1939; also Clancey 1964). In South Africa, seen hawking over rocks exposed at low tide, and also sweeping down over wet beach with obvious intention of disturbing prey (Skead 1966). In Kenya, joins other Hirundinidae and swifts (Apodidae) in feeding flocks over grass fires, typically frequenting edge of smoke cloud (Someren 1958); will travel many km to reach such a fire (Mackworth-Praed and Grant 1960). Foraging area at Amani very restricted: at one nest, all insects caught within *c.* 40 m radius (Moreau 1939). Said by Mackworth-Praed and Grant (1963) to 'fly on after dark'.

Few data on prey species, but perhaps often larger and harder-bodied than taken by most other west Palearctic Hirundinidae. 7 stomachs from Rio de Oro, April–May, contained Hymenoptera (of medium size, one of 11 mm),

flies (including *Lucilia*), beetles (Coleoptera: one of 12 mm), and other unidentified insects (Valverde 1957). 7 stomachs from Jiddah contained mostly Hymenoptera and beetles; also Hemiptera, and 1 contained many ants (Formicidae) (Bates 1936–7). Said to take mosquitoes when hunting over water, Chad (Salvan 1967–9).

No information on food of young. At Amani, rate of food delivery to nest increased steadily during first half of nestling period, then stable up to fledging. Average rate increased with brood size: for brood of 1, every 4·5 min; brood of 2, every 2·7 min; brood of 3, every 1·9 min (Moreau 1939). DJB

Social pattern and behaviour. Little information for west Palearctic, and account based largely on South African sources.

1. In central Arabia, small groups and sometimes large flocks form after breeding season, moving around to exploit local concentrations of food; largest flocks up to 300, December–January (Jennings 1980). At Ennedi (Chad), encountered singly but more often in flocks of up to 50 (Niethammer 1955*b*). Commonly associates with other Hirundinidae and swifts (Apodidae) (Serle 1939–40; Chapin 1953; McLachlan and Liversidge 1970; Frandsen 1982). BONDS. No information on mating system. Members of pair share nest-building, incubation, brooding, and feeding of young (Sclater and Moreau 1933; Serle 1955; Brooke and Vernon 1961*a*; Myburgh and Steyn 1979); at one nest, pair-members shared these activities equally (Brooke and Vernon 1961*a*). Young fed by parents for several days after fledging, in one case for at least 19 days (Brooke and Vernon 1961*a*). BREEDING DISPERSION. Usually nests in isolated pairs (Ticehurst and Cheesman 1925; Heim de Balsac and Heim de Balsac 1954); south of the Sahara, more often in colonies (Brooke and Vernon 1961*b*; Clancey 1964; MacLachlan and Liversidge 1970; Farmer 1979), e.g. in South Africa one colony of 40 pairs (MacLeod *et al.* 1952). In suburbs of Stellenbosch (Cape Town, South Africa), density *c.* 5 pairs per km² (Siegfried 1968). Site fidelity, facilitated by durability of nests, occurs. For evidence that at least one member of same pair occupied same nest in 2 successive years, see Schmidt (1964). At colony, Ile-Ife (Nigeria), one nest used for 3 successive years (Farmer 1979), though no evidence that same birds involved. In Sierra Leone, one pair nested in colony of Lesser Striped Swallow *Hirundo abyssinica* (G D Field).

ROOSTING. Nocturnal, probably most often on cliff-ledge or equivalent site. In December, Arabia, birds roosted in old nests (Ticehurst and Cheesman 1925). In winter, Cape Town, several hundred birds regularly roosted in lines along sheltered ledges on the Houses of Parliament (Gill 1936); similar roosting reported under eaves and on leeward side of buildings in East London (South Africa); apparently enters towns to roost only in cold weather (Batten 1943). In South Africa, each of 2 successive broods returned to roost and loafed in nest for some time after fledging, first brood for 23 days (Taylor 1942). Relatively active at dawn and dusk, flying in half-light (McLachlan and Liversidge 1970), and according to Frandsen (1982) more active in early morning and at dusk than at other times. Said to continue flying after dark (Mackworth-Praed and Grant 1963). 2 reports (Ticehurst and Cheesman 1925; Moreau 1939) indicate that activity typically ceases during the slightest rain; in Arabia, birds avoided rain assiduously as soon as it began, and dashed for walls of nearby building where they clustered together 'in thirties or forties' until shower stopped (Ticehurst and Cheesman 1925). Drinks by swooping low over water and dipping surface with bill (J Stannard). In Chad, drinking reported in evening at standing water (Salvan 1967-9).

2. FLOCK BEHAVIOUR. No information. ANTAGONISTIC BEHAVIOUR. In December, Arabia, 2 birds fought furiously high in the air, several times clinging together and falling *c*. 10 m before separating (Ticehurst and Cheesman 1925). When 2 pairs of Little Swift *Apus affinis* began prospecting at colony of *P. fuligula*, residents became agitated, responding to Screaming-calls of *A. affinis* intent on landing; always chased such birds and often also attacked White-rumped Swifts *A. caffer* circling too close to their nests (Brooke and Vernon 1961*b*). One breeding pair frequently chased off ♂ House Sparrow *Passer domesticus* which was apparently trying to appropriate nest-site (Winterbottom 1967). HETEROSEXUAL BEHAVIOUR. Little known. (1) Pair-bonding behaviour. In central Arabia, pairs form and prospect nest-sites in February (Jennings 1980). In western Sahara, pair-formation likewise occurs as soon as established in breeding areas (Heim de Balsac and Heim de Balsac 1951). In Israel, April, following incident considered a heterosexual encounter, though further interpretation not possible: perched bird fluttered wings and called excitedly (see 3 in Voice) as another approached in flight to *c*. 30 cm, hovered, and turned away (P A D Hollom). (2) Mating. Only one report. At Tripoli (Libya), one bird continually dived at another which was perched on a wall, eventually mounting it and copulating without ceremony before gliding away (Bundy and Morgan 1969). (3) Behaviour at nest. Nests built or repaired soon after arrival at breeding grounds, western Sahara (Heim de Balsac and Heim de Balsac 1951). At one site, *c*. 2 weeks elapsed between completion and laying (Brooke and Vernon 1961*a*). During incubation, nest-relief usually rapid and without ceremony; birds described changing over 'very neatly with a single movement' (Sclater and Moreau 1933); in same colony, at one nest, pair typically relieved each other with clutch scarcely being uncovered (Moreau 1939). RELATIONS WITHIN FAMILY GROUP. Young brooded assiduously by both parents for *c*. 3-4 days after hatching, and to appreciable extent (at least 20% of time) for about half of nestling period (Moreau 1939). At end of nestling period, young flutter out of nest to nearby perches and back again; after finally quitting nest, young fed by both parents for several days (Brooke and Vernon 1961*a*). For roosting of young, see above. ANTI-PREDATOR RESPONSES OF YOUNG. No information. PARENTAL ANTIPREDATOR STRATEGIES. When nest closely approached by man, bird swoops at and flies closely past intruder's head, producing a clapping noise, apparently with bill (Schmidt 1964).   EKD

Voice. Relatively little used (Mackworth-Praed and Grant 1963), and birds described as 'remarkably silent' (Clancey 1964). Little information for west Palearctic, and the following account based largely on extralimital sources. For clapping sound, produced apparently by bill-snapping, as response to human intrusion at nest, see

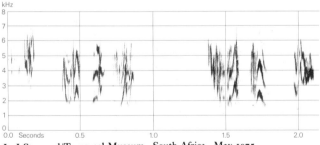

I J Stannard/Transvaal Museum South Africa May 1975

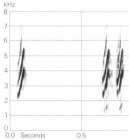

II T Harris/Transvaal Museum
South Africa March 1982

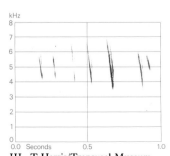

III T Harris/Transvaal Museum
South Africa March 1982

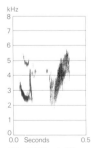

IV T Harris/Transvaal Museum
South Africa March 1982

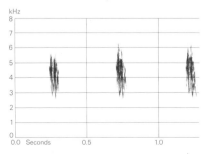

V P A D Hollom Israel April 1979

Social Pattern and Behaviour. This sound perhaps the occasionally heard 'clacking call, like a chat' (Turdidae) (Mackworth-Praed and Grant 1963).

CALLS OF ADULTS. (1) Song. Following sounds presumably song: a soft twittering (Frandsen 1982); a melodious twitter (McLachlan and Liversidge 1970); a low twitter (Mackworth-Praed and Grant 1963). In Fig I, liquid twittering song of birds drinking and feeding somewhat resembles song of Swallow *Hirundo rustica* (J Hall-Craggs). (2) Contact-call. In our recordings (all of nominate *fuligula*), sounds described by recordists as contact-calls include clipped somewhat nasal 'wit' or 'wik', rather like *H. rustica* (P J Sellar); in Fig II, 'wik wik-wik' (J Hall-Craggs); rapid high-pitched 'chi' sounds (Fig III), a series of which ends with an emphatic 'chis-wick' (Fig IV) like wagtail *Motacilla* (P J Sellar). Excepting last, these probably same as call described as a little high-pitched 'twee' (Mackworth-Praed and Grant 1963). High-pitched 'cheep-cheep-churr', repeated several times in flight (Frandsen 1982) was possibly same or closely related. (3) Other calls. Following descriptions suggest calls different from those listed above, though no details of function or context. A low burring chirp, like Sand Martin *Riparia riparia* (Sclater and Moreau 1933). A typical martin-like chortle, different from 'chirrup' of Crag Martin *P. rupestris* (Ticehurst and Cheesman 1925). Urgent rapid 'chir chir chir ...' (Fig V) by bird (*obsoleta*) perched and fluttering wings as another drew near in apparently heterosexual encounter (P A D Hollom: see Social Pattern and Behaviour).

CALLS OF YOUNG. In recording by C Haagner (South Africa), irregularly repeated 'chip chip ...' sounds, very like sparrow *Passer*, given by young; no further details.

EKD

Breeding. Little information from west Palearctic. SEASON. North-west Africa: February–April, perhaps some as early as January (Heim de Balsac and Mayaud 1962). SITE. On vertical rock surfaces, usually under overhang, mainly on cliffs, but also under bridges, on houses and other buildings, particularly under ledges and verandahs, etc. (3–)6–15 m above ground (Heim de Balsac and Mayaud 1962). Solitary. Nest: deep half-cup of mud without incorporated vegetable matter, lined with fine plant material and feathers; one nest made of 960 mud pellets (Sclater and Moreau 1933). Building: by both sexes; at one site, building continued for *c.* 40 days (Brooke and Vernon 1961*a*). EGGS. See Plate 80. Elongated, sub-elliptical, smooth and slightly glossy; white, spotted and speckled black and purple-grey, often concentrated at broad end. 18.5×13.0 mm ($17.1–19.5 \times 12.0–14.0$), $n = 15$; calculated weight 1.65 g (Schönwetter 1979). Clutch: 2–3. Up to 3 broods in Tanzania (Sclater and Moreau 1933). INCUBATION. About 17 days (16.5–18), $n = 4$ nests (Moreau 1940, 1962). In Tanzania, by both sexes in shifts averaging 2–12 min; eggs covered for

50–90% of day (Anon 1939). FLEDGING TO MATURITY. Fledging period 15(–28) days (Moreau 1940); 25–30 days, $n = 4$ nests (Brooke and Vernon 1961*a*). No further information.

Plumages (*P. f. obsoleta*). ADULT. Closely similar to Crag Martin *P. rupestris*, and sometimes indistinguishable apart from size. Upperparts and sides of head mouse-grey or pale brown-grey, paler and less brown than *P. rupestris*; feather-tips faintly fringed buff from mantle to upper tail-coverts when plumage fresh; back to upper tail-coverts often slightly paler than mantle and scapulars. Lores black-brown, contrasting more strongly with paler grey crown than in *P. rupestris*. Chin, throat, lower sides of neck, and chest pale pink-buff to isabelline-white; chin and throat unspotted (unlike *P. rupestris*), except for some brown-grey spots on cheeks and sometimes a few faint streaks on chin and sides of throat. Belly pink-buff or isabelline, like chest, but often rather more deeply coloured and with faint and fine grey shaft-streaks, gradually merging into pale brown-grey of flanks, vent, and under tail-coverts. Flight-feathers and upper wing-coverts uniform brown-grey, lesser upper wing-coverts, primary coverts, and inner webs and tips of primaries not darker than remainder of upperwing (unlike *H. rupestris*). Under wing-coverts and axillaries pale buffish-grey, coverts along leading edge of wing tipped cream-buff. Tail brown-grey; number of subterminal white or pale cream spots variable, as in *P. rupestris* (2–5 on each tail-half), but t6 more often with small spot of 2–4 mm (in 17 of 38 examined; in *P. rupestris*, in 8 out of 42), and size of spots on t2–t5 slightly smaller—length of spot on t5 8.5 (6) 6–11 in adult, 8.8 (6) 7–12 in juvenile (in adult *P. rupestris* 11.9 (12) 9–15, in juvenile 9.5 (10) 5–14). NESTLING. No information. JUVENILE. Similar to adult, but ground-colour of upperparts slightly paler and purer grey and feather-tips fringed buff to off-white along tips, widest from mantle to upper tail-coverts; underparts with deeper buff wash than adults, dark spots on cheeks absent. Tertials and greater, median, and lesser upper wing-coverts fringed buff along tips (soon bleaching to white); tips of secondaries and inner primaries narrowly edged pale buff (very fresh adult has similar edges). T6 narrower than in adult, terminal part of inner web less strongly bulging, but character useful in direct comparison only and some birds (of any age) intermediate. FIRST ADULT. Only a few examined. Similar to adult, but retains juvenile flight-feathers, tail, tertials, and greater upper wing-coverts, as in *P. rupestris*; tertials and greater coverts have traces of bleached off-white fringes, fringes of some inner greater coverts still rufous.

Bare parts. ADULT. Iris dark brown or black-brown. Bill black-brown or black, mouth pinkish-grey. Leg and foot pinkish-brown, brownish-flesh, black with slight flesh tinge, dusky brown, or black (probably more pinkish in paler races, more blackish in darker races). (Bannerman 1939; Ali and Ripley 1972; Mackworth-Praed and Grant 1973; BMNH.) NESTLING. No information. JUVENILE. Like adult, but base of lower mandible and gape-flanges pale yellow at fledging, leg and foot flesh-pink (Brooke and Vernon 1961*a*; BMNH). Not known at what age adult colours acquired.

Moults. ADULT POST-BREEDING. Starts during incubation or when feeding young, as in *P. rupestris*; sequence of moult similar. In Ahaggar (southern Algeria), starts late April to mid-May with p1, followed by scattered feathers of head and body from primary moult score 15–25, and by t1 from score *c.*

20 (ZMA). Flight-feather moult July–August in North Africa; head and body new August, slightly worn winter, heavily worn spring (Hartert 1910). In *obsoleta*, Egypt, moult starts June–July (BMNH); in Iran, moult completed or almost so late August to mid-September (Vaurie 1951*c*). In Ennedi (Chad) and Aïr (Niger), moult starts March–April (BMNH, ZFMK). POST-JUVENILE. Partial: head, body, and variable number of lesser and median upper wing-coverts. In Iran, starts early September (Vaurie 1951*b*). In April, Ahaggar, head and body fairly new but flight-feathers and tail heavily worn, 1st complete moult starting with p1 from about late April or May (ZMA).

Measurements. *P. f. presaharica* (northern Algeria), *spatzi* (Ahaggar, southern Algeria), *obsoleta* (Egypt, Sinai, Dead Sea region, and south-west Iran), and *perpallida* (Al Hufuf, eastern Saudi Arabia), combined, all year; skins (Vaurie 1951*c*, 1953; RMNH, ZMA). Tail is to tip of t4–t5, fork is tip of t1 to tip of t4–t5; bill (S) to skull, bill (N) to distal corner of nostril; exposed culmen on average 4·2 less than bill (S).

	♂		♀	
WING	119·8 (2·84; 22)	114–124	117·9 (3·89; 19)	110–125
TAIL	49·5 (2·35; 21)	47–52	49·2 (2·49; 18)	46–54
FORK	4·1 (1·08; 5)	2–5	3·2 (0·70; 7)	2–4
BILL (S)	11·0 (0·68; 21)	9·0–12·0	10·7 (0·73; 18)	9·0–12·0
BILL (N)	5·6 (0·36; 5)	5·1–6·0	5·1 (0·39; 7)	4·5–5·5
TARSUS	10·9 (0·48; 5)	10·4–11·7	10·5 (0·74; 7)	9·7–11·5

Sex differences not significant.

Wing and tail, sexes combined: (1) *presaharica*; (2) *spatzi*; (3) *obsoleta*, Egypt; (4) *obsoleta*, Dead Sea region to south-west Iran; (5) *perpallida*; (6) *buchanani* (Aïr, northern Niger) (Vaurie 1951*c*, 1953; RMNH, ZMA).

	WING			TAIL		
(1)	120·6 (2·45; 9)	117–123		50·9 (1·92; 7)	49–54	
(2)	118·7 (1·67; 8)	117–121		50·3 (1·16; 8)	49–52	
(3)	116·8 (—; 14)	110–125		48·1 (—; 14)	42–52	
(4)	121·2 (2·55; 7)	118–124		48·4 (2·17; 7)	46–52	
(5)	119·2 (—; 3)	118–121		51·3 (—; 3)	50–52	
(6)	113·2 (1·33; 6)	111–115				

Ages combined in samples above, as age of some birds unknown.

Known-age birds, Egypt and Sinai: wing, adult 121·3 (2·46; 5) 118–125, juvenile 116·3 (3·25; 6) 111–121; tail, adult 50·8 (1·04; 5) 49–52, juvenile 48·1 (1·72; 6) 46–50 (RMNH, ZMA).

Weights. *P. f. presaharica.* Figuig (south-east Morocco), April: 16·2, 24·5 (BTO). *P. f. spatzi.* Ahaggar (southern Algeria), 14–16·5 (9); Ennedi (northern Chad), ♂♂, 9, 11 (Niethammer 1955*b*).

Structure. North African populations. Similar to *P. rupestris*, except for slightly smaller size. 10 primaries: p9 longest, p8 0–3 shorter, p7 7–13, p6 17–23, p5 28–34, p4 37–43, p1 57–65; p10 reduced, 75–90 shorter than p9, 9–12 shorter than longest upper primary covert. Depth of tail-fork 2–5 mm. Length of middle toe 13·0 (12) 11–14, relatively shorter than *P. rupestris*, but remainder of leg and foot similar.

Geographical variation. Marked and complex, both in colour and size. Often split into 2 species—pale *P. obsoleta* from North Africa and Asia and darker *P. fuligula* from Afrotropics (e.g. by Mackworth-Praed and Grant 1973)—but better considered single polytypic species (following Hall and Moreau 1970) comprising 3 distinct groups: (1) *obsoleta* group (North Africa and Asia), slightly paler and smaller than *P. rupestris*, average wing 118–122; (2) *rufigula* group (West, central, and East Africa), much darker and distinctly smaller than *obsoleta* group, average wing 109–113; (3) *fuligula* group (southern Africa, north to Angola and western Moçambique), paler than *rufigula* group, but generally darker than *P. rupestris*, and larger than both of these, average wing 130–135. *Rufigula* and *fuligula* groups formerly often united in *P. fuligula*, and *obsoleta* kept separate, but in fact *obsoleta* and *rufigula* groups show complete intergradation in colour and size, while *fuligula* group apparently abruptly larger than *rufigula* group, though intergrading in colour. Position of close relatives *P. rupestris* (Eurasia) and Dusky Crag Martin *P. concolor* (India) uncertain, and sometimes all included in single species (e.g. by Meinertzhagen 1954), as *P. concolor* is dark and small like *rufigula* group and *P. rupestris* is similar in colour to north-west African races of *obsoleta* group (differing only in spotted upper throat and slightly greater size). Birds intermediate between *P. rupestris* and *obsoleta* group are known (Dorst and Pasteur 1954; Heim de Balsac and Mayaud 1962), but their breeding ranges overlap extensively in Asia, and *P. rupestris* thus treated here as separate species, following Voous (1977).

Many races have been recognized in *obsoleta* and *rufigula* groups, though some perhaps not justifiable, especially when strong influence of age, bleaching, and abrasion taken into account; Vaurie (1953, 1959) followed here. In general, colour clinally darker from north to south, and recognized races in sequence of gradually darker colour are: *perpallida*, *obsoleta*, *pallida*, *presaharica*, *spatzi*, *arabica*, *buchanani*, *pusilla*, *rufigula*, and *bansoensis*. Paler races up to *buchanani* considered to belong to *obsoleta* group, last 2 dark ones comprise *rufigula* group, *pusilla* considered to belong to either *rufigula* group (Mackworth-Praed and Grant 1973) or to *obsoleta* group (Hartert 1910); Hall and Moreau (1970) leave position of *spatzi* and *buchanani* open, considering these intermediate, and put *arabica* in *obsoleta* group. Considering also size, however, *obsoleta* group seen then to comprise *perpallida* to *spatzi* (average wing over 118), *rufigula* group comprises *buchanani*, *rufigula*, and *bansoensis* (average wing below 113), and Ennedi birds (having colour as *spatzi* from Ahaggar: Niethammer 1955*b*; ZFMK), *arabica*, and *pusilla* are intermediate. *P. f. obsoleta* from Dead Sea region slightly paler and greyer than *P. rupestris* (see Plumages), *obsoleta* from Sinai and Nile valley slightly paler (BMNH); *perpallida* from north-west Arabia and Al Faw, southern Iraq (Ticehurst 1922*b*) pale whitish-grey above, chin to chest almost pure white (Vaurie 1951*c*; BMNH); *pallida* close to *obsoleta*, but slightly darker and more sandy above; *spatzi* from Ahaggar and Fezzan similar in colour to *P. rupestris*, but upper throat uniform and throat to belly slightly deeper buff in fresh plumage; *presaharica* like *spatzi*, but upperparts slightly paler brown-grey and with more extensive greyish-buff feather fringes, appearing slightly paler and more sandy, chin to belly paler, pink-buff rather than buff when plumage fresh, whiter when worn; *buchanani* and *arabica* are both darker and more sooty-brown on upperparts than more northern races, flanks and belly to under tail-coverts darker grey-brown. For other races, see Bannerman (1939) and Mackworth-Praed and Grant (1963, 1973). CSR

Ptyonoprogne rupestris Crag Martin

Du. Rotszwaluw   Fr. Hirondelle des rochers   Ge. Felsenschwalbe
Ru. Скалистая ласточка   Sp. Avión roquero   Sw. Klippsvala

Hirundo rupestris Scopoli, 1769

Monotypic

Field characters. 14·5 cm; wing-span 32–34·5 cm. Head, body, wings, and tail all broader or bulkier than those of any other west Palearctic hirundine; 15% larger than African Rock Martin *P. fuligula*. Chunky hirundine with almost unforked tail. Essentially uniform dusky brown, with almost black under wing-coverts; at close range shows dark-mottled throat, paler and warmer buff-brown forebody, and white spots on underside of tail. Flight least energetic of west Palearctic hirundines, involving much steady gliding. Sexes similar; no seasonal variation. Juvenile separable at close range.

Adult. Upperparts brown-grey, with dusky tone in some lights but usually appearing paler than Sand Martin *Riparia riparia*; underparts smoky-buff, palest on chest; whole plumage thus more uniformly coloured than any other west Palearctic hirundine. At close range, may show off-white chin and throat mottled with dark brown and grey, but most obvious features are dark, almost black under wing-coverts (forming 'shadowed' area in any light), off-white chevrons on lateral under tail-coverts, and bold white spots below tips of tail-feathers. Tail-spots often visible even from above when tail spread. Juvenile. Plumage warmer in tone than adult due to widespread rufous feather-fringes. Throat less dark, only indistinctly mottled. At all ages, bill black, legs brown.

Apart from risk of momentary confusion with smaller, uniformly coloured *Apus* swifts, field identification of *P. rupestris* easy in Europe since *P. fuligula* restricted to desert areas of North Africa and Middle East. In these areas, however, *Ptyonoprogne* martins mingle on passage, and *P. rupestris* may even attend colonies of *P. fuligula*; thus care needed not to confuse *P. rupestris* with darker races of *P. fuligula* (see that species, p. 248). Flight quite graceful but distinctly slower, more stable, and less energetic than other west Palearctic hirundines (except *P. fuligula*); most recalls Red-rumped Swallow *Hirundo daurica* but spends even more time in slow gliding. Rarely comes to ground except to collect nest-material, using shuffling walk. Much more localized than other hirundines except *P. fuligula*, spending much time around breeding areas.

Voice rather feeble. Song a rapid throaty twittering. Commonest call consists of variations on 'prrit' or 'chwit'.

Habitat. Breeds in west Palearctic in low middle latitudes, from temperate to Mediterranean zone, mainly in warm dry continental climates, but attachment to mountainous regions involves exposure to sharply differing temperatures. Deep shadow, winds, and snow are generally avoided. In Switzerland, occurs from lowlands to above treeline, most frequently in warm, dry, and sheltered situations, even when nests must be placed on shady cool crags; altitudinal range 274–2150 m. May be displaced by adverse weather, especially in spring (Glutz von Blotzheim 1962). In USSR, breeds exclusively in steep craggy regions, avoiding wooded cliffs and arid situations. Dependent on access to streams, rivers, or springs, from near sea-level to 2500 m in Caucasus and almost 4000 m in Pamirs, where night frosts frequent even in mid-summer. In some river valleys, breeds on crags rising directly out of water. In Ala Dagh mountains (Turkey), nests up to 2700 m, and in September seen feeding up to peak at 4000 m (Gaston 1968). After breeding season, feeds over alpine meadows and farmed lands (Dementiev and Gladkov 1954*b*). In Cyprus (and elsewhere) sometimes breeds on walls or under eaves of buildings up to *c.* 1700 m, in winter descends to rocky coasts (Bannerman and Bannerman 1971). In Morocco, hundreds spend winter over swamps and a lake (Smith 1965). In India in winter likes neighbourhood of ancient hill-forts, and will rest for short periods on fort walls or ledges of rocky cliffs (Ali and Ripley 1972). Although sometimes associated with human settlements, and ready to nest on buildings (e.g. Cramp 1970*a*), infrequently has occasion to do so, and shows little sign of adapting to commensalism with man. Although highly aerial, peculiar among west Palearctic Hirundinidae (except African Rock Martin *P. fuligula*) in mainly using airspace below level of dominant neighbouring land surfaces.

Distribution. France. Former colonies in Basse Seine have disappeared (Yeatman 1976). Not seen Seine-Maritime since 1976 (Debout 1980). Yugoslavia. Has recently expanded northwards (Vasić 1985). Rumania. Expanded north since 1968 (WK). Iraq. Probably breeding in north (Chapman and McGeoch 1956; McGeoch 1963). Israel. Breeds occasionally Mount Hermon (HM). Tunisia. Has bred (Heim de Balsac and Mayaud 1962).

Accidental. East Germany, Malta.

Population. No information on population trends; colonizing buildings and motorway viaducts in some areas, especially in north.

Spain. Locally numerous, especially in north (AN).

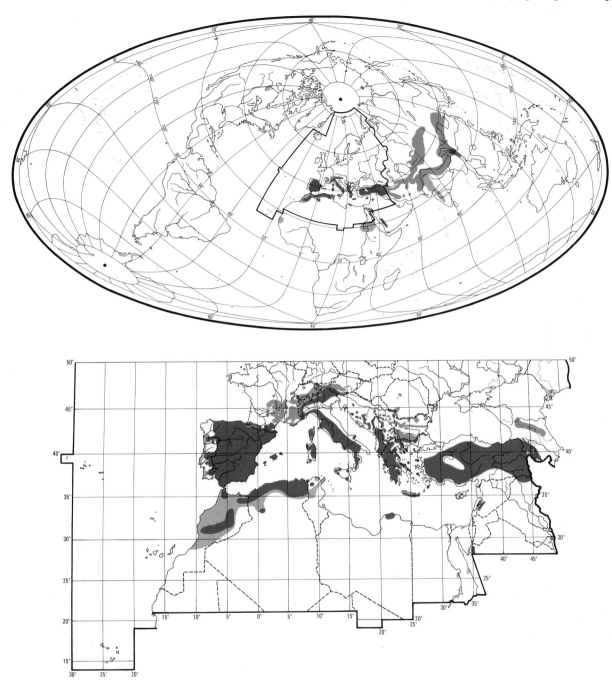

Galicia: part census 145-95 pairs, 1973-8 (Guitián Rivera *et al.* 1980). FRANCE. 1000-10 000 pairs (Yeatman 1976). WEST GERMANY. Almost extinct: over 10 pairs at 3-4 sites. Bred Bayern until 1974, but no nests 1975-80 (Bauer and Thielcke 1982). Bred Oberallgaü 1980-2 (AH). AUSTRIA. No counts; motorway viaducts recently colonized (HS, PP). ALBANIA. Scarce (AN). BULGARIA. Fairly common (TM). CYPRUS. Fairly common (Flint and Stewart 1983).

Movements. Northern populations partially migratory; mainly resident elsewhere though making altitudinal movements. Though largely migratory across southern Europe, from Greece to Spain, some winter regularly on northern side of Mediterranean basin and especially at western end. 2000-3000 birds roost on Gibraltar in winter, making conspicuous morning and evening flights to and from feeding areas in Cadiz province (Elkins and Etheridge 1974; Cortés *et al.* 1980). Birds ringed in winter on Gibraltar recovered in Spain (Cadiz, Valencia, Castellón, Mallorca), France (Pyrénées Atlantiques), and Italy (Genoa); except for winter recoveries in Cadiz, finding dates span March–October (BTO). However, other birds cross Straits of Gibraltar into Morocco in large numbers, with peak passages October and March (Cortés *et al.* 1980; Tellería 1981), contrasting with rare and irregular appearances on Malta (Sultana and Gauci 1982); this, combined with recoveries (above), suggests birds from Italy westwards tend to cross Mediterranean at narrowest point.

Winter range in Africa mainly northern parts of Morocco, Algeria, and Tunisia, though small numbers also occur in Sénégal (Morel and Roux 1973), along Nile valley as far south as northern Sudan, and on Red Sea coast and in western Highlands of Ethiopia (Meinertzhagen 1930; Cave and Macdonald 1955; Urban and Brown 1971).

Resident in Cyprus and Turkey, though leaves upland areas for winter, when found mostly in coastal and other lowland places; recorded on passage at Belen (south-east Turkey) from late August onwards (Sutherland and Brooks 1981). A summer migrant to USSR, and these presumed to winter in China, Indian subcontinent, and Middle East; scarce in central and eastern Arabia, however, especially in mid-winter (e.g. Bundy and Warr 1980, Gallagher and Woodcock 1980). In Indian subcontinent, breeding population leaves mountainous areas in autumn for lower ground, and range then (October–March) extends into peninsular India (Ali and Ripley 1972). RH

Food. Small insects. Prey mostly taken in flight, which typically involves much more steady gliding than other west Palearctic Hirundinidae except African Rock Martin *P. fuligula* (D J Brooks). Pair studied in breeding season, Italy, used different hunting behaviour at different times of day. In early morning before sunlight reached cliffs, hunted at bottom of valley along stream; when sun on cliffs, all hunting done by flying to and fro very close to rock-face; often passed one spot repeatedly, suggesting localized food source. In early afternoon when cliffs no longer illuminated, fed elsewhere. On windy days, hunted very close to cliff-faces, 'exploiting the thermals and performing true acrobatics' involving 'stoops' and 'veerings' recalling swifts (Apodidae); also hovered and performed 'rollings'. (Farina 1978.) Feeds closer to cliff-face than House Martin *Delichon urbica* (Strahm 1953). Takes insects disturbed from rock surfaces by the bird's own flight, and will pick insects from rocks as it passes (Murr 1923; Glutz von Blotzheim 1962; Watson 1964). 3 birds on ground, Aden, recorded feeding on abundant gnats and flies (Meinertzhagen 1954), and in Portugal recorded dipping to water apparently to take insects on surface (Mathews 1864). In Galicia (Spain), hunting range in winter *c.* 1·5 km², when breeding, reduced to *c.* 0·2–0·25

km² (Guitián Rivera *et al.* 1980). Birds roosting at Gibraltar, December, suspected of feeding only within 5 km of the peninsula (Finlayson 1978), considerably less than Elkins and Etheridge's (1974) suggestion of 16 km. In Switzerland, forages up to 1·5 km from colony (Glutz von Blotzheim 1962).

In Galicia, depends on a few groups of multi-generation annual insects—flies (Diptera), stoneflies (Plecoptera), caddis flies (Trichoptera), etc. 32 items found in nest and faeces, June, comprised 13 beetles (Coleoptera: Elateridae, Carabidae, Histeridae, Chrysomelidae), 11 flies (Diptera), 4 pond skaters (Gerridae), 3 butterflies (Pieridae), and 1 wasp (Vespoidea) (Guitián Rivera *et al.* 1980). In Switzerland, takes mainly dipteran flies (Brachycera, Nematocera), weevils (Curculionidae), other small beetles, Lepidoptera, ants (Formicidae), and spiders (Araneae) (Glutz von Blotzheim 1962). On Aegean islands, small moths (Lepidoptera) form large part of diet (Watson 1964). Stomachs from Aksu-Dzhabagly (Kazakhstan) contained bugs and flies (Dolgushin *et al.* 1970). 2 stomachs from Kirgiziya (USSR) contained flies and Hymenoptera (Pek and Fedyanina 1961). Stomachs from Mongolia, May, contained mostly small dung beetles (Aphodiinae) (Piechocki *et al.* 1982).

No information on differences between diets of adults and young, though studies cited above presumably include items brought to nests. Parents deliver food to nest every *c.* (1·1–)2–5 min; rate less than average in cold, dull, or windy weather (see, especially, Strahm 1956; also Hoffmann 1936, Prenn 1937, Brandt 1963, Zedler 1963), and may speed up (Strahm 1956) or slow down (Uhl 1929) when young close to fledging. DJB

Social pattern and behaviour. For major studies, see Prenn (1937), Strahm (1953, 1954, 1956, 1963), and Farina (1978). For summary, see Cramp (1970a).

1. Gregarious outside breeding season, sometimes markedly so where food locally abundant. In Gibraltar, late October, feeding flocks of up to 300–400 (Elkins and Etheridge 1974). In lower Rhône (France), March, concentration of 200 birds over water (see Schifferli *et al.* 1982). In spring, migrant flocks typically 3–7(–25), exceptionally up to 300 (Cramp 1970a). On arrival, Switzerland, up to 100 gather together (Glutz von Blotzheim 1962). At Gibraltar, does not usually associate with other species, but, on spring and autumn passage, other Hirundinidae, and sometimes Swifts *Apus apus*, occasionally join flocks of *P. rupestris* (Elkins and Etheridge 1974). On Malta, however, regularly associates in flocks with other Hirundinidae (Sultana and Gauci 1975). In spring, Austria, commonly associates with House Martin *Delichon urbica* and Swallow *Hirundo rustica*, to lesser extent with Sand Martin *Riparia riparia* (Prenn 1929, 1937). BONDS. No information on mating system. At one nest, sexes shared brooding equally (Hauri 1968), but at another, ♀ did most (Farina 1978; see Relations within Family Group, below). Both sexes feed young, which remain dependent on parents for up to 14–21 days after fledging (Strahm 1956; Hauri 1968; Guitián Rivera *et al.* 1980). Age of first breeding not known. BREEDING DISPERSION. Solitary or forming small loose colonies: e.g. in Switzerland, typically 2–5 pairs (Schifferli *et al.* 1982), and rarely more than 10 pairs

(Strahm 1953, but see below). Similar colony size typical in Bayern, West Germany (Murr 1923), and Austria (Prenn 1937). In Galicia (Spain), at 32 localities, 1–40 pairs per locality, mostly 1–6, average *c.* 4·5–6·1 (Guitián Rivera *et al.* 1980). Colonies may be larger elsewhere: e.g. in Iran, several each of *c.* 60 birds (Erard and Etchécopar 1970). In Switzerland, Hess (1920) recorded colonies of 25, 50, and over 100 pairs (mixed with smaller numbers of Alpine Swift *Apus melba*), but such sizes atypically large and perhaps refer to aggregations of smaller colonies. In various places, Switzerland, colonies 1–5 km apart, otherwise more sporadic (Glutz von Blotzheim 1962). Nests usually well-spaced, and never as close as those of *D. urbica*; in Switzerland, average separation 30·5 m (10–80), $n = 13$ (Strahm 1953, 1963). In Bayern, up to 40 m apart, closest 3 m (Murr 1923). In colony in Galicia, 6 pairs on *c.* 360–420 m² of cliff-face; distance to nearest neighbouring nest 14·4–16·2 m, $n = 5$ (Guitián Rivera *et al.* 1980). Pair defend territory around nest and close to cliff-face, but also commonly feed in neutral ground adjoining and away from colony area; 3 nest-territories respectively *c.* 200, 270, and 310 m² (Strahm 1963). Feeding area of family group, shortly after young fledged, *c.* 600 × 200 m (Hauri 1966). Several vantage points in or near territory used for courtship, nesting and preening, surveillance, and for launching expelling attacks (Strahm 1963: see below); one pair used *c.* 10 perches within a few metres of nest (Farina 1978). Colonies sometimes mixed with, or adjacent to, those of *D. urbica* (Meinertzhagen 1935; Impe 1971; Hunziker-Lüthy 1973); in Israel, sometimes mixed with Pale Crag Martin *P. fuligula* (Meinertzhagen 1954). For association with *A. melba*, see above. Nest used over several years (e.g Strahm 1963), but not known if by same pairs. One pair used same nest for 2nd brood (Flück and Flück 1984). ROOSTING. Always on cliffs, in caves, and other rocky places. During incubation, ♀ roosted on nest, ♂ elsewhere with conspecifics (Farina 1978). By day, birds frequently loaf in and beside territory on small rock outcrops or gently sloping ledges protected by overhangs (Strahm 1963; Farina 1978); these used especially for preening, most often early morning and late afternoon (Farina 1978); birds also scratch (Prenn 1937), and sunbathe by raising and stretching wing and exposing underparts (Strahm 1954; Piechocki *et al.* 1982). Birds resort to sheltered loafing-sites as soon as rain begins (Prenn 1937; Hauri 1966). At end of August, 40–50 thus assembled on ledge, perched *c.* 5 cm apart and all in area *c.* 6 × 2 m (Strahm 1956). According to Prenn (1937), after fledging family typically return to nest to roost, but pattern more variable as young grow older, e.g. young of one pair stopped returning to nest to roost 14 days after fledging (Strahm 1956). In mid-August, brood of 4, which had left nest 6 days previously, roosted in a row in horizontal crevice alongside nest; in late September, nest used by 4 birds for roosting but not known if adults or young (Hauri 1968, which see for time of roosting relative to sunset). At another nest, brood of 5 roosted in nest with 1 parent, other parent in a crevice (Hauri 1966). Young went to roost later in evening and left later in morning than nearby *D. urbica* (Strahm 1956). Roosting outside breeding season well studied. In Turkey, October, 90 together on cliff-face (Vittery *et al.* 1972). At Gibraltar, large numbers nightly occupy ledges and 'chimneys' among stalactites in limestone caves (Elkins and Etheridge 1974; Finlayson 1978). From some difference in proportions of adults and immatures occupying particular caves, Finlayson (1978) suggested adults may annex best roost-sites. From October to December, 1500–2000 birds roost nightly, decreasing in March. In fair weather, with light to moderate winds, arrival at roost begins 30–60 min before sunset, later in

December; arrival up to 2–3 hrs before sunset in dull cloudy weather, even earlier in rain and strong winds. In latter case, arrival protracted but otherwise birds arrive 'spectacularly, pouring' silently over cliff edge; flight to roost direct with no 'swarming' as in (e.g.) *H. rustica*. Most go straight to roost-sites, but on warmer evenings some feed first. Morning departure less conspicuous, birds usually not stirring until (15-)30–60 min after local sunrise, and large numbers remain on ledges until 1–1½(–3½) hrs after sunrise, sunbathing and preening in sheltered spots (Elkins and Etheridge 1974). Sunbathing before exodus especially marked in cold weather. In December, majority departed from roost within *c.* 20 min; most left in parties of up to 10 birds, but, during peak departure, up to 60(–80) together (Finlayson 1978). Birds seem loath to leave roost in cold weather (Elkins and Etheridge 1974). Heavy rain delays departure, and once, in extreme conditions, birds still departing at 15.55 hrs (Finlayson 1978). Birds often quite vocal (see 2 in Voice) for a short while, both on arrival and departure (Elkins and Etheridge 1974).

2. Said to be quite tame in breeding season, allowing approach to within 4–5 m (Brandt 1963), but see Parental Anti-predator Strategies (below). FLOCK BEHAVIOUR. Outside breeding season, at Gibraltar, aerial chasing common among flocks, birds calling (see 4 in Voice). Perched flocks show behaviour similar to 'dreads' of terns (Sternidae): group suddenly takes off and wheels away in silence, later returning to rest again. Alarm-calls (see 3c in Voice) give warning of predator. Once, in mid-winter, flying flock of 500 panicked at approach of unidentified raptor: hurriedly dived landwards, hugging slopes closely. This response elicited by Peregrine *Falco peregrinus*, but Honey Buzzard *Pernis apivorus* and Griffon Vulture *Gyps fulvus* on autumn migration usually mobbed (Elkins and Etheridge 1974). For further reactions to raptors, see Parental Anti-predator Strategies (below). ANTAGONISTIC BEHAVIOUR. Highly aggressive towards conspecifics and many other species encroaching on nest-territory (e.g. Brandt 1963, Juon 1968). One isolated pair patrolled nest-territory, flying to and fro along cliff. May dive several times at intruder, with bill open, and giving Anger-calls (see 5 in Voice) at bottom of dive, or may seize it by the nape, causing both to fall locked together; may also chase such intruders, once when within *c.* 30 m of nest (Strahm 1963; Farina 1978, 1979). In possible confrontation, 3 birds, which had been chasing, tumbled down in the air in a whirling mass, giving twittering calls (Murr 1923). Young flying close to nest-territory subjected to brief silent pursuit (Farina 1978), especially if they try to land (Strahm 1956). Following species are also expelled: *A. apus*, *D. urbica*, Robin *Erithacus rubecula*, Black Redstart *Phoenicurus ochruros*, wagtails *Motacilla*, Wallcreeper *Tichodroma muraria*, Bullfinch *Pyrrhula pyrrhula*, and Tree Sparrow *Passer montanus* (Prenn 1937; Strahm 1956, 1963; Farina 1978, 1979; see also Parental Anti-predator Strategies, below). At one nest, ♂ made 90% of attacks on *A. apus* (Farina 1978). No firm information on threat-postures, but when ♂ tried forcibly to relieve incubating mate she resisted with apparent threat-display: gaped and sleeked plumage (Farina 1978). HETEROSEXUAL BEHAVIOUR. (1) Pair-bonding behaviour. Not well understood. A few displays reported, but not all have certain relevance to pair-formation. On first arrival at colony area, much noisy aerial chasing along cliff-face (Strahm 1954; Piechocki *et al.* 1982); often 3 birds involved. At colony in Switzerland, chasing stopped as soon as pairs formed (Strahm 1954), and thus function perhaps differs from mate-guarding interpretation of aerial chasing in Sand Martin *Riparia riparia*, in which chases are most frequent after pair-formation. Dis-

playing birds settle on rocks from time to time (Lindner 1919), but no ground-display described. Once, 2 birds flew towards each other until their bills touched, but not known whether anything was transferred (Strahm 1954). (2) Courtship-feeding. Probably rare and only one definite report: ♀ at nest twice given food by visiting bird, presumably mate (Hauri 1968). In another study, incubating ♀ never fed by ♂ (Prenn 1937). (3) Mating. Only one description of behaviour suggestive of mating: on 3 occasions, at start of egg-laying (mid-April to mid-May), 2 birds flew one above the other, almost touched, with tails spread, and called (not described) loudly; usually, upper bird held the other for an instant, and for 1-2 s they flew level together (Strahm 1954). (4) Behaviour at nest. At one nest, both sexes quite often caught feathers in flight or picked them out of nest; feather often dropped (repeatedly and in rapid succession) a little way and then retrieved (Farina 1978). In similar report, bird carrying plume of feather-grass *Stipa* in bill, flew several times past nest in every direction, then dropped grass and let it fall 10-20 m before retrieving it; flew back towards nest and dropped grass again; sequence performed 3 times, then bird presented grass, bill to bill, to other bird (presumed mate) which carried it to nest; behaviour continued, with feathers or grass, until July (Strahm 1954). Nest built or repaired on arrival at colony (e.g. Hauri 1968); at one site, several weeks elapsed before pair began building (Hauri 1966). New site chosen if old nest collapsed (Strahm 1953). Nest-building sometimes occurs in intense bouts, e.g. 13 visits in 15 min (Juon 1968), but usually seems to occupy relatively little time each day, and to be protracted. In one case, only 1 member of pair building: collected grasses for lining *c.* 20-25 m from nest; got hold of grasses, then tried to tear them away from rocks; flew around warily on approach to nest; only 2 min 35 s building during 1 hr (Strahm 1953). In another case, 7 min building (in 12 stints) in ½ hr (Strahm 1954). Until completion of nest, pair quite often absent throughout day (Nitsche 1967). During incubation, ♀ continues to line nest; also makes brief excursions to feed and defecate (Strahm 1954), leaving nest for up to 20 min at a time (Hauri 1968). Meeting-ceremony at nest involves no special display, but accompanied by brief Contact-calls (Murr 1923: see 2 in Voice). ♂ at one nest often arrived giving apparent alarm-call, ♀ responding with Contact-call (Prenn 1937). At another nest, returning bird (presumed ♂) often only perched on rim without attempting to relieve mate, and soon flew off (Strahm 1954). When ♂ returns and calls, ♀ often flies off with him (Prenn 1937), though this perhaps more likely early in incubation. In another study, ♂ occasionally forced ♀ off nest, though she sometimes adopted apparent threat-posture (see above); if dislodged, ♀ immediately returned to displace ♂ (Farina 1978). RELATIONS WITHIN FAMILY GROUP. Young brooded quite closely in first few days, especially by ♀ (Prenn 1937; Strahm 1956). Change-over in brooding synchronized with visits to feed young (Hauri 1968). At one nest, young brooded irregularly for *c.* 10-11 days, steadily less each day (Farina 1978). At first, parents swallow faecal sacs but later carry them away (Strahm 1956). At one nest, 9 out of 10 faecal sacs removed by ♀ (Farina 1978). When young older, faeces left to accumulate around nest (Prenn 1937; Strahm 1956). Young beg from every conspecific bird which passes: gape, call (see Voice), and vibrate wings. Always occupy same respective positions in nest, and seem to maintain a certain individual-distance. When young older, parents take 1-2 s to feed them, perching to do so. Near fledging, transfer food while hovering momentarily in front of nest. Young now perch on rim and exercise wings. One brood fledged shortly after an alarm caused by Sparrowhawk *Accipiter*

nisus, but returned to nest after a while (Strahm 1956.) Near fledging, young may sit outside nest, where they are fed (Hauri 1966). For roosting of young, see above. After fledging, young fly, calling, towards parents, and are fed in the air (Strahm 1956; Trillmich 1968; see Bonds, above). Young of 1st brood strongly attracted by those of 2nd brood sitting on nest-rim; tried to land near nest, and would not leave even when chased by parents. On fledging, young flew *c.* 100 m and returned to cliff in rapid straight flight. When both fledglings took off, young of 1st brood pursued them (Farina 1978.) ANTI-PREDATOR RESPONSES OF YOUNG. When human intruder entered cave, young in nest became silent and withdrew deep into nest (Prenn 1937). When raptor passed near colony-area, fledged young took off with adults (Hauri 1966) and once joined adults in pursuing Kestrel *Falco tinnunculus* (Farina 1978). PARENTAL ANTI-PREDATOR STRATEGIES. (1) Passive measures. Nest attentiveness very high (Farina 1978: for details, see Breeding). During incubation, off-duty bird (almost invariably ♂) performs sentinel role on vantage point near nest (Strahm 1963; Farina 1978), giving alarm-calls (see 4 in Voice) to warn of approaching predator (Murr 1923; Stadler 1928a; Strahm 1954; Farina 1978). Alarm-calls of *D. urbica* also elicit upflight (Strahm 1956). (2) Active measures: against birds. Alerted by repeated alarm-calls, birds mount concerted mobbing attacks on *F. tinnunculus*, *A. nisus*, Buzzard *Buteo buteo*, Black Kite *Milvus migrans*, Jay *Garrulus glandarius*, and Jackdaw *Corvus monedula* (Deleuil 1913; Strahm 1954; Brandt 1963; Farina 1978). Birds launch dive-attacks from above and below (Deleuil 1913); make repeated attacks with bill open, giving Anger-calls when near target, and may pursue predator for several hundred metres (Strahm 1954; Farina 1978). Perched *F. tinnunculus* subjected to dive-attacks in which birds, with Anger-calls, brushed predator with their wings (Strahm 1954). (3) Active measures: against man. ♀ disturbed by observer when approaching nest to feed young changes into a 'swinging' or 'pendulum' flight-path in which she oscillates back and forth, perhaps indicating conflict between feeding and attack; gives 'uuii-' call (see 3c in Voice) with each oscillation (Farina 1978). Birds make bold dive-attacks as on raptors, when nest closely approached; once, when observer 12 m below nest (Strahm 1953). Give alarm-calls and dive at intruder's head (Farina 1979). At one nest, 90% of attacks made by ♀, while ♂ circled higher. Early in season, intruder escorted *c.* 100 m away with repeated attacks. ♀ also tended to attack after feeding young. Adults especially aggressive around fledging time. During incubation of 2nd clutch, both parents attacked mirror used to inspect nest; gave alarm-calls and were joined by agitated young from 1st brood (Farina 1978, which see for seasonal variation in intensity and frequency of attacks).   EKD

Voice. Freely used throughout the year, and birds possess wide repertoire (Elkins and Etheridge 1974). However, most calls quiet (Stadler 1928a; Bergmann and Helb 1982). Several calls similar to House Martin *Delichon urbica*. Some sounds evidently represent a gradation and, given this overlap, following scheme tentative.

CALLS OF ADULTS. (1) Song. A throaty and persistent rapid twittering sound, given in flight; not very conspicuous (Bergmann and Helb 1982, which see for sonagram). A short quiet mixture of all the calls, rather like *D. urbica* (Prenn 1937). Instance of probable song occurred when bird followed a series of 'zewi' sounds with a quiet chirruping, rather like *D. urbica*, but different

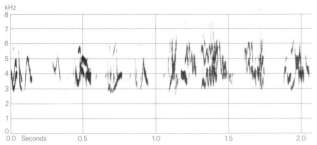

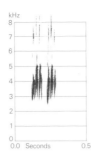

I   C Chappuis/Sveriges Radio (1972–80)   Greece   May 1967

II   C Chappuis/Sveriges Radio (1972–80)
France   August 1964

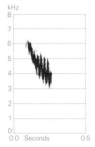

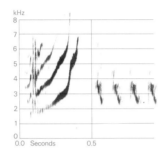

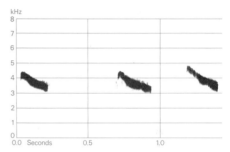

III   C Chappuis/Sveriges Radio (1972–80)
France   August 1964

IV   P J Sellar   France   May 1977

V   Roché (1966)   France   June 1965

timbre (Stadler 1928a). Frequently heard amalgam of several calls, e.g. 'widzrrk' or 'zwidi zr' or 'dewid dewid zr' (Stadler 1928a), probably therefore song. Another sequence, by birds chasing one another around colony, was rendered 'diè-diè-diè-trt-gsigi-gsigi-diè-kagsi-kagsi' (Prenn 1937). Similarly excited song in recording (Fig I) compared to sound of bicycle in which all moving parts in need of oil (J Hall-Craggs). Comparable sequence, associated with aerial chase, described by Hoffmann (1917). Subdued song, likened to Swallow *Hirundo rustica*, also heard at times in winter quarters (Elkins and Etheridge 1974). See also Calls of Young. (2) Contact-call. Variations on 'prrrit' (Elkins and Etheridge 1974); in flight, short 'pit-pit' or 'ti', sometimes a longer 'trit' or disyllabic 'pritit' (Bergmann and Helb 1982, which see for sonagrams); 'chwit' or, from agitated group of breeding birds, 'chwitit' (A M Allport). In one of our recordings, birds in flight give variously 'pri-pit' (Fig II, 2 calls), 'prip', and 'zi' (see 4a, below) (P J Sellar, E K Dunn). In another recording, a harder repeated 'drit', followed by call 4b, perhaps indicating some annoyance or anxiety. For other renderings, see Hoffmann (1917, 1936), Murr (1923), and Prenn (1937). Given singly, or often in series, with or without pauses (Stadler 1928a). Commonest call, given (e.g.) when flying around colony (Hoffmann 1936), during Meeting-ceremony (Murr 1923; Prenn 1937; Farina 1978), and between flock members. Call given when feeding young, and audible at 120 m (Murr 1923), was probably this call. At Gibraltar in autumn, 'continual hubbub' from feeding flocks (Elkins and Etheridge 1974) was presumably mostly Contact-calls. (3) Alarm- and

Warning-calls. A variety of sounds, often disyllabic, probably indicating different levels of alarm. (a) Call most commonly heard by Stadler (1928a) probably of mild alarm: sound of varying pitch, rendered 'zrr', rather like dull quiet chirruping of sparrow *Passer*; sometimes 'z-i-rr', 'dschirr', or 'd'sh-r'; sometimes harsh and higher pitched, almost like 'excitement calls' of Goldfinches *Carduelis carduelis* fighting (see call 5, below), also sometimes trisyllabic. Often has no terminal 'r', producing sound indistinguishable from 'zi' of Yellowhammer *Emberiza citrinella* (Stadler 1928a). In recording, 'dzir' (J Hall-Craggs: Fig III). This call, 'dzrji', with falling pitch, given when disturbed (Bergmann and Helb 1982, which see for sonagram). (b) Various calls, with 2 or more syllables, related: loud 'gsigsi', like 'warning-call' of *H. rustica* (Prenn 1937); 'dsjissiji', given several times in rapid succession when Kestrel *Falco tinnunculus* nearby (Hoffmann 1936); a loud 'chiupi' on sight of predator (Farina 1978). (c) Plaintive 'uuii', given mainly by ♀, often in characteristic 'swinging' flight (see Social Pattern and Behaviour) when disturbed while provisioning young, and thought to have warning function (Farina 1978). In recording (Fig IV, 1st unit), a rising 'SFOOee', rather like finch (Fringillidae) (P J Sellar). Quiet plaintive 'wheee' from perched birds on appearance of predator (Elkins and Etheridge 1974) presumably this call. Several calls, listed by Stadler (1928a), probably the same or variants: ; 'dwid' or 'zewi' or more emphatic 'zwidi'; also, more compressed 'zwid' or 'wid'; Murr (1923) reported 'ziwi', with 1st or 2nd syllable stressed, similar to *H. rustica*. (d) A pure 'siu' (Bergmann and Helb 1982); in recording (Fig V), a plaintive descending pure whistle 'eeoo' (J Hall-Craggs); also described as a loud whistling 'teeoo' or 'tsiu', repeated quickly 2–3 times, by perched birds (Elkins and Etheridge 1974); a long-drawn 'ziu', with emphasis on 2nd syllable, or 'zieh' when falcon *Falco* near nest (Murr 1923). (4) Excitement-call. In autumn, Gibraltar, during aerial chases, a rapid 'chu-chu-chu-chu-chu', similar to flight-call of Redpoll *Carduelis flammea* (Elkins and Etheridge 1974). This call possibly derived from a rapid series of call 4a; perhaps the sound roughly like 'ch-ch-ch-ch' in recording (Fig IV, last 4 units), though probably not uttered by same bird that gives preceding 'SFOOee' (J Hall-Craggs; see call 3c). (5) Anger-calls. A 'rrrr' which, during attack on intruder, increases in intensity, reaching peak when bird nearest target; given with bill open (Farina 1978). Angry-sounding 'dschrü-dschrü' given by birds mobbing unidentified falcon *Falco* (Murr 1923); apparently an emphatic variant of call 3a. (6) Other calls. Single 'pitcha' sounds (no further details) (Elkins and Etheridge 1974). For other sounds, perhaps same or variants of those listed here, see Hoffmann (1917) and Elkins and Etheridge (1974).

CALLS OF YOUNG. Food-call of small young a quiet 'silb' or 'srb'; later, also given between feeds. Call becomes harsher with age; thus, young sitting on rim of nest gave a continuous chirping 'zrb zrb' (Prenn 1937). This presumably the same as 'dsjirrrb' and 'dschjirrrb' sounds, heard at nest when adults came to feed young (Hoffmann 1936); said to be same as call 1 of adult, but somewhat higher pitched (Farina 1978). When fed, young 22–24 days old gave long-drawn screeching 'chiiii-gigigigig' sounds (Prenn 1937). Fledged young gave related 'd(s)ig', 'dsje', 'dsji(g)' or 'dsji(rk)' sounds (Hoffmann 1917). Warbling sound, given by 'yearlings' perched at colony, interpreted as song (Farina 1978).

EKD

Breeding. SEASON. Main laying period from mid-May, with 2nd clutches laid in July (Strahm 1954; Makatsch 1976). SITE. On vertical rock wall, in crevice, small hollow, or occasionally in shallow tunnel, and usually under an overhang. Frequently on building, e.g. castle, church, house (Niederfriniger 1973). Main orientation between east and south-west: of 35 nests, Switzerland, 11 faced east, 2 south-east, 4 SSW, 17 south-west, and 1 west, with 3 in tunnels (Strahm 1953). Height above ground up to 40 m. Of 38 nests, Switzerland: 0–5 m, 3%; 6–10, 29%; 11–15, 18%; 16–20, 24%; 21–25, 13%; 31–35, 5%; 36–40, 8%; Strahm 1953). Solitary or in small loose colonies. Nest: half-cup of mud, lined with feathers and plant material. Of 4 nests, Switzerland, average length 14·4 cm (12·3–17·0), width 10·0 (8·9–11·0), height 8·2 (6·5–10·5), with wall thickness at rock 2·1 mm (2·0–2·3), and at rim 1·4 (1·1–2·0) (Strahm 1954). Mud pellets used varied from 4 × 3 mm to 8 × 10 mm

(Strahm 1954). Building: By both sexes; taking 9–20 days, *n*=4 (Guitián Rivera *et al.* 1980); one nest took 14 days to complete but 6 of these cold and no building carried out (Strahm 1954). See also Social Pattern and Behaviour. EGGS. See Plate 80. Long sub-elliptical, smooth and slightly glossy; white, sparsely spotted red and grey, concentrated at broad end. 20·2 × 14·0 mm (17·2–23·2 × 12·7–15·4), *n*=105 (Schönwetter 1979). Average weight of 6 eggs 2·2 g (Strahm 1954). Clutch: (1–) 3–5. Of 29 clutches (but including some newly-hatched young), Switzerland: 1 egg, 1; 2, 9; 3, 6; 4, 8; 5, 5; average 3·2 (Strahm 1954). Normally 2 broods. INCUBATION. 13–17 days (*n*=9 nests) (Guitián Rivera *et al.* 1980). By ♀, though ♂ may occasionally relieve for short periods. In 980 min of observation at one nest, Italy, ♀ sat for 89% of time, ♂ for 5%, neither for 6%, with maximum spell by ♀ 100 min and by ♂ 24 min. On average, ♀ left nest 1–6 times per hr returning after 5–9 min; longest period of absence by both birds 28 min (Farina 1978). During 1 hr of observation, Switzerland, ♀ sat for 50 min 15 s, and was off for 9 min 45 s; longest spell of incubation recorded during study 43 min (Strahm 1954). Begins with last egg; hatching synchronous. YOUNG. Altricial and nidicolous. Cared for and fed by both parents. During 1740 min of observation from hatching to fledging, parents stayed at nest for 14% of time; ♀ brooded young for 8·8% of time and ♂ for 3·6%, with maximum spells of 2·5 min (♀) and 9 min (♂); brooding decreased towards fledging (Farina 1978). FLEDGING TO MATURITY. Fledging period 24–7 days, *n*=8 nests (Guitián Rivera *et al.* 1980). Young dependent on parents for *c.* 1–3 weeks after fledging (Strahm 1956; Guitián Rivera *et al.* 1980). Age of first breeding not known. BREEDING SUCCESS. No data.

Plumages. ADULT. Upperparts uniform brown-grey, paler and greyer than upperparts of Sand Martin *Riparia riparia*; feathers of mantle, back, and rump sometimes with narrow, faint, rufous-tinged fringe along tip when plumage fresh, giving slight sandy appearance to upperparts; fringes much less distinct than those of juvenile. Sides of head brown-grey, like upperparts, feathers in front of eye slightly darker; ear-coverts sometimes faintly mottled with paler grey. Cheeks, chin, and sides of throat mottled dark brown-grey and pale grey or buff-white, spots on cheeks sometimes forming distinct dark stripe contrasting with less heavily spotted chin and throat; cheeks, chin, and upper throat sometimes largely dark brown-grey or black-brown, faintly mottled pale buff only. Lower throat, lower sides of neck, and chest pale pink-buff (when fresh) to pale buffish-white (when worn), feather-shafts sometimes slightly brown. Sides of breast, flanks, vent, and under tail-coverts dark brown-grey or dark grey, grading through pale brown-grey on belly into buff-white of belly; in fresh plumage, breast and belly washed buff, tips of under tail-coverts fringed rufous; in worn plumage, belly bleached to pale grey-brown or off-white with pale brown mottling, under tail-coverts narrowly fringed white along tips. Central pair of tail-feathers (t1) brown-grey like upperparts, slightly darker towards tips; other tail-feathers gradually darker towards outermost, t4–t6 dull black; t2–t5 with large and rounded white spot on inner web *c.* 1 cm from tip; spots largest on t3–t5, sometimes lacking on t2(–t3);

occasionally a faint and small pale spot on inner web of t6. Length of largest spot 14·0 (1·8; 84) 10-19 mm (Elkins and Etheridge 1977). Flight-feathers, upper primary coverts, bastard wing, and lesser upper wing-coverts dull black faintly glossed green when fresh; outer webs of flight-feathers dark brown-grey (except for tips), greater and median upper wing coverts dark brown-grey, grading to paler brown-grey on innermost and on tertials; fresh greater upper wing-coverts and tertials narrowly edged white along tips. Under wing-coverts and axillaries dull black; smaller coverts along leading edge of wing fringed buff. In heavily worn plumage, feather-tips of upperparts, tertials, and belly abrade to pale grey, body appearing paler and more silvery than in fresh plumage and dark tail, wing, flanks, and vent more contrasting. NESTLING. Down short and dense, dark brownish-grey (Heinroth and Heinroth 1931). JUVENILE. Like adult, but feathers of head and body washed rufous on terminal fringes, crown and sides of head appearing brown-grey with rather narrow and indistinct rufous arcs, mantle to rump and flanks to vent brown-grey with marked rufous wash, upper and under tail-coverts brown-grey with broad rufous fringes along tips, throat, lower sides of neck, and chest washed buff-brown on grey-white ground. Spots on cheeks and chin grey-brown, paler and less contrasting than those of adult, generally not extending to throat. Tail-feathers narrowly edged buff along tips; white spots as in adult, but often less sharply defined and generally smaller, length of longest spots 11·3 (2·1; 154) 6-14 (-16) mm (Elkins and Etheridge 1977). Flight-feathers as adult, but narrowly edged buff along tips (least so on longest primaries); tertials and upper wing-coverts (except primary coverts) darker than in adult, contrastingly fringed rufous-buff along tips. In worn plumage, rufous fringes of upperparts, under tail-coverts, tertials, upper wing-coverts, tail, and flight-feathers bleach to pale buff and off-white, and rufous wash from chin to belly bleaches to buff-white; narrower fringes of crown, tail, and flight-feathers completely abraded, those of mantle to rump narrow but distinct, giving scaly effect. Also, increasing amount of juvenile feathers replaced by more uniform adult ones from September-November onwards. FIRST ADULT. Like adult, but juvenile tail, flight-feathers, greater upper wing-coverts, and variable number of other wing-coverts retained, traces of off-white fringes showing on tips of tertials, inner greater coverts, and sometimes elsewhere; tail-spots on average smaller. Adults show flight-feathers moulting up to October-November, with primaries new during winter (occasionally, moult suspended, unlike any immature); in 1st adult, virtually no flight-feathers moult until June-July of 2nd calendar year, primaries all slightly worn in autumn, distinctly worn in winter and spring. Juvenile t6 on average narrower than in adult, terminal part of inner web not as broadly rounded.

Bare parts. ADULT. Iris dark brown. Bill black, base of lower mandible grey-black or dark reddish-grey. Leg pale flesh-brown or brownish-flesh, toes slightly darker (Hartert 1910; Ali and Ripley 1972; Svensson 1984a; BMNH, ZMA.) NESTLING. Mouth yellow with pale yellow flanges. Leg and foot flesh-pink. (Heinroth and Heinroth 1931.) JUVENILE. As adult, but cutting edges of upper mandible and almost all lower mandible yellowish (generally through September: Svensson 1984a) and leg and foot reddish-flesh. Not known when adult colours obtained.

Moults. Based on Prenn (1937), Paludan (1940, 1959), Vaurie (1951c), Dementiev and Gladkov (1954b), Stresemann and Stresemann (1969), Elkins and Etheridge (1977), and specimens in BMNH, RMNH, and ZMA. ADULT POST-BREEDING. Com-

plete; primaries descendant. Starts with p1, followed by wing-coverts and body at primary moult score c. 15, tail from c. 20 (from t1 outwards, but t6 usually before t5 or t4), secondaries from 20-35 (outwards from s8, inwards from s1; s5 and s6 last); usually starts when laying, breeding, or feeding young. Moult on average starts with p1 on 15 June (mainly 10-30 June, some in late May or as late as early August); completed with regrowth of p9 10 November (mid-October to late November; those starting early August probably not until December-January, or later when moult suspended. Moult of body, wing, and tail generally completed at same time as p9 or shortly afterwards, wing-coverts usually all new by October. Only 5% of many examined on Gibraltar had moult not completed by 1st week of December; 7·6% retained old secondaries (mainly s5-s6) in mid-winter (Elkins and Etheridge 1977). POST-JUVENILE. Partial; slow and protracted. Involves head, body, and sometimes part of upper wing-coverts, but generally not flight-feathers, tail, or greater upper wing-coverts. Some start from September and in heavy moult October-November; others start November-December; moulting birds captured November still in moult when recaptured February-March. Of many examined, 70-80% still showed scaly juvenile pattern on upperparts in October, c. 30% still showed traces of scaly pattern February, less than 10% in March. Only 1 of 946 examined October-March had started primary moult (December); this perhaps only in a few early-hatched birds. (Elkins and Etheridge 1977.) 1st complete moult usually starts at same time as adult when c. 1 year old (Stresemann and Stresemann 1969).

Measurements. ADULT. Switzerland, southern Europe, and Asia Minor, all year; skins (BMNH, RMNH, ZMA). Tail is to t4-t5, fork is tip of t1 to tip of t4-t5, bill (S) to skull, bill (N) to distal corner of nostril; exposed culmen on average 4·1 less than bill (S).

WING	♂ 130·8 (2·47; 21) 127-134	♀ 130·7 (3·58; 12) 126-136
TAIL	54·5 (1·61; 14) 52-57	54·3 (1·19; 8) 52-56
FORK	4·1 (0·89; 14) 2-6	3·9 (0·98; 7) 2-5
BILL (S)	11·4 (0·50; 15) 10·7-12·2	11·5 (0·66; 8) 10·5-12·2
BILL (N)	5·5 (0·27; 15) 5·2-6·1	5·1 (0·60; 6) 4·5-5·8
TARSUS	11·7 (0·48; 15) 10·9-12·5	11·4 (0·55; 8) 10·6-12·0

Sex differences not significant.

Wing: Europe, ♂ 132·6 (2·82; 7) 129-136; Iran and Afghanistan, ♂ 131·0 (1·22; 5) 129-132, ♀ 127·9 (2·20; 9) 125-131; mountains of south-west China, ♂ 136·1 (3·56; 32) 130-143, ♀ 132·5 (2·38; 4) 130-135 (Vaurie 1951c).

JUVENILE. Wing and tail both on average c. 1·2 shorter than adult. Live birds, Gibraltar, October-December(-March) (Elkins and Etheridge 1977).

WING AD	134·0 (2·82; 391) 125-142	JUV 132·8 (2·89; 946) 124-141
TAIL	55·8 (1·26; 42) 53-59	54·6 (1·48; 73) 52-58

Wing increases with age: wing of birds of 3 years and older, 136·2 (1·98; 12) (Elkins and Etheridge 1977). Another sample from Gibraltar, mainly January-March, ages combined: wing 133·1 (2·95; 643) 125-144; tail 51·7 (1·73; 64) 48-56; bill (S) 11·6 (0·40; 34) 11-12·5; tarsus 11·8 (0·44; 33) 11-12·5 (Elkins and Etheridge 1974).

Weights. Gibraltar, averages, sexes and ages combined; total sample over 1300 (Elkins and Etheridge 1977).

	Oct	Nov	Dec	Jan	Feb	Mar
At dawn	22·6	25·6	24·5	23·8	24·2	20·4
At dusk	22·9	27·1	25·5	24·5	23·4	20·5

Juvenile usually 2–10% less than adult, sometimes 0–15%; lowest March juvenile 15·7 (Elkins and Etheridge 1977). In another sample Gibraltar: November–December, 24·9 (19·1–30·0) (*n* over 34); January, 26·3 (20·2–33·0) (*n*=*c.* 226); February, 23·9 (17·7–32·0) (*n*=*c.* 222); 1st half of March, 23·1 (20·4–30·2) (*n*=*c.* 53) (Elkins and Etheridge 1974). Eastern Turkey, Iran, Armeniya and Kazakhstan (USSR), and Afghanistan, late May to early August: ♂ 20·0 (1·30; 12) 17–22, ♀ 21·1 (2·04; 6) 19–24 (Paludan 1940, 1959; Nicht 1961; Kumerloeve 1969a; Dolgushin *et al.* 1970) Mongolia, May–July: ♂ 22·0 (0·90; 6) 21–23; ♀ 22·8 (0·96; 4) 22–24 (Piechocki and Bolod 1972; Piechocki *et al.* 1982).

Structure. Wing long and pointed, base narrow. 10 primaries: p9 longest, p8 0–3 shorter, p7 9–14, p6 18–26, p5 28–35, p4 37–46, p1 61–71; p10 reduced, narrow and pointed, hidden by greater primary coverts, 84–94 shorter than p9, 11–15 shorter than longest upper primary covert. Flight-feathers not emarginated, except for shallow notch at tip of secondaries and inner primaries; tertials short. Tail rather short, tip shallowly forked; 12 feathers, t4–t5 longest, t6 slightly shorter, t1 2–6 shorter. Bill as in Swallow *Hirundo rustica*, not as tiny as in

Sand Martin *Riparia riparia*. Leg and foot slender, bare; tarsus short, toes rather short. Length of middle toe with claw 14·7 (19) 14–16; outer toe with claw *c.* 69% of middle with claw, inner *c.* 62%, hind *c.* 58%. Claws shorter and deeper than in *H. rustica*, distinctly heavier and more strongly curved.

Geographical variation. Very slight, in size only. Birds of mountains of south-west China sometimes separated as *centralasica* (Stachanov, 1933) on account of slightly longer wing, but large overlap in size with European populations: wing of Chinese birds 135·7 (3·61; 36) 130–143 (Vaurie 1951c), of skins of European birds 131·1 (2·91; 40) 126–136 (Vaurie 1951c; BMNH, RMNH, ZMA), and of live birds from Gibraltar 133·1 (2·93; 1980) 124–144 (Elkins and Etheridge 1974, 1977). Birds from eastern Turkey, Iran, and Afghanistan perhaps slightly smaller: wing, 128·9 (2·87; 23) 121–133 (Paludan 1940, 1959; Vaurie 1951c; Kumerloeve 1969a).

Forms superspecies with African Rock Martin *P. fuligula* from Africa, Arabia, and south-west Asia east to Pakistan, and with Dusky Crag Martin *P. concolor* from India and Burma (Hall and Moreau 1970). Genus *Ptyonoprogne* sometimes included in *Hirundo*, but see Voous (1977). CSR

Hirundo rustica Swallow

PLATES 17 and 18
[between pages 232 and 233]

Du. Boerenzwaluw   Fr. Hirondelle rustique   Ge. Rauchschwalbe
Ru. Деревенская ласточка   Sp. Golondrina común   Sw. Ladusvala   N. Am. Barn Swallow

Hirundo rustica Linnaeus, 1758

Polytypic. Nominate *rustica* Linnaeus, 1758, from Europe, Asia Minor and Iraq east to Yenisey basin, western Altai mountains, Sinkiang (China) and Sikkim, and North Africa east to Libya; *transitiva* (Hartert, 1910), Lebanon, southern Syria, and Israel; *savignii* Stephens, 1817, Egypt. Extralimital: *erythrogaster* Boddaert, 1783, North America; *tytleri* Jerdon, 1864, southern Siberia from Angara basin east to Yakutsk and Olekma river, south to northern Mongolia; *gutturalis* Scopoli, 1786, eastern Asia, east from *tytleri* and nominate *rustica*, intergrading with *tytleri* between Yakutsk and coasts of Sea of Okhotsk and in Kamchatka.

Field characters. 17–19 cm, of which tail-streamers 2–7 cm; wing-span 32–34·5 cm. Most attenuated hirundine of west Palearctic; less bulky than Red-rumped Swallow *H. daurica*. Medium-sized hirundine with classic form of small bill, wide gape, streamlined body, long wings, and long forked tail with outer feathers forming streamers in adult. Upperparts shiny blue-black; underparts buff-white to red-buff, relieved by red-chestnut face and throat and blue-black chest-band and by noticeably dark undersurface of flight-feathers and white-spotted tail. Flight light and graceful, with characteristic sweeps and swoops after insects. Most vocal of west Palearctic hirundines, with distinctive song and calls. Sexes similar; some seasonal variation. Juvenile separable. 3 races in west Palearctic, adults separable in breeding plumage.

ADULT MALE. (1) West Eurasian and west North African race, nominate *rustica*. Chest-band, most of head, and rest of upperparts blue-black, noticeably shiny when fresh (except on flight- and tail-feathers) but dull or invaded with white flecks when worn and in moult.

Forehead, chin, and large area of throat above chest-band red-chestnut. Underparts including under wing-coverts and long vent buff-white, relieved by dusky-black undersurfaces of flight-feathers and tail. When spread, tail shows oval white panels near tips of all but outermost feathers. Tone of underparts variable, none pure white and darkest rufous-buff. (2) Levant race, *transitiva*. Differs from nominate *rustica* in darker, more consistently pink- to rufous-buff underparts. (3) Egyptian race, *savignii*. Differs from first 2 races in yet darker underbody and under wing-coverts, these appearing maroon to red-brown, matching tones of fore-face and throat and making whole bird appear noticeably dark. ADULT FEMALE. All races. Tail-streamers usually shorter than in ♂. Plumage pattern and tones as ♂ but underparts usually paler, with mottled, less shiny chest-band. JUVENILE. All races. Tail-streamers broader than in adult and not developed beyond 1 cm. Plumage pattern as adult but tones less vivid, with duskier, even browner upperparts and paler underparts. At close range, fore-face and throat usually orange-brown and never as intensely chestnut as adult; head appears strikingly paler and wider in front of eye. Chest-band mottled with buff. FIRST YEAR. May retain indication of juvenile plumage into 1st summer but apparent only in direct comparison with full adult. At all ages, bill and feet black; obvious yellow gape on juvenile.

In west Palearctic, unmistakable when seen well but liable to confusion with several congeners south of Sahara. Most obvious distinctions are dark head and breast contrasting with usually paler underbody and white spots on tail-feathers. Red-rumped Swallow *H. daurica* has somewhat similar form but plumage pattern differs distinctly (see that species, p. 278). Beware occasional hybrids or mutants showing pale rump or lacking chest-band; see *H. daurica*. Flight elegant and accomplished in all actions, allowing not only effective capture of flying insects through acceleration and agility but also long migration through steady speed and endurance. Action of bird on breeding grounds typically consists of bursts of easy wing-beats propelling it into long curves and swoops, jinking manoeuvres after insects, or dashing sweeps in and out of nest-sites; gliding ability marked in warm air, though not so developed as in House Martin *Delichon urbica*. On migration or between feeding areas, bird uses more constant wing-beats on more level track but action still light, with many small lifts and jinks. At nest, capable of prolonged hover and complete reverse of direction. Infrequently on ground, appearing nervous; uses shuffling walk, often with extended, flapping wings. Perches freely on wires and twigs, and clings expertly to vertical and sloping faces of cliffs and buildings. Much addicted to low-level hunting of insects near domestic and wild animal herds.

Noisiest of west Palearctic hirundines. Song a slightly spluttering but pleasing and cheerful warble with occasional rattling sounds; often has distinctive terminal 'su-seer', last note higher pitched but falling; loud at close range but hardly carrying far. Calls include clear 'witt' or 'witt-witt' and loud distinctive 'splee-plink' in alarm or excitement.

Habitat. Breeds across west Palearctic from subarctic through boreal, temperate, steppe, and Mediterranean zones in both continental and oceanic climates; missing only from arctic tundra and desert belts. Entirely dependent on constant supply of small flying insects taken in flight in lower airspace over surface, either of shallow water or clothed with low moist green vegetation. Usually avoids densely wooded, precipitous, arid, and densely built-up areas, preferring pasture grazed by large animals, meadows, and farm crops, especially where accessible open structures such as barns, sheds, stables, outhouses, and porches provide suitable nest-sites nearby, with roof-ridges, overhead wires, and bare branches or twigs as perches and places to sun and preen, and ready access

to water. Alights on ground infrequently, except to gather nest-material. Forages over open water, usually near margins. Pronounced dependence on human land uses and artefacts implies historic shift from natural habitats, presumably grasslands grazed by wild ungulates near rock exposures, hollow trees, and other original nest-sites, which would have restricted distribution. Now succeeds in almost any climate permitting adequate regular production of small flying insects. Accepts warm or chilly conditions or some sudden drops in temperature or strong winds, but hard hit by continuing frost or snowfall, or other persistently adverse conditions interfering with insect activity.

In Swiss Alps, breeds freely up broader valleys to 1775 m (Engadine, St Moritz), but avoids narrower and more shady side-valleys and exposed tops (Glutz von Blotzheim 1962). In USSR, follows human settlement to heights of 2400-3000 m (in Caucasus) and to 2000 m or more in central Asia, where water available in immediate vicinity and there are open stretches overgrown with grass or small bushes (Dementiev and Gladkov 1954*b*). Agricultural change, especially relating to livestock, can affect both breeding sites and food supply, but effects of such changes difficult to trace precisely, as is influence of climatic and land use changes in wintering areas and at intermediate staging points. Habitat similar at these seasons, but for full discussion see Moreau (1972). Will then cross seas, deserts, and other areas normally avoided, and exploit local opportunities such as bush fires.

Distribution. No evidence of any marked change.

ICELAND. Has bred *c*. 10 times (AP). FAEROES. 0–5 pairs; may breed annually (Bloch and Sørensen 1984). BRITAIN. Formerly sporadic on Orkney (Scotland), now regular (Sharrock 1976). IRELAND. No longer any gaps in distribution along west coast as 5 years ago (Sharrock 1976). MALTA. Bred 1974 (Sultana and Gauci 1982).

Accidental. Spitsbergen, Bear Island, Jan Mayen, Iceland, Azores.

Population. Fluctuating. Declines noted in Netherlands, West Germany, Denmark, Czechoslovakia, Rumania, Baltic republics (USSR), Israel, and recently Britain.

BRITAIN. 500 000 to 1 million pairs; fluctuates but no evidence of marked change in last 50 years (Sharrock 1976). Marked recent decrease shown by Common Birds Census results (Marchant 1984). FRANCE. Over 1 million pairs; fluctuates, possibly decreased (Yeatman 1976); general decrease unlikely (Hémery *et al.* 1979). BELGIUM. About 200 000 pairs (Lippens and Wille 1972). NETHERLANDS. About 150 000 pairs 1976-7 (Teixeira 1979). Declining, mainly due to habitat changes (CSR). West Germany. 0·92-4·4 million pairs (Rheinwald 1982); declining (AH). See Hölzinger (1969) for counts in Ulm area. DENMARK. Declining (Møller 1983*b*; TD); slight

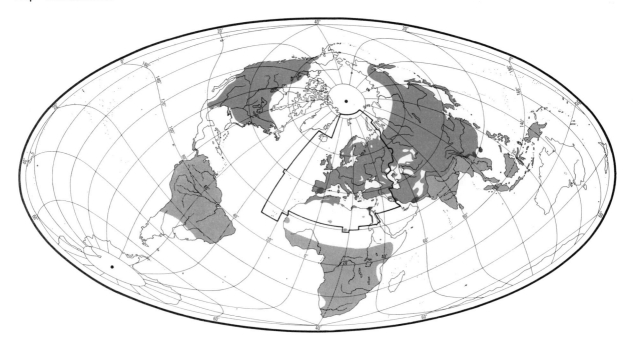

decrease 1978-82 (Vickholm and Väisänen 1984). SWEDEN. About 600 000 pairs (Ulfstrand and Högstedt 1976). FINLAND. About 300 000 pairs (Merikallio 1958). Slight decrease 1978-83 (Ulfstrand and Högstedt 1976; Vickholm and Väisänen 1984). SWITZERLAND. Census in Jura 1931-78, recent increase (Bruderer 1979). RUMANIA. Decreased (VC). USSR. Baltic republics: decreasing (HV). ISRAEL. Great decline in 1950s due to pesticides; some recovery since early 1980s (HM, UP).

Mortality. Britain: average annual mortality 63% (Lack 1949). France: average annual 1st-year mortality 73·6%, average annual 2nd-year mortality 69·3%, average annual adult mortality 43% (Hémery *et al.* 1979). France: Seine-Maritime, average annual 1st-year mortality *c.* 80% in 2nd year 47%, in 3rd year 55%, in 4th year 51%, in 5th year 87% (Jarry 1980). West Germany: in Westphalen, annual mortality 60-65% (K-H Loske). Denmark: average annual mortality *c.* 70% (Christensen 1981). Czechoslovakia: average 1st-year mortality 70%; up to 8th year, lowest mortality (33%) in 5th year (Beklová 1976). USA: annual mortality 73·4% (Mason 1953); average 1st-year mortality 68·6% (under-estimated—probably 85·7%), average annual adult mortality 42·7 ± 3·7% (Henny 1972). Oldest ringed bird 15 years 11 months (Rydzewski 1978).

Movements. Migratory. A few aberrant individuals winter every year in southern and western Europe as far north as Britain and Ireland, and recorded annually in winter in southern Spain (A Noval); small numbers winter regularly in North Africa (e.g. Smith 1965a); also 2 small resident or partly resident populations in east Mediterranean countries. Otherwise, west Palearctic birds are long-distance migrants. Similar migratory regimes apply in Asia and North America.

European and north-west Asian birds (part of nominate *rustica*) winter largely in Africa, though some in coastal Arabia and in Pakistan. In Africa, main wintering areas are south of equator, though also locally numerous in West Africa (Ivory Coast, Togo, southern Nigeria), northern Zaïre, and Uganda. Sources of birds using different parts of winter range are better known than in other European migrants, due to large-scale ringing (Drost and Schüz 1952; Davis 1965; Moreau 1972). British and Irish birds winter essentially in South Africa. Until 1962-3 winter, recoveries mainly from Transvaal, Natal, Orange Free State, and eastern Cape Province (Davis 1965), but since then a significant westerly extension of winter range into western Cape Province (Rowan 1968; Mead 1970a). In contrast, most German birds winter in Zaïre, some 1600 km further north, only minority reaching South Africa (Rowan 1968; Moreau 1972). Also 5 recoveries, December–February, of East German birds grouped in southern Ghana, perhaps indicating discrete subpopulation. However, African winter recoveries from some other European countries more widely scattered: e.g. Dutch birds in Ivory Coast and Zambia; Belgian birds in Ghana, Ivory Coast, Congo, Zaïre, and Transvaal; Swiss birds in Nigeria, Zaïre, and Cape Province. Congo basin further used by birds from Scandinavia and south-central Europe (Drost and Schüz 1952). South Africa and East Africa also important wintering regions for USSR populations of nominate *rustica*, with recoveries as far east as 92°E in Yenisey basin (Broekhuysen 1964; Rowan 1968; Britton 1980).

Passage broad-front, including large trans-desert move-

ments into and out of Africa across Sahara and Middle East; desert mortality reported more often for this than for any other species (Moreau 1961). British birds seldom recovered in extreme West Africa (Mauritania to Ivory Coast), and this apparently so for other west European countries also; hence suspected that most birds take overland route across western Sahara and West African bulge, especially in spring when more Algerian and Tunisian passage recoveries than in autumn. For fidelity of adults and young to previous breeding place, see Social Pattern and Behaviour.

Levant race *transitiva* probably a partial short-distance migrant. Winter specimens attributed to this population have been claimed (though not always reliably) from Egypt, Sudan, East Africa, and Zambia (Vaurie 1959);

not included in subsequent avifaunas from East Africa (Britton 1980) or Zambia (Benson *et al.* 1971). Egyptian race *savignii* resident. Populations of southern Palearctic Asia winter in southern Asia from India, Malaysia, and Indonesia to Japan and Melanesia. Clancey (1970) claimed small numbers of far-eastern *gutturalis* winter in south-east Africa, but identification not universally accepted (Moreau 1972).

Juvenile dispersals begin July and become oriented southwards by early August as migration begins. Autumn passage protracted, with peak exodus from north-west Europe in September and first half of October, though some birds linger there until onset of night frosts in November. Mediterranean passage and arrivals in Africa north of equator are at height mid-September to late

October, and birds become numerous in wintering regions south of equator in November. Return movement begins February, with heavy passage through Zambia by mid-February; in North Africa, Mediterranean basin, and Middle East, peak spring movement occurs mid-March to late April. Early birds return to north-west Europe in second half of March, though main arrivals mid-April to mid-May. In north USSR, arrives May and early June.

RH

Food. Almost wholly flying insects; in breeding season, especially flies (Diptera), but apparently usually fewer aphids (Aphidoidea) than taken by House Martin *Delichon urbica* (e.g. Vietinghoff-Riesch 1955, 1961), and average prey size much greater than for *D. urbica* and Sand Martin *Riparia riparia*. Prey taken almost entirely by aerial-pursuit, but many records of other methods, used especially when non-flying prey temporarily abundant or conditions unsuitable for aerial-pursuit. While actually foraging, proportion of time spent flapping recorded as $87 \cdot 5 \pm SD16 \cdot 4\%$, $n = 113$ (Bryant and Turner 1982). Feeds on ground, sometimes persistently, walking about and picking at surface (Witherby *et al.* 1938*b*; Jones 1947–8; Broekhuysen 1952; Vietinghoff-Riesch 1955; Bilby 1957; Schulze 1971; Lee and Brown 1979; King 1986). Birds wintering in south-west England often fed on ground, perching on sand and seaweed on beach (King and Penhallurick 1977). Will hover over ground or plants to pick off food items (Vietinghoff-Riesch 1955; Brooke 1956; King 1956; Hodson 1962) and recorded landing on thistle heads to take adult Diptera from them (Tubbs 1954; Currier and Howorth 1957) and on trees to take caterpillars from leaves (Sage 1954; Bilby 1957), also perching by spider's web to take insects from it (Cassidy 1971). Will take insects from water by hovering or by dipping to surface while flying slowly upwind (e.g. Würtele 1969, Ali and Ripley 1972), and seen doing so persistently on sea (King 1960*a*). Once recorded dipping to surface of water and catching stickleback (Gasterosteidae) (Jarvis 1973). In cold weather in Hungary, birds sheltered in pig-sties, perching on pigs' backs to take insects from them (Rékási 1966–7). In South Africa, often takes insects disturbed by large mammals, and recorded flying round party of Ruffs *Philomachus pugnax* to take Diptera disturbed by them from low vegetation (Taylor 1964); similar records in Vietinghoff-Riesch (1955). Recorded following a tractor ploughing field covered with tall weeds and taking insects disturbed (Schmidt 1967*b*; also Kondělka 1970), and once seen to fly along woodland edge, brushing twigs, and immediately hunt moths *Tortrix* thereby disturbed (Vietinghoff-Riesch 1955). In Sri Lanka, unlike Red-rumped Swallow *H. daurica*, apparently not attracted to grass fires to take insects (Phillips 1953). According to Bilby (1957), *gutturalis* never visits fires and nominate *rustica* only rarely. Such behaviour common in Africa, however (e.g.

Ferguson-Lees 1968*a*), and birds recorded visiting fires in southern England, though sometimes at least apparently not to feed, birds just flying leisurely backwards and forwards through smoke (Bilby 1957). Recorded stealing large butterfly (Lepidoptera) from sparrow *Passer* (Vietinghoff-Riesch 1955). When bringing 2–4 Diptera at a time to nest, will remove wings (possibly with tongue) immediately before feeding to young (Horne 1924). Pair with nest by artificial lights brought food to young by night as well as day (Semadam 1966–7; also Zucchi 1974). Near Stirling (Scotland), birds fed up to 600 m from nest, but mostly up to 250 m, average 170 m, $n = 118$ (Bryant and Turner 1982); in one other case, within 200 m of nest (De Braey 1946).

Summer diet in west Palearctic includes the following: mayflies (Ephemeroptera), damsel flies (Odonata), stoneflies (Plecoptera), grasshoppers, etc. (Orthoptera), earwigs (Dermaptera), Psocoptera, bugs (Hemiptera: Scutelleridae, Lygaeidae, Tingidae, Saldidae, Gerridae, Cixiidae, Cicadellidae, Delphacidae, Psyllidae, Aphidoidea), lacewings, etc. (Neuroptera), adult and larval Lepidoptera (Satyridae, Nymphalidae, Lycaenidae, Pieridae, Incurvariidae, Pterophoridae, Pyralidae, Tortricidae, Sesiidae, Glyphipterygidae, Coleophoridae, Tineidae, Noctuidae, Lymantriidae, Geometridae), adult caddis flies (Trichoptera), adult flies (Diptera: Tipulidae, Anisopodidae, Ptychopteridae, Psychodidae, Culicidae, Chironomidae, Ceratopogoniidae, Simuliidae, Bibionidae, Mycetophilidae, Cecidomyiidae, Stratiomyidae, Rhagionidae, Tabanidae, Therevidae, Asilidae, Empididae, Dolichopodidae, Lonchopteridae, Phoridae, Platypezidae, Pipunculidae, Syrphidae, Conopidae, Otitidae, Lonchaeidae, Lauxaniidae, Micropezidae, Psilidae, Sepsidae, Helomyzidae, Anthomyzidae, Opomyzidae, Ephydridae, Borboridae, Drosophilidae, Agromyzidae, Milichidae, Chloropidae, Tachinidae, Calliphoridae, Muscidae, Hippoboscidae), adult and larval Hymenoptera (Siricidae, sawflies Tenthredinidae, Ichneumonoidea, Cynipoidea, Chalcidoidea, Proctotrupoidea, ants Formicidae, wasps Vespidae, Sphecidae, bees Apoidea), adult beetles (Coleoptera: Carabidae, Hydrophilidae, Histeridae, Staphylinidae, Scarabaeidae, Dryopidae, Elateridae, Anthicidae, Nitidulidae, Coccinellidea, Cerambycidae, Chrysomelidae, Curculionidae, Scolytidae), spiders (Araneae), sandhoppers (Amphipoda), and stickleback (Gasterosteidae) (see especially Vietinghoff-Riesch 1955 and Kožená 1979; also Csiki 1904, Thomas 1934*a*, 1934*b*, 1937*a*, *b*, 1938, 1940, Witherby *et al.* 1938*b*, King 1956, Jarvis 1973, Smogorzhevski and Kotkova 1973). Recorded landing on ground to pick up bread and flying off with it, probably to feed young (Bell 1965). In USA will take berries from bushes—even in flocks of thousands (Vietinghoff-Riesch 1955).

Little data on food eaten by adults in summer, most information coming from studies of young (see below). In Poland, 355 items from stomachs (presumably of full-grown birds) comprised 78·0% (by number) Hymenoptera, 14·4% beetles, 5·7% Hemiptera, and 1·9% Diptera (Głowacki 1977). See also Csiki (1904) for Hungary, Florence (1914) for Scotland, and Gil (1928) for Spain.

In winter quarters, the following recorded in diet: half-grown grasshoppers, termites (Isoptera), bugs (Hemiptera), adult and larval Lepidoptera, flies (Chironomidae, Sepsidae, Tachinidae), wasps (Ichneumonoidea), winged ants, adult beetles (Scarabaeidae, Curculionidae), and sandhoppers (Broekhuysen 1952; Rudebeck 1955; Vietinghoff-Riesch 1955; Brooke 1956; Taylor 1964; Hornby 1973; Talbot 1974; Lack and Quicke 1978). Hymenoptera said to be more important food in Africa than in Britain, contributing 43·7% more to diet (see Waugh 1978).

Collar-samples from nestlings in Czechoslovakia, June–August, contained 4606 items comprising 62·3% (by number) Diptera, 28·3% Hemiptera (largely aphids and these mostly in June), 4·4% Hymenoptera, 4·0% beetles, and 1·2% others; most important Diptera were Syrphidae (hoverflies—in 55·3% of samples) and Muscidae (in 41·7% of samples); average body length of items 4·1 mm (1·5–18 mm, 56·9% 1·5–3 mm), $n=4601$ (Kožená 1979, 1980; see also for variation in diet with weather and time of day and through season). In Wales, 352 items taken from adults going to feed young comprised 91·5% Diptera, 1·7% aphids, 1·7% other Hemiptera, 0·9% ants, 1·7% other Hymenoptera, 1·4% beetles, and 1·2 % others (Thomas 1934a, 1934b, 1936, 1937a, b, 1938, 1940). See also Lyuleeva (1974) for Kaliningrad (USSR), and Ruge (1979). For comparison with diet of *D. urbica* in Moscow region (USSR), and size of prey items, see that species (p. 290). Grit found in 2·1% of 331 collar-samples from young in Czechoslovakia (Kožená 1979), and in Washington (USA) 79·9% of 159 young found dead (1–16 days old) contained grit, average 4·8 particles per bird (Barrentine 1980); particles up to 6 mm across (Vietinghoff-Riesch 1955). Bird(s) in Switzerland seen repeatedly picking up pieces of hen's eggshell, breaking up and swallowing some and carrying others away; behaviour occurred both before and during nestling period (Bürkli 1974). Adults feed young on larger prey items than they catch for themselves (Waugh 1978). When feeding young, adults at Stirling selected larger-than-average items preferentially; increase in absolute numbers of large prey available led to them forming greater proportion of diet, but no increase in proportion of smaller prey in diet when their absolute numbers increased (Turner 1982). Larger-than-average items selected also by *R. riparia*, *D. urbica*, and Swift *Apus apus*, though *H. rustica* takes larger prey on average than any of these: in decreasing order of average prey size, *H. rustica*, *R. riparia*, *D. urbica*, *A. apus* (Waugh 1979, though sources not all given); see also Table A. Large items (6·61 mg dry weight) take longer to catch and

Table A Data on food collected by adult Hirundinidae for young near Stirling, Scotland (Bryant and Turner 1982).

	Bolus dry wt (mg)	Insects per bolus	Prey wt (mg)
Sand Martin *Riparia riparia*	61·0 ± SD26·6 ($n=272$)	59·7 ± SD40·6 ($n=67$)	1·26 ± SD1·28 ($n=8080$)
Swallow *Hirundo rustica*	73·4 ± SD31·9 ($n=432$)	18·1 ± SD14·1 ($n=92$)	6·01 ± SD5·64 ($n=4960$)
House Martin *Delichon urbica*	55·2 ± SD14·8 ($n=489$)	43·8 ± SD28·1 ($n=99$)	1·88 ± SD1·49 ($n=21\,400$)

Table B Proportion of large, strong-flying Diptera in different-sized food boluses collected by adult Hirundinidae for young near Stirling, Scotland (Bryant and Turner 1982).

	Proportion of large, strong-flying Diptera (%)	
Bolus dry wt (mg)	Sand Martin *Riparia riparia*	Swallow *Hirundo rustica*
under 30	0	0
30–60	2·6	31·7
60–90	3·6	62·5
90–120	0	72·1
over 120	0	81·3

require faster flapping flight than do small items (1·06 mg dry weight) but are still more profitable (net energy gains 907 J per min and 418 J per min respectively) (Turner 1982). See Tables A–B for data on size of food boluses collected for young. Unlike in *R. riparia* and *D. urbica*, bolus weight linked more strongly to weather (and hence presumably to food availability) than to distance of food source: significantly positively correlated with prey weight, food abundance, flight speed when foraging, and average daily temperature; negatively correlated with time between nest-visits, time spent actually foraging (see below), and daily rainfall; not correlated with distance of food source; bolus weight increased significantly through the season (Bryant and Turner 1982, which see for further optimal foraging data; see also Kožená 1980). In central Scotland, 2·83 ± SD1·69 min ($n=124$) between visits, of which 2·40 ± SD1·64 min ($n=111$) spent actually foraging (Bryant and Turner 1982; see also Moreau and Moreau 1939b, De Braey 1946, Purchon 1948, Vietinghoff-Riesch 1955, Kuźniak 1967, Pröger 1979).

DJB

Social pattern and behaviour. Includes information from detailed account of study in Denmark by A P Møller. Major review by Vietinghoff-Riesch (1955). Studies on Nearctic *erythrogaster* (e.g. Bent 1942) indicate some major differences from nominate *rustica*; some of these included here.

1. Highly gregarious outside breeding season. Large numbers typically occur in pre-migratory assemblies, sometimes on migration, and at roost-sites. In Denmark, August–October, post-breeding flocks vary from a few birds to a few hundred,

average 6·9 ($n = 363$); sex-ratio in adult flocks 1:1. Juveniles tend to aggregate where adults feeding successfully, and are relatively more numerous in large flocks. (A P Møller.) Migrant flocks in Italy, April, 30-50 (see Vietinghoff-Riesch 1955). Tendency to migrate in discontinuous stream requires caution in interpreting reports of exceptionally large flocks; e.g. in Nigeria, November, 5000 birds per min for 10 min passing along coast (Elgood *et al.* 1966). In South Africa, December, birds migrated singly or in flocks of up to 125, most often 1-10 (Broekhuysen 1964). Migrant flocks sometimes associate with other Hirundinidae, less often with other species, e.g. swifts (Apodidae), Bee-eater *Merops apiaster* (Moore and Boswell 1956). For relative proportions of *H. rustica* and House Martin *Delichon urbica* in autumn migratory flocks, see Bruderer (1979). On migration, large assemblages of hundreds or thousands may occur on islands, at oases, etc. (e.g. Vietinghoff-Riesch 1955; Wallace 1982). For details of size and composition of roosts outside breeding season, see Roosting (below). BONDS. Monogamous mating system the rule, but polygyny (♂ and 2 ♀♀) sometimes occurs, and birds sometimes copulate with non-mates if they are nearest neighbours in a colony (see Breeding Dispersion, below): of 18 pairs, 6 of 11 colonial ♂♂ copulated with other ♀♀, but no solitary-nesting ♂♂ did so; similarly, of 18 paired ♀♀, 8 that copulated promiscuously all nested in colonies (A P Møller). Bigamous pairings usually synchronous, with laying by each of the 2 ♀♀ a few days apart at most. Some cases originate from 2 pairs being intent on nesting near each other; one ♂ driven off by the other who then mates with both ♀♀ and helps to raise 2 broods (Mohr 1958; Feldmann-Luternauer and Feldmann-Luternauer 1978). In one case, ♂ mated with 2 ♀♀ but ignored one brood which thus succumbed; successful ♀ disappeared after her brood fledged, and ♂ then co-operated with neglected ♀ in raising a 2nd clutch (Richardson 1956). At artificial site, bigamy occurred in 3 successive years, twice (consecutively) by same ♂; in 2nd year, young of one nest fledged first and joined brood in other nest; not known if each ♀ then fed own young. In 2 years, 2nd clutches laid in both nests, but not all raised. In 3rd year, when 1 ♀ died, ♂ paired with another ♀ who disappeared during laying. In a case of suspected bigamy, ♀ incubating in one nest helped to feed young in another nest nearby, finally deserting her own clutch. (Löhrl 1962a.) Several cases of adoption of recently-fledged young, such birds then usually being fed by foster-parents; once, 3 orphan young adopted (Weinzerl 1955). One pair simultaneously fed 5 of their own young in nest, along with 3 fledglings from another family and 1 from a 3rd (Berndt 1958). In Poland, adoption not uncommon where birds breed colonially, but unknown where only 1-2 pairs breed together (Sonnabend 1958). Pair-bond typically maintained for 2nd clutch and not infrequently for life (Boyd and Thomson 1937; Vietinghoff-Riesch 1955), although short life means that many bonds last for 1 year only. In Westfalen (West Germany), only 13 of 115 pairs bred together in 2 consecutive years, 1 pair for 3 years (K-H Loske). One ♂ returned for 11 years to same building, pairing with 2 ♀♀ for 4 consecutive years each, and with another 3 ♀♀ for 1 year each (Brombach 1982). ♂ and ♀ of pair recorded returning and breeding with new mates (Pfromm 1931), and birds that lose mate during breeding season may take new mate in same season (Vietinghoff-Riesch 1955). Of 19 pairs, 3 copulated but divorced before starting to breed; 2 of the divorced ♂♂ and 2 of the ♀♀ later took new mates (A P Møller). Due to relative fidelity of young birds to natal site (see below), inbreeding occasionally occurs: 2 cases of ♀ pairing with son of previous year (Creutz 1941), and one ♂ bred with daughter of previous year (Weinzierl 1955). Numerous reports

of hybridization with *D. urbica* (see that species, p. 291). Usually only ♀ broods. Both sexes feed young (see Relations within Family Group, below). Rarely, fledged young of 1st brood help to feed young of 2nd brood (White 1941; Williamson 1941; A P Møller). For report (doubted by Vietinghoff-Riesch 1955) of 1st brood-young also feeding incubating ♀, and helping to build 2nd nest, see Jenner (1945). For helping also by non-offspring (age unknown) in *erythrogaster*, see Myers and Waller (1977). After fledging, family bonds maintained until young 21-32 days, $n = 15$ (A P Møller). One family party stayed together for $2\frac{1}{2}$ months, and bonds may therefore continue into migration (Boley 1932). ♀♀ and most ♂♂ usually breed at 1 year, some ♂♂ not until 2: 20% of 1-year-old ♂♂ unpaired (Hémery *et al.* 1979); c. 10% of ♂♂ do not return to breed until 2 years old (Jarry 1980). BREEDING DISPERSION. Mostly solitary, occasionally colonial. Colonies usually less than 5 pairs, occasionally many more: e.g. in Poland, sometimes 20-30(-50) pairs (Sonnabend 1958). In Ulm (West Germany), 95 pairs in one stable (Hölzinger 1969). Single cases of 120 pairs at one farm, 280 nests at another (Vietinghoff-Riesch 1955, which see for other examples of up to 50 or more nests in same building). Some of these assemblies probably neighbourhood groups, e.g. 56 pairs in stable of 500 m² (Møller 1983a). *H. r. erythrogaster* typically colonial, usually 20-40 or more pairs (Bent 1942). In Nordtirol (Austria), 127 single pairs, 69 of 2 pairs, 21 of 3, 22 of 4-9 (Landmann and Landmann 1978). In Westfalen, 17 of 1-2 pairs, 9 of 3-5, 3 of 6-10, 1 more than 10 (Püttmann 1973). For annual variation in colony size, Switzerland, see Bruderer and Muff (1979). Mean number of pairs breeding together: in Manchester (England), 1·14 pairs, $n = 37$ (Cramp and Ward 1934); in Denmark, 3·1 pairs, $n = 89$ (Møller 1983a); in Finland, 2·3 pairs (Lind 1964); in Niederrhein (West Germany), 5·8 pairs, $n = 15$ (Beser 1974). In different areas of Britain, density 0·5-34·0 pairs per km² (Hollom 1930, 1985; Boyd 1933, 1936; Cramp and Ward 1934; McGinn 1979). In Denmark, 5·3-12·4 pairs per km² (Toft and Christensen 1977). In Finland, 1-3 pairs per km² (Vietinghoff-Riesch 1955; Merikallio 1958). In Uppland (Sweden), 4·9 pairs per km² (Olsson 1947). In Poland, 6·3-11·4 pairs per km² (Kuźniak 1967). In Westfalen, 8-10 pairs per km² (Püttmann 1973) and 11 pairs per km² over 26 km² (K-H Loske); in Nordrhein-Westfalen, 1·6-3·5 pairs per km² (Beenen 1970). Few data on distances between nests. In neighbourhood group, average 4·07 m ($n = 127$) between nests (Møller 1983a). Nests of bigamous ♂ 4 m apart (Richardson 1956); 2·1 m (Löhrl 1962a). Both members of pair defend territory around nest. Size of more than 100 territories usually 4-25 m² (A P Møller), once 200 m² (Møller 1974b). Territory of bigamous ♂ 12 m², that of monogamous neighbour much larger (Löhrl 1962a). Territory serves for courtship, nesting, and—in bad weather—some feeding (A P Møller). Feeding range may be shared with neighbours (De Braey 1946); for extent, see Food. Established breeders markedly faithful to previous nest-site. In Seine-et-Marne (France), 96·6% of surviving adults faithful to nest-site in successive years (Jarry 1980). In study in Oxfordshire (England), all nest-sites in wide area controlled, and ♂♂ not located in successive years presumed dead; 80% of 21 controlled ♂♂ returned to same nest-site while the rest moved to other sites nearby; no evidence of previous owner being displaced from former nest-site by another ♂. ♀♀ moved nest-site only if former mate failed to return; 9 ♀♀, known or presumed to be widows, paired again at nests 50-3050 m from former site. (C E Davies.) In Westfalen, 6·9% of 145 controlled ♂♂, and 13·3% of 173 ♀♀ changed site; ♂♂ moved on average 233 m, ♀♀ 613 m, usually to nearest available site (Loske 1982). Return of 1-year-olds to natal area much lower, e.g 0·9% of those fledging in Extremadura, Spain (Lope Rebollo 1983). In Britain, most 1-year-olds return to within 30 km of birthplace, though some found breeding up to 360 km distant (Davis 1965). In Westfalen, of 200 birds ringed as nestlings and controlled subsequently, ♂♂ more faithful to natal area than ♀♀; mean settlement distance of 146 ♂♂ from natal site 739·5 m (0–9750), of 54 ♀♀ 2472·7 m (0–6150) (K-H Loske). For similar result at Seine-et-Marne, see Jarry (1980). Pair typically uses same nest for 2nd brood; in Westfalen, 84% of 452 pairs used same nest, 16% another nest; 2% used a different room or shed (K-H Loske). In Brabant (Belgium), only *c.* 10% of pairs built another nest for 2nd clutch; 3rd clutch and replacements laid in same nest as for 1st or 2nd clutch (Brown 1924; Herroelen 1959). In Britain, proportion of nests used twice per season varied from 0 (*n* = 31) to 93% (*n* = 14), depending on area (Boyd 1935); see also Boyd (1936). In 3 years, pair with 3 nests together in a row started 2nd clutch in 2nd nest while still raising 1st-brood nestlings (Radermacher 1967). For occasional nesting in close association with *D. urbica*, see that species (p. 292). ROOSTING. In breeding season, birds roost within territory; at other times, communally in large numbers. Sometimes in spring, and typically in late summer and autumn, birds occupy communal roosts in beds of reed *Phragmites*, club-rush *Scirpus*, and reedmace *Typha*, willow-scrub *Salix*, tall grasses, and other dense vegetation in or near standing water; also in maize *Zea*, and sometimes in canopy of tall tree, or in rocks and cliffs. (Witherby *et al.* 1938*b*; Vietinghoff-Riesch 1955; Rolls and Rolls 1977; Loske 1984; A P Møller.) For similar sites in winter quarters, see below. Late-summer roosts of up to 100 000 birds occur (e.g. Kuhn 1955, Beaman 1978). Non-breeders may roost communally throughout the year, in Denmark up to a few hundred together (A P Møller). Early in breeding season, cold weather may induce communal huddling (birds sometimes on top of one another) inside buildings (Busse 1980; Bentzien 1983). Sometimes roosts thus in nests of *D. urbica*, and once 21 birds (including corpses) found in single nest (Vietinghoff-Riesch 1955). Communal roosting in buildings also reported in cold spring weather in winter quarters: e.g. in South Africa, January, in rows or radial clusters in cavities in walls, head towards centre of cluster (Broekhuysen 1961: Fig A). In breeding pairs,

A

♀ roosts on nest, ♂ nearby, often indoors (A P Møller). During nest-building, one ♂ roosted near ♀ at nest; once nest built, ♀ roosted in nest, ♂ probably in tree nearby (De Braey 1946). For roosting arrangement of bigamous trio, see Löhrl (1962*a*). During 2nd brood, ♂ may roost communally (see below). Fledglings may return to roost in nest for up to 15 consecutive nights, depending on whether further clutches started (Herroelen 1959); may roost in same building as nest for 40 days (Radermacher 1970*a*). At one nest, young of 1st brood roosted in nest with ♀ until 2nd clutch started, also (in fine weather) with ♀ in trees nearby; young of 2nd brood roosted in trees from outset (De Braey 1946). Fledglings occasionally roost in nest even when ♀ incubating 2nd clutch (Berndt and Berndt 1942; Radermacher 1967). Adults typically leave roost ½ hr

before sunrise; on cold mornings, may not begin feeding until 2–3 hrs after sunrise. Activity ends *c.* ½–¾ hr after sunset. (A P Møller.) Roosts later than *D. urbica* (Kareila 1961, which see for details). After fledging, young (also adults) typically loaf and preen on wires (etc.), in or out of nest-territory, especially around midday and again before roosting; sleep with bill in wing-coverts or more often in normal resting posture (A P Møller). In midday heat, Iran, shade sought by loafing under arches (Richter 1955*b*). Comfort behaviour often a flock activity; includes sunbathing on roofs, especially in morning (A P Møller), sometimes in company with *D. urbica* (which see for sunning behaviour, p. 292). Bathe in water by plunging (Vietinghoff-Riesch 1955), much as in Sand Martin *Riparia riparia* (see that species, p. 240). Also reported: dust-bathing (Tubbs 1954), mud-bathing (Fouarge 1971), dew-bathing on wet lawn (Staton 1950), and spray-bathing in Victoria Falls, Zimbabwe (Taylor 1954). Late-summer communal roosts may be shared with Starling *Sturnus vulgaris*, *R. riparia*, and wagtails *Motacilla*; apparently mixes freely with *R. riparia* (Vietinghoff-Riesch 1955), but not with *S. vulgaris* (Loske 1984). Marked pre-roosting activity at such roosts; following account based on several studies, all of which suggest common pattern (Giller 1955; Richard 1968; Loske 1984; see these for variation in roosting times with weather, sunset, etc.). Birds assemble over roost-area some time (e.g. 15 min to 2 hrs) before sunset. Flocks gradually bunch high over roost-site and perform spectacular manoeuvres—swirling and swerving *en masse*, and making rapid passes low over roost-site. At last light, a few birds detach themselves from flock and drop into roost, followed rapidly, and over a few minutes, by mass entry of others; enter in waves, each flock plunging, whirling and spiralling, initially in silence, but with much twittering once settled. Remaining birds bunch, passing back and forth with twittering calls before joining roost. Late arrivals drop straight in without preliminaries. Calling continues in roost for several minutes, sometimes in complete darkness. In *Phragmites*, birds typically settle where leaf joins stem, several birds to one stem (Giller 1955). Juveniles settle before adults but are more often displaced from perch sites by other birds (A P Møller). In morning, birds leave in waves after sunrise. In Nordrhein-Westfalen, communal roosting in April–May accompanied by less obvious flight manoeuvres (Loske 1984). Post-breeding communal roosts mainly juveniles: in Westfalen, July–September, 89% juveniles; 9 controlled juveniles stayed in roost for mean 18·4 days (K-H Loske). In roost of *c.* 15 000 maximum, Denmark, August–October, 69% of 423 adults were ♂♂, some of which use roost while mates attending late nestling stage of 2nd broods; 87% of 3356 birds caught were juveniles; proportion of juveniles at minimum in July, increasing to 100% by end of September or October (by which time adults have migrated). Roosting birds presumed to come from area of *c.* 500 km², and from distance of 21·4 km. (Årestrup and Møller 1980.) In winter quarters, birds reported resting on ground (Priest 1935); once, did this before roosting (Penry 1979). Communal roosts in winter are often larger than on breeding grounds, up to 1 million birds (Rudebeck 1955; Best 1977). Sometimes mixed with *R. riparia* (e.g. Becker 1974). One roost in Namibia drew birds from radius of *c.* 30–40 km (Becker 1974). Winter roost-sites typically reedbeds, elephant grass *Pennisetum* and other tall grasses, maize and other crops, and bushes; less often, trees (Verheyen 1952; De Bont 1957, 1962); in India and Malaya, also on overhead wires and buildings in well-lit areas of towns (George 1965; Medway 1973). Pre-roosting behaviour much as in late-summer roosts (e.g. Taylor 1942, Rudebeck 1955, De Bont 1957). In morning, at roost in South Africa, March, birds began calling at first light, followed about 20 min later by an 'explosive' vertical exodus of *c.* half a million birds which then departed; over next few minutes, successive waves left, but without marked upflight of first flock (Rudebeck 1955); similar description by Curry-Lindahl (1963b). Juveniles typically predominate in winter roost: e.g. 68% or more (Reynolds 1971), 80% (Best 1977), almost all juveniles (Becker 1974). At roost in Zaïre, proportion of juveniles increased with time, reaching 100% by end of winter; perhaps due to progressive arrival of juveniles or earlier spring departure of adults (De Bont 1962). Birds may use succession of roosts in winter quarters. At one roost in Namibia, 2 birds stayed at least 37 days (Becker 1974).

2. Quite bold in breeding season, and readily approachable to within 5–10 m (A P Møller). When attacked by Hobby *Falco subbuteo*, birds perched on wires dispersed with loud calls, presumably of alarm (see 7b in Voice); sometimes land on ground and crouch when raptor stoops (Vietinghoff-Riesch 1955). May also swoop upwards and climb to evade capture (Bährmann 1950; Baier 1974a). FLOCK BEHAVIOUR. Feeds sometimes in flocks, accompanied by frequent Contact-calls (see 2a in Voice). For feeding association with *S. vulgaris*, see Whitelegg (1961). During breeding season, communal displays ('song-chorus') common, though function not well understood: starting before dawn, up to 30–40 birds fly high in the air, gliding slowly or with smooth wing-beats, and sing (see 1 in Voice) continuously or give Contact-calls; display thus for up to 1 hr before dispersing (Beneden and Huxley 1951; Vietinghoff-Riesch 1955; A P Møller). Similar behaviour reported in winter quarters (Vietinghoff-Riesch 1955), and in *erythrogaster* (Samuel 1971a); see also Heterosexual Behaviour (below). For feather-'play' by flock of 30–40 birds while feeding and bathing, see Kliebe (1970). In late summer, post-breeding assemblies accompanied by much excited singing and calling, also by periodic upflights, birds dispersing to feed, chase one another (etc.), before reassembling; upflights apparently spontaneous, but sometimes initiated by Alarm-call (see 7b in Voice) of disturbed birds (Williamson 1951a). Spring pre-migratory behaviour in winter quarters evidently similar (e.g. Smith 1951). Settling of pre-migratory and roosting flocks accompanied by jockeying for favoured perch-sites; in sunbathing flock, minimum individual-distance 25 cm; in roost of *erythrogaster*, September, separation between birds mean 28 cm (Hutton 1978). Individual-distance just exceeds length of one extended wing; in roost on overhead wires, 8·6–10·7 birds per m of wire at different seasons (Medway 1973). ANTAGONISTIC BEHAVIOUR. Both members of pair defend nest staunchly against intruding conspecifics, especially after laying, but ♂ markedly aggressive from first occupation of nest-site. Rivals challenged with walking or flying approach; pecking attacks rare even when intruder sitting near nest or ♀ mate. Territorial rivals compete vocally, with Threat-display: sit side-on to each other 15–20 cm apart, carpal joints exposed, plumage sleeked, heads pointing forwards and *c.* 45° above horizontal; sing vigorously. (A P Møller.) One ♂, evidently using same display, sidled towards perched ♀, his throat plumage conspicuously ruffled (Hartley 1941). Confrontation often involves pursuit-flight, mostly silent, but sometimes accompanied by Alarm-calls, especially early in season (A P Møller). May lead to fierce fighting in which combatants sometimes fall to ground with feet interlocked: e.g. rival ♂♂ fought on and off for *c.* 8 days, often on ground, until one defeated (Mohr 1958). One ♂ with territory in stable forced intruding ♂ to stay in dark corner for almost 24 hrs (Møller 1974b). Rival ♀♀ also fight, apparently often with fatal results (Vietinghoff-Riesch 1955). Not uncommonly competes aggressively for nest-sites with other

species, notably House Sparrow *Passer domesticus*, also Spotted Flycatcher *Muscicapa striata*. Once defended nest (unsuccessfully) against *D. urbica* for 3 days; drew blood from Black Redstart *Phoenicurus ochruros*; in attack on feral Rock Dove *Columba livia*, landed on its back (Vietinghoff-Riesch 1955); buffeted *P. domesticus* with wings, and tried to land on its back and force it to ground (Baier 1974a). For fight with Blackbird *Turdus merula* after taking over its nest, see Bub (1970). In late summer and autumn, commonly chases birds of various other species, especially potential food competitors (Vietinghoff-Riesch 1955; A P Møller); perhaps an 'exaggerated social response' (Simmons 1951b). Several reports of chasing bats (Chiroptera) (e.g. Campbell 1971, Rosair 1975); possibly related behaviour. Adults sometimes chase juveniles in feeding flocks (A P Møller). HETEROSEXUAL BEHAVIOUR. (1) General. Adult ♂♂ typically first to arrive at breeding grounds, followed by adult ♀♀, then 1-year-old ♂♂ and ♀♀ (Herroelen 1959; K-H Loske). Of 73 arrivals in Westfalen, 10 April–1 May, 79·5% ♂♂ and 20·5% ♀♀ (K-H Loske). However, some birds arrive already paired (Hartley 1941; Muff 1977), indicating pair-formation starts during migration, or earlier. Thus, song heard in winter quarters, e.g. in Togo, January (Douaud 1957), and commonly on spring migration, sometimes with associated heterosexual display (Heim de Balsac and Heim de Balsac 1949–50; Verheyen 1952). In Britain, song-period late April to early October (Alexander and Alexander 1908). Most ♀♀ start laying 2–4 weeks after arrival (Herroelen 1959). (2) Nest-site selection and pair-bonding behaviour. Selection of site and attraction of mate intimately linked, and therefore treated together. ♂ takes initiative in nest-site selection. Soon after he arrives, ♂ of established pair may build nest or refurbish an old one before ♀ returns; or ♂ may wait until she arrives. Once ♀ accepts site, she typically does most building while ♂ sings nearby. (Vietinghoff-Riesch 1955.) ♂ sings in the air or perched on vantage point near nest-site; for occasional singing on ground, see (e.g.) Dawson (1976). Following account of typical Song-display and Nest-showing based on Löhrl (1962a). ♂ flies, circling around and singing, *c.* 50 m above building (i.e. nest-site). If flock approaches, ♂ sings louder, then swoops down, calling, towards site; descent sometimes stepped or rapidly spiralling. If ♀ attracted but does not approach nest-site closely, ♂ follows for some distance. Nest-showing ♂ lands at site, singing, and gives Enticement-calls (see 3 in Voice) until ♀ lands also, then he sings loudly (see 1 in Voice for variants). At end of song, ♂, with head averted, suddenly turns head towards ♀. As ♀ shows increasing interest in site, ♂ perches at site, head lowered and averted from ♀, and makes pecking movements at nest. Such Nest-showing performed at several sites, if available (Löhrl 1962a), suggesting ♀ may exercise some choice. For case of ♀ apparently choosing site without initiative of ♂, see Mohr (1958). Pair-formation complete when ♀ begins roosting at nest-site (Löhrl 1962a). In pair which arrived together on breeding grounds, ♂ made rotating movements in nest and gave loud Enticement-calls while ♀ perched nearby; ♂ removed straws from nest and took them to ♀; in the evening, ♀ roosted in nest and ♂ outside; behaved thus for a week (Vietinghoff-Riesch 1955). Communal aerial display (see Flock Behaviour, above) is perhaps heterosexual Song-display, as unpaired ♂♂ participate (A P Møller). In *erythrogaster*, resurgence of pair-bonding behaviour occurs between broods (Samuel 1971a). (3) Mate-guarding. ♂ guards mate over extended period during which neighbouring ♂♂ regularly chase and harass ♀. Guarding most intense during fertile period of ♀ (A P Møller). In presumed mate-guarding, ♂ commonly chased after ♀ with Contact-call (see 2a in Voice) whenever she left nest-site, sometimes pecking her tail; if he failed thus in making her return to nest, he coaxed her by repeatedly flying back to nest (De Braey 1946). During pursuits, either sex may give mild Alarm-calls (see 7a in Voice) (De Braey 1946; Vietinghoff-Riesch 1955). Mate-guarding more pronounced in colonial than solitary pairs, in early- than late-nesting pairs, and in 1st than 2nd clutches (A P Møller). (4) Courtship-feeding. Despite rare claims, not recorded in any of the many extensive studies (Verheyen 1947; Vietinghoff-Riesch 1955); presumably highly exceptional if it occurs at all. See also subsection 6 (below). (5) Mating. Occurs when ♀ perched near nest-site. ♂ solicits by singing, often vigorously, while ♀ nearby. ♂ flies and hovers over ♀ (A P Møller). Once, ♂ hovered with tail fanned and legs dangling, while ♀ perched or flying slowly with fluttering gliding motion; ♂ also swooped at ♀ and chased her in twisting pursuit (Nethersole-Thompson and Nethersole-Thompson 1940*b*). For suggestion that slow flight with fanned tail may have display significance outside mating context, see Hartley (1941). ♂ mounts (Fig B) and, during copulation, calls (A P

B

Møller: see 4 in Voice). Only one cloacal contact per mounting (A P Møller). One pair copulated 3 times in 5 min (Vietinghoff-Riesch 1955). Unreceptive ♀ turns head towards ♂ or pecks him; rejected ♂ flies a little way off and returns to perch by ♀ again (A P Møller). After each of several attempted copulations, one ♂ rapidly shook raised wings (Nethersole-Thompson and Nethersole-Thompson 1940*b*). Preening, especially by ♀, common after copulation. Attempted mating between pair-members equally likely to be successful within and outside fertile period of ♀, but success of promiscuous attempts markedly higher in fertile period. Most copulation occurs in morning. For 1st clutches, occurs from more than 50 days before egg-laying until 5th day of incubation, mostly from 15 days before laying until 2nd day of incubation; for 2nd clutches, from 8(-9) days before laying until 3rd(-6th) day of incubation. (A P Møller.) In *erythrogaster*, ♀ usually in nest when ♂ starts soliciting (see Samuel 1971*b*, also for associated calls). (6) Behaviour at nest. During nest-building and guarding phase, Follow-contact call (see 2c in Voice) of ♀ attracts ♂ to join her, e.g. in flights to collect nest-material. Lining (feathers) added to nest during laying and incubation, but removed after brooding phase of nestling period (A P Møller). At one nest, ♀ made periodic feeding excursions lasting 1-30 min; ♂ spent a lot of time near nest, sometimes perching there with food, but never feeding ♀ (De Braey 1946). When ♂ approaches nest closely, ♀ may leave with him, and later they return together, ♂ accompanying ♀ to nest (Vietinghoff-Riesch 1955). At nest with young, ♀ often pecked ♂; thought to stimulate him to fetch food (De Braey 1946). RELATIONS WITHIN FAMILY GROUP. ♀ sometimes helps young to hatch, e.g. one ♀ held emerging nestling by the wing and drew it out of its shell (Verheyen 1947). Young first fed, usually by ♀, *c.* 1 hr after hatching. Brooded assiduously, mostly by ♀, for first 3 days (De Braey 1946; Kuźniak 1967); steadily less until l4 days by which time young thermoregulate efficiently (Al-Rawy and Kainady 1976); thereafter, ♀ stops brooding at night and roosts on side of nest,

though parents brood occasionally up to fledging (Brown 1924; Berndt and Berndt 1942; Purchon 1948). ♂ does most feeding in early nestling stage while ♀ brooding, and takes greater share in brooding after first few days (Vietinghoff-Riesch 1955). In 7 out of 8 2nd broods, strong sex bias in provisioning developed after young 7-12 days old, in which either sex may become main provider (A P Møller). Young start begging vocally (see Voice) at 3-4 days. Each bolus of food brought by parent usually given to one nestling, not shared (A P Møller). Food-bearing parents may call (see 2a-b in Voice) to stimulate apparently sated young. Eyes open at 4-9(-11) days, but typically not until 13 days in 2nd brood. Young that fall out of nest may be fed on ground (Vietinghoff-Riesch 1955). For development of parent-offspring recognition, see Burtt (1977). Both parents remove faecal sacs until young 12-14 days old, after which young increasingly defecate over side of nest (Berndt and Berndt 1942; Radermacher 1970*a*). ♀ sometimes pecks vents of young to encourage defecation (De Braey 1946). Aggressive responses of young evident from 20-25 days (A P Møller). Commonly, brood fledges over 2 days (De Braey 1946; A P Møller). Fledged young lured away from nest by Follow-contact call (A P Møller). Newly-fledged young are fed perched on wires, tree-tops, etc. (De Braey 1946; Neumann 1978: Fig C). At one site, fed thus for 5 days, during which

C

they begged from any passing conspecific; thereafter fed almost invariably in flight, pursuing parents petulantly or flying to meet them, also accosting siblings for food (De Braey 1946). After fledging, fed by parents for average 5·3 days (3-12, *n* = 11) (A P Møller); most become self-feeding 6-8 days after fledging (Lyuleeva 1974). May remain near nest-site for up to 40 days after fledging (Radermacher 1970*a*) but usually forcibly evicted by ♂ when 2nd clutch started (De Braey 1946; Vietinghoff-Riesch 1955). For cases of exceptionally long family bonds, see Bonds (above). ANTI-PREDATOR RESPONSES OF YOUNG. When disturbed, small young gape silently, but once eyes open, retreat into depths of nest (De Braey 1946). Hide in nest when Alarm-calls given; from 30-40 days, fledglings give Alarm-calls when approached by predators (A P Møller). PARENTAL ANTI-PREDATOR STRATEGIES. (1) Passive measures. Incubating or brooding ♀ usually alerted by Alarm-calls of sentinel ♂, but does not usually leave nest until closely approached; manner of leaving varies—see subsection 4, below (A P Møller). (2) Active measures: general. Compared with *D. urbica* and *R. riparia*, gives stronger, more frequent Alarm-calls, responds faster to danger and attacks more vigorously; these differences thought due to marked territoriality of *H. rustica* (Lind 1962). (3) Active measures: against birds. On approach of raptor, Alarm-calls (see 7a in Voice) induce upflight of many birds in vicinity which pursue predator with Alarm-calls (see 7b in Voice), and mob it, making dive-attacks and sometimes apparently making physical contact (Møller 1984). Flock of 2 *H. rustica* and 2 *D. urbica* followed above and behind *F. subbuteo* for some distance, diving at its head and apparently trying to keep predator low down. Similar expelling attacks made on Kestrel *F. tinnunculus* (Baier 1974*a*) and Sparrowhawk *Accipiter nisus* (Giller 1955; Vietinghoff-Riesch 1955). Red-backed Shrike

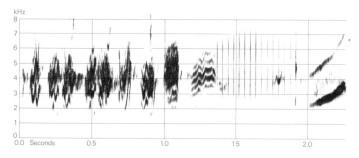

I V C Lewis England May 1978

Lanius collurio nesting near *H. rustica* vigorously mobbed at first, attacks subsiding after a week (Baier 1974*b*). For measures by *erythrogaster*, see Smith and Graves (1978) and Shields (1984). (4) Active measures: against man. Following account based on analysis of 1498 responses from 48 pairs with 1st clutches and 27 with 2nd clutches (Møller 1984). Responses varied from birds leaving nest silently and passively to giving continuous Alarm-calls and flying directly towards intruder. Pairs varied greatly, but responses of pair-members were similar. Intensity of response tended to increase with time during both 1st and 2nd clutches; thus, whereas birds left nests silently before laying, they frequently flew at observer during second half of nestling period. Both sexes showed increasing response, but response of ♀♀ more intense from the outset, and increase in response of ♂♂ restricted to second half of breeding season. Alarm-calls most frequent at peak fledging time. Birds in colonies and older birds tended to defend more actively than solitary pairs and first-time breeders. For measures by *erythrogaster* against human intruders, see Shields (1984). (5) Active measures: against other animals. Mammalian predators elicit same response as man. Cats elicit loud Alarm-calls and are collectively chased, sometimes buffeted with wings (Vietinghoff-Riesch 1955). Bird may hover over predator as if investigating it (A P Møller). Attacks on cats, dogs, weasels (Mustelidae), and other quadrupeds reported for *erythrogaster* (Bent 1942; Shields 1984).

(Figs by M J Hallam: A from photograph in Broekhuysen 1961; B from photograph in Møller 1974*b*; C from photograph in Toft and Christensen 1977.)   EKD

Voice. Freely used throughout the year. Generally more twittering than House Martin *Delichon urbica* (Witherby *et al.* 1938*b*). Bill-snapping sound, rendered 'knak-knak-knak' sometimes given by incubating birds (De Braey 1946), also (as loud as Tawny Owl *Strix aluco*) while resting at night (Vietinghoff-Riesch 1955); bill snapped rapidly 3–4 times, by either sex, during courtship (A P Møller), and may then have signal function. For detailed voice and sonagrams of *erythrogaster*, see Samuel (1971*b*). For voice of hybrid with *D. urbica*, see Elsner (1951). Following scheme based on account supplied by A P Møller. Sonagrams II–V supplied by H-H Bergmann.

CALLS OF ADULTS. (1) Song of ♂. A simple but pleasing warble, 3–15 s long or more, mixed with a throaty rattle; does not carry far (Witherby *et al.* 1938*b*). Also described as a melodious rapid twittering with many harmonic elements, ending in a fairly long clear rattle; often very short gaps between phrases (Bergmann and Helb 1982, which see for sonagram). Since phrases run together, difficult to determine which unit(s) end phrase; in recording of perched bird (e.g. Fig I), almost all rattles followed by ascending portamento (J Hall-Craggs). Song often has distinctive terminal whistle 'su-seer', with 'seer' sound falling in pitch but starting higher than 'su' (D I M Wallace); varies geographically (A P Møller). Song given perched or in flight, often near nest (Witherby *et al.* 1938*b*). Tends to include more rattles when perched than in flight (V C Lewis). Early in season, or when no ♀ nearby, rattle often omitted, but rattle prominent during pair-formation (Löhrl 1962*a*: see Social Pattern and Behaviour). Subsong a succession of not very rich short phrases (Purchon 1948). For song of juvenile, see Calls of Young. (2) Contact-calls. (a) A 'witt-witt' (Vietinghoff-Riesch 1955) or 'wid-wid' (Bergmann and Helb 1982: Fig II), also 'wic' or 'twic' or 'chwic' (Purchon 1948). More drawn-out renderings include 'wiet-wiet' (De Braey 1946), 'huit' or 'kuit' (A P Møller). In our recordings, given singly, or twice in rapid succession; in recordings by P A D Hollom, calls of perched birds mostly low and husky, those in flight higher and sharper.

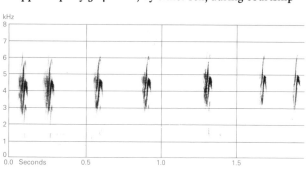

II H-H Bergmann East Germany July 1978

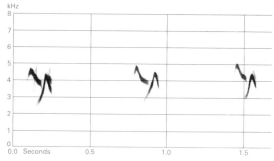

III H-H Bergmann Denmark August 1978

Given typically in communication between pair-members and between parents and young, also between members of feeding flocks (Vietinghoff-Riesch 1955). (b) Twittering variant. Simple Contact-calls are often run together in rapid series, e.g. 'tswit tswit tswit' etc. (Witherby *et al.* 1938*b*), or 'wicwicwic' etc. (Purchon 1948) to effect semblance of Subsong (Purchon 1948), though loudness varies markedly (A P Møller). Described as a long twitter, continuing for up to several minutes, and lacking final rattle of song. Given by both sexes, and so-called 'song' of ♀ is usually (perhaps always) this variant. Given inside or outside territory, often while flying to and from nest, simultaneously by pair-members while loafing near nest, also loudly as greeting; ♂♂ may also twitter thus for long periods while perched on edge of nest after feeding young, and both sexes may twitter softly to stimulate young to accept food. ♂ often progresses from twittering to song (Vietinghoff-Riesch 1955). Given throughout the year (A P Møller). (c) Follow-contact call. A 'tir-huit' given by ♀ to ♂ mate during nest-building, inducing him to accompany her to collect nest-material. Also given by ♀ when luring fledglings away from nest. (A P Møller.) (3) Enticement-call. A loud 'wi-wi-wi' given repeatedly by ♂ to attract ♀ to prospective nest-site, often inducing close approach; also used to entice young back to nest for roosting (Vietinghoff-Riesch 1955). Described as a dry continuous wheezing sound, which, in extreme form, becomes a discontinuous high-pitched 'eeeee', like a gate on a rusty hinge (Purchon 1948). (4) Courtship-call. An 'it-it-it-it' sound (A P Møller) given by unpaired ♂ when ♀ perched nearby; in *erythrogaster*, given by both sexes, and attributed a pair-bonding function (Samuel 1971*b*: 'whistle call'). (5) Copulation-call. A 'waeae waeae' given by ♂ when copulating (A P Møller). (6) Anger-call. A clear screeching 'witt titititi' given in confrontations (Bergmann and Helb 1982). (7) Warning- and Alarm-calls. Given in variety of contexts, from heterosexual chases and confrontations with mates and conspecific birds, to pursuit of and by raptors. Accordingly, variants reflect different states of alarm. Loud Contact-call and twittering (calls 2a, b) also used in alarming situations (De Braey 1946; Purchon 1948). (a) Mild Alarm-call. A 'chiir-chiir' given by both sexes when predator spotted at some distance (A P Møller); thus includes a warning function.

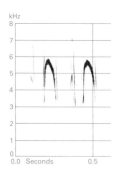

IV H-H Bergmann East Germany
July 1978

Also rendered a muted 'dschrlit' or 'tschiulit' (Bergmann and Helb 1982; H-H Bergmann: Fig III). Presumably same call as 'chirre-chirre' or 'chirre-chirre-che', sometimes given by incubating ♀ when ♂ visits nest, apparently conveying anger or annoyance (De Braey 1946); also given by ♂ driving ♀ from nest (Vietinghoff-Riesch 1955). Deep throaty sounds given from nest by ♀ apparently refusing ♂'s invitation to leave nest (Purchon 1948) perhaps include these calls. (b) Main Alarm-call. A markedly disyllabic 'ziwitt' (Bergmann and Helb 1982: Fig IV), 'zissit' (Vietinghoff-Riesch 1955, which see for variants), 'tsi-wit', 2nd syllable much higher pitched (De Braey 1946); also a distinctive 'splee-plink' (D I M Wallace). Familiar call, given repeatedly by both sexes during presence of predator (then signalling greater danger than indicated by call 7a), also by ♂ to conspecific intruders on territory (A P Møller). (c) Other Alarm-calls. A 'dschidschid' (Bergmann and Helb 1982: Fig V) in

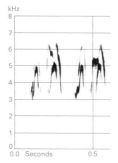

V H-H Bergmann Cyprus
April 1978

flight, signalling alarm. In imminent danger, a variety of rather muffled calls, e.g. a quiet 'dewihlik' of distress, also 'zibist', 'zetsch', 'tsätsätsä', and once, when chased by Peregrine *Falco peregrinus*, a low-pitched fluting 'flüh-flüh' (Vietinghoff-Riesch 1955). (8) Distress-call. A sharp, repeated 'weer-weer', given by both sexes, when handled, or seized by a predator (A P Møller). (9) Other calls. In recording by P A D Hollom, a rapid low-pitched bubbling or cackling sound, given by ♀ as she flew off the nest (P A D Hollom, P J Sellar).

CALLS OF YOUNG. Small young beg with a thin soft 'si-si-si', first heard on 3rd day; by 9th day, food-call rendered 'swi-swi-swi', by 14th 'swiet-swiet-swiet', gradually developing through 'twiet-twiet' to resemble adult Contact-call ('wiet') by fledging (De Braey 1946). Persistent rapidly repeated 'wid' (homologous with, e.g. 'twiet') calls of older young in nest give way to a rapid 'dsched-dsched' etc. just before being fed (Bergmann and Helb 1982); call when being fed described as explosive sizzling sound (P J Sellar). Fledged young give 'wee-wee...' calls to solicit any approaching conspecific, and 'üit-üit' calls of alarm on approach of would-be predator (A P Møller). Subsong heard from *c.* 25–30 days (A P Møller), and full song not long after, earliest August (Purchon 1948; Lunau 1952; Vietinghoff-Riesch 1955).

EKD

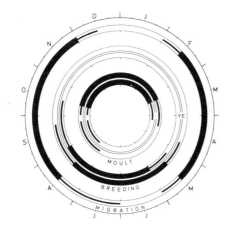

Breeding. SEASON. Britain and western Europe: see diagram. Up to 2 weeks later in Scandinavia, and 1 week earlier in southern Europe (Makatsch 1976). Southern spain and north-west Africa: first eggs March (Heim de Balsac and Mayaud 1962; P A D Hollom). SITE. On small ledge against vertical surface, e.g. beam or window-ledge in building, less often without support beneath; rarely in cave or tree. Of 192 nests, Scotland, 41% on beam, 38% in roof apex, 5% against vertical wall, 1% on wall, 15% miscellaneous (McGinn and Clark 1978). Height above ground usually 2–5 m. Of 189 nests, Scotland: 0–2 m, 5%; 2–3, 38%; 3–4·5, 41%; over 4·5 m, 16%; mean 3·3 (McGinn and Clark 1978). Usually solitary. Nest: shallow half-cup or cup of mud pellets, usually mixed with plant material, with inner lining of feathers; normal dimensions 20 cm across and 10 cm deep. Building: by both sexes, though often mostly by ♀ with ♂ helping to bring material; main nest structure completed in c. 8 days, further 2 days for lining (De Braey 1946). EGGS. See Plate 80. Elongated elliptical or oval, smooth and glossy; white, lightly marked with red-brown spots, plus some lilac and grey. Nominate *rustica*: 19·7 × 13·6 mm (16·7–23·0 × 12·3–14·8), n=250; calculated weight 1·9 g. Eggs of *transitiva*, *savignii*, and *gutturalis* perhaps smaller than these, but samples very small. (Makatsch 1976; Schönwetter 1979.) Clutch: 4–5 (2–7). Positive correlation between clutch size and nest volume (Møller 1982). Of 337 clutches, Britain: 2 eggs, 2%; 3, 9%; 4, 41%; 5, 43%; 6, 5%; 7, under 1%; mean 4·4 ± SE 0·027; slight decrease in size from south-west to south-east to north; of 93 2nd clutches, 4·05 (Adams 1957). Of 177 clutches, southern Scotland: 2 eggs, 1%; 3, 2%; 4, 34%; 5, 51%; 6, 12%: mean 4·66, with decline through season from 5·02 in May to 4·08 in August (McGinn and Clark 1978). Of 130 clutches, northern Scotland: 3 eggs, 9%; 4, 47%; 5, 38%; 6, 6%; mean 4·41, with slight decline during season (McGinn 1979). Of 86 1st clutches, West Germany: 3 eggs, 2%; 4, 25%; 5, 60%; 6, 12%; 7, 1%; mean 4·85. Of 69 2nd clutches: 3 eggs, 10%; 4, 56%; 5, 32%; 6, 2%; mean 4·25 (Löhrl and Gutscher 1973). 2–3

broods. Size of 2nd clutches and number of broods per season decrease with increasing latitude (Møller 1984b, which see for extensive review of clutch-size data). Interval of 55 days (48–65), n=11, from fledging of 1st brood to fledging of 2nd (Buxton 1946). Building of new nest starts 5 days after fledging of 1st brood (De Braey 1946). Eggs laid daily, mainly in early morning, within 2 hrs of sunrise (De Braey 1946). INCUBATION. 15·25 days (11–19), n=93 (Adams 1957). Mainly or perhaps solely by ♀. ♂ reported as taking small share (Moreau and Moreau 1939b), but only ♀ incubated in another study (De Braey 1946). In 2 periods of observation between 12.00 and 20.00 hrs, ♀ incubated for 70·5% and 72·9% of time, with 31 interruptions totalling 123 min (De Braey 1946). Begins with last egg; hatching synchronous. YOUNG. Altricial and nidicolous. Cared for and fed by both parents. FLEDGING TO MATURITY. Fledging period 19·5 days (18–23), n=110, increasing slightly through season (Adams 1957); 21·7 days ± SD1·8, n=33 (Turner and Bryant 1979). Become independent some weeks later, though earlier for 1st brood. Age of first breeding 1(–2) years. BREEDING SUCCESS. Of 1906 eggs in 441 clutches, Britain, 78·2% hatched (90·5% excluding complete clutch failures; 15% of clutches failed completely), with slightly better hatching from clutches of 4–6 than from 2–3, though difference probably not significant; of 1490 chicks hatched in 376 broods, 91·7% fledged (95·8% excluding complete brood failures); overall success 71·9%, with highest success in August, then July, then May–June (Adams 1957). Of 472 eggs laid, southern Scotland, 76·3% raised to fledging, with 8·5% lost to predation, and 7·2% to infertility (McGinn and Clark 1978). Of 45 eggs in 11 nests, Bayern (West Germany), 44 hatched and 37 fledged (82·2% overall) (Buxton 1946). Of 417 eggs laid in 86 1st clutches, West Germany, 87·3% raised to fledging; of 293 eggs in 69 2nd clutches, 87·7% raised to fledging; combining 1st and 2nd clutches, average 2·8 young fledged from 9 3-egg clutches, 3·4 young from 62 4-egg clutches, 4·4 young from 74 5-egg clutches, 5·3 young from 11 6-egg clutches, 7 young from 1 7-egg clutch; overall annual production per pair 7·3 young (Löhrl and Gutscher 1973). In Westfalen (West Germany) of 3941 eggs laid in 1st clutches, 85·9% produced fledged young; of 2651 eggs laid in 2nd clutches, 91·6% produced fledged young; of 54 eggs laid in 3rd clutches, 85·2% produced fledged young; overall success 88·2% fledged young from 6646 eggs laid (K-H Loske).

Plumages (nominate *rustica*). ADULT MALE. Forehead dark chestnut, remainder of upperparts including tertials and lesser and median upper wing-coverts dark glossy metallic-blue; white feather-bases of hindneck and mantle and partly white inner webs of inner tertial coverts occasionally visible, especially when plumage worn. Lores black, sides of head and neck from below and behind eye dark glossy metallic-blue, dull greyish-black feather-bases often just showing. Chin and throat dark chestnut, bordered below by broad band across chest;

chest-band dull greyish-black with glossy dark blue feather-tips, sometimes fully glossy blue but usually some greyish-black visible, especially when worn. Fresh feathers of central chest often narrowly tipped chestnut, bleaching to pale buff when worn, but then usually lost through abrasion or showing only as a faint pale wash; occasionally, some feathers of central chest virtually uniform chestnut, but this small patch does not fully interrupt dark chest-band. Remainder of underparts uniform (except for small blue-black patch near outer thigh), but rather variable: usually, pinkish cream-buff or pink-cream, under tail-coverts darker rufous-buff, pink-buff, or similar to belly; frequently, uniform cream-white or dingy white; rarely, uniform rufous-buff (similar to *transitiva* of Levant—so-called '*transitiva*' or '*boissonneauti*' variant) or almost pure white. Under tail-coverts usually uniform, but sometimes with black shaft-streaks, spots, or tips (Brombach 1984; RMNH, ZMA). Tail black or brown-black, slightly glossed dark blue or blue-green; t2–t5 have large rounded white or pale cream spot subterminally on inner web; t6 has similar spot, extending into a point towards narrow terminal half of feather. Flight-feathers, upper primary coverts, and bastard wing black, faintly glossed green or green-blue, rather more strongly glossed blue on tips of primaries and greater primary coverts; greater upper wing-coverts black with glossy dark blue outer webs, somewhat less deep glossy blue than median and lesser coverts. Under wing-coverts and axillaries similar to belly and vent (equally variable), often darker (similar to under tail-coverts) towards lesser coverts; small coverts along leading edge of wing spotted blue-black and buff. *In worn plumage*, upperparts and upperwing duller, greater coverts, flight-feathers, and tertials hardly glossy; some grey of feather-bases often visible on forehead, crown, and throat, some contrasting white on hindneck and mantle; chest-band duller greyish-black, feather-tips less extensively glossy or not at all; remainder of underparts whiter, pale buff, off-white, or white, less pink-buff. ADULT FEMALE. Closely similar to adult ♂, but glossy blue feather-tips on upperparts often slightly narrower, some more grey of feather-bases visible, upperparts appearing less uniform and less deep bluish, especially in worn plumage; glossy feather-tips on chest-band narrower, often entirely absent on central chest except for narrow rim below throat, chest-band appearing duller dark greyish-brown, not as extensively bluish-black as in most adult ♂♂; colour of belly on average slightly whiter than ♂, less pink, but much overlap. Best distinguishing characters are depth of tail-fork (difference from tip of t1 to tip of t6 in folded tail) and greatest length of white patch on t6, as seen from below. In ♂, fork 47–78 mm (but only 4% below 51, $n=75$); in ♀, 27–57 mm (but 90% 32–50, $n=31$). Length of patch on t6 20–34 in ♂ (but only 7% below 24, $n=43$), 13–26 in ♀ (but only 10% over 23, $n=21$). Sum of lengths of white patches on t2–t6 over 70 in ♂, less than 55 in ♀ (Bub and Herroelen 1981). Also, elongated tip of t6 often narrower in ♂: measured at 20 mm from tip, width *c.* 2·0 in ♂, 2·5–3·0 in ♀ (Svensson 1984a). NESTLING. Down long, in sparse tufts, restricted to upperparts; pale grey. A 2nd short woolly down develops between sprouting feathers at *c.* 1 week old. (Heinroth and Heinroth 1924–6.) For development and ageing, see McGinn and Clark (1978). JUVENILE. Resembles adult, differing mainly in paler forehead and throat, less deep glossy upperparts and chest, and less strongly elongated t6. Patch on forehead narrow, cream, pale buff, or rufous-buff, rarely rufous-chestnut but always narrower than adult, gradually merging through grey-brown mottling into dull black or dark brown-grey of crown; forehead-patch often extends into narrow line above lores. Upperparts less deep glossy than adult; blue duller, slightly

greener, less purplish; more dark brown-grey of feather-bases visible (greyish-white on hindneck and mantle); in particular, crown and longer upper tail-coverts often largely dull dark grey, hardly glossy; in fresh plumage, often some pale brown fringes on feather-tips of rump and upper tail-coverts. Chin and throat cream-buff, rufous-buff, or rufous-cinnamon, rarely as dark chestnut as adult, often finely spotted brown-grey at border of lower sides of neck and chest. Chest-band uniform blackish-grey, rarely with some glossy bluish-black feather-tips, mainly on sides of breast; central chest often dark brown-grey rather than blackish-grey, feather-tips often faintly fringed buff or pale brown. Remainder of underparts rather variable: on average, paler than adult ♂, usually cream-white or pale cream-pink, like adult ♀, but frequently rufous-pink or slightly rusty. Inner primaries, secondaries, and tail-feathers have narrow pale edges along tips when plumage fresh (absent in adult at same time of year). Outer tail-feathers shorter than in adult, white or cream patch on t6 sometimes shorter: length of t6 under 72, depth of fork under 31 (in adult, t6 over 75, fork over 32—under 32 in only 1 adult ♀, $n=106$); length of patch on t6 12–24 (in adult ♂, mainly 24–34, but in adult ♀ 13–26); shape of patch on t6 usually different, with extension towards feather-tip rounded rather than sharply pointed. Sometimes hard to distinguish from worn adult, which is also less glossy on upperparts, and may have chest and tail heavily abraded; apart from difference in wear, juvenile usually separable by paler rufous of forehead and throat, more rounded pale patches on t6, and broader t6 (width of t6 at point level with tip of t5 3·7–4·9 in juvenile, 2·2–3·6 in adult, $n=10$ in each). Sexes usually not separable, but chest-band of ♂ on average blacker, of ♀ browner; depth of fork in ♂ 20–30, mainly over 22, of ♀ 17–24, mainly 22 or under; patch on t6 18·0 (28) 15–24 in ♂, 15·2 (20) 12–18 in ♀. FIRST ADULT. Similar to adult; indistinguishable once last juvenile feathers (usually feathers of face, t5–t6, and outer primaries) replaced. Often recognizable up to primary moult score of 40, when worn feathers of forehead and throat paler rufous than worn feathers of adult, patch on t6 more rounded, and worn t6 broader than in adult (see Juvenile).

Bare parts. ADULT. Iris dark sepia-brown to dark brown. Bill black, cutting edges near gape often with some yellow. Leg and foot black, sometimes tinged red-brown. NESTLING. Mouth lemon-yellow, no spots on tongue; gape-flanges pale yellow or white. JUVENILE. Similar to adult, but cutting edges and lower mandible with more extensive yellow at base. Foot sometimes tinged grey or bluish-flesh.

Moults. ADULT POST-BREEDING. Complete; primaries descendant. Moult strategy varies between populations. (1) Long-distance migrants of temperate regions. Moult usually starts in winter quarters, but frequently some inner primaries, a few tertials or an occasional body feather replaced when still near breeding grounds, north of tropical winter quarters; such moult apparently suspended during migration towards tropics (but data perhaps biased, being derived in part from exhausted migrants captured during cold spells, which may have suspended moult due to physical condition, not due to migration). Moult resumed or (in bulk of birds) started upon reaching tropical regions, and apparently not suspended during movements within tropics. In England, none of adults examined in September had started (Mead 1975), but in Sweden, mid-August, 1 of *c.* 10 had shed inner 2 primaries (Persson 1979); in southern Spain, 19% had started ($n=147$), primary moult scores 1–12(–20)

(Pimm 1970); in Tien Shan (USSR), 7% had started ($n = 361$) (Gavrilov 1971). Nesting ♀, Iraq, already in moult of p1 in early June (Kainady 1976), and an apparently nesting British adult was in same stage in mid-September (Richards and Goodwin 1950). Of 541 exhausted adults during cold spell in October, Switzerland, 3% had started, primary moult scores up to 5(-10) and occasionally also tertials or a single t1 replaced; moult mainly suspended (Winkler 1975); of *c.* 530 birds (age not stated) during similar conditions in West Germany, 17 adults showed primary moult (mainly suspended), with scores 1-10(-15) (Kasparek 1976). Within Afrotropics, no marked geographical differences in timing, probably because moult started or resumed as soon as birds enter tropics and those wintering well to south are already in advanced moult on arrival there. Combining data from Uganda (Pimm 1972), Zaïre (Herroelen 1960; De Bont 1962), Zambia (Francis 1980; Ginn and Melville 1983), and South Africa (Broekhuysen and Brown 1963; Skead and Skead 1970; Mendelsohn 1973), moult starts with p1 between mid-September and mid-November, completing with p9-p10 late January to late March, with estimated duration of *c.* $4\frac{1}{2}$ months for each bird (though 185 days estimated from retraps of ringed birds). Tertials moult in sequence: middle (s8), inner (s9), outer (s7); middle shed at about same time as p1, outer at about p5. Secondaries from outer (s1, at about same time as p5) to inner (s6, completed at about same times as p9-p10); tail in sequence t1-2-3-4-6-5 or t1-2-3-6-4-5, t1 shed at about same time as p5, t5 during growth of p9, but elongated t6 often the last to grow fully (sometimes still growing on arrival in breeding area). Body starts at about same time as p1 with some feathers of back and rump, followed by remainder of body at shedding of about p2-p4; head (especially face) moulted last, completed with regrowth of p9. (Stresemann and Stresemann 1968b.) (2) Southern migratory populations. Start breeding early (e.g. in north-west Africa, Libya, perhaps Iraq and Iran, and parts of China), and primary moult generally starts in breeding area, but completed in winter quarters. Winter birds from Ivory Coast (Stresemann and Stresemann 1968b), Liberia (RMNH), and Sumatra (RMNH, ZMA) with primary moult about half-way through in September and completed November-December probably originate from such early-breeding populations. (3) Resident populations (e.g. *transitiva*, *savignii*, nominate *rustica* from Afghanistan, Pakistan, and northern India, and *gutturalis* from south-east Asia). Moult starts early, late April to June, completing September-November; moult schedule much advanced compared with passage birds through these areas (Stresemann and Stresemann 1968b; BMNH). POST-JUVENILE. Complete; primaries descendant. Sequence as in adult post-breeding, but moult on average later: 2-3 weeks later in South Africa (Mendelsohn 1973), 5-6 weeks in Zaïre (Herroelen 1960; De Bont 1962). In birds from temperate regions, does not start until arrival in tropics (though in England 2 birds September were moulting p1: Mead 1975), but in early-breeding southern migratory populations and in resident populations, where juveniles fledge April-May, moult starts near breeding area shortly after fledging. In *transitiva* and *savignii*, starts July-September, completing November-January (BMNH). Birds wintering Sumatra start immediately after arrival, on average early September, 4-5 weeks later than adults, completing about January (RMNH, ZMA); in Afrotropics, last juveniles start about early January, completing late May and June.

Measurements. ADULT. Nominate *rustica*. West and central Europe, April-October; skins (RMNH, ZMA). Tail is t6; fork is tip of t1 to tip of t6. Bill (S) to skull, bill (N) to distal corner of nostril; exposed culmen on average 4·7 less than bill (S), 1·7 more than bill (N).

		♂			♀	
WING	AD	124·4 (2·32; 64)	120-129	121·8 (2·20; 26)	118-125
	JUV	121·6 (2·00; 39)	119-126	118·5 (2·31; 28)	115-122
TAIL	AD	103·4 (8·31; 73)	89-119	86·9 (8·63; 30)	70-99
	JUV	65·9 (3·21; 28)	61-72	63·6 (2·41; 19)	60-67
FORK	AD	61·9 (8·15; 75)	47-78	44·2 (8·78; 31)	27-57
	JUV	24·1 (3·07; 28)	20-30	21·1 (2·39; 19)	17-24
BILL (S)		12·2 (0·67; 25)	11·2-12·8	12·3 (0·36; 11)	11·7-12·8
BILL (N)		5·7 (0·27; 24)	5·3-6·3	ˉ5·8 (0·28; 10)	5·4-6·2
TARSUS		10·8 (0·37; 20)	10·3-11·5	10·8 (0·45; 12)	10·2-11·4

Sex differences significant for wing, tail, and fork; for fork, see also Plumages. Juvenile wing, tail, and fork significantly shorter than adult (see also Plumages); bill and tarsus full-grown from age of *c.* 14 days (Georg and Al-Rawy 1970).

Wing, breeding birds: Balkans, 123·9 (3·06; 14) 120-130; Algeria (including Ahaggar), 122·2 (3·09; 6) 119-128; Turkey, 124·8 (5·70; 6) 119-134; Iran, ♂ 123·4 (34) 118-130; Afghanistan, 122·7 (2·55; 9) 119-128; Pakistan and northern India, 121·1 (3·35; 10) 115-126 (Vaurie 1951c; Paludan 1959; Rokitansky and Schifter 1971; Stresemann and Stresemann 1968b; RMNH, ZMA).

Wing, spring migrants: Afghanistan, ♂ 122·3 (7) 119-125, ♀ 120·3 (6) 117-126 (Paludan 1959); Morocco, live birds, ♂ 114-134, ♀ 107-131 (Ash 1969).

H. r. transitiva. ADULT. Levant, October-March; skins (BMNH, ZMA).

	♂		♀	
WING	124·9 (3·02; 11)	120-128	121·6 (2·01; 7)	119-126
TAIL	101·4 (6·12; 8)	95-112	83·8 (5·51; 6)	75-90
FORK	59·3 (6·30; 8)	48-69	39·7 (3·78; 6)	33-44

Sex differences significant.

H. r. savignii. ADULT. Nile delta (Egypt), November-May; skins (BMNH, RMNH).

	♂		♀	
WING	119·8 (2·12; 21)	117-125	116·6 (2·50; 10)	113-121
TAIL	91·2 (5·22; 21)	84-101	76·7 (4·62; 9)	72-85
FORK	49·0 (5·85; 21)	42-60	34·3 (3·91; 9)	29-41

Sex differences significant.

H. r. erythrogaster. Eastern North America, Caribbean, and northern South America, all year; skins (RMNH, ZMA). Sexed birds are adults; in juvenile, sexes not significantly different, combined.

	♂		♀		JUV	
WING	121 (8)	117-126	118 (11)	114-123	116 (26)	111-122
TAIL	88 (6)	82-93	79 (9)	70-88	60 (25)	56-66
FORK	48 (6)	41-55	39 (9)	27-45	20 (24)	15-26

H. r. gutturalis. Japan (summer) and Indonesia (winter): skins (RMNH, ZMA). Sexed birds are adults; in juvenile, sexes not significantly different, combined.

	♂		♀		JUV	
WING	116 (15)	112-122	113 (10)	110-117	113 (16)	107-118
TAIL	82 (4)	73-87	79 (6)	71-82	57 (14)	53-61
FORK	43 (4)	39-48	38 (5)	34-44	17 (14)	14-20

H. r. tytleri. Adult wing, Mongolia: 121·9 (3·16; 14) 117-127 (Piechocki and Bolod 1972).

Average increase in wing length with age, based on recaptures, West Germany: from 2nd to 3rd calendar year, 1·36 (14); in later years, 1·2 (10) per year. Average increase in depth of fork: from 2nd to 3rd calendar year, 7·9 (14); 3rd to 4th, 9·0 (4); 4th to 5th and 5th to 6th, 4·5 (6) per year. (Bub and Herroelen 1981.)

Weights. Nominate *rustica*. Breeding area. (1) West Germany. (2) East Germany. (Bub and Herroelen 1981.) (3) Netherlands, late July to mid-September (Walters 1971). (4) Turkey, Iran, and Afghanistan, April–July (Paludan 1938, 1959; Schüz 1959; Rokitansky and Schifter 1971). Netherlands: (5) August, (6) September (ZMA).

		♂				♀			
(1)	AD	18·8	(; 125)	16·1–21·4		19·4	(— ; 137)	16·0–23·7
(2)	AD	18·8	(1·89; 33)	14·0–22·5		18·6	(2·78; 21)	11·0–22·5
(3)	AD	19·4	(— ; 27)	17·4–21·3		18·7	(— ; 19)	17·6–21·5
(4)	AD	18·5	(2·17; 10)	16·0–22·0		18·6	(1·81; 4)	16·0–20·0
(5)	JUV	19·5	(1·57; 10)	17·4–22·0		19·1	(1·09; 15)	18·0–21·2
(6)	JUV	18·9	(1·55; 23)	17·4–21·8		19·1	(1·60; 9)	17·2–22·0

Unsexed breeders: Scotland, 19·0 (1·8; 52) (Turner and Bryant 1979); Iraq, May, 17·8 (16) 16·9–21·8 (Georg and Al-Rawy 1970). Juveniles, autumn: Sweden, 18·4 (0·96; 20) 16–20; Belgium, 15·2–23·5 (148) (Bub and Herroelen 1981); Netherlands, 18·4 (92) 15·2–22·0 (Walters 1971). Nestling, Scotland: on 1st day, 1·9 (0·3; 11); at peak on 14th day, 23·2 (1·9; 33); at fledging, 19·6 (1·4; 15) (Turner and Bryant 1979); peak of 23·8 on 14th day in one area, of 25·3 on *c*. 16th day in another (McGinn and Clark 1978). Nestling, Iraq: on 1st day, 1·8 (28) 1·5–2·4; at peak on 13th day (11th–15th) 20·5 (28) 17·4–24·7; at fledging, 20th–22nd day, 16·8 (28) 14–20 (Georg and Al-Rawy 1970). For development of nestling, see also De Braey (1946); for influence of weather on growth, see Rheinwald and Schulze-Hagen (1972).

On autumn migration. Cyprus, 19·4 (57) 13–24; Malta, 20·3 (147) 16–25 (Moreau 1969). Malta: 20·4 (2·5; 50) 15–25·5 (J Sultana and C Gauci). North-west Egypt: 16·1 (0·57; 8) 12·7–19·4 (Moreau and Dolp 1970).

In Afrotropics. (1) Uganda (Pearson 1971) and Zambia (Dowsett 1965a; Britton and Dowsett 1969). (2) South Africa (Broekhuysen and Brown 1963; Liversidge 1968; Mendelsohn 1973; Day 1975); maximum in March up to 27·3 (Liversidge 1968).

		(1)			(2)		
OCT		17·0	(23)	15·2–19·1	21·5	(37)	
NOV		17·7	(81)	14·5–20·8	20·8	(50)	
DEC		17·6	(100)	13·9–21·0	20·0	(99)	
JAN–FEB		17·9	(51)	16·0–20·0	19·2	(576)	
MAR		19·1	(39)	16·5–22·2	20·2	(279)	
APR		17·6	(3)	17·2–17·9	19·7	(106)	

On spring migration. Western Morocco, March: 17·9 (1·06; 10) 16–19 (Kersten *et al.* 1983). Eastern Morocco: 16·0 (2357) 11·0–28·2 (Ash 1969, which see for many details). Azraq (Jordan), 18·4 (347) 14·0–23·6; Malta, 18·3 (126) 14·3–23·5 (Moreau 1969). Malta: 19·8 (2·3; 50) 14·5–24·0 (J Sultana and C Gauci). Turkey, May: 19·4 (43) 14–23 (Vauk 1973a). Afghanistan: 18·2 (8) 17–21 (Paludan 1959).

Exhausted birds, killed during cold and rainy weather. Adult, Netherlands: 14·0 (0·89; 9) 12·9–15·3 (ZMA). South Africa. Just after arrival, November: adult, 13·0 (51) 11·3–15·4; juvenile, 12·5 (14) 10·9–14·9 (Skead and Skead 1970). April, mainly ♂♂: 13·6 (28) 11·7–15·8 (Broekhuysen 1953).

H. r. erythrogaster. New Jersey (USA), autumn: 18·8 (7) 17·0–19·8 (Murray and Jehl 1964). In moult, northern South America: ♂ 18·3 (1·39; 8) 17–20, ♀ 16·8 (0·62; 4) 16–18 (RMNH). Exhausted juveniles, autumn, after crossing Caribbean: ♂ 11·4 (0·80; 4) 10–12, ♀ 11·8 (1·29; 6) 10–13 (ZMA). For development of nestling, see Stoner (1935).

H. r. gutturalis. Average mainly 13–15, range 11–21, but average weight 17·0–17·4 in samples from late March and early April (Medway 1973).

H. r. tytleri. Mongolia, June–July: ♂ 19·0 (1·29; 7) 17–21, ♀ 20·4 (1·90; 7) 18–23 (Piechocki and Bolod 1972). Manchuria, August: ♂ 17 (8) 16–19; ♀♀ 15, 15 (Piechocki 1958).

Structure. Wing long, pointed at tip, narrow at base. 10 primaries: p9 longest, p8 (0–)1–4 shorter, p7 9–13, p6 16–22,

p5 23–33, p1 53–65. P10 reduced, narrow and pointed, hidden below primary coverts; 81–92 shorter than p9, 9–14 shorter than longest upper primary covert. Primaries rather narrow, not emarginated, but with shallow notch on tip of secondaries and inner primaries. Tertials very short. Tail rather long to long, deeply forked; 12 feathers, t1 shortest, others gradually longer towards t6; t6 longest, markedly elongated and narrow in adult ♂, less so in adult ♀ (see Plumages and Measurements). Bill short, weak, flattened dorso-ventrally; broad at base, triangular when seen from above; culmen with blunt ridge, tip slightly decurved. Short bristles along base of upper mandible. Nostril partly covered by membrane above. Tarsus very short, slender, naked; toes and claws rather short and slender, claws gently decurved. Middle toe with claw 15·1 (24) 14–16; outer toe with claw *c*. 69% of middle with claw, inner *c*. 70%, hind *c*. 68%.

Geographical variation. Slight in size, rather slight in colour (mainly of underparts). In Eurasia, largest birds in north-west (Scandinavia), gradually smaller towards east and south, ending in small *gutturalis* from south-east. Adult wing of nominate *rustica* in Scandinavia 125·0 (116–134), in East Germany 123·8 (118–131), in Netherlands, Belgium, and Britain 123·0 (114–129), in north-west Africa 122·2 (119–128); further east, 122·5 (116–130) in western USSR, 122·3 (116–130) in Iran and Afghanistan, and down to 121·1 (115–126) in Pakistan and northern India and *c*. 117 (112–123) in *savignii* from Egypt, but relatively large in Balkan countries and Turkey (see Measurements). In eastern part of species' range, wing from 121·6 (117–127) in *tytleri* from Mongolia, through 117·0 (112–121) in *tytleri* from south-east USSR and 116·4 (112–121) in *gutturalis* from Japan, to 114·5 (110–120) in *gutturalis* from China (Dementiev and Gladkov 1954b; RMNH, ZMA, and see sources in Measurements). North American *erythrogaster* has wing 119·4 (114–126), close in size to *tytleri*-like birds from north-east Asia. Chest-band completely black in nominate *rustica*, *transitiva*, and *savignii*; often some rufous feather-fringes or sometimes a few completely rufous feathers on central chest, but black not fully interrupted. In adult *tytleri*, *gutturalis*, and *erythrogaster*, rufous-chestnut of throat more or less connected through rufous to pink-buff central chest with rufous to cream-white belly, black chest-band either restricted to patches on sides of breast and broadly interrupted in middle (*erythrogaster*), or connected by a narrow dull black band across lower chest (*tytleri* and *gutturalis*). Remainder of underparts strongly variable in some populations, but hardly so in others. In general, typical nominate *rustica* from Scandinavia as well as typical *gutturalis* from eastern Asia pale cream-pink to white below, *erythrogaster* from North America saturated rusty-buff (except for white-bellied populations on islands along Gulf coast of USA, which are sometimes separated as *insularis* Burleigh, 1942); *transitiva* rusty-pink or rufous-buff, *savignii* and *tytleri* rufous-chestnut, close to colour of forehead and throat. Birds from transition zones from nominate *rustica* into *transitiva* in south-west Europe and Asia Minor, and from nominate *rustica* and *gutturalis* into *tytleri* in Western Mongolia, central Siberia, and north-east Asia are strongly variable individually, and also outside these zones some populations are paler or darker than others, while a few rusty-bellied birds occur in almost all pale-bellied populations and *vice versa*. If colour of belly scored 0 (virtually white), 1 (dingy-white or cream-white), 2 (cream or cream-pink), 3 (pinkish cream-buff with rusty tinged under tail-coverts), 4 (rufous-buff, like *transitiva*), or 5 (rufous-chestnut, like *savignii* and *tytleri*), adults from Scandinavia have average colour value 0·7 (7) 0–2,

European USSR and western Siberia 1·1 (12) 0–2, Netherlands 1·9 (46) 0–4, West Germany and France 2·3 (12) 1–4, Britain 2·3 (13) 0–3, north-west Africa 1·8 (5) 0–3, Balkans and Turkey 2·5 (23) 0–4, Iran 0·9 (51) 0–2; proportion of *transitiva*-type (score 4) everywhere below 10%, except Balkans and Turkey (30%), but here collection probably biased towards this type, as such birds uncommonly encountered in the field (Makatsch 1950). Mainly pale-bellied populations of Turkey, Iraq, and perhaps Syria suddenly replaced by true *transitiva* from coastal strip of Lebanon to Gaza, where virtually all birds score 4; otherwise not separable from occasional *transitiva*-type birds elsewhere, except for timing of moult (Stresemann and Stresemann 1968*b*). Juveniles of all races are generally paler below with less pronounced chest-band, and hence these more difficult to identify. In juvenile *gutturalis* (with virtually white belly) and *tytleri* (with rufous belly), rufous-buff of throat extends to upper chest, bordered below by more or less complete grey-brown chest-band; band not as broad and dark as most juvenile

nominate *rustica* (with belly varying between white and rufous), but chest-band of some nominate *rustica* similar, especially ♀♀; size generally smaller than nominate *rustica*. Juvenile *erythrogaster* usually more distinct from nominate *rustica*, as rusty-buff of throat extends to upper breast, dark chest-band restricted to patches at sides of breast or only indistinctly and narrowly connected by grey-brown across lower chest or (occasionally) upper chest; colour of remainder of underparts variable, between cream-white and rusty-buff (as in nominate *rustica*), but most birds tend towards saturated pink-buff (most nominate *rustica* pale cream-buff or cream-white); slightly smaller than nominate *rustica*, but much overlap.

Forms superspecies with Pacific Swallow *H. tahitica* from India and Australia to Oceania, and with Red-chested Swallow *H. lucida*, Ethiopian Swallow *H. aethiopica*, Angola Swallow *H. angolensis*, and White-throated Swallow *H. albigularis*, all from Afrotropics (Hall and Moreau 1970). CSR

Hirundo daurica Red-rumped Swallow

PLATES 17 and 18
[between pages 232 and 233]

Du. Roodstuitzwaluw   Fr. Hirondelle rousseline   Ge. Rötelschwalbe
Ru. Рыжепоясничная ласточка   Sp. Golondrina dáurica   Sw. Rostgumpsvala

Hirundo daurica Linnaeus, 1771. Synonym: *Cecropis daurica*.

Polytypic. *H. d. rufula* Temminck, 1835, southern Europe and North Africa, east through Middle East to Tien Shan mountains (USSR) and Kashmir; nominate *daurica* Linnaeus, 1771, south-east USSR (east from eastern Kazakhstan), northern Mongolia, and highlands of central China; straggler to Fenno-Scandia. Extralimital: *japonica* Temminck and Schlegel, 1847, Japan, Korea, and probably Ussuriland (USSR), south through Manchuria to lowlands of central China; 3 further races in India and *c*. 5 in Afrotropics.

Field characters. 16–17 cm, of which tail-streamers 5–6 cm; wing-span 32–34 cm. Medium-sized hirundine similar in general form to Swallow *H. rustica* but bulkier and with even stubbier bill, marginally shorter and slightly more rounded wings, and slightly shorter, blunter, and less wire-like tail-streamers (due to greater extension of 2nd and 3rd outermost feathers). Plumage differs most obviously in lack of dark chest-band and of white spots in tail, and in presence of pale rufous rump area, black (not pale) under tail-coverts, and chestnut nape. Flight rather deliberate; usually slower and with more gliding than *H. rustica*. Voice differs from *H. rustica*. Sexes similar; some seasonal variation through wear. Juvenile separable. 2 races in west Palearctic; western race, *rufula*, described here (see also Geographical Variation).

ADULT. Crown, mantle, scapulars, and upper wing-coverts blue-black, rather less shiny than *H. rustica* when fresh, dull and white-flecked when worn and in moult; wings and tail brown-black, only faintly metallic and

unmarked except for buff-white tips to tertials. Rather broad, chestnut nape less obvious in flight than combination of chestnut lower back and buff to terminally buff-white rump; rump-patch wears paler, even becoming pale orange with dull white terminal band. Forehead and lores speckled buff and black; cheeks rufous-buff. Rest of underparts to vent buff (more rufous on some but never with obvious paleness of most *H. rustica rustica*) and variably marked with narrow dark brown shaft-streaks, usually visible only at close range. Under wing-coverts buff, contrasting strongly with dusky-black undersides of flight-feathers. Tail brown-black. All but shortest under tail-coverts black, forming dark end to underbody lacking in *H. rustica* and exaggerating length of tail from below. JUVENILE. Tail-streamers shorter and blunter than in adult. Plumage pattern as adult but colours much duller and virtually unglossed above, with paler (buff) nape and upper portion of rump-patch, and paler underbody, with dark brown smudge on side of breast and shaft-streaks

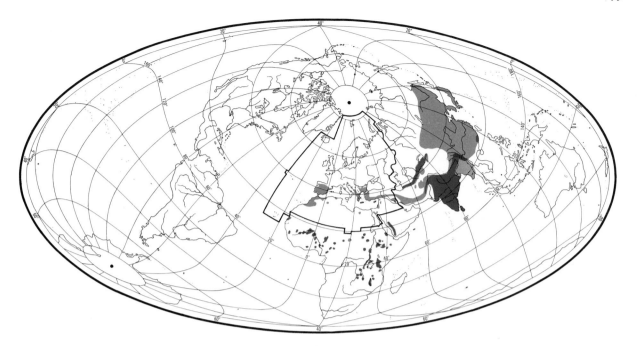

on breast only. Buff tips may show not only on tertials but on secondaries, wing-coverts, and back. At all ages, bill black; legs brown-black.

In west Palearctic, unmistakable when closely seen but requires care at long range due to (1) similarity to *H. rustica* in shape and plumage pattern, (2) similarity to House Martin *Delichon urbica* in plumage pattern and in action and height of flight, and, most confusing of all, (3) existence of plumage irregularities in those 2 species and apparent hybridization between them. Thus occurrence of (e.g.) *D. urbica* with rump stained pink, *H. rustica* with rump made pale by wear (D I M Wallace), and hybrid *H. rustica* × *D. urbica* with intermediate appearance (Charlwood 1973; Stephenson and Doran 1982) means that any vagrant *H. daurica* must be closely observed and all its characters carefully checked. In Africa south of Sahara, also subject to confusion with Mosque Swallow *H. senegalensis* and Rufous-chested Swallow *H. semirufa*. Flight almost diagnostic, being far less elegant than *H. rustica*, with flat-winged glides and almost soaring ascents, slow turns and sweeps, and far less rapid wing-beats all contributing to distinctive action which recalls Crag Martin *Ptyonoprogne rupestris* and *D. urbica* more than *H. rustica*. Hunting flight more patrolling than dashing, bird using regular route and moving rather slowly until insect prey chosen. Flight of migrants noticeably steadier than *H. rustica*. Infrequently lands on ground, and can only shuffle, but carriage there and on perch more erect than *H. rustica* and more attenuated, with more neck showing.

Less vociferous than *H. rustica*. Song a short, sweet, twittering warble, lacking vehement bursts and inter-spersed calls of *H. rustica*. Commonest call resembles *H. rustica* but more like sparrow *Passer*—'djuit'.

Habitat. In west Palearctic, breeds in lower middle latitudes in warm temperate, steppe, and especially Mediterranean zones, extending into oceanic climates, from sea-level in south-west Spain to over 1000 m in Cyprus, where sea-cliffs and caves are occupied as well as mountains (Bannerman and Bannerman 1971). In USSR, avoids sandy deserts, but ranges up to 2000–2200 m in mountains provided there is access to water, nesting even in deep dark caves, and in cracks, hollows, and niches, as well as occupied buildings and ruins, preferring those of more massive construction but also using culverts, bridges, and sites within large cities. Forages over neighbouring open areas such as meadows, pastures, and grassy glades (Dementiev and Gladkov 1954b). In India, not particularly attached to neighbourhood of water, commonly hawking over dry country, including railway goods yards, grain markets, forest clearings, and fired grass fields; rests on ledges, telegraph wires, and treetops. Occurs on plains and hills up to *c.* 1600 m (Ali and Ripley 1972). Compared with Swallow *H. rustica*, has retained closer links with primitive breeding and hunting habitats and has not yet become so committed to dependence on human structures and land uses, except in certain regions. Much more dependent on reliable warm climate, and less adaptable to sudden or drastic change. Winters in dry African grasslands, or in cultivation and forest clearings (Moreau 1972; Harrison 1982). Tends to be less confined to foraging in lowest airspace than *H. rustica*.

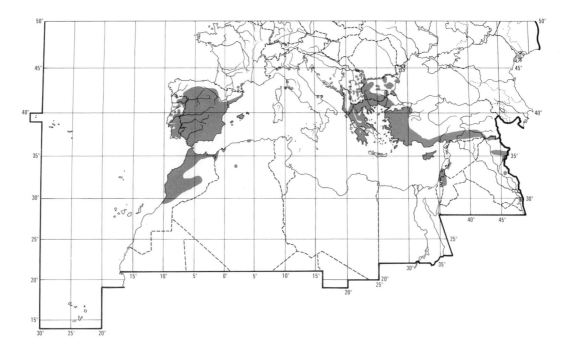

Distribution. Spreading north in Spain, France, Italy, Rumania, and Bulgaria (Wicht 1978).

FRANCE. First bred 1963 (Yeatman 1976). Has bred occasionally on Corsica since 1965 (Thibault 1983). Recently extinct Gorges de l'Ardèche (Beaufort 1983). Bred Torlard (Ardèche) 1983 (Vallée 1983). SPAIN. Spread north; before 1929 only in extreme south, but reached Ciudad Réal 1952, central Spain 1951–3, and Gerona 1960 (Lope Rebollo 1982). PORTUGAL. Spreading north (Santos Júnior 1959, 1960). ITALY. Spreading north (PB). Bred Sardinia 1965 (Schenk 1980). ALBANIA. Distribution not well known (EN). BULGARIA. Recent expansion to north (Simeonov and Michev 1980). GREECE. No changes noted (GIH). RUMANIA. Expanded north; first bred 1975 (WK). ISRAEL. More frequently nesting on houses since 1930s (HM, UP).

Accidental. Britain (recently annual), Ireland, Netherlands, Luxembourg, West Germany, Denmark, Norway, Sweden, Finland, Poland, Switzerland, Azores, Canary Islands.

Population. FRANCE. Under 10 pairs (Yeatman 1976); probably under 10 pairs (Beaufort 1983). BULGARIA. 1000–10 000 pairs (Simeonov and Michev 1980). RUMANIA. 25–100 pairs (WK).

Oldest ringed bird 7 years 9 months (Rydzewski 1978).

Movements. Northern populations migratory, though winter quarters of west Palearctic birds still unconfirmed.

African races occupying northern savanna zone and East Africa are mainly resident, though in Darfur (Sudan) there seems to be a northward post-breeding movement in April–May with corresponding return in October (Lynes 1925a; Hall and Moreau 1970). Also, birds are present only seasonally at some localities in East African highlands, and one ringing recovery showed 240 km movement (Britton 1980). Populations of Himalayas and south-east Asia are migratory, overlapping in winter in southern Asia from the Indian subregion to southern China and Indochina (Ali and Ripley 1972); vagrants to Finland (2) and Finnmark (Norway) attributed to west-central Asian race, nominate *daurica* (Haftorn 1971).

Winter range of migratory European and south-west Asian birds (*rufula*) remains unconfirmed: presumed to lie in savanna zone of northern Afrotropics, where birds inseparable in the field from resident African populations (Moreau 1972). Pronounced spring and autumn passage across Straits of Gibraltar (Heim de Balsac and Mayaud 1962; Tellería 1981). Migrants occur across whole of North Africa though are rare in Tunisia and Libya; from this, assumed that Spanish and Moroccan birds migrate towards West Africa, those from south-east Europe and south-west Asia to north-east Africa (Etchécopar and Hüe 1967). Libyan records essentially in spring (but see below), and birds occur then in Libyan Desert and Fezzan as well as coastal belt (Bundy 1976). Scattered records through Sahara show that trans-desert passage occurs (Moreau 1972); in central Chad, regular in ones and twos in autumn (Newby 1980). However, few confirmed records of *rufula* south of Sahara, and none for winter: listed as a frequent to common migrant to Ethiopia, in western highlands and Eritrea (Urban and Brown 1971); recorded (without details) as reaching Sudan (Cave and Macdonald 1955); collected in southern

Chad in October (Salvan 1967-69); acknowledged as spring migrant through Mali (Lamarche 1981) and Sénégal (Morel and Roux 1973).

Autumn passage at Straits of Gibraltar mainly September, continuing into October (Cortés *et al.* 1980; Tellería 1981). Evidently rather earlier from eastern breeding areas, since gone from Turkey by end of September (Beaman *et al.* 1975) and main Cyprus passage late August to mid-September (Flint and Stewart 1983); many fewer autumn observations than in spring, presumably due to more long-range overflying in autumn (Moreau 1961, 1972; Safriel 1968). Spring passage begins February, continuing into April, with comparable dates from regions as far apart as Morocco, Jordan and Israel, and Persian Gulf.   RH

Food. Invertebrates. Takes airborne prey by aerial-pursuit, at up to 100 m or more (Dolgushin *et al.* 1970). Hunting flight involves more steady gliding and less rapid wing-beats than Swallow *H. rustica* (D I M Wallace). In Arabia, recorded descending to camel dung to eat flies (Meinertzhagen 1954), and in Nepal seen taking termites (Isoptera) from ground in rainy weather (Diesselhorst 1968a). In Ussuriland (USSR), migrants sometimes settle on *Artemisia* plants to take insects from them (Panov 1973). In Indian subcontinent, gathers at grass fires to take insects flushed (Phillips 1953; Ali and Ripley 1972). Said sometimes to hunt in groups of 10-20 birds, unlike *H. rustica* (Fry 1970a).

♀ from Palestine, April, contained beetles (Coleoptera) (Hardy 1946). No other information on diet of adults in west Palearctic. Stomachs from Kazakhstan (USSR) contained bugs (Hemiptera), flies (Diptera), and small beetles (Dolgushin *et al.* 1970). One stomach in Kirgiziya contained beetles (Scarabaeidae) (Pek and Fedyanina 1961).

Food of 25-day-old young in nest in eastern Pyrénées (France) studied from faeces: of 255 items, 94% (by number) winged ants; also a soldier ant (unwinged), other Hymenoptera, Diptera, beetles (Carabidae, Curculionidae), and a cockroach *Loboptera*. When young 21-24 days old, 5 times fed live insects *c.* 14-20 mm long. (Prodon 1982.) Nestlings in Uzbekistan (USSR) fed on cicadas (Cicadidae) and beetles (Tenebrionidae, Coccinellidae, Chrysomelidae) (Salimov 1977). In Bulgaria, on 5th day, average 9·4 nest-visits per hr over 14 hrs; most (12·7 per hr) 08.00-15.00 hrs; little feeding 13.00-15.00 hrs. On 10th day, average 22·8 visits per hr over 15 hrs; at 15 days, average 24·5 per hr. (Simeonow 1969.) In Extremadura (Spain): at 3 days, 53 feeding visits to nest per day; at 8 days, 72; at 13 days, 138; at 18 days, 82. At 1-5 days, feeds distributed regularly throughout the day except for pause 12.15-13.30 hrs; from 16 days, sometimes 1½ hrs between feeds, with most done in early morning and at start of afternoon. (Lope Rebollo 1980.) In eastern Pyrénées, interval between nest-visits at 1 day

9·7 min, at 5 days 9·5 min, 6 days 5·6 min, 10 days 4·2 min, 18-27 days 4·2 min (no sample sizes) (Prodon 1982).

                                    DJB

Social pattern and behaviour. For important recent studies, see Lope Rebollo (1980) and Prodon (1982).

1. Gregarious outside breeding season, and though generally fewer reports of assemblages than for *H. rustica*, this may simply reflect relative abundances. In India, *nipalensis* highly gregarious in winter, *erythropygia* less so (Ali and Ripley 1972); in Sri Lanka, up to 30-40 at grass fires (Phillips 1953). Post-breeding flocks form as in *H. rustica*; flocks of 25 adults, presumably non-breeders, recorded June-July in Iraq (Sage 1960a). Post-breeding flocks typically mixed with other Hirundinidae, also swifts (Apodidae) (Dementiev and Gladkov 1954b; Nicolau-Guillaumet 1965). Migrates singly or in small parties, never in large flocks (Ivanov 1969): e.g. in Tunisia spring migrants usually alone, less often in groups of up to 4, always with other Hirundinidae (Thiollay 1977; Riols 1978); in southern Turkey, average flock size of birds passing on autumn migration 9·1 (*n* = 2093 birds) (Sutherland and Brooks 1981). See also Roosting (below). BONDS. No evidence for other than monogamous mating system; pair-bond maintained for 2nd clutch (Lope Rebollo 1980), but no information on status of bond beyond single breeding season. Brooding by both sexes, ♀ more assiduous than ♂. Both sexes feed young for up to 5-6 days after fledging (Umrikhina 1970), perhaps longer as reported being accompanied by parents for 8-9 days (Lope Rebollo 1980). Even after young capable of self-maintenance, family reunites to roost in nest nightly for 2-3 weeks after fledging (Dolgushin *et al.* 1970; Prodon 1982); see also Roosting (below). Age of first breeding 1 year (Lope Rebollo 1980). BREEDING DISPERSION. Typically solitary or in colonies of up to 3-4 pairs (Ivanov 1969). Seldom more, but one colony of 20 pairs in Turkey (Beaman *et al.* 1975) and compact colony of *c.* 50 pairs in Iraq (Sage 1960a). Nests clustered or 2-3 m apart: e.g. in one colony under a bridge, 4 nests in *c.* 1 m², minimum 29 cm between nest-entrances (Simeonow 1968). In colony, Pamir-Alay mountains (southern USSR), 5 nests (3 occupied) all touching, and nearby another 3 solitary nests (Ivanov 1969). In Kazakhstan (USSR), sometimes uses nests of *H. rustica*, building on to them (Dolgushin *et al.* 1970). No precise information on size or purpose of territory. Limited data suggest size perhaps differs with nest-dispersion: possibly confined to immediate area around nest-entrance in colonies (as in, e.g., House Martin *Delichon urbica*, but perhaps extends much further where nest solitary (E K Dunn). At one isolated nest, resident defended airspace up to 150-200 m from nest (Prodon 1982). Fidelity to previous nest-site occurs in established breeders (Vorobiev 1954; Umrikhina 1970), but few studies. Numerous reports of old nests being refurbished in successive seasons, though identity of birds usually not known; traditional sites re-used, even if nest disappears between breeding seasons, and same nest typically used for 1st and 2nd broods (Lope Rebollo 1980). One nest used for 5 successive years (*Cyprus orn. Soc.* (1957) *Rep.* 1977, 22). A measure of fidelity to natal area also occurs, but again more information needed: of 134 ringed young, 2 controlled breeding the following year 4·7 km and 8·4 km from natal site (Lope Rebollo 1980). ROOSTING. In breeding season, both members of pair roost together in nest; outside breeding season communally in reedbeds, etc. (see below). Roosting in nest continues until after young fledge (Ali and Ripley 1972; Lope Rebollo 1980). One family returned to roost in nest until 16-20 days after fledging, thereafter parents only (Prodon 1982,

which see for times of roosting and morning departure). Fledglings from 1st brood typically continue to roost in nest with parents for *c.* 15 days, by which time 2nd clutch already laid (in same nest) (Lope Rebollo 1980). Fledglings of 2nd brood tend to roost longer in nest with parents—up to 3 weeks, i.e. until autumn departure (Dolgushin *et al.* 1970); at one nest, for at least 29 days after fledging (Congreve 1945). Young return separately to nest, but roost before parents (Lope Rebollo 1980). Prior to autumn migration, independent young (presumably mainly of 1st broods) roost communally. In Spain, 2 roosts of 50 and 68 juveniles, sometimes associated with *H. rustica* and Sand Martin *Riparia riparia* (Lope Rebollo 1980). In Mali, March–April, roosts with these species at lakeside (presumably reedbeds) (Lamarche 1981). In winter in India, roosts in partially submerged reedbeds, etc.; birds assemble in early morning in hundreds or thousands on wires, sometimes also on trees or ground (Ali and Ripley 1972; see also Flock Behaviour, below). In Jordan, November, roosting on ground reported twice (Wallace 1982). In Sierra Leone, December, *c.* 300 in flock of *c.* 900 'swallows' on ground in morning after heavy rain (G D Field). In Mongolia, August, many reported resting on bushes during migration (Piechocki *et al.* 1982). Drinks and bathes in flight (Dolgushin *et al.* 1970).

2. FLOCK BEHAVIOUR. In China, post-breeding flock gathered on wires prior to migration (Fiebig 1983). Restlessness (etc.), as occurs in other Hirundinidae before migration, reported in wintering assemblies of *nipalensis* in India: birds, perched together in large numbers (see above), perform noisy upflights from time to time (Ali and Ripley 1972). ANTAGONISTIC BEHAVIOUR. Lope Rebollo (1980) saw no fighting between rivals for nest-sites or mates. Once site established, ♂ defends it against intruding conspecifics (including independent young: Lope Rebollo 1980) and other Hirundinidae. At one site, defended area within 150–200 m of nest, chasing intruders with Contact-calls, Mewing-calls, and Alarm-calls (see 2, 4, and 5 in Voice), sometimes challenging *D. urbica* more strongly than *H. rustica*; one attack on *H. rustica* accompanied by Anger-calls (see 6 in Voice) led to intruder falling into bush. (Prodon 1982.) In Zaïre, November, resident breeding birds attacked *H. rustica* among first flocks arriving on migration (Verheyen 1952). HETEROSEXUAL BEHAVIOUR. (1) General. According to Dolgushin *et al.* (1970), first birds to arrive on breeding grounds are usually single, less commonly pairs. In Extremadura (Spain), most arrive in pairs (Lope Rebollo 1980). Old birds arrive before young birds (Umrikhina 1970). Nest-building usually begins soon after arrival (e.g. Simeonow 1968), once not until 1 month after (Cotron and Prodon 1977). On arrival, established pairs spend several days in vicinity of nests, sometimes disappearing only to reappear after 1–2 days (Lope Rebollo 1980). (2) Pair-bonding behaviour. Birds form pairs a few days after arrival on breeding grounds (Umrikhina 1970). Few details of displays (etc.). Near prospective nest-site, ♂ performs display-flight in which he circles ♀, giving soft Contact-calls which ♀ sometimes reciprocates. If ♀ perches, ♂ may flutter round her and perch nearby, singing (see 1 in Voice) frequently, and sometimes spreading his tail; birds almost always use same perches. (Lope Rebollo 1980.) In display-flight, probably of this kind, described by Panov (1973), 2 birds (presumably ♂ and ♀, though not specified) fly alongside each other with frequent, synchronized wing-beats; wings fully spread, but amplitude of beats not more than 45°. Pairs performing interlaced manoeuvres high over breeding area (D I M Wallace) are probably displaying thus. In presumed heterosexual display, ♂ flew slowly, singing; from time to time, in flight, simultaneously threw head back and raised his tail forwards;

at other times, vibrated wings rapidly while holding them close to body, resulting in steep dive which ended with outspread wings. ♀ soon appeared and some calls, including Nasal-calls (see 3 in Voice) exchanged; ♂ also sang again (Nicolau-Guillaumet 1965). Song given throughout breeding season, until after fledging (Prodon 1982). (3) Nest-site selection. According to Dementiev and Gladkov (1954b), both members of pair select site. No further information. (4) Courtship-feeding. Not recorded even in lengthy study by Lope Rebollo (1980). (5) Mating. Typically preceded by singing. In 'courtship' sequence (Fig A), perhaps of pre-mating kind, ♀ perches

A

motionless, wings drooped and tail raised; ♂ perches nearby and sidles up to ♀, bill raised, singing loudly, whereupon ♀ slightly opens and vibrates wings (Panov 1973). ♂ mounts perched ♀ who acts passively; copulation lasts *c.* 2 s, followed commonly by preening and, on 2 occasions in one pair, by apparent mild Allopreening: 1–2 strokes, perhaps accidental, of bill of one bird on neck of other (Lope Rebollo 1980; F de Lope Rebollo). (6) Behaviour at nest. Incubating bird often leaves nest without immediately being relieved. On leaving nest, seeks out and rejoins mate directly, their meeting accompanied by Contact-calls. Both then return to nest together and one enters to resume incubation. Sometimes both in nest during incubation, and then ♂ often sings from inside nest (Prodon 1982). While bird incubating, mate may continue to add mud to nest (Ali and Ripley 1972). While attending fledged young of 1st brood, parents take lining material to nest in preparation for 2nd clutch (Umrikhina 1970). RELATIONS WITHIN FAMILY GROUP. Based largely on detailed study at one nest by Prodon (1982). Adults help hatching young out of egg (Lope Rebollo 1980). On day of hatching, nest attentiveness, especially by ♀, increases sharply. Food-calls (see Voice) of newly hatched young barely audible at close quarters. In succeeding days, attentiveness wanes, levelling off at 7–8 days (Prodon 1982). Eyes of young open at 10 days (Simeonow 1969). Both parents feed young, initially about equally, but from *c.* 11 days contribution of ♂ gradually diminishes to less than 40% by 20 days. Parents visit nest separately, and for *c.* 15 days carry off faecal sacs. From 16–20 days, young spend increasing amount

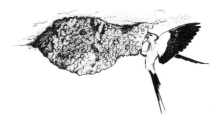

B

of time at nest-entrance, and defecate out of it (Lope Rebollo 1980). At this stage, parents typically stay outside nest to feed young (Fig B), perching briefly with outspread wings and sometimes supported by tail if staying longer, e.g. to remove faecal sac, though may enter nest up to fledging if prey item large (Prodon 1982). Near fledging, young spent a lot of time looking out of nest, especially at ground. In possible enticement, parents sometimes came to nest as if to feed, but did not perch, and gave Contact-call (which stimulates young to beg), chick at entrance responding with same call. In brood of 4, 3 fledged on same day, all returning to nest (Fig C) in early evening,

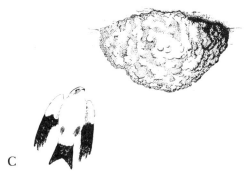

C

whereupon parents continued to provision them in nest until dusk. Next morning, brood left nest and remained airborne all day until they returned to roost at night (Prodon 1982). Initially, parents feed fledglings perched, often together, near nest (Lope Rebollo 1980), often in treetops (Dementiev and Gladkov 1954b). ANTI-PREDATOR RESPONSES OF YOUNG. On hearing Alarm-call of parents, nestlings fall silent and withdraw from nest-entrance (Prodon 1982). PARENTAL ANTI-PREDATOR STRATEGIES. (1) Passive measures. Quite wary when human intruder near nest, feeding nestlings only when intruder more than 100 m away (Isenmann 1965). Alarm-call given by either bird as soon as intruder within 20–40 m of nest; calls accompanied by rapid agitated flight. Eventually, calling induces mate to leave nest and silences young. Increasing state of alertness indicated by progression from song to Mewing-call to Alarm-call (Prodon 1982). After young fledged, parents less wary and whole family can be approached quite closely (Isenmann 1965). (2) Active measures. No information.

(Figs by M J Hallam: A based on drawings in Panov 1973; B–C after drawings in Prodon 1982.)   EKD

Voice. Relatively little used compared with other Hirundinidae (e.g. Prodon 1982). However, apparently more freely used when breeding colonially than alone (Géroudet 1962). Following account based largely on study by Prodon (1982), which see for additional sonagrams.

CALLS OF ADULTS. (1) Song of ♂. A subdued chattering twittering sound in short phrases (Bergmann and Helb 1982, which see for sonagram), given perched or in flight. Resembles *H. rustica*, but simpler, shorter, quieter, and uttered less often (Dementiev and Gladkov 1954b; Dolgushin *et al.* 1970). Distinguished from *H. rustica* mainly by shortness, structure of short phrases, and lower pitch. One ♂'s song included at least 5 different phrases, of which the most frequently used had the following features: 8–13 brief (c. 1 ms) 'clic' sounds, grouped at end of phrase to produce a dry rattle similar to *H. rustica*,

though rattle often omitted from song; rest of phrase comprised rather low-pitched, slightly nasal sounds with complex harmonic structure (in most developed form not unlike Fig IV) which, in absence of rattle, end the phrase; also contained units resembling Contact-calls (call 2). Whole phrase lasted 1·3–1·8 s including rattle, c. 1 s without it (Prodon 1982). In Fig I, phrase, at least 2·3 s long, begins with rattle, followed by 2 units resembling Contact-calls; 5th unit perhaps a faint Nasal-call (see call 3) or variant. Inclusion of Contact-calls a typical feature of song (Bergmann and Helb 1982). Song most often used in heterosexual context; when bird mildly alarmed (e.g. spots distant intruder) or frustrated (e.g. lets prey item escape), song rather agitated or aggressive (Prodon 1982). (2) Contact-call. Commonest call, quite similar to *H. rustica* (Prodon 1982), but especially like chirp of sparrow *Passer* (P A D Hollom: e.g. Fig II, 1st unit). Variable in structure and length, but always rises in pitch: rendered 'zvèit' or 'kvëit' or 'djuit' (Prodon 1982). Also described as 'tchreet' (P A D Hollom: Fig III), rather quiet 'quitsch' with distinctive complaining or pained tone (D I M Wallace), soft 'pruit-pruit' (Lope Rebollo 1980), and as softly nasal 'wüid' or 'schwüid' (Bergmann and Helb 1982, which see for sonagram). These calls much shorter than call 3 which they resemble in timbre. Single 'kwep', less sharp than *H. rustica*, given by vagrant, Ireland, in April (Browne and Harley 1953) is perhaps this call, as perhaps is call of autumn migrants: 'chirrup', rather longer, softer, and more chirrupy than call of *H. rustica* (D J Brooks). As renderings and recordings show, call given singly or twice in rapid succession, and generally repeated. Given regularly in communication between pair-members, e.g. in meeting-ceremony, and between parents and offspring; also during aggressive pursuits of other Hirundinidae near nest (Prodon 1982). (3) Nasal-call. A long-drawn nasal 'zvèèit' or 'djuui', peculiar to *H. daurica* among west Palearctic Hirundinidae (Prodon 1982, in which see Fig 4i); resembles Greenfinch *Carduelis chloris* or Rock Sparrow *Petronia petronia* (Nicolau-Guillaumet 1965). Intermediates with call 2 occur. In recording (Fig II), 2nd unit, a nasal 'tzueeee', resembles *C. chloris* (J Hall-Craggs). Fig IV shows, perhaps, a variant: a musical twanging sound (P J Sellar), rendered a buzzing nasal 'nyeeh' (E K Dunn), and likened to sound produced by a cracked tin toy trumpet (J Hall-Craggs); nasal 'queenk' of Indian *erythropygia* (Ali and Ripley 1972), perhaps the same. Nasal-calls occasionally given by ♂ in alternation with song (Prodon 1982) during pair-formation (Nicolau-Guillaumet 1965). Function not clear but perhaps homologous with Enticement-call of *H. rustica*, inviting close approach. (4) Mewing-call. A curious nasal mewing sound, c. 0·4 s long, often given 2–4 times at intervals of several seconds; like short mew of Buzzard *Buteo buteo* but diagnostic of *H. daurica* among west Palearctic Hirundinidae (Prodon 1982, in which see Fig 4d). Also

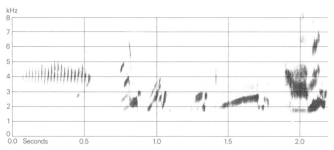

I P A D Hollom Israel April 1980

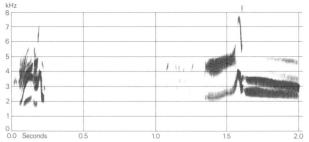

II E D H Johnson/Sveriges Radio (1972-80) Morocco March 1968

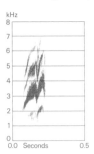

III P A D Hollom Israel April 1980

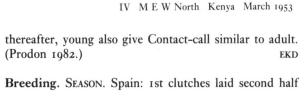

IV M E W North Kenya March 1953

thereafter, young also give Contact-call similar to adult. (Prodon 1982.)

EKD

likened to subdued mewing of a small kitten (Dementiev and Gladkov 1954b). Given only by ♂, infrequently throughout breeding season; apparently used in territorial defence (Prodon 1982). (5) Alarm-call. A brief (0·17 s) whistle, 'i(e)' or 'ki(e)' (in which 'i' pronounced 'ee'), dropping in pitch (Prodon 1982); also rendered as short sharp 'kit' (Géroudet 1961) or sharp 'kier' (Bergmann and Helb 1982). Given by either sex when intruder at nest, at intervals of less than 1 s if danger imminent (Prodon 1982; see also Social Pattern and Behaviour). (6) Anger-call. A short hard 'krr' given by ♂ near nest during expelling attack on *H. rustica* (Prodon 1982). (7) Other calls. Recording (Fig IV) shows, before and after the nasal sound, high frequency 'tik' sounds, thought to be vocal rather than instrumental (J Hall-Craggs). Function not known.

CALLS OF YOUNG. Food-call a soft repeated cheeping of variable timbre and with slightly rolling quality, 'piè-piè-piè...' or 'pjiè-pjiè...'; each unit slightly less than 0·1 s. Average pitch diminishes with age from *c.* 4 kHz at 10 days to 2-3 kHz at 19 days . Call continues to be given by fledged young when fed. Near fledging, and

Breeding. SEASON. Spain: 1st clutches laid second half of April, last in September (Lope Rebollo 1980). Greece and Bulgaria: laying from mid-May to end of July (Simeonow 1969; Makatsch 1976). North Africa: laying from end of April to July (Heim de Balsac and Mayaud 1962). SITE. Overhanging rock ledge, inside cave, under bridge or culvert, or in ruined building, etc. Of 92 nests, height above ground mainly 2-4·5 m (0·65-20), and 1·5-4·3 m when under bridge; aspect mainly south, south-east, or east (83% of 92 sites), less than 10% between west and north-east (Simeonow 1968). Nest: rounded bowl attached to overhanging, usually horizontal surface, with extended entrance tunnel also along surface. Constructed of mud pellets reinforced with plant and grass stems; lined softer vegetation, wool, and feathers. Average external dimensions of nest-chamber, length 18·1 cm (7·6-22·4), width 17·5 cm (4·7-22·1), height 10·4 cm (6·2-15·0), $n = 132$; of nest-tunnel, length 10·0 cm (4·9-14·2), width 7·1 cm (4·5-9·9), height 5·7 cm (4·3-8·2), $n = 128$; entrance, width 4·7 cm (3·3-6·5), height 4·3 cm (2·4-6·9), $n = 116$ (Lope Rebollo 1980). Average weight of nest 975 g (712-1410), $n = 19$ (Simeonow 1968). Building: by both sexes, taking 8-15 days (Simeonow 1968). EGGS. See Plate 80. Long, sub-elliptical, smooth and slightly glossy; white, occasionally with fine red-brown speckling. *H. d. rufula*, Spain: 20·3 × 14·2 mm (18·2-22·5 × 13·1-15·5), $n = 237$; weight 2·01g (1·0-2·5), $n = 332$ (Lope Rebollo 1980). No significant differences in North Africa (Schönwetter 1979). Clutch: 4-5 (2-7). Of 135 clutches, Spain: 2 eggs, 5%; 3, 15%; 4, 46%; 5, 29%; 6, 4%; 7, 1%; mean 4·13. Mean of 65 clutches, 4·51; of 56 2nd clutches, 3·8; of 14 3rd clutches, 3·6 (Lope Rebollo 1980). 2-3 broods; of 65 pairs, Spain, 86% had 2, 21% had 3 (Lope Rebollo 1980). INCUBATION. 14·5 days (13-16), $n = 21$ (Lope Rebollo 1980). 11-16 days (Simeonow 1969). Period decreases (significantly) by 0·35 days per month from April (Lope Rebollo 1980). By both sexes, though more by ♀; begins with laying of last egg or 1-2 days later (Simeonow 1969). YOUNG. Altricial and nidicolous. Cared for and fed by both parents. FLEDGING TO MATURITY. Fledging period 22-26 days (Lope Rebollo

1980); 26–27(–29) days (Prodon 1982). Age of first breeding 1 year. BREEDING SUCCESS. Of 558 eggs laid, Spain, 80·0% hatched and 74·4% raised to fledging. Of 293 eggs in 1st clutches, 77·8% hatched and 71·3% raised to fledging; of 214 eggs in 2nd clutches, 80·8% hatched and 75·2% raised to fledging; of 51 eggs in 3rd clutches, 88% hatched and all these raised to fledging. Overall fledging success varied little during year, though slightly better in April, June, July, and September than in May and August. Overall production per pair 6·38 fledged young (Lope Rebollo 1980).

Plumages (*H. d. rufula*). ADULT. Forehead and crown glossy deep metallic blue; small feathers near base of upper mandible narrowly fringed rufous or buff when fresh. Narrow supercilium and broad half-collar on rear and sides of neck deep rufous-cinnamon; lores, cheeks, and ear-coverts rufous, slightly mottled grey-brown and buff-brown; indistinct dull black patch in front of eye. Mantle and scapulars glossy deep metallic blue; crown, mantle, and scapulars similar to Swallow *H. rustica*, but slightly darker blue and less glossy; sides of feather-bases of mantle and inner scapulars pale cream-buff or off-white, hidden when plumage fresh (except sometimes on upper mantle), but showing as conspicuous streaking when plumage worn. Back, rump, and shorter upper tail-coverts rufous-cinnamon, merging into pale cinnamon-buff at sides; in worn plumage, rump, and shorter upper tail-coverts, and sides of back bleached to cream-buff, rufous-cinnamon restricted to central back. Longer upper tail-coverts black with slight blue-green gloss. Underparts rufous-buff to buff, deepest rufous on chin, lower cheeks, sides of breast, and chest, palest on central throat, belly, lower flanks, and vent; some feathers on sides of breast with indistinct dull black centres, sometimes showing as small dark patch; underparts with sharp, narrow black-brown shaft-streaks, each streak generally less than 4 mm long and less than 0·3 mm wide; vent, lower flanks, and sometimes belly unstreaked. Under tail-coverts contrastingly black, except for pale buff bases. Tail brownish-black with faint green gloss; no pale subterminal spots (unlike *H. rustica*), but sometimes a faint grey or white spot on middle of inner web of t6 or occasionally on other feathers (Hartert 1910; Bub and Herroelen 1981), especially in ♀ (Witherby *et al.* 1938*b*). Flight-feathers, tertials, and upper primary coverts brownish-black, glossed blue-green on tertials, secondaries, and greater primary coverts, faintly green elsewhere; remaining upper wing-coverts black with blue-green or dark metallic-blue gloss, deepest on lesser coverts; small coverts along leading edge of wing fringed buff. Under wing-coverts and axillaries buff or pink-buff, narrowly streaked and mottled grey on lesser coverts. *In worn plumage*, some dusky black may show on central hindneck (slightly interrupting half-collar); mantle and scapulars streaked cream; rufous rump-patch bleached to cream distally and at sides; ground-colour of underparts paler buff, small dark patch on sides of breast and narrow shaft-streaks from chin to chest more distinct; tail, flight-feathers, and upperwing brownish-black, hardly glossy. Sexes similar, even in measurements. NESTLING. Down long and soft, restricted to tufts on forehead, neck, and back; for development, see Simeonow (1969). JUVENILE. Rather similar to adult, but crown (particularly), mantle, and scapulars browner and less glossy, feather-tips narrowly fringed buff when fresh. Half-collar on sides and rear of neck paler, rufous-buff; narrow supercilium less distinct, sometimes hardly deeper coloured than lores and ear-coverts; back, rump, and shorter upper tail-coverts paler,

rufous-buff or buff, longer upper tail-coverts brown-black with pale buff tip. Ground-colour of underparts paler buff, dark shaft-streaks less pronounced and virtually absent below chest; dark patches at sides of breast slightly more distinct. Tail as adult, but duller black, less glossy, and t6 less elongated. Flight-feathers, tertials, and upper wing-coverts duller than adult, only slightly glossed green when plumage fresh; secondaries and inner primaries with faint pale edges along tips; tertials and tertial coverts with distinct rufous-buff to pale buff fringes on tips, tips of greater, median, and lesser upper wing-coverts with narrower buff fringes, soon worn off. FIRST ADULT. Similar to adult; indistinguishable when last juvenile upper wing-coverts (with traces of buff fringes) or relatively short juvenile t6 replaced.

Bare parts. ADULT. Iris brown or dark brown. Bill black or brown-black; mouth grey or pale yellow-grey. Leg and foot brown or black with red-brown or slight flesh-pink tinge. NESTLING. Bare skin flesh-pink. Bill dark grey, large gape-flanges yellow-white. Leg and foot greyish. (Neufeldt 1970.) JUVENILE. Like adult, but mouth yellow (except for tips of mandibles) and leg and foot flesh-brown or brown. (Hartert 1910; Dementiev and Gladkov 1954*b*; Ali and Ripley 1972; RMNH, ZMA.)

Moults. ADULT POST-BREEDING. Complete; primaries descendant. Few details; probably closely similar to *H. rustica*, with migratory late-nesting populations (*rufula*, nominate *daurica*, *japonica*) starting just before migration or after arrival in winter quarters, and non-migratory southern early-nesting populations starting early. In *rufula*, moult not started by late September, but all birds fresh by February (Svensson 1984*a*). In western Tien Shan, September to mid-October, 41% of 44 migrants had started primary moult, up to p2–p3 in moult (rarely p1 only or up to p5); occasionally, also some upper wing-coverts or tertials replaced (Gavrilov 1971). In USSR, single ♂ on 16 July had started with inner primaries and some feathers of body, but single ♀ 28 July less advanced (Dementiev and Gladkov 1954*b*). In Ethiopia, single ♀ *rufula* had just started primary moult in late September (BMNH). In Indian winter quarters, single *rufula* from November had primary moult score 15 with remainder of feathering old, except part of belly and lesser upper wing-coverts (ZMA). Virtually non-migratory Indian races in full moult June–August (BMNH). POST-JUVENILE. Complete; primaries descendant. A single *rufula*, India, not yet started November; another with primary moult score 27 mid-winter (t1 and body new, t3–t5 and much of head old, t2 and t6 growing) (BMNH, RMNH). 2 *japonica* in winter quarters, India, just started December, others started January; still some moult end April (Vaurie 1951*c*). Also, recorded starting late July (Witherby *et al.* 1938*b*), and hence apparently much variation in timing, as in post-juvenile of *H. rustica*, depending on date of fledging and migration strategy.

Measurements. *H. d. rufula*. ADULT. Southern Europe, Asia Minor, Cyprus, and Levant, March–July; skins (BMNH, RMNH, ZMA). Tail is t6; fork is tip of t1 to tip of t6. Bill (S) to skull, bill (N) to distal corner of nostril; exposed culmen on average 4·4 less than bill (S), 1·9 more than bill (N).

	♂		♀	
WING	124·5 (2·10; 31)	120–128	121·4 (2·48; 22)	118–127
TAIL	101·9 (3·97; 29)	94–107	95·6 (4·18; 18)	89–99
FORK	59·3 (3·32; 30)	53–64	52·0 (2·94; 18)	46–56
BILL (S)	11·2 (0·50; 13)	10·4–12·1	10·9 (1·02; 13)	9·3–12·5
BILL (N)	5·2 (0·35; 14)	4·7–5·7	5·1 (0·26; 12)	4·8–5·5
TARSUS	14·1 (0·38; 13)	13·5–14·7	13·8 (0·40; 10)	13·2–14·3

Sex differences significant for wing, tail, and fork.

Southern Spain, breeding (Lope Rebollo 1980):

WING ♂ 118·5 (2·3; 28) 114–123 ♀ 116·7 (3·3; 25) 111–125
TAIL (t6) 99·3 (4·5; 28) 90–109 92·5 (7·8; 25) 68–102
TAIL (t1) 43·8 (2·3; 21) 41–48 42·3 (3·0; 20) 39–46

Sex differences significant for tail (t6). Wing of eastern birds averages shorter: (1) southern Iran, Afghanistan, Arabia, and western Pakistan (Paludan 1959; BMNH); (2) USSR, east from eastern Turkmeniya (Dementiev and Gladkov 1954*b*).

(1) ♂ 117·8 (3·17; 13) 113–124 ♀ 118·9 (3·52; 9) 112–123
(2) 116·8 (— ; 24) 112–124 114·5 (— ; 19) 112–122

Nominate *daurica*. Central and south-east Siberia, north-west Mongolia, and mountains of central China, summer; skins (Vaurie 1955*c*; RMNH, ZMA).

WING ♂ 128·5 (2·55; 22) 124–133 ♀ 125·1 (1·89; 15) 121–128

JUVENILE. Wing 120·5 (13) 116–125; tail 77·8 (13) 71–84; fork 37·3 (13) 30–42, exceptionally 57 (BMNH, RMNH, ZMA).

Weights. *H. d. rufula*. Southern Spain, breeding adults: ♂ 22·1 (2·2; 27) 19–29, ♀ 22·4 (1·6; 25) 20–28 (Lope Rebollo 1980). West Germany, May: ♀ 19·5 (Mester and Prünte 1965*b*). Luxembourg, August: 20 (Schmitz 1980). Malta, on spring migration: 20·7 (2·6; 24) 17–26 (J Sultana and C Gauci). South-east Morocco, first half of April: 17·1 (1·79; 7) 15·2–19·9 (BTO). Greece, May: ♀ 24 (Niethammer 1943). Asia Minor, July: ♀ 19 (Kumerloeve 1968). Cyprus, April: 21·4 (Hallchurch 1981). Zagros mountains (western Iran), March: ♂ 19·4, ♀ 19·1 (Paludan 1938). Afghanistan: early June, ♂ 18; late July, ♀ 19 (Paludan 1959). Southern USSR: ♂♂ 20·5, 21·5; ♀ 17·8 (4) 15·5–19·2 (Dementiev and Gladkov 1954*b*). Kirgiziya, USSR: 18–22 (Yanushevich *et al.* 1960). Kazakhstan (USSR), May–June: ♂♂ 19, 19·8, 20; ♀ 18·5 (Dolgushin *et al.* 1970). In Bulgaria, on average 1·8 on 1st day (*n*=9); 20·9 at fledging on 20th–21st day (*n*=6) (Simeonow 1969, which see for growth curves).

Nominate *daurica*. USSR: ♀ 33 (Dementiev and Gladkov 1954*b*).

H. d. japonica. Manchuria, August: ♂ 27, ♀ 25 (Piechocki 1958).

Structure. Wing long, narrow at base; tip slightly blunter than *H. rustica*. 10 primaries: p9 longest (rarely, up to 1 shorter than p8), p8 0–4 shorter (usually 1–2), p7 6–10, p6 15–19, p5 23–27, p4 32–36, p1 55–61; p10 reduced, narrow and pointed, 80–86 shorter than p9, 6–11 shorter than longest upper primary covert. Tail long, deeply forked; 12 feathers, length of t1 39–47, depth of fork (t1 to t6) 45–62 in adult, 34–42(–57) in juvenile. T6 broader than in *H. rustica*, less attenuated, tip more broadly rounded, sometimes slightly curved inward; width of t6 (at tip of t5) 3·8 (4) 3·5–4·5 in adult ♂, 4·5 (4) 4–5 in adult ♀, 5·1 (5) 4–6 in juvenile. Bill slightly thicker and shorter than in *H. rustica*, tarsus slightly longer and thicker. Middle toe with claw 15·0 (14) 14·0–16·2. Outer and inner toe with claw both *c.* 63% of middle toe with claw, hind toe 82%; claws slightly heavier than *H. rustica*. Remainder of structure as in *H. rustica* (p. 277).

Geographical variation. Marked and complex. Involves wing length, relative length of t6, colour of underparts and rump, width of streaking on underparts and rump, and presence or absence of pale half-collar round hindneck and dark patch on sides of body near thigh. *H. d. rufula* from southern Europe and North Africa east to Iran is rather large, t6 long, tail

deeply forked; rufous collar round hindneck fully developed; rump unstreaked, rufous-cinnamon merging into buffish or cream towards rear and sides; underparts pale rufous with narrow streaks (less than *c.* 0·3 mm in width), usually not extending to lower flanks and vent; no dark patch near thighs. Ground-colour gradually paler towards east and size gradually smaller; populations from southern USSR (eastern Turkmeniya to Tien Shan mountains) and from Afghanistan and Baluchistan to Kashmir sometimes separated as *scullii* Seebohm, 1883, on account of smaller size (see Measurements), but fair degree of overlap and included in *rufula* here. Nominate *daurica*, occurring from upper Irtysh river east to Amur basin and south through Mongolia and central China to western Szechwan, larger than *rufula*, wing and t6 slightly longer (wing 121–138, t6 93–119: Hartert 1910; Vaurie 1959), fork slightly deeper; differs from *rufula* in blue-black central hindneck (collar interrupted), less extensive white or cream on feather-bases of mantle and scapulars, and slightly longer and broader streaks on underparts (on breast and belly, *c.* 0·5 mm wide and 5–6 mm long). *H. d. japonica* from Japan, Korea, and eastern China south to Yangtze Kiang river has slightly shorter wing than *rufula* (114–125: Vaurie 1959), but t6 and fork similar; collar on hindneck broadly interrupted, rump rufous with faint shaft-streaks, underparts heavily streaked (on chest, streaks *c.* 1 mm wide). Indian races smaller than more northern ones (wing 102–123); t6 short (70–82) and fork relatively shallow in races from plains of India and Sri Lanka; half-collar round neck generally well-developed; underparts vary between pale rufous with marked streaks (Himalayas) and unstreaked deep rufous-cinnamon (Sri Lanka). For Afrotropical races, see Mackworth-Praed and Grant (1960, 1973).

Forms superspecies with Striated Swallow *H. striolata* from southern Assam, southern China, and Taiwan south to Greater and Lesser Sunda Islands, but strong arguments for combining *H. striolata* and *H. daurica* in single species: *H. striolata* considered a separate species because a small and poorly streaked race of *H. daurica* north and west from Assam is replaced by a large and well-streaked race of *H. striolata* south and east from Assam; apparently no intergradation between these populations; also, tropical populations of a single species would normally be expected to be smaller than temperate ones, not larger as here (Stanford 1941). No clear morphological differences between *H. striolata* and *H. daurica*, however (possible ecological or behavioural differences not yet studied); generally, *H. striolata* larger, collar on hindneck interrupted, rump rufous with narrow streaks, underparts heavily streaked, and black patch near thighs prominent, but some races of *H. daurica* (especially *japonica*) show some or all of these characters, and some races of *H. striolata* (e.g. from Malay peninsula) show many characters of *H. daurica* (Vaurie 1951*c*). *H. d. japonica* fully intermediate between nominate *daurica* and *H. striolata*, and in fact 2 lines distinguishable: one leading from nominate *daurica* through *japonica* and populations of Taiwan and Philippines to typical *H. striolata* in Indonesia—all about similar in size (highland populations slightly larger) but with black pigmentation increasing towards south-east (gradually more extensive black on hindneck and near thigh, gradually broader streaks on rump and underparts); another line from nominate *daurica* through *rufula* and Indian races to Sri Lanka, with size gradually smaller towards south and with rufous pigmentation gradually increasing and black streaks reduced; isolated races from south-east Asia could be old derivatives from either one of these lines. CSR

Delichon urbica House Martin

Du. Huiszwaluw   Fr. Hirondelle de fenêtre   Ge. Mehlschwalbe
Ru. Городская ласточка   Sp. Avión común   Sw. Hussvala

Hirundo urbica Linnaeus, 1758

Polytypic. Nominate *urbica* (Linnaeus, 1758), Europe and North Africa east to western Siberia, Tien Shan, Kashmir, and Spiti (India). Extralimital: *lagopoda* (Pallas, 1811), central and eastern Asia from Yenisey basin, Altai mountains, and northern Mongolia east to Manchuria, Sea of Okhotsk, and Anadyrland; *dasypus* (Bonaparte, 1850), Korea, Ussuriland, Sakhalin, Kuril Islands, and Japan; *nigrimentalis* (Hartert, 1910), south-east China and Taiwan; *cashmeriensis* (Gould, 1858), Himalayas from Pakistan to south-west China.

Field characters. 12·5 cm; wing-span 26–29 cm. 10% smaller than Swallow *Hirundo rustica* (ignoring latter's tail-streamers), but head and body appear hardly less bulky due to proportionately shorter wings and tail. Medium-sized, bull-headed hirundine, with distinctly forked tail. Upperparts blue-black with broad white rump; underbody white. Flight less rapid and often at greater height than *H. rustica*. Call distinctive. Sexes closely similar; some seasonal variation. Juvenile separable at close range.

ADULT MALE BREEDING. From April, upperparts deep metallic blue-black apart from broad white rump (though tail dull black). Underbody pure white, contrasting vividly with dark upperparts, dull grey under wing-coverts, and dusky undersides to flight- and tail-feathers. ADULT FEMALE BREEDING. Underbody usually duller than ♂, washed greyish. ADULT NON-BREEDING. From August, white of rump, cheeks, throat, and flanks often mottled brownish-white. Small dark patches extend down from shoulders. Winter bird thus often lacks vivid contrast typical of summer one. JUVENILE. Duller above than adult, with marked grey-brown tone on head and noticeably brown flight- and tail-feathers. Nape often shows white flecks. At all times, bill black, feet pink-flesh; tarsi fully feathered white.

Breeding adult unmistakable but juvenile and non-breeding adult not so, inviting confusion in Europe with Red-rumped Swallow *H. daurica* (see that species, p. 278) and in Afrotropics with Grey-rumped Swallow *Pseudhirundo griseopyga* (at least 10% slighter in build, with elongated tail-streamers on adult and fully pale grey-brown rump at all times). Confusion with *H. daurica* much increased by occasional hybridization of *D. urbica* with *H. rustica* (see *H. daurica*, p. 279). Flight less fast, more steady, and much less swooping and twisting than *H. rustica*; regularly flies higher than *H. rustica* and Sand Martin *Riparia riparia*, even joining swifts *Apus*. Wing-beats less flashing, more stroked than *H. rustica*. Perches and clings; waddles on ground more easily than *H. rustica*, often raising wings and tail to do so.

Song soft and sweet, a long twitter of melodious chirps. Contact-call a hard but merry 'chirrrp'; in alarm, a shrill, plaintive 'tseep'.

Habitat. Breeds sparsely from subarctic and boreal, and more abundantly through temperate to steppe and Mediterranean zones of west Palearctic, in oceanic as well as continental climates, tolerating wet, windy, and chilly conditions but usually avoiding extremes of temperature, and vulnerable to their effects on insect prey. In suitable weather, tends to forage in airspace above lowest levels favoured by Swallow *Hirundo rustica*, but spends rather less of day in flight and, at least in north of range, seems to range less far from nest. Although transition from primitive rock-nesting to general use of buildings, bridges, and other artefacts is virtually complete over much of Europe, in some regions nesting on natural rock-faces with suitable surfaces and pitches remains locally common, e.g. in Switzerland, although one of highest sites is on a building at nearly 2200 m. As on buildings, however, protection from above by some overhang is usually sought (Glutz von Blotzheim 1962). In USSR, within west Palearctic, also nests mainly on buildings of brick or stone (sometimes of wood). Sea-cliffs also used, and some breed on rock-faces in mountains up to 2000–3000 m, feeding over glades and slopes or along rivers (Dementiev and Gladkov 1954*b*). In Cyprus, majority prefer cliffs and rocks at all elevations (Bannerman and Bannerman 1971). In western Europe, much more ready than *H. rustica* to live in large cities, not only in suburbs but, where air clean, even in centre, e.g. in London (E M Nicholson). Often mixes with other Hirundinidae, both at breeding places and in foraging, but tends to concentrate round sites often fixed by long usage, leaving wide areas unvisited. Presence of water often an attraction, although not essential. Like *H. rustica*, avoids densely wooded and arid areas, but differs in accepting mountainous and coastal as well as city habitats, and in remaining more loyal to primitive rocky nesting sites. Normally does not use reedbeds for roosting (see Social Pattern and Behaviour). General infrequency of observations of birds wintering in Africa is ascribed to their congregating at high altitudes and feeding high above ground. Commensalism with man generally accepted, and must have much increased numbers and extended range within historic times.

Distribution. Has bred irregularly in South Africa (McLachlan and Liversidge 1970).

SVALBARD. May have bred once (IB). ICELAND. Bred 1966 and 1977 (AP). FAEROES. Bred 1956, 1966 (Bloch and Sørensen 1984). BRITAIN. Inner London re-colonized, due to cleaner air (Cramp and Gooders 1967), and still spreading there (S Cramp). ISRAEL. First bred *c.* 1970 (HM, UP). LIBYA. Breeding suspected (Bundy 1976). MALTA. Bred 1981-2 (Sultana and Gauci 1982; JS, CG).

Population. Fluctuating. Marked decrease in Netherlands and (probably) Sweden; slight increase in Denmark.

BRITAIN, IRELAND. Widely reported to be decreasing but evidence unsatisfactory; recent increase reported Ireland (Parslow 1973). 300 000-600 000 pairs; population stable in study area of 1370 km² (Bouldin 1959, 1968). FRANCE. Under 1 million pairs (Yeatman 1976). BELGIUM. About 34 000 pairs (Lippens and Wille 1972). Brussels: in region to south and east 750 intact nests found; recent tendency to decrease (Walravens and Langhendries 1985). NETHERLANDS. 71 000-103 000 pairs 1966-70, *c.* 77 000 pairs 1976-7 (Teixeira 1979). Decline continues (Wammes *et al.* 1983; CSR). WEST GERMANY. 1·2-4·4 million pairs (Rheinwald 1982). See Hölzinger (1969) for counts in Ulm area and Lenz *et al.* (1972) and Witt (1984*a*) for West Berlin. DENMARK. Fluctuating; slight increase 1978-82 (Vickholm and Väisänen 1984). SWEDEN. About 400 000 pairs (Ulfstrand and Högstedt 1976); decrease in numbers ringed (Österlöf and Stolt 1982). FINLAND. About 120 000 pairs (Merikallio 1958); increased in north 1941-77 (Väisänen 1983); fluctuating. CZECHOSLOVAKIA. No change reported (KH). SWITZERLAND. Decline

Thoune area (Gunten 1957). Declined 1974-5 (Bruderer and Muff 1979). Increased 1978-83 (Vickholm and Väisänen 1984).

Mortality. West Germany: average annual mortality 36·2% (Rheinwald and Gutscher 1969); near Stuttgart, *c.* 50% (Hund and Prinzinger 1979*a*). Britain: average annual mortality 57%; double-brooded ♀♀ survive less well than single-brooded (Bryant 1979). Czechoslovakia: average mortality 65·5% in 1st year; up to 7th year, lowest mortality (55·1%) occurs in 3rd year (Beklová 1976). Switzerland: average annual mortality 67% (Gunten 1963; Menzel 1984). Oldest ringed bird 14 years 6 months (Rydzewski 1978).

Movements. Migratory. Winter records exist from Mediterranean basin and western Europe, north to Britain and Ireland, but west Palearctic and west Asian birds otherwise winter wholly in Afrotropics.

A good many passage ringing recoveries in Europe and North Africa, though few in Afrotropics (see Zink 1975). Trans-Europe passage essentially south in autumn and north in spring, without strong south-west/north-east movement via Iberia. Hence birds from North Sea countries (including Britain) recovered more often in France than Spain, Scandinavian and German birds reach Mediterranean (in autumn) via southern France and Italy, and Finnish and east Baltic birds move in direction of Greece.

Within Afrotropics, very wide scatter of observations indicates extensive size of winter range there, including humid zone north of equator, but such sightings few in relation to huge numbers of birds which must be involved (Moreau 1972). This presumably due to highly aerial

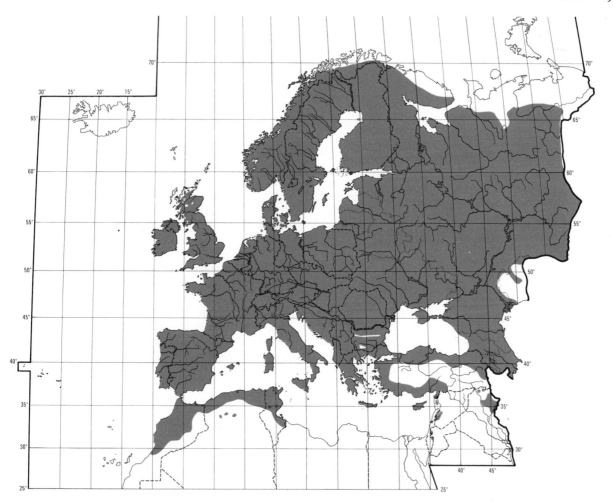

lifestyle, with high-altitude foraging like swifts (Apodidae) (Verheyen 1952). Seldom reported at passage hirundine roosts in Europe, and only one such (dubious) observation in Africa (see Social Pattern and Behaviour). Consistently common, to ground-based observers, only in highlands such as western Zimbabwe (Snell 1963), Kenya, and northern Tanzania (Britton 1980). Elsewhere, numerous at low altitude only in abnormal conditions such as when bush-fires provide good feeding opportunities (e.g. Douaud 1956, for Togo). In cold-weather mortality incidents in South Africa, more *D. urbica* picked up than Swallow *Hirundo rustica*, a reversal of relative numbers seen in flight locally (MacLeod *et al.* 1953; Skead and Skead 1970). Only 12 ringing recoveries south of 20°N: west European birds in West Africa (4), and central European and Scandinavian birds in northern Angola (1) and southern Africa (Zambia to Cape Province, 7). No recoveries in East Africa, but one ringed in Kenya found in Caucasus (USSR). Hence suggestion of longitudinal separation in Africa, but sample very small. A broad-front migrant, and appears to have same extensive trans-desert passages (over Sahara and Middle East) as *H. rustica*.

A relatively late migrant in autumn, since late broods often still in nest in August or early September. Main southerly movement through Europe September–October, with significant minority still transient in November. Return movement begins Africa in March, and first birds can reach Europe late March or early April, but main return to breeding areas in second half of April and first half of May. A few summer in Afrotropics, and breeding has occurred sporadically in Namibia and South Africa.

Central and east Siberian race *lagopoda* believed to winter Assam to Indo-China. *Dasypus* group of races (breeding Himalayas and south-east Asia north to Sakhalin) also winters in southern and south-eastern Asia. (Vaurie 1959.)   RH

Food. Almost wholly flying insects; in breeding season, especially flies (Diptera) and aphids (Aphidoidea). Prey taken almost entirely by aerial-pursuit, though many reports of birds feeding while perched on ground or trees–also (presumably while perched) from walls, rock faces, and reeds (e.g. Sage 1956b, Bilby 1957, Möhring 1958, Glutz von Blotzheim 1962, Trevor 1965, Paszkowski

1969, Schulze 1971, Harvey 1973, *Br. Birds* 1952, **45**, 69). Often hunts at greater height than Swallow *Hirundo rustica*, especially when in mixed flocks (Witherby *et al.* 1938*b*; Voipio 1970; Gatter 1976). Insect flying past normally taken from below, bird shooting steeply up with rapid wing-beats to catch it; then usually glides down to previous height (Menzel 1984). While actually foraging, proportion of time spent flapping $49 \cdot 9 \pm \mathrm{SD} 10 \cdot 4\%$, $n = 28$ (Bryant and Turner 1982). Recorded following tractor ploughing field covered with tall weeds and taking insects disturbed (Schmidt 1967*b*), and taking Lepidoptera larvae hanging from trees on threads (Paszkowski 1969). In southern Asia, rarely visits grass fires to feed on insects disturbed, unlike many other Hirundinidae; recorded visiting fires in England (Bilby 1957; Paull 1968). Injuries to prey indicate insects seized with snap of bill on abdomen or thorax (Gunten 1961). Prey for young collected in throat-pouch to form compacted bolus (see final paragraph), though long-winged insects often carried in bundles in bill (Menzel 1984). Near Stirling (Scotland), birds fed up to 2 km from nest, but mostly up to $0 \cdot 75$ km, average $0 \cdot 45$ km, $n = 45$ (Bryant and Turner 1982).

Diet in west Palearctic includes the following. Mayflies (Ephemeroptera), damsel flies (Odonata: Lestidae, Coenagriidae), stoneflies (Plecoptera), grasshoppers, etc. (Orthoptera), Psocoptera, bugs (Hemiptera: Lygaeidae, Reduviidae, Nabiidae, Miridae, Saldidae, Corixidae, Cixiidae, Cicadellidae, Delphacidae, Psylidae, Aphidoidea), lacewings, etc. (Neuroptera: Hemerobiidae, Chrysopidae), adult and larval Lepidoptera (Hepialidae, Incurvariidae, Tortricidae, Gelechiidae), caddis flies (Trichoptera), adult flies (Diptera: Tipulidae, Psychodidae, Culicidae, Chironomidae, Ceratopogoniidae, Simuliidae, Bibionidae, Scatopsidae, Mycetophilidae, Cecidomyiidae, Stratiomyidae, Rhagionidae, Tabanidae, Empididae, Dolichopodidae, Lonchopteridae, Phoridae, Pipunculidae, Syrphidae, Otitidae, Trypetidae, Lonchaeidae, Lauxaniidae, Sciomyzidae, Psillidae, Sepsidae, Helomyzidae, Anthomyzidae, Opomyzidae, Ephydridae, Drosophilidae, Agromyzidae, Milichidae, Chloropidae, Tachinidae, Calliphoridae, Muscidae), adult Hymenoptera (sawflies Tenthredinidae, Diprionidae, Ichneumonoidea, Cynipoidea, Chalcidoidea, ants Formicidae, bees Apoidea), adult beetles (Coleoptera: Cicindelidae, Carabidae, Dytiscidae, Hydrophilidae, Staphylinidae, Scarabaeidae, Elateridae, Cantharidae, Nitidulidae, Coccinellidae, Bruchidae, Chrysomelidae, Curculionidae, Scolytidae), spiders (Araneae) (Csiki 1904; Gerber 1953; Gunten 1961; Arkhipenko *et al.* 1968; Bryant 1973; Smogorzhevski and Kotkova 1973; Ko[ž]ená 1975). Takes flying insects down to $1 \cdot 1$ mm long (Menzel 1984); see also final paragraph. In September, group of 5 birds in hawthorn *Crataegus* tree recorded picking at and apparently eating berries (Miller 1972); no other records of plant material.

No detailed studies on diet of adults in breeding season, most information coming from studies of young

(see below). In Avila (Spain), 2 stomachs of adults contained beetles and Hemiptera (Gil 1927). In Hungary, several thousand birds said to have cleared large maize field of aphids in 2 days (Gerber 1953). In winter in Zaïre, takes flying ants and other insects (Chapin 1953).

Free-living adults (feeding broods of 4) in central Scotland themselves consumed average 135 kJ per bird per day, of which 95 kJ utilized (Bryant and Westerterp 1980*a*; see also Bryant and Westerterp 1980*b*).

In southern England, items in faeces of young over 7 days old (also some from adults), May–September, comprised $59 \cdot 5\%$ (by number) Diptera, $17 \cdot 8\%$ aphids, $10 \cdot 6\%$ Parasitica (Hymenoptera), $5 \cdot 0\%$ beetles, $2 \cdot 7\%$ ants, and $2 \cdot 4\%$ others; Bibionidae especially important August–September (average $42 \cdot 3\%$ of those 2 monthly totals), and aphids declined from $45 \cdot 8\%$ in May to $4 \cdot 4\%$ in August–September (sample sizes not given but apparently *c.* 1200 per month, *c.* 6000 overall); in neither of these 2 prey groups did proportions taken follow their changing availabilities from month to month; prey items tended to be larger than the average available (Bryant 1973, data corrected; see also for relative energy contribution of different food groups). In Switzerland, 97 189 food items in collar-samples from young comprised $45 \cdot 4\%$ (by number) Diptera, $33 \cdot 1\%$ aphids, $7 \cdot 2\%$ other Hemiptera, $8 \cdot 1\%$ mayflies, stoneflies, and caddis flies, $2 \cdot 6\%$ Hymenoptera, $1 \cdot 6\%$ beetles, $1 \cdot 0\%$ Psocoptera, and $1 \cdot 0\%$ others; diets of 1st and 2nd broods broadly similar (Gunten 1961; see also for variation in diet with time of day). In Krkonoše mountains (Poland), July, 3612 items in collar-samples comprised $57 \cdot 6\%$ (by number) Hemiptera (largely aphids), $31 \cdot 9\%$ Diptera, $3 \cdot 9\%$ beetles, $3 \cdot 0\%$ Hymenoptera, and $3 \cdot 6\%$ others; in August, 186 items included $48 \cdot 4\%$ Diptera and $24 \cdot 7\%$ Hemiptera; average body size of food items $3 \cdot 5$ mm (1–15), $n = 3798$; significantly fewer Diptera and more Hemiptera taken in morning than in afternoon (Ko[ž]ená 1975). In Moscow region (USSR), 1840 items brought to young comprised $53 \cdot 2\%$ (by number) Hemiptera, $17 \cdot 3\%$ Diptera, $12 \cdot 5\%$ mayflies, $8 \cdot 2\%$ beetles, $6 \cdot 0\%$ ants, $1 \cdot 6\%$ Lepidoptera, $1 \cdot 1\%$ spiders, and $0 \cdot 3\%$ Orthoptera; items mostly 2–8(-22) mm long. Diet of nearby *H. rustica* markedly different (though sample size small), involving mostly larger items: 69% Diptera, 18% ants, and 12% Hemiptera ($n = 177$); items mostly 2–14(-18) mm long. (Arkhipenko *et al.* 1968.) Grit (small stones, snail shells, earth) comprised $0 \cdot 18\%$ (by number) of 97 361 items in collar-samples from Switzerland; 2–4% of food boluses brought to young contained, or were composed entirely of, grit (Gunten 1961); found in $8 \cdot 4\%$ of 131 samples from Poland, but only from young over 6 days old (Ko[ž]ená 1975); see also Oeser (1966). When feeding young, said to take prey selectively, concentrating within each hunting flight on items of similar size, though preferred size may switch between flights (Gunten and Schwarzenbach 1962); not clear whether this result might actually be due to shifts in hunting area resulting in changed availability of the different size-classes. See Table A in *H. rustica* (p. 267) for data on size of food boluses. Over whole day, average 53 insects per bolus ($n = 1570$), but average 102 per bolus ($n = 253$) 05.00–08.00 hrs (Gunten 1961); range 1–388, $n = 618$ (Gunten and Schwarzenbach 1962); see also Kožená (1975). Unlike in *H. rustica*, bolus weight linked more strongly to distance of food source than to weather: significantly positively correlated with distance of food source, time between nest-visits, time spent actually foraging (see below), and daily rainfall; negatively correlated with maximum daily temperature; not correlated with temperature, prey weight, or food abundance (Bryant and Turner 1982, which see for further optimal foraging data). Each brood of 4, central Scotland, received average 225·5 kJ per day, of which 157·5 kJ utilized (Bryant and Westerterp 1980a). Adults tend to feed nestlings in bouts, separated by periods of *c.* $\frac{1}{2}$ hr of infrequent feeding (Lind 1960; Bryant and Westerterp 1980a). In central Scotland, 5·88 ± SD2·12 min ($n = 51$) between nest-visits, of which 4·45 ± SD2·11 min ($n = 51$) spent actually foraging (Bryant and Turner 1982; see also Moreau and Moreau 1939b). Feeding rate increases with brood size (Lind 1960).   DJB

Social pattern and behaviour. Based largely on studies by Lind (1960, 1962, 1963, 1964) in Finland. For recent review, see Menzel (1984).

1. Highly gregarious throughout the year. Flocks of several hundred common on migration and in winter quarters. In Zaïre, April, migrant flocks of 30–200 birds (Verheyen 1952). After breeding, adults and juveniles from different colonies typically form flocks prior to migration. In Britain, post-breeding flocks July–October. In southern Finland, start flocking late July, reaching maximum size (up to 400 birds) mid-August. For association with other Hirundinidae in winter quarters, see (e.g.) Chapin (1953) and Harper and Harper (1974). BONDS. Pair-bond monogamous, typically of seasonal duration or shorter. Of 152 marked pairs, 5 pairs controlled at same colony the following year, all 10 birds with new partners (Gunten 1963). Exceptionally, pair-bond renewed in subsequent years; once, same 2 birds paired in alternate years, taking different mates in intervening year (Hund and Prinzinger 1979b). Birds occasionally change mate during breeding season, sometimes after serious disturbance; of 32 pairs controlled during 1st and 2nd broods, 5 birds changed mate for 2nd brood (Gunten 1963). According to Lyuleeva (1974), regularly change mate for 2nd brood, and this likely if 1st fails (Lyuleeva 1974; Hund and Prinzinger 1979b). After mass mortality due to severe weather in June, West Germany, survivors re-paired and bred again the same season (Löhrl 1971). 3 cases of siblings paired (Rheinwald 1977; Hund and Prinzinger 1979b), and 1 of ♂ probably paired with daughter; inbreeding attributed to high site fidelity (Rheinwald 1977; see Breeding Dispersion, below). Numerous reports of hybridization with Swallow *H. rustica* (e.g. Broad 1977, Roberts 1980a, Grech 1981–3); one hybrid attempted to copulate and repaired a nest, but did not breed (Elsner 1951); for summary, and photograph of *D. urbica* copulating with *H. rustica* on ground, see Charlwood (1973). One case of hybrid with Sand Martin *Riparia riparia* (Myrbach 1975). Both sexes share brooding and feeding of young. Young

of late broods reported helping to feed young of earlier ones (Witherby *et al.* 1938b), but no such records subsequently. After fledging, young continue to receive food from parents for a few days, and usually roost with parents in nest until departure from colony (Lind 1960, but see Roosting, below). Age of first breeding 1 year (Gunten 1963). Individuals pair preferentially with others of same age (Rheinwald *et al.* 1976; Hund and Prinzinger 1985; see also Heterosexual Behaviour, below). BREEDING DISPERSION. Typically in colonies, divided into sub-colonies. Solitary pairs usually 'satellites' to main colony (Landmann and Landmann 1978); colony thus defined as all nests within 50 m of main concentration of nests (Lind 1960). In much of range, most colonies less than 5 pairs; exceptionally, more than 500 pairs, e.g. in Oxfordshire, England (Brucker 1976), and Sliwen, Bulgaria (Ernst and Thoss 1975). In following account, nests known (or presumed) to have been occupied. In east Lancashire (England), of 620 colonies over 10 years, average 3·9–4·9 nests per colony; 74·4% of 1–5 pairs, 15·2% 6–10, 5·3% 11–15, 2·9% 16–20, 1·0% 21–30, 1·3% 31–65 (Bouldin 1968, 1971, which see for influencing factors). Similar distribution reported for Nordtirol, Austria (Landmann and Landmann 1978), also for Auerbach (East Germany), where average size of 126 colonies 4·85 nests, with 37% of 1 nest, 48% 2–3, maximum 34 (Ernst and Thoss 1975). In Manchester area (England), average of 32 colonies 2·18 nests (Cramp and Ward 1934). In West Berlin, average in 2 years 19·8 pairs ($n = 47$ colonies) and 25·2 ($n = 49$), maximum 124 (Lenz *et al.* 1972). For similar results in Kiel (West Germany), see Schwarze (1975). Cliff colonies tend to be larger than those in urban areas. In Europe, colonies on cliffs mostly 10–50 nests, exceptionally 1 nest or more than 100 (Creutz 1961, which see for review); in Britain, 38 cliff colonies averaged 21 nests (6–36) (Clark and McNeil 1980). At Rügen (East Germany), 21 cliff colonies averaged 35 pairs (Plath 1977). For cliff colonies in Finland, see Lind (1960). Colonies often vary markedly in size between years: e.g. in Oxfordshire, one colony varied from 0 to 513 nests over 25 years, another from 14 to 170 over 14 years (Brucker 1976). For annual variation in Switzerland, see Bruderer and Muff (1979); for Belgium, see Rappe (1978). Density difficult to compare between studies due to variation in size of sample areas. In 11 areas of Britain, average 1·6 pairs per km² (Boyd 1936); in *c.* 10 km², Bruton (Somerset, England), *c.* 67 pairs per km² (Hollom 1930); in London, *c.* 1 pair per km² in 1949, *c.* 10 pairs per km² in 1974 (Strangeman 1975); in Oxfordshire, 10·5 pairs per km² in urban areas, 3·7 in rural areas (Alexander 1933); in Manchester 14·8 pairs per km² (Cramp and Ward 1934); over 7 years, eastern Lancashire, average *c.* 2 pairs per km², 3 if unsuitable areas excluded (Bouldin 1968). Over 3 years, Auerbach, 14–25 pairs per km² in villages, 0·6–2·1 pairs per km² over whole area (Ernst and Thoss 1975). For densities in Ulm (Bayern, West Germany), see Hölzinger (1969); for West Berlin, see Lenz *et al.* (1972). In 3 years, northern Denmark, *c.* 1 pair per km² (Møller 1974a). According to site, nests may be widely spaced or tightly packed, e.g. up to 14 clustered together in optimal sites (Plath 1977). No data on average distance between nests, but shortest distance between 2 entrance holes 8–10 cm. Defends nest-area territory of radius *c.* 10 cm around entrance, though where only 1 nest on a wall, whole length of wall may be defended, aggression waning with distance from nest (Lind 1960). Size of nest-area territory thus differs with spacing. Established breeders show marked year-to-year fidelity to nest-site, or its vicinity: in study of *c.* 65 pairs, Thunersee (Switzerland) of 65 controlled birds, 12% returned to breed in same nest, 32% on same house, 56% on a different house (Gunten 1963). ♀♀ less site-faithful than ♂♂, changing nest, colony, or village more frequently, especially after breeding failure. Pairs typically use same nest for 1st and 2nd broods, but more likely change site if 1st brood fails (Hund and Prinzinger 1979b). 1-year-olds, especially ♂♂, also relatively faithful to birthplace (Hund and Prinzinger 1979b). For settling distances from natal colony, see Lyuleeva (1974). On average over 3 years, one colony comprised 63·4% of 1-year-olds ringed as nestlings in colony, 23·5% 2-year-olds, rest 3 years or older (Gunten 1963). Many sites traditionally used over long periods (e.g. Hammer 1977). Sometimes nests alongside *H. rustica* (Balát 1964a; Schönfeld 1972); in Switzerland not uncommonly with Crag Martin *Ptyonoprogne rupestris* (Strahm 1956; Creutz 1961; Hunziker-Lüthy 1973). ROOSTING. In breeding season, roosts mainly in nest, to some extent also in trees (Fally 1984), perhaps sometimes aerially; outside breeding season, communally in various sites. Old nests claimed at start of breeding season immediately used for roosting (Lind 1960). Early in breeding season, severe weather may lead to several birds (14 recorded) in same nest—once, for 2 days continuously; eggs and young of invaded nests crushed (Oldfield 1952; Beretzk 1967). Many survive by roosting thus, but others, including adults, die (Löhrl 1971; Hund 1974). Groups of 15-40 (mostly juveniles), also recorded huddling on windowsills, sometimes on top of one another, during cold weather in Finland, late August (Lind 1960). Once paired, partners usually roost together in nest unless one excluded by aggression of other (see Heterosexual Behaviour, below). Until autumn departure, parents and offspring said usually all to roost in nest, occasionally outside in holes in walls, or in trees (Lind 1960). In mid-summer, Burgenland (Austria), after 1st broods fledged, many birds (of unspecified age) thought to roost in trees in flocks of up to 20 (Fally 1984). In Schwarzwald (West Germany), August, flocks of up to 100 apparently roosted thus (Löhrl and Dorka 1981). Young may roost in nest up to 48 days after fledging (Stremke and Stremke 1980). However, roosting as family not invariable: in 2 studies (Kareila 1961; Morath 1979), 1 parent roosted outside nest when nestlings well-grown; both parents in nest only in harsh weather (Morath 1979), and nest little used for roosting after young fledged (Kareila 1961). In Switzerland, autumn, often 10-12 birds in same nest; once 13, comprising 2 adults and 11 young thought all to be offspring of resident pair (Haller and Huber 1937). Status of birds not certain, since fledged young often roost in a neighbouring nest (Lyuleeva 1974), especially when own parents rear 2nd brood, e.g. one nest contained 2 adults, 1 offspring, and 4 strange fledglings (Stremke and Stremke 1980). Aerial roosting not proven, but thought to occur during breeding season (Buxton 1975), and 2 reports provide circumstantial evidence. In Nordrhein-Westfalen (West Germany), 31 July, c. 200-250 birds, which were known not to have roosted in colony, descended shortly after 04.45 hrs from considerable height (perhaps c. 1000 m) into colony; birds intercepted trying to get to their nests were all adult ♂♂ or juveniles; one caught was freezing cold to touch, consistent with high-altitude roosting; soon after, birds perched on wires to rest and preen (Rheinwald 1975). Also, at Voronezh (USSR), birds often observed descending to colony in morning; in fine still weather from 1 July (i.e. after 1st broods fledged) until beginning of August birds heard calling over colony at night, then flock descended slowly (for ½ hr) to colony; on completing descent, many entered 'their' nests and stayed for some time, presumably resting; these birds very cold to the touch (Semago 1974; L L Semago). In breeding season, birds go to roost well after sunset (Church 1958, which see for comparison with *R. riparia*). The following details of activity rhythm from study in Finland by Kareila (1961). Generally roosted quite late, but earlier than Swift *Apus apus*; roosted on average 23 min earlier when sky overcast. Sometimes took up to 1 hr to settle in nest, leaving tails sticking out when young present. In morning, active c. 1½ hrs before leaving nest. In July, last did not leave until 06.00 hrs, 2½ hrs after 1st; tended to leave later on cloudy days. Pair-members sometimes left nest up to ½ hr apart, the last to roost leaving first. For activity rhythm in northern Sweden, see Hoffmann (1959). For comparison with *P. rupestris* see Strahm (1956). Fledged young typically loaf and preen with adults on wires, etc. Comfort behaviour also includes water-bathing, as in *R. riparia* (Lind 1960), apparent mud-bathing (Fouarge 1971), and smoke-bathing (Pritchard 1950). Post-breeding flocks, sometimes mixed with *H. rustica*, commonly sunbathe on south-facing cliffs and roofs, on overhead wires, seldom on trees (Lind 1963). One group of birds (mostly juveniles) preened on roof with backs turned to sun, wings and tail slightly spread, and plumage ruffled (Melchior 1973). For drawings of similar posture, but with head on one side, see Prytherch (1981). Some debate as to which roost-sites most prevalent on migration and in winter quarters. Reedbeds generally little used on autumn migration (Witherby *et al.* 1938b; Lind 1960; Spencer 1978), but reportedly used in USSR (Dementiev and Gladkov 1954b; Dolgushin *et al.* 1970), and 2 reports for England (Airey 1951; Elms 1972). Autumn migrants commonly roost at *D. urbica* colonies *en route*, and in study in Leicestershire (England) most *D. urbica* nests were used by such migrants (McNeil and Clark 1977). One juvenile entered nest-hole in occupied colony of *R. riparia*, apparently to roost (Mather 1973). Spencer (1978) considered size of migrant population incompatible with exclusive use of *D. urbica* nests *en route*, and believed most roost in leafy canopy of trees; 10-15 may roost in given clump, with same trees being used for several weeks. For other reports of roosting in trees, see Adams (1966), Lyuleeva (1974), and Hughes (1980). Migrants overtaken by bad weather stop flying and cling to house walls until conditions improve (Löhrl 1971), once, 120 birds together (Azzopardi and Bonett 1980); if bad weather prolonged, use nests or other cavities (Lorenz 1932). In Morocco, November, flock of 10 rested on ground (Smith 1965a). In winter quarters, reported roosting in reedbeds (Sassi and Zimmer 1941; Verheyen 1952; Curry-Lindahl 1963a), also in leafy trees (see Verheyen 1952), and possibly on cliffs (Verheyen 1956a). In Africa, reports of *D. urbica* roosting in reedbeds thought possibly to have concerned Grey-rumped Swallow *Pseudhirundo griseopyga*; *D. urbica* never seen to roost thus, and believed to roost on the wing (R K Brooke; see also Backhurst and Allen 1974, Harper and Harper 1974).

2. Quite bold and approachable, e.g. birds collecting mud for nests allowed observer to approach to 4-5 m (Baier 1977). Much shyer during pair-formation, often flying off for no apparent reason (Lind 1960). FLOCK BEHAVIOUR. Following account based largely on Lind (1963). Forms flocks, seldom exceeding 20-30 birds, irregularly in breeding season, e.g. on cold days, when thunderstorm is approaching, after failure or disturbance of nests (see also Parental Anti-predator Strategies, below); frequent communal activities include collection of nest-material, feeding, and, from mid-summer, social flights in morning and evening (Lind 1963; Fally 1984). During nest-building, flock typically circles in morning, sometimes for 2-3 hrs, over area from which mud to be collected, before landing (Lind 1960). Main flocking activity in late summer: initially, juveniles congregate in traditional areas, travelling from colonies up to 5-6 km away in groups composed of one or more broods. After assembling, birds intersperse flying

around with comfort activities. In afternoon, flocks disperse, forming again in evening, then leave area suddenly to return to nests for night. Individual-distance is 1 body-length on wires, less on roofs and windowsills, and less in juveniles than adults (Lind 1963, which see for further details of flock dispersion and activities). In one sunbathing flock (75% juveniles), on roof, minimum 25 cm between birds (Melchior 1973). Pre-migration flocks typically restless, interspersing sudden upflights involving much high flying and calling, with comfort activities. Upflights thought to have synchronizing function and may lead directly to migratory exodus (Lind 1963). According to Lind (1963), migration in spring, unlike autumn, not communal. However, this view not supported by April observations at Upemba, Zaïre (Verheyen 1952): on arrival at staging post, migrant flocks typically 'fell' out of sky, sometimes from great height, then spent much time in rapid chasing. Periodically, flock regrouped on one or more roofs, twittering continuously, which seemed to attract others. Calling suddenly grew louder and a few took off, followed immediately by remainder. Soon after, a few broke away and returned to loafing site, remainder following suit; then another mass exodus, like first. On one day, 6 false starts before final departure, birds climbing to considerable height and flying off (Verheyen 1952). ANTAGONISTIC BEHAVIOUR. Markedly aggressive in defence of nest-area territory. Following account after Lind (1960). On first return to colony, intense competition for old intact nests, claimants defending them by diving at competitors which cling to wall or sometimes enter nests. Fighting tends to be contagious, involving up to 12 birds, and may recur for days or weeks until ownership settled. ♂ does most attacking early on, ♀ joining in more as season progresses, and sometimes defending territory alone. Fighting most vigorous at nests under construction, declining steadily through to nestling period. Satellite nesting pairs especially prone to harassment by members of main colony. Unpaired ♂♂ spend much time sitting in half-finished nests (Nest-guarding): sit with wings slightly open and drooped, flank feathers ruffled; if strange bird passes, ruffle head plumage (Threat posture: Fig A) and give Threat-calls

A

(see 4 in Voice). If intruder lands at edge of nest or nearby, occupier directs pecking movements at it, sometimes for minutes on end. If intruder attempts to enter, resident pecks, holding soft skin at base of bill, and both may fall locked together, parting before reaching ground. Resident may also hold intruder's wing-tips in bill, eliciting Distress-call (see 6 in Voice) from intruder. Resident may dive-attack which can lead to pursuit-flight—usually for 2-3 m, occasionally 25-30 m (Lind 1960). For response of unpaired Nest-guarding ♂ to intruding ♀, see Heterosexual Behaviour (below). Attacks more frequent and more vigorous on side of nest containing entrance, and birds may be prevented from building where flight paths would cross; where nests clustered, holes thus tend to point in different directions (Lind 1960). During building, disputes often arise through theft of material. Lining material may be stolen from neighbours or a deserted nest, and pair nearest deserted nest once drove off would-be looters (Lind 1960). At one colony, mud not uncommonly stolen from neighbours'

nests, provoking expelling attacks (Berndt 1982). Once, took over site being used by *H. rustica* (Urbaniak and Zatwarnicki 1979). Infringement of individual-distance in perched post-breeding flocks leads to frequent mild disputes (Lind 1963). For antagonism during pair-formation and mating, see Heterosexual Behaviour (below). HETEROSEXUAL BEHAVIOUR. (1) General. Older birds arrive at breeding grounds before 1-year-olds (Rheinwald *et al.* 1976). Most birds older than 1 year arriving at colony said to be already paired (Gunten 1963), suggesting that pairing occurs during migration. Most birds apparently form pairs before nest-building begins, but all birds, whether paired or not at outset, regularly exhibit pair-bonding behaviour at colony (Lind 1960). In Britain, song (see 1 in Voice) given mid-May to end of July (Alexander and Alexander 1908), used mainly in mate-attraction. (2) Nest-site selection and pair-bonding behaviour. Selection of nest-site and mate intimately linked, some ♀♀ evidently choosing mate by accepting site (Lind 1960, on which following account based), and therefore combined here. ♂ takes initiative in site-selection but ♀ also active. Returning birds occupy old nests first, whether whole or damaged, or, if necessary, build new ones in old sites (Lind 1960). 73% of 2- and 3-year-olds occupied old nests, 85% of 1-year-olds built new ones (Lyuleeva 1974). At first, colony visited irregularly, more regularly as time passes. Bird approaches and often hovers in front of walls, giving frequent Contact-calls (see 2a in Voice). ♂ inspects several sites and, having chosen one, keeps flying around, returning to site, defending it, etc. If threatened, visitor tends to turn away or inspect wall, or cracks in wall, this thought to identify bird as unpaired ♀ to Nest-guarding ♂ who flies behind prospective mate and, giving Contact-calls, tries to attract her back to nest-site (Nest-showing). If ♀ shows interest, ♂ lands at site and gives Enticement-calls (see 3 in Voice). ♀ flies towards ♂ and may hover in front of site, but does not land initially. ♂ follows her, trying to coax her back, and procedure repeated, sometimes for several hours. Often, ♀ hovers at different (apparently undefended) site, possibly to coax 'her' ♂ there. If ♀ finally lands beside ♂ and remains there for a while, apparently inspecting site, ♂ approaches her, singing. Mutual aggression typical of early pair-formation. If ♀ retreats, ♂ may seize her by nape or base of bill. ♀ gapes, gives Threat-calls, and may persistently peck ♂. If ♀ escapes, ♂ usually follows. Some partners remain intolerant of each other, never staying in nest together for building and roosting. Once bond strengthened, ♂ starts nest-building. In one case, ♂ and pair fought over a nest for hours; when unpaired ♂ victorious, ♀ paired anew and bred with him. (3) Courtship-feeding. Not observed by Lind (1960), who believed it plays no part in pair-formation, but is rather a response preparatory to feeding young. However, on 3 occasions during early stages of colony formation, ♂ flew to bird in the air and transferred food (Berndt 1982). Incubating birds also said to be fed by mates (Brown 1924; Niethammer 1937; Kivrikko 1947), though this not usual. (4) Mating. Following account based mainly on Lind (1960). Occurs mostly in nest; elsewhere (e.g. on roof or ground, sometimes on wires) usually unsuccessful. For Soliciting-display on wires, see Butterfield (1953); for aerial attempt, see below. ♂ typically takes initiative in soliciting, exceptionally ♀. Accounts naturally biassed towards incidents outside nest, which are more easily seen. Soliciting-display of ♂ similar to Nest-showing, but ♂ usually silent or sings in flight, rather than giving Contact-calls. Lands and crouches, singing, and runs rapidly towards ♀ in Soliciting-posture (Fig B): head lowered, throat plumage ruffled, carpal joints held slightly away from body, wings drooped, tail fanned, rump feathers ruffled; rarely, shivers wings. When ♂ runs

B

towards her, unresponsive ♀ starts pecking, retreats, or tries to escape; ♂ stops, continues singing while crouched, and starts running towards ♀ again. After several unsuccessful attempts, ♂ not uncommonly attacks ♀ with pecking movements and Threat-calls. Rejected ♂ may preen. If responsive, ♀ stays still, calls quietly, and allows ♂ to approach, sometimes looking away. In 2 cases where ♀ solicited first, she crouched horizontally, turned lowered head towards ♂, called quietly, then turned slightly away from him. When ♂ or ♀ solicits successfully, ♂ approaches ♀'s rear at an angle (Fig C), singing,

C

and pecks and then seizes her nape or crown; ♀ pushes tail aside and raises it (for drawing, see Hobbs 1950), and ♂ mounts. Copulation takes 10–30 s. In another sequence, 2 birds faced each other in nest, pecked at it frequently, and periodically raised heads to touch bills; ♂ sang and presumed ♀ occasionally shivered wings; ♂ mounted, flapping wings vigorously; pair copulated again 1 min after separating (May 1948). Gentle billing also reported in apparent Soliciting-display in which ♂ occasionally turned his back to ♀, exposing rump (Adams 1952). If ♀ tries to escape while mating, pair may flutter down to ground together. After copulating, both preen, occasionally Allopreen. Sometimes one or more conspecifics try to interfere aggressively with copulating pair (Lind 1960, which see for details). In apparent copulation attempt in the air, one bird followed another closely and landed on its back; lower bird (presumed ♀) ceased flapping and glided on bowed wings, while other, with wings aloft and fluttering, adopted an almost vertical posture, curving its tail down beneath other bird's; pair dropped gradually for a few seconds before separating (Hancock 1969). Successful copulation occurs up to 11 days before laying, mostly 3–10 days before; last seen on day 2nd egg laid; occasionally attempted when young in nest. During initial pair-bonding and nest-site selection, ♀ usually unresponsive, and copulation unsuccessful. Occurs mostly in early morning c. 2 hrs after birds 'wake up', also in evening c. 2 hrs before roosting (Lind 1960). (5) Behaviour at nest. Before laying, pair usually spend some time in nest together. In colonies, often lay in half-completed nests. Add nest-lining during laying. Nest-relief rapid, one bird usually entering nest before other leaves. Usually no calls in nest, but, if incoming bird delayed, and calls (see 2 in Voice) in flight (usually c. ½ m from nest), sitting bird, of either sex, occasionally sings quietly. Departing bird gives variant of Contact-call (see 2b in Voice), ♀ especially repeating it for first 25–150 m. After long incubation spell, sitting bird may fly off on approach of mate; after short spell, sitting bird may be reluctant to leave, and both may incubate together for several minutes. Bird thwarted in relieving mate may threaten and exceptionally attack mate (Lind 1960). Once, after mass breeding failure due to severe weather, adults removed eggs from nests (Löhrl 1971). At end of August, birds may 'play' with feathers; interpreted as residual nest-building

behaviour (Lind 1960). RELATIONS WITHIN FAMILY GROUP. From day before hatching, a twittering frequently heard from nest, thought to represent parental communication to young. One parent first brought food for young less than 2 hrs after hatching (Lind 1960). Young become audible at c. 3 days (Spencer 1975a). Faecal sacs carried away by parents for first 4 days (Lind 1960), but on c. 5th day young begin defecating from nest-entrance, and parents no longer enter nest completely to feed them (Spencer 1975a). Regularly defecate out of nest by 9th day, but adults may carry away faeces up to 14 days. Initially, young brooded closely by both parents, more by ♀. Nest-relief as for incubation, but, increasingly as young get older, without accompanying adult calls. Brooding markedly reduced when young 4–11 days old; the bigger the brood the sooner the reduction (Lind 1960). When small, young beg continuously throughout day and well into night; when older, seldom call except when parent arrives (Spencer 1975a), though P A D Hollom regularly found that well-grown young call almost throughout night. Parents stimulate begging in small young by calling or touching them with feet, bill, or breast, sometimes pecking. If young sated, parent sometimes delivers food to young of a neighbour, and one pair fed neighbouring young the day before their own hatched. Begging undirected until eyes open at 8–9 days; response to acoustic signals also well developed thereafter (Lind 1960). Young regularly poke heads out of nest-entrance from c. 9 days (Lind 1960; Spencer 1975a), initially begging whenever an adult flies past, later largely ignoring strangers. Apparently recognize parents at c. 21 days, and thereafter aggressive to intruders; parents accept strange young in nest up to 22 days, and feed only their own fledged young. Siblings sometimes beg from one another; from c. 17 days, especially when hungry, may peck at and attack one another, also peck parents. Parents usually alternate visits, but at wide entrance hole sometimes both feed young together. Parents begin enticing young to fledge from 15 days, most often 19–24, with Lure-display: with rapidly beating wings, parent flies slowly past, and hovers in front of or enters nest; not uncommonly visits other nests, or lands nearby; offers no food, and constantly utters Contact-call. Young initially silent but eventually return Contact-calls. Parent, usually ♀, lands at nest, and young lean out; parent flies off slowly with whirring wing-beats and young follow c. 0·5–3 m behind, parent and young exchanging Contact-calls (Lind 1960). Young will not leave nest unless thus enticed (Lind 1964). After first flight, in which young may land on ground, escorted back to nest (Lind 1960). In one case of luring, adult clinging by nest pulled young with it, belly to belly, as it dropped away; flew down for 3–4 m, apparently supporting young, then separated, and circled round young as it gained height (Reichholf-Riehm and Reichholf 1973). In another incident, when bird made first flight, parents frantically entered nest one after the other, as if trying to lure it back (Spencer 1975a). Up to 20 other birds, including young from previous broods, may join in luring and escorting: e.g. when 3 young fledged in rapid succession, ♀ escorted 1st chick, ♂ the 2nd, flock of 15–20 the 3rd. Young in nest may peck at these intruders (Lind 1960, 1964). For similar attraction of other birds when parents feeding young, see Baier (1977). ♀ usually takes care of fledged young while ♂ attends to those still in nest (Lind 1960). Young fly towards food-bearing parents and perform a 'tremulous' flight which often leads to aerial transfer of food. However, young never seem to follow adults very closely, and quickly achieve independence (Spencer 1975a). Once, 4 juveniles and 2 adults spent day bringing mud to a wall (Simmons 1949); evidently incipient nest-building behaviour. ANTI-PREDATOR RESPONSES

OF YOUNG. No information. PARENTAL ANTI-PREDATOR STRA-TEGIES. (1) Passive measures. In any alarming situation, e.g. when cats, horses, or human intruders nearby, birds agitated and hover by nest (Lind 1960). (2) Active measures: general. First birds to detect anything unusual give shrill Contact-call (see 2a in Voice), usually followed by mild Warning-call (see 5a in Voice) which stimulates others to fly up and increase alertness. If danger more imminent, give main Warning-call (see 5b in Voice), eliciting escape or attack, depending on circumstances. This call induces upflight of all birds in colony which, repeating call, fly in groups towards source of disturbance (Bergmann and Helb 1982). Warning-calls more frequent in small than in large colonies (Lind 1962, 1964). (3) Active measures: against birds. When raptor appears, birds fly up, gather into flock above it, and follow it some tens of metres away; from time to time, some birds mob predator (Lind 1962, 1964). Thus in Bayern, mid-July, Hobby *Falco subbuteo* attacked by 2 birds along with 2 *H. rustica*; birds tried to keep raptor low down, attacking at its head from behind, and followed it for some distance (Baier 1974a). *F. subbuteo* presenting sudden threat may also elicit escape response, birds swooping down at great speed and levelling out near ground (E K Dunn). For calls in extreme danger, see 5c in Voice. Perched juvenile Cuckoo *Cuculus canorus* mobbed by several birds which stooped at it and drove it to seek refuge in trees (White 1967b). Also attacks species attempting to usurp nests: e.g. House Sparrows *Passer domesticus* (most serious nest-site competitor, often destroying eggs and young: Lind 1961) usually subjected to flying attacks (Teidemann 1946; Summers-Smith and Lewis 1952). Vigorous defence of nest against *A. apus* led to communal pursuit by nest-owner and neighbours (Baier 1977; see also Hindemith 1972). (4) Active measures: against man. After boy smacked house wall on which birds nesting, he was always attacked on subsequent approaches (Baier 1974a). (5) Active measures: against other animals. No information.

(Figs by M J Hallam: based on photograph and drawings in Lind 1960.)

EKD

Voice. Freely used throughout the year. Generally more liquid and musical than Sand Martin *Riparia riparia* (P J Sellar). Following scheme based on Lind (1960). For voice of hybrid with Swallow *Hirundo rustica*, see Elsner (1951).

CALLS OF ADULTS. (1) Song. Very soft but sweet twittering; about twice as long as song of *R. riparia*, but given much less often (Witherby *et al.* 1938b). A short subdued chattering, basically composed of modified calls (Bergmann and Helb 1982, which see for sonagram). Our recordings vary somewhat in quality from rapid and liquid to frenzied jumble of units; may include rattles not unlike those in song of *H. rustica* except for tendency in *D. urbica* to increase or decrease rate of delivery of units. In Fig I, rattles at *c.* 1·0–1·2 s and at 1·62–1·7 s from origin show clear rallentando (J Hall-Craggs). Song often ends in series of Enticement-calls (P J Sellar: see call 3). Delivered in flight or perched, from nest, once on ground (Dawson 1976). In recording by P A D Hollom, bird in nest prior to starting clutch intermittently gives short bursts of chortling song, simpler than in Fig I. In intense encounters, ♂ may sing for up to 1 min at a time. Full song given only by ♂ but, at nest-relief,

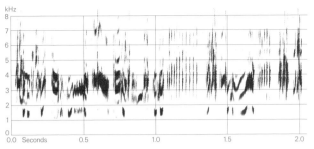

I P A D Hollom England May 1984

sitting bird of either sex occasionally responds to Contact-call (call 2a) of incoming mate by singing in subdued manner (Lind 1960). (2) Contact-calls. (a) Usual call a hard, frequently repeated 'prt' or 'pr-prt', like 2 pebbles rubbed together. Very characteristic sound at close range, but softer and more trilling (e.g. 'brüd') at distance (Bergmann and Helb 1982, which see for sonagram of 'pr-prt'). Also described as a slightly hard 'chirrrp' or 'chichirrrp', more chirruping than *H. rustica* (Witherby *et al.* 1938b). In our recordings, both 'prit' and 'pri-pit' (J Hall-Craggs: Figs II–III). Almost invariably (at

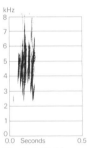

II A G Field
England July 1965

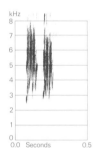

III A G Field
England July 1965

22 out of 23 nests) distinguishable between sexes (Lind 1960): in ♂, low-pitched harsh 'dridri' or 'trit'; in ♀, higher-pitched melodious 'dridri', 'drlidrli', or 'drit'. Given typically in flight in communication between pair (e.g. by incoming bird at nest-relief, by soliciting ♂); also by bird luring young to fledge, in communication with fledged offspring, and as mild expression of alarm (Lind 1960, 1962; see Social Pattern and Behaviour). Call described as an unfamiliar sound like a child's metal clicker when bird luring young to fledge (Spencer 1975a) presumably this call. Quiet calls given by soliciting ♀ (Lind 1960) probably also this call; in one precopulatory sequence, apparent ♀ gave a lively oft-repeated 'tchirrup' while soliciting ♂ (Hobbs 1950). (b) Much longer variant, typically comprising single, repeated units: in ♂, 'triit-triiit' or 'triii-trit'; in ♀, 'bliiit-bliit-bliit'; or 'pliit-pliit-pliit'. Given in flight by bird leaving nest after being relieved of incubation by mate; more persistently by ♀, usually continuing to call for 25–150 min after her departure. Wanes during brooding (Lind 1960). In our recording of bird leaving nest, 'ki-tik', more abrupt and

higher-pitched than typical disyllabic form of call 2a (J Hall-Craggs). (3) Enticement-call. Rapid 'zä zä zä...' or 'trä trä trä...', given for several seconds by Nest-showing ♂ when he lands at nest-site intent on coaxing nearby ♀ to join him there (Lind 1960). In recording, unmusical 'zi-zi-zi-zi-zi-zi...' (Fig IV) follows bout of song (J

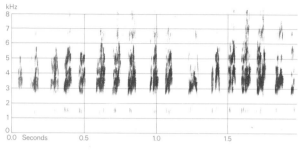

IV P A D Hollom England May 1984

Hall-Craggs). (4) Threat-call. Series of hard consonants, delivered rapidly (Lind 1960). Not further described for adult, but see Calls of Young (below). Given by either sex in confrontations with conspecifics, especially intruders during nest-site establishment, and with mate during early pair-formation (Lind 1960: see Social Pattern and Behaviour). (5) Warning- and Alarm-calls. (a) A 'trieer' or 'trieeer', given when danger mild or distant; may cause birds nearby to fly up or increase surveillance (Lind 1962); in Fig V, musical 'zieer tier', reminiscent of a

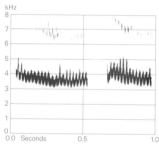

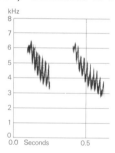

V V C Lewis England July 1975

VI P A D Hollom
England August 1984

'Trimphone' bell (J Hall-Craggs), is probably this call. (b) Main call 'tsier' or 'tsieer', stronger, shorter, and higher pitched than call 5a, given when danger more imminent (Lind 1962): in Fig VI, 'sier si-er' (J Hall-Craggs), with much steeper drop in pitch than call 5a. Although recording (Fig VI) is evidently of main Alarm-call (compare with Fig d in Bergmann and Helb 1982; 'zier' or richer 'ziürr'), call did not appear to cause upflight (P A D Hollom), and perhaps does not always do so if birds habituated to particular intruder. Unspecified Alarm-calls also given by birds disturbed from tree roost (Hughes 1980), during pre-roosting aerial manoeuvres (Spencer 1978), and during aerial chases, especially among flocks feeding high above ground prior to autumn departure (V C Lewis). (c) High-pitched, urgently

repeated 'tsitsitsitsier' or 'tsitsitsi tsi', given in extreme danger, eliciting escape or attack, depending on circumstances (Lind 1962). (6) Distress-call. A sort of hissing sound, given by intruder at nest when trapped by resident, e.g. when its wing seized in resident's bill (Lind 1960). Otherwise seldom heard. Perhaps similar to harsh hissing calls of bickering young (see below). (7) Other calls. A sort of 'chortle-rattle' (J Hall-Craggs), given probably by adult during almost continuous calling of young (see below) during night (P A D Hollom).

CALLS OF YOUNG. Food-call of small young an almost monosyllabic 'tik', repeated irregularly. When older, especially from 8 days, a louder, more persistent sawing 'zittritvitvii' (Lind 1960: for sonagram, see Bergmann and Helb 1982). Rasping cheeps or a coarse sucking sound (Fig VII) given by young c. 2 weeks old almost incessantly

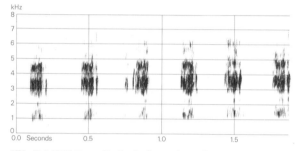

VII P A D Hollom England September 1984

throughout night, with rare breaks lasting a few seconds (P A D Hollom). From 13–15 days, young give Contact-call, rendered 'trik trik' or 'vrit' (Lind 1960); in recording by V C Lewis, sounds like 'prit' (P J Sellar); from c. 17 days, becomes clear and melodious, and, near to fledging, includes a drawn-out variant—'triiit triiit' or 'priii prtiii'; often more monosyllabic than similar Contact-call of parents luring them to leave nest, and call facilitates maintenance of family bonds outside nest (Lind 1960). Well-grown young give various calls in disputes with siblings, e.g. a harsh persistently repeated 'pshiit' (Bergmann and Helb 1982, which see for sonagram); also rendered a harsh hissing 'tschirr' or 'scherrr', perhaps usually indicating discomfort of victim (P A D Hollom). In our recordings, young near to fledging may intersperse these calls with monosyllabic and disyllabic contact-calls, e.g. in recording (Fig VIII), 'prit-prit prit', followed by

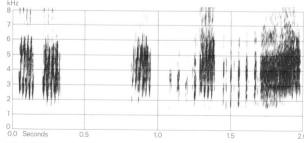

VIII V C Lewis England July 1975

extended rattle interrupted by another 'prit', and then a harsh 'scheeer'; in another recording, a long 'white noise' hiss (J Hall-Craggs). Especially after fights, 'trärrrtätätä' (Lind 1960), presumably similar to Threat-call of adult. Shortly before fledging, young not uncommonly sing, occasionally for more than 1 min; song of young distinct from adult's (Lind 1960); singing of young not heard by P A D Hollom.   EKD

Breeding. SEASON. North-west and central Europe: first eggs at beginning of May, rarely late April, main laying period begins mid-May; last young in nest until mid-October in mild year (Bryant 1975*b*; Hund 1976; Makatsch 1976; Bruderer 1979; Rheinwald 1979). Northern and north-east Europe: eggs laid from end of May or June; last young in nest in September (Lind 1960). North Africa: main laying period May–June, rarely from late March, uncommonly after July (Heim de Balsac and Mayaud 1962). SITE. Most frequent on outer walls of buildings, under eaves or other overhang; also under bridges, culverts, etc.; natural sites are on cliffs and outcrops, coastal and inland, also under overhangs; makes free use of specially designed nest-boxes. 2–3 to 50 m above ground; aspect variable. Colonial. Nest: half-cup of mud pellets, down, feathers, and other light material; cup formed against vertical wall and overhanging 'roof', with small oval entrance at top; may be contiguous with other nests, so that overall shape very variable from part-spherical to very irregular; variant nests include open-topped ones like Swallow *Hirundo rustica*, and half-spheres (against vertical walls) with side-entrances. Dimensions of 5 nests, Finland: width 17·8 cm (13·5–20·0), height 14·3 cm (13·0–18·0), depth (front to back) 10·6 cm (8·5–12·5); entrance hole 2·4 cm (2·2–2·5) high by 6·6 cm (5·2–9·2) across (Lind 1960). 10 new nests built of mean 1074 pellets (690–1495) (Lind 1960). Building: by both sexes, though ♂ more active at start, and may begin nest, with both sharing completion; 9 pairs building new nests took mean 10·4 days (8–18), while 25 pairs repairing old nests took mean 3·5 days (1–9) (Lind 1960). Pellets of wet, almost liquid, mud or clay collected from 50–150 m away, each taking 6–8 s to gather from ground; fixing pellet takes 30–60 s, accomplished with rapid, shivering movement; initial pellet fixing often over wider surface area of wall than needed for nest (Lind 1960). Lining material collected in the air or, sometimes, stolen from other nest. EGGS. See Plate 80. Sub-elliptical, smooth and slightly glossy; white, very occasionally with fine, light red spotting. North and central European populations: 19·0 × 13·1 (15·9–22·0 × 11·2–14·4), *n* = 512 (Hund and Prinzinger 1979*a*). West Mediterranean populations: 17·9 × 12·7 cm (16·0–19·6 × 12·2–13·5), *n* = 80 (Makatsch 1976). Egg size increases within clutch (Hund and Prinzinger 1979*a*), but decreases through season (Bryant 1975*b*). Fresh weight of 121 1st-clutch eggs 1·68 ± SD0·22 g, and

of 86 2nd-clutch eggs 1·64 ± SD0·21 g (Bryant 1975*b*). Clutch: 3–5 (1–7). Of 882 clutches, West Germany: 2 eggs, 4%; 3, 23%; 4, 38%; 5, 29%; 6, 5%; 7, 1%; mean 4·1, with mean of 183 known 1st clutches 4·8, and of 165 2nd clutches 3·3 (Hund and Prinzinger 1979*a*). In southern West Germany, of 111 1st clutches, 1 egg, 1%; 2, 3%; 3, 12%; 4, 28%; 5, 53%; 6, 3%; mean 4·41; of 113 2nd clutches, 1 egg, 1%; 2, 8%; 3, 54%; 4, 36%; 5, 1%; mean 3·28 (Hund 1976). In England, of 69 1st clutches, 2 eggs, 7%; 3, 18%; 4, 55%; 5, 20%; mean 3·82; of 60 2nd clutches, 1 egg, 3%; 2, 12%; 3, 72%; 4, 13%; mean 2·95 (Bryant 1975*b*). In Czechoslovakia, mean of 417 clutches 3·97 (1–6) (Kondělka 1978); of 197 clutches, 1 egg, 1%; 2, 3%; 3, 29%; 4, 47%; 5, 19%; 6, 1%; mean 3·84; mean of 37 1st clutches, 4·14 and of 38 2nd clutches, 3·34 (Balát 1974). For similar decrease from 1st to 2nd clutch, Denmark, see Møller (1974*a*). Significant decline in clutch size during season, associated with smaller 2nd clutches as noted above; additionally, early 1st clutches larger than later, and similarly for 2nd clutches, with mean clutch size of 4·71 in May declining to 3·11 in August (Kondělka 1978). Larger 1st clutches laid by older adults, but 2nd clutches of 1-year-olds larger (Rheinwald *et al.* 1976; Hund and Prinzinger 1985). 2 broods in central part of range but rare in northern and southern parts, e.g. Finland (Lind 1960) and North Africa (Heim de Balsac and Mayaud 1962). In England, 86·8% of 11 pairs attempted 2nd brood (Bryant 1975*b*). In West Germany, 35·2% of 1-year-old ♀♀ laid 2nd clutch, mean 54·5 ± SD5·7 days after 1st; 51·2% of 2-year-olds at 57·8 ± SD5·5 days; thereafter, proportion of 2nd clutches declined with age and interval increased (Hund and Prinzinger 1985). 3 broods very occasionally recorded. Eggs laid daily, around 06.00 hrs ± ½ hr (Hund and Prinzinger 1979*a*); gaps in laying of up to 4 days recorded in bad weather (Hund 1976). Interval between laying of 1st clutch and laying of 2nd 56 days (51–62), *n* = 9 (Hund 1967), and in 2 years, West Germany, 59·3 days (47–74), *n* = 80, and 54·5 days (43–68), *n* = 102 (Hund and Prinzinger 1979*a*). Interval 54·5 ± SD4·9 days, *n* = 47 (Rheinwald 1979). INCUBATION. 14–16 days (11–19). West Germany: 14·2 ± SD1·2 days, *n* = 64 (Rheinwald 1979); 15·1 days (11–19), *n* = 74 (Hund and Prinzinger 1979*a*). England: 14·6 days (12–16), *n* = 106 (Bryant 1975*b*). Switzerland: in pre-Alps, 14·4 ± SD0·7 days, *n* = 8; in lowlands 15·7 ± SD1·0 days, *n* = 13 (Bruderer 1979). Finland: 15·5 days (13–19), *n* = 43 (Lind 1960). By both sexes, ♀ for longer bouts than ♂; ♀ mean bout length 13·1 ± 0·75 min, *n* = 125, ♂ 8·7 ± 0·61 min, *n* = 168 (Lind 1960). Begins with last egg; hatching synchronous. YOUNG. Altricial and nidicolous. Cared for and fed by both parents. Brooded more by ♀ than by ♂, and small broods (1–3 young) are brooded effectively for 11 days, and large broods (6–9 young) for 5 days (Lind 1960). FLEDGING TO MATURITY. Fledging period 22–32 (–40) days. Switzerland: in pre-Alps, 23·9 ± SD2·0 days,

$n=23$; in lowlands, $27\cdot3\pm SD2\cdot2$ days, $n=45$ (Bruderer 1979). West Germany: $26\cdot2\pm SD1\cdot9$ days, $n=51$ (Rheinwald 1979). Britain: $27\cdot2\pm SD2\cdot1$ days, $n=51$ (Turner and Bryant 1979). Parents, and non-parents, attempt to lure young out of nest from $c.$ 15 days (Lind 1960, 1964). May return to nest for roosting for several days after fledging. Age of full independence not recorded. Age of first breeding 1 year. BREEDING SUCCESS. Of 1501 eggs laid Czechoslovakia, $85\cdot75\%$ hatched, of 1148 eggs hatched, 63% young reared; overall success $55\cdot8\%$ (Kondělka 1978). Of 737 eggs laid, Czechoslovakia, $84\cdot7\%$ hatched, $12\cdot3\%$ destroyed, and $2\cdot3\%$ addled; of 658 eggs hatched, $89\cdot1\%$ fledged, overall success $75\cdot4\%$ (Balát 1974). Of 440 eggs laid, England, $85\cdot8\%$ hatched; of those failing, $30\cdot6\%$ deserted, $48\cdot4\%$ infertile or died, and $21\cdot0\%$ laid in untenanted nests and not incubated. Of 377 young hatched, $94\cdot2\%$ fledged; overall success $80\cdot8\%$ (Bryant 1975b). Of 919 eggs, West Germany, $78\cdot8\%$ hatched and $91\cdot3\%$ of these raised to fledging; overall success $71\cdot9\%$ (Hund 1976). See also Table A.

Table A Breeding success of House Martin *Delichon urbica* in relation to age of each parent involved, West Germany. Figures are percentages. (Hund and Prinzinger 1985.)

Age of parent:		1	2	3	4	5	6
1ST BROODS							
Eggs hatching	♀	67·1	84·5	88·1	73·9	76·7	—
	♂	68·9	81·5	90·1	88·4	71·4	72·7
Young fledging	♀	91·3	83·1	86·8	93·0	78·2	—
	♂	81·0	84·5	86·7	89·3	—	—
Overall success	♀	61·3	70·8	76·5	68·7	60·0	—
	♂	55·8	68·9	78·1	78·9	—	—
2ND BROODS							
Eggs hatching	♀	78·2	83·6	70·2	86·5	86·7	—
	♂	86·0	80·3	87·5	73·5	62·5	66·7
Young fledging	♀	80·0	87·7	90·4	59·4	76·9	—
	♂	91·3	91·9	89·9	84·1	—	—
Overall success	♀	70·3	73·3	63·5	51·4	66·7	—
	♂	78·5	73·8	78·7	61·8	—	—

Plumages (nominate *urbica*). ADULT MALE BREEDING. Forehead, crown, hindneck, sides of neck, mantle, scapulars, and back glossy dark metallic blue; some dull grey of subterminal parts of feathers and some white of feather-bases often visible on hindcrown, hindneck, mantle, and inner scapulars, especially in worn plumage. Lores, sides of head to just below eye, and ear-coverts dull black, virtually without gloss. Rump and shorter upper tail-coverts white (exceptionally tinged rufous by soil: Steiner 1971), generally with some faint and narrow dusky shaft-streaks; longer upper tail-coverts black with slight blue gloss. Cheeks and all underparts white; chin, throat, and chest often slightly tinged pale cream, flanks slightly grey; under tail-coverts with black shaft-streaks, occasionally spotted black or dusky (Brombach 1984; probably most frequently in ♀—see below). Tail sooty-black or sepia-black, sometimes faintly edged off-white along outer edge and tip of t6. Flight-feathers, tertials, and upper wing-coverts black, slightly tinged brown on secondaries and greater coverts, slightly glossed metallic blue on lesser coverts, bastard wing, and tertial coverts; tips of secondaries faintly edged white, tip of fresh longest tertial with white fringe 0·5–1 mm wide. Under wing-coverts and axillaries dull grey; small coverts along leading edge of wing dull black with off-white tips. *In worn plumage*, grey of feather-bases visible on crown, central rump, sides of breast, and chest, white feather-bases on hindneck and mantle; new grey-brown or buff-brown non-breeding feathers may develop on rump, chin, throat, and chest; feathering of leg sometimes completely lost through abrasion. ADULT FEMALE BREEDING. Similar to ♂, but chin, throat, and chest washed pale buff-grey, flanks more extensively washed grey, under tail-coverts often with indistinct grey spots, and centre of rump sometimes spotted grey; underparts and rump not as pure white as adult ♂. In sample from Netherlands, all ♀♀ distinctly tinged buffish-grey until June, but also 4 out of 10 ♂♂; difference more difficult to see when plumage worn, buffish or greyish tinge being retained longest on chin and sides of throat (RMNH, ZMA). In sample from USSR, $c.$ 80% correctly identified by tinge of underparts, remainder intermediate or 'wrongly' coloured (Bub and Herroelen 1981). In Norway, all ♂♂ examined had under tail-coverts pure white, all ♀♀ with indistinct spots, but sample small (Schmidt 1967d). ADULT NON-BREEDING. Like breeding, but new feathers of rump, shorter tail-coverts, cheeks, chin, throat, and flanks pale brown-grey or buff-brown. Much individual variation: in some (especially ♂♂), hardly any brown obtained, in others more extensive, but these areas usually still mottled white (either because not all new feathers are brown or because moult is partial; too few examined to be certain about extent and timing). NESTLING. First down scanty, restricted to some long pale grey tufts on upperparts. In 2nd week, short and close off-white 2nd down develops between growing feathers all over body (except belly). (Heinroth and Heinroth 1924–6; ZMA.) JUVENILE. Upperparts duller than adult, sooty-brown or dark grey-brown, usually without blue gloss on crown or with traces only; tips of feathers of mantle and scapulars have variable amount of blue gloss, usually with much brown of feather-bases visible, but occasionally almost as glossy as adult. Rump and shorter upper tail-coverts white, virtually unstreaked; occasionally, much brown on feather-bases visible, especially in ♀. Longer upper tail-coverts black-brown with narrow white fringes along tips. Lores, upper cheeks, and ear-coverts grey-brown with faint paler mottling; dark crown gradually paler towards cheeks, no strong contrast between dark cap and white cheeks. Chin, throat, chest, flanks, thighs, and under tail-coverts extensively washed dark brown-grey to pale buffish-grey; lower throat often paler, belly and vent usually pure white. Grey wash to underparts rather variable in depth: in some, only slightly grey and underparts almost similar to adult (especially adult ♀), in others grey dark and extensive, resembling adult of extralimital *dasypus*. Tail shorter and fork less deep than adult (see Measurements). Flight-feathers, tertials, and upper wing-coverts browner than adult; inner primaries and secondaries narrowly edged white at tips, tertials with broad and contrasting white tips up to 2 mm wide (broader than fringes of fresh adult and not restricted to longest tertial). *In worn plumage*, crown, mantle, and scapulars browner, some off-white of feather-bases visible on hindneck and upper mantle, some grey-brown on rump; brown wash from chin to chest mottled off-white; white edges of flight-feathers abraded, white tips of tertials partly abraded; often rather similar to adult (especially to adult in partial non-breeding), but tertials still distinctly tipped white and tail-fork less deep. FIRST NON-BREEDING. Like adult non-breeding; indistinguishable when white-tipped juvenile tertials or relatively short juvenile outer tail-feathers replaced.

Bare parts. ADULT. Iris dark brown. Bill horn-black or black. Skin under feathering of leg and foot light flesh, pink-flesh with yellow tinge, or pinkish-yellow. NESTLING. Bare skin pink; bluish on future feather-tracts. Mouth uniform yellow, flanges at gape pale yellow. JUVENILE. Like adult, but basal cutting edges of upper mandible and basal half of lower mandible yellow at fledging; yellow gradually darkens to brown, and bill fully dark from late August or October. (Heinroth and Heinroth 1924-6; Witherby *et al.* 1938*b*; Ali and Ripley 1972; RMNH, ZMA.)

Moults. ADULT POST-BREEDING. Complete; primaries descendant. Frequently starts with scattered feathers of body or inner primaries when near breeding area, before reaching tropics, from (July-)August-September onwards. In Switzerland, October, 5% of 948 had started moult of tertials or inner primaries; usually s8 (middle tertial) and p1-p2 affected, exceptionally s1, s7-s9, and p3-p4; moult suspended in about half of birds which had started (Winkler 1975). In north-west Iberia, August, 3 of 5 adults showed suspended moult with p1 or p1-p2 new (Mead 1975; Mead and Watmough 1976); in Tien Shan (USSR), none of 74 August-September migrants had started primary moult however (Gavrilov 1971). Single birds had started primary moult late August (Sardinia) and mid-September (Algeria) (Vaurie 1951*c*). 8 of 22 ♀♀ from breeding colony in Belgium, 10 September, had started moult of body, but none of 9 ♂♂ (Bub and Herroelen 1981); adult ♀♀ from Netherlands, mid-September, had many new feathers on chin and throat and scattered ones on forehead, rump, and chest. Moult mainly in tropics, starting before reaching winter quarters. Data on nominate *urbica* limited; following account based mainly on wintering *dasypus* in Indonesia (RMNH), but not known whether this also applies to nominate *urbica* in Afrotropics. On arrival October-November, primary moult well advanced (score 10-30), tail old, 20-40% of body new; estimated start of primary moult early September to mid-October. Primary moult score 35-40 reached mid-December to early February. Tail starts November-December (sequence t1 to t6), at primary score 11-28, completed at same time as primaries. Ageing impossible after score 35-40; birds completing as early as late January probably adult; body virtually new at score 40-45, January-February. In South Africa, April, most birds had completed primary moult but not generally tail and rarely body (Broekhuysen 1953). In East Africa, moult from late October and November to (February-)March (Ginn and Melville 1983). ADULT PRE-BREEDING. Partial; January-April. Extent unknown; certainly grey feathers obtained early in post-breeding are replaced subsequently by white ones, and perhaps much of body replaced (Witherby *et al.* 1938*b*), but no detailed information. POST-JUVENILE. Complete; primaries descendant. Occasionally starts with body from August-October when still in temperate regions, rarely with inner primaries (but 2 out of a sample of 7 from Portugal had p1-p2 growing mid-August: Mead 1975); more usually starts on reaching tropics. In South Africa, November, 40% of 105 birds had not yet started primary moult, 17% had up to p1 new, 24% up to p2, 14% up to p3, remaining 5% up to p5(-p6) (Skead and Skead 1970). In wintering *dasypus* from Indonesia, start of primary moult more scattered than in adult post-breeding, estimated late September to early January; birds completing April-May probably all immatures. Sequence of moult as adult post-breeding, but body and tail start relatively late, from primary moult score 20-35. FIRST PRE-BREEDING. Extent unknown; perhaps all or part of body in some birds (as in adult pre-breeding), but no moult at all in

others. Birds in last stages of primary moult, late March to early May, Indonesia, either had all plumage new or had much of body very old, but not known whether new feathering 1st non-breeding or 1st breeding or whether heavily worn feathers juvenile or 1st non-breeding.

Measurements. ADULT. Nominate *urbica*, Netherlands, April-August; skins (RMNH, ZMA). Tail to tip of t6; fork is tip of t1 to tip of t6; bill (S) to skull, bill (N) to distal corner of nostril; exposed culmen on average 3·4 less than bill (S).

	♂		♀	
WING AD	111·1 (2·31; 33)	106-115	110·0 (3·10; 30)	105-116
JUV	108·4 (2·69; 15)	104-112	107·7 (3·67; 14)	102-115
TAIL AD	61·1 (1·97; 26)	58-65	60·4 (2·42; 25)	56-64
JUV	53·8 (2·16; 14)	50-57	52·8 (2·84; 13)	49-57
FORK AD	20·2 (1·53; 26)	17-23	18·6 (1·66; 26)	16-21
JUV	14·7 (1·40; 14)	13-16	13·7 (1·68; 13)	10-16
BILL (S)	9·9 (0·48; 10)	9·3-10·8	9·8 (0·61; 18)	8·7-10·7
BILL (N)	5·1 (0·28; 9)	4·7-5·5	4·8 (0·28; 17)	4·5-5·3
TARSUS	11·5 (0·45; 13)	10·9-12·3	11·4 (0·40; 18)	10·6-12·2

Sex differences not significant, except adult fork.

On average, wing increases 1 mm in length per 2·2° latitude. Adult, sexes combined: (1) Zagros mountains, Iran (Hartert 1910; Vaurie 1951*c*); (2) Algeria and Morocco (Clancey 1950*a*; Vaurie 1951*c*; RMNH, ZMA); (3) Spain, mainly Salamanca (Jordans and Steinbacher 1941; Jordans 1950; ZFMK); (4) Makedonija, southern Yugoslavia (Stresemann 1920); (5) Ravenna and Sardinia, Italy (Clancey 1950*a*; Vaurie 1951*c*); (6) Riet near Stuttgart, and (7) Bonn, West Germany (Rheinwald 1973; Bub and Herroelen 1981); (8) East Germany, mainly Halle (Bub and Herroelen 1981); (9) Hamburg, West Germany (Gruner 1977); (10) Atnasjö, southern Norway (Schmidt 1967*d*); (11) Sweden (Clancey 1950*a*).

(1)	102·5 (— ; 13)	100-106	(7) 109·9 (2·39; 238)	102-118
(2)	103·9 (— ; 32)	99-108	(8) 110·5 (2·70; 135)	104-117
(3)	104·7 (3·84; 13)	99-108	(9) 111·9 (2·43; 200)	
(4)	106·6 (3·42; 11)	102-112	(10) 115·8 (— ; 14)	112-119
(5)	107·5 (1·73; 4)	106-109	(11) 118·7 (— ; 10)	115-123
(6)	109·2 (2·45; 525)	101-118		

Wing length depends also on age. Riet (West Germany), sexes combined: 1st calendar year 108·0 (2·16; 51), 2nd 108·9 (2·27; 113), 3rd 109·5 (2·23; 47), 4th 108·8 (3·22; 32), 5th and older 107·6 (2·04; 30) (Rheinwald 1973).

Weights. Nominate *urbica*. Breeding season, Britain: ♂ 19·4 (1·3; 166), ♀ 19·7 (1·6; 227); average of ♂ rather constantly 19·4 throughout period, but 20·8 (1·3; *c.* 14) in early spring and 20·2 (1·8; *c.* 18) just before autumn migration; ♀ on average mainly 19·5 until 25th day after laying, but marked increase from 5 days before laying to peak of *c.* 22·2 at laying and weight decreasing to *c.* 18·1 at *c.* 25-35 days after laying, recovering to 20·3 (1·1; *c.* 13) before autumn migration (Bryant 1975*a*). Nestling on 1st day 1·8 (0·4; 52), reaching peak of 24·9 (2·6; 68) on 16th day, fledging with 18·3 (1·5; 54) at *c.* 27th day (Bryant 1978*a*, *b*). For growth curves of nestlings and influence of food availability on growth, see Rheinwald (1971) and Bryant (1975*b*); for growth in relation to brood size, see Bryant (1978*a*); for influence of time of year, differences between individual chicks (etc.), see Bryant (1978*b*).

Sexes and ages combined. Britain and Netherlands, combined: (1) May, (2) June, (3) July, (4) August, (5) September (BTO, ZMA). (6) Nyköping (Sweden), and (7) Weissenfels (East Germany), summer (Bub and Herroelen 1981). (8) Bonn and (9) Riet, West Germany (Rheinwald 1973). After Sahara crossing on spring migration, Morocco: (10) first half of April 1963 (BTO), (11) several seasons (Ash 1969). On migration, Malta: (12) spring, (13) autumn (J Sultana and C Gauci). (14) southern Turkey, Jordan, and Kuwait, late April and May (Kumerloeve 1970a; BTO).

(1)	18·4	(1·57; 15)	16–21	(8)	18·6	(1·22; 115)	—
(2)	18·5	(1·38; 15)	16–21	(9)	17·2	(1·32; 500)	—
(3)	17·6	(2·08; 10)	15–21	(10)	14·4	(1·07; 49)	12–17
(4)	17·2	(1·53; 11)	16–20	(11)	14·5	(— ; 252)	10–20
(5)	17·6	(1·68; 8)	16–21	(12)	17·3	(1·60; 50)	14–22
(6)	20·4	(— ; 10)	18–23	(13)	16·5	(2·00; 23)	13–21
(7)	18·1	(— ; 197)	13–23	(14)	15·9	(0·93; 5)	14–17

Riet (West Germany), 1st calendar year (juvenile) 18·0 (1·03; 51), 2nd 17·7 (1·27; 88), 3rd 17·3 (1·29; 39), 4th 17·7 (1·41; 20), 5th and older 17·5 (1·30; 25) (Rheinwald 1973).

Juveniles in winter: Ahaggar (Algeria), December, ♂ 16 (Niethammer 1963b); Zambia, January, 17·5 (Dowsett 1965a).

Exhausted birds killed during bad weather. Atnasjö (Norway), July: 13·6 (9) 13–15(–19) (Schmidt 1967d). Netherlands, May–October: 12·3 (1·29; 15) 10–14 (ZMA, RMNH). South Africa: November, mainly juvenile, 12·1 (107) 10–15 (Skead and Skead 1970); April, 13·3 (45) 12–16 (Broekhuysen 1953).

Structure. Wing rather long and narrow, bluntly pointed. 10 primaries: p9 longest, p8 (0–)1–3 shorter, p7 6–12, p6 15–21, p5 24–30, p4 28–38, p1 48–61. P10 reduced, narrow and pointed, concealed under primary coverts; in adult, 66–79 shorter than p9, 7–10 shorter than longest upper primary covert; in juvenile, 65–72 shorter than p9, 4–8 shorter than longest covert. Tail rather short, forked; 12 feathers, t6 longest, t1 16–23 (adult) or 10–16 (juvenile) shorter (see Measurements). Bill small and flattened, relatively shorter and thicker than Swallow *Hirundo rustica*. Tarsus and toes short and slender, covered by short soft feathers. Middle toe with claw 14·1 (23) 13–15; outer toe with claw *c.* 72% of middle with claw; inner *c.* 67%, hind *c.* 60%. Claws short and slender, decurved. Remainder of structure as in *H. rustica* (p. 277).

Geographical variation. Rather slight, but complex. 2 main groups discernible: *urbica* group (nominate *urbica*, *lagopoda*) in Eurasia (except Far East) south in central Asia to Kashmir and Spiti (India), and *dasypus* group (*dasypus*, *nigrimentalis*, *cashmeriensis*) from Ussuriland (south-east USSR) and Kuril Islands south through Japan, Taiwan, and southern China to Himalayas. *Dasypus* group darker and less glossy on upperparts than nominate *urbica* group; rump often more distinctly streaked black; black of lores extends along base of lower mandible to chin; throat, chest, flanks, and under tail-coverts extensively washed brown-grey; under wing-coverts blackish rather than grey; tail less deeply forked. Differences between groups in part bridged by *lagopoda* of nominate *urbica* group (tail less deeply forked) and by *cashmeriensis* of *dasypus* group (brighter glossy blue above and purer white on rump and underparts than typical *dasypus*), but *dasypus* group perhaps specifically distinct, as nominate *urbica* and *cashmeriensis* appear to meet in northern Pakistan and north-west India without interbreeding (Vaurie 1954b).

Within nominate *urbica*, variation in size (as expressed in wing length, tail length, bill length, and weight) strongly clinal, northern populations distinctly larger than southern ones. According to Clancey (1950a), 3 races separable: nominate *urbica* with wing 115–123 and tail 61–69 in Fenno-Scandia and north European USSR; *fenestrarum* (Brehm, 1831) with wing mainly 107–115 and tail 58–65 from Britain, France, and Denmark east to central European USSR and Bulgaria; *meridionalis* (Hartert, 1910) with wing 100–107 and tail 55–60 in southern Spain, Balearic islands, and north-west Africa; perhaps a 4th race in Italy, where size near *fenestrarum* but gloss of upperparts greener, less blue. Typical nominate *urbica* from southern Sweden has wing 108–117, however (Svensson 1984a), and hence scarcely larger than *fenestrarum* from western and central Europe, which thus does not warrant recognition. *D. u. meridionalis* from north-west Africa indeed smaller, but variation in size between this and more northerly populations is markedly clinal (see Measurements), with no sharp boundary; *meridionalis* thus not separated, following Vaurie (1951c). In western Asia, populations from Iraq, Transcaucasia, Iran, and lowlands from Turkmeniya east to Kazakhstan (USSR) are as small as those from north-west Africa, and birds from further north are gradually larger, as in Europe, but populations from central Asiatic mountain ranges and Kashmir are relatively large in size too. A similar cline of decreasing size towards south occurs in *dasypus* group: larger birds (*dasypus*) in Japan and smaller ones (*nigrimentalis*) in south-east China and Taiwan, with intermediate birds (*cashmeriensis*) in mountains of south-central Asia. *D. u. lagopoda* intergrades with nominate *urbica* in Yenisey basin (Hartert 1910); differs in fully white upper tail-coverts (longer ones not black) and in less deeply forked tail, fork 5–12 mm (Vaurie 1951c; Dementiev and Gladkov 1954b). CSR

Family MOTACILLIDAE pipits, wagtails

Small oscine passerines (suborder Passeres); mainly terrestrial and insectivorous. 54–58 species in 5 genera in 2 closely related groups—pipits (genera 1–3) and wagtails (4–5): (1) *Anthus* (typical pipits), 34–37 species—Old and New Worlds; (2) *Tmetothylacus* (Golden Pipit *T. tenellus*), monotypic—East Africa; (3) *Macronyx* (long-claws), 8 species—Afrotropics; (4) *Dendronanthus* (Forest Wagtail *D. indicus*), monotypic—eastern Asia; (5) *Motacilla* (typical wagtails), 10–11 species—Old World and (marginally) Alaska. Cosmopolitan but most species in Eurasia and Africa; occur in open country, often near water. Many migratory. Family represented in west Palearctic by 16 species, with 9 *Anthus* and 4 *Motacilla* breeding.

Body slender, elongated (shape often exaggerated by long tail); neck short. Sexes similar in size. Bill thin and pointed; notched. Nasal operculum and rictal bristles present. Wing medium to long; 10 primaries but p10 considerably reduced; tertials long in most species. Flight strong, undulating (especially in wagtails); song-flights, launched from ground or perch, are characteristic of both pipits and wagtails. Tail medium to long; longest in *Motacilla* which move it up and down persistently, especially during feeding. Legs medium to long, slender; hind claw often elongated—least so in forms which regularly perch in trees or live on hard, bare ground. Legs and feet rather similar to those of larks (Alaudidae) but tarsus acutiplantar with sides covered by unbroken sheath and front by fused scutes. Gait typically a walk or run. Head-scratching by indirect method. Bathe in shallow water, using typical stand-in method. Sunning behaviour poorly documented. No records of dusting. Anting reported only from *Anthus*: of direct type (ant-application).

Plumage soft; streaked and cryptic in *Anthus*, brightly and contrastingly coloured (with black, yellow, white, or orange) in other pipits and in wagtails. Many species have paler, often white, outer tail-feathers. Except in most *Anthus*, sexes often slightly different. 2 moults annually in all but some larger and tropical or sub-tropical species. Non-breeding plumage slightly different from breeding, except in most *Anthus* and some tropical species. Upperparts of nestlings covered with rather dense down, often long; mouth bright but usually without contrasting spots. Young often leave nest (on ground in most species) before capable of flight.

Family well-defined and homogeneous. Relationships to other Passeres, as indicated by traditionally accepted characters, rather obscure. Evidence from egg-white protein data seems to point to ties with Old World warblers (Sylviidae) or flycatchers (Muscicapidae) (Sibley 1970), but reduced outer primary indicates affinities with 9-primaried oscines and this supported by DNA data which place pipits and wagtails as a subfamily ('Motacillinae') within Ploceidae together with Old World sparrows ('Passerinae'), waxbills ('Estrildinae'), weavers ('Ploceinae'), and accentors ('Prunellinae') (Sibley and Ahlquist 1981*a*, 1985*a*).

Anthus novaeseelandiae **Richard's Pipit**

PLATES 19 and 27
[between pages 232 and 233]

Du. Grote Pieper   Fr. Pipit de Richard   Ge. Spornpieper
Ru. Степной конек   Sp. Bisbita de Richard   Sw. Stor piplärka

Alauda novae Seelandiae Gmelin, 1789

Polytypic. *A. n. richardi* Vieillot, 1818, USSR from Barabinskaya steppe (south-west Siberia), Semipalatinsk, and Altai mountains (eastern Kazakhstan), east to Lake Baykal and middle course of Nizhnyaya Tunguska river, accidental in west Palearctic. Extralimital: *dauricus* Johansen, 1952, USSR from Yakutiya south to Transbaykalia and neighbouring Manchuria, east to Sea of Okhotsk, Zeya river, and Great Khingan mountains; *centralasiae* (Kistiakovski, 1928), Zaysan basin and eastern Tien Shan in eastern Kazakhstan (USSR), through western and southern Mongolia to Sinkiang, Tsinghai, Kansu, and probably western Inner Mongolia (China); *ussuriensis* Johansen, 1952, lower Amur area and Ussuriland (south-east USSR) through eastern China and probably Korea south to Szechwan and Yangtze river; *sinensis* (Bonaparte, 1850), China, south of Yangtze Kiang (except Yunnan); 4 further races in southern Asia and Philippines, 12-13 in Australasia, and 8-10 in Afrotropics.

Field characters. 18 cm; wing-span 29-33 cm. Largest and most heavily built pipit known to occur in Europe (in west Palearctic, only Long-billed Pipit *A. similis* larger). Differs structurally from Tawny Pipit *A. campestris* in stouter bill, more rounded crown, broader back, up to 10% longer wings, 10-15% longer and fuller tail, 20% longer and much stouter legs, and 80% longer hind claw, and from Blyth's Pipit *A. godlewskii* in 10-15% longer and stouter bill, marginally longer wings (with folded primary-tips showing beyond longest tertial), 15% longer tail, 10% longer and stouter legs, and 30-40% longer hind claw. Noticeably large and long, rather dark, brown and buff, heavily and broadly streaked pipit, with thrush-like bill, long tail, long and strong legs, rather long toes, and exceptionally long hind claw all forming useful structural characters. Dark streaks most obvious on crown, back, and chest; wing shows pale buff double wing-bar and tertial-fringes; tail edged white. Flight powerful, with marked undulations. Commonest call a stuttered, then shouted, rasp. Sexes similar; little seasonal variation. Juvenile separable. 1, perhaps 2-3, races wander regularly to west Palearctic; difficult to separate in the field, with eastern races (*centralasiae* and *dauricus*) also closely similar to sympatric *A. godlewskii*.

ADULT. (1) Central Siberian race, *richardi*. Plumage pattern and colours recall Skylark *Alauda arvensis*. Crown, neck, mantle, and scapulars buff- to dun-brown, all streaked with black-brown (most broadly on crown and back); lower back and rump buff-brown, less streaked and appearing warmer but not obviously paler than mantle and tail; tail brown-black, with broad buff-brown edges to central feathers and bold, pure white ones to outermost. *A. n. richardi* thus has darkest, most heavily streaked upperparts of any large pipit in west Palearctic. Head well marked, with (a) pale cream-buff lore, eye-ring, and long deep supercilium creating open fore-face (like *Alauda arvensis*) and contrasting above with long dark edge to crown, (b) pale buff-brown patch on rear cheeks and distinct brown-black moustachial stripe (not penetrating into lore); both these marks and paler buff fore-cheeks bordered by (c) distinct pale cream surround extending from rear supercilium and ending at chin, and this emphasized by (d) well-streaked sides and rear of neck and long brown-black malar stripe which usually ends in patch of black streaks on side of chest-band. Chin and throat cream- to buff-white; chest and flanks buff with variable ochre or rufous tinge, chest with quite deep but usually sparsely marked band of black-brown streaks extending on to fore-flanks; rest of underparts buff-white, noticeably paler than chest and flanks. *A. n. richardi* thus has darker underparts (especially chest) than typical *A. campestris* and less richly toned underbody (particularly vent and under tail-coverts) than *A. godlewskii*. Wings buff- to dun-brown, showing (a) bold median covert panel with black-brown feather-centres and buff margins, (b) buff margins to greater coverts forming pale bar (less obvious than on median coverts), (c) broad buff edges to tertials. *A. n. richardi* thus has more obvious wing-bars than most small pipits but its median covert panel is usually not so contrasting as *A. campestris* and most *A. godlewskii*. Underwing buff-grey, noticeably darker than *A. campestris*. With wear, buff fringes to upperparts and wing bleach and become narrower and ground-colour of underparts paler; bird becomes darker and duller above, often more mottled than streaked; see Plumages for seasonal plumage sequence. (2) East Siberian race, *dauricus*, and south-central Asian race *centralasiae*. Close in size and structure to *richardi* but somewhat paler, with more contrasting streaks on upperparts and ground-colour of underparts yellower on chest and flanks. JUVENILE. All races. Even more reminiscent of *Alauda arvensis* than adult. Narrow fringes of upperpart feathers paler than adult, more grey- and buff- than brown-toned, making streaking more intense; when fresh, upperparts also relieved by buff-white feather-tips which give slightly speckled appearance. Ground-colour of underparts also paler than adult, with less buff suffusion but with broader streaks on sides of throat and chest and narrow but distinct dark brown streaks extending along whole of flanks and on to thighs. Wing pattern more distinct than adult, with a few showing off-white margins to black-centred median coverts and these birds resemble even more closely *A. godlewskii* and heaviest-marked immature *A. campestris*. FIRST WINTER. All races. With post-juvenile moult continuing at least to November, plumage often intermediate, some birds having juvenile wing pattern and adult chest pattern. At all ages, bill of *richardi* long, strong, and rather like thrush *Turdus*;

less stout in other races; dark brown horn on culmen, paler ochre-horn on cutting edges and base of lower mandible. Legs and feet long and stout; bright yellow-flesh or yellow-brown. Hind claw usually straight; up to 19 mm long, but averages 16·5 mm in *richardi* and 14·5 mm in *dauricus* (average 12 mm in *A. godlewskii*, 9 mm in *A. campestris*); length of claw usually discernible and overall length of hind toe and claw always striking, exceeding length of tarsus in some *richardi* and closely approaching it in other *richardi* and in *dauricus* and *centralasiae*.

A. novaeseelandiae is one of 4 large pipits which pose daunting problems of field identification (and hence patterns of occurrence of these species still obscure). Currently, skilled observers in Europe have in mind (1) tricky hazard of heavily streaked immature *A. campestris*, (2) long-established autumn occurrence of *richardi* (from western Siberia), and (3) proven but so far undefined vagrancy of *A. godlewskii* (from south-central Siberia), but are ignoring potential occurrence of *dauricus* (marginally sympatric with *A. godlewskii*), *centralasiae* (widely sympatric with *A. godlewskii*), *A. campestris griseus* (from south-central Asia), and *A. similis* (from Middle East). Criteria currently used in distinguishing adult and immature *richardi* from immature *A. c. campestris* are quite inadequate to exclude all other possibilities. Given similarity in plumage pattern and colour, all large pipits should be closely inspected for structural and behavioural differences. Best single character of *richardi* is long, strong, often almost orange legs and feet, with exceptionally long hind claw. On its stout legs, bird runs, walks, and even struts with obvious power and ease, occasionally even suggesting a chicken (this impression not given by any congener). *A. n. richardi* also bulkiest, darkest brown, and most heavily streaked of all large pipits, with typical immature recalling drawn-out *Alauda arvensis* much more than streaked, pale wagtail *Motacilla* suggested by similarly aged *A. campestris*. Unfortunately, *dauricus* has less bulk, rather less stout legs, and slightly shorter hind claw than *richardi* and forms with *A. godlewskii* and immature *A. campestris* a trio of birds extremely difficult to separate. Though incompletely tested, best hope of quick distinction between them (on plumage characters) may lie in face patterns (see 2nd paragraph): in *A. novaeseelandiae*, unmarked lores, long pale supercilium and cheek-surround, and heavy moustachial and malar stripes may allow ready separation from *A. campestris* (see p. 314) and *A. similis* (p. 332); for comparison with *A. godlewskii*, see that species (p. 309). Flight of *A. novaeseelandiae* strikingly powerful, with surging take-off; marked bursts of flapping wing-beats followed by obvious closure of wings against body produce long bounds and 'shooting' curves in full flight (recalling wagtail less than does *A. campestris*); flight undulations most marked of all large pipits; in landing, may first hover well above ground (in both *richardi* and *centralasiae*) before usually making characteristic terminal flutter (at least in *richardi*) during which actions—complete with dangling legs—strongly recall *A. arvensis*. Flight silhouette long and elliptical, with bulky chest and trailing tail obvious. Escape-flight rarely short (often 200–300m), and ending in dense cover when bird pressed; left quiet, often returns to point of first flush. Song-flight high and circling, with fluttering ascent and gliding or parachuting descent. Gait usually combines swift loping run and strutting walk (with large feet lifted noticeably off ground); also hops and leaps after prey. Carriage often markedly erect when alert or searching for food in rough ground but more level when moving fast, with tail held almost horizontally while running. Tail movement not fully studied but lacks pronounced wagging of *A. campestris*. Perches usually on low prominences. Wary rather than wild, allowing careful observer to approach closely. Not as gregarious as small pipits, forming scattered groups rather than flocks. Vagrants usually appear singly but may occur in scattered parties along coasts.

Song a monotonous, repeated, 5-note phrase, 'chi-chi-chee-chee-chee', with last 3 notes falling rapidly in pitch; delivered most in circling phase of song-flight but also from low bush or ground. Voice less varied than *A. campestris* but 3 types of flight-call recognized from *richardi*, *dauricus*, and *centralasiae*: (1) common call, often the only one described from European vagrants, 'r-r-ruüp', 'rreep', 'shreep', 'd-zreep', or 'sh-rout', but always explosive and harsh, with dry tone recalling particularly House Sparrow *Passer domesticus*; characterized by short, slurred, or slightly stuttered start and long, shouted, rising finish; (2) less common call, heard from anxious *dauricus*, soft 'chup' given once or twice, lacking hoarseness of *A. campestris*; (3) another less common call, heard from *richardi*, 'chirp', often given twice and apparently similar to *A. campestris*. Important also to recognize that south Asian races of *A. novae-seelandiae* (not otherwise covered in this account) have wider vocabulary, including commonly uttered soft 'chirp', thin, high-pitched 'tseep', sharp 'twit-twit-it', and 'pipit'. Apparent dichotomy of commonest call between Siberian and south Asian races may be real but possibility of Siberian bird giving unusual call needs to be considered.

Habitat. Breeds extralimitally in continental Asian middle latitudes to tropics, on mainly open lowland level or gently sloping ground of steppe, grassland, or cultivated type, warm and sunny but not arid, below 1800 m in USSR but somewhat higher and occasionally to 3000 m in Himalayan foothills. Ecologically a counterpart in east Palearctic of Meadow Pipit *A. pratensis*, geographically overlapping with Tawny Pipit *A. campestris* from which it differs in choosing more fertile, moister grassland, even marshy meadows, where it coexists with Yellow Wagtail *Motacilla flava*. Prefers, however, short not-too-dense grass, often in river valleys and sometimes in forest-steppe;

avoids taiga and other tree-grown, wetland, or broken and precipitous areas, although tolerating tussocky patches and perching freely on grass tufts and bushes. Also occurs on fallows, edges of cultivation, low dry crops, roadsides, and even stony terrain. In Zambia, breeds commonly on montane grasslands; also on dry plains and cultivation generally, and by airfields (Benson *et al.* 1971). In Zimbabwe, of general occurrence on open habitats, except denuded grassland, dry areas away from water, and some tall grassland unless mown; inhabits montane grassland (at all levels) with dense growth up to *c.* 45 cm tall and crops of lucerne up to *c.* 70 cm. Also common on other croplands where there is bare ground between plants, especially potatoes, cotton, maize, and tobacco, and on golf-courses, playing fields, grazed areas, and water-meadows. Frequents muddy shore of Lake Kariba. (Borrett and Wilson 1970.) In south-west Arabia, often on grassland among thin bushes, on which perches freely: often attends grazing livestock (Meinertzhagen 1954). Attracted to neighbourhood of water. Although a ground dweller, flies strongly and in song-flight ascends to upper airspace. On migration shows preference for ample low vegetation cover; found in front garden of bungalow above 2000 m in Baluchistan (Pakistan) and in tall marram grass in Ireland (Nicholson 1956). Vagrants on Fair Isle usually occur on rough tussocky overgrown pasture or damp ground with sedges giving plenty of low cover, in contrast to *A. campestris* which prefers very open ground with wide field of vision (Davis 1964).

Distribution. Passage visitor to many countries of west Palearctic—Britain, France, Netherlands, Denmark, Sweden, Finland, Spain (also winters in south), Italy (rare), Turkey, and Egypt (uncommon).

Accidental. Ireland, Belgium, West Germany, East Germany, Poland, Czechoslovakia, Austria, Switzerland, Yugoslavia, Greece, Bulgaria, USSR, Lebanon, Kuwait, Israel, Jordan, Cyprus, Portugal, Algeria, Morocco, Malta (almost annual).

Movements. Northern races (*richardi*, *dauricus*, *us-suriensis*, *centralasiae*) are long-distance migrants, wintering among local races from Pakistan to Indo-China and south to Malaysia (Vaurie 1959; Zink 1975). According to Lamarche (1981), *richardi* is rare but widespread in damp areas of Sahel zone in Mali, September or October to April, and birds similar in appearance to some Palearctic migrant races said to be common round Lake Chad in winter, though one collected in March had eggs forming suggesting local breeding (Hall and Moreau 1970); more information required to confirm identity of these birds. *A. n. sinensis* of south-east China apparently moves south within China during winter (Etchécopar and Hüe 1983). Other races of central and south-east Asia and Australasia are resident or largely so, some making altitudinal or short-distance movements. African races basically resident

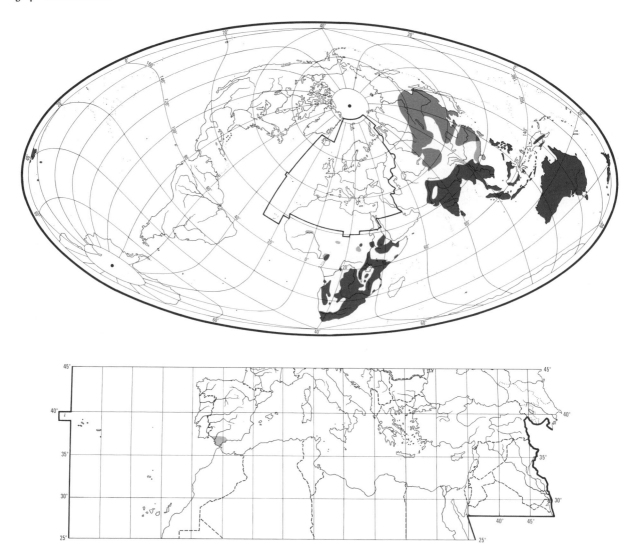

but with some confusing suggestions of local movement. Thus, East African *cinnamomeus* said to be migrant to south-west Zimbabwe and Botswana (White 1962); also to show altitudinal movement in Malawi (Benson and Benson 1977) and Eritrea (Smith 1957), though no such movement reported by Urban and Brown (1971) for Ethiopia as a whole. *A. n. lynesi*, breeding Bamenda highlands of Cameroon, has been recorded at N'Djamena (Chad) and in Nigeria at Kano and Obudu plateau, though latter might be within breeding range (White 1962; Elgood 1982).

Little information to distinguish between movements of *richardi*, *dauricus*, *ussuriensis*, and *centralasiae*. In Yakutiya district in northernmost part of range (*dauricus*), birds leave by mid-September; further south, in higher parts of Altai (*richardi*), movement appears to start early in last third of August, and by mid-September birds also absent from lower down; starts to leave Kazakhstan (*richardi*) from end of August, and exodus completed

early in last third of September (Dementiev and Gladkov 1954*a*). In India, winter visitors (presumably *richardi*) usually present October to April or early May—local and erratic, but common from Bengal eastwards (Ali and Ripley 1973*b*). Present in Sri Lanka from early October to April, with some lingering to May; unlike resident population, wintering birds move within the island, scarcely appearing in Colombo area until February with start of spring migration (Wait 1931). In Hong Kong, main influx in October but early birds arrive from late August; leave April, some remaining until May (Herklots 1967). From populations wintering in south-east Asia, stragglers move through to Malaysia (Medway and Wells 1976) and Borneo (Smythies 1981) and a few perhaps reach Australia where they would mingle with resident races and be easily overlooked (Macdonald 1973).

A feature of the movements of *richardi* is the high frequency with which birds appear well to the west of normal breeding and wintering ranges, even apparently wintering regularly in southern Spain (A Noval). Birds occur mostly as individuals but also in small parties, in western Europe from Scandinavia and Shetland to Ireland and Portugal (Zink 1975), while many reach North Africa from Morocco to Egypt (Heim de Balsac and Mayaud 1962; Etchécopar and Hüe 1967); there is a record from Lake Chad (Vielliard 1971–2) and another from the White Nile (Ogilvie-Grant 1902). On Malta, rare but almost annual in recent years, late September to late February, some birds probably overwintering (Sultana and Gauci 1982). Only 1 recovery of such a vagrant: bird ringed on autumn passage in northern East Germany and recovered Landes (south-west France) in same autumn (Zink 1975). Scale of this westward drift demonstrated by British and Irish records: over 1100 to 1982, and in every year recently (Rogers *et al.* 1983); recorded all months except June–July but most September to mid-November with a few in December; spring records far fewer but with April peak. JHE, DJB

Food. Most information extralimital. Diet mainly invertebrates, taken from among ground vegetation and from crevices in logs and rocks; sometimes takes flying insects by jumping up or in short flights (Lea and Gray 1935–6; Oliver 1955; Matthiessen 1973; McDonald 1974; Thiede and Thiede 1974; St Paul 1975; Rogers 1984).

In Zimbabwe, December–January, of 51 stomachs, 69% contained beetles (Coleoptera: Curculionidae, Chrysomelidae, Tenebrionidae, Scarabaeidae, Cerambycidae), 69% grasshoppers (Acrididae), 37% adult and larval moths (Lepidoptera), 33% Hymenoptera (bees Apoidea, Ichneumonidae, ants Formicidae, and indeterminate wasps), 31% bugs (Hemiptera: Pentatomidae, Lygaeidae, Cercopidae, Coccidae), 18% termites (Isoptera), 12% Arachnida (Araneae, Solifugidae), 12% seeds, 6% flies (Diptera), 2% cockroaches (Dictyoptera), and 2% Myriapoda (Borrett and Wilson 1970). Study from Tokomaru Bay (New Zealand) based on 57 gizzards from all months except January and July. In 80% of gizzards, invertebrates comprised 90% or more of food by volume: 77% contained beetles, 67% Hymenoptera (especially ants), 63% flies, 39% insect larvae, 25% spiders, 25% insect pupae, 18% crickets and grasshoppers, 14% moths, and 12% bugs; 1 feather louse (Mallophaga), 1 isopod (Crustacea), and 1 aquatic snail also present. Other material comprised plant seeds but in only 9% of birds was volume of seeds greater than that of invertebrates: 33% contained clovers (Leguminosae), 25% grasses (Gramineae), 23% Compositae, 23% plantains (Plantaginaceae), 9% sedges (Cyperaceae), 5% docks (Polygonaceae), 5% Labiatae, 4% Cruciferae, 2% chickweeds (Caryophyllaceae), 2% buttercups (Ranunculaceae), and 2% bedstraws (Rubiaceae). (Garrick 1981.) Study of 5 stomachs from Christchurch (New Zealand) gave similar results (Moeed 1975). In Australia, may damage young vegetables (Green and Mollison 1961) and

recorded feeding on crabs (Decapoda) (Mayo 1931). Food from stomachs of race *rufulus* from India included weevils *Myllocerus*, ants (*Philode, Cremastogaster*), termites *Termes*, bugs, spiders, plant seeds, grass leaves, and other vegetable matter (Ali and Ripley 1973*b*). Crops and stomachs of *richardi* from Transbaykalia (eastern USSR) contained mainly beetles and insect larvae (Dementiev and Gladkov 1954*a*). Also recorded in winter: adult butterfly (Lepidoptera), *Ichneumon*, and ladybird *Coccinella* (Witherby *et al.* 1938*a*), and bird in garden, south-west England, took commercial seed mix (G H Gush).

Of 48 collar-samples taken from nestlings in southern Transbaykalia, 93·7% contained Orthoptera, 8% beetles, 6% spiders, and 6% Hymenoptera; food brought to nest every 3–5 min. At another nest, where all food was delivered to young Cuckoo *Cuculus canorus*, 48% of unknown number of samples contained Orthoptera, 45% spiders, 36% beetles, 23% Lepidoptera larvae, and 6% Hymenoptera (Borovitskaya 1972). In Lena valley (eastern USSR) 149 items recovered from nestling collar-samples: 24·2% caterpillars (Lepidoptera), 19·5% flies, 16·1% beetles and larvae, 14·8% Hymenoptera and larvae, 12·1% Odonata, 4·7% Orthoptera, 4·7% spiders, 1·3% stoneflies (Plecoptera), 1·3% worms (Annelida), 0·7% lacewings (Neuroptera), and 0·7% Homoptera (Germogenov 1982). Large grasshoppers and crickets fed to young in preference to caterpillars and other small items (Murton 1970*a*). Nestlings in South Africa fed on earthworms (Oligochaeta) and flies (McDonald 1974). PJE

Voice. Probably varies geographically, but no detailed study. Call 2a is apparently that most frequently given by vagrants to Europe. Some calls confusable with Tawny Pipit *A. campestris* (see that species, p. 322); contact-calls of *lacuum* in Kenya and *malayensis* in Sri Lanka virtually indistinguishable from *A. campestris* (G Högstedt). Situation further complicated by voice of Blyth's Pipit *A. godlewskii* (see p. 311) being poorly known. For extra sonagrams, see (for Mali in winter) Bergmann and Helb (1982) and (for Siberian and Mongolian breeding grounds) Wallschläger (1984).

CALLS OF ADULTS. (1) Song of ♂. Information limited and probable differences between groups of races still to be assessed. Descriptions of song in north Palearctic races as follows. In *richardi* of western Siberia, a rather unattractive and monotonous 'tria-ia-ia-iA' given in high, circling and undulating flight, also from perch or ground; also rendered 'chi-chi-chee-chee-chee', last 3 units falling in pitch (Kozlova 1930; Witherby *et al.* 1938*a*). In Mongolia (presumably *dauricus*), birds also give simple and rather repetitive phrases: 'die-die-die-die-tia-tia-tia' (Mauersberger 1980); 'zip-zip-zip-zip-zip-zip-zip' or 'schip-schip-schip-schip-schip' (Königstedt and Robel 1983). Short phrases (4–9 units and lasting 1·05–2·1 s)

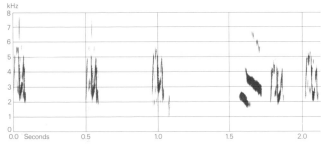

I   C Chappuis/Sveriges Radio (1972-80)   Mali   February 1969

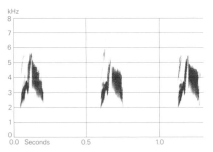

II   B N Veprintsev and V V Leonovich   USSR   May 1975

given in indulating Song-flight contain 2 unit-types: chirruping like sparrow *Passer* during ascent, followed by 4-6 ringing disyllabic 'die' sounds given while beating wings vigorously and hovering; silent descent followed by renewed ascent with song, etc.; 2-4 'die' units sometimes given as continuation of song on ground (Wallschläger 1984). Recording from Mali (race unknown), perhaps of winter Subsong, suggests continuous jumble of sparrow-like chirruping and cheeping sounds, or warble reminiscent of both lark (Alaudidae) and swallow (Hirundinidae). Largely lacking in definite patterning: 'chirrup chirrup twee-twee cheet', etc. Warbling quite sweet at times, regularly interspersed with attractive faint flutier sounds—rendered 'tjü' or 'lü' by Königstedt and Robel (1983). Delivery generally not fast, but some variation. Sequence from same recording shown in Fig I roughly 'che-ik che-ik che-ik hwee cheik cheik', 'hwee' sounding high and squeakily tonal. (J Hall-Craggs, M G Wilson.) Similar song apparently given at times also on breeding grounds. At Lake Baykal (eastern USSR), mid-July, unstructured chirruping song given only from perch or ground and perhaps by juvenile (Wallschläger 1984). In New Zealand, warbling Subsong given occasionally throughout the year (Secker 1955). Johansen (1952) likened song generally to Short-toed Lark *Calandrella brachydactyla*. For further descriptions from East Africa and New Zealand, see (e.g.) Vincent (1946-9) and Secker (1955). (2) Contact-alarm calls. (a) Call most often given by migrants. Distinctly rasping 'schreep'; typically loud, explosive, strident, and far-carrying. 1st part shorter, softer, sometimes omitted or inaudible at a distance, 2nd part louder. In timbre, reminiscent of House Sparrow *P. domesticus* or Skylark *Alauda arvensis* (Grant 1972). Similar to elements in song (see above), but can sound shorter and harsher (Wallschläger 1984). In recording (Fig II) of bird alarmed at nest, calls loud, explosive, rich, fruity, and sparrow-like, at times more buzzing or nasal: 'tree-up', 'chee-er', 'drreee', 'tsreee(p)', and more nasal 'dwee-er' (M G Wilson). Also rendered 'tsrip' or 'rrihp' with clear 'r' sound (slight trilling effect), unlike in flight-call of Tawny Pipit *A. campestris*, though confusable with some sparrow-like calls of that species (Gatter 1976). May give series of such calls if flushed suddenly: 'rrrüp rüpp' (Bergmann and Helb 1982). In New Zealand, harsh and

piercing 'tzree' given in general excitement throughout year, though often subdued in winter (Secker 1955). Recording by L Svensson, USSR, June, similarly suggests 'zreep' or 'zeek', also 'rrreee' or 'tweeep', more buzzing variants recalling Reed Bunting *Emberiza schoeniclus* (J Hall-Craggs; M G Wilson). Similar 'chreef', like Yellow Wagtail *Motacilla flava macronyx* but coarser, given by ♂ *ussuriensis* when disturbed at nest (Panov 1973). Passing migrant in northern West Germany, late October, gave a striking, drawn-out, trilling and harsh 'trruüit' *c.* 5 times every 10 s in flight (Rettig 1970). Weise (1971) noted loud, clear, markedly disyllabic 'schr-schir' once or twice each time bird flushed. (b) Recording (Fig III)

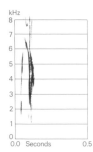

III   C Chappuis/Sveriges Radio (1972-80)
Mali   February 1969

reveals shorter and sharper 'w(h)issik' recalling *M. flava* (J Hall-Craggs; P J Sellar); 'pSICK', 'pSIT', or 'pSEEK' (M G Wilson); single short, hard 'tip' (Bergmann and Helb 1982). Highly excited birds in Mongolia and at Lake Baykal tended to change from this 'tip' to call 2a, 'tchrrip' (Wallschläger 1984). For further monosyllabic calls noted in *dauricus* and for some calls given by races of southern Asia, see Field Characters. For further descriptions, discussion, and other references, see Hakala and Tenovuo (1968); for additional renderings, also other calls given in Africa and New Zealand, see (e.g.) Vincent (1946-9), McDonald (1974), and Secker (1955).   MGW

Plumages (*A. n. richardi*). ADULT BREEDING. Like adult non-breeding (see below), but shows some contrast between fresh breeding (similar in colour to fresh non-breeding) and slightly worn retained non-breeding; usually, head, body, and (at times) some tertials and median upper wing-coverts new April-May, similar to fresh non-breeding; remaining coverts older, as described in worn non-breeding; all plumage heavily worn June-August, especially retained feathers. For variation in amount of new plumage, see Moults. ADULT NON-BREEDING. Ground-colour of upperparts buff-brown with slight rufous tinge, slightly paler on hindneck; forehead and crown marked with black streaks *c.* 2 mm wide, hindneck faintly streaked dark olive-grey; mantle and inner scapulars with broad black or olive-black feather-centres, outer scapulars and longer upper tail-coverts with ill-defined dark brown centres; rump and shorter upper tail-coverts uniform buff-brown. Long super- cilium warm buff, merging in front into pale buff of lores; narrow ring of feathers round eye pale buff; no dark stripe on lores (unlike Tawny Pipit *A. campestris*), except for dull black just in front of eye-ring and sometimes for a few dusky specks. Upper cheeks, ear-coverts, and sides of neck buff-brown with black stripes along upper ear-coverts (below rear of supercilium) and backwards from gape; rear part of ear-coverts and sides of neck faintly streaked with darker grey-brown. Lower cheeks, chin, and throat pale buff (almost white on chin), marked with distinct black malar stripe (sometimes spotted or narrow and then less distinct), which ends in black-streaked patch on upper sides of chest. Chest, sides of breast, and flanks buff or warm buff, merging into slightly paler buff on breast, sides of belly, and under tail-coverts, and into cream-white or white on central belly and vent; chest marked with distinct black spots *c.* 2 mm wide, forming band 1·5–2 cm wide, sharply divided from unmarked throat; sides of breast with faint dull black or dark grey spots, upper flanks sometimes with narrow dark shaft-streaks; remainder of underparts without dark marks. Central pair of tail-feathers (t1) dull black with poorly defined buff-brown fringes; t2–t4 black with narrow buff outer edge; t5–t6 cream-white or white; t5 with black-brown shaft, some- times with dull black outer web (except on tip), and with broad dark grey streak along inner border; remaining white on inner web forms streak 42·1 (49) 33–54 mm long, occasionally (in 6 out of 49 birds) 20–32 and then usually on one side of tail only, in only 3 birds on both sides of tail (thus as in Blyth's Pipit *A. godlewskii*); t6 with narrow dark streak along inner border. Flight-feathers dark brown-grey; outer webs of secon- daries fringed buff-brown, those of primaries more narrowly pale buff or off-white, except on emarginated parts. Lesser upper wing-coverts dull grey with broad poorly defined buff-brown fringes; median upper wing-coverts black with pale cinnamon-buff or pale buff tips *c.* 3–4 mm long; greater upper wing-coverts and tertials dull black with broad buff-brown fringes along outer webs and tips and narrower ones along inner webs (in part soon bleaching to buff or pale buff). Under wing-coverts and axillaries pale grey with cream-buff tinge; lesser coverts spotted dark grey. *In worn plumage* (about November–March), ground-colour of upperparts greyer olive- brown, less warm rufous-buff; black streaks on feather-centres more pronounced, upperparts appearing more heavily streaked except for dull greyish olive-brown rump and shorter upper tail-coverts; sides of head paler with more contrasting dark stripes; lores, supercilium, and a stripe from lores to below eye cream-buff or almost white; ground-colour of underparts paler, chest and flanks less warm buff, remainder almost white, dark spots on chest more contrasting; tips of median upper wing-coverts slightly narrower, paler buff; fringes and tips of greater coverts, tertials, and t1 bleached to pale buff, largely abraded. JUVENILE. Markedly different from adult on upperparts and upperwing, less so on underparts. All upperparts black, feathers narrowly fringed buff, slightly wider and more olive- brown only on feather-sides of outer scapulars, rump, and upper tail-coverts; forehead, crown, and hindneck thus appear black with narrow buff (when fresh) or off-white (when worn) streaks (faintly scalloped when plumage quite fresh), and mantle,

scapulars, back, and rump have narrow buff or off-white scalloping. Lesser upper wing-coverts black with contrasting pale buff fringes (unlike adult); median coverts black, as in adult but with rather evenly buff or off-white fringes *c.* 1 mm wide (in adult, feathers broadly tipped buff); greater upper wing-coverts and tertials black with buff fringe along outer web, as in adult, but fringes narrower and more sharply defined, fringe along tip paler buff or off-white. Lores, supercilium, and eye-ring cream-white or off-white, contrasting strongly with largely black crown; patch below eye and ear-coverts pale buff, finely spotted grey below eye, more densely brown and dull black along upperside and rear of ear-coverts, sometimes bordered below by dark stripe backwards from gape. Lower cheeks, chin, and throat cream-white or off-white; ground-colour of chest, sides of breast, and flanks buff (paler than adult, soon bleaching to pale buff or off-white), remainder of underparts cream-white or off-white; black malar stripe and black spots on chest and sides of breast larger and more strongly contrasting, often confluent to form a large black spot at rear of malar streak on upper sides of chest, and nearly so on upperchest, forming narrow gorget contrasting strongly with uniform throat. Black spots narrower towards upper flanks and, unlike adult, extend into dark shaft-streaks on lower flanks and feathering at thighs. Tail-feathers narrower than in adult, tips more pointed (in particular t1); pale fringes along tip of t1 narrower, paler, and more sharply defined. Flight-feathers as adult. In worn plumage, pale fringes of upperparts abraded, upperparts appearing blacker, without distinct scalloping, mottled pale buff (in particular on hindneck and upper mantle), sometimes with pale median crown-stripe; upper wing-coverts, tertials, and t1 black with narrow and sharply defined white tips and fringes (partly abraded); ground-colour of sides of head and underparts off-white. FIRST NON-BREEDING. Like adult non-breeding, but variable amount of juvenile upper wing-coverts, tertials, and tail-feathers retained (see Post-juvenile in Moults). Adult may also suspend moult during autumn migration, retaining some old coverts, tertials, or tail-feathers, but would also retain some old flight-feathers. Abraded fringes and tips of retained juvenile wing-coverts contrast strongly with black centres, unlike worn adult coverts. FIRST ADULT BREEDING. Like adult breeding, but retaining (at least) juvenile flight-feathers, greater upper primary coverts, and outer tail-feathers, these feathers more heavily abraded at tips than older birds at same time of year; often retains also a number of juvenile outer upper wing-coverts, heavily abraded. Amount of fresh breeding attained is highly variable: sometimes as much as adult, but sometimes virtually none.

Bare parts. ADULT, JUVENILE, FIRST ADULT. Iris brown or dark brown. Upper mandible and tip of lower mandible dark horn-brown or deep brown, cutting edges of both mandibles and basal and middle portion of lower mandible light brown with yellow or flesh tinge. Mouth yellow or pale flesh. Leg and foot light yellow-brown or brownish-flesh, soles chrome-yellow. (Hartert 1910; Ali and Ripley 1973*b*; BMNH, ZMA.)

Moults. ADULT POST-BREEDING. Based mainly on Dementiev and Gladkov (1954*a*) and Stresemann and Stresemann (1968*a*), with additional data from Piechocki (1958), Piechocki and Bolod (1972), and skins (RMNH, ZMA). Complete; primaries descendant. Often not completed at start of autumn migration and suspended until arrival in winter quarters; frequent irregularities in sequence of flight-feathers and tail moult

(Stresemann and Stresemann 1968a). Starts with p1 mid-July to mid-August, followed by scattered feathers of body, some wing-coverts, and tertials from primary moult score 0–10 and by tail from score 5–20; body, wing-coverts, tertials, and tail new from score c. 40, secondaries at about same time as primaries (score 50). In some birds, moult completed late August to late September in breeding area; in others, moult suspended from early September onwards, and variable amount of plumage old during migration. Moult resumed in winter quarters, completed November–December or sometimes as late as April. Due to rather irregular flight-feather moult, several scattered primaries may grow at same time, e.g. p4 and p7 or p5 and p9, with intervening ones old or new; exceptionally, even 5 feathers (p2, p4, p6, p8, p9) may grow at same time (Stresemann and Stresemann 1968a). ADULT PRE-BREEDING. Partial; March–April. Involves head and body (sometimes excluding nape, tail-coverts, throat, or belly), lesser upper wing-coverts, variable number of tertial coverts and tertials, and often t1 and median upper wing-coverts. In some birds, many upper wing-coverts, tertials, and outer tail new March–April, but these are probably non-breeding feathers replaced late in post-breeding rather than feathers of breeding plumage. Occasionally, many of these feathers are new, but head and body rather old, pointing to suppression of pre-breeding moult, with non-breeding of head and body attained earlier in moult and wing-coverts and tertials later—but this perhaps in 1st adult only. POST-JUVENILE. Partial; starts early August to late September (depending on time of hatching). Involves head and body, upper wing-coverts (usually except outer greater coverts), some or all tertials, t1, and sometimes a few other tail-feathers. Some largely in 1st non-breeding late August when still near breeding area, but most birds suspend moult or slow it down during migration, not attaining 1st non-breeding until November–February when in winter quarters. Of 20 European migrants examined (BMNH, RMNH, ZMA), second half of September to first half of November: 1 still completely juvenile; 3 juvenile except for a few new scapulars and mantle feathers; 5 had 40–80% of head and body new or growing, as well as some lesser upper wing-coverts; 2 had head and body new but wing-coverts and tertials still juvenile; remaining 9 had head and body new, as well as virtually all median upper wing-coverts, tertial coverts, some (3) or all (6) lesser upper wing-coverts, some (5) or all (4) tertials, inner (5) or virtually all (4) greater upper wing-coverts, and t1 (4) or t1–t2 (1). FIRST PRE-BREEDING. Partial; March–May. Sometimes almost as extensive as adult pre-breeding, but sometimes less extensive or almost fully suppressed. In 9 birds from USSR and China (BMNH, RMNH), March–May, head, body, wing-coverts, and often t1 and some tertials still fairly new non-breeding, outer greater coverts, upper primary coverts, other tertials, and t2–t6 or all tail still juvenile; only a limited amount of 1st breeding growing on mantle, scapulars, and sides of chest and unlikely that much more feathering would have been replaced later on (see also Adult Pre-breeding, above).

Measurements. ADULT, FIRST ADULT. *A. n. richardi.* Europe, autumn and winter (n=23), and south-central Siberia, May–August (n=11); skins (BMNH, RMNH, ZMA). Bill (S) to skull, bill (N) to distal corner of nostril; exposed culmen on average 4·3 less than bill (S). Toe is middle toe with claw; claw is hind claw. Juvenile wing includes 1st adult.

WING AD	♂ 99·9 (1·94; 9)	97–102	♀ 93·2 (— ; 2)	92–94
JUV	98·2 (2·09; 11)	94–101	93·2 (1·76; 12)	91–96
TAIL	75·5 (3·16; 20)	72–81	73·3 (2·43; 14)	69–77
BILL (S)	18·3 (0·69; 20)	17·1–19·4	17·5 (0·62; 13)	16·5–18·3
BILL (N)	10·3 (0·55; 20)	9·4–11·9	9·9 (0·47; 13)	9·0–10·6
TARSUS	31·3 (1·46; 20)	29·0–33·6	30·2 (0·80; 14)	28·7–31·3
TOE	24·7 (1·44; 7)	23·0–26·3	24·6 (1·83; 7)	23·1–27·2
CLAW	16·2 (2·60; 20)	13·3–24·5	16·8 (1·67; 13)	13·9–19·7

Sex differences significant for wing, bill, and tarsus.

A. n. dauricus. Transbaykalia (USSR), June–July, and eastern China, October–April; skins (BMNH, RMNH, ZMA).

WING	♂ 95·7 (1·96; 8)	94–99	♀ 92·4 (2·32; 8)	89–96
TAIL	73·9 (2·44; 8)	70–76	69·5 (3·21; 8)	65–73
BILL (S)	17·8 (1·17; 8)	16·6–19·8	17·8 (0·70; 8)	16·9–18·4
BILL (N)	10·3 (0·79; 8)	9·5–11·3	10·1 (0·62; 8)	9·3–10·8
TARSUS	29·9 (0·46; 8)	29·6–30·5	29·5 (1·07; 8)	28·2–31·0
TOE	24·0 (0·95; 8)	22·8–25·6	24·1 (1·23; 8)	22·9–25·6
CLAW	13·9 (1·74; 8)	11·9–16·2	14·8 (3·38; 8)	11·6–18·8

Sex differences significant for wing and tail.

A. n. centralasiae. (1) Kansu, China (Stresemann 1931), Tien Shan (USSR) and Nan Shan, China (Johansen 1952), and western and southern Mongolia (Piechocki and Bolod 1972), combined; (2) north-central Mongolia (Piechocki and Bolod 1972).

WING (1)	♂ 99·8 (1·61; 27)	97–104	♀ 95·0 (2·67; 10)	92–100
(2)	97·8 (2·14; 6)	95–101	92·5 (— ; 2)	90–95
TAIL (1)	76·1 (2·46; 19)	72–81	73·0 (1·33; 10)	71–75
(2)	74·2 (2·48; 6)	70–7	72·5 (— ; 2)	71–74
TARSUS (1)	32·2 (0·75; 6)	31–33	31·5 (1·00; 4)	31–33
CLAW (1)	15·2 (1·72; 6)	13–17	15·8 (0·96; 4)	15–17

Sex differences significant for wing and tail.

A. n. ussuriensis. USSR: ♂ 90·8 (15) 85–92·5, ♀ 87·5 (6) 85–91·5 (Dementiev and Gladkov 1954a); ♂ 90–95 (12) (Johansen 1952). North-east China: ♂ 94·5 (0·58; 4) 94–95, ♀ 89·0 (3) 88–90 (Stresemann 1931; Piechocki 1958). 5 ♂♂: wing 90–95, tail 63–70, bill (S) 18–19, tarsus 30–31 (Hall 1961).

A. n. sinensis. Southern China, 8 of each sex: wing, ♂ 87–91, ♀ 81–88; tail, ♂ 58–64, ♀ 55–62, bill (S), ♂ 17–18, ♀ 16·5–18; tarsus 27–30 (Hall 1961).

Weights. *A. n. richardi.* Fair Isle (Scotland), late September and October: 30·0 (3·52; 9) 25–36 (Fair Isle Bird Observatory, N J Riddiford). France, September: 36 (Yeatman 1967). On ship off Portugal, October: exhausted ♀, 21 (ZMA). Netherlands: September, exhausted ♂ 22·5; October, ♀ 37 (ZMA). Kazakhstan (USSR), May: ♀ 30·5 (Dolgushin et al. 1970). Switzerland, lean ♂, November: 23 (Widmer 1979; Winkler 1979).

A. n. centralasiae. (1) Kansu (China), June (Stresemann 1931). (2) Eastern Kazakhstan (USSR), June–August (Dolgushin et al. 1970). Mongolia: (3) May, (4) June, (5) July–August (Piechocki and Bolod 1972).

(1)	♂ 33·6 (1·39; 6)	31–35	♀ 37·9 (2·23; 4)	35–40
(2)	33·3 (1·38; 10)	32–37	31·6 (3·84; 4)	27–36
(3)	33·9 (2·47; 8)	30–37	28·0 (— ; 1)	
(4)	33·6 (2·07; 5)	30–35	30·7 (— ; 3)	28–33
(5)	32·8 (2·32; 6)	30–35	31·8 (3·86; 4)	28–37

A. n. ussuriensis. Manchuria (China), mainly second half of August, mainly ♀♀: 29·5 (6) 27–30 (Piechocki 1958).

Structure. Wing rather long, broad at base; tip bluntly pointed. 10 primaries: p8 longest, p7 and p9 0–1 shorter, p6 1·5–3·5, p5 9–13, p4 15–20, p3 18–23, p1 23–29; p10 reduced, narrow and pointed, 60–67 shorter than wing-tip, 10–13 shorter than longest upper primary coverts. Outer web of p6–p8 and inner of p7–p9 emarginated. Longest tertial reaches to 0–3 from wing-tip in closed wing. Tail long, tip square or slightly forked; 12 feathers, t1 2–8 shorter than t5. Bill long, rather deep and wide at base, slender near tip; straight, but culmen slightly decurved near tip. Nostrils rather large; only narrowly covered by membrane along upperside. Some long bristles near base of upper mandible. For length of middle toe with claw, and hind claw, see Measurements. Toes long and slender; hind toe without claw (measured on sole) 13–15 (9–10 in *A. campestris*, 11–12 in *A. godlewskii*). Outer toe with claw *c.* 64% of middle with claw, inner *c.* 68%; hind with claw *c.* 113%, without claw *c.* 54%. Front claws rather long and slender; hind claw very long, slender, only slightly decurved.

Geographical variation. Rather slight in Asia, mainly involving size; more marked in Australasia and (especially) Afrotropics. Migratory races of northern Asia (*richardi*, *centralasiae*, *dauricus*, *ussuriensis*) much larger than non-migratory races of southern Asia, and hence sometimes considered different species *A. richardi*, but difference bridged by intermediate *sinensis* from southern China, which intergrades with both northern and southern races. In northern Asia, variation in colour and size clinal, cline running from *centralasiae* (largest) through *richardi*, *dauricus*, and *ussuriensis* to *sinensis* (small and dark). *A. n. richardi* from Barabinskaya steppe and Altai mountains (USSR) east to Yenisey rather rufous buff-brown on upperparts in fresh plumage, streaks rather indistinct, chest and flanks rather deep buff; *centralasiae* from western and southern Mongolia and north-west China is paler buff-brown or sandy-brown with slightly narrower streaks on upperparts and paler buff tinge below in fresh plumage, inseparable when worn; birds from eastern Soviet Tien Shan mountains, Zaysan area, and probably Tuvinskaya ASSR tend slightly towards

richardi in colour. *A. n. dauricus* from Yakutiya and middle Amur and Zeya rivers south to northern Transbaykalia has ground-colour of upperparts similar to *richardi*, but streaks slightly broader and blacker and underparts rather more extensively washed buff; often indistinguishable from *richardi*, except for slightly shorter wing and tail (see Measurements); merges into *richardi* west from Nizhnyaya Tunguska and Lake Baykal; birds from southern Transbaykalia are paler than further north, nearer *centralasiae* in colour but not in size; merges into *centralasiae* in north-central Mongolia and perhaps north-east Mongolia and western Manchuria. *A. n. ussuriensis* from easternmost USSR and north-east China south to Yangtse Kiang and *sinensis* from southern China both heavily streaked black on upperparts, feather-fringes rather narrow, olive-grey; underparts extensively washed rufous-cinnamon; *ussuriensis* larger, wing of ♂ mainly 90–95, ♀ 87–92; *sinensis* smaller, wing of ♂ mainly 85–91, ♀ 81–88, tail and tarsus distinctly shorter also (see Hall 1961). For other races, see Hall (1961), Ali and Ripley (1973*b*), and Clancey (1978). For evolutionary trends, see Hall and Moreau (1970) and Clancey (1978).

Recognition. Distinguished from most other *Anthus* by large size and spotted chest. Juvenile and some adult Tawny Pipits *A. campestris* also spotted below, but upperpart-streaks of juvenile brown (not black) and adult virtually unstreaked above; both adult and juvenile show a distinct dark streak on lores (virtually absent in *A. novaeseelandiae*) and have short tarsus and (in particular) short hind claw. Blyth's Pipit *A. godlewskii* closely similar to *A. novaeseelandiae* however; see that species (p. 313). CSR

Anthus godlewskii Blyth's Pipit

PLATES 19 and 27
[between pages 232 and 233]

Du. Blyths Pieper   Fr. Pipit de Blyth   Ge. Steppenpieper
Ru. Конек Годлевского   Sp. Bisbita de Blyth   Sw. Mongolpiplärka

Agrodroma Godlewskii Taczanowski, 1876

Monotypic

Field characters. 17 cm; wing-span 28–30 cm. Close in size to Richard's Pipit *A. novaeseelandiae richardi* but less heavily built, with 10% shorter and finer-tipped bill, marginally shorter wings (with longest tertial cloaking primary-tips), 15% shorter tail, 10% shorter and finer legs, and 25–30% shorter hind claw. Large and long, rather bright, buff, streaked pipit, with general character closer to *A. novaeseelandiae* than *A. campestris* but shorter-tailed than both. In all plumages, pattern similar to *A. novaeseelandiae* and immature *A. campestris* but general tone more ochre or orange than *A. novaeseelandiae richardi*, closely matching *A. n. dauricus* and *A. n. centralasiae*, and less ochre or grey than *A. campestris*. In all plumages, indistinct cheek-surround, rather pale hindneck, prominent median and greater covert panels, and (particularly) orange-buff tone of whole underbody including under tail-coverts provide clues, but no clearly diagnostic characters yet established. Flight as *A. novaeseelandiae*. Commonest call a harsh, truncated mono-

syllable. Sexes similar; no seasonal variation. Juvenile not studied in the field.

Adult. Closely similar in appearance to *A. novaeseelandiae dauricus* and *A. n. centralasiae* but distinguished (at least in northern Mongolia where it occurs mainly with *A. n. dauricus* and intermediates between that race and *A. n. centralasiae*) by: (a) slightly smaller size (particularly of head) and shorter length (particularly of tail); (b) finer bare parts (see below); (c) less bold head pattern (see below); (d) paler, more spotted than streaked hindneck; (e) more contrasting streaking on paler mantle; (f) more distinct wing-panels, with most birds having pale sandy to white margins of median coverts contrasting vividly with black-brown centres which are almost square-cut distally (more pointed in adult *A. novaeseelandiae* and in juveniles of both species: see *Br. Birds* 1987, **80**, 50–2); creamy margins of greater coverts also obvious; (g) thin spots (rather than streaks) on chest, forming only sparse band which widens only slightly at sides and does not extend on to fore-flanks; (h) usually warm, orange-buff underwing, flanks, vent, and (importantly) under tail-coverts; (i) broader white outer margins and corners to tail; (j) basically paler legs and feet with more consistently yellow tone (less brown or orange). Head almost as well marked as in *A. novaeseelandiae* but pattern differs in : (a) interruption of pale lore by short dark brown mark in front of eye, less obvious eye-ring, and shorter supercilium which is also less well defined above due to dark edge of crown forming only short 'eyebrow'; (b) buff-brown rear cheeks, and narrow brown-black moustachial stripe joined to mark on lore; (c) less distinct pale cheek-surround due to paler (spotted rather than streaked) hindneck; (d) rather short, less distinct brown-black malar stripe (not reaching chin nor ending in patch of streaks at side of chest). In the field, structural differences from *A. n. dauricus* apparently less obvious than from *A. n. richardi* and *A. n. centralasiae* (see 1st paragraph) but field observations show that bill is finer-tipped, legs are more slender (D I M Wallace) and shorter, hind claw is shorter (Kitson 1979), and longest tertial cloaks primary-tips (Königstedt and Robel 1983). Hind claw averages 12 mm, thus falling closer to *A. campestris* than *A. novaeseelandiae*; combined length of hind toe and claw falls short of length of tarsus but not so markedly as in *A. campestris*. IMMATURE. Similar, even identical, in appearance to immature *A. novaeseelandiae*; adult's differences in head pattern and upperpart streaking likely to be suppressed by heavier markings of juvenile plumage. Thus both juvenile and 1st-winter birds have extremely confusing intermediate appearance between *A. novaeseelandiae* and *A. campestris*, with size suggesting *A. novaeseelandiae* but bare part structure, leg colour, and rather pale, tawny upperparts recalling *A. campestris*. Orange-buff tone to most of underbody provides distinction from *A. campestris* and *A. novaeseelandiae* but this variably obvious in different lights.

Best hope of diagnosis on plumage appears to be combination of face pattern and (in adult) median covert panel, both of which need acute observation. For cautionary remarks on field identification of large pipits, see *A. novaeseelandiae* (p. 302). Large pipits with appearance intermediate between *A. novaeseelandiae* and *A. campestris* occur in Britain (Grant 1980) and their assignment to a race of *A. novaeseelandiae* other than *A. n. richardi* or to *A. godlewskii* is difficult to resist. Flight of *A. godlewskii* much as *A. novaeseelandiae* but somewhat lighter on take-off and landing, lacking fluttering stall of *A. novaeseelandiae* when landing. Gait closer to *A. campestris* than *A. novaeseelandiae* and lacks strutting walk of latter. Carriage less upright than *A. novaeseelandiae*. Tail-wagging less persistent than in *A. campestris* but more frequent than in *A. novaeseelandiae*. Quite tame, particularly in montane populations.

Flight-calls include 3 types: (1) common call, 'psheeoo', too harsh and terminally stressed to cause confusion with *A. campestris* but still somewhat recalling Yellow Wagtail *Motacilla flava*; (2) common call, perhaps only of alarmed parent bird—'dzeep', with dry, anxious tone; distinguishable in Mongolia from typical rasping 'shreep' of *A. novaeseelandiae*; (3) less common call (often given once or twice before 'psheeoo'), 'chup', recalling *A. campestris* and *A. n. dauricus* or *centralasiae*.

Habitat. Breeds extralimitally in middle, lower middle, and low latitudes in continental steppe zone and in uplands and mountains. In USSR, inhabits dry rocky mountain slopes with scant vegetation (Dementiev and Gladkov 1954a). In Transbaykalia (USSR), breeding birds always associated with *Caragana* thickets in stony steppe (Dorzhiev 1983). In Mongolia, occasionally overlaps with Richard's Pipit *A. novaeseelandiae* on mountain slopes and permafrost ground and on very damp meadows, but typically differs in avoiding latter in favour of gravel or stony steppe, meadow steppe, and dry slopes, often in company with Isabelline Wheatear *Oenanthe isabellina* and Skylark *Alauda arvensis* (Königstedt and Robel 1983). Reported nesting in Assam (India) on ridge at *c.* 1600–2000 m, at least 500 m higher than *A. novaeseelandiae rufulus*, but recent confirmation lacking. On migration prefers swampy land, and in winter resorts to dry ricefields, fallow, edges of cultivation, and grassland. Recorded on migration at 6000 m on Mount Everest (Ali and Ripley 1973b). Pending further reliable data, habitat of this species remains inadequately known.

Distribution. Breeds from southern Transbaykalia and eastern Manchuria south to Tibet. Winters in India and Sri Lanka.

Accidental. England: one collected 23 October 1882 (Williamson 1977; *Ibis* 1980, **122**, 564–8). Finland: one collected 10 October 1974 (*Ornis fenn.* 1978, **55**, 84–5). Netherlands: one collected 13 November 1983 (CSR).

Movements. Migratory. Widespread and locally common in main winter quarters—Pakistan, Bangladesh, and peninsular India (Ali and Ripley 1973b).

Autumn departure starts July in Transbaykalia (USSR), birds staying *c.* 15–25 days after young leave nest (Dorzhiev 1983); August–September in eastern Tibet (Dementiev and Gladkov 1954a), with birds present until early October (Vaurie 1972). Passage extends from August to end of October in Bhutan, Sikkim, and Nepal (Diesselhorst 1968a; Ali and Ripley 1973b), with birds noted on Everest at up to 6000 m, providing evidence for broad-front southward passage through Himalayas (Martens 1971). Winter quarters occupied from early September to end of April or early May, but birds also reported in June in Sutlej valley and July in upper Assam. In Tibet (where some breed), northward passage starts early May (Vaurie 1972). Birds arrive Nerchinsk area (USSR) with considerable regularity between 7 and 15 May, though at same time other birds are still much further south moving in flocks through Tibet (Dementiev and Gladkov 1954a). Moreau (1972) reported a straggler in Africa at Lake Chad but suggested it could be a local race of Richard's Pipit *A. novaeseelandiae*.   JHE

Voice. Detailed study required to substantiate alleged differences between *A. godlewskii* and both Richard's Pipit *A. novaeseelandiae* and Tawny Pipit *A. campestris*.

CALLS OF ADULTS. Little information from outside breeding season. In Nepal, normally single call given at take-off (Fleming *et al.* 1976): harsh, with peculiarly hoarse timbre and said to be distinguishable from *A. novaeseelandiae rufulus* (Diesselhorst 1968a). Other data relate to Mongolian breeding grounds where Kitson (1979) found calls to be separable from *A. novaeseelandiae* with practice. Harsh, terminally inflected 'psheeoo' somewhat reminiscent of Yellow Wagtail *Motacilla flava*, coarseness precluding confusion with *A. campestris*, though often preceded by 1–2 soft 'chup' sounds recalling that species and *A. novaeseelandiae dauricus* (D I M Wallace). Dry, anxious 'dzeerp' probably given only by breeding birds (Kitson 1979); 'schriep' ('ie' pronounced 'ee' in English) mentioned by Mauersberger (1980) perhaps the same, though rendering suggests something very close to call 2a of *A. novaeseelandiae* (see p. 306). Very high-pitched 'drrri' also an expression of alarm in territory but considered same as flight-call given outside breeding season (see above) and similar to flight-call of *A. campestris* (Wallschläger 1984, which see for sonagram). Birds flying about in alarm near nest gave persistent 'dchip-dchip-dchip' (Königstedt and Robel 1983). Other calls mentioned by Mauersberger (1980) 'dschrill' and 'pie'. Some of the above or similar sounds, also a unit resembling that of *A. campestris*, are apparent in song, for detailed descriptions of which see Mauersberger (1980), Königstedt and Robel (1983), and Wallschläger (1984).   MGW

Plumages. Adult virtually inseparable from Richard's Pipit *A. novaeseelandiae* when plumage worn, about December–February and May–August, but distinguishable (in direct comparison) in fresh plumage September–November and March–April. Identification best made by pattern of t5 (more difficult by other plumage characters: see Field Characters), short tarsus, and rather short hind claw (see Recognition), but even these characters show some overlap with long-winged migratory races of *A. novaeseelandiae* inhabiting northern Asia (including breeding range of *A. godlewskii*). ADULT BREEDING. Fresh plumage slightly paler on upperparts with more contrasting streaking than *A. novaeseelandiae* (as in non-breeding, see below), tips and fringes of new median upper wing-coverts, tertial coverts, and tertials paler, but both species retain variable number of old non-breeding wing-coverts and tertials—these bleached, rendering identification difficult (but see Recognition). ADULT NON-BREEDING. In fresh plumage, extent and colour of upperpart streaking similar to *A. novaeseelandiae* (forehead and

crown marked with dark streaks *c.* 2 mm wide, mantle and inner scapulars with broader and more rounded dark feather-centres), but ground-colour different, sandy-buff on forehead, crown, hindneck, sides of neck, and mantle, pale olive-brown on scapulars, back, rump, and upper tail-coverts (rufescent buff-brown in *A. novaeseelandiae*); due to paler ground-colour, streaks slightly more contrasting, in particular on crown, hindneck, and mantle. Dark stripes on head similar to *A. novaeseelandiae*, though stripes also somewhat variable in extent (no dark stripe on lores, unlike Tawny Pipit *A. campestris*); ground-colour paler, however, cream rather than buff, supercilium in particular more contrastingly pale. Underparts closely similar to *A. novaeseelandiae*, extent and size of chest-spots similar (but spots often dark olive-brown or olive-black triangles instead of rounded black marks), but ground-colour slightly more saturated buff, and buff tinge more extensive, only centre of belly cream-white. For distinctive pattern of t5, see Recognition. Lesser upper wing-coverts paler than *A. novaeseelandiae* (like upperparts); centres of median and greater upper wing-coverts and tertials black as in *A. novaeseelandiae* and pale fringes equally wide, but fringes slightly paler than *A. novaeseelandiae* in equally fresh plumage, pinkish cream-buff or yellow-buff (not pale rufous-cinnamon or cinnamon-buff), contrasting slightly more with black centres. *In worn plumage*, ground-colour of head and body and pale tips and fringes of coverts and scapulars bleach to buff or pale buff in both *A. novaeseelandiae* and *A. godlewskii*, and both species then closely similar, separable only in direct comparison of series of skins. Colours of *A. novaeseelandiae* given above refer only to *A. n. richardi*; ground-colour of Mongolian race *A. n. centralasiae* paler buff-brown on upperparts than *richardi*, closer to *A. godlewskii*, tips and fringes of wing-coverts and tertials pale, as in *A. godlewskii*, underparts with paler and more restricted buff than *A. godlewskii*, but virtually inseparable. JUVENILE. Similar to juvenile *A. novaeseelandiae*, except for pattern of t5 and some measurements (see Recognition). FIRST NON-BREEDING AND BREEDING. Similar to adult non-breeding and breeding, but variable number of juvenile upper wing-coverts and tertials retained; flight-feathers, greater upper primary coverts, and t2–t6 juvenile, relatively more worn than adult at same time of year, tail-feathers narrower and more pointed. During September–November, ground-colour of upperparts and pale tips and fringes of new upper wing-coverts and tertials slightly paler and more contrasting than those of *A. novaeseelandiae*, but variable number of juvenile wing-coverts and tertials retained in both species (especially many in some *A. novaeseelandiae*), and tips and fringes of these even paler and more contrasting than those of fresh non-breeding of *A. godlewskii*, rendering identification impossible. By mid-winter, head and body worn, more similar to *A. novaeseelandiae*; new wing-coverts and tertials relatively paler-tipped than those of *A. novaeseelandiae* (again, beware of retained pale-tipped juvenile coverts of *A. novaeseelandiae*). In spring, fresh breeding plumage may develop on head and body in both species (differences as in adult breeding), but number of new (breeding), rather worn (non-breeding), and heavily worn (juvenile) upper wing-coverts and tertials strongly variable, and identification possible only by pattern of t5 and measurements (see Recognition).

Bare parts. ADULT, JUVENILE, FIRST ADULT. Iris brown or dark brown. Upper mandible and tip of lower mandible dark horn-colour, cutting edges of both mandibles and base of lower flesh-colour. Mouth bright yellow or pinkish-flesh. Leg flesh-colour, flesh-yellow, or yellowish, foot flesh-yellow or yellow-horn, soles yellow. (Ali and Ripley 1973b; BMNH, ZMA.)

Moults. ADULT POST-BREEDING. Complete; primaries descendant. In Mongolia, moult not started in 3 birds from July and early August; primary moult scores *c.* 6 and *c.* 22 on 16 August, *c.* 28 on 26 August (Piechocki and Bolod 1972). In Manchuria, plumage new but outer primaries still growing late August (Piechocki 1958). Plumage completely new on arrival in winter quarters from second week of September (BMNH). Thus, primary moult probably starts late July to mid-August and all feathering new from late August. Moult occasionally suspended during autumn migration, as in *A. novaeseelandiae* and Tawny Pipit *A. campestris*, completed in winter quarters (Hall 1961). ADULT PRE-BREEDING. Partial, with timing probably as in *A. novaeseelandiae* and *A. campestris*. Includes head, body, t1, tertials, and variable number of upper wing-coverts, as in *A. novaeseelandiae* (Hall 1961). POST-JUVENILE. Partial; extent rather variable, moult often interrupted during autumn migration. In USSR, 1 from 5 August had not yet started, 2 others from 16 and 30 August had moulted completely except upper wing-coverts (Dementiev and Gladkov 1954a); thus, probably starts between late July and late August, depending on fledging date. On arrival in winter quarters, second half of September and early October, head and body either 1st adult (except for some longer upper tail-coverts) (15 birds), or juvenile except for scattered scapulars and part of back and chest (3 birds); some lesser coverts and tertial coverts usually new. By November, some birds in full 1st adult plumage (only flight-feathers, greater upper primary coverts, and t2-t6 still juvenile), others (e.g. ♀ from Netherlands) have 1st adult head and body (occasionally except for a few scattered feathers), and juvenile wing, tertials, and tail (except for many lesser upper wing-coverts and some inner median and tertial coverts). (BMNH, ZMA). FIRST PRE-BREEDING. Apparently similar to *A. novaeseelandiae* and *A. campestris* in timing and extent, with equally marked individual variation, but only a few skins examined (BMNH).

Measurements. ADULT, FIRST ADULT. India, September–October; skins (BMNH), with additional data on wing and tail from Manchuria (Piechocki 1958) and Mongolia (Piechocki and Bolod 1972). Bill (S) to skull, bill (N) to distal corner of nostril; exposed culmen on average *c.* 4·6 shorter than bill (S). Toe is middle toe with claw; claw is hind claw.

WING	♂ 93·4 (2·36; 27)	89–98	♀ 89·1 (2·25; 29) 84–93
TAIL	68·2 (3·18; 27)	62–75	64·9 (2·21; 29) 60–69
BILL (S)	17·1 (0·61; 15)	16·2–18·1	16·6 (0·70; 12) 15·7–17·9
BILL (N)	9·6 (0·49; 15)	8·8–10·4	9·5 (0·54; 10) 8·8–10·3
TARSUS	26·9 (0·92; 16)	24·9–27·9	26·5 (1·09; 12) 24·7–28·1
TOE	22·2 (0·61; 4)	21·5–22·8	20·0 (— ; 3) 19·7–20·5
CLAW	12·7 (1·21; 15)	10·5–14·6	11·5 (1·08; 12) 9·9–13·4

Sex differences significant for wing and tail.

Measurements of vagrants in Britain (Williamson 1977; BMNH) and Finland (Rauste and Salonen 1978), both probably ♀♀, and of ♀ from Netherlands (ZMA), respectively: wing 88, 89, 93; tail 66, 65, 64; bill (S) 16·2, 16, 16·6; tarsus 27·2, 27, 27; hind claw 11·8, *c.* 13·4, 12·8.

Weights. USSR. ♂ 25·5 (Dementiev and Gladkov 1954a). Mongolia. Mid-May to early June: ♂♂ 22, 24; ♀♀ 25, 28. Late June to mid-July: ♂ 25·2 (1·79; 5) 22–26, ♀ 27·2 (1·10; 5) 26–28. August: ♂ 25·0 (3) 24–26, ♀ 24·8 (2·32; 6) 22–28. (Piechocki and Bolod 1972.) Manchuria (China), second half of August, mainly ♀♀: 26·5 (8) 25–28 (Piechocki 1958). Nepal. September: ♂ 24·7; ♀♀ 22·7, 24·9. October: ♀ 28·4 (2·35; 4) 25·1–30·5. (Diesselhorst 1968a.) Netherlands, after night in captivity, November: ♀ 20·1 (ZMA).

Structure. 10 primaries: p8–p9 longest or either one 0–1 shorter than other; p7 0·5–1 shorter, p6 1·5–3, p5 11–15, p4 15–20, p3 18–23, p1 23–27; p10 reduced, 61–68 shorter than wing-tip, 10–14 shorter than longest upper primary covert. Tail relatively slightly shorter than in *A. novaeseelandiae*, 69–78% of wing (72–83% in *A. novaeseelandiae*). Bill slightly finer than in *A. novaeseelandiae*, especially at base (but matched by many ♀♀ of northern races of *A. novaeseelandiae*); distinctly shorter than in *A. campestris*. Tarsus relatively slightly shorter than in *A. novaeseelandiae*, *c.* 29% of wing-length, as in *A. campestris* (*c.* 31% in *A. novaeseelandiae*), but proportions of toes and claws compared with tarsus near those of *A. novaeseelandiae*, toes and claws not as proportionately short as in *A. campestris*. For length of middle toe with claw, and hind claw, see Measurements. Outer toe with claw *c.* 70% of middle toe with claw, inner *c.* 67%; hind with claw *c.* 104%, without claw *c.* 54%.

Geographical variation. None.

Forms species-group with *A. novaeseelandiae* and *A. campestris* (Stepanyan 1983). Sometimes considered a race of *A. campestris* (e.g. in Dementiev and Gladkov 1954a), but breeding ranges overlap in western and southern Mongolia (Piechocki and Bolod 1972). Plumage pattern and structure nearer *A. novaeseelandiae*, but not conspecific with that species either, as breeding range overlaps completely with *A. n. centralasiae* and partly with *A. n. dauricus*. In fact, structurally rather closer to small non-migratory Indian races of *A. novaeseelandiae* than to

Table A Measurements of Blyth's Pipit *A. godlewskii*, Tawny Pipit *A. campestris*, and various races of Richard's Pipit *A. novaeseelandiae*. Data in **bold type** show no or only very slight overlap with those of *A. godlewskii*. Wedge t5 is maximum extent of white on inner web of 2nd outermost tail-feather, after examination of both sides of tail. (Hall 1961; BMNH, RMNH, ZMA.)

	A. godlewskii	*A. campestris*	*A. n. richardi*	*A. n. dauricus*	*A. n. ussuriensis* and *sinensis*	*A. n. rufulus* and *malayensis*
WING	84–96(–98)	84–98(–101)	**91–102**	(82–)92–99	81–95	**76–87**
TAIL	60–73(–75)	61–73	**69–81**	(65–)68–77	55–70	**51–59(–62)**
BILL (S)	15·7–18·1	(16·8–)17·6–20·7	(16·5–)16·9–19·4	(16·6–)17·3–19·8	16·5–19·0	15·8–18·4
TARSUS	24·7–28·1	23·9–27·6	**(28·7–)29·0–33·6**	**(28·2–)28·8–31·0**	27·0–31·0	26·0–28·9
HIND TOE (WITH CLAW)	20·5–23·3(–25·6)	**15·0–20·6(–21·4)**	**(25·4–)27·2–32·6**	(22·8–)24·3–32·1	—	—
HIND CLAW	9·9–13·4(–14·6)	**6·8–10·1**	**(13·3–)13·9–24·5**	(11·5–)13·2–19·7	12·0–14·9	(9·4–)10·4–14·5
WEDGE T5	15–30(–38)	(10–)22–42	**(20–)37–54**	(24–)35–50	—	(23–)26–40
WING/TARSUS	3·2–3·6	3·3–3·8	3·0–3·4	2·9–3·3	2·8–3·1	2·7–3·2
Sample size	27	85	34	16	more than 10	30

larger migratory Palearctic races of latter, and perhaps better considered a race of these Indian birds (under name of *A. novaeseelandiae*), while northern birds form separate species (*A. richardi*) (C S Roselaar), though position of *sinensis* then unclear, being intermediate between northern and southern birds and said to intergrade with both.

Recognition. Often hardly separable in plumage details from Richard's Pipit *A. novaeseelandiae* and juvenile Tawny Pipit *A. campestris*, but nearly always differs in measurements and often in pattern of t5. See Table A. *A. campestris* virtually always distinguishable by short hind toe or short hind claw; at all ages, differs also by dark stripe on lores (unlike *A. godlewskii* and *A. novaeseelandiae*). *A. n. richardi* and *A. n. dauricus* separable by longer tarsus and often by longer hind toe or hind claw (some *A. n. dauricus*, in particular, show short hind claw, however); also, *A. n. dauricus* usually has lower wing/tarsus ratio. White wedge on t5 always short in *A. godlewskii*, usually long in *A. n. richardi* and *A. n. dauricus*; 6 of 49 specimens of latter races showed short wedge on one side of tail, however, with other side normal, and hence both feathers must be checked (in those *A. godlewskii* examined, pattern on t5 symmetrical, both wedges short); in 3 of 49 *A. n. richardi* and *A. n. dauricus*, wedge short on both feathers, similar to *A. godlewskii*. Smaller Asiatic races of *A. novaeseelandiae* show large overlap with *A. godlewskii* in length of tarsus, hind toe, hind claw, and wedge, but *A. n. rufulus*, *A. n. malayensis*, and other races from southern Asia readily separated by short wing and tail; wedge on t5 of these races seemingly short, but not when compared with tail length and usually forms a stripe, not a triangular patch. Eastern races *A. n. ussuriensis* and (in particular) *A. n. sinensis* more difficult to identify; apart from heavier streaks on upperparts and extensively brown underparts in adult, separation from *A. godlewskii* possible only by wing/tarsus ratio, and often by shape of wedge on t5 (stripe rather than triangle).   CSR

Anthus campestris Tawny Pipit

PLATES 20 and 27
[between pages 232 and 233]

Du. Duinpieper   Fr. Pipit rousseline   Ge. Brachpieper
Ru. Полевой конек   Sp. Bisbita campestre   Sw. Fältpiplärka

Alauda campestris Linnaeus, 1758

Polytypic. Nominate *campestris* (Linnaeus, 1758), west Palearctic, western and northern Iran, and through south-west Siberia and north-west Kazakhstan (USSR) east to about Omsk; *griseus* Nicoll, 1920, eastern Iran and south-west Kazakhstan east to Tien Shan and Dzhungarskiy Alatau mountains, migrant through Middle East and Egypt. Extralimital: *kastschenkoi* Johansen, 1952, southern Siberia and north-west Mongolia from about Omsk through Altai mountains to upper Yenisey river.

Field characters. 16·5 cm; wing-span 25–28 cm. Noticeably less heavily built than Richard's Pipit *A. novaeseelandiae* and Blyth's Pipit *A. godlewskii*, with bill size and tail length intermediate, back narrower than both, wings as short as *A. godlewskii*, legs 15% shorter than *A. novaeseelandiae*, and hind toe and claw 40% shorter than *A. novaeseelandiae* and 20% shorter than *A. godlewskii*. Noticeably slimmer and shorter than Long-billed Pipit *A. similis*, with 10% shorter bill and 10–15% shorter wings and tail, but similar legs and hind claw. Noticeably long, slim pipit, with size, form, and appearance suggesting pale wagtail *Motacilla*; long, rather fine bill and legs form useful characters. Adult shows diagnostic combination of pale supercilium, virtually unstreaked chest, bold wing-covert bars contrasting with indistinct markings on rest of upperparts, and tail broadly edged white. Immature heavily streaked, resembling *A. novaeseelandiae* and *A. godlewskii*. Flight like wagtail, with less powerful action and slighter undulations than other large pipits. Commonest call a long, plaintive monosyllable, recalling Yellow Wagtail *Motacilla flava*. Sexes similar; no seasonal variation. Juvenile separable. 2, possibly 3, races occur in west Palearctic.

ADULT. (1) West Palearctic race, nominate *campestris*. Forehead and crown sandy-ochre, indistinctly mottled dark brown in centre and streaked black-brown on edge; long supercilium pale cream, contrasting with dark edge of crown and with almost black loral stripe and brown rear eye-stripe; cheeks mottled brown with dark lower border at front forming moustachial stripe; chin and throat white- to yellow-cream, with narrow indistinct malar stripe. Important to note that while *A. campestris* shows palest head of all large pipits at distance, contrast of pale supercilium and dark <-mark formed by front of eye-stripe and moustachial stripe form obvious diagnostic feature at close range. Mantle and scapulars sandy-ochre appearing uniform at distance but showing 'furrows' of small, dull brown, mottled markings at close range; rump unstreaked, appearing paler and brighter than back; upper tail-coverts as back. Folded wings sandy-brown, distinctly darker than back in ground-colour but with pale yellow-buff to buff-white fringes and tips in fresh plumage obscuring the difference; black-brown median coverts with buff tips form striking panel over forewing which thus shows more contrast than in any other large pipit; broad cream-buff fringes to tertials also noticeable. Tail brown, with broad pale buff fringes of central feathers obvious when folded and dull white to cream outer edges to 2 outermost feathers striking on take-off; effect is to give *A. campestris* pale edges to tail as broad as but less contrasting than *A. novaeseelandiae richardi*. Chest and flanks sandy-buff; breast shows a few fine, dark brown streaks on sides (sometimes also across chest) but flanks always unmarked; belly, vent, under tail-coverts and under wing-coverts yellow-cream to buff-white. Underparts thus paler and less patterned than any other large pipit. Rather yellow ground-colour of fresh upperparts may have grey tinge on crown or browner tone on back but latter usually only obvious in worn plumage (June–August) when general loss of buff fringes reduces yellow and ochre tones and increases brown or pink-brown uniformity of back and wings and makes underparts paler, with consequently sharper appearance of breast-streaks. (2) West-central Asian race, *griseus*. Averages 5% smaller than nominate *campestris*; wings and legs up to 20% shorter. Plumage patterned as nominate *campestris* but distinctly paler, overall greyer above (with ochre rump more obviously contrasting with back), and whiter, much less buff, below. Winter specimens fitting this description taken in Egypt and Arabia (Vaurie 1959) and record in October of small, grey-toned bird in England (P R Colston, R E Emmett, D I M Wallace) indicate clear possibility of *griseus* (or possibly *kastschenkoi* of central Asia) reaching west Palearctic. JUVENILE. (1) Nominate *campestris*. Differs distinctly from adult, with much more streaked appearance recalling *A. novaeseelandiae* and other large pipits and providing dangerous pitfall to inexperienced observer. Upperparts and wings basically buff-brown, with red-toned feather-centres and buff feather-margins forming distinct streaks over crown, back, and lesser wing coverts and exaggerating contrasts of centres and unworn edges of flight- and tail-feathers; isolation of median covert panel thus reduced, but pale rump and outer edges of tail more striking. Underparts basically as adult but with striking dark brown malar stripe, band of short streaks across chest, and fine striations on flanks; pattern not as marked as in *A. novaeseelandiae* but closely approaches *A. godlewskii*. Separation from *A. novaeseelandiae* safely based on lightness of leg and shortness of hind claw but distinction from *A. godlewskii* clearly more difficult and not studied in the field so far, though *A. campestris* shows more strongly marked loral stripe, less distinct streaks on upperparts, and unstreaked rump. (2) *A. c. griseus*. Differs from adult as in nominate *campestris*, but duller, less red-toned above, than juvenile nominate *campestris*. FIRST WINTER. Both races. Assumption of this plumage occurs from early July to January, this long period allowing immature autumn vagrants to appear in every stage from juvenile to apparent adult. By October, most lose fully streaked appearance of juvenile but retain, compared to adult, (a) more streaked crown, (b) at least partially streaked chest, (c) more heavily marked wings, and (d) more distinctly 'furrowed' back. At all ages, bill fine; dark brown-horn above and on tip but bright buff-flesh at base below. Legs and feet rather spindly; bright

yellow-flesh to pale ochre-flesh. Hind claw usually curved; averages 9 mm long (15·5 mm in *A. novaeseelandiae*, 12 mm in *A. godlewskii*); length of claw difficult to judge, but overall length of hind toe and claw clearly less than length of tarsus.

Except at long range, adult and fully moulted 1st winter bird distinctive; separable from *A. novaeseelandiae* and *A. godlewskii* by marked uniformity of plumage above and below, with no obvious streaks, but still subject to confusion in Middle East with *A. similis* (see that species, p. 332). Even when seen well, immature best identified by close study of structure (especially of bare parts) and behaviour (see below), since pattern of heavy streaking is shared by 3 other species in adult and/or juvenile plumage. Flight markedly undulating but lacks power of *A. novaeseelandiae* and *A. godlewskii*; strongly reminiscent of wagtail, as is slim, long-winged, and long-tailed silhouette. Escape-flight normally quite long (often exceeding 100 m when bird pressed), and usually followed by landing on open ground. Song-flight consists of fluttering ascent to considerable height and parachuting descent. Carriage most horizontal and gait most free of all large pipits, with walk, run, and alert pauses all recalling wagtail. Wags tail frequently, again like wagtail and smaller pipits and with much greater freedom than only occasional similar movement of *A. novaeseelandiae* and *A. godlewskii*. Perches freely and on higher prominences than *A. novaeseelandiae*. Can be quite tame. Migrants rarely form more than small groups on passsage and in winter range.

Song short and undeveloped, a ringing repetition of loud, tinny disyllable 'chir-ree chir-ree...'; given mainly in descent phase of song-flight but also from perch or ground. 4 flight-calls: (1) common call, loud 'tzeep', recalling Yellow Wagtail *Motacilla flava* but slightly less plaintive; (2) less common, hoarse 'chup', recalling monosyllables of House Sparrow *Passer domesticus* (but less emphatic) and juvenile Linnet *Acanthis cannabina*; (3) another less common call (perhaps used more by birds from Yugoslavia eastwards) a slurred 'chirrup' recalling monosyllables of *A. novaeseelandiae* (but less rasping) and *P. domesticus*; (4) least common or quietest call, 'tchip' or 'chirp', suggests anxiety calls of smaller pipits and usually given during take-off or landing. Other calls include single loud disyllable of song-phrase uttered singly by ♂, clear musical 'tzeeirp' given in alarm (most recalling *A. novaeseelandiae*), and quiet, metallic monosyllables from parent birds. Calls of apparent *griseus* in England 'schwup', 'chewp' (these 2 with rising inflection), 'cherrup', and 'cherrp', with tone recalling Snow Bunting *Plectrophenax nivalis* (mostly) and *M. flava*. Voice lacks explosive quality and harsh loudness of *A. novaeseelandiae*, calls not carrying as far.

Habitat. In lower middle and middle continental latitudes, from Mediterranean and steppe through temperate

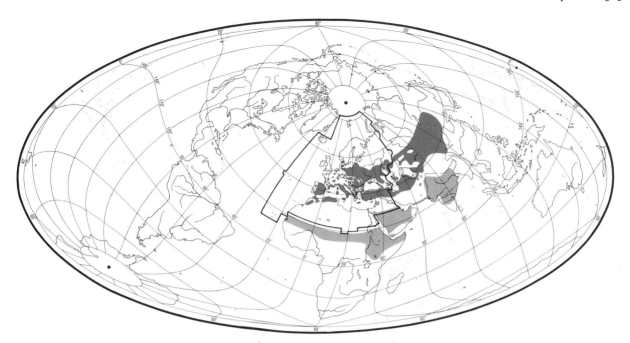

zones, preferring dry but not arid ground. Avoids steep or rocky terrain, water obstacles, tall or dense vegetation, from forest to wetland, cropland, or shrub growth. Favoured habitats tend to be more frequent in sunny continental lowlands, but locally occur at 2600 m in Armeniya (USSR), and extralimitally even to 3600 m in Balkhash region. In USSR, basically a steppe and semi-desert bird; occurs on clay, sand, and saline patches with scant vegetation (*Artemisia* and grass steppe), and in pastures and fields even of hay and cereals; locally on damp mountain heaths. Especially towards north of range, promptly occupies areas cleared of trees, and young shelterbelt strips; in Ukraine, often in sparse low growth of willows *Salix* on sands (Dementiev and Gladkov 1954*a*). In Germany, breeds on dry warm wastelands and heaths, often beside Woodlark *Lullula arborea* and Stone Curlew *Burhinus oedicnemus*, and on sandy arable fields and sandy banks of rivers or lakes; seeks water for bathing (Niethammer 1937). In Netherlands, breeds on shifting sands, sandy barren heaths, and arid soil; formerly on sand-dunes (IJzendoorn 1950). In Belgium, on most denuded parts of sandy heaths and on burnt areas or inland dunes where few other birds live (Lippens and Wille 1972). In southern Sweden, great majority of studied territories were on coastal sand-dunes dominated by heather *Calluna*, lyme-grass *Elymus*, and marram grass *Ammophila*, or in dry meadows dominated by grass *Festuca ovina* and sedge *Carex arenaria*; heath with dominant *Calluna* or *Corynephorus* also used, essential features being short vegetation cover with naked sandy patches in between and some taller growth or single trees (G Högstedt). In north-west Africa, occupies dry mountain slopes and plateaux up to 2400 m, and is

abundant in Atlas above treeline, some up to 3000 m (Heim de Balsac and Mayaud 1962). Other situations recorded include coastal sand-dunes (even close to high-water mark), dry water-courses, stony flats, roadsides, vineyards, and dry hillsides (Bannerman 1953). In southern Portugal, found feeding on dry turf with Hoopoe *Upupa epops* and Black-eared Wheatear *Oenanthe hispanica*, perching freely on rocks, less often on bushes. On Biscay coast of France, occurs in rear dunes, not fore-dunes among marram grass, running and walking through head-high vegetation on grey sand, sharing habitat with Crested Lark *Galerida cristata* and Short-toed Lark *Calandrella brachydactyla*, often perching on commanding ridges or on dead branches or twigs of planted pine at 1 m or higher. In Gotland (Sweden), seen sitting on top of low pine. (E M Nicholson.) In winter in Africa, preference for arid ground accentuated: common in coastal zone, in steppe, *Acacia* bush, or areas carrying *Suaeda*, and in barest parts of thorn-savanna, and even on edge of desert; associates with grazing cattle (Moreau 1972). Flies strongly, ascending freely above lower airspace, but is basically a ground bird.

Distribution. Range has decreased in France, Netherlands, West Germany, and Sweden.

FRANCE. Some retreat towards south (Yeatman 1976); now breeding Fontainebleau (RC). NETHERLANDS. Marked decrease in range due to cultivation; last bred on coast *c*. 1967 (Bijlsma 1978*a*; CSR). WEST GERMANY. Range decrease: ceased to breed Hamburg 1966, Saarland 1967, Rheinland 1968, and Lower Rhine 1970 (Bauer and Thielcke 1982). Irregular in Baden-Württemberg, much more widespread in 19th century (Gatter 1970).

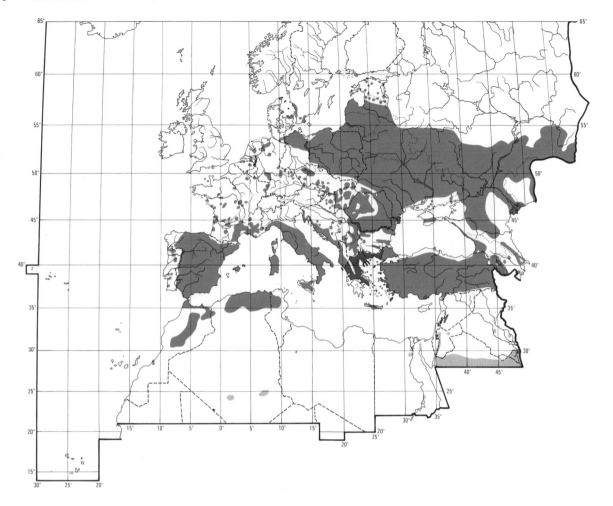

SWEDEN. In 19th century bred further north (LR); following decline, became remnant population within forest areas (Ahlén 1972). SWITZERLAND. Bred 1976 and 1981 (RW).

Accidental. Iceland, Ireland, Channel Islands, Norway, Finland.

Population. Has declined in France, Netherlands, West Germany, Sweden, Czechoslovakia, and USSR.

FRANCE. 10 000-100 000 pairs; some decline (Yeatman 1976). BELGIUM. About 150 pairs (Lippens and Wille 1972). NETHERLANDS. 75-90 pairs 1976-7 (Teixeira 1979); marked decline in 20th century (CSR). WEST GERMANY. In danger of extinction: 100-350 pairs, excluding Niedersachsen (see Bauer and Thielcke 1982 for details), and total population unlikely to be much higher (AH). DENMARK. 30-50 pairs, no recent change (TD). SWEDEN. About 100 pairs (Högstedt 1969); 100-250 pairs (Björnfors and Götmark 1981), decreased (LR); 250-300 pairs (Adolfsson 1984). POLAND. Scarce, locally very scarce (Tomiałojć 1976). CZECHOSLOVAKIA. Some decline since 1945 (KH). AUSTRIA. 20 pairs at most (HS, PP). ALBANIA.

Scarce (EN). USSR. Declined due to habitat changes (GGB).

Movements. Essentially migratory. Within breeding range, wintering occurs only in Aegean region and locally in Turkey and Levant. West Palearctic race winters in Africa and Arabia, eastern races in southern Afghanistan and India, with *griseus* also in Arabia (Meinertzhagen 1954; Moreau 1972; Ali and Ripley 1973b); however, situation in eastern races not established by ringing, and races show much overlap in appearance and measurements (see below and Geographical Variation).

WEST PALEARCTIC RACE (nominate *campestris*). In Africa, winters mainly in Sahel zone south of Sahara, but in the east, where southward movement not impeded by tree savanna or rain forest, normal range extends south almost to the equator (Urban and Brown 1971; Pearson 1972); recorded south to 3°S in Kenya (Zink 1975). Extent of southward penetration of Africa not completely established and may well fluctuate from year to year; range probably spreading south due to increased aridity. In extreme west not uncommon in Sénégal (Morel and Roux 1966) but rare in Gambia (Cawkell and Moreau 1963; Gore 1981) and no records from Sierra Leone (G D Field). Small numbers winter in Mali (Curry and Sayer 1979) and uncommon in northern Nigeria with Jos plateau (9°N) the usual southern limit, but has been recorded on coast (Rayner 1962; Elgood 1982). Only 2 winter records from Chad (Salvan 1967-9), and occurs sparsely in arid zone of Sudan south to 14°30′N (Hogg *et al.* 1984), but common in suitable habitat in Ethiopia (Smith 1960a; Urban and Brown 1971). Data suggest that major reception areas lie near Atlantic coast and in eastern Africa, but comparative paucity of observation in more central arid areas may account for apparent sparsity there. Known to winter in suitable areas within Sahara, e.g. at up to 2000 m in Ahaggar massif and Tazrouk, areas with very little vegetation (Niethammer and Laenen 1954; Blondel 1963).

Autumn passage through Europe inadequately documented to support concept that movement is divergent, mainly towards flanks of Africa; broad-front movement, at least through Mediterranean area, seems more likely. At Falsterbo (southern Sweden), autumn passage occurs late July to early October; in 1940s and 1950s, movement peaked in second half of August, but since *c.* 1970 in late August and early September (Roos 1984b). In Poland, movement occurs August to early October (Tomiałojć 1976). In Switzerland, passage occurs early August to October, but mainly September and early October (Glutz von Blotzheim 1962); small parties recorded moving through Cols de Cou and Bretolet 24 August-3 October with maximum of over 50 on 19 September (Crousaz 1961). Passage in Khar'kov region (south-central European USSR) occurs from early September, though sometimes all birds gone by end of August. In north-west Kazakhstan, passage recorded at end of July and (large numbers) during 10-13 October. Leaves northern Caucasus in mid-September. (Dementiev and Gladkov 1954a.)

Movement occurs across entire length of Mediterranean. Passage at Gibraltar is mid-August to mid-October with visible movement concentrated in mornings (Tellería 1981). On Malta, passage late August to late October with some into November (Sultana and Gauci 1982); scarce on Cyprus, early August to mid-September or rather later (Flint and Stewart 1983). Marked south-west passage down Atlantic coast of Morocco, mid-September to late October, local breeders probably moving first; stragglers to early December (Smith 1968b). Moreau (1972) referred to a late influx into Sénégal in January, inferring arrival from further north; this western-flank passage thus perhaps continues until reversed in spring. Scarce on passage through Tunisia with movement as early as late July (Thomsen and Jacobsen 1979). In Libya much less common in autumn than spring and then mainly Tripolitania (Bundy 1976). Birds moving through Egypt thought to have largely more easterly origin and may include *griseus* (Etchécopar and Hüe

1967); passage of 181 birds recorded near Suez 14 September-31 October 1981 with 50 on 18 September, maximum movement around 08.00 hrs (Bijlsma 1982). Passage noted in mid-September at Azraq, Jordan (Cameron and Cornwallis 1966).

Evacuation of African winter quarters probably starts as early as February but April records are widespread. Fairly common on passage at Dafila (eastern Morocco), mid-March to mid-April, with 85 records in 2 years, birds heading north-west to north (Smith 1968b). Northward passage through Tunisia, coastal Libya, and inland oases occurs mainly in April; birds more numerous than in autumn (Guichard 1957; Bundy 1976; Thomsen and Jacobsen 1979). Spring passage not very evident in Eritrea (Ethiopia) despite numbers wintering there (Smith 1960a). In Mediterranean area, small numbers arriving at Coto Doñana in May (Mountfort and Ferguson-Lees 1961e); passage through Balearics masked by presence of breeders but birds arrive March-April (Bannerman and Bannerman 1983). Movement through Malta mainly mid-March to mid-April (Sultana and Gauci 1982); fairly common on Cyprus, movement starting in February (Flint and Stewart 1983). Within Europe, moves through Swiss alpine passes mid-April to mid-May (Glutz von Blotzheim 1962), and arrives in Poland April-May (Tomiałojć 1976). In European USSR, arrives near Zhdanov (north of Black Sea) in late March and in Caucasus from early April, but further north not until end of April or early May (Dementiev and Gladkov 1954a).

EAST PALEARCTIC RACES (*griseus*, *kastschenkoi*). According to Ali and Ripley 1973b), *kastschenkoi* of central Asia winters only in Uttar Pradesh and Bihar (north-east India), this and remainder of winter range in India, Pakistan, and southern Afghanistan being occupied by *griseus*. Latter race said also to winter in Arabia (with nominate *campestris*) and in Somalia (Meinertzhagen 1954), though birds in Somalia listed as nominate *campestris*, not *griseus*, by Archer and Godman (1961b). Most westerly claimed record of *griseus* is at Yebjyiba in south-west Libya (Fairon 1971). Situation complicated by identification difficulties, however, and it may even be that *kastschenkoi* moves south-west to winter largely in Arabia and/or Africa.

Little data on movements of eastern races. Birds leave higher parts of Altai (south-central USSR) by end of August, but still present lower down towards mid-September (Dementiev and Gladkov 1954a). Passage at Bamian (Afghanistan) recorded early September to early October (Paludan 1959) and small parties recorded on passage in Tehran area (Iran) in mid-March and 2 at *c.* 1800 m in late April (Passburg 1959). Fairly common on passage in Oman, sometimes wintering (Gallagher and Woodcock 1980); present mid-September to mid-April (Walker 1981a, b). Arrives on central Asian breeding grounds mid-April to mid-May (Dementiev and Gladkov 1954a).

Vagrants to Britain and Ireland, 1958–72, occurred April to mid-June (15%) and late July to early November (85%), mostly late August to mid-October; average 17 per year (Sharrock and Sharrock 1976).   JHE, DJB

Food. Chiefly insects; also some seeds, mainly in winter. Feeds on ground and amongst low herbage, taking insects in stop-run-peck manner like small plover *Charadrius*, occasionally leaping up, or rarely after brief aerial pursuit (Naumann 1901; Dittmann 1925; Verheyen 1947; Dolgushin *et al.* 1970; Bijlsma 1978b). Recorded hovering briefly low over sand, apparently to locate prey (Beneden 1934). Stands on and dismembers large locusts (Acrididae) by hammering with bill; takes flying termites (Isoptera), frequently jumping in the air to catch them (Meinertzhagen 1954). Can apparently go without water for long periods (Dolgushin *et al.* 1970).

The following recorded in diet in Palearctic. Invertebrates: dragonflies (Odonata), grasshoppers (Acrididae), mantises (Mantidae), bugs (Hemiptera: Cicadidae, Coccoidea), adult and larval Lepidoptera (Pieridae, Tortricidae, Noctuidae, Lasiocampidae, Geometridae), lacewings (Neuroptera), flies and their eggs (Diptera: Tipulidae, Tabanidae, Bombyliidae, Asilidae), Hymenoptera (sawflies Symphyta, Chalcidae, ants Formicidae), beetles (Coleoptera: Cicindelidae, Carabidae, Staphylinidae, Buprestidae, Scarabaeidae, Elateridae, Dermestidae, Nitidulidae, Cucujidae, Coccinellidae, Tenebrionidae, Chrysomelidae, Curculionidae), spiders (Araneae), harvestmen (Opiliones), snails, and slugs (Pulmonata) . One record of sand lizard *Lacerta agilis* as food. (Naumann 1900; Baer 1909; Rey 1910b; Gil 1928; Verheyen 1947; Steiniger 1955; Pek and Fedyanina 1961; Averin and Ganya 1970; Dolgushin *et al.* 1970; Popov 1978; Smetana and Guseva 1981; Kostin 1983.)

In Moldavia (USSR), 41 identified items from 12 stomachs (no date) comprised 44% (by number) beetles (including 24% Curculionidae, 10% Carabidae, 10% Chrysomelidae), 19% Orthoptera, 19% Hymenoptera, 12% caterpillars, and 5% butterfly pupae; one bird taken in July contained only seeds of wild plants (Averin and Ganya 1970). In Crimea (USSR), 7 stomachs, April–July, contained only invertebrates: 78·6% of items were beetles, including 22·9% Curculionidae and 20·0% Carabidae (Kostin 1983). Extralimital study in northern Kazakhstan (USSR) by Smetana and Guseva (1981). In April, 135 items from 15 stomachs comprised 60·7% beetles (26·6% Chrysomelidae, 7·4% Nitidulidae), 35·6% ants, 3·0% Diptera, and 0·7% Hemiptera; 3 stomachs contained plant material. In August–September, 245 items from 28 stomachs comprised 60·0% beetles (34·7% Curculionidae, 12·2% Buprestidae), 14·7% Orthoptera, 11·4% Diptera, 9·8% ants, 2·4% moths, 0·8% Hemiptera, and 0·8% pupae; 3 stomachs contained plant fragments, and one some remains of a lizard.

Winter diet includes locusts, termites (in Sudan) and occasionally seeds (Meinertzhagen 1954); grasshoppers and Hymenoptera in Mali (Duhart and Descamps 1963), and assassin bugs (Reduviidae), termites, caterpillars, beetles, and wild and cultivated grass seeds in India (D'Abreu 1918).

Young fed only on invertebrates. At nest near Hannover (West Germany), young (usually 2 fed per visit) received almost exclusively caterpillars *Tortrix viridana* during infestation of them (Steiniger 1955). The only quantitative study is extralimital. In northern Kazakhstan, collar-samples from 13 nests contained 280 items: 43·2% (by number) Orthoptera, 17·5% Lepidoptera (13·9% larvae, 2·2% adults, 1·4% pupae), 11·1% spiders, 6·8% Diptera;, 5·7% Hemiptera, 5·4% Hymenoptera, 4·3% beetles, 2·5% snails, 1·8% lacewings, 1·1% dragonflies, and 0·7% others. Diet changed with age: at 1–2 days old, 30·1% Orthoptera, 20·8% adult and larval Lepidoptera, 18·9% spiders, 17·1% beetles; at 3–6 days, 58·0% Orthoptera with hard parts removed, also snails, eggshells, and sand; from 7 days, given large whole insects, 62·3% Orthoptera. Feeding rate increases with age: at 1–2 days, 30 visits per day, at 6–8 days 50, at 12 days (just prior to fledging) 80. Rate drops to $\frac{1}{3}$–$\frac{1}{2}$ of normal in very hot or rainy weather. (Smetana and Guseva 1981.)   PJE

Social pattern and behaviour. Studied in south-west Veluwe (Netherlands) by Bijlsma (1978a), in Belgium by Beneden (1934, 1938), in Denmark by Norup (1963), in West Germany by Schmidt (1960), in East Germany by Dittmann (1925, 1927) and Krüger (1980), and extralimitally in Naurzum (Kazakhstan, USSR) by Smetana (1980). Following account includes data supplied by G Högstedt from study in southern Sweden.

1. Solitary or in loose flocks outside breeding season: e.g. in winter in Mali (Duhart and Descamps 1963), and in India (Ali and Ripley 1973b). In Baden-Württemberg (West Germany), often migrates singly in spring, though mostly in twos; flocks (8–15) rare (Gatter 1970, 1976). In Veluwe, twos recorded after earlier single migrants probably pairs (Bijlsma 1978b). Loose flock of up to 30 in Orbe valley (Switzerland), late April–May, unusual (Glayre 1980), though spring flocks of 50–150 formerly occurred in southern East Germany (Dittmann 1925), but maximum 8 more recently (Rinnhofer 1970). Flock of at least 150 migrants recorded on Malta (Sultana and Gauci 1982). In Kazakhstan after arrival on breeding grounds, occurs singly, in pairs, or in small flocks, sometimes with Yellow Wagtail *Motacilla flava* (Dolgushin *et al.* 1970; Smetana 1980). Larger flocks (or concentrations) typically recorded in autumn: see (e.g.) Rinnhofer (1970) for East Germany and Beaman *et al.* (1975) for Turkey. In Taurus mountains (Turkey), flocks of up to 25 juveniles occur in August (Sutton and Gray 1972). In Naurzum, break-up of families reported in late August, birds then concentrating by roads, in clearings, and later moving to woods (Smetana 1980). Loose-knit flocks of 6–12 birds occurring in southern Sweden from early August are sometimes families, but usually random mixture from territories up to *c.* 12 km away. Such flocks resident for a few weeks before starting migration. Juveniles (9 records of ringed birds) tend to return to moulting sites in following year and may breed there if territory available (G Högstedt); at Falsterbo, most birds migrate singly in autumn, though flocks of up to 14 recorded (Roos 1984b). In Belgium, sometimes apparently migrates in

family parties: probable family of 2 adults and 4-5 young trapped together (Beneden 1938). In Veluwe, usually 2 (once 7) together; adults leave before juveniles (Bijlsma 1978*b*). Larger flocks in autumn typically for feeding and resting (Gatter 1970; Rinnhofer 1970), though loose flocks of up to 30 recorded actively migrating in Morocco by Smith (1965*a*) and reports from Kirgiz steppes and southern Kazakhstan (USSR) indicate occurrence of very large actively migrating flocks (Suschkin 1914*a*; Dolgushin *et al.* 1970). For further details, including autumn densities, see Smetana (1980), also Dolgushin *et al.* (1970). Rarely associates with other species on migration, occasionally with other *Anthus* or *M. flava* (Crousaz 1961; Gatter 1970; Bijlsma 1978*b*). BONDS. Mating system essentially monogamous (Dementiev and Gladkov 1954*a*), but polygyny not infrequently recorded; in Sweden, 11 cases ('6·7% of all breedings'—probably underestimate of true rate) all of ♂ paired with 2 ♀♀ in same territory. In 8 cases, ♂ fed young at both nests, with start of laying by 2 ♀♀ 2-5 days apart; in 3 other cases, 10, 14, and 32 days apart. May indicate occurrence of more than one type of polygyny (G Högstedt). See also Norup (1963) for probable polygyny in Denmark. No detailed information on duration of pair-bond, but site-fidelity (see Breeding Dispersion, below) perhaps encourages annual renewal of bond. In Veluwe, most birds apparently pair up before or very soon after arrival on breeding grounds; later birds all paired on arrival (Bijlsma 1978*a*), and some in Denmark also (Norup 1963). In Kazakhstan, ♂♂ said by Dolgushin *et al.* (1970) to arrive 10-12 days before ♀♀, while Smetana (1980) indicated pairs formed on migration. Birds normally pair up and breed at 1 year old; some remain unpaired or do not breed at 1 year, and older birds sometimes also fail to breed (G Högstedt). Young fed by both sexes, though some variation. In Naurzum, both sexes involved at 13 out of 18 nests, but ♀ did *c.* 3 times more than ♂ and, at other 5 nests, all (Smetana 1980). Study in Sweden showed ♀ does all feeding if young still in nest when ♂ starts moult (from *c.* 20 July) (G Högstedt). At nest where all feeding done by ♀, ♂ settled apparently to brood young at night (Steiniger 1955), though all brooding normally done by ♀ (G Högstedt). Split-brood care frequent. 1st brood also tended by ♂ if ♀ re-lays (see also Relations within Family Group, below). Independence a gradual process: young recorded still being fed at 26-29 days old, i.e. *c.* 15 days after leaving nest (G Högstedt); see also Norup (1963). In Switzerland, family unit perhaps maintained for *c.* 3 weeks after young leave nest (Glutz von Blotzheim 1981). In East Germany, presumed 1st-brood young recorded near parents apparently feeding 2nd brood (Krüger 1980). See also Relations within Family Group (below). BREEDING DISPERSION. Usually solitary and strictly territorial (Bijlsma 1978*a*; Krüger 1980; G Högstedt); for overlapping territories, see below. In Belgium in favoured dune habitat, centres of territories or presumed nest-sites (within *c.* 50 m of ♂'s song-post) usually *c.* 400-500 m apart (Beneden 1938). Pairs also *c.* 400-500 m apart where unusually high density in Causses of south-central France (Devillers and Esbroeck 1974). In Veluwe, in 1975, nests 725 m (700-750, *n*=3) apart; in 1976, 425 (300-700, *n*=5) m; in 1977, 717 (550-900, *n*=4) m (Bijlsma 1978*a*); see below. In Braunschweig area (West Germany), territories locally *c.* 900-1000 m apart (Bäsecke 1955); not clear if this is distance between edges or between centres of territories. In Naurzum, open woodland with meadow-steppe vegetation and low-lying meadows favoured; nests there sometimes 30-60 m apart and territories overlap (Smetana 1980); see Antagonistic Behaviour (below). In southern Sweden, nests at least 200 m apart, usually more. In cases of polygyny (see Bonds, above), nests of 2 ♀♀

sometimes *c.* 10-15 m apart (Norup 1963; G Högstedt). Territory used successfully by 1 pair may be taken over by different pair later in season. Most Swedish territories have short vegetation and bare patches (probably important for feeding), some taller vegetation and a few trees (used by ♂ as song-posts and look-outs). In grassy heath, territory 23±SD3·9 ha (19-28, *n*=4); in dunes 12·1±SD2·41 (9-18, *n*=14) ha; smallest in *Calluna* heath and meadows: 4·7±SD2·46 (2-11, *n*=20) ha, and 3·1±SD1·88 (1-7, *n*=12) ha respectively (G Högstedt). In Dutch study, food supply probably influenced year-to-year variation in territory size: average in 1975 27·3 ha (13-35, *n*=3); in 1976 (dry summer, food abundant), 17·0 ha (12-23), *n*=5; in 1977, 20·5 ha (15-24), *n*=4 (Bijlsma 1978*a*). Territories in Leusderheide of central Netherlands (population *c.* 8 pairs) larger at *c.* 25-50(-75) ha (Alleyn *et al.* 1971). See also Beneden (1938) and Verheyen (1947) for Belgium, and Yeatman (1976) for France. Study by Krüger (1980) indicated territories initially large, contracting up to final departure of all ♂♂, and shifts also occur (no details). Size also varied markedly with habitat—rarely larger than *c.* 3 ha in good habitat, 7·5-25 ha if less so according to Schiermann (1943). In Heller (Dresden, East Germany), 9-15 ha (Dittmann 1925, 1927). In Brandenburg (East Germany), requires 0·25-0·51 ha of wasteland, gravel pits or clear-felled area in (partially) open terrain, 1 ha in woods (Rutschke 1983). In Wallis (Switzerland), movements of ♂ suggested territory *c.* 1 km across (Glutz von Blotzheim 1981). Nest may be several hundred metres from one of ♂'s regular perches (Naumann 1900). In Naurzum, territories 0·5-2·5 ha, maximum 0·9 ha in favoured habitat as described above (Smetana 1980). Territory defended by ♂ (overlooked to large extent from his song-posts) and pair-formation and courtship take place within it (Krüger 1980; G Högstedt). Feeds both within and beyond territory, sometimes considerable distance away: e.g. in Naurzum, adults forage close to nest and up to 300-400 m away (Smetana 1980); near Braunschweig, pair-members used separate areas, generally far from nest (Bäsecke 1955; see also Schmidt (1960). In Veluwe, 2nd-brood nests averaged 50(15-75) m from 1st (Bijlsma 1978*a*); in East Germany, once *c.* 32 m (Krüger 1980). Most data on densities from East Germany. In Mecklenburg, 0·8 pairs per km² of heathland, 0·4-1·5 pairs per km² in coastal dunes (Klafs and Stübs 1977). In Brandenburg, mostly less than 10 (3-30) pairs per km² of suitable habitat (see Rutschke 1983 for more habitat details). Heller (Dresden) formerly held unusually high density of *c.* 8 pairs per km² over *c.* 1·5 km² (Dittmann 1925, 1927). Studies in former lignite mining area of East and West Germany showed 24-40 pairs per km² of restored habitat (see Kalbe 1961 and Krüger 1980 for details). In Causses, 20 birds recorded during 4-hr walk in hilly grassland with rocks. Probably highest density in France (Devillers and Esbroeck 1974). Adult site-fidelity apparently strong: of 25 colour-ringed birds, all returned to same areas of southern Sweden to breed (G Högstedt); see also Munn (1925, 1931), Beneden (1938), and Bijlsma (1978*a*). In Spree valley (East Germany), one ♀ bred *c.* 40 m from her (unsuccessful) site of previous year, and another ♀ bred *c.* 200 m (probably due to habitat change) from site she had used successfully (Krüger 1980). No information on fidelity of young to natal area. Near Dresden, large territories of *A. campestris* overlapped with smaller ones of Woodlark *Lullula arborea*, Skylark *Alauda arvensis*, and Crested Lark *Galerida cristata* (Dittmann 1925, 1927); extent of interspecific aggression not known, but see Antagonistic Behaviour (below). ROOSTING. Nocturnal and perhaps solitary though little information. On ground, in shelter of clod, grass clump, crops, in old rut, in heather *Calluna*, etc. (Naumann 1900). Pair recorded roosting on ground in heather *c*. 30 cm high in late May (G Högstedt); ♂ said by Garling (1926) always to roost on ground near nest. Report in Heinroth and Heinroth (1924-6) of birds making shallow pits in sand presumably refers to captive conditions. In Mali, from October, birds mostly in shade around noon, but also recorded moving about over burning sand (Bates 1933-4). Water-bathing rare (G Högstedt).

2. Shyness on breeding grounds varies: e.g. in coastal dunes of Sweden where used to man, less shy than (e.g.) Wheatear *Oenanthe oenanthe* and may allow approach to *c*. 10 m; especially tame and approachable when feeding young out of nest. In pre-laying period, reluctant to fly and tends to escape by running away (G Högstedt). Shyer in (e.g.) remote heaths, etc. (Bäsecke 1955), and Wadewitz (1957*a*) found it generally shy or at least wary. In Belgium, rather secretive on breeding grounds; ♂ tends to fly at *c*. 20 m, ♀ at *c*. 10 m; normally call in flight (Beneden 1938: see 3a in Voice). Migrants and vagrants may allow approach to *c*. 10-20(-40) m and typically run away, stopping to stand erect with rapidly wagging tail. May also fly about, perch on fences, etc. (Naumann 1900; Witherby *et al.* 1938*a*; Reeves 1954; Simpson 1971; Numme 1976). Bird flushed while feeding gave call 3f, followed by apparent song-fragments (Dittmann 1925). Will hide in vegetation from birds of prey (Naumann 1900). FLOCK BEHAVIOUR. Flock-members act independently when disturbed, rather like *A. pratensis*, but give loud Contact-alarm calls in flight (G Högstedt: see 3a-b in Voice). Separates from other species (see part 1) at take-off (Rinnhofer 1970). SONG-DISPLAY. Song of ♂ (see 1 in Voice) given mainly in flight, also from tree, hummock, etc.; less commonly from flat ground. Usually 60% of song given in flight, but this varies individually and with stage of breeding cycle (see subsection 1 in Heterosexual Behaviour, below) (Krüger 1980; also Witherby *et al.* 1938*a*, Dolgushin *et al.* 1970, Smetana 1980, G Högstedt). For Song-flight, takes off from perch or ground and ascends silently with fluttering wing-beats (Bijlsma 1978*a*) to *c*. 20-30(-150) m; ascent may be almost vertical (Beneden 1934, 1938; Verheyen 1947). Bergmann and Helb (1982) indicated characteristic undulations (1 song-unit per undulation) to be part of ascent. In Dutch study, birds described as flying level but with slight undulations from peak of ascent (like Song-flight of *L. arborea*)—singing, but much less vigorously than during descent (Bijlsma 1978*a*); horizontal flight with song also mentioned by Beneden (1934). Other authors referred to series of deep undulations across territory, with 1 song-unit given during descent of each undulation (Wadewitz 1957*a*; Norup 1963; Wallschläger 1984; G Högstedt). Gliding in circles prior to song during descent or in undulations mentioned only by Verheyen (1947). 2 basic types of descent described: (a) steep plummet accompanied by song (Wadewitz 1957*a*; Bijlsma 1978*a*; Smetana 1980; Bergmann and Helb 1982); (b) gliding descent resembling other *Anthus* with wings and tail raised and, in one case, hovering at *c*. 1·5 m before landing (Beneden 1934; Glutz von Blotzheim 1981); according to Beneden (1934), half gliding, half fluttering, and shallower than Meadow Pipit *A. pratensis* or Tree Pipit *A. trivialis*). Lands on perch or ground (Krüger 1980). May stay long in an erect posture on perch after Song-flight (Piechocki *et al.* 1982) or run about on ground (Voigt 1933). Song of one ♂ usually quickly elicits song in neighbour, leading to Song-duel in close proximity (Beneden 1938), though nearby ♂♂ in East German study never all sang simultaneously (Krüger 1980). In southern Sweden, song continues to sunset. No song by ♂♂ noted after 29 July. Frequent song in August from ♀♀ still feeding young (G Högstedt); see also subsection 6 in Heterosexual Behaviour and 2 in Voice (below). Song-period

in central Italy mid-April to end of July (Witherby *et al.* 1938*a*); late April to early July in Crimea, USSR (Kostin 1983), and to mid- or late July in Kazakhstan (Dolgushin *et al.* 1970). In southern East Germany, from arrival to July or early August; peaks mid-May and, after lull (presumably as prelude to 2nd brood), late June to early July (Krüger 1980). Song also noted exceptionally in autumn (Dittmann 1927). ANTAGONISTIC BEHAVIOUR. (1) General. In Veluwe, first birds (unpaired ♂♂) arrive in early April; most take up territories, late April or early May (Bijlsma 1978*a*). Near Braunschweig, some time often elapses between arrival and first territorial Song-flights (Bäsecke 1955). In southern Sweden, most territories occupied in first half of May, boundary demarcation taking 3-4 weeks up to laying (G Högstedt). In southern East Germany, birds usually move on to breeding grounds early to mid-May, but some territories occupied only in late May or early June (Krüger 1980). Fighting elicited by attempted intrusions is frequent in early stages, but usually ceases between neighbouring ♂♂ with pairing (Beneden 1938; Norup 1963). In Naurzum where nests close together in favourable habitat (see Breeding Dispersion, above), ♂♂ of pairs at same stage of breeding cycle are almost always at opposite ends of respective territories and never sing in zone of overlap (Smetana 1980). ♂ will immediately attack intruder or stuffed conspecific bird mounted near loudspeaker with playback of song. No aggression recorded between 2 ♀♀ paired with same ♂ (G Högstedt). (2) Threat and fighting. If intruder appears about to land, resident ♂ will leave perch and fly slowly towards it (Beneden 1938), giving undescribed call (Glutz von Blotzheim 1981). Attacks can be fierce, and frequently-ensuing chases are usually low and often over several hundred metres (Krüger 1980; Glutz von Blotzheim 1981). In Veluwe, in later pre-nesting phase (April-May) when aggression regular, ♂-♀ chases (see subsection 2 in Heterosexual Behaviour, below) apparently serve to advertise territorial boundary to neighbouring pair: all 4 birds not infrequently perform parallel chase along common boundary (Bijlsma 1978*a*). ♂♂ in Song-duel tend to move apart when they approach territory limit (Beneden 1938). On boundary, rival ♂♂ make only mock attacks and posture (no details) according to Krüger (1980). In close encounters, also perform Threat-gaping (mouth bright yellow in adult ♂), this sometimes leading to fights and chase during which aggressor gives call 4 (G Högstedt). In Wallis, brief fight involving 3 birds took place when ♀ solicited copulation from her mate (Glutz von Blotzheim 1981). Rather intolerant of other birds, chasing away (e.g. ♂ when ♀ nest-building) *A. arvensis*, *A. pratensis*, wagtails *Motacilla*, and (with particular hostility) *O. oenanthe* (Beneden 1938; Norup 1963). No intra- or interspecific aggression recorded outside breeding season by G Högstedt; rarely occurs (and apparently low-intensity) with *M. flava* in autumn (Beneden 1938). HETEROSEXUAL BEHAVIOUR. (1) General. In Veluwe, earlier-arriving unpaired ♂♂ attempt to attract mate by song and chasing. Such conspicuous behaviour more a feature of areas with contiguous territories; solitary pairs unobtrusive—sing and chase less. Conspicuous behaviour ceases anyway after start of breeding (Bijlsma 1978*a*). In Naurzum, about a third of birds paired by mid-April. Courtship (less protracted) also occurs mid- to late June, presumably before 2nd brood (Smetana 1980). (2) Pair-bonding behaviour. In Darfur, birds in small flocks show excitement from February, apparently as prelude to normal March departure: low chases about bush-tops accompanied by calls (Lynes 1924; see 3b in Voice). If ♀ flies up, ♂ follows and hectic chase develops: birds make rapid turns, flying close together and calling excitedly (Bijlsma 1978*a*: see 3b in Voice). In pre-laying period (including nest-building), birds mostly on ground and ♂ attends ♀ closely

(Norup 1963; G Högstedt); tend to run away if disturbed and ♂, giving an alarm-call (see 3e in Voice), will hover over ♀ still on ground (Norup 1963). (3) Courtship-feeding. Apparently does not occur. (4) Mating. In one case preceded by low and hectic chase, ♀ then gliding down to tree with wings raised and ♂ following in similar fashion to mount ♀ and copulate while beating wings (Beneden 1946). According to Norup (1963), usually takes place after flight (apparently different from chase) during which ♂ may sing, birds also flying thus after copulation. ♀ recorded wing-shivering in crouched posture to solicit copulation (Glutz von Blotzheim 1981). Courtship and probable copulation attempt noted during rearing of young (Steiniger 1955). (5) Nest-site selection. No information on selection process. ♀ at one site did most nest-building—rather feverishly, as if urged by ♂ (who also brought material). ♂ stood more erect than ♀, with head feathers ruffled; also gave short bursts of song while feeding and ♀ building; ♀ adopted more horizontal posture with sleeked plumage. Clutch complete *c.* 9-10 days after nest (Müller 1983). (6) Behaviour at nest. While ♀ incubating, ♂ guards from look-out, flies about and sings, or feeds close to nest (Norup 1963; Smetana 1980). ♀ leaves for *c.* 10 min, 1-2 times per hr (Norup 1963); for more details on stints at different stages in incubation, see Smetana (1980). ♀ will sing (see 2 in Voice) during breaks from incubation and (like ♂) later in undulating flight to and from nest when feeding young (Schmidt 1960; G Högstedt). Both birds habitually land *c.* 10 m away, sometimes dropping down from height (Bäsecke 1955), and then run to nest, tending to use one route (G Högstedt); ♂ more likely to stop occasionally, alert before proceeding. ♀ leaving nest also runs 10-20 m before flying off low (Norup 1963; also Wadewitz 1957a). In Wallis, ♂ approaching nest always flew first to root-stock *c.* 9 m below it (Glutz von Blotzheim 1981). In Naurzum, ♂ accompanied ♀ during breaks from incubation (Smetana 1980); not recorded in Danish study (Norup 1963). However, ♂ may pursue ♀ fiercely (Munn 1925, 1931), driving her back to nest (Norup 1963). RELATIONS WITHIN FAMILY GROUP. Young hatch over *c.* 8 hrs (Norup 1963) to 1 day. Raise head and open bill at any sound, more so from *c.* 2-3 days. Eyes fully open at *c.* 8-9 days (Smetana 1980). Brooded by ♀ up to *c.* 7 days (G Högstedt); in Naurzum, for long periods during first 2-3 days, for 5-15(-45) min in early morning only at 6-8 days. Shaded from sun for up to *c.* 2 hrs at a stretch when small, but not after 6-8 days. ♀ said to have watered young at 2 nests (Smetana 1980; Smetana and Guseva 1981). Wadewitz (1957a) and Steiniger (1955) recorded little shading from sun. ♀ may start to do a little feeding from 2nd day, reaching same frequency as ♂ by *c.* 6-7 days (Norup 1963). Adult often forages up to *c.* 30-40 m from nest, not flying but running to and from nest (Krüger 1980). Approach and departure otherwise much as described above, though in Naurzum, ♀ uses tree perch in both directions and, having fed young, rests there briefly, shaking feathers and bill-wiping. Sometimes sings when perched (in Swedish study, only in flight: G Högstedt). In Naurzum, ♂ much warier at nest than ♀, waiting long, moving about in tree and on ground, approaching nest by roundabout route and feeding young only after looking about long (Smetana 1980). Song of ♂ causes brooding ♀ to become more alert; larger young also on hearing song from either parent (G Högstedt). At nest near Hannover (West Germany), where ♂ accompanied ♀ on foraging trips, young moved on hearing song (probably from ♀) but did not gape; opened bill (no call given) only to receive food stuffed in by ♀ (Steiniger 1955). Normally fed whole items bill-to-bill; food-call given, often after hearing adults, from *c.* 7-8 days.

Nest cleaned at every visit, faeces being dropped *c.* 10-15 m from nest (G Högstedt); faecal sacs of young well away from nest may still be removed (Schmidt 1960). If ♂ not involved (temporarily or at all) in feeding of young, he sings (especially in morning) and guards territory (Smetana 1980). Young leave nest at 10-11 days (Norup 1963; Smetana 1980) or 13-14(-15) days (Krüger 1980; G Högstedt); able to flutter a bit at 14 days (Heinroth and Heinroth 1924-6; Norup 1963) and to escape from most ground predators at 15 days (G Högstedt). Hide in surrounding vegetation (Bijlsma 1978a). 2 brood-siblings recorded close together after leaving (Steiniger 1955); in another study, always apart, using separate hiding places (Schmidt 1960). Fed still, parents locating them by call (Smetana 1980); beg by trembling wings and may call though often silent (G Högstedt). According to Norup (1963), ♀ starts to build and re-lay before 1st brood leaves nest; perhaps refers to new ♀ or pair (see Breeding Dispersion, above). In southern Sweden, 3 2nd broods (♀♀ colour-marked) started 8, 11, and 15 days after 1st brood left nest (G Högstedt; see also Breeding). In Naurzum, stay *c.* 5-10 days in territory, then move about as family (Smetana 1980). In Veluwe, stay near nest for *c.* 2-4 weeks and near breeding grounds, singly or in loose groups, until early-September departure; family remains together longer if no 2nd brood (Bijlsma 1978a). However, Gatter (1970) and G Högstedt reported birds dispersing soon after fledging. ANTI-PREDATOR RESPONSES OF YOUNG. Crouch silent on hearing parental alarm (G Högstedt), or when threatened at *c.* 5-7 days (eyes beginning to open). Will escape at *c.* 8-9 days (Smetana 1980). 2 young out of nest crouched close together, motionless when touched, then spurted away (Steiniger 1955). Fully-fledged young may sit quiet until man *c.* 2 m away, then tend to run rather than fly (G Högstedt), though may scatter in flight when parent gives alarm-call (Smetana 1980: see 3d in Voice). PARENTAL ANTI-PREDATOR STRATEGIES. (1) Passive measures. A tight sitter (especially in rain) on eggs or young (Wadewitz 1957a; Norup 1963; G Högstedt), though Naurzum study revealed some variation during incubation: allowed approach to *c.* 20-25 m at 2-3 days, 10-12 m at 4-8 days, and *c.* 2 m from 9 days. Close to hatching, may even be caught on nest (Dolgushin *et al.* 1970) or fly up at last moment with loud wing-clap and also call (Smetana 1980: see 3d in Voice). ♀ may leave early on hearing ♂'s alarm-call (Heinroth and Heinroth 1924-6), running *c.* 15-20 m, then flying up and calling (Smetana 1980; see below). ♂ will also perform Song-flight at varying distance from nest and fly towards intruder over *c.* 400 m, or fly back and forth while singing; in one case, habitually flew up when man at *c.* 20-30 m, then moved to perch behind him (Beneden 1934, 1938; Wadewitz 1957a). In coastal Sweden, birds give few alarm-calls and simply attempt to approach nest unobserved by new route (G Högstedt). Pair with young *c.* 5-6 days old gave few if any alarm-calls when man near (Glutz von Blotzheim 1981). (2) Active measures: against birds. In Naurzum, both pair-members habitually chased off Cuckoo *Cuculus canorus*, even making dive-attacks together. Brooding ♀ attacked by Montagu's Harrier *Circus pygargus* leapt towards it, protecting offspring with wings widely spread; similarly with Kestrel *Falco tinnunculus*. When nest being plundered by such birds, parents give alarm-calls, fly about near predator in attempt to distract it, and frequently perish with brood (Smetana 1980). Corvids elicit alarm-calls and distraction-lure display (G Högstedt). (3) Active measures: against man. Slinking and wavering run typically used by ♀ leaving nest, but also by ♂ near it (e.g. Rasmussen 1921, Heinroth and Heinroth 1924-6, Durango 1936); presumably a distraction-lure display. Even from empty nest (Dementiev and Gladkov 1954a), but typically late in incubation and when with young, both sexes will perform distraction-lure display of disablement type: wings spread and quivering in apparently impeded forward movement; accompanied (after hatching according to Dolgushin *et al.* 1970) by call 3d (Smetana 1980; Glutz von Blotzheim 1981; G Högstedt); may then fly off (Bäsecke 1955). Performance typically given when not suddenly disturbed (Munn 1931), but see Verheyen (1947) for contrary view. Bird carrying food and singing while approaching nest will change to call 3d if disturbed and sometimes lands but often circles over intruder, calling constantly (G Högstedt). Simple aerial demonstration—fluttering about near nest rather like *M. flava* (Beneden 1934, 1938)—is probably similar. Recorded swooping down and calling when young threatened; will also hover over young and give contact-call in attempt to lure them away (Dittmann 1925; Schmidt 1960; G Högstedt). (4) Active measures: against other animals. Hedgehog *Erinaceus* and snake *Eryx miliaris* elicited close aerial approach and constant alarm-calls (Shnitnikov 1949; Smetana 1980).   MGW

Voice. Short units of song resemble calls, but repertoire otherwise wide and varied; this also evident in comparison with Richard's Pipit *A. novaeseelandiae* outside breeding season, though calls of *A. campestris* generally softer and less far-carrying (Grant 1972; Jonsson 1978a; Bergmann and Helb 1982). No definite evidence of geographical variation, but see Field Characters for calls from bird believed to be *griseus*. For extra sonagrams, see Bergmann and Helb (1982) and Wallschläger (1984). For exceptional case of apparent wing-clapping, see Social Pattern and Behaviour.

CALLS OF ADULTS. (1) Song of ♂. Series of complex units of 2-3(-4) sub-units delivered in rather monotonous fashion at even pace (units separated by pauses of 2-3 s) for long period or with slight initial acceleration. Units loud, clear, and metallic, rather melancholy and unmusical but not unpleasant (Voigt 1933; Witherby *et al.* 1938a; Jonsson 1978a; Bergmann and Helb 1982; G Högstedt). 3-18 units given per Song-flight (Wadewitz 1957a; Krüger 1980) or up to 25-30 (Beneden 1938). Typically has strongly modulated section, usually in middle (Wallschläger 1984). Units apparently of 2 basic types. (a) Units with last sub-unit stressed and with rising pitch, unit being rendered variously as follows: 'chivee' (Witherby *et al.* 1938a) and in recording from Iran, 'che-vee' at rate of *c.* 20 per 30 s (P A D Hollom) or 'zurCHEE' like slower buzzing Great Tit *Parus major* (M G Wilson); 'ziryh' rolled at end (Jonsson 1978a); 'zirrlüi', 'dscheri', or 'griëdlihn' (Wallschläger 1984; also Dittmann 1927, Voigt 1933); 'tschrlih' or 'trüwid' (Bergmann and Helb 1982); 'tsie-ru-flie' (Bijlsma 1978a). At Dresden (East Germany), 4-syllable units—'zieeluis' and 'zireluit' fairly often noted (Dittmann 1925). See also (e.g.) Beneden (1934, 1938), Krüger (1980), and Smetana (1980). Recording (Fig I) suggests 'tiswee' (J Hall-Craggs, P J Sellar) or 'viZEE', 'schliZEE', 'zli(t)ZEE', etc., with slightly harsh 1st part then purer-sounding but slightly buzzing 2nd part which becomes thinner, trailing away (M G Wilson). (b) Perhaps less common unit-type

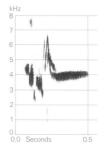

I S Palmér/Sveriges Radio (1972-80)
Sweden April 1963

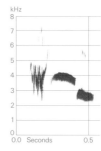

II E D H Johnson/Sveriges Radio (1972-80)
Morocco March 1963

rendered 'ziu' or 'ziurrr', with pitch descending at end (Voigt 1933; Beneden 1934); 'zrliu' in which pure-toned components follow section with frequency modulation (Bergmann and Helb 1982). Recording (Fig II) suggests 'tisle-o' or 'teesle-o'; 2nd and 3rd sub-units distinctly tonal (J Hall-Craggs, P J Sellar). Also rendered 'chlit-seeoo', with 1st sub-unit suggesting wagtail *Motacilla* and sparrow *Passer*, 2nd thin, almost sibilant, and 3rd not unlike fluting of Bullfinch *Pyrrhula pyrrhula* (M G Wilson). Suggested by Bergmann and Helb (1982) that each individual uses only one unit-type; however, in recording by P A D Hollom (Iran), 2 basic types as here described are given by same bird with marked change from simpler type to sweeter, liquid, cheerful, and attractive variant of type 1b. Apparent song noted by Schmidt (1960) in northern West Germany mid-April and early August comprised twittering (like excited mobbing-calls of White Wagtail *M. alba*), typical song-units ('zirrluih'), and 'dlit', 'dlim', and 'dieb' sounds (see below). ♂ neighbours probably distinguishable by songs (Bergmann and Helb 1982), and dialects probably exist in areas with sufficiently dense population (Wallschläger 1984). (2) Song of ♀. Identical to that of ♂ and generally given in flight to and from nest (G Högstedt): e.g. 'dzyurliy' (Smetana 1980); 'tirili' like Skylark *Alauda arvensis* said by Steiniger (1955) to be feeding-call or all-clear given by ♀; 'zürrlieh' mentioned by Bäsecke (1955) also likely to be song of ♀ or ♂. (3) Contact-alarm calls. (a) A 'tcheuk' or 'chup' (or disyllabic variants, e.g. 'serrup-serrup': Numme 1976), sometimes harsh or reminiscent of House Sparrow *Passer domesticus*, though also strongly recalling juvenile Linnet *Carduelis cannabina*. Given mainly in flight (Grant 1972). Also rendered 'sjilp', 'tschil(p)', 'zurl', 'ziurl', and 'tschrl'; given by migrants and as commonest call on breeding grounds (Bijlsma 1978a, b; also Wadewitz 1957a, Gatter 1976,

Jonsson 1978*a*, Bergmann and Helb 1982). Further renderings of this call and/or (presumably) closely related variants as follows: 'tzeuc' and 'tzi-uc' with slightly metallic timbre, 'tzic', 'tzucc', and more musical 'tsweerc' (Witherby *et al.* 1938*a*). See also Dittmann (1925, 1927) and Beneden (1934). (b) Rapidly repeated chirpy sound, sometimes on flushing or as bird lands (Grant 1972). Closely related to call 3a and perhaps only an accelerated variant: e.g. faster-than-usual 'tschilp' sounds given in ♂-♀ chase (Bijlsma 1978*a*); Bergmann and Helb (1982) noted a 'zrl zr zip...' at take-off and this apparently contains sounds resembling call 3a. Further descriptions of sounds given in rapid series: 'tscheb' or 'djep' (Gatter 1976); short 'zlip' (Rinnhofer 1970);'djip' often preceded by 'zia', 'zié', or 'ziärr' (which may be call 3c) in flight and as alarm-call when with young (Dittmann 1925, 1927); Schmidt (1960) noted similar 'di-ib' or 'zi-hip' as warning-call but likened them to 'nest-call' of Meadow Pipit *A. pratensis* and Tree Pipit *A. trivialis* (see call 3e). 'Chortling' calls mentioned by Bates (1924) probably also belong here. (c) Calls given mainly in flight and often likened to Yellow Wagtail *Motacilla flava* (this supported by sonagrams in Bergmann and Helb 1982): e.g. 'tsweep', 'tzeep', or 'zeep' (Witherby *et al.* 1938*a*; Grant 1972); 'tsiep' like Reed Bunting *Emberiza schoeniclus* (Bijlsma 1978*b*); 'tsuiit' (Numme 1976); protracted 'tsiip' (Jonsson 1978*a*); 'dieb' (Gatter 1976); 'zlie' or 'zlui' (Rinnhofer 1970); 'psja' (Bergmann and Helb 1982); 'tjya' from ♂ feeding young (Norup 1963). (d) Calls given by bird disturbed near nest, including when flushed and during distraction-lure display. A 'sree' like White Wagtail *M. alba* (G Högstedt); 'tschrieeh' said by Bäsecke (1955) to be given by only one of pair arriving at nest; 'grië' (Voigt 1933); hoarse, almost shrill with much 'r' sound (Glutz von Blotzheim 1981); squeaky or squealing sound (Smetana 1980); clear musical 'tzeeirrp' or 'chee-ip' of Witherby *et al.* (1938*a*) presumably also an alarm-call of this type. (e) A 'tji tji tji' likened to an alarm-call of *A. trivialis* (Wadewitz 1957*a*); 'tjy tjy' (Norup 1963) perhaps the same or related. Such calls noted by Glutz von Blotzheim (1981) in situation suggesting lower-intensity alarm; not always so, however: flushed birds sometimes run and call 'tsi-tsi-tsi-srië', units resembling *A. trivialis* alarm preceding change to call 3d (G Högstedt). (f) According to Dittmann (1925), units of song (see calls 1–2) may be preceded by 'tec terre tec' (as a fright-call) or 'zi zi zi' (perhaps same as call 3e). (4) Suppressed and rapid 'ti-ti-ti' given by aggressor in fights and chases (G Högstedt).

CALLS OF YOUNG. Food-call from *c.* 8–9 days like chirruping of sparrows *Passer*, especially loud after leaving nest. Like adult call 3a or 3b (Heinroth and Heinroth 1924–6; G Högstedt). Schmidt (1960) additionally noted 'dieb' (presumably adult call 3c) and, from young bird just out of nest, 'zirruih' (apparent song-unit).   MGW

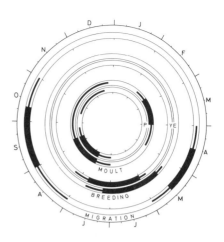

Breeding. SEASON. Western Europe: see diagram. Netherlands: 2 laying peaks, mid-May to early June, and mid-June to early July, reflecting 2 broods (Bijlsma 1978*a*). East Germany: small young still in nest early July to early August, exceptionally to mid-August (Krüger 1980). Southern Sweden; main laying period from mid-June, with extremes for 56 1st clutches of 21 May and 16 July (G Högstedt). Central and southern Europe: laying from mid-May to July (Makatsch 1976). North Africa: on low ground, laying begins second half of April, continuing to June: *c.* 2 weeks later on high ground (Heim de Balsac and Mayaud 1961). SITE. On ground in shallow hollow, often under plant tuft; bird sometimes makes scrape (Munn 1925; Dementiev and Gladkov 1954*a*). Nest: cup of grass stems and leaves, and roots, lined with finer plant material and hair. 10·5 cm outside diameter, 7 cm internal, depth of cup 5 cm (Makatsch 1976). Open side of nest usually faces east or north; this found experimentally to give shade during heat of day, and to reduce predation rate (Högstedt 1978). Building: mainly by ♀, but ♂ sometimes plays a part; takes 4–5 days (Nörup 1963; Müller 1983). EGGS. See Plate 80. Sub-elliptical, smooth and glossy; whitish, heavily marked with brown and purplish-grey spots and blotches. 21·5 × 15·7 mm (19·0–23·8 × 14·6–17·5), $n = 150$; calculated weight 2·73 g (Schönwetter 1979). Clutch: 4–5(3–6). Of 56 clutches, southern Sweden: 3 eggs, 7; 4, 29; 5, 20; mean 4·23 (G Högstedt). Of 21 clutches, East Germany: 3 eggs, 5; 4, 14; 5, 2; mean 3·85 (Krüger 1980). Of 17 clutches, Netherlands: 4 eggs, 10; 5, 6; 6, 1; mean 4·47 (Bijlsma 1978*a*). One brood, occasionally 2. Only 3 2nd broods recorded in 52 pairs, southern Sweden, with intervals between 1st and 2nd of 38, 41, and 45 days (G Högstedt). Of 12 pairs, Netherlands, 5 laid 2nd clutch (Bijlsma 1978*a*). Eggs laid daily. INCUBATION. 12 days (11·5–13), $n = 6$ (G Högstedt); 13–14 days (Makatsch 1976). By ♀ only, beginning with last egg (Smetana 1980); ♂ may help (Makatsch 1976); by both sexes (Munn

1925). YOUNG. Cared for and fed by both parents; brooded by ♀. FLEDGING TO MATURITY. Fledging period 13-14 days (G Högstedt). Independent at *c.* 4 weeks (G Högstedt), family bonds persisting for possibly 5 weeks (Glutz von Blotzheim 1981). BREEDING SUCCESS. In southern Sweden, *c.* 80-89% of eggs hatch in different years, with 60-74% fledging; up to 40% of fledged young lost through predation soon after fledging, while young noisy and poor-flying; main losses in early periods due to predation, but non-hatching responsible for 15% of losses and parasitism by Cuckoo *Cuculus canorus* for 7·7%; some variation in success in different nesting habitats, with dune habitat most successful, grass (*Corynephorus*) heath least successful, and *Calluna* heath and meadows intermediate, but samples small; strong positive correlation between fledging success and weight of young at 7-13 days (G Högstedt). In Oberlausitz (East Germany), many nests lost to weasels and martens (Mustelidae), corvids, and (especially) Marsh Harrier *Circus aeruginosus* and Buzzard *Buteo buteo*; *c.* 45% young fledged from 21 nests; more poorly concealed nests more liable to robbing (Krüger 1980). Of 56 eggs laid in 12 1st clutches, Netherlands, 91% hatched and 77% fledged; of 20 eggs laid in 5 2nd clutches, 80% hatched and 60% fledged (Bijlsma 1978*a*).

Plumages (nominate *campestris*). ADULT BREEDING. Upperparts entirely pale sandy-brown, feather-centres more pale olive-brown, fringes tinged buff; forehead and crown marked with narrow and rather indistinct olive-black shaft-streaks, mantle and scapulars with faintly darker olive shafts, back to upper tail-coverts uniform pale sandy-brown, except for narrow dark shaft-streaks on longer coverts. Supercilium long, broad, and distinct, pale cream-buff or off-white, extending from nostril to upper side of neck, bordered below by distinct dull black stripe on lores. Narrow but distinct cream or white ring of feathers round eye. Ear-coverts pale sandy-brown or olive-brown, darkest at border of supercilium, there sometimes forming distinct dull black stripe, paler and faintly streaked pale buff towards lower corner of eye, extending into pale buff patch with faint grey specks below eye. Narrow but distinct black moustachial stripe running back from gape; sometimes short, extending below eye only, but sometimes extending along lower ear-coverts to sides of neck. Sides of neck pale sandy-brown, faintly streaked and mottled buff or pale buff, especially just behind ear-coverts. Lower cheeks, chin, and throat pale buff to cream-white; a narrow but usually distinct malar stripe at sides of throat. Chest, sides of breast, and flanks buff or pale buff, remainder of underparts paler cream-buff or cream-white, palest on central belly; sides of breast with small and diffuse grey spots, sometimes extending slightly towards chest, central chest either uniform buff or with faint and narrow band of scattered tiny triangular specks on upper chest. Central pair of tail-feathers (t1) dark olive-brown, broadly fringed buff; t2-t5 dull black, t2-t4 with narrow cream outer edge, t5 with cream-buff to off-white outer web separated by dark shaft from cream or white wedge along shaft on distal part of inner web; wedge on t5 29·7 (30) 22-42 mm long, exceptionally 10 (on one tail-half only) and 17. T6 buff-white or cream-white with dark grey wedge along basal inner edge and sometimes with partly dark shaft. Flight-feathers and tertials dark grey-brown

or dark olive-brown; tertials broadly fringed warm buff or pale buff on both sides, secondaries rather broadly fringed buff on outer web, narrowly at tip, primaries narrowly edged pale buff on outer webs and tips, but only faintly on tips of longest primaries. Greater upper primary coverts and feathers of bastard wing dull black, narrowly edged buff on outer web and tip. Lesser upper wing-coverts pale sandy-brown, like upperparts; median upper wing-coverts black with broad and contrasting warm or pale buff tips *c.* 3 mm wide; greater upper wing-coverts duller black (except innermost), outer webs and tips slightly less contrastingly fringed buff or pale buff. Under wing-coverts and axillaries cream-buff to cream-white. *In worn plumage,* upperparts duller grey-brown, narrowly but distinctly streaked black-brown on forehead, crown, and mantle, traces of buff remaining mainly on nape and rump; median upper wing-coverts, tertial coverts, tertials, and t1 black-brown, pale fringes and tips worn off except for traces of off-white tips on median coverts; pattern on sides of head often less distinct, dark stripes on ear-coverts and malar stripe sometimes largely abraded, stripe below eye to shorter ear-coverts paler, off-white; ground-colour of underparts paler, chest and flanks cream, throat and belly cream-white or dirty off-white. Sexes similar, but ♀ on average more often spotted on chest than ♂, more rarely unspotted. When chest-spots scored with values between 0 (unspotted) and 5 (heavily spotted as in juvenile), ♂ scores 0·6 (14) 0-2, ♀ 1·6 (8) 0-3. ADULT NON-BREEDING. As adult breeding, except for slight differences in wear of plumage. In adult breeding, head, body, and often tertials and median upper wing-coverts new (see Moults), all flight-feathers fairly fresh to slightly worn (without suspended moult), remainder slightly worn; in non-breeding, either all plumage completely new (if moult complete), or part of plumage contrastingly old (if moult suspended during autumn migration—part of flight-feathers new, part old, unlike adult breeding). NESTLING. Down long and dense, on upperparts only; pale buff. JUVENILE. All upperparts and lesser upper wing-coverts black-brown, dark olive-brown, or rufous-brown, each feather evenly and narrowly fringed buff (when fresh) to white (when worn), giving scaly appearance; hindneck more extensively tinged buff; feather-sides of scapulars and upper tail-coverts more broadly fringed olive-buff, scaling less distinct. Supercilium and eye-ring pale buff or off-white, narrower than in adult, partly speckled brown, appearing less distinct; remainder of sides of head pale buff with indistinct dark feather-tips, latter merging into dark bar along rear of ear-coverts and sometimes into faint moustachial streak, but stripes on sides of head (including one on lores) generally less distinct than in adult. Ground-colour of lower cheeks and all underparts cream-white to off-white, chest and sides of breast slightly deeper cream-buff; malar stripe distinct, black-brown, merging into distinct gorget of short triangular spots *c.* 2 mm wide on chest and sides of breast; upper flanks with similar marks, lower flanks and sometimes sides of belly with narrower and more elongated ones. Tail and flight-feathers as adult. Median upper wing-coverts dark brown, broadly fringed pale buff to off-white, similar to lesser coverts (in adult, darker than lesser coverts); brown centre extends in blunt point towards tip (in adult, broadly rounded with narrowly protruding point). Greater upper wing-coverts as adult, but dark brown centres more sharply divided from pale fringes. FIRST ADULT NON-BREEDING. Like adult non-breeding, but variable amount of juvenile feathering retained (see Moults); in contrast, adult non-breeding either completely new or with moult suspended (retained feathers then including variable number of flight-feathers, unlike 1st non-breeding). Chest more heavily spotted than in adult non-breeding: when amount of spotting is scored

0-5 (see Adult Breeding), 1st non-breeding birds score 2·8 (15) 1-4, excluding some which retained variable amount of heavily spotted juvenile feathers (score 5). FIRST ADULT BREEDING. Like adult breeding and some indistinguishable, but juvenile flight-feathers, greater upper primary coverts, and often outer or all greater upper wing-coverts and t2-t6 retained, appearing more heavily worn than in adult. See also First Pre-breeding under Moults. Chest on average rather more heavily spotted than in adult breeding, scoring 0·8 (12) 0-2 in ♂, 1·7 (9) 0-4 in ♀.

Bare parts. ADULT, FIRST ADULT. Iris dark brown. Upper mandible and tip of lower mandible dark horn-brown, cutting edges of both mandibles and base of lower mandible pink-flesh to pale horn-brown. Leg and foot bright yellow-flesh, amber-yellow, light yellow-brown, or pale brown. (Hartert 1910; Witherby *et al.* 1938a; RMNH, ZMA.) NESTLING. Bare skin (including leg, foot, and bill) pink, future feather-tracts blue. Mouth orange or orange-red, unspotted (yellow with some dark spots in rather similar Woodlark *Lullula arborea*); gape-flanges pale yellow (Heinroth and Heinroth 1924-6). JUVENILE. Iris dark brown. Bill pink-flesh; culmen and tip pale horn, gradually darkening during post-juvenile moult. Mouth and gape-flanges as in nestling. Leg pale brown-flesh, toes brighter flesh-pink. (Heinroth and Heinroth 1924-6; RMNH.)

Moults. Based on Stresemann (1920), Stresemann and Stresemann (1968a), skins in BMNH, RMNH, and ZMA, and data supplied by G Högstedt for Sweden. ADULT POST-BREEDING. Complete; primaries descendant. Starts with p1, ♂♂ generally 2-3 weeks before ♀♀; ♂ may start before brood fledged, but only when no longer accompanying young; ♀ usually starts when young independent, starting earlier only when brood very late. In southern part of breeding range, ♂ starts from early July or perhaps late June, ♀ between late July and late August; all moult completed late August to early October. Further north in central Europe and Sweden, ♂ starts between c. 20 July and 15 August, ♀ 30 July to 1 September, but moult halted at start of migration and completed during migration stopovers or in winter quarters. Birds on autumn migration with moult suspended had primary moult scores of 5 (p1 new only; 1 bird), 10 (2 birds), 15 (3), 20 (2), 25 (3), 30 (2), and 35 (1); in Sweden, a few depart before moult started (Svensson 1984a). Birds resuming moult in winter quarters complete October-December. Timing of moult of body and tail rather variable: some migrants with suspended moult have t1 and much of body, tertials, and wing-coverts new, even at low primary score 15-25, others still largely old when suspending with score 15-25. Tail starts at score 5-20, new at score 35-40; secondaries start at score 20-25, new at same time as outer primaries. ADULT PRE-BREEDING. Partial; February to mid-April. Involves head, body (occasionally except some feathering of chin and belly), and tertial coverts, 1 or all tertials, often median upper wing-coverts, usually lesser upper wing-coverts, rarely greater coverts (never outermost); also t1 (in 9 of 24 examined: Stresemann 1920; ZMA) and sometimes all tail (Ticehurst 1923). POST-JUVENILE. Partial: head, body, median upper wing-coverts, tertial coverts, most or all lesser upper wing-coverts, some or all tertials, often t1, and rarely inner greater upper wing-coverts and all tail. Timing strongly variable, depending on fledging date. Starts at age of c. 30 days (G Högstedt) or (in captivity) when c. 2 months old (Heinroth and Heinroth 1924-6). All feathers mentioned above replaced in early-fledged birds, these completing c. 1 month after start (Dementiev and Gladkov 1954a), but late-fledged birds suspend

moult during migration and some do not start at all, these retaining some to virtually all juvenile feathers of head, body, and wing-coverts, with remaining moult carried out October-January in winter quarters. Moult starts early July to mid-September (and probably later in the few birds which do not start before autumn migration), completed early August to January. FIRST PRE-BREEDING. Mainly like adult pre-breeding. Timing as adult pre-breeding, but some birds still moulting late April and early May; extent as adult pre-breeding, but sometimes less, birds frequently retaining 1st non-breeding tertials, lesser coverts, and t1. A few birds are distinctly worn in April, except for some median coverts and tertials; these probably retain 1st non-breeding, and limited amount of new feathers probably due to late post-juvenile replacement rather than to limited 1st pre-breeding.

Measurements. Nominate *campestris*. Wing: (1) Netherlands and West Germany (RMNH, ZMA); (2) Italy, southern France, and Spain (Stresemann 1920; RMNH, ZMA); (3) Yugoslavia (Stresemann 1920; ZMA); (4) North-west Africa (RMNH, ZMA); (5) Greece and European Turkey (Makatsch 1950; Rokitansky and Schifter 1971; RMNH); (6) Asia Minor (Kumerloeve 1961, 1968, 1969a, 1970a; Rokitansky and Schifter 1971; RMNH); (7) Transcaucasia and northern and south-west Iran (Paludan 1938, 1940; Nicht 1961; RMNH). Other measurements for 1-2 (A) and 4-7 (B) combined, or all combined. Bill (S) to skull, bill (N) to distal corner of nostril; exposed culmen on average 4·3 less than bill (S). Toe is middle toe with claw; claw is hind claw.

		♂		♀	
WING	(1)	91·1 (1·88; 17)	88-94	87·3 (2·17; 11)	84-90
	(2)	91·6 (2·52; 15)	87-96	86·9 (2·81; 9)	83-91
	(3)	93·8 (1·91; 35)	90-98	88·0 (1·16; 8)	86-89
	(4)	94·5 (2·33; 9)	91-97	88·5 (3·35; 5)	84-92
	(5)	95·7 (1·79; 13)	93-98	88·6 (2·14; 4)	86-91
	(6)	95·6 (3·04; 22)	90-101	90·2 (2·98; 10)	85-94
	(7)	94·8 (1·98; 8)	92-98	88·0 (—; 2)	87-89
TAIL	(A)	67·7 (2·44; 24)	63-71	64·2 (1·67; 18)	61-67
	(B)	69·0 (2·60; 19)	65-73	64·4 (2·30; 9)	62-68
BILL (S)	(A)	18·6 (0·58; 22)	17·6-19·4	18·1 (0·66; 13)	17·0-19·0
	(B)	19·6 (0·75; 19)	18·5-20·7	18·4 (0·58; 9)	17·7-19·3
BILL (N)	(A)	10·9 (0·70; 22)	9·8-12·1	10·8 (0·52; 14)	9·9-11·7
	(B)	11·5 (0·70; 19)	10·5-12·8	10·9 (0·42; 8)	10·2-11·6
TARSUS	(A)	25·9 (1·03; 24)	24·1-27·4	25·0 (0·83; 16)	23·9-26·5
	(B)	26·7 (0·51; 19)	25·9-27·6	25·6 (1·05; 9)	24·4-27·1
TOE		19·2 (1·06; 23)	17·8-21·3	18·6 (1·23; 14)	16·8-20·4
CLAW		8·8 (1·14; 47)	6·8-11·5	8·8 (1·00; 22)	7·3-10·6

Sex differences significant, except bill (N), toe, and claw. 1st adult (with juvenile flight-feathers and often juvenile tail) and full adult combined, though wing of full adult on average 1·5 longer than 1st adult and tail on average 1·0 shorter. Hind claw mainly 7·4-10·1, only 5 birds above and 5 below this.

Weights. ADULT, FIRST ADULT. Nominate *campestris*. On migration. Kuwait and north-east Arabia, March: 20·0 (2·66; 9) 16-23 (V A D Sales, BTO). Western and northern Iran, April: ♂♂ 24, 26; ♀ 24·4 (Paludan 1938; Schüz 1959). Spring and autumn, Malta: 27·8 (2·3; 7) 26-32 (J Sultana and C Gauci). Netherlands, September: ♂♂ 24·0, 24·5, 27·2 (RMNH, ZMA). Central Afghanistan, September and early October: ♂ 23·4 (8) 20-25, ♀ 20·6 (1·78; 5) 19-24 (Paludan 1959).

In breeding area. Southern Spain, May: ♂♂ 26·5, 28·5; ♀ 21 (Niethammer 1957). Northern Greece and European Turkey, April-May: ♂ 26·8 (0·87; 4) 26-28, ♀ 25·4 (Makatsch 1950; Rokitansky and Schifter 1971). Asia Minor, Armeniya (USSR), and north-west Iran, combined: May, ♂ 28·8 (1·28; 8) 27-31,

♀ 28·8 (1·12; 4) 27-30; June, ♂ 28·3 (1·75; 6) 26-31, ♀ 33·0 (3) 31-34; July, ♂ 26·4 (1·74; 5) 23-28, ♀ 28·1 (Paludan 1940; Schüz 1959; Nicht 1961; Kumerloeve 1968, 1969a, 1970a; Rokitansky and Schifter 1971).

In winter quarters. Ahaggar, southern Algeria, December: ♀ 23 (Niethammer 1963).

A. c. kastschenkoi. Mongolia, June-August: ♂ 23·8 (1·94; 6) 21-26, ♀ 24 (Piechocki and Bolod 1972).

A. c. griseus. Afghanistan, July-September: ♂♂ 22, 25; ♀ 19 (Paludan 1959). USSR: ♂♂ 20·5, 23 (Dementiev and Gladkov 1951a).

Race unknown. Kazakhstan (USSR), April-September: ♂ 23·1 (1·38; 30) 20-26, ♀ 22·7 (2·79; 14) 19-28 (Dolgushin *et al.* 1970).

NESTLING. Nominate *campestris*. Sweden: on day 1, 2·0 (4) 1·8-2·1; day 3, 5·3 (0·75; 8) 4-6·5; day 7, 16·8 (1·80; 9) 15-20; rapid growth halted on day 10, when 21·0 (1·50; 10) 19-24; on days 11-14, 22·5 (1·10; 8) 21-24 (G Högstedt).

Structure. Wing rather long, broad at base, tip bluntly pointed. 10 primaries: p8-p9 longest, p7 0-1 shorter, p6 2-3, p5 10-14, p4 16-20, p1 24-19; p10 reduced, narrow and pointed, 59-69 shorter than wing-tip, 9-13 shorter than longest upper primary covert. Outer web of p6-p8 and inner web of p7-p9 emarginated. Longest tertials reach wing-tip in closed wing. Tail rather long, tip square or slightly forked; 12 feathers. Bill rather long, fairly deep and wide at base, slender at tip. For length of middle toe and hind claw, see Measurements. Outer and inner toe with claw both *c.* 67% of middle toe with claw; hind toe with claw *c.* 87%, without claw *c.* 48%. Hind claw short, slender, distinctly decurved.

Geographical variation. Slight, largely clinal; relatively indistinct compared with individual variation and strong influence of bleaching. 2 main lines of variation discernible: (1) northern line of rather large and sandy birds in western and central Europe leading through slightly paler and smaller birds in Volga-Ural steppes to small and rather dull grey birds in foothills of Altai; (2) southern line with large and sandy birds in North Africa leading through large but slightly greyer birds in Greece, Turkey, and Iran to slightly smaller but distinctly greyer birds in Afghanistan and Tien Shan area (Johansen 1952; C S Roselaar). Intermediate populations between both lines occur in southern Italy, Yugoslavia, Rumania, areas round Aral Sea, Lake Balkhash, and Lake Zaysan, and in Tarbagatay and southern Mongolia. As differences even between ends of clines are slight, species perhaps better considered monotypic, but 2 distinct groups recognizable among birds wintering in India and Pakistan (Hall 1961): smaller and duller grey *kastschenkoi* and larger and more sandy-grey *griseus*. Some authors recognize only *kastschenkoi* and include *griseus* in nominate *A. campestris* (Hall 1961; Ali and Ripley 1973b), or recognize only *griseus* and include *kastschenkoi* in nominate *A. campestris* (e.g. Vaurie 1954, 1959), but neither race fully similar to European nominate *A. campestris*, and to avoid further confusion both *griseus* and *kastschenkoi* recognized here. Western

and central Europe, from Iberia, northern Italy, and Hungary north to southern Sweden, inhabited by sandy nominate *campestris* with wing 87-96 (♂) or 83-91 (♀); gradually duller grey and smaller towards east, ending in dull grey *kastschenkoi* in Barabinskaya steppe, Altai, and north-west Mongolia, with wing 85-90 (♂) or 80-87 (♀). In southern line, birds from north-west Africa rather large, wing 91-97 (♂) or 84-92 (♀), but colour similar to typical European nominate *A. campestris* and difference too slight to warrant separation (though tarsus of ♂ mainly 26-28, rather than mainly 24-26 in Netherlands and West Germany). Populations from Greece, Turkey, Transcaucasia, Elburz and Zagros mountains of Iran, and western Turkmeniya (USSR) are larger and sometimes separated as *boehmii* Portenko 1960; wing of ♂ mainly 93-101, ♀ 86-94. Colour of fresh plumage in '*boehmii*' intermediate between nominate *campestris* and *griseus*: in nominate *campestris*, upperparts rather deep and uniform buff, with grey feather-centres of mantle and scapulars largely hidden underneath buff fringes; in *griseus*, upperparts light grey with rather narrow sandy feather-fringes, rump uniform light grey (not uniform buff); in populations of '*boehmii*', rather a lot of grey of feather-bases visible, feather-fringes narrower than in typical nominate *A. campestris*, wider than in *griseus*, but rather deep buff in colour, like nominate *A. campestris*, not as sandy as in *griseus*. General appearance of '*boehmii*' nearer to central European birds than to *griseus* of Tien Shan and hence included in nominate *campestris*, notwithstanding slight differences. Birds from eastern Turkey warmer sandy-brown than those from west (Kumerloeve 1968, 1969a). Sandy-grey *griseus* from central Iran eastward smaller, wing of ♂♂ from Tien Shan (USSR) 92·5 (1·86; 16) 89-96, ♀ 87·0 (1·67; 6) 84-90 (BMNH, RMNH), thus larger than *kastschenkoi* from further north; other measurements similar to typical nominate *campestris* from Europe.

Recognition. Adult with virtually unpatterned chest unmistakable, rather similar only to Long-billed Pipit *A. similis*, which differs in blunter wing (p6-p9 longest) with p5-p8 emarginated on outer web. Juvenile with spotted chest closely similar to other long-winged *Anthus*; differs from migratory races (e.g. *A. n. richardi*) of Richard's Pipit *A. novaeseelandiae* in shorter tail, tarsus, and hind claw, with hind claw more strongly curved; resident races of *A. novaeseelandiae* in Afrotropics and Indian subcontinent near *A. campestris* in size, however, though hind claw of these usually longer and straighter and upperparts more heavily streaked; adult Blyth's Pipit *A. godlewskii* very close in size and colour to juvenile *A. campestris*, but t5 blacker, usually showing small white triangle only, not longer white streak, and upperparts more distinctly streaked, not mottled brown with paler scales. Juveniles of *A. godlewskii* and of some Indian races of *A. novaeseelandiae* sometimes indistinguishable from juvenile *A. campestris* (Hall 1961). Some races of Water and Rock Pipit *A. spinoletta* unstreaked below in adult breeding, like adult *A. campestris*, but ear-coverts, upperparts, t5, and legs much darker and supercilium narrower and shorter. CSR

Anthus berthelotii Berthelot's Pipit

Du. Berthelots Pieper   Fr. Pipit de Berthelot   Ge. Kanarenpieper
Ru. Конек Бертелота   Sp. Bisbita caminero   Sw. Kanariepiplärka

Anthus Berthelotii Bolle, 1862

Polytypic. Nominate *berthelotii* Bolle, 1862, Canary Islands and Ilhas Selvagens; *madeirensis* Hartert, 1905, Madeira.

Field characters. 14 cm; wing-span 21–23·5 cm. Marginally smaller than Meadow Pipit *A. pratensis* but with form and actions recalling Rock and Water Pipits *A. spinoletta* as much as Tawny Pipit *A. campestris*. Smallest pipit of west Palearctic. Rather grey above and buff-white below, with obvious pattern on pale face, pale double wing-bar and tertial fringes, and streaked crown, mantle, and chest. Flight not markedly bounding. Sexes similar; no seasonal variation. Juvenile separable. 2 races in west Palearctic, not separable in the field.

ADULT. Upperparts brownish-grey, sharply streaked on crown, mantle, and scapulars, but unmarked on rump. Face appears pale from in front, but well marked (even striped) with obvious cream-white supercilium from base of bill to rear of ear-coverts and dark dusky line above this; rather distinct dark dusky-brown border to cheeks, all surrounded by pale area; distinct black-brown malar stripe. Wings show obvious pale bar on median coverts (with black feather-centres as striking as cream-white margins), less obvious pale buff-white bar on greater coverts, and pale buff fringes to tertials. Variation in feather-fringe colours makes some birds appear distinctly yellowish overall. Tail black-brown, outer feathers yellow-white. Underparts white, with variable pink or buff (but not yellow) suffusion on face, sides of neck and chest, and on flanks, and with distinct or obscure black-brown streaks extending from malar stripe over chest and down fore-flanks. Underwing grey- or buff-white. Effect of wear and moult not studied in the field but upperparts sometimes noticeably more uniform and chest-streaks limited, increasing suggestion of *A. spinoletta* and *A. campestris*. JUVENILE. Resembles adult but underparts always intensely streaked; tips and margins of wing-coverts and tertials rufous-buff. At all ages, bill rather long and slender; black-horn, with pale yellowish or reddish base to lower mandible. Legs yellow-brown.

The only common pipit on Canary Islands and Madeira, though confusion may arise with vagrant congeners. Of these, most likely is *A. pratensis*, but that species noticeably greener in tone, less pale and striped on face, always well streaked below, and has different voice (see p. 365]. *A. trivialis* is noticeably buffier overall, well streaked below, and again has different voice (see p. 344). Water Pipit *A. spinoletta spinoletta* and Rock Pipit *A. spinoletta littoralis* would present greatest potential confusion (see Recognition), but so far not recorded in range of *A. berthelotii*. *A. campestris* is 15% larger, with

noticeably longer tail, much less marked plumage in adult, and again different voice (see p. 313). Appearance in flight intermediate between small and large pipits, with light and rapid action lacking exaggerated 'shooting' bounds of *A. campestris*. Stance and gait closer to *A. spinoletta* than to *A. campestris*, with noticeable head movement when walking; runs quickly over flat ground, and hops and jumps nimbly over boulders; has distinctive habit of clambering over small plants when feeding. Often extremely tame. Not gregarious but forms small groups in winter.

Song a plaintive, repeated 'tsiree'. Commonest call a low-pitched, husky 'tsik', also a soft, rolled 'kree'; alarm-call a louder 'chi-iree' or 'crew-ee'.

Habitat. Only on oceanic subtropical islands of Canaries group and Madeira. Habitat varies somewhat between islands but generally in every zone from sea-level to highest ground at *c.* 1600–2100 m or even higher (see Movements). Particularly numerous on rocky or maritime plains and plateaux with scant vegetation, or on sun-baked rock-strewn hillsides covered with bushes of *Euphorbia* affording shade and nest-sites. Also common on vine-clad slopes and in coastal tomato patches, although frequently displaced by encroachment of cultivation on natural habitat. Breeding recorded on a golf course. On La Palma, the only bird found on barren lava flows. On Gomera, inhabits open hillside sloping down to sea-cliffs with immense boulders half-covered by brambles *Rubus*, thistles, coarse grass, etc. On mountains, exists in more open parts of pine forests and above them in *Retama* zone; avoids thick woods and moist ravines. A ground bird, reluctant to fly and to perch on trees, but does so on small pines in plantation on La Palma. (Bannerman 1963.)

Distribution and population. Restricted to Canary Islands and Madeira. For abundance on various islands, see Bannerman (1963) and Bannerman and Bannerman (1965).

CANARY ISLANDS. Gran Canaria: has declined due to habitat changes (Bannerman 1963); common throughout eastern islands (Shirt 1983).

Movements. Resident, with little or no evidence of even local movement. Records of birds up to *c.* 3200 m on Tenerife in winter (Wallace 1964)—higher than breeding

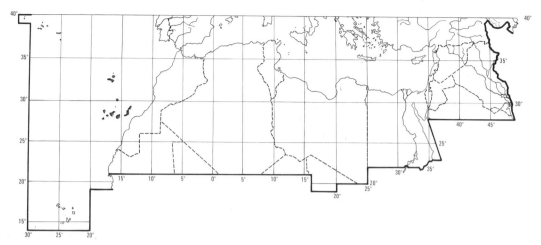

limit of *c.* 2100 m recorded by Reid (1887-8)—may suggest altitudinal movement (or that breeding occurs higher than so far proved). Despite being most common bird on dunes of north-east coast of Fuerteventura (Hooker 1958), no records from African coast only 108 km away.   JHE

Food. Insects and seeds. Feeds almost exclusively on ground, picking invertebrates while walking or (also while perched) snatching them from the air if close; sometimes makes short aerial pursuit (Koenig 1890; Ribbeck 1904; Lockley 1952; Knecht 1960; Bannerman 1963; Bergmann and Helb 1982). Stomachs analysed by Ribbeck (1904) indicated small ants (Formicidae) favourite food; also contained weed seeds, spiders (Araneae), flies (Diptera), and pupae of Lepidoptera. Other stomachs sometimes full of large plant seeds (Bolle 1857), and 3 contained numerous insects and some seeds of grasses (Gramineae) (Volsøe 1951). On Fuerteventura, recorded feeding on grasshoppers (Acrididae) (Shirt 1983).

Full-grown young fed on small crickets (Gryllidae) and caterpillars *Pieris rapae* (Bannerman 1963).   PJE

Social pattern and behaviour. Little information.

1. Usually solitary or in pairs (e.g. Volsøe 1951, Bannerman 1963), though small parties also recorded. In May–June, these probably families (Etchécopar and Hüe 1957; Ennion and Ennion 1962), but 3 or more noted in April, apparently just prior to breeding (Bannerman 1963), and up to 10 at other times (Cullen *et al.* 1952; Knecht 1960). BONDS. No information. BREEDING DISPERSION. Presumably solitary but no details. Unusually high densities recorded on favoured stony slopes and plains with sparse growth of (especially) Euphorbiaceae (Bergmann and Helb 1982). On Fuerteventura, frequently 9–10 birds noted per transect-hour (see Shirt 1983 for details). ROOSTING. No information.

2. Generally reported to be extremely tame, more so than Tawny Pipit *A. campestris* (Hüe and Etchécopar 1958). Will come to within *c.* 3 m of man standing quietly (Ribbeck 1904), and August parties of 2–10 birds allowed approach to *c.* 2 m (Knecht 1960). In contrast, Morphy (1965) found it shy and restless. Said by Bolle (1857) to crouch on stones, but overall

more likely to run off rather than fly (Godman 1872; Lockley 1952; Bannerman 1963). FLOCK BEHAVIOUR. No information. SONG-DISPLAY. Song of ♂ (see 1 in Voice) given from ground, low bush, stone or wire, also (less commonly according to Volsøe 1951) in flight. Song-flight a simple fluttering affair with undulations, and thus apparently resembling that of *A. campestris*, but no details (Cullen *et al.* 1952; Etchécopar and Hüe 1957; Bergmann and Helb 1982; Wallschläger 1984; H-W Helb). Normally takes off for Song-flight from ground (Wallschläger 1984); according to Koenig (1890), may start singing on perch, then continue in flight or conclude song during descent. On La Palma and Gomera, song not noted after end of June (Cullen *et al.* 1952). ANTAGONISTIC BEHAVIOUR. On Tenerife, ♂♂ seen fighting over ♀♀ as early as January (Koenig 1890). Chases with Bill-snapping occur (Bergmann and Helb 1982) and aerial tussles with combatants flying up breast-to-breast are frequent (Ribbeck 1904). HETEROSEXUAL BEHAVIOUR. On Tenerife, February, birds already paired and carrying nest-material (Koenig 1890); in Meadow Pipit *A. pratensis*, however (see p. 374), material sometimes carried during displays. ♂ generally nearby when ♀ building and she will discard material if ♂ gives an alarm-call (Koenig 1890: see 2 in Voice). RELATIONS WITHIN FAMILY GROUP. No information on period when young in nest. When feeding young out of nest, adults normally run to and fro rather than flying about with food (Ennion and Ennion 1962). ANTI-PREDATOR RESPONSES OF YOUNG. No information. PARENTAL ANTI-PREDATOR STRATEGIES. (1) Passive measures. According to Reid (1887-8), sitting bird tends to sneak away quietly when man approaches. Sometimes apparently sits tight (Polatzek 1908). Overall, far less demonstrative than *A. pratensis* or Rock Pipit *A. spinoletta* when with young in or out of nest (Ennion and Ennion 1962). May just run away and call (Bergmann and Helb 1982: see 2b in Voice). (2) Active measures. Sometimes performs distraction-lure display of disablement type (Shirt 1983; H-H Bergmann), though frequency of true injury-feigning not

PLATE 28 *(facing)*.
Motacilla flava Yellow Wagtail (p. 413). *M. f. flavissima*: 1 ad ♂ breeding, 2 ad ♀ breeding, 3 ad ♂ non-breeding, 4 juv. Nominate *flava*: 5 ad ♂ breeding, 6 ad ♀ breeding, 7 ad ♂ non-breeding, 8 juv. *M. f. thunbergi*: 9 ad ♂ breeding, 10 ad ♀ breeding, 11 ad ♂ non-breeding. *M. f. cinereocapilla*: 12 ad ♂ breeding, 13 ad ♀ breeding, 14 ad ♂ non-breeding, 15 juv. *M. f. feldegg*: 16 ad ♂ breeding, 17 ad ♀ breeding, 18 ad ♂ non-breeding, 19 juv. (NA)

established. Recorded fluttering and running from man with dog, then flying low in small arc back to nest after observer's departure (Polatzek 1908); in another case, bird fluttered along trailing wing, moving *c.* 15 m in and out of vegetation before flying (Bannerman 1963).   MGW

Voice. Not studied in detail, but repertoire apparently restricted. For extra sonagrams, see Bergmann and Helb (1982). For Bill-snapping, see Social Pattern and Behaviour.

CALLS OF ADULTS. (1) Song of ♂. Not loud, but quite pleasant and cheerful (Ribbeck 1904; Bannermann 1963). Similar to Tawny Pipit *A. campestris* in simplicity, lack of variety, and extreme repetitiveness; some similarity also in structure of units (J Hall-Craggs). Comprises sequence of complex units separated by pauses of *c.* 1–2 s. Individual variation apparent in structure of units: 'tschrli', 'triuit', 'zrliü' (Bergmann and Helb 1982). Presumably alternative renderings of 1st of these: 'tsiree' given 1–2 times per s (Cullen *et al.* 1952) or shrill, disyllabic 'tcher-lee tcher-lee...' given in Song-flight (Lack and Southern 1949). Recording of bird singing from ground suggests 'tiuee' or 'thiuee' (J Hall-Craggs: Fig I), 'tsirlee' at rate of *c.* 1 per s (M G Wilson). Another unit (Fig II) rather thin and forced 'slee-oo' with jingly quality (M G Wilson) or 'tuioo' with 4 distinct but connected syllables—compare, however, Fig a in Bergmann and Helb (1982) in which sub-units are discrete (J Hall-Craggs). Recording (Fig III) of bird singing from wire suggests very thin and high-pitched '(t)sillee'; another variant (Fig IV) has strained, slightly hoarse quality and different pitch pattern in final sub-unit (M G Wilson). (2) Contact-alarm calls. (a) A 'tsrl' expressing alarm near nest; barely distinguishable from song-unit, but slightly shorter and may sound strangled (Bergmann and Helb 1982). Recording by H-W Helb of bird flying up from ground suggests 'zichEE' or 'sliREE' (M G Wilson) or hoarse, quiet 'tiuee' (J Hall-Craggs). Cullen

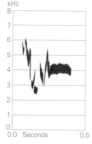

I H-W Helb
Canary Islands   March 1981

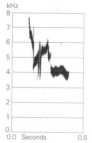

II   E D H Johnson
Canary Islands   April 1967

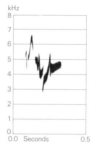

III H-W Helb
Canary Islands   April 1981

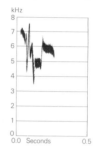

IV H-W Helb
Canary Islands   April 1981

et al (1952) also noted a 'chi-iree', apparently related to song. Feeding birds observed by Knecht (1960) gave 'tschirr' sounds; similar shrill 'tchirrt' like sparrow *Passer* said by Lack and Southern (1949) to be closer to *A. campestris* than to any other European *Anthus*. (b) Typical call a 'chup' like *A. campestris* (W R P Bourne). Other renderings of simple, mostly apparently monosyllabic calls: colourless 'tsik' commoner than call 2a (Cullen *et al.* 1952); thin 'zihp' while feeding (Knecht 1960); frequently uttered 'piet piet püit püt', sometimes in a long series, also 'trieb' in alarm (Ribbeck 1904); insignificant quiet peeping sounds given by alarmed bird moving away from nest (Bergmann and Helb 1982).

CALLS OF YOUNG. No information.   MGW

Breeding. SEASON. Canary Islands: eggs found first half of January to late May; on Tenerife, main season January–February at low altitudes, March–April higher up; Fuerteventura, April (Volsøe 1951; Ennion and Ennion 1962; Bannerman 1963; Shirt 1983). Madeira: eggs found early February to August, rarely from late January (Bannerman and Bannerman 1965). SITE. On ground, under bushes or low plants, or sheltered by stone. Nest: neat cup of plant stems and fibres, lined with hair, wool, and feathers. Building: apparently by ♀ (Koenig 1890). EGGS. See Plate 80. Sub-elliptical, smooth and glossy; pale grey, with pinkish or yellowish-brown (rarely dark grey) speckles and flecks; occasionally with dark hair-streaks at broad end. Nominate *berthelotii*: 19·7 × 14·6 mm (18·2–21·4 × 14·0–15·7), *n*=93; calculated weight 2·19 g. *A. b. madeirensis*: 19·2 × 15·2 mm (17·5–21·0 ×

PLATE 29 *(facing).*
Motacilla flava Yellow Wagtail (p. 413). Ad ♂ breeding: **1** *flavissima*, **2** *lutea*, **3** *leucocephala*, **4** *beema*, **5** nominate *flava*, **6** *simillima*, **7** *iberiae*, **8** *pygmaea*, **9** *thunbergi*, **10** *cinereocapilla*, **11** nominate *flava-feldegg* intergrade ('dombrowskii'), **12** *feldegg*.
M. f. iberiae: **7** ad ♂ breeding, **13** ad ♀ breeding, **14** ad ♂ non-breeding, **15** juv, **16** 1st ad ♂ non-breeding (1st winter).
M. f. pygmaea: **8** ad ♂ breeding, **17** ad ♀ breeding, **18** ad ♂ non-breeding, **19** juv, **20** 1st ad ♂ non-breeding (1st winter).
Variants, ad ♂ breeding: **21** 'old' *flavissima*, **22** 'brown' *flavissima*, **23** 'blue' *flavissima*, **24** '*superciliaris*'.
1st ad ♂ non-breeding: **16** *iberiae*, **20** *pygmaea*, **25** *thunbergi*, **26** *cinereocapilla*, **27** nominate *flava-feldegg* intergrade ('dombrowskii'), **28** *flavissima*, **29** *lutea*, **30** *beema*, **31** *simillima*, **32** *leucocephala*, **33** nominate *flava*, **34** *feldegg*.
Motacilla citreola Citrine Wagtail (p. 433), 1st ad ♂ non-breeding (1st winter): **35** nominate *citreola*, **36** *werae*.
Motacilla alba Pied Wagtail and White Wagtail (p. 454), 1st ad ♂ non-breeding (1st winter): **37** nominate *alba* (White Wagtail), **38** *yarrellii* (Pied Wagtail). (NA)

14·1-16·5), $n = 15$; calculated weight 2·28 g (Schönwetter 1979). Clutch: 2-5 on Canary Islands—of 35 clutches: 2 eggs, 5; 3, 3; 4, 26; 5, 1; mean 3·9 (Bannerman 1963; Shirt 1983); no information from Madeira. Prolonged season suggests at least 2 broods. No further information.

Plumages (nominate *berthelotii*). ADULT. Ground-colour of upperparts cool grey, sometimes with faint sandy suffusion; forehead and crown evenly streaked dull black, but remainder of upperparts rather uniform (especially hindneck and rump); mantle and inner scapulars with ill-defined dark grey or dull black feather-centres, longer upper tail-coverts dark grey with paler grey sides. Lores pale buff near base of bill, off-white in front of eye, broken by narrow dark grey streak from base of bill to eye. Short but distinct cream-white supercilium over eye to above ear-coverts, rather similar to supercilium of Water Pipit *A. spinoletta spinoletta*; narrow ring of cream-white feathers round eye. Upper cheeks and ear-coverts mottled dark grey, pale grey-brown and off-white; patch below eye often almost uniform off-white, sometimes with dark grey streak from gape back to lower ear-coverts and usually with dark streak on upper ear-coverts along supercilium. Lower cheeks cream-white or off-white, bordered below by dull black malar streak (not reaching base of bill, but extending down to streaked sides of breast). Chin and throat pale cream-buff to off-white, sharply divided from dull black malar stripe and evenly streaked chest and sides of breast; dark grey or dull black streaks on chest rather narrow (*c.* 2 mm), short (*c.* 1 cm), and sharply defined, those on sides of breast slightly less sharp (sometimes coalescing) and extending on to upper flanks. Ground-colour of underparts (down from chest) pink-buff, cream, or off-white, often deeper buff on chest, flanks, and thighs; unmarked below chest but flanks often partly tinged grey and sometimes with faint grey shaft-streaks. Central 4 pairs of tail-feathers sepia-black, t1 narrowly fringed grey; t5 sepia-black with narrow off-white edge and tip to outer web and long white wedge along distal part of shaft on inner web; white wedge on t5 usually 20-30 mm long, rarely 16-34, sometimes tinged grey or buffish-grey; t6 white except for dark shaft and dark grey wedge along base of inner web, distal part of outer web sometimes slightly suffused grey or buff. Flight-feathers, tertials, and upper primary coverts sepia-black or greyish-back; both webs of tertials and outer webs of secondaries broadly fringed pale buff or pale buff-grey, outer webs of primaries more narrowly and sharply edged pale buff or off-white. Greater upper wing-coverts like secondaries, showing broad but poorly defined buff fringe to outer web and tip; median coverts darker, dull black, with broad and sharp buff fringe along tip (soon bleaching to white); lesser coverts dull grey, like upperparts, fringes slightly paler grey or sandy-grey. Under wing-coverts and axillaries greyish-white or buff-white; longer coverts with slightly darker grey centres, shorter ones along leading edge of wing spotted dark grey. *In worn plumage*, upperparts dull grey with prominent black-brown streaks (least so on hindneck and none on back and rump); supercilium pronounced, white, and longer than in fresh plumage, bordered below by distinct dark grey line on lores and by black-brown upper ear-coverts; patch below eye more distinct, off-white, bordered below by more distinct dark grey stripe running backwards from gape; ground-colour of underparts paler, cream-white or off-white (sometimes stained rufous by soil); dark malar stripe and streaks on chest more sharply defined and narrower, sometimes broken into rows of triangular spots; pale fringes of t1 and tertials worn off; pale fringes on tips of median and greater upper wing-coverts

bleached to white, forming double white wing-bar, but no bars in heavily worn plumage when only traces of white remain. NESTLING. Down fairly thick, dark grey (Bannerman 1963; Harrison 1975). JUVENILE. Upperparts darker than adult, feathers with dull black or brown-black centres and rather poorly defined rufous-brown fringes (appearance is thus mottled brown and black, not as streaked on crown as adult); sides of head mottled black-brown and buff; supercilium pale buff, mottled brown-grey, short and indistinct; dusky line on lores and dark malar stripe mottled buff and not contrasting with side of head. Dark spots on chest larger, but duller and less sharply defined. Tail and wing not obviously different from adult. FIRST ADULT. Like adult and some birds probably indistinguishable; in birds examined, October-April, some retained a few or many heavily worn juvenile outer greater upper wing-coverts as well as some or all tertials or tail-feathers; these feathers distinctly more worn than neighbouring fresh 1st adult ones. See also Moults. Later in spring, all plumage heavily worn with no discernible difference in wear from adult.

Bare parts. ADULT. Iris dark brown. Upper mandible and tip of lower blackish-horn; base and middle portion of lower mandible light reddish-horn or yellowish-horn. Leg and foot pale flesh-colour to light yellow-brown, toes slightly darker. (Hartert 1910; Bannerman 1963). NESTLING. Mouth brilliant light yellow (Bannerman 1963); gape-flanges light yellow (Harrison 1975). JUVENILE. Bill greyish-flesh with dark culmen and tip; gape-flanges yellow. Leg and foot pale greyish-flesh. (BMNH.)

Moults. ADULT POST-BREEDING. Complete; primaries descendant. In full moult in June on eastern Canary Islands (Bannerman 1963), but perhaps later at high altitude in western Canary Islands and on Madeira; single bird from Madeira, mid-October, had outer 3 primaries and t5 growing, body very fresh (RMNH), but others completely new at this date. In all birds examined, plumage fresh October-December, rather fresh December-March, slightly to distinctly worn April-May: unlike many other *Anthus*, no pre-breeding moult discernible. POST-JUVENILE. Partial; head, body, and lesser and median upper wing-coverts; occasionally some tertials, t1, and variable number of inner greater upper wing-coverts; no flight-feathers or primary coverts replaced. Probably starts soon after fledging and hence timing strongly variable, as fledging occurs February-August (probably latest at highest altitudes) (Bannerman 1963; Bannerman and Bannerman 1965). Little variation in birds examined, however: all November-December birds (Madeira, Tenerife) had body fresh; in February-April (Madeira, Deserta Grande, Tenerife, Fuerteventura), body distinctly worn and retained juvenile feathers heavily worn. These samples perhaps comprise only late-fledged birds, as early-fledged birds perhaps moult tail, tertials, and wing-coverts completely and will then be indistinguishable from adult. Single March bird from Deserta Grande (Madeira group) had some median upper wing-coverts and tertials new or growing; not known whether this represented continuation of post-juvenile moult or a separate restricted pre-breeding moult.

Measurements. ADULT, FIRST ADULT. Nominate *berthelotii*. Tenerife, Lanzarote, and Fuerteventura (Canary Islands), all year; skins (BMNH, RMNH, ZMA). Bill (S) to skull, bill (N) to distal corner of nostril; exposed culmen on average 3·6 less than bill (S). Toe is middle toe with claw; claw is hind claw.

WING	♂	77·8 (1·41; 16)	75-81	♀ 73·9 (0·84; 9)	73-75
TAIL		59·5 (1·76; 16)	57-62	57·1 (2·97; 9)	53-61

BILL (S)	15·6 (0·51; 16)	14·7–16·2	15·1 (0·40; 9)	14·6–15·8	
BILL (N)	8·6 (0·43; 5)	8·0–9·0	8·8 (0·45; 6)	8·1–9·3	
TARSUS	22·7 (0·62; 16)	21·7–23·6	22·6 (0·73; 9)	21·6–23·6	
TOE	16·5 (0·57; 5)	15·9–16·7	16·4 (0·65; 6)	15·8–17·0	
CLAW	8·6 (0·61; 5)	8·2–9·7	8·4 (0·73; 6)	7·8–9·9	

Sex differences significant for wing and tail.

Birds from Ilhas Selvagens smaller than those from Canary Islands. Ranges of 2 ♂♂ and 3 ♀♀ (BMNH): wing, ♂ 75–77, ♀ 71–73; tail, ♂ 54–59, ♀ 55–56; bill (S), ♂ 14·6–15·6, ♀ 14·0–15·2; tarsus, ♂ 21·3–22·2, ♀ 17·8–22·7.

A. b. madeirensis. Madeira, Porto Santo, and Ilhas Desertas, all year; skins (BMNH, RMNH, ZMA). Exposed culmen on average 4·0 less than bill (S).

WING	♂ 78·0 (1·38; 20)	75–81	♀ 74·7 (1·03; 16)	73–77	
TAIL	58·8 (1·80; 20)	57–62	56·7 (1·76; 16)	54–60	
BILL (S)	16·5 (0·53; 20)	15·8–17·4	16·8 (0·63; 16)	15·9–17·9	
BILL (N)	9·1 (0·58; 5)	8·4–9·9	9·2 (0·35; 5)	8·6–9·5	
TARSUS	22·8 (0·55; 20)	21·9–23·6	22·7 (0·72; 16)	21·7–24·1	
TOE	17·9 (0·65; 5)	17·1–18·8	18·3 (0·72; 5)	17·6–19·2	
CLAW	9·1 (0·77; 6)	8·0–10·0	9·3 (0·85; 5)	8·5–10·6	

Sex differences significant for wing, tail, and bill (S). Bill (S) and middle toe of *madeirensis* significantly longer than nominate *berthelotii*. Ages combined, though wing of juvenile and 1st adult on average 1·4 below full adult, tail 1·1 below.

Weights. *A. b. madeirensis.* Deserta Grande, March: ♂ 17, ♀ 16 (RMNH).

Structure. Wing rather short, broad at base, tip rounded. 10 primaries: p7–p8 longest or either one 0–0·5 shorter than other; p9 and p6 both 0–2 shorter than longest, p5 5–9 shorter, p4 10–13, p3 13–16, p1 16–20. P10 reduced, hidden under reduced outermost greater primary covert; p10 7–10 shorter than longest upper primary covert, 49–56 shorter than longest primary. Outer web of p6–p8 and inner of (p6–)p7–p9 emarginated. Longest tertials reach to 2–5 mm from wing-tip in closed wing. Tail rather long, tip straight; 12 feathers. Bill long and slender; close to *A. spinoletta*; longer than (e.g.) Meadow Pipit *A. pratensis* and *A. trivialis*, more slender than (e.g.) Tawny Pipit

A. campestris. For length of middle toe and hind claw, see Measurements. Outer toe with claw *c.* 72% of middle toe with claw, inner *c.* 74%; hind toe with claw *c.* 103%, without claw *c.* 53%. Remainder of structure as in *A. pratensis* (p. 379).

Geographical variation. Slight. Bill and middle toe with claw of *madeirensis* from Madeira group of islands longer than in nominate *berthelotii* from Canary Islands and Ilhas Selvagens (see Measurements). 2 specimens from Deserta Grande (Madeira group) have pale wedge of t5 sullied grey (RMNH), but wedge pure white in other populations of *madeirensis* examined, including other birds from Deserta Grande and Chão. In *madeirensis*, populations from mountains of Madeira are the same in colour and size as coastal birds from Porto Santo and Ilhas Desertas. In nominate *berthelotii*, birds from Ilhas Selvagens are smaller than those from Canary Islands (see Measurements). Some birds (both races) are purer grey on upperparts, others slightly sandy; no apparent relation to locality, age, sex, or wear, and variation apparently individual.

Relationships with other *Anthus* not fully clear. Sometimes considered an old insular form of *A. campestris*, based on some plumage characters and voice (Lack and Southern 1949; Volsøe 1951), but general colour, pattern on sides of head, streaking on chest, structure of wing, bill, and foot, and moult suggest closer relationshihp with *A. spinoletta*, and therefore *A. berthelotii* possibly a small and pale early offshoot of that species complex (W R P Bourne, C S Roselaar).

Recognition. No other west Palearctic *Anthus* has upperparts as grey and uniform as adult *A. berthelotii*. Rock Pipit *A. spinoletta littoralis* also grey in breeding plumage, but duller and with olive tinge (not grey with slight sandy tinge) and virtually without streaks on forehead and crown. Tawny Pipit *A. campestris* has rather limited streaking, as in *A. berthelotii* (streaks narrower on crown, however), but ground-colour sandy-brown, not grey. Extent of streaking on underparts similar to *A. spinoletta littoralis* (but streaks of *A. berthelotii* slightly narrower and sharper) or to Tree Pipit *A. trivialis* (but streaks of *A. berthelotii* less broad, less sharp, less deep black, and not narrowly extending to lower flanks). Ground-colour of underparts lacks yellow tinge of many other *Anthus*.   CSR

Anthus similis Long-billed Pipit

PLATES 20 and 27
[between pages 232 and 233]

Du. Langsnavelpieper   Fr. Pipit à long bec   Ge. Langschnabelpieper
Ru. Длинноклювый конек   Sp. Bisbita campestre asiático   Sw. Långnäbbad piplärka

Anthus similis Jerdon, 1840. Synonym: *Anthus sordidus*.

Polytypic. *A. s. captus* Hartert, 1905, Levant. Extralimital: *decaptus* Meinertzhagen, 1920, Iran to southern Afghanistan and central Pakistan; *jerdoni* Finsch, 1870, eastern Afghanistan and east along Himalayan foothills to Sikkim; nominate *similis* Jerdon, 1840, hills of west peninsular India; *asbenaicus* Rothschild, 1920, Aïr (Niger); *nivescens* Reichenow, 1905, Red Sea coast of Sudan south to Somalia; *bannermani* Bates, 1930, highlands of West Africa; 1–2 further races in south-central Asia, 8–11 in Arabia and Afrotropics.

Field characters. 19 cm; wing-span 27–34 cm. Longest and bulkiest west Palearctic pipit; among larger species, has longest bill and tail but shortest hind claw; differs structurally from Tawny Pipit *A. campestris* in 15% longer bill, 10–15% longer wings, and 20% longer tail. Noticeably long, heavy-chested pipit, with large wings, rather full, dark tail, and otherwise rather dull grey-brown and ochre plumage; long, heavy bill and short hind claw diagnostic in combination. In all plumages, rather diffuse streaks of upperparts and chest, and rather dull face and outer tail-feathers form useful characters, but appearance can recall Water Pipit *A. spinoletta coutellii*. Flight as *A. novaeseelandiae*. Sexes similar; no seasonal variation. Juvenile separable.

ADULT. Crown, nape, mantle, and scapulars pale ochre-grey, dully streaked and mottled brown overall. Long but rather diffuse dull white supercilium combines with pale fore-cheek to give rather open face; dark eye prominent, set in narrow dusky eye-stripe visible across lore but only prominent behind eye; rear ear-coverts buff-brown, forming dull patch on cheeks which lack obvious surround; moustachial stripe dusky brown, usually short and smudged and not always visible in the field. Head thus has much less sharp features than *A. campestris* and other large pipits. Wings dull brown, relieved by (a) quite prominent panel formed by dull black-brown median coverts with contrasting ochre-buff margins (but much less vivid than in *A. campestris*), (b) pale ochre-buff fringes and tips to greater coverts, forming pale panel or wing-bar which further reduces obviousness of median coverts, and (c) broad pale ochre-buff fringes to otherwise dull black-brown tertials and inner secondaries. Wing thus has least contrasted pattern of all large pipits. Rump pale grey-ochre, appearing paler than back and tail but less bright than in *A. campestris*; upper tail-coverts dully streaked brown. Tail noticeably dark brown, with broad pale ochre-buff fringes to central feathers obvious when folded and rather warm buff to dun-white outer edges to 2 outermost feathers obvious on take-off (but less striking than virtually white and more extensive edges of *A. campestris*). Underparts basically dull cream, with pale grey-buff wash from sides of breast and along flanks and indistinct grey-brown streaks and smudges on chest, far less sharp than sparse marks of *A. campestris* but sufficiently obvious to show well at close range; underparts thus less bright than *A. campestris*. Overall, plumage the most uniformly pale of all large pipits, with distinctive dull, dun, ground-colour and rather shaggy quality due to lack of strong markings; hence can recall *A. spinoletta* of races *A. s. coutellii* (migrant in Middle East), and faded *A. s. spinoletta* and *A. s. littoralis*. JUVENILE. Differs less from adult than in *A. campestris* but greater colour contrast between centres and margins of streaked feathers produces generally sharper markings on upperparts, wings, and chest; face pattern more distinct, with more complete dusky eye-stripe and moustache. At all ages, bill dark brown-horn above, paler yellow-flesh at base of lower mandible, emphasizing dark tip; bill of adult noticeably long, deep, and heavy-ended, even seeming to droop at tip which has tiny terminal hook on upper mandible. Legs and feet pale yellow-brown. Tarsus and toes similar in length to those of *A. campestris* but noticeably curved hind claw averages shorter, so that overall length of hind toe and claw falls well short of tarsal length.

Identification not difficult when compared to complex problems posed by other large pipits. Race occurring in Middle East, *captus*, is confined to mountains and nearby areas of Levant, and darker, more migratory eastern races are only suspected of entering west Palearctic. Confusion with other large pipits thus normally reduced to intermingling with migrant (and in Lebanon a few breeding) *A. campestris* which is smaller and brighter, with obvious plumage marks in adult, much more distinct streaks in immature, and whiter and more extensive pale tail-sides at all times. Flight powerful and bounding, with similar bulky silhouette and actions to *A. novaeseelandiae* including surging take-off and (at least in eastern races) habit of hovering over ground cover. Escape-flight quite long but usually ends in bird perching openly on prominence; prone also to long runs after which it will skulk in cover and is then more difficult to flush than other large pipits. Song-flight a rather slow ascent-cum-hover, with flaps of wings more noticeable than in other pipits. Gait much as *A. novaeseelandiae* but does not strut. Carriage variable: stands markedly erect when alert but more level when on the move; lacks light wagtail-like appearance of *A. campestris*, looking more like long, dull *A. spinoletta coutellii* due particularly to rather short legs, relative breadth of tail, and lack of downward wag in tail movement (flicks tail up, fanning it at same time, quite unlike common action of *A. campestris*). Not gregarious.

Song rather simple compared with common small west Palearctic pipits but with distinctly longer and more disjointed phrases than *A. campestris*; notes loud and deliberate, but variable in tone and sequence. Commonest call when flushed a loud, ringing 'che-vlee', unlike commonest calls of *A. campestris*.

Habitat. Breeds in lower middle and lower latitudes, warm and largely semi-arid, in hilly and marginally in mountainous country, above 600 or often 1000 m, to *c.* 1800 m, and even to *c.* 2900 m in Indian subcontinent, but typically at low and medium elevations, on dry and sometimes steep, grassy and stony slopes with boulders, shale, or rock outcrops (Ali and Ripley 1973*b*). In Kashmir, breeds at lower levels on rough stony hillsides and on broken ground below, perching on coarse grassy tufts, stones, and rocks, but avoiding plain grassy slopes and cultivation. Moves habitually in low cover (Bates and Lowther 1952). In Baluchistan (Pakistan), occurs especially where springs emerge from rocks, up to *c.* 1000 m (E M Nicholson). In south-west Arabia, occurs at fairly high elevations on rather rocky barren ground, sometimes perching in topmost branch of a bush (Meinertzhagen 1954). In Sudan, frequents arid hilly and mountainous country, usually including places where

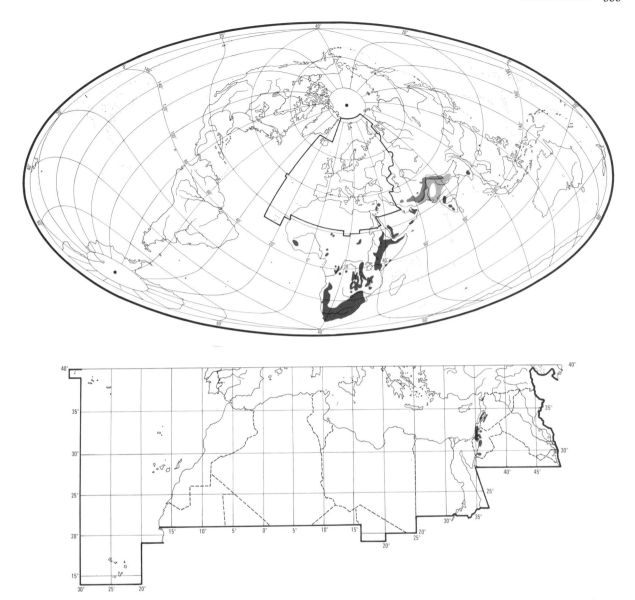

there are stretches of sand and grass in close association with stony ridges and bluffs (Cave and Macdonald 1955). In East and central Africa, most frequent in *Brachystegia* woodlands, and further north on grassy steppes, especially where there are rocky or gravel outcrops (Williams and Arlott 1980). In Southern Africa, sometimes found in same locality as Richard's Pipit *A. novaseelandiae* but prefers stony, hilly areas, including woodlands with open vegetation and montane grasslands at 2000–2250 m; perches freely on trees, bushes, and rocks (Prozesky 1970; Benson *et al.* 1971). In winter in India, descends to lower valleys and foothills, frequenting grassy plains, open low scrub jungle, dry watercourses, grassy canal banks, fallow land, wheatfields, and sand-dunes (Ali and Ripley 1973*b*). Flies apparently almost always in lower airspace. Not normally affected by man.

Distribution and population. No information on numbers or trends.

JORDAN. Probable breeding areas mapped, but so far no proof (DIMW).

Accidental. Iraq, Kuwait, Cyprus.

Movements. Most populations upland or montane residents with at most local vertical movements, but populations of Levant, Pakistan, and Himalayas are total or partial short-distance migrants.

Present all year in Levant, wintering in Syria occurring at lower altitude than breeding (A M Macfarlane), but no information on altitudinal movements in Lebanon or Israel. Presumed resident in Arabian peninsula (Bundy and Warr 1980; Jennings 1981*a*), though some local winter movement suspected in Oman (Gallagher and

Woodcock 1980). In Afghanistan, probable migrants collected late March (*decaptus*) and late May (*jerdoni*) (Paludan 1959). Pakistani *decaptus* winter in lower valleys and foothills extending to Makran coast (where scarce) and north-west India. Himalayan population of *jerdoni* also moves to lower altitude, wintering in Himalayan foothills below *c.* 900 m and in north-central and north-east India, overlapping in the west (Gujarat) with *decaptus*. Both these populations reach winter quarters early September and leave early April. Races of central and southern India resident (Ali and Ripley 1973*b*), as are birds breeding in Burma (King *et al.* 1975).

In Africa, exhibits local movements that usually involve change in altitude (Cave and Macdonald 1953; Smith 1957; White 1961; Traylor 1963; McLachlan and Liversidge 1970; Urban and Brown 1971; Benson and Benson 1977; Elgood 1982).   JHE, DJB

Food. Most information extralimital. Recorded taking berries in India (Ali and Ripley 1973*b*), though all other records are of invertebrates. Normally feeds on ground (Benson *et al.* 1971).

Bird from Israel, January, contained insects and a snail (Hardy 1946). On Socotra island (Arabian Sea), 33 birds, March–June, contained insects and remains of a scorpion (Scorpiones) (Ripley and Bond 1966). In Zimbabwe, 2 birds taken in September contained mostly grasshoppers (Acrididae) and crickets (Gryllidae), also beetles (Coleoptera), bugs (Dassidae), a mantis (Mantodea) and an ant (Formicidae); in October, 2 contained only termites *Hodotermes* and *Macrotermes*, in November 2 birds contained grasshopper nymphs, beetles (Scarabaeidae, Cicindelidae, Curculionidae), termites, spiders (Araneae), a moth (Noctuidae), and a cockroach (Blattodea); in December, 1 bird contained beetles and wasps (Braconidae, Sphecoidea) (Borrett 1973). ♂ and ♀ near Yaounde (Cameroon), November, contained insects including a large caterpillar (Germain *et al.* 1973).   PJE

Social pattern and behaviour. Poorly known; information scattered and mostly extralimital.

1. Generally solitary or in pairs (Meinertzhagen 1954; Archer and Godman 1961*b*; Ali and Ripley 1973*b*). Parties of up to 6 occur in winter (e.g. Ali 1955, Ganguli 1975, Walker 1981*b*), once up to 14 together in northern Oman (Walker 1981*a*). BONDS. No information on mating system or pair-bond. BREEDING DISPERSION. No information. ROOSTING. No information.

2. In Dhofar (Oman), rather unobtrusive outside breeding season, tending to run and crouch rather than fly (Walker 1981*b*). Said by Baker (1926) to be shyer than other *Anthus*, but sometimes quite tame in Sierra Leone (G D Field) and described as conspicuous and rather fearless in north-east Africa and Saudi Arabia (Meinertzhagen 1954; Archer and Godman 1961*b*). In Zimbabwe, 2 birds feeding in open terrain fairly approachable (Priest 1935). Disturbed birds (sometimes flushing late) may call at take-off (see 2 in Voice) and typically do not fly far, often only a few metres to perch in tree, bush, or on rock (Priest 1935; Bates and Lowther 1952; Benson *et al.* 1971). Birds flushed from crops will apparently fly up a few metres,

then hover and zigzag aimlessly before dropping down. Such behaviour said to occur before sunset without provocation. If followed, will run and hide behind grass clumps, etc. (Ali and Ripley 1973*b*). FLOCK BEHAVIOUR. No information. SONG-DISPLAY. ♂ sings (see 1 in Voice) from rock, tree, or bush or from ground, also in flight (Ali and Ripley 1973*b*; Taylor and Macdonald 1979). For Song-flight, usually ascends from perch with quivering wings—report in Ali and Ripley (1973*b*) of more flapping overall than in other *Anthus* requires confirmation. *A. s. sokotrae* ascends (while singing) to considerable height and may stay aloft for several minutes (Archer and Godman 1961*b*). Bird sometimes circles, and undulations appear to be typical as does gliding (parachute) descent with wings and tail (held at or above horizontal) widely spread. Lands on perch or ground. (Bannerman 1936; Mackworth-Praed and Grant 1960; Jennings 1981*b*, *c*; Walker 1981*b*; P A D Hollom.) Possibly sings while hovering according to Taylor and Macdonald (1979), and Gallagher and Woodcock (1980) referred to song in 'direct' as well as gliding flight. In Israel, bird sang from rock for *c.* 2½ min with only a few insignificant pauses (P A D Hollom). In Dhofar, song and courtship noted from 30 January (Walker 1981*a*); singing birds seen in Iran in early April and in northern Israel in mid-April (P A D Hollom). ANTAGONISTIC BEHAVIOUR. No information. HETEROSEXUAL BEHAVIOUR. In Kangra (Punjab, India), courtship noted from March (Whistler 1926*b*). For Dhofar, see above. While ♀ incubating, ♂ sings from perch near (often above) nest (Inbar 1977). RELATIONS WITHIN FAMILY GROUP. No information. ANTI-PREDATOR RESPONSES OF YOUNG. No information. PARENTAL ANTI-PREDATOR STRATEGIES. (1) Passive measures. Said by Baker (1934) generally to leave nest early when disturbed, although more likely to allow close approach late in incubation or after hatching. In Sierra Leone, bird sat tight on eggs, then fluttered off and perched on boulder *c.* 20 m away (Serle 1949); in Malawi, left rapidly and allegedly sang while flying to perch *c.* 50 m away (Belcher 1930). (2) Active measures. Near Simla (north-west Himalayas), bird (possibly ♀) from nest containing slightly incubated clutch performed distraction-lure display of disablement type: tumbled and fluttered about close to observers like 'rolling ball of feathers' and was chased by dogs (Dodsworth 1914).   MGW

Voice.

CALLS OF ADULTS. (1) Song of ♂. Brief descriptions available for west Palearctic race and several others indicate simply structured song probably rather uniform over most of range, though rate of delivery and timbre (sweetness) may vary. Recordings from Israel and Iran reveal simple song, though apparently with 4 well-defined unit-types compared with 2 in the similarly structured songs of Tawny Pipit *A. campestris* and Berthelot's Pipit *A. berthelotii* (J Hall-Craggs). Disjointed units loud and deliberately uttered. Some harsh, others more musical; given in a randomly varying sequence, usually 3 or more per 'phrase' (see below). Pauses between units irregular, often quite long. Overall, song quite distinct from that of *A. campestris* (P A D Hollom). According to Mackworth-Praed and Grant (1973), song of west-central and west African races usually short and unimpressive, but considerable local variation exists (no details). In Oman (*arabicus*), long series of loud, deliberate notes of 1–3 syllables: e.g. 'tewchip tschirp tswee tsup tree tushree

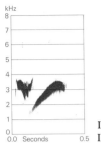 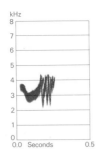 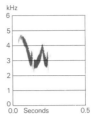 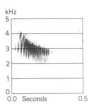

I P A D Hollom Israel April 1980

II P A D Hollom Israel April 1980

III P A D Hollom Israel April 1980

IV P A D Hollom Iran April 1971

shree' (Gallagher and Woodcock 1980); repetitious 'psweep cherup' or 'chip treep' (Walker 1981b). Description from western Saudi Arabia similarly refers to repetitive double note followed by whistle (Jennings 1981c). In Jordan, simple song quite often amplified by echo: 1st note grating, recalling bunting *Emberiza*, 2nd more musical—'chree-chewit' (Wallace 1984b). Recording of *captus* from Israel reveals disjointed song, *c.* 2–3 s between each unit; 21 units in 62 s (see below for faster rate of delivery). Rendered 'chrep shreep chew-ee', 'shreep' often rather harsh, 'chew-ee' purer and louder than other units, of which 1–3 (or more) may separate one 'chew-ee' unit from next (P A D Hollom). More detailed analysis suggests for this louder unit a piercing, wolf-whistling 'whee-ooeee' (J Hall-Craggs: Fig I) or fruity, pleasant sounding, and cheerful 'ple-WEET' or 'chew-EE(t) not unlike Woodlark *Lullula arborea*, sometimes 'psew-ee(t)' recalling wagtail *Motacilla* (M G Wilson). Same recording also contains (Fig II) a rather shrieky 'srieh' (J Hall-Craggs) or 'sweeerr', like an impure and vibrato whistle, suggesting Chaffinch *Fringilla coelebs* (M G Wilson). Fig III shows chirping or chirruping 'chirrie' recalling House Sparrow *Passer domesticus* (J Hall-Craggs); also described as a liquid and 'purred' 'sleee-oo' or 'sweer-oo', with similar sparrow-like 'chirrup' evident in another recording (from same locality) in which rate of delivery *c.* 1 unit per s and thus faster than that described above (M G Wilson). Fig IV shows a 'weez' (J Hall-Craggs); more twangy and rasping than unit shown in Fig II, almost a nasal buzz, but still like *F. coelebs*. This 'weez' apparently not in repertoire of some birds and all units probably subject to individual variation: e.g. some whistled sounds (compare Fig I) in recording from Iran rudimentary and one is more like 'chee-oo(p)' or 'psee-oo(p)' (J Hall-Craggs, M G Wilson). Phrase from same recording rendered 'jzree djep jzree jewee'. Rate of delivery *c.* 8–9 phrases per min, though composition of these erratic, and difficult to determine where one ends and another begins (P A D Hollom). Song of *sokotrae* and *asbenaicus* said to have sweet quality (Bannerman 1936; Mackworth-Praed and Grant 1960), though this less evident in *nyassae* (Mackworth-Praed and Grant 1960) and *hararensis* (Benson 1946a). For further descriptions of song given by various African races, see (e.g.) Mackworth-Praed and Grant (1960),

McLachlan and Liversidge (1970), Taylor and Macdonald (1979), and Newman (1983). (2) Contact-alarm calls. (a) Harsh 't'reep' as alarm-call (Walker 1981b); sharp 'wheet' given by alarmed *nicholsoni* at take-off (Priest 1935) presumably the same, as are perhaps the drawn-out 'tsee' given only in flight by birds in eastern Ghana (Taylor and Macdonald 1979), 'tsee-tsee' (McLachlan and Liversidge 1970), 'che-vlee' from flushed bird (D I M Wallace) and squeaky-gate noise noted from birds in Nepal by Del-Nevo and Ewins (1981). (b) Perhaps softer variants of 2a: plaintive 'peep' given by alarmed bird with young (Walker 1981b); faint 'cheet cheet' (Taylor and Macdonald 1979). (c) Metallic, sharp, clear ringing 'ki-link' or 'kilink' in alarm (Mackworth-Praed and Grant 1963; McLachlan and Liversidge 1970); soft 'plip-plip' given by some eastern races (D I M Wallace) may be softer variant. (d) Liquid 't'churrup' given in anxiety (Walker 1981b); 'djüdjep' given at regular intervals as contact-call by ♂ *leucocraspedon* (Hoesch and Niethammer 1940) is probably the same or closely related, while loud metallic chirruping described for several African races (Mackworth-Praed and Grant 1973) is perhaps also or may refer to call 2a. A 'choop' lumped with 'kilink' (see call 2c) by McLachlan and Liversidge (1970) is perhaps better placed here. (e) Constant, mournful 'chew' (Walker 1981b).

CALLS OF YOUNG. Food-call of young (apparently out of nest) a short 'tsch'b' (Walker 1981b). MGW

Breeding. SEASON. Middle East: April–July (Hüe and Etchécopar 1970). Israel: April–May (Inbar 1977). SITE. On ground, in shelter of rock or tuft of vegetation, often on slope. Nest: shallow depression with cup of vegetation, lined with finer material. Building: no information. EGGS. See Plate 80. Sub-eeliptical, smooth and glossy; whitish or grey-white, heavily marked with brown spots and freckles. *A. s. captus*: 22·5 × 16·6 (20·8–23·4 × 15·4–17·2), *n* = 22; calculated weight 2·91 g (Schönwetter 1979). Clutch: 3 (2–5). Probably 2 broods in India (Whistler 1926b). INCUBATION. 13–14 days; by ♀ only (Inbar 1977). YOUNG. No information. FLEDGING TO MATURITY. Fledging period *c.* 2 weeks (Inbar 1977).

Plumages (*A. s. captus*). ADULT BREEDING. Similar to adult non-breeding (see below), but part of non-breeding upper wing-coverts (in particular greater), tail, and sometimes tertials retained, somewhat contrasting with newer breeding. Worn non-breeding upper wing-coverts even less contrastingly patterned than in fresh non-breeding. ADULT NON-BREEDING. Upperparts rather similar to Tawny Pipit *A. campestris* with streaks narrow and indistinct, confined to forehead, crown, mantle, inner scapulars, and longer upper tail-coverts, but ground-colour different, light grey or slightly olive with sandy-buff or buff-brown tinge on feather-sides, usually not as extensively buff as *A. campestris*. (In worn plumage, buff tinge less pronounced in both species and upperparts more closely similar, but *A. similis* colder and usually purer grey, *A. campestris* more brown-grey.) Long superciliary stripe from nostril to above ear-coverts cream to white, stripe on lores dull black, both slightly narrower than similar stripes of *A. campestris*; upper cheeks and ear-coverts yellow-buff or pale buff with indistinct olive-brown streaks, not as distinctly bordered by dark stripes running back from eye and gape as in *A. campestris*; sides of neck light grey with faint cream streaks. Underparts warm cream-buff or pale rufous-buff (deeper buff than *A. campestris*), chin, throat, and central belly not much paler, vent and under tail-coverts pale tawny-cinnamon; chest and sides of breast with long, narrow, and poorly defined olive-grey shaft-streaks (in *A. campestris*, chest either uniform buff or with fine, sharp black specks, mainly restricted to upper chest); flanks washed olive. (In worn plumage, sides of head more contrastingly marked and underparts paler buff, similar to *A. campestris*, but usually a grey wash on throat, chest, and flanks, and shaft-streaks on chest and sides of breast longer and more extensive, sometimes more distinct than in fresh plumage.) Central pair of tail-feathers (t1) dull olive-grey, fringed pale sandy-grey; t2–t5 dull black with faint pale buff outer edge, t5 with narrow cream-buff wedge along shaft near tip of inner web, tip of outer web often also cream-buff; length of wedge 8–12 mm (similar wedge in *A. campestris* 22–42 mm in length and paler cream in colour, but colour affected by bleaching in both species, fading to white; outer web of t5 often mainly pale cream-buff in *A. campestris*, not largely black). T6 dull black with pale cream-buff off-white outer web and cream-buff wedge along shaft of inner web near tip (wedge 25–27 mm long; in *A. campestris*, t6 pale cream-white except for grey border along inner web). Flight-feathers and tertials dark brown-grey or greyish-black; tertials rather broadly fringed tawny-buff or pale sandy-grey on both webs, secondaries on outer web only; primaries sharply and narrowly edged pale sandy on outer webs (except emarginated parts). Lesser upper wing-coverts grey, like upperparts, longer ones with dull black centres; median and greater upper wing-coverts dark brown-grey or dull olive-brown with broad and poorly defined tawny-buff or pale greyish-buff tips and narrower duller sandy-grey fringes along outer webs (in *A. campestris*, centres deep black, sharply divided from bright pale buff fringes). Feathers of bastard wing and greater upper primary coverts dark grey or blackish-grey with faint and narrow paler grey fringes (in *A. campestris*, fringes sharp and narrow, pale buff). Under wing-coverts and axillaries warm buff or pale rufous-buff, longer coverts tinged grey, shorter coverts along leading edge of wing spotted grey. NESTLING. No information. JUVENILE. Feathers of upperparts and upper wing-coverts black-brown or dark olive-brown, those of upperparts rather evenly bordered sandy-buff or pale buff, wing-coverts with broad cream-buff or off-white tip; dark centre of median coverts terminates in sharp point towards tip, greater coverts with more evenly broad buff tip of *c.* 3 mm. Sides of head sandy-brown, supercilium broad, cream-white, most distinct behind eye. Underparts pale buff, merging into cream-white on chin and on central belly; chest and sides of breast marked with rather small triangular brown spots (2–3 mm long, 1–2

mm wide at tip). Flight-feathers and tail as in adult non-breeding. Closely similar to juvenile *A. campestris*, differing in more extensively black tail (as in adult) and wing formula (see Structure). FIRST ADULT NON-BREEDING. Similar to adult non-breeding, but part of juvenile plumage retained (some outer greater upper wing-coverts, outer tail-feathers, greater upper primary coverts, flight-feathers, and occasionally some feathers elsewhere); retained outer upper wing-coverts black with broad and sharply defined white fringe, contrasting with virtually uniform sandy-grey and olive-brown new feathers. FIRST ADULT BREEDING. Rather variable. Occasionally like adult breeding with much of head and body new but juvenile flight-feathers and greater upper primary coverts retained; these more worn than adult at same time of year. Most birds virtually suppress pre-breeding moult, retaining mixture of fairly worn 1st non-breeding plumage and heavily worn juvenile.

Bare parts. ADULT, FIRST ADULT. Iris brown or dark brown. Upper mandible and tip of lower mandible dark brown or blackish horn-brown; cutting edges of both mandibles and base of lower mandible pale flesh, brownish-white, yellow-flesh, or yellow. Leg flesh-colour, light brownish-flesh, yellowish, or flesh-yellow with paler foot. (Hartert 1910; BMNH, RMNH.) NESTLING, JUVENILE. No information.

Moults. ADULT POST-BREEDING. Complete; primaries descendant. Information limited. Single *captus* from Jordan nearing completion on 11 September (BMNH), another from Israel on 9 September (ZFMK); moult not started in late July in *decaptus* from Iran (Paludan 1940); *jerdoni*, India, in full moult August (Ali and Ripley 1973*b*). All plumage new in *captus*, *decaptus*, and *jerdoni* late September and October (BMNH, ZMA). One *decaptus* from Iran, March, had just resumed primary moult with shedding of p5 in early March, after earlier suspension with inner 4 primaries new (ZMA); thus, moult apparently sometimes suspended in autumn, as in other large Palearctic *Anthus*. ADULT PRE-BREEDING. Partial: involves head, body, tertials, tertial coverts, and lesser and median upper wing-coverts (Ticehurst 1923; BMNH, ZMA; *contra* Hall 1961, see below). Information on timing limited; single *captus* had just started moult in late December, with part of chest, some tertials, and some median upper wing-coverts growing (ZMA). POST-JUVENILE. Partial. Information on timing and extent limited. One bird from 21 August and another from early October in 1st non-breeding except for flight-feathers, greater upper primary coverts, outer tail-feathers, and outermost greater upper primary coverts (RMNH). FIRST PRE-BREEDING. Information limited. 6 out of 7 from February–April still in non-breeding (BMNH, ZFMK, ZMA) and birds like this perhaps suppress pre-breeding moult, as suggested for all age-groups by Hall (1961); one other had moult as in adult pre-breeding.

Measurements. ADULT, FIRST ADULT. *A. s. captus*. Israel and Jordan, all year; skins (BMNH, ZMA). Bill (S) to skull, bill (N) to distal corner of nostril; exposed culmen on average 4·1 less than bill (S). Toe is middle toe with claw; claw is hind claw.

	♂		♀	
WING	96·1 (1·46; 7)	94–98	90·7 (1·08; 6)	89–92
TAIL	72·6 (2·07; 5)	70–75	69·1 (2·56; 6)	67–73
BILL (S)	20·1 (0·62; 5)	19·5–21·0	19·6 (0·76; 6)	18·4–20·4
BILL (N)	11·4 (0·22; 5)	11·1–11·7	11·2 (0·46; 6)	10·7–11·8
TARSUS	26·0 (0·80; 5)	25·2–27·3	26·3 (1·30; 6)	25·0–27·8
TOE	19·6 (1·15; 4)	18·0–20·4	19·4 (0·38; 5)	19·1–20·1
CLAW	9·1 (0·85; 5)	8·0–10·0	9·4 (1·08; 6)	7·8–10·9

A. s. decaptus. Wing, ♂ 94-105, ♀ 95-101; tail, 80-91 (Ali and Ripley 1973*b*).

A. s. jerdoni. Wing, ♂ 97-105, ♀ 95-99; tail, 80-91; tarsus, 28-30 (Ali and Ripley 1973*b*).

A. s. decaptus and *jerdoni*, combined. Wing, ♂ 94-105, ♀ 92-99; tail, ♂ 71-82, ♀ 69-82; bill (S), 18-21; tarsus, 26-29; hind claw, 9-14 (Hall 1961).

Weights. *A. s. captus.* Israel, ♂♂: April, 27; September, 27 (ZFMK).

A. s. decaptus. Iran and south-west Afghanistan, March-July: ♂ 29·9 (1·71; 4) 27·7-31·5, ♀ 29 (Paludan 1938, 1940, 1959; Desfayes and Praz 1978).

Structure. Wing short, broad at base, tip rounded, 10 primaries: p7-p8 longest, p9 0-2·5 shorter, p6 0-2, p5 2-4·5, p4 8-13, p3 11-16, p1 16-21; p10 reduced, narrow and pointed, 62-70 shorter than p7-p8, 9-14 shorter than longest greater upper wing-covert. Outer web of p5-p8 and inner of (p6-)p7-p9 emarginated. Longest tertials reach wing-tip in closed wing or up to 5 shorter. Tail long, tip square; 12 rather broad feathers. Bill long and straight, rather wide and deep at base; similar in length and shape to *A. campestris* but tip of culmen slightly more decurved and distal half of bill perhaps slightly more compressed laterally. For length of middle toe with claw and hind claw, see Measurements. Outer toe with claw *c.* 75%

of middle with claw, inner 73%; hind with claw *c.* 92%, without *c.* 52%.

Geographical variation. Marked. 2 main groups: (1) virtually unstreaked races in subtropics (*captus*, *decaptus*, *jerdoni*); (2) heavily streaked races in Arabia, Socotra, and Afrotropics; races from southern India and Burma intermediate (Hall 1961). Within each group, differences usually slight, mainly involving size (as expressed in wing, tail, and bill length). *A. s. captus* from Levant rather small; upperparts rather pale grey, underparts saturated buff to greyish-buff (depending on wear) with very faint streaks on chest. *A. s. decaptus* from Zagros and southern Elburz mountains (Iran) east to southern Afghanistan and central Pakistan larger than *captus*, slightly deeper ochre-buff in fresh plumage, and with rather more well-defined chest-streaks; size of *jerdoni* from Himalayan foothills similar to *decaptus*, but upperparts darker and browner, underparts deeper rufous-buff. Isolated races of northern Niger (*asbenaicus*) and West Africa (*bannermani*) rather different from others and perhaps deserve species rank (Hall and Moreau 1970): *asbenaicus* is dusky earth-brown with sandy-buff feather-fringes on upperparts and sandy-buff on underparts, with chest virtually unstreaked; wing ♂ 93-100, ♀ 87-92; *bannermani* much darker grey-brown above and more streaked below (Bannerman 1936; Mackworth-Praed and Grant 1960, 1973).

Perhaps forms superspecies with Water/Rock Pipit *A. spinoletta* (Hall and Moreau 1970). CSR

Anthus hodgsoni Olive-backed Pipit

PLATES 23 and 27
[between pages 232 and 233]

DU. Groene Boompieper	FR. Pipit d'Hodgson	GE. Waldpieper	
RU. Пятнистый конек	SP. Bisbita de Hodgson	Sw. Sibirisk piplärka	N. AM. Olive Tree-Pipit

Anthus hodgsoni Richmond, 1907

Polytypic. *A. h. yunnanensis* Uchida and Kuroda, 1916, northern Eurasia, east to Kamchatka and south to Mongolia, Manchuria, Ussuriland (USSR), and Hokkaido (Japan). Extralimital: nominate *hodgsoni* Richmond, 1907, southern Asia from Himalayas to central and eastern China and Honshu (Japan).

Field characters. 14·5 cm; wing-span 24-27 cm. Size between Tree Pipit *A. trivialis* and Red-throated Pipit *A. cervinus*, with similar structure except for rather long forehead, receding chin, and full chest. Small but sleek pipit, with behaviour most recalling *A. trivialis*. Upperparts noticeably pale green-olive, with only faintly streaked back and plain rump; underparts noticeably clean, with ground-colour mainly white and beautifully decorated with evenly spread lines of large black spots. Face also well marked, with broad white rear supercilium and (on most) white and black rear cheek spots. Flight like *A. trivialis*. Tail movement most exaggerated of all pipits. Sexes similar; no seasonal variation. Juvenile separable.

North Eurasian race, *yunnanensis*. ADULT. Crown, cheeks, nape, back, wings, rump, and tail all basically green-olive, with bright almost iridescent sheen when fresh but duller, greyer tone when worn. Compared with other small pipits, plumage marks on head and upperparts

fewer but more obvious: (a) dark flecks on crown; (b) black brow above rear supercilium (absent on some birds); (c) supercilium pale pink-buff in front of eye, broad and white behind it; (d) black-speckled lores and narrow black rear eye-stripe; (e) white spot on upper rear cheek below end of supercilium (sometimes absent, and not diagnostic since present on some *A. trivialis*); (f) black spot on lower rear cheek, sometimes indistinct; (g) dull, faint 'furrow' lines on mantle, most obvious from behind when contrasting with unmarked nape and rump but never as striking as those of other small pipits (or of nominate *hodgsoni* of India); (h) bold white wing-bar formed by white margins to median coverts; (i) less obvious, buff, lower wing-bar formed by buff tips to greater coverts; (j) large and long tertials, again most obvious from behind; (k) white outer tail-feathers. Underparts essentially white, with delicate buff wash on throat and sides of breast and yellow-olive suffusion from rear flanks to under tail; marked by narrow black malar stripe, remarkably bold pattern of drop-like black spots, placed in lines evenly spread over breast, fore-belly, and flanks and continued as dull grey streaks under tail. Underwing yellow-buff. JUVENILE. Closely resembles adult but upperparts browner, buff and olive suffusion on rump and underparts deeper and brighter, and streaks on underparts longer and wider, lacking neatly spotted appearance of adult. At all ages, bill rather small; black-horn above, flesh- or buff-horn below. Legs flesh-pink; feet with short hind claw like *A. trivialis*.

Unmistakable when seen well but subject to confusion with both *A. trivialis* and *A. cervinus* at distance and in flight, since all 3 species have similar silhouette and wing action, commonly utter one not dissimilar call, and *A. hodgsoni* and *A. trivialis* both enter trees freely. Study of their separation far from complete and odd birds should always be approached as closely as possible. Flight free and buoyant, with action like *A. trivialis*; lacks hesitancy of Meadow Pipit *A. pratensis*. Escape-flight also like *A. trivialis*, bird taking cover in canopy if available or going into dense ground cover. Gait, posture, and carriage like *A. trivialis*; runs and walks freely (does not creep like *A. pratensis*), and will even walk confidently along branches—a skill apparently lacking in other pipits (A Cheke). Tail movement most obvious of all small pipits: often strongly pumped rather than wagged. Wary but not shy if carefully stalked. Not markedly gregarious, forming only loose groups like *A. trivialis*.

Song recalls *A. trivialis* but harsher and more sibilant. Commonest calls are 'tseep' or 'tsee', quiet and not distinctive, and 'teaze', loud and strident in full alarm and recalling Redwing *Turdus iliacus*.

Habitat. For breeding occupies large slot in upper middle and middle latitudes broadly below range of Pechora Pipit *A. gustavi* and complementing (with extensive overlap) that of Tree Pipit *A. trivialis*, which covers nearly all corresponding part of west Palearctic, from boreal through temperate zones to subtropics. Northern race, *yunnanensis*, spreads through coniferous taiga forest, mainly in its sparser sections and at its edges along river banks and on fringes of bogs and marshes, but also in birchwoods, alder thickets, and larch groves, no higher than 1000 m in north of range, but up to 3000 m in Japan and further south in USSR, where it nests in subalpine zone and even on summits, amid low shrubs and rocks, replacing Water Pipit *A. spinoletta*. In Japan, characteristic of mountains, occurring even in barren lava areas, but also nests in cultivation, and occurs on migration in orchards and on wasteland in fairly large cities (Dementiev and Gladkov 1954a). For breeding densities in different habitats, see Social Pattern and Behaviour. Apparent inconsistencies in habitat choice perhaps partly due to flexible adaptations over extensive regions offering ample food and nesting opportunities with little competition, except where *A. trivialis* competes

for more open and forest-edge situations. In Himalayas, breeds from 2700 m up to treeline and higher scrub, rising to *c*. 4500 m in Nepal, occupying glades in open forest of oak, birch, or pine, dwarf juniper or other scrub, abandoned cultivation, slopes covered with grass and bracken, and rocky ground. In winter resorts to coffee plantations, mango groves, and suitable wooded terrain (Ali and Ripley 1973b); also found on ground under trees at forest edge, on forest footpaths, and along shady highways or on outskirts of villages. When disturbed flies up to branches of trees, often quite high, but soon comes down again (Ali 1949). Available data thus suggest stronger attraction to closer stands of trees and to higher altitudes than for *A. trivialis*, but detailed study necessary of habitat difference exhibited where both species overlap in some numbers.

Distribution and population. Fairly rare in north-west of range (Dementiev and Gladkov 1954a); exact limits not known, but probable areas shown.

Accidental. Britain (annual in recent years), Ireland, West Germany, Norway, Finland, Poland, Israel, Malta.

Movements. Essentially a long-distance migrant, though birds breeding in Himalayas may winter in adjacent areas of northern India.

Main southern race, nominate *hodgsoni*, breeding from Himalayas east to Hunan (China) and Honshu (Japan), winters in Ryukyu Islands, Taiwan, Philippines, and the Indo-China countries (Vaurie 1959), with Himalayan birds moving south into almost all peninsular India and Burma (Ali and Ripley 1973b). Northern race, *yunnanensis*, breeding from Urals (USSR) to northern Japan moves to much of the same area—southern Japan, Philippines, and parts of peninsular India. Bird ringed Taiwan mid-February recovered on Sakhalin (eastern USSR) the following June (McClure 1974; Ostapenko 1981). Birds (of unknown race) are probably regular in winter in Borneo (Smythies and Harrisson 1956; Gore 1968; Smythies 1981), and a few may winter in southern Korea (Macfarlane 1963). Occurs as vagrant on Aleutian Islands, mostly May–June but also autumn (late September). Also noted on St Lawrence Island (Alaska) and in Nevada (Roberson 1980).

Timing of movements not well documented. In USSR, arrives in upper Kolyma area by mid-May, Altai mountains and Transbaykalia in late May (Dementiev and Gladkov 1954a). Departure from Kamchatka begins late August but lasts until after mid-September, while birds only reach winter quarters in southern Japan only in early November (Jahn 1942). Birds breeding in Tibet present early May to late October, but some also overwinter in deeper valleys (Vaurie 1972). Present in Malaysia late October to late April (Medway and Wells 1976) and Thailand mid-October to late April (Deignan 1945). Some records of westward vagrants to Britain regarded

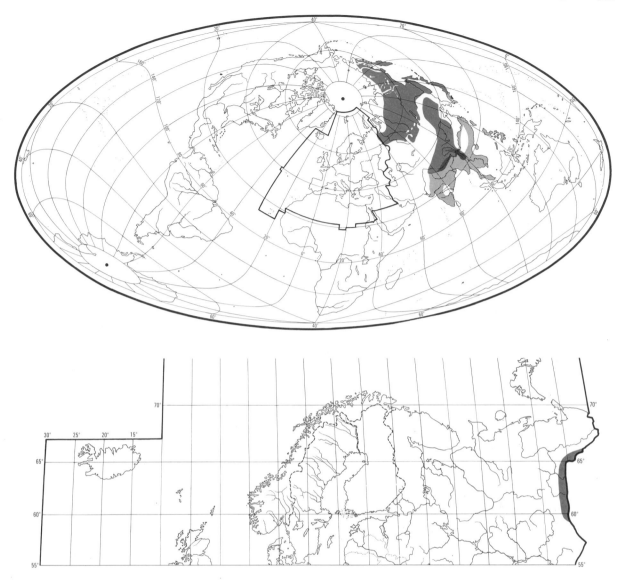

as linked to strong anticyclonic activity in central USSR (Baker 1977). Up to 1983, 27 records in Britain and Ireland: 25 occurred September–November (mostly late September and October) and 2 in spring (April–May); also, one present well inland February–March 1984 had presumably overwintered.   JHE

Food. Information almost wholly extralimital. Chiefly insects in summer and seeds in winter. Feeds on ground amongst low herbage (Matyushkin and Kuleshova 1972; Ali and Ripley 1973*b*; Mauersberger *et al.* 1982). In Nepal, visits human habitations to glean scraps (P J Ewins). In Philippines in winter, frequently feeds in pines *Pinus*, walking along branches probing for insects among needles and cones (Amadon and Jewett 1946).

Animal food includes mayflies (Ephemeroptera), dragonflies (Odonata), grasshoppers, etc. (Orthoptera), bugs (Hemiptera: Cydnidae, Aphididae), adult and larval Lepidoptera, flies (Diptera: Tabanidae, Muscidae), Hymenoptera (sawflies Symphyta, ants Formicidae), beetles (Coleoptera: Hydrophilidae, Elateridae, Curculionidae, Scolytidae), spiders (Araneae), and snails *Planorbis*. Seeds recorded from grasses (Gramineae) including millet *Panicum*, also from orache *Atriplex* and wormwood *Artemisia*. (Mason and Maxwell-Lefroy 1912; D'Abreu 1918; Gizenko 1955; Rashkevich 1965; Ogorodnikova 1979.)

Of 67 stomachs from Bihar (India), March–April, 11 contained only vegetable matter, 3 only animal food, and 53 a mixture; Curculionidae formed major part of animal food (Mason and Maxwell-Lefroy 1912). On lower Amur (eastern USSR), diet of adults in April and early May comprised seeds (78%) and beetles (14%): vegetable component disappeared by end of May, and only insects taken June–July; mayflies particularly important, 1 stomach containing 54 (Rashkevich 1965). In winter near Nagpur (India), seeds main food (D'Abreu 1918). Vagrant in southern England, February–March, fed on peanuts, seeds, and bread in garden (D Parker).

Young are fed wholly on insects (Rashkevich 1965; Ogorodnikova 1979). At 3 days old, parents bring small caterpillars, moths, and flies and at 4–5 days old also grasshoppers and ants (Rashkevich 1965).   PJE

Social pattern and behaviour. Little information and that almost entirely extralimital.

1. Only moderately gregarious. Often in flocks outside breeding season especially for feeding, but these small and generally loose-knit (Fletcher and Inglis 1924; Jahn 1942; Smith 1942; Ali and Ripley 1973b). In Krasnoyarsk (USSR) where breeds later than Tree Pipit A. trivialis, flocks, not infrequently mixed with A. trivialis, recorded from early August (Naumov and Burkovskaya 1959). Small late-summer and early-autumn flocks reported also from Mongolia (Kozlova 1930; Mauersberger et al. 1982) and western Siberia (Johansen 1952). In northern taiga of western Siberia in August (after breeding), birds mainly in clearings, low scrub, and settlements, 12–23 per km² (see Vartapetov 1984 for details); similar dispersion in north-east Altai (USSR) during autumn migration (Ravkin 1973). In Nepal, winter, feeding flocks regularly of 2–20 but never larger (P J Ewins); rarely more than 10 A. h. yunnanensis together on spring migration (Diesselhorst 1968a). BONDS. Nothing to indicate mating system other than monogamous. No information on length of pair-bond. Young fed by both sexes, though mainly by ♀; ♂ will tend young alone if ♀ lost (Rashkevich 1965). BREEDING DISPERSION. Solitary and, like other Anthus, highly territorial (Diesselhorst 1968a; Ali and Ripley 1973b). Reports of 4 nests within less than c. 440 m in Kangra, northern India (Whistler 1925b), and of 3–4 singing ♂♂ within c. 1 km east of Lake Baykal, USSR (Matyushkin and Kuleshova 1972). In taiga of western Siberia (roughly between Ob' and Yenisey rivers), northern areas hold up to 58 birds per km² on edges of bogs and shrub thickets, also on more open bogs and in pine Pinus; in middle zone, 5–7 birds per km², mainly in thickets and taiga of mixed spruce Picea and Pinus sibirica hold up to 32–34 birds per km² (see Vartapetov 1984 for details). In Krasnoyarsk, up to 7·8 pairs per km² in dense coniferous woods, c. 0·25 pairs per km² in mixed pine and broad-leaved wood (Naumov 1960). In north-east Altai, most numerous (up to 13 birds per km²) in open woods alternating with meadows and thickets of willow Salix (Ravkin 1973). For further data from Siberia, see Reymers (1966) and Izmaylov and Borovitskaya (1967). In Sikhote-Alin mountains (Ussuriland, eastern USSR), 0·4–6·1 ♂♂ per km. Favoured habitat birch Betula ermani and grassy clearings above upper limit of spruce woods. Densities fairly high on summits: 40–60 pairs per km² in krummholz birch and 20–40 pairs per km² in creeping Pinus pumila. Lower down by river mouths and coastal lagoons 10–25 pairs per km² and in dense oak Quercus woods by water 5–10 pairs per km² (Matyushkin and Kuleshova 1972); for further details in Sikhote-Alin, see Kuleshova (1976) and for lower Amur (eastern USSR), see Babenko (1984). In Kamchatka (eastern USSR), Betula ermani similarly favoured, with 10–85 birds per km² (Vronski 1977). In central and eastern Nepal, density generally low; almost exceptional for 2 ♂♂ to be within earshot. At Tongba-Luza, however, A. hodgsoni virtually the only breeding species in area of dwarf shrubs and boulders, with 1 bird every few hundred metres (Diesselhorst 1968a). ROOSTING. Bird wintering in southern England left

regular garden feeding site daily c. 30 min before dark, always departing in same direction. Also spent long periods (several hours) perched in tree during day (D Parker).

2. In Sakhalin (eastern USSR), singing ♂♂ generally shy, breaking off and disappearing at approach of man; ♀♀ unobtrusive, mostly on ground (Munsterhjelm 1922). Irish vagrant allowed approach to c. 12 m (Mullarney 1980), and another in Shetland, generally silent on ground, to c. 30 m (Dennis 1967). Said to 'sway curiously' before take-off (Fletcher and Inglis 1924). In Mongolia, migrants and breeders notably timid. When disturbed, birds fly up giving loud calls (see 2a in Voice) and scatter into tree canopy; perch on branches or even hide (Kitson 1979). Such behaviour typical (Ali and Ripley 1973b); birds perch 'rather horizontally' and perform initially fast, then gradually slower tail-pumping (Roever 1980); see also Mullarney (1980). Wintering bird in southern England (see Roosting, above) was last to fly up at any disturbance, but gave call 2a in direct flight to tree canopy (D Parker). FLOCK BEHAVIOUR. No information. SONG-DISPLAY. ♂ sings (see 1 in Voice) from top or side branch of tree or bush (e.g. Johansen 1952, Rashkevich 1965), in Japan also from overhead wires (Jahn 1942), or in flight; also from ground while feeding (Kozlova 1930). Varying reports from different parts of range, though not known if these reflect any geographical differences. Suggested by Wallschläger (1984) that frequent use of fairly dense stands of trees may lead bird to sing more often from perch than in flight. In Krasnoyarsk, no Song-flights recorded (Naumov and Burkovskaya 1959) and further west in Siberia, birds sang from perch, while suddenly and rapidly fluttering about, or while flying to another tree, but not in true ascending Song-flight (Johansen 1952); singing while flying from tree to tree also noted in Sakhalin by Munsterhjelm (1922). In contrast, Ali and Ripley (1973b) reported Song-flights much commoner than other types of performance. Song-flight closely resembling that of A. trivialis occasionally given after prolonged bout of song from perch. Ascends with rapid wing-beats to c. 10–20 m, then makes angled and gliding descent, wings and tail spread, to nearby perch (Kozlova 1930; Rashkevich 1965; Ali and Ripley 1973b). According to Matyushkin and Kuleshova (1972), 1st part of song given from perch, then further sounds during ascent. Rashkevich (1965) noted birds singing while hovering and continuing from perch after landing before renewed ascent. Earlier reports in Whistler (1925b) and Bergman (1935) and more recent comparative study of Anthus Song-display by Wallschläger (1984) indicate bird sings only from peak of ascent; see also Kapitonov and Chernyavski (1960). Song-flights occurring at peak of breeding season sometimes longer than in A. trivialis (Dementiev and Gladkov 1954a). Where ascent made from low perch, bird may fly straight down to ground. Song reduced or absent in strong wind (Matyushkin and Kuleshova 1972). In Yakutiya (eastern USSR), will sing throughout day, even at night (Vorobiev 1963); mostly in morning of fine, still days (Kapitonov and Chernyavski 1960). On lower Ob river (USSR), 53% of song in morning, 43% later in day, 4% in evening, and none at night (Podarueva 1979). In USSR, sings from arrival to mid- or late July (Portenko 1960; Vorobiev 1953); peak in Krasnoyarsk mid-June to early July (Naumov and Burkovskaya 1959); in north-east USSR, intensive song noted late May to mid-June (Krechmar et al. 1978); on lower Lena, reduced from mid-June but some to end of month (Kapitonov and Chernyavski 1960). In India, sings from mid-March to 3rd week of June (Ali and Ripley 1973b); peak in one area of Nepal early May (Diesselhorst 1968a). ANTAGONISTIC BEHAVIOUR. No information for breeding season. Wintering bird in southern England mostly indifferent

A

to other small birds in garden. Occasionally adopted (for no obvious reason) a presumed threat-posture (Fig A): head lowered, bill opened wide for a few seconds; no call heard (D Parker). HETEROSEXUAL BEHAVIOUR. Only report is that ♀ mostly on ground while ♂ sings (Munsterhjelm 1922). RELATIONS WITHIN FAMILY GROUP. Eyes of young open at 5 days. Young able to flutter about at *c*. 10 days; normally leave nest at 11–12 days and hide in surrounding vegetation; fed there by parents for at least 1–2 days, then move away (Rashkevich 1965). ANTI-PREDATOR RESPONSES OF YOUNG. Will leap out of nest and hide at *c*. 10 days (Rashkevich 1965). PARENTAL ANTI-PREDATOR STRATEGIES. (1) Passive measures. ♀ sits tightly, especially towards end of incubation (Rashkevich 1965), e.g. allowing approach to 1 m (Matyushkin and Kuleshova 1972). May fly up into tree on flushing; one bird did so each time disturbed (Whistler 1925*b*). Will also run away for some distance before flying (Rashkevich 1965) and return on foot, moving quickly through vegetation (Matyushkin and Kuleshova 1972). (2) Active measures. Recorded leaving well-incubated clutch in silent run with distraction-lure display of disablement type (Vorobiev 1963). Bird at 1 out of 4 sites in northern India fluttered down slope close to ground (Whistler 1925*b*); probably also a distraction-lure display.

(Fig by I Lewington: from drawing by D Parker.)   MGW

Voice. For extra sonagrams, see Wallschläger (1984). Full study required to determine range of calls (other than call 1) and degree of overlap. Following scheme of necessity provisional.

CALLS OF ADULTS. (1) Song of ♂. Typically comprises phrases with great variety of unit-types. In one study, birds from Mongolia and Lake Baykal area (eastern USSR) showed little repetition of units, usually no more than 2 of same type in succession; whole segments of one particular unit-type less common and tended to occur during song in flight (Wallschläger 1984). In contrast, recordings by J Boswall (China) and L Svensson (Lake Baykal) reveal much repetition of units, song strongly recalling Wren *Troglodytes troglodytes* in tendency to alternate segments of repeated units and segments of heterogeneous units. In recording of bird singing from perch, short phrases delivered in explosive manner comprise brief jingling introduction (one of 6 phrases lacks this) and coda separated by 2–5 long, loud, pure, descending 'cheeoo' or 'ptseeoo' sounds; 2nd phrase (Fig I) much like short song of *T. troglodytes* (J Hall-Craggs, M G Wilson). Similarity to Tree Pipit *A. trivialis* (often mentioned—see below) most marked in 2 unit-types: a rattle given only in flight (see below for further comparisons) and descending disyllabic 'sie' units which, when associated with Song-flight, are given only after landing. Analysis of 10 phrases given by one bird from perch (in Mongolia) revealed over 40 different unit-types, suggesting greater variation than in *A. trivialis*; however, at least 14 unit-types revealed in 3 songs of *A. trivialis* and impression of greater variety in *A. hodgsoni* perhaps due to relative brevity of its song-phrases (J Hall-Craggs). Mongolian bird typically gave (confirmed in other individuals) 2–5 particular units in regularly recurring groups (or 'blocks'); this pattern not typical of other *Anthus*. Variation basically occurred at start and end of phrase. (Wallschläger 1984.) Recording from USSR of longer and more complex song in flight contains dry rattles at various speeds, these and high-pitched 'zee-vee' units being more like Red-throated Pipit *A. cervinus* than *A. trivialis* (Svensson 1984*b*); rattles, one lasting 1·1 s and very fast at 40 'click' units per s, nevertheless equally recall *T. troglodytes* as does the song overall (J Hall-Craggs, M G Wilson). Diesselhorst (1968*a*) and Ali and Ripley (1973*b*) described song as sometimes very similar to *A. trivialis*, at other times markedly different, with harsher or harder and more wheezy or squeezed and strangled sounds; perhaps an indication of local or individual variation. In Mongolia, average length of 20 phrases given from perch 3·0 s, and connected series of phrases may last 12 s; song-phrase in flight lasts 2–5 s. Dialects apparently exist and are constant over several years: in Tereldsh (Mongolia) identical groups of units noted in 2 birds in 1979 and 1983 (Wallschläger 1984). For further descriptions and comparisons, see Jahn (1942), Johansen (1952), Dementiev and Gladkov (1954*a*), and Matyushkin and Kuleshova (1972). (2) Contact-alarm calls. Variety of transcriptions, though these suggest fairly restricted range of sounds; possibly some geographical variation, however (see below). The only available sonagrams—in study by Wallschläger (1984)—indicate that at least 2 main calls exist. (a) Likened by Wallschläger (1984), who gave no rendering, to Yellow Wagtail *Motacilla flava*. Descriptions refer to calls ranging from

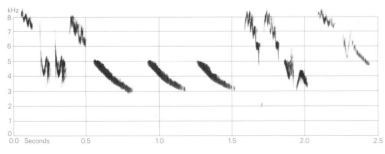

I   J Boswall   China   May 1983

faint to loud and renderings may be summarized as follows: 'tsee(p)' or 'pseep', 'tzee(p)', '(d)zeep' 'teaze', or 'tseeet'; slightly hissing 'dzeer', usually with downward inflection; also 'dzip', 'psst', and 'tsssp'. Normally given singly and intermittently in flight, also in series when disturbed (Smith 1942; Ali and Ripley 1973*b*), units getting longer as bird takes off (Wallschläger 1984). Reminiscent of *A. trivialis* or *A. cervinus*—'psst' or 'pseep', or quiet, thin 'tsssp' typically shorter than *A. cervinus* and ending abruptly (Sharrock 1980*b*; P J Ewins) and averages slightly higher pitched than *A. trivialis* (Svensson 1984*b*); in louder, more strident form, often recalls also Richard's Pipit *A. novaeseelandiae*; in close cover, and apparently as high-intensity alarm, loud, strident, and high-pitched sound ('teaze') can suggest Redwing *Turdus iliacus* (Jahn 1942; Smith 1942; Dennis 1967; Hemmingsen and Guildal 1968; Fleming *et al.* 1976; Jobson 1978; Kitson 1979; D I M Wallace). Bird released after ringing gave call similar to *A. trivialis*, but less hoarse and higher pitched (Cilia 1978). Calls ('tzeep') of birds in India and of vagrant on Shetland very like *A. trivialis*, at times practically indistinguishable. In Thailand, short, explosive 'dzeep' more like *A. cervinus*, altogether more distinctive, and separable from *A. trivialis* (Oddie 1982); evidence insufficient to support suggestion of geographical variation. Irish vagrant gave variety of similar sounding calls, overall like *A. trivialis*, frequently in flight, occasionally from ground: buzzing 'dzzz' or 'tzzz', also a more abrupt 'bzzzp' usually when flushed, or after gaining some height (Mullarney 1980). (b) A 'ziii' given as alarm-call in territory, sonagram indicating this to be a distinct call (Wallschläger 1984). A 'bee' softer than call 2a and given by perched bird (Hemmingsen and Guildal 1968) may belong here or perhaps closer to call 2c. According to Wallschläger (1984) 'ttseeee' given by migrant *yunnanensis* in eastern USSR (Panov 1973) and 'tseep' mentioned by Ali (1977) are the same. However, comparison with renderings of call 2a strongly suggests that at least the 2nd and probably both of these are closer to if not the same as call 2a. (c) Soft 'sip' given repeatedly when handled (Cilia 1978); soft, weak 'tsip' (Oddie 1982) probably the same, but context not known. A 'tizs' similar to call 2a, but much softer and 'watery'; given often during shorter, lower flights, occasionally from ground (Mullarney 1980).

CALLS OF YOUNG. No information. MGW

Breeding. SEASON. Western Siberia (USSR): June-August (Johansen 1952). SITE. On ground, in shelter of rock or tuft of vegetation. Nest: shallow depression containing cup of moss and grass, lined with fine grass and hair. External diameter 8–13 cm, internal diameter 6·5 cm, overall height 6 cm. Depth of cup 4 cm (Makatsch 1976). Building: no information. EGGS. See Plate 80. Sub-elliptical, smooth and glossy; very variable in ground-colour and pattern; dark brown, brown, or grey,

with more or less darker brown spotting and streaking. 21·1 × 15·5 mm (19·7–22·0 × 14·8–16·0), $n = 34$; calculated weight 2·6 g (Schönwetter 1979). Clutch: 4–5. Normally 2 broods (Ali and Ripley 1973*b*; Portenko 1960). Eggs laid daily (Rashkevich 1965). INCUBATION. 12–13 days, beginning with 2nd–3rd egg; hatching asynchronous—over 2, even 3, days. By both sexes according to Dementiev and Gladkov (1954*a*), by ♀ only according to Rashkevich (1965); early in incubation, ♀ leaves for several hours each day, later for only 20–30 min (Rashkevich 1965). YOUNG. Cared for and fed by both parents, more by ♀ (Rashkevich 1965). FLEDGING TO MATURITY. Leave nest at 11–12 days and fed for at least 1–2 days further, by which time flying (Rashkevich 1965).

Plumages (*A. h. yunnanensis*). ADULT BREEDING. Entire upperparts and sides of neck greenish-olive, narrowly but sharply streaked black on forehead and crown, uniform on remainder of upperparts, though usually with indistinct darker shaft-streaks on mantle. Supercilium broad and distinct, extending from nostril to upper ear-coverts; usually slightly narrower just above front of eye and broken at rear by narrow dull black line extending back from eye (pale extension of supercilium on upper ear-coverts sometimes missing); front part of supercilium deep buff when plumage fresh, rear (backwards from above eye) pale cream, sometimes with slight olive mottling; in worn plumage, front part pale buff grading to pure white above eye. Supercilium bordered above by black stripe along sides of crown (sometimes indistinct when plumage fresh). Narrow ring of cream-yellow to off-white feathers round eye, broken by dull grey in front and (more narrowly) at rear. Lores with narrow dull black line, bordered below by buff patch with faint olive mottling, extending below eye to ear-coverts; this patch often bordered below by dark olive stripe from gape backwards. Lower rear of ear-coverts with distinct black spot. For variation in pattern of sides of head, see Conder (1979), Kitson (1979), and Fairbank (1980*b*). Lower cheeks, chin, throat, chest, and sides of breast warm yellow-buff in fresh plumage (palest on chin and upper throat), virtually white in worn plumage, with some pale buff just visible on lower cheeks, lower throat, and sides of breast; buff of lower cheeks and throat separated by narrow but distinct black malar stripe, widening and breaking up into heavy black spots on lower sides of neck and sides of breast. Chest and upper flanks with broad and contrasting black streaks *c.* 2·5–3 mm wide. Lower flanks olive, remainder of underparts white, but flanks, breast, sides of belly, and under tail-coverts extensively tinged buff when plumage fresh; breast, flanks, and sides of belly contrastingly marked with black streaks 1–1·5 mm wide. Central pair of tail-feathers (t1) dull black or greyish-black, broadly bordered by poorly defined green or buff-green fringes; t2–t5 dull black with sharp and narrow green outer edge, t4 often with small white spot on tip of inner web, t5 with white wedge on distal part of inner web; length of wedge on t5 7·8 (24) 4–13 mm—no difference in length between ages, sexes, or populations. T6 white, faintly sullied olive-grey in fresh plumage; distal part of outer web olive-grey, large dark grey wedge on basal and middle portion of inner web. Flight-feathers, tertials, greater upper primary coverts, and bastard wing dull black; tertials with broad and poorly defined green-buff or green-olive fringe along outer web, a narrower one along tip; secondaries with similar but narrower fringes, greater upper primary coverts and feathers of bastard wing with narrow and sharp fringes of same colour; fringes along outer webs of primaries narrow and sharp, pale green or green-yellow, on p9 pale yellow or yellow-white. Lesser upper wing-coverts greenish-olive, like upperparts, uniform or with slightly darker olive-grey centres; median coverts contrastingly black, rather broadly (1·5–3 mm) tipped pale greenish-yellow or pale yellowish-buff, soon bleaching to white; greater coverts black with broad green-buff or green-olive fringe along outer web and yellow-buff tip, tip slightly narrower and not as pale as tips of median coverts in fresh plumage, white but narrow (if not largely lost by abrasion) in worn plumage. Under wing-coverts and axillaries yellow-buff, slightly tinged grey on longer coverts, brighter yellow on shorter coverts; coverts along leading edge of wing grey on bases. Bleaching and wear have marked influence on olive, buff, and yellow in plumage. *In very fresh plumage*, upperparts greenish-olive or olive-green, fringes of tertials, secondaries, and greater upper wing-coverts bronze-green or greenish-buff; tips of median upper wing-coverts and rear part of supercilium (from above eye backwards) pale yellow or pale buff; front part of supercilium, lower cheeks, chin, throat, chest, flanks, and under tail-coverts deep buff. *In slightly worn plumage* (about April–May), upperparts colder olive (but still uniform except for forehead and crown), fringes of tertials and secondaries olive-grey (largely worn off on tertials), tips of median and greater upper wing-coverts and rear part of supercilium pale yellow-buff or white; front part of supercilium, lower cheeks, sides of breast, lower throat, and chest pale buff or yellow; ground-colour of remainder of underparts white. *In heavily worn plumage* (about June–July), upperparts dull olive-grey with sepia-brown streaks and spots, only rump usually still uniform greenish-olive; tertials and t1–t5 black-brown, green fringes abraded; median and greater upper wing-coverts with abraded white tips; supercilium and ground-colour of underparts virtually white, only traces of some yellow remaining; pale extension of supercilium to upper ear-coverts and (often) dark malar stripe virtually lost. ADULT NON-BREEDING. Similar to adult breeding, but all plumage equally new (September–October) or slightly worn (November–December), without slight contrasts in wear among tertials and wing-coverts as in adult breeding, which has some relatively more worn non-breeding tertials and coverts retained (see Moults). NESTLING. Down dark grey and rather sparse; on upperparts only (Neufeldt 1970). JUVENILE. Like adult breeding and non-breeding, but upperparts duller brown-olive with dull black feather-centres, not as uniform and green as adult; black streaks on underparts longer and wider, but less sharply defined. Closely similar to juvenile Tree Pipit *A. trivialis*, but fringes of flight-feathers and tail bronze-green, not pale buff or yellow. (Hall 1961; Bub *et al.* 1981.) FIRST ADULT NON-BREEDING. Like adult non-breeding, but part of juvenile feathering retained (see Moults), slightly or contrastingly more worn than relatively fresh non-breeding and hence rather similar to adult breeding, but contrast in wear less marked, and timing different; about mid-winter, birds in adult breeding and in 1st non-breeding may occur together, but 1st non-breeding worn (and retained juvenile feathers distinctly worn), adult breeding new (and retained adult non-breeding still fairly fresh). FIRST BREEDING. Like adult breeding, but retained juvenile feathers (flight-feathers, greater upper primary coverts, outer tail-feathers, occasionally outer greater upper wing-coverts, and variable number of other feathers) heavily worn.

Bare parts. ADULT, FIRST ADULT. Iris ochre-brown or dark brown. Upper mandible and tip of lower mandible dark horn-brown, cutting edges of both mandibles and base of lower mandible pale flesh-colour or reddish-flesh. Leg and foot pale flesh or pink, sometimes yellowish at rear (Dennis 1967; Conder 1979;

RMNH), pale yellow with brown tinge (Gräfe *et al.* 1962; Ree 1974*b*), horny-flesh, or yellow-brown with horn-brown claws and fleshy-white soles (Ali and Ripley 1972). NESTLING. Bare skin yellowish-pink. Bill grey, darker at tip. Mouth and tongue orange, gape-flanges pale, almost white. (Neufeldt 1970.) JUVENILE. No information.

Moults. ADULT POST-BREEDING. Complete; primaries descendant. In USSR, mid-August to early October (Dementiev and Gladkov 1954*a*). Single birds from 17 July (Manchuria) and 19 August (Mongolia) had not yet started, though another on 19 August was in moult with outer 6 primaries still old; on 24–26 August and 1 September, birds from Manchuria, Mongolia, and Lake Baykal had flight-feather moult completed but some body feathers still growing; others from 1st half of September, Manchuria, had completed moult (Piechocki 1958; Piechocki and Bolod 1972; RMNH). ADULT PRE-BREEDING. Partial. (February–)March–April. Involves head, body, lesser upper wing-coverts, many or all median upper wing-coverts, often tertial coverts and most or all tertials, sometimes t1, and variable number of greater upper wing-coverts. POST-JUVENILE. Partial. Starts early August to early September, sometimes largely completed by mid-August (USSR) and fully completed in migrants from September, Manchuria. Extent about as in adult pre-breeding. (Dementiev and Gladkov 1954*a*; Piechocki 1958; RMNH.) FIRST PRE-BREEDING. Similar to adult pre-breeding, but less feathering involved in some birds and completely suppressed in at least 1 bird examined (RMNH, ZMA).

Measurements. ADULT, FIRST ADULT. *A. h. yunnanensis*. Eastern USSR, summer, and north-east China, winter; skins (RMNH, ZMA). Bill (S) to skull, bill (N) to distal corner of nostril; exposed culmen on average 3·9 less than bill (S). Toe is middle toe with claw; claw is hind claw.

	♂		♀	
WING	86·6 (1·72; 20)	84–90	83·9 (2·43; 12)	80–86
TAIL	59·8 (2·16; 19)	57–63	57·3 (1·64; 12)	55–59
BILL (S)	15·0 (0·47; 17)	14·3–15·9	14·6 (0·39; 11)	14·2–15·4
BILL (N)	8·4 (0·32; 17)	7·9–9·0	8·0 (0·23; 10)	7·7–8·4
TARSUS	21·3 (0·69; 20)	20·4–22·4	20·8 (0·61; 11)	19·8–21·7
TOE	18·8 (0·85; 12)	17·4–20·0	18·6 (0·93; 11)	17·4–20·1
CLAW	8·0 (1·02; 25)	6·7–8·8	7·7 (0·75; 14)	6·5–8·8

Sex differences significant for wing, tail, and bill (N). Hind claw exceptionally 11·7 (excluded from range).

Wing of unsexed birds, Lake Chany (south-west Siberia), on autumn migration: 85·6 (1·30; 8) 84–88 (Havlín and Jurlov 1977). Wing, Mongolia, May–August: ♂ 86·0 (3·16; 6) 81–90, ♀ 82·1 (1·96; 8) 80–86 (Piechocki and Bolod 1972). Wing, Manchuria, July–September: ♂ 85 (16) 83–88, ♀ 81·4 (8) 80–83 (Piechocki 1958).

Weights. *A. h. yunnanensis*. Fair Isle (Scotland): September 20·9, October 20·7 (Dennis 1967); late September to November 19·2 (2·37; 6) 14–21 (Fair Isle Bird Observatory, N J Riddiford). Skokholm (Wales), April: 23·5 (Conder 1979). Utsira (Rogaland, Norway), October: 17·5 (lean) (Ree 1974*b*). Helgoland (West Germany), May: ♂ 24 (Gräfe *et al.* 1962). Lake Chany, south-west Siberia, 17 August–11 September: 20·4 (2·22; 8) 17·4–24·1 (Havlín and Jurlov 1977). USSR: ♂ 22·3 (5) 20·5–25·6, ♀ 23 (Dementiev and Gladkov 1954*a*). India, winter: ♂ 19·4–24·6 (16), ♀ 17–26·3 (11) (Ali and Ripley 1972). Mongolia: May, ♂ 24·0 (3) 23–25, ♀ 21·6 (2·41; 5) 19–25; June, ♂ 20·7 (39) 18–24, ♀ 18; August, ♂ 26, ♀♀ 24, 26 (Piechocki and Bolod 1972). Manchuria, (July–)September: ♂ 22·3 (16) 21–25, ♀ 22 (8) 21–24 (Piechocki 1958).

Nominate *hodgsoni*. Nepal and India: summer, ♂ 20–22·6 (10), ♀ 20–23 (7); September, ♂♂ 23·5, 24; winter, 18–23 (8) (Diesselhorst 1968a; Ali and Ripley 1972).

Structure. Wing rather short, broad at base, tip bluntly pointed. 10 primaries: in *A. h. yunnanensis* and Japanese populations of nominate *hodgsoni*, p8 longest, p7 and p9 both 0·4 (15) 0–1 shorter, p6 2·2 (15) 1–4, p5 11·5 (15) 10–14, p4 15·9 (11) 14–18, p1 23·1 (15) 21–25; in nominate *hodgsoni* from Himalayas, p7–p8 longest, p6 and p9 both 0·9 (7) 0–2 shorter, p5 10·1 (7) 8–12, p4 14·7 (7) 13–17, p1 20·5 (7) 19–23. In both races, p10 reduced, narrow and pointed, 54–63 shorter than p8, 8–12 shorter than longest upper wing-covert. Inner web of p7–p9 and outer of p6–p8 emarginated. Longest tertial reaches to about halfway between tips of p5 and p6 in closed wing when fresh, to about p5 when worn; 5–13 mm from wing-tip (tertials relatively shorter than in *A. trivialis*). Bill rather short, thick at base; similar to *A. trivialis*. For length of middle toe and hind claw, see Measurements. Outer toe with claw *c.* 72% of middle toe with claw, inner *c.* 77%; hind toe with claw *c.* 92%, without claw *c.* 52%. Remainder of structure as in *A. trivialis* (p. 358).

Geographical variation. Rather slight. Nominate *hodgsoni* from Himalayas east to eastern China is more distinctly streaked black on upperparts than *A. h. yunnanensis*; streaks on head broader, those of mantle and scapulars rather narrow but distinct, rump and upper tail-coverts unstreaked; streaks on chest extend further down; belly more extensively streaked (Hall 1961). Wing of Himalayan populations on average *c.* 2 mm shorter, ♂ 84·4 (4) 83–88, ♀ 81·9 (4) 80–85 (RMNH, ZMA); ♂ 79–86, ♀ 77–85 (Hall 1961), tail and tarsus similar; bill to skull

slightly shorter, 14·6 (8) 13·8–15·2; wing slightly more rounded, distance of p1–p6 from tip slightly less (see Structure). Birds from Honshu (Japan) included here in nominate *hodgsoni* and those from Hokkaido in *A. h. yunnanensis*, though Hokkaido population said to be intermediate between these 2 races (Vaurie 1959). Japanese birds differ slightly from typical nominate *hodgsoni* or *A. h. yunnanensis*, however, but not enough to warrant recognition as separate race (Hall 1961): upperparts rather heavily streaked (similar to nominate *hodgsoni*), and wing and tail length and wing shape similar to *A. h. yunnanensis*, but bill to skull and tarsus slightly longer than typical populations of either race—wing ♂ 86·6 (11) 84–89, ♀ 82·0 (5) 79–83, bill (S) 15·3 (15) 14·7–16·4, tarsus 21·7 (16) 20·6–23·1 (RMNH, ZMA). Birds from south-east Tibet and Sikang (China) sometimes separated as *berezowskii* Zarudny, 1907 (e.g. Ripley 1948), but race not valid (see Hall 1961).

Recognition. Differs from all other Palearctic *Anthus* (except *A. trivialis*) in short hind claw. Only Palearctic *Anthus* with uniform greenish-olive mantle, scapulars, back, and rump, but adult non-breeding Rock Pipit *A. spinoletta littoralis* and extralimital Rosy Pipit *A. roseatus* are rather similar (though underparts of both quite different and hind claw long). In heavily worn plumage, more heavily streaked above and then difficult to separate from *A. trivialis*, especially from some eastern races of latter, which are whitish below with heavy black chest marks, quite similar to *A. hodgsoni*; short and broad white supercilium (restricted to patch above and just behind eye) then usually diagnostic, usually occurring in combination with dark patch on rear of ear-coverts (some *A. trivialis* show similar but smaller patch). CSR

Anthus trivialis Tree Pipit

PLATES 23 and 27
[between pages 232 and 233]

Du. Boompieper   Fr. Pipit des arbres   Ge. Baumpieper
Ru. Лесной конек   Sp. Bisbita arbóreo   Sw. Trädpiplärka

Alauda trivialis Linnaeus, 1758

Polytypic. Nominate *trivialis* (Linnaeus, 1758), Europe and Asia Minor, in north through western Siberia east to Lake Baykal and Yakutiya, in south to southern shores of Caspian Sea. Extralimital: *schlueteri* Kleinschmidt, 1920, mountains of west-central Asia from central and eastern Afghanistan and Tien Shan through Tekes valley (north-west China) to Dzungharskiy Alatau and Tarbagatay mountains; *haringtoni* Witherby, 1917, north-west Himalayas from Kashmir to Garhwal.

Field characters. 15 cm; wing-span 25–27 cm. Slightly bulkier than Meadow Pipit *A. pratensis*, with slimmer rear body making tail length more obvious, and slightly longer wings with narrower point. Rather small, sleek, and elegant pipit, with somewhat more attenuated form than *A. pratensis* and Red-throated Pipit *A. cervinus*. Plumage pattern typical of small pipits but differs subtly from typical *A. pratensis* and all *A. cervinus* in combination of noticeably pale eye-ring, warm but not rufous upperparts, striking wing-bars, yellow-buff, boldly spotted breast, and little-streaked flanks. At close range, quite large bill, noticeably pink legs and short hind claw are useful characters. In flight, tends to look paler than *A.*

pratensis and *A. cervinus*; flight-action as *A. cervinus*—far less hesitant and fluttering than *A. pratensis*. Habitually perches in and on trees. Song well-developed for a pipit. Sexes similar; little seasonal variation. Juvenile separable.

ADULT. Head, nape, back, and wings have noticeably warm but quite buff ground-colour, with olive or brown tone most marked in fresh plumage; overlying black-brown streaks broadest on back and most striking in worn plumage. Unmarked yellow-ochre supercilium, rather broad and complete, cream eye-ring, and brown eye-stripe and edge to cheek provide rather open-faced appearance. Folded wing shows quite striking pattern of cream margins to almost black-centred median coverts, and pale ochre tips to greater coverts and wide fringes to tertials. Face and wing pattern both more obvious than on *A. pratensis*. Rump and upper tail-coverts as rest of upperparts but almost unstreaked and thus appearing paler than back and most of tail. Tail brown, with wide buff-ochre margins on central feathers and white outer margins to darkest outer feathers forming obvious bright edges, like *A. pratensis* but less cold in tone. Chin and throat buff-white to buff, becoming faintly yellower on breast and flanks and forming warm, soft ground to fairly heavy black-brown malar stripe and wide chest-band of bold black-brown spots which extends on to fore-flanks but fades on rear flanks into much narrower, often fine and indistinct pattern; fore-underparts thus have more clustered pattern of spots than *A. pratensis*, as in Olive-backed Pipit *A. hodgsoni*, Pechora Pipit *A. gustavi*, and *A. cervinus*. Belly soft white, tinged buff when fresh, and under tail-coverts always pale buff; lower underparts thus less pure white than cleanest *A. pratensis* and never silky white as in most *A. hodgsoni*, *A. gustavi*, and winter-plumaged *A. cervinus*. Underwing buff, with grey in wing-pit. JUVENILE. Closely resembles fresh-plumaged autumn adult but usually noticeably buffier. Lacks olive or brown tones to more prominently streaked upperparts. Less yellow, more buff-toned on chest; flanks virtually unstreaked, making chest-spots more obvious. At all ages, bill larger than *A. pratensis*, averaging only slightly longer but noticeably deeper and broader; dark brown-horn, with base of lower mandible quite bright, pale buff-flesh. Legs slightly shorter than *A. pratensis*; pale flesh, with brown tone in dull light but bright pink flush in full sunlight and noticeably brighter than *A. pratensis*; hind claw a short 'hook', only 40% as long as spiky 'spurs' of *A. pratensis* and *A. cervinus* but similar in length and shape to *A. hodgsoni* and *A. gustavi*.

Most observers too ready to rely on supposedly distinctive call for identification. This actually justified only in temperate Europe in summer, for in Arctic in summer and elsewhere during migration seasons 4 other species give similar calls and/or have similar appearance: *A. hodgsoni*, *A. gustavi*, one race (*A. p. whistleri*) and some other individuals of *A. pratensis*, and *A. cervinus*. Thus *A. trivialis* is as much a key species in small pipit identification as *A. pratensis* (if not more so), and known to be joined in its movements by all 3 of the rare species above, whereas *A. pratensis* is accompanied regularly only by *A. cervinus*. Happily, *A. hodgsoni* shows distinctive green-toned plumage at close range, with noticeably more decorated face and chest, and thus no serious risk of confusion with that species if bird seen well. Because of overlaps between the 4 remaining species, most plumage characters are suspect, though streaked rumps of *A. gustavi* and *A. cervinus* separate them from plain-rumped *A. trivialis* and *A. pratensis* (and *A. hodgsoni*). Important to use calls and bare part structure to reduce confusion. Calls high-pitched and rasping in *A. trivialis* (and *A.*

hodgsoni); either close to that or short and lower pitched in *A. cervinus*; stony-toned and short in *A. gustavi*; shrill but weak, at times hysterical, in *A. pratensis*. Calls of *A. gustavi* and *A. pratensis* thus distinctive, and good view of hind claw can quickly distinguish between *A. trivialis* and *A. cervinus*. Flight-actions of *A. trivialis* and *A. pratensis* often taken to be identical but this not so: *A. trivialis* has more fluent wing-beats and rather more bounding, less fluttering progress, with slightly greater bulk and long wing-point creating distinctive silhouette (though all these features shared by *A. hodgsoni* and *A. cervinus*). Escape-flight usually quickly rising, with ascent often prolonged by half or full circle before descent to long ground cover or, when available, bush or tree. Song-flight more confident than *A. pratensis*, usually reaching greater height and containing spirals early in descent and usually starting or finishing from tree or bush. Gait rather deliberate, with high-stepping walk and short run; lacks shuffling creep of *A. pratensis* and (on occasion) *A. cervinus* and *A. gustavi*. Carriage rather upright; usually holds head higher than *A. pratensis* and often lets tail fall below line of body when perched. Wags tail with characteristic slight, gentle movement. Behaviour shy but noticeably calm, appearing more contented and less nervous than *A. pratensis*. Readily occupies bushy areas in winter and on migration; if tall vegetation absent, enters tussock grass and other high ground cover and rarely feeds in such open areas as persistently used by *A. pratensis*. Not markedly gregarious but migrants form streams, occasionally concentrated in large falls, while wintering birds can be quite densely spread through preferred habitats, forming handful-sized parties at times.

Song excels both *A. pratensis* and Rock/Water Pipit *A. spinoletta* in both quality and power, carrying much further than in *A. pratensis*: single, long, rising and falling stanza of (1) repeated monosyllables, (2) slight stuttering phrase, and (3) increasingly forceful disyllables, sequence sounding like 'chik-chik...chia-chia-wich-wich-tsee-a-tsee-a-tsee-a...'; effect is of continuous, increasingly confident but decelerating trill, with last part strongly recalling Canary *Serinus canaria* or other small finch. Commonest call, given in flight and on flushing, also much stronger than *A. pratensis* and *A. spinoletta* and audible at longer range: 'teez', 'tseep', or 'skeeze', sounding quite high-pitched and even in tone (commonest call of *A. hodgsoni* similar, even identical to some ears, but common call of *A. cervinus* less strident, with, at close range, noticeable internal pulse of volume lacking in *A. trivialis*). Alarm-call of breeding bird a slightly chinked 'zip', softer than similar call of *A. pratensis*.

Habitat. Breeds in middle and upper middle latitudes, and in Scandinavia up through subarctic to borders of Arctic; thence down through boreal and temperate to

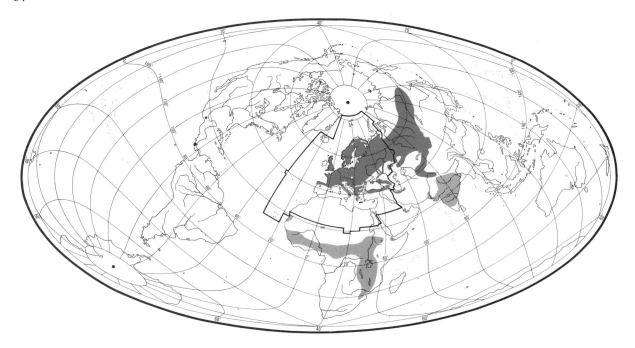

forest tundra belt, overlapping in central Palearctic with Olive-backed Pipit *A. hodgsoni*. Mainly in continental but spreading marginally into oceanic climates, between July isotherms of 10–26° C but avoiding more exposed windy and wet as well as torrid and arid conditions. Like congeners, basically a ground-feeder and ground-nester, but unique among them in west Palearctic in attachment to trees and bushes as look-outs and song-posts, no less essential in breeding territory than suitable foraging terrain and nest-sites. Accordingly shuns both open treeless and shrubless habitats and those where density of woody vegetation leaves insufficient open low herbage accessible. The remaining acceptable blends include parkland-savanna types, heathland or grassland in earlier stages of tree colonization, mature hedgerows, birch scrub, young conifer plantations 1 m or more high, open woods of oak, ash, and pine, and even closed high forest, provided there is no thick shrub layer (Yapp 1962). Newly felled woodland with some trees left may be ideal; occasionally, pylons or erected poles serve as substitute song-posts. Detailed studies in Britain have shown substantial expansions and fluctuations, partly related to transient nature of acceptable habitat conditions (Sharrock 1976; for fuller treatment see Fuller 1982). Ascends as high as trees grow, breeding in Swiss Alps even higher, in zone of dwarf conifers up to 2300 m, but avoids dense forests, orchards, cultivated land, and reedbeds (Glutz von Blotzheim 1962). In USSR, mainly in places where trees grow sparsely: forest margins, clearings, felled areas, burnt patches, shrub forests and marshes with a few taller trees, and small woods and groves. Locally in dense mixed forests, nesting in mountains up to treeline at 1000–1500 m; sometimes in bogs near Meadow Pipit *A.*

pratensis. On migration occurs in similar habitats, but also on more open steppe plains, meadows, and on sands. (Dementiev and Gladkov 1954a.) Winters in Africa among good-sized well-spaced trees, as a rule at least 10 m high with ground surface accessible. In East Africa at *c.* 1000–2700 m on areas with large *Acacia* trees and on patches of cultivation and coffee plantations shaded by tall introduced trees. In Nigeria, winters in southern more humid parts, favouring cleared strips in forest. (Moreau 1972.)

Distribution. Some range increase in Scotland.

BRITAIN. Northward expansion in Scotland in last 80 years (Parslow 1973); first bred Caithness 1968–72 (Sharrock 1976). IRELAND. Singing ♂♂ in 4 different counties 1974–6; probably now a scarce annual breeder (CDH). ITALY. Bred Sicily 1973 (PB, BM). ALGERIA. May have nested 1978 (EDHJ). PORTUGAL. Seen in north in breeding season, but nesting not confirmed (RR).

Accidental. Jan Mayen, Iceland, Kuwait, Madeira.

Population. Marked decrease Netherlands. Decrease England, north Finland and Switzerland.

BRITAIN. 50 000–100 000 pairs, fluctuating; slight decreases in England, especially in south (Parslow 1973; Sharrock 1976). FRANCE. Under 1 million pairs (Yeatman 1976). NETHERLANDS. 18 000–22 000 pairs, 1976–7 (Teixeira 1979); probable marked decrease since *c.* 1900, recent increase on Texel and other Islands (Wammes *et al.* 1983; CSR). WEST GERMANY. 520 000–620 000 pairs (Rheinwald 1982). Westfalen: in census area 1977–83 only small fluctuations (Loske 1985). SWEDEN. Over 4 million pairs (Ulfstrand and Högstedt 1976). DENMARK. Slight decrease

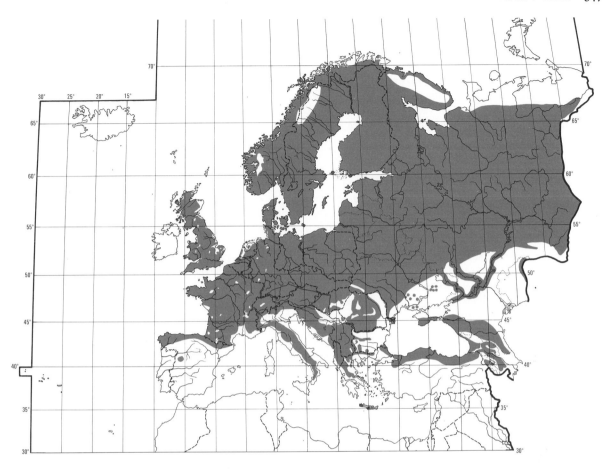

1978–83 (Vickholm and Väisänen 1984). FINLAND. About
1·65 million pairs (Merikallio 1958); decrease in north
1974–77 (Väisänen 1983); slight increase 1978–83 (Vickholm and Väisänen 1984). CZECHOSLOVAKIA. No changes
reported (KH). SWITZERLAND. Decreased (RN). GREECE.
Only a few tens of pairs (GIH).

Mortality. Belgium: average annual mortality of
young ♂♂ 65%; average annual mortality of adult ♂♂
47·5%, adult ♀♀ 66·7% (Van Hecke 1981). Oldest ringed
bird 7 years 8 months (Rydzewski 1978).

Movements. Long-distance total migrant. Winters on
Crete and islands of Aegean (G I Handrinos), in Persian
Gulf states, where scarce (Bundy and Warr 1980), and
possibly in northern Iran (Passburg 1959), but otherwise
wholly in Afrotropics and Indian subcontinent. Position
of migratory divide not known, though *schlueteri* and
haringtoni (breeding Tien Shan and north-west Himalayas, *c*. 68–80°E) winter wholly in central India, mingling
there with nominate *trivialis* which winters more extensively in the subcontinent.

Main wintering area in Africa extends across from
Guinea coast at 10°N to Ethiopia—in the west, south
only to northern edge of equatorial rain forest, but in
east extending south to Natal and Transvaal between

eastern edge of main rain forest and coastal and sub-coastal
forests bordering Indian Ocean (Mackworth-Praed and
Grant 1963; Moreau 1972; Curry-Lindahl 1981). Small
numbers encountered outside main area, particularly in
west, but no masssive concentrations recorded anywhere.
Thus, common at Nouakchott (Mauritania), *c*. 18°N (Gee
1984), and numerous in Sénégal (Morel and Roux 1966),
though few in Gambia except near Banjul (Gore 1981);
abundant in savanna woodlands and gallery forests of
Sierra Leone, even in Freetown gardens (G D Field).
Not noted in Togo until 1970 (Roo 1970) but not
uncommon in Nigeria, mainly on passage in north but
entering forest zone in clearings (Elgood *et al.* 1966;
Elgood 1982). Scarce in Niger inundation zone of Mali
(Curry and Sayer 1979). Much of western Africa south
of equator offers unsuitable terrain (rain forest or desert)
but there are a few records from Angola (Moreau 1972)
and birds occur regularly (but only in February) in
Central African Republic (Jehl 1974); in Zaïre, records
extend south to Bukavu—perhaps southern wintering
limit in central south (Lippens and Wille 1976). In
eastern Africa, major wintering area extends to just south
of Limpopo river in South Africa (Fry *et al.* 1974).
Occurs only on passage in northern Sudan, but in
southern Sudan much more numerous in autumn than

spring and some overwinter in Red Sea area (Cave and Macdonald 1955; Hogg *et al.* 1984). Common and locally abundant throughout winter in Ethiopia (Urban and Brown 1971). Not uncommon in Malawi (Benson and Benson 1977) and numerous in Zimbabwe (Irwin 1981). 2 birds recorded returning to same wintering area on Jos plateau of Nigeria in subsequent year (Smith 1965*b*).

In autumn, all populations breeding west of *c.* 15°E, plus those from Finland and even from north-west USSR east to *c.* 40°E, have been shown by ringing to move between south-west and just west of south into western Mediterranean basin and to Portugal; some from USSR after moving south-west to northern Italy then change direction to between south and south-east, at least over short distances (Zink 1975). Populations from Britain apparently fly direct to Portugal crossing Bay of Biscay, since, despite acute persecution in southern France and northern Spain, none recovered there and very few met on north coast of Spain by Snow *et al.* (1955). Most birds of Belgian origin follow Atlantic coast and are then deflected to south-west by Pyrénées, and also reach Portugal; one Belgian-ringed bird recovered in Ivory Coast (Zink 1975). Massive autumn exodus from southern Norway and Sweden varies greatly in numbers from day to day, being maximal under col conditions, with counts of over 100 birds per day occurring on some days between early September and early October; movement of 3736 birds recorded on one day at Falsterbo, Sweden (Nisbet 1957).

Occurs throughout Britain on passage, particularly on coasts and in autumn (British Ornithologists' Union 1971). Despite over 10 000 birds ringed, only *c.* 20 recoveries (Spencer and Hudson 1982): half in Portugal and singles in Norway, Netherlands, France, Morocco, and Mauritania; also a bird ringed Italy recovered in Britain. Scarce in Ireland, occurring mostly in south-east (Ruttledge 1966). Regularly met on North Atlantic weather ships (Luttik and Wattel 1979), and also noted on spring and autumn passage in Canary Islands (Bannerman 1963).

Despite major flight-path of birds from western Europe to Portugal and thence into Africa, spring and autumn passage occurs across Mediterranean from Gibraltar to Cyprus. From south-west Portugal, birds tend to move out at dusk (Owen 1958) but are few in numbers compared with other species (Moreau and Monk 1957). In Malta, autumn passage usually of single birds but up to 400 not infrequently present and exceptionally over 1000 in September–October; common on spring migration, with double figures on most days and daily maximum of over 1000; bird ringed April recovered Finland 3 weeks later (Sultana and Gauci 1982). Heaviest spring movements on Malta occur during anticyclonic weather on south-west winds (Lucca 1969). In one year at Bosporus (Turkey), late September, birds observed flying high and north-west, though this thought due to unfavourable weather (Nisbet

and Smout 1957). Spring passage there seen only in bad weather with few birds involved, thus in good weather perhaps fly too high for ground observation (Collman and Croxall 1967). In Cyprus, usually scarce on autumn passage though common in some years; common in spring, singly or up to 10 together but flocks over 100 occur (Flint and Stewart 1983). Data from North Africa indicate autumn and spring passage throughout coastal zone but of varying intensity. In Morocco, the few winter records are interpreted as passage stragglers and not as over-wintering (Heim de Balsac and Mayaud 1962). 2 birds ringed Gabès (Tunisia) in April recovered September at Pskov (north-west USSR) and Brescia (Italy); another ringed Belgium in August recovered the following March at Tafilalet (Algeria). Throughout Tunisia, more regular in spring than autumn (Thomsen and Jacobsen 1979); situation apparently similar in Libya, both near coast and at inland oases, though some bias in observation intensity perhaps responsible (Guichard 1957; Cramp and Conder 1970; Hogg 1974; Bundy 1976). Data thus suggest that higher proportion of birds overfly Mediterranean area in autumn than in spring. Nile valley and Rift Valley probably form flyway; certainly evidence that Lake Rudolf does so in spring (Fry *et al.* 1974).

Leaves Pechora river area of north European USSR September to early October; in central regions, main exodus in last third of September but as late as October from Crimea; departure from Altai mountains commences second half of August; straggler recorded in Tadzhikistan at end of November. (Dementiev and Gladkov 1954*a*.) Autumn passage through Switzerland occurs late July to late October, mainly late August to mid-September (Glutz von Blotzheim 1962), with movement through mountain passes mid-August to late October, once including massive peak of over 2000 birds on 3 September (Crousaz 1966). At Gibraltar, Tellería (1981) reported visible diurnal autumn exodus from end of August to late September with continuance on minor scale until 3rd week of October; no pattern of movement within the day. In Malta, autumn passage occurs late August to early November, with stragglers through to December (Sultana and Gauci 1982). In Cyprus, autumn passage mid-September to October, with occasional birds in November (Flint and Stewart 1983). Movement through Tunisia spans mid-September to mid-October (Thomsen and Jacobsen 1979). Visible autumn arrival recorded near Suez (Egypt) 4 September–5 November, peaking mid-September (Bijlsma 1982). In eastern Saudi Arabia, passage occurs September–November, peaking late September and early October (G Bundy). In Oman, regular on passage August–November (Gallagher and Woodcock 1980). In Iraq, uncommon in spring and autumn, with no evidence of wintering (Marchant 1963*a*). Occurs northern Iran late August to early April (Passburg 1959; Feeny *et al.* 1968).

In African winter quarters, present November–March

in Malawi (Benson and Benson 1977), numerous late October or early November to mid-March in Zimbabwe (Irwin 1981), and recorded in Eritrea (Ethiopia) 1 September–14 May (Smith 1957) and in Zaïre late September to mid-April (Lippens and Wille 1976). Occurs in Nigeria late September to early May (Elgood *et al.* 1966; Elgood 1982) and in Sierra Leone late October to mid-April; northward passage March–April, with stragglers remaining until early May (G D Field). Passage through Chad September–October and April(–May) (Salvan 1967–9). Movement in Tunisia occurs from late February to early May peaking mid-April (Thomsen and Jacobsen 1979). Passes through Malta early March to late May, mostly April (Sultana and Gauci 1982), and through Cyprus mid-March to mid-May, mainly late March to April (Flint and Stewart 1983). Further east, occurs in Oman February–April(–May) (Gallagher and Woodcock 1980), in eastern Saudi Arabia March to early May (G Bundy), and in Jordan April–May (Hollom 1959; Wallace 1984b). In Ukraine and Tadzhikistan (USSR), arrival starts late March and continues until late May, and movement through northern Caucasus occurs early April to early May (Dementiev and Gladkov 1954a). In Switzerland, passage takes place late March to mid-May, with main arrivals starting in early April and peaking in second half of month (Glutz von Blotzheim 1962).

Common on passage through most of Pakistan and western Himalayas from end of August to October and from mid-March to early May, and present in winter quarters mid-September to mid-April. *A. t. haringtoni* present in wintering areas September–April, autumn passage early September to mid-October and spring return 20 March to mid-April (Ali and Ripley 1973b). Vagrants recorded as far east as Japan (Sonobe 1982) and Alaska, June (Roberson 1980).   JHE, DJB

Food. Chiefly insects with some plant material taken in autumn and winter. Food taken mostly from ground, low herbage, and leaf litter, more rarely from twigs, branches, tree trunks, and stumps (Dementiev and Gladkov 1954a; Neufeldt 1956; Inozemtsev 1962; Ptushenko and Inozemtsev 1968); also from soil, especially during spring and autumn (Korenberg *et al.* 1972), and in summer from canopy (Averin and Ganya 1970; Van Hecke 1979b) where it walks along branches (Forbes-Watson 1983). Occasionally takes insects after short aerial pursuit from ground (Glutz von Blotzheim 1962; Van Hecke 1979b) and will flutter in front of hanging prey, e.g. small caterpillars (Van Hecke 1979b; Forbes-Watson 1983). Recorded drinking water droplets from buds and leaves (Van Hecke 1979b).

The following recorded in diet in west Palearctic. Invertebrates: grasshoppers, etc. (Orthoptera: Gryllidae, Tettigoniidae, Acrididae, Tetrigidae), earwigs (Dermaptera), bugs (Hemiptera: Cydnidae, Pentatomidae, Coreidae, Tingidae, Nabiidae, Miridae, Corixidae, Cica-

didae, Cercopidae, Cicadellidae, Psyllidae, Aphididae), lacewings and snake flies (Neuroptera), butterflies and moths (Lepidoptera: Lycaenidae, Tortricidae, Tineidae, Noctuidae, Lasiocampidae, Geometridae), caddisflies (Trichoptera), flies (Diptera: Tipulidae, Culicidae, Simuliidae, Rhagionidae, Asilidae, Muscidae), Hymenoptera (sawflies Tenthredinidae, Ichneumonoidea, Chalcidae, ants Formicidae), and beetles (Coleoptera: Carabidae, Staphylinidae, Buprestidae, Byrrhidae, Geotrupidae, Scarabeidae, Elateridae, Tenebrionidae, Coccinellidae, Cerambycidae, Chrysomelidae, Bruchidae, Curculionidae, Scolytidae), spiders (Araneae), harvestmen (Opiliones), millipedes (Julidae), and snails (Gastropoda). Plant food includes fruit of bilberry *Vaccinium myrtillus*, cranberry *V. oxycoccus*, cowberry *V. vitis-idaea* (also flowers), and elderberry *Sambucus nigra*; seeds of pine *Pinus*, birch *Betula*, violets *Viola*, chickweed wintergreen *Trientalis*, cow-wheat *Melampyrum*, Labiatae, sedges *Carex*, and grasses (Gramineae); buds of aspen *Populus tremula* and needles of spruce *Picea*. (Schuster 1930; Steinfatt 1941b; Tarashchuk 1953; Taylor 1953; Dementiev and Gladkov 1954a; Neufeldt 1956, 1961; Khokhlova 1960; Inozemtsev 1962; Ptushenko and Inozemtsev 1968; Korenberg *et al.* 1972; Popov 1978; Van Hecke 1979b; Sultana and Gauci 1982; Mal'chevski and Pukinski 1983.) Largest study done in summer in Tatarskaya ASSR (west of southern Urals, USSR) where 60 stomachs all contained some animal food: 75·0% contained Curculionidae, 41·7% bugs, 36·8% Chrysomelidae, 33·3% caterpillars, 18·3% Elateridae, 18·3% Orthoptera, 13·3% spiders, 10·0% Carabidae, 10·0% Scarabaeidae, 8·3% Staphylinidae, 8·3% Tenebrionidae larvae, 8·3% ants, 8·3% plant seeds, 5·0% small flies, 3·3% mosquitoes, 3·3% snails, and 1·7% Buprestidae (Popov 1978). In southern Kareliya (north-west USSR), 20 stomachs taken during breeding season all contained some invertebrates; 35% ants, 30% spiders, 25% unidentified beetles, 20% Chrysomelidae, 15% Coccinellidae, 15% bugs, 10% Curculionidae, 5% Byrrhidae, 5% Cerambycidae, 5% Carabidae, 5% adult moths, 5% mosquitoes, 5% other flies, and 15% plant food (Neufeldt 1961). In southern Ukraine (USSR), 14 stomachs from breeding season contained 296 items: 50% (by number) beetles (including 34·1% Curculionidae), 31·9% bugs (including 13·2% Nabidae and 12·9% Pentatomidae), 11·5% Hymenoptera (including 7·8% ants), 2·1% moths (1·4% larvae, 0·7% adults), 2·1% spiders, 1·3% millipedes, 0·7% Orthoptera, and 0·7% caddisflies (Tarashchuk 1953). Also in southern Ukraine, 6 stomachs from spring passage migrants contained 39·5% (by number) Curculionidae, 15·8% bugs, 11·8% Chrysomelidae, 10·5% moth caterpillars, 9·2% Carabidae, 3·9% spiders, 2·6% Hymenoptera, 1·3% Staphylinidae, 1·3% Tenebrionidae, 1·3% Cerambycidae, 1·3% Scolytidae, and 1·3% grasshoppers. In autumn, 16 stomachs contained 38·7% ants, 26·4% beetles, 23·0% bugs, and 11·9%♂ grasshoppers; seeds and elderberries present in

4 stomachs (Khokhlova 1960). In Moldavia (USSR), 12 stomachs, April–October, contained 56 items: 48·2% Curculionidae, 16·1% caterpillars, 14·3% Elateridae, 8·9% Buprestidae, 7·1% flies, and 5·4% Hymenoptera; also remains of Carabidae and, in 2 stomachs, plant seeds (Averin and Ganya 1970). Extralimitally, of 73 stomachs from Kirgiziya (south-central USSR), 57·5% contained Curculionidae, 21·9% Carabidae, 12·3% Scarabaeidae, 10·9% unidentified beetles, 10·9% bugs, 8·2% ants, 6·8% caterpillars, 5·5% Chrysomelidae, 5·5% flies, 5·5% grasshoppers, 4·1% Elateridae, 2·7% other Hymenoptera, 2·7% snails, 2·7% plants seeds, 1·4% Histeridae, 1·4% Coccinellidae, 1·4% Buprestidae, 1·4% Tenebrionidae, and 1·4% spiders (Pek and Fedyanina 1961); see also Kovshar' (1979).

Winter diet in West Africa includes grasshoppers, beetles, and millet grains *Panicum* (Bannerman 1951), and in Zimbabwe mainly beetles and their larvae (Curculionidae, Chrysomelidae); also moths, bugs, and termites (Isoptera) (Swynnerton 1907; Borrett 1973; Irwin 1978).

Nestlings fed exclusively on invertebrates (Steinfatt 1941b; Mal'chevski 1959; Neufeldt 1961), mainly soft items, e.g. flies, caterpillars, spiders. Few beetles fed to young, despite being major constituent of adult's diet (Inozemtsev 1962). In Voronezh region (USSR), 166 items brought to 36 nestlings in 52 visits, May–July, consisted of 37·3% flies (30·1% Muscidae), 21·1% moths and caterpillars, 18·7% bugs (18·1% aphids), 12·0% spiders, 3·0% grasshoppers, 2·4% snake flies, 2·4% beetles, 1·8% snails, 0·6% ants and 0·6% millipedes (Neufeldt 1956; see also Mal'chevski 1959). In Moscow region (USSR), 96 items from nestlings 4–8 days old in 12 nests comprised 45·8% (by number) sawfly larvae, 13·5% flies, 11·5% spiders, 11·5% moths (9·4% larvae, 2·1% adults), 9·4% bugs (5·2% cicadas, 4·2% other nymphs), 2·1% harvestmen, 2·1% snails, and 4·2% unidentified (Inozemtsev 1962). In Rominter Heide (Poland/USSR), 185 items brought to 4 young 2–9 days old, comprised 63·2% small grasshoppers, 5·4% winged insects, 3·2% caterpillars, 2·2% adult moths, 2·2% flies, 1·1% spiders, 0·5% beetles, and 22·2% unidentified (Steinfatt 1941b). In southern Kareliya, 13 collar-samples all contained animal food; 53·2% contained spiders. 38·0% mosquitoes, 38·0% moth caterpillars, 30·4% beetles, 22·8% other flies, 22·8% adult moths, 22·8% bugs, and 15·3% ants (Neufeldt 1961). Young given 2–7 feeds per hr, more frequent as they grow older; feeding very irregular when young (Steinfatt 1941b; Neufeldt 1956; Inozemtsev 1962; Ptushenko and Inozemtsev 1968). For nest of 5 young 6 days old, average for 18 hrs was 4·9 visits per hr; at 7 days old, 5·3 visits per hr; feeding rate decreased by 30–35% during rain (Ptushenko and Inozemtsev 1968). PJE

Social pattern and behaviour. Most detailed study at Kalmthout (northern Belgium) by Van Hecke (1979b). Account

includes data on extralimital *schlueteri* from study in Tien Shan (USSR) by Kovshar' (1979).

1. Often solitary outside breeding season, though small flocks and sometimes larger concentrations occur for migration, feeding, and roosting (see, e.g., Sultana and Gauci 1982 for Malta). Where birds congregate on passage, individuals show no tendency to associate closely (Steinfatt 1941b). In Leningrad and Moscow regions (USSR), flocks (up to 10) formed from mid- or late July (Ptushenko and Inozemtsev 1968; Mal'chevski and Pukinski 1983). In Kazakhstan (USSR), such flocks probably family parties; up to 10–15 occur on migration, rarely 50–100 in south (Dolgushin *et al.* 1970). For further reports of flocks prior to and during migration, see Steinfatt (1941b), Labitte (1952b), Dementiev and Gladkov (1954a), and Gatter (1976). In Antwerpen (Belgium), typically migrates singly or 2–5 together in spring, slightly larger flocks (rarely exceeding 10–12) recorded in autumn (Spaepen 1953). Dispersion similar in African winter quarters (Spaepen 1953; Serle 1957; L Grimes). In Westfalen (West Germany), roosts hold up to 100 birds, July–September; initially 85–95% juveniles, more adults from August (K-H Loske). Feeding assemblages of c. 40 recorded in Orbe valley (Switzerland), early May (Glayre 1980), and in Zaïre, 19 October (Verheyen 1956a). In Rominter Heide (Poland/USSR), shows no tendency to associate with other species (Steinfatt 1941b); in Ghana, winter, often feeds, also recorded roosting, with Yellow Wagtail *Motacilla flava* (L Grimes). BONDS. Monogamous mating system (Dementiev and Gladkov 1954a), though occasional polygamy mentioned (without details) by Norup (1963). Study in Westfalen suggested surplus of ♂♂, with c. 20–30% remaining unpaired each year (K-H Loske). At Kalmthout, marked pair remained together for 2 broods, ♀ then paired with unringed ♂ for 3rd brood; different pair maintained bond for 3 broods (Van Hecke 1979a). In Tien Shan, 2 pairs broke up after rearing 1st brood. Pair-bond probably does not normally persist beyond 1 season: none of 5 marked pairs bred together in following year (Kovshar' 1979). Observations at Bulawayo (Zimbabwe) indicate possibility of prolongation of pair-bond outside breeding season (Irwin 1978), though ♂♂ generally arrive on breeding grounds a few days before ♀♀ (Steinfatt 1941b), older ♂♂ (3–5 years) first (K-H Loske). Further comments in Van Hecke (1979b) and Kovshar' (1979) suggest pair formation takes place only after arrival on breeding grounds. For probable interbreeding with Meadow Pipit *A. pratensis*, see Rowan (1919). Young normally fed by both sexes, though some ♂♂ may do little or none and only guard nest. One ♂ fed young alone when ♀ injured (Van Hecke 1979a, b). In Tien Shan, *schlueteri* ♂♂ averaged 41% of feeding when young 1–5 days old, 46% at 6–10 days (Kovshar' 1981). ♂ having lost own brood fed 3 strange young 17 days old (Van Hecke 1979b). Pair recorded feeding young and removing faecal sacs at nest of Skylark *Alauda arvensis* during parents' absence (Cockbain 1958). Both adults also recorded near young out of nest; young probably independent c. 14 days after leaving nest (Steinfatt 1941b); according to Ptushenko and Inozemtsev (1968), fed for c. 9–10 days after leaving. Fully independent brood members may remain together for considerable time: e.g. 3 siblings controlled at roost 63 days after leaving nest (K-H Loske). See also Relations within Family Group (below). BREEDING DISPERSION. Solitary and territorial. Neighbourhood groups also reported, e.g. in Leningrad region (Mal'chevski and Pukinski 1983); probably due at least in part to specialized habitat requirements. In Moscow region, nests c. 120–200 m apart (Dementiev and Gladkov 1954a); in Tien Shan, c. 100 m apart, though 47 and 50 m recorded (Kovshar' 1979). In Bonn area (West Germany),

2 nests *c.* 80-100 m apart, 3rd *c.* 600 m from these (Lehmann 1951). Shortly after hatching or after young leave nest, territory (or part of it) may be taken over by neighbour. In one case, successful pair replaced by new pair (ringed ♀ not seen previously that season). Such shifts can lead to nests being sited very close together: e.g. 8, 13, and (built when young in neighbour's nest) 20 m apart (Van Hecke 1979*b*). Pairs also recorded nesting close together in suitable habitat of Eure-et-Loir, France (Labitte 1952*b*). Territory selected and occupied by ♂ (Steinfatt 1941*b*). In Moscow region, 'nesting territories' (see below) said to be small—usually *c.* 2800-4200 m² (rarely up to 6400 m²; borders not contiguous (Ptushenko and Inozemtsev 1968; also Geptner 1958). Most detailed information on territory in Belgian study: after arrival, birds use area of *c.* 3-4 ha; this may contract as other ♂♂ arrive. Traditionally-used neutral feeding areas (up to 0·5 ha) lie between territories and may be up to *c.* 500 m from one. Before and during incubation, birds feed in or outside (up to *c.* 300 m) territory; forage for young mostly near nest—up to *c.* 150 m, exceptionally 400 m away (Van Hecke 1979*b*). For further data on area used for feeding, see Steinfatt (1941*b*), Dementiev and Gladkov (1954*a*), Inozemtsev (1962), and Kovshar' (1979). At Kalmthout, 2nd-brood nests averaged 14·1 m (3·7-20, *n* = 13) from 1st, or 161 m (132-214, *n* = 12); where change of ♂ involved, 263 m and 370 m (Van Hecke 1979*a*, *b*). In England, once *c.* 3 m (Tracy 1925). In Eure-et-Loir, shortest distance 5-6 m, though same nest once used for 2nd brood; replacement nests also normally built close to 1st (Labitte 1952*b*). Recorded (6 cases) using same hollow for 2nd brood when 1st nest removed (Van Hecke 1979*b*). Habitat requirements clearly have major influence on density of breeding pairs: trees (etc.) important for use as song-posts, look-outs, and as perches before flying to nest; open areas for feeding also important (Lehmann 1951; Sharrock 1976). In Mecklenburg (East Germany), after some clearance in old beech *Fagus* wood (24 ha), increase from (4·2 to 42 pairs per km² over 10 years (Klafs and Stübs 1977). In Britain, census plots in semi-natural habitat hold up to 20-30 pairs per km²; newly planted conifer plantation with some taller trees left standing held 67 pairs per km². Overall, probably *c.* 25-30 pairs per occupied 10-km square (0·25-0·3 pairs per km²) (Sharrock 1976). In Sussex (southern England), over 11 territories per km² in *c.* 2·6 km² of commons, heaths, and young plantations (Hughes 1972). Along fallow slopes of Eure valley (France) *c.*15 pairs per km² over area of *c.* 40 ha (Labitte 1952*b*). In Swiss Mittelland, 54·7 ha of traditionally farmed land held average (over 3 years) 126 pairs per km²; probably lower on intensively farmed land of same area (Fuchs 1979); for further Swiss data, see Schifferli *et al.* (1982). On Kaiserstuhl (West Germany) much higher densities (up to 100 pairs per km²) recorded in larger consolidated vineyards than in older smaller ones (see Blankenagel and Seitz 1983 for further details of habitat selection). Average densities in various European countries (mainly West and East Germany) summarized by habitat type: clearings and reafforested areas 83 ♂♂ per km²; pure oak *Quercus* stands 40, semi-dry grassland, fallow and waste land and mixed orchards/meadows 38, riverine woodland 31, pine *Pinus* forest 24, beech 20, oak and hornbeam *Carpinus* 15, and fields with scattered tree clumps 9 ♂♂ per km² (Loske 1985). For further detailed data on densities in various habitats of Mecklenburg, see Klafs and Stübs (1977); for Brandenburg (East Germany), see Rutschke (1983). In Uppland (Sweden), survey of 8·67 km² revealed 5·3 pairs per km² of forest, 10 on pine moor, 26·5 on clearings, overall 5 pairs per km² (Olsson 1947). In Leningrad region, in open and moderately damp to dry pinewoods, up to 40-50 pairs per km²

(Mal'chevski and Pukinski 1983). In Moscow region, 9-12 pairs per km² of suitable open coniferous woods, 15-20 in mixed woods, over 30 pairs per km² of broad-leaved woods (Ptushenko and Inozemtsev 1968). Birds often return to birthplace to breed and also show site-fidelity in subsequent years (Spaepen 1953). In Westfalen, 11 1-year-old ♂♂ retrapped 150-1000 m from natal area, average 429 m; once, 7·2 km. Of 50 marked adult ♂♂ recovered in later years, 35 were found at exactly same site; other 15 moved average 251·7 (50-1300) m (K-H Loske; see also Kovshar' 1979). ROOSTING. Nocturnal and, outside breeding season, communal. Normally on ground in thick grass or other vegetation: e.g. migrants in potato and beetroot fields (Spaepen 1953; K-H Loske). Suggested by Van Hecke (1979*b*) that non-breeding adults in summer probably roost in trees, possibly also on ground. In Westfalen, birds roost mainly in clearings from early July to mid-September. Arrive 1-2 hrs before sunset; communal preening precedes settling on ground for roosting at dusk. Juveniles use roost on average 13 (3-26) days (K-H Loske). In Zambia and Kenya, winter, will roost in dry tussocky grass 1-1·5 m high by swamp or in forest clearing (P B Taylor). In Zaïre, birds assemble in small flocks at nightfall for roosting flight (Verheyen 1956*a*). In Leningrad region, birds leaving roost tend to feed in fields for *c.* 1 hr before moving on (Mal'chevski and Pukinski 1983). In Tien Shan, awakening (as indicated by first song or call) as follows: 05.00 hrs 10 May, 04.28 hrs 20 May and 6 June, 04.27 hrs 21 June, 04.42 hrs 6 July, 05.07 hrs 20 July (Kovshar' 1981). For further details on activity rhythms, see Song-display (below). Water-bathing recorded 5 times in Belgian study (Van Hecke 1979*b*). Anting recorded (Poulsen 1956).

2. Generally confiding at all times according to Spaepen (1953). However, ♀ especially can be very secretive on breeding grounds (Kovshar' 1979) and reports from winter quarters vary from confiding (Vrydagh 1952) to furtive and shy (Marchant 1942). Usually flies into tree when disturbed (Witherby *et al.* 1938*a*). 2 birds recorded diving down into wood to escape from Hobby *Falco subbuteo*. Sometimes gives Distress-calls (see 6 in Voice) when trapped and other bird of pair may then approach. ♀ generally gives warning-calls (see 3 in Voice) when ♂ trapped; in reverse situation, ♂ gave only few such calls but began to sing. ♀ generally quiet when handled, ♂ very restless and gives quiet Distress-calls (Van Hecke 1979*b*). FLOCK BEHAVIOUR. Migrating flocks in Belgium loose-knit and do not perform combined manoeuvres; birds give call 2 constantly however (Spaepen 1953). Larger flocks (of over 50) may be close-packed (Gatter 1976). SONG-DISPLAY. Song of ♂ (see 1 in Voice) given from perch—typically top of isolated tree in clearing or protruding branch of tree at edge of wood; also from ground, and in characteristic Song-flight. ♂ sings generally lower down in relatively strong wind (which normally reduces song) or from ground if feeding; Subsong delivered if both factors combined. Bird singing from perch (Fig A) remarkably still and exhibits

A

no special postures; little or no ruffling of throat feathers (Bergmann 1977c; Van Hecke 1979b; also Von der Decken 1972). Song-flight consists of c. 60° ascent (from perch or ground) to c. 15 m (Labitte 1952b) or up to c. 30–35 m above take-off point (Witherby et al. 1938a), followed by gliding descent (Fig B) with wings spread to form parachute, tail raised

B

and spread, and legs dangling; lands on same or adjacent perch, sometimes on ground. Usually reported to start singing near peak of ascent and continue during descent, though said by Dolgushin et al. (1970) to deliver 1st part of song from perch or at take-off, middle part during ascent, and conclusion during descent and landing. Longer burst of song normally given in flight than from perch and up to 3 phrases often delivered after landing. Performance may be repeated shortly afterwards. (Witherby et al. 1938a; Labitte 1952b; Bergmann 1977c; Bergmann and Helb 1982.) Song-flight of schlueteri (often breeding on mountain slopes) consists of arcing flight out from slope followed by return to perch; sings in shorter bursts and less persistently than nominate trivialis (Dementiev and Gladkov 1954a; Dolgushin et al. 1970). Bird will sing on return to territory after feeding, but only exceptionally sings closer than c. 10 m to nest (Van Hecke 1979b). According to Labitte (1952b), rarely moves beyond territory in Song-flight, though generally sings further away from nest during rearing of young. Song-display infectious amongst neighbours (Mal'chevski and Pukinski 1983). In Moscow region, early June, usually 8–10 times more song from perch than in flight. Most song given in 1st hour after awakening (from c. 03.00 or 04.00 hrs); 1st Song-flight performed 40–60 min after 1st song from perch, frequency of Song-flights increasing after c. 05.00 hrs (most 06.00–08.00 hrs). General reduction in activity (including song) c. 07.00–12.00 hrs, and renewed peak 15.00–17.00 hrs. In dull weather, more even pattern of activity (but rain usually inhibits Song-flights). Over 2 hrs spent singing per day. One ♂, on 7 June, spent 10 hrs 34 min on perch (sang for 2 hrs 24 min), 5 hrs 19 min on ground (Geptner 1958). In Tien Shan, during peak period (mid-May to end of June) will sing all day (c. 16 hrs); earliest 04.28 hrs, latest 20.20 hrs; at maximum 06.00–07.00 hrs, minimum 15.00–19.00 hrs. Migrants will sing—generally at low intensity and from perch (Witherby et al. 1938a; Steinfatt 1941b). In Britain, central Europe, and USSR, occurs from arrival to mid- or end of July (Witherby et al. 1938a; Labitte 1952b; Dementiev and Gladkov 1954a; Van Hecke 1979b; Mal'chevski and Pukinski 1983). In Rominter Heide, main song-period for 1st brood up to c. 25 May, then renewed from c. 10 June (Steinfatt 1941b). At Kalmthout, song also given late August to mid-September (Van Hecke 1979b). Little or no song (low intensity) given during nestling phase (Labitte 1952b), but sometimes more frequent in last few days before young leave nest, and increases abruptly after loss of eggs or small young (Van Hecke 1979b). In Tien Shan, schlueteri sings little or not at all during nest-building, more for laying and incubation, declining after hatching (Kovshar' 1979, 1981). ANTAGONISTIC BEHAVIOUR. (1) General. In Eure-et-Loir, ♂♂

take up territory on average c. 1 week after arrival (Labitte 1952b). Late arrival of 1-year-olds sometimes means some territories not occupied until June (K-H Loske). Birds may be wide-ranging initially (Van Hecke 1979b: see Breeding Dispersion, above). (2) Threat and fighting. 2–3 ♂♂ not infrequently sing close together (Mal'chevski and Pukinski 1983); may lead to chase and eviction of intruder (Labitte 1952b). Most detailed description of territory defence in Belgian study. Territory-owner seeing rival on boundary begins to sing more vigorously and clearly; may give short phrases interspersed with call 2. If other trespasses, owner performs Song-flight then chases intruder to limit if necessary. If intruder flies over giving call 2, territory-owner feeding on ground usually gives only brief song; otherwise may follow intruder for long stretch (sometimes in high undulating flight) while singing and then return to perch after eviction. Renewed intrusion immediately elicits vigorous threat-flights by owner. Aerial Song-duels also occur: may be prolonged, each bird attempting to fly above the other, and direct collisions recorded. If after several attempts intruder returns yet again, territory-owner approaches, often with excited call 2, and fight may develop, birds making contact with feet, bill, and wings. ♀♀ and strange young attempting to feed in another's territory are also evicted. Early in season, territory-owner approaches rapidly (threat-flight) on hearing playback of song; becomes increasingly restless, sings constantly, sometimes performing display-flight (no details) or landing on loudspeaker. From pair-formation onwards, some ♂♂ no longer react to playback. Noted singing simultaneously with Woodlark Lullula arborea from same perch, but A. pratensis and Linnet Carduelis cannabina driven away (Van Hecke 1979b). HETEROSEXUAL BEHAVIOUR. (1) General. Call 2 given in April is sure sign that pair-formation completed (Van Hecke 1979b). In Tien Shan (schlueteri), most pairs probably formed by 10 May (Kovshar' 1979). (2) Pair-bonding behaviour. Little information. ♀ mostly on ground while ♂ singing; ♂ flies down to her occasionally (Labitte 1952b). If ♀ attempts to leave territory when ♂ singing, ♂ chases her with short, fast turns. If chase along territory boundary, neighbouring pair may join in: ♂♂ pursue ♀♀ low over ground and give call 2 constantly, also sing briefly during pauses (Van Hecke 1979b). According to Witherby et al. (1938a), 'sexual flight' much as in finches (Fringillidae) and buntings (Emberizidae). (3) Courtship-feeding. ♂ sometimes feeds ♀ on nest (Witherby et al. 1938a; Steinfatt 1941b; Van Hecke 1979b). (4) Mating. In the few cases recorded, always in tree. In 2 cases, ♀ on ground immediately beforehand. ♂ may approach silently, sing, or give call 2 (Van Hecke 1979b). In Tien Shan, recorded once in tree c. 15 m from nest under construction, and once in early morning in tree c. 30 m from nest containing eggs 4 days from hatching; ♀ resumed incubation afterwards (Kovshar' 1979). (5) Nest-site selection. Apparently by ♀ who is accompanied by ♂ (Ptushenko

Norman Arlott.

and Inozemtsev 1968; Kovshar' 1979), also later when building (Steinfatt 1941*b*). Birds recorded carrying nest-material long before building started (Van Hecke 1979*b*); this perhaps part of heterosexual display, as described for *A. pratensis* (see p. 374). (6) Behaviour at nest. In French study, 3 cases of laying starting before nest completed (Labitte 1952*b*). ♀ normally sits facing entrance (Van Hecke 1979*b*). Twice recorded dashing off nest to catch moth (Lepidoptera). At Kalmthout, when undisturbed, ♀ normally lands in immediate vicinity of nest, fluttering briefly before landing if nest in tall grass. On departure, usually flies to tree near nest, shakes feathers and/or defecates (Van Hecke 1979*b*). In other studies, reported always to land several metres from nest and to approach on foot, also walking away some distance before flying up (Lehmann 1951; Labitte 1952*b*). From observations at 3 nests, Lehmann (1951) described bird always landing in tree near nest and waiting average *c.* 15 min, mate joining it there after minimum *c.* 7 min; hovering over nest occurred; 1 bird (possibly ♀) then moved towards nest, immediately followed by mate, birds staying only *c.* 80 s. Despite observer's claim to contrary, behaviour doubtless influenced at least to some extent by disturbance. When ♀ leaves, ♂ joins her for feeding immediately or soon afterwards. May also escort her back to nest then fly off to sing before ♀ settled, wait silently for her to settle, or sing briefly while waiting (Van Hecke 1979*b*). ♀ on nest apparently excited by presence of ♂, also when shuffling about on eggs, may give call 4 (Steinfatt 1941*b*). RELATIONS WITHIN FAMILY GROUP. Young may hatch over 2 days. Brooded for periods of *c.* 13–16 min. Still brooded at night when *c.* 3–4 days old, but no longer at 7–8 days. ♀ brooding young when ♂ approaches with food may give call 4 (Steinfatt 1941*b*). Young do not necessarily all face where parents normally approach (Lehmann 1951). Fed bill-to-bill (Witherby *et al.* 1938*a*). Faecal sacs initially eaten by parents, later carried away (Steinfatt 1941*b*); at one nest, deposited regularly on same branch (Lehmann 1951). ♂ noted singing shortly before feeding quite well-grown nestlings or young out of nest, also ascending as if for Song-flight with food in bill, but not singing. Highly likely that 2nd nest started before 1st-brood young leave; overlap proved in one case (Van Hecke 1979*a, b*). Young leave before able to fly (but see Anti-predator Responses of Young, below) at *c.* 9–13 days (Labitte 1952*b*; Dementiev and Gladkov 1954*a*; Ptushenko and Inozemtsev 1968; Kovshar' 1981), earlier (7–8 days) if disturbed; e.g. on hearing distress-calls from sibling or if parent trapped (Van Hecke 1979*b*; K-H Loske). Often, 1st leaves at 10 days,

PLATE 31 (*facing*).
Motacilla flava Yellow Wagtail (p. 413). *M. f. flavissima*: **1** ad ♂ breeding, **2** ad ♀ breeding, **3** 1st ad ♂ non-breeding (1st winter). Nominate *flava*: **4** ad ♂ breeding, **5** ad ♀ breeding, **6** 1st ad ♂ non-breeding (1st winter). *M. f. lutea*: **7** ad ♂ breeding. *M. f. feldegg*: **8** ad ♂ breeding, **9** ad ♀ breeding, **10** 1st ad ♂ non-breeding (1st winter). *M. f. thunbergi*: **11** ad ♂ breeding.
Motacilla citreola citreola Citrine Wagtail (p. 433): **12** ad ♂ breeding, **13** ad ♀ breeding, **14** 1st ad ♂ and ♀ non-breeding (1st winter).
Motacilla cinerea cinerea Grey Wagtail (p. 442): **15** ad ♂ breeding, **16** 1st ad ♂ and ♀ non-breeding (1st winter).
Motacilla alba Pied Wagtail and White Wagtail (p. 454). Nominate *alba* (White Wagtail): **17** ad ♂ breeding, **18** ad ♀ breeding, **19** 1st ad ♂ and ♀ non-breeding (1st winter). *M. a. yarrellii* (Pied Wagtail): **20** ad ♂ breeding, **21** ad ♀ breeding. *M. a. personata* (White Wagtail): **22** ad ♂ breeding.
Motacilla aguimp vidua African Pied Wagtail (p. 471): **23** ad ♂ breeding, **24** ad ♀ breeding. (NA)

last 1–3 days later; normally do not return (K-H Loske). Self-preen at *c.* 12 days (Van Hecke 1979*b*). Remain near nest or move up to *c.* 50 m 1 day after leaving; disperse and give food-calls (Steinfatt 1941*b*). In one case, *c.* 600 m away by 16 days old. Fly as well as parents at 23 days, but still spend much time in cover (Van Hecke 1979*b*). Well-grown 1st-brood young recorded approaching parents and calling; adults apparently indifferent and did not chase them away; young also seen near nest containing 2nd brood (Labitte 1952*b*). See also Bonds (above). ANTI-PREDATOR RESPONSES OF YOUNG. Probably silent for most part when parents absent or when danger threatens. When brood of 5 were touched and 1 called, others immediately abandoned nest and moved skilfully into cover, calling quietly. Older nestlings flew into tree when disturbed (Labitte 1952*b*). Skilful at hiding out of nest (Steinfatt 1941*b*). May give distress-call when handled (Van Hecke 1979*b*). PARENTAL ANTI-PREDATOR STRATEGIES. (1) Passive measures. ♀ unlikely to approach nest when man *c.* 40–60 m away; tends to wait on open perch if observer exposed, and usually calls constantly. ♂ on look-out gives call 3a when disturbed, also apparently to warn ♀ (Labitte 1952*b*; also Steinfatt 1941*b*). Birds sometimes silent even when intruder *c.* 5 m from nest, ♀ sometimes leaves when observer *c.* 0·2–3 m from nest, but often sits very tightly, especially when brooding very small young, not leaving even when vegetation over nest pulled away; once stayed even when touched with stick (Van Hecke 1979*b*). One ♀ drew grass across to screen nest when watched from hide (Steinfatt 1941*b*). (2) Active measures: against birds. Carrion Crow *Corvus corone*, Magpie *Pica pica*, Jay *Garrulus glandarius*, and (less so) Red-backed Shrike *Lanius collurio* elicit vigorous alarm-calls (see 3 in Voice) during incubation, nestling phase, and when young out of nest (Van Hecke 1979*b*); may fall silent for *G. glandarius* (Von der Decken 1972). (3) Active measures: against man. ♀ prevented from approaching nest, apart from giving alarm-calls, may come close, land, and make as if to go to nest or try to lure observer by walking away from it. ♂, also giving alarm-calls, will approach and perch conspicuously (Labitte 1952*b*; Van Hecke 1979*b*). ♀ surprised on eggs or young may stay put, or fly or flutter off silently. In fluttering distraction-flight, tail widely spread; performance similar to *L. arborea* (Van Hecke 1979*b*). According to Tucker *et al.* (1948), bird flushed gently (by tapping vegetation) almost always performs distraction-lure display of disablement type, particularly if with young, but no such display recorded by Van Hecke (1979*b*). When young bird was handled and gave distress-call, adult immediately flew in and attempted distraction (Van Hecke 1979*b*). When young left nest in alarm and flew into tree (see above) ♀ immediately came close, landed, and performed various contortions, frantically beating wing against ground. Parents soon joined young in tree (Labitte 1952*b*). (4) Active measures: against other animals. Birds silent when weasel *Mustela nivalis* nearby (Van Hecke 1979*b*), but gave alarm-calls for cat (Labitte 1952*b*), at lower intensity for deer *Capreolus* (Van Hecke 1979*b*).

(Figs by I Lewington: A from photograph by R Siebrasse; B from drawing in Bergmann and Helb 1982.) MGW

Voice. Generally silent in winter quarters (Vrydagh 1952) and song not noted there (Bannerman 1936). For extra sonagrams, see Bjerke (1971), Bergmann (1977*d*), Bergmann and Helb (1982), and Wallschläger (1984).

CALLS OF ADULTS. (1) Song of ♂. Loud and of fine quality, excelling Meadow Pipit *A. pratensis* and Rock Pipit *A. spinoletta* (Witherby *et al.* 1938*a*). Each ♂ has

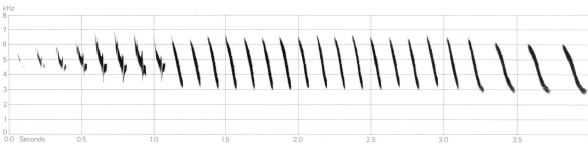

I A G Field England June 1963

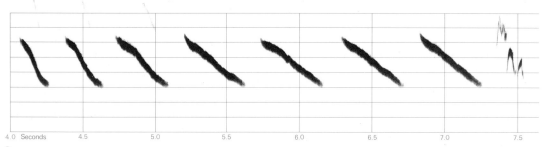

I *cont.*

repertoire of several song-types each comprising a series of contrasting segments; latter composed of repeated units—from *c.* 3-30 per segment—usually uniform but sometimes showing pronounced drift (gradation) (Fig I, segment 2); unit delivery rate from *c.* 2 per s (end of segment 2 in Fig I) to *c.* 26 per s in rattle (Fig II, segment 4). First segment tends to be constant within repertoire (Figs I and II). Song duration and delivery rate variable (J Hall-Craggs). Particularly striking are shrill, piercing, and musical disyllabic 'sIe' or 'seea' sounds like Canary *Serinus canaria* given typically towards

end of series; such sounds may rise or fall in pitch and are more fluting compared with other units (Witherby *et al.* 1938*a*; Bergmann 1977*d*; Bergmann and Helb 1982). Disturbed bird may give only truncated song (Bergmann and Helb 1982) and only 1-2 'see-you' units delivered by bird starting this segment rather low down, near ground, in Scottish birchwood (Fisher 1949). In recording of 3 songs given by 1 bird from perch, each starts with segment of 8 'teu' units (compare also Bergmann 1977*d*) at 7-8 per s and with slight accelerando. 1st song lasting 5·6 s is of 6 segment types: (i) 8 fast 'teu' units; (ii) 3

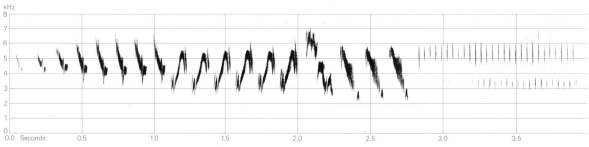

II A G Field England June 1963

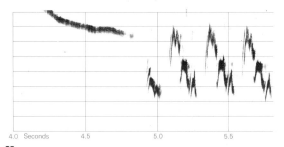

II *cont.*

slower descending 'pieu' or 'seeu' sounds; (iii-iv) short rattling and initially very fast 't-t-t-che che che che che' (presumably the bubbling trill likened to Crested Tit *Parus cristatus* by Witherby *et al.* 1938*a*); (v) 3 slowly delivered, drawn-out and rather plangent ascending 'sooeee' units; (vi) 4 'swip' sounds, also at fairly slow rate of delivery. Following after pause of *c.* 19 s, 2nd song (Fig I) comprises mainly simple units (drawn-out, descending 'seeoo' sounds prominent) and, at 7·2 s, longer than other 2 songs (mainly complex units). More complex 3rd song (Fig II) lasts 5·6 s and follows pause of *c.* 21 s: (i) after 'teu' units of 1st segment, (ii) 6 'che' followed by (iii) single 'swip', (iv) 3 descending and rather impure 'tsell' or 'tser' sounds, (v) rattle like Wren *Troglodytes troglodytes*, (vi) thin, extraordinarily high-pitched (starting at 9 kHz) and attenuated 'seee...', and (vii) final segment of 'swip' units (1st incomplete and slightly modified). In recording by P J Sellar, Sweden, series average 5·0 s long (2–8 s, *n*=6) and delivered at much faster rate with pauses 3·8 (3–5) s; 2 songs contain rattle and ensuing long high-pitched tone as described above (J Hall-Craggs, M G Wilson). At Irkutsk (Lake Baykal, USSR), bird gave songs lasting 6·8–10·5 s. Tendency noted in this population to give same segment several times in a song, but one bird gave at least 8 different segments of varying length and another bird, in East Germany, had 9 unit-types in song. For apparent use of 2 internal tympanic membranes producing typically 2-part units, see Wallschläger (1984). Norwegian study analysed 712 songs (from perch) of 7 birds: 20 different notes distinguished and 12 different song-types. Each bird had its own highly specific repertoire of song-types, apparently rather stable from day to day. 6 out of 7 birds used same introductory units through all songs. For further details, see Bjerke (1971). Song of *schlueteri* in Tien Shan (USSR) also varied, especially in ending (see above); in long bout, ♂ may alternate 3–5 variants (Kovshar' 1979). For further descriptions and renderings, see Voigt (1933), Nicholson and Koch (1936), and Dolgushin *et al.* (1970). Mimics other birds, including Chaffinch *Fringilla coelebs* (Labitte 1952*b*; Bergmann and Helb 1982; Wallschläger 1984; P J Sellar), *T. troglodytes* and Great Tit *P. major* (Briche 1962); see also Kovshar' (1979) for *schlueteri*. In Leningrad region (USSR), ♂♂ said to readily adopt individual pecularities of those first to arrive on breeding grounds, so that eventually all sing the same (Mal'chevski and Pukinski 1983). Marked dialects exist (Bergmann and Helb 1982). (2) Rasping hoarse 'teez' (Witherby *et al.* 1938*a*), nasal 'tsrieh' (Steinfatt 1941*b*; Van Hecke 1979*b*), 'psië' or 'psi' usually descending slightly in pitch and impure (Bergmann and Helb 1982). Recording (Fig III) also suggests hoarse 'tzeet' (J Hall-Craggs) or

buzzing, rather explosive 'beeez' or 'teez' (M G Wilson). Typically given in flight, notably by low-flying, diurnal migrants (Steinfatt 1941*b*), and at any time when flying some distance, more in autumn (Witherby *et al.* 1938*a*). Several calls sometimes given in succession at take-off (Bergmann and Helb 1982). Often given by ♂ during incubation period, between bouts of song, especially when mate nearby; by some ♀♀ on leaving nest or also at return, by others never. Noted once during feed and also associated with antagonistic encounters (Von der Decken 1972; Van Hecke 1979*b*). (3) Calls given mainly when disturbed. (a) Soft, rather metallic 'zip' (Witherby *et al.* 1938*a*), 'pit' or plaintive 'tip' (Labitte 1952*b*); also rendered 'tjit' (Van Hecke 1979*b*), sharp 'sritt' or 'zit' (Steinfatt 1941*b*; Bergmann and Helb 1982). Given persistently every 2–3 s by both sexes (also by migrants) and carrying up to *c.* 300 m (Labitte 1952*b*; Mal'chevski 1959; Bergmann and Helb 1982). Recording of bird disturbed near nest suggests 'tsit' or '(p)sip' at about 2 calls per s (J Hall-Craggs, M G Wilson: Fig IV), and similar 'sit' (no starting transient) in recording by V C Lewis apparently serves as contact-call to young during feed (P J Sellar). (b) High-pitched, quiet 'si si si...' given, sometimes as long series, when ground predator near nest (Van Hecke 1979*b*; Bergmann and Helb 1982). High-pitched short trill as high-intensity alarm and said to be like young Wryneck *Jynx torquilla* (Mal'chevski 1959) probably the same; short, sharp 'dzi' sounds given when flushed near nest (Steinfatt 1941*b*) may be this call or variant of call 2. Use of calls 3a and 3b detailed by Van Hecke (1979*b*): almost only call 3b given by ♂ before nest-building or laying, between bouts of song, or near feeding ♀; sometimes alternates with call 3a. From start of incubation, ♂ gives both alternately or only slow, muted variant of call 3a, this normally becoming faster and louder after hatching. Call 3b given by ♂ when ♀ brooding, kept long for ringing or when young dead in nest; given also by ♀ after hatching. Both calls used less frequently after young leave nest. (4) After return to nest, when ♂ nearby, ♀ may give quiet conversational 'pirrpipi pipipirr' audible only at close quarters (Steinfatt 1941*b*). (5) Quiet 'sriep sriep...' used, like call 4, as contact-call by ♀ to ♂; also once when ♀ came to feed young (Steinfatt 1941*b*). (6) Distress-call. Mentioned, without details, by Van Hecke (1979*b*); assumed to be a screeching sound. For contexts, see Social Pattern and Behaviour.

CALLS OF YOUNG. Cheeping food-calls of nestlings initially quiet; barely audible more than *c.* 2 m from nest (Labitte 1952*b*). In recording by V C Lewis, young being fed give thin squeals. Food- and contact-calls much louder from young out of nest; audible *c.* 150 m away (Van Hecke 1979*b*); single sharp 'siep' (Steinfatt 1941*b*); juvenile variant of adult call 2 with stressed 'r'—'tsRieh' (Van Hecke 1979*b*). In recording by D J Sutton, sharp 'pseet' or 'teep' (M G Wilson). Juveniles give adult call

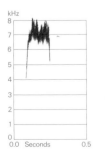

III S Carlsson Sweden
September 1983

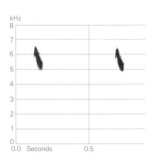

IV R W Genever
June 1966 England

3a to express alarm when near adults also calling thus. Distress-call presumably resembles that of adult (Labitte 1952*b*; Van Hecke 1979*b*). Quiet Subsong given by one of a brood of 4 *c*. 12 days after leaving nest (Took 1947).

MGW

Breeding. SEASON. North-west Europe: first eggs laid late April, with main laying season May-June; last young in nest August (Makatsch 1976; Van Hecke 1979*a*). SITE. On ground flat or sloping; in low cover, or more or less in the open. Nest: shallow depression holding substantial cup of dry grass leaves and stems, often with moss foundation, lined with finer grasses and hair. Mean inner dimensions at top, 7·9 × 6·7 cm (5·5-9·0 × 4·5-8·0), *n* = 38; mean depth 4·0 cm (2·8-6·0), *n* = 50 (K-H Loske). Building: by ♀ only, also making depression for nest (Van Hecke 1979*a*). EGGS. See Plate 81. Sub-elliptical, smooth and glossy; extremely variable in colour—often brown, grey, or reddish, but also pale blue, pink, green, or dark brown; may be evenly speckled, have dark zone at broad end, or be blotched, streaked, or spotted; some have black hair-streaks; a few are blue with no markings, or dark brown all over. 20·3 × 15·4 mm (18·0-22·5 × 14·0-16·5), *n* = 624; eggs from clutches of 4 significantly larger than from other clutch sizes; size varies with size of ♀ and through season (for details, see Van Hecke 1980). Calculated weight 2·44 g (Schönwetter 1979). Clutch: 2-6(-8). Of 288 clutches, Belgium: 2 eggs, 1%; 3, 10%; 4, 31%; 5, 51%, 6, 7%; mean 4·53. 1st clutches larger than 2nd. Of 134 1st clutches: 3 eggs, 1%; 4, 12%; 5, 78%; 6, 9%; mean 4·96. Of 57 2nd clutches: 3 eggs, 10%; 4, 63%; 5, 25%; 6, 2%; mean 4·18 (Van Hecke 1979*a*). Of 94 1st clutches, Westfalen (West Germany): 3 eggs, 5%; 4, 25%, 5, 57%; 6, 13%; mean 4·78 (K-H Loske). 1-2 broods, rarely 3. Replacements also laid. Mean 10 days (*n* = 56) between broods, with nest-building for 2nd occasionally overlapping with young of 1st still in nest (Van Hecke 1979*a*). Eggs laid daily. INCUBATION. 12-14 days. By ♀ only (Van Hecke 1979*b*). Normally leaves to feed every *c*. 45-60 min and absent for average 30 min. Over 2 days, incubation stints 2-71 min and 26-98 min, absence 1-49 and 16-48 min; 12-15 breaks from incubation per day (Steinfatt 1941). Absences also averaged *c*. 30 (15-55) min at Kalmthout (Van Hecke 1979*b*), but only a few min per hr in Tien Shan (Kovshar' 1979). YOUNG. Altricial and nidicolous. Cared for and fed by both parents (Van Hecke 1979*b*). FLEDGING TO MATURITY. Fledging period 12-14 days, but young normally leave nest 2-3 days before, earlier if disturbed, and do not return; young of same brood may leave over 2-4 days (Van Hecke 1979*b*; K-H Loske). Age of independence not known. Age of first breeding not known. BREEDING SUCCESS. Of 1324 eggs laid in 311 clutches, Belgium 73·7% hatched and 72·7% of these fledged, giving overall success of 53·6%, with mean 2·4 young fledged per clutch, and each pair rearing 3·6 young

per season (Van Hecke 1979*a*). Of 329 eggs in 69 1st clutches, Westfalen, 86·9% hatched and 82·9% of these fledged, or 72·0% overall (K-H Loske).

Plumages (nominate *trivialis*). ADULT BREEDING. Upperparts entirely olive-brown, forehead and crown rather narrowly and sharply streaked dull black, hindneck and sides of neck indistinctly dark grey; mantle, scapulars, back, and longer upper tail-coverts with rather broad but ill-defined dull black or black-brown feather-centres, rump and shorter upper tail-coverts uniform olive-brown. Supercilium narrow, buff or yellow-buff; narrowest above front of eye, faintly speckled olive above and behind eye, not sharply divided from crown. Eye surrounded by narrow, indistinct ring of buff feathers with slight olive mottling. Faint, narrow dark grey or olive-grey stripe on lores to front of eye and often also behind eye along upper ear-coverts. Upper cheeks and ear-coverts olive-brown, slightly mottled buff on upper cheeks below eye, faintly streaked buff on shorter ear-coverts, occasionally bordered below by indistinct olive stripe from gape backwards. Lower cheeks, chin, and throat warm buff to yellow-buff (palest on chin); buff lower cheeks and sides of chin separated by black malar stripe; malar stripe often distinct, sometimes curving up behind ear-coverts (but not on to ear-coverts, unlike Olive-backed Pipit *A. hodgsoni*) and extending back into distinct black patch on lower sides of neck, but stripe sometimes broken, especially at front or rear, and lower sides of neck then less heavily spotted or streaked black. Chest, sides of breast, flanks, and lower sides of belly deep buff to yellow-buff, merging into white or pale cream central belly and vent; under tail-coverts white or pale yellow-buff. Chest and sides of breast rather heavily streaked black; streaks often broader (*c*. 2 mm wide) and with lower end more rounded than those of Meadow Pipit *A. pratensis*, chest in fresh plumage often appearing rather similar to that of Song Thrush *Turdus philomelos*, not as streaky as *A. pratensis*; streaks gradually shorter and narrower towards breast and sides of belly, long and narrow on lower flanks, faint and grey on upper flanks. Central pair of tail-feathers (t1) dull black with broad but poorly defined olive-brown fringe along outer web and narrower one along inner web and tip. T2-t5 with narrow but sharply defined fringe along outer web; t4 sometimes with white spot on tip; t5 with larger white spot or small white wedge on tip of inner web, this spot or wedge 7·9 (54) 2-14 mm long, exceptionally 20-30, variation independent of age or sex. T6 white, faintly sullied grey when fresh, distal part of outer web olive-grey, basal and middle portion of inner web with dark grey wedge. Flight-feathers, tertials, upper greater primary coverts, and bastard wing dull black or sepia-black; fresh tertials have broad and rather poorly defined olive-brown fringes along outer webs, narrower ones along inner webs; more worn secondaries have narrower olive-grey or grey-buff fringes along outer webs; primaries have narrow and sharp buff or pale buff edges along outer webs; greater upper wing-coverts and feathers of bastard wing have sharp, narrow pale olive-buff fringes along outer webs and tips. Lesser upper wing-coverts olive-brown with dull black centres, like mantle and scapulars, longest having paler olive-brown, buff, or off-white tips. Median coverts black with contrasting cream or pale yellow tips *c*. 1·5-2 mm wide; some outermost (retained from non-breeding) sometimes slightly worn, black-brown with slightly narrower white tips. Greater coverts dull black with broad and ill-defined olive-brown border along outer web; tips of innermost olive-brown, of others paler buff, but variable number of outer usually retained from non-breeding, worn, black-brown with narrow white tip and grey-brown outer fringe. Under wing-coverts and axillaries pale yellow-buff to off-white, longer feathers sullied pale grey, shorter feathers (along leading edge of wing) with dark grey bases. *In worn plumage* (about May–July), upperparts slightly greyer, less warm olive-brown; dark streaks on crown, hindneck, mantle, and scapulars more distinct; pale fringes on tertials and tail worn off; supercilium paler, sometimes off-white above eye, but narrow, rear part heavily mottled brown; ear-coverts more heavily streaked and spotted dark brown, in particular towards rear; lower cheeks, chin, and throat paler, pale buff, often off-white on chin and upper throat; malar stripe narrow and mottled or virtually absent, ground-colour of underparts paler, cream or off-white, yellow-buff often restricted to upper chest, sides of breast, and upper flanks, black spots and streaks more contrasting; pale tips of median and greater upper wing-coverts and sometimes those of longest lesser bleached to white, forming 2–3 white wing-bars, but all white strongly subject to abrasion and may be virtually absent on greater and longer lesser coverts. ADULT NON-BREEDING. Like adult breeding, but all plumage equally new (though a few birds suspend flight-feather moult during autumn migration: Van Hecke 1980): greater, median, and longer lesser upper wing-coverts all uniformly tipped pink-buff, pale buff, or buffish-white; throat, chest, sides of breast, and flanks sometimes even deeper buff than adult breeding. NESTLING. Down fairly long and plentiful, dark smoke-grey; belly virtually naked (Witherby *et al.* 1938a). JUVENILE. Rather like adult non-breeding, as all plumage equally new and rather deeply coloured when plumage fresh. Ground-colour of upperparts more buff-brown, less olive. Soon affected by wear, first on sides of feather-tips, producing pale buff spots on mantle and scapulars; later on, all fringes bleached to pale buff or white; black of feather-centres of mantle and scapulars more sharply defined, ending rounded, upperparts appearing slightly scaly. Sides of head paler, more contrastingly spotted on upper cheeks and ear-coverts; ground-colour of underparts slightly paler when plumage fresh, virtually all off-white when plumage worn, spots and streaks on chest and sides of breast sometimes broader and less sharply defined. Lesser upper wing-coverts dark grey, median and greater black; tips of all coverts contrastingly buff (soon bleaching to pale buff and white), on average narrower than in adult, black of centre of covert sometimes ending in sharp point towards tip (more bluntly rounded or faintly pointed in adult, but some juveniles blunt and some adults pointed); tertials with narrower and more contrasting buff outer edges, soon bleaching to off-white. Worn juvenile often hard to distinguish from worn adult, but flight-feathers of juvenile new, those of adult heavily worn or in moult. FIRST NON-BREEDING. Similar to adult non-breeding, but variable amount of juvenile plumage retained (see Moults), contrasting with newer 1st non-breeding (in adult non-breeding, all feathering equally new; a few suspend moult during autumn migration, but these have some new flight-feathers, unlike 1st non-breeding). FIRST BREEDING. Like adult breeding, but often has more non-breeding retained (more worn and thus contrasting more with new feathers than in adult), as well as rather worn juvenile flight-feathers and often some heavily worn upper wing-coverts, part of tail, or tertials.

Bare parts. ADULT, FIRST ADULT. Iris sepia-brown to dark brown. Upper mandible and tip of lower mandible horn-brown to horn-black, cutting edges of both mandibles and base of lower mandible pale horn-brown, pale yellow with flesh tinge, pale lilac-pink, or flesh-coloured. Mouth pale pink. Leg and foot pale brown or yellow-brown with flesh tinge, lilac-pink,

pink-flesh, or pale flesh-colour. (Ali and Ripley 1973b; RMNH, ZMA.) NESTLING. Mouth orange with yellow spurs and edges to tongue and yellow spines inside upper mandible; gape-flanges pale yellow; bare skin (including leg) pink, bluish on future feather-tracts (Heinroth and Heinroth 1924–6; Witherby *et al.* 1938a). JUVENILE. Iris dark brown. Bill pink with yellow gape-flanges. Leg and foot flesh-pink. (Heinroth and Heinroth 1924–6.)

Moults. ADULT POST-BREEDING. Based mainly on Van Hecke (1980) for Belgium, Ginn and Melville (1983) for Britain, and specimens in RMNH and ZMA for Netherlands. Complete; primaries descendant. ♂ starts 3–4 weeks ahead of ♀, when some ♀♀ are still laying or incubating 2nd clutch. Moult usually starts with shedding of p1, but sometimes part of mantle, scapulars, chest, and flanks moulted first. In ♂, p1 shed late June to early August; ♀ mid-July to late August (♀ rarely from early July, perhaps failed breeders only); estimated duration of primary moult *c.* 35 days in Britain. Tail starts from primary moult score 10–20; sequence t1 to t6, but many feathers often grow simultaneously; tail regrown at moult score 35–40. All moult completed early August to mid-September in ♂, mid-August to early October (but mainly before mid-September) in ♀. Of 27 birds, late July and early August, 4 (both sexes) suspended primary moult (p1–p2 new, score 10) and 2 others probably near suspending (Van Hecke 1980); not known whether moult completed near breeding area or in winter quarters. ADULT PRE-BREEDING. Partial; January–March. Involves head, body, upper wing-coverts including tertial coverts (sometimes excluding outer median, often excluding outer greater, and occasionally no greater at all), tertials (sometimes 1–2 only), and t1 or (virtually) all tail. In southern Nigeria, January–February, 8 of 9 birds showed moult of body, some also tertials and tail; in some, only t1 replaced, in others all tail (Ludlow 1966). POST-JUVENILE. Partial. Extent and timing rather variable, depending on date of fledging. In Netherlands, birds just starting moult recorded mid-July to mid-August (mantle, scapulars, and underparts first: RMNH, ZMA); at Col de Bretolet (Switzerland), no moult after 1st 5 days of September (Jenni 1984), but in USSR young of 2nd broods start as late August (Dementiev and Gladkov 1954a), and here and elsewhere still moulting September (Stresemann 1920; Witherby *et al.* 1938a; Bub *et al.* 1981). Of 11 autumn migrants in Netherlands, all had head, body, and lesser upper wing-coverts new, 8 had tertial coverts new, 7 had median upper wing-coverts new (some others replaced a few innermost only), 6 greater upper wing-coverts (of which 3 retained some old outer ones) and most or all tertials new, and 5 had t1 new (RMNH, ZMA). Elsewhere, moult sometimes less extensive, e.g. in Sweden juvenile greater and outer median upper wing-coverts usually retained (Svensson 1984a). FIRST PRE-BREEDING. Partial. Extent and timing as in adult pre-breeding or slightly more variable. Some birds heavily worn in April and these perhaps have not moulted, retaining non-breeding.

Measurements. Nominate *trivialis*. Wing: (1) Scotland and northern England (Clancey 1950b); (2) Scandinavia (Clancey 1950b); (3) Halle and Cottbus areas, East Germany (Bub *et al.* 1981); (4) breeding birds, Westfalen, West Germany (Bub *et al.* 1981); (5) breeding birds, Netherlands, April–August (RMNH, ZMA); (6) autumn migrants, Netherlands, September (RMNH, ZMA). (7) Yugoslavia and Rumania, April–October (Stresemann 1920; ZMA). (4) live birds, others skins. Other measurements Netherlands and Yugoslavia only (RMNH, ZMA). Bill (S) to skull, bill (N) to distal corner of nostril;

exposed culmen on average 3·8 shorter than bill (S). Toe is middle toe with claw; claw is hind claw.

WING (1)	♂	86·8 (1·94; 22)	83–91	♀	84·9 (1·44; 4)	83–86
(2)		90·6 (1·71; 19)	88–94		87·0 (1·47; 11)	84–89
(3)		88·7 (— ; 37)	83–93		84·7 (— ; 27)	79–90
(4)		89·1 (4·17; 554)	84–96		85·8 (2·02; 38)	82–91
(5)		87·6 (1·38; 80)	85–90		84·6 (1·87; 30)	82–87
(6)		88·8 (2·17; 10)	85–92		86·1 (0·98; 7)	85–88
(7)		89·0 (2·41; 11)	84–92		88·0 (1·53; 7)	86–90
TAIL		59·3 (2·17; 27)	56–63		56·8 (2·02; 27)	54–60
BILL (S)		15·3 (0·58; 23)	14·5–16·3		14·9 (0·58; 25)	14·3–16·4
BILL (N)		8·5 (0·45; 17)	7·8–9·3		8·4 (0·37; 22)	7·7–9·1
TARSUS		21·1 (0·70; 26)	20·1–22·2		21·2 (0·55; 26)	20·3–22·2
TOE		19·4 (0·67; 17)	18·4–20·6		19·0 (0·82; 20)	18·0–20·4
CLAW		7·5 (0·62; 44)	6·3–8·5		7·6 (0·49; 40)	6·6–8·4

Sex differences significant for wing (2) and (4)–(6), and for tail. Hind claws of 8·8 and 9·3 (both ♂♂) excluded from range.

Wing of live birds Enskär and Nyköping (Sweden): 88·4 (2·69; 109) 83–93 (Bub *et al.* 1981). Wing of skins from Asia Minor, April–October: ♂ 89·9 (2·18; 6) 87–94, ♀ 86 (Kumerloeve 1968; RMNH).

Weights. Nominate *trivialis*. ADULT, FIRST ADULT. Sexed birds. (1) Westfalen (West Germany), breeding; (2) Belgium, breeding (Van Hecke 1980); (3) Belgium, migrants and breeding birds (Bub *et al.* 1981). (4) Netherlands, Turkey, and Iran, combined; spring and autumn migrants (Paludan 1938; Schüz 1959; Kumerloeve 1968; Rokitansky and Schifter 1971; ZMA, RMNH).

(1)	♂	21·7 (1·24; 413)	18–26	♀	25·1 (1·76; 23)	22–29
(2)		21·7 (1·17; 81)	19–25		21·3 (1·06; 78)	19–24
(3)		22·3 (— ; 125)	19–25		20·7 (— ; 98)	18–24
(4)		23·4 (2·30; 12)	20–28		22·4 (1·78; 8)	20–24

Unsexed birds. On autumn migration: (5) mainly September, Sweden (Scott 1965; Bub *et al.* 1981); (6) Malta (J Sultana and C Gauci); (7) mainly September, Lake Chany (south-west Siberia) (Havlín and Jurlov 1977). In winter quarters, Nigeria: (8) October–December (Smith 1966); (9) January–early March (Ludlow 1966); (10) 2nd half of March to early May (Ludlow 1966; Smith 1966). On spring migration: (11) south-east Morocco (Ash 1969); (12) Malta (J Sultana and C Gauci); (13) Britain (Smith 1966). Exhausted birds: (14) Netherlands, March–September (RMNH, ZMA).

(5)		22·5 (1·68; 86)	18–28	(10)	29·4 (4·72; 43)	21–38
(6)		26·3 (3·4 ; 50)	18–32	(11)	18·2 (— ; 51)	15–23
(7)		20·6 (1·87; 6)	18–23	(12)	23·3 (2·7; 52)	18–30
(8)		21·3 (1·20; 42)	18–23	(13)	20·6 (2·35; 93)	16–27
(9)		22·6 (1·78; 12)	19–25	(14)	16·0 (1·11; 6)	14–18

Race unknown. Mongolia, June to 1st half of August: ♂ 22·0 (1·11; 6) 20–24, ♀ 22·4 (2·30; 5) 19–25 (Piechocki and Bolod 1972). Ladakh (India), migrants, September: 22·1 (2·22; 13) 19–25 (Delany *et al.* 1982). Kazakhstan (USSR), March–October: ♂ 22·4 (2·05; 49) 19–28, ♀ 22·2 (1·86; 17) 19–25 (Dolgushin *et al.* 1970).

Structure. Wing rather short, broad at base, tip bluntly pointed. 10 primaries: p8 longest, p9 0–1 shorter, p7 0·5–2, p6 2·5–6, p5 11–16, p4 14–21, p1 22–29. P10 reduced, narrow and pointed; 56–65 shorter than p8, 9–13 shorter than longest upper wing-covert. Outer web of p6–p8 and inner of p7–p9 emarginated. Longest tertial reaches to about p6–p8 in closed wing, almost as long as wing-tip. Tail rather short, tip square or slightly forked; 12 feathers. Bill rather fine at tip but base wide (especially in races of central Asia) and deep. For length of middle toe and hind claw, see Measurements. Outer toe with claw *c.* 72% of middle toe with claw, inner *c.* 76%; hind with claw *c.* 86%, without claw *c.* 51%. Hind claw shorter than in other Palearctic *Anthus* (except *A. hodgsoni* and some Tawny Pipits *A. campestris*); rather heavy at base and more strongly decurved than hind claw of *A. campestris* and *A. pratensis*.

Geographical variation. Rather slight, clinal. Birds from Scotland and northern England have slightly shorter wing than typical nominate *trivialis* from Scandinavia, streaks on upperparts generally blacker, and ground-colour of throat, chest, and flanks more reddish-sandy, not as pale yellowish; thus sometimes separated as *salomonseni* Clancey, 1950, but difference very slight and much overlap. Scottish birds are at western end of cline of slightly paler colour towards east. Also a trend in size of chest-streaks: somewhat larger towards Siberia, smaller in west. Populations of eastern part of cline sometimes separated as *sibirica* Sushkin, 1925: birds from plains east of Urals on average paler and greyer on upperparts, dark streaks narrower; ground-colour of underparts paler, but size and extent of black chest-streaks variable—in some, streaks few and smaller than in nominate *trivialis*, in others streaks large (2·5–4 mm wide) and extensive (resembling those of *A. hodgsoni*), and latter group perhaps transitional towards *schlueteri*, which occurs in central Asia from Afghanistan and Tien Shan north to Tarbagatay mountains in USSR. All populations from Asia, including 'sibirica', have wing slightly blunter than typical nominate *trivialis*: p6 2–4 shorter than p8, p5 9–12, p1 20–27. Wing sometimes said to be shorter (e.g. Johansen 1952), but this not apparent in small sample of Siberian ♂♂ examined: wing 88·0 (9) 86–90. Siberian ♂♂ have rather long tail, however—60·9 (9) 58–65; other measurements similar to nominate *trivialis*. *A. t. schlueteri* also has slightly blunter wing and rather long tail; upperparts with slightly broader dark streaks than nominate *trivialis* and with slightly paler olive-grey fringes, underparts more heavily streaked black; differs from heavily streaked Siberian birds in wider and deeper bill-base. *A. t. haringtoni* from north-west Himalayas similar to *schlueteri* in wide bill-base and heavy streaks on underparts, but upperparts on average paler and streaks on underparts extend further down on flanks; size as in 'sibirica'—wing, ♂ 88·8 (9) 87–91, ♀ 85·0 (4) 84–86; tail, ♂ 60·7 (9) 59–62, ♀ 59·4 (4) 57–62.

Recognition. Differs from *A. pratensis* mainly in larger size (except for shorter hind claw) and different wing formula. Wing length of *trivialis* mainly 82–93 (*A. pratensis* mainly 73–85), bill to skull 14·3–16·4 (*A. pratensis* 13·7–15·3), bill depth at rear of nostrils 4·0–4·6 (*A. pratensis* 3·2–3·8); hind claw 6·3–8·5 (*A. pratensis* mainly 9·6–13·4, but see Measurements of both species); p7–p9 about equal, forming wing-tip, p6 2–6 shorter, p1 22–29 (in *A. pratensis*, p6–p9 longest, with p6 only 0–1·5 shorter than p7–p8, p1 17–23 shorter); outer web of p7–p8 distinctly emarginated on outer web, p6 less distinctly, with emargination *c.* 8–14 from tip of p6 (in *A. pratensis*, p6–p8 distinctly emarginated, emargination *c.* 15–19 from tip). Differences in plumages slight, not useful in practice due to marked individual variation and strong influence of bleaching and wear.

CSR

Anthus gustavi Pechora Pipit

Du. Petsjora Pieper   Fr. Pipit de la Petchora   Ge. Petschorapieper
Ru. Сибирский конек   Sp. Bisbita del Pechora   Sw. Tundrapiplärka

Anthus gustavi Swinhoe, 1863

Polytypic. Nominate *gustavi* Swinhoe, 1863, northern Eurasia. Extralimital: *commandorensis* Johansen, 1952, Komandorskiye Islands (northern Pacific); *menzbieri* Shulpin, 1928, Lake Khanka, southern Ussuriland (south-east USSR).

Field characters. 14 cm; wing-span 23–25 cm. Slightly less attenuated and less bulky than Tree Pipit *A. trivialis*, with 15% shorter tail; close in size and form to fledgling Meadow Pipit *A. pratensis*. 2nd smallest pipit of west Palearctic, with rather long and fine bill and rather short tail. Warm buff plumage sharply and copiously streaked, with no other markings showing at distance except for white belly and buff-white in outer tail-feathers; last lack cold and clean tone of tail-sides of other small pipits. Diagnostic marks apparent at close range include buff-white mantle-stripes, obvious pale double wing-bar, and fully streaked rump. Flight-action lacks strong bounds, recalling *A. pratensis*. Commonest call a distinctive short monosyllable. Sexes similar; no seasonal variation. Juvenile separable.

ADULT. Ground-colour of upperparts warm buff-brown, without olivaceous tone, and of underparts buff or white, paler on belly and under tail-coverts than both *A. trivialis* and *A. cervinus*. Upperpart pattern typical of genus but sharpness and weight of streaks on crown, back, and rump obvious and noticeably even in distribution, a feature approached only by Red-throated Pipit *A. cervinus*. Most obvious characters on upperparts are: (a) narrow but distinct, pale buff supercilium; (b) distinctly streaked nape and neck-sides; (c) almost white stripe on each side of mantle, often appearing as sharply notched line (unmatched by any pipit in west Palearctic) but less obvious when feathers ruffled or worn; (d) wide white margins to median coverts and pale buff-white tips to greater coverts, forming conspicuous double wing-bar on unworn birds, noticeably brighter than on *A. cervinus*; (e) lack of pale, unmarked rump (obvious on *A. trivialis* and *A. pratensis*); (f) if visible, warm-toned buff-white in outer tail-feathers. On underparts, the most obvious characters are (a) lack of obvious black malar stripe, with most birds showing only narrow vestige or a few tiny spots, (b) densely streaked and spotted chest, though this band narrower and more diffuse than in *A. trivialis* and *A. pratensis*, (c) well-streaked flanks, and (d) noticeably pale, almost white belly, vent, and under tail-coverts, contrasting with buff ground and streaks of chest and flanks and matched only by otherwise dissimilar Olive-backed Pipit *A. hodgsoni*. Full effect of wear in late summer not studied but likely that ground-colour becomes paler, while mantle-stripes and wing-bars become less obvious or disappear. JUVENILE. Less distinctive than adult, resembling juvenile *A. cervinus* in greater extent of streaking below. Best distinguished by voice (see below), but combination of pale buff mantle-stripes (not always matched in *A. cervinus*), white tips to median and greater coverts (buff in *A. cervinus*), and whiter ground-colour to belly and vent is strongly indicative of this species. FIRST WINTER. Following moult (from August), immature shows almost white mantle-stripes (like adult) but wear may obscure wing-bars. At all ages, bill slender, up to 17 mm long; dark horn, with pink base. Legs pink-flesh; hind claw up to 12·5 mm long—longer than *A. trivialis* but not obviously different from *A. cervinus*.

Long thought difficult to identify but well marked bird actually the most decorated of small pipits occurring in west Palearctic, with warm, heavily streaked, and bright appearance. Warm-toned outer tail-feathers likely to be noticed first; important next to check for streaked rump (shared only by *A. cervinus*) and then other characters noted above. Flight consists of erratic bursts of wing-beats interspersed with short bounds, floats, and glides; action less buoyant and confident than *A. trivialis* and *A. cervinus*, and somewhat hesitant progress thus recalls *A. pratensis*. Flight silhouette also recalls *A. pratensis*, with rather short, straight-sided tail suggesting juvenile of that species. Gait little studied; apparently as other small pipits but stance rather upright. Shy, skulking in dense cover such as long grass and centre of bush. Difficult to flush, escaping first in low flits, then in high flight.

Song of west Palearctic race consists of trilling sounds followed by guttural warbling. Commonest call a stony 'pwit' or 'p(r)it'.

Habitat. Breeds from fringe of west Palearctic eastward along a mainly subarctic band, apparently sandwiched between Olive-backed Pipit *A. hodgsoni* to south and Red-throated Pipit *A. cervinus* to north. Inhabits bushy tundra and remote taiga swamps, but not pure tundra. Overlaps with *A. cervinus* on tundra with marshy and meadow-like heaths. Data still inadequate for defining habitat separation, but appears to prefer overgrown areas with tall dense sedge, reed-grass, and plentiful shrubs or even trees, mainly in lowlands, along rivers and coasts (Dementiev and Gladkov 1954a). Extralimitally in Asia

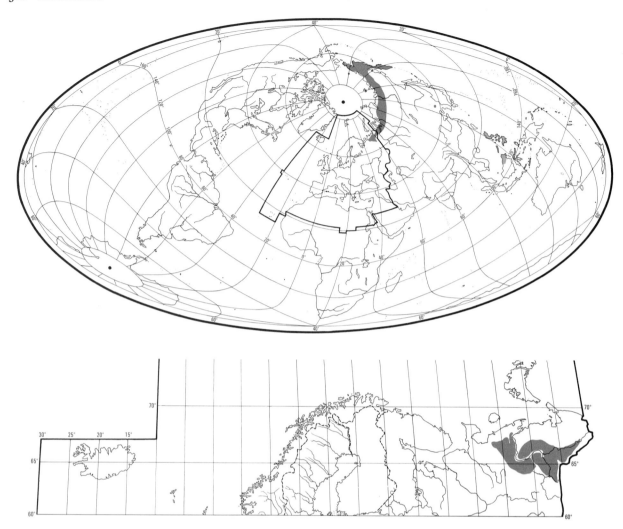

overlaps somewhat with Tree Pipit *A. trivialis*, but generally replaces it in taiga forest and shrub tundra and accompanying riverside swamps overgrown with dwarf willows and sedges. Keeping within zone of July isotherms 6-15°C, accordingly avoids such extreme low temperatures as are tolerated by *A. cervinus*, but extends to boreal as well as subarctic climates (Voous 1960). Northern limit of range corresponds with last scrub vegetation of tundra, where it is to be found in open boggy tracts with scrub, sometimes of *Pinus cembra* in patches; will at times alight on a bush. In taiga, confined to overgrown swampy glades (Bannerman 1953). For habitat of extralimital *menzbieri*, see Nazarov (1981).

Distribution and population. Reliable data scanty. No information on range changes or population trends. Not numerous (Dementiev and Gladkov 1954*a*).

Accidental. Iceland, Britain, Norway, Finland, Poland.

Movements. Migratory, wintering in East Indies.

Because of thinly spread nature of main population (nominate *gustavi*), dates of movement remain obscure: one migrant recorded on Kamchatka 6 June, and departure from Kolyma and Indigirka rivers occurs mid-September. *A. g. commandorensis* arrives on Komandorskiye islands in late May or early June and departs in second half of September (Dementiev and Gladkov 1954*a*). *A. g. menzbieri* arrives in Ussuriland from end of April to early June, nominate *gustavi* passing through at same time. *A. g. menzbieri* shows some dispersal from mid-July, migrating August to early October; nominate *gustavi* passes through Ussuriland in mid-September (Rakhilin 1979; Nazarov 1981). In Alaska (USA), recorded May-June as vagrant on Aleutians and St Lawrence Island (Roberson 1980).

In Britain (where 28 recorded up to 1985, almost all on Fair Isle, Scotland), occurs late August to mid-November (mostly late September and early October) and once (in Suffolk) in late April (Dymond *et al.* 1976; Sharrock and Sharrock 1976; Rogers *et al.* 1984).

JHE, DJB

Food. Little information available, all extralimital. Chiefly insects. Forages mainly on ground picking food from lower parts of plant stems and leaves. In late summer, young birds forage in tree branches (Nazarov 1981).

On Chukotskiy peninsula (north-east USSR), July, 5 stomachs contained 61 items, 72·1% Hymenoptera, 11·5% various larvae, 8·2% beetles (Carabidae), 3·3% flies (Diptera), 3·3% molluscs and 1·6% crane-flies *Tipula* (Portenko 1973). In Koryak highlands (north-east USSR), July–August, seen carrying dragonflies (Odonata) and crane-flies (Tipulidae) to nest (Kishchinski 1980). At Lake Khanka (eastern USSR), 1977–8, 66 samples (presumably from nestling collars) and 14 stomachs contained 32·1% (by number) flies (Culicidae, Chironomidae, Tephritidae, Tabanidae, Syrphidae), 24·2% adult and larval moths (Notodontidae, Geometridae, Lymantriidae, Pyralidae, Tortricidae, Noctuidae), 14·3% spiders (Araneae), 9·5% molluscs, 7·5% cicadas (Cicadidae), 6·3% grasshoppers, etc. (Acrididae, Tetrigidae), 4·8% beetles (Carabidae, Chrysomelidae), 0·8% sawflies (Tenthredinidae), and 0·4% caddisflies (Trichoptera). In 1977, 27·5% of samples contained spiders, 24·1% moths, and 27·5% beetles; in 1978, 27·7% contained spiders, 55·3% moths, and 67·1% flies (Nazarov 1981).   PJE

Social pattern and behaviour. Poorly known. Best study is of extralimital *menzbieri* at Lake Khanka (Ussuriland, eastern USSR) by Nazarov (1981).

1. Mostly solitary or in twos outside breeding season (e.g. duPont and Rabor 1973*a*, *b*). In Ussuriland, typically solitary on both spring and autumn migration; parties of up to 6 much less common. Nominate *gustavi* more numerous than *menzbieri* in both seasons. Independent young common by mid-July and not infrequently recorded in flocks of young Yellow Wagtails *Motacilla flava*; some broods in family unit, fed by parents, at end of July however (Nazarov 1981). BONDS. Nothing to suggest mating system other than monogamous. No detailed information on pair-bond. Young fed by both parents in nest and for indeterminate period when they leave (Nazarov 1981; Tomkovich and Sorokin 1983). BREEDING DISPERSION. Solitary and territorial though many reports in the literature of 'colonies' (some large): see (e.g.) Portenko (1939, 1960, 1973), also Gordeev (1963) who noted 'dispersed colony' along Kazym river (Tyumen', west-central Siberia). These probably more accurately described as neighbourhood groups, dictated at least to some extent by habitat availability. In one area of Chukotka (north-east USSR), 2 ♂♂ *c.* 200 m apart in willow *Salix* scrub; also in Chukotka, 6 ♂♂ recorded along 4 km of Uusenveem river (Tomkovich and Sorokin 1983). At Lake Khanka, despite lower water level in 1978, density in one river mouth up to 19 pairs per km² (twice that of previous year); attributed to increase in area of suitable breeding habitat (floating reed islands) over previous 2–3 years (Nazarov 1981). Another study at Lake Khanka indicated preference for burnt-over areas of reeds; counts of singing ♂♂ in 1978 suggested 20–30(–100) pairs per km² (Glushchenko 1981). No detailed information on territory size, though ♂ *menzbieri* sings over area of *c.* 1 ha or more (Nazarov 1981), and over 6 ha or more reported for nominate *gustavi* in eastern Chukotka (Tomkovich and Sorokin 1983), while in Pechora basin (north-west USSR), one bird evidently sang over very restricted area (Dresser 1871–81).

Common nesting associates are Red-throated Pipit *A. cervinus*, also *M. flava* and Citrine Wagtail *M. citreola* (Seebohm 1901; Portenko 1973; Tomkovich and Sorokin 1983). ROOSTING. Only report is of independent young at Lake Khanka flying out to roost at night on floating reed islands (Nazarov 1981).

2. In Philippines in winter, birds generally quiet and unobtrusive, feeding amongst dense forest undergrowth. When disturbed, will run fast and often far into cover. Occasionally flushes, though does not fly far, and tends to run into cover on landing (duPont and Rabor 1973*b*; Ripley and Rabor 1958). Migrants exceedingly shy according to Panov (1973) and solitary vagrants to Europe and America also typically furtive and quiet. In Shetland, bird flushed from cover sometimes perched on plant or fence before re-entering cover (Broad *et al.* 1973). For call sometimes given by disturbed birds, see 2a in Voice. See also Williamson (1953) and King (1980). Landing on bushes, etc., after flushing also reported from breeding grounds (e.g. Bannerman 1953, Portenko 1973); see Parental Anti-predator Strategies (below). FLOCK BEHAVIOUR. No information. SONG-DISPLAY. ♂ sings (see 1 in Voice) in flight, also (often as a continuation) from bush, etc., or ground. In Ussuriland, prolonged singing from perch occurs in strong wind (Nazarov 1981) and bird will give full-voiced yet fairly quiet song while feeding on ground (Vorobiev 1954). Following account of Song-flight compiled mainly from descriptions of nominate *gustavi* in Pechora basin and *menzbieri* at Lake Khanka. Ascends silently and fairly rapidly but not steeply to *c.* 30–70 m then flies in wide circles or smaller loops. Wings beaten evenly (not closed as in normal flight); tail slightly raised, head lowered. Then (e.g. after only 10–15 wing-beats: Nazarov 1981) hovers for a few seconds with tail spread and sings. Flies on silently for short distance and sings again whilst hovering, etc. Reminiscent of lark (Alaudidae) in wheeling and hovering flight. Nazarov (1981) reported Song-flights lasting several minutes in *menzbieri*; 20–30 min or (bird perhaps not continuously aloft) even up to *c.* 60 min recorded for nominate *gustavi* (Dresser 1871–81; Seebohm 1901; Portenko 1960; Smirenski 1979; Tomkovich and Sorokin 1983). Prior to descent, wings and fanned tail raised, body horizontal; bird then glides down in spirals or drops vertically, making sharp turns to left and right, before plummeting to land on perch or ground; *menzbieri* at least sings (1–2 bursts) during 1st stage of descent. May continue to sing from perch or ground, then ascend again. (Dresser 1871–81; Seebohm and Harvie Brown 1876; Seebohm 1901; Nazarov 1981.) For comparison with other *Anthus*, see Wallschläger (1984). In Ussuriland, sings from arrival (late April or May), less from early June and only once recorded after mid-June, though some young ♂♂ will sing in late July and early August (Nazarov 1981). Nominate *gustavi* in western and eastern parts of range noted singing in June and into early July (e.g. Gordeev 1963, Portenko 1973). At Enurmino (Chukotka), late June, song delivered 02.00–04.00 hrs; apparently less vigorous in early July (Tomkovich and Sorokin 1983). ANTAGONISTIC BEHAVIOUR. No information for breeding birds. In Ussuriland, migrants remain solitary, always driving off any conspecific birds (Nazarov 1981). HETEROSEXUAL BEHAVIOUR. Only report is of ♂ immediately interrupting Song-flight to fly down and join ♀ when she flew up from ground (Smirenski 1979). RELATIONS WITHIN FAMILY GROUP. Young hatch over a period of nearly 2 days. Eyes start to open by 3–4 days old. Young leave nest at *c.* 12–14 days old (Nazarov 1981). ANTI-PREDATOR RESPONSES OF YOUNG. No information. PARENTAL ANTI-PREDATOR STRATEGIES. (1) Passive measures. Incubating bird may allow approach to *c.* 0·4–1 m, then fly off directly and silently; once called on flushing (see 2c in Voice).

Less commonly runs off with wings half-open (perhaps a distraction-lure display—see below). Normally returns only after observer's departure, landing *c.* 10-15 m from nest and moving to it on foot (Nazarov 1981). On Chukotka, birds generally shy, some flying off, some remaining in cover and difficult to flush, others perching (mostly still concealed by foliage) in bush from where perhaps returned furtively to nest, though flushed at closer approach (Portenko 1939, 1973). (2) Active measures. If man near nest with young, ♀ will flutter about silently or call (see 2b in Voice) *c.* 5-8 m away for *c.* 1-2 min, then fly off and disappear into cover. In same situation, ♂ will ascend to *c.* 30-40 m and sing while still carrying food; not infrequently joined by neighbouring ♂♂, though all soon disappear (Nazarov 1981). Simple aerial demonstration also noted by Seebohm (1901) and Portenko (1939): erratic flight accompanied by alarm-call (see 2 in Voice); bird will fly towards intruder, then veer away (Portenko 1973). Recorded performing distraction-lure display of disablement-type, bird fluttering away on ground from nest containing eggs (Popham 1898).   MGW

Voice. Often silent outside breeding season; see Social Pattern and Behaviour.

CALLS OF ADULTS. (1) Song of ♂. Apparently varies geographically (see below) but further study required. In nominate *gustavi* (no recording available), 1st part an apparent outpouring of 'trills' like lark (Alaudidae), song of Temminck's Stint *Calidris temminckii*, or part of song of Wood Warbler *Phylloscopus sibilatrix*. This followed by low guttural warble like that given sometimes by Bluethroat *Luscinia svecica*; impression is of bird attempting to trill whilst inhaling (Dresser 1871-81; Seebohm and Harvie Brown 1876; Seebohm 1901). Loud, not very melodious, and somewhat reminiscent of *Acrocephalus* warbler (Portenko 1960, 1973). Song of *menzbieri* from eastern USSR described as a low-pitched and muted buzzing: 'ppdzhzhzhzhzhzhzhzhzheep—zhzhzheepp' or 'cheep-cheecheecheechzhzhzhzhzh cheenee'. Sounds may vary

their position in a phrase, with additional variation due to their being sometimes abbreviated or more drawn-out. Audible up to 150-200 m (Nazarov 1981). Recording of *menzbieri* suggests weird, screechy, mechanical whirring like clockwork toy (J Hall-Craggs, P J Sellar). Bird seems to work up to rapid (*c.* 40-100 'units' per s: J Hall-Craggs), dry tremolo during which there is definite impression of inhalation and exhalation. Song-phrase (Fig I) rendered 'tsi-si si werrrr tse tsi-tsi serrrrrrrrr tsi-sirrrrr tse-tsi serrrrrrrrrrr tsi-sirrrrr tse-tsi-tsi(t)'. Some tremolos short, others almost exaggeratedly attenuated; final tremolo apparently with rise in pitch and consequent change in timbre (M G Wilson). According to Johansen (1952), song of *commandorensis* much louder, more varied and more persistent than that of Tree Pipit *A. trivialis*. Recording of *commandorensis* (Fig II) reveals something in common with *menzbieri*, but tremolo (*c.* 13 units per s) more like dry rasping jingle of River Warbler *Locustella fluviatilis* or dry song of bunting (Emberizidae), also recalling Drumming of Snipe *Gallinago gallinago*; winding down after tremolo suggests 'dzee dzee zip'. Other units mostly nasal and buzzing 'dzee', but some more drawn-out, like a sucking, whirring 'wheezzzz'. Recordings thus indicate that races have distinct songs. On strength of long tremolo sections (which constitute most of the song), *menzbieri* likened to Grasshopper Warbler *L. naevia* and *commandorensis* to *L. fluviatilis*. In races for which recordings available (i.e. all except nominate *gustavi*), tremolos broken up with longer (0·1-0·3 s) structured units which generally have similar buzzing character—'chee(p)', 'dweep', 'tsee', 'dzee', etc. (J Hall-Craggs, M G Wilson.) (2) Contact-alarm calls. (a) Call apparently most frequently given outside breeding season. An unattractive but diagnostic strong, hard, forceful, and sharp-sounding 'twit' or 'pit'

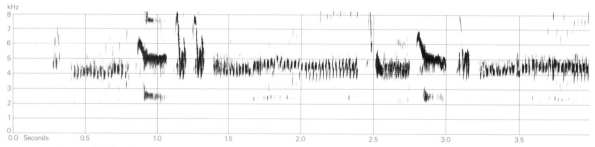

I B N Veprintsev USSR May 1977

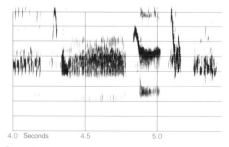

I *cont.*

II B N Veprintsev USSR May 1980

(Williamson 1973; King *et al.* 1975); stony 'pwit' or 'p(r) it' (D I M Wallace); hard, abrupt 'prrrit' quite distinct from other *Anthus* (Smythies 1981). Low-pitched and lacking sweetness or richness of tone. Often given several times in succession at take-off (Williamson 1953). Loud 'wit wit' expressing alarm on breeding grounds (Seebohm 1901) is perhaps the same or closely related. (b) ♀ will give quiet, abrupt 'tsep tsep' when alarmed by presence of man near nest with young (Nazarov 1981); not definitely distinct from call 2a. (c) One ♀ flushed from nest gave toneless hissing sound (Nazarov 1981). (d) According to Panov (1973), call-note slightly reminiscent of Water Pipit *A. spinoletta*, but harsher 'tsweep' and 'tsee-tsee-tseep'. Thin 'tzee' calls given by birds in Borneo (Smythies 1981) are probably the same, as perhaps are those given apparently infrequently by vagrants in North America—soft, difficult to hear and thus to describe, but distinguishable from Red-throated Pipit *A. cervinus* (King 1980).

CALLS OF YOUNG. Only reference is to inexpert, quiet, muttering song given by some young ♂♂ from perch or in flight (Nazarov 1981); see also Social Pattern and Behaviour.   MGW

Breeding. SEASON. Siberia: eggs laid late June and July (Makatsch 1976). SITE. On ground in low cover or in shelter of tuft of vegetation or low scrub. Nest: substantial cup of grass and other leaves, lined with finer vegetation. Outer diameter 10·5 cm, inner measurements 6 × 5 cm, 2·5 cm deep, overall height 6·5 cm (Makatsch 1976). Building: no information on role of sexes EGGS. See Plate 80. Sub-elliptical to oval, smooth and glossy; pale grey, sometimes pink-tinted or even dark red-brown, with grey speckling overall, and occasional black hair-streaks or blotches at broad end. 21·5 × 15·0 mm (20·0–24·2 × 14·5–16·2), *n* = 50; calculated weight 2·5 g (Schönwetter 1979; Firsova and Levada 1982). Clutch: 4–5(–6). Probably only 1 brood, due to high latitude (Makatsch 1976). INCUBATION. About 12–13 days. Clutch hatches over almost 2 days (Nazarov 1981). YOUNG. Altricial and nidicolous, fed by both parents (Nazarov 1981). FLEDGING TO MATURITY. Fledging period 12–14 days (Nazarov 1981). Age of independence not known.

Plumages (nominate *gustavi*). ADULT BREEDING. Forehead and crown black, forehead with fine rufous-buff or yellow-buff specks, crown with narrow and even rufous-buff or yellow-buff streaks, black on crown divided into 4 broad black streaks on centre, single narrower one on each side. Hindneck, sides of neck, and upper mantle buff with sharp black streaks (narrower than those of central crown). Lower mantle and inner scapulars black, sides of feathers with rather narrow buff borders, but sides of feathers at border between outer mantle and inner scapulars more broadly fringed cream or off-white, showing as contrasting pale V on mantle with broad black borders along both sides. Outer scapulars pale buff-brown, centres rather narrowly streaked dark brown (not as contrastingly and broadly black as on inner scapulars). Back, rump, and upper tail-coverts rufous-buff or yellow-buff, marked with sharp black streaks *c.* 2

mm wide on feather-centres; tips of feathers narrowly fringed buff when plumage fresh, but black streaks not obscured by this. Supercilium long and narrow, warm buff or yellow-buff; rather indistinct, as colour and width about similar to pale streaks on crown; supercilium sometimes paler above eye, cream-buff or pale buff, and occasionally slightly wider. Narrow but distinct ring of cream-yellow feathers around eye, broken in front. Lores and narrow patch below eye finely mottled grey and olive-buff, sometimes bordered below by narrow dull black line backwards from gape, ear-coverts ochre-buff with fine and indistinct darker streaks and spots, often bordered by brighter and more uniform buff patch at rear. Lower cheeks bright buff or yellow-buff, bordered below by mottled and thin black malar stripe (not reaching base of lower mandible), widening to large black or mottled black patch at lower side of throat. Chin and throat deep buff or yellow-buff, sometimes with a few fine black specks. Chest, sides of breast, and upper flanks deep buff or yellow-buff, broadly and contrastingly streaked black (streaks *c.* 1·5–2 mm wide); ground-colour of lower flanks usually paler buff or cream-white, but streaks (though less deep black) even broader, 2–3 mm. Breast, sides of belly, lower vent, and under tail-coverts buff or cream-buff, breast and sides of belly with narrow black streaks, under tail-coverts virtually unmarked; central belly and upper vent uniform cream or white. Central pair of tail-feathers (t1) dull black or brown-black with ill-defined olive-brown fringes along both sides; t2–t4 similar, but fringes narrower and along outer web only. T5 dull grey with pale wedge on terminal half of inner web; wedge usually 27–34 mm long, rarely 19–40 mm, once 7 mm (*n* = 20); wedge usually tinged pale cream-buff or pale isabelline, rarely (when tail worn) almost white. Outer web of t6 pale olive-grey or isabelline, inner web pale cream-buff or pale isabelline (rarely almost white) with dark grey wedge on basal half. Flight-feathers and tertials brownish-black, tips narrowly edged off-white; secondaries broadly fringed olive-brown along outer web, tertials broadly fringed olive-brown on outer web and more narrowly olive-grey on tip and inner web; primaries narrowly and sharply fringed olive-brown along outer web. Greater upper primary coverts and feathers of bastard wing brown-black, narrowly and sharply fringed olive-brown along outer web and tip. Lesser upper wing-coverts dull black with broad olive-brown fringes; tips of longest sometimes paler, off-white, forming narrow and indistinct pale wing-bar. Median upper wing-coverts black with contrasting pale greyish-buff or off-white tips, forming marked wing-bar; greater upper wing-coverts duller black, fringed olive-brown along outer web and broadly olive-grey to off-white along tip; pale tips of greater coverts form pale wing-bar, but not usually as distinct as on median coverts, (in particular) outermost and innermost greater coverts are suffused olive-grey on tips. Under wing-coverts and axillaries grey with buff suffusion; coverts along leading edge of wing largely dark grey-brown. *In worn plumage*, rufous-buff or yellow-buff streaks and fringes of upperparts and tertials bleached to paler buff and partially abraded, black of upperparts thus even more strongly prominent; ground-colour of sides of head, sides of breast, chest, and upper flanks paler buff, chin and throat cream-white or white, belly and vent more extensively white; supercilium cream-white, more distinct; broad black streaks of chest and sides of breast more contrasting; pale tips of median and greater upper wing-coverts (and sometimes those of longer lesser covers) bleached to white, wing-bars more prominent, but tips sometimes heavily abraded and then only traces of white remaining, mainly on median coverts. ADULT NON-BREEDING. Similar to adult breeding, but pale streaks on upperparts olive-buff, less bright buff; ground-colour of underparts slightly less deep buff, especially on throat. Differs mainly in degree of wear: in non-breeding, all plumage completely fresh; in breeding, head, body, lesser upper wing-coverts, tertials, tertial coverts, and t1 new, but median and greater upper wing-coverts and t2-t6 retained from non-breeding, slightly worn, and pale tips of upper wing-coverts purer white than in non-breeding, forming more distinct wing-bars. NESTLING. In nominate *gustavi*, down rather long, pale grey (Firsova and Levada 1982). In *menzbieri*, dark grey (Nazarov 1981, which see for development). JUVENILE. Like adult, but streaks on underparts less clearly defined and extending up to throat and down to belly (Hall 1961). FIRST NON-BREEDING. Like adult non-breeding, sometimes hardly distinguishable, except in those birds which retain some contrastingly worn brown juvenile feathers (one or a few outer greater coverts, some feathers of bastard wing, or an occasional tail-feather); juvenile flight-feathers and greater upper primary coverts retained, slightly more worn than those of adult at same time of year; primary coverts more pointed and frayed, black-brown with poorly defined olive-brown fringe along outer web and tip (black with sharper and narrower fringe in adult). FIRST BREEDING. Like adult breeding (though less new feathering acquired; see Moults). Retained juvenile flight-feathers and greater upper primary coverts more worn than in adult; remainder of feathering a combination of breeding and non-breeding, as in adult, but often more contrast in degree of abrasion between these plumage areas than in adult.

Bare parts. ADULT, FIRST ADULT. Iris brown or dark brown. Upper mandible dark horn-brown, cutting edges cinnamon or pale horn-brown; lower mandible light horn-brown, yellowish-flesh, or pale flesh, tip slightly darker. (Witherby *et al.* 1938a; Dementiev and Gladkov 1954a; ZMA.) Leg and foot brownish-flesh (Hartert 1910; Dementiev and Gladkov 1954a), flesh-pink (Witherby *et al.* 1938a), bright yellowish-pink (Kist and Waldeck 1961), or very pale horn-colour to pale pearl-grey (ZMA). NESTLING. In nominate *gustavi*, mouth bright yellow without dark spots (Firsova and Levada 1982). In *menzbieri*, bare skin including leg and foot orange-flesh; tip of bill brown and dark grey; gape-flanges pale yellow or yellow; mouth yellow, orange, or red-orange, corner of mouth and centre of tongue redder (Nazarov 1981). JUVENILE. No information.

Moults. ADULT POST-BREEDING. Complete; primaries descendant. 2 ♂♂ from lower Yenisey, 26 July, heavily worn, just starting moult with loss of p1 (BMNH). *A. g. menzbieri* in heavy moult 27 July and 5 August (Nazarov 1981). All plumage new on arrival in winter quarters, October. ADULT PRE-BREEDING. Partial: head, body, lesser upper wing-coverts, tertial coverts, tertials, and t1; in limited number examined, no replacement of flight-feathers, greater upper wing-coverts, t2-t6, or median upper wing-coverts. Starts from early January, completed late March or early April. POST-JUVENILE. Partial; apparently in breeding area, as moult completed on arrival in winter quarters, October. Involves all head, body, tail, and wing, except flight-feathers and greater upper primary coverts; frequently (3 out of 10 birds, October–January), some outermost greater upper wing-coverts or some feathers of bastard wing or tail retained. Timing variable in *menzbieri*, depending on hatching date; starts shortly after fledging. Fully juvenile birds encountered early July to 1 August, birds in last stages of moult between 20 July and 9 September. (Nazarov 1981.) FIRST PRE-BREEDING. Partial; apparently starts later and moult less extensive than adult pre-breeding, but only a few examined. No moult in January; in some from late February and March, Indonesia, moult had just started with scattered feathers of mantle, scapulars, and sides of breast, as well as some tertials and t1. On migration in May,

China, flight-feathers and greater upper primary coverts still juvenile, other upper wing-coverts (except sometimes lesser), all or some tertials, and tail (except sometimes t1) still non-breeding, head and body all or largely in 1st breeding.

Measurements. Nominate *gustavi*. ADULT, FIRST ADULT. Pechora and lower Yenisey (USSR), June–July, and Sulawesi (Indonesia), Basilan, and Palawan (Philippines), October–April; skins (BMNH, RMNH, ZMA). Bill (S) to skull, bill (N) to distal corner of nostril; exposed culmen on average 3·6 less than bill (S). Toe is middle toe with claw; claw is hind claw.

WING	♂ 84·5 (1·13; 12)	83–87	♀ 80·8 (1·53; 12)	78–84
TAIL	52·4 (1·92; 12)	50–56	52·3 (1·68; 12)	49–55
BILL (S)	16·3 (0·36; 12)	15·8–17·0	15·5 (0·50; 12)	14·9–16·4
BILL (N)	9·6 (0·38; 12)	9·0–10·2	9·0 (0·34; 12)	8·6–9·7
TARSUS	23·4 (0·94; 12)	21·6–24·5	22·7 (0·74; 12)	21·8–23·9
TOE	20·0 (1·04; 6)	18·6–21·0	20·4 (0·77; 9)	19·2–21·5
CLAW	10·1 (0·81; 10)	8·2–11·0	9·5 (0·72; 12)	8·2–10·4

Sex differences significant for wing and bill (S). First adult wing (with juvenile flight-feathers) combined with adult above, though wing of former on average 0·9 mm shorter.

WING. Nominate *gustavi*: ♂ 82–84, ♀ 78–81 (3) (Johansen 1952); ♂ 81–86 (8), ♀ 77–82 (6) (Hall 1961); ♂ 82–86 (9), ♀ 78–82 (9) (Bub *et al.* 1981). *A. g. commandorensis*: ♂ 83–86 (10), ♀ 79–83 (4) (Johansen 1952); ♂ 86–89 (3), ♀ 85 (1) (Hall 1961); ♂ 83–89 (9), ♀ 80–84 (5) (Bub *et al.* 1981). *A. g. menzbieri*: ♂ 76–79 (11), ♀ 74–76 (6); ♂ 80 (1), ♀ 74 (1), unsexed 73–79 (3) (Hall 1961); ♂ 79·8 (14) 77–82, ♀ 75–79 (7) (Bub *et al.* 1981).

Nominate *gustavi* (n=14): tail 47–51, bill 15–17, tarsus 20–22, hind claw 10–13. *A. g. commandorensis* (n=4): tail 51–55, bill 16–17, tarsus 24–26, hind claw 10–13. *A. g. menzbieri* (n=5): tail 45–51, bill 16, tarsus 21–23, hind claw 10–12. (Hall 1961.)

Weights. ADULT, FIRST ADULT. Nominate *gustavi*. Fair Isle (Scotland), September and early October: 23·3 (2·05; 5) 20–26 (Fair Isle Bird Observatory, N J Riddiford). Sweden, September: 22 (Tammelin 1975). *A. g. menzbieri*. Ussuriland (south-east USSR), late May to early September: 19·8 (1·23; 10) 17–21 (Nazarov 1981).

Structure. Wing rather short, broad at base; tip bluntly pointed. 10 primaries: p8–p9 longest or either one 0·5 shorter than other, p7 0–1 shorter than longest, p6 3·5–6, p5 10–12·5, p4 13·5–16·5, p1 21–26. P10 reduced, narrow and pointed, 54–59 shorter than longest primary, 9–11 shorter than longest upper primary coverts. Outer web of (p6-)p7–p8 emarginated, inner web of (p7-)p8–p9 slightly emarginated. Tertials relatively shorter than most other *Anthus*: tip of longest reaches to about halfway between tips of p5 and p6 in closed wing, 8–14 mm from wing-tip (n=30) (Svensson 1984; RMNH, ZMA). Tail rather short, tip square; 12 feathers, each distinctly pointed, even in adult. Bill rather similar to bill of Tree Pipit *A. trivialis*, being rather wide and deep at base, but longer and culmen more strongly depressed in front of nostril; thicker at base and longer than Meadow Pipit *A. pratensis* or Red-throated Pipit *A. cervinus*. For length of middle toe and hind claw, see Measurements. Outer toe with claw c. 71% of middle toe with claw, inner c. 76%; hind with claw c. 95%, without claw c. 49%. Remainder of structure as in *A. pratensis* (p. 379).

Geographical variation. Rather slight. *A. g. commandorensis* from Komandorskiye Islands (off Kamchatka) similar in colour to nominate *gustavi* (Johansen 1952) or perhaps slightly paler (Hall 1961); slightly larger, especially tarsus (see Measurements). *A. g. menzbieri* from Lake Khanka area (near Vladivostok, south-east USSR) differs more obviously: wing distinctly shorter than nominate *gustavi*, upperparts more extensively black; feather-edges narrower and paler, pale grey or off-white; ground-colour of underparts deeper yellow-buff, under tail-coverts deeper buff (Johansen 1952; Dementiev and Gladkov 1954a; Hall 1961).

Recognition. Rather similar to 1st non-breeding plumage of *A. cervinus*; both more heavily streaked on upperparts (including back and rump) than any other Palearctic *Anthus*. *A. gustavi* has dark streaks on upperparts deeper black, broader, and more sharply defined; mantle and inner scapulars mainly black, usually with large and single cream V (sometimes partly broken or open-ended); ground-colour of sides of head rich buff (most marked on supercilium, cheeks, and behind ear-coverts), contrasting with dark eye (not as olive-brown with off-white supercilium and cheeks as in *A. cervinus*); ground-colour of chest and flanks

deeper buff (cream-buff in fresh non-breeding *A. cervinus*), throat sometimes deep buff when fresh (unlike non-breeding *A. cervinus*, but sometimes almost similar to rufous-buff of some *A. cervinus* in breeding plumage); longer under tail-coverts lack black central mark or have a faint one only (usually distinct in *A. cervinus*); pale wedge on t5 nearly always longer than in *A. cervinus* (mainly 27–34 long in *A. gustavi*, exceptionally 7 only; 2–12 in *A. cervinus*); t6 and wedge of t5 sullied grey, pale buff, or isabelline (virtually white in *A. cervinus*, but some overlap in colour); feathering of tibia olive-brown, distinctly darker than lower flanks and under tail-coverts (cream-buff with some brown mottling in *A. cervinus*, rather similar in colour to lower flanks and under tail-coverts). No constant difference in upperwing (contra, e.g., Witherby *et al.* 1938a), and much overlap in extent of malar stripe and length of hind claw. See also Measurements and Structure. CSR

Anthus pratensis Meadow Pipit

PLATES 24 and 27
[between pages 232 and 233]

Du. Graspieper Fr. Pipit des prés Ge. Wiesenpieper
Ru. Луговой конек Sp. Bisbita común Sw. Ängspiplärka

Alauda pratensis Linnaeus, 1758

Polytypic. Nominate *pratensis* (Linnaeus, 1758), south-east Greenland, Europe (except Ireland and western Scotland), and western Siberia; *whistleri* Clancey, 1942, Ireland and western Scotland.

Field characters. 14·5 cm; wing-span 22–25 cm. Slightly longer and bulkier than Berthelot's Pipit *A. berthelotii* and Pechora Pipit *A. gustavi* but somewhat smaller than all other small pipits, with more rounded wings; much shorter and less bulky than large pipits, with proportionately shorter tail. Rather small, sleek but dumpy, active pipit; epitome of genus but lacking striking diagnostic field characters and much more readily identified by call than plumage. Typically olive or brown above, heavily streaked except on rump, and ochre- or grey-white below, heavily spotted and streaked on chest and flanks; striking features restricted to cold white tail-sides. At close range, rather slender bill, rather indistinct face pattern, distinct but dull wing-bars, pale brown legs, and long hind claw form useful characters. Flight fluttering, with hesitant, jerky action shared only by *A. gustavi*. Has distinctive creeping walk and uses open ground more persistently than any other small pipit. Sexes similar; little seasonal variation, but effect of wear often marked. Juvenile separable. 2 races in west Palearctic, with strongly marked birds of Irish and west Scotland race different from most nominate *pratensis*, but all populations have wide range of ground-colour tones.

(1) Main race (except Ireland and western Scotland), nominate *pratensis*. ADULT. Head, nape, back, and wings have variable, green-olive, olive-brown, or buff-brown ground-colour, with tones other than brown strongest in fresh plumage (see Plumages for seasonal plumage sequence); overlying black-brown streaks most striking

in worn plumage and heaviest on crown and back; pale fringes on sides of mantle form pair of widely spaced buff (or even buff-white) lines. Head shows indistinct grey- or yellow-white supercilium and narrow eye-ring, and dark brown eye-stripe and edge to cheek; head pattern thus dull and indistinct, usually lacking buff tone and gentle expression of *A. trivialis* or more sharply defined and darker marks of *A. gustavi*, but in worn plumage can show stronger contrasts echoing those of all other small pipits. No really striking pattern on folded wing: dull buff or buff-white margins to dark-centred median coverts, and similarly coloured but even duller tips to greater coverts and fringes to tertials. Rump and upper tail-coverts always much brighter than head, back, and tail, with more tawny or pale ochre-olive ground-colour virtually unmarked in fresh plumage and only showing streaks or blotches on upper tail-coverts when badly worn; contrast of rump stronger than in *A. trivialis* but matched by Olive-backed Pipit *A. hodgsoni*. Tail dark brown, with central feathers much as rump but outer ones noticeably darker, especially when open, and contrasting with cold white outer webs of outermost (cold tone resulting from faint dusky suffusion absent in most small congeners but present or even stronger in Rock Pipit *A. spinoletta*). Underparts have variable grey-white or ochre-white to yellow-buff ground-colour when fresh, becoming paler (especially on intensely suffused birds) when worn; unlike most *A. trivialis* and Red-throated Pipits *A. cervinus*, usually no concentration of colour on chest or warmer tone under tail, but both sometimes present. Underparts show usually distinct, narrow black malar stripe, and wide chest-band of black-brown spots and streaks which extend on to fore-flanks and, with narrower streaks, to rear flanks and lateral tail-coverts; chest less evenly spotted than on other small pipits, and more extensively marked flanks leave smaller area of lower underparts unmarked. Underwing ochre, with grey wing-pit. JUVENILE. When newly fledged, often obviously shorter tailed than adult and this structural difference if retained into autumn promotes close similarity of form to *A. gustavi*. Plumage resembles adult in pattern but dark streaks of upperparts more distinct, while all underparts tinged yellow-buff, often markedly. (2) Irish and west Scotland race, *whistleri*. ADULT, JUVENILE. In fresh plumage, from August to mid-winter, noticeably more richly coloured than nominate *pratensis*, being more rufous above and much less white below, with pink-buff (not yellow-buff) suffusion. Such birds can look startlingly different from more olive or paler brown individuals of nominate *pratensis* and may resemble *A. trivialis* and *A. cervinus* (but not *A. gustavi*). Both races at all ages have bill fairly short and slender; dark brown-horn with brown-flesh base to lower mandible. Legs quite long but often much flexed and even hidden by loose flank feathers in creeping gait; pale brown, looking bright buff in sunlight but usually not as pink as in *A. trivialis* and *A. hodgsoni*; hind claw long and little curved, forming obvious 'spur' when perched (among small pipits, shared only by *A. cervinus*).

The commonest pipit for most west European observers, and experience soon establishes general character, rather anonymous appearance, and all-important diagnostic call. Birds in typical unworn plumage less open to confusion with other small pipits than *A. trivialis*, but *whistleri*, atypical nominate *pratensis*, and any heavily worn bird (particularly when silent and at close range, with more plumage detail visible than usual) can suggest 6 other species. Important therefore to have diagnostic combination of *A. pratensis* field characters always in mind: (1) distinctive call (see below), (2) rather weak and jerky flight and frequent creeping gait (see below), (3) extensive but not bold underpart streaking, (4) unstreaked rump, (5) cold white tail-sides, (6) brown legs, and (7) long hind claw. Most frequent pitfalls are buffier individuals resembling *A. trivialis* (distinction of paler belly on *A. pratensis*, given by Witherby *et al.* 1938a, is untrustworthy) and darker birds closely resembling immature and winter ♀ *A. cervinus* (with danger extreme if rump hidden and bird silent); see *A. cervinus* (p. 380). Flight action matched only by *A. gustavi*: consists of oddly erratic bursts of wing-beats producing jerky, hesitant progress over short distance and persistent hint of flutter and recovery even over long distance; thus not bounding like large pipits nor relatively fluent like most small ones. Escape-flight sometimes low and short (with bird going quickly to ground) but usually high and erratic (with bird calling hysterically and flight ending in sweep or plummet to cover or open ground); as with all pipits, white tail-sides very obvious in such manoeuvres. Song-flight has fluttering ascent and 'plateau', and gliding descent with tail often raised and legs dangling. Gait most creeping of all pipits, with feeding bird shuffling forward on flexed legs in characteristic mouse-like posture for long periods, only occasionally walking or running in more upright pose; behaviour more typical of genus when feeding on low-flying insects and then often pounces on or makes brief aerial sortie after prey. Carriage less erect than *A. trivialis*: usually has body line almost level when perched, but more upright in alarm. Wags tail but less persistently than *A. trivialis*. Gregarious at times, and can even form flocks of thousands on migration.

Song an accelerating and then decelerating sequence of repeated, tinkling, feeble monosyllables, rising in pitch, followed by slightly more musical trill falling in pitch: 'tsee-tsee tseek tseek tsee-er tsee-er'; lacks loud terminal flourish of *A. trivialis*. Commonest call, given in flight and on flushing, a rather thin, shrill, sibilant squeak, not carrying far and weaker than most other small pipits: 'weesk', 'pheek', 'sreep', or 'ist' (only at all similar to *A. spinoletta* whose call is less squeaky and usually given singly); when repeated or yelled in full alarm, has hysterical quality lacking in all other small pipits. On breeding grounds, a shorter, less sibilant, slightly metallic 'tisp', 'chip', or 'chisk', with anxious quality. Call variants include shorter or clipped monosyllables which suggest short calls of *A. cervinus* but not hard, stony note of *A. gustavi*.

Habitat. Breeds in middle, upper middle, and upper latitudes of west Palearctic, from temperate through boreal to fringe of arctic climatic zones, and from continental to oceanic regimes, accepting rainy, windy, and chilly conditions, but avoiding ice and prolonged snow cover as well as torrid and arid areas, within rather narrow temperature range of 10-20°C on Eurasian mainland (Voous 1960). Chooses, as a ground-dweller, open areas of rather low fairly complete vegetation cover. Avoids extensive bare rock, stones, sand, soil, and close-cropped grass or herbage, and on the other hand tall dense vegetation, including woods, forests, and reedbeds. Flourishes, however, in plots of young planted trees, and perches freely on fence-posts, telegraph wires, stone walls, and other points of vantage, but, once scattered trees appear, competitive advantage seems to pass to Tree Pipit *A. trivialis*, although in some cases mingling may occur. In south of range in Switzerland, mostly confined as a breeder to moist flat bogs at 850-1080 m (formerly down to 400 m, before drainage), leaving higher terrain to Water Pipit *A. spinoletta* (Glutz von Blotzheim 1962). In Swiss Jura and in Germany, however, breeds in montane meadows up to 1500 m,

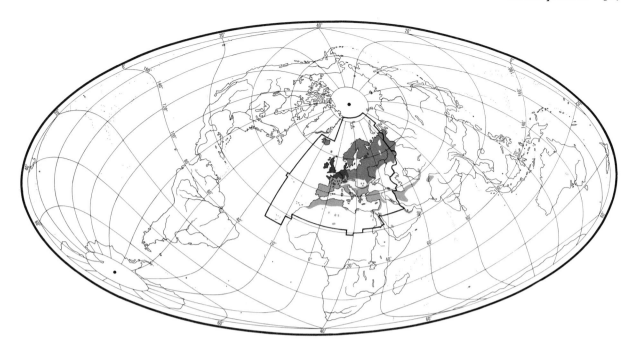

overlapping with *A. spinoletta*, as well as in moist lowland (Niethammer 1937; Biber and Link 1975). In northern USSR breeds on moss and shrubby tundra, *Sphagnum* marshes with pine and dwarf birch *Betula*, boggy hummock meadows, and marshes by inland water, or in burnt areas of forest; also on rocky patches with scree and in forest regrowth of birches up to several metres. Confined to tundra and forest zones, not extending to steppe, as a breeder. Here, as elsewhere, distribution tends to be very patchy. (Dementiev and Gladkov 1954a.) In Finland, favours wet moors, meadows, coniferous forest clearings, and treeless islets in lakes or sea, but also breeds freely in alpine zone on dry heaths with small wet patches on slopes or tops of moors, and in sub-alpine zone in open low woods of birch with bushes of juniper *Juniperus* and willow *Salix* (Pulliainen 1977). Studies in Britain indicate marked concentration on a series of distinct habitats: coastal dunes, salt-marshes (except their lower fringes), damp low-lying grasslands, lowland heaths with grass and sparse bracken, mires, grazed fens, and entire spectrum of rough heaths and moorland from those at sea-level in north, through blanket bog and gently or steeply sloping heather or grass moorland, to montane grasslands shared with Skylark *Alauda arvensis* and montane heaths avoided by it. Generally shows more catholic taste in habitats than *A. arvensis*, except on farmland. (Fuller 1982.) For densities in various habitats, see Social Pattern and Behaviour. Breeds up to summits of highest mountains in Scotland and on many inshore and offshore islands (Baxter and Rintoul 1953). Within its limits, an adaptable, enduring, and persistent species, ready to find some niche even on unfavourable terrain where choicest areas pre-empted by strong competitors,

but not immune from local declines and fluctuations of range. Winters opportunistically in terrain often similar to that used for breeding, but sometimes very dissimilar; noted in Tunisia in glades of cork oak forests near 700 m, and also on tussocky plateaux on fringe of pine forest (Bannerman 1953). Normally flies in lower airspace, even during song-flight. Encounters impact of man mainly through land reclamation.

Distribution. Some range increase in Czechoslovakia and USSR.

CZECHOSLOVAKIA. Range spread in last 70 years (Hudec and Šťastný 1979). ITALY. May breed in Alps, not confirmed (PB, BM). SPAIN. Irregular breeding in Asturias and Santander (AN). USSR. Some range increase due to habitat changes (GGB).

Accidental. Spitsbergen, Bear Island, Kuwait, Azores, Madeira.

Population. Has Increased on West German island of Mellum and in Czechoslovakia and northern Austria; decreased in Netherlands, Denmark, Poland, and part of USSR and probably southern England and Sweden.

ICELAND. Very common, with no changes noted (AP). FAEROES. 100–200 pairs (Bloch and Sørensen 1984). BRITAIN. Perhaps decreased in southern England (Parslow 1973). Probably over 3 million pairs (Sharrock 1976). FRANCE. Under 1 million pairs (Yeatman 1976). BELGIUM. About 30 000 pairs (Lippens and Wille 1972). NETHERLANDS. At least 100 000 pairs 1976–7; declined due to habitat changes (Teixeira 1979). WEST GERMANY. About 100 000–250 000 pairs (AN); large increase Mellum island (Henle 1983). DENMARK. Decrease (Møller 1982).

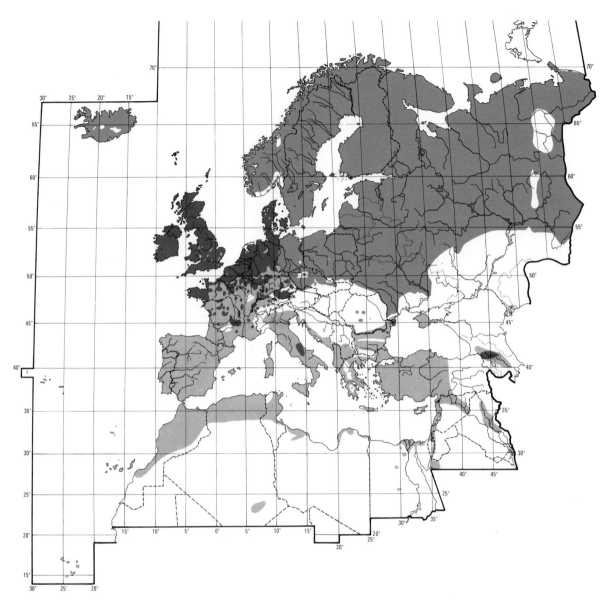

SWEDEN. Over 1·5 million pairs (Ulfstrand and Högstedt 1976). Decline in numbers of birds ringed (Österlöf and Stolt 1982). FINLAND. About 230 000 pairs (Merikallio 1958). POLAND. Decreasing due to habitat changes (AD). CZECHOSLOVAKIA. Increased over last 70 years (Hudec and Šťastný 1979). AUSTRIA. Apparently increasing in most northern part of range (HS, PP). USSR. Decreased Kremenchug area (Ukraine) after 1962 (GGB).

Mortality. West Germany: average 1st year mortality 83% (Henle 1983). Oldest ringed bird 7 years 8 months (Rydzewski 1978).

Movements. Resident or partial migrant in western Europe, but northern and eastern populations are medium-distance total migrants, though in some milder winters only extreme north completely vacated (Eber 1960). Movement almost entirely diurnal.

WESTERN POPULATIONS. Northern limits of wintering areas in western Europe not easy to define and vary with severity of season: normally, wintering occurs throughout Britain and Ireland (except uplands) and in central Europe north to Denmark and east to West Germany. Otherwise, wintering centred around Mediterranean basin. On Balearic Islands, numbers wintering thought to be declining (Bannerman and Bannerman 1983). North African wintering extends along Mediterranean coastal belt to Egypt, but penetration into Africa greatest in west, especially Morocco, with enormous numbers wintering from lowlands up to 2300 m (Smith 1965a). Most southerly occurrence is Nouakchott on coast of

Mauritania (18°N), where uncommon September–March (Gee 1984), and winters in Ahaggar massif in southern Algeria (e.g. Niethammer and Laenen 1954). In Libya, rarely south to *c*. 27°N (Bundy 1976). Not recorded in Sudan (Hogg *et al.* 1984), and identity of bird collected in southern Ethiopia (Blanchard 1969) considered unconfirmed by Urban and Brown (1971). Some reach Canary Islands and Madeira (Bannerman 1963; Wallace 1964). Birds ringed north-west Europe (Fair Isle, England, Netherlands, Belgium, West German coast) have been recovered on west coast of Morocco and Algerian coast (Heim de Balsac and Mayaud 1962).

Main passage of western populations is predominantly to southern France, Iberia, and Morocco, a south-west movement that applies to birds ringed as nestlings, as adults taken during breeding season, as migrants in autumn, and to birds ringed during winter; some, however, head east of south to Italy and Algeria. Upland breeders head randomly to adjacent lowlands but many also undertake long southerly movements (Verheyen and Grelle 1950; Eber 1960; Zink 1975; Spencer and Hudson 1978). Autumn passage in this region may begin as early as mid-August and extend to late October. At Fair Isle (Scotland), autumn departure begins late August or early September, and recoveries of birds ringed there emphasize mainly south-west movement: southern France (7), Portugal (7), Spain (8), Morocco (2), Netherlands (1), Belgium (1), and Italy (1) (Williamson 1959*b*, 1965*b*). With north-west Africa being an important wintering area much observation has been concentrated in Iberia where birds are very numerous throughout on both passages (Lletget 1945). During autumn, many birds arrive on north coast of Spain having crossed Bay of Biscay (e.g. Snow *et al.* 1955). Passage at Gibraltar on considerable scale: *c*. 4000 birds in one autumn's observation peaking mid-October to early November; visible movement maximal in early morning with subsidiary peak after 16.00 hrs (Tellería 1981). Numbers in North Africa diminish eastward: fewer birds in Algeria than in Morocco, and relatively scarce in Tunisia where most do not arrive until December (Thomsen and Jacobsen 1979). On Malta, common in autumn (from October), less so in spring (Sultana and Gauci 1982).

Populations from western Europe and Greenland extensively studied, particularly from ringing returns, with regard to direction of movement (Zang 1975; Zink 1975; Spencer and Hudson 1982). Of particular interest is flow through Greenland, Iceland, and Britain since it entails changes of direction. Birds present in eastern Greenland to early October (Bent 1950; Salomonsen 1950-1), then migrating via Iceland—birds from north of range heading nearly due south, those from south heading east. Present in Iceland to September (Hachisuka 1927), these birds joining those from Greenland and heading south-east to Faeroes and thence to Shetland where movement changes direction again (to west of

south) for Britain. Migration stream subsequently merges with populations from western Europe and changes direction again (to almost south-west) towards Iberia and Morocco. Massive exodus from southern Scandinavia observed at Utsira (south-west Norway), Lista (southern Norway), Falsterbo (southern Sweden), Ottenby (south-east Sweden), and, most spectacularly, at Blåvandshuk (western Denmark) where, over 2 weeks in September, 30 000 birds seen leaving to west, this more or less coinciding with massive arrivals on north-east coast of England (Nisbet 1957). On north-east coast of Norfolk (England), visible autumn passage occurs most days between mid-September and late November with maximum of 550 arrivals on 17 September and almost daily falls of at least 50 during 13-23 October; good correlation between visible and radar observations (Wilcock 1964). In radar counts on Scottish east coast, birds drifted north at rate of up to 3000 per day on 6-7 November with ESE wind and low visibility; also in mid-September on WNW wind, and in early November on south wind (Wilcock 1965). On Outer Hebrides (western Scotland), passage in 'full swing' by end of August, including some birds possibly from Iceland (Williamson 1959*b*); in late October, radar showed south-east departure to mainland Britain (Lee 1963). In Ireland, little evidence of emigration but birds presumed to be from Iceland have been recorded at most coastal observatories and there have been recoveries of such ringed birds in France, Spain, and Portugal (Ruttledge 1966).

Spring migration less well documented than that of autumn but ringing returns numerous enough to show that northward passage of western populations follows essentially same route as autumn, including changes in direction involved in passage via Iberia, Britain, and Iceland. Within north-west mainland Europe, including Scandinavia, movement close to north-east in almost all cases. (Zink 1975.) Spring passage in North Africa noted through Ksours mountains near northern edge of Sahara in March (Blondel 1962*b*) and passage near coast from February with large flocks mid-April (Chaworth-Musters 1939). In Tripolitania (Libya), where many overwinter, increased numbers in February suggest northward passage; also spring records from Fezzan and Libyan Desert (Bundy 1976). On Malta, 600 birds once seen arriving on 26 March; ringed birds later recovered in Italy *c*. 550 km due north (Sultana and Gauci 1982). From Cyprus departure inconspicuous but prolonged, earliest mid-February and latest mid-May (Flint and Stewart 1983). Passage throughout Iberia much greater in spring than autumn (Lletget 1945); at Coto Doñana (southern Spain) noted as late as mid-May (Mountfort and Ferguson-Lees 1961*a*). Passage recorded March–April through 4 passes in western Pyrénées (Gardner-Medwin and Murray 1958). Moves through Switzerland from end of February to early May, mainly March–April (Glutz von Blotzheim 1962), and through southern West Germany also from end of February (Rendahl 1968). In Poland, passage occurs March–April (Tomiałojć 1976), and birds arrive southern Sweden from early March, northern Sweden from mid-April (Rendahl 1968). In Britain and Ireland, passage March to mid-May (Ruttledge 1966; British Ornithologists' Union 1971), birds returning to Snowdonia (Wales) in second half of March (Walton 1979). Spring exodus from Lewis (Outer Hebrides, Scotland) mainly north-west (presumably on heading for Iceland) with a few heading east of north (presumably for Faeroes) (Lee 1963). Arrives Iceland in May (Hachisuka 1927) and eastern Greenland in mid-May (Bent 1950; Salomonsen 1950–1).

EASTERN POPULATIONS. Movements from eastern Europe and western Siberia are poorly documented but wintering areas extend east to eastern Arabia, Iran, south-central USSR, and northern Pakistan. Birds which reach Pakistan must move east of south, but lack of ringing returns means that origin of all eastern wintering birds is unknown. Zink (1975) mentioned autumn recovery east of Moscow of bird ringed in Switzerland the previous autumn showing south-west to north-east movement parallel to characteristic movement of western populations, but another bird ringed Lithuania (western USSR) and recovered in Turkey had moved well east of south. No way of assessing proportions of birds following these divergent routes nor where divisions (if any) lie between western and eastern populations.

Present October to mid-April in Gulf states of Arabia (Bundy and Warr 1980), October–March in Iraq (Chapman and McGeoch 1956), and November–April near Tehran, Iran (Passburg 1959), though small party seen moving south at south-east corner of Caspian Sea on 13 August (Feeny et al. 1968).

Autumn movement protracted and inconspicuous throughout USSR, withdrawal being mainly early September in Pechora basin, October in Moscow area, and after mid-October at Kiev. Spring arrivals occur early March in Belorussiya and mid-April in Moscow area, and extend into May around Kazan (Dementiev and Gladkov 1954a). JHE

Food. Mainly invertebrates, with some plant seeds in autumn and winter. Feeds almost exclusively on ground, walking at steady rate picking invertebrates from leaves and plant stems. Occasionally takes insects in flight which it has disturbed but never flies after them (Lebeurier and Rapine 1935b; Coulson 1956b; Glutz von Blotzheim 1962; Pedroli 1976; Prokofieva 1980; Henle 1983). One record of taking aphids in sycamore tree Acer pseudoplatanus (Robertson 1983). In Morocco, December, often seen feeding in shallow water with tail held up (Smith 1962–4). In West Germany, most insects taken less than 5 mm long (Henle 1983). For grit in stomachs, see Hågvar and Østbye (1976) and Walton (1984).

The following recorded in diet in west Palearctic.

Invertebrates: springtails (Collembola), mayflies (Ephemeroptera, Leptophlebiidae), dragonflies (Odonata: Libellulidae), stoneflies (Plecoptera), grasshoppers (Orthoptera: Acrididae, Tetrigidae), earwigs (Dermaptera), psocids (Psocoptera), bugs (Hemiptera: Acanthosomidae, Tingidae, Miridae, Cicadidae, Cercopidae, Cicadellidae, Psyllidae, Aphididae, Coccoidea), alderflies (Neuroptera: Sialidae), scorpion flies (Mecoptera: Boreidae), moths (Lepidoptera: Phalonidae, Tortricidae, Pyralidae, Coleophoridae, Noctuidae, Geometridae, Brephidae), caddis flies (Trichoptera: Limnephilidae), flies (Diptera: Tipulidae, Culicidae, Chironomidae, Bibionidae, Stratiomyidae, Tabanidae, Empididae, Rhagionidae, Dolichopodidae, Syrphidae, Psilidae, Micropezidae, Chloropidae, Tachinidae, Calliphoridae, Muscidae), Hymenoptera (sawflies Tenthredinidae, Ichneumonidae, ants Formicidae), and beetles (Coleoptera: Carabidae, Dytiscidae, Hydrophilidae, Staphylinidae, Byrrhidae, Scarabaeidae, Cantharidae, Elateridae, Coccinellidae, Chrysomelidae, Curculionidae). Also spiders (Araneae: Linyphiidae, Lycosidae, Gnaphosidae, Thomisidae, Pisauridae, Theridiidae, Tetragnathidae, Metidae), harvestmen (Opiliones), ticks (Ixodidae), centipedes (Chilopoda), woodlice (Isopoda), sand-hoppers (Amphipoda), snails (Gastropoda), earthworms (Oligochaeta), and ragworms (Polychaeta). Plant seeds include Rosaceae, Cruciferae, docks (Polygonaceae), goosefoots (Chenopodiaceae), flaxes (Linaceae), violets (Violaceae), figworts (Scrophulariaceae), heaths (Ericaceae), crowberries (Empetraceae), rushes (Juncaceae), sedges (Cyperaceae), and grasses (Gramineae). Also bread, suet, and cabbage leaves. (Newstead 1908; Rey 1910; Florence 1914; Lebeurier and Rapine 1935b; Witherby et al. 1938a; Allen 1947; Taylor 1953; Coulson 1956b; Glutz von Blotzheim 1962; Hågvar and Østbye 1976; Walton 1979; Henle 1983.)

In Snowdonia (Wales), March–October, 300 m above sea-level, 238 stomachs contained only invertebrates: of 5154 items, 48·6% (by number) flies (mostly Bibio and Dilophus), 14·0% beetles, 9·9% spiders, 8·2% larvae of moth Coleophora alticolella, 7·3% Hymenoptera, 5·6% other larvae, 4·5% bugs, 1·9% miscellaneous (Walton 1979). In southern Norway, June–September, 1200 m above sea-level, 2201 items from 88 stomachs comprised 22·4% (by number) beetles, 17·8% flies, 15·3% Hymenoptera, 13·7% Euphrasia seeds, 7·8% harvestmen, 6·6% spiders, 6·4% insect larvae, 4·7% other plant seeds, 1·3% moths, 1·0% scorpion flies, 0·8% insect pupae, 0·5% snails, 0·2% springtails, 0·1% earthworms and 0·1% ticks (Hågvar and Østbye 1976). On Mellum island (West Germany), August, 64 items from 20 stomachs comprised 31·2% (by number) beetles, 23·4% Diptera, 20·3% ants, 10·9% bugs, 9·4% plant seeds, 1·6% earwigs, 1·6% spiders, and 1·6% centipedes (Henle 1983). In France, main winter foods are plant seeds and snails: seeds of grass Poa annua especially important, found in 52·8% of 53 stomachs November–March, one containing 247 seeds; snails, mainly *Bulimus* and *Planorbis*, present in 41·9% of 62 stomachs November–April (Lebeurier and Rapine 1935*b*). In Crimea (USSR) 2 stomachs, November and January, contained beetles (Coccinellidae) and small snails, and 1 from 25 April contained spiders and small Carabidae (Kostin 1983). On Yamal peninsula (north-west Siberia), spring, unknown number of stomachs contained 19·5% (by number) Staphylinidae, 17·2% Coccoidea, 17·2% spiders, 14·9% caterpillars, 13·8% flies (Mycetophilidae), and some Carabidae, cicadas, caddis flies, sawflies, and springtails. After breeding, 29·6% Ichneumonidae, 18·5% Staphylinidae, 18·5% caddis flies, 14·8% sawflies and small numbers of alderflies, cicadas, moths, Diptera, and plant seeds. Autumn migrants' stomachs contained 41·7% Staphylinidae, 13·9% caddis flies, and some Diptera, caterpillars, Ichneumonidae, and Carabidae (Danilov *et al.* 1984).

Diet of young similar to adults'; almost exclusively invertebrates. Items brought to nest usually over 10 mm long, fed to young whole (Pedroli 1976; Prokofieva 1980). In Pennines (England) 216 items brought to 1st broods comprised 84·7% (by number) *Tipula*, 11·1% mayflies, 3·2% other adult Diptera, 0·5% Diptera larvae, 0·5% stoneflies; 122 items brought to 2nd broods comprised 41·0% *Tipula*, 30·3% mayflies, 11·5% other adult Diptera, 7·4% adult Lepidoptera, 4·9% Diptera larvae, 3·3% stoneflies, and 1·6% caterpillars (Coulson 1956*b*). In Snowdonia, 47 items fed to nestlings comprised (by number) 66·0% Diptera, 17·0% moths, 10·6% sawflies, 2·1% beetles, 2·1% harvestmen, and 2·1% springtails (Walton 1979). In Jura (Switzerland), 400 samples contained 613 items: Pyralidae (larvae), Tipulidae (adults and larvae), and Lycosidae each present in 23–24% of samples; Geometridae larvae, and adults and larvae of Syrphidae, Empididae, and Tachinidae each present in 12–16% (Pedroli 1976). In Leningrad region (USSR), unknown number of samples from 2 nests included 39·9% (by number) Diptera (mainly Tipulidae), 24·4% spiders, and also beetles, Hymenoptera, moths, bugs, grasshoppers, caddis fly larvae, earthworms, and snails (Prokofieva 1980). On Yamal peninsula, 1466 items from 209 samples included 35·4% (by number) Chironomidae, 26·6% other Diptera (mainly Tipulidae and Culicidae), 11·9% sawflies, 9·8% spiders, 7·1% moths, and 3·5% bugs. In Pennines, 10·0 (6·7–13·6) feeding visits to nest per hr for broods of 4–5, lowest in cold snowy weather, highest when warm and sunny. Rate highest after dawn declining gradually through day until sunset; average 4·87 items brought per visit (*n* = 94) (Coulson 1956*b*). On Røst (Norway), 5 10-day-old nestlings fed 68 times in 4 hrs, 4 at 3 days old 46 times in 4 hrs, and at 4 days 77 times in 6½ hrs (Wagner 1958). In Jura, ♂ made 1·4 visits per hr, ♀ 4·4; only 1 young fed per visit (Pedroli 1975). In Snowdonia, 66 stomachs from fledged young, June–October, contained 985 items: 20·6% (by number) Hymenoptera, 17·8% Diptera, 16·8% beetles, 13·1% bugs, 10·9% spiders, 8·8% insect larvae, and 4·8% larval *Coleophora alticolella*; also stoneflies, ticks, adult moths, harvestmen, snails, and seeds. Immediately after fledging (June–July), juveniles caught many more beetles than did adults; although scarcer, beetles probably more easily caught than faster-moving, more abundant flies (Walton 1979). PJE

Social pattern and behaviour. No major study but most aspects relatively well known.

Often in flocks outside breeding season, but these rather loose-knit (e.g. Witherby *et al.* 1938*a*, Ptushenko and Inozemtsev 1968). In Brittany (France), typically solitary or in small flocks (8–10) in autumn; larger but only short-lived assemblages occur for feeding (Lebeurier and Rapine 1935*b*). On Malta, wintering birds normally solitary during day, though up to *c.* 250 recorded feeding together; also gather for roosting (Sultana and Gauci 1982). Concentrations of several hundreds or even thousands recorded during spring and autumn migration (Kumerloeve 1970*c*; Turnbull 1984; D I M Wallace). Bad weather in early spring may cause breeding birds to abandon territories for short period and re-form flocks (Elfström 1979). Juvenile dispersion studied in detail in Melle area (Niedersachsen, West Germany). Flocks of independent juveniles fluctuated markedly in size and composition even during day; less true of 2–4 large flocks in which almost all juveniles concentrated at end of season; flocks larger for roosting than during day. Adults recorded only exceptionally in juvenile flocks late August and September. Small flocks of juveniles noted mainly in territories where adults still feeding young (Hötker 1982). In Kazakhstan (USSR), often migrates with other *Anthus* (Dolgushin *et al.* 1970). In Leningrad region (USSR), also usually migrates in mixed flocks: initially with Tree Pipit *A. trivialis* and Skylark *Alauda arvensis*, later with Chaffinch *Fringilla coelebs* and White Wagtail *Motacilla alba*. Sometimes feeds with buntings (Emberizidae) (Mal'chevski and Pukinski 1983), on Brittany coast also with Rock Pipit *A. spinoletta petrosus* (Lebeurier and Rapine 1935*b*). BONDS. Monogamous mating system, though polygyny recorded occasionally: in North Wales, ♂ fed young at 2 different nests (Seel and Walton 1974); at Melle, ♂ paired with 2 ♀♀ helped to rear brood of one but nest of other destroyed before hatching (Hötker and Sudfeldt 1979*a*; see also Breeding Dispersion, below); in Swedish Lapland, 2 birds incubated at nest containing 11 eggs and 3rd bird present probably ♂ (Fredriksson and Svensson 1984). Sex-ratio on Mellum island (West Germany) 1:1 and probably similar in juveniles (Henle 1983); in North Wales, that of shot birds relatively steady at 2 ♂♂: 1 ♀, March–June (Seel and Walton 1979). Few data on length of pair-bond. In North Wales, 4 pairs stayed together for 2 broods in same season. 1 pair-bond maintained for 3 successive years; 4 ♂♂ took new mates after 1 season, their mates of previous year probably having died (Seel and Walton 1979). At Melle, 1 pair remained faithful for 3 broods of 1 season; ♂ had bred in same territory with unringed ♀♀ in 2 previous years, ♀ had had 2 different mates in those years (Hötker and Sudfeldt 1978). For probable hybridization with *A. trivialis*, see that species (p. 350). ♀ does most feeding of young (Pedroli 1975)—in one study on Røst (Norway), about twice as much as ♂ (Wagner 1958). At Melle, single juveniles (not related to adults concerned) recorded as helpers at 2 nests: in one case, at 3rd-brood nest with 5 young close to fledging, helper from nest *c.* 3 km away; in other case, nest contained 4 young 5-6 days old, ♀ probably lost earlier, and copulation recorded once between adult ♂ and juvenile helper (from nest c. 3·5 km away) c. 5 days after adult ♀ last seen (Hötker and Sudfeldt 1979b). Young usually tended for c. 2-3 weeks by both parents after leaving nest, and most young have left parental territory after c. 20 days. Last broods recorded with parents 12 days after leaving territory and brood-siblings regularly stay together for some time afterwards, including within pre-departure flocks (Hötker 1982). In Belgium, 1st brood evicted from territory by ♂ after 8 days, last brood said to have remained with parents to end of October (Verheyen 1947) though this doubted by Hötker (1982); see also subsection 1 in Heterosexual Behaviour and Relations within Family Group (below). BREEDING DISPERSION. Solitary and territorial. Neighbourhood groups (of 3-4 pairs) recorded in (e.g.) Brittany where nests sometimes less than 20 m apart; once 3 within c. 100 m (Lebeurier and Rapine 1935b); see also Danilov et al. (1984) for Yamal peninsula (USSR). At Åsa (Halland, south-west Sweden), average distance between nests 110 ± SD 54·6 m (n = 32) (Elfström 1979). Study in North Wales indicated 'territorial' boundaries not sharply defined. Birds seen as inhabiting home range comprising 3 concentric zones: (a) central zone representing normal living area; (b) inner zone used by adults for feeding; (c) outer zone with perimeter marked by extreme points of foraging trips for nestling food. Considerable overlap with home ranges of adjacent pairs in outer zone (though birds more spaced in some years), moderate but sporadic in inner zone, and slight in central zone. Core area (central zone plus inner zone) averages 2·18 ± SD 0·71 ha (n = 23). In 3 pairs, little correspondence between feeding range (of both sexes) and ♂'s core range. Birds sing in c. 46% of core range (see Seel and and Walton 1979 for further details, including illustrations). At Åsa, territorial boundaries tend to consist of zone c. 20 m wide; territories square, unlike in *A. spinoletta* (see that species, p. 400) (Elfström 1979). In Brittany, territory c. 5-16 ha; most territories contain song-post. Birds also forage for young well outside territory (Constant and Eybert 1980), crossing other territories to do so if necessary (Wiprächtiger 1971). At Dieppe (Seine-Maritime, France), 2 territories each said to be c. 250 m² (c. 0·025 ha) (Ferry 1947). On Yamal, ♂ song-territories isolated throughout song-period, though pairs not infrequently settle adjacent to one another and feeding areas overlap to considerable extent (Danilov et al. 1984). 2 ♀♀ paired to one ♂ (see Bonds) used different parts of his territory without overlap; nests c. 240 m apart (Hötker and Sudfeldt 1979a). At Melle, nests for successive broods averaged 88·7 (2-320) m apart: average of 21 successful pairs 93·3 m, of 16 unsuccessful 82·8 m. In 2 cases, successful pairs changed territory and nests were 355 and 2725 m away from that previously used. Rearing of 3 broods in same territory also recorded however (Hötker and Sudfeldt 1978, 1982b). Early stages of afforestation may attract high densities: e.g. on Rhum (Scotland), 52 pairs per km² on 1-year-old ploughed and planted moorland, 18 pairs per km² on adjacent unplanted moorland; still the commonest species in 15-year-old plantations. Densities in main habitats of Britain: 1-76 pairs per km² on salt-marshes, 26-55 on chalk hill grasslands (locally absent), 28-59 on calcareous mires and lings, 18-55 on acid moors; density down to 5 pairs per km² on impoverished upland, but overall average c. 1000 pairs per occupied 10-km square (Sharrock 1976). In North Wales, annual average (4 years) 48 pairs per km² in mosaic of wet and dry grassland on sloping and level ground (Seel and Walton 1979, which see for caveats regarding methods and other references); see also Jackson and Long (1973). In Brittany, 8-64 pairs per km² (Constant and Eybert 1980).

Maximum on pasture at c. 1500 m in Swiss Jura 22 pairs per km²; on bogs and wet meadow 2-8 pairs per km² (Pedroli 1978). For further Swiss data, see Schifferli et al. (1982). In Netherlands, mostly 2-5 pairs, locally 20-30 pairs per km² (Teixeira 1979). On Mellum, breeding population in 1977 184-298 pairs, giving for green area of c. 2 km² density of 120 pairs per km²; for area of short grass actually used (c. 0·25 km²), 964 pairs per km² (Henle 1983). In Mecklenburg (East Germany), mostly 10-20 pairs per km² (Klafs and Stübs 1977, which see for differences by habitat). In Brandenburg (East Germany), large-scale censuses indicated usual maximum 10 pairs per km² even in apparently suitable habitats (Rutschke 1983). Wet meadows with sparse pine *Pinus mugo* in Krkonoše mountains (Czechoslovakia) hold up to 64 birds per km² (Klíma and Urbánek 1958). On Yamal, 2 localities held 1·6-13·5 pairs per km² over 10 years and 1·3-18·2 pairs per km² over 9 years; further details of habitat preferences in Danilov et al. (1984). In favourable habitat (mostly water meadows) of Leningrad region (USSR), up to 300 pairs per km² (Mal'chevski and Pukinski 1983). In Swiss Jura at c. 1500 m, breeds in close association with Water Pipit *A. spinoletta spinoletta* (Biber and Link 1975). At Åsa, territories overlap with those of Rock Pipit *A. s. littoralis*, nests averaging 51·8 ± SD 20 m (n = 26) apart, and c. 15 m recorded. *A. spinoletta* tolerated in zone of overlap (adult conspecific never), but repelled (like other small passerines) from within c. 5 m of nest (Elfström 1979). On Yamal, considerable overlap with territories of Red-throated Pipit *A. cervinus* (Danilov et al. 1984). Birds show strong tendency to return to natal area to breed (e.g. Henle 1983 for Mellum). In North Wales, one bird bred c. 570 m (centre of core range—see above) from natal nest; also, all returning ♂♂ used at least part of core range previously occupied (Seel and Walton 1979). Site-tenacity evident in other studies: at Melle, ♂ bred in same territory for 3 years, his mate of 3rd year having bred c. 600 m away in previous 2 (Hötker and Sudfeldt 1978). 1 Yamal ♀ bred c. 350 m away from her previous year's nest (Danilov et al. 1984). ROOSTING. Little information. Nocturnal and probably often communal outside breeding season. On ground under plants and amongst grass in the open, also in young plantations; base of osiers and hedges used mainly in severe weather (Witherby et al. 1938a; Ptushenko and Inozemtsev 1968); see also Sultana and Gauci (1982) for wintering birds on Malta. On Lyngen Peninsula (69°30'N in Norway) during continuous daylight coinciding with nestling phase, adult activity ceased for average 5 hrs 45 min per 24 hrs, 58% of this before midnight; inactivity showed closer correlation with measured minimum temperature than with measured minimum light (Hillman and Young 1977). Birds drink and bathe regularly (Henle 1983); see also King (1953). For anting, see Holt (1960) and Mester (1969); habit rare in *Anthus* (Simmons 1960).

2. For comparison and analysis of breeding-season display patterns (function, motivation, behavioural isolating mechanisms) in *A. pratensis* and *A. spinoletta littoralis*, see Elfström (1979). *A. pratensis* less shy than *A. trivialis* and *A. spinoletta* on breeding grounds according to Géroudet (1956), and typically more approachable than *A. s. spinoletta* in winter: often allows approach to c. 10 m, then habitually flies up (almost vertically) and calls (Beneden 1950: see 2 in Voice). Can be very wild, however (Lebeurier and Rapine 1935b). For example of exceptional tameness, see Allen (1947). On breeding grounds, bird faced by predator (no details) may stand erect and wag down-pointed tail, or crouch, before escaping (Elfström 1979). For escape from marauding gulls *Larus* near ship at sea, see Bub (1977). FLOCK BEHAVIOUR. Flocks, including juveniles

(Hötker 1982), generally loose-knit and when disturbed, birds tend to fly up (giving call 2a) singly or in small groups rather than *en masse* (Lebeurier and Rapine 1935*b*; Witherby *et al.* 1938*a*). Flying up to perch in trees typically occurs where several birds together during migration (Westerfrölke 1971); in Hamburg, *c.* 30 out of feeding flock of *c.* 300 perched in willows *Salix* (up to *c.* 6 m tall), with constant movement between perches and ground (Kumerloeve 1970*c*). SONG-DISPLAY. ♂ sings (see 1 in Voice) mainly in flight, though also from high or low perch (Ferry 1947) or from ground (Witherby *et al.* 1938*a*). According to latter, song from perch or ground shorter and more imperfect; at Åsa, birds more often gave 1st segment than complete song and did so in flight or on ground (Elfström 1979). Bird singing from perch appears more squat and rounded than usual: body held low, tail raised and slightly spread, wings drooped at *c.* 45% (see below). Usually gives 3–4 bursts before flying to another perch; sometimes longer series interrupted by self-preening (Ferry 1947). Song-flight serves to advertise territory and attract ♀, though also performed after pairing, thus perhaps leading to occasional polygyny (Seel and Walton 1979; see also Bonds, above). Performed after starting song from perch or ground or when disturbed. Ascends with fluttering wing-action and tail widely spread, rather like *A. arvensis*, moderately fast and not quite vertically or at *c.* 40°, to *c.* 5–35 m. Then either descends immediately or flies with markedly shallower wing-beats, sometimes circling, more or less level, for some distance. Descends in parachute style like *A. trivialis*: tail first slightly spread and horizontal, then raised and usually closed; wings bent stiffly back and raised, legs dangling (Fig A). Prior to landing (often on take-off point),

A

may fly horizontally with quivering wings; sometimes circles near ground. 1st part of song (1–3 segments) given from take-off or only after ascending a few metres, remainder during descent, part sometimes after landing. (Stadler and Schmitt 1913; Lebeurier and Rapine 1935*b*; Witherby *et al.* 1938*a*; Ferry 1947; Elfström 1979; Bergmann and Helb 1982). Bird may reascend into 2nd Song-flight without even landing (Witherby *et al.* 1938*a*) or leave several minutes between performances (Lebeurier and Rapine 1935*b*). In French study, song occasionally given in descent from tree-top. Said also to give call 2d quite often after brief silence following Song-flight (Lebeurier and Rapine 1935*b*). For illustrations, see Elfström (1979) and for schematic protrayal of Song-flights, see Ptushenko and Inozemtsev (1968). Most song given in morning and late afternoon (Ferry 1947), especially in sunshine (Lebeurier and Rapine 1935*b*); in Leningrad region, at peak (June), given more or less throughout day (Mal'chevski and Pukinski 1983). In Britain, mainly mid-March to early July, irregular but fairly frequent mid-February to mid-March and early July to early August, exceptional or only Subsong early August to mid-September (Witherby *et al.* 1938*a*); see also Seel and Walton (1979) for North Wales. In Brittany, sings from March (Lebeurier and Rapine 1935*b*), also on Mellum, but marked decline there by end of April and only occasional song in July (Henle 1983). In Moscow region, birds sing from arrival (April or early May) to mid-July (Ptushenko and Inozemtsev 1968).

ANTAGONISTIC BEHAVIOUR. (1) General. In Brittany, most pairs on territory by late February or early March. ♂♂ typically perch high and conspicuously, also perform frequent Song-flights; later, spend more time on ground (Constant and Eybert 1980); see also Seel and Walton (1979) for North Wales. Yamal territories occupied on arrival, but fights rare and occur only on days of heavy influx (Danilov *et al.* 1984). In Brittany, ♂♂ become more aggressive as season advances; early in season, ♂♂ fight over ♀♀ (Lebeurier and Rapine 1935*b*). At Åsa, most territories established before April; if later, birds normally do not breed. Birds congregating in more sheltered territory during rough weather (see part 1) tolerated by owner. Territories otherwise rarely abandoned; normally vacated after 2nd brood (Elfström 1979). At Melle, juvenile ♂ (*c.* 40 days out of nest) sang and apparently held territory for *c.* 4 days. Independent young nearly always tolerated in territories where adults still feeding young (Hötker 1982; see also part 1); not so other adults (Ferry 1947). (2) Threat and fighting. 8–10 ♂♂ noted singing fairly close together (Mal'chevski and Pukinski 1983). Territory-owner normally performs Song-flight if neighbour does so (e.g. Ferry *et al.* (1969). At Åsa, Song-duels regular, more frequent than in *A. spinoletta*, but *A. pratensis* pairs closer together (see also Breeding Dispersion, above). ♂ on territory advertises occupancy by frequent drooping of wings; may also do so when migrants overhead and may then call (see 2a in Voice) and very often (more so if unpaired) performs Song-flight. Intruders landing in territory or flying low over it evicted by owner. Walks towards intruder or flies in, sometimes to displace it directly. At close quarters (e.g. in boundary confrontation), increasing tendency to attack manifested in drooping and vibrating of wings and raising of tail, bird crouching with plumage ruffled and sometimes giving chattering call (see 2a in Voice). Threat-posture rare in such encounters, but typically used to defend individual-distance in flocks; used in breeding season mainly towards other species (including *A. spinoletta*) and in heterosexual interactions (see below). Bird faces opponent crouched low; body and tail horizontal, head forward of body though neck drawn in, bill usually closed. Bird stationary or may charge. Birds will fan and drag tail with wings drooped (Fig B); this also more typical of ♂-♀ meetings, but in

B

antagonistic interactions occasionally accompanied by stamping with one or both feet. Attack-escape conflict indicated in irregular and abrupt side-to-side movements, trampling, etc., also adoption of more upright posture (wings not drooped) with tail closed and lowered, plumage sleeked, and neck more extended. Before retreat, may also crouch low with neck drawn in, body and tail horizontal. Fights normally aerial, birds rising (exceptionally to *c.* 20 m) face to face, snapping, and clawing briefly. In chase, pursuer has tail spread as birds fly low and fast, often less than 1·5 m apart; Song-flight and Quivering-flight (slower than normal and with shallow wing-beats) may be interspersed. For details, including statistical analysis, see Elfström (1979); see also Lebeurier and Rapine (1935*b*), Morley

(1940), Ferry (1947), and Conder (1948). On Yamal, stuffed conspecific ♂ elicited aggression from territorial ♂ throughout breeding season, but not from ♀. However, ♀ attempting to solicit paired ♂ was attacked by his mate (Danilov *et al.* 1984). At Åsa, ♀♀ generally attacked less often than ♂♂ (Elfström 1979). On hearing playback of song, territorial ♂ makes erratic jerky movements, approaches, and may hover or land on loudspeaker. Commonly performs Quivering-flight (with chattering-call) and on ground will display much as described above. For details, including positive reaction to *A. spinoletta* vocalizations, see Ferry *et al.* (1969), Elfström (1979), and Vitale and Brémond (1979). HETEROSEXUAL BEHAVIOUR. (1) General. In Brittany late February to early March, ♀ often 'attended' by 2–3 ♂♂. Courtship lasts *c.* 5 weeks (Lebeurier and Rapine 1935*b*). At Melle, laying begins average 40 days (5–76, *n* = 57) after territories occupied; period shorter if occupation late (Hötker and Sudfeldt 1982*a*). Unpaired ♂ remained on territory and sang for *c.* 2 months (Wiprächtiger 1971). ♂ losing mate (e.g. if ♀ leaves with 1st brood) usually increases song and high perching; sang until early August at Åsa (Elfström 1979; Seel and Walton 1979); see also Constant and Eybert (1980). Ground display (see below) performed while feeding fledged young as prelude to 2nd brood (Wiprächtiger 1978). No conflict recorded between pair-members in study by Wagner (1958); at Åsa, Threat-posture used by ♀ to defend individual distance in early encounter with ♂ and later in season (Elfström 1979). One polygynous ♂ to some extent alternated between feeding young of one ♀ and courting other; when young of 1st ♀ reared, ♂ spent time only with 2nd ♀ (Hötker and Sudfeldt 1979*a*). (2) Pair-bonding behaviour. ♂ often approaches ♀ in Quivering-flight. ♀ may be treated initially as intruder, ♂ running at her or showing irregular movements. When near ♀ on ground, ♂ frequently droops wings and raises tail, sometimes carries grass, etc. ♂ may also stand more erect than usual, plumage neutral to sleeked. Fan-tail posture (Fig B) tends to be associated with bowing (Elfström 1979), which may serve to display white beneath tail (Nethersole-Thompson and Nethersole-Thompson 1940*a*). ♂ often flaps wings, grass-pecks, runs round ♀ while carrying material, or struts away from her in Fan-tail posture. ♀ occasionally also carries material, but often appears indifferent, feeding. Both birds may fly low giving call 2a quietly. For call particularly associated with courtship, see 3 in Voice. Chases frequent (Elfström 1979), but not sustained (Nethersole-Thompson and Nethersole-Thompson 1940*a*). In interaction witnessed by Macdonald (1968), both birds called intermittently throughout. ♂ recorded carrying material in Quivering-flight (Morley 1940). One ♀ approached by ♂ in Quivering-flight crouched slightly with head, back and tail level; ♂, stiffly erect, carpal joints slightly raised, head rigid, moved 4–5 times in semi-circle in front of ♀, also bowed and sang quietly; when ♂ left, ♀ at first crouched still, then resumed feeding (Butlin 1940). Both birds of pair recorded excitedly opening and closing raised tails at nest containing young (Nethersole-Thompson and Nethersole-Thompson 1940*a*). Strange ♀ once approached paired ♂ and adopted posture like food-begging young (Danilov *et al.* 1984; see below). (3) Courtship-feeding. ♂ regularly feeds ♀ on or just off nest, ♀ crouching and quivering wings beforehand (Nethersole-Thompson and Nethersole-Thompson 1940*a*; Elfström 1979; Seel and Walton 1979). In Swiss Jura, one ♂ fed ♀ on nest 1·6 times per hr over 14 hrs (Pedroli 1978). (4) Mating. In one case, ♂ performed Song-flight and called 10–12 times (sequences a few seconds apart), faster when ♀ approached. ♂ then sang from perch *c.* 5 m from ♀, vibrated wings, and raised tail. Copulation took place on ground after

chase. ♀ fed afterwards, ♂ (back on perch) facing her with bill open. Performance repeated after *c.* 40 min (Brown and Goodyear 1948). ♂ recorded approaching ♀ on ground while singing; during copulation lasting 2–3 s, ♂ had head and tail raised, wings drooped; chased ♀ afterwards (Conder 1948). On Røst, copulation recorded when birds had young *c.* 5 days old (Wagner 1958). (5) Nest-site selection. Probably by ♀ (see Nethersole-Thompson and Nethersole-Thompson 1943). (6) Behaviour at nest. Tends to walk to and from nest (Wagner 1958). RELATIONS WITHIN FAMILY GROUP. Young hatch over 1–2 days (Danilov *et al.* 1984). Even freshly hatched young little-brooded if weather fine, but in long stints over first 2 days if cold; ♀ then leaves only to feed herself and, as ♂'s feeding rate lower (see Food), 1–2 young may starve (Pedroli 1975). In 2 north European studies, young probably little-brooded (except at night) after *c.* 5–6 days (Davies 1958; Wagner 1958). Eyes of young open at *c.* 4–6 days (Davies 1958). Adult recorded passing food to mate (*c.* 15 m from nest) who then fed young (Wiprächtiger 1971), and on Røst, ♂ never longer than *c.* 5–12 s at nest (Wagner 1958); at Melle, both parents frequently at nest together to feed offspring (Hötker and Sudfeldt 1979*b*). Adult moving away from one nest usually gave call 2a (Steuri *et al.* 1979). Young recorded leaving nest at 8 days to escape hot sun; quite often brooded in such conditions. ♀ may start laying for 2nd brood 1–2 days before 1st-brood young leave, and these frequently still dependent at start of 2nd brood (Hötker and Sudfeldt 1982*a, b*). Young normally leave nest still unable to fly (but see below) at *c.* 11–13(–14) days (Coulson 1956*a*; Davies 1958; Pedroli and Graf-Jaccottet 1978; Constant and Eybert 1980), earlier if disturbed (Hötker 1982; Danilov *et al.* 1984). At 69°30′N, nestling period during continuous daylight 10 days (Hillman and Young 1977). On Røst, clearly able to fly on leaving at *c.* 14–15 days (Wagner 1958). At Melle, young apparently stay together for first few days after leaving, usually keeping within *c.* 100 m of nest. By *c.* 1 week after leaving, some broods still close together in territory, others (fledged) maximum 150 m apart. Shortly before independence, young may move round territory in sibling groups. 3 late broods (adults in attendance) well outside territory 12 days after leaving; crossed only unoccupied territories. Young often stay near parental territory into September (Hötker 1982). ANTI-PREDATOR RESPONSES OF YOUNG. At 10–13 days, all may leave and scatter simultaneously when disturbed (Hötker 1982). Out of nest, normally crouch and hide, also on hearing parental alarm (see 2d in Voice), though may fly up briefly when man close (Géroudet 1956). PARENTAL ANTI-PREDATOR STRATEGIES. (1) Passive measures. ♀ sometimes sits very tightly (Davies 1958), allowing approach (young *c.* 1 day old) to *c.* 0·5 m (Appert 1970). ♀ prevented from incubating will wait and call constantly, as will both adults later when feeding young (Lebeurier and Rapine 1935*b*). May not go to nest for up to 1 hr while observer present. ♂ shyer than ♀; one ♀ seen to snatch food from her hesitant mate and collect more (Wagner 1958). ♂ may sing while carrying food or give alarm-call from perch. At 2 nests, birds noticeably quieter than *A. trivialis* in similar situation (Appert 1970). (2) Active measures. All information relates to man as potential predator. Recorded shuffling and fluttering for considerable distance after shooting off nest and dropping heavily into vegetation (manoeuvre repeated). More positive distraction-lure display of disablement type performed especially if with young, though apparently less common overall than in *A. trivialis* (Witherby *et al.* 1938*a*; Tucker *et al.* 1948). Slow flapping typical of injury-feigning bird (Elfström 1979). For further descriptions, see *Br. Birds* 1947 **40**, 20, 177, 310, 344.

(Figs by I Lewington: A from photograph by H-H Bergmann; B from drawing in Elfström 1979.) MGW

Voice. More vocal than Water Pipit *A. spinoletta* (Bergmann and Helb 1982). Repertoire studied in detail by Stadler and Schmitt (1913, 1915). For extra sonagrams, see Elfström (1979), Bergmann and Helb (1982), and Wallschläger (1984).

CALLS OF ADULTS. (1) Song of ♂. Full song, normally delivered in flight, typically a long series of several different segments, each comprising a number of uniform units (Bergmann and Helb 1982). Incomplete song not infrequent (in some populations apparently more common than full song) and mostly given from perch or ground, though also in flight; often consists only of units which normally form 1st segment of full song (Stadler and Schmitt 1913; Elfström 1979; Wallschläger 1984). Such sounds rather thin, feeble, and tinkling (Witherby *et al.* 1938a), and variously rendered as follows: 'tsip' (Stadler and Schmitt 1913), 'tchip' (Bergmann and Helb 1982), 'tui' (Lebeurier and Rapine 1935b), 'tsi', 'tyi', 'tyie' (Ferry 1947), and in recording by P A D Hollom, more liquid 'tselp' or 'tlip'. Elfström (1979) noted segments of such units lasting 0·03–63·0 s. In full song, 11–200(–300) units sometimes given in irregular tempo and with brief pauses (Stadler and Schmitt 1913); while Ferry (1947) recorded 6–30 (usually 15–20) over *c.* 5–6 s when only 1st segment given. In such abbreviated songs, rate of unit delivery normally constant, 3–5 per s depending on structure of unit (Wallschläger 1984). Full song always contains segment of 'tsip' sounds which normally start quietly and become louder and faster. Typically associated with this change is a rise in pitch and consequent change in timbre (pitch sometimes first falls then rises); sounds become quieter again and following long and high-pitched tremolo segment often quiet to inaudible. Other segments (whole song of about 5 or more) mainly of tremolo type and vary in length, pitch, and timbre. Short 'i' (pronounced 'ee' in English), 'si' or tonal 'jü' sounds typically follow short rattle towards end of song (Stadler and Schmitt 1913, which see for detailed analysis and musical notation; Ferry 1947; Bergmann and Helb 1982; Wallschläger 1984). Short 'i' sounds, or soft, melancholy 'ti-u' or 'si-u' likened to Tree Pipit *A. trivialis* (Stadler and Schmitt 1913; Ferry 1947). Song lasts 10–23 s (Witherby *et al.* 1938a) or, with 5 different segment types given in regular order, 15–25 s (Ferry 1947). In recording of song lasting 21 s, bird starts with segment of 52 thin, rather squeaky or strained 'psee' or 'psip' sounds at roughly equal time intervals and lasting 11 s (Fig I shows 4 of these sounds); followed by 7 similar but lower-pitched 'whee' sounds. First of 3 rattle (tremolo) segments almost attains rich tone of thrush *Turdus* or Nightingale *Luscinia megarhynchos*, next is much thinner and drier like a mechanical jingling, and 3rd a richer sound again and recalling Sedge Warbler *Acrocephalus schoenobaenus*. Pen-

ultimate segment comprises 8 buzzing 'zzeep' sounds suggesting Skylark *Alauda arvensis* and song ends with coda of 9 rich, pure, gentle, fluting tones similar to first 2 segments but lower pitched than either. In recording by R Savage, 2nd segment following fragmented 1st (bird makes hesitant start to song from perch before ascending into Song-flight) strongly reminiscent of 'tick-tock' of Snipe *Gallinago gallinago* and perhaps an imitation (J Hall-Craggs; M G Wilson); incorporation of mimicry held to be typical by Wallschläger (1984). Comparison of various recordings, and sonagrams in sources detailed above, indicate some variation in units. Elfström (1979) found most marked variation between individuals in 1st segment, while units in later segments (2nd and 3rd 'phrases' of a 3-phrase song) had same basic form in all individuals of population, thus creating dialect. Wallschläger (1984) noted constancy of abbreviated (1st-segment) songs over several years and similarly assumed existence of dialects. (2) Contact-alarm calls. (a) In flight and especially at take-off, a thin, feeble, squeaky, normally high-pitched 'tsiip', 'psiip', or sharp 'iss(t)'. Frequently a double call or, when startled, several in succession: 'ississ issississississis iss' (Stadler and Schmitt 1915, which see for fuller discussion and variants; Witherby *et al.* 1938a; Bergmann and Helb 1982). Louder calls of high-intensity alarm have shrieked, almost hysterical quality (D I M Wallace) and quieter variants also occur. Less hissing, purer-sounding and higher-pitched than equivalent call of *A. spinoletta* (Bergmann and Helb 1982). Given mostly outside breeding season (Stadler and Schmitt 1915). In recording (last 4 units of Fig II), squeaks have slightly distorted quality and first 2 are close enough together to sound like double call (P J Sellar, M G Wilson). Elfström (1979) illustrated (but not described in detail) a chattering call. (b) In recording by P A D Hollom of October flock, birds (seldom flushed or moving far) give quite rapid series of lower-pitched 'chew' or 'piu' sounds somewhat reminiscent of finch (Fringillidae) and combined with quieter variants of call 2a (M G Wilson): 'choo-choo-choo-choo-tseep-tseep'; number of calls per burst, speed, and pitch vary. Such combinations typical of birds on migration (P A D Hollom, P J Sellar). Rough 'ch-tt utt utt utt...' of Jonsson (1978a) probably also a low-pitched sequence of this sort. (c) On breeding grounds and usually on ground or when perched, a 'tsipp' different (at least in timbre) from unit in introductory segment of song (Stadler and Schmitt 1915). Also described as a chirping 'chutt' or 'chitt'; used to maintain contact rather than to express anxiety (Jonsson 1978a; P J Sellar). Recording suggests a 'trip', with 'r' to indicate just audible break in each unit (J Hall-Craggs: Fig III); also rendered 'tlit' or 'tsit' and some units more obviously disyllabic—'zlip-ip' or 'tlip-ip'—so bird perhaps alternating or intergrading with call 2d (M G Wilson). (d) Alarm-call near nest a persistent 'stitt-itt' (Jonsson 1978a); 'tuip', 'tilic', or 'tuluc', sometimes in

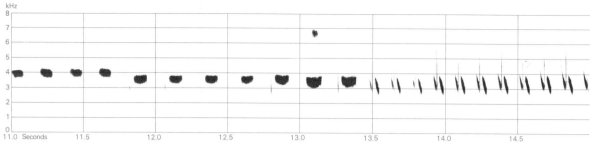

I V C Lewis Wales June 1978

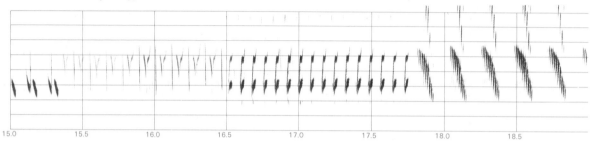

I *cont.*

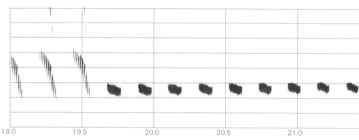

I *cont.*

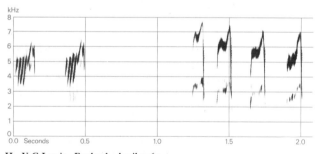

II V C Lewis England April 1962

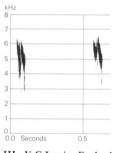

III V C Lewis England
May 1971

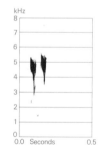

IV P J Sellar
Scotland June 1974

V P Radford Scotland July 1977

long series (Lebeurier and Rapine 1935*b*). Typically a disyllabic call (P J Sellar) and recording suggests 'ti-ti' (J Hall-Craggs: Fig IV); also rendered 'chit-ip' or 'fit-it' with squeaky, almost metallic quality and strongly recalling cricket *Gryllus* (M G Wilson; also Appert 1970). In recording of bird alarmed at presence of Kestrel *Falco tinnunculus*, sounds sharper, higher pitched, and more pronounced starting transient suggests 'kit-it' (J Hall-Craggs, M G Wilson: Fig V). Alarm-calls documented by Bergmann and Helb (1982): 'pi' or just disyllabic 'tli'; at high intensity also trisyllabic 'pititi' and quite long series of 'tli' sounds when disturbed at nest. See also Stadler and Schmitt (1915). (3) A buzzing, fruity '(p) rrreep' or 'prrreet' resembling well-grown young House Martin *Delichon urbica*, given singly or as fairly rapid series and combined with call 2a (M G Wilson: first 2 units of Fig II). Used in courtship (V C Lewis). Trilling 'tee-ee-ee' (Morley 1940) probably the same. (4) Other calls. (a) Bright ringing 'tisitsi' recalling Dunnock *Prunella modularis* (Jonsson 1978a). (b) High-pitched but muted 'errrr' given rarely by flushed bird (Stadler and Schmitt 1915).

CALLS OF YOUNG. Food-call a discordant 'stih' (Jonsson 1979); shrill call of fledglings eventually develops into adult call 2a (see Elfström 1979). Only other information relates to age at which song first given: in 2 wild juveniles at latest 40-42 days old (Hötker 1982); captive bird at 2 months (Heinroth and Heinroth 1924-6). MGW

Breeding. SEASON. Onset of laying affected by temperature in last third of March (Hötker and Sudfeldt 1982a), becoming later with increasing altitude and latitude (Coulson 1956a). Central and western Europe: first eggs from first half of April, last eggs laid beginning of August (Pedroli 1978; Hötker and Sudfeldt 1982a). Britain: from second half of April with peak 2nd week of May in south and 3rd week in north, also delayed by altitude, *c.* 1 day per 45 m (Coulson 1956a). Swedish Lapland: laying begins mid-June to early July (Davies 1978). SITE. On ground, usually concealed in vegetation. Of 277 nests, West Germany, 81·2% on slopes (sides of ditches, banks, etc.), 18·8% on flat ground; 95·2% in slight hollow, 92·5% sheltered from above by vegetation; proportion on slopes declined through season (Hötker and Sudfeldt 1982b). Nest: cup of grasses and other plant material, lined finer vegetation and hair. Building: by ♀. EGGS. See Plate 80. Sub-elliptical, smooth and glossy; quite variable, usually brown, grey, or reddish, spotted or mottled, sometimes finely streaked brown, black, or grey. 19·5 × 14·4 mm (17·2-21·6 × 13·5-15·8), *n* = 300; calculated weight 2·06 g (Schönwetter 1976). Clutch: 3-5 (2-6). Of 115 clutches, West Germany: 2 eggs, 1%; 3, 3%; 4, 31%; 5, 63%; 6, 3%; mean 4·64, declining through season (Hötker and Sudfeldt 1982a). Of 75 clutches, southern Norway: 4 eggs, 3; 5, 36; 6, 35; 7, 1; mean 5·42. Of 28 clutches, Finnmark (northern Norway); 4 eggs, 2; 5, 5; 6, 19; 7, 2; mean 5·75 (Coulson 1956a). 2 broods normal in central and western Europe, but rare in north (Coulson 1956a). 3 broods recorded from 2 ♀♀, West Germany (Hötker and Sudfeldt 1978). Eggs laid daily. INCUBATION. 13 days (11-15), *n* = 19 (Coulson 1956a). Normally by ♀ only (Seel and Walton 1979); occasionally by ♂ according to Ryves (1943). Begins with last egg, occasionally next to last (Makatsch 1976). YOUNG. Altricial and nidicolous. Cared for and fed by both parents. FLEDGING TO MATURITY. Fledging period 12·5 days (10-14), *n* = 36 (Coulson 1956a). Young often leave nest before fully capable of flight. Young tended by parents and fed, though decreasingly with time, for mean 13 days (4-39, *n* = 78) after leaving nest (Hötker 1982). Age of first breeding 1 year. BREEDING SUCCESS. Of 221 eggs laid, Switzerland, 21% predated, 3% infertile, and 76% hatched; further 7% young dead from starvation within 48 hrs, 4% thereafter, and 9% predated, giving 56% overall success up to fledging (Pedroli 1978). Of 534 eggs laid, West Germany, in 115 clutches, 65% hatched and 71% of these fledged, giving 46% overall success, with 4·45 young reared per pair per year (Hötker and Sudfeldt 1982a). In West German study of *c.* 240 pairs producing 1095 young up to mid-August, estimated 69% of eggs hatched and 78% of these fledged, giving 54% overall success (Henle 1983). In Swedish Lapland, of 112 eggs laid, in 23 clutches, 88% hatched and 88% of these fledged, giving 78% overall success (Davies 1958).

Plumages (nominate *pratensis*). ADULT BREEDING. Similar to adult non-breeding (see below), but variable amount of non-breeding retained and hence rather variable in appearance. In March-April, head and body usually new, as in fresh non-breeding; some tertials, tertial coverts, t1, and variable number of median upper wing-coverts usually new also, olive-green or pale olive-brown fringes contrasting with bleached off-white fringes of retained non-breeding wing-coverts and tertials. Plumage heavily worn by about June-July (more so than worn non-breeding in December-March), upperparts dull olive-grey with heavy black-brown marks; sides of head heavily mottled, without distinct pattern except for white eye-ring; underparts off-white with contrasting black-brown spots from rear cheek down to sides of breast, chest, and flanks; pale tips of median and greater upper wing-coverts largely abraded. ADULT NON-BREEDING. Ground-colour of upperparts rather variable: usually greenish-olive, sometimes more yellow-brown (with golden tinge) or greenish- or yellowish-olive-brown. Forehead and crown rather narrowly and sharply streaked black, hindneck and sides of neck more narrowly and faintly; mantle, inner scapulars, and back with distinct and well-defined black feather-centres; rump uniform or with feather-centres faintly darker olive-grey, longer upper tail-coverts with poorly defined dull grey centres. Lores pale buff or cream-yellow, bordered below by dull black line from base of upper mandible to front of eye; short supercilium from just above eye over ear-coverts pale yellow, finely speckled with some olive above ear-coverts. Distinct though narrow ring of pale yellow feathers round eye. Upper cheeks and ear-coverts greenish-olive to olive-brown, indistinctly streaked and spotted pale yellow just behind and below eye (sometimes showing as pale patch), faintly streaked dark olive-grey on upper ear-coverts and on lower cheeks (sometimes showing as faint dark stripes from eye backwards below supercilium and from gape backwards). Chin and throat pale cream-buff or cream-white, this colour sometimes extending upwards on sides of neck along ear-coverts; small but distinct dull black spots on side of throat join to form more or less complete malar stripe; stripe widens into large spotty or streaky dull black patch on sides of chest. Ground-colour of chest, sides of breast, and flanks pale yellow-buff to cream; central belly, vent, and under tail-coverts usually paler cream or buff-white, but entire underparts sometimes fairly deep yellowish-buff; chest, upper breast, sides of breast, and flanks with well-defined black streaks, *c.* 2 mm wide on chest, sides of breast, and lower flanks, slightly narrower on breast and remainder of flanks, often extending as fine shaft-streaks on to sides of belly and feathering near thighs; throat sometimes finely spotted grey. Central pair of tail-feathers (t1) dark grey or olive-black with rather narrow and poorly defined green-olive fringe all round; t2–t4 black with narrow yellow-green outer edge, t4 sometimes with small white spot on tip; t5 dark grey with pale yellow outer edge and short white wedge on tip of inner web, this wedge 9·9 (61) 3–15 mm in length, exceptionally 21; t6 white with grey wash on distal part of outer web and with dark grey base extending into grey wedge on middle portion of inner web. Flight-feathers and tertials dull black; tertials with rather broad olive-green or pale olive-brown fringe along both webs, secondaries with broad but ill-defined olive-green or olive-brown fringe along outer web, primaries with narrow and sharp yellow-green or pale yellow outer edge, nearly white on p9; all fringes and edges, particularly of tertials and primaries, strongly affected by bleaching and wear, soon changing to yellowish-white or off-white. Lesser upper wing-coverts olive-green or olive-brown (like upperparts), poorly defined feather-centres duller grey, tips of longer lesser coverts sometimes slightly paler, yellow-green; median coverts black with contrasting olive-green or pale olive-brown tips 1·5–2 mm wide; greater coverts dull black with less contrasting olive-green or pale olive-brown fringe along outer web and tip. Greater upper primary coverts and feathers of bastard wing dull black with narrow olive-green or yellowish edge along outer web and tip. Under wing-coverts and axillaries dull grey, broadly fringed pale yellow-green to yellow-white. *In worn plumage*, about December–March, ground-colour of upperparts duller greyish-olive, less yellowish-green or brown, fringes to sides of mantle-feathers and inner scapulars sometimes partly greyish-white, showing as pale spots or broken streaks, black streaks of upperparts more marked; yellow or buff tinge of sides of head and underparts bleached so that short supercilium, eye-ring, and mottled patch below eye off-white, lower cheeks, chin, and throat off-white with malar stripe more pronounced (but sometimes partly abraded and indistinct), and remainder of underparts cream or yellow-white with black streaks and spots on chest, sides of breast, and flanks more contrasting; fringes of lesser upper wing-coverts and t1 and those of outer webs of greater coverts and secondaries abraded, bleached to olive-grey; tips of median and greater upper wing-coverts and sides of tertials bleached to off-white. NESTLING. Down rather long and dense, restricted to upperparts; brown-grey (Heinroth and Heinroth 1924–6; Witherby *et al.* 1938a). JUVENILE. All plumage equally new or evenly worn, as in adult non-breeding, without contrast between fresh (olive-tipped) and worn (white-tipped) upper wing-coverts and tertials (unlike adult breeding and 1st adult); flight-feathers and tail new when those of adult are worn or moulting. Ground-colour of upperparts more buff-brown, not olive-green or olive-brown as adult and 1st adult; dark streaks broader, purer black and more sharply defined, in particular crown, mantle, and outer scapulars appearing darker. Ground-colour of underparts between buff and off-white, depending on wear; spots on feathers of chest, sides of breast, and flanks rather variable in size and extent, spots sometimes broad and deep black, sometimes narrow, duller black, and less sharply defined; spots usually longer and more triangular than those of adult, not rounded on side near feather-tip. FIRST ADULT NON-BREEDING. Like adult non-breeding, but variable amount of juvenile feathers retained (see Moults). Many birds retain juvenile median and greater upper wing-coverts and tertials (except sometimes for tertial coverts

or an odd tertial); bleached fringes and tips of these distinctly paler than neighbouring fresh outer scapulars (in adult, all feathers equally new, fringes and tips closely similar in colour to those of scapulars). In some birds, retained juvenile coverts and tertials still new and hardly paler than scapulars; in others, many wing-coverts and all tertials replaced (usually except outer greater coverts), and these birds hard to age without experience. No constant difference in pattern of black on retained juvenile median upper wing-coverts (*contra* Svensson 1984a). FIRST ADULT BREEDING. Like adult breeding, and many indistinguishable; retained juvenile flight-feathers, greater upper primary coverts, and t2–t6 on average more heavily worn than those of adults.

Bare parts. ADULT, FIRST ADULT. Iris dark brown. Bill brown-black or horn-black, cutting edges of both mandibles and sometimes base of lower mandible yellow-flesh, pale flesh-yellow, or flesh-pink. Leg and foot yellow-brown, light brown-flesh, pale flesh-yellow, or bright flesh-pink; joints slightly darker or tinged grey, claws pale horn, pink-horn, or (recorded in summer) black. NESTLING. Bare skin (including bill and foot) flesh-pink. Gape-flanges yellow, mouth carmine-red or orange-red. JUVENILE. Iris brown or dark brown. Bill pink-flesh, grey-brown, or flesh-brown, gradually darkening to blackish from tip towards head during post-juvenile moult. Leg and foot pink-yellow, flesh-pink, or flesh-brown. (Heinroth and Heinroth 1924–6; Witherby *et al.* 1938a; RMNH, ZMA.)

Moults. ADULT POST-BREEDING. In Britain, starts with p1 mid-June to mid-August (sometimes early June to late August), completed with regrowth of p9–p10 late August to first days of October (sometimes mid-August to mid-October); estimated duration of primary moult *c.* 50 days (Ginn and Melville 1983). Of 12 birds from Netherlands, July, none had started (ZMA, RMNH). In Finland, starts July, primary moult lasting *c.* 42 days (Haukioja and Kalinainen 1969, 1972). In USSR, flocks with moult not completed occur in late September (Dementiev and Gladkov 1954a). ADULT PRE-BREEDING. Partial. Starts late December to March, completed March–April. Usually involves head, body, tertials, tertial coverts, lesser and median upper wing-coverts, and t1, sometimes a few outer tail-feathers or all tail or some inner greater upper wing-coverts; frequently, all upper wing-coverts retained (except tertial coverts), as well as some inner tertials or all tail. In May–June, birds from Netherlands more worn than those from Scotland and Scandinavia, perhaps due to earlier moult in Netherlands or to more-frequent partial or complete suppression of moult. In USSR, starts mid-March; moult intense in second half of March, completed late April (Dementiev and Gladkov 1954a); in Asia Minor, birds moult in March, but many depart before moult started (some of these perhaps not moulting at all) and others perhaps leave while moulting (Weigold 1914). POST-JUVENILE. Partial. Start highly variable, depending on fledging date; 1st birds just starting moult encountered late May, last bird in full juvenile plumage early September; birds in active moult examined late May to early October; immatures with moult completed seen from early August onwards and some probably complete before this. Extent rather variable: birds retain at least juvenile flight-feathers and greater upper primary coverts, and often t2–t6 and some outer greater upper secondary coverts, but much variation in number of tertials, tertial coverts, and median upper wing-coverts replaced. FIRST PRE-BREEDING. Partial; extent and timing variable, as in adult pre-breeding. In some birds, no moult at all (see e.g. Stresemann 1920), or moult restricted, involving some tertials and wing-coverts only (then probably merely a continuation of post-juvenile); in others, all head, body, lesser and median upper wing-coverts, tertials, tertial coverts, and t1 replaced. Scattered feathers of upperparts and all non-breeding upper wing-coverts more often retained than in adult.

Measurements. ADULT, FIRST ADULT. Nominate *pratensis*. Wing of (1) Netherlands, breeding, May-August; (2) Netherlands, autumn migrants, second half of September to November; (3) southern Europe and north-west Africa, winter; other measurements combined for these periods; skins (RMNH, ZMA). Bill (S) to skull, bill (N) to distal corner of nostril; exposed culmen on average 3·7 less than bill (S). Toe is middle toe with claw; claw is hind claw.

WING (1)	♂ 81·3 (1·86; 31)	78-85	♀ 77·6 (2·36; 10)	74-81
(2)	82·0 (1·57; 24)	80-86	77·6 (1·72; 22)	74-81
(3)	81·0 (1·37; 18)	79-84	76·0 (1·41; 13)	73-78
TAIL	58·0 (1·73; 27)	55-62	54·7 (2·57; 15)	52-59
BILL (S)	14·7 (0·40; 25)	14·0-15·3	14·2 (0·50; 15)	13·7-15·1
BILL (N)	8·5 (0·36; 14)	8·1-9·1	8·2 (0·31; 10)	7·8-8·5
TARSUS	21·4 (0·71; 26)	20·4-22·7	20·9 (1·04; 15)	19·8-22·4
TOE	18·2 (0·91; 11)	17·2-20·2	18·6 (0·54; 10)	17·8-19·6
CLAW	10·8 (1·97; 28)	9·2-14·5	11·1 (1·03; 21)	9·7-13·2

Sex differences significant for wing, tail, and bill (S). 1st adult (with juvenile wing and often tail) and full adult combined above, though average wing of adult 1·7 longer than juvenile, average tail 2·1 longer. Hind claw mainly 9·6-13·4; only 4 birds below 9·6 (including one breeding adult with claws 3·5 only, excluded from table), only one over 13·4.

WING. Timan tundra (north-west USSR): ♂ 82·0 (8) 79-84, ♀ 77·4 (7) 74-79 (Gladkow 1941).

Weights. Skokholm (Wales). Adult: spring, 18·8 (1·25; 97) 15·2-21·8; summer and autumn, 18·4 (1·05; 228) 13·9-23·4. Juvenile and 1st adult: summer and autumn, 18·4 (675) 13·9-23·4; averages 17 June-7 July 17·9 (68), 8-28 July 18·7 (48), 29 July-18 August 18·2 (110), 19 August-8 September 18·8 (98), later on 19·2 (62). (Browne and Browne 1956.) For monthly variations, Wales, see Seel and Walton (1979). Netherlands, May-June: ♂ 17·3 (0·92; 5) 16-18 (ZMA). West Germany, summer: Ruhrtal, ♂ 19·0 (1·46; 23) 15-22, ♀ 19·5 (1·07; 15) 16-22; Osnabrück-Bielefeld area, ♂ 18·6 (1·07; 24) 17-20, ♀ 18·5 (1·41; 27) 17-19 (Bub *et al.* 1981). Iceland, April-May: ♂ 19, ♀ 17 (Timmermann 1938-49). Averages of adults, Norway: June, 18·4 (c. 1·4; 13); July, 17·5 (c. 2·3; 15); August, 17·5 (c. 2·3; 17); September, 19·9 (c. 3·0; 10) (Skar *et al.* 1972). Yamal peninsula (USSR), summer: ♂ 19·1 (18-21), ♀ 21·1 (18-27) (Danilov *et al.* 1984).

Netherlands, during autumn migration, September-November: ♂ 18·9 (1·45; 17) 16-21, ♀ 17·7 (2·06; 12) 15-21 (RMNH, ZMA). Malta, autumn and winter: 17·5 (113) 14·5-22 (J Sultana and C Gauci). On spring migration, Netherlands

and East Germany, April: ♂ 24·4 (3) 24-25 (Bub *et al.* 1981; ZMA).

Exhausted birds, mainly killed during bad weather. Skomer (Wales), January: 12·7 (12) 8-16 (Harris 1962). Netherlands, all year: ♂ 13·2 (0·31; 6) 12-14, ♀ 11·6 (1·36; 12) 9-13 (ZMA). Iceland, early May: ♂ 14 (15) 12-17, ♀ 12·5 (5) 11-15 (Timmermann 1938-49).

Nestling on 1st day after hatching, Switzerland, 1·8 (18); strong increase to 15·6 (17) on 9th day; less marked increase to 18·3 (13) on 12th day, when leaving nest (Pedroli and Graf-Jaccottet 1978). See also Seel and Walton (1979).

Structure. Wing rather short, broad at base, tip bluntly pointed. 10 primaries: p7 and p8 longest or either one 0-1 shorter than other; p6 and p9 0-1·5 shorter than longest, p5 8-12, p4 12-17, p3 14-20, p1 17-23. P10 reduced, narrow and pointed, hidden under outer greater primary covert; 51-61 shorter than wing-tip, 8-11 shorter than longest upper primary coverts. Outer web of p6-p8 and inner of p7-p9 emarginated. Tertials (s7-s9) long; tip of longest (s7) reaches to 2-6 from wing-tip when wing closed. Tail rather long, tip square or slightly forked; 12 feathers, each rather pointed at all ages, t1 3-6 shorter than t5-t6. Bill fine and slender, straight, length slightly less than half head length; culmen slightly concave in front of nostrils, slightly decurved towards fine and sharp tip. Some short fine bristles along base of upper mandible, projecting obliquely down. Tarsus rather long and slender, laterally compressed (not rounded at rear as in larks Alaudidae); tibia feathered. Toes long and slender; for length of middle toe with claw, see Measurements; outer toe with claw c. 71% of middle with claw, inner c. 74%, hind with claw c. 112%, without c. 52%. Claws slender, decurved; front claws rather short, hind claw long (see Measurements).

Geographical variation. Slight and clinal. *A. p. whistleri* from Ireland and western Scotland on average deeper and redder olive-brown above, and with slightly heavier black streaks; streaks on underparts similar to typical nominate *pratensis* from Scandinavia, but ground-colour of chest, sides of breast, and flanks markedly deeper, cinnamon-buff (Clancey 1943a, 1948; BMNH). East from western Scotland, populations gradually paler, and birds from USSR on average paler and greyer than Scandinavian ones, but difference very slight (Meinertzhagen 1933). Birds from Iceland and Greenland similar to Scandinavian birds, not to *whistleri*. Population from western Ireland on average slightly darker than those from western Scotland and hence sometimes separated as *theresae* (Meinertzhagen, 1953), e.g. by Vaurie (1954), but differences slight and variation within Britain and Ireland gradual, thus *theresae* does not warrant recognition (Clancey 1961; P A Clancey, K H Voous).   CSR

Anthus cervinus Red-throated Pipit

DU. Roodkeelpieper   FR. Pipit à gorge rousse   GE. Rotkehlpieper
RU. Краснозобый конек   SP. Bisbita gorrirrojo   SW. Rödstrupig piplärka

Motacilla cervina Pallas, 1811

Monotypic

Field characters. 15 cm; wing-span 25–27 cm. Close in size to Tree Pipit *A. trivialis*, appearing noticeably plumper than Meadow Pipit *A. pratensis* in foreparts. Rather small, sleek, but quite robust pipit, with bulk approaching Olive-backed Pipit *A. hodgsoni* and (unlike all other west Palearctic pipits except Water Pipit *A. spinoletta*) colourful, changing plumage patterns. Combination of little-marked face, broadly streaked upperparts and (in winter) buff underparts with large-spotted chest and flanks creates darker, richer plumage than *A. trivialis* and most *A. pratensis*. In summer, ♂ less streaked below with variable pink or red-buff suffusion on face, throat, and even chest, producing diagnostic appearance; summer ♀ may resemble ♂ but usually has less pink and more streaks below. At close range, rather short bill, usually brown legs, and long hind claw form useful characters. Flight like *A. trivialis*. Sexes usually dissimilar in breeding plumage; marked seasonal variation in most birds, especially ♂♂. Juvenile separable.

ADULT MALE BREEDING. Ground-colour of crown, nape, back, and wings usually noticeably dark, rich buff-brown, though some birds have these areas greyer, more rufous, or even slightly purple; on all, however, broad black-brown streaks on feather-centres reduce obviousness of paler margins even when fresh, making ♂ *A. cervinus* typically much darker above than any other small pipit. Mantle may show double V lines formed by paler margins but most obvious features are (a) buff-white or yellow-buff margins to black-centred median coverts and similarly coloured tips to brown-centred greater coverts, forming distinct upper and less striking lower wing-bars, and (b) similarly coloured fringes to tertials, obvious from side and (particularly) behind, contrasting with dark rump. Rump and upper tail-coverts as back, with broad, virtually black streaks obvious when wings relaxed (among small west Palearctic pipits, pattern shared only by Pechora Pipit *A. gustavi*). Tail black-brown, fringed buff-brown on central feathers and with cold white outer webs to outermost ones forming obvious margins, like *A. pratensis*. Face essentially buff-pink with supercilium, pale lore, and eye-ring contrasting with dark crown; browner rear eye-stripe and lower and rear edges to cheek less obvious; eye thus prominent, and front or whole of face looks open. Chin and throat uniform with face but chest and flanks variably pink-buff, orange-buff, or even red-buff, contrasting with or merging into buff or buff-white lower underparts. Some birds have throat

and chest (except lower sides) unmarked, but others have a few faint streaks across chest, and a few have as many streaks as in winter; all, however, have regular, heavy streaks from lower sides of chest along flanks. Front of underparts thus looks noticeably darker than rear (this pattern only faintly suggested by richest-coloured *A. pratensis* and markedly dissimilar from summer Water Pipit *A. spinoletta* which has marked supercilium and usually only clouded breast with few streaks on flanks). Underwing buff-grey. ADULT FEMALE BREEDING. Variation in summer ♂♂ and ♀♀ makes sexing of many paler birds dubious, but ♀ usually less warmly coloured on front of underparts, with chin and throat buff or cream and chest less pink, these areas retaining more nearly complete pattern of winter streaking. Difference between sexes most obvious in July when many ♀♀ have almost white throats. ADULT MALE NON-BREEDING. From August, loses pink suffusion of face, throat, and chest, and underparts become heavily marked by partial black malar stripe and obvious chest-band of boldly splashed black spots which extend more markedly down flanks; ground-colour and markings of underparts thus darkest of all small pipits in winter except Rock Pipit *A. spinoletta*. Fresh upperparts show paler feather-margins, these sometimes forming obvious single or double V lines on mantle. ADULT FEMALE NON-BREEDING. Again closely resembles ♂ but face pattern more pronounced, with buff-white lores and supercilium contrasting more with dark crown and brown cheeks, and chin and throat not showing any pink tone, being buff-white. JUVENILE. Closely resembles winter ♀ but with pale margins to upperpart feathers narrower, forming sharp buff V lines on mantle; whole underparts suffused yellow-ochre, and throat as well as chest and flanks heavily streaked. This plumage thus darkest of any small pipit in west Palearctic except Rock Pipit *A. spinoletta*. FIRST YEAR. Autumn migrants notoriously variable in tone of upperparts: some as dark as adult but others paler and duller, with stronger grey, or even faintly olive or tawny, ground-colour to upperparts on which pale lines on mantle may appear as marked as on *A. pratensis* (important to recognize that these markings not diagnostic in any pipit). On pale birds, underparts may also be generally whiter (like *A. pratensis*) but under tail-coverts always obviously buff (unlike nearly all *A. pratensis* except Irish and Scottish race *A. p. whistleri*). Not known whether full adult plumage attained in 1st spring: dull, streaked ♂♂ and ♀♀ with little pink on throat seen on passage in April–May and on breeding grounds in early summer are presumably either part of individual variation between adults or have undergone incomplete assumption of breeding plumage. At all ages, bill size between *A. trivialis* and *A. pratensis* in thickness but shorter than either; dark brown-horn with flesh-buff base to lower mandible. Legs yellow- to brown-flesh, pink-red in spring; hind claw a long and little-curved 'spur', as in *A. pratensis*, far longer than in *A. hodgsoni*, *A. trivialis*, or *A. gustavi*.

Typical breeding adult virtually unmistakable in west Palearctic, with Rosy Pipit *A. roseatus* safely confined to Himalayas and further east, and only the most rufous individuals of Irish and Scottish *A. pratensis whistleri* providing pitfall in western Europe; winter adult and immature usually distinctive but their variable plumage and calls promote regular or potential confusions with (1) *A. hodgsoni* (sharing similar call, but having markedly distinct plumage colours and pattern; see p. 337), (2) *A. trivialis* (sharing similar call and buff ground to plumage, but with unstreaked rump and short hind claw; see p. 344), (3) *A. gustavi* (sharing rich plumage tone to upperparts and streaked rump, but with strikingly white belly and distinctive call; see p. 359), and (4) *A. pratensis* (sharing similar plumage tones in *A. p. whistleri* and having same long hind claw, but with unstreaked rump and distinctive call; see p. 365). Flight action as *A. trivialis* but silhouette somewhat bulkier, with rather broader wing-tip. Escape-flight as *A. pratensis*, in clear reflection of similar use of open habitat. Song-flight well developed; higher than *A. pratensis* and often sustained longer before parachuting descent. Gait and carriage combine those of *A. trivialis* and *A. pratensis*, including latter's creeping walk.

Song between that of *A. trivialis* and *A. pratensis* in quality, consisting of shrill, loud monosyllables followed by bubbling trill, then hissing 4-syllable sounds repeated but not in pronounced decelerating flourish: 'twee-twee trrrrr twizz-wizz-wizz-wizz...'. 2 common calls, differing in tone and form. (1) In migration flight and in alarm, a quite loud call of 1–2 syllables—hissing and rasping at a distance but piercing at close range: monosyllabic form 'skeeze', 'sii(s)', or 'teez'; disyllabic form 'skee-eaz', 'sssii', 'tee(ee)', 'teez-eez', 'teaz-p', or 'pee-ez'; extended disyllabic form 's(k)ee-eze'. Distinguished (at close range) from similar notes of *A. trivialis* and *A. hodgsoni* by having quite marked pulse in volume, thus 'ss-se-EEZ-zeez-sss' (though this transcription suggests too long a call). (2) From wintering birds in flock and some migrants, a full, rather abrupt but less high-pitched 'teu', 'chup', 'chwit', or 'chit', uttered both singly and in rapid series and somewhat similar to shorter, quieter notes of *A. pratensis* but more musical and less clipped than monosyllable of *A. gustavi*. Alarm-call of breeding bird 'tsweep', rather hoarse and shriller than 2nd call described above.

Habitat. Arctic and subarctic, between July isotherms of 2–15°C, north of forest limits and mainly on shrubby or mossy tundra (Voous 1960), although locally in Scandinavia up to 1000 m, and near water in swamps of willow and birch. Near settlements will adapt to drained and cultivated land, as well as damp grassy flats (Bannerman 1953). Overlaps in northern Europe with Meadow Pipit *A. pratensis* and Tree Pipit *A. trivialis*, both of which are less adapted to arctic conditions. In USSR, favours hummocky humid tundra and boggy levels overgrown with sedge and other herbage, including osier thickets (Dementiev and Gladkov 1954a). For breeding densities in various habitats, see Social Pattern and Behaviour.

Winters, in Egypt, on irrigated ground, but associated then almost everywhere with short grass produced by grazing animals and above all with areas having very shallow water, as in cattle-tramped mud. Other African winter habitats include mudflats around Rift Valley lakes, newly-turned ploughland on coast, dumps, and cattle yards, high-altitude moorlands, *Artemisia* scrub, and even bare rocky outcrops up to 2500 m. (Moreau 1972.)

Distribution. No major range changes reported.

Norway. Bred in south 1975 (VR).

Accidental. Faeroes, Iceland, Ireland, Netherlands, Spain, Albania.

Population. Sweden. About 1000 pairs (Ulfstrand and Högstedt 1976); estimate unreliable (LR). Finland. Rough estimate 300–500 pairs; in danger of extinction (Merikallio 1958). Marked fluctuations in Lapland (OH). USSR. Abundant in suitable localities (Dementiev and Gladkov 1954a).

Movements. Migratory, wintering largely in tropics of Africa and south-east Asia. Migration thus longer than that of any other *Anthus* except Pechora Pipit *A. gustavi* (p. 360). Position of migratory divide within breeding range not known. For additional data, see Eber and Szijj (1960) and Zink (1975).

In Europe, winters regularly only in one region of south-east Italy (P Brichetti). Also regular in scattered areas of southern Turkey, Middle East, and North Africa, and abundant in Nile valley of Egypt (Etchécopar and Hüe 1967). In Morocco, small numbers recorded wintering near western coast (Smith 1965a). Winters all across Sahel zone, though uncommon further west than Sudan and few birds west of 5°E (e.g. Cave and Macdonald 1955, Curry and Sayer 1979). Scarce but regular in Gambia (Cawkell and Moreau 1963; Gore 1981), and regular in Ghana but only at coastal saltpans (Macdonald 1978). In Nigeria, winters mainly in north but some records from coast at Lagos (Elgood 1982). In eastern Africa, common to locally abundant in Ethiopia (Urban and Brown 1971), widespread in Kenya (especially

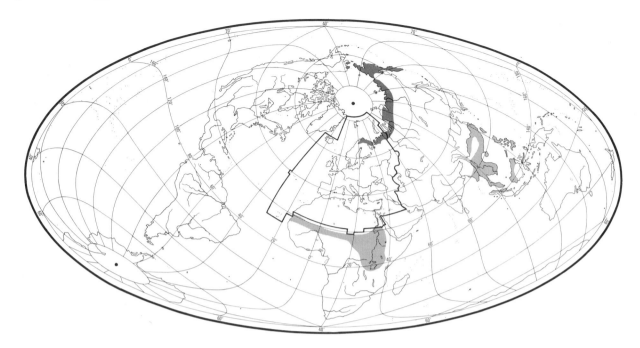

common in Rift Valley area), and regular south to north-east Tanzania (occasionally to 7°S), but scarce in Uganda (Britton 1980). 2 records (southernmost) from Ndola, Zambia (Taylor 1979, 1980b).

In accordance with eastern Africa being more important wintering area than western, the most westerly populations in northern Fenno-Scandia move mainly east of south in autumn to pass east of Baltic into central Europe, but there is also significant flow via Falsterbo, southern Sweden (Rudebeck 1947). Records scanty through west-central Europe: e.g. scarce but widespread in Poland, more numerous in autumn than spring (Tomiałojć 1976). Accidental or uncommon in Iberia (Lletget 1945; Henty 1961) and very few recorded on passage at Gibraltar (Lathbury 1970)—only 2 in an autumn's observation (Tellería 1981). Rare also on Balearic islands: 3 records in spring and 1 in autumn (Bannerman and Bannerman 1983). Large numbers occur on passage through central and east Mediterranean and Middle East. Fairly common in spring on Malta (mostly single birds but large influxes occur, once over 800 birds), though smaller numbers occur in autumn (Sultana and Gauci 1982). Uncommon in both spring and autumn in Greece but occurs in all areas (Lambert 1957; Raines 1962). Common on both passages in Cyprus (Bannerman and Bannerman 1958; Flint and Stewart 1983). Scarce in Tunisia, occurring mostly in eastern coastal areas (Thomsen and Jacobsen 1979), and seen less often on passage through desert oases of Libya than other Motacillidae (Hogg 1974). Passage through Lebanon occurs chiefly near the coast (Vere Benson 1970). At Azraq (Jordan), spring passage flocks reach peak of over 500, arriving from SSE and departing on same line

(Ferguson-Lees 1969). Common on spring passage through most of Arabia, but uncommon then in North Yemen; in Gulf states much less abundant in autumn than in spring, but not so in Oman (Meinertzhagen 1954; Bundy and Warr 1980; Gallagher and Woodcock 1980; Jennings 1981a; Cornwallis and Porter 1982). In Iraq, more common in spring than autumn (Marchant 1963a). Only small-scale spring passage through Iran (Passburg 1959) and very scarce in Afghanistan, Pakistan, and India, most records coming from Pakistan and north-west India from Gilgit to Baroda (Paludan 1959; Ali and Ripley 1973b). In Eritrea (Ethiopia), migrates through all altitudes, with marked spring movement along coast and across plateaux (Smith 1957, 1960a).

Leaves tundra of USSR usually between late August and early October (Dementiev and Gladkov 1954a). Passes Falsterbo early September to mid-October with peak during 3rd week of September (Rudebeck 1947); similar timing in Denmark (Huber 1947; Palm 1951). In Austria, passage at Lake Neusiedl occurs mid-September to mid-October (Huber 1954a). Occurs on Malta mid-October to November (Sultana and Gauci 1982) and on Cyprus October–November (Bannerman and Bannerman 1958; Flint and Stewart 1983). Passes through Tunisia and Libya late September to October (Bundy 1976; Thomsen and Jacobsen 1979), and passage at Elat (Israel) lasts from late August into December (Safriel 1968). Passage through Gulf states of Arabia occurs mid-September to November (Bundy and Warr 1980; G Bundy). Present in Sudan and Eritrea (Ethiopia) from September (Cave and Macdonald 1955; Smith 1957, 1960a; Hogg et al. 1984), and East Africa from late October (Britton 1980). Wintering birds recorded in

Morocco from mid-October (Smith 1965a), but not in Ghana and Nigeria until January (Macdonald 1978; Elgood 1982).

Not recorded in Morocco after early March (Kersten *et al.* 1983), though present in most of Afrotropical winter range until April–May (Cave and Macdonald 1955; Smith 1957, 1960a; Macdonald 1978; Britton 1980; Elgood 1982; Hogg *et al.* 1984). Strong passage March–April along shores of Lake Turkana (Kenya) with maximum of 40 birds per day in April (Fry *et al.* 1974). Passes through Tunisia late March to mid-May (Thomsen and Jacobsen 1979), Libya April–May (Bundy 1976). Passage through Gulf states of Arabia occurs mid-March to mid-May, peaking in April (Bundy and Warr 1980; G Bundy). At Elat, spring passage protracted: begins early February

and may continue until June (Safriel 1968). At Azraq, passes mid-April to early May (Ferguson-Lees 1969). Occurs Cyprus March–May (Bannerman and Bannerman 1958; Flint and Stewart 1983) and Malta late March to mid-May (Sultana and Gauci 1982). In Austria, passage at Lake Neusiedl occurs during first half of May (Huber 1954a), and birds arrive on tundra of USSR in late May (Dementiev and Gladkov 1954a).

Westward vagrants move with Meadow Pipit *A. pratensis* in both spring and autumn, perhaps reflecting shared migration route between north-west Europe and north-west Africa, and often found in mixed flocks or parties (D I M Wallace). Pattern of records in Britain and Ireland fits well enough with that for migration in western Europe (Ruttledge 1966; Ferguson-Lees 1969;

British Ornithologists' Union 1971), birds occurring mainly late August to early November and (fewer) in May (Sharrock and Sharrock 1976).

Movement into southern Asia is on same massive scale as into eastern Africa. Present in Hong Kong late September to early May, mainly December to mid-April (Herklots 1967). Birds breeding Alaska (USA) move in autumn into Siberia via Aleutian islands, and spring passage on this route even more conspicuous. Recently, increasing number of records down west coast of North America: 2 mid-September records from Washington State (USA), and *c*. 20 records (involving 66 birds) from Californian coast between mid-September and mid-November, most in late October; southernmost American record at *c*. 30°N in Mexico. (Bent 1950; Roberson 1980.) JHE, DJB

Food. Chiefly insects, also small water snails and a few seeds. Feeds on ground by pecking and probing amongst vegetation; on seashore, probes amongst washed up seaweed (Coopman 1919; Collett and Olsen 1921; Blair 1936; Bannerman 1953; Korenberg *et al.* 1972; E Jaakkola). After capture, largest prey items (*Tipula* larvae *c*. 2 cm long) are vigorously pounded on ground before swallowing. In summer feeds most actively from late morning to late afternoon. (E Jaakkola.)

The following recorded in summer diet in Palearctic. Mayflies (Ephemeroptera), dragonflies (Odonata), crickets and grasshoppers (Orthoptera), bugs (Hemiptera: Cicadidae, Coccoidea), alder flies (Neuroptera: Sialidae), larval and adult butterflies and moths (Lepidoptera), caddis flies (Trichoptera), larval and adult flies (Diptera: Tipulidae, Culicidae especially *Chaeborus*, Mycetophilidae, Chironomidae, Syrphidae, Muscidae), Hymenoptera (sawflies Tenthredinidae, Ichneumonidae), beetles (Coleoptera: Staphylinidae, Scarabaeidae, Cantharidae, Chrysomelidae, Curculionidae), spiders (Araneae), harvestmen (Opiliones), centipedes (Lithobiidae), snails (Mollusca), small worms (Annelida), and grass seeds. Outside breeding season, diet includes insects, snails, worms, and grass seeds. (Collet and Olsen 1921; Witherby *et al.* 1938a; Meinertzhagen 1954; Krechmar 1966; Kostin 1983; Danilov *et al.* 1984; E Jaakkola).

At Karigasniemi (north-west Finland), late May to early June, most important prey are spiders, Chironomidae, and larvae of Tipulidae up to 2 cm long; from mid-June, Tipulidae and mosquitoes (Culicidae) become major constituents of diet, and later still larvae and adults of Hymenoptera, Lepidoptera, and beetles taken (E Jaakkola). On Yamal peninsula (USSR), May–June, 31 stomachs contained 20·5% (by number, no sample size) beetles (Staphylinidae), 15·9% spiders, 12·3% Lepidoptera larvae; 10·3% scale insects (Coccoidea), and a few Diptera (Mycetophilidae, Syrphidae), sawfly larvae, ichneumon flies, beetles (Cantharidae, Chrysomelidae), Cicadidae, caddis flies, and grass seeds; in August, 27

stomachs contained 24·7% Staphylinidae, 20·6% caddis flies, 15·1% ichneumon flies, 5·5% adult sawflies, also cicadas, alder flies, adult Lepidoptera, Diptera, and grass seeds (Danilov *et al.* 1984).

In Chukotka (north-east USSR), July, 8 stomachs contained 121 identifiable items comprising 39·7% (by number) Hymenoptera, 26·4% Lepidoptera larvae, 15·7% beetles (including 9·9% Chrysomelidae), 14·9% unidentified insect larvae, 2·5% spiders, 0·8% Diptera, and 0·8% Hemiptera (Portenko 1973). In western Taymyr, early June, stomachs contained many centipedes (Lithobiidae), and larvae of Chrysomelidae and Scarabaeidae (Krechmar 1966).

On spring passage, April, 5 birds from Crimea (USSR) contained only insects and their larvae, mainly ants 55% (by number, no sample size) and beetles 25%; migrant in October contained 9 eggs of Orthoptera and 1 adult (Kostin 1983). April stomachs from Hadda (Saudi Arabia) were crammed with small water snails; birds also take (mostly) insects, worms, and small seeds (Meinertzhagen 1954). Winter diet includes freshwater molluscs and grass seeds (Witherby *et al.* 1938a).

Nestling collar-samples from Karigasniemi contained similar prey to that taken by adults: mainly Diptera (Tipulidae, Culicidae, Chironomidae, Muscidae), Hymenoptera (mostly larvae), caterpillars, and spiders; when nestlings very young, smaller prey favoured (E Jaakkola). On Yamal peninsula, collar-samples from 16 broods contained 4457 items comprising 28·1% (by number) Chironomidae, 11·1% Muscidae, 10·2% Tipulidae, 7·1% adult Lepidoptera, 6·3% spiders, and 1–5% each of Cicadidae, Culicidae, beetles, and caddis flies; also dragonflies, ichneumon flies, alder flies, and a single orthopteran, mayfly, and snail (Danilov *et al.* 1984). Observations at one nest at Karigasniemi over 5 days (young 3–7 days old) revealed definite feeding rhythm: 01.00–06.00 hrs 13–16 visits per hr, 06.00–14.00 hrs 8–14 per hr, 14.00–21.00 hrs 11–17 per hr, 21.00–01.00 hrs 0–6 per hr (E Jaakkola). On Yamal peninsula, feeding visits started 03.00 hrs (01.00–05.00), stopping at 21.00–23.00 hrs; 40–308 visits per day, with rate peaking at 7–10 days but not affected by brood size (Danilov *et al.* 1984). PJE

Social pattern and behaviour. No major study. For review, see Ferguson-Lees (1969). Following account includes material supplied by E Jaakkola from study at Karigasniemi (north-west Finland).

1. Gregarious outside breeding season, with mostly loose-knit flocks of varying size occurring for migration, feeding, and roosting; single birds and 'pairs' (see Heuglin 1869 for north-east Africa) also reported. On Mindoro (Philippines), January–February, birds generally in flocks; on presumed roosting flight, flocks of 2–8 (Temme 1974). In winter in Kenya, large flocks unusual even where relatively common (P B Taylor). Flocks or concentrations of up to several hundred during spring or autumn migration widely reported: see (e.g.) Bannerman (1953) for Sweden, Danilov *et al.* (1984) for Yamal peninsula and

Kostin (1983) for Crimea (USSR), Kumerloeve (1954) for Turkey, Sultana and Gauci (1982) for Malta, Butler (1905) for Egypt, Ferguson-Lees (1969) for Jordan, Smythies (1953) for Burma. Average flock size at Suez (Egypt), October–November, 7·8 (1–25), $n = 56$ (Bijlsma 1982). At Myslowice (Polish Silesia), up to *c.* 100 recorded roosting with Meadow Pipits *A. pratensis* in early October (Natorp 1929). Although Ferguson-Lees (1969) referred to occasional occurrence of thousands together during migration, report of at least 10 000 near Sofia (Bulgaria) in early October (Klein 1933) considered dubious by Niethammer (1958a). At least some flocks (up to *c.* 20–30) occurring on breeding grounds from early August are of juveniles only (see Grote 1935b, Tomkovich and Sorokin 1983; also Krechmar 1966 and Kishchinski 1980). On Yamal, break-up of families reported from end of July (Danilov *et al.* 1984), juveniles apparently migrating before adults (Bannerman 1953). Recorded associating in winter with Yellow Wagtail *Motacilla flava* in Nigeria (Serle 1957); similar association noted in wet habitats of Kenya, though birds will also associate with other *Anthus* and larks (Alaudidae) in drier terrain (Reynolds 1974; P B Taylor). On spring migration especially, associates with wagtails *Motacilla* (particularly *M. flava*), sometimes with *A. trivialis*; in autumn, including for roosting (see above), more with *A. pratensis* (e.g. Natorp 1920, 1925a, Ivanov 1952, Moore and Boswell 1957, Castan 1958, Mester and Prünte 1965a). In northern Taymyr (USSR), late August, mixed flocks formed with *M. flava* and Citrine Wagtail *M. citreola* (Krechmar 1966). BONDS. Nothing to suggest mating system other than monogamous. In Finnish study, involving some colour-ringed birds, no indication of polygamy, nor of pair-bond lasting beyond one season (E Jaakkola). For records of birds apparently already paired on spring migration, see Reiser (1905) and Dharmakumarsinhji (1976); see also subsection 2 in Heterosexual Behaviour (below). Young fed by both sexes, including for *c.* 2 weeks after leaving nest (Pulyakh 1977; E Jaakkola). At Karigasniemi, marked birds first bred at 1 year old; some ♂♂ (probably 1-year-olds) sometimes failed to get a mate and thus did not breed (E Jaakkola). BREEDING DISPERSION. Solitary and territorial according to E Jaakkola, though territories evidently not always exclusively defended (see below). On Yamal, evenly dispersed in all suitable habitats of south, isolated groups of 2–4 pairs occurring further north (Danilov *et al.* 1984). In Timanskaya tundra (north-west USSR), where *A. cervinus* one of most numerous species, nests *c.* 80–150 m apart (Gladkow 1941). Locally in western Taymyr, pairs every *c.* 200–300 m (Krechmar 1966). At Karigasniemi, territory *c.* 0·4–0·5 ha; defended by ♂ and used for pair-formation, courtship, and nesting. Territory expanded during incubation (see also Antagonistic Behaviour, below) to become more like home-range where food collected for nestlings. Smaller 'satellite' feeding areas several hundred metres from nest used especially by ♂ when foraging for young (E Jaakkola). On Yamal, territories initially discrete, but considerable overlap as more birds arrive. Neighbouring pairs apparently use same area for feeding; ♂♂ use much larger area, 2–3·5(–9) ha, than ♀♀ (see illustrations in Danilov *et al.* 1984). Pair-formation normally takes place within territory, but some territories apparently occupied by already formed pair (Danilov *et al.* 1984). For aggression towards and tolerance of other species in territory, see Antagonistic Behaviour (below). In Timanskaya tundra, territory *c.* 1 ha (Gladkow 1941). At Karigasniemi, maximum density in bog habitat 27 pairs per km^2 over *c.* 1·25 km^2 (E Jaakkola). In Bol'shezemel'skaya tundra (north-west USSR), 30–40 pairs per km^2 over 2 years (Pulyakh 1977). Most detailed data from Yamal study: at Kharp, density over 10 years 9·7–38·1 pairs

per km^2; in 1974 (when numerous), locally 105 pairs per km^2 in sparsely wooded area, up to 72 pairs per km^2 in tundra and up to 29 pairs per km^2 in bogs and meadows; in 1973 (low numbers), 13 pairs per km^2 in sparse woods, 16 pairs per km^2 in tundra. At Khadyta, 11·7–77·9 pairs per km^2 over 11 years. For further details, see Danilov *et al.* (1984). For densities in various tundra habitats of Chukotka (north-east USSR), see Tomkovich and Sorokin (1983), and, for taiga of western Siberia, Vartapetov (1984). On Yamal, ♂ (1 out of 18 ringed adults) returned to occupy territory *c.* 700 m from that of previous year (Danilov *et al.* 1984); adult site-fidelity probably not uncommon, as in other *Anthus*. ROOSTING. Nocturnal and, outside breeding season, communal. At Myslowice, in early October, birds roosted communally with *A. pratensis* in reedbed, perched on plant stems, also apparently lower down, perhaps on ground, in grass and other vegetation; arrived at nightfall, later than *A. pratensis* (Natorp 1929). Communal roosting with *M. flava*, also in early October and in similar habitat, reported from North Yemen by Phillips (1982), and birds wintering in Kenya apparently have similar roosting habits (P B Taylor). On Yamal, bird recorded roosting in hollow prepared for nest (Danilov *et al.* 1984; see also Breeding). At Karigasniemi, ♀ generally roosts in nest immediately after its completion, ♂ in tree nearby (E Jaakkola). Captive bird roosted on perch, unlike *A. trivialis* and Water Pipit *A. spinoletta* which did so communally on ground (Coopman 1919). At Karigasniemi, when actively singing (see part 2), ♂♂ feed mostly from late morning to late afternoon and for shorter period in evening before roosting (E Jaakkola); for ♀'s activity rhythms, see subsection 6 in Heterosexual Behaviour (below).

2. Can be difficult to observe on ground, staying in cover or moving into it when disturbed (e.g. Grote 1925), and flushing late (e.g. when observer at *c.* 15 m). Silent or gives contact-alarm call (see 2a in Voice) on flushing and often does not fly far, dropping into cover again or sometimes flying to bush or tree. Relative shyness perhaps varies with wind and weather (Tischler 1917; Natorp 1925a, b; Melcher 1951; Mester and Prünte 1965a; Wilson 1975b; Taylor 1979). Sometimes allows approach to *c.* 15–25 m, or even to 1 m (Leser and Schnorr 1960; Peitzmeier and Westerfrölke 1960; Bock *et al.* 1961; Mester and Prünte 1965a; Lenz 1970); thus tamer than *A. pratensis* (Tischler 1917) or same as that species if associating with it (Mester and Prünte 1965a). In Voronezh region (USSR), early May, 1 migrant ♂ (out of *c.* 12–15) allowed approach to within *c.* 5 m and sang (see 1 in Voice), apparently 'at observer', in posture similar to that shown in Fig C (M G Wilson). At Myslowice, October, 1 of several flushed when Short-eared Owl *Asio flammeus* flew up followed and then mobbed it, giving constant contact-alarm calls; joined by *A. pratensis*, *M. flava*, and Skylarks *Alauda arvensis* (Natorp 1925a). FLOCK BEHAVIOUR. Feeding flocks at Myslowice show varyingly close association with *A. pratensis*, sometimes keeping separate as discrete flock. Similarly when moving to roost; on arrival there, drop down vertically into reedbed (Natorp 1925a, 1929). SONG-DISPLAY. On breeding grounds, ♂ sings (see 1 in Voice) mainly in flight (Bannerman 1953), though also from perch (Grote 1925). Song-flights performed occasionally by spring migrants (Jung 1967). In spring especially, but also in autumn (Natorp 1925a), more likely to give quieter Subsong (see 1b in Voice) while stationary (on mound) or moving about and feeding on ground, sometimes from perch (Tischler 1917; Čtyroký 1958; Mester and Prünte 1965a; Jung 1967). Song-flight resembles that of *A. pratensis*, but more complex and with well-defined pattern according to E Jaakkola; also often higher (see below) and longer according to Witherby *et al.* (1938a) and Johansen

(1952). Takes off from ground (early in season from thawed patch in tundra) or perch (at least in spring migrants: Jung 1967) and ascends at an angle, with rapid fluttering wing-beats, to *c.* 15 m (spring migrants: Jung 1967) or *c.* 20-30 m; according to Portenko (1973), to several tens of metres and thus higher than other tundra species. May descend (while singing) immediately after reaching peak of ascent (only Grote 1925 reported birds hovering), but usually (perhaps typically later in season) first glides horizontally on stiffly extended wings for *c.* 50 m (Kapitonov and Chernyavski 1960), sometimes moving in wide arc (Portenko 1973). Descends in parachute-style, wings half-spread, tail spread and slightly raised, landing on ground or tree (Bent 1950; Kapitonov and Chernyavski 1960; Portenko 1973; E Jaakkola). Portenko (1973) reported birds turning sharply to land facing take-off point. In Song-flights with horizontal gliding phase, first notes of song given during ascent; this followed by silent glide, 2nd part of song being given at start of descent and remainder during it (E Jaakkola; see also Kapitonov and Chernyavski 1960, Portenko 1960, Wallschläger 1984). Silent and incomplete Song-flights recorded in north-east USSR in late June (Portenko 1973). In Siberian tundra, will sing at any time of day or night (Grote 1935*b*; Johansen 1952). In western Taymyr, 44% of song given in morning, 28% later in day, 4% in evening, and 24% at night (Podarueva 1979). In Bol'shezemel'skaya tundra, if overcast and wet, starts at *c.* 09.00-10.00 hrs, ending by *c.* 22.00 hrs. If clear, will sing almost throughout, with break of $1-1\frac{1}{2}$ hrs at 22.00-24.00 hrs (Pulyakh 1977). In Yakutiya (eastern USSR), late May, wind and cold down to -12°C caused no reduction in song (Vorobiev 1963), and Song-flights by spring migrants in West Germany also continued in strong wind (Jung 1967); on lower Lena (eastern USSR), however, song ceased in strong wind (Kapitonov and Chernyavski 1960). Song-period in USSR normally from arrival until July, thus continuing in some cases after fledging and sometimes renewed before departure (Johansen 1952; Portenko 1960). In north-east USSR, intense song noted for *c.* 1 week after arrival (Portenko 1973); on lower Lena, wanes by mid-June, but sometimes continues to mid-July at night (Kapitonov and Chernyavski 1960). See also Krechmar (1966), Kishchinski (1980), and Kishchinski *et al.* (1983). ANTAGONISTIC BEHAVIOUR. (1) General. Yamal territories established and occupancy marked with Song-flights soon after arrival. Fights fairly common during first 3-5 days, but rare thereafter (Danilov *et al.* 1984). At Karigasniemi, territorial antagonism of ♂♂ declines during incubation (E Jaakkola). (2) Threat and fighting. 2 ♂♂ associating with 1 ♀ in Niedersachsen (West Germany) in spring recorded only once in brief fluttering fight when call 3 also given; otherwise showed no rivalry (Jung 1967); see also subsection 2 in Heterosexual Behaviour (below). In Threat-posture described in Finnish study, ♂ has bill raised to show off rufous throat and breast; wings raised, spread and slowly flapped. This posture used to drive off conspecific birds and also (e.g.) *A. pratensis*. ♂-♂ chases also occur (E Jaakkola). On Yamal, birds threaten at close quarters in posture (Fig A) similar to Threat-posture of *A. pratensis* (see p. 373), though this not recorded by E Jaakkola. ♂ neighbours seen to make combined attack on intruders attempting to sing in their territories, though no aggression shown towards feeding birds,

nor towards stuffed conspecific birds (♂ and ♀) set up in territory. ♀♀ not recorded participating in fights (Danilov *et al.* 1984). However, at Karigasniemi, ♀♀ seen to drive off (e.g.) Lapland Bunting *Calcarius lapponicus* (E Jaakkola). In Finnmark (northern Norway), no interspecific aggression noted with *A. pratensis*: e.g. pair allowed to forage within *c.* 5 m of incubating ♀ *A. cervinus* (Haftorn 1959). Migrants sometimes aggressive towards other birds: e.g. in Niedersachsen, ♂ attacked and chased (over *c.* 150 m) Little Ringed Plover *Charadrius dubius*; also attacked, but did not chase, Serin *Serinus serinus* (Jung 1967); in central England, early May, bird persistently chased Swifts *Apus apus* and, at least once, Linnet *Carduelis cannabina* (Fowler *et al.* 1984). HETEROSEXUAL BEHAVIOUR. (1) General. On Yamal, nest-building begins 1-5 days after pair-formation. Period between arrival of first birds and start of laying in year with early spring 15-20 days; in another year, 1-12 days (Danilov *et al.* 1984). (2) Pair-bonding behaviour. ♂ may chase ♀ as in other *Anthus* (e.g. Portenko 1973): ♀ flies low and with slow fluttering action; tail spread and white outermost feathers conspicuous (E Jaakkola); call 4 given (Witherby *et al.* 1938*a*), though not known if by ♂ only or by both sexes. In Niedersachsen, bout of courtship occurred twice between migrants in early May. While presumed ♀ passive in grass clump, ♂ adopted Courtship-posture: tail steeply raised, wings slightly drooped and shivered, head held stiffly up to reveal rufous throat and breast, bill slightly open. Moved with light dancing steps in semi-circles before ♀ for *c.* 30 s. Performance repeated *c.* 30 min later, but ♂ then carried grass (see *A. pratensis*, p. 374, and subsection 5, below). Finally, after short pause, ♂ ran on in normal posture, taking no more notice of ♀. No further sign of any bond between either of 2 ♂♂ and ♀ (Jung 1967). Finnish study confirmed use of Courtship-posture or similar one (Fig B) by ♂ on breeding grounds; ♂ calls (see

B

4 in Voice) and rapidly quivers wings and tail while standing face to face with ♀. ♀ adopts similar posture: bill slightly raised, but wing-quivering less marked, though slightly raised tail is flicked. ♀ gives call 5a (E Jaakkola). Pair-formation also includes display by ♂ to ♀ on ground (Fig C) and in the air (Fig D) (Danilov *et al.* 1984). (3) Courtship-feeding. As in *A. pratensis*,

A

C

D

♀ typically fed by ♂ on or off nest (Congreve 1936) during incubation. At Karigasniemi, took place mainly around midday; in case of re-nesting, occurred between building of new nest and incubation (E Jaakkola). Detailed observations in Finnmark and additional data from Finnish study reveal following pattern: on seeing ♂ carrying food or in response to his contact-call, ♀ gives loud contact-call (see 5b in Voice) from nest, then flies to ♂ as soon as he moves to traditional spot *c.* 10–20 m from nest; ♀ adopts Begging-posture (crouches with head and tail raised, shivers wings below line of back), and gives Begging-call (see 5c in Voice). Afterwards, ♀ returns immediately to nest or feeds nearby for a few minutes (Haftorn 1959; E Jaakkola). (4) Mating. Sometimes follows bout of courtship as described above (E Jaakkola). (5) Nest-site selection. No information on selection process. At Karigasniemi, when pair feeding near nest under construction and ♀ building occasionally, ♂ suddenly picked up nest-material and ascended into Song-flight (E Jaakkola). On Yamal, ♀ normally lays 1–2 days after nest completed; in 2 cases, eggs laid in unfinished nest (Danilov *et al.* 1984). (6) Behaviour at nest. In Finnmark, ♀ tended to leave nest directly in flight, but ran last few metres on return (Haftorn 1959). Traditional route may be used, leading to creation of corridor (Dementiev and Gladkov 1954*a*). On Yamal, ♀♀ left nest 13–71 times per day, usually for break of 2–5 min, and never exceeding 25 min (Danilov *et al.* 1984). In Norwegian study, ♀ spent 153½ min away from nest in 24 hrs, including 61 min of active feeding; maximum break 9 min. Accompanied off nest 4 times by ♂ who made last feeding visit at 18.09 hrs, ♀ then remaining on nest until short break to self-preen at 01.51 hrs; ♂'s feeding of ♀ showed marked peak 02.27–05.00 hrs (Haftorn 1959). At Karigasniemi, ♀ feeds away from nest mainly in early morning and in evening before roosting (E Jaakkola). RELATIONS WITHIN FAMILY GROUP. Hatching recorded over average 1·4 days, *n* = 18 (E Jaakkola), or *c.* 3 days. Eyes of young open at *c.* 3–4 days (Dementiev and Gladkov 1954*a*). ♀ broods young for long periods during first few days and does little feeding. Later, however, young not brooded in poor weather and this, combined with little or no feeding, may lead to high mortality (Pulyakh 1977). Young fed bill-to-bill (Witherby *et al.* 1938*a*). Leave nest at 11·6 days (9–14), early if brood large (Danilov *et al.* 1984), or 12·6 days (11–15), *n* = 42 nestlings (E Jaakkola); then hide in nearby vegetation (Dementiev and Gladkov 1954*a*). Can flutter short distance at this stage, but run well according to Pulyakh (1977); see also Vorobiev (1963). Apparently stay close to nest for certain period (Tomkovich and Sorokin 1983). Age of independence not known exactly, but Kapitonov and Chernyavski (1960) recorded young able to fly well and still with parents; see also Bonds (above). ANTI-PREDATOR RESPONSES OF YOUNG. Will leave nest early if threatened (Vorobiev 1963), running off in different directions (Kishchinski 1980). PARENTAL ANTI-PREDATOR STRATEGIES. (1) Passive measures. ♀ generally sits tightly, flushing at last moment, even allowing herself to

be caught when just leaving. Will run about close to man by nest and even return, settling to incubate with man *c.* 2 m away (Dresser 1871–81; Coopman 1919; Congreve 1936). ♀ can be furtive and retiring like *A. pratensis* (Bannerman 1953); one ♀ left eggs when man at *c.* 5 m, then stayed far away (Portenko 1973). ♂ typically noisy and demonstrative, less skulking than *A. pratensis* (Blair 1936; Congreve 1936). (2) Active measures. All information relates to man as potential predator. Both sexes will flutter near or high above intruder, giving constant alarm-calls (see 2d–e in Voice), even for *c.* 20 min continuously; also perch conspicuously, or land briefly on ground (Bannerman 1953; Dementiev and Gladkov 1954*a*; D J Brooks); in north-east USSR, only ♂ recorded flying high, and birds, although highly excited, did not come as close as *M. flava* (Portenko 1973). Distraction-lure display of disablement type said by Bannerman (1953) to be less frequent than in *A. pratensis*. ♀ flutters off nest and flies low so that spread tail brushes vegetation (E Jaakkola). Williams (1941) compared performance to that of Reed Bunting *Emberiza schoeniclus*; bird fluttered along ground briefly, then flew to nearby fence, both adults continuing to give alarm-calls.

(Figs by I Lewington: B from drawing by E Jaakkola; others from drawings in Danilov *et al.* 1984.) MGW

Voice. Wintering birds watched in Morocco by Smith (1965*a*) gave only call 2a; this and call 2b noted in Kenya (P B Taylor). Variety of calls (including song) given by migrants (see below). For extra sonagrams, see Bergmann and Helb (1982) and Wallschläger (1984). Some calls difficult to classify and following scheme therefore provisional.

CALLS OF ADULTS. (1) Song of ♂. (a) Full, loud song superior in quality to that of Meadow Pipit *A. pratensis* (Witherby *et al.* 1938*a*; Haftorn 1971) and closer to Tree Pipit *A. trivialis* (Blair 1936) or, in volume and harsher tone, Water Pipit *A. spinoletta* (Coopman 1919). In timbre and composition differs from other *Anthus* according to S Palmér. 3 main components distinguished by Witherby *et al.* (1938*a*): shrill and prolonged 'twee' sounds given about 4 times, somewhat like 'see-a' of *A. trivialis*. but less musical and higher pitched; bubbling trill; more sibilant 'twizz' and 'wizz' sounds. Full song (from perch) thus rendered 'twee-twee-twee-twee tyrrrrrrrrr twizz-wizz-wizz-wizz twizz-wizz-wizz-wizz twizz-wizz-wizz-wizz'. In flight, song longer and more varied (Witherby *et al.* 1938*a*), with minor warbling or twittering; musical, lively, pleasing and like Canary *Serinus canaria* (Williams 1941; Bent 1950). Generally includes species-specific 'spiss' (see call 2a) or (not all birds) drawn-out 'tiii' or 'tiie' (Haftorn 1971, which see for further renderings of song). More detailed sonagraphic analysis by Bergmann and Helb (1982) and Wallschläger (1984) showed song to be structurally similar to *A. trivialis*., with quite long series of about 8 segments of varying length, each comprising a number of uniform units. Fig I shows song with 10 segments, as follows. (i) 9 attractive, descending 'psee-a' or 'psee-oo' units—pure-sounding, with only very slight buzzing effect. (ii) Slow, loud rattle (tremolo) of 9 'tip-it' sounds separated by brief pause

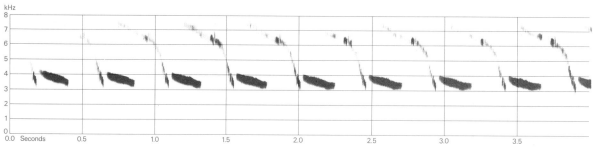

I S Palmér/Sveriges Radio (1972–80) Norway July 1963

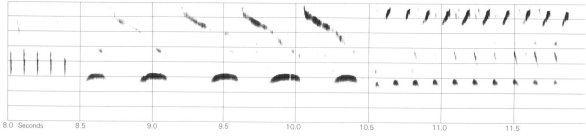

I *cont.*

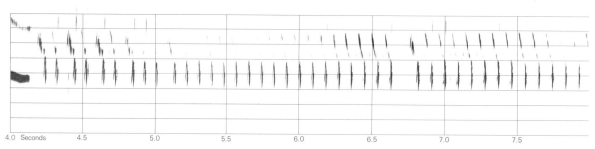

I *cont.*

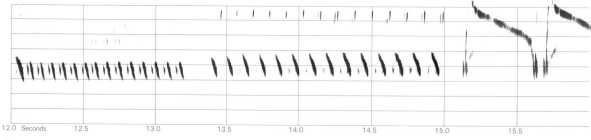

I *cont.*

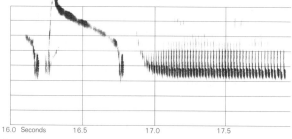

I *cont.*

from (iii) segment of 18 similar but perhaps slightly more liquid ('tip-a') units which show crescendo and accelerando. (iv) After longer pause, segment of same unit-type but with more or less steady volume. (v) 5 very high-pitched, squeaky, descending 'quee-a' units with simultaneously-sounding tones below them at 3 kHz (diads). (vi–viii) Segments of 10, 15, and 14 units respectively, all of rattle (tremolo) type: these vary in speed, pitch, and timbre. Quite musical middle segment of the 3 recalls finch (Fringillidae) or, in another song by same bird, Yellowhammer *Emberiza citrinella*. End of song rather remarkable: (ix) penultimate segment comprises 3 very thin, attenuated, almost strained 'pseeeoo' sounds evidently closely related to call 2a but here (and even more in another song by same bird) characterized by peculiar 'dissonance', comprising 2 portamento tones sung in contrary motion, i.e. one (loud) starts above 8 kHz and descends, while the other (quiet) starts at *c.* 4 kHz and ascends. The lower voice fades rapidly on the sonagram and is easier to hear than to see. This is in contrast to the diads in 1st segment of this song where the loud lower notes overpower the start of each high tone. In 5th segment, however, the simultaneous tones are both easy to hear. In each of these diads it may be assumed that there is independent but simultaneous use of the 2 internal tympanic membranes (J Hall-Craggs; see also Bergmann and Helb 1982; Wallschläger 1984); (x) concluding segment a remarkably full-sounding, buzzing rattle, as in Bergmann and Helb (1982) Fig a. (J Hall-Craggs, M G Wilson.) Other 2 songs in same recording similar, both starting with segment of 'quee-a' sounds, though length of this and other segments apparently varies, as does order of segments; full, loud rattle may be omitted (as noted by Portenko 1973). (b) Subsong given in Westfalen (West Germany) in spring a rather monotonous trilling or twittering at more or less steady pitch and audible *c.* 30 m away (Bock *et al.* 1961). In Oman from 21 March, a thin, scratchy, unmusical warble, interspersed with call 2a (Walker 1981*a*). In Niedersachsen (West Germany), something closer to full song given in flight and (quietly) on ground: began with 'psi' calls (like call 2a), followed by a trill and variations; occasional clear whistling notes (probably 1st segment as described above). Overall like *A. pratensis* but more varied and higher pitched. Lasted from a few seconds up to (usually) 1–2 min, sometimes up to 10 min with only short interruptions (Jung 1967). Another spring migrant in West Germany gave similar quiet twittering and trilling song with 'dja' sounds at start (Mester and Prünte 1965*a*). See also Tischler (1917) and Natorp (1925*a*). (2) Contact-alarm calls. (a) In typical form an excellent aid to identification. Long, drawn-out hissing 'psssss', starting loudly but trailing off (Sharrock 1980*b*). Characteristically high pitched, with distinctive thin tone and normally descending in pitch at end (Rogers 1981*b*), though pitch sometimes steady (Bergmann and Helb 1982). Also rendered as follows: 'psiih', 'bsiii', 'bi-is', 'bie-e', 'tsiih', or sharp, high-pitched, drawn-out 'zjiiieh' recalling Reed Bunting *Emberiza schoeniclus* (Chernel 1919; Coopman 1919; Melcher 1951; Leser and Schnorr 1960; Jung 1967; Melchior 1975: Fig II from recording analysed at half-speed playback); see also Dementiev and Gladkov (1954*a*). Loud and penetrating at close quarters, slightly sharper and louder than *E. schoeniclus* (Natorp 1925*a*). Sharp, drawn-out, and high-pitched 'psiëh' or 'psüe', slightly impure in timbre—'electric'; (or metallic: Taylor 1979), and also likened to Penduline Tit *Remiz*

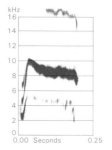

II   P A D Hollom
Egypt   April 1979

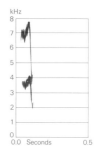

III   H-H Bergmann
Greece   April 1979

pendulinus (Bergmann and Helb 1982). Longer and more plaintive than equivalent call of *A. pratensis* (e.g. Goulliart *et al.* 1968) and may sound slightly disyllabic, as indicated in some of the above renderings (M G Wilson); see also Field Characters. Typically given in flight (Bergmann and Helb 1982), especially in longer flights by migrants (Jung 1967), but also from ground (Natorp 1925*a*). Normally given singly rather than the double note of *A. pratensis* (Tischler 1917; Coopman 1919), though several calls may be given in succession when flying over (Melcher 1951); see also call 2c. (b) Presumed variant of call 2a closer to call 2 of *A. trivialis*: e.g. one of several calls in recording by P A D Hollom of spring migrants in Egypt a harsher 'wheeez(p)' or 'zeez' (M G Wilson). Coarse 'tzeez' harsher than *A. trivialis* (Chapman and McGeoch 1956) and 'skeez' of Witherby *et al.* (1938*a*) may refer to this variant or to typical call 2a, but (e.g.) Tischler (1917) and Appert (1951) found call 2a to be typically softer (less harsh) and more drawn-out than *A. trivialis*. In Kenya, winter, birds frequently give call 2a and variants such as harder 'tseez' (P B Taylor). In recording by P A D Hollom of birds released after being handled, call perhaps only slightly harsher than typical 2a (M G Wilson). (c) Quiet, softer and shorter 'spie' (sometimes as a series—'spie spie spie'—with varying emphasis), 'dji', or 'djie-e' given in shorter, lower flights by migrants (Tischler 1917; Natorp 1925*a*; Jung 1967). A 'su' given by migrant from perch (P A D Hollom) probably of similar type. In recording of flushed flock, calls much like *A. pratensis*: 'peep', 'tijk', 'pseet', or 'psit' (Fig III), 'psip'; also a double 'pseet-eet' and a lower-pitched conversational series of 'tweep' sounds: one such series rendered 'psi-psit-psi-psi' (Bergmann and Helb 1982; M G Wilson). See also Field Characters. Coopman (1919) mentioned a 'ty-töitt-tjt' given by captive bird especially in evening and apparently in response to wild conspecific birds flying over; difficult to judge from rendering but perhaps also a conversational series. Full, musical, rather abrupt 'chüp' given in flight, when flushed, or when perched; used, like call 2a (or 2b), also by migrants (Witherby *et al.* 1938*a*); probably also belongs here, but difficult to understand why these authors attributed greater importance to it than to call 2a. (d) A short 'pit' or 'djik' given when disturbed (Bergmann and Helb 1982). Short, repeated 'te' or 'dytt' given frequently on breeding grounds, e.g. in flight over intruder (Haftorn 1971) clearly the same. Recording (Fig IV) suggests short

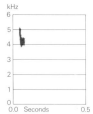

IV S Palmér/Sveriges Radio (1972–80) Norway July 1963

'peep', harder than in *A. pratensis* (P J Sellar), or quite hard, rather explosive 'tip', 'tyip', 'chip', or slightly metallic 't(y)eep' (M G Wilson). (e) Alarm-call at nest said by Witherby *et al.* (1938a) to be a rather hoarse, shrill 'tsweerp'; confirmation required as not mentioned by other authors and call 2d is at least one alarm-call given on breeding grounds (and away from them). However, possible explanation of this in Congreve (1936) who claimed that alarm-calls of sexes differ, ♂'s resembling noisy alarm-call of *A. trivialis* and ♀'s being faint squeaky note like ♀ *A. trivialis*. (3) A hard 'tschr' given occasionally by migrants (Bergmann and Helb 1982). Recording suggests bubbling 'prrrrt' (P J Sellar); sonagram (Fig V)

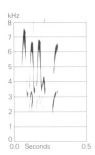

V P A D Hollom Egypt April 1979

closely similar to Fig e in Bergmann and Helb (1982). Whispering sound difficult to describe but roughly 'spr(i) r(i)r(i)t' associated with antagonistic interactions between migrants (Jung 1967) perhaps the same or a related call. (4) Courtship-call of ♂. A harsh trill (E Jaakkola). A 'tsrrrrrup' given during ♂-♀ chases (Witherby *et al.* 1938a) perhaps the same. (5) Calls given by ♀ in contact with ♂. All probably only variants of calls 2a–c. (a) Long 'spiii' given by ♀ during courtship (E Jaakkola). (b) Loud 'psii psii' like Yellow Wagtail *Motacilla flava* given from nest in contact with ♂ (Haftorn 1959). (c) Begging-call. Rapid, excited 'psi-psi-psi' given in Begging-posture prior to Courtship-feeding (Haftorn 1959). (d) Muted 'psy' given before return to nest after being fed by ♂ (Haftorn 1959).

CALLS OF YOUNG. No information.   MGW

Breeding. SEASON. At southern extent of range, first eggs laid end of May; further north, laying from early or mid-June to early July; last young, from late or repeat layings, fledge up to mid-August (Ferguson-Lees 1969; Makatsch 1976; Danilov *et al.* 1984). SITE. On ground, in side of hummock or bank, or sheltered by low scrub; sometimes at end of short 'tunnel' in mossy hummock. Nest: hollow in moss or ground, filled with cup of grass leaves and stems, with some moss and dead leaves in base; minimal lining of finer grass, hair, and some feathers. External diameter 7–11 cm, internal diameter 6–8·5 cm, depth of cup 3–4·7 cm (Pleske 1928; Danilov *et al.* 1984). Building: ♂ makes hollow in moss (10–15 cm deep, 6–10 cm across); only ♀ builds, filling hollow with grasses (etc.), brought initially by both birds then only by ♂ (Danilov *et al.* 1984). EGGS. See Plate 80. Sub-elliptical, smooth and glossy; variable, mainly grey, buff, olive, or pinkish, with fine spots, speckles, or blotches of brown, grey, or red-brown; sometimes with fine black hair-streaks. 19·2 × 14·2 mm (17·1–21·0 × 13·4–15·5), $n = 150$; calculated weight 2·01 g (Schönwetter 1979). Weight 2·09 g (1·85–2·40), $n = 26$ (Danilov *et al.* 1984). Clutch: 5–6 (2–7). Of 63 clutches, Finland: 4 eggs, 3; 5, 31; 6, 26; 7, 3; mean 5·46 (E Jaakkola). Of 18 clutches, Norway: 5 eggs, 3; 6, 14; 7, 1; mean 5·9 (Thome 1926). At different sites on Yamal peninsula (USSR): mean 5·32 (3–7), $n = 143$; 5·17 (2–7), $n = 52$; 5·35 (5–6), $n = 14$ (Danilov *et al.* 1984). One brood; replacements laid after egg loss (Ferguson-Lees 1969). INCUBATION. 12·4 days (11–14), $n = 11$ (E Jaakkola); 11·5 days (10–13), $n = 28$ (Danilov *et al.* 1984). Begins with 2nd–3rd egg; hatching over mean 1·8 days (1–3). By ♀ only; mean time on nest per day 21·0 hrs (18·5–22·7), $n = 34$. (Danilov *et al.* 1984.) YOUNG. Cared for and fed by both parents. FLEDGING TO MATURITY. Fledging period 12·6 days (11–15), $n = 42$ (E Jaakkola). Age of independence and first breeding not known. BREEDING SUCCESS. Of 162 eggs laid, Finland, 68·5% hatched and 62·9% fledged, with main losses due to desertion and non-hatching (E Jaakkola). On Yamal peninsula, hatching success 83·4%, with 91·8% of young reared to fledging, sample sizes not given (Danilov *et al.* 1984). From 66 eggs in 14 nests, Chukotka (north-east USSR), 40 young fledged giving 60·6% overall success (Tomkovich and Sorokin 1983).

Plumages. ADULT MALE BREEDING. Entire upperparts, including rump and upper tail-coverts, streaked buff-brown and black or brown-black: black streaks rather narrow but sharp on forehead and crown; indistinct on hindneck; broad (3–4 mm wide) on mantle, scapulars, and back; rather broad but indistinctly defined on rump and upper tail-coverts (especially towards flanks, where brown rather than black), on rump sometimes partly concealed by buff-brown feather-tips; distal parts of lateral feather-fringes of mantle and inner scapulars sometimes bleached to pale grey-brown or buff, showing as pale short streaks, occasionally joining to form 4 streaks or double open-ended V, but not as marked as single V of Pechora Pipit *A. gustavi*. Upper sides of neck and ear-coverts buff-brown; remainder of sides of head and neck, chin, and throat pink; deepest rusty-pink on broad supercilium and on cheeks, slightly paler pink-buff on lores and on narrow ring of feathers round eye; often an indistinct grey-brown or buff-brown line from gape to lower ear-coverts. Chest deep buffish-pink or rusty-pink, merging into buff on sides of breast and into pale buff, pink-buff, or cream-buff on remainder of underparts (including under tail-coverts and feathering of tibia). Underparts down from chest streaked black or black-brown to variable extent. Usually, broad streaks (*c.* 2 mm wide) on sides of chest; narrower and shorter ones on chest (not joining into rounded gorget, unlike non-breeding and some adult ♀ breeding); rather broad and long ones on flanks; rather short and triangular spots on breast; often fine streaks on sides of belly; broad streaks on longer under tail-coverts (but these sometimes mottled or concealed, occasionally absent). Some birds (perhaps 1st adult) have heavier streaks on chest and flanks, with streaks of breast and sides of belly more extensive, and ground-colour of chest less pink, rather more deep buff; others have chest virtually uniform saturated buffish-pink, with black streaking restricted to rather narrow streaks on sides of breast and lower flanks. Central pair of tail-feathers (t1) dull black, bordered by poorly defined buff-brown fringe along both sides and tip; t2–t5 black with sharp narrow buff-brown edge to outer web, t5 with short off-white wedge on tip of inner web, 6·0 (25) 2–10 mm long, rarely absent or up to 13 long; t6 white, distal part of outer web suffused grey-brown or buff-brown, base of inner web with dull black or sepia-brown streak (remaining white on inner web reaching to 30–40 mm from tip along shaft), white sometimes faintly sullied grey or buff. Flight-feathers black-brown, tips narrowly edged off-white, outer webs of primaries narrowly and sharply edged buff, outer web of secondaries more broadly and indistinctly fringed pale buff-brown. Tertials and greater upper wing-coverts black with broad greyish-buff fringes; fringes poorly defined from black near base of feathers, more sharply at tip (where buff often bleached to pale grey-buff or off-white). Median upper wing-coverts black with sharply defined pink-buff or cream tips (soon bleaching to off-white), forming wing-bar 2–3 mm wide. Lesser upper wing-coverts dull grey or grey-black with poorly defined buff-brown fringes, rather like upperparts. Tips of worn greater upper wing-coverts may bleach and then show as 2nd wing-bar, but tips of both greater and median coverts also subject to abrasion and largely disappear through wear (though not as completely as in 1st breeding). Upper primary coverts and bastard wing black-brown to dull black, narrowly and sharply fringed pale buff-brown along outer webs, faintly along tips. Under wing-coverts and axillaries buff or cream-buff, sometimes deeper buff along leading edge of wing, where spotted dull grey. *In worn plumage*, black streaking of upperparts heavier, contrasting more with bleached pale buff of feather-sides; pink of sides of head and from chin to chest bleached to pinkish-buff, not as saturated shiny pink; black streaks on remainder of underparts more pronounced; ground-colour of underparts from breast downwards off-white; pale tips of median and greater upper wing-coverts and pale wedge on tip of t5 largely worn off. ADULT FEMALE BREEDING. Like adult ♂ breeding, but pink of sides of head and underparts less deep and saturated, more buff; supercilium, lores, and chest buff or pale buff, often virtually without pink; chest more heavily streaked, similar to streaking in non-breeding, black markings on upper chest sometimes joining to form broken and rounded gorget. *In worn plumage*, chin and throat virtually without pink, pale buff predominating. Some ♀♀ resemble typical ♂♂ and *vice versa*. ADULT MALE NON-BREEDING. Upperparts similar to breeding; upper wing-coverts and tertials similar, but all equally new, with no contrast between fresh inner and lesser coverts and slightly worn outer

coverts and greater primary coverts. Sides of head and neck and underparts rather variable: always distinctly less extensively pink-buff than adult ♂ breeding, but some birds similar to some adult ♀ breeding. Supercilium buff or warm buff, sometimes pink-buff, lores pale buff; ear-coverts buff-brown as in breeding, but more often with dull black stripe from gape to lower ear-coverts; sides of neck buff-brown with indistinct grey streaks. Lower cheeks, chin, and throat pink-buff or warm buff, sometimes as deep rusty buff as in breeding, but buff bordered by rounded black-spotted gorget below instead of extending to chest; some birds with paler buff throat show distinct black malar stripe, extending to side of breast, more rusty birds lack this. Sides of breast, chest, flanks, and longer under tail-coverts heavily streaked black on buff or cream-buff ground, sides of belly less heavily streaked. Tail and flight-feathers as adult ♂ breeding. ADULT FEMALE NON-BREEDING. Upperparts and underparts from breast downwards similar to adult ♀ breeding and adult ♂ non-breeding, underparts heavily streaked; wing-coverts and tertials all equally new, as in adult ♂ non-breeding. Supercilium, lores, lower cheek, and chin pale buff; patch below eye indistinctly mottled pale buff and buff-brown, bordered below by buff-brown or dull black line from gape to lower ear-coverts; pale buff of lower cheeks and chin separated by rather distinct black-brown or black malar stripe; throat usually rather deep buff or pink-buff, more rarely uniform pale buff (deep buff or pink-buff not as extensive as in adult ♀ breeding or adult ♂ non-breeding). NESTLING. Down fairly long and plentiful, dark grey-brown; underparts naked (Witherby *et al.* 1938*a*). JUVENILE. Similar to adult ♀ non-breeding, but throat pale buff or cream-buff (not pink-buff); pale fringes along sides of feathers of upperparts narrower, buff or pale buff, upperparts appearing darker black-brown, less tawny; black streaks on underparts longer and broader, but duller and less sharply defined. (Witherby *et al.* 1938*a*; Hall 1961.) FIRST ADULT NON-BREEDING. Upperparts, upperwing, and tail as adult, usually with some contrast in colour and abrasion between fresh non-breeding (body, lesser and inner upper wing-coverts, some or all tertials, and sometimes t1) and retained juvenile (outer greater upper wing-coverts, greater upper primary coverts, and remainder of tail); adult non-breeding shows no such contrast, but adult breeding shows contrast between fresh breeding and old non-breeding (see Moults for timing and extent). Sides of head and neck and underparts as in adult ♀ non-breeding; supercilium, lores, lower cheeks, chin, and throat pale buff, usually without deeper buff or pink-buff on throat (supercilium sometimes partly mottled brown, indistinct); black malar stripe distinct, extending to heavily streaked sides of breast, chest, and flanks. Sexes similar. FIRST ADULT BREEDING. Like adult breeding, but sometimes separable by slight differences in extent of moult (see Moults). Juvenile greater upper primary coverts and sometimes retained outer greater coverts more heavily worn than those of adult; median and greater upper wing-coverts usually retained non-breeding, distinctly abraded.

Bare parts. ADULT, FIRST ADULT. Iris dark brown. Bill dark brown or horn-brown, base of lower mandible yellow-horn or yellowish flesh. Leg and foot brownish-flesh or yellowish-flesh, bright pink-red in spring. (BMNH, RMNH.) NESTLING. Mouth flesh-red or vivid crimson; gape-flanges yellow or very pale yellow (Witherby *et al.* 1938*a*; Ferguson-Lees 1969). JUVENILE. Iris dark brown. Bill flesh-pink with darker culmen and pale yellow gape-flanges. Leg and foot flesh-pink with darker joints and toes. (BMNH.)

Moults. ADULT POST-BREEDING. Complete; primaries descendant. Moults on breeding grounds July–August; completed before start of migration, September–October. ADULT PRE-BREEDING. Partial. Starts November–January, face and throat first, followed by remainder of head and body late January to early March. Completed March, when head and body new (though ♀ sometimes retains scattered old feathers), as well as t1 (occasionally whole tail), tertials, tertial coverts, lesser upper wing-coverts, some or all median upper wing-coverts, and sometimes a few inner greater upper wing-coverts. (D J Pearson, BMNH, RMNH.) POST-JUVENILE. Partial. Starts shortly after fledging, early July to mid-August; in some, completed early August when still in or near breeding area, others have some feathers still growing in September during migration. Involves head, body, tertials, tertial coverts, lesser upper wing-coverts, variable number of median and greater upper wing-coverts, and sometimes t1; some outer longer lesser and greater upper wing-coverts as well as an occasional tertial sometimes retained. In some birds (perhaps those fledged late), moult restricted to head, body, and some median upper wing-coverts and tertial coverts. FIRST PRE-BREEDING. Partial; rather variable in timing and extent. Some start January, others not until March; completed April or early May. Extent sometimes as adult pre-breeding, but usually more restricted. Usually involves head, body, tertial coverts, some or all tertials, and t1, but in ♀ sometimes only face, throat, chest, and some scapulars and tertials, and occasionally no moult at all.

Measurements. ADULT, FIRST ADULT. Mainly north-west USSR (summer) and Balkans, Middle East, and Egypt (on migration), some from elsewhere in western part of range; skins (RMNH, ZMA). Bill (S) to skull, bill (N) to distal corner of nostril; exposed culmen on average c. 3·8 less than bill (S). Toe is middle toe with claw; claw is hind claw.

WING	♂ 87·7 (1·64; 18)	85–90	♀ 84·4 (1·64; 5)	82–86
TAIL	58·8 (2·52; 18)	56–63	56·2 (1·64; 5)	54–58
BILL (S)	14·7 (0·58; 18)	13·8–15·7	14·5 (0·88; 5)	13·7–15·3
BILL (N)	8·2 (0·47; 18)	7·5–8·9	8·4 (0·16; 5)	8·1–8·5
TARSUS	21·7 (0·59; 18)	20·7–22·5	20·7 (0·52; 5)	20·2–21·4
TOE	19·0 (1·03; 12)	17·6–20·2	18·9 (0·40; 5)	18·4–19·5
CLAW	10·4 (0·98; 18)	8·8–11·9	9·9 (0·96; 6)	8·9–10·9

Sex differences significant for wing and tarsus.

China and Taiwan, winter and spring; skins (BMNH, RMNH, ZMA).

WING	♂ 86·1 (1·43; 8)	84–88	♀ 84·0 (2·08; 10)	81–87
TAIL	58·2 (2·12; 8)	55–61	57·2 (1·60; 10)	55–60
BILL (S)	14·7 (0·39; 8)	14·2–15·2	14·0 (0·31; 9)	13·6–14·6
TARSUS	21·6 (0·88; 6)	20·8–23·2	21·7 (0·36; 9)	21·2–22·1

Sex differences significant for wing and bill (S). 1st adult (with juvenile flight-feathers and often tail) similar to adult, combined above.

WING. (1) Breeding birds from Timanskaya tundra, north-west USSR (Gladkow 1941). (2) Autumn migrants, Manchuria (Piechocki 1958).

(1)	♂ 87·5 (1·79; 56)	82–91	♀ 83·5 (2·00; 29)	77–86
(2)	85·3 (—; 13)	83–87	82·9 (—; 9)	82–85

Wing of adult ♂♂ from breeding area: northern Europe, 86·5 (14) 82–91; north-east Siberia, 84·5 (18) 80–89 (Vaurie 1954).

Weights. ADULT, FIRST ADULT. Fair Isle (Scotland): May, ♂ 21; June, 16·5; September, 18·5; October, 21·5 (Fair Isle Bird Observatory, N J Riddiford). Summer: north-west USSR, ♂ 20·5 (29) 17–24, ♀ 20·3 (24) 17–24; eastern USSR, ♂ 21·2 (4) 20–23 (Dementiev and Gladkov 1954a). Yamal peninsula

(USSR), summer: ♂ 21·1 (20) 20–24, ♀ 20·8 (17) 20–23 (Danilov *et al.* 1984).

On autumn migration. Malta: 19·8 (1·7; 12) 16–22·5 (J Sultana and C Gauci). North-east Turkey, October: ♂♂ 21, 24 (Kumerloeve 1968). Kazakhstan (USSR), September–October: ♂♂ 19·3, 22·5, 24·2; ♂♂ 24, 24 (Dolgushin *et al.* 1970). Manchuria, mainly 1st half of September: ♂ 20·4 (13) 17–24, ♀ 19·6 (9) 18–21 (Piechocki 1958).

In winter quarters. Ghana, November–February: 20·1 (0·57; 4) 19·3–20·6 (BTO). Nigeria, February–March: 20·0 (0·92; 6) 19–21 (Ward 1963). Central Kenya: October to early March, 19·3 (1·3; 59) 17–23; late March and early April, 21·8 (2·9; 15) 17–28 (D J Pearson).

On spring migration. Western Morocco, early March: 19·6 (Kersten *et al.* 1983). South-east Morocco, early April: 17·5 (BTO). Malta: 21·0 (2·1; 30) 18–27 (J Sultana and C Gauci). Kuwait, Saudi Arabia, Turkey, and northern Iran, April and early May: ♂♂, 19, 23·5; ♀ 21·5; unsexed, 15·7, 18, 20·5, 23 (Schüz 1959; Kumerloeve 1963; BTO). Kazakhstan (USSR), April–May: ♂♂ 19·3, 19·5; ♀♀ 17, 19·5 (Dolgushin *et al.* 1970). Exhausted ♂, Netherlands, September: 16 (ZMA).

NESTLING. At hatching, c. 1·5; peak of 19·5–20·6 reached on c. 11th day, weight stable until fledging on 13th day (Dementiev and Gladkov 1954a).

Structure. Wing rather short, broad at base, tip bluntly pointed. 10 primaries: p9 longest, p8 0–1·5 shorter, p7 0–2·5, p6 1·5–5, p5 11–16, p4 16–21, p1 23–28. P10 reduced, narrow and pointed, 56–64 shorter than p9, 11–12 shorter than longest upper primary covert. Outer web of p6–p8 and inner of p7–p9 emarginated. Tip of longest tertial 0–6 mm from wing-tip (n=48) (Svensson 1984a). Tail rather long, tip square; 12 feathers. Bill slender, relatively fine at base, as in Meadow Pipit *A. pratensis*. For length of middle toe and hind claw, see Measurements. Outer toe with claw c. 75% of middle toe with claw, inner c. 79%; hind with claw c. 110%, without claw c. 53%.

Geographical variation. Slight. Populations from western part of breeding range (west from Taymyr peninsula) on average larger than those from north-east Siberia and western Alaska, feather-fringes on upperparts warmer brown-olive with pale spots (less cold green-grey) and underparts more heavily streaked (Johansen 1952); these western birds sometimes separated as *rufogularis* Brehm, 1842, but overlap in size large and difference in colour slight compared with individual variation and influence of wear and bleaching, not warranting separation (Vaurie 1954).

Recognition. Rather similar to *A. pratensis*; juvenile and adult ♀ in worn breeding stated to be inseparable from that species (Johansen 1952). In fresh plumage, *A. cervinus* has rump almost as heavily streaked as mantle and back (in *A. pratensis*, rump much less streaked), upperparts contrastingly streaked pale buff and dark rufous-brown (in *A. pratensis*, less contrastingly olive-brown and dark brown); in worn plumage, both species heavily streaked dark brown above, including rump, but *A. cervinus* always has some orange-buff on throat. In *A. cervinus*, 80% of c. 150 birds had contrasting brown-black central streak or patch (sometimes concealed) on longer under tail-coverts (in *A. pratensis*, under tail-coverts white, sometimes partly suffused grey-brown, unstreaked). Under wing-coverts and axillaries buff-white or (sometimes) rusty-brown, never yellowish (in *A. pratensis* grey-white or cream-white, virtually always with pale yellow tinge) (Svensson 1984a). Measurements of both species overlap extensively, except in wing length, which is useful as an identification character in sexed birds only (in *A. cervinus*, ♂ 84-91, ♀ 81-87; in *A. pratensis*, ♂ 78-86, ♀ 74-81). Longer wing of *A. cervinus* mainly due to relatively longer outer primaries, hence greater distance between tips of inner primaries and wing-tip in folded wing: p6 1-5 shorter than wing-tip (in *A. pratensis* 0-2), p5 11-16 (8-12), p4 16-21 (12-17), p1 23-28 (17-23).

CSR

Anthus spinoletta Rock Pipit and Water Pipit

PLATES 21, 22, and 27
[between pages 232 and 233]

Du. Oeverpieper/Waterpieper Fr. Pipit maritime/Pipit spioncelle Ge. Strandpieper/Wasserpieper
Ru. Горный конёк Sp. Bisbita ribereño costero/Bisbita ribereño alpino
Sw. Skärpiplärka/Vattenpiplärka

Alauda Spinoletta Linnaeus, 1758

Polytypic. (1) *Petrosus* group (Rock Pipit). *A. s. petrosus* (Montagu, 1798), Britain and Ireland (except Outer Hebrides and St Kilda), north-west and western France, and Channel Islands; *meinertzhageni* Bird, 1936, Outer Hebrides; *kleinschmidti* Hartert, 1905, Faeroes, Shetland, Fair Isle, and (perhaps this race) Orkney and St Kilda; *littoralis* C L Brehm, 1823, Fenno-Scandia (including Denmark) and north-west USSR. (2) *Spinoletta* group (Water Pipit). Nominate *spinoletta* (Linnaeus, 1758), mountains of central and southern Europe from Iberia to Balkans and (perhaps this race) north-west Turkey; *coutellii* Audouin, 1828, eastern Turkey, Caucasus, northern Iran, and Kopet-Dag (Turkmeniya, USSR); *blakistoni* Swinhoe, 1863, central Asia from north-east Afghanistan east to Transbaykalia and Nan Shan (extralimital). (3) *Rubescens* group (Water Pipit). *A. s. rubescens* (Tunstall, 1771), north-east Siberia east from Taymyr peninsula south to Nizhnyaya Tunguska and Baykal'skiy mountains, east to Sea of Okhotsk, Kamchatka, and Chukotsk peninsula, and northern North America east to western Greenland, accidental in west Palearctic; *japonicus* Temminck and Schlegel, 1847, eastern Siberia south of *rubescens*, east to Kuril Islands, Sakhalin, and Sikhote Alin, overlapping with *blakistoni* in southern Transbaykalia, accidental in west Palearctic; *c.* 2 further races in North America.

Field characters. Rock and Water Pipit groups separable in adult breeding plumage but often not so at other times.

(A) **Rock Pipit.** 16·5-17 cm; wing-span 22·5-28 cm. 10-15% larger than Meadow Pipit *A. pratensis*, with more bulk, longer and stronger bill and legs, and fuller tail; close in size to Tawny Pipit *A. campestris* but less slim and less obviously long-tailed. Quite large pipit with more robust form than smaller pipits. Shares Water Pipit's mottled (not streaked) upperparts, and dark bill and legs, but differs in grey and white or smoky outer tail-feathers, dusky underwing, and (in spring and summer) retention of well-streaked underparts. Flight confident, lacking hesitancy of *A. pratensis* and involving characteristic sweep before landing. Call distinctive. Sexes closely similar; marked seasonal variation in northernmost race. Juvenile not easily separable. 4 races in west Palearctic, 1 (*littoralis*) separable from other 3 in the field.

ADULT BREEDING. (1) North European race, *littoralis*. Upperparts basically dark olive-grey with soft darker streaking; less brown or olive than the 3 western races (except on rump and upper tail-coverts). Marked by obvious buff to off-white supercilium and paler grey to off-white wing-bars and tertial-fringes. Tail largely brown-black; outer web of outermost feather pale grey to dusky white, forming obvious (but not clean) edge when tail opened. Underparts basically pale cream to buff, with chin and upper throat almost white and lower throat, chest, and flanks deeper buff or pink; on birds with deepest chest colour, streaks usually absent on breast and faint on flanks, but birds with paler buff chest usually have obvious streaking across chest and down flanks. Most variable of all races, inviting confusion particularly with Nearctic race of Water Pipit *rubescens*. (2) British and north-west French race, *petrosus*. Darker, more olive above than *littoralis*, with indistinct, often broken or spotted dirty cream supercilium, and dull buff-brown wing-bars and tertial-fringes. Tail has pale grey-brown outermost feather, looking smoky and forming only dull edge when tail opened. Dirtier, more olive-toned below than *littoralis*, with copious, diffuse, dark brown streaks covering all underparts except chin, upper throat, and vent; thus appears as dark below as above (unlike *littoralis* or any race of Water Pipit). A few birds show faint pink-buff tone on lower throat. Prior to autumn moult, wear reduces uniformity of plumage, with mottling on upperparts becoming distinct, tertial-fringes bleaching, and lower underparts wearing paler. (3) Faeroes race, *kleinschmidti*. Marginally largest of all races of Water or Rock Pipit in west Palearctic. Closely resembles *petrosus* but with yellower tone to feather-fringes of upperparts and ground-colour of underparts; thus appears streaked darker brown above and much more heavily streaked below. Face also suffused yellow, unlike *petrosus* and *meinertzhageni*. Outer tail-feathers brown-white. (4) Outer Hebrides race, *meinertzhageni*. Resembles *petrosus* but upperpart markings almost black, rump and fringes of wing feathers green-olive, and ground-colour of underparts yellower. ADULT NON-BREEDING. Olive tone of upperparts and streaking of underparts increased; bird assumes generally dusky olive appearance, particularly in poor light. (1) *A. s. littoralis*. Incompletely studied in the field but most retain resemblance to *rubescens*, looking creamier or yellower below than *petrosus* and showing more obvious supercilium and wing-bars. Heavy dark grey or brown streaks on chest and flanks more extensive. (2) *A. s. petrosus, kleinschmidti*, and *meinertzhageni*. All show less seasonal difference than any other race of Rock or Water Pipit, merely becoming darker, with streaks on underparts olive-brown and much diffused. Separation of these races in winter not studied, but *kleinschmidti* apparently retains more marked yellow tone above and below than others. JUVENILE. All races. Resembles winter adult but shows browner upperparts, with black feather-centres more obvious (creating distinct mottling or dull streaking) and pale feather-tips more distinct, and somewhat cleaner underparts, with malar stripe and chest- and flank-streaks sharper and slightly darker. In all races and at all ages and seasons, bill black-horn, with (in winter) only faint dun or yellowish base to lower mandible; legs noticeably dull or dark, from dirty flesh-brown through black-brown to lead-grey.

Dark bare parts (except in east Asian *japonicus*) and characteristic call are distinctive features of both Rock and Water Pipits, but complex racial differences and similarity of appearance to smaller and larger pipits (for larger pipits, see Water Pipit, below) present many pitfalls to inexperienced observer. With smaller pipits, most confusion arises between Rock Pipit and *A. pratensis* or Red-throated Pipit *A. cervinus* (which can breed in areas abutting or even overlapping with Rock Pipit); these 2 other species may both appear dark on upperparts and well-streaked below but are distinctly smaller and have white outer tail-feathers, pale legs, and different calls. Separation of Rock Pipit from Water Pipit and, especially, of their component races is often difficult and still incompletely studied; compounded by overlaps of winter ranges caused by transatlantic vagrancy of *rubescens* and altitudinal migration in remainder of Water Pipit group, by little-understood coastal and over-sea movement of Rock Pipit, and by overlaps in winter habitat preferences. Important to remember that migrant nominate *spinoletta* Water Pipits move north-west and east as well as south to sheltered coastal marshes and inland waters, and may share estuaries and their grassy edges with at least some migrant *littoralis* Rock Pipits, but rarely occur on rocky sea-shores inhabited by mainly sedentary *petrosus* Rock Pipits. Of all races of Rock and Water Pipits, *littoralis* is the rogue since it can resemble at least 3 others and occurs in all the species' winter habitats, though preferring flooded meadows to nearby river or sea-shores on spring migration (Flamborough Ornithological Group, D I M Wallace). Flight of both Rock and Water Pipit easy and confident, lacking hesitant flutter of *A. pratensis*, but action does not have urgent bounds of larger pipits, since wing-beats more frequent than theirs over short distance and produce noticeably steady progress on migration; over breeding habitat and along shores in winter, flight often ends in long, curving sweep (on shore, bird then immediately disappearing among seaweed or rocks). Flight silhouette quite bulky, being fuller-bodied and broader-tailed than smaller pipits and far less long-tailed than large ones. Rock Pipit appears somewhat heavier in flight than Water Pipit, though this probably an illusion due to darker underparts. Escape-flight usually low and short. Song-flight like *A. pratensis*, but usually from and to rock or cliff edge; final 'parachute' phase slower. Gait less creeping than *A. pratensis* but lacks persistent long runs of large pipits; hops, walks, and runs both on short grass and among rocks and seaweed. Rarely perches on wires or trees. Not gregarious. Fairly tame.

Song recalls *A. pratensis* in phrasing, but much louder, with notes both more musical and more tinkling, and terminal trill stronger. Flight- and general alarm-call 'phist', 'feest', or 'weesp', less of a cheep than *A. pratensis* with less squeaky, fuller, yet still sibilant tone; usually given singly but in excitement repeated in more slurred trisyllabic form. Alarm-call of parent bird an insistent 'tchip', distinctly metallic.

(B) Water Pipit. Size and structure as Rock Pipit except for slightly longer tail and legs and slightly shorter bill; bare part colours and flight as Rock Pipit. Plumage differs in less-streaked or plain underparts in spring and summer, white outer tail-feathers, and almost white underwing. Sexes closely similar; marked seasonal variation. Juvenile easily separable. 4 races occur in west Palearctic; 3 covered here, separable in breeding plumage but at other times subject to confusion with each other and with 1 race of Rock Pipit (*littoralis*); for 4th (*japonicus*), see Geographical Variation.

ADULT BREEDING. (1) Central and south European race, nominate *spinoletta*. Upperparts basically dull grey, with brown tone slight on crown and nape but often pronounced on back and wings; strikingly marked by quite deep and long, white or pink-white supercilium (more contrasting than in any other race of Rock and Water Pipit), pale grey-brown margins to median wing-coverts and tips to greater coverts (forming double wing-bar more obvious than in any other race), and pale grey-buff tertial-fringes (often overlapping to form pale panel). Tail largely brown-black; outer web of outermost feather white, combining with white wedge on inner web and white tip to next-to-outermost feather to form obvious edge and corner when tail opened. Underparts basically dull white, with strong (♂) or faint (♀) pale buff-pink wash from throat to mid-belly; flanks may show fine streaks, but only ♀ retains marks strong enough to suggest pattern of winter plumage on chest. Prior to autumn moult, wear reduces colour contrasts in both upper- and underparts, with wing-bars becoming less obvious, tertials noticeably paler, and pink flush much reduced. (2)

Caucasus race, *coutellii*. Plumage appears faded, being paler grey above than nominate *spinoletta*, with sandy (not brown) tone and rather more obvious markings, and paler buff below. Chest mottled or obsoletely streaked in both ♂ and ♀. Supercilium broad but dull. (3) Nearctic race, *rubescens*. Browner above than nominate *spinoletta* but ashier on crown and mantle than north European Rock Pipit, *littoralis*; supercilium less contrasting than in nominate *spinoletta*, buff or pink, usually showing small streaks. Much buffier below than nominate *spinoletta*, with little or no contrast below chest and vent and always some (often much) obvious streaking on chest and flanks. Looks white-throated, unlike nominate *spinoletta* and *coutellii*. ADULT NON-BREEDING. Grey tone of upperparts and full pink tinge of underparts lost; all races become browner above and streaked below. Separation from Rock Pipit becomes difficult, even impossible. (1) Nominate *spinoletta*. Upperparts essentially brown, lacking pure olive tone. Distinct white supercilium retained. Underparts basically dull white, well-streaked dark brown on chest and flanks, hardly so on belly, and not at all on vent and throat. (2) *A. s. coutellii*. Not studied in the field but apparently slightly paler brown above and less streaked below. (3) *A. s. rubescens*. Slightly more olive- or grey-toned above than nominate *spinoletta* and (importantly) remaining buff below, especially on under tail-coverts—never white as in *spinoletta* nor sullied olivaceous as in Rock Pipit. Streaks on underparts noticeably narrow and almost black. JUVENILE. All races differ from adult as in Rock Pipit. In all races and at all ages and seasons, bare parts as in Rock Pipit (except that legs of east Asian *japonicus* are rather pale brown).

See Rock Pipit (above) for separation from that group and for separation of component races. Among other pipits, mistakes possible between Water Pipit and *A. campestris* or Long-billed Pipit *A. similis* breeding in uplands. Of last 2, *A. campestris* easily distinguished by tawny plumage (streaked only in juvenile), form and flight (like wagtail *Motacilla*), pale legs, and loud plaintive call; *A. similis* much more confusing, with sandy-grey plumage particularly suggesting *coutellii*, but still separable on greater length and size, pale bare parts, and loud disyllabic call. Important also to remember that *A. spinoletta* very unlikely to breed in arid habitat of *A. campestris* and *A. similis*. Water Pipit has flight, gait, etc., as Rock Pipit, though appears lighter in flight (see above). Nominate *spinoletta* also walks over watercress and other floating vegetation, and *rubescens* accompanies gait with much wagtail-like tail-wagging (this habit less obvious in other races, except in alarm). Usually shy.

Voice as Rock Pipit, though common call said to be less loud.

Habitat. Sharply divided into 2 groups which on ecological grounds would merit specific separation. Common features limited to tolerance of extreme conditions and often sunless climates, and liking for neighbourhood of water and rocks.

(A) Rock Pipit. Coastal. Ranges from middle to upper latitudes in temperate, boreal, and arctic zones, rarely penetrating more than a short distance inland and almost entirely attached to rocky sea-cliffs and crags, rarely much higher than 100 m and often down to shore level (Sharrock 1976). Avoids totally exposed situations, preferring sheltered gullies or inlets, and islands, even far offshore; not troubled by high and gusty winds or prolonged heavy rain and salt spray.

(B) Water Pipit. Montane. Breeds in west Palearctic in middle and lower-middle latitudes at considerable elevations, in Switzerland infrequently below 1400-1800 m and thence up to 2600 m or even higher. Habitat here is on cool levels or moderate slopes, clear of treeline, birds foraging in fine weather on short grass, but when wet rather on dwarf heath. Prefers moist or wet meadows with pools, watercourses, or snowmelt, or fresh or permanent snow cover. (Glutz von Blotzheim 1962.) Prefers zone of stunted trees with sparse ground cover or moist meadows, often close to glaciers and on steep nearly bare crags, even above snowline (Niethammer 1937). In Bayern (West Germany), breeds at *c.* 615-2230 m; in Carpathians at 1200-2400 m; in mountains of south-east Europe to almost 3000 m. Breeds mainly on grassy, bouldery, swampy ground between dwarf krummholz growth and persisting snow patches, and on moist stony or moorland areas with more or less low plant cover, tolerating wide altitudinal and climatic range (Pätzold 1984). In Armeniya (USSR) breeds at 2000-3150 m. Habitat in USSR varies, especially extralimitally where arctic birch thickets and stone screes with rhododendron thickets and rocky tundra may be occupied. Descends in winter to lower ground, or banks of mountain streams, occurring in spring on boggy lowlands with shrubs, sandy lowlands, and arable land. (Dementiev and Gladkov 1954*a*.) In western Europe, descends to flooded or damp meadows, watercress beds, estuaries and seashores, including mudflats (Bannerman 1953).

Distribution. Some spread in Denmark, Norway, and Finland.

BRITAIN, IRELAND. No evidence of any widespread change in distribution; for minor changes, see Parslow (1973) and Sharrock (1976). WEST GERMANY. Has bred Hessen (Witt 1972). DENMARK. Spread in islands in Kattegat (TD). NORWAY. Some range extension (VR). FINLAND. First bred Valassaaret islands 1962 (OH). EAST GERMANY. First bred Harz 1964 (SS). ALBANIA. May breed but no proof (EN). PORTUGAL. Probably breeds but no proof (RR).

Accidental. Spitsbergen, Jan Mayen, Iceland, Malta, Canary Islands.

Population. Increasing Finland.

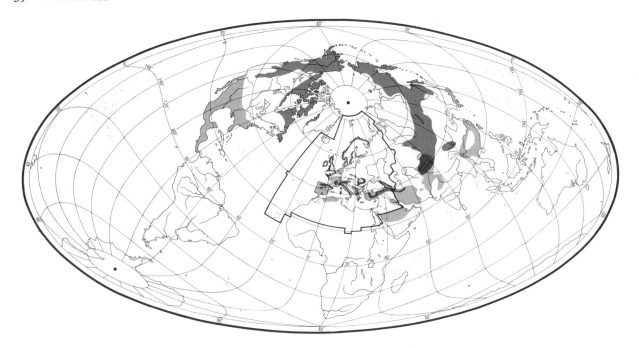

FAEROES. 2000-2500 pairs (Bloch and Sørensen 1984). BRITAIN, IRELAND. Over 50 000 pairs; no evidence of any widespread change (Parslow 1973; Sharrock 1976). FRANCE. 10 000 to 100 000 pairs (Yeatman 1976). WEST GERMANY. Bayern: 2000-6000 pairs (Bezzel *et al.* 1980). Baden-Württemberg: 15-25 pairs (Hölzinger *et al.* 1970). DENMARK. About 100 pairs (Kiis 1981). SWEDEN. About 100 000 pairs (Ulfstrand and Högstedt 1976). FINLAND. Rough estimate 500 pairs (Merikallio 1958). Increasing; hardly exceeds 1000 pairs (OH). POLAND. 2000-3000 pairs in Tatry mountains (P Profus). CZECHOSLOVAKIA. About 200 pairs (Witt 1982). No changes reported (KH). ITALY. Sardinia: 10-30 pairs (Schenk 1980). GREECE. Only a few tens of pairs (GIH).

Oldest ringed bird 8 years 10 months (Rydzewski 1978).

Movements. Migratory, partially migratory, dispersive, or resident in different parts of range, races differing considerably in extent and direction of movement (Zink 1975; Pätzold 1984). Along ice-free shores Rock Pipits are mainly sedentary, and most European Water Pipits are resistant to winter conditions, only retreating slightly from severe frost—either south or to lower altitudes (Voous 1960). Difficulty in assigning individuals to a race (within Rock Pipit and Water Pipit groups) obscures movement patterns.

(A) **Rock Pipit**. Faeroes population of *kleinschmidti* winters almost exclusively within breeding range (Pätzold 1984), though most 1st-years from Fair Isle leave in autumn (Williamson 1965). Thus, 2 ringed there as nestlings recovered south-east Scotland, both over 300 km south, and (as only proof that birds of British

origin may reach European continent to winter) another recovered in Netherlands 700 km SSE (Zink 1975). Bird ringed January on coast of north-east England was recovered in May, at sea just north of Faeroes (Zink 1975).

Populations of mainland Britain and Ireland, *petrosus*, are basically resident with local dispersive movements, birds appearing away from breeding areas from September (Zink 1975). No long-distance recovery from over 2000 ringed on Skokholm, Wales (Evans 1966a). Bird from Lundy Island (south-west England) recovered 210 km NNW in south-east Ireland is further proof that movements may extend beyond local level. Numbers along east coast of Britain increase in winter but this at least partly due to immigration of *littoralis* from Norway (as shown by one recovery in south-east Scotland of a bird ringed in south-west Norway). Irish birds move only locally (Ruttledge 1966). Many Rock Pipits (perhaps mostly *petrosus*) move via Straits of Gibraltar (Lathbury 1970; Tellería 1981) into north-west Africa, mainly Morocco (Heim de Balsac and Mayaud 1962) and also Algeria (Etchécopar and Hüe 1967). Some may overwinter on northern edge of Sahara, e.g. oasis of Ghardaia in Algeria (Gaston 1970).

Baltic and northern populations of *littoralis* vacate breeding areas in winter: thus, bird ringed as nestling in extreme north-west USSR was recovered in December on Vesterålen islands of north-west Norway over 600 km west (Zink 1975); also, many recoveries show movements between WSW and south to Britain, Portugal, and Sicily, and birds may occur anywhere within area enclosed by these extreme recoveries with regular passage records from central Europe. Also occurs rarely on Malta (Sultana

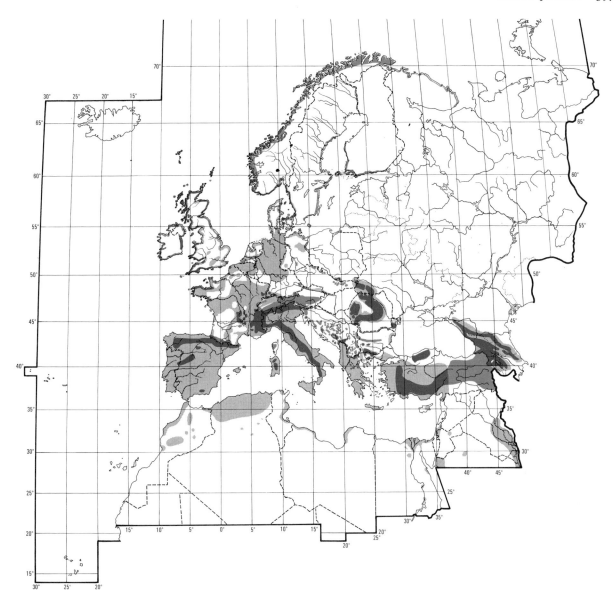

and Gauci 1982). Winters in small numbers on Baltic coast of East Germany but rarely completely absent; single birds winter on larger lake shores in Mecklenburg; only a vagrant further inland (Pätzold 1984). Within Britain, occurs on east coast (south from Shetland) and south coast, also western Ireland; mainly October–November and late March and early April (British Ornithologists' Union 1971).

(B) **Water Pipit**. Populations from mountains of central and southern Europe (nominate *spinoletta*) recorded in winter almost throughout inland central Europe north-west to Tongres, Belgium (Christiaens and Nelissen 1970), and western Netherlands (Koning 1982), rather rarely on coast (Pätzold 1984). Many thus move north in autumn, though movement in Alps often purely altitudinal, birds descending from mountains to adjacent lakes and rivers (Crousaz 1966; Zink 1975). At some places (e.g. Berlin area), numbers at roosts peak October–November and in February, perhaps indicating onward movement (Witt 1983). Considerable proportion of this race make long movements, mostly to south-west as with birds ringed in western Czechoslovakia and recovered on Ebro river in north-east Spain, *c.* 1500 km away (Zink 1975); others from same source have moved south into northern and southern Italy, distances of 500–1200 km. Although not supported by ringing recoveries, large numbers of nominate *spinoletta* move into North Africa (Blondel 1962*b*; Zink 1975), although birds (presumably of this race) are scarce in Libya (Bundy 1976), as on Crete (Stresemann 1956*a*). No evidence that populations of Spanish mountains make other than local altitudinal movements. Examples of this race show faithfulness to winter quarters with 6 out of 23 ringed birds recaptured in same area the following year (Peltzer 1967*b*).

A. s. coutellii, breeding in Turkey, Caucasus, and northern Iran, makes altitudinal or longer movements, presumably accounting for birds wintering in Middle East, including Egypt (Etchécopar and Hüe 1967). There is evidence that birds breeding in Elburz mountains of northern Iran descend to lower levels in winter, with small flocks noted at Veramin and Pahlevi in February (Passburg 1959). In Cyprus, fairly common March-April, less so mid-October to November, some also wintering (Flint and Stewart 1983). Near Suez, 321 birds arrived 21 October-5 November (Bijlsma 1982). Arrives northern Caucasus late March but does not move up to breeding grounds until April-May (Dementiev and Gladkov 1954*a*).

Populations of north-east Siberia, northern North America, and western Greenland (*rubescens*) are long-distance migrants wintering in south-east Asia, southern USA, and Central America. Autumn movement in North America almost south-west in the east, east of south in the west, with some possible evidence of loop migration, e.g. abundant in New England in autumn but that region avoided in spring. Spring passage relatively slow, starting late March in the east with birds in Washington at beginning of April and Newfoundland in early May, but not reaching extreme north, with snow-melt, until early June. (Bent 1950). Leaves western Greenland late August or September, returning May or early June (Salomonsen 1950-1). Individuals of this race (presumably from North America) have been recorded in Scotland on St Kilda and Fair Isle, and twice in south-east Ireland (British Ornithologists' Union 1971); also 3 records from Helgoland (Pätzold 1984) and 1 from Italy (Moltoni 1952).

A. s. japonicus of east-central Asia winters in Japan and south-east Asia (Pätzold 1984), but recorded west to Pakistan (Ali and Ripley 1973*b*), Israel (C Persson), and Italy (*Riv. ital. Orn.* 1962, **32**, 51-4). Birds wintering in south-east China present February-May (Caldwell and Caldwell 1931), suggesting slow autumn passage and late swift return in spring. JHE, DJB

Food.
(A) Rock Pipit. Mainly invertebrates. Feeds on ground among tide wrack and rocks, rarely making short pursuits to catch insects in flight. *A. s. littoralis* frequently wades in sea water, following waves as they retreat (Gibb 1956; Pätzold 1984). Will feed among household waste, and recorded trying to take meat hung up to dry (Collett and Olsen 1921). On Faeroes, October, pair of *kleinschmidti* recorded feeding among leaves and twigs of bush or tree (Williamson 1947*b*). Recorded persistently associating with tractor moving pebbles on beach and uncovering animal food (Kendrick 1985).

The following recorded in diet in west Palearctic. Aphids (Aphididae), adult and larval flies (Diptera: Chironomidae, Coelopidae, Sphaeroceridae, Calliphoridae), ants (Formicidae), adult and larval beetles (Coleoptera: Scarabaeidae, Carabidae, Curculionidae), amphipods (Amphipoda), isopods (*Idotea*, *Ligia*), small marine molluscs (young dog-whelks *Nucella lapillus*, small periwinkles *Littorina*), crabs *Cancer*, terrestrial snails *Helix*, slugs *Arion*, small worms (Annelida: Polychaeta, Oligochaeta), small fish (rockling *Gaidropsarus vulgari*, sand-eels Ammodytidae), and seeds (Dresser 1871-81; Newstead 1908; Collett and Olsen 1921; Elton 1928; Witherby *et al.* 1938*a*; Gibb 1956; Barnard 1963; Feare 1970; Duncan 1981; Pätzold 1984; Kendrick 1985).

In Norway in spring, several stomachs contained seeds, though summer food mainly Diptera and their larvae from rotting seaweed, rarely crustaceans; one stomach, July, contained amphipods and beetles (mainly *Aphodius* and their larvae). Winter diet in Norway mostly crustaceans; in December, one bird full of *Littorina*, also some amphipods. (Collett and Olsen 1921.) At Porthcew (Cornwall, England) in winter, main food was *Littorina neritoides*, eaten whole and singly—during 386 counts took $32.9 \pm SD\ 11.9$ per min; also took larvae of Chironomidae (during 187 counts, $34.1 \pm SD\ 14.9$ per min), kelp flies *Coelopa* (especially larvae, during 29 counts $8.2 \pm SD\ 5.4$ taken per min), and *Idotea* ($2.0 \pm SD\ 1.1$ taken per min, no sample size). Amphipods were rarely taken in Cornwall or on Skokholm Island (Wales) but were an important constituent of diet at Cape Clear (Ireland) and Loch Sunart (Scotland) where about 20 per min taken. Total daily dry weight intake in December estimated at 25 g periwinkles (of which 19.75 g inorganic) and 0.9 g of midge larvae (Gibb 1956).

On St Kilda (Scotland), adults seen carrying fish to young taken from around colony of Puffins *Fratercula arctica*—rocklings of 20-25 mm and sand-eels of 50-60 mm (Duncan 1981). No further information on diet of young.

(B) Water Pipit. Mainly invertebrates; also some plant material. Feeds mainly on ground, but occasionally catches insects in flight by making short leaps or flying from perch. In cold spells in high mountains during breeding season, feeds around burrow-entrances of marmots *Marmota*. (Glutz von Blotzheim 1962; Pätzold 1984.) The following recorded in summer diet in west Palearctic. Invertebrates: springtails (Collembola: *Isotoma*), stoneflies (Plecoptera), crickets and grasshoppers (Orthoptera: Acrididae, Tettigoniidae), bugs (Hemiptera: Pentatomidae, Cicadidae, Aphididae), lacewings and snake flies (Neuroptera), scorpion flies (Mecoptera: *Boreus*), butterflies and moths (Lepidoptera: Nymphalidae, Noctuidae, Geometridae), caddis flies (Trichoptera: Phryganeidae), flies (Diptera: Tipulidae, Culicidae, Bibionidae, Empididae, Rhagionidae, Syrphidae, Calliphoridae, Muscidae), Hymenoptera (sawflies Tenthredinidae, ants Formicidae), beetles (Coleoptera: Carabidae, Dytiscidae, Hydrophilidae, Staphylinidae, Histeridae, Scarabaeidae, Elateridae, Coccinellidae, Chrysomelidae, Curculionidae), spiders (Araneae: Philodromidae, Phalangidae, Ageliniidae especially *Argyroneta aquatica*), harvestmen (Opiliones), centipedes and millipedes (Myriapoda), snails (Gastropoda, Bivalvia), and worms (Annelida). Plant food includes alga *Ulothrix zonata*, and berries of *Vaccinium* and *Empetrum*. (Gil 1928; Witherby *et al.* 1938*a*; Glutz von Blotzheim 1962; Pätzold 1984.) Winter food includes freshwater snails (especially *Limnaea*), amphipods *Gammarus*, small Diptera, remains of a small dead fish, and plant material including seeds (Witherby *et al.* 1938*a*; Johnson 1970).

In Krkonoše mountains (Czechoslovakia/Poland), July-August, stomachs contained large quantities of alga *Ulothrix zonata* which formed up to 75% of diet by volume even when insects superabundant (Pätzold 1984). Those living near snowfields take so-called 'snow insects' (springtails *Isotoma saltans*, *I. nivalis*), scorpion fly *Boreus hyemalis*, and in North America members of order Grylloblattodea (Pätzold 1984). Extralimitally in Tien Shan mountains (USSR) in summer, 3 stomachs of *blakistoni* contained 60% (by number) Curculionidae; also centipedes (Lithobiomorpha), spiders, harvestmen, and caterpillars of Noctuidae (Zlotin 1968; see also Pek and Fedyanina 1961). In USA, significant amount of plant material taken only in autumn and winter, maximum *c.* 40% by volume (in mid-winter), $n = 285$ stomachs (Martin *et al.* 1951). In Utah, October, 103 stomachs of *rubescens* contained 3462 items of which 62·1% (by number) Hemiptera (including 45·8% *Nysius*, 2·7% leafhoppers, 2·6% aphids), 16·2% beetles (including 7·5% *Sitona*, and 4·6% Curculionidae), 13·9% weed seeds, 3·1% adult and larval Lepidoptera, 1·7% Hymenoptera, 0·9% Diptera, 0·9% Orthoptera, 0·6% springtails, 0·3% spiders, and 0·3% bristle-tails (Thysanura), thrips (Thysanoptera), and lacewings (Knowlton 1944).

Diet of young similar to adults, mainly insects. In Riederalp (Switzerland), 18 collar-samples contained 230 identifiable items: 59·1% (by number) Diptera (including 20·9% Rhagionidae), 20·0% Lepidoptera (including 16·5% Geometridae), 7·8% spiders and harvestmen, 5·7% Orthoptera, 3·9% beetles, 1·7% Hymenoptera, 1·3% Hemiptera, and 0·5% snails (Glutz von Blotzheim 1962). In Talasskiy Alatau mountains (Tien Shan, USSR), items in 120 collar-samples included 22·4% (by number) Lepidoptera, 22·7% cicadas, aphids, and psocids (Psocoptera), 16·0% Diptera, 9·0% Orthoptera, 9·0% beetles; also other Hemiptera, Hymenoptera, Myriapoda, ticks (Ixodidae), and, less commonly, dragonflies (Odonata), earwigs (Dermaptera), lacewings, and snake flies. Young fed 5–18 times per hr according to age. On day of fledging, brood of 5 fed average 6 times per hr 07.00–11.00 hrs (Kovshar' 1979). At Cape Whittle (north-east USA), 6 *rubescens* nestlings, 2–9 days old, fed average

3·6 times per hr over 13·5 hrs (Johnson 1933). In Wyoming (USA), nestlings fed 6–7·6 times per hr at 1 day, 9·0 at 2 days, 11·2 at 5 days, 16·0 at 8 days (Verbeek 1970; see also Pickwell 1947).   PJE

Social pattern and behaviour.

(A) Rock Pipit. Most important but by no means comprehensive studies on *petrosus* in Britain by Gibb (1956) and on *littoralis* at Åsa (Halland, south-west Sweden) by Elfström (1979). Account includes material from study of *littoralis* on islands of Nidingen and Malö (south-west Sweden) supplied by C Askenmo.

1. Normally territorial throughout the year (Gibb 1956; Elfström 1979; Tucker 1981), at least in parts of range. *A. s. petrosus* not all that gregarious but often in small parties in winter; these nomadic and probably young and/or immigrants. Rarely in regular flocks except on migration (Witherby *et al.* 1938*a*; Gibb 1956). Large flocks of *littoralis* said to occur in southern Norway in winter (Dresser 1871–81). In south-west England, resident *petrosus* defend individual territories in winter. Such a territory normally resumed in September and individual bird may hold same territory for several (perhaps up to 3) years; winter territory (or closely similar area) probably used later for breeding but this not proven. Territory used for feeding, though birds will occasionally feed in neighbouring territory or temporarily abandon territories (especially in bad weather) and then congregate at favourable food source: (Elfström 1979; C Askenmo); see also Antagonistic Behaviour (below). Territory of 'usual size' (see Breeding Dispersion, below) even when density almost twice normal (Gibb 1956). Sometimes gregarious for roosting, but *littoralis* normally less so than Water Pipit (Pätzold 1984). In Noord-Holland (Netherlands), winter sex-ratio probably about equal, unlike in Water Pipit (Koning 1982). Few reports of flock sizes: in Kent (south-east England), where *petrosus* only an occasional breeder, maximum 60 during autumn passage; in winter, regularly 15-30, exceptionally up to 70-120 (Taylor *et al.* 1981). Flock of 48 *littoralis* recorded inland and *c.* 50 on coast of West Germany (Ringleben 1953; Witt 1982). Feeding birds sometimes associate with Meadow Pipit *A. pratensis*, less commonly with Skylark *Alauda arvensis* and Snow Bunting *Plectrophenax nivalis* (Dresser 1871-81). BONDS. *A. s. petrosus* not studied in detail, but nothing to indicate mating system other than monogamous. At Åsa, only monogamous pairs of *littoralis* recorded (T Elfström). However, strong evidence of polygyny in colour-marked *littoralis* populations on Nidingen and Malö: inferred from records of more than 1 (maximum 3) nests within a territory, broods being fed by 1 ♀ or several broods being fed by same ♂; on Nidingen, over 4 years, 12-40% of 40-50 ♂♂ polygynous; in 1984, 9% of 32 ♂♂ on Malö (C Askenmo). Little information on duration of pair-bond. In *petrosus*, dissolved with break-up of territory in July (Gibb 1956). If both members of a *littoralis* pair return, they usually re-pair in same territory. If ♀ lost, ♂ always retains old territory and takes new mate. Widowed ♀ more likely to move to another territory some distance away. Some 1-year-old ♂♂ fail to acquire territory and thus a mate, but may establish very small territory late in season (C Askenmo). Young fed by both parents (Witherby *et al.* 1938*a*). Polygynous ♂ tends to concentrate on earliest brood and feeds those of secondary ♀♀ little or not at all. After fledging, brood frequently split betwen parents (C Askenmo). BREEDING DISPERSION. Solitary and highly territorial (Gibb 1956). At Åsa, average distance between nests $310 \pm SD108$ m, $n = 25$ (Elfström 1979). On Isle of May

(Scotland), 2 nests exceptionally c. 10 m apart (Southern 1937). At Asa, territory elongated along shore, with 1 or 2 distinct borders consisting (as in *A. pratensis*) of zone c. 20 m wide (Elfström 1979). On small islands, birds also occupy central part so that some territories separated from shore by others, or connected by narrow flight corridor. On Nidingen, density unusually high and territories often small (C Askenmo); larger on mainland at Åsa (see map in Elfström 1979). In Morbihan (France), birds use c. 75 m of rocky beach and sward extending behind to parallel road (Ferry *et al.* 1969). In Britain, in favourable habitat, territory c. 200-250 m long on average; some evidence of longer territories where density low. Where territories abut, boundaries remarkably precise, though not landward or seaward. Territory defended by both sexes, though more by ♂; ♀ apparently attacks other ♀♀. Mates of polygynous ♂♂ (see Bonds, above) may have partly separated home-ranges within ♂'s territory and conflicts between such ♀♀ quite common (C Askenmo). Nest normally sited within territory, but birds often feed far outside it and only small proportion of nestling food obtained within it (Gibb 1956). On Swedish islands, most nestling food collected within territory; adults most likely to forage outside territory in cold weather (C Askenmo). Detailed information on densities only from Britain. On Cumbrae island (Strathclyde, Scotland), estimated 39 territories, markedly clumped, along 18·6 km of coast; density 2·1 territories per km (2·5 excluding harbour where none), 4·3 in favoured stretch holding 14 territories. Up to 10 pairs per km in part of Calf of Man. On Skokholm (Wales), 4·8-6·0 per km; in Cornwall (south-west England) 3·6-6·0 per km of optimal habitat, 3·0-4·2 on less indented coastline; on Isle of May, 4·8-5·4; at Clevedon (south-west England) 4·2; at Loch Sunart (Scotland), 3·0-4·2; on Cape Clear Island (Ireland), 3·5; on Boreray (St Kilda, Western Isles), 2·5—commonest breeding passerine, at least 17 pairs in 77 ha (Duncan 1981); in Sussex 1·5; on Soay (Western Isles) 0·9. Average from these figures 3·48 territories per km, but biased in favour of well-studied and presumably well-occupied areas. Overall, probably 50 pairs per occupied 10-km square. On St Kilda, c. 150 pairs in c. 6 km², about half on cliffs, half inland. (Sharrock 1976; Tucker 1981 and sources cited therein.) For nesting association with *A. pratensis*, including territorial overlap, see that species (p. 372). No proof that residents hold same territory for more than 1 season, but likely. On Åland Islands (Finland), where *littoralis* probably exclusively migratory, ♀ recorded breeding on same islet c. 20 m from where ringed as nestling 3 years previously (Grenqvist 1935); see also Bonds (above) and Water Pipit (below). ROOSTING. Nocturnal and apparently not infrequently solitary or only a few together (Pätzold 1984). The race *petrosus* roosts chiefly in rock clefts and gullies (Witherby *et al.* 1938a). On Isle of May, mid-July, 3-4 birds roosted communally in cover of thistles (Compositae); about 12 used rock fissure c. 1·5 m wide (Southern 1937). In Wageningen (Netherlands), *littoralis* uses groynes by Rhine; in few cases where roost shared with Water Pipit, *littoralis* in dry terrain (Bijlsma 1977). Pätzold (1984) also mentioned roosting by *littoralis* in short-grass meadows. Winter activity rhythms studied in south-west England. In December, birds feed most intensively and rest for average only 35-39 min daily; pattern changes by March when rest for average 3 hrs daily. Routine modified by food supply: e.g. when food plentiful 28 December, birds fed for 6½ hrs, rested and self-preened for 1¾ hrs and defended territory for ¾ hr (Gibb 1956). *A. s. littoralis* reported to go to roost slightly later (c. 10 min) than Water Pipit (Pätzold 1984). Anting recorded in *petrosus* (Barnard 1963).

2. Comparative behavioural study at Åsa revealed displays and postures identical to those of *A. pratensis* (Elfström 1979). Not especially shy outside breeding season (Dresser 1871-81) and on shore, among rocks, typically allows closer approach than Water Pipit (Pätzold 1984). If disturbed, will flutter short distance; special Fluttering-flight (flying with rapid and shallow wing-beats) occurs in this situation according to Elfström (1979). Will give call 2a (2-3 times if flushed) and flick wings and tail nervously on landing (Naumann 1900). On Helgoland (West Germany), *littoralis* usually flies c. 15-20 m and often up to rock ledge, allowing observer to pass before flying down to resume feeding (Gätke 1891). In Faeroes, *kleinschmidti* tame and confiding outside breeding season; frequently perches in trees, especially when alarmed (Williamson 1947b). Can be wary, almost shy on breeding grounds like Water Pipit (Pätzold 1984). Typically crouches and hides from raptors; does not freeze afterwards, but especially vigilant when resuming activity. Occasionally (perhaps inexperienced juveniles) crouches and hides from Cormorant *Phalacrocorax carbo* and Snipe *Gallinago gallinago* (Gibb 1956). FLOCK BEHAVIOUR. No information. SONG-DISPLAY. ♂ sings (see 1 in Voice) in flight also from perch or from ground. Territorial ♂ often gives abbreviated variant of song from perch and then has wings drooped and tail raised (C Askenmo). Song-flight resembles that of *A. pratensis*. According to Dresser (1871-81), *petrosus* sings during ascent at c. 50° to great height and descends silently at same angle, or may sing throughout flight. In *littoralis*, slower average wing-beat frequency (9·4 per s) than *A. pratensis* (13·6 per s) during ascent, and during phase with rapid shallow wing-beats (6·1 and 13·2 per s). Song-flight probably serves to attract mate as more frequently performed by unpaired bird, but also apparently stimulates neighbouring ♂♂ to sing (Elfström 1979; C Askenmo). For relative frequency of Song-duels at Åsa, see *A. pratensis* (p. 373). In Faeroes, early to mid-April, subdued variant of full song given while feeding (Williamson 1947b). In Britain, main song-period late March to early July; irregular, but fairly frequent song mid-March and rest of July to early August, exceptional or only Subsong earlier in March (Witherby *et al.* 1938a); February, September (probably not uncommon in fine weather) and October song noted by Meiklejohn (1948a). At Åsa, ♂♂ losing mate and not re-pairing sing throughout summer to early August (Elfström 1979). ANTAGONISTIC BEHAVIOUR. (1) General. At Åsa, most interactions in March when territories being established. New territories rarely established after April. Later in year, 3 ♂♂ moved between their families in a communal feeding area and territory to display: 1-4 visits per hr for 5-24 min (n=53) (Elfström 1979). In *petrosus*, territories break up immediately after breeding season, in July (Gibb 1956). (2) Threat and fighting. Extremely aggressive towards conspecific birds (except mate: C Askenmo) where many gather at favourable food source; if this in a territory, owner may be almost constantly occupied with chasing—e.g. up to 17 (maximum 45) times per hr. Aggression manifested throughout year primarily in chases and (more in spring and breeding season) in threat-displays (see below). Chases used for territorial defence in summer (also in defence of mate) and winter, including within flocks (i.e. fighting over food), and against other species of comparable size. Owner always dominant in territory and expels all conspecific birds (except mate), including those attempting to feed; in late summer, also trespassing juveniles of early broods (Gibb 1956). However, at Åsa, independent juvenile *littoralis* usually tolerated (Elfström 1979). On Isle of May, 2 pairs nesting close together (see Breeding Dispersion, above) but with young at different stages had evidently developed unusual tolerance, though frequent intrusions by juveniles led to fierce fights, and other pairs

highly territorial as normal (Southern 1937). In breeding season, ♂ spends much time on look-out from which will also sing, sometimes also ascending into Song-flight. Will patrol territory in flight, including in winter (Gibb 1956; C Askenmo). Antagonistic postures and their use in particular situations much as in *A. pratensis*, though Wing-drooping apparently less common overall in *littoralis* and raising of tail more so (for full details, see Elfström 1979). Study of *petrosus* indicated bird attacked in flock always retreats when challenged, other then desisting. Close encounter on breeding grounds (boundary dispute or territory-owner trying to chase another too far) as follows. Birds tense, more erect than usual and with plumage sleeked, face up at less than 1 m or occasionally up to 2 m apart. Advance and retreat constantly, sometimes turning back on rival, apparently in attack-escape conflict. Often ends after a few seconds or up to 2 min with birds casually starting to mock-feed. Fight ensued when birds only c. 23 cm apart and was preceded by wing-flicking like that of Dunnock *Prunella modularis* (and thus perhaps different from wing-trembling common in *Anthus*). Both attacked simultaneously in flight (chases and fights almost exclusively aerial and as in *A. pratensis*: Elfström 1979) and Bill-snapped, then threatened as previously before one started to mock-feed (Gibb 1956). Territorial ♂♂ will perform parallel-walks along boundary; this seldom leads to fight and birds usually return to centre of territory (C Askenmo). When paired ♂ approached to within c. 1 m of pair on territory, both ♂♂ called (see 4 in Voice). All 3 birds crouched, then territorial ♂ chased intruder which later returned only to be attacked by ♀ (Gibb 1956). At Åsa, one closely watched ♂ more often aggressive towards conspecific birds than towards *A. pratensis* but chased latter more from feeding than from nesting area. Species apparently able to recognize each other by plumage and song (Elfström 1979). Will chase wagtails *Motacilla* (Gibb 1956) and recorded chasing Wheatear *Oenanthe oenanthe* in 'song-flight posture' (Conder 1948). Adopts Threat-posture (in breeding season mainly heterosexual as in *A. pratensis*; see that species (p. 373) in interspecific confrontations if attacked bird stands its ground: head held low and forward, bill open, wings slightly raised, tail fanned and lowered to ground (Gibb 1956). Other elements apparently here combined with Threat-posture typical of *littoralis* and *A. pratensis* (see p. 373), but differences insignificant. Higher arching of wings in Threat-posture of ♂ (whose ♀ nest-building) elicited by Cuckoo *Cuculus canorus*; in flying attack on *C. canorus*, ♂ had bill open and feet thrust forward (Gibb 1956). For experiments with playback of voice, see references in *A. pratensis* (p. 374). HETEROSEXUAL BEHAVIOUR. (1) General. In Cornwall, pair-formation takes place in early spring. Pair usually fed close together in March, synchronizing their activities (Gibb 1956). ♂ attends ♀ closely up to and during laying also when she forages outside territory (C Askenmo). ♂ also displays briefly (less than 1 week) late in season, even after 2nd brood (Elfström 1979). (2) Pair-bonding behaviour. ♀ treated as intruder initially but rarely attacked (Elfström 1979). Shortly after pairing, ♂ may chase ♀ briefly; she retreats but does not quit territory (Gibb 1956). Other low, fluttering chases accompanied by calls (see 3 in Voice) from both birds and, in some cases, ♀ apparently pursues ♂ rather than vice versa; some chases fast and with sharp turns (C Askenmo). Other behaviour in heterosexual interactions much as in *A. pratensis* (which see): includes frequent Song-flights by ♂, also song on ground when accompanying ♀, Quivering-flight, Wing-drooping (most commonly) and raising of tail, Fan-tail posture (mainly sexual and associated with pair-formation, including ♂'s first approach (C Askenmo); expression of high-intensity excitement), wing-

vibrating (recorded only once at Åsa), and symbolic carrying of nest-material (see, especially, Elfström 1979, also Williamson 1947b, Gibb 1956). ♂ and ♀ seen gaping at each other when close together; significance unknown (C Askenmo). ♂ losing mate late in season (May-July) performs more Song-flights and perches prominently much like Water Pipit in Alert-posture (Elfström 1979; see below). (3) Courtship-feeding. No information. (4) Mating. Fan-tail posture (as in *A. pratensis*, p. 374) associated with copulation attempt (Elfström 1979). (5) Nest-site selection. ♂♂ recorded behaving as if inspecting possible nest-sites or guiding ♀ to them. Carrying of material by ♂ (see above) sometimes stimulates ♀ to build (C Askenmo). No further information on heterosexual behaviour. RELATIONS WITHIN FAMILY GROUP. Young fed bill-to-bill (Witherby *et al.* 1938a). Juvenile *littoralis* recorded on another islet c. 200 m from nest 25 days after leaving it (Grenqvist 1935). ANTI-PREDATOR RESPONSES OF YOUNG. No information. PARENTAL ANTI-PREDATOR STRATEGIES. (1) Passive measures. ♀ sits tightly on eggs or small young, sometimes leaving only when (nearly) touched (Naumann 1900; Gibb 1956). (2) Active measures: against man. Like Water Pipit, often performs simple aerial demonstration with constant calls (Dresser 1871-81). In distraction-lure display of disablement type, ♀ fluttered c. 2 m silently from nest with head held back, bill closed and slightly raised, wings arched above back, tail fanned and lowered to ground. Waddled thus stiffly to and fro for c. 1 min. Stayed closed and called (see 2b in Voice) while man at nest (Gibb 1956). Wing-flapping typical during such display (Elfström 1979).

(B) Water Pipit. Most important studies of nominate *spinoletta* by Catzeflis (1978) at Col de Balme (Switzerland/France border) and monograph by Pätzold (1984). North American *rubescens* studied by (e.g.) Pickwell (1947), *alticola* by Verbeek (1970), and *blakistoni* in Tien Shan mountains (USSR) by Kovshar' (1979).

1. Solitary or gregarious outside breeding season. Flocks occur for roosting and (usually loose-knit) for feeding: often in groups of 2-5, sometimes 20-60 or up to 200 or more. Tendency to flock more pronounced in late winter and spring (Huber 1954b; Peltzer 1967a; Johnson 1970; Pätzold 1984). For reports of large flocks, see Bijlsma (1977, 1982) and Witt (1982, 1983). Very large flocks (of up to several thousand) sometimes occur in North America (e.g. Knowlton 1944). Normally solitary during autumn migration, rarely 2(-3) together (Pätzold 1984). At Halle (East Germany), typically solitary during day, flocks forming for roosting (Tauchnitz 1977). Shows fidelity to winter quarters where (in central Europe and Macedonia at least) ♂♂ apparently predominate (Diesselhorst 1957; Mester 1957c; Koning 1982; Witt 1982; Pätzold 1984), though not always markedly so: trapping at Halle roost showed sex-ratio of 11♂♂: 7♀♀ (Tauchnitz 1977). In Britain, birds reported to use same area (even same few square metres) for feeding over several weeks (Johnson 1970); not clear to what extent this defended but see Antagonistic Behaviour (below) and Rock Pipit (above). Break-up of flocks in spring very much dependent on snow cover: e.g. at Col de Balme in year with late spring, roving flocks (3-5 birds) recorded in late May (Catzeflis 1978; see also part 2). Breeding territories also frequently abandoned if poor weather persists (Verbeek 1970), birds moving down even after laying in late June (Catzeflis 1978). See Kovshar' (1979) for similar pattern in *blakistoni*. At Col de Balme, birds recorded using communal feeding area early in June: e.g. 12 in 100 m² (Catzeflis 1978); see Breeding Dispersion (below). Nomadic flocks of independent 1st-brood young (e.g. in Vitosha mountains, Bulgaria, up to 30) joined by young of 2nd broods, then

in late August or September by adults after moult (Pätzold 1984). At Col de Balme, not infrequently up to *c.* 150 (95% juveniles), once 200 in late August. Ringing showed that birds used different feeding sites from day to day. During one week in mid-September of 1973, estimated 405-570 juveniles probably from *c.* 12 km² (calculated from density; see below). Last big flocks (*c.* 50-130) recorded late September, but if no snow cover, flocks may remain high up in mountains until October (Catzeflis 1978). Further reports of flocking after breeding season in Naumann (1900), Černy *et al.* (1970), and Verbeek (1970). Will associate with Rock Pipit *A. s. littoralis* (Pätzold 1984), but in Wageningen (Netherlands) used separate feeding and roosting habitat (Bijlsma 1977); see also Roosting (below). BONDS. Monogamous mating system, though Schifferli (1969) reported nest with 8 eggs assumed to be from 2 ♀♀. Little information on duration of pair-bond. In marked pair of *blakistoni*, bond dissolved immediately after 1st brood fledged. Both birds re-paired with other birds for 2nd brood (Kovshar' 1979). For loose association between apparent pair-members in winter, see subsection 1 in Heterosexual Behaviour (below). Young fed by both parents, more by ♂ during 1st week when ♀ broods for *c.* 17-20% of time, parents then taking equal share (Kovshar' 1979). ♂ may tend 1st brood alone if ♀ re-lays (Meiklejohn 1930; Witherby *et al.* 1938a). BREEDING DISPERSION. Solitary and territorial. In favoured habitat of low shrubby vegetation (e.g. *Pinus pumila*) alternating with marshy open areas, streams, and rocks, pairs barely 100 m apart (Naumann 1900); in Switzerland, *c.* 40 m between nests recorded (Bannerman 1953). In Tien Shan, normal minimum *c.* 100 m, but less than 50 m recorded where density very high (Kovshar' 1979). At Col de Balme, territory (as marked by repeated circular Song-flights of ♂: see part 2) averages 4900 m² (2000-12 000), *n* = 25. No other conspecific bird apart from mate tolerated within it; some feeding done outside (Catzeflis 1978). In Tien Shan, territory *c.* 2000-3000 m². Defended mainly by ♂, especially against other conspecific singing ♂♂. Feeding area larger; food for young not infrequently collected *c.* 100-200 m from nest (Kovshar' 1979). In Wyoming, 5 territories of *alticola* averaged *c.* 1810 (1580-3355) m². On Bylot Island (North-West Territories, Canada), average size of *rubescens* territory 5600 m², each pair feeding along *c.* 100 m of 4-5 valleys (Drury 1961; Verbeek 1970); see also Pickwell (1947). Territory boundaries are often snowfields, marshy areas, or high rocks (Pätzold 1984). In Tien Shan, in case of re-pairing for 2nd brood (see Bonds, above), ♂ bred *c.* 70 m from 1st-brood nest, ♀ *c.* 200 m (Kovshar' 1979). Densities in various parts of range as follows. At Col de Balme in alpine zone (including some unsuitable habitat), general average *c.* 16 pairs per km². Probably *c.* 15-20 pairs per km² as a rule in alpine zone of central Alps, though *c.* 30 pairs per km² recorded in most favourable meadows (Catzeflis 1978). For further Swiss data, see Schifferli *et al.* (1982). In Krkonoše mountains (Czechoslovakia), density generally higher in wetter places. Highest (44·8 birds per km²) on rocky slopes and in basins; vegetated rock-fields hold 36·1 birds per km², *Pinus mugo* thickets 23·0, and meadows with sparse *P. mugo* only 8·9 birds per km² (Klíma and Urbánek 1958). Also in Krkonoše, 8 pairs recorded in 6 km²—1·3 (maximum 4) pairs per km². In Lower Tatra (Czechoslovakia), 26 pairs per km² (Pätzold 1984). In Malá Fatra (Czechoslovakia), transect counts of singing ♂♂ on different days in a season showed 29, 32, and 59 pairs per km² (Černy *et al.* 1970; Pätzold 1984). At Sasashi (Svanetia, Georgian SSR), 9-10 pairs of *coutellii* per km² (Zhordania and Gogilashvili 1976). Data from North America include *c.* 180 pairs per km² on Bylot Island (Drury 1961; Pätzold 1984) and *c.* 50 pairs per km² on alpine

tundra of Wyoming (Verbeek 1970); see also Pickwell (1947). For *japonicus* in eastern USSR, see Kishchinski (1980). Fidelity to breeding area over several years recorded (Pätzold 1984). Of 102 *alticola* (including 9 adult ♀♀) ringed one year, 1 adult ♀ returned the next year to breed *c.* 400 m from her previous site. 4 ringed juveniles also returned to their natal area (Verbeek 1970). ROOSTING. Nocturnal and (outside breeding season) communal, though birds may be rather dispersed in roost (Sunkel 1952: Frost 1972). Shows preference for damp or wet habitat with dense vegetation (mostly *Phragmites*, *Typha*, or *Carex*), birds often roosting on plant stems over water of varying depth. Bushes and trees such as willow *Salix* and alder *Alnus* apparently important as perches used before moving to roost proper, but birds will also roost on *Salix* branches overhanging water (Witt 1982, 1983; Pätzold 1984). In Egypt, *coutellii* similarly roosts in reedbeds and long grass bordering swamps (Witherby *et al.* 1938a). In Wageningen, prefers wetter habitat than Rock Pipit *A. s. littoralis* (Bijlsma 1977), though at Halle up to 50 birds continued to use reedbed when it dried out (Tauchnitz 1977). Will share roost with (e.g.) Reed Bunting *Emberiza schoeniclus* and White Wagtail *Motacilla alba* (Mester 1957c; Frost 1972). Birds probably travel from feeding grounds up to 5(-9) km away (Mester 1957c; Witt 1983). Most birds arrive 1-2 hrs before sunset and have entered roost by sunset; those arriving late (up to *c.* 30 min after sunset) usually enter roost immediately (see also Flock Behaviour, below). Morning exodus normally 30 min before sunrise at earliest. Leave fairly bunched, by sunrise at latest, most heading directly to feeding grounds (Witt 1983); for further details of activity rhythms in winter, see Witt (1984b). In Sauerland (West Germany), arrival at roost normally completed *c.* 30 min before sunset, apparently not influenced by weather. Birds roost in a horizontal posture, bill pointed forward. On 23 February, birds begin to leave around sunrise (Mester 1957c). On breeding grounds, roosts on ground in grass (Pätzold 1984). At Col de Balme, birds still roosting communally in early April use *Alnus* and rhododendron at 1550-1650 m (Catzeflis 1978); presumably roost on ground in such thickets. Move higher up in morning from 06.30-07.30 hrs in small flocks (3-4) and recorded moving down at *c.* 15.30 hrs. In late June and early July, active (vocally) from *c.* 04.00 hrs, *c.* 1 hr later in August. On 1 June, last song noted at 20.30 hrs; on 29 August, last calls at 19.15 hrs. In late September, active from *c.* 07.30 hrs and in roost by *c.* 18.30 hrs. Communal roosting by independent juveniles recorded from mid-August in 2 years; in one year, used rhododendron and *Vaccinium* at 2100 m, in other, meadow at *c.* 2160 m (Catzeflis 1978). Naumann (1900) noted lull in activity on breeding grounds 13.00-15.00 hrs and considered birds roost at this time. Comfort behaviour recorded in the wild includes water-bathing (rare), sand- and dust-bathing, sunning (not uncommon), and (only in sun) anting; for further details, including observations on captive birds, see Mester (1969) and Pätzold (1984).

2. Rather shy and not easily approached outside breeding season (e.g. Johnson 1970 for Britain). Shyer than Rock Pipit (Ringleben 1949) and *A. pratensis* (Stadler 1928b; Beneden 1950), e.g. inland wintering birds flying at *c.* 40-50 m from man (Pätzold 1984), though exceptionally allowing approach to *c.* 5-6 m (Ringleben 1957). Typically flies up in rather zigzag fashion and gives Contact-alarm call (see 2a in Voice) some time after take-off (Stadler 1928b); longer flights quite high (Pätzold 1984). May also fly away low or freeze (even for several minutes) *c.* 40-50 m from man. 2 birds flushed more or less simultaneously sometimes chase and call (Beneden 1950). At roost in Netherlands, birds extremely shy in open, but

stayed put despite disturbance after entering roost (Berg 1975); see also Flock Behaviour (below). In Derbyshire (north-central England), birds tamest at roost in March–April, allowing approach to *c.* 20 m (Frost 1972). Will freeze in presence of Kestrel *Falco tinnunculus* and once hid behind stone from Sparrowhawk *Accipiter nisus* (Beneden 1950); roosting flock recorded pursuing *A. nisus* for considerable distance (Mester 1957*c*). Somewhat less shy on breeding grounds than in winter, but still wary, especially when with young. ♂ (much less commonly ♀) will adopt Alert-posture (Fig A) with wings

A

drooped and tail slightly raised (Pätzold 1984). May call (see 2c in Voice) each time accompanied by deep downward movement of tail, or may fly up giving call 2a, land, and then fly off far to return later (Naumann 1900). Birds on migration usually fly up silently but may call after gaining height (Pätzold 1984). FLOCK BEHAVIOUR. In large post-breeding flocks (see part 1), several birds flying up together probably family members. If one bird flies round giving Contact-alarm call, whole flock usually takes off, birds soon landing, typically in trees where adopt Alert-posture and may hold it for several min (Naumann 1900). Birds arrive at roost singly or in small flocks, often approaching at some height (e.g. 20–40 m, even in wind: Mester 1957*c*). In Sauerland, drop down to roost directly or settle in trees and give call 2a from there. Early arrivals especially circle roost for *c.* 5–10 min, calling constantly before dropping down (Mester 1957*c*). In West Berlin, circling ('group-flight') usually evolves out of reconnoitring flights by small flocks on arrival; these may be joined by birds in trees or already in roost so that all birds often participate; some then break away to enter roost or to perch (Witt 1983). Sometimes circle with *A. pratensis* (Berg 1975). At Halle, birds initially silent and are widely dispersed in vegetation and by pools; fly up again *c.* 10 min before dark, some perch, then move to roost in flocks of 10–25 for noisy circling. Silent immediately in roost after sudden plunge (Tauchnitz 1977). Birds occasionally call from roost in response to later arrivals. Singing (see 1b in Voice) in roost noted in early March (Mester 1957*c*). Morning departure heralded by Contact-alarm calls (Witt 1983). SONG-DISPLAY. ♂ sings (see 1 in Voice) from perch (rock, tree, bush) or in flight (Pätzold 1984); according to Naumann (1900), mainly in flight, less commonly from perch or also while running about on ground. Occasional song from ground also noted in *alticola* (Verbeek 1970). In Krkonoše, May to early June, birds sang (incomplete song) predominantly from perch; assumed Alert-posture and moved tail down (5–10°) in time with song-notes. Bout of song lasting *c.* 5–8 min followed by pause of up to 30 min (Pätzold 1984). In Canada, *rubescens* recorded running about nervously with plumage sleeked, then singing briefly from rocks before ascending into Song-flight (Drury

1961); singing before ascent frequent in *rubescens* (Pickwell 1947) and also noted in nominate *spinoletta* (e.g. Schmidt 1908). Of 100 Song-flights in Krkonoše, bird took off from bush in 88% and from ground or rock in 12% (Pätzold 1984). Song-flight basically as in *A. pratensis* according to Stadler (1928*b*). In nominate *spinoletta*, bird uses rapid, shallow wing-beats, ascending at *c.* 30–45°, rarely spiralling, to *c.* 10–30 m. From peak of ascent, circles or flies in long arc, often crossing territory, to perform return flight to starting point later; may thus fly 40–60(–80) m (Pätzold 1984). Not known to what extent (if at all) gliding involved before descent phase, but does not hover (Pätzold 1984). Slow, gliding descent similar to that of Tree Pipit *A. trivialis*: wings widely spread and well forward, tail raised and partly spread. In study by Pätzold (1984), birds descended at 45–60(–80)° and in 72% of 100 cases landed on bush. May also fly from top of one tree to another (Zang 1972). Description in Naumann (1900) indicates initially spiralling descent after circling, then long, rather flat glide to landing point; see also Walther (1972). Rare plummeting descent reported only by Schmidt (1908). Starts to sing *c.* 2 s after take-off (only from peak of ascent according to Wallschläger 1984), then continues throughout Song-flight (see illustration in Pätzold 1984 and Voice for details) or ends flight silently (Stadler 1928*b*). Usually silent for *c.* 4–5 s after landing, but not infrequently gives incomplete song (as typically when perched) or continues full song immediately after landing. (Naumann 1900; Schmidt 1908; Stadler 1928*b*; Walther 1972; Pätzold 1984.) Variant Song-flight described by Pätzold (1984) apparently serves to mark territory boundaries. Used by territory-owner when neighbour performs true ascending Song-flight (rarely, both perform thus). Territory-owner flies low from bush to bush or rock to rock by boundary and sings (usually only 1st segment), continuing also during brief pauses on perch. One bird continued thus for 7 min, finally changing to Song-flight when neighbour silent and continuing when neighbour started again. Especially long (over several hundred metres) and high (up to 130–170 m) parabolic Song-flights described for *rubescens* which may also ascend steeply and in zigzags, then descend almost to take-off point. After landing, may give call 2a or sing with neck extended and plumage sleeked (Pickwell 1947; Drury 1961; Woodell 1979). See also Verbeek (1970) for *alticola* and for *blakistoni* Kovshar' (1979). Little song before sunrise or during afternoon. Will sing (from perch) in mist and low cloud, but not in rain (Naumann 1900). In Krkonoše, 5 June, one ♂ sang 54 times 09.00–15.00 hrs (43% from perch, 22% in ascending Song-flight, 35% in variant—see above). In Bulgaria at 1900 m, 19 June, song started at 04.20 hrs (Pätzold 1984). In Harz (East Germany), 23 May, birds sang with pauses of varying length 06.00–14.00 hrs (Walther 1972). In Tien Shan, *blakistoni* starts *c.* 30 min before dawn, then lull around midday and renewed song in evening (Kovshar' 1979). Incomplete song noted in Belgium in winter (Beneden 1950), also in mid-March in west Caspian lowlands (USSR) before birds moved up to breeding grounds (Schüz 1959). Complete and incomplete song noted from migrants in April, especially when passing through in some numbers (Stadler 1928*b*). In Europe, song normally ceases or at least wanes from about mid-July. At Col de Balme, maximum intensity in one year 25 June–5 July coinciding with laying and start of incubation. After marked decline from mid-July, occasional song noted during August (Catzeflis 1978). See also Kovshar' (1979) for *blakistoni*. ANTAGONISTIC BEHAVIOUR. (1) General. In Tien Shan, over a week may elapse between arrival on breeding grounds and first song (Kovshar' 1979). At Col de Balme, in year with late spring, only 50% of territories occupied by late May; *c.* 80% by mid-May of another year. Territories provisionally established above 1900 m from early April, birds singing irregularly and moving down to roost each evening. Snow-free patches occupied first. Birds will also sing lower down with onset of spring (Catzeflis 1978). See also Verbeek (1970) for *alticola*. (2) Threat and fighting. ♂ marks occupancy with Song-flights, usually delineating territory by circling or traversing it repeatedly from limit to limit, and, after landing (often on elevation: see Song-display, above), by adopting Alert-posture even when no rival visible (Pätzold 1984). In early phase when territorial behaviour strongest, ♂ *alticola* (but not ♀) often walks about with tail raised and wings drooped (Verbeek 1970); see *A. pratensis* (p. 373) for similar posturing. Rival ♂♂ may approach to within 1 m of each other on territory limit (e.g. snowfield) (Pätzold 1984). In *alticola*, often does so casually while feeding; birds then have wings drooped and move tail up and down. Move back and forth while feeding until one makes flying attack at the other, not infrequently chasing it far beyond territory limits (Verbeek 1970; Pätzold 1984). Birds may threaten at close quarters with wings drooped and head up, apparently to show off throat (as in Red-throated Pipit *A. cervinus*, p. 386). In subsequent chase, both give accelerated calls (Bijlsma 1977). On Bylot Island, silent *rubescens* sometimes trespassed with impunity while feeding, at other times were vigorously attacked; no detectable relation with breeding cycle. Trespasser once gave normal flight-call (see 8 in Voice) and territory-owner responded with Trill-call (see 6 in Voice), followed by diving attack from high look-out. Apparently antagonistic chases sometimes involved 3 birds, including at least 1 ♀ (Pickwell 1947; Drury 1961). Nominate *spinoletta* ♀ remains apparently indifferent, continuing to feed and not taking part in antagonistic interactions (Pätzold 1984). On return from chase, territory-owner usually lands on perch and assumes Alert-posture, sometimes mock-preening. May remain in Alert-posture and call (see 6 in Voice) when another ♂ flies over or take off to chase it beyond limit (Verbeek 1970; Pätzold 1984). No fights recorded by Pätzold (1984), but 2 *rubescens* ♂♂ once fought long and desperately on ground and in the air (Pickwell 1947). In Britain, early winter, will chase off conspecific birds and other species; 2 *A. spinoletta* at a watercress bed tend to stay at opposite ends. Markedly aggressive towards *A. pratensis* and *M. alba* (Johnson 1970). In Belgium in winter, *A. pratensis* chased persistently on ground and in the air (Beneden 1950) and on Swiss breeding grounds, bird carrying food chased singing ♂ *A. pratensis* (Gunzinger 1983). At Col de Balme, birds also chased juvenile Golden Eagle *Aquila chrysaetos*, Cuckoo *Cuculus canorus*, and Nutcracker *Nucifraga caryocatactes*; short chases with either species as initiator also often involved Wheatear *Oenanthe oenanthe*, and less frequently Skylark *Alauda* arvensis (Catzeflis 1978). HETEROSEXUAL BEHAVIOUR. (1) General. In Belgium, winter, probable pair-members apparently maintain loose contact vocally and by following each other in longer flights. When one in danger, other said to approach rapidly (Beneden 1950). Pair-formation takes place in some cases before arrival on breeding grounds (presumably in flocks at lower altitudes), sometimes only after ♂ has taken up territory (Pätzold 1984). (2) Pair-bonding behaviour. Chases, perhaps sexual, recorded at one roost from late February (Mester 1957c). Ground display much as in other *Anthus* and certain Alaudidae. ♂ has tail raised more steeply than in Alert-posture, wings drooped to brush grass and vibrated, head up and bill slightly open; may circle ♀ thus (Pätzold 1984). See also Fig B which shows pair in courtship-display. In Switzerland, late January, 2 birds called loudly, moved sideways (while facing) with dancing steps

for *c.* 30 s, fluttered up bill-to-bill (but with no contact), flew off and, after landing, one circled other (which rotated to follow its movements) with dancing steps for *c.* 1 min before both flew off (Geissbühler 1949). In Austria, late May, 2 birds performed simultaneous counter-circling 'courtship dance' (Gressel and Petersen 1981). In *alticola*, both birds recorded carrying nest-material and giving undescribed call; ♂'s behaviour rather frenzied (Verbeek 1970; see *A. pratensis*, p. 374). (3) Courtship-feeding. At Col de Balme, ♂ feeds ♀ near nest about once per hr (Catzeflis 1978). ♂ gives call 4 once or twice, ♀ then responding with call 5 and running or flying to ♂. ♀ begs by shivering wings. Flies back (or walks and feeds: Verbeek 1970), landing *c.* 1-3 m from nest and walking to it (Pätzold 1984). Further details on *alticola* in Verbeek (1970). In Tien Shan, some *blakistoni* ♂♂ feed ♀ fairly frequently on nest, others not at all (Kovshar' 1979). (4) Mating. No details. Courtship-sequence on ground (see subsection 2) not followed by copulation (Pätzold 1984). (5) Nest-site selection. No information on selection process. ♀ seen making scrape with rocking movements of body. ♂ waited in Alert-posture *c.* 5 m from scrape to which ♀ took material (Pätzold 1984). In Tien Shan, ♀ begins laying 1-3 days after nest completed (Kovshar' 1979). (6) Behaviour at nest. While ♀ incubating, ♂ generally on look-out nearby and sometimes ascends into Song-flight (Dresser 1871-8). ♀ off nest longer (8·1 compared with 4·3 min) when not fed by ♂ (Verbeek 1970). RELATIONS WITHIN FAMILY GROUP. Young brooded by ♀ for *c.* 6 days (Johnson 1933) or regularly for 4-5 days, then only sheltered from rain or sun (Verbeek 1970). ♀ *rubescens* frequently interrupted brooding to rearrange young in nest (Johnson 1933). Eyes of young open at *c.* 4-5(-6) days (Verbeek 1970; Catzeflis 1978; Pätzold 1984). In *alticola*, reacted to human voice at 3 days, hand movement at 6 days (Verbeek 1970). In *rubescens*, young initially aroused by call 7; by day 8, heard (wing-noise) or recognized parent (Johnson 1933). For further details of physical development, see Pätzold (1984). Young fed bill-to-bill (Witherby *et al.* 1938a). During period of regular brooding (see above), most food brought by ♂ who passes all or part of it to ♀ who then feeds young. ♂ will feed young direct if ♀ absent and always does so after brooding has stopped (Verbeek 1970). In *rubescens*, ♀ often examined food brought by ♂ and rarely ate some; food most often divided and both parents fed young. ♀ sometimes called off nest by ♂. Reluctant to rise and allow ♂ to feed young in late evening. Later in brooding period, ♀ sometimes foraged near nest and would dart at ♂ (apparently with hostile intent) if he approached (Johnson 1933). Pickwell (1947) recorded adult walking long distance to and from nest, frequently starting to collect food without flying. Faecal sacs initially consumed, later carried away mostly by ♀ (Johnson 1933) or in early days more by ♂ (Verbeek 1970). Young stand in nest and exercise wings, also self-preen in sun 1-2 days before leaving (Verbeek 1970; Pätzold 1984) at (data on various races) *c.* 13-15 days, earlier if disturbed. Some can

B

C

fly a bit on leaving. Beg as shown in Fig C. Stay mainly hidden in vegetation and near nest for *c.* 1 week (*alticola*) though one *rubescens* brood frequently recorded with one or both parents within *c.* 300 m from nest up to 18 days after leaving; entered another territory but no intermingling occurred there. Adult sometimes gave call 7 to summon young for feeding (Johnson 1933). Fully independent *c.* 28–30 days after leaving nest. (Naumann 1900; Verbeek 1970; Kovshar' 1979; Pätzold 1984). ♀ probably starts building again before or soon after 1st brood leaves nest (Meiklejohn 1930). ANTI-PREDATOR RESPONSES OF YOUNG. Nestlings fall silent on hearing parental alarm (Pätzold 1984). No longer react to human sounds or movements from *c.* 7–8 days and crouch deeper in nest from 9–10 days (Verbeek 1970). Call when handled at 7 days, then attempt to move into cover when placed on ground (Pätzold 1984). If handled or suddenly threatened from 12 days old, tend to leave nest. Can be caught at this stage, but then run off, wing-flapping. Out of nest, otherwise normally hide, allow close (*c.* 0·5 m) approach, then fly up (Verbeek 1970; Pätzold 1984). PARENTAL ANTI-PREDATOR STRATEGIES. (1) Passive measures. Incubating ♀♀ allow varyingly close approach (1–20 m). One left regularly on foot when observer at *c.* 20 m; no change through habituation. Reluctant to return to eggs or to feed young while man still present: e.g. waited alertly at *c.* 40 m, finally swallowed food and flew off (Pätzold 1984; also Meiklejohn 1930, Woodell 1979). May approach young out of nest in zigzag run (Géroudet 1957a). For fairly rapid habituation to man using hide, see Pickwell (1947) and Pätzold (1984). (2) Active measures: against man. Adult will call (see 2c in Voice), each time moving tail down, when man some distance (e.g. 100 m) from nest or young. Perches conspicuously in (exaggerated) Alert-posture, plumage ruffled. May fly back and forth between bushes, then gradually come closer, circling intruder and calling constantly. Changes to higher-intensity alarm-calls (see 2d–f in Voice) as danger becomes more acute (Naumann 1900; Walther 1972; Pätzold 1984). On Bylot Island, walked about calling (see 2a in Voice) and moving tail; also mock-fed while approaching nest (Drury 1961). When young *rubescens c.* 1 week old, both parents nearby very agitated, ♂ also performed Song-flights (snatches of song). Distraction-lure display of disablement type—fluttering along and dragging one wing—recorded only at first visit. Bird afterwards (and on subsequent visits) called from rock while moving tail (Woodell 1979). Pätzold (1984) noted distraction-display only on day of hatching: bird flitted *c.* 2 m from nest, tail slightly spread. (3) Active measures: against other animals. Will flutter over dog, giving alarm-calls and land so close as to be in danger of being caught (Naumann 1900).

(Figs by I Lewington: A and C from photographs in Pätzold 1984; B from drawing in Géroudet 1957a.)   MGW

Voice.

(A) **Rock Pipit**. Less well studied than Water Pipit (see below). For analysis of song and some calls of *littoralis*, also comparison with Meadow Pipit *A. pratensis*, see Elfström (1979). For further sonagrams, see also Pätzold (1984) and Wallschläger (1984).

CALLS OF ADULTS. (1) Song of ♂. (a) Resembles *A. pratensis*, but louder, notes fuller, rather more musical, and with more pronounced terminal trill (Witherby *et al.* 1938a). Said by Ferry *et al.* (1969) to be similar to *A. pratensis* in virtually all respects—slightly deeper toned but this evident only in direct comparison. Recording of *petrosus* (Fig I) reveals harsher, more chirruping song than in *A. pratensis*, generally lacking sweet quality of that species (P J Sellar); comprises 5 segments as follows. (i) After 2 single 'cheep' sounds, 27 double and quite harsh 'cheep-a' units (11 shown in Fig I) with quality of sparrow *Passer* or Chaffinch *Fringilla coelebs* and showing accelerando. (ii) 16 thin, but rather musical 'ge' or 'gee' units with rise in pitch, accelerando, and recalling part of song of Linnet *Carduelis cannabina*. (iii) 12 higher-pitched 'psee' sounds. (iv) 10 lower-pitched but still musical units. (v) 3 distinctive dry rattles recalling *Acrocephalus* warbler (M G Wilson). Song (after 2 introductory units) lasts 18 s. Units in segments ii–iv show use of 2 internal tympanic membranes (J Hall-Craggs). Like Water Pipit (see below) and also *A. pratensis*, frequently gives abbreviated variant(s). In *littoralis*, songs of 1st-segment units only lasted 0·05–33·5 s (Elfström 1979); individual variation, dialects (etc.), as in *A. pratensis* (p. 375). Recording of *littoralis* by T Elfström mostly of 1st-segment song though some units recalling *C. cannabina* also given (see above). Unlike in *petrosus* (Fig I), 1st-segment units monosyllabic; rather harsh, impure, buzzing 'tleet', 'treep', 'chree', or 'tsree'; mechanical, like motor warming up. In view of individual variation, probably not significantly different from *petrosus* (M G Wilson). According to Stadler (1928b), probably no major differences between songs of Water and Rock Pipit, though Pätzold (1984) and Wallschläger (1984) found some differences in small amount of material examined, and further detailed sonagraphic analysis undoubtedly required. (b) Subdued song of *kleinschmidti* (Faeroes) ventriloquial, with sweet, whispered quality and audible only over a few metres (Williamson 1947b). (2) Contact-alarm calls. (a) A 'tsüp' or 'phist' more metallic than *A. pratensis*, otherwise differing as Water Pipit (Witherby *et al.* 1938a; see below). Calls of *littoralis* ('tsip' and variants—for classification, see *A. pratensis*, p. 375) more drawn-out, with higher top frequencies and more frequency modulation than in *A. pratensis*, but differences minute and not easy to appreciate by ear according to Elfström (1979). In recording of *petrosus* (Fig II), calls less thin and delicate than 'squeal' of *A. pratensis* (P J Sellar); quite full-sounding, varyingly harsh, slightly strained 'pseeit' or 'psee-er', sometimes several

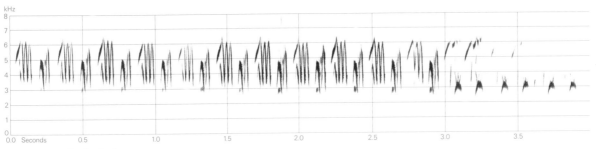

I   P J Sellar   Scotland   June 1975

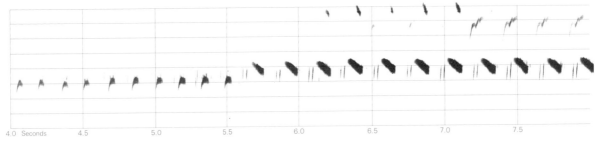

I   *cont.*

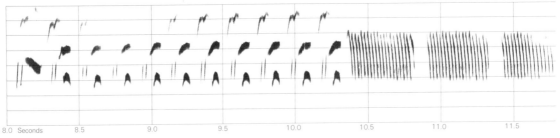

I   *cont.*

in fairly rapid succession. See below for possible slight differences from Water Pipit. Calls in Fig II show rising pitch—unlike in Fig c in Bergmann and Helb (1982) which shows call of Water Pipit having steady or slightly descending pitch; however, not known if this is a constant and significant difference (M G Wilson). (b) Alarm-call a high-pitched, shrill 'chip' or 'chick' (Witherby *et al.* 1938*a*); single 'ssit' (Jonsson 1978*b*). In recording, bird with young out of nest gives long and fairly rapid series of 'pee(t)', 'bii(t)' or 'seet' sounds. Some calls have more cheeping, chirruping quality, slightly reminiscent of

sparrow and more clearly disyllabic—'trreet', some harder and more explosive 'prrit' (M G Wilson). Fig III shows 'pit' followed by 'tcheoo' (J Hall-Craggs). Chipping calls given by ♀ after distraction-lure display (Gibb 1956) probably also belong here. May be given during whole of approach to nest, thus perhaps serving also as contact-call with young (P A D Hollom). (3) In hetero-sexual chases, both birds give an excited 'tss-ss-ss-ss' (Gibb 1956); perhaps an accelerated variant of call 2a. (4) Soft chipping call from ♂ in antagonistic interactions (Gibb 1956).

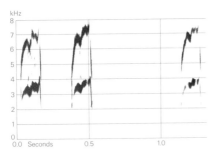

II   P J Sellar   Scotland   June 1975

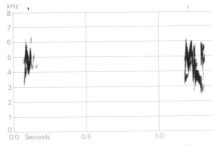

III   P J Sellar   Scotland   June 1975

CALLS OF YOUNG. In recording by P A D Hollom, *petrosus* nestlings being fed give rapid and quite loud 'psee' sounds with squeaky quality (some perhaps slightly husky) much as in adult call 2a (P J Sellar, M G Wilson). Captive *littoralis* gave mild form of adult call 2a with hoarse quality audible only at close quarters; sometimes disyllabic 'wie wis' or as a series 'sisisisisi' resembling Common Sandpiper *Actitis hypoleucos*. Calls an octave lower than those of Water Pipit, though varied in pitch and timbre (Stadler 1928*b*) much as in Water Pipit (Pätzold 1984; see below).

(B) Water Pipit. Repertoire much as in Meadow Pipit *A. pratensis*. Studied in detail by Stadler (1928*b*). Highly vocal on breeding grounds, using wide range of calls when disturbed. Outside breeding season, less vocal than *A. pratensis*, mostly giving call 2a which is less commonly uttered on breeding grounds (Naumann 1900; Pätzold 1984). Account includes some data from studies of Nearctic races; extent to which these differ from west Palearctic races unknown. For sonagrams, see Bergmann and Helb (1982), Pätzold (1984), and Wallschläger (1984).

CALLS OF ADULTS. (1) Song of ♂. (a) Intermediate between Tree Pipit *A. trivialis* and *A. pratensis*; confusable especially with *A. pratensis* (similarity confirmed by sonagraphic analysis: Wallschläger 1984), but generally superior and equalling in abandon *A. trivialis*, though drawn-out sounds inferior (Stadler 1928*b*; Witherby *et al.* 1938*a*). More melodious and louder than *A. pratensis* (Walther 1972). Rattling and tinkling song made up of *c.* 4–5 segments each comprising 1 unit-type, though each segment may be incorporated more than once (Wallschläger 1984). Song from perch (often also that given throughout Song-flight) normally only of unit-type given in 1st segment of less common full song. In Krkonoše mountains (Czechoslovakia), over 5 days, 52% of songs of 1st segment only, 8% full (Pätzold 1984); see also below. 1st-segment units usually metallic, slightly sharper, purer-sounding, and more metallic than in *A. pratensis* and *A. trivialis* (Pätzold 1984). Timbre also described as noisy and impure, sounds recalling Yellowhammer *Emberiza citrinella* or Chaffinch *Fringilla coelebs*. Units (apparently of 1–2 syllables) variously rendered 'dwis', 'wiss-wiss', 'sib', 'zi-zi', 'bit', 'dwi' (Stadler 1928*b*; Pätzold 1984); see also Schmidt (1908) who noted marked individual differences and listed many transcriptions perhaps not all of 1st-segment units as here defined. In song from perch, bird usually gives 8–12 units in 4–5 s, often 15–30 bursts with pauses of 10–12 s. Accelerando typical, volume about even. Full song (lasting 9·3–12·0 s: Wallschläger 1984) proceeds as follows: (i) 1st segment of 8–12 units, rarely with irregular pauses but with even volume; (ii) brief (1–2 s), more drawn-out 'bieth-bieth' often rising in pitch and often omitted; (iii) 6–10 'zillit' or 'zillitr' units normally given over peak of trajectory (for further division of this segment and

apparent simultaneous use of 2 internal tympanic membranes, also variants, see Stadler 1928*b*; (iv) last segment often omitted—3–6 'zim-zim' units lasting *c.* 2 s with pitch steady or slightly descending (Stadler 1928*b*; Pätzold 1984). Whole song rendered 'vitt vitt vitt-vitt-vitt...vietvietvietviet...tritritritritri...tsiatsiatsia' (Géroudet 1957*a*). Rendering of final (3rd) part of song as a rattling 'trrr' or 'rirrt' (Schmidt 1908) suggests something closer to final rattles in song of Rock Pipit (see above), and some Water Pipit songs certainly contain such rattles (H-H Bergmann). Further details in, especially, Stadler (1928*b*), also Naumann (1900) and Schmidt (1908). (b) Quiet Subsong given while running about on ground has hoarse, hissing, rasping quality recalling Swallow *Hirundo rustica* and Siskin *Carduelis spinus*; higher- and lower-pitched variants of call 2a interpolated (Naumann 1900). In similar situation, *alticola* may give 'tjwee' sounds in faster sequence than normal in ascent of Song-flight (Verbeek 1970). Quiet Twitter-song at roost comprises 'bsrieh' sounds (Mester 1957*c*); perhaps also variants of call 2a. Subdued twittering song sometimes given by ♀ (Pätzold 1984) probably similar. (2) Contact-alarm calls. (a) A 'tsiip' appreciably fuller, less feeble and squeaky, rather more grating than *A. pratensis* (Witherby *et al.* 1938*a*). Further descriptions: 'psri', 'psiet', or 'psscht', less pure and lower pitched than in *A. pratensis* (Bergmann and Helb 1982); generally sharp 'hiss(t)' or '(w)iss(t)', though timbre varies (Stadler 1928*b*), with vowel between 'i' and 'e' (Pätzold 1984); incisive 'dzip', harder and louder than *A. pratensis* (Devillers 1964), audible up to *c.* 300 m (Beneden 1950). See also (e.g.) Schmidt (1908). Presumed equivalent in *rubescens* 'tseep' and 'zzeeep' (alarm) in various combinations (see Drury 1961). Call 2a given mainly outside breeding season, rather uncommonly on breeding grounds for which Witherby *et al.* (1938*a*) mentioned variants 'tsiip-iip-iip' and (possibly call 2c) 'tsip'. Normally given singly (unlike in *A. pratensis*) and several seconds after take-off (suggesting contact- rather than alarm-call: Pätzold 1984); usually with perceptible pauses if repeated, but excited bird will also give (rapid) series (pitch then falling) and this not so easily distinguished from *A. pratensis* (Alexander 1924; Stadler 1928*b*; Meinertzhagen 1930; Witherby *et al.* 1938*a*; Beneden 1950; Devillers 1964; Bergmann and Helb 1982; Pätzold 1984). Said by Ringleben (1949) to differ from equivalent call of Rock Pipit *littoralis*, but no details. Wintering birds in Britain variously described as giving harsher or less harsh call than Rock Pipit *petrosus*; individual variation probably involved (Johnson 1970); perhaps longer in Water Pipit (Stewart 1971), slightly less loud and less shrill (Alexander 1924) or quieter, but harsher (Nicoll 1906). According to Lack (1932), 'pheest' of Water Pipit readily separable from 'phist' of *A. s. petrosus*. Berg (1975) found 'ieps-ieps' to be identical to equivalent call of Rock Pipit (*littoralis* and *petrosus*); see also further above. (b) A 'ierrp-ierrp' with pronounced 'r' sound considered distinct from Rock Pipit (Berg 1975); perhaps also in repertoire of Rock Pipit (and yet possibly the same) is short, rolling 'drrrt' easily distinguished from *A. pratensis* (Jonsson 1978b). (c) Best-known alarm call. Clear, short 'bit' or 'bsi' (Pätzold 1984), weak, metallic 'zi' (sometimes 2 close together) (Stadler 1928b), 'pi' (Bergmann and Helb 1982); persistent 'zit' or 'zip', sometimes more like 'zrr' or 'trr' (Schmidt 1908). Virtually indistinguishable from equivalent call of *A. pratensis*. Given persistently with short pauses, though speed varies—generally faster, the nearer the danger; up to 100-120 calls per min. Usually given from perch, but may continue as bird flies off (Schmidt 1908; Pätzold 1984). Agitated chirping sounds (Johnson 1933) and 'zing-zing' (Drury 1961) are presumably the equivalent in *rubescens*. (d) At higher intensity, changes to harder sounds recalling ticking of Blackbird *Turdus merula*. May be combined with song fragments (Pätzold 1984; also Naumann 1900, Stadler 1928b). (e) When man *c*. 3-4 m from young, parent often gives screeching 'srie' (Pätzold 1984); drawn-out, slightly hoarse 'sijt' (Bergmann and Helb 1982) perhaps the same or related. (f) If young directly threatened, highest-intensity alarm expressed in yapping 'wö-wö-wö', dog-like especially if strengthened by echo (Pätzold 1984). (g) ♀ may give soft, fluting 'jübb-jübb' in contact with trapped young and/or to express alarm (Pätzold 1984). (3) A 'gick gerick' and similar *Passer*-type calls perhaps serve as all-clear (Naumann 1900). (4) A 'tiijet' or 'tujet' given by ♂ to call ♀ off nest for Courtship-feeding (Pätzold 1984). In *alticola*, 'tjueet' (Verbeek 1970). (5) In response to call 4 from ♂, ♀ gives shorter, quieter and soft 'wiet' or 'wiejet' (Pätzold 1984). In *alticola*, 'peet' from nest and 'wee-wee-wee-wee' while flying to ♂ (Verbeek 1970). (6) ♂ may give hard 'titititii' (in *alticola*, 'pitititii': Verbeek 1970) or, at higher intensity, yapping 'twoi-twoi-twoi', apparently as threat towards overflying conspecific (Pätzold 1984). Trill-call of *rubescens* about to attack (Drury 1961) perhaps the same. (7) Low gurgling chirp given by *rubescens* approaching nest to feed young and (probably similar) twittering chirp as feeding-call for young out of nest (Johnson 1933). Situation and related calls not studied in west Palearctic races. (8) A 'gege' or 'tete' usually as a series and resembling Linnet *Carduelis cannabina* said to be contact-call (Schmidt 1908). Excited twittering of ♂ *rubescens* to ♀ (Johnson 1933) perhaps the same as is 'dik-dik dik-dik dik-dik dik' resembling Redpoll *C. flammea* given by *rubescens* pair flying about and feeding and apparently different from normal 'chip-chip chip-chip-chip' flight-call (Drury 1961); see also Pickwell (1947).

CALLS OF YOUNG. Nestlings normally quiet (Stadler 1928b). Give a buzzing sound when fed, this becoming louder with age. When handled at *c*. 7 days, give thin cheeping sounds. Fledged young give persistent 'psip' (or 'spieb' expressing fear: Naumann 1900) or 'biesitt',

more vigorously and faster if hungry; captive birds gave up to 40 per min (Pätzold 1984). Quiet and softer than adult call 2a, sometimes in rather attractive series: 'wieswitt wieswi witt' or 'wies wi wi witt' (Naumann 1900). See also (Stadler 1928b). Captive juvenile (perhaps ♀) sang in subdued fashion with bill closed from *c*. 6 weeks old (Pätzold 1984).   MGW

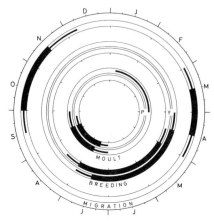

Breeding.

(A) **Rock Pipit.** SEASON. Britain and Ireland: see diagram. Scandinavia: from mid-May in south but not before June in north; *petrosus* nests up to 2 weeks earlier than *spinoletta* in same area (Makatsch 1976). SITE. In hole or hollow in cliff, from near base to top, or in bank or under thick vegetation, never far from shore. Nest: as Water Pipit but with inclusion of seaweed as material. Building: by ♀, taking 4-5 days (Kovshar' 1979). EGGS. See Plate 81. Sub-elliptical, smooth and glossy; grey-white, finely but heavily spotted olive-brown and grey, sometimes with cap at broad end. *A. s. kleinschmidtii*: 22·2 × 16·1 (19·0-24·0 × 14·2-17·0), *n*=100; calculated weight 2·95 g. *A. s. petrosus*: 21·6 × 16·0 mm (17·8-24·2 × 14·1-17·2), *n*=250; calculated weight 2·85 g; *littoralis* not significantly different (Schönwetter 1979). Clutch: 4-6(-9); in Tien Shan (USSR), mean 5·45, sample size not given (Kovshar' 1979). 2 broods in south of range, 1 in north. INCUBATION. 14-15 days. By ♀ only; occasionally by ♂ (Ryves 1943). YOUNG. As Water Pipit (see below). FLEDGING TO MATURITY. Fledging period 16 days (Witherby *et al.* 1938a). No further information.

(B) **Water Pipit.** SEASON. Southern Europe: eggs laid end of April to early July (Makatsch 1976). SITE. In side of steep bank or hollow, well concealed by overhanging vegetation; sometimes at end of short tunnel. In Wyoming (USA), most nests oriented between north-west and north-east, to avoid prevailing wind causing snow drift, and worst heat from sun (Verbeek 1981). Nest: cup of grass stems and leaves, and moss, with slight lining of finer leaves and a few hairs; outer diameter 14 cm (12-17), internal diameter 6·8 cm (6·5-7·2), overall height 4·5 cm (4·3-5·0), depth of cup 2·0 cm (1·5-2·5) (Pätzold

1984). Building: by ♀, though ♂ may bring some material (Verbeek 1970). EGGS. See Plate 81. Sub-elliptical, smooth and glossy; grey-white, heavily mottled brown and grey, sometimes with dark zone or cap at broad end, and occasionally with black hair-streaks. Nominate *spinoletta*: 21·4 × 15·7 mm (19·0–24·0 × 14·9–16·5), n = 100; calculated weight 2·73 g (Schönwetter 1979). Clutch: 4–6(–8). Of 59 clutches, Tien Shan (USSR): 3 eggs, 2; 4, 6; 5, 59; 6, 2; mean 4·9, but mean for larger area 5·33 (Kovshar' 1979). Increase with latitude in North America. Thus, of 104 clutches, 35–45°N: 3 eggs, 7; 4, 25; 5, 64; 6, 8; mean 4·7. Of 40 clutches, 65–75°N: 3 eggs, 1; 4, 1; 5, 17; 6, 19; 7, 2; mean 5·5 (Verbeek 1970). 2 broods. Eggs laid daily, most in early morning (Pätzold 1984). INCUBATION. 14–15 days. By ♀ only, from last or penultimate egg. YOUNG. Cared for and fed by both parents; brooded by ♀ for first few days, while ♂ does feeding; thereafter both feed and ♀ only broods during rain or hot sun (Verbeek 1970). FLEDGING TO MATURITY. Fledging period 14–15 days (Pätzold 1974). Age of independence and first breeding not recorded. BREEDING SUCCESS. Of 241 eggs laid in 51 nests, Tien Shan (USSR), 90·0% hatched, with 5% infertile and 5% dead in shell; young fledged successfully from 47% of 79 nests, all others lost, with 39% of nests predated; early nests much more vulnerable because of lack of nesting cover, with main predators stoat *Mustela erminea* and snakes (Kovshar' 1979). In Wyoming, of 386 eggs laid in 84 nests, 74% hatched and 77·6% of young hatching survived to fledging; overall success 66·8%, with 80·4% of nests successful (Verbeek 1981). Of 99 eggs laid, Switzerland/France border, 76% hatched and 58% fledged (Catzeflis 1978).

Plumages (Rock Pipit *A. s. littoralis*). ADULT BREEDING. Entire upperparts dark grey, slightly tinged olive (especially on mantle and scapulars), rump and upper tail-coverts slightly olive-brown; forehead and crown uniform or with very faint dark streak on feather-centres, mantle and scapulars usually with more distinct though ill-defined dark grey-brown feather centres. Lores finely mottled dark brown and cream, almost similar in colour to faint buff-and-grey mottled stripe above lores and to cream-white and grey mottled front part of cheeks. Narrow and broken eye-ring cream, rather contrasting; distinct cream or off-white supercilium from just above eye to above ear-coverts, where merging into grey of nape. Ear-coverts grey-brown, shorter with narrow off-white shaft-streaks, those along supercilium darker, forming faint dark stripe. Chin cream or off-white; cheeks and throat pink-buff to pale pink-cinnamon, sometimes with fine spots along side of throat, forming faint malar stripe; pink of throat gradually merges into warm pink-buff or vinous-cinnamon chest and breast. Chest and breast usually uniform in centre, but with ill-defined olive-brown spots or short streaks at sides; occasionally centre with faint dusky spots or narrow streaks also. Flanks pink-cream or pale cream-yellow with ill-defined olive-brown streaks; belly, vent, and under tail-coverts pink-cream to cream-white, under tail-coverts sometimes with some olive-grey of bases visible. Tail dark brown to brown-black; central pair (t1) with broad and ill-defined olive-brown or olive-grey fringes, t2–t5 with narrow olive outer

edges; t5 usually with small pale grey spot or wedge on tip of inner web at shaft if not too worn; outer web of t6 pale grey or pale olive-grey, slightly darker olive-grey towards tip, inner web dusky grey with rather ill-defined pale grey or pale brown-grey wedge on tip; length of this wedge 25·5 (65) 22–27 mm, rarely 19–21 or 28–31. Flight-feathers black-brown or greyish-black, darkest on outer webs and tips of primaries; basal inner webs with ill-defined dull grey borders; narrow, well-defined, yellow-white edges along outer webs, sometimes also on tips of primaries; broader and less clear-cut olive-grey fringes along outer webs of secondaries. Lesser upper wing-coverts dark grey with paler grey or olive tips, median coverts greyish-black or brown-black with rather broad and contrasting pale olive-grey or (if bleached) off-white tips, greater like median but dark centres largely hidden under broad ill-defined olive-grey fringes and pale fringes on tip partly or fully worn off. Tertials greyish-black or brown-black with broad but ill-defined olive-grey or olive-brown fringes along outer web and tip. Greater upper primary coverts greyish-black with narrow ill-defined olive fringe along outer web. Under wing-coverts and axillaries dusky grey, most feathers faintly and narrowly washed pale yellow along borders when not too worn. Marked individual variation, in part caused by variable extent of pre-breeding moult. Birds with full pre-breeding moult as described above retain only non-breeding upper wing-coverts (not tertial coverts), some tertials, and t2–t6; tips of median coverts bleached to brown-white, those of greater coverts largely worn off; these birds rather scarce. More commonly, also scattered non-breeding feathering of upperparts (mainly from lower mantle to tail-coverts) and underparts (mainly down from flanks and belly) retained; this feathering rather worn and appearing more sharply streaked brown; feathers of upperparts more olive-brown, less grey; on underparts off-white, less pinkish or cream. Some birds have pre-breeding moult rather extensive, but character of new feathering intermediate between breeding and non-breeding; new feathers of upperparts rather olive-brown, less pure grey; supercilium restricted to patch above eye, cream with olive spots; throat and breast pink-buff or cream-pink, all or partly with ill-defined olive-brown spots or short streaks. Finally, pre-breeding moult sometimes largely or fully suppressed (may occur in up to half of local breeding population; however, includes 1st adults which more often may suppress moult): plumage largely worn non-breeding, sometimes except for some new uniform pink-buff or vinous-cinnamon feathers on underparts; worn upperparts very heavily streaked brown or almost uniform dark grey-brown; pale supercilium absent; underparts dirty white with heavy brown spots on chest, sides of breast, and flanks. ADULT NON-BREEDING. Upperparts dark olive-grey, very faintly streaked with paler olive on forehead, crown, and mantle, slightly brighter greenish olive-brown on rump and upper tail-coverts; some darker olive-brown of feather-bases often visible on lower mantle and scapulars; feathering along sides of crown with faint olive-yellow specks, forming very indistinct supercilium. Lores and front part of upper cheek finely mottled off-white and grey; narrow and broken eye-ring olive-yellow; rear of upper cheek and ear-coverts brownish-olive; lower cheeks and lower sides of neck dark olive-grey or dark brown-olive with some ill-defined pale cream-yellow spots backwards from gape and at rear of ear-coverts. Chin and throat pale cream-yellow to off-white, chin faintly mottled dusky olive, throat more distinctly. Chest, sides of breast, and flanks pale cream-yellow with heavy but ill-defined dark olive-grey or dusky olive-brown spots and streaks; streaks slightly narrower on flanks but almost coalescent on chest and sides of breast, upper chest sometimes with narrow rounded gorget of sharper and blacker spots. Remainder of underparts cream-yellow to cream-white, sides of belly sometimes with narrow ill-defined olive streaks. Tail, flight-feathers, wing-coverts, and tertials as adult breeding, but all upper wing-coverts and tertials equally new; median and greater coverts with pale greyish tips 2-3 mm wide, all tertials and tertial coverts with broad olive-brown fringes along outer web and tip. *In worn plumage*, about mid-winter, upperparts more olive-brown, blackish feather-bases of mantle and scapulars visible; mottled olive and cream supercilium sometimes more pronounced, mainly behind eye; ground-colour of underparts dirty cream-white, spots on sides of head and neck and on chest, sides of breast, and flanks slightly darker, browner, and sharper; pale tips of greater upper wing-coverts and olive fringes of tertials partly or largely worn off. NESTLING. Down long and dense, on upperparts only; light grey on 1st day, later on dark grey or brown-grey (Heinroth and Heinroth 1931; Witherby *et al.* 1938a; Pätzold 1984). For development, see Pätzold (1984). JUVENILE. Rather like adult non-breeding, but plumage looser, in particular upperparts, vent, and tail-coverts; upperparts like adult, but slightly browner, less olive-grey; mantle and scapulars with long but ill-defined dull black streaks; sides of head darker than in adult, olive-brown and black-brown except for indistinct eye-ring and short mottled cream supercilium; ground-colour of underparts pale cream-buff to buff-white, marked with heavy but ill-defined black-brown or dark olive-brown spots on sides of throat, chest, sides of breast, faint spots on flanks, and narrow streaks on upper belly. Remainder of plumage as adult non-breeding. FIRST ADULT NON-BREEDING. Like adult, but part of juvenile feathering retained. Colour and pattern of retained juvenile virtually similar to adult non-breeding, differing mainly in being more worn: greater and sometimes outer or all median coverts more worn than tertial coverts, 1-2 tertials often more worn than others, t1 sometimes newer than t2-t6 (in adult, all feathering equally new or slightly worn). Tips of primaries, greater upper primary coverts, and tail feathers narrower, more pointed, and more worn than those of adult at same time of year, less broadly and smoothly rounded. See also Svensson (1984a). FIRST ADULT BREEDING. Like adult breeding, but pre-breeding moult more often suppressed (see Moults); tips of retained juvenile primaries, greater upper primary coverts, and tail-feathers more pointed and worn than those of adult, but abrasion strong at any age once nesting started and ageing then impossible.

Bare parts. ADULT, FIRST ADULT. Iris brown, dark brown, or black-brown, sometimes warm brown or red-brown. Bill dark horn-brown or black-brown in autumn and winter, base of lower mandible yellowish, yellow-horn, pink-flesh, or yellow-flesh; fully black-brown or black in spring and summer. Leg and foot rather variable. In *petrosus* and *littoralis*, pale red-brown, flesh-brown, horn-brown, or dark brown in late summer and autumn (occasionally up to at least November; perhaps in 1st adult only), 1st adult sometimes with yellow soles; dark grey, dark brown, black-brown, or grey-black in winter, spring, and early summer. In nominate *spinoletta* and *blakistoni*, brown or dark brown in autumn and winter; blackish flesh-brown, deep brown with blackish toes, or fully blackish in spring. In *rubescens*, dusky purplish-grey, black-brown, or blackish. In *japonicus*, rather pale brown, never blackish. NESTLING. Bare skin (including bill, leg, and foot) flesh-grey or pink-flesh. Mouth reddish-flesh, reddish-orange, carmine-red, or orange-red; gape-flanges deep yellow or very pale yellow. JUVENILE. Iris brown. Bill dark horn-brown, cutting edges and lower mandible yellow-flesh or yellow-brown. Leg and foot

flesh-brown, grey-brown, brown, or dull grey with flesh tinge. (Hartert 1903-10, 1921-2; Heinroth and Heinroth 1931; Witherby *et al.* 1938a; Dementiev and Gladkov 1954a; Ali and Ripley 1973b; Harrison 1975; Pätzold 1984; RMNH, ZMA.)

Moults. ADULT POST-BREEDING. Complete; primaries descendant. In Britain, *petrosus* starts July, p1-p2 first, completed with p9-p10 and inner secondaries mid-August to mid-September, average duration of primary moult *c.* 42 days (Ginn and Melville 1983). On Fair Isle, *kleinschmidti* moults between mid-July and early September; tail starts when p1-p3 new, secondaries when primary moult halfway through; some *meinertzhageni* on Outer Hebrides finish mid-October (Williamson 1965b). In montane race *alticola* of *rubescens* group, western USA, moult starts with mantle and chest before p1 shed; tertials shed with p3-p5 (sequence: s8, s9, s7), tail rapidly shed with p5-p6 (sequence: t1 to t6), feathers growing almost simultaneously; ♂ starts *c.* 14 days before ♀, primary moult lasting *c.* 1 month, mid-July to late September (Verbeek 1973). Some *blakistoni* start early August; some *japonicus* in heavy moult late July and completed by late August or late September (Dementiev and Gladkov 1954a). Nominate *spinoletta* moults July-September; one from Czechoslovakia, 19 August, had not started (RMNH, ZFMK, ZMA); in French Alps, some in moult August and early September, others near completion mid-August or fully completed early September (Frelin 1983). ADULT PRE-BREEDING. Partial, but extent highly variable, and sometimes no moult. In all races, mainly January-March, sometimes December-April; moult rapid, but much variation in starting time (Stresemann 1920). Usually involves head, body, 1-2 tertials (occasionally all), t1, and often tertial coverts and some median upper wing-coverts. Situation further complicated by fact that some birds moult into a dull and streaked plumage similar to non-breeding rather than into brighter and more uniform breeding plumage—especially ♀♀ of *spinoletta* group and both sexes of *petrosus* group. Moult most often limited or absent in birds from coastal Norway and Atlantic islands (including Britain), less often in Brittany and Baltic, and only occasionally in *spinoletta* group (Stresemann 1920; Mayaud 1952c; Williamson 1965b; RMNH, ZFMK, ZMA). POST-JUVENILE. Partial: head, body, many or all lesser and variable number of median upper wing-coverts, tertial coverts, 0-3 tertials, and sometimes t1 or variable number of greater upper wing-coverts. Starts shortly after fledging at age of *c.* 1 month and hence timing rather variable; starts early July to late August, completed mid-August to late September. In French Alps, many nominate *spinoletta* fully juvenile in August; many moulting late August and early September and only a few after 10 September (Frelin 1983). FIRST ADULT PRE-BREEDING. Like adult pre-breeding, but moult more often limited or absent; in some birds, head, body, tertial coverts, tertials, and t1 involved, others retain non-breeding rump, tail-coverts, and belly, and non-breeding or juvenile wing-coverts, tertials, and tail; sometimes only scattered feathers of throat and chest new. Moult mainly February-April. (RMNH, ZMA.)

Measurements. ADULT, FIRST ADULT.
(A) Rock Pipit. *A. s. littoralis.* Wing: (1) Scandinavia, summer (slightly worn), (2) western Europe, winter; other data combined; skins (RMNH, ZMA). Bill (S) to skull, bill (N) to distal corner of nostril; exposed culmen on average 4·1 less than bill (S). Toe is middle toe with claw; claw is hind claw.

WING (1)	♂ 90·8	(1·72; 18)	88-93	♀ 83·5 (1·48; 6)	82-85
(2)	91·2	(1·58; 46)	89-96	85·1 (2·06; 42)	81-90
TAIL	62·2	(1·78; 35)	59-66	57·3 (1·95; 21)	54-60

BILL (S)	18·0 (0·60; 32)	17·2–19·2	17·3 (0·42; 21)	16·5–17·9
BILL (N)	10·8 (0·62; 27)	9·8–12·0	10·4 (0·48; 19)	9·7–11·2
TARSUS	23·0 (0·78; 35)	21·7–24·4	22·4 (0·65; 20)	21·4–23·4
TOE	19·6 (0·95; 27)	18·2–21·7	19·2 (0·62; 16)	18·3–20·1
CLAW	10·3 (0·89; 22)	8·6–11·5	10·2 (0·67; 15)	9·1–11·2

Sex differences significant, except claw. 1st adult with retained juvenile wing and tail combined with older birds, as tail similar and 1st adult wing only 0·7 mm shorter than wing of full adult.

A. s. petrosus. Southern Scotland and England, all year; skins (RMNH, ZMA).

	♂		♀	
WING	90·6 (1·90; 10)	88–94	83·5 (2·98; 8)	80–88
TAIL	63·8 (2·26; 10)	60–67	57·6 (3·09; 7)	55–63
BILL (S)	17·9 (0·48; 10)	17·2–18·6	17·8 (0·78; 7)	16·9–18·8
BILL (N)	10·7 (0·48; 10)	9·8–11·4	11·0 (0·44; 6)	10·6–11·7
TARSUS	23·2 (0·78; 10)	22·2–24·4	23·1 (0·37; 7)	22·3–23·4

Sex differences significant for wing and tail. Average toe 19·4, average claw 10·25, hence as in *littoralis*. Wing, western France: ♂ 84–92 (24), ♀ 78–85 (16) (Mayaud 1952c).

A. s. meinertzhageni. Outer Hebrides, 10 ♂♂: wing 87–94, bill (S) 16–17 (Witherby *et al.* 1938a) Wing, ♂ 88–95, ♀ 83–89; bill (S), ♂ 18–19, ♀ 19 (Hall 1961).

A. s. kleinschmidti. Faeroes, ♂: wing 93 (6) 92–96 (Vaurie 1959); wing 93–96(–99) (Vaurie 1954).

(B) Water Pipit. Nominate *spinoletta.* Europe from Netherlands and Czechoslovakia south to Spain and Yugoslavia, all year; skins (RMNH, ZMA; wing includes data from Stresemann 1920).

	♂		♀	
WING	91·5 (1·83; 61)	88–96	85·4 (1·77; 36)	82–90
TAIL	65·1 (2·15; 34)	61–69	60·3 (1·99; 23)	57–64
BILL (S)	16·9 (0·56; 34)	16·0–17·8	16·4 (0·54; 21)	15·3–17·1
BILL (N)	9·9 (0·47; 33)	8·9–10·8	9·7 (0·39; 22)	8·9–10·0
TARSUS	24·2 (0·58; 34)	23·0–25·3	23·7 (0·50; 23)	22·8–24·6
TOE	19·6 (1·14; 13)	17·8–21·1	19·5 (—; 3)	18·4–20·2
CLAW	10·6 (0·71; 25)	9·2–12·0	10·8 (0·82; 17)	9·3–12·1

Sex differences significant, except bill (N), toe, and claw.

A. s. coutellii. Caucasus and Iran, all year; skins (Hartert 1921–2; Stresemann 1928; Paludan 1938; Schüz 1959; RMNH, ZFMK, ZMA).

	♂		♀	
WING	87·5 (1·60; 16)	84–90	80·2 (0·91; 5)	79–82
BILL (S)	16·8 (0·46; 5)	16·2–17·4	16·1 (—; 3)	15·6–16·5
TARSUS	24·5 (1·54; 4)	23·4–25·8	23·4 (—; 3)	22·6–25·1

Sex differences significant for wing.

A. s. blakistoni. Tien Shan to Lake Baykal, USSR; all year (RMNH, ZFMK, ZMA); wing includes data from Mongolia (Piechocki and Bolod 1972).

	♂		♀	
WING	92·5 (2·32; 21)	90–98	88·8 (4·16; 7)	84–94
BILL (S)	16·7 (0·63; 14)	15·9–17·6	16·6 (0·66; 5)	15·6–17·3
TARSUS	23·1 (0·91; 7)	21·9–23·9	22·7 (—; 3)	22·1–23·2

Sex differences significant for wing.

A. s. rubescens. Western Greenland, Labrador, and north-east USA, all year; skins (RMNH, ZFMK, ZMA).

	♂		♀	
WING	87·6 (1·35; 8)	85–90	82·2 (2·64; 10)	78–86
TAIL	63·4 (2·22; 8)	60–67	59·0 (1·58; 10)	56–61
BILL (S)	16·0 (0·61; 8)	15·3–16·8	15·5 (0·46; 10)	14·8–16·1
BILL (N)	9·5 (0·40; 8)	9·3–10·0	9·2 (0·34; 10)	8·7–9·7
TARSUS	22·7 (0·60; 8)	21·8–23·4	21·7 (0·74; 10)	20·6–22·8

Sex differences significant, except for bill. Middle toe with claw 18·0 (0·53; 9) 17·3–18·8; hind claw 9·5 (0·85; 15) 8·3–10·0, exceptionally 11·3 (RMNH, ZMA).

Weights.
(A) Rock Pipit. *A. s. petrosus.* Skokholm (Wales): adult 24·6

(7) 22·7–25·7; juvenile and 1st adult non-breeding 24·5 (228) 19·6–30·5 (Browne and Browne 1956, which see for details of juvenile weight).

A. s. littoralis. Norway, March–May: 21–26 (5) (Haftorn 1971). Migrants Helgoland: 23·5 (19) 19·5–27·5 (Weigold 1926). Netherlands: mainly October, 22·4 (1·82; 59) 18–26·5 (Koning 1982); October–March, ♂ 25·1 (3·51; 7) 21·5–32·5, ♀ 21·8 (1·14; 4) 20·8–23·3, exhausted birds 15·0 (1·95; 5) 13–17 (RMNH, ZMA).

(B) Water Pipit. Nominate *spinoletta.* Czechoslovakia, August: 22·1 (1·67; 101) 18·4–26·5. Braunschweig (West Germany), winter: 23·9 (2·10; 100) 19·5–24·0. Westfalen (West Germany), February–March: ♂ 26·4 (1·61; 12) 23·5–29·2. (Bub *et al.* 1981). Southern West Germany, winter: ♂ 24·9 (1·46; 7) 22·7–27·0, ♀♀ 21·0, 25·1 (Diesselhorst 1957). Netherlands, October–December: 24·1 (2·04; 25) 20–27 (Koning 1982); ♂ 23·4 (2·61; 5) 20–27, ♀♀ 22, 23 (RMNH, ZMA). French Alps, August to 1st half of October: 23·4 (1·95; 1407) (Frèlin 1983, which see for many details). Malta, winter: 21·8 (1·3; 11) 19·5–23 (J Sultana and C Gauci). For growth of nestling, see Pätzold (1984).

A. s. coutellii. Iran, late March and early April: ♂ 22·0 (2·15; 4) 19–24, probable ♀ 19·2 (Paludan 1938; Schüz 1959). This race or intermediate with nominate *spinoletta*, north-east Turkey, October: ♂♂ 24, 25 (Kumerloeve 1968).

A. s. blakistoni. Kazakhstan (USSR) and Mongolia, combined (Dolgushin *et al.* 1970; Piechocki and Bolod 1972); migrants may include some *rubescens*.

	♂		♀	
MAR–APR	20·7 (1·35; 4)	19–22	20·0 (—; 1)	—
MAY–JUN	21·1 (1·81; 18)	17–27	21·8 (3·68; 8)	17–27
JUL–OCT	21·9 (2·05; 12)	19–27	20·5 (2·14; 5)	18–24

A.s. rubescens. Northern Alaska and neighbouring north-west Canada, May to early August: ♂ 21·3 (41) 19–26, ♀ 20·1 (8) 19–23 (Irving 1960). Alabama (USA), ♂♂: January, 24·7 (3) 24–25; April, 34, 37 (Stewart and Skinner 1967). Ohio (USA): 23·2 (5) 19–24 (Stewart 1937).

Structure. Wing rather long, broad at base, tip bluntly pointed. 10 primaries: p7–p8 longest or either one 0·5 shorter than other, p9 0·5–3 shorter, p6 0·5–1·5, p5 9–13, p4 14–19, p3 16–22, p1 19–25; no significant difference between *petrosus*, *littoralis*, and nominate *spinoletta* in small samples tested. P10 reduced, tiny and pointed; 56–70 shorter than wing-tip, 9–15 shorter than longest upper primary covert. Outer web of p6–p8 and (slightly) inner of p6–p9 emarginated. Tertials long; longest *c.* 3–6 shorter than wing-tip in closed wing. Tail rather long, tip square or very slightly forked; 12 feathers. Bill rather long, straight, tip sharply pointed. Nostrils oval. A few fine bristles at base of upper mandible. Tarsus rather long, toes long and slender. For length of middle toe with claw and hind claw, see Measurements. Outer toe with claw *c.* 70% of middle with claw, inner *c.* 73%; hind with claw *c.* 94%, without claw *c.* 49%.

Geographical variation. Marked and complex. 3 main groups: (1) *petrosus* group (Rock Pipit) in western and northern Europe in which breeding plumage generally heavily streaked, supercilium absent, and wedge on outer tail-feather grey; (2) *spinoletta* group (Water Pipit) from central and southern Europe east to central Asia in which breeding plumage generally uniform vinous on underparts, supercilium distinct, and wedge on outer tail-feathers white; (3) *rubescens* group (Water Pipit) from eastern Asia, North America, and western Greenland which is rather like *spinoletta* group but chest-marks sharply defined and tail with more white. There has recently been a tendency among Russian authors to separate *rubescens* group (as a distinct species) from *spinoletta* and *petrosus* groups (e.g. Stepanyan 1978*a*), mainly on account of marginal overlap (without apparent hybridization) between *rubescens* and *blakistoni* (of *spinoletta* group) in Khentei-Chikoy uplands in southern Transbaykalia (Nazarenko 1978). On the other hand, European authors tend towards separation of *petrosus* group (as a species) from *spinoletta* and *rubescens* groups, mainly on account of differences in breeding plumage, habitat, voice, and some other characters. However, *littoralis* of *petrosus* group intermediate in plumage between *spinoletta* group and typical *petrosus*, and differences between these 2 groups are mainly due to differences in extent of moult; hence preferable to combine both into a single species (Williamson 1965*b*). The 3 groups are apparently incipient species, and choice of whether any one has reached status of separate species is (on current knowledge) arbitrary; splitting into 2 species (in whatever combination) probably unjustified, whole assemblage best split into 3 or combined into 1; pending further research (e.g. on behaviour and song), classical concept of single species followed here.

Racial characters of each group as follows (for geographical ranges, see p. 393). (1) *Petrosus* group. Recognition of races and their characters largely based on Williamson (1965*b*). For *littoralis*, see Plumages; birds from western Norway, sometimes considered to belong to *petrosus*, are actually inseparable from *littoralis* (Pethon 1968). *A. s. petrosus* closely similar to *littoralis* in non-breeding plumage, though fringes of mantle and scapulars perhaps slightly more pure olive. Rather different in breeding plumage, however: though pre-breeding moult almost as extensive as in *littoralis*, new breeding feathers rather similar to non-breeding, showing hardly any reduction of spotting on chest. Upperparts olive-grey, greyer than in breeding, but not as pure dark grey as breeding *littoralis*; supercilium less noticeable than *littoralis*, short and spotted; upper chest sometimes vinous, but much less extensive than in many *littoralis*, mainly restricted to fringe below whitish throat; pale wedge on t6 very slightly darker grey than in *littoralis*, length similar—23·3 (10) 19–31 mm. See also Johnson (1970). In western France, many *petrosus* acquire *littoralis*-like breeding plumage (Mayaud 1952*c*). *A. s. meinertzhageni* similar to *petrosus* in non-breeding plumage, but underparts slightly yellower (like *kleinschmidti*, but upperparts darker); in breeding plumage, feather-centres of head and body blackish, not brown; feather-fringes of mantle and scapulars greenish-olive, rump uniform greenish-olive, brighter than *petrosus*; ground-colour of underparts creamy-buff, paler than *petrosus*, but not as bright as *kleinschmidti*; many birds have a few vinous feathers below lower throat; sides of head and tail as in *petrosus*. *A. s. kleinschmidti* has virtually no supercilium and pale wedges of outer tail-feathers brown-white; in non-breeding more yellow-olive on upperparts than other races of *petrosus* group, feather-centres darker (but not as black as *meinertzhageni*), underparts more heavily suffused with deeper brighter yellow, flanks heavily washed olive, and chest-spots large, often coalescing; breeding plumage similar to non-breeding except sometimes for a few vinous feathers below throat and yellow-brown sides of head (grey-brown in *petrosus* and *meinertzhageni*). (Williamson 1965*b*.)

(2) *Spinoletta* group. Size as in *petrosus* group (except for rather small *coutellii*), but tail (in all races) and tarsus (in nominate *spinoletta* and *coutellii*) relatively longer and bill relatively shorter. In non-breeding, nominate *spinoletta* differs from *petrosus* group in decidedly brown, greyish-brown, or dark olive-brown upperparts (not as pure olive or dark olive-grey), virtually uniform or with slightly darker feather-centres visible on mantle and scapulars (as in *petrosus* group); often a whitish supercilium, faintly mottled grey, indistinct or absent above lore, more pronounced just above eye and ear-coverts (usually less distinct in *petrosus* group, except for some *littoralis*; occasionally, faint in nominate *spinoletta* also); ground-colour of underparts pale cream-buff, pale pink-buff, or off-white, not as yellowish as in *petrosus* group (though yellow of *littoralis* often pale and rather tinged with cream); underparts less profusely spotted and spots narrower and more sharply defined, olive-brown rather than greyish-olive, not as coalescent as in *petrosus* group, not as sharp as in *rubescens* group; streaks on flanks narrower; pale tips of lesser and median upper wing-coverts grey-brown, not as olive-grey as in *petrosus* group (in all groups, tips off-white when plumage worn); wedge on t6 white, sharply defined, 29·7 (43) 24–36 mm long, tip of t5 with small white wedge 8·6 (28) 4–15 mm long; wedge occasionally grey, however, as in *petrosus* group (recorded in Czechoslovakia, Bulgaria, and Yugoslavia: Königstedt 1980; Königstedt and Müller 1983; ZMA); axillaries and under wing-coverts more broadly fringed white or cream-buff, appearing paler than in *petrosus* group. In breeding plumage, forehead, crown, and nape grey with slight olive-brown tinge, mantle and scapulars down to upper tail-coverts more decidedly brown or olive-brown, generally not as greenish-olive as in *petrosus* group, but some *littoralis* similar; supercilium and narrow broken eye-ring vinous-pink, pink-buff, or white, sometimes indistinct above lore, but usually more distinct than in non-breeding plumage and generally more distinct than in breeding plumage of *petrosus* group (except for some *littoralis*); lower cheeks, lower sides of neck, and chin down to breast and flanks deep vinous-pink to pale vinous-buff, often with faint and narrow dusky streaks on sides of breast and flanks and frequently (especially ♀) with some diffuse grey spots on cheeks, lower sides of neck, and chest; belly to under tail-coverts pink-buff to off-white; underparts much less streaked and spotted than in *petrosus* group, but some little-marked *littoralis* are close to relatively heavily marked nominate *spinoletta*, though flanks of nominate *spinoletta* usually less heavily marked than any *littoralis* and belly to under tail-coverts pinkish or whitish, not yellowish. Tail and wing as in non-breeding. In worn plumage, upperparts dark brownish-grey (similar to worn *littoralis*); supercilium and eye-ring still conspicuous (faint or absent in *petrosus* group); vinous of underparts bleached to dirty buff-white and largely worn off, some grey of feather-bases exposed, but dark streaks and spots on cheeks, sides of neck, chest, and flanks generally much less distinct than in *petrosus* group; outer web of t6 often bleached to white, or (in contrast to this) pale wedges on outer tail-feathers sometimes grey due to contamination by soil. Birds east to north-central Asia Minor similar to nominate *spinoletta* (Kumerloeve 1961), but those from Bayburt in north-east Turkey intermediate and those from Varsambeg and Rize (further east in Turkey) inseparable from *coutellii* (Kumerloeve 1968). *A. s. coutellii* from Caucasus area and *blakistoni* from central Asia are both similar to nominate *spinoletta*, but upperparts in non-breeding slightly paler brown, marked with slightly more pronounced dark olive-brown feather-centres; streaks on underparts narrower and more restricted, often confined to malar stripe, band across chest, and some fine streaks on upper flanks; in breeding plumage, forehead, crown, and hindneck paler grey than nominate *spinoletta*, mantle and scapulars down to upper tail-coverts paler sandy olive-brown with more pronounced dark streaks on mantle and scapulars; supercilium broader, upper cheeks paler; underparts paler, pink-buff or sandy-buff, much less vinous; tail as nominate

spinoletta. A. s. blakistoni sometimes included in *coutellii*, but *coutellii* distinctly smaller than *blakistoni* (and nominate *spinoletta*), and *blakistoni* somewhat paler all over than *coutellii* and with even finer marks on underparts.

(3) *Rubescens* group. Structurally near *spinoletta* group, but *rubescens* (not *japonicus*) smaller. In all races, tail whiter, and spots on underparts sharper. Non-breeding plumage of *rubescens* dark olive-brown with some dark grey tinge on upperparts, like nominate *spinoletta* (less pale and streaky than *blakistoni*); supercilium and eye-ring distinct, sometimes less so above lore (thus as in *spinoletta* group); ground-colour of underparts (especially under tail-coverts) often warmer cream-buff than in *spinoletta* group; marks on underparts sharp and narrow, darker than in *spinoletta* group, olive-black or black-brown, restricted to malar stripe, lower side of neck, chest, side of breast, and flank (rather as in nominate *spinoletta*, more spots than in *blakistoni*); t6 white except for large dark wedge 36·4 (29) 30–42 mm long on basal inner web, and sometimes dusky smudge on tip of outer web; t5 with fairly extensive white wedge 16·1 (25) 8–24 mm long on tip of inner web (and sometimes partly white outer web), t4 with small white spot. In breeding plumage, upperparts slightly greyer than in non-breeding; supercilium longer and more distinct; underparts warm pink-buff, marked with fine and sharp dark spots on malar stripe, sides of breast, chest, and flanks (hence, not as deep vinous as nominate *spinoletta*, not as pale pink-buff or sandy as *blakistoni*, and not virtually unspotted or marked with a few diffuse spots as in both these races). *A. s. japonicus* similar to *rubescens* in non-breeding, but upperparts darker greyish olive-brown with slightly more pronounced black feather-centres visible on mantle and scapulars; ground-colour of underparts similar, but spots much heavier, black, sharply defined, and often more extensive, especially on breast and flanks; tail as *rubescens*; underparts rufous-cinnamon or pink-cinnamon in breeding, marked with sharp and distinct black spots on cheeks, chest, and flanks. Races of *rubescens* group with sharp black spots below, much white in tail, and (in *japonicus*) heavy markings and pale legs rather dissimilar from races of *petrosus* group (though linked by *spinoletta* group with intermediate characters); rather more easily confused with familiar smaller species of *Anthus* than with *petrosus* group, e.g. with Tree Pipit *A. trivialis* or Meadow Pipit *A. pratensis*. CSR

Motacilla flava Yellow Wagtail

PLATES 28, 29, and 31
[facing pages 328, 329, and 353]

Du. Gele Kwikstaart   Fr. Bergeronnette printanière   Ge. Schafstelze
Ru. Жёлтая/Желтолобая трясогузка   Sp. Lavandera boyera   Sw. Gulärla

Motacilla flava Linnaeus, 1758

Polytypic. LUTEA COMPLEX. *M. f. flavissima* (Blyth, 1834), Britain and locally also on continental coast of north-west Europe; *lutea* (S G Gmelin, 1774), basin of lower Volga river north to Kazan' and Perm', east through plains of Kazakhstan to Lake Chany and Lake Zaysan; *taivana* (Swinhoe, 1863), eastern Siberia from basins of Vilyuy and Vitim rivers east to Magadan area, Sea of Okhotsk, and Sakhalin (extralimital race). FLAVA COMPLEX. (1) *Flava* group. Nominate *flava* Linnaeus, 1758, Europe from western France east to foothills of Urals and lower Volga, north to southern Sweden, Leningrad, Moscow, Gor'kiy, and Kazan' areas of European USSR (intergrading with *thunbergi* in broad zone from central Sweden and southern Finland east to Vologda, Kirov, and Perm' areas of European USSR and with *beema* in steppes along lower Volga and Ural rivers), south to central France, Alps, northern Yugoslavia, central Rumania, and northern Ukraine (intergrading with *iberiae* just north of Pyrénées, with *cinereocapilla* in southern Austria, northern Dalmatia, Bosnia-Hercegovina, and Serbia, and with *feldegg* in inland Montenegro, Serbia, southern Rumania, and southern Ukraine); *beema* (Sykes, 1832), lower Volga east to Sayan mountains (near Lake Baykal), north in south-west Siberia to *c.* 60°N, south to northern Kazakhstan and to northern foothills of Altai (apparently also Ladakh), partly overlapping with *lutea*; *leucocephala* (Przevalski, 1887), north-west Mongolia and neighbouring parts of northern Sinkiang (China) and Tuvinskaya oblast (USSR), straggling to Middle East; *cinereocapilla* Savi, 1831, Italy including Sicily and Sardinia, and Istra (north-west Yugoslavia), intergrading with *iberiae* on Corsica, in Mediterranean France, and perhaps Algeria and Tunisia; *iberiae* Hartert, 1921, south-west France, Iberia, and north-west Africa; *pygmaea* (A E Brehm, 1854), Egypt. (2) *Feldegg* group. *M. f. feldegg* Michahelles, 1830, southern and eastern Yugoslavia (north to southern Dalmatia and Serbia), Bulgaria, southern Ukraine, Crimea, western shores of Caspian Sea, Iran, and Afghanistan, south to Greece, Turkey, Levant, and Iraq; *melanogrisea* (Homeyer, 1878), from Volga delta and eastern shores of Caspian Sea east to Tarbagatay and Ili basin, grading into *feldegg* in Turkmeniya and Uzbekistan; mainly south of *beema* but partly overlapping *lutea*. (3) *Thunbergi* group. *M. f. thunbergi* Billberg, 1828, Norway east to Gydanskiy peninsula in northern Siberia, north of nominate *flava* and *beema*, with which it intergrades; *plexa* (Thayer and Bangs, 1914), northern Siberia from Taymyr peninsula to Kolyma basin, south to *c.* 60°N (extralimital); *simillima* Hartert, 1905, shores of north-east Sea of Okhotsk, Kamchatka, and northern Kuril Islands, straggling to Europe; *c.* 4 further races from eastern Siberia (south and east of *plexa* and *simillima*) and Alaska, south to Altai, northern Mongolia, and Manchuria.

Field characters. 17 cm; wing-span 23–27 cm. Somewhat smaller than Pied Wagtail *M. alba*, with sleeker form and 20% shorter tail; noticeably less attenuated than Grey Wagtail *M. cinerea*, with at least 25% shorter tail. Smallest, most compact of west Palearctic wagtails, with form and silhouette more like pipit *Anthus* than any of the others. Wags tail constantly when on ground, but this less pronounced than in *M. alba* and *M. cinerea*. Plumage of both adult and 1st-winter basically yellow below and on patterned edges of wing-feathers. Adult breeding ♂♂ of the many races differ in head pattern: various combinations of yellow, white, bluish, grey, and black. Sexes dissimilar in summer, less so in winter. Seasonal variation most marked on head (of ♂ only) and chest. Juvenile and 1st-winter bird separable from adult of same race. 10 races breeding in west Palearctic, 2 others accidental: 2 yellow-headed, 1 white-headed, 2 blue-headed with complete supercilium, 5 grey-headed with or without supercilium, and 2 black-headed. Adult ♂♂ of some races easily identifiable in the field but others less so; situation also complicated by frequent occurrence of (1) racial hybrids, (2) unusually pale birds, and (3), as part of the species' normal variability, occasional birds which resemble races far distant from those individuals' actual geographical origins (at least 4 west Palearctic races subject to this last problem). Most ♀♀, immatures, and non-breeding ♂♂ not racially identifiable in the field. Only widespread nominate *flava* described in full here, followed by main characters of ♂♂ and well-marked ♀♀ of other races.

Western and central mainland European race, nominate *flava*. ADULT MALE BREEDING. Crown and nape blue-grey, complete supercilium white; lore and cheek grey, darker than crown but often flecked white on lore and under eye; chin usually white. Back and rump yellowish-green, wearing to olive and green-brown. Wing-coverts striking, with almost black ground-colour and distinct greenish-yellow fringes and bold tips forming obvious double wing-bar on median and greater coverts. Flight- and tail-feathers almost black, with fringes to tertials yellow- or buff-white and outermost tail-feathers white. Underparts bright, full yellow, with greenish patches on breast-sides and suffusion along upper flanks. Under wing-coverts mostly white. ADULT FEMALE BREEDING. Lacks full head pattern and colour contrasts of ♂, with duller grey or grey-brown crown, and less contrast between duller off-white supercilium and browner (less grey) cheeks; back browner (less green) and underparts less uniform yellow, showing buffy chest and dark-brown-spotted 'necklace' (or indication of it). ADULT MALE NON-BREEDING. Full pattern and immaculate colours of breeding plumage lost from August. Head pattern duller, with green-brown tips to new feathers of crown, browner cheeks, and yellower supercilium. Back browner but rump yellower; wing markings less contrasting. Restricted white chin virtually invisible and rest of underparts less uniform,

with buff chest marked by indistinct, dark necklace and clear yellow area most noticeable from rear flanks to under tail-coverts. ADULT FEMALE NON-BREEDING. Duller than non-breeding ♂, with buff or dull yellow tone to supercilium, and paler underparts with mottled necklace and/or central spot on chest. JUVENILE. Noticeably darker than even dullest ♀♀, having earth-brown upperparts with only faint olive tone; blackish 'frown' lines on edges of crown obvious at close range. Face noticeably dark, with buff supercilium, brown cheeks, and buff chin and throat all tending to be mottled. Malar stripe and distinct broad necklace dark brown, recalling swarthy, dusky marks of juvenile *M. alba*. Also paler below than adult ♂, with foreparts buff and hindparts buff-yellow. FIRST WINTER. Juvenile plumage shed by September, then resembles adult ♀. ♂ somewhat greener on back and less strongly marked on chest than ♀; some ♂♂ also show cleaner face pattern, with early indication of greater contrast between crown, supercilium, and cheeks. FIRST SUMMER MALE. As adult male breeding but often retains some white- (not yellowish-) tipped greater coverts. FIRST SUMMER FEMALE. As adult female breeding or intermediate between it and non-breeding. At all times (and in all races), bill and legs black.

Other races. ADULT MALE BREEDING. (1) Iberian race, *iberiae*. Rather small. Compared to nominate *flava*, crown and nape darker, with ashy (not blue) tone; supercilium narrower, thus less obvious and sometimes vestigial in front of eye; chin and throat white, extending to upper breast; mantle darker green; breast-sides tinged olive. (2) Italian race, *cinereocapilla*. Resembles *iberiae* but supercilium absent or restricted to streak behind eye, cheeks darker, white throat more restricted or washed yellow, and rest of underparts slightly paler yellow than both nominate *flava* and *iberiae*. (3) Egyptian race, *pygmaea*. Smallest race, 10% smaller than *cinereocapilla* or nominate *flava*. Both crown and mantle greener than in *cinereocapilla*, contrasting less; supercilium usually absent or, at most, a trace behind eye. (4) Central Palearctic race, *beema*. Compared to nominate *flava*, crown paler, very pale blue to pale grey, depending on light; pronounced white supercilium and streak below eye. Chin and throat sometimes white. (5) North European race, *thunbergi*. Crown and nape slate-grey; supercilium white, often broken, vestigial, or even lacking; cheeks almost black or distinctly slatier than crown; chin and throat usually yellow or with white restricted to chin (unlike *iberiae* and *cinereocapilla*); sides of breast tinged olive, often joined by indistinct necklace. (6) Kamchatka race, *simillima*. Crown paler than *thunbergi*, supercilium white, quite broad and complete; cheeks grey, paler than *thunbergi*. Thus extraordinarily similar to nominate *flava* except for less blue in tone of crown and darker back (see Geographical Variation for additional structural characters). (7) Race of south-east Europe, Turkey, Levant, and Iraq, *feldegg*. Visibly the largest race, with rather long bill. Top, rear, and sides of head velvet-black, enhancing bulky look of bird and contrasting sharply with olive mantle and wholly yellow underparts; fringes and tips of wing-coverts and tertials noticeably bright, emphasizing wing-bars and tertial-lines; sides of breast and fore-flanks olive, even dusky. For separation from *thunbergi*, see Berg and Oreel (1985). (8) North and east Caspian race, *melanogrisea*. Resembles *feldegg* at distance, but close view allows separation based on brighter, more yellowish back and white stripe under forepart of black head. (9) British race, *flavissima*. Differs distinctly from all preceding races in having forehead often yellow, crown green-yellow or yellow-green, supercilium yellow, cheeks yellow-green, and upperparts sometimes washed yellow. (10) Race of central European USSR and central Kazakhstan, *lutea*. Differs from *flavissima* in having forehead, forecrown, and even all of crown yellow, therefore lacking supercilium; with only faintly green cheeks, bird can appear wholly yellow-headed. (11) North-west Mongolian race, *leucocephala*. Differs even more distinctly from preceding races in having white head, except for faint grey wash on rear crown and cheeks. ADULT FEMALE BREEDING. Racial identification in spring and summer best made by reference to ♂ of pair but compared to *flava*, ♀♀ of 4 races or racial pairs reasonably distinctive on their own, with (a) *feldegg* and *melanogrisea* showing swarthy, black-mottled or -patched head and bright wing marks as in ♂, (b) *iberiae* and *cinereocapilla* showing darker cheeks and whiter chin and throat than nominate *flava*, (c) *thunbergi* showing darker crown, cheeks and upperparts and less obvious supercilium than nominate *flava*, and darkest necklace, and (d) *flavissima* typically darker, browner above than *flava*, with only 2-toned head pattern and paler yellow underparts.

RACIAL HYBRIDS and birds resembling distant races are not fully understood. Important to note that (e.g.) British observations on breeding birds have indicated that (1) overshooting of at least nominate *flava* into range of *flavissima*, with subsequent pairing, can produce ♂♂ resembling both *beema* and *leucocephala*, and (2) that nominate *flava* and *flavissima* (at least) include brown and blue-grey birds that lack substantial yellow pigment, while some ♂♂ of *flavissima* resemble *lutea*. In Europe, some zones of intergradation sufficiently wide to give rise to populations of apparently distinct and constant appearance, e.g. *flava* × *feldegg* hybrids of Rumania ('*dombrowskii*') which are numerous enough to be recognized both in winter quarters and on migration (Someren 1934; Wallace 1955, 1984*b*). These and other birds offer field observers an interesting but as yet not meaningful pursuit; see Geographical Variation for further details.

A bird of vast complexity in systematics and morphology and thus subject to both confusion with other species and doubts in racial identifications. Unlikely to be mistaken for *M. alba*, except as juvenile when the 2 species are closest in appearance. In Britain, some juvenile *flavissima* approach swarthiness of juvenile *M. alba yarrellii*, and separation best based on basic upperpart tone, and on cleaner chest and usually yellow vent of *flavissima*. Grey variant *M. flava* lacking yellow vent may suggest strange *M. alba* but is distinguished by lack of chest-band. *M. flava* unlikely to be mistaken either for *M. cinerea*, since, although that species is yellow-bodied, it is also yellow-rumped and has predominantly grey head and mainly black wings with long white wing-band in all plumages (in addition to structural differences, most obviously very long tail). Much more serious risk of confusion between *M. flava* (particularly yellow-headed races in adult plumage, any race in pale sub-adult plumage, or pale variant) and *M. citreola* (in both adult and sub-adult plumage). Important to recognize that adult *M. citreola* although yellow-headed and yellow-bodied, is much duskier on upperparts, particularly on sharply demarcated nape which is black on ♂. Among adults, risk of confusion with this species thus lies between ♂ or ♀ *flavissima* and ♀ *M. citreola*; latter differs most distinctly in almost white fringes and tips to median and greater wing-coverts and tertials, less patterned face and crown, dusky nape (less sharply demarcated than in ♂ *M. citreola*), dusky upperparts, and paler vent. Among sub-adults, confusion less restricted, since pale variants of several races of blue- or grey-headed *M. flava* may be closely similar to immature *M. citreola* (see that species, p. 434). *M. citreola* also has reasonably distinctive call. *M. flava* has bounding flight but not as 'shooting' as in *M. alba* and *M. cinerea*—even appearing to stall occasionally. Flight silhouette less attenuated than *M. alba* and *M. cinerea*, with obviously shorter tail; recalls longer-tailed pipits *Anthus* and confusion can occur in brief or distant view. Gait as *M. alba* but stance on ground less horizontal; much addicted to following cattle and tolerant of denser ground cover than *M. alba* when foraging.

Song a brief warble, interspersed with call notes; given intermittently in flight or from perch. Common call (loudest in *feldegg*) often the first signal of bird's presence: shrill, drawn-out, quite musical but plaintive 'tsweep', with slight terminal accent.

Habitat. Breeds in west Palearctic from lower middle to high latitudes, near July isotherm of 10°C, in arctic tundra and subarctic, boreal, temperate, steppe, and Mediterranean zones, mainly continental but marginally oceanic, largely on level or gently sloping lowlands (Voous 1960), but in Caucasus and elsewhere in damp meadows on river banks and lake shores up to 2000-2500 m (Dementiev and Gladkov 1954*a*). In west of range (e.g. East and West Germany, Switzerland) confined however to lowlands, avoiding mountains and broken, arid, sandy, stony, or bare ground, forests, and enclosed landscapes (Niethammer 1937; Glutz von Blotzheim

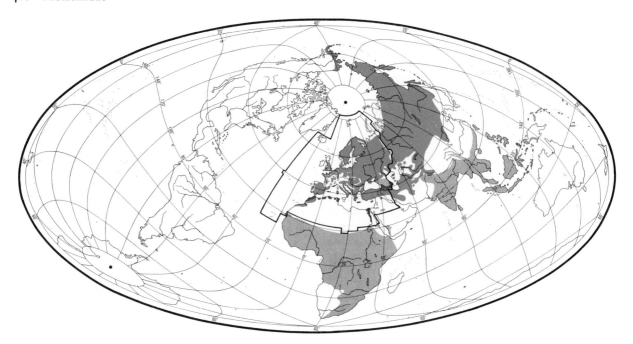

1962). In breeding season, occupies fringes of wetlands, such as riversides, lakesides, upper levels of salt-marshes, floodlands, moist pastures, water-meadows, excavations, subsidences, grazed fens with scattered small trees, sedge marshes, slacks within sand-dunes, and similar habitats both natural and artificially managed, especially where herbage low, dense, luxuriant, moist, and preferably near shallow surface water. Also requires low perches on bushes, wires, fences, banks, or walls. In England, breeding population strongly concentrated within *c.* 20 km of coastline and on lower reaches of slow-flowing rivers, in water-meadows, grazed semi-marshland, and cattle pastures; comparable artefact areas such as sewage farms and reservoir margins are also attractive. Abandonment of breeding areas or population declines is apparently sometimes due to changes in land use or management, especially where denser or taller growth of vegetation during breeding season has resulted (Smith 1950). In far north, occupies *sphagnum* bogs with sparse stunted pine trees, hummocky peat-bogs covered with moss and overgrown with shrubs, and occasionally arable land or areas by villages or among trees such as pine or birch, finding suitable nest-sites in very small parcels of marshy ground (Dementiev and Gladkov 1954*a*). In some parts of range will also breed on relatively dry farmland, heath, moor, and upland areas, although these generally support much lower densities than damper terrain (Sharrock 1976; Fuller 1982). Considering conspicuous racial plumage differences, variations in habitat choice appear inconsiderable in relation to wide geographical and climatic range. In winter in Africa, as to some extent in breeding season, exploits a niche left open by native species, in close association with large herbivores, both wild and domestic. Range of winter habitats is wide, extending from residual rain-pools and ricefields to parched acacia steppe, and up to 2200 m; essential requirement is for unobstructed ground, whether short grass or cultivation, including cassava plots and inside of a banana grove (Moreau 1972). In Sierra Leone, occurs on drying swamps, mountain plateaux, games fields, and tarred roads; also on coastal mudflats among waders (Charadrii), and treetops and roofs (G D Field). In winter in India, frequents pastures and moist grassy places along riversides and pond margins, foraging on ground, commonly in attendance on grazing cattle, or on irrigated lawns in urban areas. Runs about and feeds on wet mud in squelchy newly-cut paddyfields around coastal backwaters, sometimes coming close to the reapers at work. Will fly for 30 km or more to roost communally in reedbeds and sugarcane fields (Ali and Ripley 1973*b*). Evidently highly successful in finding an ample ecological niche for large population forced to vacate immensely extensive Palearctic breeding grounds for at least 6 months in each year and to compete with indigenous avifauna. Flies freely, and on migration ascends to upper airspace in crossing mountains such as Himalayas, but primarily a ground bird. Experiences few problems in relation to human activities.

Distribution. No longer breeds regularly in Ireland, and some contraction of range in northern Britain, France, and West Germany.

BRITAIN. *M. f. flavissima* is main race. Marked contraction since 1930s. Bred formerly in Cornwall, and once bred as far north as Aberdeen. Nominate *flava* breeds occasionally, with proved or suspected records as far

north as Aberdeen and as far west as Isles of Scilly. Birds resembling *beema* have bred in southern England, and one resembling *leucocephala* bred Kent 1908. (Smith 1950; Milne 1959; Parslow 1963; Sharrock 1976). IRELAND. *M. f. flavissima* is main race. Two large colonies Mayo/Galway and around Lough Neagh became extinct by 1941 (Parslow 1967; Sharrock 1976). Since then, has bred 1957-8, 1965-8, 1981, and probably 1980 (CDH). Bird resembling *cinereocapilla* bred near Belfast 1956 and nominate *flava* Belfast 1963 (Parslow 1967; Sharrock 1976). FRANCE. Apparently some range contraction since 1936 (Yeatman 1976). Nominate *flava* is main race, but *flavissima* breeds in north coastal zone from Somme to Finistère, and *iberiae* (with a tendency towards *cinereocapilla* in appearance) in Pays Basque and on Mediterranean coast. Birds resembling *cinereocapilla* bred

Alsace 1948. (Yeatman 1976; RC.) Birds resembling *feldegg* bred Seine-et-Marne 1980 (Siblet and Tostain 1984). NETHERLANDS. Nominate *flava* is main race. *M. f. flavissima* overlaps with nominate *flava* on coast, but no interbreeding, preferring tulip-fields rather than grasslands (Dijk 1975). WEST GERMANY. Some contraction of range. Nominate *flava* is main race, but *flavissima* breeds sporadically on North Sea islands, and *cinereocapilla* occurs, probably with hybridization in south-west. (AH.) DENMARK. Nominate *flava* is main race. *M. f. flavissima* first bred 1970; few records since, mostly west coast of Jutland (TD). NORWAY. Some expansion (VR). Nominate *flava* is main race, but under 50 pairs of *flavissima* also breed (Paulsen 1978). FINLAND. *M. f. thunbergi* predominates over nominate *flava*, interbreeding frequently (Merikallio 1958). CZECHOSLOVAKIA. Nominate *flava* is main race, but birds resembling *cinereocapilla* breed in small numbers in central Slovakia. *M. f. thunbergi* bred northern Bohemia 1968; *feldegg* bred 1919, 1965, and 1974–5. (Joń 1970; Davola 1972; KH.) AUSTRIA. Nominate *flava* is main race, intergrading with *cinereocapilla* in south. *M. f. feldegg* has bred almost annually since 1969 in Neusiedlersee area and elsewhere (HS, PP). SWITZERLAND. Has spread (Géroudet 1983). ITALY. *M. f. cinereocapilla* is main race, *feldegg* local, nominate *flava* rare (PB, BH). MAURITANIA. Banc d'Arguin: breeds Nair and probably other islands (Roux 1960; Naurois 1969a).

Accidental. Bear Island, Iceland, Faeroes, Azores, Madeira.

Population. Some decline in Britain, Netherlands, West Germany, southern Sweden, and Czechoslovakia. Has increased northern Sweden and northern Finland.

BRITAIN. About 25 000 pairs. Marked decrease in Scotland; in England and Wales, perhaps a decline in south and slight increase in north; few changes since 1950s. (Smith 1950; Parslow 1963; Sharrock 1976.) FRANCE. 10 000–100 000 pairs (Yeatman 1976). NETHERLANDS. Nominate *flava*: 17 000–25 000 pairs 1976–7 (Teixeira 1979); marked decrease since 1960s with habitat changes (Wammes *et al.* 1983; CSR). *M. f. flavissima*: 150–200 pairs 1976–7 (Teixeira 1979). WEST GERMANY. 45 000–87 000 pairs (Rheinwald 1982). Declining (AH). SWEDEN. About 400 000 pairs (Ulfstrand and Högstedt 1976). Decreasing in south due to habitat changes, increasing in north in clear-felled forest areas (LR). Numbers of autumn migrants at Falsterbo doubled since 1940s (Roos 1983). FINLAND. About 400 000 pairs (Merikallio 1958). Has increased in north (Väisänen 1983). EAST GERMANY. Moderately abundant (SS). CZECHOSLOVAKIA. Marked decline since 1960, now often rare (KH). AUSTRIA. Has declined considerably due to habitat loss (HS, PP).

Oldest ringed bird 8 years 5 months (Rydzewski 1978).

Movements. Most populations migratory, wintering in Afrotropics, India, and south-east Asia. Winters also between Black and Caspian Seas and at a few sites in Turkey. Egyptian race, *pygmaeus*, largely resident, and some parts of breeding range in north-west Africa and southern Spain occupied through the winter, with possibility that some individuals are resident.

Several factors make this a particularly well documented migrant: large populations; conspicuous (mostly diurnal) movement; use of huge communal roosts, both on migration and in winter, facilitating ringing; assumption by ♂♂ of racially distinct breeding plumage shortly before spring migration. On the other hand, confusion can arise through racial intermediates and disjunct pattern of geographical variation (see final paragraph).

Main Afrotropical winter range bounded in north by the line Sénégal to bend of Niger river, Lake Chad, Khartoum, and northern Eritrea; to the east, absent roughly beyond a line from Berbera (Somalia) to mouth of Zambezi; to south-west, largely absent between Zaïre river estuary south to Namibia but occurs in extreme south as far west as Cape Town; a few winter in Saharan oases and January records from North Africa suggest some overwinter from Morocco to Tunisia (Zink 1975). Nominate *flava* the most widespread race (occurring from 10°W to 50°E), consistent with its breeding range and mainly north–south broad-front migration.

East to Cameroon, nominate *flava* most abundant, with *iberiae*, *cinereocapilla*, *thunbergi*, and *feldegg* locally numerous, *feldegg* tending to favour coastal zone; *flavissima* recorded certainly only east to Mount Nimba (Liberia), and *lutea* occurs east of Liberia (Douaud 1957; Curry-Lindahl 1964; Elgood *et al.* 1966; Vielliard 1971–2; Gore 1981; Louette 1981; Elgood 1982; G D Field); for racial distribution within Nigeria, see Wood (1975). In central Africa (Chad, Zaïre, Zambia), nominate *flava* and *thunbergi* occur throughout, with also *iberiae*, *beema*, and *melanogrisea* in Chad, and *lutea* and *feldegg* in Zaïre, but *lutea* apparently the only additional race in Zambia (Salvan 1967–9; Benson and Benson 1977; Curry-Lindahl 1981). East from Sudan, Uganda, and Tanzania south to Malawi, Zimbabwe, and the Cape, nominate *flava* and *thunbergi* are numerous, as are *lutea* and *beema* (both breeding mainly east of Volga); *leucocephala* from Mongolia (wintering mainly in India) has occurred as vagrant, and several claims for British *flavissima* though these more probably refer to *lutea*; only nominate *flava* reaches South Africa in any numbers (Cave and Macdonald 1953; Smith 1960a; Dowsett 1965b; Urban and Brown 1971; Moreau 1972; Benson and Benson 1977; Pakenham 1979; Irwin 1981; Hogg *et al.* 1984). Individuals commonly faithful to wintering site from year to year (see Social Pattern and Behaviour).

Populations from east of Urals and Caspian Sea winter mainly in Indian subcontinent and from Burma and south-east China through south-east Asia to western New Guinea.

Precise wintering areas of the various races are not well established but in the main lie between south-east and south-west of respective breeding areas. *M. f. lutea*, *melanogrisea*, and *beema*, breeding in south-central USSR, all winter in both eastern Africa and India, but not known whether there is a clear migratory divide somewhere in western Asia. *M. f. leucocephala*, breeding largely in north-west Mongolia, moves west of south to winter in north-west India (Smith 1950; Ali and Ripley 1973b). Populations from eastern Siberia and central and eastern Asia move between south and south-west to winter in south-east Asia. *M. f. tschutschensis* of Alaska and north-east Siberia moves mainly west of south to winter in Indonesia and Philippines, rarely (probably this race) reaching northern Australia (Macdonald 1973; Pizzey 1980). Birds leaving Alaska move first west via Aleutian Islands to east coast of Asia (Bent 1960), though a few straggle south-east along American west coast as far south as California (Roberson 1980).

Movement broad-front in both spring and autumn, with numerous sightings of migrants at sea in all areas (Mills 1956; Serle 1956; Paige 1960; Fry 1961; Long 1961; Warham 1964; Casement 1966; Cheke 1967). West and central European populations move between WSW and SSW in autumn (Zink 1975). *M. f. flavissima* from Britain first moves south into extreme south-west France, then changes direction to nearer west to reach coast of Portugal, then redirects again to head nearly south into Morocco. Some perhaps head across Bay of Biscay direct to Iberia, though Snow *et al.* (1955) saw only a single bird come in from the sea west of Santander, while at La Coruña eastward coastal passage included *flavissima* (this perhaps a reverse movement by birds which had been heading west along north Spanish coast and turned back after striking west Iberian coast). Numerous recoveries around Bilbao suggest easternmost section of Biscay generally overflown (Zink 1975). Birds from Scandinavia move between south-west and south: those leaving via Falsterbo tend to move through Iberia, while of those leaving further east (via Ottenby or Öland) over half travel south through northern Italy. Birds from Finland have been recovered in Greece on a heading well east of south (Zink 1975). In USSR, northernmost races leave first: thus *thunbergi* leaves Murmansk area from early August, nominate *flava* leaves between August and early September, *lutea* has vacated Irgiz river area by late August, while *feldegg* (southernmost race in European USSR) leaves mainly during September (Dementiev and Gladkov 1954a).

Autumn passage in Switzerland has been noted as early as late July but main passage begins second half of August and peaks through September usually to end abruptly in early October, though individuals have been noted still passing in first third of November (Glutz von Blotzheim 1962).

In south-west Portugal, strong evening easterly movements recorded at Quarteira towards Straits of Gibraltar (Henty 1961) where steady passage of nominate *flava* and *iberiae* occurs (Lathbury 1970); extends from early August to early November peaking mid-September, but minor peaks in early and later August probably indicate passage of different populations; passage heaviest in early morning and again around 18.00 hrs (Tellería 1981). Passage through Balearic Islands obscured by local breeding of *iberiae*, but nominate *flava*, *thunbergi*, and *cinereocapilla* pass in numbers, though *flavissima* is scarce (Bannerman and Bannerman 1983). Occurs on Malta in flocks of up to 1000, mostly September and early October; birds retrapped up to 12 days later having made considerable fat deposition; mostly nominate *flava* but with significant number of *cinereocapilla* and *feldegg*, small numbers of *flavissima* and *thunbergi*, and possibly also *lutea* and *iberiae* (Sultana and Gauci 1982). The small numbers of

thunbergi (breeding right across northern Europe from Norway to east of Urals) suggest that most overfly Mediterranean. In Cyprus, passage mainly mid-August to mid-October, involving flocks of up to 1000, peaking mid-September to early October; nominate *flava* commonest, *feldegg* less so, and a few *thunbergi* also occur (Flint and Stewart 1983).

Arrival on North African coast not well documented and most birds perhaps overfly. Scarce in Libya, September–October, though once recorded widely in south-west in October (Bundy 1976). Nominate *flava* common on passage along Atlantic coast of Morocco, early September to mid-October; also a few *flavissima* recorded and some possible *cinereocapilla* (Smith 1965). Arrives in Afrotropics mainly during October, rapidly occupying the whole region. Thus, at Sokoto (northern Nigeria), first recorded 29–30 September in 2 successive years (Mundy and Cook 1972), while first arrival of flocks of nominate *flava* at Ibadan (southern Nigeria) over 4 successive years was 2–4 October (Elgood 1959). Arrives towards end of month in southern parts, e.g. Zimbabwe (Irwin 1981). By the time birds arrive in Afrotropics, racial identification almost impossible, and thus virtually no reliable data on arrival dates of the different races. In east Mediterranean region, passage through Turkey (*thunbergi*, nominate *flava*, *feldegg*, *lutea*) occurs early August to mid-October, peaking in first half of September (Nisbet and Smout 1957; Smith 1960b; Ballance and Lee 1961; Sutherland and Brooks 1981; Porter 1983). Main movement in September also through Jordan (Cameron and Cornwallis 1966) and Israel (Safriel 1968), and occurs August to early October in Iraq (*feldegg* commonest, also *thunbergi*, nominate *flava*, *lutea*), with at least some westward movement and birds making stopovers of a few days or more (Chapman and McGeoch 1956; Marchant 1963a). On south Caspian coast, occasionally seen arriving off sea, but movement largely along coast, both east and (mostly) west; movement declined after mid-September, and races identified were *lutea*, nominate *flava/beema*, *thunbergi*, and *feldegg/melanogrisea*. Also, steady passage noted over Elburz mountains (northern Iran) in early September (Feeny *et al.* 1968), and southward movement recorded near Tehran 2 September–20 October (J B Heigham). Common on passage in eastern Arabia, mid-August to November (Bundy and Warr 1980; Gallagher and Woodcock 1980).

Over the 2–3 weeks before spring departure from Africa, fat build-up amounts to *c.* $\frac{1}{3}$ of winter weight (Ward 1964; Smith and Ebbutt 1965; Fry *et al.* 1970), and Wood (1982) deduced that birds preparing to leave northern Nigeria had acquired enough fat for 60–70 hrs continuous flight, sufficient to cross Sahara non-stop if no head-winds; final take-off then needs to be from desert's southern edge, typically in late afternoon, entailing 3 nights and 2 days on passage. Movement is broad front, with little evidence for use of Nile valley as major flyway.

Evacuation is progressive from south, starting March in extreme south with slow progress to edge of Sahara. In Zaïre, races come into breeding plumage and prepare for departure in sequence: southernmost-breeding *feldegg* going first, followed by mid-latitude-breeding nominate *flava*, and northern-breeding *thunbergi* last (Curry-Lindahl 1958); similarly in Nigeria (Wallace 1969). Over 8 years, last departure (mainly nominate *flava*) from Ibadan (southern Nigeria) was 28 April-9 May (Elgood 1959; Elgood *et al.* 1966). Birds ringed in Nigeria recovered north of Sahara as follows: Italy 26, Malta 14, Tunisia 5, Finland 5, USSR 5, Algeria 4, Libya 4, Greece 3, Sicily 2, and 1 each Morocco, Germany, Austria, Yugoslavia, and Crete (Sharland 1979; Elgood 1982). Spring return movement of best-studied western races, *flavissima* and nominate *flava*, often by different route from that used in autumn, tending to lie to eastward: thus, British *flavissima* migrate further east through north-west Africa and south-west Europe than in autumn (ringing recoveries in Portugal only in autumn; in eastern Morocco, western Algeria, and eastern Spain only in spring; similarly with nominate *flava*, birds trapped Belgium also found on average further east on spring migration than in autumn, and spring recoveries of birds ringed in autumn in Germany are again further east—in Malta, Tunisia, and Adriatic coast of Italy). Birds from eastern Europe, however, seem to show opposite tendency with birds from Finland moving south-east in autumn to Balkans and Greece and returning in spring further west via Italy. (Zink 1975; Dittberner and Dittberner 1984.)

Spring passage in North Africa much more obvious than autumn. At Defilia oasis (Figuig, eastern Morocco), abundant on spring passage with arrival throughout the day irrespective of weather; birds departed mostly between north and west; mainly nominate *flava*, *iberiae*, and *cinereocapilla*, though *thunbergi*, *flavissima*, and *feldegg* also identified, suggesting generally east of north destination (Smith 1968b); birds delayed 4 days (6–9 April) by very bad weather showed no significant weight change (Ash 1969). In Tunisia, passage late March to end of May, peaking early May; *flavissima* regarded as accidental (Deleuil 1954; Blanchet 1955; Thomsen and Jacobsen 1979). Daily passage at Kufra oasis (Libya) through April (Cramp and Conder 1970), and departure from Tripolitanian coast extends from mid-March to mid-May; birds abundant through this period in descending order nominate *flava*, *feldegg*, *iberiae*, *cinereocapilla*. and *flavissima/lutea* (Guichard 1957; Bundy 1976; Willcox and Willcox 1978). At Elat (Israel), passage (of nominate *flava* and *feldegg*) occurs early March to late May (Safriel 1968), and migrants common late February to May in Gulf states of Arabia (Bundy and Warr 1980).

♂♂ reach breeding grounds before ♀♀ (Smith 1950; Spaepen 1957). In Britain, *flavissima* arrives late March to mid-May (Hollom 1962), and breeding grounds in most of central Europe reoccupied from late March

(Spaepen 1957; Zink 1975). In southern Sweden, nominate *flava* arrives from mid-April while *thunbergi* arrives in north from early May (Rendahl 1967). In Switzerland, earliest arrivals 10 March but usually begin late March or early April and continue well into May with peak in second half of April; mostly nominate *flava* but also *cinereocapilla* which suggests that reverse migration must follow to take birds south to their breeding area (Italy); northern *thunbergi* notably later and *flavissima* noted only seldom (Glutz von Blotzheim 1962). In France, passage of *thunbergi* noted only in May (Mayaud 1952a). In USSR, *feldegg* arrives in south mid-March to early May; nominate *flava* arrives Crimea early April and Moscow area mid-April; *thunbergi* keeps pace with nominate *flava* at first but does not reach Lapland until end of May or early June; *lutea* arrives Armeniya in April, and *beema* arrives Altai region mid-May (Dementiev and Gladkov 1954a).

Many records of birds resembling a particular race well outside that race's normal range, but some (at least) of these are part of the species' normal variability (see Geographical Variation) and do not necessarily indicate vagrancy. Birds showing the characters of several races have been recorded in Britain, for example, mainly in spring and sometimes well outside their normal range: continental nominate *flava* occurs regularly and has bred occasionally (and probably hybridizes with *flavissima*), chiefly near east and south coasts; *thunbergi* occurs most years; *feldegg* has been seen at various localities, including during breeding season, though most records of this race in western Europe probably relate to dark-headed variants of *thunbergi* (Berg and Oreel 1985); birds identical to *simillima* from Far East have been collected at Fair Isle in autumn (and sighted elsewhere); birds resembling *beema*, *leucocephala*, and *cinereocapilla* have bred (British Ornithologists' Union 1971). JHE

Food. Small invertebrates. 3 main foraging techniques (for more detailed breakdown, including use of high flight, see Wood 1976). (1) Picking. Picks items from ground or water surface while walking. (2) Run-picking. Makes quick darting run at prey, picking it up either from surface or as it takes off. (3) Flycatching. Makes short flight from ground or perch, catching prey in mid-air—either in bill or by knocking it down with wings. (Smith 1950; Davies 1977.) Occasionally takes insects from plants in hovering flight (Glutz von Blotzheim 1962), or flies low over water snatching insects from surface (Kishchinski 1980). Tail assists balance when run-picking and flycatching; especially important when turning rapidly in flight (Davies 1977). Usually no pursuit made if attempt to seize prey fails (Dolgushin *et al.* 1970). Often feeds in association with grazing cattle and sheep, taking insects disturbed by animals or blood-sucking species from animals themselves. Also known to hunt insects emerging from earth turned over by ploughs in fields (Schmidt 1967b). Use of various feeding techniques varies according to whether birds feeding in flocks or alone. At Oxford (England) in April, flock birds (feeding at shallow pools) employed picking almost exclusively (99·5%), while single birds (feeding at dung pats) used greater variety of techniques—picking 83·8%, run-picking 8·7%, flycatching 7·5% (Davies 1977). Prey swallowed whole, but larger items often beaten against ground or rock first (Dolgushin *et al.* 1970); grasshoppers and crickets (Orthoptera) shaken vigorously causing head and appendages to fall off before swallowing (Wood 1976). Smaller species usually fed to young complete with wings; larger species have hard and inedible body parts and wings removed; large molluscs usually given to young in broken shells (Prokofieva 1980). At Oxford, single birds generally feed more slowly than those in flocks, with flock birds in April taking 35·8±SE1·2 items per min, solitary birds 8·7±SE1·1 items per min. However, at midday when peak number of prey available at dung pats, rate of energy intake higher for solitary birds at dung pats (285 J per min) than for birds feeding in flocks at pools (196 J per min). Despite great differences between flock and solitary sites in feeding methods and prey eaten (see below), birds switched rapidly from one to the other; switches sometimes resulted in maximization of energy intake but at other times not, probably because feeding efficiency constrained by need to engage in other activities. On arrival at Oxford in April, birds all fed in flocks at pools, but pattern changes gradually so that 2 weeks later all fed singly at dung pats; during period of change, tended to feed in flocks at pools in early morning and evening, singly at dung pats at midday. (Davies 1977.) In Afrotropics, overall food consumption does not alter markedly through course of winter, but food increasingly difficult to obtain in late winter as dry season progresses. In Nigeria, measure of difficulty determined by number of paces between capture of successive items: in late November, average 6·5 paces; in March, average 22·6 paces (Wood 1976). During winter, only early in season is there a marked diurnal pattern of feeding: highest feeding rates after dawn and towards dusk, lower during heat of midday. Later in winter, feeding rates almost constant through the day (Wood 1976; see also Fry *et al.* 1972).

The following recorded in diet in west Palearctic. Larval and adult mayflies (Ephemeroptera), dragonflies (Odonata: Coenagriidae, Lestidae, Libellulidae, Gomatidae, Corduliidae), stoneflies (Plecoptera), grasshoppers, etc. (Orthoptera: Acrididae, Tetrigidae), earwigs (Dermaptera), termites (Isoptera), bugs (Hemiptera: Aphididae, Delphacidae, Miridae, Psyllidae), alder flies, etc. (Neuroptera), larval and adult butterflies and moths (Lepidoptera: Noctuidae, Pieridae, Lymantriidae), larval and adult caddis flies (Trichoptera), larval and adult flies (Diptera: Chironomidae, Drosophilidae, Scatophagidae, Sphaeroceridae, Sepsidae, Chloropidae, Calliphoridae,

Mycetophilidae, Lonchopteridae, Syrphidae, Bibionidae, Tipulidae, Agromyzidae, Opomyzidae, Dryomyzidae, Empididae, Asteiidae, Tephritidae, Muscidae, Asilidae, Conopidae, Culicidae, Limoniidae, Tabanidae, Tachinidae, Rhagionidae, Therevidae, Hippoboscidae, Hypodermatidae), Hymenoptera (sawflies Symphyta, Ichneumonidae, Braconidae, Cynipidae, Pteromalidae, ants Formicidae, wasps Vespidae), adult and larval beetles (Coleoptera: Curculionidae, Scarabaeidae, Carabidae, Chrysomelidae, Byrrhidae, Dascillidae, Dytiscidae, Anthicidae, Elateridae, Geotrupidae, Staphylinidae, Buprestidae), spiders (Araneae), harvestmen (Opiliones), Myriapoda, small molluscs, and worms. (Baer 1909; Rey 1910b; Drost 1948; Kovačević and Danon 1952; Smith 1950; Mester 1959; Glutz von Blotzheim 1962; Grosskopf 1968; Ptushenko and Inozemtsev 1968; Brosset 1971b; Szabó 1976; Davies 1977; Popov 1978; Prokofieva 1980; Kostin 1983; Dittberner and Dittberner 1984.) The following recorded in Afrotropical winter quarters: stick and leaf insects (Phasmida), cockroaches (Dictyoptera), bugs (Pentatomidae), larvae of moths (Sphingidae), beetles (Cerambycidae), crustaceans, berries of *Salvadora*, and small seeds. (D'Abreu 1918; Fry *et al.* 1972; Wood 1976.)

At Oxford, April–May, analysis of faeces from birds feeding in flocks at pools showed diet almost wholly Diptera: in April, 1335 items comprised 85·9% (by number) Chironomidae, 4·8% Scatophagidae, 3·5% Sphaeroceridae, 1·4% Chloropidae, 1·2% beetles, 3·2% others (mostly Diptera); in May, 2862 items comprised 44·1% Drosophilidae, 34·8% Chironomidae, 5·4% Sphaeroceridae, 4·3% Aphididae, 3·4% Ichneumonidae, 2·0% Scatophagidae, 1·4% Coleoptera, and 4·6% other Diptera; in May, Diptera selected preferentially (especially Drosophilidae—17·6% in trap samples but 44·1% in faeces), while Ichneumonidae selected against (30·1% in trap samples but only 3·4% in faeces). Birds feeding singly at dung pats also took almost wholly Diptera: 667 items comprised 43·6% (by number) Sphaeroceridae (1–4 mm long, mostly 3–4 mm), 35·1% Scatophagidae (5–10 mm), 6·4% beetles (2–3 mm), and 14·9% others (mostly Diptera); large Sphaeroceridae (3–4 mm) selected preferentially (comprised 10·1% of total available, but 41·3% in faeces) and Scatophagidae selected against (77·1% of total available but only 35·1% in faeces). (Davies 1977.) In Moscow region (USSR), June, 6 stomachs contained 168 items comprising 83% (by number) Psyllidae, 7% beetles (mostly Chrysomelidae), 2·3% Diptera, 2·3% Cynipidae, 1·8% caterpillars, and 2·6% others (Ptushenko and Inozemtsev 1968). In Crimea (USSR), April, contents of 6 stomachs comprised 62% (by number) beetles (Chrysomelidae, Curculionidae, Carabidae), 30·5% ants, and 7·3% earwigs and Diptera (Kostin 1983). In Tatarskaya ASSR (east-central European USSR), of 56 stomachs, many contained beetles: 41% contained Curculionidae, 15% Chrysomelidae, 11% Buprestidae, 4% Elateridae, 4% Staphylinidae, 11% Carabidae, 16%

spiders, 2% Myriapoda, 4% Odonata, 11% Orthoptera, 11% cicadas, 14% aphids, 21% other Hemiptera, 23% Hymenoptera, 25% small Diptera, 18% mosquitoes, 32% caterpillars, 7% aquatic invertebrates; of 10 further stomachs from same location, 5 contained Curculionidae, 3 Geotrupidae, 3 Elateridae, 2 Staphylinidae, 2 Chrysomelidae, 1 Buprestidae, 2 Orthoptera, 1 cicadas, 1 other Hemiptera, 1 caterpillars, 2 aquatic invertebrates (Popov 1978). For further quantitative data from west Palearctic, see Baer (1909), Rey (1910b), Kovačević and Danon (1952), and Averin and Ganya (1970); for east Palearctic, see Rustamov (1958), Pek and Fedyanina (1961), and Dolgushin et al. (1970).

In winter, Nigeria, items in 89 stomachs comprised 45·3% (by number) Hemiptera, 22·9% beetles, 14·9% Hymenoptera, 14·0% Orthoptera, 1·9% Diptera, and 1·0% others; also takes termites when swarming at start of rains (Wood 1976). See also Fry et al. (1972). For elsewhere in Afrotropical region (mainly Uganda), see Owen (1969).

Food of young similar to adults'. In Tatarskaya ASSR, diet during first 3 days of life dominated by small delicate aphids and Psyllidae; after 4 days, bigger items and insects with hard chitinous skins eaten (dragonflies, Orthoptera, butterflies) (Popov 1978). In Leningrad region (western USSR), observations at 3 nests showed Diptera comprised 44·8% (by number) and Ephemeroptera 24·2% of nestling diet; others items mainly dragonflies, beetles, caddis flies, and molluscs (Prokofieva 1980). In Hungary, observation of 1 nest showed 90% of items were grasshoppers, adult and larval Lepidoptera, and beetles (Szabó 1976). For comparison with diet of Citrine Wagtail M. citreola, see that species (p. 437). Most food brought morning and evening, less during heat of midday (Ptushenko and Inozemtsev 1968). Frequency of feeding varies with age of young, availability of food, weather, nearby crops, and distance to be travelled; rate initially increases with nestling age (Smith 1950). In West Germany, over 29 hrs of observation, at 5 days old 15 feeds per hr, at 6 days 19 per hr, at 7 days 29 per hr, at 8 days 37 per hr, at 9 days, 34 per hr, at 10 days 34 per hr, at 11 days 21 per hr (Ringleben 1935). In Oka valley, (USSR), June, 6 young 3 days old fed from 03.50–21.30 hrs with 7·3 visits per hr; at 6 days, 03.40–21.35 hrs with 9·8 visits per hr (Ptushenko and Inozemtsev 1968).   ILG, CAT

Social pattern and behaviour. For major reviews see Smith (1950) and Dittberner and Dittberner (1984), concerning respectively flavissima and (mostly) nominate flava. No detailed study on differences between races.

1. Gregarious outside breeding season. In winter quarters, dispersion varies with food supply, but majority form small flocks. At Vom (Nigeria), most in flocks, often associated with cattle, though some ♂♂ defend feeding-territories throughout winter in richest feeding areas adjacent to streams (Wood 1978, 1979). Average length of stream within 6 territories 82·5 m (60–130), average area 0·25 ha (0·15–0·36) (Wood 1976). Similar

dispersion reported in Gabon where one territory c. 600 m² and flocks typically 3 birds (Brosset 1971b). Will also defend water troughs (Wood 1976). In Zaïre, usually territorial, sometimes in flocks of 4–6 (Verheyen 1956a; Ruwet 1965). In East Africa, seldom forms flocks except for roosting and migration (Reynolds 1974). Fidelity to winter quarters may be marked, e.g. of 6 birds ringed in territories, Freetown (Sierra Leone), 2 returned to occupy them the following year (G D Field). For fidelity to at least general area, see Ashford (1970), Moreau (1972), Reynolds (1974), and Wood (1976). For communal roosting, see below. Migrates in small parties: usually less than 20, sometimes 60–80 (Crousaz 1961); in spring often in ones and twos (Smith 1950); in autumn, southern Turkey, average 1·8 (n = 379 birds) (Sutherland and Brooks 1981). Much larger numbers, often thousands comprising several races, may assemble at major stopovers on spring migration (Meinertzhagen 1954; Williamson 1955b): e.g. at Khartoum (Sudan), February–March, huge flocks associated with White Wagtails M. alba and Red-throated Pipits Anthus cervinus (Butler 1905). On arrival at breeding grounds in southern England, flocks disband after 2 weeks (Davies 1977). After breeding season, family groups may form larger flocks, of up to c. 40 (Smith 1950). BONDS. Monogamous mating system the rule, and pair-bond of seasonal duration (Drost 1948; Dittberner and Dittberner 1984). For possible case of bigamy (♂ and 2 ♀♀), see Drost (1948). For interbreeding with Citrine Wagtail M. citreola, see that species (p. 437). Some birds arrive on breeding grounds already paired. Not known if this arises from pairing in winter quarters or on spring migration, e.g. pairs evident in spring migrating flocks of thunbergi, which then show 1:1 sex-ratio. Also possible that pair-bonds may develop in previous year, as juveniles show incipient courtship-behaviour in late summer and form 'engagement' pairs (see Heterosexual Behaviour, below) prior to autumn migration. Pair-bond typically maintained for 2nd clutch, even if 1st fails (Dittberner and Dittberner 1984). Pair-bond 'looser' for 2nd brood, and ♂ may pair anew for 2nd brood if 1st mate dies (Drost 1948). Both members of pair brood and feed young. Family bonds maintained for some time after leaving nest: young self-feeding at 17–18 days (Dittberner and Dittberner 1984) but remain with parents for several weeks thereafter (Smith 1950). Age of first breeding 1 year (Dittberner and Dittberner 1984). BREEDING DISPERSION. Solitary and territorial, though tends to form neighbourhood groups in which territories overlap, especially when feeding young (Dittberner and Dittberner 1984). Few data on size of territory; one ♂ kept to area 300 × 300 m (Schifferli 1968). Distances between nests in neighbourhood group often 30–60 m, sometimes only 20 m (Dittberner and Dittberner 1984). On an island, Helgoland (West Germany), 3 nests in a triangle 17, 18, and 22 m apart, a 4th 30 and 32 m from 2 of these, overall c. 4 pairs per ha; on another island, 3 nests 10, 50, and 60 m apart, another 150–200 m away (Drost 1948). In England, average c. 1 pair per 5 km of occupied waterway (Sharrock 1976). On farm in Cheshire (England), over 7 years 2–4 pairs on c. 4 ha (Smith 1950). From review by Dittberner and Dittberner (1984) of different habitats in East Germany, West Germany, Netherlands, and USSR, density 0·03–33·0 pairs per 10 ha, average 2·92 (n = 37 densities), though average 1·52 when 2 highest densities excluded. In Netherlands, 2–3 pairs per km² on arable, 10–25 on undisturbed grassland, exceptionally 36 on certain polders (Teixeira 1979). In southern Finland, 1·7 pairs per km² in cereal fields, 16·4 pairs per km² in abandoned fields (Tiainen and Ylimaunu 1984). In Hemishofen-Ramsen (Switzerland), 10 pairs in 2 km² (Schifferli et al. 1982). In Asturias (Spain), c. 0·66 pairs per ha in 30 ha (Noval 1974). For nesting association with *M. citreola*, see that species (p. 437). Territory serves for courtship, nesting, and some feeding, but residents typically also feed far outside territory, e.g. birds commuted between 2 islands 2 km apart when feeding young (Drost 1948). Builds new nest (17-100 m from 1st) for replacement clutches and for 2nd clutches (30-50 m from 1st), incurring some changes in territorial boundaries (Dittberner and Dittberner 1984). Residents remain in territory and its vicinity until autumn departure (Schümperlin 1984). Fidelity of ♂ to previous territory often marked from year to year (Dittberner and Dittberner 1984, which see for examples). On Helgoland, 6 ♂♂ returned to breed twice, 2 bred for 3 years, 1 for 5 years; some young also returned to breed and in 5 years 31-80% of birds breeding had been ringed on the island in previous years (Drost 1948). ROOSTING. In breeding season, in territory or communally; outside breeding season, communally, often in large numbers. On arrival at breeding grounds, some ♂♂ (first arrivals) roost singly under tussock (etc.) in reclaimed territory. Others, perhaps most, roost communally, typically in *Phragmites* or willow *Salix* scrub, once in rhododendron with *M. alba* (Witherby et al. 1938a; Smith 1950). Roost usually in lower ⅓ of cover, sometimes quite close to water surface (Dittberner and Dittberner 1984, which see for account of habitat, extent, and associates in roosts, central Europe). Communal roost may be occupied throughout breeding season: e.g. roost in East Germany first occupied mid-April, mostly by ♂♂; in May-June by unpaired ♂♂ and some pairs; size of roost increased in late June as family groups joined, numbers maximum August-September. Adults used roost up to 51 days, average 19·5, juveniles for up to 30 days, average 13. Some birds markedly faithful to site, returning for up to 4 successive years. (Dittberner and Dittberner 1984.) Communal roosting in spring not uncommon in Britain, e.g. in Kent, April, 40-50 in reedbed (Manser and Owen 1949; Burton 1970, which see for mixture of *flavissima* and nominate *flava*). Initial preponderance of ♂♂ reported in Cheshire, April, where c. 15 ♂♂ in reedbed with hundreds of *M. alba* (Boyd 1949). Following account of roosting behaviour based on Dittberner and Dittberner (1984). Birds start arriving at roost 1-2 hrs before sunset in fine weather, peak arrival just before sunset. Fly in at considerable height in small groups, calling (see 2a in Voice) constantly; first arrivals may circle silently, later birds drop steeply down. Typically assemble first near roost at open site, e.g. fields, meadows, or shore (Smith 1950), where they preen, feed, sing, etc. Then enter roost directly *en masse* before dark; late arrivals forego prior assembly and go straight to roost. At peak migration, some flocks enter roost 20-40 min after sunset. In breeding pairs, ♀ roosts near nest during building, ♂ elsewhere in territory or perhaps in communal roost. When young hatch, ♀ stays on nest at night and, once young no longer need brooding, she usually roosts near nest in long grass. (Dittberner and Dittberner 1984.) Once young fledge, family may join communal roost, or parents roost in territory, e.g. crouched in scattered ones or twos in rough tussocky pasture (Smith 1950). 2 pairs thus occupied territory until autumn departure, and once, flock of 60-80 roosted overnight in their territory (Schümperlin 1984). Whereas communal roosts in breeding grounds may contain 100 or more birds, those in winter quarters may contain tens of thousands: e.g. at Kano (Nigeria) up to 57 000 (Ashford 1970); at Lake Chad, evening roosting flights up to 50 000 (Ash et al. 1967; Fry et al. 1972). In West Africa, numbers diminish markedly in March-April (Fry et al. 1972; Wood 1979). Sites often traditional: include *Typha*, tall grasses (including sugar-cane) in dry and marshy situations, and trees, especially fig *Ficus*

dekdekena (Smith 1955a; Sharland and Harris 1961; Smith and Ebbutt 1965; Fry et al. 1972; Reynolds 1974; Wood 1976, 1978). At Lake Chad site, spring, bulk roosted in *Phragmites*, a small proportion in clumps of woody vegetation along shore (Fry et al. 1972). From March onwards at Asmara (Eritrea), *M. flava* arrived *en masse* with *M. alba* but the species roosted in separate trees (Smith 1955a). At Vom, catchment area of roost c. 270 km² in November, 825 km² in February (Wood 1976). At Ibadan (Nigeria), movement into roost lasted 28 min, birds arriving in flocks of 2-60, average 20 (Broadbent 1969). Roosting behaviour very similar to that described (above) for communal roosts in breeding grounds; for detailed accounts, see Smith and Ebbutt (1965), and Wood (1976). When nearly all birds assembled, calling from roost reaches a peak, audible at considerable distance; sustained for c. 10 min. In morning, birds depart in waves (Wood 1976, which see for times of arrival and departure). Territory-holders also join communal roosts (Vrydagh 1952; Verheyen 1956a); at Vom, typically arrived earlier and left later than flock birds (Wood 1976). Individual birds often change roost-sites during winter (Smith and Ebbutt 1965; Reynolds 1974; Wood 1976). At Lake Chad, birds seek shade or frequent wetter places during hottest part of day (Fry et al. 1972). Often drink around midday, and in evening before flying to roost (Wood 1976). For illustrations of comfort behaviour, and for bathing, see Dittberner and Dittberner (1984).

2. In breeding season, may crouch if surprised by Hobby *Falco subbuteo* (Dittberner and Dittberner 1984). At Lake Chad, readily takes refuge in *Phragmites* and other tall vegetation if alarmed (Fry et al. 1972). For raptors causing alarm at winter roost, see Wood (1976). During migration, tends to flee when approached (Dittberner and Dittberner 1984, which see for raptors eliciting rapid escape from roost). For other responses when threatened, see Parental Anti-predator Strategies (below). FLOCK BEHAVIOUR. In August-September, imminent emigration evident in restlessness of roosting flocks: groups fly around in circles giving Contact-calls, then settle again. Restlessness increases in September, then main parties depart by night (Smith 1950). Calling and conspicuous tail movements said to help maintain cohesion of feeding flocks (Wood 1976). Different races mingle freely in winter (Moreau 1972; Wood 1976). On arrival at breeding grounds, Oxford (England), birds which defended dung-pat feeding-sites by day, and then maintained individual-distance of 500-1200 m, switched to feeding in flock at flooded pool in early morning and evening, and then fed amicably, often less than 1 m apart (Davies 1977). SONG-DISPLAY. ♂ sings (see 1 in Voice) from perch and often in display-flight. ♂ *flavissima* typically sings April-July (Witherby et al. 1938a). Begins Song-display from elevated perch, with body somewhat hunched, breast puffed out, neck drawn in, head tilted back, wings drooped, and tail lowered. With first unit of song, head sometimes jerked back. Suddenly, ♂ launches himself into the air and ascends steeply for a few metres; then, with tail fanned and slightly raised, breast puffed out, legs dangling and toes clenched, descends with rapidly fluttering wings (Fig A), giving trilling variant of song (see 1b in Voice). Just before reaching ground, bird swoops upwards again but does not call, followed by another fluttering descent with song, then another ascent, and so on; Song-flight thus a series of long wavering undulations. (Smith 1950.) Display of nominate *flava* differs little, if at all, from *flavissima* (Witherby et al. 1938a). Dittberner and Dittberner (1984) considered Song-flight of *flava*, involving 3 ascents to c. 4-5 m, to be exceptionally long, and shorter undulating flights more typical; ♂ *flava* may also approach ♀ in slow hesitant flight (Dittberner

A

and Dittberner 1984). Song-flight of *pygmaea*, as described by Simmons (1952*b*), not obviously different. Song-display wanes when raising 1st brood, but resurgence occurs before 2nd clutch (Bürkli 1977; Dittberner and Dittberner 1984). For variant when young out of nest, see Parental Anti-predator Strategies (below). Song-display tends to be directed at rival ♂♂, and also attracts ♀♀. If ♂ fails to attract ♀ with Song-flight, switches to perched Courtship-song (see 1b in Voice); unpaired ♂♂ may sing thus for lengthy periods (Dittberner and Dittberner 1984). ANTAGONISTIC BEHAVIOUR. Territory-owners, especially ♂♂, highly aggressive to trespassing conspecifics and numerous other passerines. Onset of aggression coincides with arrival of ♀♀ on spring migration. ♂♂ then contest territorial claims vocally (see Song-display, above), by threat and by fighting. Typical sequence, described by Smith (1950), as follows. Rival ♂♂ begin by facing each other in Upright-threat posture (Fig B, left): sitting back on its tail, bird thrusts breast forward and ruffles plumage, tilts head back until nape almost tucked into scapular feathers, and droops wings. Rivals held Upright-threat posture for *c.* 10 s, then resident crouched forward, breast still puffed out, gaped (Fig B, right), and gave

B

a hissing-call (see 5 in Voice). Rival responded by strutting about in front of crouched bird and swaying from side to side in Upright-threat posture, this provoking crouched bird to spring forward and attack; in fierce fight that followed, combatants struck with bill and claws, rolled over and over, then fluttered upwards face to face and tumbled down interlocked. Fighting alternated with rapid twisting and turning pursuit-flights for *c.* 2 hrs around territory, with resident usually in pursuit. Trespasser pressed his claim much less strongly the next day, and was absent the day after that. Aggression especially evident after birds paired, since unpaired ♂♂ from neighbouring territories continually invade those of paired birds. When pairs meet, usually ♂♂ fight, and (especially early on) ♀♀ remain passive in a tense crouched posture, occasionally Upright-threat posture. Sometimes, however, ♀♀ confront each other in boundary disputes (Smith 1950), notably during nest-building; usually, in disputes between pairs, ♂ fights ♂

and ♀ fights ♀ (Dittberner and Dittberner 1984), but sometimes a mêlée develops with birds skirmishing indiscriminately (Smith 1950). For expelling attacks on other (mostly passerine) trespassers, see (e.g.) Harvey (1923), Thönen (1948), and Schwarz (1949). Threat, chasing, and attacking also reported among territorial ♂♂ in winter quarters (Vrydagh 1952; Brosset 1971*b*), but not by 1-year-olds; territorial ♂♂ give monosyllabic Warning-call (see 5 in Voice) to warn off other birds approaching their territory; regularly patrol boundaries, adopting Forward-threat posture (not seen in breeding season) towards neighbours: bird crouches, leans forward, gapes (Fig C), and calls inter-

C

mittently; rushes at intruders and occasionally fights. Also highly aggressive towards Red-throated Pipit *Anthus cervinus* (Wood 1976). In Zaïre, territory-owners attacked conspecific trespassing flocks (Verheyen 1956*a*). In late summer, juvenile ♂♂ show aggression in flocks, sing, and occupy 'pseudo-territories' (Dittberner and Dittberner 1984). HETEROSEXUAL BEHAVIOUR. (1) General. ♂♂ arrive on breeding grounds 1–2 weeks before ♀♀ (Smith 1950). Some birds arrive already paired (Dittberner and Dittberner 1984: see Bonds, above, and also subsection 2, below). (2) Pair-bonding behaviour. Song-display (see above) evidently important in attracting mate. Following account of subsequent ground-display after Smith (1950). ♂ runs towards ♀ and when *c.* ½ m away, adopts Advertising-posture (Fig D,

D

left): ruffles breast, nape, and mantle plumage, lowers and half-opens wings, and drags tail; giving Sreeze-call (see 4b in Voice), runs thus around ♀. At greater intensity, ♂ leaps into the air and hovers over ♀, his breast puffed out, tail widely spread and lowered. When ♀ *flavissima* first arrived on territory, ♂ and ♀ exchanged disyllabic Contact-calls (see 2a in Voice); ♂ then left his song-perch, flew to ♀, and performed ground-display, as described above. ♀ responded with quiet calls, but was otherwise passive. ♂ then returned to perch, to sing and call. Display repeated after *c.* 1 hr, and occurred regularly thereafter until laying. (Smith 1950.) Initially, ♂ often pursues ♀ (Schwarz 1949); e.g. ♂ lands from Song-flight and then displays, ♀ flies up, and ♂ follows (Dittberner and Dittberner 1984). Pair-formation usually complete *c.* 1 week after ♀ arrives in ♂'s territory, egg-laying starting *c.* 10 days after that (Smith 1950). Incipient pair-bonding behaviour of juveniles occurs at pre-roosting assemblies, July–August; 'engagement' pairs thus formed, and partners often roost together, e.g. one pair perched and billed together at roost

(Dittberner and Dittberner 1984). Little hard information on pair-formation of races other than *flavissima* and nominate *flava*. Ground-display of *pygmaea* thought to be more striking (postures more extreme) than *flavissima* (Simmons 1952*b*), but sample too small to prove difference. In Camargue (France), spring, ♀♀ of nominate *flava* are attracted to song of ♂ *iberiae* and *cinereocapilla* (Schwarz 1956). (3) Courtship-feeding. 2 reports of ♂ feeding incubating ♀ (Meier 1954; Ptushenko and Inozemtsev 1968), but in lengthy study by Smith (1950), never seen at any time, and evidently exceptional. (4) Mating. Accounts by Smith (1950) and Dittberner and Dittberner (1984) appear to differ somewhat. According to Smith (1950), usually occurs after Advertising-display of ♂, as described above; if ♀ then solicits by crouching with raised tail, ♂ mounts. Sometimes, ♀ solicits by crouching flat, ruffling plumage, half-opening and drooping wings, and raising tail vertically (Soliciting-posture: Fig E); may hold posture or begin rotating

E

on the spot several times, whereupon receptive ♂ mounts. According to Dittberner and Dittberner (1984), copulation usually occurs with few preliminaries, ♂ flying directly on to back of ♀ perched on ground or branch; once, perched ♀ solicited by shivering wings, more intensely as ♂ (perched nearby) flew to join her, then mounted; after copulating, ♂ flew off and ♀ shook herself and preened, later flying off calling followed by ♂. Little information on calls associated with copulation, but presumed Subsong (see 1d in Voice) occurred once, apparently also Bill-snapping after copulation (P J Sellar). Copulation occurs at various times of day 3–4 days before nest-building, also during nest-building. Pairs copulate regularly prior to 2nd clutch, when 1st brood leaving nest. ♂♂ intruding on resident's territory not uncommonly copulate with resident's mate (Dittberner and Dittberner 1984, which see for several cases of promiscuous copulation). (5) Behaviour at nest. ♀ chooses nest-site: runs around inspecting various sites (Dittberner and Dittberner 1984). On finding prospective site, she behaves as if soliciting copulation, and thus makes shallow scrape (Smith 1950). One ♀ made several scrapes before choosing one (Dittberner and Dittberner 1984). ♂ plays no part, other than running around with puffed out breast. ♀ builds nest and performs virtually all incubation, while ♂ mostly guards. No record of incubation by ♂ nominate *flava* (Dittberner and Dittberner 1984) but ♂♂ of *flavissima* and *iberiae* sometimes incubate when ♀ long absent (Smith 1950; Glutz von Blotzheim 1955). In *flavissima*, nest-relief very discrete: relieving bird lands *c.* 50 m from nest and begins giving Contact-calls. Sitting bird leaves nest, running considerable distance before flying up and reciprocating Contact-calls. Relieving bird makes slow, leisurely approach to nest, calling intermittently as it goes (Smith 1950). RELATIONS WITHIN FAMILY GROUP. For detailed ontogeny of physical development, voice, alarm responses (etc.), see Dittberner and Dittberner (1984). First food delivered to

one brood *c.* 1½ hrs after apparent time of hatching (Dittberner and Dittberner 1984), but Smith (1950) believed parents start feeding young the day after hatching. Young begin giving food-calls (audibly at least) at 5–6 days; eyes open at 4–6 days (Smith 1950; Dittberner and Dittberner 1984). At 1 week, peck at insects on rim of nest (Dittberner and Dittberner 1984). Up to 4 days, faecal sacs sometimes swallowed by parents but more often carried away; parents stimulate young reluctant to defecate by prodding with bill around cloaca, or tugging at down on spinal tract, occasionally even seizing chick's head and shaking it. From 11–12 days, several of brood may walk short distance from nest, exercise wings, and return; near fledging, may leave nest to meet food-bearing parents (Smith 1950). Finally quit nest early in morning (Smith 1950), lured by Enticement-call (Dittberner and Dittberner 1984: see 2b in Voice). 4 young dispersed 1 m from nest by the day after leaving; hopped and fluttered a few metres further in the next few days; first flew 5 days after leaving nest (Bürkli 1977). After leaving nest, chick nearest it fed by ♀, others *c.* 70 m away fed by both parents (Ringleben 1935). When 1st brood being tended outside nest, ♂ tends to spend more time in territorial advertisement, performing Song-flights preparatory to 2nd clutch (Dittberner and Dittberner 1984). One pair fed young at least 11 days out of nest, and family bonds maintained another 6 days after that (Meier 1954), though family may stay together considerably longer (Smith 1950). ANTI-PREDATOR RESPONSES OF YOUNG. From *c.* 7 days, sometimes in response to a Warning-call (see 5 in Voice), young cower in nest when disturbed, bills resting on rim (Smith 1950), but from *c.* 8 days may be enticed from nest by parents (Dittberner and Dittberner 1984). If alarmed near time when ready to leave nest, scramble out of nest, run, and hide; immediate stimulus is often parental Warning-call (Dent 1907; Dittberner and Dittberner 1984: see 5 in Voice). When handled or trapped, young give a distress-call (see Voice) which instantly stimulates siblings to run and hide, and attracts parents and other conspecifics in vicinity (Dittberner and Dittberner 1984). PARENTAL ANTI-PREDATOR STRATEGIES. (1) Passive measures. For discrete nest-relief, see above. On approach of potential predator, sentinel ♂ alerts sitting mate with Alarm-calls (Meier 1954: see 5 in Voice). ♀ usually leaves quietly when disturbed, less often sits tight, exceptionally allowing herself to be touched; flushed ♀ slow to return to nest, and then often performs 'dancing flight' (presumably hovering) over nest (Dittberner and Dittberner 1984). (2) Active measures: against birds. Following based on Dittberner and Dittberner (1984). If threatened by raptors, ♀ leaves nest and joins ♂ in chasing them, sometimes aided by neighbours. Dive-bombs Cuckoo *Cuculus canorus*, but not during chick-rearing; once, *C. canorus* chased by up to 8 ♂♂ and 3 ♀♀. May hover over and flutter around Short-eared Owl *Asio flammeus* on ground. During aerial demonstration, may give Threat-song (see 1c in Voice). (3) Active measures: against man. Response varies, partly with stage of breeding. Usually, after hatching, adults fly towards intruder, then circle his head, calling excitedly (se 1c and 5 in Voice) as nest approached; when intruder at nest, birds run about nearby, giving constant Alarm-calls; no such demonstration during incubation (Smith 1950). ♂ may also perform Song-flight; when young out of nest, Song-flight apparently includes steeper ascent (resembling Tree Pipit *Anthus trivialis*) than when directed at conspecifics (Thönen 1948). Singing ♂ may be joined by ♀ and neighbours. Rarely, ♀ performs distraction-lure display of disablement type (Dittberner and Dittberner 1984). When nestlings being ringed, ♀ once came up to within a few metres and shuffled along with drooped wings and fanned tail; ♂ nearby did not display, but both uttered Distraction-calls (Buxton 1947: see 7 in Voice). (4) Active measures: against other animals. No information.

(Figs by C Rose: A from drawing in Schwarz 1949; B, D, and E from illustrations in Smith 1950; C from drawing in Wood 1976.)   EKD

Voice. Freely used in breeding season, less so at other times. More varied than often indicated, and more information is needed on differences between races (Dittberner and Dittberner 1984, which see for review). Generally lower pitched further south in range (Chappuis 1969). Unless otherwise indicated, statements by Dittberner and Dittberner (1984), below, refer to nominate *flava*, those of Smith (1950) to *flavissima*. When handled, or when jockeying for perch-sites at roost, Bill-snapping not uncommon, evidently signalling anger or threat (Dittberner and Dittberner 1984); in recording by R Margoschis, England, sounds following copulation appear to include Bill-snapping (P J Sellar). In following scheme, based largely on Dittberner and Dittberner (1984), several calls closely related, and divisions therefore to some extent arbitrary.

CALLS OF ADULTS. (1) Song of ♂. Rarely heard outside breeding season, and not reported from winter quarters; sometimes given when perched at autumn roost, and then quieter and shorter than usual (Witherby *et al.* 1938*a*; Dittberner and Dittberner 1984). 3 types of song are distinguished by Dittberner and Dittberner (1984). (a) Territorial-song. Not very loud, but often prolonged, rapid sequence of twittering sounds, comprising loosely connected syllables of Contact-call-type (call 2), given from ground or elevated perch. Marked individual variation, e.g. 'srie srühsrisiep srühsrieh srühsisiep siep...' (Dittberner and Dittberner 1984). Also described as a rhythmically repeated sequence of rather inconspicuous short phrases of 1–5 'zier' units in which syllables may be run together (Bergmann and Helb 1982, which see for sonagrams). One *flavissima* gave a repeated, rather musical, high-pitched 'tsee-wee-sirr tsee-wee-sirr' with 'tsweep' (Contact-call) interspersed (Smith 1950). Phrases similar to these evident in recording: e.g. phrase shown in Fig I which has forced squeaky character and, in final unit, audible descent (J Hall-Craggs). (b) Courtship-song. Given from elevated perch, often in Song-flight. From descriptions in Dittberner and Dittberner (1984), most

obvious differences from 1a are in simpler, more rhythmic repetition of Contact-call-type units and a distinctive final unit, e.g. 'srie srie sriehä' or 'sriep sriep sriep zier', or a twittering at end. In Song-flight, only twittering variant given (during fluttering descent); in *flavissima*, this rendered a rapidly repeated, sweet, musical trilling 'sree-sree-sree', not unlike Robin *Erithacus rubecula* (Smith 1950). In *pygmaea*, an unmusical hard rattling trill, 'trrizzz' (Simmons 1952*b*). If ♂ nominate *flava* fails to attract ♀ thus, song reverts to series of up to 60 or more calls, given usually in groups of 3(–4), e.g. 'sriep sriep sriep' (Dittberner and Dittberner 1984); musical repetition of 'szree-sree' (Smith 1950). (c) Threat-song. Louder and faster than 1a, with 's' and 'z' sounds dominant. Given when chasing conspecific rivals and would-be predators in flight (Dittberner and Dittberner 1984). (d) In recording by R Margoschis, England, liquid warbling (presumed Subsong) given during copulation (P J Sellar). (2) Contact-alarm calls. (a) Most commonly heard call differs slightly between races. In nominate *flava*, a sharp monosyllabic 'psie' (Bergmann and Helb 1982, which see for sonagrams), also rendered a monosyllabic 'psüip' with emphasis at end, or a disyllabic 'psiib', sometimes descending in pitch and similar to but sharper than Reed Bunting *Emberiza schoeniclus* (Dittberner and Dittberner 1984). Recording of flock of nominate *flava* in Egypt, April, suggests 'psssrt' (P J Sellar). In *flavissima*, a rather prolonged, shrill musical 'twee' (J Hall-Craggs: Fig II), often disyllabic (Witherby *et al.* 1938*a*, which see for variants), e.g. 'tsi-weep' (Fig IIIb); recordings show that given bird uses both monosyllabic and disyllabic variants during a bout of calling (P J Sellar). According to Czikeli and Knötzsch (1979, which see for sonagrams), *iberiae*, *cinereocapilla*, and *feldegg* all harsher than nominate *flava* ('trie' instead of 'psijip'), while *thunbergi* gives a slight variant of *flava*. Wallschläger (1983, which see for sonagrams) confirms that *thunbergi* and *flava* very close in timbre and structure. Call of *thunbergi* rendered 'rssli rsliu' (Schüz 1956); for regional differences in *thunbergi* within Finland, see Sammalisto (1956). Other comparisons between races as follows. Compared with nominate *flava* and *flavissima*, *cinereocapilla* more grating (Witherby *et al.* 1938*a*), and *feldegg* harsher, e.g. 'shrreep' (Gibb 1946) or 'psrii' in

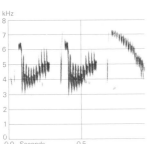

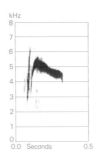

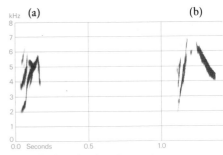

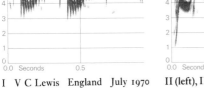

I V C Lewis England July 1970   II (left), III (right) V C Lewis England June 1971   IV J Gittins Sweden June 1961

which 'r' sound marked compared with other races (see Dittberner and Dittberner 1984); for other renderings of *feldegg*, see Niethammer (1937) and Bergmann and Helb (1982); for sonagrams, see Wallschläger (1983). *M. f. iberiae* gives shrill 'sreee' or 'shreee', quite different from *flavissima* (Smith 1962-4). Call used, perched or in flight, as communication between pair and between flock members: e.g. ♂ uses call to summon ♀ off nest, or sitting ♀ calls thus to bring ♂ to nest (Dittberner and Dittberner 1984, which see for harsher variants). (b) Enticement-call. A quiet 'psiehp' to lure young from nest after *c.* 8 days old (Dittberner and Dittberner 1984); see also call 5. (c) Other quiet calls given between pair-members and audible only at close quarters, e.g. 'djup djiup djup' by ♂ *flava* (Dittberner and Dittberner 1984); short 'chchch' of *iberiae* at nest-relief (not known by which bird) possibly a greeting-call (Glutz von Blotzheim 1955). For other weak calls, perhaps related, see Bergmann and Helb (1982). (3) Feeding-call. Series of low sweet sounds, 'syew syew syew', often given to nestlings (Smith 1950); perhaps not different from call 2b. (4) Excitement-calls. (a) ♂ nominate *flava* flying toward ♀ in courtship-display calls 'dürrr quürrr djürrr' (Dittberner and Dittberner 1984). (b) Musical 'sreeze' given by ♂ *flavissima* circling ♀ during pair-formation (Smith 1950). (5) Warning- and Alarm-calls. Effectively high-intensity Contact-calls. A loud 'psiehp' given by ♂ nominate *flava* to warn mate during nest-site selection or building—also (given sharply) to draw near-fledged young away from danger (Dittberner and Dittberner 1984). Sharp penetrating 'psieee' by *thunbergi* when disturbed feeding young (Fig IVb) is probably this call. Attacks by either sex on rivals and predators in breeding territory accompanied by loud sharp 'psrieh' (♂) and 'psiih' (♀) (Dittberner and Dittberner 1984); 'psie' (Bergmann and Helb 1982). Low guttural hiss given in Forward-threat posture (Smith 1950) perhaps this call. Also given in territorial disputes in winter quarters (B Wood). When young in nest, similar 'prüip psiehp' ('prüip' short and emphatic) given by parents flying high over nest; on hearing call, small young crouch in nest, older young leave it (Dittberner and Dittberner 1984). In *flavissima*, musical 'sree sripp sripp', in which 'sripp' units staccato and lower-pitched than 'sree', and given when human intruder at nest (Smith 1950), evidently same or closely related. In *thunbergi*, short 'quip' calls (Fig IVa) not unlike House Sparrow *Passer domesticus*, and interspersed with more frequent 'psieee' calls (J Hall-Craggs) are perhaps related. (6) Distress-call. A short sequence of drawn-out 'psiehp' sounds (Dittberner and Dittberner 1984); clearly a high-intensity Warning-call. (7) Distraction-call. In intense excitement, 'schrüü schrüüh' or 'chrä chrä chrä' (by ♂ when young being ringed, and especially when young gave a distress-call; see below). Also given by either sex during aerial demonstration ('distraction-flight') against predators (Dittberner and Dittberner 1984).

Harsh grating call given by ♀ *flavissima* performing distraction-display, and by ♂ nearby, when human intruder at nest (Buxton 1947), evidently this call; also 'krrr krrr krrr' by ♀ *iberiae* leaving nest (Glutz von Blotzheim 1955).

CALLS OF YOUNG. Food-call initially a rather weak wheezy rendering of adult Contact-call (Smith 1950). From 5-11 days, rather thin high-pitched 'sissisi'; also rendered 'psieh', weaker and thinner than adult. From 1 week, young beg with increasing intensity when parent still some distance from nest, reaching maximum on arrival. From 12 days up to fledging, food-call a loud 'srieh srieh srieh', very like adult Contact-call, and accompanied by wing-shivering (Dittberner and Dittberner 1984). For sonagraphic comparison with adult (nominate *flava*), see Wallschläger (1983). Up to 12 days, young often give quiet 'psiehp' calls, like Enticement-call of adult, to locate one another outside nest (Dittberner and Dittberner 1984). Distress-call (e.g. when trapped or handled) a loud 'tschirrrp', causing well-grown siblings to leave nest in panic and hide in vegetation; juveniles give quiet 'quiehp quähp' or shrill 'psriip psriip', e.g. when trapped at roost, attracting others to the spot. From independence, juvenile ♂♂ give a quiet twittering song, especially at roost-sites (Dittberner and Dittberner 1984).

EKD

Breeding. SEASON. Northern Scandinavia: most eggs laid June. Southern Scandinavia: laying from last week of May. Britain and Ireland: first eggs mid- to late April, main laying period May, last eggs found early August. Southern and south-east Europe: from end of April to early June. North Africa: from end of April to end of May, with most in first half of May (Heim de Balsac and Mayaud 1962; Makatsch 1976; Mason and Lyczynski 1980). SITE. On ground, in side of tuft of vegetation; 52% of 329 nests in Britain and Ireland were close to water (Mason and Lyczynski 1980). Nest: cup of grass leaves and stems placed in shallow scrape, lined with hair, wool, or fur; external diameter up to 10·5 cm, internal diameter 5·5 cm, depth of cup 4·5 cm (Smith 1950; Makatsch 1976). Building: by ♀ only, though usually accompanied by ♂. Scrape formed by turning and pressing with breast; material brought from up to 1 km away; takes 4 days to 3 weeks (Smith 1950). EGGS. See Plate 81. Sub-elliptical, smooth and glossy; grey-white to buff, densely spotted various shades of brown, often with dark hair-streak. Nominate *flava*: $18·5 \times 14·0$ mm ($17·0-120 \times 12·8-15·3$), $n = 200$; calculated weight 1·8 g. *M. f. flavissima*: $19·2 \times 14·2$ mm ($17·0-21·5 \times 12·7-15·5$), $n = 120$; calculated weight 1·9 g. Measurements of *iberiae*, *cinereocapilla*, *thunbergi*, and *feldegg* not significantly different from nominate *flava* (Schönwetter 1979). Clutch: 4-6 (3-8). Of 135 clutches, Britain and Ireland: 3 eggs, 8%; 4, 9%; 5, 41%; 6, 41%; 7, 1%; mean 5·2. No variation in Britain and Ireland with latitude or altitude, but limited data suggest cline of increasing clutch size across range from south to north, e.g. mean 5·7 ($n=87$), Finland (Smith 1950; Haartman 1969*a*; Makatsch 1976; Mason and Lyczynski 1980). One brood in nominate *flava*, rarely 2; 2 more regular in *flavissima* (Smith 1950; Makatsch 1976; Mason and Lyczynski 1980). Eggs laid daily, mostly early in morning (Smith 1950). INCUBATION. 12·4 days (11–13), $n=33$ (Smith 1950; Mason and Lyczynski 1980). By both sexes, though mostly ♀ takes larger share. At 4 nests in 3 different years, 10.00–18.00 hrs, ♀ sat for 67% of time, during 1st week of incubation, ♂ for 17%, and eggs uncovered for 17%; during 2nd week, ♀ 70%, ♂ 20%, uncovered 10%; incubation spells lasted 1–2 hrs (Smith 1950). Begins with last egg; hatching synchronous. YOUNG. Altricial and nidicolous. Cared for and fed by both parents. FLEDGING TO MATURITY. Fledge at *c.* 16 days, but leave nest at 11·1 days (10–13), $n=14$, with disturbance an important factor in determining nest exodus (Smith 1950). Stay with parents for several weeks after fledging, possibly departing together on migration (Smith 1950). Age of first breeding 1 year. BREEDING SUCCESS. Of 99 eggs laid in 19 clutches, England, 75·7% hatched and 88% of these fledged, giving 66·6% overall success (Smith 1950). Of 808 eggs in 157 clutches, Britain and Ireland, 67·6% hatched and 75·8% of these fledged, giving 51·3% overall success; mean 0·88 young reared from 9 clutches of 3 eggs, 2·5 from 18 of 4, 3·04 from 68 of 5, and 2·98 from 55 of 6; of 65 cases, 30·8% of losses due to desertion, 29·2% predation, 15·4% agricultural operation, 13·8% weather, 7·7% trampling by cattle, 1·5% infertility, 1·5% miscellaneous (Mason and Lyczynski 1980).

Plumages (nominate *flava*). ADULT MALE BREEDING. Forehead, crown, hindneck, and upper sides of neck medium bluish-grey, often slightly paler on forehead, sometimes with some partly olive-green feathers on crown or hindneck. Mantle, scapulars, and back yellowish olive-green, rump and upper tail-coverts brighter yellow-green, but longest coverts with duller olive centres. Supercilium white, long and distinct though rather narrow extending from just behind nostril back to above rear of ear-coverts; narrow eye-ring white, broadly broken by dark grey in front and at rear. Remainder of sides of head rather variable: in 70 breeding birds from Netherlands, lore, upper cheek, and ear-coverts either uniform dark grey, usually except for some faint white spots below eye extending backwards to lower short ear-coverts (55%), or medium bluish-grey with more pronounced white stripe from just below eye backwards (45%); in birds with darker cheeks, chin and throat either uniform yellow (and these birds often also show most extensive olive-green tinge on crown and hindneck and sometimes on front part or rear of supercilium) or (about as often) chin and narrow partly interrupted stripe along lower cheek white, separating yellow of throat from grey of cheek; in birds with paler cheeks, chin and narrow stripe along cheeks white (stripe sometimes wider than in dark-cheeked birds), remainder of throat yellow; chin and throat all-yellow in only 2 birds, and yellow then also invading pale stripe from below eye backwards as well as part of supercilium; in 1 bird, throat fully white.

Apart from these variants an exceptional grey morph occurs (2 out of 70), in which mantle, scapulars, and back olive-grey (only rump tinged green) and underparts white with pale yellow tinge on chest and belly (no absolute schizochroism, as yellow pigment not fully absent; see also Goethe 1977). Remainder of underparts down from throat and lower sides of neck bright deep yellow, sometimes with orange tinge; sides of breast olive-green, flanks slightly tinged olive-green, chest sometimes with some diffuse olive-green spots. Central 4 pairs of tail-feathers (t1–t4) black with narrow yellow-green edge along outer web; t5–t6 largely white, both with broad greyish-black wedge on base and middle portion of inner web; base and variable part of shaft of t5 also dark. Flight-feathers dark grey or greyish-black; narrow edges along tips of secondaries and inner primaries greyish-white, narrow edges along outer webs of primaries pale yellow or yellow-white, broader and less sharply defined fringes along outer webs of secondaries pale yellow or pale greenish-yellow. Tertials dark grey-brown or greyish-black, shorter ones with very broad pale green-yellow fringe along outer web, longest with narrower and sharper yellow-white fringe. Greater upper primary coverts greyish-black, tip narrowly edged grey, outer web narrowly pale green-yellow. Lesser upper wing-coverts yellowish olive-green, like upperparts but often with some black of bases visible; median coverts dull black or greyish-black with pale green-yellow or pale yellow tips 2–4 mm long; greater coverts greyish-black or dark brown-grey with less sharply defined pale green-yellow or yellow-white tips and outer fringes (widest and more greenish on tertial coverts). Colour of coverts depends on extent of pre-breeding moult: in some, most or all median and some inner greater coverts black with pale yellow-green tips, old outer greater more brown-grey with paler yellow-white tips, but others have many coverts old (though tips never as white and abraded as in retained juvenile coverts of some 1st adults). Under wing-coverts and axillaries white, marginal coverts tinged yellow; longer and marginal coverts have variable amount of grey on feather-bases. *In worn plumage* (from about late May), forehead and crown slightly paler, mantle and scapulars duller green with variable amount of brown-grey of feather-bases visible; white of supercilium sometimes partly worn off, especially above lore; yellow of underparts paler; pale tips and fringes of flight-feathers, tertials, and wing-coverts largely worn off, upperwing wholly dark brown-grey except for traces of whitish tips on median and greater coverts. ADULT FEMALE BREEDING. Forehead, crown, and hindneck either medium-grey (slightly duller, less uniform, and less bluish than adult ♂ breeding), or brownish or greyish olive-green. Mantle, scapulars, and back olive-green, usually slightly greyer or browner than adult ♂ breeding, less yellowish; rump and upper tail-coverts brighter green, not as yellow-green as ♂. Sides of head as adult ♂ breeding, but supercilium often less clear-cut, sometimes tinged pale buff; dark stripe on lore less distinct; ear-coverts often more olive-brown, less grey. Underparts yellow, but paler than in adult ♂ breeding and often much white of feather-bases visible on chin and throat; sometimes a few dark olive spots on lower sides of neck or chest. Tail and wing as adult ♂, but often a greater number of upper wing-coverts old, brown-grey with frayed pale yellow fringes and tips. *In worn plumage*, underparts paler than in adult ♂ breeding, often largely white except for yellow wash on chest and belly. Sometimes hard to separate from adult ♂ breeding, but generally less contrast between grey hindneck and green upperparts; mantle and scapulars duller green; less contrasting pattern on sides of head; and paler yellow underparts. ADULT MALE NON-BREEDING. Rather like adult ♂ breeding and (in particular) adult ♀ breeding; forehead and crown medium grey (not ashy) with variable amount of olive-green tinge; mantle, scapulars, and back rather dull olive-green, rump and upper tail-coverts slightly brighter green (less yellowish than adult ♂ breeding); side of head as adult ♂ breeding, but supercilium less clear-cut and sometimes partly suffused pale buff or yellow; upper cheek and ear-coverts brown-grey, less uniform; underparts usually paler yellow than adult ♂ breeding (occasionally, almost as deep yellow), uniform except for whitish chin and whitish border along brown-grey cheek, often with golden-buff tinge on chest; sometimes a few dark olive spots on lower sides of neck and occasionally on chest. Tail and wing as adult breeding, but all upper wing-coverts equally fresh; median and greater coverts and tertials greyish-black with broad pale green-grey tip and with slightly more yellowish fringe along outer web. ADULT FEMALE NON-BREEDING. Upperparts dull brownish olive-grey, except for slightly brighter olive-grey rump and upper tail-coverts. Supercilium pale buff or buff-white, not sharply divided from brown-grey forehead and from poorly defined brown-grey stripe on lore. Upper cheek and ear-coverts brown-grey or brownish olive-grey, mottled pale buff just below eye and on lower ear-coverts. Chin, throat, and lower sides of neck pale buff or pale yellow-buff, sometimes with some dusky grey or olive spots on sides of neck or sides of chest; remainder of underparts pale yellow with some white of feather-bases showing through, chest and sometimes upper flanks tinged pale buff. Tail and wing as adult ♂ non-breeding, but tips and fringes of all feathers often less yellow-green, more pale brownish-grey. Differs from adult ♂ non-breeding in less contrast between crown and mantle, buffier supercilium, more buff-and-brown upper cheek and ear-coverts, and pale buff underparts, appearing pale yellow on belly to under tail-coverts only, not largely yellow. NESTLING. Down fairly long and plentiful, on upperparts only; yellow, greyish-yellow, buff-white, buff, or yellow-brown, dependent on age (Witherby *et al.* 1938*a*; Dittberner and Dittberner 1984). For development, see Dittberner and Dittberner (1984). JUVENILE. Upperparts variable, pale buff-brown with darker feather-centres—olive-brown, dark grey-brown, or dull olive-grey; side of crown with black stripe. Lore mottled grey and buff, often with a dark spot in front of eye. Supercilium from above eye backwards and narrow eye-ring pale buff or off-white. Upper cheek and ear-coverts brown or dull grey with fine buff or off-white specks and streaks. Broad stripe running backwards from gape pale buff or off-white, bordered by distinct black malar stripe below; black of malar stripe extends up into short black band at rear of cheek and down into boldly spotted black band across upper chest. Underparts pale yellow-buff, pale yellow, buff-white, or dirty white. Tail, flight-feathers, tertials, and greater upper primary coverts virtually as adult. Lesser upper wing-coverts like upperparts; median and greater coverts dull black with pale yellow-buff tips which soon bleach to white (coverts blacker than in adult, pale tips more sharply demarcated; tips buffier or whiter, less yellowish). Some birds rather similar to juvenile White Wagtail *M. alba*, but fringes of tail-feathers and tertials yellow (not whitish), malar stripe distinct, and band across upper chest sharp, narrow, and black (not a diffuse grey band or patch). FIRST ADULT MALE NON-BREEDING. Sometimes similar to adult ♂ non-breeding, differing only in retained juvenile outer greater upper wing-coverts and some or all tail-feathers and tertials (white tips of coverts contrasting in colour and wear with neighbouring fresh yellow-tipped 1st adult feathers), but others much less bright grey and green above and less deep yellow below, closely similar to adult ♀ non-breeding (but again, part of juvenile

coverts retained, not all equally fresh as non-breeding adult); latter birds duller and more olive-grey above than adult ♂ non-breeding, without contrast in colour between crown and mantle; supercilium and underparts pale buff or cream, lower sides of neck and chest often more strongly tinged buff or (at sides) olive; yellow tinge restricted to belly, vent, and under tail-coverts; tips and fringes of new median and inner greater upper wing-coverts pale brownish- or olive-grey, not yellow-green as in some brighter 1st adult ♂♂ and all adult ♂♂. FIRST ADULT FEMALE NON-BREEDING. Like adult ♀ non-breeding and some less bright 1st adult ♂ non-breeding, but usually duller grey-brown on upperparts, rump less green than adult ♀ non-breeding; underparts sometimes white with buff tinge on lower cheeks, throat, and chest, pale yellow restricted to lower vent and under tail-coverts, but sometimes as extensively pale yellow as adult ♀ non-breeding. Part of juvenile feathers retained, as in 1st adult ♂ non-breeding; juvenile greater upper wing-coverts tipped white, contrasting in colour and wear with more brownish- or olive-grey tips of other new coverts. FIRST ADULT MALE BREEDING. Like adult ♂ breeding, but underparts slightly less deep yellow March–April (similar once adult and 1st adult both slightly worn); some juvenile outer greater upper wing-coverts often retained, tip white, strongly abraded (in adult, greater coverts usually a mixture of breeding and non-breeding, but even tips of worn non-breeding coverts not as white and frayed as juvenile coverts of 1st adult at same time of year); tips of flight-feathers and greater upper primary coverts more worn and pointed than those of adult at same time of year. FIRST ADULT FEMALE BREEDING. Rather variable: some birds have grey crown, green mantle and scapulars, distinct pattern on sides of head, and extensive yellow underparts, like adult ♀ breeding, but others similar to adult ♀ non-breeding or intermediate, both due to new feathering obtained in pre-breeding moult being similar to non-breeding, as well as due to partial retention of old non-breeding plumage. Part of juvenile feathering retained (see First Adult Male Breeding). As in 1st adult ♂ breeding, many indistinguishable, especially from late May onwards when plumage worn.

Bare parts. ADULT, FIRST ADULT. Iris brown to black-brown. Bill dark horn-brown, horn-black, greyish-black, or black, paler greyish or (perhaps in 1st autumn only) yellowish at base of lower mandible. Mouth yellowish-pink. Inside of mandible and tongue perhaps grey in adult, flesh to orange in 1st autumn (Bub *et al.* 1981). Leg and foot dark slate-grey, brown-black, or black. NESTLING. At hatching, body including bill and leg yellow-red, gape-flanges pale yellow, bill-tip dark; later on, body flesh-coloured, gape-flanges deeper yellow, bill grey-yellow, mouth yellow, leg greyish-flesh, and soles pale yellow (Dittberner and Dittberner 1984, which see for details). Mouth orange-yellow or orange, 2 brown spots on base of tongue (Witherby *et al.* 1938*a*). JUVENILE. Iris dark brown. Bill slate-brown or horn-brown with yellowish base and gape-flanges. Leg and foot flesh-grey, purplish-grey, slate-grey, or slate-black. (Hartert 1903-10; Witherby *et al.* 1938*a*; Ali and Ripley 1973*b*; Dittberner and Dittberner 1984; RMNH, ZMA.)

Moults. ADULT POST-BREEDING. Complete; primaries descendant. In Britain, *flavissima* starts with p1 between *c.* 1 July and *c.* 7 August (mainly 10-31 July, average 19 July), completed with p9-p10 *c.* 15 August to *c.* 25 September (mainly 20 August-10 September, average 31 August); duration of primary moult 43-45 days (Hereward 1979; Ginn and Melville 1983). Bird may start autumn migration before moult finished. Tail moult usually centrifugal (t1 first, t6 last) or t1 and t6 shed at same time and others later on; moult less often simultaneous (all at same time) or irregular. (Hereward 1979.) Small sample of *thunbergi* from northern Finland started about June–July, completed mid-August; in southern Finland, estimated duration of primary moult *c.* 40 days; tail moult started mainly at primary moult score 1–20, completed at 35–45 or slightly after p9; secondary moult mainly from score 22–35, completed with regrowth of p9 or slightly later (Haukioja 1971). In Netherlands and East Germany, timing of moult in nominate *flava* same as *flavissima*; duration of primary moult 40 days; moult of some body feathers occasionally late June, main body and tertial moult at primary score 10–35 (Dittberner and Dittberner 1984; RMNH, ZMA). Timing elsewhere in central and southern Europe apparently similar to *flavissima* and nominate *flava*; in USSR, *feldegg* starts with some body feathers late June, flight-feathers and tail from mid-July; completed mid-August (Dementiev and Gladkov 1954b); moult August–September in Yugoslavia (Stresemann 1920). In *melanogrisea*, Afghanistan, moult had started 5 August (Paludan 1959); in *leucocephala*, Mongolia, small sample moulted on average slightly earlier than average British *flavissima* (Piechocki and Bolod 1972). Egyptian *pygmaea* perhaps start early (as in post-juvenile, see below), but no information. ADULT PRE-BREEDING. Partial: head, body, tertial coverts, often some or all tertials, and highly variable number of other wing-coverts and tail-feathers. Some or all median upper wing-coverts usually replaced, but occasionally all upper wing-coverts except for some outer greater, or none at all except for tertial coverts. At least t1 replaced, but sometimes many or all tail-feathers (least often t4–t5). Timing rather variable: said to start January–February, completed March–April, lasting 12–13 weeks (Ginn and Melville 1983; Dittberner and Dittberner 1984), though in north-east Africa starts (October–)November (Bub *et al.* 1981), and in Indonesia body moult of *taivana* and *simillima* starts (late September–) October, followed by tertials December–January and tail January–March (Voous 1950). In Kenya, moult of head and part of mantle from November–December, but main moult of head, body, and many wing-coverts starts January or early February, completed after *c.* 50 days in March or early April; no difference between nominate *flava*, *beema*, and *lutea* (D J Pearson). In Nigeria, tertials and tail replaced 15 January–20 February, head and body 10 February to early April (Wood 1978); moult sometimes still not completed on arrival in Europe (Curry-Lindahl 1963b). In north-west India, ♂ *melanogrisea* had finished moult early March, when ♀ still moulting (Ticehurst 1923). POST-JUVENILE. Partial: head, body, lesser and median upper wing-coverts, often some or all tertial coverts, sometimes 1–2(–3) tertials, and highly variable number of inner greater upper wing-coverts and tail-feathers. Starts at *c.* 6 weeks old, taking *c.* 6 weeks. In Europe, moult mainly mid-June to first half of September, but birds occasionally still fully juvenile in early September, leaving for winter quarters with part of body and all tail and upper wing-coverts juvenile (Heinroth and Heinroth 1924–6; Dittberner and Dittberner 1984; RMNH, ZMA). In Sweden, over 95% of birds retained all old greater upper wing-coverts and some retained all other wing-coverts also (Svensson 1984a); in Britain, 48% retained all old greater coverts, but 41% had 1–4 inner ones new, 5% had 5–7 new, and 7% had all 9 new (*n* = 582), though much variation throughout season and between years (Hereward 1979). Some spring birds show worn non-breeding inner greater coverts and contrastingly new non-breeding outer coverts, and hence moult apparently sometimes continued in winter quarters; on the other hand, birds retaining some heavily worn juvenile greater, median, or lesser coverts and some tail-feathers occur frequently.

Sample of *pygmaea*, Egypt, from April contained a juvenile bird, moulting birds, and a bird in fresh non-breeding (ZFMK); moult thus earlier than in Europe. FIRST PRE-BREEDING. Like adult pre-breeding, but more variable in extent; part of non-breeding plumage on head and body sometimes retained (after moult, upper wing-coverts and tail a mixture of heavily worn juvenile, worn non-breeding, fairly fresh non-breeding, and fresh breeding plumage, proportion of each plumage variable). In Kenya, timing as adult pre-breeding (D J Pearson), but in Indonesia moult on average *c.* 1 month later (Voous 1950).

Measurements. ADULT, FIRST ADULT. All samples from breeding area, April–September, unless otherwise noted; skins (RMNH, ZFMK, ZMA). In all races, 1st adult with retained juvenile flight-feathers and occasionally tail combined with older birds: tail length similar, but wing of 1st adult on average 1·6 mm shorter than fully adult. Bill (S) to skull; exposed culmen on average 3·9 less, bill to distal corner of nostril on average 6·6 less (from 6·2 in *thunbergi* to 7·2 in *feldegg*).

M. f. flavissima. Britain and western Netherlands.

WING	♂ 82·6 (2·76; 25)	79–89	♀ 77·7 (2·02; 7)	75–81
TAIL	70·2 (2·96; 25)	66–76	68·1 (3·13; 7)	65–72
BILL (S)	16·4 (0·62; 24)	15·3–17·1	15·6 (0·51; 7)	14·9–16·3
TARSUS	23·9 (0·61; 23)	22·9–25·1	23·6 (0·37; 6)	23·2–24·2

Sex differences significant for wing. Hind claw 9·4 (0·71; 26) 7·9–10·6, rarely up to 12·4.

M. f. lutea. Mainly Volga–Ural steppes, USSR; wing includes migrants northern Iran (Schüz 1959).

WING	♂ 81·6 (1·60; 21)	79–85	♀ 80·0 (1·83; 4)	78–82
TAIL	67·8 (1·80; 12)	65–71	65·5 (— ; 2)	64–67
BILL (S)	16·1 (0·61; 14)	15·1–17·0	16·1 (— ; 2)	16·0–16·2
TARSUS	23·8 (0·66; 12)	22·8–24·7	23·8 (— ; 2)	23·1–24·5

Sex differences not significant. Hind claw 9·8 (1·10; 14) 8·7–10·6, rarely up to 12·0.

Wing of live birds, Kenya: adult ♂ 84·3 (1·68; 200) 80–88, 1st adult ♂ 82·7 (1·68; 200) 78–88 (D J Pearson).

Nominate *flava*. Mainly Netherlands.

WING	♂ 81·8 (2·06;127)	77–86	♀ 78·5 (2·01; 67)	74–82
TAIL	69·4 (2·58; 51)	65–76	67·1 (2·54; 33)	62–73
BILL (S)	15·9 (0·41; 34)	15·1–16·5	15·6 (0·55; 21)	14·7–16·2
TARSUS	24·0 (0·54; 36)	23·3–25·0	23·5 (0·76; 26)	22·5–24·8

Sex differences significant. Hind claw 9·4 (1·12; 58) 8·6–10·5, occasionally 7·1–14·9.

M. f. beema. Volga—Ural steppes east to south-west Siberia.

WING	♂ 82·1 (1·52; 34)	79–85	♀ 78·7 (1·35; 5)	76–80
TAIL	67·4 (1·99; 28)	63–72	64·8 (1·57; 5)	62–67
BILL (S)	16·0 (0·53; 28)	15·2–17·1	15·8 (0·65; 5)	15·1–16·7
TARSUS	23·9 (0·86; 28)	22·4–25·3	23·6 (0·65; 5)	23·0–24·4

Sex differences significant for wing and tail. Hind claw 9·6 (1·07; 33) 7·9–10·7, rarely up to 12·9.

M. f. leucocephala Mongolia (Piechocki and Bolod 1972).

WING	♂ 81·6 (2·41; 14)	77–87	♀ 80·0 (—; 3)	79–81

Sex differences not significant.

M. f. cinereocapilla. Italy.

WING	♂ 83·0 (1·73; 49)	79–87	♀ 79·6 (1·57; 10)	77–82
TAIL	71·9 (2·11; 32)	67–76	67·8 (3·53; 9)	62–72
BILL (S)	16·2 (0·67; 21)	15·2–17·2	15·7 (0·66; 9)	15·0–16·6
TARSUS	24·1 (0·97; 21)	22·8–25·2	23·7 (0·80; 9)	22·7–24·9

Sex differences significant for wing and tail. Hind claw 9·2 (0·86; 30) 8·3–10·7.

M. f. iberiae. Spain and Portugal.

WING	♂ 80·6 (1·90; 13)	78–84	♀ 75·1 (1·16; 6)		73–77
TAIL	69·1 (3·20; 13)	64–74	64·7 (1·99; 6)		62–67
BILL (S)	16·2 (0·49; 13)	15·4–16·9	16·1 (0·74; 6)		15·2–16·9
TARSUS	24·3 (0·51; 13)	23·7–25·1	23·4 (1·58; 6)		22·7–24·1

Sex differences significant, except bill (S). Hind claw 9·4 (1·04; 18) 8·6–10·5, rarely down to 6·8.

Mallorca, wing: ♂ 77·9 (1·96; 9) 75–82 (Stresemann 1920); ♂ 77–82, ♀ 73–78 (Jordans 1924). Banc d'Arguin, Mauritania: wing ♂♂ 78, 80; ♀ 76 (Roux 1960).

M. f. pygmaea. Egypt.

WING	♂ 76·2 (1·84; 10)	74–79	♀ 72·4 (1·75; 5)		70–75
TAIL	62·6 (1·28; 7)	61–65	59·6 (1·70; 4)		57–61
BILL (S)	15·5 (0·30; 7)	15·1–16·0	15·1 (0·23; 5)		14·9–15·5
TARSUS	23·3 (0·57; 7)	22·6–24·0	23·2 (0·44; 5)		22·6–23·7

Sex differences significant for wing and tail. Hind claw 9·0 (0·90; 11) 7·2–10·0.

M. f. feldegg. Mainly Greece and Turkey, some Yugoslavia, Transcaucasia, and Turkmeniya (USSR).

WING	♂ 83·1 (2·23; 20)	80–87	♀ 80·0 (1·95; 6)		77–82
TAIL	69·9 (2·37; 20)	66–73	67·4 (1·08; 5)		66–69
BILL (S)	16·8 (0·59; 20)	16·0–17·9	16·6 (0·53; 5)		16·2–17·6
TARSUS	24·2 (0·78; 19)	23·2–25·5	24·0 (0·70; 6)		23·1–24·7

Sex differences significant for wing. Hind claw 9·5 (0·77; 23) 8·6–10·2, rarely 7·6–11·3.

Southern Yugoslavia, wing: ♂ 81·9 (1·66; 60) 78–85, ♀ 77·6 (2·06; 18) 74–82; ♂ exceptionally 74 (Stresemann 1920). Turkey, Iran, and Armeniya (USSR), wing: ♂ 84·1 (2·59; 13) 81–90, ♀ 79·3 (3) 78–82 (Stresemann 1928; Paludan 1940; Schüz 1959; Nicht 1961; Kumerloeve 1963, 1969a; Rokitansky and Schifter 1971).

M. f. thunbergi. Fenno-Scandia and (on spring migration) Netherlands.

WING	♂ 83·4 (1·67; 42)	80–87	♀ 80·2 (1·78; 27)		77–84
TAIL	69·8 (1·76; 32)	66–72	67·1 (1·80; 25)		63–70
BILL (S)	15·9 (0·48; 33)	15·1–16·6	15·5 (0·56; 25)		14·5–16·3
TARSUS	23·8 (0·57; 31)	22·9–25·1	23·4 (0·64; 25)		22·2–24·3

Sex differences significant. Hind claw 9·5 (0·85; 55) 8·2–10·5, rarely up to 11·5.

Weights. *M. f. flavissima.* Skokholm (Wales): 18·3 (8) 15–20 (Browne and Browne 1956). Southern Norway, July: ♂ 15 (Haftorn 1971).

M. f. lutea. Northern Iran, April: ♂ 18·0 (1·32; 7) 16–20, ♀♀ 16·5, 18 (Schüz 1959). In Kenya, average early morning weight of adult ♂ 16·4 (1·1; 100) in October, gradually increasing to 17·0 (1·2; 100) in January and to 17·5 (1·6; 100) in first half of March, strongly increasing to 18·6 (2·6; 100) in first half of April, with individual maximum of 26·4; in 1st adult ♂, average 15·8 (0·9; 50) October, 16·4 (1·3; 50) January, 16·9 (1·0; 50) early March, 17·7 (1·8; 50) early April; in adult ♀ (probably mainly of this race), 15·3 (1·2; 100) October, 15·7 (1·1; 100) January, 16·0 (1·1; 100) early March, 16·9 (1·9; 100) early April, with individual maximum of 22·6; in 1st adult ♀, 14·6 (1·0; 50) October, 15·5 (1·6; 50) January, 15·7 (1·2; 50) early March, and 16·1 (1·7; 50) early April (D J Pearson).

Nominate *flava.* Sweden, July–August: adult 17·0 (1·03; 13) 15–19; juvenile and 1st adult 17·5 (1·03; 53) 16–20 (Bub et al. 1981). East Germany: ♂ 17·7 (113) 14–27, ♀ 17·2 (49) 14–21; see Dittberner and Dittberner (1984) for monthly averages, which go up to 22 (♂) and 21 (♀) by October. West Germany: April, ♂ 16·8 (14) 12–21, ♀ 15·4 (8) 14–17; August, ♂ 16·6 (66) 13–20, ♀ 15·6 (49) 12–18; October, ♂ 20·3 (5) 18–25·2

(Mester 1959). Netherlands: April–May, ♂ 16·8 (1·37; 21) 15–20, ♀ 14·8 (0·70; 5) 14–16; August–September, ♂ 16·4 (1·43; 8) 15–18, ♀♀ 14·6, 17 (RMNH, ZMA). Belgium: August 16·4 (28), September 17·1 (147) (Spaepen 1957, which see for details). Col de Bretolet, Switzerland: until 10 September 15·6 (94), 11–25 September 16·4 (114), after 25 September 16·7 (72) (Crousaz 1961, which see for daily variations). Morocco, spring: ♂ 15·1 (57) 12·3–20·4 (Ash 1969). For growth of nestling, see Dittberner and Dittberner (1984).

M. f. beema. Migrants Iran and Afghanistan: April, ♂ 16·3 (3) 16–17, ♀ 19; September, ♂ 17·1 (11) 16–19, ♀ 16·0 (12) 14–18 (Paludan 1959; Schüz 1959). Lake Chany, south-west Siberia: mainly adults in wing moult, July, ♂ 16·5 (1·25; 13) 15–19, ♀ 15·6 (0·92; 7) 14–17; full-grown, mainly August, 15·7 (1·09; 59) 13–18 (Havlín and Jurlov 1977).

M. f. leucocephala. Mongolia, mainly June: ♂ 17·6 (1·64; 14) 14–20, ♀ 18·7 (3) 18–20 (Piechocki and Bolod 1972).

M. f. cinereocapilla. Morocco, spring: ♂ 14·9 (37) 12·6–18·2 (Ash 1969).

M. f. iberiae. Morocco. March: 16·1, 16·2 (Kersten et al. 1983). Spring migrants: ♂ 14·3 (29) 12·4–16·5 (Ash 1969). Spain, May: ♂♂ 17, 18; ♀ 15·5 (ZFMK).

M. f. feldegg. Greece, Turkey, Armeniya (USSR), and Iran, combined, April–July: ♂ 18·6 (1·98; 13) 15–22, ♀ 18·5 (3) 14·6–24 (Paludan 1940; Niethammer 1943; Schüz 1959; Nicht 1961; Kumerloeve 1963, 1969a; Rokitansky and Schifter 1971).

M. f. melanogrisea. Afghanistan, March–August: ♂ 17·2 (0·84; 5) 16–18 (Paludan 1959).

M. f. thunbergi. Migrants Helgoland, May and early June: ♂ 16·6 (15) 14–19·3, ♀ 15·0 (3) 14–17 (Krohn 1915; Weigold 1926). Migrants northern Iran, April: ♂ 18·5 (1·91; 4) 16–20 (Schüz 1959). Exhausted ♀, at sea off Canary Islands, May: 14 (ZMA).

Various races, combined or unidentified. Malta: spring 16·8 (1·9; 53) 13·5–22, autumn 18·5 (2·7; 67) 14–24·5 (J Sultana and C Gauci). Morocco, spring: ♀ 13·9 (61) 11·2–17·9 (Ash 1969). For tables of monthly weights in Afrotropics, see Ward (1964) for north-east Nigeria, February–April; Smith and Ebbutt (1965) for central Nigeria, October–April; Pearson (1971) for Uganda, October–April; for graphs of monthly weights of nominate *flava* and *thunbergi* for Zaïre, Camargue (France), and northern Sweden, December–June, see Curry-Lindahl (1963b); for central Nigeria, November–March, see Wood (1978); see also Fry et al. (1970, 1972) for north-east Nigeria and Fry et al. (1974) for northern Kenya. Maximum of nominate *flava*, Zaïre, February: ♂ 28 (Curry-Lindahl 1963b).

Structure. Wing rather short, broad at base, tip bluntly pointed. 10 primaries: p8 longest, p9 0–1 shorter, p7 0·5–1·5, p6 3–7, p5 10–14, p4 14–18, p1 23–27. P10 reduced, narrow and sharply pointed; 53–60 shorter than wing-tip, 7–11 shorter than longest upper primary covert. Outer web of p7–p8 emarginated, sometimes slightly p6 also; inner web of p8 with slight notch, p8 and sometimes p7 with very faint notch. Tertials long; longest about equal to wing-tip in closed wing (4 shorter to 2 longer when fresh, slightly shorter when worn; however, up to 8 shorter in fresh juvenile). Tail rather long (relatively shorter than in Pied Wagtail *M. alba* and Grey Wagtail *M. cinerea*), tip square, slightly forked or rounded; 12 feathers, t6 1–4 shorter than t2–t3. Bill rather long, straight; rather slender at base; tip of culmen slightly decurved, sharp at end. Nostril small, oval; partly covered by membrane above and by frontal feathering at rear. Tarsus, toes, and claws rather long and slender (relatively longest in races of eastern Asia). Middle toe with claw 19·1 (38) 17–20·5; outer toe with claw *c.*

72% of middle with claw, inner *c.* 75%; hind with claw *c.* 90%, without claw *c.* 48%. Claws slender and sharp, slightly decurved; front claws rather short, hind claw long (see Measurements).

Geographical variation. Marked and complex; mainly involves colour of ♂ breeding plumage, less so other plumages; also (but scarcely) size. Division as followed here derived from theories of Domaniewski (1925), Sushkin (1925), Stresemann (1926), Johansen (1946), Sammalisto (1961), and Svensson (1963): 2 complexes recognized, often considered separate species: (1) *lutea* complex (*lutea, flavissima, taivana*); (2) *flava* complex (all other races). Every member of *lutea* complex overlaps partly or fully with *flava* complex, apparently with limited interbreeding, though some gene-flow between the complexes occurs (Sammalisto 1971), and members of *lutea* complex are perhaps not closely related to each other as measurements and structure of each are closer to the neighbouring member of *flava* complex than to other members of *lutea* complex; each member of *lutea* complex perhaps a separate early offshoot of part of the *flava* complex, as suggested by Vaurie (1957, 1959). *Flava* complex here subdivided into 3 groups: grey-headed *thunbergi* group in north, blue-headed nominate *flava* group in mainly temperate latitudes, and black-headed *feldegg* group in south, from Balkan countries to eastern Kazakhstan (USSR). Each of these groups sometimes considered a separate species also, but as they are connected by hybridization zones of variable width, better combined into a single highly polytypic species. For status of groups, see also Mayr (1956), Rendahl (1967), and Dijk (1975); for races (including extralimital ones) Hartert (1903-10, 1921-2), Hartert and Steinbacher (1932-8), Smith (1950), Grant and Mackworth-Praed (1952), Vaurie (1959), Sammalisto (1961), Sharrock and Dale (1964), Beregovoy (1970), and Dittberner and Dittberner (1984).

Variation in size small, especially compared with variation in colour. *M. f. pygmaea* and (less markedly) *iberiae* are smaller than other races (see Measurements); *feldegg* has rather long bill (especially compared with *thunbergi*, which is otherwise sometimes similar); all eastern races (*taivana* and all races of *thunbergi* group east from *plexa* and *simillima*) rather large, especially in bill, tarsus, toes, and claws (in 40 birds of various eastern races, bill 15·8-17·4, tarsus 23·7-26·4, hind claw 10-13), but birds from Bering Sea area smaller.

Breeding ♂♂ of all races readily separable in colour, apart from birds of unstable local populations in hybridization zones; occasionally, some individuals from a hybridization zone are confusingly similar to birds of another race without being stragglers from elsewhere: see, e.g., Stresemann (1920), Witherby *et al.* (1938a), Smith (1950), Sammalisto (1968a), Reichholf (1968), Voous (1969), and Hausmann (1983) for individual variation. Non-breeding ♂♂, ♀♀, and juveniles often more difficult to separate, though general characters (e.g. depth of colour of crown and ear-coverts, presence of supercilium, extent of yellow on throat) similar to breeding ♂. For variation in blue-headed nominate *flava* of *flava* group, see Plumages. Breeding ♂ *beema* paler ash-grey on crown, lores, and ear-coverts than ♂ nominate *flava*, supercilium often wider, white stripe from just below eye to lower ear-coverts (present also in some nominate *flava*) more pronounced; as in nominate *flava*, throat either fully yellow or with broad white stripe along lower cheek, but in palest *beema* chin and throat white, grey on side of head restricted to lores and upper ear-coverts and such birds resemble Mongolian *leucocephala*. In fact, continous cline from western Europe to Mongolia, with head gradually paler from

nominate *flava* through *beema* to *leucocephala*, and even as far west as Netherlands some birds inseparable from *beema*; white-headed birds occasionally reported in Europe and Middle East more probably pale variants of *beema* than true *leucocephala*. ♀ and non-breeding ♂ *beema* resemble nominate *flava*, but crown and ear-coverts sometimes slightly paler and pale stripe below eye to lower ear-coverts more pronounced. *M. f. beema* intergrades with members of *thunbergi* group in north and east of breeding range, resulting in birds appearing similar to nominate *flava*. Breeding ♂ *cinereocapilla* similar to breeding ♂ nominate *flava*, but forehead, crown, and hindneck on average darker bluish-grey; lore, upper cheek, and ear-coverts often darker plumbeous-grey, without white below eye or with a few white specks only; supercilium absent or (occasionally) faintly present just behind eye, rarely present as narrow line in front of eye; chin and throat white, but centre of throat frequently has yellow wash and occasionally whole throat yellow. ♀ and non-breeding ♂ *cinereocapilla* generally have darker ear-coverts than nominate *flava*, no supercilium or a narrow one (narrower than nominate *flava*), and throat white (but sometimes slight yellow or buff tinge, especially on cheek and lower throat, though not as yellow as ♀ *thunbergi*); often some dark olive-brown or grey spots on sides of throat and (occasionally) chest, but not a full collar as some *thunbergi*. Breeding ♂ *iberiae* similar to breeding ♂ *cinereocapilla* in dark crown, lore, upper cheek, and ear-coverts and in largely or fully white chin and throat, but supercilium usually long, narrow, and distinct. ♀ and non-breeding ♂ *iberiae* like *cinereocapilla*, but supercilium present, on average longer and wider than those *cinereocapilla* which show supercilium; upper cheek and ear-coverts darker than in nominate *flava*, chin and throat whiter. Breeding ♂ *pygmaea* closely similar to breeding ♂ *cinereocapilla*, differing mainly in smaller size; supercilium absent or present as a trace behind eye, but occasionally long and narrow as in *iberiae*; crown extensively tinged green, mantle and scapulars rather dark olive-green; centre of throat usually yellow, whole throat sometimes white; much olive on feather-bases of chest and flanks, showing as a dark patch on chest when plumage worn. ♀ and non-breeding ♂ *pygmaea* have short and narrow supercilium and usually some yellow on centre of throat. Breeding ♂ *feldegg* of *feldegg* group distinct: top, rear, and sides of head black (faintly tinged green when plumage fresh), sharply divided from yellowish olive-green mantle and from yellow chin, throat, and lower sides of neck; yellow of underparts rather deep, often with pronounced olive-green patch on side of breast (but ♂♂ of other races sometimes similar). ♀ and non-breeding ♂ *feldegg* highly variable: lore, area round eye, and ear-coverts usually black, often with slight green tinge, upper cheek sometimes spotted pale yellow; black often extends to forehead or crown; remainder of crown and hindneck olive-green or olive-grey, like mantle and scapulars; no supercilium or a faint olive-yellow one behind eye only; depth of yellow on underparts variable, but extent of yellow as in breeding ♂. Breeding ♂ *melanogrisea* similar to *feldegg*, but upperparts brighter yellowish-green and white stripe runs from chin below black cheek; ♀ and non-breeding ♂ similar to *feldegg*, but chin and sides of throat white instead of yellow. In breeding ♂ *thunbergi* of *thunbergi* group, top and sides of head and upper body resemble *cinereocapilla*: top of head dark grey, lores, upper cheeks, and ear-coverts slate-grey to blackish; mantle, scapulars, and head slightly duller olive-green (less yellowish) than in breeding ♂ nominate *flava*, rump green (less yellow-green); tips and fringes of upper wing-coverts and tertials often yellowish-green (not as yellow-white as nominate *flava, cinereocapilla*, or *feldegg*—but much variation, mainly due to variable number of coverts involved in pre-breeding moult); supercilium absent or very faint and narrow; unlike *cinereocapilla*, chin and throat yellow, occasionally except for white upper chin or a very faint white line running below cheek; often many olive-green spots on lower side of neck, extending to chest in 23 of 40 birds examined, forming full collar in 5, but virtually absent in 2. Some ♂♂ have forehead or part of crown black and these closely similar to *feldegg*, differing in slightly paler yellow underparts and usually in presence of dark spots on lower side neck and sometimes chest. ♀ and non-breeding ♂ *thunbergi* differ from nominate *flava* in darker crown, duller green upperparts, and greyish-black ear-coverts; usually no supercilium, or present just in front of or behind eye only, but occasionally (6 of 25 examined) long, pale buff, yellow, or white (though narrower than in nominate *flava*); lower sides of neck always have dark olive or blackish spots, like some *cinereocapilla* or nominate *flava*, but (unlike those races) spots extend across upper chest in 18 of 25 examined, forming broken or full collar (often more pronounced than in breeding ♂); yellow of underparts either extends up to chin, or (especially in 1st adult non-breeding), up to dark collar only. M. f. *simillima* closely similar to nominate *flava*, but upperparts slightly darker and bill, leg, and foot slightly longer; in ♂, bill to skull 16·8 (10) 16·1-17·5, tarsus 25·0 (10) 24·0-26·1, hind claw 11·1 (10) 9·6-13·1 (ZMA). Breeding ♂ *flavissima* of *lutea* group has forehead greenish-yellow; crown yellowish olive-green (like mantle and scapulars), but forecrown occasionally greenish-yellow, hardly greener than supercilium, and then resembling breeding ♂ *lutea*; supercilium and broken eye-ring bright yellow; lore mottled dusky grey and yellow; upper cheek and lower ear-coverts yellow with olive-green mottling, upper ear-coverts yellowish olive-green; underparts entirely bright yellow, except for olive-green patch on side of breast and sometimes slight olive suffusion on chest. Breeding ♀ *flavissima* closely similar to breeding ♀ nominate *flava*, but supercilium usually has distinct yellow tinge; upperparts duller greyish olive-brown than breeding ♂, rump less yellowish, supercilium and underparts slightly to distinctly paler yellow. Non-breeding ♂ and ♀ *flavissima* similar to nominate *flava*, but supercilium, lower sides of neck, and chest usually deeper buff or yellow-buff; ageing as in nominate *flava*. M. f. *lutea* similar to *flavissima* but breeding ♂ usually differs in having bright yellow forehead, forecrown, broad supercilium, upper cheek, and lower ear-coverts; lore, upper ear-coverts, rear crown, and nape green-

yellow, paler than breeding ♂ *flavissima*; head sometimes entirely yellow (then resembling Citrine Wagtail M. *citreola*, but with upperparts olive-green, not grey or black), or (especially in 1st breeding ♂) crown, hindneck, and upper ear-coverts sometimes olive-green (then similar to *flavissima*, though supercilium usually wider and upper cheek less mottled olive); frequently some blackish spots on lower sides of neck or (more rarely) across chest; ♀ and non-breeding ♂ indistinguishable from *flavissima*; see also Pearson and Backhurst (1973).

For overlap or hybridization between *thunbergi* and nominate *flava* in Fenno-Scandia, see Sammalisto (1956, 1958, 1968a), Rendahl (1967), and Vepsäläinen (1968); similar zone extends east through central European USSR. For contact area of *flavissima* and nominate *flava* in western Europe, see Mayaud (1952a) and Dijk (1975). For individual variation in hybridization area of *cinereocapilla* and *iberiae* in Camargue (France), see Schwarz (1956); similar variation occurs on Corsica, while populations of Algeria and Tunisia are more similar to *iberiae* in colour but like *cinereocapilla* in size. For contact zone of *cinereocapilla* and nominate *flava*, just west and north of Alps, see Kinzelbach (1967). For Yugoslavia, where *cinereocapilla*, nominate *flava*, and *feldegg* come into contact, see Kroneisl-Rucner (1960) and Matvejev and Vasić (1973). In Balkan countries, wide hybridization zone between nominate *flava* and *feldegg* extends from southern Rumania to southern Ukraine, with some *feldegg* influence north to southern Poland; birds from this area sometimes separated as '*dombrowskii*' (Tschusi, 1903), showing dark grey crown and blackish ear-coverts in combination with broad white supercilium. Breeding range of *feldegg* is expanding, however, apparently with very little hybridization; for situation in Balkans and neighbouring parts of south-east Europe, see Kleiner (1936, 1939), Keve (1958, 1978), Leisler (1968), Sammalisto (1968b), and Radu (1975). Situation in plains of lower Volga and Ural rivers complex, as birds intermediate between nominate *flava* and *beema* hybridize with birds intermediate between *feldegg* and *melanogrisea*, while influence of nearby *lutea* (in south-east) and *thunbergi* (in Urals) also apparent; some birds black-headed with white or yellow supercilium of variable extent, and these sometimes separated as *superciliaris* (Brehm, 1854), others grey-headed with yellow supercilium (similar to hybrids of nominate *flavissima* and nominate *flava*, named *perconfusus* Grant, 1949), but many more combinations occur.   CSR

Motacilla citreola Citrine Wagtail

PLATES 26, 29, and 31
[between pages 232 and 233, and facing pages 329 and 353]

Du. Citroenkwikstaart   Fr. Bergeronnette citrine   Ge. Zitronenstelze
Ru. Желтоголовая трясогузка   Sp. Lavandera cetrina   Sw. Citronärla

Motacilla citreola Pallas, 1776

Polytypic. Nominate *citreola* Pallas, 1776, northern USSR from Kanin peninsula to central Siberia, and in central Asia east of Tomsk, Kuznetskiy Alatau, Sayan mountains, and Lake Baykal, south to Mongolia, intergrading with *werae* from Zaysan basin east to Sayan mountains and north through Altai to Kuznetskiy Alatau; *werae* (Buturlin, 1907), plains of southern USSR, from central European USSR east to Tomsk and foot of Altai mountains, intergrading with *calcarata* in Kirgizskiy mountains, Issyk-Kul basin, and valley of Tekes river. Extralimital: *calcarata* Hodgson, 1836, Iran, Afghanistan, and Pamir mountains, east through Himalayas to Ordos (China) and north to western Tien Shan (USSR), Tibet, and Koko Nor (China).

Field characters. 17 cm; 24-27 cm. Only marginally larger than Yellow Wagtail *M. flava* but with 10% longer tail usually obvious; less attenuated than Grey Wagtail *M. cinerea*, with 15% shorter tail; less deep-chested than White Wagtail *M. alba*. Most constant marks are slate-grey upperparts, and striking double white wing-bar and white fringes to tertials. Breeding ♂ shows fully yellow head and underparts; contrast of yellow head with black neck-shawl diagnostic. Breeding ♀ suggests dusky-backed *M. flava*, and juvenile and 1st winter birds recall pale *M. alba*; all show noticeably grey flanks. Sexes dissimilar; seasonal variation in ♂. Immature separable. 2 races in west Palearctic but only breeding ♂♂ (scarcely) separable.

(1) North Siberian race, nominate *citreola*. ADULT MALE BREEDING. Head and underparts rich yellow, some grey on upper flanks. Hindneck dusky-black, forming contrasting shawl which reaches sides of breast (and exaggerates upright carriage of head). Back and rump slate-grey, with paler (bluer) or duller (browner) tones in some lights; rear upper tail-coverts, wings, and tail almost black. Upperparts boldly relieved by double white wing-bar (formed by long tips to median and greater coverts), obvious white fringes to tertials, and flashing white outer feathers to rump and tail. Underwing pale grey-buff, noticeably divided by wide pale central panel formed by white tips to rear coverts. ADULT MALE NON-BREEDING. Loses most of yellow head and black neck-shawl; assumes more ♀-like pattern of brown-grey rear crown and cheeks (hence retaining yellow only on forehead and supercilium) and paler, more primrose-yellow belly and vent. Upperparts also browner in tone. ADULT FEMALE. Readily distinguished from breeding ♂ by head pattern, since most of crown and cheeks grey and neck-shawl hardly evident. Rest of plumage patterned as ♂ but grey upperparts less clean in tone and yellow underparts paler, with more grey on upper flanks. Somewhat resembles ♀♀ of *M. flava flavissima*, *M. f. lutea*, and *M. f. taivana* (of eastern Siberia) but readily distinguished from typical birds of these races by greyer upperparts, white wing markings, and paler vent. JUVENILE. Not studied in the field. Differs from adult ♂ in being browner above and lacking any intense yellow below. Plumage pattern and colours recall *M. alba* more than *M. flava*, with deep brown-grey flanks strongly suggesting *M. alba* but lack of strong chest-necklace pointing to *M. flava*. At close range, shows unmistakable set of characters: pale yellow, buff, or brown forehead, wide supercilium, mainly buff chest-band (with isolated 'stickpin' marks and dots usually not forming necklace), very pale, yellow- or buff-washed belly and vent, and bold wing marks. FIRST WINTER. Differs from juvenile in being cleaner (bluer or greyer) above and even paler below. Recalls *M. flava* (particularly pale individuals of *M. f. thunbergi* and grey mutants) from in front, *M. alba* from behind. In British and Swedish vagrants, no fully

yellow tones visible in the field, and pale buff colours (tinged pink or primrose) restricted to forehead, fore-supercilium, chest-band, and belly. Compared to aberrant *M. flava*, safest characters are (a) pale cream forehead exaggerated by yellow or buff fore-supercilium, (b) cream or white rear-supercilium, (c) blue- or slate-grey upperparts, often looking very dark in dull light, (d) bold white wing-bars, (e) extensive grey sides to breast and flanks, and (f) lack of yellow vent. Compared to *M. alba*, lack of full necklace, more distinctly patterned face, and buff chest-band allow ready distinction. FIRST SUMMER. Adult plumage not acquired in 1st spring; ♂♂ may have plumage intermediate between 1st winter and adult ♀. At all ages, bill and legs dark brown or black, base of lower mandible paler in immature; hind claw long (short in *M. alba*). (2) West-central USSR race, *werae*. ADULT MALE BREEDING. 5-10% smaller and paler than nominate *citreola*. As nominate *citreola*, including black neck-shawl, but back always grey, lacking any dusky suffusion; underparts weaker yellow, with less grey flanks.

All plumages except adult ♂ breeding overlap in appearance with commoner *M. alba* and *M. flava*, so identification of *M. citreola* needs care. Some 1st winter birds with dark cheeks and narrower supercilia have noticeably 'capped' heads but not known if this results from individual or racial variation or from hybridizaton with *M. flava*; one spring bird (on Scilly, England) showed characters that indicated either hybridization or only partial assumption of adult ♀ plumage (D I M Wallace), and other such birds known. Thus important to study suspect *M. citreola* fully for form, voice, and behaviour; all offer useful additional clues. Flight silhouette recalls *M. flava*, but more robust, with slightly longer tail. Flight action again like *M. flava* but slightly more powerful wing-beats give more even acceleration into bounds (and avoid hint of stalling so obvious in *M. flava*). Gait rather more sedate than other wagtails, with noticeably less tail-wagging than *M. alba* and probably less than *M flava*. Stance, perching, and feeding behaviour closer to *M. flava* than to *M. alba*.

Commonest call (often given in flight) 'dzzip', 'dzzeep', 'sweeip', or 'zreep'; recalls *M. flava*, but harsher and perhaps shriller and shorter. Song a repetition of similar sounds. Alarm-call (of pressed vagrant) more disyllabic, 'ze-reep'; in extralimital *calcarata* also a metallic 'pzeeow'.

Habitat. Breeds from arctic and subarctic through boreal and temperate dry continental zones, and extralimitally to subtropics, ranging from sea-level to upper level of meadow vegetation at above 4500 m in Pamirs, where densest population occurs at 3500 m. In tundra belt, inhabits osier thickets on coast and on islands in large river deltas, perching often on bushes. Also by lakes on marshy-shrubby tundra, on wet sections of mountain tundra, and among willow bushes on tussocky mountain meadows. In south in upper part of forest belt and in

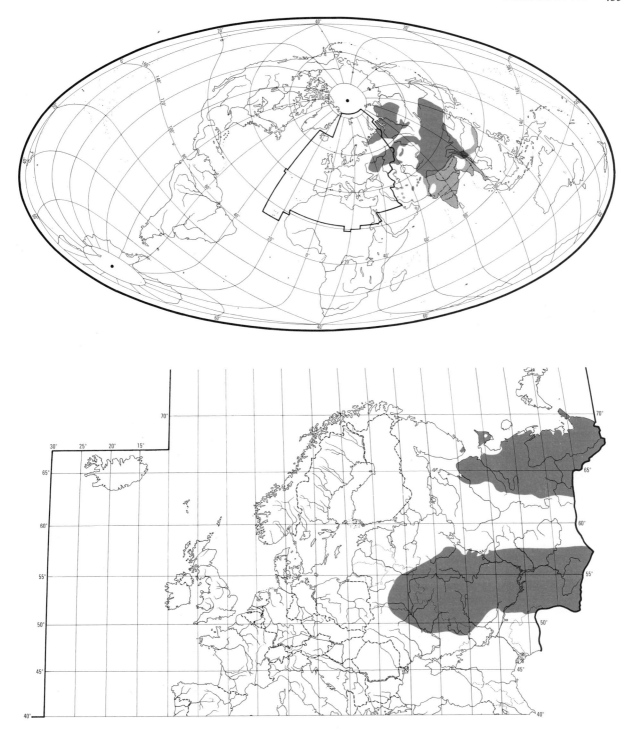

mountains, often in very damp places; also in river valleys in peaty hummocky bogs or marshy meadows covered with sparse low shrubs. (Dementiev and Gladkov 1954a.) In Afghanistan, breeds in rather dry grass-clad areas along rivers, or in fields round villages at *c.* 1000-3100 m (Paludan 1959). In Kashmir, frequents wet low-lying grassland and weedy marsh from *c.* 1600 m, breeding in hundreds on wet spongy ground covered with soft grass of varying length interspersed with little bushes, extending also to neighbouring ricefields and cut-down reedbeds; at higher elevations, by tarns with wet grassy margins and streams running through marshy patches below glaciers (Bates and Lowther 1952). Frequents sedges and grassy margins of lakes, streams, water-meadows, bogs, and irrigated ploughed land; essentially a marsh-inhabiting or water bird, breeding mostly between 3000 and 4600 m in Indian subcontinent. In winter frequents marshes, squelchy grassy margins of ponds, and irrigated ricefields. (Ali and Ripley 1973b.) Also occurs by rivers, canals, drains, swamps, pools, flooded areas, and reedbeds, never away from water, usually close to its edge or on surface of half-submerged aquatic plants (Hutson 1954).

Distribution. Signs of westward spread in USSR; isolated breeding recently in Finland and Czechoslovakia.

BRITAIN. Adult ♂ feeding 4 young 1976; ♀ not seen (Cox and Inskipp 1978). SWEDEN. Adult ♂ feeding 3 young 1977; ♀ not seen and young not identified (*Vår Fågelvärld* 1979, **38**, 47) so breeding not proven (LR). FINLAND. ♂ bred with ♀ Yellow Wagtail *M. flava* in south-east 1983–4 (Mikkola 1984). CZECHOSLOVAKIA. Bred northern Moravia 1977 (Kondělka 1980). USSR. Signs of spread west and south-west, including isolated nesting (Wilson 1979, 1984; Shapoval 1982).

Accidental. Iceland, Ireland, Britain (now annual), Netherlands, West Germany, Norway, Sweden, Finland (now annual), Poland, Czechoslovakia, Austria, Switzerland, Italy, Greece, Rumania, Israel, Jordan, Syria, Kuwait, Turkey, Cyprus.

Population. USSR. Very common in tundra where there are osier thickets; not numerous further south (Dementiev and Gladkov 1954a).

Movements. Migratory, wintering mainly in India and south-east Asia.

The most northerly breeding population, from Siberia south to *c.* 60°N, winters from south-east of Caspian Sea and southern Iran to India (Voous 1959; Cox and Inskipp 1978). Birds from central European USSR and Siberian steppes, Tomsk area, and east to the Altai migrate through eastern Iran, Turkestan, and Afghanistan to winter through most of Indian subcontinent (Ali and Ripley 1973b). Population breeding north-west Manchuria through northern Mongolia to the Altai and Chinese Turkestan winters in southern China (Caldwell and Caldwell 1931; Etchécopar and Hüe 1983). Within Indian subcontinent, *calcarata* winters only in north, birds breeding in northern Pakistan, Kashmir, and southern Tibet moving only short distances into adjacent plains of Pakistan, northern India, Bangladesh, and Assam (Ali and Ripley 1973b).

Rather little information on timing of movements. Autumn departure from USSR starts mainly in early September (Dementiev and Gladkov 1954a), and passage in Afghanistan occurs mid-September (Paludan 1959). Northern birds arrive India in September and depart

mostly March–April, extreme dates 5 September–17 May (Ali and Ripley 1973b). Noted in extreme northern Thailand (nominate *citreola*) early November to early May (Deignan 1945). Spring passage in Afghanistan early May (Paludan 1959), and birds observed crossing southern borders of USSR late March, with intensive movement in first half of April and late arrivals to mid-May (Dementiev and Gladkov 1954a).

Occurrence in Middle East evidently peripheral to main movement: scarce winter visitor and passage migrant, October–March, to eastern Saudi Arabia and United Arab Emirates (Bundy and Warr 1980; G Bundy); uncommon and irregular in Oman, September–April (Gallagher and Woodcock 1980).

Most vagrants to western Europe occur September–November; *c.* $\frac{1}{3}$ of total on Fair Isle (Scotland). Apart from isolated breeding records (see Distribution), very few spring records (Finland, West Germany, Poland, Austria). (Cox and Inskipp 1978.) JHE, DJB

Food. Information almost exclusively extralimital. Diet invertebrates, often aquatic. Usually forages in or near wet habitat (see Habitat). 3 foraging techniques. (1) Picking. Picks items from ground or water surface while standing or walking on ground or wading in shallow water—even up to belly, long legs being well adapted to this (Meinertzhagen 1938); will perch on twig above water to take prey on surface (M G Wilson), and walk on floating leaves of water plants. Hunts prey flushed by grazing animals, bird walking around and between legs of cattle, etc. (N van Swelm). (2) Immersion. In a development of picking, bird plunges head into water to catch insect larvae (Chernyshov 1981). (3) Flycatching. Snatches insects flying past with brief upward flutter (Dolgushin *et al.* 1970). Hovering, as used by other *Motacilla*, not recorded in *M. citreola* (Chernyshov 1981).

The following recorded in diet. Adult and larval dragonflies (Odonata), grasshoppers, etc. (Orthoptera: Acrididae, Tetrigidae), mantises (Dictyoptera: Mantidae), bugs (Hemiptera: Corixidae, Notonectidae, Cicadidae), adult and larval butterflies and moths (Lepidoptera, including Noctuidae), larval caddis flies (Trichoptera), adult and larval flies (Diptera: Tipulidae, Culicidae, Chironomidae, Stratiomyidae, Empididae), larval ants, wasps and bees (Hymenoptera), pupal and adult beetles (Coleoptera: Carabidae, Dytiscidae, Staphylinidae, Geotrupidae, Scarabaeidae, Elateridae, Tenebrionidae, Chrysomelidae, Curculionidae), spiders (Araneae), small molluscs (e.g. *Lymnaea*), small fish, and one record of unidentified plant material (D'Abreu 1918; Pek and Fedyanina 1961; Dolgushin *et al.* 1970; Popov 1978; Chernyshov 1981; Kostin 1983; Danilov *et al.* 1984).

In Tatarskaya ASSR (east-central European USSR) in breeding season, 11 stomachs contained beetles (Curculionidae, Elateridae, Geotrupidae, Carabidae, larvae of Tenebrionidae), spiders, Orthoptera, Hemiptera, mosquitoes, and other small flies, caterpillars, and aquatic invertebrates (Popov 1978). In Kirgiziya (south-central USSR) 62 stomachs contained mainly Carabidae, Curculionidae, and Diptera; 3 stomachs contained spiders and 1 contained unidentified plant material (Pek and Fedyanina 1961). On Yamal peninsula (USSR), 6 stomachs contained mainly Staphylinidae by number and Tipulidae larvae by weight; other items included Chironomidae, Dytiscidae, small Carabidae, and Empididae (Danilov *et al.* 1984). In Nagpur (India), March, 1 stomach contained 5 water bugs (Corixidae), 5 small Hymenoptera, 4 small beetles (2 Curculionidae), and 1 fly; another, April, contained 4 Chrysomelidae, 2 grasshoppers (Tetrigidae), 1 small Elateridae, 1 beetle pupa, and 1 spider (D'Abreu 1918).

Diet of young basically similar to adults', though adult beetles may be less important (Chernyshov 1981). In comparative study of nestling diet in *M. citreola* and Yellow Wagtail *M. flava* in northern Kazakhstan and Baraba (Novosibirsk, USSR), differences in species taken found only in items of secondary importance; of 450 items recorded in visual and stomach samples, 85% found in diet of both species. Most important were dragonflies (larvae and adults) and larvae of aquatic beetles, also spiders and flies. However, some differences between the species: *M. citreola* given mainly dragonfly larvae, *M. flava* mainly adult insects; of beetles, only *M. citreola* fed larger larvae (*Acilius sulcatus*). Aquatic items accounted for 67% (by volume) of diet of *M. citreola*, 20·1% for *M. flava*. (Chernyshov 1981.) No information on feeding rates.

ILG, CAT

Social pattern and behaviour. Relatively little studied.

1. Quite gregarious during breeding season, markedly so at other times. At end of breeding season, young form small nomadic flocks prior to migration (Dementiev and Gladkov 1954a; Potapov 1966), not uncommonly with other *Motacilla*. Migrates in flocks of up to 10–15, also singly within flocks of Yellow Wagtail *M. flava* (Dolgushin *et al.* 1970). In winter quarters, occurs in scattered pairs or groups, sometimes in large flocks, usually in association with other *Motacilla* (Ali and Ripley 1973b). At Quetta (Baluchistan), April, thousands together along river edge (Meinertzhagen 1920a). On arrival at breeding grounds, often initially in flocks, e.g. in Kirgiziya (USSR), April, flocks of 8–15(–25) (Yanushevich *et al.* 1960). BONDS. No hard information on mating system or duration of pair-bond. However, pair-bond probably of seasonal duration in most cases since birds not firmly attached to any particular partner on arrival at breeding grounds (Günther 1972). In Finland, at same site in 2 successive years, ♂ bred with ♀ *M. flava* (Mikkola 1984), and breeding with *M. flava* suspected elsewhere (Wilson 1976; Cox and Inskipp 1978; Edqvist 1979). Both members of pair care for young (Günther 1972; Ali and Ripley 1973b), and both brood. Family bonds maintained for some time after leaving nest, until young fully grown (L L Semago). Will breed in immature plumage and age of first breeding thus probably 1 year (e.g. Whitehead 1909). BREEDING DISPERSION. Solitary and territorial, but typically in neighbourhood groups (Dolgushin *et al.* 1970). In Ladakh (Kashmir), estimated 31 territories, none overlapping, in *c.* 25 ha; average

size (calculated from illustration of territorial dispersion: E K Dunn) *c.* 0·4 ha (0·12–1·5, *n* = 25); territories tended to be smaller in patches of sea buckthorn *Hippophae rhamnoides*, larger in main areas of cultivation (Delany *et al.* 1982). In Moscow region, territories usually smaller when first occupied, larger once boundaries finalized (Günther 1972). In neighbourhood group, nests not uncommonly 50 m apart (Potapov 1966; Günther 1972). Average distance between 4 nests in Kazakhstan (USSR), 29·5 m (20·5–33·5); between 24 nests, Baraba (Novosibirsk, USSR), 30·1 m (13·5–64·0) (Chernyshov 1981, which see for distances from nests of *M. flava*). In Moscow region (USSR), 4 pairs 200, 300, and 400 m apart (Ptushenko and Inozemtsev 1968). On Yamal peninsula (USSR), 0·7–9·0 pairs per km² locally up to 5 pairs in 4–6 ha (Danilov *et al.* 1984). In northern Kazakhstan, 70–300 pairs per km² over 2 years; in Novosibirsk, 8–310 pairs per km² over 3 years (Chernyshov 1981). In Pechora basin (north-west USSR), 0·2–35·6 birds per km², according to habitat (Estafiev 1981). Nests usually near other Motacillidae, especially *M. flava* (Dolgushin *et al.* 1970). Territories overlap little or not at all with those of conspecifics, but may overlap considerably with those of *M. flava* (R Günther, L L Semago; see also Antagonistic Behaviour, below). Territory used for courtship, nesting, and some feeding, residents typically also feeding outside territory (Delany *et al.* 1982). ROOSTING. Outside breeding season, communally in reedbeds, etc. (Whitehead 1909; A Gretton); recorded roosting with *M. flava* (P J Ewins).

2. FLOCK BEHAVIOUR. No information, but for confrontations in flocks, see Antagonistic Behaviour (below). SONG-DISPLAY. Not unlike pipit *Anthus*. From start of breeding season, ♂ advertises with song (see 1 in Voice) from perch, flies up (apparently singing throughout: Dementiev and Gladkov 1954a) and, with vigorous wing-beats, hovers, and then suddenly descends at an angle on to elevated perch; repeats display a few minutes later (Grote 1925). Pair with young in nest flew a few metres into the air and from time to time hovered for a few seconds over some spot, not only near nest but also some distance away; this interpreted as possible Song-display preparatory to 2nd brood (Günther 1972), but perhaps a response to human intruder (E K Dunn; see Parental Antipredator Strategies, below). ANTAGONISTIC BEHAVIOUR. ♂ defends territory vigorously against conspecific ♂♂, much less so against ♀♀; residents attack and expel rival ♂♂ in pursuit-flights up to 100 m long (Günther 1972; R Günther). In Voronezh (USSR), apparent territory-holder adopted a presumed threat-posture: 'tensed and flattened' before driving off intruding ♂ (M G Wilson). Rarely (Chernyshov 1981) or occasionally (Günther 1972; R Günther) attacks *M. flava* in breeding season; one ♂ regularly entered territory of *M. flava* and frequently attacked resident ♂ (L L Semago). In Essex (England), ♂ *M. citreola* raising brood alone chased off ♀ *M. flava* which landed nearby (Cox and Inskipp 1978). Occasionally attacks Whinchat *Saxicola rubetra* (Günther 1972; R Günther). In May, 2 ♂♂ seen to fly up face to face in brief aerial tussle (M G Wilson). In Kashmir, late summer, birds in flocks described as being as quarrelsome as White Wagtails *M. alba*, chasing one another in erratic pursuit-flights (N van Swelm). HETEROSEXUAL BEHAVIOUR. (1) General. ♂♂ arrive on breeding grounds before ♀♀ (Kozlova 1930; Potapov 1966). ♂♂ tend to display much less than *M. flava*, and generally much less conspicuous (e.g. less vocal) than that species (L L Semago). (2) Pair-bonding behaviour. Begins soon after arrival on breeding grounds, e.g. in Moscow region, some ♂♂ defend territories *c.* 1 week after arrival (R Günther). Advertising ♂ commonly sits on exposed perch and gives Contact-calls (see 2

in Voice), with tail movements as in other Motacillidae (Wällschlager 1983). Apart from Song-flight (above), no other displays yet described. (3) Courtship-feeding. None known. (4) Mating. Occurs during nest-building (Semago *et al.* 1984). Only one description: ♂ began by adopting a posture apparently the same as, or similar to, Advertising-posture of *M. flava*: fluttered, then crouched horizontally, back feathers ruffled, tail spread and depressed. After copulating, ♂ ran *c.* 1 m away from ♀ and flew off (Mauersberger *et al.* 1982). (5) Behaviour at nest. No information on nest-site selection. ♀ builds nest; begins soon after arrival in territory, once the day after arrival (L L Semago). No further information. RELATIONS WITHIN FAMILY GROUP. Little known. Young in nest give incessant food-calls (see Voice) when parents away (Grote 1925). Both sexes feed young but ♀ does more (R Günther). At nest where no ♀ present, ♂ raised brood to fledging (Cox and Inskipp 1978). Parents continue to feed young for some time out of nest (L L Semago). ANTI-PREDATOR RESPONSES OF YOUNG. No information. PARENTAL ANTI-PREDATOR STRATEGIES. (1) Passive measures. Accounts of boldness at nest vary. Described by Osmaston (1925) as very wary when intruder near nest, but not shy according to Phillips (1949). Sentinel ♂ calls (probably 3 in Voice) to warn ♀ of approaching danger (Seebohm and Harvie Brown 1876). At one nest, incubating *calcarata* did not flush until almost trodden upon (Baker 1926). In 2 cases, ♀♀ flushed when intruder 6 m from nest, in 3 cases ♂♂ at 12 m; during disturbances, ♀♀ usually stayed nearer nest than ♂♂ (Günther 1972; R Günther). When young in or out of nest, response to man varies: some very wary and do not feed young as long as intruder nearby; others carry on bringing food to nest (L L Semago). (2) Active measures: against birds. Once, 2 birds attacked Magpie *Pica pica* (R Günther). (3) Active measures: against man. Bird flushed from nest flies up, calling excitedly (Günther 1972: see 3 in Voice), and one or both parents then demonstrate vocally—perched or, more often, flying around. Flight described as fluttering (N van Swelm), typically in tight circles (Denby and Phillips 1976). In Pechora basin (USSR), when neighbourhood group of breeders disturbed, birds perched on top of bushes (often in peculiar 'bunched-up' posture), flew around intruder in circles, or hovered overhead, calling; sometimes dozens together, often approaching from a distance, and following intruder until he left (Seebohm and Harvie Brown 1876).   EKD

Voice. Freely used in breeding season, less so at other times; birds said to be less vocal in breeding season than Yellow Wagtail *M. flava* (L L Semago). No information on differences, if any, between races. For other sonagrams, see Bergmann and Helb (1982).

CALLS OF ADULTS. (1) Song of ♂. (a) In recording of perched bird, song a sequence of phrases composed of units resembling Contact-calls (see 2a below), variously combined together—notably 'cheeoo ee' (Fig I) and 'ch ch chk'; in one recording, *c.* 6 phrases in 20 s (E K Dunn). (b) In Song-flight, a rapid repetition of units resembling Contact-calls, but softer and more delicate (Grote 1925; Dementiev and Gladkov 1954a). A rapid chattering sequence containing many Contact-call units, frequently given in Song-flight (Bergmann and Helb 1982). (2) Contact-calls. Most frequently heard calls, given perched or in flight, and used in communication

between pair, and between flock-members. (a) Main type. Often monosyllabic, sometimes disyllabic, and any one bird utters both. Monosyllabic call widely agreed to be harsher, and perhaps shriller and shorter than *M. flava* (Svensson 1977; Cox and Inskipp 1978; Edqvist 1979). To the practised ear, thus readily distinguishable from *M. flava* (Cox and Inskipp 1978), though *calcarata* apparently almost indistinguishable from eastern races of *M. flava* (Lekagul *et al.* 1985). Rendered 'sreep', 'drreep', or 'sweeip' (Svensson 1977); 'dzzip' or 'dzzeep', constant in pitch and with rasping quality reminiscent of Tree Pipit *Anthus trivialis* and Reed Bunting *Emberiza schoeniclus* (Cox and Inskipp 1978). Recording of ♀ in flight (Fig II) suggests 'psreee' (P J Sellar). Disyllabic variant described as buzzing 'wizzzippp' (M G Wilson), 'chiz-zit' (Ali and Ripley 1973b), or 'zeet-zeet' (Walker 1981a). Studies and sonagrams presented by Bergmann and Helb (1982) and by Wallschläger (1983) suggest considerable variation in repertoire of individuals: one ♂ gave mainly 'zilit', also 'slit'; one ♀ gave series of 'zilit' and 'schlit' sounds, all apparently somewhat different from one another (Wallschläger 1983, which see for details). (b) Apparently subdued variants of call 2a not uncommon,

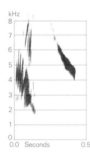

I   P A D Hollom   Iran   May 1977

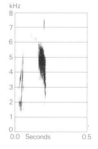

II   P A D Hollom   Iran   May 1977

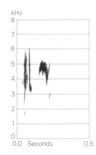

III   Svensson (1984)   USSR   June 1983

e.g. a short 'chitt' or quite soft 'chirrtt' (M G Wilson). (3) Alarm-call. When disturbed by human intruder at nest, gives calls described as excited (Günther 1972), loud (Denby and Phillips 1976), and high-pitched (N van Swelm); evidently a variant of call 2. Recording (Fig III) suggests strident 'tssik' or 'pssseee', occasionally interspersed with lower-pitcheed 'twitup' (P J Sellar); 'tschiit' (Bergmann and Helb 1982, in which see Fig c) is evidently this call. (4) Other calls. In *calcarata*, a metallic 'pzeeow', slightly higher pitched and less rasping than more frequently heard 'dzeep' (call 2a); said to be diagnostic of *M. citreola* (Lekagul *et al.* 1985), but more information needed on affinities with other races and other calls.

CALLS OF YOUNG. Little information. Food-calls of young in nest an incessant twittering (Grote 1925). Contact-call not described, but contact-calls of fledged young suspected to be *M. citreola* × *M. flava* hybrids were similar to adult *M. citreola* (Cox and Inskipp 1978; Edqvist 1979).   EKD

Breeding. SEASON. Northern USSR: eggs laid mid-June. Southern USSR: eggs laid late April to June (Makatsch 1976). SITE. On ground in hollow in bank, or under thick vegetation or stone. Nest: cup of moss and plant leaves and stems, lined with hair, wool, and feathers, thicker in north of range. External diameter 11 cm, height 5·5 cm, internal diameter 6·5 cm, depth of cup 3·5 cm Makatsch 1976). Building: by ♀ (Semago *et al.* 1984). EGGS. See Plate 81. Sub-elliptical, smooth and glossy; buff or pale grey, finely speckled grey or grey-brown; sometimes mottled light brown. Nominate *citreola*: 19·5 × 14·4 mm (18·0–21·3 × 13·8–15·3), $n = 85$; calculated weight 2·0 g. *M. c. calcarata*: 21·0 × 15·3 mm (18·5–22·5 × 14·0–16·0), $n = 125$; calculated weight 2·42 g (Schönwetter 1979). Clutch: 4–6(–7). 1–2 broods. 2 broods probably regular in Moscow region, USSR (Günther 1972). INCUBATION. 14–15 days (Yanushevich *et al.* 1960). By both sexes (Ali and Ripley 1973*b*). FLEDGING TO MATURITY. Fledging period 13–15 days (Yanushevich *et al.* 1960). According to Gladkov (1962), 11–12 days, this perhaps premature departure. Fed by parents for some time after fledging (Semago *et al.* 1984). Age of first breeding probably 1 year (see Social Pattern and Behaviour). No further information.

Plumages (nominate *citreola*). ADULT MALE BREEDING. Entire head bright yellow or golden-yellow, feathers of forehead, crown, and nape narrowly tipped dusky when fresh, hindcrown and nape sometimes slightly washed dull olive-green; ear-coverts sometimes slightly tinged olive. Hindneck and upper mantle black, forming contrasting black band, extending down to upper sides of breast. Lower mantle and scapulars dark olive-grey (when fresh) or dark grey with variable olive tinge (when worn), grading to dark slate-grey on rump and to black on longer upper tail-coverts; outer webs of outer tail-coverts broadly fringed white. Underparts bright yellow, but sides of breast and flanks washed olive-grey and under tail-coverts white

with pale yellow tinge; occasionally some black spots on sides of chest, forming incomplete collar. Central 4 pairs of tail-feathers (t1–t4) black, t1 with narrow pale yellow or white fringe all round, t2–t4 with fringe along outer web only (soon worn off); t5–t6 mainly white, each with black wedge on basal part of inner web, extending into point along inner border up to $\frac{2}{3}$ of tail length on t5, to *c.* $\frac{1}{2}$ on t6. Flight-feathers and greater upper wing-coverts dark brown, ground-colour of tertials and longest feathers of bastard wing virtually black; outer web and tips of flight-feathers narrowly edged white, of primary coverts pale grey; outer web and tip of shorter 2 tertials with broad and rather poorly defined pale yellow-grey fringe *c.* 4–5 mm wide (soon bleaching to greyish-white), longest tertial with more well-defined white fringe along outer web (*c.* 3 mm wide at middle). Lesser upper wing-coverts dark olive-grey (like upperparts), longer lesser with darker grey-brown centres; median coverts dull black with contrastingly white tips 3–6 mm wide (widest on central coverts; partly washed pale yellow when plumage fresh); greater coverts dull black with broad white tip and outer fringe (5–8 mm wide on central coverts, narrower and with pale grey-brown tinge on outermost); white tips of median and greater coverts occasionally very broad, black at bases concealed. Under wing-coverts and axillaries pale grey-brown, longer coverts broadly tipped white; lesser coverts along leading edge of wing dark brown with white tips. *In worn plumage*, yellow of head and underparts slightly less bright and deep, upperparts duller and greyer, pale fringes and tips of tertials largely worn off. ADULT FEMALE BREEDING. Similar to adult ♂ breeding, but hindcrown and hindneck dark olive-grey or dull grey, like mantle and scapulars, sometimes partly mottled yellow-green or yellow; yellow of forehead, forecrown, and ear-coverts sometimes mottled and washed olive-green (isolating broad yellow supercilium); black band across mantle absent or indicated by black mottling only, mainly at sides; yellow of head sometimes slightly paler, of underparts often slightly paler, under tail-coverts virtually pure white; white tips of median and greater upper wing-coverts on average narrower, on median 3–5 mm wide, on greater 3–6 mm; pale fringes of tertials slightly narrower, those of inner 2 often grey, not as pale as adult ♂ breeding. ADULT MALE NON-BREEDING. As adult ♀ breeding, but white tips of upper wing-coverts and pale fringes of tertials average wider. ADULT FEMALE NON-BREEDING. Central forehead, crown, hindneck, and sides of neck dark olive-grey or dark grey, similar to mantle and scapulars, tinged yellow-green when fresh, in particular on forehead and crown; no black on upper mantle; remainder of upperparts as adult ♀ breeding, with dark grey rump merging into dull black upper tail-coverts. Sides of forehead, broad supercilium, and remainder of sides of head pale yellow (paler than adult ♀ breeding), ear-coverts mottled and streaked olive. Underparts pale yellow, but sides of breast and flanks pale ash-grey and under tail-coverts virtually white. Wing as adult ♀ breeding. Similar to adult ♂ non-breeding, but forehead and forecrown less extensively yellow, no black on upper mantle, head and underparts paler yellow, and pale tips and fringes on wing narrower. NESTLING. Down long and dense; grey (Günther 1972). JUVENILE. Forehead buff, merging into dull olive-brown crown; entire upperparts dull olive-brown, feathers narrowly and faintly fringed buff or paler olive-brown when plumage fresh; upper tail-coverts black, outer with narrow pale buff or off-white outer fringe. Supercilium and narrow eye-ring pale buff, widest and almost white above ear-coverts. Lores, upper cheeks, and ear-coverts dull olive-brown, narrowly streaked white or pale buff below eye and on shorter coverts. Underparts pale buff or cream-buff, virtually white on chin, upper throat, vent, and under tail-coverts, washed olive-brown on sides of breast and flanks; a narrow dull olive-brown collar extends down from sides of neck, often broken into spots or entirely absent on central chest; sometimes an indistinct brown malar stripe. Tail as adult; wing as adult, but shorter lesser upper wing-coverts faintly fringed buff, longer lesser contrastingly tipped pale buff or pale yellow; white or pale yellow tips of median coverts 2-3 mm wide, on greater 2-5 mm, pale outer fringes of tertials *c.* 2 mm. FIRST ADULT NON-BREEDING. Differs from adult non-breeding in cream or white ground-colour of head and underparts, virtually without yellow. Central forehead, crown, hindneck, sides of neck, and remainder of upperparts medium grey, in fresh plumage slightly tinged buff on forehead and olive on remainder of upperparts, on rump merging into black upper tail-coverts; lateral upper tail-coverts with white outer fringes. Sides of forehead, broad supercilium, band round rear of ear-coverts, lower cheek, chin, and throat white or pale cream, sometimes with bright yellow spot on supercilium above front of eye. Lore, upper cheek, and ear-coverts grey, narrowly streaked or mottled white or pale cream below eye and on shorter coverts. Sometimes a narrow area of dark spots from sides of neck to chest. Sides of breast and flanks medium grey, remainder of underparts cream-white, often slightly deeper cream-buff on chest and purer white on vent and under tail-coverts. Wing and tail as in adult non-breeding, but juvenile flight-feathers, upper primary coverts, usually outer greater coverts, often a variable number of tail-feathers and tertials, and sometimes a few outer median and lesser coverts retained, plumage not as uniformly new as adult at same time of year; often a sharp contrast between fresh inner greater coverts (with broad white tips) and neighbouring worn outer (with narrower tips), and sometimes a similar contrast in median coverts; tips of central median and greater upper wing-coverts pure white, not partly pale yellow as in fresh adult. Sexes similar. FIRST ADULT BREEDING MALE. Like adult ♂ breeding, but hindcrown and hindneck often mottled or streaked dark grey or dull black, forecrown sometimes mottled dusky, not as pure yellow as adult; yellow of head and underparts sometimes slightly paler; flight-feathers and greater upper primary coverts still juvenile, more worn than those of adult at same time of year; heavily worn juvenile outer greater coverts sometimes contrast with newer inner ones (but adult often also shows contrast between old outer non-breeding coverts and fresh inner coverts). FIRST ADULT BREEDING FEMALE. Rather variable. Some are indistinguishable from adult ♀ breeding except sometimes for relatively more worn juvenile flight-feathers and greater upper primary coverts; usually no black on upper mantle. Others similar to adult ♀ non-breeding, with dark forehead and crown and pale yellow supercilium and underparts. Others still mainly like 1st non-breeding, with supercilium and underparts mainly white and crown dark, but some pale yellow may show on forehead, supercilium, cheek, or throat.

Bare parts. ADULT. Iris dark brown. Bill black. Leg and foot black or black-brown. NESTLING. No information. JUVENILE, FIRST ADULT. At about fledging, iris dark brown, bill flesh with greyish culmen, mouth pink-red, leg and foot greyish-flesh (Swelm 1980). Later on, bill dark horn-brown or black-brown, base of lower mandible flesh-brown or grey; gape-flanges pale yellow until early autumn. Mouth yellow, orange-yellow with pink centre, or yellowish-pink. Leg and foot horn-brown or black-brown. (Williamson 1955a; Ali and Ripley 1973b; RMNH, ZMA.)

Moults. ADULT POST-BREEDING. Complete; primaries descendant. In *werae* and nominate *citreola*, starts with p1 in early or mid-July (Dementiev and Gladkov 1954a; Piechocki and Bolod 1972; RMNH), soon followed by shedding of t1-t2 and some feathers of forehead and sides of chest; in full moult August (Havlín and Jurlov 1977), completed mid- or late August (Dementiev and Gladkov 1954a; Piechocki 1958). In *calcarata*, starts late July or August (Dementiev and Gladkov 1954a) or *c.* 5-20 August (Delany *et al.* 1982); by late August, primary moult score 25-45 reached; all moult completed (score 50) about mid- or late September (Delany *et al.* 1982). Tail starts with primary score 0-10 (sequence approximately 1-2-6-3-4-5: RMNH, ZMA), secondaries at score *c.* 20, all completed at about same time as p9-p10 (Delany *et al.* 1982; RMNH). ADULT PRE-BREEDING. Partial; February-March. Involves head, body, tertials (1-2 occasionally excluded), all upper wing-coverts (except for 1-6 outer greater coverts, average 3·3, n = 18), and all or part of tail (in particular t1 and t6, but occasionally none). In ♀, moult on average less extensive than in ♂, ♀ more often retaining all tail and occasionally back, rump, and outer lesser and median upper wing-coverts. POST-JUVENILE. Partial; starts soon after fledging and timing hence strongly variable. Some in first non-breeding mid-July, others with part of body still juvenile in September or (rarely) October; in late-fledged birds, moult continued in winter quarters, especially wing-coverts and tail. During autumn migration, some completely in non-breeding except for flight-feathers and greater upper primary coverts, but usually at least some juvenile outer greater coverts and tail-feathers retained, and occasionally many other wing-coverts, tertials, and part of body also; by early spring, often hardly any juvenile remaining except for flight-feathers and primary coverts. FIRST ADULT PRE-BREEDING. Partial; timing as in adult pre-breeding, but extent highly variable. In some, moult as extensive as adult pre-breeding, but many birds replace only head, body (often excluding part of back or rump and sometimes lower belly or vent), some or all median and tertial coverts, 0-2 tertials, and a few tail-feathers (mainly t1 and t6). Plumage (still fairly fresh) acquired late in post-juvenile moult is retained (e.g. many greater or outer median coverts, some tertials, t4-t5).

Measurements. ADULT, FIRST ADULT. Nominate *citreola*. Lower Pechora river (north European USSR), summer: skins (BMNH). Bill (S) to skull, bill (N) to distal corner of nostril; exposed culmen on average *c.* 4·2 less than bill (S).

	♂		♀	
WING	87·9 (1·49; 17)	85-90	82·3 (1·65; 7)	80-85
TAIL	76·6 (2·15; 17)	74-81	72·6 (1·90; 7)	70-76
BILL (S)	17·5 (0·60; 17)	16·7-18·8	16·8 (0·66; 7)	16·1-17·4
BILL (N)	10·2 (0·31; 17)	9·7-10·7	9·8 (0·44; 7)	9·1-10·2
TARSUS	26·3 (0·79; 17)	25·3-27·7	24·6 (0·70; 7)	23·8-25·6

Sex differences significant.

Tien Shan and Altai mountains (USSR) and neighbouring China, summer; skins (BMNH, RMNH, ZMA). Toe is middle toe with claw, claw is hind claw.

	♂		♀	
WING	85·2 (0·96; 11)	84-87	80·7 (1·91; 6)	77-82
TAIL	71·5 (2·34; 11)	69-75	69·0 (2·02; 6)	66-72
BILL (S)	17·5 (0·41; 11)	16·9-18·1	16·8 (0·56; 7)	16·1-17·6
BILL (N)	10·6 (0·34; 11)	10·0-11·3	10·0 (0·37; 6)	9·5-10·6
TARSUS	25·8 (0·78; 11)	24·7-26·8	25·0 (0·38; 6)	24·4-25·2
TOE	20·3 (0·71; 9)	19·5-21·8	20·4 (0·87; 4)	19·7-21·6
CLAW	11·0 (1·30; 9)	8·2-12·6	12·0 (1·64; 6)	10·1-13·6

Sex differences significant for wing and bill.

Transbaykalia, Mongolia, and Manchuria, summer and autumn;

skins (Piechocki 1958; Piechocki and Bolod 1972; RMNH, ZMA).

WING	♂ 91.2 (1.36; 21)	89–93	♀ 83.4 (3.89; 13)	77–88

M. c. werae. South European USSR, April–June; skins (BMNH, RMNH, ZMA).

WING	♂ 82.4 (1.47; 18)	80–84	♀ 78.5 (2.50; 5)	76–82
TAIL	69.3 (2.13; 18)	66–71	66.6 (1.95; 5)	65–69
BILL (S)	16.8 (0.41; 18)	16.0–17.3	16.6 (0.46; 5)	16.0–17.3
BILL (N)	10.0 (0.52; 18)	9.2–10.8	9.7 (0.21; 5)	9.4–9.9
TARSUS	24.1 (0.61; 18)	23.0–25.1	23.5 (0.47; 5)	23.1–24.2
TOE	19.1 (0.92; 6)	18.3–20.6	17.9 (— ; 2)	17.3–18.5
CLAW	10.3 (0.86; 18)	9.1–11.8	10.1 (1.15; 4)	9.5–11.9

Sex differences significant for wing and tail.

Wing: European USSR, ♂ 82.1 (2.52; 20) 78–85 (Gladkow 1941); all USSR, ♂ 79.3 (114) 74–83, ♀ 77.0 (35) 74–81 (Bub *et al.* 1981), ♀ 74.4 (5) 70–77 (Dementiev and Gladkov 1954a). Tail: ♂ 68–75, ♀ 65–72 (Sushkin 1925).

JUVENILE. Though retained juvenile wing of 1st adult on average 1.8 shorter than full adult, all ages combined in tables above.

Weights. Nominate *citreola*. Kazakhstan, USSR (Dolgushin *et al.* 1970), Mongolia (Piechocki and Bolod 1972), and Manchuria, China (Piechocki 1958), combined: March–April, ♂ 20.5 (1.87; 4) 18–23; May–June, ♂ 22.2 (1.32; 13) 20–24, ♀ 21.1 (2.10; 8) 18–25; July–August, ♂ 19.9 (2.11; 13) 17–23, ♀ 20. Northern USSR, summer: ♂ 20.7 (14) 19–23, ♀ 19.9 (11) 17–23 (Dementiev and Gladkov 1954a). Yamal peninsula (USSR), summer: ♂♂ 19.0, 20.4, 22.1; ♀♀ 20.6, 23.8 (Danilov *et al.* 1984). Britain: September, 16.8–18.8; early October, 15.4 (Williamson 1955a); October, 18.6 (Axell *et al.* 1965).

M. c. werae. Kazakhstan: March–April, ♂ 18.7 (1.22; 8) 17–21; May–June, ♂ 19; ♀♀ 14.4, 19.1 (Dolgushin *et al.* 1970). Turkey, May: ♀ 15 (Vauk 1973a). Lake Chany (south-west Siberia), mainly August: adult 16.7 (0.97; 5) 15.6–18.2; juvenile 16.1 (1.19; 52) 13.4–18.1 (Havlín and Jurlov 1977).

M. c. calcarata. India, March–April: 18.1 (13) 15–21 (Ali and Ripley 1973b). Afghanistan and Kansu (China), June–July: ♂ 18.5 (2.13; 7) 15–22, ♀♀ 18, 23 (Stresemann *et al.* 1937; Paludan 1959). Ladakh (India), mainly August: 18.5 (1.41; 32) 15–22 (Delany *et al.* 1982).

Structure. Wing rather short, broad at base, tip bluntly pointed. 10 primaries: p7–p9 longest or any one 0–0.5(–1) shorter than others; p6 1–2.5 shorter than longest, p5 8–12, p4 12–16, p3 15–19, p1 20–24. P10 reduced, narrow and pointed; 52–62 shorter than wing-tip, 7–12 shorter than longest greater upper wing-covert. Outer web of p6–p8 and (faintly) inner of (p7–)p8–p9 emarginated; emargination on outer web of p8 18–24 from tip of p8, equal to tip of about p2 in closed wing (sometimes halfway between p2 and p3 or near p1). Longest tertial reaches tip of p5–p8 in closed wing. Tail rather long, tip square or slightly rounded; 12 feathers, t6 1–4 shorter than t2–t3. Bill long and slender, similar to Yellow Wagtail *M. flava*, but tip relatively slightly longer. Tarsus, toes, and claws relatively longer than in *M. flava* (see Measurements), except in *werae*. Outer toe with claw *c.* 73% of middle with claw, inner *c.* 75%, hind about similar (of which 57% is claw). Hind claw somewhat less decurved than in *M. flava*.

Geographical variation. Marked; involves colour of ♂ breeding plumage and size, but differences partly obscured by strong individual variation. Breeding ♂ of nominate *citreola* has upperparts dark grey with olive tinge, broad black band across upper mantle and down to upper sides of breast, and olive-grey flanks. ♂ *werae* similar but upperparts slightly paler grey, black band on average narrower, flanks and sides of breast less extensively washed with paler olive-grey, and yellow of head and underparts slightly less deep yellow; often indistinguishable unless compared directly with nominate *citreola* (but see Measurements). Birds from Zaysan basin through Altai to Sayan mountains are rather pale, near *werae*, but size similar to nominate *citreola*, and these sometimes separated as *quassatrix* Portenko, 1960; however, individual variation marked, with many birds indistinguishable from nominate *citreola* from further north and east, and hence included in nominate *citreola* here. Breeding ♂ *calcarata* from Iran to central China has mantle and scapulars wholly black, back and rump sometimes black also, head and underparts on average slightly deeper yellow than nominate *citreola*, black on t5 sometimes slightly more extensive; however, a few nominate *citreola* also wholly black above (Johansen 1952; Vaurie 1959; Mauersberger 1982). Races hard to distinguish in juvenile, non-breeding, or ♀ plumage; in all these plumages, *werae* has less grey wash on flanks than nominate *citreola*, breeding ♀ usually lacking black on mantle, and head and underparts paler yellow than in nominate *citreola*; *calcarata* averages darker grey on upperparts and flanks, sometimes with more black in tail, juvenile having more pronounced dark stripe along side of crown and on cheek, and 1st breeding ♀ having ground-colour of side of head and underparts usually white rather than yellow as in most ♀♀ of nominate *citreola* and *werae*. Size also helpful (especially when sex known). *M. c. werae* smaller than nominate *citreola*, with tarsus particularly short. Wing and tail of *calcarata* similar to nominate *citreola*, but bill slightly longer, tarsus, middle toe with claw, and hind claw markedly longer: tarsus of ♂ *calcarata* mainly over 26.5, of ♀ over 25.5; middle toe with claw of ♂ mainly over 21.5, of ♀ over 20.5; hind claw mainly over 11.5 (nominate *citreola* mainly below these values) (RMNH, ZMA). Sum of tarsus and middle toe with claw over 49 in *calcarata*, below 47 in nominate *citreola* (Stresemann *et al.* 1937).

Recognition. Sometimes hard to distinguish from Yellow Wagtail *M. flava* or White Wagtail *M. alba*, though breeding ♂ with yellow head and underparts unmistakable (upperparts grey, not green as in *M. f. flavissima* and *M. f. lutea*). Other plumages closely similar to *M. alba* on upperparts, showing similar gradual darkening from grey of back to dark grey on rump and black on upper tail-coverts (underparts of *M. alba* quite different, with distinct dark gorget, widest in centre). Sides of head and underparts rather like *M. flava*, but supercilium wider, especially above ear-coverts, extending as pale band behind ear-coverts down to lower throat, and tips of median and greater upper wing-coverts and fringes of tertials and flight-feathers broad and white (slightly pale yellow on tips of coverts in fresh plumage), not narrow and yellow or grey-yellow as in *M. flava* (juvenile *M. flava* has tips of coverts pale yellow or white, but narrower than in *M. citreola*, and fringes of flight-feathers and tertials always tinged yellow, not white). Nominate *citreola* and *calcarata* also separable from *M. flava* by large size (especially of bill and tarsus) and all races have different wing-formula: p6–p8 emarginated on outer web in *M. citreola* (and *M. alba*), p7–p8 in *M. flava* (p6 sometimes slightly emarginated, but not as marked as p7–p8); length of emarginated part of outer web of p8 18–24 in *M. citreola* (and *M. alba*), 15–20 in *M. flava*; emargination on p8 near tip of p2 in closed wing in *M. citreola* (sometimes between p2 and p3), between p3 and p4 in *M. flava* (and *M. alba*, or sometimes equal to p3); p6 1-2·5 shorter than wing-tip in *M. citreola*, 3-6 in *M. flava* (and *M. alba*) (*n* = 10 in each), or, according to

literature, p6 1·9 (25) 0·5-4 shorter in *M. citreola*, 3·0 (25) 2·5-8 in *M. flava* (Williamson and Ferguson-Lees 1955) or 1-4(-5) in *M. citreola* and 3-6 in *M. flava* (Svensson 1984*a*).

CSR

Motacilla cinerea Grey Wagtail

PLATES 26 and 31
[between pages 232 and 233, and facing page 353]

Du. Grote Gele Kwikstaart   Fr. Bergeronnette des ruisseaux   Ge. Gebirgsstelze
Ru. Горная трясогузка   Sp. Lavandera de cascada   Sw. Forsärla

Motacilla Cinerea Tunstall, 1771

Polytypic. *M. c. patriciae* Vaurie, 1957, Azores; *schmitzi* Tschusi, 1900, Madeira; *canariensis* Hartert, 1901, Canary Islands; nominate *cinerea* (Tunstall, 1771), north-west Africa and Europe, east to Caucasus and Iran; *melanope* Pallas, 1776, Asia from Ural mountains and Afghanistan east to middle Amur river. Extralimital: *robusta* (Brehm, 1857), eastern Asia from Kamchatka and Okhotsk Sea region south to north-east China and Japan.

Field characters. 18-19 cm; wing-span 25-27 cm. Longer and more attenuated than any other west Palearctic wagtail, with slimness of rear body enhanced by exceptionally long tail (up to 35% longer than in Yellow Wagtail *M. flava*). Very graceful, lithe, slim wagtail, with almost constantly moving tail so long that it 'whips'. Plumage essentially grey above and yellow below; wings largely black (but with obvious white bar showing in flight) and tail black with white outer feathers. ♂ has black bib in summer. Flight action exceptionally bounding. Call distinctive. Sexes dissimilar; seasonal variation marked in ♂. Juvenile separable. 5 races in west Palearctic; race of Europe and north-west Africa easily distinguished from 3 isolated island races (see Geographical Variation for west Asian race).

ADULT MALE BREEDING. (1) European and north-west African race, nominate *cinerea*. Upperparts (except for olive-yellow rump) basically medium grey, with olive, blue, or slate tone according to angle of light; head has black lore, narrow white supercilium and eye-ring, and complete, quite broad white border to cheek, below which black chin and throat form striking bib. Wings olive-grey on coverts, becoming virtually black elsewhere; folded wing shows narrow yellow to white fringes to tertials and narrow white line across bases of inner secondaries. Tail black, with very obvious white outer feathers. Underparts from bib to under tail-coverts, and rump and upper tail-coverts, lemon-yellow, with greyish suffusion on flanks. In flight, white bases to secondaries and inner primaries form obvious wing-bar which together with yellow rump and long black tail form 3 striking characters on fleeing bird; underwing grey-white, with white, almost translucent, band along centre. (2) Azores race, *patriciae*, and Madeiran race, *schmitzi*. Darker grey above, particularly on cheeks; white supercilium less obvious. Area of white on outer tail-feathers reduced; *patriciae* noticeably long-billed. (3) Canary Islands race, *canar-*

iensis. Less dark above than *patriciae* and *schmitzi* but more vividly coloured than nominate *cinerea*, with underparts canary-yellow or even orange-yellow and white marks on face and wings more contrasting (wing markings in particular far more noticeable). ADULT FEMALE BREEDING. All races. General appearance as ♂ but readily separated by lack of discrete bib: chin and throat buff-white, merely mottled black towards and on chin. ADULT MALE NON-BREEDING. All races. Supercilium becomes buff-white and indistinct; black bib lost, whole of throat buff-white. Breast buff-yellow, contrasting with pale throat but merging with yellow belly and vent. ADULT FEMALE NON-BREEDING. All races. Closely resembles ♂ non-breeding, but in direct comparison buffier chest and paler yellow underparts may allow separation. JUVENILE. All races. Resembles winter ♀ but at close range separable by green-brown tone to upperparts, buff face markings (much less distinct than adult's), buffier chest with grey-black mottling on breast-sides (occasionally forming broad chest-band), wide grey flanks, very pale yellow-white belly, buff fringes and tips to wing-coverts, and dull grey fringes to tertials. FIRST WINTER. All races. Closely resembles winter ♀ but buffier on chest. Sexes not separable. FIRST SUMMER. As adult breeding, but ♂ often has some white in throat-bib and ♀ often has chin and throat all-white. At all ages, bill grey-black (with paler base in juvenile); legs brown-flesh, distinctly paler than any other west Palearctic wagtail.

Unmistakable. Bird in atypical surroundings may suggest *M. flava* momentarily but no other west Palearctic wagtail is as slim or long-tailed, or shows combination of yellow rump, single white bar on both upper- and underwing, and pale legs. Flight markedly bounding at height, flitting and darting along streams and among boulders. Action as *M. alba*, but wing-beats looser, with translucent panel in midwing remarkably obvious at times, and 'shooting' curve or fall exaggerated by marked

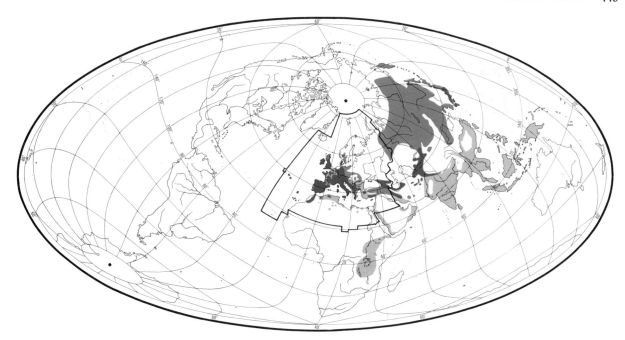

acceleration and 'whipping' or 'streaming' tail. Flight silhouette much more attenuated than any other west Palearctic wagtail. Gait similar in action to *M. alba* but noticeably delicate as bird picks its way along and hops over obstructions; walk, run, and jump all accompanied by almost incessant, nervous wagging of tail which may cause feathers to separate and fan. Stance on ground most horizontal of all west Palearctic wagtails, with tail even held up as bird dashes about. Perches often, with length and slight wagging of tail always obvious; body attitude more angled than on ground but tail rarely drooped as much as in *M. alba*. Not gregarious except at winter roosts; usually migrates singly.

Song in flight involves trill with opening notes recalling Wren *Troglodytes troglodytes*; from perch, a staccato, then more melodious warble, with opening notes effectively an extension of commonest call—a high-pitched, metallic disyllable, 'tzitzi', staccato or stuttered, and shorter or more clipped than *M. alba*.

Habitat. In west Palearctic, occurs mainly in temperate middle and lower-middle latitudes, overlapping sparingly into boreal and Mediterranean. Typical breeding habitats include combination of: (1) fresh water, especially fast-running streams and rivers, but also canals, lowland streams, and margins of lakes, both oligotrophic and eutrophic; (2) rock slabs, boulders, vertical rock faces, shingle or gravel stream beds, or artefacts such as sluices, weirs, locks, culverts, walls, or roofs; (3) sheltering trees, shrubs, or dense herbage; (4) holes, ledges, or hollows for nesting. Upland and mountain streams often provide ample choice of such requirements; lowland streams, lakes, and reservoirs generally less; lowland canals,

woodland pools, habitations with artificial tanks or drains, and other peripheral types of site are occupied locally or infrequently. Territories accordingly often linear in pattern, ascending to limit of suitable waters, in Switzerland above 1600 m (above treeline) (Glutz von Blotzheim 1962); in Morocco to 3000 m (Heim de Balsac and Mayaud 1962) and in Georgia (USSR) commonly at 600–800 m but also met at 2500 m or even 3000 m. In USSR, mainly in forest zones wherever suitable waters occur, but sometimes above treeline to foot of glaciers; also in human settlements, although much less so than White Wagtail *M. alba* (Dementiev and Gladkov 1954a). Avoids waters with dense marginal or emergent vegetation inhibiting easy movement, but will roost in reedbeds. In winter, shifts generally to lowlands, estuaries, coasts, and artificial situations such as sewage farms, retaining more marked attachment to water than congeners. Has a range of winter quarters in Africa from ornamental water in Egypt to mountain streams on Kilimanjaro, Tanzania (Moreau 1972). Flies readily and widely in lower-middle as well as lower airspace, reconnoitring alternative habitats. Little affected by disturbance or changes in land use, which it can often turn to advantage.

Distribution. Became established in central Europe shortly after 1850 (Voous 1960). Has since bred Netherlands, Sweden, Finland, and spread in Norway and Poland.

BRITAIN. Expanded in eastern and southern England in 1950s; in Scotland irregularly includes Orkney and Outer Hebrides (Parslow 1973; Sharrock 1976). NETHERLANDS. First bred *c.* 1915 (Voous 1960); probably colonized 1850–1900 (CSR). DENMARK. First bred 1923

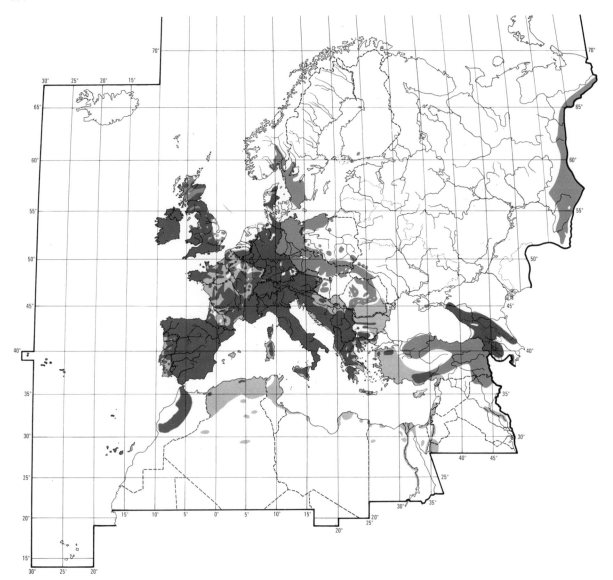

(Jørgensen 1970) and spread (TD). NORWAY. Some range expansion (VR). SWEDEN. First bred 1916 (LR). FINLAND. First bred 1976; now regular Karkkila, occasionally elsewhere (OH). POLAND. In mid-19th century only in mountainous areas; spread to lowlands from 1865 and bred there in 20th century (Tomiałojć 1976). TURKEY. Very thinly distributed except in eastern Black Sea mountains (Vittery *et al.* 1971).

Accidental. Iceland, Faeroes.

Population. Fluctuates with hard winters. Has increased in Scandinavia, but decreased recently in Netherlands and Poland.

BRITAIN, IRELAND. 1000-10 000 pairs; fluctuates markedly with hard winters. Increased in eastern and southern England in 20th century, especially in 1950s; decreased after 1960 with hard winters (Sharrock 1964, 1969; Parslow 1973). About 25 000-50 000 pairs, with recovery during late 1960s (Sharrock 1976). FRANCE. 10 000-100 000 pairs (Yeatman 1976). BELGIUM. About 1500 pairs (Lippens and Wille 1972). NETHERLANDS. 110-125 pairs 1975-6, 130-150 pairs 1977, 150-175 pairs 1978 (Teixeira 1979); perhaps numbers double these in 1940s, with decline due to hard winters and water pollution (CSR). WEST GERMANY. 20 000-50 000 pairs (Rheinwald 1982). DENMARK. Probably 60-75 pairs (Jørgensen 1970); now a few hundred pairs (TD). NORWAY. Probably increasing (VR). SWEDEN. About 500 pairs (Ulfstrand and Högstedt 1976); numbers now somewhat higher though fluctuating with hard winters (LR). FINLAND. 5-10 pairs. POLAND. Recent decline in lowlands, except Pomerania (AD). EAST GERMANY. Widespread but scarce (SS). HUNGARY. About 70 pairs (LH). USSR. Estonia: breeds occasionally

(HV). GREECE. Several hundred pairs (GIH). AZORES. Abundant. Not persecuted (G le G).

Movements. Mainly a partial migrant, but wholly migratory or resident in some parts of range. Found in winter throughout most of European breeding range but some move to Africa as far south as southern Malawi (Benson and Benson 1977). Winter range also includes Middle East, and birds from central and eastern Asia move to India and south-east Asia as far as New Guinea, with single record from northern Australia (Macdonald 1973; Slater 1974). Populations of Azores (*patriciae*), Madeira (*schmitzi*), and Canary Islands (*canariensis*) are resident (Bannerman 1963; Bannerman and Bannerman 1965, 1966).

In autumn, movements within Europe of over 200 km (as shown by ringing) have been mainly between north-west, south-west, and south-east in central and western Europe, and mainly around south-west in Britain. Thus a nestling from Belgium recovered in England and an autumn migrant from Kärnten (Austria) recovered in Dalmatia (Yugoslavia); even longer, westward, movements (*c.* 800 km) are Denmark to South Wales, and East Germany to south-east England. There are also cases of roughly eastward movement in England and north-west France including a movement of 240 km ENE from Brittany to Channel Islands which could be regarded as reverse migration. (Zink 1975; Jørgensen 1976; Tyler 1979.)

Most move south-west to Portugal, Italy, or north-west Africa (Lathbury 1970; Heim de Balsac and Mayaud 1962; Zink 1975). Winter recoveries of birds ringed in central Europe are relatively evenly distributed from the Rhine south-west to Iberia though with large gap in north-east Spain (Zink 1975). Recoveries of birds ringed Germany and recovered in Africa as follows: Morocco 8, Algeria 3, Tunisia 1, Mauritania 1, Mali 1, Sénégal 6, Ghana 2 (Schloss 1982). 2 Czechoslovakian-ringed birds recovered in Algeria (Jørgensen 1976).

In Britain, mainly resident with local dispersion: only 25% of 68 recoveries were over 100 km with only 3 birds crossing English Channel (Zink 1975). However, coastal autumn passage occurs, most marked at southern headlands and inshore islands and almost synchronous throughout Britain and Ireland, peaking mid-September (Sharrock 1969). Bird ringed in Portugal, October, recovered in Yorkshire in April (British Ornithologists' Union 1971); see also above.

Largely resident from western France to southern Europe but a few from France are recovered later in Spain (Zink 1975). In Switzerland spring passage begins inconspicuously in February and continues into April though very few observations; in autumn, birds recorded far from breeding habitat as early as July but main movement not much before mid-August, peaks late September into early October then declining sharply at end

of month; wintering regular in lowlands but some winter at higher altitude than breeding grounds (Glutz von Blotzheim 1962). Recoveries on north coast of Spain indicate that Bay of Biscay may be overflown though Snow *et al.* (1955) saw only one actually arrive off sea in late October; conspicuous diurnal movements recorded along coast were perhaps to and from roost. Autumn passage at Straits of Gibraltar occurs late August to early November, mainly mid-September to mid-October; over 200 birds recorded in a single autumn (Tellería 1981). Common on Malta, early September to late March; autumn arrival peaks October–November; numbers increase again late February to early March (Sultana and Gauci 1982). In north-west Africa winters widely from coast to first oases in Sahara but on small scale (Smith 1968*b*). Also regular in small numbers at Nouakchott, Mauritania (Gee 1984). Breeders in Haut Atlas of Morocco make altitudinal movements (Chaworth-Musters 1939). Scarce in Tunisia, usually present October to late March (Thomsen and Jacobsen 1979). Passage recorded near Suez, 14 September–16 October; 39 birds of total 840 *Motacilla* (Bijlsma 1982). In Libya a common winter visitor to the coastal zone of Tripolitania but less numerous Cyrenaica; paradoxically only accidental on passage inland in west but regular and not uncommon at oases in eastern Libyan desert (Cramp and Conder 1970; Hogg 1974; Bundy 1976). The oases probably supplement main Nile valley flyway into equatorial eastern Africa. A common winter visitor in suitable areas of Egypt (Zink 1975), but a sparse double passage migrant in Sudan with very few spring records, and not recorded west of the Nile flyway (Hogg *et al.* 1984). Common winter visitor in Uganda, Kenya, and Tanzania (Mackworth-Praed and Grant 1960); Nyanga highlands in Zimbabwe (Williams 1984) is southernmost record.

Recoveries do not suggest that spring and autumn migration routes differ, nor those of adults and 1st-year birds (Zink 1975).

Present all year in Iran, and subject to local altitudinal movements in Elburz mountains (Norton 1958; Passburg 1959). Occurs in small numbers in Gulf states of Arabia and in Oman, both on passage and with some overwintering, mid-August to mid-May (Bundy and Warr 1980; Gallagher and Woodcock 1980).

Northern and central Asian populations (including those of Urals) are migratory, presumably wintering in India, south-east Asia, and beyond, though perhaps also in Middle East or even Afrotropics. Recorded as vagrant to islands off Alaska, mostly in spring (Roberson 1980).

Marked birds wintering in South Wales stayed in same area throughout winter (D Hansford), and birds may return to same area in successive years (Tyler 1979). In winter on Malta, 12 birds re-trapped at place of ringing after 1 year, 7 after 2 years, 1 after 3 years, and 1 after 4 years (Sultana and Gauci 1982). JHE

Food. Largely insects. 2 main foraging techniques. (1) Picking. Bird walks or runs, repeatedly picking up small items or chasing more mobile prey, with tail wagging and snapping up, down, or to one side, apparently to flush insects; may also wade in shallow water picking up tadpoles or lunging for small fish. (2) Flycatching. Flies from perch or ground; if from ground, flight a steep or near-vertical fluttering leap to maximum c. 6 m. (Schifferli 1961a; S J Tyler.) In addition, may hover (intermittently or continuously) to obtain flying insects or prey from leaves or tree crevices (Ruttledge 1973), and may take prey in aerial-pursuit using zigzag flight, bird tumbling and circling, tail apparently acting as rudder (Schifferli 1961a). Technique employed depends on availability, type, and mobility of prey; picking most common in winter, while flycatching may increase in summer (S J Tyler). Repeatedly beats or bangs certain prey against rocks or ground before swallowing: in this manner extracts caddis fly larvae from cases and kills or immobilises dragonfly nymphs and small fish. Food carried to young in bill, many items at once. May remove wings from insects before feeding to nestlings; banks of stream beside several nests in Britain littered with dismembered wings of damsel flies *Agrion virgo*. (S J Tyler.) Noted extralimitally as feeding near foraging animals (Dolgushin *et al.* 1970). Little quantitative data on hunting success. In Britain in winter, 11–36 pecks per min recorded, but success rate undetermined (S J Tyler). In Britain in summer, 20 items per min recorded in morning, 30 in mid-afternoon, decreasing towards dusk; estimated 19 800 items consumed per day (B Wood). Most food for nestlings gathered within 100 m of nest (Jørgensen 1977).

The following recorded in diet in west Palearctic. Adult and nymphal mayflies (Ephemeroptera), stoneflies (Plecoptera), damsel flies and dragonflies (Odonata: Agriidae, Cordulegasteridae), lacewings (Neuroptera), larval moths (Lepidoptera: Geometridae, Noctuidae), larval caddis flies (Trichoptera), flies (Diptera: Tipulidae, Culicidae, Chironomidae, Stratiomyidae), adult and larval ants (Formicidae), beetles (Coleoptera: Carabidae, Byrrhidae, Scarabaeidae, Elateridae, Curculionidae). Also known to take small molluscs (e.g. mussels Lamellibranchia), small crustaceans (e.g. amphipods *Gammarus*), spiders, small amphibia (e.g. tadpoles Anura), and small fish (e.g. minnows *Phoxinus* and sticklebacks Gasterosteidae); also recorded as picking at dead toad (Anura) and taking vegetable matter (potato) in winter. (Baer 1909; Gil 1927; Sátori 1942; Kovačević and Danon 1952; Harley 1955; Schifferli 1961a; Glutz von Blotzheim 1962; McCluskey 1972; Tyler 1972; Ruttledge 1973; Kostin 1983.)

Little quantitative data. In East Germany, late March and early April, 3 stomachs contained 3 beetles (2 Curculionidae, 1 Byrrhidae) and remains of caddis fly larvae and lacewings (Baer 1909). In Crimea, 2 stomachs (no date), contained *Gammarus* and moth larvae (12 Geometridae, 2 Noctuidae) (Kostin 1983). Extralimitally, in south-east Kazakhstan (USSR), April–August, of 11 stomachs 10 contained only insects with up to 8 different families per stomach and 1 stomach contained c. 10 small seeds (Dolgushin *et al.* 1970). In Kirgiziya (south-central USSR), of 26 stomachs (no date), 7 contained beetles (Carabidae) (Pek and Fedyanina 1961). In India, September–November, 3 stomachs contained 28 flies, 10 Hemiptera, 7 Orthoptera, 6 beetles, 3 dragonflies, and 1 spider (D'Abreu 1918).

Diet of young similar to adults' but dominated initially by soft-skinned larvae of Ephemeroptera, Plecoptera, Trichoptera, and Diptera, progressing with age to animals with hard chitinous skins and wings (Roshardt 1927). Variations in nestling diet also correlated to weather, available foraging areas and time of breeding season (early to late) (Schifferli 1961a). Quantitative data on nestling diet solely extralimital. In southern Lake Baykal area (USSR), collar-samples from 36 nestlings aged 3–9 days contained 352 invertebrates including 10 orders of insects, as well as spiders and terrestrial molluscs; mainly Ephemeroptera (23·9% by number), Plecoptera (12·8%), Diptera (11·3%, including 6·8% Culicidae, 4·5% Tipulidae), and Trichoptera (6·2%). From end of June to early July Ephemeroptera, Plecoptera, and Trichoptera accounted for 85% (by number) of sample; at end of July, this group accounted for only 5·4%, with Diptera then most important at 32·8%. Invertebrates 5–16 mm (2·5–30); portion fed to each nestling typically comprised 1 large item to 18 smaller items (Sonin and Anuchina 1979). In Tien Shan (USSR), 16 nests received great variety of items but predominantly Ephemeroptera, Plecoptera, and Diptera; 18 families of Diptera recorded, but at 13 nests Muscidae most important, accounting for 12·6% (by number) of all items (for details see Kovshar' 1979). In Switzerland, at nest of 5, feeding rate averaged 17·4 visits per hr with total 280 feeds per day; feeding occurred 04.15–20.00 hrs, with slight decrease in rate around 09.00 hrs and another at midday, but after 14.00 hrs rate exceeded morning rate (Schifferli 1961a). Number of feeds per day increases with nestling age until 2–4 days before fledging. Broods of similar size fed less frequently in June than July, yet weights of young remained similar due to July feeds being smaller and of poorer quality (Schifferli 1972). ILG, CAT

Social pattern and behaviour. Based largely on material supplied by S J Tyler. Account includes some data on *melanope* from Tien Shan, USSR (Kovshar' 1979).

1. Outside breeding season, dispersion varies from solitary to relatively gregarious, depending on food supply. After fledging, juvenile brood-members remain together for up to 2–3 months, after joining with others at favourable feeding sites, notably sewage works and watercress beds (Tyler 1970, 1972; S J Tyler). Flocks usually small, probably of family groups (Boase 1952; Palmer and Ballance 1968). At some rich feeding sites, large numbers congregate but each defends individual-distance, birds thus appearing scattered (S J Tyler); more than

50 birds recorded at such sites (Taverner 1976). Along watercourses in Hampshire (southern England), wintering birds occupy individual home-ranges; no defence seen, but density relatively low (Tyler 1972). In Ethiopian highlands, October–March, birds defended (against conspecifics) territories of 100–200 m along streams and rivers: territory often within territory of Mountain Wagtail *M. clara*, but no interspecific aggression (S J Tyler). Apparent lack of interspecific competition reported elsewhere (Moreau 1972; Williams 1984). However, in Zaïre, bird defended territory against *M. clara* and wintering Yellow Wagtails *M. flava* (Verheyen 1956a). BONDS. Monogamous mating system, pair-bond lasting only for duration of breeding season. Owing to high mortality, most pairs last one season only, but ringing evidence suggests that, where both partners survive, they may pair up the next season (S J Tyler). In one case of clutch of 11 eggs, polygamy suspected (Cormack 1954). For case of 2 ♂♂ associated with one nest, see Witt (1976). Occasionally hybridizes with White Wagtail *M. alba* (Dornbusch 1968; Buchet and Jougleux 1979). Both members of pair brood and feed young; continue to tend them for 1–2(–3) weeks after fledging, longer for last broods. Pair may split fledged brood, each tending (e.g.) 2–3 young, but if ♀ re-lays soon after 1st brood fledges, ♂ alone cares for 1st brood for a few days or more (S J Tyler.) Thus, one ♂ stayed with brood for 3 weeks after fledging, then joined ♀ to help incubate 2nd clutch (Tyler 1970). In case where 3 broods raised in same nest, ♂ left rearing of 3rd brood almost entirely to ♀ (Roshardt 1927). Age of first breeding 1 year (S J Tyler). BREEDING DISPERSION. Solitary and territorial. Territories often not contiguous, and size therefore difficult to assess. In study in Denmark, shortest distance between breeding pairs usually 300–400 m, once 100 m (Jørgensen 1977). On upper River Monnow (Wales), pairs 700–800 m apart, and territories contiguous; on some other tributaries of lower Wye (Wales), pairs 400–500 m apart, minimum 150 m (S J Tyler). In Tien Shan, minimum distance between nests 40–60 m (Kovshar' 1979). In Britain, dispersion widely studied. Density varies markedly within and between river systems, e.g. on Wye, averge 4·1 pairs per 10 km, but 22·2 pairs per 10 km over one section of 4·5 km (RSPB River Survey). Density higher in fast-flowing shallow water, lower in sluggish deeper water (Tyler and Tyler 1972). Density increased with altitude, gradient, and presence of rapids or 'riffles' (Marchant and Hyde 1980; Round and Moss 1984). Average density of different rivers in Wales 2·38–4·67 pairs per 10 km (Round and Moss 1984). In New Forest (southern England), average c. 3 pairs per 10 km (1·5–25); lowest on acid streams, highest on base-rich streams. On River Wylye (Wiltshire, southern England), average 6·25 pairs per 10 km on fast-flowing, shallow upper stretch, but apparently nil on more sluggish, deeper lower reaches (Tyler and Tyler 1972). In Sussex (England), average 3·8 pairs per 100 km² on clay, 30·8 pairs per 100 km² on greensand (Merritt *et al.* 1970; which see for other densities in southern England, also Palmer 1983). In Switzerland, 14–30 pairs per 10 km of waterway (Schifferli *et al.* 1982; see also Schifferli 1961a, and Schifferli 1972). In River Our and tributaries (Belgium, Luxembourg, West Germany) c. 5·4 pairs per 10 km (Wiesemes 1982). In Harz mountains (West Germany), 0·6–25 pairs per 10 km, highest on main rivers (Oelke 1975, which see for densities elsewhere in West Germany; see also Damm 1976, Hannover 1977). In East Germany, 2–10 pairs per 10 km in Mecklenburg (Klafs and Stübs 1977), 1·7–2·5 pairs per 10 km in Brandenburg (Rutschke 1983); see also Kolbe (1963). Territory serves for pair-formation, nesting, and feeding (S J Tyler). Birds feed mostly within 100 m of nest (Jørgensen 1977), typically inside

territory, but birds breeding in farm buildings, etc., also fly up to ½ km to feed at streams or rivers (S J Tyler). In Tien Shan, commonly flies 100–200 m or more to feed, this thought to be often outside territory (Kovshar' 1979). Successive clutches may be laid in same nest, or, more often, a different one in territory. In New Forest, 3 out of 12 pairs used same nest twice (Tyler 1970). Rearing of 3 broods in same nest reported by S J Tyler; see also Bonds (above). In 73% of 52 cases, 2 nests of pair less than c. 6 m apart (in 35% less than c. 2 m), in 15% 6–30 m, 6% 3–100 m, 6% over 100 m; some well-separated nests possibly belonged to different pairs (Tyler 1972). In Tayside (Scotland), average distance between successive nests 110 m (n=9); only 1 pair used same nest twice (Nicoll 1980). Nest-sites highly traditional, typically used over many years, but no good information on fidelity of given pairs to given site (Tyler 1970, 1972). Some evidence from ringing, however, that birds faithful to breeding territories in successive years, also that young faithful to natal area (Tyler 1979). In Tien Shan, of 5 ♂ and 1 ♀ controlled over 2 years, 3 ♂♂ bred not more than 50 m from former nest-site, 2 ♂♂ 200–400 m away; ♀ moved c. 2 km upstream; of 78 ringed nestlings, only 1 recorded subsequently, 2 years later, c. 300 m from where born, but not breeding. (Kovshar' 1979.) In New Forest, 5 of 37 nests were close to occupied nests of *M. alba*, and 1 less than 5 m away (Tyler 1970). Because of similar choice of nest-site, often breeds close to Dipper *C. cinclus*. Rarely aggressive towards *C. cinclus* or *M. alba*, though former occasionally chased (S J Tyler). One pair ousted Robin *Erithacus rubecula* from its nest, laid, and reared 3 of their own young and 1 *E. rubecula* (Cohen 1963). ROOSTING. In breeding season, in territory; ♀ typically on nest, ♂ nearby, perhaps in tree or ledge under bridge (S J Tyler). In one case, pair with nestlings both roosted away from nest (Schifferli 1961a). Outside breeding season, at least in some areas, roosts communally in reedbeds, trees, bushes, or other thick vegetation near water (Naumann 1900), not uncommonly on city buildings, e.g. wall of bomb-damaged building (C Lynch). According to Naumann (1900) reedbeds used less than by other *Motacilla*. In Britain, roosts usually of less than 50 birds, often 10 or less (S J Tyler). Often associated with *M. alba* (S J Tyler). In exceptionally large roost in Hampshire (England), January, over 180 with 100 *M. alba* (Taverner 1979); also in Hampshire, January, 81 with 42 *M. alba* in reedbed (Truckle 1968). Traditional reedbed site at Slapton Ley (south-west England) held up to c. 100 birds (*Devon Bird Rep.* 1948). In Beirut (Lebanon), 20–25 in small tree; birds assembled each night with sparrows *Passer*, but roosted separately from them (Cawkell 1947a). At regular lakeside roost of 40–50 in Cheshire (England), birds arrived singly or in groups of 2–3 (R J Raines). In New Forest, wintering birds—whether on home-ranges or feeding with others at sewage works (etc.)—left in evening to roost, arriving back on feeding sites c. ½–1 hr after dawn. Loafing behaviour includes preening, bathing, and sunning. Sunning bird crouches on pebbles or rocks, spreads wings, cocks head back, and raises tail slightly. (S J Tyler.)

2. Outside breeding season, birds readily flush when man approaches to c. 15–20 m, and usually give 1 or more mild Alarm-calls (see 3 in Voice). Fly to cover when disturbed by predators such as Sparrowhawk *Accipiter nisus*. In spring, bird flew towards and chased Kestrel *Falco tinnunculus*; mobbed Little Owl *Athene noctua*. At winter feeding site, birds disturbed by cat crouched low and stayed still until danger passed—but in same incident flock of *M. alba* gave alarm-calls and flew up to nearby hedgerow (S J Tyler). FLOCK BEHAVIOUR. Little information. Forms small flocks; members well-spaced and give

Contact-calls (see 2a in Voice) to facilitate flocking in flight (S J Tyler). SONG-DISPLAY. Song-period mainly March–July, occasionally February–October (Boase 1952; S J Tyler). In Rome (Italy), sings throughout winter, though little in December and early January (Alexander 1917). Song continues during nest-building, and between fledging of 1st brood and start of 2nd clutch (S J Tyler). ♂ sings (see 1a in Voice) both perched and in Song-flight. Sings repeatedly from elevated perch (e.g. tree, rock), sometimes quivers wings and ruffles rump feathers (S J Tyler). Song-flight often compared to Tree Pipit *Anthus trivialis*: typically, ♂ descends parachute-fashion from high perch, wings outspread and held steady or fluttering, tail raised slightly, and rump exposed. Gives trilling song (see 1b in Voice) during descent, then Contact-calls before landing on ground or low perch (Ruthke 1938; S J Tyler). ♀ commonly nearby during ♂'s Song-flight; when ♀ once flew on a little way, ♂ followed her, performing Song-flight as he went (Ruthke 1938). In sequence preceding mating (see Heterosexual Behaviour, subsection 3, below), singing ♂ flew around above ♀, sometimes quite high (Eggebrecht 1939). In Tien Shan, displaying ♂ may raise tail almost vertically giving repeated Contact-calls between flights (Kovshar' 1979). Song-flights sometimes performed by adults and juveniles in autumn (Naumann 1900). ANTAGONISTIC BEHAVIOUR. In breeding season, residents (especially ♂♂) highly aggressive towards rivals, particularly where territories contiguous (Jørgensen 1977). Birds often reported attacking their own images in window-panes, etc. (e.g. Moreau 1965, Schulze and Schulze 1977). ♂ *robusta* challenges rivals with Threat-posture (Fig A): raises head, thus

A

displaying throat, then may fly, with same inclination of head (Fig B), towards rival (Panov 1973). Intruders into territory readily expelled in pursuit-flight. Resident ♂ pursues trespassers thus for up to *c*. 30 m from nest (Kovshar' 1979). Pursuit may lead to fierce mid-air clashes, rivals flying up and confronting each other, one then usually flying off, other pursuing (S J Tyler). 2 cases of bird killing or drowning another (Bassett 1953; *Devon Bird Rep.* 1976): in one incident, ♂ used his bill to force head of young intruder under water (Bassett 1953). In repeated boundary disputes between pair and ♂ of another pair, fights alternated with lengthy bouts of singing; in longest fight, pair tussled on ground for *c*. 2 min (Faris 1937).

B

HETEROSEXUAL BEHAVIOUR. (1) Pair-bonding behaviour. Frequent chasing occurs. Ground-display typically interspersed with Song-display. Often after Song-flight, ♂ lands near ♀ and runs towards her, strutting around in Advertising-posture: stretches up, head and bill pointing up to present throat and breast to ♀; may also spread tail (S J Tyler) and droop wings (Kleinschmidt 1931; Vollbrecht 1939: Fig C). If ♀ receptive,

C

ground-display of ♂ may lead to mating (see below). (2) Courtship-feeding. None reported. (3) Mating. On ground, ♀ often preens beforehand (S J Tyler). Commonly, ♀ solicits while ♂ is singing (often in Song-flight) nearby. Adopts Soliciting-posture (Fig D), not unlike Advertising posture of ♂.

D

Crouches with tail almost vertical, head thrown back, plumage ruffled, and wings shivering; posturing thus, ♀ may constantly turn to face displaying ♂ who hovers over ♀, lands beside her, mounts, and copulates briefly; also reported mounting directly from hovering descent (Eggebrecht 1939; Vollbrecht 1939). ♂ may take off from song-post after a bout of song, pursue ♀ in flight a short way, and mate with her on ground, both then flying off together (S J Tyler). In one instance after mating, ♀ shivered wings for some time while ♂ ran about in front of her, his upper tail-coverts and lower back feathers markedly ruffled (Vollbrecht 1939). Once immediately after copulation, ♂ faced ♀ and sang in apparent Threat-posture (Kleinschmidt 1931, which see for drawing). In the following account, ♂ apparently solicited passive ♀, his display mainly a variant of Advertising-display: ♂ arrived near ♀ in fluttering flight, body plumage ruffled, landed nearby and ran towards her with head lowered and thrust forward, wings arched and fluttering, rump plumage raised, and tail lowered and partly spread; finally fluttered up and tried to mount ♀ who turned away, rebuffing him. ♂ landed and continued displaying but with wings closed and drooped; called (not described) during display (Boase

1952), perhaps Contact-lure call (Vollbrecht 1939) (see 2b in Voice). Little information on frequency of mating. In one pair, 2 copulations and 3 attempted copulations seen in 2 days (Vollbrecht 1939). (4) Behaviour at nest. Details of nest-site selection only for *melanope*. Occurs immediately after pair-formation and initiated by ♂ who, on finding suitable site, calls up ♀. Giving repeated Contact-calls, ♂ goes in and out of chosen site until ♀ joins him. If ♀ flies on further, ♂ follows and repeats inspection procedure if he then finds another site. Sometimes ♂ tries to attract ♀ by performing Song-flight which ends at prospective nest-site. If territory occupied early in season, nest-site selection may be prolonged: e.g. once lasted 2 weeks during which ♂ carried material to 3 sites; after starting to build at 3rd site, pair rejected it in favour of earlier choice. Nest for 2nd clutch often built at a site inspected earlier in season. (Kovshar' 1979.) While ♀ nest-building, ♂ *robusta* sometimes adopts a bowed-posture with wings drooped rather (Panov 1973: Fig E); significance not known. Nest-relief

E

typically rapid and furtive, as follows. Relieving bird gives brief Contact-call but not closer than *c.* 20-30 m to nest; makes discreet approach, sitting on perch usually 10-15 m from nest, and looking all around for some time before flying or running quietly to nest. Sitting bird leaves in silence, calling only when airborne, and relieving bird sits immediately (Eggebrecht 1939; S J Tyler.) Once, incubating ♀ drove ♂ from nest (Kleinschmidt 1931). See also Parental Anti-predator Strategies (below). RELATIONS WITHIN FAMILY GROUP. Young said not to be fed on day of hatching (Roshardt 1927). At first, brooded assiduously by both sexes, typically more by ♀ (e.g. Roshardt 1927); said to be brooded only by ♀ at night (Buxton 1961). At various nests on Tayside (Scotland), ♂ found incubating or brooding on 28% of visits (Nicoll 1980). Food-begging conspicuous up to 7-8 days (S J Tyler). Both parents feed young and practise nest-sanitation: take faecal sac from nest and fly short distance to drop it in running water, landing first on rock or at water's edge (S J Tyler), and may tap it a few times to sink it (Schifferli 1961*a*). One pair which bred *c.* 1-1·5 km from nearest stream habitually dropped faecal sacs in nearby water trough (Schücking 1963). At one nest, parents ate faeces during 1st day after hatching, thereafter removed them (Schifferli 1961*a*). On leaving nest, young initially seek suitable cover nearby, e.g. brood remained in ivy thicket for 2-3 days (Schulze and Schulze 1977). In *melanope*, fledged young return to general area of nest nightly for *c.* 1 week but do not roost in nest. Start self-feeding 4-5 days after leaving nest (Kovshar' 1979), though parents continue to feed them for up to 1 week afterwards: young follow parents around, running up to them and begging (S J Tyler). For brood-division and duration of family ties, see Bonds (above). ANTI-PREDATOR RESPONSES OF YOUNG. Well-grown young may leave nest prematurely if disturbed, e.g. at 9-11 days (Kovshar' 1979; S J Tyler). When danger threatens, alarm-calls (see 3 in Voice) of parents induce juveniles to crouch and remain motionless (S J Tyler). PARENTAL ANTI-PREDATOR STRATEGIES. Account by S J Tyler. (1) Passive measures. Bird usually sits tight, leaving nest only when danger imminent, e.g. human intruder within a

few feet; even then, may stay motionless until hand reaches out towards nest. (2) Active measures: against birds. No information. (3) Active measures: against man and other animals. On approach of human intruder, birds become very agitated and give Alarm-calls during nest-building stage, but respond little during first 10 days of incubation. Most responsive around fledging time: parents then fly frantically to and from human intruders and dogs, giving repeated intense Alarm-calls (see 3b in Voice). Either sex often performs distraction-lure display of disablement type towards man or dog, almost invariably when brooding small young or when young about to fledge or just fledged: lands in front of intruder and, by combination of walking, fluttering, and flying short distances, evidently tries to lead intruder away; intense Alarm-calls given throughout. Distraction-display also performed around hatching time.

(Figs by C Rose: C from drawing in Kleinschmidt 1931; D adapted from drawing in Eggebrecht 1939; others from drawings in Panov 1973.)   EKD

Voice. Freely used in breeding season, less so at other times. Described as the only west Palearctic *Motacilla* with really pronounced territorial song; no differences between recordings of songs (see 1, below) from Europe, Siberia, and Mongolia (Wallschläger 1984, which see for sonagrams). Also, recorded calls (see 2a and 3b, below) of *patriciae* (Azores) apparently not different from those of nominate *cinerea* (Knecht and Scheer 1971, which see for sonagrams). Performs Bill-snapping, probably in situations when excited or agitated, as in other *Motacilla*. In recording by R Tassell, England, fledged brood give rapidly repeated Bill-snapping sounds: ticking 'prik', like snapping of dried twigs (P J Sellar); apparently given in response either to approaching parent or its Warning-call.

CALLS OF ADULTS. (1) Song of ♂. Given mainly from perch, less often in Song-flight. For song-period, see Social Pattern and Behaviour. (a) Most commonly heard ('simple') song a repetition of phrases, each of 3-10 uniformly pitched shrill units resembling call 2; song may end with units usually of lower pitch; duration of phrases 0·65-1·3 s (Wallschläger 1984, which see for variants). In our recordings, song shows crescendo, but no accelerando; all units consistently *c.* 0·1 s apart (J Hall-Craggs). Song shown in Fig I apparently similar to

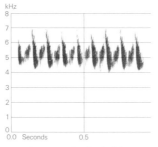

I V C Lewis England May 1972

that described by S J Tyler as 'tee tee tee tee tee tee' or 'tchee tche tche tche tche tche tche' (this 7-unit phrase common). Fig II shows 'double' song-phrase: 2nd part,

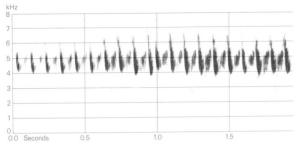

II   V C Lewis   England   May 1972

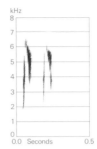 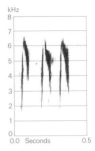

III (left), IV (right)   V C Lewis   England   June 1977

continuous with 1st, starts at 11th unit, with renewed crescendo from 11th to 13th unit (J Hall-Craggs). For other renderings and sonagrams, see Bergmann and Helb (1982). (b) After short pause, simple sequence (as above) may be followed by trill, especially in parachute-like descent of Song-flight, e.g. 'chu chu chu chu tre-tre-tre-tre-tre...', commonly given 3–4 times (S J Tyler: see Social Pattern and Behaviour). This variant described in one bird as resembling opening notes of song of Wren *Troglodytes troglodytes* ('gee' units), followed by short trill ('zee' units): 'gee-gee-gee-gee-zee-zee-zee-zee' (Witherby *et al.* 1938a, which see for other renderings). ♂ gave 'zississississi' during ascent in Song-flight and shrill 'zier zier zier' during parachuting descent (Ruthke 1938); for similar renderings, see Kleinschmidt (1931) and Boase (1952). (c) Also possesses a chattering song, with fluent trills and pure tones, to some extent with mimicry, and especially associated with courtship (Bergmann and Helb 1982). In 2 of our recordings, perched birds sang thus: a frenzied rollicking sound made up of units highly varied in length, pitch, and timbre (E K Dunn), resembling in quality song of Canary *Serinus canaria* (P J Sellar). Difficult to determine, in these recordings, where phrases begin and end, but one section suggests something like 'tweeeee twe-twe-twe titititiii treeteri trretrretrre tweetweetweetweetwe   tweetew   tweeeeeee   tsee tweetweetweetwee-tisew...' (E K Dunn); perhaps used as intense self-advertisement, since given by ♂ of site-prospecting pair when another bird (presumed ♂) intruded on them. Low warbling Subsong (Witherby *et al.* 1938a) perhaps also of this type. (2) Contact-calls. (a) Common call, given when perched or in flight, a single 'tchee', 'chee', or 'tchu'; also disyllabic 'che-sic', 'chissik', 'tscheesik' (S J Tyler), or 'chee-chee' (Fig III); in Fig IV, 'chee-chee-chee' (J Hall-Craggs). Thinner and higher pitched than Pied Wagtail *M. alba* (S J Tyler). For similar renderings, see Boase (1952). In flight, sharp and shrill 'ziss-zississ', of 1–4 syllables (Bergmann and Helb 1982); in *patriciae* feeding on ground, 'di-di-did', etc. (Knecht and Scheer 1971). Single call commonly serves in communication between pair-members, notably when nest-building and when approaching or leaving nest; disyllabic variant especially prevalent in flight (S J Tyler). (b) Contact-lure call. High-pitched 'füid' given by ♂ to invite close approach of ♀, e.g. prior to copulation

(Vollbrecht 1939). (3) Warning- and Alarm-calls. (a) Plaintive, drawn-out, rising '(t)weee' (E K Dunn, J Hall-Craggs); 'züih' (Bergmann and Helb 1982, which see for sonagram); 'chweeet' or 'wee-eet' (S J Tyler), though our recordings show no end transient, i.e. no 't' (J Hall-Craggs). (b) As alarm mounts, bird gives rapid and increasingly agitated succession of strident Contact-call-type units. In *patriciae* alarmed at nest, 'zi-zi...' (Knecht and Scheer 1971), sonagram resembling Fig III. Calls 3a and 3b freely combined: e.g. '(t)weee che-che' (Fig V), repeated *ad lib* when disturbed feeding young;

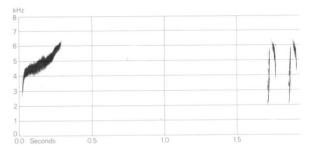

V   V C Lewis   England   May 1972

single 'che' units alternate with disyllabic variant (E K Dunn, J Hall-Craggs). Other sequences: a loud, harsh repeated 'tchissik' or 'tchutchu tchutchu tchutchu' or 'totu totu tu tu tu chweet'; when flying in alarm or performing distraction-display near human intruder, an intense 'che-chee che-chee che' or 'cherchee cherchee chee' (S J Tyler). (4) Other-calls. (a) In recording by R Tassell, England, high-pitched penetrating 'syit syit syit' given repeatedly by adults to well-grown juveniles (P J Sellar). Function not known. (b) 'Complex calls', e.g. 'zritzridrü' (Bergmann and Helb 1982, which see for sonagram).

CALLS OF YOUNG. Food-call of small young a very quiet cheeping, twittering sound (Kleinschmidt 1931). Contact-call of juvenile a 'chip', or 'chit', duller than adult's call. For Bill-snapping, see 1st paragraph.   EKD

Breeding. SEASON. North-west Europe: first eggs laid in last few days of March; main laying period April–May, last eggs laid early August (Tyler 1972; Makatsch 1976). Central and eastern Europe; first eggs second half of April (Makatsch 1976). North Africa: laying from late

March to May (Heim de Balsac and Mayaud 1962). Canary Islands: early March to June (Bannerman 1963). Azores: laying from late May (Bannerman and Bannerman 1966). SITE. In hole or crevice in wall or bank, under bridge, or among tree roots. Of 150 nests, Denmark: 34% in hole in wall, 23% in hole in bank or ground, 11% under bridge, 10% in nest box (Jørgensen 1977). Of 699 nests, Britain: 32% in hole in bank, etc., 25% under bridge, 23% in hole in wall (Tyler 1972). Normally close to water; only 3% of 673 nests, Britain, more than 30 m (up to several hundred metres) from water (Tyler 1972). Of 150 nests in Denmark 39% 0-1 m above water surface, 47% 1-2 m, 10% 2-4 m, 2% 4-6 m, 2% over 6 m, maximum 10-15 m Jørgensen 1977). Of 617 nests in Britain, 62% 0-1·65 m above water, 29% 1·65-3·3 m, 9% over 3·3 m (Tyler 1972). Nest: cup of grass, roots and small twigs, often with moss, lined hair; size variable, shaped to fit crevice. Building: both sexes gather material and probably build, but ♀ may collect lining and complete nest (S J Tyler). EGGS. See Plate 81. Sub-elliptical, smooth and glossy; whitish, cream or grey-buff, faintly marked grey or grey-buff; eggs of *patriciae* said to be paler (Bannerman and Bannerman 1966). Nominate *cinerea*: 19·0 × 14·3 mm (17·0-21·7 × 13·0-15·5), n = 200; calculated weight 1·91 g (Schönwetter 1979). *M. c. patriciae*: 19·5 × 14·8 mm (17·5-20·5 × 14·0-15·6), n = 44 (Bannerman and Bannerman 1966); calculated weight 2·1 g (Schönwetter 1979). Clutch: 4-6(3-7). Of 636 clutches, Britain: 3 eggs, 3%; 4, 19%; 5, 61%; 6, 17%; 7, 1%; mean 4·9; no significant variation through season, but clutches possibly larger towards north (Tyler 1972). Of 126 clutches, Denmark: 2 eggs, 1%; 3, 5%; 4, 4%; 5, 33%; 6, 55%; 7, 2%; mean 5·4 (Jørgensen 1977). 2 broods, occasionally 3. Replacements laid after egg loss, including 3rd clutch after loss of 2nd (S J Tyler). Interval between fledging of 1st brood and laying of 2nd clutch c. 2 weeks (1-30 days), n = 54 (Tyler 1972; S J Tyler). May use same or different nest. Eggs laid daily (Boase 1952). INCUBATION. 11-14 days. By both sexes. In one study, ♀ in periods of 20-90 min, ♂ 20-68 min (Buxton 1961). Begins with last egg; hatching synchronous. YOUNG. Cared for and fed by both parents. FLEDGING TO MATURITY. Fledging period 13-14 days (11-17) (Tyler 1972). May return to nest for a few days after fledging, when still fed by parents; still dependent on parents 2 weeks after fledging, including when 2nd clutch being laid (Boase 1952; Tyler 1972). Age of first breeding 1 year. BREEDING SUCCESS. In Britain, estimated 54% of 162 nests found before or during laying produced at least 1 fledged young; from 338 eggs laid in 68 such nests, 42% young fledged; hatching success of 246 eggs, 94%; of 396 young hatching in larger sample of nests, 82% fledged (Tyler 1972). In Denmark, of 590 eggs laid in 112 clutches, 72% hatched, and 79% of young fledged, giving overall success of 56·8%, with mean 3·0 young reared per pair per year (Jørgensen 1977).

Plumages (nominate *cinerea*). ADULT MALE BREEDING. Forehead, crown, hindneck, sides of neck, mantle, scapulars, and back medium-grey, feather-tips fringed olive-green when fresh, upperparts then showing distinct olive-green tinge; purer bluish-grey (in particular on crown and hindneck) when worn, only slightly tinged olive on forehead, lower mantle, and scapulars. Rump and upper tail-coverts contrastingly olive-green, almost pure bright yellow on lateral coverts. Lores dull black, upper cheeks and ear-coverts medium grey, separated from crown by long and narrow white supercilium (narrowest and sometimes indistinct in front of eye). Narrow white eye-ring, broken at front and rear; cheek sometimes with some white mottling below eye. White stripe from base of lower mandible backwards, widening towards rear of lower cheek. Chin and throat black, feathers narrowly tipped white when fresh. Remainder of underparts bright yellow, often deepest on under tail-coverts, usually slightly paler and with slight grey wash on sides of body, occasionally tinged orange-buff on chest and upper flanks. Central pair of tail-feathers (t1) black with broad olive-green or olive-grey fringe at sides; t2-t3 black (t3 sometimes with some white at tip of outer web); t4 white with black outer web (except for distal ¼) and sometimes with narrow black fringe along base of inner web (exceptionally, black of outer web extends just across shaft to inner web, and black fringe along inner web broad); t5 white with black outer web (except distal third and extreme base; exceptionally fully white); t6 fully white (rarely, shaft dark near base); bases of outer webs of t2-t4(-t5) tinged yellow-green. Flight-feathers, tertials, upper primary coverts, and greater and median upper wing-coverts black; outer webs of primaries and greater coverts narrowly fringed olive-green, of tertials and tertial coverts more broadly green-yellow (yellow-white or white on longest tertial); tips of median and greater upper wing-coverts and of greater upper primary coverts rather narrowly and indistinctly fringed olive-grey; basal halves of secondaries white, showing as narrow white band beyond tips of upper greater coverts; inner webs of primaries with white base. Lesser upper wing-coverts medium grey, like upperparts; longest with slightly darker centres. Under wing-coverts and axillaries pale grey, longest coverts tipped white. In worn plumage, May-July, upperparts uniform dull bluish-grey with contrasting greenish-yellow rump; some white of feather-bases sometimes visible on upper cheeks, chin, and upper throat; flanks paler with more extensive grey wash on sides of body; pale fringes and tips of flight-feathers and upper wing-coverts worn off, wing mainly black except for grey lesser upper wing-coverts and distinct white band across bases of secondaries. ADULT FEMALE BREEDING. Like adult ♂ breeding, but chin and throat either white, variably mixed with some wholly or partly black feathers (especially on sides of throat and on lower throat), or (more rarely) chin and throat mainly black, like adult ♂ breeding, but white tips and bases of feathers slightly more extensive, chin and throat appearing grizzled white when fresh, slightly mottled white when worn, strongly mottled white when heavily abraded (much white of feather-bases showing). Ear-coverts often slightly paler grey than in ♂, faintly washed green in fresh plumage; supercilium and lower cheeks faintly washed buff. ADULT MALE NON-BREEDING. Like ♂ breeding, but chin and throat white, no black (first black feathers may appear from December). All plumage equally new until November-December, forehead, crown, mantle, and scapulars tinged olive-green; distinct grey tips to upper wing-coverts and pale yellow fringes to tertials; supercilium narrow, white; ear-coverts faintly streaked white; chin and throat white, tinged buff at border of deep yolk-yellow chest; remainder of underparts bright yellow. ADULT FEMALE

NON-BREEDING. Like adult ♂ non-breeding, but supercilium, ear-coverts, cheeks, chin, and throat slightly washed buff; breast and belly on average slightly paler yellow. NESTLING. Down rather long and dense, on upperparts only, pale greyish-yellow or golden-buff. Fully feathered at fledging at *c.* 2 weeks, but tail and flight-feathers continue growing until 3rd–4th week. (Heinroth and Heinroth 1924–6; Witherby *et al.* 1938a; RMNH.) JUVENILE. Upperparts grey-brown or mouse-grey, slightly tinged olive; rump and upper tail-coverts olive-green, not as strongly contrasting as adult and 1st adult, only lateral feathers purer pale yellow. Sides of head and neck grey, like upperparts; short supercilium and narrow broken eye-ring off-white, rather indistinct; grey of cheeks gradually merges into off-white chin and throat; occasionally a faint and mottled grey malar stripe. Chest and upper flanks pale yellow-buff, sometimes with dark patch on uppersides of chest; vent and under tail-coverts pale yellow; remainder of underparts white with slight yellow or buff tinge. Tail as adult, but fringes along basal part of feathers buffish-grey or grey, less yellowish. Wing as adult, but fringes along tertials yellow-white (soon bleaching to white), fringes along tips of median upper wing-coverts pale grey, buff-grey, or buff (not olive-grey). FIRST ADULT NON-BREEDING. Like adult non-breeding, but flight-feathers, variable number of tail-feathers, outer greater upper wing-coverts, and sometimes a few tertials or other wing-coverts retained, contrasting in colour and wear with fresh neighbouring feathers (in adult non-breeding, all plumage equally fresh); tips of worn outer greater coverts white, abraded, contrasting with new grey and smooth-edged inner coverts. Supercilium and narrow broken eye-ring buff, not as white as in adult; chin and throat buff (almost white on central throat), merging into yellow-buff of chest and sides of breast (chin and throat of adult whiter, contrasting more with yolk-yellow chest); exceptionally, a broken black collar across lower throat; under tail-coverts and often vent and central belly bright yellow, but remainder of underparts silky-white or yellow-white (not as bright as adult); occasionally, entire underparts white except for pale buff chest and sides of throat and pale yellow under tail-coverts. *In worn plumage* (about December–February), upperparts dull grey; supercilium, chin, throat, and belly dirty white, chest and sides of breast pale grey-buff, yellow virtually restricted to under tail-coverts. Sexes indistinguishable. FIRST ADULT BREEDING. Like adult breeding and often indistinguishable; flight-feathers and greater upper primary coverts still juvenile, thus browner and more heavily worn than in adult at same time of year; occasionally, some heavily worn juvenile outer greater wing-coverts retained. Chin and throat of ♂ often have more white on feather-bases than in adult ♂ breeding, more extensively mottled white in worn plumage; chin and throat of ♀ often completely white or with a few dull grey feathers only.

Bare parts. ADULT, FIRST ADULT. Iris dark brown. Bill greyish-black. Mouth greyish-pink. Leg and foot pale brown, pink-brown, or dark brownish-flesh; soles pale grey or yellow-grey (in 1st autumn perhaps deep horn-grey). (Hartert 1910; Witherby *et al.* 1938a; Bub *et al.* 1981; RMNH, ZMA.) NESTLING. Bare skin (including bill, leg, and foot) yellowish-flesh; mouth bright yellow, gape-flanges pale yellow (Heinroth and Heinroth 1924–6; Witherby *et al.* 1938a). JUVENILE. Iris dark brown. Bill horn-black with flesh tinge on base and yellow flanges. Leg and foot greyish-flesh or pink-brown, toes sometimes flesh-grey. (RMNH, ZMA.)

Moults. ADULT POST-BREEDING. Complete; primaries descen-

dant. In Britain, starts with p1 between late June and late July, rarely August; completed with regrowth of p9–p10 late August to mid-September (rarely mid-October); estimated duration of primary moult *c.* 65 days (Ginn and Melville 1983). In Netherlands and West Germany, starts mid- or late July; p7–p10 and secondaries still old early August, tertials moulting; all moult completed late September (Bub *et al.* 1981; RMNH, ZMA). In southern Yugoslavia, in full moult August, completed from mid-September onwards (Stresemann 1920). On Canary Islands, starts late June to mid-July (BMNH). In USSR, starts mid- or late July, completed late August or September (Dementiev and Gladkov 1954a). ADULT PRE-BREEDING. Partial. Mainly February–March, in tropics and on Atlantic islands sometimes from December (exceptionally also in Europe: Hartert 1910; Stresemann 1920). Involves head, body, some or all tertials and tertial coverts, sometimes lesser and median upper wing-coverts, rarely inner greater coverts, sometimes t1, occasionally also t6 or t2, and in tropics sometimes many tail-feathers. Occasionally, part of old head and body retained, in particular part of chin and throat of ♀, sometimes parts of hindneck, rump, vent, or tail-coverts. POST-JUVENILE. Partial; starts shortly after fledging, sometimes when tail and outer primaries still growing. Timing and extent rather variable, depending on fledging date; some largely in 1st non-breeding by mid-July, others not until late September. Involves head, body (some juvenile body feathers sometimes retained until mid-winter), 1–2 or all tertials and tertial coverts, all lesser and median upper wing-coverts, frequently t1, and sometimes all tail (in particular in tropical winter quarters). (RMNH, ZMA.) 2–6 outer greater upper wing-coverts usually retained; sequence of tail-feather replacement t1-6-2-5-4-3 (Bub *et al.* 1981); sometimes all plumage replaced except flight-feathers and greater upper primary coverts (Ginn and Melville 1983; RMNH, ZMA). FIRST PRE-BREEDING. Partial. In tropical winter quarters and on Canary Islands and Madeira, sometimes from December; in Europe, mainly late February to late April. Extent as in adult pre-breeding, but frequently less extensive, and occasionally part of non-breeding chin, throat, hindneck, rump, or flanks retained, especially in ♀.

Measurements. Nominate *cinerea*. Western and central Europe, all year; skins (RMNH, ZMA). Juvenile wing and tail include data from 1st adult with retained juvenile flight-feathers and tail. Bill (S) to skull, bill (N) to distal corner of nostril; exposed culmen on average 4·4 less than bill (S). Toe is middle toe with claw; claw is hind claw.

	♂			♀		
WING AD	85·0	(2·00; 13)	82–89	83·8	(2·11; 10)	80–86
JUV	83·6	(1·59; 25)	81–86	82·7	(1·80; 16)	80–86
TAIL AD	97·9	(3·63; 14)	94–104	95·8	(2·94; 10)	92–99
JUV	94·7	(3·00; 17)	91–101	93·9	(2·40; 14)	90–98
BILL (S)	16·1	(0·51; 29)	15·1–16·9	15·9	(0·55; 24)	15·1–16·7
BILL (N)	9·5	(0·46; 25)	8·5–10·1	9·4	(0·42; 22)	8·8–10·2
TARSUS	20·8	(0·67; 35)	19·4–21·8	20·6	(0·53; 31)	19·7–21·6
TOE	16·3	(0·84; 10)	15·4–17·6	15·8	(0·60; 13)	14·8–16·8
CLAW	6·5	(0·65; 20)	5·5–7·4	6·4	(0·55; 24)	5·4–7·3

Sex differences not significant.

M. c. patriciae. Azores, all year; skins (BMNH).

	♂			♀		
WING	83·7	(2·39; 19)	80–88	81·2	(1·68; 11)	78–84
TAIL	87·6	(2·37; 19)	84–92	87·2	(1·76; 11)	85–90
BILL (S)	17·5	(0·62; 12)	16·8–18·3	17·8	(0·59; 10)	16·9–18·8

M. c. schmitzi. Madeira, all year; skins (BMNH, ZMA).

	♂			♀		
WING	82·5	(1·71; 13)	79–86	81·3	(0·82; 6)	80–83
TAIL	89·2	(3·33; 12)	84–94	89·1	(2·35; 6)	86–92
BILL (S)	16·3	(0·70; 11)	15·4–17·4	16·2	(0·26; 6)	15·7–16·4

M. c. canariensis. Canary Islands, March–July (BMNH, RMNH, ZMA).

WING	♂ 83·5 (2·04; 11)	81–88	♀ 81·7 (2·73; 6)	77–84	
TAIL	92·0 (2·13; 11)	88–95	92·4 (3·34; 6)	88–95	
BILL (S)	16·0 (0·65; 11)	15·1–16·8	16·2 (0·41; 6)	15·4–16·6	

Wing and tail, sexes combined. Nominate *cinerea*: (1) western and central Europe, (2) Balkans, (3) Asia Minor and northern Iran. *M. c. melanope*: (4) Afghanistan, (5) Tien Shan mountains and neighbouring parts of central Asia, (6) Mongolia. *M. c. robusta*: (7) Japan and (in winter) Taiwan. (Stresemann 1920, 1928; Paludan 1940; Vaurie 1957; Schüz 1959; Kumerloeve 1961; Piechocki and Bolod 1972; BMNH, RMNH, ZMA.)

(1)	WING	83·7 (1·94; 64)	80–89	TAIL	95·7 (3·63; 62)	90–104
(2)		82·4 (1·97; 55)	79–86		97·0 (3·72; 39)	90–105
(3)		83·1 (1·97; 14)	81–87		92·6 (4·03; 10)	91–96
(4)		81·1 (1·83; 12)	78–84		88·6 (3·63; 12)	83–95
(5)		81·2 (1·82; 28)	78–85		87·4 (3·59; 28)	82–93
(6)		82·9 (2·19; 12)	80–86		91·4 (3·75; 11)	84–99
(7)		83·0 (2·00; 45)	79–88		89·4 (2·81; 51)	83–95

In nominate *cinerea*, 4 out of 62 birds from western and central Europe had tail below 91·5, 3 out of 39 from Balkans, and 2 out of 10 from Turkey and Iran; in *melanope* and *robusta*, 3 out of 12 from Afghanistan had tail over 91, 4 out of 28 from Tien Shan, and 8 out of 51 from Japan and Taiwan, but 6 out of 11 from Mongolia.

M. c. melanope. Central Asia, skins (BMNH, RMNH, ZMA).

WING	♂ 81·8 (1·55; 18)	80–85	♀ 80·1 (1·81; 10)	78–83
TAIL	87·9 (3·39; 18)	82–93	86·4 (3·95; 10)	82–91
BILL (S)	16·5 (0·38; 12)	15·9–17·2	16·3 (0·95; 5)	15·6–17·2
TARSUS	19·8 (0·56; 12)	18·9–20·5	19·4 (0·38; 5)	19·0–19·8
CLAW	6·5 (0·36; 12)	5·9–7·1	5·9 (0·44; 5)	5·5–6·3

Sex differences significant for wing and claw.

Weights. ADULT, FIRST ADULT. Nominate *cinerea*. Sexed birds from whole geographical range (Paludan 1940; Niethammer 1943, 1950, 1957; Schüz 1959; Kumerloeve 1961, Rokitansky and Schifter 1971; Desfayes and Praz 1978; Bub *et al*. 1981; Hallchurch 1981; RMNH, ZMA): (1) March, (2) May–June, (3) July–August. (4) Switzerland and West and East Germany, date not stated, probably spring and summer (Bub *et al*. 1981).

(1)	♂ 18·1 (—; 22)	16–21	♀ 18·4 (4·47; 4)	15–25
(2)	17·4 (—; 19)	15–20	17·3 (1·86; 5)	14–19
(3)	18·7 (—; 26)	15–22	17·6 (1·64; 6)	15–20
(4)	18·0 (1·41; 87)	15–22	17·2 (1·42; 26)	14–20

Sexes combined; mainly West Germany (Bub *et al*. 1981), some from other areas and authors, as cited above:

JUN	17·5 (—; 14)	15–20	SEP	19·2 (1·01; 49)	16–21
JUL	18·1 (1·53; 56)	15–22	OCT	18·3 (1·60; 31)	15–21
AUG	18·6 (2·16; 97)	16–22	NOV–FEB	18·4 (2·15; 10)	14–21

Malta, autumn and winter: 17·5 (1·3; 50) 15–22 (J Sultana and C Gauci). Exhausted birds: Norway, ♂ 12 (Willgohs 1955); Netherlands, ♀ 13·3 (ZMA).

M. c. melanope. Afghanistan and Kazakhstan (USSR), combined: March–May, ♂ 15·6 (1·60; 10) 14–19, ♀ 15·4 (1·28; 7) 14–18; June–July, ♂ 15·7 (0·93; 10) 14–18, ♀ 15·5 (1·32; 5) 14–18; August, ♂♂ 18·5, 21; ♀♀ 16·2, 18·4 (Paludan 1959; Dolgushin *et al*. 1970). Manchuria (China), mainly July–August: 16·5 (18) 14–18 (Piechocki 1958). Mongolia, May–July: ♂ 15·8 (1·64; 5) 14–18, ♀ 18·5 (1·05; 6) 17–20 (Piechocki and Bolod 1972).

M. c. robusta. Taiwan, mainly October–December: ♂ 15·9 (1·86; 7) 15–20, ♀ 16·7 (2·57; 9) 14–22 (RMNH).

NESTLING, JUVENILE. On 1st day, 2·1 (7) 1·5–2·5; on 9th, 16·7 (5) 15·5–18·5; on 14th, shortly after fledging, 17·1 (7) 15·5–18 (Schifferli 1972).

Structure. Wing rather short, broad at base, tip bluntly pointed. 10 primaries: p8 longest, p7 and p9 both 0–1·5 shorter, p6 5–7, p5 11–15, p4 17–20, p3 20–24, p1 25–29. P10 reduced, narrow; 56–63 shorter than p8, 8–13 shorter than longest greater upper wing-covert. Outer web of p6–p8 and inner web of (p6–)p7–p9 emarginated. Longest tertials reach wing-tip in closed wing. Tail long, tip square; 12 feathers. Bill straight and slender; a few fine bristles along base of upper mandible. Tarsus and toes short and slender. For length of middle toe with claw and hind claw, see Measurements. Outer toe with claw *c*. 72% of middle toe with claw, inner *c*. 75%, hind *c*. 79%. Front claws short and fine; hind claw rather short, slender, decurved.

Geographical variation. Slight; mainly involves tail length, depth of colour of body, extent of supercilium, and amount of black on tail. Nominate *cinerea* from Europe large with proportionately long tail (usually over 91·5: see Measurements). Populations from Asia Minor, Caucasus, and northern Iran have slightly shorter tail and hence sometimes separated as *caspica* (Gmelin, 1774), but tail mainly over 91·5 and therefore included in nominate *cinerea*. *M. c. melanope* from central Asia slightly smaller than nominate *cinerea*, especially tail (see Measurements); tarsus shorter, but bill slightly longer. No information on Urals population; provisionally included in *melanope*. *M. c. robusta* from eastern Asia similar to nominate *cinerea* in wing and tarsus length, but tail distinctly shorter (see Measurements) and hind claw and bill to skull longer (on average, hind claw 7·0, bill 16·9). *M. c. patriciae* (Azores), *canariensis* (Canary Islands), and *schmitzi* (Madeira) all slightly smaller than nominate *cinerea*, but tarsus and hind claw proportionately short in *canariensis* and bill long in *schmitzi* and (in particular) *patriciae* (see Measurements). Nominate *cinerea* medium grey on upperparts, bright yellow below, tail with much white; some individual variation, some birds richer and darker yellow below and some with distinct black border along inner web of t4 and dark shaft to t6 (in particular in Iberia and north-west Africa, but occasionally elsewhere in Europe or Turkey). Most birds of *canariensis* deeper yellow below and with more black on tail than average nominate *cinerea*, but difference slight and bridged by North African and Iberian populations, and *canariensis* considered a poor race. *M. c. schmitzi* is distinctly darker slate-grey on upperparts and ear-coverts than nominate *cinerea*; white supercilium reduced to narrow stripe behind eye, white stripe on lower cheeks narrow and indistinct; black of throat deeper; underparts often deep yellow, tail has dark streak along inner web of t4 and partly dark shaft to t6. *M. c. patriciae* similar to *schmitzi*, but dark streak on inner web of t4 broader (t4 sometimes wholly dark except for short white wedge on tip of inner web), t5 has narrow dark streak along inner web (unlike most *schmitzi*), and shaft of t6 more extensively dark (see Vaurie 1957 for details). *M. c. melanope* similar in colour to nominate *cinerea*, but usually some black on inner web of t4 and on basal shaft of t6 (like *canariensis*). *M. c. robusta* similar to *melanope*, but upperparts slightly darker slate-grey, tending towards *schmitzi* and *patriciae*, but slightly more olive, less pure grey.

Forms superspecies with Mountain Wagtail *M. clara* from Afrotropics (Hall and Moreau 1970).   CSR

Motacilla alba Pied Wagtail and White Wagtail

(Pied Wagtail refers to *yarrellii*, White Wagtail to all other races)

Du. Rouwkwikstaart/Witte Kwikstaart Fr. Bergeronnette d'Yarrell/Bergeronnette grise
Ge. Trauerbachstelze/Bachstelze Ru. Белая трясогузка
Sp. Lavandera enlutada/Lavandera blanca Sw. Engelsk sädesärla/Sädesärla

Motacilla alba Linnaeus, 1758

Polytypic. Nominate *alba* Linnaeus, 1758, continental Europe, Iceland, Faeroes, Asia Minor, and Levant, grading into *dukhunensis* in Urals, lower Volga basin, western Caucasus, and eastern Turkey; *yarrellii* Gould, 1837, Britain, Ireland, and locally on coasts of western continental Europe; *subpersonata* Meade-Waldo, 1901, Morocco; *dukhunensis* Sykes, 1832, south-east European USSR and central Siberia, east to Altai mountains and Yenisey basin, south to Transcaucasia, northern Iran (north of Elburz mountains, east to Gorgan), and lowlands of northern Uzbekistan and Kazakhstan; *persica* Gould, 1876, western and central Iran, east from Zagros mountains, north to southern slopes of Elburz, in winter west to Iraq; *personata* Gould, 1861, eastern Iran and Turkmeniya (USSR), east through mountains and neighbouring plains of central Asia to western Kun Lun, Pamirs, Dzhungaria, north-west Mongolia, and western Sayan, north to lower Angara river; occasionally occurs west to Egypt. Extralimital: *ocularis* Swinhoe, 1860, north-east Siberia, east of *dukhunensis*, from about Yenisey east to Chukotsk and central Kamchatka; *baicalensis* Swinhoe, 1871, southern Siberia, east of *personata* and south of *ocularis* from about Lake Baykal and northern Mongolia to middle Amur river; *lugens* Gloger, 1829, southern Kamchatka and southern shores of Sea of Okhotsk south to Japan and Ussuriland; *leucopsis* Gould, 1838, southern Mongolia, northern and eastern China, and eastern Amurland; *alboides* Hodgson, 1836, Kashmir through Himalayas and southern Tibet to south-west China.

Field characters. 18 cm; wing-span 25–30 cm. Somewhat bulkier and markedly less attenuated than Grey Wagtail *M. cinerea*, with 25% shorter tail; slightly larger than Yellow Wagtail *M. flava*, with 20% longer tail. Ground-haunting, active insectivore, sharing form and actions of pipit *Anthus* but differing markedly in mainly pied plumage and long tail, often wagged. Flight action generically diagnostic, with bounding progress most marked of all small west Palearctic passerines. Adult essentially black (or grey) and white, ♂♂ of the various races differing mostly in back colour (black or grey) and more or less well marked face. Racial distinctions much less clear in ♀♀, even absent in immatures. Sexes dissimilar in all but juvenile plumage but distinctions subject to overlap within and between races; marked seasonal variation on foreparts. Juvenile and 1st-year separable from adult of same race. 6 races occur in west Palearctic, with widespread group of 5 grey-backed races (comprising 2 with unmarked face-sides, 2 with masked faces, 1 variable), and 1 (mainly British and Irish) black-backed race. Only widespread nominate *alba* described in full here, followed by main characters of ♂♂ of other races.

Race of continental Europe except south-east, nominate *alba*. ADULT MALE BREEDING. Crown from above eye, nape, chin, throat, chest, ground-colour of wing (including tertials), and all but outermost tail-feathers black. Mantle, scapulars, rump, upper tail-coverts, and suffusion on flanks clean grey, darkest on upper mantle, palest on scapulars. Forehead, forecrown, lores, cheeks, patch on lower sides of neck, fringes and tips of median and greater wing-coverts, fringes of tertials, underbody, and edges of outer tail-feathers white. Bird thus has strikingly white face (contrasting with black of rear head and of throat and chest), striking double white wing-bars (formed by tips of median and greater coverts), and bold white lines along tertials and edges of tail. In flight, black areas increased by greater exposure of wing but paleness of back and rump remains striking. Underwing white. Wear dulls both black and grey plumage and reduces white feather-fringes. ADULT FEMALE BREEDING. Resembles ♂ but plumage markings and contrasts usually less striking, with black mottling on forehead, black crown invaded by grey, duskier mantle and rump, white-mottled chin, and duskier flanks. ADULT MALE NON-BREEDING. Forehead mottled black, crown partly grey. Chin and throat white (occasionally cream or tinged yellow), leaving black chest-band; otherwise as breeding ♂ but contrasts somewhat subdued. ADULT FEMALE NON-BREEDING. Loses white forehead, whole of crown becoming grey, and white lores and cheeks also invaded by dusky or olivaceous mottling; underparts as non-breeding ♂, with dull white chin and throat and black chest-band, though overall effect of moult is to 'close' top of head and 'open' lower face, quite unlike ♂. JUVENILE. At fledging, smaller and shorter-tailed than adult. Lacks bold pattern and clean tones of breeding adult and noticeably swarthier than even non-breeding ♂, with dusky or brown tinge to all grey plumage and dirty appearance to all white plumage. Lack of contrast most obvious in (1) darker head, with fully coloured crown and indistinct buff-white supercilium, (2) splashed rather than patched chest pattern, with smoky or broken malar stripe and crescentic chest-band often resembling those of strongly marked juvenile *M. flava*, (3) dusky-brown flanks, usually particularly visible behind chest-band, (4) indistinct buffy fringes and tips to wing-coverts, forming only dull wing-bars, and (5) dull white fringes to tertials. FIRST WINTER. Plumage usually acquired by September. ♂ resembles adult ♀ non-breeding but crown blacker. ♀ similar but grey crown duller, tinged olivaceous. Both sexes retain juvenile flight- and tail-feathers. FIRST SUMMER. By March, plumage predominantly as adult but wing pattern sometimes incomplete. Distinguished by dull brown flight-feathers. At all ages, bill and legs black.

Other races. ADULT MALE BREEDING. (1) Race of Britain, Ireland, and (locally) nearby continental coast, *yarrellii*. Lacks any pure grey plumage, being dusky- or jet-black on back and rump; thus truly piebald, with starker contrast in plumage pattern than other races. White markings noticeably pure and broad on fringes and tips of wing-coverts and along fringes of tertials, almost forming panels on best-marked birds. Dusky flanks obvious, particularly when fluffed out. (2) Moroccan race, *subpersonata*. Lacks unmarked face-sides of *alba* and *yarrellii*, with half-masked appearance created by black cheeks and smaller white patch on neck-sides. Also less white over and around eye, with white on lore reduced to a streak below black fore-eye-stripe. Back dusky and less clean than nominate *alba* but paler than *yarrellii*. (3) West and central Iranian race, *persica*. Variable: some show face pattern of *personata* (see below) but others resemble birds of the *alba/dukhunensis* cline. (4) Race of south-east European USSR and west-central Asia, *dukhunensis*. Head pattern as nominate *alba*, but back paler grey, wing-bars and tertial-edges broader and whiter. (5) Race of eastern Iran to north-west Mongolia, *personata*. Head more extensively black than in any other race occurring in west Palearctic, with white markings restricted to forehead, eye-ring, and supercilium. White wing-bars and tertial fringes wide, often broad enough to form solid panel (suggesting African Pied Wagtail *M. aguimp* but that species always white-throated).

Specific identification not difficult, with only *M. aguimp* (in west Palearctic) providing closely matching appearance and yet easily distinguished by striking differences in forepart and wing patterns. Racial identification of adult ♂♂ varies from easy to difficult (or, in *persica*, suspect due to unstable morphology): in western Europe, separation of nominate *alba* from British *yarrellii* on characters given above usually dependable but occasional intermediates not assignable. Racial identification of ♀♀ varies from fairly easy (but not completely trustworthy) to impossible: again, separation of nominate *alba* and *yarrellii* often practical but care needed to establish fully grey rump of nominate *alba* and to avoid trap of merely grey-backed 1st-summer or dusky-backed adult *yarrellii*; field study of ♀♀ of other races incomplete (but see Geographical Variation). Racial identification of juveniles and of 1st-winter migrants largely impractical, dogged not only by markedly reduced racial differences but also by convergent appearance of some aberrant juvenile *M. flava* and juvenile Citrine Wagtail *M. citreola*; field studies incomplete but for discussion of interspecific confusion between nominate *alba* and *M. flava* and *M. citreola*, see those species (pp. 415 and 434). Faced with birds other than adult ♂♂, important to remember that *M. alba* has structure and voice which, though broadly overlapping with *M. aguimp*, do allow fairly ready separation from *M. flava* and *M. citreola*. Most useful characters are (1) apparently higher carriage and more constant movement of head, (2) apparently lower-slung chest, showing both deep and wide markings in all plumages, (3) longer, closer-folded and thus often apparently thinner tail, and (4) disyllabic form of common calls. Risk of confusion with *M. flava* and (particularly) *M. citreola* thus avoidable by close attention to general character, even in cases of troublesome juvenile and 1st-winter plumages. Flight noticeably free and bounding, consisting of alternating short bursts of rapid wing-beats and long, 'shooting' curves or falls with wings closed; action surprisingly constant even in restricted space, when often accompanied by noticeable tail-fanning. Flight silhouette essentially diamond-shaped, with extension of long, thin tail which does not 'whip' like that of Grey Wagtail *M. cinerea*. Appearance in flight recalls longer-tailed pipits *Anthus* but action and silhouette both sufficiently distinctive to allow instant separation at distance; most marked differences are greater length of curving flight trajectory and narrower tail. Gait remarkably free: nimble walk, short or long run, hops, and leaps all employed when foraging. Especially when walking, head moves backwards and forwards; walking and, in particular, the end of a run often accompanied by obvious tail-wagging which makes white outer tail-feathers 'flash'. Has angled stance when walking, noticeably horizontal when running, with head held well above chest; more variable on perch, though tail often hangs down. Migrant assemblies smaller than those of *M. flava*.

Song simple but jaunty and twittering; nominate *alba* sings more freely than *yarrellii*, even doing so on passage. At least 3 regular calls: commonest flight-call 'tschizzik' (said to be softer in nominate *alba*), but one contact-call, 'tzeurp', lacks disyllabic form and can suggest commonest monosyllable of *M. flava* and *M. citreola* though less shrill and more rippled; alarm-call a sharp 'chick'.

Habitat. From highest mainland latitudes to middle oceanic as well as continental zones, Arctic and subarctic, boreal, temperate, steppe, Mediterranean, and even desert fringes, from July isotherm of 4°C to subtropics, penetrating into cold windy rainy regions and into arid and hot climates. Correspondingly adaptable as between wide variety of waterside habitats: lakes, rivers, streams, canals, estuaries, and sea coasts; also others distant from water, especially where agriculture, pastoralism, human settlements, roads, tracks, airfields, parks, gardens, gravel-pits,

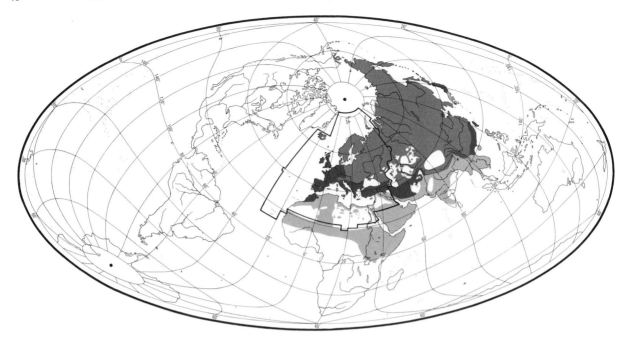

or other human intervention have provided essential bare spaces or very low vegetation cover. Differs from Yellow Wagtail *M. flava* in avoidance of tall or dense vegetation, except for roosting, then using reedbeds, bushes, and palm-groves and also artefacts such as large urban buildings, sugar-cane plantations, tree-lined urban streets, and horticultural glasshouses. To some extent shares attachment to grazing animals, and extends this to livestock in farmyards and small pens, especially where water provided. Differs also in readiness to breed in enclosed situations, including gardens with lawns or urban open spaces, or contrastingly where there are extensive bare tracts either of arable land or waste, even semi-desert. Differs from Grey Wagtail *M. cinerea* in preference for flat especially lowland sites, and in avoidance of rushing watercourses and dense stands of trees. Will however ascend even above treeline in association with dwellings or mountain huts, up to 2000 m or more in Switzerland (Glutz von Blotzheim 1962) and in Caucasus to 3000 m. In USSR, also nests in rocky parts of pine forest, by timber yards, on burnt patches, and on tundra at forest margins (Dementiev and Gladkov 1954a). Has evidently adapted and colonized extensively in recent times in wake of changing land use and practices, most of which have been turned to good account, although some have lately proved locally adverse. Although largely a ground bird, flies freely in lower and middle airspace, especially on migration and in winter, when it occupies similar habitats.

Distribution. Expansion in western Ireland and south-west Morocco.

JAN MAYEN. Bred 1908 (Seligman and Willcox 1940).

BRITAIN, IRELAND. Spread in western Ireland, e.g. Galway and Mayo (Sharrock 1976). Bred Isles of Scilly early 20th century (Parslow 1973). A few nominate *alba* breed occasionally, mainly Shetland (Scotland) (Sharrock 1976). CHANNEL ISLANDS. Bred 1952 and 1960 (*yarrellii*) and 1980 (nominate *alba*) (Long 1981b). FRANCE. *M. a. yarrellii* breeds occasionally, mainly coast of Pas de Calais and Somme; mixed pairs frequent (RC). Breeds irregularly on Corsica (Thibault 1983). NETHERLANDS. 4-10 pairs of *yarrellii* breed, mainly along west coast (Teixeira 1979). WEST GERMANY. *M. a. yarrellii* is a rare breeder in north (AH). SYRIA. No recent proof of breeding (AMM). ALGERIA. May breed occasionally; no proof (EDHJ). MOROCCO. Marked expansion in south-west (GT).

Accidental. Spitsbergen, Bear Island, Jan Mayen.

Population. Increased Ireland, decreased eastern England and Rumania.

ICELAND. Common in lowlands; no known change (AP). FAEROES. 0-5 pairs (Bloch and Sørensen 1984). BRITAIN, IRELAND. Under 1 million pairs. No evidence of marked widespread change, but has increased Ireland and perhaps decreased slightly in parts of England; earlier decrease in Scotland now arrested (Parslow 1973). About 500 000 pairs, decreased eastern England (Sharrock 1976). FRANCE. Under 1 million pairs (Yeatman 1976). BELGIUM. About 70 000 pairs (Lippens and Wille 1972). NETHERLANDS. 50 000-75 000 pairs (Teixeira 1979). WEST GERMANY. 750 000-1 300 000 pairs (Rheinwald 1982). SWEDEN. Not less than 500 000 pairs (Ulfstrand and Högstedt 1976); counts of migrants at Falsterbo suggest possible increase since 1942-4 (Roos 1978). FINLAND. About 430 000 pairs (Merikallio 1958). Increased in north

1941–77 (Väisänen 1983). Decreased 1979–83 (Vickholm and Väisänen 1984). CZECHOSLOVAKIA. No changes reported (KH). RUMANIA. Decreased (VC).

Survival. Germany: of 134 young ringed, 79·8% recovered in 1st year of life, 9·7% in 2nd year, 8·2% in 3rd year, 1·5% in 4th year, and 0·8% in 5th year (Drost and Schüz 1940). Finland: annual overall mortality 48% (Haukioja 1969). East Germany: 1st-year mortality 74% (Ölschlegel 1985). Oldest ringed bird 9 years 11 months (Rydzewski 1978).

Movements. Varies from wholly migratory to more or less resident. Most northern populations migrate south to Mediterranean area, tropics and subtropics of Africa, peninsular India, and south-east Asia.

Populations of nominate *alba* breeding in area south of Denmark within 50–55°N and 6–12°E winter from Garonne valley (France) south through central Iberia to Morocco; found in inland areas, thus not overlapping to any extent with *yarrellii*. Probably many nominate *alba* winter in favourable areas well north into breeding zone: certainly true of Lake Geneva basin even though rest of Switzerland vacated (Glutz von Blotzheim 1962). Birds from northern Europe generally (all nominate *alba*) have been recovered wintering round entire Mediterranean coastline though very many more from African shore, particularly Algeria to Tunisia and also (largely birds from Finland) from Israel to Nile delta. One bird ringed at 69°N in extreme northern Norway recovered in winter in northern Turkey (Zink 1985). Nominate *alba* winters

September–April also throughout Arabia and Sinai, even in quite arid areas (Meinertzhagen 1954); also some records of *dukhunensis*, breeding east of Urals, from eastern Arabia.

Nominate *alba* winters through much of Africa north of equator (Heim de Balsac and Mayaud 1962; Moreau 1972; Curry-Lindahl 1981) with considerable numbers north of Sahara and at oases within desert (Bundy 1976). In extreme west, common south to Mauritania (Gee 1984), Sénégal (Morel and Roux 1966), and Gambia (Gore 1981), but becoming uncommon in Sierra Leone (G D Field). Not uncommon in Niger inundation zone of Mali (Curry and Sayer 1979). In Nigeria winters commonly in extreme north, less commonly south to *c.* 9°N; occasional claims south to coast regarded as dubious because of possible confusion with African Pied Wagtail *M. aguimp* (Elgood 1982). Further east, numerous in Sudan (Hogg *et al.* 1984) and Ethiopia (Urban and Brown 1971), and a few cross equator with stragglers reported south to Malawi (Benson and Benson 1977) and probably to Livingstone in Zambia (White 1961), though sight records from adjacent Victoria Falls regarded as unlikely (Benson *et al.* 1971). *M. a. dukhunensis* from east of Urals (wintering mostly in India and south-east Asia) has been found occasionally in Ethiopia (Urban and Brown 1971) and in north-east Zaïre (Lippens and Wille 1976).

Populations from Britain and Ireland (*yarrellii*) largely resident, progressively so southward, but with many upland areas, especially in north, deserted in winter (British Ornithologists' Union 1971); thus, northern Scotland largely vacated (Baxter and Rintoul 1953) but too little data to show whether birds wintering in Ireland and southern Britain are mainly of local origin or to what extent they originate from further north. Short-distance movements (5–80 km, $n = 226$) appeared random; movements over 80 km within Britain ($n = 38$) orientated between north-west and north or between south-east and south. Ringing shows that main wintering area of *yarrellii* abroad lies in Brittany, western Iberia, and Morocco. 124 overseas recoveries were from Norway (1), West Germany (1), France (39), Spain (21), Portugal (60), and Morocco (2), mostly of 1st-year birds; both severe winters and favourable breeding seasons lead to increased movement. These long-distance migrants keep to southerly direction as far south as French/Spanish border on Biscay but then turn between SSW and south-west to reach western Iberia. One bird (alleged to be *yarrellii*) ringed in autumn in Morocco was recovered 3 months later in Tunisia, *c.* 1800 km east. (Davis 1966; Zink 1985.)

Races breeding east of Urals winter as follows. *M. a. dukhunensis* breeding largely between Urals and Yenisey river winters mainly in northern India but also west to Afghanistan with some in eastern Arabia and eastern Africa (see above) suggesting broad-front migration essentially southward; *personata* of west-central Asia winters

from Iran east to northern India, suggesting a more south-west heading in autumn (Ticehurst 1926–7); *baicalensis* from south-central Siberia winters widely from Iran east through India to south-west China and south to Thailand; east Asian *lugens* may winter as near breeding grounds as southern Japan and Taiwan but also in north-east China (e.g. Ostapenko 1981); *ocularis* of north-east Siberia winters from north-east India through Thailand (McClure 1974) and south-west China and as far east as Philippines; *leucopsis* of east-central Asia winters in southern China and Thailand (Delacour and Mayr 1946; Bent 1950; Vaurie 1959; Ali and Ripley 1973*b*; Etchécopar and Hüe 1983).

Movements of *subpersonata* from Atlas mountains of Morocco, *persica* from mountains of Iran, and *alboides* from eastern Himalayas through to south-west China are largely altitudinal, though some *persica* winter in Iraq and some *alboides* move south to Thailand.

The few nominate *alba* which breed most years in eastern Greenland are said to move almost due east to Iceland in autumn (Bent 1950) and then join with numerous Iceland breeders to proceed between SSE and south-east across Ireland (Ruttledge 1966), Britain, and western Iberia to winter in West Africa; Icelandic birds recovered south to Sénégal (Zink 1985). Populations of nominate *alba* breeding in Norway may move south-west into Britain or move nearer south at first into Denmark, all these west European birds then heading south-west into extreme south-west France where large numbers recovered (Zink 1985), especially October–November, most having been taken north of Loire though heavy persecution in western France could be introducing bias. Birds from central Europe move mainly south-west, with a few between SSW and south but one bird ringed as nestling in Switzerland had moved south-east to southern Italy, though perhaps via Mediterranean coast of France (Zink 1985), as recorded in Blackcap *Sylvia atricapilla* (Zink 1977). Appears to be migratory divide just east of Denmark at *c.* 10–12°E, birds west of this line heading well west of south to western France, Iberia, and western Africa, those to east heading east of south towards eastern Mediterranean (Zink 1985).

Movement of birds from south of Baltic not well known, though recoveries showing south-west movement have occurred; recoveries of birds from south-east Baltic area have been to SSW (from Vistula estuary to Tunisia) and SSE (from Lithuania to Bulgaria) (Zink 1985). The only cases of recoveries from further east concern spring movement of birds ringed in Kutch (India) and recovered in south-west European USSR, inferring previous autumn movement to east of south-east. Passage through Swiss passes late September to late October with movement showing marked diurnal rhythm with peaks at *c.* 08.00 hrs and 17.30 hrs and only very few between 13.00 and 15.30 hrs (Crousaz 1961).

Autumn passage occurs across entire length of Mediterranean. Strong passage through Gibraltar (Lathbury 1970). Birds occurring on Balearic islands mainly 1st-years and include some *yarrellii* (Bannerman and Bannerman 1983); on Malta, much more numerous in autumn than spring, with some overwintering—no *yarrellii* recently though reported in the past (Sultana and Gauci 1982); also mostly in autumn on Cyprus, occurring late September to mid-November, peaking in second half of October (Flint and Stewart 1983).

Passage of Icelandic nominate *alba* through Britain and Ireland occurs mostly August-October (Ruttledge 1966; British Ornithologists' Union 1971), though passage of Icelandic birds on Fair Isle may start in late July with stragglers to November but peaks late August to early September (Williamson 1965). In southern Finland, passage begins late August and peaks mid-September with only stragglers in October (Zink 1985); near Helsinki, conspicuous reverse (northward) migration occurs up to time of peak southward movement when it ceases abruptly (Koskimies 1947). Timing and reverse movement similar at Ottenby, southern Sweden (Bergman 1948). In Switzerland, autumn departure generally begins *c.* 10 September, peaks mid-October, but continues regularly well into December (Glutz von Blotzheim 1962). In European USSR (nominate *alba*), departure may start in late August, but more often in early September; main exodus through September and completed by mid-October; movement tends to follow river valleys, being concentrated in morning and evening in good weather but extending throughout day in poor conditions.

Further east (*dukhunensis*), movement starts early September in small flocks, but temperature drop produces larger parties moving particularly along Yenisey river; in Pamirs, continues to late October or early November. Little data for other Asiatic races: *baicalensis* moves September to early October; *ocularis* leaves Anadyr and Yakutsk in late September; *lugens* leaves Kamchatka late August to early September; *personata* leaves Krasnoyarsk from mid-August but present until early October. (Dementiev and Gladkov 1954a.) Autumn passage of *dukhunensis* in northern Baluchistan, Northwest Frontier Province, and from Gilgit and Kashmir east to Sikkim is from end of September (earliest 13 September) to October; *personata*, some of which may have bred at no great distance, arrive in Himalayan foothills at end of August; *leucopsis* common on passage at high altitudes in southern Tibet and eastern Himalayas from mid-August to mid-September; in Nepal, passage of *ocularis* occurs during October, of *baicalensis* rather earlier (Ali and Ripley 1973b). Earliest arrival in Malaysia 8 September (*ocularis*) (Medway and Wells 1976).

Little to suggest that birds moving to breeding areas in spring do not mainly retrace autumn routes (in contrast to considerable evidence for loop migration of western populations in *M. flava*). There are, however, 2 spring recoveries of British *yarrellii* from Belgium, from which

area there are no known autumn recoveries; also, nestling from near Hamburg recovered on spring migration 2 years later on Bornholm (Denmark): these data indicate possibility of some spring looping to east (Zink 1985).

Abundant on spring passage through Haut Atlas of Morocco, March to mid-April (Chaworth-Musters 1939), and spring passage well documented through Mediterranean area (Cortés *et al.* 1980; Sultana and Gauci 1982; Flint and Stewart 1983), though mostly reported as being on smaller scale than in autumn. Apart from winter mortality, disparity could be due to overflying (particularly of the central islands) rather than to any change of route.

Little data on spring arrival times for partially migratory *yarrellii*, even for northern areas which are completely vacated in winter. Arrival of nominate *alba* over wide areas of central Europe may be as early as February but mainly March-April, while arrival in southern Scandinavia is late March and in northern Scandinavia around mid-April (Rendahl 1967). On Fair Isle, passage may start mid-March, but is usually early April to early May with stragglers some years into June (Williamson 1965). In Switzerland, first arrivals have been observed in early February but more usually at end of February or in first third of March, peaking late March and first half of April (Glutz von Blotzheim 1962).

In USSR west of Urals, nominate *alba* arrives as river edges become ice-free but large-scale passage later, in southern areas mainly late March, in middle latitudes April, but not until May in Lapland and Timan tundra. East of Urals, *dukhunensis* appears in March in south, and in last third of April north-east of Aral Sea, but arrival continues until late May in north of that race's range. Arrival of other Asiatic races similarly dependent on latitude with *lugens* arriving in Kamchatka late April to early May while *personata* arrives on Pamir plateau in March when area still snowbound. (Dementiev and Gladkov 1954a.) Spring passage of Asiatic races through Indian subcontinent, though not well documented, seems to follow same routes as in autumn (Ali and Ripley 1973b). In Thailand, races recorded as follows: *alboides* 11 October-12 February, *ocularis* 28 October-5 April, *baicalensis* 26 September-9 March, *leucopsis* 19 September-29 March (McClure 1974).

In west of range, nominate *alba* recorded once on west coast of Greenland (Bent 1950) and even as far west as Fort Chimo on Hudson Bay (Canada)—2 adults with 2 juveniles (Todd 1963). In east, records of both *ocularis* and *lugens* in coastal western Alaska, and vagrants (probably all *lugens*) recorded along west coast of America south to Los Angeles, mostly in autumn but with single spring inland record from Oregon (Roberson 1980). JHE

Food. Small invertebrates. 3 main foraging techniques. (1) Picking. Picks items from ground or water surface while walking; will also walk on floating vegetation. (2)

Run-picking. Makes quick darting run at prey, picking it up either from surface or as it takes off. (3) Flycatching. Makes short flight from ground, catching prey in mid-air. (Davies 1977.) Will also take food from water while hovering (Glutz von Blotzheim 1962), and recorded hovering repeatedly for periods of $c.$ 5 s, $c.$ 10 cm up, to take small swarming insects (Cope 1985). Tail assists balance when run-picking and flycatching; especially important when turning rapidly in flight (Davies 1977). In winter, feeds at edges of puddles, lakes, and rivers, taking items washed up (Glutz von Blotzheim 1962; Davies 1976; Davies and Houston 1981; Mal'chevski and Pukinski 1983). Also recorded as hunting insects emerging from earth in field turned over by plough (Schmidt 1967b), and will take invertebrates from backs of pigs (Glutz von Blotzheim 1962). Feeds both singly and in flocks (see Social Pattern and Behaviour). Prey usually swallowed whole, but larger items often beaten against ground or rock first (Dolgushin et al. 1970). At Oxford (southern England), flock birds (feeding at shallow pools) employed picking exclusively, while single birds (feeding at dung pats) used greater variety of techniques—picking 67·4%, run-picking 13·6%, flycatching 19·0%; single birds generally fed more slowly than those in flocks, with flock birds taking $73·2 \pm SE3·3$ items per min, solitary birds $10·1 \pm SE0·6$ items per min. Despite great difference between flock and solitary sites in behaviour used and prey eaten (see below), birds switched rapidly from one to the other; switches sometimes resulted in maximization of energy intake but at other times not, probably because feeding efficiency constrained by need to engage in other activities. (Davies 1977.) In some instances, shared defence of feeding territory (see Social Pattern and Behaviour) leads to increased feeding rates (Zahavi 1971; Davies 1980). *Contra* Broom et al. (1976), no good evidence that roosts serve as food information centres, allowing searching birds to accompany those who have located food; see also Fleming (1981).

The following recorded in diet in west Palearctic. Adult and larval mayflies (Ephemeroptera), damsel flies and dragonflies (Odonata: Agriidae, Aeshnidae), adult and larval grasshoppers, etc. (Orthoptera: Acrididae), earwigs (Dermaptera), bugs (Hemiptera: Cicadellidae, Hydrometridae, Gerridae, Delphacidae, Aphididae), larval lacewings (Neuroptera), adult and larval butterflies and moths (Lepidoptera), adult and larval caddis flies (Trichoptera: Phryganeidae), adult, pupal, and larval flies (Diptera: Chironomidae, Drosophilidae, Muscidae, Sphaeroceridae, Chloropidae, Sepsidae, Bibionidae, Ephydridae, Tipulidae, Culicidae, Tabanidae, Tachinidae, Syrphidae, Calliphoridae, Mycetophilidae, Lonchopteridae, Agromyzidae, Opomyzidae, Dryomyzidae, Empididae), adult and larval Hymenoptera (sawflies Symphyta, Ichneumonidae, ants Formicidae), adult and larval beetles (Coleoptera: Carabidae, Staphylinidae, Haliplidae, Silphidae, Scarabaeidae, Curculionidae, Chry-

somelidae, Cerambycidae, Elateridae, Byrrhidae), small crustaceans (e.g. amphipods *Gammarus*), spiders, small fish (e.g. minnows *Phoxinus*, fry of perch *Perca fluviatilis*), seeds and herbaceous plant material, and bread and other food put out in gardens (Baer 1909; Rey 1910b; Gil 1927; Elton 1928; Kovačević and Danon 1952; Summers 1953; Neufeldt 1961; Glutz von Blotzheim 1962; Fincher 1963; Ptushenko and Inozemtsev 1968; Averin and Ganya 1970; Popov 1978; Glue 1982; Mal'chevski and Pukinski 1983; see also review by Ölschlegel 1985). Outside west Palearctic, recorded also taking crabs, small molluscs, earthworms *Lumbricus*, and berries (D'Abreu 1918; Dolgushin et al. 1970; Ali and Ripley 1973b; Kovshar' 1979).

At Oxford, March, 4654 items in faeces of birds feeding in flocks at pools comprised almost wholly Diptera: 96·7% (by number) Chironomidae, 1·2% beetles, and 2·1% others (mostly Diptera). Birds feeding singly at dung pats also took mostly Diptera: 779 items comprised 37·5% Sphaeroceridae (1-4 mm long, mostly 3-4 mm), 35·5% Scatophagidae (5-10 mm), 12·5% beetles (2-3 mm), and 14·5% others (mostly Diptera); large Sphaeroceridae (3-4 mm) selected preferentially (comprised 10·1% of total available, but 35·9% in faeces) and Scatophagidae selected against (77·1% of total available but 35·9% in faeces); prey size smaller than that generally available (6-8 mm, rather than 7-9 mm) and feeding experiments showed that this size range produced greatest rate of energy intake (larger items required longer handling time). (Davies 1977.) In Leningrad region (USSR) (no date), over 50% of food samples from young and adult stomachs (no number) contained Diptera, primarily Tipulidae, Culicidae, Tabanidae, and Tachinidae (Mal'chevski and Pukinski 1983). This predominance of Diptera not found in other studies of adult diet. In southern Karel'skaya ASSR (western USSR), 10 stomachs contained invertebrates found mainly by water: mayfly larvae in 8 stomachs, beetles (Hydrometridae, Gerridae, Chrysomelidae) in 3 (Neufeldt 1961). In Moscow, June, 65 stomachs contained 30% (by number) dragonflies (mainly larvae), 30% beetles (Carabidae, Chrysomelidae), 15% caterpillars, 20% other insects (mainly Diptera and Hymenoptera), and 5% shoots of herbaceous plants (Ptushenko and Inozemtsev 1968). In Moldavia (southwest USSR), 22 stomachs (no dates) contained mainly beetles (Curculionidae, Chrysomelidae, Cantharidae, Carabidae); also larval and adult Diptera, Ephemeroptera, Orthoptera, Lepidoptera, and spiders (Averin and Ganya 1970). In Tatarskaya ASSR (east-central European USSR), 35 stomachs (no dates) contained many beetles: Curculionidae in 51%, Carabidae in 23%, Elateridae 17%, Chrysomelidae 11%, Geotrupidae 6%, larval Tenebrionidae 3%, Dytiscidae 3%, Staphylinidae 3%, Buprestidae 3%, also Hemiptera in 31%, Hymenoptera in 23%, caterpillars 17%, mosquitoes 12%, other small Diptera 23%, spiders 11%, Orthoptera 6%, Cicadellidae 3% (Popov 1978). In Kazakhstan (south-central USSR),

September, 1 stomach contained 30% (by number) Orthoptera and 15% small seeds (Dolgushin *et al.* 1970); for further data from east Palearctic (showing diet mainly beetles), see Yanushevich *et al.* (1960), Pek and Fedyanina (1961), and Dolgushin *et al.* (1970).

No data on differences between diets of adults and young. For study including some stomachs of young, see above; for only other data on typical diet of young in west Palearctic, see Smogorzhevski and Kotkova (1973). In Tien Shan (south-central USSR), collar-samples from young contained mostly Diptera (Calliphoridae 23·1%, Muscidae 10·8%, Tipulidae 6·1%) and Ephemeroptera; also Lepidoptera (mainly Noctuidae), Orthoptera, Hemiptera (Miridae, Tingidae), and molluscs; at 11 nests at same site, Diptera seen to be brought on 45% of 704 occasions (Tipulidae 36·8%), larval and adult Lepidoptera 10·5%, Orthoptera 4·8%, and others (including beetles, dragonflies, spiders, earthworms, and breadcrumbs) 2·8% (Kovshar' 1979). Adults recorded bringing small fish to young; on one occasion, *c.* 4 cm long and picked up dead or dying (Brown 1948; Chappell 1949). No real difference in feeding rates between 1st and 2nd broods (Kovshar' 1979). In Moscow region in nest with 6 young 9 days old, feeding occurred 03.40-21.10 hrs; 330 visits averaged 18·9 per hr (3-28), 3·3 per hr per nestling, each nestling receiving 55 portions during the day. Feeding rates increased with age of young: 140 visits per day at 3 days, 204 at 5 days, 330 at 9 days. (Ptushenko and Inozemtsev 1968.) At 7-8 days, recorded receiving average 15 visits per hr (Mal'chevski and Pukinski 1983). In Tien Shan, usually 2-5 items brought to nest per visit, fed to 2-3 young; not infrequently single items brought repeatedly, or large balls containing 10-15 items (Kovshar' 1979).   ILG, CAT

Social pattern and behaviour. Well studied, especially winter dispersion and roosting. For recent review, see Ölschlegel (1985).

1. Dispersion outside breeding season varies from gregarious to solitary, even territorial, depending on food supply. Migrates mostly in small flocks (Dementiev and Gladkov 1954a); e.g. in eastern Scotland, spring migrants in flocks of 2-3(-20) (Evans 1901). However, in late autumn, USSR, birds (usually juveniles) typically migrate singly (Dementiev and Gladkov 1954a). In winter, Oxford (England), most birds in flocks, e.g. of 5-20 on ploughed fields; however flock dispersion highly flexible from day to day, even hour to hour, as food distribution changed, and territorial birds periodically joined flock (Davies 1976, 1982: see Flock Behaviour, below). In Surrey (England), exceptionally large feeding flock of 150-175 reported (Weller 1947). In Egypt, nominate *alba* occurs singly, in small parties, or in flocks of up to thousands (Witherby *et al.* 1938a), latter perhaps for roosting. In Oxford, some adult ♂♂ held contiguous feeding territories along river banks October-March; 7 territories along 3·2 km; 2 territories subsequently used (by same winter residents) for breeding, other 5 abandoned (Davies 1976). Average length of territory *c.* 300 m (*n*=21) (Davies 1982; Houston *et al.* 1985, which see for discussion of optimum territory size). Further south in range, winter territorialism

more common (Greaves 1941; Goodwin 1950; Zahavi 1971; Broom *et al.* 1976). In Israel, territorial where food dependable and localized, in flocks where food more temporary and dispersed; territories established mainly in autumn, and dispersion quite stable after mid-November. Territory size varied from less than 10 m in diameter on rubbish tip to more than 100 m in city, reflecting food supply. (Zahavi 1971.) Throughout range, winter territory established by ♂ (resident) but he may form temporary association, of varying length (see Bonds, below), with a 1st winter bird or adult ♀ ('satellite'). Satellite helps to defend territory, and may be admitted to several different territories during winter; tolerated on territory when food supply is such that benefits to owner of help with defence outweigh costs of sharing food. In Oxford, residents accepted satellites (typically 1st-winter birds) for periods of up to 7 days. (Davies 1980, 1982; Davies and Houston 1981.) In Japan, similar dispersion in *lugens*: of 10 territories, 3-4 occupied by temporary 'pairs', rest by single ♂♂ (Higuchi and Hirano 1983). In Israel, ♂♂ solitary in small territories, but in large territories often admitted ♀ satellite, pair thus formed frequently persisting for weeks or months; some ♀♀ consorted simultaneously with 2 or more neighbouring ♂♂, and so gained access to their territories (Zahavi 1971). For displays facilitating association of resident and satellite, see Antagonistic Behaviour (below). In Egypt, territories likewise often held by ♂ and ♀ (Goodwin 1950; Simmons 1965), sometimes by ♂ and 2 ♀♀ (Greaves 1941). In Israel, most residents retained same territory in successive winters: of 27 marked birds, 11 occupied same territory next autumn; flocks also contained high proportion of same birds in successive winters (Zahavi 1971). Fidelity to wintering areas also established in Malta (Sultana and Gauci 1982). BONDS. No evidence for other than monogamous mating system. Pair-bond usually lasts only for duration of breeding season, though may occasionally form on winter territory (see above); thus, in 2 out of 7 territories, ♂-satellite association established pair-bond for subsequent breeding (Davies 1976). In Israel, ♂♂ and ♀♀ sharing winter territories (see above) arrived on and left territory at different times, therefore presumably not migrating or breeding together (Zahavi 1971). Occasionally, breeding pairs faithful over 2 seasons (Leinonen 1974a; see also Breeding Dispersion, below). Several reports of nominate *alba* hybridizing with *yarrellii* (e.g. Champernowne 1908, Sueur 1982, Mayer 1983). For hybridization with Grey Wagtail *M. cinerea*, see that species (p. 447). Almost all brooding by ♀. Both members of pair feed young; tend them for 4-7 (-11) days after fledging (Leinonen 1973a). ♀ stays longer than ♂ with 2nd brood (Persson 1977; for further details of parental roles, see Relations within Family Group). Age of first breeding 1 year; some 1-year-olds do not breed (Leinonen 1973b). BREEDING DISPERSION. Solitary and territorial. Territories often not contiguous, and size therefore difficult to determine. In Finland, density 1·1-36·7 pairs per km², average *c.* 2-3 pairs per km² (Leinonen 1974a). In Uppland (Sweden), 2·1 pairs per km² (Olsson 1947). In Netherlands, pairs per km² near Leiden (Teixeira 1979). In Switzerland, average 1-5 pairs per km² in open country, 10-20 pairs per km² in villages (Schifferli *et al.* 1982). Territory serves for pair-formation, nesting, and feeding; in 2 territories, Oxford, most food collected in vicinity of nest, but pair often moved up to 1 km to feed (Davies 1976). Of 64 pairs which lost nest (at any stage), 77% abandoned territories; remainder built repeat nests in same territories, usually in different place from 1st nest (Leinonen 1973a). In pairs that raise 2 broods, little information on siting of nests for 2nd brood: one hybrid (*alba* × *yarrellii*) pair built nest for 2nd clutch *c.* 5 m from 1st (Mayer 1983). Territories typically used year after year. In central Finland, a proportion of adults (both sexes) return to breed in same territory (but at different nest-site) though territories often have different occupants from year to year. Of 38 ♀♀ ringed as adults, 10 bred in same area at least twice; of these, 7 used same territory, 3 a different one; 5 bred in same territories for 2 consecutive years, 1 each for 4 and 5 years. Of 13 ♂♂ ringed as adults, 5 returned to same area; of these 2 used same territory, 3 a different one. Of 436 ringed young, at least 4·1% returned to breed in same area; though none bred in natal territory, 11 bred within 2 km of it. (Leinonen 1974a.) In Sweden and Finland, not uncommonly breeds in occupied nests of Osprey *Pandion haliaetus* and White-tailed Eagle *Haliaeetus albicilla* (Durango 1949, which see for references). ROOSTING. Widely studied. In breeding season, in territory, less often communally. Outside breeding season, all roost communally, territorial birds joining flock birds (Zahavi 1971; Davies 1976). Communal sites varied: often in emergent marsh vegetation (e.g. *Phragmites*, *Typha*, *Glyceria*) or in thick foliage of bushes, shrubs, and small trees (e.g. willows *Salix*, gorse *Ulex*, rhododendron, holly *Ilex*, ivy *Hedera*, and other evergreens); also inside or on roofs of buildings (notably inside greenhouses) and at sewage works, birds even roosting on slowly rotating arms of treatment apparatus (Witherby *et al.* 1938a; Johnston *et al.* 1943; Rappe 1960; Boswall 1966a, b; Broom *et al.* 1976; Busche and Meyer 1978; Chandler 1979; Engelen 1979). Urban sites often well illuminated and noisy (e.g. Brown *et al.* 1976). In London, 20 roosts located in 1978-9, of which 11 in reedbeds; of 70 cases 1936-78, 46% in reedbeds, 33% in trees and bushes, 11% in buildings (including greenhouses), 10% miscellaneous (sewage farms, etc.); though most sites are reedbeds, these are occupied mainly in autumn, and each typically by not more than *c.* 150 birds; majority of birds use other sites, especially later in winter (Chandler 1980). Of 33 urban roosts in Britain, 52% in roadside trees, 18% on glass roofs, 9% on other buildings, 9% in shrubs and bushes, 6% inside building, 6% undescribed (Boswall 1966b). In winter quarters elsewhere in south of range, communal sites include reeds *Arundo donax* and trees *Ficus* (Sultana and Gauci 1982), palms (Meinertzhagen 1940), sugar cane (Greaves 1941), and river barges (G D Field). Many sites traditional, e.g. small plane *Platanus* trees, Dublin (Eire) used consistently for 37 years (Boswall 1966b). Sometimes roosts with other species including Yellow Wagtail *M. flava* (see that species, p. 423), Tree Sparrow *Passer montanus* (Rappe 1960), Spanish Sparrow *P. hispaniolensis* (Curmi 1977), Blackbird *Turdus merula* (Goethe 1934), Swallow *Hirundo rustica* and Sand Martin *Riparia riparia* (Bircher 1980), and Raven *Corvus corax* (Meinertzhagen 1940). Roosts may comprise tens, hundreds, or sometimes thousands of birds. Roost in reedbed, Kent (England), contained *c.* 5000 birds in September (Boswall 1966c, which see for review of other roosts in Britain exceeding 1000 birds; see also Bergman 1948, Kennedy *et al.* 1954, Busche and Meyer 1978, Chandler 1979, Sultana and Gauci 1982). At Casablanca airport (Morocco), November, *c.* 6000-7000 (I R Hepburn). Size of given roost varies seasonally with various factors. Roosts of *yarrellii* occur throughout the year, most often August-March when resident and migrants often roost together. Some occupied only in autumn, or spring and autumn, probably mostly by migrants (Boswall 1966b; Chandler 1980). Roosts often continue to be used in breeding season by smaller numbers of birds, including ♂♂ and ♀♀ breeding locally, and later by juveniles (Sales 1972; Broom *et al.* 1976, which see for review), though ♂♂ may predominate in summer (Meiklejohn 1937; Chandler 1979). Sequence at roost in Kent (England) probably typical of *yarrellii* roosts which accommodate migrants:

roost first occupied by juveniles in June; numbers increased in September, evidently by migrants, and at maximum January-February, minimum April-May. Another roost, late August to April, showed no marked fluctuations, indicating use by residents only. (Chandler 1979.) At Leopoldsburg (Belgium), nominate *alba* arriving in spring roosted communally from late March, left to breed in May; roost reoccupied in June, followed by big influx of adults and juveniles in July; all departed *en masse* mid-October (Rappe 1960). Similar sequence reported by Busche and Meyer (1978). At Falsterbo (Sweden), 1st-brood juveniles arrived first, together with a few adults, from mid-summer onwards, followed in July by bulk of ♂♂, and finally ♀♀ and 2nd-brood juveniles (Persson 1977). Roost size also affected by changes in site, in turn influenced by weather (Fleming 1981). In autumn, roosts may break up for birds to seek more favourable sites, e.g. birds in reedbed disbanded to power station (Boswall 1966b). In London, October, small roosts may be abandoned for bigger ones (Chandler 1980), and elsewhere same process may account for build-up in large roosts, February-March (Broom *et al.* 1976; Chandler 1979). Catchment area of roost in Berkshire (England) *c.* 12 km radius (Sales 1972; Broom *et al.* 1976); in Leopoldsburg, *c.* 5 km (Rappe 1960). Entry to and departure from communal roosts well studied (e.g. Moffat 1931, Greaves 1941, Boswall 1966c; following account based largely on Broom *et al.* (1976) and Chandler (1979, which see for seasonal variation in roosting times). Birds start arriving typically *c.* 1-1½ hrs before sunset (earlier if overcast) in small flocks, and assemble on any suitable nearby space (e.g. roof, fence, tree, road). Then preen and feed communally; may also bathe (Rappe 1960; Marshman 1977), and squabble (Rappe 1960; Cortés 1977). Almost all birds may gather thus for *c.* 1 hr before moving to roost (Broom *et al.* 1976). In June at Kent roost, juveniles arrived earlier and entered roost earlier than adults (Chandler 1979). Shortly before exodus from assembly site to roost, groups of birds periodically make apparently spontaneous wide circular flights; some return to starting point, others land elsewhere or enter roost. Upflights most frequent, and by biggest flocks, at times of year when numbers at roost relatively high or changing rapidly (Chandler 1979). Group departures may be stimulated by 'invitation flights' and calling of 1-2 birds (Cortés 1977). Pattern of entry differs with site, e.g. at sewage works, birds did not reconnoitre site, but entered over *c.* ½ hr in groups of up to 50, giving special call (not described, but different from typical Contact-call: see 2 in Voice), which apparently attracts others. At reedbeds, however, birds often fly over site before entering, and tend to enter *en masse* (Broom *et al.* 1976; see also Greaves 1941). At some roosts, no prior assembly occurs: birds may arrive in flocks, often at considerable height, giving Flight-contact calls (see 2a in Voice), and circle briefly before plummeting straight into roost (Mayes 1935; Boswall 1966c). Variety of entry patterns may also occur at given roost, depending on conditions: e.g. at one reedbed, birds sometimes entered directly, at other times made circling flights beforehand (Fleming 1981). Last birds may enter roost well after sunset. As birds settle in roost, a chorus of twittering arises, in spring also snatches of song (see 1 in Voice), subsiding once all birds have entered. In roost occupied June-May, chorus occurred September-April, coinciding with occupation by migrants (Chandler 1979). On entering roost, birds jostle for position; competition for sites not uncommonly accompanied by Bill-snapping sounds and mild Alarm-calls (see 5 in Voice). Individual bird sometimes occupies same perch-site several nights in succession. Birds do not huddle together, even in cold weather: median distance between birds 17 cm ($n = 200$), minimum 7 cm (Broom *et al.* 1976). In the morning, chorus of calling may continue until all birds leave (Boswall 1966*c*). Birds leaving roost may assemble nearby before dispersing; from *c.* ½ hr before sunrise (winter), birds joined assembly silently in ones and twos, and spent 10–15 min there, many preening, before departing—often calling—in larger groups (Broom *et al.* 1976). Some roosts disperse directly, without prior assembly (e.g. Boswall 1966*c*, Fleming 1981). In mid-winter, territorial ♂♂ spent average 64·6% of each 24 hrs sleeping, 31·9% feeding, 2·2% nesting and preening, and 1·3% defending (Davies 1982); on territory, October–November, spent 90·1% of daylight feeding, 6·2% resting and preening, 3·7% defending (Davies 1976).

2. Not particularly shy. Birds disturbed at sewage-works roost all flew up and landed in compact flock in flat area, otherwise more widely dispersed (Sales 1972). At same roost in summer, assembling birds mobbed Kestrel *Falco tinnunculus* and Carrion Crows *Corvus corone* (Broom *et al.* 1976). Birds settling at roost intensified calling when Sparrowhawk *Accipiter nisus* flew nearby, and flew up when same raptor made a close pass (Boswall 1966*c*; Fleming 1981). FLOCK BEHAVIOUR. Feeding flocks quite well studied in winter. Dispersion of food and birds closely linked. Flock feeding at edge of small pool of floodwater was rather dense (1–2 m between birds), moving as a unit in same direction with little aggression; at large pool, flock more dispersed, birds feeding individually, 10–30 m apart; when food concentrated by wind, birds fed singly and defended temporary territories. (Davies 1976, 1977.) In feeding flocks in Israel, birds, *c.* 20–30 cm apart, sometimes drove one another away, but rarely chased further. When food artificially concentrated, spacing increased from *c.* 50 cm to 5–10 m (Zahavi 1971). In feeding flock of 12–14, apparently 1-year-olds, birds moved together over ground, those at rear periodically flying to catch up with rest (Scherner 1982). In flight, birds in flock regularly give Flight-contact calls, e.g. when approaching or leaving roost (see above). SONG-DISPLAY. ♂ sings (see 1 in Voice) rather little, and then mostly from elevated perch, often a roof-top, or from ground. In Egypt, February, ♂ sang while displaying on ground near another bird, presumed ♀ (Hartley 1946*a*: see Heterosexual Behaviour, below). More often attracts ♀ with series of Contact-calls (Boase 1926; Witherby *et al.* 1938*a*; Bergmann and Helb 1982). Song given in a relaxed posture, and thought by Bergmann and Helb (1982) to have no territorial significance. Song-flight seldom described, and evidently little used compared with some other *Motacilla.* 2 accounts, however, indicate ritualized Song-flight, similar to *M. cinerea*: in March, loudly singing ♂ made slow hovering descent from *c.* 15 m, tail raised vertically; stopped singing once landed on roof (Fisher 1978). In similar account, bird made near-vertical parachute-like descent (Hornbuckle 1978). On completing similar descent, ♂ copulated with ♀ (Venables 1981). For silent parachute-like descent of ♂ to join ♀, see Richardson (1948). In Italy, song given January–June and September–December (Alexander 1927). In Egypt, continues almost throughout winter (October–March), with gap of *c.* 2 weeks in January (Hartley 1946*a*, which see for diurnal rhythm). ANTAGONISTIC BEHAVIOUR. Birds, especially ♂♂, highly aggressive towards rivals in defence of territory or food. Often reported attacking their own images in window-panes, etc. (e.g. King 1954, Simpson 1970, Borrmann 1974). Territorial defence similar in summer and winter; following account by Zahavi (1971) for wintering birds. In typical boundary dispute, rivals walk with tense gait, calling (see 5a in Voice) and Head-bobbing, along either side of boundary; Fig A shows Head-bobbing birds with head retracted (left) and stretched (right). At higher

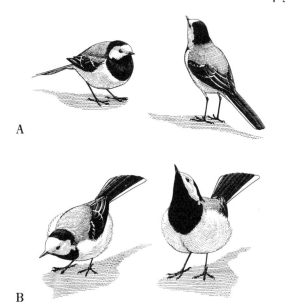

A

B

intensity, breast plumage ruffled, tail somewhat spread, and Head-bobbing gives way to Bowing-display (Fig B). For illustration of 2 rival *ocularis* bowing face-to-face, see Panov (1973). Rivals periodically jump *c.* ½ m into the air. Threat-display may be prolonged, sometimes developing into fight in which combatants flutter into the air, trying to grasp each other with bills and feet (Zahavi 1971). Several reports, in and out of breeding season, of apparent Threat-displays associated with 1 or more birds performing Bouncing-flight: thus, in Norfolk, December, ♂ flew towards another bird with exaggerated undulating flight, body upright, tail closed and straight down, wings shut during each descent (Fig C); other bird (perhaps

C

♀) then also performed Bouncing-flight to land further on (Richardson 1948). In May, ♂ and 2 ♀♀ once seen to perform Bouncing-flight after Head-bobbing (Nisbet and Eltis 1951). Same display, thought to be heterosexual, also observed in Egypt, November–December (Hartley 1946*a*). Significance of Bouncing-flight not clear; interpreted here as antagonistic but heterosexual function cannot be ruled out. Short, steady hovering flights with body vertical or at 45° perhaps a related display: recorded from 7 adult and 1st-winter ♂♂ in flock of 18 birds, October (Jeffrey and Wallace 1956), and from 2 birds (apparently ♂ and ♀), February (D J Brooks). In winter, Israel, trespassing ♀♀, intent on sharing territory with resident ♂, were initially rebuffed (Zahavi 1971). ♂ darts at ♀ with snapping bill (Goodwin 1950: see Voice). ♀ progressively appeases owner,

D

first by low-intensity Appeasement-display (Fig D): retreats slowly, tilting bill slightly upwards, and giving Appeasement-call (see 4 in Voice). ♂ may or may not respond by abating attack, but tendency to attack often wanes over subsequent days as ♀ intensifies Appeasement-display (Fig E): raises bill and tail,

E

and droops and quivers wings. Attack and appeasement most likely when birds first meet on territory each morning (see also Goodwin 1950), or when one lands by the other during day. ♀ helps to defend territory, usually against other ♀♀. (Zahavi 1971.) In winter, Oxford, satellite (see part 1) sharing ♂'s territory rarely chased intruders from territory. Resident ♂♂ answered Flight-contact calls of would-be trespassers with Advertising-call (see 3 in Voice), which was often sufficient to deter intruder. Neighbouring residents trespassed silently, typically when owner absent. Residents also chased other insectivorous passerines when they trespassed. (Davies 1976, 1980, 1981a, 1982; Davies and Houston 1981.) Rival passerines also chased from bird-table (Blake 1980; Patten 1982), and attacked by flock-feeding birds (Tiainen 1981). Highly aggressive towards *M. cinerea* nesting nearby (S J Tyler). ♂ birds seen making dive-attacks on Dipper *Cinclus cinclus* and forced it to submerge (Perray 1978). HETEROSEXUAL BEHAVIOUR. (1) General. In central Finland, ♂♂ arrive on breeding grounds before ♀♀ (Leinonen 1973a), and adult ♀♀ before 1-year-old ♀♀. Pair-bonding behaviour occurs in flocks during spring migration (Leinonen 1973b). According to Dementiev and Gladkov (1954a), nominate *alba* apparently arrives on breeding grounds in pairs. Flocks disperse rapidly, most birds occupying territories 2 weeks after arrival (Leinonen 1973a). Resurgence of courtship occurs between fledging of 1st brood and start of 2nd clutch (Schweinsteiger 1938). (2) Pair-bonding behaviour. For vocal attraction of ♀ by ♂, see Song-display (above). Once ♀ nearby, ♂ performs elaborate ground-display, more complex than in other *Motacilla*. Following account, based on Boase (1926, which see for variants), describes interaction between ♂ and ♀ *yarrellii*, although more than 1 ♂ may participate. Initially, ♂ chases ♀; on landing, ♂ typically adopts crouched Upright-advertising posture, with bill pointed skywards exposing bold throat pattern. Pursued ♀ mostly passive, flying further on. If ♀ receptive, ♂ approaches her, performing apparent variant of Head-bobbing or Bowing-display (see

above): zigzags towards her, jerking head down and forward, briefly fluttering wings as he retracts head. In apparently higher-intensity approach, ♂ performs Tilted-advertising display: tilts head toward ♀, spreads wing on side nearest her, and spreads and twists tail towards her to expose upper surface; both wing and tail brush ground. In nominate *alba*, display similar but ♂ reported tilting away from ♀, raising wing nearest her (Witherby *et al.* 1938a; Brusewitz 1980: Fig F). Display

F

obviously same or similar in *personata*, *lugens*, and *leucopsis* (Dementiev and Gladkov 1954a; Panov 1973, which see for drawings). (3) Courtship-feeding. Not recorded, but see Relations within Family Group (below). (4) Mating. Takes place on ground, sometimes on a roof. ♂ appears usually to take initiative, and several pre-copulatory displays reported, including those listed in subsection 2 (above). ♂, performing apparent high-intensity Appeasement-display (Fig E), may approach ♀ (Boase 1926; Cooke 1951); one singing ♂ approached ♀ apparently thus in Egypt, February, but did not mate with her (Hartley 1946a). With or without this preliminary, ♂ may face ♀, bowing in spasmodic, springy fashion by bending his legs (Richards 1945); this perhaps different from variant Head-bobbing described above. With each bow, rises on toes and lowers bill to ground. Then runs in circles round ♀ in ♂-soliciting posture: holds head low, spreads and lowers tail, spreads and quivers wings, and ruffles plumage on wings and back (Boase 1926; Boyd 1945). In variant, ♂, bowing deeply so that bill alternately stretched upwards and touched ground, ran around ♀ in ever-decreasing circles (Witherby *et al.* 1938a). Before copulating, one ♂ performed Tilted-advertising display, dragging himself towards ♀ (Kirkman 1913). Receptive ♀ adopts ♀-soliciting posture, evidently the same as, or very similar to, high-intensity Appeasement-display on winter territory (see above): raises tail almost vertically, holds wings out from body and slightly drooped, and quivers them. Copulation follows (Witherby *et al.* 1938a; Boyd 1945). In perhaps low-intensity variant, ♀ crept or shuffled round ♂ with wings and tail somewhat spread and lowered (Boase 1926). One ♂ copulated with ♀ on completing descent from Song-flight; during descent, ♀ gave a soliciting-call (Venables 1981; see 4b in Voice). Another ♂ flew from a distance straight on to back of ♀ soliciting on roof and made pecking movements at her nape (M G Wilson). Unreceptive ♀ may dash at and peck ♂ (Witherby *et al.* 1938a). Descriptions of mating in nominate *alba* suggest behaviour, especially of ♀, possibly different from *yarrellii*. In one account, ♀ adopted ♀-soliciting posture (as in *yarrellii*), accompanied by quiet calls (see 4a in Voice). ♂ ran in ever-tighter circles around ♀ who then stopped calling, sleeked plumage, and lay flat on ground, bill and tail horizontal. Circling ♂ then raised head, ruffled throat and breast, dragged tail, and gave apparent Contact-call every couple of steps. ♂ finally approached ♀ obliquely from rear and flew on to her back (Steinbacher 1948). Similar description by Portig (1942). In another account, ♂ tripped round ♀, moving head up and down (perhaps no different from *yarrellii*), opening and closing bill in time with movement. After some time, ♀ turned to face ♂, trembling tail and rear body. ♂ continued circling ♀, then approached her from behind and attempted to grasp her nape. ♀ flew off, followed by ♂, into bushes, where copulation thought to have occurred (Sudhaus 1970). For intense pre-copulatory display of captive nominate *alba*, see Teschemaker (1913). According to Niethammer (1937), pair usually copulate several times in succession. (5) Behaviour at nest. Not clear which sex selects nest-site. Both look for site, ♀ apparently more actively. Select site several days before nest-building, but sometimes begin building in 2–3 places before making final choice (Leinonen 1973a). Once, ♀ started preparing new nest before 1st brood fledged (Schweinsteiger 1938). Nest-relief usually silent unless sitting bird cannot see out, and then relieving bird usually calls briefly before entering. Usually runs to nest. After 10th day of incubation, sitting bird, of either sex, sometimes reluctant to yield to relieving bird; may gape at or drive off mate, calling (Löhrl 1957b; Leinonen 1973a: see 5b in Voice). Often flies directly from nest (Löhrl 1957b) or, if nest on ground, runs some distance then flies. 2 days before hatching, parents at one nest began pecking at bottom of nest, apparently searching for eggshells (Leinonen 1973a). RELATIONS WITHIN FAMILY GROUP. Following account based largely on Leinonen (1973a). ♀ usually broods during hatching period, ♂ paying little attention to nest at this stage. During first few days, ♀ spends most time brooding, steadily less from 2nd–6th day, after which broods only at night. Both parents feed young, beginning on day of hatching. Although ♂ does not usually feed young on 1st day, feeding shared about equally over whole nestling period. ♂ feeds brooding ♀: during first 2 days, usually gives all food to ♀ who eats part of it and feeds rest to young; gradually, ♂ begins feeding young directly, and, after 6th day, both parents feed young independently. Young start begging as soon as ♀ rises to greet incoming ♂, or when approaching parent calls. If young reluctant to feed, food-bearing parent calls (see 2b in Voice) to stimulate them (see also Löhrl 1957b). During first 2 days, parents eat faeces; on 3rd day, carry them away about half of the time, and, from 4th day, regularly remove them. Parents peck at cloacal region of young to stimulate defecation. From 10 days, young defecate over rim of nest. Both parents present when young fledge. When 1st young fledges, both parents follow it, calling, to where it lands, then one parent returns to nest and calls as if encouraging others to leave. Young usually fledge at short intervals, sometimes over 1 day. Brood usually stays near nest for 1st day after fledging, then disperses further. In the evening, brood initially reassembles away from nest for roosting. 1st brood fledglings fed mostly by ♂, probably because ♀ preparing for 2nd clutch. Similar post-fledging care by ♂ described by Schweinsteiger (1938). According to Persson (1977), however, ♀ stays longer in territory with 2nd brood than ♂. ANTI-PREDATOR RESPONSES OF YOUNG. When disturbed by human intruder, young in nest near fledging markedly timid; crouched, pulled wings half-way across eyes, and gaped (Schweinsteiger 1938). Young may leave nest prematurely, i.e. at (11–)12–13 days, if disturbed; departure stimulated by alarm-call (not described) from one or more young (Leinonen 1973a). For premature departure of hole-nesting broods, see Leinonen (1974b). PARENTAL ANTI-PREDATOR STRATEGIES. (1) Passive measures. No information. (2) Active measures: against birds. Once, when Great Tit *Parus major* looked into nest-hole, sitting bird gaped vigorously (Löhrl 1957b). Pursues raptors, e.g. *A. nisus*, with variant of song (Bergmann and Helb 1982: see 1 in Voice). (3) Active measures: against man. Human intruder near nest stimulates vigorous

vocal demonstration (see 5 in Voice); at high intensity, may be accompanied by distraction-display which apparently takes various forms. When disturbed by human intruder, sitting ♀ sometimes flew silently from nest and, with shivering wings, ascended to some height, then dropped to ground and performed apparent distraction-lure display: 'toddled' in a stiff posture with wings raised obliquely and shivering (Leinonen 1973a). When young recently fledged, parents nearby very agitated by human intruder: performed distraction-lure display of disablement type: ran low immediately in front of intruder on bent legs with wings and tail dragging, as if injured (Leinonen 1974b). In recording by P Radford, Scotland, approach of recordist to nest with nearly fledged young elicited increasingly intense Alarm-calls from ♂ parent, accompanied by audible wing-beating sounds, presumably a distraction-lure display of disablement type.

(Figs by C Rose: C after drawing in Richardson 1948; F after painting in Brusewitz 1980; others after drawings in Zahavi 1971.)   EKD

Voice. Freely used throughout the year, more often in breeding season. Differences, some documented here, occur between races, but no comprehensive study. For additional sonagrams, see Bergmann and Helb (1982). As in other *Motacilla*, Bill-snapping appears to indicate threat, anger, excitement, etc. Recording by J Fisher contains chorus of Bill-snapping given by birds assembling and apparently squabbling at roost, sounding like small twigs catching fire (P J Sellar); same noise described as a crackling, like crumpling cellophane (Emley 1985). ♂ in Egypt, winter, described Bill-snapping at ♀ in expelling-attack from his territory (Goodwin 1950). For wing-beating sounds, see Social Pattern and Behaviour.

CALLS OF ADULTS. (1) Song of ♂. In *yarrellii*, a simple but lively warbling twitter, consisting largely of repeated slurred Contact-calls (call 2) with variants and modulations (Witherby *et al.* 1938a: Fig I). Similar description for nominate *alba* (Bergmann and Helb 1982). Song given seldom by *yarrellii* but apparently more freely by nominate *alba*, not uncommonly by migrants (Witherby *et al.* 1938a). In Egypt, winter, bird kept up a soft gabbling song as it ran and displayed behind another bird (Hartley 1946). Apparent Subsong given in recording of nominate *alba* by R Goodwin. A variant, so-called 'excitement song' (not further described), regularly given when pursuing raptors (Bergmann and Helb 1982). (2) Contact-calls. (a) Main type. In nominate *alba*, a monosyllabic 'zit', or 'psit', sometimes 2–3 syllables, e.g. 'ziti', 'zilipp', or 'zittip', less sharp than Grey Wagtail *M. cinerea*; given especially in flight, at rate of 1 call per undulation, but also when perched (Bergmann and Helb 1982). Detailed sonagraphic analysis shows that call in nominate *alba* usually comprises at least 2 sub-units, of which 1st consistently descends in pitch from 7·8 kHz to 3·6 kHz and shows rather pronounced frequency modulation; 2nd sub-unit varies markedly between individuals. Little variation in nominate *alba* throughout range. (Wallschläger 1983.) Call of nominate *alba* tends to be softer than

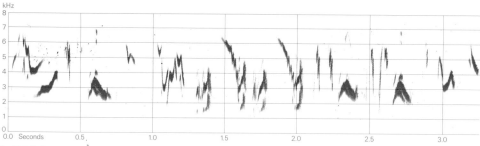

I P J Sellar Sweden May 1973

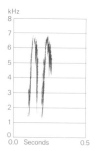

II P J Sellar
Sweden May 1973

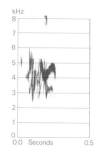

III V C Lewis
England April 1978

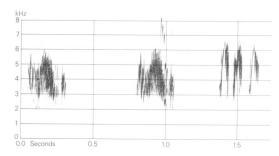

IV P A D Hollom Scotland June 1980

shrill 'tschizzik' of *yarrellii* (Witherby *et al.* 1938a; B E Slade); call of *yarrellii* also rendered 'tissick' (J Hall-Craggs: Fig II), 'chissik', and monosyllabic 'chisk' (e.g. Boase 1926). Comparison of sonagrams reveals differences among other races: perched *baicalensis* calls 'ziti', very like nominate *alba* (see above); *lugens* superficially similar to nominate *alba* but sub-units reversed (Wallschläger 1983). In Iran, *personata* calls 'pii-tup' or 'piiup', very like sparrow *Passer* and apparently very different from nominate *alba* (Erard and Etchécopar 1970). (b) Variants. In nominate *alba*, quiet rather toneless 'tp', 'tptp', 't'tp', or 'zlit' (Bergmann and Helb 1982). Pair near nest exchanged 'zlipp' like *Passer*; also used to stimulate young to accept food (Löhrl 1957b). Twittering 'tisup' sounds, softer and sweeter than usual call, given by *yarrellii* settling in roost—harsher and more intensely when raptor appeared (Boswall 1966c). (3) Advertising-call. A highly distinctive musical 'chee-wee' or 'che-wee' (Fig III), given by territory-owners (not by other birds) as principal means of advertising occupancy of territory and to deter intruders from landing in it (see Social Pattern and Behaviour) often given in response to disyllabic Contact-calls of would-be intruder (Davies 1976, 1981a, 1982). Also rendered 'tzi-wirrp', more musical than disyllabic Contact-call (Witherby *et al.* 1938a). Evidently same call, described as a persistent loud 'tschrli', illustrated by Bergmann and Helb (1982). Bird with territory commonly gives long series of this call, often interspersed with Contact-calls and variants. Variant of such a sequence perhaps 'zrrrip zrrrip chi-chi-chi' (Fig IV) in which 'zrrrip' a hard, almost harsh buzzing sound, ending in an audible click (J Hall-Craggs), and reminiscent of Flight-call of Skylark *Alauda arvensis* (P J Sellar); Fig

IV shows part of a long series, mostly of 'zrrrip' sounds. (4) Soliciting- and Appeasement-calls. (a) A repeated, quiet 'zirrlih' given by ♀ nominate *alba* inviting copulation (Steinbacher 1948). (b) A continuous shrill piping 'pee pee pee' given by ♀ during ♂'s descent from Song-flight, culminating in copulation (Venables 1981). Rapidly repeated high-pitched call (not further described) given by ♀ during appeasement-displays to facilitate her access to winter territory of ♂ (Zahavi 1971) is possibly call 4a or 4b. (5) Alarm- and Anger-calls. (a) Emphatic repeated 'quizzick', evidently a more clipped and harder variant of disyllabic Contact-call, given in increasingly rapid succession as danger mounts. Shrill or harsh disyllabic calls given in territorial disputes (Dosseter 1944; Zahavi 1971) are presumably the same or similar. (b) Incisive 'chick', given especially by breeding birds (Witherby *et al.* 1938a). In recording (Fig V, last 4 units), a hard,

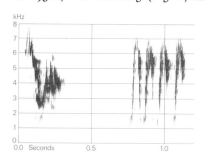

V A P Radford
Scotland July 1973

urgently repeated 'ki-ki-ki-kik' (J Hall-Craggs) given by ♂ when human intruder near nest. In recording by P A D Hollom, staccato 'tic' sounds given by food-bearing parent approaching nest when intruder nearby; at higher intensity, a rapidly delivered 'tic-tic-tic' (P J Sellar).

Sitting ♂ nominate *alba* resisting nest-relief by incoming ♀ called 'diuck'; not heard in any other context (Löhrl 1957*b*); presumably related to above. (c) Emphatic 'cheeooEE' calls (Fig V, 1st unit) and 'queezew' calls supplement predominant calls 5a and 5b. In recording by J Fisher of birds at roost, similar calls rendered 'tcheeoo', like *Passer*; these also accompany rasping sounds, like Starling *Sturnus vulgaris*, and Bill-snapping (P J Sellar). (6) Other calls. (a) Special call, not described but different from disyllabic Contact-call, given by birds leaving pre-roost assembly for roost; apparently attracts others to departing group (Broom *et al.* 1976). (b) Hard 'sticknick' call given by ♂ nominate *alba* performing pre-copulatory display (Steinbacher 1948).

CALLS OF YOUNG. In recording by J Burton and C Johnson, food-calls of nestlings a high-pitched needling 'sisisisisi...', given more rapidly when fed. Food-calls of fledged young similar but stronger and lower-pitched: rendered a long series of sharp 'zississi...' sounds (Bergmann and Helb 1982, which see for sonagram). Premature departure from nest stimulated by alarm-call (not described) of 1 or more young (Leinonen 1973*a*).

EKD

Breeding. SEASON. North-west Europe: first eggs laid beginning of April, main period late April to mid-May, last eggs early August (Makatsch 1976; Mason and Lyczynski 1980). Iceland and northern Scandinavia: laying begins early June (Makatsch 1976). Central Finland: first eggs in 2nd week of May; main period mid-May to early June, last eggs 2nd week of July; up to 1 week earlier in southern Finland; closely correlated with temperature (Leinonen 1973*a*). South and south-east Europe: late April to mid-July (Makatsch 1976). North Africa: eggs found from late May to June (Heim de Balsac and Mayaud 1962). SITE. Hole or crevice in wide variety of natural and artificial sites, including building, wall, bank, cliff, pile of debris, dense bush, old nest of other species; from ground level upwards. Of 309 nests in Finland, 82·5% in artificial sites (57·2% in buildings, 11·5% in stone walls, 5·6% in wooden structures, 7·9% in heaps of material, 0·3% in metal structures) and 17·5% in natural sites (13·6% in rock or stone crevices, 3·9% among tree roots and in other holes) (Leinonen 1973*a*). In Britain, 58·8% of unknown number of nests were at 0–1·5 m, 24% at 1·5–3 m, 17·2% over 3 m, with highest 15·2 m (Mason and Lyczynski 1980). Nest: cup of twigs, grass stems and leaves, roots, and moss, lined with hair, wool, and feathers. In artificial crevices (not in natural sites) may have substantial base of twigs, before cup with lining added; dry weight of 120 nests in artificial sites, Finland, 108 g (47–164), and of 10 nests in natural sites 44 g (13–74) (Leinonen 1973*a*). Building: much variation in role of sexes; of 27 nests, ♂ took part in building *c.* ¼, and 6 built entirely by ♂; most usual for ♂ to assist in early stages, with ♀ completing nest, including lining,

by herself; takes 4–7 days (Leinonen 1973*a*). EGGS. See Plate 81. Sub-elliptical, smooth and glossy; whitish, blue-white or grey, with fine, even freckling of grey-brown or grey, occasionally with brown spots. Nominate *alba*: 20·2 × 15·1 mm (18·00–22·0 × 14·2–16·3), $n = 250$; calculated weight 2·3 g. *M. a. yarrellii*: 20·5 × 15·3 mm (18·6–22·2 × 14·2–16·5), $n = 140$; calculated weight 2·35 g (Schönwetter 1979). Clutch: 5–6(3–8). Of 697 clutches, Britain: 3 eggs, 3%; 4, 17%; 5, 48%; 6, 32%; 7–8, 1%; mean 5·10 ± SE0·03, declining through season (Mason and Lyczynski 1980). Of 210 clutches, central Finland: 4 eggs, 5%; 5, 26%; 6, 68%; 7, 1%; mean 5·7 ± SE0·04, declining through season; clutch size increasing with latitude, with mean for mainly southern Finland 5·4 ± SE0·05 (Leinonen 1973*a*). 2 broods, rarely 3, in south of range, mainly one in north. Replacements laid 11–47 days after loss, $n = 7$ (Mason and Lyczynski 1980). Interval between broods mean 9·2 days (3–23), $n = 16$ (Leinonen 1973*a*). Eggs laid daily, in early morning (Leinonen 1973*a*). INCUBATION. 12·6 days (11–16), $n = 69$, becoming shorter by 0·03 days per day during season (Leinonen 1973*a*). By both sexes, with ♀ sitting throughout night and taking greater part during day; in 210 hrs daytime observation, ♀ sat for 61·3%, ♂ 24·0%, no sitting bird 14·7%; ♀ sat in spells of mean 31·9 min (1–100), $n = 146$, ♂ for 21·3 min (1–50), $n = 43$ (Leinonen 1973*a*). Begins with last egg; hatching synchronous. YOUNG. Altricial and nidicolous. Cared for and fed by both parents. Brooded by ♀ for first 5 days, then at night (Leinonen 1973*a*). FLEDGING TO MATURITY. Fledging period 13·7 days (11–16), $n = 44$ (Leinonen 1973*a*). Fed by parents for 4–7 days after fledging, rarely to 11 days (Leinonen 1973*a*). Age of first breeding 1 year. BREEDING SUCCESS. Of 1147 eggs laid in 207 clutches, central Finland, 78·9% hatched; of 832 young in 153 broods, 77% fledged (Leinonen 1973*a*). Of 5747 eggs laid in 1186 nests, Britain, 63·8% hatched and 82·6% of these fledged, giving overall success of 52·7%; clutches of 3 eggs produced mean 1·59 young, 4 eggs 2·32, 5 eggs 2·80, 6 eggs 3·29; of 463 nest-failures, 49·9% due to predation, 33·5% desertion, remainder due to weather and miscellaneous causes (Mason and Lyczynski 1980).

Plumages (nominate *alba*). ADULT MALE BREEDING. Forehead, forecrown, and sides of head and neck white. Hindcrown and nape black, usually with slight blue gloss. Remainder of upperparts medium grey, contrasting sharply with black of nape; lower rump and shorter upper tail-coverts usually slightly browner or duller grey, longer upper tail-coverts black, outer with broad white outer fringe. Chin, throat, and chest deep black; black of throat often extends slightly upwards to lower rear of ear-coverts, black of chest has rounded lower margin. Sides of breast and flanks medium or light grey, usually separated from black of chest by narrow white band, merging rather gradually into white of remainder of underparts; thighs mixed black, black-brown, and off-white. Central pair of tail-feathers (t1) black with narrow white outer fringe; t2–t4 black, t4 sometimes with partly white edge along outer web;

t5–t6 white with long and narrow dark wedge along basal and middle portion of inner border. Flight-feathers and greater upper primary coverts black or greyish-black, outer webs and tips narrowly edged pale grey (except tips and emarginated parts of outer primaries); fringes along outer webs of secondaries often wider and less sharply defined, those of outer primaries narrower, sharper, and almost pure white. Tertials black or brownish-black, longest one with broad white outer fringe, shorter ones with pale drab-grey outer web merging into white along outer border. Lesser upper wing-coverts medium grey, like upperparts; median coverts black with white tips *c.* 2–3 mm long; greater coverts greyish-black with broad pale drab-grey fringe and tip merging into white along border; some innermost and outermost median and greater coverts usually paler greyish-black or grey-brown and with greyish rather than white fringes and tips. Bastard wing black, shorter feathers narrowly edged white. Under wing-coverts and axillaries white, shorter coverts along leading edge of wing with dull black bases. *In worn plumage*, about July, white of face and sides of head less sharply defined; grey of upperparts slightly browner; flight-feathers, tertials, and upper wing-coverts browner, white fringes and tips largely worn off. In very fresh plumage, about March, white of face and sides of head sometimes slightly tinged cream or yellow. Occasionally, grey of mantle, scapulars, and sides of breast partly suffused black; this mainly in western Europe and then probably due to hybridization with British black-backed race *yarrellii*, but some birds from central Europe far from breeding range of *yarrellii* show it, too. Exceptionally, some grey feathers on crown and white feathers on chin retained from non-breeding plumage. ADULT FEMALE BREEDING. Like adult ♂ breeding, but white of forehead and forecrown usually with some black or dark grey mottling (in 12 of 14 examined), often some grey on central crown or nape and nape then less sharply defined from grey of upper mantle (in 11 of 14 examined); black of chin sometimes mottled white (in 2 of 14 examined), and pale tips and fringes of median and greater upper wing-coverts less extensive and less pure white; some indistinguishable from adult ♂ breeding. ADULT MALE NON-BREEDING. Upperparts and sides of head like adult ♂ breeding, but forehead regularly mottled black (8 of 14 examined) and black of crown and nape often mixed grey (11 of 14 examined), less sharply defined from mantle; underparts rather different, chin and throat white (sometimes slightly tinged cream or pale yellow), white of throat separated from white sides of neck by solid or mottled black stripe extending from rear of lower cheek to sides of chest; chest black, forming broad black gorget, but lower border less smoothly and roundly ending as in breeding; belly to under tail-coverts often slightly suffused pale cream-buff; tips of median upper wing-coverts and outer webs and tips of greater often slightly less pure white. ADULT FEMALE NON-BREEDING. Forehead mainly grey, variably mottled white, rarely as white as adult ♀ breeding or adult ♂ non-breeding; forecrown to nape grey, often with slight olive tinge (like mantle and scapulars), usually mixed with some black feathers, especially on sides of crown and nape. Lores, supercilium, ear-coverts, and upper cheeks less pure white than adult ♀ breeding and adult ♂ non-breeding, tinged cream-yellow and faintly speckled olive-grey; rear of ear-coverts sometimes largely olive-grey, forming dark bar connecting grey nape with black stripe at rear of lower cheek. White on sides of neck tinged cream and partly suffused grey. Remainder of upperparts and all wing and tail as in adult ♀ breeding; underparts as in adult ♂ non-breeding, mainly off-white with bold black gorget ending in mottled point on central breast and with grey sides of breast and flanks. NESTLING. Down rather long and dense,

on upperparts only; grey-white to smoke-grey (Heinroth and Heinroth 1924–6; Witherby *et al.* 1938a). JUVENILE. Entire upperparts pale brown-grey, tinged olive-brown on forehead, crown, rump, and sometimes elsewhere, usually with faint darker brown stripe at side of crown; upper tail-coverts black-brown. Sides of head without contrasting pattern, except sometimes for dirty white or pale brown supercilium; lores, upper cheeks, and ear-coverts closely mottled grey and cream-buff or cream-yellow, rear of ear-coverts sometimes with dark brown bar; lower cheeks and sides of neck more uniform pale brown-grey. Chin and throat pale cream-buff or dirty white, sometimes with faint grey mottling. Chest with ill-defined dark grey or brown-grey patch, bordered by buff below and at sides; depth of colour and extent of patch rather variable, sometimes restricted to short bar on upper chest, sometimes much broader and narrowly extending laterally into dark stripe along lower side of neck (as gorget of adult non-breeding, but grey rather than black). Sides of breast and flanks pale brown-grey, remainder of underparts white with variable cream or pale buff suffusion. Tail as adult ♂ breeding, but fringes along fresh t1 pale buff or yellow, not white. Flight-feathers and tertials as adult ♂ breeding, but slightly less deep black, greyer than adult when fresh, browner when worn; fringes along outer webs slightly narrower and more yellow when fresh (in adult, broader and whiter) but similar when worn. Lesser upper wing-coverts brown-grey with olive-brown or buff-brown tips. Median coverts dull black with pale cream-yellow or buff-white tip (black usually extending into a point at shaft; in adult, white more extensive at shaft, forming shallow notch); greater coverts greyish-black with brown-grey fringe along outer web and white one along tip (fringes less extensively white than in adult). *In worn plumage*, upperparts paler brown-grey without olive tinge, ground-colour of underparts off-white. FIRST ADULT MALE NON-BREEDING. Like adult non-breeding, but upperparts olive-grey, merging into medium grey on rump (in adult non-breeding completely grey with slight olive tinge at most). Forehead olive-yellow, mixed olive-grey to variable extent; crown often mottled with some black in front and at sides and then rather like adult ♀ non-breeding, but central crown mixed olive instead of grey (in adult ♂ non-breeding, crown and nape largely or fully black). Sides of head and underparts as in adult non-breeding (showing distinct black gorget extending down from rear of cheeks), but sides of head, chin, throat, and chest just below gorget strongly suffused cream-yellow or pale yellow; as in some adult ♀ non-breeding, upper cheeks, ear-coverts and sides of neck sometimes tinged olive-grey, contrasting with pale yellow supercilium and eye-ring. Tail and tertials as adult non-breeding, but often a mixture of old and new feathers (all new in adults which have completed post-breeding moult). Wing as adult, but juvenile flight-feathers, greater upper primary coverts, and a variable number of outer greater upper wing-coverts retained; flight-feathers and primary coverts browner and with more frayed tips than adult at same time of year, outer greater coverts usually distinctly browner and more worn than neighbouring fresh inner coverts (see also Broom *et al.* 1976 and Svensson 1984a). *In worn plumage*, about December–February, upperparts greyer, less olive, but never as uniform cold medium grey as adult; sides of head and throat less yellow, but not as white and uniform as adult. FIRST ADULT FEMALE NON-BREEDING. Like 1st adult ♂ non-breeding and often indistinguishable; black on crown restricted to sides or fully absent (on average, less black on crown than 1st adult ♂ non-breeding, but overlap complete). FIRST ADULT MALE BREEDING. Like adult ♂ breeding and sometimes indistinguishable. Some birds (6 of 46 examined) retain a few grey non-breeding feathers mixed in black of crown or nape or some white feathers on chin; retained juvenile flight-feathers often distinctly browner than in adult, less black, showing contrast between brown inner secondaries and new black longest tertial; tips of outer primaries more frayed than adult at same time of year; a variable number of brown and worn juvenile outer greater upper wing-coverts often retained, contrasting with blacker non-breeding or 1st breeding inner coverts. FIRST ADULT FEMALE BREEDING. Some like adult ♀ breeding and indistinguishable except for browner and more worn flight-feathers or outer greater coverts (see First Adult Male Breeding, above); others retain much worn non-breeding on head and body: crown and nape heavily mixed grey, with black often restricted to narrow stripe at sides of crown and narrow bar on forecrown, or even fully absent; chin occasionally mottled white; exceptionally, entire 1st non-breeding plumage retained throughout spring and early summer. See also Leinonen (1973*b*).

Bare parts. ADULT, FIRST ADULT. Iris dark brown or black-brown. Bill, leg, and foot black or horn-black. Mouth dark grey to black. NESTLING. Bare skin, including bill and foot, flesh-pink; mouth orange-yellow, tongue reddish, gape-flanges pale yellow. JUVENILE. Iris dark brown. Bill dark horn-brown or black-brown with flesh or pale brown tinge on base of lower mandible, gape-flanges pale yellow. Mouth yellowish-red. Leg and foot grey with flesh tinge, greyish-black, or black. (Hartert 1910; Heinroth and Heinroth 1924–6; Witherby *et al.* 1938*a*; Bub *et al.* 1981; RMNH, ZMA.)

Moults. ADULT POST-BREEDING. Complete; primaries descendant. In a non-migratory population of *yarrellii*, southern England, average duration of primary moult estimated at 76 days, with average start 16 July, average completion 30 September; moult of secondaries started with shedding of p5–p6, completed 3–25 days after regrowth of p10; tertials moulted between shedding of p2 and p8, taking 8–9 weeks; body started with shedding of about p2 (lesser upper wing-coverts, mantle, scapulars, and rump first), completed with p10 or slightly later (Baggott 1970). In a migratory population, southern Scotland, moult started with p1 between mid-June and early August, on average 5 July; average completion estimated at 16 September (range late August to late September), after *c.* 73 days; tail shed with p3–p4, regrown during growth of p7–p8, tail moult lasting *c.* 60 days (Galbraith 1977). In Netherlands, a migratory population of nominate *alba* shed p1 4 July–3 August, with a few up to late August, on average 16 July; moult completed with regrowth of p10 after 73 days (average of population), *c.* 68 days (according to recaptured birds), or perhaps down to 53 days in late-starting birds; tail started at primary moult score 3–10, completed at about regrowth of p10, taking *c.* 59 days, sequence t1–6–2–5–4–3; secondaries started at primary score 20–29, completed *c.* 1 week later than p10; moult sometimes not completed before birds depart in early October (Jukema and Rijpma 1984). In southern Sweden, birds usually start in first half of July when gathered in roosts after breeding, but some (especially ♀♀) with late-fledged juveniles start in 2nd half of July when still in own territory (Persson 1977). In northern Finland, starts 4–20 July (average 8 July), taking *c.* 46–48 days; tail starts at primary score 1–20, secondaries at (10–)20–26, all completed at about regrowth of p9–p10 (Haukioja 1971). In Yugoslavia, nominate *alba* moults August–October (Stresemann 1920); in USSR, nominate *alba* late July to September, occasionally November, *personata* completing August (Dementiev and Gladkov 1954*a*). Single *dukhunensis*

from northern Iran had just started 29 July (Paludan 1940). See also Snow (1965). ADULT PRE-BREEDING. Partial: head, body, lesser upper wing-coverts, tertial coverts, and 1–2(–3) tertials, usually many or all median upper wing-coverts, occasionally a variable number of inner greater coverts, often t1 or t6, and sometimes a variable number of other tail-feathers. Moult of ♀ on average less extensive than in ♂. Starts December–February, completed late February to early April. POST-JUVENILE. Partial. Starts at age of *c.* 6 weeks, when tail just full-grown (Heinroth and Heinroth 1924–6); hence, timing of moult highly variable, starting between about mid-June and late September. In Britain, duration of moult in *yarrellii c.* 7 weeks, but late-fledged birds (starting late) moult more rapidly and may complete in *c.* 6 weeks; average starting date 2 August, average date of completion 19 September; most birds had finished before 4 October (Baggott 1969, 1970, 1973, which see for sequence of feather replacement and variation in extent). In Netherlands, moult of head and body of nominate *alba* starts late June to September, tail from late July; moult halted mid-September irrespective of stage reached: early-hatched birds replace all head, body, upper wing-coverts, tertials, and tail, late-hatched part of head and body only or virtually nothing at all (Jukema and Rijpma 1984). In all populations, relatively few birds replace all upper wing-coverts, and most birds retain at least 2–3 outer greater coverts. FIRST ADULT PRE-BREEDING. Like adult pre-breeding but more variable in extent, moult in some birds much less extensive. Usually, all head, body, lesser upper wing-coverts, and tertial coverts replaced, as well as most or all median coverts, 1–2 tertials, t1, t6, or (occasionally) some greater coverts, but occasionally no moult at all, especially in ♀.

Measurements. ADULT, FIRST ADULT. Nominate *alba*. Northern and central Europe, mainly March–October; skins (RMNH, ZMA). Juvenile wing and tail refer to retained juvenile feathers of 1st adult. Bill (S) to skull, bill (N) to distal corner of nostril; exposed culmen on average 4·9 less than bill (S). Toe is middle toe with claw; claw is hind claw.

WING AD	♂ 91·6	(2·06; 44)	87–96	♀ 88·2 (2·02; 18)	85–92
JUV	89·5	(2·24; 67)	84–94	85·6 (2·29; 37)	80–90
TAIL AD	85·0	(2·56; 80)	80–90	82·0 (2·44; 42)	77–87
JUV	83·4	(3·57; 22)	76–88	81·1 (2·81; 11)	76–85
BILL (S)	16·2	(0·60; 40)	15·2–17·2	16·0 (0·71; 21)	14·9–16·9
BILL (N)	9·0	(0·48; 18)	8·5–9·8	9·2 (0·58; 14)	8·5–10·1
TARSUS	23·8	(0·83; 38)	22·7–25·0	23·2 (0·75; 23)	22·2–24·4
TOE	17·3	(0·47; 13)	16·6–18·4	17·6 (0·81; 7)	16·8–18·9
CLAW	6·6	(0·71; 22)	5·5–8·1	6·6 (0·74; 15)	5·8–7·8

Sex differences significant for wing, adult tail, and tarsus. Exceptionally short tarsus of 20·7 (♂) and 21·6 (♀) excluded from range. Southern Yugoslavia (Stresemann 1920):

WING AD	♂ 91·5	(1·54; 20)	89–95	♀ 87·2 (0·75; 6)	86–88
JUV	90·1	(1·93; 12)	87–93	85·2 (1·49; 8)	83–87

No difference in size between birds from Iceland, Scandinavia, central Europe, and southern Europe (BMNH, RMNH, ZMA).

M. a. yarrellii. Mainly Britain and Ireland, some western and south-west Europe, all year; skins (RMNH, ZMA).

WING AD	♂ 92·0	(1·45; 12)	90–94	♀ 87·8 (2·14; 8)	87–92
JUV	89·3	(2·36; 25)	84–92	85·1 (2·19; 15)	81–89
TAIL AD	84·9	(2·92; 35)	78–90	81·2 (2·27; 20)	77–86
BILL (S)	16·0	(0·38; 34)	15·1–16·5	15·6 (0·45; 16)	14·9–16·2
BILL (N)	9·2	(0·35; 27)	8·4–9·8	8·9 (0·43; 9)	8·3–9·5
TARSUS	23·8	(0·80; 34)	22·4–25·2	23·2 (0·71; 15)	22·2–24·4

Sex differences significant, except bill (N). Wing, live birds: adult ♂ 86–96 (50), 1st adult ♂ 84–95 (80), adult ♀ 83–91 (36), 1st adult ♀ 82–89 (54) (Svensson 1984*a*).

M. a. dukhunensis. Eastern Turkey (Kumerloeve 1969a), northern Iran (Paludan 1940; Schüz 1959), Caucasus, and western Siberia, ages combined; skins (RMNH, ZMA).

WING	♂ 92·6 (1·78; 15)	90–96	♀ 87·7 (2·36; 10)	84–92
TAIL	87·8 (1·94; 9)	84–91	83·1 (1·98; 6)	80–85

Sex differences significant.

Migrants, Afghanistan (Paludan 1959):

WING	♂ 90·0 (—; 22)	86–96	♀ 86·2 (—; 11)	84–89
TAIL	86·5 (—; 22)	82–94	83·9 (—; 10)	80–88

M. a. personata. Tien Shan (USSR) and Sinkiang (China), summer; skins, ages combined (RMNH, ZMA).

WING	♂ 95·9 (1·60; 14)	92–99	♀ 89·8 (1·71; 8)	87–92
TAIL	91·5 (2·62; 14)	89–96	86·2 (2·00; 8)	83–90
BILL (S)	17·2 (0·38; 14)	16·6–17·8	16·4 (0·54; 8)	15·8–17·2
TARSUS	24·2 (0·43; 14)	23·6–24·9	23·5 (0·49; 8)	23·0–24·1

Sex differences significant.

North-east Iran: wing, ♂♂ 96, 97, ♀ 92; tail, ♂♂ 100, 104, ♀ 95. Altai: wing, ♂ 92·4 (11) 89–95, ♀ 87·3 (3) 86–89; tail, ♂ 91·5 (11) 86–96, ♀ 89·3 (3) 87–92 (Stresemann 1928).

Afghanistan, summer (Paludan 1959).

WING	♂ 94·2 (1·33; 16)	92–96	♀ 88·8 (1·97; 14)	86–91
TAIL	92·4 (3·12; 16)	87–100	86·4 (2·41; 10)	83–90

M. a. persica. Iran. Wing, ♂♂ 93, 96, 97, 97; ♀♀ 87, 90 (Paludan 1938, 1940; Schüz 1959).

Weights. Nominate *alba*. Norway: May, 19·6–21·9 (4); July–September, 20–23·6 (8) (Haftorn 1971). Södermanland (Sweden), autumn: 21·0 (2·13; 93) 17·3–25·0 (I Nord). East Germany, spring to autumn: ♂ 21·6 (1·50; 27) 19–25, ♀ 19·9 (1·80; 23) 16–23 (Bub *et al.* 1981). Netherlands: March–May, ♂ 20·8 (2·71; 18) 17–26, ♀ 20·7 (2·08; 7) 18–24; July–September, ♂ 20·4 (1·15; 9) 18–22, ♀ 20·0 (1·05; 9) 18–22 (RMNH, ZMA). Belgium: adult ♂ 20–24·6 (25), adult ♀ 17·6–21·9 (16), laying ♀ 24·2–27·9 (4) (Bub *et al.* 1981). Malta, autumn and winter: 21·8 (1·8; 50) 18·0–27·5 (J Sultana and C Gauci). Greece, November: ♂♂ 24, 25 (Makatsch 1950). USSR: ♂ 21·5 (68) 17·7–24·1, ♀ 19·2 (56) 16·4–23·4 (Bub *et al.* 1981). Netherlands: adult, late June and July, moult not or just started, 20·3 (164); August, in full moult, 19·8 (107); September, moult nearing completion, 20·0 (48); juvenile, July, moult not or just started, 19·7 (43); August, in full moult from juvenile to 1st adult, 19·6 (30); late September, post-juvenile moult completed, 19·3 (39) (Jukema and Rijpma 1984). Exhausted birds: 14·8 (1·31; 5) 13·5–16·1 (Timmermann 1938–49; ZMA). For growth of nestling, see Leinonen (1973a).

M. a. yarrellii. Skokholm (Wales): 23·6 (11) 20·6–27·0 (Browne and Browne 1956). Helgoland (West Germany), March: ♂ 19, ♀♀ 19, 20, 23 (Krohn 1915). Exhausted ♂, January: 14 (Harris 1962).

M. a. dukhunensis. Turkey and northern Iran (including some intermediates between nominate *alba* and *dukhunensis*), March–July: ♂ 23·2 (1·95; 8) 19·5–26; ♀♀ 22, 23; unsexed 21, 25 (Paludan 1940; Schüz 1959; Kumerloeve 1961, 1969a, 1970a). On migration, Afghanistan: March–April, ♂ 24·5 (4) 23–27, ♀ 22·1 (4) 19–25; September–October, ♂ 22·2 (16) 19–25, ♀ 20·8 (6) 20–23 (Paludan 1959). India, winter: 20·2 (17) 18–24 (Ali and Ripley 1973b).

M. a. persica. Iran, May–July: ♂ 24·8 (3) 24·6–25·0; ♀♀ 19·4, 23 (Paludan 1938, 1940; Schüz 1959).

M. a. personata. Afghanistan: March, ♂ 24, ♀ 22·3 (3) 22–23; May–July, ♂ 24·2 (1·40; 12) 21–26, ♀ 24·7 (2·00; 9) 22–28; September, ♂♂ 28, 29, ♀♀ 25, 26 (Paludan 1959).

Structure. Wing rather short, broad at base, tip bluntly pointed. 10 primaries: p8–p9 longest or (occasionally) either one 0–1 shorter than other; p7 (0–)0·5–1 shorter than longest, p6 3–6, p5 12–16, p4 17–22, p1 25–31; p10 reduced, narrow and sharply pointed; 55–65 shorter than longest primary, 9–12 shorter than longest upper primary covert. Outer web of p6–p8 and (rather faintly) inner of p7–p9 emarginated. Tertials long, longest reaching to about tip of p6–p7 in closed wing. Tail long, tip square or slightly forked; 12 feathers. Bill rather long, straight, sharply pointed; rather wide at base, laterally compressed at tip; tip of culmen slightly decurved. Nostrils small, oval, partly covered by membrane above and by frontal feathering at rear. Some fine hair-like bristles at base of upper mandible. Tarsus and toe rather long and stout; claws rather short, decurved. For length of middle toe and hind claw, see Measurements. Outer toe with claw *c.* 72% of middle with claw, inner *c.* 76%, hind with claw *c.* 81%, without claw *c.* 40%.

Geographical variation. Marked and complex.

M. alba and close relatives in Afrotropics and Asia form group of fairly large to large wagtails with variable black-and-white head pattern, grey or black upperparts, black chest, and white belly. In view of partial overlap without interbreeding between some members of this group some of the larger black-backed populations are considered as separate species, viz. African Pied Wagtail *M. aguimp*, Large Pied Wagtail *M. maderaspatensis*, and Japanese Wagtail *M. grandis* (see *M. aguimp*, p. 474), but status of some other members (here still included in *M. alba* following long tradition) far from settled. In particular, large black-backed *lugens* from eastern Asia seems to overlap with smaller black-backed *leucopsis* from China with only limited amount of hybridization and may have reached species level, while *personata* from central Asia and *dukhunensis* from further north behave as separate species in area of contact in USSR, though hybridizing freely elsewhere. For taxonomic problems of the *M. alba* complex, see Johansen (1952), Nazarenko (1968), and Mauersberger (1980, 1982, 1983).

Differences between races most pronounced in ♂ breeding, less so in ♀ breeding and ♂ non-breeding, and immature birds often difficult to identify. Splitting of *M. alba* into groups (or even species) should be done according to head pattern of ♂ breeding rather than (as advocated by Vaurie 1959) according to upperpart colour; 4 main subspecies-groups distinguishable, as follows. (1) Races in which head, neck, and chest black except for contrasting white forehead and supercilium: *personata* (upperparts grey) and *alboides* (upperparts black—*M. grandis* similar, but chin white). (2) Races with forehead and sides of head and neck white except for black stripe on lore and through eye to side of nape: *lugens* (upperparts black) and *ocularis* (upperparts grey). (3) Races with forehead and sides of head and neck fully white, but chin and throat black: *yarrellii* (upperparts black) and nominate *alba* and *dukhunensis* (upperparts grey). (4) Races similar to group 3 but with chin and upper throat white: *leucopsis* (upperparts black) and *baicalensis* (upperparts grey). Group 1 appears closely related to group 2, and group 3 to group 4, but groups 1–2 more distant from groups 3–4 (C S Roselaar). *M. a. persica* and *subpersonata* more difficult to allocate. *M. a. persica* is a variable hybrid population between *personata* and nominate *alba* or *dukhunensis*; usually shows white forehead and side of head, and often some white on side of neck (like nominate *alba*, but with black band down side of upper neck). *M. a. subpersonata* similar to *persica* but lores and ear-coverts black (lower cheeks and sides of neck white, unlike *personata*, and much less white on tips of upper wing-coverts); perhaps an old offshoot of *M. aguimp* (Smith 1968a) or a stabilized hybrid population between nominate *alba* and *M. aguimp*.

M. a. yarrellii of Britain and Ireland differs from nominate *alba* mainly in much darker upperparts; white tips of wing-coverts and tertials slightly wider but this often not noticeable; difference in size negligible. In adult ♂ breeding, upperparts black, sometimes with some grey diffusion (especially on rump), occasionally with part of old grey non-breeding plumage retained, showing as grey mottling; sides of breast blackish-grey or black, separated from black chest by narrow white bar; flanks dark grey (darker than in nominate *alba*). Adult ♀ breeding similar to adult ♂ breeding, but forehead often speckled black, mantle, scapulars, and back often with pronounced grey grizzling or mottling, and sides of breast and flanks slightly paler; frequently, some grey non-breeding feathers retained on upperparts. 1st breeding plumages similar to adult breeding, but juvenile flight-feathers and variable number of wing-coverts or tail-feathers retained, as in nominate *alba* (see Plumages); forehead of 1st adult ♂ usually white, as adult ♂, but upperparts often partly mixed or washed grey, as adult ♀; 1st adult ♀ often shows more retained grey feathers on upperparts than adult ♀. Adult ♂ non-breeding similar to adult ♂ breeding, but mantle, scapulars, back, and sometimes nape variably mixed grey, and chin and throat white with yellow tinge. Adult ♀ non-breeding and 1st adult ♂ and ♀ non-breeding as variable as in nominate *alba* (see Plumages for ageing) and often difficult to separate from nominate *alba*: grey of mantle, scapulars, sides of breast, and flanks on average darker than in nominate *alba*, more extensively dark on flanks, mantle and scapulars sometimes partly suffused black, and (most importantly) rump black or greyish-black rather than medium grey. Juvenile as juvenile nominate *alba* and often indistinguishable: olive-grey or grey of upperparts and flanks on average slightly darker, dark stripe at side of crown and dark gorget on average blacker and more pronounced. For sexing and ageing, see also Svensson (1984a). *M. a. dukhunensis* similar to nominate *alba* in all plumages, but grey of upperparts on average slightly paler and white tips of median and greater upper wing-coverts wider; populations from

Caucasus area, northern Iran, and lower Volga north to Saratov have upperparts darker grey, like nominate *alba*, but tips of coverts wide, like *dukhunensis* (Hartert 1921-2; RMNH); on the other hand, some birds from north-east Turkey are as pale as *dukhunensis* above, but wing-coverts like nominate *alba* (Kumerloeve 1961). Populations of Zagros mountains and of southern slopes of Elburz mountains in Iran, separated as *persica*, are variable intermediates between *personata* and nominate *alba* or *dukhunensis*. *M. a. personata* markedly different from nominate *alba* in breeding plumage: head, neck, and chest black, but forehead, supercilium, and eye-ring white; white fringes along median and greater upper wing-coverts and along tertials very wide, often completely hiding black feather-bases, especially in ♂; mantle to back, lesser upper wing-coverts, sides of breast, and flanks grey, like nominate *alba* (darker than *dukhunensis*); sexes similar, but contrast between black nape and grey mantle often less marked in ♀, and some ♀♀ (especially 1st adults) retain some grey non-breeding feathers on crown or nape and white ones on chin. In non-breeding plumage, adult ♂ and sometimes adult ♀ and 1st adult ♂ similar to breeding but central crown and nape washed grey or olive-grey, chin white, and throat variably mottled black and white, white of chin separated from white of sides of forehead by black line on lore (rather like *M. grandis*); other non-breeding plumages have uniform grey upperparts (except for yellow-white or pale buff forehead and supercilium), dark grey stripe extending from lores to olive-grey ear-coverts, white chin and variable amount of white on throat, and less extensively white upper wing-coverts (but still much more white than any nominate *alba*); juvenile has chin to chest almost uniform dull grey, not white with dark gorget as in nominate *alba*, *dukhunensis*, and *baicalensis*. Isolated *subpersonata* from Morocco easy to identify in breeding plumage, having forehead, eye-ring, lower cheek, and side of neck white, remainder of head and neck black; remainder of body, and size, as nominate *alba*; in non-breeding, adult ♂ and some adult ♀♀ and 1st adult ♂♂ similar to breeding, but chin and throat white or mottled white, in other non-breeding plumages probably similar to 1st adult non-breeding nominate *alba*, except for dark grey lores; juvenile indistinguishable from juvenile nominate *alba* (BMNH, ZFMK). CSR

Motacilla aguimp African Pied Wagtail

PLATES 30 and 31
[facing pages 352 and 353]

Du. Afrikaanse Witte Kwikstaart   Fr. Bergeronnette pie   Ge. Witwenstelze
Ru. Африканская трясогузка   Sp. Lavandera blanc africana   Sw. Afrikansk sädesärla

Motacilla aguimp Dumont, 1821

Polytypic. *M. a. vidua* Sundevall, 1850, Afrotropics, except extreme south. Extralimital: nominate *aguimp* Dumont, 1821, basins of Orange and Vaal rivers (South Africa and neighbouring Namibia), merging into *vidua* in eastern Cape Province, Natal, and northern and eastern Transvaal.

Field characters. 18·5-19 cm; wing-span 26-31 cm. Larger than White Wagtail *M. alba*, though overlapping in most measurements. Greater size less evident than plumper body, enhanced by plumage pattern of head and forebody. Essentially black and white, differing from *M. alba* most in head, chest, and wing pattern. Face pattern more linear, with long white supercilium, completely black lores and cheeks, and long white throat. Bold white

transverse panel on folded wing becomes even more obvious in flight when large white bases of flight-feathers exposed. Sexes closely similar; little seasonal variation. Juvenile separable.

ADULT MALE BREEDING. At a glance, velvet-black and white plumage may recall ♂ of British race of White Wagtail *M. alba yarrellii* but differs distinctly from that or any other race of *M. alba* as follows: (a) black of

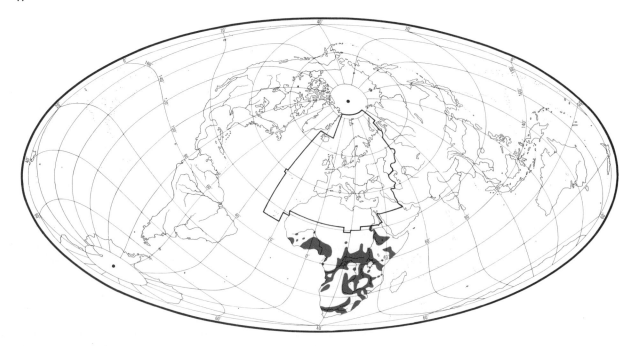

crown reaching forward to bill; (b) bold white supercilium; (c) wholly black lores and cheeks, joining hindneck and with downward extension from rear cheeks to chest-band; (d) permanently white chin, throat, and upper chest; (e) rather narrow but bulging black chest-band; (f) broad white blaze across folded wing, formed by white wing-coverts, broad fringes to secondaries and tertials, and bases to primaries (in flight, large area of white even more obvious). ADULT FEMALE BREEDING. Dark plumage areas less intense than ♂. White on wing-coverts often less extensive. ADULT MALE NON-BREEDING. Black on upperparts becomes mostly greyish. ADULT FEMALE NON-BREEDING. Black on upperparts becomes mostly dark olive-grey, paler than ♂ non-breeding. JUVENILE. Dark plumage less intense than adult, dark grey-brown. Head and chest pattern less distinct, with narrow supercilium (smudged around eye) and mottled chest-band. Wing marks less bold: pale fringes and tips buff and off-white. At all ages, bill and legs black.

Adult unmistakable; juvenile confusable with darker juveniles of *M. alba* but easily distinguished by large area of white on flight-feathers and coverts. Plumper form obvious to observer long familiar with *M. alba* and this visible even in flight, with silhouette less narrow-tailed and action rather less bounding. Gait and behaviour as *M. alba*.

Song superior even to that of *M. alba alba*, recalling Canary *Serinus canaria* in both rhythm and phrase. Commonest disyllabic call 'chizzit', more slurred and less divided than equivalent in *M. alba*.

Habitat. Tropical and subtropical. Largely a lowland bird but adaptable to variety of habitats, often along rivers; these include sand-banks, rapids, rocky perennial streams, and beds of large rivers, but birds avoid strongly tidal water and creeks (G D Field). A tame village bird, often away from water in vicinity of human habitations, including rest houses up to 2300 m—even roads running through woodlands. Also occurs on coastal lagoons. Likes feeding along water's edge, particularly by larger rivers and especially those which are rocky. (Prozesky 1970; Benson *et al.* 1971; Serle and Morel 1977.)

Distribution and population. No information on range changes or population trends.

EGYPT. Probably still breeds regularly in extreme south along Lake Nasser (PM, WCM).

Accidental. Israel.

Movements. Resident, particularly in areas modified by man. Occasional non-breeding visitor to Zanzibar and Pemba, Tanzania (Pakenham 1979), so subject to some local movement. Post-breeding dispersal probably sufficient to account for rapid appearances at new reservoirs and irrigation and drainage channels. Communal roosting produces local diurnal movement. Away from human habitation in Sierra Leone, where birds inhabit banks of larger rivers subject to seasonal flooding, the flooding necessitates local movement (G D Field).   JHE

Voice. See Field Characters.

Plumages (*M. a. vidua*). ADULT MALE BREEDING. Centre of forehead, crown, and remainder of upperparts deep black; often duller and slightly greyish on rump, broad white fringes along outer webs of upper tail-coverts. Long and broad white supercilium, extending backwards from nostril and side of forehead, bordered below and behind by black stripe from lores and gape backwards over upper cheeks and ear-coverts and up from rear of ear-coverts to rear of crown. Chin, throat, and sides of neck white; throat bordered below by broad and rounded black gorget on chest, laterally extending upwards almost to rear of ear-coverts; remainder of underparts white, washed grey on flanks, mottled black on thighs. Central 4 pairs of tail-feathers (t1–t4) black, t1 narrowly fringed white along outer web when fresh, t2–t4 sometimes with faint white outer edge; t5–t6 white, t5 with black border along basal and middle portion of inner web. Flight-feathers and tertials black; secondaries, inner primaries, and (faintly) tertials narrowly fringed white along tips; secondaries and tertials with broad and contrasting white fringe along basal and middle portions of outer web (up to tip on tertials); bases of p1–p7(–p8) contrastingly white, middle portion of inner web of outer primaries with white wedge. Lesser upper wing-coverts, greater upper primary coverts, and bastard wing black, median upper wing-coverts and shorter upper primary coverts white (some black at bases of inner webs, usually hidden). Greater upper wing-coverts white, but innermost (tertial coverts) black with white outer fringe and others with partly black inner webs (largely hidden, not extending to tips and bases of coverts). Under wing-coverts and axillaries white. ADULT FEMALE BREEDING. Like adult ♂ breeding, but mantle, scapulars, and back sooty-grey or dull greyish-black, and rump often dark grey, not as deep black as adult ♂; black of hindcrown often contrasts with slightly paler mantle; median and greater upper wing-coverts often more extensively black than in adult ♂, median coverts with black shaft and often with some black showing on inner web, greater coverts often extensively black on both webs, but tips white. ADULT MALE NON-BREEDING. Like adult ♂ breeding, but crown, mantle, scapulars, and back dull dark grey or greyish-black; forehead, sides of crown, upper tail-coverts, lores, and ear-coverts still black; rather like adult ♀ breeding, but wing usually more extensively white. ADULT FEMALE NON-BREEDING. Like adult ♀ breeding, but black on upperparts replaced by dark olive-grey (paler than adult ♂ non-breeding) except on centre of forehead and forecrown. Sides of head, underparts, and upper wing-coverts patterned black-and-white as adult ♀ breeding (marks on sides of head and on chest not as dull and ill-defined as juvenile; wing-coverts more contrastingly marked). JUVENILE. Central forehead, crown, and remainder of upperparts dark grey-brown with slight olive tinge, pale grey of feather-bases sometimes partly showing on mantle, scapulars, back, and rump. Supercilium narrow, white, widening into larger white patches on sides of forehead and above ear-coverts; sometimes broken by olive-brown above front of eye, partly freckled dusky, or partly washed pale olive. Lores, ear-coverts, and upper cheek dull greyish-black or olive-black bordered behind by white side of neck. Underparts white, tinged cream on chin, chest, and flanks; ill-defined dark grey-brown or dull black gorget extends from below ear-coverts across lower throat and upper chest, partly hidden by olive or buff feather-tips in fresh plumage. Tail as adult; flight-feathers as adult, but white of bases slightly less extensive. Tertials olive-black (darkest on longest) with white fringe along outer web and olive-buff fringe along tip. Lesser upper wing-coverts and tertial coverts dark brown-grey, like upperparts, tips narrowly fringed buff when fresh. Median upper wing-coverts white with distinct black shaft-streak and ill-defined dull black base to inner web; greater coverts white, middle portion of inner web or both webs dull black, merging into dusky freckling (not as clear-cut as adult). FIRST ADULT NON-BREEDING. Like

adult and sometimes indistinguishable. Juvenile flight-feathers, greater upper primary coverts, and sometimes outer greater upper wing-coverts or some tail-feathers retained, relatively more worn than adult at same time of year, and black on greater upper wing-coverts less sharply defined. Upperparts of ♂ grey, black restricted to forehead, sides of crown, upper tail-coverts, lores, and ear-coverts; in ♀, black virtually restricted to lores and front part of ear-coverts. FIRST ADULT BREEDING. Like adult breeding, but part of juvenile plumage retained, as in 1st non-breeding. Part of 1st non-breeding sometimes retained (especially in ♀), upperparts showing mixture of dull grey non-breeding and black breeding plumage.

Bare parts. ADULT, JUVENILE. Iris dark brown. Bill black; some yellow at base of both mandibles in juvenile shortly after fledging. Leg and foot slate-black or black. (Bannerman 1936; BMNH, RMNH, ZMA.)

Moults. ADULT POST-BREEDING. Complete; primaries descendant. Probably starts soon after breeding, but as breeding season strongly variable in Afrotropics (depending on local rains), no information on timing, as the few moulting birds examined were from widely scattered localities. In Sudan, moult completed August (BMNH). ADULT PRE-BREEDING. Partial: head, body, lesser upper wing-coverts, tertials, tertial coverts, and all or part of tail and median upper wing-coverts. Timing varies with locality; one from southern Egypt in heavy moult of body and tail 11 February (Meinertzhagen 1930). Single ♂ from Sudan just starting 19 October, others had moult completed February (BMNH). POST-JUVENILE. Partial: head, body, and usually all tail, tertials, and upper wing-coverts (a few tail-feathers or greater upper wing-coverts sometimes retained), but no flight-feathers or greater upper primary coverts. Starts soon after fledging, completed August in birds examined from Sudan. FIRST PRE-BREEDING. No moult up to January–February in birds examined from Sudan, but 2 in moult mid-March. Moult less extensive than in adult pre-breeding, ♀ in particular retaining much non-breeding.

Measurements. ADULT, FIRST ADULT. *M. a. vidua.* Southern Egypt, Sudan, and Ethiopia, all year; skins (BMNH, RMNH, ZMA). Bill (S) to skull, bill (N) to distal corner of nostril; exposed culmen on average *c.* 4·3 less than bill (S).

WING	♂ 96·4 (3·07; 10)	93–102	♀ 91·3 (2·39; 11) 88–96
TAIL	90·0 (3·52; 10)	86–92	87·1 (3·50; 11) 83–93
BILL (S)	18·3 (0·48; 10)	17·6–19·1	17·9 (0·75; 11) 16·9–18·5
BILL (N)	10·8 (0·46; 9)	10·0–11·4	10·8 (0·43; 10) 10·1–11·5
TARSUS	25·2 (0·90; 9)	23·9–26·4	24·2 (0·94; 11) 23·2–25·6

Sex differences significant for wing and tarsus. 1st adult with retained juvenile wing and sometimes tail similar to adult, combined above.

Birds from Liberia smaller: wing, ♂ 91·8 (3) 91–93, ♀ 85·5 (1) (RMNH). South Africa: wing, ♂ 89–98 (16), ♀ 85–98 (11) (McLachlan and Liversidge 1970).

Weights. *M. a. vidua.* Zambia: 27·0 (55) 23–30 (Dowsett 1965a).

Structure. 10 primaries: p7–p8 longest, p9 0·5–3 shorter, p6 1–2, p5 6–10, p4 12–16, p3 15–19, p1 20–24; p10 reduced, 57–65 shorter than wing-tip, 6–10 shorter than longest upper primary-covert. Outer web of p6–p8 emarginated. Bill relatively slightly longer and heavier than in Pied Wagtail *M. alba.* Middle toe with claw 17·1 (6) 16–19. Hind claw 5·9 (8) 5·3–6·4. Remainder of structure as in *M. alba.*

Geographical variation. Slight in size (see Measurements), pronounced in colour. *M. a. vidua* from all Afrotropics except south has broad white supercilium, much white on side of neck, white flanks, and rather narrow black bands from upper cheeks to ear-coverts and across chest. Nominate *aguimp* from western and central South Africa and neighbouring Namibia distinctly darker: white supercilium slightly narrower, only a small spot of white on side of neck, broad black chest-band, and broad black band over cheeks to ear-coverts (merging into black of hindcrown and of upper sides of chest), and black flanks. Birds examined from eastern South Africa intermediate between *vidua* and nominate *aguimp* or nearer *vidua*: black bands over cheeks backwards and (in particular) across chest rather broad, but flanks pale grey or white.

Forms superspecies with *M. alba*, Large Pied Wagtail *M. maderaspatensis* from Indian subcontinent, and Japanese Wagtail *M. grandis* from Japan (Hall and Moreau 1970). *M. aguimp* sometimes included in *M. alba* (see, e.g., Hartert and Steinbacher 1932–8, Irwin 1960), but differs from it to about the same degree as do *M. maderaspatensis* and *M. grandis*, and (though not overlapping in breeding range with it as those 2 species do) is thus considered a separate species (Hall and Moreau 1970).

Recognition. Differs at all ages from *M. alba* in white bases of flight-feathers and broad white fringes along outer webs of secondaries.   CSR

Family PYCNONOTIDAE bulbuls

Small to medium-sized oscine passerines (suborder Passeres); many arboreal but some terrestrial to greater or lesser extent; most frugivorous and insectivorous, some also taking nectar and pollen. About 120 species (some known as greenbuls, brownbuls, bristlebills, or leafloves) in *c.* 14 genera of which largest are: (1) *Pycnonotus*, 47–51 species—Africa and especially southern Asia; (2) *Phyllastrephus*, 19 species—Afrotropics and Madagascar; (3) *Hypsipetes*, 20 species—Madagascar and southern and eastern Asia; (4) *Criniger*, 10 species—Afrotropics and Asia. Found in Old World only, mainly in forest and parkland: tropics of Africa and Asia, north to Middle East and Japan, east to Philippines and Indonesia. About equal number of species in Afrotropics and southern Asia, but most genera (13) in Afrotropics. Mostly sedentary. Family represented in west Palearctic by 3 species of *Pycnonotus*, all breeding.

Body moderately slender; neck short. Sexes of similar size in most species but ♂ larger in some. Bill short to medium in length and rather curved; often slender and pointed, sometimes hooked, with or without a notch. Rictal bristles usually well-developed; nasal operculum present to greater or lesser extent. Wing rather short, broad and rounded. Flight relatively weak-looking in

most species, but swift and agile in some (if only over short distances); flycatching sallies common. Tail medium to long; tip usually round, square, or graduated. Legs short, toes weak. Gait a hop. Head-scratching by indirect method. For information on bathing, sunning, and anting in Common Bulbul *Pycnonotus barbatus*, see p. 487. No indication of dusting.

Plumage soft and, especially on rump, dense; often a patch of hair-like feathers on nape, sometimes concealed. Colour mainly sombre—brown, olive, or green—but often with bright or contrasting head markings and/or red, yellow, orange, or white under tail-coverts. Some species crested. Sexes alike. At all ages, a single complete annual moult shortly after breeding season. Nestling naked at hatching; mouth without contrasting spots.

A clearly defined family (Delacour 1943). Rather resemble babblers (Timaliidae) in appearance but closer relationship to cuckoo-shrikes (Campephagidae) suggested by some anatomical studies (Beecher 1953; Bock 1962), though most superficial characters of the two differ (Sibley 1970), and DNA evidence places Pycnonotidae in Old World warbler group (superfamily 'Sylvioidea') not far from swallows (Hirundinidae) and warblers (Sylviidae) (Sibley and Ahlquist 1985a).

Pycnonotus leucogenys **White-cheeked Bulbul**   PLATES 32 and 34
[between pages 544 and 545]

Du. Witoorbuulbuul   Fr. Bulbul à joues blanches   Ge. Weissohrbülbül
Ru. Белощекий булбуль   Sp. Bulbul cariblanco   Sw. Vitkindad bulbyl

Brachypus leucogenys Gray, 1835

Polytypic. *P. l. mesopotamiae* Ticehurst, 1918, Iraq, Arabia, and southern Iran. Extralimital: *leucotis* (Gould, 1836), lowlands of Pakistan and northern India, intergrading with *mesopotamiae* in south-east Iran, southern Afghanistan, and western Pakistan; nominate *leucogenys* (Gray, 1835), foothills and mountains from eastern Afghanistan along Himalayas to Assam, intergrading with *leucotis* in hills of northern Pakistan.

Field characters. 18 cm; wing-span 25·5–28 cm. 10% smaller than Common Bulbul *P. barbatus* and Yellow-

vented Bulbul *P. xanthopygos*. Bulbul of similar form to *P. barbatus* and *P. xanthopygos* but with diagnostic black,

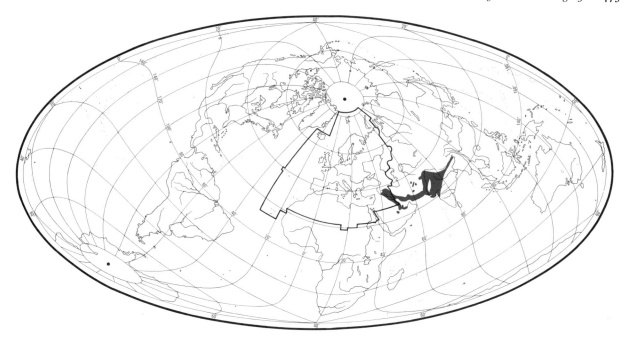

white-cheeked head. Sexes similar; no seasonal variation. Juvenile separable.

ADULT. Head including throat black, with bold white blaze below eye, over cheeks, and on to side of neck (recalling Great Tit *Parus major*). Upperparts dull grey, with darker flight-feathers and black-brown tail whose feathers show terminal white spots. Chest pale dusky-brown, underbody pale grey-buff, almost grey-white on rear flanks. Vent rich yellow. Eye red-brown; narrow yellow eye-ring. JUVENILE. Head browner than adult, with white cheeks less clear-cut. Vent paler. At all ages, bill and legs black.

Unmistakable in west Palearctic. Behaviour, flight, and other actions much as *P. barbatus*; even more associated with human settlements, becoming tame pest.

Song a fruity, musical 4-syllable unit, frequently repeated, suggesting speeded-up Golden Oriole *Oriolus oriolus*. For calls, see Voice.

Habitat. Mainly subtropical, in warm dry areas from coastal mangroves to *c.* 1800 m in hills of Baluchistan (Pakistan) and up to *c.* 2400 m in Nepal, but prefers broader valleys to side valleys and hills, where a climatic barrier is reached (Bates and Lowther 1952). A bird of open country not of forest, and of bushes rather than trees. In eastern Saudi Arabia, frequents palm groves (especially with rank undergrowth), bush cultivation, reedbeds, tamarisk scrub, and gardens (G Bundy). In Kashmir, common round hill villages with cultivation, in gardens, along roadsides, and on house boats, flying up on to tables in use to share food with people. Often feeds on ground but also on bushes, perching on their tops and on trees (Whistler 1941). Nests not only in bushes but in low pollarded willows growing out of water and flanking lake channels in Kashmir, where also will winter inside occupied houseboats. Otherwise found in light forest and secondary growth, amongst hedgerows and bushes (Bates and Lowther 1952). Elsewhere in India overlaps with Red-vented Bulbul *P. cafer* but prefers semi-desert to richer deciduous aspects, though will occupy gardens and groves (Ali and Ripley 1971). A thicket-lover around Delhi, favouring berried bushes *Capparis* and *Salvadora*; also *Zizyphus* (Hutson 1954). Flourishes through capacity for adaptation and commensalism, keeping pace with human developments. Flies mainly in lowest airspace.

Distribution and population. IRAQ. Widespread and common (Allouse 1953); has spread up Euphrates valley (Chapman and McGeoch 1956).

Movements. Essentially resident, though populations of Himalayan foothills seem subject to some local upward altitudinal movement following breeding (Ali and Ripley 1971): in Pech valley of eastern Afghanistan, birds seen during 9 days in early April but not during next 25 may have been migrants (Paludan 1959); said also to be migrant visitor to Quetta valley April–November, but recorded there all months except January–February (Ticehurst 1926–7). JHE

Food. Mainly insects, fruit, and berries; also seeds, buds, and nectar. In India commonly catches insects on ground (Whistler 1949). Also hawks flying insects (Dewar 1923; Whistler 1949; Meriwani 1973), usually launching pursuit from prominent perch; most common in evening. In

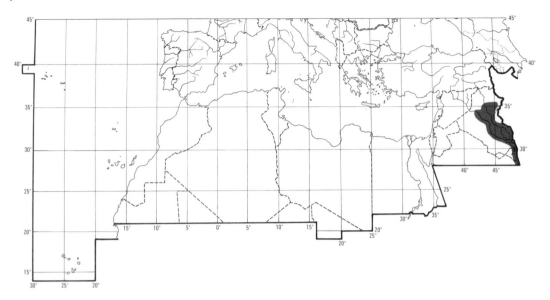

Kashmir feeds throughout daylight hours, although rather less active 12.00-15.00 hrs in July-August (A Gretton).

In Iraq, ripe dates *Phoenix* a favourite item, also unripe figs *Ficus*, and occasionally mantis *Mantis religiosa* (Sage 1960a); also in Middle East, takes mulberries *Morus* and pomegranates *Punica* (Hüe and Etchécopar 1970). In Indian subcontinent, takes wide range of fruit and berries, including mistletoe *Viscum*, barberry *Berberis*, neem *Melia*, caper *Capparis aphylla*, peeloo *Salvadora persica*, *Rubus*, *Lantana*, *Zizyphus*, and pea *Pisum* from gardens. Buds, nectar, and seeds (all unspecified) also recorded as food items. Breadcrumbs occasionally taken, from the hand in Iran, and off tables in Kashmir. Insects commonly taken, including larval and adult butterflies (Lepidoptera), ants (Formicidae, especially winged), and beetles (Coleoptera). Also spiders (Araneae). (Ali 1949; Whistler 1949; Sage 1960a; Diesselhorst 1968a; Erard and Etchécopar 1970; Hüe and Etchécopar 1970; Ali and Ripley 1971.)

Insects form higher proportion of diet when adults feeding young (Ali 1949).   AG

Social pattern and behaviour. Little studied, especially in breeding season.

1. Outside breeding season, in pairs or small flocks of 5-6, sometimes more at favourable food source (Whistler 1941); occasionally in large flocks (Ali and Ripley 1971). At Basrah (Iraq), October, *c.* 20 together (Moore and Boswell 1956). In Nepal, flocks of 5-6 early March, but in pairs by end of April (Diesselhorst 1968a). Flocks likely to include paired birds (see below). BONDS. No evidence for other than monogamous mating system. In Iraq, however, associations of 3 birds commonly occur at nests under construction (Marchant 1963b), raising possibility of helpers, as in Common Bulbul *P. barbatus* (see p. 487). Pair-bond maintained all year (Ganguli 1975). Apparently constant companionship of pair-members (Phillips 1950) and intimate roosting behaviour (see below) lend support to notion of strong, enduring pair-bond. Hybridization with Red-vented Bulbul *P. cafer* reported by Sibley and Short (1959)

and Bundy and Warr (1980). No further information. BREEDING DISPERSION. Solitary and territorial. In Iraq, territories *c.* 4 ha or more; defended also against Magpies *Pica pica* and Hooded Crows *Corvus corone* (Meinertzhagen 1954). No further information. ROOSTING. Outside breeding season, forms communal roosts of 5-20 birds (Gadgil and Ali 1976). At rest, captive pair usually leaned against each other (Short 1964). No further information.

2. Bold and confiding (Phillips 1950), even entering houses to feed (Ali and Ripley 1971). Mobs owls (Strigidae) vigorously and noisily until they move away (Meinertzhagen 1954). FLOCK BEHAVIOUR. No information. SONG-DISPLAY. Song (see 1 in Voice) given from ground, perch, or in short flight. Song-period at least April-October (Moore and Boswell 1956). Vocalization (apparently song) given from perch accompanied by continual bowing and posturing (Ali and Ripley 1971). ANTAGONISTIC BEHAVIOUR. Study by Short (1964) of captive pair indicated rich repertoire of postures and displays, interpreted as appropriate to different intensities of attack-flee conflict. For calls associated with displays, see 3 in Voice. At low intensity, the following performed: (a) Crest-raising; (b) abrupt and momentary Tail-spreading; (c) quick upward Tail-flicking followed by slower lowering, which may be accompanied by side-to-side tail movement and slightly ruffled plumage; (d) slight outward Wing-flicking. At higher intensity: (a) Wing-spreading—distinct spreading and moderate raising of wings, evidently an escalation of Wing-flicking, and threatening imminent attack; (b) Vent-showing display—raises and spreads under tail-coverts so that they stand out vertically and laterally, heightening effect of colour patch (display associated with Tail-flicking and Crest-raising); (c) Gaping-display—directs open bill at opponent for up to 5 s (evidently pronounced threat). In presumed antagonistic display, but of uncertain intensity, bird performs slow side-to-side Head-turning with white cheek plumage apparently ruffled. Bird confronted by Gaping-display retreats and/or adopts appeasing Crouched-posture: faces opponent, lowers body, and sleeks plumage. (Short 1964.) Various accounts document most of these postures and displays in the wild: Crest-raising and incessant Wing-flicking regularly performed by birds moving restlessly about (Ali and Ripley 1971), presumably indicating general wariness. At sunset often Wing-flicks while giving apparent Anger-calls (Phillips 1950: see 3

in Voice). Tail-flicking (high above level of head) with side-to-side movements reported by Silsby (1980). In autumn flock, Iraq, some birds chased one another with Anger-calls (see 3a in Voice); in one dispute, bird flew up, circled, then landed and evidently performed Tail-flicking and Vent-showing display (Moore and Boswell 1956). HETEROSEXUAL BEHAVIOUR. Virtually nothing known. In Iraq, flocks break up into pairs late January (Marchant 1961). Captive pair often Allopreened (Short 1964). RELATIONS WITHIN FAMILY GROUP, ANTI-PREDATOR RESPONSES OF YOUNG. No information. PARENTAL ANTI-PREDATOR STRATEGIES. In India, distraction-lure display of disablement type recorded when human intruder near nest: bird flew to ground short way in front of intruder, feigning injury in one wing, then flew with laboured flight to bush; gave Warning-call (see 4b in Voice) when approached (Cumming 1902). No further information. EKD

Voice. Rich and varied repertoire, freely used. Some sounds confusable at distance with Bee-eater *Merops apiaster* (Ali and Ripley 1971). No information on differences between races. Account based partly on study by Short (1964) of 2 captive nominate *leucogenys*.

CALLS OF ADULT. (1) Song. In eastern Saudi Arabia, a fruity, musical 4-syllable unit, frequently repeated, suggesting speeded-up Golden Oriole *Oriolus oriolus* (D J Brooks). In Iraq, songs include bubbling, chuckling, and warbling sounds, also a protracted series of 'chip-chop' phrases (Moore and Boswell 1956). Melodious phrases of 3–4 units in apparently endless variety of combinations (Phillips 1950). 2 common phrases rendered 'tea for two' and 'take me with you' (Ali and Ripley 1971). In Nepal, renderings include: 'we-did-de-dear-up', 'whet-what', 'who-lik-lik-leer', 'three-thirty', and 'take-it-eber' (Fleming *et al.* 1976). In recording (Fig I), a brief liquid burbling phrase difficult to render but something like 'what are you DOing THEN' (J Hall-Craggs) or 'whe-whor-o-whor-up', repeated at intervals of *c.* 4–11 s; in another recording (Fig II) of simpler song, interval 2–3 s; in a 3rd (Fig III), phrases 7–11 s apart and longer, and units delivered faster than in Figs I–II (E K Dunn, J Hall-Craggs). (2) Contact-call. A 'pit' or 'pit-pit' or 'pit-pit-pit' (Short 1964). Probably this call the busy, rather squeaky, 'rusty' tonal sounds, rendered 'k-zee zee kr-zer ze' (J Hall-Craggs: Fig IV), rather like high-pitched sparrows *Passer* (P J Sellar), given by undisturbed birds in tree. In flight, 'plee-plee-plee' (Fleming *et al.* 1976)

perhaps the same or related. (3) Anger-calls. Evidently modified Contact-calls; several variants occur, as follows, in confrontations with conspecifics. (a) 'PIT-pit PIT-pit' or 'PIT-a PIT-a' (Short 1964). Calls rendered a harsh angry 'tchak tchak' (Moore and Boswell 1956), and 'wik-wik-wik-wiker' (Fleming *et al.* 1976) are probably the same or similar. (b) At higher intensity, 'PIT-lo' or 'PIT-pit-lo'; variants contain more 'pit' units; also rarely 'PIT-a-lo' or 'PIT-it-lo' (Short 1964). (c) Slow chattering 'PIT-pit-it-it-it-it' or a rapid chattering 'pitititit' (Short 1964). (4) Alarm- and Warning-calls. (a) Sharp 'peep' (Short 1964). (b) Anxious soft clear low note, like human trying to whistle 'you', given by *leucotis* performing distraction-display, and presumed to warn mate (Cumming 1902).

CALLS OF YOUNG. No information. EKD

Breeding. Little information from west Palearctic. SEASON. Iraq: eggs found April to mid-July, mainly mid-April to mid-May (Marchant 1963*b*). Eastern Saudi Arabia: eggs laid early March to July (R J Connor). Afghanistan, India: eggs found April–June (Hüe and Etchécopar 1970; Ali and Ripley 1971). SITE. In low bush or sometimes in branches of low tree, 0·3–3·5 m above ground; average 1·36 m (0·8–3·2) for 20 nests (Marchant 1963*b*, which see for sites); also in creepers, thatch, or building; usually well concealed but can be exposed; occasionally on ground (Dementiev and Gladkov 1954*a*; Ali and Ripley 1971). Nest: substantial cup of grass stems and leaves, roots, and thin twigs, lined with finer rootlets, lichens, and grass; *c.* 7·5 × 5 cm (Marchant 1963*b*). Building: probably only ♀ brings material (Marchant 1963*b*); bird shot while building was ♀ (Paludan 1959). EGGS. See Plate 81. Sub-elliptical, smooth and glossy; pinkish-white, heavily marked with spots, blotches, and streaks of red, with underlying small purple spots. *P. l. mesopotamiae*: 23·8 × 16·9 mm (22·6–25·1 × 15·8–18·1), $n = 22$; calculated weight 3·48 g (Schönwetter 1979). Clutch: 3 (2–5) (Ali and Ripley 1971). Of 28 clutches, Iraq: 2 eggs, 4; 3, 17; 4, 7; mean 3·1 (Marchant 1963*b*). Eggs laid at daily intervals. After one nest was deserted, replacement clutch started, in new nest, 5–9 days later. 2 (possibly 3) broods per season. (Marchant 1963*b*; see also Ticehurst *et al.* 1921–2.) 3 broods proved

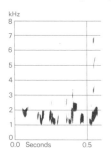

I P A D Hollom
Iran April 1972

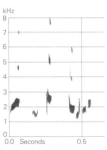

II P A D Hollom
Iran April 1972

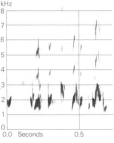

III P A D Hollom
Iran April 1972

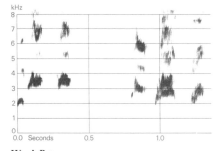

IV A Paterson
India September 1982

in eastern Saudi Arabia (R J Connor). INCUBATION. About 12 days (*c.* 10½–12½); probably does not usually begin until clutch complete (Marchant 1963*b*). FLEDGING TO MATURITY. Fledging period 9-11 days (Marchant 1963*b*). No further information. BREEDING SUCCESS. Of 48 eggs, Iraq, 12 hatched and maximum 8 young fledged; only 3 nests out of 19 produced fledged young; except for 2 infertile eggs, all failures due to total destruction of nest (Marchant 1963*b*).

Plumages (*P. l. mesopotamiae*). ADULT. Forehead, crown, and lores deep glossy black, remainder of upperparts, upper wing-coverts, and tertials uniform grey-brown or drab-grey; feathers of hindneck and sides of neck narrowly tipped black, rump and upper tail-coverts slightly paler and greyer, fringes along tips of upper wing-coverts slightly paler. Chin and throat black, like crown; black of lower throat extends up along sides of neck to sides of hindcrown, isolating long white patch from ear-coverts to lower cheeks behind gape. Chest, sides of breast, and flanks grey-brown or drab-grey, like upperparts or slightly paler, merging gradually into grey-white of central belly and vent. Under tail-coverts bright yolk-yellow or orange-yellow. Tail brownish-black, slightly paler drab-grey towards base; fringes along tip of central pair (t1) dusky grey; t3-t6 with 7-11 mm of tip white (most extensive on inner webs), sharply demarcated; t2 with less clear-cut dusky-white tip of *c.* 4 mm. Flight-feathers, greater upper primary coverts, and bastard wing dark grey-brown to black-brown with indistinct drab-grey fringes along outer webs. Under wing-coverts and axillaries pale grey-brown, longest coverts fringed white. Bleaching and wear have strong influence: by May-June, black of head faded to black-brown, drab-grey of upperparts, chest, and flanks to pale grey-brown, fringes and tips of flight-feathers and tail-feathers to pale grey. NESTLING. Naked (Harrison 1975). JUVENILE. Black of head duller, tinged brown, not extending as far back on hindcrown or as far down on lower throat as in adult; white patch on side of head smaller, less clear-cut; rump and upper tail-coverts slightly paler; feathers of underparts softer and looser than in adult, breast to vent cream-white with some pale grey-brown of feather-bases visible; under tail-coverts paler yellow; basal and middle portions of tail-feathers slightly paler, greyer, feathers narrower, *c.* 9-12 mm wide near tip of t1 and t6 (12-15 mm in adult); less white on tips of t3-t6 (3-8 mm), white less sharply defined, and less square-cut than in adult; flight-feathers slightly paler grey-brown; tips of tertials and upper wing-coverts less smooth, softer, bordered by faint and ill-defined grey-buff fringes. FIRST ADULT. As adult; indistinguishable when last juvenile feathers shed.

Bare parts. ADULT. Iris red-brown, brown, or dark brown. Narrow bare ring round eye: orange-yellow, yellow, or dull lemon-yellow in *mesopotamiae*, black in other races. Bill dark horn-brown or black. Mouth yellow. Leg and foot greyish-black, dark horn-brown, or black; soles white to dark grey. (Hartert 1921-2; Ripley 1951; Ali and Ripley 1971; BMNH, RMNH.) NESTLING. Bare skin dark purple (Harrison 1975). JUVENILE. At about fledging, bill pink-flesh, leg and foot greyish-flesh. During post-juvenile moult, iris dark brown; bill, leg, and foot brown-black; ring round eye dull orange. (BMNH.)

Moults. ADULT POST-BREEDING. Complete; primaries descendant. In *mesopotamiae*, moult starts with shedding of p1 mid-July to late August; completed with regrowth of p9-p10

October-November. Moult of body, tail, and secondaries starts at primary moult score *c.* 20, but sometimes some body moult before primaries start; moult completed with outer primaries (score 50); face, outer tail-feathers, and middle secondaries last. (BMNH, RMNH.) In *leucotis*, complete moult about September (Ali and Ripley 1971). POST-JUVENILE. Complete; primaries descendant. Sequence as in adult, but timing more variable, depending on hatching period. Starts with p1 June-August, completed with p10 September-November. (BMNH, RMNH.)

Measurements. ADULT. *P. l. mesopotamiae*. Wing (1) Iraq, (2) northern Arabia; other data Iraq only, all year; skins (BMNH, RMNH.) Bill (S) to skull, bill (N) to distal corner of nostril; exposed culmen on average 4·0 less than bill (S).

			♂			♀	
WING (1)	♂	93·2 (2·11; 19)	91-98	♀	89·1 (1·38; 12)	87-92
(2)		92·2 (2·73; 9)	88-97		87·7 (1·72; 5)	86-90
TAIL		85·2 (3·04; 13)	80-90		81·4 (1·93; 9)	78-84
BILL (S)		18·0 (0·42; 14)	17·5-18·8		17·4 (0·36; 9)	16·9-18·0
BILL (N)		9·3 (0·29; 13)	8·9-9·8		8·9 (0·21; 9)	8·5-9·2
TARSUS		22·4 (0·44; 13)	21·7-23·1		21·4 (0·73; 9)	20·4-22·4

Sex differences significant.

P. l. leucotis. India: wing ♂ 81-91, ♀ 80-88; tail, ♂ 66-84, ♀ 73-80; bill to skull, ♂ 15-18, ♀ 14-16; tarsus, ♂ 21-22, ♀ 20-23 (Ali and Ripley 1971).

Nominate *leucogenys*. Northern India, all year; skins (RMNH, ZMA).

				♀		
WING	♂	93·7 (2·89; 6)	90-97	♀	90·1 (2·29; 7)	88-94
BILL (S)		18·9 (0·74; 5)	18·2-19·9		18·4 (0·97; 5)	17·3-19·4

Sex differences significant for wing. Tail and tarsus similar to *mesopotamiae*.

JUVENILE. Wing on average *c.* 6 shorter than adult, tail *c.* 12 shorter.

Weights. *P. l. mesopotamiae*. Kuwait, March-June, unsexed: 31·1 (1·75; 4) 29-33 (V A D Sales, BTO). Southern Iraq, February, unsexed: 35·4 (BMNH).
P. l. leucotis. South-west Afghanistan, early April: ♂ 31, ♀ 30 (Paludan 1959). India, unsexed: 23 (10) 18-28 (Ali and Ripley 1971).
Nominate *leucogenys*. Eastern Afghanistan, early April: ♂ 31, ♀ 32 (Paludan 1959). India, unsexed: 34-38 (5) (Ali and Ripley 1971).

Structure. 10 primaries: in adult *mesopotamiae* and nominate *leucogenys*, p6 longest, p5 and p7 both 0-1·5 shorter, p8 3-5, p9 15-21, p10 38-42, p4 2-3, p3 5-7, p2 8-12, p1 10-15; in juvenile, p10 more broadly rounded at tip and relatively longer, 27-38 shorter than wing-tip. Tail rather long, tip square or slightly rounded; 12 feathers, t6 4-9 shorter than t2-t3. Bill short and thick in *mesopotamiae*, more slender in *leucotis* and nominate *leucogenys*. Feathers of forehead and crown slightly elongated in *mesopotamiae* and *leucotis*, crown appearing slightly peaked when feathers raised; feathers markedly lanceolate in nominate *leucogenys*, *c.* 2-3 cm long, forming curled crest. Toes short; middle toe with claw 18·1 (4) 17-20. Remainder of structure as in Common Bulbul *P. barbatus*.

Geographical variation. Rather slight throughout much of range, but nominate *leucogenys* from foothills of Himalayas markedly different, sometimes considered separate species (e.g. by Vaurie 1958, 1959). *P. l. mesopotamiae* from central Iraq drab-grey on upperparts, chest, and flanks; populations from southern Iraq south through Kuwait to Bahrain and Al Hufuf on average paler in worn plumage (probably due to stronger bleaching), but closely similar when fresh; these slightly paler birds sometimes separated as *dactylus* Ripley, 1951 (Ripley 1951; Gallagher and Rogers 1978). *P. l. leucotis* from lowlands of Pakistan and north-west India similar to *mesopotamiae*, but narrow eye-ring black (not yellow), plumage around eye more extensively black, bill more slender, underparts paler, belly more extensively white, and wing and tail shorter; intergrades with *mesopotamiae* in south-east Iran. Nominate *leucogenys* markedly different from *mesopotamiae* and *leucotis*, and latter 2 races therefore often considered to form a separate species White-eared Bulbul *P. leucotis*. Feathers of forehead and crown of nominate *leucogenys* elongated and pointed, dark brown with narrow grey fringes, bordered by long and narrow white supercilium at sides; upperparts, upper wing-coverts, tail-base,

and fringes of flight-feathers olive-green (not drab-grey); bill slender, white ear-patch small, belly and vent extensively yellowish-white, and under tail-coverts bright yellow (not as orange as *mesopotamiae*), but these differences from *mesopotamiae* bridged by *leucotis*; size large, as in *mesopotamiae*. As nominate *leucogenys* intergrades into *leucotis* in narrow zone in foothills of northern Pakistan, all forms treated here as single species, following Ripley (1958), Sibley and Short (1959), Ali and Ripley (1971), and Voous (1977).

Considered by Hall and Moreau (1970) to form superspecies with Red-vented Bulbul *P. cafer* from southern Asia with which it occasionally hybridizes (Sibley and Short 1959), with Yellow-vented Bulbul *P. xanthopygos*, with Common Bulbul *P. barbatus* and its relatives (see p. 489), and some other species from southern Asia.   CSR

Pycnonotus xanthopygos Yellow-vented Bulbul

PLATES 32 and 34
[between pages 544 and 545]

Du. Arabische Buulbuul   Fr. Bulbul des jardins   Ge. Vallombrosabülbül
Ru. Желтопоясничный булбуль   Sp. Bulbul árabe   Sw. Levant bulbyl

Ixus xanthopygos Hemprich and Ehrenberg, 1833

Monotypic

Field characters. 19 cm; wing-span 26·5–31 cm. Close in size to north-west African race of Common Bulbul *P. barbatus barbatus*. Similar in appearance and behaviour to *P. barbatus*, differing most markedly in blacker head, grey-white eye-ring, and yellow vent. Sexes similar; no seasonal variation. Juvenile difficult to separate.

ADULT. Basic plumage pattern as *P. barbatus*, but rather more contrasting, with largely black head and tail standing out from pale grey-brown body (slightly paler below than above). Vent lemon-yellow (off-white in *P. barbatus*). Dark eye made obvious by grey-white eye-ring. JUVENILE. Slightly paler above than adult, with duller head and loosely feathered vent.

Unmistakable when diagnostic yellow vent visible, but not so in brief view when general appearance so similar to *P. barbatus* that instant identification quite impossible. Confusion unlikely, however, since both species are sedentary and have separate breeding ranges.

Voice much as *P. barbatus*.

Habitat. Mediterranean and subtropical, but habitat otherwise scarcely distinguishable from that of Common Bulbul *P. barbatus*, and broadly similar to White-cheeked Bulbul *P. leucogenys*. In Arabia, a bird of gardens, palm groves, and fairly thick bush in wadi beds (Meinertzhagen 1954). Ranges from sea-level to summit of Jebel Suda above 2000 m on moist southern highlands (Jennings 1981a). In Lebanon, in moister valleys with trees by rivers at no great altitude, and in orange and banana plantations as well as in gardens, orchards, groves, and

thickets, chiefly in coastal strip, including Beirut and its suburbs (Vere Benson 1970).

Distribution and population. No information on range changes or population trends.

EGYPT. Sinai: common in oases (SMG, PLM, WCM).

Movements. Apparently sedentary.   JHE

Food. Mainly fruit, seeds, and insects; occasionally leaves, flowers, and nectar. Insects often taken in flight, especially at dusk, by swooping up like bee-eater *Merops* from tree perch; also by pecking from foliage, sometimes in brief hover (Gallagher and Woodcock 1980). In Lebanon, gathers in large flocks to feed on flying ants (Formicidae) (Mackintosh 1944). Also noted hunting among sticks, etc., on ground (Bark Jones and Hartley 1945). Recorded chasing Hoopoes *Upupa epops* and stealing mole-crickets (Gryllotalpidae) from them (O Hasson).

Diet mainly fruit, including peaches and plums *Prunus*, strawberries *Fragaria*, and tomatoes *Solanum*. In Israel, fruit pulp found in one December bird, and leguminous flowers, perhaps beans, in one from February. In North Yemen, takes fruit of prickly pear *Opuntia*. Also recorded: leaves, seeds, nectar, insect larvae, flies (Diptera), moths (Lepidoptera), flying ants, bees, and wasps (Hymenoptera), mole-crickets and locusts (Orthoptera), worms (presumably earthworms Lumbricidae), and snails (Gastropoda) (Mackintosh 1944; Hardy 1946; Meinertzhagen

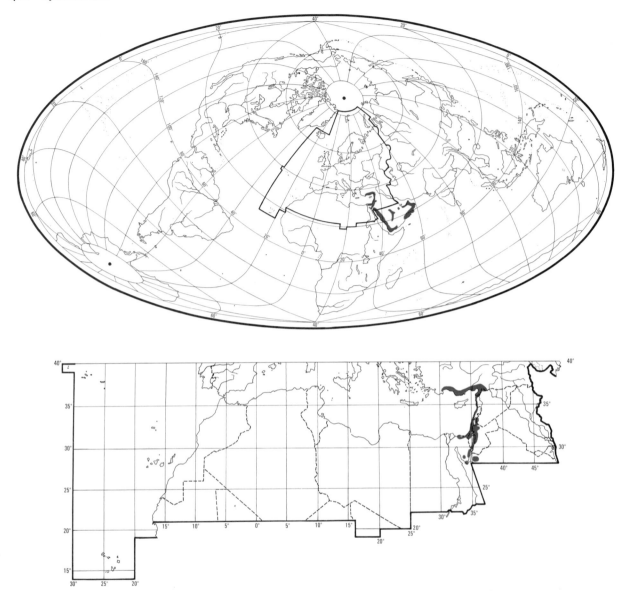

1954; Arnold and Ferguson 1962; Gallagher and Woodcock 1980; Phillips 1982; O Hasson, Y Yom-Tov).

Young fed entirely on insects and spiders on 1st day,
more fruit subsequently until entirely fruit near fledging
(O Hasson).   AG, EKD

Social pattern and behaviour. Major study in Israel by
Hasson (1983), and account includes further material from O
Hasson. Otherwise little known.

1. Gregarious throughout the year when food locally abundant, otherwise mostly in pairs or 'duos'. Duo comprises 2
siblings of same or different sex, associating closely (Hasson
1983; see also Bonds, below). In Jordan, April–May, flocks of
25–50 recorded (Wallace 1984*b*). In winter, Lebanon, feeding
flocks of *c.* 40 'pairs' (Mackintosh 1944). Where density high,
Israel, flocks comprise a few hundred birds or even thousands;
after breeding season, 67–83% 1st-years, depending on nest-

predation (O Hasson). Some birds (duos or pairs) hold territories
all year but will vacate them temporarily to join feeding flocks
(Hasson 1983). BONDS. Monogamous mating system (Inbar
1977). Pair-bond maintained all year (Arnold and Ferguson
1962; Inbar 1977; Hasson 1983), and though often lasts
for several years, not necessarily life-long; in 12 cases one
pair-member (10 ♀♀, 2 ♂♂) replaced by another bird, perhaps
initiated by unpaired ♀ soliciting paired ♂ (see Heterosexual
Behaviour, below) or after territorial disputes with intruding
duos (see Antagonistic Behaviour, below). In several cases birds
which disappeared from territories were known to be alive
elsewhere, e.g. ♀ left her mate and became paired with ♂ in
neighbouring territory. Young dependent on parents for *c.*
30 days (O Hasson). When they leave parental territory,
brood-members remain closely bonded in duos, or often trios.
In trios, one bird gradually severs bonds with siblings, and
associates with another (unrelated) bird. From data on 2–3
broods, members of duos (apparently at *c.* 4–6 months old) take

new partners, forming ♂-♀ pairs of non-siblings and begin to seek a territory. (Hasson 1983; O Hasson.) For apparent pairing with Common Bulbul *P. barbatus*, see Meinertzhagen (1930). Both members of pair feed young (Inbar 1977). Age of first breeding sometimes 1 year, but more often 2 or more depending on how long it takes to establish a territory (Hasson 1983). BREEDING DISPERSION. Solitary and territorial. Territories usually contiguous in favourable habitat. In botanic gardens in Tel-Aviv (Israel), average of 15 territories 0·13 ha (0·05–0·19); much larger in desert areas (O Hasson). Territory serves for courtship, nesting, and feeding, although birds readily seek food outside their territories, at high density almost inevitably in another's territory (Hasson 1983.) ROOSTING. Nocturnal, in dense foliage. Territorial birds roost in territory; non-territorial birds typically trespass into occupied territories at dusk to seek a suitable site for roosting alone. One record of apparent communal roosting, several birds entering tree. Young probably roost with parents for a while after fledging. Members of duos or pairs sleep snuggled up together; for similar daytime loafing, see Fig A. Comfort activities include bathing. Feeding activity

A

of flocks (see below) begins often before sunrise if food source is insects; for vegetable food, activity begins typically shortly after sunrise, sometimes 1 hr or more after, and ends usually in late afternoon, up to *c.* ½ hr before dark. (Hasson 1983; O Hasson.) For diurnal song activity, see below.

2. Readily approachable (Tristram 1865). Confiding near habitation, but shy elsewhere (Gallagher and Woodcock 1980). FLOCK BEHAVIOUR. Cohesion very loose, flock representing a temporary aggregation of birds continuously joining and leaving. Feeding flocks noisy (see Voice) and animated, and, since they usually infringe territorial space, provoke disputes with residents (Hasson 1983). In Lebanon, small noisy parties sometimes form, usually soon after dawn (Vere Benson 1970). SONG-DISPLAY. Song (see 1 in Voice) given by territorial ♂ from an elevated (but not necessarily exposed) perch in territory. May also be given in snatches while bird feeds and preens in treetops, rarely in flight (Hasson 1983.) On 2 occasions, given by member of a small flock, performer perching higher than the others (Bark Jones and Hartley 1945). Song-duels occur (Bark Jones and Hartley 1945). In Israel, sings irregularly November–March, song fully developed by end of March and maintained until October, declining thereafter (O Hasson). Sings all year round in Lebanon (Vere Benson 1970). In Israel, sings from before sunrise, most intensely in first hour of activity, declining during day, with some increase towards evening (O Hasson). ANTAGONISTIC BEHAVIOUR. Account based on Hasson (1983): territories defended by ♂ or ♀ or both together. Pair capable of ousting 2 or 3 duos trespassing to feed, but if a large flock invades, residents given Alarm-threat calls (see 4 in Voice); at best, resident ♂ may attack individuals in flock. Residents call similarly when they detect birds trespassing to roost. By day, attack single intruders in flight or immediately after landing.

Resident threatens rival by standing opposite, this often enough to induce retreat. Threat greater if pair-members stand close together facing intruder(s). If intruder holds ground, resident (usually ♂) performs Forward-threat display (Fig B): spreads tail and moves it slowly up and down; sometimes lowers wings and ruffles head plumage. At higher intensity, spreads tail wider, accentuating motion, increasingly ruffles head, especially crown, and gives ever louder Alarm-threat calls. If intruder still persists, resident attacks. Just before, or during fight, if rival challenges at close quarters, both birds perform Upright-threat display (Fig C): face each other, stretch necks,

B   C

point bills upwards, and sleek plumage; maintaining this posture, make quick opening and closing movements of their wings (Wing-flicking) and tails. Residents fight with neighbours and other intruders, usually both resident pair-members joining in. If intruder, or more often duo, slow to retreat, and if resident ♂ slow to threaten or attack, ♀ mate begins to Allopreen his neck. Allopreening apparently painful to recipient and thought, therefore, to irritate and provoke in this context. Allopreened ♂ invariably becomes more aggressive, gives increasingly intense Alarm-threat calls, and eventually attacks. Once, when territorial pair invaded adjacent territory, resident ♂ was absent and fighting continued as intruding ♂ chased resident ♀ inside her own territory. Intruder's mate followed him and Allopreened his neck during lulls in fighting. When confrontation ceased, ♀ mate of intruder returned to her own territory (and nest). For *c.* 1 hr thereafter, ♂ intruder alternated between the 2 territories, attending both ♀♀ in turn; in adjacent territory, consorted with resident ♀ as if she were his mate, even helping her to expel another neighbour. No permanent change of partner resulted. HETEROSEXUAL BEHAVIOUR. Account based on Hasson (1983). (1) Pair-bonding behaviour. Pair-formation probably occurs in various ways; sometimes promiscuity of territorial pair-members, or paired or unpaired non-territorial birds soliciting paired territorial birds, suggest ways in which new pairs may form, though no hard evidence that these lead to permanent changes of mate. Possible sequences as follows: (a) Most commonly, duos (see Bonds) become breeding pairs once they succeed in establishing territories. Sometimes, mostly among young birds, duos form momentarily (for several minutes), with associated close snuggling (see below), apparently to gauge compatibility. (b) Initially, aggressive encounter between 2 duos may lead to formation of new duo. Thus, if fight between 2 members of 2 duos is prolonged, there may be a gradual change from violent pecking to Allopreening, in this context thought to be less provocative (see below), indicating the possibility of a new relationship. (c) In 3 cases, a non-territorial bird (presumed ♀) tried to solicit a territorial ♂ while his mate was absent; ♀ tried to approach ♂, fluttering her wings (see Subdued Meeting-ceremony, below). ♂ threatened but did not attack, and tried to avoid ♀. Each time, a 3rd bird, probably partner of soliciting ♀, was present but showed much less interest in the territorial ♂ and was once attacked by him. In each case, encounter ended quickly as soon as rightful ♀-mate returned to territory; resident pair-members snuggled up to each other, performed Subdued Meeting-ceremony (see below), and the interlopers retired. (d) Once, a territorial ♂ consorted with a neighbouring territorial ♀ when her mate was absent (see Antagonistic Behaviour). Formation of duos, and eventually of breeding pairs, characterized by the following behaviour: (a) Relinquishing of individual-distance. Only members of the same pair or duo, even if only of ephemeral duration, snuggle close to each other. See also Roosting. (b) Consorting together. Duos are together during most of the day and night, and only rarely do territorial pair-members separate during the day for more than ½ hr. (c) Meeting-ceremonies. Of 2 kinds, depending on length of intervening separation. In Subdued Meeting-ceremony, usually performed after relatively brief separation, both partners, or the one being approached, lightly flutter wings, spread tail, and give quiet Greeting-calls (see 2a in Voice). During the meeting, or immediately after, they usually snuggle up to each other. Noisy Meeting-ceremony, one bird spreads its wings (Fig D) and gives loud Greeting-calls (see

D

2b in Voice). Noisy Meeting-ceremony performed by territorial birds after a long separation and in confrontations, immediately before or after bird lands close to its mate. Sometimes followed by Subdued Meeting-ceremony. (d) Allopreening. Mutual Allopreening. Common in duos and pairs. Although apparently unpleasant, even painful to recipient, especially young bird, Allopreening believed to be important in bonding. (e) Begging. ♂ does not feed ♀, but occasionally prevents her from eating, whereupon she apparently begs: rapidly flutters her wings and, occasionally, slightly lowers her head and sometimes opens and closes her bill, while frequently calling (see 3 in Voice). Such behaviour not always confined to duos and pairs. (f) Hunched-display. A bird turning away from its partner while moving from place to place may perform Hunched-display (Fig E), perhaps indicating appeasement: perches some distance away (at least 1 m) from partner, usually hunches and ruffles back, lowers head, points bill upwards, and opens and closes it; lowers and sometimes spreads tail, and slowly moves drooped

E

wings in and out in exaggerated manner. (2) Mating. No information. (3) Behaviour at nest. ♀ builds, accompanied by ♂. 4-7 days between start of building and 1st egg (O Hasson). RELATIONS WITHIN FAMILY GROUP. First fed immediately after hatching (O Hasson). Until c. 5 days old, young beg by gaping, swaying heads, and giving food-calls (see Voice). Same response elicited by touching nest. At c. 5 days, eyes begin to open, and young cease calling thereafter (Aharoni 1928). Nest-sanitation occurs throughout nestling period (O Hasson), though older young also defecate over side of nest (Aharoni 1928). Juvenile out of nest begged with trembling wings (P A D Hollom). If only 1 fledgling, new nest started c. 1 week after it fledges; probably later with larger broods. ♂ continues feeding 1st brood while ♀ builds new nest (O Hasson). ANTI-PREDATOR RESPONSES OF YOUNG. No information. PARENTAL ANTI-PREDATOR STRATEGIES. Alarm-calls (see 4 in Voice) given when raptor nearby. Regularly mobs and attacks Jay *Garrulus glandarius*, especially when close to nest. *G. glandarius* or man near nest elicits distraction-display (O Hasson), but no details.

(Figs by J P Busby: from photograph and drawings in Hasson 1983.)   EKD

Voice. Freely used. Repertoire rich and varied. No detailed information on regional differences, but see call 1.

CALLS OF ADULTS. (1) Song. Monotonous repetition of single phrase which, in Israel, varies somewhat between regions (Hasson 1983). Composed of short phrases of 2-8 syllables, variously stressed (Bark Jones and Hartley 1945). Sometimes rich and flute-like, sometimes like contralto whistle (but lower) of Blackbird *Turdus merula*, sometimes husky and deep, nearly always in short disjointed snatches. Occasionally mimics other birds, especially (in Lebanon and Jordan) Thrush Nightingale *Luscinia luscinia*, Golden Oriole *Oriolus oriolus*, and *T. merula* (Vere Benson 1970). In Israel, wild birds not heard to mimic but caged birds do so readily (O Hasson). According to Bark Jones and Hartley (1945), pitch usually constant but may rise on last note; this not supported by other sources, however. Presumed song-units and phrases reported by King (1978) include descending 'wheeew'; 'pheeew-wheeew' of which 2nd unit descending; 'whee-too-too', the 'too' notes lower; 'too-too-too-too'; 'took-whee-too', 2nd unit lowest, 3rd highest. In our recording, main phrase (Fig I) a mellow 'twur-tu-TWEE-teeroo' (E K Dunn) or 'teu-strip-(s)tree-tu' (P A D Hollom); in recording, this phrase repeated 4 times at intervals of c. 3 s, then 5th phrase (Fig II) a more elaborate 'twur-tu-TWEEtwer-u-weri' (E K Dunn). Given such variation within and between individuals, quality and overall structuring of song are more instructive for identification than specific renderings. (2) Greeting-calls. Vary according to intensity of Meeting-ceremony (see Social Pattern and Behaviour for definitions). (a) Thin, faint, chirping, whistling sounds given in Subdued Meeting-ceremony (Hasson 1983). 'Bubbling courtship call' (Gallagher and Woodcock 1980) is probably the same (O Hasson). (b) Sharp, robust sounds, something like 'MINE, you're MINE, you're MINE, you're

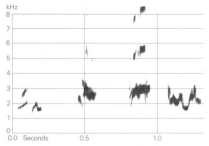

I P A D Hollom Israel April 1980

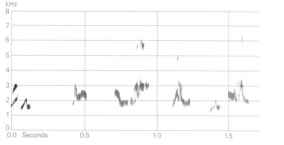

II P A D Hollom Israel April 1980

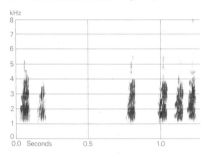

III P A D Hollom Israel April 1979

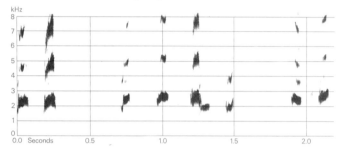

IV P A D Hollom Israel April 1979

MINE', given in course of Noisy Meeting-ceremony (Hasson 1983; O Hasson). (3) Begging-call. A 'weak sound' (not further described) given frequently by ♀ apparently soliciting ♂ (mate) for food, or at least access to food (Hasson 1983). (4) Alarm-threat call. A sharp, burring, scolding 'weck' (King 1978). At growing intensity, a rapid repetition of one syllable, producing an increasingly loud high-pitched chatter; serves to threaten conspecifics in confrontations, also used when mobbing (Hasson 1983). In Fig III, harsh 'tscheck' sounds, rather like Magpie *Pica pica* (J Hall-Craggs). Slowly repeated 'krik krik' or 'prip prup' (Gallagher and Woodcock 1980) perhaps low intensity variant. Loud sharp call of 1-2 notes given when Hobby *Falco subbuteo* or Sparrowhawk *Accipiter nisus* nearby (O Hasson). (5) Other calls. (a) Strong, sharp 'wit-wit-wit...', also 'teewit' and 'tew' sounds (different from call 5) given by non-territorial birds, especially in flock, and by territorial birds when outside their own territory; function not known (O Hasson). (b) A 'tchink-tchink-tchink kutchi-kutchi-kutchi tiwinktiwinktiwink' (Gallagher and Woodcock 1980); sequence not readily reconciled with calls listed above.

CALLS OF YOUNG. Food-call of small young a piping cheeping sound (Aharoni 1928; see also Social Pattern and Behaviour); of juvenile, nasal 'pee' and 'zee' sounds (Fig IV), not unlike contact-calls of White-cheeked Bulbul *P. leucogenys* but a little more liquid and less wheezy (E K Dunn, J Hall-Craggs: compare Fig IV of that species, p. 477).   EKD

Breeding. SEASON. Middle East: eggs laid end of May to beginning of July (Hüe and Etchécopar 1970). Israel: April–August (Arnold and Ferguson 1962). Oman: nest with eggs and young found late April (Walker 1981a). Gulf states of Arabia: eggs found March (Bundy and Warr 1980). SITE. In bush or low palm. Nest: small cup of thin twigs, grass stems, moss, and leaves, sometimes also string and wool; lined with hair, shredded bark and rootlets (Müller 1879; Inbar 1977). According to Aharoni (1928), nest unlined. Building: no information on role of sexes. EGGS. See Plate 81. Sub-elliptical, smooth and glossy; light violet to pinkish-white, well marked with violet- or red-brown and grey spots and speckles. 24·3 × 16·9 mm (22·0–27·2 × 15·0–17·8), $n = 17$; calculated weight 3·55 g (Schönwetter 1979). Clutch: 3 (2–4). 2–3 broods (Arnold and Ferguson 1962). INCUBATION. About 14 days; by both parents according to Inbar (1977), ♀ only according to O Hasson. YOUNG. Fed by both parents (Inbar 1977). FLEDGING TO MATURITY. Fledging period 13–15 days (Inbar 1977). BREEDING SUCCESS. No information.

Plumages. ADULT. Forehead, crown, lores, ear-coverts, cheeks, and chin deep black with slight bluish gloss, shading to brown-black on hindcrown, rear of ear-coverts and cheeks, and on lower chin and to dark brown on throat. Entire upperparts grey-brown or pale drab-brown, slightly tinged olive-yellow in fresh plumage (in particular on rump and upper tail-coverts). Chest to belly and flanks pale grey-brown or drab-grey (slightly paler than upperparts), gradually darker towards upper sides of chest and sides of mantle. Vent and under tail-coverts contrastingly bright yellow. Tail black, central pair of feathers (t1) tinged drab-brown; all feathers with slightly paler drab-grey tips, but this poorly defined and restricted, hardly visible. Flight-feathers, tertials, and upper wing-coverts dark grey-brown, slightly darker than upperparts (darkest, brown-black, on inner webs of outer primaries and on longest feathers of bastard wing); flight-feathers and greater upper primary coverts with narrow pale grey-brown or drab fringes along outer webs. Under wing-coverts and axillaries pale grey-brown or drab-grey, similar to chest and flanks; small coverts along leading edge of wing tinged yellow. *In worn plumage*, black of face duller, slightly browner; upperparts slightly paler and browner; underparts pale drab (less greyish), grading to grey-white on central lower belly; vent and under tail-coverts paler yellow; tail, upper wing-coverts, and flight-feathers darker and duller brown, but frayed tips and outer edges of feathers bleached to paler grey-brown; when plumage heavily worn, upperparts bleached to drab-grey and underparts (down from chest) to cream-white. Sexes similar (but see Measurements). NESTLING. Entirely naked. JUVENILE. Differs from adult in same characters as juvenile Common Bulbul *P. barbatus* differs from adult (see p. 488); under tail-coverts yellow, loose and fluffy. FIRST ADULT. As adult; indistinguishable once juvenile p10 shed (longer, broader, and with more rounded tip than in adult).

Bare parts. ADULT, FIRST ADULT. Iris brown or dark brown. Bare and warty ring round eye pale flesh-grey, greyish-white, bluish-white, or white. Bill plumbeous-black, darkest on culmen and tip. Leg and foot slate-blue to black, soles dark flesh-grey, dark grey, or greyish-black. (Hartert 1910; BMNH, RMNH.) NESTLING. Bare skin black, gape red (Aharoni 1928). JUVENILE. No information.

Moults. ADULT POST-BREEDING. Complete; primaries descendant. Starts with a few feathers on body from late May onwards, followed by p1 early June to early August. Moult completed with p9, p10, some middle secondaries, or outer tail-feathers September to early November. POST-JUVENILE. Complete; primaries descendant. Timing and sequence as adult post-breeding, but some birds finish as late as November or early December. (BMNH, RMNH, ZMA.)

Measurements. ADULT. South-central Turkey, Lebanon, Israel, and Jordan, all year; skins (BMNH, RMNH, ZMA).

Bill (S) to skull, bill (N) to distal corner of nostril; exposed culmen on average *c.* 3·7 less than bill (S).

	♂		♀	
WING	99·4 (1·81; 18)	96-103	92·8 (1·42; 19)	90-95
TAIL	89·1 (3·31; 15)	84-95	84·6 (2·07; 15)	80-88
BILL (S)	21·0 (0·74; 15)	19·7-22·0	20·3 (0·55; 15)	19·3-21·2
BILL (N)	10·8 (0·52; 15)	9·9-11·7	10·4 (0·42; 15)	9·8-11·2
TARSUS	23·6 (0·65; 14)	22·7-24·9	22·8 (0·63; 15)	21·8-23·7

Sex differences significant except for bill (N).

JUVENILE. Wing and tail on average *c.* 8 shorter than in adult.

Weights. Israel: 44 (56) 35-46 (Inbar 1977).

Structure. Closely similar to north-west African race of Common Bulbul *P. barbatus barbatus*, but wing slightly shorter, tail slightly less rounded at tip, bill longer and more slender, less deep at base, more sharply pointed at tip; also, feathers of crown slightly longer, forming short and broad rough crest when erected (less markedly so in north-west African *P. b. barbatus*, but some Afrotropical races of *P. barbatus* similar); narrow ring of bare skin round eye, covered with small warts. 10 primaries: p6 longest, p5 and p7 0-1(-2) shorter, p8 2-5, p9 14-17, p10 36-40, p4 1-3, p3 3-7, p2 7-11, p1 10-16. 12 tail-feathers; t3-t4 longest, t1 2-5 shorter, t6 4-8. Middle toe with claw 18·9 (5) 18-20.

Geographical variation. None. All populations virtually identical in colour in fresh plumage, but more different when worn, some bleaching more rapidly than others: birds from Asia Minor still fairly dark in spring, those of lowlands of southern Jordan and western Arabia markedly paler.

Sometimes considered a race of *P. barbatus* (e.g. by Vaurie 1959), or some Afrotropical races of *P. barbatus* are included in *P. xanthopygos* (e.g. by Mackworth-Praed and Grant 1960, 1963). See, however, *P. barbatus* (p. 489) and Hall and Moreau (1970). CSR

Pycnonotus barbatus Common Bulbul

PLATES 32 and 34
[between pages 544 and 545]

DU. Grauwe Buulbuul   FR. Bulbul commun   GE. Graubülbül
RU. Обыкновенный бульбуль   SP. Bulbul naranjero   SW. Trädgårdsbulbyl

Turdus barbatus Desfontaines, 1789

Polytypic. Nominate *barbatus* (Desfontaines, 1789), north-west Africa; *arsinoe* (Lichtenstein, 1823), Egypt (not Sinai) south to Red Sea coast and 8°N in Sudan, west through Sahel zone to inland Mauritania, north to Tibesti (Chad) and Aïr (Niger). Extralimital: *schoanus* Neumann, 1905, Eritrea and central Ethiopia to south-east Sudan; *inornatus* (Fraser, 1843), Sénégal to Ghana and Mali; *nigeriae* Hartert, 1921, central Nigeria south to Gabon, east to central Cameroon; *goodi* Rand, 1955, northern Nigeria to south-west Chad; 8-11 further races in southern Afrotropics.

Field characters. 19 cm; wing-span 26·5-31 cm. Slightly larger than Corn Bunting *Miliaria calandra* and White-cheeked Bulbul *P. leucogenys*; 15-20% smaller than any *Turdoides* babbler. Medium-sized passerine, with high-crowned head, rather broad wings, and long, slim body and tail; general character most recalls small, long-tailed thrush *Turdus* or sober shrike *Lanius*. Plumage sombre dusky-brown, with paler greyish underparts. Sexes similar; no seasonal variation. Juvenile difficult to

separate. 2 races in west Palearctic, distinguishable by size and plumage.

ADULT. (1) North-west African race, nominate *barbatus*. Upperparts and wings essentially dusky (even olive-toned) brown, relieved only by noticeably darker, umber-brown head (almost black in front of eye) which contrasts with paler grey-buff chest and even paler grey-ochre or white underbody and vent; area below head may be more or less mottled and blotched with dusky patches or diffuse

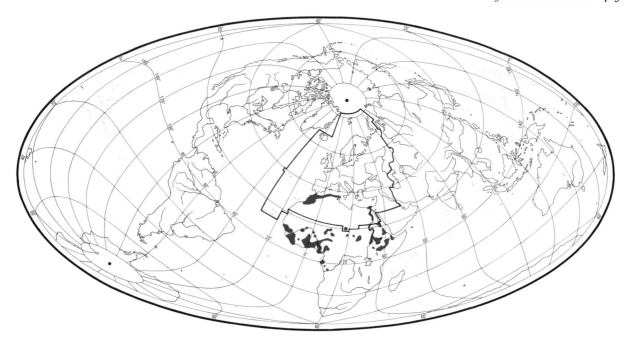

lines. Dull black tail contrasts little with rump but strikingly with pale vent. (2) Saharan and Egyptian race, *arsinoe*. Distinctly smaller than nominate *barbatus*, with 10% shorter wings. Upperparts paler, exaggerating contrast of head (itself darker due to blacker tone which extends to crown and throat) and tail. JUVENILE. Both races. Duller than adult about face but doubtfully distinguishable except in family party. At all ages, bill and legs grey-black.

Liable to confusion only with Yellow-vented Bulbul *P. xanthopygos*, when much blacker head and yellow vent of that species not seen. Flight said to be weak but actually no less strong than Jay *Garrulus glandarius*; bird capable of rapid escape and dramatic plunge through cover; hint of instability stems mostly from erratic wing-beats which give fluttering action to broad wings, exaggerated by apparent waving and spreading of long tail. Gait includes hopping, leaping, and clambering. Often gregarious, groups of birds moving about together.

Given to sudden, staccato outbursts of noise, with both calls and song rich and fluting. Song essentially a series of clear, melodious whistles.

Habitat. Tropical and subtropical; in west Palearctic, also breeds marginally in warm Mediterranean zone. In north-west Africa, above all a plains bird, but ascends freely to 700–900 m and locally in Haut Atlas to 2300 m. Always in green and fertile places, including wooded streams, gorges, and oases, as well as gardens and orchards (Heim de Balsac and Mayaud 1962). A counterpart to Blackbird *Turdus merula*, being characteristic of gardens, orchards, and palm groves, but also sometimes occurring in wild and desolate regions (Etchécopar and Hüe 1967).

In Sudan may be found almost anywhere with trees and bushes, especially in gardens and cultivation, but also up to *c.* 2500 m on Jebel Marra (Cave and Macdonald 1955). In West Africa, on similarly wide range of habitats, although absent from closed forests (Serle and Morel 1977). In Sierra Leone, quick to colonize forest clearings, sometimes penetrating into main forest along logging trails, or entering it to feed on fruiting lianas; in northern savanna and on mountain plateaux, most abundant bird even where remote from man (G D Field). In East Africa, also in coastal scrub, thickets and open forest, as well as old cultivation and gardens (Williams and Arlott 1980).

Distribution. EGYPT. Strongly extended range in Nile delta and valley this century (Koenig 1924; Borman 1928; Meinertzhagen 1930). Parts of Suez Canal area colonized since 1945 (PLM, WCM).

Population. EGYPT. Common (PLM, WCM). CHAD. Very common where vegetation plentiful (Simon 1965).

Oldest ringed bird 9 years 11 months (Rydzewski 1978).

Movements. Resident and (often at least) sedentary (Heim de Balsac and Mayaud 1962; Etchécopar and Hüe 1967; G D Field), but being associated with man-made habitats moves relatively quickly into newly modified areas such as exploited forest reserves (Okia 1976) and even isolated new developments such as the cattle ranch and hotel at Obudu plateau (Nigeria), *c.* 17 km from and *c.* 1200 m above Obudu town (Elgood 1982; J H Elgood).

JHE

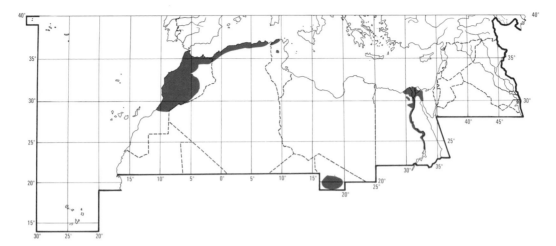

Food. Mainly fruit and insects; also seeds, flowers, young leaves, nectar, and even crystallized gum. Much of the following information from outside west Palearctic. Insects commonly caught on ground by scouring vegetation, although aerial feeding, particularly on winged termites, is common. Bird may launch pursuit from perch or ground, sometimes flying almost vertically up; returns to perch or ground to eat prey (Bannerman 1936; Brooke 1973; Markus 1974). In Uganda will 'dart and tumble' in aerial-pursuit of insects, particularly winged termites (Isoptera) (Stanfield 1973). Aerial-pursuit occurs particularly on overcast days, and even after sunset. Opportunistic, taking advantage of caterpillar outbreaks even if not in the bird's typical habitat, and takes termites from broken galleries that are accidentally exposed, usually by the the bird landing on a branch with active galleries (Brooke 1975; Vernon 1975). Fruit normally taken from trees, but also when fallen. Berries swallowed whole, but larger fruits pecked into; will make small hole in side of orange *Citrus* while still on tree and clean out flesh inside (Irby 1875; Hartert 1902). Probes flowers, particularly *Aloe*, for nectar.

In west Palearctic recorded feeding on apricot *Prunus*, dates *Phoenix*, figs *Ficus*, oleander *Nerium*, pomegranate *Punica*, citrus fruits *Citrus*, *Zizyphus*, asphodel (Liliaceae), strawberry tree *Arbutus*, oleaster *Elaeagnus*, mastic tree *Pistacia*, and jasmine *Jasminus fruticans*. Mainly takes fruit, but also fleshy petals of oleander and asphodel; also seeds. Takes insects, including 'white ants' (presumably termites Isoptera). Outside west Palearctic, the following recorded (mainly from Zimbabwe and adjacent countries): wide variety of fruit (over 50 species) including figs (especially preferred), *Asparagus*, avocado *Persea*, blackberry *Rubus*, spinous cucumber *Cucumis*, *Duranta* berries, guava *Psidium guajava*, gourds (Cucurbitaceae), Persian lilac *Melia*, peach *Prunus persica*, red pepper *Capsicum*, and Virginia creeper *Parthenocissus quinquefolia*; flowers of *Acacia*, *Aloe aculeata*, bottlebrush *Calothamnus quadrifidus*, *Eucalyptus*, *Pinus*, *Protea*, mesquite *Prosopis*

juliflora, and butter-tree *Bassia*; young leaves of *Acacia* and *Prosopis*; nectar from *Aloe* and *Greyia sutherlandi*. Also feeds from sugar bowls and 2 records of crystallized gum taken from *Acacia* (no insects trapped in gum). Animal food as follows: many records of termites (mainly winged) and winged ants (Formicidae), also grasshoppers, etc. (Orthoptera), mantises (Mantidae), greenfly (Aphididae), adult and larval Lepidoptera, larval flies (Diptera), a beetle (Coleoptera), and 1 record of geckos (Geckonidae). (Dekeyser 1956; Horváth 1959b; Archer and Godman 1961b; Brosset 1961, 1971a; Heim de Balsac and Mayaud 1962; Etchécopar and Hüe 1967; Oatley and Skead 1972; Stanfield 1973; Brooke 1975; Masterson 1975; Pettett 1975; Vernon 1975; Thévenot 1982.)

In Zimbabwe, termites 67% (by number) of insect prey, and figs 24% (by number) of fruit taken (calculated from Brooke 1975). In Sénégal, 2 ♂♂, August–September, contained unidentified black berries; 2 ♂♂ from February contained *Zizyphus* fruits (Dekeyser 1956). In Sudan, feeds on almost any kind of sweet fruit approaching maturity and may cause considerable damage where birds numerous (Schmutterer 1969).

No information on food of young in west Palearctic. In Kenya, brooding ♀ will swallow berries brought by ♂, but passes much of insect food to young; young receive only insects until 1 week old, then also fruit. Fed 8–12 times per hr, much less at midday (Someren 1956). In Gabon, nestlings fed 60 times in 20 hrs (Brosset 1971a).

AG

Social pattern and behaviour. Little studied and most information extralimital; most detailed study by Someren (1956) for *tricolor* in Kenya.

1. Gregarious outside breeding season, occurring in small parties, sometimes larger aggregations at (e.g.) favourable feeding sites (Pineau and Giraud-Audine 1979). In Ghana, flocks of up to 30 common, November–January (Grimes 1972). In Zimbabwe, September, flock of more than 70 birds reported (Vernon 1975). At least some birds within such flocks are presumably pairs (see below). BONDS. No evidence for other than monogamous mating system, and pair-bond probably

maintained all year: e.g. in Kenya, *tricolor* presumed to pair for life (Someren 1956); in Gabon, *nigeriae* 'permanently' in pairs (Brosset 1971*a*). In Sierra Leone, evidence from colour-ringing shows that at least sometimes helpers attend nest and assist with feeding of young (G D Field). For apparent pairing with Yellow-vented Bulbul *P. xanthopygos*, see Meinertzhagen (1930); for hybrids with Red-eyed Bulbul *P. nigricans* in southern Africa, see Irwin (1958) and Markus (1963, 1967). ♂ and ♀ share nest-duties (Someren 1956; Brosset 1971*a*). Parents tend young for some time after fledging (Someren 1956). BREEDING DISPERSION. Solitary and territorial. No information on size of territory, but in Sierra Leone 2 nests *c.* 15 m apart (G D Field). In Morocco, average 1·6 pairs per km² in undisturbed woodland, 14·8 pairs per km² in degenerate woodland, *c.* 21-24 pairs per km² in semi-arid scrub extending up into high valleys (Thévenot 1982). In Morocco, many pairs appeared not to breed (Brosset 1961); these perhaps pre-breeding pairs or 'duos', as in *P. xanthopygos* (p. 480). In Sierra Leone, ringed birds tended to stay in same garden year after year (G D Field), suggesting established birds site-faithful. No further information. ROOSTING. Little information for breeding season: in Gabon, pair spent every night for 2 months in same clump of foliage (Brosset 1971*a*). Outside breeding season, communally in trees (Grimes 1972). In Morocco, in dense foliage of orange *Citrus*, fig *Ficus*, etc. (Brosset 1961). In West Africa, leaves forest in evening to roost in plantations and gardens (Bannerman 1936). One of the first birds active in the morning, and the last to retire at night, doing so with much 'twitter and fuss'. Loafs and preens at midday and sunset (Someren 1956). At nest with young, ♀ preened, then sunbathed: lay on side, spread a wing and tail, raised crest fully, and ruffled rump plumage (Someren 1956). In Zimbabwe, anting behaviour as follows: birds crouched, spread wings and tail, and tucked head under breast; one brushed underside of wings and belly with jerking head movements, then flew to branch and preened breast and outspread wings in vigorous, highly agitated manner; ants found on ground below perch (Brown and Newman 1974). Readily drinks and bathes in rivers, streams, gutters, etc. (Brooke 1975); typically under cover of overhanging thicket or stem dipping into water (Bannerman 1936). Not uncommonly bathes gregariously, especially in early morning or evening (Someren 1956). In Tanzania, often rain-bathes by flopping about on leaves of trees, especially *Cassia nodosa* (Archbold 1971).

2. Typically bold and animated (Bannerman 1936). Once flushed, no longer seeks cover, but perches conspicuously (Gurney 1871), giving repeated Alarm-calls (see 3 in Voice), raising crest, twitching wings, and spreading tail when agitated (Someren 1956). Readily and noisily mobs predators, including snakes, cats, and other small carnivores (Bannerman 1936; Vincent 1946-9; McLachlan and Liversidge 1970). FLOCK BEHAVIOUR. Winter flocks described as loose (Grimes 1972; Vernon 1975), though 'coveys' of 8-10 flushing together (Gurney 1871) suggests greater cohesion. SONG-DISPLAY. Little information, and few sources clearly distinguish singing (see 1 in Voice) from calling. Sings typically from elevated perch, sometimes raising wings (see also Heterosexual Behaviour, below); sings all day, especially at dawn and dusk (Serle and Morel 1977). Starts just before dawn, and continues for *c.* 2 hrs before feeding; in Kenya, frequently sings or calls at night, especially when moonlit (Someren 1916, 1956), but in Gabon, not heard to sing at night (Brosset 1971*a*). ANTAGONISTIC BEHAVIOUR. Quarrels with conspecifics and other species when feeding, drinking, or bathing (e.g. Brooke 1975). Not markedly aggressive in breeding season, however, though nesting pair

said to resent close approach of conspecifics (Someren 1956). In Morocco, some competition for food with Blackbirds *Turdus merula* (Meise 1959). Up to mid-June, Morocco, groups of 3-5 frequently 'in battle' (Lynes 1925*b*), and chases during this period (Mathey-Dupraz 1926) perhaps antagonistic. Chasing through trees also reported by Bannerman (1936). HETERO-SEXUAL BEHAVIOUR. Little known. Established pairs evidently almost inseparable: e.g. when feeding young, search for food together and return to nest together. Between feeds, bird may fly up into tree and is at once joined by mate who is greeted with Greeting-call (see 2 in Voice) accompanied by half-raised wings (Meeting-ceremony); pair then sit side by side (Someren 1956). RELATIONS WITHIN FAMILY GROUP. Young brooded constantly for first few days, when ♂ often feeds brooding ♀. ♂ also broods, and both sexes feed young. At midday, when feeding sporadic, parents sit on edge of nest and shade young with outspread wings. (Someren 1956, 1958.) Adults recorded apparently swallowing faecal sacs at nest (Irvine and Irvine 1974). On leaving nest, young do not fly strongly, and hide in bushes for a few days before able to follow parents around (Someren 1956). ANTI-PREDATOR RESPONSES OF YOUNG. No information. PARENTAL ANTI-PREDATOR STRATEGIES. (1) Passive measures. When bird sitting on nest, mate usually mounts guard nearby, and calls if danger approaches (Brosset 1971*a*). (2) Active measures. Birds often demonstrate with Alarm-calls (see 3 in Voice) when human intruder near nest (Bannerman 1936) but may also be remarkably tolerant: one ♀ continued attending young in nest while intruder in full view close by (Someren 1958). See also introduction to part 2 (above). EKD

Voice. Rich repertoire freely used. Readily mimics great variety of other species (Reichenow 1904-5). Following account includes extralimital information. Spacing, sequence, and intonation of calls vary with race and locality (Someren 1956, which see for examples, mainly *tricolor*, East Africa) but no details for west Palearctic. Rather little information available about calls of adult, but some calls of fledged young (see below) presumably part of adult repertoire.

CALLS OF ADULTS. (1) Song. Fluting sonorous phrases of abrupt, somewhat clipped syllables (Etchécopar and Hüe 1967). In Morocco, markedly varied: commonly a deep-throated 'quoquodéh', last syllable stressed or higher pitched, mixed with other motifs; also 'teterih' somewhat reminiscent of, but more guttural than, Song Thrush *Turdus philomelos* or Blackbird *T. merula*; 2 birds sang identical phrase 'quoquí reröri quii' (Meise 1959, which see for variants). Other descriptions from Morocco include 'pwit pwit quiterà quiterà' with timbre resembling *T. merula* (Irby 1875), and 'huit huit huit hwitera hwitera' (Hartert 1902); also 'vigouro vigouro', 'wee-te-treeou', 'whit-ti-tre', and 'tree-ou tree-ou tree-ou tree-ou' (P A D Hollom). For other renderings of song from North Africa, see Dresser (1871-81) and Whitaker (1896). Recording of nominate *barbatus* (Fig I) suggests 'tu-TWEE-twur-tu-TWEE' (E K Dunn) or 'pik WEE qut tyitWEE', low (unaccented) units liquid and throaty, accented ones fluting; units delivered in staccato manner. Average interval between phrases 8 s (*n* = 7); in recording of *tricolor*, rendered 'TWEE-tu-tereu', interval *c.* 2·5 s

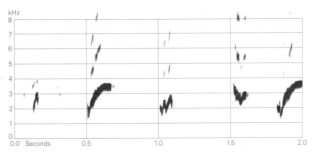

I L Grimes Ghana March 1974

(E K Dunn). For West Africa, Serle and Morel (1977) described short piping 'tooee-tee-tee-teeo', pleasant and ringing, often followed by some harsh chattering notes; see also Williams and Arlott (1980) for East Africa. (2) Greeting-call. A low 'cheedle cheedle cheedlelit', given by bird (*tricolor*) of unspecified sex on arrival of mate near nest (Someren 1956). In recording (Fig II), bird gives jumble of warbling ('cheedling') sounds, resembling song of Swallow *Hirundo rustica*; 3 overlapping notes near end (at 1·3–1·7 s) probably given by mate arriving at nest, whereupon cheedling song subsides (J Hall-Craggs). (3) Anger- and alarm-calls. In *tricolor* (Zaïre), a scolding chatter in presence of predator, attracting other conspecific birds to the spot (Vincent 1946–9)—'chit-chit' or 'gwit-gwit', repeated persistently (Mackworth-Praed and Grant 1963; McLachlan and Liversidge 1970). Fig III shows 'chit' sounds somewhat like alarm-call of *T. merula*, but less metallic (P J Sellar); compared with *T. merula*, longer, rather lower pitched, and possessing marked frequency modulation (J Hall-Craggs). In recording made when cat nearby, calls more liquid and, as alarm increased, run together to produce almost a bubbling sound (P J Sellar). In nominate *barbatus* (Tunisia), loud repeated shrill 'tit tit tit' (Erlanger 1899) probably this call. More complicated sequence in *tricolor* (Kenya) an oft-repeated 'tit-twit ti-twat-twut-twut tii-oo to-wit' (Someren 1956, which see for another sequence given by pair chivvying snake).

CALLS OF YOUNG. The following are calls of recently fledged captive *tricolor* (Kenya). Food-call 'tweet-tweet'. When alone, a piping whistle, apparently of distress. When nestling into handler, a near purr of contentment. When raptor sitting in tree, an angry scolding twitter. (Preston 1975.)   EKD

Breeding. SEASON. North-west Africa: eggs from mid-May to August (Heim de Balsac and Mayaud 1962). SITE. In tree or bush, in variety of positions. Nest: cup of grass stems and leaves, lined hair. Building: no information on role of sexes. EGGS. See Plate 81. Sub-elliptical, smooth and glossy; pinkish-white to white, heavily spotted and speckled with red-brown and lilac. Nominate *barbatus*; 23·8 × 16·8 mm (22·1–26·0 × 14·0–18·8), n = 35; calculated weight 3·44 g. *P. c. arsinoe*: 21·7 × 15·5 mm (20·3–23·0 × 14·0–16·4), n = 20; calculated weight 2·08 g (Schönwetter 1979). Clutch: 2–3(–4). INCUBATION. In Kenya, 12–13 days, almost exclusively by ♀ (Someren 1956); in Gabon, 14 days, by both sexes; eggs laid daily (Brosset 1971a). FLEDGING TO MATURITY. Fledging period 12–14 days. No further information.

Plumages (nominate *barbatus*). ADULT. Face fuscous-brown, almost black on lores, on feathering round eye, and near gape, merging into dark brown on crown, ear-coverts, cheeks, and throat, and gradually paler further down. Some white of feather-bases sometimes visible just behind ear-coverts, showing as faint pale spot or bar. Entire upperparts dark grey-brown, often slightly paler drab-brown on rump. Chest dark brownish-grey, breast and flanks paler brownish-grey, merging into pale drab on belly and white on vent and under tail-coverts; longer under tail-coverts with indistinct grey-brown shaft-streaks and often with faint pale yellow feather-sides. Tail dull black, central pair of feathers (t1) and outer webs of others slightly tinged drab-brown; tips of feathers faintly paler grey. Flight-feathers, tertials, and upper wing-coverts dark grey-brown, like upperparts or slightly darker; longest feather of bastard wing and inner webs of flight-feathers slightly paler drab-brown. Under wing-coverts and axillaries pale brown-grey, like flanks; small coverts along leading edge of wing greyish-white or white. Bleaching and wear have some influence: face slightly browner; upperparts duller and browner, less grey; chest slightly browner; underparts down from belly dirty white; tail, flight-feathers, tertials and upper wing-coverts have bleached and frayed grey-brown tips and outer fringes (Someren 1956; Harrison 1975). JUVENILE. Closely similar to adult; feathers softer and looser, especially on rump, vent, and under tail-coverts. Face less deeply coloured, duller grey-brown; upperparts slightly paler grey-brown, under tail-coverts uniform white (no yellow and grey-brown, unlike adult). Tail-feathers narrow, rather pointed, c. 8–12 mm wide (in adult, tip broadly rounded, c. 12–17 wide); p10 longer and with more broadly rounded tip than adult, 23–35 mm shorter than wing-tip (in adult, shorter and narrower, 32–44 mm shorter than wing-tip). FIRST ADULT. Indistinguishable from adult once relatively long and broad juvenile p10 replaced.

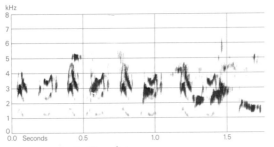

II M E W North Kenya August 1953

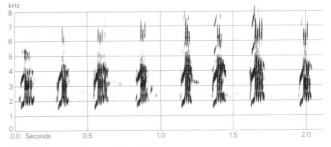

III L Grimes Ghana March 1974

Bare parts. ADULT, FIRST ADULT. Iris red-brown, brown, or dark brown. Bill horn-black or black. Mouth orange-yellow. Leg and foot dark horn-brown, horn-black, or black; soles yellow-horn or pale horn-brown. (BMNH, RMNH.) NESTLING. Bare skin dark flesh-brown on upperparts, slightly paler below; mouth yellowish-pink, gape yellow (Someren 1956). JUVENILE. Iris dark brown. Bill black-brown, gape-flanges yellow. (BMNH.)

Moults. ADULT POST-BREEDING. Complete; primaries descendant. The following based on only 10 moulting birds, Morocco, Algeria, and Egypt. Starts with p1 early August to mid-September; completed with p9-p10 early October to mid-November (BMNH, RMNH, ZMA). Body, upper wing-coverts, and tertials mainly old at primary moult score c. 20, largely new at 30-35; moult of tail and secondaries starts at score c. 20, completed at about same time as outer primaries (score 50) or slightly later. POST-JUVENILE. Complete; primaries descendant. Timing and sequence as in adult post-breeding. A few spring birds from north-west Africa had body, wing-coverts, and tail rather fresh, but flight-feathers (especially outer primaries) and greater upper primary coverts rather worn and pointed, more so than others of same date, and these partly worn birds perhaps 1st adults with some juvenile plumage retained.

Measurements. ADULT. Nominate *barbatus*. North-west Africa, all year; skins (BMNH, RMNH, ZMA). Bill (S) to skull, bill (N) to distal corner of nostril; exposed culmen on average c. 3·7 less than bill (S).

	♂		♀	
WING	103·4 (2·11; 17)	100-107	96·8 (1·96; 18)	93-100
TAIL	90·2 (2·75; 16)	87-95	85·6 (2·28; 14)	83-89
BILL (S)	20·2 (0·92; 15)	18·7-21·9	19·5 (0·74; 14)	18·5-21·1
BILL (N)	10·6 (0·52; 16)	9·8-11·6	10·1 (0·46; 14)	9·4-11·1
TARSUS	23·5 (0·76; 16)	22·5-25·1	22·8 (0·63; 13)	21·8-23·7

Sex differences significant, except bill (S).

P. b. arsinoe. Egypt and northern Sudan, all year; skins (BMNH, RMNH).

	♂		♀	
WING	93·6 (1·53; 21)	91-97	88·7 (1·84; 12)	86-91
TAIL	85·5 (2·09; 11)	82-89	79·2 (3·15; 8)	75-83
BILL (S)	19·1 (0·83; 10)	18·1-20·3	17·8 (0·59; 7)	17·1-18·8
BILL (N)	9·8 (0·39; 11)	9·2-10·4	9·3 (0·52; 7)	8·8-10·1
TARSUS	21·9 (0·85; 12)	20·6-23·1	21·8 (0·81; 7)	20·4-22·6

Sex differences significant for wing, tail, and bill (S).

Weights. *P. b. arsinoe*. Aïr (Niger), early February: ♂ 34-41 (4), ♀ 26. Ennedi (Chad), early April: ♂ 26, ♀ 28. (Niethammer 1955*b*).

No information on nominate *barbatus*. Populations similar in wing length to nominate *barbatus* occur in eastern South Africa; these generally 39-42 on average (Crowe *et al.* 1981, which see for details of many Afrotropical populations).

Structure. Wing short, broad, and rounded. 10 primaries: in adult, p6 longest, p5 and p7 0-1·5 shorter, p8 3-6; in nominate *barbatus*, p9 14-20 shorter, p10 37-44, p4 1-4, p3 4-7, p2 6-11, p1 10-14; in *arsinoe*, p9 12-17 shorter, p10 32-42, p4 1-4, p3 3-5, p2 6-9, p1 9-13. Juvenile p10 relatively longer and broader, 23-35 shorter than wing-tip. Outer web of (p4-)p5-p8 and (slightly) inner web of p6-p9 emarginated. Tertials very short, not exceeding p1. Tail rather long, tip rounded; 12 feathers: t6 7-12 shorter than t1-t2 in nominate *barbatus*, 3-8 shorter in *P. b. arsinoe*. Bill rather short and strong; culmen decurved, cutting edges slightly decurved (more strongly near bill tip). Nostrils oval, partly covered by membrane above.

Strong stiff bristles along base of upper mandible, finer ones near nostril and on chin. Leg and foot short and slender. Front toes partly fused at base. Middle toe with claw 19·6 (8) 19-21; outer toe with claw c. 74% of middle with claw, inner c. 68%, hind c. 79%. Claws short, rather slender, decurved.

Geographical variation. Rather slight in west Palearctic, more pronounced in Afrotropics; involves size (as expressed in wing length), contrast in colour between head and body, and (in Afrotropics) occurrence of fleshy wattle or warts round eye and colour of under tail-coverts; situation complex in Afrotropics, where several slightly differentiated populations overlap without interbreeding in some areas, but interbreed elsewhere; not clear which of these should be included in *P. barbatus*. In North Africa, *P. b. arsinoe* from Egypt and northern Sahel zone (Atar in Mauritania east to central and northern Sudan) smaller than nominate *barbatus* from north-west; face more extensively and deeper glossy black, reaching crown, ear-coverts, cheeks, and chin, merging into dark brown on hindcrown, sides of neck, and throat; upperparts as nominate *barbatus* (both strongly variable due to bleaching and wear), but rump slightly paler and greyer; small silvery-white spot or bar behind ear-coverts (faintly present in a few nominate *barbatus*) slightly larger and more often visible; chest extensively dark grey-brown, contrasting rather with pale brown flanks and cream-white (if fresh) or dirty white (if worn) remainder of underparts; under tail-coverts more rarely with some yellow at sides. Thus, *P. b. arsinoe* smaller than nominate *barbatus*, with head darker and body paler (in particular underparts from breast downwards), contrasting more with head. Yellow-vented Bulbul *P. xanthopygos* from Levant and Arabia closely similar to *arsinoe* in colour of head and upperparts, but size and paler grey-brown chest similar to nominate *barbatus*, and differs from both in fully bright yellow under tail-coverts and pale eye-ring; sometimes included in *barbatus* (e.g. by Vaurie 1959), but treated as separate species here following Rand (1958) and Hall and Moreau (1970); geographical ranges touch in northern Sinai and some hybrids known (Meinertzhagen 1930).

P. barbatus assemblage treated here as single species comprising 4 subspecies-groups, but some authors separate these as 4 species comprising a superspecies. Some of the subspecies-groups hybridize locally, though elsewhere may overlap without interbreeding. The 4 groups are: (1) *barbatus* group (nominate *barbatus*, *arsinoe*, *schoanus*, *inornatus*, *nigeriae*, and *goodi* in West and central Africa south to Gabon, Cameroon, Chad, northern Sudan, and central Ethiopia), with white under tail-coverts and uniform mantle and chest; (2) *somaliensis* (Harar area of Ethiopia and northern Somalia), with white under tail-coverts, pronounced white spot behind ear-coverts, and scaly mantle and chest; (3) *dodsoni* (eastern Somalia, south-east Ethiopia, and eastern Kenya) with scaly mantle and chest and white ear-spot (like *somaliensis*), but with yellow under tail-coverts; (4) *tricolor* group (central and southern Afrotropics, except western and central part of southern third, south of *barbatus* and *dodsoni* groups), similar to *barbatus* group, but with yellow under tail-coverts. In arid south-west of Afrotropics, *P. barbatus* replaced by Red-eyed Bulbul *P. nigricans* (similar to *tricolor* group, but iris red instead of brown and with naked orange ring round eye), and in western Cape Province (South Africa) by Cape Bulbul *P. capensis* (mainly uniform dark brown with yellow under tail-coverts and naked whitish ring round eye).

P. barbatus considered to form superspecies with *P. nigricans*, *P. capensis*, and *P. xanthopygos*; perhaps also with some species from southern Asia (Hall and Moreau 1970). CSR

Pycnonotus cafer (Linnaeus, 1766) **Red-vented Bulbul**

FR. Bulbul culrouge   GE. Russbülbül

Widespread resident in Pakistan, India, Sri Lanka, Borneo, and western Yunnan (China). Successfully introduced into Fiji, Tonga, and Samoan and Hawaiian Islands, unsuccessfully to Australia and New Zealand (Long 1981). In Kuwait, probably bred 1983–5 and breeding confirmed 1986–7; 5 pairs; perhaps now established (C Pilcher).

Family BOMBYCILLIDAE waxwings, hypocoliuses

Medium-sized oscine passerines (suborder Passeres) in 2 sub-families: Bombycillinae (waxwings) and Hypocoliinae (single species, Grey Hypocolius *Hypocolius ampelinus*), both represented in west Palearctic.

2 further groups (both confined to New World) sometimes included as additional subfamilies—silky fly-catchers *Ptilogonys*, *Phainopepla*, and *Phainoptila* and Palmchat *Dulus dominicus*—but here treated as distinct (see Voous 1977). These and Bombycillinae similar to one another in skeletal characters (Arvey 1951) and undoubtedly more closely related to one another than to other Passeres; though they obviously diverged long ago—*Dulus* having markedly different egg-white proteins from others (Sibley 1970)—it is a matter of opinion whether all should be retained in single family or split into 3 families. DNA data, however, place all within Bombycillidae and point to close relationship with Cin-clidae (dippers), an enlarged Turdidae (chats, Old World flycatchers, thrushes), and an enlarged Sturnidae (see Mimidae, p. 542), these together forming superfamily 'Turdoidea' within the 'Muscicapae' (Sibley and Ahlquist 1985a). Relationships of *Hypocolius* not clear (see Hypocoliinae, p. 502). Solitaires *Myadestes* provisionally placed with silky flycatchers in family Ptilogonatidae by Voous (1977), though generally considered genuinely turdid.

Subfamily BOMBYCILLINAE waxwings

Comprises a single genus *Bombycilla* of 3 species confined to temperate and subarctic Northern Hemisphere: Wax-wing *B. garrulus* (northern Eurasia and North America), Cedar Waxwing *B. cedrorum* (North America), and Japanese Waxwing *B. japonica* (north-east Asia). Ar-boreal; insectivorous and frugivorous. Migratory and nomadic, with periodic irruptions. Only *B. garrulus* occurs in west Palearctic, breeding.

Body plump-looking; neck short. Sexes of similar size. Bill short, thick, and broad at base; slightly hooked and notched. Wing long and pointed; 10 primaries (p10 minute). Flight strong and direct; flycatching sallies from perch not uncommon. Tail short and square, 12 feathers; coverts elongated. Leg short but toes strong, middle and outer united at base; claws long. Front of tarsus scutellated. Often forages on ground, with hopping gait. Foot not used in feeding. Head-scratching by indirect (overwing) method. *B. cedrorum* recorded anting in captivity, using direct method (ant-application).

Plumage soft, dense, and silky; a delicate vinous-brown with paler yellow or whitish belly and contrasting velvet-black face-mask and throat-patch, partly bordered by white lines. Lores and narrow frontal band, including nostrils, covered with short, dense, plush-like feathers. Crest of short dense feathers present in all species. Flight-feathers of *B. garrulus* and *B. japonica* have contrasting white or yellow marks; those of *B. cedrorum* largely uniform. Red, drop-like, waxy appendages present on tips of secondaries in *B. garrulus* and *B. cedrorum* but these lacking in *B. japonicus* which just has tips of feathers themselves red. Tail grey-black with contrasting yellow or red tip. Sexes nearly alike. Juvenile plumage much duller; streaked, especially on underparts, and lacks black on throat; crest shorter. A single (complete) annual moult in adult; a partial autumn moult in juvenile. Nestling naked at hatching; mouth bright red with contrasting coloured patch on each side of palate.

For nearest relatives in Passeres, see Bombycillidae (above).

Bombycilla garrulus Waxwing

Du. Pestvogel   Fr. Jaseur de Bohême   Ge. Seidenschwanz
Ru. Свиристель   Sp. Ampelis europeo   Sw. Sidensvans   N. Am. Bohemian Waxwing

Lanius Garrulus Linnaeus, 1758

Polytypic. Nominate *garrulus* (Linnaeus, 1758), Fenno-Scandia to western Siberia. Extralimital: *centralasiae* Poliakov, 1915, central and eastern Siberia; *pallidiceps* Reichenow, 1908, North America.

Field characters. 18 cm; wing-span 32–35·5 cm. Size and general form similar to Starling *Sturnus vulgaris*. Medium-sized, vinaceous-brown passerine, unique in west Palearctic in having both bold crest on head and (usually) bright 'waxy' appendages on secondaries. At close range, black bib (of adult), contrasting grey rump and rich brown vent, and yellow terminal band on black tail obvious. Flight and social behaviour recall *S. vulgaris* but essentially arboreal. Call distinctive. Sexes closely similar; no seasonal variation. Juvenile separable.

ADULT MALE. Crest, head, back, wing-coverts, chest, and flanks all essentially vinaceous-brown, with overtones varying from chestnut on face to pink on crest, nape, chest, and flanks. Most obvious plumage features are: (a) narrow black forehead and eye-stripe; (b) white-edged black bib over chin and throat; (c) long, ash-grey rump, contrasting with brown back and (d) blackish tail with broad yellow tip; (e) warm sienna-brown vent, contrasting with buff or yellowish belly; (f) remarkably decorated pattern of wing feathers—primaries essentially black and secondaries grey-black, but primary coverts and secondaries have obvious white tips, secondaries have bright wax-red appendages on ends of all but 3 innermost feathers, and primary-tips have bright V-shaped margins (white on inner webs and yellow on outer). Underwing pale grey. ADULT FEMALE. Closely resembles ♂ but duller in direct comparison, with greyer back, less black bib, and both tail and wing markings less sharp. Red appendages on secondary-tips usually markedly smaller and fewer. JUVENILE. Basically brown like adult but lacks its bold contrasts and is thus essentially much drabber and less immaculate in appearance. Differs in: (a) much shorter crest; (b) diffuse, dull white supercilium and upper cheeks, making face look pale; (c) paler, brown-streaked throat and underbody; (d) duller rump, lacking grey tone; (e) duller vent, only ending in chestnut under tail-coverts and not forming bold warm patch; (f) undeveloped decorations to flight-feathers, with red appendages on secondaries very small and no white margins to inner webs of primaries. FIRST WINTER AND SUMMER. Head and throat pattern as adult but markings on primary-tips as juvenile. At all ages, bill and legs black.

Unmistakable at close range. Cedar Waxwing *B. cedrorum* of North America, a potential escape and vagrant, is smaller and has white vent and pale yellow belly, flight-feathers show only wax-red appendages to secondaries, other flight-feather decorations of *B. garrulus* being absent. Potential confusion with *S. vulgaris*, at middle and greater distance (particularly in flight), stems from similarity of basic form, flight action, and perched silhouette. Important to note that plumage tones of *B. garrulus* vary widely according to light intensity and background colour—can look as dark in silhouette as *S. vulgaris*. Flight silhouettes and actions strongly recall *S. vulgaris*, but shorter bill, fuller head (due to laid-back crest), and bulkier body combine to produce less angular form, while actions are somewhat slower, with less rapid wing-beats (even in migratory flight) and noticeably more floating glides and turns; capable of hovering. Wing-beats make rattling sound. Flocks fly in compact formations, again recalling *S. vulgaris*, but rarely comprise more than hundreds of birds. Agile in tree and bush foliage, with feeding actions and gait reminiscent of large tit *Parus* or crossbill *Loxia*. When feeding on ground, gait restricted to hops and shuffles, markedly less free than *S. vulgaris*. Often astonishingly tame and frequently 'lazy', sitting around in groups almost immobile. Best known for irregular eruptions from breeding range in years of food scarcity, during which bird forsakes its taiga home and may appear on berry-bearing plants almost anywhere.

Commonest call a feeble though distinctive high, sibilant trill, 'sirrrrr'; audible at close range from bird perched or in flight. Song essentially a variation of this.

Habitat. Breeds in west Palearctic in upper middle latitudes in subarctic and boreal zones up to 10°C July isotherm, stopping short of treeline, in belt of dense tall taiga, especially of spruce *Picea* and pine *Pinus*, sometimes mixed with broad-leaved species such as birch *Betula*. Occurs largely in lowlands and valley forests, but also in uplands, although apparently not in mountains. Prefers for breeding old stunted conifers festooned with hanging witch-hair lichen *Usnea*; dense forest interiors and fringes by peat swamps or dwarf heath are both acceptable sites (Bannerman 1953; Dementiev and Gladkov 1954a). On switching in autumn to diet of berries, often confronts choice between finding adequate supplies for winter in native forests or launching eruptive movements to alternative supply sources in temperate lands, seeking profuse crops of fruits of rowan *Sorbus*, rose *Rosa*, or other trees and shrubs, including introduced garden varieties. At this season occurs on roadsides, in parks

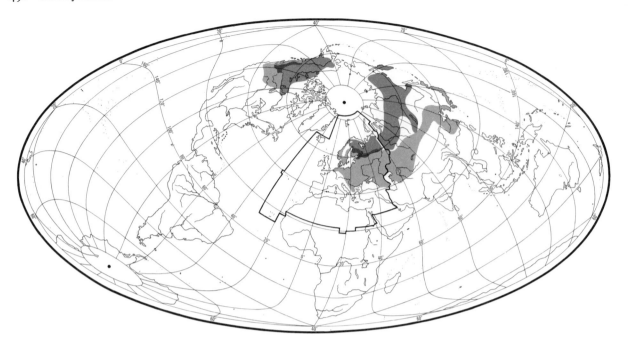

and gardens, along hedgerows, and wherever berries can be found, regardless of human presence, and abandoning any special attachment to conifers. Occurs in Nepal in mixed spruce and rhododendron forest at *c*. 3650 m (Ali and Ripley 1972).

Habitat in Nearctic appears identical with that in Palearctic. In Canada breeds in open coniferous or mixed forest, burnt areas, and muskegs (Godfrey 1966). In Alaska, dead stubs of small black spruces on wet boggy shrubby terrain were noted as attractive. In Mount McKinley National Park breeds in spruce on wooded slope (Gabrielson and Lincoln 1959; Murie 1963).

Although so markedly arboreal, hardly ever straying far away from trees or shrubs, appears less thoroughly attached to them than other arboreal species such as tits (Paridae) or treecreepers (Certhidae).

Distribution. Little information on range changes. Moves further south than mapped in irruption years (see Movements).

SWEDEN. Breeds irregularly further south (LR). USSR. Estonia: breeds occasionally (HV).

Accidental. Iceland, Faeroes, Ireland, France, Spain, Belgium, Albania, Greece, Turkey, Israel, Cyprus, Malta, Algeria.

Population. SWEDEN. About 500 pairs (Ulfstrand and Högstedt 1976), doubtfully accurate (LR). FINLAND. About 5000 pairs, numbers variable; some decline since 19th century (Merikallio 1958). USSR. Probably not rare (Dementiev and Gladkov 1954*b*).

Oldest ringed bird 12 years 7 months (Rydzewski 1978).

Movements. Partial migrant, often making eruptive movements. In northern Europe regularly overwinters within breeding area and also annual limited movements to southern Sweden and Denmark with recent extension to north-central Europe (Zink 1985); occurs almost annually in East Germany (Klafs and Stübs 1977; Rutschke 1983) but less often in West Germany, which suggests that invasion from further east than northern Scandinavia occurs in East Germany. Major eruptions not directly attributable to severe winters sometimes occur, with some evidence of 10-year cycle (Zink 1985). These eruptions do not necessarily coincide with movements of other eruptive species, and do not always coincide in the countries involved (Lack 1954*a*). A further general feature is that individual birds may travel great distances: e.g. bird ringed Poland in winter recovered the following winter in eastern Siberia, *c*. 5500 km away (Voous 1960).

Breeding populations (nominate *garrulus*) of northern Fenno-Scandia, and probably from further east, move both south-west to Britain and western Europe and also south to south-east to central and eastern Europe and there are occasional records of more westerly movement to Iceland (Timmermann 1938–49), the Faeroes, and eastern Greenland (American Ornithologists' Union 1957), and North Atlantic weather ships (Luttik and Wattel 1979). Few recoveries of birds ringed as nestlings: 2 ringed northern Fenno-Scandia recovered in same year in Yugoslavia and Czechoslovakia, another recovered in subsequent year near Leningrad, USSR (Zink 1985).

Invasions of Britain occur October–March, with maximum numbers in mid-winter, and occasionally extend to April, rarely to May, and 2 invasions in July. In most

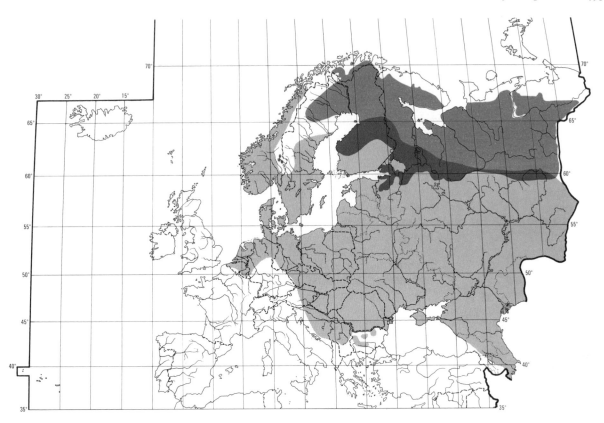

years only a few birds occur, but irregularly there are large numbers, mostly in eastern areas from Shetland to Kent; invasion has occurred more occasionally in western England and western Scotland, rarely in Wales and north-west Scotland. Scarcer in Ireland, occurring mostly in north and east, but not every year; 1965-6 notable with flocks of 300 (Witherby *et al.* 1938a; Baxter and Rintoul 1947; Gibb 1948; Cornwallis 1961; Ruttledge 1966; Everett 1967; British Ornithologists' Union 1971).

Best-documented eruption is that of 1965-6. In autumn 1965, acute imbalance between population size and food supply in Fenno-Scandia led to large-scale eruption (Cornwallis and Townsend 1968). Invasion of Britain began earlier than usual, with party of 13 in Argyll (Scotland) on 22 September; by mid-October, flocks of over 100 were at 3 areas in Scotland and one of 230 in Norfolk; birds occurred in the east from Shetland to Kent with visible diurnal movement down coast. Invasion and southward movement continued to mid-November followed by westward movement to Wales and Ireland. Gradual withdrawal January-March with stragglers through April to 24 May. (Cornwallis and Townsend 1968.) In Scotland, 1500 records covering 10 000 birds (Everett 1967). Cornwallis and Townsend (1968) also detailed scale of 1965-6 irruption throughout Europe with enough data to trace significant southward movements in north-west USSR, in September-October. In north-central areas, irruptions most spectacular in Britain, with

others in Ireland, Denmark, southern Sweden, and Poland, all October-November; further south similar irruptions occurred in West Germany, eastern France, and Czechoslovakia in November-December; smaller numbers present in southern France and Mediterranean coast December-February, and birds reached southern Italy and Sicily in January, Turkey and Greece in December. Major irruptions such as 1965-6 are associated with southward extension of normal breeding range into southern Fenno-Scandia (Svärdson 1957).

Invasions through Europe have been well documented (e.g. Schüz 1933, Bernis 1965). 1965-6 irruption involved 34 000 birds in Baden-Württemberg (West Germany), the largest invasion climaxing in mid-December after which very large flocks dispersed into smaller groups (Neub 1967). Ringing recoveries for that year can be interpreted as showing birds moving initially in a 'preferred direction', replaced later by random movements in search of food (Cornwallis and Townsend 1968). Data for Finland (Siivonen 1941), Netherlands (Van Oordt 1950), France (Mayaud 1941-5), West Germany (Zink 1969; Hölzinger 1972), East Germany (Creutz 1959b), Hungary (Keve 1950), and Italy south to Sicily (Moltoni 1969) supplement the survey for 1965-6 of Cornwallis and Townsend (1968). In Switzerland, Glutz von Blotzheim (1966) acquired massive data detailing all recoveries since 1903-4 irruption in western and central Europe and more detailed information on invasion progress in Switzerland in 1963-4 and 1965-6. Movements found to have 3 phases: (1) early widespread mass invasion, (2) regionally restricted immigration on smaller scale, (3) withdrawal characterized by scattered occasional birds in eastern and central areas (or by complete absence). Ringing recoveries in irruption years show widespread scatter of both headings and destination areas, birds moving from WSW through south to ESE; bird ringed south-west Finland in October recovered at Alma-Ata (Kazakhstan, USSR) in January, *c.* 4000 km ESE (Zink 1985). Birds ringed in Switzerland were recovered in Czechoslovakia, USSR, Rumania, Germany, Sweden, and France (Schifferli 1965). The 1970-1 invasion of Italy was on large scale for anywhere in southern Europe and extended south to Puglia with evidence that birds in this area had flown on heading south or west of south across Adriatic (Moltoni 1971); in autumn-winter of both 1974-5 and 1975-6, numerous occurrences in northern Italy from Liguria to Abruzzi with ♀♀ outnumbering ♂♂ and juveniles outnumbering adults (Moltoni and Brichetti 1976).

On Malta, recorded November 1903 (3), January 1913 (2), January 1948 (1), November 1965 (1), and January 1971 (1)—all years of large irruptions in Europe (Sultana and Gauci 1982). One record (5 birds) on Cyprus, January 1966, during massive European irruption (Flint and Stewart 1983). Also one for North Africa: quite large numbers near Blida (Algeria), late autumn 1841 (Heim de Balsac and Mayaud 1962).

Populations breeding between Fenno-Scandia and Urals also subject to eruptive movements but less well documented. Large numbers in some years arrive northern Poland mid-October to early November and depart March-April or even early May; mostly flocks of 'several score' but up to 400 and once 700 in Warsaw; fewer in mid-winter than at passage times (Tomiałojć 1976). In USSR, ringing has shown movements to Poland and Hungary (and from Poland to eastern Siberia mentioned above); movements have extended as far south as Crimea with northward return from mid-March, though some remain until April (Dementiev and Gladkov 1954a). Only accidental in Indian subcontinent; recorded January-March in north-west Pakistan, Kashmir, and Nepal (Ali and Ripley 1972). Noted in north-east Tibet, mid-March to mid-April (Vaurie 1972). In China, moves south to Hopeh and Shantung; more occasional west to Kansu and on lower Yangtse Kiang (Etchécopar and Hüe 1983). In Korea substantial invasions occurred in 1964-5, 1965-6, and 1967-8, mainly December-February; in severe winters birds move further south and remain until April, once to mid-May (Gore and Pyong-Oh 1971).

In North America, populations of *pallidiceps* breeding in Alaska and north-west Canada east to Hudson Bay move to winter throughout southern Canada east to Nova Scotia suggesting most fly ESE; other western birds move south to southern California, southern Arizona, central New Mexico, and Texas. Between these 2 routes, many move to central Ohio and central Indiana and more birds to the east to Massachusetts, Connecticut, and Pennsylvania. These movements vary in penetration and intensity from year to year and also in main direction of movements (between south and ESE). (Bent 1950; American Ornithologists' Union 1957.)   JHE

Food. In summer, mainly insects, especially mosquitoes (Culicidae); in winter, chiefly fruit, also buds and flowers. Change from insects to fruit (and back again) occurs gradually according to weather and abundance of insects (Greschik 1934; Hölzinger 1972). Pátkai (1966) noted birds in April flycatching when sunny but returning to fruit when sky became overcast. Insects caught mainly by flycatching from tops of trees, shrubs, telegraph poles, etc. (Schüz 1933; Baxter and Rintoul 1947); flights are fluttering and interspersed with glides (Ivanov 1969). Rokitansky (1959) described 3 methods: steep rise, returning to same or nearby tree; flight through gentle arc, gliding to distant tree; fluttering bat-like flight-path. In winter may hunt insects in crevices (Baxter and Rintoul 1937). Fruit taken mainly from tree but sometimes from ground below it. Berries picked with slight stooping motion, held briefly in bill and swallowed with quick toss of head. Sometimes bird clings to underside of branch and eats berry from below (Baxter and Rintoul 1937). Occasionally taken in hovering flight (Creutz 1974). Where size permits (e.g. rowan *Sorbus aucuparia*, hawthorn *Crataegus*), berry swallowed whole, but with others (e.g. *Cotoneaster*, *Viburnum*) seed and pulp taken and skin discarded or left attached to branch (Baxter and Rintoul 1937; Gibb 1948). In winter in Bayern (West Germany) feeds mainly 11.00-14.00 hrs (Reichholf-Riehm 1972). Usually makes short visits to food, feeding intensively. May consume 2-3 times own body weight of berries per day (Gibb and Gibb 1951; Bagnall Oakeley 1961; Borowski 1966; Glutz von Blotzheim 1966) with concomitant high production of faeces (Gibb 1948; Borowski 1966). These described as being like a string of pearls—undigested seeds with a gelatinous covering (Schoennagel 1972). One record from Denver (USA) of forcible regurgitation of large quantity of clear fluid containing undigested seeds (Bergtold 1917). Between meals, large quantities of water or snow taken (Baxter and Rintoul 1937; Gibb 1948; Glutz von Blotzheim 1966). Birds also noted flycatching snowflakes (Warga 1939b; Pátkai 1966). Willow, poplar, and cherry bark taken by birds may be roughage rather than food (Thielcke 1965; Bezzel 1966). Stones and earth taken possibly to aid digestion (Thielcke 1965). Birds also recorded pecking at white-washed walls possibly for salt, taking chalk from horse manure (Glutz von Blotzheim 1966), and pecking at other birds' dried faeces (Schüz 1934).

On breeding grounds, food predominantly dipteran flies, in particular mosquitoes and midges (Culicidae,

Chironomidae, Ceratopogonidae, Bibionidae). Also may-flies (Ephemeroptera), stoneflies (Perlidae), and caddisflies (Phryganeidae) (Greschik 1934; Dementiev and Gladkov 1954*b*). In southern Kareliya (USSR), of 26 stomachs taken May–September, 30·7% contained fruit of cran-berry *Vaccinium oxycoccus*, 30·7% bilberry *V. myrtillus*, 19·2% rowan, 15·3% raspberry *Rubus idaeus*, 3·8% bog bilberry *V. uliginosum*, 7·7% leaf buds of aspen *Populus tremula*, and 3·8% flower buds of willow *Salix*; also 30·8% beetles (Curculionidae, Chrysomelidae, Elateridae, Cerambycidae), 7·7% spiders (Araneae), and 3·8% sawfly larvae (Tenthredinidae); predominance of berries due probably to abundant crops; insects taken mainly in June (Neufeldt 1961). Stomach of one ♂ from Yakutskaya ASSR (eastern USSR) contained crane-flies *Tipula*, grasshoppers *Tetrix*, and cranberries (Vorobiev 1963).

In winter, main food is berries of rowan and hawthorns *Crataegus monogyna*, *C. laevigata*, and *C. prunifolia*, but takes wide range of others including many Rosaceae: whitebeam *Sorbus aria*, wild service tree *S. torminalis*, roses *Rosa canina*, *R. arvensis*, bramble *Rubus fruticosus*, blackthorn *Prunus spinosa*, bird cherry *P. padus*, wild strawberry *Fragaria vesca*, apples *Malus*, pears *Pyrus*, plums, etc. *Prunus*, cotoneaster *Cotoneaster*, firethorns *Pyracantha*, amelanchiers *Amelanchier ovalis*, *A. canadensis*, medlar *Mespilus germanica*, stransvaesia *Stransvaesia davidiana*; also bilberry, etc. *Vaccinium*, bearberry *Arctostaphylos uva-ursi*, prickly heath *Pernettya mucronatta*, guelder rose *Viburnum opulus*, wayfaring tree *V. lantana*, honeysuckle *Lonicera periclymenum*, elder *Sambucus nigra*, snowberries *Symphoricarpos rivularis*, *S. racemosus*, hackberry *Celtis occidentalis*, sugarberry *C. australis*, Japanese pagoda tree *Sophora japonica*, crow-berry *Empetrum nigrum*, currants *Ribes*, mistletoes *Viscum album*, *Loranthus europaeus*, privets *Ligustrum vulgare*, *L. japonica*, barberries *Berberis*, buckthorn *Rhamnus catharticus*, alder buckthorn *Frangula alnus*, sea buckthorn *Hippophae rhamnoides*, juniper *Juniperus communis*, Vir-ginia creepers *Parthenocissus*, vine *Vitis vinifera*, wild vine *Ampelopsis hederacea*, holly *Ilex aquifolium*, ivy *Hedera helix*, Russian-olives *Elaeagnus*, spindle *Euonymus europaeus*, dogwood *Cornus sanguinea*, Cornelian cherry *C. mas*, asparagus *Asparagus officinale*, lily-of-the-valley *Convallaria majalis*, mulberries *Morus*, yew *Taxus baccata*, Japanese persimmon *Diospyros kaki*, date-plum *D. lotus*, ameisans *Ameisan*, and *Lycium*. Seeds also taken from alder *Alnus glutinosa*, birch *Betula pendula*, ash *Fraxinus excelsior*, lilac *Syringia vulgaris*, sycamore *Acer pseudo-platanus*, field maple *A. campestre*, black poplar *Populus nigra*, aspen, limes *Tilia*, tree of heaven *Ailanthus altissima*, spruces *Picea*, plane *Platanus hybrida*, sun-flowers *Helianthus*, and false acacia *Robinia pseudacacia*. In late winter and spring, buds are eaten from hawthorns, willows, rowan, elms *Ulmus*, beech *Fagus sylvatica*, ash, black poplar, pear, buckthorn, limes, pines *Pinus*, and chickweed *Stellaria media*. Also flowers of *Crocus*, *Fuschia*,

jasmine *Jasminum*, broom *Sarothamnus scoparius*, raspberry, primrose *Primula vulgaris*, alder, birches, willows, osier *Salix viminalis*, sycamore, poplars, aspen, cherries, mulberries, and potato *Solanus tuberosum*. Shoots and leaves of chickweed, larch *Larix decidua*, cedar *Thuja occidentalis*, limes, and false acacia, and sap from silver birch and maple are consumed, together with moss *Orthotrichum* and lichens *Physcia dubia*, *P. ascendens*, and *Xanthoria parietina*. Animal items taken in winter include snails (*Chondrula tridens*, *Theba carthusiana*), shield bug *Lecanium corni*, aphids (Aphidoidea), flies *Bibio*, and beetles (*Phoriden*, *Aphodius prodromus*, *A. contaminatus*, *Omalium rivulare*, and *Dorytomus longimanus*). In hard conditions also takes scraps from bird tables—bacon rind, bread, cold porridge, orange, banana, carrot. (Schuster 1930; Schüz 1933, 1934; Greschik 1934; Baxter and Rintoul 1937; Warga 1939*b*; Géroudet 1942; Gibb 1948; Keve 1950; Holmberg 1952; Dementiev and Gladkov 1954*b*; Dittberner and Dittberner 1959; Rokitansky 1959; Filipăscu 1964; Bezzel 1966; Dorka 1966; Glutz von Blotzheim 1966; Everett 1967; Hölzinger 1972; Creutz 1974; Tyrer and Moran 1977; Krechmar *et al.* 1978; Popov 1978.)

In Britain, main foods are berries of hawthorn (Baxter and Rintoul 1947; Gibb 1948) and cotoneaster (Everett 1967). In winter 1946–7, 75% of 2650 records were of birds eating hawthorn, 5·5% *Rosa*, and 5·3% cotoneaster (Gibb 1948) while in 1965–6 winter in Scotland, of 954 records, 40·8% came from birds eating cotoneaster, 11·1% hawthorn, 9·2% *Berberis*, 8·4% *Rosa*, 7·1% apples, 4·7% yew, 3·9% honeysuckle, 3·4% elder, 2·7% rowan, 2·3% bramble, 2·1% pears, 1·5% scraps, 1·2% holly, 0·3% *Pyrancantha*, 0·3% juniper, 0·2% plums, 0·2% privet, 0·2% currants, 0·1% ash, 0·1% elm, and 0·1% willow buds (Everett 1967). In northern and central Europe, rowan berries most important food (Schüz 1933; Corn-wallis 1961; Glutz von Blotzheim 1966). In 1970–1, Baden-Württemberg (West Germany), 34·4% of 26 373 records were of apples, pears, plums, and quinces, 18·1% rowan, 12·4% *Rosa*, 7·5% privet, 7·2% cotoneaster, 6·9% guelder rose, 3·5% sea buckthorn, and 2·5% hawthorn (Hölzinger 1972). In East Germany, 1970–1, 39·8% of 206 records were berries of rowan, 18·9% hawthorn, 16·5% *Rosa*, and 9·7% mistletoe (Creutz 1974). In 1965–6 in the Alps and Jura (Switzerland), birds took rowan almost exclusively; only when these were completely stripped did they take hawthorn and occasionally *Berberis* and blackthorn; in lowlands, guelder rose 36% of *c.* 470 records, cotoneaster 29·4%, apples and pears 21·1% (Glutz von Blotzheim 1966). In Hungary, main winter food is berries of *Loranthus europaeus* until February and then *Sophora japonica*. Below these in decreasing order of frequency come *Celtis occidentalis*, Virginia creeper, juniper, hawthorn, *Elaeagnus*, *Rosa*, privet, *Prunus*, snow-berries, buds of chickweed, flowers, and insects (Greschik 1934). In Volga-Kama region (USSR), 60% of stomachs contained rowan berries, 40% lily-of-the-valley, 6% *Rosa*, and 5% hawthorn (Popov 1978). Unusual records include birds drinking sap from maple in April near Berlin (East Germany) and from birch in March in Glasgow, Scotland (Dittberner and Dittberner 1959; Tyrer and Moran 1977).

Diet of young similarly mixed. In Scandinavia, 5-day old nestlings fed with insects and crowberries (Haftorn 1957). In USSR, nestlings given buds of bilberry, previous year's berries of cranberry and cowberry, beetles (Chrysomelidae), mosquitoes (Culicidae), flying ants (Formicidae), small dragonflies, caddisflies, and other insects (Novikov 1975). Fledglings in July seen taking berries of blue-berried honeysuckle *Lonicera caerulea* (Bergman 1935). PJE

Social pattern and behaviour. Poorly known, especially on breeding grounds. Following account includes material on North American *pallidiceps*. Better-known and behaviourally similar Cedar Waxwing *B. cedrorum* of North America studied by (e.g.) Putnam (1949).

1. Highly gregarious, especially outside breeding season. Compact flocks of varying size occur for migration, feeding, and roosting. Nomadic flocks often small, but sometimes of up to several hundred or even thousands (Bernard 1904; Tschusi zu Schmidhoffen 1905-10; Warga 1929, 1939a; Peters 1959; Glutz von Blotzheim 1966; Krauss 1972; Kolunen and Vikberg 1978; for North America, see (e.g.) Goodwin (1905), Lincoln (1917), and Rathbun (1920). Flocks usually larger in early winter, breaking up more by spring (Goodwin 1905; Bezzel 1966). In Hungary, winter flocks reported to contain up to *c*. 80% juveniles, and ♂♂ also apparently slightly more numerous than ♀♀ (Warga 1939a); see also Tricot (1965) for Belgium. One ♂ recorded returning to winter in Budapest after 5 years (Warga 1939a). Birds move about in small flocks (6-8) after arrival on breeding grounds (Wahlstedt 1970). In Finnmark (northern Norway), parties of usually 2-3 (once over 12) pairs recorded 'feeding and playing together' during nest-building, sometimes even later (Blair 1936; Bannerman 1953); see also Bonds (below). ♂♂ form flocks for feeding while ♀♀ incubating (L Raner). Family parties move about on breeding grounds after fledging, then aggregate into larger flocks during August (Stegmann 1936; Johansen 1952; Dementiev and Gladkov 1954b; Järvinen 1980). For post-breeding densities in taiga of Siberia, see Vartapetov (1984). Winter flocks will associate for feeding, occasionally for roosting, with thrushes *Turdus*, especially Fieldfare *T. pilaris* (Tschusi zu Schmidhoffen 1905-10, 1922; Keve 1952, Creutz 1974); see also part 2. BONDS. Little information, but nothing to suggest mating system other than monogamous. Pairs can be obvious in winter flocks (e.g. Pohl 1922); Mal'chevski and Pukinski (1983) recorded this in November-December, but considered pair-formation took place when birds closer to breeding grounds. In West Berlin, November-March, courtship (see part 2) involved birds of various ages including 1st-winter birds (J Herrmann). Age of first breeding not known, but some at least do not breed at 1 year as nomadic flocks of non-breeding immatures occur on breeding grounds (Blair 1936; Bannerman 1953; Leonovich 1977). Both sexes feed young—at one nest in southern Norway, about equally (Haftorn 1957; Bub *et al.* 1984). Family party apparently remains together for considerable time after fledging (Järvinen 1980). BREEDING DISPERSION. Solitary (Dementiev and Gladkov 1954b). Captive birds apparently defended only

immediate vicinity of nest (Meaden and Harrison 1965). No information on territoriality of wild birds, though these frequently reported to be almost 'colonial' when numerous: e.g. in northern Norway, 'little colony of 7 pairs', or up to dozen pairs 'in small area', though not as close together as *T. pilaris*; of 6 nests by Pasvik river, none more than *c*. 50 m apart (Blair 1936; Bannerman 1953). Such groups reported also in Canada (Anderson 1915). Data on densities from USSR only. In Pechora basin area (including northern Urals), 0·8-5·2 birds per km² in spruce *Picea* with green moss, 1·2-12·8 birds per km² in pine *Pinus* with sphagnum moss, 5·6-35·6 birds per km² in willow *Salix* (Estafiev 1981). In northern taiga between Ob' and Yenisey rivers, 7-9 birds per km² in preferred dense riverine taiga and burnt-over areas by pinewoods, 1-2 birds per km² in pinewoods and by raised bogs, 0·1-0·6 birds per km² in other habitats. Much less common overall in middle taiga belt, mainly in pinewood clearings and tall thickets (6-8 birds per km²), few in pines and mixed riverine woods; 0·1-0·4 birds per km² in most other habitats (see Vartapetov 1984). ROOSTING. Nocturnal and, outside breeding season, communal; no information for breeding grounds. Mostly in trees, including woods or plantations, also in hedge. In Niederösterreich (Austria), birds used oak *Quercus* still with some foliage, but avoided nearby pines (Tschusi zu Schmidhoffen 1905-10). In Vienna (Austria), up to 1000 birds used *Thuja* trees, December-January (Peters 1959). In Scotland, roosting on gables recorded (Baxter and Rintoul 1937). During 3 days of snow, birds roosted in window recesses of woodland hut (Naumann 1901). Hard frost may cause birds to clump together, e.g. recorded perching one on top of another in tree-fork (Kirikov 1974). In northern Kazakhstan (USSR), birds roosted on ground or on low branches under cover of matted vegetation (Krivitski 1962). In French Alps, *c*. 200 birds roosted only in highest branches of larches *Larix*, sometimes 8-10 in a row on one branch. Regular daily routine in January: arrived on feeding grounds (altitude 1500 m) at *c*. 08.30 hrs, leaving *c*. 16.30-17.30 hrs for roost 2-3 km away and at 1730 m (Besson 1966). In south-east Bayern (West Germany), December-January, birds fed in villages mostly 11.00-14.00 hrs, thereafter making roosting flight to riverine wood (Reichholf-Riehm 1972). Birds assembled at another Bayern roost from *c*. 16.00 hrs; left in morning before fully light (Bezzel 1966). In Savonlinna (Finland), late October, most flying activity from *c*. 20-40 min before sunrise (*c*. 07.30 hrs) to 08.15 hrs. Evening movement 16.00-16.30 hrs. Birds assembled in tall birches *Betula* before flying to roost in woods (Grenquist 1947). For further details of activity rhythms in Hungary, see Tschusi zu Schmidhoffen (1905-10) and Warga (1939a). Water-bathing apparently regular in winter (Warga 1939a); in Moscow (USSR), birds bathed in melt-water puddles then moved to warm ground (pipes underneath) to dry (Wunderlich and Vietinghoff-Scheel 1975). Sun-bathe and self-preen on high perches (Creutz 1974).

2. Feeding birds in winter well known to be fearless and easily approached: e.g. to *c*. 8 m (Besson 1966) or, with care, to *c*. 1-2 m (Creutz 1974). Excited or alarmed bird will adopt an erect posture with crest fully raised (Bannerman 1953). In French Alps, some members of feeding flock would adopt posture with neck and bill extended almost vertically when approached closely (Besson 1966); perhaps also expresses anxiety. At Vienna roost, birds tended to leave quietly at approach of observer despite dark (Peters 1959). Constant contact-calls (see 2a in Voice), nervous and restless behaviour with sleeking of feathers recorded prior to migration in early April (Dathe 1975). Paired birds on breeding grounds varied between wary and very approachable (Blair 1936); family parties

not shy but more restless than birds in winter (Stegmann 1936). Flocks of *pallidiceps* tend to hide from raptors (Cameron 1908) while nominate *garrulus* recorded fleeing from Sparrowhawk *Accipiter nisus* (Warga 1939*a*). 2 birds attacked by Merlin *Falco columbarius* flew straight up into the air with much chattering (Bannerman 1953: see 2e in Voice). In East Germany, Blackbird *Turdus merula* flying past quite often caused birds to flee (Creutz 1974); in North America, birds harried by American Robin *T. migratorius* flew short distance then resumed feeding (Goodwin 1905). FLOCK BEHAVIOUR. Study of captive birds showed them to be sociable, but to maintain individual-distance of *c.* 2·5–5 cm (Meaden and Harrison 1965); see, however, Roosting (above) and Heterosexual Behaviour (below). Flocks remain compact, members giving contact-calls; increase in calls (and longer calls than usual) heralds take-off (Creutz 1974; Bergmann and Helb 1982; Mal'chevski and Pukinski 1983; L Raner). Birds in trees often attract down flock or may cease activity to fly after passing flock (Warga 1939*a*). When disturbed, nearly all flock-members take off and circle with much calling until they coalesce, then all land in co-ordinated fashion. One flock of *c.* 2000 *pallidiceps* flew up *en masse*, circled, and then broke up into smaller flocks for departure (Rathbun 1920). At Vienna roost, birds arrived quietly and rather inconspicuously; came in fairly high, then plummeted into trees. Call 2a given only occasionally from roost and fresh arrivals elicited no calls (Peters 1959). In French Alps, birds changed trees frequently before settling to roost (Besson 1966). SONG-DISPLAY. On breeding grounds, ♂ sings (see 1 in Voice) from tree-top near nest (Bannerman 1953; L Raner). Birds of both sexes in winter flocks will sing from perch in chorus (Bub *et al.* 1984). In eastern England, February–March, one bird sometimes sang while perched for up to *c.* 10 min (Turner 1914). Quiet song also noted in East Germany, primarily on mild spring days, but in September and November–December as well (Creutz 1974). No special posture recorded for singing but throat movement noted in captive bird (Meaden 1964). In Novosibirsk (USSR), 55% of song given in morning, 33% later in day, 12% in evening, none at night (Podarueva 1979). ANTAGONISTIC BEHAVIOUR. (1) General. *B. g. pallidiceps* said to have a gentle disposition and to show no aggression towards conspecific birds (Goodwin 1905). Lack of reports of aggressive behaviour in wild nominate *garrulus* perhaps confirms this as typical, but the only detailed studies relate to captive birds. Latter show some upsurge of antagonistic behaviour (not apparent at other times) in May; sociability does not break down completely during nesting but ♂♂ apparently compete for attention of ♀♀ (Meaden and Harrison 1965). However, other captive ♂♂ showed marked territoriality, their aggression reaching peak during nest-building and laying (Herrmann 1985–6). (2) Threat and fighting. Some brief apparently antagonistic chases occurred between pairs in Hungary, mid- to late March (Warga 1939*a*). Most other information relates to captive birds. ♀ ruffling feathers close to pair in courtship-display (see below) threatened by ♀ of pair. ♂ will defend nest where ♀ incubating. Usually adopts Upright-threat posture close to intruder: body upright, plumage sleeked and crest depressed, so that bird appears tall and thin; dark throat-patch prominent as bill slightly raised. May then perform Threat-gaping with bill opened wide; this recorded when 1 bird lands by another's nest. Will also open and close bill repeatedly, often producing loud Bill-snapping while leaning towards opponent in Forward-threat posture. Any attack directed at head, but actual pecking probably does not occur (Meaden and Harrison 1965). Following threat, dominant bird may perform bouts of wing-raising with simultaneous tail-raising and swaying of fore-body (Herrmann

1985–6). Single *pallidiceps* in flock of *c.* 12 *B. cedrorum* neither elicited nor initiated any aggression (Griscom and Harper 1915). 2–3 birds in flock of *T. pilaris* recorded chasing tits *Parus* and Bullfinch *Pyrrhula pyrrhula* (Hennemann 1921). Sometimes chases away thrushes (Baxter and Rintoul 1937), but more likely to be chased by such species (Creutz 1974). HETEROSEXUAL BEHAVIOUR. (1) General. Elements of courtship behaviour frequently recorded in winter: e.g. in West Berlin on 20 days, 21 November–28 March (J Herrmann) (see also below). In northern Norway, courtship continued in one year to 3rd week of June (Bannerman 1953). On Malaya Sos'va river (Tyumen', USSR), pair-formation takes place from end of April (Dementiev and Gladkov 1954*b*). (2) Pair-bonding behaviour (including Courtship-feeding). If ♀ breaks away during bout of courtship, ♂ will pursue her in fast chase, both birds giving call 4a or, less frequently, call 2e, or ♂ may fly to tree-top to sing (Bannerman 1953). Apparently heterosexual chases (accompanied by call 2a) recorded in East Germany (Creutz 1974) and North America (Anderson 1909). In West Berlin, early February, 1 bird (perhaps ♂) of apparent pair repeatedly flew back and forth between top (where mate perched) and lower branches of tree with peculiar shallow wing-beats and with plumage ruffled (J Herrmann). In Leningrad region (USSR), ♂ of pair also recorded performing display-flight (no details) in mid-May (Mal'chevski and Pukinski 1983). Most detailed study of courtship behaviour done on captive birds. Courtship-posture (Hump-posture) antithesis of Upright-threat posture as body feathers ruffled and body axis horizontal. ♂ may hop towards ♀, call (see 4b in Voice) and display only a few cm from her (birds even touching) in Hump-posture (Figs A–B): tail depressed, feathers of lower

A   B

back, rump, and upper tail-coverts ruffled, simultaneously also those of belly and under tail-coverts, bird appearing much larger than usual and short-legged. More marked ruffling of feathers (to some degree those of mantle and breast also) in higher-intensity variant (Fig C, left) makes bird appear almost spherical and, from front, laterally flattened as wings held close to body. ♂ has fore-body pivoted slightly towards ♀, head turned slightly away. Crest raised almost vertically. As head

C

turned, closed tail bent laterally towards ♀. Responsive ♀ will ruffle feathers similarly, this then leading to silent and solemnly deliberate Gift-passing ceremony (Figs D-E). Birds may face

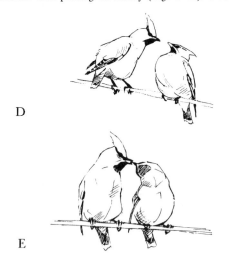

D

E

same or opposite directions. ♂ then has neck extended and places edible or inedible item (picked up before approaching ♀) into ♀'s open bill; ♀ slightly crouched with bill raised. With ♂ sometimes offering item to ♀ several times, performance appears highly ritualized (J Herrmann). Item may be passed back and forth up to 14 times. After moving apart, birds remain motionless with plumage somewhat less ruffled, but hop together again for repeat performance after a few seconds. (Meaden and Harrison 1965; see also Meaden 1964.) In another study of captive pair, birds gave call 2a faintly before performance and (slightly louder) while both holding items. Sometimes only 1 object passed back and forth, however. Description also refers to opening and simultaneous quivering of wings and to measured circular dance on tree trunk (Cube 1950). Drooping and quivering of wings, repeated raising and lowering of crest and ecstatic fluttering about each other while giving call 2a incessantly reported for wild birds by Bannerman (1953). Several authors described 2 birds keeping apart from rest of winter flock and Billing (in some cases without contact) accompanied by frequent raising of crest (e.g. Tschusi zu Schmidhoffen 1905-10, Warga 1939a, Gibb 1948, Creutz 1974). In wild *pallidiceps*, much symbolic Billing ('sham-feeding') occurred though plant bud passed at least once; bill and head motions of 2 birds resembled actions of Budgerigars *Melopsittacus undulatus* and performance accompanied by call 4a (Höhn 1951). According to Meaden and Harrison (1965), 'true courtship-feeding', which occurs fairly often during breeding season, not linked with any display. ♂ also feeds ♀ on nest during incubation (Dementiev and Gladkov 1954b) and (in captive birds) in early stages after hatching (Meaden 1970). As ♀ sometimes takes and eats item proffered by ♂ in courtship-display as described above, perhaps difficult to draw clear line between Courtship-feeding and Gift-passing ceremony. On breeding grounds, ♀ begs from ♂ by wing-quivering (L Raner). Other reports of interactions (in winter) involving ♀ being fed by ♂ referred to ♂ sidling up to ♀, jumping from one side of her to the other, prancing in circle and Billing with ♀, wing-rattling faintly, once hovering in front of ♀ to pass berry and sometimes feeding her twice in quick succession; call 2a given loudly by ♂ or both birds (see Berndt 1933, Gibb 1948). *B. garrulus* recorded accepting food from *T.*

pilaris (Wahlstedt 1970). (3) Mating. In Britain, winter, after Billing, excited ♂ hopped around on branches behind motionless ♀, but no copulation took place (Gibb 1948). ♂ *pallidiceps* repeatedly hovered over crouched ♀, brushing her rump, but did not copulate (Höhn 1951). Crouched ♀ in presumed soliciting-posture held wings away from body and tail slightly raised (Berndt 1933). Captive birds copulated rather infrequently, perhaps only following 1 out of 10 courtship displays (Meaden and Harrison 1965). Copulation recorded in flocks, late March to early April (Tschusi zu Schmidhoffen 1905-10); on breeding grounds in northern Norway, in one year as late as 17 June (Blair 1936). First copulation noted in one study only after ♂ had fed ♀ with insect rather than with berry (L Raner). (4) Nest-site selection. In captive birds, apparently by ♂ (see Herrmann 1985-6). Soon after arrival on breeding grounds, ♂ will pick up and toss aside twigs and moss, but in 3 cases, ♀ apparently stimulated to start building only after receiving insect food from ♂ (L Raner). According to Nilsson (1944), building ♀ accompanied by ♂; some evidence of building by both sexes, however (e.g. Haartman 1969a; see also Herrmann 1985-6). (5) Behaviour at nest. No information for wild birds, but see Herrmann (1985-6) for study in captivity. RELATIONS WITHIN FAMILY GROUP. At nest in southern Norway, young c. 5 days old still brooded for long periods by ♀. ♂ brought food at intervals of c. 30-120 min. Food regurgitated from crop, ♀ taking some from ♂'s mouth to feed young, but most feeding done by ♂ direct. Faecal sacs swallowed by adults who left together after young fed, ♀ returning with food after c. 4½-13 min (Haftorn 1957). In Västerbotten (Sweden), adult arriving at nest always called ('whistled': presumably contact-call) until mate arrived and only then were young fed (Hägglöf 1954). Young recorded refusing plant food after period of bad weather; insects presumably important in diet (L Raner). Young leave nest at 14-15 days or up to 3-4 days later in some cases, perhaps due to bad weather (Haartman 1969a; Järvinen 1980; L Raner). Apparently stay together as family for considerable time (Järvinen 1980). In northern Baykal area (eastern USSR), late July, young able to fly well still fed by parents (Stegmann 1936). Young *pallidiceps* said by Cameron (1908) to have been fed in flock up to end of January. For study in captivity, see Herrmann (1985-6). ANTI-PREDATOR RESPONSES OF YOUNG. Brood of 5 were bolt upright in nest at approach of man; all left when nest touched. Gave undescribed call when handled (Newton 1861). PARENTAL ANTI-PREDATOR STRATEGIES. (1) Passive measures. ♀ generally sits very tightly, even early in incubation; after flushing (when tree shaken or only when nest touched), may remain (also ♂) perched as close as c. 1 m while observer present. ♀ fed and brooded young only c. 10 min after being handled. Another pair continued to feed young in nest which had been removed and placed on ground. (Blair 1936; Nilsson 1944; Dementiev and Gladkov 1954b; Haftorn 1957; Wahlstedt 1970.) Some birds rather wary, disappearing when disturbed at nest (Bannerman 1953). (2) Active measures. In western Canada, birds nesting on lake islands flew c. 100 m towards approaching boat, then circled back, calling (see 2d in Voice) constantly (Anderson 1915). May fly about silently amongst trees near nest without approaching closely or give contact-alarm calls (Anderson 1909; Blair 1936; Bannerman 1953; Wahlstedt 1970).

(Figs by J P Busby: A and C-E from photographs by J Herrmann; B from photograph in Meaden and Harrison 1965.)

MGW

Voice. Repertoire rather poorly known apart from calls 1-2a. Not very vocal on breeding grounds (Blair 1936).

Wing-rattling at take-off and landing audible up to *c.* 30 m away and said to be as characteristic as most familiar call 2a (Blair 1936; Bannerman 1953). For Wing-rattling in heterosexual context and for Bill-snapping in threat, see Social Pattern and Behaviour. For extra sonagrams, see Bergmann and Helb (1982). No evidence that Nearctic *pallidiceps*, for which some material included below, differs significantly from nominate *garrulus*.

CALLS OF ADULTS. (1) Song. (a) Song of ♂ rather poorly developed. In winter, plaintive and subdued and given with bill closed. Sustained 'trill' not unlike Redpoll *Carduelis flammea* (clearly similar to, if not identical with, call 2a), mixed with drawn-out wheezing like Greenfinch *C. chloris* (Turner 1914) or twittering like Swallow *Hirundo rustica* (Tschusi zu Schmidhoffen 1905-10); see also Naumann (1901). Song at nest also described as a warbling of which basic component is a slightly varied and prolonged call 2a (Bannerman 1953; L Raner). Recording of captive (and injured) bird suggests series of tinkling tremolos (see call 2a) varying in length and pitch (some fruitier than others) and interspersed with a rather pebbly scraping sound preceded by 'tik' sub-units. Some harsher units rather like abbreviated Craking-call of Corncrake *Crex crex*. Length of pauses between units varies considerably. Fig I shows tremolo followed by 2 units with harsher, scraping quality: 1st of these 'tik tik tik scrape', 2nd 'tik tik scrape', but 'tik' sub-units compressed so that both units sound more like 'tik scrape' (J Hall-Craggs, M G Wilson). Harsher unit also described as smacking 'zi-tschret' (Bergmann and Helb 1982). (b) Song of ♀ apparently approaches ♂'s in quality. ♀♀ may sing in chorus with ♂♂ but quieter (Bub *et al.* 1984). (2) Contact-alarm calls. (a) Most familiar call. Clear, tinkling tremolo, typically high pitched, though some variation, also in length (P J Sellar: Fig II). Thin tinkling sounds not infrequently likened to part of song of *C. flammea* (see call 1), but more silvery (Bannerman 1953). Further descriptions and renderings. High-pitched 'sirr' or 'srii' or rapid 'dididi' (Bergmann and Helb 1982); 'tsee-ee-ee' rather like food-call of nestling passerine (Meaden and Harrison 1965); shrill 'shree-e' (Baxter and Rintoul 1937); 'zrrr' or 'zirrr' (Warga 1939*a*). Rendering 'zir-r-r-r' for *pallidiceps* (Cameron 1908) considered not to express decidedly sibilant quality of each syllable (Griscom and Harper 1915, which see also for comparison

with Cedar Waxwing *B. cedrorum*). See also (e.g.) Hammling (1914), Wright (1921), Davis (1932), and Garnett (1932). Given in flight and when perched and sometimes linked with heterosexual display (see Social Pattern and Behaviour). Higher-intensity variant shriller and given by bird alarmed by raptor or when nest threatened (Hutchinson 1932; Bannerman 1953; Meaden and Harrison 1965). (b) Low 'brrt brrrt' given by excited bird and accompanied by crest-raising (Rehberg 1970); perhaps a low-pitched variant of call 2a. (c) A 'zieh' like Redwing *Turdus iliacus* given by single birds in winter (Baxter and Rintoul 1937). (d) In fights with Starling *Sturnus vulgaris*, screeching sounds as if from 'miniature' Magpie *Pica pica* (Stokes 1947). Short series of high-pitched screaming notes from disturbed *pallidiceps* in flight (Anderson 1915) perhaps the same or a related call. Brevity of descriptions makes it difficult to judge whether such screeches are only higher-intensity variant of call 2a. (e) Short, loud chatter given during ♂-♀ chase and (presumably similar) chattering given by birds escaping from raptor (Bannerman 1953). (f) Sharp 'zick' like Grey Wagtail *Motacilla cinerea* given when highly excited at nest (L Raner). Perhaps the same or a related call is ticking like Robin *Erithacus rubecula* given most frequently (apart from call 2a) by small mid-March flock (Swaine 1932). (3) Low 'chup-chup' given by captive birds of both sexes during breeding season, especially when feeding young; audible only at close quarters (Meaden 1964). (4) Calls associated with ♂-♀ encounters. (a) Sharp 'tweek tweek' given by both birds during chase (Bannerman 1953). Soft squeaking sounds completely unlike call 2a given by *pallidiceps* during courtship-display (Höhn 1951) probably the same or closely related. (b) Soft 'zutt' closely resembling call 3 given by captive ♂ prior to courtship-display (Meaden and Harrison 1965); soft fluting 'düdü' (Naumann 1901) perhaps the same. (5) Nest-site call. Not described, but given by captive birds within *c.* 60 cm of nest (Meaden 1964). (6) Other calls. (a) Churr like alarm-call of Whitethroat *Sylvia communis* followed by long wheezing sound as if cork being drawn (Gibb 1948); perhaps elements of song (see above). (b) Occasional 'peep' or 'wheep' given by members of small winter flock (Swaine 1932). (c) Soft whistling 'soo', not unlike 'seep' of *E. rubecula* given in winter, especially in late afternoon (Baxter and Rintoul 1937).

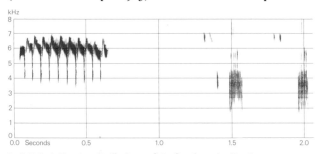

I   S Palmér/Sveriges Radio (1972-80)   Sweden   April 1963

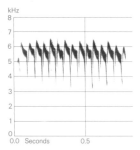

II   P J Sellar   Scotland   November 1970

CALLS OF YOUNG. Captive small nestlings give a quiet 'sirrit' or 'sirr't'; 'tirr' when older (Herrmann 1985-6). Fledged young give a soft 'sit sit' (L Raner).　MGW

Breeding. SEASON. Northern Scandinavia: laying normally begins mid-June, but from 1st week in early seasons, or even last few days of May (Blair 1936; Makatsch 1976; L Raner). SITE. In tree, 3-15 m above ground; in low pine or scrub, usually close to stem, but in taller tree, often out on branch (L Raner). Nest: cup with base of thin twigs 2-15 cm long, then grass and reindeer moss, with lining of dry grass and sometimes fine lichens (Blair 1936; L Raner). Nest: Building: by both sexes (L Raner). EGGS. See Plate 81. Sub-elliptical to oval, smooth and glossy; grey-blue to pale blue, sometimes buffish, lightly spotted black and grey. 23.8×17.5 mm $(21.0-28.3 \times 15.7-18.8)$, $n=140$; calculated weight 3.8 g (Schönwetter 1979). Clutch: 5-6 (4-7). Of 19 clutches, Sweden: 5 eggs, 15; 6, 4; mean 5.2 (L Raner). One brood. INCUBATION. 14-15 days. By ♀ only. YOUNG. Altricial and nidicolous. Cared for and fed by both parents. FLEDGING TO MATURITY. Fledging period 14-15 days, up to 17 days in bad weather (L Raner). No further information.

Plumages (nominate *garrulus*). Sexing only possible in known-age birds (i.e. adult or 1st adult) due to overlap in characters between (especially) adult ♀ and 1st adult ♂. ADULT MALE. Feather-tuft covering nostril and narrow line on forehead along base of upper mandible velvet-black, widening in front of eye and extending back to just below eye and over eye to above rear of ear-coverts. Forehead and side of crown above eye rufous-chestnut, merging into drab-brown with vinous-pink tinge on central crown and on elongated feathers of hindcrown. Hindneck and side of neck slightly paler drab-brown and with less vinous tinge; lower mantle, scapulars, and upper back drab-brown, merging into grey-brown lower back and ash-grey rump and upper tail-coverts. Triangular patch at base of lower mandible white, extending into narrow and largely hidden stripe running backwards to below eye and another running along sides of black throat. Upper cheek and ear-coverts pale vinous-drab, patch at rear of lower cheek rufous-cinnamon. Chin and throat deep black, often sharply divided from vinous-drab of chest, side of breast, and upper flanks. Central belly, vent, thighs, and lower flanks pale drab-grey or pale brown-grey, sometimes with some greyish-white wash; under tail-coverts contrastingly deep rufous-chestnut. Tail dark grey, gradually darkening towards black distally; tips of all feathers contrastingly bright yellow; shafts near tips sometimes red, occasionally extending laterally into narrow glossy waxy plates. Flight-feathers black, merging into pale grey on base of inner webs; outer webs of inner and middle primaries (about p1-p6) with bright yellow margin at tip, joining narrowing white margin to tip of inner web (except on p1) to form contrasting pale V on each feather-tip; V reduced and yellow paler or white on (p6)p7-p8, virtually absent on p9. Secondaries dark grey, tips of outer webs with oblong white patch, subterminally bordered by blackish-grey; shafts with bright red waxy oblong appendages at tips, projecting beyond vane of outer web; tertials (s8-s10) uniform drab-brown, s8 usually with tiny red waxy appendage at tip. Greater upper primary coverts dull black with bold and contrasting white tips; remaining upper wing-coverts and bastard wing drab-brown, like upperparts. Under wing-coverts and axillaries light grey, longest coverts tinged drab on tips, shorter ones along carpal edge spotted dull black. ADULT FEMALE. Similar to adult ♂. Main distinguishing characters (in order of decreasing importance) as follows. (1) Length of longest appendage on secondaries (usually on s6-s7). In sample from Yugoslavia, ♂ 5-10 mm (mainly 7-9), ♀ 5-8 (mainly 5-6) ($n=48$) (Cvitanić 1962); in Netherlands, ♂ 6.5-9, ♀ 5-7 ($n=20$) (RMNH, ZMA). (2) Width of yellow tips to tail-feathers, measured at shaft (excluding shaft itself). In Yugoslavia, width on t1 in ♂ 6-10 (mainly 6-8), in ♀ 4-7 (mainly 5-6) (Cvitanić 1962); in Netherlands, width on t6 in ♂ 7-10, in ♀ 5.5-7.5 (RMNH, ZMA); on t1, ♂ 5.5-8.5, ♀ 4-6; on t5, ♂ 7-11, ♀ 5-8 (Svensson 1984a). (3) Extent of pale V on primary-tips. In ♂, pale V usually present on p2-p7(-p8), white margin on tip of inner web broad and extending 6-11 mm from shaft on p4-p5; in ♀, V present on p3-p6 (but sometimes p2-p8, as in ♂), white margin on inner web narrower and extending only 4-8 mm along tip on p4-p5. (4) Number of red waxy appendages on secondaries. See Table A. (5) Presence of red shafts and appendages on tips of tail-feathers. In Hungary, 45% of 190 ♂♂ showed red on tail-tips, 12% of 196 ♀♀ (Warga 1939a, b). (6) Depth of yellow of tail-tip. In Hungary, tip deep yellow in 25 of 27 ♂♂, but in only 3 of 13 ♀♀; in remainder pale yellow (Warga 1939a); depth of yellow of tail- and primary-tips depends partly on age (Bub 1963). (7) Length and number of elongated crest feathers. Longer and more numerous in ♂ (see Structure). (8) Colour and contrast of black throat-patch. Patch glossy in 23 of 27 ♂♂ and in 5 of 13 ♀♀, duller and partly tinged drab or grey in remainder; patch sharply demarcated in 19 of 27 ♂♂ and in 6 of 13 ♀♀ (Warga 1939a). (9) Colour of forehead and under tail-coverts. On average, ♂ deeper chestnut, ♀ more rufous. NESTLING. No information. JUVENILE. Entire upperparts, upper secondary coverts, and much of sides of head and neck pale grey-brown or olive-brown (in adult, more saturated drab-brown with vinous crown and grey rump); feathers of forehead with whitish bases, central rump faintly streaked paler grey. Narrow line along forehead, lores, and over eye dull black, not as well-defined as in adult; small whitish patch behind eye; cheeks pale olive-grey, almost white. Chin and throat dull white, chin with indistinct dark stripe at side. Chest, sides of breast, and flanks deep olive-grey, faintly streaked off-white or pale buff; belly and vent dirty white or buff-white with faint darker olive

Table A　Number of red waxy appendages on tips of secondaries and tertials of Waxwings *Bombycilla garrulus* of different ages and sexes. Figures are percentages for each age/sex class (+: less than 1%) (Warga 1939a, b; Cvitanić 1962; RMNH, ZMA).

No. of appendages	Adult ♂	Adult ♀	1st adult ♂	1st adult ♀
0	0	0	0	3
1	0	0	0	3
2	+	0	+	10
3	0	0	1	30
4	+	+	17	41
5	1	14	36	9
6	34	62	27	2
7	36	18	16	1
8	27	6	3	+
9	1	0	0	+
Sample size	229	225	887	702

shaft-streaks. Under tail-coverts pale chestnut or vinous-cinnamon. Tertials and upper primary coverts virtually as adult; tail and flight-feathers as 1st adult. FIRST ADULT. Like adult, but differs in pattern on primary-tips; also, many other slight differences—compare description of sex characters for 1st adult (below) with those for adult (above). In both sexes, pattern on primary-tips restricted to short yellow (middle primaries) or white (innermost and outermost primaries) margin on outer web, not extending to inner web and thus not forming pale V on feather-tip (unlike adult). Some adult ♀♀ have part of V on inner web rather narrow, short, or restricted to p3-p6, but pale margin of outer web never stops abruptly at shaft, as it does on juveniles; a very few juvenile ♂♂ show some white on tip of inner web of a few primaries, but this not connected with pale margin of outer web. Sexes separable as follows (characters listed in sequence of decreasing importance). (1) Width of yellow tail-band. In Yugoslavia, width on t1 in ♂ 4-11 mm (mainly 5-8), ♀ 3-8 (mainly 4-5) (*n*=122) (Cvitanić 1962); in Netherlands, width on t6 in ♂ 6·5 and over, ♀ 6·5 and less (*n*=64); also, division between yellow and black straight in ♂, often a shallow V in ♀ (RMNH, ZMA); width on t1 in ♂ 5-8, ♀ 2-5; width on t5 in ♂ 7-10·5, ♀ 3-6 (Svensson 1984a). (2) Length of longest appendage on secondaries. In Yugoslavia, in ♂ 2-6 mm (mainly 4-6), ♀ 0-6 (mainly 2-4); in Netherlands, ♂ 3-7 (mainly 3·5-5·5), ♀ 0-4 (mainly 1·5-3). (3) Colour and contrast of black throat-patch. Usually deep black and sharply demarcated in ♂, drab-grey on lower throat and merging into colour of chest in ♀. (4) Number of appendages on secondaries. See Table A. (5) Presence of red shafts or appendages on tail-feather-tips. In Hungary, 12% of 782 ♂♂ had some red, but only 0·6% of 621 ♀♀ (Warga 1939a, b). (6) Colour of tail-band. Deep yellow in about half of ♂♂ examined, mainly pale yellow in ♀♀ (Warga 1939a, RMNH, ZMA). (7) Length and number of elongated crest feathers. See Structure. (8) Colour of forehead and under tail-coverts. Deeper in ♂, as in adult.

Bare parts. ADULT. Iris dark brown. Bill black with blue-grey base. Leg and foot black. NESTLING. Mouth deep cherry-red with violet-blue patch on each side of palate and on lower mandible; tongue vinous-red. (Pycraft 1909.) JUVENILE. Iris brown. Bill black-brown with flesh-grey base. Leg and foot yellow-brown or flesh-brown. (Hartert 1910; RMNH, ZMA.)

Moults. ADULT POST-BREEDING. Complete; primaries descendant. August-November. Sometimes not completed before autumn migration: of 87 migrants through Baltic states of USSR, second half of October and November, 45 moulted flight-feathers and many also tail (Schüz 1933). Moult occasionally suspended in autumn, perhaps completed later in winter or spring (Svensson 1984a; Ginn and Melville 1983). However, of many examined Hungary, moult had been completed upon arrival late November and December (Warga 1939a); a few from Netherlands had some feathers of body old upon arrival, October (RMNH, ZMA). POST-JUVENILE. Partial: involves head, body, and upper wing-coverts; no tertials, greater primary coverts, tail, or flight-feathers. Starts August (Stegmann 1936) or later; plumage sometimes still fully juvenile mid-October (Schüz 1933) or in full moult late October (RMNH). Most birds examined from late October and first half of November, Netherlands, had scattered growing feathers (mainly on underparts), but no remaining juvenile body feathers, while a few still had some growing feathers in late November and early December (RMNH, ZMA).

Measurements. Fenno-Scandia, Netherlands, and Hungary, mainly autumn and winter; skins (RMNH, ZMA). Juvenile wing and tail refer to 1st adult (retained from juvenile plumage). Bill (S) to skull, bill (N) to distal corner of nostril; exposed culmen on average 7·2 less than bill (S).

	♂		♀	
WING AD	118·0 (3·25; 11)	114-125	118·1 (2·50; 9)	114-122
JUV	116·3 (2·52; 33)	113-123	116·0 (2·93; 31)	111-121
TAIL AD	62·4 (2·42; 11)	60-67	60·6 (2·17; 8)	57-64
JUV	59·5 (2·27; 32)	55-64	58·5 (2·22; 30)	55-63
BILL (S)	17·8 (0·89; 28)	16·8-19·5	17·8 (0·64; 24)	17·0-18·9
BILL (N)	7·5 (0·56; 25)	6·8-8·6	7·1 (0·34; 18)	6·6-7·7
TARSUS	20·8 (0·58; 25)	19·9-21·8	20·4 (0·58; 25)	19·4-21·3

Sex differences not significant.

Weights. (1) Norway, October-January (Haftorn 1971). (2) East Germany (Bub *et al.* 1984). (3) Netherlands, late October-February (RMNH, ZMA). Hungary, December-April: (4) adult, (5) 1st adult (Warga 1939a).

	♂		♀	
(1)	63·9 (— ; 13)	59-74	65·8 (— ; 14)	58-75
(2)	63·4 (6·88; 21)	50-75	61·4 (7·66; 16)	50-83
(3)	59·7 (6·38; 15)	52-68	62·6 (7·52; 14)	51-83
(4)	56·0 (— ; 27)	45-65	59·6 (— ; 13)	50-70
(5)	54·5 (— ; 38)	34-69	52·5 (— ; 22)	38-64

Yamal peninsula (USSR), summer: ♂ 55; ♀♀ 45, 47; unsexed 43, 45 (Danilov *et al.* 1984). Finland, autumn: 62·1 (9·14; 50) (Pulliainen *et al.* 1981). Norway, October-November: 65·2 (44) 50-80 (Haftorn 1971). Baltic states (USSR): October, 57·0 (273); November-December, 55·1 (375) (Schüz 1933). Hungary: 56·4 (5·63; 455) 34-72 (Warga 1939a). Belorussiya (USSR): ♂ 66 (9) 54-73, ♀ 69 (6) 66-73 (Fedyushin and Dolbik 1967). Maximum of 1st adult ♂♂, France: 85 (Goulliart *et al.* 1965). Exhausted birds, Netherlands: 42, 43, 46 (ZMA). Berlin: November-December, adult 62·2 (5) 54-67, 1st adult 58·1 (18) 49-65; January-March, adult 57·8 (36) 49-68, 1st adult 56·0 (76) 47-64 (J Herrmann).

Structure. Wing rather long, broad at base, tip pointed. 10 primaries: p8-p9 longest or either one 0-2 shorter than other, p7 3-8 shorter than longest, p6 11-15, p5 18-22, p4 24-31, p1 39-48. P10 reduced, narrow and pointed, 72-85 shorter than wing-tip, 15-20 shorter than longest upper primary coverts. Tips of secondaries with oblong glossy wax-like appendages, number varying with sex and age (see Table A); similar appendages sometimes on outermost tertials or on 1-10 tail-feathers. Tail short, slightly rounded; 12 feathers. Crown with dense crest of long and soft decurved feathers; length of longest feathers 49-70 in adult ♂, 40-56 in adult ♀, 40-60 in 1st adult ♂, 40-54 in 1st adult ♀ (Warga 1939a); 13-16 elongated feathers in adult ♂, 5-9 in adult ♀, 7-11 in 1st adult ♂, 7-10 in 1st adult ♀ (Bub 1963). Plumage of body soft and dense. Bill short and rather stubby, wide at base; straight, but tip of culmen decurved and gonys curved upwards; slight notch inside tip of upper mandible. Nostril oval, covered by tuft of rather soft bristles. Tarsus and toes short, but rather strong; front of tarsus and uppersurface of toes covered by large scutes. Middle toe with claw 21·5 (20) 20-23; outer toe with claw *c.* 70% of middle with claw, inner *c.* 67%, hind *c.* 70%. Claws short but strong, decurved.

Geographical variation. Slight, in general colour only. *B. g. centralasiae* from central and eastern Siberia slightly paler vinous-brown on mantle and scapulars, paler grey on rump and upper tail-coverts, and paler on belly and vent, cream-coloured rather than grey. However, nominate *garrulus* from northern Europe similarly pale in summer and many *centralasiae* wintering in China and Japan not separable from nominate *garrulus* (Hartert 1921-2); perhaps as many as 50% of *centralasiae* similar in colour to nominate *garrulus* (Hartert and Steinbacher 1932-8) and hence recognition of *centralasiae* doubtful. *B. g. pallidiceps* from North America also slightly paler than nominate *garrulus*, but mantle and scapulars dull drab-grey, not as vinous drab-brown as *centralasiae*, and flanks grey with slight olive tinge, less vinous-drab. CSR

Subfamily HYPOCOLIINAE hypocoliuses

Comprises a single species: Grey Hypocolius *Hypocolius ampelinus*, breeding Turkmeniya (USSR) to Iraq. Marginal breeder in west Palearctic.

Closely resembles Bombycillinae (waxwings) in many aspects of anatomy and habits, so far as known (see, e.g., Delacour and Amadon 1949), with rather similar short and stubby bill, black face-mask, thick crest, contrastingly patterned flight-feathers and tail, and soft plumage, though colour more subdued—grey and buff predominating and red absent. Tail much longer and wing shorter, with more rounded tip. ♀ duller than ♂, with no face-mask and less obvious wing and tail pattern. Juvenile plumage rather similar to ♀'s; not streaked as in Bombycillinae. For further information, see *H. ampelinus*.

Relationships to other Passeres, even Bombycillinae, still uncertain. According to Lowe (1947), *Hypocolius* shows chat-like anatomical features which place it with Turdidae but this not confirmed by what little is known of behaviour and general biology. No data yet from analysis of egg-white proteins or from DNA-DNA hybridization.

Hypocolius ampelinus Grey Hypocolius

PLATES 33 and 34
[between pages 544 and 545]

Du. Zijdestaart   Fr. Hypocolius gris   Ge. Seidenwürger
Ru. Свиристелевый сорокопут   Sp. Empelis gris   Sw. Grå Palmfågel

Hypocolius ampelinus Bonaparte, 1850

Monotypic

Field characters. 23 cm; wing-span 28-30 cm. Almost as long as but slighter than Great Grey Shrike *Lanius excubitor*, with tail proportionately longer and not graduated, and bill without hook. Unique, sleek, long-tailed passerine, with somewhat shrike-like appearance but with behaviour recalling both Waxwing *Bombycilla garrulus* and babbler *Turdoides*. Plumage essentially pale grey above and pale isabelline-buff below, with black primaries tipped white and tail tipped black. ♂ has black face mask extending to nape. Sexes dissimilar; no seasonal variation. Juvenile separable.

ADULT MALE. Crown dull isabelline, becoming grey at rear; narrow frontal band, lores, cheek-patch, and complete but narrower collar black; chin and throat pale isabelline; head markings thus convey shrike-like pattern but without typically sharp contrasts. Mantle, scapulars, wings, rump, upper tail-coverts, and tail dull to light blue-grey depending on light, with markings restricted to broad white tips to black primaries and broad black terminal band on upper- and underside of tail. Breast dull grey, washed isabelline; rest of underparts pale isabelline. Underwing cream. ADULT FEMALE. Somewhat smaller than ♂, and plumage more uniform, lacking black head markings and appearing pale grey-brown at distance, but isabelline face, throat, and vent obvious at close range and dark bill and eye stand out. Most of primaries grey-brown, with black restricted to subterminal bands behind dull white tips. Tail duller than ♂ but with similar black terminal band, though less obvious below. JUVENILE. Plumage entirely pale sandy-brown, though ♂ has dull black tips to tail-feathers. At all ages, bill dark black-horn; legs yellow-flesh.

Unmistakable when seen well, but in brief view may suggest shrike or babbler, with direct flight into dense cover particularly recalling babbler's escape behaviour. Tail-band and primary markings (of adult) diagnostic. Flight strong, direct or circling, without undulations of shrike, and often at considerable height; action whirring and laboured when climbing, fast-flapping over distance, and gliding in descent and before landing. Usually unobtrusive, keeping to thick palm scrub, etc., but if disturbed from there flies off for up to 800 m. Gait and carriage not precisely recorded but apparently adopts both upright and level postures; hops and clambers deliberately through dense cover; said to cock tail on landing though keeps it lowered when perched upright.

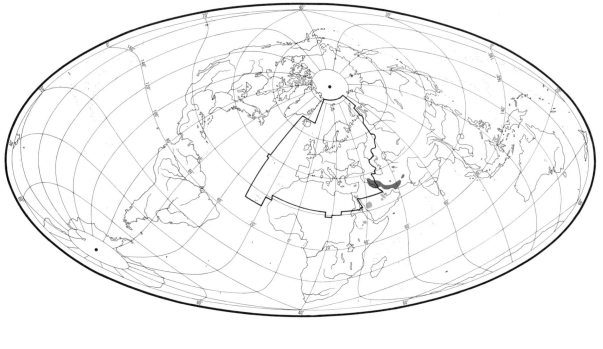

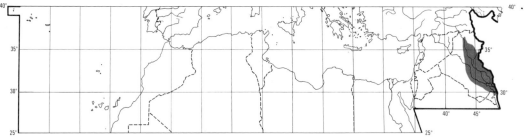

Apparently becomes tame in close association with human habitat, sitting amazingly tight in dense cover, but birds in natural habitat shy, flying off when disturbed.

Voice includes a variety of mewing or whistling sounds, and a fairly low-pitched and harsh monosyllable, 'chirr' or 'kirr'; also a higher-pitched but less harsh chirp when scolding.

Habitat. Subtropical and tropical, in more vegetated belts within arid lowlands, level or gently undulating; often in river valleys fringing desert or semi-desert with patchy or thin scrub, open broad-leaf scrub, groups of trees, irrigated areas, gardens, or palm groves (Harrison 1982). In Iraq, a bird of date palms, frequenting young palms and running up and down leaf-stems; also frequents low scrub jungle, almost impenetrable in places, of acacia, tamarisk, poplar, and willow, poplar being preferred. Usually keeps to outside of bushes when searching for insects; fond of sitting up conspicuously at top of a bush, which may be thorny and leafless; may roost in thorn bushes. Courtship flight rises to *c.* 30–50 m (Ticehurst *et al.* 1921-2). During winter in eastern Saudi Arabia

frequents thick palm scrub in oases and cultivated areas, often near settlements; in April, when presumably on migration, occurs occasionally in other types of scrub (G Bundy). In winter in India occurs occasionally in semi-desert and open deciduous scrub jungle; chiefly arboreal, feeding in trees and bushes but sometimes dropping to ground briefly to pick up an insect; perches low on branches (Ali and Ripley 1972). In Afghanistan, party in April in low tamarisk scrub among fields at *c.* 520 m perched in top of bushes but also settled on ground; in May on small wooded islands in river at *c.* 900 m (Paludan 1959*a*). Habitat vulnerable to loss by destruction of scrub in recent decades. Flies usually in lowest airspace.

Distribution and population. Little information on range changes and none on population trends.

IRAQ. Breeds widely but locally (Ticehurst *et al.* 1921-2; Allouse 1953). USSR. First nested 1979 (Peklo and Sopyev 1980).

Accidental. Egypt, Kuwait.

Movements. Most birds apparently short-distance migrants. Recorded on breeding grounds in winter only in Karun district of western Iran where small numbers occur; much more numerous there in summer (Ticehurst et al. 1921-2). Data available so far suggest it winters largely in Saudi Arabia, being uncommon but locally numerous in western areas from Hejaz mountains northwards, in central Arabia (Jennings 1981a), and in oases of the Eastern Province (G Bundy). Possibly winters for short periods on Bahrain, where recorded November-December (Hammonds 1984), and occurs irregularly mid-March to April in Kuwait. Rare further south and east in Arabian peninsula, with 2 March records in United Arab Emirates (Bundy and Warr 1980) and 3 in Oman (Dhofar and Masirah island), November and March (Gallagher and Woodcock 1980). Wanders irregularly to southern Afghanistan (Paludan 1959), and rare vagrant to Pakistan and north-west India, occurring November-April (Ali and Ripley 1972). Not proved to winter regularly in Africa: in northern Sudan, 3 records from Gebel Elba (Etchécopar and Hüe 1967; S M Goodman), also a more dubious one (Cave and Macdonald 1953); one old record from Ethiopia, at Massawa on the coast (Urban and Brown 1971).

Apparent migrants seen in eastern Saudi Arabia in April (G Bundy), and birds arrive at Baghdad (Iraq) in mid-April (Marchant 1963b); arrives Bushire (Iranian coast) in March (Cumming 1899), and recorded in Turkmeniya (USSR) May-October (Flint 1962).

JHE, DJB

Food. Fruit and some insects. Searches for food among trees, rarely descending to ground (Cumming 1899; Moore and Boswell 1957). Noticeably deliberate in feeding movements when perched on bush, stretching and balancing to reach berries with, at times, tail angled well downwards (P A D Hollom). Chews fruit, rejecting skin and stones, though small stones and pips swallowed and later excreted (Meinertzhagen 1949c, 1954). Will also fly down from perch like shrike Lanius apparently to take insects on ground (Vorobiev 1973) and recorded flying up to c. 3-4 m to catch insect, then returning to perch (Bunni and Siman 1979).

In Iraq, seen eating ripening dates, and gizzards of other birds contained berries of Lycium europaeum; some insects—chiefly beetles (Coleoptera)—also taken (Cumming 1899; Ticehurst et al. 1921-2). In Iraq, fruit in diet included Lycium, mulberry Morus, fig Ficus, and date (Ticehurst 1926), this apparently the preferred sequence, with many Morus alba taken in summer, according to Meinertzhagen (1954). Also in Iraq, berries, and occasionally leaves, of Lycium barbarum (Siman and Bunni 1978; Bunni and Siman 1979), and shoots of tamarisk Tamarix pentandra (S Marchant). In Saudi Arabia, winter, only berries recorded: Solanum incanum, Withania somnifera, Lycium arabicum, Ochradenus baccata (Resedaceae) (Meinertzhagen 1949c, 1954). In Tedzhen and Murgab valleys of southern Turkmeniya (USSR), main food is berries of Lycium turcomanicum, e.g. 4 stomachs from Murgab, May-June, were full of these (Flint 1962; Vorobiev 1973, 1980; Peklo and Sopyev 1980). Stomachs in India contained only berries, of Lanata aculeata, Salvadora persica, and Zizyphus (Ali and Ripley 1972).

In Iraq, regurgitated food of unknown composition given to small young. Older nestlings fed with berries, fruits, other vegetable matter, and larval and adult insects (Siman and Bunni 1976b). In Murgab valley, young given berries of L. turcomanicum (2-4 per visit; see also Social Pattern and Behaviour) and various insects—grasshoppers (Acrididae), ant-lions (Myrmeleontidae), and cicadas (Cicadoidea), larger insects being broken up before feeding to young (Peklo and Sopyev 1980; Sopyev 1981). Hand-reared birds ate wet clay (but not sand), presumably to aid digestion (Cumming 1899). PJE, MGW

Social pattern and behaviour. Studied in some detail Umm Al-Khanazeer Island in Tigris river, Baghdad, Iraq (Siman and Bunni 1976a, 1978; Bunni and Siman 1979). For recently discovered breeding population in Murgab valley (southern Turkmeniya, USSR), see Peklo and Sopyev (1980) and Sopyev (1981).

1. Gregarious, especially outside breeding season when forms mostly small flocks of 5-10 birds (Ticehurst et al. 1921-2; Moore and Boswell 1957; Paludan 1959; Ali and Ripley 1972). Larger flocks or gatherings occasionally reported: e.g. c. 40-50, mostly ♀♀ or juveniles, in Iraq, early October (S Marchant); up to 120 in some small areas of palm scrub in central or eastern Saudi Arabia, October-April, in years when more than usual cross Persian Gulf (Bundy 1985). Separate-sex parties recorded in Saudi Arabia in winter (Meinertzhagen 1949c); in Iraq, all-♂ and mixed-sex flocks noted in late March (Bunni and Siman 1979); in Iran, late April, flocks of 5-7 (or up to 12 for feeding) apparently contained few adult ♂♂ (P A D Hollom). In southern Turkmeniya, mid-May, flock of 16-18 perhaps actively migrating; other birds paired but frequently formed groups of 2-3 pairs (Flint 1962); later study during same period showed gathering of c. 35 (mostly paired birds), also a few single birds and threes (Peklo and Sopyev 1980); flock of 12 noted in late October (Flint 1962). Tendency of several breeding pairs to form flock also reported from Iraq by Ticehurst et al. (1921-2); general increase in numbers there, flocks containing many juveniles, from late July (Moore and Boswell 1957); juvenile-only flocks of 6-20 birds also recorded (Siman and Bunni 1976b). BONDS. Mating system evidently monogamous, but no details. Pair-formation apparently takes place after arrival on breeding grounds and bond strong once established, though no evidence of duration beyond one season. All nest-duties shared by both sexes. Family remains intact for indeterminate period after fledging (Cumming 1899; Bunni and Siman 1979). BREEDING DISPERSION. In Iraq, normally in small loose colonies with nests 4-15 m apart (Ticehurst et al. 1921-2; Siman and Bunni 1978a). On Umm Al-Khanazeer (c. 120 ha), no solitary pairs recorded. Largest colony of 5 pairs in area only c. 15 m across. Each pair (both sexes) defends small nest-site territory (nest and immediate vicinity). Food usually collected away from nesting area, adults flying further when feeding young (Siman and Bunni 1978). Nests sometimes more

dispersed in Iraq: 2 *c*. 50 m apart, 3rd *c*. 100 m from there (Marchant 1963*b*). June counts in Murgab valley indicated *c*. 40 pairs in 200 ha of dense riverine thickets (Peklo and Sopyev 1980; Sopyev 1981). Replacement nest usually sited close to that previously used: once, only 2·5 m away (Siman and Bunni 1978). Roosting. Nocturnal and, outside breeding season, communal. In trees and (thorn) bushes: in Iraq, 'scores' in old orchard, September–October of one year and large roosts recorded in late August of another year (Ticehurst *et al.* 1921–2; Marchant 1961, 1962). Either sex may roost on nest during incubation (Siman and Bunni 1976*a*).

2. After landing on top of bush, etc., may remain there a few seconds and adopt Alert-posture: body vertically erect, tail down, head and neck extended. Excited bird will also ruffle nape feathers to form small crest and both flick and rotate tail in manner of shrike *Lanius*. Commonly then dives into thick cover, sometimes after hearing alarm-call (see 2d in Voice), and waits for danger to pass (Ticehurst *et al.* 1921–2; Flint 1962; Sopyev 1981; Bundy 1985). In Iran, party of 5 birds *c*. 4 m up in tree allowed slow, quiet approach to *c*. 8 m (P A D Hollom). In Iraq, generally wary rather than shy; may fly about wildly, but tame when perched (S Marchant): e.g. continued to feed 'within a few feet' of observer and stayed even when 2 shot (Meinertzhagen 1954). Can be approached closely on occasions also in Saudi Arabia (Bundy 1985), though earlier study emphasized its extreme shyness there in open scrub habitat, birds flying at *c*. 300–400 m from man or not allowing approach closer than *c*. 100 m. In thick bush habitat or when surprised in isolated tree or bush, will remain hidden, silent and frozen. Sometimes retreats further into bush if closer approach attempted. 1 bird remained on perch and only called (see 2a–b in Voice) as observer moved hand closer; finally made rapid 'explosive' escape as typically when subjected to extreme disturbance—stones, shaking of bush, etc. (Meinertzhagen 1949*c*, 1954). Birds engaged in courtship will break off immediately if disturbed (Bunni and Siman 1979). Flushed birds usually fly up high and move far away (P A D Hollom, S Marchant). Flock Behaviour. In Iran, late April, flocks often flew about at *c*. 50 m, giving frequent flight-calls (P A D Hollom: see 3 in Voice); small winter flocks recorded circling thus for up to *c*. 20 min (Bundy 1985). May wheel several times like sparrows *Passer* when coming down to bush (Ticehurst *et al.* 1921–2). Birds in excitable August flocks followed one another into bushes like babblers *Turdoides* (Moore and Boswell 1957). Large early-October flock in Iraq clustered at top of bare bush, also crept about within it; tail movements when settling and perched, also behaviour when flushed, as described above (S Marchant). Feeding birds tend to keep fairly close together, rather like Waxwing *Bombycilla garrulus*; can be noisy, giving frequent Mew-calls (see 1 in Voice; Peklo and Sopyev 1980; P A D Hollom). Song-display. Nothing described; see, however, Voice and Heterosexual Behaviour (below). Antagonistic Behaviour. (1) General. Antagonism intense among ♂♂ after pair-formation, and territorial defence most marked early in season, especially during nest-building (Siman and Bunni 1978; Bunni and Siman 1979). (2) Threat and fighting. ♂ landing *c*. 50 cm from paired ♀ immediately chased away by her mate. Also defends nesting territory against conspecific birds and others such as Iraq Babbler *T. altirostris* and Graceful Prinia *Prinia gracilis*, though doves *Streptopelia* generally ignored. Incubating bird normally indifferent to these other species, also to White-cheeked Bulbul *Pycnonotus leucogenys*, but ♀ once left nest briefly to chase *Phylloscopus* warbler that had come within *c*. 10 cm of nest. Most defence thus undertaken by off-duty

bird, but mate will help to drive away conspecific intruder approaching close to nest. Attacks sometimes with wings and bill and gives call 2e, both participants calling thus in fights (Siman and Bunni 1976*a*, 1978; Bunni and Siman 1978, 1979). Heterosexual Behaviour. (1) General. ♂♂ apparently arrive on breeding grounds in Iraq a few days before ♀♀; in 2 years, first pairs noted 12 and 5 days after first ♀. Courtship-dance (see below) recorded early in season after pair-formation and during nest-building, mainly April–May; Courtship-feeding takes place during building and laying, Billing also a feature of early season, common during building, less so during laying. Some similarity in courtship to Cedar Waxwing *B. cedrorum* (Bunni and Siman 1979) and thus to *B. garrulus* (see that species, p. 497). In southern Turkmeniya, birds paired by mid-May (Flint 1962). (2) Pair-bonding behaviour (including Courtship-feeding). In early spring, unpaired ♂ will give repeated Mew-calls (see 1 in Voice) from high perch; such calls (perhaps a simple song) apparently important for pair-formation and unpaired ♀ will usually give Mew-call in response to ♂. Several ♂♂ sometimes fly about close to bushes when ♀♀ present; may fly up *c*. 3–4 m, then turn abruptly to land in same bush. Black-and-white wing-patch prominent in such flights which apparently serve to attract ♀. Courtship-display flight performed by pairs early in season, especially during nest-building: birds ascend to *c*. 20 m (or 30–50 m: Ticehurst *et al.* 1921–2) and, keeping 1–2 m apart, describe circles *c*. 15–20 m across. Kirr-call (see 3 in Voice) given loudly during performance lasting *c*. 2–5 min. Once landed afterwards *c*. 5 m from nest on bush (Bunni and Siman 1978, 1979). Billing takes place near nest and usually after ♂ returns from chasing intruder. Once proceeded as follows: ♂ perched, bill open, *c*. 60 cm from ♀ in (incomplete) nest; when ♂ approached to *c*. 30 cm, ♀ hopped towards him and inserted bill into his, birds then Billed (Bill-rubbed) for a few seconds before ♂ moved to nest and ♀ away (see also subsection 4, below). More complex ♂–♀ interaction is Courtship-dance in which ♂ (usually with food in bill) Hop-circles ♀, describing vertical circles *c*. 1–2 m across. While Hop-circling, ♂ frequently shakes one or both slightly drooped wings (perhaps more of a shivering or trembling movement) and sometimes spreads tail slightly; ♂ usually Hop-circles 2–4 times; no calls noted. ♂ coming close (*c*. 15 cm) to ♀ may half-open wings (more on side nearer ♀) and circle ♀ twice in *c*. 2 min. ♂ will eventually Hop-sidle up to ♀ until touching, birds parallel and facing same direction. ♀ then assumes Begging-posture with bill raised vertically and breast-feathers of both birds ruffled where they touch. ♂ then turns head to present food to ♀ who normally accepts it, then Hop-circles ♂, though without spreading wings or tail. Whole performance lasts 2–5 min, ending with food being eaten by either bird or dropped. ♀ may perhaps pass food back to ♂ (see *B. garrulus*, p. 498). During laying (but not subsequently), ♂ feeds ♀ on nest, e.g. twice in *c*. 25 min; ♀ extends neck up to ♂ (on nest-rim) with open bill; ♂ flies off giving Kirr-call. (Bunni and Siman 1979.) (3) Mating. Rarely recorded in Iraq study; probably takes place within thick bushes. Once, when ♀ on horizontal branch *c*. 1 m above ground, ♂ Hop-circled her (circles *c*. 1 m across, but not clear whether horizontal or vertical), gave loud Mew-calls, also wing-shaking and tail-spreading. ♂ then ceased calling, mounted ♀, both facing same direction. Balanced with slight opening of wings for *c*. 3–4 s on ♀'s back, then moved sideways and down for cloacal contact, ♀ remaining quiet and motionless. After dismounting, ♂ resumed Circle-dance, birds copulating 4 times in *c*. 3 min. ♂ finally moved to perch *c*. 3 m from ♀ who shook feathers, then started to feed (Bunni and Siman 1979). (4) Nest-site selection and behaviour at nest. No information on selection process. During building and laying, pair maintain close contact. Birds often collect material together, spending *c*. 30 s to 6 min away from nest. Give low, continuous Mew-calls while collecting material or when they meet near nest (Siman and Bunni 1976*a*, 1978; Bunni and Siman 1978). May fly in high together and call (probably Kirr-call: Bunni and Siman 1978), landing directly near nest (Marchant 1963*b*). 1 bird (more often ♂) may sit in nest, take material from mate and add it to structure, but most often change-over takes place and is accompanied by faint and continuous Mew-calls, incoming bird then adding material after taking over. Kirr-call given on leaving nest. Most building done in early morning and nest completed in *c*. 4 days normally before laying, though material occasionally added after start of laying. ♀ will sit on nest on day before laying which normally takes place *c*. 05.00–08.00 hrs (Siman and Bunni 1976*a*, 1978; Bunni and Siman 1979). ♀ at or on nest during laying and ♂ also tends to stay within *c*. 5 m, birds maintaining contact with Kirr-call; ♂ may also pay brief visits to nest and sometimes accompanies ♀ off nest, when Billing or Courtship-feeding occur. Incubating bird may pant with open bill and move tail up and down; will rise to rearrange eggs, also occasionally to take small insect or plant food within reach. Off-duty bird usually perched *c*. 2–7 m from nest: self-preens, feeds, rests, and guards nest. On approach for nest-relief gives Kirr-call (1–3 times); sitting bird makes 'shaking movements', departs and also gives Kirr-call from nearby branch, then flies off, mate moving in to take over. If sitting bird fails to respond immediately, incoming bird will return to perch and wait for mate to leave. Faint Mew-calls sometimes given as birds pass for change-over (as earlier during building). Usually follow same route to and from nest, ♀ especially moving rather furtively. Off-duty bird moving away from regular perch tends to give Kirr-call, same call also announcing change-over on return (Siman and Bunni 1976*a*). RELATIONS WITHIN FAMILY GROUP. When hatching imminent, adult excited, restless, rising and settling, looking down and turning eggs. Hatching takes place over up to 73 hrs. Eggshells removed (Siman and Bunni 1976*a*). Young brooded by both parents (Cumming 1899), for *c*. 5–7 days (Siman and Bunni 1976*b*). In Murgab valley, late June, young 2–4 days old not brooded during period of most intense feeding 05.00–07.30 hrs, but then brooded for period of 3–32 min by each parent after feeds, 07.30–10.30 hrs. Nearly always 1 parent perched on nest-rim 10.30–18.00 hrs to shade young from sun (Peklo and Sopyev 1980; Sopyev 1981). Eyes of young almost open at 4 days (Siman and Bunni 1976*b*). Adult will give Kirr-call when feeding young, but normally main Feeding-call (see 4 in Voice) given in approach to nest (occasionally accompanied by jerky tail-twisting) and continued until young gape to receive food (Bunni and Siman 1978) which (for small young) parent regurgitates (Siman and Bunni 1976*b*). In Murgab valley, young given only berries during hottest time of day: perched near nest or on rim, adult mandibulates berries, thus squeezing juice and fleshy pulp into mouth of nestling; remains dropped into nest and/or eaten (later) by adult (Peklo and Sopyev 1980; Sopyev 1981). Faecal sacs swallowed by parents. Young move on to nest-rim and eventually leave nest at *c*. 13–14 days (Siman and Bunni 1976*b*). ANTI-PREDATOR RESPONSES OF YOUNG. Immediately freeze on hearing adult call 2d. Give food-call when handled (Bunni and Siman 1978). PARENTAL ANTI-PREDATOR STRATEGIES. (1) Passive measures. Disturbed birds may desert during building and especially laying (Sharpe 1886; Siman and Bunni 1976*a*), perhaps even destroying their own nest (Marchant 1963*b*). Incubating bird may stand in nest when alarmed, crest raised and bill open,

looking about. ♂ once sat tight when man only *c*. 1 m from nest; often allows approach to *c*. 3–4 m, then leaves quickly and quietly, hopping *c*. 1–2 m before flying away low. Incubating bird crouched on nest when Magpie *Pica pica* near. Showed no reaction to Scops Owl *Otus scops c*. 8 m from nest, though owl being mobbed by bulbuls and babblers (Siman and Bunni 1976*a*). (2) Active measures: against birds. *T. altirostris* and Carrion Crow *Corvus corone* elicit a variety of alarm-calls (Bunni and Siman 1978: see 2a–c in Voice). (3) Active measures: against man. Off-duty bird will give alarm-calls when man *c*. 12 m from nest and bird leaving at closer approach (see above) will fly to bush and call similarly or join mate (sometimes other conspecific birds) in flying above intruder or hopping about excitedly in tree or bush, crest raised and cheek feathers ruffled (Siman and Bunni 1976*a*). Call 2d given apparently as warning to young, call 2e when threat more acute; if young handled, parent will perform distraction-lure display of disablement type (no details) *c*. 2 m from man and give call 2f (Bunni and Siman 1978). (4) Active measures: against other animals. Dogs and jackals *Canis*, also cats, elicit alarm-calls (see 2a–c in Voice), but bird will leave nest only if these come close. Birds with eggs or young give call 2e on sighting snake. ♂ recorded 'displaying both wings' and calling when snake *c*. 30 cm below nest; bulbuls and babblers eventually joined in mobbing which continued for several minutes before snake chased away by observer (Siman and Bunni 1976*a*; Bunni and Siman 1978). In Murgab valley, snakes elicited alarm-calls from ♀, also from Great Reed Warbler *Acrocephalus arundinaceus* and Booted Warbler *Hippolais caligata*; birds flew over and dive-attacked snake which nevertheless succeeded in taking 1–2 young (Peklo and Sopyev 1980; Sopyev 1981). MGW

Voice. Not very vocal outside breeding season (Meinertzhagen 1954; Bundy 1985); flocks can be noisy, however, calls being then audible over some distance (Ticehurst *et al.* 1921–2). Calls typically throaty and mellow (P A D Hollom). Detailed study by Bunni and Siman (1978) who distinguished 7 adult calls (scheme modified in following account) based on associated activities of birds, but noted no song (see, however, call 3 and Social Pattern and Behaviour).

CALLS OF ADULTS. (1) Mew-calls. Variety of mewing or whistling sounds: e.g. mewing 'meee' often given repeatedly and loudly; 63–85 calls per min (Bunni and Siman 1978). May resemble Buzzard *Buteo buteo* (Moore and Boswell 1957): thus pleasing, soft whistling 'pieur' (S Marchant) or 'peeeooo' with descending pitch and occasionally an almost fluting 'p-uuu' (Bundy 1985); low, soft 'wheooo' given as pair-contact during nest-building, occasionally later during incubation (Bunni and Siman 1978). Constant change in timbre and pitch noted in birds from southern Turkmeniya (USSR), with shrill 'peeu' and also 'pyau' (Flint 1962). Recordings by P A D Hollom confirm existence of mewing and whistling sounds, varying in length, pitch, and loudness: 'peeu', 'mee(w)', 'quee', etc.; many calls rather plaintive, wavering, or more like brief murmur. Attractive 'wheeoo' resembles Whistle-call of ♂ Wigeon *Anas penelope* (M G Wilson: Fig I); resemblance less marked in some shorter and less purely tonal sounds (unit 1 in Fig II), while

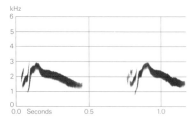

I P A D Hollom Iran April 1971

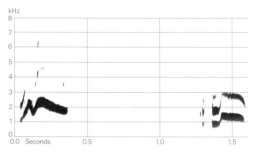

II P A D Hollom Iran April 1971

2nd unit in Fig II suggests 'au-kyee', slightly reminiscent of start of Long-call in Herring Gull *Larus argentatus*, though shorter (J Hall-Craggs). Mew-calls given by both sexes throughout breeding season, also in flocks at other times (e.g. constantly when moving about in bush), from perch, at take-off, in flight, and on landing (Bunni and Siman 1978); no 'wheeoo' calls noted from flying birds in Iran (P A D Hollom). (2) Calls given mainly when disturbed. Birds of both sexes disturbed at nest or with young give mostly separate call-types as listed below, but sometimes a variable mixture (descriptions from Bunni and Siman 1978 unless stated otherwise). (a) A 'meew meew meew mevit mevit veekveek...kirr' and (b) 'meew meew meew kirrrkirrru kirrrek...kirrr'. Major component of calls 2a-b is intense, high-pitched 'meew' given 2-3 times; 1-3 'kirrr' sounds may be interpolated in either phrase; *c.* 60-80 phrases noted per min. Given when disturbed at nest with or without eggs. Scolding chirp, high pitched and round, also harsher, lower-pitched warning 'chirr' (Meinertzhagen 1954) probably also call 2a-b variants. (c) Single 'meew' given towards other birds, also various animals (see Social Pattern and Behaviour); intensity increases after hatching and peaks when young leave nest. Loud and continuous but otherwise identical 'meew' given by ♂ near ♀ prior to copulation. (d) Abruptly ending 'me' given 1-2 times by alarmed parent or flock member. (e) Loud 'veek' given repeatedly by both sexes when young threatened, also used in fights. (f) A 'kegh' given by both sexes as high-intensity alarm. (3) Kirr-call. Loud, trilling 'kirrr' in several variants: 'kirrrek'; 'kirrrkirrru kirrru'; 'kirrrkirrrkirrru'; 'kirrrek' given 1-3 times, others 3-4 times; 'kirrr' may sound more like 'kirrrl' in compound calls. Given in flight by both sexes; loud, continuous 'kirrrkirrrkirrr' (termed 'song' without further discussion

by Bunni and Siman 1979) is feature of Courtship-display flight; also serves as contact-call, particularly during nest-building, nest-relief, and when feeding young (Bunni and Siman 1978). Recording of perched birds suggests soft, liquid 'turrerer'; entirely lacking in harshness, syllables compressed to give rich, brief trill effect (Fig III). Shorter and crisper calls from birds in flight may suggest Bee-eater *Merops apiaster* (P A D Hollom): rich, fruity chirruping 'queeru(p)', etc. (M G Wilson: Fig IV,

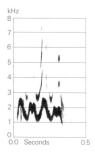

III P A D Hollom Iran April 1971

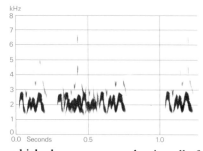

IV P A D Hollom
Iran April 1971

which shows some overlap in calls from 2 birds). Short melodious trill, reminiscent of *M. apiaster* but softer, also according to Paludan (1959), Flint (1962), and Vorobiev (1973). Other descriptions: liquid, high-pitched 'quieeur' (Moore and Boswell 1957); weak 'quee', sometimes vaguely disyllabic and with reedy quality, as flight-call from flocks (Bundy 1985). Description in King (1978*b*) presumably represents summary of calls 1-3: varied, sonorous, rolling whistled sounds 'wheew a-wheeew whee-di-du whee-du di-di-du du-du-du'; 'wheew' and 'whee' usually strongly inflected upward, 'di' and 'du' with chuckling quality. (4) Feeding-call. Soft, faint 'wit' or 'vit' given by either sex to encourage young to gape (Bunni and Siman 1978).

CALLS OF YOUNG. Peeping sounds given from egg *c.* 2 hrs before hatching (Siman and Bunni 1976*a*). Food-call from day of hatching 'vis'. Out of nest, a louder 'visk', faster when begging and sometimes interspersed with single 'veeio'—probably the plaintive double note from young just out of nest mentioned by Cramp (1970*b*) and this or another closely similar to an adult Mew-call (Bunni and Siman 1978).   MGW

Breeding. SEASON. Iraq: eggs from early May to June or July (Marchant 1963*b*); during 2 years, first eggs laid 11 May, last on 5-12 July (Siman and Bunni 1976*a*).

SITE. In bush or low tree, 1–4 m above ground, often well hidden in densest part of bush but may be more exposed in tree. Nest: base of small twigs, with cup of grass and tufts of vegetable down, lined with more down, sometimes wool and hair. Loosely colonial. Building: by both sexes (Marchant 1963b). EGGS. See Plate 81. Sub-elliptical to oval, smooth and rather glossy; white to very pale grey, sometimes tinged green when fresh, usually with zone of lead-grey to grey-brown blotches round broad end, sometimes forming band, occasionally scattered or sometimes unmarked. 20·6 × 26·8 (18·5–21·6 × 24·5–28·6), n = 68 (Siman and Bunni 1976a). Calculated weight 5·1 g (Schönwetter 1979). Clutch: 3–4(–5). Of 47 clutches, Iraq: 3 eggs, 5; 4, 26; 5, 16; mean 4·2, declining through season with mean 4·6 (n = 25) before 15 June, and 3·8 (n = 22) after 15 June (Siman and Bunni 1976a). Of 11 clutches, Iraq: 3 eggs, 3; 4, 7; 5, 1; mean 3·8 (Marchant 1963b). Re-lays after egg loss (Marchant 1963b). Eggs laid daily, occasionally a day missed; takes place in morning (mostly 05.00–08.00 hrs) immediately after nest completed, though some material added during first 2 days of laying. 2 broods. (Siman and Bunni 1976a.) INCUBATION. 14·2 days (13·9–14·5, n = 4), measured from laying of last egg to hatching of 1st egg, though incubation begins with 1st egg, and hatching asynchronous, with mean interval between eggs in same clutch 20.2 hrs (5–49, n = 15). By both sexes, though ♀ takes greater share (Siman and Bunni 1976a). YOUNG. Cared for and fed by both parents, more by ♀; brooded for 5 days after all eggs hatched (Siman and Bunni 1976a).

Plumages. ADULT MALE. Forehead and forecrown cream-pink, merging into pale bluish-grey on hindcrown; narrow black line runs from base of upper mandible over lore and round eye, widening behind eye and over ear-coverts into broad black patch extending across nape (where partly hidden by slightly elongated feathers of hindcrown). All upperparts including tertials and upper wing-coverts pale bluish-grey. Chin and upper throat buffish-cream or pink-cream, merging into pale grey of chest and sides of breast; flanks pale grey; central belly, vent, and under tail-coverts cream-pink, sometimes with vinous or mauve tinge, merging into pale grey of chest and flanks. Tail pale blue-grey, each feather with black triangle on tip (apex towards feather-base); length of triangle at shaft c. 30–35 mm on t1, 10–20 mm on t6, but only 5–10 mm at side of each feather. Primaries deep black with straight and clear-cut white tip; tip c. 12 mm wide on p1, gradually larger towards p7–p9 (25–30 mm), but c. 30–35 mm on inner web of p9; tip of p9 suffused grey-brown. Secondaries pale blue-grey, inner webs and bases all or partly dark grey or dull black (not sharply defined). Greater upper primary coverts black with partly white edge. Longest feathers of bastard wing black, 2nd longest pale blue-grey with black suffusion on outer web. Axillaries pale blue-grey, under wing-coverts pale cream-buff or white. Some variation in colour of crown: in some, mainly pale buff or salmon-buff, in others largely blue-grey, independent of wear. In worn plumage, upperparts duller grey, less bluish, underparts dirty white with grey tinge on chest and flanks. ADULT FEMALE. All upperparts, sides of neck, tertials, and upper wing-coverts uniform pale buff-brown or drab; no black on sides of head or nape. Chin, throat, and lower cheeks cream-buff; chest, sides of breast, and flanks pale buff-brown or drab, gradually merging into pale cream-buff or pink-buff of central belly, vent, and under tail-coverts. Tail buff-brown or drab, like upperparts, t1–t5 with ill-defined black tips c. 10–15 mm wide (not as broad and clear-cut as in ♂; virtually no black on t6); black tips more distinctly defined on pale grey undersurface of tail. Secondaries and bastard wing buff-brown or drab, like upperparts and upper wing-coverts; inner webs of secondaries dusky grey towards bases (not as dark as adult ♂). Flight-feathers and greater upper primary coverts buffish-grey or dull drab-grey; tips of p1–p8 with black tips c. 8 mm wide, contrastingly bordered distally by white fringe 2–3 mm wide; black of tips not sharply divided from grey of remainder of primaries; greater primary coverts with ill-defined blackish tips. Axillaries and under wing-coverts buff or drab, like flanks. In worn plumage, upperparts duller, drab-grey, underparts paler buff or dirty white. NESTLING. Down white or pale grey, on upperparts only. Eyes open on c. 4th day; well-feathered by 10–12th day. (Siman and Bunni 1976b; Peklo and Sopyev 1980.) JUVENILE MALE. Closely similar to adult ♀; feathers of head and body narrower, looser; entire upperparts and upper wing-coverts uniform yellow-buff (no black on head), underparts isabelline. Tail as adult ♂, but black of tips duller, browner, and less extensive, distally bordered by pale buff fringe. Tertials fringed pale buff along tips. Flight-feathers as adult ♂, but white tips of primaries have partial grey-brown suffusion; secondaries browner, less extensively dark on bases. Greater upper wing-coverts brown or black-brown, fringed off-white along tip and along base of outer web. JUVENILE FEMALE. Head and body like juvenile ♂, paler on upperparts and more isabelline below than adult ♀; differs from adult ♀ and juvenile ♂ in uniform drab flight-feathers, slightly duskier brown towards tip, faintly fringed pale buff on secondaries and inner primaries (primaries lack adult ♀'s subterminal black blotch and distinct white fringe along tips; no broad white tip with grey suffusion and black subterminal bar as in juvenile ♂); greater upper primary coverts virtually uniform drab-grey or buff (no black on tips as in adult ♀); tail uniform buff-brown or drab, indistinctly paler buff along tip and sometimes slightly darker brown subterminally (some black on tip in adult ♀ and juvenile ♂); greater upper wing-coverts paler than in adult ♀, faintly edged pale buff on tip. FIRST ADULT. Like adult, but juvenile flight-feathers, greater upper primary coverts, t2–t6 or all tail-feathers, and variable number of outer greater upper wing-coverts retained, showing characteristic juvenile pattern.

Bare parts. ADULT, FIRST ADULT. Iris brown or red-brown in breeding season, brown or dark brown during rest of year. Bill black with horn-coloured base; lower mandible flesh-coloured at base in 1st autumn and winter. Leg and foot reddish-flesh in breeding season, yellowish-flesh or brownish-flesh during rest of year (Ali and Ripley 1972; BMNH.) NESTLING. Bare skin (including bill and leg) pink or flesh-colour (Siman and Bunni 1976b). Gape-flanges pale yellow, mouth red-pink (Peklo and Sopyev 1980). JUVENILE. Iris dark brown. Bill dark horn-brown, base of lower mandible flesh-colour. Leg and foot brownish-flesh. (BMNH.)

Moults. ADULT POST-BREEDING. Complete; primaries descendant. Information limited. In late June, Iraq, plumage heavily worn but moult not started. One in full moult on 1 August, Iraq, with primary moult score c. 17 and all tail-feathers growing (BMNH); moult completed or nearly so in 6 birds from

Afghanistan, late October (Vaurie 1958). Thus, moult probably usually late summer and early autumn, but one from early March, Afghanistan, had outer primaries still growing (score 43) (BMNH). POST-JUVENILE. Partial: head, body (sometimes not all lower belly, vent, or tail-coverts), lesser and median upper wing-coverts, some or all tertials and tertial coverts, and sometimes t1 or a few to many greater upper wing-coverts. No information on timing, but probably starts within a few weeks after fledging; birds examined from November and later had moult completed (BMNH).

Measurements. ADULT, FIRST ADULT. Whole geographical range, all year; skins (BMNH, RMNH). Bill (S) to skull; exposed culmen on average *c*. 5·8 less; bill to distal corner of nostril *c*. 10·4 less.

	♂		♀	
WING	102·1 (2·68; 15)	97–106	101·3 (2·35; 17)	97–105
TAIL	110·5 (5·51; 14)	109–122	103·7 (3·80; 17)	96–109
BILL (S)	19·9 (0·51; 14)	19·1–20·6	19·7 (0·79; 15)	18·3–21·0
TARSUS	23·9 (0·63; 14)	23·4–24·9	24·3 (0·78; 14)	23·2–25·4

Sex differences significant for tail.

JUVENILE. Wing of 1st adult with retained juvenile flight-feathers on average *c*. 2·9 shorter than adult, tail *c*. 6·2 shorter, but both combined with adult in table above.

Weights. ADULT, FIRST ADULT. Iraq: 2 ♂♂ and 1 ♀ 2 oz (*c*. 57 g) each (Ali and Ripley 1972). South-west Afghanistan, mid-April: ♂ 48; ♀♀ 49, 50, 55 (Paludan 1959).

Structure. Wing rather short, broad at base, tip bluntly pointed. 10 primaries: p8 longest, p7 and p9 both 0·5–2 shorter, p6 4–6, p5 8–10, p4 11–15, p1 21–25. P10 reduced, narrow and pointed; 59–63 shorter than wing-tip, 4–8 shorter than longest upper primary covert. Outer web of p6–p8 and inner web of p7–p9 emarginated; outer web of p8 and inner of p9 slightly broadened in middle. Tail long, tip rounded; 12 feathers, t6 10–18 shorter than t1–t2. Bill remarkably similar to Waxwing *Bombycilla garrulus*: short and blunt, upper mandible rather bulbous (not laterally compressed as in shrikes *Lanius*); nostrils bare, except for narrow membrane above (not covered by short bristles as in *Bombycilla*); often a minute notch inside tip of upper mandible. Plumage soft, as in *Bombycilla*, but feathering slightly shorter and less dense. Leg and foot rather short and heavy; scutellations heavier than in *Bombycilla*, claws slightly longer. Middle toe with claw 21·6 (3) 20–23; outer toe with claw *c*. 73% of middle toe with claw, inner *c*. 64%, hind *c*. 76%.   JW, CSR

Family CINCLIDAE dippers

Quite small to medium-sized oscine passerines (suborder Passeres); mainly terrestrial and aquatic—being unique among Passeres in living in close contact with water, foraging largely below surface of flowing streams and rivers for invertebrates and, to lesser extent, fish. 5 closely similar species in single genus *Cinclus*: Dipper *C. cinclus* (Eurasia, North Africa), Brown Dipper *C. pallasii* (Asia), American Dipper *C. mexicanus* (North and Central America), White-capped Dipper *C. leucocephalus* (South America), and Rufous-throated Dipper *C. schulzii* (Argentina)—last two sometimes considered conspecific. Occur along fast-moving hill and mountain streams. Mainly sedentary with some altitudinal or southerly movement, depending on severity of winter. 1 species in west Palearctic, breeding widely but locally.

Though expert divers, no special morphological adaptations for life under water apart from dense plumage and broad membrane above nostrils which can be closed when head submerged; oil-gland, however, larger than that of most other Passeres. Body short and rotund, robust-looking; neck short. ♂ larger than ♀ in most species. Bill slender, appearing slightly upcurved; compressed laterally and slightly hooked and notched. Nostrils narrow. Wing short and broad, tip rounded; concave beneath, fitting sides of body closely. 10 primaries, p10 short but well developed. Flight direct and rapid with quick wing-beats; wings used also for underwater swimming. Will drop into or dive under water from flight at times, though more usually from rock, etc. High display-flights (with or without song) and flycatching sallies reported. Tail short, square or nearly so; 12

feathers, stiff. Legs long and stout, with strong toes and claws (middle claw sometimes pectinated); tarsus smooth except at extreme base. Feet not webbed but used for swimming on surface as well as for foraging under water (moving against current) and wading in shallow water. Gait a walk. Head-scratching by indirect method. Bathe in typical passerine stand-in manner. No information on sunning. Dusting not recorded and unlikely. Anting observed in both Old World species; of direct type (ant applied in bill). The name 'dipper' refers to characteristic habit of bobbing whole body up and down; often accompanied by conspicuous blinking of pale eyelid and whitish nictitating membrane (see *C. cinclus*, p. 516).

Plumage soft, long, and dense with thick layer of underlying down. Sombre in colour, largely brown or black, with or without white on head and chest; some species polymorphic. Sexes alike. Juvenile like adult but spotted or barred below in forms with white on underparts. A single (complete) annual moult in adult. Peculiar among Passeres in shedding innermost 5 primaries simultaneously, followed by gradual descendant loss of outer ones when most of others again fully grown; secondaries moulted late in comparison with primaries. Partial post-juvenile moult shortly after fledging. Nestling has upperparts covered with long, dense down; mouth orange-yellow, no spots.

Dippers resemble Troglodytidae (wrens) in shape of body and relative length of tail and, accordingly, are traditionally placed near them, but many other morphological characters point to position nearer Turdidae (chats, thrushes) (e.g. Stejneger 1905, Ripley 1952). Egg-white protein data also indicate much closer relationship to Turdidae than to Troglodytidae (Sibley 1970; Sibley *et al.* 1974), while DNA data place Cinclidae next to Turdidae and Sturnidae (starlings) in superfamily 'Turdoidea' and far from Troglodytidae, which is put in another superfamily entirely—though within same assemblage ('Muscicapae').

Cinclus cinclus Dipper

PLATE 35
[between pages 544 and 545]

Du. Waterspreeuw   Fr. Cincle plongeur   Ge. Wasseramsel
Ru. Оляпка   Sp. Mirlo acuático   Sw. Strömstare

Sturnus Cinclus Linnaeus, 1758

Polytypic. Nominate *cinclus* (Linnaeus, 1758), Fenno-Scandia east to Pechora river and south to Baltic states and Denmark; also central France, Pyrénées, north-west Iberia south-east to Sierra de Guadarrama, Corsica, and Sardinia; *hibernicus* Hartert, 1910, Ireland, Outer Hebrides, and west coast of Scotland; *gularis* (Latham, 1801), Orkneys, Scotland (except west), England, and Wales; *aquaticus* Bechstein, 1803, central and southern Europe from Belgium, West Germany, and southern Poland south to Greece, mainland Italy, Sicily, and eastern France, also eastern and southern Spain; *minor* Tristram, 1870, north-west Africa; *olympicus* Madarász, 1903, formerly Cyprus, now extinct; *caucasicus* Madarász, 1903, Turkey, Caucasus, Transcaucasia, and Iran north from Elburz mountains, east to Khorasan (north-west Iran); *persicus* Witherby, 1906, Zagros mountains (south-west Iran), intergrading with *caucasicus* in north-west Iran (Azarbaijan and Hamadan) and perhaps south-east Turkey; *rufiventris* Tristram, 1884, Lebanon; *uralensis* Serebrovski, 1927, Ural mountains region. Extralimital: 3–4 further races from Afghanistan to central China and eastern Siberia.

Field characters. 18 cm; wing-span 25·5–30 cm. Size between wheatear *Oenanthe* and small thrush *Turdus*; bulk shows in deep chest and belly, made more obvious by short wings and half-cocked tail. Medium-sized, rotund passerine, with shape suggesting huge Wren *Troglodytes troglodytes*, restricted to hill streams in which it both walks and swims. Plumage generally black-brown, relieved (in west Palearctic races) by broad white throat and chest in adult and white-mottled underparts in juvenile. Flight fast and direct. Sexes similar; no seasonal variation. Juvenile separable. 10 races in west Palearctic, differing mainly in colour of belly; only 2 described here (see also Geographical Variation).

(1) North European race, nominate *cinclus*. ADULT. Head and sides of neck dark brown, rest of upperparts, wings, and tail dark slate (with feathers broadly margined black-brown and appearing dully mottled at close range), and rear underparts black-brown (with variable but usually small suffusion of chestnut on mid-breast). Underwing black-brown. Dark plumage sharply relieved by white chin, throat, foreneck and breast. In summer, wear reduces or removes feather-margins and bird becomes paler on head and greyer elsewhere. JUVENILE. All upperparts dark-slate, with black-brown feather-margins providing obvious mottling; underparts basically white under head to dull grey towards tail, copiously spotted, scaled, and closely barred with dusky and black-brown feather-tips as far back as rear flanks, but only faintly marked by dull pale scaling on under tail-coverts. Secondaries fringed white. Underwing shows white feather-margins. (2) Mainland British race, *gularis*. ADULT. Differs distinctly from nominate *cinclus* in having strongly chestnut area below white breast, this usually extending on to fore-flanks and mid-belly. JUVENILE. Darker above than nominate *cinclus*, with more heavily marked underparts including rufous-buff under tail-coverts. All races. FIRST YEAR. Resembles adult but distinguished at close range by white tips to outer greater wing-coverts and (in winter) by grey-white areas just beside or below white chest. Bill and legs black-brown. White eyelid visible at close range when bird blinks.

Unmistakable, there being no other congener in west Palearctic, but racial identifications between 'chestnut-bellied' and 'black-bellied' forms are specious. In Britain, occurrences of 'black-bellied' birds traditionally assumed to stem from movements of northernmost nominate *cinclus*, but, in rest of Eurasia, geographical variation highly complex, some populations even varying within same mountain range (see Geographical Variation). Flight direct and rapid along straight runs of stream or river but more swinging, even jinking, among boulders and round bends. Fast and regular wing-beats produce constant action, but this varied by sideways tilts of head and body; rarely rises much above water level but will do so when pressed or in display. Gait involves walking, running, and occasional hopping, but most obvious movement is characteristic bobbing of body accompanied by downward flick of tail and blinking of white eyelid. Enters and submerges in water freely, walking over stream bed or swimming upstream, with flicks of wings propelling

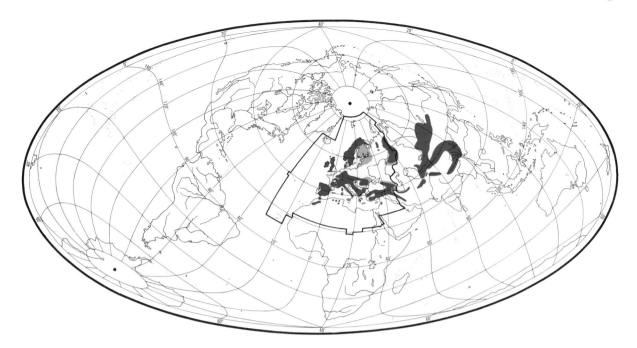

it downward as well as forward in order to counteract buoyancy. In areas of wide climatic extremes, birds leave higher streams to winter along shallower, slower rivers, a few using atypical waters such as millponds and canals.

Song a sweet, rippling trill, recalling *T. troglodytes*; calls include sharp, metallic 'clink-clink' in flight (often first indication of bird's presence) and loud 'zit zit zit'.

Habitat. Unique in west Palearctic: beside, on, and under swift-running streams and rivers of mountainous and hilly regions and dispersed over middle, higher, and lower middle latitudes, continental and oceanic, between July isotherms of 10-22° C in boreal temperate Mediterranean and steppe climatic zones (Voous 1960). Will descend locally and infrequently to lowlands where fast-flowing water created, even artificially, and sometimes occurs by lakes or on seashore (Sharrock 1976). Typical habitat contains plenty of rock faces and boulders but mainstay of foraging habitat is shallow water, often with gravel bottom, which apparently needs to total *c.* 4000 m^2 per territory (Shooter 1970). Typical habitats tend to have little aquatic or bankside vegetation, while other neighbouring vegetation appears to have only slight significance for the species, which is one of the few to be totally confined to rivers over greater part of range. Where lakes are adopted they are generally oligotrophic, with sparse vegetation (Fuller 1982). Despite specialized habits and habitat, has become largely dependent for nest-sites on artefacts such as bridges, culverts, weirs, walls, and other structures related to exploitation of watercourses, which are often more frequent and accessible than natural crevices or waterfalls (Sharrock 1976). In Switzerland, breeds at all levels from 200 m to

well over 2000 m; presence of forests, villages, and even cities along banks is immaterial (Glutz von Blotzheim 1962). In USSR, ascends to *c.* 2500 m, but will winter wherever open rivers or ice-free stretches occur, even at air temperatures of -40° C; moves away from breeding habitat only so far as weather conditions dictate (Dementiev and Gladkov 1954*b*). Increases in stream acidity are apparently detrimental (*Freshwater Biol.* 1986, **16**, 501-7). Runs and flies well, but is only secondarily an aerial or terrestrial species, owing to success in exploiting special aquatic habitats for which there is little interspecific competition.

Distribution. Range has decreased in Britain (though slight increase in parts of England) and France. Extinct on Cyprus.

BRITAIN, IRELAND. Before 1950, bred on Isle of Man and Orkneys (Parslow 1973). Slight extension of range in Midlands and south-central England (Sharrock 1976). FRANCE. Formerly bred Normandy (Yeatman 1976), and in Brittany to 1930s at least (Guermeur and Monnat 1980). NETHERLANDS. Bred 1910-15 and 1933 (CSR). TURKEY. Rather sparsely distributed (Vittery *et al.* 1971). CYPRUS. Extinct, last recorded 1945 (Flint and Stewart 1983).

Accidental. Faeroes, Iraq, Tunisia, Malta.

Population. Declining East Germany and Poland.

BRITAIN, IRELAND. No evidence of any widespread change, with 1000-10 000 pairs (Parslow 1973), but *c.* 30 000 pairs according to Sharrock (1976). Increased north-west Ireland 1972-82 (Perry 1983). FRANCE. 1000-10 000 pairs (Yeatman 1976). Alsace: 100-200 pairs

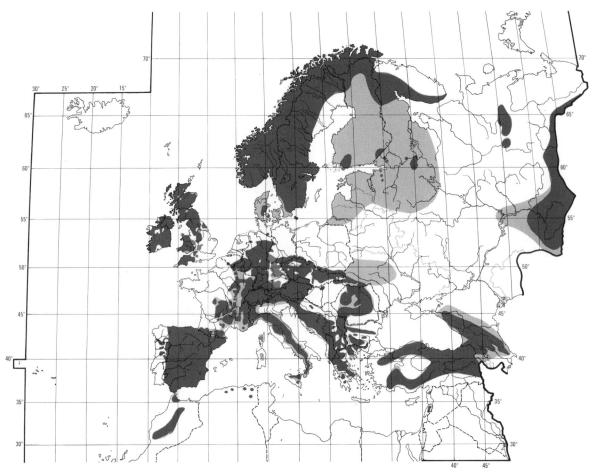

(Kempf 1976). BELGIUM. About 1200 pairs (Lippens and Wille 1972). WEST GERMANY. 5700-6600 pairs (Rheinwald 1982); 7300-32 200 pairs (AH). DENMARK. 6-8 pairs (TD). SWEDEN. Estimated 10 000 pairs (Ulfstrand and Högstedt 1976). FINLAND. Estimated 400 pairs (Merikallio 1958); some hundreds of pairs (OH). EAST GERMANY. Scarce, declining (SS). POLAND. Scarce (Tomiałojć 1976). Some recent declines, especially in Sudeten (AD, LT). CZECHOSLOVAKIA. No change in numbers reported. AUSTRIA. 46 pairs in Kamp river system, Lower Austria (Dick and Sackl 1985). GREECE. Only a few tens of pairs (GIH). BULGARIA. Quite common in upland areas (TM). USSR. Common in west and Caucasus (Dementiev and Gladkov 1954b). Estonia and Latvia: breeds occasionally (HV).

Survival. Britain: average annual 1st-year mortality 59%; average annual adult mortality 40-45% (Galbraith and Tyler 1982). England: average annual overall mortality 64·4% (Robson 1956). Oldest ringed bird 7 years 10 months (Rydzewski 1978).

Movements. Most populations resident but undertake local post-breeding dispersal movements, often involving altitudinal change. However, north European populations

of *cinclus* are subject to medium- or long-distance partial migration, some birds staying in breeding areas as long as water continues to flow, even beneath ice perforated by air-holes (Voous 1960).

Little information on movements of southern populations of nominate *cinclus* (breeding central France, Pyrénées, north-west Iberia, Corsica, and Sardinia), but a proportion of main northern populations of this race (breeding Fenno-Scandia and north-west USSR), and apparently also of *uralensis* from Urals, vacate breeding areas to make journeys of up to 1000 km (at least) in case of better-documented western populations (Dementiev and Gladkov 1954b; Zink 1981). Many recoveries of birds from Norway wintering Sweden, Denmark, Finland, and Baltic states (Andersen and Wester 1976; Zink 1981; Galbraith and Tyler 1982), and nominate *cinclus* now annual in winter in Netherlands (C S Roselaar). Movement is mainly south-east or even ESE, birds crossing Gulf of Bothnia into Finland and Gulf of Finland into Estonia and Latvia rather than south across Baltic to Denmark and East Germany, though some birds from southern Norway head south-east across Skagerrak to winter in Denmark (Zink 1981). Wintering birds arrive late October to early November, departing March to mid-April (Zink 1981). A few Norwegian birds head west to east coast of Britain, especially Shetland; recorded most years October–April, with peak in Shetland late March and early April suggesting birds have wintered elsewhere and are on return passage (British Ornithologists' Union 1971). To the east, records of nominate *cinclus* extend south only a little beyond lower Elbe to Berlin and in Poland reach only northernmost provinces (Zink 1981). Birds from Baltic area east to Urals can winter at -40°C provided some ice-free water remains (Dementiev and Gladkov 1954*b*).

In central and western Europe (*aquaticus*) fledglings leave parental territories at 6–7 weeks to become nomadic (Zink 1981). In Harz mountains (Germany), birds (probably mainly juveniles) occur both above and below breeding limits of 265–865 m (Zang 1981). Movements usually follow watercourses but some cross watersheds (Jost 1969). Almost all recoveries less than 50 km from ringing site (Zink 1981), though 160 km (juvenile) recorded (Glutz von Blotzheim 1962) and bird of unknown origin on Malta must have travelled at least this distance (Sultana and Gauci 1982). Such movements apparently regular and not caused by weather (Haller 1934; Creutz 1966; Zink 1981). In Switzerland in winter (September or October to February or March) occurs at lower altitudes away from breeding areas, notably at lake-sides, but also found above treeline at up to 2600 m (Glutz von Blotzheim 1962; Schifferli *et al.* 1982). Largely resident in Poland, but may descend in winter to Lublin and even Warsaw (Tomiałojć 1976). In Rila mountains of Bulgaria, noted breeding at 1500 m by fast streams that freeze in winter, so probably makes altitudinal movements (Mountfort and Ferguson-Lees 1961*b*).

British and Irish populations (*hibernicus*, *gularis*) sedentary apart from local post-breeding dispersal, affecting juveniles more than adults and ♂♂ more than ♀♀ (Galbraith and Tyler 1982). In winter in Wales and elsewhere, mountain streams regularly vacated in favour of lower reaches; some birds cross watersheds and in severe weather move to coasts and estuaries (S J Tyler). Some winter immigration to Britain from Europe—mainly nominate *cinclus* (probably mostly from Norway), also 2 possible records of *aquaticus* (British Ornithologists' Union 1971).

North-west African population (*minor*) is essentially sedentary but some individuals make local altitudinal movements (Chaworth-Musters 1939; Etchécopar and Hüe 1967).

Caucasus population winters below zone of deciduous forests, though some said to remain at 1500–1600 m (Dementiev and Gladkov 1954*b*), and some dispersal occurs in Turkey (*Orn. Soc. Turkey Bird Rep.* 1966–75), but little other evidence of movement in populations of south-east Asia (*caucasicus*, *persicus*, *rufiventris*), and those from further east appear to make only local or altitudinal wanderings.   JHE

Food. Large invertebrates of stream beds, especially larvae of caddis flies (Trichoptera). Feeds predominantly while submerged, walking on stream bed. In calm water, wings held to side; in rougher water, tail spread and bird progresses by use of wings. Leg movements continue as if on land. To surface, spreads tail, holds wings open and accelerates rapidly up. Actions underwater thus a combination of swimming, walking, and 'flying' (Penot 1948; Glutz von Blotzheim 1962; Sudhaus 1972), wing movements allowing extreme manoeuvrability (Ingram *et al.* 1938; Heim de Balsac 1949–50; Jones and King 1952). Will submerge like grebe *Podiceps* while swimming or 'belly floating' on surface (Penot 1948; Glutz von Blotzheim 1962). In calm conditions will walk directly into water and submerge without recourse to swimming (Horsfield 1915; Eggebrecht 1937; Rankin and Rankin 1940; Penot 1948; Glutz von Blotzheim 1962; Robson 1972). Also jumps into water from rocks or directly from flight, and submerges immediately (Vader 1971). Prior to dive, or from edge, frequently dips head 2–3 times below surface of water (Jones and King 1952; Lüdtke 1966; Vader 1971; Sudhaus 1972). In all cases of submersion, enters water facing current (Rankin and Rankin 1940; Penot 1948; Lüdtke 1966). Will up-end like swan *Cygnus* (Rankin and Rankin 1940; Jost 1975). Mean duration of dives 3·2 s (2·4–7·4, *n*=197), increasing with depth of water (Efteland 1975). Other measurements: 5–7 s (Eggebrecht 1937); 3–5 s (Holmes 1939; Sudhaus 1972); 5–10 s (Rankin and Rankin 1940); 5–7 s (Glutz von Blotzheim 1962); 3–4 s (Robson 1972); rarely longer than 10 s (Rankin and Rankin 1940; Efteland 1975), 23 s recorded in the wild and 30 s in captivity (Vader 1971; Efteland 1975). Maximum rate of dives 11·5 per min (Efteland 1975); frequently remains on surface for less than 1 s before diving again (Eggebrecht 1937; Sudhaus 1972; Efteland 1975). May repeat dives in same place, swim on, or fly upstream and dive again (Eggebrecht 1937; Penot 1948). Will dive in water up to 1·5 m deep but most often in less than 1 m (Rankin and Rankin 1940; Efteland 1975). Main feeding technique under water is to move stones and feed on items exposed underneath (Sudhaus 1974). Swallows smaller items under water, others (e.g. fish, larvae of caddis flies Trichoptera) brought to surface (Richter 1955*a*). Carries up to 5 items crosswise in bill when feeding young (Rankin and Rankin 1940). Deals with caddis fly larvae by pressing tube in bill and beating it on rock; grips head of larvae and finally removes tube by beating and tossing; small tubes may be eaten whole (Klaas 1952; Jost 1972). Also feeds on land, following stream banks and turning over stones, leaves, and debris (Eggebrecht 1937; Glutz von Blotzheim 1962; Sudhaus 1974), and uses similar technique in shallow water (Sudhaus 1974). Catches flying insects using short flights, seizing prey from below (Eggebrecht 1937; Glutz von Blotzheim 1962; Creutz 1966). Can spend up to 66% of day feeding (Creutz 1966); most active early morning and late afternoon prior to roosting (Efteland 1975). Pellets of undigested material are cylindrical with rounded ends, light to dark brown (Smith 1979): 6-20 × 3-10 mm (Jost 1975); 16 × 15 mm (Richter 1955a); average 16 × 6 mm, n = 30 (Smith 1979).

Diet includes the following. Invertebrates: springtails (Collembola), Protura, two-pronged bristle-tails (Diplura), silverfish (Thysanura), adult and nymphal mayflies (Ephemoptera, particularly Baetidae, Ecdyonuridae, Ephemeridae), dragonflies (Odonata), adult and nymphal stoneflies (Plecoptera: particularly Nemouridae and Perlidae), adult crickets and grasshoppers (Orthoptera), earwigs (Dermaptera), adult and nymphal bugs (Hemiptera), lacewings (Neuroptera), adult and larval caddis flies (Trichoptera: Rhyacophilidae, Sericostomatidae, Hydropsychidae, Phryganeidae, Glossosomatidae, Limnephilidae, Odontoceridae), butterflies and moths (Lepidoptera), adult and larval flies (Diptera), adult and larval wasps and ants (Hymenoptera), adult and larval beetles (Coleoptera: Dryopidae, Dytiscidae, Curculionidae, Carabidae, Haliplidae, Staphylinidae, Dascillidae, Hydrophilidae, Scarabaeidae, Cantharidae, Tenebrionidae, Chrysomelidae, Dryopidae, Ipidae, Gyrinidae), spiders (Araneae), shrimps (Amphipoda), isopods (Isopoda), daphnia (Phyllopoda), crayfish Astacus, millipedes (Diplopoda), centipedes (Chilopoda), leeches (Hirudinea), earthworms (Oligochaeta), flatworms (Platyhelmenthes), roundworms (Nematoda), snails (Gastropoda), bivalves (Bivalvia). Also fish (minnows Phoxinus phoxinus, loach Cobites barbatulatus, bullhead Cottus gobio) and eggs of salmon Salmo salar. Plant material (not identified) taken irregularly. (Vollnhofer 1906; Jones and King 1952; Sage 1964b; Creutz 1966; Jost 1975; Smith 1979; Ormerod 1985a; Ormerod et al. 1985.)

In Wye catchment area (England), April–June, faecal analysis (72 samples, 814 items) indicated diet dominated by mayfly nymphs (56·7% by number, mostly Baetidae) and caddis fly larvae (25·3%); also 7·5% stoneflies, 2·8% Diptera larvae, 2·5% beetles; by dry weight, mayfly nymphs 75·9%, caddis fly larvae 56·8%, stonefly nymphs 7·2%, Diptera larvae 1·7%; fish in 7% of faeces and aerial insects in 18% (Ormerod 1985a; see also Smith and Ormerod 1986). In Lancashire (England), April, 393 items from faeces comprised estimated percentage contributions 75·9% (by weight) mayfly nymphs, 16·2% caddis fly larvae, 4·4% Diptera, 3·4% stonefly nymphs; dry weight of items in pellets estimated 1·3-4·0 mg per item (Ormerod 1985b). In north-west Ireland, April, 510 items in faeces comprised 24·6% mayfly nymphs (mostly Baetidae), 10% stonefly nymphs, 63·2% (by weight) caddis fly larvae (largely Hydropsychidae), and 2·0% others (Ormerod and Perry 1985). In West Germany, March–August, caddis fly larvae present in 92·6% of 337 stomachs (in some stomachs up to 800 larvae tubes), stoneflies in 21·4%, Gammarus in 17·2%, molluscs in 8·0%, beetles in 7·4%, plant material in 6·5%, fish in

6·2%; during September–February, caddis fly larvae present in 89·7% of 145 stomachs, stoneflies in 24·1%, molluscs in 13·8%, fish in 13·8%, beetles in 11·7%, and plant material in 7·6% (Vollnhofer 1906). In Wye catchment area, January–February, 4850 items in faeces and pellets included Diptera (34·4% by number), mayflies (18·0%), and stoneflies (11·3%); by weight, fish 63·7%, caddis fly larvae 19·4%, molluscs 6·1%, mayflies 3·2%, shrimps 3·2%, Diptera 3·1%, and stoneflies 1·3% (Ormerod and Tyler 1986). In Yorkshire (England), July–November, 353 items in regurgitated pellets comprised 48·4% (by number) caddis flies, 26·1% beetles, 9·3% Gammarus, 1·4% Diptera, 1·4% Hymenoptera, 0·3% Lepidoptera; up to 17·0% of items non-aquatic (Smith 1979; E Smith). Seasonal analyses show predominance of mayflies, stoneflies, and caddis flies during breeding season (Creutz 1966; Jost 1975; Ormerod 1985a, b; Ormerod and Perry 1985) with shrimps, molluscs, beetles, and fish becoming more important during winter when much larger proportion of aquatic prey taken (Vollnhofer 1906; Creutz 1966; Jost 1975; Ormerod and Tyler 1986; E Smith). Figures calculated from Jost (1975) show caddis fly larvae in over 80·0% of pellets March–May, declining to 10% December–March; Gammarus increased from 10% June–August to 80% December–March but declined to zero March–May; beetles most frequent March–May but still less than 10%. Requires up to 79 g of food per day (Pastukhov 1961).

Diet of young similar to adult's: largely insect larvae supplemented by adult insects, crustaceans, molluscs, and fish (Jost 1975; Ormerod 1985a, b). Diversity and size of prey items brought to nest increases with nestling age, as does weight of each meal. At 5 days old, young receive up to 20 items per meal, older young receive a few larger items per meal (Jost 1975). Mayfly nymphs predominate up to 9 days with proportion of caddis fly larvae increasing up to day 15 (Ormerod 1985a). Overall, caddis fly larvae are the most numerous prey in diet of young (Jost 1975; Ormerod 1985a). Diet of young up to 15 days old in Wye catchment area as found from 50 faeces, comprised 82·5% (by weight) caddis fly larvae, 14·2% mayfly nymphs, and 2% stonefly nymphs (Ormerod 1985a). In Lancashire, 140 items in faeces from 4 young 15-17 days showed diet comprised 62·5% (by weight) caddis flies, 33·9% fish, 2·2% stoneflies, 1·3% mayflies, and 0·1% Diptera (Ormerod 1985b). In north-west Ireland, 246 items from young 7-15 days old showed diet comprised 95·3% (by weight) caddis flies, 3·8% mayflies, and 0·9% stoneflies (Ormerod and Perry 1985). In West Germany, 950 items in collar-samples from 128 nestlings aged 0-13 days comprised 63·5% mayflies, 10·9% caddis flies, 10·6% Diptera, 6% stoneflies, 1·5% Hemiptera; mayflies almost sole food given on day 1, with caddis flies introduced from day 3 (Jost 1975). In Scotland, collar-samples from 4 nestlings comprised 28·7% (by weight) adult caddis flies, 25·1% mayfly and stonefly nymphs, 24·3% caddis

fly larvae, 20·2% adult mayflies and stoneflies, and 1·6%
adult mosquitoes *Anopheles*; snails, beetles, and leeches
also recorded; minimum length of items 7·5 mm (Shaw
1979*a*). Young fed at a nest in Ireland every 7–15 min,
rate not affected by age of young (Rankin and Rankin
1940). Nest in France visited every 2–16 min (Penot
1948). In West Germany, young found to receive almost
33% more food at 23 days than at 9 days (Eggebrecht
1937).   TAS

Social pattern and behaviour. Well studied; see especially
Eggebrecht (1937), Balát (1962), and Creutz (1966).
1. Territorial throughout the year (Shaw 1979*b*). Dispersion
outside breeding season varies. Usually solitary, established
breeders of either sex defending individual territories (Vogt
1944; Robson 1956; Preuss 1959; Shaw 1979*b*). However, some
birds remain paired during the winter, both occupying same
territory (Horst 1941; Richter 1953; Robson 1956). Whether
birds single or paired, winter territory may be separate from
breeding territory (Balát 1962), but usually not far away.
Nominate *cinclus* migrating (apparently from further north) to
winter in northern Sweden showed site-fidelity, 17% of ringed
birds returning to same winter stream next season (Lundberg
et al. 1981). Unpaired ♂♂ more likely to move to another
territory outside breeding season. 2–3-year-old ♂♂ more
sedentary, not uncommonly waiting near an occupied territory
for it to fall vacant (Richter 1953, 1956). In autumn, Northumberland (England), 1st-year ♀ may challenge and displace
adult ♀ from breeding territory. 1st-year ♀♀ failing to displace
resident in autumn breed next spring on inferior sites nearby
until better ones become available. 1st-year ♂♂ likewise
challenge adult ♂♂, rarely successfully. Territories vacated by
birds at end of breeding season may be occupied by another
pair for duration of winter (Haller 1934). Winter territory
usually smaller than breeding territory (Sudhaus 1972). Territories of solitary birds minimum 100 m of waterway (Schuster
1953), typically 100–200 m (Balát 1962), 150–200 m (Sudhaus
1972), 200–500 m (Vogt 1944), average 350 m (220–630); of
pairs, average 440 m (220–830) (Holmbring and Kjedemar
1968). In Lübeck area (northern West Germany) up to 7 birds
along 300 m (Arndt 1979). For winter densities in various river
systems, West Germany, see Schuster (1953) and Kramer
(1968). When streams freeze, birds (irrespective of status)
readily seek open water elsewhere (Schuster 1953; Balát 1962),
often on lower reaches or even on coast, and small aggregations
may then occur, individuals defending individual-distance rather
than territories (G Shaw). In severe conditions, 2 ♂♂ recorded
feeding amicably at same ice-hole (Alder 1966). 'Concentrations'
of 21–26 birds reported in Harz mountains, West Germany
(Haensel 1977). BONDS. Monogamous mating system the rule.
Pair-bond usually lasts only for duration of breeding season,
but may continue throughout year (see above); maintained
outside breeding season in migratory nominate *cinclus* (Sudhaus
1972). Exceptionally, ♂ bigamous (for interactions preceding
and promoting bigamy, see Heterosexual Behaviour below). In
2 cases, ♂ simultaneously attended young in 2 nests in
contiguous territories (Mork 1975: see also Breeding Dispersion,
below). For similar case, see Galbraith (1979). In 2 of 3 cases
recorded by Sacher (1980*a*), the 2 nests attended by ♂ were
asynchronous by 1 and 2 weeks. Pairs reported breeding
together for up to 4 consecutive years (Creutz 1966; Fuchs
1970; Andersson and Wester 1976). However, mate-fidelity not
invariable (Fuchs 1970, which see for examples). When pairs

break up, partners often re-pair with neighbours (G Shaw: see
also Breeding Dispersion, below). ♀ alone broods but both
sexes feed young until independent (G Shaw). ♂ said to take
greater share of feeding 2nd broods (Balát 1962). ♀ who lost
mate failed to raise brood on her own; re-paired within 14
days (Baake 1982). In another case, when ♀ disappeared, ♂
nest-building with new ♀ 9 days later (Fuchs 1970). Fledged
broods may be attended by only one parent (Kovshar' 1979).
In 2 studies (Fuchs 1970; Sudhaus 1974), ♀ fed fledged young
more than did ♂. Parents continue to feed young until *c.*
(5–)7(–11) days after fledging (Fuchs 1970; Sudhaus 1972).
According to Balát (1962, 1964*b*), young not independent until
11–18 days after leaving nest. Reports of lengthy dependence
perhaps influenced by young sometimes remaining in territory
during period of juvenile moult; 3 out of 5 young which thus
stayed more than 2 weeks did not latterly beg for food (Fuchs
1970). ♀♀ usually first breed at 1 year, but ♂♂ may remain
unpaired at 1 year or older; of 7 ♂♂, 4 first bred at 1 year, 1
at 2 years, 2 at 3 years (Richter 1953, 1956). BREEDING
DISPERSION. Solitary and territorial. Territories often separated
by unoccupied length of waterway, but sometimes abut (Robson
1956). No estimates of area occupied by territory. Length
thought to be governed by area of shallow water suitable for
feeding; width 1·5–15 m (Shooter 1970) and length greater
where stream narrower (Robson 1956). During fledging period,
area defended smaller than at beginning of season (Balát 1962).
Length varies from 110 to 1500 m, depending on quality
(Eggebrecht 1937; Richter 1953; Balát 1964*b*; Görner 1971).
In Tatra mountains (Poland), 1000–3000 m (Sokołowski 1964).
Density highest in shallow, fast-flowing streams (Jost 1975),
and thus tends to increase with gradient, altitude, and presence
of rapids (Round and Moss 1984). In central-eastern Europe,
density mostly 0·7–10 pairs per 10 km (Oelke 1975). On 3 river
systems in Wales, 1·15–2·67 pairs per 10 km (Round and Moss
1984). In Cumbria (England), 2·1 pairs per 10 km on limestone,
6·2 on sandstone; average territory 430 m (110–640), *n* = 23
(Robson 1956). In Derbyshire (England), 5·4 pairs per 10 km
on gritstone, 5·8 on limestone (Shooter 1970). In Perthshire
(Scotland), up to 9·7 pairs per 10 km, in Galloway (Scotland)
up to 7·7 pairs per 10 km (G Shaw). On 402 km of Esk river
system (southern Scotland), 1·4–2·2 pairs per 10 km (Cowper
1973). Over 140 km, south-west Norway, 1·4–2·2 pairs per 10
km (Efteland and Kyllingstad 1984). In West Germany: 2·4
pairs per 10 km (1·1–3·7) on 522 km of waterway, eastern
Hessen (Jost 1975); also in Hessen, *c.* 2·7 pairs per 10 km on
River Kinzig (Klein and Schaack 1972); *c.* 4 pairs per 10 km
in Harz mountains (Oelke 1975); for details of density on River
Neckar, see Horst (1941). On 196 km of Kamp river system
(Lower Austria), 2·3 pairs per 10 km (G Dick and P Sackl).
Nest may be built in any suitable site within territory; distance
between nests of neighbouring pairs 290–1850 m (Balát 1964*b*),
closest 100 m (Kovshar' 1979). In Norway, neighbouring nests
reported *c.* 50 m and *c.* 100 m apart, once 8 m, suggesting
possible bigamy (Efteland 1983). In 2 cases of bigamy, ♂
simultaneously attended nests 700 and 800 m apart (Mork
1975), in 2 other cases 2 km and 3·5 km apart (Sacher 1980*a*).
Territory serves for pair-formation, nesting, and feeding (G
Shaw). Bird may leave territory (even crossing watershed) to
feed, roost, etc. (J Alder). Exceptionally, nest placed away from
water, once more than 100 m (Moon 1923); one pair had to fly
more than 2·8 km to feed, due to local pollution (Efteland
1983). Pairs markedly faithful to territory, though not uncommonly switch to (presumably preferred) neighbouring one
if opportunity arises (Richter 1953, 1956). When pair breaks
up, one partner, of either sex, typically retains territory (G

Shaw). Likewise, if either partner dies, other retains territory and re-pairs (Balát 1962). Nest-sites traditional; one said to have been used for 123 years (Gladstone 1910). Birds more faithful to sites than nests; may renovate old nests but, at 24 sites, only 6 used for 3 consecutive years (Robson 1956). Often builds new nest alongside old one (Schuster 1953; Sudhaus 1972). Same nest, but re-lined, typically retained for 2nd clutch (Brook 1912; Diesselhorst 1938). In 26 of 27 cases, 2nd clutch laid in same nest as 1st (Shaw 1978). Young ♂♂ relatively faithful to natal area: ♂♂ from 11 broods ringed as nestlings settled on average 2·5 km (1·3-3·6) from natal site (Richter 1956). Territories not uncommonly overlap with those of Grey Wagtail *Motacilla cinerea*, Pied Wagtail *M. alba*, and Wren *Troglodytes troglodytes* (Mawby 1961). ROOSTING. Nocturnal. Mostly solitary throughout year, less often communal and then mainly outside breeding season. Even where pair or small group roost together, they never huddle close together, exceptionally as little as 10 cm apart (G Shaw). Solitary roost-sites in territory typically sheltered concealed places over running water, similar to those used for nesting, e.g. old nest (sometimes in cold weather: Hewson 1969), cavity under bank, behind waterfall, under bridge, inside culvert (G Shaw, S Andersson). In Småland (southern Sweden), communal roosting in winter widespread (S Andersson). In Scotland and northern England, communal winter roosts at traditional sites, mainly in roofed cavities, on girders (etc.) under bridges (Hewson 1969; Shaw 1974, 1979b; Moss 1975; J Alder). Average distance between 5 roosts, all bridges, 1897 m (896-3109). Communal sites usually within an occupied territory, but roosting birds tolerated by resident (Hewson 1969)—though resident ♂ increasingly aggressive towards others as breeding season approaches (Keicher 1983; see also Antagonistic Behaviour, below). Not more than 7 birds at given roost (J Alder). Total numbers at 4 roosts varied nightly from 13 to 27, highest in worst weather (Shaw 1979b). Sites used in most months, though tend to be vacated March-June; juveniles begin using roosts in July, but most users adult. Average collective total at 5 roosts 4·0 in April, 11·0 August, 4·6-8·5 December-March (Hewson 1969). For similar seasonal patterns, see Shaw (1979b) and Keicher (1983). Birds approach roost singly or often several together, at dusk (Hewson 1969), sometimes flying exceptionally high (Richter 1953). Before roosting, typically assemble at suitable perch-sites nearby to feed and perform comfort activities; assembly often also accompanied by marked social interaction: singing (see 1 in Voice), calling, Dipping, Blinking (for details, see part 2, below), running, and flying up and down stream (Hewson 1969; Keicher 1983). Entry into, and morning exit from roost typically silent, direct, and in poor light (Hewson 1969; Shaw 1979b). When territory-owner aggressive in spring, birds about to roost may wait until he is absent, or under cover (Keicher 1983). For daily activity rhythm, see Prato (1981) and Keicher (1983); latter also for details of roosting times relative to light intensity. First birds to enter roost are juveniles, then adult ♀♀, then adult ♂♂ (i.e. dominant birds last); this order reversed in morning. Dominants also regularly supplanted subordinates at favoured perch-sites, both at roost and pre-roosting assembly (Keicher 1983). Individuals relatively faithful to given roost, and to perch-sites within roost, from season to season (Hewson 1969; Keicher 1983). Pair-members regularly roost relatively near each other, but not physically close, January-March (Keicher 1983; G Shaw). In breeding season, ♂ usually roosts near nest, sometimes on rock in water; ♀ roosts in nest until young half-grown, thereafter near nest (Hewson 1969; G Shaw). For some days after leaving nest, young show some tendency to return to nest for roosting (Shaw 1978). On rocks (etc.),

bird sleeps with head tucked under wing, plumage ruffled and largely covering legs and feet. For loafing, ♂ has one or more preferred rocks in stream near nest-site (G Shaw). Also reported loafing daily for up to 2½ hrs at a time in recesses in river bank (Berthet 1947). Comfort behaviour includes bathing, especially before roosting (Hewson 1969; Keicher 1983), and sunbathing (Arndt 1979). Also performs anting behaviour, accompanied by much Dipping (for details see part 2, below) and preening (Roskell 1982). For other descriptions of anting, see Creutz (1952) and Angus (1980). For anting-like behaviour in which fish (held in bill) rubbed into plumage, see Johnston (1985).

2. Rather shy, usually allowing approach to 20-30 m (Vogt 1944); in town, to 2-4 m (Pfeifer 1974). On close approach typically flies off or dives (Arndt 1979), giving Contact-alarm calls (Horst 1941: see 2 in Voice). Escape-flight usually short, and bird alights within sight of observer; repeated disturbance eventually elicits longer flight, sometimes returning over intruder at height of 10-20 m, but always along line of river (G Shaw). Occasionally, if followed, doubles back, flying as high as 30-60 m (Witherby *et al.* 1938b). In breeding season, birds can seldom be driven beyond their territorial boundaries (Penot 1948). Bird pursued by Sparrowhawk *Accipiter nisus* flies low in zigzag path over water and, if pressed, drops into water or hides under vegetation (Creutz 1966; G Shaw). Overflying *A. nisus* induced Dipping (see below) and Contact-alarm calls (Vogt 1944). Birds handled at communal roost give repeated Alarm-calls (G Shaw: see 2 in Voice). In almost any situation, bird signals excitement, mild alarm, threat, or mixture of these, by repeated Dipping and Blinking, rate of both increasing when approaching nest, mate, or when disturbed by intruder. Dipping a jerky bobbing motion due to flexing of legs: may be performed with little change in posture, but at most intense bird extends wings with tips almost touching ground, tail lowered, whole body rising and falling (G Shaw). In heterosexual encounters, ♂ usually Dips more deeply than ♀, and the longer the intervals between Dips, the deeper the Dip (Eggebrecht 1937). At same time, bird Blinks white upper eyelids, this almost certainly having signal function (Alder 1957; Creutz 1966). Blinking always accompanies Dipping but not vice versa (Rankin and Rankin 1940; G Shaw). Dips on average *c.* 50 times per min, minimum 38; Blinks 36-54 times per min at rest, at faster rate when Dipping (Rankin and Rankin 1940). For tail-raising in response to disturbance, see Antagonistic Behaviour (below). FLOCK BEHAVIOUR. None observed. No obvious cohesion in communal roosts or other occasional small gatherings (G Shaw). SONG-DISPLAY. Song (see 1 in Voice) given by both sexes; according to Keicher (1983), ♂ sings more, but in another study ♂ and ♀ sang equally (H Galbraith). Sings throughout year except when moulting (July-August), and resurgence from September associated with establishment of winter territory (Hewson 1967). Also sings little during chick-rearing (Diesselhorst 1938; Penot 1948). In Renfrewshire (Scotland), no singing May-July; maximum September-October and January-March, latter coinciding with dissolution of winter territories held by unpaired birds, and establishment of breeding territories (H Galbraith). In Württemberg (West Germany), song of ♂ peaks February-March (Keicher 1983). Song given at virtually any time of day, but mostly morning and evening; little around midday (Eggebrecht 1937). From May to August, birds sing regularly morning and evening at roost-site (Keicher 1983). In winter, one ♂ sang just before sunrise, prior to leaving roost, little by day, more at sunset when challenging intruders *en route* to roost (H Galbraith). Both sexes sing perched or in flight, in almost any part of territory. Lone bird usually sings quietly—almost

A

C

invariably from ground, generally on rock or water's edge. Near nest-site, particular rocks usually favoured. Singing in flight frequent in presence of mate. Facing mate, or conspecific rival, perched bird, of either sex, sings loudly in Advertising-display (Fig A, left): stands upright, throws head back, pointing bill upwards, and turns head slowly from side to side with white breast plumage ruffled; lowers and fans tail, and half-spreads wings, pointing them sideways or upwards; wings may be stationary, quivered, or flicked, depending on context (H Galbraith). For Advertising-display combined with Blinking in a threat context, see Antagonistic Behaviour (below). For other variants, associated with Antagonistic or Heterosexual context, see below. ♀ also sings, often in subdued manner, in normal relaxed posture (G Shaw); longest bouts of uninterrupted song 17 s by ♂, 13½ s by ♀ (Eggebrecht 1937). Bouts of singing can last 6 min (Rankin and Rankin 1940). Occasionally, in heterosexual encounter, song given in High-flight. In March, 2 birds thus flew close behind each other, *c.* 30 m above ground, apparently both singing continuously. In April, one pair displayed similarly *c.* 15 m above ground over distance of at least 200 m (Hewson 1967), and High-flight of similar length and altitude seen by H Galbraith. In January, pair Dipping at stream edge suddenly flew downstream and then to considerable height and circled twice in wide arc, one ahead of other; trailing bird sang continuously then made 2 solo circuits, still singing (Moody 1955). Similar display reported, March and June, by pair with young in nest (Robson 1956). ANTAGONISTIC BEHAVIOUR. Highly aggressive to conspecifics during most of year, less so during fledging period when juveniles (and to some extent also adults) from neighbouring territories allowed to trespass (Balát 1962). Rivals threaten each other with Advertising-display (see above), accompanied by loud song. Aggressive display sometimes involves fewer wing movements than similar display to mate: in 12 displays (9 by ♀) against

territorial intruders, only 3 involved wing-quivering, 3 were accompanied by spasmodic wing-flicking, and 6 by no wing-movements (H Galbraith). Threat-display common between pair-members (see also below) and between territorial rivals, notably between ♂♂ in autumn (J Alder). For aggression of pair to rival ♀, see Heterosexual Behaviour (below). Birds display face to face, usually 1-2 m apart, typically for 5-15 s (G Shaw). Aggressor, often 1st year ♂ (Fig B, left) stands erect, flicking wings forward and Blinking. Resident (Fig B, right), also upright, sleeks plumage, keeps wings still, and sings. Aggressor may adopt an oblique threat-posture (Fig C) or, after approaching to *c.* 1 m, a low threat-posture (Fig D, left). Resident becomes more erect (Fig D, right), and sings more

D

loudly, breast puffed out, tail spread, head moving from side to side. Aggressor attacks, then pursues fleeing resident and may attempt to catch resident's tail, or grapple in water; may give Contact-alarm calls or sing. In response to rival trespassing in flight, resident at nest stood fully erect, sang loudly, raised tail markedly (Fig E) then lowered and raised it again ; may

B

E

respond similarly to sudden sharp sound or disturbance. Tail-raising also reported by Eggebrecht (1937), Manuel (1949) and Sudhaus (1972). Chased bird may withdraw or drop into water or to water's edge (G Shaw). In one dispute, September, 2 ♂♂ kept fluttering a few cm above water, singing and calling (Penot 1948). Boundary disputes between ♂♂ usually brief and mild, but if both land on solid ground or in water fierce fight may follow, birds pecking and wing-cuffing with feet interlocked; exceptionally, one killed (Bunn 1963; G Shaw); in autumn, much foot injury results (J Alder). If swimming bird enters another's territory, it is immediately attacked and expelled. One bird, presumably a trespasser, submerged with only its head above water for nearly 1 min when another bird was nearby (Preuss 1959). At start of breeding season, one bird with communal roost in its territory challenged roosting birds by running, swimming, or flying towards them, sometimes running around singing in Advertising-display; chased birds (of either sex) up to 150 m (Keicher 1983). Paired ♂ and ♀ are not uncommonly aggressive to each other when they suddenly meet during breeding season (G Shaw): commonly, when sitting ♀ leaves nest, then encounters mate, he displays at her (as in Figs B–C), causing her to retreat, then pursues her back to nest (J Alder). Early in breeding season, aggressive bill-fencing not uncommon (S Efteland and K Kyllingstad). When population density high, fierce disputes between neighbours, of either sex, may lead to bird ejecting young from nest of rival (J Alder). ♂ often chases *M. cinerea* and Kingfisher *Alcedo atthis* (Eggebrecht 1937, 1939) from vicinity of nest; less often, *T. troglodytes* and Song Thrush *Turdus philomelos* (Creutz 1966). HETEROSEXUAL BEHAVIOUR. (1) General. Pair-formation can apparently take place at any time of year, but mainly in spring (Richter 1953; Fuchs 1970). Maximum activity March–April, waning with onset of incubation; resurgence occurs in June, when 1st brood almost fledged, probably preliminary to starting 2nd, also in autumn after moult completed (Creutz 1966). Pair-bonding behaviour reported in winter territories (Diesselhorst 1938; Schuster 1953; Balát 1962; Creutz 1966), but since display between prospective or established pair-members is almost indistinguishable from confrontations between rivals (H Galbraith), some caution required in interpreting behaviour in winter and indeed at any time (E K Dunn). Unexpected encounters, perhaps early in pair-formation, may be more antagonistic than heterosexual (G Shaw; see below). Where pairs remain together for several years (see Bonds, above) on breeding and winter territories, pair-bonding behaviour inconspicuous, sometimes almost absent (Creutz 1966). (2) Pair-bonding behaviour. For vocal attraction of pair-members, see Song-display (above). Pair-formation may begin in territory with one partner, initially usually ♂, approaching other. At first, ♀ coy, often flying off, but gradually she becomes bolder. Whenever she approaches ♂, he runs (sometimes swims) towards her, singing the while in Advertising-display (Creutz 1966). At close quarters, at times bill to bill, ♂ runs with tripping gait around ♀, commonly Dipping and Blinking, sometimes wing-quivering and giving Rattling-call (see 3 in Voice). Dipping and Rattling-calls appear to be important elements of Meeting-ceremony (Rankin and Rankin 1940; G Shaw). In 5 out of 7 Advertising-displays between ♂ and ♀ during winter, rapid wing-quivering accompanied display, this apparently more frequent than when display hostile (see above); in 5 out of 7 incidents, displaying partner was ♀ (H Galbraith). Dipping and/or Advertising-display often followed by departure of ♀, ♂ giving chase, singing. Pursuit may lead to High-flight (see Song-display, above). On landing, pair display again (Creutz 1966). In established pairs especially, pursuer or pursued may land directly in water with a splash, and swim ashore (Hewson 1967). In early spring, unpaired ♀ (Fig A, front), from nearby territory presents herself, passively, on nest-site of pair (Fig A, rear) who then challenge intruder; resident ♀, singing loudly, flies at rival who retreats, chased by singing pair, towards her territory, thus signalling to ♂ that she is unpaired territory-owner. Much chasing and lengthy boundary dispute

may follow, sometimes resulting in bigamy by ♂ (J Alder). Intensity of display between pair-members increases as season advances. As ♀ becomes more receptive, she increasingly takes initiative, approaching ♂ and performing Advertising-display, typically inducing same in ♂ (Creutz 1966). (3) Courtship-feeding. On alighting near ♂, ♀ may perform Soliciting-display (Fig F, right): crouches slightly, holds wings partly open and

F

quivers them rapidly; frequently accompanies display with Gaping, song, or Rattling-calls (S Efteland and K Kyllingstad). ♀ continues to display thus into incubation period. ♂ may respond by feeding ♀ (G Shaw), often singing and giving Rattling-calls as he does so (Eggebrecht 1937). Tug-of-war over food may develop, sometimes resulting in one bird, wings raised and quivering, toppling back into water (Rankin and Rankin 1940). Once ♂ offers food freely, and ♀ accepts it, pair-formation complete (Creutz 1966). ♂ continues feeding ♀ during incubation at or near nest (Eggebrecht 1937). (4) Mating. Takes place anywhere in territory; sometimes near nest and probably sometimes in it (Eggebrecht 1937; Creutz 1966; G Shaw). ♀♀ solicit (as described for courtship-feeding) from middle of January to end of March (J Alder). At first, ♂ may ignore soliciting ♀ (Diesselhorst 1938), or more often ♀ rebuffs ♂, lunging at him with raised wings (Creutz 1966). On 2 occasions, singing ♀ apparently refused copulation by continually turning to face ♂, who kept trying to get behind her (Eggebrecht 1937). Soliciting-display of ♀ stimulates ♂ to mount directly, or after proferring food. Without further ado, ♂ jumps on ♀'s back, or sometimes flies and lands straight on her back. While copulating, both partners balance with beating wings (Eggebrecht 1937; Creutz 1966). No calls heard. 2 copulations lasted 2 s and 7 s (Eggebrecht 1937). ♀ may follow Soliciting-display with Invitation-display (Fig G): crouches, raises wings higher than

G

in Soliciting-display and quivers them, then turns abruptly, presenting somewhat raised tail to ♂ and turns head slightly towards him: ♀, sometimes followed by ♂, then flies quickly to nest where mating may take place. Invitation-display very like begging posture of young, and ♂ recorded mounting offspring displaying thus. (J Alder.) No marked post-copulatory behaviour: pair may do nothing (G Shaw), or ♂ may stay around,

Dipping and Blinking, while ♀ shakes herself and preens (Creutz 1966); once, pair went off and fed close together (Eggebrecht 1937). (5) Behaviour at nest. Both members of pair seen investigating potential nest-sites but not known which sex makes final choice (G Shaw). ♂ often visits ♀ while she is building inside nest; flies up and clings to nest, singing and giving Rattling-calls (Eggebrecht 1937, which see for diurnal rhythm of nest-building). Between bouts of building, ♂ sings intermittently near nest (Rankin and Rankin 1940; Manuel 1949). When ♀ incubating, ♂ often perches near nest (Görner 1971). On returning after absence, ♂ may sing with wing-quivering or (especially near laying) give Rattling-calls and variants (Eggebrecht 1937; Rankin and Rankin 1940; see 3 in Voice). After building, bird often dives straight into water, ♀ doing likewise during breaks from incubation (Schuster 1953; Görner 1971). Birds may swim under water to approach nest when entrance covered by flood (Alder 1963). Bird, usually ♀, sometimes covers eggs with leaves (Purvey 1985; J Alder), thick layer apparently more often indicating something amiss in nesting attempt than (e.g.) strategy to conceal clutch. ♀ typically removes lining from successful nest after young fledged: of 150 nests, 80 successful were de-lined, 10 probably successful were de-lined or partly so, 60 unsuccessful were left with lining intact. (J Alder.) RELATIONS WITHIN FAMILY GROUP. Young brooded by ♀ only until 12-13 days (G Shaw), once apparently until 23 days (Eggebrecht 1937). At one nest, ♀ brooded assiduously for spells of 20-65 min, once 113 min; brooded progressively less as young aged, not at all after 13 days, at which time she also stopped roosting overnight on nest. Both parents feed young, until 7 days old, approaching nest with Rattling-calls (Rankin and Rankin 1940); also periodically call thus during transfer of food (Eggebrecht 1937). Small young fed, inside nest-chamber, on single items, older young a food-ball (G Shaw); usually all food given to one chick (Eggebrecht 1937; G Shaw). A 'carousel' system of feeding, as in *Alcedo atthis*, operates, chick just fed retreating to rear and allowing another to occupy nest-entrance (Sokołowski 1964). From *c.* (7-)11-12 days, young lean out of entrance, gape, and beg loudly; are fed there by visiting parents clinging briefly (G Shaw). Young start calling when incoming parent up to 50 m from nest (Sokołowski 1964); sibling rivalry thought to operate at this stage (G Shaw). Up to 9 days, parents carry faecal sacs from nest; ♀ twice seen removing soiled leaves; after 9 days, young defecate out of nest (Eggebrecht 1937). Both parents now collect faeces (unless they sink) from below nest, usually dropping sacs in flight into water not far from nest (G Shaw). In conditions of prolonged flood, cold, or food shortage, parents may eject young (usually small), dead or alive, from nest, which they may subsequently re-use (J Alder). Close to fledging, young tend to flutter from nest if accessible perches available nearby, then return to it (Shaw 1978). For diving (etc.) of young, see below. Brood commonly fledges over more than 1 day, e.g. one brood of 5 took 3-4 days (Fuchs 1970). Fledglings beg by gaping, giving food-calls and wing-quivering (G Shaw). Start Dipping as soon as they fledge; also spend a lot of time bathing, swimming, and preening (Sudhaus 1972). Practise feeding skills, initially with little success (Eggebrecht 1937). ANTI-PREDATOR RESPONSES OF YOUNG. From *c.* 7 days before fledging, young capable of diving and swimming under water, and readily abandon nest prematurely in this way (Hewson 1969; G Shaw) in response to parental Alarm-calls (Balát 1964b). Immediately after leaving nest, prematurely or not, young seek refuge in vegetation, roots (etc.), on river-bank (Eggebrecht 1937; G Shaw); one chick hiding thus gave hissing sound when man 1·5 m away, and swam off, later crouching

with head drawn in and remaining motionless when touched (Süss 1970). Fledglings fall silent and fly when approached by man (Eggebrecht 1937). PARENTAL ANTI-PREDATOR STRATEGIES. (1) Passive measures. ♀ sits very tightly, not flushing until nest closely approached or touched. Some departing birds dive straight into water; may give Alarm-calls. Sentinel ♂ gives Contact-alarm calls if danger approaches, eliciting no response from incubating ♀ (G Shaw). (2) Active measures: against birds. When Magpie *Pica pica* entered territory, bird gave Contact-alarm calls (Rankin and Rankin 1940). No further information. (3) Active measures: against man. Sometimes, mainly when well-grown young in nest, demonstrative when nest disturbed, bird flying directly at and around intruder, calling. When nest approached during building (not after-wards), ♂ bold and aggressive, flying at intruder in peculiar upright manner, prominently displaying white breast, and giving Contact-alarm calls (Rankin and Rankin 1940). No distraction-displays seen in lengthy study by G Shaw, but one record of possible distraction-lure display alleged to deceive intruder about location of nest-site: adult flew rapidly about in spirals, finally perching on branch, then both parents flew repeatedly to hollow rootstock away from nest (Horst 1941).

(Figs by J Alder: from his original drawings, but incorporating elements from material supplied by I Byrkjedal and R Roalkvam.) EKD

Voice. Relatively simple repertoire used throughout the year. All vocalizations given by both sexes. Non-vocal sounds occur, though display significance, if any, not known: wing-whirring (Bergmann and Helb 1982, which see for sonagram), and rapid patting or slapping sounds during nest-building process, evidently associated with working material into fabric (P A D Hollom).

CALLS OF ADULTS. (1) Song. A very sweet rippling warble, given equally by ♂ and ♀ (Witherby *et al.* 1938b). Song of ♂ comprises a variety of notes in apparently any order and repeats short phrases and units, louder as ♀ (or other conspecific) approaches; song of ♀ a less sweet series of whistles and disconnected units (Rankin and Rankin 1940), usually easily distinguishable from ♂ by being more scratchy and less melodious (G Shaw). Song difficult to render, but one sequence (by bird of unspecified sex) described as 'zi zi kep kep töp tja tja zerb srit srit tsarrrr trurr ziii titja' (Creutz 1966). In song (Fig I), some quite harsh churring units occur, this part of lengthy song rendered '(k)err ti(k) ti(k) ti(k) ti(k) (s)-zeek (s)-zeek chik-chik ti ti chu chu che zik zik zik zik k-cheek' (J Hall-Craggs). Song given perched, often quietly when bird alone, or in flight; occasionally at night (Witherby *et al.* 1938b). Sexes sing alone, but usually in display, almost invariably when approaching each other, sometimes with food in bill (Rankin and Rankin 1940); at such times, song may be combined with call 3 (see below, and Fig IV). Song reported in High-flights and expelling pursuit-flights (see Social Pattern and Behaviour, also for song-period). Sustained sweet Sub-song also given (Witherby *et al.* 1938b; see also Fig IV). (2) Contact-alarm calls. Commonest call a loud high-pitched 'zit' (Fig II), given usually 2(-4) times in

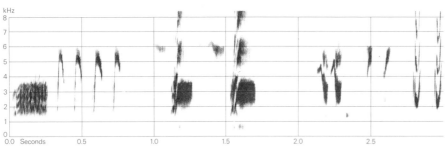

I   V C Lewis   England   November 1971

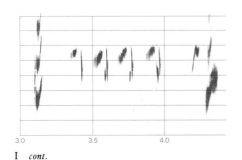

I   *cont.*

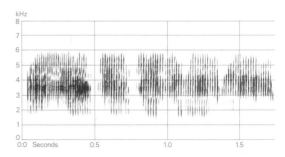

II   P A D Hollom
Scotland   May 1981

III   P A D Hollom   England   February 1985

succession (G Shaw) at rate of 75-100 per min, when excited or alarmed (Rankin and Rankin 1940), also commonly on approaching nest (e.g. Alder 1963), perched or more often in flight (P A D Hollom). Also rendered 'tze' and, in confrontations, a rapid rasping 'tze-tze-tze' (Keicher 1983). Variants include: 'zrik' and perhaps 'zerrb' (Bergmann and Helb 1982); at times of high stress e.g. in chase, or when handled, or flushed from nest, a shriller metallic 'zlint' (G Shaw), or high musical 'zlink', given usually in flight (Rankin and Rankin 1940). In flight, rate of calling increases with mounting agitation (Eggebrecht 1937). (3) Rattling-call. A loud rattling 'r-r-r-r' and, at other times, a short, low, rolling 'zur-r-r-r' (Rankin and Rankin 1940). In recording at nest nearing completion, arrival of bird when mate already present gave rise to variety of 'churr' sounds. At high intensity, sounds may be very long and reeling, given in rapid sequence (Fig III), at lower intensity brief and strangled, e.g. Fig IV shows 2nd of 2 churrs (2nd unit here) which merge into quiet song, perhaps Subsong; 1st, low-pitched, unit apparently contributed by mate (J Hall-Craggs).

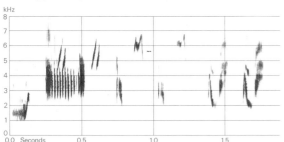

IV   P A D Hollom   England   February 1985

Mixture with song-units said to denote high arousal (Eggebrecht 1937, which see for rendering). Rattling-calls given in greeting between pair-members (Meeting-ceremony), also by ♀ performing Soliciting-display (G Shaw), often at or near nest. In Witherby *et al.* (1938*b*), Rattling-call assigned to *aquaticus*, but well developed in *gularis* also. (4) Other calls. (a) 'gri gri' given by bird emerging from roost and flying downstream (Keicher 1983). (b) Low-pitched 'gö', 'zö', 'kep' or 'kröp' sounds (Creutz 1966). Not clear if these different from, or variants of, calls listed above.

Calls of Young. Food-calls of young in nest shrill and squeaking (G Shaw), rendered 'zi-zi' or 'zü-zü-zü' (Horst 1941). Fledged young beg with loud, high-pitched 'zip-zip-zip...' (Bergmann and Helb 1982), or 'wi-wi-wi', very like Contact-alarm call of Common Sandpiper *Actitis hypoleucos* (Sudhaus 1972); same comparison made by Horst (1941) who rendered call 'die-die-die-die-die'. Series of 'gi-gi' calls, given rarely by independent young, June–October, when conspecifics flying overhead (Keicher 1983), presumably same or closely related. Renderings 'flui' (Manuel 1949), and 'fluii' (Penot 1948) indicate disyllabic interpretation possible. Fledged young also give Contact-call very like call 2 of adult (Sudhaus 1972), rendered 'zj-zj-zj' (Horst 1941). Juveniles begin singing in August, just before completing moult (H Galbraith).   EKD

Breeding. Season. British Isles: eggs laid last week of February to mid-June, with peak late March to late April; little regional variation (Shaw 1978). North-west and central Europe: laying begins mid-March, but delayed by cold springs, and becomes later with increasing altitude, e.g. by 5·92 days per 100 m in Harz mountains, West Germany (Balát 1964b; Makatsch 1976; Zang 1981). North Africa: laying from mid-March to May (Heim de Balsac and Mayaud 1962). Scandinavia: from early May in southern Norway and southern Finland, but from mid-May to early June in northern Finland (Haartman 1969; Andersson and Wester 1975). SITE. In cavities or on ledges above, often overhanging water; much use made of artificial sites where available; also uses nest-boxes. Of 1159 sites in Britain, 50·6% natural (15·4% on ledge on rock face, 10·4% under bank, 7·7% in hole or crevice, 6·4% among tree roots, 5·6% under waterfall, 2·8% in hollow stump, 1·4% on rock, 0·9% in vegetation), and 49·4% artificial (17·3% in cavity under bridge, 17·2% on ledge under bridge, 6·9% in wall, 3·1% in pipe or culvert, 2·5% on sluice, 1·6% on waterwheel, 0·4% in building); only 1% more than 1 m (horizontally) from water, 2 furthest 4 m and 7 m; of 1276 nests, 36·4% 0–1 m above water, 43·5% 1–2 m, 10·6% 2–3 m, 4·8% 3–4 m, 2·4% 4–5 m, 2·4% over 5 m (Shaw 1978). Of 84 nests in Norway, 56% on rock walls, 15·5% under overhangs, 28·5% in artificial sites (Andersson and Wester 1975). Of further 60 nests in Norway, 45% under bridge, 25% in rock face, 16% on rock in stream, 14% under waterfall or in building (Efteland and Kyllingstad 1984). Of 66 nests in Hungary, 35% on rocks, 33% under bridges, 17% in nest-boxes (Balát 1964b). Nest: domed structure of moss and grass stems and leaves, with wide entrance, usually pointing down towards water, with inner cup of stems, rootlets, leaves, and hair. Dimensions vary according to site; range of 8 nests, length 17–26 cm, width 13–22 cm, height 16–23 cm, entrance 5–7 cm, internal diameter 10 cm, depth 4·5–7 cm (Balát 1964b). Mean dry weight of cavity nest 159 g ($n=5$), and of ledge nest 264 g ($n=4$) (G Shaw). Building: by both sexes, with ♀ completing lining; moss gathered 10–30 m from nest, wetted if not already wet; first stops up crevices before nest proper begun at base, then builds back and sides, followed by front and entrance, finally inner cup and lining (Balát 1964b). After 1st brood fledged, lining renewed for 2nd (Balát 1962). Construction takes 18 days (9–28), $n=18$ in Harz mountains, West Germany (Zang 1981), where breeding season short, but 28 days (25–32), $n=40$ in Britain (Shaw 1978). EGGS. See Plate 81. Sub-elliptical, smooth and glossy; white. Nominate *cinclus*: 26·0 × 18·7 mm (22·0–30·6 × 17·0–20·1), $n=60$; calculated weight 4·6 g. *C. c. aquaticus*: 25·6 × 18·8 mm (23·4–28·4 × 16·5–20·1), $n=200$; calculated weight 4·55 g; *hibernicus* identical. *C.c. gularis*: 26·4 × 18·4 mm (22·0–29·5 × 16·5–19·9), $n=110$; calculated weight 4·75 g (Schönwetter 1979). Clutch: 4–5(1–8). Of 705 clutches, Britain: 1 egg, 1%; 2, 2%; 3, 9%; 4, 33%; 5, 51%; 6, 4%; 7, less than 1%; mean 4·42 ± SD0·88 (Shaw 1978). Of 91 clutches, West Germany: 1 egg, 1%; 3, 7%; 4, 14%; 5, 60%; 6, 18%; mean 4·85, decreasing by 0·17

eggs per 10 days of season, and 0·20 eggs per 100 m of altitude (Zang 1981). Of 74 clutches, Norway: mean 5·08 ± SD0·74 (Efteland and Kyllingstad 1984). 1–2 broods, rarely 3. 11% of 56 pairs in Scotland had 2 broods (G Shaw), and 14% of 70 pairs in Hungary, mainly in years when season started early (Balát 1964b). In 2 years, Norway, 5·3% of 38 pairs had 2 broods, and 6·5% of a further 62 pairs (Mork 1975). Re-lays after egg loss. 2nd clutch laid mean 9·9 days (1–25), $n=27$ after fledging of 1st brood (Shaw 1978). Eggs laid daily. INCUBATION. Mean 16 days (12–18), $n=26$ (Shaw 1978). By ♀ only but ♂ recorded covering eggs during her absence and may incubate rarely (Balát 1964b). For details of incubation spells, see Eggebrecht (1937). Begins with last, or perhaps penultimate egg; hatching synchronous. YOUNG. Altricial and nidicolous. Cared for and fed by both parents. FLEDGING TO MATURITY. Fledging period 22 days (20–24), $n=15$ (Shaw 1978). Length of period apparently dependent on size of nest and brood, longer where nest and brood larger (Baake 1982). Young fed for up to 18 days after fledging before becoming independent; more often for *c*. 1 week after fledging (see Social Pattern and Behaviour); fed by ♂ only if ♀ starting 2nd clutch (Balát 1962, 1964b; Hewson 1967). Age of first breeding 1 year (G Shaw). BREEDING SUCCESS. Of 1986 eggs in 455 nests, Britain, 61·0% hatched, and 84·1% reared to 14 days, equals 51·4% overall success; main losses from human interference (30·4%), desertion (25·4%), floods (19·6%), and predation (18·4%); omitting complete clutch losses, overall success 83·1% (Shaw 1978). In West Germany, of 258 eggs laid in 66 clutches, 80% fledged, mean 3·91 young per nest; 5 2nd broods produced mean 3·0 young (Zang 1981). In Hungary, of 178 eggs laid in 38 nests, 50·6% young reared to fledging (Balát 1964b). In Norway, data from 74 nests indicated 56% of nests started and 51% of eggs laid produced fledged young (Efteland and Kyllingstad 1984).

Plumages (*C. c. gularis*). ADULT. Forehead and crown down to upper cheek and ear-coverts, hindneck, and sides of neck dark chocolate-brown to dark drab-brown, contrasting with remainder of upperparts and lesser and median upper wing-coverts, which are dark slate-grey with broad deep black fringe to each feather; grey feather-centres largely hidden under black fringes on mantle and lesser coverts, appearing mainly black with limited grey spotting; scapulars, back, rump, and median coverts grey with distinct black scaling, upper tail-coverts largely grey. Chin, throat, and chest white (occasionally, with some faint black feather-tips when plumage fresh). Broad band across breast chestnut-brown or rufous-brown, merging into dark brown and black-brown on central and lateral belly; feathers faintly and narrowly tipped off-white in fresh plumage. Sides of breast, flanks, vent, and under tail-coverts dark slate-grey, suffused brown at border of belly; longer under tail-coverts narrowly tipped pale rufous or off-white. Tail dark slate-grey with black shafts. Flight-feathers, tertials, greater upper wing-coverts, upper primary coverts, and bastard wing brown-black, tinged slate when fresh; outer webs of tertials, secondaries, and greater coverts paler slate-grey along fringe;

tertials, secondaries, and inner primaries sometimes narrowly tipped white in fresh plumage; greater primary coverts, greater coverts, and longest feather of bastard wing narrowly and indistinctly fringed grey along tip. Under wing-coverts and axillaries dark brownish-grey, longest coverts faintly and narrowly tipped off-white. *In worn plumage*, brown of head and neck paler, dull grey-brown, less deep chocolate-brown; black feather-fringes on upperparts partly abraded, more grey of feather-bases visible; brown of breast paler rufous-brown or dull tawny-cinnamon; flight-feathers and greater coverts browner, without white tips. NESTLING. Down long and dense, dark smoke-grey; on upperparts only. JUVENILE. Upperparts entirely slate-grey, each feather with black or black-brown fringe along tip; fringes not as clear-cut as adult, often slightly narrower; forehead, crown, and hindneck appear grey with dark mottling, only inner scapulars, back, and rump with scaly pattern as adult. Lores, ear-coverts, upper cheeks, and sides of neck dull dark grey or brown-grey with faint paler grey or grey-buff mottling or streaking; some traces of white just above and below eye. Chin and throat white, part of feathers narrowly tipped black in fresh plumage; chest, breast, and central belly white, sometimes suffused cream or buff, each feather with black tip, variable in width, giving slightly mottled to heavily scaled appearance; sides of breast, flanks, vent, and under tail-coverts slate-grey, under tail-coverts with ill-defined pale brown tips. Tail and flight-feathers as adult, but slightly browner, less slaty; tips of tertials, secondaries, and inner primaries broadly fringed white when fresh (not narrowly as in some adults). Upper wing-coverts grey, as in adult, but duller and less slaty; lesser and median hardly tipped black; greater coverts and inner greater primary coverts narrowly tipped white (unlike adult); tips of greater upper primary coverts rather pointed, slightly tinged brown (in adult, broadly and smoothly rounded, indistinctly fringed grey). Under wing-coverts grey with white tips, primary coverts largely white. Some variation in colour of sides of head (buff or grey), width of black bars on chest and breast, and ground-colour of chin to belly (cream-white or buff), in part probably sex-linked: ♀ on average buffier and more heavily barred than ♂ (Balát 1964b; RMNH, ZMA). FIRST ADULT. Similar to adult, but breast and belly with pale brown to off-white feather-tips and chest sometimes with narrow black feather-tips (usually worn off by late winter or spring); juvenile flight-feathers, tail, greater upper primary coverts, and many or all greater upper wing-coverts retained, browner than in adult and somewhat contrasting in colour and wear with newer neighbouring 1st adult feathers; outer greater coverts and inner primary coverts with narrow white tips (absent in adult, but may occasionally wear off in 1st adult in spring and summer; Balát 1964b; Andersson and Wester 1973; Görner 1978); greater upper primary coverts narrower and with more pointed and frayed tip than in adult (tip not broadly rounded nor with smooth edges).

Bare parts. ADULT, FIRST ADULT. Iris deep brown, rich brown, or deep red-brown in adult, grey-brown, dull brown, or dark brown in 1st summer and autumn. Eyelids white, except for narrow black rim round eye; lower often just visible as a narrow short stripe at rest, upper conspicuous only when eye closed, showing as contrastingly white spot ('winking' or 'blinking') (Alder 1957). Bill horn-black, brown-black, or black, base of lower mandible sometimes tinged slate. Leg and foot dark brown, black-brown, or black, front of tarsus and toes leaden-grey or whitish-slate, soles pale grey or horn-brown. (Hartert 1910; Witherby *et al.* 1938b; Ali and Ripley 1973b; Parsons and Reid 1975; Bub *et al.* 1984; RMNH.) NESTLING. Mouth pink, bright

yellow, or orange-yellow. Bare skin orange-flesh. JUVENILE. Iris grey-brown or dark brown. Bill horn-black. Leg and foot dark horn-brown, sometimes with flesh tinge. (Heinroth and Heinroth 1924–6; Witherby *et al.* 1938b; Bub *et al.* 1984; BMNH.)

Moults. ADULT POST-BREEDING. Complete; primaries partly simultaneous, partly descendant. Starts with p1, in Britain between late May and late July, mainly first half of June. Inner 5 primaries (p1–p5) shed simultaneously; p6 lost when p1–p3 full-grown and p4–p5 growing, mainly about mid-July; other primaries then follow in descendant order, each shed when previous feather *c.* ⅓ grown; moult completed with regrowth of p9–p10 in August (mainly mid-August), occasionally September (Galbraith *et al.* 1981). Body moults at same time as primaries, but occasionally some moult of mantle and scapulars occurs April–May, perhaps representing partial pre-breeding moult (C S Roselaar). Moult of secondaries starts with shedding of s1 at about same times as loss of p7, about late July; s2–s6 follow about August at about shedding of p9, sequence s2 to 26, but s6 often lost out of order, mainly directly after s3, and occasionally s5 also; last feathers still growing when primaries already completed. Tertials (s7–s9) replaced in irregular sequence, sometimes 2 or all 3 simultaneously; mainly early July to early August. Tail replaced in irregular sequence or, more often, partly or wholly simultaneously; mainly early July to mid-August. (Galbraith *et al.* 1981.) Estimated average duration of primary moult in Britain *c.* 70 days (48–98 days) (Galbraith *et al.* 1981) or *c.* 64 days (Ginn and Melville 1983); elsewhere *c.* 2 months (Richter 1954) or a few days less (Balát 1960). See also Stresemann (1920), Vaurie (1951b), Dementiev and Gladkov (1954b), Richter (1954), Balát (1960), and Stresemann and Stresemann (1966). Adults accompanying dependent juveniles start 4–5 weeks later than solitary ones (Richter 1954). POST-JUVENILE. Partial: head, body, lesser and median upper wing-coverts, and sometimes a few inner greater upper wing-coverts. Starts at *c.* 60 days old, *c.* 4 weeks after fledging; as fledging occurs March–September (Fuchs 1972), timing of moult also strongly variable; mainly June–September (as in adult post-breeding), occasionally April–October (RMNH, ZMA).

Measurements. ADULT, FIRST ADULT. *C. c. gularis*. England and southern Scotland, all year; skins (BMNH, RMNH, ZMA). Bill (S) to skull, bill (N) to distal corner of nostril; exposed culmen on average *c.* 5·8 shorter than bill (S).

	♂		♀	
WING	95·3 (2·00; 18)	92–99	87·5 (2·23; 16)	82–91
TAIL	52·0 (2·64; 11)	48–56	48·8 (2·94; 13)	46–54
BILL (S)	22·5 (0·74; 11)	21·1–23·4	21·4 (0·85; 13)	19·6–22·4
BILL (N)	12·6 (0·71; 11)	11·8–13·5	12·0 (0·53; 13)	11·0–12·5
TARSUS	30·2 (0·75; 10)	29·4–31·3	28·1 (0·69; 12)	27·3–29·3

Sex differences significant. Live birds, Scotland: wing, ♂ 96·0 (1·87; 89) 91–100, ♀ 88·1 (2·11; 81) 80–92 (Galbraith and Broadley 1980).

Nominate *cinclus*. Fenno-Scandia all year; skins (RMNH, ZMA).

	♂		♀	
WING	97·8 (2·00; 13)	95–101	88·9 (1·46; 7)	87–91
TAIL	53·4 (2·37; 13)	50–57	48·1 (2·85; 7)	46–53
BILL (S)	22·3 (0·91; 13)	20·9–23·5	20·9 (0·70; 7)	19·8–21·8
BILL (N)	12·3 (0·62; 13)	11·4–13·3	11·4 (0·51; 7)	10·8–12·0
TARSUS	29·5 (0·79; 13)	28·6–30·7	27·4 (0·66; 7)	26·6–28·7

Sex differences significant.

Live birds, Norway and Sweden: wing, ♂ 98·4 (1·86; 337) 94–104, ♀ 89·9 (1·86; 432) 84–97; tarsus, ♂ 29·8 (0·85; 143) 25·5–31·5, ♀ 27·8 (0·81; 185) 25·3–30·1 (Andersson and Wester

1971, 1975; Lundberg *et al.* 1981). Corsica and Sardinia: wing, ♂ 92·8 (1·71; 8) 91–96, ♀ 87·2 (0·29; 4) 87–88 (RMNH, ZFMK, ZMA). North-west Spain: wing, ♂ 95·4 (1·92; 15) 92–99, ♀ 86·8 (1·33; 6) 85–89 (RMNH, ZFMK, ZMA).

C. c. aquaticus. Wing: (1) West and East Germany and northern Czechoslovakia (RMNH, ZMA), (2) southern Austria and mainland Italy (RMNH, ZMA), (3) eastern Hungary, Rumania, southern Yugoslavia, Greece, and European Turkey (Stresemann 1920; Makatsch 1950; Rokitanksy and Schifter 1971; RMNH, ZMA). Other measurements for all areas combined, all year; skins (RMNH, ZMA).

WING (1)	♂ 95·1 (1·72; 16)	91–98	♀ 86·5 (1·98; 8)	84–90
(2)	94·7 (0·98; 6)	93–96	88·1 (1·72; 7)	84–90
(3)	94·1 (1·87; 20)	91–98	86·1 (2·17; 8)	82–89
TAIL	50·6 (1·77; 26)	48–54	46·9 (1·64; 17)	44–50
BILL (S)	22·2 (0·73; 25)	21·0–23·5	21·2 (0·60; 16)	20·2–22·4
BILL (N)	12·3 (0·61; 25)	11·5–13·6	11·9 (0·61; 15)	11·0–12·9
TARSUS	29·5 (0·92; 29)	27·9–31·3	27·7 (0·99; 18)	26·8–28·9

Sex differences significant, except bill (N).

Wing of live birds: West Germany, ♂ 94·7 (2·33; 139) 91–102, ♀ 86·9 (1·71; 146) 83–93 (Bub *et al.* 1984); East Germany, ♂ 94·3 (80) 90–101, ♀ 86·8 (71) 84–89 (Görner 1981).

C. c. minor. North-west Africa: wing, ♂ *c.* 98, ♀ *c.* 82–90 (Hartert 1910); bill to skull, ♀ 24 (7) 22–26 (Vaurie 1959).

C. c. caucasicus. (1) Taurus mountains, southern Turkey (Kumerloeve 1961), and north-east Turkey (ZFMK). (2) Caucasus and Transcaucasia, (3) Elburz mountains and north-east Iran, (4) Azarbaijan and Hamadan, north-west Iran (Vaurie 1951b; RMNH, ZFMK, ZMA). Bill measured to skull.

WING (1)	♂ 95·1 (1·68; 11)	93–98	♀ 87·9 (3·11; 6)	84–92
(2)	94·2 (2·30; 13)	91–98	86·7 (3·23; 9)	83–93
(3)	94·7 (1·98; 7)	93–99	87·2 (2·23; 6)	84–90
(4)	97·6 (2·05; 10)	95–101	92·0 (1·41; 6)	90–94
TAIL	52·6 (—; 27)	48–59	47·4 (3·34; 19)	42–53
BILL (2)	22·4 (0·72; 8)	20·9–23·0	21·0 (1·08; 7)	19·7–22·5
(3)	22·2 (—; 6)	21·5–22·5	21·4 (—; 8)	20·5–22·5
(4)	22·8 (—; 10)	21·5–23·5	21·8 (0·47; 6)	21·2–22·5
TARSUS	28·7 (—; 3)	28·1–29·2	27·2 (0·39; 5)	26·6–27·5

C. c. rufiventris. Lebanon: wing, ♂ *c.* 92·5, ♀ 84 (Hartert 1910; Vaurie 1959).

C. c. persicus. Zagros mountains, Iran: wing, ♂ 100·5 (1·90; 11) 98–104, ♀ 92·1 (1·64; 8) 90–94; bill (S), ♂ 22·2 (11) 21–23, ♀ 21·9 (8) 21·2–22·5 (Vaurie 1951b).

JUVENILE. Wing not full-grown until 6–8 weeks old, 3–5 weeks after fledging (Fuchs 1972). First adult with retained juvenile wing and tail combined with older birds in tables above, though wing on average 0·6 shorter and tail 2·8 shorter.

Weights. *C. c. gularis.* Scotland, ♂ 69·1 (3·0; 124) 60–76, ♀ 58·1 (3·3; 101) 50–67 (Moss 1975). Average of ♂ *c.* 70 January, decreasing to *c.* 68 in May when nesting, but high level of *c.* 70 reached July–August when in moult, followed by minimum of 66 September–October. In ♀, *c.* 59 December–February, increasing to *c.* 67–68 March–April when nesting, decreasing again to *c.* 60 June–August when moulting, and followed by minimum of *c.* 57–58 September–October. (Galbraith and Broadly 1981.)

Nominate *cinclus.* Norway, May–June: ♂ 66·4 (3·28; 32) 60–73, ♀ 57·4 (4·82; 52) 50–72 (Andersson and Wester 1975). South-east Sweden, November–April: ♂ 69·2 (4·20; 286) 58–84, ♀ 58·4 (3·78; 369) 49–72; slightly lower weight on arrival in about first half of November, average of ♂ 67·8, ♀ 57·8, constant throughout winter (♂ mainly 69–70, ♀ 58–59),

increasing to average 71·1 (♂) or 61·1 (♀) shortly before departure in second half of March and first half of April (Andersson and Wester 1972). Northern Sweden, November–April: ♂ 73·1 (4·4; 61) 59–84, ♀ 59·7 (4·2; 27) 51–66 (Lundberg *et al.* 1981). North-west Spain, April: ♂♂ 62, 67; ♀ 56 (ZFMK).

C. c. aquaticus. West and East Germany: ♂ 64·2 (3·75; 178) 53–76, ♀ 55·4 (5·54; 180) 46–72 (Görner 1981; Bub *et al.* 1984).

Averages throughout year (Görner 1981; Bub *et al.* 1984).

JAN–MAR	♂ 65·9 (28)	♀ 57·0 (21)	AUG–SEP	♂ 66·4 (104)	♀ 54·9 (87)
APR–MAY	64·3 (38)	56·7 (40)	OCT–DEC	65·2 (47)	53·4 (65)
JUN–JUL	65·3 (53)	54·7 (22)			

Sierra Nevada (Spain), May: ♂ 62 (Niethammer 1957). European Turkey, May: ♀ 46 (Rokitansky and Schifter 1971).

C. c. persicus. South-west Iran, April: ♂ 61·2, ♀♀ 74·2, 56·2 (ill) (Paludan 1938).

NESTLING. At fledging, 51·5 (4·49; 9) 45–59 (Görner 1981).

Structure. Wing short and broad, tip rounded. 10 primaries: p8 longest, p9 1–2·5 shorter, p7 0–1·5, p6 2–4, p5 6–10, p4 11–14, p1 19–26. P10 rather short and narrow, reduced; 39–49 shorter than p8, 3 shorter to 5 longer than longest upper wing-covert. Outer web of (p6–)p7–8 and inner web of (p7–)p8–p9 emarginated. Tertials very short; shorter than secondaries in closed wing. Tail short, tip square or slightly rounded; 12 feathers. Bill short and slender, tip laterally compressed; culmen slightly decurved near tip, cutting edges straight or curved slightly up, gonys distinctly curved up. Tip of upper mandible forms small hook. Nostril narrow, protected by broad membrane above. No bristles near base of bill. Feathering of head and body dense. Tarsus and toes rather long, heavy; rear of tarsus with sharp ridge. Middle toe with claw 25·6 (39) 24–28 in ♂, 24·6 (22) 23–26 in ♀; outer toe with claw *c.* 69% of middle with claw, inner *c.* 66%, hind *c.* 71%. Claws short and strong, decurved.

Geographical variation. Marked and complex; involves colour of head and nape and width of dark feather-fringes on remainder of underparts (both strongly affected by bleaching and wear), colour of breast and belly (often with marked individual variation), and (slightly) size. Polymorphism occurs in central Asia, all-dark birds living alongside white-chested ones. In west Palearctic, breast and belly often either rufous-brown or blackish-brown, but distribution of each of these colour-groups not contiguous, range of rufous ones broken by blackish ones and *vice versa*. Some populations fairly uniform in appearance, others highly variable (notably in Switzerland, Pyrénées, Sierra de Guadarrama, and Carpathian basin). To avoid having races with non-continuous geographical ranges within each of dark-breasted and rufous-breasted groups, many races formerly recognized. However, all populations of both groups closely similar and excessive splitting not warranted, in spite of patchy distributions (Greenway and Vaurie 1958). In Czechoslovakia, birds of steep, damp, and well-vegetated ravines have darker underparts, those of wider, warmer, and more open valleys are more rufous below (Balát 1961), and same may apply to whole west Palearctic, dark-breasted birds inhabiting areas with cooler and wetter climates, rufous-breasted warmer and drier areas. Geographical variation may be obscured by ♂ being darker on average than ♀, and some darkening occurs with age (Richter 1954; Balát 1961).

C. c. gularis from Britain is rather dark above, crown and nape rather dark greyish-brown, drab-brown, or chocolate-brown, feather-fringes on remainder of upperparts brown-black, 2–3 mm wide; underparts rather dark, but with pronounced chestnut-brown breast and central belly, merging into brownish-black vent. *C. c. hibernicus* from Ireland and western Scotland darker above than any other west Palearctic race; crown and nape dark chocolate-brown; feather-fringes blacker, *c.* 3 mm wide in fresh plumage; rufous of breast duller, darker, and more restricted, belly more extensively brownish-black; darkest birds similar to nominate *cinclus* below, but distinctly darker above. Nominate *cinclus* from northern Europe has upperparts similar to *gularis*, but breast and belly completely blackish-brown; some rufous-brown tinge often visible along border with white chest, but not as much as in *gularis*. Populations from western and central France and Pyrénées ('*pyrenaicus*' Dresser, 1892) and from Sardinia and Corsica ('*sapsworthi*' Arrigoni, 1902) close to nominate *cinclus* and (for simplicity) included in it here, though crown and nape on average very slightly paler, feather-fringes on remainder of upperparts faintly narrower, and breast sometimes has slightly more extensive rufous-brown tinge; populations from north-west Iberia ('*atroventer*' Floericke, 1926) also included in nominate *cinclus*, being similar except for slightly deeper brown-black belly and vent. Size of these south-western populations of nominate *cinclus* near *gularis* and *hibernicus*, not as great as typical nominate *cinclus* from Fenno-Scandia. No specimens examined of *uralensis* from east European USSR; upperparts stated to be intermediate between north European nominate *cinclus* and central European *aquaticus*, breast chocolate-brown, paler than nominate *cinclus*, less rufous than *aquaticus* (Hartert and Steinbacher 1932-8); considered a valid race by Greenway and Vaurie (1958), but included in nominate *cinclus* by Dementiev and Gladkov (1954b), where however strongly different *caucasicus* and *rufiventris* are also included. *C. c. aquaticus* from central Europe (south to north-west France, West Germany, and western Czechoslovakia) rather pale above and bright rufous-chestnut or deep chestnut on breast and belly; crown and nape rather pale grey-brown (like '*pyrenaicus*' and *caucasicus*), feather-fringes of remainder of upperparts rather narrow (slightly narrower than in *gularis* and nominate *cinclus*, rather similar to *caucasicus*), breast and belly brighter and paler rufous than in *gularis*; smaller than typical nominate *cinclus* from northern Europe, but similar in size to populations of southern and western Europe. Populations from Austria, mainland Italy, Sicily, and south-east France ('*meridionalis*' Brehm, 1856) on average slightly paler than *aquaticus* from central Europe, but many virtually indistinguishable. Birds from Balkans and Greece ('*orientalis*' Stresemann, 1919) slightly darker than *aquaticus* on average, breast and upper belly deeper chestnut-brown; in particular, populations from southern Yugoslavia and Greece darker than average *aquaticus* (perhaps grading into *caucasicus* of northern Turkey: Bub *et al.* 1981), but populations of Carpathians, Rumania, central Yugoslavia, and Hungary individually strongly variable and similarly dark birds occur occasionally within typical *aquaticus* from Germany. Populations from eastern and southern Spain inseparable from *aquaticus* (Greenway and Vaurie 1958). *C. c. minor* from north-west Africa near *aquaticus* in colour and size of wing, but bill longer (see Measurements). *C. c. caucasicus* from Caucasus, Transcaucasia, and Iran north from Elburz mountains and east to Khorasan differs from other west Palearctic races in colour of upperparts; dull grey-brown of crown and nape extends to mantle and scapulars, mantle without distinct grey feather-centres, only rump and upper tail-coverts grey with dark feather-fringes (similar in this respect to extralimital races of central Asia); breast and belly uniform dull grey-brown, rather like crown and nape, sometimes slightly tinged rufous on breast and occasionally rather similar to nominate *cinclus* below, but colour greyer, less deep brown. Birds from Taurus mountains (southern Turkey) are close to Caucasus birds in appearance and also included in *caucasicus*, though breast slightly paler brown and less greyish (Kumerloeve 1961). Birds from northern Turkey sometimes separated as *amphitryon* Neumann and Paludan, 1937; usually included in nominate *cinclus* (Greenway and Vaurie 1958; Vaurie 1959), but size and underparts similar to *caucasicus* and upperparts intermediate between *caucasicus* and *aquaticus*, and hence included here in *caucasicus*; birds from Ulu Dag (north-west Turkey) tend more strongly towards *aquaticus*, but still nearer *caucasicus* (ZFMK). Position of now-extinct *olympicus* from Cyprus obscure: apparently close to *caucasicus* (Greenway and Vaurie 1958), perhaps especially to Taurus birds and perhaps better combined with that race, but *olympicus* maintained here in the absence of direct comparison between birds from Cyprus and Taurus and for the sake of stability of nomenclature (*olympicus* is an older name than *caucasicus*). Populations of Iranian Azarbaijan similar in colour to *caucasicus*, but distinctly larger, like *persicus* from further south; birds from Hamadan area tend somewhat towards *persicus* in colour (Vaurie 1951b). Typical *persicus* from south-west Iran larger and paler than any west Palearctic race; breast and belly pale rufous; *rufiventris* from Lebanon similar, but smaller, and breast and belly slightly darker rufous (but only 2 specimens known) (Hartert 1910; Vaurie 1951b). Extralimital races either rather like *caucasicus*, or all-dark, or largely white below; see Vaurie (1959) and Stepanyan (1977).

CSR

Family TROGLODYTIDAE wrens

Tiny to fairly small oscine passerines (suborder Passeres); insectivorous, living mainly in scrub close to ground or on ground close to scrub—though some found in marshes or in rocky areas. 52-60 species in up to 14 genera, including: (1) *Campylorhynchus* (cactus wrens), 12 species; (2) *Salpinctes* (rock wrens), 2 species; (3) *Cistothorus* (marsh wrens), 4 species; (4) *Thryothorus*, 16-21 species; and (4) *Troglodytes*, 5-6 species. Found mostly in the Americas, only a single species—Wren *Troglodytes troglodytes*—in Eurasia (including many islands of North Atlantic and North Pacific). Most species sedentary.

Body short and rotund; neck short. ♂ slightly larger than ♀ in many species. Bill slender and somewhat curved, with sharp tip; usually rather short but fairly long in some terrestrial species. Nostril partly covered by small membrane. Bill used by some species for turning over stones, leaves, and other ground debris when searching for food. Wing short and rounded; 10 primaries, p10 well developed. Flight direct and rapid with fast, whirring wing-beats. Tail short to long, often cocked up; feathers soft and often narrow, number variable (12 in *Troglodytes*). Leg and toes strong, with long claws; front toes partly fused at bases. Tarsus scutellate. Usually gait a hop but some species can run. Head-scratching by indirect method. Bathe in shallow water in usual passerine stand-in manner, but reported using in-out method at times. Dusting a characteristic habit. Adopt lateral and spread-eagle postures when sunning. No anting records.

Plumage thick and soft; generally closely barred or streaked in various shades of brown though underparts often partly uniform buff to white; often with some contrasting spots on wing and tail; some species have distinct pale supercilium. Sexes alike. Juvenile plumage closely similar to adult. Adults usually have single full moult annually in autumn but in some species also partial 2nd moult in spring; post-juvenile moult partial, autumn. Nestling down rather short and scanty; mouth brightly coloured, without spots.

Relationships usually thought to be closest with Mimidae (mockingbirds) on morphological and anatomical grounds, Certhiidae (treecreepers) being not far distant. Egg-white protein patterns, however, are quite unlike those of Mimidae and Turdidae (chats, thrushes), but rather similar to those of *Certhia* treecreepers and *Parus* tits (Paridae) (Sibley 1970). DNA data support egg-white protein data, indicating close relationship with *Certhia* and New World gnatcatchers *Polioptila*; these all placed in single family by Sibley and Ahlquist (1985*a*), close to Paridae within superfamily 'Sylvioidea'—which also includes Hirundinidae (swallows), Pycnonotidae (bulbuls), and Sylviidae (Old World warblers)—far removed from Turdidae and allies in superfamily 'Turdoidea', though all are members of same larger assemblage ('Muscicapae').

Troglodytes troglodytes **Wren**

PLATE 36
[between pages 544 and 545]

Du. Winterkoning	Fr. Troglodyte mignon	Ge. Zaunkönig
Ru. Крапивник	Sp. Chochín	Sw. Gärdsmyg	N. Am. Winter Wren

Motacilla Troglodytes Linnaeus, 1758

Polytypic. Nominate *troglodytes* (Linnaeus, 1758), continental Europe from Fenno-Scandia and Urals south to Pyrénées, north-west Spain, Portugal, mainland Italy, and Greece, grading into *indigenus* in central and southern England, into *kabylorum* along Mediterranean coast of France and in central Spain and Sicily, into *hyrcanus* in eastern Rumania and perhaps Ukraine, Crimea, and northern Asia Minor, and into *cypriotes* in western and southern Asia Minor; *indigenus* Clancey, 1937, Britain and Ireland, north to Inner Hebrides and Orkneys; *fridariensis* Williamson, 1951, Fair Isle; *hirtensis* Seebohm, 1884, St Kilda; *hebridensis* Meinertzhagen, 1924, Outer Hebrides; *zetlandicus* Hartert, 1910, Shetland; *borealis* Fischer, 1861, Faeroes; *islandicus* Hartert, 1907, Iceland; *koenigi* Schiebel, 1910, Corsica and Sardinia; *kabylorum* Hartert, 1910, north-west Africa, Balearic Islands, and southern Spain; *juniperi* Hartert, 1922, north-west Cyrenaica (Libya); *cypriotes* (Bate, 1903), Crete, Rhodos, Cyprus, and Levant; *hyrcanus* Zarudny and Loudon, 1905, eastern Asia Minor, Caucasus, Transcaucasia, and northern and western Iran. Extralimital: *c.* 25-30 further races in central and eastern Asia and North America.

Field characters. 9-10 cm; wing-span 13-17 cm. Not the smallest but the shortest bird in west Palearctic, due to habit of holding tail erect; appears about half size of Dunnock *Prunella modularis*. Tiny, restless, and pugnacious passerine, usually seen at or near ground level and at any distance appearing warm brown overall. Most obvious characters are rather long, thin bill, pale buff supercilium, barred wings and flanks, and often cocked tail. Flight confident, with rapid whirring action. Sexes similar; no seasonal variation. Juvenile separable at fledging. 13 races in west Palearctic, with some isolated forms markedly distinct but most markedly clinal; only continental European race, nominate *troglodytes*, described here (see also Geographical Variation).

ADULT. Essentially russet-brown above and paler buff-brown to white below, barred (or tending to be) all over. Obvious plumage features restricted to: (a) cream-buff supercilium; (b) white-buff chin and throat; (c) obvious, even white-based, barring on folded primaries, under tail-coverts, and flanks; (d) brighter (most rufous) tone to rump and tail-base; (e) tiny white spots on tips of median coverts. In flight, no characters show and bird becomes just a compact, round-winged, warm brown, almost bee-like creature. JUVENILE. Often shows pale yellow gape in first weeks after fledging. Plumage even more warmly coloured than adult, but with less obvious markings, notably lack of white tips to under tail-coverts and median wing-coverts. At all ages, bill dark horn with paler, yellower, lower mandible and base; legs usually light brown.

Unmistakable in west Palearctic, with all other members of Troglodytidae confined to Nearctic. Even juvenile *P. modularis* much larger and longer-tailed and has different behaviour. Flight actions all rapid and whirring—difficult to distinguish, except in brief glide before landing when rounded wings and spread tail present 'parachute'

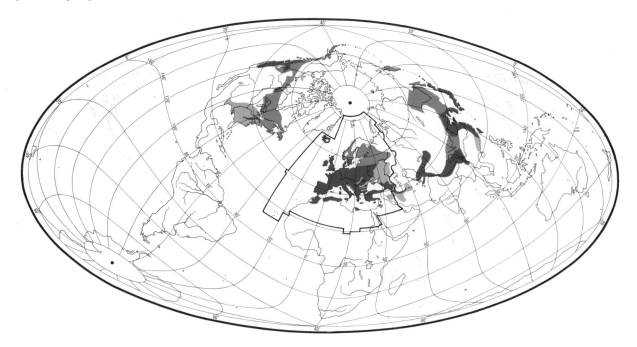

appearance. Gait rather mouse-like, slowly or quickly creeping, hopping, and climbing in constant search of ground and low plant cover. Behaviour combines quiet skulking with frequent, noisy and irascible squabbles with fellows or other passerines; actions include obvious bobbing, tail-wagging, and body-swivelling, and amazing vibrating of whole bird during singing. Not social during day but packs into roosts in hard weather.

Song a well-structured, shrill (but not unsweet), pulsing, warbling phrase delivered as if bird trying to burst lungs; usually lasts 4–6 s. Calls include hard, clicking trisyllable, and churring or rattling in alarm.

Habitat. Breeds in west Palearctic throughout middle latitudes and in parts into higher latitudes, from Mediterranean through temperate to boreal zones, with oceanic rather than continental tendency, between July isotherms of *c.* 10–22°C, thus avoiding extremes of cold and heat. In suitable situations will nest above treeline (in Switzerland, to 2000 m or even 2400 m), but mainly in lowlands from sea-level; also a successful colonist of both inshore and offshore oceanic islands. Within predominantly moist mild climatic range, suitable habitat offered by wide variety of low cover and foraging opportunities, including herb and field layers of plant growth (within or outside woodland), crops and aquatic vegetation, fallen trees and branches or heaps of brash, hedgerows, gardens, parks, and shrubberies. Is attracted to earthen banks, stone walls, outhouses and other free-standing structures, and natural crags, fissures, sea-cliffs, and other faces or slopes providing cavities, crevices, and interstices which can be profitably explored or used for roosting or nesting.

Apart from wide open tracts of bare rock, snow, ice, sand, gravel, mud, soil, or water, and also forests without undergrowth, extensive farmlands, and closely built-up urban areas, suitable habitat exists as discontinuous layer of low natural cover or artefacts over much of remaining landscape, irrespective of how it might otherwise be classed. The most favourable habitats, however, are damp deciduous or mixed woods with much small growth and thick cover, overgrown valleys with streams, untended garden-woodland areas, and sheltered deep tree-fringed lanes. Arid areas are avoided and there is a preference for nesting near brooks and lakes, and thus for mountain areas in southern parts of range, e.g. in North Africa most breed in stream valleys or forests at 1200–1800 m, rising to 3000 m in Haut Atlas. In southern USSR, breeds in forest zone of mountains in woods with heavy underbrush or in thickets along mountain streams, running amid rocks and cliffs with piles of stones grown with dense low vegetation. In Carpathians, at *c.* 1350–1800 m in beech *Fagus* and spruce *Picea*. In north, breeds in taiga and mixed forests or open woodlands and alder thickets, provided these contain dense clusters of shrubs, nettles, or other undergrowth and heaps of dead vegetation or dry branches (Dementiev and Gladkov 1954*b*). In Britain, during build-up of population after heavy reductions in hard winter, woodland and streamside habitats reoccupied first, followed, as these become saturated, by gardens, and finally by farmland hedgerows. The most abundant and widely distributed bird species in British woodlands, both in summer and winter. One of the 3 most widespread species in Britain, and one of the most adaptable, found in almost every habitat from seashore to highest mountain boulder-fields. Among other

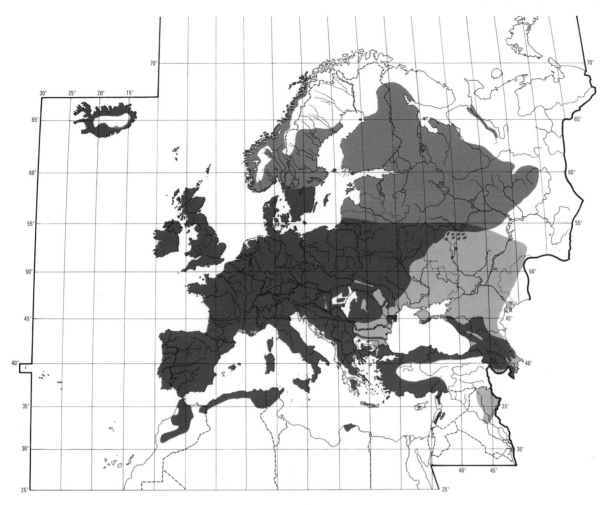

habitats in Britain are rocky coastlands with breeding sites in fissures and caverns, lowland heaths where deep heather is nest-site, chalk downland with yew and oak scrub, willow *Salix* scrub, high marsh with dense tall vegetation, and steeply sloping heather or grass moorlands at over 500 m (Sharrock 1976; Fuller 1982). Near coast, prefers dense cover of bramble *Rubus* or bracken *Pteridium*, in most sheltered sites (Hawthorn 1975*b*); rarely feeds in intertidal zone (Hawthorn 1975; Armstrong and Whitehouse 1977), though Dixon (1885) stated that *hirtensis* often does. For preferred feeding sites inland, see Food. After breeding season, with removal of constraints of access to nest-sites and feeding young, catholicity of occurrence in even wider range of habitats is so much increased that a list of those on which the species will never occur might be easier to compile than the converse, but would risk soon becoming falsified. (Nicholson 1951; Armstrong 1955; Voous 1960; Glutz von Blotzheim 1962.)

Distribution. No evidence of any marked range changes.

ISRAEL. First bred 1970s (HM, UP).

Accidental. Madeira.

Population. Marked fluctuations after hard winters.

ICELAND. Claims of decline not substantiated. See also Bengtson (1970). FAEROES. 250–500 pairs (Bloch and Sørensen 1984). BRITAIN, IRELAND. Over 1 million pairs, but suffers major losses in hard winters (Parslow 1973). 10-fold increase shown by Common Birds Census index 1964–74 after 2 hard winters; *c.* 10 million pairs at peak (Sharrock 1976). FRANCE. Over 1 million pairs (Yeatman 1976). BELGIUM. About 200 000 pairs (Lippens and Wille 1972). NETHERLANDS. Estimated 350 000 pairs; marked fluctuations after hard winters (Teixeira 1979). WEST GERMANY. Estimated 1·8–2·9 million pairs (Rheinwald 1982). SWEDEN. Estimated 500 000 pairs (Ulfstrand and Högstedt 1976); marked fluctuations after severe winters, e.g. 10-fold decrease 1975–9 (LR). FINLAND. Estimated 6000 pairs (Merikallio 1958); very marked fluctuations (OH). CZECHOSLOVAKIA. No change in numbers reported (KH). POLAND. Fairly numerous (Tomiałojć 1976).

Survival. England: average annual overall mortality *c.* 63% (Hawthorn and Mead 1975); average annual adult mortality 58% (Innes 1981). Oldest ringed bird 6 years 7 months (Dejonghe and Czajkowski 1983).

Movements. Migratory, partially migratory, and resident. Continental populations exhibit long- and short-distance movements, either on basically north–south axis or altitudinal, with many in southern parts of range sedentary. In Europe and probably in Asia, apparently unable to endure winter further north than *c.* -7°C January isotherm (Armstrong 1955).

Some birds of Icelandic race, *islandicus*, move to coastal zone (Armstrong 1955) or come near farms in winter (Timmermann 1938-49), but races endemic to Faeroes (*borealis*), Shetland (*zetlandicus*), Fair Isle (*fridariensis*), and St Kilda (*hirtensis*) not known to undergo even local movements as these smaller islands have no areas far enough from sea to be significantly colder (British Ornithologists' Union 1971). In Outer Hebrides (*hebridensis*), local movement to coast in winter probably induced by deterioration of inland cover and may lead to birds occupying intertidal zone (Hawthorn *et al.* 1976).

Populations of *indigenus* from mainland Britain have been subject to fairly intensive ringing, almost 170 000 having been ringed up to 1981 (Spencer and Hudson 1982). Despite low recovery rate (less than 0·6%), returns show 2 types of movement: directions random for distances up to 50 km, but longer movements (mostly 50-250 km) highly orientated towards south in autumn and north in spring, while 3 ringed at Dungeness were recovered in Bordeaux and Marseilles, southern France (Hawthorn and Mead 1975). Movement essentially nocturnal and many migrants attracted to lighthouses (Patten 1912; Henty 1980). Movement may begin as early as August (Armstrong 1955) and movement from headlands noted regularly in autumn, e.g. at Dungeness mid-September to early November; smaller numbers arrive in spring, mid-March to late April (Parslow 1969). Within Britain, local movement (especially of 1st-year birds) to reedbeds provides adequate winter shelter except in the most severe conditions (Hawthorn 1975a). Hilbre Island (Cheshire, England) has no breeding population but small numbers winter there regularly, November–March (Hawthorn 1975b). Double passage and occasional overwintering occurs on Isle of May, Scotland (Baxter and Rintoul 1918). In Ireland (*indigenus*), movements masked, as in mainland Britain, by majority being sedentary or undertaking only short-distance random movements, though clear seasonal departures and influxes occur at headlands and lighthouses; many arrive November, probably to winter, origin unknown (Ruttledge 1966). Has been recorded on North Atlantic weatherships (Luttik and Wattel 1979).

In western Europe, movements of nominate *troglodytes* again masked by large proportion of birds wintering within breeding range but large-scale ringing has shown that individuals make substantial journeys (Zink 1981). General direction between south and south-west, with birds ringed Germany recovered western France; movements mostly less than 1000 km but a few in excess of 2500 km, e.g. Gotland (Sweden) to southern Spain. Such movements apply to birds ringed as nestlings and as breeding-season adults, though some of the latter move more to west, e.g. Belgium to Channel Islands. Same direction of movement shown by birds ringed autumn or winter and recovered before or during the following breeding season; one ringed in autumn in Lithuania (western USSR) recovered in December near Bordeaux, *c.* 2000 km south-west. Nominate *troglodytes*, probably from Scandinavia, is frequent at Fair Isle (Williamson 1965). Ringed bird seen in northern Iceland, where no birds had been ringed, may well have been nominate *troglodytes* from Scandinavia (Armstrong and Whitehouse 1977). Populations of northern Europe, Fenno-Scandia, and western USSR show great fluctuations in numbers from year to year reflecting severity of winters since so many individuals attempt overwintering and perish (Hildén 1974). In USSR, smaller numbers seen in winter, and March influx apparent in many areas (Dementiev and Gladkov 1954b). Northward spring passage inconspicuous but movement at Danish lighthouses noted 20 March-20 May (Tischler 1941; Preuss 1978) and movement through Sweden is slow; arrival at Värmland (southern Sweden) not before 12 April, and at Västerbotten (northern Sweden) not before mid-May (Rendahl and Vestergren 1959). In Poland, spring passage March–April; slow autumn withdrawal occurs September–December, peaking November, and movement clearly heavier than in spring; overwintering occurs more in west than east (Tomiałojć 1976). In mountainous areas further south (e.g. Alps) movement altitudinal only.

Populations in North Africa (*kabylorum* in north-west Africa, *juniperi* in Libya) are resident though subject to altitudinal movement in Atlas mountains (Blanchet 1955; Heim de Balsac and Mayaud 1962; Etchécopar and Hüe 1967; Thomsen and Jacobsen 1979). Immigration into North Africa would seem possible, but no evidence for it. Thus, resident on Gibraltar (Lathbury 1970), and no mention of passage occurring there (e.g. Tellería 1981). Resident in Balearics (Bannerman and Bannerman 1983). On Malta, where no breeding birds present, a scarce winter visitor October to late March, with up to 12 birds in a season; ringed birds retrapped up to 4 months later in same winter (Sultana and Gauci 1982). Resident on Cyprus, with only local movement to low ground in winter (Flint and Stewart 1983).

Relatively little known about movements of Asian populations. Possible influx recorded late January along Caspian coast of northern Iran (Passburg 1959). Altitudinal movements occur in Himalayas (Armstrong 1955; Ali and Ripley 1973b), China (Etchécopar and Hüe 1983), and Korea (Gore and Pyong-Oh 1971).

In North America, most continental populations migratory, moving south as far as Gulf of Mexico in winter. In autumn, moves through Massachusetts in October, and southward progress traced by early arrival dates:

Maryland 10 September, Washington 25 September, Florida 20 October, New Orleans 24 October, Oklahoma 5 November. Birds reaching extreme south start north as early as February, and arrivals in southern Canada and Newfoundland occur from early April, rather later in north-west of range. (Bent 1948; Paynter 1957.) Record from Azores (Van Vegten and Schipper 1968) not considered acceptable by G Le Grand.   JHE

Food. Chiefly insects, especially beetles (Coleoptera); also spiders (Araneae). Mostly taken from surface of leaves (upper- and underside), twigs, and bark or from crevices in bark, rocks, walls, and leaf litter. Usually feeds within 2 m of ground, lower in winter than in summer (Armstrong 1955). Territorial ♂♂ feed higher than ♀♀ (Armstrong 1955), occasionally even in tree canopy in spring (Glutz von Blotzheim 1962). Feeds lower in exposed sites. In scrub habitat in Britain, 51% of feeding observations on ground, 42% in herb layer, 53% in shrub layer; in oak *Quercus robur* woodland, 13% on ground, 36% in herb layer, 43% in shrub layer (Beven 1964). In scrub, feeding location shows preference for hawthorn *Crataegus monogyna* (66%), rose *Rosa* (10%), and blackthorn *Prunus spinosa* (10%) (Beven 1964). *T. t. hirtensis* feeds largely in bracken *Pteridium* and dock *Rumex obtusifolius* (Harrisson and Buchan 1934). Will take insects in flight (Glutz von Blotzheim 1962). Occasionally enters water to secure prey, immersing head (Forbush 1929) or rarely whole body (Haslam 1844). Small fish taken from water surface (Bagnall-Oakeley 1968). In winter, will follow badgers *Meles meles* to obtain food from disturbed ground (D G C Harper). Most prey items less than 1 cm long (Armstrong 1955). Small prey swallowed whole, larger items (over 7 mm) may be struck on perch or stone to kill them before being eaten or fed to young (Bagnall-Oakeley 1968; Pullinger 1971). Caddis fly larvae (Trichoptera) beaten on twigs before case removed (Rope 1889). Froglets (Anura) broken apart before being fed to fledged young (Czikeli 1975). Wings of large flies (Diptera) and dragonflies (Odonata) may be removed (Armstrong and Whitehouse 1977). Seldom seen drinking (Armstrong 1955). Nestlings eject indigestible remains as pellets (Armstrong and Thorpe 1952).

The following recorded in diet in west Palearctic. Invertebrates: bristle-tails (Thysanura), springtails (Collembola), mayflies (Ephemeroptera), small dragonflies (Odonata), stoneflies (Plecoptera), grasshoppers (Orthoptera), earwigs (Dermaptera), bugs (Hemiptera), lacewings (Neuroptera), scorpion flies (Mecoptera), larval and (less often) adult moths and butterflies (Lepidoptera, especially Noctuidae and Geometridae), caddis flies (Trichoptera), adult and (less often) larval and pupal flies (Diptera, especially Tipulidae, Culicidae, Chironomidae, Empididae, Syrphidae, Muscidae), Hymenoptera (sawflies Tenthredinidae, Chalcidoidea, adults and eggs of ants

Formicidae), adult and (less often) larval beetles (Coleoptera: in winter, especially Histeridae, Staphylinidae, Cryptophagidae, Chrysomelidae, Curculionidae especially *Sitona*; in summer, also Carabidae, Scarabaeidae, Elateridae, Scolytidae, Platypodidae), small spiders (Araneae), mites (Acarina), harvestmen (Opiliones), millipedes and centipedes (Myriapoda), woodlice (Isopoda), small littoral crustaceans (especially *Gammarus pulex*), and small snails (Mollusca). Vertebrate food taken includes tadpoles, young frogs, and small fish and their fry (e.g. minnow *Phoxinus phoxinus*, goldfish *Carassius auratus*). Plant food recorded includes fruit of elder *Sambucus* and raspberry *Rubus idaeus*, seeds of bilberry *Vaccinium myrtillus* and wood sorrel *Oxalis acetosella*, and fragments of seaweed (Phaeophyceae). (Thompson 1849; Rope 1889; Csiki 1908; Newstead 1908; Baer 1909; Rey 1910a; Florence 1914; Clarke 1915; Collinge 1924-7; Gil 1928; Schuster 1930; Harrisson and Buchan 1934; Kleiner 1937; Hyde-Parker 1938; Timmermann 1938-9; Huxley 1949; Kovačević and Danon 1950; Armstrong 1955; Bagenal 1958; Glutz von Blotzheim 1962; Prokofieva 1962; Bagnall-Oakeley 1968; Ptushenko and Inozemtsev 1968; Pullinger 1971; Czikeli 1975; Armstrong and Whitehouse 1977; Hamilton 1978; Kostin 1983.)

10 stomachs from Crimea (USSR) throughout the year, contained only invertebrates: 40% Hemiptera, 15·7% snails, 12·9% beetle larvae and adults, 11·4% ants, 8·6% spiders, 5·7% Lepidoptera larvae, 5·7% Lepidoptera adults (Kostin 1983). In Hungary, 4 stomachs from April-August contained 14 items: 71% beetles (Cryptophagidae, Scarabaeidae, Curculionidae), 29% ants *Lasius*; 19 from October-March contained 121 items comprising 72·7% beetles (Curculionidae, mostly *Sitona tibialis*, 43·8% of total items), 9·9% Diptera (especially Muscidae), 8·3% snails, 5·7% Hymenoptera, 1·6% Hemiptera, 1·6% spiders (Csiki 1908). One stomach from Sachsen (East Germany), January, contained spiders and small beetles (Baer 1909). A stomach from Ireland, January, contained only small beetles (Thompson 1849). Unspecified series of stomachs from Iceland in winter contained many beetles (Staphylinidae), also adult and larval flies (Timmermann 1938-49). Stomach from Bayern (West Germany), October, contained several small spiders and 1 fly *Sepsis* (Rey 1910a). Stomach of ♀ from Scotland, January, contained beetles, spiders, and 1 fly larva; 3 ♂♂ from same site in April contained many small Crustacea (possibly *Gammarus pulex*) and traces of seaweed and beetle larvae (Florence 1914). Adult ♂ from Spain, May, contained variety of beetles (especially Elateridae, Curculionidae, and Cynipidae) and bird in June contained ants, *Sitona*, larvae of Noctuidae and Geometridae, adults of Tipulidae, Neuroptera, and Aphididae, and small seeds (Gil 1928). 4 stomachs from Britain in summer contained 58% (by volume) Geometridae larvae, 22% Tipulidae larvae, and 20% aphids; also seed fragments (Collinge 1924-7). Stomachs from Switzerland, July-August, con-

tained Orthoptera, aphids *Rhynchota*, Neuroptera, ant pupae and adults, mites, woodlice (Isopoda), and myriapods; in November, also small seeds (Glutz von Blotzheim 1962). 2 stomachs from Sachsen (East Germany), April, contained mostly small beetles (Chrysomelidae and Curculionidae) and spiders (Baier 1909). 5·6% of 90 stomachs contained snail remains (Kleiner 1937). 3 ♂♂ from Yugoslavia, August, contained mostly beetles (especially Scolytidae), Lepidoptera, ants, and Hemiptera (Kovačević and Danon 1950-1). 10 stomachs from Moscow region (USSR), April–July, contained 101 items: 15% Hemiptera (Aradidae), 22% Homoptera, 12% Coleoptera (mostly Curculionidae), 8% Diptera, and 33% Araneae; in August–September, observed taking aphids and scale insects (Coccoidea) from lower surface of nettle leaves (Ptushenko and Inozemtsev 1968). 20 stomachs from Kirgiziya (USSR), April–July, contained only insects: many small beetles (especially Curculionidae), scorpion flies (Boreidae), bugs (Pentatomidae and Naucoridae), Diptera, and Lepidoptera larvae (Pek and Fedyanina 1961). Bird once seen to pick up slug *Arion hortensis* and reject it (Armstrong 1955).

Diet of young similar to adult's but generally lacking in Coleoptera. Prey size increases with age of chicks (Armstrong and Whitehouse 1977). In Leningrad area (USSR), collar-samples from 2 broods, June–August, contained 1525 items: 41·2% (by number) spiders, 13·1% Hemiptera (Pentatomidae, Tingitidae, Reduviidae, Aphididae, Cicadidae), 12·4% Lepidoptera (mostly Noctuidae and Geometridae, also Satyridae, Pyralidae, Tortricidae, Lycaenidae), 9·4% Diptera (Empididae, Tipulidae, Chironomidae, Mycetophilidae, Culicidae, Tachinidae, Syrphidae, Limoniidae, Chloropidae, Leptidae), 2% Hymenoptera, 2% beetles, 1·3% snails, and 1% seeds; also Trichoptera, Mecoptera, Neuroptera, Plecoptera, Ephemeroptera, Collembola, Opiliones, Myriapoda, Isopoda, Acarina, and 15% unidentified insects (Prokofieva 1962). In Britain, one brood received moth larvae: Geometridae, Noctuidae, and *Mamestra brassicae*; also small larvae of *Callimorpha jacobaeae*, though larger larvae rejected (Newstead 1908). Brood in southern England received mosquitoes *Theobaldia zebrina* on 84% of visits—usually sole prey brought, several at a time; also 5 moths, 2 small caterpillars, 1 fly, 1 hoverfly (Syrphidae), 1 dragonfly, and 1 spider (Armstrong and Whitehouse 1977). 8 nestling stomachs contained only Tipulidae larvae (Diptera) and 1 Geometridae larvae (Collinge 1924-7). On St Kilda (off western Scotland), July–August, brood received mostly Noctuidae larvae, some larval Geometridae, and few Myriapoda, earwigs and spiders; parents ate beetles at this time; young in a cliff site received more Diptera (Harrisson and Buchan 1934). Brood by a trout hatchery in England was fed almost entirely on fry of trout *Salmo trutta* (Huxley 1949). Additional food noted by Kovshar' (1979) in Tien Shan mountains (south-central USSR) includes Orthoptera,

Tineidae (Lepidoptera), Larvivoridae (Diptera), and Licoridae (Araneae). Rate of feeding visits to nest increases as chicks grow; higher rate with bigger broods (Armstrong 1955). On St Kilda, parents bring food *c*. 03·15-21·30 hrs (Bagenal 1958). Frequency highest in morning and evening with a lull at midday; as chicks near fledging, rate lower in afternoon and evening peak reduced (Harrisson and Buchan 1936; Armstrong 1955; Bagenal 1958; Prokofieva 1962; Armstrong and Whitehouse 1977). In Britain, 3 broods of 4 fed on average 106·3 times on day 1, increasing to 256 during 15 hrs on day 15. Brood of 6 had 397 visits in 1 day, 2 days before fledging, and ♀ working alone made peak of 442 visits per day to another brood of 6. Visit rates by ♂♂ vary between 9·5 per hr (*n*=2) in *zetlandicus* to 31 (*n*=2) in nominate *troglodytes*. (Armstrong 1955; Bagenal 1958.) Brood of 5 in West Germany received on average 19·3±2·1 visits per hr during 17.00-19.00 hrs (Goertz 1960).   AGG

Social pattern and behaviour. Well studied. See especially Kluijver *et al.* (1940), Armstrong (1955), Kentish (1976), Armstrong and Whitehouse (1977), and Garson (1978). For *hirtensis*, see Harrisson and Buchan (1934, 1936), Armstrong (1953b), Williamson (1958b, 1959), and Waters (1964); for *hebridensis*, Armstrong (1953a); for *islandicus*, Armstrong (1950); for *borealis*, Williamson (1947c), these referred to below collectively as 'island races'. Nearctic races not considered here. Unless otherwise stated, account refers to nominate *troglodytes* and/or *indigenus*.

1. In sedentary population, ♂♂ territorial for most of year, less so when chick-rearing, moulting, and in winter (Armstrong 1955, 1956). Territories of *hirtensis* appear to break down when family parties start moving around; in winter, 'territories' contract markedly, birds confining themselves to well-sheltered parts (Waters 1964). In autumn, when territorial defence less and population high, given area may hold twice as many ♂♂ as in spring (Armstrong 1955). Migrants show marked fidelity to winter quarters: e.g. ringing at reedbed in Berkshire (England) showed that at least 40% of birds caught one winter returned the following winter (Hawthorn *et al.* 1971); since annual mortality *c*. 60% (see Population), presumably great majority of birds surviving between years returned to same wintering area. No flocking reported, but in hard weather outside breeding season, ♂♂ virtually abandon territories and readily feed together (Kluijver *et al.* 1940). In winter, reported feeding close together on shore (Timmermann 1935). See also Roosting (below). BONDS. Mating system varies with race and, in nominate *troglodytes/indigenus*, with quality of habitat. Island races thought to be typically monogamous (Armstrong 1955), though status of *islandicus* uncertain (Armstrong and Whitehouse 1977). In western Europe, *c*. ½ ♂ nominate *troglodytes/indigenus* monogamous, rest successively polygynous with 2-3(-4) ♀♀ such that broods overlap (Kluijver *et al.* 1940; Armstrong 1955; Garson 1978; Lovaty 1985). According to P J Garson, monogamous ♂♂ behave as would-be polygamists but fail to attract more than 1 mate. In 1937 at Stein (Netherlands), of 12 ♂♂, 6 polygynous; of these, 2 involved in breeding attempts with 3 ♀♀ at same time, while 4 bred with 2 ♀♀; in only 3 cases did laying periods overlap for 2 ♀♀ in a given territory. In 1938, 4 out of 7 ♂♂ polygynous; in 1939, 3 out of 6 (Kluijver *et al.* 1940). In primeval forest, Białowieża (Poland), on average less than 20% of ♂♂ polygynous, never breeding

with more than 2 ♀♀ per season, and then only in optimal (ash and alder) habitat (Wesołowski 1981, 1983). For factors affecting mate choice, see Heterosexual Behaviour (below). Pair-bond, even in cases of monogamy, never close, ♂ and ♀ associating only for courtship and mating; ♀♀ typically mate with different ♂♂ for successive broods, and between seasons; exceptionally ♀ lays more than 1 clutch for same ♂ within a season. ♀♀ thus probably range over a wide area in search of ♂♂ (Kluijver *et al.* 1940). Role of sexes in brood-care varies with mating system. In island races, ♂ apparently takes about equal share in feeding young from a few days after hatching (Armstrong 1955; Armstrong and Whitehouse 1977), though, in some nests of *islandicus*, apparently only ♀ feeds young in nest (Armstrong 1950). In nominate *troglodytes/indigenus*, ♂ often plays little or no part in feeding young in nest, though role varies with time of year, degree of polygyny (etc.). At all nests in Wytham, Oxfordshire (England), ♀ fed young until fledging, with only token help from ♂ (Garson 1978). According to Armstrong (1955), however, ♂ gave appreciable assistance in *c.* 40% of breeding attempts. Kluijver *et al.* (1940) found negligible contribution by ♂ in 14 out of 19 breeding attempts, and ♂♂ gave substantial help only at end of breeding season; 2 which helped early in season were monogamous. After fledging, however, ♂ starts helping ♀ to escort and feed young, sometimes taking sole charge if ♀ starting new clutch (Armstrong 1955). One ♂ alternated in escorting 2 broods which fledged same day (Kluijver *et al.* 1940; see also Roosting, below). Dallmann (1977) found that although both sexes escorted young, ♂ did not feed them much. Pair continue to feed young typically for *c.* 9 days after fledging, exceptionally up to 18 days (Armstrong 1955). Siblings may continue to associate after independence: one brood found roosting together 17 days after fledging (Armstrong and Whitehouse 1977). In another brood, groups of 2 and 3 roosted together at separate sites 26 days after fledging (Kluijver *et al.* 1940). Age of first breeding not known. BREEDING DISPERSION. Solitary and territorial. Territories contiguous or dispersed, depending on uniformity of habitat (e.g. Wesołowski 1983). Boundaries commonly topographical, e.g. break in cover (Armstrong 1955; Garson 1978) and, in *hirtensis*, remarkably constant from year to year (Waters 1964). Size of territory varies with quality of habitat. In Białowieża, also varies markedly within given habitat: 0·67–3·74 ha in ash-alder, 2·24–8·07 in oak-hornbeam (Wesołowski 1983). These territories generally larger than in man-modified habitats in western Europe: e.g. at Wytham, average 0·37–0·85 ha (0·15–1·16) over 3 years (Garson 1978, 1980*b*). In gardens at Cambridge (England), 0·48–1·6 ha (Armstrong and Whitehouse 1977). In garden at Stein, 0·3–1·7 ha (Kluijver *et al.* 1940). In island races, size of territory requires further study, but thought to be relatively large compared with nominate *troglodytes/indigenus*. In *islandicus*, 2 territories perhaps *c.* 7·5 ha and 33 ha (Armstrong 1950), and territories of *hebridensis* and *zetlandicus* thought to be similar (Armstrong 1952, 1953*a*). In *hebridensis*, however, territory much smaller, estimated *c.* 1–1·9 ha (Armstrong 1955; Williamson 1958*b*; Waters 1964). According to Cody and Cody (1972), *hebridensis* 0·37 ha (*n* = 4), *zetlandicus* 0·20 (*n* = 6), *islandicus* 0·23 (*n* = 4); these perhaps underestimates. Density of nominate *troglodytes/indigenus* widely studied. Density in man-modified habitats of western Europe often exceeds 150 territories per km²; however in primeval forest, Poland, densities up to 50 territories per km² in optimal habitats, less than 20 per km² in suboptimal ones (Wesołowski 1981, 1983; Tomiałojć *et al.* 1984, which see for details). Within western Europe, density varies with habitat. In moorland, Orkney, 2·4 territories per km² (Lea and Bourne 1975); in New

Forest (southern England), 16·2 per km² (Glue 1973); in farmland, Suffolk (England), 20·7 per km² (Benson and Williamson 1972); in young conifer plantation, Anglesey (Wales), 63–99 per km² (Insley and Wood 1972); in sessile oak, Gwynedd (Wales), 107 per km² (Gibbs and Wiggington 1973). At Wytham, 232 per km² (British Trust for Ornithology); in mixed deciduous wood with rhododendron understorey, Somerset (England), 380 per km² (Parsons 1976). For densities in oak and spruce, Ireland, see Batten (1976). In central Italy, 182 per km² (Lambertini 1981). In Switzerland, from 10 per km² in spruce-pine to 77 per km² in mixed montane beech (Schifferli *et al.* 1982). In Morocco, 23 'pairs' per km² in degenerate woodland, up to 74 per km² in humid forest (Thévenot 1982). In spruce forest, southern Kareliya (north-west USSR), 4–16 per km² (Zimin 1972). For reduction after hard winter, see (e.g.) Brouwer and Daalder (1982). For increase after forest clearance in East Germany, see Schönfeld (1975*a*). Outer part of territory frequented much less than inner part (Garson 1978). Territory defended by ♂ only; serves for display, nesting, and feeding of ♂ and young; ♀♀ breeding with resident ♂ feed in his territory but also range more widely (Kluijver *et al.* 1940; Armstrong 1955). ♂♂ faithful to territory in successive seasons (Kluijver *et al.* 1940; Garson 1978). One ♂ occupied same territory for 5 years, often attempting to extend it (Armstrong and Whitehouse 1977). ♀♀ nesting simultaneously in ♂'s territory typically do so far apart, never close together (Armstrong 1955). On average, ♀♀ moved *c.* 200 m between successive nesting attempts (Kluijver *et al.* 1940). Territorial ♂♂, especially polygynous nominate *troglodytes/indigenus*, build numerous nests, though some do so without attracting any ♀♀, and thus fail to breed (Kluijver *et al.* 1940). In Cambridge, ♂♂ completed up to 5 nests per season, and started up to 3 others (Armstrong 1955). In 3 years, average of 5·8, 8·14, and 5·0 nests per ♂, overall average 6·3 (*n* = 25), several nests incomplete. One ♂ built 15, 10, and 9 nests in 3 years (Kluijver *et al.* 1940). In Bayern (West Germany), 1–5 nests per ♂, average 2·5 (*n* = 88); ratio of nests used for breeding to those not used varied from 1·6:1 to 1:6, depending on region (Dallmann 1977). In Allier (Switzerland), 26 nests in 4 territories; one territory contained 11 nests, of which 5 used for breeding (Lovaty 1985). In Białowieża, ♂ usually builds more than 1 nest, but situation complicated by use of nests from previous season (Wesołowski 1983, which see for frequency of use of available nests in different habitats). In island races, ♂ thought to build usually not more than 2 nests per season (Armstrong 1955); perhaps up to 3 in *zetlandicus*, but no hard evidence (Armstrong 1952), and 1 or more 'auxiliary' nests in *islandicus* (Armstrong 1950). In *hirtensis*, average 1·5 nests per ♂ per season (Waters 1964). Exceptionally, same nest used twice per season, perhaps only when others destroyed or already in use (Armstrong 1955). Occasionally ♂ repairs predated nest for re-use (Garson 1978). Good sites often traditional, used for several years (Armstrong 1955; Waters 1964). One nest used for 4 successive years (Hoffmann 1976). ROOSTING. Nocturnal. Mostly solitary, but communal outside breeding season in hard weather-especially in freezing conditions but also in wind and rain (Armstrong and Whitehouse 1977). In breeding season, uses sites only moderately well insulated, e.g. ivy, hollow log. ♂♂ occasionally roost in own nests, typically so in *hirtensis* (Armstrong 1955). Breeding ♀ roosts in chosen nest, sometimes from before laying (Owen 1919) until after fledging (see below). In late summer, *hebridensis* roosted in heather banks and scrapped cars (Hawthorn *et al.* 1976). From January onwards, 2–3 birds not uncommonly roost together (Dunsheath and Doncaster 1941; Armstrong 1955: for possible breeding association, see Heterosexual Behaviour, below). Just before breeding season, ♂♂ and ♀♀ tend to adopt separate solitary roost-sites, this pattern continuing after breeding season until harsh weather intervenes. Roosting assemblages may then contain adult ♂♂, ♀♀, and juveniles (Kluijver *et al.* 1940; see also below). Communal sites typically well insulated, usually 1·5-5 m above ground, notably nests of own species or of other passerines (not uncommonly House Martin *Delichon urbica*), nest-boxes, or holes in trees, walls, or under roofs (Armstrong 1955). For other sites, see Labitte (1937), Dunsheath and Doncaster (1941), and Hart (1984). Nest recorded being built in January and then used for communal roosting (Sharrock 1980a). Seldom more than *c.* 10 roost together, though 14 reported as early as October (Moore 1960). In severe conditions, one nest-box roost contained 61 birds which took *c.* 15-30 min to enter at night and *c.* 20 min to disperse in morning (Flower 1969). 96 birds once roosted in loft (Haynes 1980). 30 birds removed from cavity in thatch returned there when released (Dewar 1902); may thus develop strong attachment to site, despite disturbance (Armstrong 1955). Sites often traditional within and between seasons, drawing birds from up to 2 km. Nightly establishment of roost seems to be initiated by a 'leader', probably resident ♂, whose activity, especially vocalizations (see 1-2 in Voice) near site, and outflights from site, apparently attract others. (Armstrong 1955; Armstrong and Whitehouse 1977.) Birds squat in roost up to 2-3 layers deep in bigger assemblages, heads facing inwards, tails towards entrance or sides. By end of January, when weather improving, much squabbling occurs at roost, and up to 4-5 birds seen falling in a bunch as they tried to evict one another (Armstrong 1955). At end of December, of 18-25 birds competing to enter roost, only 7 finally entered (Sudhaus 1964). Resident ♂, while accepting other ♀♀, may station himself at roost-entrance and ward off other ♂♂, but insistent ♂ may gain entry, sometimes usurping resident and associated ♀♀ and forcing them to go elsewhere. Association of ♂ and ♀♀ in roost appears not to lead to subsequent breeding between them (Armstrong and Whitehouse 1977). In breeding season, birds first arrive at roost-site about sunset, somewhat later in winter (Armstrong and Whitehouse 1977). Usually leave roost 30-40 min before sunrise (Clark 1949). For details of diurnal rhythm, see Armstrong (1955) and Whitehouse and Armstrong (1953). In Tübingen (West Germany), January-March, average time of entry 19 min after sunset, varying with the weather; average departure 29 min before sunrise, irrespective of weather (Ammermann 1975). In Kiel (West Germany), average entry 26 min (8-46) after sunset, depending on weather and light intensity (Sudhaus 1964). ♂♂ leave roost about same time in relation to 'dawn twilight' throughout the year; while incubating and rearing young, ♀♀ leave later than ♂♂. ♀ also tends to roost earlier than ♂, though not when feeding well-grown young in nest (Armstrong 1955). Recently fledged young roost in one of ♂'s unoccupied nests or in disused nest (exceptionally, occupied) of another species; especially on 1st day out of nest, may also use some convenient cavity (Whitehouse and Armstrong 1953, which see for other examples). Brood may be distributed between 2 nests; on 2 occasions 'parents' said to have hastily constructed extra nests to accommodate roosting young (Cox 1922). For *c.* 2 weeks after fledging (usually longer for 2nd broods), young escorted to roost by ♂, less often ♀ (Kluijver *et al.* 1940; Whitehouse and Armstrong 1953), sometimes by both parents (Armstrong and Whitehouse 1977). If ♂ is escort, he moves towards roost-site with series of short flights, calling and singing; indicates site by singing outside, entering and leaving. ♀ may give Whisper-song (see 1c in Voice). If disturbed,

♀ will, during evening, conduct young to alternative site, and yet another if disturbed again. If young roost sufficiently early, ♀ may feed them in roost; may roost with young up to and occasionally after fledging. (Kluijver *et al.* 1940; Armstrong 1955.) One ♀ allowed fledged 1st brood to roost in nest where she was incubating her 2nd clutch (Mächler 1947). Comfort activities include preening, dust-bathing (Schmidt 1966; Stainton 1978; Edwards 1980; Hohl 1981), bathing in dew-covered grass (Carman 1973) or wet foliage; once seen jumping in and out of shallow stream (Armstrong 1955). Sunbathes by stretching head forward, fanning tail, spreading wings (sometimes until tips almost meet in front of bill), and ruffling back (Harrisson and Buchan 1936; Armstrong 1955). In December, bird seen nibbling at bill of another, then climbing on to its back, and pecking at head and neck; thought to be possibly grooming for parasites (Schmidt 1966). Anting not recorded (Armstrong 1955).

2. Not especially shy, often allowing close approach. When excited or alarmed, typically cocks tail and may bob body (Armstrong 1955). On Sark (Channel Islands), sudden disturbance by human intruder often elicits curtailed song (J Hall-Craggs: see 1 in Voice). Alarmed *hebridensis* seeks refuge in small gullies in peat, heather (etc.), and squats there before finally moving away through cover; bird flushed from refuge will flit away and squat again (Armstrong 1953a). Disturbed *hebridensis* or *zetlandicus* also crouches low with tail partly spread and lowered; possibly a threat-posture (Armstrong 1955). Loud explosion caused nominate *troglodytes* to 'freeze' on branch in an upright posture, bill pointing vertically; remained motionless for *c.* 6 s before gradually relaxing (Radford 1981). May respond to cat by squatting motionless and silently on branch for up to 2-3 min until danger passes; this response specially typical before roosting, perhaps to prevent predator discovering roost-site. In winter, roosting birds at first squat when discovered, then usually 'explode' from site (Armstrong 1955). In unusual threat-display, juvenile confronted water vole *Arvicola terrestris* by squatting flat, fanning wings until tips almost met in front of gaping bill, ruffling rump, and holding tail vertically up or down, or flicking it up and down; several times bird sprang *c.* 5 cm into the air, remaining silent throughout (Barrett 1947). Bird, apparently ♂, gave a Grating-call (see 4b in Voice) as it flew into cover to escape Sparrowhawk *Accipiter nisus*. Readily joins parties of other birds mobbing owls (Strigiformes), moving around excitedly giving Alarm-calls (see 7b in Voice), standing on tip-toe, and flicking wings in and out (Wing-quivering) constantly (Armstrong 1955). Once, 5 birds, including 2 resident neighbouring ♂♂, mobbed weasel *Mustela nivalis* (Garson 1978). FLOCK BEHAVIOUR. None recorded. SONG-DISPLAY. For general accounts, see Kluijver *et al.* (1940), Armstrong (1944, 1955), Clark (1949), Kentish (1976), and Garson (1978). Given mainly by ♂, occasionally by ♀ (see 1c in Voice). following account refers mainly to Territorial-song (see 1 in Voice) of nominate *troglodytes*/ *indigenus*. Song given throughout the year, except when moulting (Kreutzer 1974a); also sings less from December-January. Frequency increases February-March, initially mainly in morning and to lesser extent before roosting. In late March and April, sings regularly in afternoon also. Singing often most intense in June, declining July (though juveniles also start then), dropping abruptly almost to nil in August during adult moult; some resurgence thereafter. (Armstrong 1955.) *T. t. hirtensis* sings February-October, mainly April-July; *zetlandicus* similar (Waters 1964). Rate of Territorial-song shows strong diurnal rhythm: ♂♂ start singing as soon as they leave roosts, and song-rate then maximum; at dawn ♂♂ commonly sing for

up to 20 min per hr, giving *c.* 130 songs per hr (Garson 1978), sometimes over 200 per hr (Armstrong 1955). Sing less after cold nights (Garson 1978; Garson and Hunter 1979). Song declines both in rate and length until noon, staying sporadic thereafter, though bird may sing briefly before roosting (Clark 1949; Garson 1978). Singing ♂ weaves head from side to side (J Hall-Craggs); sometimes said that tail cocked also, but this not always so; song given from near ground, up to 10–15(-20) m on a tree or building. High vantage points especially favoured early in breeding season (Armstrong 1944, 1955); *hirtensis* used higher perches as season progressed (Waters 1964). ♂ often flies from ground to elevated perch and sings immediately after neighbour has sung, frequently leading to rivals holding Song-duel for several minutes. Usually give a few songs from one perch, then fly to another, sometimes singing as they go. Singing in flight apparently more common in island races, especially in Nest-showing ceremony (Armstrong 1953*a*, 1955; see Heterosexual Behaviour, below). Residents rarely sing against each other less than 15 m apart; interval between songs significantly shorter against rival (average 8·49 s) than when singing alone (10·79 s). Rivals usually reply within 1 s of each other, sometimes overlapping. ♂♂ thus patrol their territories, singing repeatedly. (Garson 1978.) For song of ♂ during nest-building, courtship, and Nest-showing, see below, and 1b in Voice. ANTAGONISTIC BEHAVIOUR. Territorial ♂♂ highly aggressive to other ♂♂ for most of the year, especially breeding season (see part 1). ♀♀ do not defend territories and pass freely between ♂-territories, though some ♀♀ may defend nest-area against potential intruders. ♂ threatens rivals mostly by Song-display (see above). ♂ may trespass into neighbour's territory by remaining silent (Armstrong 1955). According to Garson (1978), however, resident ♂♂ rarely trespass, *contra* Armstrong (1955) and Kentish (1976). Song-display combined with postural display: main threat is Wing-quivering display: bird rapidly twitches wings which may be held slightly out from body or, *in extremis*, raised vertically (Wings-raised display: Fig A) (Armstrong 1955). In prolonged skirmish (see

A

below), perched singing rivals Wing-quivered intensely, rapidly vibrating wings through a small arc, while frequently flicking their unfanned tails backwards and forwards (Garson 1978). Rivals Wing-quiver as they gradually approach each other through the undergrowth, flitting from perch to perch. Bird may also adopt Upright-threat posture: stands up at full stretch, cocks tail high, and erects plumage over most of body. One bird sang at another with bill and tail held high and periodically made Bill-snapping sounds (Armstrong 1955: see Voice). In

response to playback of song, ♂ perched close to source, drooped and quivered wings, and moved cocked tail from side to side (Bergmann 1977*a*, which see for drawing). Pursuit-flights unusual, and overt fighting rare (Armstrong 1955), but brief confrontations occur late winter, early spring, involving chases, singing, and Anger-calls (Kluijver *et al.* 1940; Garson 1978: see 5 in Voice). Exceptionally, rivals confront each other physically, sparring in the air, interlocking feet, pecking, and calling (Williamson 1947*c*; Armstrong 1955). In one lengthy dispute, rivals pursued each other for more than 4 hrs, with much singing and flying from bush to bush in disputed territory; performed intense threat-display (described above). In this case, non-neighbour was trying to usurp resident. Residents also try to expand territories at expense of neighbours. In contests between residents and intruders, intruders usually win. (Garson 1978.) Fledged broods usually trespass with impunity into neighbouring territories (Armstrong and Whitehouse 1977). In *hirtensis* territorial defence breaks down once young leave nest (Williamson 1958*b*; Waters 1964). From mid-July onwards, when juvenile ♂ nominate *troglodytes/indigenus* start singing, they apparently settle at the edge of established territories and are chased by the resident ♂ (Garson 1978). Few reports of aggression to other species. For possible competition with Cetti's Warbler *Cettia cetti* in Kent (England), see Hollyer (1975). For other threat-displays and aggression, see introduction to part 2, and also Roosting (above). HETEROSEXUAL BEHAVIOUR. (1) General. Pair-formation in nominate *troglodytes/indigenus* usually begins once ♂ has nests available, continuing until 2–3 weeks after last clutches laid. Occasionally, desultory courtship between ♂ and ♀ roosting together outside breeding season, especially late winter, may result in mating the following breeding season, but not necessarily so (Armstrong 1955). Study of mate choice, especially by ♀, pioneered by Kluijver *et al.* (1940) and elaborated by Garson (1978, 1980*b*) who believes ♀♀ choose mates rather than nests or territories; ♀♀ tended to mate more readily with territorial ♂♂ possessing greatest number and density of available nests. At Białowieża, no relationship found between territory size, availability of nests, and occupancy rate of ♀♀, i.e. degree of polygyny; ♂♂ arrive, establish territories, and build nests *c.* 1 month before ♀♀ arrive (Wesołowski 1983). (2) Pair-bonding behaviour. ♂ and ♀ do not form strong pair-bonds; courtship and copulation only activities which bring ♂ and ♀ together, and then for relatively short periods. Resident ♂♂ intercept and court any ♀ entering their territory. ♂ typically first attracts ♀ by Territorial-song. Soon after ♂ begins singing, ♀ may be heard giving Chattering-call (see 2b in Voice) and approaches ♂. Courtship behaviour involves following displays, some of which also used in threat context (Armstrong 1955). Unless otherwise stated, descriptions refer to nominate *troglodytes/indigenus*. (a) Sexual-chasing. ♂ flies closely behind ♀. When chase ends, he may deliver Courtship-song (see 1b in Voice) as he flits about above where ♀ has landed, or else flies off to feed and give Territorial-song. Often he rushes at ♀ (Pouncing-display) where she has landed, and, during brief mêlée, Excitement-calls (see 4b in Voice) given, followed by excited snatch of song from ♂. Pouncing-display frequently occurs without prior chasing, especially when ♀ feeding newly fledged young, and may represent ritualized attempted copulation, intended to stimulate ♀. In Fluttering Pursuit-flight (Fig B), usually in early morning, ♀ flies near ground and just ahead of ♂; both fly rather slowly with fluttering moth-like action and tails fanned, rendering rounded wings and plumage pattern conspicuous. (Armstrong 1955.) (b) Wing-quivering display. According to P J Garson, similar to, but movement more spasmodic than, that described

B

for Song-display against rival ♂♂ (see above). In presence of
♀, ♂—often with animated movements—extends and quivers
drooped or half-raised wings (as in Fig C, left), once producing

C

audible wing-rattling sounds (Armstrong 1955: see Voice).
Fanned tail may be moved from side to side, or forwards and
backwards (Kluijver *et al.* 1940). Display often accompanied by
Courtship-song and, later in season, by brief calls. Occasionally
Wing-quivering leads to bird becoming airborne. Display
thought to excite ♀; also performed by *hirtensis* and *islandicus*
(Armstrong 1955). (c) Meeting-ceremony. Island races, also
some nominate *troglodytes/indigenus*, perform Meeting-ceremony
in which one or both sexes spread and quiver wings, ♂
sometimes also singing, or both calling (Armstrong 1952, 1955).
(d) Variants of Wings-raised display. Sometimes, in interludes
during chasing flights, or when ♂ perched near hidden ♀, ♂
raises wings high, as in threat-display (see Fig A, and
Antagonistic Behaviour) but posture maintained for *c.* 1 s. ♂
may repeat display several times, changing perch between
performances (Kluijver *et al.* 1940; Armstrong 1955). ♂ *borealis*
also displays thus (Williamson 1947*c*). ♂ *zetlandicus* performs
apparent variant, beating raised wings rapidly in whirling
manner ('windmill display': Armstrong 1955). In aerial Wings-
raised display, ♂ nominate *troglodytes/indigenus* concludes flight
from higher to lower perch by floating daintily down with
wings uptilted at 45°, giving a few wing-beats before landing
(Armstrong 1955). Various components of displays (a)-(d) may
be combined with collection of nest-material (Armstrong 1955).
One ♂ carried material with neck markedly extended, Wing-
quivering, and singing (Witherby *et al.* 1938*b*). (3) Courtship-
feeding. Not recorded (e.g. Armstrong 1955, Garson 1978), but
see subsection 6, below. (4) Nest-site selection. Account based
on Armstrong (1955). Site initially chosen by ♂, but ♀ exercises
choice of his completed nests. Exceptionally, ♀ builds nest and
may then choose site alone. ♂ shows nest to ♀, indirectly
by intermittent bursts of song during building, directly by
Nest-showing display: flits attentively around and ahead of ♀
and gives Courtship-song, trying thus to lure ♀ to nest. On
nearing nest, ♂ increasingly animated, Wing-quivering, hopping
from perch to perch as he sings (Fig C, left). If ♀ approaches, ♂
suddenly darts into nest for period varying from a few seconds
to several minutes, then flies out giving Excitement-calls (see
4a in Voice); continues Wing-quivering and singing as ♀
approaches gingerly and looks into nest. If she now enters,
♂'s excitement intense: flies in and out of nest, bobbing here
and there. If ♀ accepts nest, ♂ later spends longer spells with

her in nest. Nest-showing display of *zetlandicus* apparently
more overt and vocal, ♂ singing strongly and repeating display
several times: *hebridensis* apparently similar (Armstrong 1952,
1953*a*). (5) Mating. Commonly occurs near nest in which ♀
will lay (Kluijver *et al.* 1940). ♀, and to some extent also ♂,
solicits copulation. In typical sequence, ♀ sits on low branch
in undergrowth and performs ♀-soliciting display (Fig D):

D

partly extends wings and quivers them rapidly, also fans and
quivers tail, accompanied by Soliciting-calls (see 3 in Voice).
May display thus up to 20 times, at intervals of a few seconds
to a minute or more, occasionally shifting perch, initially with
no response from ♂ (Armstrong 1955). ♀ moves back and forth
on perch, and, usually after some Soliciting-calls, cocks tail
(Kluijver *et al.* 1940). If ♀ fails to solicit ♂ after repeated
efforts, she leaves his territory. Interested ♂ periodically
descends to ♀ who responds by moving away. Receptive ♂ may
move his tail slowly from side to side. One ♂ ran towards
soliciting ♀, flapping his wings vigorously (Armstrong 1955).
In apparent Soliciting-display, ♂ may approach ♀ in posture
(Fig E) similar to upright 'freezing' posture (see introduction

E

to part 2); in one incident, ♂ gave prolonged loud song, and
approached ♀ with neck well extended, 'not unlike Bittern
Botaurus stellaris in alarm posture'; ♀ then adopted same posture
and ♂ attempted copulation (Nelder 1948). Another ♂ adopted
same posture, but with tail dragging on ground, in presence of
♀ and, in another incident, 2 birds faced each other *c.* 1 m
apart with bills raised at 45°, and each bowed several times to
its partner: in neither incident, however, was mating said to
have followed (Marples 1940). Copulation rapid, ♂ flying down
('pouncing': Armstrong 1955), mounting, and returning to
perch (Kluijver *et al.* 1940). Copulation takes *c.* 2-5 s, ♂ balancing
with whirring wings. During presumably successful copulation,
subdued Excitement-calls (see 4b in Voice) given. Mating most
common in April, though induced experimentally early March,
and exceptionally occurs until November. ♂ may visit nest and
copulate with ♀ completing clutch or incubating. Once, ♂
attracted by alarm-calls of newly fledged young tried to
mount one of them. (Armstrong 1955.) (6) Behaviour at nest.
Nest-building activity of ♂ greater after recent rain (Kluijver

et al. 1940; Armstrong 1955; Garson 1978). When she has accepted nest (see above), ♀ begins lining (Kluijver *et al.* 1940), though she may line nest in which she does not lay (Garson 1978); may continue lining nest after incubation (Armstrong and Whitehouse 1977). ♂ takes little interest at this stage, and visits nest seldom during incubation and nestling periods (Kluijver *et al.* 1940); tends to stay away until young hatch (Garson 1978). Once, ♀ drove ♂ from nest just after eggs hatched (Marples 1940). Visits by ♂ tend to be early in morning (Whitehouse and Armstrong 1953), not uncommonly when incubating ♀ off feeding (Armstrong and Whitehouse 1977). Visiting ♂ sings or calls outside nest, or occasionally looks in, incubating ♀ responding with Chattering-call (Armstrong 1955). Incubating ♀ very sensitive to disturbance, readily deserting (Sacher 1980*b*). RELATIONS WITHIN FAMILY GROUP. From hatching, ♀ broods assiduously; ♀ appears to brood for *c.* ½ daylight hours on day of hatching, gradually less over next week (Armstrong 1955; Armstrong and Whitehouse 1977). ♀ roosts with young sometimes until 9 days old (Kluijver *et al.* 1940), exceptionally until fledging (Armstrong 1955). ♂ nominate *troglodytes/indigenus* does not brood, but ♂ *hirtensis* once contributed near fledging. Young probably fed within 1 hr of hatching. Apparently do not beg vocally for first 2 days (Armstrong 1955; Waters 1964), food-calls (see Voice) intensifying with age. ♀ may give Whisper-song or call softly when feeding nestlings. For role of sexes in feeding, see Bonds (above). Eyes open at *c.* 5 days in *hirtensis*, *c.* 7 in nominate *troglodytes/indigenus* (Armstrong 1955). Parents do not usually remove faecal sacs until *c.* 4th day, regularly thereafter. At *c.* 1 week, parents perch outside nest entrance to feed young. After being fed, chick turns round and voids faecal sac for parent to collect, then retires and another chick occupies front position. Some parents retrieve faecal sacs dropped on the ground. Young usually fledge in the morning; ♀ may possibly lure young to fledge by combination of Wing-quivering, calling (see 3 in Voice), and cocking tail. Fledged young huddle together on branch (etc.) for 1–2 days, giving frequent contact-calls (see Voice), then become more scattered as they

approach independence (Armstrong 1955). Both ♂ and ♀ feed and escort fledged young; ♂ typically now stops singing, but gives Contact-alarm calls almost continuously (Garson 1978). For roosting, see Roosting (above). ANTI-PREDATOR RESPONSES OF YOUNG. At 2 nest-boxes, tapping side of box induced cowering when broods 4 and 7 days old. Well-grown young fledge 'explosively' if unduly disturbed, e.g. if nest approached too closely; young then able to make short flights and give alarm-calls (see Voice) not heard in nest, attracting parents (Armstrong 1955). PARENTAL ANTI-PREDATOR STRATEGIES. (1) Passive measures. No information. (2) Active measures: against birds. No information, but see introduction to part 2 (above). (3) Active measures: against man. On approach of human intruder to nest or young, parent may give Reeling-call (see 7b in Voice), though less often than for cat (see below) and hops and flits agitatedly. Escape may contain element of laboured flight, wing action appearing retarded and more fluttering than usual. Bird disturbed at nest may perform apparent distraction-lure display similar to Meeting-ceremony (see above); observed in nominate *troglodytes/indigenus* (Armstrong 1955) and *hebridensis* (Armstrong 1953*a*). More elaborate distraction-displays, e.g. injury-feigning, not recorded, but ♂ nominate *troglodytes/indigenus* disturbed while nest-building gave snatch of song and momentarily draggled wings as he fluttered away over ground (Armstrong 1955); *hirtensis* behaved similarly (Harrisson and Buchan 1934). ♂ *zetlandicus*, alarmed by intruder, gave an Excitement-call (see 4b in Voice) and, after feeding young in nest, hopped up rock with tail fanned and wings drooped (Armstrong 1955). (4) Active measures: against other animals. On approach of ground predator, especially cat, less often dog, rat *Rattus*, and stoat *Mustela erminea*, bird gives Reeling-call, and moves about excitedly. Once, Reeling-call apparently summoned young away from stoat and towards roost-site. ♂ with fledged brood highly demonstrative to cat, apparently performing simple distraction-lure display: darts from perch to perch near predator, calling intensely, and sometimes thus induces cat to follow him. (Armstrong 1955.)

(Figs by C Rose: from drawings in Armstrong 1955.) EKD

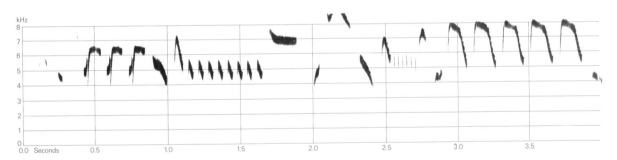

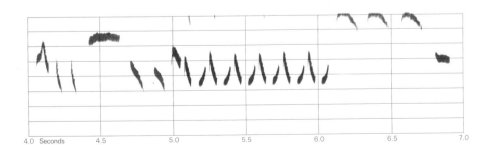

I A G Field England
May 1965

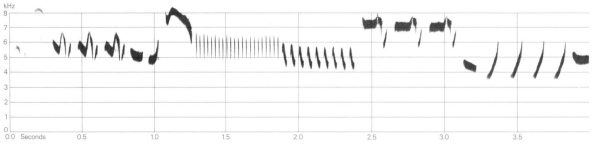

II I Baker Scotland June 1977

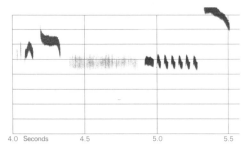

II *cont.*

Voice. Freely used throughout the year, especially by ♂. Vocalizations readily intergrade, making division of calls somewhat arbitrary, and following scheme based on, but more condensed than, Armstrong's (1955) elaborate treatment. Sonagrams of our recordings suggest most calls based on 'tik', 'tek', or 'tut' sounds, delivered with variety of speeds and patterns (J Hall-Craggs). Differences, if any, between nominate *troglodytes* and *indigenus* not known. Compared with nominate *troglodytes/indigenus*, calls of *islandicus* different, those of *hebridensis*, *hirtensis*, and *zetlandicus* apparently similar (Armstrong 1955). Differences in songs of island races require further study, but some major distinctions, based largely on unpublished material supplied by J Hall-Craggs, are described here. For Nearctic races, see Kroodsma (1980). Non-vocal sounds are occasionally made by nominate *troglodytes/indigenus*, and perhaps other races: ♂ threatening rival may accompany song with Bill-snapping; handled bird may also Bill-snap; Wing-quivering display of ♂ once produced audible 'snip' sounds of vibrating wing feathers (Armstrong 1955). For additional sonagrams, see Bergmann and Helb (1982).

CALLS OF ADULTS. (1) Song. Widely studied, but with conflicting results, especially concerning racial differences. Strongly modifiable to suit purpose and context (Armstrong 1944, 1955). Given mainly by ♂, but also by ♀. (a) Territorial-song of ♂. A well-structured rattling warble of clear shrill notes, delivered with remarkable vehemence (Witherby *et al.* 1938b). J Hall-Craggs distinguishes 4 segment types in songs of nominate *troglodytes/indigenus* and island races as follows. Type A: series of heterogeneous units characterized by lack of uniformity in duration and acoustic structure of adjacent

units, and in time intervals between them (e.g. *c.* last 10 units in Fig IV); such segments comprise about half of song, and usually begin and end it. Type B: series of repeated units which, in contrast to type A, show marked uniformity of unit structure and duration, and of interval between units (e.g. in Fig I, 5 Λ-shaped units towards end of song). Type C: series of rapidly repeated clicks sounding, overall, something like a rattle; units appear as fine vertical lines on sonagrams, and are repeated 1–30 times at rate of *c.* 28–40 units per s (e.g. rattle of 22 units at *c.* 1·3 s in Fig II). Type D: single continuous unit, 0·1–0·5 s long, frequency 4–5 kHz (or higher or lower), sounding like a buzz (less stereotyped than rattle of type C), with frequency modulation at over 180 per s, e.g. Fig II at 4·5 s; apparently confined to populations on or near cliffs. Fig I shows song for nominate *troglodytes*. Full songs last *c.* 4–6 s, curtailed songs (during aggressive interactions) *c.* 3–5 s (Garson 1978). Curtailed songs have a rattle near or at end, but full song always ends in type-A segment (J Hall-Craggs). Curtailed songs (*c.* 7% of total) typical of Song-duels between rival ♂♂ (Clark 1969), also when suddenly disturbed (J Hall-Craggs: see Social Pattern and Behaviour). At high intensity, song stifled, congested, and somewhat harsh, and, in close encounters, combined with call 5 (Armstrong 1944, 1955). Exceptionally, long songs produced by running 2–3 normal songs together (Clark 1949). For singing rate, see Social Pattern and Behaviour. Each ♂ has 4–7 (median 5) different phrases in his repertoire, and typically gives same phrase 10–20 times before switching to another. Local populations share same phrase types (i.e. collectively a dialect) (Kreutzer 1973, 1974a, b, which see for major quantitative study of song structure and geographical variation; see also Chappuis 1976). Pitch markedly deeper in south than in north of geographical range (Chappuis 1969). For renderings of *juniperi* in Libya, see Hartert (1923). In northern Europe, island races differ in quality and delivery. Interpretation of verbal descriptions (e.g. Armstrong 1950, 1953a, 1955, Waters 1964) commonly complicated by lack of synonymy of terms, e.g. 'trill' used indiscriminately to embrace apparently rattle, buzz, and rapid reiteration in type-B segments (see above). Some descriptions contradictory or subjective, and only partly supported by the following analysis made by J Hall-Craggs. Song tends to get longer the more northerly the race; variety within songs also greater in island races than in nominate *troglodytes/indigenus*; thus in island races 64% of repeated units comprised 2 or more sub-units, whereas in nominate *troglodytes/indigenus* 68% consisted of single continuous notes. Songs, especially of *zetlandicus*, also of *hebridensis*, *hirtensis* (Fig II), *indigenus* on Orkney and on islands off Co Kerry (Ireland) (P G H Evans), and of nominate *troglodytes* on Sark (Channel Islands), contain pronounced buzzes; otherwise songs similar in structure and delivery to mainland nominate *troglodytes/indigenus*. Songs of *borealis*, *islandicus* (Fig III), and *fridariensis* (Fig IV) lack buzzes, and are delivered more slowly, thus sounding less urgent, less vehement, and more musical; *borealis* and *fridariensis* show low repetition rate in type-B segments while *islandicus* shows long-duration type-A segments, yielding longest and perhaps most musical song of any race (see also Armstrong 1952 for apt description of *islandicus*). Full song of *fridariensis* invariably contains, near beginning and end, distinctive type-B segments sounding like 'weedle-weedle-weedle' (Sellar 1974); in Fig IV, these

the somewhat M-shaped units, of which 4 near beginning and 4 near end. Subsong divided by Armstrong (1955) according mainly to context; 2 kinds (b and c) recognized here. (b) Courtship-song of ♂. Compared with Territorial-song, usually much abbreviated, softer, sweeter, more warbling, and rapidly repeated (Kluijver *et al.* 1940; Armstrong 1944, 1955). If ♂, singing thus, approached by ♀, his song gets progressively softer, remaining abbreviated until, when ♀ nearby, he sings very few introductory notes (normally heard in Territorial-song), but maintains a high song rate. Courtship-song given specifically to attract ♀, e.g. in sexual chases, in Wing-quivering display, when nest-building; during Nest-showing display, even more refined variant ('Invitatory Song') given (Armstrong 1944, 1955). (c) Whisper-song. Given mainly by ♀, rarely ♂. Resembles distant song of Swallow *Hirundo rustica*, varying from occasional 'whit' sounds to sustained twittering, sometimes also subdued but harsh 'skreek' sounds. Given by ♀ at nest, occasionally when incubating, more often after hatching, i.e. when brooding, feeding young, leading

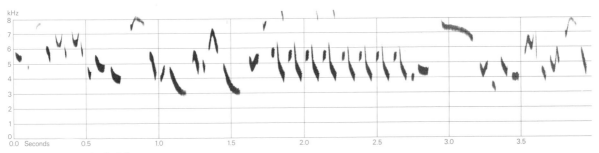

III P J Sellar Iceland July 1972

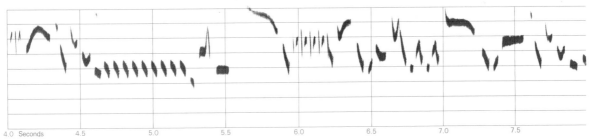

III *cont.*

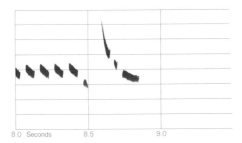

III *cont.*

young from nest or into roost (Armstrong 1955). (2) Contact-alarm calls. (a) Ticking-call. Commonest call, a single, double or repeated sharp note, like large pebbles knocked together (Armstrong 1955). Rendered a hard, clicking, slightly tremulous 'tic-tic-tic' (Witherby *et al.* 1938*b*); a hard noisy series of 'tek tek' sounds (Bergmann and Helb 1982: Fig V). Given to communicate presence, mainly by ♂ throughout the year, in almost any context, including presence of predators; loud harsh 'chrrt chrrt' in presence of cat possibly this call (E K Dunn). Variable in pitch and rate of delivery: often a higher pitched 'tick', like beads knocked together, given by both sexes

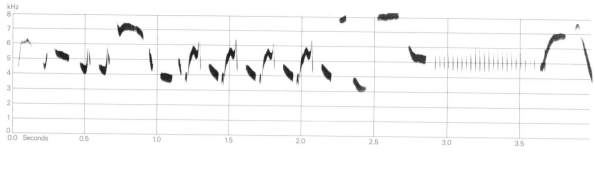

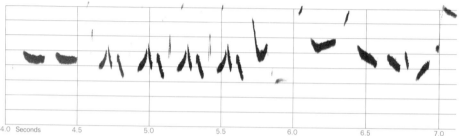

IV P J Sellar Scotland
June 1974

throughout the year (Armstrong 1955); 'titititititicc' (Witherby *et al.* 1938*b*) presumably this call. Another variant a richer deeper 'teeek' or 'tuuuk' (Kluijver *et al.* 1940). ♀ *zetlandicus* typically gave disyllabic 'chittick' or 'chittuck' (Armstrong 1952). (b) Chattering-call. Loud rapid succession of low-pitched, rather harsh sounds, typically given by ♀ greeting visits of ♂ to nest, also sometimes when approaching singing ♂, or when singing ♂ approaching her. Rarely given by ♂. Higher intensity signal than 1b, expressing a measure of excitement (Armstrong 1955). (c) Chleep-call. A 'chleep' or slightly grating 'chit' sound, exchanged by birds, of either sex, approaching winter roost (Armstrong 1955). (3) Soliciting-call of ♀. Rapid series of urgent, somewhat squealing sharp 'weeech-eech-eech...' sounds, given by ♀ soliciting copulation (Armstrong 1955); 'tseer' of ♀ in courtship (Kluijver *et al.* 1940) perhaps this call. Similar call once used by ♀ apparently to lure young from nest, described as a subdued single note resembling contact-call of Swallow *Hirundo rustica*, and rather similar to, but louder than, the succession of calls or Whisper-song (see

1c above) given at nest when feeding small young (Armstrong 1955). (4) Excitement-calls. (a) Churring-call. Short rapid succession of ticking or churring sounds given by ♂ in intense heterosexual encounter, e.g. when Nest-showing, also when escaping, or being released from, capture. Rarely given by ♀ (Armstrong 1955). (b) Grating-call. Harsh, low-pitched sound, resembling House Sparrow *Passer domesticus*, given mainly by ♂, seldom ♀, in intense sexual excitement, e.g. when 'pouncing' on ♀ in attempted copulation, also during copulation. Similar sound, not unlike Sedge Warbler *Acrocephalus schoenobaenus*, signalling some anger or aggression, given by ♂ fleeing Sparrowhawk *Accipiter nisus*, and by ♂ *hirtensis* alarmed by human intruder at nest (Armstrong 1955). (c) Excited ♂ *zetlandicus* gave a series of 'scrape-purrs' (not further described), only once heard from nominate *troglodytes/indigenus* (Armstrong 1952). (5) Anger-calls. Variously a wheeze, wheezy squeal, spitting hissing noise, squeaking sound like something (e.g. cork) rubbed against glass, given in confrontations between ♂♂, especially outside breeding season; once, in a fight,

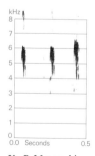

V R Margoschis
England July 1979

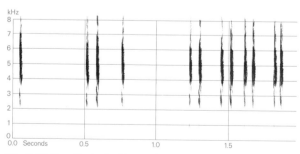

VI P J Sellar Scotland May 1976

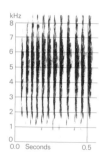

VII A P Radford England
August 1968

described as a shrill squeak like loud shrew *Sorex* (Armstrong 1955). (6) Distress-call. Soft, high-pitched squeak given by either sex when captured; may be accompanied by Bill-snapping (Armstrong 1955: see Introduction, above). (7) Alarm-calls. (a) Chitter-call. Short rapid succession of 'chit' sounds (Fig VI), higher-pitched than call 4a. Given in variety of situations, commonly by ♀ with brood to attract ♂ when predator present (Armstrong 1955); 'tr-tr-tr' (Bergmann and Helb 1982) probably this call. (b) Reeling-call. A regular and rapid succession of chittering sounds, each series given at intervals of 1–3 s, and rendered 'kreeee' (Armstrong 1955: Fig VII); also 't-r-r-r-r-r-t' when cat near nest (E K Dunn), 'trrr' (Bergmann and Helb 1982). Call rendered 'stirrrrrrrrrup' (Witherby *et al.* 1938*b*) is perhaps the same. Given in presence of ground predators, especially cat, also man and owls (Strigiformes) (Armstrong 1955). (c) Alarmed ♀ *zetlandicus* gave 'chilp' call.

CALLS OF YOUNG. Food-call of nestlings and fledglings a high-pitched squeak (Armstrong 1955), rendered 'kree' in well-grown *zetlandicus* (Armstrong 1955). Food-calls at first faint and barely audible, latterly raucous. In nest of young, shrill, rasping, thin, but powerful calls differed in pitch and purity between individuals, rendered variously 'see(p)', 'swee', and 'tswee' (P A D Hollom). Out of nest, food-calls also serve for contact between brood-members, and with parents. Alarm-call a chittering sound, like call 7a of adult but less vigorous; given only after fledging, e.g. on approach of human intruder (Armstrong 1955). Song given by juveniles from mid-July onwards, initially consisting of long warbling phrases (Garson 1978); develops from a few scratchy splintered notes into irregular succession of loud jangles and sweet Subsong (Armstrong 1955).   EKD

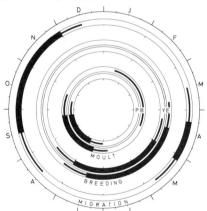

Breeding. SEASON. Britain and north-west Europe: see diagram. Central Europe: up to 1 week later (Makatsch 1976). European USSR: main laying period second half of May, though in Caucasus region begins in May (Dementiev and Gladkov 1954*b*). Cyprus: eggs found early April to May (Bannerman and Bannerman 1958). General tendency to get later with increasing latitude.

SITE. Very variable, but essentially a hollow, crevice, or hole at ground level or at up to *c.* 10 m high, average height increasing during season; in side of wall, tree, or steep bank, often inside building, and recorded in numerous artefacts. Of 170 sites, England, 42% in brambles, 32% in fork, 10% against trunk, 5% in box or basket, 4% in mixed bramble and bracken, 2% in branches, 5% other sites (Garson 1978). Nest: domed structure of leaves, grass, moss, and other vegetation, lined with feathers and hair. Mean dimensions: length 14·5 cm, breadth 13·0 cm, height 11·3, *n* = 116; internal height 5·6 cm, diameter 6·2 cm, *n* = 27 (Dallmann 1977). ♂ builds 5–8 (2–12) nests from which his ♀(♀) each select one for lining and use. Of 11 ♂♂, Netherlands, 1 nest built by 1, 4 by 3, 5 by 2, 7 by 2, 8 by 2, and 10 by 1; mean 5·4. Island races more usually build 1–2. (Armstrong 1955.) Building: main structure by ♂, lined by ♀, occasionally by ♂; ♀ very occasionally helps build main nest; takes 1 to 4–5 days (Armstrong 1955). EGGS. See Plate 81. Sub-elliptical, smooth and glossy; white, sometimes with speckling of black or brown at broad end. Nominate *troglodytes/indigenus*: 16·4 × 12·6 mm (14·7–18·4 × 11·6–13·8), *n* = 150; calculated weight 1·32 g. *T. t. hirtensis*: 18·6 × 13·9 mm (17·0–20·3 × 13·0–15·0), *n* = 100; calculated weight 1·84 g; *zetlandicus* and *borealis* very similar, samples of other races too small for comparison (Schönwetter 1979). Clutch: 5–8 (3–9); up to 16 recorded, though in view of polygamy (see Social Pattern and Behaviour), laying by 2 ♀♀ cannot be ruled out. Of 1115 clutches, Britain: 2–3 eggs, 4%; 4, 9%; 5, 25%; 6, 43%; 7, 13%; 8, 4%; 9, 2%; mean 5·7 (Garson 1980*a*); declined through season but no relation with latitude; ♀♀ breeding with more polygynous ♂♂ lay bigger clutches than those with less polygynous ♂♂, irrespective of time of year (Garson 1978). Of 44 clutches, Poland: 4 eggs, 1; 5, 2; 6, 9; 7, 32; 8, 4; mean 6·84 (Wesołowski 1983). Of 90 clutches, Netherlands: 2 eggs, 1; 3, 1; 4, 5; 5, 27; 6, 45; 7, 11; mean 5·6. Increase with latitude in Europe. (Armstrong 1955.) Eggs laid daily, mainly in early morning. Re-lays after clutch loss, 5–8 days later. 2 broods, 2nd started 10–14 days after fledging of 1st (Armstrong 1955). INCUBATION. 16·02 days (12–20), *n* = 43. By ♀ only, beginning gradually during laying; hatching takes place over 1–3 days. Incubates for 10–15 min at a time, alternating with periods of 6·7–9·0 min off nest; pattern varies during day. (Armstrong 1955.) YOUNG. Cared for and fed by both parents, brooded more or less continuously for first few days. FLEDGING TO MATURITY. Fledging period 17·3 days (14–19), *n* = 42 (Armstrong 1955). Become independent 9–18 days after fledging, occasionally not until later (Armstrong 1955). Age of first breeding not known. BREEDING SUCCESS. In 825 clutches, Britain, 71% of eggs hatched and 92% fledged, giving 65% overall success (Garson 1980*a*). In West Germany, 76 pairs reared average 3·7 young per pair per year (Berthold 1977).

Plumages (nominate *troglodytes*). ADULT. Entire upperparts rufous-brown, often duller and sometimes tinged dull greyish-olive or dark fuscous on forehead, crown and hindneck, often brighter rufous on lower rump and upper tail-coverts; scapulars, back, rump, and upper tail-coverts finely and closely barred dull black; bars usually rather ill-defined, but sometimes blackish and more strongly contrasting, occasionally extending to mantle; feather-centres on rump sometimes with white spot. Eye-ring and long narrow supercilium pale buff or off-white, partly mottled with fine brown spots; supercilium sometimes rather indistinct in front of eye. Lores, upper cheek, and ear-coverts spotted olive-brown or dark brown and off-white. Chin to chest and sides of neck and breast pale buff-brown or pale rufous-brown; feather-centres usually paler, those of central throat and sometimes chest off-white. Central belly uniform buff-white, remainder of underparts closely and narrowly barred black, rufous-brown, pale buff, and cream-white; dark bars usually rather faint on upper flank, blacker and more contrasting on lower flank and vent, but much individual variation in colour and extent. Tail, tertials, secondaries, and inner primaries rufous-brown, narrowly and closely barred black; outer primaries closely barred black and cream-buff or white; inner web of flight-feathers uniform dull black, merging into pale grey on basal border. Upper wing-coverts rufous-brown, narrowly and rather irregularly barred black; median coverts with fine white speck at tip and often with fine white shaft-streak. Axillaries and under wing-coverts pale grey-brown or light grey, sometimes slightly suffused buff and often with dark brown spots and arcs. *In worn plumage*, upperparts slightly duller and browner; dark bars of scapulars, back, and tertials slightly more contrasting; ground-colour of sides of head and underparts paler, grey, less tinged with brown, supercilium white, spots on outer primaries whiter. NESTLING. Down rather short and scanty, restricted to head and upperparts; dark grey or greyish-black (Heinroth and Heinroth 1924–6; Witherby *et al.* 1938*b*). JUVENILE. Upperparts similar to adult, but ground-colour slightly paler and duller rufous-brown, more faintly and extensively barred dark grey-brown. Supercilium indistinct, pale specks on sides of head less well-defined. Chin and throat duller grey than in adult, faintly mottled dusky; chest and breast with extensive but ill-defined brown fringes along feather-tips. Flanks, vent, and under tail-coverts extensively tinged rufous-brown, faintly barred with darker brown (ground-colour less pale than in adult, and under tail-coverts lack white tips). Tail, tertials, and flight-feathers as adult (but see First Adult); upper wing-coverts as adult, but ground-colour paler rufous-brown, dark bars narrower and closer, and median coverts without white spots. FIRST ADULT. Like adult, but juvenile flight-feathers, longer feathers of bastard wing, and variable number of tertials, greater coverts, and tail-feathers retained. Differences in pattern of largest feather of bastard wing and number of light spots on outer web of p7 considered useful age characters in Britain by Hawthorn (1971, 1974), but apparently not valid for populations of (e.g.) Switzerland (Jenni and Winkler 1983), Sweden (Svensson 1984*a*), or Netherlands (RMNH, ZMA). Best ageing character is retention of juvenile outer greater upper wing-coverts, contrasting in colour, length, wear, and pattern with new inner greater coverts. In adult, all greater coverts uniform medium brown or grey-brown, similar in colour to median coverts; some greater often with pale or white tips. In 1st adult, 3–6 (1–8) inner greater coverts new and like adult's; outer old, more rufous than neighbouring new median and greater coverts, often shorter, and without pale tip. Also, longest feather of bastard wing of adult usually has white outer margin, unlike most 1st adult. (Jenni and Winkler 1983;

Svensson 1984*a*.) Birds with all juvenile coverts retained difficult to age (Jenni and Winkler 1983).

Bare parts. ADULT. Iris dark brown. Upper mandible horn-black or black-brown, cutting edges of upper mandible paler horn or flesh-brown, lower mandible pale horn-brown or yellow-brown, sometimes with dusky tip and yellowish or flesh base. Leg and foot dull brownish-flesh, reddish-grey, light brown, pale yellow-brown, or greyish-brown; in *juniperi*, dark coffee-brown, sometimes with yellow soles. NESTLING. Bare skin pink. Mouth bright yellow. Gape-flanges pale lemon-yellow. Leg and foot flesh-pink. JUVENILE. Like adult, but lower mandible pale flesh-pink, flesh-yellow, or orange-yellow; gape-flanges pale yellow, disappearing at *c.*1 month old. (Heinroth and Heinroth 1924–6; Witherby *et al.* 1938*b*; Stanford 1954; Sage 1956*a*; RMNH, ZMA.)

Moults. ADULT POST-BREEDING. Complete; primaries descendant. In Britain, starts with p1 between early July and mid-August, completed with regrowth of p10 early September to mid-October; estimated duration of primary moult *c.* 60 days (Hawthorn 1974; Ginn and Melville 1983). ADULT PRE-BREEDING. 25 of 78 birds examined from Europe, North Africa, and Iran showed partial moult mid-January to late April, involving all or part of face, crown, and throat (Vaurie 1951*b*). POST-JUVENILE. Partial: head, body, lesser and median upper wing-coverts, and variable number of tertials, greater upper wing-coverts, and tail-feathers. Birds from early broods start about July and tend to renew tail, some greater coverts (usually 5), and sometimes tertials; tail, tertials, and greater coverts often all shed simultaneously; total duration of moult 45–60 days (Hawthorn 1974; Ginn and Melville 1983). Birds starting later replace fewer tail-feathers and no tertials; those starting as late as September-October replace head, body, and a few tertial coverts only (Hawthorn 1974; RMNH). Tail-feathers shed simultaneously at *c.* 10 weeks old; completely regrown in *c.* 18 days (Heinroth and Heinroth 1924–6). Moult of body starts *c.* 1 week before tail and greater coverts, in Britain between about mid-July and late September; some feathers still growing when tail and coverts new, early September to mid-October (Hawthorn 1974). Probably much geographical variation in number of greater coverts replaced, and more juvenile coverts retained in northern Europe than in south (Jenni and Winkler 1983).

Measurements. ADULT, FIRST ADULT. Nominate *troglodytes*. Netherlands, all year; skins (RMNH, ZMA). Bill (S) to skull, bill (N) to distal corner of nostril; exposed culmen on average 3·1 less than bill (S).

	♂		♀	
WING	49·8 (1·31; 73)	47–52	46·9 (1·15; 30)	45–50
TAIL	31·5 (1·40; 30)	29–34	30·4 (1·55; 22)	28–32
BILL (S)	14·0 (0·42; 26)	13·2–14·7	13·3 (0·62; 21)	12·2–14·2
BILL (N)	8·2 (0·38; 26)	7·6–8·8	7·8 (0·40; 18)	7·3–8·5
TARSUS	17·6 (0·42; 29)	17·0–18·3	16·9 (0·43; 24)	16·2–17·6

Sex differences significant. Wing of one ♂ 54 (breeding adult, RMNH).

Sweden: wing, ♂ 49·5 (10) 47–52, ♀ 46·0 (4) 45–47; bill (S), ♂ 14·1 (10) 13·5–15 (Vaurie 1951*b*, 1955*b*). Rumania and southern Yugoslavia: wing, ♂ 49·2 (1·08; 9) 48–51, ♀ 46·5 (1·13; 8) 45–48; bill (S), 14·0 (1·27; 6) 13–16 (Stresemann 1920; RMNH, ZMA).

T. t. indigenus. England and southern Scotland, all year; skins (RMNH, ZMA).

	♂		♀	
WING	50·0 (1·03; 10)	48–52	46·3 (1·13; 11)	45–48

BILL (S) 14·0 (0·54; 9) 13·3–14·8   13·2 (0·43; 10) 12·5–13·8
TARSUS 18·5 (0·57; 10) 17·6–19·1   17·4 (0·55; 11) 16·6–18·3
Sex differences significant.

Live birds Berkshire (southern England), August–April: wing of adult 49·5 (1·94; 515) 46–54; juvenile and 1st adult significantly shorter, 48·6 (1·76; 653) 45–53; range of adult ♂ 50–54, adult ♀ 46–50, immature ♂ 49–53, immature ♀ 45–49 (Hawthorn 1975a).

T. t. koenigi. Sardinia, all year; skins (RMNH, ZMA).
WING ♂ 49·0 (1·00; 12) 48–51   ♀ 45·5 (0·84; 5) 44–47
BILL (S) 13·8 (0·66; 12) 12·8–14·7   13·4 (0·80; 5) 12·5–14·3
TARSUS 17·4 (0·50; 13) 16·7–18·1   16·9 (0·59; 6) 16·3–17·8
Sex differences significant for wing. Bill (S), ♂ 15·2 (15) 14·5–16 (Vaurie 1959).

T. t. kabylorum. Algeria and southern Spain, all year; skins (RMNH, ZMA).
WING ♂ 48·3 (0·80; 6) 47–49   ♀ 45·4 (1·45; 6) 43–47
BILL (S) 14·1 (0·82; 6) 13·4–15·4   13·3 (0·36; 6) 12·9–13·8
TARSUS 17·7 (0·46; 6) 17·3–18·4   16·9 (0·26; 6) 16·5–17·2
Sex differences significant for wing and tarsus. Bill (S), ♂ 14·5 (20) 13·5–15·5 (Vaurie 1959).

T. t. fridariensis. Fair Isle. Wing 46–53, bill (S) 13–15(–16), tarsus 18–20(–21) (*n* = 20) (Williamson 1951b).
T. t. hirtensis. St Kilda. Wing, ♂ 51–55, ♀ 48–52; bill (S) ♂ 13·5–16; tarsus, ♂ 18–21 (Salomonsen 1933; Witherby *et al.* 1938b; Vaurie 1959).
T. t. hebridensis. Outer Hebrides. Wing, ♂ 46–53; bill (S), ♂ 12·5–13·5; tarsus, ♂ 17–19 (Salomonsen 1933; Witherby *et al.* 1938b). Bill (S), 14·5 (5) 14–15 (Vaurie 1955b).
T. t. zetlandicus. Shetland. Wing, ♂ 50–54, ♀ 47–51; bill (S) (14–)15–16; tarsus 17–20 (Hartert 1910; Salomonsen 1933; Witherby *et al.* 1938b; Williamson 1951b; Hawthorn 1980). Wing 52·5 (9) 52–54, bill (S) 15·9 (9) 15–16·5 (Vaurie 1955b).
T. t. borealis. Faeroes. Wing 54 (8) 50–56 (Vaurie 1959). Wing, ♂ 51–56, ♀ 51–53; bill (S), ♂ 15–17, ♀ 15–16; tarsus, ♂ 19·5–21, ♀ 19–20·5 (Salomonsen 1933). Wing, ♂ 49–55, tarsus 18–20 (Hartert 1910).
T. t. islandicus. Iceland. Wing, ♂ 58·2 (13) 56–61, ♀ 53–59; bill (S), ♂ 15–17, ♀ 15–17; tarsus, ♂ 20–21·5, ♀ 19·5–21 (Hartert 1910; Salomonsen 1933; Timmermann 1938–49; Vaurie 1959; ZMA).
T. t. juniperi. Cyrenaica (Libya). Wing, ♂ 46–48 (15); bill (S), ♂ 16 (9) 15·5–16·5 (Hartert and Steinbacher 1932–8; Stanford 1954; Vaurie 1959).
T. t. cypriotes. Crete. Wing, ♂ 46–52 (6), ♀ 47 (1) (Jordans and Steinbacher 1943).
Race unknown. Northern Asia Minor. Wing, ♂ 48·0 (0·93; 8) 46–49, ♀ 45·3 (3) 45–46; bill (S), 13·5–14·5 (Jordans and Steinbacher 1948; Rokitansky and Schifter 1971; Kumerloeve 1970a).
T. t. hyrcanus. Caucasus, Transcaucasia, and northern and south-west Iran (Stresemann 1928; Vaurie 1951b; Schüz 1959).
WING ♂ 50·0 (—; 51) 47–53   ♀ 47·3 (—; 22) 46–49
BILL (S) 14·9 (—; 37) 13·2–16·0   14·3 (—; 21) 13·2–15·0

Weights. Nominate *troglodytes.* Netherlands: (1) May–October; (2) November–April (RMNH, ZMA). East Germany: (3) all year (Bub *et al.* 1984).
(1) ♂ 10·0 (0·47; 12) 9·0–11·0   ♀ 7·8 (0·79; 8) 7·0–9·0
(2) 10·4 (1·49; 10) 8·9–12·5   8·4 (1·16; 6) 6·8–10·0
(3) 9·5 (—; 14) 8·0–11·7   9·5 (—; 4) 7·6–11·3
Southern Sweden, September: 8·9 (0·68; 130) (Scott 1965). West Germany: 9·3 (1·10; 272) 6–12 (Bub *et al.* 1984). France:

9·1 (12) 7·6–10·4 (Mountfort 1935). Central Italy, October–March: 9·4 (0·9; 407) (Ioale' and Benvenuti 1982, with curves of diurnal and monthly variations). Portugal, late July to September: 8·8 (0·89; 13) 7·1–10·2 (C J Mead). Greece, June–July: ♂♂ 10, 10 (Niethammer 1943). Exhausted birds: 7·6 (0·42; 11) 6·6–8·1 (Bub *et al.* 1984; ZMA, RMNH).
T. t. indigenus. Southern England: adult, ♂ 10·3–11·0, ♀ 8·9–9·5; juvenile and 1st adult, ♂ 10·1–10·8, ♀ 8·8–9·3 (Hawthorn 1975a); April–July, 9·2 (70) 7·4–11·7 (Harris 1962). Skokholm (Wales): 9·5 (26) 8·4–11·7 (Browne and Browne 1956). Isle of May (Scotland): 9·3 (18) (Williamson 1951b). Orkney, September–October: 10·2 (0·97; 23) 8·5–11·7 (J A B King; BTO). Exhausted birds, southern England and Wales, winter: 7·0 (12) 5–8·6 (Harris 1962; Ash 1964).
T. t. fridariensis. Fair Isle. 12·2 (50) (Williamson 1951b).
T. t. hirtensis. St Kilda. 12·5, 12·8 (BTO); ♀♀, 12·5, 14·5 (Waters 1964). On 1st day out of nest, 10·6 (1·07; 6) 9·5–12·5 (Waters 1964).
T. t. hebridensis. Outer Hebrides: average of 39 summer birds *c.* 1 higher than *indigenus* from England (Hawthorn *et al.* 1976).
T. t. zetlandicus. Shetland. ♂ 12–15, ♀ 10·5–13·5 (Hawthorn 1980).
T. t. islandicus. Iceland, November–March, ♂♂ 16·5, 20; ♀ 15 (4) 13·5–16·4 (Timmermann 1938–49).
T. t. cypriotes. Cyprus, October: 8·0, 9·2 (Hallchurch 1981).
Race unknown. Malta, winter: 8·9 (0·9; 15) 7·0–11·5 (J Sultana and C Gauci). Asia Minor, April–July: ♂♂ 8·5, 10; ♀ 10 (Kumerloeve 1970a; Rokitansky and Schifter 1971). USSR: 9·0 (31) 7·9–10·9 (Bub *et al.* 1984).

Structure. Wing short, broad, tip rounded. 10 primaries: p6–p7 longest, p8 0–2 shorter, p9 4–7, p10 16–22, p5 0–1, p4 2–4, p3 3–5, p1 6–7. Outer web of p5–p8 and (slightly) inner of p6–p9 emarginated. Tail short; 12 soft feathers with rounded tips, t6 much shorter than others. Bill long and fine, culmen and cutting edge of upper mandible slightly decurved, lower mandible straight; tip laterally compressed. Nostrils narrowly oval, partly covered by membrane above. Leg and foot strong, especially in some island races. Middle toe with claw 14·7 (21) 14–16 in nominate *troglodytes*, 16–18 in some large races like *islandicus*; outer toe with claw *c.* 74% of middle with claw, inner *c.* 72%, hind *c.* 89%. Claws strong, sharp, decurved; hind claw particularly strong.

Geographical variation. Marked and complex. Mainly clinal in continental Europe, but obscured by marked individual variation; some isolated island races more strongly differentiated. Variation involves depth of ground-colour of body, extent and width of barring, size (as expressed in wing and tail lengths and body weight), and relative size of bill, tarsus, and foot. In continental Europe and Mediterranean area, body very slightly smaller towards south; *koenigi* (Sardinia and Corsica), *cypriotes* (Crete to Levant), and *hyrcanus* (Caucasus to Iran) virtually similar in wing and tail length to nominate *troglodytes* from northern and central Europe, *kabylorum* (southern Spain, Balearic islands, and north-west Africa) and *juniperi* (Cyrenaica) slightly smaller; however, bill length relatively larger towards south, and though this hardly noticeable in *koenigi, kabylorum* from southern Spain, and in birds from northern Asia Minor, bill markedly longer in north-west African *kabylorum*, in *cypriotes, hyrcanus,* and (particularly) *juniperi*. No marked variation in size of leg in south of west Palearctic. Another cline runs from southern England (where measurements similar to nominate *troglodytes* from central Europe) north to Iceland: *indigenus* (Ireland, northern England, and Scotland north to

Inner Hebrides and Orkneys) similar in size to nominate *troglodytes*, but tarsus longer and foot heavier; *hebridensis* (Outer Hebrides) similar, but bill rather short; wing, tail, bill, tarsus, and foot gradually longer towards north in sequence *fridariensis* (Fair Isle), *hirtensis* (St Kilda), *zetlandicus* (Shetlands), and *borealis* (Faeroes), ending in large *islandicus* from Iceland (in particular, bill of *zetlandicus* and *borealis* proportionately long). In nominate *troglodytes*, colour of upperparts generally rufous-brown, barred dusky on scapulars, tertials, back, rump, and upper tail-coverts; underparts grey-white or buff-white, with deeper rufous and well-barred flanks, vent, and under tail-coverts; populations of humid western Norway (sometimes separated as *bergensis* Stejneger, 1884) have upperparts slightly duller and more saturated brown, less rufous, but difference too slight to warrant recognition. Towards south, populations of arid Mediterranean France, central Spain, southern Italy, Sicily, and Greece are more diluted rufous-brown or greyish-brown above, purer white below with more contrasting dark bars, tending towards *kabylorum*, *koenigi*, and *cypriotes*. *T. t. koenigi* from Corsica and Sardinia rather greyish olive-brown above, with bars often extending up to mantle; heavy and contrasting bars on underparts often extend higher up belly or breast than in nominate *troglodytes*. *T. t. kabylorum* pale, like *koenigi*; extent of barring rather variable, sometimes restricted to tertials and tail-coverts, sometimes as extensive as in *koenigi* (though bars often narrower). *T. t. juniperi* from Cyrenaica (Libya) similar to *kabylorum*, but upperparts slightly greyer, foot darker, and bill longer (Stanford 1954). *T. t. cypriotes* not well differentiated; usually well barred, like *koenigi*, but upperparts greyer, less brown, and bill longer; several populations united in single race here (following Vaurie 1955*b*), but differ slightly and sometimes split into *stresemanni* Schiebel, 1926 (Crete), *seilerni* Sassi, 1937 (Rhodos), and *syriacus* Meinertzhagen, 1933 (Levant). *T. t. hyrcanus* from Caucasus to Iran close in colour to nominate *troglodytes*, but barring slightly more extensive and bill longer. Populations from northern Asia Minor problematical; either considered inseparable from Rhodos birds (Jordans and Steinbacher 1948) or intermediate between nominate *troglodytes* and *hyrcanus* (Kumerloeve 1970*a*). *T. t. indigenus* from Ireland, northern England, and Scotland slightly darker above than nominate *troglodytes*, less bright and rufous, but birds south of about Lancashire and Yorkshire variably intermediate between *indigenus* and nominate *troglodytes* (see Clancey 1943*b*, Vaurie 1955*b*). On Atlantic islands north of *indigenus*, 3 lines of upperpart colour development, each with increasing barring towards north (1) from *indigenus* through *hebridensis* to *zetlandicus* (upperparts progressively darker rufous-brown to sooty-brown); (2) from *indigenus* through *fridariensis* to *hirtensis* (upperparts greyer, less rufous); (3) from *indigenus* through *borealis* to *islandicus* (upperparts duller, less rufous, more earth-brown or grey-brown); variation in underparts different, grading from buffish *hebridensis* through gradually paler and more contrastingly barred *fridariensis*, *zetlandicus*, *borealis*, and *islandicus* to whitish well-barred *hirtensis* (Williamson 1951*b*; see also Salomonsen 1933). For extralimital races, see Vaurie (1951*b*, 1955*b*, 1959) and Ridgway (1901–11).   CSR

Family MIMIDAE mockingbirds

Medium-sized oscine passerines (suborder Passeres); most species terrestrial or living in scrub or low trees close to ground, feeding on ground-living invertebrates, fruits, and seeds. 29–31 species in *c.* 11–13 genera (6–10 monotypic), including: (1) *Mimus* (typical mockingbirds), 8–9 species; (2) *Toxostoma* (typical thrashers), 10 species; (3) *Oreoscoptes* (Sage Thrasher *O. montanus*), monotypic; (4) *Dumetella* (Gray Catbird *D. carolinensis*), monotypic; and (5) *Melanotis* (blue mockingbirds), 2 species. Confined to New World, with greatest diversity in subtropics and tropics. Most species non-migratory, except those in higher latitudes. 2 species accidental in west Palearctic.

Body rather elongated and thrush-like in most species. Sexes similar in size. Bill medium to long, strong; straight to sharply decurved. Used for flicking over leaves and other ground debris when feeding, also for digging in some species. Rictal bristles few but always present. Wing short in most species, broad, and rounded; 10 primaries. Flight rather slow and laboured, with deliberate wing-beats; typically of short duration. Wings flicked up and down repeatedly in characteristic manner by some species, especially when feeding. Tail long. Legs and toes strong; tarsus distinctly scutellated, rather long in most species. Middle and outer front toes fused at base. Gait a walk or run. Head-scratching by indirect method. Bathe in typical stand-in manner. Anting recorded in at least 3 species, including *D. carolinensis* (frequently) and Brown Thrasher *T. rufum*; of direct type (ant-application).

Plumage somewhat variable in colour but generally rather uniform grey or brown, sometimes more rufous; underparts either similar or paler grey-brown to white, often with dark spots; under tail-coverts or tip of tail sometimes contrastingly coloured. Dark malar stripe in several species. Wing frequently has contrasting bars. Sexes virtually alike. Juvenile closely similar to adult but underparts usually faintly spotted. Adult has single complete moult annually in autumn; juvenile has partial moult in autumn.

Though at times placed near Troglodytidae (wrens) or Timaliidae (babblers), Mimidae now often considered closely related to Turdidae (chats, thrushes). Serological and egg-white protein evidence, however, shows closer resemblance to Sturnidae (starlings) than to Turdidae (Stallcup 1961; Sibley 1970), while DNA data point to such a close relationship between Mimidae and Sturnidae that both united (in 'Sturnidae') by Sibley and Ahlquist (1984), though in same superfamily ('Turdoidea') as Turdidae within the 'Muscicapae' assemblage.

Toxostoma rufum **Brown Thrasher**

PLATE 37
[between pages 544 and 545]

Du. Rosse Spotlijster   Fr. Moqueur roux   Ge. Rote Spottdrossel
Ru. Коричневый пересмешник   Sp. Trillador   Sw. Rödbrun härmtrast

Turdus rufus Linnaeus, 1758

Polytypic. Nominate *rufum* (Linnaeus, 1758), eastern North America, west to eastern Texas and western parts of Arkansas, Missouri, Iowa, and Minnesota. Extralimital: *longicauda* (Baird, 1858), Great Plains area of North America, west of nominate *rufum*.

Field characters. 24 cm; wing-span 29–31 cm. Approaches Mistle Thrush *Turdus viscivorus* in size, but form quite unlike any indigenous Palearctic passerine (except babblers *Turdoides*) due to rounded tail being almost half total length of bird. Plumage thrush-like, with red-brown upperparts and black-spotted, white underparts, but also double white wing-bar. Bill sharp and decurved. Sexes similar; no seasonal variation. Juvenile separable.

ADULT. Upperparts red-brown, with chestnut tone on crown, rump, and tail; relieved most obviously by bold wing-bars formed by white tips (buffy in autumn) and blackish subterminal marks on median and greater coverts; also by dusky forecrown, black-brown centres to white-fringed tertials, and dusky tips to primaries. Face essentially dusky, with obvious paler-toned supercilium and brown mark on rear cheeks; dominated by pale eye (iris yellow). Underparts cream, with long black-brown malar stripe, obvious lines of black-brown spots extending from sides of neck and breast to flanks and belly, and buff tones on chest, flanks, and under tail-coverts. Bill quite long and slightly decurved, ending in spiky tip; dusky horn or black. Legs grey horn. JUVENILE. Face pattern less clear, and upperparts dully spotted black. Iris even paler (white) and bill and legs have pink tone.

Plumage pattern may recall spotted thrush *Turdus* and attitudes even Rufous Bush Robin *Cercotrichas galactotes*, but rather large size, overall length, wing-bars, and bushy tail combine into appearance unique in west Palearctic. Flight and gait recall small, nimble Magpie *Pica pica* with all actions pronounced and powerful: flight-action includes bursts of beats from short wings, and bird hops and runs quickly, and raises and waves tail. Skulks in cover but will feed in the open, hammering at nuts with great force. Cocky and robust.

Calls given by vagrant to Britain included loud, harsh, throaty 'tschek' or 'chip', recalling short monosyllable of *P. pica*.

Habitat. Breeds in Nearctic middle temperate latitudes in lowlands and sparsely to *c.* 1700 m on elevated plains, in dense tangled underbrush along prairie stream bottoms (Niedrach and Rockwell 1939). Frequents open brushy woods, scattered patches of brush, small trees in open areas, and also shelterbelts, copses, and planted shrubberies in suburbs, where dense thickets offering suitable nest-sites occur (Johnsgard 1979). While overlapping with Gray Catbird *Dumetella carolinensis* differs in often preferring drier and more open situations such as brushy pastures and new second growth (Pough 1949). Habitat choice varies regionally, however. In New England, prefers remoteness from human habitations, utilizing bushy pastures, briar patches, tangles along fences, dry thickets, brushy hillsides, and edges of woodland. In Tennessee, has nested near college buildings on a campus, and in Kansas occurs in city parks and gardens and on farm fields. Trees needed as song-posts (Bent 1948). In Canada, prefers deciduous thickets (Godfrey 1966). Despite attachment to dense woody vegetation, spends much of time on ground, frequently dusting-bathing on dirt roads. Makes only short low flights, apparently even on migration (Forbush and May 1939).

Distribution. Breeds in central and eastern North America from southern Canada to eastern Texas and Florida. Winters eastern and southern USA.

Accidental. England: 1, Dorset, November 1966 to February 1967 (British Ornithologists' Union 1971).

Movements. Migratory over northern two-thirds of range, with breeding and wintering ranges overlapping in southern USA, north on east coast to Massachusetts, and in Mississippi valley north to Illinois (American Ornithologists' Union 1983; Farrand 1983). Departure on autumn migration progresses from late September in Montana and Saskatchewan, to early October in northern USA, and late October in Pennsylvania and Virginia. Ringing recoveries suggest breeders generally move south, but some eastern birds are forced westward by Gulf coastline. Spring arrivals average late March in North Carolina and Virginia, April in northern USA, and early May in Montana and Saskatchewan.

Accidental records in North America include Point Barrow (Alaska), southern Hudson Bay, Newfoundland, western USA, northern Mexico, western Cuba, Bahamas, and Bermuda (Peterson and Chalif 1973; Wingate 1973; American Ornithologists' Union 1983).   TL-E

Voice. See Field Characters.

Plumages (nominate *rufum*). ADULT. Entire upperparts bright tawny-rufous, forehead often tinged pale olive-brown, crown sometimes deeper, rufous-chestnut, rump and upper tail-coverts usually paler rufous-cinnamon. Lores, area round eye, ear-coverts, and upper cheek indistinctly mottled olive-brown or rufous-brown and pale buff (usually palest below eye), ear-coverts with fine pale buff shaft-streaks; lower cheek behind gape often with slightly darker (dull black) mottling, forming indistinct and mottled moustachial stripe. Sides of neck rufous-brown with olive or grey tinge and sometimes indistinct pale buff streaks. Lower cheek, chin, and throat pale cream or off-white with distinct mottled black or dark grey malar stripe extending down sides of throat to chest. Ground-colour of chest, flanks, and under tail-coverts cream-buff; sides of breast often darker, pale cinnamon-buff, belly and vent paler cream-yellow or cream-white; chest, sides of breast, and upper flanks boldly marked with dull black rounded-triangular spots (black of each spot often partly tinged dark rufous-brown, in particular on sides of breast), breast and sides of belly with smaller dull black spots, lower flanks with short and broad dull black streaks. Thighs olive-brown or buff, mottled dusky. Tail-feathers rufous-brown; narrow and indistinct fringes along tips of feathers and faint edge along outer web of t6 rufous-buff or off-white (soon wearing off). Flight-feathers dull tawny-rufous on outer webs (often brighter cinnamon-rufous on bases of primaries and rufous-buff on tips), dull black on inner webs. Tertials tawny-rufous; innermost ones narrowly tipped rufous-buff or pale buff, subterminally bordered by black. Upper wing-coverts tawny-rufous, median and greater with contrasting cream-buff or white tips *c.* 2–3 mm wide (narrower and more rufous on tertial coverts), subterminally bordered by black (not sharply divided from rufous). Longer feathers of bastard wing black with rufous base and white fringe along distal half of outer web. Under wing-coverts and axillaries pale cream-buff (like flanks), shorter coverts finely spotted dull black, longest tinged pale rufous-grey. *In worn plumage*, upperparts less deep rufous, slightly tinged olive; sides of head and neck dull grey with faint dark olive-brown and pale grey mottling; ground-colour of entire underparts dirty white or pale grey (faintly cream on chest and under tail-coverts); tips of inner tertials and median and greater upper wing-coverts bleached to white, but partly worn off. JUVENILE. Forehead and crown uniform pale olive-brown with buff tinge (in adult, deep rufous, only forehead slightly olive), merging into buff-brown on hindneck and sides of neck and dull rufous-cinnamon on upperparts; mantle to upper tail-coverts and lesser upper wing-coverts marked with indistinct dull black spots (not uniform deep rufous). Sides of head indistinctly mottled buff and brown; dark malar stripe faint, composed of fine spots. Feathers of underparts narrow and loose; ground-colour of chest, sides of breast, flanks, and under tail-coverts yellow-buff (not as deep cream-buff as adult), of remaining underparts cream-white or dirty white; spots on chest and sides of breast dull black (in adult, partly rufous), more triangular and less clear-cut than adult; dull black streaks on flanks narrower, *c.* 2 mm wide (in adult, 3–4 mm). Tail-feathers narrower than in adult; width of t6 (at widest part, sub-terminally) 12–13 mm (in adult, 15–16 mm), t1 *c.* 15 (in adult, 17–18). Longest feathers of bastard wing dark olive-brown (in adult, tip largely black), pale outer border narrowly reaches shaft at tip (in adult, tip near shaft fully black). Greater and median upper wing-coverts rather indistinctly tipped buff, subterminal black marks less distinct (in adult, tips contrastingly cream-white); tertials with narrow and rather indistinct pale rufous fringes (in adult, tips of innermost more contrastingly

black with cream fringe). FIRST ADULT. Like adult, but juvenile flight-feathers, greater upper primary coverts, bastard wing, outer or all tail-feathers, and sometimes a few tertials or outer greater upper wing-coverts retained; these contrastingly more worn than adult at same time of year. Flight-feathers and rather narrow juvenile outer tail-feathers slightly worn in autumn (in adult, tips smooth and rounded), heavily in spring (adult still quite fresh); by May, ground-colour of underparts often dirty greyish-white (in adult, still bright yellowish-cream).

Bare parts. ADULT, FIRST ADULT. Iris bright lemon- or sulphur-yellow. Bill dark greyish-horn, slate-black, or black, basal half of lower mandible flesh-pink. Leg and foot dusky yellowish-flesh or grey. JUVENILE. Iris pale grey or almost white. Bill dull pink-buff; gradually darkening from base of upper mandible and from middle of lower; tip light horn. Leg and foot dull pink-buff or pale greyish-blue. (Ridgway 1901–11; Bent 1948; Roberts 1955; RMNH.)

Moults. ADULT POST-BREEDING. Complete; primaries descendant. July–August (Bent 1948). POST-JUVENILE. Partial: head, body, and wing-coverts, but no flight-feathers, greater upper primary coverts, or outer tail-feathers; sometimes also retains some tertials or outer greater upper wing-coverts and often all tail (RMNH, ZMA). Late July and August (Bent 1948); still many feathers growing mid-September (BMNH).

Measurements. Nominate *rufum*. ADULT, FIRST ADULT. Eastern North America, April–October; skins (BMNH, RMNH, ZMA). Bill (S) to skull, bill (N) to distal corner of nostril; exposed culmen on average 5·3 less than bill (S).

WING	♂ 105·4 (2·82; 24) 100–112	♀ 103·2 (1·78; 16) 99–106
TAIL	122·7 (5·20; 18) 116–132	120·3 (4·42; 10) 112–126
BILL (S)	30·5 (1·25; 17) 28·9–32·6	27·9 (1·90; 10) 25·2–30·8
BILL (N)	20·2 (1·27; 16) 18·4–22·5	18·1 (1·74; 9) 16·2–20·8
TARSUS	34·7 (1·14; 18) 33·5–36·4	34·1 (0·76; 11) 33·2–35·1

Sex differences significant for wing and bill.

JUVENILE. 1st adult with retained juvenile flight-feathers and tail included in table above, though wing on average 2·2 shorter and tail 4·4.

Weights. Ohio (USA). Adult: April–May, ♂ 73·0 (6), ♀ 68·8 (20); June, ♂ 64·8 (5), ♀ 61·9 (27); July–September, ♂ 74·2 (2), ♀ 68·1 (10). Juvenile: June, 63·4 (18); July, 59·7 (34). (Baldwin and Kendeigh 1938.) Kentucky (USA), late April and early May: ♂♂ 63·1, 64·2 (Mengel 1965). New Jersey (USA), on autumn migration: 65·3 (*c.* 3·8; 78) 52–75 (Murray and Jehl 1964). Alabama (USA), December–January: ♂ 81, ♀ 79 (Stewart and Skinner 1967). Dorset (England), November: 81 (Incledon 1968).

Structure. Wing rather short, broad at base, tip rounded. 10 primaries: p6 and p7 longest, p8 2–4 shorter, p9 11–15, p10 37–41 (adult) or 34–38 (juvenile), p5 0–2, p4 2–5, p3 6–10, p2 9–14, p1 12–17. Outer web of p5–p8 and (faintly) inner of p6–p9 emarginated. Tertials very short. Tail long, tip rounded, 12 feathers; t1–t2 longest, t3 0–5 shorter, t4 2–12, t5 8–18, t6 18–30. Bill long, almost equal to head length; rather heavy; straight for basal half, but distal half of culmen and of cutting edges distinctly decurved, ending in sharp tip; gonys virtually straight, not concave as in other *Toxostoma*. Nostrils oval, surrounded by narrow membrane except below. Distinct bristles along base of upper mandible near gape. Tarsus and toes strong, covered with distinct scutes in front; bases of middle and outer toes

PLATE 34. *Pycnonotus leucogenys mesopotamiae* White-cheeked Bulbul (p. 474): **1-2** ad. *Pycnonotus xanthopygos* Yellow-vented Bulbul (p. 479): **3-4** ad. *Pycnonotus barbatus* Common Bulbul (p. 484): **5-6** ad. *Bombycilla garrulus* Waxwing (p. 491):**7-8** ad ♂. *Hypocolius ampelinus* Grey Hypocolius (p. 502): **9-10** ad ♂, **11** ad ♀. (NA)

PLATE 35. *Cinclus cinclus* Dipper (p. 510). *C. c. gularis*: **1-3** ad fresh (autumn), **4** ad worn (spring), **5** juv. Nominate *cinclus*: **6** ad fresh (autumn), **7** NW Iberian form ('*atroventer*') ad fresh (autumn). *C. c. aquaticus*: **8** ad fresh (autumn). *C. c. caucasicus*:·**9** ad fresh (autumn). *C. c. rufiventris*: **10** ad fresh (autumn). (HB)

PLATE 36. *Troglodytes troglodytes* Wren (p. 525). Nominate *troglodytes*: **1–3** ad fresh (autumn), **4** ad worn (spring), **5** juv. *T. t. islandicus*: ad fresh (autumn). *T. t. zetlandicus*: **7** ad fresh (autumn). *T. t. hirtensis*: **8** ad fresh (autumn). *T. t. hyrcanus*: **9** ad fresh (autumn). *T. t. kabylorum*: **10** ad fresh (autumn). (HB)

PLATE 37. *Empidonax virescens* Acadian Flycatcher (p. 43): **1** ad breeding fresh (spring), **2–3** 1st autumn. *Toxostoma rufum* Brown Thrasher (p. 543): **4–6** ad fresh (autumn). *Dumetella carolinensis* Gray Catbird (p. 545): **7–9** ad fresh (autumn). (PJKB)

PLATE 38. *Prunella modularis* Dunnock (p. 548). Nominate *modularis*: **1** ad ♂ fresh (autumn), **2** ad ♂ worn (spring), **3** ad ♀ fresh (autumn), **4** juv. *P. m. occidentalis*: **5** ad ♂ fresh (autumn). *P. m. obscura*: **6** ad ♂ fresh (autumn). *P. m. hebridium*: **7** ad ♂ fresh (autumn). (NA)

PLATE 39. *Prunella montanella* Siberian Accentor (p. 560): **1** ad fresh (autumn), **2** ad worn (spring), **3** juv.   *Prunella ocularis* Radde's Accentor (p. 565): **4** ad fresh (autumn), **5** ad worn (spring), **6** juv.   *Prunella atrogularis* Black-throated Accentor (p. 568): **7** ad fresh (autumn), **8** ad worn (spring), **9** juv. (NA)

PLATE 40. *Prunella collaris* Alpine Accentor (p. 574). Nominate *collaris*: **1** ad ♂ fresh (autumn), **2** ad worn (spring), **3** ad ♀ fresh (autumn), **4** juv. *P. c. subalpina*: **5** ad ♂ fresh (autumn). *P. c. montana*: **6** ad ♂ fresh (autumn). (NA)

PLATE 41. *Prunella modularis* Dunnock (p. 548): **1–2** ad ♂ fresh (autumn). *Prunella montanella* Siberian Accentor (p. 560): **3–4** ad fresh (autumn). *Prunella ocularis* Radde's Accentor (p. 565): **5–6** ad fresh (autumn). *Prunella atrogularis* Black-throated Accentor (p. 568): **7–8** ad fresh (autumn). *Prunella collaris* Alpine Accentor (p. 574): **9–10** ad ♂ fresh (autumn). (NA)

PLATE 42. *Cercotrichas galactotes* Rufous Bush Robin (p. 586). Nominate *galactotes*: **1** ad fresh (autumn), **2-3** ad worn (spring), **4** juv. *C. g. syriacus*: **5-7** ad worn (spring). *C. g. familiaris*: **8** ad worn (spring). *Cercotrichas podobe podobe* Black Bush Robin (p. 595): **9** ad. (HB)

PLATE 43. *Erithacus rubecula* Robin (p. 596). Nominate *rubecula*: **1** ad fresh (autumn), **2** ad ♂ worn (spring), **3** juv. *E. r. tataricus*: **4** ad ♂ worn (spring). *E. r. melophilus*: **5** ad ♂ worn (spring), **6** 1st ad fresh (1st autumn). *E. r. superbus*: **7** ad ♂ worn (spring). *E. r. caucasicus*: **8** ad ♂ worn (spring). (HB)

PLATE 44. *Luscinia luscinia* Thrush Nightingale (p. 616): **1** ad fresh (autumn), **2** ad worn (spring), **3** juv.   *Luscinia megarhynchos* Nightingale (p. 626). Nominate *megarhynchos*: **4** ad fresh (autumn), **5** ad worn (spring), **6** juv. *L. m. hafizi*: **7** ad fresh (autumn). (HB)

PLATE 45. *Luscinia cyane* Siberian Blue Robin (p. 661): **1** ad ♂, **2** ad ♀, **3** 1st ad ♂ non-breeding (1st winter).   *Tarsiger cyanurus* Red-flanked Bluetail (p. 664): **4** ad ♂ bright morph fresh (autumn), **5** ad ♂ bright morph worn (spring), **6** ad ♀ worn (spring), **7** 1st ad ♂ worn (1st spring), **8** juv. (HB)

Hilary Burn

partly fused. Middle toe with claw 28·6 (12) 27–30; outer toe with claw *c.* 68% of middle with claw, inner *c.* 63%, hind *c.* 73%. Claws strong, short, decurved.

Geographical variation. Slight. *T. c. longicauda* from Great Plains area larger, wing on average *c.* 7 longer, tail *c.* 8·5 (Ridgway 1901–11); upperparts slightly paler rufous, underparts

and pale tips of wing-coverts slightly purer white; inner webs of flight-feathers and shafts of tail-feathers darker brown (Bent 1948).

Forms superspecies with Long-billed Thrasher *T. longirostre* from southern Texas (USA) and eastern Mexico and with Cozumel Thrasher *T. guttatum* from Cozumel Island off Yucatan, Mexico (Engels 1940). CSR

Dumetella carolinensis Gray Catbird

Du. Katvogel   Fr. Moqueur chat   Ge. Katzenvogel
Ru. Птица-кошка   Sp. Pájaro gato   Sw. Kattfågel

Muscicapa carolinensis Linneus, 1766

Monotypic

PLATE 37
[between pages 544 and 545]

Field characters. 18·5 cm; wing-span 24–26 cm. Approaches Redwing *Turdus iliacus* in overall length but has different structure, with rounded tail almost as long as body. Plumage essentially dark grey, with black cap and rusty under tail-coverts. Often flicks tail. Sexes similar; no seasonal variation. Juvenile separable at close range.

ADULT MALE. Dull slate-grey, with sooty to black cap, paler grey but hardly contrasting supercilium, slightly paler, ash grey underparts, blacker tail, and rufous-chestnut under tail-coverts. ADULT FEMALE. Usually duller than ♂, with sooty cap, faint contrast between chest and paler throat and belly, and some grey in under tail-coverts. JUVENILE. Duller than adult ♀, with brown tone to underbody, faint dark spots on chest and flanks, and (on some) paler under tail-coverts. At all ages, bill and legs black-brown.

Unmistakable. Has structure and habits of other Mimidae but none has such uniformly dark plumage. Flight chat-like, with bursts of strong wing-beats (from fairly short wings) allowing dashes into and out of cover, and flicking and spreading of tail obvious in tight manoeuvres. Hops from branch to branch, and uses balancing tail movements on ground. Although reputedly a great skulker, often appears on edge of cover and not really shy.

Calls likely from vagrants are cat-like mew and harsh 'kak kak kak'.

Habitat. In Nearctic middle temperate latitudes, in continental lowlands. Breeds in thickets, woodland edges, shrubby marsh borders, orchards, parks, and similar habitats combining dense vegetation with a vertical or horizontal 'edge' (Johnsgard 1979). Favourite wild environment is dense shrubbery and vine tangles near streams, ponds, and open alder swamps; also breeds in brushy cut- or burnt-over lands, hedgerows, field borders, and ornamental shrubbery arising from human land-uses (Pough 1949). Usually nests in low dense thickets, tangles of vines or small bushy trees, often bordering marshes, streams, or forest; sometimes in spruce or pine, and will use bushes overhanging lakes (Bent 1948). Ventures into the open only where there is immediate access to dense cover. In mountain regions does not extend beyond foothills at *c.* 1700 m, e.g. in Colorado Rockies (Niedrach and Rockwell 1939). In Canada, breeds in low dense deciduous (and occasionally coniferous) thickets along streams, ponds, roadsides, and woodland edges, as well as in gardens (Godfrey 1966).

Distribution. Breeds in North America from southern Canada, eastern Oregon, central Arizona, and north-east Texas south to central part of Gulf states. Winters in eastern USA (south from New England) south to Panama and Caribbean islands.

Accidental. Channel Islands: 1, Jersey, mid-October 1975 (Long 1981*a*). West Germany: 1 collected Helgoland, October 1840 (Niethammer 1937). East Germany: 1, Leopoldshagen, May 1908 (Makatsch 1981).

Movements. Migratory over most of range, with some overlap of breeding and wintering ranges on east coast and in deep south of USA; resident on Bermuda (Wingate 1973). This abundant autumn migrant departs breeding grounds September to early October arriving in southern USA mid- to late September, and in Honduras, Nicaragua, and Panama in late October (Bent 1948). Transient and wintering along east coast of USA south from Massachusetts (where rare in winter) and through Central America (except north-west Mexico) to Panama; also on

PLATE 46 (*facing*).
Luscinia calliope Siberian Rubythroat (p. 638): 1 ad ♂ worn (spring), 2 ad ♀, 3 1st ad ♂ fresh (1st autumn), 4 1st ad ♀ fresh (1st autumn), 5 juv.
Luscinia svecica Bluethroat (p. 645). Nominate *svecica*: 6 ad ♂ breeding, 7 ad ♀ breeding, 8 ad ♂ non-breeding, 9 1st ad ♀ non-breeding fresh (1st autumn), 10 juv. *L. s. cyanecula*: 11 ad ♂ breeding, 12 ad ♀ breeding, 13 1st ad ♂ non-breeding fresh (1st autumn). *L. s. pallidogularis*: 14 ad ♂ breeding, 15 1st ad ♂ fresh (1st autumn). *L. s. magna*: 16 ad ♂ breeding. (HB)

Bahamas and Cuba—though rare on islands further south and east, e.g. Hispaniola, Anguilla (Bond 1961; Rappole *et al.* 1983).

Leaves Honduras, Nicaragua, and Panama in April, reaching Texas and Louisiana in early April, Florida mid-April to early May, and northern and western limits of breeding range *c.* 1 month later; present in North Dakota and British Columbia by mid-May (Bent 1948).

Particularly in years of abundant berry crops, stragglers of this frugivorous species attempt to winter north to South Dakota and southern Ontario. Accidental records in Nearctic in Newfoundland and James Bay (Canada), western Oregon, California, Nevada, Lesser Antilles, and as far south as northern Colombia (American Ornithologists' Union 1983).   TL–E

Voice. See Field Characters.

Plumages. ADULT MALE. Centre of forehead and all crown deep black; crown sharply divided from long grey supercilium and from dark grey hindneck. Entire upperparts from mantle backwards dark plumbeous-grey. Sides of head often darker and duller grey than supercilium, in particular lores and upper ear-coverts, merging into dark grey on sides of neck and plumbeous-grey on sides of breast and flanks. Underparts from chin to vent and thighs slate-grey, slightly paler and less bluish than upperparts; under tail-coverts rufous-chestnut. Tail black, bases of outer webs slightly tinged plumbeous-grey along borders. Lesser and median upper wing-coverts dark plumbeous-grey, like upperparts; remainder of upperwing, flight-feathers, and tertials dull greyish-black (darkest on inner webs of primaries and on bastard wing), tinged slate-grey on outer webs; fringes along outer webs of primaries paler slate-grey or ash-grey. Under wing-coverts and axillaries pale slate-grey (slightly paler than flanks), shorter coverts along leading edge of wing tinged plumbeous, longer coverts slightly tinged rufous on borders. *In worn plumage* (spring), little different from fresh autumn plumage: crown slightly duller, often tinged brown towards rear; remainder of upperparts and upper wing-coverts slightly browner, less deep plumbeous; throat and belly slightly paler grey; flight-feathers and tertials slightly browner, but tips of primaries still smooth and unfrayed until about June. ADULT FEMALE. Closely similar to ♂, and not always distinguishable. Crown more sooty-black, not as deep black as ♂, merging more gradually into grey of hindneck; forehead tinged grey to variable extent; mantle and scapulars with slight olive tinge, not as dark and pure plumbeous-grey as ♂; chin, throat, belly, and vent slightly paler grey, underparts often showing some contrast between darker grey chest and flanks and paler grey throat and belly (♂ virtually uniform); shorter under tail-coverts sometimes more extensively grey on centres, not as uniform chestnut as ♂. *In worn plumage* (spring), sexes differ in same characters as in fresh plumage, but sexing more difficult due to individual variation in wear: worn ♂ duller on crown, slightly olive on upperparts, and slightly paler grey on throat and belly, thus rather similar to fresh ♀; some birds of both sexes still fresh March–May, others more worn, and hence slightly worn ♂♂ occur alongside fresh ♀♀ and these are indistinguishable in colour; in breeding pairs, ♂ usually darker than ♀, however. JUVENILE. Rather similar to adult ♀; feathers of body narrower and shorter, rather loose, especially on rump and vent. Forehead and crown dull greyish-black; mantle, scapulars, back to upper tail-coverts, and lesser upper wing-coverts dull grey with slight

brown tinge; underparts dull brownish-grey with faint dull black spots or bars on chest, breast, and flanks and sometimes with buff tinge on throat and flanks; under tail-coverts rufous-cinnamon or rufous-fawn (as in adult or slightly paler), soft and loose. Tertials, median and greater upper wing-coverts, greater upper primary coverts, and tail with narrow rufous edges (sometimes olive and less distinct; occasionally virtually absent). FIRST ADULT MALE. Similar to adult ♂, but black of crown contrasts slightly less with dark grey hindneck, central forehead often tinged with variable amount of grey, and upperparts often faintly tinged olive. Best distinguished by presence of retained juvenile feathers: flight-feathers, greater upper primary coverts, tail, some or all greater upper wing-coverts, and often some or all tertials juvenile; in particular, juvenile coverts contrastingly browner than fresh plumbeous neighbouring ones; coverts and tail often have faint and narrow rufous or pale edges along tips when not too worn. FIRST ADULT FEMALE. Similar to adult ♀, but hindcrown more extensively tinged grey, no contrast between dull black crown and dark grey hindneck, and centres of under tail-coverts slightly more extensively grey retained, as in 1st adult ♂. Part of juvenile feathering retained, as in 1st adult ♂.

Bare parts. ADULT. Iris brown. Bill black. Inside of upper mandible dark slate. Leg and foot dark horn-brown, horn-black, or black, toes darkest. JUVENILE. Iris grey or grey-brown. Bill dusky pink-buff to dusky horn-brown. Leg and foot dusky buff-brown, dusky pink-buff, dusky purplish-brown, or black-brown, soles grey. (RMNH, ZMA.) FIRST ADULT. Like adult, but colour of gape pinkish at least until autumn migration (Raynor 1979; ZMA).

Moults. ADULT POST-BREEDING. Complete; primaries descendant. August–September, start depending on completion of breeding (Raynor 1979). Moult completed in birds examined from mid-September and later. POST-JUVENILE. Partial; August–September. Start apparently dependent on fledging date; completed by mid- or late September (Raynor 1979). In birds examined, moult completed from mid-August onwards; of 17 August–January birds: 1 retained many median upper wing-coverts; 2 retained tertial coverts; 9 retained all tertials, 5 1–2 tertials; 7 retained all greater upper primary coverts, 9 2–5 outer ones only; 16 retained all tail, 1 t4–t6 only; all 17 retained flight-feathers and greater upper primary coverts. No pre-breeding moult, but post-juvenile sometimes continued in winter quarters, as a few greater upper wing-coverts or tertials are sometimes contrastingly newer than others in spring.

Measurements. ADULT, FIRST ADULT. Eastern USA, April–October, and Mexico, October–April; skins (BMNH, RMNH, ZMA). Bill (S) to skull, bill (N) to distal corner of nostril; exposed culmen on average *c.* 4·6 less than bill (S).

WING	♂ 91·4 (2·21; 31)	88–96	♀ 88·6 (1·95; 14)	86–92
TAIL	90·2 (3·07; 29)	85–97	87·5 (2·61; 13)	85–92
BILL (S)	20·8 (0·76; 28)	19·3–21·9	20·4 (0·44; 14)	19·7–21·2
BILL (N)	11·1 (0·54; 28)	10·3–12·0	11·1 (0·43; 14)	10·7–12·0
TARSUS	28·4 (0·81; 29)	26·9–29·8	27·8 (1·27; 14)	25·9–29·4

Sex differences significant for wing and tail.

Wing of live birds, New York (USA): 91·0 (2·91; 2582) 80–102 (Raynor 1979).

JUVENILE. Retained juvenile wing of 1st adult on average *c.* 3·0 shorter than full adult and tail *c.* 3·6 shorter, but both combined with adult in table above.

Weights. Ohio (USA). Adult: May, ♂ 35·3 (24), ♀ 38·4 (43);

June ♂ 34·2 (25), ♀ 36·3 (39); July, ♂ 32·7 (21), ♀ 34·7 (20). Juvenile and 1st adult: July, 34·9 (75); August, 34·7 (38); September, 35·5 (6). (Baldwin and Kendeigh 1938.) New York, May–October: adult ♂ 38·3 (3·52; 126) 32–49; adult ♀ 40·9 (4·38; 79) 32–54; juvenile and 1st adult 39·9 (2·99; 1767) 27–50; all birds 39·9 (3·15; 2584) 27–54 (Raynor 1979). On migration, Illinois, mainly September: ♂ 40·2 (7), ♀ 38·2 (11) (Graber and Graber 1962). On migration, Georgia, early October: adult ♂ 38·0 (16) 34–43; adult ♀ 36·4 (6) 33–39; 1st adult ♂ 37·5 (6) 35–39 (Johnston and Haines 1957). On migration, New Jersey, autumn: 35·2 (2·86; 591) 23–45 (Murray and Jehl 1964). Panama, mainly October: 31·3 (2·4; 100) (Rogers and Odum 1966). Southern Mexico and Cayman islands: October–January, ♂ 37·7 (5·82; 8) 32–49, ♀ 34·0 (2·35; 5) 30–36; April, ♂ 51 (Olson *et al.* 1981; RMNH). On spring migration, Louisiana: 35·7 (0·27; 46) (Rogers and Odum 1966).

Structure. Wing rather short, broad at base, tip rounded. 10 primaries: p6–p7 longest or either one 0–1 shorter than other, p5 and p8 both 2–4 shorter, p9 10–15, p10 35–42, p4 6–10, p3 9–13, p1 13–18. Outer web of (p5–)p6–p8 and inner of (p6–)p7–p9 emarginated. Longest tertials reach to p1–p3 in closed wing. Tail long, tip rounded; 12 feathers, t2 longest, t1 and t3 1–3 shorter, t4 3–5, t5 5–9, t6 14–19. Bill straight, slender; distal half of culmen decurved; cutting edge of upper mandible has slight notch just behind tip; gonys straight or slightly convex. Nostrils oval, partly covered by membrane above. Distinct bristles project obliquely above gape. Tarsus and toes rather long, slender; tarsus scutellated, not smooth as in chats and thrushes (Turdidae); bases of middle and outer toe fused. Middle toe with claw 22·9 (12) 20–24; outer toe with claw *c.* 71% of middle with claw, inner *c.* 64%, hind *c.* 68%. Claws short but strong, decurved.

Geographical variation. Slight and clinal, involving size only. Northern birds slightly larger than southern ones: wing of birds from British Columbia (Canada) on average 3·8 mm longer than wing of birds from Mississippi valley, with birds from north-east USA intermediate between these and birds from Bermuda perhaps even smaller than Mississippi population (Ridgway 1901–11).

Allocation of this species to Mimidae sometimes disputed, as some characters nearer Turdidae (see, e.g., Pence and Casto 1975); however, all Mimidae probably close to Turdidae (Sibley and Ahlquist 1984).   CSR

Family PRUNELLIDAE accentors

Small oscine passerines (suborder Passeres); mainly terrestrial, often feeding close to scrub or boulders; food mainly insects in summer, but seeds and fruits also important in winter. 13 species in single genus *Prunella*, widespread in Eurasia, with greatest density of species in mountains of central Asia (Marien 1951). Some species sedentary, others migratory or move altitudinally in winter. 5 species in west Palearctic, all breeding.

Body compact; neck short. Sexes of similar size. Bill peculiar: of medium length, tapering rather evenly to point; hard, wide at base with laterally swollen appearance; culmen rounded in cross-section. Used for flicking over leaves and other debris when feeding. Nostril free of feathering, covered by membrane (operculum). Wing short to moderately long, rounded or rather pointed; 10 primaries, p10 very short but not vestigial. Flight strong and rapid in most species, straight or undulating; usually sustained over short distances only and made at low height. Some species perform song-flights. Tail moderately long in most species, short in others; tip square or notched; 12 feathers. Leg fairly short with strong toes; hind claw the longest. Tarsus scutellated in front, some scales being more or less fused. Usual gait a hop or shuffling walk, typically while moving close to ground with legs well bent; some species run. Head-scratching by indirect method. Bathe in typical stand-in manner. Sunning common (in Dunnock *P. modularis* at least): performed in lateral posture, often characteristically with one wing raised; spread-eagle posture of most other passerines not recorded. Dusting reported in *P. modularis* but strong likelihood of confusion with other behaviour and confirmation needed. No anting records.

Plumage thick and rather coarse. Generally stone-coloured, tinged with grey, brown, or rufous; often streaked above; uniform dull grey or buff below, with contrasting black or black-and-white spotted or barred throat-patch and with bright rufous or deep buff chest or flanks; many species have uniform grey patch at side of neck; often a pronounced supercilium. Wing in some species contrastingly patterned rufous-brown and black, marked with whitish spots. Sexual differences slight. Juvenile similar to adult but underparts streaked. Adult has single complete moult annually in autumn, juvenile a partial one in autumn. Nestling down dark grey, brown-black, or black; mouth brightly coloured with contrasting dark spots.

Relationships to other oscine families long considered obscure. Bill shape and jaw musculature rather thrush-like but tarsus scutellated and young not really spotted as in Turdidae, while 'turdine thumb' structure of syringeal musculature absent. Presence of crop, muscular gizzard, and operculum suggests affinities with finches (Fringillidae) and buntings (Emberizidae), though these similarities usually considered due to convergence, while structure of egg-white proteins points to closer relationship with Sylviidae (Old World warblers) and Muscicapidae (flycatchers) (Sibley 1970). DNA data, however, place *Prunella* close to Passeridae (sparrows), Ploceidae (weavers), Estrildidae (waxbills), and Motacillidae (pipits, wagtails), these together forming an enlarged 'Ploceidae' within superfamily 'Fringilloidea' fairly remote from thrushes and allies, though all members of same larger assemblage ('Muscicapae') (Sibley and Ahlquist 1981b; see also for general review of taxonomic position of Prunellidae).

Prunella modularis Dunnock

PLATES 38 and 41
[between pages 544 and 545]

Du. Heggemus   Fr. Accenteur mouchet   Ge. Heckenbraunelle
Ru. Лесная завирушка   Sp. Acentor común   Sw. Järnsparv

Motacilla modularis Linnaeus, 1758

Polytypic. *P. m. hebridium* (Meinertzhagen, 1934), Ireland and Outer and Inner Hebrides; *occidentalis* (Hartert, 1910), eastern Scotland, England, and Wales, grading into *hebridium* in western Scotland and into nominate *modularis* in western France; nominate *modularis* (Linnaeus, 1758), central and northern Europe, from central France, Netherlands, and Norway east to Ural mountains, south to Alps, central Yugoslavia, and central Rumania; *mabbotti* Harper, 1919, Iberia, Pyrénées, south-central France, Italy south from Apennines, and (perhaps this race) Greece; *meinertzhageni* Harrison and Pateff, 1937, southern Yugoslavia and Bulgaria; *fuscata* Mauersberger, 1971, mountains of Crimea (USSR); *euxina* Watson, 1961, north-west Asia Minor; *obscura* (Hablizl, 1783), eastern Turkey and Iran north through Caucasus to lower hills of Crimea.

Field characters. 14·5 cm; wing-span 19–21 cm. Same length as House Sparrow *Passer domesticus* but less bulky, with fine bill, evenly domed head, and slimmer rear body and tail. Sleek but plump, ground-creeping passerine, with warbler-like bill, rather round head, and constant, seemingly nervous wing-twitching. Plumage essentially grey on head and chest, brown elsewhere (except for whitish belly), copiously streaked on upperparts and flanks; lacks any obvious character. Sexes rather similar; little seasonal variation. Juvenile separable. 8 races in west Palearctic, forming cline across Europe; Caucasus race the most distinct.

ADULT MALE BREEDING. (1) North and central continental European race, nominate *modularis*. Head, hindneck, throat, and breast basically smoky-grey, with brown-black streaks over top of head and neck, brown suffusion and indistinct grey-white streaks on lores and cheeks (most obvious behind eye), and hoary chin. Mantle, scapulars, and smaller wing-coverts warm brown, profusely streaked brown-black; rump and upper tail-coverts dull brown, unstreaked and contrasting slightly with black-brown tail. Greater coverts and secondaries black-brown, with warm brown fringes and small buff tips forming dull bar across coverts and spots on tertials; primaries black-brown. Underwing grey. Sides of breast and flanks buff-brown, streaked black- and rufous-brown; below this area and on upper breast, smoky-grey of foreparts fades into grey-white and then almost pure white on centre of belly, which looks strikingly pale when exposed. Under tail-coverts buff, strongly dappled brown. From about September to January, upperpart feathers have wider fringes and tips, making plumage less obviously streaked; grey of face and foreparts more washed brown. (2) Irish and Hebridean race, *hebridium*. Upperparts much darker than nominate *modularis*, with rich rufous tone to brown plumage and streaks blacker; head and underparts darker grey, with almost purple tone on head; belly less white. (3) British race, *occidentalis*. Intermediate in appearance between *hebridium* and nominate *modularis* but variable with (e.g.) ground-colours of isolated Scilly population much less grey, buffier brown. Upperparts lack rich brown of *hebridium* and underparts lack white belly of nominate *modularis*, but old ♂♂ can have dark purple-grey on head of *hebridium*. (4) Caucasus, eastern Turkey, and Iran race, *obscura*. Noticeably paler and duller brown above, less smoky and duller grey below, with greater invasion of hoary feather-tips on chin, throat, and below breast. (NB Important to recognize that plumage variable within races and susceptible to changes in tone according to light; racial identification of migrants thus hazardous—clarity of white on belly the only useful distinction between nominate *modularis* and western extreme of *hebridium/occidentalis* cline, and dull buff ground-colour of upperparts and flanks the best indication of *obscura*.) ADULT FEMALE. All races. Particularly in spring, less richly coloured than ♂: head and foreparts paler grey (and much suffused brown when fresh); upperparts less darkly streaked, with wider brown feather-fringes. JUVENILE. All races. Somewhat resembles browner ♀ but whole plumage more ochre or buff in ground-colour, lacking any clear grey tone except on throat and having much more prominent striations over almost all underparts (with only narrow belly-centre showing dull white). Pale buff tips to greater coverts form quite striking wing-bar. At all ages, bill short and fine; black-horn with pink-brown base to lower mandible; legs warm pink to red-brown; iris noticeably red-brown.

Though without striking markings, has distinctive character (and voice). Little risk of confusion with congeners: Alpine Accentor *P. collaris* much larger and more boldly patterned, and Asian species all heavily marked on head, with paler underparts. In brief view can also be mistaken for other small ground-hugging passerines (e.g. juvenile Robin *Erithacus rubecula*, small bunting *Emberiza*, dark warbler Sylviidae). Flight normally low, quite fast and whirring, with rather round wings and quite long, full tail obvious as bird ducks into cover; in autumn, migrants may make fluttering ascent, partially circling attempt at departure, and then high departure flight, with less whirring action and more sustained, slightly undulating progress. Gait includes characteristic mouse-like shuffle or creep with small hops and mincing walk (legs almost hidden), as well as more active hopping with simultaneous flicking of wings and occasional jerking of tail. Carriage usually rather horizontal when creeping, with head and tail often above level of body, but more upright when hopping and particularly so when resting or when singing from perch. General behaviour restless, busy yet unobtrusive; keeps close to cover but not really shy. Seeks food at close range, like Wren *Troglodytes troglodytes*; unlike small chats (Turdidae), does not regularly pounce out of and back into cover.

Song a flat little warble; high-pitched and somewhat recalling short section of song of *T. troglodytes* but rather formless and lacking that species' pulsing vehemence— 'treep titti-treep titti-treep titti-treep' or 'weed sissi-weed sissi-weed sissi-weed'; usually delivered from low cover. Commonest call a shrill, piping, explosive monosyllable, 'pseek' or 'tseep', more shrieked and drawn out in alarm.

Habitat. In upper and middle latitudes, mainly in temperate but marginally in subarctic, boreal, and Mediterranean zones (and in south especially in montane zone), between July isotherms 13–26°C (Voous 1960). Apparently, like congeners, *P. modularis* evolved in scrub and stunted coniferous arctic-alpine and wooded tundra habitats which are still occupied in Switzerland, southern USSR, and elsewhere. In Switzerland, prefers cool temperate climates with relatively high precipitation, reaching highest density in mountain forests of spruce *Picea* and larch *Larix*, up to *c.* 2200 m, choosing glades and edges away from tall trees and among scrub coniferous growth, dwarf heath, alder *Alnus* or willow *Salix* bushes; sometimes in mixed woodland but absent from lowland parks and gardens and only exceptionally in broad-leaved woods without conifers (Glutz von Blotzheim 1962). In Germany, chooses similar habitats, except in west where it has become a common park and garden bird in lowlands and on coast (Niethammer 1937). In USSR in summer, occurs up to treeline at 2500–2600 m in Caucasus, descending in winter sometimes only to 1000 m; frequents thickest tangles of subalpine vegetation, including birch *Betula*, maple *Acer*, juniper *Juniperus*, rhododendron, and bramble *Rubus*. In Transcarpathia, breeds up to 1500–1700 m mainly in spruces and pines *Pinus* but more sparsely in beeches *Fagus*. In north of range, still mainly in spruce but also in mixed and broad-leaved woodlands, especially along rivers and streams. On migration and in winter, inhabits forests and orchards with bushes, scrub, or hedgerows, and even reeds and wetland plants (Dementiev and Gladkov 1954*b*). In Białowieża forest (Poland), occurs most densely in swampy stands of alder with ash *Fraxinus* or spruce; more thinly in mixed oak *Quercus* and hornbeam *Carpinus*, and in spruce and pine mixed with oak and birch (T Wesołowski). In Belgium, departure from montane and coniferous habitats much more marked, although young conifer plantations and conifers in parks and gardens still used, but also thickets, brambles, hedges, wooded marshes, and edges of large forests, main population being in lowlands (Lippens and Wille 1972). In Scotland, breeds up to 500 m or more among juniper but elsewhere in Britain shift away from montane and coniferous habitat is almost complete, having occurred apparently *c.* 200 or more years ago (Montagu 1831). Now a pioneer species, commonly invading a wide variety of scrub-grown situations, it has adapted to coppice woodland with vigorous ground vegetation and widely spaced tall standard trees, and then to field hedgerows, farms, railway embankments and cuttings, churchyards, parks, gardens, and vacant urban land, as well as many semi-natural bushy and shrubby areas, including both inshore and offshore islands where high winds and exposure inhibit forest growth. Even rocky scrubby coastal cliffs are often occupied, as are sand-dunes. In contrast, however, to the equally abundant Robin *Erithacus rubecula* and Wren *Troglodytes troglodytes*, it is common in woods only of pedunculate oak *Quercus robur*, and to lesser extent in those of native pine and in young coniferous plantations. (Yapp 1962; Sharrock 1976; Fuller 1982.) While continuing to benefit from remoteness and undisturbed nature of primitive habitats, has also adapted to changing land use and to management practices artificially maintaining growth stages which in nature could be only transitory. This contrasts with other *Prunella*.

Distribution. Expanded south-east in Sweden and in Bavaria.

BRITAIN. Colonized Outer Hebrides and Orkneys (Scotland) in second half of 19th century (Parslow 1973). Bred Fair Isle (Shetland) 1974 (Sharrock 1976). WEST GERMANY. Expanded range in southern Bayern (Reichholf 1984). SWEDEN. Expanded range to south-east 1900 to 1950s (LR). TURKEY. Thinly distributed (Vittery *et al.* 1971). ALBANIA. May breed in north (EN). ALGERIA. May have bred recently but no proof (EDHJ).

Accidental. Iceland, Faeroes, Egypt, Libya, Algeria, Morocco.

Population. Increased Netherlands, Finland, Poland, and Czechoslovakia, declined Denmark and Rumania.

BRITAIN, IRELAND. Over 1 million pairs; no evidence of marked widespread change but probably increased rather than decreased (Parslow 1973). Estimated 5 million pairs (Sharrock 1976). Marked increase in inner London (Cramp and Tomlins 1966). FRANCE. Under 1 million pairs (Yeatman 1976). BELGIUM. About 120 000 pairs (Lippens and Wille 1972). NETHERLANDS. 125 000– 170 000 pairs (Teixeira 1979); increased, now common, even in city of Amsterdam (CSR). WEST GERMANY. Estimated 2·1–2·3 million pairs (Rheinwald 1982). DEN-MARK. Some decline 1978–82 (Vickholm and Väisänen

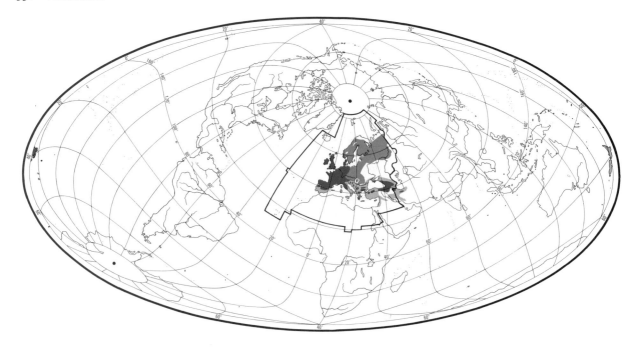

1984). SWEDEN. Estimated 1·7 million pairs (Ulfstrand and Högstedt 1976). FINLAND. Estimated 8000 pairs (Merikallio 1958); rare until 1950s, huge increase since (Hildén and Koskimies 1984; OH); almost stable 1978–83 (Vickholm and Väisänen 1984). POLAND. Slowly increasing (Tomiałojć 1976; AD, LT). CZECHOSLOVAKIA. Increased after 1960, especially at lower elevations and in urban habitats (KH). GREECE. Only a few tens of pairs (GIH). RUMANIA. Decreased (VC).

Survival. England: average annual adult mortality 51% (Innes 1981). Oldest ringed bird 9 years (Rydzewski 1978).

Movements. Resident, partial migrant, and, in northern and central Europe, total migrant.

Populations of Ireland and western Scotland (*hebridium*) mainly sedentary (Zink 1975). Populations of eastern Scotland, southern Britain, and western France (*occidentalis*) for the most part make only short dispersive movements, rarely over 30 km (British Ornithologists' Union 1971; Zink 1975). Main continental populations (nominate *modularis*), especially those breeding in northern areas (Fenno-Scandia, northern East and West Germany, Poland, and northern USSR east to Urals) and to lesser extent those from southern areas (central France to Corsica, Sardinia, and central Italy), move to winter in south-west Iberia, Mediterranean area, southern USSR, and Turkey (Vaurie 1959). Considerable passage to islands of western and central Mediterranean: common on Minorca and Mallorca though not on Ibiza or Formentera (Bannerman and Bannerman 1983); common on Malta October–March, with notable fresh influx February–March suggesting passage from Africa (Sultana

and Gauci 1982). Apparently rather rare in North Africa, however, and only at all regular in northern Tunisia (e.g. Willcox and Willcox 1978, Thomsen and Jacobsen 1979, Goodman and Watson 1983). Southernmost records (race not determined) at Elat (southern Israel), January–February (Safriel 1968).

Recoveries of birds ringed as nestlings, breeding adults, and migrants show movement on wide front and close to a north-east/south-west axis; thus, birds ringed Norway to Poland recovered in winter in south-west Spain, Balearics, Sardinia, and northern half of Italy; marked absence of recoveries in north-east Spain (Zink 1975). In USSR, rarely winters well north (e.g. Kaliningrad at 55°N on Baltic) but majority move south; further south (e.g. in Ukraine), degree of movement probably varies with severity of winter (Dementiev and Gladkov 1954b). Autumn migration occurs at Kaliningrad and Minsk (west-central USSR) September–November, and probably quite general at this time. Most easterly recorded wintering on Aral Sea, where small flocks seen in 2 successive winters (Rashek 1965). Poland vacated in winter with most birds moving south-west, some to Spain (Zink 1975); prominent passage throughout the country March–April and September to early November, with some birds regularly moving within the country to milder west, especially to Baltic coast area, Kielce area, and Mazurian Lakes (Tomiałojć 1976). In Switzerland, winter visitors disappear from late February while upland breeding birds arrive late March (or even early April); in autumn, movement away from upland breeding areas starts in August, but regular migration not until mid-September with peak in first half of October, stragglers until late November (Glutz von Blotzheim 1962). Ap-

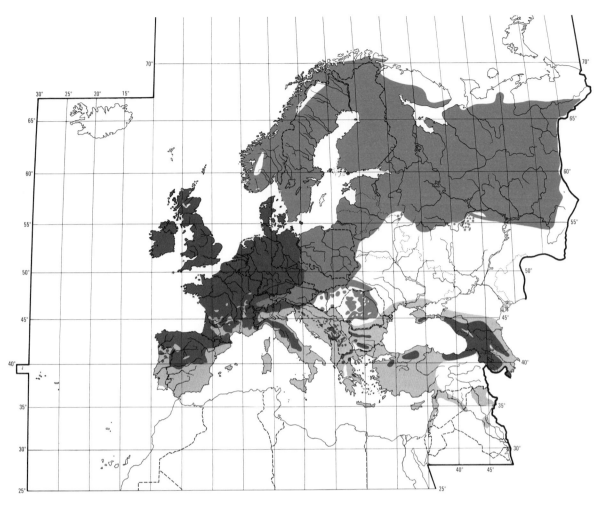

parently little wintering in Britain by continental birds, though migrant nominate *modularis* regular on Fair Isle and fairly regular on east coast (British Ornithologists' Union 1971; see also final paragraph); movements by birds of unknown origin have been noted over London, one pronounced westward autumn movement peaking in early October (Grant 1968).

For birds breeding in Iberia (*mabbotti*), no clear indications of any (even local) seasonal movement (Zink 1975). Populations breeding in north-west Turkey (*euxina*) and in Crimea, Caucasus, north-east Turkey, and Iran (*obscura*) are resident or only partially migratory; North Caucasian birds make local altitudinal movements, while at least some from south Caucasus descend to Black Sea coast as early as late July or early August (Dementiev and Gladkov 1954*b*; Beaman 1978). However, specimens of *obscura* known from Egypt (Goodman and Watson 1983), and these birds had perhaps travelled via Cyprus where wintering may also occur, but few records there and race not determined (Flint and Stewart 1983).

Individuals may winter in quite different areas in successive winters: bird ringed November, northern

France (and presumed to have wintered there or possibly to south-west), was wintering November–January the following year in Isle of Man; another from same area of France was recovered the following December at Cannes on Mediterranean coast (Zink 1975). This contrasts with evidence of 3 successive years' wintering by an individual in Malta (Sultana and Gauci 1982).   JHE

Food. Largely insects, plus significant proportion of seeds in winter. Predominantly a ground feeder, spending much of time in cover. Feeds under bushes, hedges, young conifers, and piles of twigs, and amongst roots and leaf litter (Glutz von Blotzheim 1962; G Bishton). In summer feeds more often in vegetation, up to 8 m above ground (G Bishton). In winter moves nearer human habitation and regularly seen in farms, gardens, and on bird-tables and window-ledges (Glutz von Blotzheim 1962; Glue 1982). Major foraging technique is a steady hop along ground with body horizontal, accompanied by ceaseless pecking movements; always moves forward and does not retrace steps (Naumann 1823; Witherby *et al.* 1938*b*; Müller 1974; G Bishton). Also, especially on flat ground, will shuffle along with belly nearly touching ground, legs well bent (Witherby *et al.* 1938*b*; G Bishton). Less commonly forages while standing still, picking food directly from vegetation, and turning over leaves, or probing soil to depth of 1 cm (Müller 1974; G Bishton). Also recorded flycatching (G Bishton), dislodging seeds from seed heads then picking up those fallen on ground (Harrison 1962), and picking items from faeces of Blackbird *Turdus merula* (Simmons 1985*e*). Highest proportion of time spent feeding in January (94%), decreasing to 54% in May; feeding activity rises again once young hatch (G Bishton).

The following recorded in diet in west Palearctic. Invertebrates: springtails (Collembola), stoneflies (Plecoptera), small grasshoppers (Orthoptera), earwigs (Dermaptera), bugs, etc. (Hemiptera), adult and larval butterflies and moths (Lepidoptera), larval scorpion flies (Mecoptera), flies (Diptera), ants (Formicidae), sawflies (Symphyta), and beetles (Coleoptera: Carabidae, Curculionidae, Scarabaeidae, Elateridae, Throscidae, Hydrophilidae, Chrysomelidae, Staphylinidae, Leiodidae). Regularly takes spiders (Araneae) and, less commonly, harvestmen (Opiliones), false scorpions (Pseudoscorpiones), small snails (Gastropoda), and earthworms (Oligochaeta). Plant food mainly seeds, predominantly taken in latter half of summer and during winter: particularly favours docks (Polygonaceae), legumes (Leguminosae), cranesbills (Geraniaceae), nightshades (Solanaceae), and figworts (Scrophulariaceae); also pines (Pinaceae), birches (Betulaceae), nettles (Urticaceae), goosefoots (Chenopodiaceae), amaranthuses (Amaranthaceae), pinks (Caryophyllaceae), buttercups (Ranunculaceae), poppies (Papaveraceae), Cruciferae, legumes, spurges (Euphorbiaceae), mallows (Malvaceae), primroses (Primulaceae), willowherbs (Onagraceae), Labiatae, plantains (Plataginaceae), thistles (Compositae), and sedges (Cyperaceae). Takes berries of heaths (Ericaceae), crowberry *Empetrum*, bramble *Rubus*, elder *Sambucus*, and holly *Ilex*. In winter will take cereals, particularly wheat *Triticum* and barley *Hordeum*. At bird-tables will eat crumbs, potatoes, meat, fat, and fruit. (Naumann 1823; Baer 1909; Florence 1912, 1914; Gil 1928; Lebeurier and Rapine 1936*b*; Witherby *et al.* 1938*b*; Dementiev and Gladkov 1954*b*; Glutz von Blotzheim 1962; Harrison 1962; Inozemtsev 1963; Ptushenko and Inozemtsev 1968; Glue 1982; Kostin 1983; Mal'chevski and Pukinski 1983; Bishton 1985; G Bishton.) Grit regularly found in stomachs and faeces, *c.* 1 mm diameter (G Bishton), but recorded up to 2·2 mm (Rey and Reichert 1908). Grit most common in winter months, when seeds form major part of diet (G Bishton).

In Shropshire (England), May-October, 95 samples of faeces had large invertebrate content. Of all invertebrates 62% (by number) beetles, 29% spiders, 4% earthworms, 3% Diptera, 2% snails, 1% springtails. No seeds present in faeces May-July, but in August-October nettle *Urtica*

dioica made up 46·4% of total faecal volume, grass *Holcus lanatus* 34·4%, thistle *Cirsium* 0·1%. During November-April, from 107 samples of faeces, beetles 73% of total invertebrates, spiders 12%, snails 7%, springtails 4·5%, earthworms 3·5%; seeds more common over winter: nettle 35% of total faecal volume, elder *Sambucus nigra* 18·4%, dock *Rumex* 6·5%, willowherb *Epilobium angustifolium* 0·4%, and grass *Poa annua* 0·1%. Over whole year, Curculionidae 56·7% by number of total beetles (particularly *Ceuthorhynchus* and *Otiorhynchus singulatus*—*Leiosoma deflexum* the only curculionid taken throughout winter and spring), Staphylinidae 29·9% (taken in all months except April-June, most frequently *Stenus clavicornis*, *Tachinus signatus*, and *Anotylus sculpturatus*), Carabidae 9·8% (taken July-December), Leiodidae 2·4%, and Chrysomelidae 1·2%. Also took spiders, harvestmen, false scorpions, snails (mostly in February), and (in June) some Lepidoptera. Over whole year, average volume of seeds 52%; maximum 90·4% in February, declining to 0 in April. *Urtica* most important: 29·1% of total faecal volume, *Holcus* 8·6%. Grass seeds commonly taken July-August, *Sambucus* and *Rumex* in late winter. (Calculated from data provided by G Bishton.) In Scotland, 4 stomachs from May-September all contained seeds (especially grass *Aira*) and 1 also contained Curculionidae; of 12 from November-March, all contained seeds (knotweed *Polygonum* in 10, chickweed *Stellaria media*, clover *Trifolium*, and *Rumex* also common) and 2 also contained beetles (Florence 1912, 1914). For other British studies see Newstead (1908) and Collinge (1924-7). In Spain, June, 4 stomachs contained seeds, especially *Polygonum*, spurrey *Spergula*, and *Stellaria*; invertebrates mainly beetles (especially Curculionidae); also earwigs, larvae of moth *Tortrix*, and spiders (Gil 1928). In Sachsen (East Germany), April-May, 3 adult ♂♂ contained only insects, especially beetles (Curculionidae most common), also ants and Diptera larvae (Baer 1909). In Brittany (France), 35 stomachs taken during winter contained 804 seeds of which 30% by number *Polygonum*; no other single species predominant although *Spergula* (18%) important during November, and mouse-ear *Cerastium* (8%) during January; stomachs mostly contained mixture of above and gorse *Ulex*, broom *Sarothamnus*, trefoil *Lotus*, and vetch *Vicia* (Lebeurier and Rapine 1936*b*). In Crimea (USSR), December-March, 9 stomachs contained 97·2% (by number) seeds (31·6% *Amaranthus*, 22·3% *Setaria*), and 2·8% invertebrates (largely molluscs and beetles); 7 stomachs from March contained 92·4% seeds (55% *Amaranthus*, 5·7% Cruciferae, plus buckwheat *Fagopyrum*, Chenopodiaceae, and grasses) and 7·6% invertebrates (4·7% molluscs, plus beetles, moth larvae, and ants) (Kostin 1983).

Diet of young similar to summer diet of adults: largely insects and occasionally seeds (Mal'chevski and Pukinski 1983; Bishton 1985). In Shropshire, 105 faecal samples from 35 nestlings May-July, contained the following.

Beetles in 75% of samples: 73% contained Curculionidae (mostly *Ceuthorhynchus*), 26% Staphylinidae (most frequently *Tachinus signatus*), 11% Carabidae, 11% Chrysomelidae, and 6% beetle larvae. Diptera in 75% of samples: 29% contained Muscidae, 22% *Calliphora*; remainder mostly Sepsidae and Mycetophilidae. Arachnids (spiders, harvestmen, and false scorpions) in 51% of samples, Lepidoptera larvae in 27% (mainly winter moths *Operophtera fagata* and *O. brumata*), Hemiptera in 22% (mostly Aphidoidea and Miridae), snails 20%, earthworms 11%, Hymenoptera 6%, adult sawflies 4%, and scorpion fly larvae 2%. Beetles eaten throughout breeding season; amongst Diptera, Muscidae common up to early July (57·3%), declining to 1·5% for rest of month, and *Calliphora* 7% up to early July, 35% for remainder. Arachnids taken throughout, while Lepidoptera larvae peaked in mid-June (83%) compared to 14% over rest of period; Hemiptera, snails, and earthworms only common in July. Seeds (mostly rape *Brassica napus*, present in 9% of samples) eaten in July only. Lepidoptera larvae eaten more commonly by nestlings 0–5 days old; older nestlings fed more snails and earthworms (Bishton 1985). In Rominter Heide (Poland/USSR), of 65 items brought in bill to nest 58% flies or fly-like insects, 15% spiders, 14% Lepidoptera larvae, 6% Lepidoptera adults, and 6% unidentified. Seeds also fed to young (Steinfatt 1938). In Leningrad region (USSR), 31 food samples obtained directly from young in July comprised 108 seeds of spruce *Picea abies* and 385 invertebrates: 34% spiders, 17% Lepidoptera larvae, 12% Curculionidae, 8% Cicadellidae larvae, 8% snails, and 4% aphids. 73 samples from 2 nests in Moscow region (USSR), July, comprised 27% spiders, 60% aphids, 2% larval sawflies (Tenthredinidae), 2% larval hoverflies *Syrphus*, 5% Lepidoptera (mainly adult Tortricidae), 2% snails, and 1% beetles (Inozemtsev 1963). Young fed invertebrates bill-to-bill, and seeds by regurgitation following softening in adult's crop (Niethammer 1937; Steinfatt 1938). 20–30 insects per portion brought to nest (Mal'chevski and Pukinski 1983). Feeding rates from 8 nests in Oxfordshire (England) with total 12·3 visits per hr (3·8–20·8): ♀ 5 per hr (2·2–8·7), ♂ 5·2 per hr (1·3–13·5), ♂ helper (see Social Pattern and Behaviour) 1·5 per hr (0·2–4·9); for remaining 0·6 visits per hr, bird not identified. Total 4·1 visits per hr per nestling (0·9–8·4); after leaving nest, feeding rate rises to 7·8 times per hr per fledgling (4·2–12·2) (calculated from Karanja 1982). Rates per brood range from 10 to 19 feeds per hr for broods of 2–5 fed by 2 parents, and from 6·5 to 20 feeds per hr for broods of 1–5 fed by 2 parents with 1 helper (Davies 1985). Near Moscow, rates per brood of 5 ranged from 4·2 feeds per hr (at 3 days old) to 11·2 per hr (9 days old); ♀ made 59–62% of visits, ♂ 38–41%; ♂ brought largest items (Ptushenko and Inozemtsev 1968); *c.* 12 visits per hr at steady rate throughout the day, though increasing to 13–15 per hr

during 06.00–11.00 hrs (Inozemtsev 1963). In Rominter Heide, hourly rate increases with age of young: on days 1–2, 3·9 feeds per hr (♀ 2·35, ♂ 1·55); days 8–9, 5·2 feeds per hr (♀ 2·4, ♂ 2·8); days 11–12, 6·3 per hr (♀ 3·1, ♂ 3·2); feeding intervals 2–37 min (Steinfatt 1938).

TAS

Social pattern and behaviour. Well studied. Account based on Davies (1983, 1985), Davies and Lundberg (1984), and Snow and Snow (1982, and unpublished) except as indicated.

1. Essentially solitary outside breeding season, occupying individual home-ranges, though local feeding aggregations may give appearance of gregarious behaviour: e.g. 180–200 congregated in upland area with low dense vegetation, 3 October, Switzerland (Lehner 1975); at least 80 at mound of rotting potatoes, October–April, England (Cheke 1985). Records of smaller concentrations in weedy areas along field edges, etc., probably due to locally good foraging conditions. Home-ranges of ♂ and ♀ are independent, and occupants of home-ranges are dominant over intruders. Ranges of 2–6 birds (of either sex) may overlap locally and all may congregate temporarily at rich feeding patches within areas of overlap. Birds (of either sex) leave home-ranges at times to visit good foraging areas nearby, where ♂♂ dominant over ♀♀ (this dominance reversed in breeding season). Competition between individuals for food during hard weather (but perhaps only at artificially clumped food supplies) may be intense and lead to death (Everett and Hammond 1975). Social organization of migrants in winter quarters not studied, but in Italy occasional song in late autumn and more frequent song in early spring indicates some degree of territoriality perhaps occurs (Alexander 1917). In late winter and early spring, ♂♂ begin to set up exclusive territories within winter home-ranges. BONDS. Typically, mating system essentially monogamous or involves polyandrous trio of birds (♀ plus 2♂♂); in addition, polygyny regular (♂ plus 2, occasionally 3, ♀♀), and 'polygynandry' occasional (2–3 ♂♂ plus 2–4 ♀♀). Trios arise when ♀ frequents territories of 2 ♂♂ (see Heterosexual Behaviour, below), and these territories then tend to coalesce; conflict between the 2 ♂♂ gradually declines and one (the 'α-♂', usually the older) becomes dominant over the other ('β-♂'), though both use same perches for singing. Less often, trio formed by ♂ persistently intruding on territory of monogamous pair until he is accepted. Both α- and β-♂♂ (though mostly α-♂) copulate with polyandrous ♀ who regularly tries to 'lose' escorting α-♂ in order to copulate with β-♂ or (as also do monogamous ♀♀) with neighbouring ♂♂. Generally, ♂♂ of a trio that have copulated with a ♀ feed her young, and those that have not do not. Indirect evidence suggests that β-♂♂ that have not copulated with ♀ may destroy her eggs and young chicks, so inducing her to lay again (Davies 1985). These observations, and others described in part 2, indicate conflicting strategies for ♂ and ♀: ♂ tries to maximize his paternity, while ♀ tries to maximize her breeding success by copulating with several ♂♂ in order that these will all feed her young. Bonds (such as they are) do not persist outside breeding season, though some contact often occurs then between birds that have bred together (see Roosting, below, and Voice). Incubation, and brooding of young, by ♀ only. Young fed by ♀ alone or by ♀ and one or more ♂♂ (see above); share of feeding by β-♂ generally less than by α-♂, but increases with brood size and with proportion of copulations achieved. Young fed for up to 17 days after leaving nest. Age of first breeding normally 1 year (Davies and Lundberg 1984). BREEDING DISPERSION. Solitary and territorial, though modified

by complexities described under Bonds (above). Highest recorded overall density 640 breeding adults per km² (Cambridge Botanic Garden) (Davies and Lundberg 1984). In 9·3 ha of woodland and scrub in southern England, overall densities 230 birds per km², but close to 1000 birds per km² on basis of fraction of area actually used. Much lower breeding densities recorded in woodland and other areas with sparse or localized undergrowth. Territory size highly variable, depending mainly on cover and foraging quality; territory of ♂ larger, mostly 0·15–0·3 ha, that of ♀ as little as 800 m² (Snow and Snow 1982). In farmland with hedges, territories more or less linear along hedges; in one study, one territory per 188 m (Williamson 1971). Territory and nest-site fidelity of ♀♀ well-marked; ♂♂ more likely to move from year to year, depending on opportunities to occupy more favourable habitat and feeding areas. ROOSTING. Typically 1–2 m above ground in holly *Ilex* or hawthorn *Crataegus* in or near hedge (G Bishton). Generally 2 birds together. From October to mid-January, just prior to roosting, common for 2 individuals (probably ♂ and ♀) to contact each other with frequent Seep- and Trill-calls (see 3–4 in Voice), often from high tree perches, then fly off together. In spring, ♂ and ♀ may roost together in same bush, in territory. Bathing usually in shallow water, but dew-bathing and bathing in heavy rain also recorded. Dust-bathing apparently rare (Bishton 1984). Sun-bathing frequent, March–September; peculiar among passerines, bird lies on side and raises one wing so that sun strikes underside then turns and repeats, raising other wing, each position held for about ½ min (Teager 1967).

2. Uses Alarm-call (see 5 in Voice) in presence of Kestrel *Falco tinnunculus*, fox *Vulpes*, or cat outside breeding season. Interactions between adults intelligible only in light of complex territorial and mating system (see part 1). SONG-DISPLAY. Song (see 1 in Voice) usually delivered from exposed perch with wide range of heights depending on vegetation structure; occasionally given in flight. Main song-period from early spring to end of breeding season; song normally by ♂ only, occasionally by ♀. Especially at beginning of breeding season, interactions between ♂♂, in course of establishment of dominance relations and territory boundaries, involve Song-duels at a distance and, at closer quarters, low-volume song, postures, and displays. In Song-duels, song typically loud and full-length when ♂♂ well apart, becoming shorter and of lower volume, and given at shorter intervals as they approach one another (see also Antagonistic Behaviour). ANTAGONISTIC BEHAVIOUR. At close quarters, aggressive display may involve the following elements: ruffling of plumage, especially of rump, with body held horizontal; flicking of wings, either rapid, or slower ('wing-waving') with wings held up for appreciable fraction of second, either both wings together or alternately (Fig A);

A

occasional bill-wiping on perch, with body pivoting between each bill-wipe; subdued songs at very short intervals. Aggressive displays usually accompanied by frequent shifts of position, often with 4 or even more individuals involved and close together, sometimes continuing for minutes on end. Fighting

generally rare (sometimes occurring at good food sources—probably only at artificial ones—outside breeding season) but common between α- and β-♂♂ competing for same ♀ (see Bonds, above) and can lead to death. Competitive interactions over food normally settled by withdrawal of bird subordinate in local dominance hierarchy. Outside breeding season (September–February), both sexes advertise presence in home-range by uttering high-pitched Seep-call (see 3 in Voice), usually in series, most often before or after short movements within home-range. HETEROSEXUAL BEHAVIOUR. (1) Pair-bonding behaviour. In early spring, ♂♂ frequently chase ♀♀ for periods of up to 10 min, following them closely; when ♀ perches, ♂ perches by her and may sing. Extent of ♀'s wanderings at this time helps to determine type of mating system which develops (see Bonds, above). ♂ and ♀ of former pairs or trios may answer each other's Seep-calls and then move towards one another. ♂ and ♀ also communicate by Trill-call (see 4 in Voice). During ♂'s song-period, Trill-calls are mainly by ♀♀, sometimes in response to ♂'s song. ♂ often accompanies foraging ♀, singing from perches above her. In last 4–5 days before 1st egg laid, ♂ (α-♂ in case of trio) begins to follow ♀ closely, mainly staying within 5 m of her (mate-guarding) for up to 40 min in each hour. (2) Courtship-feeding. None known. (3) Mating. During mate-guarding period, neighbouring ♂♂ often intrude and try to copulate, but most intense competition comes from β-♂, and α-♂ may spend much time keeping him away (see above). β-♂ nevertheless often succeeds in copulating if α-♂ temporarily out of sight or ♀ has 'escaped' from him (see Bonds, above). Copulation attempts frequently interrupted in trios, infrequently (by neighbouring ♂♂) in monogamous pairs. Pre-copulation display elaborate. ♀ crouches, ruffles body feathers, shivers wings, and quivers tail, raising it to expose cloaca (Fig B, upper). ♂ hops from side to side behind her (Fig B, middle) and pecks cloaca (Fig B, lower) for up to 2 min before mating.

B

Copulation itself extraordinarily brief: ♂ appears to jump over ♀ (and action has been so interpreted by some observers), cloacal contact lasting for fraction of second. During pecking by ♂, ♀'s cloaca becomes pink and distended and from time to time makes strong pumping movements, sometimes ejecting a droplet which contains a mass of sperm; ♂ looks intently at such droplets and copulation follows immediately after. Pecking of ♀'s cloaca more prolonged when another ♂ has spent more time with her. This process interpreted by Davies (1983) as part of ♂'s strategy of attempting to increase his share of paternity. Copulation occurs first when nest-cup complete and ♀ beginning to line it; continues until beginning of incubation. (4) Behaviour at nest. During incubation, ♀ leaves nest to forage without stimulus from ♂. ♂ seldom comes to nest during incubation; may remain for up to 1 min on nest-edge, and stick bill deep into nest-cup; according to Steinfatt (1938), occasionally feeds ♀ on nest. RELATIONS WITHIN FAMILY GROUP. Nestlings brooded (by ♀ alone) for much of day when first hatched, decreasingly as they grow. Adults feed nestlings from bill (insects) or from crop (small seeds); typically food from bill given first, then a number of items from crop (Steinfatt 1938). Faecal pellets of nestlings eaten by parents in first few days, then carried away. Young leave nest at 12 (11–15) days; skulk in low cover and continue to be fed by parents for 14–17 days. ANTI-PREDATOR RESPONSES OF YOUNG. From *c.* 9 days old, young may leave nest if disturbed and usually will not remain in nest if put back. Fledged young become silent on hearing parents' Alarm-call (see 5 in Voice). PARENTAL ANTI-PREDATOR STRATEGIES. If human intruder approaches nest, incubating or brooding ♀ slips off silently. Both parents utter repeated Alarm-calls if intruder approaches nest with large young, or recently fledged young. No further details.

(Figs by J P Busby: from original drawings.) DWS

Voice. Most calls high-pitched, and all except song of simple structure. Following account largely from Snow and Snow (1983, and unpublished).

CALLS OF ADULTS. (1) Song of ♂. Short, undistinguished, rather high-pitched warble. Consists of a number of more or less sharply modulated units, many of them peaked in sonagrams (Fig I); some very brief, others more prolonged, louder, and less sharply inflected. More prolonged and louder units nearly always in frequency range 5–6 kHz, rarely 4·5–6·5 kHz; briefer, sharply modulated units 3·5–7·5 kHz. Loud, prolonged units never contiguous but always separated by one or more of briefer units. Also, 1–4 rapid trills usually interspersed: consist of 3–6 (7–8) repeated notes, usually 0·03–0·04 s apart (interval tends to be shortest when trill has many notes); trills never contiguous. Apart from the notes of a trill, and occasional repeated units (2–3 in sequence, less rapid than true trill), all units in a typical song (Fig II) are different from one another. Song mostly 2–3·5 s long, occasionally up to 4·5 s; longer songs, up to *c.* 5 s, may be produced by repetition of a section at the end. Songs delivered in bouts of usually 2–8, sometimes much more; average number of songs per bout *c.* 8 at height of song-period in southern England, decreasing to 3–4 at end of main song-period (July). Each ♂ has repertoire of up to 6 (perhaps more) different

songs, each usually completely different in all units from others; these song-types retained from year to year, usually with minor and occasionally major modifications. Neighbouring ♂♂ (and α- and β-♂♂ sharing a territory: see Social Pattern and Behaviour) have strong tendency to incorporate into their songs sequences exactly copied from one another, young birds probably copying songs of older birds; entire songs may consist of 'copies' of sections of other birds' songs, hence strong tendency for local dialects to occur. For modification of song during Song-duels between ♂♂, see Social Pattern and Behaviour. When closely escorting ♀, ♂ may utter very low-volume songs usually while perched just above her; only one analysed, and that different in structure from any of that ♂'s loud songs. (Snow and Snow 1983; B K and D W Snow.) Subsong by solitary ♂♂ regular, usually given from low thick cover; very different in structure from loud song; typically more or less continuous, with greater frequency range and phrasing much less well defined (Thorpe and Pilcher 1958). (2) Song of ♀. Infrequent but probably regular in some individuals. Shorter and less elaborate than ♂'s though improving in form and complexity on repetition; uttered by paired ♀ in breeding season as a contact signal when temporarily separated from ♂. (Bronsart 1954; Bartlett 1970; B K and D W Snow.) (3) Seep-call. Most familiar call outside breeding season. A loud single 'seep' (Fig III) lasting *c.* 0·12 S (0·10–0·16), with fundamental frequency of 5–5·5 kHz. In sonagrams, usually appears more or less markedly 'humped', and may begin with brief downward inflection, this part of call comparatively weak. Sometimes 2(–3) calls given in quick succession, intervals only a little longer than calls themselves. This call used almost exclusively outside breeding season (September–February) by both sexes. Given most often before or after short movements within home-range, long sequences given especially after periods of absence from home-range. Serves to advertise presence of calling bird within home-range, and to maintain contact between members of a (former) pair, who may answer each other's calls and move towards one another. (B K and D W Snow.) (4) Trill-call. Rapid 'ti-ti-ti' or 'he-he-he' of 3–4(–5) notes, each 0·05–0·07 s, with fundamental frequency mainly 4·6–5·9 kHz. In sonagrams individual notes have characteristic form (Fig IV) with frequency falling from beginning to end of each (but with hump in middle), and also slightly from one note to the next. Much less loud than Seep-call, and falling inflection gives plaintive sound. Fundamentally a contact-call, uttered by both sexes, often alternately by paired ♂ and ♀. Often immediately precedes flight, and may be continued in flight and after landing; the usual pre-roosting call, uttered from high perch before bird flies off to roosting site; also characteristic of birds in state of migratory restlessness (Steinfatt 1938; Grant 1968). In resident English populations, becomes more frequent in autumn, with peaks at dawn and dusk and

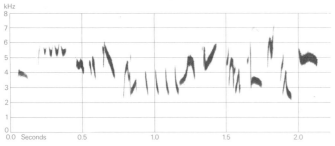

I P J Sellar England March 1983

III D W Snow England
March 1979

IV D W Snow England
March 1981

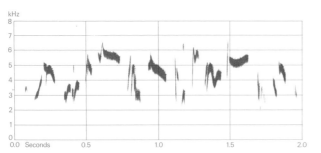

II P A D Hollom England March 1981

V P J Sellar England August 1982

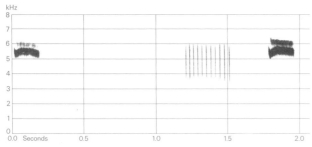

VI A P Radford England May 1982

few calls between 10.00 hrs and shortly before sunset. In spring and summer, use by any one pair variable, depending on circumstances; more often uttered by ♀—often in response to ♂'s song or when bigamous ♂ leaves for his other ♀'s territory. During period when ♂ is closely escorting ♀, one or both may give faint trills audible only at close quarters. (B K and D W Snow.) (5) Alarm-call. Loud single note of same duration as Seep-call (see 3) but somewhat higher-pitched, fundamental frequency mainly 5·2–6 kHz; sonagram (Fig V) differs in showing little tendency to humped shape; generally flatter, but may fall in frequency towards end, to produce more plaintive sound. Uttered by both sexes when there is threat to nest containing large young (usually at least 6 days old) or to fledged young for about a week after leaving nest, causing young to keep silent. Varies in volume and rate of repetition: at most intense, loud and piercing, uttered with bill wide open. Tends to be given at regular intervals, up to rate of nearly 1 per s when bird highly alarmed. Relatively infrequent outside breeding season but may be uttered in long

sequences in response to presence of Kestrel *Falco tinnunculus*, fox *Vulpes*, or cat. May also be given (while fleeing) when attacked by Robin *Erithacus rubecula* at artificial feeding site, and once heard (from loser, as it flew off) after fight between 2 *P. modularis*. (B K and D W Snow.) Repeated churring call (Fig VI, middle unit) given (in addition to usual Alarm-call) by bird in presence of Cuckoo *Cuculus canorus* near nest with egg (Radford 1985); apparently seldom used.

CALLS OF YOUNG. Nestlings utter high-pitched, sharply modulated food-calls which begin with rapid rise in pitch (starting at *c.* 4 kHz and peaking at 6·9–7·0 kHz) then end suddenly. Essentially the same food-call continues for first few days after leaving nest, becoming longer and more varied in form and frequency. On leaving nest, faint version of adult Seep-call (see 3, above) is uttered as location call; becomes adult-sounding (including double 'seep') at 18 days after leaving nest. Older, independent juveniles frequently give 'seep' of typical adult form but slightly higher pitched. Adult Trill-call (see 4, above) first heard from young *c.* 6 weeks out of nest. First song attempts may be made soon after leaving nest (Alabaster 1946), and tentative, unformed song regular in 1st autumn; definitive song-types (see Calls of Adults) not developed until territory taken up in late winter or early spring. (B K and D W Snow.)   DWS

Breeding. SEASON. Britain and north-west Europe: see diagram. Leningrad region (USSR): laying begins mid-May (Dementiev and Gladkov 1954*b*). Sweden: eggs laid late April to June (Makatsch 1976). SITE. In bush, hedge, or low tree, 0·5–3·5 m above ground, sometimes in side of bank; normally well concealed. Occasionally uses old

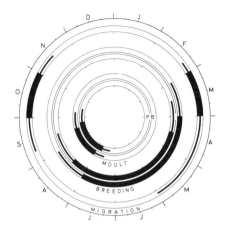

nest of another bird. Nest: quite substantial cup of twigs, leaves, stems, roots, and other plant parts, lined with wool, hair, moss, and sometimes feathers. Outer diameter up to 15 cm, inner 6–8 cm, depth of cup 4–5 cm. Building: by ♀ only (Davies 1985; D W and B K Snow). EGGS. See Plate 82. Sub-elliptical, smooth and glossy; bright blue, rarely with a few reddish spots. Nominate *modularis*: 19·2 × 14·5 mm (17·5–21·2 × 13·0–15·5), n = 170; calculated weight 2·13 g. *P. m. occidentalis*: 19·8 × 14·7 mm (18·4–22·5 × 14·0–15·5), n = 100; calculated weight 2·28 g (Schönwetter 1979). Clutch: 4–6 (3–7). Of 22 clutches, Poland: 4 eggs, 5; 5, 10; 6, 7; mean 5·1 (T Wesołowski). In polyandrous trios, clutch larger (mean 4·11) if both ♂♂ have copulated with ♀ than if only α ♂ (see Social Pattern and Behaviour) has done so (mean 3·36) (Davies 1985). 2 broods, occasionally 3. Eggs laid daily. INCUBATION. 12–13 days. By ♀ only, beginning with last egg (Makatsch 1976) or sometimes morning after (Davies 1985). YOUNG. Cared for and fed by both parents. FLEDGING TO MATURITY. Fledging period 11–12 days, sometimes leaving nest 1 day before able to fly (Steinfatt 1938). Become independent 14–17 days later. Age of first breeding 1 year. BREEDING SUCCESS. In 2 years, hatching success 84·8% (n = 372 clutches) and 81·2% (n = 260); fledging success 79·7% (n = 302 clutches) and 80·2% (n = 227); average brood size 3·8 (n = 256 broods) and 3·3 (n = 211) (BTO). Monogamous pairs fledged on average 2·1 young per attempt (n = 41). Where 2 ♂♂ involved, 1·3 young per attempt (n = 35) if only the dominant ♂ copulated and fed young, 3·2 (n = 33) if both ♂♂ contributed; ♀♀ thus achieved better breeding success by securing 2 cooperating ♂♂. (N B Davies.)

Plumages (nominate *modularis*). ADULT MALE. Forehead, crown, and hindneck dull olive-grey with rather narrow and poorly defined black-brown shaft-streaks; some uniform dull blue-grey of feather-bases usually visible. Mantle and scapulars bright rufous-brown, feathers with black shaft-streaks *c.* 2–3 mm wide restricted to tips; feather-bases more olive-brown. Back, rump, and upper tail-coverts olive-brown, often indistinctly streaked dusky on back. Sides of head back to ear-coverts grey-brown, slightly purer grey on long but in-

distinct supercilium, finely peppered off-white on upper cheek, narrowly streaked white on ear-coverts; longer ear-coverts tipped rufous-brown or olive-brown. Sides of neck blue-grey, contrasting somewhat with spotted hindneck and duller and less bluish grey of throat. Chin finely mottled pale grey and off-white; some dark feather-bases sometimes visible on sides of chin, forming poorly defined and indistinct malar stripe. Lower cheeks, throat, and chest medium grey, feathers faintly tinged olive-brown on tips and faintly fringed off-white when plumage fresh. Sides of breast, flanks, and thighs olive-brown, marked with ill-defined dull black or black-brown streaks on flanks (extending as indistinct dull olive-brown spots up to sides of breast). Central belly, vent, and under tail-coverts greyish-white, under tail-coverts with bold olive-brown arrow-marks and streaks. Tail black-brown with slight olive tinge; feather-sides narrowly fringed rufous-brown. Flight-feathers dull black, outer webs fringed rufous-brown, except along emarginated parts of primaries; rufous-brown deepest near bases of outer secondaries. Tertials black, fringed rufous-brown along both webs. Lesser upper wing-coverts olive-brown, like rump; median coverts olive-brown with black wedge on centre of each covert and with paler olive-grey fringe along tip; greater coverts rufous-brown with black shaft-streak, tip of each covert with cream-buff fringe along outer web, more rufous fringe on inner web. Greater upper primary coverts dark olive-brown, grading to black on tip; outer web fringed olive-brown. Axillaries and upper wing-coverts dull grey with paler grey or olive tinge near tips. *In worn plumage*, about February–July, black-brown and olive-brown of feather-tips of forehead, crown, hindneck, and ear-coverts largely worn off, entire head and neck appearing dull plumbeous-grey, slightly tinged olive-brown on crown and ear-coverts, finely speckled white on lores, upper cheeks, and behind eye, slightly purer blue-grey on supercilium and sides of neck. Rufous-brown of mantle, scapulars, tertials, tail, and wing slightly tinged olive-grey, black of feather-centres slightly browner. Chin to breast uniform deep grey, merging into greyish-white on central belly and flanks; sides of breast and flanks olive-brown, streaked dull black or darker olive-brown on flanks. ADULT FEMALE. Like adult ♂ and often indistinguishable. Throat to breast rather deep grey (as in adult ♂), but chin on average more extensively mixed off-white, chest-feathers with broader olive-brown fringes along tips, and tips of breast feathers more extensively fringed off-white. *In worn plumage*, chest and breast not as uniform grey as ♂, retaining olive fringes on chest and faint pale arcs on breast; also, belly then on average more extensively and purer white. NESTLING. Down rather long but scanty, on head and upperparts only; black-brown or dull black. For development, see Heinroth and Heinroth (1924–6) and Mächler (1975). JUVENILE. Forehead and crown olive-brown, indistinctly mottled black on feather-tips; hindneck similar, but ground-colour often paler, buff. Mantle and scapulars buff-brown, not as deep rufous-brown as adult, narrowly streaked black; rump to upper tail-coverts buff-brown with faint darker mottling. Sides of head finely mottled olive-brown; indistinct supercilium (back from eye) and sides of neck buff or pale buff with some dusky specks. Underparts quite different from adult: ground-colour pale buff-brown or buff, often slightly greyish on chin and throat and nearly white on central belly and vent; chin and throat finely mottled dusky grey; chest, breast, and flanks more distinctly marked with dull black triangular shaft-streaks. Feathering of body narrow and loose, especially on rump, underparts, and tail-coverts. Tail, flight-feathers, and primary coverts as adult, but tail-feathers often narrower and more pointed at tips. Lesser upper wing-coverts buff-brown with dull black tips; median coverts olive-brown ending in square black spot on tip, laterally bordered by buff spots (in adult, black ends in narrow point subterminally and tip fringed narrowly and evenly pale greyish-buff); greater coverts as adult, but pale spot on tip of outer web larger and more rounded, cream or white (not a narrow cream-buff fringe), and a similar spot sometimes on tip of inner web. FIRST ADULT MALE. Like adult ♂, but juvenile flight-feathers, greater upper wing-coverts, primary coverts, and usually tail and tertials retained; due to abrasion and individual variation in pattern, juvenile character of retained feathers often hard to detect and ageing often difficult. Tips of outer webs of greater coverts and tertials often have larger and more contrastingly white spots than adult (in adult, still rufous-buff in spring); tertials and tail-feathers slightly worn in autumn, heavily worn in spring, but all or partly new in some birds (in adult, tips unfrayed until April–May); also, feathers of crown and hindneck more broadly tipped olive and black-brown, less uniform dull grey in worn plumage; feather-tips of chest and breast more extensively fringed off-white, chest more extensively tinged olive-brown, and belly more extensively white; underparts not as uniform deep grey as adult ♂, but closely similar to adult ♀ (chin of 1st adult ♂ on average greyer, however, not as whitish as adult ♀). See also Bare Parts. FIRST ADULT FEMALE. Similar to 1st adult ♂, but chin and throat heavily mottled off-white or pale buff; ground-colour of chest and breast less deep grey, tinged pale olive-brown or grey-brown, feathers broadly fringed buff-white or pale olive-buff; belly extensively off-white, white often extending up to breast. In worn plumage, underparts olive-grey from throat to chest, not deep grey from throat to breast as in adult ♀ and 1st adult ♂ or down to mid-belly as in adult ♂.

Bare parts. ADULT. Iris reddish-brown, brick-red, or bright red; occasionally (at least in some ♀♀ from November–February and perhaps also at other times of year) light brown or warm brown. Bill bluish-black, base of upper mandible and much of lower mandible pale horn-pink; in breeding season, bill black-brown or black. Leg and foot light red-brown to flesh-red, sometimes pale straw-brown. NESTLING. Bill flesh-pink with duller grey tip; gape-flanges light pink. Mouth orange with black spot on each side of base of tongue. Leg and foot flesh-pink. JUVENILE. At fledging, iris dark grey, grey-brown, or greyish-olive; bill flesh-pink with dark horn-grey tip; leg and foot pale greyish-flesh or brownish-flesh. After completion of post-juvenile moult, iris greyish-olive, olive, olive-brown, or sepia-brown; bill horn-black with pale yellow-brown base (most extensively so on lower mandible); leg and foot light yellow-brown. Adult iris and bill colour gradually acquired during mid-winter, but some variation in timing, in part depending on fledging date and perhaps on sex. Iris occasionally red-brown by October–November, but usually still olive-brown or sepia-brown in mid-winter (though bill then usually fully black-brown, like adult, and leg and foot light brown or pale red-brown), acquiring red-brown or brick-red colour in spring; some birds have iris olive-brown or cinnamon-brown throughout summer of 2nd calendar year, however. (Heinroth and Heinroth 1924–6; Poulding 1969; Mächler 1975; Shapoval 1981; Bub et al. 1984; Svensson 1984a; RMNH, ZMA.)

Moults. ADULT POST-BREEDING. Complete; primaries descendant. In Britain, starts with shedding of p1 between 10 July and 20 August (on average, 30 July), completing with regrowth of p10 5 September–5 October (on average, 22 September); a few birds start as soon as early July and complete from mid-August, while others do not start until late August and complete up to mid-October. Estimated duration of moult 54 days (based on average speed of moult) or 60 days (based on retraps of ringed birds). Secondaries start from primary moult score 18–30(–35), when p1–p2 about full-grown; completed from score 42 or sometimes after completion of primary moult (score 50). Moult of tail-feathers usually centrifugal; start at primary score 5–20 (sometimes 0–25), completed at score 39–48 (sometimes 36–50). Tertials and body start at about same time as tail, completed at primary score 23–35 (some body feathers sometimes later). (Ginn 1975; RMNH, ZMA.) Scattered data from Netherlands (RMNH, ZMA), USSR (Dementiev and Gladkov 1954b), Balkans (Stresemann 1920), and Iran (Marien 1951) fit within British pattern. POST-JUVENILE. Partial: head, body, and lesser and median upper wing-coverts; in early-fledged birds, also tertials, tertial-coverts, and often all or part of tail. Starts at age of c. 5 weeks; moult mainly July to early September, occasionally mid-June to mid-October. (Dementiev and Gladkov 1954b; Stresemann 1920; Witherby et al. 1938b; Ginn and Melville 1983; Bub et al. 1984; Jenni 1984; ZMA.)

Measurements. ADULT, FIRST ADULT. *P. m. occidentalis.* England and southern Scotland, all year; skins (BMNH, RMNH, ZMA). Bill (S) to skull, bill (N) to distal corner of nostril; exposed culmen on average 3·2 less than bill (S).

WING	♂ 70·3 (1·61; 27)	68–74	♀ 68·7 (1·40; 30)	66–71
TAIL	57·9 (2·64; 18)	54–62	57·5 (2·24; 23)	53–61
BILL (S)	14·7 (0·49; 17)	13·9–15·4	14·6 (0·61; 22)	13·7–15·6
BILL (N)	8·1 (0·60; 15)	7·3–9·1	8·0 (0·55; 20)	7·2–8·8
TARSUS	21·5 (0·75; 18)	20·3–22·6	21·5 (0·80; 23)	19·9–22·4

Sex differences significant for wing.

Nominate *modularis.* Netherlands, all year; skins (RMNH, ZMA).

WING	♂ 71·1 (1·25; 78)	69–74	♀ 68·6 (1·56; 48)	65–72
TAIL	58·0 (2·04; 30)	55–60	55·7 (1·60; 25)	53–59
BILL (S)	14·6 (0·49; 27)	13·9–15·5	14·4 (0·57; 23)	13·6–15·6
BILL (N)	7·9 (0·34; 21)	7·4–8·5	7·9 (0·35; 21)	7·4–8·6
TARSUS	20·6 (0·58; 29)	19·8–21·7	20·3 (0·51; 26)	19·4–21·2

Sex differences significant for wing and tail. 1st adult (with retained juvenile flight-feathers and tail) combined with full adult above, though 1st adult wing on average 1·3 shorter than full adult and tail on average 1·5 shorter: e.g. wing ♂, adult 72·1 (17) 71–74, 1st adult 70·8 (61) 69–73; wing ♀, adult 69·6 (12) 68–72, 1st adult 68·3 (36) 65–71 (RMNH, ZMA).

P. m. obscura. Iran, all year; skins (Marien 1951).

WING	♂ 69·8 (1·71; 9)	67–73	♀ 68·1 (—; 8)	67–69
TAIL	58·4 (—; 8)	56–60	57·8 (—; 8)	54–61
BILL (S)	13·6 (0·46; 9)	12·5–14	13·4 (—; 9)	13–14

Wing; skins, except where stated. (1) *P. m. hebridium* and *occidentalis,* Scotland; (2) *occidentalis,* England; (3) nominate *modularis,* West Germany (Niethammer 1971). Nominate *modularis:* (4) East Germany; (5) Rheinland, West Germany, live birds (Bub et al. 1984); (6) southern Urals, USSR (Dolgushin et al. 1972). *P. m. mabbotti:* (7) Italy (Mauersberger 1971; RMNH, ZMA). *P. m. obscura:* (8) Caucasus, all year; (9) Levant and south-west Iran, winter (Mauersberger 1971). *P. m. meinertzhageni:* (10) southern Yugoslavia (Stresemann 1920); (11) Bulgaria (Mauersberger 1971). *P. m. euxina:* (12) Ulu Dag, north-west Asia Minor (Watson 1961b). *P. m. fuscata:* (13) Crimea (Mauersberger 1971).

(1)	♂ 68·9 (1·43; 31)	65–72	♀ 66·9 (1·8; 20)	64–70
(2)	69·1 (2·12; 31)	65–74	67·7 (1·89; 26)	63–70
(3)	69·7 (1·7; 33)	65–74	68·0 (—; 9)	66–69

(4)	71·1 (3·74; 28)	66–75	67·8 (2·32; 28)	62–72	
(5)	70·4 (4·79; 105)	67–74	69·4 (1·89; 68)	65–73	
(6)	(—; 6)	67–70	64·5 (—; 2)	64–65	
(7)	70·3 (0·80; 8)	69–71	67·2 (1·50; 4)	66–69	
(8)	69·1 (1·13; 34)	67–71	67·5 (1·52; 34)	65–70	
(9)	69·6 (0·55; 5)	69–70	67·7 (0·76; 7)	67–69	
(10)	70·1 (1·41; 14)	69–72	67·7 (1·68; 11)	65–70	
(11)	69·4 (3·74; 7)	64–74	67·0 (—; 3)	65–68	
(12)	68·8 (—; 2)	68–69			
(13)	70·0 (1·15; 7)	69–72			

Average of tail *c*. 59·7 in ♂ of *occidentalis* and *hebridium*, *c*. 58·0 in ♀; significantly shorter in nominate *modularis* from West Germany: ♂ 57·6, ♀ 55·9 (Niethammer 1971).

Weights. ADULT, FIRST ADULT. Nominate *modularis*. Sexed birds. (1) East Germany (Bub *et al.* 1984). (2) West Germany (Niethammer 1971). Netherlands: (3) March–May, (4) June–September, (5) October–February (RMNH, ZMA). Southern Urals (USSR): (6) October–November (Dolgushin *et al.* 1972).

(1)	♂ 19·8 (6·57; 13) 14·5–25·0	♀ 18·1 (1·63; 6) 14·7–20·0
(2)	20·5 (—; 10) 19·0–22·0	18·9 (—; 4) 16·0–23·5
(3)	21·4 (1·98; 10) 18·5–25·0	21·1 (2·00; 6) 19·1–24·5
(4)	18·5 (1·43; 6) 16·9–20·3	18·7 (1·15; 5) 17·3–20·0
(5)	20·7 (2·14; 7) 18·3–23·0	18·4 (1·77; 5) 16·0–20·0
(6)	22·1 (2·49; 5) 19·5–25·1	18·5 (—; 1)

Unsexed birds. Finland: (7) spring, (8) autumn (Hildén 1974). Southern Sweden: (9) autumn (Scott 1965). North Jutland, Denmark: (10) spring. Helgoland, West Germany: (11) spring, (12) autumn. West Germany: (13) all year. (Bub *et al.* 1984.) Northern France: (14) all year (Mountfort 1935).

(7)	18·5 (2·4; 33)	15–24	(11)	21·0 (—; 34)	14–21·5
(8)	19·0 (1·9; 47)	15·5–23·5	(12)	19·2 (—; 29)	15–25
(9)	19·1 (1·18; 37)	—	(13)	19·7 (1·92; 527)	14–26
(10)	19·0 (1·60; 133)	16–24	(14)	21·0 (—; 5)	20–22

P. m. occidentalis. (1) Britain, all year (Marples 1935). (2) Dorset (Ash 1964). (3) Kent (Scott 1965). (4) Skokholm Island, Wales (Browne and Browne 1956). Lincolnshire, late August and early September: (5) 05.00–11.00 hrs, (6) 11.00–20.00 hrs (Gordon 1962).

(1)	21·7 (—; 69)	17–26	(4)	20·1 (—; 8)	18–23
(2)	20·8 (2·5; 50)	13–26	(5)	20·2 (—; 86)	17–24
(3)	20·0 (—; 52)	—	(6)	22·1 (—; 51)	17–26

P. m. meinertzhageni. Bulgaria, June: ♂ 19·5 (Niethammer 1950). Perhaps this race, northern Greece, January: ♂ 22 (Makatsch 1950).
P. m. euxina. North-west Turkey, late April: ♂♂ 21, 22 (Watson 1961b).
P. m. obscura. Southern Turkey, December–March: ♂ 20, ♀ 22 (Kumerloeve 1970a). Northern Egypt, November: ♂ 17·8 (Goodman and Watson 1983). Northern Iran, March: ♂♂ 26, 25; unsexed 22 (Schüz 1959).
Race unknown: either wintering or migrating nominate *modularis* or *mabbotti*. (1) Southern France, October–February (G Olioso). (2) Central Italy, October–March (Ioale' and Benvenuti 1982, which see for diurnal and monthly variation curves). (3) Malta, autumn and winter (J Sultana and C Gauci).

(1)	19·2 (1·43; 33)	17–22	(3)	19·5 (1·8; 50)	16–24·5
(2)	20·8 (1·6; 201)				

Exhausted birds. Nominate *modularis*. Netherlands, all year: ♂ 16·0 (1·01; 5) 15–17·5, ♀ 13·9 (1·18; 5) 12–16 (ZMA). *P. m. occidentalis.* Southern England, winter 16·2, 16·4 (Ash 1964).
NESTLING, JUVENILE. For growth curves, see Birkhead (1981).

Structure. Wing rather short, broad at base, tip either bluntly pointed or more rounded. 10 primaries: in nominate *modularis*, p7 longest, p8 0–1·5 shorter, p9 3–6, p6 0–2, p5 1·5–4, p4 6–9, p1 11–15·5 (*n* = 42; Netherlands, East and West Germany, and Scandinavia); in *occidentalis*, p6–p7 longest, p8 0–2 shorter, p9 4–7, p5 1–3, p4 5–7, p1 10–12·5 (*n* = 54; Britain). In both races, p10 reduced, narrow; 34–44 shorter than wing-tip, 3 shorter to 4 longer than longest upper primary covert. P9 2 longer to 2 shorter than p4 in *occidentalis*, 1–5 longer in nominate *modularis*. (Scott 1962a; RMNH, ZMA.) In *occidentalis*, south-east England, p9 on average 0·48 longer than p4 (*n* = 36); in nominate *modularis*, southern Sweden, 3·20 longer (*n* = 20) (Scott 1965). Wing formula independent of age and sex (Nitecki 1969; ZMA). More pointed wing with longer p9, as in nominate *modularis*, occurs also in *meinertzhageni*, *euxina*, *fuscata*, and *obscura*, more rounded wing with relatively shorter p9, as in *occidentalis*, occurs also in *hebridium* and *mabbotti*, though some local variation (Hartert 1910; Meinertzhagen 1948; Vaurie 1959; Watson 1961b; Scott 1962a, 1965; Nitecki 1969; Mauersberger 1971). Outer web of p5–p8 and (slightly) inner of p6–p9 emarginated. Tertials short; longest reach to about tip of p1–p2 in closed wing. Tail rather short, tip square or slightly forked; 12 feathers. Bill short; rather broad and flattened near base, laterally compressed near tip; culmen straight or slightly concave, except for slightly decurved tip; lower mandible straight but gonys slightly kinked upwards. Nostrils narrow, partly covered by membrane above. A few fine bristles along base of upper mandible. Leg and foot rather short and strong; front of tarsus with large scutes. Middle toe with claw 17·7 (18) 17–19 mm; outer toe with claw *c*. 70% of middle with claw, inner *c*. 68%, hind *c*. 76%.

Geographical variation. Rather complex, involves general colour (see below), wing formula (see Structure), and (scarcely) size (see Measurements). Colour variation within all populations large, mainly due to differences in age, sex, bleaching, and abrasion; races often separable only in direct comparison between series of birds of same sex and with similar stage of plumage wear. Treatment of races based mainly on Vaurie (1959) for western, central, and northern Europe, and on Mauersberger (1971) for south-east part of geographical range. Vaurie (1959) described all trends in colour but did not name them all and perhaps recognized too few races; Mauersberger (1971) named all trends and perhaps separated too many races, but as no south-eastern races (except *obscura*) examined for present study, no judgement can be made here about their validity. *P. m. occidentalis* from England, Wales, and eastern Scotland differs from nominate *modularis* from central and northern Europe in more extensively dusky olive-brown crown and hindneck, marked with slightly longer and broader black-brown streaks; mantle and scapulars slightly duller rufous-brown; underparts on average slightly darker grey, especially throat and chest of ♂; sides of breast and flanks slightly darker olive-brown, less conspicuously streaked; see also Structure. *P. m. hebridium* from Inner and Outer Hebrides heavily mottled black-brown on crown and hindneck; mantle, scapulars, and tertials have broad and poorly defined black feather-centres and rather narrow dull brown (not rufous-brown) feather-sides; rump and upper tail-coverts rufous olive-brown; grey of underparts slightly darker than in *occidentalis*, extending further down belly. Birds from western Scotland (sometimes separated as *interposita* Clancey, 1943) average slightly paler than *hebridium*, tending towards *occidentalis*; birds from Ireland (sometimes separated as *hibernicus* Meinertzhagen, 1934) slightly paler and less heavily marked on underparts than typical *hebridium* and upperparts rather deep rufous-brown; however, both these populations included here in *hebridium*, as difference slight and much overlap in colour (Vaurie 1955*b*). Populations from Wales and north-west England more or less intermediate between *occidentalis* and Irish birds, and birds from western France about intermediate between *occidentalis* and nominate *modularis* (Meinertzhagen 1948; Vaurie 1959). *P. m. mabbotti* from Pyrénées and south-central France is a rather poorly differentiated race; rather close to nominate *modularis*, but wing-tip blunt, as in *occidentalis*; upperparts greyer, less rufous-brown than *occidentalis*, darker and greyer than nominate *modularis*. Populations from Iberia similar to typical *mabbotti* from France, but upperparts gradually more rufous towards north-west Portugal (rather like *occidentalis*, but underparts paler), and birds from that area sometimes separated as *lusitanica* Stresemann, 1928. Birds from central Italy are greyish, like typical *mabbotti*, but a few are more rufous, like '*lusitanica*' (Mauersberger 1971; RMNH, ZMA). Race of birds breeding in Sicily and Greece not established. *P. m. meinertzhageni* from southern Yugoslavia (Matvejev and Vasić 1973) and Bulgaria rather grey above, like *mabbotti*, but spots on upperparts heavier and blacker, on mantle and scapulars sometimes up to 4 mm wide; feather-fringes of mantle and scapulars paler and greyer than in nominate *modularis*, but underparts darker grey (Watson 1961*b*; Mauersberger 1971). *P. m. euxina* from north-west Asia Minor has lighter grey crown than nominate *modularis*, marked with paler brown streaks; streaks on mantle and scapulars brown (not black), ill-defined; rump and upper tail-coverts grey rather than olive-brown; underparts darker and more extensively grey; flanks with strongly reduced streaks (Watson 1961*b*). *P. m. fuscata* from montane coniferous forests of Crimea close to *euxina* from same habitat in Asia Minor; upperparts slightly darker with slightly better-defined brown streaks; underparts deeper grey, showing hardly any white except for some on vent; differs from *meinertzhageni* in browner and more diluted streaks on upperparts, and in darker, less pure grey underparts; differs from nominate *modularis* in greyer upperparts with less sharply defined streaks and in extensively grey underparts (Mauersberger 1971). *P. m. obscura* from Caucasus area, Iran, and eastern Turkey browner than other races, less greyish above than *euxina* and *fuscata*; underparts more extensively washed olive-brown on paler grey ground than any other race; ground-colour of mantle and scapulars duller rufous-brown than nominate *modularis*, streaks on upperparts browner, less black; sides of head and neck washed pale olive-brown or buff; throat to breast extensively marked with off-white or pale buff feather-tips, forming scaly pattern; birds from lower hills of Crimea perhaps slightly paler (Mauersberger 1971; ZMA).

Forms superspecies with Maroon-backed Accentor *P. immaculata* from Himalayas and Japanese Accentor *P. rubida* from Japan (Marien 1951; Mauersberger 1971).   CSR

Prunella montanella Siberian Accentor

PLATES 39 and 41
[between pages 544 and 545]

Du. Bergheggemus   Fr. Accenteur montanelle   Ge. Bergbraunelle
Ru. Сибирская завирушка   Sp. Acentor de Pallas   Sw. Sibirisk järnsparv

Motacilla montanella Pallas, 1776

Polytypic. Nominate *montanella* (Pallas, 1776), north-west Siberia from Ural mountains east to Khatanga and from Altai and Lake Baykal east to Stanovoy and Sikhote Alin mountains. Extralimital: *badia* Portenko, 1929, north-east Siberia east from lower Lena and south to northern and western shores of Sea of Okhotsk.

Field characters. 14·5 cm; wing-span 21–22·5 cm. Close in size to Dunnock *P. modularis*; marginally smaller than Black-throated Accentor *P. atrogularis*. Of similar form to *P. modularis* but with much more distinctive plumage: bold pale supercilium, black cheek-patch, little-streaked rufous-brown upperparts, and ochre-buff chest splashed there and along flanks with black and rufous-brown. Sexes similar; little seasonal variation. Juvenile separable.

ADULT MALE. Crown dusky-brown with long black edge; supercilium long and (behind eye) broad, pale buff to cream; cheek-patch dusky-black, with narrow pale buff border at rear; chin and throat pale buff. Nape, back, and wings essentially brown, with rufous or chestnut tone strongest on mantle, scapulars, and fringes of wing-coverts; indistinct streaks show far less than on any other west Palearctic congener. Wings relieved by almost white tips to median and greater coverts which form narrow double wing-bar (obvious only at close range). Rump greyer and more uniform than back; tail grey-brown. Underwing ochre-buff. Breast buff, with wash of similar colour or bright ochre over flanks and sides of belly; well marked with diffuse black streaks on breast and sharper rufous-brown ones along flanks. Centre of belly and vent cream, vent streaked grey. ADULT FEMALE. As ♂, but black lower edge to crown narrower, upperparts less rufous, and breast less boldly marked black. JUVENILE. Not studied in the field. Less rufous above than adult, and face pattern duller and less contrasting. Underparts dull isabelline with large brown spots on chest, breast, and sides of throat. At all ages, bill black-horn, with ochre base to lower mandible; legs bright yellow-brown.

Easily distinguished from typical *P. atrogularis* by lack of black or largely black throat, more uniform upperparts (streaked rufous rather than blackish), rufous-brown fringes to larger wing-feathers, and rufous-brown (not grey-brown) flank-streaks. Distinguished from Radde's Accentor *P. ocularis* by smaller size, less contrasting face-pattern (*P. ocularis* has pure white supercilium), and more uniform upperparts. Flight, gait, and behaviour apparently as congeners.

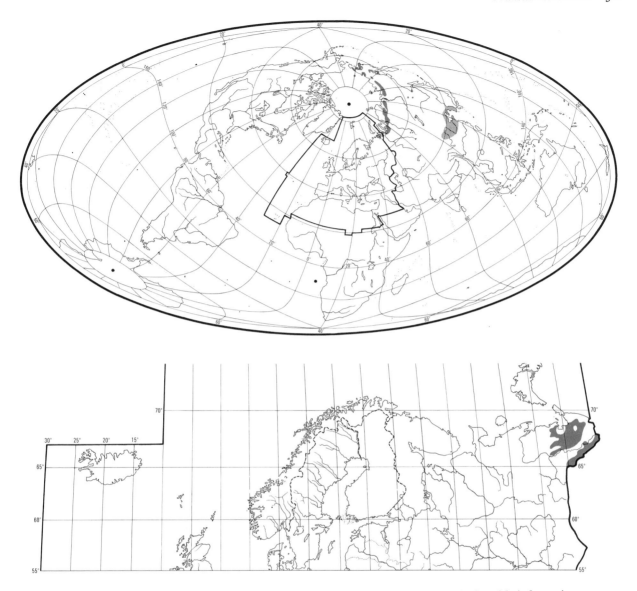

Song resembles *P. modularis*; more melodious than *P. atrogularis*. Commonest call a quiet 'dididi'.

Habitat. In boreal and subarctic zones of continental upper latitudes. Breeds along river banks and valleys and also on mountains to treeline, where it occurs in stunted spruces and birch crowns in sparse woodlands; also in spruce taiga or recumbent cedar patches. Also found on low ground in tangles of shrubbery along river banks, chiefly willows. Often sings from top of bush or tree. In winter prefers river bank thickets and streamside tangles but also occurs in various kinds of forests and scrub. Replaces Dunnock *P. modularis* from Urals eastward, but has not adapted correspondingly to temperate lowland habitats or, except in winter, to human settlements. (Dementiev and Gladkov 1954*b*.)

Distribution and population. No information on range changes or population trends.

USSR. Not abundant (Dementiev and Gladkov 1954*b*). For distribution in northern Urals, see Estafiev (1979).

Accidental. Sweden, Finland, Czechoslovakia, Austria, Italy, Lebanon.

Movements. Migratory, whole population wintering in Korea and eastern China.

In north of range, vacates northern Urals early September, and in Anadyr' (eastern Siberia) wandering begins early August, birds leaving by mid-September. In southern part of breeding range, present until mid-October or even mid-November. Main spring passage through Mongolia from late March, and through Soviet Maritime Territories April–May. Southern breeding grounds apparently reoccupied from early May, northernmost not until early June. (Dementiev and Gladkov 1954b.)

Has straggled to western Europe, once to Kashmir in April (Ali and Ripley 1973b), to Japan in winter (Sonobe 1982), and to Alaska in autumn (Roberson 1980).

DJB, JHE

Food. All information from outside west Palearctic. Diet largely insects; in winter, also seeds (Portenko and Vietinghoff-Scheel 1976). Forages on ground, pecking at soil, dead leaves, and grass (Kozlova 1975; Fiebig 1983); also regularly in trees and bushes (Stresemann *et al.* 1937; Černý 1944; Kozlova 1975; Kishchinski 1980; Fiebig 1983). Often seen in areas where snow has blown away or thawed (Panov 1973). In winter will approach houses, gardens, and urban areas (David and Oustalet 1877; Panov 1973; Fiebig 1983).

Little information on details of diet. 2 stomachs from Yamal peninsula (USSR), contained 3 adult craneflies (Tipulidae) and a few seeds of bistort *Polygonum bistorta* (Danilov *et al.* 1984). Stomach of ♀ from north-east USSR, September, contained seeds of crowberry *Empetrum* (Portenko 1973). Stomach of ♀ from Yakutskaya ASSR, September, contained only seeds (Vorobiev 1963). Seeds of Amaranthaceae and Betulaceae also recorded (David and Oustalet 1877; Portenko and Vietinghoff-Scheel 1976; Kishchinski 1980).

Diet of young similar to that of adults. Largely insects, especially beetle larvae (Dementiev and Gladkov 1954b; Kozlova 1975). Several items carried in bill to nest at each visit (Portenko 1939).

TAS

Social pattern and behaviour. Little information and that almost entirely extralimital.

1. Normally solitary or in small loose parties outside breeding season: e.g. in Beijing (China), November–March (Fiebig 1983); similarly during migration (Kozlova 1930; Vorobiev 1963; Panov 1973). In Koryak mountains (north-east USSR), late summer and autumn, singly or in twos (perhaps pairs) for feeding (Kishchinski 1980). From mid-August in Anadyr' (north-east USSR), more commonly in pairs than in families; little association with other species (Portenko 1939). 2 vagrants on Tsushima (Japan) associated fairly closely with Siberian Meadow Bunting *Emberiza cioides* (Thiede 1975). BONDS. Nothing to suggest mating system other than monogamous; no details, however. In Ussuriland (eastern USSR) where common on passage and rarely winters, two recorded together through February and into early March perhaps a pair (Panov 1973); may indicate prolongation of pair-bond outside breeding season or early pair-formation. Young fed by both sexes (Morozov 1984) and ♂ recorded apparently continuing alone after ♀ lost (Zasypkin 1981). BREEDING DISPERSION. Solitary or in small groups (Etchécopar and Hüe 1983). In *badia*, pairs normally far apart (Dementiev and Gladkov 1954b). In one part of Yamal peninsula (USSR), nominate *montanella* recorded in groups of 2–6 pairs, with nests close together (Danilov *et al.* 1984). *P. m. badia* collects food for young near nest (Dementiev and Gladkov 1954b); no further information on size, use, or defence of territory. Densities in Bol'shezemel'skaya tundra (north-west USSR) in July: in fragmented and not very tall spruce *Picea*

on slopes and river terraces, 30·8 pairs in 1 km²; in relatively dense riverine spruce 16·7 pairs per km² (Estafiev 1979). Other data refer to eastern USSR: in lower Amur region, June, average 5 birds per km² in *Pinus pumila* (Babenko 1984); 3 territorial ♂♂ recorded along 0·5 km of tributary of Kanchalan river (Chukotsk region), but otherwise scarce and local (Kishchinski *et al.* 1983). Marked site fidelity noted in Yamal study: *c.* 30% of marked birds returned to previous year's nesting area (Danilov *et al.* 1984). ROOSTING. No information.

2. Generally rather secretive, remaining in cover for feeding, etc. (Seebohm 1901; Stegmann 1936; Thiede 1975; Fiebig 1983). Singing ♂♂ said by Johansen (1955) to be exceptionally shy, disappearing rapidly into cover if approached. Vagrant in Bohemia (Czechoslovakia), flew about low when disturbed, soon dropping into vegetation, but avoiding trees (Černý 1944); in Mongolia, however, feeding bird flew up into tall trees when flushed (Kozlova 1930). FLOCK BEHAVIOUR. No information. SONG-DISPLAY. Song of ♂ (see 1 in Voice) normally given from top of bush or tree (Portenko 1939; Dementiev and Gladkov 1954b). Migrants will sing while moving about on ground and feeding; tend to do this more than flying to tree or rock especially to sing (Kozlova 1930). Sings from arrival on breeding grounds in various parts of USSR: e.g. from early June on Yamal (Danilov *et al.* 1984), continuing well into July—e.g. 6 June–22 July, most vigorously in last week of June, on Putorana plateau (Krasnoyarsk) (Morozov 1984). On lower Lena river, maximum song-output in early morning (Kapitonov and Chernyavski 1960). See also Vorobiev (1963) and Kishchinski *et al.* (1983). ANTAGONISTIC BEHAVIOUR. In southern Yakutiya (USSR), autumn-passage birds rather aggressive; respond to contact-call (see 2 in Voice) from conspecific birds, but fights often occur if birds meet (Noskov and Gaginskaya 1977). HETEROSEXUAL BEHAVIOUR. On Yamal, early June, courtship and copulation attempt (no details) recorded in pair near empty nest; ♀ sat briefly in nest then flew out (Danilov *et al.* 1984). In Anadyr', ♂ seen chasing ♀, apparently to drive her back to nest (Portenko 1939). RELATIONS WITHIN FAMILY GROUP. No information. ANTI-PREDATOR RESPONSES OF YOUNG. In Anadyr', nestlings extended their necks and gaped when molested (Portenko 1939). PARENTAL ANTI-PREDATOR STRATEGIES. (1) Passive measures. Will sit tightly, allowing close approach. If startled, usually flies off immediately, and returns furtively (Danilov *et al.* 1984). (2) Active measures. Will fly about silently from tree to tree, staying in view and very rarely giving alarm-call (Danilov *et al.* 1984: see 2 in Voice). Also recorded moving about but staying mainly in cover—e.g. when small young in nest; both birds of one pair very agitated, ♂ sometimes singing (Portenko 1939). In Yakutiya, ♀ attacked man vigorously when he touched her unfinished nest (Vorobiev 1959). MGW

Voice. Overall like Black-throated Accentor *P. atrogularis* (Dementiev and Gladkov 1954b). In China, winter, birds give only Contact-alarm call (David and Oustalet 1877).

CALLS OF ADULTS. (1) Song of ♂. Close to Dunnock *P. modularis*, more melodious than *P. atrogularis* or extralimital Brown Accentor *P. fulvescens* (Dementiev and Gladkov 1954b), and quite powerful (Etchécopar and Hüe 1983). Of fine quality and likened to Rustic Bunting *Emberiza rustica*, and with some similarity fo *Sylvia* warbler though squeakier and higher pitched, not as rich and powerful as (e.g.) Blackcap *S. atricapilla* (Portenko 1939). Kapitonov and Chernyavski (1960) also likened it to *P. atrogularis* and Garden Warbler *S. borin*. Recording

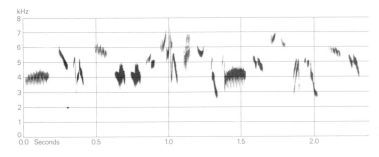

I B N Veprintsev and V V Leonovich USSR June 1978

apparently indicates existence of 2 song-types: short phrases recall *P. modularis* but in Fig I more buzzy sounds exhibiting rapid frequency and/or amplitude modulation and recalling Skylark *Alauda arvensis*; last loud buzz followed by 'si-zik' unit. Other song-type (Fig II) less buzzy and sweet warbling renders it closer to *P.*

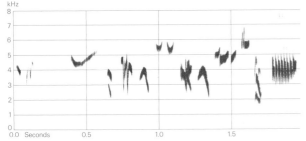

II B N Veprintsev and V V Leonovich USSR June 1978

modularis, though last unit is pronounced and loud buzz as culmination of crescendo (probably not typical) throughout phrase; final buzz can be very quiet, however. Duration of song 1·9-2·2 s; frequency 1·5 to over 7·0 kHz (J Hall-Craggs, M G Wilson). (2) Contact-alarm call. Few available descriptions all refer to trisyllabic call, variously rendered as follows: quiet 'dididi' as contact-call from wintering birds in China (Fiebig 1983); 'tsee-ree-see' or 'see-r-see' as characteristic call given frequently throughout stay in Anadyr', USSR (Portenko 1939); short, quiet 'teeseesee' from birds alarmed at nest (Danilov *et al.* 1984); 'til-il-il' resembling tit *Parus* (Seebohm 1901). CALLS OF YOUNG. No information.   MGW

Breeding. SEASON. Siberia: June–July (Makatsch 1976). SITE. In thick shrub or fork of low tree, close to main stem, 0·4-8 m from ground (Danilov *et al.* 1984). Nest: compact cup of twigs, plant stems and leaves, and moss, lined finer material and hair. Outer diameter 12 cm, inner 6 × 5 cm, height 7 cm, depth 3·5 cm (Makatsch 1976). Building: no information on role of sexes. EGGS. See Plate 82. Sub-elliptical, smooth and glossy; deep blue-green. 18·6 × 13·7 mm (17·0-20·6 × 12·6-14·4), $n=35$; calculated weight 1·86 g (Schönwetter 1979). 19·1 × 14·1 mm (17·8-20·0 × 13·2-14·8), $n=13$; weight 1·95 g (1·7-2·3) (Danilov *et al.* 1984). Clutch: 4-6. INCUBATION. 10 days at one nest; apparently by ♀ alone (Danilov *et al.* 1984). No further information.

Plumages (nominate *montanella*). ADULT. Central forehead and crown black-brown, feathers faintly tipped greyish-olive or rufous-olive when fresh. Long black stripe running from base of culmen along side of forehead and crown to side of hindneck and sometimes faintly across hindneck; stripe bordered below by long and broad supercilium (widest backwards from eye), rather deep buff when plumage fresh, cream or off-white when worn. Mantle and scapulars rufous-chestnut, feathers rather narrowly bordered olive-grey or buff at sides and with black spot on shaft at tip, upperparts appearing chestnut with rather narrow and poorly defined buff or olive streaks and black spots. Back olive-brown with rufous-brown spots; rump and upper tail-coverts uniform pale olive-cinnamon or olive-grey with slight buff tinge. Lores, upper cheeks, and ear-coverts black or brown-black, usually bordered by narrow buff bar behind, extending down from supercilium; patch at sides of neck ash-grey with slight buff or olive suffusion. Chin, throat, chest, breast, and flanks ochre-buff or pale cinnamon-buff; chest and breast with boldly black feather-centres, largely hidden when plumage fresh (least so on breast), showing clearly when worn; sides of breast with bold rufous-chestnut blotches, gradually merging into sharper and narrower rufous-brown or olive-brown shaft-streaks on flanks and sides of belly. Central belly, vent, and under tail-coverts cream or off-white, longer under tail-coverts with broad olive-grey shaft-streaks. Tail dark olive-brown, feathers with narrow deeper rufous-olive outer edges. Flight-feathers and greater primary coverts dull sepia-black, outer webs rather narrowly edged pale rufous-brown to pale olive-grey (paler on outermost primaries), inner webs edged off-white on basal and middle portions. Lesser upper wing-coverts dark olive-grey with broad rufous-chestnut fringes along tips; median coverts dark grey with contrasting cream-buff or yellow-white fringe along tip (often slightly spotted rufous at shaft); greater coverts and tertials with dark grey centres extending into narrow dull black point at tip, outer webs of coverts and both webs of tertials broadly fringed rufous-brown, inner webs of coverts broadly olive-grey, fringes of both webs ending in contrasting cream-buff or yellow-white spot at tip; pale spots on tips of median and greater coverts form narrow double wing-bar. Axillaries and under wing-coverts ochre-buff, like flanks. Some individual variation in width of rufous-chestnut streaks on mantle, scapulars, and flanks, in width and depth of colour of olive-grey or buff fringes on feather-sides of mantle and scapulars (sometimes completely absent), in extent of black spots on mantle and scapulars, and in colour of rump and upper tail-coverts (more rufous, olive-grey, or buff). Variation in part dependent on bleaching and wear, and probably on age, but in part also sexual (though sexes often indistinguishable): in ♂, black lateral crown-stripe wider and central crown and nape blackish-brown (in ♀, more rufous-olive-brown), rufous-chestnut streaks of mantle and scapulars more extensive, and black spots on chest and breast on average larger and deeper black. NESTLING. Down brownish-black, on upperparts only (I A Neufeldt). JUVENILE. Upperparts earth-brown, feathers of mantle and scapulars with large brown spots on centres. Supercilium dirty white with brown specks. Side of head pale brown, central ear-coverts tinged buff. Underparts dirty isabelline with large brown triangular spots on chest and breast, smaller spots on sides of throat. (Hartert 1910). FIRST ADULT. Similar to adult, but juvenile flight-feathers, tail, greater upper primary coverts, and variable number of outer greater coverts retained, these feathers contrasting slightly in wear and colour with neighbouring fresh ones (e.g. with tertials, tertial coverts, or median upper wing-coverts; in adults, all plumage equally new); primaries, primary coverts, and tail relatively more worn than in adult at same time of year.

Bare parts. ADULT. Iris yellowish-brown, light brown, or dark brown. Bill black or black-brown, base of lower mandible brown, pale horn-brown, beige, or yellowish. Leg and foot yellowish-brown, flesh, or light brown-red with dirty-brown soles. (Hartert 1910; Dementiev and Gladkov 1954b; Ali and Ripley 1973b; Lindell et al. 1978; BMNH, RMNH.) NESTLING. Mouth bright red (Portenko 1939); tip of tongue dark, base with 2 dark spots (I A Neufeldt). JUVENILE. No information.

Moults. ADULT POST-BREEDING. Complete; primaries descendant. July–September (Bub et al. 1984). POST-JUVENILE. Partial: head, body, lesser and median upper wing-coverts, tertials, tertial coverts, and variable number of inner greater upper wing-coverts. July–September (Bub et al. 1984). Several specimens had just started in first half of August, when others had just fledged; one in advanced moult on 25 July (Dementiev and Gladkov 1954b). Intense moult on 10 August in Ural mountains (Portenko and Vietinghoff-Scheel 1976). Birds with moult completed examined from second half of September (BMNH, RMNH).

Measurements. ADULT, FIRST ADULT. Nominate montanella. West and central Siberia (Yenisey to Lake Baykal: BMNH, RMNH, ZMA), Czechoslovakia (Černý 1944), and Mongolia (Piechocki et al. 1982), combined, all year; skins. Bill (S) to skull, bill (N) to distal corner of nostril; exposed culmen on average 3·2 less than bill (S).

WING	♂ 74·6	(1·68; 13)	72–78	♀ 71·7 (1·32; 8)	70–73
TAIL	64·0	(2·39; 12)	60–68	61·2 (2·71; 8)	58–65
BILL (S)	13·5	(0·51; 12)	12·9–14·3	13·2 (0·59; 8)	12·5–14·0
BILL (N)	7·9	(0·53; 12)	7·2–8·6	7·3 (0·45; 8)	6·7–7·8
TARSUS	19·6	(0·61; 12)	18·8–20·6	19·2 (0·69; 8)	18·2–19·9

Sex differences significant for wing and tail. 1st adult with retained juvenile flight-feathers and tail similar to full adult.

WING. (1) Nominate montanella; (2) badia (Dementiev and Gladkov 1954b).

(1)	♂ 72·6	(8) 70–74	♀ 69·3	(12) 65–73
(2)	69·2	(7) 67–72	68·7	(6) 63–70

Weights. Finland, October: 15·5 (Eriksson et al. 1976). USSR: ♂ 17·0, 17·3, 17·8; ♀ 17·5 (Dementiev and Gladkov 1954b). Kazakhstan (USSR), January: ♂ 18·0 (Dolgushin et al. 1972). Mongolia, late September: ♂ 18·5 (Piechocki et al. 1982). China: January, ♂♂ 16·5, 18·4 (Stresemann et al. 1937); March, 18·3 (4) 17·2–20 (Rensch 1924).

Structure. Wing rather short, broad at base, tip bluntly pointed. 10 primaries: p7 longest, p6 and p8 both 0–1 shorter, p9 3–5, p5 2–4, p4 7–9, p1 14–17; p10 reduced, narrow and pointed, 38–43 shorter than p7, 2 shorter to 1 longer than longest upper primary covert. Outer web of p5–p8 and inner of (p6–)p7–p8 emarginated. Middle toe with claw 16·6 (5) 15·5–17·5; outer toe with claw c. 75% of middle with claw, inner c. 70%, hind c. 80%. Remainder of structure as in Black-throated Accentor P. atrogularis (see p. 574).

Geographical variation. Rather slight in colour of upperparts, slight in size. P. m. badia from north-east Siberia has darker crown than nominate montanella from further west and south; in fresh plumage, mantle and scapulars more intensively rufous-chestnut, with less grey or buff at feather-sides and paler red-brown spots on feather-tips (less blackish); rump and underparts tinged deeper ochre-buff, rump less olive, flanks less yellow-buff; streaks on flanks often more rufous-brown, less olive; in worn plumage, crown often virtually uniform black, mantle and scapulars uniform rufous (Dementiev and Gladkov 1954b; BMNH, RMNH). For size difference, see Measurements.

Forms species-group with P. atrogularis, Kozlov's Accentor P. koslowi, Radde's Accentor P. ocularis, and Yemen Accentor P. fagani (see P. ocularis, p. 568).

Recognition. Differs from P. atrogularis in rufous-chestnut streaks on mantle and scapulars, rufous-brown fringes along tertials, flight-feathers, and greater upper wing-coverts, uniform buff throat without black, larger black feather-centres on chest and breast (often clearly visible on breast), rufous-brown rather than grey-brown streaks on flanks, and slightly more distinct pale spots on tips of median upper wing-coverts. A few P. atrogularis show slight rufous tinge to feather-centres on mantle and scapulars, however, and black patch on chin and throat of P. atrogularis sometimes small, and (if plumage fresh) sometimes largely hidden under broad buff feather-fringes; in worn plumage, some black of feather-bases also visible on breast of P. atrogularis and pale spots on tips of coverts abraded in both species. For difference from P. ocularis, see that species (p. 568). CSR

Prunella ocularis Radde's Accentor

PLATES 39 and 41
[between pages 544 and 545]

Du. Steenheggemus Fr. Accenteur de Radde Ge. Steinbraunelle
Ru. Пестрая завирушка Sp. Acentor alpino perso Sw. Kaukasisk järnsparv

Accentor ocularis Radde, 1884

Monotypic

Field characters. 15·5 cm; wing-span 22–23 cm. Noticeably larger than Dunnock *P. modularis* and Siberian Accentor *P. montanella*. Rather bulky accentor, with basic plumage pattern most like *P. montanella* but colours differing in bold white supercilium, black (not rufous) streaks on upperparts, and greyer rump and wings; usually shows broken malar stripe. Sexes closely similar; little seasonal variation. Juvenile separable.

ADULT MALE. Head pattern recalls Whinchat *Saxicola rubetra*: crown brown-black; supercilium pure white; obvious, almost diamond-shaped, cheek panel dusky-black; chin and throat white; usually a mottled dusky malar stripe. Hindneck grey, streaked dusky; mantle and scapulars pale olive- to sandy-brown, quite heavily streaked black; rump and upper tail-coverts olive-grey to olive-brown; rear upperparts thus greyest and best-marked of black-faced accentors. Wings essentially grey-brown: 2 wing-bars, formed by cream margins to median coverts (obvious at close range) and off-white fringes to greater coverts (less obvious); tertials black-brown with buff fringes. Underwing buff. Tail sepia. Breast and flanks warm buff, with sides of breast (when worn) mottled dusky and flanks (at all times) streaked black-brown; at distance, only flank-streaks show and buff tone appears to extend down to belly. Centre of belly and vent cream-white, producing marked pale area. ADULT FEMALE. At close range, shows duller, grey-streaked crown, and paler, less buff underparts. JUVENILE. Resembles adult but plumage generally warmer in tone, with markings less contrasting; in particular, supercilium less distinct and throat less clean. At all ages, bill black-horn with pale brown base to lower mandible; legs dull red-brown.

Similar in plumage pattern and colour to *P. montanella* but noticeably larger and more distinctly streaked above, with cleaner supercilium. Widely separated from it geographically at all times of year, however, though much less so from similar Brown Accentor *P. fulvescens* which occurs as close to west Palearctic as central Afghanistan. Flight low and flitting, even whirring. Hops on straight legs (not bent like *P. modularis*); also walks with twitching and hopping movements. Fairly upright when on the move, less so when feeding. Recalls *P. modularis* in general character and restriction to dense cover and adjacent open ground.

Short twittering song resembles *P. modularis*. Call like *P. modularis* but quieter and thinner.

Habitat. In lower middle and lower continental latitudes, in warm arid zone. In Transcaucasia, breeds at altitudes between 2000 m and 3000 m on rocky or stony mountain slopes carrying scrub vegetation, chiefly junipers; generally stays in bushes or shade of rocks (Dementiev and Gladkov 1954*b*). In the Ala Dagh (Turkey), with dry Mediterranean climate, found breeding in area of sheep and goat pastures and gullies with scattered bushes of barberry *Barbarea* between 2400 m and 2600 m; feeds among bushes and low vegetation or (after breeding) on rocky slopes, and sings from tops of bushes (Gaston 1968). Winters at lower altitudes in shrubby growth bordering mountain streams (Harrison 1982).

Distribution and population. No evidence of range or population changes.

USSR. Numbers extremely small (Dementiev and Gladkov 1954*b*).

Accidental. Israel.

Movements. Apparently resident in Turkey and Caucasus (*Orn. Soc. Turkey Bird Rep.* 1970–5; G G Buzzard), though no winter records from Turkey at least. Populations in Iran winter south of breeding ranges, at lower altitude (D A Scott). Thus, present only in winter (October–April) in several lowland areas of Iran (Marien 1951). Not recorded from Gulf states of Arabia, but one November record from Masirah island off Oman (Gallagher and Woodcock 1980) (presumably this relates to this species and not to Yemen Accentor *P. fagani*, formerly considered conspecific).   JHE, DJB

Food. In south-west Iran, November–December, 3 stomachs contained small seeds; 1 also contained 1–2 small winged insects (L Cornwallis). No further information on diet. Field observations suggest foraging done largely on ground among snow, rocks, scrub, and low vegetation, and alongside streams (Suschkin 1914*b*; Meiklejohn 1948*b*; Gaston 1968). Nest observations in Turkey showed feeding rates of *c.* 9–11 visits per hr during 06.00 hrs–10.00 hrs, and 4–6 per hr during 10.00–16.00 hrs, then declining to nil by *c.* 19.00 hrs (Gaston 1968).

For detailed study of (nestling) diet and feeding methods in Brown Accentor *P. fulvescens*, see Kovshar' (1979).   TAS

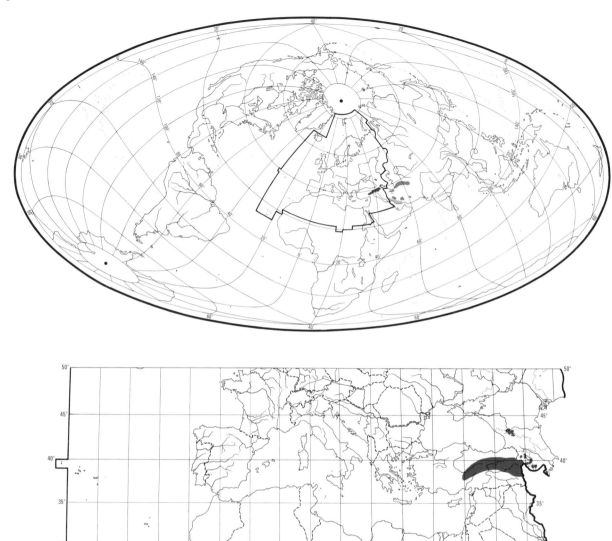

Social pattern and behaviour. Very poorly known.

1. Very little information on dispersion outside breeding season. Loose-knit parties recorded in southern Iran after breeding (Desfayes and Praz 1978). Does not form large flocks in autumn (Dementiev and Gladkov 1954b). In Ala Dagh mountains (southern Turkey), feeding birds recorded associating with Black Redstart *Phoenicurus ochruros* (Gaston 1968). BONDS. No information on mating system or duration of pair-bond. Both parents feed nestlings and fully fledged young (Meiklejohn 1948b; R P Martins and C R Robson). BREEDING DISPERSION. Solitary (Dementiev and Gladkov 1954b). In Ala Dagh mountains, favourite song-post of 1 ♂ at least 100 m from nest, suggesting large territory (Gaston 1968). ROOSTING. No information.

2. In Caucasus (USSR/Turkey), birds stay mostly in juniper *Juniperus* thickets and emerge only to re-enter cover quickly or to fly off (Suschkin 1914b). Varying reports of relative shyness from Iran: singing ♂ flew from rock into cover when disturbed (Stresemann 1928); shy and not easily approached (P A D Hollom); bird seen daily in Lar valley, 9-12 July, frequented thick vegetation, but perched openly when disturbed (Meikle-

john 1948b); 1 of 2 in Elburz mountains, early August, was exceptionally tame, allowing close approach (Norton 1958). Recorded wing-flicking (while feeding), horizontal tail-flicking, up-and-down tail-jerking, and tail-quivering, much as in Dunnock *P. modularis*; perhaps performed only when alert and anxious (Meiklejohn 1948b; Norton 1958; Desfayes and Praz 1978; P A D Hollom). FLOCK BEHAVIOUR. No information. SONG-DISPLAY. ♂ sings (see 1 in Voice) from top of bush or rock: e.g. in Iran, early May, bird gave 10 songs in 60 s from prominent little ridge. In Turkey, early June, first song noted at 03.25 hrs (P A D Hollom; also Gaston 1968). Will sing during nestling phase (R P Martins and C R Robson). ANTAGONISTIC BEHAVIOUR. No information. HETEROSEXUAL BEHAVIOUR. In Turkey, early June, display associated with copulation involved some wing-raising (perhaps wing-waving) like *P. modularis* (P A D Hollom). RELATIONS WITHIN FAMILY GROUP, ANTI-PREDATOR RESPONSES OF YOUNG. No information. PARENTAL ANTI-PREDATOR STRATEGIES. Adult once seen to carry faecal sac of young *c.* 250 m away (R P Martins and C R Robson).

MGW

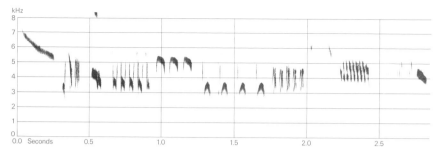

I P A D Hollom Iran April 1972

Voice. CALLS OF ADULTS. (1) Song of ♂. Quiet, sweet, and gentle, with twittering or clear bubbling quality and closely resembling Dunnock *P. modularis* rather than Alpine Accentor *P. collaris* (Stresemann 1928; Meiklejohn 1948*b*; Gaston 1968; Desfayes and Praz 1978; P A D Hollom). Phrases last *c.* 3–4 s (Gaston 1968). Descriptions of song given in Iran, early May: 'swee swee trurrerer' and 'turrer che turrer churrerer'; also 'slee vit chur chur tui', clear and loud in still air with rock-face background (P A D Hollom). In recording, short stereotyped phrases begin with quiet, high-pitched but descending portamento, and end in faint, attractive flourish 'vee vee-veet' or 'vee-vee vee-veet'. Longer phrase (Fig I) has similar portamento sound at start but overall more warbling with gentle quality, purring rattles prominent (J Hall-Craggs, M G Wilson). (2) Contact-alarm calls. (a) A 'trill' of 3–4 syllables, similar to that of other Asiatic Prunellidae (Suschkin 1914*b*). (b) Call-note like *P. modularis*, but quieter, thinner, and less penetrating (P A D Hollom), with piping quality (Norton 1958); presumably the equivalent of Seep-call in *P. modularis*, rather than same as call 2a.

CALLS OF YOUNG. No information.   MGW

Breeding. SEASON. Southern USSR: nests found June–August (Dementiev and Gladkov 1954*b*). SITE. In low bush, in one case nearly 1 m above ground (R P Martins and C R Robson). Nest: cup of twigs, leaves, and stems, with lining of finer material, including hair, wool, grass, and moss; one nest *c.* 16 cm across, cup *c.* 6 cm deep (R P Martins and C R Robson). Building: no information. EGGS. Sub-elliptical, smooth and glossy; blue. No measurements available. Clutch: 3–4. INCUBATION. No information. YOUNG. Fed by both parents (R P Martins and C R Robson). No further information.

Plumages. ADULT MALE. Forehead and crown dull black or brown-black; hindneck brown-black with faint dull grey streaks. Mantle and scapulars pale olive-brown or pale sandy-buff (darkest when plumage fresh), each feather with black central streak 2–3 mm wide; feather-sides paler and greyish when plumage worn. Back similar, but black streaking less pronounced; rump and upper tail-coverts uniform olive-grey or pale olive-brown. Broad and contrastingly white supercilium, extending backwards to sides of nape, usually widest behind eye; lores, upper cheeks, and ear-coverts dull black, forming distinct black mask; rear part of ear-coverts speckled buff in fresh plumage. Sides of neck ash-grey, sometimes faintly speckled dusky. Lower cheeks, chin, and upper throat cream-buff (if fresh) or white (if worn), lower cheeks sometimes with dusky specks, forming reduced malar stripe. Cream of lower throat gradually merges into warm cinnamon-buff on chest and sides of breast; feather-bases of chest dull black, showing as black mottling when plumage heavily worn (from about May). Remainder of underparts cream-buff, pale buff, or cream-white, palest on central belly and vent; flanks and sometimes upper belly and sides of belly with rather broad but ill-defined dark olive-brown or dull black streaks. Tail greyish-black or dark sepia, both webs of central pair (t1) and outer webs of others narrowly edged pale buff-brown (if fresh) or off-white (if worn). Flight-feathers like tail; sharp and rather narrow outer edges pale buff to cream-white (not extending to emarginated parts of outer primaries), edges broader and darker olive-brown or rufous-brown towards base of secondaries and inner primaries. Tertials dull black with broad buff-brown, rufous-brown, or pale buff fringes. Lesser upper wing-coverts greyish-buff, some dull black of feather-centres sometimes visible; median coverts black with contrasting cream-buff or cream-white tip of *c.* 1·5–2 mm; greater coverts black with broad rufous-brown, olive-brown, or greyish-buff fringes along outer webs, merging into clear-cut white fringe at tip. Greater upper primary coverts black with narrow olive-brown outer fringe merging into off-white at tip of outer web. Axillaries and under wing-coverts buff, some dark grey of feather-bases showing on longer coverts and on small coverts along leading edge of wing. ADULT FEMALE. Like adult ♂, differing mainly in less uniformly black crown, which is streaked black-brown and dull olive-grey; lower cheeks and neck slightly more extensively mottled dusky; ground-colour of chest and sides of breast slightly paler buff, throat and belly more cream-white. NESTLING. Down greyish (R P Martins and C R Robson). JUVENILE. No information. FIRST ADULT. Like adult, but juvenile flight-feathers, tail, greater upper primary coverts, and outer or all greater upper wing-coverts retained; these do not show constant difference in pattern from new 1st adult feathers, but in particular tail, outer primaries, and primary coverts more worn than adult at same time of year: in autumn, adult fresh and smoothly tipped, in 1st adult tips slightly frayed; in spring, adult primaries and coverts still virtually smooth at tips, tail slightly worn; in 1st adult, primaries and coverts distinctly abraded, tail heavily.

Bare parts. ADULT. Iris brown. Bill black-brown with yellow-brown or horn-brown base of lower mandible and basal cutting edges of upper mandible. Leg and foot reddish-brown. (Dementiev and Gladkov 1954*b*; BMNH). NESTLING. Mouth red; tip of tongue black, base with 2 black spots (R P Martins and C R Robson). JUVENILE. No information.

Moults. ADULT POST-BREEDING. Complete; primaries descendant. No moult in many birds examined October–April (Marien 1951). Adults with body fresh but outer primaries still growing encountered mid-August; moult completed 20 August, except for some growing feathers on crown (Dementiev and Gladkov 1954b). POST-JUVENILE. Partial: head, body, lesser and median upper wing-coverts, tertials, tertial coverts, and variable number of inner greater upper wing-coverts. Birds in fresh juvenile plumage and not yet moulting encountered 2–22 July and mid-August; others in moult mid-August (Dementiev and Gladkov 1954b).

Measurements. ADULT, FIRST ADULT. Eastern Turkey and northern Iran, spring; skins (BMNH, ZFMK). Bill (S) to skull, bill (N) to distal corner of nostril; exposed culmen on average 3·2 less than bill (S).

WING	♂ 78·1 (0·66; 6)	77–79	♀ 73·5 (3)	73–74
TAIL	64·6 (1·67; 5)	62–66	60·5 (3)	59–62
BILL (S)	14·2 (0·30; 6)	13·7–14·5	14·3 (3)	14·0–14·7
BILL (N)	8·0 (0·13; 6)	7·8–8·2	8·0 (3)	7·9–8·2
TARSUS	22·0 (0·36; 6)	21·5–22·4	20·8 (3)	20·4–21·0

Sex differences significant, except bill. 1st adult with retained juvenile wing and tail combined with adult, though juvenile wing on average 0·6 shorter than adult and tail 1·2 shorter (samples very small).

Iran (mainly Zagros mountains), October–April; skins (Marien 1951).

WING	♂ 76·7 (2·22; 24)	74–80	♀ 74·1 (10)	71–78
TAIL	67·1 (— ; 23)	62–73	64·4 (10)	61–72
BILL (S)	13·7 (0·24; 23)	13–14·5	13·3 (10)	12–14

Weights. Eastern Turkey, May: ♂ 25 (Kumerloeve 1968). Southern Iran, June: ♂ 20 (Desfayes and Praz 1978).

Structure. Wing rather short, broad at base, tip rounded. 10 primaries: p7 longest, p6 and p8 both 0–0·5 shorter, p9 3–7, p5 0·5–2, p4 4–6, p3 7–9, p1 11–15. P10 reduced; 38–46 shorter

than wing-tip, 0–4 longer than longest upper primary covert. Outer web of p5–p8 and inner of p6–p9 rather faintly emarginated. Bill strong, straight, broad and heavy at base, tip sharp; similar to bill of Black-throated Accentor *P. atrogularis huttoni*. Middle toe with claw 17·5 (7) 16·5–18·5; outer toe with claw *c.* 72% of middle with claw, inner *c.* 71%, hind *c.* 78%. Remainder of structure as in *P. atrogularis* (p. 574).

Geographical variation. None.

Sometimes considered a race of Brown Accentor *P. fulvescens* of central Asia (e.g. by Hartert 1910, 1921–2), but shows also many features in common with Black-throated Accentor *P. atrogularis*, to which it appears more closely allied and with which it may constitute a superspecies (Marien 1951). As geographical ranges of *P. fulvescens* and *P. atrogularis* overlap widely, these 2 cannot comprise a single superspecies, and all 3 forms better considered full species within a species-group (Stresemann 1928; Vaurie 1955b). Yemen Accentor *P. fagani* sometimes considered a race of *P. ocularis* (Dementiev and Gladkov 1954b), but better kept separate (Hartert and Steinbacher 1932–8) as it appears equally closely related to other members of the *ocularis-atrogularis-fulvescens* group (Vaurie 1955b); hence considered another full species of this group, as is Siberian Accentor *P. montanella* and Kozlov's Accentor *P. koslowi*.

Recognition. Closely similar to *P. atrogularis* and *P. montanella* and easily confused with either of these (e.g. Boswell and Naylor 1957). Differs from both in pure white supercilium; unlike *P. atrogularis*, no black on throat; unlike *P. montanella*, no rufous streaks on upperparts; crown of ♂ more uniform black than ♂ *P. atrogularis*; flanks rather more heavily streaked than *P. montanella* and *P. atrogularis huttoni* (*P. a. atrogularis* virtually unstreaked), but streaking less extensive and lower cheeks less heavily spotted black than in *P. fagani*. Tarsus slightly longer than in other west Palearctic *Prunella* (except for much larger Alpine Accentor *P. collaris*). CSR

Prunella atrogularis Black-throated Accentor

PLATES 39 and 41
[between pages 544 and 545]

DU. Zwartkeelheggemus FR. Accenteur à gorge noire GE. Schwarzkehlbraunelle
RU. Черногорлая завирушка SP. Acentor gorginegra SW. Svartstrupig järnsparv

Accentor atragularis Brandt, 1844

Polytypic. Nominate *atrogularis* (Brandt, 1844), north-west USSR. Extralimital: *huttoni* (Horsfield and Moore, 1854) (synonym: *lucens* Portenko, 1929), central Asia.

Field characters. 15 cm; wing-span 21–22·5 cm. Slightly larger than Dunnock *P. modularis* and Siberian Accentor *P. montanella*. Accentor of similar form to *P. montanella* but with more heavily marked head, black on chin and throat, black-streaked brown back, and rather pale brown rump. Sexes similar; little seasonal variation. Juvenile separable.

ADULT. Crown and nape olive-brown, heavily streaked and bordered black; large dull black cheek-panel, bordered by long buff to cream supercilium and narrow cream malar stripe; narrow cream eye-ring and grey patch

on side of neck; black chin and throat appear as distinct bib when plumage worn (spring and summer) but as speckled patch when fresh (winter). Mantle and scapulars buff-brown, streaked black-brown (recalling *P. modularis*, not *P. montanella*), back and rump buff- to olive-brown. Tail dark olive-brown. Wings olive-brown, with white spots on tips of median and greater coverts (forming indistinct wing-bars), and buff-brown fringes to black-brown tertials. Underwing buff. Border to bib buff-white; deep breast and flanks rich, bright buff, becoming cream on belly and vent; breast unmarked except for a few

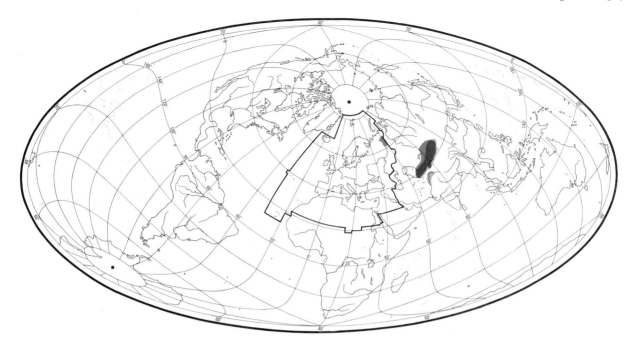

grey- or olive-brown spots which become streaks on flanks and sides of vent. When plumage fully worn, all black-marked areas intensify and underparts become paler, with more spots on sides of and even across chest. JUVENILE. Not studied in the field. Duller above and more streaked below than adult. Head pattern more diffuse, lacking grey patch on side of neck. At all ages, bill black with horn base to lower mandible, legs pink- to brown-flesh.

Birds with full black bib of breeding season unmistakable, but those showing only speckled throat less distinctive, suggesting both *P. montanella* and Radde's Accentor *P. ocularis*. Of the 3 species, *P. montanella* and *P. atrogularis* show vagrancy patterns to south and south-west though *P. atrogularis* yet to reach western Europe. Observer faced with dark-cheeked accentor must concentrate on colour of and heaviness of streaking on both upper- and underparts and on throat pattern. Flight, gait, and behaviour much as *P. modularis*, but a more robust bird and less shy, perching freely in the open even when disturbed.

Song a simple piping resembling *P. modularis* and *P. montanella*. Calls include quiet 'teeteetee'.

Habitat. Breeds in upper and middle continental latitudes, in north in subalpine belt in clumps of stunted spruce shrubs, while in central Asia it inhabits tall conifer forests; also nests in broad-leaved forests and in scrub with plenty of juniper, or in impassable thickets. Avoids open areas, living in low dense and often thorny bushes or on ground, but sings from top of shrub or small tree, hiding in thicket when alarmed. On passage occurs in osiers, oleander, or tamarisks in river valleys. In winter

may enter orchards, or stay in bottomland forests, bramble thickets, and reeds or dense grasses on banks of irrigation canals (Dementiev and Gladkov 1954*b*). In India, winters in hills up to *c.* 2500 m, but mostly below 1800 m, in scrub jungle, tea gardens, orchards, and bushes near cultivation. In plains, occurs in sandy semi-desert near cultivation and among grass tussocks, feeding on ground, but on being disturbed perches freely on bushes and in lower branches of trees (Ali and Ripley 1973*b*).

In Afghanistan, seen in late winter at *c.* 500 m in dune country with tamarisks, and in cultivated valley at *c.* 1100 m (Paludan 1959). In north, breeding range overlaps Siberian Accentor *P. montanella*; available data do not permit definition of ecological distinctions between the two, although *P. atrogularis* far more concentrated in more southerly highlands rather than boreal lowlands and valleys.

Distribution and population. No evidence of range or population changes.

USSR. Numbers quite low (Dementiev and Gladkov 1954*b*).

Accidental. Israel.

Movements. Urals population (nominate *atrogularis*) migratory, central Asian population (*huttoni*) partially so, also moving altitudinally. Combined winter quarters lie in Afghanistan, Pakistan, Kashmir, and mountains of central Asia, races apparently mixing over much of this range, though nominate *atrogularis* largely to south and west, *huttoni* largely to north and east. Thus, nominate *atrogularis* recorded mostly from Afghanistan, though also in northern Pakistan and apparently east to north-west

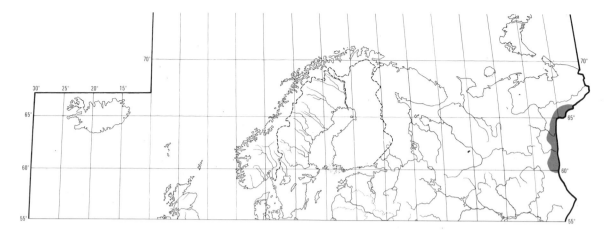

China, and *huttoni* mostly in central Asia and Kashmir, though also west to northern Afghanistan (Dementiev and Gladkov 1954*b*; Paludan 1959; Ali and Ripley 1973*b*; Etchécopar and Hüe 1983).

Autumn departures from Urals begin August, birds subsequently appearing scattered widely to south and south-east. Reaches Orenburg (southern Urals) in October and southern Turkmeniya in November (Dementiev and Gladkov 1954*b*). Arrives northern Pakistan and India usually mid-October, leaving by 3rd week of March (Ali and Ripley 1973*b*). Present to mid-March in Afghanistan (Paludan 1959). In central Asia, birds recorded wintering at Gul'cha (Kirgiziya) November–December, and present further north in Talasskiy Alatau mountains (presumably on breeding grounds) from late February (Portenko and Vietinghoff-Scheel 1976). Nominate *atrogularis* reaches *c.* 50°N (Orenburg, Barnaul) in April (Dementiev and Gladkov 1954*b*).　　　　　　　DJB, JHE

Food. All information from outside west Palearctic. Diet largely insects, supplemented by seeds (mainly in winter). Will also take other small arthropods. Feeds on ground, favouring woods, dense shrubs, grassy clearings, and stream banks; in winter, normally in patches of weeds, reeds, bushes, and alongside ditches (Ivanov 1969). Grit found in stomachs of adults (Ivanov 1969; Dolgushin *et al.* 1972).

Predominantly small insects, including springtails (Collembola), two-pronged bristle-tails (Diplura), adult and larval grasshoppers (Orthoptera), adult and larval earwigs (Dermaptera), adult and larval bugs (Heteroptera), lacewings (Neuroptera), adult and larval butterflies and moths (Lepidoptera), caddisflies (Trichoptera), adult and larval flies (Diptera) and adult and larval beetles (Coleoptera: Carabidae, Curculionidae, Chrysomelidae, Staphylinidae, Dermestidae, Bruchidae, and Tenebrionidae). Also takes mites (Trombidoidea), spiders (Araneae), crustaceans (Crustacea), earthworms (Oligochaeta), and small snails (Mollusca). Plant food mainly seeds of grasses and herbaceous plants, predominantly taken in late summer and winter: Amaranthaceae, Chenopodiaceae, Compositae, Gramineae, Leguminosae, Polygonaceae, Ranunculaceae, Rosaceae, Umbelliferae; less commonly, seeds of trees—Pinaceae, Cupressaceae, Betulaceae (Yanushevich *et al.* 1960; Pek and Fedyanina 1961; Ivanov 1969; Dolgushin *et al.* 1972; Kovshar' 1979; Piechocki *et al.* 1982).

Little information on proportions of different items in diet. In Kazakhstan (USSR), 36 stomachs taken throughout the year contained grasshoppers, psyllids (Psyllidae), aphids (Aphidoidea), bugs, flies, ants (Formicidae), adult and larval butterflies and moths, beetles, and seeds of Chenopodiaceae, Leguminosae, Gramineae, *Amaranthus*, bristle-grass *Setaria*, and spruce *Picea*; of 23 stomachs taken September–April all contained seeds and 10 contained animal matter; of 13 taken May–August, all contained insects and 4 contained seeds (Dolgushin *et al.* 1972). 16 stomachs from the Pamirs (USSR), spring, autumn, and winter, contained mainly spiders and insects (grasshoppers, earwigs, butterflies and moths, flies, ants, and beetles) with some holding up to 80 ants or 25 Curculionidae; very few seeds (Ivanov 1969). 25 stomachs from Kirgiziya (USSR), all year, contained beetles, especially *Aphodius prodromus* (Tenebrionidae), Chrysomelidae, and Carabidae; also seeds (Polygonaceae, Leguminosae, Gramineae, Rosaceae, Compositae, Umbelliferae, spruce *Picea*, juniper *Juniperus communis*) and mollusc shells (Pek and Fedyanina 1961).

Food of young similar to that of adults. Largely insects supplemented by seeds. Of 40 collar-samples in Kazakhstan (USSR), June–July, 39 contained insects, 11 contained spiders, 6 contained molluscs, and 10 contained seeds (Dolgushin *et al.* 1972). 239 neck-collar samples from 18 nests in Zailiyskiy Alatau mountains (Kazakhstan) comprised 56·1% (by number) aphids (adults and larvae), 18·5% flies (adults and larvae), 5·3% butterflies and moths (adults and larvae), 5·2% spiders, 3·7% beetles (adults and larvae), 3·2% springtails (adults), 2·8% ants (adults and larvae), 1·9% grasshoppers (adults and larvae), 1·2% snails, 0·8% earwigs (adults and larvae),

0·5% bugs (adults and larvae), 0·2% lacewings (adults and larvae), 0·2% mites, 0·1% crustaceans, 0·1% earthworms; also a few two-pronged bristle-tails and caddisflies (Kovshar' 1979). Stomach of fledgling Oriental Cuckoo *Cuculus saturatus* being fed by *P. atrogularis* contained grasshoppers and berries (Irisov 1967). Grit sometimes fed to young (Dolgushin *et al.* 1972). Young fed by both parents, each bringing one large or several small items per visit. Usually given to one nestling only although 2–3 may be fed per trip. Adults appear at nest every 5–10 min (Dolgushin *et al.* 1972). In Tien Shan, nestlings 2–4 days old fed 2–5 times per hr; at 8–13 days, 11–16 times per hr; of 422 visits, 38% by ♂, 62% by ♀ (Kovshar' 1979).   TAS

Social pattern and behaviour. West Palearctic race, nominate *atrogularis*, poorly known; extralimital *huttoni* studied in Tien Shan (southern USSR) by Gavrilov (1973) and Kovshar' (1979).

1. Nominate *atrogularis* solitary or in small parties outside breeding season; an old record of *c.* 150 together (Dementiev and Gladkov 1954*b*). *P. a. huttoni* occurs singly or in pairs, with loose-knit parties (usually under 10 birds) formed only outside breeding season; no true flocks, though occasionally up to 20 during post-breeding dispersal and up to a few dozen recorded during migration (Dementiev and Gladkov 1954*b*; Ivanov 1969; Dolgushin *et al.* 1972; Abdusalyamov 1973); overall more gregarious than Dunnock *P. modularis* (Ali and Ripley 1973*b*). In Terskey-Alatau mountains (Tien Shan), 100–110 birds per km² of creeping *Juniperus* belt, August–October, 20 birds per km² in later winter (Vtorov 1967). Birds wintering in mountains apparently remain within relatively small area; bird ringed as nestling 25 August controlled *c.* 200 m away 15 December (Kovshar' 1979). In USSR, little true association with other species, though *huttoni* frequently recorded together with Red-mantled Rosefinch *Carpodacus rhodochlamys* (Dementiev and Gladkov 1954*b*). In winter in India, *huttoni* sometimes associates with Rufous-breasted Accentor *P. strophiata* (Ali and Ripley 1973*b*). BONDS. Nothing to indicate mating system other than monogamous, though no detailed study. Not known when pair-formation takes place, perhaps only after arrival on breeding grounds (Dolgushin *et al.* 1972). No information on duration of pair-bond. Both sexes feed young and perform nest-sanitation (Gavrilov 1973). Young tended (by ♂ alone if ♀ re-lays) for *c.* 7–9 days after leaving nest. With transition to independence, broods break up, young birds then solitary or, less commonly, in twos (Abdusalyamov 1973; Gavrilov 1973; Kovshar' 1979). BREEDING DISPERSION. Solitary and strictly territorial. In Zailiyskiy Alatau (Tien Shan), pairs usually 100–300 m apart (Gavrilov 1973). However, another study in same area indicated minimum distance between nests in *Juniperus* 40–65 m (*n*=4); at upper limit of spruce *Picea* wood, singing ♂♂ commonly (30–)40–50 m apart. Territory smaller than in Brown Accentor *P. fulvescens*. Food collected mostly within *c.* 50–70 m of nest, though not infrequently up to 100–150 m away. 2nd-brood nest *c.* 50–60 m from 1st (Kovshar' 1979); nests *c.* 10–30 m from 1st-brood nest in same territory also assumed to be for 2nd brood, rather than indication of 2nd pair moving in to occupy vacated territory (Gavrilov 1973). In Kumbel'-Say (Tadzhikistan, USSR), territories 2–3 ha, or up to 10–12 ha, of suitable habitat (Abdusalyamov 1973). Little information on densities. On western slopes of northern Urals in Pechora basin (north-

west USSR), 1 bird per km² of spruce with green moss; 20·4 birds per km² in spruce with herb layer. 8·8–24·0 birds per km² of birch *Betula*, larch *Larix*, and willow *Salix* but such habitat apparently used mostly for feeding rather than nesting (Estafiev 1981). In Terskey Alatau (Tien Shan), 70 birds (*huttoni*) per km² of creeping juniper belt (Vtorov 1967). Some evidence of adult and juvenile (natal) site fidelity in study by Kovshar' (1979): of 154 (11 ♂♂, 8 ♀♀, 122 nestlings, 13 older young) ringed 1971–5, 5 (3 adult, 2 juveniles) recovered (including birds nesting) 30–2000 m from ringing locality. Minimum distance between nests of *P. atrogularis* and *P. fulvescens* in Tien Shan *c.* 30 m (see also Antagonistic Behaviour, below). Often nests close to Himalayan Rubythroat *Luscinia pectoralis* (Kovshar' 1979). ROOSTING. No information.

2. Not shy and can be watched as close as 2–3 m, but often skulks in cover, especially when disturbed (Whistler 1919; Dolgushin *et al.* 1972). Birds feeding on ground may give alarm-call (see 2 in Voice) and fly up into tops of trees (Sushkin 1938; Shnitnikov 1949). Report in Ali and Ripley (1973*b*) apparently indicates disturbed birds sometimes perch more openly. Will fly away strongly if pressed (Fleming *et al.* 1976). Wing-flicking occurs as in other Prunellidae (Whistler 1926*a*; Jones 1947–8). FLOCK BEHAVIOUR. No information. SONG-DISPLAY. ♂ sings (see 1 in Voice) typically from elevated perch—top of tree or bush or protruding side branch, less commonly lower down on stump or rock (Shnitnikov 1949; Gavrilov 1973; Kozlova 1975). May fly to high perch, sing once or several times, then disappear into bushes; sometimes sings for up to *c.* 15 min from same perch (Gavrilov 1973). Will sing at any time of day: in Tien Shan, June, 04.07–20.30 hrs. Maximum intensity on 13 April of one year: 142 songs in 15 min around 07.00 hrs; 202–285 songs per hr recorded during nest-building (mid-April); 96–298 per hr during laying (early May). Will sing also during nestling period—see Relations within Family Group, below (Kovshar' 1979). In northern Urals, song of nominate *atrogularis* noted up to at least 11 July (Johansen 1955). In Tien Shan, birds sang from late January in one year, late March in another; song-period extends to late July or early August, though some juvenile song later than this. Increase in song in 2nd half of June presumably prelude to 2nd brood (Kovshar' 1979). ANTAGONISTIC BEHAVIOUR. Song apparently plays major role in deterring potential intruders; important for a species which leads rather secretive life (Gavrilov 1973). No further information on intraspecific territoriality. In Tien Shan, always subordinate to more powerful *P. fulvescens*. The 2 species once recorded fighting over food, but *P. atrogularis* does not normally attack *P. fulvescens* intruding on its territory, tending rather to accompany it until it leaves. ♂♂ of 2 species once sang alternately from perches *c.* 1 m apart (also *c.* 20 m from nest of *P. atrogularis*). When *P. fulvescens* left, ♂ *P. atrogularis* immediately occupied vacated perch and sang more loudly than before (Kovshar' 1979). HETEROSEXUAL BEHAVIOUR. (1) General. In Kazakhstan (USSR), birds apparently paired up to *c.* 1 month before breeding starts and ♂ sings vigorously during this period (Dolgushin *et al.* 1972). No information on Pair-bonding behaviour except for possible indication of mate-guarding in subsection 4 (below). (2) Courtship-feeding. ♂ once recorded feeding ♀ on nest (Dolgushin *et al.* 1972). (3) Mating. No information. (4) Nest-site selection and Behaviour at nest. Observation in Tien Shan, mid-April, suggests nest-site selected by ♀: ♀ flew about from bush to bush, ♂ accompanying her and singing. When ♀ stayed longer than usual in scrub, ♂ broke off song and followed her in. ♀ builds, collecting material *c.* 5–15 m from nest; again, ♂ remains close and sings. Only once did ♂ sit briefly in empty

nest (Kovshar' 1979). In Tien Shan, laying recorded 8-11 days after nest completed; for 2nd brood, ♀ may start laying even before nest finished. ♀ leaves nest every *c.* 30-60 min to feed for *c.* 5-30 min (Gavrilov 1973). RELATIONS WITHIN FAMILY GROUP. Hatching usually more or less synchronous though sometimes extends over 2-3 days. Small (unfeathered) young brooded by ♀ for long periods after feeds, but not once feathers start to appear. Young 2 days old brooded 49·4% of day; at 4 days old, for 70·4% of morning (Gavrilov 1973; Kovshar' 1979). Adults usually arrive at nest separately and alternately, though sometimes appear together. Approach cautiously, flying short distances from bush to bush, then hopping last few metres. After feeding young, ♂ may fly to tree or bush and sing briefly before flying off to collect more food (Gavrilov 1973). Young leave, barely able to flutter far (Dolgushin *et al.* 1972), at 11-14(-15) days; 1 out of brood of 4 left at 10 days and stayed on ground nearby for rest of day; single nestling once remained in nest for 16 days. As a rule, stay in or near nesting territory for long period, though 1 brood recorded *c.* 800 m from nest after 20 days. (Gavrilov 1973; Kovshar' 1979). ANTI-PREDATOR RESPONSES OF YOUNG. No information. PARENTAL ANTI-PREDATOR STRATEGIES. Warier than *P. fulvescens* when man near nest, tending to dive silently into cover and remaining hidden; only rarely gives alarm-call (Kovshar' 1979). May become habituated to man quite quickly, however: e.g. continuing to feed and tend nestlings when photographer at *c.* 5-7 m (Dolgushin *et al.* 1972). MGW

Voice. Only call 3 noted from extralimital *huttoni* in Indian winter quarters (Ali and Ripley 1973*b*).

CALLS OF ADULTS. (1) Song of ♂. Closely resembles Dunnock *P. modularis* and Siberian Accentor *P. montanella* (Sushkin 1938, which see also for comparison with extralimital Brown Accentor *P. fulvescens*; Michaelis 1977; J Hall-Craggs). Pleasant but simple and piping with 'tsi' and 'kri' sounds (Johansen 1955 for nominate *atrogularis*). Rather monotonous song of *huttoni* rendered 'teets-reetee-teets-reetee-teets-reetee-tsee-feetee' (Gavrilov 1973). Recording of *huttoni* (Fig I) reveals short phrases much like *P. modularis*. Sounds sweet but thin and brittle, almost splintery (sizzling jumble: P J Sellar); buzzes also occur and a prominent and attractive feature is relatively long and low-pitched but ascending tone ('hwee')—5th unit from last in Fig I. Compare also songs of *P. montanella* (p. 562) and Radde's Accentor *P. ocularis* (p. 567) (J Hall-Craggs, M G Wilson). (2) Contact-alarm call. Of 3-4 syllables (Sushkin 1938; Johansen 1955). Quiet 'teeteetee' when alarmed at nest (Kovshar' 1979); gentle and tremulous 'zee-zee-zee-zee' (Dolgushin *et al.*

1972); also rendered 'tseerree' or 'tseerreesee' (Abdusalyamov 1973). (3) A soft 'trrt' from *huttoni* in winter (Ali and Ripley 1973*b*); significance not known.

CALLS OF YOUNG. No information on calls of nestlings. Will sing rather primitive song from August after moult (Gavrilov 1973; Kovshar' 1979). MGW

Breeding. Except where indicated, data from Kovshar' (1979) for extralimital *huttoni* in Tien Shan (south-central USSR). SEASON. Urals: eggs found June (Makatsch 1976). Tien Shan: eggs laid from beginning of May to end of July. SITE. On branch of tree or shrub. Of 171 nests, 56% in spruces, 40% in juniper shrubs, 4% in deciduous shrubs. 50% of nests by trunk, 26% from 0·5 to 1 m from trunk; in juniper shrubs nearly half nests in centre, not higher than 2 m. Of all nests, 77% 0·3-3 m above ground, highest 13 m. Nest: cup of twigs and moss, with some grass leaves and stems, lined with finer material and hair. Building: by ♀, taking 13-18 days for early nests, but 7-8 days for later ones. EGGS. See Plate 82. Sub-elliptical, smooth and glossy; deep blue-green. 19·6 × 15·0 mm (18·2-20·5 × 14·5-15·4), $n=13$; calculated weight 2·27 g (Makatsch 1976). Clutch: 3-5 (1-6). Of 123 clutches: 2 eggs, 3%; 3, 12%; 4, 62%; 5, 22%; 6, 1%; mean 4·04; clutch size increases at first during season, later declines. Eggs laid 5-6 (1-10) days after nest completed in May, but 0-2 days after in June. 2 broods, possibly 3. INCUBATION. 12·3 days (11-14), $n=15$. Begins with last egg. By ♀ only. YOUNG. Cared for and fed by both parents. FLEDGING TO MATURITY. Fledging period 12·5 days (11-14), $n=6$. Young fed for at least 1 week after fledging. Age of first breeding not known. BREEDING SUCCESS. Average 55·7% of 140 nests successful, with most losses due to predation.

Plumages (nominate *atrogularis*). ADULT. Forehead, central crown, and hindneck olive-brown, broadly streaked black-brown; uniform black-brown stripe at side of crown, bordering supercilium. Mantle and scapulars evenly streaked black-brown and pale buff-brown, rather similar to mantle and scapulars of Dunnock *P. modularis*, but dark streaks slightly browner and with slight olive tinge and pale ground-colour slightly more buffish, tinged pink when fresh, slightly olive-grey when worn, not as rufous-cinnamon as in *P. modularis*. Back to upper tail-coverts uniform buff-brown or olive-brown, except for indistinct dusky streaks on back; slightly paler and often more buffy than dull olive-brown or grey-brown rump of *P. modularis*.

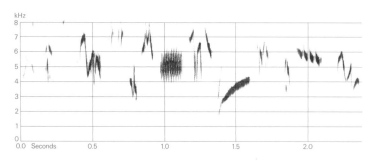

I Svensson (1984) USSR June 1983

Pale supercilium long and distinct, extending from nostril to side of hindneck; buff in fresh plumage, cream-buff or off-white when worn. Lores finely mottled pale buff and dull black; narrow eye-ring cream or off-white; upper cheeks and ear-coverts black-brown or dull black (feathers faintly tipped brown when plumage fresh). Side of neck behind and below ear-coverts buff, bordered behind by ash-grey patch with faint darker spots between side of mantle and side of chest. Chin and upper throat dull black or brown-black, feathers tipped buff when fresh, separated from black of upper cheek by narrow cream malar stripe running backwards from lower mandible; occasionally, face completely black, cream malar stripe and cream front part of supercilium absent. Black throat-patch bordered below by cream-buff or cream-white gorget, merging into deep buff or cinnamon-buff of sides of breast, chest, flanks, and thighs. Flanks and thighs with rather indistinct and ill-defined dull grey-brown or olive-brown shaft-streaks. Belly, vent, and under tail-coverts pale buff or cream-white, marked with grey-brown or olive-brown streaks on sides and coverts. Tail dark olive-brown, narrowly fringed pale buff along both webs on central feathers and along outer web and tip on outer feathers. Flight-feathers dark grey-brown, narrowly but sharply fringed buff or pale buff along outer webs and tips; bases of outer webs of secondaries more widely buff or pale olive-brown; tertials dull black or black-brown with broad and sharp buff-brown or pale olive-brown fringes. Upper wing-coverts buff-brown or pale olive-brown; lesser coverts with indistinct dusky centres, median with broad and contrasting dull black centres, greater with dull black shaft-streak near base which widens into broad area at tip, tip bordered at sides by paired buff to off-white spots formed by narrow broken fringe. Greater upper primary coverts and feathers of bastard wing dark grey-brown to dull black, narrowly fringed buff or off-white. Axillaries and under wing-coverts buff, partly mottled by grey of feather-bases. Abrasion and bleaching have marked influence. *In fresh plumage*, upperparts evenly streaked buff and brown, except for dark lateral crown-stripes and uniform back to upper tail-coverts; supercilium, side of head behind and below ear-coverts, and gorget below black throat-patch buff, pink-buff, or cream-buff; black of chin and throat speckled buff; chest uniform deep buff. *In worn plumage*, forehead and crown almost uniform olive-black or black, mantle and scapulars heavily streaked black; chin and throat virtually uniform dull black; supercilium, bar behind ear-coverts, malar stripe, and gorget below black throat-patch cream-white or white; sides of breast, chest, and flanks paler buff; often many rounded black feather-centres showing as large spots on chest and central breast; buff fringes of tail and flight-feathers bleached, partly abraded; sides of tips of greater coverts and tertials bleached, showing as row of white spots. Sexes largely similar; ♂ on average has more black on crown, narrower buff fringes to feathers of chin and throat, and more black on feather-bases of chest and central breast, but these characters strongly subject to change by abrasion and some ♀♀ may show more black than some less-worn ♂♂. NESTLING. Down dark smoke-grey, almost black; on upperparts only (I A Neufeldt). JUVENILE. Upperparts like adult, but duller, less sharply streaked. Supercilium narrow, but distinct; off-white, finely speckled brown. Side of head and neck finely mottled dark brown, buff-brown, and off-white; no ash-grey patch on side of neck. Chin and throat dark grey-brown, finely speckled pale buff, not sharply divided from cheeks and chest (blacker and more contrasting in adult). Chest, sides of breast, and flanks buff or greyish-buff, marked with ill-defined dark brown shaft-streaks; belly and vent cream or off-white, faintly and narrowly streaked brown at sides. Tail

and flight-feathers as adult. Lesser upper wing-coverts like mantle, remainder of wing-coverts as adult. Differs from juvenile *P. modularis* in paler brown upperparts with less blackish shaft-streaks, distinct pale supercilium, darker throat, more distinct streaks on chest and flanks, and pale brown (not dark rufous-brown) edges along flight-feathers, tail-feathers, and greater coverts (Bub *et al.* 1984). FIRST ADULT. Like adult; juvenile flight-feathers, tail, greater upper primary coverts, and variable number of greater upper wing-coverts retained, coloured as in adult but relatively more worn. In autumn, all plumage new in adult, wing and coverts slightly worn in 1st adult, tail moderately worn; in spring, adult wing and coverts slightly worn, tail moderately worn, juvenile wing and coverts distinctly worn, tail heavily worn. Colour differences from adult slight, depending partly on sex: in spring, adult ♂ the most contrastingly coloured, 1st adult ♀ least so, 1st adult ♂ and adult ♀ intermediate. In adult ♂ in spring, crown and hindneck blackish, contrasting with mantle; in 1st adult ♀, central crown and hindneck similar to mantle. Ear-coverts, chin, and throat of adult ♂ deep black, most pale feather-tips worn off by spring, pale malar stripe narrow; in 1st adult ♀, ear-coverts, chin, and throat dull brown, narrow pale buff or off-white feather-tips retained in spring, malar stripe wider; adult ♀ and 1st adult ♂ intermediate. Sexing and ageing often impossible, especially in spring and summer.

Bare parts. ADULT, FIRST ADULT. Iris brown or dark brown. Bill blackish, base of lower mandible paler horn-brown, yellow-brown, or flesh-brown. Leg and foot light brown, yellowish-brown, or flesh-brown. (Hartert 1910; Dementiev and Gladkov 1954*b*; Ali and Ripley 1973*b*; BMNH.) NESTLING. At hatching, bare skin (including bill, leg, and foot) orange-pink; mouth and tongue orange, tongue with 2 distinct rounded spots on base and with dark tip, latter gradually paler towards fledging date; gape-flanges narrow, orange-pink with faint white edge (Neufeldt 1970). JUVENILE. No information.

Moults. ADULT POST-BREEDING. Complete; primaries descendant. August (Marien 1951) or July–September(–October) (Bub *et al.* 1984). In USSR, early July, several adult *huttoni* and single nominate *atrogularis* heavily worn but moult not yet started; one nominate *atrogularis* had just started primary moult on 19 August; moult in both races completed late August to late September (Dementiev and Gladkov 1954*b*). POST-JUVENILE. Partial; starts late July to late August, completed from early September, depending on hatching date and altitude of breeding area (Dementiev and Gladkov 1954*b*; Portenko and Vietinghoff-Scheel 1976; Bub *et al.* 1984; BMNH, ZMA). Involves head, body, lesser and median upper wing-coverts, and usually tertials, tertial coverts, and variable number of inner greater upper wing-coverts; in some birds from early September, tertials and tertial coverts still juvenile, but these replaced later on, as mid-winter birds had them new or rather new (BMNH, RMNH, ZMA).

Measurements. ADULT, FIRST ADULT. Nominate *atrogularis*. Iraq, Transcaspia (USSR), and north-west India, winter; skins (Ludlow and Kinnear 1933; BMNH, ZMA); wing includes summer data from Yamal peninsula, USSR (Danilov *et al.* 1984). Bill (S) to skull, bill (N) to distal corner of nostril; exposed culmen on average 2·6 shorter than bill (S).

	♂		♀	
WING	74·9 (1·54; 7)	72–76	72·2 (1·55; 7)	70–74
TAIL	60·5 (1·22; 4)	59–62	59·2 (— ; 3)	57–61
BILL (S)	12·6 (0·57; 6)	12·0–13·5	12·3 (0·43; 5)	11·9–13·0

BILL (N)   7·0 (0·24; 4)   6·7-7·3   6·9 (— ; 3)   6·6-7·2
TARSUS   20·2 (1·10; 4)   19·0-21·6   18·9 (— ; 3)   18·6-19·3
Sex differences significant for wing and tarsus.

Iran and Afghanistan, winter: bill (S), ♂ 12·1 (11) 11-14, ♀ 12·2 (10) 11-13 (Marien 1951). Bill to nostril rarely over 7·5 (Dementiev and Gladkov 1954b).

P. a. huttoni. Kirgiziya, September-May; skins (RMNH, ZMA). Exposed culmen on average 3·2 shorter than bill (S).

WING   ♂ 75·4 (1·86; 35)   72-80   ♀ 72·1 (1·76; 12)   69-75
TAIL   63·1 (2·51; 34)   59-68   60·9 (1·87; 12)   57-64
BILL (S)   14·2 (0·66; 37)   13·1-15·6   14·2 (0·60; 12)   13·2-14·8
BILL (N)   7·8 (0·54; 28)   7·0-8·7   7·9 (0·38; 12)   7·4-8·4
TARSUS   20·9 (0·61; 35)   19·6-21·7   19·1 (0·51; 12)   18·8-20·7

Sex differences significant for wing, tail, and tarsus. Adult and 1st adult combined, though retained juvenile flight-feathers and tail of 1st adult significantly shorter than adult, e.g. in ♂: wing, adult 77·0 (1·92; 9) 74-80, 1st adult 74·6 (1·38; 22) 72-77; tail, adult 64·9 (2·32; 9) 62-68, 1st adult 62·3 (2·06; 21) 60-66. Central Asia: bill (S), ♂ 13·8 (0·34; 24) 13-15, ♀ 13·5 (10) 12-15 (Marien 1951). Bill to nostril usually over 7·5 (Dementiev and Gladkov 1954b).

Weights. Nominate *atrogularis*. Yamal peninsula (USSR), summer: ♂ 19 (Danilov et al. 1984). Probably this race, Lake Chany (south-west Siberia), 20 September: 14·6, 16·0 (Havlín and Jurlov 1977).
    P. a. huttoni. Kazakhstan, USSR (possibly including some wintering nominate *atrogularis* in October-April): (1) May-August, (2) October-November, (3) December-April (Dolgushin et al. 1972).

(1)   ♂ 19·0 (1·07; 6)   17·5-20·1   ♀ 18·8 (2·14; 5)   15·5-20·7
(2)   18·5 (2·44; 11)   14·7-23·6   18·6 (2·74; 4)   14·5-20·5
(3)   19·5 (2·04; 8)   16·5-22·5   18·2 (1·19; 4)   16·5-19·0

South-east Kazakhstan, May-June: ♂ 20·2 (9) 19·1-21·4; juvenile ♂ 17·4; juvenile ♀♀ 15·7, 17·9 (Bub et al. 1984). Afghanistan, March: ♂ 20; unsexed 19, 21 (Paludan 1959). India, November-December: ♂ 17·8-20 (5); ♀♀ 17·1, 19·4 (Ali and Ripley 1973b).

Structure. Wing rather short, broad at base, tip bluntly pointed. 10 primaries: p6-p8 longest, equal; in nominate *atrogularis*, p9 and p5 both 2-4 shorter, p4 7-8, p1 14-16; in *huttoni*, p9 3-5 shorter, p5 1-3, p4 6-8, p1 13-16; in both races, p10 reduced, narrow and pointed, 40-44 shorter than wing-tip, 1 shorter to 3 longer than longest upper primary covert. In nominate *atrogularis*, p9 often equal to or longer than p5; in *huttoni*, p9 usually shorter than p5 (Hartert and Steinbacher 1932-8; Ludlow and Kinnear 1933; Dementiev and Gladkov 1954b; Vaurie 1955b; Bub et al. 1984). Outer web of p5-p8 and inner of p6-p9 emarginated. Tail rather short, tip square or slightly forked; 12 feathers. Bill as in *P. modularis* (see p. 559). Tarsus and toes rather short and slender; middle toe with claw 15·7 (2) 15-16·5 in nominate *atrogularis*, 17·5 (12) 16-18 in *huttoni*; outer toe with claw c. 70% of middle with claw, inner c. 73%, hind c. 76%.

Geographical variation. Rather slight; mainly involves bill size, wing formula, and colour pattern of throat. Nominate *atrogularis* from north-west USSR has shorter bill (see Measurements), often more pointed wing-tip (see Structure), and virtually always a whitish gorget round black of throat, separating black from buff of chest and rear of cheeks. *P. a. huttoni* from mountains of central Asia has longer bill, more rounded wing-tip, and black of throat directly bordered by buff extending from chest. Also, tail of *huttoni* on average slightly longer; crown browner (less black than nominate *atrogularis*); dark shaft-streaks on upperparts and flanks slightly darker and broader; ground-colour of upperparts slightly darker; black throat-patch slightly larger; chest deeper and more extensively cinnamon-buff; flanks more distinctly streaked. Birds from Altai mountains slightly paler than typical *huttoni* from Tien Shan and sometimes separated as *menzbieri* Portenko, 1919, tending slightly toward nominate *atrogularis*, but differences very slight and recognition not warranted (Hartert and Steinbacher 1932-8; Vaurie 1955b).
    Forms superspecies with Kozlov's Accentor *P. koslowi* from Mongolia and neighbouring parts of China (Portenko and Vietinghoff-Scheel 1976); *P. atrogularis* and *P. koslowi* form species-group with Siberian Accentor *P. montanella*, Radde's Accentor *P. ocularis*, and Yemen Accentor *P. fagani* (see *P. ocularis*, p. 568).

Recognition. See *P. montanella* (p. 564) and *P. ocularis* (p. 568).   CSR

Prunella collaris Alpine Accentor

PLATES 40 and 41
[between pages 544 and 545]

Du. Alpenheggemus   Fr. Accenteur alpin   Ge. Alpenbraunelle
Ru. Альпийская завирушка   Sp. Acentor alpino   Sw. Alpjärnsparv

Sturnus collaris Scopoli, 1769

Polytypic. Nominate *collaris* (Scopoli, 1769), north-west Africa and western and central Europe, in east south to Italy and northern Yugoslavia (Slovenija and north-east Serbia), east to Carpathian mountains; *subalpina* (C L Brehm, 1831), south-east Europe, from Croatia, southern Serbia, and Bulgaria south to mainland Greece and Crete, also southern Asia Minor east to at least central Taurus; *montana* (Hablizl, 1783), northern and eastern Turkey, Caucasus area, Iran, and Kopet-Dag in Turkmeniya (USSR). Extralimital: *rufilata* (Severtzov, 1879), Afghanistan, Pamirs, and Kun Lun mountains north to Tien Shan; *whymperi* (Baker, 1915), Himalayas from Kashmir to Kumaon; 3-6 further races in central and eastern Asia.

Field characters. 18 cm; wing-span 30-32·5 cm. At least 25% larger and much more robust than any other accentor. Quite strong-billed, bulky accentor, with form somewhat recalling lark *Alauda* or pipit *Anthus*. Plumage patterned as Dunnock *P. modularis* but more colourful, being basically dull blue-grey, with mottled brown back, white-edged black panel across forewing, white-speckled throat, rufous-splashed flanks, and white-tipped dark brown tail. Flight much more free than *P. modularis* and carriage erect. Sexes similar; little seasonal variation. Juvenile separable. 3 races in west Palearctic, becoming greyer eastward across Europe to Asia Minor, but differences slight.

ADULT. Head, breast, and belly ashy-grey, tinged blue or brown according to angle of light, palest on belly and vent and little marked except for black and white speckling of chin and throat. Mantle basically grey-brown, with black-brown feather-centres forming broad streaks; scapulars edged rufous; lower back and rump grey, tinged buff, but upper tail-coverts brown with rufous tinge, streaked black and slightly scaled with white. Tail dark brown, with all but central feathers showing broad white tips on inner webs which appear as terminal band or line of spots in flight. Wings strongly patterned, with (a) grey-brown lesser coverts, usually hidden, (b) black median coverts with small buff-white tips forming thin upper wing-bar, (c) mainly black greater coverts with conspicuous white spots on tips forming prominent dark transverse panel and striking lower wing-bar, (d) black tertials with rufous edges and white tips, (e) black-brown secondaries with rufous edges, forming pale panel when folded, and (f) dark brown primaries with buff tips. Underwing rusty-grey, with faint black and white marks on feathers giving mottled appearance. Flanks and sides of breast bright chestnut, with feathers towards rear of flanks increasingly margined white and darker-centred in oval-scaled pattern. Under tail-coverts black-brown, margined white in conspicuous arrowhead-scaled pattern. JUVENILE. Plumage pattern basically as adult but duller, less ashy-grey above and more uniform below, with buff-grey, unspeckled throat and rest of underparts brown, scaled buff and lacking 2-tone contrast of adult. Both wing-bars buffier, making dark transverse panel of greater coverts less contrasting. FIRST WINTER AND SUMMER. Some retain dull wing pattern of juvenile. Bill blackish, base yellow in adult. Legs red-brown.

Unmistakable in close view, differing distinctly from *P. modularis* in larger size, greyer plumage, and strongly patterned and coloured wings, flanks, tail, and under tail-coverts. Much less distinctive at long range, with pale-speckled throat and chestnut flanks difficult to see; overall pattern and colour then suggest dark Skylark *Alauda arvensis* or pipit *Anthus*, this appearance enhanced by undulating flight, winter habit of feeding in flocks, and rather lark- or pipit-like calls. Important therefore to exclude other sympatric montane passerines by clear view of plumage pattern and colours. Flight more free than any west Palearctic congener, with fluent action and undulating progress. Song-flight short, with upward flutter, brief hover, and downward glide also recalling

lark. Escape-flight short among rocks but longer from sward or scree and directed towards rocks or over ridge. Gait a quick walk varied with little runs and hops. Carriage more erect than *P. modularis*, with head usually held higher, but also adopts lower postures when feeding without quite assuming such mouse-like attitudes as other accentors. Flutters and jerks wings and tail like *P. modularis* or *Saxicola* chat. Rarely perches on low plants but often sings from top of rock. Unobtrusive in breeding season but often obvious and gregarious below snowline in winter.

Song resembles *P. modularis*, but lower pitched and more musical. All calls tend to recall *A. arvensis*: commonest (e.g. when flushed) rather quiet and oddly ventriloquial—a rippled 'tchirririp'; others include a more tinkling variant of this and a shorter, huskier 'churrp' or 'teurrp'.

Habitat. Breeds exclusively in mountain ranges of middle latitudes, from *c.* 1800-2000 m up to snowline; normally up to *c.* 2600-3000 m in Switzerland and in Caucasus, but extralimitally to 4000 m in central Asia and even higher in Himalayas—seen at nearly 8000 m on Mount Everest. Encounters accordingly low temperatures and exposed conditions, although apparently choosing sunniest available sites: patches of alpine grassland strewn copiously with boulders or large stones, often on flat plateaux but sometimes on slopes or scree, and almost always well above treeline and relatively free of shrub growth. Does not normally choose, and may actively avoid, precipitous or broken terrain and neighbourhood of shrubs or stunted trees, but is at home close to old snow patches, and even close to abandoned or occupied mountain huts or places where picnic remains have been left by skiers. A ground-feeder and ground-nester, occasionally perching on low bush or plant-stem, and singing from rock or ground or in low song-flight. In Indian subcontinent, also occurs on cliffs and moraines. In winter, sometimes stays in breeding area but usually shifts to lower slopes, remaining commonly above 1800 m, sometimes entering villages or visiting chalets, hotels, ski-lifts, or stables. Rarely and locally shifts to rocky and scrubby ground in lowlands, or migrates further afield.

Distribution. No evidence of range changes.

FRANCE. Status in Massif Central imprecise (Yeatman 1976). PORTUGAL. May breed, but no proof (RR). TURKEY. Sparsely distributed (Vittery *et al.* 1971). ALGERIA. First proved breeding 1978 (EDHJ).

Accidental. Britain, Netherlands, West Germany, Denmark, Norway, Sweden, Finland, Hungary, Israel, Malta.

Population. Little information on population trends.

FRANCE. Under 10 000 pairs (Yeatman 1976). WEST GERMANY. 700-1200 pairs (Bezzel *et al.* 1980). POLAND.

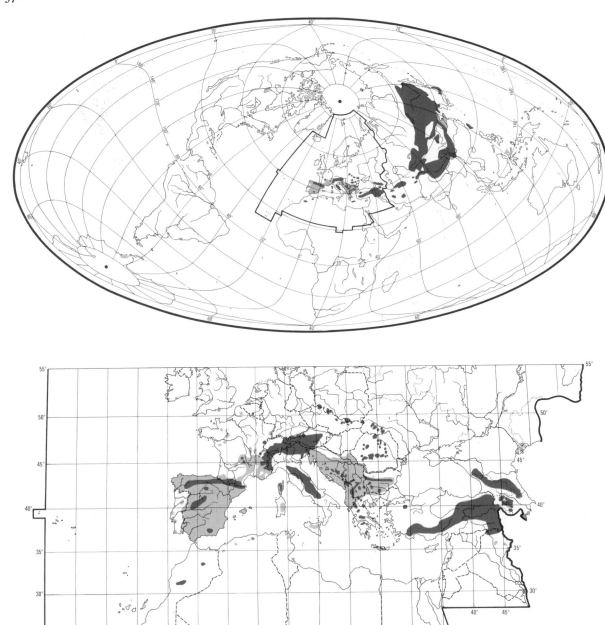

Estimated 230–380 pairs (P Profus). CZECHOSLOVAKIA. Estimated 100 pairs; no changes reported (KH). RUMANIA. Decreased (VC). USSR. Not abundant (Dementiev and Gladkov 1954*b*).

Movements. Resident or subject to local altitudinal or more distant movements.

Most descend in winter below snowline or seek snow-free patches. Wintering in lowland southern Europe apparently widespread though birds presumably sparsely distributed and occurrences perhaps irregular, but no details available. In southern Poland, populations largely move from some mountain systems (Sudety, Babia Gora,

Tatry) to other ranges (Swietokrynskie, Ikerskie) from which absent in summer (Tomiałojć 1976). In France, birds from Alps reach Basse-Provence and Côte d'Azur (Yeatman 1976); thus, Swiss bird ringed as a nestling was found southern France in October (Zink 1975). In Switzerland, many remain in high Alps through winter but most move to low ground, especially near areas of human activity, with local movement due to weather; upland breeding areas start to be fully reoccupied from mid-March, but mainly April. Some birds visit snow-line both before and after breeding, and weak but distinct southerly movement occurs through several passes in October, notably at Col de Bretolet where maximum daily count 18 birds on 12 October (Glutz von Blotzheim 1962). Bird ringed December at Bergamo (northern Italy) recovered 49 km WSW 8 weeks later (Zink 1975). In Greece, birds noted as low as *c.* 1000 m on Mount Ida by early September (Bateson and Nisbet 1961). Regular in autumn on Gibraltar (though birds seldom seen on consecutive days) and occurrence thought to be linked with severe weather in southern Spain (Lathbury 1970). Rare and irregular on Malta (though up to 6 birds recorded together), occurring mainly October-November(-April) (Sultana and Gauci 1982). Scarce but regular winter visitor and passage migrant on Balearic islands, November to March or early April (Bannerman and Bannerman 1983). Birds breeding in Atlas mountains of Algeria and Morocco descend to lower altitudes in winter where numbers probably augmented by birds from Europe (Chaworth-Musters 1939). Scarce but regular, late October to late March, in Tunisia (Thomsen and Jacobsen 1979; A Blanchet). Movements of birds breeding central and eastern Asia not well known but apparently also mainly altitudinal with occasional wandering (Dementiev and Gladkov 1954*b*; Ali and Ripley 1973*b*).

Vagrants occur west and north to Britain and Fenno-Scandia. Less regular now in Britain than previously: e.g. only 1 record in period 1958-72, though total of *c.* 30 before that; mostly August-January, with a few March-June (Sharrock and Sharrock 1976).   JHE, DJB

Food. Largely insects, plus significant proportion of plant seeds. Food taken principally on ground, bird moving with quick small hops or walking quietly. Sometimes makes aerial sallies and will chase active prey on foot. Can hover for short periods. Clings to rock faces like Wallcreeper *Tichodroma muraria* (Corti 1961). While foraging often 'freezes' like Blackbird *Turdus merula*. May clean bill by rubbing either side on vegetation (Dyrcz 1976; A Dyrcz). Forages among rocks, stones, lichens, moss, grassy vegetation, snow fields, and streams (Géroudet 1957*a*). Particularly favours edge of melting snow and pecks actively in water or sodden ground. Splits off moss and lichens from rocks to find insects (Dyrcz 1976; A Dyrcz), and can use broad-based bill to extract food from cracks in rock (Richard 1938). In winter may move to lower altitudes and regularly feeds at farms, barns, gardens, bird tables, balconies, roads, and woodland edges—any place where snow melt is rapid (Corti 1965; Dick and Holupirek 1978). In hard weather will search for seeds in horse manure (Géroudet 1957*a*; Dyrcz 1976; A Dyrcz). Grit regularly taken by adults (Naumann 1823; Fatio and Studer 1907; Jordania 1970).

Favours small insects including adult damselflies (Odonata), grasshopper adults and eggs (Orthoptera), earwigs (Dermaptera), bugs (Heteroptera), adult and larval butterflies and moths (Lepidoptera), adult and larval flies (Diptera), ants (Formicidae), and beetles (Coleoptera: Carabidae, Curculionidae, Scarabaeidae, El-

ateroidea, Cantharoidea, Tenebrionidae). Less commonly takes spiders (Araneae), small snails (Mollusca), round-worms (Nematoda), and earthworms (Oligochaeta). Plant food mainly seeds of alpine plants and grasses, predominantly taken in latter half of summer and during winter: seeds of pines Pinaceae, Scrophularia, primrose *Primula*, and Cornaceae recorded; Leguminosae, Compositae, and Gramineae recorded from outside west Palearctic (Yanushevich *et al.* 1960; Maruyama *et al.* 1972). Also takes cereals, particularly oats *Avena* and wheat *Triticum*. Will eat small leaves and berries. In winter recorded taking berries of mistletoe *Viscum album*. At bird-tables and around human habitation will eat crumbs, scraps, and fat (Naumann 1905; Gil 1928; Witherby *et al.* 1938*b*; Dementiev and Gladkov 1954*b*; Géroudet 1957*a*; Glutz von Blotzheim 1962).

Little information on proportions of different items in diet. 7 stomachs from Italian Alps, April-September, contained berries, small seeds, grain, fly larvae, moth larvae, and many nematode worms *Gordius aquaticus* (Corti 1965). Female from Spain, June, contained mostly grass seeds and roots; also several beetles (*Aphodius*, *Dasytes*) and parts of various grasshoppers (Acrididae) (Gil 1928). 3 stomachs from Switzerland, October and November, contained seeds of Cornaceae and *Scrophularia*, oats, small leaves, beetles, earwigs, ants, and spiders (Fatio and Studer 1907). Also in Switzerland, June-July, stomachs contained small Carabidae, Curculionidae, and Elateridae; 4 adults taken in August contained many spiders. Only one contained seeds—of *Pinus*. 1 juvenile, August, contained seeds of bistort *Polygonum bistorta* and of primrose *Primula minuta* (Naumann 1905). In the central Caucasus (USSR), 4 December-February stomachs contained high proportion of seeds and terrestrial molluscs (Boehme 1958). Two studies from outside west Palearctic give a more detailed break-down of diet. 8 stomachs from Kirgiziya (USSR), August, contained beetles, grasshoppers, bugs, fly larvae and adults, moth larvae, and ants; winter stomachs contained higher proportion of seeds (Leguminosae and Polygonaceae) plus spruce seeds *Picea* and terrestrial molluscs (Yanushevich *et al.* 1960).

Diet of young similar to that of adults: largely insects supplemented by seeds ground up in adult's crop (Géroudet 1957*a*). Nest observations in Switzerland indicated predominance of caterpillars in diet of young nestlings (Géroudet 1957*a*). In Italian Alps, Diptera important, especially crane-flies (Tipulidae) (Corti 1965). Insects predominate in food brought to nests in Karkonosze mountains (Poland): comprised especially dung-flies (Scatophagidae) and crane-flies (Tipulidae); also winged ants, adult and larval moths, crumbs, and berries (Dyrcz 1976; A Dyrcz). Similar results obtained outside west Palearctic. In Yakutskaya ASSR (eastern USSR), nestlings fed exclusively insects (including adult Lepidoptera) and spiders (Vorobiev 1963). 2 well-feathered young, Switzerland, visited on average every 24 min (9–32) over 4-hr period (Praz 1976). 3 11-day-old nestlings, Poland, fed by both parents and 1 helper: visited on average every 9 min over 10½ hrs; of 71 visits, ♀ made 21, ♂ 25, helper 7, 18 unknown). 2 13-day-old nestlings, Poland, fed by both parents and 1 helper: visited every 10·5 min over 9¾ hrs; of 56 visits, ♀ made 44, ♂ 6, helper 2, 4 unknown (calculated from Dyrcz 1977). TAS

Social pattern and behaviour. Most detailed studies by Dyrcz (1976) in Karkonosze mountains (Poland) and by Praz (1976) in Val d'Hérens (Valais, Switzerland). Some references to Dyrcz (1976) in following account include additional data supplied by A Dyrcz.
1. More gregarious than Dunnock *P. modularis* (Witherby *et al.* 1938b). In Karkonosze, frequently 2–4 together at start of breeding season, birds also congregating for feeding some distance from breeding area, then splitting up mainly into pairs on return. Some birds recorded temporarily in small flocks (3–4 birds) during breeding season (see also Bonds and part 2, below). Then, as typically also in other parts of range, family parties and larger flocks occur for feeding in post-breeding period (Dyrcz 1976). Feeding flocks loose-knit, birds not really sociable. Small nomadic flocks may move higher up in late summer (Géroudet 1957a). On west and south-west slopes of Val d'Hérens, 'flock' stable throughout winter, with up to 80 recorded together in favourable feeding areas around cattle sheds, by garden walls, forest edge, etc. Some birds solitary or in twos (perhaps pairs) or threes during day but may rejoin flock at end of day. (Praz 1976.) For Switzerland, see also Schifferli *et al.* (1982). In Rumanian Carpathians, generally in groups of 4–6, but up to 80–120 noted feeding together in late September (Kalabér 1978). Average winter flock size in mountains of Cuenca Etremeña de Guadiana (central Spain) 3 (1–10), *n* = 29 (Ferrero Cantisán *et al.* 1983); see also Anon (1977). In Kirgiziya (USSR), in pairs spring and summer, mostly solitary in autumn, then flocks of 5–12 in winter (Kuznetsov 1962); for densities outside breeding season in creeping juniper *Juniperus* belt of Tien Shan (USSR), see Vtorov (1967). For further reports of flocks (mostly small, though in some cases up to 40–50), see Portenko and Vietinghoff-Scheel (1974), Schmidt (1973) for Budapest (Hungary), Dementiev and Gladkov (1954b), Boehme (1958), and Dolgushin *et al.* (1972) for USSR, Porter *et al.* (1969) for Turkey, Passburg (1966) for Iran, and Smith (1962–4) and Juana and Santos (1981) for Morocco. In Mallorca (Spain), winter, feeding birds associated with Thekla Lark *Galerida theklae* and Chaffinch *Fringilla coelebs* (Watkinson undated); in Iraq, mid-March, with rock sparrows *Petronia* and various Fringillidae (Moore and Boswell 1957). BONDS. Mating system probably essentially monogamous and pair-bond of seasonal duration (Dyrcz 1976; Praz 1976). However, in Switzerland, numerous non-breeders associated intimately with breeding pairs (Praz 1976; Schifferli *et al.* 1982) and, in Karkonosze, ♀♀ recorded copulating promiscuously in flock of 4 birds (Dyrcz 1976). Young fed by both sexes, including for several days after leaving nest. In Karkonosze, 1 additional adult helper recorded at 2 out of 3 nests where birds colour-ringed. 1-year-old ♂ helper (also colour-marked) at nest containing 3 11-day-old young had been ringed as nestling nearby and was perhaps therefore offspring of pair it helped. Other nest with helper contained 2 young 13 days old. Frequent occurrence of threes and fours in breeding season suggests co-operative breeding perhaps more widespread (Dyrcz 1976, 1977). See

also Food for helper's (small) contribution. Families generally stay together rather a long time, e.g. up to late August if not double-brooded (Géroudet 1957a), and one 1st-year bird in Japan stayed with parents until April (Maruyama *et al.* 1972). BREEDING DISPERSION. Solitary and territorial. Pairs in Val d'Hérens several tens of metres apart (Praz 1976). In USSR, solitary pairs of *montana* widely dispersed (Dementiev and Gladkov 1954b). Loose 'colonies' mentioned by (e.g.) Schifferli *et al.* (1982) perhaps better termed neighbourhood groups and, according to Géroudet (1957a), such dispersion probably due to snow conditions which are highly variable from year to year. Boundaries not distinct so that difficult to calculate territory size, but see illustrations in Maruyama *et al.* (1972). In Karkonosze, ♂ apparently has more or less permanent song-posts, but will also move up to several hundred metres to sing. Food for young collected mostly near nest, but sometimes well outside territory—up to several hundred metres away or even more (Dyrcz 1976). General pattern, according to Bergmann and Helb (1982), is for birds to defend small nesting-territory and to feed in large home-range. See also Praz (1976). Little information on densities. In Val d'Hérens, counts include 14 birds (including 5 active pairs) along *c.* 2 km at 2300–2700 m altitude, mid-June, and 18–19 birds (at least 9 active pairs) along *c.* 2 km at 2700–3070 m in early July (Praz 1976). More recent information for Switzerland suggests *c.* 6–8 pairs per km² (Schifferli *et al.* 1982). In creeping juniper belt of Tien Shan, 10 birds per km² (Vtorov 1967). ROOSTING. Nocturnal and, outside breeding season, apparently sometimes communal. In central Spain, winter, flocks and single birds arrived late in evening to roost in holes in walls of buildings (Ferrero-Cantisán *et al.* 1983). At Mount Gellért (Budapest, Hungary), February-March, 4 birds left regular feeding site for (undiscovered) roost *c.* 1 hr before dark at 17.00 hrs; back in area from *c.* 09.30 hrs next day (Szijj 1954). On Fichtelberg (1214 m, East Germany), winter, single bird roosted in sheltered site on building, also resting there during day in snow and wind (Dick and Holupirek 1978). Young return to nest to roost for up to several days after leaving. Birds spend much time self-preening; sun-bathing also recorded, bird lying on side and raising one wing then, after bout of feeding, in similar fashion the other (Dyrcz 1976). Frequent water-bathing by young birds recorded in Nepal, late July (Diesselhorst 1968a).
2. Not especially shy and often allows close approach: e.g. in Nepal, birds running just a few paces from observer and flushing only at closer approach (Diesselhorst 1968a); see also (e.g.) Naumann (1905), Whistler (1926a), Ludlow and Kinnear (1933), Stegmann (1936), and Schmidt (1973). ♂♂ will sing within a few metres of man (Wilson 1887; Géroudet 1957a) and birds can be tame in villages in winter (Schifferli *et al.* 1982). Bird wintering around ski-lift station, East Germany, usually allowed approach to *c.* 3–5 m; when pressed, flew higher up in building or away to small wood (Dick and Holupirek 1978). In Iraq, mid-March, isolated birds also not shy and flushed only short distance; flock of 5 more nervous and rapidly flew out of sight (Moore and Boswell 1957). May move down slope when disturbed, then up again stealthily (Jordania 1970; Mau 1972). In contrast to high-density breeding areas, solitary pairs quiet and secretive, difficult to observe (Praz 1976). Wing-flicking occurs, much as in *P. modularis* (Bannerman 1954), and alarmed bird flying from rock to rock will also flick or twitch tail (Dementiev and Gladkov 1954b; Dolgushin *et al.* 1972). Pair apparently disturbed by observer during bout of courtship moved away, both mock-pecking rapidly and excitedly perhaps as an alarm-reaction (Praz 1976). FLOCK BEHAVIOUR. In Val d'Hérens, large flock split up following attacks by

Sparrowhawks *Accipiter nisus* and Goshawk *A. gentilis*, numerous smaller parties then moving about more and further than hitherto (Praz 1976). Small flock in flight said to 'play' in upcurrents (Naumann 1905). In central Spain, small winter flocks usually dispersed in a line a certain distance apart for feeding (Ferrero Cantisán *et al.* 1983). SONG-DISPLAY. ♂ sings (see 1 in Voice) mainly from rock, stone, edge of rocky shelf, sometimes from low bush or other plant or flat ground, also in flight (Witherby *et al.* 1938*b*; Dyrcz 1976; Bergmann and Helb 1982). Bird singing from rock may adopt more erect posture than normal (Naumann 1905) and move head about in lively fashion (Géroudet 1957*a*). Prior to Song-flight may quiver tail and give Ripple-calls (see 2 in Voice) from rock, also at take-off; then makes steep ascent, giving further Ripple-calls, to *c.* 20–30 m, closes and opens wings, hovering briefly (*c.* 1 s), then descends (while singing) to land silently and slightly higher than take-off point, resting briefly on pile of snow, etc. before ascending again. In Caucasus (USSR), evening Song-flights may continue thus indefatigably until sunset (Stadler 1931; Jordania 1970, 1974; Zhordaniya and Gogilashvili 1976). According to Géroudet (1957*a*), occasional Song-flights short and direct, lacking ascent like lark (Alaudidae); however, other reports refer to high, soaring and hovering flight, with ascent continuing (bird also singing) after soaring and hovering (Stadler 1931; Bergmann and Helb 1982) and some gliding occurs with wings held stiffly up in steep V (Dejonghe 1984). Descent may resemble that of pipit *Anthus* (Stresemann 1943). Sings mostly in early morning and before sunset (see Fig I in Dyrcz 1976) on sunny days, less persistently in bad weather. In fine weather in June starts *c.* 1½ hrs before sunrise and stops *c.* 1 hr after sunset—thus vocally active over *c.* 18 hrs (Dyrcz 1976). Will sing in winter, but main song-period in Europe February to July, with some also in September (Géroudet 1957*a*), e.g. calmly perched in sun (Corti 1959). Song during breeding season includes that from members of small flocks; feeding birds will also give Subsong in October (Dyrcz 1976). Vagrant to East Germany sang from late January (Dick and Holupirek 1978) and in northern Baykal area (eastern USSR) song continues to late July or early August when young out of nest (Stegmann 1936). Song over by end of July also in Nepal (Diesselhorst 1968*a*). In Mongolia, vigorous song given in any weather in late September, even dull, cold, windy conditions; sings only in quiet, sunny and warm weather in October–November, then normally silent in coldest period, resuming from March (Kozlova 1930). In Kazakhstan (USSR), sings in September, also in fine weather at any time in winter (Dolgushin *et al.* 1972). ♀ will also sing (Bannerman 1954), but context not known; one definite record in Polish study (Dyrcz 1976). ANTAGONISTIC BEHAVIOUR. (1) General. In Japan, late August to early November, antagonism perhaps arose from individual-distance encroachments (Maruyama *et al.* 1972); quarrelling in feeding parties also noted by Géroudet (1957*a*). (2) Threat and fighting. Song and fighting recorded between 2 ♂♂ early May, one turning attention to ♀ after other left (Corti 1959); also in Swiss Alps, ♂ seen feeding with pair, without territorial rivalry, though ♀ once chased another ♀ (Bannerman 1954). In Val d'Hérens, ♀ of pair engaged in courtship fled when 3rd bird arrived; after long chase involving 2 ♂♂, presumably 1st of these returned to ♀. Apparent pair (♀ carrying food accompanied by singing ♂) flying over well-populated area caused much excitement: joined in flight by 3rd bird and provoked louder song in 4th (Praz 1976). In Karkonosze, when conspecific intruder appears near nest, 1 of the owners (also giving Subsong) usually makes quick, short attack-flight, driving it off; pursuit normally brief. Chases often

involve 3 birds and pursuer in aerial chase frequently has tail spread. When trio lands, 1 bird will approach another in apparent threat-posture: neck extended forward, neck feathers ruffled, calm song given (Dyrcz 1976). HETEROSEXUAL BEHAVIOUR. (1) General. In Val d'Hérens in fine spring, birds gradually spend more and more time near breeding grounds. In one year, much song and birds apparently already paired by 19 May, while intense activity with pursuits, song, display, and copulation attempts noted on 13 June, so that start of breeding probably well synchronized (Praz 1976). Also in Swiss Alps, mid-June, only occasional Song-flights and ♂♂ ignored soliciting ♀♀ (Bannerman 1954); ♂♂ probably physically ready for copulation only later in season (Aichhorn 1969). Timing very much dependent on snow cover (and late snowfalls): on same slope, early July, some birds preparing for 2nd brood, others 300 m higher up at same stage for 1st (Praz 1976). In Kirgiziya (USSR), pair up in April but no breeding before June (Kuznetsov 1962). In Făgăras massif (Rumania), 2 out of 9 pairs engaged in courtship-display (see below) as late as mid-August (Kalabér 1978) and in Nepal, apparent soliciting behaviour (no response from bird displayed at) noted in small feeding flock, late July to early August (Diesselhorst 1968*a*). (2) Pair-bonding behaviour. 2 birds (perhaps pair) may perform distinctive undulating display-flight and call (see 6 in Voice). Single bird in similar performance made occasional sudden turns and flew in 'closed curve' (Dyrcz 1976). Aerial chases accompanied by calls (see 4 in Voice) and song noted on Crete (Greece), mid-June (Stresemann 1943), were possibly sexual. In Val d'Hérens, 19 May, behaviour much as in winter, but bouts of foraging interrupted by song, only 1 bird singing in trio. Probable pair associated closely later (Praz 1976); pair-members frequently feed together in territory early in season (Dyrcz 1976) and ♂ may chase ♀ (or vice versa) on the ground (Dejonghe 1984). By mid-June in Val d'Hérens, more excited behaviour includes display and chases, with single birds also joining in (Praz 1976). Courtship-display proceeds as follows: ♂ gives mixture of song and Ripple-calls and, with wings drooped and held slightly away from body, plumage ruffled, hops about or circles ♀, body trembling. ♀ may move with short steps, calling quietly to ♂ who (still in posture as described) follows her closely and also calls (Praz 1976; Kalabér 1978; Dejonghe 1984). Courtship resumed after (presumably) same ♂ returned from chasing away interloper and landed close by ♀ (Praz 1976). (3) Courtship-feeding. Nothing recorded. (4) Mating. Much in common with *P. modularis*. Usually takes place on ground and near future nest-site. May occur while pair feeding, ♀ running in front of ♂ and adopting Soliciting-posture (Dyrcz 1976) or follows courtship sequence as described above (Corti 1959; Kalabér 1978). In Soliciting-posture, ♀ flattens and crouches, wings half-spread and quivering; may move thus with short steps and give quiet Ripple-calls (Praz 1976; Kalabér 1978); holds tail closed and steeply raised, moving it rapidly from side to side; ruffles vent feathers, exposing bright rosy-red cloaca which is constantly in motion; may also defecate occasionally (differs from Soliciting-posture of *P. modularis* only in that tail more steeply raised and wings less widely spread). ♂, just behind ♀, stops feeding and stands motionless for several seconds, looking at her (Dyrcz 1976; Fig A); may make jerky head movements back and forth, but *contra* some reports (and unlike *P. modularis*) does not peck at ♀'s cloaca (Aichhorn 1969). ♂ suddenly leaps at ♀'s tail and ♀ usually also leaps away immediately: analysis of film of captive birds showed that ♂ clings to ♀ for only *c.* 0·15 s, one foot on her back, other on her belly, and beating wings for balance. As ♂ pushes off, ♀ almost tips over forward. (Aichhorn 1969; Dyrcz

A

1976.) Extreme rapidity of copulation perhaps facilitated by ♂'s greatly elongated and sinuous sperm duct which hangs down as egg-shaped sac (12 × 8 mm) either side of cloacal aperture and part of whose contents may be expelled immediately when pressed against ♀ (Aichhorn 1969). ♀ quite often rushes forward then doubles back to come in front of ♂ and again to solicit copulation which may take place 3 times in succession and perhaps up to 100 times per day (Aichhorn 1969). ♂ often sings (Subsong mixed with full song) after copulation. Maximum recorded in Polish study 19 copulations in 30 min (Dyrcz 1976); in Rumania, at 3–5 different sites over 17–20 min, each time preceded by courtship-display (Kalabér 1978). In Karkonosze, copulation recorded 19th to 3rd day before laying, also in ♀ still feeding 1st-brood young 19 days old (Dyrcz 1976). (5) Nest-site selection and Behaviour at nest. No information on selection process. ♀ collects material usually several metres from nest, occasionally up to c. 20 m away. In Karkonosze, on 13th day of incubation, ♀ incubated for 392 out of 557 min, feeding away from nest during other 165 min. Longest incubation stint 51 min, longest period off nest 17 min (Dyrcz 1976). Relations within Family Group. In one case, young hatched over 16 hrs—16.00 to 08.00 hrs next day (Dyrcz 1976). Brooded (not known by which parent) for c. 10–15 min per hr during feeding, also at night (Kalabér 1978), but not known for how long overall. In Val d'Hérens, ♀ collecting food joined by ♂ who accompanied her (also singing) when she flew off, but carried no food (Praz 1976). According to Witherby et al. (1938b), young fed (by both adults) with food from crop; not clear, however, to which stage this refers and insects certainly brought in bill (see Food). Young call irregularly between feeds, then noisily when adult arrives (Praz 1976). React at 1–2 days to lights and sounds, raising head, opening bill and calling (Kalabér 1978), also when gape flanges touched (Diesselhorst 1939); otherwise stay flat, crouched low in nest. Faecal sacs carried away some distance (Kalabér 1978). Leave nest at c. 16 days before able to fly far (Géroudet 1957a; Dyrcz 1976). According to Kalabér (1978), can walk and run at 10 days and may perhaps leave earlier than 16 days; normally run off and hide. 1 (out of 2) leaving nest perhaps prematurely hid under stone in grass nearby (Praz 1976). Fed intensively especially during first few days; in one case, up to 12 days after leaving, but also doing some self-feeding at that stage. Begged from adults at 39 days, but not known whether actually fed then (Dyrcz 1976). Anti-predator Responses of Young. Nestlings very lively at c. 10 days, but freeze in any danger (Kalabér 1978); may evacuate nest when disturbed (Richard 1938), but not known at what age. Young out of nest will hide under bushes, amongst rocks; sometimes simply crouch flat and, at c. 16 days, fly off only when about to be seized (Kalabér 1978; also Praz 1976). Parental Anti-predator Strategies. (1) Passive measures. Incubating bird may allow very close approach and can sometimes almost be touched; quick to return if flushed, even if observer c. 5 m away (Kalabér 1978). Relative shyness of birds feeding young varies: e.g. slow to lose shyness and using concealed approach to nest when men 10–15 m away; disappearing at approach of observer and not coming to nest while observer visible at c. 50 m, or waiting motionless on rock until observer hid; recorded feeding young when man c. 1·5–3 m away and staying within c. 1 m while young handled. (Diesselhorst 1939; Dolgushin et al. 1972; Lanz and Wigger 1976; Praz 1976; Kalabér 1978.) (2) Active measures: against man. In Val d'Hérens, ♀ attempting to feed young out of nest moved about c. 6–7 m from observer and called frequently; joined by ♂ who was silent and carried no food. Birds then moved up to c. 100 m away. At a different site, adult (probably ♂) gave alarm-calls (not specified, but see 7 in Voice) from rock c. 30 m away and also later when feeding young after observer had entered hide (Praz 1976). In distraction-lure display of disablement type, ♀ typically flies from nest, lands some distance away, and occasionally extends one or both wings, quivering them as though injured, and also calling (see 7a in Voice). In Savoie (France), bird allowed progressively closer approach, also landed closer to nest, through incubation to 10 days after hatching. Some variation: a) simply landed; b) one or both wings extended; c) 2nd variant combined with rapid opening and closing of bill, bird sometimes calling. Progressed from 1st to 3rd variant with advance of incubation to hatching and beyond. (Barash 1975.) In Switzerland, when young handled for ringing escaped, adult made apparent dive-attack on man, calling (probably shrieking alarm-call: see 7b in Voice) and flying close to his shoulder, then landed and performed distraction-lure display, fluttering about with tail spread until young bird caught; then returned to roof of hut where moved about but silent (Lanz and Wigger 1976). (3) Active measures: against other animals. Stoat *Mustela erminea* elicits frequent alarm-calls (Dolgushin et al. 1972).

(Fig by J P Busby: from photograph in Aichhorn 1969.)

MGW

Voice. Noisy on breeding grounds where density high, also in autumn feeding flocks (Praz 1976). Song (when amplified by echo), also certain calls when heard close to, can be astonishingly loud (Naumann 1905; Géroudet 1957a). Repertoire studied in some detail by Stadler (1931), but multiplicity of other renderings and, in many cases, lack of contextual details make it difficult to assess its true extent; some calls may be duplicated below and scheme therefore provisional. For extra sonagrams, see Bergmann and Helb (1982). Account includes material supplied by A Dyrcz.

Calls of Adults. (1) Song of ♂. Phrases of varying length or several in succession giving a varied, well-sustained, continuous chattering warble at moderate speed; slightly slower than Dunnock *P. modularis* and generally superior to that species, being more developed and musical, lacking its silvery timbre (lower pitched), though sonagrams otherwise quite similar (Stadler 1931; Géroudet 1957a; Bergmann and Helb 1982; A Dyrcz, D I M Wallace). Often likened to Skylark *Alauda arvensis*, but this only a very rough similarity according to Bergmann and Helb (1982); Stadler (1931) found general tonal quality to resemble *A. arvensis*, especially at a distance, but noted narrower frequency range in *P. collaris*. Characteristic components are relatively hard and low-pitched trills or ripples—'brürrr', 'tjürr', 'dire', etc.;

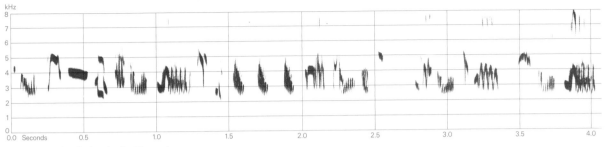

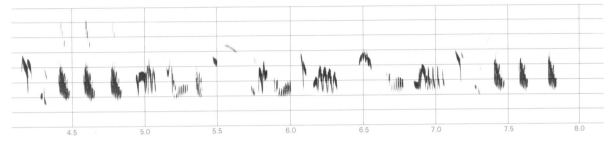

I   C Chappuis   Switzerland   May 1964

I   *cont.*

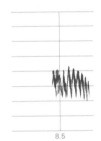

I   *cont.*

such sounds (evidently call 2 variants, some resembling *Stenobothrus* grasshopper) often given at start (overall very variable) of song, sometimes in rapid succession, sometimes separated by pauses, and frequently interpolated later, but generally absent from fragmented songs. Mixed with some fine, more tonal, and to some extent reiterated components (Naumann 1905; Stadler 1931; Bergmann and Helb 1982); such clear, fluting sounds may recall (in volume and timbre) Crested Lark *Galerida cristata* (Naumann 1905, which see for further details, including renderings). End of song rather arbitrary, with sudden break-off in middle of a motif according to Stadler (1931, which see for details, also musical notation). However, typical song-phrase in Nepal said by Fleming *et al.* (1976) to be of 6 notes slurring upward to a final 'tee-dee'; not clear whether this represents regional variation, but resembles, at least in part, rendering of 1 section—'ide ide ildü iii iii i'—by Stadler (1931). Recording confirms song as continuous rich warble of clear, fluty whistles (some thin, melancholy, and trailing away reminiscent of *A. arvensis*) mixed with purring 'prrreerrr', 'chirrr', etc. Fig I shows 3 sections of song,

each ending in group of 3 'tschk' (J Hall-Craggs) or slightly buzzing 'chuch', 'chich' or 'dsche' sounds (M G Wilson) then, after longer pause, call 2 ('chirriree'). Some passages recall *P. modularis*, but richer. For further descriptions and comparisons, see (e.g.) Müller (1929) and Dolgushin *et al.* (1972). Weak Subsong given with bill closed (A Dyrcz); for contexts, see Social Pattern and Behaviour. Quiet variant of song mentioned by Müller (1929) and Géroudet (1957a) perhaps same. Early-September song noted in Austria contained mainly 'trrüi-trrrüill' sounds (Corti 1959). Weak song also given by ♀ (Bannerman 1954). (2) Ripple-call. Commonest and most characteristic call typically rolled or rippling. Often loud, clear and pleasant sounding, though sometimes quieter and, with oddly ventriloquial effect, can be difficult to locate. Number of syllables variable and renderings include 'tchirririRIP', 'truiririp', 'tritritri', 'chirrup', 'datürr', 'tijürr', 'drürr', 'türrr', etc. At times rather like penny whistle but lower pitched (Naumann 1905; Stadler 1931; Witherby *et al.* 1938b; Géroudet 1957a; Jonsson 1982; D I M Wallace). May resemble (in timbre) *A. arvensis* (Stadler 1931), more so at distance (Witherby *et al.* 1938b). Recording (Fig II) suggests

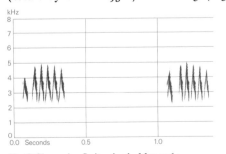

II   C Chappuis   Switzerland   May 1964

'tirrirree', 'prreerrrrr', etc., with individual sub-units delivered very rapidly to give attractive rolled or purred effect; difficult to render satisfactorily, however (M G Wilson), as rolled 'r' runs simultaneously with 'ee' sounds (J Hall-Craggs). Typically given to signal take-off or when flushed (Witherby *et al.* 1938*b*; Géroudet 1957*a*), often in long rapid series, with or without pauses (Stadler 1931). See above for inclusion in song. (3) Renderings such as 'trrui', 'brrui' or 'truit', also 'dri' or 'driet' (mostly given during rearing of young) apparently treated as variants of call 2 by Naumann (1905) and Géroudet (1957*a*). However, 'trüi' starts at higher pitch than remainder of unit, giving it characteristic sharp 't' sound at beginning (this feature shared with call 4—see Fig III—which is deemed closer in type to call 3 than to call 2); also, call 2 usually starts lower than, or on level with, remainder of unit (J Hall-Craggs); 'trüi' not linked with typical call 2 ('drürr') by Bergmann and Helb (1982) which see also for sonagraphic evidence of differences. (4) Very noisy (i.e. not tonal) and coarse, chirruping sounds like sparrow *Passer*: 'zerre', 'zerreze', 'derre dedede' or 'derre dididi', 'schirr', 'dschri dschap'; all listed as flight-calls by Stadler (1931). Ringing disyllabic flight-call noted from vagrant in southern England (James and James 1955) may refer to one of these. A 'veerr-veer', something like very loud chirp of Tree Sparrow *P. montanus*, is 'call-note' (A Dyrcz), but may belong here, though rendering suggests possible relationship with call 2. Following calls separated from call 2 by timbre and (in some cases) brevity and probably best included here: 'tchy' and 'pyrrt' (sometimes combined) more piercing than call 2 (Jonsson 1982); more metallic or huskier 'churrp', 'teurrp' or 'chichurrchurrp' (Witherby *et al.* 1938*b*; D I M Wallace). In recording (Fig III), presumed calls (perhaps part of song) rather dry, quite harsh and explosive 'chich', 'tschk', or 'dsche' sounds (see call 1); patterned in pairs with 1st unit louder than 2nd and 3 unit-pairs per group. Sounds reminiscent of flight-calls of Redpoll *Carduelis flammea*, also when combined (in an apparent chorus) with call 2, though not unlike chattering and chirruping of sparrows (J Hall-Craggs, M G Wilson). Same pairs of units rendered 'zak-zak' by Bergmann and Helb (1982) who also noted a 'tr-TSCHI-zak' like a short song-motif; similarly, apparent combination of calls 2 and 4—'dscherre-dürr'—typically given during chases (Stresemann 1943); in recording, 'dsche' or 'tschk' (Fig IV) is followed by call 2. See Stadler (1931) for song-motifs as calls, and other combinations. (5) (a) A 'zije' or 'züjü', sometimes quite like Bullfinch *Pyrrhula pyrrhula* and far-carrying, given when perched (Stadler 1931). (b) Other (apparently similar) calls separated from call 5a by Stadler (1931): 'tjü', 'tü', 'tjüjü', or 'tüte'; 'tju-tju-tju' slightly like Linnet *C. cannabina* (Jonsson 1982) probably the same, as are perhaps the soft 'thiup' and 'thiuyup' given when perched or on ground (Géroudet 1957*a*). (6) A 'tee-drit tee-drit' given during display-flight (A Dyrcz).

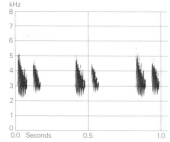
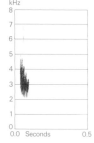

III C Chappuis Switzerland May 1964

IV C Chappuis Switzerland May 1964

(7) Calls given when disturbed. (a) Very weak 'pee-pee-peee-pee-pee' expresses slight astonishment or anxiety (A Dyrcz). High-pitched repetitive call given during distraction-lure display (Barash 1975) probably the same or related. (b) A shriek something like an alarm-call of Blackbird *Turdus merula*, but much weaker is main alarm-call (A Dyrcz), though Fleming *et al.* (1976) noted a high-pitched 'tsip' in this context in Nepal. (8) Feeding-calls. A 'tchick-tcheeck-tchick' or 'pli-pli-pli-pli' very like weak Bubbling-call of ♀ Cuckoo *Cuculus canorus* (A Dyrcz). (9) Mild 'grweck-grweck-grweck' used by ♀ in contact with fledged young (A Dyrcz). (10) Other calls. (a) Impure 'doide' given in flight (Stadler 1931). (b) An 'iredI' and 'id' like *A. arvensis* (Stadler 1931), according to same author, 'dridlit' mentioned by Michel (1917) probably the same; see, however, call 6.

CALLS OF YOUNG. Very weak twitter given in first few days of life (A Dyrcz);' quiet 'tschib' when being fed (Michel 1917) probably the same. Develops into sharp, chirping 'tcheet-tcheet' given by older nestling or fledgling on seeing adult with food. Loud single 'tcheerrit' also noted (A Dyrcz), but context not known. Young rather noisy generally and calls amplified by rocks carry far (Géroudet 1957*a*). Twittering song of bird less than 1 year old contained distinct 'trüi' sounds (Dick and Holupirek 1978).   MGW

Breeding. SEASON. Pyrénées and Alps: see diagram. Little apparent variation across range (Dementiev and Gladkov 1954*b*; Makatsch 1976). SITE. In crevice in cliff or rocks, or between boulders. Nest: loosely made of grass leaves and stems, with neat inner cup of moss lined with hair and feathers. Base of nest 17 × 12 cm, height 8 cm, inner diameter 8 cm, depth of cup 5·3 cm (Richard 1938). Building: ♀ reported collecting material (Dyrcz 1976); no further information. EGGS. See Plate 82. Sub-elliptical, smooth and glossy; uniform pale blue. Nominate *collaris*: 23·0 × 16·6 mm (20·5-26·7 × 15·0-17·1), $n = 70$; calculated weight 3·38 g. *P. c. subalpina*: 23·6 × 16·8 mm (22·7-26·4 × 16·4-18·4), $n = 12$; calculated weight 3·57 g (Schönwetter 1979). Clutch: 3-4 (-6). 2 broods. Eggs laid daily. INCUBATION. (11-)14-15 days. By both sexes. (Makatsch 1976; Kalabér 1978.)

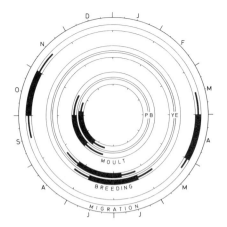

YOUNG. Cared for and fed by both parents. FLEDGING TO MATURITY. Fledging period *c*. 16 days or more; young leave nest before able to fly far. No further information.

Plumages (nominate *collaris*). ADULT. Forehead and crown dull grey, feathers slightly tinged olive-brown and with indistinct dull black shaft-streaks; hindneck, sides of neck, and upper mantle purer grey, more faintly streaked dusky. Lower mantle, scapulars, and back grey with distinct olive-brown or olive-buff tinge and broad but ill-defined dull black feather-centres, outer webs of outer scapulars extensively tinged rufous; rump like back, but feather-centres faintly tinged dusky only, largely olive-grey or olive-buff; upper tail-coverts greyish olive-brown with black shaft-streak, broad white tip, and narrower black subterminal bar. Sides of head dull grey, finely speckled off-white on lores, cheeks, and round eye, tinged olive-brown on ear-coverts; indistinct superciliary stripe from behind eye slightly purer grey. Chin and throat white with narrow black bars, pattern gradually merging into speckled cheeks, but sharply divided from uniform grey chest; amount of barring on chin and throat rather variable: in some, each feather with double black bar, throat appearing fully and closely barred; in others, mainly white with single terminal bar or spot only, showing as broken bars or irregular spotting. Chest and sides of breast dull grey with slight olive-brown tinge, some feathers sometimes having faint dusky arc at tip; sides of breast and lower chest partly tinged rufous, merging into rufous-cinnamon on flanks and lateral upper belly. Feathers of flanks broadly fringed white at sides, rufous centres grading to rufous-brown and olive-black on lower flanks. Central belly grey or buffish-grey, each feather with white tip and dull black or grey subterminal arc; vent and thighs pale grey or off-white with indistinct dusky marks. Under tail-coverts white with contrasting olive-black centres. Tail-feathers black-brown with narrow grey edges; tips of outer webs with indistinct grey-brown spot, tips of inner webs with large and contrasting pink-buff or white semi-circular spot (less distinct on innermost pair). Primaries greyish-black, gradually paler grey towards bases of inner webs; outer webs of inner primaries narrowly edged rufous-brown, of outer primaries pale buff. Secondaries and tertials black, base of outer web broadly fringed rufous, fringes narrower distally but widening to form rufous patch on tip; inner webs of secondaries with broad and poorly defined buff fringe, of tertials with sharper deep rufous fringe. Lesser upper wing-coverts dull olive-grey; median coverts contrastingly black with narrow white tip, tending to form triangular spot on shaft; greater coverts black with rufous olive-grey base of outer web

(most extensive on inner coverts) and triangular white spot on tip of outer web. Lesser upper primary coverts spotted black-and-white, greater primary coverts and feathers of bastard wing black with short white streak on tip of outer web, coverts with olive-grey outer fringe. Under wing-coverts and axillaries grey with buff tinge, feathers with white tip and often narrow dusky subterminal arc. *In worn plumage*, upperparts duller, browner, and more heavily streaked; lower mantle and scapulars often largely black-brown; black bars or spots on chin and throat partly abraded, chin and throat sometimes mainly white with poorly defined grey mottling; rufous of breast and flanks bleached to cinnamon-buff, dusky arcs on central belly abraded; rufous fringes and tips of secondaries and tertials bleached and largely abraded; white spots on tips of upper tail-coverts, tail-feathers, and median and greater upper wing-coverts, and short streaks on greater upper primary coverts smaller due to abrasion, sometimes almost absent. Sexes similar, but ♀ slightly duller and less contrastingly coloured, rufous on underparts on average slightly paler and less extensive; see also Measurements. NESTLING. Down long but scanty, on upperparts and thighs only; dark grey. JUVENILE. Forehead, crown, hindneck, and sides of head and neck dull grey with brown tinge on feather-tips; not unlike adult, but feathers softer, browner, and less streaked. Mantle, scapulars, and back to upper tail-coverts streaked cinnamon-buff and black or black-brown, outer scapulars slightly deeper rufous-cinnamon and more sharply streaked, rump and upper tail-coverts less distinctly streaked. Underparts quite different from adult: basically cinnamon-buff with narrow and poorly defined grey shaft-streaks; flanks slightly deeper rufous-cinnamon and more contrastingly streaked with black; central belly, vent, and under tail-coverts paler cream-buff, streaks pale grey and very faint. Tail and flight-feathers as adult. Lesser upper wing-coverts like upperparts; median coverts and tertials black-brown with smaller rufous-buff spot on tip of inner web and larger one on tip of outer web (soon partly bleaching to white); greater coverts similar, but spot on tip of inner web reduced or absent. FIRST ADULT. Like adult, but juvenile tail, flight-feathers, primary coverts, and greater coverts retained. Tail, flight-feathers, and primary coverts similar to adult, but more worn than adult at same time of year (tips of adult primaries and tail usually have smooth edges up to about March–April, those of juvenile frayed by November–December). Greater upper wing-coverts mainly as in adult, but differ in shape, size, and colour of pale spots on tips of outer webs: in adult, spots small and white, often ending in sharp point along shaft; in 1st adult, spots larger and more rounded, cream but tinged buff or rufous at border with black. In not-too-worn nominate *collaris*, length of spot (measured along shaft of *c*. 3rd outermost covert) 2·4 (7) 2·2–2·8 mm in adult, 3·6 (10) 3·0–4·5 mm in 1st adult; in *subalpina*, *c*. 1 mm longer at all ages.

Bare parts. ADULT. Iris hazel-brown, brown, or red-brown. Upper mandible and tip of lower mandible brown-black or horn-black, basal cutting edge of upper mandible and basal and middle portion of lower mandible yellow. Leg and foot pink-brown or reddish-brown, sometimes with yellowish soles. NESTLING. Mouth bright red (Kalabér 1978); 2 oval black spots on base of tongue. Gape-flanges white. JUVENILE. Iris grey-brown or brown. Bill dark horn, paler below, pinkish at gape. Mouth orange with 2 oval black spots on base of tongue. (Hartert 1910; Witherby *et al.* 1938b; Dementiev and Gladkov 1954b; Ali and Ripley 1973b; RMNH.)

Moults. ADULT POST-BREEDING. Complete; primaries descendant. In nominate *collaris*, August–November (Witherby *et al.* 1938*b*); in *subalpina*, August–September (Stresemann 1920). In *montana*, no flight-feather moult in July, but one had just started (with body) on 16 July, USSR (Dementiev and Gladkov 1954*b*), another (with primaries) on 28 July on Ulu Dag, north-west Turkey (ZFMK). Most birds in moult August, with last ones starting mid-August; tail moult rapid; all plumage new in October and thus moult probably completed September (Dementiev and Gladkov 1954*b*). In *whymperi*, no moult until mid-August, but moult well-advanced in one from 14 September (Marien 1951). Further east, in Nepal, no moult by 18 August, but one had started 24 August (Diesselhorst 1968*a*); in Mongolia, moult well-advanced in 2 birds from 2 August (Piechocki and Bolod 1972). POST-JUVENILE. Partial; juvenile flight-feathers, tail, greater upper primary coverts, and some or all greater coverts retained, but apparently no tertials or tertial coverts. July–October (Witherby *et al.* 1938*b*) or July–August (Stresemann 1920); in specimens examined, first bird with moult completed was from 6 September, last fully juvenile one late August. In 8 birds from north-east Turkey, 25 August–23 September, head and neck in heavy moult but body largely completed (ZFMK). Moult apparently sometimes continued in winter, as 2 birds from February–March had middle greater coverts contrastingly newer than innermost, only 3–4 outermost still juvenile. (RMNH, ZMA.)

Measurements. ADULT, FIRST ADULT. Nominate *collaris*. Spain, Switzerland, Austria, and Italy (RMNH, ZFMK, ZMA), and southern Poland (Dyrcz 1976), all year; skins. Bill (S) to skull, bill (N) to distal corner of nostril; exposed culmen on average 4·1 less than bill (S).

WING	♂ 104·6 (2·85; 18)	101–110	♀ 97·4 (1·60; 10)	95–100
TAIL	65·1 (2·65; 17)	60–70	58·7 (0·93; 7)	55–64
BILL (S)	17·4 (0·38; 13)	16·8–18·0	17·4 (0·14; 4)	17·3–17·6
BILL (N)	9·6 (0·46; 14)	9·0–10·3	9·7 (0·54; 4)	9·4–10·5
TARSUS	25·4 (0·68; 15)	24·4–26·6	24·8 (0·44; 4)	24·4–25·4

Sex differences significant for wing and tail.

Range of wing in ♀ 95–103 (Hartert 1910). Wing, Pyrénées and north-east Spain: ♂ 98–107 (7), ♀ 95–104 (Jordans 1933). Wing, Sierra Nevada (Spain): ♂ 99–105 (8), ♀♀ 95, 100 (Niethammer 1957). Wing, Algeria: ♂ 104, ♀ 97 (Hartert and Steinbacher 1932–8).

P. c. subalpina. Yugoslavia, Bulgaria, Greece, and Crete, all year; skins (Stresemann 1920; BMNH, RMNH, ZFMK, ZMA).

WING	♂ 103·1 (2·20; 17)	100–107	♀ 97·6 (1·96; 17)	94–100
TAIL	63·6 (3·08; 9)	61–69	60·2 (3·31; 12)	55–66
BILL (S)	17·7 (0·64; 9)	17·4–18·3	17·7 (0·34; 9)	17·4–18·3
BILL (N)	9·9 (0·47; 9)	9·4–10·4	9·8 (0·46; 9)	9·0–10·3
TARSUS	25·8 (0·40; 9)	25·3–26·4	25·6 (0·65; 9)	24·6–26·6

Sex differences significant for wing and tail.

Taurus mountains (Turkey), January, ♀♀: wing 95·7 (3) 93–98, bill (S) 17·5 (3) 17·4–17·6, bill (N) 9·8 (3) 9·4–10·2, tarsus 24·6 (3) 24·4–24·7 (BMNH).

P. c. montana. Caucasus, north-east Turkey, northern Iraq, and northern Iran, all year; skins (Stresemann 1928; BMNH, RMNH, ZFMK, ZMA).

WING	♂ 103·4 (2·14; 18)	100–107	♀ 97·1 (2·04; 16)	93–102
TAIL	65·8 (3·26; 13)	61–70	62·3 (2·87; 14)	55–66
BILL (S)	17·1 (0·44; 9)	16·3–17·6	16·4 (0·40; 9)	16·0–17·2
BILL (N)	9·8 (0·40; 9)	9·1–10·3	9·0 (0·41; 9)	8·5–9·5
TARSUS	25·0 (0·60; 9)	24·3–25·9	24·2 (0·46; 9)	23·6–24·8

Sex differences significant.

Ulu Dag (north-west Turkey), July, ♀: wing 97·4 (3·20; 4) 94–101, bill (S) 16·9 (0·48; 4) 16·5–17·3, bill (N) 9·5 (0·28; 4) 9·2–9·8, tarsus 24·0 (0·76; 4) 23·3–24·8 (ZFMK).

P. c. rufilata. Afghanistan (Paludan 1959) and USSR (Marien 1951; Dementiev and Gladkov 1954*b*; RMNH, ZMA).

WING	♂ 100·1 (—; 11)	95–105	♀ 95·5 (—; 8)	89–102

P. c. whymperi. Kashmir and Punjab (Marien 1951; ZMA).

WING	♂ 97·9 (2·66; 12)	93–102	♀ 92·6 (1·12; 8)	91–94
TAIL	64·8 (2·95; 13)	61–69	60·6 (1·11; 7)	59–62
BILL (S)	15·7 (0·54; 13)	15·0–16·8	15·6 (0·46; 8)	15·0–16·3

Weights. Nominate *collaris*. Switzerland: 32·9–40·6 (16) (Bub *et al.* 1984). Southern Poland, May–June: ♂♂ 40, 43, ♀ 43 (Dyrcz 1976). Sierra Nevada (Spain), May: ♂ 38·5–43 (8), ♀♀ 36·5, 40 (Niethammer 1957). For growth curve of nestlings, see Dyrcz (1976): maximum of 38–40 reached on 12th day.

P. c. subalpina. Greece, June: ♂♂ 39, 40, ♀ 42·5 (Niethammer 1943).

P. c. montana. South Iran, early summer: juvenile ♂ 33 (Desfayes and Praz 1978).

P. c. rufilata. Eastern Afghanistan, June: ♂♂ 29, 32; ♀♀ 28, 29 (Paludan 1959). Kazakhstan and Kirgiziya (USSR), all year: ♂ 31·1 (3·43; 9) 25–35, ♀ 30·0 (3·62; 5) 25–34 (Yanushevich *et al.* 1960; Dolgushin *et al.* 1972).

Structure. Wing rather long, broad at base, tip bluntly pointed. 10 primaries: p7–p8 longest, p9 1–4 shorter, p6 1–3, p5 3–8, p4 11–17, p3 16–21, p1 22–28. P10 reduced, rather narrow; 59–69 shorter than wing-tip, 1–6 shorter than longest upper primary covert. Tertials short. Outer web of p5–p8 and inner of (p6–)p7–p9 emarginated. Tail rather short, tip square or slightly rounded; 12 feathers. Bill short, straight, tapering abruptly to sharp point; base wide and rather flattened, tip laterally compressed. Nostrils slit-like, covered by membrane above. A few fine bristles along base of upper mandible and on chin. Leg and toes rather short, but strong. Middle toe with claw 21·2 (11) 20–22; outer toe with claw *c.* 73% of middle with claw, inner *c.* 72%, hind *c.* 78%. Claws rather strong, decurved.

Geographical variation. Rather slight in size, pronounced in colour (but slight in west Palearctic). Generally, smaller towards east, with *whymperi* of Kumaon (India) and birds from eastern Asia smallest, but difference slight. In west Palearctic, bill and tarsus of *subalpina* distinctly larger than *montana*; nominate *collaris* intermediate. Colour gradually darker towards east. *P. c. subalpina* from Balkans close to nominate *collaris*, but upperparts purer and paler grey, less tinged olive and buff, narrower and more sharply streaked dusky on mantle and scapulars, paler rufous or olive-brown on outer scapulars; chest paler and purer grey; flanks more narrowly streaked with paler rufous; throat-patch less heavily barred, chin often almost uniform white. *P. c. montana* from north-east Turkey, Caucasus area, and Iran rather olive-grey on upperparts, with distinctly buff mantle and scapulars; like nominate *collaris*, but crown and hindneck more uniform olive-grey and streaks on mantle and scapulars olive-brown rather than black, less contrasting; rufous on flanks deep, as in nominate *collaris*, but more extensive, invading sides of breast, lateral belly, lower flanks, and under tail-coverts. Birds examined from Ulu Dag (north-west Turkey) too heavily worn to be certain of colour, but nearer *montana* in size and hence provisionally included in that race; see also Kumerloeve (1961). Birds from Taurus mountains (southern Turkey) inseparable in colour and size from *subalpina*, and birds from Crete also similar to *subalpina*. Of extralimital races, *rufilata* from western mountains of central Asia similar to *montana*, but rufous on underparts more uniform, less streaked white; *whymperi* from western Himalayas is like *rufilata*, but grey of head, upperparts, and chest distinctly darker, black-and-white throat contrasting more with cheeks and chest. Birds from eastern Himalayas darker still, and those from eastern Asia combine dark grey head, upperparts, and chest with deep chestnut upper tail-coverts, outer scapulars, and belly; throat-patch rather narrow. Racial allocation of some west Palearctic populations not certain. Birds from Algeria

separated as *nigricans* Heim de Balsac, 1925, and considered close to *subalpina* by Hartert and Steinbacher (1932–8), but included here in nominate *collaris* following Vaurie (1955*b*). Confirmation also required of position within nominate *collaris* for birds from Morocco, Corsica, and central and southern Italy, and within *montana* for birds from south-west and southern Iran.

Forms species-group with Himalayan Accentor *P. himalayana*, these 2 sometimes separated from other Prunellidae in genus *Laiscopus* on account of larger size and different plumage pattern.   CSR

Family TURDIDAE chats, thrushes

Small to medium-sized oscine passerines (subfamily Passeres); of several adaptive types, feeding on invertebrates (often spiders, worms, and snails as well as insects) and fruits (often berries). About 300 species, most occurring in woodland, parkland, and scrub, though many feed on ground and some are entirely terrestrial; a few closely attached to water. Almost cosmopolitan in distribution but with largest diversity in tropical Africa and Asia. Many species migratory. Well represented in west Palearctic.

Turdidae here treated as comprising 2 subfamilies: Turdinae (chats, thrushes) and Enicurinae (forktails). Latter wholly extralimital and not considered further. Following Voous (1977), solitaires *Myadestes* excluded from family (but see further, below). Within Turdinae, chats sometimes considered to form separate tribe ('Erithacini' or 'Saxicolini') from true thrushes ('Turdini').

Chats placed in *c.* 33 genera, including: (1) *Cercotrichas* (bush robins), 10 species; (2) *Brachypteryx* (shortwings), 6 species; (3) *Erithacus* (typical robins), 3 species; (4) *Luscinia* (nightingales, bluethroats, etc.), 12 species; (5) *Tarsiger* (bush robins), 4 species; (6) *Cossypha* (robin chats), 13 species; (7) *Copsychus* (magpie-robins, shamas), 8 species; (8) *Irania* (White-throated Robin *I. gutturalis*), monotypic; (9) *Phoenicurus* (redstarts), 11 species; (10) *Sialia* (bluebirds), 3 species; (11) *Cercomela* (rock chats), 9 species; (12) *Saxicola* (stonechats), 11 species; (13) *Myrmecocichla* (ant chats), 9 species; (14) *Oenanthe* (wheatears), 19 species; and (15) *Saxicoloides* (Indian Robin *S. fulicata*), monotypic. Majority of chats occur in Old World (87 species in Eurasia, 77 in Afrotropics), sole endemic representatives in New World (if *Myadestes* excluded) being bluebirds *Sialia*—though Bluethroat *L. svecica* and Wheatear *O. oenanthe* have now extended their breeding ranges into northern North America. 35 species in west Palearctic: 4 accidental, 1 migrant only (Eversmann's Redstart *P. erythronotus*), and remainder breeding.

Thrushes placed in *c.* 15 genera, including: (1) *Monticola* (rock thrushes), 12 species; (2) *Myiophoneus* (whistling thrushes), 7 species; (3) *Zoothera* (mountain thrushes,

ground thrushes), 26 species; (3) *Hylocichla* (Wood Thrush *H. mustelina*), monotypic; (4) *Catharus* (nightingale-thrushes), 10 species; and (5) *Turdus* (typical thrushes), 59–62 species. Thrushes not as exclusively in Old World as chats: *c.* 55 species in Eurasia, 21–25 in Afrotropics, *c.* 46 in Americas, and a few on oceanic islands. 20 species in west Palearctic: 10 accidental, rest breeding.

Body compact in most species; neck short. Sexes generally of similar size. Bill usually slender but fairly stout in some species; of medium length. Used by some species (e.g. Blackbird *T. merula*) for flicking aside debris when searching for food. Nasal and rictal bristles usually present though nasal bristles often poorly developed. Wing shape varied, from short and rounded to long and pointed; 10 feathers, p10 often short. Flight varied, from weak to strong, straight to undulating. Tail of medium length to fairly long in many species but short in some (e.g. *Brachypteryx*) and very long in others (e.g. some *Copsychus*); mostly square or rounded; typically 12 feathers. Leg and foot usually strong; tarsus typically quite long and 'booted' (with scutes fused into single smooth sheath). Foot used in feeding by a few species at least (e.g. *T. merula*, which scratches backwards with one foot to uncover food hidden in earth or ground litter). Gait a run and/or hop. Head-scratching by indirect method. Bathe in shallow water, using typical stand-in method. Sunning frequently observed, birds adopting lateral and spread-wing postures at higher intensities. No reports of dusting. Anting recorded from several species, both chats (*Erithacus*, *Luscinia*, *Cossypha*, *Copsychus*), using direct method (ant-application), and thrushes (*Myiophoneus*, *Hylocichla*, *Catharus*, *Turdus*), using both direct method and indirect (ant-exposure); also from *Myadestes*.

Plumage generally cryptic and often spotted below in larger thrushes *Zoothera*, *Hylocichla*, *Catharus*, and *Turdus*; much brighter in some thrushes (e.g. *Monticola*) and in many chats, especially on side of head, underparts, and tail where often a contrasting pattern. Occurrence of sexual dimorphism highly variable: within many genera, sexes alike in some species, dissimilar in others, with ♀

the duller, more cryptic sex—though tendency for some adult ♀♀ to resemble 1st adult ♂ in many dimorphic species. Polymorphism occurs, in both sexes of some *Oenanthe*. Juvenile plumage always cryptic and typically spotted. Adults usually have single complete moult annually in autumn, but several species have additional partial moult in spring—occurrence varying greatly between species, even within same genus. Juveniles moult in autumn, usually partially but completely in some tropical species. Nestling typically covered in long, scanty down but naked in a few species; mouth without contrasting spots.

Chats and thrushes share a distinctive feature of syringeal musculature—so-called 'turdine thumb'—found otherwise, so far as known, only in Muscicapidae (Old World flycatchers) (Ames 1975). Most members of Turdidae also well-characterized by booted tarsus, spotted juvenile plumage, nasal bristles, and double humeral fossae. One or other of these features, however, absent in some species traditionally included in Turdidae and, conversely, shared by species of other families, e.g. Muscicapidae, Timaliidae (babblers), and Sylviidae (Old World warblers). Some thrush-like species thus difficult

to allocate. In view of possible close relationship of these and 9 other similar taxa, all sometimes united (as subfamilies) within an enlarged 'Muscicapidae' comprising *c.* 1430 species (e.g. Bock and Farrand 1980). Egg-white proteins of Turdidae, however, rather different from those of Sylviidae and Timaliidae, while the two latter taxa appear to be closely related (Sibley 1970)—as already suggested by Delacour.(1946). This position also now supported by DNA data which indicate that, within assemblage 'Muscicapae', Turdidae and Muscicapidae are closely related to each other (in superfamily 'Turdoidea') but not to Sylviidae and Timaliidae (in superfamily 'Sylvoidea'), and that the enlarged 'Muscicapidae' of earlier authors—as endorsed by (e.g.) Hartert (1910) and Mayr and Amadon (1951)—is a polyphyletic assemblage including members of all 6 superfamilies of Passeres as defined by the DNA-DNA hybridization method (Sibley and Ahlquist 1985a). DNA data also indicate that, within an enlarged 'Turdidae', chats (including *Myadestes*) are more closely related to Muscicapidae than to true thrushes (Sibley and Ahlquist 1980).

Cercotrichas galactotes Rufous Bush Robin

PLATE 42
[between pages 544 and 545]

Du. Rosse Waaierstaart   Fr. Agrobate roux   Ge. Heckensänger
Ru. Тугайный соловей   Sp. Alzarola   Sw. Trädnäktergal

Sylvia galactotes Temminck, 1820. Synonyms: *Agrobates galactotes*, *Erythropygia galactotes*.

Polytypic. Nominate *galactotes* (Temminck, 1820), south-west Europe, North Africa, Sinai, Israel, Jordan, and southern Syria; *syriacus* (Hemprich and Ehrenberg, 1833), south-east Europe, west and central Turkey, and Levant south to Lebanon; *familiaris* (Ménétries, 1832), Transcaucasia and Iraq east to Iran, Afghanistan, and Kazakhstan (USSR), in west intergrading into *syriacus* in western Transcaucasia, south-east Turkey, and probably north-east Turkey and northern Syria; *minor* (Cabanis, 1850), southern Sahara from Sénégal to northern Somalia, north to central Mauritania, Tibesti (Chad), and perhaps Ahaggar (Algeria). Extralimital: *hamertoni* (Grant, 1906), eastern Somalia.

Field characters. 15 cm; wing-span 22-27 cm. 10% smaller than Nightingale *Luscinia megarhynchos*, with rather slimmer form ending in long, fan-shaped tail. Medium-sized, strong-billed, long-tailed, and sprightly chat, with posture frequently recalling Wren *Troglodytes troglodytes*. Plumage essentially bright rufous- to grey-brown above and buff-white below, with obvious pale supercilium, double wing-bar, and diagnostic orange-rufous tail tipped black and white. Flight chat-like in action but silhouette recalls large warbler. Sexes similar; no seasonal variation. Juvenile separable. 4 races in west Palearctic, forming 2 easily distinguishable pairs.

ADULT. (1) Iberian, North African, and Israeli race, nominate *galactotes*. Upperparts and wings rufous-brown, becoming more rufous on rump and even brighter red-chestnut on tail, of which all but central feathers show subterminal black marks and (towards sides) increasingly obvious and striking white tips above and below. Head conspicuously marked with long, cream supercilium and eye-ring, brown eye-stripe, cream fore-cheeks, and pale brown rear cheeks. Wings marked by 1 faint wing-bar (formed by narrow pale rufous tips to median coverts) and 1 strong one (broad rufous-white ends to greater coverts), dark brown tertials fringed rufous to cream, and dark brown wing-point. Chin and throat almost white but rest of underparts and underwing sandy-white, with slightly more pink- or rufous-brown wash on breast and flanks. (2) South Saharan race, *minor*. Resembles nominate *galactotes* in colour but upperparts pink-brown rather than rufous-brown; also distinctly smaller, with 10% shorter wings and bill. (3) Races of south-east Europe, Turkey, and Lebanon eastwards, *syriacus* and *familiaris*.

Differ markedly from first 2 races in grey-brown upperparts and less warmly suffused underparts, with fully rufous (and thus more contrasting) rump and tail shown by *syriacus* but becoming paler in eastern race (*familiaris*). At close range, both these races also show (a) increased contrast between whiter eye-ring and supercilium and dark stripe on edge of crown and darker eye-stripe and (b) deeper black subterminal band on tail. JUVENILE. All races. Closely resembles adult, but paler (sandier or greyer) above and on flanks, with faintly sand-speckled or olive-mottled throat, breast, and fore-flanks. Indistinguishable from adult after autumn moult. At all ages, bill brown-horn, with dull flesh base to lower mandible; legs bright pale brown to grey-flesh.

Unmistakable in west Palearctic but liable to confusion with several congeners in African winter quarters. Flight rather less flitting than more typical chats (e.g. nightingales *Luscinia*, Robin *Erithacus rubecula*), with flutter and dash more reminiscent of large warbler (e.g. Great Reed Warbler *Acrocephalus arundinaceus*). However, any confusion with warbler quickly dispelled by instant assumption when landed of postures and wing and tail movements strongly recalling *Luscinia* and even *T. troglodytes*: most characteristically, cocks tail right up over back, producing almost U-shaped bird set upon rather long legs; in excitement also bobs head and swivels body, and other actions include drooping and forward-flicking of wings, and tail-fanning and -wagging. Gait hopping, accompanied by tail-jerks—or may keep tail cocked while hopping. Though often skulking, not really shy, feeding in the open near cover and perching in exposed situations to sing. Often forms loose groups on migration and when newly arrived on breeding grounds.

Song rich, varied, and beautiful, but with phrases disjointed: combines series of clear ringing notes recalling lark, pulsing notes suggesting Nightingale *Luscinia megarhynchos*, and murmuring warble; usually delivered from elevated perch or butterfly-like, parachuting display-flight. Commonest call a hard 'teck teck' or 'chack chack'.

Habitat. Breeds in dry middle and lower middle latitudes, in Mediterranean, steppe, and desert fringe zones, above 25°C July isotherm, mainly in lowlands (Voous 1960). In north-west Africa, only natural habitat in uplands is in tamarisks *Tamarix* and vegetation bordering wadis; very common in Cañon du Dra (Morocco) in jungle of tamarisks and other shrubs growing beside stagnant water, and fairly common in wooded wadi beds. Not attracted to natural maquis and forest, and avoids both mountains and bare plains. More attracted by man-made habitats such as parks, orange groves, gardens, and groups of prickly pear *Opuntia*; in steppes, favours areas planted with bushes and trees, especially *Pistachia*, *Ziziphus*, *Rhus*, and *Retama*; on edge of and within Sahara, uses marginal cultivation and oases (Heim de

Balsac and Mayaud 1962). Often associated with man in North Africa, living near houses, e.g. on cultivated ground enclosed by dead brushwood (Etchécopar and Hüe 1967). In Lebanon, found in coastal strip, lower hills, and river gorges, in bushes, plantations, groves, and arid scrub (Benson 1970). At Azraq (Jordan) breeds in oasis and in border zone of tamarisk, spiky grass, sedges, and halophytes (Nelson 1973). In USSR, considered aptly named in Russian as 'river forest nightingale' from preference for valleys with bottomland forests, shrubs, and bulrush beds. Breeds, however, not only near water but in arid areas, even in desert sand-dunes with scattered shrubs; also in saline or alkaline places, on steppes with shrub growth, and locally beyond lower belt of foothills to juniper zone, as high as 1950 m in Armeniya. Sometimes breeds in orchards and vineyards. Perches freely on bush-tops, and feeds on ground under bushes and trees as well as in bushes (Dementiev and Gladkov 1954*b*). In Indian subcontinent occurs on migration in dry scrub jungle, tamarisks, and broken stony country (Ali and Ripley 1973*b*). In Africa, winters in dry *Acacia* steppe, sometimes around human settlements on coastal plain but also in dry uplands below about 1000 m (Moreau 1972).

Distribution. No evidence of major range changes.

FRANCE. No proof of former alleged breeding in south (RC). ALBANIA. Almost certainly breeds in lowlands (AN). BULGARIA. May have bred, but no proof (TM). TURKEY. Formerly bred Ankara (Vittery *et al.* 1971). LEBANON. Probably bred near Jounieh 1974–5 (AMM). KUWAIT. Singing ♂♂ in summer 1978 but no proof of breeding (PRH). LIBYA. Distribution in east uncertain (GB). ALGERIA. May breed Ahaggar (EDHJ).

Accidental. Britain, France, West Germany, Norway, Switzerland, Bulgaria, Rumania, Malta, Canary Islands.

Population. Little information on population trends.

GREECE. A few hundred pairs (GIH). SPAIN. Locally numerous (AN). USSR. Generally common (Dementiev and Gladkov 1954*b*). ISRAEL. Major decline due to pesticides in 1950s; slow recovery from early 1970s but former population size not recovered (UP). EGYPT. Locally common (SMG, PLM, WCM). LIBYA. Tripolitania: locally common (Bundy 1976).

Movements. Eurasian and North African populations are migratory, wintering in northern Afrotropics. Sub-Saharan breeding populations (including those of Tibesti in Chad) are resident.

Nominate *galactotes* (breeding Iberia, North Africa, and Levant north to southern Syria) is strictly a summer visitor to breeding range. Presumed to winter in Sahel zone of West Africa (Heim de Balsac and Mayaud 1962), though confirmed records there inadequate to account for numbers that must be involved, probably due to overlap with resident race *minor* (Moreau 1972). At species level, a common passage migrant and winter visitor to Mauritania (Gee 1984) and a regular migrant through Mali in September–October and April (Lamarche 1981). Nominate *galactotes* collected Sénégal, Mali, Niger, and northern Nigeria (Morel and Roux 1966, 1973; Moreau 1972; Elgood 1982). Migrants cross Libya in spring at least, including records from Fezzan and Libyan Desert (Bundy 1976), though Newby (1980) and Lynes (1925a) recorded only resident *minor* in central Chad and western Sudan respectively; hence doubt about winter range of

nominate *galactotes* in north-central Africa. However, in eastern Sudan this race reported as non-breeding visitor, October–March, along Nile and east to Red Sea (Cave and Macdonald 1955). Autumn passage everywhere is poorly documented, but exodus from breeding areas occurs September to early October. More North African records in spring, when birds arrive early April in southern Algeria and Tunisia and later in April or into early May further north (Heim de Balsac and Mayaud 1962); reaches southern Iberia in late April and early

May, but not until late May in central Spain and apparently some not until early June in central Portugal (Bannerman 1954; Beven 1970).

Eastern races *syriacus* (breeding Balkans, Asia Minor, northern Syria, and Lebanon) and *familiaris* (breeding Transcaucasia and Iraq to central Asia) migrate through east Mediterranean area and Middle East to winter (overlapping) in north-east Africa. Eastern birds cross north-west India in autumn, but not seen in spring. In Ethiopia, *syriacus* perhaps more common than *familiaris* (Urban and Brown 1971), though in about equal numbers in Somalia (Ash and Miskell 1983). Winter range extends south to Kenya, where locally common, and probably to extreme north-east Tanzania (Britton 1980), but only single records from Uganda and north-east Zaïre. No confirmed records of these populations from Sudan. Some early Kenyan skins assigned to *syriacus*, though all fresh-plumaged spring birds handled recently have shown characters of *familiaris* (Britton 1980). Sole Zaïre record, December, attributed to *syriacus* (Chapin 1953).

In Soviet Asia, most leave breeding range in second half of August, minority lingering to mid-September (e.g. Dolgushin *et al.* 1972); passage across Arabia mid-August to mid-October (e.g. Bundy and Warr 1980). Leaves Turkey August–September, a few occurring late September on south coast (*Orn. Soc. Turkey Bird Rep.* 1966–75); passage through Cyprus August to early October (Flint and Stewart 1983). Begins to arrive Somalia in September (Ash and Miskell 1983), though main passage into Kenya not until November (Britton 1980). Spring departure from Kenya March–April. Passage through Somalia, Arabia, Middle East, and Cyprus mid-March to mid-May, and main arrival in Turkey in mid-May. First birds reach Kazakhstan at end of April or early May, immigration continuing throughout May (Dolgushin *et al.* 1972). RH

Food. Mostly insects and earthworms, often rather large; occasionally fruit. Feeding method varies with prey. Pursues ants, Orthoptera, etc., on ground. Takes small Diptera and Hymenoptera from flowers, sometimes hovering to do so. Locates earthworms by probing in soft ground, throwing earth aside with bill once worm found; also hunts worms like Blackbird *Turdus merula* by using fast run followed by pause with head cocked to one side and quick jab with bill. Takes Lepidoptera in flight. (Sage 1960*b*.) May wipe bill after feeding (Lambert 1965). Grasshoppers for delivery to young sometimes have legs removed (Beven 1970).

Diet includes the following. Invertebrates: dragonflies (Odonata), grasshoppers, etc. (Orthoptera: Gryllotalpidae, Acrididae), earwigs (Dermaptera), mantises (Mantidae), bugs (Hemiptera: e.g. cicadas, Pentatomidae), larvae of ant-lions (Myrmeleontidae), adult and larval Lepidoptera (Lycaenidae, Pyralidae, Alucitidae, Sphingidae), adult and larval flies (Diptera: Culicidae, Taban-

idae, Asilidae), adult and larval Hymenoptera (Ichneumonoidea, ants Formicidae, wasps Vespidae, bees Apoidea), adult and larval beetles (Coleoptera: Carabidae, Staphylinidae, Scarabaeidae, Buprestidae, Elateridae, Tenebrionidae, Coccinellidae, Curculionidae, Scolytidae), spiders (Araneae), millipedes and centipedes (Myriapoda), earthworms (Oligochaeta). Plant material: fruits, e.g. of *Nitraria*, orange. (Dementiev and Gladkov 1954*b*; Mountfort 1958; Sage 1960*b*; Watson 1964; Dolgushin *et al.* 1972; Mambetzhumaev and Abdreimov 1972; Sagitov and Bakaev 1980.)

In southern Morocco, May, one adult contained 5 large Lepidoptera larvae and 2 small beetles; in June, an adult contained 4 fruits of *Nitraria*, and a juvenile contained 2 fruits of *Nitraria* and insects (Valverde 1957). In Greece, takes beetles, small Orthoptera, larval Lepidoptera, and earthworms (Watson 1964). In Iraq, takes Hymenoptera (especially ants), Orthoptera (up to *c.* 5 cm long), mantises, and small Lepidoptera (Sage 1960*b*). Near Amu Dar'ya river (south-central USSR), 49 stomachs contained largely beetles, Diptera (adults and larvae), and larval hawkmoths (Sphingidae) (Mambetzhumaev and Abdreimov 1972). In Uzbekistan, 17 stomachs contained largely ants and Curculionidae (Sagitov and Bakaev 1980). In Turkmeniya, May–June, stomachs contained ants, adult and larval beetles, and adult and larval Lepidoptera (Dementiev and Gladkov 1954*b*); ant larvae may comprise up to 82·8% of diet (Dolgushin *et al.* 1972).

Nestlings in Portugal seen to be fed on beetles, small dragonflies, Lepidoptera larvae, flies, grasshoppers, earthworms, and berries (Beven 1970). Near Amu Dar'ya, young fed largely on ants, larval Lepidoptera (mostly Sphingidae), beetles, Orthoptera, and cicadas (Mambetzhumaev and Abdreimov 1972). In Turkmeniya, young may be given up to 74% Acrididae (Dolgushin *et al.* 1972). In Iraq, young fed on Orthoptera and Lepidoptera larvae (Ticehurst 1926), and may be given worms in large numbers (Sage 1960*b*). DJB

Social pattern and behaviour. Important studies by Sage (1960*b*), Beven (1970), and López Iborra (1983). Account includes notes supplied by E N Panov for Turkmeniya (USSR).

1. Little information on dispersion outside breeding season. At end of breeding season, Alicante (Spain), family groups of 4–5 not uncommon, and thought possibly to migrate together (López Iborra 1983). In late April, Tiouine (Morocco), over 40 in 'quite small area', suggesting some may have been migrants (Sage and Meadows 1965). BONDS. No evidence for other than monogamous mating system. Both sexes tend young for *c.* 3 weeks after fledging, longer for 2nd broods (López Iborra 1983: see Relations within Family Group, below). After fledging, brood of 1 fed by both parents but brood of 2 divided such that apparently 1 chick always fed by 1 adult, other chick by other adult (Sage 1960*b*). Pair-bond maintained for 2nd brood (Brosset 1961; López Iborra 1983). Age of first breeding not known. BREEDING DISPERSION. Solitary and territorial, contiguous in suitable habitat (Heim de Balsac and Heim de

Balsac 1954). Territory serves for courtship, nesting, and feeding. When feeding young, adults moved up to 800 m from nest to find food (Cano 1960). In Alicante (Spain) average area of territories 2-8 ha ($n = 5$); density 2·8 pairs per km² (López Iborra 1983). In western Sahara, singing ♂♂ c. 200 m apart (Valverde 1957). In Kazakhstan (USSR), singing ♂♂ as close as 80–100 m apart (Dolgushin et al. 1972). In south-west Uzbekistan (USSR), nests mostly 140–160 m (60–250) apart (Sagitov and Bakaev 1980, which see for transect counts). In northern Sahara, 1 April–30 May, 8 counts gave average of 2 birds in 24 km of wadi (Blondel 1962a). In 2 cases of 2nd broods, 2nd nest built at least 1·5 m from 1st (Brosset 1961). ROOSTING. Little known. In captivity, readily dust-bathes (Beven 1970). Young start dust-bathing around time they become independent of parents (López Iborra 1983).

2. In presence of man, bold and easily approachable (Pitman 1921). Inquisitive, readily approaching stationary human intruder to investigate (Aplin 1896; Laenen 1949–50). Most striking features are tail and wing movements, accentuated when alarmed, but conspicuous even when mildly excited. Moves tail slowly up and down, or lowers and fans it, often cocking it vertically or laying it almost horizontally along back; to accommodate tail movements, flicks wings forward or partly spreads and droops them until tips nearly touch ground (Tucker 1947; Beven 1970). These postures and movements are elements of the more emphatic Antagonistic and Advertising displays (see below). For calls accompanying alarm, see 2–3 in Voice. FLOCK BEHAVIOUR. None reported. SONG-DISPLAY. ♂♂ start singing (see 1 in Voice) almost immediately upon arrival on breeding grounds (Meise 1959; Ivanov 1969; Nelson 1973). ♂ sang little near nest during building, but more often after 1st egg laid (Ponomareva 1974). In Israel, song conspicuous early summer, declining late June (Pitman 1921). Apparently does not sing in winter in Kenya (D J Pearson). ♂ sings from exposed perch, also not infrequently in Song-flight. Perched song often from favoured vantage point, e.g. top of bush, tree, or overhead wires (Witherby et al. 1938), from 0·5–8 m above ground (Valverde 1957); during perched Song-display, tail conspicuously immobile (López Iborra 1983). ♂ sometimes gives Subsong on ground, accompanied by slowly lowering moderately spread tail, then raising it (E N Panov: Fig A). In Song-flight,

A

bird flies up from bushes and glides down (Dolgushin et al. 1972) or horizontally with wings raised and tail fanned (Beven 1970; Jonsson 1982; Bergmann and Helb 1982). Following details of Song-flight from E N Panov: flight not very high, usually in a straight line, sometimes in an arched trajectory. Bird flies slowly with gentle, shallow wing-beats. When changing direction, sometimes turns on motionless slightly raised wings, and also postures thus prior to landing on bush. On landing, the fanned tail is moved briefly up and down 2–3 times and then held horizontally (Fig B). Once, an unpaired ♂ performed a short flight on irregular path, rather like display-

B

flight of wheatears Oenanthe. For report of ♀ singing (subsequently shot and dissected), Afghanistan, early May, see Meinertzhagen (1938). ANTAGONISTIC BEHAVIOUR. ♂ vigorously defends territory until young independent; ♀ (mate) takes no part (Sage 1960b). When conspecific bird trespasses, resident ♂ turns or flies towards it and, when nearby, performs Upright-threat display (Fig C): stands upright, head horizontal, bill

C

sometimes gaping; fans tail and constantly moves it up and down; when raised, tail begins to close as it passes beyond the vertical, and is fully closed as it arcs forwards to almost touch the head. Meanwhile, partially spreads and lower wings until they nearly touch ground, flicks them well forwards and, with undersides facing forwards, holds them there for 1–2 s. A few moments later, folds wings and closes and lowers tail (Sage 1960b; Beven 1970). Drawings by E N Panov suggest less intense display, in which wings slightly drooped and upright tail moved through smaller arc either side of vertical. Occasionally, defending resident dips head or stretches it out until bill almost on ground (Beven 1970). Thereafter, with lowered head and puffed out plumage, resident typically runs backwards and forwards in front of intruder, then makes quick dash at intruder, repeats Upright-threat display, and makes another forward dash. Occasionally intruder flies off, but occasionally adopts Forward-threat posture (Fig D): squats low, slightly opens and droops wings, fans tail and lowers it or, less commonly, raises it vertically; stretches head forwards and gapes widely, exposing orange-yellow interior. Posture may somewhat delay, but not

D

inhibit, forward dash of defending ♂. Intruder usually retreats without a fight (Sage 1960*b*). On border between territories, 2 ♂♂ may come as close as 1 m and perform Upright-threat display at each other, sometimes leading to brief fight, accompanied by Excitement-calls (E N Panov: see 3a in Voice). HETEROSEXUAL BEHAVIOUR. (1) General. In south-west Uzbekistan (USSR), ♂♂ arrive before ♀♀ on breeding grounds (Sagitov and Bakaev 1980). (2) Pair-bonding behaviour. Self-advertisement of ♂ similar to Upright-threat display (see above) but ♂ either turns bodily away from ♀ or, if turned towards her, keeps bill closed and averts head (Beven 1970, which see for photographs). Pair-formation at Sokoto (Nigeria) also included chasing (Mundy and Cook 1972). (3) Courtship-feeding. None reported prior to laying, but see subsection 5, below. (4) Mating. No information. (5) Behaviour at nest. Not known which sex chooses nest-site. Ceremony, interpreted by E N Panov as possible nest-site selection, as follows: in early May, around sunset, perched ♂ alternately lowered closed tail and raised it, shivering drooped wings when tail upright (Fig E); also made some mock-pecking movements and gave

E

Excitement-calls. Then ♂ raised half-spread tail high, as in Fig A, and flew to ♀. Pair stood together for some time, then flew to foot of a thick bush and stayed there, unseen, for some time before ♂ flew out and glided, singing, to the foot of another bush. Then ♂ moved higher into bush and sang with tail spread and lowered. When ♂ and ♀ together at base of bush, Excitement-calls (see 3a in Voice) and (once) Alarm-calls heard. Both sexes build (Ticehurst 1926), ♀ continuing on day 1st egg laid (Ponomareva 1974). Approaches nest silently when building (Ponomareva 1974) and, once incubating, sits tight and motionless (Pitman 1921). At start of incubation (i.e. during laying), ♀ sits only at night, later on also by day, being fed on nest by ♂ (Dementiev and Gladkov 1954*b*). In morning, sitting ♀ leaves nest for relatively long intervals, incubating more assiduously in the afternoon (López Iborra 1983). While ♀ incubating, ♂ adopts sentinel role nearby (Sagitov and Bakaev 1980). RELATIONS WITHIN FAMILY GROUP. Eyes of young open at 6 days (Sagitov and Bakaev 1980). Until 8–9 days, young respond with food-calls to extraneous noises and vibrations near nest (López Iborra 1983). Fed in nest by both parents (Sage 1960*b*). According to Sagitov and Bakaev (1980), leave nest at *c.* 10 days but this perhaps premature, and 12–13 days (Dolgushin *et al.* 1972) seems more plausible. Well-grown chick which fell from nest was fed on ground (Cano 1960). On first leaving nest, young perch separately and give food-calls until parents return with food. Parent announces arrival with apparent Warning-call (see 4 in Voice), to which young respond with food-call. As they grow, young become increasingly mobile, but more often perch together. By 25 days, tirelessly follow parents, begging with food-calls and shivering wings; often invade periphery of a neighbouring territory. (López Iborra 1983.) In fledged brood of 2, where parents divided the young strictly between them, these followed respective parents closely;

young in brood of 1, fed by both parents, waited, calling incessantly for parents to return or occasionally followed one or other parent, running close behind, wing-shivering and jerking its tail every time parent halted (Sage 1960*b*). At *c.* 3 weeks after leaving nest, young begin self-feeding, often coincident with hatching of 2nd clutch to which parents now devote all their time. Parents then largely ignore 1st brood, and respond aggressively to entreaties for food. Parents more attentive to 2nd brood, and 1 family party together for at least 4 weeks. (López Iborra 1983.) ANTI-PREDATOR RESPONSES OF YOUNG. From 7 days, young show fear response if touched: withdraw into nest, retract their heads into their shoulders, partly open their wings, often raise their back plumage, sometimes raise the tail and gently deposit a faecal sac on rim of nest. Newly fledged young seek refuge in vegetation surrounding nest. (López Iborra 1983.) According to Dolgushin *et al.* (1972), newly fledged brood escaped from danger mainly by running. PARENTAL ANTI-PREDATOR STRATEGIES. (1) Passive measures. When human intruder first enters territory, sentinel bird gives Warning-call (López Iborra 1983). If human intruder in vicinity of nest, ♀ at first sits tight; with progressive approach of intruder, ♀ first flies off, then, if nest imminently threatened, returns to nest-area (Sagitov and Bakaev 1980). According to Dementiev and Gladkov (1954*b*), ♀ allows approach to within 2 m before flushing. In Almeria (Spain), when human intruder near nest, 1 parent stayed on guard, giving Alarm-calls (see 3b in Voice), while other fed young, albeit at reduced rate and by modifying usual route to nest (Cano 1960). (2) Active measures: against man. In Israel, birds disturbed sitting on nest in thick cover stretched neck and twisted it about in a snake-like fashion in direction of intruder; if this failed to deter intruder, bird flew off (Pitman 1921). Parents often highly demonstrative when nest discovered, behaviour containing strong elements of distraction-lure display: follow closely, and sometimes hover near, intruder, giving Alarm-calls (Ali and Ripley 1973*a*); hover over or near nest, dive madly into bushes, and hop into the open with up-jerked tail (Meinertzhagen 1930). In Iraq, bird moved away from nest with fanned tail, said to be distraction-display (Moore and Boswell 1956).

(Figs by J P Busby: A, B, and E from drawings by E N Panov; C and D from drawings and photographs in Beven 1970.) EKD

Voice. Freely used in breeding season. Little information for other times of year.

CALLS OF ADULT. (1) Song. Given by ♂; once reported for ♀ (Meinertzhagen 1938), this presumably exceptional. Commonly described as rich and musical (Beven 1970), sweet, pure and warbling (Vere Benson 1970; Gallagher and Woodcock 1980); loud, sweet, and rather twittering, somewhat like Robin *Erithacus rubecula*, and delivered in thrush (*Turdus*)-like jerks (Aplin 1896). For comparisons with other passerines, see Steinfatt (1954), Meise (1959), and Vere Benson (1970). Comprises short phrases, liberally punctuated by pauses (Jonsson 1982). Delivery varies from distinctly phrased to continuous; phrases begin with loud, drawn-out, pure-toned, fluting, and (to some extent) trilled units, followed by quieter short sequences somewhat reminiscent of *E. rubecula*; this chattering part of song may be quite protracted and may include mimicry. Each ♂ has several types of phrase-beginnings in his repertoire. (Bergmann and Helb

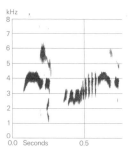

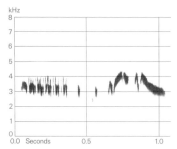

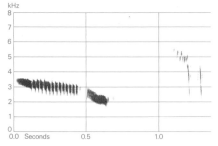

I   E D H Johnson   Morocco
April 1966

II   E D H Johnson   Morocco   April 1966

III   P A D Hollom   Iran   April 1971

1982, which see for sonagrams.) Our recordings likewise demonstrate rich repertoire of phrases in song of a given bird. Figs I and II depict two phrases from discontinuous, rather lazy, dry song, perhaps even Subsong: Fig I shows a musical 'seet t-rrreee', Fig II a tremolo followed by 'si si wee-choo' representing terminal flourish, not unlike Chaffinch *Fringilla coelebs*, of a phrase *c.* 1 s long. 2nd recording a more continuous, fulsome, sweeter, song, with much longer phrases: one phrase (Fig III) rather like Blackcap *Sylvia atricapilla*, with introductory tremolo; another phrase (Fig IV) shows terminal flourish comparable to that in Fig II. (J Hall-Craggs.) Renderings of other song-phrases (Meise 1959) include 'essobibibIt', last syllable higher-pitched and emphasized, then, after pause, a slightly descending 'gögögögö'; another rendered 'tobebIt dih dededö'; a 3rd quite long, quiet, and chattering, rendered 'örrr dedede örr örr dididide debIt'. Song often includes a 'teviu' sound, like Song Thrush *Turdus philomelos* (Jonsson 1982). On spring migration, a quiet warble, like Redstart *Phoenicurus phoenicurus*, interspersed with occasional rising notes (Walker 1981*a*). Subsong (see Social Pattern and Behaviour) more hurried than full song (Moore and Boswell 1956). (2) Contact-alarm call. A sibilant 'sseeep' or 'tseeeet', sometimes shortened to 'zip', 'zetk', 'tsip' (etc.), given in winter (D J Pearson), on spring passage (Gallagher and Rogers 1980; Gallagher and Woodcock 1980), and (rendered 'prriiii') among family members prior to autumn migration (López Iborra 1983). Other renderings include a plaintive but melodious 'tsee tsee' (Ripley 1951; Beven 1970), or 'dziii' (E N Panov). (3) Excitement- and Alarm-calls. (a) Rapidly repeated variant of call 2, rendered 'zi-zi-zi...' (E N Panov), or sharp squeaky 'pzeet-pzeet-pzeet', signals

excitement or agitation. Also described as a sometimes harsh 'chi-chi-chi', recalling Great Tit *Parus major* but weaker and less metallic (Cano 1960); a rapid series of 'prriiii' (López Iborra 1983) or 'triii' sounds (Bergmann and Helb 1982). Given with bill wide open, often in confrontations with neighbours, and in display interactions with ♀ (E N Panov). Also when human intruder near nest or young (López Iborra 1983). (b) A hard harsh 'teck' (Meise 1959; Beven 1970; Bergmann and Helb 1982); a scolding 'tack tack' like warbler (Sylviinae) (Vere Benson 1970). (4) Warning-call. In low-intensity danger, a short, soft, sweet, fluty 'piu'. Tends to be first call given when human intruder enters territory. Also used to contact hidden young, e.g. by parent returning with food (López Iborra 1983). Whistle or pipe recalling Bullfinch *Pyrrhula pyrrhula* or Nightingale *Luscinia megarhynchos*, heard regularly in captivity (Beven 1970), also 'twee' like *Phylloscopus* warbler (Moore and Boswell 1956), presumably the same or related. (5) Other calls. A hoarse 'tchiep', a harder 'tiut', and a low rolling 'chrrr' (Jonsson 1982). Short, abrupt 'tchok' given by ♂ approaching ♀ mate (E N Panov). 'Sing' sound like Dunnock *Prunella modularis*, and (once) 'gew-gaw' (Moore and Boswell 1956).

CALLS OF YOUNG. Food-calls in nest described as soft, almost inaudible little peeping sounds. Newly-fledged young give a whistling 'pri-pi-pi-pi-pi-pi' (López Iborra 1983). From 4–14 days after fledging, food-call a loud chirrup like 'a toy canary whistle' (Beven 1970). Sound rendered a vibrant 'prrrr prrrr', accompanied by wing-shivering (López Iborra 1983), presumably this call. Audible up to *c.* 40 m (Sage 1960*b*).   EKD

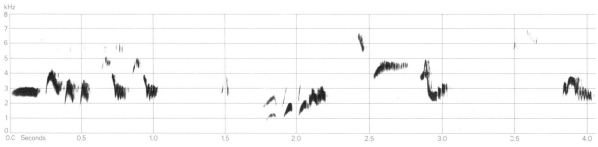

IV   P A D Hollom   Iran   April 1971

Breeding. SEASON. Iberia, Greece, and North Africa: main laying period second half of May and early June (Witherby *et al.* 1938*b*; Bannerman 1954; Makatsch 1976). Iraq: eggs laid early or mid-May to late June (Tucker 1947; Sage 1960*b*; Marchant 1963*b*). Southern USSR: main laying season late May and June (Dementiev and Gladkov 1954*b*). SITE. In thick bush or low tree, often near trunk; mean height above ground 2·12 m (1·7-2·5), *n* = 9 (López Iborra 1983). Occasionally uses old nest of another species (Ali and Ripley 1973*a*). Nest: loosely-constructed untidy structure of fine twigs, grasses, and rootlets, lined with vegetable down, wool, hair, and feathers, and (in southern parts of range) often a piece of snake skin (90% of 70 nests, Palestine) (Pitman 1921; Beven 1970). Outside diameter 13-20 cm, inner 7-9 cm, height 6·5-10·0 cm, depth of cup 4·0-6·5 cm (Dementiev and Gladkov 1954*b*; Makatsch 1958; López Iborra 1983). Building: by both sexes (Dementiev and Gladkov 1954*b*). EGGS. See Plate 82. Sub-elliptical, smooth and fairly glossy; white or very pale grey, sometimes tinged blue or green, heavily marked with brown, purplish-brown, or purplish-grey spots, speckles, and small streaks. Nominate *galactotes*: 22·5 × 16·3 mm (19·5-26·0 × 14·3-18·0), *n* = 125; calculated weight 3·2 g. *C. g. syriaca*: 21·7 × 16·0 mm (18·5-22·7 × 14·3-17·3), *n* = 90; calculated weight 2·98; *familiaris* not significantly different. Clutch: 4-5 (2-6). Of 68 clutches, Palestine: 2 eggs, 9; 3, 35; 4, 23; 5, 1; mean 3·2 (Pitman 1921); mean of 16 clutches, Iraq, 3·8 (Marchant 1963*b*). 2 broods. Eggs laid daily (Sagitov and Bakaev 1980). INCUBATION. 13 days (Beven 1970; Sagitov and Bakaev 1980; López Iborra 1983). By ♀ only, beginning with last egg (Sagitov and Bakaev 1980); suggested in Dementiev and Gladkov (1954*b*) that in early stages incubation only carried out at night, but this needs confirmation. YOUNG. Cared for and fed by both parents (Sage 1960*b*). FLEDGING TO MATURITY. Fledging period 12-13 days (Dolgushin *et al.* 1972); for one brood 15 days (Cano 1960), but may leave at 11 days (López Iborra 1983), or *c.* 10 days (Sagitov and Bakaev 1980). Fed after fledging but period of dependence not known; single young fed by both parents, but brood of 2 may be split between parents (Sage 1960*b*). BREEDING SUCCESS. Of 36 eggs in 9 clutches, south-west Uzbekistan (USSR), 1 egg unaccounted for, 1 infertile, 10 destroyed by reptiles or mammals, and 24 hatched from which 16 young fledged, giving overall success of 44% (Sagitov and Bakaev 1980).

Plumages (nominate *galactotes*). ADULT. Entire upperparts rufous-brown, slightly tinged vinous-brown when fresh, but rump and upper tail-coverts brighter, pale rufous-cinnamon. Long but rather narrow supercilium and eye-ring cream-white; eye-ring broken in front and behind. Stripe on lores dark brown-grey; dark brown stripe running backwards from eye over upper ear-coverts. Cheeks cream-white, faintly mottled grey, merging into pink-brown ear-coverts. Chin and throat cream-white; sides of neck and chest cinnamon-pink, pink-buff,

or greyish-cream (depending on wear); sides of breast rufous-cinnamon or rufous-brown; flanks paler, pink-brown. Belly, vent, and under tail-coverts cream-pink, almost white on centre of belly and vent. Tail bright rufous-cinnamon, like rump and upper tail-coverts or slightly deeper rufous; central pair of feathers (t1) uniform or with ill-defined dusky olive-brown spot on tip; t2 with larger and more distinct dark ochre-brown or black band on tip, 7-12 mm wide; t3-t5 with similar band subterminally, distally bordered by pale cream or white tip, 2-4 mm wide on t3 to 10-14 mm on t5; white tip on t6 15-20 mm wide, often more extensive along outer edge, black subterminal band narrower, *c.* 5 mm wide, not or scarcely extending to outer web. Flight-feathers and greater upper primary coverts dark brown-grey, outer webs with narrow cinnamon-rufous fringe widening towards feather-base; tips of secondaries and inner primaries narrowly and sharply fringed pale buff-grey or off-white; reduced p10 virtually uniform dark brown-grey. Tertials dark brown-grey, sides extensively washed rufous-brown (especially on inner tertials), tips more narrowly. Lesser upper wing-coverts rufous-brown, like upperparts; median coverts similar but centres slightly darker brown; greater coverts dark brown with ill-defined rufous-brown fringe along outer web merging into off-white fringe at tip. Bastard wing dark brown-grey or blackish-brown, feathers narrowly fringed off-white along outer web and tip. Under wing-coverts and axillaries cream-pink or white. *In worn plumage*, upperparts slightly paler rufous-brown, less vinous; underparts paler, off-white with pale buff or pink-buff chest, sides of breast, and flanks; dark bands on tail less deep black, more olive-brown, those on central feathers sometimes less contrasting; fringes along tips of tertials and median and greater upper wing-coverts bleached to greyish-white. NESTLING. Entirely naked (Witherby *et al.* 1938*b*). JUVENILE. Closely similar to adult, not always distinguishable. Feathering of body markedly shorter and looser, tips fluffy. Upperparts slightly greyer, less saturated rufous-brown, more sandy-buff than adult when plumage worn; rump and upper tail-coverts slightly paler pink-cinnamon, markedly loose. Sides of neck, sides of breast, and flanks pale cream-buff to cream-white, similar to remainder of underparts (in adult, distinctly darker, rufous-cinnamon to pink-brown); chest with indistinct olive-grey feather-tips, showing as indistinct mottling. Tail like adult, but black subterminal bands reduced, and white on tips less extensive; black absent from wholly rufous t1, absent or restricted to spot on inner web on t2, usually larger on t3-t6 but restricted to rather irregular and narrow patch on inner web (in adult, t3-t5 have fully black band across both webs); occasionally, no black on tail except for faint spot on inner web of t4 or t5; pale tail-tips often tinged pink-buff (sometimes rufous in birds with restricted amount of black), 3-6 mm wide on t3-t4(-t5) to 13-16 mm on t6. Flight-feathers like adult, but pale fringes along tips on average wider but less sharply defined; tips of outer primaries (including p10) and greater upper wing-coverts often narrowly fringed buff-white. FIRST ADULT. Like adult; juvenile flight-feathers, greater upper primary coverts, and perhaps other feathers retained, but these hardly separable from adult (see juvenile). In autumn, tail and flight-feathers fresh (in adult, heavily worn or primaries moulting); in spring, primaries, greater primary coverts, and tail distinctly worn, tips frayed, tips of coverts rather pointed (in adult, tips of primaries, tail-feathers, and coverts broad and smoothly edged until April-May; some birds intermediate and difficult to age).

Bare parts. ADULT, JUVENILE, FIRST ADULT. Iris brown. Bill dark horn-brown or black-brown, base of lower mandible paler horn-brown, flesh-brown, greyish-flesh, bluish-white, or whitish. Mouth orange-yellow. Leg and foot pale brown, flesh-brown, light grey-brown, greyish-flesh, purple-flesh, or pink-flesh; claws horn-colour. (Hartert 1910; Witherby *et al.* 1938*b*; Sage 1960*a*; Lambert 1965; Ali and Ripley 1973*a*; BMNH, ZFMK.) NESTLING. Bare skin black. Mouth orange-yellow, no tongue-spots; gape-flanges white. (Witherby *et al.* 1938*b*.)

Moults. ADULT POST-BREEDING. Complete; primaries descendant. Data on timing conflicting: either before autumn migration or immediately afterwards; perhaps sometimes suspends moult during migration. In nominate *galactotes*, moult in autumn, occasionally from July (Witherby *et al.* 1938*b*). In *syriacus*, moult July-September (Stresemann 1920), but in full moult in Syria in December (Witherby *et al.* 1938*b*); late autumn and winter (Svensson 1984*a*). In *syriacus* or *familiaris*, Kenya, flight-feathers heavily worn up to November, moult starting November or early December (Pearson and Backhurst 1976). In *familiaris*, USSR, usually no moult in June, though one just starting 29 June; many start in first half of July, but some not yet moulting early August; one bird nearing completion 26 August (Dementiev and Gladkov 1954*b*). POST-JUVENILE. Partial or perhaps occasionally complete: head, body, some or all tertials, central, all, or no tail-feathers, and many or all wing-coverts. Timing rather variable, depending on hatching; June-October in nominate *galactotes* and *syriacus* (Stresemann 1920; Witherby *et al.* 1938*b*; RMNH, ZMA); in *familiaris*, one in 1st adult 25 June, another 13 August (Dementiev and Gladkov 1954*b*). Usually also a partial moult in 1st winter, but some *familiaris* replace entire plumage then except for a few inner primaries and some may moult completely (Svensson 1984*a*). In *minor*, Sudan, complete moult in winter, one starting with p1 in January, another half-way through moult in February (BMNH, ZFMK).

Measurements. ADULT, FIRST ADULT. Nominate *galactotes*. Spain and north-west Africa, April-September; skins (RMNH, ZMA). Bill (S) to skull, bill (N) to distal corner of nostril; exposed culmen on average 3·8 less than bill (S).

WING	♂ 87·7 (2·11; 17)	84-92	♀ 86·1 (2·19; 16)	81-89
TAIL	68·9 (2·85; 18)	64-73	67·5 (1·67; 16)	65-70
BILL (S)	20·2 (0·84; 16)	19·0-21·6	19·2 (0·66; 14)	18·1-20·1
BILL (N)	11·8 (0·83; 11)	10·8-13·2	11·4 (0·63; 12)	10·8-12·2
TARSUS	27·0 (0·83; 17)	25·7-28·4	26·6 (1·07; 16)	24·9-28·1

Sex differences not significant, except bill (S). 1st adult with retained juvenile flight-feathers and tail combined with adult above, though juvenile wing on average 1·8 shorter than adult, tail 1·2 shorter.

Wing of (1) *syriacus* (Stresemann 1920; RMNH, ZFMK, ZMA) and (2) *familiaris* (Paludan 1938, 1959; RMNH, ZFMK, ZMA), April-September; skins. Other data combined for both races.

WING (1)	♂ 88·3 (1·89; 17)	85-92	♀ 85·6 (0·92; 6)	84-87
(2)	87·9 (2·12; 13)	84-92	85·7 (— ; 3)	83-87
TAIL	64·2 (1·47; 12)	62-66	62·0 (1·67; 8)	59-64
BILL (S)	19·4 (0·63; 12)	18·4-20·4	19·5 (0·43; 8)	18·9-20·2
BILL (N)	11·5 (0·44; 12)	10·8-12·1	11·4 (0·48; 8)	10·7-12·2
TARSUS	25·8 (0·72; 12)	24·9-27·0	25·8 (0·99; 7)	24·6-27·2

Sex differences significant for wing and tail.

C. g. minor. Sénégal, northern Chad, and northern Sudan, all year; skins (RMNH, ZFMK).

WING	♂ 82·1 (2·05; 7)	79-85	♀ 79·2 (1·84; 4)	78-81
TAIL	66·8 (4·13; 6)	61-71	65·1 (0·48; 4)	64-66
BILL (S)	18·9 (0·85; 7)	17·6-19·7	18·0 (0·66; 4)	17·4-18·9

BILL (N)	11·5 (0·85; 6)	10·2-12·2	10·6 (0·51; 4)	10·3-11·4
TARSUS	25·9 (0·57; 6)	25·2-26·7	24·1 (0·71; 4)	23·6-25·1

Sex differences significant for wing and tarsus.

Weights. Nominate *galactotes*. South-east Morocco, mainly first half of April: 22·7 (1·42; 9) 21-25 (BTO). This race or *syriacus*, on spring migration, Malta: 24·4 (1·1; 6) 22-26 (J Sultana and C Gauci).

C. g. syriacus. Greece and Turkey, May-June: ♂♂ 19, 21, 23·5, 24, 29·5; unsexed 28 (Niethammer 1943; Kumerloeve 1963; Rokitanksy and Schifter 1971; Vauk 1973*a*; ZFMK).

C. g. syriacus or *familiaris*. Kenya, October-January: 20·3 (26) 17·5-24 (Pearson and Backhurst 1976). On migration Azraq (Jordan): second half of April, 21·3 (1·80; 22) 18-25; first half of May, 22·3 (1·73; 21) 19-26 (BTO).

C. g. familiaris. Bahrain (Ripley 1951), Sjarjah and north-east Saudi Arabia (BTO), Iran (Paludan 1938; Desfayes and Praz 1978), USSR (Nicht 1961; Dolgushin *et al.* 1972), and Afghanistan (Paludan 1959), combined: April, ♂♂ 22, 28, unsexed 19·2 (2·01; 4) 17-22; May and early June, ♂ 22·3 (1·15; 9) 20-24, ♀ 22; July, ♂♂ 21, 21; unsexed 18, 21; September, ♂♂ 16, 18. USSR: ♂ 21·8 (5) 20-25, ♀♀ 22·5, 27 (laying) (Dementiev and Gladkov 1954*b*). India, winter: 21·7 (11) 19-27 (Ali and Ripley 1973*a*).

C. g. minor. Chad and Niger: ♂♂ 17, 19, 21; ♀♀ 19, 21 (Niethammer 1955*b*; ZFMK). Northern Nigeria, November-March: 18·4 (0·9; 6) 17-19·5 (Fry 1970*b*).

Structure. Wing rather short, broad at base, tip rounded. 10 primaries: in nominate *galactotes*, p7-p8 longest, p9 3-5·5 shorter, p6 0-2, p5 3·5-5·5, p4 8-10, p1 14-19; in *syriacus* and *familiaris*, p8 longest, p9 1·5-3 shorter, p7 0-2, p6 1-4, p5 4-8, p4 9-13, p1 17-21. In all races, p10 reduced, narrow, tip narrowly rounded; in nominate *galactotes*, 40-47 shorter than wing-tip, 2-5 (adult) or 4-8 (juvenile) longer than longest upper primary covert; in *syriacus* and *familiaris*, 45-49 shorter than wing-tip, 2 shorter to 2 longer than longest primary covert in adult, 1-3 longer in juvenile. Outer web of (p5-)p6-p8 and inner web of (p6-)p7-p9 emarginated; emarginations sometimes indistinct on outer webs. Tertials short, longest equal to secondaries. Tail long, tip rounded; 12 rather broad feathers with rounded tips; t1 longest, t6 5-10 shorter in nominate *galactotes*, 2-6 in *syriacus* and *familiaris*. Bill rather long, slender; wide at gape, laterally compressed at tip; basal half straight, culmen and cutting edges of distal half slightly decurved. Nostrils oval, partly covered by membrane above. Tarsus long and slender. Toes rather long and slender; middle toe with claw 18·6 (28) 17-20; outer toe with claw *c*. 68% of middle with claw, inner *c*. 65%, hind *c*. 70%. Claws short, decurved.

Geographical variation. Marked; 2 main groups separable, differing in general colour, wing formula, and relative tail length. In all races, rump to tail rufous-cinnamon; in western group of races (nominate *galactotes*, *minor*, *hamertoni*), colour of upperparts rufous-brown, rather like rump and tail, but in eastern group (*syriacus* and *familiaris*) upperparts grey-brown or drab-brown, contrasting markedly with rufous-cinnamon rump and tail; tail of eastern group slightly shorter relative to wing length; wing-tip slightly more pointed (see Structure). *C. g. minor* from belt south of Sahara differs from nominate *galactotes* from western Mediterranean area in warmer pink-brown upperparts, smaller size (see Measurements), relatively longer tail, and more rounded wing-tip; some local variation in colour of upperparts (see Vaurie 1955*a*). Nominate *galactotes* from Egypt and southern part of Levant similar in size and colour to populations from Iberia and north-west Africa, but tail on average slightly longer and wing more rounded, tending slightly toward *minor* in structure. Apart from colour of upperparts and structure, *syriacus* from Balkans, Turkey, and northern part of Levant differs also from nominate *galactotes* in whiter supercilium, more distinct dark grey-brown stripes at sides of crown, through eye, and backwards from gape, whiter patch below eye, greyer (less sandy or creamy) chest and flanks, broader and more extensive black subterminal band on tail, and slightly less white on tail tips (8–15 mm on t6). *C. g. familiaris* from Transcaucasia and Iraq to eastern Kazakhstan (USSR) and Iran closely similar to *syriacus*, with which it intergrades in eastern Turkey, northern Syria, and western Transcaucasia; crown, mantle, and scapulars on average paler and greyer, rufous of rump to tail slightly paler, and underparts almost entirely white; much local variation in colour—in particular, birds inhabiting Turkmeniya (USSR) and eastern Iran markedly pale grey above and these sometimes separated as *deserticola* (Buturlin, 1908), but darker and paler birds often occur side by side and hence recognition not warranted (Hartert 1921–2; Dementiev and Gladkov 1954*b*). Birds from Bahrain richer and darker brown than *familiaris* (nearer *syriacus*), and smaller (Ripley 1951). Extralimital *hamertoni* from eastern Somalia is even smaller than *minor*, with darker upperparts; sometimes considered a separate species (Meinertzhagen 1949*b*).

Forms species-group with White-winged Scrub Robin *C. leucophrys*, Brown-backed Scrub Robin *C. hartlaubi*, Kalahari Scrub Robin *C. paena*, and Karroo Scrub Robin *C. coryphaeus*, all from Afrotropics (Hall and Moreau 1970). For survey of systematic position of genus *Cercotrichas*, see Heim de Balsac and Mayaud (1951) and Sage (1960*a*).   CSR

Cercotrichas podobe Black Bush Robin

PLATE 42
[between pages 544 and 545]

Du. Zwarte Waaierstaart   Fr. Merle podobé   Ge. Russheckensänger
Ru. Чёрный тугайный соловей   Sp. Alzacola negro   Sw. Svart trädnäktergal

Turdus podobe P L Statius Müller, 1776

Polytypic. Nominate *podobe* (P L Statius Müller, 1776), Africa; *melanoptera* (Hemprich and Ehrenberg, 1833), Arabia.

Field characters. 18 cm; wing-span 23–28 cm. Close in size and structure to Rufous Bush Robin *C. galactotes* but with 40% longer tail and 10% longer legs. Sprightly, bush-haunting chat with similar general character to *C. galactotes* but all-black except for white tips to under tail-coverts and bold white tips to graduated outer tail-feathers, last always conspicuous on cocked tail. Sexes similar; no seasonal variation. Juvenile separable. African race, nominate *podobe*, described here.

ADULT. Lead- or sooty-black, relieved only by (a) conspicuous white tips to 4 outermost pairs of tail-feathers, (b) white tips to under tail-coverts (forming chequered ventral pattern obvious from behind), (c) rufous under wing-coverts, and (d) rufous inner webs (not tips) of primaries and secondaries sometimes visible on folded wing and in flight. (Arabian race, *melanoptera*, usually has rufous areas black.) JUVENILE. As adult but duller, more sooty-brown than lead-black, and pale tips to under tail-coverts narrow or indistinct. At all ages, bill black, legs black-brown.

Unmistakable. Flight, gait, and behaviour strongly reminiscent of *C. galactotes* but spends more time on ground. Pugnacious in defence of territory.

Song consists of fairly short but pleasantly varied, babbling phrases, with predominantly sweet, fluted notes interspersed with more scratchy or throaty ones. Call thrush-like.

Habitat. Tropical and marginally subtropical, in arid mainly lowland regions from fringe of desert through shrub and acacia savanna, especially on sandy soils into which it probes with bill. Avoids close stands of trees and banks of rivers, using shrubs rather than trees as song-posts and foraging on more-or-less bare ground, but often under bushes. For nesting, needs crevices, which may be in tree-trunks, or between leaves of date palm, or even in roof of disused small building (Bannerman 1936; Serle *et al.* 1977). In Saudi Arabia, prefers dry bush vegetation and gardens for breeding; avoids summits of mountains and juniper zone (Jennings 1981*a*).

Distribution. Breeds in Africa in dry region south of Sahara from Sénégal to Red Sea, and in western and southern Arabia.

Accidental. Algeria: 2 on 8 February 1968 at Tamanrasset (23°N), and another almost certain in nearby Ahaggar massif on 12 February 1968 (Gaston 1970). Israel: at Elat, 1 in April 1981, 3 in April–May 1985, and 3 in March–May 1986 (*Dutch Birding* 1986, 8, 85–8).

Movements. Apparently largely or wholly resident, in both Africa and Arabia (Meinertzhagen 1954; Newby 1980; Elgood 1982). However, a very uncommon non-breeding visitor to Somalia, November–March (Ash and Miskell 1983); also to Ethiopia according to Archer and Godman (1961*b*), though listed as resident in Ethiopia by Urban and Brown (1971). Capacity for dispersal shown by spread into central Saudi Arabia since 1973, following increased irrigation (Jennings 1980).

JHE, DJB, RH

Voice. See Field Characters.

Plumages (nominate *podobe*). ADULT. Head, body, and lesser and median upper wing-coverts entirely dark slate-grey or plumbeous-black, more slaty when plumage fresh, slightly brown-black when worn; upper tail-coverts almost black, under tail-coverts black with broad white tips. Tail black, slightly plumbeous when fresh, brown-black when worn; t2 usually with narrow white tip; t3-t6 more broadly tipped white, 15-19 mm of tip of t6 white. Tertials, greater upper wing-coverts, upper primary coverts, and bastard wing dark brown or blackish-brown, feathers tinged plumbeous on outer webs and tips when plumage fresh; bastard wing and small coverts along leading edge of wing sometimes narrowly fringed white on tips when fresh. Flight-feathers sooty-black, basal $\frac{3}{4}$ of inner webs contrastingly rufous-cinnamon (less extensively on p9 and virtually absent on p10 and innermost secondary). Under wing-coverts and axillaries plumbeous-black. JUVENILE. Rather like adult, but much duller, fuscous-brown; feathering of body loose and fluffy (in particular, on back and tail-coverts); greater upper wing-coverts and tertials partly tinged rufous; white tips of under tail-coverts narrow or absent, 0-2 mm (in adult, 4-5 mm); tail-feathers narrower, less white on tips; t1-t3 fully black, t4 with ill-defined tip 2-4 mm wide, white tip of t6 10-12 mm wide; flight-feathers more extensively rufous, rufous extending to base of p10 and to bases of outer webs of p1-p8; p10 often longer and broader (Heim de Balsac and Mayaud 1951). FIRST ADULT. As adult.

Bare parts. ADULT. Iris hazel-brown, umber-brown, or brown. Bill black. Leg and foot dark horn-brown, dark slate-brown, or black. (Bannerman 1936; BMNH, RMNH, ZMA.) JUVENILE. Iris hazel-brown. Bill horn-brown with paler tip. Leg and foot brown, paler than bill (Heim de Balsac and Mayaud 1951; BMNH.)

Moults. ADULT POST-BREEDING. Complete; primaries descendant. Starts shortly after nesting, July-August, varying somewhat due to local differences in timing of breeding; moult completed August-October (Bannerman 1936; Heim de Balsac and Mayaud 1951; Mackworth-Praed and Grant 1973; BMNH, RMNH, ZMA). POST-JUVENILE. Complete; starting soon after fledging, about June-August. None in moult examined, but no possible 1st adults retained feathers with juvenile characters (Heim de Balsac and Mayaud 1951; BMNH, RMNH, ZMA).

Measurements. Nominate *podobe*. ADULT. Sudan and Ethiopia, all year; skins (BMNH, RMNH, ZMA). Bill (S) to skull, bill (N) to distal corner of nostril; exposed culmen on average 4·5 shorter than bill (S).

WING	♂ 93·1 (2·79; 15)	90-102	♀ 88·7 (2·29; 15) 85-92
TAIL	109·9 (5·12; 12)	104-119	101·4 (2·63; 10) 97-105
BILL (S)	20·0 (0·78; 13)	19·0-21·3	18·9 (0·89; 11) 17·8-20·1
BILL (N)	11·9 (0·67; 13)	11·1-12·9	11·4 (0·51; 11) 10·5-12·1
TARSUS	29·3 (0·83; 13)	28·1-30·4	28·1 (0·95; 11) 26·7-29·4

Sex differences significant. Ennedi (Niger) and Aïr (Chad), wing: ♂ 92·2 (3) 90-94, ♀ 88·3 (3) 87-89 (ZFMK).

JUVENILE. Wing on average 6·6 shorter than adult, tail on average 10·4 shorter (BMNH).

Weights. Ennedi (Niger) and Aïr (Chad): ♂ 25·7 (3) 24-27, ♀ 25·0 (4) 24-26 (Niethammer 1955*b*).

Structure. Wing rather long, broad at base, tip rounded. 10 primaries: p7 longest, p8 0·5-2 shorter, p9 6-9, p6 0-1, p5 1-3, p4 3-7, p1 12-16. P10 somewhat reduced; in adult, 35-40 shorter than p7, 13-20 longer than longest upper primary covert, total length 28-33, width 4-5; in juvenile, slightly longer and with more rounded tip, (25-)32-35 long, 5-6 wide (Heim de Balsac and Mayaud 1951; RMNH, ZMA). Outer web of p5-p8 and inner of p6-p9 emarginated. Tail long, tip graduated; 12 broad feathers with rounded tips; t1 longest, t6 26-32 shorter. Bill as in Rufous Bush Robin *C. galactotes*, but tip slightly finer. Leg and foot as in *C. galactotes*, but toes proportionately slightly longer and heavier. Middle toe with claw 19·2 (5) 18-20·5; outer toe with claw *c.* 72% of middle with claw, inner *c.* 68%, hind *c.* 77%. Remainder of structure as in *C. galactotes*.

Geographical variation. Rather slight. Arabian race *melanoptera* differs from nominate *podobe* of Africa in lacking rufous-cinnamon on inner webs of flight-feathers, these being sooty-black or (occasionally) with faint indication of rufous (Vaurie 1959); some populations show rufous more often than others, but amount generally less than in nominate *podobe* (Heim de Balsac and Mayaud 1951; BMNH). CSR

Erithacus rubecula Robin

PLATES 43 and 47
[between pages 544 and 545, and facing page 652]

Du. Roodborst   Fr. Rougegorge familier   Ge. Rotkehlchen
Ru. Зарянка   Sp. Petirrojo   Sw. Rödhake

Motacilla Rubecula Linnaeus, 1758

Polytypic. *E. r. melophilus* Hartert, 1901, Britain and Ireland, grading into nominate *rubecula* (see below); nominate *rubecula* (Linnaeus, 1758), Azores, Madeira, western Canary Islands (La Palma, Gomera, Hierro), continental Europe east to Urals and south to Spain, Italy, Greece, Asia Minor, and *c.* 50°N in central European USSR, grading into *melophilus* in belt from Denmark through Netherlands, south-east England, and western France to north-west Spain and Portugal; *superbus* Koenig, 1889, Gran Canaria and Tenerife (Canary Islands); *witherbyi* Hartert, 1910, Tunisia and Algeria, grading into nominate *rubecula* in Morocco and southern Spain and on Corsica and Sardinia; *tataricus* Grote, 1928, western Siberia east from Urals; *valens* Portenko, 1954, Crimea; *caucasicus* Buturlin, 1907, Caucasus and Transcaucasia; *hyrcanus* Blanford, 1874, south-east Transcaucasia and northern Iran.

Field characters. 14 cm; wing-span 20–22 cm. 15–20% smaller than Thrush Nightingale *Luscinia luscinia* and Nightingale *L. megarhynchos*. Small, robust chat, with rather large head and fat chest. Grey to brown above; face and chest diagnostically drenched orange-buff to rufous-chestnut. Sexes similar; little seasonal variation. Juvenile separable. 8 races in west Palearctic, differences clinal and obvious only in fresh adult and 1st-winter plumages.

ADULT. (1) Race of continental Europe and Asia Minor, nominate *rubecula*. Upperparts largely olive-brown, washed grey on back and wings; rump grey-brown subtly contrasting with warm brown upper tail-coverts and tail; in winter, often a wing-bar formed by buff tips to greater coverts. Forehead, surround to eye, fore-cheeks, chin, throat, and chest orange, with rufous tinge in some lights but no full red tone. Orange face and chest separated from olive-brown upperparts by quite wide band of delicate blue-grey feathers, noticeably loose on sides of chest and by upper flanks. Flanks bright, warm buff-olive; belly, vent, and under tail-coverts white, with pale buff suffusion most marked under flanks and tail. Underwing buff. In flight, retreating bird looks noticeably uniform but warmer tail and pale buff-white lateral tail-coverts noticeable, particularly when bird turns. Plumage fresh and fully saturated in autumn; old, worn, and often greyer above and paler below by spring, becoming even patchily abraded by summer. (2) British and Irish race, *melophilus*. Upperparts warmer and darker tawny-brown than nominate *rubecula*, chest deeper rufous-orange. (3) North-west Asian race, *tataricus*. Forms end of eastward cline of decreasingly saturated plumage, first noticeable in central European populations and quite marked beyond Moscow. Distinctly paler, with bluer-grey wash above and markedly paler below, with chest appearing 'faded' orange in spring and noticeably white belly and vent (but worn nominate *rubecula* can also look like this). (4) Race of Gran Canaria and Tenerife (Canary Islands), *superbus*. Abruptly different in appearance from birds of other Canary Islands (nominate *rubecula*): conspicuously darker, grey to mud-brown above and richly red-chestnut on face and chest, contrasting noticeably with wider deep blue-grey surrounding fringe and almost white belly and vent. (5) Algerian and Tunisian race, *witherbyi*. As dark above as *melophilus* but smaller, with chest colour of nominate *rubecula*. (6) Crimean race, *valens*. Greyer above than nominate *rubecula* but less blue than *tataricus*. Differs from all preceding races in having upper tail-coverts and tail distinctly rufous. (7) Caucasus race, *caucasicus*. Darker above and on chest than *valens* but with lower rump and tail still more rufous. (8) North-west Iranian race, *hyrcanus*. Forms end of south-eastward cline of increasingly saturated plumage. At least 10% longer-billed than other races. Differs from *caucasicus* in rich red-chestnut breast and reddest upper tail-coverts and tail. JUVENILE. All races. Quite unlike adult, lacking uniform upperparts and orange chest. Essentially brown above and buff below, with copious pale buff spots on feathers of rear crown, mantle, and scapulars and dark brown crescents and pale buff spots on lower face, chest, and flanks; wings show buff-spotted lesser and median coverts and fuller wing-bar on greater coverts. Underwing and belly buff. FIRST WINTER. All races. Juvenile plumage replaced from August in migratory races, by late September in sedentary ones; then doubtfully distinguishable from adult, though immature has less saturated plumage and fuller wing-bar (see Plumages). In all races at all ages, bill dark brown; legs brown.

Unmistakable in good view but liable to suggest other small chats (e.g. nightingales *Luscinia*, Red-flanked Bluetail *Tarsiger cyanurus*) in brief glimpse or in spotted juvenile plumage; tail pattern and tail action important to resolution of such latter confusion (see discussions under other species). Flight usually low and flitting, with easy, rapid wing-beats, and quick turns and dives amongst and into cover; action often contains half-cock of tail, producing jerky progress which over longer distance also rises and falls as if bird uncertain how to sustain lift. Shows well-balanced form in flight, with broad head and rump, quite rounded wings, and full tail. Stance usually rather upright, with bold head carriage, prominent chest, and frequently flicked wings and tail making ♂ especially appear full of character; excited or suspicious bird bows and bobs, even swaying on legs with cocked tail. Uses both short and long hops in quick succession, movement characteristically interrupted by pauses and wing- and tail-flicks. Perches freely on all sorts of natural and artificial materials but may also skulk in thick cover for long periods; has characteristic 'drop' from lower perches, bird falling forward with flick of tail and loosening of wing-tips; also darts after food from hedge or bush. Pugnacious. Essentially a bird of dense vegetation; British race *melophilus* tame (seemingly addicted to human habitation and affection), but continental races noticeably wilder and more skulking, this behaviour persisting on migration and in winter quarters.

Song bitter-sweet in quality (particularly in autumn): short, liquid (even bubbling) warble interspersed with shriller notes and lasting usually less than 3 s; given by ♂ in all months except July and also by ♀ in autumn and winter. Commonest call a scolding 'tic tic' confusable with notes of many other species, e.g. *Tarsiger cyanurus* and Little Bunting *Emberiza pusilla*.

Habitat. Breeds in upper and especially middle latitudes of west Palearctic, in boreal, temperate and Mediterranean zones, in oceanic and continental mainly humid lowlands and wooded mountains to treeline (up to 2200 m in Switzerland) between July isotherms 13–23°C (Voous 1960). Preferred habitat includes elements of cool shade, moisture, cover of at least medium height and not more than medium density, patches or fringes of open ground

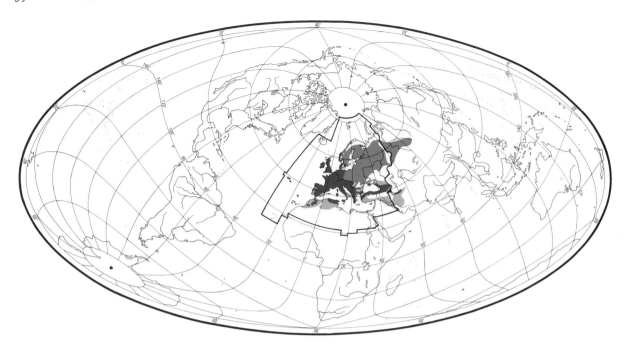

(free of tall or obstructive vegetation, stones, ridges, or marked irregularities) and song-posts giving adequate view without undue exposure. Avoids hard dry surfaces of all types; prefers those lately disturbed or lending themselves to light probing for foraging. Accordingly avoids all kinds of wide open spaces, natural or cultivated, but also wetlands unless overgrown with woody vegetation. In some parts of range occupies coniferous forest, especially with moist mossy floor and ample dead wood, but in others is attached to broad-leaved or mixed woodland and, especially in Britain and parts of western Europe, to parks, gardens with trees and shrubs, road verges, burial grounds, and other humanly managed and disturbed habitats. In Britain, commonly shares such habitat with Dunnock *Prunella modularis* (except where trees become too tall and dense for that species) and Wren *Troglodytes troglodytes* (so long as opportunities for ground feeding are adequate). Although largely a ground-feeder, seeks wide variety of raised perches, even at some height, and values proximity of some water, and of walls, banks or rock faces, especially when clothed in vegetation. Spread in parts of range in modern times from forest to farmed and settled habitats may have arisen from early association with man in woodland management operations (Sharrock 1976). An arboreal and ground species, flying only in lower airspace.

Distribution. FAEROES. Bred 1960, 1964, and 1966 (Bloch and Sørensen 1984). BRITAIN, IRELAND. Colonized Orkney in late 18th century (Parslow 1973). Has bred sporadically Tiree and Coll (Scotland) (Sharrock 1976). FINLAND. Spread in recent decades (OH).

Accidental. Jan Mayen, Iceland, Jordan.

Population. Fluctuates, but has decreased Denmark and increased Finland.

BRITAIN, IRELAND. Over 1 million pairs; no evidence of marked widespread change, except for short-term fluctuations, often caused by hard winters (Parslow 1973). Estimated 5 million pairs; fluctuates, especially after hard winters (Sharrock 1976). FRANCE. Over 1 million pairs (Yeatman 1976). WEST GERMANY. 1·9–5·5 million pairs (Rheinwald 1982); probably too low (AN). DENMARK. Fluctuating; some decrease 1978–82 (Vickholm and Väisänen 1984). SWEDEN. Not less than 5 million pairs (Ulfstrand and Högstedt 1976). BELGIUM. About 260 000 pairs (Lippens and Wille 1972). NETHERLANDS. 120 000–170 000 pairs; stable (Teixeira 1979). FINLAND. 410 000 pairs (Merikallio 1958). Increase in recent decades (OH). CZECHOSLOVAKIA. No changes reported (KH). TURKEY. Numbers rather small (Vittery *et al.* 1971).

Survival. Europe: annual mortality 58–62%, lowest in 1- and 2-year-olds (Paevski 1977). Britain: annual adult mortality 62%; annual 1st-year mortality 72% from 1 August (Lack 1965). Finland: annual overall mortality 76% (Haukioja 1969).

Movements. Most populations partially migratory, with ♂♂ more sedentary than ♀♀; totally migratory in north-east of range and probably largely sedentary in extreme south. Winters south to Saharan oases and Middle East. Movements in western Europe documented by ringing recoveries (see Erard 1966, Speek 1972, Mead 1984*a*), but few reliable data for east and south of range. A nocturnal migrant, though some local movements occur by day.

Main wintering area lies from Ireland and Britain

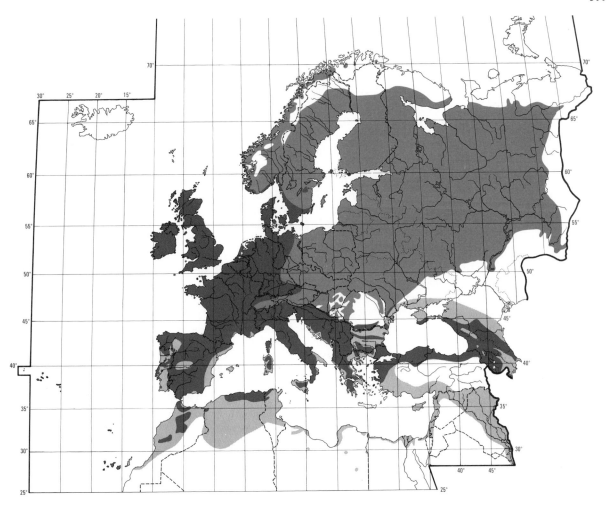

south to Morocco and south-east through Europe with few birds north-east of line from southern Denmark to Bosporus (Turkey), though some December records as far north-east as Västerås, Sweden (58°31′N 16°47′E), and Masuria district, north-east Poland (Tomiałojć 1976; J Ekberg). High ground of central Europe largely vacated. Notable concentrations in Mediterranean basin, including areas where breeding does not occur, e.g. parts of Iberian and Yugoslavian coasts, southern Turkey, and most Mediterranean islands. In North Africa, scarce south of coastal strip and records below 32°N rare. Reports from Ahaggar massif (Algeria) at 24°N in January (P E Smith) require confirmation. Often abundant near North African coast: densities in Algeria range from 62 per km² in hillside maquis to 250 per km² in mixed woodland (Rooke 1947). Similar densities near Casablanca (Morocco) with 130 per km² in mixed rural habitats (J Wilson). Scarcer around Gulf of Sirte, but moderately common along Cyrenaican coast of Libya into Egypt (Meinertzhagen 1930; Bundy 1976). Up to 50 per km² in coastal scrub in western Sinai (B Little). In east Mediterranean, scarce away from coast. Winters on Black Sea coasts north to

Odessa, and east to Caspian Sea, avoiding high ground (Dementiev and Gladkov 1954b). At least sometimes winters in south-west Turkmeniya near Krasnovodsk (Dementiev and Gladkov 1954b). South-east limit in Middle East poorly defined. Regularly winters in small numbers in Kuwait and more irregularly east along south Persian Gulf to Dubai (Bundy and Warr 1980), exceptionally to Oman (D G C Harper). Status to north of Gulf requires investigation, though common in Zagros mountains (Iran) and Basra marshes, Iraq (Dementiev and Gladkov 1954b; D G C Harper). Wintering grounds of different races poorly defined except for *melophilus* and western populations of nominate *rubecula*.

British and Irish populations (*melophilus*) largely resident, but some birds migrate SSW at least as far as southern Iberia. Most recent review of ringing data by Mead (1984a). Juvenile dispersal can begin in May, but few more than 1 km from natal site before early July. Peak dispersal by juveniles varies from early July to late September in different populations (Harper 1984a; D G C Harper). Few juveniles move more than 5 km (Mead 1984a); late dispersers within population tend to move furthest (Harper 1984a). Proportion of adults leaving breeding site variable. In most populations ♂♂ tend to be sedentary: in Enniskillen (Ireland) 75%, n = 24 (Burkitt 1926); Devon (England) 70%, n = 20 (Lack 1965); Cambridge (England) 77%, n = 74 (Harper 1984a); Oxfordshire (England) 69%, n = 36 (D G C Harper). ♀♀ less so, however, with corresponding figures for above studies 36% (n = 25), 33% (n = 33), 30% (n = 78), 3% (n = 31). Many sites abandoned by both sexes, including some dense woodland (Lack 1965; D G C Harper) and upland areas (D G C Harper, M B Horan). Timing of adult dispersal usually bimodal with birds leaving before or after moult. Relative proportions in peaks variable, ♂♂ rarely leaving before moult. At some sites, dispersal delayed until up to 2 months after moult (D G C Harper). Ringing recoveries suggest that most birds move less than 5 km (Mead 1984a); notable influxes into marginal breeding habitats ranging from reedbeds (D G C Harper) to suburban areas (Swann 1975; East 1981b). Some birds (probably less than 5% of ♀♀ and very few ♂♂: Mead 1984a; Lack 1986) emigrate, with recoveries along Atlantic coast from Netherlands south to c. 37°N in Spain (Mead 1984a). All recoveries within 500 km of Atlantic, except for 1 in Switzerland (found in April and so may have wintered even further south). Field identifications of *melophilus* south to Malta (Sultana and Gauci 1982) and Gibraltar (T Bacon) require confirmation; inspection of material at BMNH suggests *melophilus* and nominate *rubecula* only separable using large series of specimens (D G C Harper). Small numbers of *melophilus* emigrating means passage largely obscured by larger movements of nominate *rubecula*, especially on east coast of Britain (see below). Autumn passage through Ireland and south-west Britain mid-August to late October (exceptionally late November), peaking early October, appears to involve mainly *melophilus*. Return passage much less marked, especially in west where noted late February to May, peaking early March. (Ruttledge 1975; Dymond 1980; D G C Harper.) Racial identification of birds on passage in Hebrides, Orkney, and Shetland dubious; here treated under nominate *rubecula* (see below). Arrivals at breeding sites occur mid-December to early May, with large variations between populations; earliest arrivals presumed to involve short-distance movements. At 2 suburban sites, most ♀♀ arrived by early February, pairing at about same time as sedentary ♀♀ (Jackson 1958; Harper 1985c). In 4 rural studies, migrant ♀♀ arrived from late February, pairing later than sedentary ♀♀ (Burkitt 1926; Lack 1965; Beven 1976; East 1981b). Upland areas in Scotland sometimes not occupied until April or even May (D G C Harper).

Continental European populations of nominate *rubecula* vary from totally migratory in north to (probably) sedentary in south. Scandinavia, USSR, Poland, and West and East Germany largely vacated in winter, with a few birds in southern Sweden, mainly on coast. Birds wintering in Poland and East Germany do so especially in urban areas (Tomiałojć 1976; K Metz). All except a few birds, mainly ♂♂, said by Drost (1935) and Niethammer (1937) to leave Germany, but proportion wintering in West Germany probably exceeds 30% (C Hoth). Estimated 50% of Belgian population migratory (Verheyen 1956b), and this proportion assumed to decrease in more southerly populations (Erard 1966), with most Iberian populations believed largely sedentary (C M Herrera). Uplands of central Europe largely abandoned with only c. 10% of Swiss and c. 25% of Czech population resident (Schifferli 1961b; K L Béres), most of Swiss residents probably ♂♂ (Glutz von Blotzheim 1962). Autumn passage on broad front. Passage on Polish Baltic coast mid-August to early November, peaking late September and October; passage in south appears to occur slightly later than in north (Erard 1966; compare Schifferli 1961b and Speek 1972), and also slight tendency for adults to pass later than juveniles (Lebreton 1962; Pettersson 1983; but see Murillo and Sancho 1969). Departs from Urals (USSR) late September, but rarely vacates Moscow area until mid-October; exceptional record of 2 in 12-15°C of frost on 5 November near Moscow (Dementiev and Gladkov 1954b). Northern Norway largely vacated by early October (S Rjukan), and passage in Belgium ceases by mid-November (Lippens and Wille 1972). Arrivals at wintering sites in Andalucia (Spain) occur from early September, peaking in October (C M Herrera), and North Africa reached by late September (exceptionally late August: M Pickers), with largest influxes during October (Rooke 1947; Bundy 1976). Most records from Saharan oases in period late December to March; one in November at c. 30°N (Bundy 1976). Wintering range and passage best known for Scandinavian populations (see Erard 1966, Mead 1984a): most winter from Baltic coast of Germany south-west through Low Countries, western France, and Iberia into Morocco and from there east along northern and southern coasts of Mediterranean; no ringing recoveries in Africa east of Gulf of Sirte (Libya). Scandinavian birds probably responsible for most of passage by nominate *rubecula* through Britain (possibly Ireland also: Ruttledge 1975), especially on east coast where most falls occur during periods of north-east winds (Taylor 1984). However, continental birds from as far east as Poland sometimes reach Britain (Erard 1966). Few nominate *rubecula* winter in Britain (Mead 1984a; D G C Harper); further data on distribution required. Racial status of birds on passage and wintering in Hebrides, Orkney, and Shetland controversial, but most are probably nominate *rubecula* (D G C Harper). Some nominate *rubecula* usually reach north-west to Faeroes on autumn passage, occasionally in large numbers (Bloch and Sørensen 1984). Birds breeding in eastern Scandinavia tend to winter further east than those from west (Erard 1966), though birds caught on autumn passage at single location scatter widely, e.g. birds at Christiansö (Denmark) recovered between Iberia and Greece (Rabøl 1982). However, no Scandinavian breeding bird or young recovered east of Italy (see Mead 1984*a*). Possible that birds ringed on passage in Scandinavia and recovered as far east as Cyprus (Erard 1966) had straggled west of normal passage route from USSR and subsequently re-oriented (one bird ringed in Sweden in October recovered next August in Estonia, USSR: Kastepold and Kabal 1982). Otherwise, recoveries in east Mediterranean suggest south-east orientation by minority of Scandinavian birds, or undetected secondary movement eastward through Mediterranean. Wintering range of USSR breeders poorly defined, but likely to be somewhat to east of Scandinavian birds; unlikely that many, if any, nominate *rubecula* from USSR winter in Caucasus or Tashkent (Dementiev and Gladkov 1954*b*) since ringing recoveries tend to be towards SSW or south-west (Erard 1966). Birds ringed at Rybachiy (Kaliningrad, USSR) recovered south-west to Iberia, south to Algeria, and east through Mediterranean to Turkey (Erard 1966). Populations of western Europe winter in same general area as Scandinavian birds (see Erard 1966). Birds ringed March–September in Netherlands recovered further west in winter than those ringed October–February (Speek 1972). Possible that many of latter were on passage from more north-easterly populations; same-winter movements from Netherlands into West Germany (Speek 1972) might indicate re-orientation by such birds. Wintering range of central European breeders indicated by concentration of recoveries in south-east France and Iberia; also singles from Switzerland, southern West Germany, and Czechoslovakia to Algeria, and from Hungary to Yugoslavia (Erard 1966). Extent of movement by Mediterranean populations not known, though in Balkans and Turkey probably partially migratory with vertical displacement in upland areas (Erard 1966), migrants probably reaching North Africa. One recovery in Greece from Malta (Sultana and Gauci 1982) possibly of bird from more northerly population. Passage across Bosporus (Turkey) from late September to early November, peaking in late October (Porter 1983). Arrivals in Cyprus from early October onwards, in force by early November (Flint and Stewart 1983). Birds ringed Cyprus in October recovered in Lebanon in spring (Flint and Stewart 1983).

First signs of return passage in early February with increasing numbers along coasts of Algeria and Morocco (Rooke 1947; B Little). By late February, birds leaving Cyprus (Flint and Stewart 1983) and arriving in Switzerland and southern West Germany (Schifferli 1961*b*; C Hoth), exceptionally southern Norway (S Rjukan). Peak passage throughout most of range in March, with exceptional records as far north as 69°N in Norway by end of March (S Rjukan). Although Baltic coast of USSR reached in mid-March, birds do not reach Urals until early May (Dementiev and Gladkov 1954*b*). Egypt vacated

from early March or late April, and all North Africa by end of April, probably earlier in west than in east (Rooke 1947; Bundy 1976; M Pickers). Apparent increase in numbers during March in southern Greece (T Köhl), and slight passage on Lebanese coast in late March (P Steele). Cyprus largely vacated by mid-April, with stragglers until early May (Flint and Stewart 1983). After April, few records from Mediterranean islands where breeding does not occur, though exceptional summer records from Malta (Sultana and Gauci 1982). Wintering sites in Andalucia (Spain) vacated between mid-February and late April (J Halcide). Some sites in Alps not occupied until early May (Schifferli 1961*b*), about time northern Scandinavian sites fully occupied (H Christiansen), though this sometimes further delayed by late thaw (S Rjukan).

E. r. tataricus, breeding east from Urals, totally migratory. Recorded in winter from south-east Kazakhstan (USSR) and Iran (Vaurie 1959). No data on timing.

Caucasian and Transcaucasian races, *caucasicus* and *hyrcanus*, appear to be mainly sedentary with vertical displacement in upland areas. However, in north of breeding range, North Ossetia (USSR) almost entirely vacated from September or November (Dementiev and Gladkov 1954*b*). Migrants may move south to Zagros range (Iraq) and north Persian Gulf (Dementiev and Gladkov 1954*b*). 4 out of 70 wintering birds examined in Cyprus were probably *hyrcanus* (Flint and Stewart 1983). Return passage little known; arrives in North Ossetia late February or early March and wintering sites in south-west Turkmeniya (USSR) vacated during March (Dementiev and Gladkov 1954*b*).

Populations of Morocco (nominate *rubecula*) and of Algeria and Tunisia (*witherbyi*) are virtually restricted to montane forest. No evidence for vertical displacement and presumed sedentary (Rooke 1947; Moreau 1966; J Wilson).

Populations of Azores, Madeira, and western Canary Islands (nominate *rubecula*) and of central Canary Islands (*superbus*) are presumed sedentary. Absence of distinctive *superbus* outside breeding range suggests residency; not known to breed on Lanzarote, but April record there may indicate breeding rather than dispersal (Trotter 1970). On Tenerife, *superbus* virtually restricted to montane forests and no evidence for vertical displacement, possibly because movement to lower altitudes might incur competition with resident Blackcap *Sylvia atricapilla* (Lack and Southern 1949). The species is regularly noted in eastern Canary Islands between late September and November and from February to early May; heaviest passage in late spring while *superbus* nesting, and birds moving are therefore presumably nominate *rubecula* of unknown origin (Bannerman 1919; K Lockwood). 2 October records of nominate *rubecula* on Ilhas Selvagens between Madeira and Canary Islands (Araújo and Rufino 1981) might indicate dispersal from Madeira or extreme straggling by birds arriving in North Africa.   DGCH

Food. Invertebrates, especially beetles and (in southern Spain at least) ants; also fruit and seeds in winter. Uses 2 main methods to locate prey: (1) perching on bush or low branch, flying down to eat prey, then returning to perch (may fly up to c. 20 m to pick up tiny insect); (2) hopping on ground, but usually not turning over leaves, etc. In woodland, thus depends primarily on animals moving on surface of litter layer, but also sometimes takes prey from branch or leaf, and occasionally catches items in the air. (Lack 1948a.) In experimental study, 5 of 10 wild birds readily took conspicuous immobile food, but only 1 took any cryptic immobile items (Lawrence 1985a). In Sussex (southern England) in winter, both sexes found to increase proportion of time spent ground foraging and to decrease hunting from perch at lower temperatures (Table A); also, ♂♂ significantly decreased lengths of ground foraging bouts and (when hunting from perch) time between perch changes and time between capture attempts (Table B). In southern Spain, October–February, 212 observations indicated hunting from perch used most in November and February (during 82% of observations in each month), least December–January (53%); ground foraging varied from 4–6% (November and February) to 42–48% (December–January); other methods 7–13% (Herrera 1977; see also Beven 1959). When ground frozen, will accompany Pheasant *Phasianus colchicus*, mole *Talpa*, man, etc., to feed at areas where hard surface broken (Lack 1948a); also (when no frost) will feed on ground disturbed by mole tunnelling beneath (Lack 1965; Sharrock 1982); well-known habit of accompanying digging gardeners presumably an extension of this (Lack 1948a). Will take items from surface of streams, and recorded catching small fish from a drying stream and by plunging into pond (Lack 1948a). Once seen persistently wading briefly in stream up to c. 7·5 cm deep to pick items off bottom; also hovering briefly over surface before plunging in, immersing head and shoulders (Goodwin 1953b). Not infrequently feeds between tidemarks on seashore, and seen to plunge head into saltwater pool and to take crustaceans from mud of a saltwater creek. Will take berries which other birds in bushes have let fall to ground, and sometimes picks growing fruits off bushes itself, often hovering near them. (Lack 1948a.) Recorded holding small snail in bill and repeatedly hitting it against fence, like Song Thrush *Turdus philomelos* (Pryce-Jones 1972). Bird catching bee *Apis* appeared to pick off and discard sting, then kill bee by crushing its head before swallowing it whole (Suffern 1965b). During thick snow, one bird came to artificially provided food on average every 19 min (6–30). Large eyes perhaps assist in feeding in poor light as, among diurnal woodland birds, it is often the latest active at dusk and earliest at dawn. (Lack 1948a.) Will feed after dark near artificial lighting at all times of year (e.g. King 1966, England 1978a), though most records refer to birds feeding young or to periods of cold weather (D G C Harper). Recorded catching flying insects around street lamp, and moths (Lepidoptera) at lighted window (Green 1978a).

Diet in west Palearctic includes the following. Invertebrates: damsel flies and dragonflies (Odonata), crickets, etc. (Orthoptera: Gryllidae, Tettigoniidae), earwigs (Dermaptera), bugs (Hemiptera: Anthocoridae, Tingidae, Reduviidae, Cicadellidae), thrips (Thysanoptera), ant-lion larvae *Myrmeleon*, adult and larval Lepidoptera (Pieridae, Noctuidae, Geometridae), adult and larval flies and their eggs (Diptera: Tipulidae, Trichoceridae, Bibionidae, Borboridae, Muscidae), Hymenoptera (sawflies Tenthredinidae, Cynipoidea, ants Formicidae, bees Apoidea), adult and larval beetles (Coleoptera: Carabidae, Gyrinidae, Silphidae, Staphylinidae, Lucanidae, Scarabaeidae, Elateridae, Dermestidae, Cantharidae, Coccinellidae, Cerambycidae, Chrysomelidae, Curculionidae), spiders (Araneae), mites (Acari), woodlice (Isopoda), sandhoppers (Amphipoda), millipedes (Diplopoda), centipedes (Chilopoda), small molluscs (including slugs), earthworms (Oligochaeta). Small fish (minnow *Phoxinus*, roach *Rutilus*) and lizards *Lacerta*. Plant material mainly fruits and seeds: of juniper *Juniperus*, yew *Taxus*, spurrey *Spergula*, strawberry *Fragaria*, apple *Malus*, cherries *Prunus*, Cotoneaster, rowan *Sorbus*, bramble *Rubus*, vine *Vitis*, wild vine *Ampelopsis*, *Phytolacca*, privet *Ligustrum*, mistletoe *Viscum*, currants *Ribes*, elder *Sambucus*, *Viburnum*, snowberry *Symphoricarpos*, oak *Quercus*, dogwood *Cornus*, *Lonicera*, *Rubia*, bilberry *Vaccinium*, strawberry-tree *Arbutus*, *Myrtus*, oleander *Olea*, *Osyris*, *Phillyrea*, pistachio *Pistacia*, spindle *Euonymus*, sea buckthorn *Hippophae*, buckthorn *Rhamnus*, *Smilax*, *Tamus*, mezereon *Daphne*,

Table A   Sex and temperature differences in proportions of different foraging techniques used by Robins *Erithacus rubecula* in Sussex (south-east England) in winter. Figures are percentages; total observation time 860 min. (East 1980.)

Sex	Temp (°C)	Ground foraging	Perch hunting	Leaf gleaning	Bark pecking	Fly-catching
♂	0–5	54	34	0	12	0
♂	6–10	37	45	5	11	2
♀	0–5	72	25	0	3	0
♀	6–10	49	34	4	11	1

Table B   Sex and temperature differences in winter foraging behaviour of Robins *Erithacus rubecula* in Sussex (south-east England). Figures are median times in seconds. (East 1980.)

Sex	Temp (°C)	Length of ground-foraging bouts	Time between perch changes	Time between capture attempts	n
♂	0–5	18	14	26	20
♂	6–10	9	20	71	40
♀	0–5	24	16	32	17
♀	6–10	19	14	30	33

nightshade *Solanum*, cereal grain; occasionally herbage and once a gilled fungus *Melanoleuca*. Will eat carrion, especially in cold weather—once, meat in butcher's shop; also table scraps, especially fat. (Florence 1914; Gil 1928; Schuster 1930; Witherby *et al.* 1938*b*; Lack 1948*a*; Dementiev and Gladkov 1954*b*; Suffern 1965*b*; Flint 1976; Herrera 1977, 1984*a*; Kiss *et al.* 1978; Radford 1978, 1984; Kostin 1983; Mead 1984*a*.)

In experimental study with 30 captive birds, found not to prefer plant food to animal food at any season, and no preference for any particular sort of fruit; birds fed all-fruit diet lost weight or died (see also below), but addition to diet of 3 g of beetle larvae per day sufficient to maintain body weight (Berthold 1976). For experimental study, with unreliable conclusions, on food colour preferences, see Radford (1970). Seen to catch and drop (apparently deliberately) butterflies—a white (Pieridae) and a small tortoiseshell *Aglais urticae* (Lack 1948*a*). According to Herrera (1978*b*), individuals with short tarsus or long bill take significantly greater variety of prey sizes, but analysis based on only 12 stomachs and inadequately controlled. Food items often tiny: observer 1 m away may not be able to see what bird is taking (Lack 1948*a*).

Of 197 food items from 13 birds (place and time of year not given), 42·1% (by number) adult beetles (mostly Chrysomelidae), 18·3% seeds, 7·1% fruits, 6·1% beetle larvae, 5·6% earwigs, 4·1% Hemiptera, 3·6% larval Lepidoptera, 3·6% larval Diptera, 3·0% adult Diptera, 2·5% spiders and mites, 1·0% ants, 1·0% other Hymenoptera, 1·0% molluscs, 0·5% earthworms, and 0·5% thrips (Lack 1948*a*). In breeding season in Crimea (southern USSR), stomachs contents included 39·4% beetles (mostly Curculionidae), 16·1% Hymenoptera (including 8·8% ants), 14·7% millipedes, 8·7% Hemiptera, 5·8% Lepidoptera larvae, and 0·7% plant material; in winter, stomachs contained insects and Arachnida, also berries and seeds (Kostin 1983). In holm oak *Quercus ilex* woodlands in southern Spain, October–February, stomachs contained 16·5% (by volume) plant material in October (*n* = 17), rising to 78·4% in December (*n* = 16), falling again to 39·1% in February (*n* = 22); plant material was largely remains of acorns left by Nuthatches *Sitta europaea* and Great Tits *Parus major*; of 1905 invertebrates present, 75·9% (by number) ants, 12·2% beetles, 4·3% insect larvae, 2·3% earwigs, and 5·1% others; ants and beetles largely 2–8 mm long, larvae up to 20 mm or more. In farmland of southern Spain, November–January, 22 stomachs contained 26·0–42·3% (by volume, monthly averages) plant material, entirely berries and other pulpy fruit; these birds had lower body weights than those from woodland and had weight negatively correlated with proportion of fruit in stomach (positively correlated in woodland birds, which had been eating acorns instead of soft fruit); invertebrates component similar to woodland birds, 311 items comprising 72·7% (by number) ants,

15·1% beetles, 5·5% insect larvae, and 6·7% others. (Herrera 1977.) At 1150 m in Sierra del Pozo (southern Spain) over 4 winters, faeces showed fruit in diet comprised 43·9% (by number) *Pistacia lentiscus*, 31·0% *Phillyrea latifolia*, 19·3% *Viburnum tinus*, and 5·8% others (*n* = 487); at El Viso (altitude 100 m) over same period, 88·8% *Pistacia lentiscus*, 11·2% others (*n* = 133) (Herrera 1984*a*, which see also for fruit taken by Blackbird *Turdus merula* at these sites). Other data as follows. Britain: Florence (1914). USSR: Tarashchuk (1953), Neufeldt (1961), Ptushenko and Inozemtsev (1968), Ganya *et al.* (1969), Averin and Ganya (1970), Prokofieva (1972*a*). Rumania: Kiss *et al.* (1978). Hungary: Csiki (1909). Yugoslavia: Kovačević and Danon (1952). Spain: Gil (1928).

For data on energy intake and utilization, see Korodi Gál (1965) and Korodi Gál and Nagy (1965).

For data on food of young, see Ptushenko and Inozemtsev (1968) and Mal'chevski and Pukinski (1983).

DJB

Social pattern and behaviour. British and Irish populations (*melophilus*) well known; main studies by Burkitt (1924*a*, *b*, 1925*a*, *b*, 1926), Lack (1939, 1940*a*, *b*, 1948*a*, 1965), Jackson (1958), East (1980, 1981*a*, *b*, 1982), and Harper (1984*a*, *b*, 1985*a*, *b*, *c*, *d*). Other populations little known; Adriaensen and Dhondt (1984) provide the only details for main European race, nominate *rubecula*.

1. Solitary and strongly territorial for most of year, defence of territory typically relaxed only during severe winter weather and moult (Lack 1965). Outside breeding season, both sexes defend individual territories; exceptionally, pair-bond maintained (D G C Harper; see Bonds, below). A few birds hold more than one territory simultaneously for up to 6 weeks: these are normally adult ♂♂ which move (up to 500 m) between breeding and winter territories, and juveniles establishing first territories in autumn (Adriaensen and Dhondt 1984; D G C Harper). Timing of territory establishment by juveniles variable. In Cambridge (England), they set up territories in early July when adults start to moult and cease to defend territories: most (62%, *n* = 106) of the juveniles and some (17%, *n* = 82) adults were evicted during fights as adults completed moult (Harper 1984*a*). In Oxfordshire (England) and Enniskillen (Ireland), most juveniles do not establish territories until late September or October after adults have done so (Burkitt 1926; D G C Harper). Median area of non-breeding territory variable both within and between habitats: mixed rural, Devon (England), 0·30 ha (0·07–0·53, *n* = 33) (Lack 1965); suburban, Cambridge, 0·27 ha (0·08–0·63, *n* = 108) (Harper 1984*a*); wood, Oxfordshire, 0·41 ha (0·11–1·09, *n* = 62); downland scrub, Sussex (England), 0·73 ha (0·32–1·34, *n* = 18) (D G C Harper). At Antwerp (Belgium), 17 autumn territories 0·2–0·7 ha (Adriaensen and Dhondt 1984). Similar densities (up to 300 birds per km²) found in Mediterranean wintering sites (Rooke 1947; C M Herrera, D G C Harper). No evidence that ♀♀ defend smaller territories than ♂♂, nor that within-site variation in territory area is related to food density (Harper 1984*a*). Fidelity to wintering sites high in sedentary populations, with most ♂♂ and some ♀♀ defending same site throughout life (Burkitt 1926; Lack 1965). Migrants often return to same wintering sites in successive years (Bannerman and Bannerman 1971; Herrera and Rodrigues 1979; Finlayson 1980), but proportion doing so

not known. Typically migrates singly, occasionally in small parties of 2-10 birds; exceptional flocks of 50-60 perhaps result from concentration of movements by topographic features (D G C Harper). Often territorial at migratory stop-overs (e.g. Szuk-Olech 1964), even while resting on boats (Nelson 1907; D G C Harper). Territories defended while on autumn passage on Lundy Island (England) 0·08-0·20 ha, $n = 13$ (D G C Harper). BONDS. Mating system typically monogamous, but 6% of 71 ♂♂ in Cambridge practised simultaneous bigamy (Harper 1985a) and other accounts suggest low frequency of bigamy typical (Lack 1965; Jackson 1958). Substantial number of ♂♂ fail to pair: 20%, $n = 65$ (Lack 1965); 15·9%, $n = 63$ (East 1981b); 20%, $n = 71$ (Harper 1985c). Bereaved ♀♀ usually move elsewhere unless caring for eggs or young (Lack 1965; East 1981b) and then usually attract unpaired ♂ within days or even hours (D G C Harper). Pairs usually form at breeding site (Lack 1965), but 3% of 92 pairs at Cambridge moved to new site after pairing (Harper 1985c) and at sites not used for wintering at least 50% of 18 pairs arrived already paired (D G C Harper). Pair-bond of seasonal duration, usually terminated immediately before (Harper 1985c) or after (Burkitt 1926) moult; 2 cases of pair sharing territory all year (D G C Harper). Frequency of mate-desertion typically low: in Devon, 12% of 35 pairs broken by ♀ deserting ♂ (Lack 1965); none of 31 ♀♀ in Sussex deserted (D G C Harper); highest rate recorded in Cambridge where 23-37% of 92 ♀♀ deserted and 1% of 92 ♂♂ (only known record). Rarity of desertion by ♂♂ probably due to loss of territory which this would involve. Desertions tend to occur before 1st clutch—either within days or hours of nearby ♂ losing ♀ or while pair caring for last brood of year (Harper 1985b). When both ♂ and ♀ return to same general area to breed in successive years, they typically re-pair: in 2 British studies, this occurred in 70% of 21 cases (Lack 1965; Harper 1985b). Mate-fidelity seems to result from extreme site-fidelity, not vice versa (Harper 1985b; see Breeding Dispersion, below). In woodland site where management caused large changes in habitat, site-fidelity of ♂♂ remained high but only 1 out of 10 ♀♀ with chance to re-pair did so (D G C Harper). Age of first breeding 1 year (Burkitt 1926), but many ♂♂ do not breed in any given year (see above) and these perhaps include higher proportion of 1st-years. Only ♀ broods. Both parents feed young, ♂ taking larger share, especially if subsequent breeding attempt made, when often responsible for nearly all care of fledglings (Lack 1965; East 1981b; Harper 1985b). If ♀ loses mate while caring for young, and re-pairs, new ♂ does not feed young and often attacks them (East 1981b; D G C Harper). Brood-division not uncommon, especially for final brood (Harper 1985b). 2 instances of 3 adults feeding nestlings: in 1st, extra bird was crippled; in 2nd, it was unpaired ♂ from adjacent territory (Harrison 1952; D G C Harper). Young become independent 16-24 days after leaving nest, period decreasing with season in Cambridge but increasing in Sussex (D G C Harper). Aggression extremely rare between nestlings, but common among fledglings, often resulting in peck-order (D G C Harper). BREEDING DISPERSION. Solitary and strongly territorial; at low density, territories often more aggregated than expected from habitat (L Workman, D G C Harper). Median territory area defended by pairs varies both within and between habitats; tends to be about twice the area of winter territory at same site (compare with above): Devon 0·55 ha (0·18-1·47, $n = 29$) (Lack 1965); Cambridge 0·56 ha (0·25-0·83, $n = 63$) (Harper 1984c); Oxfordshire 0·80 ha (0·30-1·12, $n = 53$) (D G C Harper); Sussex 1·44 ha (0·72-2·10, $n = 18$) (D G C Harper). At Antwerp, spring territories 1·2-2·3 ha, $n = 14$ (Adriaensen and Dhondt 1984); on Outer Canary Islands

(*superbus*), 0·9-1·5 ha, $n = 7$ (J Pearson). Unpaired ♂♂ tend to lose ground to neighbouring pairs and are sometimes evicted (Lack 1965; Harper 1984a). Maximum breeding densities in Britain 250-300 pairs per km², average for woodland 30 pairs per km², for farmland 30 pairs per km² (Mead 1984a). In primaeval forest of Białowieża (Poland), density ranged from 48 pairs per km² in mixed coniferous-deciduous to 7 pairs per km² in mixed ash *Fraxinus* and alder *Alnus* (Tomiałojć *et al.* 1984). In Morocco, 25·8 pairs per km² in forest, 1·5 in degenerate forest (Thévenot 1982). Even at low densities, territories rarely exceed 2 ha, even if surplus suitable habitat available (D G C Harper). Territory used for all activities, but birds frequently trespass on other territories while feeding (Lack 1965; Harper 1984a); at low density, nearly all foraging may occur off territory (Burkitt 1926). For defence of territory against other species, see Antagonistic Behaviour (below). Resident ♂♂ typically faithful to one territory for life (Burkitt 1926) but occasionally move up to 500 m (Lack 1965; D G C Harper). Migrant ♂♂ often return to same breeding sites in successive years (Burkitt 1926; Lack 1965; D G C Harper), but proportion unknown. ♀♀ also show strong site-fidelity: at one site, 88% of 16 ♀♀ present in successive years paired on same territory, and 71% of 14 1-year-olds paired in areas which they had defended as juveniles (Harper 1985c). Typically builds new nest for 2nd brood (e.g. Lack 1965) but exceptionally uses same one (Shepperd 1953). ROOSTING. Nocturnal. Usually roosts singly on territory in dense cover 1-3 m above ground, especially in ivy *Hedera* or conifers; less often uses crevices and even holes, including nest-boxes. Occasionally enters buildings. Minority of pairs roost together during breeding season, except for period between nest-building and cessation of brooding when ♀ usually roosts in or close to nest. Mean times of arrival and departure from roost at Cambridge 18 min after sunset and 22 min before sunrise, but sometimes active much later and earlier (D G C Harper). Communal roosts occasional outside breeding season, usually of less than 10 birds, exceptionally 35 (D G C Harper). Communal roosts near Aberdeen (Scotland), occupied September-March, considered to be used by winter visitors defending territories up to 1 km away (Swann 1975). However, communal roosts at 2 English sites, occupied July-April, contained non-territorial birds (D G C Harper). Catchment areas of communal roosts overlap and can each cover 3 km² or more. Some roosts less than 500 m apart. 2 cases of roosts used in successive years, although vacated during breeding season. Birds often enter and leave communal roosts while completely dark, and considerable chasing and vocalization occur in roost (Swann 1975; D G C Harper). Newly fledged young typically roost singly in dense cover, closer to ground than adults; however, occasionally one or more found roosting at nest-site. Independent juveniles often join communal roosts (D G C Harper). Occasionally sings by night (e.g. Jones 1955), more frequently if artificial light available (e.g. Hollom 1946, King 1966) or if disturbed, e.g. by gunfire (White 1967a) or conspecific song (D G C Harper). Sometimes feeds by night (see Food). Comfort behaviour includes sunning, preening, bathing, and anting. Sunning bird fans wings and tail and ruffles body feathers, especially on back which is typically directed at sun (D G C Harper). Bathing usually in shallow pool or slow moving water, but occasionally in damp vegetation (Denny 1952) or snow (D G C Harper). Even in cold weather, often bathes immediately before roosting (Kennedy *et al.* 1954; Hancock 1965; D G C Harper). Active anting with *Formica* and *Lasius* ants occasional (Simmons 1966; D G C Harper), but more frequently uses millipedes, especially *Iulus* (Goodliffe 1969; D G C Harper). Holds millipede or ant (exceptionally up to 4) in bill and rubs it on underside of extended and raised wing, with spread tail thrust sideways and forward to press behind. Sometimes adopts anting posture in dense smoke. (D G C Harper.)

2. In Britain and Ireland typically confiding (sometimes even entering buildings), especially outside breeding season, but elsewhere often extremely secretive (Lack 1965). During primary moult, becomes extremely skulking. On detecting avian predator, plunges into dense cover or presses body against ground or perch, as if to hide orange breast (D G C Harper). SONG-DISPLAY. Over 99% of 2736 song-bouts (see 1 in Voice) delivered from perch, 71% 1–3 m above ground, exceptionally up to 13 m; remaining song-bouts given in flight, mainly while chasing intruders or potential mates (D G C Harper). For vocal and behavioural responses of ♂ and ♀ to playback of ♂ and ♀ song, see Hoelzel (1986). Perches typically concealed and often same perch used repeatedly despite numerous alternatives (Mead 1984*a*). Both sexes sing outside breeding season while defending individual territories; in 2 exceptional cases when pair shared winter territory only ♂ sang (D G C Harper). In some populations, wintering ♀♀ sing less than ♂♂ (Lack 1965; East 1982), exceptionally singing rarely or not at all (D G C Harper); however, most-detailed study found no significant difference (Harper 1984*a*). On average, ♀♀ use lower perches than ♂♂ (East 1982; Harper 1984*a*), but this probably related to sexual differences in foraging (East 1980, 1982; see Food). Prior to pairing, ♂ (but not ♀) shows following abrupt changes: sings more, and uses higher, less concealed perches more centrally positioned on territory than before (East 1982; D G C Harper). In Britain, change occurs between late December and March (exceptionally May) with marked variation within local population; coincides approximately with switch from autumn to spring song (see 1 in Voice) (D G C Harper). Within hours of pairing, equally abrupt reversal of change occurs (except spring song retained); if mate lost, reversal of change occurs again within hours (Harper 1984*a*, 1985*c*). Paired ♀♀ rarely sing, except during fights or while predator near nest (Lack 1965; East 1982; Harper 1984*a*). Song very rare during primary moult, but ♂♂ otherwise sing all year. Seasonal peaks of song coincide with territory establishment and pairing (Lack 1965). Marked diurnal rhythm in song output: pronounced dawn chorus and weaker one at dusk, when Tic-calls (see 2 in Voice) much more prominent (Lack 1965; Harper 1984*a*). Role of song in territorial behaviour discussed by Lack (1965); see below. ANTAGONISTIC BEHAVIOUR. Usually extremely aggressive towards both conspecific birds and some other small passerines throughout year except during primary moult. Aggression most frequent and violent when territories being established or boundaries changing during pairing period. (Lack 1965.) Serious injury to eyes and legs during intraspecific fighting sometimes common and in high-density population in Cambridge at least 10% of 98 adult ♂♂ and at least 3% of 86 adult ♀♀ died as result. Vast majority of encounters resolved without physical contact; serious escalations nearly always related to territorial establishment and extremely rare between neighbours of more than a few days. (Harper 1984*a*.) Outside breeding season, both sexes aggressive to intruders regardless of sex, though some neighbours of opposite sex tolerate each other. These nearly always mates of previous year or parent and offspring, each bird defending part of ♂'s former breeding territory; likely to pair in following year (Harper 1985*c*). When either seriously challenged by conspecific, other often gives support, so that apparent pairs can be seen driving off intruders (Harper 1984*c*). During breeding season, ♂ more aggressive than ♀; both more likely to attack members of same sex, and ♀

often avoids ♂ intruders (Lack 1965; D G C Harper). Intruder typically retreats as soon as it detects conspecific (Lack 1965), even if another intruder (D G C Harper), although ♂ intruders attracted towards nesting ♀♀ (Harper 1984*a*). When owners detect intruder they often merely make presence known by moving on to conspicuous perch or vocalizing (usually song or Tic-calls). Deterrence of intruders by presence of owner may explain regular return visits to territory by owners feeding elsewhere (D G C Harper). Foraging intruders often not evicted (Lack 1965). However, if intruder stands ground, advances, or gives territorial vocalizations (song or repeated Tic-calls), owner usually approaches singing or giving Tic-calls. Sometimes intruder replies, resulting in Song-duel in which opponents tend to match type of vocalization used and rate of delivery. Song-duels often prolonged, although usually interrupted by foraging bouts if they last longer than 30 min. (D G C Harper.) Intruders more likely to enter into Song-duel early in day, if weather mild, or if non-territorial themselves (Harper 1984*a*). Song- and Tic-call duels are common between neighbouring territory-holders, but decrease in length and frequency as birds become more familiar with each other (D G C Harper). As rivals approach, display of orange breast becomes more important. Threat-display highly variable, influenced in part by relative heights of rivals (Fig A, in which upper bird

A

is above rival, middle bird is level with rival, and lower bird is below rival); thus, breasts directed at each other prior to launching attack, body usually held stationary with bill above horizontal, and plumage sleeked (Harrison 1973); otherwise, plumage is often ruffled, especially on throat and crown. Position of wings and tail especially variable; loser tends to flick them more than rival. Tail sometimes held below horizontal and fanned, more frequently raised towards vertical. Wings occasionally extended and raised. Slow lateral swaying of body common; occasionally, legs rapidly flexed with body bobbing up and down. (D G C Harper.) Variation between postures seems to be continuous (D G C Harper, *contra* Harrison 1973), but quantitative analysis much needed. Threat-display sometimes given in silence; typical vocalizations are song and Tic- and Antagonistic-calls (see 5 in Voice), with rivals tending to use same vocalizations. Song often has curious strangled quality (see 1 in Voice), and output positively correlated with length of display (Chantrey and Workman 1984). Sigh- and Hiss-calls (see 13-14 in Voice) more typical of actual fights, but sometimes given if rival approaches caller. Repeated ground-pecking occasionally occurs during Threat-display (Radford 1967b); more frequent when rival is a dummy (D G C Harper). Juveniles rarely give or receive Threat-display until post-juvenile moult initiated, their few aggressive interactions prior to this being dominated by Tic-call duels and physical conflict. This supports contention that orange breast acts as releaser for aggression (Lack 1965), though not the sole one (Chantrey and Workman 1984): thus, stuffed Great Tit *Parus major* is usually vigorously attacked when presented together with recording of *E. rubecula* song (D G C Harper). Considerable variation in response to dummies: Lack (1965) found bundles of orange feathers attacked, and claims of display directed at Mead's (1984a) orange beard contrast with repeated failure in some populations to evoke response with anything other than stuffed or freeze-dried specimen (D G C Harper). Very crude dummy often attacked in one population (Chantrey and Workman 1984) but not even approached in another (D G C Harper). Fights occurred in 13% of 1067 encounters between intruders and owners (Harper 1984a). Physical contact usually starts abruptly, often preceded by vocalizations (see below). Initial attack often involves striking opponent single blows with wings and/or feet, or bowling it off perch. As fight develops, birds begin to roll over and over on ground; then flutter up face to face, while striking with legs, before tumbling to ground interlocked. While interlocked, attempt to pin each other to ground; victor rains blows down on other's head, especially around eyes, occasionally blinding or killing it. Bill-snapping common during fights, and typical vocalizations include song (for fighting variant, see 1 in Voice) and Tic-, Antagonistic-, Sigh-, and Hiss-calls. Most fights last less than 1 min, before loser (almost always intruder) flees; some encounters continue sporadically for up to 1 hr, exceptionally several days. During such long contests, fighting alternates with chasing (owner usually in pursuit), Song- and Tic-call duels, and Threat-displays, as well as bouts of foraging. Some fights develop without gradual escalation described above, and in one exceptional case intruder killed within 80 s of being detected by owner. Loser usually terminates encounter by rapid retreat; no submissive posture recorded and probably no specific call (but see 5 in Voice and below). Trespasser intent on foraging often studiously ignores owner, turning its back towards it, as if to hide orange breast. Brown-breasted juveniles (and occasionally adults) often give repeated Contact-calls as if begging for food (see 6 in Voice) if approached by aggressive conspecific, regardless of latter's age or sex. Single Contact-calls also recorded from losing birds during antagonistic encounters throughout year, suggesting possible submissive function. (D G C Harper.) Other small passerines (rarely larger Greenfinch *Carduelis chloris*) frequently attacked, but rarely excluded from territory (Lack 1965). Sometimes attacks small mammals at food sources (Smith 1960c; D G C Harper). Most victims seem simply to get in way of foraging bird (D G C Harper) and role of mistaken identity (Lack 1965) seems small. Most intense and consistent aggression directed at potential food competitors,

especially redstarts *Phoenicurus*, wheatears *Oenanthe*, and *Saxicola* chats. Occasionally, aggression with Redstart *P. phoenicurus* results in interspecific territoriality (D G C Harper) and stuffed *P. phoenicurus* sometimes attacked (Palmer 1895), even at sites where they do not breed (D G C Harper). Also a report, however, of sharing nest, incubating side by side with *P. phoenicurus* (Booth 1967). Dunnock *Prunella modularis* also victim of much aggression, possibly because ground foraging leads to competition for undisturbed ground. Other species rarely attract Threat-display or song regardless of whether they have orange in plumage or not (D G C Harper, *contra* Lack 1965). HETEROSEXUAL BEHAVIOUR. (1) General. Where most ♂♂ resident, commonest pairing strategy (78% of 89 cases) is for ♀ to join ♂ on his winter territory; otherwise, ♂ moves to join ♀ on her territory (10%) or birds with adjacent territories fuse them (12%) (Harper 1985c). Situation in migratory populations unclear with unknown proportion arriving already paired; unpaired ♂♂ tend to arrive a few days ahead of unpaired ♀♀, but considerable overlap (D G C Harper). (2) Pair-bonding behaviour. For role of song in mate attraction by ♂♂, see Song-display (above). Migrant ♀♀ wander for several days (exceptionally up to 17 days recorded), interacting with succession of ♂♂ before pairing; resident ♀ leaving winter territory to pair usually joins ♂ within 1 day (Harper 1985c). While approaching ♂, ♀ often gives Contact-calls. Early in season, ♂ often highly aggressive, but later treats any silent (even freeze-dried ♂) or Contact-calling bird as if ♀, approaching excitedly and singing loudly. As ♀ draws closer, ♂ often pecks at ground before retreating while singing, and such repeated approaches by ♀ and retreats by ♂ (Song-and-following ceremony) result in hectic chase of ♂ by ♀ (easily misinterpreted as territorial dispute), both birds singing during sporadic rests. Interaction often terminated by ♀ leaving territory or starting to forage, though then often followed closely by ♂, flying excitedly to and fro and flicking wings and tail while at rest. While off territory, ♂ generally silent and makes repeated attempts to drive ♀ back; unusually likely to fight owner if detected intruding with ♀. Most ♂♂ give up and return to territory to sing loudly within a few minutes of ♀ leaving, but in low-density population ♂ may follow ♀ for up to 2 hrs, up to 900 m from territory. Increasingly, ♀ spends most of her time with one ♂, and once bouts of Song-and-following become interrupted by pair foraging together on territory rather than by ♀'s departure, ♀ is then unlikely to pair elsewhere and Song-and-following usually ceases within hours (exceptionally 5 days). Song-and-following also prominent, when birds from adjacent territories pair, with each following other around other's territory. If ♂ moves to join ♀, pairing (initiated by extreme aggression) is inseparable from attempted eviction by ♂; ♀'s behaviour suddenly switches from resistance to participation in Song-and-following. (Harper 1984a; D G C Harper.) In one study, ♂♂ with large territories more likely to pair and paired earlier than those with small ones (Harper 1985c), but this not always so (Lack 1965; D G C Harper). Although behaviour of ♀ strongly suggestive of active mate choice (East 1981b), this difficult to demonstrate (Harper 1985c). Aggressive response of ♀ to ♂ and to any new ♀ usually prevents a 2nd (bigamous) pairing. Song-and-following much reduced in intensity during formation of such a bigamous pair-bond. In contrast to pair-formation earlier in the year, Courtship-feeding (see below) prominent if pair formed during breeding season, regardless of whether ♂ already has a mate or not (D G C Harper). Interval between pairing and nest-building lasts up to 4 months in sedentary populations; pair spend little time together and no obvious pair-bonding behaviour occurs (Lack 1965). Especially in early spring, pair-members often intensely aggressive towards each other, even giving Threat-display (Lindsay-Blee 1939; Teager 1939). Pairs of previous year less aggressive towards each other than new pairs, regardless of age (Harper 1984a). Much aggression related to foraging, and artificial provision of food reduces it (Harper 1984a); during severe weather pairs often separate at least temporarily (Lack 1939; Colquhoun 1940; D G C Harper). Once nest-building begins, aggression unusual, though one ♀ blinded in one eye by mate during egg-laying period (D G C Harper). ♀♀ of bigamous ♂ usually extremely aggressive towards each other, dividing territory between them (Lack 1965; Harper 1985c), though in 2 out of 9 cases shared territory amicably (D G C Harper). Occasionally nest less than 1 m apart (Shepperd 1953) and exceptionally share nest (Forges 1959; D G C Harper). (3) Courtship-feeding. Rarely observed before nest-building unless pairing occurs during breeding season (D G C Harper); ♀ starts giving Contact-calls regularly (up to 20 times per min) within 3 days of completing nest (East 1981b) or before lining it (88% of 66 nests: D G C Harper). Continues to call until eggs hatch, exceptionally longer, even if nesting attempt fails (East 1981b; Harper 1985a; D G C Harper). Initially ♂ makes no response to persistent calling; ♀ sometimes approaches ♂ giving Feeding-call (see 7 in Voice) or even attacks him (D G C Harper). Within a few days (exceptionally 15: D G C Harper), ♂ starts to collect food for ♀, while giving infrequent Contact-calls. ♂ sometimes utters Subsong as he approaches ♀, who stoops, quivering her slightly opened wings rapidly against her body, and accelerates rate of Contact-calls. As food passed from bill to bill (Fig B), ♀ gives

B

Feeding-calls. Food typically consists of large items or bundles of 4–6 items (East 1981b; D G C Harper, *contra* Lack 1940a, c). Contact-calling rate highest while ♀ off nest during incubation, typically 10–15 per min, compared with 5–10 per min during egg-laying (East 1981b), although if time spent on nest included (c. 66% of 938 daylight hours during incubation) positions slightly reversed (D G C Harper). Courtship-feeding rate roughly constant at 30–50 feeds per day from c. 2 days before 1st egg until hatching. Both Contact-calling and Courtship-feeding rates positively correlated with clutch size during egg-laying but not during incubation. (East 1981b; D G C Harper.) Although most Courtship-feeding not followed by copulation, 65% of 1489 copulations followed Courtship-feeding (*contra* Lack 1965). During incubation, ♂ often collects food and then approaches nest giving Contact-calls or Subsong to attract ♀. In some pairs, ♀ only leaves nest when called off by ♂ but in others ♂ never collects food until approached by begging ♀. Occasionally ♂ feeds ♀ on nest; rate of such behaviour highly variable between pairs. (D G C Harper.) In captivity, ♀♀ recorded offering food to adults of both sexes, though only ♂♂ accepted (Lack 1940a). Tame ♀♀ and exceptionally ♂♂ will give Contact-calls to humans and give Feeding-call when fed. (4) Mating. Usually no obvious preliminaries by either sex except that ♀ crouches as ♂ mounts

from ground or nearby perch (Lack 1965). However, 18% of 939 copulations preceded by display by one or both sexes (D G C Harper). Most often (10% of cases), crouching ♀ extends and slightly raises wings which are then flapped slowly and asynchronously as she sways body from side to side (Fig C);

C

head usually directed away from ♂. 4 records of ♂ adopting similar posture before mounting. In other display used by ♀ (7% of cases), she holds tail almost vertical (Fig D); height of

D

head variable but usually directed away from ♂. Display by ♂ (8% of cases) involves slowly hopping around, ♀ c. 1 m away, with tail fanned, wings held close to ground, and bill above horizontal (Fig E); usually sings quietly. When displaying on

E

perch, movements more limited. Immediately before or during mounting, ♀ usually (89% of cases) gives Copulation-calls (see 8 in Voice). Copulation takes place mainly on ground (80% of cases), lasting c. 1(–4) s with single cloacal contact. Starts up to 15 days before 1st egg, peaks just before or during laying, and rare after more than 2 days of incubation. Highest rate at dawn and also smaller peak around dusk. Rate during egg-laying apparently highly variable: at Cambridge, 2·5 per hr over 195 hrs; in Oxfordshire, 0·56 per hr over 43 hrs; in Sussex, 0·16 per hr over 25 hrs (D G C Harper). No records of ♂ forcing copulation on ♀ he is not paired to, but voluntary copulation (preceded by Courtship-feeding) occurs occasionally between birds not paired to each other, though ♀ usually avoids strange ♂♂ whether or not they are carrying food (D G C Harper). (5) Behaviour at nest. Nest-site chosen by ♀ after inspecting several possibilities. Frequently squats in potential site giving subdued Contact- and/or Churring-calls (see 11 in Voice) and occasionally collects pieces of material without building (juveniles of both sexes behave similarly while establishing 1st territory). In spring, behaviour of ♂ variable. Many do not accompany prospecting ♀, but others follow closely; a few follow ♀ into sites, utter Subsong, and exceptionally collect material. (D G C Harper.) ♀ builds nest (Lack 1965), and few reports of ♂ building (Frankum 1955; Olivier 1959) though occasionally carries material without building (D G C Harper). Most ♀♀, even if hand tame, extremely discreet while building; if watched too closely will ostentatiously carry material to false site, then sneak to real nest up to 50 m away. Similarly careful when leaving or returning to nest; during incubation, ♀ often perches close to nest for up to 3 min before returning (D G C Harper). For ♂'s role in calling ♀ off nest (which might reduce predation), see Courtship-feeding (above). Until hatching, ♀ often drives ♂ away from nest (Harper 1984a). RELATIONS WITHIN FAMILY GROUP. ♀ helps chicks out of shell (D G C Harper). Brooded by ♀ only, usually up to 7 days after hatching (East 1981b). Both parents feed young; of 8 broods, all fed within 2 hrs of hatching, although ♂♂ took up to 7 hrs and in all cases ♀ fed young before ♂ (Harper 1984a). Young receive more feeds from ♂ than from ♀ (East 1981b) and ♂ carries more food per feed. Young beg by calling (audible even before hatching) and gaping (directed vertically up, even if nest in cavity with entrance below cup). Eyes open at 4-6 days, after which begging directed more towards parent. While young being brooded, ♂ often transfers food to ♀ either on or close to nest. Captive ♀ fed by observer divided each bundle of food on average between 1·8 chicks (n=317) out of brood of 4; less likely to divide bundle as young grow older. (D G C Harper.) Both parents remove faecal sacs; sometimes eaten, especially by ♀, but usually carried away, especially after day 5; sacs sometimes dropped on another territory. Parents stimulate reluctant young to defecate by pecking around cloaca or tugging at spinal feather-tract. Young leave nest voluntarily before capable of sustained flight. Of 39 broods, all young fledged within 18 hrs of siblings (Harper 1985b); 78% fledged between sunrise and 12.00 hrs (D G C Harper). During fledging, ♀ tends to feed nestlings, while ♂ tends to feed fledged young (Harper 1985b). Most young leave nest without enticement, though parents sometimes perch nearby holding food and giving Contact-calls if last 1-2 slow to leave; once, ♀ pushed chick out of nest. For first 4-5 days after leaving nest, chicks fly weakly and spend most of time within 1 m of ground. Typically move little, except for rapid dispersal from nest-site, during which parents lure chicks by carrying food and giving Contact-calls. Young fledglings usually widely dispersed on territory, but sometimes remain as tight group. As powers of flight improve, become much more mobile, but tend not to follow parents. Family usually remains on territory, though if bigamy practised one ♀ often moves her brood to undefended ground up to 500 m away. Similar movements sometimes made by whole families in low-density populations. (D G C Harper.) When ♀ starts another nesting attempt, ♂ takes over most of feeding of young (Lack 1965; East 1981b; Harper 1985b, d); proportion of feeds by ♀ in such cases increases with brood size (D G C Harper). Brood-division not uncommon: stable for periods of days, typically developing 5-10 days after leaving nest, exceptionally soon after fledging (Harper 1985b). Less frequent if brood followed by another nesting attempt (38% of 55 cases, compared with 82% of 45 cases: D G C Harper), incubating ♀ then feeding only 1 chick (Harper 1985b). In 77% of 35 cases, parents fed chicks of opposite sex (D G C Harper; see also Harper 1985b). Function of brood-division not known (but see Harper 1985b). Newly fledged young repeatedly mandibulate pebbles, leaves, etc., but attention increasingly turns towards potential prey and self-feeding usually evident within 8 days of leaving nest, increasing greatly during next week. Most

foraging attempts initially on ground, feeding in vegetation and in flight becoming more frequent as flight improves. Siblings often aggressive to one another, and stable hierarchy often develops with dominants interfering with parental feeding of subordinates. Adults rarely aggressive to own young, but often attack those of other pairs; ♀♀ of bigamous ♂ can be extremely violent towards each other's young (Harper 1985d). Occasionally feed strange young, mainly when own young recently fledged; ♂♂ more prone to this than ♀♀, and one ♂ permanently adopted strange chick though ♀ ignored it (Harper 1985b, d). Parents feed young for 16-24 days after leaving nest, frequency decreasing sharply after c. 10 days (Harper 1985b). Timing of independence appears partly influenced by young, adults sometimes carrying food while giving Contact-calls after young have dispersed; subordinate young tend to leave natal site before dominants (D G C Harper). In Cambridge, independent juveniles rarely moved more than 1 km from nest before early July (Harper 1984a). Initially lead roving existence, occasionally aggregating in large numbers at communal roosts (see Roosting, above) or good feeding site (exceptionally 32 in 0·03 ha of scrub). For establishment of territories in 1st autumn, see above. Family bonds typically broken at independence, though 2 siblings recorded associating closely for 2 months (D G C Harper). ANTI-PREDATOR RESPONSES OF YOUNG. From c. 5 days after hatching, young fall silent and cower in nest in response to parents' Alarm-calls (see 4 in Voice), and become increasingly likely to do so also if disturbed. After c. 10 days, liable to scramble out of nest and hide in nearby cover if disturbed. Older nestlings liable to give adult Hiss- or Distress-calls (see 15 in Voice) when handled and Distress-call usually prompts siblings to leave nest. Recently fledged young fall silent and cower whenever disturbed, or in response to Alarm-calls. As young become more mobile, tend to seek cover before freezing. About 10 days after fledging, become likely to give Tic- and Alarm-calls when conspecifics do so and when they encounter predators. Cuddly toy, previously eliciting no response, drew strong mobbing response from 15-day-old fledglings shown it holding dead conspecific. (D G C Harper.) PARENTAL ANTI-PREDATOR STRATEGIES. (1) Passive measures. Incubating and brooding ♀♀ often sit extremely tightly, perhaps explaining their higher mortality during breeding season than ♂♂ (Harper 1985c). During incubation, ♂ (rarely ♀) sometimes gives Alarm-calls as predator approaches nest; once young hatched, both sexes nearly always do so (East 1981a), change occurring within a few hours of hatching (D G C Harper). (2) Active measures: against birds. Responses extremely variable, tending to be strongest around fledging. Birds larger than Mistle Thrush *Turdus viscivorus* near nest, and all birds very close to it, usually provoke vocal demonstration, but often perfunctory (D G C Harper). Both Tic- and Alarm-calls usually used together, bird predators eliciting more Alarm-calls than mammals (East 1981a; D G C Harper). Diverting attacks rarely delivered except on Corvidae and all species at nest; even so, many pairs do not attack predators at nest. Owls often closely approached. Little Owl *Athene noctua* often attracts distraction-flights and mock-attack flights; once repeatedly buffeted by ♂ while 20 m from nest. Tawny Owl *Strix aluco* sometimes mobbed by more than one pair, though in 3 out of 11 such cases resident ♂ chased intruding conspecifics (D G C Harper). Raptors (Accipitriformes, Falconiformes) usually avoided. When chased by Sparrowhawk *Accipiter nisus*, one ♂ gave repeated Tic-calls (East 1981a), but this seems unlikely to represent distraction behaviour to protect family since similar behaviour observed also outside breeding season. Willow Warbler *Phylloscopus trochilus* landing at nest-entrance containing nestlings was attacked by ♂, losing feathers about head. Cuckoo *Cuculus canorus* sometimes provokes persistent Alarm-calls, but usually ignored except during period from nest-building to hatching when occasionally chased relentlessly and dive-attacked by both sexes. (D G C Harper.) Stuffed *C. canorus* sometimes attacked (Smith and Hosking 1955; D G C Harper). (3) Active measures: against man. Detailed study by East (1981a). For use of Hiss-call by sitting ♀, see 14 in Voice. Vocal demonstration rare before hatching, but then increases dramatically until shortly after fledging (East 1981a; D G C Harper). Both Tic- and Alarm-calls used, with Tic-call increasing in relative frequency, especially after fledging. Humans attract more Alarm-calls than most mammals (D G C Harper). Rarely approaches intruder closely before hatching, but later likely to fly towards intruder and even circle head calling loudly. When intruder at nest, some birds will mount repeated diverting attacks while giving rattling bursts of Tic-calls; others remain at distance mainly using Alarm-calls. A few, mainly hand tame, strike intruder especially about hands and face, even if rest of torso bare, giving Hiss-call and/or Bill-snapping (Witherby 1920; D G C Harper). One tame bird perched on observer as nestlings were ringed (D G C Harper). On average, ♂♂ more aggressive than ♀♀, but great variation. (4) Active measures: against other animals. Snakes and slow worms *Anguis* cause great agitation, parents making emphatic vocal demonstrations, interrupted by much apparent displacement behaviour (e.g. preening). Birds avoid close approach; sitting ♀♀ usually flush as soon as they detect reptile. Response to mammals variable; intensity changes through nesting cycle as for humans (see above). Small mammals up to size of stoat *Mustela erminea* ignored unless close to nest or fledged young, and then subjected to vocal demonstration sometimes escalating into diverting attacks; dead pygmy shrew *Sorex minutus* found under incubating ♀ after commotion at nest suggests direct attack. Defence sometimes results in death of parent: weasel *Mustela nivalis* found with 2 freshly dead adults by nest containing live nestlings. Response to larger mammals extremely variable. Fox *Vulpes vulpes* and badger *Meles meles* often attract vocal demonstration as soon as detected on territory from start of incubation until young independent. However, most other mammals ignored unless close to nest, when treated like humans (see above), except that Alarm-calls rarely as common. Cats occasionally struck; once, ♂ fluttered repeatedly into cat's face as cat toyed with fledgling. Responses to predators at night virtually unknown. When approached by hedgehog *Erinaceus europaeus* in dark, 2 incubating ♀♀ sat tight until intruder *c.* 50 cm away; one gave Hiss-call and Bill-snapped for *c.* 30 s before flushing, while other slipped off nest without calling; neither heard to call again and both deserted. (D G C Harper.)

(Figs by N McCanch: A after illustrations in Lack 1965 and Harrison 1973; B after drawing in Lack 1965; others after sketches by D G C Harper.)   DGCH

Voice. Highly vocal throughout year, except during primary moult when virtually silent. Song distinctive and familiar, but extensive and variable repertoires require further investigation. Some calls (perhaps most) not context-specific. Bill-snapping frequent during fights and while attacking predators (D G C Harper).

CALLS OF ADULTS. (1) Song. A melodic varied warble; for sexual differences, see below. Autumn song (given by both sexes from moult until pairing period) differs from that used in spring (given from pairing period until moult,

but only exceptionally by ♀), though no detailed analysis: autumn song softer and more wistful than spring song and tends to contain longer phrases (D G C Harper). Most comprehensive quantitative analysis of song structure (based on recordings made throughout the year) by Bremond (1966, 1967, 1968). Typically a succession of phrases each lasting 1–3 s (exceptionally 5·5 s: D G C Harper) and given at rate of 10(–13) per min. Each phrase contains 4–6(–11) different motifs, most lasting less than 0·45 s (D G C Harper), average 4·79 motifs per phrase ($n = 740$ phrases from 28 birds: Bremond 1966). Song highly variable within and between individuals: overall *c.* 1300 motifs known to be used. Motifs not repeated within phrases, and successive phrases different (Bremond 1966, 1967, 1968). Individuals have extensive repertoires of phrases and motifs (275 motifs recorded from one bird: D G C Harper) but difficult to quantify consistently due to apparently continuous modification by singer, and no adequate data on average repertoire size. Individuals seldom share songs or even motifs (Bremond 1966, 1968): each of 3 ♂♂ and 1 ♀ used at least 100 songs and, despite 2 of them being close neighbours, the 3 ♂♂ shared only 0–3 songs (Hoelzel 1986). Mimicry often suspected but difficult to assess due to apparent modification of motifs derived from model: birds mimicking songs of Great Tit *Parus major* and Blackcap *Sylvia atricapilla* gradually modified 'adopted' motifs (D G C Harper). Fig I (spring song) shows

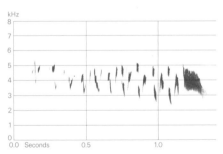

I P J Sellar Ireland April 1976

good mimicry of song of Chaffinch *Fringilla coelebs* (J Hall-Craggs). Overall frequency range 2–9 kHz (Hoelzel 1986). Phrases usually begin with relatively high-pitched motifs (nearly always above 4 kHz, typically above 6 kHz) consisting of discrete notes. There are marked changes in tempo, also in pitch which, in successive motifs, typically alternates above and below *c.* 4 kHz; this alternation important in species-recognition. Phrases tend to start and end quietly, and in 90% of cases, finish in a low-pitched motif. (Bremond 1966, 1967, 1968.) Fig II (spring song) shows typical switching between high and low frequencies early in phrase but not in latter section (whose units are characterized by unusually strong 2nd harmonics); this phrase also lacks a feature prominent elsewhere in both this and an autumn recording, namely the termination of most phrases with long diads or triads

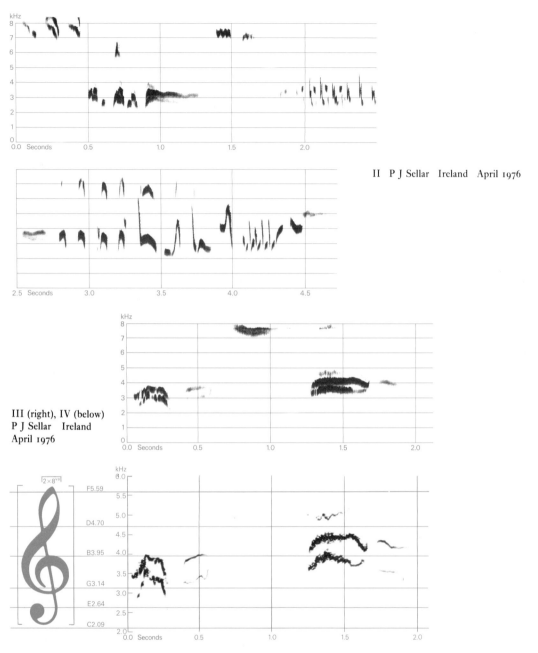

II   P J Sellar   Ireland   April 1976

III (right), IV (below)
P J Sellar   Ireland
April 1976

(simultaneous use of 2–3 'voices') of low to mid frequency. In spring song, of 14 phrases, 6 ended in diads, 5 in triads; exceptional phrase shown in Fig III comprises entirely diads except for penultimate triad; (Fig IV shows same phrase with use of log scale magnifier to clarify structure—see p. 29). In autumn song, no triads but 4 of 6 phrases ended in diads, e.g. Fig V shows 3 short high-pitched diads preceding a long terminal one of lower pitch. (J Hall-Craggs.) ♂'s song of greater duration, diversity, and repetition rate than ♀'s; average duration of ♂'s 1·86 s, of ♀'s 1·14 s (Hoelzel 1986, which see for details). Pitch of song tends to decrease from northern

France to Morocco (Chappuis 1969); further information required. Song of some populations tends to be much simpler than just described. Spring song of *superbus* (Gran Canaria and Tenerife, Canary Islands) comprises short phrases (1–2 s) lacking complex changes in tempo and pitch (D G C Harper, P J Sellar). Near Kirkenes (Norway), similar spring song noted: short phrases (1–1·5 s) repeated at unusually high rate of 15–20 per min (D G C Harper). During antagonistic encounters, so-called 'Fighting Song' has curious strangled, almost slurred, quality (Lack 1965), with phrases often lasting 5 s (D G C Harper): victor in fight often gives loud

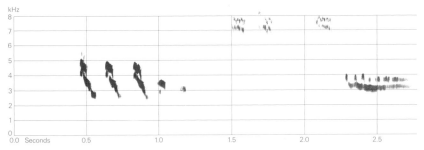

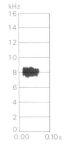

V P J Sellar England October 1973

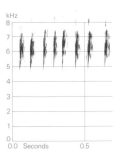

VI P J Sellar England
September 1978

VII I Baker England
April 1981

emphatic song as rival retreats (Lack 1965); unlike in normal song, successive phrases often similar or even identical (D G C Harper). Subsong, given by both sexes in variety of circumstances, is a subdued, complex and rambling warble containing discordant notes. Structurally more variable than full song and phrases not always apparent; when present, successive phrases often differ greatly in length (most 1–15 s). Mimicry more obvious than in full song: has included songs of Blackbird *Turdus merula*, *S. atricapilla*, and *F. coelebs* (P J Sellar, D G C Harper). (2) Tic-call. Short sharp 'tic' (3–6 kHz) mainly used in territorial defence, but also frequently as alarm-call. Very variable rate of delivery and tonal quality; as urgency increases, call repeated more rapidly and sound more deep-throated (D G C Harper, P J Sellar). Recording (Fig VI) shows Tic-calls ranging from more than 4 kHz to more than 8 kHz (as described also by Bremond 1966), and given at rate of *c*. 12–14 per s. More urgent Tic-calls in same recording are lower pitched (3–7 kHz) with similar but less regular rate of delivery. (J Hall-Craggs.) When challenging conspecific, call is often repeated monotonously like clockwork, but sometimes given with irregular tempo to form repeated phrase. Prior to intra- and interspecific attacks, often delivered as explosive rush of notes, impossible to count by ear (D G C Harper). Mammalian predators attract more Tic-calls than Alarm-calls (see call 4, below) (East 1981*a*). (3) Contact-alarm call. High-pitched (*c*. 7 kHz), thin 'tswee', said to resemble Spotted Flycatcher *Muscicapa striata* (Witherby *et al.* 1938*b*), but not as thin or grating; almost like Penduline Tit *Remiz pendulinus* (D G C Harper). Heard in variety of contexts and often intergrades with other high-pitched calls (calls 4–10). In particular, intermediate

between Alarm- and Contact-calls (see calls 4 and 6 below). Compared to Alarm-call, less penetrating, softer, and sounds less plaintive; more distinct from Contact-call, being much sharper but less abrupt. Typically given singly, or calls at least 5–10 s apart. Heard in circumstances suggesting mild alarm: bird wandering off territory, entering trap, encountering novel object, or being closely approached by observer. Also given when birds approach each other in non-antagonistic contexts (when often quiet and conversational; see call 9) and during territory establishment. (D G C Harper.) (4) Alarm-call. High-pitched, sharp, and generally thin 'tseep' notes (*c*. 7 kHz), sometimes an elongated 'tseeeee...' up to 0·8 s long which typically sounds plaintive and appears to fade away. Sonagrams (see Bergmann and Helb 1982) suggest shorter calls consist of 2 frequency bands separated by *c*. 0·7 kHz, while longer calls contain only 1 (possibly corresponding to upper band of short calls) (D G C Harper). Given in presence of predators: typically repeated and often interspersed with other calls, especially Tic-calls. Avian predators attract more Alarm-calls than Tic-calls (East 1981*a*). (5) Antagonistic-calls. High-pitched (*c*. 7·5 kHz) 'see', varying in tonal quality. Possibly 2 calls involved: soft sibilant note lasting 0·4 s given by subordinates and a piercing loud call lasting up to 0·8 s given by dominants with bill slightly open. However, not all calls can be assigned to this dichotomy (D G C Harper). (6) Contact-call. Soft, high-pitched (*c*. 7 kHz) 'dzeep' (Gillham 1955), 'zit' (Harper 1985*a*), or 'tsip' (J Pearson), repeated at regular intervals by ♀ begging for food, when grades into Feeding-call (see 7, below). Also heard in other contexts, including ♀ searching for mate and from subordinate in antagonistic encounters. (7) Feeding-call of ♀. An often hissing, high-pitched sound (*c*. 6–7 kHz) given as ♀ fed by ♂, difficult to describe: 'sweez-eez-eez...' (Witherby *et al.* 1938*b*), 'jeer-jeer...' (Gillham 1955), or 'zheet-zheese-eese-eese...' often accelerating in tempo (D G C Harper). Described as churring by East (1981*b*), but see call 11. (8) Copulation-call of ♀. Slurred rapid repetition of Contact-call (see 6, above) given during mounting (East 1981*b*), sometimes reminiscent of Feeding-call: 'zlee-zlee-zlee...' or 'zlee-zeep-zeep...' (D G C Harper). Sometimes interrupted with Churring-calls (call 11). (9) Foraging-calls. Thin metallic 'peep' calls (Fig VII, mostly 8 kHz) heard from foraging member of pair (I Baker) appear not to be Contact-calls and possibly identical with quiet conversational 'büb' notes (*c.* 7 kHz), inaudible beyond 10-15 m, frequently heard from foraging birds of both sexes, especially when close to mate or (in case of tame birds) to human (D G C Harper). (10) Flight-call. Repeated soft, sibilant, high-pitched (6-7 kHz) 'sip', 'sissip', or similar, mainly heard from nocturnal migrants (D G C Harper). (11) Churring-call. Variable churring notes, often interspersed with other calls, heard in many circumstances; usually 3-6 kHz (D G C Harper). Call of adult tending nestlings (V C Lewis) sounds similar to calls heard in autumn rendered 'ti-churr' (with emphasis on 2nd, lower-pitched, unit: Fig VIII) and a faster less

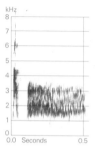

VIII P A D Hollom England
September 1983

powerful 'ti-chuk-chuk' (P A D Hollom). Low rattling 'chi-chuc-chuc...' of variable length, decreasing in pitch and tending to fade away, sometimes heard as birds approach each other in non-antagonistic contexts (D G C Harper). Fast, staccato 'cha-cha-cha', resembling alarmed Sardinian Warbler *Sylvia melanocephala*, but less emphatic, heard from foraging bird (J Pearson). (12) Brooding-calls. Very subdued and abrupt notes, often resembling other calls (especially calls 6 and 11) given during incubation and brooding. Bursts of 5-26 curt 'chup' notes given by captive ♀ turning eggs and as egg hatched (D G C Harper). (13) Sigh-call. Plaintive, drawn-out (typically longer than 1 s), and staccato whine, sounding as if breath being sucked in (Walpole-Bond 1938), given during fights and when cornered by predator, but not once seized (D G C Harper). (14) Hiss-call. Short (typically less than 1 s) explosive hiss (up to *c.* 7 kHz) resembling spit of cat (Witherby *et al.* 1938*b*); sounds like forced exhalation and given with bill slightly open during fights (D G C Harper). ♀♀ and older nestlings sometimes give longer, more sibilant hissing when disturbed in nest and unable to escape (Walpole-Bond 1938; D G C Harper). (15) Distress-call. Shrieking scream, with rapid changes in pitch and tempo, given when cornered or seized by predator, resembling high-pitched (*c.* 5-7 kHz) version of call of Song Thrush *Turdus philomelos* in similar circumstances. Less likely to call than latter species; one study found 22% of 270 netted *E. rubecula* did so (Greig-Smith 1982*d*; Inglis *et al.* 1982); another found calling by 35% of dependent fledglings (*n*=89), 15% of independent juveniles (*n*=117), and only 4% of adults (*n*=226), with no

differences between sexes or seasons (D G C Harper).

CALLS OF YOUNG. Call from chick in egg and young nestlings a soft piping 'peeb' (D G C Harper) or 'piub' (J Pearson). This gradually changes, with age, into adult Contact-call (call 6, above), so that by fledging very similar to adult call, though tends to be more scratchy. Fledglings often give Contact-calls with characteristic tempo, allowing observer to recognize individuals (D G C Harper). Newly hatched chicks sometimes give quiet hissing 'pez-pez-pez...' while being fed; by fledging, this appears to develop into loud, more abrasive adult Feeding-call (see call 7, above). Once feeding complete, but while ♀ still present, chicks often continue begging or give whining, soft 'siib' calls which seem to stimulate brooding by ♀ (D G C Harper). Nestlings give variety of other notes; further information required. By *c.* 10 days old, chicks being handled are likely to give adult Hiss- or Distress-calls (calls 14-15, above), and fledglings heard to use all adult calls except full song and calls 8, 10, and 12. Subsong once heard 12 days after leaving nest, but rarely used until independent of parents. Tic-calls (call 2, above) not heard from nestlings and not freely used by fledglings until *c.* 5 days after leaving nest, although sometimes given by chicks exploding from nest due to predator. Possibly develops from short, sharp 'te' or 'de' given by chicks in mild distress, e.g. when sibling sitting on head (D G C Harper).   DGCH

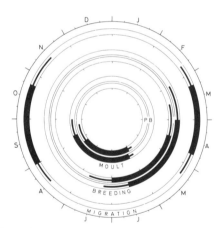

Breeding. SEASON. Britain and Ireland: see diagram; rarely, nests also found in all other months of year (Lack 1946*a*, 1948*b*). Canary Islands: from mid- or late March to early June (Bannerman 1963). Madeira: eggs laid April-June (Bannerman and Bannerman 1965). Northwest Africa: eggs laid mid-April to late May (Heim de Balsac and Mayaud 1962). Southern Europe: laying begins April. Central Europe: laying from end of April to late July (Makatsch 1976). Scandinavia: from very end of April in south but not before mid- or late May in north (Haartman 1969*a*). USSR: from mid-May to mid-June in north; from mid-April in south (Dementiev and Gladkov 1954*b*). SITE. Natural hollow in tree-stump, bank, among tree-roots, rock-crevice, or hollow tree, from ground level to *c.* 5 m up, rarely to 10 m. Makes much use of artificial holes in wide variety of man-made objects, often in or attached to buildings; uses nest-boxes. Nest: bulky structure of dead leaves forming base, on which cup of moss, grass, and leaves is built, lined with finer material including hair, vegetable fibre, and occasionally feathers; average dimensions, outer diameter 13 cm, inner diameter 5 cm, height 4·5 cm, depth of cup 3 cm, weight 16 g (Makatsch 1976). Building: by ♀, taking *c.* 4 days, a few hours each day (Lack 1965). EGGS. See Plate 82. Sub-elliptical, smooth but not glossy; in nominate *rubecula* and *melophilus*, white or faintly bluish, variably marked with sandy-red freckles and small blotches, often sufficient to colour entire egg uniform reddish, but can be very sparse or (rarely) absent, or gathered towards broad end; in *superbus*, whitish-pink or creamy-white, thickly flecked and spotted with rust-red, especially towards broad end, with underlying purplish blotches often forming girdle at broad end. Nominate *rubecula*: 19·6 × 15·0 mm (17·0–22·2 × 13·8–16·1), *n*=200; calculated weight 2·4 g; *melophilus* and *superbus* not significantly different (Makatsch 1976; Schönwetter 1979). Clutch: 4–6 (2–8); increases from west to east and from south to north (see, especially, Lack 1946a, 1948b). Of 1091 clutches, England and Wales: 2 eggs, less than 1%; 3, 2%; 4, 19%; 5, 55%; 6, 22%; 7, 1%; 8, 1%; mean 5·0. Varies with date with 4·6 in March (*n*=97), 5·0 in April (*n*=792), 5·2 in May (*n*=473), and 4·8 in June–July (*n*=136). Other means: Canary Islands 3·5 (*n*=17), North Africa 4·2 (*n*=15), Azores 4·7 (*n*=22), Iberia 4·9 (*n*=8), north-east Scotland 5·5 (*n*=76), west-central Europe 5·9 (*n*=99), Scandinavia 6·3 (*n*=132). 2 broods, rarely 3, except in north of range where normally 1. 2nd clutch started 1–3 days before 1st brood fledges to 1–12(–21) days after fledging (Lack 1948b). Eggs laid daily, occasionally with gap of 1 or more days (Lack 1948b). INCUBATION. 13·7 days (12–21), *n* = 105; slightly longer in March–April (14·0 days), than in May–June (13·5 days) probably because of delay in start of incubation after laying of last egg in earlier clutches (Lack 1948b). By ♀ only. In 56-hr watch, 67·4% of time spent on nest in spells of 38·0±SE 5·13 min (*n*=35), with periods off nest of 13·8±SE1·38 min (*n*=40) (East 1981b). Normally begins with last egg, but may be a delay of 1 day, or may, in late and/or large clutches, start with penultimate egg (Lack 1948b). Hatching occurs typically during night or early morning (15 out of 17 cases) (Harper 1984a). YOUNG. Cared for and fed by both parents; brooded for 7 days after hatching (East 1981b). ♂ continues to feed alone when ♀ starts 2nd brood (Lack 1965). FLEDGING TO MATURITY. Fledging period 13·4 days (10–18), *n*=125; slightly shorter after mid-May than before, and longer for broods of 4–5 than for smaller or larger broods (Lack 1948b). Become independent 16–24 days after fledging. Age of first breeding 1 year. BREEDING SUCCESS. In England and Wales, of 1426 eggs laid, 71% hatched and 77% of these fledged, giving 55% overall success. Of 282 nests with newly-laid eggs, 54% hatched full clutches, 18% lost 1 egg, 6% lost 2 or more eggs, and 22% failed completely. Of 407 nests with newly-hatched young, 73% fledged all young, 4% lost 1, 3% lost 2 or more, 20% failed completely. Overall success 53% in March–April, 61% in May, and 46% in June–July. (Lack 1948b.)

Plumages (nominate *rubecula*). ADULT. In fresh plumage, entire upperparts from crown backwards brownish olive-green, sometimes with slight grey-green tinge; front and sides of crown partly tinged ash-grey, upper tail-coverts usually washed cinnamon on tips. Ash-grey stripe running backwards from above eye extends along side of neck to side of breast; ash-grey often slightly mottled olive at upper side of neck. Ear-coverts olive-buff with orange shaft-streaks. Forehead, lores, ring round eye, cheeks, chin, throat, chest, and breast bright deep orange. Belly and central vent silky-white, extending up to mid-breast; some ash-grey of feather-bases sometimes showing through. Flanks and sides of vent pale olive-buff; thighs duller olive-grey. Under tail-coverts pale buff or cream-white. Tail greyish-black, t1 often tinged olive-brown; outer webs of all feathers fringed pale olive-brown or cinnamon-brown, widest near bases. Flight-feathers and greater upper primary coverts greyish-black, outer webs fringed pale olive-brown (except for p9–p10 for emarginated part of outer primaries), inner webs with pale yellow-buff wedge or fringe, widest near base. Tertials olive-brown with duller blackish basal inner webs, brighter brownish olive-green fringes along outer webs, and narrow olive-brown or slightly rufous fringe along tip. Lesser and median upper wing-coverts brownish olive-green, like upperparts. Greater upper wing-coverts similar to lesser and median, but inner webs dull black; tips either uniform olive (35 of 59 examined), or with narrow rufous-cinnamon spot at shaft; these spots usually narrow, ill-defined, merging into olive on tip, or restricted to middle coverts (in juvenile and 1st adult, spots large, sharply defined, triangular or rounded, and present on middle and outer coverts); hence, presence of spots no indication of age. Under wing-coverts and axillaries pale buff or cream-buff; small coverts along leading edge of wing grey with orange-buff tips or sometimes fully orange. Some variation in colour of upperparts (some more olive-brown, others more greyish-olive) and depth of orange of face and chest (some slightly reddish, others paler orange), but variation due to bleaching and abrasion much more marked. *In worn plumage* (in some from February, in others not until April–May), upperparts distinctly greyish-olive or dull grey with limited olive tinge, warmer greyish olive-brown restricted to rump, upper tail-coverts, and fringes of tail-feathers; some white of feather-bases sometimes visible on rump; ash-grey supercilium more extensive, reaching sides of forecrown, but contrasting less with greyish upperparts than in autumn; deep orange face and chest still bright in spring, but face and throat bleached to orange-buff or yellow-orange by June–July; olive-buff of flanks and sides of vent more restricted; upper belly washed with grey, lower belly to under tail-coverts dirty white. Sexes virtually similar, though face and chest of ♀ on average slightly paler orange, especially in spring and summer. NESTLING. Down grey and rather scant at hatching, sooty-black, long and rather dense at age of *c.* 1 week; on upperparts only (Heinroth and Heinroth 1924–6; Witherby *et al.* 1938b). For development, see Heinroth and Heinroth (1924–6), Lack and Silva (1949), and Balcells and Ferrer (1968). JUVENILE. Upperparts and sides of head pale olive-brown or olive-grey, each feather with poorly defined buff subterminal spot (palest at shaft) and blackish rim at tip; rump and upper tail-coverts virtually uniform buff-brown or cinnamon. Chin and throat buff to buff-white; chest, breast, and flanks deeper buff or pale buff-brown with black feather-fringes, appearing scaly; remainder of underparts pale buff or cream-white with traces of dull grey feather-fringes. Tail as adult, but feathers usually pointed at tip (in adult, feathers broader with squarer tips). Flight-feathers, tertials, and greater upper primary coverts as adult, but tips of tertials and coverts usually less broadly rounded; tertials sometimes with buff spot on tip (occasionally also secondaries). Lesser and median upper wing-coverts olive-brown with large bright buff subterminal spot on tip and often with narrow black rim, like feathers of upperparts. Greater upper wing-coverts dull black with olive-brown outer fringe and large triangular or rounded deep buff spot on tip (soon bleaching to pale buff), innermost usually with black fringe along tip. FIRST ADULT. Like adult, but (in fresh plumage) upperparts slightly duller and greyer, less olive-green; forecrown and sides of crown slightly less extensively suffused grey (often no grey at all in ♀); face and throat on average less deep orange, in particular chin and throat yellowish-orange, in ♀ sometimes entirely orange-yellow. Juvenile flight-feathers, greater upper primary coverts, variable number of greater outer secondary coverts, and tail retained; on average, these feathers more worn than on adult at same time of year (especially in spring). Juvenile tail-feathers more pointed than those of adult (see Svensson 1984a). Spots on tips of juvenile greater secondary coverts usually larger than those present in some adults, rounded-triangular, less narrow, usually present on all retained coverts, not only on middle ones as in most adults which have them; however, some birds difficult to age thus. *In worn plumage*, colour strongly changed by bleaching and wear (as in adult); upperparts become duller olive-grey, sometimes almost pure dull grey; face and chest paler orange (often paler and less saturated than in spring or early-summer adult, ♀ sometimes particularly pale orange-buff); buff spots on greater coverts bleached and abraded, making ageing more difficult, but tips of tail-feathers usually still more sharply pointed and more heavily abraded than those of adult. For details of ageing criteria, see Karlsson *et al.* (1986).

Bare parts. ADULT, FIRST ADULT. Iris dark brown or black-brown. Bill dark olive-brown, dark horn-brown, or black-brown; base of lower mandible paler brown or flesh-grey. Inside of upper mandible and palate grey or grey-black in adult, yellowish or greyish-white in 1st adult in autumn; grey gradually spreads on inside of mandible in 1st winter, but most birds still show some yellow on palate in spring and summer of 2nd calendar year (Svensson 1984a). Leg and foot flesh, greyish-flesh, flesh-brown, brown, dark brown, or black-brown. NESTLING. Mouth yellow, unspotted; gape-flanges pale yellow. JUVENILE. At fledging, iris dark brown, bill dark horn-brown with flesh-coloured cutting edges and base of lower mandible; gape-flanges pale yellow; inside of mouth yellow; leg and foot pale flesh, greyish-flesh, or flesh-brown; soles pale yellow. 1st adult colours obtained during post-juvenile moult. (Heinroth and Heinroth 1924-6; RMNH, ZMA).

Moults. ADULT POST-BREEDING. Complete; primaries descendant. In Britain, starts with shedding of p1 early June to early August, completed with regrowth of p10 late July to late September; estimated duration of primary moult 60 days (Ginn and Melville 1983). Tail moults centrifugally; t1 shed shortly after p1, all feathers regrown at primary moult score 30-40.

Tertials and body start approximately at shedding of p3-p4; tertials, flanks, and breast largely new when primary moult halfway through; head moulted last, completed before p9 full-grown. Secondaries start with s1 at shedding of p5, last feather (s6) regrown at same time as p9-p10 or shortly after. (Ginn and Melville 1983; RMNH, ZMA.) In 24 captive birds, moult started with p1 between about 1 July and 25 August, completed with regrowth of p9-p10 (25 August-)15 September-10 October; p10 shed 7 weeks after p1, p9-p10 regrown 80-85 (75-90) days after shedding of p1; tertials (sequence: s8-7-9) and secondaries (sequence: s1 to s6) started 3 weeks after p1, tertials regrown at *c.* 60 days, secondaries at about same time as p9-p10 or slightly later; tail (sequence: t1 to t6) completely renewed within 7 weeks after shedding of p1 (t1 shed with p1); body started *c.* 18 days after p1, completed at *c.* 11 weeks (sequence: belly-chest-upperparts-throat-crown-flanks) (Rogge 1966). POST-JUVENILE. Partial: head, body, lesser and median upper wing-coverts, usually tertial coverts, variable number of greater upper wing-coverts (1-3 innermost in 129 of 150 examined, none in 13, 4-5 in 7, all 6 in 1, but perhaps more often all in southern Europe), and occasionally a few tertials. In captivity, started at age of *c.* 45 days (Heinroth and Heinroth 1924-6), in the wild at 6-7 weeks (Ginn and Melville 1983); completed at age of 3 months (Heinroth and Heinroth 1924-6) or *c.* 55 days after start (Ginn and Melville 1983). Starts late May to mid-August, completed late July to early October (Witherby *et al.* 1938b; Dementiev and Gladkov 1954b; Jenni 1984; RMNH, ZMA).

Measurements. ADULT, FIRST ADULT. Nominate *rubecula*. Wing: (1) Sweden, (2) Netherlands, breeding, April-August, (3) Netherlands, late September to March (RMNH, ZMA); (4) Balkans, Greece, and Turkey (Stresemann 1920; Niethammer 1943; Watson 1961b; RMNH, ZMA); other measurements for Sweden and Netherlands combined; skins. Bill (S) to skull, bill (N) to distal corner of nostril; exposed culmen on average 3·8 less than bill (S).

WING (1)	♂ 73·2 (1·31; 8)	72-74	♀ 70·6 (1·93; 7)	68-73
(2)	72·8 (1·81; 17)	70-76	70·5 (1·25; 10)	68-73
(3)	72·8 (1·70; 66)	70-77	71·4 (1·62; 23)	68-74
(4)	72·6 (1·55; 30)	70-76	70·5 (1·89; 19)	68-74
TAIL	56·4 (1·87; 99)	53-60	55·1 (2·39; 38)	51-59
BILL (S)	14·4 (0·48; 92)	13·7-15·6	14·2 (0·46; 35)	13·5-15·1
BILL (N)	7·6 (0·48; 21)	6·9-8·2	7·1 (0·30; 10)	6·7-7·5
TARSUS	25·1 (0·78; 96)	23·9-26·6	25·2 (0·70; 37)	24·0-26·7

Sex differences significant for wing and tail. 1st adult (with retained juvenile flight-feathers and tail) combined with adult in table above, though juvenile wing and tail slightly shorter: e.g. wing, adult ♂ 73·7 (16) 71-77, 1st adult ♂ 72·6 (67) 70-76; tail, adult ♂ 56·6 (16) 54-60, 1st adult ♂ 56·3 (67) 53-59 (Netherlands: RMNH, ZMA).

Sardinia, August-November, skins (RMNH, ZMA).

WING	♂ 73·5 (1·51; 14)	72-76	♀ 70·9 (1·79; 8)	69-73
TAIL	57·5 (2·37; 14)	54-60	55·9 (2·15; 8)	53-59
BILL (S)	14·5 (0·44; 13)	13·9-15·2	14·3 (0·89; 8)	13·4-15·1
TARSUS	25·6 (0·51; 13)	24·8-26·4	25·4 (0·87; 8)	24·6-26·6

Sex differences significant for wing.

E. r. melophilus. Britain, all year; skins (RMNH, ZMA).

WING	♂ 74·5 (1·81; 22)	72-78	♀ 71·4 (1·42; 21)	69-74
TAIL	57·4 (2·09; 21)	53-61	55·2 (1·42; 20)	52-57
BILL (S)	14·6 (0·50; 20)	13·9-15·6	14·7 (0·55; 18)	13·9-15·5
TARSUS	26·4 (0·71; 22)	25·7-27·7	25·9 (0·70; 20)	24·6-26·8

Sex differences significant, except bill (S).

E. r. superbus. Gran Canaria and Tenerife (Canary Islands), all year; skins (RMNH, ZFMK, ZMA).

WING	♂ 71·2 (1·21; 26)	69–73	♀ 68·5 (1·25; 21)	65–70	
BILL (S)	15·2 (0·40; 17)	14·7–15·9	15·2 (0·51; 13)	14·5–16·2	
TARSUS	25·2 (0·49; 17)	24·3–25·8	25·2 (0·52; 13)	24·1–26·1	

Sex differences significant for wing.

E. r. hyrcanus. South-east Transcaucasia and northern Iran, summer, and Iraq, winter (BMNH, ZMA).

WING	♂ 74·2 (1·84; 12)	72–78	♀ 72·0 (2·07; 10)	70–76	
BILL (S)	16·5 (0·50; 12)	15·7–17·2	15·9 (0·32; 10)	15·3–16·3	
TARSUS	26·9 (0·86; 11)	25·6–27·8	26·1 (0·94; 10)	24·9–27·4	

Sex differences significant for wing and bill (S).

E. r. witherbyi. North-west Africa, wing: ♂ 69·5 (10) 66–73, ♀ *c.* 67·5 (Hartert 1910; Vaurie 1959). Southern Portugal, wing: ♂ 72·5 (4) 71–74, ♀ 69·9 (4) 67–72 (RMNH, ZMA).

Weights. ADULT, FIRST ADULT. Nominate *rubecula*. (1) Norway, July–October (Haftorn 1971). (2) South-west Norway, late September and early October, on autumn departure; (3) Fair Isle, late September and early October, on arrival (Nisbet 1963). (4) Southern Sweden, September (Scott 1965). (5) Denmark, late March to May (Petersen 1972). (6) France (Mountfort 1935). (7) Central Italy, October–March (Ioale' and Benvenuti 1982, which see for details). Portugal: (8) September, (9) early October (C J Mead). Gibraltar: (10) autumn, (11) spring (Finlayson 1981). (12) Western Morocco, March (Kersten *et al.* 1983). (13) Malta, winter (J Sultana and C Gauci).

(1)	17·8 (— ; 37)	15–21	(8)	15·1 (1·0 ; 17)	13–17	
(2)	17·7 (0·98; 20)	—	(9)	15·3 (0·7 ; 15)	14–16	
(3)	14·9 (1·42; 24)	—	(10)	15·5 (0·15; 62)	—	
(4)	15·8 (0·87; 282)	—	(11)	16·5 (1·44; 28)	—	
(5)	16·7 (1·19; 1040)	—	(12)	16·1 (0·7 ; 9)	—	
(6)	18·8 (— ; 10)	16–23	(13)	17·7 (1·5 ; 50)	14–22	
(7)	18·2 (1·9 ; 694)	—				

Sexed birds. Netherlands: (14) late September–October, (15) November–February, (16) April–July, (17) exhausted birds, autumn and winter (ZMA). (18) Greece and Turkey, February–May(–July) (Niethammer 1943; Watson 1961b; Rokitansky and Schifter 1971).

(14)	♂ 16·9 (1·14; 21)	15·5–20·5	♀ 15·8 (1·61; 17)	13·5–18·5	
(15)	19·0 (2·31; 12)	16·8–24·5	18·3 (1·63; 7)	15·8–20·0	
(16)	16·6 (1·07; 11)	15·1–18·0	18·5 (3·74; 5)	15·2–25·0	
(17)	12·2 (1·37; 10)	10·5–13·8	11·2 (1·42; 4)	10·0–12·3	
(18)	17·1 (1·63; 5)	14·8–18·7	18·5 (3·03; 5)	15·2–22·0	

Heaviest ♀♀ in samples above were egg-laying.

E. r. melophilus. Britain, all year: 19·7 (109) 15·5–25·3 (Marples 1935). Skokholm Island (Wales), all year: 17·8 (59) 13·8–22·2 (Browne and Browne 1956). Southern England, all year: 18·2 (0·5; 50) 14·2–22·5; exhausted ♂ 13·4 (Ash 1964). Britain, all year: ♂ 20·7 (249) 17–26, ♀ 19·9 (259) 16–24; ♀ exceptionally 26 (Lees 1949, which see for diurnal and monthly variations).

E. r. witherbyi. Sierra Nevada (southern Spain), May: ♂ 15·5 (Niethammer 1957).

E. r. caucasicus. North-east Turkey, October–November: ♂ 17·0 (1·83; 4) 15–19, ♀ 17 (Kumerloeve 1968).

E. r. hyrcanus. Southern Iraq, February: ♂ 14·2 (BMNH). Northern Iran: March, ♂ 19·5; July, ♀ 19·4 (Paludan 1940; Schüz 1959).

E. r. tataricus. Kazakhstan (USSR), October–February: ♂ 16·3 (10) 14·2–20·2; ♀♀ 15·8, 16·9 (Dolgushin *et al.* 1970).

NESTLING, JUVENILE. *E. r. melophilus*. At hatching, 1·85 (10).

Averages of large samples: on day 1, 2·2–3·2; day 7, 13·1–13·9 (individual range 8·6–17·8); day 10, at end of rapid weight increase, 17·8 (individual range 11·6–21·9); day 13, when about to fledge, 18·2–18·5 (individual range 12·2–23·9). (Lack and Silva 1949; Lees 1949.) Fledging weight maintained until end of post-juvenile moult (Lees 1949).

Structure. Wing short, broad at base, tip rounded. 10 primaries; in nominate *rubecula* (except Sardinia), *caucasicus*, and *hyrcanus*, p6 longest, p7 equal in length but occasionally 0–1 shorter, p8 2–4 shorter, p9 9·5–13·5, p5 1–2, p4 6–8·5, p1 12–15. Nominate *rubecula* from Sardinia similar, but p4 4–6·5 shorter and p1 11–14; birds from Iberia and Italy perhaps similar. In *melophilus*, p9 9–14 shorter, p4 4–7·5, p1 11–15; in *superbus*, p9 8–14 shorter, p5 0·5–1·5, p4 4–7, p1 10–12. In all races, p10 reduced, narrow, 30–40 (nominate *rubecula*) or 28–36 (*melophilus* and *superbus*) shorter than wing-tip, 5–13 longer than longest upper primary covert. Outer web of p5–p8 and inner of (p6–)p7–p9 emarginated. Tertials short, longest equal to secondaries. Tail rather short, tip square; 12 feathers, tips slightly pointed (especially in juvenile and 1st adult: see Svensson 1984a), tail-tip appearing slightly forked when spread. Bill short and straight, rather strong (especially in *melophilus*); tip of culmen slightly decurved. Nostrils oval, partly covered by membrane above. Some fairly long bristles along base of upper mandible. Tarsus long and slender. Toes rather long, slender; middle toe with claw 17·8 (21) 17–19 in nominate *rubecula*, 18·4 (15) 17·5–19·5 in *melophilus*; outer toe with claw *c.* 72% of middle with claw, inner *c.* 63%, hind *c.* 77%. Claws short, decurved.

Geographical variation. Slight; strongly clinal except for isolated *superbus* from Gran Canaria and Tenerife (Canary Islands). Involves colour of upperparts, depth of rufous on face and chest, presence of rufous on upper tail-coverts and tail-base, and (to slight extent) size. Variation in colour of upperparts strongly affected by abrasion (brighter and more olive in autumn, duller and greyer in spring; age also has some influence, 1st adults often abrading more rapidly than older birds); depth of rufous of face and chest shows marked individual variation, with tendency to deeper rufous-orange in adult ♂, paler yellow-orange in 1st adult ♀, intermediate in adult ♀ and 1st adult ♂. In addition to *superbus*, populations of nominate *rubecula* from Scandinavia, of *melophilus* from Britain and Ireland, and of *hyrcanus* from south-east Transcaucasia and northern Iran are the only well-characterized forms; all other races and populations either similar to one of these (differing only in measurements) or more or less intermediate. Nominate *rubecula* from Scandinavia characterized by brownish olive-green upperparts and orange face and chest. *E. r. melophilus* from Britain differs in fresh plumage by warmer dark olive-brown upperparts, more rufous-brown on upper tail-coverts and tail-base, deeper orange-rufous face and chest, and deeper buff-brown flanks; in worn plumage, upperparts duller brownish-grey, tail-base still rufous-brown when not too worn; face and chest less deep orange-rufous than in fresh plumage, but on average darker and less yellowish-orange than nominate *rubecula*; difference often visible only in comparison between series of birds of same age and sex; in autumn and winter, colour of upperparts is best character for identification, in spring and summer colour of chest. See also Measurements. *E. r. hyrcanus* from northern Iran browner on upperparts than nominate *rubecula*, rather like *melophilus* but less warm olive; face and chest rufous-orange like *melophilus*, flanks rather pale, like nominate *rubecula*. Differs from both *melophilus* and nominate *rubecula* in rufous-cinnamon to rufous-chestnut upper tail-coverts and tail-base and in long bill. In worn plumage, similar to *melophilus* except for rufous tail-base and long bill. *E. r. superbus* from Gran Canaria and Tenerife dark greyish-olive on upperparts, darker and greyer than nominate *rubecula*; wider band of ash-grey on forecrown and from side of crown down to side of breast, ash-grey tinge often extending to central crown and across upper mantle; face and chest even deeper rufous-chestnut than in *melophilus*; belly and vent whiter; small. Variation clinal throughout Europe; trends described below usually visible in fresh plumage only; for details, see also Lack (1946–7, 1951, 1965) and Vaurie (1955*a*). Birds from West and East Germany and France on average warmer olive-green than Scandinavian ones, less cold and greyish in colour; birds from Corsica, Sardinia, Italy, and north-east Spain (sometimes separated as *sardus* Kleinschmidt, 1906) similar but face and chest on average slightly deeper orange; however, all these best included in nominate *rubecula*. Populations from western continental seaboard, from Denmark to Brittany (France), on average darker olive-brown than those from central Europe, tending towards *melophilus*, but face to chest pale like nominate *rubecula*; sometimes separated as *armoricanus* Lebeurier and Rapine, 1936, but many indistinguishable from nominate *rubecula*. Western and southern part of Iberian peninsula as well as north-west Africa inhabited by birds closely similar to *melophilus*, agreeing in saturated colours, though not as brown above; north-west African populations distinctly smaller than *melophilus* (warranting recognition as *witherbyi*), Iberian

populations intermediate in size and, as not well-differentiated from '*armoricanus*' or '*sardus*', included in nominate *rubecula*. Birds from Balkans, Greece, and western Turkey named *balcanicus* Watson, 1961; said to be greyer on upperparts than nominate *rubecula*, but grey coloration mainly caused by abrasion and nominate *rubecula* in equally worn plumage similar (Stresemann 1920; Lack 1951; RMNH, ZMA). *E. r. tataricus* of Urals and western Siberia recognized here following Vaurie (1955*a*, 1959), though difference from nominate *rubecula* slight: upperparts paler and greyer; face and chest paler orange (but worn 1st adult ♀♀ of nominate *rubecula* similar). Crimea and Caucasus inhabited by 2 poorly-differentiated races which are transitional between nominate *rubecula* and *hyrcanus*: *valens* from Crimea slightly paler above than nominate *rubecula*, differing mainly in rufous upper tail-coverts and tail-base; *caucasicus* from Caucasus and Transcaucasia close to nominate *rubecula*, but upperparts slightly browner, less olive, face and chest slightly deeper orange-rufous, and upper tail-coverts and tail-base rufous, like *valens* and *hyrcanus*; differs from *hyrcanus* in slightly less bright and extensive rufous on tail-base, browner upperparts, slightly paler chest, and slightly shorter bill.

Forms species-group with Japanese Robin *E. akahige* from Japan, southern Sakhalin, and southern Kuril Islands and with Ryukyu Robin *E. komadori* from Ryukyu Islands. Genus *Erithacus* sometimes merged with *Luscinia* and *Tarsiger*, but all 3 maintained here following Lack (1954*b*) and Voous (1977).

CSR

Luscinia luscinia Thrush Nightingale

PLATES 44 and 47
[between pages 544 and 545, and facing page 652]

Du. Noordse Nachtegaal   Fr. Rossignol progné   Ge. Sprosser
Ru. Обыкновенный соловей   Sp. Ruiseñor ruso   Sw. Näktergal

Motacilla Luscinia Linnaeus, 1758

Monotypic

Field characters. 16·5 cm; wing-span 24–26·5 cm. Similar in size to Nightingale *L. megarhynchos*, but similar in structure only to easternmost race *L. megarhynchos hafizi*, having more pointed wings and marginally longer tail than European *L. m. megarhynchos*. Has similar form, general appearance, and behaviour to *L. megarhynchos*, differing in drabber, more olive-grey tone to plumage, mottled or spotted chest, and even richer voice. From above and behind, best distinction is reduced contrast of dull rufous-brown tail with dull brown upperparts; from below and in front, shows pale throat and mottled chest. Sexes similar; no seasonal variation. Juvenile separable.

ADULT. Upperparts typically dark olivaceous grey-brown, lacking marked rufescence of *L. megarhynchos* and with warm tone only on rufous-tinged upper tail-coverts and dark, dull rufous-brown tail; bird thus appears dull-rumped with little contrast between tail and rest of upperparts (*L. megarhynchos* usually shows brighter rump as well as tail). Underparts typically dull white,

with clean throat more sharply demarcated than in *L. megarhynchos* by dusky-brown malar stripe and similarly coloured lower cheeks and upper chest; chest and flanks mottled or diffusely spotted dusky or brown, with somewhat moth-eaten appearance; centre of belly unmarked but vent and under tail-coverts pale buff, sometimes with faint brown bars or mottling. On well-marked birds, chest completely suffused brown, dividing throat and lower body in pattern quite unlike *L. megarhynchos*. Some variation in plumage tones and underpart markings occurs, and birds seen breeding near Moscow (USSR) in May were astonishingly more rufous above and far less mottled on chest than any seen in central Europe or Middle East (D I M Wallace); these probably indistinguishable on plumage from duller *L. megarhynchos* (Hollom 1960). JUVENILE. Typically darker than *L. megarhynchos*, with paler, more contrasting spots on tips of tertials and wing-coverts, but much variation and individuals perhaps not safely identifiable in the field.

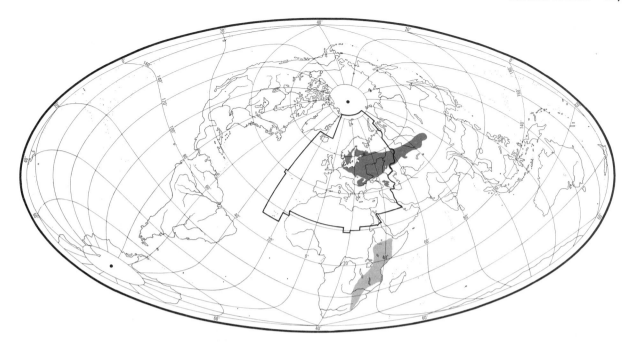

FIRST WINTER AND SUMMER. Distinguished from adult as in *L. megarhynchos* (see p. 627). At all ages, bare parts as *L. megarhynchos*.

Separation from *L. megarhynchos* most studied in migrants, becoming fairly easy in circumstances where differences in plumage tone, tail contrast, and fore-underpart pattern fully visible, but more rufescent birds with indistinctly mottled chests perhaps not identifiable in the field if not singing. Flight and other behaviour apparently identical to *L. megarhynchos* but usually shyer, often escaping by running and crouching. Not gregarious but forms small concentrations on passage. Prefers even damper areas for breeding than *L. megarhynchos*.

Song immediately recalls *L. megarhynchos*: louder, no less beautiful, and even more impressive, but more often lacks magnificent crescendo; has solemn, somewhat religious quality, with staccato phrases and pure bell-like peals dominant, interspersed with rasping 'dserr' notes; usually delivered from dense cover by day or night. Calls rather more clipped and higher pitched than *L. megarhynchos* with monosyllabic 'whit' recalling *Ficedula* flycatcher rather than *Phylloscopus* warbler, and a more grating croak.

Habitat. Close counterpart of Nightingale *L. megarhynchos*; see that species (p. 627) for a comparative analysis of habitat, which, like geographical range, approaches and sometimes overlaps with it, but maintains clear specific separation. Inhabits more continental, easterly and northerly temperate breeding grounds, overlapping boreal and steppe zones in middle latitudes between July isotherms 17–25 °C. Essentials of habitat—comprising deep soft humus with (usually) ground cover of dead leaves, tall and dense but patchy herbage, and plenty of tall bushes, shrubs, or low trees forming thicket or open woodland—are typically found along river banks or near standing water, normally in river valleys or on lowland plains. In central and east-central Europe, breeds in carrs of alder *Alnus*, small moist depressions, pond margins, and excavations with scrub of willow *Salix*, ash *Fraxinus*, birch *Betula*, poplar *Populus*, and hazel *Corylus* above rank stands of nettles *Urtica* and bramble *Rubus*; also on cleared patches of forest with burgeoning regrowth, and, much less frequently, in orchards, parks, and cemeteries (Niethammer 1937; Voous 1960). Presence of tall trees, even forming forests, appears acceptable, provided canopy is not so close as to inhibit scrub developing at lower levels, as would normally be the case with (e.g.) conifers. Where interspecific competition for breeding sites occurs with *L. megarhynchos*, that species normally confines itself to drier types within its broader habitat spectrum. Sharp and enduring boundaries of range observed by both species plainly imply critical limits of acceptability for habitat, in which climate and nutritional demands for young may prove no less determinant than character and form of vegetation. Apparently *L. luscinia* is even more averse to elevated or broken terrain than *L. megarhynchos*, but in USSR breeds in overgrown defiles (Dementiev and Gladkov 1954*b*). In African winter quarters, geographic separation perhaps again more important than habitat distinctions. Highly localized in dense thickets, sometimes shared with *L. megarhynchos*, mostly below 1500 m; in southern Africa, will occupy small patches of dense cover such as thick hedges of *Euphorbia* or *Lantana* (Moreau 1972).

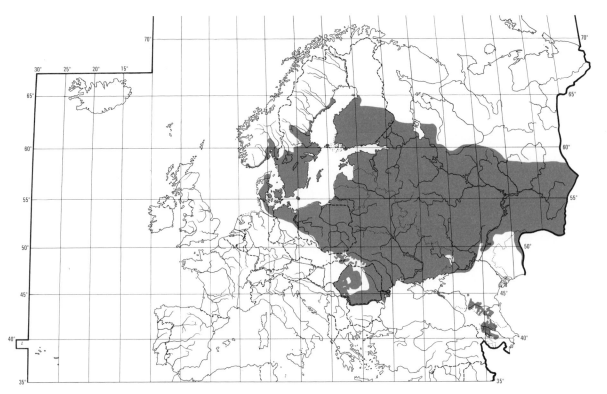

Distribution. Has spread west in West Germany, Denmark, Finland, and Poland, and south-west in East Germany; marked recent expansion in Norway and (after large range contraction) in Sweden.

WEST GERMANY. Expanding to west (Orr 1976). DENMARK. Spreading west (TD). NORWAY. Spreading south and north (VR). SWEDEN. In 18th century bred over much of southern and central Sweden, but declined in 19th and early 20th centuries, when found only on south coast; then expanded again and present range larger than in 18th century (LR; see also Karvik 1952, Bjärvall 1965, Elofson 1968, Holmbring 1972). FINLAND. Spread to west and north (OH); see map in Haartman (1973) POLAND. Spreading to west (AD). EAST GERMANY. Range expanded to south-west in recent years (SS). AUSTRIA. Bred in east until early 19th century (LS). YUGOSLAVIA. Bred 1900, 1921, 1950, and 1972 (VV). USSR. Leningrad region: expanded north-east since 1900 (Mal'chevski and Pukinski 1983).

Accidental. Britain (now annual), France, Belgium, Spain, Netherlands, Switzerland, Italy, Kuwait, Libya, Malta.

Population. Increasing Sweden, Finland, Leningrad region of USSR, and probably Poland; decreasing Rumania.

WEST GERMANY. 10–100 pairs (AH). NORWAY. Estimated 200–500 pairs 1984; increasing since 1960s (VR). SWEDEN. About 15 000 pairs; marked increases (Ulfstrand and Högstedt 1976; LS). FINLAND. About 200 pairs

(Merikallio 1958); huge increase to c. 8000 pairs (Hildén and Koskimies 1984; P Koskimies). POLAND. Scarce (Tomiałojć 1976); probably increasing (AD). CZECHOSLOVAKIA. No changes reported (KH). See also Kux and Weisz (1977). RUMANIA. Decreased (VC). USSR. Leningrad region: recent increase (Mal'chevski and Pukinski 1983).

Oldest ringed bird 8 years 11 months (Rydzewski 1978).

Movements Migratory, wintering entirely in Africa—largely south of equator with some north to southern Ethiopia (Ash 1973). In Kenya, locally common at 500–1500 m, east of highlands, mainly in Meru and Kitui districts to Voi–Taita area; numbers variable according to extent of rains in December–February (Pearson 1984). Few records west of Rift Valley and none from Uganda. Recorded in Tanzania west to Lake Manyara, Tabora, and Rukwa (Britton 1980). Probably common in Zambia (Benson et al. 1971) mainly in extreme north, east of Muchinga Mountains, the copper belt, and south of 15°S (D R Aspinwall). Common in southern Malawi (Benson and Benson 1977). Nowhere numerous in Zimbabwe (Irwin 1981), recorded on Zambezi in Mozambique (Vincent 1935), on Okavango river in Namibia, Botletle river in Botswana, and southwards into Transvaal and Natal, South Africa (Clancey 1980). 2 records in Nigeria, September–October, well to west of normal range (Elgood 1982).

South-east autumn passage necessary for west Baltic

population before southward heading into north-east Africa (Zink 1973). In East Germany, passage almost entirely east of Spree and Neisse rivers (Creutz 1980). Common migrant Bulgaria and Rumania (Stresemann 1948) and Black Sea coast of Turkey (*Orn. Soc. Turkey Bird Rep. 1968–75*). Passage occurs across eastern Mediterranean—thus, common on Cyprus (Flint and Stewart 1983) but only a vagrant on Malta (Sultana and Gauci 1982)—and recorded on passage in Iraq (Allouse 1953; Sage 1960*a*) and south of Caspian Sea, Iran (Feeny *et al.* 1968). Uncommon in Gulf states of Arabia (Bundy and Warr 1980), scarce in Oman (Gallagher and Woodcock 1980), and rare in North Yemen (Cornwallis and Porter 1982). Enters Africa east of 25°E, sometimes in large numbers (Archer and Godman 1961*b*; Zink 1973), western Baltic population thus crossing at least 20° of longitude. Common on passage in eastern Egypt (Watson 1973*b*). Many enter Africa across northern Red Sea, e.g. 2nd most common migrant on Sudanese coast in August-September (Nikolaus 1983). Very common in Sudan on Nile system but absent in west (Hogg *et al.* 1984). Only one record in southern Sudan (Nikolaus and Pearson 1982) but widespread on autumn passage in Ethiopia mainly in and west of Rift Valley (Ash 1973). Passage into Kenya therefore from west of north, birds migrating east of Lake Turkana and Aberdares most common in central and southern areas with eastern edge of passage well defined, occasionally recorded east to Garissa but none from coast, lower Tana or north-east Kenya (Pearson 1984).

Return passage less protracted and obvious through Kenya, with more birds probably overflying and movement generally further east (sometimes common on coast, unlike in autumn), probably heading for central and eastern Ethiopia, and possibly Somalia; no spring records in Sudan (D J Pearson). In Egypt, much less common than in autumn, and scarce in Arabia (e.g. Bundy and Warr 1980). Most overfly, many probably making landfall in Middle East. Common at Azraq in Jordan (Wallace 1982) and in Lebanon and Syria (Meinertzhagen 1935), and northward passage occurs across Iraq (Allouse 1953; Sage 1960*a*), Iran (Witherby 1907; Passburg 1959), and eastern Mediterranean. In Cyprus may be more common than supposed with many sight records of Nightingale *Luscinia megarhynchos* probably referable to *L. luscinia*, e.g. of 357 limed nightingales in spring only 112 *L. megarhynchos* (Flint and Stewart 1983). Scarce in Turkey in spring (*Orn. Soc. Turkey Bird Rep.* 1968–75).

Leaves breeding areas mainly from early August, although first passage in Crimea by late July and present in some areas of USSR until mid- or late September (e.g. Dementiev and Gladkov 1954*b*, Berger 1967). Passage on Turkish Black Sea coast from mid-August to late September, up to late September on Bosporus, and from early August on central plateau (Porter 1983; *Orn. Soc. Turkey Bird Rep.* 1968–75). Passage south of Caspian Sea occurs from end of August to mid-September (Feeny *et al.* 1968) and through Middle East from late August to early October, peaking mid-September (Meinertzhagen 1920*b*; Wallace 1982). Passes through Cyprus from mid-August to October with peak in late August and September (Flint and Stewart 1983), through Egypt mid-August to mid-October with peak late August to mid-September (Watson 1973*b*), and through central Sudan August-November with peak from late September (Hogg *et al.* 1984). Arrives on Sudanese Red Sea coast from 10 August to 12 October, mainly 25 August-20 September, and main passage through Ethiopia from 2nd week of September to early November (D J Pearson). Movement into Kenya begins end of October; peak at Ngulia 10 November-10 December with some passage until late December, this timing indicating step migration with some birds probably spending up to 2 months in green areas in eastern Sudan and western Ethiopia (Nikolaus and Pearson 1982). Birds make stopovers in large numbers in east Kenyan bushlands only after vegetation has come into leaf, usually late November to late December (Pearson 1984). Main arrival in southern Africa from late November (Benson and Benson 1977; Clancey 1980), but high weights of some birds in southern Malawi in December and early January suggested these still on passage (Hanmer 1979).

Leaves winter quarters in March, and exodus complete by early April (Benson and Benson 1977; Irwin 1981). Passage through eastern Zaïre occurs mid-March (Curry-Lindahl 1981) and through Kenya from late March to 3rd week of April with peak in 1st week of April (D J Pearson). Wintering population leaves southern Ethiopia in early April but passage continues until end of April (Ash 1973). Present in Jordan, north-east Israel, Syria, and Lebanon from mid-April to early May (Meinertzhagen 1935; Wallace 1982; P V Hayman), and late April to mid-May in Iraq and Iran (Witherby 1907; Allouse 1953; Sage 1960*a*). Passes through Cyprus in late March and April (Flint and Stewart 1983) and central plateau of Turkey from mid-March, passage peaking late April to early May (*Orn. Soc. Turkey Bird Rep.* 1968–75). Occurs mid-April to mid-May on European coasts of Black Sea (Blaszyk and Steinbacher 1954; Moreau 1972), and reaches south-west of breeding range (Rumania) from mid-April (Portenko and Wunderlich 1977). In European USSR, arrives from late April in south of range to early or mid-May in north; average first arrival at Tomsk (south-central Siberia) 17 May over 16 years (Dementiev and Gladkov 1954*b*; Johansen 1955). Arrives at Hiddensee (East Germany) in first 10 days of May and whole population usually present by 20 May (Berger 1967). Arrives in Sweden in first half of May (Rendahl 1960*a*). Spring records in Britain 6 times more frequent than autumn records, possibly linked to range expansion (Jenni 1981).

In East German study, ♀♀ arrive on breeding grounds

8–10 days later than ♂♂ (Berger 1967). No evidence of age separation on autumn migration with adults and 1st-years arriving together in both Sudan and Kenya (D J Pearson).

Birds recovered in Egypt from Sweden (3) and Denmark (1); Danish autumn recovery from eastern Austria; Swedish spring recoveries on Arabian Red Sea coast and in Poland; Danish 1st-year recovered in Natal (South Africa) on 20 November. 2 atypical recoveries: Swedish bird recovered in autumn in Italy (considered by Zink 1973 to require confirmation), and 1st-year from Norway trapped 12 days later (1099 km away) in southern England (Hilprecht 1954; R Hudson). Birds ringed November–December at Ngulia (Kenya) recovered in Syria (April), Lebanon (April), Ukraine (July), and Finland (June) (G C Backhurst; D J Pearson).   MGK

Food. Arthropods and some fruit. Feeds largely on ground, hopping around and disturbing leaves to search; particularly during nestling phase also feeds in herb and shrub layers and even recorded foraging in crown of tree (Neumann 1943; Meinertzhagen 1954; Neufeldt 1956; Emmrich 1971); see below for proportions of diet taken from different foraging sites. Will also take flying insects in brief aerial-pursuit (Emmrich 1971), and recorded clinging to tree-trunk like tit *Parus* (Neumann 1943).

Summer diet in west Palearctic includes the following. Invertebrates: damsel flies or dragonflies (Odonata), grasshoppers, etc. (Orthoptera: Tettigoniidae, Acrididae), earwigs (Dermaptera), Psocoptera, bugs (Hemiptera: Pentatomidae, Cercopidae, Psyllidae, Aphidoidea), snake-flies (Neuroptera: Rhaphidiidae), adult, pupal, and larval Lepidoptera (Tortricidae, Notodontidae, Arctiidae, Noctuidae, Lymantriidae, Geometridae, Lasiocampidae), caddis flies (Trichoptera), adult and larval flies (Diptera: Culicidae, Muscidae), adult and larval Hymenoptera (sawflies, ants Formicidae), adult, pupal, and larval beetles (Coleoptera: Carabidae, Elateridae, Tenebrionidae, Lagriidae, Cerambycidae, Curculionidae), spiders (Araneae), woodlice (Isopoda), millipedes (Diplopoda), small gastropod molluscs. Plant material mainly fruits: of elder *Sambucus*, currants *Ribes*, bramble *Rubus*, mulberry *Morus*, vine *Vitis*, *Amelanchier*; occasionally also seeds. (Neufeldt 1956; Emmrich 1971; Kostin 1983; Mal'chevski and Pukinski 1983.)

On Hiddensee island (East Germany), before and after breeding, invertebrates in stomachs of adults (sample size not given) comprised 52·7% Hymenoptera (largely ants), 24·5% beetles (largely adults), 8·8% Hemiptera, 4·4% earwigs, and 9·6% others; 52·8% of items were of species living on or under the ground, 33·6% from herb or shrub layer, 13·6% not determined; said also to take berries readily when available, and sometimes also seeds (Emmrich 1971). In Crimea (USSR), May and August, 92 items in 5 stomachs included 67% (by number) ants, 15% millipedes, and 2% seeds (Kostin 1983).

Bird on passage through Lebanon in May had eaten more soft fruit than insects (Meinertzhagen 1954). One bird in Kenya, November, contained ants and beetles (Lack and Quicke 1978).

Diet of young on Hiddensee markedly different from adults', though this perhaps partly due to different times of season at which studied (see above): 514 items from collar samples comprised 38·2% (by number) beetles (mostly larvae), 20·0 % Lepidoptera (mostly larvae), 12·7% Diptera (mostly adults), 11·1% woodlice, 9·4% Arachnida, 3·3% Hymenoptera (mostly adults), 2·9% Hemiptera, and 2·4% others; 53·6% of items were of species living on or under the ground, 43·4% from herb or shrub layer, 3·0% not determined; items were markedly smaller than those eaten by adults themselves (Emmrich 1971). In Voronezh region (USSR), 133 items in collar-samples comprised 29·3% Lepidoptera (mostly larvae), 20·3% larval snake-flies (Rhaphidiidae), 18·0% spiders, 15·8% beetles (mostly adults), 8·3% Hemiptera, 3·8% small molluscs, 3·0% adult Diptera, and 1·5% Orthoptera (Neufeldt 1956; see also Mal'chevski 1959). Independent young take berries not infrequently (Mal'chevski and Pukinski 1983).   DJB

Social pattern and behaviour. Major studies at Hiddensee (East Germany) by Berger (1960), and in European USSR by Simkin and Steinbach (1984). For review, and comparison with Nightingale *L. megarhynchos*, see Hilprecht (1954). Account includes material supplied by M Pryl and J Sorjonen (Finland).

1. Mostly solitary. In autumn, Budapest region (Hungary), migrates singly or in small loose flocks (Farkas 1954; Schmidt 1975); sampled flock birds always juveniles, suggesting young (perhaps brood-siblings) associate for migration (Farkas 1954). In Kenya, autumn passage migrants commonly occupy territories for up to 2-3 weeks; once settled in winter quarters, typically sedentary and territorial (Pearson 1984). In Zimbabwe, 3 territorial birds, allegedly ♂♂, c. 100 m apart from one another, a 4th c. 40 m away (Smith 1951). In Zimbabwe, commonly associates with Red-backed Scrub Robin *Erythropygia leucophrys* (Smith 1951). BONDS. Monogamous mating system (M Pryl, J Sorjonen); once ♂ paired with 2 ♀♀ (J Sorjonen). Pair-bond breaks down once young independent (Steinfatt 1939b; see below), and birds typically take new mate each year (J Sorjonen). One report of hybridization with *L. megarhynchos* in captivity, producing young (Nöhring 1943), but no record of hybridization in the wild (Orr 1976). Only ♀ broods, while both members of pair feed young (J Sorjonen). Brood-division after fledging may occur, as in *L. megarhynchos*: in reported case, each adult seemed to be tending 2 young (Neumann 1943). Young fed for 2–3 weeks after leaving nest (Steinfatt 1939b; Neumann 1943; J Sorjonen). In perhaps exceptional case, ♂ fed 30-day-old young of 1st brood while ♀ incubated 2nd clutch (Hilprecht 1954). Some ♂♂ breed first at 1 year, but ♀♀ and most ♂♂ not until later years (J Sorjonen). BREEDING DISPERSION. Solitary and territorial. In European USSR, typically forms neighbourhood groups of 3-9 (2-21) pairs (Simkin and Steinbach 1984, which see for details of dispersion). In favourable habitat, density relatively high. In European USSR, apparent territory size contracts when bulk of birds arrive, often initially from more than 4 ha to 0·25 ha (Simkin and Steinbach 1984; see also Ptushenko and Inozemtsev 1968).

Size of territory typically *c.* 1 ha (0·2-2·0); at Siikalahti (south-east Finland), nests 30-100 m apart (J Sorjonen). At Helsinki (Finland), average distance between nests *c.* 50 m (*n* = 15), shortest *c.* 20 m (M Pryl). In East Germany, nearest distance between nests in 4 cases 75 m, 106 m, 109 m, and 168 m (Hilprecht 1954). Around Matsalu Bay (Estonia, USSR), singing ♂♂ *c.* 50-100 m apart (Dementiev and Gladkov 1954*a*). Along edge of reservoir, Voronezh (USSR), at least 44 singing ♂♂ in *c.* 8 km (Wilson 1976). For other densities in USSR, see Ptushenko and Inozemtsev (1968). On Hiddensee (East Germany), 140 ♂♂ per km² (Klafs and Stübs 1977, which see for other densities; see also Berger 1960). In lower Oder valley (East Germany), 6 pairs per km² (Rutschke 1983, which see for other densities). In Schleswig-Holstein (West Germany), over several years, *c.* 4-9 *L. luscinia* together with *c.* 3-6 *L. megarhynchos* in 11 ha (Orr 1976). In ash-alder, Białowieża (Poland), 24 pairs per km² (Tomiałojć *et al.* 1984). In Warsaw (Poland), *c.* 100 ♂♂ per km² in thickets of willow *Salix* along River Vistula, 4-25 ♂♂ per km² in parks (Luniak 1969). At Siikalahti, overall 4·3 pairs per km², varying with height of coppice: at 3-8 m 0·38 per ha, at 8-12 m 0·82, at 12-16 m 0·46, over 16 m 0·24 (Sorjonen 1983*a*). In one case, nest for 2nd brood 10 m from 1st (Hilprecht 1954). Territory serves for pair-formation, nesting, and raising young. In one territory, food for young collected mainly within 50 m of nest, at most 100 m away (Steinfatt 1939*b*). At another nest, a few days after hatching, ♂ flew 50-100 m to collect food (Neumann 1943). Early arrivals (mostly old ♂♂) usually reoccupy former territory, or one nearby; some ♀♀ also site-faithful (Berger 1960). ROOSTING. Presumably solitary, in dense scrub, during breeding season. Comfort activities include sunning early in cold mornings, and bathing (J Sorjonen). On autumn migration, Budapest, flocks roost preferably near water, favouring broom *Cytisus*, willow (especially if mixed with elder *Sambucus*), nettle *Urtica*, and bramble *Rubus* (Farkas 1954). For interactions with *L. megarhynchos* in roosts, see Antagonistic Behaviour (below). After leaving nest, young typically do not return to it, but fledged runt found in nest on 2 evenings, apparently roosting there (Neumann 1943).

2. Though favouring dense thickets, not especially shy. In breeding season, approachable to 3-5 m when singing from dense cover, flushing earlier if song-post more exposed (J Sorjonen). Outside breeding season, boldness varies apparently between individuals and circumstances. In winter quarters, Zimbabwe, tame and confiding when in full view, especially shortly after sunrise when ♂♂ sing from exposed perches (Smith 1951). On autumn migration, Egypt, some shy and skulking, others staying more in the open and closely approachable (Bodenham 1945). On passage in Arabia, reluctant to fly when disturbed, often running long distances under cover of bushes before taking off (Meinertzhagen 1954). See also Flock Behaviour (below). In various situations when suspicious or excited, bird raises tail and fans it, sometimes repeatedly, to one side or the other, then lowers it (Fig A). Accompanies

tail-fanning by wing-twitching, wing usually lower on the side on which tail fanned (Berger 1960). FLOCK BEHAVIOUR. In Budapest, birds in loose flocks disturbed at autumn stopovers gave characteristic twist of the tail, then flew off giving Contact-alarm call (Schmidt 1975: see 3a in Voice). In March, Zimbabwe, a few birds, congregated in *c.* 0·2 ha, flew from one clump of shrubs to another with excitement-calls (see 3b in Voice), behaviour possibly indicating premigratory restlessness (Smith 1951). SONG-DISPLAY. ♂ sings typically from dense cover, but often from near edge, at average 4·6 m (1-15) above ground (Sorjonen 1983*b*, which see for influence of environmental conditions on transmission of song). Usually sings from higher than does *L. megarhynchos* (Berg 1980). Especially before pair-formation, some ♂♂ have 2 regular song-posts up to 100 m apart (Mal'chevski and Pukinski 1983). Main song-post usually well removed from nest-area (Berger 1960; Simkin and Steinbach 1984) For evidence that song serves to prevent rivals invading territories, see Göransson *et al.* (1974). In full song, bird's body vibrates and tail shivers (Orr 1976). Song comprises mainly Courtship-song, delivered, often by day, from dense cover near or on ground a few metres from ♀, and Territorial-song, typically delivered at night from rather higher perch (Sorjonen 1977, 1983*b*: see 1 in Voice). For song during territorial disputes, see Antagonistic Behaviour (below). Courtship-song given many times daily, usually before noon, 5-10 days before incubation (Sorjonen 1974). Most singing activity is at night, except in cloudy weather when activity more evenly spread (Sorjonen 1977, which see for references). On arrival on breeding grounds, USSR, ♂♂ said to sing at first only nocturnally, later throughout the day, especially at dawn and at dusk (Dementiev and Gladkov 1954*a*). In south-east Finland, song heard throughout 24 hrs, peaking around midnight; song during daylight hours most prevalent early in breeding season, and especially characteristic of non-breeding birds. Nocturnal song began after sunset at beginning and end of song-period (see below), and before sunset in middle of song-period. (Sorjonen 1977, which see for individual variation.) At Pappilanniemi (Finland), June, nocturnal Territorial-song began before sunset and stopped slightly before sunrise; 1 ♂ sang most frequently from 21.00-04.00 hrs, least 04.00-05.00, 12.00-14.00, 16.00-21.00; bird sang for *c.* 50% of 24 hrs, longest bout (i.e. pauses not exceeding 5 s) 2 hrs 16 min, longest break 1 hr 19 min; no detectable effect of weather on song output (Piiparinen and Toivari 1958). For diurnal pattern, Sweden, see Göransson and Karlsson (1978). According to Orr (1976), *L. luscinia* less likely than *L. megarhynchos* to give up singing in bad weather. In winter quarters, Zimbabwe, sings from sunrise until midday, only sometimes in the evenings; sings in all weathers, often in considerable heat (Smith 1951; see also below). Some song given on spring migration (Simkin and Steinbach 1984) but mostly ♂♂ start singing shortly after arrival on breeding grounds, e.g. in Kazan (USSR), *c.* 1 week after (Dementiev and Gladkov 1954*a*). Song-period short, and similar throughout breeding range. In Schleswig-Holstein, begins 2nd week of May, finishes before end of June (Orr 1976). Similar pattern reported from south-east Finland; there, breeding ♂♂ started singing *c.* 1 week before non-breeding birds but, at end of season, stopped singing *c.* 12 days before non-breeders. Breeding ♂♂ sang mostly before laying, less during incubation, and not at all after hatching. (Sorjonen 1977.) Steinfatt (1939*b*) found little singing after hatching, none after fledging. In Sweden, song begins early May, declining abruptly end of June but continuing until early July (Göransson and Karlsson 1978). In Leningrad (USSR), singing heard exceptionally until early

A

August (Mal'chevski and Pukinski 1983), this probably from juveniles which begin singing about this time (Dementiev and Gladkov 1954a; Hilprecht 1954). ♂♂ sing regularly in winter quarters (Orr 1976). In Zimbabwe, sing—intermittently—at least from early January, frequently February–March (Smith 1951). In Zambia, Song-duels (see below) reported (M G Kelsey). ANTAGONISTIC BEHAVIOUR. Newcomers tend to arrive in evening, trespass on occupied territories, remain silent until dawn, then begin to sing (Simkin and Steinbach 1984), eliciting alarm-calls (Berger 1960; Simkin and Steinbach 1984: see 5 in Voice) or strangled song from resident who may then chase rival along ground or in low erratic flight (Berger 1960). In typical sequence, intruder, singing, flies towards resident, leading to lengthy Song-duels, also sometimes pursuit-flight which may result in repeated fights, each 3–5 min, on the ground, punctuated by chasing and song. In succeeding days, intense aggression gives way to Song-duels as—usually—intruder adopts a peripheral territory. (Simkin and Steinbach 1984.) Disputes mild at start of season, more vigorous when ♀♀ arrive; fights rare once ♀♀ incubating (Berger 1960). Occasionally, resident also chases other species, notably *Sylvia* warblers (J Sorjonen) and other passerines near nest (M Pryl). One resident ♂ chased other birds, especially Icterine Warbler *Hippolais icterina*, out of tree-tops near nest, and once fought with White Wagtail *Motacilla alba* (Neumann 1943). Typically aggressive in contact with *L. megarhynchos*. In communal autumn roosts, quarrelled noisily with *L. megarhynchos* (Farkas 1954). In shared winter quarters, apparently competes with *L. megarhynchos* (Pearson 1984). HETEROSEXUAL BEHAVIOUR. (1) General. ♂♂, especially older birds, arrive on breeding grounds some days before ♀♀ (Stresemann 1948; Berger 1960; Pryl 1980), e.g. on Hiddensee, 8–10 days before (Berger 1967). (2) Pair-bonding behaviour. Initial contact vocal. ♀ may apparently attract ♂ (with call 4a) or vice versa (call 4c). ♀ calling on ground variously draws head in, ruffles plumage, droops and trembles wings, trembles body slightly. Attracted ♂ lands and adopts Advertising-posture: somewhat droops wings, fans tail; once raised and repeatedly fanned tail to one side, and moved off with big hopping leaps (Advertising-display) followed hesitantly by ♀. (Berger 1960.) ♂ sometimes follows Courtship-song by chasing ♀ low over the ground (Sorjonen 1977; J Sorjonen). In 2 cases (2 pairs, both in the morning), ♂ rapidly pursued ♀ back and forth through thicket. When ♀ landed, ♂ flew around her giving excited Courtship-song, then stopped singing and performed intense Advertising-display: shivering his drooped wings and with throat feathers ruffled and tail fanned and raised (Fig B, left), ♂ dashed along the ground

B

towards ♀ who drooped her wings, otherwise remained passive (Fig B, right). ♂ then leaped backwards away from her to a distance of 0·5 m (Fig C), and made another forward dash, sequence repeating itself several times. Finally ♀ flew off, and ♂ chased her excitedly back and forth again. The nests of the 2 pairs were 10 and 20 m from display-site. (M Pryl.) Pair-formation observed in several pairs from 11 days before laying until 1st day of incubation (Sorjonen 1977). ♀ begins nest-building *c.* 4 days after pair-formation (Stresemann 1948). (3) Courtship-feeding. None reported. (4) Mating. Preceded

C

by chasing and Advertising-display, as described above. Occurs on the ground. (J Sorjonen.) One sequence as follows: after chase, ♀ sat wing-shivering, tail-fanning, and calling (perhaps 4a in Voice) 3 m from ♂. Then, with rapid head-tossing, and giving call 4b, ♀ approached ♂ on the ground; continuing to tail-fan and wing-shiver, and crouching low, ♀ moved on to some twigs, closed tail, and moved it sideways. ♂, on ♀'s right, nestled close to her, apparently giving quiet purring sounds (see 4d in Voice) and mounted ♀. Pair copulated 4 times, ♂ mounting from ♀'s right each time. Both then flew off, ♂ with his head well back. (Berger 1960.) After mating, ♂ may start singing. Copulation occurs just before and during laying. (Sorjonen 1977, J Sorjonen.) (5) Behaviour at nest. ♂ may perform variant of Advertising-display (see above)—perhaps in courtship or to some extent as a Meeting-ceremony—when ♀ carries material to nest, or when both birds arrive at nest at same time to feed young: ♂ faces ♀, ruffles throat feathers, bows with drooped and shivering wings, and raises and lowers fanned tail. One ♂ displayed thus to a nest-building ♀ 4 times in 1 hr, and, the following day, 5 times in 1 hour. (Pryl 1980; M Pryl.) Similar behaviour described by Steinfatt (1939b): when ♂ near nest, he slightly raised and opened wings, and moved spread or partly spread tail from side to side in manner of Red-backed Shrike *Lanius collurio*. Neumann (1943) reported similar display by ♂ meeting ♀, while, after hatching, ♀ responded by crouching slightly and shivering her wings. During incubation, ♀ leaves nest by running in a low posture, and returns to nest indirectly and discreetly (Mal'chevski and Pukinski 1983; M Pryl). RELATIONS WITHIN FAMILY GROUP. Only ♀ broods young, most assiduously after hatching (Pryl 1980, J Sorjonen). At one nest, ♀ brooded for 27% of the time from 02.00–09.00 hrs, 18% from 09.00–15.00 hrs, 15% from 15.00–22.00 hrs (Pryl 1980). Brooding spells 1–19 min, average 10·4 (Steinfatt 1939b). ♂ does not feed brooding ♀, so she leaves nest occasionally and in silence to self-feed (Steinfatt 1939b). In hot summer, young not brooded after 4th day (Berger 1960). Both members of pair feed young, ♂ doing most during first few days until ♀ relieved of need to brood young; ♂ sometimes passes food to ♀ at nest for transfer to young. Young beg by gaping silently or with food-calls. Eyes of young open after 7 days. For first few days, one or both parents swallow faecal sacs, later carry them away. (Steinfatt 1939b; Neumann 1943; Berger 1960.) Young typically leave nest at *c.* 9–10 days (Steinfatt 1939b; Pryl 1980), before able to fly properly (Dementiev and Gladkov 1954a; Pryl 1980). Spend next 4–5 days in low cover (Ptushenko and Inozemtsev 1968). Disperse from nest quite rapidly, begging incessantly (Steinfatt 1939b; Neumann 1943). Parents continue to tend young for up to 3 weeks after fledging, and brood-members may continue to associate for some time after independence (J Sorjonen; see also introductory paragraph, section 1). ANTI-PREDATOR RESPONSES OF YOUNG. If danger threatens, young at least 4 days old crouch in nest (Berger 1960). Leave nest 1–2 days early if disturbed

(Ptushenko and Inozemtsev 1968). Fledged young, often warned by parents (see 5 in Voice), press themselves silently and motionless on the ground (Steinfatt 1939*b*; J Sorjonen): press body forwards, and tuck head in such that bill points vertically upwards, rendering eyes invisible from the front. On day after leaving nest, perched young fluttered clumsily away when approached. (Neumann 1943.) PARENTAL ANTI-PREDATOR STRATEGIES. (1) Passive measures. ♀ sits very tightly, not flushing until intruder right beside nest, and even then sometimes reluctant to leave nest (Dementiev and Gladkov 1954*a*; Orr 1976; M Pryl, J Sorjonen). Sentinel ♂ warns (see 5a in Voice) sitting ♀ of approaching danger, and both parents call thus after hatching, starting when intruder tens of metres from nest (M Pryl). ♂ comes quickly to nest if ♀ gives an alarm-call (Berger 1960). When ♀ disturbed from nest, ♂ interrupted his song to chase her back (Hilprecht 1954). (2) Active measures: against birds and man. No clear distinction. When nestlings or recently fledged young threatened, parents fly around giving alarm-calls (J Sorjonen). Call 5b gives way to call 5a on closer approach of predator (Berger 1960). Usually fly towards predator, land *c*. 1 m away, call, and if predator about to attack, fly off instantly; this thought to confuse predator (Mal'chevski and Pukinski 1983). Sometimes, when nestling being handled, ♀ launched direct attack, giving intense Alarm-calls (J Sorjonen: call 5a). ♀ regularly drove *L. collurio* from nest (Berger 1960). Twice, in presence of human intruder, ♀ gave warning-call while crouching and wing-shivering (Neumann 1943); no further details, but possibly a mild distraction-display. Parent sometimes lands and hops away, as if trying to distract predator (Berger 1960). (3) Active measures: against other animals. Rarely, feigns injury in presence of predator, e.g. cat (Mal'chevski and Pukinski 1983). No further information.

(Figs by M J Hallam: A from photograph in Knystautas and Liutkus 1982; B and C from drawings by M Pryl.)   EKD

Voice. Widely studied. Very like Nightingale *L. megarhynchos* (Orr 1976), but song distinguishable to practised ear. Calls not easily differentiated by function and several seem to convey a mixture of contact, excitement, and mild alarm, according to emphasis, timbre, etc. For detailed study of song types and dialects in European USSR, see Simkin (1981). For relationship between song and establishment of neighbourhood groups, see Simkin and Steinbach (1984). Account based partly on notes supplied by J Sorjonen. For additional sonagrams, see Bergmann and Helb (1982) and Sorjonen (1983*b*).

CALLS OF ADULTS. (1) Song of ♂. Typical sequences of low sobbing 'tschok' or 'jug' units, rattles, and plaintive 'pew' sounds; given both in and out of breeding season (see Social Pattern and Behaviour). Sometimes includes mimicry of *L. megarhynchos* and other species (Hilprecht 1954; Stadler 1959, which see for list; Bergmann and Helb 1982). For musical notation, see Stadler (1959) and especially Sotavalta (1956). Quantitative analysis by Sotavalta (1956), also Sorjonen (1983*b*) on which the following based. Usually all phrases within a song are different, and phrases are composed of units usually repeated twice or more. Introductory phrase seems to vary greatly within and between repertoires of ♂♂. Among the most stereotyped phrases are 'castanet' phrase of 1–12 loud units (1–4 kHz), and rattling phrase of 11–80 units and wide frequency range. ♂ gives 2 kinds of song (see Social Pattern and Behaviour for functions). (a) Territorial-song: Given mostly at night and consists of 4–10 phrases, successive phrases alternating in pitch between intermediate (1–4 kHz) and higher (5–6 kHz) frequencies (Sorjonen 1983*b*); strangled variant given in daytime confrontations. (b) Courtship-song: often given by day, near ♀; rather weak, and high-pitched units more common than in Territorial-song (Sorjonen 1983*b*). Recordings and sonagrams of Territorial-song analysed by J Hall-Craggs indicate that all typical phrases of *L. luscinia* (and *L. megarhynchos*: see below) have 3-part structure as follows. (a) Introductory section: comprises relatively long, sometimes very quiet tones, rendered

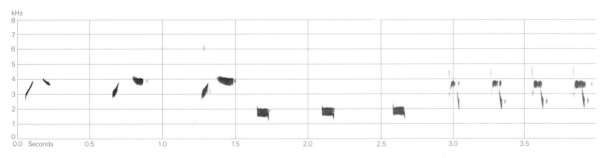

I   P J Sellar   Finland   May 1971

'pew' and variants (e.g. 'peeoo'or 'pooee'); in one rather long phrase (Fig I), lasting 6·4 s, introduction consists of 3 'pooee' units, with marked crescendo. (b) Middle section: one or more short series of rather slowly repeated units of varying timbre, some tending to be tonal, others noisy; in Fig I, middle section begins with 3 low-frequency tonal sounds, then 5 hybrid units (tonal with some noise). (c) Terminal section: a coda made up of noisy units (most often a rattle, variable in speed and volume, or a castanet sequence: occasionally, as in Fig I, both); coda typically preceded and followed by a quite short high note; in Fig I, terminal rattle (artificially boosted in sonagram to clarify units) flanked by brief notes. Comparison of overall structure with *L. megarhynchos* as follows. Phrases of *L. luscinia* likely to be up to twice as long (typically *c.* 5 s rather than *c.* 2·5 s), except when *L. megarhynchos* starts phrase with series of long 'pew' sounds. Whereas *L. luscinia* begins all phrases of normal Territorial-song with quiet low 'pew' sounds, *L. megarhynchos* begins only *c.* 20% of phrases thus; *L. megarhynchos* usually and *L. luscinia* sometimes makes pronounced crescendo through 'pew' sequence (*contra* Orr 1976 and Berg 1980 who stated that *L. luscinia* lacks crescendo). In middle section, *L. megarhynchos* usually gives rapid trills, tremolos or rattles; these numerous and very variable in timbre and delivery rate for given bird; middle section in *L. luscinia* mostly shorter figures of notes or units repeated relatively little. Terminal sections of *L. megarhynchos* brief—a few, usually repeated notes and a flourish, occasionally only one or the other. Terminal figures of *L. luscinia* almost always rattles; these vary in volume, rate, and duration but are consistently noisy, not tonal, in available recordings, and end with a very short flourish less complex than that of *L. megarhynchos*. (J Hall-Craggs.) For other comparisons with *L. megarhynchos*, see Hammling (1909), Orr (1976), Berg (1980), Bergmann and Helb (1982), and Sorjonen (1983*b*). As these not all in agreement with our recordings, degree of difference between the species may depend on individuals and geographical variation. Our recordings support Bergmann and Helb's (1982) contention that speed of delivery of *L. luscinia* somewhat slower and units clearer than *L. megarhynchos*. *L. luscinia* also said to have trisyllabic 'chiddy-ock' sequence, lacking in *L. megarhynchos* (Orr 1976); this feature not readily identifiable in our recordings, but 'chi-di-di-ock' (J Hall-Craggs: Fig II, repeated 4 times), is perhaps the same or similar. (2) Song of ♀. According to Berger (1960), incubating ♀ typically responds to Courtship-song of arriving mate with abbreviated melodious song which, though variable in length, usually contains 5 syllables. Renderings of 2 songs: 'tschie-tschie-tet-tet-schere', and 'tschi-tschi-tschorr-tscherrschick'. (3) Contact-alarm and Excitement-calls. (a) Drawn-out, low rattling sound (Fig III, end) variously rendered 'drrr' (J Sorjonen), 'karr' (Steinfatt 1939*b*; Neumann 1943; Berger 1960), a hard sonorous 'krr' (Bergmann and Helb 1982). Familiar call,

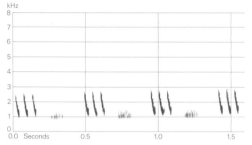

II   B N Veprintsev   USSR   June 1959

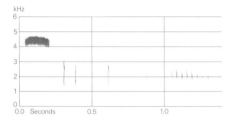

III   P J Sellar   Denmark   June 1961

often indicating suspicion (Berger 1960), mild agitation (etc.), given when disturbed on territory (Steinfatt 1939*b*), or on passage (Schmidt 1975). For more intense alarm variant see call 5a. (b) A 'tack' or 'tacktack' (Steinfatt 1939*b*), a quiet muffled 'dak' or 'dakdak' (Stadler 1959); in winter quarters, apparently prior to spring migration, low excited 'tuc tuc' (Smith 1951). Like call 3a, seems to signal usually low intensity excitement. Also combined with other calls, however, in alarm, e.g. in Fig III, 'whee (call 5b) tuk tuk tuk krrrrrrrrr', probably by both parents (J Hall-Craggs). In alarm, 'klack' (Berger 1960) perhaps variant. (4) Courtship-calls. (a) Quiet drawn-out 'thied' or 'thiet' heard just prior to ground courtship, once definitely from ♀ (Berger 1960); apparently summons ♂ and perhaps invites close approach or solicits copulation. Presumably a variant of call 5b which is also heard in courtship (J Sorjonen). (b) ♀ soliciting copulation gave 'karr', like *L. megarhynchos* (Berger 1960). (c) 'Sirr-sirr' and 'tik-tik' calls, resembling food-calls of small young, and given during courtship (J Sorjonen); evidently this call given near ♀ by advertising ♂, who may also give a peculiar buzzing sound (Simkin and Steinbach 1984). (d) Purring sounds, thought to be given by ♂, but possibly ♀, just before ♂ mounted ♀ (Berger 1960). (5) Alarm- and Warning-calls. (a) In extreme alarm, a harsh low 'kärrr ärrr', e.g. when demonstrating at human intruder near nest with young (J Sorjonen), also during territorial disputes (Berger 1960; J Sorjonen). 'Koarrk-tack' (Berger 1960) probably a combination with call 3b. See also 5b, below. (b) 'wied' sound, varying in quality; sometimes quiet, soft, and anxious, given with bill closed, by both members of pair as warning when predator near nest (Neumann 1943), also during territorial disputes. Rendered 'ieth ieth' (pronounced 'eet eet'), given in a steady series (Berger 1960), also rather high-pitched 'wheet' or 'seep' rather like Chaffinch *Fringilla coelebs* (Orr 1976). For other renderings see Stadler (1959) and Bergmann and Helb (1982). Also given loudly and sharply, like rusty hinge, with bill open, to warn young in or out of nest (Neumann 1943). This variant rendered a loud short whistling 'wit' (J Sorjonen); 'hüit' or a fluting 'whit', sharper and higher-pitched than *L. megarhynchos* (Stresemann 1948; Hilprecht 1954). See also call 4a. Call 5b commonly combined with calls 3a or 5a, thus 'wied karr' (Neumann 1943), 'i-krrr' (Mal'chevski and Pukinski 1983), 'zikrr' (Bergmann and Helb 1982). May also be combined with call 3b, e.g. 'i tak' (Stadler 1959), or 'siep tack' (Steinfatt 1939*b*). In Fig III 'whee' precedes 3 'tuk' sounds and 1 'krrrrrrrr' sound (J Hall-Craggs). (c) Pursued bird gave loud 'raät raät' when rival resident caught up with it and made physical contact; during preceding chase, 'krapp krapp krapp' (Berger 1960), perhaps variant. (d) Once young hatched, ♂ gave a startling low bell-like sound when human intruder bent over nest (Neumann 1943). (6) Other calls. 'Djup' sounds, given singly or repeated up to 3 times (Piiparinen and Toivari 1958).

CALLS OF YOUNG. In nest or recently fledged, young beg with 'dsirt-tak' or 'dsirt-tak-tak' when parents off collecting food (J Sorjonen). Food-call of fledglings also rendered a sharp loud 'tschi' or 'tschitsche', often followed by 'tack' sounds (Neumann 1943; see adult call 3b). Also described as a smacking 'tschi-tscha-tscha' (Bergmann and Helb 1982). When parents near, an excited 'dsirt-sit-sit-si-si' (J Sorjonen). In our recording (Fig IV) of family party, calls of young a slow knocking

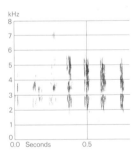

IV P J Sellar Denmark June 1961

rattle rendered 'chi chi chi...' (J Hall-Craggs). Sequence for older fledglings 'tsi-tsi-karr' (Neumann 1943), indicating addition of adult call 3a. For development of song, see Social Pattern and Behaviour.　EKD

Breeding. SEASON. Laying begins mid- to late May over most of range for which there is information (Steinfatt 1939*b*; Dementiev and Gladkov 1954*b*; Dittberner and Dittberner 1973; Makatsch 1976; Orr 1976; Pryl 1980). SITE. On ground, among dead branches, roots, or thick leaf litter, frequently in heavily shaded position. Nest: loose and bulky structure, with basal pad of leaves and

cup of grass leaves and stems, lined with finer material and hair. Outer diameter 13·2 cm (10–15·1), *n* = 18; inner diameter 6·6 cm (6–8), *n* = 19; height 8·3 cm (5–10), *n* = 15; depth of cup 5·3 cm (3–6·5), *n* = 17; weight 36·1 g (21–60), *n* = 19 (Pryl 1980). Building: by ♀ only; mainly in early morning, taking about 4 days (Pryl 1980). EGGS. See Plate 82. Sub-elliptical, smooth and slightly glossy; variable, buff, olive, greenish, or grey-blue, with reddish markings normally giving tinge to whole egg, occasionally just to broad end; often with irregular chalky-white marks. 21·8 × 16·2 mm (20·0–24·2 × 15·3–17·1), *n* = 80 (Schönwetter 1979). Fresh egg weight 3·18 ± SE0·05 g, *n* = 39 (Pryl 1980). Clutch: 4–5. Of 47 clutches, western Finland: 4 eggs, 9; 5, 36; mean 4·8. One brood. Re-lays after clutch loss; once, on day following loss on day 8 of incubation. (Pryl 1980.) 13·3 ± SE0·14 days, *n* = 12. By ♀ only. Not fed on nest, but leaves nest for mean 9·3 min once or twice per hr. (Pryl 1980.) YOUNG. Brooded by ♀; fed by both parents (Pryl 1980). FLEDGING TO MATURITY. Fledging period 9·6 ± SE0·2 days, *n* = 8 (Pryl 1980). Independent 2–3 weeks after leaving nest. Some ♂♂ breed first at 1 year, ♀♀ and most ♂♂ later. (J Sorjonen.) BREEDING SUCCESS. Of 216 eggs in 45 nests, western Finland, 86% hatched and 89·2% of these fledged, to give 76·8% overall success (Pryl 1980).

Plumages. ADULT. Closely similar to Nightingale *L. megarhynchos*, but upperparts, upper wing-coverts, and tertials duller, dark olive-brown or earth-brown, not as rufous or umber-brown as *L. m. megarhynchos*, not as pale greyish-olive as *L. m. hafizi*. Upper tail-coverts and tail tinged rufous, as in *L. megarhynchos*, but rufous tinge less bright and t1 often hardly rufous, mainly deep brown; however, much variation in tail colour and many birds indistinguishable on this character. Pale buff eye-ring often more contrasting than in *L. megarhynchos*, lores and cheeks often more heavily mottled olive-brown, ear-coverts more distinctly streaked pale buff and olive-brown, but much individual variation in both species. Often a mottled dark olive-grey or olive-brown malar stripe (unlike *L. megarhynchos*); chin and throat slightly mottled or washed olive-grey (uniform dirty white or cream in *L. megarhynchos*). Feather-centres on chest and sides of breast olive-brown, tips pale grey-buff, sometimes darker olive again towards fringes; chest and sides of breast thus show distinct dark mottling or appear olive-brown with indistinct pale scaling (uniform pale olive-grey, grey-buff, or dirty white in *L. megarhynchos*; a few *L. megarhynchos* are sometimes contaminated by soil in spring, however, this showing as dark scaling). Flanks, sides of vent, and thighs darker and more extensively pale olive-brown than *L. megarhynchos*, less uniform, sometimes extending as indistinct olive mottling or wash to central belly. Under tail-coverts often have some olive bars at sides (unlike *L. megarhynchos*). Flight-feathers and greater upper primary coverts as *L. megarhynchos* (but see Structure); colour of outer fringes variable, sometimes distinctly rufous and contrasting with olive-brown upperparts and upper wing-coverts. Under wing-coverts and axillaries duller and more olive-grey than in *L. megarhynchos*, more extensively barred or marked olive-brown. NESTLING. As *L. megarhynchos*. JUVENILE. Similar to *L. megarhynchos*, and rather variable in both ground-colour and in contrast and extent of pale spotting. Separable only when flight-feathers well-enough grown to see emargination on inner web of p8: in *L. luscinia*, not emarginated or with a faint notch 5–8 mm from tip; in *L. megarhynchos*, usually a distinct notch 11–14 mm from tip. FIRST ADULT. Like adult, differing as in *L. megarhynchos*. Pale buff spots on tips of retained greater upper wing-coverts or tertials often more distinct than in 1st adult *L. megarhynchos*, but occasionally hard to see; greater coverts (especially innermost) occasionally have traces of 1–2 dull black subterminal bars.

Bare parts. ADULT, FIRST ADULT. Iris dark brown. Bill dark brown, dark grey-brown, or black-brown; cutting edges and base of lower mandible pale brown, flesh-grey, pink-brown, or pinkish. Mouth yellow. Leg and foot pale brown, brown-flesh, pale flesh, or pink-brown with front of tarsus purple. (Witherby *et al.* 1938*b*; Davis 1958*b*; Davis and Hope Jones 1958; Robson 1967; RMNH.) NESTLING, JUVENILE. As in *L. megarhynchos*.

Moults. ADULT POST-BREEDING. Complete; primaries descendant. In East Germany, starts within 1 week after young fledge, with p1 generally shed between 25 June and 5 August (mainly about 10 July). Primary moult very rapid, involving up to 6 feathers simultaneously, rendering bird almost flightless; primaries completed with regrowth of p9–p10 after 32–35 days. Tertials and tail shed at about dropping of p4–p5; sequence of tertials s8–9–7 and tail moulted centrifugally, but many feathers and sometimes all may grow simultaneously. Secondaries start with s1 after shedding of p7–p8; s2–s6 shed during growth of p9 and often not fully completed when all primaries new. (Berger 1967.) In USSR, starts in second half of July, completed in second half of August (Dementiev and Gladkov 1954*b*). POST-JUVENILE. Partial: all head and body and most wing-coverts, but not flight-feathers, tail, or greater upper primary coverts, and only 0–5 inner greater upper secondary coverts. Starts within a few days after fledging, before flight-feathers and tail full-grown; largely completed after 18 days. Sequence approximately: back–sides of breast–belly–mantle and scapulars–chest–neck–crown–forehead, sides of head, and ear-coverts. (Berger 1967.)

Measurements. ADULT, FIRST ADULT. All geographical range, all year; skins (BMNH, RMNH, ZMA). Bill (S) to skull, bill (N) to distal corner of nostril; exposed culmen on average *c.* 4·4 less than bill (S).

WING	♂ 90·8	(2·31; 17)	87–94	♀ 87·9 (1·89; 18)	84–90
TAIL	65·6	(2·83; 14)	62–70	63·1 (1·78; 18)	60–67
BILL (S)	17·1	(0·58; 14)	16·3–17·8	16·6 (0·48; 17)	15·7–17·3
BILL (N)	8·9	(0·57; 14)	7·9–9·6	8·6 (0·54; 18)	7·8–9·4

TARSUS	27·2 (0·60; 14)	26·3–28·3	27·3 (0·72; 18)	26·2–28·3

Sex differences significant for wing, tail, and bill (S). 1st adult with retained juvenile wing and tail similar to full adult.

Ethiopia, live birds: wing, 88·9 (2·4; 28) 82–93 (Ash 1973).

Weights. Belorussiya (USSR), summer: ♂ 27 (5) 25–30 (Fedyushin and Dolbik 1967). East Germany, late May and early June: ♂ 24·5 (1·61; 6) 23–27, ♀♀ 25, 26 (Stresemann 1948). Near Budapest (Hungary): May 27·0 (3) 26–28, July 22·0, August 26·0 (3·33; 44), September 30·2 (4·59; 16) (G L Lövei and T Csörgo 1970). Kazakhstan (USSR), May–September: ♂ 24·0 (4·54; 6) 18–31, ♀ 24·0 (2·95; 5) 20–27 (Dolgushin *et al.* 1970). Cyprus, autumn: 27·2 (16) 25–32 (Moreau 1969). Egypt, September–October: 24·4 (2·24; 33) 21–30 (Moreau and Dolp 1970). Ethiopia: October–November, 23·9 (24) 21–28; December–February, 19·0 (3) 18–20; early April, 34·1 (Ash 1973). Kenya, October–January: 23·3 (87) 19–29 (Pearson and Backhurst 1976). Kenya and Zambia: September–January, 22·1 (2·75; 5) 19–25; March, 28 (Dowsett 1965; RMNH). Malawi (data approximate, read from graph): December, 24·0 (27) 22–28; January, 24·5 (20) 22–28; March, 27·5 (7) 24–33; April, 29 (Hanmer 1979). Kenya, mid-April: 28·2 (2·37; 19) (Britton and Britton 1977). Azraq (Jordan), spring: 21·3 (1·37; 11) 20–25 (Moreau 1969; BTO). Stragglers, western Europe: spring, 19·7, 21·6; autumn, 29·6 (3·61; 9) 23–35 (Davies 1958*b*; Davis and Hope Jones 1958; Robson 1967; Smeenk 1969; Nicolau-Guillaumet 1972; Koch 1977; Jenni 1981; Schelbert 1984).

Structure. Closely similar to *L. megarhynchos*, but markedly different in wing formula. 10 primaries: p8 longest, p9 2–6 shorter, p7 3–5, p6 6–10, p5 9–12, p4 12–16, p1 21–25; tip of p9 equal to p7 ($n=4$), slightly longer than p7 ($n=7$), or slightly shorter ($n=3$), halfway between p6 and p7. P10 more strongly reduced, stiff and narrow; 53–63 (adult) or 51–59 (juvenile) shorter than p8, 8·2 (6) 6–11 (adult) or 5·7 (7) 3–8 (juvenile) shorter than longest upper primary covert. Only 1 primary outer web (p8) emarginated (p7–p8 in *L. megarhynchos*); inner web of p9 distinctly emarginated; inner of p8 sometimes slightly emarginated, 5–8 mm from tip (p8 distinctly in *L. megarhynchos*, 11–14 from tip). Tail rather short; t6 4–11 shorter than t1. Bill proportionately shorter than in *L. megarhynchos*. Middle toe with claw 20·0 (10) 19–22. Remainder of structure as in *L. megarhynchos*.

Geographical variation. None.

Forms superspecies with *L. megarhynchos*, breeding ranges having limited overlap in narrow zone from Denmark to Balkans.   CSR

Luscinia megarhynchos Nightingale

PLATES 44 and 47
[between pages 544 and 545, and facing page 652]

Du. Nachtegaal   Fr. Rossignol philomèle   Ge. Nachtigall
Ru. Южный соловей   Sp. Ruiseñor común   Sw. Sydnäktergal

Luscinia megarhynchos C L Brehm, 1831

Polytypic. Nominate *megarhynchos* C L Brehm, 1831, Europe and North Africa, in southern USSR east to Crimea and Taman peninsula (Krasnodar), in Middle East to central Turkey and Levant; *africana* (Fischer and Reichenow, 1884), Caucasus area and eastern Turkey, east to Iran and Kopet Dag (south-west Turkmeniya, USSR); *hafizi* Severtzov, 1872, from Aral Sea and eastern Turkmeniya east to Mongolia.

Field characters. 16·5 cm; wing-span 23–26 cm. 15% larger than Robin *Erithacus rubecula*, with proportionately smaller head and longer tail giving well-balanced form suggesting small thrush *Turdus*. Medium-sized, graceful chat, with alert, rather upright carriage, noticeably uniform plumage, and skulking habits. Russet-brown above, warmest on tail; dull brown-grey below, with paler throat and vent; pale eye-ring emphasizes gentle expression. Flight flitting and low. Song rich and fluty. Sexes similar; no seasonal variation. Juvenile separable. 3 races in west Palearctic, with ends of cline distinguishable.

ADULT. (1) European race, nominate *megarhynchos*. Upperparts brown, usually tinged russet and becoming brighter, more chestnut on lower rump and upper tail-coverts and fully chestnut-brown on tail. Extension of brown over cheeks emphasizes rounded shape of head; dark face relieved only by paler brown-buff lore and buff eye-ring. Centres of tertials, primary-tips, and central tail-feathers dark brown. Underparts basically dull cream-white, with clouded grey to brown suffusion on sides of throat, over breast, and along flanks; vent and under tail-coverts brighter, cream-buff (often noticeable when tail cocked). Underwing buff-white. Wear makes whole bird rather paler, reducing rufous tone on back. (2) Races of Caucasus and eastern Turkey eastward, *africana* and *hafizi*. These races are eastern end of cline towards paler plumage and longer wings and tail: *hafizi* up to 15% longer-winged and -tailed than nominate *megarhynchos*, and distinctly paler—sandy-grey-toned above and whiter below. JUVENILE. All races. Ground-colours as adult but distinctly and liberally spotted buff and dark brown above, and dark brown on lower throat, chest, and flanks; pale tips to median coverts create obvious panel but pale brown tips of greater coverts form only a dull bar; upper tail-coverts and tail chestnut-brown, as adult. FIRST WINTER AND SUMMER. Spotted juvenile plumage mainly replaced by September but tips of tertials and outermost greater coverts retain pale spots, these marks sometimes persisting through 1st summer. At all ages, bare parts as Thrush Nightingale *L. luscinia*: bill dark brown-horn with paler base to lower mandible (pale area more extensive in juvenile); legs pale brown or pale flesh (always pale in juvenile).

The only breeding nightingale in western and central Europe and around Mediterranean, and thus readily identifiable—only slight possiblity of confusion with similarly coloured species such as Rufous Bush Robin *Cercotrichas galactotes* and juvenile *E. rubecula* or White-throated Robin *Irania gutturalis*. Across central Europe (from Germany south-east to Black Sea) in breeding season and around North Sea in migration periods, closely similar *L. luscinia* also occurs (see that species, p. 616). Flight recalls that of *E. rubecula*, being performed at low level and consisting of bursts of wing-beats, leading into sweeping glides, sudden turns, and dives into cover; action thus flitting but rather less so than *E. rubecula*, with longer wings and especially longer, fuller tail allowing more floating progress. Moves on ground by short or long hops on rather long legs, usually with rather erect carriage (head and tail both held up, wing-tips drooped below line of tail). Movements accompanied by frequent flicks of wings and tail; cocks tail when excited. Like *E. rubecula*, has endearing habit of holding head on one side when foraging. Ground-hugging but not always skulking, feeding around ground cover as much as within it when undisturbed. Not gregarious but occurs in scattered falls on migration.

Song remarkably rich and varied and extremely vigorous: most distinctive parts are rapid, staccato 'chock chock...' and slow, fluting 'pioo pioo...' in crescendo; also introduces throaty, musical chuckles and contrasting croaks; delivered by day and night, usually from dense cover but occasionally from open perch. Calls also varied: most commonly, 'hweet' like *Phylloscopus* warbler, 'tack tack' like *Sylvia* warbler or wheatear *Oenanthe*, and croaking 'krrr'. Also has harsh and grating scolding call, 'tschaaa', accompanied by bowing threat posture.

Habitat. Breeds in west Palearctic in middle and lower-middle latitudes, with some oceanic bias, in mild and warm temperate, Mediterranean, and steppe climatic zones between July isotherms 17–30°C (Voous 1960). Differs from Thrush Nightingale *L. luscinia* in more southerly, westerly, and generally somewhat warmer breeding range, less restricted to lowlands, valleys, and neighbourhood of water in most regions, and more ready to inhabit drier sandy soils and sunny hillsides. In Switzerland, breeds exceptionally up to 1100 m although only locally above 600 m, and in canton Bern almost exclusively associated with standing or running water in mild situations (Glutz von Blotzheim 1962; Lüps *et al.* 1978). In England, rarely above 200 m (Simms 1978).

Occupies 3 distinguishable habitat types, of which 1st closely resembles that of *L. luscinia*, in thickets or woods near water: e.g. in England, 19 out of 40 territories adjoined streams and remainder were near small pools of stagnant water, although woods were mainly avoided in favour of open sunny thorn thickets with patches of tufty grass in between, the rest being in roadside spinneys or large gardens (Harthan 1934). Breeding distribution in England lies almost entirely south of 19°C June isotherm (Norris 1960), and records over a long period show that this pattern has persisted with little variation and with regular sharply defined cut-off lines, ruling out both cooler and wetter regions, but still including substantial areas unoccupied presumably through lack of suitable cover or nesting or foraging conditions. In England, favours coppice-with-standards woods cut every 12–15 years, years 5–10 being the best; also occurs along tall overgrown hedgerows and farm tracks edged by bushes or woodland, especially with standing oaks *Quercus* or fallen trees. In one area, mature yew *Taxus* with oak and

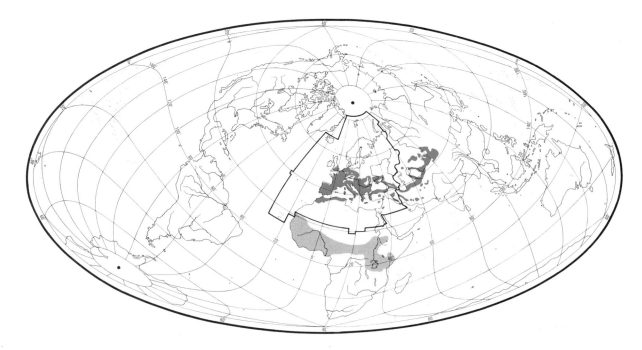

shrub growth on a chalk hillside attracts a high density. Elsewhere, rough meadowland with small fields and high thick thorn hedges especially favoured. 2nd distinguishable habitat type is thus a mixture with drier soil, no surface water, and thicket, scrub, or managed open woodland including various open spaces, not necessarily flat, and offering unobstructed feeding ground with ample leaf-litter and sunny as well as shady conditions. Edges of woods of oak and beech *Fagus sylvatica* are preferred to depths of woodland, and disturbance avoided as far as possible, although large quiet gardens may be occupied (Yapp 1962; Sharrock 1976; Fuller 1982). In Belgium and Netherlands, also inhabits bushy coastal sand-dunes (IJzendoorn 1950; Lippens and Wille 1972). Sandy, higher and drier situations also chosen in central Europe in area of overlap with *L. luscinia*; see that species (p. 617). In southern USSR breeds in cover along river banks, in open woodland of hornbeam *Carpinus*, among oaks, beeches, and alders *Alnus*, in thickets of sea buckthorn *Hippophae rhamnoides*, hawthorn *Crataegus*, and other shrubs, and sometimes in orchards. Central Asian race *hafizi* occurs densely in orchards devoid of undergrowth and ascends to *c.* 2600 m, thus representing in extreme form the 3rd main habitat type, characteristic of Mediterranean region, on dry and warm hillsides or valleys, often near human settlements, and even involving use of telegraph poles and rooftops for singing (Simms 1978). In Coto Doñana (Spain), breeds in shrubbery of bramble *Rubus*, heath *Erica*, and *Pistacia* in pinewoods, and in cork oak *Quercus suber* savanna heath (Valverde 1958). On Mallorca, inhabits rich maquis and thin woodlands (Parrack 1973). In Camargue (France), breeds in riverain woodland, degenerate oak wood, and shrubs

below juniper *Juniperus phoenicea* on old sand-dunes (Hoffmann 1958).

In winter in tropical Africa, frequents savanna woodland, thorny scrub, river gallery edges in mountains, humid forest edges and clearings, low second growth, and tangles of small trees, bushes, and rank herbage fringing watercourses, sometimes occurring in tall grass (Moreau 1972; Serle and Morel 1977; G D Field). At all seasons a ground and perching bird, fond of cover and not flying further or higher than essential. Avoids disturbance by man where possible, but often indirectly affected by reclamation or unsuitable management of habitats, which however may as often be rendered untenable by ecological succession.

Distribution. Range has decreased in Britain and USSR. Some spread in East Germany.

BRITAIN. Some slight decline at periphery of range (Parslow 1973; Hudson 1979). EAST GERMANY. Rapidly colonized lowland areas in Oberlausitz after 1950 (SS). USSR. Formerly bred further north in Ukraine (GGB).

Accidental. Iceland, Ireland, Norway, Finland, Sweden, Kuwait.

Population. Decreased Britain, France, East Germany, and Rumania, and probably Netherlands and Poland. Some recent increase (following decrease) in Czechoslovakia.

BRITAIN. 1000–10 000 pairs; declined since 1950 (Parslow 1973). About 10 000 pairs according to Sharrock (1976), but this estimate believed by Hudson (1970) to be wrong—probably near 4000 singing ♂♂. About 4770 singing ♂♂ 1980 (Davis 1982). FRANCE. Over 1 million

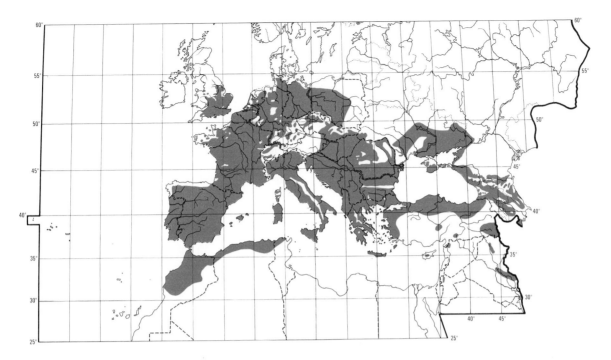

pairs; declined (Yeatman 1976). WEST GERMANY. 33 000–39 000 pairs (Rheinwald 1982); declining (Yeatman 1976; AH). Increased upper Main 1972–80 (Kortner 1981). BELGIUM. About 3000 pairs; decreasing (Lippens and Wille 1972). NETHERLANDS. 4500–5000 pairs (Teixeira 1979); estimate probably too low (CSR). Decrease, not linked with Sahel drought (Wammes *et al.* 1983). POLAND. Fairly numerous (Tomiałojć 1976); probably slight decrease (AD). CZECHOSLOVAKIA. Decreased 1900–50, then slow increase (KH). RUMANIA. Decreased (VC). USSR. Decreased northern and central Ukraine and Caucasus (GGB).

Oldest ringed bird 7 years 11 months (Rydzewski 1978).

Movements. Migratory, wintering in Afrotropics and one area of Iraq (H Siman).

WESTERN POPULATIONS, nominate *megarhynchos* (breeding in Europe, western Turkey, and north-west Africa), winter between Sahara and rain forest from West Africa east to Uganda (Vaurie 1959; Portenko and Wunderlich 1977). European breeding birds leave in autumn between end of July and September, with peak passage through British bird observatories on south and west coasts in late August (Riddiford and Findley 1981). Movement through Europe broadly south-west, with birds occurring throughout Mediterranean region though commonest in west. In Malta, passage is from mid-August to mid-October (Sultana and Gauci 1982), and on Cyprus from late August to mid-October, though rather scarce (Flint and Stewart 1983). In Turkey, apparently leaves August–September (*Orn. Soc. Turkey Bird Rep.* 1966–75); peak passage at Bosporus late July to early August (Porter 1983).

Zink (1973) documented recoveries of birds ringed in Europe (mainly central Europe to 15°E). Predominant movement of all populations is south-west, with concentrations of recoveries in south-west France August–September and in south-west Iberia August–October. 2 September recoveries near north-west Moroccan coast (ringed West Germany and southern France) are the only ones in North Africa in autumn, where few birds are seen, though occurring mid-August to early October (Heim de Balsac and Mayaud 1962). Exceptions to general pattern are recoveries in Italy and south-east France of some birds ringed in East Germany, and 2 movements east of south, from Germany to south-east Italy (recovered September) and from Hungary to Greece (recovered February). Of the relatively few spring recoveries, none are in Iberia and only 2 in south-west France, but 2 well inland in Morocco. Portenko and Wunderlich (1977) and others have suggested a more easterly route taken in spring, but recoveries from Tunisia (April) to Hungary and from Malta (March) to Czechoslovakia (May) show that movement is still on south-west/north-east axis. One bird ringed September in Malta and recovered the following April in northern Italy, and 2 ringed in Tunisia in April and recovered in northern Italy in June and October indicate similarity of spring and autumn routes. 3 out of 1326 ringed at Gabès (Tunisia) have been retrapped on subsequent spring passages (Arnould *et al.* 1959). The relative scarcity in much of North Africa and also Middle East in autumn (e.g. only 1 record for Libya, in September: Bundy 1976) suggests Mediterranean and

Sahara normally crossed in one continuous flight. In Egypt, however, scarcer in spring than in autumn (Etchécopar and Hüe 1967).

Detection in Africa often relatively easy as birds sing on winter territory and also sometimes on passage. First arrivals south of Sahara in August (e.g. Morel and Roux 1966, Lamarche 1981), though arrival in (e.g.) Sénégal not complete until October (Morel and Roux 1966), and evidence then of slow southward spread during autumn. In Gambia, present (uncommonly) only from mid-November (Gore 1981). Few observations in Mali and perhaps occurs only on autumn passage (Lamarche 1981). In Nigeria, first arrivals in north from early September, though main passage at Kano is in second half of October, and a few stay to winter; in south, however, not present until early November, with wintering birds at Ibadan mid-November to early April (Elgood *et al.* 1966, 1973). Bird ringed in Tunisia in spring recovered in Nigeria at Akure. Further east, less information on this race, but common in some autumns in Ouadi Rimé (Chad) with extreme dates 19 September-27 March (Newby 1980) and common on passage and in winter in Sudan (Moreau 1972) with autumn passage in Darfur 7 September-27 October (Dementiev and Gladkov 1954b).

Some present in Sénégal until early May (Morel and Roux 1966), but spring passage through Nigeria concentrated in late March and early April with arrivals in North Africa and southern Europe at this time. Unlike autumn, many records in spring along North African coast and on Mediterranean islands and even commonly inland in Algeria and Libya, so passage obviously on broad front. Peak passage at Gabès (Tunisia) 31 March-22 April but birds occur throughout North Africa on passage until early May. On Malta, spring passage occurs mid-March to mid-May, numbers usually in low double figures, but with daily maximum of 60. In Cyprus, 112 limed one spring compared with 245 Thrush Nightingales *L. luscinia* (Flint and Stewart 1983). Stresemann (1948) gives peak spring arrival dates on breeding grounds progressively later further north ranging from 30 March at Tanger (Morocco) to 28-29 April at Lübeck, West Germany (54°N). Arrival may be delayed by weather, with birds arriving in waves between spells of bad weather during one cold, wet spring (Horstkotte 1971). Peak arrival time at south coast observatories in Britain is late April, with late May arrivals of spring overshoots on east coast (Riddiford and Findley 1981).

EASTERN POPULATIONS, *africana* (breeding from Caucasus and eastern Turkey to Iran) and *hafizi* (breeding east from Aral Sea), winter mainly in Kenya and Tanzania, though movements and wintering areas confused by difficulties of separating the 2 races. *L. m. hafizi* said by Vaurie (1959) to be migrant through north-east Africa, Iraq, Iran, and Arabia, and *africana* said to have 'migration and winter quarters the same as in *hafizi*'. *L. m. africana* arrives east Transcaucasia during mid-April

and Georgia by early May (Dementiev and Gladkov 1954b). *L. m. hafizi* leaves breeding grounds not before end of August (Johansen 1955). Scarce on passage in northern Oman late August to October and in May, and common in southern Oman with specimens collected late September and early October (Gallagher and Rogers 1980). Presumably this is the race referred to by Erard and Etchécopar (1970) as common in Iran at Bandarabbas, 28-30 April and less so round Zahedan 9-13 May. On Red Sea coast of Sudan, Nikolaus (1983) caught 22 between 3 and 10 September (race not stated), and much migration into and out of eastern Africa clearly occurs through Arabia. Birds in Uganda are mainly nominate *megarhynchos*, though *hafizi* has occurred at Kampala. In Kenya, the species occurs mainly in east, centre, and south-east; *hafizi* predominates on coast, with *africana* increasingly inland. Earliest arrival 22 October, but mainly during November (suggesting several weeks spent further north in north-east Africa), with song on winter territories December-March; small onward passage at Ngulia in south-east, continuing into December, with 464 ringed 1969-84 (compared with 12 595 *L. luscinia*); 60% *africana*, 40% *hafizi*. Also some evidence of spring passage in Kenya, with occasional concentrations on coast in early April. (Pearson 1984.) Further north, although only 2 old records each from Somalia and Ethiopia, now found by Ash and Miskell (1983) to be locally common in winter south of 2°N in southern Somalia (*hafizi* or intermediates). Also fairly common in southern Ethiopia on passage in October, with lack of ringing retraps indicating continuing passage in November and with December-February records demonstrating overwintering. One bird ringed in January 1971 was retrapped at same site on 20 and 23 November 1971. One December bird weighed 50% more than the average, presumably in preparation for onward passage within Africa. One in April was similarly heavy, and such pre-migratory fattening in Ethiopia is consistent with scarcity in Arabia in spring. So far, only *hafizi* identified in Ethiopia, though all 3 races thought to occur (Ash 1973).   RAC

Food. In breeding season, terrestrial invertebrates, especially beetles (Coleoptera) and ants (Formicidae); in late summer, also berries. Feeds on ground, taking food mostly from litter layer but also from bare ground and from leaves or twigs or while gripping bark. Moves on ground by long hops with frequent pauses but will also drop on to prey from perch and catch insects in flight (Witherby *et al.* 1938b; Horstkotte 1965; Rustamov 1982).

In Crimea (USSR) in breeding season, 11 stomachs contained 89·3% ants (Formicidae) and 8·9% beetles (Coleoptera); also 1 mollusc, 1 millipede (Julidae), and 1 woodlouse (Isoptera) (Kostin 1983). Other animal food recorded: small Lepidoptera, caterpillars, flies (Diptera), ant pupae and larvae of other Hymenoptera, adult and larval beetles (Carabidae, Scarabaeidae, Elateridae,

Curculionidae), spiders (Araneae), earthworms (Oligochaeta) (Rey 1908; Gil 1927; Witherby *et al.* 1938*b*; Dementiev and Gladkov 1954*b*; Hilprecht 1954). Fruits taken largely in second half of summer; include strawberry *Fragaria*, bird cherry *Prunus padus*, currants *Ribes*, alder buckthorn *Rhamnus*, *Amelanchier*, elder *Sambucus*, and dogwood *Cornus* (Schuster 1930).

In winter, eastern Nigeria, all of many stomachs examined contained only insects (Serle 1957).

Young studied in Thüringen (East Germany) given 767 items comprising 25·5% adult and larval beetles, 24·5% adult, larval, and pupal Diptera, 11·6% adult and larval Hymenoptera, 11·1% adult, larval, and pupal Lepidoptera, 8·1% Hemiptera, and 7·7% Arachnida; remaining 11·5% comprised springtails (Collembola), stoneflies (Plecoptera), grasshoppers, etc. (Orthoptera), lacewings, etc. (Neuroptera), caddis flies (Trichoptera), isopod crustaceans, millipedes (Diplopoda), and gastropod molluscs. Young at one nest fed mainly on larvae of moth *Tortrix viridana*. (Emmrich 1971.) In Turkmeniya (south-central USSR), 151 items from collar-samples comprised 61% insects, 17% woodlice, 15% berries, and 7% spiders; berries given towards end of nestling period. Further sample of 53 items comprised 32% (by number) insect larvae, 30% ants, 15% spiders, 13% Orthoptera, 8% berries, and 1 earthworm. In another study, young given no ants, most important groups being Orthoptera (20%), mosquitoes (12%), and caterpillars (12%). (Rustamov 1982.)   DJB

Social pattern and behaviour. Major studies in Westfalen (West Germany) by Horstkotte (1965), and at Marchegg (Austria) by Grüll (1981). For review, and comparison with Thrush Nightingale *L. luscinia*, see Hilprecht (1954).

1. Mostly solitary. Local concentrations occur on spring migration in Kenya (Pearson 1984). Early ♂♂ arrive on breeding grounds alone; later ♂♂ in small groups, sometimes with ♀♀ (Horstkotte 1971). At end of breeding season, family parties may associate loosely (Horstkotte 1965): e.g. in West Berlin, mid-August, loose assembly of 39 birds in 2·7 ha persisted for some days, briefly 20 in 0·6 ha (Elvers 1972). At Marchegg, ♂♂ stayed in their territories for some time after young independent; in Westfalen, all birds left after young independent (Horstkotte 1965). Territorial in winter quarters (e.g. Cawkell and Moreau 1963, Nicolai 1976; see also Antagonistic Behaviour, below); birds in Kenya commonly sedentary for weeks or months (Pearson 1984). In Sierra Leone, one resident per territory, but territories may be close together (G D Field). In Togo, winter territories apparently small, 3-4 birds often singing *c.* 10 m apart in same thicket (Douaud 1957). BONDS. Monogamous mating system the rule. A few reports of bigamy (♂ plus 2 ♀♀): in one case, overlapping clutches laid by 2 ♀♀ in ♂'s territory (Davis 1975); in 2 similar cases, ♂ attended both nests (Clodius 1894). Pair-bond typically breaks down at end of breeding season around time when family bonds dissolve and moult begins (Stresemann 1948). Of 107 pairs, only 6 re-formed in following season, though others settled near to former partners, sometimes in neighbouring territory, raising possibility of re-pairing to former mate at subsequent stage; of the 6 pairs which re-formed, the members of 4 pairs had belonged to

different pairs in the past, or changed mates in a later season (Hilprecht 1954). Where given pair starts 2nd clutch, ♂ mate not uncommonly departs with 1st brood while ♀ starts incubating. In 2 cases, ♀ then raised 2nd brood herself (Horstkotte 1969; Grüll 1981). In 2 other cases, ♀ paired and bred with a neighbouring bachelor ♂ soon after her mate departed with 1st brood (Horstkotte 1969). Only ♀ broods; both parents feed young. If one parent disappears, other may continue to raise young (Hilprecht 1954). After young able to fly (within 5 days of leaving nest), parents typically divide brood between them (Hilprecht 1954; Horstkotte 1965). According to Horstkotte (1965), family may re-unite later, *contra* Hilprecht (1954). In one case, ♀ stayed in territory with 1 young, ♂ having left earlier with another (Horstkotte 1965). Young fed for 14-15 days after leaving nest (Horstkotte 1965), 15-20 days (Rustamov 1982). Once, ♀ still feeding young 23 days after fledging. At least some birds breed at 1 year old (Grüll 1981). BREEDING DISPERSION. Solitary and territorial. At Marchegg, average size of territory, calculated from perimeter song-posts, 0·67 ha (0·13-1·9) (Grüll 1981). In Magdeburg (East Germany), 0·2-0·4 ha, nests often 50 m apart (Hilprecht 1954); in Westfalen, 0·35 ha (0·31-0·38) (Horstkotte 1965). In Ashkhabad (Turkmeniya, USSR), average size of 15 territories 0·39 ha (0·13-0·78), average 67 m (20-94, *n* = 13) between nests; slight increase in territory size after hatching (Rustamov 1982). At start of breeding season, progressive arrival of birds leads to contraction of territory size, e.g. from 0·8 ha to 0·55 ha, though late arrivals may extend territories at expense of neighbours (Grüll 1981). Density, based usually on singing ♂♂, varies greatly, maximum locally *c.* 1 pair per ha, e.g. in Magdeburg 7 pairs in 6·35 ha (Stein 1968). In Niedersachsen (West Germany), locally 7 ♂♂ in 7 ha, more widely 1·8 per km² (Flade 1979); in various towns, Niedersachsen, 0·2-2·9 pairs per km² (Scherner and Wilde 1972). In Hamburg area (West Germany) over 6 years, 0·04-0·05 pairs per km² (Leuschner 1974). In optimal habitat, Karben (Hessen, West Germany), 100-150 territories per km² (Heerde 1982). In Switzerland, 40-60 pairs per km² at Nyon, up to 80 in Allondon Valley (Schifferli *et al.* 1982, which see for transect counts). Over *c.* 80 ha of predominantly oak *Quercus* wood, Gloucestershire (England), 1927-60, 5-30 ♂♂ per km² (Price 1961). In Surrey (England), 1957-71, 0·06 territories per km², maximum in any year 0·52 (Wheatley 1973). For Kent and south Worcestershire (England), see Stuttard and Williamson (1971), and Harthan (1934). In Morocco, 4·1 pairs per km² in woodland, 18·3 in degenerate woodland (Thévenot 1982). Replacement nests built in same territory (Horstkotte 1965); once, nest for 2nd brood 17 m from 1st (Horstkotte 1969). Territory serves for pair-formation, nesting, and most feeding; parents with fledged young may feed beyond territory, though never trespassing on neighbouring territories (Grüll 1981, which see for territory structure), or family may vacate territory altogether (Horstkotte 1965). Territories traditional, and within a given area the same (best) ones are always occupied first. ♂♂ markedly site-faithful, ♀♀ less so. At Marchegg, ♂♂ older than 1 year showed strong site-fidelity unless they bred unsuccessfully the previous year. Thus, over 3 years, 76% of ♂♂ occupied territory of previous year or an adjacent one, rest moved further away; on average, site-fidelity of ♂♂ 48·5%, of ♀♀ 22%. (Grüll 1981.) In Magdeburg, of 746 ♂♂ ringed as adults, 19·2% returned to former breeding area (same park) compared with 9·6% of 428 ♀♀. 2·4% of ♂♂ shown to have moved to another park in town, 3·7% of ♀♀; on average ♀♀ also moved further. Majority of young ♂♂, but few ♀♀, return to natal area to breed. (Hilprecht 1954.) For Hamburg area, see Leuschner (1974). ♂♂ recorded moving up to 1120 m within breeding season after failing to attract ♀ (Hilprecht 1954). ROOSTING. Little known. One ♀ bathed regularly during incubation and brooding (Horstkotte 1965). For aggression with *L. luscinia* in communal autumn roosts, see that species (p. 622).

2. Secretive but not especially shy, especially outside breeding season: e.g. on spring migration, Iraq, will sing from scantiest of cover (Moore and Boswell 1956). In Khartoum (Sudan), September, very tame, keeping mainly to ground and, when approached, hopping away under cover rather than flying (Butler 1905). In various situations where suspicious or excited (e.g. when approaching nest to feed young: Ticehurst 1912), flicks up tail and lowers it more slowly; unlike *L. luscinia*, does not fan tail in this context (Berger 1960). FLOCK BEHAVIOUR. In loose assembly in West Berlin, mid-August, a number of birds were singing (see 1 in Voice) softly, and aggressive chasing not uncommon (Elvers 1972). No further information. SONG-DISPLAY. ♂ sings chiefly from low undergrowth, also low branches of trees, sometimes fully exposed. Occasionally in flight (Witherby *et al.* 1938*b*; see Antagonistic Behaviour, below). By day, sings from several perches, often in the open, regularly changing perch with start of next song; nocturnal song given mainly from one perch, used several nights in succession. Diurnal song mainly an interaction with approaching rivals, nocturnal song more long-distance advertisement to ♀♀. (Hultsch 1981.) Nocturnal song thus associated mainly with pair-formation (Hilprecht 1954), and, in Austria, given for only *c.* 15 days by ♂♂ which paired, much longer by those (all 1-year-olds) which failed to pair (Grüll 1981). Horstkotte (1965) distinguishes 3 types of song: (1) Territorial-song, given in a stiff upright posture, wings slightly open, near boundary, until hatching of young. Pale throat plumage conspicuous when singing (Bergmann and Helb 1982). For Song-duels (etc.), see Antagonistic Behaviour, below. (2) Courtship-song, directed at ♀, associated more with nest (see subsection 5, below); ceases with hatching. (3) Contact-song, rather fragmentary and short, for contact with ♀ in territory. In addition to these song types, Subsong given, notably in winter quarters (Nicolai 1976; G D Field). After independence, young ♂♂ sing from time to time, but imperfectly. Young ♀♀ also sing to some extent, never as loudly or as clearly as ♂♂ (Hilprecht 1954). For detailed differences between song types, see Voice. In Westfalen, song most vigorous in evening until midnight, then lull until 02.00 hrs. Silent *c.* 05.00–06.00 hrs. Vigorous thereafter for 2 hrs. Sings with short pauses during day, most vigorously in morning. (Horstkotte 1965.) 1 ♂ sang regularly by day until young 5–6 days old, also intermittently at night until fledging (Ticehurst 1912). In winter quarters, sings from dense cover. In Zaïre, rarely sang at night (Chapin 1953), and diurnal song most commonly reported elsewhere, e.g. in Sierra Leone, sings throughout the day, usually Subsong, sometimes almost full song (G D Field). Winter song differs from summer song mainly in being less loud and powerful (e.g. Douaud 1957, Nicolai 1976); said by Nicolai (1976) to be as structured and persistent as summer song, by Walker (1981*b*) to be less rich and varied. Sings throughout winter (e.g. Douaud 1957, Serle 1957, Morel and Roux 1966, Pearson 1984), and regularly on spring passage (Witherby *et al.* 1938*b*). In central Europe, song-period late April until end of June (Stadler 1957), in Mediterranean until end of July (Stresemann 1948). In Austria, little song at outset; ♂♂ begin to sing at night once territories established; song builds up over following week to peak at end of April, coinciding with arrival of ♀♀; ♂♂ which arrive at this time start singing straight away. Song declines before mid-May (laying), then a brief resurgence (end of laying), but little from

end of May/beginning of June, although some resurgence 5–10 June and again after fledging; almost all ♂♂ stop singing early July, coinciding with break-up of families and onset of adult moult. (Grüll 1981.) ANTAGONISTIC BEHAVIOUR. (1) General. Little aggression on first arrival on breeding grounds, increasing markedly when ♀♀ arrive, waning once incubation begun (Hilprecht 1954). Main defence is by Territorial-song, though varies with context. Full song given only away from territorial boundary, before or after dispute; when intruder detected at edge, resident sings loudly, approaches giving strangled song or alarm-calls (see 4 in Voice) (Hilprecht 1954; Grüll 1981). Accompanies bowing threat-posture with 'tschaaa' call (D I M Wallace: see 4d in Voice). Rivals approach each other, singing the while. Song-duel may lead to chasing, sometimes to treetops and back to ground again, accompanied by song and twittering calls. Fights, in which bills used, occur. (Hilprecht 1954.) Winter territorial defence likewise achieved by song, alarm-calls, and chasing (Douaud 1957; Nicolai 1976; L G Grimes), sometimes in competition with *L. luscinia* (Pearson 1984; see also Roosting, above). At end of breeding season, parents may attack strange independent offspring (Grüll 1981). Near laying, ♂ was increasingly aggressive to Redstart *Phoenicurus phoenicurus*, less so to Chiffchaff *Phylloscopus collybita* and Robin *Erithacus rubecula* trespassing in breeding territory (Horstkotte 1965). For call of bird chasing Cuckoo *Cuculus canorus*, see 4d in Voice. For aggression to ♀ mate during pair-formation, see Heterosexual Behaviour (below). HETERO-SEXUAL BEHAVIOUR. (1) General. ♂♂ arrive a few days before ♀♀ on breeding grounds (Stresemann 1948; Horstkotte 1969, 1971). Oldest ♂♂ evidently arrive first: in Austria, 1-year-old ♂♂ did not occupy territories until at least 9 days after older birds. Up to 2 weeks may elapse between initial arrival and establishment of territories. (Grüll 1981.) Pair-formation (of given ♂ and ♀) seems to occur within 1 week (Horstkotte 1965). ♀ builds nest soon after pair-formation (Stresemann 1948). (2) Pair-bonding behaviour. ♂ on ground near ♀ gives quiet Courtship-song; occasionally flies up to a branch, singing, fans and raises tail, keeping it raised sometimes for several seconds (Hilprecht 1954; Horstkotte 1965). May also droop wings, and exchange harsh Alarm-calls with ♀ (Hilprecht 1954; see 4d in Voice). In pauses, when feeding, ♂ performs peculiar hopping jumps (Horstkotte 1965), presumably elements of Advertising-display (see below). If ♀ moves away, ♂ pursues her through the undergrowth giving repeated bouts of Courtship-song, typically accompanied by tail movements (as described above: Hilprecht 1954). Also pursues her with Bleating-calls (see 3a in Voice) in whirring flight, gliding down and landing by her side. Then, ♂ may perform intense Advertising-display: with tail fanned and wings spread, 'danced' in front of ♀: as the 2 birds moved along, ♂ jumped towards ♀, not closer than 50 cm, then back again, several times in rapid succession (Hilprecht 1954); this evidently similar to Advertising-display of ♂ *L. luscinia* (p. 622). In perhaps less intense variant, ♂, after flying to ♀'s side, tripped around her in a rather stiff posture with wings drooped, ♀ remaining passive. Next day, ♀ followed ♂. She landed on ground, and performed apparently appeasing Scraping-display: pressed down near ♂ with her wings spread; ♂ performed semi-circular dance a couple of times, almost dragging his fanned tail. At other times, when ♀ suddenly appeared, ♂ called aggressively (Horstkotte 1965). Display of ♂ near ♀ described by Géroudet (1954) as follows: quivers wings, moves fanned tail up and down, and lowers head to the level of his perch, perhaps indicating some threat (see above). Strengthening of pair-bond indicated by increasing contact-calls (see 2b in Voice) between pair, these progressively replacing Courtship-song of ♂. (Horstkotte 1965.) (3) Courtship-feeding. None reported, but see subsection 5, below. (4) Mating. ♂ repeatedly flew up to a branch above ♀, fanned his tail, and made rather deliberate movements toward ♀ in a stiff posture. ♂ suddenly flew down and, after a long chase, accompanied by Bleating-calls (of ♂), copulated with ♀ (Horstkotte 1965). In Fig A, birds

A

shown immediately prior to copulation: ♂ (left) calls, holding tail fanned and raised, wings half-spread, quivering, while ♀ (right) crouches somewhat (Géroudet 1954). In chases leading to copulation, calls (see 3b in Voice) given by both sexes; after copulation, ♂ returned to song-perch to sing, ♀ remaining on the ground (Rustamov 1982). (5) Nest-site selection and behaviour at nest. ♀ selects site, accompanied by—and perhaps stimulated by—♂ (Grüll 1981). ♂ gave Courtship-song while ♀ collected nest material, also fanned and lowered his tail, similar to pre-copulatory movement. Pair moved around territory. ♂ now behaved as ♀ did when following ♂ during pair-formation (see above): ♂ flew to ground, nestled down, and made rotating movements, beating wings vigorously against the ground (Scraping-display). ♂ repeated display when ♀ moved further away. Many scrapes found in vicinity and display possibly stimulates ♀ to seek nest-site, though in this instance site had already been chosen. During incubation, ♂ used Courtship-song to summon ♀ off nest (for feeding together), and to lead her back to nest, this ceasing after 6th day of incubation when bond apparently weaker, pair-members then feeding independently. From 13th day of incubation until small young in nest, ♂ occasionally fed ♀ on nest. (Horstkotte 1965.) This also reported by Hosking and Newberry (1949) and Rustamov (1982); see Fig B. As ♂ commonly delivers food to

B

♀ for young (see below), primary intention possibly not to provision ♀. RELATIONS WITHIN FAMILY GROUP. Young beg immediately upon hatching. Brooded regularly by ♀ for first 4 days, less from day 5, by which time eyes fully open. Thereafter brooded for a few days at night, and by day if weather bad. (Horstkotte 1965.) While ♀ brooding, ♂ does most feeding,

often giving food to ♀ for transfer to young (Hilprecht 1954; Rustamov 1982). At one nest, possibly suspicious ♂ regularly gave food to ♀ a few m from nest (Ticehurst 1912). If ♀ absent, ♂ feeds young directly and may brood them briefly (Horstkotte 1965). Once young no longer brooded, both parents feed them. Approach nest gradually and quietly, flying from perch to perch, giving a feeding call (not described) from side of nest (Rustamov 1982). ♂ often gave a burst of song on arrival with food. Both sexes perform nest-sanitation. At one nest, when young small, parents ate faeces from bottom of nest, but, from 6 days collected faeces from rim or as they were voided, then swallowed or carried them away. (Ticehurst 1912.) Young leave nest at 10-12 days (Horstkotte 1965), dispersing into surrounding cover until able to fly 3-5 days later (Hilprecht 1954; Horstkotte 1965). For brood-division after fledging, see Bonds (above). Young start self-feeding at 8-9 days after leaving nest, but are partly fed by parents for typically another 6 days (Horstkotte 1965). Until independent, parents may lead young to feeding area outside territory (Horstkotte 1965). Independent young disperse and lead nomadic existence (Stresemann 1948). ANTI-PREDATOR RESPONSES OF YOUNG. From 4-7 days, young give intense rattling-calls (see Voice) when handled; from 7-8 days, young crouch (Horstkotte 1965), pressing against sides of nest and gripping with feet when lifted (Rustamov 1982). Well-grown young threat-gape, give rattling-calls, then may freeze; seek cover on leaving nest and fall silent on hearing warning-calls (see 4b, c in Voice) of parents. From 8-9 days after leaving nest, fly off when approached by human intruder. (Horstkotte 1965.) PARENTAL ANTI-PREDATOR STRATEGIES. (1) Passive measures. Incubating ♀ leaves nest if closely approached; sits low and tightly in nest from day 9; near hatching, may allow herself to be touched, peck intruder's hand, and flush only a short distance, this continuing when brooding small young. In alarm, when ♀ off nest, ♂—with extreme tail-fanning—tended to lead her back to the nest (Hilprecht 1954; Horstkotte 1965); described as chasing her back in fluttering flight (Hilprecht 1954). (2) Active measures: against birds. When Magpie *Pica pica* threatens nest, parents give excited and persistent alarm-calls (Hilprecht 1954). (3) Active measures: against man. Parents mostly silent during incubation, regularly give alarm-calls after hatching (Hilprecht 1954), most intensely when young just out of nest (Horstkotte 1965). Mild alarm- or Warning-calls (see 4b in Voice) typically give way to harsh alarm-calls (4a, d in Voice), or a mixture of the two, as man nears nest (Rustamov 1982). When ♀ touched on nest, ♂ fluttered around giving loud excited Warning-calls (Hilprecht 1954). On 2 successive days, when intruder at nest, ♀, giving apparent Warning-calls, hopped away with wings markedly drooped, then returned and continued calling; same ♀ once approached intruder to within 2 m; became agitated and aggressive, and—still calling—flew at and mobbed intruder. At another nest, ♀ performed more explicit distraction-lure display: drooped her wings and 'fluttered like a bat' in the leaf litter, not moving far. (Horstkotte 1965.) After fledging, when brood divided, usually each parent warns the young in its care (Hilprecht 1954). Parents continue warning young until independence, especially in evening before going to roost. Warning given from perches near song-perches (Grüll 1981). (4) Active measures: against other animals. Apparently responds to cat as for man (Hilprecht 1954; Rustamov 1982). One account indicates distraction-display towards cat (Pröschl 1981). Once chased and drove off squirrel *Sciurus* (Hilprecht 1954).

(Figs by M J Hallam: A from drawing in Géroudet 1954; B from photograph in Hosking and Newberry 1949.)   EKD

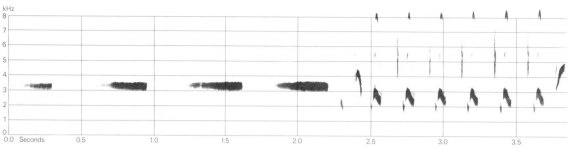

I   V C Lewis   England   May 1964

Voice. Widely studied. Repertoire and quality similar to Thrush Nightingale *L. luscinia* (for differences, see that species, p. 623). For musical notation, see Schmitt and Stadler (1918), Voigt (1933), and Stadler (1957). For rate of learning of song patterns, see Todt *et al.* (1979). For additional sonagrams, see Bergmann (1977*c*) and Bergmann and Helb (1982).

CALLS OF ADULTS. (1) Song of ♂. Remarkable for its richness, variety and vigour (Witherby *et al.* 1938*b*). A succession of phrases 2–4 s long, separated by pauses of about same length (Bergmann and Helb 1982). Many units have clear, rich, liquid, bubbling, or piping quality, other sequences a kind of musical chuckle, and even rather toneless or unmusical notes and variants of alarm-call (4a, below) occur. Most striking are the rapid loud sequence of 'chooc' units, and the fluting, much higher pitched 'pioo' repeated rather slowly in striking crescendo (Witherby *et al.* 1938*b*). Capable of mimicry (Stadler 1957, which see for extensive list), including song of *L. luscinia* (Hilprecht 1954). Song described by Bergmann and Helb (1982) as follows: full-sounding phrases with quiet introductory units of call 4b, e.g. 'hüid'; followed by loud variable phrases such as 'tjuk-tjuk-tjuk...' (same as 'chooc' above) and most characteristically the 'dü dü...' sounds ('pioo' above) which are pure tones often with increase in volume (crescendo) and rate of delivery. Systematic analysis reveals 3-part structure of song, shared by *L. luscinia* (J Hall-Craggs); in analysis by Todt (1970*a*), 4-part structure derived from subdivision of 1st section into quiet and louder units. See *L. luscinia* (p. 623) for detailed explanation and comparison. Notable contrast with *L. luscinia* is that in 5 recordings (115 phrases) of *L. megarhynchos*, only *c.* 20% of phrases begin with introductory 'pew' (or 'pioo',

'd'ü', etc.) units which are therefore by no means diagnostic. In Fig I, 4 long introductory 'pew' units, followed by 6 repeated units (middle section), then characteristic terminal flourish. Fig II (same bird as Fig I) shows phrase lacking introductory 'pew' sequence: begins with 5 notes of different structure, pitch, and duration, followed by a 12-unit rattle, then 4 disyllables (like a slow short trill—the well-known 'chooc' or 'tjuk' sequence), finally a loud flourish. (J Hall-Craggs.) Similar regular structure described by Todt (1970*a*) in analysis of song of bird with repertoire of at least 250 phrases: song comprised more than 600 different unit-types, 5–20 per phrase; bird delivered 400 phrases per hr, and main difference between phrases was in introductory units. Phrase types used in regular cyclic succession (Todt 1971, which see for details). 34 ♂♂ each had repertoire of 160–280 phrase-types (Hultsch and Todt 1981, which see for relationship between repertoire-sharing and territorial dispersion). For repetition rate of phrases, and further analysis of structure, see Panov *et al.* (1978). Typical daytime songs differ from nocturnal song in being shorter overall, in phrase and pause length, also less variable (fewer phrase-types, repeated more often); characteristic whistling phrases much less common by day (Hultsch 1981; see also Todt 1971). 3 kinds of song distinguished by Horstkotte (1965). (a) Territorial-song: given loudly and continuously in long bouts, especially at night when it serves as long-distance advertisement, also by day in interactions with rivals. (b) Courtship-song: typically quieter than Territorial-song and directed at ♀, notably in vicinity of nest. (c) Contact-song: rather fragmentary and short, in communication with ♀ anywhere in territory. For diurnal and seasonal variation in song, also for winter song and Subsong, see Social Pattern and

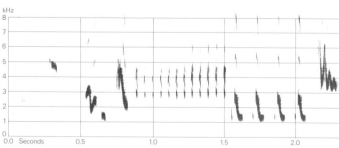

II   V C Lewis   England   May 1964

Behaviour. (2) Contact-alarm calls. (a) As in *L. luscinia*, softer variant of call 4a (below), often described as a churr, may serve as a contact-alarm call (V C Lewis). (b) A hard 'tacc tacc' or softer 'tucc tucc' (Witherby *et al.* 1938*b*). In Fig III, low 'tuk tuk' (like Blackbird *Turdus*

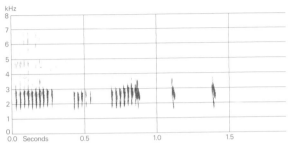

III V C Lewis England April 1964

merula), preceded by 3 'rrrr' units (call 4a) given by ♀ (P J Sellar). Exchanged between pair members (Horstkotte 1965), sometimes in loosely connected series (Bergmann and Helb 1982). (c) Food-bearing parents hovering over concealed fledged offspring gave 'errp' calls (Horstkotte 1965), perhaps contact-calls; young responded similarly. (3) Courtship-calls. (a) Bleating-call of ♂. A 'mä-ä-ä-ä-ä' or 'hä-hä-hä-hä' given in whirring aerial pursuit of ♀ prior to ground display and copulation (Horstkotte 1965); this perhaps the whinnying sound as a coda to song-phrases in recording by J W E Douglas. (b) Twittering call of ♂ chasing ♀ through bushes, and—in more subdued meeting—a 'whispering' call (Hilprecht 1954) perhaps courtship calls. 'trr-tse-tse-tse' given by both sexes during chases (Rustamov 1982) perhaps same or related. (4) Alarm- and Warning-calls. (a) In alarm, a croaking frog-like 'krrrrr' (Aplin 1916; Witherby *et al.* 1938*b*); a hard sonorous 'karr' repeated at (e.g.) 1 s intervals, not uncommonly in combination with call 4b, also 2b: in Fig III, 1st 3 units. Also given by residents in winter territories (e.g. Moreau and Moreau 1928; Douaud 1957). (b) A soft whistling 'hweet' very like *P. collybita* and *P. trochilus*, but often louder (Witherby *et al.* 1938*b*). Also rendered a clear 'hüid', louder and harder than *P. collybita*, given repeatedly (Bergmann and Helb 1982: Fig IV), often with call 4a, e.g. 'fid-krr', lower-

IV E Simms/BBC France May 1954

pitched, less sharp and penetrating than *L. luscinia* (Hammling 1909). Given when disturbed on breeding grounds or migration (Bergmann and Helb 1982), also in winter quarters (Douaud 1957). (c) At higher intensity,

more drawn out and higher pitched (Horstkotte 1965). In recording by V C Lewis, sequence rendered 'see (or 'sih') see see' (J Hall-Craggs) by alarmed ♀; perhaps given only by ♀♀ (V C Lewis). ♀ gave 'prärrr, ziep-ziep-ziep-ziep' (the 'ziep' plaintive), also 'fiep-fiep-fiep', possibly distraction-calls, when intruder at nest (Horstkotte 1965). This call perhaps the long loud 'uninflected' squeaks given by 2 birds, often in alternation for several minutes, in Egypt, winter (Moreau and Moreau 1928). (d) A harsh 'raäk' or 'praäk' (etc.), like Jay *Garrulus glandarius*, given in extreme alarm, but also signalling anger and annoyance, e.g. by ♂ confronting ♀ mate (Hilprecht 1954; Horstkotte 1965). Also rendered a drawn-out 'rräd' noise (Bergmann and Helb 1982). Harsh grating 'tchäää' (Witherby *et al.* 1938*b*; Fig V), given

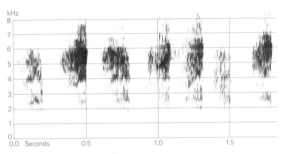

V L Koch/BBC 1935-8

by bird attacking Cuckoo *Cuculus canorus*. (5) Other calls. (a) In Zaïre, winter, a harsh dry 'kh-k-k-k-k-k' (Chapin 1953), perhaps an alarm-call. (b) Parents arriving at nest with food for young gave a 'feeding call' (Rustamov 1982); not known if additional to calls listed above.

CALLS OF YOUNG. Food-calls of nestlings rendered as follows: 'zizzits' (Heinroth and Heinroth 1924-6), 'tzezezezezezeze' (Hilprecht 1954). In recording by P A D Hollom, food-calls of nearly-fledged young a series of urgent, rapidly repeated 'tik' sounds (Fig VI), also

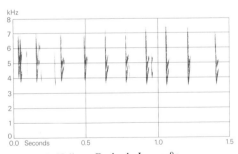

VI P A D Hollom England June 1985

'chituk' or 'chidik' (J Hall-Craggs). Fledged young beg with rattling calls rather like Blackbird *Turdus merula* but less raucous, accelerating when food brought (P A D Hollom); also similar to young Redstarts *Phoenicurus phoenicurus*: 'raäk' and variants frequently before, during, and after feeds (Horstkotte 1965). Whistling, and other sounds, commonly given; thus, one fledgling gave 'prärrr-hi-hie-trrrrr' (Horskotte 1965; compare sequence from ♀ in call 4c, above), also 'fiedt', sometimes combined with a low 'gor' (Hilprecht 1954).   EKD

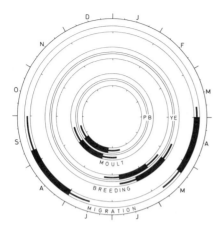

Breeding. SEASON. See diagram for southern Europe. Little variation over whole range, with similar dates in southern England, central Europe, south-west USSR, and North Africa (Dementiev and Gladkov 1954b; Heim de Balsac and Mayaud 1962; Horstkotte 1969; Makatsch 1976; Morgan 1982). SITE. On ground or slightly above, in twigs and undergrowth below scrub, or in dense herbage. Of 283 nests in England, 0·7% below ground level, 27·2% at ground level, 26·1% 0–15 cm above, 35·3% 15–30 cm, 3·9% 30–45 cm, 4·6% 45–60 cm, 2·1% over 60 cm; mean 20 cm (Morgan 1982). Nest: bulky cup of dead leaves and grass, lined with finer material and feathers, occasionally domed. Building: by ♀, in one case taking 3 days (Horstkotte 1965). EGGS. See Plate 82. Sub-elliptical, smooth and very slightly glossy; pale blue, green, or green-blue, finely speckled and mottled pale red-brown, often no more than a general reddish tinge; sometimes forming more marked zone at broad end. 20·9 × 15·5 mm (18·2–23·0 × 13·9–16·9) n = 200; calculated weight 2·65 g (Schönwetter 1979). Clutch: 4–5 (2–6). Of 138 clutches, England: 2 eggs, 1%, 3, 1%; 4, 25%; 5, 71%; 6, 3%; mean 4·75 (Morgan 1982). Of 32 clutches, West Germany: 4 eggs, 4; 5, 25; 6, 3; mean 4·9 (Horstkotte 1969). 2 broods in south of range, 1 in north. Re-lays after clutch loss. INCUBATION. 13 days (Witherby et al. 1938b). By ♀ only. Begins with 1st egg. YOUNG. Cared for and fed by both parents. Where double-brooded, ♀ may start to build 2nd nest before young fledged, when presumably only ♂ feeds them (Horstkotte 1969). FLEDGING TO MATURITY. Fledging period 11 days (Witherby et al. 1938b). Young independent 2–3 weeks after leaving nest. At least some birds breed at 1 year (Grüll 1981). BREEDING SUCCESS. In England, 23% of nests fail completely during incubation, 10·5% during fledgling period; 69% produce at least 1 fledged young;

mean clutch size at laying 4·75 (n = 138); mean brood size at hatching 4·39 (n = 72), and at 8 days 4·19 (n = 86) (Morgan 1982). In West Germany, 14·8% of 332 broods totally lost—due to predation (7·5%) and desertion (7·2%) (Hilprecht 1954).

Plumages (nominate *megarhynchos*). ADULT. In fresh plumage, entire upperparts deep brown, varying somewhat in appearance: in some birds, dark umber-brown with slight grey cast, in others, warmer russet-brown or dark olive-brown; rump and upper tail-coverts usually brighter dark rufous-brown, but sometimes scarcely more rufous than remainder of upperparts. Lores and upper cheeks pale grey, cream, or buff-white, finely mottled dark grey; in some, contrastingly paler than forehead, in others hardly so; narrow ring of buff-white feathers round eye, sometimes indistinct. Ear-coverts umber-brown or russet-brown, like upperparts but usually slightly paler, and faintly streaked pale buff; bordered above and at rear by uniform dull ash-grey or grey-brown stripe; greyish lore and area above ear-coverts sometimes form indistinct and broken supercilium. Lower cheeks, sides of neck, chest, and sides of breast pale brown-grey, sometimes slightly tinged buff or ochre, merging into dull white chin and throat and off-white, cream, or greyish-white belly and vent. Flanks grey-brown or pale olive-brown; under tail-coverts pale cinnamon-buff or pale pink-buff. Tail deep chestnut-brown, central pair sometimes tinged blackish-brown on centre and tip. Flight-feathers blackish-grey, outer webs fringed chestnut-brown or rufous-brown (widest near base), inner webs fringed pink-grey on base and middle portion; tertials and greater upper primary coverts deep brown or black-brown with rufous-brown fringe. Lesser, median, and greater upper wing-coverts brown, like upperparts; fringes along tips paler rufous-brown or olive-brown. Under wing-coverts and axillaries buff or cream-buff; longest coverts, primary coverts, and small coverts along leading edge of wing have much grey showing on bases (especially when worn). *In worn plumage*, upperparts, upper wing-coverts, and fringes of flight-feathers and tertials usually slightly paler, dark olive-brown or rufous-brown, but much individual variation, as in autumn—some birds almost chestnut-brown, others more greyish olive-brown; grey supercilium sometimes more marked; underparts paler, throat and belly purer white, lower cheeks, chest, and sides of breast pale grey, scarcely darker than throat and belly, but sometimes grey-brown and more contrasting; tail usually uniform rufous-brown. NESTLING. On 1st day, down scanty, rather short, grey; at c. 1 week, longer and more plentiful, grey-black; on upperparts only (Heinroth and Heinroth 1924–6; Witherby et al. 1938b). JUVENILE. Upperparts, sides of head, and lesser and median upper wing-coverts rufous-brown or dark olive-brown, each feather with golden-buff subterminal spot at centre and black fringe along tip; rump brighter rufous, upper tail-coverts uniform rufous-chestnut. Chin and throat pale buff or off-white, finely speckled dusky grey at sides; sides of neck, chest, sides of breast, and flanks buff or yellow-buff, each feather with blackish arc at tip, appearing scaly; breast, belly, and vent similar, but ground-colour paler buff and dark scaling more widely spaced; under tail-coverts uniform buff. Tail and flight-feathers as adult, but tail sometimes with narrow pale spot on tip of some feathers; tertials and greater upper wing-coverts as adult, but tips with pale rufous-brown fringe; fringe paler rufous-buff and more spot-like towards inner coverts, narrowly bordered black distally on tertial coverts. *In worn plumage*, spots on upperparts paler buff and ground-colour of underparts virtually white. Closely similar to juveniles of

Robin *Erithacus rubecula* and Redstart *Phoenicurus phoenicurus*, but tail and fringes of flight-feathers uniform chestnut-brown or rufous-brown (in *E. rubecula*, olive-brown; in *P. phoenicurus*, tail paler rufous with black central pair and fringes of flight-feathers buffish); fewer emarginated primaries than both of these. FIRST ADULT. Like adult, and sometimes hard to distinguish. Juvenile flight-feathers, tail, tertials, and greater upper primary coverts retained, often similar to those of adult, especially when worn (rufous fringes on tips of tertials and primary coverts and small buff spots on tips of some tertials and tail-feathers soon wear off). Variable number of juvenile greater upper secondary coverts retained; innermost of these usually have distinct pale buff spot on tip, readily visible even when worn (though sometimes completely abraded, feather-tip then showing notch), but in some birds only outermost coverts retained, or tips of juvenile coverts only slightly washed paler rufous, without distinct pale spot; no marked difference in wear of primaries and tail between adult and 1st adult in spring.

Bare parts. ADULT, FIRST ADULT. Iris dark brown. Bill dark brown or black-brown, cutting edges, base, and middle portion of lower mandible brownish-flesh, flesh-grey, or livid-flesh. Leg and foot pale brown, brownish-flesh, buff-grey, pale grey-brown, or pale livid-flesh. NESTLING. Inside of mouth orange, no tongue spots. Gape-flanges white, shading to pale yellow at angle of gape. JUVENILE. Iris dark brown. Culmen and tip of bill dark brown or brown-black, remainder pale flesh, dull flesh, or greyish-flesh; gape-flanges pale yellow. Leg and foot pale flesh, flesh-yellow, flesh-grey, or greyish flesh-brown (Hartert 1910; Heinroth and Heinroth 1924-6; Witherby *et al.* 1938*b*; RMNH, ZMA.)

Moults. ADULT POST-BREEDING. Complete; primaries descendant. In Britain, starts with p1 late June to late July, completed with regrowth of p9-p10 late July to late August or early September; duration of primary moult *c.* 45 days (Ginn and Melville 1983). In Netherlands and France, some not started by 15-17 July, some completed 20-21 August (RMNH, ZMA). In Balkans, moult July-August (Stresemann 1920). Single ♂ *africana*, north-east Turkey, in last stages of primary moult on 8 September (ZFMK). *L. m. hafizi* from USSR encountered in various stages of moult mid-July to mid-August (Dementiev and Gladkov 1954*b*). POST-JUVENILE. Partial: head, body, lesser and median upper wing-coverts, often tertials, sometimes inner greater upper wing-coverts, and occasionally a few tertials. Starts at age of 5-6 weeks, completed at *c.* 3 months (Heinroth and Heinroth 1924-6); in birds from Netherlands, France, Spain, Yugoslavia, and Greece, starts late June to early August, completed late July to early September, mainly mid-August (Stresemann 1920; RMNH, ZMA). *L. m. hafizi* just started mid-June in Tadzhikistan (Dementiev and Gladkov 1954*b*); not yet started or in full moult 11-12 July in Afghanistan (Paludan 1959).

Measurements. ADULT, FIRST ADULT. Nominate *megarhynchos*. Netherlands and France, April-August; skins (RMNH, ZMA). Bill (S) to skull, bill (N) to distal corner of nostril; exposed culmen on average 4·3 shorter than bill (S).

WING	♂ 83·9	(1·67; 28)	81-87	♀ 81·8 (2·02; 23)	78-85
TAIL	64·5	(2·06; 29)	62-68	62·1 (2·08; 20)	58-65
BILL (S)	17·4	(0·53; 26)	16·7-18·4	16·8 (0·42; 16)	16·3-17·5
BILL (N)	8·8	(0·35; 26)	8·3-9·6	8·8 (0·34; 16)	8·2-9·2
TARSUS	27·5	(0·71; 25)	26·6-28·6	26·9 (0·72; 20)	25·8-27·8

Sex differences significant, except bill (N). 1st adult with retained juvenile flight-feathers and tail combined with older birds above, though juvenile wing on average 1·4 shorter and tail 0·9 shorter.

Wing, live breeding birds, southern France: 83·5 (2·31; 14) 80-88 (G Olioso).

Wing. Nominate *megarhynchos*: (1) Balkans, Greece, and western Turkey. April-September (Stresemann 1920; RMNH, ZMA); (2) Portugal, breeding (RMNH, ZMA). *L. m. hafizi*: (3) Mongolia (Piechocki and Bolod 1972; Piechocki *et al.* 1982) and Altai (RMNH).

(1)	♂	86·2 (1·75; 62)	83-90	♀ 83·2 (1·30; 9)	81-85
(2)		82·8 (1·92; 6)	81-85	77·0 (— ; 1) ·	—
(3)		94·7 (2·26; 10)	91-99	92·0 (— ; 1)	—

Data from large samples, measured with slightly different technique (wing and tarsus on average 0·6 shorter than comparable birds in RMNH and ZMA, tail 3·6 longer): (1) nominate *megarhynchos* (mainly Balkans, Ukraine, and Crimea), (2) *africana* (Caucasus and south-west Turkmeniya), (3) *hafizi* (Asiatic USSR); bill to nostril (Loskot 1981).

WING	(1)	♂ 85·1 (2·65; 66)	79-91	♀ 83·1 (2·73; 32)	76-88
	(2)	86·2 (2·22; 93)	82-92	84·4 (1·65; 18)	80-87
	(3)	91·2 (2·74; 120)	85-98	90·1 (2·69; 50)	84-97
TAIL	(1)	68·9 (3·26; 66)	62-76	67·5 (2·54; 32)	62-75
	(2)	74·1 (2·89; 93)	67-85	74·5 (1·74; 18)	70-76
	(3)	81·2 (2·63; 120)	74-88	80·3 (2·69; 50)	76-86
TARSUS	(1)	26·7 (0·78; 66)	25-29	26·4 (1·15; 32)	23-28
	(2)	26·6 (0·96; 93)	24-30	26·4 (0·81; 18)	25-28
	(3)	27·3 (0·88; 120)	25-29	26·9 (0·85; 50)	25-29
BILL	(1)	9·1 (0·50; 66)	8-10	8·8 (0·72; 32)	8-10
	(2)	9·1 (0·48; 93)	8-10	9·1 (0·30; 18)	8-10
	(3)	9·3 (0·55; 120)	8-11	9·1 (0·27; 15)	8-10

Weights. ADULT, FIRST ADULT. Nominate *megarhynchos*. In breeding area. Netherlands, France, and Turkey, April-June: ♂ 20·9 (1·73; 9) 17-23, ♀ 19·4 (3) 17-24 (Kumerloeve 1970*a*; Rokitansky and Schifter 1971; RMNH, ZMA). Southern France, May-July: 20·5 (0·60; 15) 19-21 (G Olioso). Exhausted birds: ♂♂ 15, 16; ♀ 16 (RMNH, ZMA). For weight increase before migration, see Clafton (1971).

On autumn migration. (1) Southern France, August-September (G Olioso). (2) Portugal, August and early September (C J Mead). (3) Gibraltar (Finlayson 1981). (4) Malta (J Sultana and C Gauci). (5) Inland Algeria, late September (Bairlein *et al.* 1983). (6) Lake Chad, October-November (Fry 1970*b*). (7) Lake Chad (Dowsett and Fry 1971).

(1)	23·7	(4·6; 21)	18-36	(5)	28·0 (1·9; 4)	25-30
(2)	22·3	(2·7; 8)	18-27	(6)	22·2 (0·7; 6)	21-24
(3)	24·0	(2·2; 18)	—	(7)	19·4 (— ; 19)	16-22
(4)	28·0	(3·4; 50)	19-35			

In winter quarters. Nigeria and Ghana: (8) November-January, (9) late February to early May (Ludlow 1966; Smith 1966; BTO).

(8)	19·3	(1·1; 10)	17-21	(9)	23·1 (2·4; 4)	21-26

On spring migration. (10) Lake Chad (Dowsett and Fry 1971). (11) Eastern Morocco (Moreau 1969). (12) Malta (J Sultana and C Gauci). (13) Gibraltar (Finlayson 1981). (14) Southern Spain, April (Mester 1971). (15) Southern France, April (G Olioso). (16) Southern England (Clafton 1971).

(10)	29·1	(— ; 8)	22-33	(14)	19·7 (— ; 22)	15-24
(11)	18·3	(— ; 276)	12-25	(15)	20·1 (0·9; 15)	18-22
(12)	21·8	(2·3; 50)	17-29	(16)	21·4 (— ; 15)	18-23
(13)	20·0	(2·1; 55)				

L. m. africana. Northern Iran, April–May: ♂♂ 21, 21·5 (Schüz 1959). Kenya, November–December: 19·8 (1·65; 207) 16–25 (D J Pearson). Probably this race, Azraq (Jordan), April–May: 19·9 (3·06; 4) 16–23 (BTO).

L. m. hafizi. Iran, Afghanistan, Kazakhstan (USSR), and Mongolia, combined, May–July: ♂ 25·5 (2·45; 14) 21–30, ♀ 23·8 (2·58; 9) 21–28 (Paludan 1959; Dolgushin *et al.* 1970; Piechocki and Bolod 1972; Desfayes and Praz 1978; Piechocki *et al.* 1982). Kenya, November–December: 20·8 (1·84; 116) 16–26 (D J Pearson).

Race unknown. North-west Egypt, on autumn migration: 22·6 (0·4; 24) 20–27 (Moreau 1969; Moreau and Dolp 1970). Ethiopia: November–January, 20·1 (1·9; 22) 16–25, one 30·0; one from 1 April, 31·3 (Ash 1973). *L. m. africana* or *hafizi*, Kenya, late March and April: 27·3 (6·5; 6) 21–39 (D J Pearson).

Structure. Wing rather short, broad at base, tip bluntly pointed. 10 primaries: p8 longest, p9 4–7 shorter, p7 0–2, p6 3–6, p5 6·5–10·5, p4 10–14, p1 16–22. P10 reduced, 42–49 (adult) or 40–45 (juvenile) shorter than p8; 2 shorter to 4 longer than longest upper primary coverts in adult (on average 1·6 longer, $n = 17$), 3·2 (16) 1·5–5·5 longer in juvenile. In nominate *megarhynchos*, tip of p9 usually between tips of p5 and p6 or equal to p6 (rarely slightly longer than p6 or equal to p5); in *hafizi*, usually equal to p5 (Hartert 1910). Outer web of p7–p8 and (more distinctly) inner of (p7–)p8–p9 emarginated. Tertials short, longest about equal to secondaries. Tail rather short, but relatively longer in eastern races; tip rounded; 12 feathers, t1 longest, t6 3–9 shorter. Bill straight; rather heavy at base, slender and laterally compressed at tip; distal part of culmen slightly decurved. Nostrils oval, partly covered by membrane above. Some fine bristles in feathering of lores, along gape, and on chin. Legs long and rather slender; toes rather long and slender. Middle toe with claw 20·0 (39) 19–22; outer toe with claw *c.* 70% of middle with claw, inner *c.* 64%, hind *c.* 75%. Claws rather short, strong, decurved.

Geographical variation. Marked, involving general colour, and length of wing and tail. Strongly saturated umber-brown, with chest clouded olive-buff, in populations from Britain, western France, north-west Spain, and Portugal; gradually paler eastward, terminating in pale brown *hafizi* of central Asia. Wing and tail shortest in populations from north-west Africa, Iberia, and Corsica, gradually longer eastwards, ending in large Mongolian populations of *hafizi*; gradual increases in wing and tail length not strictly parallel, as wing in birds of Caucasus and Iran (*africana*) rather short, but tail long. Racial division followed here is that of Loskot (1981), though boundaries between races not as sharp as suggested there. Splitting into 8–10 races as advocated by Eck (1975a, b) not warranted, such a treatment giving too much importance to minor differences in colour and size, which are easily bridged by individual variation and influence of bleaching and wear in a single population. These dubious races include: *corsa* Parrot, 1910 (Corsica—upperparts like nominate *megarhynchos*, underparts strongly tinged brown-grey; wing, ♂ 83–85, ♀ 79–82); *luscinioides* von Jordans, 1924 (Mallorca—upperparts dark grey-brown, no rufous tinge; wing, ♂ 82–87, ♀ 81–84); *caligiformes* Clancey and von Jordans, 1950 (Britain—upperparts duller and less rufous-brown than nominate *megarhynchos*; wing 81–88); *tauridae* Portenko, 1954 (Crimea—lighter and more rufous than nominate *megarhynchos*, wing longer); *baehrmanni* Eck, 1975 (Balkans, Greece, Cyprus, and Turkey—upperparts greyer than nominate *megarhynchos*, wing, tail, and wing-tip longer, wing 81–91) (Stresemann 1920; Jordans 1924; Clancey and Jordans 1950; Eck 1975a, b). Characters of all these 'races' too slight to warrant separation and all included here in nominate *megarhynchos*. *L. m. hafizi* differs markedly in colour from nominate *megarhynchos*: upperparts greyer, chest sandy-buff, remainder of underparts virtually white; whitish lore and rather distinct supercilium; edges of flight-feathers sandy-grey, under wing-coverts and axillaries cream. *L. m. africana* from Caucasus and Iran rather intermediate between *hafizi* and nominate *megarhynchos*; upperparts duller brown-grey and sometimes darker than nominate *megarhynchos* (less rufous; not sandy as in *hafizi*), underparts rather pale with grey-brown chest. CSR

Luscinia calliope Siberian Rubythroat

PLATES 46 and 47
[facing pages 545 and 652]

Du. Roodkeelnachtegaal   Fr. Calliope sibérienne   Ge. Rubinkehlchen
Ru. Соловей-красношейка   Sp. Gargarita roja   Sw. Rubinnäktergal

Motacilla Calliope Pallas, 1776

Monotypic

Field characters. 14 cm; wing-span 22·5–26 cm. Marginally larger than Robin *Erithacus rubecula*; close in size and structure to Bluethroat *L. svecica* but with 10% longer tail. Rather small but sturdy, long-legged, ground-haunting chat, with elegant character most recalling *L. svecica*. Essentially brown above and brown-grey below, with strikingly pale vent and under tail-coverts obvious when dark brown tail cocked; ♂ has striking face pattern, with white supercilium and submoustachial stripe and red throat; ♀ has white throat and buff-white supercilium. Sexes dissimilar; no seasonal variation. Juvenile separable.

ADULT MALE. Upperparts faintly olivaceous, light umber-brown, darkest on crown and cheeks; in May–June, strongly suffused rufous on fringes of dark brown flight-feathers and base of dusky-brown tail. Face and throat beautifully decorated, with long, white supercilium, almost black lore, white submoustachial stripe and deep, iridescent, bright ruby-coloured throat, which is narrowly bordered by slate-black malar stripe and necklace across throat. Sides of neck and band below necklace strongly suffused grey, but lower chest and flanks less intensely coloured, buff-grey to buff. Centre of belly and under

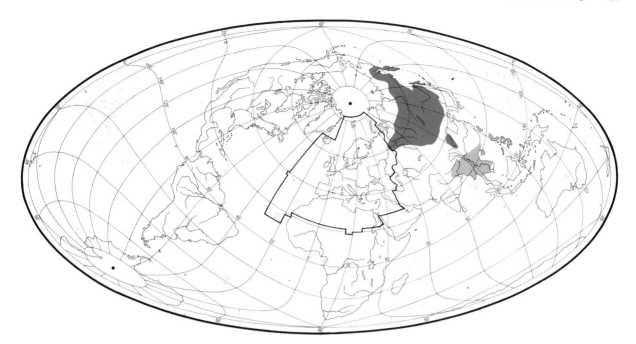

tail-coverts conspicuously buff-white. Underwing pale grey, mottled buff-brown. ADULT FEMALE. Resembles ♂ except in face pattern, most birds having only buff supercilium (indistinct behind eye), indistinct buff-white submoustachial stripe, faint slate-spotted malar stripe, and white throat (all these marks vary in intensity and contrast); chest and flanks uniform brown-buff. JUVENILE. Ground-colour and pattern as ♀ but liberally spotted pale buff above and dark brown below; difficult to distinguish from *E. rubecula* and Thrush Nightingale *L. luscinia*. FIRST YEAR. Resembles adult ♂ and ♀ but distinguished by pale spots on tips of retained juvenile greater coverts; face and throat pattern often duller or less sharp. At all ages, bill brown-horn, with paler flesh base to lower mandible; legs rather long, lead-brown.

Adult ♂ unmistakable; escaped Himalayan Rubythroat *L. pectoralis* much more heavily marked on face and throat (with almost black crown, white forehead, and wide black border and breast around smaller, crimson throat-patch) and has white-based and -tipped outer tail-feathers. Adult ♀ and 1st year far less distinctive, inviting confusion from behind with *E. rubecula*, *L. svecica* (if tail pattern not visible), *L. luscinia* (if smaller size not apparent), and *L. pectoralis*; easily separated from 1st 3 in good side or front view, but sight of throat (no wide grey surround) and outer tail-feathers (no white tips) necessary to eliminate *L. pectoralis*. Juveniles may be inseparable in the field from juveniles of *E. rubecula* and *L. luscinia*. Vagrants occur in damp ground cover or dense coastal thickets. Commonest call a loud whistling 'tiuit-tiuit', less plaintive than Nightingale *L. megarhynchos* and *L. svecica* and more drawn-out than *L. luscinia*; alarm-call a harsh, rattled 'churr'.

Habitat. Approximately replaces Robin *Erithacus rubecula* in breeding niche eastwards from Urals region across upper middle and middle continental latitudes in boreal and cool temperate zones (Harrison 1982), mainly in lowlands from flat sea coast up river valleys, but also occurring above treeline in krummholz and subalpine shrub growth. Although widespread in taiga, tends to avoid dense coniferous stands, preferring thickets of bird cherry *Prunus*, birch *Betula*, and willow *Salix*, but also larch *Larix* and pine *Pinus* with fallen trees, heaps of broken branches, tall grass, small bogs, and regrowth after burning; sometimes in open meadows or clearings, especially near rivers or coast. Especially numerous where bushy thickets in glades abut mixed or broad-leaved stands of trees (Naumov 1962); see also Social Pattern and Behaviour. Although singing from treetops, feeds mostly on ground beneath, or with immediate access to, dense cover (Dementiev and Gladkov 1954b).

In winter in India resorts to dense scrub near water, hedges near villages, underbrush along sides of country roads, long grass, sugarcane, reeds, and sometimes tea plantations, up to 1500 m (Ali and Ripley 1973a).

Distribution and population. No information on range or population changes.

USSR. Common, locally abundant (Dementiev and Gladkov 1954b).

Accidental. Iceland, Britain, France, Denmark, Italy.

Movements. Long-distance migrant. Passes through Mongolia, China, Korea, and Japan, to winter in southeast Asia from Philippines and southern China to Assam (eastern India) and Bangladesh. As in other Siberian

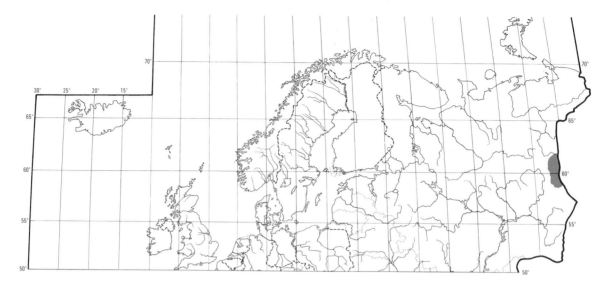

species which winter essentially in south-east Asia, there is probably an easterly passage in autumn (reverse in spring) within USSR to pass north of major mid-Asian mountain systems before turning southwards. Hence rarity in Soviet Central Asia and winter absence from western parts of Indian subcontinent. However, recently found to be scarce but fairly regular migrant to Bharatpur in north-central India (Ali and Ripley 1973*a*); significance of this as yet unknown.

Autumn departures begin late August; exodus complete by mid-September in northern breeding areas (e.g. Yakutsk), though not until early October further south (e.g. Transbaykalia). Passage through China and into wintering regions at height late September to November, returning during May. Mid latitudes of breeding range re-occupied mid-May, though not until late May or early June in north (e.g. Yenisey basin, Yakutsk) and in mountainous areas (e.g. Altai). (Grote 1934; Vorobiev 1954, 1963; Johansen 1955; Dolgushin *et al.* 1970.)   RH

Food. Feeds mostly on ground, but also among low bushes and reeds; runs rapidly in short spurts (Ali and Ripley 1973*a*). Stomachs from eastern USSR contained insects (largely beetles) and their larvae (Dementiev and Gladkov 1954*b*). In eastern USSR, takes amphipods *Gammarus* on seashore in spring and early autumn on Sakhalin (Gizenko 1955), and on Kuril Islands, August–September, seen taking flies and their larvae near fishing stations (Bergman 1935).

In Lena valley (eastern USSR), 124 items fed to young comprised 10 adult and 25 larval Hymenoptera, 17 adult and 6 larval beetles (Coleoptera), 7 adult and 7 larval flies (Diptera), 15 spiders (Araneae), 12 molluscs, 11 adult and 1 larval Hemiptera, 7 adult stoneflies (Plecoptera), 1 adult and 2 larval Lepidoptera, 1 adult dragonfly (Odonata), 1 Myriapoda, and 1 unidentified adult insect (Germogenov 1982).   DJB

Social pattern and behaviour. Information scanty and entirely extralimital. For more detailed studies of closely related Himalayan Rubythroat *L. pectoralis*, see Gavrilov and Kovshar' (1970) and Kovshar' (1979).

1. Usually solitary in winter (Ali and Ripley 1973*a*), though 'parties' recorded in Burma by Smythies (1953); see also Song-display (below). On migration also normally solitary or in small (probably rather loosely associated) flocks of 2–8 birds (Pleske 1889–94; Kozlova 1930; Grote 1934; Nechaev 1969; Moskvitin 1972; Panov 1973). In north-east Altai (USSR), during post-breeding period, 1–40 birds recorded per km², mainly in river valleys (Ravkin 1973). Small gatherings occur at time of departure in September (Portenko 1954). BONDS. Nothing to suggest mating system other than monogamous, but no detailed information on this or other aspects. Young apparently fed by both sexes (see Dolgushin *et al.* 1970). BREEDING DISPERSION. Solitary and territorial. No information on distance between nests nor on territory size, though ♂ apparently sings often from one song-post or changes frequently but still remains within one small area, only occasionally moving 'quite far' (Grote 1925, 1934; Naumov 1962); perhaps indicates relatively small territory. Territory apparently defended by ♂, but no further information on its use. Data on densities in USSR as follows. In middle Siberian broad-leaved woods 15–26 (locally up to 40) pairs per km² (Naumov 1962). In depths of northern taiga and pinewoods between Ob' and Yenisey rivers, mainly by water—2–5 pairs per km of bank; further south shows preference for mixed woods, there averaging 6–9 pairs per km² (Moskvitin 1972). See also Naumov (1960), Berman and Kolonin (1967), Puzachenko (1968), Panteleev (1972), and Vartapetov (1984). In Salairskiy Kryazh (mountains, east of Novosibirsk), 31–73 birds per km² in depths of coniferous and broad-leaved forest, 53–61 per km² near glades, 18 in depths of *Pinus-Picea-Abies*, and 31 birds per km² of riverine willow *Salix* thickets (Chunikhin 1965); see also Ravkin (1973) for north-east Altai. Along Angara river (west of Lake Baykal), 4–32 birds per km² in *Populus-Betula-Pinus*, glades near *Populus* being favoured; 12–19 birds per km² of taiga with bogs (Ravkin 1984). Further east, in Koryakskiy khrebet (mountains, north of Kamchatka), rather unevenly dispersed, with large areas of apparently suitable habitat unoccupied; locally up to 3 pairs (feeding young) in 0·5 ha of willow thickets (Kishchinski 1980);

see also Kishchinski *et al.* (1983). Same habitat in Kamchatka holds up to 5 pairs per km² (Lobkov 1983). On Kunashir island (south Kuril Islands), 2–3 singing ♂♂ per km of coastal thickets, 3–4 birds along 1 km in meadows in south of island (Nechaev 1969). Numerous on east coast of Moneron island (off Sakhalin): 300–400 pairs per km² (Gizenko 1955). ROOSTING. No information on roosting habits; for nocturnal song, see part 2. Fledged young bathe readily (E N Panov).

2. Restless and shy (Dementiev and Gladkov 1954*b*), staying mainly in cover and difficult to observe (e.g. Pukinski 1975), though more in the open high up in mountains (Dementiev and Gladkov 1954*b*). One of the shyest birds of Japan (Jahn 1942), ♀ more so than ♂ (Austin and Kuroda 1953). Less wary than other *Luscinia* of middle Siberia according to Naumov (1962). If threatened, tends to head for cover immediately, slipping mouse-like into thickets, etc. (Grote 1934; Unfricht and Unfricht 1983). Singing bird may fall silent at once when disturbed (Grote 1925) or continue to sing and allow (noisy) approach to *c.* 8–10 m or even less; rarely flies off, tending rather to move furtively lower down in same bush, singing again there or further away (Grote 1934; Naumov 1962; Moskvitin 1972). Breeding birds very secretive during incubation, less so when feeding young (Kishchinski 1980) and relatively more approachable during autumn migration than in summer (Portenko 1939). Alert bird has tail raised and slightly fanned (Unfricht and Unfricht 1983) and will flick it over back (Ali and Ripley 1973*a*); tail-raising occurs in many other situations apart from in territorial conflict (see Fig A and below), and not confined to breeding season, being performed also by migrants (E N Panov). FLOCK BEHAVIOUR. No information. SONG-DISPLAY. ♂ sings (see 1 in Voice) often from within dense bush, not uncommonly low down, but may move about constantly within bush or tree canopy, or at least within very small area (see Breeding Dispersion, above); sometimes uses more open perch on bush, tree, or fence-post (Munsterhjelm 1922; Grote 1925; Jahn 1942; Austin and Kuroda 1953; Portenko 1954; Nechaev 1969). Some variation: in Tomsk area of Siberia, favours for morning song dead branches of fallen trees, stumps, and dead branches of shrub in sun, rarely sings from top of dead tree (Moskvitin 1972); also in Siberia, Naumov (1962) rarely noted birds singing in open and never from tree-top. Latter perhaps used as song-post more in mountains (Grote 1934; Dolgushin *et al.* 1970). See also Unfricht and Unfricht (1983). Singing bird (Fig A, left) has tail steeply

A

raised and wings drooped and whole body may tremble. Throat feathers ruffled and ruby patch visible over considerable distance; feathers seem to rise in waves in powerful vibrating movement so that white basal parts revealed (Munsterhjelm 1922; Jahn 1942). In USSR, sings mainly in (early) morning and evening, sometimes at night, though may continue throughout night (Grote 1934; Bergman 1935; Dementiev and Gladkov

1954*b*; Vorobiev 1963); in Japan, also mostly in early morning, but not uncommonly at night (Jahn 1942). High up in mountains (where nights cold) tends to sing only after midday, continuing to dusk (Grote 1934). In Novosibirsk, 45% of song in morning, 30% later in day, 21% in evening, 4% at night (Podarueva 1979). See also Naumov (1962) and Mauersberger *et al.* (1982). In Koryakskiy khrebet, ♀ (in ♂-type plumage) noted singing 9 July (Kishchinski 1980); no contextual or other details, however. Song-period in USSR extends more or less from arrival (in mid- or late May to June) to early or mid-July (e.g. Gizenko 1955, Johansen 1955, Naumov 1962, Vorobiev 1963), in Tomsk area (Moskvitin 1972) and on Kunashir island (Nechaev 1969) sometimes to early August. Vigorous song noted in Koryakskiy khrebet early to late June (Kishchinski 1980) and on Moneron island 6–20 July (Gizenko 1955). Will sing also on migration (Moskvitin 1972) and in winter quarters: e.g. in Hong Kong, November–December, perhaps also in January (Dove and Goodhart 1955; Herklots 1967) and also starts in Burma before April departure (Smythies 1953). ANTAGONISTIC BEHAVIOUR. Near Tomsk, territories occupied ('by pairs') 5–7 days after arrival in early June (Moskvitin 1972). Several ♂♂ may sing fairly close together (e.g. Dresser 1871–81); 3 did so in one bush in Hong Kong, 1 December (Herklots 1967). ♂♂ sing regularly during territorial disputes, also flirting tail as described (E N Panov: see Fig A). HETEROSEXUAL BEHAVIOUR. Little information. ♂♂ arrive before ♀♀ in Sakhalin (Gizenko 1955). Increased ♂ activity noted near Tomsk 18 June included probable early pair-formation in the form of sexual chases (Moskvitin 1972). ♂ generally remains close to nest or brood (Naumov 1962) and may sing close to nest throughout night (Dresser 1871–81). RELATIONS WITHIN FAMILY GROUP. Adults (especially ♀) tend to approach nest very cautiously, mainly on ground (Stresemann *et al.* 1937; Dolgushin *et al.* 1970). Food-begging young raise and tremble both wings (sometimes one wing higher than the other) and make slight bobbing movement while giving food-calls (E N Panov). Faecal sacs carried away or swallowed by adults (Dolgushin *et al.* 1970). After leaving nest (captive birds did so before able to fly: Unfricht and Unfricht 1983) young stay mostly on ground in cover and also fed there (Naumov 1962); apparently reluctant to fly—e.g. running along paths on Moneron island (Gizenko 1955). Family stays together not more than 2 weeks according to Naumov (1962); also suggested by Moskvitin (1972) that break-up comes fairly quickly. ANTI-PREDATOR RESPONSES OF YOUNG. May leave nest prematurely, scattering and hiding (Dementiev and Gladkov 1954*b*; Kishchinski 1980) at which extremely skilful (Johansen 1955). Young out of nest may crouch until almost trodden upon and loath to fly (Naumov 1962). PARENTAL ANTI-PREDATOR STRATEGIES. (1) Passive measures. Incubating ♀ tends to sit tightly, even allowing herself to be seized on nest according to Dresser (1871–81). In Altai, ♀ (with only slightly incubated clutch) allowed approach to *c.* 4 m, then slipped away and disappeared (Sushkin 1938); see also Tomek (1984). Slow to return (Dresser 1871–81). More likely to show agitation close to hatching and when with young and may then remain within 3–10 m of nest after flushing (Naumov 1962; Pukinski 1975). (2) Active measures. ♂ not infrequently first to give alarm-call (see 2 in Voice) and will come close when ♀ disturbed (Naumov 1962). ♀ with young may show herself and perform apparent distraction-lure display, excitedly throwing up wings (see photograph in Pukinski 1975). Highly agitated ♀ near young out of nest gave quiet alarm-call (Dolgushin *et al.* 1970).

(Fig by M J Hallam: from drawing by E N Panov.) MGW

Voice. Calls 1–2 given both in breeding season and in winter quarters.

CALLS OF ADULTS. (1) Song of ♂. Strikingly beautiful, rich, varied, melodious, and sustained warbling with some hard 'squeezed' notes. Often loud and delivered with great vigour and intensity (Bergman 1935; Jahn 1942; Yamashina 1982; Svensson 1984*b*); almost as powerful as Nightingale *L. megarhynchos* or Thrush Nightingale *L. luscinia* according to Dementiev and Gladkov (1954*b*). Said by Grote (1934) almost to equal *L. luscinia* in richness of sounds, and Seebohm (1901) considered it more melodious than Bluethroat *L. svecica* and very little inferior to *L. megarhynchos*. Johansen (1955) likened it to a fine thrush *Turdus* song and available recordings confirm close affinity with unusually varied (little straight repetition of simple sounds) and enterprising Song Thrush *T. philomelos* (J Hall-Craggs). In Japan, said to have faint resemblance to Red-flanked Bluetail *Tarsiger cyanurus* (Yamashina 1982) and (part of) song rendered 'chil chil chil-li chilli' (Austin and Kuroda 1953). Song given in short phrases separated by quite long pauses or is long and continuous (E N Panov). In recording by R Naumov, 3 songs last *c.* 5–8 s, with pauses of 3 s and 4 s; 3 songs recorded by L Svensson last *c.* 5–10 s and separated by pauses of *c.* 3 s and *c.* 2 s. Recording by R Naumov reveals bright whistles, richer and throatier sounds (some close to *L. megarhynchos*), tinkling, twangy, chortling, and hollow, xylophone-like sounds; ventriloquial effect apparent, but may result simply from considerable differences in loudness in this recording. Twangy sounds also evident in another recording (Fig I), but overall, song quieter, sweeter, more restrained and at times not unlike Starling *Sturnus vulgaris* (J Hall-Craggs, M G Wilson). Not infrequently mimics other birds: e.g. song may start with copy of Chaffinch *Fringilla coelebs* (Dementiev and Gladkov 1954*b*) and near Krasnoyarsk (USSR), astonishingly accurate imitation of Scarlet Rosefinch *Carpodacus erythrinus* song is especially common (Naumov 1962). Recording by L Svensson contains much expert mimicry, e.g. song of Yellow-breasted Bunting *Emberiza aureola* (introductory units of Fig I) and calls of wide range of bird species (Svensson 1984*b*). For song (no description) of ♀, see Social Pattern and Behaviour. (2) Contact-alarm calls. Most authors mention 2 basic types (not infrequently combined); some evidence, though few details, for existence of a 3rd (see below). (a) Loud, melodious, drawn-out, and rather plaintive whistle. Sonagram from E N Panov shows 2 tonal portamenti, each call *c.* 0·3 s long and at first steady in pitch, then descending from *c.* 4·4 to 3·2 kHz; rendered 'peeeeoo' (Panov 1973). Other renderings apparently indicate various pitch patterns occur: 'tiuit tiuit', very loud when startled (Grote 1934; Dementiev and Gladkov 1954*b*); thin 'puri-puri' (Yamashina 1982); in Burma, winter, rather plaintive 'chee-wee' given several times while moving about in cover and difficult to locate (Smythies 1953); plaintive whistle given by ♂ in winter (Munn 1894) almost certainly this call rather than attempt to sing, *contra* Ali and Ripley (1973*a*); loud, wailing 'ürü' apparently with falling pitch, also an apparently rising 'huü' noted by Hemmingsen and Guildal (1968) presumably also belong here. For comparisons with other *Luscinia*, see Field Characters and relevant accounts. (b) For many authors, the typical call given when disturbed, but calls 2a and certainly 2c are also given in that context. Harsh and sometimes loud 'tacking' or 'chacking', etc., typical of Turdinae: e.g. 'chakh' or 'chokh' (Panov 1973) and sonagrams by E N Panov show close affinity with 'duck', 'djuk', or 'dak' of various *Turdus* thrushes, and 'tak' of certain *Luscinia* (J Hall-Craggs, M G Wilson).

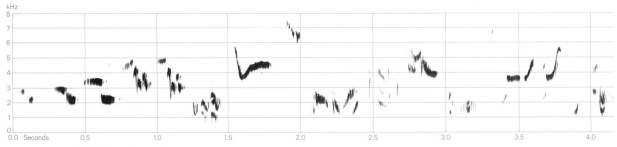

I Svensson (1984) USSR June 1983

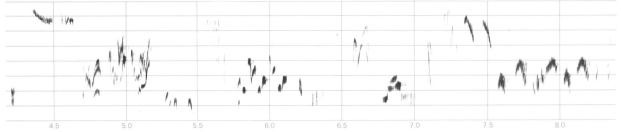

I *cont.*

Further descriptions as follows: 'chot chot' when alarmed (Yamashina 1982); in Nepal, 'guck' like *T. cyanurus* (Fairbank 1980*a*) or metallic 'chep chep' given at intervals of a few seconds (Rand and Fleming 1957; Fleming *et al.* 1976); harsh 'ké' or 'ché', sometimes combined with song, is call-note (Ali and Ripley 1973*a*). See also (e.g.) Dove and Goodhart (1955) and Dolgushin *et al.* (1970). Apparent combination of calls 2a and 2b: 'hüit tack tack' as contact-call (Grote 1934); 'zäck zäck-tschwu-hwiu tschu-hwiu' as warning-call (Stresemann *et al.* 1937). (c) Presumably distinct from call 2b is loud rattle or harsh churr mentioned as alarm-call by Smythies (1953) and Ali and Ripley (1973*a*); no further details, but such a call certainly exists in repertoire of some other *Luscinia*.

CALLS OF YOUNG. Food-call of captive young a quiet hissing sound according to Gräff (1975). Sonagram from E N Panov shows relatively long (longer than 0·5 s) squeaking food-call with rapid frequency and amplitude modulation (perhaps producing the hissing quality already alluded to); of 3 units shown, 1st relatively steady, others ascend in pitch. Another sonagram of a food-call from E N Panov depicts reiterated and structured descending ripples (J Hall-Craggs, M G Wilson). Quiet squeaking or cheeping sounds given at 2–3 weeks old while preening are slightly reminiscent of subsong (E N Panov): apparently (from sonagram by E N Panov) a half-tonal, half-trilling and rather high pitched (up to *c.* 6·75 kHz) sound (J Hall-Craggs). MGW

Breeding. SEASON. USSR. In Tomsk area, laying begins early June (Johansen 1955), and main fledging period 10–20 July (Moskvitin 1972); eggs recorded near Krasnoyarsk mid-June to end of July (Naumov 1962). SITE. On ground in thick tussock or under thick low bush, occasionally just above ground. Nest: loosely constructed of extremely fine stems and vegetable fibres, slightly lined with hair and plant fluff; external diameter *c.* 9·0 cm, internal *c.* 6·5 cm (Dementiev and Gladkov 1954*b*); often spherical with side entrance (e.g. Naumov 1962, Moskvitin 1972). EGGS. See Plate 82. Sub-elliptical, smooth and glossy; pale blue, finely marked with red-brown speckles and mottles, giving overall greenish tinge, sometimes gathered at broad end. 20·6 × 15·6 mm (19·0–22·4 × 14·6–16·5), *n* = 30; calculated weight 2·65 g (Schönwetter 1979). Clutch: 5(4–6). One brood (Dementiev and Gladkov 1954*b*). INCUBATION. Period unknown. By ♀, not known whether ♂ helps (Dementiev and Gladkov 1954*b*). No further information.

Plumages. ADULT MALE. Upperparts entirely dark brownish-olive, slightly brighter olive or olive-brown on rump and upper tail-coverts, often deeper and browner on forehead and crown. Narrow but distinct white supercilium, almost reaching base of culmen, extending back to above rear corner of eye. Lores and upper cheek to just below eye black; narrow eye-ring black, but partly white at border of supercilium and with some tiny white spots on lower rear. Ear-coverts deep brown-olive or

grey-olive, sides of neck paler olive-grey. Broad and distinct white stripe running backwards from lower mandible, separated from bright scarlet chin and throat by narrow black malar stripe. Fresh feathers of chin and throat narrowly tipped white (worn off by October–November), white of feather bases generally not visible. Narrow black malar stripe extends right round lower throat, forming narrow black gorget 1–3 mm wide (sometimes dark grey on lower throat). Rear of cheeks and upper chest dark ash-grey, slightly tinged olive-brown when plumage fresh, merging into olive-brown on sides of breast and lower chest. Belly and central vent off-white, not sharply divided from pale olive-brown or buff-brown flanks and sides of vent; under tail-coverts cream-buff or almost white. Tail dark brownish-olive or dull black with olive tinge, sides of feathers sometimes slightly brighter olive-brown towards bases. Flight-feathers and greater upper primary coverts dull black, outer webs narrowly margined deep brown or dark olive-brown, except on p9–p10 and on emarginated parts of outer primaries; inner webs with narrow pale grey-buff border along basal and middle portions. Lesser and median upper wing-coverts dark brownish-olive, like upperparts; greater coverts and tertials dark grey, merging into dark olive-brown on sides and tips, where narrowly and evenly fringed paler olive-brown. Under wing-coverts and axillaries pale grey with broad pale buff-brown tips; smaller coverts along carpal joint largely white. *In very fresh plumage* (about September), some fine olive spots may show on supercilium and on white stripe on lower cheek; chest to vent tinged olive. *In worn plumage* (spring), no marked changes from fresh autumn plumage described above, but narrow black gorget often more distinct, chest broadly dark ash-grey, sides of neck purer ash-grey, and fringes of flight-feathers and tertials paler olive-brown or olive-grey. ADULT FEMALE. Rather variable, ranging from little-marked like 1st adult ♀ (see below) to almost as contrastingly coloured as adult ♂. Upperparts as adult ♂ or slightly brighter dark olive-brown. Supercilium white (but usually shorter and less sharply defined than adult ♂) or buff with some olive mottling. Pale eye-ring more fully developed than in adult ♂, white or pale buff, usually broken in front and at rear. Lores uniform greyish-black (less deep black than adult ♂) or dark olive-brown with pale buff mottling. White stripe on lower cheek smaller and less sharply defined than adult ♂, partly mottled olive, or entirely absent; cheeks olive-brown or dark olive-grey like ear-coverts and sides of neck. Chin and throat basically white, not surrounded by black gorget and hence less sharply defined than in adult ♂ (sometimes some traces of black at sides of chin); almost always tinged with scarlet-red or pinkish-red; in some, red restricted to indistinct wash on central throat, in others, entire chin and throat scarlet-red, but red appearing less deep because white feather-bases show through, and feather-tips more broadly white or slightly olive when plumage fresh. Rear of cheeks and upper chest either dark ash-grey (as adult ♂, but less extensive) or olive-brown. Remainder of underparts, tail, and wing as in adult ♂. *In heavily worn plumage*, red on throat bleaches considerably and sometimes completely worn off. NESTLING. Down smoky-grey; sparse and long (Neufeldt 1970). JUVENILE. Exceedingly similar to juveniles of Robin *Erithacus rubecula* and Thrush Nightingale *Luscinia luscinia*, showing similar spotted upperparts, uniform rufous upper tail-coverts, and buff underparts with dark scaling, but differs from *E. rubecula* in heavier bill and larger foot and from *L. luscinia* in wing formula (3 primaries emarginated instead of 1). FIRST ADULT MALE. Like adult ♂, but juvenile flight-feathers, tail, tertials, greater upper wing-coverts, and greater upper primary coverts retained. Fringes along tips of greater upper wing-coverts and tertials paler buff-brown, tending towards pale buff at shaft; pale buff may form contrasting spot, especially on inner coverts and inner tertials and when plumage worn, but sometimes difficult to see, in particular on outer coverts and longer tertials and when plumage fresh. Tips of some tail-feathers and greater upper primary coverts sometimes have narrow pale buff spot at shaft. Plumage otherwise similar to adult ♂, but supercilium and white stripe at lower cheek often finely mottled olive when fresh, black below eye finely spotted white or buff; feather-tips of chin and throat rather more extensively white, taking longer to wear off; shade of scarlet on chin and throat usually similar, but occasionally paler pink-red or orange-red; black malar stripe does not extend across lower throat or scarcely so; chest more extensively olive, less grey. Closely similar to some adult ♀♀, but differs in retained juvenile greater coverts and tertials. First Adult Female. Upperparts entirely dark olive-brown, darkest on crown, often slightly rufous on tail-coverts. Supercilium and eye-ring pale buff, mottled olive-brown, often indistinctly demarcated; lores dark olive-brown or black-brown. Lower cheeks, chin, and throat white, partly mottled olive-brown (in particular on cheeks and sides of throat), white not clear-cut and on cheeks sometimes virtually absent; no red tinge on throat. Chest and flanks buff-brown or olive-brown, less olive-grey and less clear-cut than in adult ♀. *In worn plumage* (spring), white of supercilium, eye-ring, stripe on cheek, and throat more clear-cut, chest greyer, and occasionally faintly pink on throat, but head pattern often still less clear-cut than adult ♀. Part of juvenile plumage retained, as in 1st adult ♂; juvenile characters as in 1st adult ♂.

Bare parts. Adult Male. Iris brown or dark brown. Bill bluish-black with paler grey base to lower mandible; cutting edges near gape pale blue-grey; whole bill almost black during breeding. Mouth dark. Leg and foot dull flesh-grey, pale plumbeous, grey-brown, or horn-brown. Adult Female. As ♂ but bill dark horn-brown or dark grey-brown with paler flesh-grey base to lower mandible. Nestling. Mouth yellow; gape-flanges pale yellow, almost white (Neufeldt 1970). Juvenile. Iris dark brown. Bill horn-black, cutting edges and lower mandible flesh-pink. Gape-flanges and mouth yellow-pink. Leg and foot dull flesh-pink with grey joints. First Adult. Like adult, but bill dark horn-brown to plumbeous-black with pink-grey or flesh-brown base to lower mandible; mouth pink-buff or pink-white; leg pink-purple with flesh-coloured soles. (Hartert 1910; Ali and Ripley 1973a; Lowe 1979; Svensson 1984a; BMNH, ZMA.)

Moults. Adult Post-breeding. Complete; primaries descendant. August to mid-September (Dementiev and Gladkov 1954b). Post-juvenile. Partial: head, body, lesser and median upper wing-coverts and tertial coverts; no flight-feathers, tail, greater upper wing-coverts, and greater upper primary coverts, and usually no tertials. Starts late July to early September, completed late August to late September (Dementiev and Gladkov 1954b; Piechocki and Bolod 1972; ZMA).

Measurements. Adult, First Adult. Japan, China, and Taiwan, all year; skins (RMNH, ZMA). Bill (S) to skull, bill (N) to distal corner of nostril; exposed culmen on average 4·7 less than bill (S).

WING	♂	79·7 (2·98; 36)	73–85	♀	75·8 (1·74; 13)	73–79
TAIL		60·4 (2·53; 26)	56–64		56·2 (2·25; 12)	52–59
BILL (S)		17·4 (0·65; 25)	16·5–18·5		17·1 (0·56; 13)	16·2–17·9
BILL (N)		9·2 (0·46; 25)	8·6–9·8		8·9 (0·49; 13)	8·4–9·6
TARSUS		30·5 (1·01; 26)	28·3–32·1		29·2 (0·96; 13)	27·8–31·2

Sex differences significant for wing, tail, and tarsus. 1st adult with retained juvenile flight-feathers and tail combined with older birds above, though juvenile wing on average 0·8 shorter than adult, tail 0·6 shorter.

Wing. (1) Kamchatka (USSR), summer; (2) Japan, mainly migrants; (3) Taiwan, winter; (4) eastern China, mainly winter; (5) north-east Tsinghai and Kansu (China), summer; (6) Mongolia, May–August; (7) western and central USSR (Stresemann *et al.* 1937; Dementiev and Gladkov 1954b; Nechaev 1969; Piechocki and Bolod 1972; RMNH, ZMA).

(1)	♂	80·7 (— ; 16)	77–85	♀	76·6 (— ; 7)	74–78
(2)		83·0 (1·26; 7)	80–85		75·8 (2·18; 6)	73–79
(3)		80·8 (1·75; 8)	78–83		75·7 (— ; 3)	74–77
(4)		77·6 (3·25; 11)	73–82		75·9 (1·08; 4)	73–78
(5)		77·1 (— ; 10)	75–80		74·0 (— ; 4)	72–75
(6)		78·5 (1·05; 6)	77–80		—	
(7)		74·5 (— ; 31)	71–78		70·5 (— ; 13)	67–75

Weights. Kamchatka (USSR), summer: ♂♂ 24·3, 29·4; ♀ 22·5 (3) 21·8–23·2; juvenile 23·3, 24·1 (Nechaev 1969). Hong Kong, June: ♂ 21·2 (ZMA). Mongolia, May: ♂ 22·6 (1·82; 5) 20–25 (Piechocki and Bolod 1972). Kazakhstan (USSR), May–June: ♂♂ 23·5, 25 (Dolgushin *et al.* 1970). Migrants, Hopeh (China): ♂ 19 (10) 17–27, ♀ 18 (10) 16–22 (Cheng 1963). Late August and early September, Manchuria (China): ♂ 22·7 (3) 20–25 (Piechocki 1958). Mongolia, August: moulting juvenile ♂ 25 (Piechocki and Bolod 1972). Lake Chany (south-west Siberia), August–September: ♂ 24·1 (Havlín and Jurlov 1977). Taiwan, eastern China, Nepal, and India, November–March: ♂ 21·4 (1·42; 9) 19–23, ♀ 20·2 (2·95; 4) 16–23; exhausted birds, ♂ 16·5 (3) 15–18, ♀ 16·4 (Diesselhorst 1968a; Ali and Ripley 1973a; RMNH, ZMA). Britain, October: 20 (Lowe 1979).

Structure. Wing rather long, broad at base, tip bluntly pointed. 10 primaries: p7 longest, p8 0–1 shorter, p9 6·5–11·5, p6 0·5–3, p5 3·5–7, p4 6–11, p1 13–19; tip p9 about equally often between p4–p5 and between p3–p4 (*contra* Stresemann *et al.* 1937). P10 reduced; 31–38 shorter than p7, 4–10 longer than longest upper primary covert. Outer web of p6–p8 and inner of p7–p9 emarginated. Tertials short; longest about equal to secondaries. Tail rather short, tip rounded; 12 feathers, t6 6–11 shorter than t1. Bill rather long, rather heavy and wide at base; closely similar to bill of Nightingale *L. megarhynchos*. Tarsus and toes long and slender. Middle toe with claw 21·3 (17) 19·5–23·5; outer toe with claw *c.* 70% of middle toe with claw, inner *c.* 66%, hind *c.* 76%. Claws rather short, decurved.

Geographical variation. Slight, clinal. Western birds smaller than those from eastern Siberia: wing of ♂♂ in west mainly 71–78 (Hartert and Steinbacher 1932–8), in Kamchatka 81–85 (Nechaev 1969); Kamchatka birds hence sometimes separated as *camtschatkensis* Gmelin, 1789, but no sharp boundary between western and eastern populations, and birds of central Siberia (from where the species was described) are intermediate, hence recognition not warranted. Isolated population in north-west Tsinghai and Kansu (China) sometimes separated as *beicki* Meise, 1937, and said to differ in relative tail length and wing formula (Stresemann *et al.* 1937), but many Siberian birds similar (Vaurie 1955a; RMNH, ZMA), and species therefore considered monotypic.

Forms superspecies with Himalayan Rubythroat *L. pectoralis* from central Asia. CSR

Luscinia svecica Bluethroat

PLATES 46 and 47
[facing pages 545 and 652]

Du. Blauwborst   Fr. Gorgebleue à miroir   Ge. Blaukehlchen
Ru. Варакушка   Sp. Pechiazul   Sw. Blåhake

Motacilla svecica Linnaeus, 1758

Polytypic. Nominate *svecica* (Linnaeus, 1758) (synonyms *gaetkei*, *robusta*), Scandinavia east through northern Siberia to western Alaska, south to *c.* 60°N in European USSR, to *c.* 57°N in western Siberia, to Altai and Sayan mountains in central Siberia, and to *c.* 55°N along shores of Sea of Okhotsk, also some isolated populations in mountains of central Europe; *namnetum* Mayaud, 1934, western France; *cyanecula* (Meisner, 1804), central Europe from eastern and northern France and Netherlands east to Carpathian basin, north-west Ukraine, Smolensk, and Leningrad (USSR), also Spain; *volgae* (Kleinschmidt, 1907), central European USSR from north-east Ukraine north to *c.* 57°N and east to middle Volga river, intergrading with *cyanecula* in west and with nominate *svecica* further north; *pallidogularis* (Zarudny, 1897) (synonyms *saturatior*, *altaica*), south-east European USSR (east from Volga) and lowland steppes of south-west Siberia south to Turkmeniya, east to foothills of Altai mountains and upper Yenisey, intergrading with *volgae* in west and with nominate *svecica* in north and east; *magna* (Zarudny and Loudon, 1904), Caucasus area, eastern Turkey, and Iran. Extralimital: *tianschanica* (Tugarinov, 1929), Pamir and Tien Shan mountains; 3 further races in central Asia and western Himalayas.

Field characters. 14 cm; wing-span 20-22·5 cm. Marginally smaller than Robin *Erithacus rubecula*, with markedly slimmer form and proportionately longer legs. Small, graceful, elegant chat, with noticeably erect carriage and characteristic cocking and fanning of tail. Bright chestnut patches at bases of outer tail-feathers diagnostic; rest of plumage essentially dark brown above, silky buff-white below with dusky flanks; breeding ♂ has blue throat, bordered below with black-white-chestnut bands, ♀ has white throat and black-splashed necklace. Marked racial variation in tone of upperparts and colour and shape of spot on ♂'s throat (white in west and south, red in north and east). Flight like *E. rubecula* but ends in characteristic low sweep into cover. Sexes dissimilar; marked seasonal variation in ♂'s foreparts. Juvenile separable. 6 races in west Palearctic, ♂♂ of 4 distinguishable but ♀♀ and immatures hardly so. 4 races described here (see also Plumages and Geographical Variation).

ADULT MALE BREEDING. (1) North Eurasian race, nominate *svecica*. Upperparts and wings dark brown, with only faint, greyer fringes to longer wing-coverts and tertials. Face little marked, with dark streaks on crown forming almost black line above buff fore-supercilium and grey rear supercilium which contrasts less with brown, buff-flecked lore and cheek, but throat and chest are beautifully patterned, with metallic blue throat and upper breast containing red-chestnut spot and bordered below by black, white, and red-chestnut bands. With wear, supercilium becomes whiter and chestnut breast-band narrower. Underparts buff-white, with flanks washed dusky. Underwing tawny-brown on axillaries and along leading edge, grey elsewhere. Tail strikingly patterned, with bright rufous-chestnut bases to outer feathers forming obvious square panels at base of otherwise black-brown feathers which thus produce ⊥ pattern, like

wheatear *Oenanthe*; upper tail-coverts also chestnut on bases, with warm tone showing when worn. (2) Central European race, *cyanecula*. Differs distinctly from nominate *svecica* in having throat-spot white or lacking; upperparts average darker. (3) South-central USSR race, *pallidogularis*. Foreparts more immaculate in appearance than nominate *svecica*, with paler blue throat, paler chestnut throat-mark (a broad bar or triangle rather than a spot), and less intensely chestnut and less obvious breast-band; upperparts distinctly greyer, underparts noticeably silky-white. (4) Race of Caucasus, eastern Turkey, and Iran, *magna*. Largest race, with 5-10% longer wings than other west Palearctic races. Resembles *pallidogularis* in appearance but with even paler blue throat, no throat-spot or only a small white one, and obvious chestnut breast-band. ADULT FEMALE BREEDING. All races. Upperparts, wings, tail, and rear of underparts as ♂. Below cheeks, differs distinctly from ♂ in usually lacking noticeably blue throat or discrete throat-spot, having instead obvious brown-black moustache joined to brown-black necklace, broken up along lower edge, interspersed with rufous-buff and fusing with dusky-brown wash on flanks. In more intensely marked birds, a diffuse blue band above necklace, and blue feathers show in moustache and over chin; a few red or white tips may echo throat-spot of ♂. Individual variation makes separation of races virtually impossible other than by tone of upperparts (hard to judge in the field). ADULT MALE NON-BREEDING. All races. By September, discrete pattern of throat and breast lost in growth of less colourful and more variegated plumage: chin and upper throat buff-white, initially bordered with brown-black lines and (with wear) invaded later by black bars or patches above pale patch (rather than spot) and metallic blue, white-tipped patch or band; breast-bands less obvious, with black line under blue band either marked or indistinct, white one broken and obvious only on sides of breast, and red-chestnut one deeper but fringed and tipped white. Supercilium noticeably dull; rest of underparts noticeably creamier, becoming buff at rear. Note that even white-spotted races may grow some chestnut feathers in centre of throat and that only the most clearly marked birds can be separated. ADULT FEMALE NON-BREEDING. All races. As ♀ breeding but rarely with any blue on chin, around moustache, or even on lower throat. JUVENILE. All races. Tail and most flight-feathers as adult but rest of plumage brown-black above and buff below, liberally scaled and/or spotted pale buff on upperparts and dark brown on underparts. Pale cream-buff tips to tertials more obvious than dull buff tips to median and greater coverts, which hardly form wing-bars. FIRST-WINTER MALE. All races. Like non-breeding ♂ but throat less well marked, with spot diffuse and usually paler in red-spotted races, blue less pronounced and chestnut breast-band much paler. Outermost greater wing-coverts (retained from juvenile plumage) show buff tips and these often persist through 1st summer. FIRST-WINTER FEMALE. All races. Like adult ♀ but chin and centre of throat whiter; presence of blue feathers quite exceptional. Buff-tipped outermost greater wing-coverts retained as ♂. At all ages, bill dark brown-horn; legs ochre-brown.

Unmistakable; no other small west Palearctic chat has similar tail pattern, except for mountain- and desert-haunting Red-tailed Wheatear Oenanthe xanthoprymna chrysopygia which also has rufous rump and lacks blue throat or necklace. Racial identifications other than between nominate svecica and cyanecula not traditionally attempted, but a pale immature resembling pallidogularis has appeared with other Siberian birds in eastern England (R E Emmett, D I M Wallace), and heavily marked ♀♀, unlike nominate svecica or cyanecula, pass through Jordan in spring (D I M Wallace). Flight free but nearly always close to ground, with flitting action and characteristic terminal flat glide or sweep into base of cover; flight silhouette recalls E. rubecula but with apparently broader rump and tail. In the open, gait like E. rubecula, with bold, upright carriage and long (even bounding) hops, varied by fast, short runs; when in cover, creeps furtively like mouse and capable of moving through densest vegetation. Frequently holds tail up and then looks remarkably long-legged, especially when bobbing in anxiety or excitement; also frequently flirts and fans it, rufous basal patches flashing. Migrants utilize ditches and swamp edges rather than hedgerows and other drier sites.

Song loud, sweet, and exceptionally varied; reminiscent of Nightingale Luscinia megarhynchos but not so rich or full in tone, having merrier, more tinkling quality; characteristic phrases include throaty 'torr-torr-torr-torr' and metallic, ringing 'ting-ting-ting', recalling sounds of both metal triangle and bell of alarm clock; often strongly imitative, cyanecula particularly. Sings from dense cover or (more often) from open perch or in song-flight like that of smaller pipits Anthus. Calls include hard 'tacc tacc', lower croaked 'turrc turrc', and plaintive 'hweet' like L. megarhynchos and Phylloscopus warbler.

Habitat. Breeds from arctic and boreal upper latitudes to temperate and steppe middle latitudes and montane regions, continental and mainly cool. Patchiness in south of range suggests approach to relict status. Best adapted to regions intermediate between forest and open plains or valleys, such as wooded tundra with marshy glades among spruce Picea, dwarf willows Salix, and junipers Juniperus, woods of birch Betula, and shrubby wetlands, ascending from sea-level to high Scandinavian fjells. Also on floodplains and banks of rivers and lakes in dense but low woody vegetation, sometimes with reeds and rushes or willows and alders Alnus, predominantly in moist or wet situations with ample cover, and on damp alpine meadows, in USSR up to 4000 m (Dementiev and Gladkov 1954b); sometimes in tall grass or near habitations. South European race cyanecula differs ecologically in stronger preference for bushy sites by water, including flood levels with alders and reedbeds (Niethammer 1937), while southern populations in Spain at up to 2000 m frequent dry stony slopes covered with Spanish broom Spartium junceum, singing from topmost twigs (Witherby 1928). Apparently therefore critical habitat requirement is for copious low dense vegetation c. 1–2 m tall, interspersed with open ground in patches and free of mature trees, especially conifers, and of continual disturbance. Surface water of whatever type often favours growth and maintenance of such habitat, but probably not in itself an essential. Compared with Robin Erithacus rubecula, Nightingale L. megarhyncos, and Thrush Nightingale L. luscinia, has no apparent forest affiliations. Extralimitally in Afghanistan found breeding in mountain valley at about 2700 m in dense well-grown scrub of willow and buckthorn Hippophae along river; on autumn migration, also in a plantation of poplar Populus, in lucerne fields, among tussocks along a small stream, and among rocks at c. 2900–3000 m (Paludan 1959). In Indian subcontinent in winter, frequents reeds, clumps of grass, maize, sugar-cane, tamarisks and other bushes near water, bushes around well-watered cultivation, and dry tidal mudflats with salt-marsh vegetation; roosts with swallows Hirundo and other birds in reedbeds (Ali and Ripley 1973a). In tropical African winter quarters, occupies edges of watery places, as a rule swamps, even very small ones, but also moist marshes with clumps of shrub; in Ethiopia, in rank herbage by running streams at over 1900 m (Moreau 1972). Where development occurs, habitat apt to be target for clearance and drainage, producing tendency to widen gaps between southern montane breeding grounds and range of nominate svecica in boreal and arctic zones. Secretiveness, however, safeguards the species against significant direct persecution, except to some extent on

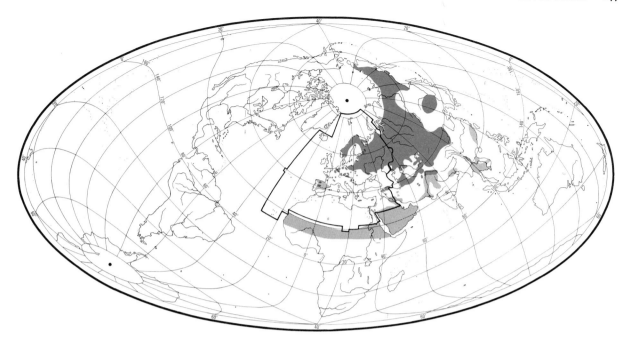

passage. In this regard it ranks as one of the least aerial passerines when not on migration, apart from making brief song-flights in breeding season, and sometimes flycatching in equally brief sorties. Although much on ground, mostly uses places under good cover, even including potato or turnip fields and patches where gorse has been burnt.

Distribution. Range decreased in Netherlands; recent occasional nesting in Scotland, Switzerland, and Italy.

BRITAIN. Scotland: bred 1968 (Greenwood 1968), singing ♂♂ present 1979, probably bred 1983 (Spencer *et al.* 1986). FRANCE. In east, more common in 19th century near ruins (RC). For recent expansion in Brittany, see Guermeur (1977). NETHERLANDS. Marked decrease in range since *c.* 1930 (CSR). NORWAY. Some range expansion (VR). EAST GERMANY. Breeds occasionally outside mapped area. POLAND. Nominate *svecica* recently bred in Tatra (Z Głowaciński and P Profus). SWITZERLAND. Bred 1937 (Hess 1927), 1980 (Wartmann 1980), and 1983 (*Br. Birds* 1984, 77, 586-92). AUSTRIA. Nominate *svecica* breeds at 6 sites, first proved 1976 (Krösche 1976). About 12 sites for *cyanecula*; decreasing and rare (HS). ITALY. 3 pairs bred in Alps 1983 and 2 pairs 1984 (PB). RUMANIA. Bred 1927 and 1967-8 (WK).

Accidental. Iceland, Faeroes, Ireland, Greece.

Population. FRANCE. 1000-10 000 pairs (Yeatman 1976). In north at least 30 pairs of *cyanecula* in 1977 (Godin and Loison 1978); *c.* 850 pairs of *namnetum* in Morbihan and Loire-Atlantique (Guermeur 1977). BELGIUM. About 900 pairs (Lippens and Wille 1972). NETHERLANDS. Estimated 900-1200 pairs 1976-7 (Teixeira 1979); marked decrease since *c.* 1930 (CSR). WEST GERMANY. About 600-1050 pairs (see Bauer and Thielcke 1982 for details, also Rettig 1983). Marked decline Niederrhein (Hubatsch 1983). SWEDEN. Estimated not less than 300 000 pairs (Ulfstrand and Högstedt 1976). FINLAND. Estimated 27 000 pairs (Merikallio 1958); increased 1941–73 (Väisänen 1983). EAST GERMANY. About 170-190 breeding pairs (SS). POLAND. Marked decline, especially in south; recently extinct Silesia (Tomiałojć 1976; AD). CZECHOSLOVAKIA. About 20 pairs of *cyanecula* in south-west Slovakia and southern Bohemia; decreased in south-west Slovakia since 1960 due to drainage. 10-15 pairs of nominate *svecica* in Krkonoše mountains (KH). HUNGARY. About 20 pairs (LH). AUSTRIA. *L. s. cyanecula* rare; decreasing. 10-15 pairs of nominate *svecica*. (HS.)

Movements. Mainly migratory, west Palearctic populations having extensive wintering area extending from Mediterranean basin south to northern Afrotropics, and east to Indian subcontinent. Eastern races (not considered further here) winter in Indian subcontinent and south-east Asia.

Northern race, nominate *svecica* (breeding Scandinavia east across northern USSR), is found in winter throughout Mediterranean and over entire African winter range of the species (Zink 1973). 44 recoveries of Scandinavian-ringed birds (Staav 1975) indicate 2 migration routes: (1) south into Africa, with recoveries in northern Italy in September and Algerian Sahara (27°N`1°E) in April; (2) south-east, probably to winter in Pakistan and north-west India (no winter recoveries, however), with passage birds found in Turkestan near Afghanistan border, in Tien Shan mountains, and 4 others showing clear south-east

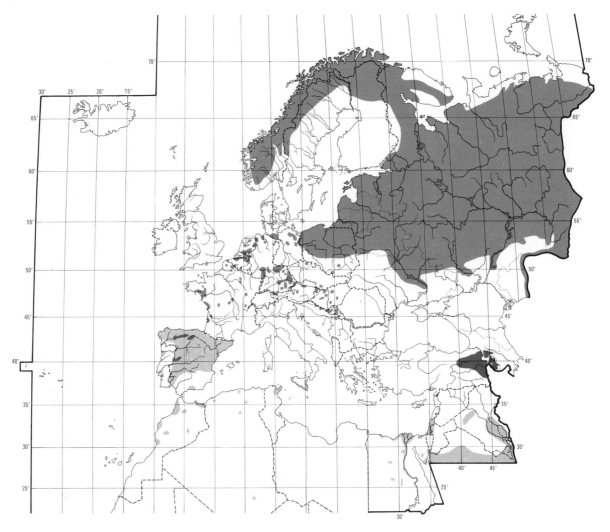

orientation including one on Caspian east coast on 10 October; considering low chance of recovery in these regions, this must be a major route. Birds leave north Scandinavian breeding grounds in late August, with peak numbers in south-east Sweden in 1st week of September (Staav 1975, 1978). Many remain off-passage in marshy areas here and some show site-faithfulness (e.g. 1 ringed 8 September 1971 was retrapped 30 August 1973, 31 August 1975, and 3 September 1977). Similarly, bird ringed 21 September 1966 at Slapton (south-west England) was retrapped there on 14 September 1968, and an adult ♂ on 17 May 1958 was retrapped there on 5 May 1963, giving further evidence of regular SSW orientation. Sharrock (1970) showed an early September peak in Britain, with arrivals coinciding with big falls of other migrants (notably in early September 1965 when over 100 were recorded in a massive fall in East Anglia); this also suggests a regular small-scale orientation towards SSW. Since that time, British autumn records have declined, but spring overshoots on east coast remain

relatively common, with more than half in Shetland. Apparently few nominate *svecica* migrate across central or eastern Europe, and these mainly in spring. Thus Schmidt (1970) summarized 33 Swiss records, mainly in April. This race seldom identified on Malta, but is the most common migrant race in Cyprus, with autumn passage mid-October to early November and some wintering (Bannerman and Bannerman 1971; Flint and Stewart 1983). Recoveries of Scandinavian birds show that some are still in Turkestan and Kazakhstan in April; arrive in southern Sweden in 2nd week of May. Immigration is almost all to islands and peninsulas of Baltic coast, confirming that main migration route is from south-east. Arrivals peak in mid-May, and ♂♂ arrive on average a few days earlier than ♀♀ (Staav 1975, 1978). Dolgushin *et al.* (1970) gave extensive details of passage timing through Kazakhstan, though little distinction was made between the various races (mainly nominate *svecica* and *pallidogularis*). First arrivals immediately north of the mountains (e.g. Tashkent, Chimkent) are in first half of

March, but do not reach the north until early April. Passage very extended, with birds recorded in all areas (especially river valleys) until mid-May, suggesting a mixture of local and northern breeding birds; 3 nominate *svecica* have been taken at Panfilov in the second half of April. ♂♂ move through much earlier than ♀♀. Dispersal of local birds occurs from early August and breeders have left foothills around Alma-Ata by early September. An extended autumn passage includes birds in west and central Kazakhstan in mid- and late October which are probably in part from Scandinavia and north-west Russia. Johansen (1955) gave broadly similar passage dates for USSR east of Urals, with arrival in far north (Gydanskiy peninsula) as late as 18 June. More eastern populations of nominate *svecica* are thought to migrate through eastern China, to winter south-east China and Indonesia.

South-west races, *cyanecula* (breeding Spain and central Europe) and *namnetum* (breeding western France). Recoveries of birds ringed in west, central, and eastern Europe, including German and Swiss recoveries of birds from western USSR, are all evidence of autumn orientation between west and south, with concentrations then in southern Spain and western Portugal (Zink 1973). Most recoveries there are September–November. Recoveries December–February in south-west France (1), Spain (7), and Algeria (1, from Bayern). 7 birds ringed on Campania coast of southern Italy in December 1982 and retrapped in January 1983 demonstrate overwintering there also (Scebba and Vitolo 1983), but only a few winter in Camargue (France), and majority of European population must winter in Africa. Also winters occasionally in western two-thirds of Turkey, but race not mentioned (*Orn. Soc. Turkey Bird Rep.* 1966–75). Main spring passage is early. Over 50% pass through Neusiedlersee (Austria) in first half of April, ♂♂ slightly ahead of ♀♀; peak spring passage in southern Germany is 31 March to 3 April, and in Switzerland 4–7 April (Schmidt 1970). Of 44 *cyanecula* caught in Schwäbische Alb (southern West Germany) 1964–70, peak was first 10 days of April, with ♂:♀ ratio 2:1; weights varied from 14·5 g to 23 g, indicating that onward passage was occurring from the site (Ullrich 1972). Peak period for *cyanecula* in southern England is end of March (Sharrock 1970).

L. s. volgae (breeding central European USSR) winters in Middle East (including Iraq) and Egypt. Occurs on migration in south-east Ukraine and from lower Volga to lower Don and lower Kuban (Vaurie 1959). Meinertzhagen (1930) found them wintering at regular localities in Egypt (including Sinai), 18 November–22 March. Conder (1981) estimated 50–100 wintering at Azraq (Jordan), all red-spotted birds. Returns to northern Caucasus and Crimea by early April and to Moscow by end of April or early May, leaving in second half of September. ♂ caught on 2 April at Pingarati (eastern Rumania) may indicate some passage west of Caspian to

wintering areas in eastern Mediterranean (Munteanu 1966).

Caucasian and Iranian race *magna* has occurred in winter in Iraq (Meinertzhagen 1930); also a scarce passage migrant in Eritrea, 31 August to late October and again late March to 9 May (Smith 1957), and occurs in winter in Ethiopia and Sudan (Moreau 1972). *L. s. pallidogularis* (breeding south-east European USSR and Soviet Central Asia) winters from Syria, Iraq, and southern Transcaspia through Iran to northern India. Few data, however, on wintering within west Palearctic. Passage through Gissarskaya basin (south-west of Pamirs) occurs mid-March to mid-May and late August into November (Ivanov 1969).

WINTERING IN AFRICA. The distribution and relative abundance of the 2 main European races (*cyanecula* and nominate *svecica*) in Africa is confused. In general, *cyanecula* is more common in the west, *svecica* in the east, but both occur throughout. In western Morocco, Smith (1965a) found only *cyanecula*, with 7 on 22 December in one area, though at La Moulouya (eastern Morocco) 6 *namnetum* and 2 nominate *svecica* were caught December–January (Brosset 1959); passage birds of both races were also caught there. Nominate *svecica* also winters in Tunisia. Heim de Balsac and Mayaud (1962) and Zink (1973) stated that records in North Africa are mainly in passage periods and that the few winter records are mainly of *cyanecula*, including mid-winter dates in Algerian and Libyan Sahara. Return passage starts early, with 30–40 seen in western Morocco on 28 February thought to be migrants (Smith 1965a); spring records in north-west Africa are more frequent than in autumn. In West Africa the species is common in Sénégal; Morel and Roux (1966) were not sure of the racial proportions, but of 47 caught 1969–72, all 30 ♂♂ were *cyanecula* (Jarry and Larigauderie 1974), the only recent record of nominate *svecica* being at Richard-Toll on 7 March 1973 (Morel and Roux 1973). Birds are present in Sénégal from mid-September to end of March, exceptionally end of May. In Mali, *cyanecula* is described as fairly common north of 17°N in Niger and Bari, arriving October–November and leaving in March. Distribution of nominate *svecica* here is thought to be the same though evidence is poor; caught regularly at Bamako November–December (Lamarche 1981). Uncommon in Nigeria, occurring mainly in north, and these early December to March, mainly nominate *svecica*; described as common north of Sokoto in January (Elgood 1982). In Chad, *cyanecula* arrives in north of country in October, nominate *svecica* (less common) in November; wintering range may be moving south due to increasing aridity (Newby 1980). Further east, Friedmann (1962) found birds at Fada (800 km north-east of Lake Chad), which provides a link to known wintering area in Darfur (Sudan). Majority of *cyanecula* probably pass through Iberia or cross Mediterranean direct to winter in tropical Africa from Mauritania and Sénégal through

Niger inundation zone discontinuously as far east as western Sudan, but only a small and perhaps decreasing proportion of nominate *svecica* do so, with records of that race fragmented and sometimes contradictory. RAC

Food. Largely terrestrial invertebrates, mostly insects; in autumn, also some seeds and fruits. Feeds on ground, hopping, running briefly, and pausing (Witherby *et al.* 1938b; Ali and Ripley 1973a); also takes items from low vegetation (Ptushenko and Inozemtsev 1968) and will catch insects in the air (Klimmek 1950). Searches for food by turning over leaves and soil (Meinertzhagen 1954). Adult catching caterpillar for its own consumption always shakes it to expel gut contents (L L Semago).

Diet includes the following. Invertebrates: larval mayflies (Ephemeroptera), dragonflies, etc. (Odonata), stoneflies (Plecoptera), grasshoppers, etc. (Orthoptera: e.g. Tettigoniidae), earwigs (Dermaptera), bugs (Hemiptera: Cicadellidae, Reduviidae), larval Lepidoptera (Pieridae, Noctuidae, Lymantriidae, Geometridae), caddis flies (Trichoptera), adult and larval flies (Diptera: Tipulidae, Chironomidae, Culicidae), Hymenoptera (Tenthredinidae, Ichneumonidae, ants Formicidae), adult and larval beetles (Coleoptera: Carabidae, Dytiscidae, Staphylinidae, Scarabaeidae, Elateridae, Cantharidae, Tenebrionidae, Bruchidae, Chrysomelidae, Curculionidae), spiders (Araneae), sandhoppers (Amphipoda), shrimps *Crangon*, small snails (Gastropoda), earthworms (Lumbricidae), young frogs *Rana*. Plant material: seeds or fruits of docks and knotweeds (Polygonaceae), strawberries *Fragaria*, blackberries *Rubus*, bird-cherry *Prunus padus*, Leguminosae, alder buckthorn *Rhamnus*, elder *Sambucus*, and wheat; once, greenstuff recorded in stomach. (Gil 1928; Schuster 1930; Witherby *et al.* 1938b; Mayaud 1938; Dementiev and Gladkov 1954b; Pek and Fedyanina 1961; Scott 1962b; Ptushenko and Inozemtsev 1968; Godin *et al.* 1977; Danilov *et al.* 1984.)

Most detailed study of adult diet from Kirgiziya (south-central USSR), where 47 stomachs most commonly contained ants (in 25), Lepidoptera larvae, and beetles (mostly adults, especially Curculionidae and Carabidae); few seeds (Pek and Fedyanina 1961). In Voronezh region (central European USSR), apparently indifferent to berries (L L Semago), though taken in autumn in Moscow region (Ptushenko and Inozemtsev 1968). For further, mostly sparse, data see Gil (1928) for Spain, Mayaud (1938), for France, Baer (1909) for East Germany, and Dementiev and Gladkov (1954b), Ptushenko and Inozemtsev (1968), and Danilov *et al.* (1984) for USSR.

On Yamal peninsula (USSR), food in 42 collar-samples from 5 broods comprised mainly beetles, spiders, and larval Tenthredinidae (19–20% by number each); 103 samples from a further 3 broods included 23·3% larval Tenthredinidae, 23·3% Tipulidae, 18·8% flies, and 14·7% Lepidoptera (Danilov *et al.* 1984). Young in one nest in northern Sweden fed on caterpillars (including

those of moth *Oporinia*) and other insects (Peakall 1956). Prey brought to nest in Belgium was mainly earthworms, also Tipulidae, small crustaceans (apparently), and 2 young frogs *Rana* (Godin *et al.* 1977). In Voronezh region, young fed on caterpillars and Orthoptera (L L Semago). DJB

Social pattern and behaviour. No major study, but most aspects moderately well known. For summary, see Schmidt (1970).

1. Normally solitary outside breeding season. Small concentrations (up to 10) reported from Ghana in winter, when apparently territorial (Walsh and Grimes 1981). Essentially solitary on migration, though concentrations do occur (e.g. Staav 1978); see also Antagonistic Behaviour (below). Territoriality poorly developed or absent on migration stopovers of up to 6 days in southern West Germany, though 2–3 birds sometimes close together (Schmidt-Koenig 1956; Ullrich 1972). Solitary also on arrival on breeding grounds (Danilov *et al.* 1984), though ♂♂ said by Portenko (1954) to arrive in small flocks. Juveniles solitary after break-up of family (Ptushenko and Inozemtsev 1968) and birds remain thus or in small parties in autumn (Bannerman 1954). Can show fidelity to autumn resting area: 1 trapped at same locality in Sweden in 4 different years (Staav 1978); same perhaps true of winter quarters (Cheesman and Sclater 1935). BONDS. Mating system probably essentially monogamous, but 2 cases of polygyny in Finnish study: ♂ recorded feeding young at 2 nests *c.* 30 m apart (Peiponen 1960). In Yamal (USSR), ♂, apparently paired with 3 ♀♀ who started laying 4–6 days apart, guarded all 3 nests and fed all 3 broods, but mainly in first few days after hatching, i.e. devoted most attention to nest where young just hatched (Danilov *et al.* 1984). On breeding grounds at Kilpisjärvi (Finland), ♂:♀ ratio 1·4:1 (but see, e.g., Fraine 1982 for sexual behavioural differences possibly influencing this). Additional ♂ once recorded associating with pair from laying, staying mostly near ♀, but later displaying (see part 2) at both pair-members, also occasionally visiting nest (Peiponen 1960). In northern West Germany, at least 2 surplus ♂♂ visited one nest and one brought food but neither fed young (Klimmek 1950). At other nests, additional single ♂♂ recorded helping to feed 3 well-grown young (Krösche 1979) and feeding young of another pair when his own had fledged (Theiss 1973). Birds (of either sex, sometimes unpaired, but including widowed ♀) will feed strange young out of nest (Peiponen 1960; Theiss 1973). In Yamal, after marked ♂ died, another ♂ (see above) moved in, guarded nest, and fed young until fledging (Danilov *et al.* 1984). Brooding normally by ♀ alone (e.g. Klimmek 1950, Peakall 1956, Arheimer 1982), but at one nest in eastern USSR ♂ brooded young more than ♀ (Kishchinski 1980). Young fed by both sexes about equally (Schmidt 1970; Arheimer 1982), more by ♂ (e.g. Klimmek 1950), or by ♀ (Witherby 1928; Tadić 1975). Where double-brooded, most feeding by ♂ while ♀ re-building, but ♀ recorded feeding 1st brood during laying (Theiss 1972) and incubation of 2nd clutch (Tuschl 1985). Family apparently breaks up on breeding grounds prior to departure (Kishchinski 1980). ♀ breeds first at 1 year old, ♂ probably also (see Müller 1982). BREEDING DISPERSION. Solitary and territorial. Groups of pairs not infrequently reported (e.g. Danilov *et al.* 1984); perhaps due primarily to habitat constraints. In Vendée (France), nests sometimes only 25–30 m apart (Labitte and Languetif 1962), also by Leda river (Ostfriesland, northern West Germany) (Klimmek 1950), though birds using polder ditches in Ostfriesland normally *c.* 600 m

apart (Blaszyk 1963). In central Spain, at *c.* 1935-2170 m altitude, common on slopes with broom *Cytisus purgans*—e.g. 4 ♂♂ within *c.* 50 m of observer, territories contiguous, and 2 singing ♂♂ frequently *c.* 20 m apart (Witherby 1928). In Dischmatal (Switzerland), 4 territories (pairs with nest or young) at *c.* 1200 m were *c.* 280-720 m apart (Koch 1983). Locally in northern and north-east USSR, pairs or singing ♂♂ 50-100 m apart (Krechmar 1966; Kishchinski 1980) or 150-250 m apart (Portenko 1973). Important study of clay-pit population at Mannheim showed territory-centres far apart, up to 400 m. Much mobility in areas between centres. In one year, 9 ♂♂ very close together; spent most time on or near song-posts, but highly mobile, using others' song-posts, and thus no clear boundaries. 1 ♂ used area *c.* 700 × 30-150 m, another *c.* 400 × 400 m; more restricted 'home-ranges' (apparently not due to outside pressure) *c.* 50 × 70 m and *c.* 40 × 60 m (see also Antagonistic Behaviour, below). 'Home-range' and territory-centre (core area) larger in vocally and otherwise more active ♂♂. Same song-posts, excursion zones (areas visited outside core), and escape-routes used by birds year after year (see also below for site-fidelity). Some ♂♂ apparently stayed in one area for several days, then bred *c.* 500-800 m away, but shift perhaps (partly) due to disturbance (Schmidt-Koenig 1956); see also Schmidt (1970) for Hungary. Further examples of 'territory' size (where measured from song-posts, perhaps to be viewed with caution following findings of Schmidt-Koenig 1956) indicate similar variability to that described above, estimates varying from 0·45 to 25 ha (see Radetzky 1970, Schmidt 1970, Theiss 1973, Dittberner and Dittberner 1979). In Yamal, song(-display) territories typically mobile; boundaries initially strict, but soon after groups established (see above), neighbours will sing in partly (sometimes considerably) overlapping territories (Danilov *et al.* 1984). Territory established and defended by ♂; used for nesting, though in Yamal, when water unusually high, 4 nests built outside ♂'s song-territory (Danilov *et al.* 1984). ♂ sometimes moves far from nest (Fredriksson *et al.* 1973). In various studies in both lowland and montane habitats, food collected 20-90 m from nest (Schmidt 1970; Theiss 1972; Wartmann 1980), or even several hundred metres (Blaszyk 1963). Densities from various parts of range as follows. In Savoie (France), 60±5 pairs per km² or 80±5 pairs if only area actually occupied is counted (Tournier 1973). At least 15 pairs in *c.* 10 × 1·5 km in Demer valley of Diest-Aarschot, Belgium (Peeters 1979). Up to 7·5 pairs per km² in one part of Drenthe (Netherlands), 0·2-4 pairs per km² otherwise (Dijk and Os 1982). In Ostfriesland, formerly *c.* 40-50 ♂♂ or pairs in 20 km², 1 pair per 40-50 ha or along 1·2 km of suitable ditches. Declined to *c.* 15 ♂♂/pairs in 20 km²; only concentration is 7 ♂♂ in 1·5 km² (Blaszyk 1963; Rettig 1974). At one lake in Mecklenburg (East Germany), 6·2-6·8 pairs per km² in 2 years (Klafs and Stübs 1977), and at 3 localities in Brandenburg 1·3-2·7 pairs per km² (Dittberner and Dittberner 1979). One of 7 most numerous species in willow *Salix* along middle Vistula, Poland; for details, see Luniak (1971). On Nord-Fugløy island (northern Norway), 13 pairs in 22 km² (Lütken 1964). At Kilpisjärvi, average 10 pairs per km² for whole area, but up to *c.* 89 pairs per km² in rich meadow/birch *Betula* forest patches in some years (Järvinen and Pryl 1980); see also Järvinen *et al.* (1980). Various habitats of Pechora basin including adjoining Urals (north-west USSR) hold 1·6 (in bogs) to 96 (in montane larch *Larix*) birds per km² (see Estafiev 1981). In Yamal, favours willow thickets near water: 4·2-7·7 and 1·3-9·1 pairs per km² at 2 sites over 9 years; other densities listed include up to 60 pairs per km² in willow and 50 in alder *Alnus* (Danilov *et al.* 1984). For further data from extralimital USSR, see (e.g.)

Naumov (1960), Kishchinski (1980), and Vartapetov (1984); for Mongolia, see Piechocki *et al.* (1982). In one case, 2nd-brood nest sited *c.* 6 m from 1st (Theiss 1972). ♂♂ show marked site-fidelity: at Mannheim, some occupied exactly same core area over 2(-3) years; one returning for 3rd year occupied partly new core area apparently due to pressure from many new neighbours; yet another chose new core area within former overall 'territory' (Schmidt-Koenig 1956). 3 marked ♂♂, but no ♀♀, returned to one Yamal breeding area (Danilov *et al.* 1984), and strong evidence of fidelity (up to 6 years) in Belgium (Godin and Loison 1978). ROOSTING. In India, winter, *pallidogularis* roosts with swallows (Hirundinidae), wagtails *Motacilla*, etc., in reeds and bushes by water (Ali and Ripley 1973*a*). Roosting at least partly nocturnal and presumably solitary on breeding grounds. In Finnish Lapland, incubation and feeding rhythm closely follow temperature changes; sustained incubation spell (and thus restraint from feeding) *c.* 24.00-04.00 hrs coinciding with time of daily minimum temperature (Peiponen 1970), though Schmidt (1970) indicated ♀'s main rest period 23.00-07.00 hrs and that of parents feeding young 02.00-03.00 to 06.00-07.00 hrs. Average length of nocturnal rest during feeding of young in Swedish Lapland 4½ hrs; starts 22.00-00.25 hrs, ends 03.10-05.15 hrs, this period matching low activity of insects serving as food of young (Arheimer 1982). For similar results from detailed observations at 2 nests with young in Swedish Lapland, see Lennerstedt (1973) who noted slightly reduced but more or less continuous activity shortly before hatching. Diurnal rhythm in Swedish Lapland perhaps further indicated by number of birds trapped at different times: low activity apparently 04.00-06.00 hrs and *c.* 16.00-17.00 hrs (Fredriksson *et al.* 1973). *L. s. cyanecula* generally less active in late morning and afternoon, more so again from evening (Schmidt-Koenig 1956; Schmidt 1970). For nocturnal song, see part 2. Water-bathes readily, several times daily in any weather and also at night (Naumann 1905; L L Semago).

2. Outside breeding season generally rather shy, or at least unobtrusive and skulking, staying close to ground and moving about furtively in thick cover (e.g. Witherby *et al.* 1938*b*, Schmidt-Koenig 1956, Lütken 1964). Alarmed bird will slink away with head low, body tilted forward, and tail raised (see below), then straighten up to look about before spurting further (Ali and Ripley 1973*a*). When flushed, flies off low and for short distance. Will feed in the open if undisturbed (Witherby *et al.* 1938*b*; Fry 1966*b*), and sometimes easier to approach than above descriptions indicate (Naumann 1905; Knorr 1927; Lorenz 1954). Can become relatively tame on breeding grounds—e.g. near human settlements in Arctic, but even where birds used to man, ♀ still secretive and shy (♀ *cyanecula* flies little until hatching: Fraine 1982), ♂ often bold and demonstrative (Bannerman 1954; Radetzky 1970); see also Antagonistic Behaviour (below). ♂ flushed on territory will fly off low, then up to another look-out (Schmidt 1967*c*). Both sexes overall less skulking when with young (e.g. Ticehurst and Ticehurst 1902, Lütken 1964). Fledged young fairly confiding (unlike adults) in Spain, mid-August (Ern 1966), though not so in Switzerland when able to fly well (Koch 1983). When on ground or perch, frequently raises tail (up to *c.* 90° to body) and flirts it (moving it laterally and up and down with simultaneous fanning). Tail-spreading on landing, and bobbing (like Robin *Erithacus rubecula*), also typical. For calls given by alarmed bird, see 2 in Voice. Red spot of nominate *svecica* or white spot of *cyanecula* briefly twice normal size in vigorous song and thus probably an important accompanying signal in territory advertisement; throat and

breast pattern as well as red marks at base of tail similarly have important signal function in other antagonistic and heterosexual displays (Christoleit 1928, 1929; Bergmann 1977b). FLOCK BEHAVIOUR. No information. SONG-DISPLAY. ♂ sings (see 1a–b in Voice) from within undergrowth or, when more excited, from exposed perch up to c. 6–10 m high, or in flight (Witherby et al. 1938b; Armstrong and Westall 1953; Schmidt 1967c, 1970; Tournier 1973; Wartmann 1980). In central Spain, each ♂ used 3–4 favourite perches (Witherby 1928). In Oberfranken (West Germany), ♀ twice reported singing from perch in late March and early April (Theiss 1973), and both sexes reported to sing in winter (Ferguson-Lees 1968b). However, all detailed information on Song-display refers to ♂. When singing from perch, droops (and trembles) wings, and raises and slowly opens and closes tail; throbbing movements of throat cause spot to change size and shape and, especially in white-spotted cyanecula, apparently to twinkle (Natorp 1928; Garnett 1930; Ptushenko and Inozemtsev 1968; Bergmann 1977b, with drawing; L L Semago). Will move about in this posture, then down to ground (Johansen 1955). Bird (in some cases juvenile) singing in autumn may quiver tail, which is raised only slightly; movement of throat feathers often the only indication that bird is singing (L L Semago: see 1c in Voice); see also Fig A which

A

shows bird giving Subsong. Takes off for Song-flight from perch or ground. Sometimes flutters up briefly, while singing and with tail spread, to c. 2–5 m, then dives into cover. Will also ascend at angle (to just a few metres) or spiral up, briefly hover (perhaps even loop the loop), describe one or several circles or zigzag, then make gliding descent, singing all the while, wings (which may be quivered) and tail widely spread and angled like pipit Anthus, legs stiffly extended down; often lands near ♀, continuing to sing. Glides sometimes over c. 20–30 m and will perform display repeatedly, moving from perch to perch, though some ♂♂ do not perform Song-flights at all. (Witherby 1928; Armstrong and Westall 1953; Bannerman 1954; Schmidt-Koenig 1956; Corley Smith and Bernis 1956; Géroudet 1963; Ptushenko and Inozemtsev 1968; Schmidt 1970; Theiss 1973; Tuschl 1985.) In Hungary, cyanecula sings well before dawn, most intensely 05.00–07.00 hrs, then again at dusk; at peak, some sing with pauses throughout night. Little singing around midday, especially if hot (Schmidt 1970); see also Mayaud (1938) and Schmidt-Koenig (1956). Nominate svecica sings more at night than other Arctic passerines. Song noted 21.00–03.00 hrs (pause 01.00–02.00 hrs), including Song-flights 24.00–01.20 hrs, also later in morning (Palmgren 1935; Armstrong 1954). In arctic Norway will sing at all hours (pauses generally around midnight and midday). Further south in Norway, ♂♂ sing almost only at night from start of incubation; silent when colder or darkness more intense, but sing again for 1–2 hrs after daybreak (Bannerman 1954). In Novosibirsk (c. 55°N in USSR), 36% of singing in morning, 11% later in day, 30% in evening, 23% at night (Podarueva 1979). See also Natorp (1928), Kishchinski (1980), and Tuschl (1985). In Moscow region, increased song (dawn to dusk) and frequent

Song-flights after arrival of ♀♀ (Ptushenko and Inozemtsev 1968). Song inhibited by cold, wet, wind, or fog, but strongly infectious—e.g. all singing in sun after snow—and typically more singing where birds clustered than from ♂ of isolated pair (Schmidt-Koenig 1956; Schmidt 1970; Theiss 1973). Birds (in various parts of range) sing from arrival or from up to c. 1 week later (Schmidt 1970). In arctic Norway, song-period c. 6 weeks (Bannerman 1954), up to early July (Lütken 1964), ceasing with hatching (Ticehurst and Ticehurst 1902). In Yamal, output reduced after pair-formation, i.e. during laying (or building); quite a lot of singing to about mid-July (Danilov et al. 1984). In various other studies, singing peaked around time of nest-building and laying, waning thereafter (Mayaud 1938; Kapitonov and Chernyavski 1960; Krechmar 1966; Portenko 1973; Tournier 1973; Kishchinski 1980; Wartmann 1980; Müller 1982); e.g. in northern West Germany, ceases with start of incubation (Klimmek 1950). In Oberfranken, ♂ (even of isolated pair) will sing until young leave nest, others even later (Theiss 1972), and singing frequent in Spain in June during feeding of young (Witherby 1928). Resurgence of singing noted for 2nd brood (Tuschl 1985; L L Semago). In Brandenburg, adults sing to July, juveniles in August–September (Dittberner and Dittberner 1979). Other records of late-summer or early-autumn song (some by adults; see 1c in Voice) from Spain at 1600 m altitude in mid-August (Ern 1966), in Yakutiya (eastern USSR) in mid-August (Vorobiev 1963), and in Voronezh (L L Semago). Will sing also in mid-winter (e.g. Moreau and Moreau 1928, Walsh and Grimes 1981). Migrants sing, but usually not full song nor in flight (Géroudet 1963). ANTAGONISTIC BEHAVIOUR. (1) General. Apparent territorial behaviour in Ghana, winter, includes frequent calling, chases, and song (Walsh and Grimes 1981). In Afghanistan, late April, where large numbers congregating on passage, birds highly pugnacious, frequently fighting (Meinertzhagen 1938); such behaviour typical in these circumstances (Naumann 1905). ♂♂ on territory and displaying in France from late March; ♂-♂ chases occur early April (Mayaud 1938). Some territorial disputes recorded in montane population of central Spain, mid-June (Witherby 1928). Older, well-established ♂♂ are first to arrive at Mannheim (19–24 March), and sing from 1st day; arrival and occupation of territories continues to mid-April. Quiet and secretive birds probably bound for breeding grounds further north; others show low-intensity territorial behaviour (quiet singing, some high perching, tail fanning, but not Song-flights) and may eventually take up territory in same general area (see also Breeding Dispersion, above). Loud and excitable ♂♂ (not of any particular age group) occupy large 'home-range' and more extensive core area than others; especially likely to perform Song-flight when approached (Schmidt-Koenig 1956; also Schmidt 1970). Yamal

Hilary Burn

PLATE 48. *Irania gutturalis* White-throated Robin (p. 671): **1** ad ♂ fresh (autumn), **2** ad ♂ worn (spring), **3** ad ♀ worn (spring), **4** 1st ad ♂ fresh (1st autumn), **5** juv. (HB)

PLATE 49. *Phoenicurus erythronotus* Eversmann's Redstart (p. 678): **1** ad ♂ worn (spring), **2** ad ♀ fresh (autumn).   *Phoenicurus ochruros gibraltariensis* Black Redstart (p. 683): **3** ad ♂ worn (spring), **4** ad ♀ worn (spring).   *Phoenicurus phoenicurus* Redstart (p. 695). Nominate *phoenicurus*: **5** ad ♂ worn (spring), **6** ad ♀ worn (spring). *P. p. samamisicus*: **7** ad ♂ worn (spring).   *Phoenicurus moussieri* Moussier's Redstart (p. 708): **8** ad ♂ worn (spring), **9** ad ♀ worn (spring).   *Phoenicurus erythrogaster* Güldenstädt's Redstart (p. 712): **10** ad ♂ worn (spring), **11** ad ♀ worn (spring). (VR)

PLATE 50. *Phoenicurus ochruros* Black Redstart (p. 683). *P. o. gibraltariensis*: **1** ad ♂ fresh (autumn), **2** ad ♂ worn (spring), **3** ad ♀ fresh (autumn), **4** ad ♀ worn (spring), **5** 1st ad ♂ worn (1st spring), **6** juv. *P. o. aterrimus*: **7** ad ♂ worn (spring). *P. o. semirufus*: **8** ad ♂ worn (spring). *P. o. phoenicuroides*: **9** ad ♂ worn (spring), **10** ad ♀ worn (spring), **11** 1st ad ♂ fresh (1st autumn). (VR)

PLATE 51. *Phoenicurus phoenicurus* Redstart (p. 695). Nominate *phoenicurus*: **1** 1st ad ♂ fresh (1st autumn), **2** ad ♂ worn (spring), **3** ad ♀ worn (spring), **4** 1st ad ♂ fresh (1st autumn), **5** 1st ad ♀ fresh (1st autumn), **6** juv. *P. p. samamisicus*: **7** ad ♂ worn (spring), **8** ad ♀ worn (spring), **9** 1st ad ♂ fresh (1st autumn). (VR)

territories in riverine woods are occupied 3–10 days before those in tundra bushes. Old and young ♂♂ continue arriving in numbers when other birds already nesting; sing briefly, then disappear. Hierarchy apparently exists where territories overlap: subordinate ♂ will sing there only if dominant ♂ absent. ♀♀ do not participate in territorial conflicts (Danilov *et al.* 1984). (2) Threat and fighting. Territory-owner may fly immediately to song-perch when ♂ neighbour approaches (Klimmek 1950). Bird attempting to establish territory may at first feed unobtrusively on edge of another's territory; freezes if owner displays (see below) and is attacked on moving, this leading to chase from which owner returns in long Song-flight to perch. Persistent intrusions may eventually allow bird to establish itself, occupancy being then advertised by loud singing, including in flight. At Mannheim, reciprocal visits by territorial ♂♂ regular: gradually move closer, while singing excitedly (Schmidt-Koenig 1956: see 1b in Voice). In Hungary, neighbouring ♂♂ sometimes perched a few metres apart without hostility (Schmidt 1970). Study at Kilpisjärvi (including experiments with stuffed birds) revealed movements and postures of ♂ similar to *E. rubecula*. 3 main postures (with much overlap), depending on relative positions of opponents, as follows. (a) Commonest is Head-up posture (Fig B) adopted by

B

attacker if both on same level; bird appears unusually tall and short-bodied, with closed tail at 60–90° to body, neck extended; no ruffling of feathers, nor (*contra* some reports) are wings drooped (see, however, Heterosexual Behaviour, below). (b) Attacker below opponent extends neck to show off throat and breast pattern; raises tail to 45°–60° and sways fore-body from side to side; no ruffling of feathers. (c) Attacker above opponent (and especially prior to chase) holds head forward and fans and lowers tail (Fig C). (Peiponen 1960.) Rival ♂♂ will hop about rather stiffly at close quarters, glide to ground with tail

C

PLATE 52 (*facing*).
Phoenicurus erythronotus Eversmann's Redstart (p. 678): **1** ad ♂ fresh (autumn), **2** ad ♂ worn (spring), **3** ad ♀ fresh (autumn), **4** 1st ad ♂ fresh (1st autumn).
Phoenicurus moussieri Moussier's Redstart (p. 708): **5** ad ♂ worn (spring), **6** ad ♀ worn (spring), **7** 1st ad ♂ fresh (1st autumn), **8** juv ♂.
Phoenicurus erythrogaster Güldenstädt's Redstart (p. 712): **9** ad ♂ worn (spring), **10** ad ♀ worn (spring), **11** 1st ad ♂ fresh (1st autumn), **12** juv ♀. (VR)

fanned and mock-feed with body feathers sleeked. Adopt Head-up posture, flicking tail back and forth (less commonly fanning it) and strut around each other. Both disappear silently or one attacks and chases other, presumed victor returning in Song-flight towards territory-centre or breaking off chase to perch and sing excitedly while fanning tail. Disputes sometimes involve 3 ♂♂—probably provoked and increased in intensity by presence of ♀♀ (Schmidt-Koenig 1956; Peiponen 1960). Pair feeding young were restless and ♀ called frequently (see 2c in Voice) when strange ♂ near, but paired ♂ attacked this more brightly plumaged of 2 strangers only when young starting to leave nest (Klimmek 1950). Further evidence in experimental study that aggression declines with advance of breeding season, though unpaired ♂♂ always ready to intrude into territory to display at model; for further details of display, attacks, attempted copulation, etc., see Peiponen (1960). Highly excited ♂ will sometimes threaten man at close quarters (Peiponen 1960), showing off striking breast pattern and orange gape (Gamble 1952). In Hungary, generally tolerant towards other passerines, though ♂ sometimes attacks Yellow Wagtail *M. flava*, Linnet *Carduelis cannabina*, and Reed Bunting *Emberiza schoeniclus*; Little Owl *Athene noctua* ignored (Schmidt 1970; see also Theiss 1973). HETEROSEXUAL BEHAVIOUR. (1) General. No evidence that birds pair up on migration, and Peiponen (1960) considered that pair-formation takes place on breeding grounds when ♀ enters ♂'s territory. Normal for ♂♂ to arrive on breeding grounds before ♀♀: e.g. *c*. 5–7 days in Yamal where a further 3–5 days elapse between arrival of ♀♀ and start of building (Danilov *et al.* 1984); in Oberfranken, nest-building from *c*. 16–18 days after arrival, depending on weather (Theiss 1973); in France, courtship noted during first 2–3 weeks after arrival (Mayaud 1938). At Kilpisjärvi, early in season when still few ♀♀ present or later between unpaired ♂♂, nominate *svecica* ♂ sometimes shows ambivalent behaviour, i.e. performing ♀ role in pair-formation; see also subsection 5 (below) for nest-scraping by ♂ in presence of another. Pair-formation appears aggressive, inseparable from typical threat. ♂ continues to direct threat-posture at ♀ at least until start of incubation (Peiponen 1960); this is presumably the posture illustrated and described by Vaughan (1979) who noted that bird (also during nestling period) would briefly flick tail while partly opening wings when approached by mate (Fig C); see also drawing in Barber (1975). (2) Pair-bonding behaviour. ♀ enters ♂'s territory, apparently first attracted by his song. If she leaves, ♂ will chase her rapidly; if they enter another territory, ♂ will return to his own and continue Song-flights. Whole process repeated if ♀ approaches again (Peiponen 1960). ♂ landing from Song-flight recorded beating bush with wings and tail and chasing ♀ when she flushed from there (Portenko 1973). Although ♀ approaches ♂ for pair-formation, only ♂ displays—♂ sometimes on perch, but both usually on ground, often in cover. ♂ adopts Head-up posture: tail raised to at least 45° (not so steeply as in ♂–♂ encounter: Theiss 1973) and may be flirted or fanned (Fig D); head thrown back, sometimes nearly

D

touching tail; wings perhaps slightly drooped (see, however, Antagonistic Behaviour, above) or, according to Naumann (1905), may even brush ground; ♂ sways from side to side while circling ♀ and gives short bursts of quiet, strangled song; captive ♀ responded with short, loud singing (Barber 1975). ♂ may ascend into Song-flight, then land and continue display. ♀ remains motionless, watching ♂, ground-pecks (captive ♀), or runs about under bushes, apparently indifferent. (Natorp 1928; Reboussin 1928; Mayaud 1938; Witherby *et al.* 1938*b*; Peiponen 1960; Géroudet 1963; Theiss 1973.) (3) Courtship-feeding. At Kilpisjärvi, incubating ♀ fed about once every 2 hrs by ♂, especially at night (Järvinen and Pryl 1980). No feeding on or off nest recorded by Peiponen (1960), but noted in France (on nest) by Reboussin (1928) and Mayaud (1938); see also Relations within Family Group (below). (4) Mating. ♂ who had previously sung from perch glided down (tail spread) to ground where ♀ solicited with wings drooped and trembling, and copulated (Klimmek 1950). Copulation may also follow chase as part of pair-formation (Peiponen 1960). (5) Nest-site selection and behaviour at nest. No information on selection process. 2 ♂♂ once seen *c.* 30 cm apart in territory, one scraping and rotating while the other watched (Peiponen 1960); significance unknown. ♀ builds nest mostly in morning (over *c.* 3–6 days), collecting material generally close by; ♂ accompanies her, or sings, calls, or feeds nearby, and may briefly visit unfinished nest. 1–4 days between completion of nest and laying. ♂ recorded chasing ♀ from unfinished nest during disturbance. Contact-call (see 2a–b in Voice) or song from ♂ may be signal for incubating ♀ to leave; ♂ will also move about near nest, fan tail when perched, etc. ♀ leaving nest initially cautious, then moves away in swift, curving flight (see also Relations within Family Group, below, for adults' approach to nest). ♂ will accompany ♀ while feeding, chase her back to nest or visit latter perhaps to encourage her return. (Klimmek 1950; Bannerman 1954; Peakall 1956; Theiss 1973; Dittberner and Dittberner 1979; Wartmann 1980; Järvinen and Pryl 1980; Danilov *et al.* 1984.) RELATIONS WITHIN FAMILY GROUP. Young *cyanecula* in northern West Germany brooded (by ♀) for *c.* 6 days (Klimmek 1950); in north-east USSR (nominate *svecica*) for at least 8 days (Kishchinski 1980). At nest in Swedish Lapland (68°21′N), ♀ brooded young assiduously for 2–3 days, being fed then by ♂; ♀ frequently visited young *c.* 4–5 days old without feeding them, though sometimes brooded them momentarily. ♂ passed food for young to brooding ♀, also when both arrived at nest together (Peakall 1956). In Hungary, ♀ regularly passed food to ♂ while crouching and rapidly wing-shivering (Radetzky 1970). Adults will land on ground and run to nest or fly low through cover (Kapitonov and Chernyavski 1960; Radetzky 1970; Dolgushin *et al.* 1970); may fly to particular perch before and after visiting nest (Theiss 1972). Young fed bill-to-bill (Witherby *et al.* 1938*b*). Captive birds gave call 3a, apparently to encourage gaping (Barber 1975). Eyes of young start to open from *c.* 5–6 days (Kapitonov and Chernyavski 1960; Kishchinski 1980), fully open at 8 days (Klimmek 1950). Both adults clean nest, initially swallowing faecal sacs, later (from *c.* 3–4 days) carrying them away (see Peiponen 1959); ♂ recorded passing faecal sac to ♀ (Klimmek 1950). Data from northern Fenno-Scandia and USSR south to Moscow indicate young leave nest, still unable to fly, at *c.* 10–15 days; shorter period perhaps due to disturbance in some cases (Dunaeva and Kucheruk 1941; Peakall 1956; Ptushenko and Inozemtsev 1968; Lennerstedt 1973; Järvinen and Pryl 1980; Danilov *et al.* 1984). In Swedish Lapland, nestling period 12–14 days (*n*=6 nests); in 3 cases, whole brood left within 24 hrs, in other 3 within 24–48 hrs (Arheimer 1982). May be

well dispersed for feeding after leaving nest (Koch 1983) or stay together for first few days, hiding or hopping and running about in vegetation, then dispersing when able to fly short distance, not returning to nest (Theiss 1973), though elsewhere reported to disperse initially to *c.* 25 m, whole brood then returning to roost in nest (Peeters 1979). Call constantly to maintain contact with parents (Schmidt 1970), but adults will also feed strange young out of nest (see Bonds, above). In Moscow region, young fed for 7–9 days out of nest, but family apparently splits up only when adults start moult (Ptushenko and Inozemtsev 1968). In Oberfranken, ♀ recorded feeding otherwise-independent 1st-brood juvenile *c.* 1 month after it left nest. Another 1st-brood juvenile seen perched on 2nd-brood nest *c.* 3 weeks after leaving natal nest (Theiss 1972). In Switzerland, independent young attempted (unsuccessfully) to associate with families of Black Redstart *Phoenicurus ochruros* and Whinchat *Saxicola rubetra* (Bürkli 1983). ANTI-PREDATOR RESPONSES OF YOUNG. Freeze in nest at approach of danger—e.g. predatory frogs *Rana* in Hungary (Schmidt 1970). May leave nest prematurely at approach of man (Gladkov 1962; Müller 1982). Out of nest, normally fall silent and freeze on hearing parental alarm-call (see 2 in Voice), refusing to move even when touched (Watzinger 1914; Witherby 1928), but recorded running, hopping, and finally fluttering into bush to escape from man (Krösche 1979; Koch 1983). PARENTAL ANTI-PREDATOR STRATEGIES. (1) Passive measures. ♀ normally a tight-sitter and may even allow herself to be seized (Ticehurst and Ticehurst 1902; Dementiev and Gladkov 1954*b*; Theiss 1973), but likely to leave quickly if disturbance repeated (Schmidt 1970). (2) Active measures: against man. Recorded (repeatedly) performing apparent distraction-lure display of rodent-run type on leaving nest (Bannerman 1954), once moving thus *c.* 4 m before giving alarm-calls (Wartmann 1980). ♂ usually sings when man enters territory (Schmidt 1970), and will do so even from bush where young hidden (Hostie 1937); some Song-flights at this time are in form of direct attack, bird flying at intruder and veering away or repeatedly passing him (Schmidt-Koenig 1956; Lütken 1964; Theiss 1973). One ♀ with eggs gave alarm-calls (see 2c in Voice) *c.* 3 m from man (Wartmann 1980). When feeding young, both sexes will perch prominently and call persistently while flicking wings and flirting tail (Guichard 1937; Ingram 1955; Lütken 1964); ♂ sometimes calls or moves about silently in cover, ♀ also but may be bolder, coming close, fluttering around observer, and even trying to peck him (Watzinger 1914; Portenko 1939, 1973; Bannerman 1954; Krösche 1979). Variety of alarm-calls (see 2a–b, 3c in Voice) given by ♀ flying about near nest with young (Kapitonov and Chernyavski 1960). (3) Active measures: against other animals. When adder *Vipera berus* near nest with eggs, ♀ flew about, called (see 2c in Voice), fanned tail, and beat wings; both birds called similarly and dive-attacked snake after hatching (Wartmann 1980).

(Figs by M J Hallam: A–B from drawings by E N Panov; C from photograph in Vaughan 1979; D from drawing in Barber 1975.)   MGW

Voice. Most outstanding feature is remarkable mimetic song, snatches of which are also given in winter (by both sexes), though call 2 more common then (Moreau and Moreau 1928; Smythies 1953; Moore and Boswell 1956; Ferguson-Lees 1968). Repertoire apparently the same in the 2 best-studied races—nominate *svecica* and *cyanecula*. For extra sonagrams, see Bergmann (1977*b*) and Bergmann and Helb (1982).

CALLS OF ADULTS. (1) Song. Full song by ♂ only, but ♀ gives quiet and loud but normally short bursts (no detailed description). (a) Normal territorial or advertising song (also used in Song-duels) loud, strongly imitative, and thus varied; sometimes likened to Marsh Warbler *Acrocephalus palustris* (Czajkowski 1973; Wartmann 1980). Comprises phrases separated by distinct pauses. In bird (*cyanecula*) singing at night, Neusiedler See (eastern Austria), late April, phrases 8–21 s long, pauses of 2–19 s. Song-phrase typically starts with initially slow and hesitant, short whistled units (e.g. 'djip-djip...'), accelerando and crescendo then leading to variety of other motifs. In Oberösterreich (Austria), mostly began with quiet 'üd üd üd üd üd' then accelerando and crescendo and final loud 'sueci sueci trutsitrutsitrut' (Watzinger 1914). Much repetition of units within a phrase is typical, but sweet and musical notes alternate with harsher, discordant hissing sounds. Some sounds recall Nightingale *L. megarhynchos*, but never attain richness and fullness of that species. Liquid trills or muted churr, or bell-like sounds are characteristic, especially a fine, full-toned 'torr-torr-torr-torr'; clear, silvery, metallic 'ting-ting-ting' (like note struck on metal triangle or small, fine-toned gong) often given near end of song. (Blair 1936; Witherby

et al. 1938*b*; Bannerman 1954; Géroudet 1963; Schmidt 1970; Portenko 1973; Bergmann 1977*b*; Bergmann and Helb 1982.) Considerable individual variation but apparently no fundamental differences between nominate *svecica* and *cyanecula* (Natorp 1928; Witherby *et al.* 1938*b*). According to Portenko (1939), nominate *svecica* from Anadyr (eastern USSR) includes in song sounds more suggestive of *Acrocephalus* warbler than does European birds; song of nominate *svecica* from Bratsk (north-west of Lake Baykal, eastern USSR) similar to that of Fenno-Scandian birds (Svensson 1984*b*). In central Spain, rather gentle and quiet song of *cyanecula* at times like Dunnock *Prunella modularis*; perhaps a copy (Witherby 1928). An outstanding mimic, apparently not modifying sounds but producing accurate copies of calls or songs of other bird species, crickets (Orthoptera), tree frogs *Hyla*, and such mechanical sounds as scythe-whetting and locomotive whistles. 50 mimicked bird species (mostly passerines) listed by Schmidt (1970) based on observations in Hungary and other data from literature; see also Blair (1936), Bannerman (1954), Schmidt-Koenig (1956), Corley Smith and Bernis (1956), Czajkowski (1973), Tournier (1973), Wallschläger (1978), and Dittberner and Dittberner (1979). Bird will even mimic

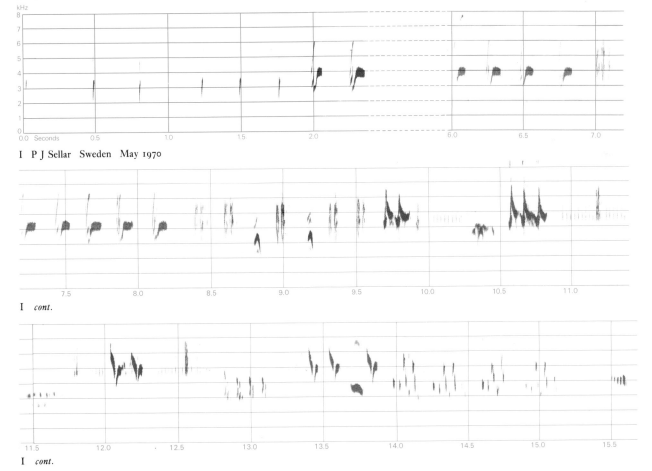

I   P J Sellar   Sweden   May 1970

I   *cont.*

I   *cont.*

whole chorus of sounds from flock, e.g. of sparrows *Passer* (Schmidt-Koenig 1956). Species mimicked early in season mainly those sharing same general nesting area; for examples associated with various habitats, see Schmidt (1970). At Velencei Tó (lake, Hungary), ♂♂ incorporate sounds of virtually all other birds arriving in or passing through study area. New sound usually mimicked immediately and for several days, then replaced by another, though those mimicked earlier may recur later in season (Schmidt 1970). Older birds in particular apparently retain copies over winter, i.e. mimic certain species before they have arrived or ahead of a particular stage in their breeding cycle. Phrase of a particular species may be fashionable in a given year, birds apparently learning from one another. Confused jumbles of sounds also given; 8–10 or more species may be mimicked in a burst of song and each ♂ normally has 1–2 favourites (L L Semago). Recording of 7 songs from nominate *svecica* reveals rich mixture of light, tinkling sounds, buzzes, 'dweep', 'zweep', and 'stip-ip' sounds sometimes strongly suggesting cricket, and attractive 'hoo-EET'. Typically harsher towards end with sound becoming noticeably strained, almost strangled, and hisses, nasal whines, rattles, and scratchy

sounds sometimes producing effect like Black Redstart *Phoenicurus ochruros*. Fig I shows 6 'deep' sounds followed by excellent mimicry of Chaffinch *Fringilla coelebs* 'pink' calls and some little knocking rattles (Fig I omits song section 2·4–5·6 s which contains only such units), then irregular accelerando in mixture of tones and long rattles with rapid reiteration of units at *c.* 40–50 per s. Final brief squeak preceded by 4 exquisite figures, almost but not quite identical, separated by pauses of less than 0·1 s: like fragment of recording of droplets falling from fountain into pool; a remarkable liquid bubbling sound or, at ¼-speed playback, like tintinnabulation of tiny bells. Total length of song just under 16 s. In another song (*c.* 18–19 s) by same bird, 27 distinctive introductory buzzes have rather sudden accelerando after 6th unit. Fig II shows last 7 of these buzzes with some irregularity or hesitation. Buzzes followed by 10 mild 'kyow' units, ripple of 6 notes, single 'kyow' and 3-note ripple, then 4 more 'kjow', 3 'peechu', and 4 'tyoot' units. Alternating squeaks and rattles (rather 'zizzing' because of reiteration at over 40 per s) lead to final coda 'wheet oi-ti-tiu'. (J Hall-Craggs, M G Wilson.) (b) So-called 'Excitement-song' a strangled and rapid series of motifs, usually with

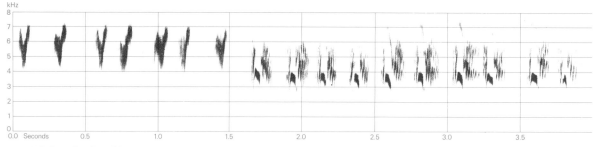

II P J Sellar Sweden May 1970

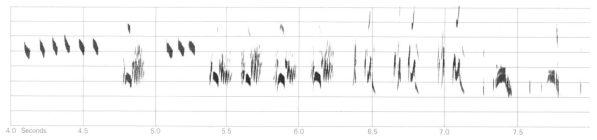

II *cont.*

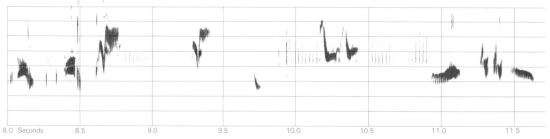

II *cont.*

stereotyped repetition. Used in close encounter with rival and may be reduced to barely audible chirping. Given in Song-flight (also often before or after this), and a variant (no details) during courtship (Schmidt-Koenig 1956); see also Armstrong and Westall (1953) and Géroudet (1963). (c) Song given in early autumn with bill closed is barely audible but still contains mimicry (L L Semago). Warbling in low undertone presumed by Witherby *et al.* (1938*b*) to be Subsong; ♂ *cyanecula* singing in eastern England in spring was inaudible at *c.* 20–30 m (Garnett 1930). In sonagrams of Subsong from recording by E N Panov (Uzbekistan, USSR), rattles apparently predominate. Song in winter may be sweet and brilliant (Moreau and Moreau 1928) or subdued, but musical (Walsh and Grimes 1981). (2) Contact-alarm calls. (a) Call given (sometimes in series) when disturbed typically a hard, guttural, and clicking chat-like 'tak' or 'tk' (Moreau and Moreau 1928; Witherby *et al.* 1938*b*; Kapitonov and Chernyavski 1960; Géroudet 1963; Bergmann and Helb 1982: Fig III). Further descriptions:

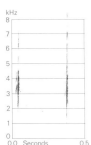

III P J Sellar Sweden July 1965

'tch-tch-tch-tch-tch...' (Hostie 1937); 'teck' (Theiss 1972); 'chat-chat' from ♀ (Greenwood 1968); soft 'dhuck-dhuck' in Iraq, winter (Moore and Boswell 1956); 'k' sounds also serve as ♂ contact-call with ♀ (Klimmek 1950). In *abbotti* of Himalayas (and apparently also in nominate *svecica* and *pallidogularis*), 'tick tick' of alarm distinguished from 'call note'—'chuck chuck' resembling Blyth's Reed Warbler *A. dumetorum* (Ali and Ripley 1973*a*). (b) Weak, rather plaintive whistling sounds. Variety of transcriptions, perhaps indicating that 2 or more variants actually exist. Renderings as follows: 'hweet', 'huid', 'ouit', 'fyueet', or 'fid', resembling *Phylloscopus* warbler or redstart *Phoenicurus* (Watzinger 1914; Witherby *et al.* 1938*b*; Bannerman 1954; Kapitonov and Chernyavski 1960; Géroudet 1963); Theiss (1972) noted both 'huid' and, from alarmed ♂ near young out of nest, 'isst'. Not clear whether the following are significantly different from 'huid', etc., of other authors: weak whistling 'sii' (Bergmann and Helb 1982); high-pitched 'seep' from alarmed ♀ (Greenwood 1968). Similar weak and drawn-out 'seeeep seeeep' noted (in addition to 'fyueet') by Kapitonov and Chernyavski (1960). Calls 2a–b often combined: 'tsi tchak-tchak...' as contact-call, sometimes interpolated into song (Jonsson 1979); see also Watzinger

(1914), Uhl (1930), Guichard (1937), Hostie (1937), and Corley Smith and Bernis (1956). (c) Various churring or croaking sounds: quite soft 'trr trr' (Géroudet 1963) or rather soft, somewhat croaking 'turrc turrc' (Witherby *et al.* 1938*b*); harsh churr given after hatching (Ticehurst and Ticehurst 1902); unmusical dry vibrato (Corley Smith and Bernis 1956). Rattling 'chraä' given by alarmed birds in Switzerland (Wartmann 1980) perhaps also belongs here. A 'tschrr-teck-teck-tschrr' given by ♀ disturbed by strange ♂ (Klimmek 1950) is clearly combination of calls 2a and 2c and same perhaps true of 'ti ti zir tit' noted in Hungary by Bergmann and Helb (1982). (3) Other calls. (a) Low 'chuck-chuck' given by captive birds apparently to elicit gaping in young (Barber 1975); probably closely related to or same as call 2a. (b) A 'chii' shown in sonagram by Bergmann and Helb (1982) but no further details. (c) Melodious, bubbling 'churrree-churrree', apparently in alarm (Kapitonov and Chernyavski 1960); perhaps related to call 2c. (d) Musical bell-like note once given by alarmed ♂ near nest (Bannerman 1954).

CALLS OF YOUNG. Food-call a sharp 'psirr', rising in pitch (Bergmann and Helb 1982). Similar, perhaps identical, 'srieeh' sounds described by Theiss (1973): given during last few days of nestling period and also form start of short song-phrases given *c.* 3 weeks after leaving nest.   MGW

Breeding. SEASON. Central Europe: laying begins late April. Scandinavia: eggs laid from late May. Finland: laying at 69°N normally takes place during 2 weeks beginning 9 June ± 4 days (Makatsch 1976; Järvinen and Pryl 1980). SITE. On ground in dense vegetation, in tussock, under bush, or in hollow in low bank. Nest: cup of grass stems and leaves, with roots and moss, lined with hair and finer vegetation. Mean internal diameter of 65 nests 6 cm (5·5–6·5), depth of cup 6·6 cm (5·5–8·0) (Järvinen and Pryl 1980). Building: by ♀, taking 4–5 days (Theiss 1973); sometimes helped by ♂ according to Mayaud (1938). EGGS. See Plate 82. Sub-elliptical, smooth but only slightly glossy; pale blue, green, or blue-green, finely marked red-brown, often indistinct, giving rusty tinge to shell, occasionally more heavily mottled. Nominate *svecica*: 18·4 × 14·2 mm (16·9–20·7 × 12·8–15·9), n = 110; calculated weight 1·98 g. *L. s. cyanecula*: 18·8 × 14·2 mm (17·1–20·5 × 12·5–15·6), n = 150; calculated weight 2·02 g. *L. s. namnetum*: similar to nominate *svecica*, but sample small. (Schönwetter 1979.) Mean weight of 58 eggs (nominate *svecica*) 1·95 g (Järvinen and Pryl 1980). Clutch: 5–6 (4–8). Of 32 clutches, northern Finland: 5 eggs, 6; 6, 14; 7, 12; mean 6·19 (Järvinen and Pryl 1980). Eggs laid daily. One brood in north of range, 2 in south. INCUBATION. 13–14 days (Haartman 1969*a*). Usually by ♀ only (Haartman 1969*a*); ♂ helps occasionally (Schmidt 1970; Tuschl 1985). Begins with last egg (Makatsch 1976). In Swedish Lapland, clutches of 6 eggs hatched over 24–36(–48) hrs, of 7 eggs over 60–72 hrs (Arheimer

1982); see also Peakall (1956) and Kishchinski (1980). YOUNG. Cared for and fed by both parents. FLEDGING TO MATURITY. Fledging period 14 days, but may leave nest 1–2 days earlier (Haartman 1969a). Age of first breeding 1 year, probably for both sexes (Müller 1982). BREEDING SUCCESS. Of 120 eggs laid in northern Finland, 77·5% hatched, and 74·2% of these fledged, giving overall 57·5% success, with mean 4·45 young fledging per nest (Järvinen and Pryl 1980).

Plumages (*L. s. cyanecula*). ADULT MALE BREEDING. Forehead buff-brown, feathers with rounded black centres. Crown streaked black and dark olive-brown (more grey-brown towards rear); *c.* 3 black streaks on central crown, but these often rather irregular in shape; 1 broader black stripe on each side of crown, sometimes faintly tinged blue. Line from nostril backwards pale cream-buff, widening into broad cream-white or off-white supercilium above eye and ear-coverts. Lore mottled pale cream-buff and black, patch in front of eye largely black. Narrow ring round eye pale buff. Upper cheek and ear-coverts streaked rufous-buff and dull black, merging into grey-brown on longer ear-coverts. Occasionally, part of forehead, lores, and upper cheek replaced in pre-breeding moult, resulting in largely black forehead, whiter front part of supercilium, and bright blue lore and front of cheek. Mantle, scapulars, and back dark grey-brown or drab-brown, slightly purer grey when plumage fresh, duller brown when worn; side of neck paler brown-grey (palest at rear of lower cheek); rump olive-brown; upper tail-coverts dark grey-brown, feathers with variable amount of deep rufous-cinnamon on bases (often concealed, but sometimes forming conspicuous patch, especially when plumage worn). Chin, lower cheek, throat, and upper chest bright and glossy cobalt-blue; contrasting rounded spot or short bar on central lower throat silky-white, sometimes with a few rufous feather-tips at border or partly hidden under bright blue feather-tips. Occasionally, spot absent, throat entirely blue (so-called 'wolfi'-variant, occurring in 8% of 90 birds examined from central Europe). When throat freshly moulted, February–March, blue feathers often faintly tipped or washed grey. Lower border of blue throat-patch black, forming gorget *c.* 3 mm wide across lower chest; feather-tips at lower gorget white, forming narrow and often indistinct white bar below black, this in turn bordered below by deep rufous-cinnamon breast-band 1–1·5 cm wide. Sides of breast dark grey-brown, merging into greyish-buff flanks. Remainder of underparts off-white (variably tinged pale buff or pale olive-grey), under tail-coverts tawny-cinnamon to cream-buff or dirty white. Central pair of tail-feathers (t1) dark grey-brown or greyish-black (some hidden rufous at extreme base), t2–t6 with distal 40% dull black or brown-black and remainder deep rufous-cinnamon or rufous-orange; rufous tail-base often largely hidden by upper tail-coverts when these not heavily worn and tail not fully spread. Flight-feathers, tertials, and all upper wing-coverts dark grey-brown or drab-brown, like upperparts; feather-centres slightly darker, black-brown, narrow fringes along outer edges and tips paler, grey, but easily worn off. Axillaries, inner and shorter under wing-coverts and lesser under primary coverts tawny-buff or slightly rufous with some grey-brown at bases, other under wing-coverts pale brown-grey (like undersurface of flight-feathers), narrowly tipped pale buff. ADULT FEMALE BREEDING. Upperparts and sides of head and neck down to upper cheek basically similar to adult ♂ breeding; black streaks on crown on average shorter and less distinct (especially central ones), general tinge of mantle and scapulars sometimes slightly more olive-brown, less dull greyish drab-brown; no blue on lore or upper cheek. Lower cheek and chin to breast usually markedly different, though much individual variation: ground-colour of lower cheek, chin, and throat pale cream or off-white, with broad solid black stripe running down each side of throat, connected at lower end by solid black band across upper chest; black stripes connected by fine black spots on lower chin and at base of bill, isolating whitish chin, whitish rear of lower cheek, and whitish rounded or triangular spot on lower throat. Lower chest dull black with variable dull blue tinge and white feather-tips, bordered below by grey-brown breast which shows variable amount of rufous-cinnamon on feather-bases. In some birds (probably 2-year-olds), dull blue restricted to some mottling on lower cheek and chin and to chest, and rufous on breast pale, limited in extent, and partly concealed by grey-brown streaking, but in others (probably birds of 4th calendar year and older) lower cheek and chest largely dull blue or bright blue, chin and upper throat mottled blue, and rufous on breast deep and extensive; latter differ from adult ♂ by grey-mottled chin, traces of black stripes on sides of throat and of black feather-bases on chest, and some dark spots on less uniform rufous breast; spot on central throat either silky-white or (equally often) pale rufous (perhaps surprisingly for white-spotted race). Remainder of underparts, tail, and wing as adult ♂ breeding, but sides of breast and flanks sometimes with faint grey-brown streaks. ADULT MALE NON-BREEDING. Differs from adult ♂ breeding mainly in lower cheek, chin, throat, and chest, as these parts of plumage replaced twice a year; remainder of plumage as in breeding, but fresh instead of worn. Upperparts slightly brighter dark olive-brown, especially scapulars, back, and rump, feather-fringes of back, rump, and upper tail-coverts slightly rufous-brown; forehead and crown less distinctly streaked black, only lateral crown-stripe distinct, remainder faintly spotted. Supercilium less distinct, especially rear part; front part from nostril to above eye and narrow broken eye-ring pink-buff or cream-buff, rear part above ear-coverts dull olive-grey. Lores, upper cheek, and ear-coverts dark brown with buff mottling and narrow buff streaks, bordered by grey at sides of neck. Lower cheek pale blue, feathers narrowly tipped buff; malar stripe dull black, broken by pale buff feather-tips, widening towards sides of chest and not reaching base of bill. Chin and upper throat pale buff, often partly mottled pale blue and dull black or almost entirely pale blue; central lower throat usually slightly rufous, feathers with concealed silky-white bases. Band across upper chest pale blue with narrow white feather-tips, bordered by dull black band with broader white feather-tips below, and this band in turn bordered below by bright rufous-cinnamon band across lower chest. Sides of breast and remainder of underparts as adult ♂ breeding, but breast and belly tinged cream and lower flanks and under tail-coverts slightly deeper pink-cinnamon. Tail and wing as in adult ♂ breeding, but tips of all upper wing-coverts and outer webs and tips of tertials with faintly paler olive-brown fringes. ADULT FEMALE NON-BREEDING. Like adult ♀ breeding, but plumage fresh, upperparts, underparts down from breast, tail, and wing as adult ♂ non-breeding. Distinct black malar stripe, as in adult ♂ non-breeding, but hardly any blue on cheeks and chest; lower cheek pale buff or cream, like chin and throat, sometimes slightly mottled pale blue or dull black; band across chest dull black, sometimes with very slight blue tinge, each feather broadly fringed off-white, sometimes appearing scalloped; chest pale greyish-buff, often with some rufous-cinnamon wash on feather centres, not as extensive and deep rufous as adult ♂ non-breeding. NESTLING. Down long and scanty at hatching, grey; dark slate-grey and denser after a few days; on upperparts only (Heinroth and Heinroth 1924-6; Witherby *et al.* 1938*b*). For development, see Peiponen (1962). JUVENILE. Upperparts, lesser and median upper wing-coverts, and chest brown-black, feathers with narrow buff shaft-streak ending in rounded spot on tips of longer feathers and narrowly bordered black terminally. Rump and upper tail-coverts bright rufous-cinnamon. Supercilium and narrow eye-ring indistinct, pale buff with fine dusky specks; lores, upper cheek, and ear-coverts black-brown with fine buff spots and streaks. Chin, throat, sides and centre of breast, and upper flanks buff, each feather with broad black fringe along tip, appearing heavily scalloped; central lower throat sometimes almost uniform cream-white. Remainder of underparts pale buff or cream with some dusky suffusion. Tail and flight-feathers as adult; greater upper wing-coverts and tertials black-brown with narrow tawny or rufous-buff fringes along outer webs and tips, tending to form triangular spot on tips. Under wing-coverts and axillaries pale buff with much dull black of feather-bases visible. Darker than juveniles of other small Turdidae; approached by some Stonechats *Saxicola torquata*, but chest blacker, tail-base rufous, and greater coverts, tertials, and flight-feathers browner with less contrasting broad bright rufous fringes. FIRST ADULT MALE NON-BREEDING. Like adult ♂ non-breeding, but juvenile flight-feathers, tertials, tail, greater upper wing-coverts (often except tertial coverts), and greater upper primary coverts retained; flight-feathers as adult; tail and primary coverts as adult, but tips often narrower and more sharply pointed, less broadly rounded, primary coverts sometimes with small buff spot on tip; tertials and greater upper wing-coverts as adult, but tips usually with rufous-buff to tawny-buff triangular spot, sometimes narrowly bordered black terminally on inner coverts and (if retained) tertial coverts when not too worn; in worn plumage, spots bleach to off-white and partly wear off, showing as small pale triangles mainly on inner greater coverts only; when very heavily worn, spots fully worn off, feather-tips then often showing small triangular indentation. Remainder of plumage as adult ♂ non-breeding, but lower cheeks, chin, and throat often with less pale blue suffusion and dull black spots, mainly pale buff, especially on chin and throat; rufous-cinnamon of chest sometimes slightly paler; often close to adult ♀ non-breeding, which has less cinnamon on chest and no juvenile wing-coverts and tertials. FIRST ADULT FEMALE NON-BREEDING. Like adult ♀ non-breeding, but part of juvenile feathers retained, as in 1st adult ♂, which see for juvenile characters. Unlike adult ♀ non-breeding, no blue or rufous on cheeks, chin, throat, and chest; lower cheeks, chin, and throat pale buff or cream-white with some dusky specks on sides of chin and on upper throat; heavy black malar stripe along sides of throat and black band across upper chest, black partly concealed by pale grey or off-white feather-fringes; lower chest grey (like sides of breast and upper flanks), sometimes with some tawny at centres or sides of feathers and usually with some dull black on central tips. FIRST ADULT MALE BREEDING. Like adult ♂ breeding, but part of juvenile feathers retained, as in 1st adult non-breeding; retained greater coverts usually with trace of pale buff or white spot on tip. Unlike many other small Turdidae, no marked difference in wear between age-groups: many full adults have tail- and primary-tips as heavily worn as 1st adults. Remainder of plumage as adult ♂ breeding, separable only when series of skins compared: blue of chin and throat often slightly paler and less intense, rufous chest-band often narrower, less than 1 cm wide, slightly paler orange-rufous, often more distinctly separated by narrow white bar from black on upper chest. FIRST ADULT FEMALE BREEDING. Like 1st adult ♀ non-breeding: part of juvenile feathers retained, and (unlike

adult ♀ breeding) generally no blue on cheek or throat and no rufous on chest. Chin, lower cheek, and central throat cream or off-white; malar stripe and band across upper chest heavy, but less uniformly black than adult ♀ breeding, narrow pale feather-fringes partly concealing black.

Bare parts. ADULT, FIRST ADULT. Iris hazel-brown or dark brown. Bill black or brown-black, base slightly paler and with blue tinge; bases of cutting edges yellow or orange. Mouth bright lemon-yellow or orange-yellow. Leg and foot dark brown, black-brown, or black, sometimes with faint flesh tinge; soles sometimes yellow. NESTLING. Mouth orange, no spots. Gape-flanges pale yellow. JUVENILE. Iris dark brown. Bill bluish-black, cutting edges and base of lower mandible pink-flesh; gape-flanges pale yellow. Leg and foot greyish-flesh-brown. (Heinroth and Heinroth 1924-6; Witherby *et al.* 1938*b*; Ali and Ripley 1973*a*; BMNH, ZMA.)

Moults. ADULT POST-BREEDING. Complete; primaries descendant. Usually starts in breeding area immediately after nesting, but some may not start until arrival in winter quarters: one on 19 September, Morocco, was just starting body moult and remainder of plumage old (Ginn and Melville 1983). In nominate *svecica*, northern Finland, starts with shedding of p1 about mid-July, completed with p9-p10 after 42-45 days in second half of August (Haukioja 1971, which see for details). In USSR, nominate *svecica* starts about late July to late August, completed from late August onwards, tail-feathers growing simultaneously; in *volgae*, heavy moult by July (Dementiev and Gladkov 1954*b*). In *cyanecula*, starts mid-July to early August, completed from mid-August onwards (RMNH, ZMA). In Ladakh (India), starts with p1 July to early August, completing after *c.* 50 days late August to late September; tail moult rapid, starting at about same time as p1, completed when primary moult about half-way through; secondaries start at primary moult score *c.* 20-30, completed after *c.* 30-35 days at same time as p9-p10 or slightly later (Delany *et al.* 1982). ADULT PRE-BREEDING. Partial: in ♂, lores, ear-coverts, cheeks, chin, and throat (thus excluding underparts from rufous breast-band downwards), and sometimes forehead or some other body feathers; in ♀, often less feathering involved (usually part of cheeks and throat only), occasionally none at all. February-April, mainly completed by late March. POST-JUVENILE. Partial: head, body, and wing-coverts, but not flight-feathers, tail, greater upper primary coverts (often except 1-2 tertial coverts), tertials, or part of bastard wing. Starts at 6-8 weeks old, completed at 3 months (Heinroth and Heinroth 1924-6; RMNH, ZMA). For relation between moult and weight increase, Sweden, see Lindström *et al.* (1985). In *cyanecula*, starts between early July and late August, completed early August to late September (RMNH, ZMA), as in nominate *svecica*, *volgae*, and *pallidogularis* (Witherby *et al.* 1938*b*; Dementiev and Gladkov 1954*b*). FIRST PRE-BREEDING. Like adult pre-breeding, but sometimes less extensive in ♂ (occasionally some old feathers in cheeks or central throat retained); in ♀, often restricted to a few feathers of cheeks and throat or no moult at all.

Measurements. ADULT, FIRST ADULT. Nominate *svecica*. Wing and tail of (1) breeding birds from northern Sweden, (2) migrants Helgoland (West Germany) and Netherlands; other measurements combined for these; skins (RMNH, ZMA). Bill (S) to skull, bill (N) to distal corner of nostril; exposed culmen on average 4·3 less than bill (S).

WING (1) ♂ 76·6 (1·47; 11) 74-79 ♀ 73·8 (0·84; 5) 73-75
(2) 78·0 (1·64; 16) 76-81 75·9 (1·07; 6) 74-77

TAIL (1)	52·8 (1·35; 11)	51–55	51·7 (1·15; 5)	50–53
(2)	53·9 (2·12; 16)	51–57	53·0 (0·84; 6)	52–54
BILL (S)	15·9 (0·46; 28)	15·2–16·8	15·6 (0·54; 10)	15·1–16·1
BILL (N)	8·8 (0·37; 25)	8·2–9·5	8·5 (0·51; 10)	8·0–9·1
TARSUS	27·5 (0·80; 27)	26·5–29·2	26·6 (0·55; 11)	25·8–27·3

Sex differences significant for wing and tarsus. Migrant sample virtually all 1st adult birds; wing of older birds on average 0·8 longer, tail similar.

L. s. cyanecula. Netherlands, March–September; skins (RMNH, ZMA). Exposed culmen on average 3·9 less than bill (S). Juvenile wing and tail include those of 1st adult.

WING AD	♂ 75·4 (1·41; 17)	73–78	♀ 71·6 (1·38; 10)	70–73
JUV	74·2 (1·43; 34)	71–77	70·7 (1·24; 12)	69–72
TAIL AD	54·0 (1·78; 15)	51–57	50·2 (2·14; 10)	48–53
JUV	53·0 (1·86; 24)	50–57	49·7 (1·60; 12)	47–52
BILL (S)	15·7 (0·60; 45)	14·7–16·9	15·3 (0·52; 20)	14·5–16·1
BILL (N)	8·6 (0·46; 34)	7·8–9·4	8·3 (0·38; 20)	7·7–8·9
TARSUS	26·6 (0·92; 49)	24·6–28·2	26·3 (0·76; 22)	24·6–27·4

Sex differences significant, except tarsus. Spain (mainly Sierra de Gredos), April–June, ♂♂: wing 78·4 (1·87; 8) 77–82; bill (S) 16·8 (0·33; 8) 16·4–17·4; tarsus 27·3 (0·73; 8) 25·9–28·3 (BMNH).

L. s. namnetum. Western France: wing, ♂ 67–72 (29), ♀ 64–68 (6); tail, ♂ 46–54 (29), ♀ 45–51 (6) (Hartert and Steinbacher 1932–8).

L. s. volgae. Ukraine (USSR), May–July, ♂♂: wing 74·2 (1·11; 13) 72–76, bill (S) 15·9 (0·46; 11) 15·2–16·6 (RMNH, ZFMK, ZMA).

L. s. pallidogularis. Volga steppes (USSR) to foot of Altai mountains, summer (RMNH, ZMA), and spring migrants Middle East (RMNH) and Afghanistan (Paludan 1959).

WING	♂ 72·2 (1·30; 16)	70–75	♀ 69·2 (0·96; 4)	68–70
TAIL	54·1 (2·53; 7)	52–57	50·1 (0·63; 4)	49–51
BILL (S)	15·7 (0·88; 11)	14·8–16·6	15·5 (0·61; 4)	14·7–16·2
TARSUS	26·9 (0·56; 7)	26·3–27·7	25·8 (0·35; 4)	25·4–26·2

Sex differences significant for tail and tarsus.
Live birds, Lake Chany (south-west Siberia), August–September: wing, ♂ 70·7 (2·35; 128) 67–79, ♀ 68·2 (2·53; 59) 62–76 (Havlín and Jurlov 1977).

L. s. magna. Eastern Turkey and Iran, summer, and Kuwait and Iraq, on migration March–September (Kumerloeve 1968; BMNH).

WING	♂ 81·9 (1·32; 17)	79–84	♀ 78·5 (2·14; 6)	76–81
BILL (S)	17·1 (0·62; 14)	16·4–18·3	17·1 (0·84; 5)	16·4–18·0
TARSUS	28·0 (0·60; 13)	27·2–29·0	27·5 (0·89; 5)	26·6–28·5

Sex differences significant for wing.

Weights. Nominate *svecica*. Yamal peninsula (USSR), summer: ♂ 18·2 (108) 14·8–25·5, ♀ 18·3 (34) 14·7–21·6 (Danilov *et al.* 1984). On migration, West Germany: ♂ 20·3 (14) 17–22, ♀ 18·8 (6) 17–22 (Weigold 1926). On migration, Netherlands: September, ♂ 18·4 (4) 17–19, ♀ 18; April–May: ♂♂ 19, 20; ♀ 17 (RMNH, ZMA). Average of juveniles, northern Sweden, 31 July–25 August: ♂ 17·1 (50), ♀ 16·6 (21); fat accumulation in southern Sweden in September (Lindström *et al.* 1985, which see for details). For weight increase of nestlings, see Peiponen (1962) and Arheimer (1982). See also Stolt and Mascher (1962).

L. s. cyanecula. Portugal: late August and first half of September, ♂ 16·0 (1·64; 24) 13–20, ♀ 15·3 (1·57; 21) 13–18; second half of September, ♂ 16·9 (2·31; 8) 13–19, ♀♀ 16·1, 20·0; October, ♂ 17·1 (1·58; 4) 16–19 (C J Mead). Baden-Württemberg (West Germany), on migration late March–April: average of ♂ 19·3 (27), ♀ 18·6 (14); range 14·5–23 (Ullrich 1972, which see for variations throughout day). Naples area

(Italy), October–March: ♂ 17·9 (1·19; 9) 16·8–19·5, ♀ 16·5 (1·34; 9) 14·2–18·5 (G L Lövei and S Scebba). South-east Morocco, April: 13·6 (0·83; 12) 12–15 (BTO).

L. s. namnetum or probably that race. Western France, late July and August: 14·7 (1·43; 7) 13–17 (Marion 1977).

L. s. volgae. Ukraine (USSR), April–July: ♂ 19·7 (3) 19–21 (ZFMK).

L. s. pallidogularis. Lake Chany (south-west Siberia), August–September: ♂ 16·1 (1·06; 132) 14–19, ♀ 15·6 (1·19; 64) 13–20, juvenile 15·9 (1·19; 23) 14–20 (Havlín and Jurlov 1977). South-west Afghanistan, migrants late March: ♂ 16·2 (1·64; 5) 14–18 (Paludan 1959).

L. s. magna. Eastern Turkey, May: ♂ 20, ♀ 22 (Kumerloeve 1968). Probably this race, Kuwait, March–April: 17·1 (1·45; 9) 16–20 (V A D Sales; BTO).

L. s. tianschanica. Central Afghanistan, migrants September–October: ♂ 16·0 (11) 15–17, ♀ 14·6 (5) 14–16 (Paludan 1959).

Race undetermined. Nominate *svecica* or *cyanecula*. Malta: autumn 19·3 (2·7; 25) 14–24; spring 18·0 (2·3; 17) 14–23 (J Sultana and C Gauci). North Nigeria, December: 15·5, 15·5 (Fry 1970b). *L. s. volgae* or *pallidogularis.* Cyprus, October: 16·9 (12) 14–18 (Hallchurch 1981). *L. s. pallidogularis, tianschanica*, and possibly some migrant nominate *svecica*, combined, Kazakhstan (USSR): April–June, ♂ 16·0 (1·39; 24) 13–18, ♀ 18·0 (0·71; 4) 17–19; July–September, ♂ 15·9 (0·93; 23) 14–18, ♀ 16·7 (1·96; 11) 14–21; October–November, ♂ 17·8 (2·00; 4) 15–19, ♀ 17·6 (3) 16–19 (Dolgushin *et al.* 1970).

Structure. Wing rather short, broad at base, tip bluntly pointed. 10 primaries: p8 longest, p9 3–8 shorter, p7 0–1·5, p6 1–3, p5 3·5–6, p4 7–10, p1 12–16; no significant difference between *cyanecula* and nominate *svecica*. P10 reduced, narrow; 33–42 shorter than p8, 2–7 longer than longest upper primary covert. Outer web of (p5–)p6–p8 and inner of (p6–)p7–p9 emarginated. Tertials short; longest about equal to secondaries. Tail rather short, tip slightly rounded; 12 feathers, t6 3–6 shorter than t1. Bill rather short, straight, sharp, and slender; rather fine at base, laterally compressed at tip. Nostril small, oval; partly covered by narrow membrane above and by frontal feathering at base. Some fine hairs along base of upper mandible. Tarsus and toes long and slender. Middle toe with claw 18·4 (27) 17·5–19·5; outer toe with claw *c.* 70% of middle with claw, inner *c.* 65%, hind *c.* 78%. Claws rather long, slender, and sharp, decurved.

Geographical variation. Marked and complex; much individual variation locally, especially in apparent zones of secondary intergradation, such as central European USSR where *cyanecula* meets *pallidogularis*; also, clinal gradation from *pallidogularis* towards north and north-east into nominate *svecica*, towards south into *tianschanica*, and towards south-east into pale Mongolian races; large *magna* from Caucasus area, eastern Turkey, and Iran and small but long-billed race from western Himalayas are more isolated and distinct, though latter shows zone of apparently secondary intergradation into *tianschanica* in Kashmir and western Kun Lun mountains. *L. s. cyanecula* from Belgium and eastern France east to Carpathians and approximately to Smolensk, Novgorod, and Leningrad in western USSR is rather large, and ♂ has rounded silky-white spot or bar on lower throat in breeding plumage, only rarely absent (see Plumages). *L. s. namnetum* from western France similar, but smaller in size (see Measurements). Birds from mountains of northern and central Spain (Witherby 1928; Ern 1966; Schmidt 1970) usually included in *cyanecula*, but in fact intermediate between this race and similarly isolated southern mountain race *magna* from Caucasus area or even nearer latter: large in size (see Measurements) and throat-spot often absent; of 7 ♂♂ examined, 2 had small white bar like *cyanecula*, 3 had limited amount of white concealed under blue, and 2 had no white at all (BMNH); of 17 birds seen in the field, 12 fully blue, 3 had white spot, and 2 a large reddish-buff spot (Corley Smith and Bernis 1956). *L. s. magna* from Caucasus and eastern Turkey to Iran large (see Measurements), throat of breeding ♂ entirely blue or with some concealed white, occasionally with small white spot; ♀ and non-breeding plumages inseparable from *cyanecula*, except for size. Birds east from Ukraine, Smolensk, Novgorod, and Leningrad through plains of USSR south of *c.* 60°N to Tien Shan, Mongolia, and west of Lake Baykal are generally smaller than populations further north and west; many races described from this area, differing only slightly in general colour, and situation not fully established (for contrasting opinions see, e.g., Kozlova 1945, Vaurie 1959, Stepanyan 1978*a*); of these southern birds, *pallidogularis* (synonyms *saturatior*, *altaica*) is a well-marked race, occurring from Volga and Turkmeniya east to Altai and upper Yenisey; breeding ♂ has lower cheeks, chin, and throat glossy pale cerulean-blue (much paler than in *cyanecula*) and spot on lower throat rufous-cinnamon, shaped as a large broad bar or broadly triangular spot; rufous breast-band rather narrow and pale; upperparts, sides of breast, and flanks paler brownish-grey, less drab-brown; ♀ and non-breeding plumages paler on upperparts and with narrower black malar stripe and narrower black band across upper chest. Area from Ukraine and central European USSR east to Volga, between ranges of *pallidogularis* and *cyanecula*, inhabited by highly variable *volgae*: throat of breeding ♂ deep blue and spot on lower throat small, like *cyanecula*, but spot rufous with narrow white border all round or on lower edge only, more rarely fully white or rufous (spot paler rufous and smaller than in nominate *svecica*); rufous breast-band rather narrow; ♀ and non-breeding plumages inseparable from *cyanecula*. Northern Eurasia and Alaska inhabited by nominate *svecica*, which intergrades with *cyanecula*, *volgae*, and *pallidogularis* within a zone from Leningrad eastwards along *c.* 60°N. Nominate *svecica* (synonyms *gaetkei*, *robusta*) characterized by large and rather triangular deep rufous throat-spot in breeding ♂ (not a narrower bar or more rounded spot as in most southern races); no constant difference from *cyanecula* in colour of upperparts or depth of blue on throat. ♀ and non-breeding plumages generally indistinguishable from *cyanecula* except sometimes when series of skins compared: malar stripe of ♀ *cyanecula* averages heavier and black band across upper chest broader than in ♀ nominate *svecica* (throat-spot either white or rufous in both races); adult ♂ non-breeding *cyanecula* has lower throat usually silky-white with narrow rufous feather-tips, nominate *svecica* more cream-white with slightly broader rufous tips. Isolated populations of red-spotted birds occur in mountains of central Europe within range of *cyanecula* (Krösche 1979; Vit 1979; Wartmann 1980; Müller 1982); these apparently glacial relics of nominate *svecica*. CSR

Luscinia cyane Siberian Blue Robin

PLATES 45 and 47
[between pages 544 and 545, and facing page 652]

Du. Blauwe Nachtegaal Fr. Rossignol bleu du Japon Ge. Blaunachtigall
Ru. Синий соловей Sp. Coliazul siberiano Sw. Blånäktergal

Motacilla Cyane Pallas, 1776

Polytypic. Nominate *cyane* (Pallas, 1776), southern Siberia from Altai to Sea of Okhotsk and (perhaps this race) eastern Mongolia, northern China, and Korea. Extralimital: *bochaiensis* (Shulpin, 1928), middle and lower Amur basin and Sakhalin south to Ussuriland and Japan.

Field characters. 13·5 cm; wing-span 20-21 cm. Close in size to Robin *Erithacus rubecula* and Red-flanked Bluetail *Tarsiger cyanurus* but with 10-15% shorter wings, 25-30% shorter tail, and 10% longer legs. Rather small but compact, robust chat, with relatively large bill and head, rather short wings and tail, and long legs; shape, and habit of running on ground, recall small crake *Porzana*. ♂ intensely coloured, dark blue above, with deep black face-mask, and white below; ♀ and immature dull, olive-brown above, fulvous-brown and white below. Legs pale flesh. Sexes dissimilar; no seasonal variation. Juvenile separable.

ADULT MALE. Upperparts basically slate-blue, brightest on forehead and crown, dullest on flight-feathers, and darkest on tail. Lores, most of cheeks, and area down side of neck (to shoulder) jet-black, merging with bluer rear cheeks and rest of neck but contrasting sharply with silky white throat and foreneck. Sides of chest and flanks washed dull dusky blue, looking shadowed above silky white underbody and vent. Underwing dark grey, mottled black. ADULT FEMALE. Upperparts olive-brown, often with green tone (in bright light) and tinged rufous on wings, upper tail-coverts, and tail. Pale buff eye-ring offset by olive-brown cheeks which contrast with buff-white throat. Lower sides of neck, breast, and flanks tinged olive-buff, and breast mottled and scalloped with pale brown; rest of underparts dull white, lacking silkiness of ♂. Underwing pale buff-brown. Rump often blue, tail usually slightly bluish. JUVENILE. Not studied in the field (see Plumages). FIRST-WINTER MALE. Resembles adult ♀ but often shows patches or tones of blue above, particularly on rump, upper tail-coverts, and tail. At all ages, bill dark black-horn, paler, even pink-white on lower mandible; legs pale flesh.

Adult ♂ unmistakable in west Palearctic but could suggest several other 'blue robins' in India and south-east Asia. ♀ and immature liable to confusion with *T. cyanurus* and several other 'blue robins' in main winter range, and with *T. cyanurus* as vagrant in west Palearctic. Distinction from *T. cyanurus* best based on (1) different form (particularly long legs), (2) pale legs, (3) lack of orange flanks, and (4) lack of strongly blue-toned tail. Flight light, but fast wing-beats yield rather fluttering action, recalling *E. rubecula* before other chats; agile in flight, quickly ducking in and out of cover and sweeping low across gaps in it. Gait remarkably free, incorporating short and long hops, high-stepping run, and frequent pauses. Holds head well up, with wings and tail drooped, as in other chats. Constantly quivers short tail (unlike *T. cyanurus*). Shy and skulking, rarely leaving ground cover and then only briefly. Not gregarious.

Alarm-call a rapid 'chuck-chuck-chuck'; not known if migrants call.

Habitat. Breeds in middle and lower-middle latitudes of continental and oceanic east Palearctic, characteristically in deep taiga of spruce *Picea*, fir *Abies*, birch *Betula*, aspen *Populus*, and other (mainly coniferous) forest trees, with dense shady canopy and fallen trees but no undergrowth, often by riversides and near meadows with tall herbage. Locally also in well-lit oakwoods *Quercus*. Mostly in lowlands and river valleys but also on mountain foothills and near coasts. In Honshu (Japan), occurs from *c.* 700 to *c.* 1800 m (Yamashina 1982). In Borneo in winter, occupies primary and secondary forest, especially along streams, from sea level to *c.* 1800 m (Smythies 1981). Secretive and largely terrestrial. (Dementiev and Gladkov 1954*b*; Ali and Ripley 1973*a*.)

Distribution. Breeds in southern Siberia from Altai east to Amurland and Sakhalin, south to Manchuria, Korea, Japan, and northern China. Winters in south-east China west to Burma, Philippines, Borneo, and Sumatra.

Accidental. Channel Islands: 1 probably 1st-year ♀, Sark, 27 October 1975 (Rountree 1977).

Movements. Migratory. Winters in southern Asia from southern China, Indochina, Thailand, and southern Burma south to Philippines, Borneo, Malaya, and Sumatra, straggling to eastern India (Bengal, Manipur). Passage seems to be essentially through Mongolia and China, i.e. to east of Himalayas and associated mountain systems. Several ringed at Pasoh, Malaya, in winter returned to same locality in subsequent season (Medway and Wells 1976). A nocturnal migrant, subject to casualties at radio towers and lighthouses while crossing populated regions.

Leaves USSR breeding areas late August and September, only a few individuals lingering to end of September (Dementiev and Gladkov 1954*b*); crosses China in September and early October (e.g. Hemmingsen and Guildal 1968), and passes through Malaya mid-September to mid-November (Medway and Wells 1976). Spring passage in Malaya during April, through China early May to early June. Heavy passage at Lake Khanka

(Ussuriland) in second half of May (Vorobiev 1954), and USSR breeding areas reoccupied in early June.   RH

Voice. See Field Characters.

Plumages (nominate *cyane*). ADULT MALE. Upperparts and upper sides of neck entirely deep blue, darker and more plumbeous than (e.g.) cerulean-blue of crown, wing, and tail of Blue Tit *Parus caeruleus*. Lore velvet-black, widening towards front of eye and extending into black stripe from cheek across lower side of neck to side of breast; not sharply divided from blue of upper side of neck, but contrasting markedly with underparts; black often finely mottled blue below eye. Entire underparts silky-white; flanks tinged pale blue-grey, throat and chest sometimes slightly washed buff and with faint dark scaling when plumage fresh, particularly at sides. Tail dull black, central pair of feathers (t1) and outer webs of others extensively tinged blue. Flight-feathers, tertials, and greater upper primary coverts dull black, outer webs extensively tinged blue (both webs on tertials); blue sometimes paler and more greyish than on upperparts, in particular on outer primaries and when feathers worn. Lesser and median upper wing-coverts deep blue, like upperparts; greater coverts dull black or greyish-black with blue wash on outer web and tip. Under wing-coverts and axillaries dark grey with blue tips. *In worn plumage*, upperparts slightly deeper and more glossy blue, but some grey of feather-bases visible when heavily abraded in mid-summer. ADULT FEMALE. Upperparts and sides of neck greyish olive-brown, sometimes slightly brighter olive on forehead, rump often (but not always) contrastingly blue. Rather indistinct supercilium (not extending back from eye) and more distinct eye-ring orange-buff to cream-buff, often finely mottled olive-brown on supercilium. Lores and upper cheeks mottled olive-brown and buff or cream-white, ear-coverts olive-brown with fine pale buff or cream-white streaks. Lower cheek, chin, and throat white, feather-tips washed buff and spotted olive-brown—in particular, lower cheek, side of throat, and lower throat show fine dark spots or bars. Feathers of chest and side of breast white with olive-brown tips and buff wash subterminally; olive-brown virtually contiguous on sides of breast and sometimes across upper chest (forming narrow dark chest-band), but much white and buff visible on central chest, olive-brown tips forming dark scaling. Remainder of underparts white, washed pale olive-brown on flanks and thighs. Tail dull olive-brown, usually slightly tinged blue on outer webs. Flight-feathers, greater upper primary coverts, and tertials dull black, outer webs extensively tinged olive-brown (both webs of tertials); olive-brown often brighter and less greyish than on upperparts. Upper wing-coverts greyish olive-brown, like upperparts; fringes along tips and outer webs of greater coverts often slightly brighter olive-brown. Under wing-coverts and axillaries pale buff-brown with greyish bases. JUVENILE. Upperparts brown, finely streaked buff on forehead and crown, more coarsely spotted buff or rufous-buff on mantle, scapulars, and back; longer feathers narrowly fringed black on tip, upper tail-coverts uniform rufous-buff. Underparts buff or cream-buff, throat, chest, sides of breast, and flanks with ill-defined brown scaling; central belly off-white. Tail as adult, tinged blue in ♂, virtually or completely without blue in ♀. Flight-feathers, tertials and greater upper primary coverts as adult ♀, but outer webs of ♂ often (not always) partly tinged pale grey-blue or slate-blue; tertials and primary coverts sometimes have faint rufous fringe along tip. Lesser and median upper wing-coverts brown with buff spots, like mantle and scapulars; greater upper wing-coverts as adult ♀, but tips narrowly though contrastingly fringed rufous; greater coverts of ♂ often duller grey-brown than in adult ♀ and partly tinged blue on outer webs. FIRST ADULT MALE NON-BREEDING. Strongly variable. Differs from adult ♂ and ♀ in retaining juvenile greater upper wing-coverts (with rufous tips), tertials, and flight-feathers. Upperparts olive-brown with limited amount of blue on tail or with more extensive blue from rump to tail (both like adult ♀), but some birds have upperparts and upperwing olive-brown with much blue tinge on all feather-tips, and a few are as extensively and deeply blue as adult ♂, differing only in rufous fringes on otherwise blue greater upper wing-coverts. Side of head and neck like adult ♀, but side of head sometimes tinged blue and with traces of black down side of neck. Underparts either like adult ♀, or mainly white as adult ♂, though extensively washed buff from chin to chest and cream on flanks, sides of belly, vent, and under tail-coverts; sides of throat and all chest with narrow olive-grey arcs, narrower and less regular than scaling of adult ♀, sometimes showing as dusky mottling only. FIRST ADULT FEMALE NON-BREEDING. Like adult ♀, but juvenile flight-feathers, tail, most greater upper wing-coverts, and often tertials retained (see Juvenile); greater coverts and sometimes tertials have trace of rufous or buff fringes on tips. Upperparts generally olive-brown, sometimes with faint blue tinge on tail-base, rarely some blue on rump and upper tail-coverts. Never blue on upperwing (unlike some 1st adult ♂ non-breeding). FIRST ADULT MALE BREEDING. Like adult ♂, but retains juvenile flight-feathers, tail, usually tertials, and variable number of median and greater upper wing-coverts—usually distinctly duller and browner than blue upperparts and new lesser and tertial coverts; greater coverts have traces of rufous tips. Scattered olive-brown non-breeding feathers sometimes retained on upperparts (particularly back and rump), and some buff feathers with brown fringes sometimes retained on chest and flanks. FIRST ADULT FEMALE BREEDING. Like 1st adult ♀ non-breeding; occasionally, part of non-breeding retained, bleached and worn, and underparts in particular then pale and with limited dark scaling.

Bare parts. ADULT, FIRST ADULT. Iris dark brown. Bill black, bluish-black, or dark horn-brown, base of lower mandible livid white, flesh-grey, pale horn, or brown. Inside upper mandible dark in adult, pink with blue cast in 1st adult. Leg and foot pale flesh-white, pink, greyish-flesh, or flesh-brown. (Hartert 1903–10; Ali and Ripley 1973a; Rountree 1977; Svensson 1984a; RMNH.) JUVENILE. No information.

Moults. ADULT POST-BREEDING. Complete; primaries descendant. Starts with p1 mid- or late July; in early August, moult intense (flight almost impossible), with many tail-feathers sometimes growing simultaneously (Dementiev and Gladkov 1954b). ADULT PRE-BREEDING. Partial. In ♂, involves head, body, lesser and median upper wing-coverts, and tertials. Starts November–December, generally completed mid-January. No ♀♀ in moult examined, but those from March fresh or slightly worn and hence moult probably as in ♂. POST-JUVENILE. Partial: head, body, lesser and median upper wing-coverts, tertial coverts, and perhaps sometimes tertials and tail. In USSR, starts second half of July, completed from mid-August; perhaps rather earlier in Korea and Japan (Dementiev and Gladkov

1954b). FIRST PRE-BREEDING. Partial, but extent rather variable. Starts January–March, completed April. In some, extent as in adult pre-breeding; in others, more restricted, mainly confined to sides of head and neck and chin to breast, with scattered feathers on rest of body. Some spring birds have tail and tertials new, but not known whether these replaced in post-juvenile or in 1st pre-breeding. Juvenile flight-feathers, outer greater upper wing-coverts, and greater upper primary coverts retained.

Measurements. China (on migration) and Malay peninsula and Borneo (winter); skins (BMNH, RMNH, ZMA). Bill (S) to skull, bill (N) to distal corner of nostril; exposed culmen on average 3·8 less than bill (S). Juvenile wing refers to retained juvenile wing of 1st adult.

WING AD	♂ 76·2 (1·98; 17)	74–81	♀ 76·1 (2·46; 7)	73–79
JUV	74·1 (2·52; 23)	71–79	71·9 (1·43; 7)	70–74
TAIL	47·1 (1·29; 15)	46–50	46·2 (2·00; 10)	43–49
BILL (S)	15·6 (0·59; 16)	14·8–16·4	15·2 (0·39; 11)	14·6–15·7
BILL (N)	8·4 (0·37; 16)	7·8–9·0	8·2 (0·26; 11)	7·6–8·7
TARSUS	25·7 (0·86; 15)	24·4–27·2	26·2 (0·72; 12)	25·0–27·4

Sex differences significant for juvenile wing. Juvenile wing significantly shorter than adult wing.

Weights. USSR: ♂♂ 14, 16 (Dementiev and Gladkov 1954b). Kuril islands: ♂ 15·5 (Nechaev 1969). Mongolia, May: ♂ 16 (Piechocki and Bolod 1972). Manchuria (China), May: ♂ 16 (Piechocki 1958). Hopeh, China: ♂ 15·8 (7) 14–18, ♀ 13·5 (4) 11–15 (Cheng 1963). Korea, mid-May: ♀ 20 (Tomek 1984). Malay peninsula: October 13·6 (1·05; 19); winter 14·8 (1·15; 52) (Nisbet 1968). Channel Islands, October: probable ♀ 15·8 (Rountree 1977).

Structure. Wing rather short, broad at base, tip rounded. 10 primaries: p7–p8 longest (p7 rarely 0–0·5 shorter than p8; p8 sometimes 0–2·5 shorter than p7), p9 6–11 shorter, p6 (0–)1–2, p5 5–9, p4 9–12, p1 14–19. P10 reduced, narrow; 37–43 shorter than wing-tip, 2–7 longer than longest upper primary covert. Outer web of p6–p8 and inner of (p6–)p7–p9 emarginated. Tertials very short; longest about equal to secondaries. Tail short, tip slightly rounded; t6 3–5 shorter than t1. Bill rather short, slender; rather wide at base, laterally compressed at tip. Nostrils oval, partly covered by membrane above. Some short fine bristles along base of upper mandible and on chin. Tarsus and toes long and slender. Middle toe with claw 18·4 (10) 17–19·5; outer toe with claw c. 68% of middle, inner c. 66%, hind c. 74%. Claws rather short, decurved.

Geographical variation. Slight. Adult ♂ of *bochaiensis* from middle Amur river east to Sakhalin and south to Japan darker blue on upperparts than nominate *cyane* from further west, less greyish; ♀ darker above (Vaurie 1959); also, wing of *bochaiensis* shorter (in ♂, 67–72; in ♂ nominate *cyane* 70–76: Hartert and Steinbacher 1932–8). Difference very slight, however, and both darker and paler birds occur in Japan (at least during migration), and no relationship between colour and size in those examined (BMNH, RMNH, ZMA); separation of *bochaiensis* thus perhaps not warranted.

Forms superspecies with Indian Blue Robin *L. brunneus* from Himalayas.

CSR

Tarsiger cyanurus **Red-flanked Bluetail**

PLATES 45 and 47
[between pages 544 and 545, and facing page 652]

Du. Blauwstaart   Fr. Rossignol à flancs roux   Ge. Blauschwanz
Ru. Синехвостка   Sp. Coliazul cejiblanco   Sw. Blåstjärt

Motacilla cyanurus Pallas, 1773

Polytypic. Nominate *cyanurus* (Pallas, 1773), northern Eurasia. Extralimital: *rufilatus* (Hodgson, 1845), Afghanistan to north-central China.

Field characters. 14 cm; wing-span 21–24·5 cm. Close in size and structure to Robin *Erithacus rubecula*, though slightly shorter-billed and longer-tailed. Small, fairly compact chat, with general character recalling both *E. rubecula* and redstart *Phoenicurus*; behaviour often suggests flycatcher *Ficedula*. Typical ♂ intensely coloured, dark sheeny blue above, white and grey below, with long splash of orange along flanks. ♀ and dull ♂ olive-brown above, with pale eye-ring and darker, blue-washed rump and tail; dull white below, with brown chest and orange flank-panel. Sexes dissimilar; no seasonal variation. Juvenile separable.

Siberian and Japanese race, nominate *cyanurus*. ADULT MALE. In worn plumage (spring and summer), head including cheeks and side of neck, back, wings, and tail essentially dark blue, with sheen producing paler, sky-blue tones and 'flashes' on forehead, smaller wing-coverts, rump, and on fringes of flight-feathers. Lore almost black; supercilium white, running from bill to mid-point of ear-coverts and becoming pale blue behind eye. Chin and throat cream-white, contrasting with dusky-blue sides of chest and fore-flanks; middle and rear flanks bright orange (variably visible, as in Redwing *T. iliacus*). Rest of underbody cream-white (without dusky or grey suffusion of south Asian race, imported to Europe as cagebird). Underwing buff. In fresh plumage (from August or September), blue feathers of upperparts fringed olive-brown (especially on head) and sides of breast mottled buff. Dull morph ♂♂ resemble ♀, but variable. ADULT FEMALE. Head, back, and wings dark olive-brown (with greener wash in some lights); lore pale buff, abutting distinct buff-white eye-ring but not forming supercilium. Rump, upper tail-coverts, and tail dull olive- to grey-blue, with brighter pale blue sheen strongest on fringes of outer tail-feathers. Underparts well-patterned with large, pale cream throat contrasting with olive-brown cheeks, sides of neck, and joined breast-patches; fore-flanks olive grey-brown and rest of flanks orange, tinged olive. JUVENILE. Basic pattern and colour as adult ♀ but liberally spotted with cream and scalloped dark brown on head and back and mottled dark on chest. Flanks patchy, with cream-white spots and dull brown tips hiding orange panel. FIRST YEAR MALE. May resemble adult ♀ or assume intermediate plumage, with patches of blue above and more cream-white below. Bill black-horn; legs red-brown.

Unmistakable given view of orange panel on flanks, but ♂ from behind may suggest ♂ Siberian Blue Robin *Luscinia cyane* and ♀ and immature can recall *E. rubecula* and ♀ *L. cyane*. Despite earlier belief, contrast of white throat with cheeks, side of neck, and/or chest not a prime character, for this and eye-ring also shown by *L. cyane*. Important therefore to recognize that *T. cyanurus* combines shape of foreparts, gait, and perching behaviour of *E. rubecula* with shape of rear parts and flight of *Phoenicurus*, pouncing hunt of *Saxicola* chat, and tail movement and fly-catching of *Ficedula*, being often high in trees and never as persistently on ground as *L. cyane*. Flight free and light, with fluent wing-beats and rather loose-tailed appearance. Gait essentially hopping, whether along branches, in foliage, or over ground. Carriage generally upright, like *E. rubecula* and *Phoenicurus*, but noticeably level and mouse-like when searching coniferous foliage for food. Constantly flicks wings and tail open, like *Saxicola*. Although shy in breeding habitat, ♂ perches openly on tree-top when singing; becomes less wild in winter, perching on posts or wires near dense cover.

Song (in southern Siberia) consists of rather short phrases but is clear and far-carrying, suggesting a pure-voiced thrush *Turdus* as much as a chat: 'whew-wee-whew-wee-wee-wellu-it', with slight dip in first 4–5 notes and distinct fall at end. Commonest call of breeding birds 'weep' or 'peep', often repeated. Call from vagrants a short 'teck-teck', recalling *E. rubecula*.

Habitat. Breeds in upper-middle and marginally in upper continental latitudes, exclusively boreal and montane, in thick mossy conifer forest, especially taiga, on moist soil, generally with undergrowth, and with July temperatures of 15–24°C (Voous 1960). Also mixed forest with birch *Betula* and rhododendron. In Far East, more often in birchwoods, even up to 3000 m in Japan, where trees no more than 2–3 m high (Dementiev and Gladkov 1954b). Widely separated population in Himalayas inhabits forests of spruce *Picea*, pine *Pinus*, and birch with little undergrowth, or dense rhododendrons in open forest, but avoids pure broad-leaved stands and scrub above treeline, although breeding freely up to 4000 m (Ali and Ripley 1973a). Less of a ground bird than relatives such as Thrush Nightingale *Luscinia luscinia*, not only singing but foraging freely in trees, and also flycatching. Sometimes found in gardens and orchards. Autumn straggler in Shetland (Scotland) kept flitting about low from pool

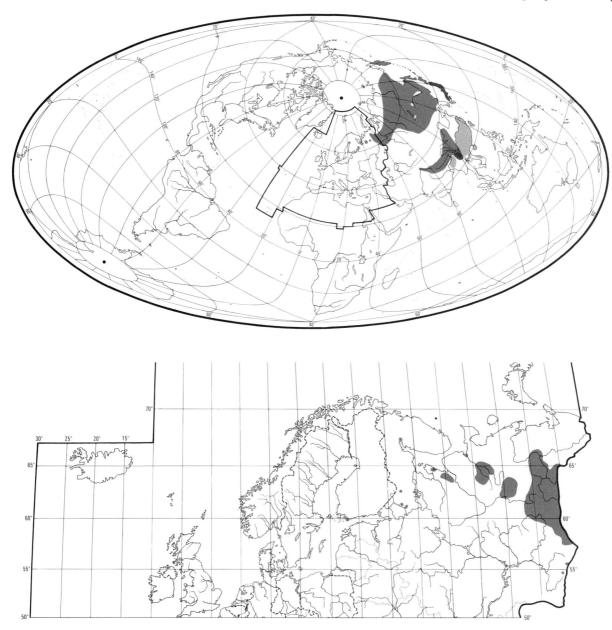

to pool catching insects on barren hills, but such habitat choice probably atypical and compelled by circumstances (Bannerman 1954). While mainly associated with moist forests appears less attached to watercourses, ponds, and swampy terrain than (e.g.) *L. luscinia*.

Distribution. Westward expansion into Finland apparently halted, but has spread further west into Estonia 1980.

FINLAND. First recorded 1949 then apparently spread west, with most reports from Kuopio, Kuhmo, and Kuusamo (Merikallio 1958). Westward expansion now apparently stopped, and only a few ♂♂ recorded in recent years, not annual (OH). USSR. Bred Estonia 1980 (Rootsmäe 1981).

Accidental. Britain, Channel Islands, Netherlands, West Germany, Denmark, Norway, Sweden, East Germany, Czechoslovakia, Italy, Lebanon, Syria, Cyprus.

Population. FINLAND. Estimated 500 pairs (Merikallio 1958); this figure probably too high as only 1 nest found (OH); see also Distribution. USSR. Common on taiga; more frequent in Arkhangel'sk region since 1938 (Dementiev and Gladkov 1954*b*).

Movements. Northern populations are long-distance

migrants (see below). Southern race *rufilatus* (breeding Himalayan region and western China) mainly shows short-distance altitudinal movements, but some birds reach Assam, Burma, and northern Thailand (Ali and Ripley 1973a); not considered further here.

Japanese population of northern race, nominate *cyanurus* (breeding USSR, Mongolia, Manchuria, and Japan), is resident except in northern island of Hokkaido (Sonobe 1982); other populations migrate through China, Korea, and Japan to winter in southern China, Taiwan, southern Korea, Indochina, Thailand, and Burma. Hence USSR birds, especially those from European parts, must make long easterly movements (in autumn) within USSR, passing north of major central Asian mountain systems, before turning south through Mongolia and China (Helminen 1958). Several autumn migration records, late September and early October, from Kazakhstan (Dolgushin *et al.* 1970), but birds do not cross Turkmeniya, and sole record of nominate race in Indian subcontinent was from Cachar in Assam (Ali and Ripley 1973a).

Autumn migration in USSR begins early September; northern edge of range deserted by mid-September, though some linger in southern Siberia into early October (occasionally later). Crosses Mongolia and China during October. Return passage begins April, vanguard reaching southern Siberia (e.g. Lake Khanka) in second half of April. Spreads north and west during May, reaching Arkhangel'sk region around 20 May-4 June. (Kozlova 1930; Dementiev and Gladkov 1954b; Vorobiev 1954, 1963.) RH

Food. Insects; also fruits and seeds outside breeding season. Feeds in low trees, shrubs, and on ground. Catches insects by hopping about on ground, by perching and flying down to take items located, and by brief aerial-pursuit like flycatcher (Muscicapidae) (Dementiev and Gladkov 1954b; Ali and Ripley 1973a). Several birds seen tearing off growing fruits of spindle *Euonymus* like flycatcher (Omel'ko 1979).

In Pechora area (north European USSR), presumably in summer, 11 stomachs contained mostly adult beetles (Coleoptera), also beetle larvae, caterpillars, and a spider (Araneae) (Dementiev and Gladkov 1954b). Said by Grote (1934) to eat berries in late summer, and in autumn in far-eastern USSR recorded eating fruits of buckthorns, *Aralia*, and *Euonymus* (Dementiev and Gladkov 1954b; Omel'ko 1979). In eastern China (presumably in winter), stomachs contained small beetles (Carabidae), ground bugs (Lygaeidae), and flies (Diptera); often also berries in autumn (Kolthoff 1932). In Himalayas, south Asian race eats insects, including caterpillars (Ali and Ripley 1973a); 2 stomachs from eastern Himalayas contained ants (Formicidae), wasps (Chrysididae), and *Polygonum* seeds (Morioka and Sakane 1981). Wild-caught captive

bird took adult Lepidoptera, though apparently preferring small ones (Bezzel and Löhrl 1972). DJB

Social pattern and behaviour. Most aspects poorly known.

1. Solitary in winter (Yamashina 1982); in Japan, both sexes then defend territory against conspecific birds (Jahn 1942), though no information on territory size nor is it known how long territory held. In Nepal, usually recorded singly or in (loose) pairs (Fleming and Traylor 1964; Fleming *et al.* 1976). Mostly solitary also on spring migration (e.g. Hemmingsen and Guildal 1968, Panov 1973); in Nepal, mid-June, birds apparently newly arrived included flock of 4-5 (Diesselhorst 1968a). In USSR, after breeding, may move about in loose feeding flocks (Grote 1934). Density of 8 birds per km² recorded in dense conifers along rivers of western Siberia in late August (Vartapetov 1984). In Ussuriland (eastern USSR), usually solitary also for autumn migration, sometimes in twos or rarely in loose-knit flocks of 3-5 (Panov 1973). Larger autumn flocks (15-20) reported from Yenisey area of Siberia (Grote 1925). BONDS. Nothing to indicate mating system other than monogamous, though no details—nor on duration of pair-bond. Young fed by both sexes (Diesselhorst 1968a); at one nest in Finland, more by ♀ (Skoog 1981). Tended for long time after leaving nest (Jahn 1942), and apparent family recorded in Sakhalin (eastern USSR) as late as mid-October (Munsterhjelm 1922). ♂♂ breed at 1 year old (Bergman 1935; Dementiev and Gladkov 1954b); not known whether same true of ♀♀. BREEDING DISPERSION. Solitary and territorial (Dementiev and Gladkov 1954b); in Nepal however, only rarely 2 pairs so close that territories contiguous (Diesselhorst 1968a). In Finland, 3 singing ♂♂ recorded *c.* 300 m apart (Helminen 1958). At Kuusamo (Finland), ♂'s song-post *c.* 300-400 m from nest (Mikkola 1973; Skoog 1973); not known, however, if this gives even an approximate idea of territory size, though one ♂ reacted aggressively to playback of song *c.* 250 m from nest (Skoog 1981); see also part 2 (below). Movements of singing ♂♂ in Nepal suggested 'fairly large' territory; in one case, at least 50 × 50 m (Diesselhorst 1968a). Territory defended by ♂ (Portenko 1954). Feeding may be done quite a long way from nest (Vorobiev 1963). Information on densities restricted, especially from small west Palearctic range. In Pechora area of north-east European USSR numbers fluctuate: in year when many in most favourable habitat along river valley, calculated density 120 ♂♂ along 100 km; far fewer in less favourable habitat (Dementiev and Gladkov 1954b); in middle Pechora basin, Estafiev (1981) recorded only 0·4 birds per km², though 7 birds per km² in adjacent northern Urals. In pine *Pinus* forest with clearings in western Siberia, 1-2 birds per km²; average over summer in southern taiga of Ob' valley 3 birds per km² (Vartapetov 1984). In Siberia, fir *Abies* forest with glades may hold 6-20 birds per km²; densities generally lower in other habitats of western and central Siberia, though data for mixed woods along upper Lena river in Reymers (1966) indicate *c.* 48 birds per km² (Ravkin 1984). For Yenisey taiga, see Rogacheva (1962). In north-east Altai (south-central USSR), mainly in pure *Pinus* and birch *Betula* (7-8 birds per km²); fewer in sparse woods and most dense taiga of middle altitudes (see Ravkin 1973). Highest density in Sikhote-Alin' mountains (eastern USSR) in *Betula ermani*—116 birds per km²; 2-3·5 birds per km² in other habitats (Kuleshova 1976). Not especially numerous in Kamchatka (eastern USSR): e.g. 7 singing ♂♂ along 27 km (Dementiev and Gladkov 1954b). On Kunashir (southern Kuril Islands, eastern USSR), 5-6 ♂♂ recorded along 1 km in woods of *Picea* and *Abies* Nechaev 1969). In

open montane *Abies* woods of Khumbu (Nepal) not much above 1 pair per 3–4 ha and often less, although quietness of birds early in season may cause average density to be considerably underestimated (Diesselhorst 1968a). ROOSTING. No information on roosting habits. For nocturnal song, see part 2. Water-bathing recorded (Kozlova 1930).

2. Some variation (in part regional or seasonal) in reports of relative shyness. On breeding grounds can be shy and secretive, ♀ especially so (e.g. Sushkin 1938, Skoog 1973). In Finland, singing ♂ may retreat when approached; one was difficult to approach to within *c.* 50 m and constantly moved on, often up to *c.* 100 m; always flew into canopy and re-emerged only near new landing place (Sovinen 1952); typically flies close to ground between perches (Skoog 1981). Also in Finland, Skoog (1981) noted that shyness varies individually and same bird may show different degrees of shyness on different occasions. For further comments regarding restless, shy, and skulking habits, see (e.g.) Munsterhjelm (1922), Grote (1934), and Ali and Ripley (1973a). In Nepal, breeding birds very quiet prior to hatching (Diesselhorst 1968a). On Kuril Islands, not shy and easily approached (Dresser 1871–81; Bergman 1935). More confiding on migration than in summer (Portenko 1954; Vorobiev 1954). Tame in winter (Yamashina 1982). In Nepal, November–December, not especially shy, but generally keeps to deep forest undergrowth or thick scrub at forest edges (Fairbank 1980); in China, birds also tend to stay in dense thickets in autumn, but more in open in spring (Kolthoff 1932). Vagrant in Shetland shy (Bruce 1948); another on Mellum island (West Germany) hard to flush but when perched openly allowed approach to *c.* 6 m (Dohle *et al.* 1957). See also Alerstam (1974) and Svensson (1974) for Sweden. Constantly flicks tail open and shut (Whistler 1926a; Ali and Ripley 1973a), or first spreads it then twitches it up at intervals of 3–4 s (Fleming *et al.* 1976), or spreads it slightly after lowering it (Panov 1973). Also twitches wings constantly (as flight-intention movement: Bergmann and Helb 1982) and this not infrequently combined with a bobbing or curtseying movement of whole body, sometimes accompanied by call 2b (Bezzel and Löhrl 1972; Ali and Ripley 1973a; Panov 1973). FLOCK BEHAVIOUR. In Nepal, mid-June, small flock moved about in open, birds calling constantly (Diesselhorst 1968a: see 2d in Voice). SONG-DISPLAY. ♂ sings (see 1 in Voice) usually from tree-top or other prominent perch, sometimes lower down and half in cover (Grote 1934; Portenko 1954; Johansen 1955; Mikkola 1973; Bergmann and Helb 1982; Yamashina 1982). In Nepal, singing ♂♂ restless, not staying long on one perch, but moving about through crowns of low trees (Diesselhorst 1968a); perhaps due at least in part to disturbance. Singing bird adopts a rather erect posture, back curved, tail (which may be quivered) pointed straight down, throat feathers ruffled (Sovinen 1952). In Finland, sings mainly in early morning and late evening; twice noted around midday (see Mikkola 1973 for details). Will also sing at night: at Kuopio (Finland), mainly 23.00–01.00 hrs; for further details of evening and morning song, see Sovinen (1952). In Tibet (Ludlow and Kinnear 1944) and USSR (Dementiev and Gladkov 1954b) also sings in morning and around sunset (see also Grote 1934 and Sushkin 1938). In Japan, starts at crack of dawn and continues through day and (later than other species) to late dusk (Jahn 1942); will sing in rain, even in stormy weather (Yamashina 1982). Song-period rather long. In Finland, 12 May–20 July (Mikkola 1973); see also Skoog (1973). In Ussuriland, song noted during peak passage in late April (Panov 1973). In European USSR (Dementiev and Gladkov 1954b) and Siberia (Grote 1934; Johansen 1955), sings from arrival on breeding grounds—song initially subdued or

full and loud (Reymers 1966)—to mid- or late July, though in northern Lake Baykal area (eastern USSR), ♂♂ had stopped by end of June/early July, when young still in nest (Stegmann 1936), but continued to mid-August on Kuril Islands (Nechaev 1969). In Nepal, song wanes very rapidly after start of incubation; some ♂♂ singing very persistently mid-June are probably unpaired; very little song after *c.* 20 June (Diesselhorst 1968a). Will sing also in autumn (e.g. in Pechora area, 21 September) but song usually quiet and incomplete (Dementiev and Gladkov 1954b), though full song noted as late as 9 October in Ussuriland (Panov 1973). ANTAGONISTIC BEHAVIOUR. In Japan, birds move on to breeding grounds and establish territories (by song) from March (Jahn 1942). On Kuril Islands, pairs recorded on territory and frequently driving off conspecific birds 1 week after arrival at end of April (Bergman 1935). In Kangra (northern India), birds not pugnacious, ♂♂ meeting in same territory without fighting (Whistler 1926a). Possibly antagonistic chases by 2 birds in Finland accompanied by call 2c (Skoog 1981). In an apparently aggressive posture elicited by playback of song in territory, ♂ had wings slightly drooped, tail raised at *c.* 35° and slightly spread, orange flank feathers ruffled (see illustration in Skoog 1981). Call 2b may also be given by ♂ in response to playback of song (Skoog 1973), ♂ moving (sometimes hopping) around tape-recorder on ground. Also noted flying close and giving call 2d in similar situation (Skoog 1981). HETEROSEXUAL BEHAVIOUR. At Beijing (China), ♂♂ apparently arrive *c.* 2 weeks ahead of ♀♀ (Hemmingsen and Guildal 1968). Near Krasnoyarsk (USSR), pair-formation in one year around 12–13 May (Dementiev and Gladkov 1954b); apparently quite long gap between arrival and start of breeding in Altai (Sushkin 1938). For Kuril Islands, see Antagonistic Behaviour (above). ♂ will chase ♀ and call (Nechaev 1969: see 2c in Voice) and apparently also feeds her on nest (Dementiev and Gladkov 1954b; Ali and Ripley 1973a). RELATIONS WITHIN FAMILY GROUP. No information on nestling phase. Young out of nest spend most time on ground (Bergman 1935; Dementiev and Gladkov 1954b); may indicate they leave before able to fly. In Altai, brood of full-grown (and moulting) young dispersed, but still fed and guarded by adults (Sushkin 1938). ANTI-PREDATOR RESPONSES OF YOUNG. Tried to escape when handled perhaps *c.* 5–6 days before would normally have left nest. Large nestlings at another site started to leave as observer approached (Vorobiev 1963). Show little fear of man when out of nest and on ground (Bergman 1935). PARENTAL ANTI-PREDATOR STRATEGIES. (1) Passive measures. ♂ recorded freezing on perch, bill pointed up and remaining thus *c.* 3–4 m from man searching for nest (Ali and Ripley 1973a). ♀ generally a tight sitter (Dresser 1871–81). Birds may approach closely, but tend not to visit nest while observer near according to Whistler (1926a); ♀ of pair feeding young more skulking than ♂ (Skoog 1973). (2) Active measures: all information relates to man as potential predator. In Nepal, change in adults' behaviour coincided with hatching: both gave loud warning-calls (see 2a in Voice) as soon as man entered nesting area (Diesselhorst 1968a). Such calls may increase in volume as man gets closer to nest and other bird species thus attracted will join in mobbing (Whistler 1926a). One ♂ in Finland often gave persistent warning-calls (see 2a–b in Voice) from regular perch (Skoog 1973); report in Dresser (1871–81) that ♂ will come close in attempt (usually successful, but if not flies off) to distract man near nest where ♀ sitting perhaps no more than this type of vocal demonstration. Birds will also perch prominently, fly or hop about nearby when nest or young threatened and (when with young) make apparent attack-flights (see Davidson 1898, Sushkin 1938, Dementiev and Gladkov

1954*b*, and Vorobiev 1963). One ♀ apparently performed distraction-lure display on leaving nest containing young (Sushkin 1938), but no details.   MGW

Voice. In north-east China, migrants rather quiet (Hemmingsen and Guildal 1968). For extra sonagrams, see Bezzel and Löhrl (1972), Mikkola (1973), and Bergmann and Helb (1982).

CALLS OF ADULTS. (1) Song of ♂. Short (*c.* 1·5–2 s, or less; see Figs I–III) but distinctive and often loud phrases audible up to several hundred metres (*c.* 500 m: Skoog 1981) and of several different types, though true extent of individual and/or geographical variation yet to be assessed (see further below). A bird will sing several different song-types but sometimes gives one repeatedly with short pauses over long period: e.g. 5 sonagrams from song of Finnish bird showed no variation. Captive bird (presumed nominate *cyanurus*) would sing one type (especially in morning during full-intensity period) then switch to other (normally sung during day) for similar period. High whistled sounds may have rather melancholy quality and sometimes given in cadences (descending pitch) like Red-breasted Flycatcher *Ficedula parva* or Redwing *Turdus iliacus*. In Finland (nominate *cyanurus*), 2 distinctly different song-types. One type clear and melodious and, particularly in loudest shrill (silvery) sounds in middle, resembling thrush *Turdus*; quietest at start, slightly louder at end where like Wood Sandpiper *Tringa glareola*. Rendered variously as follows: 'itrU-tU-tititit', 'tetee-teeleee-tetete', 'titi TIILTYYL-TIILTYYL titi', 'titi TRILIYY-TILIYY lilili' or 'tritri TRYYTYY-TRIITYY tritritri'. (Sovinen 1952; Bezzel and Löhrl 1972; Mikkola 1973; Skoog 1981; Bergmann and Helb 1982.) Other song-type vibrant and with rolling final notes: 'itrU-tU-ri-hrrrr' (Skoog 1981); presumably same type also described as stereotyped and repeated 'joji joji yrrr' or 'tru-tri tru-tri trrr' (Svensson 1974); see also discussion of captive bird (above). Songs given in fairly rapid sequence; when singing most actively, pause between phrases 4–7(–10) s (Sovinen 1952; Bezzel and Löhrl 1972; Mikkola 1973). Recording reveals song reminiscent in tonal quality and pitch of Mistle Thrush *T. viscivorus*, though more repetitive, with little variation, only last note occasionally omitted. Sweet throaty warbling, clear and ringing: 'te tloo tlee tee-ti-ti' (J Hall-Craggs, P J Sellar: Fig I). In recording of nominate *cyanurus* from Mongolia (Schubert 1982), rich but repetitive whistles of 2 ♂♂ in apparent bout of counter-singing recall *T. iliacus*, roughly 'whee whee whee whee whee whee-oo' (M G Wilson). In another recording from Finland (Fig II), song thinner, much less throaty and, after sweet beginning suggestive of *Galerida* lark, final flourish is slow rattle, recalling Redstart *Phoenicurus phoenicurus*: after inaudible 1st note, 'si-swoo swee swep-sip-sip-sip-sip'; compare also Fig III which shows a song almost identical in pitch and time pattern, but delivery rate much slower. 1st note slightly stronger than in Fig

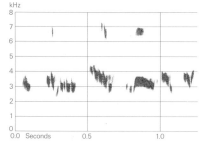

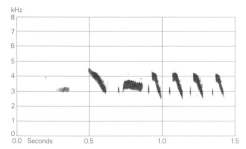

I B Hörnfeldt/Sveriges Radio (1972–80)   Finland   June 1968

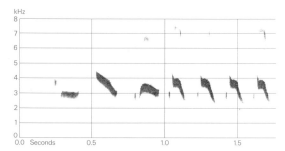

II G Alenius/Sveriges Radio (1972–80)   Finland   June 1968

III J Paatela/Sveriges Radio (1972–80)   Finland   July 1969

II but still faint (strong in Fig I) and phrase rendered 's-si-swoo swee swep sip-sip-sip-sip'. Overall patterning similar between all 3 songs, but 3rd note from end in Fig I (5th from end in Figs II–III) most clearly marks its identity. (J Hall-Craggs, M G Wilson.) In USSR (nominate *cyanurus*), song one of characteristic sounds of Siberian taiga: simple alternation of high- and low-pitched sounds gradually fading away and fitting in well with gloom of dense forest. See also Dresser (1871–81), Grote (1934), Portenko (1937), and Johansen (1955). For song of nominate *cyanurus* in Japan, see Jahn (1942), Mikkola (1973), Skoog (1981), and Yamashina (1982). Distinctive short territorial song of *rufilatus* in Nepal rendered 'tü trri tü trri' (like that noted by Johansen 1955 for Siberia) or 'tsy türr türr' (Diesselhorst 1968*a*); see also Ali and Ripley (1973*a*) and Fleming *et al.* (1976). (b) Low song (perhaps Subsong) given in China in spring said by Kolthoff (1932) to differ from that given on breeding grounds; no details, however. (2) Contact-alarm calls. (a) Given commonly on breeding grounds as contact- or excitement-call, particularly after hatching, between family-members, also apparently as alarm-call by vagrants. A 'fid fid' like Black Redstart *P. ochruros* (similarity confirmed sonagraphically by Bergmann and Helb 1982) or sharp 'fit' or 'feet' like Willow Warbler *Phylloscopus trochilus* (Grote 1934; Sushkin 1938; Johansen 1955); 'heet', occasionally like Thrush Nightingale *Luscinia luscinia*, variable in volume and rate of delivery, depending on excitement (Skoog 1973, 1981); see also Svensson (1982). Recording by G Alenius (Finland) suggests series of thin 'swi' or 'seep' sounds not unlike Meadow Pipit *Anthus pratensis* (P J Sellar). (b) Not infrequently follows call 2a. Short, low-pitched, harsh and quiet sound often given in (rapid) burst of 2–3: 'dagdag' (Bezzel and Löhrl 1972); 'krak-krak' (Panov 1973); 'tacktacktack' as alarm given near nest by nominate *cyanurus* in Japan and *rufilatus* in India (Jahn 1942; Ali and Ripley 1973*a*). Despite renderings, the following probably also this call or closely related: quiet 'cheg-cheg' (Potorocha 1972); 'tschk-tschk' roughly like *Sylvia* warbler (Dohle *et al.* 1957) or 'tongue-clicking' 'trk trk' (Hemmingsen and Guildal 1968); in Nepal, winter (race unknown), throaty 'guk' or 'gug' resembling Siberian Rubythroat *Luscinia calliope* (Fairbank 1980*a*). On breeding grounds in Finland, calls 2a–b combined in quite long series: 'hiit hiit hiit kack-kack hiit hiit hiit'; pauses *c*. 1 s between 'hiit' sounds, 'kack' muffled and barely audible even close to (Skoog 1973, 1981). (c) Hard and noisy (i.e. not tonal) 'tke-tke' sounds (Bergmann and Helb 1982) distinct from call 2b. Skoog (1981) noted call resembling Spotted Flycatcher *Muscicapa striata* from ♀ and an irritated chattering during chase; these, as well as apparent ticking from ♂ pursuing ♀ (see Nechaev 1969) perhaps also of this type. For other descriptions, presumably of this call, see Galloway *et al.* (1961), Hemmingsen and Guildal (1968), Ali and Ripley (1973*a*), Svensson (1974), Fleming *et al.* (1976), and Yamashina (1982). (d) Rattle-call. Given on breeding grounds and outside breeding season. A 'zerrr' resembling an alarm-call of *T. iliacus* given by ♂ in Finland (Skoog 1981; see also Bezzel and Löhrl 1972, Alerstam 1974, Svensson 1974, Fleming *et al.* 1976, Yamashina 1982). May follow call 2c (Kolthoff 1932) or 2a (Vorobiev 1963). In Nepal, small flock of *rufilatus* gave persistent, hard 'siep siep' combined with 'trrt trrt' (Diesselhorst 1968*a*); first of these presumably call 2a, second likely to be Rattle-call. Whistled 'trweet' sometimes given by alarmed *rufilatus* after call 2c (Ali and Ripley 1973*a*) perhaps a similar combination. (3) Other calls. (a) A 'tchay-tchay-tchay...' given in slow sequence in autumn (Munsterhjelm 1922); low 'sesese' given when disturbed in spring (Kolthoff 1932) perhaps the same or related. (b) Ringing 'ping' (Hemmingsen and Guildal 1968). CALLS OF YOUNG. Chirping sounds given when being fed similar to young redstarts *Phoenicurus* (Jahn 1942). A 'chak chak' given by fledged young, apparently as an alarm (Portenko 1937). Song apparently fully developed at 1 year old (Bergman 1935). MGW

Breeding. SEASON. Pechora basin USSR: 1st broods on the wing in late June, 2nd in mid-August (Dementiev and Gladkov 1954*b*). SITE. On ground in hollow among tree roots, or in hole in bank, or slightly above ground in stump or fallen log. Nest: cup of moss, grass, and roots, lined with softer grass, wool, hair, and sometimes pine needles. Building: no information on role of sexes. EGGS. See Plate 82. Sub-elliptical, smooth and slightly glossy; white, sometimes lightly marked with brownish blotches, usually at broad end. Nominate *cyanurus*: 17·9 × 13·4 mm (16·6–19·0 × 13·34–15·0), *n* = 27; calculated weight 1·85 g (Schönwetter 1979). Clutch: 5–7. Sometimes 2 broods in Pechora basin. (Dementiev and Gladkov 1954*b*.) INCUBATION. Period unknown. By ♀ only (Dementiev and Gladkov 1954*b*). YOUNG. No information. FLEDGING TO MATURITY. Fledging period 15 days (Harrison 1975). ♂ breeds at 1 year old (Dementiev and Gladkov 1954*b*). No further information.

Plumages (nominate *cyanurus*). ADULT MALE. BRIGHT MORPH. In fresh plumage, upperparts, cheeks, ear-coverts, and sides of neck dark blue, each feather with olive-brown fringe on tip, partly covering dark blue, in particular on forehead, ear-coverts, and hindneck, least so on sides of crown, rump, and upper tail-coverts. Short and broad cream supercilium running from nostril to above eye. Lores, front part of cheeks, and broken ring round eye dull blue, partly speckled off-white. Chin, throat, and chest cream-buff; some white of feather-bases visible on throat, some dull blue-grey sometimes on chest. Sides of breast dark blue, feathers broadly fringed buff. Flanks bright rufous-orange. Remainder of underparts (including thighs) buff or olive-buff, merging to virtually white on central belly and longer under tail-coverts. Tail blue, like upperparts; inner webs of t2–t5, both webs of t6, and undersurface of all feathers dull dark grey. Flight-feathers and tertials greyish-black; tertials with much dark blue tinge, especially on outer webs; rather narrow fringes of outer webs of flight-feathers paler cerulean-blue or (particularly secondaries) partly olive-brown. Lesser upper wing-coverts glossy bright blue; median and tertial coverts dark blue (like upperparts) with some dull black visible on centres; greater coverts dull black with much dark blue tinge on outer webs and often with olive-brown fringe along tip. Greater upper primary coverts and longer feathers of bastard wing dull black, outer webs tinged blue or olive-brown with cerulean-blue admixed. Under wing-coverts and axillaries orange-buff or cream-buff, shorter coverts with dull grey bases. *In worn plumage* (about January–May), upperparts uniform dark blue (brightest on rump, some traces of olive-brown feather-fringes sometimes visible). Broad short supercilium cream-white, bordered above by some bright blue on sides of crown and merging into bright blue supercilium behind eye (sometimes partly speckled white). Lores, cheeks, ear-coverts, and sides of neck dull dark blue, somewhat brighter blue towards sides of breast. Chin and throat cream-white, chest buff. Remainder of underparts, tail, and wing as in fresh plumage. *In heavily worn plumage* (about June–August), much dull grey of feather-bases visible on upperparts; white and bright blue of supercilium partly abraded, less distinct; underparts off-white with some grey of feather-bases visible, but sides of breast dull blue-grey and flanks often still with much orange; fringes along outer webs of flight-feathers bleached to off-white and pale bluish-olive. DULL MORPH. Largely like adult ♀ (see below), but rather variable. Generally, variable amount of blue tinge visible on feather-bases of olive-brown upperparts; front part of supercilium often distinct, cream-white, rear part and upper border tinged bright blue; rufous-orange on flanks deeper and more extensive than in adult ♀, remainder of underparts paler cream-buff or off-white; frequently much blue tinge on upper wing-coverts, including some glossy bright blue on lesser coverts (unlike ♀). ADULT FEMALE. Upperparts rather dark olive-brown; somewhat paler buff-brown near base of upper mandible, blue with some olive-brown feather-tips on rump and upper tail-coverts. Short and ill-defined olive-buff or cream-buff supercilium, running from nostril to just above eye. Distinct cream-white eye-ring. Sides of head and neck (including rear-part of supercilium) olive-brown, often slightly paler and more olive than upperparts, faintly speckled or streaked cream near bill-base and just below and behind eye. Chin and throat pale buff or cream-white, often rather sharply defined from olive-brown or olive-grey of cheeks, chest, and sides of breast. Flanks rufous-orange, slightly paler and less extensive than in adult ♂. Olive-brown or olive-grey of chest and orange of flanks rather gradually merges into white of central belly; vent and under tail-coverts cream. Tail blue, like adult ♂ but on average slightly paler and more cerulean. Lesser and median upper wing-coverts olive-brown, like upperparts; remainder of wing including flight-feathers and tertials as adult ♂, but all fringes olive-brown, deepest and slightly rufous along outer edges of flight-feathers. *In worn plumage*, upperparts, sides of head and neck, and chest duller and more greyish olive-brown or brown-grey, short supercilium whiter but still poorly defined and hardly extending behind eye; chin, throat, and belly whiter; rump to tail paler and more greyish-blue; outer edges of tertials and flight-feathers paler olive-brown or buff. NESTLING. No information. JUVENILE. Upperparts and sides of head dark grey-brown, each feather with buff or ochre central spot bordered by variable amount of dull black on tip. Chin to chest buff with black-brown arcs or streaks, remainder of underparts almost uniform pale buff to off-white. Tail as in adult (darker blue in ♂, on average paler cerulean-blue in ♀). Wing as in adult ♀, but tertials and greater coverts fringed olive-buff along tips, tending to form paler buff spot at shaft. FIRST ADULT MALE. Like adult ♀, showing olive-brown upperparts and chest in fresh autumn plumage, more grey-brown in spring. Blue restricted to rump, upper tail-coverts, and tail, like adult ♀, but some 1st adult ♂♂ have some bright blue on lesser upper wing-coverts, sometimes largely hidden, and a few show variable amount of blue tinge on centres of new tertials, tertial coverts, or scapulars. Differs mainly from adult ♀ (and dull morph ♂) by retention of juvenile flight-feathers, tail, greater upper primary coverts, greater upper secondary coverts, and often all or part of tertial coverts and tertials; these more worn than adult at same time of year and relatively more worn than neighbouring fresh feathers. Also, greater upper wing-coverts and (if retained) tertials show olive-buff or rufous fringes, often with paler buff spot at centre (in adult ♀, fringes uniform olive-brown), but sometimes hardly distinguishable, especially when worn. FIRST ADULT FEMALE. Like adult ♀, but part of juvenile feathers retained, as in 1st adult ♂. Closely similar to 1st adult ♂, but no blue on lesser upper wing-coverts, tertials, tertial coverts, or scapulars; rump to tail paler cerulean-blue, not as dark blue as 1st adult ♂, but some overlap in colour between sexes and difference hard to see in spring; orange of flanks on average paler and less extensive than in 1st adult ♂.

Bare parts. ADULT. Iris brown or dark brown. Bill brown-black or black; inside of upper mandible dark brown-grey, dark grey, slate-grey, or black. Leg and foot horn-brown or black. Bill, inside upper mandible, and leg apparently darkest in bright morph of adult ♂. NESTLING. No information. JUVENILE. Iris dark brown. Bill, leg, and foot greyish-brown, gape-flanges and bill-tip yellow-white. FIRST ADULT. Iris dark brown. Bill dark horn-brown; inside of upper mandible pink-yellow. Leg and foot light brown, grey-brown, or dark horn-brown; soles whitish or pale grey. (Hartert 1910; Dementiev and Gladkov 1954b; Mainwood 1972; Sandeman 1978; Skoog 1981; Svensson 1984a; BMNH, ZMA.)

Moults. ADULT POST-BREEDING. Complete; primaries descendant. In USSR, one bird starting late July, some in moult August, one nearing completion mid-September (Dementiev and Gladkov 1954b). In Japan, none in moult up to mid-July (RMNH, ZMA). Captive 1-year-old ♂, West Germany, started late July (Bezzel and Löhrl 1972). Moult starts late and not clear whether it occurs shortly before or just after autumn migration; some birds very worn in mid-August, with moult not started; one migrant from Korea freshly moulted 19 October (Svensson 1984a). POST-JUVENILE. Partial: head, body, lesser and median upper wing-coverts, sometimes tertial coverts, and occasionally a few tertials or inner greater upper wing-coverts. In USSR, early-fledged birds start from mid-July; some nearing completion about 20 August, but others still fully juvenile then (Dementiev and Gladkov 1954b). Moult completed in migrants examined late September and October (BMNH, RMNH, ZMA).

Measurements. ADULT, FIRST ADULT. Nominate *cyanurus*. Whole geographical range, all year; skins (BMNH, RMNH, ZMA). Bill (S) to skull, bill (N) to distal corner of nostril; exposed culmen on average 4·1 less than bill (S).

	♂		♀	
WING	80·0 (1·77; 37)	77–84	76·4 (1·80; 20)	73–79
TAIL	58·5 (1·99; 21)	55–62	55·9 (1·44; 14)	53–58
BILL (S)	13·9 (0·73; 21)	12·7–14·9	13·7 (0·47; 12)	13·0–14·3
BILL (N)	6·8 (0·46; 20)	6·2–7·5	6·8 (0·37; 11)	6·1–7·3
TARSUS	22·5 (0·98; 20)	21·1–24·0	22·6 (1·02; 11)	21·4–24·1

Sex differences significant for wing and tail. No difference between eastern and western birds; no difference between 1st adult (with retained juvenile flight-feathers and tail) and older birds.

Weights. Nominate *cyanurus*. USSR: ♂ 13·8 (10) 12·2–15·2, ♀ 15·1 (6) 12·0–17·8 (Dementiev and Gladkov 1954b). Kuril islands: ♂♂ 13·8, 14·1; ♀ 13·8 (Nechaev 1969). Lake Chany (south-west Siberia), late September: probable ♀♀ 12·2, 13·7 (Havlín and Jurlov 1977). Hopeh, China: ♂ 11·9 (13) 11–13, ♀ 12·4 (12) 10–16 (Cheng 1963). Taiwan, February: ♂ 11 (RMNH). Exhausted ♂, Hong Kong: 8·7 (ZMA). Adult ♂, Scotland, May: 14 (Mainwood 1972). First adults, western and central Europe: September, East Germany, ♂ 12·1 (Weber 1973); October, West Germany, ♀♀ 8·6 (after 1 night captivity), 10 (Dohle *et al.* 1957; Vauk 1973b); October, Netherlands, fat ♂ 14·1 (ZMA); October, Scotland, unsexed 12 (Sandeman 1978).

T. c. rufilatus. See Diesselhorst (1968a) and Ali and Ripley (1973a).

Structure. Differs markedly from *Erithacus* and *Luscinia* in rather long wing and tail, short and fine bill, and short and slender leg and toes. Wing rather long, broad at base, tip bluntly pointed. 10 primaries: p6–p7 longest, p8 2–4 shorter, p9 10–14, p5 1·5–3, p4 7–11, p3 12–16, p1 16–20; tip of p9 about equal to p3–p4 in closed wing. P10 reduced, narrow, 34–40 shorter than wing-tip, 7–14 longer than longest upper primary covert. Outer web of p5–p8 and inner of p6–p9 emarginated. Tertials short; longest about equal to secondaries. Tail rather short, tip slightly forked; 12 feathers, t1 2–5 shorter than t3–t4, t6 1–4 shorter. Bill short, straight; rather wide at base; tip of culmen slightly decurved. Nostrils oval, partly covered by membrane above. Some long and rather strong bristles along base of upper mandible. Leg and toes short and slender. Middle toe with claw 17·3 (10) 16·5–18·5; outer toe with claw *c.* 72% of middle, inner *c.* 65%, hind *c.* 77%. Claws rather short, slender, decurved.

Geographical variation. Slight within northern Eurasia: populations from eastern part of range (sometimes separated as *ussuriensis* Stegmann, 1929, or *pacificus* Portenko, 1954) on average slightly darker and bluer than typical nominate *cyanurus* from further west, but all differences very slight (Vaurie 1959; RMNH) and recognition not warranted. *T. c. rufilatus* from central Asia more markedly different: tail and tarsus longer (hence, differing in tail/wing and tarsus/wing ratio: Stresemann

et al. 1937), wing-tip more rounded (p6 longest, p1 14–18 shorter, p9 14–16, with tip of p9 about equal to tip of p2 in closed wing). Upperparts of adult ♂ usually brighter blue than in nominate *cyanurus*; supercilium paler blue, usually without white on front part or with white restricted to feather-bases; throat-patch narrower and purer white, not broad and buffish; ♀ and 1st adult ♂ as nominate *cyanurus*, but throat and belly on average whiter. In some areas, many adult ♂ are dull morph, apparently retaining ♀-like plumage throughout life, and bright morph adult ♂♂ rare (Ali and Ripley 1973a). Colour of adult ♂ from north-central China rather similar to nominate *cyanurus*, though size and structure near typical *rufilatus*, and these birds sometimes separated as *albocoeruleus* Meise, 1937 (Stresemann *et al.* 1937), but colour rather variable in many populations of *rufilatus* and recognition not warranted. ♀ and dull morph ♂ of populations of western Himalayas paler olive-grey or olive-brown on upperparts than typical *rufilatus* from further east and hence western birds sometimes separated as *pallidior* (Baker, 1924), but difference very slight.

Forms species-group with Rufous-bellied Bluetail *T. hyperythrus* from central and eastern Himalayas.   CSR

Irania gutturalis White-throated Robin

PLATES 47 and 48
[between pages 652 and 653]

Du. Perzische Nachtegaal   Fr. Iranie à gorge blanche   Ge. Weisskehlsänger
Ru. Соловей-белошейка   Sp. Petirrojo turco   Sw. Vitstrupig näktergal

Cossypha gutturalis Guérin-Méneville, 1843

Monotypic

Field characters. 16·5 cm; wing-span 27–30 cm. Nearly 20% larger than Robin *Erithacus rubecula*; close in size to nightingales *Luscinia*, with similar form and structure except for slightly longer wings and tail. Quite robust and bulky chat, more recalling robin-chats *Cossypha* of Africa than Palearctic relatives. Diagnostic combination of rather long black tail, white vent, and rufous-buff flanks. ♂ striking, with black face-mask contrasting with white throat, narrow white supercilium, dark blue-grey upperparts, and rich rufous-orange chest. ♀ much less colourful, with brown-grey head and back. Sexes dissimilar; little seasonal variation. Juvenile separable.

ADULT MALE. Crown, nape, rear cheeks, back, wing-coverts, and fringes to tertials and secondaries blue- to lead-grey (depending on angle of light), duller when worn; flight-feathers grey-black; rump and upper tail-coverts grey-black, merging with black tail. Side or head on, shows striking pattern of narrow white to grey-white supercilium, black lores, fore-cheeks, and sides of neck, and pure white chin and throat. Lower throat sometimes shows black necklace. Breast rich rufous-orange, this colour extending on to fore-flanks, becoming orange-buff on lower and rear flanks and towards centre of belly, and merging with buff-white vent and almost white under tail-coverts. A few birds have underparts wholly cream. Under wing-coverts rufous-orange. All birds become

paler with wear. Bill and legs black. ADULT FEMALE. Wing (above and below), rump, and tail much as ♂ but head and back brown-grey, with face pattern restricted to faint, buff fore-supercilium and eye-ring and diffusely bordered, dull white throat. Underparts basically buff-white but strongly washed rufous on breast and along flanks; some birds also show fulvous on breast. JUVENILE. Ground-colours much as adult ♀ but with small buff spots on crown and cheeks, and large buff or dull cream spots on back, rump, tertials, and wing-coverts, with those on median and greater coverts forming pale disjointed bars; also a mixture of buff spots and dark mottling on chest. Pale buff supercilium more distinct than on ♀ but rufous tone to underparts restricted to patch on side of chest and wash along flanks, thus usually paler-bellied. Bill and legs brown. FIRST-WINTER MALE. Duller than adult, with noticeably paler, buffier, less rufous-orange underparts and brown tone to upperparts.

♂ unmistakable in west Palearctic but not so in winter range of East Africa, where several robin-chats *Cossypha* have similarly coloured plumage and may cause brief confusion when their rufous rumps and tails not seen. ♀ and immature puzzling, suggesting several other chats, e.g. Thrush Nightingale *L. luscinia* and ♀ Redstart *Phoenicurus phoenicurus* until black tail visible (a character shared only with Blackstart *Cercomela melanura* but that

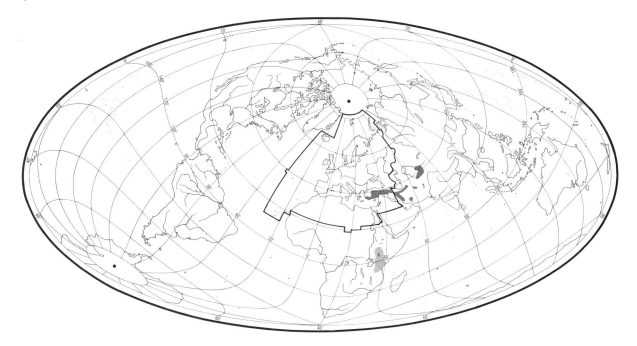

species 15% smaller, with more uniform, paler grey or brown plumage). Flight free and light, with action and escape behaviour as nightingales *Luscinia*; due to relatively long wings and tail, flight silhouette can suggest a *Turdus* thrush as much as a chat. Gait and stance as nightingale *Luscinia*, moving with long, quick hops and frequently drooping wings; when in the open or perched and excited, often cocks tail high and may wave it. Inveterate skulker in ground cover and difficult to flush. Not gregarious.

Song a short phrase of clear bell-like notes, of similar quality and volume to those of nightingale *Luscinia*. Calls include 'tirric', 'churr', and 'chick', also recalling *Luscinia*.

Habitat. Breeds in warm dry continental lower-middle latitudes, in USSR largely in stony arid uplands, on more or less steep slopes or in ravines of mountain streams and narrow stony gullies, often with scattered junipers *Juniperus*, *Zygophyllum*, almonds, and other shrubs, or rank grasses on edge of mountain steppes, especially at 1000–2200 m up to subalpine zone where habitat overlaps with Shore Lark *Eremophila alpestris*. Perches in bushes or tall weeds, dropping to ground at times, but often perches openly on stones or runs from one to another (Dementiev and Gladkov 1954*b*). In Afghanistan, found in scrub at 1800–2100 m in a side valley below stony steep slopes after breeding season, but a pair seen at around 2550 m in late May (Meinertzhagen 1938; Paludan 1959). In Lebanon, occurs in bushes, thickets, tangled undergrowth in mountains and ravines, and similarly in Iraq in fairly dense scrub on rocky hillside (Benson 1970). Also in Iraq noted as hopping on a stack of brushwood and occurring on spring passage in willows *Salix* and scrub (Ticehurst *et al.* 1921–2). Breeds in Iran

at up to *c.* 2850 m in oak *Quercus* steppe, in forest with bush layer, or on hillsides where oaks had been removed but bushes remain. In East African winter quarters described as a skulker in thick bush, preferring ravines and hill slopes, but also occurs in ground-water forest; also found in patchy thickets interspersed with open ground, into which it emerges to feed, and in Serengeti (Tanzania) in *Commiphora*-*Acacia* open woodland (Moreau 1972). Thus resembles Bluethroat *Luscinia svecica* and Rufous Bush Robin *Cercotrichas galactotes* rather than other west Palearctic members of robin-nightingale group in becoming most nearly dissociated from forest habitats, and also from water or humid soils, although still dependent on some woody cover and perches.

Distribution. Little information on range changes.

LEBANON. Bred formerly (Kumerloeve 1969*d*) and may still do so, but no recent proof (AMM).

Accidental. Britain, Norway, Sweden, Greece, Jordan, Cyprus, Egypt.

Population. No information on population trends.

USSR. Not abundant (Dementiev and Gladkov 1954*b*).

Movements. Migratory, wintering in rather restricted area of East Africa.

Winters in Kenya in plateau country north and east of highlands and in Tanzania mainly in north-east and dry interior (Pearson 1984), with recent record as far south as Mbarali 8°35′S, 38°40′E (D J Pearson). Small numbers also winter, apparently regularly, in Zimbabwe (Parnell 1976). Arrives in Turkey from mid-April, with continuing arrivals into May (*Orn. Soc. Turkey Bird Rep.*

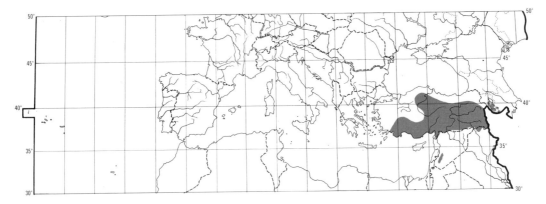

1966-75*). Arrival apparently synchronous across entire range. Dolgushin *et al.* (1970) gave arrival date of 24 April in Ugam valley and Karatau (Kazakhstan), and Ivanov (1969) last week in April in Pamir Alay on eastern edge of breeding range. Most have left USSR breeding grounds by end of August. Nowhere common on passage. Uncommon migrant to all areas of Saudi Arabia (Jennings 1981*a*). Passage records from North Yemen include 2 on 18 April and a late bird on 2 November 1979 (Cornwallis and Porter 1982). In Iraq, more common in spring than autumn (Kumerloeve 1964), and in Iran Erard and Etchécopar (1970) saw only 2 migrants, on 30 April and 1 May at Bandar-abbas. Further west, even less common. Has occurred 2-29 April at Azraq, Jordan (Wallace 1982); Kumerloeve (1964) gave 3 passage records, near Jerusalem August 1885, in Syrian desert on 3 March 1905, and north of Saïda (Syria) on 3 October 1958; only 2 records (16 March and 1 April) for Cyprus (Flint and Stewart 1983), only 1 from Egypt (in Sinai) (Goodman and Storer 1985), and 3 from Sudan (Nikolaus 1984). This scarcity perhaps indicates that east Mediterranean and north-east Africa usually overflown (Moreau 1972) or that birds take more easterly route and enter Africa through Eritrea. Autumn passage in Eritrea recorded 28 August-23 September, being common above *c.* 2000 m but with none in spring (Smith 1957). Known as passage migrant to Ethiopia from 18 August to 17 October and again 2 March-13 April (mainly 19 March-9 April (Ash 1980), and in Somalia west of 46°E in September-October and again in March, but uncommon (Ash and Miskell 1983). Southward movement is apparently gradual, with birds not reaching Kenya until November.

Progressive southward movement occurs through early part of winter, with numbers in Tsavo (south-east Kenya) peaking late November to early January, birds apparently moving on as bush dries out. Some do spend all winter as far north as Kenya, and one ringed near Nairobi in December was re-trapped in same area the following February. Good evidence of return spring passage mid-March to mid-April along eastern edge of highlands (Leuthold 1973; Lack 1983; Pearson 1984). Here they presumably fatten to cross north-east Africa without alighting, accounting for paucity of spring records in this area.   RAC

Food. In breeding season at least, mainly insects. Feeds mainly on ground, turning over leaves to search; also in trees and bushes (Meinertzhagen 1954; Adamyan 1963*b*), and in North Yemen seen feeding on insects in prickly pear (Phillips 1982).

In Armeniya (USSR), diet almost exclusively insects (mainly beetles Coleoptera) in spring and summer, partly changing to plant food in autumn; 58 items from stomachs comprised 47% (by number) adult beetles (including Carabidae, Scarabaeidae, Tenebrionidae, Chrysomelidae), 21% ants (Formicidae), 9% bugs (Hemiptera), 9% beetle larvae (Dermestidae), 5% grasshoppers (Acrididae), 3% moth (Noctuidae) larvae, 3% wasps (Hymenoptera), and 3% spiders (Araneae) (Adamyan 1963*b*). In autumn, said to feed mainly on honeysuckle berries (Dementiev and Gladkov 1954*b*).

Young in Armenian study fed on larval Noctuidae (57% of 49 items), *Acronycta* larvae (31%), grasshoppers (8%), and spiders (4%) (Adamyan 1963*b*). At one nest in northern Iraq, young brought exclusively insects, mainly small caterpillars (Čtyroký 1972).   DJB

Social pattern and behaviour. Some aspects poorly known; fullest study by Adamyan (1963*b*) in Armeniya (USSR).

1. Not very sociable (Suschkin 1914*b*). Generally solitary on migration (e.g. Abdusalyamov 1973). By early August in western Caucasus (Turkey/USSR), families had already broken up and birds mostly solitary, less commonly in twos (Suschkin 1914*b*). Frequently quite large numbers together in winter according to Hüe and Etchécopar (1970), though recent observations in East Africa suggest birds territorial, individuals being recorded in same thicket over many weeks (D J Pearson); see also part 2 (below). BONDS. Nothing to suggest mating system other than monogamous; no information on duration of pair-bond. In Armeniya, ♂♂ first to arrive on breeding grounds and pair-formation takes place there (Adamyan 1963*b*). ♀ broods young which are fed by both parents, including for some time after leaving nest (Adamyan 1963*b*; Čtyroký 1972); see part 2. BREEDING DISPERSION. Solitary and territorial. At Beyşehir Gölü (Turkey), pairs or singing ♂♂ in population of 6-7 or 8-10 pairs at least 150-200 m apart, only once 60-70 m (Kumerloeve 1964). In northern Iraq, 4 nests *c.* 300-1000 m

apart (Čtyroký 1972). Near Erevan (Armeniya), 11 nests in gorge 80-200 m apart (Adamyan 1963b). In Khodzhai-Bekhob (Tadzhikistan), pairs 800-900 m apart (Abdusalyamov 1973). In Aksu-Dzhabagly reserve (Kazakhstan, USSR), highest density at least 5 pairs in juniper *Juniperus* along 0·5 km of 2 narrow gorges (Ivashchenko 1982). Scree with scattered bushes and trees at 1300-2300 m altitude on southern slopes of Gissarskiy mountains (Tadzhikistan) held 1-3 pairs per 1-km stretch (Leonovich 1962). Territory occupied on arrival in Armeniya *c.* 100-400 m²; area later defended most vigorously is that immediately adjacent to nest (Adamyan 1963b); see also Parental Anti-predator Strategies (below). In gorge near Erevan, population more or less stable from year to year and nests sited near those of previous year (Adamyan 1963b); perhaps indicates same territories used from year to year, though no proof that same birds involved. In northern Iraq, ♂ collected food for well-grown young close to nest, sometimes within *c.* 10 m (Čtyroký 1972). Nest for one replacement clutch sited 6 m away from unsuccessful 1st clutch (Adamyan 1963b). ROOSTING. No information.

2. Usually described (breeding birds, those in winter quarters, and vagrant) as shy or at least elusive, spending much time skulking in dense bushes, etc. Not so shy if approached cautiously, though normally dives into cover from which can be virtually impossible to flush. Sometimes flies low to another bush, up into top of solitary tree (disturbed while feeding), or flies off far (Witherby 1903; Ticehurst *et al.* 1921-2; Stresemann 1928; Lyaister and Sosnin 1942; Kumerloeve 1964; Hüe and Etchécopar 1970; Sutton and Gray 1972; Cederwall and Svenaeus 1973; Beaudoin 1976; Jonsson 1982). ♀ altogether more secretive than ♂ on breeding grounds, spending most time in cover and only rarely calling (Adamyan 1963b; Dolgushin *et al.* 1970). ♂ will perch in full view at top of bush to sing (Jonsson 1982), but does not tolerate close approach, tending to break off and quickly enter cover (Kumerloeve 1964). Birds engaged in courtship oblivious of man (Witherby 1903) and fairly easy to see during building and feeding of young, though much more secretive during incubation and after young leave nest (Leonovich 1962; Dolgushin *et al.* 1970; Abdusalyamov 1973). Trapped ♂ 'very wild' and gave presumed distress-call (Čtyroký 1972: see 4 in Voice). When alarmed, will adopt an erect posture, legs extended and head raised so that white throat-patch prominent; at same time, wings held out to side or drooped and tail slowly raised to *c.* 45° (sometimes 2-3 times in succession) and spread like Rufous Bush Robin *Cercotrichas galactotes* or some *Luscinia*, and slowly lowered. When wings held out to side, tail sometimes moved horizontally to one side (Dementiev and Gladkov 1954b; Moore and Boswell 1956; Dolgushin *et al.* 1970; Erard and Etchécopar 1970; E N Panov); in Iran, ♀ collecting nest-material frequently flicked tail up (P A D Hollom). FLOCK BEHAVIOUR. No information. SONG-DISPLAY. ♂ sings (see 1 in Voice) from an exposed and elevated perch—top of bush or tree—sometimes from middle of tree, or in flight (Wadley 1951; Dolgushin *et al.* 1970; Cederwall and Svenaeus 1973; P A D Hollom). In Fars (south-west Iran), both sexes noted singing in flight and when perched (Witherby 1903), but other authors refer only to song by ♂, and this record perhaps relates to ♂♂ with dull underparts (see Plumages). In East Africa, winter, usually sings from perch low down in bush. 2 performed apparent Song-duel *c.* 30 m apart; one watched closely had head extended, back flattened, wings drooped, and tail spread while singing (D J Pearson: Fig A). On breeding grounds in Armeniya during courtship period, singing ♂ has tail raised vertically (and wings drooped) (Adamyan 1963b). In Iran, early May, also noted singing during

A

brief pauses between changing perch and dropping down to feed (P A D Hollom). May ascend into Song-flight after singing from perch. Detailed description of Song-flight lacking, but a number of variants (or perhaps simply phases) occur. May ascend while singing, then, continuing to sing loudly, glide slowly or rapidly with wings and tail widely spread and rigid for several tens of metres, to land (tail widely spread) on bush or rock. Several Song-flights sometimes performed in quick succession, bird thus moving from perch to perch across territory (Leonovich 1962; Adamyan 1963b; Ivanov 1969; Dolgushin *et al.* 1970; Jonsson 1982). Other authors referred to a striking performance with tail spread, but wings fluttered (Moore and Boswell 1956; Cederwall and Svenaeus 1973). Alternately gains and loses height (E N Panov). In Iran, bird once performed Song-flight over *c.* 25 m between perches; wing-beats very slow—slower than Greenfinch *Carduelis chloris* in Song-flight; in Turkey, Song-flight once performed over *c.* 50 m (P A D Hollom). Paludan (1938) likened Song-flight to that of pipit *Anthus*, bird making angled descent with wings half-open. Description from Fars referred to bird 'flying down' (not clear whether this an abbreviated Song-flight from perch to ground, or just descent phase) while quivering wings and singing. On landing, will raise and fan tail (Witherby 1903), doing this repeatedly before moving into bush (Leonovich 1962). In Armeniya, starts to sing usually from *c.* 06.00 hrs; song wanes during hottest part of day, followed by resurgence in evening, but pauses between songs then longer at *c.* 10-20 min. One ♂ sang vigorously 06.00-13.00 hrs, then only 3 times over *c.* 3½ hrs (Adamyan 1963b); see also Dolgushin *et al.* (1970). In Tadzhikistan, birds most active 10-31 May in one year and sang even during hottest part of day (Leonovich 1962). Will sing also in bad weather—wind, etc. (Kumerloeve 1974). Vagrant in Sweden June-July mostly gave full loud song, once Subsong (Cederwall and Svenaeus 1973; see 1b in Voice). At Beyşehir Gölü, ♂♂ still in song during feeding of young in early June (Kumerloeve 1964). In Armeniya, song intensity and persistence gradually increase from time of pair-formation, but decline markedly from start of incubation. Significantly less song after hatching and ceases when young leave nest: e.g. 3 June in one year, though later starters may continue to mid-June. Song in 2nd half of June associated with 2nd breeding cycle (Adamyan 1963b). In Tadzhikistan, sings from mid-April, occasionally still late May (Abdusalyamov 1973); only some ♂♂ sing in Kazakhstan from arrival in late April (Dolgushin *et al.* 1970). Song noted in East African winter quarters in December and, quite strongly, January-March (*Scopus* 1979, **2**, 105-25; D J Pearson). ANTAGONISTIC BEHAVIOUR. Fairly peaceable toward conspecific and other birds according to Suschkin (1914b). However, in Armeniya, fights not infrequent during period of territory establishment soon after arrival on breeding grounds and territorial ♂♂ recorded chasing off Cuckoo *Cuculus canorus* and sparrows *Passer* (Adamyan 1963b). Will also sing at an intruder (Moore and Boswell 1956); see also Song-display (above). ♀ attacking large caterpillar (Lepidoptera) adopted apparently aggressive posture with crown feathers ruffled and wings half-open (Meinertzhagen

1954). HETEROSEXUAL BEHAVIOUR. General. ♂♂ first to arrive on Armeniyan breeding grounds, ♀♀ following several days (e.g. on 25 April, 6 days) later. Nest-building takes place during 1st half of May (Adamyan 1963*b*). In 2 areas of Tadzhikistan, birds paired and on territory by late April or early May; building and laying during 2nd half of May (Leonovich 1962; Vorobiev 1968; Abdusalyamov 1973). (2) Pair-bonding behaviour. In Fars, ♂-♀ chases, with birds singing, occurred continually in April (Witherby 1903). In Kazakhstan, such chases tend to take place after sunset when ♂ has stopped singing. Birds' behaviour rather phrenetic: fly fast from bush to bush or briefly land on stones, then take off and chase again, etc., both birds calling (Dolgushin *et al.* 1970: see 3 in Voice). (3) Courtship-feeding and mating. ♀ not fed by ♂ on nest. Mating usually follows (brief) aerial chase as described above, ♂ driving ♀ to open spot and copulating, or ♂ may chase ♀ rapidly and agilely through branches. Birds probably copulate several times during breeding season; recorded in Armeniya, 4 and 10 May (Adamyan 1963*b*) and in Iran, also in early May (P A D Hollom). (4) Nest-site selection and behaviour at nest. No information on selection process. ♀ builds nest, sometimes (or often) accompanied, at slight distance, by ♂. Material normally collected within *c.* 50 m of nest. Eggs normally laid at night or in early morning (Adamyan 1963*b*). In Kazakhstan, 2 records of nest being completed only after 1st egg laid (Ivashchenko 1982). ♀ incubates for *c.* 20-35 min, then flies off to feed for *c.* 5-10 min; usually accompanied off nest by ♂. ♀ will throw out any damaged eggs (Adamyan 1963*b*). RELATIONS WITHIN FAMILY GROUP. Hatching may take place over 3 days. Young brooded by ♀ for total of *c.* 5-6 days by which time ♀ almost entirely occupied with feeding young. Young able to beg *c.* 30 min after hatching; give muted barely audible food-calls at 1-2 days old and able to stand well and beg at 3 days. First fed (always by ♀) within *c.* 3-4 hrs of hatching, ♂ bringing food and passing it to ♀ who (rarely) eats it herself or passes it to young. ♂ tends to use regular route to nest, landing on particular branch of tree where nest sited. Little collecting of food done by ♀ during first few days (when mostly brooding), but she will fly off to feed herself regularly every *c.* 25-35 min; ♂'s feeding rate increases as young grow and he will feed young himself if ♀ absent. At 6 days old, young quiet in nest and clump together. Eyes fully open by 8th day (Adamyan 1963*b*, which see for further details of physical development). In Iraq, brood *c.* 10 days old defecated on to nest-rim and faecal sacs removed by adults. Young probably leave nest at *c.* 12 days (Čtyroký 1972). However, in Armeniya, leave (not yet able to fly) at 9-10 days old and move about in thickets. Able to flutter *c.* 10-15 m at 13-15 days; land clumsily and rush to hide under stones. Fly at 16-18 days, but still fed and guarded by parents at this age and up to *c.* 25-30 days old (Adamyan 1963*b*). ANTI-PREDATOR RESPONSES OF YOUNG. At 3 days old, crouch in nest at approach of man (Adamyan 1963*b*). For hiding out of nest, see above. PARENTAL ANTI-PREDATOR STRATEGIES. (1) Passive measures. ♀ a tight sitter, almost allowing herself to be touched before leaving (Ramsay 1914); tight-sitting also recorded once in ♂ (Leonovich 1962). In Turkey, ♀ left nest containing small young at approach of observer. Neither adult came into view or called while observer present (Kumerloeve 1964; at another nest, adults skulked and flitted about (Ramsay 1914). In Iraq, one ♀ relatively tame, continuing to feed young, but ♂ moved about in trees nearby, reluctant to visit nest while observer in hide (Čtyroký 1972). (2) Active measures: all information relates to man as potential predator. Nest defended by both sexes from start of incubation. After hatching, brooding ♀ may adopt threat-posture (Fig B):

B

head thrown back, bill opened wide; freezes thus for a few seconds (Adamyan 1963*b*); similar behaviour noted in extralimital White-throated Rock Thrush *Monticola gularis* (Neufeldt and Sokolov 1960). Birds feeding young in nest will move about in trees and on ground, raise tail vertically and give constant alarm-calls (presumably call 2b and/or 2c). Such calls also given intensely by 2 pairs when disturbed at night, as known to occur in Nightingale *Luscinia megarhynchos* (Čtyroký 1972). Parents also very agitated when man near young just out of nest: flew in close, called, and constantly opened wings and raised tail (Lyaister and Sosnin 1942).

(Figs by M J Hallam: A from drawing by D J Pearson; B from drawing in Adamyan 1963*b*.) MGW

Voice. Rather quiet on breeding grounds (Dementiev and Gladkov 1954*b*), particularly at end of season, in August (Suschkin 1914*b*) and ♀ more so than ♂ (Adamyan 1963*b*). Sings also in winter quarters, but no information on other calls given there.

CALLS OF ADULTS. (1) Song. Majority of descriptions are of song given by ♂; Witherby (1903) noted song by both sexes (see Social Pattern and Behaviour) and gave no indication of differences. (a) Song comprises short phrases—e.g. in Iran, main phrase 2-3 (rarely 7-8) s, though may be preceded by quieter, brief trilling notes (P A D Hollom); longer, full, rich and loud songs tend to be delivered in flight (Moore and Boswell 1956; Hüe and Etchécopar 1970). Typically fluid, melodious, clear and quite loud, but slow, warbling phrases limited and without much variety (Cederwall and Svenaeus 1973; P A D Hollom). Characteristically mixes mellow, fluting whistles with lower-pitched, harsher warbling (Leonovich 1962; Hüe and Etchécopar 1970) or bubbling—'as if too excited to sing properly'—(Witherby 1903). Such rich bursts mixed with grating passages typically given when perched according to Moore and Boswell (1956). Another description emphasized unmusical quality and likened it to coarser version of song of Rüppell's Warbler *Sylvia rueppelli*. Frequently includes chattering 'tyrr-r'r'r'' (Jonsson 1982) which may be trill of bell-like quality mentioned as typical by Dementiev and Gladkov (1954*b*) and Dolgushin *et al.* (1970). In Turkey, song more melodious than Rufous Bush Robin *Cercotrichas galactotes*, less jerky than Nightingale *Luscinia megarhynchos* and, in volume and tonal quality, more like Orphean Warbler *S. hortensis* (Kumerloeve 1964; Cederwall and Svenaeus 1973) or Rock Thrush *Monticola saxatilis* (Hüe and Etchécopar 1970). For further comparisons with other *Sylvia* and *Luscinia*, see Dolgushin *et al.* (1970). Recording confirms song as mixture of attractive whistles (including 'wheet-er wheet-a') and rather mechanical,

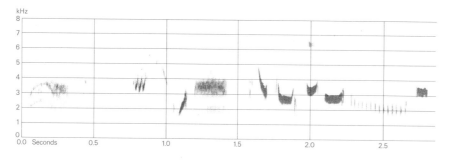

I   P A D Hollom   Iran   May 1977

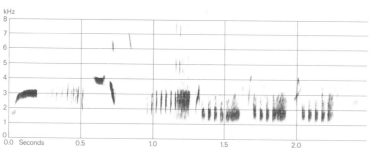

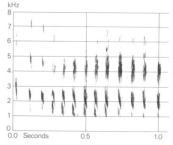

II   P A D Hollom   Iran   May 1977

III   P A D Hollom   Iran   May 1971

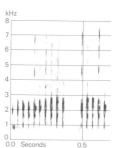

IV   P A D Hollom   Iran   May 1971

throaty, half-suppressed or squeezed-out rattles (at times reminiscent of distant football fan's rattle), often at start of phrase; accelerating series of such rattles—'turrerr'—prominent in Fig I (M G Wilson). Overall like *L. megarhynchos*, but with throttled quality (P J Sellar). Fig II shows phrase beginning with corvine rasp (rather like quiet Jay *Garrulus glandarius*), followed by clear and rich whistles which are preceded (in middle) and followed (penultimate unit) by rattles recalling Wren *Troglodytes troglodytes* (J Hall-Craggs; M G Wilson). Sometimes mimics other bird species (Hüe and Etchécopar 1970): in Tadzhikistan (USSR), Golden Oriole *Oriolus oriolus* (at start of song), *S. hortensis* (see above), Pied Wheatear *Oenanthe pleschanka*, etc. (Leonovich 1962). (b) Low warbling Subsong given by Swedish vagrant (Cederwall and Svenaeus 1973). (2) Contact-alarm calls. (a) Noisy 'tji-tjytt' is contact-call (Jonsson 1982); 'tzi-lit' reminiscent of White Wagtail *Motacilla alba* given by Swedish vagrant (Cederwall and Svenaeus 1973), also 'tirric' (D I M Wallace) probably the same, and 'tchryk' given in flight (E N Panov) perhaps also. (b) Dry, hard 'tec', 'chick' or chacking sound recalling nightingale *Luscinia* and given singly or as series by both sexes, but rather rarely by ♀

(Dolgushin *et al.* 1970; Hüe and Etchécopar 1970; D I M Wallace). Recording suggests 'chek' or 'chack' like powerful *Sylvia* warbler. At least once, bird gives double 'chek-chek' and calls also develop into dry, rattling chuckle, descending in pitch (P J Sellar, M G Wilson). Fig III shows 2 isolated 'chek' units then rattling series of 'chk' sounds (J Hall-Craggs). (c) Dry 'churr' or 'trrr' given by both sexes and similarly reminiscent of nightingale *Luscinia* (Hüe and Etchécopar 1970; Desfayes and Praz 1978); also rendered 'tchaurr' and 'tcharrtcharr' (E N Panov). In recording (Fig IV) 'krrrr' has slightly wooden timbre and is distinct from longer (chuckling rattle) variant of call 2b, being richer and suggesting Magpie *Pica pica*. Calls 2b-c probably have alarm function (M G Wilson). (d) Alarm-call of ♀ in Iran a thin 'eet...eet' or 'zut...zut' occasionally interspersed with call 2c (Desfayes and Praz 1978); such a combination rendered 'een een rrrr' recalling nightingale *Luscinia* (E N Panov); in Turkey, ♀ carrying food similarly gave a piping like Bullfinch *Pyrrhula pyrrhula*, but less deep and mellow (P A D Hollom). (3) Bell-like call given by ♂ during courtship chase. ♀'s call in same context said by Dolgushin *et al.* (1970) to be different; no details however. P A D Hollom noted chattering or chirruping sounds, perhaps from both sexes, during chase. (4) 'Crying' of trapped ♂ (Čtyroký 1972) presumably a distress-call.

CALLS OF YOUNG. Rhythmic sounds given from pipped eggs (Adamyan 1963*b*); apparently linked with development of breathing organs (Mal'chevski 1959). Call given on sighting ♀ with food like that of young Robin *Erithacus rubecula* (Čtyroký 1972); quiet whistling from 4-day-old nestlings (Adamyan 1963*b*) perhaps the same. Apparently independent juveniles in Turkey gave 3 calls:

(a) low scraping sound; (b) 'churr'; (c) 'whit' (Sutton and Gray 1972). Alarm-call (presumably adult-type) given from 13-15 days (Adamyan 1963*b*) is perhaps 2nd of these.   MGW

Breeding. SEASON. Armeniya (USSR): eggs laid from first half of May (Adamyan 1963*b*). SITE. In lower part of shrub or small tree, on stump, or in tree crevice. 15-125 cm above ground (Adamyan 1963*b*). Nest: cup of dry grass leaves, twigs, and bark, lined with vegetable down and hair; often some feathers, bits of rag, paper, sheep's wool, etc. (Adamyan 1963*b*). External diameter 9-14 cm, internal diameter 6·5-7 cm, height 7·5-11·5 cm, depth of cup 5·2-5·8 cm (Dolgushin *et al.* 1970; Abdusalyamov 1973). Building: by ♀, over 5-6 days (see Adamyan 1963*b* for details). EGGS. See Plate 82. Sub-elliptical, smooth and fairly glossy; greenish-blue, with yellowish or rusty-brown spots usually coalescing at broad end. 20·7 × 14·5 mm (19·5-22·5 × 13·0-16·0), $n = 10$; calculated weight 2·66 g (Schönwetter 1979). Clutch: (3-)4-5. Eggs laid daily. Replacements laid (Adamyan 1963*b*). INCUBATION. 13 days: first 3 hatch on day 13, 4th on day 14, 5th on day 15. Starts with laying of 3rd egg. By ♀, though ♂ may take over when ♀ off nest (Adamyan 1963*b*; Abdusalyamov 1973). FLEDGING TO MATURITY. See Social Pattern and Behaviour. BREEDING SUCCESS. In Armeniya (USSR), 18 young reared from 47 eggs; 7 eggs and 15 young destroyed by snakes, 4 eggs addled, 2 overheated, and 1 crushed (Adamyan 1963*b*).

Plumages. ADULT MALE. Entire upperparts, ear-coverts, and sides of neck medium grey, contrasting with black tail and (less so) with dark grey-brown tertials; rather similar to Blackstart *Cercomela melanura*, but grey darker, less light and ashy, and upper tail-coverts grey (black in *C. melanura*). Rather narrow supercilium white, running from nostril backwards, but not extending far behind eye, where merging into light grey above ear-coverts. Lores and lower cheeks black, forming bold black stripe downwards from base of bill along sides of throat, contrasting markedly with white front part of supercilium and with white chin and throat, more gradually merging into grey on rear of upper cheeks and on lower sides of neck. Black of lower sides of throat sometimes connected by black bar across upper chest (in 6 of 80 examined); this bar usually narrow (2-3 mm wide, partly mottled rufous), exceptionally 10-15 mm wide. Colour of remainder of underparts rather variable: usually deep rufous-cinnamon, but pale orange-buff or cream-yellow in 10 of 80 examined; in all, colour gradually paler towards belly and lower flanks, vent and under tail-coverts pale cream or white, thighs dull grey, mottled white. Basal sides of black tail-feathers fringed medium grey. Flight-feathers and tertials dark grey-brown; outer webs of tertials tinged grey, outer webs of flight-feathers fringed grey (widest towards bases of secondaries, absent on p9-p10 and on emarginated parts of outer primaries). Lesser and median upper wing-coverts medium grey, like upperparts; greater coverts similar but inner webs black except for tips. Longest feather of bastard wing black, rather contrasting with remainder of upperwing; remainder of bastard wing and greater upper primary coverts greyish-black with broad but ill-defined medium grey fringes along outer webs and tips. Under wing-coverts and axillaries rufous-cinnamon to

cream-yellow, like chest and flanks; small coverts along leading edge of wing and greater and outer under primary coverts dark grey with paler grey tips. *In worn plumage*, upperparts and upper wing-coverts slightly duller, sometimes with olive-brown tinge, not as bluish medium grey as in fresh plumage; rufous or cream of chest, breast, and flanks slightly bleached (less so on axillaries and under wing-coverts); grey outer fringes of tertials and flight-feathers bleached and partly abraded. ADULT FEMALE. Upperparts and sides of neck grey with slight brown tinge, purer grey on rump and upper tail-coverts. Short stripe above lore pale buff, forming short and ill-defined supercilium, not extending above and behind eye; narrow eye-ring pale buff. Lore, front part of cheek, and sides of chin finely spotted grey and pale buff or white, merging gradually into grey on rear of cheeks, sides of throat, and sides of neck. Ear-coverts grey, partly with slight rufous tinge. Central chin and central throat white, rather contrasting with grey sides of chin and throat and with grey band across upper chest and sides of breast (chest faintly suffused buff). Lower chest and flanks bright orange-cinnamon, paler towards breast and sides of belly, merging into white on lower belly, vent, and under tail-coverts; some grey of feather-bases often showing on breast. Tail black, contrasting with grey upper tail-coverts. Flight-feathers dark grey or greyish-black, secondaries with pale buff-grey outer fringe and inner primaries with narrow buff-grey outer edges. Lesser and median upper wing-coverts grey with slight brown tinge, like upperparts; greater coverts purer grey, tertials darker grey with faintly paler buffish-grey outer fringes and tips. Greater upper primary coverts dark grey, faintly and narrowly edged paler grey along outer web and tip. Under wing-coverts and axillaries bright orange. *In worn plumage*, upperparts purer bluish ash-grey, orange of underparts faded. Marked individual variation, probably homologous with cream- and rufous-bellied 'morphs' of ♂ (no real morphs, as intermediates occur). ♀ with rufous underparts, as described above, comparable with rufous-bellied ♂. Some ♀♀, probably comparable with cream-bellied ♂, have underparts (including under wing-coverts) white with faint cream-buff tinge, but sides of throat and chest grey; feathers of chest pale subterminally, grey tips showing as indistinct scaling or barring. Intermediate birds combine grey-scaled chest with slight rufous-cinnamon suffusion on flanks. NESTLING. Down dark grey, fine and sparse, slightly denser on back (Adamyan 1963*b*). JUVENILE. Rather like pale-bellied adult ♀, but upperparts with triangular pale buff spots, size varying from *c.* 3 mm across on crown to *c.* 5 mm on lower scapulars and tertials; rump mixed grey and buff; upper tail-coverts black-brown with olive-brown tips. Sides of head mottled dark brown and pale buff or cream-white; cream supercilium long but ill-defined. Underparts pale cream-buff, merging into white on vent and under tail-coverts; lower throat, chest, sides of breast, and upper flanks with rather faint black-brown scales or spots. Tail black, tips with traces of rufous. Flight-feathers as adult, but tips of secondaries and inner primaries sometimes with pale rufous spots. Upper wing-coverts brownish-grey, each with distinct rufous (when fresh) or pale buff (when worn) triangular spot on tip; triangles largest (*c.* 4 mm long) on median and tertial coverts, smallest (*c.* 2 mm) on outer greater coverts and greater upper primary coverts. Under wing-coverts and axillaries deep rufous or pale cream. See also Čtyroký (1972). FIRST ADULT MALE. Like adult ♂, but upperparts washed olive-brown, less pure grey; cheeks, sides of neck, and chest dark grey-brown, feathers with rather broad buff or cream tips, appearing scaled; sides of breast and flanks washed rufous orange, merging into cream-white on central belly and vent. Juvenile flight-feathers, tertials, tail, greater upper primary coverts, and often some outer greater upper wing-coverts retained, coverts often showing pale buff or off-white triangular spots on tips. In spring, similar to adult ♂ (perhaps after a partial pre-breeding moult); some birds retain a number of pale-spotted greater coverts or primary coverts, others only some old tail- or flight-feathers (see Moults, and Svensson 1984a). FIRST ADULT FEMALE. Like adult ♀, but cheeks, sides of neck, and chest barred or mottled grey and cream and part of juvenile pale spotted feathers retained (see 1st adult ♂). In spring, like adult ♀, but variable amount of juvenile feathers still present, as in 1st adult ♂.

Bare parts. ADULT, FIRST ADULT. Iris dark brown or dark chocolate-brown. Bill, leg, and foot black in adult ♂, black or black-brown in ♀ and 1st adult ♂. Mouth flesh-coloured in adult ♀. (Dementiev and Gladkov 1954b; Cederwall and Svenaeus 1973; BMNH.) NESTLING. Bare skin grey, tinged orange-yellow on underparts. Gape-flanges pale yellow. At 3 days, bill, leg, and foot pale brown; at 8 days, iris dark brown. (Adamyan 1963b.) JUVENILE. Iris dark brown. Bill, leg, and foot brown (BMNH).

Moults. ADULT POST-BREEDING. Complete; primaries descendant. Perhaps usually a complete moult late June to September on breeding grounds, but quite a few birds suspend moult during autumn migration, retaining 1–4 secondaries which are probably replaced in winter quarters (Svensson 1984a). In mid-July, moult of body feathers started, and in some birds also flight-feathers (Paludan 1959; Rokitansky and Schifter 1971); single ♀ from south-east Turkey, 26 June, had just started wing, body, and tail, and another from USSR, July, had already completed moult (BMNH). In USSR, moult of body starts late June; moult completed about mid-August, except for some flight-feathers (Dementiev and Gladkov 1954b). In Armeniya (USSR), moult starts with p1 from 11 June; primaries half-way through moult late June to July, completed late July or early August. Greater upper primary and secondary coverts replaced at same time as primaries, but moult of secondaries delayed until after leaving breeding area. Body moult starts with shedding of p1–p2, completed from 20 July; tail moults centrifugally, t1 shed at about same time as p4–p5. (Adamyan 1963b.) Moult nearing completion on arrival Kenya, November (D J Pearson). POST-JUVENILE. Partial, perhaps occasionally complete. Starts within 10–15 days of fledging, with head, body, and lesser and median upper wing-coverts (Adamyan 1963b); moult of body feathers completed early July to late August (Dementiev and Gladkov 1954b; Adamyan 1963b; BMNH). Moult suspended during migration, and continued in winter quarters with tertials, 1–3 outer secondaries, greater

upper wing-coverts (except sometimes for some outer), occasionally some tail-feathers, up to 8 primaries and greater upper primary coverts, and some feathers of bastard wing (Svensson 1984a). FIRST PRE-BREEDING. In winter quarters; extent not fully established, but some birds replace sides of head and underparts at least (BMNH, ZFMK).

Measurements. ADULT, FIRST ADULT. Whole geographical range, all year; skins (Paludan 1938, 1940, 1959; Nicht 1961; Vauk 1973a; BMNH, RMNH, ZFMK, ZMA). Bill (S) to skull, bill (N) to distal corner of nostril; exposed culmen on average 4·4 less than bill (S).

WING	♂ 96·2	(2·32; 20)	92–101	♀ 94·6 (1·94; 15)	91–99
TAIL	70·2	(2·49; 14)	67–73	68·9 (3·02; 11)	64–73
BILL (S)	19·0	(0·76; 15)	17·9–20·2	19·0 (0·78; 11)	17·8–20·3
BILL (N)	10·6	(0·58; 15)	9·7–11·4	10·8 (0·59; 11)	10·1–11·8
TARSUS	25·9	(0·64; 14)	25·8–26·7	25·5 (0·86; 11)	24·3–26·8

Sex differences not significant. Wing of 1st adult with retained juvenile flight-feathers on average 5 shorter than in older birds. Wing ♂ (88–)91–102(30), ♀ 87–98(21) (Svensson 1984a).

Weights. (1) April and early May, Kuwait (BTO). (2) April–May, (3) June–July, USSR (Nicht 1961; Dolgushin *et al.* 1970), Turkey (Kumerloeve 1964, 1969a; Rokitansky and Schifter 1971; Vauk 1973a), Iran (Paludan 1938, 1940; Desfayes and Praz 1978), and Afghanistan (Paludan 1959), combined.

(1)	♂ 22·3	(1·73; 8)	19–24	♀ 20·4 (2·64; 8)	16–23
(2)	23·1	(1·69; 7)	21–26	22·8 (1·09; 4)	21–24
(3)	24·5	(1·61; 7)	22–27	24·3 (— ; 2)	23–25

1st adult ♂, Sweden, June: 23·5 (Cederwall and Svenaeus 1973). Sudan, early September: ♂ 19, increased to 21·5 on recapture after 9 days (Nikolaus 1984). Kenya, November–December: 21·9 (18·7–25·2) (Pearson and Backhurst 1976).

Structure. Wing rather long, broad at base, tip bluntly pointed. 10 primaries: p8 longest, p9 4–7 shorter, p7 0–1, p6 1·5–4, p5 6–11, p4 10–15, p1 18–23. P10 reduced, rather narrow and pointed; 47–52 shorter than wing-tip, 0–5 longer than longest upper primary covert. Outer web of p6–p8 and inner of p7–p8(–p9) emarginated. Tail rather long, tip slightly rounded; 12 feathers, t6 4–10 shorter than t2–t4. Bill rather long, straight, slender; wide at base, laterally compressed at tip; tip of culmen slightly decurved. Nostrils oval, partly covered by narrow membrane above. Some fine bristles at base of both mandibles and near nostrils. Tarsus rather long and slender; toes rather short, slender. Middle toe with claw 18·9 (4) 17·8–20·0; outer toe with claw *c.* 65% of middle with claw, inner *c.* 64%, hind *c.* 72%. Claws rather short, slender, decurved. CSR

Phoenicurus erythronotus Eversmann's Redstart

PLATES 49 and 52
[between pages 652 and 653]

Du. Eversmann's Roodstaart   Fr. Rougequeue d'Eversmann   Ge. Sprosserrotschwanz
Ru. Красnoспинная горихвостка   Sp. Colirrojo de Eversmann   Sw. Altairödstjärt

Sylvia erythronota Eversmann, 1841

Monotypic

Field characters. 16 cm; wing-span 25·5–27 cm. 10–15% larger and bulkier than Redstart *P. phoenicurus*.

Rather large redstart, with proportionally larger head, less slim body, and slightly shorter tail than smaller congeners. ♂ unusually patterned, having fully rufous fore-underparts and back, and 2 white patches on wing. ♀ fawn-coloured, with prominent eye-ring and whitish vent. Flirts tail upwards but does not quiver it. Sexes dissimilar; seasonal variation in ♂. Juvenile separable.

ADULT MALE. From mid-winter to summer (when plumage worn), forehead and indistinct supercilium grey-white, crown and nape ash-grey; lores, cheeks, and broad stripe on side of neck black, joining mostly black wing-coverts. Inner wing-coverts white, creating panel below scapulars extended by buff-white fringes of inner secondaries; white primary coverts form 2nd patch; flight-feathers brown-black. Back, rump, outer tail, and underparts from chin to sides of belly dark orange-chestnut; lower underbody and vent cream to white; central tail-feathers dark brown. From September to early winter (when plumage fresh), colourful pattern subdued by presence of broad brown feather-fringes on crown, nape, and back, and white tips on fore-underparts. ADULT FEMALE. Fawn grey-brown above and on chest and fore-flanks, dull white on throat, belly, and vent. Rather uniform but for striking buff-white eye-ring, obvious fulvous-white fringes to tertials and wing-coverts, buff rump, and bright rufous upper tail-coverts and tail. JUVENILE. Brown with dull yellow spots above, cream with black mottling (on foreparts) below. Wings and tail much as adult ♀. FIRST-WINTER MALE. Resembles winter adult but with feather-fringes even broader; black areas duller. At all ages, bill and legs black.

♂ unmistakable but ♀ and immature far less distinctive, recalling *P. phoenicurus* but lacking any full orange-buff below and having paler buff rump, duller, less bright-sided tail, more conspicuous eye-ring, and paler fringes on wing. Flight and behaviour not well studied but apparently as *P. phoenicurus*, though tail movement consists of distinct upward flirt without nervous quiver. Perches on rocks and bushes. Solitary.

Song a quiet but persistent babbling chatter. Commonest call a soft, slurred croaking monosyllable; alarm-call a croaking 'gre-er'.

Habitat. Breeds in lower-middle latitudes, extralimitally, in temperate continental montane regions up to 5400 m at upper limit of spruce *Picea*, but mainly much lower in sparse woodlands, often of stunted trees, broad-leaved or coniferous, or of tall shrubs, on dry stony rather than moist soil. Avoids tall dense coniferous stands and low shrub tangles, preferring well-lit glades or forest margins, and neighbourhood of dry slopes or scree. After breeding, descends to lower levels with trees and high bushes, to forests beside lower reaches of rivers, and also orchards (Dementiev and Gladkov 1954*b*). In winter in Indian subcontinent, usually below 2100 m in arid country: waste land, scrub jungle, olive groves, orchards, dry river beds, wooded compounds, and avenues and groves of *Acacia*, *Prosopis*, and similar trees. Usually perches on a

stone or lower branch of a tree or thorn bush, descending momentarily to pick up an insect and return (Ali and Ripley 1973*a*). In Afghanistan found in spring in oak scrub in steep valley at 1500–1700 m, and in a hotel garden at *c.* 700 m; in autumn in high open valley at 2550 m (Paludan 1959).

Distribution. Breeds in Tien Shan, Tarbagatay, Altai, and from north-west Mongolia to west and south of Lake Baykal. Winters in southern USSR, Afghanistan, Iran, northern Pakistan, and western Himalayas.

Accidental. USSR, Turkey, Iraq, Kuwait.

Movements. Vary from altitudinal displacement to short- or medium-distance migration, with a few birds penetrating south-west as far as Persian Gulf. The following account based mainly on Kovshar' (1979) and especially Dathe and Neufeldt (1983).

Wintering areas vary from year to year, even month to month, as direct result of weather and food availability; birds often forced to move well away from breeding grounds to find suitable conditions. True migrant in north-east of breeding range (Altai region, *c.* 50–55°N), leaving high mountains for adjacent lowlands during second half of September, vacating these in turn during October; in spring, vanguard reappears in second half of March, with immigration continuing into early May. In Dzungaria (*c.* 45–50°N), where also largely migratory, some remain in lower mountains until advent of snow and frost in early November, or even into December around Alakol'; on lower hills of Dzungarian Alatau, present in autumn mid-September to mid-December. Return movement in spring can begin late February, but usually 10 March to mid-April. In Tien Shan system, in south-west of range (*c.* 40–45°N), birds use 3 different strategies: (1) typical altitudinal movement within breeding range; (2) short- to medium-distance dispersals, to lower mountains and valleys on periphery of summer range; (3) seasonal migration away from breeding grounds.

Birds from different breeding regions (Altai, Dzungaria, Tien Shan) not separable in wintering areas, in absence of ringing. Autumn migration oriented south-westwards, across western Mongolia, Chinese and Soviet Turkestan, Transcaspia, and Afghanistan; some winter (locally common) in Soviet Central Asia (southern Kazakhstan, Turkmeniya, Tadzhikistan), but others winter south to Himalayas (Nepal westwards), hills and plains of northern Pakistan, southern Afghanistan, and much of Iran. Also occurs occasionally or in very small numbers further west: in Azerbaydzhan (USSR), north-west Iran, Iraq, and along Arabian side of Persian Gulf from Kuwait to northern Oman. Passage occurs across southern Kazakhstan from early October to early November and early March to mid-April, and through Afghanistan in October and March; wintering areas further south are occupied November–March. ♂♂ leave breeding grounds 1 month

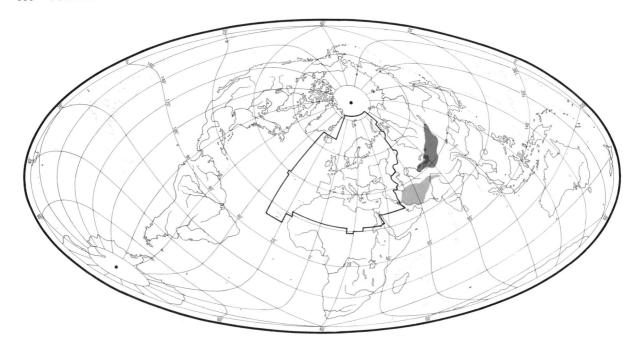

later than ♀♀ in autumn, and return 1-2 weeks earlier in spring. (Dementiev and Gladkov 1954*b*; Vaurie 1959; Dolgushin *et al.* 1970; Ali and Ripley 1973*a*; Bundy and Warr 1980; Walker 1981*a*; Dathe and Neufeldt 1983.)

RH

Food. Largely fruit and seeds in mid-winter, largely insects at other times. Feeds by (1) picking items from ground (sometimes rummaging in leaf-litter or clearing away thin snow layer) or from vegetation (including bushes), (2) flying on to prey from low perch, or (3) taking aerial prey in brief flight. Also recorded fluttering by wall in attempt to catch insect. (Ivanov 1969; Dolgushin *et al.* 1970; Ali and Ripley 1973*a*.)

Breeding season diet largely insects, especially adult and larval beetles (Coleoptera) and larval Lepidoptera (see Yanushevich *et al.* 1960, Pek and Fedyanina 1961, Dolgushin *et al.* 1970, Kovshar' 1979).

Outside breeding season, more fruit and seeds eaten, especially in mid-winter when these may be taken more or less exclusively. Thus, in autumn in Zailiyskiy Alatau mountains (Kazakhstan, USSR), stomachs contained only invertebrates. In October, December, and March in Talasskiy Alatau mountains (Kazakhstan), 12 stomachs contained mainly invertebrates—dragonflies, etc. (Odonata), grasshoppers (Acrididae), bugs (Pentatomidae), fly larvae, ants (Formicidae), beetles (Dytiscidae, Scarabaeidae, Staphylinidae, Tenebrionidae, Curculionidae), spiders (Araneae), and centipedes, etc. (Myriapoda); only 3 of these stomachs also contained seeds, though stomachs from January contained only seeds and *Berberis* berries. Near Alma-Ata (Kazakhstan), one stomach from late December contained only seeds of bryony, seeds and

stalks of *Rhamnus*, and seeds of *Berberis*; one from February contained mainly insects—dipteran flies (Dolichopodidae), a bug (Hemiptera), 2 beetles (Chrysomelidae, Curculionidae), and Myriapoda, also seeds of *Convolvulus* and green plant material. (Dolgushin *et al.* 1970.) In Kirgiziya, August, some stomachs contained berries of sea buckthorn *Hippophae rhamnoides* and seeds of Leguminosae; in winter, 9 contained berries of *Hippophae* and *Berberis*, 2 contained seeds of docks, etc. (Polygonaceae), 1 contained rose *Rosa* hips, and 4 contained plant material and beetles (Yanushevich *et al.* 1960). On lower Vakhsh (Tadzhikistan) in winter, berries of *Eleagnus* are main, perhaps only, food up to late January when birds start to move away from thickets (Ivanov 1945). Birds from Uzbekistan (USSR) in November–December and March contained insects and spiders with very little plant material (Ivanov 1969).

DJB

Voice. For detailed study of calls (not song), comparison with other *Phoenicurus*, and sonagrams, see Böhme and Böhme (1986).

CALLS OF ADULTS. (1) Song of ♂. Fine, cheerful and varied (Johansen 1955); not like typical *Phoenicurus* according to Sushkin (1938). In Tien Shan (southern USSR), sings rather little and quietly: purling or babbling chatter more like subsong than true song. The most characteristic feature is persistence: 5–6 (or more) bouts follow one another almost without interruption. Rarely includes mimicry of other birds: e.g. in 2 ♂♂ song started with accurate copy of Willow Tit *Parus montanus songarus*. Uncertain and quiet song sometimes given by juveniles before completion of moult, mid- to late August. Normally sings from tree or bush, less commonly from rock or stone, not infrequently also in flight (for further details, see Kovshar' 1979). (2) Contact-alarm calls. (a) Distinctive growling rattle—'dr-r-r-r' (Gallagher and Woodcock 1980); soft (muted) 'trr' (Flint *et al.* 1984); resembles 'karr' of nightingale *Luscinia* (Sushkin 1938). Volume and timbre apparently vary, and other descriptions, presumably of same call, as follows: peculiar croaking 'gre-er' or creaking 'greer' as alarm-call, also in winter quarters (Whitehead 1909; Whistler 1920; Jones 1947-8); in Nepal, loud, nasal and rasping 'chaaaaaaan' (Fleming *et al.* 1976). (b) Softer, slurred variant of call 2a said by Whistler (1920) to be the 'ordinary call'. (c) Loud whistling 'few-eet' (Flint *et al.* 1984); short, abrupt whistle or squeak given by ♂ apparently as warning to ♀ feeding on ground, but not heard at other times (Kovshar' 1979) is perhaps the same or a related call.

MGW

Plumages. ADULT MALE. In fresh plumage (about late August to December), forehead, crown, and hindneck pale grey-brown, mantle and scapulars pale grey-brown with some deep rufous of feather-bases visible on mantle and inner scapulars, some black on outer scapulars. Back rufous with some grey feather-tips, rump and upper tail-coverts uniform deep orange-rufous. Lores, narrow ring round eye, and stripe just below and behind eye to upper ear-coverts black, faintly speckled pale buff, bordered by narrow and rather indistinct pale ash-grey supercilium above. Lower cheeks, rear part of ear-coverts, and sides of neck pale grey-brown with some black of feather-bases visible. Chin, throat, chest, and upper flanks pale pinkish-grey, usually with some deep orange-rufous of feather-bases visible (in particular from chin to central chest). Sides of belly, lower flanks, and under tail-coverts orange-buff or pale buff, merging into white on central belly and vent. Central pair of tail-feathers (t1) black with narrow rufous outer edge; other tail-feathers deep orange-rufous, t2-t5 often with dusky wash and dark shaft at tip (sometimes hardly visible, occasionally forming more contrasting dull black tip 3-5 mm wide), t6 with dusky streak on distal part of outer web. Flight-feathers dark grey, almost black on innermost tertials; inner webs bordered off-white at bases; outer webs of primaries narrowly edged off-white (except p9 and emarginated parts), outer webs of secondaries and outer webs and tips of tertials more broadly fringed tawny or pale buff (widest and virtually white on outer webs of tertials and on middle portions of outer webs of inner secondaries, forming pale panel on closed wing). Shorter lesser upper wing-coverts, outer median coverts, most greater coverts, and bastard wing deep black, outer and greater coverts rather broadly fringed white or pale buff on tips; longer lesser coverts, inner median coverts, and tertial coverts contrastingly white (faintly edged pale brown-grey on tips when fresh), forming large and conspicuous wing-patch. Greater upper primary coverts white (forming another prominent patch) with 3-5 mm of tip black; black has rather square-cut basal edge and is narrowly edged pale grey-buff on outer web and distally. Under wing-coverts and axillaries white; axillaries, some longer coverts, and marginal coverts have partly black bases. *In worn plumage*, markedly different, as broad pale grey-brown fringes on upperparts and pink-white fringes on underparts gradually wear off, revealing bright bases. From about February-March, forehead, crown, hindneck, and upper mantle bluish ash-grey (almost white on

forehead and sides of crown, especially when worn, June-July); mantle, inner scapulars, and back contrastingly deep and glowing rufous-cinnamon; narrow bar across base of bill, lores, ring round eye, upper cheeks, and ear-coverts black, forming contrasting mask extending back over sides of neck and outer scapulars to inner tertials; sometimes a narrow black bar between blue-grey hindneck and rufous mantle. Rump to tail as in fresh plumage, slightly paler orange-rufous than mantle and back. Chin to breast and upper flanks deep rufous-cinnamon (some traces of white feather-fringes usually present on central breast and upper flanks), contrasting with cream-white or off-white belly and vent. Flight-feathers and wing-coverts as fresh plumage, but pale fringes on outer webs of flight-feathers and on tips of black outer and greater upper wing-coverts largely worn off, white bar from carpal joint to inner tertial coverts and white patch on bases of primary coverts more contrasting, but white fringes of outer webs of tertials and inner secondaries narrower (in particular in summer, when heavily worn), hardly forming pale patch in closed wing. ADULT FEMALE. In fresh plumage (about late August-January), upperparts and sides of neck grey-brown, slightly paler and more sandy than upperparts of ♀ Redstart *P. phoenicurus* and plumage markedly denser and looser; rump buff-cinnamon, merging into deep orange-rufous of upper tail-coverts (not as uniform rufous-cinnamon as *P. phoenicurus*). Narrow bar across base of upper mandible, lores, upper cheeks, and ear-coverts pale buff with slight dusky mottling; small patch in front of eye and upper ear-coverts usually mottled and streaked dark brown (unlike *P. phoenicurus*). Narrow pale buff eye-ring (as in *P. phoenicurus*). Lower cheeks, sides of breast, chest, and upper flanks pale brownish-grey, merging into cream-white or buff-white on chin, throat, central belly, and vent, and into buff or pink-buff on lower flanks and under tail-coverts; some deeper buff or rusty-buff of feather-bases often visible on chest and breast. Tail as adult ♂, often with slightly dusky tips. Flight-feathers and tertials as adult ♂, but pale fringes more sandy-brown, in particular those of tertials less white and less contrasting. Lesser and median upper wing-coverts dull black with broad pale sandy-grey or pale buff fringes (fringes almost concealing black, especially on lesser coverts); greater coverts dull black with broad sandy-buff fringes along outer webs and tips. Fringes of outer webs and tips of tertial coverts often whitish (sometimes also those of median coverts), forming trace of pale wing-bar like adult ♂; occasionally, tertial and median coverts largely white, forming more complete bar. Bastard wing dull black; greater upper primary coverts dull black with broad pale grey-buff fringe along outer web and narrower one on tip (no white on base, unlike adult ♂). Under wing-coverts and axillaries as adult ♂, but partly tinged buff. *In worn plumage*, upperparts duller and greyer; underparts paler with more buff, rusty-buff, or grey-brown of feather-bases visible (less uniform than in fresh plumage), but change in colour of body slight compared with adult ♂; change in wing more marked, however, black of bases of median and greater upper wing-coverts more exposed and sandy-buff of outer webs of tertials and of tips of median and greater upper wing-coverts bleached to off-white, forming double wing-bar. Pale tips of coverts wear off during nesting (sometimes even from April), wing-bars disappearing; in some birds, buff fringes abrade before bleaching, and thus no conspicuously white wing-bars present. JUVENILE. Upperparts brown with rusty-yellow spots and dark brown feather-fringes; rump rufous with black feather-edges; upper tail-coverts uniform rufous; underparts cream with black arcs, belly almost uniform pale cream. Lesser and median upper wing-coverts black with buff tips, remainder of wing and tail as adult (but see First

Adult, below). Sexes separable: greater upper primary coverts of ♀ dull black with paler fringes, ♂ with variable amount of white on bases. FIRST ADULT MALE. Like adult ♂, but grey-brown fringes of upperparts and sides of head and pink-white fringes of underparts broader, more fully concealing bright colours of feather-bases in autumn and taking longer to wear off; bright plumage acquired March or later. Juvenile flight-feathers, tail, tertials, greater upper wing-coverts, and greater upper primary coverts retained; flight-feathers and tail relatively more worn than adult at same time of year and with less broad and smooth tips; black of tertials and wing-coverts browner; fringes on tips of greater upper wing-coverts broader on outer web than on inner web (about equal on both webs in adult), but this hard to see when plumage worn. Best ageing character is greater upper primary coverts: tip more broadly black-brown than in adult, and black on outer web often runs submarginally as mottled line towards base on at least some (sometimes all) coverts, white of base sometimes largely concealed; dark line on outer web sometimes absent, but then black of tip rather unequally distributed over both webs, unlike rather square-cut division of adult. FIRST ADULT FEMALE. Like adult ♂, but part of juvenile feathers retained, as in 1st adult ♂; retained feathers virtually similar to adult ♀, differing only in relatively greater wear; tips of greater and median upper wing-coverts and outer webs of tertials sooner bleached to white (from about November–December) and sooner worn off completely.

Bare parts. ADULT, FIRST ADULT. Iris brown or dark brown. Bill, leg, and foot black; mouth yellowish or flesh (Dementiev and Gladkov 1954b; Ali and Ripley 1973a; ZMA). JUVENILE. No information.

Moults. ADULT POST-BREEDING. Complete; primaries descendant. Probably starts early July or earlier, as moult completed in birds from 16–19 August (Dementiev and Gladkov 1954b; RMNH, ZMA). POST-JUVENILE. Partial: head, body, lesser and median upper wing-coverts, and occasionally some inner greater upper wing-coverts; no flight-feathers, tertials, tail, greater upper primary coverts, and longer feathers of bastard wing. In USSR, moult started in one from 24 July; some completed late August, when others still fully juvenile; in Mongolia, some start mid-August, others then already in fresh 1st adult; head, scapulars, and chest replaced last (Dementiev and Gladkov 1954b). Moult completed in birds from Tien Shan (USSR) from mid- and late August (RMNH, ZMA).

Measurements. ADULT, FIRST ADULT. Mainly Turkmeniya and Kirgiziya (USSR), some elsewhere USSR, China, and Iran, August–April; skins (RMNH, ZMA). Bill (S) to skull,

bill (N) to distal corner of nostril; exposed culmen on average c. 4·8 mm less than bill (S).

	♂		♀	
WING	87·7 (1·83; 51)	84–92	85·1 (1·47; 13)	83–89
TAIL	67·6 (1·92; 41)	64–71	65·8 (2·66; 13)	62–69
BILL (S)	15·5 (0·44; 27)	14·8–16·3	15·5 (0·60; 13)	14·7–16·2
BILL (N)	7·7 (0·30; 28)	7·3–8·2	7·9 (0·43; 13)	7·3–8·5
TARSUS	23·6 (0·57; 29)	22·6–24·7	23·3 (0·58; 13)	22·4–24·2

Sex differences significant for wing and tail. 1st adult (with retained juvenile flight-feathers and tail) combined with older birds above, though juvenile wing on average 1·8 shorter than adult and tail on average 0·5 shorter: e.g. wing, adult ♂ 88·6 (1·34; 25) 86–91, 1st adult ♂ 86·8 (1·83; 26) 84–92.

Weights. (1) November–March, Kazakhstan, USSR (Dolgushin et al. 1970) and Afghanistan (Paludan 1959), combined. (2) April–May, Mongolia (Piechocki et al. 1982) and Kazakhstan. (3) June–August, Mongolia (Piechocki and Bolod 1972) and Kazakhstan. (4) October, Kazakhstan and Afghanistan. (5) Undated, Kirgiziya (Yanushevich et al. 1960).

	♂		♀	
(1)	18·7 (1·53; 18)	15–22	17·8 (1·51; 5)	17–21
(2)	18·4 (2·06; 7)	16–22	18·0 (— ; 1)	—
(3)	18·7 (1·62; 13)	16–22	—	
(4)	18·0 (0·95; 5)	17–19	17·2 (— ; 2)	17–18
(5)	— (— ; 19)	15–22	— (— ; 12)	16–22

Structure. Wing rather short, broad at base, tip rounded. 10 primaries: p7 longest, p8 0·5–3 shorter, p9 11–15, p6 0–0·5, p5 1–3, p4 7–11, p3 10–15, p2 14–18, p1 16–20. P10 reduced, 39–44 shorter than p7, 9–13 longer than longest upper primary covert. Outer web of p5–p8 and (slightly) inner of p6–p9 emarginated. Tertials short. Tail rather short, tip square or slightly forked; 12 feathers, t1 and t6 1–5 shorter than t3–t4. Bill short, rather slender; straight, but tip of culmen decurved. Nostril oval, covered by membrane above. Long fine bristles near base of upper mandible, shorter bristles on forehead and chin. Tarsus and toe rather long and slender. Middle toe with claw 16·9 (30) 15·5–18 mm; outer toe with claw c. 68% of middle with claw, inner c. 66%, hind c. 72%. Claws rather short, slender, slightly decurved.

Geographical variation. None.

Forms superspecies with Alashan Redstart P. alaschanicus from eastern Kunlun Shan mountains east to Ho-lan Shan mountains (central China), which is sometimes included in P. erythronotus, latter then forming single polytypic species; for reasons for separating them, see (e.g.) Vaurie (1972). P. alaschanicus differs from P. erythronotus in relatively longer and more strongly rounded tail and (in ♂) in white central belly and in absence of black mask on side of head—blue-grey of crown reaches down to lores and ear-coverts and rusty-orange of underparts up to cheeks (Hartert 1910; BMNH).   CSR

Phoenicurus ochruros Black Redstart

Du. Zwarte Roodstaart   Fr. Rougequeue noir   Ge. Hausrotschwanz
Ru. Горихвостка-чернушка   Sp. Colirrojo tizón   Sw. Svart rödstjärt

Motacilla Ochruros S G Gmelin, 1774

Polytypic. *P. o. gibraltariensis* (J F Gmelin, 1789), western and central Europe east to Latvia and Crimea, south to northern Spain, Sicily, and Greece; also north-west Africa and (probably this race) western Asia Minor; *aterrimus* von Jordans, 1923, Portugal and southern and central Spain north to Valladolid; *semirufus* (Hemprich and Ehrenberg, 1833), Levant; nominate *ochruros* (S G Gmelin, 1774), mountains of Turkey east from *c.* 32°E, Caucasus, Transcaucasia, and northern Iran east to Mazandaran; *phoenicuroides* (Horsfield and Moore, 1854), Tien Shan through mountains of north-central Asia to western Sayan, south to Fergana basin and northern Mongolia; *rufiventris* (Vieillot, 1818), Transcaspia from Bol'shoy Balkhan and Kopet-Dag (Turkmeniya) and north-east Iran east through Himalayas north to Turkestanskiy and Alayskiy Khrebet, Kun Lun Shan, eastern Tsinghai, and Ningsia (China); this race and/or *phoenicuroides* occurs in Middle East and Egypt outside breeding season. Extralimital: *xerophilus* Stegmann, 1928, Astin Tagh (Sinkiang) to western Kansu and Tsinghai, China.

Field characters. 14·5 cm; wing-span 23–26 cm. Marginally larger than Redstart *P. phoenicurus*, with 10% longer wings and 10–15% longer tail; 20% smaller than Güldenstädt's Redstart *P. erythrogaster*. Redstart of closely similar form to *P. phoenicurus* but differing distinctly in less colourful plumage (in western races), less attenuated form, ground-hugging behaviour, and running (as well as hopping) gait. Dominant colour in all plumages dusky-black in ♂, dusky grey-brown in ♀ and immature; in all plumages, striking chestnut rump and tail and (in western adult ♂♂) white wing-panel. Eastern races have underparts below chest increasingly saturated deep chestnut, suggesting *P. phoenicurus* and *P. erythrogaster*. Sexes dissimilar; seasonal variation in ♂. Juvenile separable. 5–6 races occur in west Palearctic, with ♂♂ of westernmost races (the only truly 'black' redstarts) easily separable from others.

ADULT MALE. (1) Race of Europe (except south and central Iberia) and north-west Africa, *gibraltariensis*. In spring (when plumage worn), head, back, wing-coverts, and underparts to rear flanks and thighs dusky slate, blackest on head, mantle, and chest; wings brown-black with dull but distinct ochre-grey to off-white wing-panel formed by basal fringes to tertials and inner secondaries. Underwing dusky grey, merging with flanks. Centre of belly grey-white, vent and under tail-coverts orange-buff, these areas combining into pale vent obvious only at close range. Rump and tail bright chestnut, with central tail-feathers and tips of outer dark brown. From August, fresh plumage copiously marked with pale feather-tips: tips olive-brown on upperparts, grey on sides of head and throat, and ochre or white on underparts; thus crown, lower mantle, and middle of flanks look 'hoary' (but less so than in *P. phoenicurus*) and ventral area paler, even markedly white; wing-panel more prominent. (2) Central and south Iberian race, *aterrimus*. Intensely black on head, mantle, and chest, with larger and more

contrasting wing-panel always appearing white. Underwing dusky-black. (3) Race of central Turkey and Caucasus east to north-west Iran, nominate *ochruros*. Variably intermediate between *gibraltariensis* and west-central Asian races, but most individuals darker above than former and chestnut below breast like latter; wing-panel variable. Underwing dusky-black, sometimes chestnut. (4) Levant race, *semirufus*. Back blacker than *gibraltariensis*; underwing and body below breast rich chestnut; wing-panel indistinct. (5) West-central Asian races, *rufiventris* and *phoenicuroides*. Resemble *semirufus* but often have grey or white forehead, blue-grey crown and wings, and more bright orange-chestnut on underbody (reaching upper breast) and underwing; wing-panel absent. ADULT FEMALE. (1) Western races, *gibraltariensis* and *aterrimus*. Upperparts grey-brown, noticeably paler than ♂, but distinctly smokier and duskier than ♀ *P. phoenicurus*; eye-ring less distinct than in ♀ *P. phoenicurus*. Underparts brown, also with smoky or dusky wash from throat to flanks but brown-white on centre of belly and vent, becoming orange-buff on under tail-coverts. Fringes of tertials and inner secondaries pale brown, forming indistinct wing-panel when bleached. Rump and tail less bright chestnut than ♂, with central tail-feathers paler brown and less contrasting. (2) Eastern races. *P. o. phoenicuroides* distinctly less dusky overall and much paler brown below than western races; other races intermediate. JUVENILE. All races. Resembles adult ♀ but more uniform dusky-brown, lacking any grey tone above; liberally flecked and barred dark brown—on upperparts down to rump, and on chest and flanks (not pale-spotted like *P. phoenicurus*). FIRST WINTER AND SUMMER. All races. From August, ♂ assumes ♀-like or partial adult ♂ plumage; in partial adult ♂ plumage, eastern races already show contrast between dark chest and dull chestnut underbody. ♀-like 1st-year ♂ often has greyer upperparts, chest, and flanks, and paler wing-panel than 1st-year ♀. At close

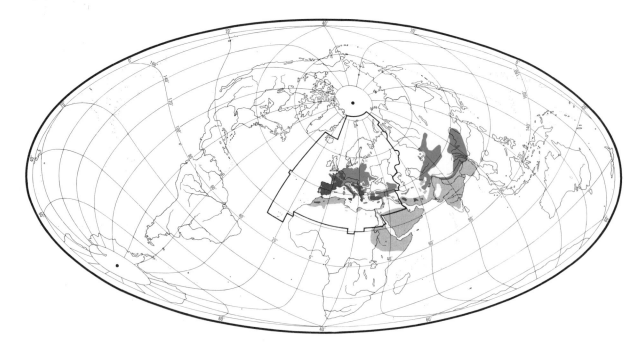

range, both sexes in either plumage distinguished from adult ♀ by retained dull juvenile wing-coverts and fringes to tertials. At all ages, bill and legs black.

Western races *gibraltariensis* and *aterrimus* unmistakable in all plumages, with uniform dusky appearance quite distinct from any *P. phoenicurus*. Adult ♂♂ of eastern races with chestnut underbody do recall *P. phoenicurus* but are quickly separable on blacker back and lack of large pale vent; ♀♀ and 1st-years of eastern races more likely to be confused with *P. phoenicurus* but differences in form and gait should prevent confusion, while lack of any warm rufous or buff on underbody above vent is apparently trustworthy. Flight like *P. phoenicurus*, with similar fluent action but rather less flickering wing-beats and much less loose-tailed appearance, since flight silhouette more compact, recalling Robin *Erithacus rubecula*. Gait a brisk hop like *P. phoenicurus* but, unlike that species, often also runs quickly like wagtail *Motacilla*; pauses in alert, upright posture. Less shy than *P. phoenicurus* and markedly more terrestrial, perching mainly on rocks and stone artefacts though using hedgerows on migration. Tail movement as *P. phoenicurus* but less constant. Mainly solitary.

Song a rather rushed warble, less full-toned and quieter than *P. phoenicurus*, with highly individual subterminal phrase suggesting rattle of ball-bearings shaken together or poured into bottle; often ends in louder burst of rushed, ringing notes which are also included in earlier part of song. Commonest call a short 'tsip' (which does not recall *Phylloscopus* warbler, unlike common call of *P. phoenicurus*); this note often introduces scolding 'tucc-tucc'; in alarm, gives a rapid rattle 'tititicc'.

Habitat. Breeds in west Palearctic in middle latitudes, oceanic as well as continental, mainly in sunny warm or mild temperate, Mediterranean, and steppe or montane climates, avoiding persistently wet or humid situations and dense vegetation of any height (Voous 1960). Favours rocky, stony, boulder-strewn broken or craggy terrain, including cliffs, right up to snowline, but in Switzerland infrequently higher than 2400 m (Glutz von Blotzheim 1962). In Carpathians, chiefly on slopes of mountains covered with junipers *Juniperus*, as well as scree and boulders; in Altai on dry mountain slopes or piles of stones covered with shrubs, but these not essential (Dementiev and Gladkov 1954b). Avoids forests and meadows; is attracted by shady ravines and river banks, and by huts and dwellings in settlements. Frequency and convenience of nest-sites in walls or roofs of buildings, outbuildings, and wide variety of other structures have evidently led to evolution of close commensalism with man, spreading from montane to lowland regions and facilitating extensive northward spread across plains and valleys. This has extended even into large cities where absence or scarcity of water, trees, shrubs, and grasslands more than compensated for by presence of waste patches colonized by weed species, with many bare disturbed areas of soil, and a choice of commanding song-posts and cavities suitable for nesting. Suburbs, parks, and landscaped areas are therefore less often colonized than industrial sites, railways, warehouses, large public buildings, and (especially) churches with towers, provided there is immediate access to open ground with adequate food for rearing brood and space for display which involves wide-ranging flight. Distinguished among small

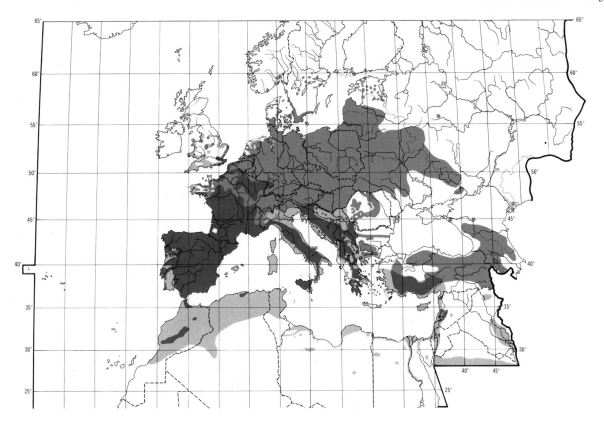

ground-feeding passerines by partiality for conspicuous high song-posts, often at 20 m or more, on hard perches, natural or artificial, avoiding trees and shrubs. In winter, occurs at lower altitudes but in similar habitats.

Distribution. Has spread in north-west Europe, including Scandinavia and USSR.

BRITAIN. Bred Durham 1845, Sussex 1909 (perhaps), Sussex 1923, and Cornwall for some years, then annually since 1939 mainly in south-east England. Has bred in most English counties, but still predominantly in south-east (Parslow 1973; Sharrock 1976; Morgan and Glue 1981). FRANCE. Some spread, e.g. Maine *c.* 1950 and interior of Normandy 1958 (Yeatman 1976). NETHERLANDS. Some spread in range (CSR). WEST GERMANY. Bred in north since 19th century (Yeatman 1976). DENMARK. Colonized mid-19th century (Dybbro 1976). NORWAY. First bred 1944 (VR). SWEDEN. First bred 1910 (Morgan and Glue 1981). FINLAND. Has bred regularly since 1970s (OH). USSR. Spreading in north (GGB). Lithuania: first bred 1939 (HV). Latvia: first bred 1923; range increase after 1945 (HV). (HV). Ukraine: now breeding in centre (Belik 1977).

Accidental. Faeroes, Iceland, Madeira.

Population. Has increased in north-west Europe, including Scandinavia.

BRITAIN. From 1942 to 1979 increased from under 20 to 104 singing ♂♂, with highest figures in 1975 and 1977-8 (Morgan and Glue 1981; see also Fitter 1965, 1971, 1976). FRANCE. 110 000 to 1 million pairs (Yeatman 1976). NETHERLANDS. 3000-4000 pairs 1970-8 (Teixeira 1979). Increased (CSR). WEST GERMANY. 380 000 to 1·3 million pairs (Rheinwald 1982). DENMARK. About 200 pairs (Dybbro 1976); *c.* 150 pairs (*Br. Birds* 1979, **72**, 275-81). Increasing (TD). NORWAY. 20-30 pairs (VR). SWEDEN. About 300 pairs (Ulfstrand and Högstedt 1976). FINLAND. 5-10 pairs (OH). POLAND. Scarce (Tomiałojć 1976). Stable (AD). CZECHOSLOVAKIA. No changes reported (KH). USSR. Lithuania: fairly common by 1959 (HV). Voronezh: increased (L L Semago).

Oldest ringed bird 8 years 5 months (Rydzewski 1978).

Movements. Resident, partial migrant, or migrant in different parts of range, main wintering area for west Palearctic breeders being Mediterranean basin, with a few on west coast of Europe and some southward extension to north-west and (especially) north-east Africa, and east through Middle East and Arabia to India and Burma. Populations breeding Balkans, Italy, and Spain thought to be sedentary (Erard and Yeatman 1967). Vaurie (1959) stated that birds breeding Portugal and central and southern Spain (*aterrimus*) are resident. Birds breeding in plains and low valleys of northern Spain are sedentary, but mountain breeders winter in neighbouring valleys (Erard and Yeatman 1967). In Haut Atlas of Morocco, mountain populations descend in January to 2000 m or below, but still seen on 30 November between 2100 m and snow-line (Heim de Balsac and Mayaud 1962).

P. o. gibraltariensis (breeds in most of European range east to Ukraine and Crimea). A few winter in Britain and west and central France; regular in winter in Belgium, and occasional in Switzerland and Germany. Northern limit of wintering varies from year to year. Southern limit is northern edge of Sahara with occasional records further south, e.g. regular in Algeria at Beni Abbes 30°N, and 2 seen at Tamanrasset in Ahaggar massif at 23°N (Erard and Yeatman 1967). Further south still, recorded on 7 April at $20\frac{1}{2}$°N (Atar, Mauritania), north of Kanem (Chad) on 7 October (race not specified), and at Bamako (Mali) in December (Lamarche 1981). However, bulk winter in Mediterranean basin, being common on coasts and islands in western Mediterranean. Further east, on Malta, occurs on passage mid-October to early November and March, and present during winter in relatively small numbers (Sultana and Gauci 1982); common along coastal zone of Libya from October to mid-April with majority in west, some birds at inland oasis towns, and a few records further south in desert (Bundy 1976). Also winters in Egypt, with spring passage in late February (Dementiev and Gladkov 1954b). Common winter visitor to Cyprus, late October to February, with 2 birds re-trapped in same area in subsequent winters (Flint and Stewart 1983); evidence of passage influxes mid-October to early November and early to mid-March (Hubbard 1969). This race is also the commonest wintering in Middle East. As suggested by dates of passage and wintering in Mediterranean, movement is quite late in autumn and early in spring, at least in western Europe. On southern edge of Swiss Jura, juveniles first to leave breeding grounds, dispersing July–August, when 98% of 1935 birds trapped were 1st-years; this movement apparently not orientated, and Zink (1981) listed 3 movements of 100–290 km between ENE and NNE from natal area in July of birds ringed Belgium and Germany. Very few 1st-years seen in Jura after mid-August, and from mid-September passage of adults starts, with only 3% of birds trapped being in juvenile plumage; peak passage c. 10 October, with rapid decline after 20 October. (Biber 1973.) A few have reached wintering areas by early September, but most not before second half of October. Birds are back on breeding grounds in Germany in early March, in Kiev (USSR) by late March, and Carpathians by late March or early April (Dementiev and Gladkov 1954b). On north-west limit of breeding range, passage in Britain peaks in late March in west, but mid-April in east (Langslow 1977); birds breeding at Dungeness (Kent) arrive mainly late April and are largely immatures (Riddiford and Findley 1981).

Zink (1981) gave detailed ringing recovery maps, and prior to that Hempel and Reetz (1957) and Erard and Yeatman (1967) also interpreted European ringing recoveries. Birds ringed west of c. 13°E migrate in autumn between west and south, mainly to Mediterranean coasts, with few in western Portugal and none in north-west Spain. For example, nestlings ringed in Belgium and Luxembourg have been recovered October–January in Portugal (1), southern Spain (12), eastern Spain (3), Balearic islands (3), and North Africa (7), with also an April recovery in southern Spain and another in March on coast of south-east France. Birds raised in Switzerland also travel to Spain and Balearics, with recoveries in France mainly during migration periods, including 2 autumn recoveries on Atlantic coast almost due west of ringing sites. The many recoveries in North Africa are mainly coastal, but there are 3 well inland in Morocco; most are south to south-west orientations of birds almost all ringed west of 13°E, though there is one from West Germany to Egypt. Zink (1981) split the large numbers of recoveries of birds ringed at 8–13°E into a number of sub-samples, showing October recoveries on south and south-east coast of France and also northern Italy for birds ringed at 10–13°E and *en route* to wintering areas in western Mediterranean. Fewer recoveries in Spain by January than earlier in winter, and in southern France a February–March concentration of birds on spring migration occurs. For ringing sites at 13–18°E there is increasing evidence of a migratory divide with recoveries in winter in both east and west Mediterranean—though according to Erard and Yeatman (1967) migration takes place on broad front throughout Mediterranean, birds freely crossing mountains. However, recoveries show that east of 18°E most move south or south-east to Balkans or elsewhere in east Mediterranean, although one ringed eastern Poland was recovered in central Italy in winter. November–February recoveries in Cyprus to 1978 are from Czechoslovakia (3), Poland (2), and Dublyany, USSR (50°N 24°E). There are 2 recoveries almost due east of central Europe on north-east shore of Caspian, 1 possibly storm-blown (Erard and Yeatman 1967). A few birds ringed west of 13°E in Germany have also moved south-east with recoveries from Oberbayern to Rhodes (Greece) in February and from Westfalen to Turkey in October. Birds ringed on passage in Britain and Ireland are recovered mainly to east at 2–12°E in central Europe in summer (apart from 1 in central France and 1 in Switzerland), and in Iberia in winter (Langslow 1977).

Nominate *ochruros* (breeding eastern Turkey and Caucasus east to northern Iran) makes only short movements to winter in western Zagros mountains of Iran and in north-west and western Iraq; present on breeding grounds from end of March or early April to end of September (Dementiev and Gladkov 1954b; Vaurie 1959). Norton (1958) suggested there is an initial (August) movement to higher altitudes after breeding. Little information from Iraq: Meinertzhagen (1924*b*) found it common in Kurdistan and at Mosul in December with a few at Baghdad and Nasiriyeh in January, and Ticehurst (1921-2) found both nominate *ochruros* and *phoenicuroides* present late October to mid-March. Flint and Stewart (1983) gave 3 records for Cyprus: 1 in spring, 2 in November.

P. o. semirufus (breeding in highlands of Levant) apparently moves little (though presumably small scale altitudinal migrations occur), extending in winter only to Cyprus and upper Euphrates according to Vaurie (1959), though no records for Cyprus listed by Flint and Stewart (1983). Also said by Etchécopar and Hüe (1967) to winter from Egypt southwards.

P. o. phoenicuroides (breeding in mountains of central Asia) winters in central and west Indian plains, southern Iran, Arabia, Somalia, and Ethiopia, and occasionally in Egypt and Sudan (Dementiev and Gladkov 1954*b*; Vaurie 1959). Meinertzhagen (1924*b*) found it common at Baghdad (Iraq), November-January, but only small numbers winter in Arabia, though Cornwallis and Porter (1982) found 31 *P. ochruros* at 5 localities in North Yemen in March 1982, the 3 adult ♂♂ all being *phoenicuroides*. A fairly common visitor, October-March, north of $9\frac{1}{2}$°N in Somalia, mainly at 1300-2000 m (Ash and Miskell 1983). This wintering at high altitude contrasts with *gibraltariensis* and may account for shortage of information on winter distribution. 'Rather uncommon' in Sudan (Moreau 1972). Migratory route into Africa unclear. Mann (1971) cited several on Dahlak Islands August-September, which seems early to be so far south, and only one has been seen on mainland of Eritrea (Moreau 1972). Departs from high mountains early, leaving Altai during late August or early September, though at Semirech'e (south-east Kazakhstan) not until late October, returning there at end of March. Arrives in southern Altai by 3 April, Chulyshman mid-April, and northern Altai and Pamirs at end of April. (Dementiev and Gladkov 1954*b*; Johansen 1955.)　　RAC

Food. Small or medium-sized invertebrates and fruit. Feeds on ground, hopping or running (and recorded digging 2-4 cm into hard surface to get at larvae), by flying from perch on to ground prey (and usually returning quickly to perch), and by taking aerial prey in short flight; also takes items from walls, foliage, etc., by hovering nearby (Witherby *et al.* 1938*b*; Weber 1970*b*; Gallagher and Woodcock 1980; Menzel 1983). Recorded associating with Turnstones *Arenaria interpres* on beach, taking items exposed (King 1983), and feeding at night under lights, taking flying insects attracted (Gwinner 1958; Allinson 1979). Squashes large insects before feeding them to small young (Menzel 1983).

Diet includes the following. Invertebrates: grasshoppers, etc. (Orthoptera: Tettigoniidae, Acrididae), earwigs (Dermaptera), cockroaches (Blattidae), bugs (Hemiptera, including Aphidoidea), adult and larval Lepidoptera (Nymphalidae, Pieridae, Noctuidae), flies (Diptera: Tipulidae, Culicidae, Asilidae, Tabanidae, Calliphoridae, Muscidae), Hymenoptera (ants Formicidae, wasps Vespidae, bees Apoidea), adult and larval beetles (Coleoptera: Cicindelidae, Carabidae, Silphidae, Staphylinidae, Scarabaeidae, Elateridae, Tenebrionidae, Coccinellidae, Cerambycidae, Chrysomelidae, Curculionidae, Scolytidae), spiders (Araneae), woodlice (Isopoda), millipedes (Diplopoda), small molluscs, earthworms (Oligochaeta). Plant material mainly fruits, also seeds: of juniper *Juniperus*, docks and knotweeds (Polygonaceae), Cruciferae, strawberry *Fragaria*, bramble *Rubus*, hawthorn *Crataegus*, cherry *Prunus*, Leguminosae, alder buckthorn *Frangula*, dogwood *Cornus*, elder *Sambucus*, currant *Ribes*, *Lycium*, mulberry *Morus*, olive *Olea*, *Daphne*, *Parthenocissus*, honeysuckle *Lonicera*, ivy *Hedera*, bilberry *Vaccinium*, strawberry-tree *Arbutus*. (Erard 1959; Pek and Fedyanina 1961; Herrera 1978*a*; Kostin 1983; Menzel 1983.) Also recorded feeding on nectar of *Aloe* during winter in Gibraltar (Cortés 1982).

In holm oak *Quercus ilex* woodlands in southern Spain, October-February, 32 birds contained only invertebrates: of 709 items, 54·8% ants, 31·7% beetles, 3·5% earwigs, 3·2% various larvae, and 6·8% others; items mostly 2-8 mm long (Herrera 1978*a*). In Kirgiziya (south-central USSR), 50 stomachs contained Hemiptera, ants, and beetles (Pek and Fedyanina 1961). Fruits taken mainly in late summer and autumn (Niethammer 1937). For other (scant) data on adult diet, see review by Menzel (1983); also King (1983) for England, Erard (1959) for France, Radermacher (1982) for Germany, Kovačević and Danon (1952) and Tutman (1962) for Yugoslavia, Gil (1928) for Spain, Kovshar' (1962), Yanushevich *et al.* (1960), and Kostin (1983) for USSR, Hardy (1946) for Palestine, and Piechocki *et al.* (1982) for Mongolia.

Food at one site in London (England) with abundant midges (Diptera) nearby (food presumably destined largely for young—deduced from observations and nest-remains) comprised *c.* 33% Diptera (largely midges), *c.* 17% seeds and fruits, *c.* 6% Hemiptera (largely aphids), *c.* 6% Lepidoptera, *c.* 4% ants, and *c.* 33% unknown (sample size not given) (Meadows 1969). See also Mey (1974). Fruit said to be given to young from 8 days old (Niethammer 1937; Meitz 1972).　　DJB

Social pattern and behaviour. Main references: Nesenhöner (1956), Henze (1958), Senk (1962), and Menzel (1983).

1. Outside breeding season, solitary, in pairs, occasionally in small groups, probably only loosely associated (Moreau and Moreau 1928; Witherby *et al.* 1938*b*; Anon 1959; Géroudet 1963; Menzel 1983). Exceptional gathering of *c.* 800 birds on small island near Mallorca (Spain), late October (Goethe 1933). Same ♂♂ apparently occupy same area through winter (Blondel 1962), perhaps returning there year after year (Moreau and Moreau 1928) though no reports of territories being held. BONDS. Mating system probably usually monogamous. Bigamy

sometimes occurs, ♂ feeding 2 broods simultaneously within short distance of each other; ♀♀ may then tolerate each other, though if nests are close they never bring food simultaneously (Hoehl 1941; Gebhart and Sunkel 1954; Gutscher 1958; Parker 1960; Glutz von Blotzheim 1962). In one case of apparent bigamy involving 2 nests 34·5 m apart, one built later than other, ♂ mainly attended 1st nest until young left, then moved to 2nd nest (Frost *et al.* 1982). 2 records of ♂ feeding 2 1st and 2nd broods, i.e. each of the 2 ♀♀ reared 2 broods (Hoehl 1941; Gutscher 1958). One case of 2 ♀♀ tending same clutch and young, alternating incubation of eggs though often sitting together on nest, tolerating one another; recently hatched young were fed first only by ♂ and 1 ♀ while 2nd ♀ brooded young, often preventing 1st ♀ from feeding them; later, all 3 shared in feeding (Menzel 1983). Trio of 2 ♀♀ and 1 ♂ also recorded feeding same brood at another site (Stephan 1978). 2nd ♂ may be present on territory (see Antagonistic Behaviour, below). Several records of hybridization with Redstart *P. phoenicurus*. In one hybridizing pair, ♂ *P. phoenicurus* mainly gave typical song of *P. ochruros*, but warning-calls of both species; 1st brood reared had mixture of characters of the 2 species, 2nd were all typical *P. ochruros* (Böhm and Strohkorb 1964). ♂ *P. phoenicurus* recorded associating with pair of successfully breeding *P. ochruros* and pairing with ♀ (3 weeks after young fledged) when her mate disappeared (Menzel 1983). In another case, ♂ *P. phoenicurus* took over territory of ♂ *P. ochruros*, driving him away, and paired with ♀ after she had abandoned 1st nest; young raised looked like *P. ochruros*; during raising of next brood, former mate returned and was driven off again but finally managed to help feed young, with open hostility confined to nest area (Robert and Toulon 1984). Andersson (1963) gave 3 records of successful hybridization, each with ♀ being *P. phoenicurus*. Reports of hybrid ♂♂ by Kleinschmidt (1908), Ringleben (1948), and Dathe (1950). One ringed pair of *P. ochruros* bred together in 2 successive years (Drost and Desselberger 1932). 18 out of 23 birds returned to breed in same locality (10 ringed as adult, 6 as juveniles, 2 age unknown). One young bird returned to nearby site, 4 found 8–14 km away: one bird ringed as adult and one ringed as juvenile returned to same site after 2 years, one ringed as adult returned after 4 years, and another bred at same site for 5 consecutive years (Drost and Desselberger 1932). In monogamous single-species pairs, pair-bond (sometimes, at least) lasts through whole season for 3 broods (Jenny 1946; Löhrl 1957a). In one case, ♂ took new mate after 1st brood, though still fed young of that brood until 8 days after fledging; new mate used same nest-box as former mate (Senk 1962). Young brooded by ♀ alone, but fed by both sexes, including for some weeks after leaving nest (see Relations within Family Group, below). ♂ takes over main responsibility if ♀ re-lays (Nesenhöner 1956; Menzel 1983); for variations where up to 3 broods reared, see (e.g.) Henze (1958) and Hui-Früh (1975). ♂ recorded feeding young alone after loss of mate, but soon re-paired and reared young successfully with new partner (Turcek 1965). In one case, 2nd-brood young fed 3rd brood in and (without assistance from parents) also outside nest (Menzel 1983). Many (perhaps all) ♂♂ breed at 1 year old (Currie 1949; Pflugbeil 1951; Verbeek 1984); in one case, ♀ 1 year old at first breeding (Becker 1984). BREEDING DISPERSION. Solitary and territorial. 6 West German territories ranged from 2·9 to 7·4 ha (Nesenhöner 1956); in France, most 0·35–1 ha, probably rarely exceeding 1·5 ha (Géroudet 1963); smallest East German territory recorded by Menzel (1983) less than 1 ha. Pairs recorded nesting 6–13 m (or less) apart, though this leads to constant disputes (Géroudet 1963); one case, however, of 2 pairs nesting 60 cm

apart in building without aggression (Greiner 1953). Birds typically range up to *c.* 200 m from nest for foraging (Géroudet 1963). Territory may contract after pair-formation (Senk 1962) but may expand again shortly after young of 1st brood have hatched (Marchant 1938). In 2 studies (Nesenhöner 1956; Menzel 1983), territory never completely surrounded by others. Up to 80 pairs per km² in alpine zone of Switzerland, but considerable variation lower down, despite many suitable nest-sites: 0–150 pairs per km² though with concentrations in some areas, e.g. 3 pairs in one barn (Glutz von Blotzheim 1962). 200–300 pairs per km² at site above 1800 m in Svanetia, Georgian SSR, USSR (Zhordania and Gogilashvili 1976). In Ladakh (Kashmir), 40 territories per km² over *c.* 25 ha of marginal cropland (Delany *et al.* 1982). Territory almost always delimited basically by song-posts or by buildings (Menzel 1983). Pair may forage together within territory (Rayfield 1941). ♂ alone defends territory, establishing it immediately after arrival (Pflugbeil 1951); remains within it after breeding (or returns to it if family has previously been led away), maintaining it until final departure, though defence apparently wanes during moult (Nesenhöner 1956; Senk 1962; Menzel 1983). Subsequent nests usually built close by (e.g. Nesenhöner 1956, Henze 1958, Menzel 1983): in 15% of 108 cases, consecutive nests 0·2–10 m apart; in 22%, same nest used for 2nd brood (Pflugbeil 1951; see also Rentsch 1975)). Löhrl (1962b) and Menzel (1983) recorded occasional autumn nest-building by ♀; nest sometimes completed but no eggs laid. ♂ may use same nest-site in successive years (Senk 1962). ROOSTING. Nocturnal and usually solitary, though migrants sometimes loosely communal (Anon 1959) and fledged young recorded roosting clumped together on flat top of stone fence support (Menzel 1983). Other sites used include exposed rock or wall ledges, bushes, and stones (migrants), also holes and other sheltered places or inside buildings; one wintering bird roosted by warm-air vent (Anon 1959; Heer 1978; Klose 1978). Adult roosts on nest only when incubation started (Menzel 1983). Sunbathing bird lay on ground with wings outstretched for *c.* 3 min (Gliemann 1976). Water- and dust-bathing also recorded (Anon 1959).

2. Not especially shy during winter, allowing approach to 30 m (Güth 1956; Ludwig 1973). Bird disturbed at roost dropped to ground and ran away (Anon 1959). FLOCK BEHAVIOUR. No information. SONG-DISPLAY. Song (see 1 in Voice) considered by Blume (1967) to attract ♀, intimidate rivals, maintain contact between pair members, and assist with sexual synchronization. ♂ usually sings from high, exposed perch (Oelke 1960; Menzel 1983), sometimes in flight (Witherby *et al.* 1938b), but no details, and most likely exceptional. Birds sing in early spring from nest-building, through incubation, and as soon as young leave nest (but not whilst in nest) for 1st brood (Menzel 1967), and then until 2nd brood hatched. Last regular song 20 and 29 July at 2 British sites (Rayfield 1941; Ashby 1942). All singing birds identified by Gnielka (1969) were fully adult plumaged ♂♂, though in Pamir mountains (USSR) birds in ♀-type plumage (presumably young ♂♂) also sing (Potapov 1966). Quite a long period of intermittent autumn song (September–October), usually only after moult (Ashby 1942; Dare 1953; Senk 1962; Bergmann and Helb 1982), which may be directed at ♀ (Homann 1960) though without copulation, usually while still on breeding territory (Fitter 1946); may be heard in winter quarters (Moreau and Moreau 1928; Gibb 1946; Ludwig 1973; Klose 1978). Most singing done in morning and evening up to 21.00 hrs (Dare 1953); occurs at night, mainly near artificial light or on moonlit nights (Heinroth and Heinroth 1924–6; Meinertzhagen

1935; Gwinner 1958; Kaiser 1961; Zucchi 1974; Martelli 1976). Song output high during pair-formation at *c.* 5000 songs per day (Ruthke 1941); in early April, 5650 songs 04·27–19·07 hrs (reduced to 920 when cool). Considerable increase in intensity after sunrise, with 10 songs per min; no really long pauses throughout period though some interruptions later in morning when moving about on roof or feeding; wanes slightly around midday and early afternoon, with resurgence by evening (Nesenhöner 1956). ♂ may have more than one regular song-post within territory; song-posts shifted around territory during summer (Fitter 1944, 1947), and may even be sited outside it (Dare 1953). ANTAGONISTIC BEHAVIOUR. (1) General. All other ♂♂ usually driven out by territory-owner (Senk 1962; Potapov 1966; Menzel 1983), but in one study, 2nd ♂ tolerated by territory-holding ♂, though resident affected by its presence and birds sometimes followed each other (not necessarily chasing) (Ashby 1942). (2) Threat and Fighting. For threat posture adopted by ♂ in inter- and intraspecific disputes within territory, see Fig A; for associated calls, see 4b and 6b in

A

Voice. Little fighting, however (Nesenhöner 1956), and records of 2 ♂♂ tolerating each other when song-posts 60 m apart. Trio of 2 ♀♀ and 1 ♂ (see Bonds, above) showed mutual tolerance and united to expel intruding ♂ (Stephan 1978). Generally little interest shown in other species, even *P. phoenicurus* (even when nesting amongst up to 5 pairs of *P. phoenicurus* in 1·3 ha of timber yard), and no attacks recorded (Menzel 1983). Will attack ♂ Güldenstädt's Redstart *P. erythrogaster* (Potapov 1966). Some aggression (including dive-attack) recorded towards Great Tit *Parus major* nesting nearby and entering *P. ochruros* nest-box (Wennrich 1979). HETEROSEXUAL BEHAVIOUR. (1) General. In Britain, ♂♂ arrive on breeding grounds 3–4 weeks before ♀♀ (Fitter 1948). In East Germany, at 2 sites, ♂♂ arrive 1–2 weeks before ♀♀, though ♂♂ arrive over period of *c.* 2 weeks at one of these sites; ♂♂ take up territories immediately after arrival (Pflugbeil 1951; Menzel 1983). In Bayern (West Germany), ♂♂ arrived only 4–7 days before ♀♀ (Henze 1958). (2) Pair-bonding behaviour. Apparent courtship chases (see below) and incipient nest-building recorded in Egyptian winter quarters (Moreau and Moreau 1928). Once ♀ arrives in ♂'s territory, pair-formation takes a number of days with ♂ showing readiness for mating by singing, though ♀ takes more active part (Nesenhöner 1956; Menzel 1983). Courtship involves fast chases by ♂ after ♀, ♂ singing constantly; ♀ settles and ♂ flies about in front of her, singing (see 3 in Voice), and twisting from side to side; further chasing follows (Harrison 1943). ♂ once perched *c.* 1 m from ♀, held body almost vertical, stretched head up, half-opened bill, ruffled feathers, fanned tail, drooped wings slightly, and moved slowly up and down through 30–40°; vertical movement performed 3

times; when ♀ flew off, ♂ abruptly relaxed posture, and gave chase (Gliemann 1976). In one interaction, ♂ called (see 6a in Voice) but ♀ remained silent during chase, though call (see 7 in Voice) apparently given by ♀ when ♂ flew around her, wing-shivering and pecking at her cloaca; no copulation took place then (Nesenhöner 1956). (3) Courtship-feeding. Not recorded. (4) Mating. Incident in which courtship led to mating described by Nesenhöner (1956) as follows. ♂ arriving where ♀ was quietly perched immediately sang excitedly, fluttering on roof. ♀ then fluttered or hopped to ♂, who moved towards her, as if begging (calling quietly), and beating his wings with head up. ♀ grabbed his bill in apparent defence but with extended bill-contact. ♂ flew back a bit then immediately approached and mounted her. ♀ lay on roof with head pressed down and tail raised and twisted to right. Copulation lasted *c.* 4 s, ♂ beating wings vigorously and clinging on to either side of ♀'s body with feet; afterwards, momentarily held wings still and outspread, then flew off. In another case, ♂ sang (see 3 in Voice), then apparently called (see 5 and 6a in Voice) including during brief pursuit of ♀ on ground prior to copulation, after which birds flew off calling (see 4b in Voice), and ♂ later sang (Laferrère 1953). (5) Nest-site selection. Apparently, ♀ normally selects site (Geyr von Schweppenburg 1940, 1942; Löhrl 1957*a*), although Geyr von Schweppenburg (1940) gave record of at least one pair where ♂ obviously participated and checked a number of suitable sites, and Senk (1962), on basis of 3 pairs, considered that ♂ chooses site. No published observation of Nest-showing display as well known in *P. phoenicurus* (Menzel 1983). (6) Behaviour at nest. ♂ often in close attendance during incubation by ♀ (Rayfield 1941). Both parents commonly perch near nest when bringing food, and after feeding (Ashby 1942), using these perches regularly (Heyder 1980). RELATIONS WITHIN FAMILY GROUP. For first 7 days, pattern of brooding (by ♀ alone) more or less as during incubation, though brooding stints gradually shortened; from day 8 onwards, young brooded only at night and in morning and evening. From day 11, no daytime brooding on warmer days, and one ♀ stopped brooding at night from day 9 (Nesenhöner 1956), another from day 7 (Menzel 1983). ♂ may bring food for young a day or so before hatching (Henze 1958). Eyes begin to open on 5th day, fully open two days later. Up to day 6, young beg by gaping with head held vertically up, then change to gaping towards arriving adult. Young restless, clumped together at first, then become less restless after some feather growth. (Menzel 1983.) Usually both parents feed young about equally even when ♀ doing a lot of brooding, and 1st brood sometimes fed exclusively by ♀ (unless overlapping with 2nd clutch, when ♂ may feed young alone) (Ashby 1942; Henze 1958; Menzel 1983). In captivity, ♂ may feed ♀ as well as young (Teschemaker 1912*b*), though in the wild Menzel (1983) recorded ♂ initially passing food to ♀ for feeding to young; after first few days, ♂ would announce arrival with calls (unspecified), and ♀ would leave nest. Unlike in *P. phoenicurus*, ♂ may remain at nest for a while after passing food (Nesenhöner 1956; Menzel 1983). Ashby (1942) reported ♂ always removing faeces, but Comte (1928) recorded only ♀, while no sexual differences noted by Menzel (1983). Apparently unpaired ♂ occupying same territory may also remove faeces (Ashby 1942). During first few days, tend to swallow faecal sacs; later drop them up to 20 m from nest or sometimes deposit them on branch. Nest kept very clean until a few days before fledging, and ♀ cleans it thoroughly if to be used again. (Menzel 1983.) Dead young (4·8 g) recorded being carried 14 m from nest (Koch 1948–9). Near to fledging, young may sit or roost on rim of nest (Menzel 1967*b*). Fledging may extend over 3 days though usually within 1 day. Disturbance causes young to fledge slightly prematurely, while rainy, cold weather delays it. Sometimes leave nest when not fully able to fly. (Menzel 1983.) Recently fledged young seek safety (or are led by parents) amongst rocks, vegetation or other cover and may remain there much of day and night (Henze 1958; Potapov 1966; Menzel 1983). After fledging, may roost nearby or return to nest to roost until 4 days after fledging (Menzel 1967), though generally do not return to nest at all (Menzel 1983). At one site, ♂ took brood away whilst ♀ incubated next clutch, ♂ returning to feed young of next (3rd) brood regularly only when young 2 days old (Heyder 1980). However, Menzel (1983) recorded 1st brood always fed within nesting territory up to complete independence. ♀ feeding fledged young of 3rd brood may drive off young of 2nd brood although ♂ may feed both (Henze 1958). However, according to Nesenhöner (1956), ♀ usually tolerates young of previous brood near nest. Fledglings apparently do not wander far from nest whilst being fed by parents (e.g. Potapov 1966). Young may be independent by 11 days after fledging (Senk 1962), or family may stay together for 3–4 weeks (Menzel 1983). ANTI-PREDATOR RESPONSES OF YOUNG. Young in nest in pipe scrambled out of reach when approached (Fitter 1948). When hand placed over nest, young will gape up to day 6, but crouch from day 7 (Nesenhöner 1956). PARENTAL ANTI-PREDATOR STRATEGIES. (1) Passive measures. No details. (2) Active measures: ♂ reacted to Little Owl *Athene noctua* not by mobbing but by calling (see 4–5 in Voice) and bobbing up and down (Dare 1953). Call 4 at least given when Red-backed Shrike *Lanius collurio* and cat near nest or fledged young (Henze 1958; Bergmann and Helb 1982).

(Fig by C Rose: from drawing in Blume 1967.)

PGHE, MGW

Voice. As well as in breeding season, song given intermittently in autumn and occasionally in winter (see Social Pattern and Behaviour). No difference between nocturnal and diurnal song (Martelli 1976). Call 4 given throughout year, though generally rather quiet in winter (Dare 1953). For additional sonagrams, see Bergmann and Helb (1982), Menzel (1983), and Böhme and Böhme (1986).

CALLS OF ADULTS. (1) Song of ♂. Short (c. 1–2 s), scratchy, stuttering, falsetto sounding (squeezed or strangled) phrases apparently of 2 basic types given separately with distinct pause, or type 2 often follows type 1 after short pause (e.g. 0·3–0·7 s in recordings: J Hall-Craggs) to form compound song-phrase lasting c. 4 s (Witherby *et al.* 1938b) and rendered 'tetetetet-kschsch tetet' (Voigt 1933; Bergmann and Helb 1982). Further descriptions: thin sweet notes often followed by short metallic chattering 'tsi-tsi-tsi-tsi', the combination being given many times in rapid succession (Gibb 1946); quick warble rather like Redstart *P. phoenicurus* but less rich and loud (often used alone), then remarkable sound like handful of small metal balls shaken together (or peculiar spluttering or crackling like wireless atmospherics: Dare 1953) introduced near end and followed by a couple of double notes or confused warble louder and more musical than opening part (Witherby *et al.* 1938b). More detailed analysis showed 1st type, after inconspicuous beginning, to be like 'jirr-titititi...', often ending on a rising note

like an open question; 2nd type (part of song) consists of scratching-spluttering (hissing) introduction and a connected phrase or motif: 'krchch-titütili' or 'krchch-tütitititi', ending with slight descent like answer to question (Bergmann and Helb 1982). However, in recording by E D H Johnson, song not unlike Chaffinch *Fringilla coelebs* has slow, measured delivery and apparently presents opposite pattern (i.e. answer then question), especially when upward flick (rising note) given at end of 2nd part of song (P G H Evans, J Hall-Craggs). ♂♂ have several variants of each song-phrase type in repertoire (Bergmann and Helb 1982); variation evident in recordings includes degree to which both phrase-types are extended by repetition of final units (see Figs I–II). Although sometimes too quiet to print, most striking feature in recordings after loud introductory phrase is gravel-crunching sound (P J Sellar) already alluded to (above) which starts 2nd phrase; this followed by crescendo and loud terminal jumble of notes not unlike Robin *Erithacus rubecula* (P G H Evans, J Hall-Craggs: Figs I–II). In Pamir mountains (southern USSR), song (of *phoenicuroides*) less melodious than at lower altitudes: e.g. at c. 4200 m altitude 'chi-ko...chi-chi-chi-chi-chi-ko'; 2 main variants, however— 'vi-te-re-re...vi-te' and 'rri-chi-chi-chi-chi-tiu', then gravelly '...chri-chi-chiu', song ending loud and sharp or as a quiet trill (Potapov 1966); songs of *phoenicuroides* described here probably well within range of variation suggested for European birds (see above). Other bird species mimicked in song given in and outside breeding season include Oystercatcher *Haematopus ostralegus*, Chiffchaff *Phylloscopus collybita*, Great Tit *Parus major*, Blue Tit *P. caeruleus*, Starling *Sturnus vulgaris*, *F. coelebs*, Greenfinch *Carduelis chloris*, and Linnet *C. cannabina* (Dare 1953; Gundelwein 1955; Moore and Boswell 1956; Gliemann 1976; Weibull 1979). (2) Occasional quiet twittering reported in Menzel (1983) perhaps Subsong. (3) Courtship-song. Similar to full song, but more wheezy and grating; given by ♂ flying close to perched ♀, apparently also during chase (Harrison 1943). Stuttering confused sounds difficult to render satisfactorily; roughly 'gr-gr-grr-sch-schvididi', followed by low and hurried call 4b —'ket-ket-ket' (Laferrère 1953). (4) Contact-alarm calls. 2 basic sounds (renderings suggesting some variation in timbre within each) frequently combined, especially when alarmed, though also given separately. (a) A 'sip', 'sit' or 'tsip' (Rayfield 1941; D I M Wallace). Only call given in winter, often as a series, e.g. 'sit-sit-sit-sit'; normally soft, though once clear, loud and arresting; usually given when perched, occasionally in flight (Dare 1953), not necessarily expressing alarm. Differs from equivalent call of *P. phoenicurus* (see that species, p. 704) in not resembling *Phylloscopus* warbler (Voigt 1933; D I M Wallace); see also call 5. (b) Harder 'tuc', 'tk', 'chuc-chuc' recalling Wheatear *Oenanthe oenanthe* and sometimes apparently expressing aggression (Dare 1953)

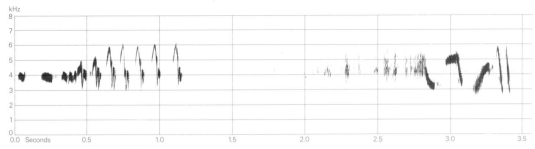

I   P A D Hollom   France   May 1973

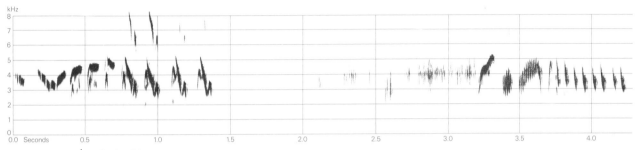

II   E D H Johnson   Spain   May 1959

or, in recording by P A D Hollom, like quiet distant ticking of *E. rubecula* (J Hall-Craggs, P J Sellar); also rendered 'tzi' or hard rapid 'kkkkk kkk' (Voigt 1933), 'ket-ket-ket' (Laferrère 1953). Examples of calls 4a–b combined: usual alarm-call a toneless 'fid-tk-tk...', sounds sometimes given in rapid alternation, and at high intensity (e.g. when fledged young threatened), units given at ever faster rate, forming continuous series (Bergmann and Helb 1982); 'ui-tic' or 'ui ui-chuc' recalling Stonechat *Saxicola torquata* (Dare 1953); 'sip tititic' or, from highly agitated ♀, 'sip sip sip' running into 'sippr-wit sip sip tututuc' (Rayfield 1941). (5) Slow 'fü-i fü-i' closely resembling call of Bonelli's Warbler *P. bonelli* and given (apparently by ♂) during courtship (Laferrère 1953). Clear 'tui-tui' sounds of varying volume given by ♂ in breeding season, also (in addition to call 4a–b) when alarmed by Little Owl *Athene noctua* (Dare 1953); perhaps the same, but confirmation required. (6) Rattling, purring and related sounds. (a) A 'kr-kr-kr-kr' or not unlike distant 'kru-kru-kru' Flight-call of Black Woodpecker *Dryocopus martius* given (apparently by ♂) during courtship-chase on ground (Laferrère 1953). Short purring sounds of varying volume given by ♂ during courtship (Nesenhöner 1956) are probably the same—as perhaps is a short, quick rattle, low, harsh and purring, given probably by ♂ in presence of ♀, perhaps as a greeting (Ashby 1942) perhaps also. (b) Bergmann and Helb (1982) noted (and illustrated sonagraphically) 'tre...' and 'tre-tre-tre...' given especially during fights with conspecific birds. Relationship (if any) with call 6a not known. (7) Quiet melodious whistling sounds given probably by ♀ during courtship (Nesenhöner 1956).

CALLS OF YOUNG. Food-call shrill and mouse-like 'tsitt tsitt' during first few days, becoming louder, stronger, and harsher and persisting up to *c.* 1 week after leaving nest (Rayfield 1941; Ashby 1942; Menzel 1983); harsh urgent chattering from young out of nest (Ashby 1942) and 'tzi-ditt tzi ditt...' given when begging from adult Crag Martin *Ptyonoprogne rupestris* nesting close by (Strahm 1956). Repeated ticking sounds (adult call 4b) in form of rattle given from a few days after leaving nest (Ashby 1942; Menzel 1983).   PGHE, MGW

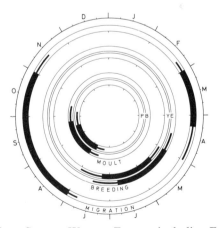

Breeding. SEASON. Western Europe, including England: see diagram. Eastern Europe: up to 2 weeks later. Southern Europe and North Africa: similar to western Europe (Heim de Balsac and Mayaud 1962; Makatsch 1976). SITE. On ledge in cave or building, or in hole or crevice in rock or wall. Up to 45 m above ground, mostly

1–4 m (Menzel 1983). Nest: loose cup of grass leaves and stems, moss, and other plant material, lined with wool, hair, and feathers. Outer diameter 12–15 cm, inner 6 cm, overall height 7–10 cm, depth of cup 4–5 cm (Menzel 1983). Building: by ♀, taking 5–8 days (Senk 1962). EGGS. See Plate 82. Sub-elliptical, smooth and glossy; in *gibraltariensis*, white, rarely tinged blue or with faint brownish spotting; in nominate *ochruros*, pale blue to blue-green. *P. o. gibraltariensis*: 19·4 × 14·4 mm (17·0–21·5 × 13·3–15·9), *n* = 200; calculated weight 2·16 g; nominate *ochruros* not significantly different (Schönwetter 1979). Clutch: 4–6 (2–8). Of 469 clutches from various areas of Europe: 2 eggs, less than 1%; 3, 2%; 4, 16%; 5, 71%; 6, 8%; 7–8, less than 1%; mean 4·9; little variation through season (Menzel 1983). 2 broods, occasionally 3 (e.g. Henze 1958, Becker 1984). Eggs laid daily (Senk 1962). INCUBATION. 13–17 days. By ♀ only, beginning with last egg (Senk 1962; Menzel 1983). YOUNG. Cared for and fed by both parents. FLEDGING TO MATURITY. Fledging period 12–19 days (Menzel 1983); independent from 11 days after (Senk 1962), perhaps sometimes much later (see Social Pattern and Behaviour, also for age of first breeding). BREEDING SUCCESS. 1 ♀ laid 62 eggs over 5 seasons, rearing 49–51 young to fledging (7–14 per year) (Becker 1984).

Plumages (*P. o. gibraltariensis*). ADULT MALE. In fresh plumage (autumn), narrow strip along base of upper mandible black, rump and upper tail-coverts deep rufous-cinnamon or almost rufous-chestnut, remainder of upperparts medium grey, each feather with ill-defined olive-brown tip (widest on lower mantle, inner scapulars, and back, narrower elsewhere, showing as slight brown tinge on mainly grey feathering); feather-centres of lower mantle and scapulars sometimes with some black; sides of crown above and behind eye slightly paler and purer grey. Sides of head down from just above eye and ear-coverts, chin, and throat black, feathers faintly and narrowly tipped grey on lores and chin, more broadly on ear-coverts, rear of cheeks, and lower throat, where black sometimes almost hidden. Chest and sides of breast pale grey, sometimes with slightly olive-brown tinge, variable amount of black of feather-centres visible. Breast, belly, flanks, and vent grey, palest on vent, often washed buff-brown on flanks, and sometimes with some darker grey of feather-bases visible on breast and belly. Under tail-coverts cinnamon-rufous. Central pair of tail-feathers (t1) dull black or brown-black with rufous fringe along base of outer web; t2–t6 bright rufous-cinnamon, often slightly tinged brown on tip of outer web of t2 and t6, and occasionally on tips of all feathers. Flight-feathers and tertials dull black, greyish-black, or brownish-black; outer webs of primaries narrowly edged off-white; outer webs of secondaries and outer tertials broadly fringed white (except on tips), forming obvious white wing-patch. Lesser and median upper wing-coverts grey, some black on centres sometimes visible, especially on median; greater coverts and greater upper primary coverts dull black with medium grey or ash-grey fringe along outer web and tip (tips sometimes partly washed buff or pale olive-brown, like those of tertials). Axillaries and under wing-coverts black (shorter feathers) or dull grey (longer ones), tipped pale grey or off-white. *In worn plumage* (spring), olive-brown fringes of upperparts and white or pale grey ones of sides of head and

underparts largely or fully worn off, upperparts medium grey except for black rim along base of upper mandible and cinnamon-rufous rump and upper tail-coverts (some brown usually visible still on back); sides of head, chin to chest, and sides of breast uniform black, flanks grey, breast black with traces of grey feather-fringes, gradually merging into grey-white belly and vent; grey fringes of upper wing-coverts partly or fully worn off, lesser and median coverts with much black of centres exposed; olive-brown or buff on tips of greater coverts and tertials (if any) worn off; white of wing-patch usually partly worn off, but patch usually well-visible on bases of inner secondaries and outer tertials up to June–July. Some individual variation in colour of body: of 50 birds in worn plumage examined from central Europe (RMNH, ZFMK, ZMA), 2% had pronounced white patch on forehead; 8% had belly and vent tinged rufous instead of grey-white (rather like nominate *ochruros*, but much less pure and extensive than *phoenicuroides* and *rufiventris*); 58% had upperparts uniform medium-grey (though often tinged darker grey on mantle and scapulars), 36% had much black visible on feather-centres of lower mantle and scapulars (or almost completely black), 6% had hindneck, mantle, and scapulars virtually black (and breast and upper belly more extensively black also), resembling Iberian *aterrimus*. ADULT FEMALE. In fresh plumage (autumn), upperparts dark brown-grey, often slightly tinged olive-brown on mantle and scapulars; lower back paler buff-brown, merging into rufous-cinnamon on rump and upper tail-coverts. Sides of head and chin to chest and flanks like upperparts, but slightly paler and less saturated; lores, chin, and rim along front of and below eye pale buff-grey, rim sometimes extending over eye and forming indistinct broken eye-ring. Brown-grey of chest and flanks gradually merges into grey-white of central belly and vent; under tail-coverts rufous-cinnamon to pale orange-buff. Tail like adult ♂. Flight-feathers and tertials dark brown or dark greyish-brown, outer webs and tips narrowly edged off-white (except tips of outer primaries); basal outer edges of secondaries and outer edges of tertials usually wider, forming indistinct pale wing-patch. Lesser upper wing-coverts dark brownish-grey, like upperparts, fringes slightly paler grey; median and greater coverts and greater upper primary coverts dark brown or dark greyish-brown, outer webs and tips narrowly fringed grey. Under wing-coverts and axillaries pale grey-brown or pale buff-grey, shorter coverts with dark brown centres. *In worn plumage* (spring), upperparts and upper wing-coverts slightly greyer and duller, less brown; tail-coverts paler rufous-cinnamon; slightly darker brown shafts sometimes visible on chest and breast, appearing obscurely streaked. NESTLING. Down rather long and dense, dark grey; on upperparts only. For development, see Menzel (1983). JUVENILE. Rather like adult ♀, but head and upperparts darker and duller grey with slight olive-brown tinge, feathers with very faint dark fringe along tip; upper tail-coverts paler, cinnamon or pale orange-buff; lores dull brown-grey with fine buff mottling; ear-coverts with faint pale shaft-streaks. Chin to breast and flanks like upperparts, but slightly paler and browner, dark feather-tips slightly more distinct. Breast grey-brown, merging into buff on central belly and vent, feather-tips indistinctly fringed dull black. Under tail-coverts pale cinnamon or rufous-buff. Tail as adult, but feather-tips more often tinged brown or olive-brown. Flight-feathers and tertials like adult, but pale outer edges narrower, pale brown; fringes along outer edges and tips of tertials buff; fringes of secondaries and tertials not as pale as adult ♀ and not as broad and white as adult ♂, not forming pale wing-patch. Lesser and median upper wing-coverts like scapulars or slightly darker grey-brown; greater coverts and greater upper primary coverts dark grey-brown or black-brown, outer webs and tips of greater coverts rather broadly fringed pale grey-brown, pale olive-brown, or buff-brown, of primary coverts more narrowly pale grey-brown or buff-grey. Underwing dull grey with pale buff feather-tips. Entire plumage softer, looser, and shorter than in adult ♀, in particular back, vent, and tail-coverts. FIRST ADULT MALE. Rather variable; some birds in bright or 'paradoxus' morph resemble adult ♂, others in dull or 'cairii' morph resemble adult ♀; others intermediate. In all birds, juvenile flight-feathers, tail, many or all greater upper wing-coverts, all greater upper primary coverts, and usually all tertials retained, fringes browner (less greyish) and tips narrower and relatively more worn than those of adult at same time of year. In bright morph, fringes of retained greater coverts and tertials distinctly browner and more worn than greyer ones of new neighbouring coverts, feather-centres less black; no white wing-patch, but rarely some tertials new, and these have white outer border as in adult, creating small patch. Head and body like adult ♂, but olive-brown feather-fringes of upperparts and upperwing wider, largely concealing grey when plumage fresh; off-white fringes on tips of black feathers of sides of head, chin, and throat wider, lores, upper cheeks, chin, and throat mottled black and white. When plumage fresh, ear-coverts and chest largely grey-white or pale olive-grey; remainder of underparts more heavily suffused pale olive-brown or buff-brown; cinnamon of tail-coverts less deep. *In worn plumage* (spring), upperparts still with traces of olive-brown (not as uniform medium grey or black-and-grey as adult); lores, ear-coverts, throat, and chest still with many narrow off-white feather-fringes; retained juvenile wing-feathers brown and heavily abraded, less black than lesser and median upper wing-coverts. In dull morph, plumage almost identical to adult ♀; fringes of retained juvenile feathers slightly browner than more greyish ones of new lesser and median coverts when plumage fresh, but difference hard to see when plumage worn. When series of skins examined, upperparts, chest, and flanks duller grey than ♀, only slightly tinged brown (less olive-brown), fringes of lesser upper wing-coverts rather broad and pale ash-grey (in ♀, brown-grey), of median coverts grey (in ♀, brown), but these differences hard to see in spring when plumage worn. Best ageing characters are tail-tips (relatively narrower and more worn than adult at same time of year) and greater upper primary coverts (browner with buffier fringes and with more heavily worn and sharply pointed tips than in adult at same time of year). Generally, dull morph more common than bright morph (see, e.g., Stresemann 1920, Biber 1978, Verbeek 1984), though some local variation in proportion of each. FIRST ADULT FEMALE. Like adult ♀, but part of juvenile plumage retained, as in 1st adult ♂. Fringes of upper wing-coverts slightly browner than those of adult (in particular, lesser coverts less greyish), but ageing often difficult, especially in spring when plumage worn; differs in shape and wear of tail-tips and greater upper primary coverts (see 1st adult ♂, above). Similar to dull morph of 1st adult ♂, but in series slightly more olive-brown on upperparts, chest, and flanks, less deep grey (see 1st adult ♂, above).

Bare parts. ADULT. Iris dark brown or black-brown. Bill black. Leg and foot black. NESTLING. Inside mouth rich yellow, no spots on tongue; gape-flanges ivory-colour. JUVENILE, FIRST ADULT. Like adult, but bill, leg, and foot dark grey, dark horn-brown, or black-brown, sometimes with paler grey or yellowish soles; gape-flanges pale yellow at fledging. Adult colours obtained in late summer and early autumn of 1st

calendar year. (Witherby *et al.* 1938*b*; Menzel 1983; RMNH, ZFMK, ZMA.)

Moults. ADULT POST-BREEDING. Complete; primaries descendant. In England, *gibraltariensis* usually starts with p1 during about second half of July, finishing (with p9-p10) after *c.* 50 days, mainly in first half of September; however, some birds start as late as *c.* 15 August, not finishing until early October (Ginn and Melville 1983). Tail, mantle, and sides of breast start from primary moult score 5-10, secondaries from *c.* 30 (ZFMK, RMNH). Same picture as for England applies to *gibraltariensis* in central Europe (Stresemann 1920; RMNH, ZFMK, ZMA), and apparently also to nominate *ochruros* from Caucasus, Transcaucasia, and north-east Turkey, which is in heavy moult August (Dementiev and Gladkov 1954*b*) or starts approximately mid-August, completing mid-September (ZFMK). In *phoenicuroides*, USSR, some start mid-July but most in August (Dementiev and Gladkov 1954*b*), while 2 late-July birds from Mongolia had started but 2 from first half of August had not (Piechocki and Bolod 1972). 2 *rufiventris* from Afghanistan, September, had primary moult completed but some body feathers still growing (Paludan 1959). POST-JUVENILE. Partial: head, body, lesser and median upper wing-coverts, occasionally 2-6 inner greater upper wing-coverts and tertial coverts, and rarely tertials. Timing highly variable, depending on hatching date. Starts on average 89 ± 6.9 days after hatching; moult completed mid-August to mid-October, on average 163 ± 11.7 days after hatching (Berthold 1983). In Netherlands, first birds with moult completed encountered late July, others still fully juvenile up to mid-September; most birds from first half of September had moult completed, however (RMNH, ZMA). In Switzerland, moult about mid-August to mid-September (Biber 1973). In *phoenicuroides*, USSR, some nearing completion late July, when others just fledged; generally, completed by late August or early September (Dementiev and Gladkov 1954*b*). In *rufiventris*, one from Afghanistan still moulting 4 September (Paludan 1959). In nominate *ochruros*, north-east Turkey, 3 birds from late August and early September had completed moult (ZFMK).

Measurements. ADULT, FIRST ADULT. *P. o. gibraltariensis.* Western and central Europe (mainly Netherlands), all year; skins (RMNH, ZMA). Juvenile wing and tail include retained juvenile wing and tail of 1st adult. Bill (S) to skull, bill (N) to distal corner of nostril; exposed culmen on average 4·2 mm less than bill (S).

WING AD	♂ 87·6 (1·58; 28)	85-91	♀ 85·4 (2·44; 7)	83-90
JUV	85·4 (1·98; 39)	82-89	82·3 (1·72; 28)	79-85
TAIL AD	60·8 (1·72; 6)	58-63	58·4 (2·39; 7)	55-61
JUV	59·0 (2·37; 10)	56-63	56·9 (1·39; 17)	55-60
BILL (S)	15·2 (0·48; 40)	14·3-16·3	14·9 (0·50; 23)	14·1-15·8
BILL (N)	8·0 (0·46; 15)	7·4-8·5	7·9 (0·35; 23)	7·5-8·6
TARSUS	23·4 (0·50; 16)	22·6-24·2	23·3 (0·82; 22)	21·8-24·6

Sex differences significant for wing and tail. Juvenile wing and tail significantly shorter than adult. No significant difference between measurements of 1st adult ♂ in bright and dull morph, e.g.: wing, bright morph 85·2 (2·24; 14) 83-88, dull morph 85·5 (1·85; 25) 82-89.

P. o. aterrimus. Ages combined. Portugal and central and southern Spain, April-June; skins (RMNH, ZFMK, ZMA).

WING	♂ 89·0 (2·07; 14)	86-94	♀ 82·9 (1·44; 4)	81-84
BILL (S)	16·0 (0·50; 9)	15·2-16·7	14·5 (— ; 3)	14·3-14·8

Nominate *ochruros*. Ages combined. Eastern Turkey and Caucasus, April-August; skins (ZFMK, ZMA).

WING   ♂ 84·4 (1·48; 11)   83–87   ♀ 82·6 (2·53; 4)   79–86
BILL (S)   15·1 (0·64; 11)   14·3–15·8   15·1 (— ; 3)   14·4–15·4

P. o. phoenicuroides. Ages combined. Wing: (1) Tien Shan, USSR (RMNH, ZMA), (2) Mongolia (Piechocki and Bolod 1972; Piechocki et al. 1982); other data Tien Shan only.

WING (1)   ♂ 82·8 (1·44; 15)   80–85   ♀ 80·2 (— ; 2)   78–82
   (2)   82·9 (1·52; 10)   81–85   80·7 (1·97; 6)   78–83
BILL (S)   15·5 (0·77; 15)   14·2–16·6   15·3 (— ; 2)   15·0–15·6
TARSUS   23·9 (0·71; 9)   22·9–25·0   23·2 (— ; 2)   22·9–23·5

P. o. rufiventris. Ages combined. Wing: (1) Afghanistan (Paludan 1959), (2) Kashmir and Ladakh (RMNH, ZMA); other data Kashmir and Ladakh only.

WING (1)   ♂ 83·8 (2·64; 14)   77–87   ♀ 79·3 (— ; 3)   76–82
   (2)   87·8 (2·05; 12)   84–91   83·7 (3·22; 6)   79–87
TAIL   60·9 (1·35; 12)   59–63   59·3 (2·66; 7)   56–62
BILL (S)   15·8 (0·48; 12)   15·0–16·6   15·4 (0·63; 6)   14·6–16·0
BILL (N)   8·6 (0·52; 8)   7·8–9·1   7·8 (— ; 3)   7·7–8·0
TARSUS   24·7 (0·61; 11)   23·7–26·1   24·6 (0·73; 6)   23·3–25·5

In all samples, sex differences significant for wing.

Weights. *P. o. gibraltariensis.* Sexes combined. (1) Skokholm, Wales (Browne and Browne 1956). (2) Southern France, September–November (G Olioso), (3) Malta, winter (J Sultana and C Gauci). Gibraltar: (4) autumn, (5) spring (Finlayson 1981).

(1)   16·1 (— ; 8)   14·8–17·4   (4)   15·5 (1·15; 30)   —
(2)   16·2 (1·30; 10)   14·0–18·0   (5)   17·0 (1·43; 16)   —
(3)   16·5 (1·5 ; 50)   13·0–20·0

Europe, all year: ♂ 15·8 (1·75; 6) 13–18, ♀ 16·2 (2·10; 4) 14–19 (Niethammer 1943; Makatsch 1950; BTO, RMNH, ZMA). For fluctuations throughout year, see Berthold (1983); average reaches peak of c. 23·8 in early December. Exhausted ♂♂: off Norway, April, 12 (Willgohs 1955); Netherlands, October, 9·2 (ZMA).

P. o. aterrimus. Sierra Nevada, Spain: May, ♂ 17·5 (0·61; 5) 17–18·5; ♀♀ 15·5, 16 (Niethammer 1957; ZFMK).

Nominate *ochruros.* Turkey: June, ♂ 16 (Kumerloeve 1970a). Northern Iran: March, ♀ 18 (Schüz 1959).

P. o. semirufus. Mt Hermon (Levant): June–July, ♂ 13, ♀ 15 (ZFMK).

P. o. phoenicuroides and *rufiventris.* Saudi Arabia and Kuwait (BTO), Iran (Desfayes and Praz 1978), Afghanistan (Paludan 1959), Kazakhstan, USSR (Dolgushin et al. 1970), India and Pakistan (Ali and Ripley 1973b), and Mongolia (Piechocki and Bolod 1972; Piechocki et al. 1982), combined.

MAR–APR   ♂ 13·8 (0·44; 5)   13·4–14·5   ♀ 17·9 (— ; 2)   17·5–18·3
MAY–JUL   15·1 (0·87; 24)   13·0–16·0   16·4 (2·44; 7)   14·0–21·0
AUG–OCT   15·6 (1·67; 13)   14·0–18·0   14·9 (1·44; 11)   12·8–17·0
NOV–DEC   16·4 (— ; 2)   15·5–17·3   15·7 (— ; 3)   15·0–17·0

Autumn (mainly September), Ladakh: 15·7 (0·95; 52) 13·7–19·0 (Delany et al. 1982).

Structure. Wing rather long, broad at base, tip bluntly pointed. 10 primaries: p7 longest, p8 0–2·5 shorter, p9 8–13, p6 0–1, p5 2–4, p4 8–12, p3 11–16, p1 16–21 (*gibraltariensis* and *aterrimus*) or 13–20 (nominate *ochruros* and *phoenicuroides*). P10 reduced; 33–47 shorter than p7, 4–11 longer than longest upper primary covert. Outer web of p5–p8 and (rather faintly) inner of p6–p9 emarginated. Tertials short; longest equal to inner secondaries. Tail rather long, tip square or slightly rounded; 12 feathers, t6 1–5 shorter than t2–t4. Bill rather long, straight, slender; rather wide at base, laterally compressed at tip; tip of culmen decurved towards sharp tip. Nostril small, oval; partly covered by membrane above. Some fine bristles along base of upper mandible. Tarsus and toes rather long and slender. Middle toe with claw 16·9 (20) 16–19; outer toe with claw c. 69% of middle with claw, inner c. 65%, hind c. 71%. Claws rather short, sharp, decurved.

Geographical variation. Marked: in ♂, involves extent of black and grey on upperparts, colour of underparts and underwing, presence of white on forehead, and amount of white on wing; in ♀ and juvenile, mainly involves general colour; in both sexes, size (as expressed in wing length). 2 main subspecies-groups discernible: (1) *gibraltariensis* group with *gibraltariensis* and *aterrimus*, occurring Europe and North Africa east to Crimea and (probably) western Turkey; (2) *phoenicuroides* group with *phoenicuroides*, *rufiventris*, *xerophilus*, and *semirufus*, in central Asia, west to Turkmeniya, north-east Iran, and Levant. Nominate *ochruros* from eastern Turkey, Caucasus, and northern Iran combines characters of both main groups, apparently as a result of secondary intergradation; much individual variation, but in general nearer to *phoenicuroides* group. Within each main group, races mainly separated on account of differing amount of black on upperparts of ♂, but individual variation in this character marked and some races appear not to be well-differentiated. ♂♂ of *gibraltariensis* group characterized by uniform grey forehead and crown, black breast merging into grey-white central belly and vent, and large white wing-patch formed by broad white outer fringes of secondaries and outer primaries; ♀ and juvenile generally dark brownish-grey. ♂ *gibraltariensis* usually has upperparts uniform medium grey, but lower mantle and inner scapulars frequently black (see Plumages); ♂ *aterrimus* has hindneck, mantle, scapulars, and upper back largely black in 50% of 14 birds examined (and these usually show largely black belly and wing-coverts also), but in 29% only lower mantle and scapulars black and in 21% upperparts fully grey, and these inseparable from *gibraltariensis*, and hence separation of *aterrimus* perhaps doubtful, especially considering that 6% of 50 ♂♂ *gibraltariensis* examined were as black as darkest *aterrimus* and ♀♀ inseparable. Typical ♂ of *phoenicuroides* from Tien Shan mountains east to Mongolia differs markedly from *gibraltariensis* by fully deep rufous-cinnamon sides of breast, belly, flanks, vent, axillaries, and under wing-coverts, sharply contrasting with black chest; forehead often white, contrasting with black rim along base of upper mandible (white sometimes concealed, especially in fresh plumage); white wing-patch absent, fringes of tertials and secondaries grey (similar to upper wing-coverts or slightly paler ash-grey); upperparts usually grey with slight black suffusion on mantle, but sometimes as black as darkest *aterrimus*; ♀ and juvenile paler than *gibraltariensis*, upperparts (except rufous rump and upper tail-coverts) paler grey-brown with sandy suffusion, underparts pale buff-brown or buffish-grey, belly, vent, and flanks usually pale buff to off-white (but sometimes warmer sandy-buff or pale rufous-cinnamon), and often a distinct buff eye-ring; both sexes smaller than *gibraltariensis* (see Measurements). Following Stepanyan (1978), populations from Turkmeniya (USSR) east through Pamir-Alay mountains to Himalayas separated as *rufiventris*; typical ♂ *rufiventris* from Himalayas characterized by largely black upperparts and rather deep rufous underparts, ♀ by rufous tinge to underparts, and both by large size; dark colour shared by some *phoenicuroides*, however, and some *rufiventris* are pale, and large size not true for populations of (e.g.) Turkmeniya and Afghanistan (Zarudnyi and Bil'kevich 1918; Paludan 1959), which are small like *phoenicuroides*; hence distribution and characters of both races require confirmation, as does position of *xerophilus* from China, described as being pale like *phoenicuroides* and large like *rufiventris* (Stegmann 1928; Hartert and Steinbacher 1932–8). *P. o. semirufus* from Levant small, like *phoenicuroides*; ♂ similar to *phoenicuroides*, but upperparts usually with much black (sometimes up to crown and down to back); black of chest reaches slightly further down; sides of breast, axillaries, and under wing-coverts either rufous or mixed rufous and grey; wing has white patch, like *gibraltariensis*; ♀ pale, like *phoenicuroides*, or intermediate in colour between *phoenicuroides* and *gibraltariensis*. Nominate *ochruros* small, like *phoenicuroides*; colour of ♀ and juvenile pale or intermediate; ♂ from southern and eastern Turkey and Transcaucasia rather like *gibraltariensis*, but grey-white of belly and vent replaced by rufous, grey of sides of breast, axillaries, and under wing-coverts mixed with variable amount of rufous, and no white wing-patch; in northern Iran, belly of ♂ deeper rufous and more sharply divided from black of breast, rather like *semirufus*, but black on upperparts more restricted (usually on mantle only), breast more extensively black, underwing mixed grey and rufous, and usually no white wing-patch. Population of Caucasus highly variable: some as extensively rufous as birds from northern Iran, others similar to *gibraltariensis* (especially in western Caucasus), but sometimes without white wing-patch or with partly rufous belly (see Stegmann 1928 for details).

Recognition. ♂♂ of *gibraltariensis* group (*gibraltariensis* and *aterrimus*) easily separated from Redstart *P. phoenicurus phoenicurus* by grey forehead, grey or grey-and-black upperparts, black belly, and white wing-patch (*P. p. phoenicurus* has white forehead, grey upperparts, rufous belly, and no wing-patch). ♀♀ of *gibraltariensis* group distinctly darker and greyer than ♀ *P. p. phoenicurus*. In races of Caucasus and central Asia, characters become confusingly mixed, races of *phoenicuroides* group of *P. ochruros* having white forehead, rufous belly (but black chest, rufous not extending as high up as in *P. phoenicurus*), and no wing-patch (but white patch present in *semirufus* and some nominate *ochruros*), while *P. p. samamisicus* from eastern Turkey, Caucasus, and Iran has large white wing-patch (often extending to outer webs of primaries, unlike *P. ochruros*) and sometimes (in *incognita* morph of northern Iran and neighbouring Turkmeniya) largely black upperparts; ♀♀ of *phoenicuroides* group distinctly paler brown-grey than ♀♀ of *gibraltariensis* group, and very similar to ♀ *P. phoenicurus*, but chest and flanks usually paler brown-grey (not rufous-brown) and belly often more tawny-rufous, less white. Both species readily identifiable by wing-formula (see Structure) and by differences in measurements: tail of *P. phoenicurus* virtually always shorter than in any *P. ochruros*, and wing, bill, and tarsus often shorter.　　　　　CSR

Phoenicurus phoenicurus Redstart

PLATES 49 and 51
[between pages 652 and 653]

Du. Gekraagde Roodstaart　　　Fr. Rougequeue à front blanc　　　Ge. Gartenrotschwanz
Ru. Обыкновенная горихвостка　　　Sp. Colirrojo real　　　Sw. Rödstjärt

Motacilla Phoenicurus Linnaeus, 1758

Polytypic. Nominate *phoenicurus* (Linnaeus, 1758), Europe, northern Asia, and north-west Africa, east to Lake Baykal, south to Balkan countries and Ukraine; *samamisicus* (Hablizl, 1783), Crimea, Caucasus, eastern and southern Turkey, and Levant, east to Iran, intergrading with nominate *phoenicurus* in Balkans, western Turkey, and northern Caucasus.

Field characters. 14 cm; wing-span 20·5–24 cm. Similar in length to Robin *Erithacus rubecula* but with much more attenuated form, most obvious in flatter crown, longer wings, and slim rear body extending into rather long tail; marginally smaller and less robust than Black Redstart *P. ochruros*. Small, elegantly dressed, and graceful chat, with fine bill, rather long wings and tail, and slim rear body. Brilliant rufous-chestnut rump and tail always eye-catching whether flirted in flight or characteristically quivered on ground. ♂ blue-grey above, with white forehead and supercilium; deep rufous-orange to white below, with black face and throat. ♀ brown-grey above, buff to white below, with pale eye-ring and characteristic demure expression. Sexes dissimilar; much seasonal variation in ♂. Juvenile separable. 2 races in west Palearctic, with ♂♂ well differentiated at extremes of cline.

ADULT MALE. (1) European, north Asian, and north-west African race, nominate *phoenicurus*. From mid-winter to July (when plumage worn), forehead (and rarely forecrown) white, extending over eye to form usually short supercilium; crown, back, and scapulars blue-grey; wings black-brown, with narrow buff-grey fringes and tips on tertials and larger coverts (not forming obvious pattern). Rump, upper tail-coverts, and tail bright chestnut with darker brown central tail-feathers. Narrow frontal band over bill, lores, lower face (including sides of neck), throat, and upper breast black, contrasting with orange-chestnut breast and flanks, these merging with paler orange rear flanks, white belly and rufous-buff under tail-coverts; under wing-coverts pale chestnut. (2) South-west Asian race, *samamisicus*. Usually differs distinctly from nominate *phoenicurus* in having conspicuous white wing-panel formed by broad white fringes to tertials and secondaries and variably extended by narrow white fringes to primaries; some show black tinge to mantle and richer, orange-red underparts. ADULT FEMALE. (1) Nominate *phoenicurus*. Head and back grey-brown with faint fawn tone, relieved only by buff-white eye-ring; wings duller than ♂, with no grey fringes. Underparts distinctly paler, with almost white throat, pale orange-buff chest and flanks, cream to almost white belly, and buff vent. Rump and tail as ♂, though rump less bright. Some birds more intensely marked and coloured, resembling 1st-winter ♂. (2) *P. p. samamisicus*. Pale wing-panel indicated on a few, and eye-ring more obvious, but otherwise no constant difference from paler, greyer individuals of ♀ nominate *phoenicurus*. From August (when plumage fresh), loses immaculate appearance: grey upperparts marked with brown or red-brown feather-tips, underparts with white; bird thus appears 'hoary', with less clear forehead pattern, less black on face, duller upperparts, paler underparts, and much paler-edged tertials and larger wing-coverts. JUVEN-ILE. Both races. Upper tail-coverts and tail as adult ♂ but rest of plumage essentially brown above and buff-cream below, liberally spotted ochre-buff on back and wing-coverts and densely barred dark brown on throat, chest, and flanks. Spots on median and greater coverts form obvious bars; pale buff fringes of tertials form faint panel on folded wing. ♂ *samamisicus* shows white on wing. FIRST-WINTER MALE. Both races. From August, usually distinct from late juvenile and 1st-winter ♀: resembles winter adult but lacks grey-toned back and crown, obvious white supercilium, and hoary upper breast, having close black bars on cheeks and grey-white bars on throat. FIRST-SUMMER MALE. Abrasion removes most brown fringes from feathers but, in nominate *phoenicurus* at least, appearance remains less than im-maculate, with less blue-toned crown and back, white flecks on face and throat, and noticeably dull wings. At all ages, bill and legs black.

Far from unmistakable, with (1) adult ♀ subject to confusion particularly with eastern races of *P. ochruros* (see that species, p. 683), (2) immature ♂ on passage in Middle East confusable with Eversmann's Redstart *P. erythronotus* (see that species, p. 678), and (3) ♀ and juvenile confusable with 6 other species of *Phoenicurus* of similar sex or age. Faced with this barrage of pitfalls, important to realize that *P. phoenicurus*, while by far the commonest and most ubiquitous of genus on migration, shows markedly different habitat preferences in breeding season from *P. ochruros*, and habitat does not more than marginally overlap that of Güldenstädt's Redstart *P. erythrogaster* (which is much larger), Moussier's Redstart *P. moussieri* (distinctly shorter and more compact), or *P. erythronotus* (larger and bulkier). Best marks of *P. phoenicurus* are (1) slim, elegant form, (2) blue tone of ♂'s crown and back, (3) clean tone of ♀'s underparts, (4) almost constant tail-quivering (shared only by *P. ochruros*), (5) distinctive calls, and (6) arboreal behaviour. Flight fluent and agile, with slight undulations (obvious from behind) and terminal sweeps and turns into cover; beats wings rapidly, but length of wings and tail, and apparently loose joint between tail and body produce less whirring, more flitting action than in *Saxicola* chats, most recalling *E. rubecula* or longer-tailed *Luscinia* chat. Carriage usually half-upright on open perch, more upright on ground but much less so in foliage; length of tail usually obvious and enhanced by almost constant, neurotic quivering movement (strictly up and down, except in display). Gait a light quick hop, interspersed with tail-quivering. Perches freely on wide range of objects but tends to feed from hidden position. Shy rather than wild.

Song sweet, rather melancholy in tone and weak in volume, with repetition of main phrases suggesting *E. rubecula* but having characteristic feeble, squeaky, or mechanical jangle at end. Calls include loud, plaintive, *Phylloscopus*-like 'hweet'; rather liquid, slightly explosive 'tuick' often run together with tremulous, scolding 'hwee-tucc-tucc'.

Habitat. Breeds in west Palearctic from upper to middle latitudes, mainly continental and lowland, in boreal, temperate, steppe, and Mediterranean zones between July isotherms 10–24° C. Requires sheltered but fairly open wooded or parkland areas with access to dry secure nest-holes in trees, rocks, walls, banks, or other places and without too dense or tall unbroken undergrowth or herbage. At least in west of range prefers broad-leaved or mixed trees, but in some parts occupies open pine-woods, and is adapted to woodland edges, streamside and roadside trees, orchards, and gardens in human settlements (Voous 1960); also heaths and commons with scattered trees or copses, pollard willows *Salix* along streams or ditches, open hilly country with loose stone walls, old ruins, quarries, and rocky places. Much more arboreal than Robin *Erithacus rubecula* but less shade-loving; perches freely on exposed branches of trees at any height, as well as on bushes, fences, walls, and buildings, but also at times skulks in foliage (Witherby *et al.* 1938*b*). For densities in various habitats, see Social Pattern and Behaviour. In Switzerland, widespread between 600 and 1000 m, but breeds sparingly up to *c.* 2000 m, in mountains favouring human neighbourhoods, but also ruins and rocky places, and open woods of larch *Larix* and Arolla pine *Pinus cembra*. Presumed original ground-breeding status now almost entirely superseded by more elevated nest-sites, artificial as well as natural (Glutz von Blotzheim 1962). In Berlin, density of breeding in zone of gardens, cemeteries, and parks is up to 14 or more times greater than in forest (Bruch *et al.* 1978). In USSR, nests to northern boundary of tree growth, and in south mainly in hills, extending locally into stunted krummholz towards tundra and on mountains, but generally prefers broad-leaved and mixed forests, less frequently pinewoods, while dense spruce *Picea* avoided (Dementiev and Gladkov 1954*b*). For discussion of supposed original habitat on pine-heath or in pinewoods, see Buxton (1950) and Yapp (1962)—also for comment on peculiar anomalies of occupancy and neglect of apparently suitable areas, and unaccountable fluctuations in range, possibly linked with vulnerability to climatic and other factors on migration. Moreau (1972) reviewed

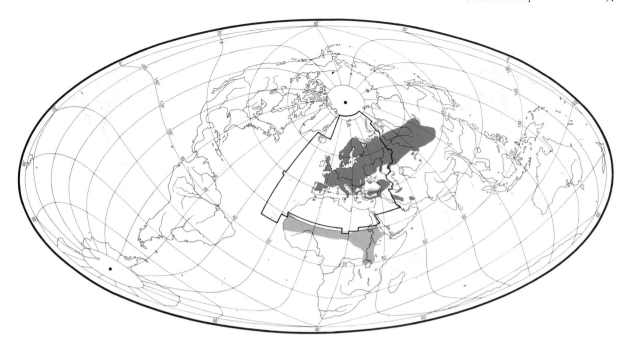

situation in African winter quarters. In west, in Sénégal inhabits both dry *Acacia* steppe and more lush thickets of *Acacia nilotica*; in Gambia found on edges of cultivated ground. In Eritrea, winters in every woodland except dry *Acacia*, nominate *phoenicurus* at higher altitude (mainly above 1300 m) than *samamisicus*. Evidently both an adaptable and a vulnerable species, even though benefiting from human impacts in breeding range.

Distribution. No evidence of any marked range changes.
BRITAIN. Recent slight expansion to Isle of Man and eastern Cornwall (Sharrock 1976). IRELAND. Scattered breeding records since 1895, now probably annual (CDH); more breeding records 1968-72 than previously in 20th century (Sharrock 1976). IRAQ. May breed in north (Moore and Boswell 1956). CYPRUS. May have bred (Flint and Stewart 1980). TUNISIA. Breeds Cape Bon (Heim de Balsac and Mayaud 1962); no recent evidence (MS).
Accidental. Iceland, Chad, Madeira.

Population. Has decreased Britain, France, Netherlands, Norway, Sweden, Finland, East Germany, Poland, Czechoslovakia, Switzerland, and Estonia, due in part to drought in Sahel winter quarters, but some increase in Denmark.
BRITAIN. 10000-20000 pairs; marked decline in southern, eastern, and central England and to lesser extent in Scotland (Alexander and Lack 1944). No evidence of further decrease, and some increase in southern England (Parslow 1973). 50000-100000 pairs; recent decrease due to drought in Sahel (Winstanley *et al.* 1974; Sharrock 1976). FRANCE. Recent decline due to Sahel drought

(Yeatman 1976). BELGIUM. About 45000 pairs (Lippens and Wille 1972). NETHERLANDS. 50000-60000 pairs (Teixeira 1979); declined due to habitat changes and Sahel drought (Wammes *et al.* 1983; CSR). WEST GERMANY. 84000-450000 pairs (Rheinwald 1982). DENMARK. Increased, now a few hundred pairs (Vickholm and Väisänen 1984; TD). NORWAY. Decrease in lowland population especially in south-east (VR). SWEDEN. About 1·5 million pairs; decreasing for at least 20 years (LR); numbers of birds ringed declined 1960-79 (Österlöf and Stolt 1982). Increase in south (Johnsson 1981; Wirdheim 1981; Hildén and Koskimies 1984; Vickholm and Väisänen 1984). FINLAND. About 67000 pairs; marked decrease in recent decades (Järvinen 1981). EAST GERMANY. Marked decline in recent decades (SS). In study areas, increase 1927-40, marked decrease 1955-78 (Berndt and Winkel 1979). POLAND. Scarce (Tomiałojć 1976). Marked decline since 1960s (AD). CZECHOSLOVAKIA. Strong decrease since 1960 (KH). AUSTRIA. Decreasing (AS, PP). SWITZERLAND. Marked decline during late 1950s (Bruderer and Hirschi 1984); some recovery in 1970s (RW). USSR. In west, increase 1920-40, sharp decline since (GGB). Leningrad region: sharp decrease since 1960s (Mal'chevski and Pukinski 1983). Estonia: decrease in recent decades (HV).
Survival. Netherlands: annual 1st-year mortality *c.* 79%; annual adult mortality *c.* 62% (Buxton 1950). Finland: annual overall mortality 51% (Haukioja 1969). Oldest ringed bird 9 years 6 months (Dejonghe and Czajkowski 1983).

Movements. Migratory. Movement mainly nocturnal, with broad-front trans-desert passages across Africa and Middle East. 2 distinct populations: nominate *phoenicurus*

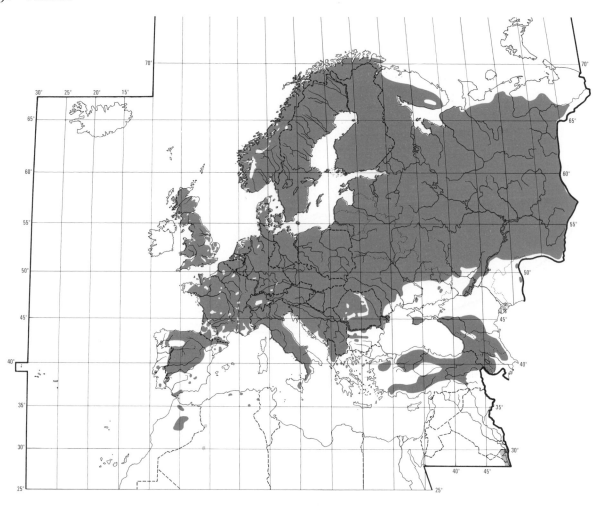

breeds Europe, Siberia, and north-west Africa, and winters across Afrotropics north of equator; *samamisicus* breeds around Black and Caspian Seas and in northern Middle East, and winters in Arabia, Sudan, and Ethiopia.

A good many December–February ringing recoveries in west Mediterranean basin, from Iberia and Morocco to Italy (mapped in Zink 1981); also a few other winter observations north to latitudes of Britain and East Germany. However, main wintering zone is in scrub-savanna belt of Africa at 9°30′–15°30′N, with southward extension to *c*. 2°N in eastern Zaïre and Uganda; also occurs sparsely in Kenya, but only 1 Tanzanian record (Moreau 1972; Britton 1980; Zink 1981). Isolated records (4 birds) claimed from Namibia in December 1957 (Sauer and Sauer 1960). With preference for arid and semi-arid latitudes in winter (some remaining all season even in Saharan oases), population declines noted during current cycle of Sahelian droughts (e.g. Marchant 1982).

Autumn movement through Europe mainly south-westward. Leaves breeding grounds in second half of August, with peak passage through north-west Europe in early September and numbers there diminishing gradually during October. Hempel and Reetz (1957) mapped recoveries (mainly ringed Germany), and found movements mainly to south-west France and north-east Spain in August–October, with further concentrations of recoveries in southern Spain and Portugal September–November; also a few autumn recoveries in Italy and Algeria, showing that minority took more south-easterly route. More recent review of recoveries in Zink (1981). Birds from west and central Europe (east to 15°E) all moved SSW–WSW towards Iberia in autumn, presumably pausing there to gain weight for next stage of journey. A few Iberian recoveries as early as August, but most in September–October; 3 main groupings there—birds found just south of Pyrénées (mostly in September), in northern Portugal, and in south-west Spain/Algarve (mainly September–October). Relatively few autumn recoveries in North Africa, suggesting tendency for Mediterranean and northern Sahara to be overflown then, though falls of migrants (some large) have been noted there. Nestlings ringed northern Scandinavia (north of 62°N) and Finland showed suggestion of migratory divide, with autumn recoveries in Iberia and in Yugoslavia/

Italy/Algeria, while birds ringed Poland and Soviet Baltic States produced more autumn recoveries in northern Italy than elsewhere. First arrivals south of Sahara in first half of September, but not common there until mid-October (e.g. Morel and Roux 1966, Newby 1980). Wing lengths in Sénégal, at western end of winter range, suggested that birds there were from eastern as well as western Europe (Blondel 1967; Moreau 1972).

Spring passage in Africa more conspicuous than in autumn. In Sénégal, successive waves of birds appear, fatten, and vanish, with first full weights about mid-April (G Morel). In Nigeria, birds at 9°50'N gained weight from late March, prior to departure (Smith 1966), but at 11°N wintering birds departed without weight gains and presumably fattened up further north (Fry 1970b; Moreau 1972). In Mali, there can be hundreds within a few ha of scrub near water in March–April (Lamarche 1981). In north-east Africa also, passages of nominate *phoenicurus* and *samamisicus* are more marked (e.g. in Eritrea) in April than in autumn (Smith 1957). In contrast to autumn, there are many recoveries of European-ringed birds from March to early May in North Africa (especially Morocco and Algeria), in keeping with larger spring (than autumn) falls of migrants generally in North Africa. Few spring recoveries in southern or western Iberia, though more in north-east Spain (Pyrénées region) and on passage through France. Vanguard arrives northern Europe in first half of April; main arrivals there mid-April to mid-May, with passage declining into June (e.g. Bruns and Heinrich 1968, Vauk and Schröder 1972, Riddiford and Findley 1981). RAC, RH

Food. On breeding grounds at least, largely insects (especially adult and larval Lepidoptera and Coleoptera) and spiders. 4 main feeding methods. (1) Picks items from ground; apparently does not probe for worms (etc.) and rarely searches in leaf-litter, though this recorded in Nigeria (Serle 1939–40). (2) Feeds in trees and other vegetation (perhaps especially when feeding young: Buxton 1950), picking items from trunks, branches, and leaves, including by hovering near foliage, etc. (3) Flies from perch on to prey on ground, normally returning to perch to eat it. (4) Takes aerial prey in brief flight from perch. (Buxton 1950; Menzel 1971.) In Norwegian birch *Betula* forest, late May to mid-July, foraging done largely on ground, in tree foliage, and in flight (each category *c.* 25–30% of total 643 foraging records); also in herb layer (*c.* 12%) and on trunks and branches of trees (*c.* 8%); ♂♂ fed significantly more often in flight and less often on ground than ♀♀, and hovered more often at tips of vegetation (Hogstad 1977; see also Schmidt 1967a). This difference between sexes in foraging methods apparently widespread in breeding season (Buxton 1950). In winter in Sierra Leone, feeds mainly on ground (preferably open, not grassy), occasionally by flutter-flights to take

prey from vegetation (G D Field). Sometimes beats prey against perch, and normally removes (presumably only large) insects' wings before eating body (Buxton 1950); removes legs of mole-crickets *Gryllotalpa* (Menzel 1971). Rarely drinks (perhaps normally never), though captive birds do so readily (Buxton 1950; Menzel 1971).

Diet in west Palearctic includes the following. Invertebrates: mayflies (Ephemeroptera), damsel flies (Agriidae), grasshoppers, etc. (Orthoptera: Gryllotalpidae, Gryllidae, Tettigoniidae, Acrididae), earwigs (Dermaptera), Psocoptera, bugs (Hemiptera: Aradidae, Scutelleridae, Pyrrhocoridae, Nabiidae, Aphidoidea), lacewings, etc. (Neuroptera: Raphidiidae, Hemerobiidae), adult and larval Lepidoptera, including hairy larvae (Nymphalidae, Pieridae, Cossidae, Tortricidae, Coleophoridae, Noctuidae, Lymantriidae, Thyatiridae, Geometridae, Sphingidae), caddis flies (Trichoptera), adult flies (Diptera: Tipulidae, Tabanidae, Therevidae, Asilidae, Empididae, Dolichopodidae, Syrphidae, Trypetidae, Tachinidae, Muscidae), adult, larval, and pupal Hymenoptera (Pamphilidae, sawflies Tenthredinidae, Argidae, Cimbicidae, Diprionidae, Ichneumonoidea, ants Formicidae, wasps Vespidae, Pompilidae, bees Apoidea), adult and larval beetles (Coleoptera: Carabidae, Staphylinidae, Scarabaeidae, Elateridae, Cantharidae, Meloidae, Tenebrionidae, Coccinellidae, Cerambycidae, Bruchidae, Chrysomelidae, Curculionidae, Scolytidae), spiders (Araneae), harvestmen (Opiliones), mites (Acari), woodlice (Isopoda), millipedes (Diplopoda), small molluscs, earthworms (Oligochaeta). Plant material mainly fruits: of juniper *Juniperus*, yew *Taxus*, pear *Pyrus*, cherries *Prunus*, rowan *Sorbus*, brambles *Rubus*, *Amelanchier*, dogwood *Cornus*, elders *Sambucus*, currants *Ribes*, crowberry *Empetrum*, strawberry-tree *Arbutus*, privet *Ligustrum*, buckthorn *Rhamnus*, alder buckthorn *Frangula*. (Data mainly from Menzel 1971 and Dornbusch 1981; also Schuster 1930, Buxton 1950, Ganya *et al.* 1969, Kostin 1983.)

Of 601 items in adult diet, Hiddensee (East Germany), 51·7% (by number) Hymenoptera (almost all adults), 23·5% beetles (largely adults), 7·2% Hemiptera, 7·1% adult Diptera, and 10·5% others (Emmrich 1975). In Crimea (USSR), items from 16 stomachs, April–May and August–September, comprised 43·6% Hymenoptera (mostly ants), 37·9% beetles, 8·1% Hemiptera, and 10·4% others (Kostin 1983). In Moldavia (USSR), 50 stomachs contained mainly invertebrates, with plant material (7·4% of total) eaten only from July up to departure in September–October (Ganya *et al.* 1969). For other data, see review by Menzel (1971); also Rey (1908) for Germany, Csiki (1908) for Hungary, and Mal'chevski (1959), Khokhlova (1960), Pek and Fedyanina (1961), Ptushenko and Inozemtsev (1968), Averin and Ganya (1970), Prokofieva (1972a), and Kovshar' (1979, 1981) for USSR.

In Ahaggar massif (southern Algeria), commonly eats

ants (Laferrère 1968). In Sénégal, February, one bird contained termites (Isoptera) (Dekeyser 1956).

Collar-samples from young in East Germany contained 1143 items comprising 27·0% adult and larval Lepidoptera, 22·0% Arachnida, 20·0% beetles (mostly adults), 11·0% Hymenoptera (mostly adults), 8·4% Diptera (mostly adults), 5·7% Orthoptera, and 5·9% others; proportion of beetles rose as young grew, proportion of Arachnida fell (Bösenberg 1960). Further 1256 items from collar-samples in East Germany comprised 20·0% beetles, 18·9% Lepidoptera, 17·8% Arachnida (mostly spiders), 17·4% Diptera, 10·7% Hemiptera, 7·4% Hymenoptera, and 7·8% others (Dornbusch 1981). At Frankfurt am Main (West Germany), items in collar-samples comprised 59·3% (by number) Lepidoptera, 11·6% beetles, 8·1% spiders, 6·2% Diptera, and 14·8% others (Pfeifer and Keil 1959). See also Mikkola and Mikkola (1976) for Fenno-Scandia, Ruiter (1941) for Netherlands, Emmrich (1975) for East Germany, Pruska (1980) for Poland, Korodi Gál and Györfi (1958) for Rumania, and Neufeldt (1956), Ganya et al. (1969), Averin and Ganya (1970), and Kostin (1983) for USSR.   DJB

Social pattern and behaviour. Main references: Boubier (1925), Ruiter (1941), Buxton (1950), and Menzel (1971).
    1. In winter, apparently usually solitary or in pairs (e.g. Mundy and Cook 1972). On migration, occurs singly or in parties of up to 50-60 or more, exceptionally in company with other species (Buxton 1950). BONDS. Probably mainly monogamous, though several cases of bigamy (♂ plus 2 ♀♀) (Kierski 1934; Mühl 1958; Creutz 1959a, 1962; Burri 1960; Sutter 1960; Haartman 1969b; Menzel 1971; Male 1975; Löhrl 1976). Other cases of supposed bigamy (Warga 1926, 1939c; Ruiter 1941; Rosenson 1946) could have resulted from ♂ helping a 2nd ♀ after her 1st mate was lost. ♂ usually extends territory size on becoming bigamous. In 9 out of 10 cases, the 2 nests of a bigamous ♂ were over 75 m apart (Ruiter 1941); other recorded distances 8-150 m (Creutz 1959a, 1962; Menzel 1971). One pair reported driving off neighbouring pair after their own young fledged, and new ♀ incubated their eggs; ♂ of that ♀ then fed not only her but also young of a nest in a 3rd site where ♀ had probably lost her mate; eggs of his original mate did not hatch (Warga 1926). Extent of care by bigamous ♂ varies. In 4 cases of apparent bigamy, ♂ fed and tended young of both ♀♀ equally (Warga 1939c); in another, ♂ not seen to feed one brood at all and the other not very assiduously (Burri 1960). When bigamy successive, ♂ feeds young of 1st ♀, later moving to 2nd ♀, and may then return to feed 2nd-brood young of 1st ♀ (Creutz 1959a, 1962; Menzel 1971); young of final brood first fed by ♂ when 7 days old (Menzel 1971). Record of 10 eggs incubated by 2 ♀♀ (Berndt 1938) and of 16 eggs in 2 layers (probably laid by 2 ♀♀ though only 1 ♂ and 1 ♀ ever seen) (Schönfeld 1972). A number of records of hybridization with Black Redstart P. ochruros; see that species (p. 688). If ♀ disappears during season, ♂ may re-pair—after 5 days in one case (Althen 1950-1). Same birds sometimes re-pair in subsequent years, due to year-to-year site-fidelity (Ruiter 1941; Menzel 1971); see Breeding Dispersion (below). Of 64 pairs, only 12 re-formed in following year, in 6 cases using same nest-box; 3 bred within 75 m of previous year's site, other 3 76-150 m away (Ruiter 1941). Warga (1939c) recorded only 2

cases of fidelity over 2 years. No cases of pair-fidelity of over 3 years in either study. Warga (1939c) recorded ♂ frequently stealing mate and nest-site from another ♂. Both sexes may breed at 1 year old (Ruiter 1941; see also breeding). 91% of 54 1-year-olds in Netherlands that returned to study area settled more than 225 m from natal site, possibly because of exclusion due to their later arrival on breeding grounds (Ruiter 1941). 25% of 1-year-old ♂♂ and 58% of 1-year-old ♀♀ said not to breed (Ruiter 1941), though these estimates questioned by Buxton (1950). Both parents care for young both in nest and after fledging although if another clutch laid, ♂ feeds fledged young whilst ♀ lays next clutch (Ruiter 1941; Buxton 1950; Menzel 1971). Young remain together in care of parents until 10-14 days after fledging (Henze 1958). BREEDING DISPERSION. Solitary and territorial. study area in Niederlausitz (East Germany), territories 1400-5000 m² (Menzel 1971). One territory in West German village 1600 m² (Blume 1966). Average territory size in Netherlands 1 ha (Ruiter 1941). For details of territory distribution, see Menzel (1971). Territory size contracts as more ♂♂ arrive (Menzel 1971). Bigamous ♂♂ generally have larger territories than other ♂♂. In Britain, 5 Welsh oakwood Quercus sites contained average 67 territories per km²; 2 broad-leaved woods in northern England contained average 26 territories per km²; 3 broad-leaved woods in New Forest (southern England) contained average 58 territories per km² (BTO). At one site in East Germany, density 266 pairs per km² (Menzel 1971). Lowest recorded densities 0·4 pairs per km² (Kriwanek 1965) and 0·7 pairs per km² (Bruns 1957), both in spruce Picea woods. In Switzerland, up to 120 pairs per km² in parks and gardens; also common in open deciduous woods, orchards at lower altitudes with 60 and sometimes 100 pairs per km²; above 800 m, only 10 pairs per km² (Glutz von Blotzheim 1962). Along Tisza river (Hungary), 110 pairs per km²; 1 pair per 80-100 m of pollarded willows Salix (Schmidt 1967b). Other densities: in West Germany, 50 pairs per km² in bomb-damaged city (Drost 1949) and 90 pairs per km² in urban park (Drost 1949); in Poland, 1 pair per km² (Tomiałojć et al. 1984); in East Germany, 0·4 pairs per km² in pine forest and 1·5 pairs per km² in marshy broad-leaved woodland (Schiermann 1934); in Sweden, 0·7 pairs per km² in forest and 2·9 pairs per km² on cultivated ground (Olsson 1947); for Finland, see Palmgren (1930), Siivonen (1935), Soveri (1940), Perttula (1945), and Merikallio (1946). Nest-box studies over 11 years in Swedish Lapland indicated 4·0 pairs per km² in meadow birch Betula forest, 5·6 pairs per km² in heath birch forest; in same areas before nest-boxes installed, c. 10 pairs per km² (Enemar 1980). In Britain, highest density in oak Quercus/hazel Corylus and oak/birch coppice in western Scotland was 49 pairs per km², with 20 pairs per km² in alder Alnus woods and open Scots pine in Wester Ross, although oak wood in same region held twice that density (Sharrock 1976). Territory defended by ♂ (Menzel 1971). Several records of mixed broods in same nest: e.g. with Pied Flycatcher Ficedula hypoleuca, fledging successfully (Ushakova and Ushakov 1976; Sundkvist 1979); sharing nest-box with Robin Erithacus rubecula, incubating either singly or together with no aggression (Booth 1967); see also Amann (1949) and Mackenzie (1954). Replacement clutches often only a few metres from 1st, within same territory; furthest 63 m (Menzel 1971). 2nd brood also close to 1st: in same nest or up to 180 m away (Ruiter 1941). 23% of surviving young birds returned to natal area in one Dutch study (Ruiter 1941), and 2 1-year-olds recovered 100 km and 360 km away in East German study (Menzel 1971). In Netherlands, site-fidelity stronger in ♂♂: 73% of adult ♂♂ (n=60) returned to their territory, and 55% of ♀♀ (n=66) to within 150 m of their nest-site; 32% of ♂♂ and 21% of ♀♀ returned to breed at same nest-box, and only 3% of ♂♂ bred more than 350 m away (Ruiter 1941). One ♂ recorded occupying same garden for 6 consecutive summers until its death (Burri 1960). ROOSTING. Solitary and nocturnal. In holes in trees, masonry, or rocks (Naumann 1905). ♀ said by Ruiter (1941) to start roosting in nest after 1st egg laid, though in 15 pairs studied by Menzel (1971) not until clutch half-complete; may roost on nest-rim (Ruiter 1941) or in nest (Menzel 1971).

2. Often confiding and easily approached, at least in mainland Europe. When disturbed may give alarm-call (see 6 in Voice); normally ceases characteristic tail-shivering and makes for cover (Buxton 1950). FLOCK BEHAVIOUR. No information. SONG-DISPLAY. ♂ (occasionally ♀) sings (see 1 in Voice) from high and exposed perches, same ones used throughout breeding season (Buxton 1950; Menzel 1971). Circling Song-flight also recorded: a few metres above ground and *c.* 5 min long (Link and Rutter 1973). Stresemann (1910) noted ♂ in whirring display-flight, then singing from perch. Little song for *c.* 2 days after arrival on breeding grounds but then intense, particularly if other ♂♂ in vicinity (Menzel 1971), even when ♀ not yet present (Diesselhorst 1968*b*). About 2 weeks after arrival, ♂ suddenly switches from full song to strangled, monotonous song (see 1b in Voice) given near one favoured nest-site, at same time performing Nest-showing display (Diesselhorst 1968*b*); see subsection 5 in Heterosexual Behaviour (below). During nest-building, ♂ often becomes silent (often taking cover in deeper undergrowth) but then starts singing persistently during incubation, though usually silent when ♀ comes off nest to feed (Buxton 1950; Ward 1956; Kostin 1983). Song given mainly from a few days after arrival on breeding grounds to hatching of 2nd (or last) brood; wanes through season and may cease altogether when chicks hatch and during rest of season unless 2nd brood raised (Buxton 1950; Ward 1956; Diesselhorst 1968*b*; Menzel 1971). Singing occurs through night in Finland (Heyder 1934); in central Europe, usually starts *c.* 1 hr before sunrise (Menzel 1971). Generally no singing during moult; only a few cases of song in August and September, and then only a little in morning (Menzel 1971). No singing recorded on migration in Sénégal (Morel and Roux 1966) nor on wintering grounds in Africa (Stoneham 1931; G D Field), although Buxton (1950) recorded occasional song by migrants and once in mid-March in Nigeria. ANTAGONISTIC BEHAVIOUR. (1) General. Territory-holding ♂ shows aggression to intruding ♂♂ (and to other species such as *P. ochruros*: e.g. Turrian 1980) by extending neck, sleeking plumage, fanning tail slightly, opening wings slightly, and calling (see 7 in Voice); may also sway body from side to side; intruder sings briefly in response (see 1b in Voice) before chase ensues, and owner may call (see 6b in Voice) after intruder has left (Buxton 1950; Menzel 1971; Böhme and Böhme 1986). Brief contests may occur on territory boundaries with ♂ singing, chasing off other ♂, then returning and singing again, but if territory intrusions more persistent the 2 birds call at each other; contests generally cease before incubation begins (Buxton 1950). Contests also occur at any time with apparently unpaired ♂♂: involve close pursuit over greater distance, sometimes ending in scuffle, in which ♀ might join; these intrusions do not evoke song in resident bird (Buxton 1950). Up to 4 ♂♂ recorded fighting on ground (Menzel 1971). Birds of both sexes recorded fighting own image in window pane (Mühl 1955; Menzel 1971). Potential nest-site users (e.g. Wryneck *Jynx torquilla*, Starling *Sturnus vulgaris*, Blue Tit *Parus caeruleus*, Spotted Flycatcher *Muscicapa striata*) may be driven away (Buxton 1950; Diesselhorst 1968*b*). One ♂ which fed nestling Short-toed Tree-

creepers *Certhia brachydactyla* attacked their parents if they met at nest, driving them away (Gass 1975). When young hatch, ♂ recorded as very aggressive to other birds (Ebbutt 1947). 2 ♀♀ paired with same ♂ may fight and chase one another, though chases generally cease after incubation (Mühl 1958). On wintering grounds, may be aggressive to other Palearctic migrants, though sharing bush amicably with Didric Cuckoo *Chrysococcyx caprius* (G D Field). HETEROSEXUAL BEHAVIOUR. (1) General. ♀♀ generally arrive on breeding grounds a few days or up to 2 weeks after ♂♂ (Buxton 1950; Menzel 1971; Löhrl 1976; Bergmann and Helb 1982). Among 12 pairs in Oberlausitz, average 7 days (4-12) between pair-formation and laying (Menzel 1971). (2) Pair-bonding behaviour. After successful attraction of ♀ on to territory (sometimes only after several weeks: e.g. Blume 1966, Menzel 1971), ♂ adopts Greeting-posture (Fig A) in front of her:

A

crouches slightly with wings spread and tail fanned (Buxton 1945; Menzel 1971). ♂ no longer gives call 7 (linked with aggressive behaviour) but instead call 4a with bill wide open (Buxton 1950). During courtship usually silent (though accompanied by more excited chirruping and singing from ♂ as season progresses), ♂ following ♀ around trees in vicinity of future nest; both birds hold tail slightly or widely fanned (Buxton 1945, 1950). Both Buxton (1950) and Diesselhorst (1968*b*) described display in which ♂ flies to ♀, perching nearby and beating wings deeply (prior to nest-building at same site by ♀). Courtship displays and mating may continue through egg-laying and into incubation, though last full display on day after laying of last egg (Buxton 1945). (3) Courtship-feeding. Not recorded other than by Wolff (1941) who noted ♂ frequently feeding ♀ off nest. During courtship, ♂ may, after singing and on approach of ♀, go through the motions of feeding her but without food in bill; ♀ crouches and ♂ mounts but copulation unsuccessful. On one occasion, ♂ with bill full of food was approached by ♀ giving soft call (no details), and both flew off; food transfer not observed. ♀ also recorded begging food from ♂, giving call like food-call of young (presumably call 4a) and craning up her neck and opening her bill. (Buxton 1945.) (4) Mating. Copulation generally starts only when ♀ starts nest-building (Menzel 1971), though attempts recorded the day before nest-building (Buxton 1945). One detailed description by Alpers (1942): ♂ landed *c.* 0.5 m from ♀ on fence, gradually moving towards her with tail slightly raised, wings drooped, whole body trembling markedly; ♀ crouched as ♂ approached, and copulation took place (twice in quick succession) only after ♂ had advanced and retreated 3 times; both birds called throughout (see 5 in Voice). ♀ also described as fanning tail (apparently indicating readiness for copulation) while ♂ flattens body and raises wings, quivering them, until they probably touch above his back; holds tail half-depressed (though sometimes raised) and widely fanned, lowers head, stretches neck, and presses body down on to branch, moving towards ♀ and calling all the time (see 4a in

Voice) with mouth wide open. ♀ also gives this call, and copulation follows. ♂ then flies off in wild rapid flight, giving a short burst of loud song (see 2 in Voice), sometimes chasing ♀ thus. Copulation may also be preceded by the 2 birds making small vertical jumps into the air whilst facing each other. (Buxton 1945, 1950; also Fleischmann 1977.) Copulation may occur quite high in tree or lower (Menzel 1971). (5) Nest-site selection. Generally ♂ selects nest-site (though one case of ♀ doing so) and often visits nest-holes even when no ♀ present (Menzel 1971). Ruiter (1941) recorded ♂♂ entering nest-holes only just before beginning of nest-building and always before 08.00 hrs, though this not found by Blume (1966) or Menzel (1971). ♂ generally goes to a number of different sites and offers these to ♀ (Blume 1966). Buxton (1945, 1950) described 3 main behaviour patterns of ♂ associated with Nest-showing: (a) flies in, looks out of entrance again, and occasionally sings with head out of hole, displaying white forehead (Fig B); (b)

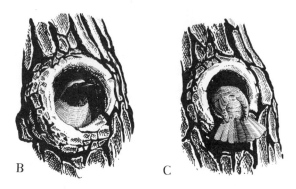

B   C

flies in and out (sometimes very frequently), showing rufous tail and rump, clinging to nest-box just below hole, occasionally with brief song or ticking call (see 1b and 6a in Voice)—or may appear to enter but not do so, holding tail fanned (Fig C); (c) performs gliding Song-flight with tail and wings spread (Fig D), but probably only when other 2 types unsuccessful

D

(see also Diesselhorst 1968b). 2 additional types described by Blume (1966), though Menzel (1971) considered these only variants: (d) sings from entrance hole with feet on hole (Fig E); (e) one ♂ sang while perched by nest, wings slightly drooped and fanned and tail raised (Fig F), then quickly went into box and flew out again, sequence repeated several times. As soon as ♀ inspects hole and is ready to take in nest-material, ♂'s Nest-showing stops (Menzel 1971). Diesselhorst (1968b), when describing Nest-showing by one ♂, distinguished between Bill-lowering apparently to show white forehead to ♀ above him (♂ frequently giving strangled variant

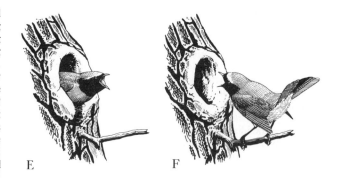

E   F

of song: see 1b in Voice), and Bill-raising apparently to show off black throat; these displays followed by ♀ joining ♂ in nest then emerging with some old nest-material and perching quietly by box; ♂ flew around silently, then both birds showed interest in another nest-box (Diesselhorst 1968b). (6) Behaviour at nest. Usually 1-2(-3) days between completion of nest and start of laying (Menzel 1971). ♂ visits nest only occasionally during laying and rarely (though sometimes regularly) feeds ♀ on nest (Koefoed 1935; Ruiter 1941; Buxton 1950; Schönfeld 1962; Menzel 1971). ♂ generally announces arrival at nest by calling (see 6e in Voice), and ♀ then usually leaves (Menzel 1971). Once when ♀ left nest to feed, ♂ approached, sang, then on no response from ♀ chased her until she returned to nest (Passow and Passow 1913). RELATIONS WITHIN FAMILY GROUP. Eggs hatch over 1-2(-3) days (Menzel 1971). Eyes start to open on 6th day, fully open at 9 days; for further details of physical development, see Bussmann (1940) and Menzel (1971). Young brooded by ♀ only (Menzel 1971), usually continuously throughout 1st day but with frequent though brief breaks from 2nd day (Bussmann 1940; Buxton 1950; Kostin 1983). Brooding during day continues until young c. 5-7 days old, during night until 4-10 days old (Bussman 1940; Ruiter 1941; Menzel 1971). ♂ may bring food for young even before hatching and start to feed them as soon as they hatch (Buxton 1950; Doerbeck 1966). Young fed by both sexes, usually more by ♀ (Salmen 1930; Ruiter 1941; Menzel 1971), but ♂ in one study fed young 3 times as much as ♀ though with smaller items (Ward 1956); both Bussmann (1940) and Buxton (1950) observed more feeds by ♂, though with less extensive data. In some cases ♂ does little or no feeding (Menzel 1971). Food-bearing ♂ calls (see 6e in Voice) close to nest, usually causing ♀ to leave, then feeds young; however, ♂ will occasionally (Bussmann 1940; Menzel 1971) or regularly (Ward 1956) give food to ♀ for transfer to young. If one parent disappears, other parent will attempt to rear young alone though usually without much success (Salmen 1930; Ruiter 1941; Beer-Heinzelmann 1956; Doerbeck 1966; Menzel 1968, 1971; Schönfeld 1975b). Especially during first 5 days, feeding parent calls (see 6b in Voice) to encourage young to gape (Menzel 1971); in captive pair, ♀ seen pecking at gape-flange of chick for same purpose (Fleischmann 1977). Young (several fed per visit) given very small items initially and adult recorded squashing insects in bill immediately prior to feeding them to young. Young react to any noises caused by arrival of parents from c. 5th day, gaping immediately (Menzel 1971). Up to 7 days old, young usually gape silently with head and neck straight up, then gradually change to gaping in direction of arriving adult (Buxton 1950; Menzel 1971). During last 2 days in nest, capable of standing and taking food at entrance. Young cluster in nest and are generally restless when ♀ absent, though less so from

6-8 days (Menzel 1971). Faecal sacs generally swallowed by parents until young *c.* 4 days old, afterwards carried 5-50 m away (Menzel 1971). Nest-sanitation by ♂ only up to *c.* 6 days, about equal thereafter (Bussmann 1940). Probably not possible for adults to remove all faeces after *c.* day 13-14 (Bussmann 1940). When young almost ready to fledge, disturbance may cause premature departure. May move to rim of nest when near to fledging, if site allows it. Young leave singly, mainly in morning although full departure may take half a day or longer. In 45 cases, young left over 1 day, in 17 cases over 2 days. No cases of young returning to nest. (Menzel 1971.) According to Menzel (1971) young able to fly pretty well on fledging, but Schmidt (1967*b*) stated that young normally leave before able to fly. Either or neither parent may initiate departure of young from nest (Boubier 1925; Bussmann 1940; Buxton 1950). Young remain together after fledging, being fed for between 6 days and 3 weeks (Henze 1943; Buxton 1950; Menzel 1971); family breaks up soon after (Menzel 1971) or not until well into August (Garling 1944). Young usually fed by both sexes after leaving nest unless ♂ taking care of 2 parallel broods, or when broods overlap (Ward 1956; Menzel 1971; see below). In study of one pair in southern West Germany, fledged young remained in cover on or near ground for several days, coming into open to be fed only after 4th day; self-feeding from 9th day, though still regularly fed by parents 12(-13) days after fledging; family bonds loose after *c.* 14 days, but parents continue to guard, giving alarm-calls (see 6 in Voice) on approach of man (Löhrl 1976). After fledging, young greet any adults arriving by beating wings rapidly, giving food- and contact-calls (Menzel 1971). 1st brood may stay 2-3 days near nest after fledging while ♀ starts repairing nest; ♂ then feeds both incubating ♀ and fledged young of 1st brood, whereas ♀ chases 1st brood away from nest and, when 2nd brood hatches, 1st brood is chased out of territory (Kostin 1983). Buxton (1950) recorded young independent at age of *c.* 1 month, though still keeping together and roosting together. Broods sometimes overlap: in one case, ♀ had begun nest-building during nestling period of 1st brood, and both parents continued to feed 1st brood when fledged (Menzel 1971); in another case, ♀ stopped feeding earlier brood as soon as 1st egg laid (Menzel 1971); see also Löhrl (1976). ANTI-PREDATOR RESPONSES OF YOUNG. From 9 days old, crouch on hearing any strange noises. During last few days of nestling period, remain quiet and crouch on hearing warning-calls of adults (Nice 1943; Menzel 1971). After fledging, young give adult alarm-calls (Buxton 1950). PARENTAL ANTI-PREDATOR STRATEGIES. (1) Passive measures. No details. (2) Active measures: Passow and Passow (1913) considered ♂ to be more excitable than ♀ on approach of potential predator, coming quite close and giving alarm-calls. When human approached fledged young, ♀ flew to a post giving alarm-calls, then flew *c.* 10 m to trunk of tree, running vertically up it with wings held away from body, exposing rufous rump and spreading tail and pressing it against trunk; followed by flight to high branch and continuous calling; sequence repeated once (Briggs 1984). Similar behaviour noted towards Sparrowhawk *Accipiter nisus* (Buxton 1950). Recorded mobbing Barn Owl *Tyto alba*, calling (D S Bunn: see 6 in Voice). In defence of nest and young, will attack vigorously any intruder (e.g. *J. torquilla*, *S. vulgaris*, tits *Parus*) approaching too close, even pecking at it until it flies off (Buxton 1950).

(Figs by C Rose: A-D from drawings in Buxton 1950; E-F from drawings in Blume 1966.)   PGHE

Voice. Song given occasionally by migrants in Africa (see Social Pattern and Behaviour, but no detailed

information on use of other calls outside breeding season. Song and calls deeper in south of range than in north (Chappuis 1969). Calls lower pitched than *Phoenicurus* of montane habitats (Böhme and Böhme 1986, which see for sonagrams and detailed comparative analysis). For further sonagrams, see Blume (1967) and Bergmann and Helb (1982). ♂ paired with ♀ Black Redstart *P. ochruros* frequently gave song and calls of that species (Böhm and Strohkorb 1964); another ♂ singing like *P. ochruros* was perhaps offspring of hybrid pair (Brzozowski 1984); *P. phoenicurus* has considerable mimetic ability, however (see below).

CALLS OF ADULTS. (1) Song. (a) Main song of ♂ comprises 2-part phrases of *c.* 2 s duration, but highly variable. 1st part normally species-specific, relatively pure-toned or frequency-modulated, and consisting of a high-pitched, slightly drawn-out tone followed by 2-4 lower-pitched or also higher-pitched, slightly shorter units: 'ji-gjü gjü gjü...' or 'jü-jik jik jik...'. 2nd part contains different-sounding, partly clicking or rattling, partly pure-sounding passages which contain rich mimicry (see below), each ♂ having several introductory motifs and numerous imitations (Bergmann and Helb 1982). Attractive, sweet, but rather melancholy and rapid tinkling tones recall Robin *Erithacus rubecula*, but generally harsher, more strangled, less liquid, though can be fuller and more liquid with repeated units; ends in characteristic weak, squeaky or mechanical trill or jangle (Menzel 1971; P G H Evans, D I M Wallace). Figs I-II clearly show easily audible frequency and amplitude modulation of units 2-3 following high-pitched 1st tone. Same 3 introductory units present in all 8 songs recorded by P A D Hollom; one song consists of these units alone. In Fig I, units 3-4 (separated by pause of 150 ms) also easily audible, higher-pitched units 6-7 less so and, with only *c.* 60 ms between them, less easily perceived as discrete; last 2 units attractive 'twhooee twhooee', pronounced starting transient making it easy (despite pause of less than 60 ms) to hear when 2nd unit starts. Last 6 units of Fig II are good copy of 2 3-unit segments of Great Tit *Parus major* song. In recording by S Palmér, each of 66 songs begins with reiteration like Chaffinch *Fringilla coelebs* (perhaps influenced by *F. coelebs* singing nearby), though unit pattern changes between 1st and 6th songs; likely mimicry of *F. coelebs* song flourishes at, and towards, end of song in Fig III (J Hall-Craggs). For detailed treatment of song-phrases of 8 birds, see Thimm (1973) and Thimm *et al.* (1974). Mimics a wide range of other bird species (Bergmann and Helb 1982): over 30 (mainly passerines) listed by Menzel (1971); see also Saxenberger (1905), Stresemann (1910), and Jouard (1926). Lesser Whitethroat *Sylvia curruca* apparently mimicked in 2nd part of song (see sonagram in Bergmann and Helb 1982), though rattle like *S. curruca* (but not necessarily mimicry) may also start song (V C Lewis). (b) Song variant with strangled, harsh sounds reminiscent

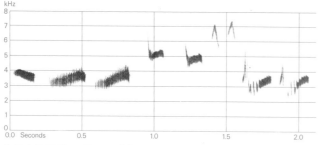

I   P A D Hollom   France   May 1973

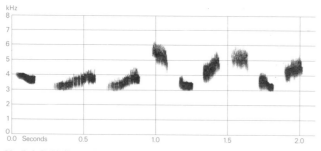

II   P A D Hollom   France   May 1973

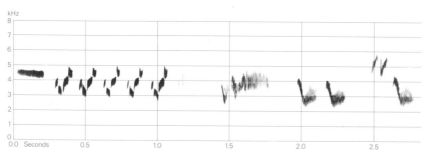

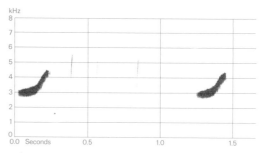

IV   P A D Hollom   Wales   July 1979

V   V C Lewis   England   May 1961

III   S Palmér/Sveriges Radio (1972–80)
Sweden   May 1957

(and perhaps copy) of Whitethroat *S. communis* warble and rendered 'dzi dzi dzi dzi dzi' given by ♂ during antagonistic interactions, Nest-showing, and (an abbreviated form) after fledging (Buxton 1950; Diesselhorst 1968*b*). (c) Occasional song given by ♀ resembles full song of ♂ (Buxton 1950); see also Calls of Young. (2) Courtship-song. Hurried jumble of liquid notes rather like Dipper *Cinclus cinclus* given by ♂ during copulation (P J Sellar), but also associated with courtship and soliciting of copulation (P G H Evans). (3) Courtship-call. ♂ in so-called 'copulation flight' gave sound similar to Bleating-call of Nightingale *Luscinia megarhynchos* (Horstkotte 1965; see that species, p. 634); presumably distinct from calls 1b and 4. (4) Pre-copulatory calls. (a) Very high-pitched, soft hissing sound not unlike nestling food-call given by both sexes in courtship and prior to copulation (Alpers 1942; Buxton 1950). (b) In captive pair, frequent characteristic and throaty 'tok tok' calls (perhaps only different rendering or variant of call 6a) given by ♂ attempting to approach ♀ for copulation (Fleischmann 1977). (5) Copulation-call. Cheeping sounds given by both sexes during copulation (Alpers 1942). (6)

Contact-alarm calls. (a) ♂ gives an ascending 'hueee' (sometimes in isolation—see also 6d) followed by varying number of 'tic' or 'tk' sounds: e.g. 'hueee tk tk tk hueee tk' (P G H Evans, J Hall-Craggs: Fig IV). In north-east Greece, 1st unit rather different, resembling local *F. coelebs* call ('hiid') and 'sih' of Thrush Nightingale *L. luscinia*, so that whole rendered 'fid-tek-tek' (Bergmann and Helb 1982). Repeated fairly consistently and given in a variety of situations involving contact, including arrival at nest to feed young (see also call 6e) and even during soliciting of copulation by ♂, also in alarm whilst mobbing predator (Buxton 1945). (b) Sharp scolding 'tchuk' not unlike Blackcap *Sylvia atricapilla* associated with antagonistic interactions, e.g. given by territorial ♂ after intruder has left (Buxton 1950), and, in recording by V C Lewis, probably same rapid (at times) stony ticking sounds of high-intensity alarm in presence of fledged young, are presumably variant of 'tk' sound; see 6a for discussion of variants. (c) High-pitched 'see' given by ♂ excited by intruder (e.g. *J. torquilla*) at nest (Buxton 1950); perhaps resembles 1st unit of call given by bird in Greece (see above). (d) Call of ♀ similar to 'hueee' of ♂, but given more powerfully and often in long series: recording (Fig V) suggests 'hueet' with steeper rise in pitch and more abrupt ending than in call of ♂; recalls Willow Warbler *P. trochilus*, but less delicate (P G H Evans, P J Sellar). Ticking sounds apparently given at least occasionally by ♀ (see Buxton 1950); further study required. (e) Strangled sound given by ♂ announcing arrival at nest and probably similar quiet strangled sounds given by either sex apparently to encourage young to gape (Menzel 1971). (7) Rasping sounds linked with antagonistic interactions (Buxton 1950).

Calls of Young. Food-calls at *c.* 4 days very weak (faint unrhythmical chirpings), becoming louder up to 9th day. Bursts of rasping sounds given in last days of nestling period and after fledging (Buxton 1950; Menzel 1971); in recording by C Fuller, apparently shorter and more explosive when begging, longer rasps when being fed. Contact-call given by recently fledged young (perhaps also in nest) 'tic tic' or 'tchack-tchack' (Buxton 1950); short strangled sound sometimes also given as contact-call by fledged young (Menzel 1971). Young captive ♂♂ gave quiet rudimentary song like *S. atricapilla c.* 1 week after leaving nest; apparently stimulated by father who would give contact-calls until juvenile started singing again; young ♀ sang even louder than 2 juvenile ♂♂ (Fleischmann 1977). Other captive young started to sing at 18 and 19 days (Nice 1943). PGHE

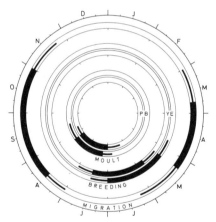

Breeding. Season. North-west Europe: see diagram. Southern Europe: up to 2 weeks earlier (Makatsch 1976). Northern Finland: late May to late June (Järvinen 1978). Site. Hole in tree, rocks, or building, less often in bank, among tree-roots, or heap of stones; readily uses nest-box. Mean height above ground of 97 nests (excluding boxes) in Finland 1·6 m (0–8·0), with 26% below 1 m, 32% 1–2 m, 26% 2–3 m, and 16% over 3 m (Haartman 1969a). Nest: loose cup of grass, moss, and other vegetation, lined with wool, hair, and feathers. Building: by ♀, taking *c.* 5 days (Ruiter 1941). Eggs. See Plate 82. Sub-elliptical, smooth and glossy; pale blue. 18·7 × 13·8 mm (16·6–21·5 × 12·3–15·2), *n*=250; calculated weight 1·9 g

(Schönwetter 1979). Clutch: 5–7 (3–10). Increases with latitude, but declines through season, with older birds laying larger clutches than younger. Of 157 clutches, Czechoslovakia: 3 eggs, 3%; 4, 11%; 5, 24%; 6, 37%; 7, 22%; 8, 3%; mean 5·7. Of 206 clutches, Netherlands, 1-year-olds laid mean 6·04 ± SE0·17 (*n*=23), and older birds 6·51 ± SE0·07 (*n*=61). Of 233 clutches, northern Finland: 4 eggs, 3%; 5, 11%; 6, 32%, 7, 47%; 8, 6%; 9, 1%; mean 6·4. (Ruiter 1941; Menzel 1971; Pulliainen *et al.* 1982.) 2 broods over most of range, but only 1 in north. Mean of 10 2nd broods, Netherlands, 5·0 (Ruiter 1941). Eggs laid daily, from sunrise to *c.* 08.00 hrs (Ruiter 1941). Incubation. 12–14 days (10·5–17) (Ruiter 1941); 12·8 days (11–15), *n*=22 (Järvinen 1978). By ♀ only. Continuously at night, and in spells of 8–25 min during day. Begins with last egg, though will sit on rim of nest from 1st egg. Hatching mainly synchronous, but can be asynchronous. (Ruiter 1941; Haartman 1969a.) Young. Cared for and fed by both parents, though ♀ takes larger share (Ruiter 1941). Fledging to Maturity. Fledging period 14–15 days (13–17) (Buxton 1950); 13·0 days (12–15), *n*=5 (Järvinen 1978). Young independent 2–3 weeks after fledging (Buxton 1950). Age of first breeding mainly 1 year, but some not until 3 or even 4; among 50 pairs studied over 6 years in Netherlands, 59% of birds bred at 1 year, with 75% of ♂♂ and 42% of ♀♀ (Ruiter 1941). Breeding Success. Of 479 eggs laid in 76 nests, northern Finland, 81·2% hatched, and 92·3% of these fledged, giving 74·9% overall success with mean 4·7 young fledged per nest. Overall success very similar for different-sized clutches of 4–7 eggs, increasing very slightly from 75 to 79·8%, but only 50% for 8-egg clutches; however, number of young fledged per nest increased from 3·0 to 5·6 for 4- to 7-egg clutches, declining to 4·0 for 8 eggs (Järvinen 1978).

Plumages (nominate *phoenicurus*). Adult Male. In fresh plumage (August–November), narrow strip of feathers along base of upper mandible, lores, ring round eye, upper cheek, and ear-coverts deep black, usually with some narrow and faint white feather-tips at rear of cheek and rather narrow pale brown or grey-brown feather-tips on rear ear-coverts. Narrow line along forehead and above lore white, merging into pale grey supercilium above and behind eye. Remainder of forehead and all crown, hindneck, mantle, scapulars, and back medium bluish-grey, feathers with ill-defined olive-grey tips on top of head and with warmer olive-brown tips on mantle, scapulars, and back; olive-brown tips almost coalesce on lower mantle, lower scapulars, and back, hiding grey, but remainder of upperparts appear grey with slight olive wash only. Rump and upper tail-coverts bright rufous-cinnamon, longest coverts often almost rufous-chestnut. Lower cheek, chin, throat, and side of neck black, each feather with white fringe along tip; fringes generally less than 1–1·5 mm wide on throat and even less or absent elsewhere; area thus appears mainly black, sometimes fully black except for narrow white band at border of rufous chest. Underparts from chest and sides of breast downwards deep rufous-cinnamon, gradually paler towards pale cinnamon-buff under tail-coverts; feathers of chest and breast have white tips 1-2 mm wide (but rufous-cinnamon still prevailing), more white on flanks and sides of belly, central belly and vent largely white. Central pair of tail-feathers (t1) dull black or brown-black, narrowly fringed rufous-chestnut round tip, broadly so at base; t2-t6 deep rufous-cinnamon or rufous-chestnut, outer web of t6 and tips of some other feathers sometimes slightly tinged olive-brown. Flight-feathers, tertials, and greater upper primary coverts greyish-black, outer web of flight-feathers narrowly and sharply edged grey or pale grey-buff; tips of primary coverts narrowly pale grey-buff, outer webs of primary coverts narrowly edged grey, outer webs and tips of tertials more broadly edged pale cinnamon-buff, often with partial grey tinge towards bases of tertials. Lesser and median upper wing-coverts bluish-grey or ash-grey, some black of feather-bases visible along leading edge of wing, some dark grey on median coverts; some median coverts tipped rufous or pale cinnamon-buff, especially outermost; greater coverts greyish-black with grey fringe along outer web and rufous-cinnamon or rufous-buff fringe along tip. Longest feather of bastard wing sometimes contrastingly black. Under wing-coverts and axillaries bright rufous-cinnamon or orange-rufous. *In worn plumage* (from mid-winter, sometimes from November), olive fringes of upperparts and white ones of underparts worn off; broad white band on forehead reaches back to above front corner of eye; crown, hindneck, mantle, scapulars, back, and lesser and median upper wing-coverts uniform pure bluish-grey, sometimes with indistinct paler grey or whitish supercilium; frontal feathering, sides of head and neck, chin, and throat uniform deep black (rarely, some faint fringes on throat until April); underparts rufous-cinnamon with rather sharply demarcated white central belly and vent and buff thighs and under tail-coverts, traces of white fringes remaining only on lower breast and on lower and rear flanks; pale edges along flight-feathers, tertials, primary coverts, and greater upper wing-coverts bleached to pale buff or off-white and partly abraded, but usually still distinctly pale grey on outer webs of greater primary coverts and greater upper wing-coverts. ADULT FEMALE. In fresh plumage, upperparts down to scapulars and back grey-brown or pale olive-brown, tinged or spotted buff on forehead and along sides of forecrown; top of head usually more greyish, scapulars and back more brownish. Rump and upper tail-coverts as adult ♂ or sometimes slightly paler rufous-cinnamon. Lore mottled pale buff; narrow but distinct eye-ring pale buff. Ear-coverts brown with faint dusky mottling and pale buff streaking, usually slightly darker than crown and hindneck and distinctly darker than cheek. Front part of cheek, chin, and sometimes central throat pale cream-buff or dirty white, merging into pale brown-grey on rear cheek, side of throat, and side of neck. Chest, side of breast, and flanks rufous-cinnamon, feathers with ill-defined pale brown-grey tips on chest and sides of breast and broader white tips on flanks. Some birds have underparts largely rufous-cinnamon (paler and less saturated than ♂) with white central belly and vent and tawny-buff under tail-coverts; others have chest, sides of breast, and flanks pale brown-grey with very limited cinnamon tinge and central breast, belly, and vent more extensively white or cream-pink. Tail as adult ♂. Flight-feathers, tertials, and greater upper primary coverts browner than in ♂, narrow and even edges along outer webs of flight-feathers and primary coverts pale brown or pale grey-buff; broader fringes along outer webs and tips of tertials slightly deeper cinnamon-buff. Lesser and median upper wing-coverts like mantle and scapulars or slightly greyer; greater coverts dark greyish-brown with cinnamon-buff outer fringes and tips, like tertials. Under wing-coverts and axillaries pale rufous-cinnamon or tawny-buff (distinctly paler than in adult ♂),

usually with much grey of feather-bases visible. Strong individual variation, in particular in head and underparts, most marked when plumage worn. *In worn plumage*, little variability among typical pale birds described above: upperparts duller grey-brown than in fresh plumage, eye-ring paler and more distinct, usually a rather distinct buff stripe from nostril to above eye when not too worn, and often a contrast between dirty buff-white throat and cinnamon-rufous or pale cinnamon-brown chest, sides of breast, and flanks. Frequently, however, dark grey or dull black of feather-bases exposed on rear cheek and lower throat, showing as dark mottled grey-and-buff collar, and birds with this usually have a mottled grey-brown and cream-white forehead-patch and extensively rufous underparts (paler than adult ♂). More rarely (6 out of 183 ♀♀ in study in Netherlands: Ruiter 1941) upperparts dull grey with olive-brown tinge (like 1st adult ♂), forehead shows some white mottling (less white than any ♂), sides of head (from lores and ear-coverts downwards), chin, and throat dull black with much buff mottling (much less black than 1st adult ♂), and underparts rather deep rufous; amount of black on throat usually increases with age (Ruiter 1941). NESTLING. Down fairly long, on upperparts only; dark grey. For development, see Meidell (1961). JUVENILE. Upperparts down to back olive-brown with distinct black borders along feather-tips and paler buff or golden-rufous feather-centres, appearing spotted or scalloped. Rump and upper tail-coverts rufous-cinnamon, feathers partly tipped black. Sides of head closely mottled and streaked buff, olive-brown, and dark grey; no distinct eye-ring. Underparts buff, deepest on chest and sides of breast and grading to tawny on vent and under tail-coverts; feathers of chest, breast, and flanks have distinct black fringe along tip, appearing scalloped; throat, belly and vent less distinctly fringed black. Lesser and median upper wing-coverts and tertial coverts dull dark grey with contrasting rufous-buff triangular tip; tips of median and tertial coverts and sometimes others fringed black. Sexes similar. For tertials, greater upper primary coverts, and underwing, see First Adult. FIRST ADULT MALE. In fresh plumage (autumn), like fresh adult ♂, but olive-brown feather-tips of upperparts and lesser and median upper wing-coverts much broader, largely or fully concealing brown-grey feather-bases; white line across forehead and above lore and eye partly or fully concealed; narrow eye-ring buff (in adult ♂, black); black of lore, upper cheek, and ear-coverts largely concealed by olive-brown feather-tips (least so on shorter ear-coverts); black of lower cheek, chin, and throat largely (sometimes fully) concealed— pale fringes broader than in adult (1·5-3 mm wide on throat) and partly or entirely pale brown (all white in adult); underparts slightly more extensively washed white than adult; axillaries and under wing-coverts pale buff to rufous-cinnamon. Juvenile flight-feathers, tertials, tail, greater upper primary coverts, greater upper wing-coverts, and sometimes tertial coverts retained, but these hardly more worn than in adult at same time of year; tail as adult; flight-feathers as adult, but middle and sometimes longer primaries have ill-defined pale fringe along tip (hardly so in adult), fringes along outer webs and tips of tertials, greater upper primary coverts, and greater upper wing-coverts uniform buff or cinnamon-buff (in adult, partly or largely pale grey), and inner greater coverts and (if retained) tertial coverts have traces of black fringes along tips. By mid-winter, fringes still prominent over all head and body (in adult, largely worn off). *In worn plumage* (spring and summer), white forehead-patch usually smaller and less clear-cut than in adult ♂; upperparts duller grey with distinct olive-brown wash (in adult, pure blue-grey); black of sides of head, chin, and throat less deep black, with traces of pale fringes present on throat up to May–June; eye-ring black (perhaps after a partial pre-breeding moult) or mottled black and buff. In worn plumage at all ages, flight-feathers, tertials, and greater coverts browner, with pale edges and fringes bleached and worn off, but worn 1st adult usually has pale brown or buff fringes along outer webs of greater upper wing-coverts and greater upper primary coverts, thus contrasting more with grey of adjacent median coverts (in adult, greater and primary coverts greyer and less contrasting). FIRST ADULT FEMALE. Like adult ♀ and hardly distinguishable, especially in autumn. Feathers of forehead always without buff or cream-white subterminal spotting, those of lower throat without any black or dark grey subterminally. Chest, sides of breast, and flanks grey-brown or pale buff-brown with ill-defined off-white feather-fringes, not as deep and extensively rufous as many adult ♀♀; under tail-coverts sometimes paler pink-buff. Part of juvenile plumage retained, as in 1st adult ♂, but difference from adult ♀ slight: middle primaries often have pale buff fringe along tip (absent in adult ♀, or very narrow and pale grey); tertials more pointed, with tip broadly fringed buff (more square-tipped in adult, with much narrower or almost no buff fringe at tip); inner greater upper wing-coverts and (if retained) tertial coverts have traces of black fringe along tip. *In worn plumage*, throat, chest, and sides of breast grey-brown or buff-brown with indistinct darker grey-brown mottling, with hardly any rufous tinge (in adult, throat either whiter or blacker, usually contrasting with pale rufous-cinnamon chest and sides of breast). Worn juvenile feathers generally indistinguishable from adult.

Bare parts. ADULT, FIRST ADULT. Iris dark sepia-brown or dark brown. Bill horn-black or black. Gape and mouth yellow. Leg and foot dark horn-brown, horn-black, or black, soles sometimes paler horn-brown or yellow-brown. NESTLING. Mouth pale orange or orange-yellow, no spots. Gape-flanges yellow-white. JUVENILE. Iris dark brown. Bill, leg, and foot yellowish-flesh; culmen dusky horn-black. At *c.* 14 days old, bill, leg, and foot darken to adult colour and mouth changes from yellow to pink. (Heinroth and Heinroth 1924–6; RMNH, ZMA.)

Moults. ADULT POST-BREEDING. Complete; primaries descendant. In Britain, starts with p1 between mid-June and late July, mainly early July; finishes with p9–p10 and inner secondaries after *c.* 40 days, in late July to early September, mainly mid-August (Ginn and Melville 1983). Data from Netherlands, West Germany, and Mongolia basically similar; a few start early June, and moult generally completed by last week of August; tail moults centrifugally, starting at primary moult score 10–15, and many or all feathers grow simultaneously, finishing at primary score 35–40; body moult mainly at primary scores 25–40, late July and early August (Piechocki and Bolod 1972; RMNH, ZMA). In USSR, moult completed late August (if single-brooded) or September (if double-brooded) (Dementiev and Gladkov 1954b). POST-JUVENILE. Partial: head, body, lesser and median upper wing-coverts, and often some or all tertial coverts. Moult occurs mainly at 5–7 weeks old and timing thus strongly dependent on date of hatching: starts with some feathers of breast, flanks, and mantle between mid-June and late August, completed early July to mid-September (Heinroth and Heinroth 1924–6; RMNH, ZMA). In Switzerland, many birds in juvenile plumage until mid-August, some up to mid- or late September (Jenni 1984). Some moult sometimes occurs in winter quarters, December–March (Fry 1970b).

Measurements. ADULT, FIRST ADULT. Nominate *phoenicurus*. Netherlands, West Germany, and migrants from north-west Africa; all combined, as differences between breeders and migrants of different areas not statistically significant; skins (RMNH, ZMA). Bill (S) to skull, bill (N) to distal corner of nostril; exposed culmen on average 4·2 less than bill (S). Juvenile wing and tail include retained juvenile wing and tail of 1st adult.

	♂		♀	
WING AD	80·8 (2·05; 20)	77–84	78·3 (2·05; 11)	75–81
JUV	79·7 (1·99; 47)	76–84	78·1 (1·78; 39)	74–82
TAIL AD	55·6 (2·03; 18)	52–58	53·9 (2·05; 11)	51–57
JUV	55·3 (1·92; 45)	51–59	54·1 (2·17; 30)	50–58
BILL (S)	14·7 (0·57; 54)	13·6–15·9	14·5 (0·56; 48)	13·4–15·8
BILL (N)	7·9 (0·45; 41)	7·2–9·0	7·7 (0·39; 33)	6·9–8·5
TARSUS	21·8 (0·66; 64)	20·5–23·2	22·0 (0·64; 49)	20·8–23·3

Sex differences significant for wing. For wing of live migrants of different sexes and ages in spring and autumn in Camargue (France), see Blondel (1967).

Wing of ♂♂. Live birds or data from skins recalculated to be comparable with live birds: (1) Britain, spring and summer; (2) France, Belgium, Netherlands, and Luxembourg, summer; (3) northern and central part of West Germany, East Germany, and Poland, summer; (4) southern part of West Germany, Italy, Hungary, and Yugoslavia, summer; (5) Rumania, Bulgaria, and south European USSR, summer; (6) Fenno-Scandia, spring to autumn; (7) north European USSR, spring to autumn; (8) Asiatic USSR, spring to autumn; (9) *samamisicus* (Blondel 1967).

(1)	77·2 (1·9; 112)	73–83	(6)	81·7 (1·9; 109)	76–86
(2)	77·8 (2·2; 87)	73–83	(7)	81·2 (2·0; 159)	76–86
(3)	80·1 (2·1; 141)	74–85	(8)	82·0 (1·7; 128)	77–86
(4)	80·3 (2·6; 48)	75–86	(9)	79·9 (2·3; 60)	—
(5)	81·1 (2·0; 72)	77–86			

Weights. ADULT, FIRST ADULT. Nominate *phoenicurus*. Sexed birds. (1) Norway, July–August (Haftorn 1971). (2) Sweden, September (Scott 1965). (3) South-west Siberia and Kazakhstan, USSR, August–September (Dolgushin *et al.* 1970; Havlín and Jurlov 1977). (4) Poland, autumn (Busse 1972). (5) Netherlands, August–October (RMNH, ZMA). (6) Dungeness (Kent), Britain, autumn (Scott 1965). (7) Northern Nigeria, October–March (Fry 1970b). (8) Ad Damman and Sjarjah, north-east Arabia, March to early May (BTO). (9) Cyprus, 1st half of April (BTO). (10) Netherlands, April–May (RMNH, ZMA). (11) Denmark, spring (Petersen 1972). (12) USSR and Mongolia, April–August (Dolgushin *et al.* 1970; Piechocki and Bolod 1972; Piechocki *et al.* 1982; Danilov *et al.* 1984). (13) Exhausted birds, Netherlands (RMNH, ZMA).

	♂		♀	
(1)	15·9 (—; 7)	14–18	15·9 (—; 7)	15–18
(2)	14·4 (0·54; 20)	—	14·3 (0·74; 28)	—
(3)	14·8 (1·08; 21)	13–17	14·4 (1·76; 15)	13–16
(4)	14·7 (1·36; 1016)	10–19	14·5 (1·43; 1282)	9–21
(5)	15·6 (1·78; 21)	13–20	15·3 (2·15; 18)	12–20
(6)	14·5 (—; 12)	—	15·0 (—; 30)	—
(7)	14·5 (—; 14)	13–16	13·4 (—; 10)	12–15
(8)	13·8 (1·42; 25)	11–17	12·5 (0·94; 8)	11–14
(9)	14·3 (2·74; 7)	12–19	15·2 (1·19; 4)	14–16
(10)	16·1 (1·30; 11)	14–20	15·0 (0·68; 6)	14–16
(11)	15·9 (—; 143)	13–19	15·4 (—; 90)	13–18
(12)	15·2 (2·06; 15)	10–18	15·3 (1·51; 11)	12–17
(13)	10·9 (1·08; 9)	9–12	10·1 (0·30; 4)	10–11

Unsexed birds. Autumn: (1) Gibraltar (Finlayson 1981); (2) interior Algeria (Bairlein *et al.* 1983); (3) Malta (J Sultana and C Gauci); (4) Cyprus (Moreau 1969); (5) Egypt (Moreau and

Dolp 1970); (6) Lake Chad (Dowsett and Fry 1971). Spring: (7) central Nigeria (Smith 1966); (8) Egypt, (9) Azraq (Jordan), (10) Cyprus (Moreau 1969); (11) Malta (J Sultana and C Gauci); (12) Morocco (Moreau 1969); (13) Gibraltar (Finlayson 1981); (14) Britain (Smith 1966).

(1)	15·0 (1·26; 13)	—		(8)	11·9 (— ; 79)	9–16
(2)	16·0 (3·96; 4)	12–20		(9)	12·9 (— ; 37)	11–17
(3)	16·8 (2·7 ; 50)	11–24		(10)	15·3 (— ; 65)	12–20
(4)	16·8 (— ; 58)	12–21		(11)	15·7 (2·0 ; 50)	11–21
(5)	16·9 (2·40; 60)	11–21		(12)	13·2 (— ;110)	11–17
(6)	13·7 (— ; 14)	12–16		(13)	14·0 (1·38; 24)	—
(7)	15·7 (2·24; 14)	12–20		(14)	13·9 (1·22;275)	10–20

P. p. samamisicus. Turkey and Iran, combined. Adult, March–May: ♂ 14·4 (1·05; 6) 13·6–16·5, ♀ 14·4. Juvenile ♂♂, June–July: 15, 16. (Paludan 1938, 1940; Schüz 1959; Kumerloeve 1969a, 1970a.)

NESTLING, JUVENILE. For growth, see Ruiter (1941) and Meidell (1961).

Structure. Wing rather long, broad at base, tip bluntly pointed. 10 primaries: p7–p8 longest or either one 0–1 shorter than other; p9 and p5 5–10 shorter than longest, p6 1–3, p4 10–14, p1 17–22. No difference in wing shape between breeders from Scandinavia, Netherlands, Balkans, Spain, and North Africa; rather variable in all populations. P10 reduced, 37–45 shorter than wing-tip, 2–8 longer than longest upper primary covert. Outer web of p6–p8 and inner of p7–p9 emarginated. Tertials short. Tail rather short, tip square or very slightly rounded; 12 feathers, t6 1–4 shorter than t2–t3. Bill short, straight, but tip of culmen slightly decurved towards sharp tip; wide at base. Nostril rather small, oval, partly covered by membrane above. Some fine bristles along base of upper mandible. Tarsus and toes rather short, slender. Middle toe with claw 16·5 (43) 15–

19; outer toe with claw *c*. 70% of middle with claw, inner *c*. 62%, hind *c*. 72%. Claws short and sharp, decurved.

Geographical variation. Slight in size (see Measurements), more marked in colour of flight-feather fringes. ♂ *samamisicus* from Crimea, eastern Turkey, and Levant east through Iran and Turkmeniya to southern Uzbekistan, south-west Tadzhikistan, and probably adjacent north-east Afghanistan differs from nominate *phoenicurus* in broad white fringes along outer webs of flight-feathers, forming conspicuous white area—rather variable in extent, sometimes consisting of whole outer webs of tertials and secondaries (except tips) and rather broad fringes along primaries, sometimes formed by rather narrow fringes on secondaries only; in worn plumage, fringes partly wear away and white area sometimes less obvious. White on wing also present in juvenile ♂ and 1st adult ♂, and sometimes represented by broad cream-buff or pale grey fringes in adult ♀. No other differences from nominate *phoenicurus*. A morph ('*incognita*') with more extensive black at base of upper mandible and lores and largely black mantle, scapulars, and lesser upper wing-coverts occurs locally within range of *samamisicus*, in particular in northern Iran. Exact boundaries between *samamisicus* and nominate *phoenicurus* difficult to establish: in northern Caucasus (but not in Crimea) many breeders inseparable from nominate *phoenicurus*, others intermediate, showing small wing-patch (Stegmann 1928); in Turkey, *samamisicus* occurs west to at least Bolu in north and Middle Taurus mountains in south, but birds with *samamisicus* characters occur occasionally west to eastern Bulgaria (Makatsch 1950) and southern Yugoslavia (Matvejev and Vasić 1973), indicating wide intergradation zone.

Recognition. For separation from Black Redstart *P. ochruros*, see that species (p. 695). CSR

Phoenicurus moussieri Moussier's Redstart

PLATES 49 and 52
[between pages 652 and 653]

DU. Diadeemroodstaart   FR. Rubiette de Moussier   GE. Diademrotschwanz
RU. Белобровая горихвостка   SP. Colirrojo diademado   SW. Diademrödstjärt

Erithacus Moussieri Olph-Galliard, 1852. Synonym: *Diplootocus moussieri*.

Monotypic

Field characters. 12 cm; wing-span 18·5–20·5 cm. 15% shorter than Redstart *P. phoenicurus*, with relatively shorter wings and tail and sturdy body. Smallest redstart of west Palearctic, with rather compact form recalling *Saxicola* chat (as does ♂'s plumage). ♂ blackish above with large white wing-patch and huge white circlet round crown. ♀ resembles ♀ *P. phoenicurus*, but underparts have much rufous. Sexes dissimilar; seasonal variation in ♂. Juvenile separable.

ADULT MALE. From February to July (when plumage worn), head down to cheeks, back, and wings black, strikingly marked by (1) white circlet running across fore-crown, becoming supercilium, then continuing as line down rear cheeks and widening into wide white wedge on sides of nape, (2) bold white panel on bases

of inner primaries and secondaries, and (3) bright, warm orange-rufous rump and tail (central tail-feathers dark brown). Underparts orange-rufous, with paler cream-buff rear belly and rufous-buff underwing. From August (when plumage fresh), head, body, and wing-coverts marked with broad pale brown fringes, reducing contrasts of pattern. ADULT FEMALE. Head and back brown-grey, with buff supercilium (usually indistinct) and no obvious eye-ring. Wings brown-black, with buff fringes to tertials and greater coverts. Rump and tail pale orange-rufous, lacking full warmth of ♂. Underparts sometimes brown-buff but usually orange-rufous on sides of throat, breast, and flanks; belly and vent cream. JUVENILE. Suggests ♀ but dully spotted above and scaled on fore-underparts (though markings weaker than on most young chats). ♂

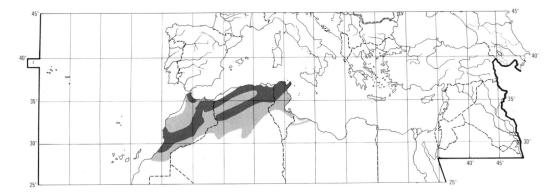

has white wing-panel indicated. Bill and legs black in adult, paler in juvenile.

♂ unmistakable; ♀ and immature less distinctive, but small size, compact form, and warm rufous underparts soon learnt. Most serious risk of confusion occurs with 1st-winter ♂ *P. phoenicurus* which may show faint circlet on forehead (but has black speckles on throat). Flight easy but rapid; rather fluttering wing-beats and shorter-tailed silhouette recall Stonechat *S. torquata* more than typical *Phoenicurus*. Usually stays close to ground level. Hops. Shivers tail like other *Phoenicurus*. Not shy, ♂ occasionally inquisitive.

Song a short simple warble. Commonest call a loud 'wheet' or 'beezp', usually followed by a rasping 'tr-rr-rr'.

Habitat. In lower-middle latitudes in warm dry Mediterranean climate, at all elevations from sea-level to 1900 m in Algeria and Tunisia and to 3000 m in Moroccan Haut Atlas. In east of range, in broken maquis, on dry grassy, stony, or rocky slopes, and in old or degraded forests on broken terrain. In Atlas region, occurs at forest base and on stony summits, forest summits, and denuded plateaux, and at higher elevations among bushes and xerophytic plants (Heim de Balsac and Mayaud 1962). In winter, shifts to lower ground along fringe of Sahara, in bushes in wadis, and in *Zizyphus* scrub on plains (Smith 1965a). Often regarded, on account of habitat and mannerisms, as combining characteristics of a more typical *Phoenicurus* with those of Stonechat *Saxicola torquata*. Virtually unaffected by human impacts.

Distribution and population. No information on range changes or population trends.

Libya. Breeding suspected in Tripolitania (Bundy 1976).

Accidental. Italy, Malta.

Movements. Resident, dispersive, and perhaps migratory over relatively short distances, normally staying within North Africa. Extent of movement within general breeding area impossible to quantify in absence of ringing recoveries—apart from nestling ringed near Tetouan (nor-

thern Morocco) and recovered the following December c. 12 km north-west. Evidently widespread in Morocco outside breeding season (Heim de Balsac and Mayaud 1962; Smith 1965a; Pineau and Giraud-Audine 1977), and numbers move into valley of Moulouya, north-east Morocco (Brosset 1961). Winter distribution thus suggests general movement out of Moyen and Haut Atlas to lower altitudes. Movements to south and east of breeding range are better documented. Blondel (1962a, b) noted December–January influx (mainly of ♂♂) into semi-desert round Monts des Ksours (Algeria), supplementing local population; during February, local birds returned to mountains and migrants also left. Further east in Atlas Saharien, Arnault (1929) also noted departure from mountains round Djelfa in October to avoid snow (♂♂ preceding ♀♀ by several days) and linked this to large numbers found in winter round Laghouat. However, few move more than 70 km south of Laghouat (Niethammer and Laenen 1954). Southern limit in Morocco is c. 28°N (J D R Vernon), and in Algeria 30–32°N (Geyr von Schweppenburg 1918; Serle 1943; Dupuy 1966). Similar degree of movement in Tunisia (Thomsen and Jacobsen 1979). In Libya, where not proved breeding, sometimes almost common in winter in north-west, from Sabratha to Wadi Kaam and in Jabal Nafusah, and easternmost record is one at Benghazi (Bundy 1976). 2 old records in Italy (Moltoni 1945) and 5 recent ones (March to mid-May) in Malta (Sultana and Gauci 1982). RAC

Food. Takes insects on ground (e.g. seen feeding for long period on ants Formicidae) (Arnault 1929) and frequently digs with bill, looking around after each dig (Gurney 1871). Makes brief flights from ground in pursuit of flying prey, e.g. large black flies (Sage and Meadows 1965). In central Morocco, stomachs contained mostly beetles (Coleoptera), grasshoppers (Orthoptera), and larvae; 1 contained only small ants (Meinertzhagen 1940). Stomach of ♀ from Tunisia, December, contained stone, stem, and flesh of an olive *Olea* (Blanchet 1951).

Adults recorded feeding young Cuckoo *Cuculus canorus* in nest on larvae of moth *Thaumetopoea* (Chavigny and Le Dû 1938). DJB

Social pattern and behaviour. Poorly known.

1. Apparently solitary outside breeding season (Niethammer and Laenen 1954). Birds seen in central Morocco in autumn were predominantly ♂♂ (Meinertzhagen 1940). BONDS. Mating system not studied, but no evidence to indicate other than monogamy. Pair-formation probably takes place soon after arrival on breeding grounds (see Arnault 1929). No information on length of pair-bond or age of first breeding. Both parents seen to feed young in nest (M G Wilson), and young remain with parents after leaving nest, even when self-feeding (Arnault 1929). BREEDING DISPERSION. Solitary and territorial, though no details of dispersion or territory size. ROOSTING. No information.

2. Confiding, allowing approach to 6–8 m, but ♀ more shy than ♂, both in and outside breeding season (Meinertzhagen 1940; Meise 1959). Shivers tail constantly like any other *Phoenicurus*, and flicks wings like Stonechat *Saxicola torquata* (Smith 1965a; M G Wilson). Alarm-calls (not confined to breeding season) given by both sexes on approach of human (Snow 1952; Smith 1962–4; P A D Hollom: see 3 in Voice). FLOCK BEHAVIOUR. No information. SONG-DISPLAY. Song (see 1 in Voice) delivered by ♂ at least from February (northern Tunisia) from tree, bush, rock, or ground (Meise 1959; M G Wilson). In Algeria, 2 ♂♂ seen in apparent Song-duel, mid-May (Kleinschmidt 1908). During courtship, ♂ gives soft thin song (Arnault 1929), including while chasing ♀ (Smith 1962–4). ANTAGONISTIC BEHAVIOUR. No information. HETEROSEXUAL BEHAVIOUR. No information. RELATIONS WITHIN FAMILY GROUP. See Bonds (above). No further information. ANTI-PREDATOR RESPONSES OF YOUNG. No information. PARENTAL ANTI-PREDATOR STRATEGIES. ♀ extremely shy and usually leaves when human 20–30 paces from nest. If flushed from nest, ♀ may give alarm-call (see 3 in Voice), then remain in vicinity, continuing to call thus (Koenig 1895). Both adults will come close and give alarm-calls when man near nest with young, but ♀ perhaps overall less tolerant of humans than ♂ and more vocal when young threatened (Kleinschmidt 1908; P A D Hollom, M G Wilson). PGHE, MGW

Voice. CALLS OF ADULTS. (1) Song of ♂. Thin reedy warble recalling Dunnock *Prunella modularis* (P G H Evans); alternates between rustling 'ir' sounds like Serin *Serinus serinus* and more vocal 'svisvi' (Jonsson 1982). Song given during courtship quite pleasant but feeble, carrying only a few metres (Arnault 1929). Length of song-phrase varies: 2–2·5 s (P A D Hollom), *c.* 4 s (Snow 1952) or just under 5 s to just over 6 s in recordings by E D H Johnson (J Hall-Craggs). Usually 8–10 (5–12) repetitions of 2–4-syllable motifs 'zrätä', 'zízerä', 'zízeräze', sounding sharp and strangled. Typical full phrase rendered 'zízräzrize zerízerízer zeräzerîzera', usually with 2 unstressed sounds as introduction (Meise 1959; see also Kleinschmidt 1908); also rendered 'd(z) er-dzee-der-dzer' with buzzy quality (P A D Hollom). In recording (Fig I), song begins quietly, building up to full sound pressure level in first second. 4 introductory notes followed through rest of (apparently complete) song by virtually regular alternation between notes (tonal sounds) and short rasps or rattles with varying (but rapid) rates of delivery of constituent units; often so fast as to sound like buzz. Usually 2–3 tones between each rasp. (P G H Evans, J Hall-Craggs.) In another recording by E D H Johnson, song often prefixed by call like finch (Fringillidae) (P J Sellar). (2) Subsong in recording by P A D Hollom shows same alternating pattern of narrow-band and broad-band units (tones and rattles); rapid quiet jumble of low whistles, chirrups and chucks (P G H Evans; J Hall-Craggs). (3) Contact-alarm calls. According to Jonsson (1982), thin 'hih' like Nightingale *Luscinia megarhynchos* is contact-call, while a variety of stone-crunching 'trrrr' sounds function as warning. Division probably not clear-cut however and in recording

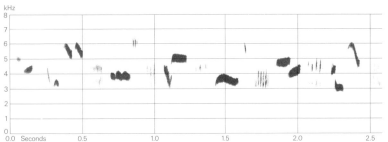

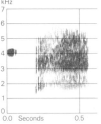

II P A D Hollom Morocco March 1978

I E D H Johnson Morocco April 1973

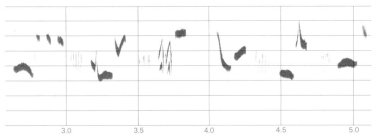

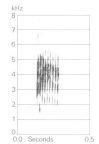

I *cont.*

III (left), IV (right) E D H Johnson Algeria April 1981

(Fig II), ♂ near nest gives 2 sounds in conjunction—'pi chirrr' (J Hall-Craggs), rasping 2nd unit also sounding like 'shirr' or 'schwirr' (P A D Hollom) or, in recording (also of ♂) by T C White, low rasp like putting out line from fishing reel (P G H Evans), while brief, high-pitched whistling 'hee', or 'heet' or 'hweet' resembling squeaky hinge (slightly longer than 1st unit of 'pi-chirrr' from ♂) is given by ♀. In another recording, 'hee' sound (Fig III) much as described, but descends slightly in pitch; grating rattle (Fig IV, also from ♂) so different from rasp shown in Fig II (and evident in recording by T C White) that it may well have different function (no further proof, though) (P G H Evans, J Hall-Craggs). Some indication in recordings of possible sex difference in calls (see also Kleinschmidt 1908 and Redstart *P. phoenicurus*, p. 704), with ♀ giving only 'hee' sound, ♂ this as well as rasp and rattle. However, further study required to determine whether difference constant, as Koenig (1895) noted combination of whistles and rasping or rattling calls ('tsit tsit-karr karr') from ♀ in vicinity of nest.

CALLS OF YOUNG. No information. PGHE

Breeding. SEASON. Algeria and Tunisia: eggs laid 1st week of April to mid-June. Moroccan Sahara: laying from mid-March (Heim de Balsac and Mayaud 1962). SITE. On ground, under tussock or low bush, also in hole in tree or wall. Nest: loose cup of light vegetation, lined with feathers and hair. Building: no information on role of sexes. EGGS. See Plate 82. Sub-elliptical, smooth and glossy; white or very pale blue, proportions of different coloured eggs varying with region (Heim de Balsac and Mayaud 1962). 18·2 × 14·0 mm (16·8–20·3 × 13·4–15·8), *n* = 83; calculated weight 1·9 g (Schönwetter 1979). Clutch: 4–5(3–6). Of 143 clutches, Algeria and Tunisia: 3 eggs, 3%; 4, 76%; 5, 18%; 6, 3%; mean 4·2 (Heim de Balsac and Mayaud 1962). One brood, possibly 2. YOUNG. Cared for and fed by both sexes (M G Wilson). No further information.

Plumages. ADULT MALE. In fresh plumage, about August–December, crown, hindneck, mantle, scapulars, and back black, feather-tips broadly fringed pale grey-brown or pale rufous-brown, partly concealing black. Crown bordered in front and at sides by white diadem, consisting of white bar across forehead, extending laterally into stripe over eye widening into large white patch on sides of neck; tips of white feathers slightly mottled grey. Rump and upper tail-coverts deep orange-rufous. Narrow line above base of upper mandible, lores, narrow line over eye and broad band below eye black, extending into black band over ear-coverts to lower sides of neck; some pale grey-brown mottling on ear-coverts and sides of neck. Entire underparts (including thighs) deep orange-rufous, feather-tips narrowly fringed white, but vent and shorter under tail-coverts cream-pink or white. Central pair of tail-feathers (t1) black with rufous edges at sides; other feathers deep rufous-cinnamon, sometimes slightly paler orange-rufous outwards; t6 with black streak on distal part of outer web. Flight-feathers and tertials deep black (slightly paler, dark sepia-grey, on inner webs and tips of secondaries and inner primaries), basal halves of outer webs of inner 5–6 primaries and all of secondaries contrastingly white, forming conspicuously white wing-patch; basal inner webs of tertials with some white (mainly concealed); tertials with faint and narrow pale buff fringes along tips. Upper wing-coverts, greater upper primary coverts, and bastard wing deep black, tips of median and greater coverts and of primary coverts narrowly fringed white. Under wing-coverts and axillaries deep rufous-cinnamon, longest coverts dark grey. *In worn plumage* (from about February onwards), brownish fringes of upperparts, tertials, and sides of head and white fringes of underparts and wing-coverts worn off, upperparts, sides of head, and upperwing completely deep black except for contrastingly white diadem and wing-patch and for rufous rump and upper tail-coverts; underparts uniform deep orange-rufous except for paler orange-white or cream-white vent. ADULT FEMALE. In fresh plumage, upperparts, ear-coverts, and sides of neck pale brown-grey or pale olive-grey, similar to upperparts of adult ♀ Redstart *P. phoenicurus*; rump orange-buff, merging into deep orange-rufous on upper tail-coverts. Narrow bar at base of upper mandible, lores, cheeks, and chin pale buff, usually faintly mottled dusky grey (in particular in front of eye and on lower cheeks); narrow eye-ring pale buff or cream-white (sides of head similar to adult ♀ *P. phoenicurus*). Throat pale buff with faint grey-buff or grey mottling. Chest, breast, and flanks rather variable: either greyish-brown with some faint rufous-brown on feather-bases, or light orange-brown with grey wash on chin, throat, and chest, or rather deep rufous-cinnamon with faint grey feather-fringes (similar to but less intense than adult ♂). Central belly, vent, and under tail-coverts pink-buff or cream-white, merging into rufous-brown of flanks and breast. Tail rufous-cinnamon; central pair (t1), often outer border of t6, and sometimes some outer borders of other feathers dark olive-brown or black-brown. Flight-feathers, tertials, greater upper wing-coverts, greater upper primary coverts, and bastard wing black-brown; outer webs of flight-feathers narrowly fringed pale grey-buff (except emarginated parts of primaries), outer webs and tips of greater primary coverts narrowly fringed pale grey, outer webs and tips of tertials and greater wing-coverts more broadly fringed pale brown-grey or pale olive-grey. Lesser and median upper wing-coverts like mantle and scapulars or sometimes slightly greyer. Under wing-coverts and axillaries orange-buff, cream-buff, or pale buff. *In worn plumage*, upperparts duller and more greyish-olive-brown (less olive in ♀♀ with much rufous on underparts); sides of head more heavily mottled grey-brown and buff, eye-ring sometimes hardly visible; underparts rather variable, in some largely rufous-cinnamon except for cream-buff chin, throat, vent, and under tail-coverts, in others dull rufous-brown or dull greyish-brown except for buff central throat and pale buff vent and under tail-coverts; chest, sides of breast, and upper flanks sometimes faintly streaked dusky brown. NESTLING. No information. JUVENILE. Upperparts grey-brown with slight rufous tinge, feathers faintly tipped black; rump and upper tail-coverts rufous-cinnamon. Eye-ring indistinct, pale buff. Cheeks, ear-coverts, sides of neck, chin to chest, and upper flanks pale buff-brown (paler buff on chin, throat, and upper flanks), feathers with dark grey fringes along tips, appearing scaled; remainder of underparts uniform yellow-buff, merging into cinnamon-buff on under tail-coverts. Tail and flight-feathers as adult. Lesser and median upper wing-coverts dark grey-brown with ill-defined rufous-buff fringes on tips (fringes grey in adult ♀); greater coverts similar but fringes extending along outer web. Greater upper primary coverts black-brown with ill-defined buff fringes. Axillaries and under wing-coverts buff. Sexes mainly similar, but ♂ with white bases of outer webs of inner primaries, showing white wing-patch (like adult ♂). Rather similar to juvenile Robin *Erithacus rubecula*, but more uniform and tail mainly rufous; paler and less heavily scaled than juveniles of *P. phoenicurus* and Stonechat *Saxicola torquata*. FIRST ADULT MALE. Like adult ♂, but juvenile tail, flight-feathers, greater upper primary coverts, and variable number of greater upper wing-coverts retained. Tail and flight-feathers as adult, but more heavily worn than those of adult at same time of year (difference most marked in spring); retained dark brown outer wing-coverts and upper primary coverts have bleached buff fringes and contrast strongly with new black median and inner greater wing-coverts. In heavily worn plumage, pale fringes worn off, but greater upper primary coverts distinctly browner, more pointed, and more worn than black and rounded ones of adult. White diadem sometimes less broad than adult, partly speckled black or interrupted on centre of forehead. FIRST ADULT FEMALE. Like ♀ and sometimes hard to distinguish. Part of juvenile feathers retained, as in 1st adult ♂, but these hardly contrasting in colour with new feathers; best distinguished by greater wear of retained feathers, in particular greater upper primary coverts dark brown, pointed, and with traces of buff fringes (in adult, brown-black with rounded tips and narrower more even grey or grey-buff fringe). Variation in colour of underparts as in adult ♀.

Bare parts. ADULT, FIRST ADULT. Iris bright dark brown. Bill, leg, and foot black. (Hartert 1910; BMNH.) NESTLING. No information. JUVENILE. Iris dark brown. Bill flesh-coloured with dark horn-brown culmen and tip; gape-flanges pale yellow. Leg and foot dull flesh-grey with darker joints and toes. (BMNH.)

Moults. ADULT POST-BREEDING. Complete; primaries descendant. Starts with p1 about mid-June to mid-July. 1-year-old ♂ from Algeria, 21 June, had primary moult score 6, and 4 central pairs of tail-feathers simultaneously shed, but head and body still old (ZMA); single ♂ from 29 June had not started moult, others from 29 June and 7 July had shed p1-p2. Moult completed in 7 adults examined October. (BMNH.) POST-JUVENILE. Partial:

involves head, body, lesser and median upper wing-coverts, and inner or all greater upper wing-coverts, but no flight-feathers, tertials, tail, greater upper primary coverts, and longer feathers of bastard wing. Starts shortly after fledging, about late May to early August; completed August–September, perhaps occasionally earlier. (BMNH, RMNH, ZMA.)

Measurements. North-west Africa, all year; skins (BMNH, RMNH, ZMA). Bill (S) to skull, bill (N) to distal corner of nostril; exposed culmen on average 4·0 mm shorter than bill (S).

WING	♂ 67·6 (1·28; 20)	65–70	♀ 66·0 (1·99; 15)	62–71
TAIL	47·0 (1·72; 20)	44–51	45·9 (1·78; 15)	44–49
BILL (S)	14·8 (0·42; 17)	14·3–15·6	14·9 (0·44; 15)	14·2–15·7
BILL (N)	8·1 (0·35; 17)	7·6–8·8	8·1 (0·37; 15)	7·5–8·8
TARSUS	24·3 (0·78; 19)	23·2–25·4	23·9 (0·67; 16)	22·8–25·1

Sex differences significant for wing. First adult with retained juvenile wing and tail combined with full adult, though 1st adult wing on average 1·4 shorter than adult and tail on average 1·2 shorter.

Weights. Algeria, November: ♂♂ 14·5, 15; ♀♀ 15, 15 (ZFMK).

Structure. Wing short, broad at base, tip rounded. 10 primaries: p6-p7 longest, p8 and p5 0·5-2 shorter, p9 6-9, p4 3-5, p3 5-8, p2 7-11, p1 8-13. P10 reduced, 24-33 shorter than wing-tip, 6-11 longer than longest upper primary covert. Outer web of p5-p8 and inner of p6-p9 emarginated. Tertials short. Tail short, shorter than in other *Phoenicurus*, more like *Saxicola* chats; tip square or slightly rounded; 12 feathers, t6 0-4 shorter than t1. Bill slender as in *P. phoenicurus*, not as heavy as in *Saxicola*; some fine bristles along base of upper mandible, as in other *Phoenicurus*, not as strong as in *Saxicola*. Tarsus long and slender; toes rather short and slender. Middle toe with claw 15·7 (20) 14·5-16·5; outer toe with claw *c.* 69% of middle with claw, inner *c.* 63%, hind *c.* 78%. Claws short and rather slender.   CSR

Phoenicurus erythrogaster **Güldenstädt's Redstart**

PLATES 49 and 52
[between pages 652 and 653]

Du. Witkruinroodstaart   Fr. Rougequeue de Güldenstädt   Ge. Bergrotschwanz
Ru. Краснобрюхая горихвостка   Sp. Colirrojo siberiano   Sw. Bergrödstjärt

Motacilla erythrogastra Güldenstädt, 1775

Polytypic. Nominate *erythrogaster* (Güldenstädt, 1775), Caucasus area. Extralimital: *grandis* (Gould, 1850), central Asia.

Field characters. 18 cm; wing-span 28-30 cm. Close in length to Rock Thrush *Monticola saxatilis*; nearly 30% larger than Redstart *P. phoenicurus*, with proportionately longer wings and shorter tail. Biggest redstart of west Palearctic, with rather long-crowned head and more robust appearance than smaller congeners. ♂ has diagnostic combination of white crown and nape and large white wing-patch, but otherwise recalls ♂ Black Redstart *P. ochrurus* of eastern races. ♀ suggests ♀ *P. phoenicurus* but has rather uniform dusky-buff underparts and duller

tail. Sexes dissimilar; slight seasonal variation in ♂. Juvenile separable.

ADULT MALE. Forehead and crown white, becoming silver-grey on nape. Face, throat, breast, back, and wings black with slaty tone, marked by bold white wing-patch formed by deep white bases to secondaries and inner primaries and narrowing white bases to outer primaries; in flight, patch becomes long panel on both surfaces of wing. Underparts below breast, rump, tail, and under wing-coverts deep maroon-chestnut; central tail-feathers

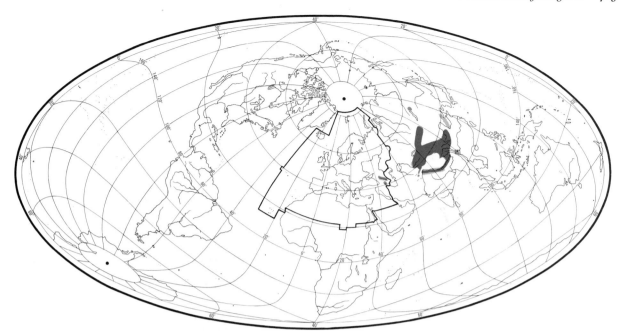

rather browner. After autumn moult, plumage briefly duller with ashy fringes to back and breast and greyer crown and nape. ADULT FEMALE. Head, back, and scapulars brown-grey; eye-ring buff-white. Lower face and breast pale dull grey, becoming ochre-buff on flanks and buff-white on belly. Wings dark brown-grey, without striking panel of ♂. Underwing pale brown-grey. JUVENILE. Resembles ♀ but head, back, and fore-underparts obscurely mottled; median coverts tipped yellow-buff. ♂ shows dull white wing-panel. FIRST-YEAR MALE. As adult but slightly duller. At all ages, bill and legs black.

♂ unmistakable. ♀ less distinctive but compared to congeners in west Palearctic, larger size, duller underparts, and less bright chestnut tail soon obvious to experienced observer. Flight fluent and floating, with rather long wings producing less fluttering action than in smaller congeners. Gait and behaviour typical of genus. Shy.

Calls include a weak 'lik', and a harder 'tek' in alarm.

Habitat. Breeds mainly extralimitally in elevated rugged middle latitudes of central Asia up to perennial snow and glacier line, at *c.* 5000 m, in severe climates with summer snow, hail, and sleet on plateaux and mountain peaks, stonefields, mountain tundra, steep crags with scree, stony patches with scattered bushes, and alpine meadows. Nominate *erythrogaster* in west Palearctic inhabits uppermost belts of mountains and narrow defiles traversed by rapid mountain streams; also scree and detritus of glacial moraines. In winter, descends to river valleys, occupying thickets of Siberian buckthorn *Hippophae*, and sometimes perching on trees (Dementiev and Gladkov 1954*b*). In Tibet breeds up to 5500 m; along northern

Himalayas favours zone at 3900-4800 m, and always above *c.* 3600 m, in dry barren alpine terrain above dwarf scrub zone, on boulder-strewn meadows and slopes or river-beds. In winter descends sometimes to 900 m, but rarely below 1500 m and will occur at 4800 m, resorting to rocky moraines and to hillsides near streams and river-beds, especially where thickets of *Hippophae* grow in valley bottoms. Flies from rock to rock, hawking flying insects in the air, or perches on boulders or scrub, making brief descents to pick insects from ground. Soars in display flight from prominent rock (Ali and Ripley 1973*b*). Thus, even surpasses accentors (Prunellidae) in primitive attachment to most exposed high altitudes, and contrasts with extent of adaptation by Redstart *P. phoenicurus* to boreal habitat in sunny lowlands, retaining only slightest vestiges of evolution from montane origins.

Distribution and population. No information on range changes or population trends.

USSR. Numbers not high (Dementiev and Gladkov 1954*b*).

Accidental. Bulgaria, Kuwait.

Movements. Mainly a short-distance altitudinal migrant, many descending to foothills, valleys, and plains for winter months, while some birds disperse further. In very severe winters, descends to 500 m in Kabardino-Balkarskaya ASSR (Baziev and Chunikhin 1963). Outside breeding season, Caucasian birds disperse both northwards (as far as Kislovodsk in Stavropol' oblast) and southwards into Georgia and Armeniya (Dathe and Neufeldt 1984). Birds from Pamirs range west in winter to Samarkand, via Zeravshan valley (Ivanov 1969). In

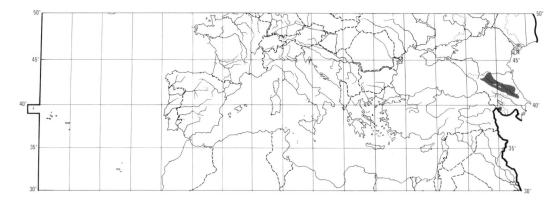

northern India, scarce south of main range even in winter, when recorded south to Kashmir and Chamba (Ali and Ripley 1973a). Believed to be a rare winter visitor to north China plain (e.g. Hopei), such birds possibly from Transbaykalia population (Walters 1981). Has straggled to northern Saudi Arabia, and possibly to Red Sea hills of northern Sudan (Nikolaus 1981).

A hardy species which often remains at high altitude, especially ♂♂, even during severe weather; ♀♀ entirely desert high summer grounds in winter (Ali and Ripley 1973a). Timing of movement varies geographically, and from year to year according to weather severity. In Caucasus, harshest areas vacated by late September but others not until October or November (Baziev and Chunikhin 1963); young birds descend first, and adults may not complete exodus until mid-December (Boehme 1958). Spring return in Caucasus is in March, but movement gradual and dependent on thaw. In contrast, descent begins earlier and return is later in harsher regime of Soviet Asian mountains (e.g. Pamirs), where main movements occur mid-September to mid-October and April to early May (Ivanov 1969; Dolgushin et al. 1970).   RH

Food. In summer, insects, especially beetles (Coleoptera); berries in winter. Feeds by picking items from ground, ice, water, or bank of stream, hopping or running about— even feeds inside animal carcasses, though presumably taking insects attracted to them; also flies on to prey from low perch, sometimes pursuing it along ground, or takes aerial prey in brief flight (Ludlow and Kinnear 1933; Stegmann 1936; Kozlova 1952; Drozdov and Zlotin 1962; Ali and Ripley 1973a; Piechocki et al. 1982).

Diet includes the following. Invertebrates: grasshoppers, etc., and their eggs (Orthoptera: Tettigoniidae, Acrididae), bugs (Hemiptera: e.g. Cicadellidae), adult and larval Lepidoptera (Noctuidae, Satyridae), adult and larval flies (Diptera: e.g. Chironomidae), Hymenoptera (Ichneumonoidea, ants Formicidae), adult and larval beetles (Coleoptera: Carabidae, Histeridae, Scarabaeidae, Elateridae, Tenebrionidae, Chrysomelidae, Cur-

culionidae), Arachnida, centipedes (Chilopoda), earthworms (Oligochaeta). Plant material mainly fruits and seeds: of *Juniperus*, docks and knotweeds (Polygonaceae), buttercups (Ranunculaceae), barberry *Berberis*, Leguminosae, sea buckthorn *Hippophae*, *Ephedra*, sedges (Cyperaceae), and grasses (Gramineae); also leaves. (Pek and Fedyanina 1961; Baziev and Chunikhin 1963; Zlotin 1968; Ali and Ripley 1973a; Fleming et al. 1976; Walters 1981; Piechocki et al. 1982.)

In summer in Kirgiziya, takes exclusively insects, largely adult and larval beetles (Yanushevich et al. 1960; see also Pek and Fedyanina 1961), though summer stomachs analysed by Kuznetsov (1962) often contained small seeds and leaves. In spring and autumn, Tien Shan, 11 stomachs contained exclusively insects, mainly beetles (Zlotin 1968).

During January–February in Caucasus, midges (Chironomidae) can be important, birds taking large numbers from river banks or water surface (Drozdov and Zlotin 1962), but winter diet seems to consist typically of berries, especially *Hippophae* which may be taken more or less exclusively (Boehme 1959; Yanushevich et al. 1960; Baziev and Chunikhin 1963; Zlotin 1968; Ali and Ripley 1973a), even in late April when insects available (University of Southampton 1982).

No information on food of young.   DJB

Social pattern and behaviour. Poorly known.

1. In winter, birds that move away from breeding areas occur in loose flocks (Portenko 1954; Boehme 1958; Kydyraliev 1959). In south-east Kazakhstan, said to stay in pairs in winter (Shnitnikov 1949). No noticeable flocks or large gatherings recorded in Caucasus though densities in buckthorn *Hippophae* thickets may number 377 birds per km² (Drozdov and Zlotin 1962; Baziev and Chunikhin 1963). In winter, frequently seen in loose company with other species, e.g. Eversmann's Redstart *P. erythronotus* (Kuznetsov 1962; Abdusalyamov 1973; Ali and Ripley 1973a). ♂:♀ ratio 7:1 near Dargavs, Caucasus (Drozdov and Zlotin 1962). In Tadzhikistan (USSR) groups of 5–6 ♂♂ seen in early October (Abdusalyamov 1973). BONDS. No information on mating system. Pair-formation and establishment of territories within c. 10–15 days of arrival (Kydyraliev 1959; Potapov 1966; Ivanov 1969; Dolgushin et al. 1970;

Abdusalyamov 1973). Both sexes feed young before and after fledging (Yanushevich *et al.* 1960; Baziev and Chunikhin 1963; Abdusalyamov 1973). Young fed for at least 7 days after fledging (Potapov 1966). No information on duration of pair bond or age of first breeding. BREEDING DISPERSION. Solitary and territorial; territories usually large. In Pamirs, pairs usually 900–1000 m apart in typical habitat (Abdusalyamov 1973), occasionally 300–400 m apart (Potapov 1966). Densities in certain places in Nepal may be sufficient for territories to be contiguous (Diesselhorst 1968*a*). Observed collecting food for young 200–300 m from nest (Potapov 1966; Ivanov 1969). Density 1 pair per 1·5 km on rocky slopes near watershed ridges in Pamirs (Potapov 1966); 8 pairs in 2·5 km² in one area of crag and scree in Artysh range, Tuvinskaya ASSR (Zabelin 1976). Densities generally lower than these however: e.g. 3 pairs in 3 km² and 1 pair in 4 km² in two localities in Caucasus (Baziev and Chunikhin 1963). Territory used for pair-formation, nesting, and feeding. ROOSTING. May roost under crags or in rodent holes during heat of day, in bad weather, or at night (Abdusalyamov 1973).

2. Outside breeding season, both sexes generally shy and secretive (Abdusalyamov 1973; Ali and Ripley 1973*a*), rarely allowing approach closer than *c.* 30 m (Potapov 1966), though ♀ also described as more secretive than ♂, hiding in centre of *Hippophae* bushes whilst ♂ perches more prominently on top (Shnitnikov 1949). Actions and behaviour typical of *Phoenicurus*, flying from rock to rock or perching on boulders or scrub, frequently tail-flicking (Ali and Ripley 1973*a*; Walters 1981). FLOCK BEHAVIOUR. No details. SONG-DISPLAY. ♂ sings (see 1 in Voice) from prominent perch, with tail spread or slightly lowered; also in Song-flight (rare according to Schäfer 1938), gliding with quivering wings and tail widely spread (Dolgushin *et al.* 1970; Ali and Ripley 1973*a*). However, generally rather silent with song rarely heard even during courtship period (Dolgushin *et al.* 1970; Walters 1981); heard singing only 6 times in 2 summers by Potapov (1966); but see Schäfer (1938) for Tibet. Only report of song heard outside breeding season from Severo-Osetinskaya ASSR (Caucasus) when ♂♂ noted singing loudly on sunny days in January–February (Drozdov and Zlotin 1962). ANTAGONISTIC BEHAVIOUR. Frequent furious fights occur between ♂♂ during territory demarcation at beginning of season, also with Black Redstart *P. ochruros* and Desert Wheatear *Oenanthe deserti* (Potapov 1966; Abdusalyamov 1973). Disputes between territorial ♂♂ include chases accompanied by Excitement-call (see 4 in Voice). Postures and displays during confrontations are similar to those of other *Phoenicurus* (Böhme and Böhme 1986). HETEROSEXUAL BEHAVIOUR. ♂♂ arrive on breeding grounds before ♀♀ (Potapov 1966; Abdusalyamov 1973). No description of courtship or mating behaviour, but ♂ recorded chasing and fighting with ♀ and giving Excitement-call (Boehme 1982). ♀ apparently selects nest-site and incubates eggs but with ♂ in close attendance (Potapov 1966). RELATIONS WITHIN FAMILY GROUP. No information on behaviour in nest. After fledging, young remain close to nest for 1–3 weeks (Potapov 1966; Chernikin 1976). Nest sanitation by both parents, faecal sacs discarded 30–200 m from nest (Dolgushin *et al.* 1970). ANTI-PREDATOR RESPONSES OF YOUNG. Alarm-call given from day 13 (Boehme 1958). Reluctant to fly when approached for first 2 days after leaving nest, then more cautious on 3rd day with parents in constant attendance (Chernikin 1976). PARENTAL ANTI-PREDATOR STRATEGIES. (1) Passive measures. Very secretive during incubation and rearing of young (Osmaston 1925; Chernikin 1976). Birds lose usual shyness at nest with young (Potapov 1966; Abdusalyamov 1973), though ♂ far more cautious than ♀, ready to flee in silence on seeing danger (Abdusalyamov 1973; see also Johansen 1955, Walters 1981). Potapov (1966), however, noted ♂♂ continuing to feed young when observer 4–5 m away. (2) Active measures. When young in nest, ♂ (who apparently does all guarding of nest) may attack potential predators, e.g. perching on back of chipmunk *Tamias sibiricus* and pecking at it (Chernikin 1976). After fledging, parents give Alarm-calls if man approaches, and ♀ recorded almost landing on camera lens; 3 days after fledging, no close flying by parents though still some Alarm-calls (Chernikin 1976). PGHE

Voice. Poorly known. Both Alarm-calls and song noted outside breeding season (Boehme 1958; Drozdov and Zlotin 1962). Repertoire smaller than in other *Phoenicurus* of USSR (Böhme and Böhme 1986, which see for sonagrams and analysis of calls, but not song).

CALLS OF ADULTS. (1) Song. Short, clear, melancholy whistles, somewhat reminiscent of Blackbird *Turdus merula* in tone, with twittering or ticking sounds interpolated (Potapov 1966; Ali and Ripley 1973*a*). Singing bird will also give sound like falling stones, characteristic of Black Redstart *P. ochruros* song (Potapov 1966). (2) Contact-call. A weak 'lik' (Fleming *et al.* 1976); less clear than that of Redstart *P. phoenicurus* with never more than 2 syllables (Sushkin 1938; Walters 1981). (3) Alarm-call. 'Tsee-tek tsee-tek-tek' given by adults in vicinity of nest or young, but also in winter quarters (Boehme 1958; Potapov 1966). (4) Excitement-call. 'Drrr' given when 2 ♂♂ chase each other in territorial disputes (Fleming *et al.* 1976). Probably same as given when ♂ pursues or fights with ♀ (Boehme 1982).

CALLS OF YOUNG. Food-call of fledged young 'tsee-kre tsee-kre-kre'; as in *P. ochruros*, resembles adult's Alarm-call; only call given at 8–9 days old (Potapov 1966). Alarm-call uttered from day 13. At 40 days, ♂ gives muttering sounds. During moult (at 2 months old) young gave only Alarm-calls and an anxiety call developed from food-call (Boehme 1982). 'Display' sounds (adult Excitement-call) start at 4 months old, and by 6 months have developed into unstructured song (Boehme 1982). See also Böhme and Böhme (1986). PGHE

Breeding. SEASON. Altai mountains (USSR): eggs laid from June (Dementiev and Gladkov 1954*b*). SITE. On ground, in hole or crack in rocks, on building (e.g. under roof), and also recorded using nest-box at 6 m (Dolgushin *et al.* 1970; Chernikhin 1976). Nest: bulky cup of grass stems and leaves, lined with wool, hair and feathers. Building: mainly by ♀ (Potapov 1966). EGGS. Sub-elliptical, smooth and slightly glossy; blue (nominate *erythrogaster*: Dementiev and Gladkov 1954*b*) or white with pale reddish freckles (*grandis*: Ali and Ripley 1973*a*). No measurements available for nominate *erythrogaster*. *P. e. grandis*: 22·0 × 15·7 mm (21·0–22·4 × 15·0–16·9), *n* = 6; calculated weight 2·1 g (Schönwetter 1979). Clutch: in *grandis*, 4 (3–5) (Dementiev and Gladkov 1954*b*). INCUBATION. 12–16 days (Abdusalyamov 1973), by ♀

(Kydyraliev 1959). YOUNG. Cared for and fed by both sexes (Dolgushin *et al.* 1970). FLEDGING TO MATURITY. Fledging period 14 days (Potapov 1966; Ivanov 1969), 21-22 days (Kydyraliev 1959; Yanushevich *et al.* 1960). Fed by parents for at least 7 days after fledging (Potapov 1966). No further information.

Plumages (nominate *erythrogaster*). ADULT MALE. Forehead, crown, and hindneck white. Mantle, scapulars, and back sooty-black, sometimes with slight plumbeous tinge on mantle. Rump and upper tail-coverts deep rufous-chestnut. Sides of head and neck black (from narrow band over base of upper mandible and from just above eye downwards), grading to plumbeous-black on chin, throat, and upper chest. Remainder of underparts uniform deep rufous-chestnut. Tail deep rufous-chestnut; central pair of feathers (t1) and fringe along tip of outer web of others often tinged dusky olive-brown or bronze; at times, only tip of t1 tinged olive-brown and occasionally dusky tinge entirely absent. All upper wing-coverts, bastard wing, and tertials sooty-black; tertials, outer greater coverts, inner greater primary coverts, and longest feather of bastard wing narrowly edged white on tip when plumage fresh; bases of tertials extensively white, but this largely hidden at rest. Flight-feathers black, slightly more grey-brown on inner webs; secondaries and inner primaries narrowly edged pale grey on tips; basal $\frac{2}{3}$ of secondaries contrastingly white (except for narrow horn-brown shaft-streak), showing as 2·5-3 cm wide bar along upperwing; innermost primaries (p1-p2) similar to secondaries, but white strongly reduced towards outer primaries, p5-p7 with *c.* 0·5 cm of white beyond upper primary coverts, p8-p10 with restricted white visible from below only. Under wing-coverts and axillaries deep rufous-chestnut; smaller coverts along leading edge of wing and primary coverts black. Unlike other west Palearctic *Phoenicurus*, abrasion has hardly any influence on plumage. *In fresh plumage* (autumn), white of forehead and hindneck has slight grey wash on feather-tips, soon wearing off; *in worn plumage*, chin to chest slightly deeper black, remainder of underparts slightly paler chestnut, and narrow white edges on tips of some flight-feathers, tertials, and upper wing-coverts worn off. ADULT FEMALE. Sides and top of head, mantle, scapulars, and back pale brown-grey or pale drab-grey, similar to upperparts of ♀ Redstart *P. phoenicurus* or slightly paler and greyer; lores slightly mottled grey and off-white, forming indistinct pale stripe; eye-ring pale buff-grey or cream-white, often indistinct. Drab-grey of upper rump merges into pale cinnamon or cinnamon-buff on lower rump; upper tail-coverts contrastingly deep rufous-chestnut (paler than in ♂). Chin to chest pale drab-grey (greyer and more uniform than in ♀ *P. phoenicurus*, throat less cream, chest less pale cinnamon); breast, belly, flanks, and vent pale sandy-grey, variable amount of pale cinnamon or deep rufous-cinnamon of feather-bases showing through; under tail-coverts rufous-chestnut, sometimes contrasting with remainder of underparts. Tail as adult ♂, but t1 and terminal part of outer web and tip of others more extensively tinged bronze-brown or olive-brown, t1 often entirely brown. Flight-feathers, tertials, greater upper wing-coverts, greater upper primary coverts, and bastard wing dark brownish-grey, outer webs and tips rather narrowly bordered by ill-defined pale drab-grey fringes (except tips of outer primaries); less blackish than fresh feathers of ♀ *P. phoenicurus* and fringes wider and less clear-cut. Lesser and median upper wing-coverts pale brown-grey or pale drab-grey, like upperparts; centres of median coverts often slightly darker brown. Under wing-coverts and axillaries pale brown-grey,

partly with darker brown bases. *In worn plumage*, upperparts and upperwing duller grey-brown, underparts pale grey-brown with variable amount of pale cinnamon-brown or rufous-cinnamon tinge on chest and belly; pale fringes of flight-feathers and wing-coverts bleached to off-white in spring, worn off in summer. NESTLING. Down dark brown, on upperparts only, including eyelids (Neufeldt 1970). JUVENILE MALE. Head and body rather like adult ♀, though more variable. Crown to back pale brown-grey, like ♀, but feathering more fluffy and often some pale grey of feather-bases visible; feather-tips either uniform pale brown-grey, or with narrow darker brown fringes (crown, mantle, and scapulars showing faint dark scaling), or more extensively dull brown with ill-defined buff or off-white spots. Tail-coverts cinnamon or pale rufous, as in adult ♀. Sides of head, chin, throat, and chest pale buff, feather-tips slightly darker, grey-brown, showing as indistinct scaling (tips sometimes largely grey-brown, buff feather-centres showing as ill-defined pale spots); belly, vent, and flanks pale cream-buff. Lesser and median upper wing-coverts dull brown-grey with broad yellow-buff tips. Tail as adult ♂; amount of dusky tinge rather variable. Flight-feathers, tertials, greater upper wing-coverts, greater upper primary coverts, and bastard wing as adult ♂ (black with large white wing-bar) but differs in more extensive grey-buff or pale buff fringes to secondaries, inner primaries, primary coverts, and longer feathers of bastard wing, and (in particular) in rather broad buff, pink-buff, or buff-white fringe on tip of outer web of greater upper wing-coverts, sometimes extending narrowly to tip of inner web and often forming shallow triangle (apex at shaft). JUVENILE FEMALE. Head and body as juvenile ♂. Tail and wing as adult ♀, wing browner than juvenile ♂ and without white wing-bar. Flight-feathers, tertials, greater upper wing-coverts, and greater upper primary coverts with slightly wider and more buffy tips than in adult ♀, in particular on tips of outer webs of outer greater coverts. FIRST ADULT MALE. Like adult ♂, but forehead, crown, and hindneck rather more extensively tinged grey, white of hindneck less sharply divided from black of mantle; mantle and scapulars slightly greyer; underparts less uniform deep rufous-chestnut, tinged orange on central belly and vent. Juvenile flight-feathers, tertials, outer or all greater upper wing-coverts, greater upper primary coverts, and longer feathers of bastard wing retained; relatively browner and more worn than adult; pale tips of outer webs of outer greater coverts bleached to buff-white or white, often partly worn off (showing as indentations), but usually obvious until July of 2nd calendar year. FIRST ADULT FEMALE. Like adult ♀, but part of juvenile feathering retained, as in 1st adult ♂; often hardly distinguishable from adult ♀, apart from relatively more worn primaries and greater upper primary coverts.

Bare parts. ADULT, FIRST ADULT. Iris brown to black-brown. Bill black. Inside of upper mandible black in adult, yellow or orange in 1st adult (at least during 1st autumn). Leg and foot brown-black to black. NESTLING. No information. JUVENILE. Iris dark brown. Bill horn-black, gape-flanges yellow. Leg and foot horn-black, soles grey. (Hartert 1910; Ali and Ripley 1973a; Delany *et al.* 1982; RMNH.)

Moults. ADULT POST-BREEDING. Complete; primaries descendant. ♂ starts before ♀ (Dementiev and Gladkov 1954b); in Mongolia, late August, ♂ of pair with dependent young had primary moult score 9, ♀ not yet started (Piechocki and Bolod 1972). Starts with p1 early July to early August, completed August-September with outer primaries and some secondaries. In Caucasus, one ♂ completely new 14 August; in Tadzhikistan, some completed by mid-August; in Altai, *c.* 21 August. Some body feathers and secondaries still growing on arrival winter quarters, Ladakh, late September and October. (Dementiev and Gladkov 1954*b*; Ali and Ripley 1973*a*; Delany *et al.* 1982; RMNH.) POST-JUVENILE. Partial: head, body, lesser and median upper wing-coverts, and occasionally some inner upper wing-coverts; no flight-feathers, tertials, outer greater coverts, or greater upper primary coverts. Starts soon after fledging; some feathers of sides of breast and lesser wing-coverts first, early July to late August, followed by mantle, scapulars, belly, and chest, and then by remainder of head, body, and wing-coverts. Completed August-September (or early October). (Dementiev and Gladkov 1954*b*; Delany *et al.* 1982; RMNH.)

Measurements. Nominate *erythrogaster*. Caucasus, August-April; skins (RMNH, ZFMK, ZMA). Bill (S) to skull, bill (N) to distal corner of nostril; exposed culmen on average 4·8 mm less than bill (S).

	♂			♀		
WING	102·8 (1·04; 47)	100-108		98·4 (1·76; 13)	95-101	
TAIL	69·3 (2·00; 47)	66-75		67·6 (2·09; 8)	64-70	
BILL (S)	16·7 (0·44; 45)	15·9-17·5		16·4 (0·63; 13)	15·5-17·3	
BILL (N)	8·7 (0·36; 43)	8·1-9·5		8·6 (0·36; 7)	8·1-9·0	
TARSUS	25·9 (0·53; 44)	24·9-26·9		25·8 (0·59; 8)	24·9-26·5	

Sex differences significant for wing and tail. Wing and tail of 1st adult (with retained juvenile flight-feathers and tail) significantly shorter than older birds, but combined in table above; e.g.: wing, adult ♂ 105·9 (1·18; 4) 105-108, 1st adult ♂ 102·5 (1·62; 43) 100-106; tail, adult ♂ 72·8 (2·22; 4) 69-75, 1st adult ♂ 69·0 (1·77; 43) 66-73 (RMNH, ZMA).

P. e. grandis. Tien Shan (USSR), all year, skins (RMNH, ZMA).

	♂			♀		
WING	103·7 (1·73; 25)	101-108		98·3 (—; 3)	97-100	
TAIL	70·5 (2·10; 15)	68-74		68·3 (—; 3)	67-69	

Wing. Ladakh (India), September-November, live birds: (1) adult, (2) 1st adult (Delany *et al.* 1982). (3) Tibet, skins (Vaurie 1972).

	♂			♀		
(1)	107·1 (2·36; 77)	101-113		102·1 (2·36; 102)	96-108	
(2)	105·4 (2·46; 240)	99-112		100·2 (2·38; 150)	94-106	
(3)	108·0 (—; 99)	92-115		102·3 (—; 45)	95-107	

Weights. ADULT, FIRST ADULT. *P. e. grandis* (similar in size to nominate *erythrogaster*). (1) February, north-central China (Stresemann *et al.* 1937); (2) March-April, Kazakhstan, USSR (Dolgushin *et al.* 1970); (3) July-September, Nepal (Diesselhorst 1968*a*), Mongolia (Piechocki and Bolod 1972), and Kazakhstan, combined; (4) late September to November, Ladakh, India (Delany *et al.* 1982); (5) undated, Kirgiziya, USSR (Yanu-shevich *et al.* 1960).

	♂			♀		
(1)	— (—; 4)	30-32		— (—; 5)	25-31	
(2)	27·3 (3·27; 6)	24-31		25·1 (3·64; 4)	23-31	
(3)	28·6 (1·94; 8)	25-31		26·3 (3·35; 11)	18-30	
(4)	25·2 (2·26; 367)	22-29		23·9 (1·34; 346)	21-28	
(5)	— (—; 8)	22-31		— (—; 11)	22-33	

Structure. Wing rather long, broad at base, tip rounded. 10 primaries: p6-p7 longest, p8 1-4 shorter, p9 9-16, p5 1-3, p4 6-10, p3 11-15, p1 17-22. P10 reduced, 44-54 shorter than p6-p7, 7-12 longer than longest upper primary covert. Outer web of p5-p8 and inner of (p5-)p6-p9 deeply emarginated. Tertials short. Tail rather short, tip square or slightly rounded; 12 rather broad feathers, t6 2-5 shorter than t1. Bill short and rather slender; tip of upper mandible decurved. Nostrils oval, partly covered by membrane above. Short fine bristles along base of upper mandible and on chin. Tarsus rather short and slender; toes short and rather strong. Middle toe with claw 18·4 (19) 17·5-19·5; outer toe with claw *c.* 74% of middle with claw, inner *c.* 69%, hind *c.* 75%. Claws short, rather strong, decurved.

Geographical variation. Rather slight. *P. e. grandis* from central Asia differs in ♂ by dark slate-grey mantle, scapulars, chin, throat, and upper chest (more greyish in 1st adult, darker plumbeous-black in older birds, but not as black as nominate *erythrogaster*), paler rufous-chestnut rump, tail-coverts, under-parts, and tail (1st adult with narrow pale sandy feather-fringes on underparts in fresh plumage, unlike nominate *erythrogaster*), and more extensive white on primary-bases, forming bar 1-1·5 cm wide (excepting p9-p10); pale tips of greater upper wing-coverts in 1st adult ♂ often broader than in 1st adult ♂ nominate *erythrogaster*. Forehead, crown, and hindneck extensively tinged grey in fresh plumage in *grandis* from Tien Shan (especially in 1st adult), unlike nominate *erythrogaster*, but top of head of *grandis* from western Himalayas apparently paler. ♀♀ of both races rather closely similar, but upperparts and chin to breast of *grandis* paler and more sandy-grey, less cold drab-grey; belly and vent more extensively cream-white, tip of tail with less dusky suffusion. *P. e. grandis* from west-central Asia similar in size to nominate *erythrogaster*, but birds from eastern Himalayas, Tibet, and central China on average larger and sometimes separated as *vigorsi* (Horsfield and Moore, 1854) or *maximus* Kleinschmidt, 1924, but dif-ferences too small and too gradual and overlap in size too large to warrant this: wing of ♂ in western Himalayas 101-109 (23), in Tien Shan 99-109 (24), in Tibet 102-111 (16), in central China (Kansu) 106-111 (4) (Ludlow and Kinnear 1944); in Kansu, adult ♂ 108·9 (10) 105-111 (Stresemann *et al.* 1937).

CSR

Cercomela melanura **Blackstart**

Du. Zwartstaart   Fr. Traquet de roche à queue noire   Ge. Schwarzschwanz
Ru. Чернохвостка   Sp. Colinegro real   Sw. Svartstjärt

Saxicola melanura Temminck, 1824

Polytypic. Nominate *melanura* (Temminck, 1824), Dead Sea depression south to Sinai and north-west and central Saudi Arabia; *lypura* (Hemprich and Ehrenberg, 1833), south-east Egypt, eastern Sudan, and north-east Ethiopia; *airensis* Hartert, 1921, Aïr (northern Niger), Chad (north to Tibesti), and western and central Sudan. Extralimital: *neumanni* Ripley, 1952 (synonym: *erlangeri* Neumann and Zedlitz, 1913), Arabia south from Mecca and Hadhramawt; *aussae* Thesiger and Meynell, 1934, eastern Ethiopia and northern Somalia; *ultima* Bates, 1933, Mali and southern Niger.

Field characters. 14 cm; wing-span 23–27 cm. Close in size to Redstart *Phoenicurus phoenicurus*. Small, rather long-billed and noticeably slim chat with rather uniform grey plumage relieved by black rump and tail. Tail flicked open constantly. Sexes similar; no seasonal variation. Juvenile separable. 3 races in west Palearctic, easternmost distinguishable in the field from other 2.

ADULT. (1) Levant and north Arabian race, nominate *melanura*. Head, chest, and upperparts essentially bluish ash-grey, with darker wings and black rump and tail. Faint face-pattern caused by paler surround to dusky cheeks. Belly white to grey-white. Underwing pearl-grey. ♀ somewhat browner above than ♂ in direct comparison. (2) North African races, *lypura* and *airensis*. Browner than nominate *melanura*. JUVENILE. All races. Resembles adult (lacking spotted appearance of most chats) but browner, with pale throat and paler fringes and tips to wing-coverts and tertials. FIRST WINTER AND SUMMER. All races. Resembles adult, but retained juvenile tertials and greater coverts often distinctly browner than neighbouring feathers. At all ages, bill and legs black.

Unmistakable. No other west Palearctic passerine is virtually uniform grey but for black rump and tail. Flight noticeably light and floating, recalling that of *P. phoenicurus* but with shorter tail evident. Gait and stance like Black Redstart *P. ochruros* or small wheatear *Oenanthe*. Flicks tail open constantly, accompanied by flicking of drooped wings.

Song a pleasant but little-varied warble. Commonest call a soft, fluty 'cher-u'.

Habitat. Breeds in warm arid lower latitudes, temperate, subtropical, and tropical. In Sudan, occurs on arid rocky hills and mountains where there is some scattered scrub but keeping mainly to rocks, and replaced by extralimital Brown-tailed Rock-chat *C. scotocerca* where scrub thicker (Cave and Macdonald 1955). Shows preference for thorny bushes in rocky ravines; often shares habitat with White-crowned Black Wheatear *Oenanthe leucopyga* (Etchécopar and Hüe 1967). In Saudi Arabia, prefers dry rocky desert areas; occurs on mountains but avoids summits and juniper zone in Hejaz and Asir (Jennings 1981a). Usually sits in and not on top of bush (Mei-

nertzhagen 1954). In Tibesti (Chad), seen entering crevice high up vertical rock face (Guichard 1955). Normally found on steep terrain rather than on level areas.

Distribution. No reported changes in range.

SYRIA. Seen Yarmuk valley in winter; may be resident but no proof of breeding (AMM).

Accidental. Syria, Kuwait.

Population. No information on population trends.

EGYPT. Sinai: locally common (SMG, PLM, WCM). CHAD. Tibesti: common in all vegetated areas (Simon 1965).

Movements. Resident or even sedentary throughout range, e.g. Gallagher and Woodcock (1980) and Jennings (1981a) for Arabia, Meinertzhagen (1920b) for Palestine, Lamarche (1981) and Ash and Miskell (1983) for sub-Saharan Africa. Some dispersal might be expected in view of arid habitats occupied, and Newby (1980) saw 3 birds of uncertain status in Ouadi Rime in central Chad, south of known breeding range. In general, however, virtual absence of records anywhere outside breeding range suggests little long-distance movement occurs. RH

Food. In Palestine, March and September–October, stomachs contained beetles, caterpillars, other insects, and seeds; also takes ants (Hardy 1946). In Niger, August–September, stomachs contained insects (Villiers 1950). Recorded eating berries in South Yemen (Gallagher and Rogers 1980), and in eastern Saudi Arabia once seen to eat a berry of *Lycium shawii* (J Palfery). Feeds mainly by perching on rocks, trees, and bushes, dropping on to prey on ground; also by searching vegetation and occasionally in brief hover or aerial-pursuit (Gallagher and Woodcock 1980).

In Oman, seen carrying small caterpillars and winged insects, presumably for young (Bundy 1986).   DJB

Social pattern and behaviour. Poorly known.

1. Mainly solitary outside breeding season (Arnold and Ferguson 1962; Gallagher and Rogers 1980; Jennings 1980).

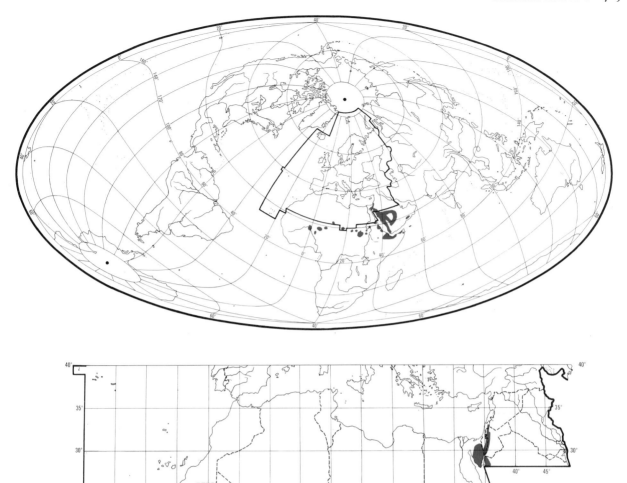

Not known whether breeding territory maintained all year. BONDS. Presumably mainly monogamous. Duration of pair-bond and age at first breeding not known. Not known if both parents feed young in nest but young beg food from parents several weeks after leaving nest-hole (Bundy 1986). BREEDING DISPERSION. Solitary and territorial. Densities vary depending on availability of suitable habitat (e.g. bushes of *Acacia* and *Euphorbia*). In Oman, in only one area (a narrow ravine) were 2 pairs detected in close proximity. In desert wadis, 6 pairs found in 1·5-km stretch and 4 pairs in another 1-km stretch; 3 pairs present along 500 m of cliff bordering a small settlement. South Arabian *neumanni* apparently has larger territories than north Arabian nominate *melanura*, though no comparative data. (Bundy 1986.) ROOSTING. No information.

2. Fearless (Tristram 1867; Meinertzhagen 1954), with restless flitting between rocks or small trees, frequently fanning wings and conspicuous black tail, interspersed with occasional singing (Tristram 1867; Lynes 1925a; Bates 1933-4; Gallagher and Woodcock 1980: see 1 in Voice). Constant opening and closing of wings and tail occurs in a measured way—not like quick flick of Wheatear *Oenanthe oenanthe*: tail expanded to *c*.

45° each side, primaries well fanned but inner wing only slightly extended (Lynes 1925a). Wings may be drooped slightly (Fig A) like Rufous Bush Robin *Cercotrichas galactotes* (Arnold and

A

Ferguson 1962). Tail-spreading and singing rarely observed in late summer and autumn during moult (Bates 1933-4). FLOCK BEHAVIOUR. No information. SONG-DISPLAY. Song delivered often from tree-top or other perch (Bates 1933-4; Gallagher and Woodcock 1980). Sings particularly in morning and evening (Niethammer 1955b). Song probably uttered during most of year, since noted October, December, and March-May (Hardy

1946; Niethammer 1955*b*; Cornwallis and Porter 1982; C Chappuis, E D H Johnson), though not heard in February in Arabia (Bates 1936–7) and in Niger very little in late summer and autumn (Bates 1933–4). ANTAGONISTIC BEHAVIOUR. 'Territorial competition' by individuals reported regularly among nominate *melanura* (Bundy 1986). No further information. HETEROSEXUAL BEHAVIOUR. At start of breeding season, aerial chases occur, birds calling continually (Jennings 1980). Birds frequently hold wings drooped and tail fanned at this time (Lynes 1925*a*; Meinertzhagen 1954; L Cornwallis). Both members of pair take part in nest-building (Cornwallis and Porter 1982). No further information. RELATIONS WITHIN FAMILY GROUP. No information on behaviour in nest. Young attended by both parents after fledging (L Cornwallis). ANTI-PREDATOR RESPONSES OF YOUNG. No information on responses when in nest, but once fledged will run to safety in rock crevices on approach of potential predator (Bates 1933–4). PARENTAL ANTI-PREDATOR STRATEGIES. (1) Passive measures. No information. (2) Active measures. ♀ with unfledged young performs distraction-display with wings fanned and open and projected forward, tail fanned and depressed, and head held stiffly out and downwards; gives harsh call (Walker 1981*b*: see 4c in Voice). Will attack and mob predators such as snakes (R Frumkin).

(Fig by C Rose: from drawing in Arnold and Ferguson 1962.)   PGHE

Voice. Poorly known. Song perhaps varies geographically (see below).

CALLS OF ADULTS. (1) Song of ♂. Series of loud, well-spaced notes 'chree chrew chitchoo chirri chiwi...' (Gallagher and Woodcock 1980); 'chee-yu chee-yu' or 'chee-yu-chwe' or 'chee-oo-chuk', with 'yu' and 'oo' sounds lower pitched (King 1978); quiet 'chup t'chee chup ter tutcher' (Walker 1981*b*); occasional chirps recall House Sparrow *Passer domesticus* (Hartert 1921; Lynes 1925*a*). In song of *erlangeri*, emphasis on first and last syllables—'CH-lulu-WE'; perhaps slightly more trilling than in nominate *melanura* whose 3–4 syllables are equally emphasized—'ch-we-we' (Bundy 1986). Recording of *aussae* from Ethiopia reveals monotonous repetition of slow warbling phrases: 'chi koo chri-ki chiu-teoo' (P G H Evans: Fig I); thus apparently similar to other songs described above. 3 songs of *ultima* from Niger longer (1·28–1·54 s) than in *aussae* (0·9–0·96 s). Song of *ultima* (Fig II) significantly different from *aussae*: quite pleasant 'weedley' song, comprising liquid notes and some scratchy

sounds (J Hall-Craggs); typical of wheatear *Oenanthe* (Chappuis 1975; see also Lynes 1925*a*) or a bit like Corn Bunting *Miliaria calandra* (P J Sellar). Slightly more piercing than *aussae*, presumably due to higher-pitched components (not unlike squealing food-calls of fledged Blackbird *Turdus merula*) and, unlike *aussae*, contains some repetition of units or short figures which are not always precise but easy enough to hear. Variation occurs between songs of *ultima* in variety of unit structure, rate of vibrato, sound pressure level, and frequency range. (J Hall-Craggs.) (2) Subsong. In recording of *ultima* by C Chappuis, typical of Turdinae, wheezy and hollow-sounding (P J Sellar); protracted jumble of low warbling notes with low, harsh, strangled units (P G H Evans). (3) Contact-calls. Loud, liquid 'chura lit' (Walker 1981*b*) or loud 'tyootrit' like fragment of song shown in Fig I or one unit of Song Thrush *T. philomelos* song (J Hall-Craggs, P J Sellar: Fig III); minimum *c*. 3 s between successive calls (P G H Evans). Probably at least closely related calls noted in Israel rendered 'chairee', 'churee', or 'cheerarie', and, in Jordan, soft, fluty 'cher-u' (P A D Hollom). (4) Alarm-calls. (a) High-pitched whistling 'feefee' (L Cornwallis); 'whee' (P A D Hollom) perhaps the same. (b) Alarmed bird carrying food gave a rasping 'skirrrr' and other probably related sounds also noted in Jordan include throaty 'hrate', harsher, less far-carrying 'tschirr' and rather harsh, nasal 'shirp' (P A D Hollom). (c) Harsh 'tweep dra' from ♀ performing distraction-lure display (Walker 1981*b*); perhaps a combination of sounds described in 4a–b.

CALLS OF YOUNG. No information.   PGHE

Breeding. SEASON. Southern Sahara: March–June (Heim de Balsac and Mayaud 1962). Probably similar in Egypt and Israel (Hüe and Etchécopar 1970). Oman: from mid-February (Bundy 1986), continuing locally until end of September (Walker 1981*b*). SITE. Crevice in rock, up to 0·5 m from entrance (Jennings 1980). Nest: cup of grass stems and leaves, lined with hair and finer vegetation. May have platform of small pebbles. Building: no information on role of sexes. EGGS. See Plate 82. Sub-elliptical, smooth and fairly glossy; very pale blue, finely speckled red to red-brown, speckles sometimes

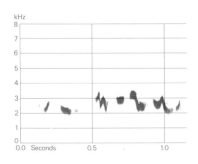

I   E D H Johnson   Ethiopia   December 1969

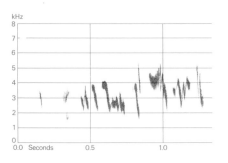

II   C Chappuis   Niger   October 1971

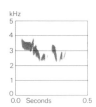

III   P A D Hollom   Israel April 1979

concentrated at broad end. 19·6 × 14·7 mm (18·0–21·5 × 14·0–16·5), n = 26; calculated weight 2·26 g (Schönwetter 1979). Clutch: 3–4 (Hüe and Etchécopar 1970). No further information.

Plumages (nominate *melanura*). ADULT. Entire upperparts backwards to rump uniform bluish ash-grey, feathers faintly fringed brown along tips in fresh plumage. Upper tail-coverts black, shorter ones at sides tipped ash-grey or white. Lores indistinctly mottled ash-grey and pale brown, sometimes slightly paler than forehead, sometimes slightly darker; black spot just in front of eye, extending in narrow line just below eye and sometimes faintly backwards along upper ear-coverts. Ear-coverts pale greyish-brown. Cheeks, chin, throat, and chest pale ash-grey, feather-tips slightly washed pale buff when plumage fresh; remainder of underparts cream-white or off-white, merging into pale grey on flanks; belly and vent slightly tinged pink or pale cinnamon in fresh plumage. Tail black. Flight-feathers, tertials, greater upper primary coverts, and longest feathers of bastard wing dark brown-grey; outer webs of flight-feathers narrowly and sharply edged pale grey or white (except emarginated parts of primaries), tips of secondaries and inner primaries narrowly edged white; tips and outer webs of tertials and greater upper primary coverts more broadly fringed ash-grey. Upper wing-coverts ash-grey, median with slightly duller and browner shafts, greater with duller and browner centres. Under wing-coverts white, lesser with some grey of feather-bases visible; axillaries pale grey. *In worn plumage*, upperparts paler ash-grey but with much dull grey of feather-bases visible, chin to breast and flanks pale grey, heavily mottled by dark grey feather-bases; pale edges and fringes of flight-feathers and wing-coverts partly worn off. Face and underparts sometimes contaminated with soil of environment, tinged rufous, brown, or dirty grey. NESTLING. No information. JUVENILE. Like adult, but feathers of head and body looser, more brown-grey or sandy-grey (less bluish ash-grey), upper tail-coverts black with brown tips, breast and belly cream-buff. Flight-feathers and greater upper primary coverts as adult, but off-white edges broader and less clear-cut; broad, rather clear-cut white fringes along outer webs and tips of greater upper secondary coverts (in adult, grey and less clear-cut). FIRST ADULT. Like adult, but juvenile flight-feathers, tertials, tail, greater upper primary coverts, and many outer or all greater upper wing-coverts retained; these have whiter fringes than in adult and are more subject to abrasion. Retained tertials and greater upper wing-coverts often distinctly browner and more worn than neighbouring median coverts and (if new) tertial coverts (in adult, all equally new and grey); greater upper wing-coverts and greater upper primary coverts distinctly worn January–March (in adult, still new and with smooth edges), heavily worn May (in adult, slightly worn); when bleached, sepia flight-feathers contrast with black tail.

Bare parts. ADULT, FIRST ADULT. Iris brown, dark brown, or sooty brown. Bill greyish-black, brown-black or black. Leg and foot black, soles sometimes pale horn-brown. NESTLING. No information. JUVENILE. Iris dark brown. Bill, leg, and foot brown. (Hartert 1921–2; Bannerman 1936; BMNH, RMNH.)

Moults. ADULT POST-BREEDING. Complete; primaries descendant. Data limited. In nominate *melanura*, 3 from late May and early June heavily worn, but not yet moulting; 10 from September completely new, but one from 8 September had p7-

p10 growing (score 40), one from 27 September had p9 still growing (BMNH, RMNH, ZFMK, ZMA). In Oman, nominate *melanura* in body and tail moult July–August, but *neumanni* perhaps not until later, October–November (Bundy 1986). In *airensis*, all birds heavily worn September, primary moult started; completely new mid-October (Bannerman 1936). POST-JUVENILE. Partial: head, body, lesser and median upper wing-coverts, and variable number of inner greater upper wing-coverts; flight-feathers, tertials, tail, outer or all greater coverts, and greater upper primary coverts retained. Of 18 birds from Sinai, September, 5 retained juvenile tertial coverts and all greater coverts, 11 retained greater coverts but not tertial coverts, 4 retained 1–5 outer greater coverts only; 1 had some tertials new (ZFMK). In Dead Sea area and Jordan valley, mid-July, some largely in juvenile plumage, others completely in 1st adult, but some feathers still growing. Moult completed in most birds from early September. (BMNH, RMNH, ZMA.) In *airensis*, moult perhaps complete; birds starting moult encountered mid-July (BMNH, ZFMK, ZMA).

Measurements. ADULT, FIRST ADULT. Nominate *melanura*. (1) Dead Sea area and Jordan valley, (2) Sinai; skins (BMNH, RMNH, ZFMK, ZMA). Bill (S) to skull, bill (N) to distal corner of nostril; exposed culmen on average 4·0 less than bill (S).

WING (1)	♂ 83·1	(1·33; 16)	81–86	♀ 78·3	(1·84; 12) 76–82
(2)	85·7	(1·78; 19)	81–89	81·0	(1·76; 11) 79–84
TAIL (1)	61·2	(2·62; 11)	57–65	56·7	(1·60; 10) 54–59
BILL (S) (1)	17·0	(0·70; 12)	15·8–18·1	16·9	(0·56; 9) 16·1–17·6
(2)	17·6	(0·50; 19)	16·8–18·4	17·0	(0·43; 11) 16·4–17·6
BILL (N)(1)	9·9	(0·73; 11)	8·9–11·1	9·9	(0·45; 10) 9·3–10·6
TARSUS (1)	24·2	(0·86; 11)	23·4–25·6	23·1	(0·89; 10) 21·8–23·9

Sex differences significant, except bill (S) (1) and bill (N). Retained juvenile tail of 1st adult similar to tail of full adult, but retained juvenile wing on average 1·1 shorter than adult wing; 1st adult combined with full adult in table above.

C. m. airensis. Northern Niger and Sudan, range of 9 ♂♂ and 8 ♀♀: wing, ♂ 78–82, ♀ 74–77, tail, ♂ 57–58, ♀ 53–55; exposed culmen, 12–13; tarsus, ♂ 23–24, ♀ 22–23 (Bannerman 1936). Ennedi (northern Chad) and Aïr (northern Niger): wing, ♂♂ 80·5, 81·5, ♀♀ 76·8 (1·19; 4) 76–78; bill to skull, ♂♂ 15·7, 16·8, ♀ 16·3 (0·38; 4) 16·0–16·8 (ZFMK). Single ♂ from Tibesti (northern Chad) rather large: wing 84, tail 59, bill to skull 16·1, tarsus 23·8 (BMNH).

Weights. *C. m. airensis*. Ennedi (northern Chad) and Aïr (northern Niger): ♂ 14–15 (4); ♀♀ 15, 15 (Niethammer 1955b; ZFMK).

Structure. Wing rather short, broad at base, tip rounded. 10 primaries: p7 longest, p8 0–2 shorter, p9 5–10, p6 0–1, p5 2–4, p4 7–9, p3 10–13, p2 12–16, p1 14–18. P10 reduced, 37–45 shorter than p7, 4–9 longer than longest upper primary covert. Outer web of p5–p8 and inner of p6–p9 emarginated. Tertials short. Tail rather short, tip slightly rounded; 12 feathers, t6 2–5 shorter than t1. Bill rather short and slender; straight but tip of culmen decurved towards sharply pointed tip and cutting edges slightly decurved. Nostrils oval, partly covered by membrane above. Rather long and strong bristles at base of upper mandible, finer bristles along frontal feathering and on chin. Tarsus and toe rather long and slender. Middle toe with claw 15·9 (10) 15–17; outer and inner toe with claw both c. 68% of middle with claw, hind c. 80%. Claws short and rather strong, decurved.

Geographical variation. Rather slight, clinal, involving general colour of body only. Nominate *melanura* (Levant) and *neumanni* (southern Arabia) rather pale and grey, *lypura* and *aussae* (western and southern Red Sea coasts) pale and brown. Colour gradually darker and browner through Sahel zone towards west: *airensis* intermediate, *ultima* darkest and strongly brown. Variation in size very slight: nominate *melanura* from Dead Sea depression and Sinai on average apparently slightly larger than others, *airensis* from Ennedi (Chad) slightly smaller. Colour of upperparts of nominate *melanura* bluish ash-grey; underparts pale ash-grey, merging into white on central belly, vent, and under tail-coverts. *C. m. lypura* from Red Sea coasts of Jebel Elba (south-east Egypt) to Eritrea (Ethiopia) pale sandy grey-brown on upperparts, only lesser upper wing-coverts purer grey (paler than in nominate *melanura*); chest and flanks dirty pale brown-grey or pale sandy-grey, merging into pale cream on central belly, vent, and under tail-coverts. Upperparts of *airensis* from eastern part of Sahel zone (northern Niger to central Sudan, north to Tibesti) light grey-brown or drab-brown; chin and throat buffish-grey; cheeks, chest, and flanks light sandy-brown, merging into cream-buff on central belly and to cream-white on lower vent and under tail-coverts (BMNH); birds from Tibesti and Ennedi (northern Chad) darker than those from Aïr in northern Niger (Niethammer 1955*b*; BMNH, ZFMK); may form a separate race. For extralimital races, see Bannerman (1936), Meinertzhagen (1954), Mackworth-Praed and Grant (1960), and Bundy (1986).

Forms species-group with Red-tailed Chat *C. familiaris*, Brown-tailed Rock-chat *C. scotocerca*, and Sombre Rock-chat *C. dubia*, all of north-east Africa, *C. familiaris* also extending south to southern Africa (Hall and Moreau 1970).   CSR

Saxicola rubetra **Whinchat**

PLATES 54 and 55
[between pages 736 and 737]

Du. Paapje   Fr. Traquet tarier   Ge. Braunkehlchen
Ru. Луговой чекан   Sp. Tarabilla norteña   Sw. Buskskvätta

Motacilla Rubetra Linnaeus, 1758

Monotypic

Field characters. 12·5 cm; wing-span 21–24 cm. Close in size to Stonechat *S. torquata* but somewhat slimmer, with less rounded head (appearing so partly due to face pattern) and noticeably longer wings. Small but quite robust, rather heavy-billed chat, with compact outline when perched and broad cruciform silhouette when wings extended in flight. Adult shows diagnostic combination of striking white line across inner wing-coverts and dark oval or diamond-shaped cheek-panel between long pale supercilium and pale lower border. ♂ vividly patterned, with black-brown crown, white supercilium, black-brown cheeks with white lower border, and warm buff underparts. ♀ similarly patterned but colours subdued, with supercilium and cheek-border buffier. Streaked brown and black above, with white bases to sides of tail. Immature duller than ♀, with even more subdued head pattern and no white line on wing-coverts. Sexes dissimilar; quite marked seasonal variation. Juvenile separable.

ADULT MALE BREEDING. From March, head black-brown with long, deep, white supercilium (reaching nape) and long, deep, white lower border to cheek ending in upturned patch at rear; pattern gives bird apparently rather flat-crowned head. Back dark brown, with black feather-centres appearing as spotted lines on rufous-brown background at close range; rump paler than back, with rufous-brown fringes to feathers dominant and dark centres reduced to small streaks and bars. Wing-coverts largely black but with bold white bar across centre and on tertials. Flight-feathers brown-black, with rufous-brown fringes to inner secondaries and small white patch on bases of outer primaries. Underwing basically grey-black, with buff hue due to fringes of axillaries and larger coverts. Tail almost black, with buff fringes to feathers and all but central pair showing white on bases (and along outer web of outermost feathers), creating small white triangles on sides of tail-base. Underparts from chin to rear flanks noticeably uniform orange-chestnut; vent and under tail-coverts pale buff. Spring migrants in East Africa and Middle East often strikingly drabber and greyer above and duller below than birds occurring in Europe. ADULT FEMALE BREEDING. Resembles ♂ but plumage less immaculate, with duller, rougher supercilium, browner or mottled cheeks, paler, more buff (less orange) underparts, and less distinct white line over inner wing-coverts and duller white triangles on sides of tail. ADULT MALE NON-BREEDING. From August, fresh plumage has striking contrasts reduced: supercilium tinged cream, cheeks flecked brown, mantle feathers and scapulars tipped buff or white, white line on coverts subdued, and orange underparts tipped and fringed buff, finely spotted and streaked dark brown on chest, while wing shows paler, buff fringes to tertials and narrow buff bar across tops of greater coverts. ADULT FEMALE NON-BREEDING. Closely resembles non-breeding ♂ but white line on coverts usually restricted to greater coverts. JUVENILE.

Much less distinctive than adult, inviting confusion with juvenile *S. torquata*. Dull, dark brown above, with black feather-centres and noticeably broad, pale buff or rufous-brown tips to mantle and scapulars; upper tail-coverts rufous-buff, little marked and appearing as small pale patch above tail. Face lacks full supercilium (being often indistinct in front of eye and never paler than buff behind it) and has cheeks much marked with rufous and buff feather-margins. Wing-coverts streaked and spotted rufous on outer and buff on inner, but no white line. Underparts basically buff, darkest on chest and liberally speckled or mottled dark brown on throat and chest. FIRST WINTER. From August, ♂ and ♀ resemble non-breeding adults except for partially retained wing-coverts and flight-feathers; thus no white line on wing-coverts. FIRST SUMMER. From March, resemble breeding adults but coverts much browner with white line only partially indicated. At all ages, bill and legs black.

Distinctive but not unmistakable. Dark pale-lined face, dark upperparts, and orange underparts of breeding bird present easy target, but much buffier, even sandy and rather uniform look of 1st-winter and non-breeding birds liable to invite confusion with eastern races of *S. torquata* (see that species, p. 737), and juvenile can closely approach appearance of juvenile of any race of *S. torquata*. Identity of juvenile confirmed by well-marked supercilium (at least behind eye), wide rufous-buff fringes on upperparts (suffusing back with sandy tone), and pale underparts (particularly throat); also, tail shows white triangles on sides of base (though some eastern races of *S. torquata* have white tail-base). Separation of juveniles of eastern forms of both species not studied in the field. Flight usually low, level, and fast, with noticeably faster wing-beats than wheatears *Oenanthe* creating whirring action and rather long, broad wings dominating silhouette which ends in rather short, square tail (occasionally appearing to have narrow base due to pattern). Looks strongly coloured in flight, with conspicuously warm glow above in non-breeding plumages, dark underwing usually obvious, and uniform underparts striking; white marks on wing of adult show as oblique panel across coverts and tiny mirrors against primary coverts and at all ages white bases to outer tail-feathers show when tail spread. Stance on ground typically half-upright but adopts more level posture when perched, rarely suggesting 'tiny guardsman' as does *S. torquata*. Gait a rapid hop, with occasional run. Less nervous than *S. torquata* but flicks wings and tail in excitement, bobbing in alarm. Perches at low level, preferring taller spray among ground cover of otherwise uniform height. Noticeably active at dusk. Can be sociable, with family parties obvious in early autumn, but solitary in winter.

Song variable in length, phrase, and quality, in part recalling *S. torquata*, Wheatear *Oenanthe oenanthe*, Redstart *Phoenicurus phoenicurus*, and even Bluethroat *Luscinia svecica*; always contains harsh introductory phrase, skirling notes, and silvery warble, with rather ventriloquial quality; warbling sequence can even suggest Whitethroat *Sylvia communis* or include jangling run like that of Corn Bunting *Miliaria calandra*. Usually sings from perch (often high, e.g. telegraph wire or tree-top) but also in brief fluttering flight. Commonest call a disyllabic, somewhat scolding 'tick-tick', like *S. torquata* but softer and often preceded by more fluent, even musical syllable—hence 'tuee-tick-tick'; also a more buzzing but still clipped 'tza' and a short, hard, rattling 'churr' in alarm.

Habitat. In west Palearctic, breeds largely to north of Stonechat *S. torquata*, in boreal and temperate but only marginally in steppe and Mediterranean zones of middle and upper middle latitudes, with more continental bias than *S. torquata*, although favouring moister and less rough habitats. General separation of range suggests avoidance of interspecific competition with *S. torquata* and greater use of seasonal growth of green vegetation such as bracken *Pteridium* and tall grasses or herbage in preference to permanent woody shrubs; where the 2 species breed near each other, *S. torquata* apparently dominant (Phillips 1970). Accepts sparser and less robust perches than *S. torquata*, often using posts, fences, or tall weeds; accordingly less dependent on heaths and moors and on coastal situations (except open dunes) and more attracted to grassy areas, including some farmland types, young conifer plantations with grassy ground cover, railway embankments, verges of quiet roads, fringes of wetlands, and grassy uplands to *c.* 500 m in Britain. (Sharrock 1976; Fuller 1982.) In Canton Bern (Switzerland), prefers level or gently sloping moist or wet terrain, such as meadows, depressions, and reedbeds free of flooding, but avoids cornfields or other crops, and meadows where hay cut before 1 July; thus, up to altitude of *c.* 670 m, concentrated in marshy areas (Lüps *et al.* 1978). On Swiss mountains, breeds up to *c.* 1800 m, exceptionally to 2000–2200 m in zone of scattered juniper *Juniperus* and other alpine shrubs (Glutz von Blotzheim 1962). In USSR, also breeds in glades, clearings, and burnt patches in forests, even where dense and unbroken all around; in mountains of Armeniya breeds up to 2230 m (Dementiev and Gladkov 1954*b*).

In winter in Africa, widespread in vegetated areas wherever open places occur with suitable perches and access to ground: on savanna north of Congo especially in cultivation; in Uganda up to 2300 m, especially in tall-grass areas, but in Kenya and Tanzania in fairly short grass with bushes (Moreau 1972). On coast, occurs in gardens and on farms; also in forest clearings and up to barest hilltops, and on open areas where crops have been cleared or bush recently cut (Bannerman 1951). In Sierra Leone, occurs on any open site, especially in grassland and grass woodland, even entering more open

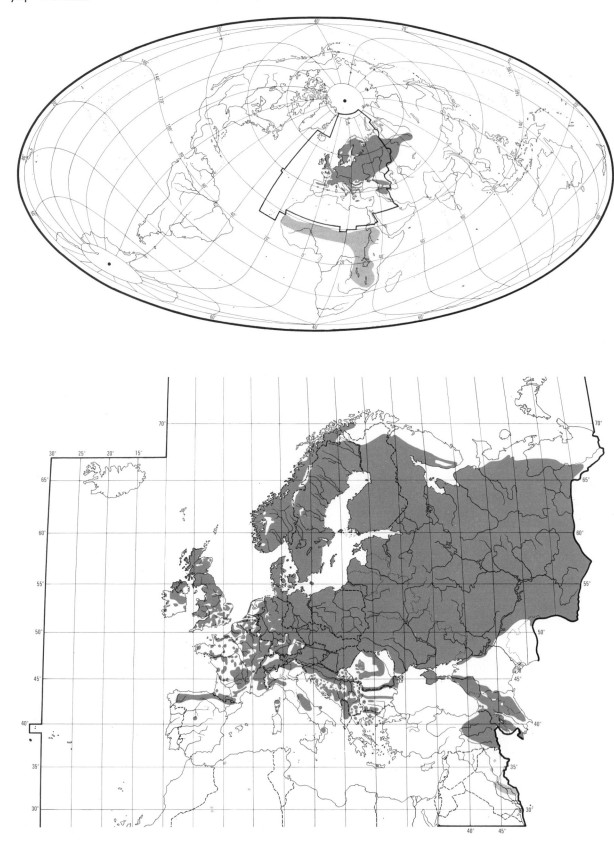

forest areas; once seen on top of tree over 25 m tall at end of logging trail; occasionally among smaller mangroves (G D Field).

Distribution. IRELAND. Breeds sporadically (Sharrock 1976). NETHERLANDS. Marked decrease in range since 1920-40 (CSR).

Accidental. Spitsbergen, Bear Island, Iceland.

Population. Declined markedly in Netherlands, Belgium, and France, less so in West Germany and Czechoslovakia. Recent increase after decline in Britain and perhaps Ireland.

BRITAIN. 10 000-100 000 pairs; declined in Midlands and south-east of England in last 50-60 years, probably partly due to habitat loss, but some increase in upland Britain due to increase in young conifer plantations (Parslow 1973). 20 000-40 000 pairs; some increase since early 1960s (Sharrock 1976). FRANCE. 10 000-100 000 pairs (Yeatman 1976); decline in north (Mouton 1984). BELGIUM. About 1000 pairs (Lippens and Wille 1972). NETHERLANDS. About 1000-1100 pairs (Teixeira 1979); marked decline caused by habitat changes and Sahel drought (Wammes *et al.* 1983; CSR). WEST GERMANY. 55 000-60 000 (Rheinwald 1982), perhaps too high (AH). Declining in some areas; for detailed counts see Bauer and Thielcke (1982). SWEDEN. About 350 000 pairs (Ulfstrand and Högstedt 1976). FINLAND. About 300 000 pairs (Merikallio 1958). Catastrophic decline in south in last 2-3 years (OH), but some increase in north (Väisänen 1983); decreased 1978-83 (Vickholm and Väisänen 1984). CZECHOSLOVAKIA. Decreased in 20th century, especially in lowlands; at higher altitudes, small increase after 1966 (Hudec 1957; KH). SWITZERLAND. Some decline (Géroudet 1983; Mouton 1984). GREECE. A few tens of pairs (GIH).

Survival. West Germany: annual adult mortality (breeding birds) 54%; annual 1st-year mortality 67% (Schmidt and Hantge 1954).

Movements. Essentially a trans-Saharan migrant, wintering in tropical Africa, though also regularly in Algeria (E D H Johnson) and Iraq (H Y Siman); other wintering records north of Sahara are exceptional but widely scattered through Mediterranean basin and western seaboard of Europe north to Britain. In West Africa, winters in wider latitudinal belt than most Palearctic migrants, occurring not only in savanna zone but also in cleared cultivated areas in other vegetation belts further south. Wintering range extends from Sénégal through Nigeria and Zaïre to Uganda, and uncommonly in Kenya and Tanzania. Small numbers reach Malawi, and in Zambia there were about 21 records to 1972, but including at least 6 at one site near Lusaka in January 1972 so probably winter regularly there (Tucker 1972). Also recorded a few times from Namibia and eastern South

Africa, but perhaps only a vagrant this far south. Zink (1980) summarized winter sightings in Africa and suggested that there are two main wintering regions: (1) from Sénégal to Cameroon, and (2) from north-east Zaïre and Uganda to Zambia. There is a gap of 10° longitude in Central Africa between these 2 regions where there have been only passage records, e.g. rare passage migrant in Chad (Newby 1980). Apparently also a gap between Kenya and Zambia, in which there have been very few records.

Birds leave north European breeding grounds in late August and September, with peak numbers on passage in western Europe in early September. Passage peaks at British west- and east-coast bird observatories and in continental Europe are simultaneous, indicating general departure over wide area at this time; suggestion of double peak of passage on west coast of Britain could refer to departures of 1st and 2nd broods (Riddiford and Findley 1981). Birds ringed in autumn in western and central Europe have moved south-west to western Iberia. Recoveries are in 2 main areas: (1) western France and northern Spain on the south-west edge of the Pyrénées, and (2) western Iberia, especially Portugal. Majority of recoveries are in September in western France and September-October in western Iberia. Very few recoveries in North Africa in autumn, but many in spring. For example, British-ringed birds have been recovered in autumn in France (8), Spain (17), and Portugal (16) with none in North Africa, but in spring in Morocco (6), Algeria (4), Tunisia (1), Spain (2), and France (2); French recoveries were in May but remainder in April (Zink 1973, data updated to 1983 inclusive). This suggests that birds cross Sahara in single flight from Iberia in autumn, but frequently make landfall in North Africa in spring.

Birds ringed as nestlings in Finland may move either through Iberia (15 recoveries) or Italy (10); main recovery period again September-October but 2 had reached northern Italy by August. 2 nestlings ringed in south-west Poland also recovered in northern Italy in August. The only autumn recovery in North Africa of a bird ringed in Europe was in north-west Morocco in October, ringed in eastern Poland. (Zink 1973.)

Fairly common in much of central and east Mediterranean in autumn, though 700 at 2 localities in Malta in late September 1969 was exceptional (Sultana and Gauci 1982). There is also a mid- to late-September peak on Cyprus (Flint and Stewart 1983). Widespread on passage in Turkey (*Orn. Soc. Turkey Bird Rep.* 1966-75), but only a few are seen at Azraq (Jordan) or in Iraq, and relatively scarce also in Arabia (e.g. Bundy and Warr 1980, Cornwallis and Porter 1982), so presumably this entire region is normally overflown from southern USSR. In north-east Africa, there is a rapid passage through Eritrea (Ethiopia) in late August and first half of September (Smith 1957), while birds are scarce in Somalia, with only 2 records in autumn and 6 in spring (Ash and Miskell 1983); hence this region also may normally be overflown.

First arrivals at wintering sites are in mid- or late September (e.g. Elgood *et al.* 1966, Morel and Roux 1966, 1973, L G Grimes). In Gambia, first record is 13 September, but mainly a passage migrant here in October–November and February–March, with the last date 9 May (Gore 1981). Recoveries of ringed birds from Nigeria to Tripoli (Libya) and to Poland, along with a bird ringed in April in Tunisia and found wintering in Togo (Elgood *et al.* 1966; Zink 1973) suggest that birds wintering in this region derive from central Europe. Distribution, abundance, and origins of birds wintering in central and East Africa are not well documented. Birds are relatively common in Uganda and as far east as Nairobi (Kenya), but scarce east of this, and Kenya-Tanzania rift appears to mark the normal eastern limit (Britton 1980). At Gabès (Tunisia), 0·84% of migrants ringed were subsequently recaptured there (Moreau 1972). RAC

Food. Invertebrates and some seeds. Hunts from perch, flying to and taking prey mainly from ground or in vegetation, sometimes in flight like flycatcher—e.g. in Sierra Leone in winter when tall grass hides ground on first arrival (Prokofieva 1980; G D Field). In experimental study in Sweden: foraging time per unit area decreased with increasing distance from nest; took higher proportion of food available near nest than distant from it; foraging area decreased when food density increased (Andersson 1981). In Scotland, pairs nesting close to pair of Stonechats *S. torquata* foraged further from nest; diet apparently similar, but included more adult Lepidoptera (Phillips 1970). Will forage in groups on migration: in Belgium, time between sallies decreased with increasing flock size (Draulans and Vessem 1982). Handles hairy caterpillars as does *S. torquata* (see p. 742), though handling time not recorded.

The following recorded in diet in west Palearctic. Invertebrates: mayflies (Ephemeroptera), dragonflies (Odonata), grasshoppers, etc. (Orthoptera), earwigs (Dermaptera), bugs (Hemiptera: Pentatomidae, Cicadellidae, Aphidoidea), adult and larval Lepidoptera (Noctuidae, Geometridae, Arctiidae), larval caddis flies (Trichoptera), flies (Diptera: Tipulidae, Stratiomyidae, Tabanidae, Rhagionidae, Asilidae, Empididae, Syrphidae, Tachinidae), Hymenoptera (sawflies Tenthredinidae, Braconidae, ants Formicidae, bees Apoidea), adult and larval beetles (Coleoptera: Cicindelidae, Carabidae, Dytiscidae, Silphidae, Scarabaeidae, Byrrhidae, Elateridae, Cantharidae, Tenebrionidae, Coccinellidae, Chrysomelidae, Curculionidae), spiders (Araneae), woodlice (Isopoda), centipedes and millipedes (Myriapoda), snails (Gastropoda), earthworms (Oligochaeta). Plant material: seeds and blackberries *Rubus*. (Rey 1908; Steinfatt 1937;

Witherby *et al.* 1938*b*; Kovačević and Danon 1952; Schmidt and Hantge 1954; Scott 1962*b*; Prokofieva 1980; Steinbacher 1981; Kostin 1983.)

Near Leningrad (USSR), July, 2 stomachs contained 34 items comprising 14 ants, 10 aphids (Hemiptera), 3 Curculionidae, 3 other beetles, 1 sawfly larva, 1 bee *Bombus*, 1 Braconidae, and 1 spider (Prokofieva 1980). In Crimea (USSR), April, 3 stomachs contained 21 items comprising 10 ants, 7 beetles (including 6 Tenebrionidae), 2 small bees, 1 shield bug (Pentatomidae), and 1 caterpillar; in September, 9 stomachs contained 27·8% beetles, 25% Formicidae, 16·7% Hemiptera, 2·8% Orthoptera, 8·3% unidentified insects, and 19·4% Solanaceae seeds (Kostin 1983). In Ural river floodplains (USSR), sometimes feeds exclusively on beetles but sometimes Orthoptera and bugs in addition to spiders, snails, and worms (Dementiev and Gladkov 1954*b*). In East Germany, May, 2 stomachs contained only beetles; 2 in September contained 7 beetles and 3 Lepidoptera (Rey 1908). In Yugoslavia, September–October, of 5 stomachs, all contained beetles, 2 contained ants, 1 Pentatomidae, and 1 spiders (Kovačević and Danon 1952).

Young in 5 nests near Leningrad given mainly whole medium-sized insects, including hard beetles, though small young given softer prey, e.g. small spiders and sawfly larvae. Beetles predominantly adult *Phyllopertha horticola* (Scarabaeidae) (13·5%); also Diptera, adult and larval Tenthredinidae, ants, mayflies, Orthoptera, Odonata, larval Trichoptera, leaf hoppers (Cicadellidae), millipedes, and molluscs (Prokofieva 1980). See also Steinfatt (1937) and Schmidt and Hantge (1954). JL

Social pattern and behaviour. Detailed studies by Schmidt and Hantge (1954) near Heidelberg (West Germany) and Géroudet (1957*b*) in Switzerland; summarized by Frankevoort and Hubatsch (1966).

1. Solitary in winter quarters. Territorial according to Serle (1949), though apparently not in Ghana (Grimes 1972). Flocks of up to *c.* 30 occur on migration (Harber 1948; Draulans and Vessem 1982), but usually less than 6–8 (Walpole-Bond 1938). BONDS. Monogamous mating system. Pair-bond sometimes renewed in successive years, but not always stable through season, 42% of 33 replacement nests involving change of partner; during nestling or fledgling stage, either partner may desert mate and re-pair with another bird, leaving former mate to rear young alone. One record of siblings pairing. Unpaired ♂ may join breeding pair, occasionally bringing food and defending young with calls (see below); only rarely chased off by breeding ♂. (Schmidt and Hantge 1954.) ♀ alone broods young (Groebbels 1950; Schmidt and Hantge 1954; Géroudet 1957*b*). ♂ accompanies ♀ collecting nest-material, driving her back at territory boundary (Schmidt and Hantge 1954). Both sexes feed young (Hellmich 1983). Parents accompany young after leaving nest, often dividing brood between them (Frankevoort and Hubatsch 1966). As soon as they can fly (17–19 days) family leaves territory and breaks up when young *c.* 26–28 days out of nest. Age of first breeding 1 year. (Schmidt and Hantge 1954.) BREEDING DISPERSION. Territorial and mostly solitary, though neighbourhood groups of pairs form around returning

breeders (Schmidt and Hantge 1954). Density highly variable: high on alpine meadows in Switzerland, e.g. 80 pairs per km² over 15 ha at 1000 m and *c.* 100 pairs per km² at 2000 m (Oggier 1979), and 62·5 pairs per km² over 16 ha (Manuel and Beaud 1982); low densities include 2·5 per km² over 78·5 ha in Alsace, France (Kempf 1982), 6–12 pairs per km² in Brecon, Wales (Massey 1972), 1·2 pairs per km² in Uppland, Sweden (Olsson 1947), 3–4 pairs per km² in Ayrshire, Scotland (Gray 1974). Density depends on habitat: e.g. in Vantaa (Finland), 2·1 pairs per km² in drained cereal fields, but 94·5 pairs per km² in abandoned fields (*n* = 113 pairs) (Tiainen and Ylimaunu 1984). Distances between nests generally small—less than 100 m in neighbourhood groups in West Germany (Schmidt and Hantge 1954), average 99 m (60–126), *n* = 7, in more or less linear habitat in Scotland (Gray 1973), though Horstkotte (1962) recorded average 330 m. Territory size less than *c.* 1 ha in Ayrshire, Scotland (Gray 1973), at least 0·75 ha in West Germany (Schmidt and Hantge 1954), average 0·43 ha in Netherlands (Frankevoort and Hubatsch 1966). Areas contract after spring boundary disputes, but may be greater late in season, when density lower (Schmidt and Hantge 1954). Few nestlings return to breed in natal area: none of 79 ringed in Switzerland (Géroudet 1957*b*) and 6·5% in West German study, only one such bird (♀) breeding in parents' territory, others settling over 250 m away. Of 54 breeding ♂♂, 47% returned to same area after 1 year, 15% for 2 years, and 4% for 3 years; in almost half such cases (for both ♂♂ and ♀♀) birds faithful to territory of previous year. (Schmidt and Hantge 1954.) Territory used for pairing, feeding, and nesting (Schmidt and Hantge 1954), though Gray (1973) saw food gathered outside, on hayfields. Andersson (1981) found most food was collected within 100–150 m of nest; one pair in Ayrshire, Scotland gathered food within *c.* 25 m (Phillips 1970). ROOSTING. Nocturnal. In vegetation on ground, at base of bushes (etc.); sometimes in family parties in autumn, but singly in winter quarters. Rather crepuscular, remaining active until dark. (Witherby *et al.* 1938*b*.) In more or less continuous daylight of arctic summer, adults feeding young rested for 3–4 hrs from midnight, similar to other species in same habitat (Lennerstedt 1973). Comfort behaviour includes bathing and sunbathing (Frankevoort and Hubatsch 1966).

2. Alarm-calls (2 in Voice) given in presence of birds of prey (e.g. Kestrel *Falco tinnunculus*) outside breeding season (P W Greig-Smith). FLOCK BEHAVIOUR. In flocks formed during migration stopovers, birds move independently, 7–10 m apart; higher rate of prey capture attempts in larger groups (Draulans and Vessem 1982). SONG-DISPLAY. Song of ♂ (see 1 in Voice) usually given from elevated perch, occasionally in flight. Sometimes sings while feeding (P W Greig-Smith). ♂ starts singing after pair-formation, with lull after laying (Frankevoort and Hubatsch 1966). Géroudet (1957*b*) noted song up to hatching (*c.* 20 July) in Switzerland. Song starts 85–100 min before sunrise, intermittent through day, occasionally after dark (Frankevoort and Hubatsch 1966). Song-flight similar to Stonechat *S. torquata*: bird flies up, singing, with body erect and tail spread, making tail pattern conspicuous; followed by drop back to perch (Frankevoort and Hubatsch 1966; Bergmann and Helb 1982). ANTAGONISTIC BEHAVIOUR. ♂ defends territory by Song-display (see above) and flight-displays (Schmidt and Hantge 1954): may fly at intruder with undulating flight, followed by 'excitement' variant of song (see 1 in Voice) given when perched, with wings opened and tail fanned (Fig A); or may fly slowly towards intruder, spiralling upwards on nearing it, and giving 'excitement' song. Either version may be followed by chase and physical fight during which screeching calls (5 in

A

Voice) may be given. After interaction, ♂ often peforms 'somersaults and rolls' in the air on returning to mate (Gray 1973). Fights particularly vigorous between ♂ arriving late and hoping to take up his previous year's territory and ♂ which has already occupied it (Schmidt and Hantge 1954). ♂ may or may not tolerate other adults near young (Groebbels 1950). Does not retaliate against chasing by *S. torquata* or Wheatear *Oenanthe oenanthe* (Greig-Smith 1982*e*), but dominant over Meadow Pipit *Anthus pratensis* and Reed Bunting *Emberiza schoeniclus* (Phillips 1970). HETEROSEXUAL BEHAVIOUR. (1) Pair-bonding behaviour. At start of season, ♂ follows ♀ entering his territory, singing quietly; ♀ reacts only weakly (Schmidt and Hantge 1954). Display by ♀ involves holding body horizontally with wings out and beaten rapidly (Fig B), but

B

never seen immediately before copulation (Frankevoort and Hubatsch 1966). No evidence of further pair-bonding behaviour before replacement clutch laid (Schmidt and Hantge 1954). Interactions between mates include gaping. (2) Courtship-feeding. Schmidt and Hantge (1954) twice recorded ♂ taking food to incubating ♀, and Groebbels (1950) saw ♂ feed ♀ brooding young 1–2 days old. (3) Mating. Copulation occurs on ground or in low bush, preceded by short chase (P W Greig-Smith); or ♀ may approach ♂ after a short, simple song (Thalmann 1981). Copulation may occur several times with short pauses between (Horstkotte 1962). (4) Behaviour at nest. ♀ chooses nest-site, though Eccles (1955) saw ♂ chase ♀ with nest-material, causing her to change site. Incubating ♀ leaves nest on hearing alarm-calls of ♂ (Schmidt and Hantge 1954). ♂ and ♀ bringing food to nest for young do so quietly; if undisturbed, call only rarely (Steinfatt 1937); show anxiety at such times by slow tail-cocking (Fig C). RELATIONS WITHIN FAMILY GROUP. Brooding continues for 2–3 days after hatching (Géroudet 1957*b*). Chicks open eyes on 4th or 5th day (Groebbels 1950; Frankevoort and Hubatsch 1966), and beg by stretching up and calling (Steinfatt 1937). ♂ and ♀ take

C

more or less equal share in feeding young and removing faecal sacs (Hellmich 1983); ♀ may visit without food, to take faecal sac (Groebbels 1950). Young leave nest before able to fly (Steinfatt 1937), then remain on ground for a few days; seen on bushes 2 weeks after leaving nest (Géroudet 1957b; Gray 1974), and continue to be fed by parents for 2–3 weeks after leaving (Géroudet 1957). Brood-division often occurs (Frankevoort and Hubatsch 1966). ANTI-PREDATOR RESPONSES OF YOUNG. Nestlings crouch when disturbed; after leaving nest, disperse under vegetation (Géroudet 1957b; Frankevoort and Hubatsch 1966). Intense parental warning-calls (see below) may cause young to leave nest early (Schmidt and Hantge 1954). Recently fledged young, when disturbed, may adopt upright, elongated posture with bill 60° above horizontal (Andrews 1981). PARENTAL ANTI-PREDATOR STRATEGIES. (1) Passive measures. Incubating ♀ sits very tight, more so as incubation proceeds, and can be caught on nest after day 9 (Schmidt and Hantge 1954; Géroudet 1957b). (2) Active measures. During incubation, ♂ and later both parents give warning-calls (see 2 in Voice) when human intruder c. 100 m from nest; call-rate increases with addition of Contact-calls (see 3 in Voice), as intruder approaches (Steinfatt 1937; Schmidt and Hantge 1954). Calling probably has same functions of warning young and distracting predator as equivalent calls of *S. torquata* (Greig-Smith 1980). Pair chased and mobbed Red-backed Shrike *Lanius collurio* near nest (Groebbels 1950), and Cuckoo *Cuculus canorus* near nest was vigorously attacked and driven away (Horstkotte 1962: Fig D). Exceptionally, ♂ may strike human at nest (Géroudet 1957b).

(Figs by D Rees: A adapted from photograph in Frankevoort and Hubatsch 1966; B from photograph in England 1978b; C from photograph in Hellmich 1983; D based on photograph in Edwards *et al.* 1949.)   PWG-S

D

Voice. Highly vocal in breeding season. Quieter in winter quarters, though occasional songs heard in Sierra Leone (Almond 1956; G D Field), Ghana (L G Grimes), and Nigeria (Marchant 1953). See Stadler (1952) for musical notation of song and calls. Additional sonagrams given by Bergmann and Helb (1982) and Schwager and Güttinger (1984); see latter also for detailed analysis of 56 songs.

CALLS OF ADULTS. (1) Song. Given by ♂ only; consists of short phrases (0·6–1·8 s long), given in lengthy bouts. Frequency range *c.* 2–8 kHz, slightly broader than song of Stonechat *S. torquata*, and includes many more low-pitched sounds (Schwager and Güttinger 1984), giving a fluting, harmonious, and pleasant-sounding character (Heinroth and Heinroth 1924–6; Stadler 1952). Phrases comprise units of very variable structure; each ♂ has extensive repertoire of units (over 100), largely different from neighbours (Schwager and Güttinger 1984, on which following analysis based). Units grouped into 12 types, including the following and intermediates: rattling, rasping sounds given in rapid succession; discrete whistles of rather uniform pitch, whistles of more variable pitch comprising 2 or more sub-units; units with rapid frequency modulation; harsh sounds of varied and complex form; unit (diads) indicating simultaneous use of 2 internal tympanic membranes; noises. Whistles comprise a high proportion of repertoire. In recording (Fig I), song shows, near start, a long continuous whistle, with pitch change, then chattering and rattling in middle section, and short tonal sounds at end except for penultimate rasping unit. A later song by same bird (Fig II) starts with pronounced rasps, ends with a rattle, and has distinctive portamento-type tones in between. (J Hall-Craggs.) Bird rarely repeats 'song-types' within singing bout (less so than *S. torquata*), and switches units so that changes in pitch are marked (more so than *S. torquata*), producing richly varied song, in contrast to more monotonous song of *S. torquata* (Schwager and Güttinger 1984). Notable for song mimicry of wide range of bird species (Groebbels 1950; Gray 1974; Thalmann 1981; Bergmann and Helb 1982). Fig III shows perfect mimicry of 'duple' call of Bullfinch *Pyrrhula pyrrhula* (1st 2 units), followed by long terminal flourish of Chaffinch *Fringilla coelebs* song. In Fig IV, first 1·4 s shows mimicry of Corn Bunting *Miliaria calandra*, remainder perhaps of low-frequency notes of Robin *Erithacus rubecula*. (J Hall-Craggs.) ♂♂ in one locality appeared to learn imitations from one another: thus, calls of Swift *Apus apus* and Nightingale *Luscinia megarhynchos* copied by one ♂ were in repertoire of neighbour a few days later, and in half the population after a week (Schmidt and Hantge 1954). Variants of normal song include Subsong, a simple quiet 'ziwüziwü' given before copulation (Thalmann 1981), and an 'excitement' song, comprising rapid repetition of rattling and scratchy motifs, in interactions with other ♂♂ (Schmidt and Hantge

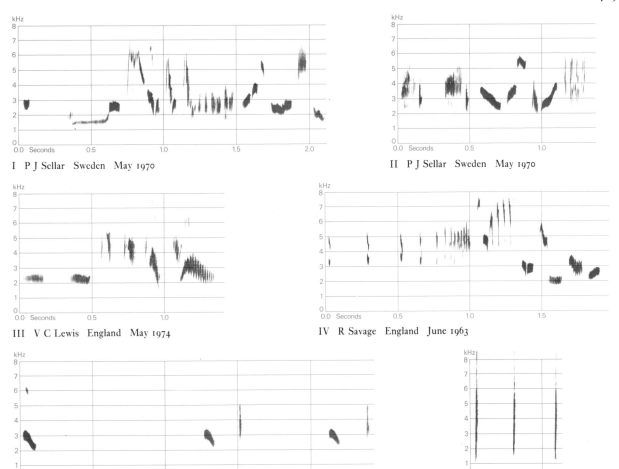

I   P J Sellar   Sweden   May 1970

II   P J Sellar   Sweden   May 1970

III   V C Lewis   England   May 1974

IV   R Savage   England   June 1963

V   P A D Hollom   Wales   July 1979

VI   V C Lewis   England   May 1960

1954). (2) Warning-call. Brief, whistled 'fiu' (Chappuis 1975), 'jüf', 'djü' (Bergmann and Helb 1982), or 'fu' (Groebbels 1950), weaker and lower-pitched than equivalent call of *S. torquata* (Chappuis 1975). Warns of (e.g.) approach of intruder towards nest. Given in alternation with call 3: e.g. in recording (Fig V) of ♀ with recently fledged young, 'fu(or'fiu') fu-tk fu-tk'; in this recording, 'fu' given alone (1st unit) apparently differs slightly in structure from 'fu' followed closely by 'tk' (J Hall-Craggs). (3) Contact-call. Harsh 'tec tec' (Chappuis 1975: Fig VI), 'tack', 'teck' (Groebbels 1950), 'tk tk' (Fig V: see call 2 for explanation), or 'zk zk' (Bergmann and Helb 1982); not as sharp as in *S. torquata*. Draws attention to bird, in distraction of predators and other contexts. (4) Anxiety-call. Ruttledge (1961) described a 'chup', 'chip', or 'tuup', considered unlike any call of *S. torquata*; used in situations of great anxiety. (5) Schmidt and Hantge (1954) described excited screeching calls associated with fights between ♂♂.

CALLS OF YOUNG. Incessant 'dscherr' food-calls given by nestlings when parents bring food (Bergmann and Helb 1982).

PWG-S

Breeding. SEASON. North-west Europe: see diagram. Little variation across range. SITE. On ground in vegetation, usually well hidden. Of 476 sites in Britain, 58% in open grass, 14·3% bracken *Pteridium*, 11·0% mixed low vegetation, 6·9% heather *Erica*, 4·8% gorse *Ulex*, 4·8% other low scrub, 0·2% stone wall; of 239 nests, 97% at or near ground level (0–15 cm), with highest 38 cm (Fuller and Glue 1977). Nest: cup of grass stems and leaves and moss, lined with finer material and hair. Building: by ♀ (Schmidt and Hantge 1954). EGGS. See Plate 82. Sub-elliptical, smooth and glossy; pale blue, with very fine speckling of red-brown (sometimes sparse, or concentrated towards broad end) giving rusty appearance. 18·9 × 14·3 mm (16·6–21·5 × 13·3–15·4), $n=250$; calculated weight 2·06 g (Schönwetter 1979). Clutch: (2–)4–7. Of 152 clutches, Britain: 2 eggs, 1%; 4, 8%; 5, 20%; 6, 63%; 7, 8%; mean 5·66; decline during

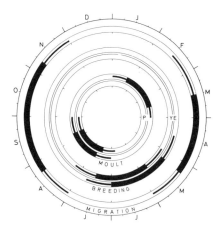

season from mean $5.88 \pm SE0.08$ ($n = 84$) in April–May to $5.31 \pm SE0.12$ ($n = 58$) in June–July (Fuller and Glue 1977). Of 111 clutches, Finland: 4 eggs, 4%; 5, 13%; 6, 45%; 7, 32%; 8, 6%; mean 6·2 (Haartman 1969a). One brood, occasionally 2. Eggs laid daily. INCUBATION. 12–13(–15) days. By ♀ only. Begins with last egg (Makatsch 1976). YOUNG. Cared for and fed by both parents (Schmidt and Hantge 1954). FLEDGING TO MATURITY. Leave nest at 12–13 days if undisturbed, starting to fly at 17–19 days. Independent at 28–30 days. Age of first breeding 1 year. (Schmidt and Hantge 1954.) BREEDING SUCCESS. In Britain, data from nest record cards suggest 72% hatching success and 94% fledging success, giving 86·5% overall (Fuller and Glue 1977). For West Germany, see Schmidt and Hantge (1954).

Plumages. ADULT MALE BREEDING. Entire upperparts sandy-buff, each feather with black central streak or spot rounded towards tip where narrowly fringed buff or off-white; when plumage fresh, March–April, forehead and crown appear black with buff scalloping, mantle and scapulars down to upper tail-coverts buff with rounded spots; nape often slightly greyer with less distinct dull black marks; fringes in some birds tinged pink-cinnamon (especially when plumage quite fresh), in others pale yellow-buff or greyish-buff (especially if worn). Very distinct broad white supercilium running from base of nostril to side of nape, sometimes continued as uniform buff line across upper nape. Narrow eye-ring white, broadly broken by black in front, more narrowly at rear. Sides of head rather variable; in some birds, lore, cheek, and ear-coverts fully black except for some olive-brown specks on ear-coverts and for distinct white stripe from base of lower mandible backwards to below eye (bordered above by black lore and below by narrow black malar stripe); more often, central ear-coverts extensively olive-brown with some black spots, rear part of ear-coverts mottled buff, and no white on cheek except for some white mottling near gape. Chin and side of throat white, forming conspicuous white stripe below black of cheek, extending into white patch at rear of cheek and on side of neck. Central throat, chest, sides of breast, and flanks uniform rufous-cinnamon to pale orange-buff (deepest when plumage fresh), paler tawny-yellow towards rear of flanks; central breast, belly, and vent cream or off-white, under tail-coverts pink-buff, pale buff, or buff-white. Central pair of tail-feathers olive-black with

indistinct grey fringes and some hidden white at base, basal $\frac{2}{3}$ of other tail-feathers white, distal $\frac{1}{3}$ olive-black (least extensively on inner web; inner web of t6 olive-black for 14–23 mm: Svensson 1984a); shafts often partly black, t6 with narrow white outer edge. Flight-feathers greyish- or olive-black; fringe along outer web of secondaries rufous-cinnamon (widest near base), narrow edge along tip of secondaries and along outer web and tip of inner primaries pale buff-grey or pale buff; edge along inner web of flight-feathers off-white (widest near base), basal $\frac{1}{4}$ of outer web of inner primaries white or buff-white, reaching c. 6 mm beyond tip of inner upper primary coverts. Central and inner median and longer lesser upper wing-coverts white, outer coverts and shorter lesser coverts black with faint buff fringe; tertial coverts (gc7–gc9) fully white, greater upper wing-coverts and tertials black with cinnamon-buff fringe along outer web and tip (soon bleaching to pale buff and partly worn off), inner web of inner greater coverts (gc5–gc6) sometimes white, basal half of outer web of tertials white. Greater upper primary coverts white, tip of outer coverts black for up to 3·5–6·5 mm, inner with less black on tip or (perhaps in birds older than 3 years: Schmidt and Hantge 1954) fully white; black of tips narrowly edged off-white. Under wing-coverts and axillaries cream or white with some black or grey on bases, greater primary coverts grey. In worn plumage, June–July, buff fringes on upperparts worn off, except for feather-sides of nape, inner scapulars, and lower mantle to upper tail-coverts, which appear streaked black and pale buff; forehead, crown, and sides of head almost fully black, supercilium even more markedly contrasting than in fresh plumage; throat and chest bleached to tawny-cinnamon or buff, breast and flanks to pale buff or dirty cream. ADULT FEMALE BREEDING. Upperparts as adult ♂ breeding. Supercilium long and broad, as adult ♂ breeding, but pale buff or cream, less white and slightly less contrasting. Narrow cream eye-ring. Lores and upper ear-coverts grey-brown or dark brown, lower ear-coverts buff with dusky spots and fine pale shaft-streaks. Cheek and side of neck buff with ill-defined dusky grey feather centres; front part of cheek sometimes largely cream bordered by black at lower rear; side of neck just behind cheek often with cream or white spot. Upper chin pale buff, gradually merging into warmer cinnamon-buff on chest and flanks (paler than adult ♂ breeding and without contrasting white stripe along lower cheeks); feathers at sides of breast partly suffused dull black; remainder of underparts pale buff to cream-white. Tail as adult ♂, but olive-black tips not sharply defined from white base, white tinged buff or pink subterminally of black (this less clear-cut division forming best sexing character at any age). Flight-feathers as adult ♂ breeding, but inner borders cream instead of white and pale base of outer web of inner primaries tinged buff and more restricted. Lesser and median upper wing-coverts like upperparts, but inner median and longer lesser coverts with triangular white tips (in adult ♂, largely or fully white). Greater upper wing-coverts and tertials as adult ♂ breeding, but ground-colour black-brown rather than black, and buff fringes slightly wider; tertials with restricted amount of pale buff on basal outer web; tertial coverts often partly white (see Svensson 1984a), forming pale patch on inner wing, but generally much less white than adult ♂ breeding. Greater upper primary coverts dull black or olive-black with narrow off-white fringe and some uniform cream or off-white on base. In worn plumage, upperparts blacker (but less so than in adult ♂ breeding, and hardly any black on sides of head), underparts bleached to pale buff. ADULT MALE NON-BREEDING. Upperparts as adult ♂ breeding, but fringes warmer rufous or cinnamon (less buff) and narrow fringe on each feather-tip of mantle, scapulars, upper tail-coverts, and sometimes crown white, giving spotted appearance. Sides of head as adult ♀ breeding, but buff supercilium and narrow eye-ring less contrasting. Underparts as adult ♀ breeding, but chest, sides of breast, and flanks deeper rufous-cinnamon, each feather with slightly paler fringe and contrasting black spot (spots occasionally narrow or virtually absent). Tail as adult ♂ breeding. Flight-feathers, tertials, and greater upper primary coverts like adult ♂ breeding, tertials with partly white bases and primary coverts with much white; remainder of upper wing-coverts rather like adult ♀ breeding, central and inner median and longer lesser coverts with variable amount of white on tip or base, tertial coverts white with tip partly or fully black (see Svensson 1984a). ADULT FEMALE NON-BREEDING. Upperparts and sides of head like adult ♂ non-breeding (upperparts thus warmer cinnamon than in adult ♀ breeding and with white spots); underparts either like adult ♀ breeding or chest deeper rufous-cinnamon (like adult ♂ non-breeding), with black spots or (sometimes) without these. Tail as adult ♀ breeding, pattern less contrasting than in adult ♂ non-breeding. Flight-feathers, tertials, and greater upper primary coverts like adult ♀ breeding, uniform pink-buff on bases restricted; upper wing-coverts as in adult ♀ breeding, white on inner coverts more restricted than in adult ♂ non-breeding or virtually absent, but variable in both and much overlap (see Svensson 1984a). NESTLING. Down fairly long and plentiful, on upperparts only; dark smoke-grey with paler bases (Witherby *et al.* 1938b). JUVENILE. Forehead, crown, lower mantle, scapulars, and back olive-brown or buff-brown with black spots, each feather with narrow contrasting pale buff or cream-white shaft-streak. Nape, sides of neck, and upper mantle pale buff with indistinct dusky mottling. Rump and upper tail-coverts rufous-buff with black spots and paler buff feather-tip. Supercilium indistinct, extending from just above eye backwards; pale buff or cream, like colour of crown-stripes. Lore, upper cheek, and ear-coverts buff-brown, finely spotted black on lore and upper ear-coverts, finely streaked cream on remainder of ear-coverts. Entire underparts including under wing-coverts pale buff or cream-buff, slightly deeper buff on throat, chest, and sides of breast, where feather-tips irregularly mottled dull black, sometimes showing as faint scalloping. Tail as adult (black tips clear-cut in ♂, less well-defined in ♀). Lesser and median upper wing-coverts and tertial coverts dull black with fine white shaft-streak and triangular white or cream spot on tip; greater coverts black (♂) or black-brown (♀) with broad rufous-cinnamon fringe along outer web and tip. Flight-feathers, tertials, and greater upper primary coverts mainly like adult non-breeding, but see 1st adult non-breeding below. FIRST ADULT MALE NON-BREEDING. Like adult ♂ non-breeding, but white on outer web of inner primaries restricted (reaching *c.* 3 mm beyond tip of inner primary coverts) and tinged buff, tertial coverts and bases of outer webs of tertials with less white (see Svensson 1984a), greater upper primary coverts white with black tip 7–11 mm long or largely black with buffish base hidden under longest feathers of bastard wing; tips of primary coverts with broader buff fringes than adult. Very similar to adult ♀ non-breeeding, but black of tail-tip contrasting. See also Bare Parts. FIRST ADULT FEMALE NON-BREEDING. Very similar to adult ♀ non-breeding (and to 1st adult ♂ non-breeding, except for less contrast in tail); virtually no uniform buff at base of greater upper primary coverts; tertial coverts black with buff fringes, without white except sometimes for broad fringe along tip (see Svensson 1984a). FIRST ADULT MALE BREEDING. Like adult ♂ breeding, but tertials, outer webs of inner primaries, and greater upper primary coverts with less white or pale buff on bases, as in 1st adult ♂ non-breeding (for

primary coverts, see Labhardt 1984 and Svensson 1984a; however, at least 5 of 30 1st adults examined had replaced all primary coverts and some birds inner primaries also and these feathers then similar to adult ones); white tertial coverts often partly or fully tipped black; some outer greater upper wing-coverts often retained from juvenile plumage, distinctly browner than neighbouring inner greater coverts or median coverts (in adult, all black or only slightly contrasting); chest rarely spotted black. FIRST ADULT FEMALE BREEDING. Like adult ♀ breeding and often indistinguishable; tertial coverts on average less white than adult ♀ (see Svensson 1984a).

Bare parts. ADULT, FIRST ADULT. Iris red-brown or dark brown. Bill black. Inside of upper mandible slate-grey to black in adult; pink-grey or yellow-white in 1st autumn, some darkening in October; sometimes still a little pink-grey in spring of 2nd calendar year or light grey in 2nd autumn (Svensson 1984a). Leg and foot black. NESTLING. Mouth pale orange shading to chrome-yellow, no spots. Gape-flanges pale yellow shading to ivory-white or cream-white. JUVENILE. Iris dark brown. Bill dark horn or greyish-black, cutting edges and base of lower mandible flesh-yellow, gape-flanges yellow. Leg and foot black-brown or slate-black. (Heinroth and Heinroth 1924–6; Witherby *et al.* 1938b; Harrison 1975; ZMA.)

Moults. ADULT POST-BREEDING. Complete; primaries descendant. In Britain ($n = 89$), starts with shedding of p1 between early July and 1st week of August, exceptionally from late June or up to *c.* 20 August; moult completed after *c.* 50 days with regrowth of p9–p10, mid-August to late September, before migration starts (Ginn and Melville 1983). Small samples from Netherlands, West and East Germany, Finland, and Yugoslavia fit British pattern (Stresemann 1920; Ginn and Melville 1983; RMNH, ZFMK, ZMA). Moult of body, tertials, and tail mainly between primary moult scores 10 and 40; head and secondaries during last stages of primary moult. ADULT PRE-BREEDING. Partial: head, body, lesser upper wing-coverts, most or all median upper wing-coverts, and tertial coverts; exceptionally some inner greater upper wing-coverts and tertials. January–March. POST-JUVENILE. Partial; extent rather variable. Some birds (probably early-hatched ones) replace head, body, lesser and median upper wing-coverts, and tertial coverts; others suspend moult at start of migration (late August and September) and these often retain juvenile tertial coverts, some (occasionally all) median and longer lesser upper wing-coverts, and sometimes part of body feathering (some scapulars or some feathers of belly, head, or neck). Moult starts soon after hatching, at age of *c.* 5 weeks, June–August; sides of breast, flanks, scapulars, and lesser upper wing-coverts first; completed at age of 9–10 weeks, from late July (Heinroth and Heinroth 1924–6; ZMA). FIRST PRE-BREEDING. Like adult pre-breeding, but less extensive: tertial coverts usually replaced, but variable number of 1st non-breeding outer median and longer lesser upper wing-coverts often retained, and occasionally most wing-coverts as well as part of body still in worn 1st non-breeding on arrival in breeding area. At least 5 of 30 1st adult ♂♂ had all greater upper primary coverts replaced, at least 2 had also replaced inner 4 primaries, and t1 sometimes replaced (ZMA); this moult perhaps in ♀ also, as single ♀ from January, Ivory Coast, had inner primaries replaced (Moreau 1972).

Measurements. ADULT, FIRST ADULT. Wing (1) Britain, Netherlands, central Europe, and migrants through north-west Africa (ZMA), (2) Dalmatia and Makedonija, Yugoslavia

(Stresemann 1920; RMNH, ZMA); other measurements combined, April to early October; skins. Bill (S) to skull, bill (N) to distal corner of nostril; exposed culmen on average *c.* 4·1 less than bill (S).

WING (1)	♂ 77·2 (1·88; 42)	74–81	♀ 75·9 (1·40; 20)	73–78
(2)	77·0 (1·83; 16)	74–81	75·6 (1·40; 13)	74–78
TAIL	44·9 (1·75; 46)	42–49	43·1 (1·42; 17)	41–46
BILL (S)	14·8 (0·59; 43)	13·8–15·5	14·8 (0·36; 18)	14·2–15·4
BILL (N)	7·8 (0·44; 37)	7·0–8·5	7·9 (0·36; 18)	7·1–8·5
TARSUS	22·2 (0·64; 45)	21·0–23·4	22·4 (0·38; 18)	21·8–23·1

Sex differences significant for wing and tail. 1st adult with retained juvenile flight-feathers and tail combined with older birds, though wing (not tail) of 1st adult slightly (not significantly) shorter; e.g. wing, adult ♂ 78·0 (16) 76–81, 1st adult ♂ 76·7 (42) 74–80; adult ♀ 76·7 (7) 75–78, 1st adult ♀ 75·5 (13) 73–78 (Stresemann 1920; ZMA). Live birds, Switzerland: wing, adult ♂ 76·3 (1·4; 112) 73–80, 1st adult ♂ 74·9 (1·0; 61) 70–80, ♀ 73·4 (1·8; 106) 68–77 (Labhardt 1984). Wing, Sweden: ♂ 73–83 (87), ♀ 71–78 (54) (Svensson 1984a).

Weights. ADULT, FIRST ADULT. Sexed birds, combined for whole geographical range: (1) April–June, (2) August–September (Makatsch 1950; Schüz 1959; Kumerloeve 1963, 1969a; Dolgushin *et al.* 1970; Rokitansky and Schifter 1971; Havlin and Jurlov 1977; RMNH, ZMA).

(1)	♂ 16·5	(1·54; 14)	14–19	♀ 16·7 (1·72; 6)	14–19
(2)	16·9	(1·92; 16)	15–22	15·8 (—; 2)	15·6–16

In Switzerland, ♂ on average *c.* 16 during May, decreasing to *c.* 15·4 mid-June to early July, increasing to 16·7 when migrating Col de Bretolet; ♀ 15·5 (11·1; 26) on arrival, 20·3 (1·9; 22) at laying, 17·4 (0·7; 8) during incubation, 15·4 (0·9; 18) when young small, 15·4 (1·3; 16) when young just fledged, 16·1 (59) on migration Col de Bretolet (Labhardt 1984; which see for graphs and many details). Minima of exhausted birds, Netherlands, August: ♂ 12·8, ♀♀ 10·1, 10·8 (ZMA); maximum of laying ♀ 23·5 (Labhardt 1984). Eastern Turkey, October: ♂ 27 (Kumerloeve 1968).

Unsexed birds. (1) Portugal, September (C J Mead). (2) Malta, autumn (J Sultana and C Gauci). Central Nigeria: (3) late October and November, (4) 15 March–15 April, (5) 2nd half of April (Smith 1966). (6) Northern Nigeria, April (Fry 1970b). (7) Gibraltar, spring (Finlayson 1981). (8) Malta, spring (J Sultana and C Gauci). (9) Southern England and Wales, April–May; (10) Fair Isle (Britain), May–early June (Smith 1966).

(1)	19·4 (2·73; 5)	15–22	(6)	19·2 (—; 11)	15–23
(2)	18·7 (2·8 ; 50)	13–24	(7)	16·0 (1·94; 10)	—
(3)	14·5 (0·81; 8)	—	(8)	16·7 (1·9 ; 50)	13–21
(4)	18·7 (2·67; 24)	—	(9)	16·6 (1·52; 72)	13–22
(5)	20·6 (2·87; 33)	13–26	(10)	14·6 (1·64; 27)	11–18

Structure. Wing rather long, broad at base, tip bluntly pointed. 10 primaries: p8 longest, p9 1·5–4·5 shorter, p7 0–1·5, p6 1·5–4, p5 5–9, p1 17–22. P10 reduced, 38–47 shorter than p8, 2 shorter to 3 longer than longest upper primary covert. Outer web of p6–p8 and inner of p7–p9 emarginated. Tertials short. Tail rather short, tip square, 12 feathers. Bill short, straight; rather stout at base, tip of culmen slightly decurved. Nostril small, oval; partly covered by membrane above and by feathering at base. Some long stiff bristles at base of upper mandible. Tarsus rather short and slender. Toes rather long and slender. Middle toe with claw 18·6 (25) 18–20; outer toe with claw *c.* 66% of middle with claw, inner *c.* 62%, hind *c.* 75%. Claws rather long, slender and sharp, decurved.

Geographical variation. Slight, clinal; colours gradually less saturated towards south-east and east, but some exceptions. Differences too slight and individual variation too marked for recognition of any races, though many proposed. In populations from Britain and Ireland, sometimes separated as *hesperophila* Clancey, 1949, from dark western end of cline, breeding ♂ has purer and more extensive black feather-centres on crown, mantle, and ear-coverts and more rufous feather-fringes on upperparts; juvenile rather dark brown (Clancey 1949); however, many inseparable from Scandinavian birds (Vaurie 1959). In south-west and southern Yugoslavia, dark feather-centres on upperparts on average slightly narrower and feather-fringes slightly paler yellow-brown, and these populations sometimes separated as *spatzi* (Erlanger, 1900), but again many birds inseparable from those further north (Stresemann 1920); size similar to birds from western and central Europe (see Measurements). Birds from Caucasus and Siberia, at eastern end of cline, sometimes separated as *noskae* (Tschusi, 1902), showing rather pale feather-fringes like *spatzi*, but dark feather-centres broader and size stated to be larger, wing up to 82 (Hartert 1910); however, large size not apparent in small Siberian sample examined (wing of 5 ♂♂ 76–79), and *noskae* not recognized, following Hartert (1921–2), Hartert and Steinbacher (1932–8), and Johansen (1954). Breeders from Van Gölü area (south-east Turkey) separated as *sengueni* Kumerloeve, 1969; described as showing rather broad dark feather-centres on upperparts (like *noskae*; broader than in central Europe), bordered by sandy-brown feather-fringes (Kumerloeve 1969a), but sample very small and, pending further research, not recognized here.

Recognition. In all plumages, separable from Stonechat *S. torquata* by wing formula and usually by longer wing; wing-tip formed by (p7–)p8 (in *S. torquata*, p6–p8), p9 1·5–4·5 mm shorter, p5 5–9, p1 17–22 (in *S. torquata*, p9 4·5–8 shorter, p5 1–3, p1 10·5–14); reduced p10 3 mm longer to 2 mm shorter than longest upper primary covert (in *S. torquata*, 4·5–10 mm longer); outer web of p6–p8 emarginated (in *S. torquata*, p5–p8). Broad pale supercilium and white tail-base do not occur in *rubicola* group of *S. torquata*, but are matched by some races of *maura* group of latter; these, however, show virtually unstreaked white or rufous-buff rump and upper tail-coverts, unlike *S. rubetra*.   CSR

Saxicola dacotiae Canary Islands Stonechat

PLATES 54 and 55
[between pages 736 and 737]

Du. Canarische Roodborsttapuit   Fr. Traquet des Canaries   Ge. Kanarenschmätzer
Ru. Канарский чекан   Sp. Tarabilla canaria   Sw. Kanariebuskskvätta

Pratincola dacotiae Meade-Waldo, 1889

Polytypic. Nominate *dacotiae* (Meade-Waldo, 1889), Fuerteventura; *murielae* Bannerman, 1913, Alegranza and perhaps Montaña Clara (off northern Lanzarote), now apparently extinct.

Field characters. 12·5 cm; wing-span 19-20·5 cm. Closely approaches size of Stonechat *S. torquata*, with similar form and with appearance strongly recalling its paler, eastern races. Endemic to eastern Canary Islands. Spring ♂ dark black-brown above, with narrow supercilium, broad white collar extending from throat, and white patch on inner wing-coverts; buff-white to pink-white below, with orange-rufous bib on breast. ♀ drab brown above, greyer white below, with indistinct supercilium, narrow white patch on inner wing, and only faintly buff chest. Pure white throat diagnostic; tail fully black. Sexes dissimilar; marked seasonal variation in ♂. Juvenile separable. 2 races (1 apparently extinct), probably not distinguishable in the field.

ADULT MALE. From about January to May (when plumage worn), crown, cheeks, and nape dark brown-black, with narrow white supercilium from base of bill to behind eye (but not reaching end of ear-coverts) and faint brown tips on crown. White chin, throat, upper breast, and collar contrast noticeably with crown, mantle, and chest-patch. Rest of upperparts dark brown, mottled and streaked pale grey; grey-white upper tail-coverts usually obvious, contrasting with brown-black tail. Wings black-brown, with paler, buff fringes to flight-feathers not obvious but white inner median and greater coverts forming oblique panel. Underwing white. Diffuse, roughly triangular orange-chestnut patch on chest, fading into buff-white flanks and white vent. From July (when plumage fresh), feather-fringes make crown, nape, mantle, and scapulars more uniform pale grey-brown, supercilium longer and wider but more diffuse, and white collar dirtier and interrupted at rear; chest and belly more uniformly coloured and with vinaceous tone. ADULT FEMALE. Much duller than ♂, suggesting flycatcher (Muscicapidae). Upperparts browner than ♂, particularly lacking black head, and chest only faintly tinged dull yellow. No white collar, and dull white supercilium and wing-panel less obvious. From July (when plumage fresh), more uniform pale grey-brown above and more buff below. JUVENILE. Resembles ♀ but throat dirty white, head and mantle spotted buff-white, chest speckled and barred brown, and tertials broadly fringed buff. FIRST-YEAR MALE. As adult, but black of head shows many brown fringes (even when worn) and underparts whiter. FIRST-YEAR FEMALE. As adult, but underparts whiter. At all ages, bill black-horn; legs black-brown.

The only small *Saxicola* chat breeding in Canary Islands (and never recorded elsewhere), but subject to confusion there with *S. torquata* (fairly regular on passage and in winter) and Whinchat *S. rubetra* (regular on passage). Best distinguished from *S. torquata* by white throat and narrow white supercilium, from *S. rubetra* by pale or only orange-patched underparts and all-black tail. Flight, carriage, and general behaviour apparently similar to *S. torquata* (including nervous tail-jerking), but runs as well as hops. Not shy, and can be inquisitive.

Song resembles that of *S. torquata*: a rather scratchy 'bic-bizee-bizeeu', etc.; also sings in flight with a repeated 'liu-liu-liu-screeiz', 1st part mellow, 2nd rasping. Commonest call a sharp 'chep' like *S. torquata*.

Habitat. Now restricted to Fuerteventura, an oceanic and subtropical island. Climate very warm with night temperatures exceeding 32°C, but windy on west coast which is mainly avoided. Widely distributed from mountains to seashore, mainly on steep, stony, and sparsely vegetated ground, frequenting both open hillsides and secluded shallow valleys or ravines. Found also on a lava stream or slope, on barren volcanic terrain, and low hills and stony plains, although not generally a plains bird. Sometimes in cultivation, fields, gardens, or verdant valley bottoms with tamarisks, but often favours stony sides of valleys or barren flats, and well-developed desert scrub, which offer such nest-sites as holes in walls or hollows under rocks or bushes, including *Euphorbia*. (Bannerman 1963; Collins 1984; Phillips 1986.)

Distribution. No information on range changes. Now found only on Fuerteventura (Collins 1984; Phillips 1986).

CANARY ISLANDS. Bred Alegranza until at least 1913, when an apparent family party was seen also on Montaña Clara (Bannerman 1914); not recorded on either island since.

Population. CANARY ISLANDS. 750 ± 100 pairs (Phillips 1986). Possibly increasing (Bannerman 1963).

Movements. Apparently sedentary, with no confirmed records away from breeding islands, though the only birds ever seen on Montaña Clara were not proved to have bred there (Bannerman 1963).   DJB

Food. Invertebrates. For flying insects, flutters from perch on (e.g.) rock or tree, usually returning to new perch. Also flies down to catch prey on ground: usually hops but will also run, pausing on low perch to swallow prey and look around (Collins 1984). Food for young sought mainly on ground (Phillips 1986).

Important prey items include Lepidoptera larvae, ants and ichneumons (Hymenoptera), flies (Diptera), and centipedes (Chilopoda) (Collins 1984); also beetles (Coleoptera) and spiders (Araneae) (Pérez Padrón 1983). Favourite prey said to be a large fly (Thanner 1905).

Food items brought to young were mainly terrestrial insects (D R Collins); included grasshoppers, etc. (Orthoptera), adult Lepidoptera, and a fly (Muscidae) (D A Hill).   JL

Social pattern and behaviour. Poorly known, though behaviour appears generally similar to Stonechat *S. torquata*. Account based largely on material supplied by D R Collins.

1. Solitary or in pairs or loose groups outside breeding season (F J S Jones). BONDS. Mating system apparently monogamous, but pair perhaps sometimes helped by extra ♂ which is not chased from nest (D R Collins) and may help parent(s) to escort fledged young (see below). Pair-bond lasts at least for duration of single breeding season (D R Collins). Some recently fledged parties of young were each accompanied by parents and additional ♂, or by 2 ♂♂ in adult plumage, or by parents and additional ♀ (C J Bibby and D A Hill). Only ♀ broods young (D R Collins), though parents take equal share in feeding them (D A Hill). Young of 1st brood tolerated in territory by parents with 2nd brood (D R Collins). Age of first breeding not known. BREEDING DISPERSION. Solitary and territorial. In suitable habitat (dry watercourses) reaches 4 pairs per km² (D R Collins); 9 adults plus 9–10 juveniles in 2-km strip (Shirt 1983); *c.* 8 pairs per km² in Vallebron area (D A Hill). Neighbouring nests 100–400 m apart (*n*=9); most more than 200 m (D R Collins), but may be as close as 40 m (D A Hill). Successive nests of one pair 25 m apart (Collins 1984). Territory defended only early in breeding season; later on, birds wander near neighbours' nests (D R Collins). ROOSTING. No information.

2. Not particularly wary. May call (2–3 in Voice), with wing- and tail-flicking, on approach of human intruder (Bannerman 1963). FLOCK BEHAVIOUR. In loose non-breeding groups, birds *c.* 30 m apart (F J S Jones). SONG-DISPLAY. Only ♂ sings (see 1 in Voice). No song heard in August (F J S Jones), and performances rare and incomplete in December, though increases sharply after first rain of winter (D R Collins). Most song given from perch (see call 1a). In Song-flight (see call 1b) bird flies up singing, may bob up and down briefly, then drops suddenly to boulder or bush (Collins 1984). ANTAGONISTIC BEHAVIOUR. (1) One ♂ chased own juvenile (age unknown) within territory, and another chased Spectacled Warbler *Sylvia conspicillata* near incubating ♀ (D R Collins). HETEROSEXUAL BEHAVIOUR. Courtship sequence observed twice, ♂ flicked tail slowly then flew at ♀, bowed, and presented his nape to her. One ♂ seen feeding ♀ on nest on day of hatching. (D R Collins.) No further information. RELATIONS WITHIN FAMILY GROUP. No conflict between siblings (D R Collins). Young leave nest after 16–18 days, then remain hidden in scrub for several days, giving food-calls. (Collins 1984; D R Collins.) ANTI-PREDATOR RESPONSES OF YOUNG. No information.

PARENTAL ANTI-PREDATOR STRATEGIES. (1) Passive measures. Incubating ♀ flushes on 1st alarm-call by ♂, or when human intruder very close to nest (D R Collins). (2) Active measures. ♂ and ♀ give persistent alarm-calls (see 2–3 in Voice), and may approach human intruder to within 1 m, wing- and tail-flicking. Also give alarm-calls in presence of Great Grey Shrike *Lanius excubitor* and ground squirrels *Atlantoxerus*; these mobbed if close to eggs or young. (Collins 1984; D R Collins.)   PWG-S

Voice. Very like Stonechat *S. torquata*. Quite vocal, during breeding season and when disturbed.

CALLS OF ADULTS. (1) Song of ♂. (a) When perched, a scratchy 'bic-bizee-bizeeu' with variants such as 'bizee-beeu'. (b) Song different in flight, containing mellow, lark-like 'liu' and loud rasping 'screeiz'. (Collins 1984.) (2) Contact-alarm call. Short (*c.* 20–40 ms in our recording), harsh 'chut' (Meade-Waldo 1889: Fig I, call

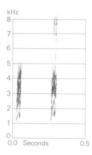

I   E D H Johnson   Canary Islands   April 1967

of ♂ as recordist approached nest with young), 'chep', 'chup', or 'tcheck' (D R Collins), given by both sexes, singly or in runs of 2–3 calls. Louder and sharper than call of *S. torquata* (Meade-Waldo 1889), and slightly louder than Berthelot's Pipit *Anthus berthelotii* in same habitat (F J S Jones). (3) Thin, high-pitched alarm call, 'seit' or 'suit', intermixed with contact-alarm calls (D R Collins). (4) In extreme anxiety, uses rasping sound similar to call 2 but quieter and more prolonged (D R Collins).

CALLS OF YOUNG. After leaving nest, fledglings give thin wheeze 'zirr' as food-call (D R Collins).   PWG-S

Breeding. SEASON. Earliest eggs in January, but most mid-February to late March, with start perhaps governed by onset of winter rains (Bannerman 1963; Collins 1984). SITE. On ground among stones and rocks or low down in wall, not more than 0·5 m above ground, often sheltered by overhanging stone or low bush, but can be exposed (Collins 1984). Nest: firmly built cup of plant stems and roots, including much *Salsola*; lined with goat hair; inner diameter *c.* 5 cm (Collins 1984). Building: by ♀ only. EGGS. See Plate 82. Sub-elliptical to short sub-elliptical, smooth and glossy; light green-blue, with fine speckling or mottling of red-brown, usually thicker towards broad end. 17·8 × 14·1 mm (16·0–18·8 × 13·5–14·7), *n*=14; calculated weight 1·88 g (Schönwetter 1979). Clutch: 4 (2–5) (Bannerman 1963). 3 complete clutches of 3 eggs recorded (Collins 1984). 2 broods recorded. Eggs laid daily. INCUBATION. 13–15 days. By ♀ only, beginning with last egg (Collins 1984). YOUNG. Cared for and fed by both parents. FLEDGING TO MATURITY. Fledging period 16–18 days (Collins 1984). BREEDING SUCCESS. Of 47 broods, mean size at fledging 2·3 ± SE0·2 (Phillips 1986).   DRC

Plumages (nominate *dacotiae*). ADULT MALE. In fresh plumage (about July–November), forehead, crown, and central hindneck pale grey-brown or greyish-olive, some black of feather-bases sometimes visible; mantle, scapulars, back, and rump similar, but feather-bases browner and less contrasting, sometimes visible as dark streaking; upper tail-coverts white with broad tawny-rufous tips. Broad white patch above lore; long but narrow white supercilium; rear part of supercilium often mottled dusky grey and pale grey-brown, less distinctly defined; due to dull black mottling, supercilium sometimes broken above front corner of eye. Lore pale grey-brown, feathering in front of and just below eye as well as shorter ear-coverts spotted white and dull black, upper and rear ear-coverts black with faint grey-brown tips. Chin and throat white, extending up into white band along sides of neck; white band partly mottled grey-brown and vinous-brown. Chest, sides of breast, belly and flanks vinaceous-cinnamon, darkest on chest, merging into white on lower belly, vent, and under tail-coverts. Tail greyish-black, feathers narrowly edged sandy-grey to off-white (widest on tips and along outer web of outermost feather). Flight-feathers, tertials, and greater upper primary coverts greyish-black, outer webs and tips narrowly edged pale sandy-brown or grey. Lesser and median upper wing-coverts black with pale grey-brown tips, greater greyish-black with pale sandy-grey fringes along outer webs and tips; inner longer lesser coverts and all tertial coverts largely white, forming contrastingly white wing-patch (often hidden under scapulars at rest, however). Under wing-coverts and axillaries white, some dull black of feather-bases partly visible. *In worn plumage* (about January–May), rather marked changes in colour due to abrasion of feather-tips. Forehead, crown, central hindneck, upper cheeks, and ear-coverts black, uniform or with limited amount of brown of feather-tips remaining; mantle, scapulars, back, and rump streaked dark brown and pale grey-brown; upper tail-coverts grey-brown with much white showing on shorter and lateral feathers (but white sometimes completely absent). Supercilium narrower, more sharply defined, and usually more restricted than in fresh plumage, often reduced to small white spot above lore and short white stripe above eye, occasionally (2 of 20 examined) entirely absent. Spot in front of eye and narrow ring round eye black, latter finely spotted white at lower border. Band along sides of neck distinct, fully white; sometimes hardly interrupted by dark grey-brown on central hindneck. Vinaceous-cinnamon of underparts bleached to cream or off-white on belly and flanks and to orange-cinnamon on chest; chest more contrasting with remainder of underparts than in fresh plumage. Pale fringes and edges of tail, flight-feathers, tertials, and upper wing-coverts largely worn off, ground colour of these feathers more brownish-black. ADULT FEMALE. In fresh plumage, entire upperparts pale grey-brown or pale greyish-olive, except for more tawny lower rump and upper tail-coverts; dark brown of feather-centres usually partly visible, showing as dark streaks. Lore, eye-ring, and front part of cheek buff, finely mottled grey just in front of and below eye; ear-coverts and sides of neck and breast grey-brown with indistinct buff streaking or spotting; some off-white of feather-bases sometimes visible along sides of neck. Chin and throat

cream-white; chest and upper flanks cream-buff, vinaceous-cream, or vinaceous-buff, merging into paler cream-pink on lower belly, vent, and under tail-coverts. Tail, flight-feathers, and wing-coverts as in adult ♂, but white wing-patch usually more restricted, tertial coverts partly black-brown. *In worn plumage*, upperparts closely streaked dark brown and grey-brown; upper tail-coverts hardly tawny. Lore and ring round eye pale buff, remaining sides of head, and sides of neck and breast grey-brown with faint paler mottling, broad supercilium slightly paler. Chest and upper flanks bleached to cream-yellow, belly to cream-white. Pale edges and fringes of flight-feathers, tail, and upper wing-coverts largely worn off. NESTLING. Down long and sparse, on head and back only; grey (D R Collins). JUVENILE. Upperparts dull brown-grey, marked with narrow and faint off-white spots on forehead and crown and with slightly larger but faint and poorly defined off-white or pale buff spots on mantle, scapulars, and back, there partly edged black along tip; sides of feathers of mantle and scapulars slightly darker brown; rump and upper tail-coverts uniform pale greyish-tawny, buff-brown, or pale rufous-brown. Narrow but distinct off-white supercilium, extending from near nostril to just above eye, but hardly over ear-coverts and sometimes absent above lores; distinct white or pale cream eye-ring. Lores and cheeks off-white with faint dusky grey mottling; upper ear-coverts brown-black, others dark brown-grey with narrow white shaft-streaks; sides of neck brown-grey with faintly darker feather-tips. Chin and throat off-white (faintly mottled grey at sides); chest and sides of breast pale rufous-cinnamon, buff, or cream-buff with narrow brown feather-tips, appearing scaly; remainder of underparts pale cream or off-white, slightly mottled grey on flanks. Tail, flight-feathers, and tertials as adult; tail greyish-black with narrow pale grey or sandy-buff edges along outer webs and tips, tertials with broad rufous-buff or pale buff fringes along outer webs and tips. Lesser upper wing-coverts like scapulars; median dark grey or dull black with ill-defined pale buff tips or triangular spots; greater dark grey or black-brown with larger rufous-buff (if fresh) to off-white (if worn) tips of 2–3 mm; shape of pale tips more triangular and with narrow pale shaft-streak on inner and tertial coverts; tertial coverts with variable amount of white. Greater upper primary coverts greyish-black with ill-defined pale grey or off-white fringes along outer webs and tips. Under wing-coverts and axillaries cream-white. FIRST ADULT MALE. Like adult ♂, but forehead, crown, and sides of head less uniform dull black in worn plumage, showing many brown fringes even if heavily worn; underparts less uniform vinaceous-cinnamon in fresh plumage, showing more contrast between cinnamon chest and cream or off-white belly; in worn plumage, entire underparts sometimes off-white, cinnamon of chest worn off. Juvenile flight-feathers, tail, tertials, greater upper primary coverts, and usually variable number of outer greater upper wing-coverts retained, these distinctly browner and more worn than those of adult at same time of year, especially tail sepia-brown rather than greyish-black, heavily abraded in spring. Retained juvenile coverts browner and off-white fringes along tips wider than in neighbouring blackish 1st adult ones. FIRST ADULT FEMALE. Like adult ♀, but entire underparts pale pink-buff in fresh plumage (in adult, chest, deeper vinaceous-buff or cream-buff, belly cream-pink) and virtually uniform off-white when worn (in adult, chest still buffish); on average, less white on tertial coverts, occasionally virtually absent; retained juvenile feathers as in 1st adult ♂, but no marked colour difference between juvenile and 1st adult feathers and hence more difficult to age; retained juvenile feathers more worn than neighbouring fresh 1st adult feathers, and retained juvenile greater upper primary coverts and in particular tail more heavily abraded than those of adult at same time of year.

Bare parts. ADULT, FIRST ADULT. Iris brown or dark hazel. Eye-lids pale orange. Bill dark horn-colour or black. Leg and foot dark brown or black. (Hartert 1910; Bannerman 1963; BMNH.) NESTLING. Bill dark horn with pale yellow cutting edges; gape-flanges pale straw-yellow (D R Collins). JUVENILE. Bill, leg, and foot blackish (D R Collins).

Moults. ADULT POST-BREEDING. Complete; primaries descendant. Of 26 birds from March, none moulting, though 1-year-olds in particular heavily worn. Single 1-year-old ♂ from 5 May had shed innermost primaries; some lesser coverts and t1 growing. Adult ♂ from 12 May had not yet started, though heavily worn; 1-year-old ♂ from 12 May had primary moult score 22 (inner 3 primaries new, p4–p6 shed or growing), but tail and body old except for some growing feathers on sides of breast. Single ♂ from 17 June had almost completed moult: score 43, and tail and body new except for some growing feathers. These scanty data indicate moult starts late April or May, completed June–July. POST-JUVENILE. Partial: head, body, lesser and median upper wing-coverts, tertial coverts, and variable number of greater upper wing-coverts (some outer usually retained). Starts shortly after fledging (mid-February to late April: Collins 1984); birds in full juvenile plumage examined from March and mid-May. Single bird from 30 May had body in 1st adult, but head and neck still largely juvenile; 2 from 17 and 28 June retained only limited amount of juvenile on head and neck, many new feathers growing; another from 28 June had moult completed. (BMNH, RMNH, ZFMK, ZMA.)

Measurements. ADULT, FIRST ADULT. Nominate *dacotiae*. Fuerteventura (Canary Islands), March–June; skins (BMNH, RMNH, ZFMK, ZMA). Bill (S) to skull, bill (N) to distal corner of nostril; exposed culmen on average 3·9 shorter than bill (S).

WING	♂ 62·8 (1·23; 23)	60–66	♀ 61·7 (1·34; 12) 60–64
TAIL	45·7 (1·58; 23)	43–49	45·2 (1·79; 12) 43–48
BILL (S)	15·9 (0·44; 23)	15·1–16·6	15·8 (0·43; 12) 15·0–16·4
BILL (N)	9·1 (0·41; 23)	8·4–9·9	8·9 (0·38; 12) 8·3–9·4
TARSUS	22·7 (0·49; 23)	21·8–23·5	22·7 (0·48; 12) 21·8–23·3

Sex differences significant for wing. 1st adult with retained juvenile flight-feathers and tail combined with adult, though wing and tail of 1st adult on average 1·2 shorter than in older birds.

S. d. murielae. Alegranza and Montaña Clara (Canary Islands). Wing, ♂ 63·5–67, ♀ 60–63; tail, ♂ 48–49, ♀ 48; exposed culmen, ♂ 11–13, ♀ 11–12·5; tarsus, ♂ 21–22, ♀ 21·5–23 (Bannerman 1963).

Weights. No information.

Structure. Wing short, broad at base, tip rounded. 10 primaries: p7 longest, p8 0·5–1·5 shorter, p9 4–7, p6 0–0·5, p5 0·5–1·5, p4 2·5–4, p3 5–7, p1 9–10. P10 reduced, 24–29 shorter than wing-tip, 7–11 longer than longest upper primary covert. Outer web of p5–p8 and inner of p6–p9 emarginated. Tertials short. Tail short, tip square; 12 feathers. Bill as in Stonechat *S. torquata*. Tarsus and toes rather long, slender. Middle toe with claw 16·7 (5) 16–17·5; outer toe with claw *c.* 73% of middle with claw, inner *c.* 68%, hind *c.* 79%. Claws rather short; sharp and decurved.

Geographical variation. Slight. *S. t. murielae* (apparently extinct) from small islet off northern Lanzarote differs from nominate *dacotiae* of Fuerteventura in more uniform rufous-cinnamon underparts without white on belly; crown slightly lighter, more red-brown, underparts more uniform vinaceous-buff in fresh plumage, chest-patch less sharply defined (Bannerman 1914; Hartert 1921–2; Vaurie 1959). However, nominate *dacotiae* also uniform vinaceous-cinnamon from chest to belly when plumage fresh, chest hardly darker than remainder of underparts; though contrast between chest and whitish remainder of underparts marked in heavily worn plumage, chest to belly still fairly uniform when slightly worn (March), especially in adult (1-year-olds more contrastingly coloured then); also, colour of crown feathers rather variable, and hence *murielae* rather a poor race.

Closely related to Stonechat *S. torquata*. Forms superspecies with *S. torquata*, Réunion Stonechat *S. tectes* (synonym: *S. borbonensis*) from Réunion in southern Indian Ocean, and White-tailed Stonechat *S. leucura* from India (Hall and Moreau 1970).   CSR

PLATE 53 (*facing*).
Saxicola torquata Stonechat (p. 737). *S. t. rubicola*: **1** ad ♂ worn (spring), **2** ad ♀ worn (spring), **3** 1st ad ♂ fresh (1st autumn), **4** 1st ad ♀ fresh (1st autumn), **5** juv. *S. t. hibernans*: **6** ad ♂ worn (spring), **7** ad ♀ worn (spring), **8** 1st ad ♂ fresh (1st autumn), **9** 1st ad ♀ fresh (1st autumn). *S. t. variegata*: **10** ad ♂ worn (spring), **11** 1st ad ♂ fresh (1st autumn). *S. t. armenica*: **12** ad ♂ worn (spring), **13** ad ♀ worn (spring), **14** 1st ad ♂ fresh (1st autumn). *S. t. maura*: **15** ad ♂ worn (spring), **16** 1st ad ♂ fresh (1st autumn). *S. t. stejnegeri*: **17** ad ♂ worn (spring), **18** 1st ad ♂ fresh (1st autumn).
Saxicola caprata rossorum Pied Stonechat (p. 752): **19** ad ♂ worn (spring), **20** ad ♀ worn (spring), **21** 1st ad ♂ fresh (1st autumn). (VR)

PLATE 54. *Saxicola rubetra* Whinchat (p. 722): **1** ad ♂ breeding, **2** ad ♀ breeding, **3** 1st ad ♂ non-breeding (1st winter), **4** juv. *Saxicola dacotiae* Canary Islands Stonechat (p. 733): **5** ad ♂ worn (spring), **6** ad ♀ worn (spring), **7** 1st ad ♂ fresh (1st autumn), **8** juv. (VR)

PLATE 55. *Saxicola rubetra* Whinchat (p. 722): **1** ad ♂ breeding, **2-3** 1st ad ♂ non-breeding (1st winter). *Saxicola dacotiae* Canary Islands Stonechat (p. 733): **4** ad ♂ worn (spring), **5-6** 1st ad ♂ fresh (1st autumn). *Saxicola torquata* Stonechat (p. 737). *S. t. rubicola*: **7** ad ♂ worn (spring), **8-9** 1st ad ♂ fresh (1st autumn). *S. t. maura*: **10-11** 1st ad ♂ fresh (1st autumn). *S. t. stejnegeri*: **12** 1st ad ♂ fresh (1st autumn). *Saxicola caprata rossorum* Pied Stonechat (p. 752): **14** ad ♂ worn (spring), **15-16** 1st ad ♂ fresh (1st autumn). (VR)

PLATE 56. *Cercomela melanura* Blackstart (p. 718). Nominate *melanura*: **1** ad ♂ worn (spring), **2** ad ♀ worn (spring), **3** juv. *C. m. lypura*: **4** ad ♂ worn (spring). *C. m. airensis*: **5** ad ♂ worn (spring). *Myrmecocichla aethiops* Ant Chat (p. 754): **6** ad ♂ worn (spring), **7** 1st-autumn ♀, **8** juv. (NA)

PLATE 57. *Oenanthe leucopyga* White-crowned Black Wheatear (p. 876): **1** ad ♂ worn (spring), **2** ad ♀ worn (spring), **3** 1st ad ♂ fresh (1st autumn), **4** 1st ad ♂ moulting into ad (1st summer), **5** juv. *Oenanthe leucura leucura* Black Wheatear (p. 885): **6** ad ♂ worn (spring), **7** ad ♀ worn (spring), **8** 1st ad ♂ fresh (1st autumn), **9** juv. (NA)

Saxicola torquata Stonechat

Du. Roodborsttapuit   Fr. Traquet pâtre   Ge. Schwarzkehlchen
Ru. Черноголовый чекан   Sp. Tarabilla común   Sw. Svarthakad buskskvätta

Motacilla torquata Linnaeus, 1766

Polytypic. *S. t. rubicola* (Linnaeus, 1766), western and southern Europe (except for range of *hibernans*), north-west Africa, Turkey (except mountains of east), and Caucasus and Transcaucasia east to a line from Groznyy to Tblisi, intergrading with *hibernans* in north-west France and western Belgium and Netherlands; *hibernans* (Hartert, 1910), Britain, Ireland, western Brittany, and west coast of Iberian peninsula; *variegata* (S G Gmelin, 1774), steppes of lower Volga and from mouth of Ural river south along west Caspian Sea to eastern Caucasus; *armenica* Stegmann, 1935, mountains of eastern Turkey, southern and eastern Transcaucasia, northern Iran east to Elburz and south through Zagros mountains to Kerman; *maura* (Pallas, 1773), northern and eastern European USSR and south-east shore of Caspian Sea east to middle Yenisey river, Lake Baykal, and north-west Mongolia, south to Altai, Tien Shan, Pamir, Kashmir, Afghanistan, and north-east Iran. Extralimital: *stejnegeri* (Parrot, 1908), eastern Siberia east from middle Yenisey and Lake Baykal, south to eastern Mongolia, Manchuria, Korea, and Japan, perhaps accidental in west Palearctic; *indica* (Blyth, 1847), Himalayas, grading into *maura* in eastern Kashmir and northern Punjab; *przewalskii* (Pleske, 1889), central and south-west China, west to eastern Kun Lun mountains and Tibet; *felix* Bates, 1936, south-west Arabia; *c.* 15 further races Afrotropics, including Madagascar and Comoro islands.

Field characters. 12·5 cm; wing-span 18–21 cm. Close in size to Whinchat *S. rubetra* but with 10% shorter, more rounded wings and noticeably round head (due partly to plumage pattern). Smallest widespread chat of west Palearctic, with virtually diagnostic character: large, round, busby-like head, compact form, upright stance, often bold behaviour (of ♂), constant nervous twitching of wings and tail, and whirring flight. When perched, recalls guardsman on sentry post; in flight, huge bumble bee. Adult ♂ has wholly black head (lacking pale throat of other *Saxicola* chats in breeding plumage) and chestnut breast. ♀ and (particularly) immatures far less distinctive, with most characters subject to complex geographical variation: most marked in decreasing saturation of fresh plumage (east to central Siberia), differences in underpart and tail patterns, and increase in facial contrasts in eastern races, which have pale supercilium and throat. Sexes dissimilar; much seasonal variation. Juvenile separable. 5 races occur in west Palearctic, separable in the field into 3 groups.

ADULT MALE. (1) Western races (Europe, south-west Asia, and north-west Africa), *hibernans* and *rubicola*. Distinguished by usually dull white, grey or buff underwing (not looking wholly black in the field), small, white-mottled, dull rufous or deep orange rump and upper tail-coverts, rather uniform, russet underparts,

noticeably dark brown or black head, and solidly brown-black or black throat. Spring and fresh autumn plumages not dissimilar, with wear producing evenly black head and upperparts, large isolated white panels on sides of neck, white rump streaked and mottled black, white centre of belly and under tail-coverts, darker underwing, and sharply defined white panel on inner median and greater coverts. (1a) Race of Britain, Ireland, western Brittany, and western coastal Iberia, *hibernans*. Darkest of all races, with least contrast between upper- and underparts; fringes of upperpart feathers and whole underparts markedly rufous, with collar and pale areas on rump and vent all restricted. (1b) Main European, Turkish, and west Caucasus race, *rubicola*. Averages paler than *hibernans*; fringes of upperpart feathers less rufous but not pale buff. Rump often whiter, especially on sides and in eastern birds, but not forming bold, uniform patch. (2) Caspian races (except south-east), *variegata* and *armenica*. No intergradation with easternmost *rubicola*, differing distinctly in larger size (with wings up to 15% longer), white base to tail extending up to half visible length (even more on some *variegata*), larger and whiter rump, huge white wing-panel, and white collar broken only on nape; less differentiated from Siberian races (and intergrading with *maura*), but size difference still marked (as *maura* is even smaller than *rubicola*). Spring and fresh autumn plumages dissimilar: from August, noticeably fringed buff on head, back, and wings and tipped buff on underparts. (2a) North Caspian race, *variegata*. Most variegated of west Palearctic races, with huge pale rump (only rarely showing orange spots), white tail-base (up to half visible length of feathers and sometimes extending even further on outer feathers), huge white shoulder-patch appearing to form collar, and pale (only rufous) breast. (2b) South Caspian race, *armenica*. Largest race; differs

PLATE 58 *(facing)*.
Oenanthe isabellina Isabelline Wheatear (p. 756): **1** ad ♂ (spring), **2** ad ♀ (spring), **3** 1st ad ♂ fresh (1st autumn).
Oenanthe oenanthe Wheatear (p. 770). Nominate *oenanthe*: **4** ad ♂ breeding, **5** ad ♀ breeding, **6** 1st ad ♂ non-breeding (1st winter), **7** 1st ad ♀ non-breeding (1st winter), **8** 1st ad ♂ breeding (1st spring), **9** juv. *O. o. leucorhoa*: **10** ad ♂ breeding, **11** ad ♀ breeding, **12** 1st ad ♂ non-breeding (1st winter), **13** juv. *O. o. seebohmi*: **14** ad ♂ breeding, **15** ad ♀ breeding, **16** juv. (NA)

from *variegata* in having only extreme base of tail white (not exceeding a quarter of visible length of outer feathers) and deep chestnut chest and flanks darker even than *hibernans* and contrasting with pure white belly. (3) Northern USSR, central Asian, and south-east Caspian races, *maura* and *stejnegeri* ('Siberian Stonechat'). Widely separated from *rubicola* but intergrading with *armenica* and *variegata*. Differ from other groups mainly in broad orange or partly orange rump, pale underparts, prolonged retention of streaked upperparts and pale fringes to inner secondaries and tertials, and black underwing. Spring and fresh autumn plumages markedly dissimilar: from August, conspicuously fringed sandy-buff or buff-white and creating much paler, lined or panelled appearance on back and wings; contrast of wing-panel much reduced and full contrast between upperparts and underparts not attained until June. (3a) West Siberian and south-east Caspian race, *maura*. Averages smaller than *rubicola* and *armenica*. No white visible on tail base. Breast much paler rufous, grading into pink-white vent; neck-panels divided by dark nape; back often with dull stripes, caused by retention of pale buff fringes to feathers. Rump large, usually white or pale orange, not streaked. (3b) East Siberian race, *stejnegeri*. Smaller and deeper coloured than *maura* (not larger and paler, *contra* Robertson 1977), with longer and broader-based bill, but intergradation with *maura*, particularly in worn plumage, makes field separation dubious at best. ADULT FEMALE. (1) Western races, *hibernans* and *rubicola*. Lack sharply demarcated pattern of ♂: head much mottled brown and thus only spotted and lined black, sometimes with vestigial pale buff supercilium, pale brown flecks under eye, and pale buff chin; white patch on neck much smaller; back and wings more olive, mottled and streaked black on mantle, with narrower white wing-panel; rump and upper tail-coverts dull, not contrasting with tail; underparts more uniform, paler russet. Spring and fresh autumn plumages not dissimilar, but wear produces marked loss of pale fringes to head, back, and wing-feathers (making them blacker, particularly throat), increases size and contrast of white neck-patch and wing-panel, and makes breast rather brighter. Racial differentiation untrustworthy in the field, but decreasing saturation of feather-fringes and feather-centres noticeable from darkest *hibernans* to palest *rubicola*; *hibernans* has most russet underparts. (2) Caspian races, *variegata* and *armenica*. Differ from western races as in ♂♂. Differ from ♂♂ of own races in smaller size of wing-panel and collar, somewhat paler (*armenica*) or much paler (*variegata*) upperparts (caused by less rufous, more yellow-buff or even pale sandy fringes and tips on head and back), and paler, more uniform underparts. In breeding season, buff fringes to upperparts much reduced and plumage thus noticeably darker though still lacking full contrasts of ♂, and ♀ *variegata* remaining much paler than ♀ *rubicola*. (3) Siberian races, *maura* and *stejnegeri*. Differ from western races as in ♂♂, and from western

races and Caspian races in having pale throat, additional pale wing-panel (caused by broad tertial-fringes), and pale supercilium (though this indistinct on many *maura*). Differ from ♂♂ of own races in having paler brown, sandier, or even buff-white fringes to upperparts, pale supercilium, throat usually uniform with rest of under-parts, duller but still contrasting rump, and noticeably pale sandy-buff margins to tail. Wear makes spring plumage noticeably darker in *maura* (which recalls *armenica* and paler eastern *rubicola*) and darker still in *stejnegeri* (which recalls *hibernans*). JUVENILE. All races. Resembles dull ♀, but has only speckled or pale throat and no obvious collar, being instead liberally spotted and streaked buff on head and back, washed with rufous on rump, and finely spotted and barred dark brown on chest and flanks. ♂ shows hint of white wing-panel, white bases to tail-feathers (in races possessing these), and somewhat darker underparts. FIRST WINTER. All races. From August or September, resembles autumn adult, though close view of wing-coverts may reveal retained dull juvenile feathers. Prominence of supercilium, pale throat, and tertial-panels in Siberian races probably tells of immaturity. At all ages, bill and legs black-horn.

Adult ♂ unmistakable, no other small ♂ chat being wholly black-headed. Adult ♀♀ of western and Caspian races also unmistakable, but adult ♀♀ of Siberian races and all juveniles subject to marked overlaps of characters. Also, ♀ and juvenile *maura* and *stejnegeri* resemble ♀ and juvenile *S. rubetra* in face and body pattern, and ♀♀ and juveniles of Siberian races and north Caspian race *variegata* may even suggest a wheatear *Oenanthe*. Separation of races discussed by Robertson (1977), but his criteria for *stejnegeri* now known to be wrong (see above and Geographical Variation). See Movements for racial allocation of eastern vagrants to western Europe. On Canary Islands, confusion possible with Canary Islands Stonechat *S. dacotiae* (see that species, p. 733). Confusion with ♂ Pied Stonechat *S. caprata* impossible but with ♀ all too easy (see that species, p. 752). Flight more whirring than flitting, with rapid, jerky bursts of wing-beats producing flat and direct flight; will hover; turns with great agility, ducking into cover or bouncing on to perch, immediately flicking wings and tail on landing. Flight silhouette noticeably short, with blob-like head, rather short and round wings, wide rump, and square tail. Gait a bouncing hop, but spends less time on ground than redstarts *Phoenicurus* or wheatears *Oenanthe*, feeding mostly from perch. Carriage most erect of all west Palearctic chats: unlike *S. rubetra*, often seems totally upright (especially when tail drooped and head stretched up) and rarely less than half-upright; uprightness emphasized by large head on plump body. Restless, inquisitive, and pugnacious in breeding season, with ♂ frequently working himself into 'fury' at disturbance, first flirting open wings and tail and then jerking tail up and down and constantly scolding; ♀ usually more

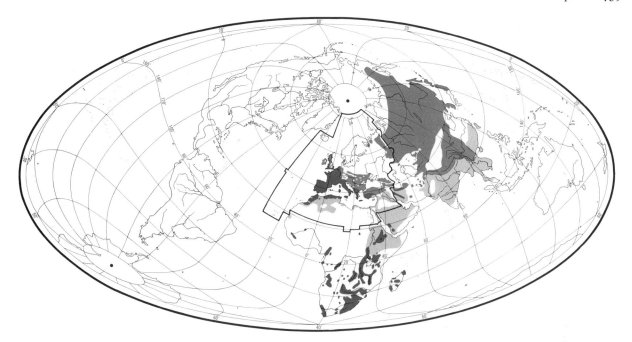

secretive and less aggressive. Behaviour less overt on passage and in winter: still perches prominently but prone to long escape-flights followed by skulking; this behaviour particularly marked in vagrants of eastern races to western Europe.

Noisy in breeding season. Song mostly a repetition of one clear and sharp and one lower-pitched and more throaty note, rising and falling rhythmically and recalling Dunnock *P. modularis* in overall pitch and volume. Calls include a plaintive 'hweet' and a most characteristic, insistent, hard 'tsak tsak' (particularly from scolding ♂); often run together into 'hwee-tsak-tsak' or 'hwee-tsak hwee-tsak', harsher notes sounding like clash of 2 pebbles. At least one other note recalls *S. rubetra* (see Voice for fuller discussion) but *S. torquata* far less noisy than that species on passage and in winter, many birds remaining silent even under pressure.

Habitat. Breeds in west Palearctic in middle and lower middle latitudes, in temperate, steppe, and Mediterranean zones, except for disjunct range of eastern race *maura* which extends into boreal zone of north-west of the region. Absent from high-altitude mountainous regions in north of range, and from high forest, wetlands, and open expanses which are bare or have only sparse or low vegetation. Not affected by lack of standing or running water. Within these limitations, inhabits wide variety of dry plains and hillsides, often submarginal for agriculture, characterized by scattered bushes, shrubs, stones, walls, or fences 1 m or more high, used as look-outs or song-posts commanding lower heathland, grassland, or bare patches. In Britain, prefers rough coastal areas where gorse *Ulex*, heather *Calluna* or *Erica*, and bracken *Pteridium* are interspersed with close-cropped grass; where grass grows tall and dense, Whinchat *S. rubetra* usually takes over, but *S. torquata* apparently dominant in shared habitat (Phillips 1970). Also uses young conifer plantations where ground cover is heather (Sharrock 1976), and breeds on open stages of sand-dunes and open shingle beaches where vegetation suitable, as well as scrub-clad slopes above low sea-cliffs (Fuller 1982). In Belgium, frequents borders of railways and canals and industrial sites (Lippens and Wille 1972). In Germany, found on sunny, more or less stony and dry hills and bushy slopes, sandy forest clearings, heaths, wasteland, and even on quite moist moorland or in osier stands (Niethammer 1937). In Switzerland, breeds mostly below *c.* 600 m, but nest recorded at 1450 m; favours sunny eroded terrain with low herbage and suitable perches, including vineyards, and scree at foot of mountains (Glutz von Blotzheim 1962). Siberian races *maura* and *stejnegeri* differ markedly in preference for moist meadows with plentiful grass and tall herbage and marshes with shrubs, as well as dry pines on stony mountain slopes, ascending to 3000 or even 4000 m (Dementiev and Gladkov 1954b). Winter habitat similar; in India, resorts to tamarisk thickets, reedbeds, bushes on dry mudflats along tidal creeks, and sand-dunes by sea (Ali and Ripley 1973b). Although restless and active, confines flight to lower airspace and usually to short distances, preferring it to terrestrial movement. Sometimes perches on treetops or telegraph wires, but usually on lower vantage points, with unobstructed field of vision.

Distribution. BRITAIN. Some recent spread inland (Sharrock 1976). NETHERLANDS. Some recent spread to

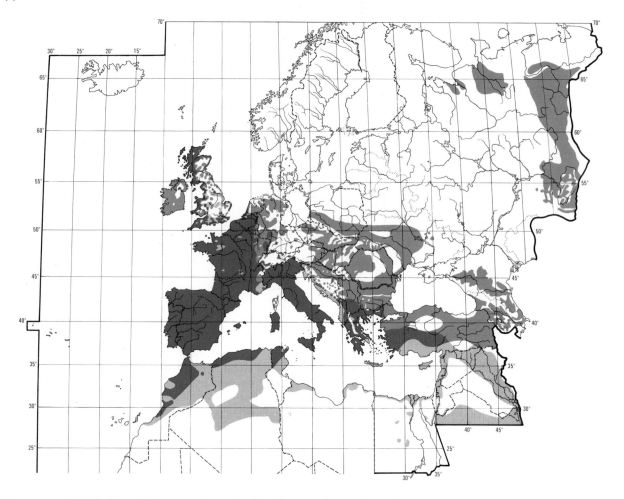

dune areas (CSR). EAST GERMANY. Irregular breeder. Bred Brandenburg 1961 and 1966, Thüringen 1963–71, and Oberlausitz 1960 (SS). NORWAY. Bred 1973–9, with up to 14 sites (Ree 1977; Munkejord 1981). TURKEY. Thinly distributed (Vittery *et al.* 1971).

Accidental. Iceland, Faeroes, Sweden, Madeira.

Population. Has declined Britain, Netherlands, West Germany, Czechoslovakia, and probably France.

BRITAIN. 1000–10 000 pairs (Parslow 1973). Marked fluctuations after hard winters and decline with habitat changes due to human activity (Magee 1965). Long-term decline in eastern England and locally elsewhere (Parslow 1973). 30 000–60 000 pairs; susceptible to hard winters (Sharrock 1976). FRANCE. 100 000–1 million pairs; some decline suspected (Yeatman 1976). NETHERLANDS. 2400–2600 pairs 1973–7 (Teixeira 1979). Declined (CSR). WEST GERMANY. 12 000 pairs (Rheinwald 1982); much too high, recent rapid decline (AH). Nordrhein-Westfalen: declined 65% mid-1960s to end of 1970s (Mildenberger 1984); declined 60% 1970–82 (Flinks and Pfeifer 1984). See also Bauer and Thielcke (1982) for recent estimates. DENMARK. 2–5 pairs (TD). CZECHOSLOVAKIA. Decreased

since 1960 and especially since 1970 (KH). SWITZERLAND. About 230 pairs 1977–9 (RW). Decreased 1972–6 after earlier increase (Biber 1984).

Oldest ringed bird 6 years (Dejonghe and Czajkowski 1983).

Movements. Varies from migratory to resident in different parts of range, being sensitive to cold winter weather.

European races, *rubicola* and *hibernans*, winter in south of breeding range, with notable concentrations in southern and eastern Spain, Balearic islands, and Algerian coast. Mediterranean populations apparently resident, North African probably so (Heim de Balsac and Mayaud 1962), though no evidence from ringing. Ringing at one site in Tunisia produced 6 retraps from one winter to next, but these were probably of winter visitors. *S. t. hibernans* (breeding Britain, Ireland, north-west France, and western Iberia) often stated to be resident (e.g. Vaurie 1959), though many British populations are at least partially migratory. Study in coastal Ayrshire (Scotland) showed that many, including some 1st-year birds, wintered within study area; 19 known movements from natal to wintering site varied from 1·5 to 17·5 km, average 8 km (Phillips 1976). On Jersey (Channel Islands), Johnson (1971*b*) found *c.* 14 winter territories regularly occupied in his study area, comprising in part the resident population; many 1st-years moved out in winter and there was an early spring return passage. Birds ringed as nestlings or during breeding season in Britain have been recovered in winter in western and southern France (3), Spain and Balearic islands (7), Portugal (1), and Algeria (1). Others leave inland sites to winter on south and west coasts. Recent colonization of western Norway thought to be by Scottish birds overshooting on spring migration (Ree 1977). Remaining European populations are mainly migratory, wintering in Mediterranean basin, with some extension south in North Africa. Autumn movements take place from September to early November and wintering areas generally vacated by end of March, although one record in Algerian Sahara on 12 April (Heim de Balsac and Heim de Balsac 1954). Between 1937 and 1964, there were only 0–12 known wintering sites in Belgium (Van Hecke 1965*a*). Majority of October–February recoveries from Belgium and Netherlands are in Spain (especially south-west), Algeria (6), and Morocco (3) (Van Hecke 1965*b*). A group of recoveries on western edge of Pyrénées in October and March suggests that main passage route may skirt the mountains. European ringing recoveries have been mapped by Zink (1973) and Van Hecke (1965*b*). Recoveries of birds ringed central Europe (Switzerland, eastern France, Germany) are all in same general area, but with increased tendency to winter on North African coast east to Tunisian border. One from Hungary travelled south-west to Italy by 9 September, and 2 from Czechoslovakia also moved south-west to Italy in September and October (Matoušek 1960; Kaňuščák and Kubán 1969). Of 26 recoveries in North Africa, all but 2 were near coast; recoveries in east were 2 from southern Germany and 2 from northern Italy (Zink 1973). Only a very few remain to winter in central Europe (e.g. Dobler and Stadelmann 1975, Dathe 1983). There is some extension of wintering into northern Sahara. In Algeria, Johnson (1971*b*) found the species occurring regularly to northern limits of Sahara. Birds were also present on winter territories at oases surrounding Grand Erg Occidental and west of Grand Erg Oriental (Oued Mya) in Algeria as far as 500 km into Sahara and south to about 29°N. They are also found to *c.* 28°N on Atlantic coast of Morocco (Zink 1973) and occur fairly regularly in western Canary Islands and as vagrant to Madeira (Etchécopar and Hüe 1967). However, no satisfactory evidence for migration across western Sahara, and the often-quoted occurrence in Sénégal (e.g. Dementiev and Gladkov 1954*b*, Etchécopar and Hüe 1967) presumably refers to local breeding race (Jarry and Larigauderie 1971). Other Saharan sightings documented by (e.g.) Blondel (1962*b*) and Germain (1965). Birds wintering commonly in central and eastern Mediterranean

are presumably from populations breeding in south-east Europe and Turkey, but no ringing recoveries. Dementiev and Gladkov (1954*b*) gave wintering area of birds breeding in Ukraine, Crimea, and western Caucasus as Mediterranean, Levant, Egypt, and Yemen; perhaps also Iraq (Meinertzhagen 1924*b*), though *maura* may be more common there (Sage 1960*a*). In Turkey, most inland breeding areas vacated and small numbers winter mainly in coastal areas. There are late February and March concentrations on south coast (e.g. 34 at Karataş, 9–12 March 1972) and there is also a September passage (*Orn. Soc. Turkey Bird Rep. 1970–5*). This race is a common winter visitor to Malta, where many retraps demonstrate winter site-fidelity (Sultana and Gauci 1982); also common on Cyprus, with occasional October to early November and late February to early March passage influxes of up to 50 birds (Flint and Stewart 1983). Common on Libyan coastal belt October–April, with occasional records in inland desert regions (Bundy 1976).

S. t. variegata (breeding in north Caspian area) is migratory, passing through eastern Transcaucasia (in September and March: Dementiev and Gladkov 1954*b*), Iraq, and Arabia (Vaurie 1959), and winter range extends to eastern Sudan and Ethiopia. Widespread in Eritrea where there are bushes, and recorded 9 September to end of March (Smith 1957). In Somalia, only one old record of 2 birds in September in north-west (Ash and Miskell 1983).

S. t. armenica (breeding in south Caspian area) apparently winters in Iraq (though not seen by Sage 1960*a*), western Arabia, Egypt, Sudan, and northern Ethiopia (Vaurie 1959). This race listed by Urban and Brown (1971) as probably an uncommon visitor to highland grasslands of Rift Valley and western highlands of Ethiopia.

S. t. maura (breeding western Siberia and south-east Caspian) is long-distance migrant, present on breeding grounds in south and centre of its range by mid-April to early May and beginning to leave Altai and other mountainous regions by August, but mainly in September, with stragglers until early October (Dementiev and Gladkov 1954*b*). Majority winter in northern India and north-west Burma. In Iraq, regular on passage (Moore and Boswell 1956; Sage 1960*a*), but winter status uncertain; Meinertzhagen (1924*b*) found none north of Baghdad November–January. Status in Arabia even less clear. One of this race was trapped on Cyprus in December and pale birds with white rumps (of uncertain race) are often seen (Flint and Stewart 1983). Clearly confusion elsewhere as well, and no mention of *maura* in recent East African literature, with only *variegata* and *armenica* reaching as far south as Ethiopia.

S. t. stejnegeri (breeding eastern Siberia) winters north-west India and Burma east to Indonesia and southern China. Arrives on breeding grounds in Baykal area early May, Ussuriland (easternmost USSR) in late April, Sakhalin mid- to late May, and Yakutiya mid-May, ♂♂ a few days ahead of ♀♀. Most leave Yakutiya during second half of August, and passage in Japan is late September and October with return in mid- to late April (Dementiev and Gladkov 1954*b*; Vorobiev 1954, 1963).

Both *maura* (mostly) and *stejnegeri* are vagrants to western Europe, and Robertson (1977) listed 25 records of these 2 races of which 16 in Britain and 10 in 1974; 24 in period 9 September–7 November. Apparent now, however, that at least some records of *stejnegeri* are incorrect, as (relatively) large and pale birds formerly ascribed to *stejnegeri* are probably *variegata*, *armenica*, or intergrades of these 2 races with *maura* (C S Roselaar); see Field Characters and Geographical Variation for identification criteria. *S. t. variegata* recently recorded in Norway in June (T Oxelsen and T I Bjønnes). RAC

Food. Small and medium-sized insects and other invertebrates. Locates terrestrial prey from elevated perch (e.g. bush), then flies, glides, or hops to ground, picking prey up on landing or while standing on ground; may return to same perch or new one (Moreno 1984*b*). In East Sussex (England), average perching height 1 m in spring, 1·6 m in summer with ♂♂ in spring tending to perch higher and closer to top of vegetation than ♀♀, and selecting higher perches than average available; spring perch preferences corresponded to those allowing highest chance of capture per visit to perch, but no perch-height preference shown in summer. Time on perches averaged 25 s with no differences between sexes or between spring and summer, but tended to be longer on higher perches (Greig-Smith 1983). In northern Spain, summer, median time on perch 11·7 s ($n=287$) if capture attempt followed, 13·9 s ($n=416$) if no capture attempt, though capture probability stayed about constant for at least 30 s; higher perches enabled searching of larger area—time spent searching from each perch (before moving to new one) and distance moved between perches both positively correlated with perch height; eventual flight to prey likely to be longer if searching time longer (Moreno 1984*b*). In Spain, most common technique in 368 capture attempts was hunting from perch for ground prey (80% of observations); also used aerial hawking (9%: prey pursued and captured in flight), hover-gleaning (6%: prey picked off substrate during brief hover), and flutter-pursuit (5%: locates prey from distance, flies rapidly to it, then flutters after it along ground or in air as it attempts to escape) (Moreno 1984*b*). In spring in East Sussex, only diving-to-ground technique used; in summer, also uses flycatching and snatching prey off foliage; change in behaviour linked to appearance of new prey types (Greig-Smith 1983). Also recorded repeatedly hovering for 2 s just above water and picking food from surface (Pounds 1944), hovering to feed on fish fry (H J Boyd), and dropping into water, submerging feet and

belly, apparently to pick food from water surface (Hodgson 1978). Once seen unsuccessfully to hammer snail on wall (Fisher 1979). Swallows hairy caterpillars whole, but first softens them up for 5–10 min by running them through bill from end to end and beating them on ground with forward or side-to-side head movements (King 1978*a*). Will feed near man, perching on newly turned drains, turves, or gardening tools following gardening activity (Bottomley 1978; Campbell 1980; Martin 1980). Foraging occurs either near or far from nest (Greig-Smith 1984). Pair in Scotland nesting close to pairs of Whinchat *S. rubetra* foraged mostly in heather *Calluna* and closer to nest—when young large, mostly within *c*. 25 m (Phillips 1970).

Diet in west Palearctic includes the following. Invertebrates: damsel flies (Odonata), earwigs (Dermaptera), bugs (Hemiptera: e.g. Pentatomidae), lacewings (Neuroptera), adult and larval Lepidoptera (e.g. Lycaenidae, Noctuidae, Arctiidae), adult and larval flies (Diptera: e.g. Syrphidae), Hymenoptera (e.g. sawflies Symphyta, Parasitica, ants Formicidae), adult and larval beetles (Coleoptera: Carabidae, Hydrophilidae, Staphylinidae, Scarabaeidae, Lampyridae, Tenebrionidae, Coccinellidae, Curculionidae), spiders (Araneae), woodlice (Isopoda), sandhoppers (Amphipoda), snails (Gastropoda), earthworms (Oligochaeta) and their eggs. Vertebrates: small fish, lizards up to 8 cm long. Plant material: seeds, blackberries *Rubus*. (Witherby *et al.* 1938*b*; Scott 1962*b*; Johnson 1971*a*; Bibby 1979; Fisher 1979; Martin 1980; Greig-Smith and Quicke 1983; H J Boyd, C J Colthrup, G Olioso.)

Few studies of adult diet, but for extralimital analysis of 60 stomachs from Kirgiziya (south-central USSR), see Pek and Fedyanina (1961). In Crimea (USSR), March, 2 stomachs contained 27 caterpillars, 12 adult and larval beetles (Carabidae, Tenebrionidae, Scarabaeidae), and 2 Hemiptera (Kostin 1983). For other data, see Baer (1909), Labitte (1944), Kovačević and Danon (1952), Hudec (1957), Ziegler (1966), and Kuz'menko (1977). Diet in autumn and winter said to include snails and exceptionally seeds (Dementiev and Gladkov 1954*b*): in Andalucia (southern Spain), ants comprised 54·3% of diet in autumn and winter (Herrera 1984*b*), and 2 stomachs from Jersey, January, contained exclusively Formicidae (Johnson 1971*a*).

Food of young in East Sussex studied from faeces. Of 121 faeces, May–August, 49% contained weevils (Curculionidae), 68% other beetles, 15% Syrphidae, 46% other Diptera, 40% shieldbugs (Pentatomidae), 11% other Hemiptera, 35% Parasitica (Hymenoptera), 18% bees and wasps, 8% ants, 42% larvae of Lepidoptera and Symphyta, 39% adult Lepidoptera, 15% earthworms, and 6% spiders; also earwigs, lacewings, damsel flies, and some plant material; proportion of weevils in diet declined from mid-July due to reduced availability. Adults thus took prey varying from cryptic immobile animals on ground or vegetation to large fast-flying flies and moths captured in flight. Most prey types occurring in habitat were taken but shieldbugs and caterpillars taken in much higher proportion than their abundance in habitat, and other Hemiptera and spiders in lower proportion; only 2 ladybirds (Coccinellidae) recorded in food but many warningly-coloured Hymenoptera. Large broods fed fewer adult Lepidoptera and more hoverflies (Syrphidae) than small broods. (Greig-Smith and Quicke 1983.) In Jersey, young fed small Diptera, unidentified larvae, Lepidoptera, and larvae of glow-worm *Lampyris noctiluca* (Johnson 1971a). Also recorded being given woodlice (Isopoda) (Bibby 1979) and small lizards (E L Turner).   JL, DJB

Social pattern and behaviour. Main studies of *hibernans* by Johnson (1961, 1971a) in Jersey (Channel Islands) and Greig-Smith (1980, 1981, 1982a, b, c, 1983, 1984, 1985) in Sussex (England); and of *rubicola* by Agatho (1960-1) in Netherlands, and Ziegler (1966) in Westfalen (West Germany). No obvious differences between races.

1. Mostly solitary or in pairs outside breeding season, though occasional feeding aggregations occur, e.g. *c.* 20 birds with *c.* 30 Linnets *Acanthis cannabina* (Liedekerke 1970). In Belgium 64% of wintering birds paired (Van Hecke 1965a), in England 62% (Greig-Smith 1979a); in both studies, ♂♂ more often solitary than ♀♀. Pairs which do not disperse after breeding (i.e. especially those on coasts) defend same or similar territories through winter (Labitte 1944; Parrinder and Parrinder 1944-5; Johnson 1971a; Hughes 1975; Phillips 1976; Bibby 1978; Phillips and Greig-Smith 1980). Local density sometimes high on migration routes and winter quarters: e.g. 3-20 birds per ha in Gibraltar (J C Finlayson); many contiguous winter territories less than 0·25 ha in south-east Spain (Johnson 1961). Usually, however, winter territories larger than breeding: 1·2-7·0 ha (*n* = 14) in Jersey (Johnson 1971a), 2-7 ha in Ayrshire, Scotland (Phillips 1970). During breeding season, parties of independent juveniles form wandering groups (*c.* 5-10) between territories (P W Greig-Smith). Mixed-species flocks form around solitary birds, especially in breeding season, taking advantage of greater vigilance of *S. torquata* (Greig-Smith 1981); such associations brief, involving 1-5 other birds, e.g. Meadow Pipit *Anthus pratensis*, Whitethroat *Sylvia communis*, Willow Warbler *Phylloscopus trochilus*, *A. cannabina*, and Reed Bunting *Emberiza schoeniclus* (Tallowin and Youngman 1978; P W Greig-Smith). BONDS. Mating system essentially monogamous, though small proportion of ♂♂ polygynous (Johnson 1961, 1971a; Frankevoort and Hubatsch 1966) or associate temporarily with 2 ♀♀ at start of breeding season (P W Greig-Smith). One record of polyandry, ♀ associating with 4 ♂♂ in succession, of which 3 cared for young (Ziegler 1966). Both sexes breed at 1 year old. Pair-bond generally persists through breeding season, though Ziegler (1966) noted mate changes between 1st and 2nd broods; ♂♂ failed to replace mates killed after start of breeding (Greig-Smith 1982c). Johnson (1971a) recorded breakdown of territoriality and bonds in Jersey over 3-4 days in March, as migrants arrived, with subsequent mate changes; not seen elsewhere, though immigrant ♂ may displace territory owner, retaining ♀ (P W Greig-Smith). Mates often change between years, even if both partners alive (Agatho 1960-1; Johnson 1971a; Greig-Smith 1982a). May remain paired outside breeding season (Greig-Smith 1979a). Pair-formation perhaps occurs on spring migration stopovers as *c.* 50% of birds arriving in

Westfalen (West Germany) were paired, including 1st-years (Ziegler 1966). Only ♀ broods young though both parents feed them, in nest and after fledging, taking equal shares on average, but this varies between pairs (Greig-Smith 1982c). Young usually fed by parents for *c.* 5 days after fledging (Johnson 1971a), though may be up to 30 days (Freitag 1943; Agatho 1960-1). See also Relations within Family Group (below). Young remain in territory until next brood hatch, and may leave unprompted (Freitag 1942, 1943) or be driven out by ♂ (Ziegler 1966; Johnson 1971a; P W Greig-Smith). BREEDING DISPERSION. Solitary and territorial. Heathland in Sussex (England) held 2·8, 5·6, and 9·6 pairs per km² in 3 years (Greig-Smith 1979a); 7·3 pairs per km² recorded in Dorset (Bibby 1978), and 7 pairs per km² in Jersey (Johnson 1971a). Mildenberger (1950) found a maximum 6 pairs per km² on heathland, but 8-12 pairs per km² in grassland in Rheinland (West Germany). Much higher densities also recorded: e.g. 33 pairs per km² in Asturias, Spain (Noval 1974), 31·5 pairs per km² over 40 ha of suitable habitat in Cornwall, England (Parrinder and Parrinder 1944-5), 179 pairs per km² over 14 ha of waste ground in Maastricht, Netherlands (Agatho 1960-1). Also low densities: e.g. 0·5-5·2 pairs per km² on Cape Clear Island, Ireland (Sharrock and Bromfield 1968). Breeding pairs on heaths in Sussex defend ranges of 0·8-*c.* 4 ha, *n* = 45 (Greig-Smith 1979a), but less on coastal cliffs, 0·3-1·0 ha, *n* = 15 (Parrinder and Parrinder 1944-5), and more on un-cultivated strips within farmland in Netherlands, 1·0-4·8 ha, excluding arable feeding areas (Agatho 1960-1). Though territories may adjoin, nests of neighbouring pairs usually more than 150 m apart (P W Greig-Smith); 400-750 m in study in south-west Veluwe, Netherlands (Bijlsma 1978c), *c.* 300 m in West Germany (Mildenberger 1950). Territory used for pairing, display, breeding, and feeding; nest-material often gathered outside. Portion used changes through season, due to distances between successive nests, which can be 900 m apart (Frankevoort and Hubatsch 1966) though usually less—in Sussex, up to 270 m, median 95 m, *n* = 70 (Greig-Smith 1982b), and in West Germany 15-150 m in Westfalen (Ziegler 1966), 50-180 m in Wetzlar, Hessen (Freitag 1943). Singing by ♂ concentrated near centre and at boundary (Greig-Smith 1982a). Year-to-year site fidelity low in Sussex, probably due to mortality, though Johnson (1971a) found frequent territory changes by survivors (and see Ziegler 1966). In Netherlands, 37% of both ♂♂ and ♀♀ returned to same area, but some moved territories (Agatho 1960-1; see also Frankevoort and Hubatsch 1966). ROOSTING. Nocturnal and solitary or in family group on territory (or in nearby reedbed: Freitag 1943). Breeding ♀ roosts on nest (Johnson 1971a; P W Greig-Smith), otherwise sites as in Whinchat *S. rubetra* (Witherby *et al.* 1938b: see that species, p. 727). ♂ may call brood to roost, gathering young singly or in twos (Ziegler 1966; M B R Eagles).

2. Highly vigilant due to high proportion of time spent on elevated perches (Greig-Smith 1983; Moreno 1984b); usually takes flight when human at *c.* 55 m (Greig-Smith 1981). Reacts to flying predators (Kestrel *Falco tinnunculus*, Hobby *F. subbuteo*, Sparrowhawk *Accipiter nisus*, Hen Harrier *Circus cyaneus*, Buzzard *Buteo buteo*, Great Grey Shrike *Lanius excubitor*) with 1-5 Warning-calls (see 2 in Voice) and hides in vegetation (Greig-Smith 1980). Johnson (1971a) observed ♂ freeze silently for 10 s as *A. nisus* flew past. FLOCK BEHAVIOUR. In flocks of juveniles (see part 1), birds chase and supplant one another, though usually keep at least 2 m apart (P W Greig-Smith). SONG-DISPLAY. ♂ sings (1 in Voice), perched on top of tree or bush, or (3% of phrases: Greig-Smith 1982c) in flight. Song-posts higher (median 1·8 m, *n* = 700) than perches

used while feeding (0·9 m, *n* = 353); therefore these activities rarely mixed (Greig-Smith 1979*a*). On perch, ♂ stands erect, head often raised and white wing-coverts exposed (Fig A).

A

Song-flights include phrases of typical song plus buzzes and whistles (see 4–5 in Voice); flight slow and jerky, with shallow wing-beats and periodic brief hovers, bird sometimes rising and falling several times without forward movement; keeps forebody slightly raised, tail lowered, and feet dangling, white rump and neck- and wing-patches conspicuous (Johnson 1971*a*; Panov 1973; Bergmann and Helb 1982; P W Greig-Smith: Fig B).

B

Mechanical whirring or rattling of wings described by Freitag (1943) and Bergmann and Helb (1982). Flights usually last less than 15 s, at heights of 10–25 m (Agatho 1960–1; Johnson 1971*a*; P W Greig-Smith); ♂ generally flicks wings repeatedly on landing (Greig-Smith 1982*e*). Most singing is in isolation or near to mate; rarely in territorial interactions, though Song-duels with other ♂♂ occur occasionally (Greig-Smith 1982*c*). Song heard March–July in England, occasionally up to mid-October (Frankevoort and Hubatsch 1966); starts in February in Jersey (Johnson 1971*a*), March in Brittany, France (Lebeurier and Rapine 1936*a*). Individual paired ♂♂ have initial song-period (up to 30 days) starting just after pairing and ending with start of incubation; later in season, shorter song-periods occur just before ♂'s own or neighbour's next clutch is laid (Greig-Smith 1982*a*). No obvious variation through the day (Greig-Smith 1982*a*). Unpaired ♂♂ sing at high rates, near to neighbouring ♀♀, but stop in mid-season and leave area (Greig-Smith 1982*c*). ANTAGONISTIC BEHAVIOUR. In early-season groups of ♂ with 2 ♀♀, most aggression occurs between ♀♀ (P W Greig-Smith). Territorial ♂♂ tolerant of intrusion before pairing (Frankevoort and Hubatsch 1966) and defence subsequently reaches peak near hatching (Ziegler 1966). Average rate of interactions in Sussex fell from 1 incident per 5 hrs in March to 1 per 20 hrs in May (Greig-Smith 1982*a*). Territorial ♂ chases intruding ♂♂ and physical contact sometimes occurs, accompanied by calls (4, 5, or 7 in Voice). Display includes wing-flicking 2–3 times after landing on perch,

especially at intermediate heights; also exposure of white wing patches, both slow and fast tail-flicking, and fanning of tail-feathers. All most frequent at start of season (Greig-Smith 1982*e*). Only rarely aggressive to intruding ♀♀ (Ziegler 1966). ♀♀ keep behind mates during interactions (Agatho 1960–1; Frankevoort and Hubatsch 1966; P W Greig-Smith). ♂♂ (and occasionally ♀♀: Zucchi 1973) aggressive to other species. This infrequent in heathland populations studied by Mildenberger (1950) and Greig-Smith (1982*f*), occurring with (e.g.) *A. pratensis*, *P. trochilus*, and *E. schoeniclus* (also see Agatho 1960–1); however, fierce and persistent against *S. rubetra* and Wheatear *Oenanthe oenanthe* on spring (but not autumn) passage. Aggression more frequent at upland site in North Wales (average 5 incidents per hr): dominant over *A. pratensis* and *S. rubetra* but subordinate to *O. oenanthe* (Greig-Smith 1982*f*, also Phillips 1970). Stuffed Cuckoo *Cuculus canorus* by nest was always violently attacked, though *S. torquata* is rarely parasitized (Agatho 1960–1). HETEROSEXUAL BEHAVIOUR. (1) General. Migrant ♂♂ and ♀♀ arrive on breeding grounds together, or ♀♀ up to 1 week later (Mildenberger 1950; Agatho 1960–1; Ziegler 1966; Greig-Smith 1979*a*). (2) Pair-bonding behaviour. Song-display (see above) probably important. Agatho (1960–1) and Frankevoort and Hubatsch (1966) described aerial 'dance' in which ♂ flies backwards and forwards up to 20 times over the back of ♀ on ground. ♂ also chases ♀ rapidly and persistently, sometimes making physical contact (Freitag 1943; Agatho 1960–1; Panov 1973; P W Greig-Smith). Chase sometimes followed by ♂ hovering above ♀, calling (see 2 in Voice) and displaying white patches on body, or by ♂ approaching ♀ on ground, with body low and wings drooped, occasionally 'bowing' to her (Freitag 1943; Mildenberger 1950; Agatho 1960–1; Frankevoort and Hubatsch 1966; Johnson 1971*a*). ♂ may fan tail, slowly raising and lowering it (Fig C). (3)

C

Courtship-feeding. Occasional records of ♂ feeding ♀, sometimes on nest (Mildenberger 1950; Linsenmair 1960; Panov 1973; Johnson 1971*a*: Fig D) but never seen by Agatho (1960–1) or Ziegler (1966). Frankevoort and Hubatsch (1966) saw ♂ pass food to brooding ♀ for transfer to young. ♂ sometimes flies to touch ♀'s bill, without food (P W Greig-Smith). (4) Mating. ♂

D

displays on ground, bowing with body low, head back, and wings and tail spread, displaying white patches (Freitag 1943; P W Greig-Smith). ♀ then solicits by crouching and shivering wings, giving quiet calls (Mildenberger 1950; Frankevoort and Hubatsch 1966: see 6 in Voice). Bill-touching once seen just before copulation (P W Greig-Smith). Mounting lasts a few seconds; usually occurs on ground or on perch up to 1 m high (Freitag 1943; Mildenberger 1950). Pre-copulatory displays occur from a few days before egg-laying (P W Greig-Smith). (5) Behaviour at nest. Agatho (1960-1) and Frankevoort and Hubatsch (1966) stated that ♀ chooses nest-site, but Doughty (1970, with captive birds) saw initial inspection by ♂, and ♂ once seen placing first piece of nest-material before ♀ took over building (P W Greig-Smith). Other records of ♂♂ occasionally with nest-material (Lebeurier and Rapine 1936a; Mildenberger 1950; Agatho 1960-1), and ♂ twice seen chasing ♀, apparently trying to snatch material from her (P W Greig-Smith). Incubating ♀ often leaves nest at call from ♂ (e.g. 2-3 in Voice) or even a song phrase (Mildenberger 1950; Frankevoort and Hubatsch 1966; Johnson 1971a; P W Greig-Smith). ♀ returns to nest from feeding place in a single low flight, often approaching covertly, hidden by vegetation; ♂ may then fly to nest, perching above. When ♀ sitting, ♂ occasionally hovers briefly in front of entrance (Ziegler 1966; P W Greig-Smith). RELATIONS WITHIN FAMILY GROUP. Nestlings fed on day of hatching (Agatho 1960-1; Frankevoort and Hubatsch 1966; P W Greig-Smith) and ♀ broods them for *c.* 5 days thereafter (Mildenberger 1950; Ziegler 1966), time on nest declining from 80% on 1st day to 50% on 4th day (Greig-Smith 1979a). Eyes open on 5th day (Groebbels 1950; Ziegler 1966). In nest, begging involves gaping, calling, and rearing up towards parent arriving at nest (Greig-Smith 1980). After leaving nest, beg by gaping and wing-shivering. Nestlings not aggressive, but in large broods compete for rear of nest where they lie partly on backs of siblings in front, enabling easier access to incoming parents, leading to faster growth (Greig-Smith 1985). ♂ and ♀ bring food equally on average (Agatho 1960-1; Greig-Smith 1982c); ♂ may delay feeding for a few days after hatching (Johnson 1971a), or may take larger share than ♀ for first few days (Greig-Smith 1979a). If both arrive at nest together, ♀ may give way to ♂ (M B R Eagles). ♂ and ♀ remove faecal sacs (Parrinder and Parrinder 1944-5), carrying them up to 50 m away, sometimes to habitual place (P W Greig-Smith). ♂ usually takes over feeding and defence of 1st or 2nd brood young while ♀ starts next clutch (Plucinski 1956; Agatho 1960-1; Ziegler 1966; Johnson 1971a; Greig-Smith 1980)—though not in study by Mildenberger (1950). ♂ may even continue to tend 1st brood while ♀ cares for 2nd-brood young (Ziegler 1966). In 3rd brood, ♂ may leave, ♀ rearing young alone (P W Greig-Smith). Young stay in nest up to 17 days if undisturbed (P W Greig-Smith), though usually leave after 12-13 days (Freitag 1943; Mildenberger 1950; Johnson 1971a), scattering under ground vegetation, being led by parents to nearby refuge under bush or hedge, average 5·5 m (*n* = 22) from nest (Ziegler 1966). They remain hidden for 4-5 days, parents bringing food, then start to follow parents. Brood-division may occur, if young are widely scattered (Agatho 1960-1; M B R Eagles). After *c.* 12-15 days (Johnson 1971a; P W Greig-Smith), ♂ and occasionally ♀ (Guldi 1965) may be aggressive to young; they leave territory after *c.* 6 weeks (38-54 days) out of nest (Freitag 1943). Dispersing young may move only just out of territory or further—e.g. on Sussex heath often 500-1000 m in first 3 months (P W Greig-Smith). ANTI-PREDATOR RESPONSES OF YOUNG. Parental Warning-calls (see 2 in Voice) inhibit begging, both in and out of nest (Greig-Smith 1980). If disturbed, older

nestlings may leave nest, crouch head-down nearby, and return after intruder gone (Groebbels 1950; Ziegler 1966; Johnson 1971a; P W Greig-Smith). When handled, nestlings give shrill calls, once causing ♂ to fly at intruder, and may defecate (P W Greig-Smith). PARENTAL ANTI-PREDATOR STRATEGIES. (1) Passive measures. Incubating or brooding ♀ may leave nest early if alerted by ♂'s Warning-calls (see below) or sit tight on nest until almost touched (P W Greig-Smith). (2) Active measures. The following based on Greig-Smith (1980). Reactions to birds of prey (see above) more likely if pair have fledged young; vocal response also well-developed towards terrestrial intruders (humans, dogs, etc.) and to corvids (Carrion Crow *Corvus corone*, Magpie *Pica pica*), but none made to grass snake *Natrix natrix*. Long series of Warning- and Distraction-calls (see 3 in Voice), often alternating, starts when intruder 100-120 m away, both calls increasing in rate as intruder approaches; as intruder withdraws, Warning-calls decline though Distraction-calls continue, combined with frequent short flights and wing- and tail-flicking. Calling rates change through nesting cycle—low during egg-laying and incubation, rising through nestling stage to peak (*c.* 40-50 calls per min) *c.* 2-3 days after young leave nest; in some pairs, also small temporary increase at start of incubation. ♂ and ♀ call equally on average, but pairs may show sex difference (Greig-Smith 1982c). ♂ approaches slightly closer to intruder at nest, responds earlier, and perches higher. More Warning-calls given if brood large, but overall level of distraction not related to brood size. Pairs calling at lower than average rates were more liable to predation, but high rates could also attract corvids.

(Figs by D Rees: A based on drawing in Bergmann and Helb 1982; B and D from drawings in Panov 1973; C from photograph by E Hosking in *Orn. Mitt.* 1965, 17, 146.)

PWG-S

Voice. Highly vocal during breeding season, notably in territorial advertisement (♂) and nest defence (♂ and ♀). Calls (mostly alarm) rarer outside breeding season, and birds almost silent in winter. For additional sonagrams, see Greig-Smith (1980, 1982c), Bergmann and Helb (1982), and Schwager and Güttinger (1984) who analyse structure of over 100 songs; Stadler (1952) gives musical notation of song and calls. Wing-whirring recorded as part of Song-display (Freitag 1943; Bergmann and Helb 1982: see Social Pattern and Behaviour).

CALLS OF ADULTS. (1) Song of ♂. Full song a variable, rather melancholy warbling sequence of phrases, sometimes delivered in 'scrappy' short bouts with unfinished phrases, but also unbroken sequences of several dozen phrases (P W Greig-Smith). Varies from *c.* 3-7 kHz. Compared to Whinchat *S. rubetra*, shrill (fewer low-pitched sounds) and monotonous due to (i) repetition of phrases (up to 12 times in succession) leading to smaller repertoire within singing bout; and (ii) smaller pitch changes between successive units. Frequency-modulated notes also used more than by *S. rubetra*. (Schwager and Güttinger 1984.) For further comparison with *S. rubetra*, see description of Fig II (below). Bird gives average 6·3 phrases before moving perch (P W Greig-Smith). Phrases 0·8-1·5 s long, separated by gaps of at least 2·0 s (Greig-Smith 1982c; Schwager and

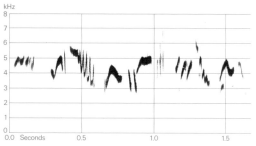

I R Savage England June 1982

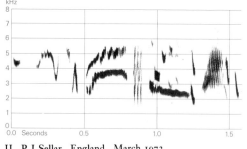

II P J Sellar England March 1973

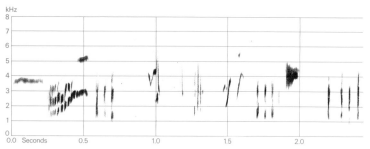

III V C Lewis Wales March 1973

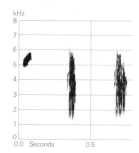

IV P J Sellar Scotland July 1977

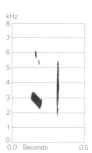

V P J Sellar Scotland July 1977

Güttinger 1984). Phrases divided into complex units, *c*. 8 per phrase (Bergmann and Helb 1982; Greig-Smith 1982*c*; Schwager and Güttinger 1984). Individual ♂♂ have large repertoires of units (up to 100 types), largely different from neighbouring ♂♂, though recordings from Germany and France show some identical units. Units grouped into 12 categories (for summary list, see *S. rubetra*, p. 728, which shares same unit classification). 'Whistles' form significant proportion of repertoire, producing melodious component of song. (Schwager and Güttinger 1984.) Songs given in flight (see Social Pattern and Behaviour) contain more whistles and are more varied than perched song (Frankevoort and Hubatsch 1966; P W Greig-Smith). Mimicry of other passerines occurs, less commonly than in *S. rubetra*. Thonnerieux (1980) heard ♂ copying Great Tit *Parus major* song on approach of human intruder; 2 ♂♂ in Sussex (England) copied nearby Wren *Troglodytes troglodytes*, Yellowhammer *Emberiza citrinella*, Meadow Pipit *Anthus pratensis*, Jay *Garrulus glandarius*, and *P. major* (P W Greig-Smith). In Fig I, song *c*. 1·6 s long, musical and much like Dunnock *Prunella modularis* (J Hall-Craggs; P J Sellar). In Fig II,

the 2 longest and most complex units are structured diads (indicating simultaneous use of 2 internal tympanic membranes); although there is a short rattle in the middle of the phrase, and a sort of rasping buzz just before the end, song conspicuously lacks the enthusiastic rasping and rattling of *S. rubetra* song. However, Subsong (Fig III), perhaps based partly on calls, contains 4 loud clear rattles, descending at times to less than 1 kHz, and is thus lower pitched than full song. (J Hall-Craggs.) Quiet incomplete courtship-song heard at dusk by Freitag (1943); see also Mildenberger (1950). (2) Warning-call. A short note (less than 100 ms): 'whit' (Johnson 1971*a*; Greig-Smith 1980), 'uit' (Lebeurier and Rapine 1936*a*), 'fid' (Bergmann and Helb 1982), 'sit' (Groebbels 1950), 'wiet' (Frankevoort and Hubatsch 1966). Frequency begins at *c*. 4 kHz, rises rapidly by *c*. 1 kHz, falling slightly before end of note (Greig-Smith 1980). In one pair, when human intruder near fledged young, ♂ called 'wheet', as above (Fig IV, 1st unit), but ♀'s call (Fig V, 1st unit) sounded more like 'fiu' of *S. rubetra*—*c*. 2–3 kHz, and descending throughout (J Hall-Craggs); not known if this a regular sexual difference. Used by both sexes, mainly to warn young when terrestrial predators present; also used by ♂ to call ♀ off nest, in alarm at flying birds of prey, and in courtship (see Social Pattern and Behaviour). Often mixed with Distraction-call (call 3) in long sequences (Greig-Smith 1980), as in Figs IV–V (see below). (3) Distraction-call. Harsh 'chack' (Johnson 1971*a*; Greig-Smith 1980), 'teck' (Groebbels 1950), 'trat', 'kr(r)' (Bergmann and Helb 1982), or 'träk' (Lebeurier and Rapine 1936*a*). Duration *c*. 100 ms, over broad range of frequencies between *c*. 0·5 and 7 kHz, with little or no modulation (Greig-Smith 1980; Bergmann and Helb

1982). In Fig IV (♂), this call the last 2 units, rendered 'chat'; in Fig V (♀), terminal 'tek' (J Hall-Craggs). Draws attention of intruder, and, with wing- and tail-flicks and frequent perch-changes, distracts intruder away from nest (Greig-Smith 1980). Also used in courtship and territorial interactions (Johnson 1971*a*; P W Greig-Smith). (4) Brief, hoarse, rattling 'krrrr', occasionally used in Song-flights (Panov 1973; P W Greig-Smith); less often by perched ♂ in interaction with neighbour near start of breeding season. (5) Drawn-out 'weee-eee', up to *c.* 1·5 s; used occasionally in courtship, territorial interactions, and other social contexts (Agatho 1960-1; Johnson 1971*a*; P W Greig-Smith). (6) Soliciting-call. Quiet 'sisisisisi' given by ♀ soliciting copulation with wing-shivering (Freitag 1943). (7) Aggressive-call. In fight between territorial ♂♂, one gave 'chee-chee' when chase ended in physical contact (P W Greig-Smith).

CALLS OF YOUNG. Food-call of nestlings an excited, continuous 'shee-shee-shee-shee'; increases in volume with age (Greig-Smith 1980). When handled, nestlings may give short scream of distress, causing higher calling rates by parents. Contact-alarm call of fledged independent young is a harsh, slightly trilled 'prrrt', rather similar to adult call 3; its use ceases before 1st winter (P W Greig-Smith). PWG-S

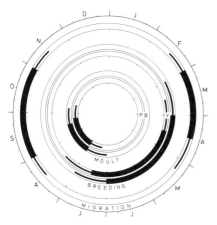

Breeding. SEASON. North-west Europe: see diagram. From mid-April in south and central Europe (Makatsch 1976). Northern USSR: from mid-May (Dementiev and Gladkov 1954*b*). SITE. On or close to ground in dense vegetation, at base of bush, in tussock, or low down in thick scrub. Of 523 nests, Britain: 33·5% in gorse *Ulex*, 21·6% open grass, 20·8% heather *Calluna*, 10·3% bracken *Pteridium*, 8·8% other low scrub, 4·4% mixed low vegetation, 0·6% stone walls (Fuller and Glue 1977). Usually facing north, or entrance sheltered from sun. Mainly near ground with 86·1% of 474 nests in Britain at 0-15 cm, 9·1% 15-30 cm, and 4·8% over 30 cm, with maximum 122 cm (Fuller and Glue 1977). Nest: loose, unwoven cup of dry grass stems and leaves, lined with hair and feathers sometimes with wool; *c.* 8 cm across

and 5 cm deep; reached via horizontal or vertical tunnels through vegetation, *c.* 25 cm long (Johnson 1971*a*). Building: only ♀ builds, though ♂ once seen carrying material (Parrinder and Parrinder 1944-5); done mainly in early morning, taking 4-10 days (Johnson 1971*a*). EGGS. See Plate 82. Sub-elliptical, smooth and moderately glossy; pale blue to green-blue, variably marked with red-brown—often very finely but sometimes more heavily and with cap round broad end. *S. t. hibernans*: 18·7 × 14·4 mm (16·5-21·3 × 13·2-15·5), *n* = 148; calculated weight 2·0 g. *S. t. rubicola*: 18·1 × 14·3 mm (16·0-20·0 × 13·2-15·4), *n* = 115; calculated weight 1·97 g. *S. t. maura*: 17·0 × 13·5 mm (15·4-18·8 × 12·3-14·6), *n* = 100; calculated weight 1·65 g. (Schönwetter 1979.) Clutch: 4-6 (2-7). Of 197 clutches, Britain: 2 eggs, 1%; 3, 3%; 4, 10%; 5, 63%; 6, 22%; 7, 1%; mean 5·06. Some variation during season, with mean 4·92 ± SE0·06 (*n* = 126) from mid-March to end of April, 5·49 ± SE0·10 (*n* = 51) in May and first half of June, and 4·80 ± SE0·15 (*n* = 20) from mid-June to early August (Fuller and Glue 1977). Of 101 clutches, southern England: 2 eggs, 1%; 3, 3%; 4, 25%; 5, 46%; 6, 25%; mean 5·3 (Greig-Smith 1979*a*). Of 63 clutches, Jersey (Channel Islands): 2 eggs, 1; 3, 1; 4, 7; 5, 26; 6, 27; 7, 1; mean 5·27. Of 27 1st broods, mean 5·22; of 21 2nd broods, 5·48; of 13 3rd broods, 5·15; of 2 4th broods, 4·5 (Johnson 1971*a*). 2-3(-4) broods, 4th perhaps mainly after earlier egg loss, but 4 successful broods also recorded (Plucinski 1956; Guldi 1965). Resident birds start earlier and regularly have 3 broods; migrants, with later start, have 2 (Parrinder and Parrinder 1944-5; Johnson 1971*a*). Of 37 pairs, England, 100% had 2nd attempt, 84% 3rd attempt, 8% 4th attempt (Greig-Smith 1979*a*). Only 4-5 days between broods (Johnson 1971*a*). In Surrey (southern England), of 30 pairs laying 1st clutch, 97% attempted 2nd, and 50% a 3rd (Morgan and Davis 1977). Eggs laid daily, at 06.00-07.00 hrs (Johnson 1971*a*). INCUBATION. 13-14 days. By ♀ only. Begins with last egg, or occasionally penultimate or even earlier one (Johnson 1971*a*). YOUNG. Cared for and fed by both parents. ♀ broods intensively in first 24 hrs (gradually decreasing), and may feed exclusively for first 4-5 days before ♂ joins in (Johnson 1971*a*). FLEDGING TO MATURITY. Fledging period 13·5 days (12-16), *n* = 22, with earliest departures before capable of flight (Johnson 1971*a*). Fed by both parents for 4-5 days after fledging, then ♀ starts to build nest for next brood, while ♂ continues to feed 1st brood for further 5-10 days (Johnson 1971*a*). Age of first breeding 1 year (Johnson 1971*a*). BREEDING SUCCESS. Of 332 eggs laid in 63 clutches, Jersey, 81% hatched (made up of 90% of 141 1st clutch eggs, 74% of 115 2nd, 73% of 67 3rd, and 89% of 9 4th), and 79% fledged (78% of 127 1st brood, 67% of 85 2nd, 100% of 48 3rd, and 100% of 8 4th), giving overall success of 64% (70% for 1st broods, 50% for 2nd, 73% for 3rd, and 89% for 4th). 72% of unsuccessful eggs infertile, 17% lost to predation (Johnson 1971*a*).

Of 122 eggs laid, England, hatching success 93% and fledging success 97%, 91% overall (Parrinder and Parrinder 1944–5). Of 153 eggs laid, England, fledging success 60% (Phillips 1968). Nest record cards, Britain, indicated hatching success 82% and fledging success 96% or 79% overall (Fuller and Glue 1977). Of 475 eggs laid, southern England, 80·8% hatched; main losses from infertility (7·8%) and addling (4·8%); 72·9% of young survived to fledging, giving overall success of 59·0%, with young lost mainly through predation (13%) (Greig-Smith 1979a). Success linked with nest cover which thickened well in mid-season (e.g. bracken moorgrass *Molinia*) compared with that which did not (e.g. heather, gorse *Ulex*) (Greig-Smith 1984). Nests for successive broods further apart after predation, but also after good rearing success; least closest after low rearing success for reasons other than predation (Greig-Smith 1982b).

Plumages (*S. t. rubicola*). ADULT MALE. In fresh plumage (autumn), upperparts from forehead down to rump deep black, each feather broadly fringed pale buff-brown or grey-buff with slight olive tinge, fringes largely concealing black (least so on hindneck, mantle, and outer scapulars). Upper tail-coverts white, longest with black spot on tip, others fringed rufous-cinnamon, white often visible on sides only. Side of head down from eye, chin, and throat black, feathers narrowly tipped sandy-buff to off-white; fringes much narrower than on top of head, largely black side of head often contrasting with largely grey-buff of pale buff-brown forehead and crown. Side of neck down to side of breast white; white partly concealed by black or cinnamon feather-tips on rear and lower side of neck and on side of breast, uniform white only just behind cheek. Chest, breast, and upper flank rufous-cinnamon, feather-tips with vinous-pink tinge; gradually paler buff-cinnamon and tawny-buff towards sides of belly and rear flanks, pink-buff on belly and under tail-coverts, cream-white on vent. Tail dull black, feathers narrowly edged sandy-grey along outer web and tip; outer web of outer pair (t6) narrowly fringed off-white. Flight-feathers and greater upper primary coverts dull black, greyish-black, or brownish-black, secondaries and inner primaries narrowly fringed sandy-buff along outer web and off-white along tip, outer primaries very faintly edged pale buff along outer web, primary coverts narrowly and evenly fringed off-white along outer web and tip; basal inner web of primary coverts occasionally white, usually concealed. Upper wing-coverts and tertials deep black, evenly fringed sandy-buff; fringes often partly off-white, especially towards bases, basal parts of outer web of tertials and sometimes of s6 white, greater tertial coverts fully white, shorter tertial coverts and 2 inner greater coverts partly white; white forms contrasting patch on inner wing, though often hidden under scapulars in closed wing. Under wing-coverts and axillaries black or greyish-black, tipped white. *In worn plumage*, from about February, upperparts largely black and brown-black, pale feather-fringes bleached to sandy and partly worn off (least so on mantle and scapulars), upper tail-coverts white (except for black spot on longest ones); side of head, chin, and throat fully black, patch on side of neck fully white, contrasting sharply with side of head; chest deep rufous, contrasting with white sides of breast, sides of belly and flanks gradually paler rufous-orange to orange-buff towards rear; central belly, vent, and under tail-coverts virtually white;

upperwing largely black, flight-feathers slightly browner, white patch on inner wing more strongly contrasting, pale fringes bleached to pale buff or pink-buff, largely worn off (least so on greater coverts and tertials). By June–July, all pale fringes of upperparts and upperwing worn off, but upperparts not uniform black as some grey of feather-bases often visible; black spots on tips of upper tail-coverts worn off, but sometimes white also, uppertail appearing dark; white of sides of neck and breast partly worn off, chest still rufous-cinnamon, but breast and flanks bleached to dirty buff and remainder of underparts dirty white, often with much grey of feather-bases visible. ADULT FEMALE. Upperparts dull olive-grey, usually with slight sandy-pink tinge, especially on mantle, scapulars, and back; some ill-defined black of feather-bases partly visible, especially on hindneck, mantle, and scapulars; rump and upper tail-coverts slightly brighter buff-brown or pale cinnamon-brown, tail-coverts with narrow black feather centres; fringes of upperparts often greyer and always wider and less clear-cut than in fresh adult ♂, black feather-bases duller and more restricted, upper tail-coverts without white. Side of head and neck like upperparts, supercilium and narrow eye-ring only slightly paler buff-grey and ear-coverts sometimes slightly darker brown; side of neck appears sometimes slightly paler due to concealed white of feather-bases slightly showing through. Lower cheek, chin, and throat cream-buff, black of feather-bases sometimes slightly shining through and occasionally just visible on lower throat. Remainder of underparts down from chest and sides of breast virtually uniform pink-cinnamon, feather-tips slightly paler when freshly moulted, central belly and vent slightly paler pink-cream. Tail and flight-feathers as adult ♂; greater upper primary coverts as adult ♂, but narrow, even, and clear-cut fringes pale buff-grey rather than off-white. Upper wing-coverts and tertials black-brown (less deep black than adult ♂), rather broad and clear-cut fringes pink-cinnamon or pale sandy-cinnamon; tertial coverts contrastingly white (innermost sometimes partly black), white on bordering greater coverts, tertial bases, and shorter tertial coverts often more restricted than in adult ♂. Under wing-coverts and axillaries pink-buff or white, grey bases largely concealed except on longest feathers. *In worn plumage*, from about mid-winter, fringes of upperparts bleached and partly abraded, upperparts appearing streaked dull black or brown-black and greyish-sandy, upper tail-coverts still mainly pale rufous-cinnamon, supercilium sometimes more distinct, in particular just above and behind eye, grey-white; some white of feather-bases often visible at rear of cheek; much black usually visible on central throat (but still concealed on chin and lower cheek); underparts paler and less extensively rufous-cinnamon, some dull black visible on side of breast; central belly, vent, and under tail-coverts cream-white; fringes of upperwing and tertials bleached to greyish-sandy and partly abraded. When heavily worn, June–July, head and upperparts (including upper tail-coverts and upperwing) black-brown and throat almost black, all finely mottled brown or buff (in particular forehead, lore, chin, and ear-coverts), rather similar to adult ♂ but less uniform deep black; side of neck often with mottled white patch (smaller and less uniform than adult ♂, not extending to side of breast); chest and flanks dirty tawny-buff, remainder of underparts pale buff to dirty white with some dark grey of feather-bases visible; white patch on inner upper wing usually smaller than in adult ♂, under wing-coverts still mainly dirty buff, less blackish. NESTLING. Down rather long but scarce; on upperparts only; brown-grey (Witherby *et al.* 1938b). For development, see Greig-Smith (1985). JUVENILE. Upperparts black-brown or black, each feather with buff to off-white central mark, forming narrow streaks on crown, wider triangular drops on mantle and scapulars where each mark narrowly bordered black; lower rump and upper tail-coverts tawny-buff or rufous-cinnamon, partly spotted black. Side of head closely mottled dull black and pale buff, lore and supercilium (backwards from eye) sometimes almost uniform buff. Side of neck, lower cheek, chin, and throat pale buff or pale grey-buff, finely speckled dark grey; underparts down from chest and sides of breast warmer buff or cinnamon-buff, chest and sides of breast spotted black, flanks and upper sides of belly faintly mottled grey (underparts not scalloped, unlike many other juvenile thrushes Turdinae). Lesser and median upper wing-coverts greyish-black with large triangular buff tips, greater coverts dull black with broad pink-cinnamon or rufous tips. Sexes similar, but tertial coverts of ♂ white with buff tips, outermost one black with broad white shaft-streak, innermost often with partly black inner web; tertial coverts of ♀ generally with less white, mainly black with broad buff tip and narrow off-white shaft-streak. For tail, flight-feathers, tertials, and greater upper primary coverts, see First Adult. FIRST ADULT MALE. Like adult ♂, but juvenile flight-feathers, tail, tertials, many or all greater upper wing-coverts (not tertial coverts), and greater upper primary coverts retained. In fresh plumage, pale feather-fringes of upperparts, ear-coverts, cheek, chin, and throat slightly wider than in adult ♂ in equally fresh plumage, deep black of feather-bases slightly more concealed but still visible (least so on crown, mantle, inner scapulars, and central throat). Tail as adult, but pale fringes along feather-tip slightly broader and less clear-cut; black at centre of tip of outer feathers ending pointed, less rounded. Tertials as adult ♂, but bases sometimes with less white, fringes often buff (less white), slightly less clear-cut; innermost tertial covert sometimes partly black; tips of secondaries and inner primaries with broader, greyer, and less clear-cut fringes than adult; tips of greater upper primary coverts with broader, less clear-cut, and more frayed white fringes than adult, outer web with narrower and more buffy fringe (fringes in adult narrow, of even width, whitish, smooth at tip); retained juvenile outer greater upper wing-coverts sometimes browner than neighbouring median and inner greater coverts, less deep black. See also Bare Parts. *In worn plumage*, retained juvenile feathers browner and more worn than neighbouring fresh feathers and than those of adult at same time of year, but difference hard to see when heavily worn. FIRST ADULT FEMALE. Like adult ♀, but part of juvenile feathers retained, as in 1st adult ♂. Characters of retained feathers as in 1st adult ♂, but often hardly any difference in colour from neighbouring feathers. Innermost tertial covert usually with some black; white or cream at base of outer web of tertials sometimes very restricted (more so than in 1st adult ♂), but some adult ♀♀ similar.

Bare parts. ADULT. Iris dark brown. Bill, leg, and foot black. Inside of upper mandible slate-grey. (RMNH, ZMA.) NESTLING. Mouth yellow, no spots; gape-flanges whitish-yellow (Witherby *et al.* 1938b). JUVENILE. Iris brown. Bill flesh-pink with dark horn-grey culmen and tip. Leg and foot flesh-brown, slate-brown, or slate-black. FIRST ADULT. Like adult, but inside of upper mandible partly flesh-pink or yellow-pink to at least September and sometimes October; many darken to slate-grey in October (Svensson 1984a; ZMA).

Moults. ADULT POST-BREEDING. Complete; primaries descendant. In Britain, *hibernans* starts with shedding of p1 about 10 July to late August, completing with regrowth of p9–p10 early September to mid-October; on average from c. 1 August to c.

22 September, with estimated duration of moult c. 50 days (Ginn and Melville 1983). Moult of *rubicola* in Netherlands similar to *hibernans* (RMNH, ZMA), but moult may start earlier in southern Europe and North Africa, completed early September in Yugoslavia (Stresemann 1920). Moult of *maura* and *stejnegeri* apparently earlier than in *hibernans*, but data available for southern populations only, viz. southern USSR (Dementiev and Gladkov 1954b), Turkmeniya (RMNH), Mongolia (Piechocki and Bolod 1972), and Manchuria (Piechocki 1958): starts late June to mid-July, halfway through moult in July and early August, completed early August to early September. POST-JUVENILE. Partial: head, body, lesser and median upper wing-coverts, tertial coverts, often 1–2 tertials, and 0–6 greater upper wing-coverts. Starts at age of 4–6 weeks and hence timing strongly variable; in all races, birds in full juvenile plumage encountered early May to late August, moulting birds June–September. Moult completed at age of 2–3 months, from early July in south (North Africa, Turkey) and from mid- or late July further north; some feathers of neck and head grow last, occasionally up to early October. (BMNH, RMNH, ZFMK, ZMA.)

Measurements. ADULT, FIRST ADULT. *S. t. rubicola.* Wing and bill to skull of (1) Netherlands, Belgium, central France, and northern Italy; (2) north-west Africa; (3) southern Yugoslavia, Greece, and Turkey; other measurements combined, all year; skins (Stresemann 1920; RMNH, ZFMK, ZMA). Bill (N) to distal corner of nostril; exposed culmen on average 4·5 less than bill to skull.

		♂		♀	
WING	(1)	66·4 (1·31; 26)	64–68	65·4 (1·33; 18)	63–68
	(2)	67·7 (1·47; 6)	66–70	67·2 (1·20; 4)	65–68
	(3)	64·8 (1·44; 58)	63–69	63·9 (1·24; 28)	62–66
TAIL		45·2 (1·25; 27)	43–48	44·8 (1·37; 18)	42–48
BILL	(1)	14·9 (0·56; 22)	13·7–15·6	15·1 (0·51; 17)	14·0–15·8
	(2)	15·6 (0·35; 6)	15·1–16·0	16·0 (0·52; 4)	15·2–16·4
	(3)	15·0 (0·75; 7)	14·1–16·2	14·7 (0·23; 4)	14·3–14·9
BILL	(N)	8·1 (0·38; 18)	7·6–8·7	8·2 (0·40; 13)	7·5–8·8
TARSUS		22·7 (0·47; 25)	21·9–23·2	22·9 (0·33; 15)	22·3–23·5

Sex differences significant for wing (1) and (3). 1st adult with retained juvenile flight-feathers and tail combined with older birds, as juvenile wing on average only 0·8 shorter than full adult and tail 0·6 shorter. Wing of live birds, Balearic islands: ♂ 67·5 (1·84; 21) 64–71, ♀ 65·7 (9) 64–68 (Mester 1971); data of 9 ♂♂ from western Caucasus: wing 63·1 (61–67), bill to skull 15·0 (13·5–15·7), tarsus 21·7 (21·0–22·5) (Buturlin 1929).

S. t. hibernans. Wing and bill to skull of (1) Britain, (2) western Portugal; other data combined, all year; skins (BMNH, RMNH, ZFMK, ZMA).

		♂		♀	
WING	(1)	67·4 (1·52; 11)	65–70	66·2 (1·30; 10)	64–68
	(2)	66·9 (1·44; 12)	65–69	65·4 (1·43; 7)	64–67
TAIL		46·5 (1·64; 17)	44–49	45·1 (1·55; 8)	43–47
BILL	(1)	14·6 (0·55; 9)	14·0–15·3	14·9 (0·49; 10)	14·2–15·6
	(2)	15·8 (0·49; 11)	15·0–16·4	15·6 (0·40; 6)	15·3–16·2
TARSUS		23·1 (0·70; 20)	22·2–24·4	23·0 (0·46; 13)	22·3–23·9

Sex differences not significant.

S. t. maura. Kopet Dag, Tien Shan, and Altai mountains, USSR, mainly summer; skins (RMNH, ZFMK, ZMA). Bill (S) to skull.

	♂		♀	
WING	69·9 (1·69; 27)	68–74	69·2 (1·04; 6)	67–71
TAIL	46·7 (2·24; 13)	43–49	45·9 (0·82; 5)	45–47
BILL (S)	14·2 (0·49; 27)	13·7–15·1	14·2 (0·78; 5)	13·4–15·0
BILL (N)	7·6 (0·53; 13)	7·0–8·4	7·3 (0·49; 5)	6·8–8·0
TARSUS	21·8 (0·73; 27)	20·5–23·1	21·4 (0·39; 5)	20·9–21·8

Sex differences not significant.

S. t. stejnegeri. Japan and eastern China, mainly winter; skins (RMNH, ZFMK, ZMA).

WING	♂ 70·6 (1·71; 21)	68–73	♀ 68·7 (1·32; 13)	67–71
TAIL	48·0 (2·20; 18)	44–52	47·5 (2·15; 10)	44–51
BILL (S)	14·4 (0·40; 19)	13·8–15·2	14·4 (0·57; 13)	13·7–15·3
BILL (N)	7·5 (0·27; 16)	7·1–8·0	7·5 (0·13; 11)	7·3–7·7
TARSUS	21·7 (0·42; 19)	20·9–22·3	21·4 (0·64; 13)	20·3–22·4

Sex differences significant for wing.

S. t. variegata. Volga–Ural steppes, summer, and Sudan, winter; skins (RMNH, ZFMK).

WING	♂ 71·9 (0·46; 7)	70–74	♀ 70·2 (– ; 3)	68–72
BILL (S)	14·2 (0·91; 5)	13·7–15·0	14·1 (– ; 3)	13·7–14·4
TARSUS	21·1 (0·45; 5)	20·4–21·5	21·1 (– ; 3)	20·8–21·4

Wing of ♂: 72 (10) 69–75 (Vaurie 1959).

S. t. armenica. Eastern Turkey and Iran, summer: wing, ♂ 73·4 (1·96; 9) 71–77, ♀♀ 72, 73·5; bill (S), ♂ 14·5 (0·82; 5) 13·5–15·5; tarsus, ♂ 21·7 (3) 21·3–22·1 (Paludan 1938; RMNH, ZFMK). Zagros mountains, south-west Iran: wing, ♂ 76 (15) 74–81 (Vaurie 1959).

Weights. ADULT, FIRST ADULT. *S. t. rubicola.* Sexed birds. Netherlands (ZMA), Greece (Niethammer 1943; Makatsch 1950), and Turkey (Kumerloeve 1968; Rokitansky and Schifter 1971), combined, all year: ♂ 14·2 (1·60; 10) 11·5–17, ♀ 14·8 (1·38; 5) 13·5–17. Unsexed birds. Southern France: (1) July–August, (2) September–October, (3) November–February (G Olioso). Gibraltar: (4) autumn (Finlayson 1981). Balearic islands, summer: (5) adult, (6) juvenile (Mester 1971). Malta: (7) winter (J Sultana and C Gauci). Cyprus: (8) October (Hallchurch 1981).

(1)	15·2 (1·17; 6)	13–16	(5)	14·6 (– ; 16)	13–16
(2)	14·6 (0·97; 11)	13–17	(6)	13·7 (– ; 27)	11–16
(3)	15·1 (1·31; 5)	13–16	(7)	15·3 (1·4 ; 50)	13–19
(4)	16·5 (1·94; 31)	–	(8)	14·5 (– ; 9)	13–16

S. t. hibernans. Skokholm (Wales): 15·5 (8) 14–17·4 (Browne and Browne 1956). Portugal, July: 15·2 (1·10; 8) 13·4–16·7 (C J Mead).

S. t. variegata. Migrants northern Iran, March: ♂♂ 13, 15 (Schüz 1959).

S. t. armenica. North-east Turkey (Kumerloeve 1968, 1969a), Armeniya (Nicht 1961), and Zagros mountains, Iran (Paludan 1938): March, ♂♂ 13·8, 14·3; May, ♂♂ 14·6, 15, ♀ 12·5; June, ♂ 16, ♀♀ 14, 15. Perhaps this race (wing 71–74) Kuwait: ♂♂ 11, 11, 12·2; ♀♀ 11·5, 12·5 (V A D Sales, BTO).

S. t. maura. (1) March, northern Iran and Afghanistan (Paludan 1959; Schüz 1959). (2) May–June, Kazakhstan (USSR) and Afghanistan (Paludan 1959; Dolgushin et al. 1970). (3) August and a few September, Lake Chany, south-west Siberia (Havlín and Jurlov 1977). (4) September and early October, Afghanistan (Paludan 1959).

(1)	♂ 15·5 (1·87; 6)	13–18	♀ 14·0 (1·83; 4)	12–16
(2)	13·4 (– ; 4)	11–15	14·2 (2·45; 4)	11–17
(3)	13·6 (0·98; 14)	12–15	13·6 (1·25; 7)	12–15
(4)	12·7 (– ; 11)	11–15	12·1 (– ; 10)	11–13

S. t. stejnegeri. Mongolia: May–June, ♂ 14·4 (2·07; 5) 12–17, ♀♀ 12, 15 (Piechocki et al. 1982); August, ♂ 14·8 (0·98; 6) 14–16, ♀ 13·8 (1·50; 4) 13–16 (Piechocki and Bolod 1972). Manchuria, July–September: 15·4 (9) 15–16 (Piechocki 1958). Hopeh, China: ♂ 14·7 (10) 13–16, ♀ 15·0 (10) 14–16 (Cheng 1963).

NESTLING, JUVENILE. For growth curves, see Greig-Smith (1985).

Structure. Wing short, broad at base, tip rounded. 10 primaries: p7 longest; in *rubicola* and *hibernans*, p8 0–1 shorter, p9 4·5–8, p6 0–0·5, p5 1–3, p4 4–6·5, p1 10·5–14; in *maura* and *stejnegeri*, p8 0–1 shorter, p9 4–7·5, p6 0–0·5, p5 1–2, p4 3·5–6·5, p1 11–14 (and hence closely similar to west and central European races); in *armenica*, p1 up to 16·5 shorter than wing-tip. In all races, p10 reduced, 4·5–10 longer than longest upper primary covert, 28–35 shorter than wing-tip (but up to 38 in *armenica*). Outer web of p5–p8 and inner of p6–p9 emarginated. Tertials short. Tail rather short, tip square or very slightly rounded; 12 feathers, t6 2–5 shorter than t2–t3. Bill short, straight; rather heavy and wide at base, pointed at tip; tip of culmen decurved, projecting over tip of lower mandible. Nostrils small, oval; partly covered by membrane above and by frontal feathering at base. Some long and stiff bristles at base of upper mandible. Tarsus rather long and slender, toe long and slender. Middle toe with claw of *rubicola* and *hibernans* 17·5 (15) 16·5–18·5, of *maura* and *stejnegeri* c. 1 mm less; outer toe with claw c. 67% of middle with claw, inner c. 65%, hind c. 74%. Claws rather long and slender, sharp, decurved.

Geographical variation. Marked; involves general proportions, size (as expressed in wing and bill length), general colour (especially head and upperparts), and amount of white on upper tail-coverts and tail-base. In Palearctic, 2 main groups discernible: (1) *rubicola* group in west, with *rubicola* and *hibernans*, and (2) *maura* group in eastern Europe and Asia, with *maura*, *stejnegeri*, *variegata*, *armenica*, *indica*, and *przewalskii*. Both groups come into contact in Caucasus, Transcaucasia, and eastern Turkey without apparent intergradation; some overlap may occur, but both inhabit different levels, *rubicola* in lowlands and hills, *armenica* at high level in mountains, though apparently not such a difference in Caucasus where *rubicola* and *variegata* meet. In *rubicola* group, average wing mainly 64–68, in *maura* group slightly longer (69–72) and up to average of c. 75 in mountain races *armenica* and *przewalskii*. Bill rather short in northern populations of *rubicola* group (average 14·6–15·1) but longer in populations of Iberia, north-west Africa, Greece, Crete, and Turkey (not western Caucasus), where average 15·6–16·0; some populations sometimes separated on account of this long bill, e.g. *desfontainesi* Blanchet, 1925, north-west Africa, and *graecorum* Laubmann, 1927, Greece, but differences too slight. Bill in *maura* group generally shorter, on average 14·1–14·5. Tarsus rather long in *rubicola* group, averages ranging from 22·7 to 23·1 (longest in west of range), with individual values mainly 22·3 and over; shorter in *maura* group, averages 21·1–21·8, individual values mainly below 22·3. Also, toes in *maura* group slightly shorter, middle toe 15–17 (16·5–18·5 in *rubicola* group). Within *rubicola* group, *hibernans* from Britain, Ireland, Brittany, and coastal Portugal darker than *rubicola* from most of Europe, north-west Africa, western Caucasus, and western, central, and northern Turkey: in fresh plumage, feather-fringes of head and upperparts darker rufous-brown (in *rubicola*, pale buff-brown to sandy olive-grey), feather-centres of ♀ deeper black (♂ deep black in both races); chest, breast, and flanks darker, rufous-chestnut rather than rufous-cinnamon; in worn plumage, *hibernans* and *rubicola* usually not separable, as feather-fringes on head and upperparts wear off completely in ♂♂ and bleach to sandy-buff in ♀♀ of both races; difference in colour of feather-centres of upperparts of ♀♀ hard to detect; underparts in both sexes of *hibernans* on average darker than in *rubicola*, but some overlap and much individual variation. Some autumn birds from coastal dunes of Netherlands and Belgium (and perhaps also north to Denmark and south to north-west France) similar to *hibernans*; not known whether these are local birds or migrants from Britain, as breeders from these areas are too worn to assign to any race. Best characters for identifying ♂ *maura* from east European USSR and Turkmeniya east to Lake Baykal and north-west Mongolia are fully and extensively white lower rump and upper tail-coverts, without black spots on longer coverts (but rump and coverts extensively tinged orange-buff when plumage fresh, as in *rubicola*); ♀ has virtually unstreaked rufous-buff rump and does not show black on throat in worn plumage, unlike ♀ *rubicola*. Amount of white on tail-base of ♂ not a good identifying character: of 30 *maura* examined (Altai, Tien Shan, Uzbekistan, Turkmeniya), most had tail fully black and only a few had up to 10 mm of base of t2-t5 white, hidden under coverts. However, no birds from Urals and western Siberia examined, and these perhaps show more white: of 17 ♂♂ from western Siberia, only 3 were mainly black, 14 had 11-17 mm of tail-base white (Ticehurst 1938), thus basal $\frac{1}{4}$-$\frac{1}{3}$ of tail; this perhaps a result of intergradation of *variegata* (with much white on tail-base), which occurs north to at least mouth of Ural river. Adult ♂ *maura* in fresh autumn plumage has crown, hindneck, mantle, scapulars, chin, and throat black with sandy-buff feather-fringes 2-3 mm wide, sides of head with narrower fringes (1-2 mm), lores and shorter ear-coverts fully black, rump and upper tail-coverts white with rufous-buff feather-tips 3-4 mm long, underparts pale rufous-cinnamon with brighter cinnamon or rufous-chestnut chest, and axillaries and under wing-coverts black with narrow white fringes; thus closely similar to adult ♂ *rubicola*, but black of chin and throat more concealed, rump-patch more extensively pale, and underwing blacker; in worn plumage, from about mid-winter, pale fringes worn off, and then similar to *rubicola*, except for whiter rump and upper tail-coverts and sometimes larger white patches on sides of neck and inner upper wing. 1st adult ♂ *maura* in fresh plumage is rather different from adult ♂ *maura* and 1st adult ♂ *rubicola*, as hardly any black visible on head and upperparts: crown down to upper rump, chin, and throat sandy olive-brown or buff-brown with black feather-bases largely concealed, only forehead, lores, front of cheeks, and ear-coverts have some black; uniform buffy chin and throat contrast with rufous-cinnamon or pink-cinnamon chest, flanks, and belly; feathers at sides of neck and breast, on lower rump, and upper tail-coverts white, but this hidden under rufous tips; less white on inner upperwing than adult ♂ (see Svensson 1984a), but buff fringes along secondaries and tertials wider, forming pale panel. Plumage worn from March (thus later than in adult ♂) and then similar to adult ♂, but black of mantle, scapulars, chin, and throat still has fairly broad buff fringes which do not wear off until early summer. ♀ *maura* in fresh plumage sandy-grey to buff-brown on upperparts, forehead and crown with black feather-centres, feathers of nape, mantle, scapulars, and back with ill-defined dull black central streaks; rump and upper tail-coverts rufous-buff, longer coverts with narrow dark shaft-streaks, shorter feathers sometimes partly white. Lores, supercilium, and narrow eye-ring buff, cheeks and ear-coverts mottled dull black and buff. Chin and throat pale buff, contrasting with pale rufous-cinnamon or tawny-buff chest and flanks. Under wing-coverts and axillaries dark grey with broad buff fringes. Often differs from 1st adult ♂ only in absence of (largely hidden) white on sides of neck and absence of (largely hidden) black on cheek and throat. Tail-base always black. Differs from ♀ *rubicola* in slightly paler sandy-buff upperparts with more contrasting pale rufous-buff to buff-white rump and upper tail-coverts, more distinct supercilium, pale buff (if fresh) to cream (if worn) chin and throat, and wider buff fringes along secondaries and tertials. Ageing of ♀ difficult: adult has narrower pale greyish-buff fringes of even width along outer webs and tips of greater upper primary coverts; 1st adult has buff fringes, wider at tip than at sides. See also Robertson (1977). *S. t. variegata* from eastern Caucasus and shores of north-west and northern Caspian Sea similar to *maura*, but feather-fringes of upperparts in fresh plumage paler sandy-buff or yellowish-sandy, and rump-patch larger and whiter; ♂ differs conspicuously in having much white on tail-base, $\frac{1}{3}$-$\frac{1}{2}$ of tail length white (except on t1); in those examined, black on tip of t1 extended for 32-35 mm, minimum extent of black on t2-t5 5-25 mm (mainly 10-15 mm on outer web, rather more on inner; in one bird, white extended along shaft to tail-tip), maximum extent of black on outer web of t6 10-35 mm (thus, tail pattern like wheatear *Oenanthe*); ♀ usually has lateral tail-base buff or cream for *c.* 1 cm (unlike ♀ *maura*); ♂ has large white patches on sides of neck and on inner upperwing, and deep chestnut chest. *S. t. armenica* close to *variegata*, intergrading with it in eastern Transcaucasia (Stegmann 1935): apart from slightly longer wing, differs mainly in blacker tail of ♂, which has *c.* $\frac{1}{4}$ of base white; in those examined, 32-42 mm of t1 black, minimum 20-37 mm of t2-t5 black, at least 28 mm of outer web of t6 black, sometimes up to base; chest of ♂ deep chestnut, belly and flanks with much white; intergrades with *maura* in north-east Iran. *S. t. stejnegeri* from eastern Siberia to Korea and Japan close to *maura* and by some considered to be inseparable (e.g. Svensson 1984a). In fresh plumage, *stejnegeri* from Japan has feather-fringes of upperparts and general tinge of underparts darker and more rufous than in *maura* (colour of *stejnegeri* comparable with *hibernans*, *maura* with *rubicola*); in worn plumage, both virtually identical, but base of bill distinctly wider in *stejnegeri*; tail-base as south Siberian *maura*, fully black or (occasionally) with base of t2-t5 up to 1 cm white. In winter birds from eastern China, attributed to *stejnegeri* and probably originating from eastern Siberia, bill-base narrow (near *maura*) and both upperparts and underparts slightly paler rufous-buff, intermediate between *maura* and Japanese *stejnegeri*; *maura* and *stejnegeri* probably intergrade over wide area of central and eastern Siberia and Mongolia. Extralimital *indica* similar to south Siberian *maura*, but smaller; *przewalskii* like *stejnegeri* but larger and more deeply and extensively rufous below. For survey of other races in Arabia and Afrotropics, see Deignan *et al.* (1964).

Forms superspecies with Canary Islands Stonechat *S. dacotiae* and Réunion Stonechat *S. tectes*. CSR

Saxicola caprata Pied Stonechat

PLATES 53 and 55
[between pages 736 and 737]

Du. Zwarte Roodborsttapuit   Fr. Tarier pie   Ge. Mohrenschwarzkehlchen
Ru. Чёрный чекан   Sp. Tarabilla negra   Sw. Svart buskskvätta

Motacilla Caprata Linnaeus, 1766

Polytypic. *S. c. rossorum* Hartert, 1910, Transcaspia and Iran, east to Syr Darya, Pamirs, and north-west Pakistan; accidental in Iraq. Extralimital: *bicolor* Sykes, 1832, central Pakistan and northern India east to Nepal and northern Bangladesh, grading into *rossorum* in south-east Iran, Baluchistan (Pakistan), and Kashmir; nominate *caprata* (Linnaeus, 1766), northern Philippines; 13 further races in south-east Asia and New Guinea area.

Field characters. 13·5 cm; wing-span 21–23 cm. 10–15% larger than Stonechat *S. torquata*. Robust chat with typical *Saxicola* character but very dark plumage, relieved on ♂ only by white wing-patch, rump, and vent, and on ♀ by rusty or buff on rump and vent. Sexes dissimilar; seasonal variation in ♂. Juvenile separable.

ADULT MALE. Sooty-black except for white panel on wing (beside scapulars), white rump and upper tail-coverts, and white belly and under tail-coverts. From July, brown fringes to fresh feathers of head and body create mottled and dull-scaled appearance. ADULT FEMALE. Essentially brown above, with buff lore and eye-surround, buff edges to wing-coverts, tertials, and inner secondaries, and pale rusty rump and upper tail-coverts contrasting with dull black tail. Fulvous-grey below, with rusty tinge to breast and belly, pale buff vent, and cream under tail-coverts. JUVENILE. Resembles ♀ but feathers of upperparts paler-spotted. At all ages, bill and legs black.

♂ unmistakable but ♀ and immature suggestive of darker races of *S. torquata*, and this difficulty not studied in the field. Flight agile; like *S. torquata*, will hover. Gait and behaviour much as *S. torquata*; ♂ perches openly, with twitching tail movements conspicuous.

Song brisk and whistling, 'chip-chepee-chewee chu'; calls include repeated, plaintive 'chep chep-hee' or 'chek chek trweet' (accompanied by tail-jerk) used against territorial intruder, and sharp, scolding 'chuh' in strong alarm.

Habitat. In continental lower middle latitudes, in plains and hills, but avoiding mountains and also forests and steppes, preferring low scrub, often on stony hillsides, moist places with thickets near reedbeds and coarse grass, especially beside rivers, canals and ponds, and where tamarisk clumps, willows, and grass alternate with cultivation; also orchards, gardens, cultivation or semi-cultivation and damp meadows, especially near water. In Kashmir, recorded exceptionally to *c.* 2400 m (Bates and Lowther 1952), and in Baluchistan (Pakistan) to 2000 m (E M Nicholson); in Afghanistan, however, only common to *c.* 1200 m (Paludan 1959) and in much of India only locally above *c.* 1700 m (Whistler 1941). A ground feeder, flying down from vantage point on top of bush, weed-stalk,

etc.; nests on ground or in bank or wall; will sing from treetop or telegraph wire; flies only briefly and in lower airspace. Fledglings seem to gather in beds of rushes and tall grasses (Hutson 1954). Favours lower elevations, open terrain, and proximity to water. Needs plenty of raised look-outs commanding some bare ground, but not interested in cover except for nesting, and indifferent to neighbourhood of man.

Distribution. Breeds from Transcaspia and eastern Iran through Oriental region to Indonesia (except Borneo), Philippines, and New Guinea. Northern races migratory, sedentary in tropics.

Accidental. Iraq: reported at Fao in November and March (Allouse 1953).

Movements. Largely migratory in west of breeding range, but eastern tropical races sedentary. Although most movement apparently nocturnal, single birds and small parties also seen moving by day (Paludan 1959).

Populations of Transcaspia, eastern Iran, and Afghanistan (*rossorum*) move south, with most individuals wintering from south-east Iran along Makran coast into plains of north-west Pakistan. However, a few stragglers reach west into western Iran and Iraq. Little information on autumn passage; Gissar valley (Tadzhikistan, USSR) vacated by 22 September. Spring passage reaches Turkmeniya in early April, but higher ground of Tadzhikistan often not occupied until early May. Often suggested that ♂♂ arrive first, but extremely skulking behaviour of ♀♀ make assessment of such claims difficult.

Populations of north Indian peninsula from Pakistan east to Burma (*bicolor*) are partially migratory, regularly wintering south of breeding range to *c.* 14°N and less frequently to *c.* 12½°N. Post-breeding dispersal starts in early August, but few details on timing of either passage. ♂ recorded on ship off Laccadive islands (*c.* 10°N 71½°E), 6 October (Simpson 1984). Other races primarily sedentary, although there are winter records from western Yunnan which seem to refer to *burmanica*, breeding in central India. (Dementiev and Gladkov 1954*b*; Ali and Ripley 1973*b*.)   DGCH

Voice. See Field Characters.

Plumages. ADULT MALE. In fresh plumage, upperparts, chin to chest, sides of breast, and flanks black; feathers of mantle and scapulars with indistinct brown tips (*c.* 1 mm wide), chin to chest and sides of breast with more distinct buff feather-tips. Rump and upper tail-coverts white, feathers of rump slightly tipped buff, longer upper tail-coverts tipped black. Central breast and belly, vent, lower flanks, and under tail-coverts white. Tail black. Flight-feathers, tertials, greater upper wing-coverts, and longest feather of bastard wing black; inner webs and tips of flight-feathers slightly tinged grey-brown; narrow and indistinct fringes along tips of tertials, secondaries, and inner primaries off-white, faint fringes along tips of greater upper primary coverts and longest feather of bastard wing brown. Remainder of upperwing deep black, but lesser upper wing-coverts (except outermost and small ones along leading edge of wing) and tertial coverts contrastingly white, forming distinct patch on wing. Under wing-coverts and axillaries black. *In worn plumage* (from about mid-winter), plumage entirely black except for contrastingly white rump, lower flanks, central breast to under tail-coverts, and wing-patch; sometimes traces of off-white feather-fringes visible on chest and usually faint brown fringes on greater upper primary coverts; flight-feathers appear black-brown from about May. ADULT FEMALE. Upperparts and sides of head dark grey-brown or brown; rump rufous-cinnamon, merging into deep rufous on longer upper tail-coverts. Lower cheeks, chin, and throat grey-brown with variable amount of sandy-buff suffusion. Chest, sides of breast, and flanks rather variable (independent of age), rufous-brown to dull brown or greyish-brown; gradually merging into buff on belly and to cream on vent and under tail-coverts. Tail uniform dull black. Tertials and flight-feathers black-brown, narrowly fringed grey-buff on outer webs of flight-feathers (except emarginated parts of outer primaries) and all round on tertials. Greater upper primary coverts dark brown or black-brown with narrow grey-buff edges along outer webs and tips. Upper wing-coverts dark grey-brown, tips narrowly fringed buff, tips of greater more broadly fringed rufous-buff; some inner lesser and shorter tertial coverts sometimes suffused white, giving slight indication of white wing-patch. Under wing-coverts and axillaries sandy-buff, some dark brown bases visible on lesser coverts. *In worn plumage*, generally duller and browner, buff fringes largely worn off, wing appearing black-brown; lores, cheeks, chin, and ear-coverts speckled buff and brown; chest, belly, and flanks sometimes with faint dark shaft-streaks. JUVENILE. Upperparts and lesser upper wing-coverts dark grey-brown with small and narrow pale buff spots on feather-centres of crown and (indistinctly) hindneck; large cream or pale buff spots on mantle, scapulars, and lesser coverts, only narrowly fringed dusky along feather-tips. Lower rump and upper tail-coverts uniform cinnamon-buff or rufous-buff. Sides of head dark grey-brown with fine cream-buff specks; narrow cream ring round eye. Underparts pale cinnamon-buff (if fresh) or dirty pale buff (if worn) with narrow dark brown scaling; scaling faint on chin and throat, distinct on chest, sides of breast, and thighs, absent on central belly and vent. Under tail-coverts uniform cream-white. Tail black with faint and narrow rufous-buff fringes. Flight-feathers as adult (blacker in ♂, browner in ♀). Tertials black with broad buff fringes. Greater upper primary coverts black with rather narrow but sharp buff fringe along outer web and tip. Median upper wing-coverts black with buff shaft-streaks and broad rufous-buff tip; greater coverts black with broad rufous-buff tip or triangular spot on tip. Axillaries and under wing-coverts sandy-buff. Sexes similar, but ♂ usually has innermost median or lesser upper wing-coverts white with partial buff tinge. FIRST ADULT MALE.

Like adult ♂, but black duller, less deep; feathers of upperparts (except rump) broadly fringed grey-brown in fresh plumage, traces of fringes remaining even in worn plumage, showing as brown wash. Chin, throat, chest, sides of breast, and upper flanks with broad pale buff feather-fringes; those of chin, throat, and central chest gradually wear off, but those elsewhere still prominent in spring and early summer; white of central belly not as sharply divided from breast and flanks as adult. White of wing-patch often more restricted, outer lesser coverts dull black with narrow buff fringes rather than white. Under wing-coverts and axillaries black with buff fringes. Juvenile flight-feathers, tail, greater upper primary coverts, 1–6 outer greater upper wing-coverts, and often tertials retained, distinctly browner and more worn than those of adult at same time of year, and contrasting with fresh black neighbouring feathers; retained greater coverts with buff outer edge and often large triangular pale buff or off-white spot on tip, greater upper primary coverts with broad sandy fringes along outer webs and tips (widest on innermost). FIRST ADULT FEMALE. Like adult ♀, but part of juvenile plumage retained, as in 1st adult ♂. Colour of retained plumage as in adult ♀, differing only in wear, and hence ageing often difficult. In some birds (not all), outer greater wing-coverts with large buff triangular spots, unlike adult; greater upper primary coverts sometimes more pointed at tip than in adult and with broader buff fringes.

Bare parts. ADULT, FIRST ADULT. Iris dark brown. Bill black in ♂, dark brown to brown-black in ♀. Mouth pale pink, yellow-pink, grey-pink, or brown-pink, apparently varying with age. Leg and foot black-brown or black. (Dementiev and Gladkov 1954*b*; Ali and Ripley 1973*b*; BMNH.) JUVENILE. No information.

Moults. ADULT POST-BREEDING. Complete; primaries descendant. In USSR, moult prolonged, in second half of July and August, but locally some moult mid-June; completed or nearly so by early August (Dementiev and Gladkov 1954*b*). In Afghanistan, starts about early July, completed late August; body largely new at primary moult score 35–40; tail starts at score 15–20, completed at same time as outer primaries (BMNH). POST-JUVENILE. Partial: head, body, lesser and median upper wing-coverts, often some tertial coverts and inner greater upper wing-coverts, and occasionally some tertials; juvenile flight-feathers, tail, greater upper primary coverts, and 1–6 outer greater upper wing-coverts retained. In USSR, in heavy moult early and mid-August (Dementiev and Gladkov 1954*b*); in Afghanistan and Pakistan, starts about mid-July and one bird from mid-September just completed (BMNH).

Measurements. ADULT, FIRST ADULT. *S. c. rossorum*. Turkmeniya (USSR) and Afghanistan, February–September; skins (BMNH, RMNH). Bill (S) to skull, bill (N) to distal corner of nostril; exposed culmen on average *c.* 4·4 less than bill (S).

	♂	♀	
WING	76·5 (2·07; 14)	73–79	75·4 (1·21; 8)	74–77
TAIL	52·8 (1·39; 9)	51–56	50·9 (1·68; 8)	48–53
BILL (S)	15·2 (0·38; 9)	14·6–15·6	14·7 (0·58; 8)	14·0–15·4
BILL (N)	8·3 (0·46; 9)	7·6–9·0	7·9 (0·41; 8)	7·4–8·4
TARSUS	22·3 (0·68; 9)	21·3–23·1	22·4 (0·61; 8)	21·4–23·2

Sex differences significant for tail. Retained juvenile wing of 1st adult on average 2·7 shorter than older birds, tail 2·1 shorter, but all ages combined in table.

Wing, Afghanistan: adult ♂ 76·1 (1·76; 9) 74–79; juvenile ♂ and 1st adult ♂ 73·3 (1·12; 9) 71–75; ♀ 72·6 (1·90; 7) 70–76 (Paludan 1959).

Weights. *S. c. rossorum.* Afghanistan. April: ♂ 14·5 (0·58; 4) 14–15, ♀ 15·3 (3) 14–17. June–July: ♂ 19·0 (4·34; 6) 15–25, ♀ 12. September and early October: ♂ 16·6 (4·00; 8) 13–26, ♀ 16·0 (3) 15–18. (Paludan 1959.)

Structure. Wing rather short, broad at base, tip rounded. 10 primaries: p7 longest, p6 and p8 each 0–0·5(–1) shorter, p9 6–8, p5 1–2, p4 3–5, p3 6–8, p1 9–12. P10 reduced, 29–35 shorter than wing-tip, 8–12 longer than longest upper primary covert. Outer web of p5–p8 and inner of p6–p9 emarginated. Tertials short. Tail rather short, tip square or slightly rounded; 12 feathers, t6 2–4 shorter than t2–t3. Bill relatively short but wider and deeper at base than in Stonechat *S. torquata*, more numerous and heavier bristles at base of lower mandible. Leg slightly heavier than in *S. torquata*, foot larger. Middle toe with claw 18·1 (6) 17·5–19; outer toe with claw *c.* 70% of middle with claw, inner *c.* 66%, hind *c.* 76%. Remainder of structure as in *S. torquata*.

Geographical variation. Relatively slight in view of large geographical range; mainly involves size, amount and distribution of white on adult ♂, and depth of colour of adult ♀. *S. c. rossorum* rather large and ♂ has much white on belly, often extending up to middle of breast; *bicolor* of northern India slightly smaller (wing, ♂ 68–74 (20), ♀ 64–72 (20): Vaurie 1959), white of underparts of ♂ restricted to under tail-coverts and vent, ♀ on average slightly darker brown and with darker rufous upper tail-coverts. Nominate *caprata* from Philippines is smaller, and white of ♂ restricted to wing-patch, rump, and under tail-coverts. For names and distribution of other races, see Deignan *et al.* (1964).

Recognition. ♂ *rossorum* easy to identify, but ♀ and juvenile rather similar to ♀ and juvenile *S. torquata*. Generally darker on upperparts than *S. torquata maura* (which occurs in same general area), more like *S. t. rubicola* (with which hardly likely to overlap), though throat less blackish. Larger than any race of *S. torquata* (except *armenica*, which is pale), especially in wing and tail length; wing more rounded. CSR

Myrmecocichla aethiops Ant Chat

PLATES 56, 64, and 65
[between pages 736 and 737, and facing pages 761 and 856]

Du. Bruine Miertapuit   Fr. Traquet-fourmilier brun   Ge. Russschmätzer
Ru. Африканский чекан   Sp. Hormiguero   Sw. Svart termitskvätta

Myrmecocichla aethiops Cabanis, 1850

Polytypic. Nominate *aethiops* Cabanis, 1850, Sénégal to Lake Chad and northern Cameroon; *sudanensis* Lynes, 1920, western and central Sudan and (perhaps this race) neighbouring northern Chad; *cryptoleuca* Sharpe, 1891, East Africa.

Field characters. 17 cm; wing-span 28·5–32 cm. About 15% smaller than Rock Thrush *Monticola saxatilis* but of similar form, being noticeably bulkier than any other west Palearctic chat. Large, rather short-tailed, stumpy, ground-loving chat, with almost wholly brown-black plumage relieved only by striking white panel on primaries visible in flight. Sexes similar; no seasonal variation. Juvenile separable.

ADULT MALE. Sooty to brown-black, mottled brown on fore-face, throat, and chest and fully black only on tail and under tail-coverts. White inner webs of primaries (⅔ of length from bases) show in flight, forming panel like that of Magpie *Pica pica*. ADULT FEMALE. As ♂ but usually browner in direct comparison. JUVENILE. More uniformly coloured than adult, with little mottling. At all ages, bill and legs black.

Unmistakable in west Palearctic (but subject to confusion with ♀ and juvenile Sooty Chat *M. nigra* in Afrotropics). Flight weak, slower than wheatears *Oenanthe* and with noticeably more flapping wing-beats. Walks, runs, hops, and leaps, frequently cocking and spreading tail and drooping wings. Perches on plants and telegraph wires; much attracted to hollows.

Vocal, with pairs and family parties frequently calling, sometimes in unison. Song a whistled warble, based on piping 'tee-chu'. Bird often droops wings as it calls.

Habitat. Tropical dry lowlands from border of desert through very open bush (intermediate between orchard-bush and thorncountry) to cultivation and seashore—will occupy a derelict village and even enter a town amidst a level dusty plain. A ground feeder, but less terrestrial than wheatears *Oenanthe*; often perches on walls or even telegraph wires. For breeding habitat, an essential appears to be presence of unlined wells or large artifical holes in ground, suited to excavation within them of horizontal shafts or hollows to serve for nesting and roosting (Bannerman 1936). Frequents open farmland, villages, clay and laterite quarries, road cuttings, and wells in arid belt of West Africa (Serle and Morel 1977). In East Africa, common along roadsides and in open country with scattered bush and trees; also in open *Acacia* woodland in highlands and Rift Valley (Williams and Arlott 1980). Occurs in montane grassland of Tanzania crater highlands at 1800–2700 m, among small trees and bushes, and in degraded savanna of Nigeria showing all stages of regeneration from ex-farmed land, or more or less dense low scrub with frequent large shade trees (Moreau 1972). In Sudan, reaches northern limit well within tropics, ascending to *c.* 2700 m; a weak flier (Cave and Macdonald 1955). Peculiar nest-site requirements presumably a limiting factor on spread over otherwise suitable habitat.

Distribution. Breeding resident from Sénégal to Niger and Lake Chad and northern Nigeria to northern Cameroon and Sudan in semi-arid country, and in highlands of Kenya and northern Tanzania.

Accidental. Chad: 1 collected Tibesti 1954 (Simon 1965).

Movements. Little information; probably largely sedentary. Tame and conspicuous behaviour suggests unlikely to be regularly overlooked in well-watched areas. No evidence that any congeners migrate.

No evidence for seasonal movements, nor for movements between 3 disjunct breeding areas: West Africa (nominate *aethiops*), central western Sudan (*sudanensis*), and highlands of Kenya and Tanzania (*cryptoleuca*). Specimen collected in Tibesti (Chad), originally identified as *sudanensis*, but probably not safely separable from other races (Malbrant and Receveur 1955; Louette 1981; D G C Harper).

DGCH

Voice. See Field Characters.

Plumages (*M. a. sudanensis*). ADULT. Largely dark chocolate-brown or black-brown, feather-tips of head and neck rather narrowly and indistinctly fringed brown, those of chin, throat, and chest broader and more distinct, paler buff-brown, throat and chest appearing closely mottled black and brown. Under tail-coverts and tail black. Flight-feathers black (browner when worn), basal $\frac{2}{3}$ of inner webs of inner primaries and basal half of inner webs of outer primaries contrastingly white. Greater upper wing-coverts and greater upper primary coverts black; lesser and median upper wing-coverts brown like upperparts. Axillaries and under wing-coverts black-brown. *In worn plumage*, feather-fringes of head, neck, and chest bleached to pale sepia-brown and partly abraded. JUVENILE. Like adult, but head, neck, and chest black-brown, like remainder of body, feather-tips only very indistinctly fringed dark brown. FIRST ADULT. Indistinguishable from adult.

Bare parts. ADULT. Iris dark brown. Bill, leg, and foot black. (BMNH.) JUVENILE. Iris dark blue-grey. Bill, leg, and foot blackish. Gape pale yellow. (Bannerman 1936.)

Moults. ADULT POST-BREEDING. Complete; primaries descendant. In Sudan, moult starts with p1 mid-October to mid-February, and completed with p9-p10 January-May. POST-JUVENILE. Complete, primaries descendant. No further information, as plumage not separable from adult once body moult started. Wide scatter in start and termination of primary moult in Sudan probably due to different age-groups being involved, and some evidence of 2 moulting groups: one starting October-November and completing January-February, another starting January-February and completing April-May. Age of these groups unknown, however, though perhaps correspond to post-juvenile moult of 1st and 2nd broods. (BMNH.)

Measurements. ADULT. *M. a. sudanensis*. Western and central Sudan, all year; skins (BMNH). Bill (S) to skull, bill (N) to distal corner of nostril.

WING	♂ 106·8 (1·93; 8)	104-109	♀ 101·0 (2·67; 8)	98-105
TAIL	63·6 (1·69; 8)	61-66	60·4 (2·53; 8)	57-64
BILL (S)	20·8 (0·59; 8)	19·8-21·4	20·6 (1·59; 8)	20·0-21·6
BILL (N)	12·5 (0·51; 8)	11·8-13·3	12·6 (0·78; 8)	11·8-13·6
TARSUS	32·0 (1·00; 8)	30·8-33·5	30·9 (0·74; 8)	29·8-32·0

Sex differences significant, except bill.

Nominate *aethiops*. West Africa, ranges of 15 ♂♂ and 8 ♀♀: wing, ♂ 110-115, ♀ 108-115; tail, ♂ 67-74, ♀ 70-76; exposed culmen, ♂ 17-19, ♀ 17-18; tarsus, ♂ 34-36, ♀ 33-35 (Bannerman 1936).

M. a. cryptoleuca. East Africa: wing 107-122 (Mackworth-Praed and Grant 1960).

Weights. No information on nominate *aethiops* or *sudanensis*. In rather large *cryptoleuca*, Kenya, average of 5 birds *c.* 58 (Leisler *et al.* 1983).

Structure. Wing short, broad at base, tip rounded. 10 primaries: p6-p7 longest, p5 and p8 1-2 shorter, p9 8-11, p4 2-5, p3 4-8, p1 11-15. P10 short, somewhat reduced; 41-47 shorter than p6-p7, 13-21 longer than longest upper primary covert (at all ages). Outer web of p5-p8 and inner of p6-p9 emarginated. Tertials short. Tail short, tip slightly rounded; 12 feathers, t6 4-8 shorter than t1. Bill rather short, but heavy; culmen distinctly decurved, cutting edges slightly decurved. Nostrils small and rounded, partly covered by membrane above. Some strong bristles at base of upper mandible. Leg and foot markedly strong. Middle toe with claw 21·6 (8) 20·5-22·5; outer toe with claw *c.* 73% of middle with claw, inner *c.* 71%, hind *c.* 83%. Claws strong, rather short, sharply decurved.

Geographical variation. Rather slight; involves colour (see below) and size (see Measurements). *M. a. sudanensis* is smallest and darkest race. Nominate *aethiops* is larger and mainly dark brown rather than black-brown; fringes on head, neck, and chest wider and paler grey-brown. East African *cryptoleuca* largest; ground-colour dark, like *sudanensis* or even blacker, fringes of head, neck, and chest rather pale.

Forms superspecies with Southern Ant Chat *M. formicivora* of southern Africa (Hall and Moreau 1970).

CSR

Oenanthe isabellina Isabelline Wheatear

Du. Isabeltapuit   Fr. Traquet isabelle   Ge. Isabellsteinschmätzer
Ru. Каменка-плясунья   Sp. Collalba isabel   Sw. Isabellastenskvätta

Saxicola isabellina Temminck, 1829

Monotypic

Field characters. 16·5 cm; wing-span 27–31 cm. On average, larger than Greenland race of Wheatear *O. oenanthe leucorhoa*, with the following subtle differences in appearance and form: longer, slightly hooked bill; more bull-headed look; basally broader but still long, though slightly less pointed wings; rather shorter and broader, more rounded tail; more exposed thigh and tarsus. Largest wheatear in west Palearctic, with long, strong bill, long, evenly domed head, rather long body often held markedly upright, large wings, broad tail, and long strong legs. Palest *Oenanthe*, with less contrast between upperparts and underparts than even ♀♀ of most other species; essentially isabelline-brown above and buff-white below, marked only by black lores (in ♂), darker wings (especially when pale fringes worn off), broad white rump, and wide black band on tail. In flight, wings show diagnostic pattern of dark rim above and below, with wholly pale under wing-coverts. Sexes closely similar; little seasonal variation except from wear. Juvenile separable.

ADULT MALE. Plumage noticeably dense, wetting easily. Upperparts sandy-brown with isabelline suffusion; sometimes a grey hue on scapulars when fresh, yellow when worn; in dull light, often pink or mauve tint on mantle. Wings brown; look pale when fresh (except for black-brown bastard wing and tertial-centres) with broad cream or pale isabelline-buff fringes and tips to coverts, tertials, and secondaries; lesser and median coverts dark with brown centres, forming speckled patch at shoulder above dark bastard wing; margins of median and tips of greater coverts form 2 pale wing-bars; fringes of pale inner secondaries form pale panel. Wings when worn look darker, contrasting with back and body and showing few pale fringes and tips or even none. Secondaries cloak more of primaries than in *O. oenanthe*, making wing-point appear short. Face least marked of any ♂ *Oenanthe*, but at close range shows pattern reminiscent of several ♀♀ wheatears, especially pale *O. oenanthe leucorhoa*: quite deep well-defined grey-white supercilium in front of and above eye, black-speckled or (when worn) black lore, cream-white crescent under large brown-black eye, and buff to pale red-brown rear cheeks. Underparts basically cream, with buff throat, sandy- to isabelline- or pink-buff breast and flanks, and pale buff under tail-coverts; thus bird often looks no more than warm-chested or pale-bellied, with head, breast, back, and (if unworn) wings appearing uniform at distance. Rump and upper tail-coverts white, making rump look typically equal in depth to mainly black tail. White bases to outer tail-feathers cover *c.* ½ their length, thus forming only small square white patches on sides of tail-base and allowing only short upward extension of dark terminal band; this pattern usually insufficient to produce classic ⊥ of most *Oenanthe* and black tail-band 25–30% deeper than that of *O. oenanthe*. Narrow pale buff-white tips on fresh tail-feathers visible at close range. Underwing strikingly pale, with diagnostic pattern: axillaries and under wing-coverts white to buff, showing faint speckling along leading edge and lines in centre of wing-coverts only when worn; flight-feathers broadly fringed almost white on basal ¾ of inner webs and looking pale silvery-grey-brown over whole surface except for dark tips; overall effect heightened by breadth of wing at base. Pale inner webs of flight-feathers also discernible from above when wing fully extended, making brown tips form darker edge to end and rear of rather plain wing (but this character apparently reduced or lost through wear).

ADULT FEMALE. Closely resembles ♂ and often indistinguishable. Usually slightly smaller and duller, with narrower grey or black speckled lores, more diffuse and always cream supercilium, paler cheeks, and browner flight and tail-feathers. JUVENILE. Closely resembles dullest ♀♀ but somewhat buffer above, with even less obvious face pattern, buffer supercilium, and faint overlay of paler streaks and dark tips on upperparts and of dark brown freckles on sides of throat and breast. Bill long and strong, with tiny hook at end of culmen and broad, deep base (but not out of proportion to rather long, evenly domed head); black-horn at all ages. Legs and feet long and strong, often with lower thigh clear of body; black at all ages.

Apparently distinctive, but dogged by confusion with pale ♀ and 1st-year *O. oenanthe leucorhoa*, and reliable distinction is still disputed. However, the following characters of *O. isabellina* are important. Although no constant overall difference in customary measurements, form and structure most characterized by: (1) bill plus head length well over ½ length of folded wing (½ or less in *O. o. leucorhoa*); (2) rather long, often evenly elliptical, not full-breasted body shape in upright attitudes; (3) long, narrow but also broad-based wings, with folded or extended points slightly shorter and blunter than in *O. oenanthe leucorhoa* and usually leaving more than ½ visible tail exposed (½ or less in *O. oenanthe leucorhoa*); (4) long, often exposed thighs, adding to overall leg length and relative stoutness of strong legs and feet; (5) relatively short and slightly rounded tail; (6) exceptionally wide spread of wings and tail in flight, (not matched by *O. oenanthe leucorhoa* which has evenly balanced outline). Although no marked difference in plumage pattern and colours, appearance most characterized by: (1) limited face markings, with particular lack of long rear supercilium and usually dusky-brown lores and cheeks of typical *O. oenanthe leucorhoa*; (2) sandy-pink rather than grey-brown tones in ground-colour of upperparts, with speckled lesser and median wing-coverts and dark bastard wing and tertial-centres more contrasting than on *O. oenanthe leucorhoa*; (3) greater breadth of both white rump and dark brown or black tail-band, producing striking white patch followed by black 'blob' and with tail only rarely showing large ⊥ of *O. oenanthe leucorhoa*; (4) when unworn, rather plain sandy-brown wings, with eye-catching dark rim above and below and remarkably pale underwing, not shown by even palest immature *O. oenanthe leucorhoa*; (5) when worn, dark brown wings, lacking blackness above of *O. oenanthe* and still retaining pale and virtually unmarked under wing-coverts (within *O. oenanthe*, *leucorhoa* has darkest worn underwing); (6) sandy-pink rather than orange-russet tones in ground-colour of underparts, with belly usually creamier than in *O. oenanthe leucorhoa* and producing subtle contrast with breast and flanks. Again, no absolute differences in behaviour, etc., from *O. oenanthe leucorhoa*, but the following are characteristic: (1) often more upright, noticeably taller stance; (2) much more developed loping run on open, level ground; (3) slower, more hesitant (but still rapid) bob of head in alarm; (4) jerkier wagging and flagging of tail (Tye and Tye 1983); (5) more flapping, less flitting flight action, with noticeable 'parachute' appearance due to large area of wings and tail being strikingly displayed on landing. Little chance of confusion with other *Oenanthe* except for ♀ and immature Desert Wheatear *O. deserti* (much smaller, with less obvious, dull buffish rump and all-black tail) and Red-rumped Wheatear *O. moesta* (smaller, with rufous rump and largely black tail). Flight powerful; able to glide and sweep over longer distance than other *Oenanthe*. Escape-flight often long (in keeping with normally open habitat), frequently towards elevated rock or ridge; also runs away from disturbance, taking flight less quickly than *O. oenanthe*. Flight silhouette less cruciform than in other *Oenanthe*, with breadth of wing-base cloaking body and rump and making tail appear even shorter and broader. Gait essentially as *O. oenanthe*, with loping run not diagnostic but used much more habitually and for longer distances, interrupted by brief hopping most often on broken ground. Carriage variable but, at least when foraging, more constantly erect than other *Oenanthe*, even at times apparently almost upright, with length of legs creating impression of stilt; at other times or against high wind, adopts less erect, rounder form or even horizontal

posture. Quite shy, often keeping infuriatingly long distance between itself and observer; best chance of close view comes to observer hidden near favoured food source. Vagrants to western Europe have favoured short grass swards and coastal sand dunes.

Song loud, rich, and varied, with croaks, pipes, and whistles flung out with abandon and interspersed with mimicry of other ground birds, especially *Melanocorypha* larks. Subsong quieter, with whistles recalling Starling *Sturnus vulgaris*, chuckles, and a grating 'cher-we-do'. Calls usually loud, with rich, piped 'weep' or 'dweet' from breeding bird and rather high-pitched, disyllabic 'wheet-whit' and quiet 'tcheep' or 'cheep' from spring and autumn vagrants, sometimes accompanied by wing-flick and head-bob.

Habitat. Breeds in west Palearctic in lower middle and middle continental latitudes on plains and plateaux up to 3500 m in warm arid climate. Prefers level or gently sloping terrain, open but with sufficient isolated shrubs or large rocks, and with clay or sandy soil but not loose sand or surface gravel—although occurs on scree in Caucasus. Found locally on river banks with rich grass cover, and even, in passing, on mown lawns, but prefers very short, sparse vegetation with ample bare patches. Accordingly, largely a steppe and steppe-desert bird, dependent on opportunities for nesting in burrows (Dementiev and Gladkov 1954*b*), but also occurs in forest-steppe in USSR, though avoiding forest and wetlands. Prefers perching on stones or banks to bushes, and likes old ploughland (Bannerman 1954). In Baluchistan (Pakistan), common on plateaux at 1200–2200 m; fond of irrigation banks and excavations, and attracted to vicinity of water; also on roadsides and near houses, though wary (E M Nicholson). At Azraq (Jordan), occurs in wadis in limestone area and much given to perching on bushes, but this probably related to birds on passage (Hemsley and George 1966; Nelson 1973). In Afghanistan, occurs in breeding season at all altitudes up to 3000 m in open flat desolate areas, avoiding mountains, fertile valleys, and normally even margins of cultivation (Paludan 1959). In detailed study in south-west Iran of large sample of breeding territories, flat ground, sometimes also with slope, universally preferred to anything steeper; only Desert Wheatear *O. deserti* showed stronger insistence on flatness and stronger rejection of boulders and rock outcrops. While overlapping freely in habitat, *O. deserti* differed in strong liking for scattered bushes and dislike of stony surfaces. *O. isabellina* normally chose undeveloped steppe, but occasionally used abandoned fields or edge of cultivation—predominantly silt surfaces on wide open plains with widely spaced shrublets, subject to spring flush of herbaceous plants transiently surrounding them but not obstructing free movement. Ill-adapted to flimsy perches but freely resorts to clods, boulders, and firm plants to locate prey and (if shade not accessible) escape

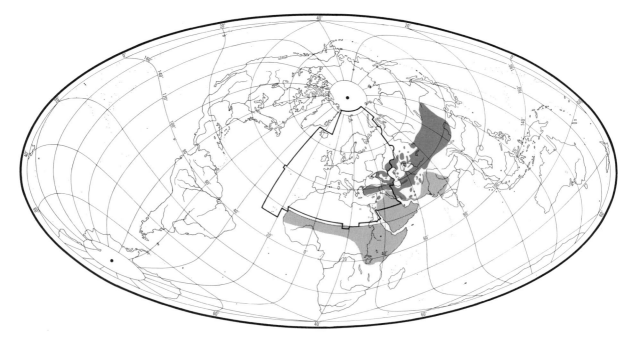

unduly high ground temperatures. Habitat conditions preclude common use of rock sites for nesting, rodent burrows being normal substitute; their availability perhaps influences extent to which otherwise suitable habitat is exploited. Some indication also that pre-emption of steeper and more rocky terrain by other more dominant *Oenanthe* might be partial cause of concentration on less desirable flatlands, and that this might explain apparently atypical occupancy of scree in Caucasus where such competition does not occur (see above). In any event, existing habitat clearly stable and of long standing, associated with conspicuous adaptations to role of ground-dweller hunting over open spaces. (Cornwallis 1975.) In winter in Africa, ranges from coastal flats to 2500 m, mainly in drier and more barren open places in East Africa, but also in short grass at over 1500 m; in West Africa, especially favours burnt ground, in Sénégal sharing habitat with Wheatear *O. oenanthe*, and Black-eared Wheatear *O. hispanica*; in Chad, favours flat sandy bare ground; in Kenya occurs on Nairobi golf course; in Eritrea (Ethiopia), at all altitudes in open country (Moreau 1972). In Red Sea Province of Egypt, spring, occurred along shoreline, in hotel and monastery gardens (even under palm trees and bushes), at a nomad camp, and commonly along arid coastal plain and on flat stretches among mountains, but rarely in open desert (P A D Hollom and E M Nicholson). Winters to extreme west of Sahelian savannas (Heim de Balsac and Mayaud 1962). Mainly a ground bird, flying whenever necessary, normally in lowest airspace.

Distribution. Range increases in USSR and Turkey.
BULGARIA. Bred 1972 and 1981 (TM). RUMANIA. Bred 1975-8 (WK). USSR. Some spread south in 20th century (GGB) and west of Don (Oeser and Martin 1983); spread north in Voronezh region (Wilson 1976). TURKEY. Apparently spreading west and north-west (Kumerloeve 1975). IRAQ. Position unclear. Seen near Amara in July and thought likely to breed in salt deserts (Ticehurst *et al.* 1921-2), also near Habbaniyah in August (E A Chapman and J A McGeoch), but no proof of breeding in Iraq (Marchant 1963*b*). JORDAN. Not found nesting Azraq (Wallace 1983*a*).

Accidental. Britain, Norway, Sweden, Finland, Spain, Malta, Morocco, Madeira.

Population. No information on trends, but range expansion in USSR and Turkey (see Distribution) suggests possible increase.
GREECE. Under 50 pairs (GIH).

Movements. Migratory.
Winters up to northern edge of Sahel zone in West Africa, as far west as south-west Mauritania and northern Sénégal. Common in Mali between $13\frac{1}{2}°$ and 17°N and the commonest *Oenanthe* there north of 16°N (Lamarche 1981; see also Bates 1933-4). Winter distribution in central West Africa may be discontinuous (Moreau 1972); though regular in Chad, rare west of 19°E (Newby 1980). Winters in some areas of Egypt (Goodman and Ames 1983) and common in Sudan (Cave and Macdonald 1955); abundant in Ethiopia (Ash 1980), including Eritrea (Smith 1957), and in Somalia (Ash and Miskell 1983). Occurs regularly in north-east Zaïre (Lippens and Wille 1976; Curry-Lindahl 1981). Common in Kenya, eastern Uganda, and northern Tanzania, but very occasional in

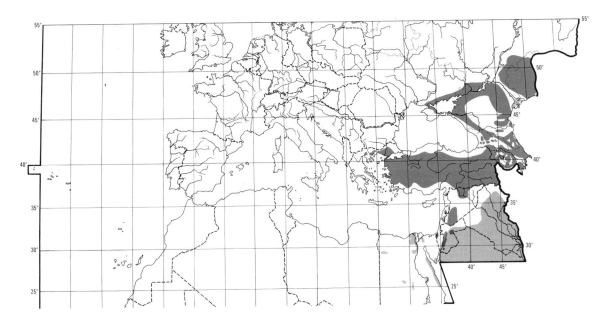

southern and western Uganda; few records south of 4°S in Tanzania (Britton 1980). 3 records in northern Zambia (Aspinwall 1977). Generally rather common through Arabia (e.g. Jennings 1981a). Recorded mid-January at Azraq, Jordan (Conder 1981), and in December and early January in Syria (Macfarlane 1978). Widely distributed winter visitor in Mesopotamia, Iraq (Allouse 1953); uncommon in lowland Iran (Cornwallis 1975). In some years, a few (mainly ♂♂) winter in southern Turkmeniya (USSR) at *c.* 40°N (Bel'skaya 1965a; Panow 1974), though Turkmeniyan population wholly migratory according to Ivanitski (1981c); for earlier detailed discussion, see Rustamov (1954). Fairly common in Pakistan and north-west India in winter (Ali and Ripley 1973b).

Migration protracted; signs of movement as early as late January and late July, but some birds remain in wintering areas furthest west until mid-March, and in breeding areas furthest east until mid-October. Passage more conspicuous in spring than in autumn in Mediterranean areas furthest west. Assessment of movement particularly in Middle East and Turkey complicated by overlap of passage birds with both winter and summer visitors. Wintering areas of individual breeding populations not known. ♀ which bred in south-west Iran (ringed May 1971), recovered in Qatar 27 January 1972, may already have been on passage (Cornwallis 1975).

Passage largely nocturnal (Hogg *et al.* 1984) and on broad front, probably mainly in a WSW or south-west direction in autumn. Most of Ukrainian population probably migrates east of Black Sea, although records in Rumania and Bulgaria suggest some movement to west (see Robel 1981). Rather widespread migrant in Turkey, especially on Black Sea and southern coasts, and in western Anatolia (*Orn. Soc. Turkey Bird Rep.* 1966–75).

Regular both spring and autumn in Cyprus, and fairly common on plains (Bannerman and Bannerman 1971). Occurs in very small numbers on Sicily and Malta, chiefly in spring (Moltoni and Brichetti 1978; Sultana and Gauci 1982). In Tunisia, recorded on passage mainly on south-east central plain, occasionally in autumn, rather more regularly in spring (Thomsen and Jacobsen 1979). For scattered records of passage across Algeria, see Ledant *et al.* (1981). Regularly migrates through Libya; inland records both spring and autumn, but on coast of Tripolitania and in northern Cyrenaica occurs only in spring (Bundy 1976). Generally common or fairly common on passage through Middle East—e.g. Marchant (1963a) for Iraq, Meinertzhagen (1954) for Arabia, and Nelson (1973) for Azraq (Jordan). Recorded from ships in Red Sea and on Socotra in Gulf of Aden (Moreau 1938). In East Africa, small numbers regularly on passage in Lake Victoria basin and coastal lowland Kenya, which lie outside main wintering areas (Britton 1980). Late autumn movement in November demonstrated by presence in night falls of migrants at Ngulia (south-east Kenya); such further southward movement may represent 'step migration' or late arrival (Curry-Lindahl 1981). Further east, common on passage in Baluchistan (Christison 1941) and movement noted in Ladakh into upper Indus valley (Delany *et al.* 1982). These records probably concern passage into southern Asian wintering area.

Leaves breeding grounds from August (sometimes late July) in Mongolia, beginning of September in Tien Shan (south-central USSR). Departure rather protracted with stragglers present in the Gobi until mid-October and in Kazakhstan until early November (Dolgushin *et al.* 1970; Portenko and Vietinghoff-Scheel 1971), while late-October departure from Turkmeniya apparently normal (Panow 1974). In south-west Iran, juveniles and ♀♀ leave the breeding area once young are independent (end of May); ♂♂ remain, most leaving by early September; passage birds occur August to November (Cornwallis 1975). Elsewhere in Middle East, passage spans similar dates: described as August to mid-November in Persian Gulf states (Bundy and Warr 1980), and 2 years' records in Iraq spanned end of July to early November, chiefly September (Marchant 1961, 1962). Movement in Turkey most conspicuous August–September (*Orn. Soc. Turkey Bird Rep. 1966–75*); in Cyprus, mainly September, spanning mid-August possibly to mid-November (Bannerman and Bannerman 1971). Isolated records in Tunisia from late August to late November (Thomsen and Jacobsen 1979); 4 inland records, Libya, September–October (Bundy 1976). Passage across Red Sea recorded until early November (Moreau 1938). Arrives in Eritrea from late August (Smith 1957). Present from September in Sudan (Cave and Macdonald 1955) and Somalia (Ash and Miskell 1983). Occurs in East Africa from October (exceptionally mid-September), with southward passage still evident in November (Britton 1980). Arrives in Chad October–November (Newby 1980), and in Mali September–October (Lamarche 1981).

Departure from wintering areas chiefly February–March. Recorded until mid-March in northern Sénégal (Morel and Roux 1973). Leaves Mali March–April (Lamarche 1981). Last dates in eastern Zaïre early March, and most have left East Africa by end of March although some remain until mid-April (Britton 1980; Curry-Lindahl 1981). Passage recorded chiefly mid-February to late March in Tunisia (Thomsen and Jacobsen 1979) and Libya (Bundy 1976). In Malta, recorded mid-March to mid-April (Sultana and Gauci 1982). Passage through Cyprus early March to late April (Bannerman and Bannerman 1971); normally first *Oenanthe* to arrive in Turkey, with records from early March (*Orn. Soc. Turkey Bird Rep. 1966–75*). In Red Sea Province of Egypt, widespread passage noted early and mid-March (P A D Hollom and E M Nicholson). Migrants occur late January to early May in eastern Saudi Arabia, peaking late February and March (G Bundy); other records indicate this is general pattern for Middle East. Return to breeding grounds in south-west Iran from early February (for fidelity to breeding sites, see Social Pattern and Behaviour); later in February, occurs on passage in quite large numbers (Cornwallis 1975). Arrives in Turkmeniya mid- to late February (Panow 1974; Ivanitski 1981c), in Armeniya (USSR) from early March (Portenko and Vietinghoff-Scheel 1971), southern Kazakhstan from early March, further north up to mid-April, with ♂♂ 5–7 days before ♀♀ (Dolgushin *et al.* 1970). A few arrive in Mongolia as early as end of March and peak arrival in extreme north-east of breeding range mid-April (Portenko and Vietinghoff-Scheel 1971).

In Indian subcontinent, passage in Ladakh and Kash-mir recorded August–October (Delany *et al.* 1982) and March–April. Usually present on wintering grounds in Pakistan and north-west India until March (Ali and Ripley 1973b). Stragglers recorded as far east as Japan and southern Kuril Islands (Portenko and Vietinghoff-Scheel 1971). MGK, DFV

Food. Information mostly from outside west Palearctic. Diet mainly invertebrates; ants and beetles particularly important. Usually forages by making quick dashes along ground after prey (Cornwallis 1975). Sometimes uses perch (e.g. bush, stone) to watch for prey, drops down to ground, and eats item before flying up to same or new perch (Dolgushin *et al.* 1970). Always uses firm perch (Cornwallis 1975). Digs in soil with bill to extract invertebrates, especially in early spring when few on surface (Panow 1974; Ivanitski 1980). Seeks prey in rodents' burrows, particularly in early spring or when ground covered by snow (Bibikov and Bibikova 1955). Recorded catching grasshoppers (Acrididae) by leaping (Mauersberger *et al.* 1982). Also said often to locate prey by hovering (Jonsson 1982), though this unusual in Iran (L Cornwallis). In strong winds in Grampian (Scotland), ran quickly forward in horizontal position and suddenly half-squatted, legs slightly splayed, to capture prey (Beaman and Knox 1981). In Tsavo (Kenya), 94% of 268 food items taken on ground or in herbage (mostly by running about, 8% in sallies from perch), 4% in the air, and 1% in bushes (Lack 1980). Large prey such as cockroaches (Blattidae) and scorpions (Scorpiones) dragged to and fro across ground and eaten piecemeal (Panow 1974). Mole-cricket *Gryllotalpa* killed by repeatedly seizing and throwing to one side (Bel'skaya 1961). Will also hammer large prey with bill and swallow whole. Recorded unsuccessfully attacking toad-headed lizard *Phrynocephalus*. (E N Panov.) When water available will sometimes drink, but not always (Bel'skaya 1965a). In winter quarters, will feed in recently burned areas where prey such as termites and ants unaffected and easily seen (Dittami 1981). Relatively long legs well suited to cursorial foraging, and relatively short tail to upright searching stance (Cornwallis 1975).

PLATE 59 (*facing*).
Oenanthe pleschanka Pied Wheatear (p. 792). Nominate *pleschanka*: 1 ad ♂ fresh (autumn), 2 ad ♂ worn (spring), 3 ad ♀ worn (spring), 4 1st ad ♀ fresh (1st autumn), 5 juv. *O. p. cypriaca*: 6 ad ♂ worn (spring).
Oenanthe hispanica Black-eared Wheatear (p. 806). Nominate *hispanica*: 7 ad ♂ pale-throated morph worn (spring), 8 ad ♂ black-throated morph worn (spring), 9 ad ♀ worn (spring), 10 1st ad ♂ fresh (1st autumn), 11 1st ad ♂ worn (1st spring), 12 juv. *O. h. melanoleuca*: 13 ad ♂ black-throated morph worn (spring), 14 ad ♂ pale-throated morph worn (spring), 15 ad ♀ worn (spring), 16 1st ad ♂ fresh (1st autumn).
Oenanthe finschii Finsch's Wheatear (p. 831): 17 ad ♂ worn (spring), 18 ad ♀ worn (spring), 19 1st ad ♂ fresh (1st autumn), 20 juv. (NA)

Norman Arlott.

PLATE 60. *Oenanthe deserti* Desert Wheatear (p. 820). *O. d. homochroa*: **1** ad ♂ worn (spring), **2** ad ♀ worn (spring), **3** 1st ad ♂ fresh (1st autumn), **4** 1st ad ♂ worn (1st spring), **5** juv. *O. d. atrogularis*: **6** ad ♂ worn (spring), **7** 1st ad ♂ fresh (1st autumn). Nominate *deserti*: **8** ad ♂ worn (spring). (NA)

PLATE 61. *Oenanthe moesta* Red-rumped Wheatear (p. 841): **1** ad ♂ worn (spring), **2** ad ♀ worn (spring), **3** 1st ad ♂ fresh (1st autumn), **4** juv ♂. *Oenanthe xanthoprymna* Red-tailed Wheatear (p. 847). *O. x. chrysopygia*: **5** ad worn (spring), **6** ad heavily worn (summer), **7** 1st ad ♂ fresh (1st autumn), **8** juv. Nominate *xanthoprymna*: **9** ad ♂. (NA)

PLATE 62. *Oenanthe picata* Eastern Pied Wheatear (p. 858). *Opistholeuca* morph: **1** ad ♂ worn (spring), **2** ad ♀ worn (spring), **3** 1st ad ♂ fresh (1st autumn), **4** juv. *Picata* morph: **5** ad ♂ worn (spring). *Capistrata* morph: **6** ad ♂ worn (spring). *Oenanthe alboniger* Hume's Wheatear (p. 872): **7** ad ♂ worn (spring), **8** ad ♀ worn (spring), **9** 1st ad ♂ fresh (1st autumn), **10** juv. (NA)

PLATE 63. *Oenanthe lugens* Mourning Wheatear (p. 858). *O. l. halophila*: **1** ad ♂ worn (spring), **2** ad ♀ worn (spring), **3** 1st ad ♂ fresh (1st autumn), **4** juv ♀. Nominate *lugens*: **5** ad ♂ worn (spring). *O. l. persica*: **6** ad ♂ worn (spring). *Oenanthe monacha* Hooded Wheatear (p. 867): **7** ad ♂ worn (spring), **8** ad ♀ worn (spring), **9** 1st ad ♂ fresh (1st autumn), **10** juv. (NA)

The following recorded in diet. Dragonflies (Odonata), grasshoppers, etc. (Orthoptera: Gryllotalpidae, Acrididae), web-spinners (Embioptera), cockroaches and mantis pupae (Dictyoptera: Blattidae, Mantidae), termites (Isoptera), bugs (Hemiptera), Neuroptera (adult and larval ant-lions Myrmeleontidae, lacewings Chrysopidae), adult and larval Lepidoptera (Noctuidae, Lymantriidae, Sphingidae), caddis flies (Trichoptera), adult and larval flies (Diptera: Tabanidae, Stratiomyidae, Asilidae, Muscidae), Hymenoptera (Ichneumonoidea, Scoliidae, ants Formicidae, wasps Vespidae, Pompilidae, bees Apoidea), adult (mostly) and larval beetles (Coleoptera: Cicindelidae, Carabidae, Silphidae, Staphylinidae, Scarabaeoidea, Elateridae, Cantharidae, Meloidae, Tenebrionidae, Coccinellidae, Cerambycidae, Chrysomelidae, Curculionidae), mites (Acarina), spiders (Araneae), harvestmen (Opiliones), *Galeodes araneoidis* (Solifugae), Scorpiones, millipedes (Diplopoda), woodlice (Isopoda), small lizards (Sauria), molluscs, and seeds. (Morrison-Scott 1937; Hollom 1959; Bel'skaya 1961, 1965a; Pek and Fedyanina 1961; Kovshar' 1966; Zlotin 1968; Dolgushin *et al.* 1970; Panow 1974; Cornwallis 1975; Panchenko 1976; Ivanitski 1980; Kostin 1983.)

In Crimea (USSR), June–August, stomachs of 10 birds contained 59·8% (by number) ants, 7·5% Curculionidae, 7·5% other beetles, 7·5% adult and (mostly) larval Lepidoptera, 6·5% bees and wasps, 5·6% Hemiptera, 4·7% spiders, and 0·9% Orthoptera (Kostin 1983). At Jiddah (Saudi Arabia), January–April, ants occurred in all of 5 stomachs analysed; also beetles (in 4), larval Lepidoptera (2), mites (2), adult and larval Diptera (1), larval Myrmeleontidae (1), Hemiptera (1), Blattidae (1), and seeds (3) (Morrison-Scott 1937). In south-west Iran, 84% of 37 stomachs collected throughout the year contained ants (61% of 2347 items), 78% contained beetles, 30% Hemiptera, 30% spiders, 22% termites, 3% bees, 19% other Hymenoptera, 11% Orthoptera, 11% woodlice, 5% Diptera, 5% Embioptera, 3% Lepidoptera, and 3% millipedes; insect larvae occurred in 16%, insect pupae in 3%, seeds in 3%, and other plant material in 16%; field observations showed lizards, scorpions, and dragonflies occasionally taken; of 2323 prey items measured, 56% 2–5 mm long, 40% 5–10 mm, 4% 10–15 mm, and less than 0·5% 15–20 mm (Cornwallis 1975). In Kirgiziya (south-central USSR), 75 stomachs contained principally ants (in 20%) and beetles (some larval), especially Curculionidae (in 37%) and Carabidae (in 17%); small numbers of adult and larval Diptera, adult and larval Lepidoptera, Hemiptera, and Orthoptera occurred, and occasionally caddis flies, bees, spiders, plant fragments, and seeds (Pek and Fedyanina 1961). In Tien Shan (south-central USSR), 8 stomachs contained chiefly Curculionidae and dung beetles, also adult and larval Lepidoptera, larval flies, and harvestmen (Zlotin 1968). In detailed study in Turkmeniya (south-central USSR), stomachs of 218 birds collected throughout the year contained 3411 invertebrates including 37·9% (by number) ants and 36·2% termites; 56% of stomachs contained ants, 52·3% beetles (chiefly Curculionidae, Carabidae, and adult and larval Tenebrionidae), 27·6% termites, 24·3% caterpillars, 16% Acrididae, and 14% Hemiptera; lacewings, adult and larval ant-lions, cockroaches, Hymenoptera (chiefly spider-hunting wasps Pompilidae), and Diptera occurred in small numbers; also spiders, harvestmen, mites, woodlice, Myriapoda, molluscs, 2 small lizards *Eremias*, and 41 seeds of Leguminosae and *Ceratocarpus turkestanicus*. Ants occurred in diet with almost equal frequency at all seasons (slightly more in winter), beetles more frequently in autumn (68%) and winter (100%) than in spring (43·6%) and summer (49·2%). Diversity of prey taken increased when unusually mild, dry winter followed by unusually wet, cold spring. (Bel'skaya 1965a.)

Diet of young less varied than that of adults, with greater proportion of caterpillars. In Turkmeniya, on one day in April, 06.00–18.00 hrs, adults recorded bringing 404 items on 308 visits to 6 7-day-old nestlings; prey 61·3% (by number) caterpillars (Sphingidae and Noctuidae), 24·8% Myriapoda, 5·2% beetles, 5% spiders, 2·8% adult Lepidoptera, 0·7% Acrididae, 0·2% ants; on average, 1·7 items brought per visit in morning, 1·1 in middle of day, and 1·0 in evening. Individual young

PLATE 64 (*facing*).
Cercomela melanura melanura Blackstart (p. 718): **1–2** ad ♂ worn (spring).
Myrmecocichla aethiops Ant Chat (p. 754): **3–4** ad ♂ worn (spring).
Oenanthe isabellina Isabelline Wheatear (p. 756): **5–6** ad ♂ worn (spring).
Oenanthe oenanthe oenanthe Wheatear (p. 770): **7–8** ad ♂ worn (spring).
Oenanthe pleschanka pleschanka Pied Wheatear (p. 792): **9–10** ad ♂ worn (spring).
Oenanthe hispanica hispanica Black-eared Wheatear (p. 806): **11–12** ad ♂ pale-throated morph worn (spring).
Oenanthe deserti homochroa Desert Wheatear (p. 820): **13–14** ad ♂ worn (spring).
Oenanthe finschii Finsch's Wheatear (p. 831): **15–16** ad ♂ worn (spring).
Oenanthe moesta Red-rumped Wheatear (p. 841): **17–18** ad ♂ worn (spring).
Oenanthe xanthoprymna chrysopygia Red-tailed Wheatear (p. 847): **19–20** ad ♂ worn (spring).
Oenanthe picata Eastern Pied Wheatear (p. 858): **21–22** ad ♂ *opistholeuca* morph worn (spring).
Oenanthe lugens halophila Mourning Wheatear (p. 858): **23–24** ad ♂ worn (spring).
Oenanthe monacha Hooded Wheatear (p. 867): **25–26** ad ♂ worn (spring).
Oenanthe alboniger Hume's Wheatear (p. 872): **27–28** ad ♂ worn (spring).
Oenanthe leucopyga White-crowned Black Wheatear (p. 876): **29–30** ad ♂ worn (spring).
Oenanthe leucura leucura Black Wheatear (p. 885): **31–32** ad ♂ worn (spring). (NA)

received average 33 feeds per day at 5 days, 51·3 at 6 days, and 54·5 at 9 days. Proportion of beetles increased with age of young. Prey often presented live. Young c. 1 week after fledging seen to run up to prey timidly, flapping their outspread wings before catching prey or even backing away. (Bel'skaya 1961, 1965a.) In Ukraine (south-west USSR), fragments collected from 10 nests after young had fledged comprised chiefly beetles, also 2 moths (Noctuidae) and 1 bee (Panchenko 1976). In Iran, June, flying young recorded being fed black ants (Buxton 1921).   DFV

Social pattern and behaviour. Major studies in Badkhyz (southern Turkmeniya, USSR) and Tuvinskaya ASSR (south-west of Lake Baykal, eastern USSR) by Ivanitski (1978, 1981b, c, d, 1982); further important data and comparisons with other *Oenanthe* in Panow (1974) and Panov (1978). Account includes information supplied by E N Panov.

1. Normally solitary and territorial in winter, though apparent pairs recorded towards end of winter and during migration when birds sometimes loosely gregarious, also associating with other *Oenanthe* (Bates 1933-4; Archer and Godman 1961b; Duhart and Descamps 1963; Bel'skaya 1965a; Cornwallis 1975). In south-west Iran, ♂♂ and ♀♀ hold separate individual territories (L Cornwallis). In Kenya, birds are attracted to recently burnt areas, both sexes there establishing separate territories (c. 2 ha) defended against conspecific birds of both sexes and certain other species (including *Oenanthe*); in two cases, 'pairs' defended combined territory of larger-than-average size (Dittami 1981; Leisler et al. 1983); for further details, see Pied Wheatear *O. pleschanka* (p. 797) and Antagonistic Behaviour (below). In northern Iran, late March, c. 15 birds well spaced along c. 3 km of Caspian shore and no concentrations (Schüz 1959). Loose groups of 6-8 spring migrants (birds c. 20-30 m apart) occur in USSR, but normally solitary then and birds staying 1-4 days at migration stopovers are territorial (Ivanitski 1982). On Karpathos (Greece), apparent migrant pair occupied oval territory c. 60 × 40 m for at least 1 week (Kinzelbach and Martens 1965). On breeding grounds, young birds forced out to edge of parental territory and there establish (at c. 1½ months) initially small home-range which develops into territory (Ivanitski 1978, 1982); see also part 2. BONDS. Mating system apparently varies between areas. Monogamous in Iran (L Cornwallis). In Tuva (Tuvinskaya ASSR), population evidently wholly monogamous, unpaired ♂♂ rare and no unpaired ♀♀ noted. In Badkhyz, up to c. 30-40% of ♂♂ bigamous (Ivanitski 1978, 1981c). Apparently more ♂♂ than ♀♀ in various parts of Turkmeniya: in adults (n = 338, February-November) ratio 1·6:1; in nestlings and fledglings (n = 62, April-July), 2:1. However, this counterbalanced by (apparently) later maturation of ♂♂: 47% of ♂♂ and 33% of ♀♀ did not breed in March, 35·3% of ♂♂ and 10% of ♀♀ in April. Birds of either sex not breeding at start of season are probably (mostly) from late broods of previous year, and as the ♀♀ then mature on average sooner than the ♂♂, this creates the potential for simultaneous or successive polygyny. 1st (resident) ♀ tends to react aggressively towards new ♀ only during nest-building and copulation and an unpaired ♀ usually settles in a territory only when ♀ there already incubating; little aggression shown by 1st ♀ if she has young out of nest (and, in USSR populations studied by E N Panov, ♀ *O. isabellina* anyway less aggressive overall than *Oenanthe* with strict territorial system; for Iran, see below). ♂ will also attack 1st ♀ in defence of new ♀ (see

part 2). Non-breeding but territory-holding young ♂♂ may display only when ♀♀ appear; tend to attack and chase off ♀♀. Unpaired ♀♀ non-territorial, initially nomadic within 'home-range' embracing territories of 3-5 pairs and unpaired ♂♂. Some tolerance shown by ♂ territory-owner towards ♂ intruders (which may display at resident ♀) and this might favour promiscuity, but no promiscuous copulations by paired birds recorded by Ivanitski (1978); see, however, subsection 4 in Heterosexual Behaviour (below) for indication that such copulation is attempted. In northern and mountain populations, 1 synchronous breeding cycle is normal and all birds breed at 1 year old; no polygyny recorded. Little information on duration of pair-bond. In Turkmeniya, sexes arrive more or less at same time (♂♂ sometimes a few days earlier) and pair up immediately; in some cases, perhaps already paired on arrival—see discussion of dispersion outside breeding season, above. (Bel'skaya 1965a, 1979; Ivanitski 1978, 1981b, c, 1982; E N Panov.) In south-west Iran, ♂ whose ♀ went missing at start of breeding season paired with another ♀ the same season. By 2nd year of study of 2 neighbouring pairs, one ♂ was missing and his mate then moved to join and pair with the other ♂ whose mate had also disappeared. Pair-members normally separate when young disperse. (Cornwallis 1975.) Young normally fed by both sexes, but in southern Turkmeniya at least, more by ♀. Normal for 2nd brood to begin before 1st-brood young independent and ♂ feeds these alone if ♀ re-lays (Bel'skaya 1965a, 1979; Ivanitski 1978). No information on amount of feeding of each ♀'s brood done by bigamous ♂. One case of brood-division recorded in south-west Iran: each parent fed 2 young in different halves of territory (L Cornwallis). Young fed for c. 12-14 days after leaving nest and, when close to independence (at c. 25-30 days) are attacked first by ♀ (if starting to re-lay) then by ♂ and forced to stay on edge of territory. With development of aggression and territoriality, these young may attack offspring of later broods. In Badkhyz, 1st-brood young sometimes stay long in parental territory, even until 2nd brood fledges, or longer; in Tuva, break-up of family comes when adults and young move away c. 1-1½ weeks after single brood have left nest. (Bibikov and Bibikova 1955; Bel'skaya 1965a; Ivanitski 1978, 1981c.) BREEDING DISPERSION. In Iran, solitary and strictly and aggressively territorial (L Cornwallis). In USSR, solitary and only loosely territorial. In open plains with many rodent colonies (see below), pairs occupy large, widely scattered territories and rarely come into contact (Panow 1974; Ivanitski 1978; E N Panov). In Zeravshan valley (Tadzhikistan, USSR), pairs 1·5-2 km apart (Abdusalyamov 1973). Territories sometimes clustered. In Tuva, nests average 168 ± 19 m apart, minimum 32 m (n = 13); 136 ± 11 m, minimum 80 m (n = 13), from nests of Wheatear *O. oenanthe*, and 80 ± 10 m, minimum 56 m (n = 9), from those of Desert Wheatear *O. deserti* (Ivanitski 1980). In Badkhyz, c. 30 m between nests recorded (Ivanitski 1978). In south-west Iran, average distance between neighbouring nests 83·4 m (20-170), n = 7 (L Cornwallis, E K Dunn). Territory established and defended by ♂; used for courtship, copulation, nesting (new nest normally built for 2nd clutch), rearing of young, and at least some feeding, adults foraging up to c. 80 m from nest (Bel'skaya 1965a). In Turkmeniya, territory may be used almost twice as long as in Tuva and can hold up to 12-14 birds—adults and 2 broods (Ivanitski 1981c). Size not easy to estimate owing to overlap zones c. 20-50 m wide. In one study, 48% of all conflicts between neighbouring ♂♂ took place in peripheral zones, others arose during frequent intrusions by ♂♂ into neighbour's territory (see Antagonistic Behaviour, below). According to Rustamov (1954), ♂ defends only c. 10-15 m radius of nest.

Average territory size in south-west Iran 2·2 ha (0·52-6·9), $n = 27$ (L Cornwallis, E K Dunn). In Badkhyz, territory averages 2·7 ha (1·1-6·4), $n = 12$ (Ivanitski 1978); in later study (Ivanitski 1981c), territory of 1st-year ♂♂ 0·8±0·1 ha ($n = 6$), of older monogamous ♂♂ 1·5±0·2 ha ($n = 16$), and of bigamous ♂♂ 3·2±1·2 ha ($n = 6$). Territories of 1st-year ♂♂ usually located in less favourable areas; defended against others of same age and against older ♂♂ though latter always dominant. In 2 areas of Tuva, density of local neighbourhood groups higher than in Turkmeniya, and territory in Tannu-Ola foothills 0·6±0·1 ha ($n = 13$) and on Tsagan-Shibetu 1·3±0·2 ha ($n = 15$). 2 ♀♀ paired with same ♂ show weak territorial division of ♂'s territory, each chasing off the other from her own area, or one may dominate the other in larger part of shared territory. (Ivanitski 1981c, 1982.) All information on densities is from USSR where major density-determining factor said to be availability of rodent burrows for nesting (Panow 1974), though other habitat factors perhaps also involved: e.g. in Turkmeniya, *c.* 10 birds per 10-km stretch where 4·7 colonies of gerbil *Rhombomys opimus* per ha, just over 1 bird per 10 km where only 2 such colonies per ha; infilling of unoccupied rodent burrows by wind-blown sand also reduces density (Babaev 1967). Where *R. opimus* numbers reduced due to ploughing, *O. isabellina* correspondingly declines: in Kopet-Dag south of Ashkhabad (Turkmeniya) in 1960, 20 birds along 10 km, in 1966-9 *c.* 1 (Bel'skaya 1979). For further data, see Bel'skaya (1965a) who noted 190 nests per km² over 32·5 ha of Badkhyz. In Ukraine, overgrazing by sheep has apparently created habitat favourable to rodents and hence to *O. isabellina*: on stony steppe 140 pairs per km² over 25 ha, and in pasture adjoining dirt road 1250 pairs per km² over 2 ha; density lower in *Stipa-Festuca* steppe—e.g. 6·25 pairs per km² over area 8 km × 100 m (Panchenko 1976). Also in Ukraine, 4 pairs recorded along 1 km of coastal lagoon on Sea of Azov (Oeser and Martin 1983). In Azerbaydzhan, 15 birds per km² in valley with tamarisks *Tamarix* (Drozdov 1965). Favourable habitat with many rodent holes in Volga-Ural sands holds up to 300-400 pairs per km² (Shevchenko *et al.* 1969; see also Shevchenko 1969). Further data from Caspian region in Poslavski (1963, 1974) and Neruchev and Makarov (1982). In south-east Kazakhstan and Kirgiziya, 1-15 birds recorded along 10-km stretches (Bibikov and Bibikova 1955) and similar transects at 1500-2000 m in Tuva showed 4-16 pairs per 10 km (Ivanitski 1980, which see for habitat preferences). Fidelity to breeding territory recorded in south-west Iran: 4 out of 7 marked ♂♂ occupied same territory in 2 successive years; other 3 missing in 2nd year (Cornwallis 1975). ROOSTING. Nocturnal and normally in rodent hole which is also used for sheltering from bad weather (Swerew 1927) and excessive heat (Rustamov 1954); see below. Whole family will use same hole (probably in some cases the nest-hole): e.g. in Ukraine, late May, 1st chick entered hole to roost 15 min after sunset, others by 15 min later, parents 15 min after that (Panchenko 1976). According to Ivanitski (1982), brood sometimes splits into 2, each group roosting apart. Later, and outside breeding season, solitary roosting probably typical. In Turkmeniya, active from dawn to after dark—up to 22.00-23.00 hrs on moonlit night in March (Rustamov 1954; Bel'skaya 1965a). For diurnal activity rhythms when feeding young in south-west Iran, see Cornwallis (1975). Wintering birds in Mali feed mainly in evening (Duhart and Descamps 1963). Young birds bathe readily in captivity and, unlike other *Oenanthe*, also fond of sand-bathing (Panow 1974).

2. Vagrants more approachable than *O. oenanthe* (Nicolau-Guillaumet 1971; Rogers 1981), these (as well as apparent migrants in Greece) allowing approach to *c.* 6-20 m (Kinzelbach

and Martens 1965; Bernis *et al.* 1973; Lubián and Moreno 1974; Rogers 1981). Excited bird may have ruffled plumage and typically bobs head and fore-body (as in *O. oenanthe* but slower, see p. 781), often accompanied by tail-raising and wing-flicking; constant tail-wagging like wagtail *Motacilla* is characteristic (not just of alarmed bird: Catley 1981), and tail sometimes fanned then (Raethel 1955). Perhaps crouches briefly when approached (Schüz 1959), but normally runs off fast, body held low, when disturbed, then stopping alert in a high erect posture. (Nicolau-Guillaumet 1971; Panow 1974; Beaman and Knox 1981; Rogers 1981; Tye and Tye 1983.) Pursued by Hobby *Falco subbuteo*, bird dived down and took refuge under bicycle, close to man (Passburg 1966). FLOCK BEHAVIOUR. No information. SONG-DISPLAY. ♂ sings Territorial-song (see 1a in Voice) from low perch (pile of stones, bush, etc.) or in flight (Panow 1974; Bergmann and Helb 1982). In Turkey, sometimes fluttered only 1-2 m into the air (P A D Hollom) or fluttered up calling (see 4 in Voice), suddenly half closed wings and dived vertically to rock, sometimes running then to ♀ (Wadley 1951). In fuller (rather variable) Song-flight, ascends steeply with fluttering wing-beats, tail widely spread and lowered (colour pattern conspicuous), to *c.* 10-20 m or, according to Sarudny and Härms (1926), 'several hundred feet'. Body and head may be horizontal or (perhaps nearer peak of ascent), head raised. Tail rarely raised during ascent; feet may dangle. Hovers for a few seconds at peak, sometimes also briefly once or twice during ascent and 1st part of descent. Descent is otherwise gliding (sometimes in spirals) with wings and tail spread, and/or plummeting and still accompanied by loud song which continues after landing, sometimes near ♀, on perch. (Swerew 1927; Ludlow and Kinnear 1933; Wadley 1951; Raethel 1955; Paludan 1959; Bel'skaya 1965a; Panow 1974; Bergmann and Helb 1982; P A D Hollom; E N Panov.) Will also sing from ground—either Territorial-song or louder variant of Subsong (Panow 1974: see 1b in Voice). Song-flights performed frequently by ♂ when ♀ feeding young; more by monogamous than by polygamous ♂♂, though latter will increase rate if both their ♀♀ incubating simultaneously. At peak of ♂'s self-advertisement phase, Song-flights (with tail fanned) last 15·0±1·2 s ($n = 56$); at low intensity, 8·2±1·0 s ($n = 18$) and tail usually closed (Ivanitski 1978, 1982). ♀ occasionally also sings, including in flight (see subsection 2 in Heterosexual Behaviour, below). In Kara-Kum (Turkmeniya), ♂ sings mostly in morning (e.g. in April from *c.* 05.00-06.00 hrs), less so in evening, but will sing (from perch or in flight) even during hottest time of day; few if any 'desert' (steppe) species sing as much (Rustamov 1954; Panow 1974). Will sing also at night (Bel'skaya 1965a; Kozlova 1975). In Nakhichevanskaya ASSR (south of Caucasus, USSR), in cold, late spring, birds arrived from 4 March and sang actively from 11 March (E N Panov); in south-east Kazakhstan, display and song noted from *c.* 5-7 days after late-March arrival (Bibikov and Bibikova 1955). In Soviet Central Asia and Mongolia, song continues to mid-June (Pleske 1889-94). Winter Subsong given from perch (Smith 1971) or, in Egypt, throughout winter from ground with bill closed (Moreau and Moreau 1928). ANTAGONISTIC BEHAVIOUR. (1) General. In Turkmeniya, spring migrants on temporary territories show some aggression and interest in burrows. Local ♂♂ initially not very demonstrative, not performing Song-flights, but stay in (and defend) small area usually centred on rodent colony. Few contacts initially as density low, but conflicts more likely as ♂'s mobility and size of territory increase; in Badkhyz, conflicts and Song-flights extend over *c.* 3 months (Ivanitski 1978, 1982). In south-east Kazakhstan, territories occupied *c.* 5-7 days after arrival in late March

(Bibikov and Bibikova 1955). On Cyprus, migrants very aggressive towards other birds, especially *Oenanthe* (Mason 1980). In Tuva, strict linear hierarchy among sympatric *Oenanthe*, with *O. isabellina* dominant over *O. oenanthe* (even migrant *O. isabellina* feed with impunity in *O. oenanthe* territory), *O. deserti*, and *O. pleschanka*. *O. isabellina* involved in more and longer conflicts than other species and shows more extravagant display postures. In other parts of range, apparently dominated by *O. finschii*, Eastern Pied Wheatear *O. picata*, Hume's Wheatear *O. alboniger*, and Mourning Wheatear *O. lugens* (Ivanitski 1980; also Cornwallis 1975 for Iran), though ♂ *O. finschii* chasing ♀ *O. isabellina* will be attacked and driven away by her mate. ♀ generally less aggressive than other ♀ *Oenanthe* but sometimes drives from nest conspecific birds of both sexes, occasionally also ♀ *O. finschii* (Panow 1974). Will drive off Crested Lark *Galerida cristata* (Rustamov 1954; Ivanitski 1980; L Cornwallis). 2 birds in Mongolia approached to within *c.* 1 m of 2 Demoiselle Cranes *Anthropoides virgo* and attacked them vigorously (Mauersberger *et al.* 1982). In winter quarters, also, attacks more than is attacked, but hierarchy evidently less strict than on breeding grounds as *O. isabellina* sometimes attacked by *O. oenanthe* (Leisler *et al.* 1983). (2) Threat and fighting. In Kenya, winter territories demarcated mainly by threat (no details) from hummock; only one Song-flight recorded (Leisler *et al.* 1983). At Massawa (Eritrea) birds drove off conspecific birds and certain other species from artificial food source: 2 ♂♂ in conflict faced up on ground, hovered bill-to-bill at 'a few feet', landed and repeated the action, then both chased presumed ♀ (Smith 1971). On breeding grounds, birds with adjacent territories sometimes perform low-intensity parallel run with wing-flicks, tail movements, mock-preening, and mock-feeding. With time, increased territorial advertisement expressed in Song-flights and, on ground, aggressive bird will fan and raise or lower tail, adopting Head-up posture (usually indicating attack-intention) with neck extended, bill raised, and plumage sleeked. In apparent Threat-gaping, tail widely fanned and raised or lowered; bill opened wide to reveal black gape against pale face (see drawings in Panow 1974 and Ivanitski 1981*d*), and effect further enhanced by presence (in most ♂♂) of equally black lores, thus extending mask back to eyes (L Cornwallis). Bobbing, wing-shivering, Dancing-flight (see Heterosexual Behaviour, below), slow mincing gait, and long slow Quiver-flights (shallow wing-beats, tail spread) also recorded in antagonistic interactions. Conflicts sometimes continue for hours, though not always with much demonstrative behaviour. Fierce fight can occur at any moment, even in low-intensity dispute; may be preceded by mock-feeding (birds picking up small twigs, etc.) and attack then ensues, attacker also calling (see 7 in Voice); birds fly up, breast-buffeting, and chase may follow (or birds alternately rest and fight), more active bird returning immediately afterwards to own territory. Extreme aggression (typically with loud Subsong) usually arises only when attempt made to settle in centre of another's territory. However, unpaired ♂♂ and paired ♂♂ whose ♀♀ are incubating constantly intrude into core of neighbour's territory, especially during pair-formation and copulation. ♀ spending much time near territory limit during 1st copulation phase (see Heterosexual Behaviour, below) shows no aggression to other birds of either sex, but intrusions by her can cause conflicts and other ♂♂ will generally perform more Song-flights and stay near limit of a territory where pair-formation is taking place. Intruder usually attacked and returns to own territory after short chase; sometimes, pair and ♂ intruder fly far beyond territory limit, resident ♂ performing Song-flight on his return. Intruding ♂ sometimes succeeds in displaying at resident ♀

who may be neglected by her mate in favour of a 2nd ♀. Unpaired ♀♀ constantly intrude into territories of pairs (see also Bonds, above). In early stages 1st (resident) ♀ hostile and ♂ makes no attempt to stop her, sometimes even attacking new ♀ himself. Later, when ♂ displays to new ♀, 1st ♀'s attacks on new ♀ elicit aggression from ♂ who tends to chase 1st ♀ back to nest, so that new ♀ may eventually be able to settle in territory; in such a case, long and fierce fights occur between ♂ and 1st ♀, but no hostility between them outside new ♀'s centre of activity. For posture adopted by ♂ threatening ♀, see Fig A. 1-year-old ♂♂ subordinate to older ♂♂ and generally less

A

demonstrative even in conflicts (most fights in any case involve only older ♂♂), rarely adopting Head-up posture, flirting tail, etc.; also perform fewer Song-flights. (Pleske 1889–94; Panow 1974; Ivanitski 1978, 1980, 1981*b*, *c*, *d*, 1982; E N Panov.) HETEROSEXUAL BEHAVIOUR. (1) General. Formation of breeding population in Badkhyz takes place over *c.* 1½ months from mid- or late February; in Tuva, process complete *c.* 2 weeks after main arrival late April to early May, pair-formation, nest-building, and incubation being relatively synchronous. In both populations, pair-formation completed in *c.* 1½–2 weeks; *c.* 8–12 days between start of building and incubation (Ivanitski 1978, 1981*b*, *c*, 1982). On Mangyshlak peninsula (east Caspian, USSR), pair-formation noted late March to early April, nest-building and copulation late April (E N Panov); timing of pair-formation in Mongolia similar (Kozlova 1933). Where ♂♂ and ♀♀ arrive together on breeding grounds, early contact between sexes apparently accidental, ♀♀ moving about where ♂♂ establishing territory; later, ♂'s Song-flights are important marker. Some ♀♀ quick to pair up and breed, others stay long (sometimes weeks) unpaired (Ivanitski 1982). (2) Pair-bonding behaviour and nest-site selection. General scheme of pair-formation as in other *Oenanthe*, but rather variable. Some interactions at burrow described below may refer only to nest-site selection rather than pair-formation, though probably no fundamental differences. When ♀ appears in territory, ♂ typically does not approach immediately but starts loud song, raises (closed or fanned) tail, and may make little steps; shows round-headed appearance due to ruffled feathers. May approach ♀ in this Tail-up posture (wings also slightly drooped), but more likely to run (Fig B) or fly to burrow in Song-flight or

B

low, slow Quiver-flight (with loud song), standing there in Tail-up posture (Fig C, right) before entering. Typically gives short Territorial-song at this time and Rasp-call (see 6 in Voice) when in burrow. Flamboyant and varied behaviour by ♂ at burrow more likely to attract ♀ and to lead to pair-formation. After hesitant approach, ♀ (head feathers typically sleeked) may

C

freeze in bowed posture over burrow (as may ♂), then enter, sometimes followed by ♂ who usually exits first (in Tail-up posture) and will bow towards entrance (Fig C, left) while ♀ inside. ♂ also reported to move with light dancing steps around burrow, alternating with adoption of a high erect posture (bill horizontal); or half-enters hole so that tail still visible, comes out backwards, runs about, approaches ♀, then runs away from her. ♂ sometimes takes bit of nest-material into burrow or comes out with some object, running or flying short distance with it. May fly to another burrow for repeat performance, but ♀ tends to stick to only 2–3 burrows demonstrated by ♂ in first hours of pair-formation. When not nest-showing, ♂ will sing from mound or in flight. Short fights at burrow between ♂ and ♀ (initiated by either sex) not uncommon in early stages. Other features of ♂'s behaviour apparently associated with pair-formation: Head-up posture (sometimes held for several minutes while performing various activities); Looking-down (body erect, head bowed, bill tucked against breast) and another posture in which body horizontal, bill pointed down, wings opened and held forward, and closed tail moved up and down (for illustrations, see Panow 1974); moving towards ♀ in zigzag leaps (tail fanned, wings open) or (perhaps only preceding copulation) in zigzag flight; walking around ♀ while singing with wing-tips brushing ground, tail fanned, head down. (Bel'skaya 1965a; Panow 1974; Ivanitski 1982; E N Panov.) Ivanitski (1981b) distinguished 2 phases in which copulation may take place. (i) Over 3–8 days during nest-building, with ♂ (typically showing low-intensity postures—no tail-fanning, tail-raising, bill-raising, or sleeking of head feathers) following ♀ closely (presumably mate-guarding as in *O. oenanthe*, see p. 784). ♀ highly active, with much movement about territory including Quiver-flight, Song-flight, and frequent and long visits to territory limit (despite ♂'s attempts to prevent this); ♀ also frequently approaches ♂ (sometimes in Quiver-flight) and quivers wings while on ground. If ♂ adopts an erect posture and attempts to approach ♀, she will move away in slow mincing gait, still wing-quivering. Quiver-flight by ♀ induces ♂ to follow similarly (though tail closed), this leading eventually to fast chase, ♂ close behind ♀. Such chases characteristic of whole pair-formation and -consolidation period (including 2nd phase—see below); usually ends with ♀ hiding in burrow around which ♂ performs Dancing-flight (see subsection 4, below) and other ostentatious displays as illustrated by Ivanitski (1981b). (ii) Towards end of nest-building, ♀ much less mobile while ♂ now very demonstrative, frequently involved in territorial conflicts, displays to other ♀♀, Song-flights, and Quiver-flights. Tail-flirting and sleeking of plumage typical. ♂ spends much time on display-ground (e.g. boulder or edge of

hollow), flying there in Quiver-flight, singing. Postures adopted there highly variable: wing-shivers, moves about in slow mincing gait, runs, enters holes, and sings—loud Subsong mainly on ground, loud Territorial-song in flight, landing always on display-ground. If 2nd ♀ settles in territory, 1st ♀ reverts to behaviour typical of 1st phase, ♂ in his courtship of new ♀ likewise. (Ivanitski 1981b; E N Panov.) (3) Courtship-feeding. Not known to occur; incubating ♀ not fed on nest by ♂ (Bel'skaya 1965a), though at one nest with young one adult would wait by nest and beg food from mate on its return: extended neck, opened bill, and quivered wings (Bel'skaya 1961); not known whether food ever passed. (4) Mating. On breeding grounds, copulation recorded over *c.* 10–15 days from *c.* 3 days after start of nest-building to *c.* 5 days after start of incubation. During 1st phase (see above), copulation may follow chase. Pre-copulatory behaviour highly variable or nil; preliminaries reduced late in season or where 2nd ♀ has settled in territory. Post-copulatory behaviour highly stereotyped. ♂ mounts, settles, bends tail under ♀'s for cloacal contact. Flies away quickly afterwards, typically swinging from side to side and giving either call 5 (E N Panov) or long series of call 2b (for up to several minutes: see Voice); call 2b also given from perch or ground after landing (Ivanitski 1981b). ♂'s departure in straight Quiver-flight with song but no calls apparently indicates copulation unsuccessful. Birds may copulate again within *c.* 40–50 min. During 2nd phase, ♀ flies (Quiver-flight) then runs to ♂ at his display-ground. ♂ then performs Dancing-flight (typically associated with copulation, being performed *c.* 1–10 s beforehand: Ivanitski 1978): moves over ♀ in series of arcing leaps (Fig D) of radius *c.* 1·5 m, irregularly

D

beating wings and brushing ground with breast (or wing: Bel'skaya 1965a); 2–50 leaps per series (*c.* 1 per s), short series (up to 5 leaps) sometimes separated by pauses of a few seconds. Rolling-display of ♂ (Fig E), unique to *O. isabellina*, also

E

occurs. ♂ may eventually land from Dancing-flight close to (wing-quivering) ♀ and creep towards her with wings spread and quivering (Fig F, right); crawls thus on to ♀ and copulates,

F

flying off in swinging fashion as described above. Most contacts at this stage do not lead to copulation which anyway takes place only after many unsuccessful attempts (e.g. c. ¼ of all attempts frustrated by intervention of other ♂♂). ♂ recorded copulating with his 1st mate c. 20 min after series of fights caused by arrival of new ♀. ♂ neighbour will use a display-ground in its owner's absence and attempt to copulate with that bird's mate there. When ♀ incubating, ♂ may attempt (with all preceding ceremony) to copulate with lump of earth, etc. (Ivanitski 1981b; also Panow 1974, E N Panov.) (5) Behaviour at nest. ♀ starts to build soon after suitable site found, taking material from near entrance, later from further away; tends to return in Quiver-flight (Ivanitski 1981b, which see for further details). ♂ stays close, at least in early period, and will sing (various song-types) while bowed over nest when ♀ inside. Later, rarely comes close when ♀ arrives (Panow 1974). RELATIONS WITHIN FAMILY GROUP. No information on brooding. At one nest in Mongolia, both adults usually arrived with food together, ♀ always feeding young first; if flying in, landed c. 1–2 m from nest (Mauersberger et al. 1982). For begging posture of young and posture indicating refusal to take food, see Panow (1974), also Voice. Nest apparently cleaned by both adults (Swerew 1927), faecal sacs being swallowed or dropped close to nest (Bel'skaya 1961). Young emerge when fully feathered, but stay close to burrow entrance for c. 5–6 days, although will move about within complex tunnel system of rodent colony, thus straying far from nest which they finally leave at c. 13–15 days, still unable to fly; fed for c. 2 weeks more though capable of self-feeding within c. 5–6 days (Rustamov 1954; Bel'skaya 1965a; Panow 1974; Panchenko 1976). ♀ may re-lay when 1st brood still dependent; independent 1st-brood young may be attacked furiously by ♂ if they approach 2nd nest (Panow 1974). ANTI-PREDATOR RESPONSES OF YOUNG. After first emerging from nest, disappear immediately into burrow if threatened. Will also hide in burrow (not necessarily nest-hole) after fledging (Swerew 1927; Bibikov and Bibikova 1955; Bel'skaya 1965a). PARENTAL ANTI-PREDATOR STRATEGIES. (1) Passive measures. After feeding young, parent tends to stick head out of hole and look about before flying off; will retreat inside if disturbed (Mauersberger et al. 1982). In Greece, birds ran from man and bolted down burrow (Watson 1961a); likewise for birds of prey (Swerew 1927). (2) Active measures. In Mongolia, ♂ apparently directed Song-flight at man during nest-building (Piechocki et al. 1982). Highly agitated birds, especially when young out of nest, will give alarm-calls (see 2 and 5 in Voice) for man, dog, or raptor, often then retreating into hole (Swerew 1927; Johansen 1954; Panchenko 1976). Recorded mobbing Little Owl Athene noctua in Iran (Cornwallis 1975). Suslik Citellus or marmot Marmota approaching nest similarly elicits loud clicking alarm-calls (see 2 in Voice) and is subjected to repeated dive-attacks, usually resulting in expulsion; afterwards, bird will run about with wings drooped and tail fanned before returning to nest (Swerew 1927; Bibikov and Bibikova 1955).

(Figs by K H E Franklin: from drawings in Panow 1974 and by E N Panov.)   MGW

Voice. Outside breeding season uses Subsong and calls 2 and (especially) 5 (e.g. Nicolau-Guillaumet 1971, Beaman and Knox 1981, Rogers 1981a, Leisler et al. 1983). Apparently silent after breeding, during migration, and also in early period after arrival on breeding grounds, though vocal there later, loud and astonishingly mimetic song being the most striking feature (Bibikov and Bibikova 1955; Robel 1981; Königstedt and Robel 1983). For additional sonagrams, see Bergmann and Helb (1982).

CALLS OF ADULTS. (1) Song. Differs distinctly from other Oenanthe. Improvisatory, with (in both main song-types) much expert mimicry of bird and other sounds, and usually lacking any kind of stereotypy. Roughly divisible into 2 main types (based on structural organization). Given mainly by ♂, but ♀ sings occasionally (probably both main types), including in Song-flight. (a) Territorial-song. Given in flight or from ground or perch. Of fine quality ('nightingale of the desert'); loud, liquid, melodious, and varied, incorporating notes with overtones, (series of) calls 2, 5, and 6 (sometimes song consists only of such calls), clear and masterly imitations, and sometimes part of loud Subsong (see below). Comprises short phrases of c. 1–2 s or often longer sequences (4–30 or more units) comprising segments of similar sounds or sound combinations, sometimes modified, e.g. 3–4 such segments during a Song-flight. Loud, piercing whistles often given at end of song or may precede it (Panow 1974; Panov 1978; E N Panov; also Bergmann and Helb 1982, Jonsson 1982); in recording (Fig I), 9 'weeoo' units preceded by 'whee weee wee-er ee-oo' (J Hall-Craggs); in recording by S Taylor (Greece), such whistles feature prominently all through the song. Recording by P A D Hollom reveals remarkable jumble of sounds in rather disjointed bursts: 'tzuk' (call 2a), rather nasal and quiet churring, and much louder hard rattles like Mistle Thrush Turdus viscivorus; piping like Oystercatcher Haematopus ostralegus as well as other squeaks ('kewk' or 'keek'), whistles, including the errand-boy 'whee-ou-whit' (see call 5), and gurgles at times apparently ventriloquial (perhaps due to simultaneous use of 2 internal tympanic membranes which is regular in song: Panov 1978); harsh barking like mating call of ♂ fox Vulpes and a gentler yapping or laughing sound; short and explosive units, some harsh and almost metallic, and dry fizz not unlike gravelly sound of Black Redstart Phoenicurus ochruros; explosive, spluttering rattles and loud rasps; bleating like young goat; harsh

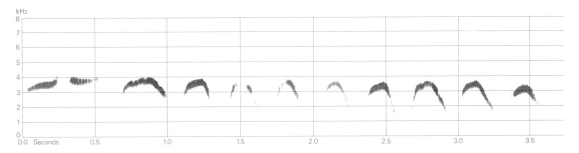

I E D H Johnson Turkey April 1972

'chack' sounds like Fieldfare *T. pilaris*; mimicry of larks (Alaudidae), probably including Calandra Lark *Melanocorypha calandra*, also Kiew-call of Little Owl *Athene noctua*. In sonagrams from this recording, Fig II shows pleasant whistle similar to but lower pitched than 'weeoo' (see above) followed by attractive warbling, and Fig III (first 3 units) is mimicry of Quail *Coturnix coturnix* (P A D Hollom, M G Wilson). *O. isabellina* has few rivals as a mimic, copying remarkable range of bird, mammal, and mechanical sounds, including bellow of camel and, according to Dementiev and Gladkov (1954b), whole noise of passing caravan (Radde and Walter 1889;

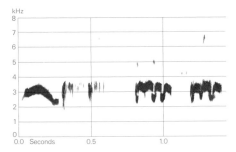

II P A D Hollom Iran May 1977

Pleske 1889-94; Sarudny and Härms 1926; Swerew 1927; Shnitnikov 1949; Raines 1962; Panow 1974; Panov 1978; E N Panov.) (b) Subsong. Loud variant typically given during antagonistic interactions, sometimes also during pair-formation and prior to copulation. Random sequence of very varied sounds—bleating, wailing, clucking, hissing, hoarse whistles, call 4, etc. Contains unusual combinations of extraordinarily high-pitched and quite low-pitched sounds. Brief description from Turkey by Wadley (1951) perhaps relates to this type. Typical quiet Subsong normally given when alone. (Panow 1974; Panov 1978; E N Panov.) In Oman, a mixture of whistles like Starling *Sturnus vulgaris*, grating 'cher we do', chuckles, and 'dweet' sounds (Walker 1981b). (2) Clicking calls. (a) Muffled clicking 'tchok' typically alternating with short whistle-calls (call 5), also with call 2b (as in many other *Oenanthe*), when disturbed at nest or with fledged young. Such combinations sometimes form a short song-type sequence (Panov 1978; E N Panov). (b) A characteristic clear loud click given in various situations: during conflicts, when disturbed (see above) or on sighting high-flying raptor, also incorporated in song; after copulation, rapid sequence (5-6 clicks per s, slowing after 8-10 s to 2-3 per s given during first few seconds of ♂'s flight (see also call 5). Apart from that associated with copulation, fastest rate of delivery is from adult disturbed when tending young (Ivanitski 1981b). No details of how calls 2a and 2b differ, but see Panov (1978). Recording suggests '(t)zuk' (P A D Hollom) or (2nd unit of Fig IV) 'chik' or 'chk' (J Hall-Craggs); in Fig V ('tuk tuk tik tuk see tuk'), some clicks lower pitched (J Hall-Craggs), the 2 types perhaps corresponding to 2a-b as defined above; both are shorter than a nevertheless obviously related grating 'tschr' illustrated by Bergmann

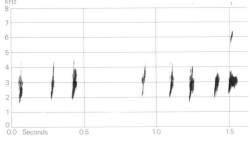

III P A D Hollom Iran May 1977

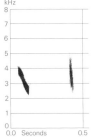

IV P A D Hollom Iran May 1977

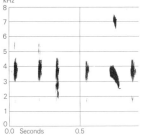

V P A D Hollom Iran May 1977

and Helb (1982). (3) Threat-call of ♀. Harsh, apparently low-pitched clicks given in rapid succession to form short rattle; in sonagram from E N Panov, 9-12 clicks per rattle. Given by ♀ when approached by a strange ♂ (also sometimes by her own ♂) and in ♀-♀ conflicts (Panov 1978; E N Panov). (4) High-frequency short 'zik' given by ♂ in interactions with conspecific birds of either sex (after fight, before copulation), when alone on display-ground (see Social Pattern and Behaviour), also in loud and quiet Subsong (E N Panov). (5) Whistle-calls. Several variants, all apparently descending in pitch. Often combined with clicking-calls when alarmed, given also by ♂ during pair-formation and, according to E N Panov, always after copulation (see, however, call 2b). Recording (Fig IV, 1st unit) suggests 'cheup' or 'teup' (P A D Hollom); not noticeably disyllabic, rather like flight-call of Chaffinch *Fringilla coelebs*, but loud, pure-sounding, and well-spaced units given in quite long series, sometimes accelerated with pauses shortened (M G Wilson). Further descriptions of this variant: 'tjip' (Bergmann and Helb 1982; Jonsson 1982); '(t)cheep' like domestic chick (Wadley 1951), loud 'weep' (Wallace 1984a), though sometimes subdued and rather soft, and infrequently given by vagrants (Beaman and Knox 1981; Rogers 1981a); see also Fischer (1974) and Robel (1981). Recording also reveals a 'see' (Fig V, penultimate unit) which lacks sharper cut-off of 'teup' and perhaps for that reason sounds higher pitched; recording by E D H Johnson contains 'sip', close to but different from both (J Hall-Craggs). Often mentioned as part of song, but apparently also with independent function as calls, are 'weeoo' (most typical call according to Raines 1962, though only infrequently given according to Königstedt and Robel 1983) and loud, high-pitched, slightly drawn-out 'wheet whit' (Wadley 1951; Rogers 1981a) said by Raethel (1955) to serve as contact-call. (b) Rasp-call. Like rasping hunger- or contact-call of young (see below). Frequently given by ♂ during pair-formation (Panov 1978; E N Panov). Sharp croaking 'trill' given in territorial conflicts (Panow 1974) is perhaps series of clicks forming rattle or rasp.

CALLS OF YOUNG. Contact-call a short (*c.* 0·25 s), rather low-pitched rasp. Adult call 2a also serves as contact-call. Food-call apparently typically longer than the rasp—higher pitched, more like a rattle (Panow 1974; Panov 1978; E N Panov); hard 'trr' and 'trrai' distinguished and illustrated by Bergmann and Helb (1982).   MGW

Breeding. SEASON. USSR: in Transcaspia, laying begins end of March, though sometimes considerably earlier in south, with (e.g.) building in Turkmeniya from 18-24 February and hatching recorded 12-18 March; clutches recorded up to July in northern Transcaspia (Shevchenko 1979); in southern Transcaucasia, nest-building noted (in one early spring) mid-April, fledged young 3-4 May; in lowland Kazakhstan, most young fledge late May, in steppes between Volga and Ural rivers early June, at *c.* 2000 m in Altai mountains *c.* 20 June (Panow 1974; Ivanitski 1978; E N Panov). SITE. Normally in burrow of rodent (see Panow 1974 for list of species) or sometimes of bee-eater *Merops*, occasionally in natural hole or crevice in ground or rock. Nest: bulky cup of dried grass, roots, and hair, lined with hair, wool, and feathers. Outer diameter 10·7-13·4 cm, depth of cup 3·0-5·7 cm, height 5·0-6·3 cm, thickness of wall 1·7-3·4 cm, n = 15 (Bel'skaya 1965a). Distance from entrance of burrow mean 208 cm (95-310), n = 22 (Bibikov and Bibikova 1955); see also Bel'skaya (1965a). Building: by ♀, taking at least 5 days (Panow 1974); hollow in nest-chamber perhaps made by bird (Bel'skaya 1965a). EGGS. See Plate 83. Sub-elliptical, smooth and glossy; pale blue, rarely with faint reddish specks. 21·7 × 16·2 mm (19·0-24·0 × 15·2-17·3), n = 100; calculated weight 3·0 g (Schönwetter 1979). Clutch: 5-6(4-7); according to Shevchenko (1979), range 3-9, with mean 6·9 (n = 107) in April, 6·0 (n = 84) in May, 4·9 (n = 47) in June; mean brood size 5·3 (n = 155). 2(-3) broods in south of range, normally 1 in north (Panow 1974; Ivanitski 1981c); in northern Transcaspia, 35-40% of ♀♀ double-brooded (Shevchenko 1979). Eggs laid daily (Bel'skaya 1965a). INCUBATION. 12 days, starting with 4th egg. By ♀ only (Bel'skaya 1965a). YOUNG. Cared for and fed by both parents (Dementiev and Gladkov 1954b), though ♂ apparently does little feeding (see Social Pattern and Behaviour). FLEDGING TO MATURITY. Young leave nest at 13-15 days before able to fly, and are fed by parents for at least 2 weeks further (Panow 1974); see also Social Pattern and Behaviour, including for age of first breeding. BREEDING SUCCESS. In Voroshilovgrad (Ukraine, USSR), main predator *Mustela eversmanni* which destroyed 5 out of 35 occupied nests, 3 others being lost to unknown cause (Panchenko 1976). In northern Transcaspia, loss of nests much greater if occupied rather than unoccupied rodent burrows used; some also fail due to flooding and predation by ants (Formicidae); total loss of eggs and young 57% (Shevchenko 1979). In Badkhyz (Turkmeniya), 82·3% of 29 eggs hatched; 62% of eggs laid produced fledged young (Bel'skaya 1979).

Plumages. ADULT. Upperparts (backwards to upper rump) and side of neck uniform pale sandy-brown, crown and hindneck often slightly greyish, rump at border of white brighter pink-cinnamon; freshly moulted feathers with distinct vinous-brown tinge, but this rapidly fading to sandy-brown. Lower rump and upper tail-coverts cream-white or white, together with white tail-base usually forming narrower white band than in Wheatear *O. oenanthe*. Supercilium rather variable: in fresh plumage, consists of narrow cream-white line from nostril backwards, widening to large off-white patch above lore, narrowing again above eye, and gradually merging into buff above ear-coverts, where hardly contrasting with sandy-brown crown; in worn plumage, supercilium whiter, broader, and more distinct, also behind eye, though extension over ear-coverts or extension above lore occasionally faint or absent. Lore deep black, forming contrasting dark stripe, widening in front of eye; black stripe on average wider in ♂, narrower in ♀, but occasionally faint or absent in any sex or age, especially in fresh autumn plumage, when sometimes dark grey and partly hidden under pale buff or white feather-tips. Narrow eye-ring cream-buff or white, broken in front and behind, usually distinct. Front part of cheek pale buff or off-white (sometimes extending into narrow white stripe below eye), merging into warmer buff or pink-buff on rear of cheek and lower ear-coverts and this in turn into dark rufous-brown or buff-brown (often with faint pale streaking) on upper ear-coverts; upper border of ear-coverts immediately below supercilium dark grey, forming short extension of black loral stripe behind eye, but dark grey often partly or fully hidden. Central chin and throat pale cream-buff or white, merging into warm pink-cinnamon (if fresh) or deep buff (if worn) on side of throat, chest, side of breast, and upper flank; remainder of underparts including under tail-coverts pale cream-buff (if fresh) to off-white (if worn); newly-moulted feathers of underparts markedly bright vinous-pink, but this colour rapidly fading. Tail-base white, tail-tip dark olive-brown to black (narrowly fringed off-white terminally when fresh); maximum extent of black on central pair (t1) 39·2 (16) 35–42 in ♂, 36·0 (17) 32–39 in ♀; minimum extent on t3–t5 23·2 (20) 19–27 in ♂, 20·6 (20) 18–24 in ♀; maximum extent on t6 26·0 (20) 22·5–30 in ♂, 23·5 (21) 21–26 in ♀; thus, band on tail-tip *c.* 20–26 wide, with extension of black on t1 often partly or largely hidden by white upper tail-coverts, resulting in black block of uniform width, not unlike tail of Desert Wheatear *O. deserti*. Flight-feathers dark brown or dark grey-brown, darkest towards tips, fringed pale greyish-buff or off-white along tips, brighter pink-cinnamon along outer webs of secondaries, and more narrowly pale buff along basal parts of primaries; basal inner webs of flight-feathers with sharp white or isabelline border (unlike *O. oenanthe*, which is uniform dark grey). Greater upper primary coverts dark grey-brown with even pink-buff or off-white fringe along outer web and tip. Lesser and median upper wing-coverts pale sandy-brown, like mantle and scapulars or with slightly darker centres; greater coverts, shorter feathers of bastard wing, and tertials dark grey-brown or dark brown with broad pink-cinnamon fringe along outer web and tip. Longest feather of bastard wing dull black or brown-black with narrow off-white fringe along tip, distinctly darker than remainder of wing except tips of flight-feathers. *In worn plumage*, flight-feathers, tertials, greater coverts, and greater upper primary coverts duller dark grey-brown, fringes bleached to pale buff or off-white and partly abraded, longest feather of bastard wing browner and sometimes less contrasting. Axillaries and under wing-coverts white, slightly tinged buff on lesser coverts; some grey of feather-bases usually visible on longer feathers and marginal coverts. NESTLING. Down long; pale grey (Panow 1974). JUVENILE. Rather like adult, but upperparts less saturated sandy-brown, more buff, each feather with ill-defined pale buff or off-white centre (most distinct on forehead, crown, nape, and mantle) and with narrow ill-defined dark terminal rim (most distinct on mantle and scapulars), appearing mottled; rump buff, sometimes with black scalloping, upper tail-coverts white with some buff on tips. Side of head like adult, rather variable, but without black loral stripe (sometimes a faintly grey one) and ear-coverts with faint black scalloping. Underparts like adult, but feathering looser and shorter (especially on chin, vent, and under tail-coverts), side of throat, chest, breast, and flank buff, sometimes with narrow but distinct black scalloping (occasionally extending to chin and belly), but this occasionally virtually absent. Tail, flight-feathers, tertials, and greater upper primary coverts as adult, but fringes along tips broader, less sharply defined, and deeper pink-buff, rufous-buff, or cinnamon. Lesser and median upper wing-coverts pale grey-brown with broad buff tips, greater coverts dark grey-brown with broad and ill-defined cinnamon fringe along outer web and tip. Longest feather of bastard wing blackish, as adult. Axillaries and under wing-coverts pale buff or isabelline-white. FIRST ADULT. Like adult and sometimes hardly distinguishable. Juvenile flight-feathers, tail, greater upper primary coverts, and some or all tertials and greater upper secondary coverts retained, relatively more worn than neighbouring fresh feathers, and browner and more worn than adult at same time of year (especially tail-tips and primary coverts); fringes along tips and outer webs of primary coverts and greater coverts broader and less clear-cut than those of adult, but difference hard to see when plumage worn.

Bare parts. ADULT, FIRST ADULT. Iris dark brown. Bill, leg, and foot black. Mouth black. (Hartert 1903–10; Witherby *et al.* 1938*b*; Panow 1974; ZMA.) NESTLING. Bill, leg, and foot yellowish-grey. Mouth and gape-flanges lemon-yellow. (Panow 1974; E N Panov.) JUVENILE. Iris dark brown. Bill, leg, and foot flesh-grey, flesh-brown, or brown. (Hartert 1903–10; RMNH, ZFMK, ZMA.)

Moults. ADULT POST-BREEDING. Complete; primaries descendant. Mainly June–August, occasionally to October (Witherby *et al.* 1938*b*). In Turkmeniya (USSR), early June to late July; in mountains of Soviet central Asia completed mid-August or late August (Dementiev and Gladkov 1954*b*). In Afghanistan, one just started 27 July, others well advanced or almost completed by 8–13 August (Vaurie 1949); some from late July and early August almost completed (Paludan 1959). In Iran, one from 6 June had started; moult completed in birds from 30 July and late August (Vaurie 1949). ADULT PRE-BREEDING. Partial: head, body, occasionally a few lesser or median upper wing-coverts, rarely 1–2 tertials or tertial coverts, frequently t1; (November–)December–February (Witherby *et al.* 1938*b*; RMNH, ZFMK, ZMA). POST-JUVENILE. Partial: head, body, lesser and median upper wing-coverts, tertial coverts, variable number of greater upper wing-coverts (sometimes all), frequently 1–2 tertials, occasionally t1. Timing strongly dependent on hatching date: with fledging period from second half of April to late July, moult starts early June to early August and completed early July to early September (Vaurie 1949; Dementiev and Gladkov 1954*b*; Piechocki 1958; Paludan 1959; RMNH, ZFMK, ZMA). In captivity, moult takes place in 2nd month of life (E N Panov). FIRST PRE-BREEDING. Partial. Perhaps occasionally as extensive as adult pre-breeding, but usually more restricted, involving all or part of chest, breast, and flanks only, as well as sometimes a few feathers elsewhere; occasionally, no moult at all. In captivity, starts in 5th month; involves head, body, tail, and some wing-coverts and tertials (E N Panov).

Measurements. ADULT, FIRST ADULT. Whole geographical range, all year; skins (RMNH, ZMA). Bill (S) to skull, bill (N) to distal corner of nostril; exposed culmen on average 4·6 shorter than bill (S).

	♂		♀	
WING	101·7 (2·49; 30)	97–106	96·1 (1·75; 24)	93–100
TAIL	57·8 (2·34; 17)	54–62	53·7 (2·23; 19)	50–57
BILL (S)	20·6 (0·77; 26)	19·4–21·9	19·5 (0·90; 19)	18·5–20·8
BILL (N)	11·5 (0·39; 21)	10·9–12·1	11·0 (0·59; 17)	10·2–11·9
TARSUS	32·4 (0·89; 20)	31·3–34·1	29·8 (1·28; 19)	28·1–31·8

Sex differences significant. 1st adult with retained juvenile flight-feathers and tail combined with older birds above, though juvenile wing on average 0·8 shorter than adult and tail 2·6 shorter. Juvenile bill and tarsus not full-grown until completion of post-juvenile moult.

Wing. (1) Turkey (Kumerloeve 1961, 1970a); (2) Iran and Afghanistan (Paludan 1940, 1959); (3) Mongolia and Manchuria (Piechocki 1958; Piechocki and Bolod 1972); (4) Tien Shan, USSR (RMNH, ZMA).

(1)	♂ 96·3 (— ; 11)	92–99	♀ 94·0 (— ; 7) 91–96
(2)	98·5 (2·24; 12)	96–104	95·3 (3·44; 6) 89–99
(3)	99·8 (3·20; 8)	96–106	95·8 (1·98; 8) 92–98
(4)	103·1 (1·86; 14)	101–106	98·6 (1·70; 4) 96–100

Weights. Whole geographical range. (1) March–April; (2) May; (3) June; (4) July–August (Paludan 1940, 1959; Piechocki 1958; Nicht 1961; Dolgushin et al. 1970; Rokitansky and Schifter 1971; Piechocki and Bolod 1972).

(1)	♂ 30·3 (2·28; 7)	28–34	♀ 27·6 (— ; 3) 22–32
(2)	28·5 (3·15; 6)	25–34	30·0 (4·35; 6) 25–38
(3)	30·1 (1·86; 9)	28–34	27·3 (1·80; 9) 25–31
(4)	31·0 (2·28; 6)	28–33	30·5 (3·62; 8) 26–36

Kuwait: January–March, 26·8 (6·38; 8) 20–40; August–September, 28·8 (6·74; 4) 22–35 (BTO). Kirgiziya (USSR): ♂ 27–38 (20), ♀ 22–36 (15) (Yanushevich et al. 1960). Afghanistan, early October: ♀♀ 25, 39; sex unknown 39 (Paludan 1959).

Structure. Wing rather long, broad at base, tip bluntly pointed. 10 primaries: p8 longest, p9 2–5 shorter, p7 0–1·5, p6 2–6, p5 9–13, p4 13–17, p1 24–30. P10 reduced, 50–64 shorter than p8, 0–6 shorter than longest upper primary coverts. Outer web of p6–p8 and inner of p7–p9 emarginated. Tail short, tip square or slightly rounded; 12 feathers. Bill rather long, straight, rather heavy at base, laterally compressed at tip; tip of culmen slightly decurved, sometimes forming fine hook. Nostril rather small, oval, partly covered by membrane above. Some stiff bristles at base of upper mandible, many finer ones at nostril and on chin. Tarsus relatively and absolutely longer than in other Palearctic *Oenanthe*; toes rather long and slender. Middle

toe with claw 18·9 (14) 17·5–20·5; outer toe with claw c. 72% of middle with claw, inner c. 65%, hind c. 76%. Claws rather short, strong, slightly decurved.

Geographical variation. Very slight. Some populations slightly darker and browner, others paler and greyer or more sandy, and also some variation in size (see Measurements), but variation irregular and some populations highly variable (Vaurie 1949, 1959; RMNH, ZMA).

Forms superspecies with Red-breasted Wheatear *O. bottae* of northern Afrotropics and southern Arabia, and perhaps with Capped Wheatear *O. pileata* of southern Afrotropics (Meinertzhagen 1954; Hall and Moreau 1970; Portenko and Vietinghoff-Scheel 1971). Forms species-group with Wheatear *O. oenanthe*, both differing from other Palearctic *Oenanthe* in short tail and relatively long to very long tarsus.

Recognition. Closely similar to ♀ and 1st non-breeding ♂ of *O. oenanthe*, differing in pale sandy-brown upperparts (not darker drab-brown), paler ear-coverts (not contrasting with cheeks), buff and brown flight-feathers, tertials, and upper wing-coverts, with only longest feather of bastard wing blackish (all-black or brown-black in *O. oenanthe*, with more contrasting cinnamon-buff fringes), white or pale buff axillaries and under wing-coverts (in *O. oenanthe*, dull black with rather narrow pale fringes), and (most importantly, as valid at all ages) sharp pale border along basal inner webs of flight-feathers (inner webs wholly dark in *O. oenanthe*), long tarsus (longer than in most *O. o. oenanthe*, but not *O. o. leucorhoa*), and broader black tail-band, minimum extent of black on t3–t5 18–27 mm (mainly 16–20 in *O. o. oenanthe*, but 18–22 in *O. o. leucorhoa*), maximum on t6 21–30 (14–23 in *O. o. oenanthe*, but 20–25 in *O. o. leucorhoa*). Length of bill, amount of white on lower rump, contrast, extent, and colour of supercilium, and presence of black loral line sometimes useful characters (e.g. Königstedt and Röbel 1983, Tye and Tye 1983, Svensson 1984a, Alström 1985), but less exact and thus less valuable than those cited above, at least in the hand. CSR

Oenanthe oenanthe Wheatear

PLATES 58, 64, and 65
[facing pages 737, 761, and 856]

Du. Tapuit   Fr. Traquet motteux   Ge. Steinschmätzer
Ru. Обыкновенная каменка   Sp. Collalba gris   Sw. Stenskvätta   N. Am. Northern Wheatear

Motacilla Oenanthe Linnaeus, 1758

Polytypic. *O. o. leucorhoa* (Gmelin, 1789), north-east Canada, Greenland, and Iceland; nominate *oenanthe* (Linnaeus, 1758), Europe, western and northern Siberia, and Alaska, east from Faeroes and Ireland, south to Pyrénées, Alps, Yugoslavia, central Rumania, southern Urals, and Yakutia; *libanotica* (Hemprich and Ehrenberg, 1833), Spain south of Pyrénées, Balearic islands, eastern Rumania, Crimea, Greece, Turkey, Levant, Iran, and from Kazakhstan (USSR) and Afghanistan east to Altai, Transbaykalia, and Mongolia; *seebohmi* (Dixon, 1882), north-west Africa.

Field characters. 14·5–15·5 cm; wing-span 26–32 cm. Eurasian race, nominate *oenanthe*, 10% longer than Robin *Erithacus rubecula*. Structure and bulk noticeably variable, with Greenland race *leucorhoa* up to 15% larger and longer-winged than nominate *oenanthe*, approaching Isabelline Wheatear *O. isabellina* in size and even exceeding

it in wing length; otherwise, usually larger and longer-winged but shorter-tailed than other *Oenanthe*. Bold, bouncy, ground-loving chat, epitome of *Oenanthe* and occurring more widely than any other. White rump and white tail with black ⊥ make striking flight character (shared by 8 other *Oenanthe*). Specific characters most obvious in spring and summer, with fully blue-grey crown, nape, and back of ♂ diagnostic, and always pale or clean throat and breast of ♀ helpful. Sexes markedly dissimilar in breeding plumage, less so in winter. Juvenile separable. 4 races in west Palearctic, ♂♂ separable in the field.

ADULT MALE BREEDING. (1) Main Eurasian race, nominate *oenanthe*. Crown, nape, mantle, scapulars, and back pale blue-grey, contrasting with white frontal band and supercilium, black lores, cheek-patch, and wings, and broad white rump and upper tail-coverts. Tail white with broad black terminal band and central feathers, forming bold ⊥-shape. Chin, throat, and breast pale pink-buff; rest of underparts buff-white to white. Under wing-coverts and axillaries black, fringed and tipped white; appear dappled grey in glimpses. Above description fits most birds in Europe with plumage worn but not abraded; important to note however that (a) some from Britain and north-west Europe are generally paler, with silvery-grey crown and back and much whiter, even wholly white underparts, and (b) many become paler with heavy abrasion, which increases white tones and makes upper-wing drab black-brown and under wing-coverts black-flecked rather than grey-mottled. (2) East Canadian, Greenland, and Icelandic race, *leucorhoa*. Usually noticeably larger than nominate *oenanthe* (bill, wing, and tail 15-20% longer, legs 20-30% longer), with bold, full-chested, and usually taller appearance. Plumage pattern as nominate *oenanthe* but most show less blue and apparently never any silvery or white tones on grey crown and back, and much richer and darker, more uniformly coloured underparts. Uniform underparts, when combined with large size, virtually diagnostic of race. Under wing-coverts always black-speckled and black ⊥ on tail even more pronounced. Above description applies to east Canadian and Greenland birds; Icelandic birds intermediate in size and appearance between these and nominate *oenanthe*. (3) Spanish, south-east European, and central Asian race, *libanotica*. Resembles pale individuals of nominate *oenanthe*, with pale grey upperparts and cream or white underparts; bill slightly longer and black tail-band narrower. (4) North-west African race, *seebohmi*. Similar to *libanotica* but with distinctive fully black face and throat, almost black under wing-coverts, and even narrower black tail-band. ADULT FEMALE BREEDING. (1) Nominate *oenanthe* and *leucorhoa*. Basic plumage pattern as ♂ but (except for rump and tail) all features less clear-cut. Crown and back grey-brown and wings black-brown, all noticeably dirtier in tone than in ♂. Supercilium variable, cream to pale buff, not extending to frontal band or to rear cheeks. Cheek-patch on some no more than a dark brown eye-stripe, on others a brown patch (darkest on rear cheeks), and on a few black-brown like winter ♂. Underparts usually buffier than ♂ but sometimes almost white. Underwing as ♂ but buffier. With heavy wear, grey tones of back and paleness

of underparts increase but none ever as contrasting in pattern as ♂. (2) *O. o. libanotica*. Often remarkably similar to ♂; differs only in slightly duller upperparts, some pale fringes on wing-coverts and tertials, and slightly darker, often tawny chest. (3) *O. o. seebohmi*. Usually less black on throat than in ♂, or none. Upperparts average paler and sandier than nominate *oenanthe* and *libanotica*. ADULT MALE NON-BREEDING. All races. Bold contrasts and immaculate tones of breeding plumage lost in July and August. New upperpart feathers have grey, buff-brown, and brown tips masking basic pattern: crown and back become grey-brown, rear cheeks (and lower throat in *seebohmi*) hoary with indistinct borders, and wings scaly with all but smallest feathers boldly edged and tipped buff-white to buff and creating wing-bar across greater coverts and obvious pale panel on longest tertials and inner secondaries. Underparts darker (buffier) and more uniform, with pale rear belly difficult to see. Thus more like ♀, but sexing still possible on shorter but still clean supercilium, black lores and area under eye, blacker wings, and variable patches of grey above. Differentiation of nominate *oenanthe* from *leucorhoa* less easy but *leucorhoa* generally browner above (especially on feather-fringes and -tips) and usually deeper buff on chest. *O. o. seebohmi* still shows fully black chin and upper throat (and upperparts tend towards grey-sand). ADULT FEMALE NON-BREEDING. All races. Appearance less changed than in ♂ due to more uniform breeding plumage. Crown and back dusky- to buff-brown; face pattern less contrasting; wings with obvious buff scaling on small coverts and broad red-buff fringes to greater coverts and inner flight-feathers (creating noticeably pale marks on dark brown ground at close range); underparts wholly buff and thus usually closer in tone to upperparts. Thus looks much more uniform than in spring. Racial differentiation dubious, but at least some *leucorhoa* darker buff below than nominate *oenanthe*. JUVENILE. All races. Differs distinctly from adult, having basic plumage pattern and colours of ♀ but copiously spotted buff and/or scaled dark brown on crown, back, lesser wing-coverts, sides of neck, and breast. Racial differentiation impractical. FIRST WINTER. All races. From October, usually indistinguishable from adult ♀ non-breeding, though some ♂♂ show hint of full face pattern and all have somewhat duller, less brightly fringed wings than adult by November. FIRST SUMMER. All races. Both sexes closely resemble adult but some young ♂♂ noticeably less clean and contrasting, with indistinct frontal band, duller face pattern, browner back, and cream rather than white ground-colour to body. Some spring ♂♂ with brown wings may be this age. At all ages, bill and legs black.

The most widespread and best known *Oenanthe*, but only adult ♂ (and, in *libanotica*, adult ♀) easily identified. Potential pitfalls as follows: (1) adult ♂ breeding of nominate *oenanthe* (pale individuals) and *libanotica* may resemble pale-throated morph of Black-eared Wheatear

O. hispanica; in north-west Africa, black-throated *seebohmi* provide further possible confusion, even with Mourning Wheatear *O. lugens halophila*; (2) adult ♂ non-breeding (i.e. without full grey upperparts) liable to suggest several other *Oenanthe*, with pale sandy back of some *seebohmi* again providing particularly dangerous confusion with black-throated morph of *O. hispanica*; (3) adult ♀ lacks any single, easily visible diagnostic character and has appearance approached by several other *Oenanthe* (though these usually show blacker under wing-coverts than all races of *O. oenanthe* except *seebohmi*); (4) juvenile has no diagnostic character and thus liable to suggest several other *Oenanthe* in south of range; (5) 1st-year less distinctive than adult, with ♀ lacking diagnostic character. Identification of non-breeding birds best done on (1) tail length (proportionately shorter than any congener except *O. isabellina*), (2) tail pattern (bolder than *O. hispanica*, less so than *O. isabellina*, matching Pied Wheatear *O. pleschanka* and Finsch's Wheatear *O. finschii*), (3) under wing-coverts (distinctly darker than *O. isabellina* but, except in *seebohmi*, paler and less uniformly dark than other species), (4) face pattern (rather distinct, with better-formed supercilium than in other *Oenanthe*), (5) upperpart tones (generally greyest in genus), and (6) lower underpart tones (generally buffiest and darkest in genus, especially in non-breeding plumage). For more detailed comparison with most easily confused species, see *O. isabellina* (p. 756), *O. pleschanka* (p. 792), *O. hispanica* (p. 806), and *O. finschii* (p. 831); all other non-black *Oenanthe* in west Palearctic show obvious differences in tail length and rump and tail pattern. Flight like that of all chats but silhouette fuller across wing-bases, rump, and tail; action essentially flitting but progress rapid and direct, usually just above ground. Actions include characteristic upward sweep to perch, stretched-wing stall before landing on ground, and rolling of body angle in flight (allowing glimpses of underwing). Escape-flight usually short and directly away from disturbance, ending in bird landing on prominent perch, but sometimes jinking, with rapid disappearance into (e.g.) gulley or rock pile. On ground, usually makes series of hops broken by pauses or brief perching on raised ground; also uses loping run on flat surfaces; always accompanied by frequent flicks of wings and tail, and tail also spread at times. Stance usually half-upright, with fully erect posture when inquisitive or alarmed, and more squatting, horizontal carriage on linear perch or when facing strong wind. Given to bowing whole body and bobbing head and forebody. Note that these general actions of *O. oenanthe* not specific, being shared by all congeners.

Song melodious and enthusiastic, lacking fitful quality of Stonechat *Saxicola torquata* and Whinchat *S. rubetra*; main phrase a repeated, short, pleasing, lark-like warble linked by creaking and rattling notes and usually containing some mimicry of other nearby breeding birds.

Commonest call a harsh 'chack-chack' or (in mild alarm) 'weet-chack-chack'. Alarm-calls of breeding ♀ *libanotica* include loud 'weet', dangerously suggestive of *O. isabellina*.

Habitat. Breeds from high and low Arctic through boreal and temperate zones to steppe, Mediterranean, and subtropical arid zones, from July isotherms of 3°C to over 32°C, and from extreme continental to extreme oceanic climates, reaching Nearctic tundra from both European and Asian distribution areas. Much of this expansion must have occurred since the last glaciation, and far surpasses that of other *Oenanthe* with which, however, it shares constraints of requiring ready-made rock or burrow nest-site immediately neighbouring seasonally insect-rich bare patches or short swards for easy foraging. Has exploited stony and shrub tundra, rocky slopes, scree, and alpine meadows above treeline in mountains. Also high rocky rolling ridges either in foothills or well up mountains below permanent snowline in Alaska, (Gabrielson and Lincoln 1959; Godfrey 1966). European habitat also open but even more diversified: flat, sandy, sparsely vegetated arctic tundra at sea-level, sand-dunes, coastal islands, shingle, cliff-tops, heaths and downland closely grazed by rabbits or sheep, roadsides with short grass, moors, steppe, bogs, clearings in forests, riverside bluffs, embankments, walled fields, rocky alpine meadows and defiles, and sparsely vegetated mountain-top plateaux up to 3000 m, even among snowfields, and extralimitally in USSR even to 3500 m; also above 3000 m in Iran (Cornwallis 1975). Largely unaffected by degree of slope from flat to steep, or substrate whether of rich fertile soil, clay, sand, limestone, small stones, or exposed bedrock. Equally adaptable to dry hot conditions, incessant rain, snow, mist, or wind. Does not however tolerate closed situations such as forests, wetlands, gardens, or orchards, and requires ample bare or unobstructed spaces to function as a ground bird, although readily accepting and perching on points of vantage such as banks, projecting rocks, walls, fences, shrubs, overhead wires, and (infrequently) trees. Has not adapted well to cultivation, although often frequenting its margins and foraging readily on bare ploughland or fallow. Not a house bird, but tolerates neighbourhood of human settlements in some localities. More a lowland species in northern and middle range, becoming much more montane towards south, and thus reconciling its status as the most widespread *Oenanthe* with that of being rarely in competition with congeners on same ground, except on migration or in winter. In Swiss Alps, breeds mainly at 2000-2500 m, favouring sunny open short-grass slopes with stone walls, heaps of stones, detritus, and boulders; also moraines, rocky meadows, and neighbourhood of mountain huts. Avoids narrow, shady valleys, damp places, and neighbourhood of woods, but occurs in open shrubland (Glutz von Blotzheim 1962). Analysis of

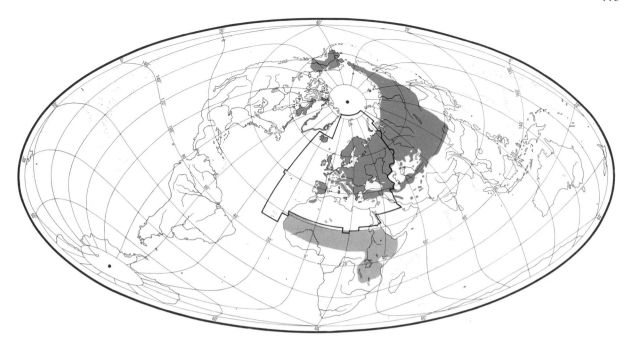

breeding distribution on British uplands showed occupancy of paramaritime moorland and bog, heather or grass moorland (both well-drained gently sloping and steeply sloping), montane heath, grassland, and fell-fields; blanket bog was only habitat type unoccupied. Among upland passerines, only Skylark *Alauda arvensis* and Meadow Pipit *Anthus pratensis* were more widespread. (Fuller 1982.) In south-west Iran, prefers conjunction of flat areas with rocky bases of hills providing outcrops and piles of boulders, and some bare ground among open or dense herbage; breeds markedly higher than other *Oenanthe* which favour flat sites (Cornwallis 1975). North African race *seebohmi* nests on stony plateaux at 1700–2300 m (Heim de Balsac and Mayaud 1962). Greenland race *leucorhoa* also has distinctive habitat, preferring drier and warmer interior to cool coastal belt; frequents sloping hillsides with stretches of talus, stony moraines, ravines, valleys, dry riverbeds, low rolling heathland with scattered boulders or large rocks, and even dense shrub growth with stony clearings, sometimes perching on twigs of birch *Betula* and willow *Salix*, although more often on rocks (Salomonsen 1950–1). On migration, usually occurs in lowlands and on managed grasslands or arable. In African winter quarters found on bare soil from sea-level to over 3000 m, favouring hillsides and rocky outcrops (Williams and Arlott 1980). Normally flies in lower airspace like other *Oenanthe*.

Distribution. Some range decrease in England and France.

SVALBARD. Bred 1954, 1973–4, and probably 1937 (Løvenskiold 1964; IB). JAN MAYEN. Probably bred 1924 (Bird and Bird 1935). BRITAIN. Has disappeared from some southern counties of England (Parslow 1973; Sharrock 1976). FRANCE. Disappeared as breeder from Bocages, Normandy, Autunois, and Hérault (Yeatman 1976). IRAQ. May breed in north (Allouse 1953).

Accidental. Bear Island, Chad, Madeira, Cape Verde Islands.

Population. Has decreased Britain, Ireland, France, Belgium, Netherlands, West Germany, Finland, East Germany, Czechoslovakia, and USSR.

ICELAND. No known changes (AP). FAEROES. Estimated 200–300 pairs (Bloch and Sørensen 1984). BRITAIN. 1000–10 000 pairs; marked decrease in southern half of England, disappearing or becoming rare in several counties where formerly locally common, mainly due to habitat changes (Parslow 1973). About 80 000 pairs (Sharrock 1976). IRELAND. Recent decline in inland counties (Sharrock 1976). FRANCE. 10 000–100 000 pairs; decreased (Yeatman 1976). BELGIUM. About 650 pairs; marked decrease (Lippens and Wille 1976). NETHERLANDS. 1400–1600 pairs; major decrease since *c*. 1900 due to habitat changes (Teixeira 1979; CSR). WEST GERMANY. About 7000 pairs (Rheinwald 1982); figure too high, and numbers falling in many areas (AH). For local counts, see Bauer and Thielcke (1982). SWEDEN. About 500 000 pairs (Ulfstrand and Högstedt 1976). FINLAND. Estimated 300 000 pairs (Merikallio 1958). Moderate decline recently in many areas (OH), though increased in north 1941–77 (Väisänen 1983). EAST GERMANY. Decreased (Kneis 1982). CZECHO-SLOVAKIA. Marked decline since 1960 (KH). ITALY. Sardinia: 75–150 pairs (Schenk 1980). USSR. Has decreased in west (GGB). Baltic republics: recent decrease (HV).

Oldest ringed bird 7 years (Rydzewski 1978).

Movements. Migratory, though North African race *seebohmi* probably only partially so.

Apart from birds wintering regularly in Tigris and Euphrates valleys of Iraq (H Siman), entire world population, including birds breeding in Nearctic, winters in Africa—in broad belt south of Sahara from West African coast to Indian Ocean, and south in eastern Africa to northern Zambia. Although winter records do exist from eastern North America south to West Indies, and small wintering population therefore suggested (see Salomonsen 1934), these records can probably be attributed to stragglers (Snow 1953). Overwinters exceptionally in Europe, e.g. 10 birds in Lincolnshire

(England), 1958-9 (Posnett and Bates 1960). Migration mainly nocturnal, although diurnal movement recorded, e.g. at Falsterbo (Sweden). Large landfalls sometimes occur at coastal localities in western Europe in autumn: in September 1965, *c.* 8000 present on 4 km of coastline in Suffolk (England) during 'fall' conditions (Davis 1966).

EURASIAN AND ALASKAN POPULATION (nominate *oenanthe* and *libanotica*) is common over most of winter range in western Africa: see (e.g.) Browne (1982) for Mauritania and Lamarche (1981) for Mali; uncommon in Gambia, with evidence of passage further south (Gore 1981); common in Nigeria, chiefly in north, occasionally reaching coast at Lagos (Elgood 1982). In Central African Republic, recorded south to Bamingui National Park (7°30′N), where frequent (Green 1983). Very common in Sudan, especially west of Nile and on coastal plain (Hogg *et al.* 1984); common in Ethiopia (Ash 1980), though no evidence of wintering in Eritrea (Smith 1957); locally common in southern Somalia (Ash and Miskell 1983). Uncommon in south and south-west Uganda, and sparse in northern and eastern Kenya; common elsewhere in East Africa, occurring as far south as *c.* 8°S in Tanzania (Britton 1980). Regular in small numbers in north-east Zaïre (Lippens and Wille 1976), and a few records from Burundi (Gaugris *et al.* 1981). Very scarce in northern Zambia, with records in (e.g.) Isoka region (Aspinwall 1977), but some confusion may occur with juvenile Capped Wheatear *O. pileata* (Benson *et al.* 1971). Occasionally recorded on Mashonaland plateau in Zimbabwe and rarely south to 18°S (Irwin 1981).

Ringing recoveries show European birds moving SW-SSW in autumn, extreme western populations probably moving south or even SSE initially (see Zink 1973). 111 autumn recoveries up to 1982 of birds ringed as adults in Britain showed heading between south and south-west with 34 recoveries in France (mainly Brittany and south-west France), 38 in Spain, 7 in Portugal, and 30 in Morocco. 22 autumn recoveries of young ringed in Britain showed similar pattern: France (11), Spain (6), Portugal (2), Morocco (3); also 39 spring recoveries including Morocco (20), Algeria (6), Spain (6), and France (2). (BTO.) SW-SSW heading also found in birds ringed in Norway: 12 recoveries were from Denmark (3), West Germany (2), France (1), Spain (5), and Algeria (1); one from extreme north-east Norway also found to SSW, in Finland (Holgersen 1982). Recoveries of birds (mostly full-grown) ringed in West and East Germany up to 1977 showed general south-west heading for southern recoveries—France (7), Italy (3), Spain (2), and Morocco (1), and NE-NNE heading for northern recoveries—Denmark (1), Norway (1), Sweden (1), and Finland (1) (Zang 1979).

Passage occurs on broad front across southern Europe, Mediterranean, and full length of North African coast. Regarded by Moreau (1961), in review of Mediterranean-Saharan migration, as outstanding example of near equality of abundance in both seasons, in contrast to many passerines; impression of high numbers more or less evenly distributed along North African coast is consistent with strong westerly component in route of populations furthest east. Described as very common on passage in Malta (Sultana and Gauci 1982); less so in autumn in Cyprus (Flint and Stewart 1983) and Crete (Moreau 1961), suggesting many overfly. Recorded from ships in eastern Mediterranean (Horváth 1959a). Very numerous on passage in eastern Morocco (Brosset 1956); in Algeria, passage particularly heavy on Hauts Plateaux (Ledant *et al.* 1981); see also Gouttenoire (1955) for Tunisia, Bundy (1976) for Libya, and Moreau (1927) for Egypt. Widespread and common on passage throughout Turkey (*Orn. Soc. Turkey Bird Rep. 1966-75*), abundant in

autumn in adjoining area of northern Iraq (Moore and Boswell 1956). Heavy passage in Turkmeniya (Dementiev and Gladkov 1954b), but only a straggler through Pakistan and north-west Himalayas (Ali and Ripley 1973b). Records further south, presumably outside main migration route, generally in small numbers with passage more conspicuous in spring: in south-west Iran, passage widespread in small numbers in spring, rare in autumn (Cornwallis 1975); in Gulf states of Arabia occurs regularly but with lesser numbers in autumn (Bundy and Warr 1980); generally uncommon elsewhere in Arabia, including North Yemen (Cornwallis and Porter 1982), though said to be common in Aden in autumn (Browne 1950). Passage occurs across Red Sea, with records from Dahlac archipelago (Mann 1971). Movement into western Tanzania takes place mainly east of Lake Victoria (Britton 1980). Some evidence for 'step migration', with late arrival in eastern Zaïre and then decrease in numbers present towards end of year (see Curry-Lindahl 1981).

Migration seasons notably protracted. Birds leave breeding grounds chiefly from August, although on Faeroes some dispersal occurs from end of July (Williamson 1947a), and in England, some movement southward noted from mid-July, with passage continuing until *c.* 3rd week of October, and stragglers into November (Witherby *et al.* 1938b). Most breeding birds leave Skokholm Island (Wales) by late August or early September; passage well under way by mid-August, continuing (including *leucorhoa*) to mid- or late October (*Skokholm Bird Rep. 1936-70*; Boswall 1952). In northern Denmark, records span early August to early November, with peak departure late August (Møller 1978b). Passage recorded on West German coast over 12 weeks, but most occurs over 4-5 single days (Hantge and Schmidt-Koenig 1958); in Camargue (southern France), extreme dates 19 July-20 October (Glegg 1941). At La Dombes (east-central France), however, heavy passage concentrated within *c.* 10 days between mid-August and mid-September, with little annual variation (Vaucher 1955). Passage across Mediterranean and North African coast spans mid-August to November, chiefly September and early October (see, e.g., Heim de Balsac and Mayaud 1962, Sultana and Gauci 1982, Flint and Stewart 1983). In south-east Algeria, recorded mid-September to early December, with peak in early October (Laferrère 1968). Passage through Pskov (western USSR) starts early August with peak late August or early September; still present at Minsk in October with 1 record from mid-November; in Khar'kov, passage occurs early September to early October (Dementiev and Gladkov 1954b). Present in Greece up to late October (Watson 1964). In Turkey, passage from early August to mid-October in west and to late October on central plateau and in east; has been recorded up to late November (*Orn. Soc. Turkey Bird Rep. 1966-75*).

In north-east Siberia, autumn passage starts late

August. Similar further south in USSR, but through-passage of northern birds produces protracted migration period. Birds leave Anadyr' (north-east USSR) late August, Gydanskiy peninsula (north-central USSR) early September, Krasnoyarsk (central USSR) late September to early October. Birds at higher altitudes leave earlier: e.g. mid-September in Altai, where present at lower altitudes up to late September; mountain regions of Tien Shan vacated by end of September. In Kazakhstan, departure from north over by end of October; further south (e.g. northern Aral Sea, Mangyshlak) departure starts mid-September. In Turkmeniya, however, passage recorded from first half September. (Dementiev and Gladkov 1954*b*; Dolgushin *et al.* 1970.) Recorded September–October in Pakistan and north-west India (Ali and Ripley 1973*b*). Present in mountains south of Caspian Sea (Iran) up to mid-September (Feeny *et al.* 1968), and on breeding grounds in south-west Iran up to end of August (Cornwallis 1975). Movement at Elat (Israel) chiefly late August to early October, with 1 record in mid-December (Safriel 1968). Passes through Gulf states of Arabia late August to November, with some records in early August and early December (Bundy and Warr 1980). Recorded in Ethiopia (where some winter) from late August (Ash 1980). Arrives in winter quarters chiefly from late September and early October: e.g. early October in Sénégal (Morel and Roux 1966), mid-September in Chad (Newby 1980), late September in East Africa and Nigeria (Britton 1980; Elgood 1982).

Departure from winter quarters protracted, probably especially in west, with passage noted from late January in southern Morocco (Heim de Balsac and Heim de Balsac 1951), and records from mid-February to May in Algeria (Ledant *et al.* 1981). Morel and Roux (1966) also noted prolonged departure from Sénégal. Early departures imply birds may reach northern edge of desert *c.* 6 weeks before arrival on breeding grounds (Moreau 1961); stopovers noted (e.g.) in northern Morocco early March (Brosset 1957); see also Social Pattern and Behaviour for both spring and autumn stopovers. Present in East Africa until late March, with occasional summering records (Britton 1980). Occurs up to mid-May in Ethiopia (Ash 1980); recorded in May also in Sudan and Somalia (Cave and Macdonald 1955; Ash and Miskell 1983). Passage across North African coast and Mediterranean chiefly March–April, tailing off to mid-May (e.g. Sultana and Gauci 1982, Bundy 1976).

In north-west Europe, a notably early spring migrant. Thus, often the first passerine to reach Britain, where sometimes recorded early March (exceptionally late February) (Witherby *et al.* 1938*b*), but more usually from mid-March with peak in early April, usually *c.* 1–2 weeks later in north than in south; arrival peaks variable, usually related to calm anticyclonic weather with clear nights in Iberia and France (e.g. Williamson 1962). First arrivals in Netherlands mid-March (Commissie voor de Nederlandse

Avifauna 1970). In Norway, arrives in south from mid-March but not present in arctic regions until mid-May (Blair 1936; Haftorn 1971). In central Sweden, average arrival date 23–24 April (Lundberg and Edholm 1982). In Aegean (Watson 1964) and south coast of Turkey, passage early March to early May, and from early April in Thrace and on Black Sea coast; present in eastern Turkey from early May (*Orn. Soc. Turkey Bird Rep.* 1966–75). Passage through Middle East from late February to mid-May: on Arabian peninsula, recorded from early March on Red Sea coast of Saudi Arabia (Bates 1936–7); crosses Gulf states late February to May with peak in March (Bundy and Warr 1980). Recorded in Pakistan and north-west India chiefly in April (Ali and Ripley 1973*b*). First arrivals in Iran late March (Schüz 1959; Cornwallis 1975). In Kazakhstan (USSR), present in south and south-west by end of March, but not in north and north-east until end of April or early May (Dolgushin *et al.* 1970). Present in Altai from late March, Transbaykalia from mid-April, and from early May on Pechora river (European USSR, 65°N) and Yakutsk (eastern USSR, 62°N). Arrivals in far north of USSR (including north-east) not until late May or early June. (Dementiev and Gladkov 1954*b*.) First arrivals in Alaska in May (Murdoch 1885).

♂♂ arrive on breeding grounds before ♀♀. Of those caught on passage by bird-limers in Cyprus, 40% of ♂♂ had been caught by 2 April, 50% by 4 April; only 10% of ♀♀ had been caught by 2 April, 50% not until 10 April (Horner and Hubbard 1982). In study south of Caspian Sea, first ♂ arrived 25 March, first ♀ on 7 April (Schüz 1959). Arrival later at higher altitudes: e.g. west of Alma-Ata (Kazakhstan), present in foothills from early April but not seen at 2400–2500 m until 30 April (Dolgushin *et al.* 1970). Up to 33% extra weight put on prior to northerly Sahara crossing: specimens collected near Lake Chad in March had range of weights 20–33 g, of which 2–33% fat (Ward 1963). For site-fidelity to breeding grounds, see Social Pattern and Behaviour.

ICELAND, GREENLAND, AND EAST CANADIAN POPULATION (*leucorhoa*) winters from Sénégal and Sierra Leone east to Mali; birds presumed to be *leucorhoa* occasionally reach Gulf of Guinea coast in Togo (Douaud 1957). Autumn migration involves south-east crossing of North Atlantic, and frequency of records from ships south-east of Greenland is clear evidence that large numbers fly non-stop from Greenland to western Europe (Snow 1953). Under strong cyclonic conditions, flight may be 2400 km, lasting 30 hrs. Following such conditions, birds arriving on Fair Isle (Scotland) sometimes weigh as little as 20–25 g (probably 40 g before departure); under weaker cyclonic conditions, arrivals on Fair Isle heavier, averaging 30 g (see Williamson 1961).

Breeding grounds vacated August–September, with ship records in North Atlantic mainly from end of August to late September (Snow 1953), though birds still present in southern Iceland in early October (Gudmundsson 1970). Influx of passage birds on Faeroes in late summer swells local population (Williamson 1947*a*). Radar observations from Isle of Lewis (Scotland) showed broad-front passage with early-September movement heading south-east, presumably from Iceland, and later (but overlapping) September movement heading ESE, presumably from Greenland; birds apparently flew lower at night, probably to see land more easily (Lee 1963). Arrival peaks generally after 'drift' conditions, otherwise often protracted. Large falls occur, especially on outlying islands: over 1000 arrived on Fair Isle on 20-21 September 1959 following cyclonic conditions (Williamson 1959). Recorded mainly on west coasts of Norway, Britain, and Ireland, and October records in Sweden, Denmark, northern West Germany, and Netherlands. Regular in France on coasts of Brittany and the south-west. Passage through western Europe from mid-August, though mainly September-October.

Ringing recoveries confirm that Greenland breeders migrate south-east to western Europe and suggest that after making landfall they then head south-west. Up to 1974, 10 foreign recoveries of birds ringed in Greenland, of which 5 were in autumn (September-October) in France (4) and Portugal (1) (Salomonsen 1967, 1971, 1979). Of *leucorhoa* ringed on passage in Britain, 5 autumn recoveries: Morocco (3) in September, October, and December, France (1) in October, and Spain (1) in September; all on SSW-SW heading indicating direction change described above (BTO). Unlikely that birds regularly fly directly to Iberia (Williamson 1961), records in Spain and Portugal probably resulting from south-west passage from western Europe. Main passage into North Africa through Morocco and Algeria west of 3°E (Heim de Balsac and Mayaud 1962). Passage through North Africa mainly September-October, with arrival on wintering grounds from October, usually later than nominate *oenanthe* (e.g. Lamarche 1981).

Recorded in winter quarters until April but most birds probably leave in March since passage evident in western Europe in April; noted on passage in western Sahara in February (Heim de Balsac and Heim de Balsac 1951). Route taken probably similar to that in autumn, although spring records in Algeria extend to 5°40′E (Moreau 1961).

Up to 1974, 5 spring foreign recoveries of Greenland-ringed birds: Channel Islands (1), Belgium (1), and Britain (3) (Salomonsen 1967, 1971, 1979). British-ringed passage birds recovered in Portugal (1) and Algeria (1) (BTO). Records from ships and ringing recoveries may indicate tendency to more northerly overland route in spring initially, or may be drift effect (see Snow 1953, Zink 1973). Passage through western Europe later than nominate *oenanthe*. Earliest record in Netherlands 14 April (Commissie voor de Nederlandse Avifauna 1970), most records from end of April. Passage through Skokholm

late April to late May (*Skokholm Bird Rep. 1936-70*). Records from ships in North Atlantic from early April to early June (Snow 1953), and protracted passage occurs through outlying Scottish islands with birds present on St Kilda until late May and arrival peaks occurring during drift conditions (e.g. Williamson 1962). Occurs on Surtsey (off southern Iceland) from mid-April but mainly in May (Gudmundsson 1970). Arrives in Greenland mainly from May (Meltofte 1975).

In Surtsey study, peak arrivals on 30 April in one year, birds probably having left Britain on 28-29 April. Most very fat (47·8% of ♂♂ over 31·8 g, 17·9% over 28·2 g) and thus able to make journey without greatly depleting reserves. 11·9% of ♂♂ had no fat reserves and may have experienced longer journey, more adverse weather conditions, or perhaps were individually less able to put on fat (Gudmundsson 1970). Birds recorded on ships in weak condition recover rapidly when fed (Elkins 1983). Individuals in May in north-east Greenland often in poor condition, probably as result of bad weather when migrating or shortage of available food on arrival (Meltofte 1975).

NORTH AFRICAN POPULATION (*seebohmi*) migratory, although sections may show only altitudinal movement, as wintering recorded in central and south-west Morocco: several ♂♂ in lowlands at Ouarzazate *c.* 50 km from known breeding areas noted by Thévenot *et al.* (1982), and *c.* 20 (possibly *seebohmi*) in 2 successive years by Smith (1965*a*). Occurs commonly in south-west Mauritania (chiefly east of 14°30′W), and this presumably the main wintering area (Browne 1982); fairly common across Sénégal border at Podor (Morel and Roux 1973), and recorded occasionally elsewhere in northern Sénégal (Morel and Roux 1966) and in Sahel zone of Mali (Lamarche 1981).

Leaves breeding grounds September-October, returning March-April (Thévenot *et al.* 1982). Recorded at Menaka (south-east Mali) in September, and at Tamanrasset (southern Algeria) late March (Ledant *et al.* 1981); 2 trapped in Sénégal in early April carried large fat reserves (Morel and Roux 1966). Some ♂ *seebohmi* reported among other *O. oenanthe* at Kufra oasis in Libyan Desert, early April (Cramp and Conder 1970).

DFV, MGK

Food. Chiefly insects; also spiders, molluscs, and other small invertebrates, supplemented by berries. Normally locates prey visually, chiefly on ground or in low vegetation. 2 main foraging techniques, which may be used in same area. (1) Running: in flat areas of short turf, runs (or sometimes hops) short distance, stops to pick up item or to scan ground ahead, and then runs on (Brooke 1981; Moreno 1984*b*). (2) Perching: in areas of scattered perches (e.g. stones, low bushes), uses these to scan ground nearby, drops down for item, and then returns to perch or moves to new one; tends first to scan area closest to perch, then progressively further. Frequently flies (or from very low elevation runs or hops) to new perch, whether or not prey capture attempted. (Moreno 1984b.) In northern Spain, height of perches usually less than 20 cm and very few above 100 cm (Moreno 1984b). Perching technique often noted in heavy rain (Tye 1982). Flying insects sometimes caught in fluttering pursuit from ground, or by sallying from perch, or snapped at from ground (Mildenberger 1943; Kneis and Lauch 1983; Moreno 1984b). Occasionally also leaps up for aerial prey, or to pick prey from vegetation. May pursue elusive grasshopper (Acrididae) with series of leaps and bounds. (Tye 1982.) Young studied by Moreno (1984a) learned first to forage along ground (hopping rather than running), but will also rapidly learn to snap at flying insects (Nicholson 1930, for *leucorhoa*). Occasionally hovers like Kestrel *Falco tinnunculus* at height of up to *c.* 10 m where grass too long for usual techniques, diving or gliding down for food sighted on ground, then returning to hovering position, often at different height. (Gordon 1942; Pounds 1942; Kishchinski 1980.) Also forages by probing, striking ground with sideways scoops of bill to loosen soil (Nicholson 1930; Panow 1974); perhaps sometimes locates prey by sound, as recorded holding head on one side before darting forward to extract larva (Nicholson 1930, for *leucorhoa*). Forages intensively among animal droppings and dung-heaps (Roberts 1934; Posnett and Bates 1960). On migration, *leucorhoa* seeks flies (Diptera) among algae along seashore (Mayaud 1932). When tackling large or unfamiliar prey, may slowly open wings and move them forward (Panow 1974). Large or hard items may be hammered or beaten on ground (Meinertzhagen 1954; Panow 1974; Tye 1982). One record in Baltic area of lizard *Lacerta agilis* taken, probably by *leucorhoa*, which hopped around, jabbing at lizard, withdrawing then renewing attack until tail detached; this taken aside, broken up, and consumed with difficulty; attack renewed until bird disturbed (Grössler 1959). Recorded catching by the head, and immediately swallowing, bumble bees *Bombus* landing on flowers (King 1968); also seen to strike bees on ground, presumably to remove sting (Tye 1982). Hairy caterpillars may be run through bill from end to end, and beaten on ground with forward or side-to-side head movements before being swallowed whole (King 1978a; Tye 1982). Grasshoppers beaten on ground, shedding some legs and wings, before being swallowed (Tye 1982). Seen to take blackberries *Rubus fruticosus* by leaning forward and downward to pull at them, at times beating wings to maintain position (Veysey 1961). In winter quarters, will exploit recently burnt areas where prey more readily visible (Dittami 1981). May forage late in evening, even in near-darkness (Nicholson 1930; Berck 1961). Hard and indigestible parts of prey regurgitated as pellets (Panow 1974; Tye 1982). Flexibility of jaw muscles assists in utilizing both

large and small prey (Potapova and Panov 1977). For feeding rates on Skokholm island (Wales) of migrants and of ♀ before and during incubation, see Brooke (1979).

The following recorded in diet in west Palearctic. Invertebrates: springtails (Collembola), dragonflies (Odonata), grasshoppers, etc. (Orthoptera: Gryllidae, Acrididae), earwigs (Dermaptera), bugs (Hemiptera), snake flies (Neuroptera: Raphidiidae), adult and larval butterflies and moths (Lepidoptera: Pieridae, Nymphalidae, Zygaenidae, Arctiidae, Noctuidae, Geometridae), flies (Diptera: adult and larval Tipulidae, Tabanidae, Calliphoridae, Muscidae), Hymenoptera (small sawflies Symphyta, Ichneumonidae, ants Formicidae, bees Apoidea), adult (mostly) and larval beetles (Coleoptera: Cicindelidae, Carabidae, Gyrinidae, Hydrophilidae, Staphylinidae, Geotrupidae, Scarabaeidae, Elateridae, Tenebrionidae, Cerambycidae, Chrysomelidae, Curculionidae), spiders (Araneae), centipedes (Chilopoda), woodlice (Isopoda), small land molluscs (*Helix*, *Clausilia*), and earthworms (Oligochaeta). One record of lizard *Lacerta agilis* as prey of *leucorhoa*. Plant food mainly berries: blackberry *Rubus fruticosus*, rowan *Sorbus aucuparia*, red currant *Ribes rubrum*, and elderberry *Sambucus nigra*, also seeds; plant fragments include moss, lichen, leaves of bog bilberry *Vaccinium uliginosum*, and freshwater algae. (Bailly 1853; Clarke 1902; Morbach 1934; Roberts 1934; Witherby *et al.* 1938b; Axell 1954; Grössler 1959; Posnett and Bates 1960; Berck 1961; Neufeldt 1961; Veysey 1961; Glutz von Blotzheim 1962; Scott 1962b; Menzel 1964; King 1968; Ptushenko and Inozemtsev 1968; Conder 1969; Averin and Ganya 1970; Popov 1978; Brooke 1981; Tye 1982; Kostin 1983; J Moreno.) In Greenland, additional items recorded in diet of *leucorhoa* include caddis flies Trichoptera, larval Lepidoptera (Lymantriidae), and spider mites (Acarina); also bog bilberry *Vaccinium uliginosum* and crowberry *Empetrum nigrum*. (Longstaff 1932; Witherby *et al.* 1938b; Tinbergen 1961.)

Analyses show frequent preponderance of beetles in diet. In Volga-Kama region (USSR), all of 14 stomachs analysed contained beetles (especially Curculionidae in 78%), 36% contained Hymenoptera, 29% Lepidoptera, 14% Orthoptera, 14% Hemiptera, and 14% Diptera; larval insects (except Elateridae) found in 21%, Elateridae in 36%; also unidentified fruit in 14% (Popov 1978). In Moldavia (south-west USSR), adult insects in 25 stomachs comprised 70·2% (by number) beetles (38% of them Curculionidae), 25% ants, and 4·8% grasshoppers (Averin and Ganya 1970). In Crimea (USSR), March–August, stomachs of 23 birds contained 44·4% (by number) Hymenoptera, 36·1% beetles, 7·2% adult and (mostly) larval Lepidoptera, and small numbers of other insects; also 3·9% spiders and 1·7% centipedes (Kostin 1983). In Karel'skaya ASSR (north European USSR), adult stomachs contained mainly beetles; also recorded were ants, snake flies, small sawflies, and Ichneumonidae (Neufeldt 1961). In Breckland (eastern England), analysis of 193 faecal samples suggested beetles and ants taken most often, and grasshoppers probably important also; direct observation indicated spiders, bees, mites *Erythraeus phalangoides*, and woodlice as other possibly frequent prey (Tye 1982). Of 5 birds collected in different parts of central and south-east Iceland, June–August, adult Lepidoptera occurred in 3 stomachs, a Lepidoptera larva once, and beetles in 2; 13 Ichneumonidae found in 4 stomachs and 5 Diptera (*Scatophaga*) in 2; fragments of moss and lichen occurred in 2, and freshwater algae and leaf of bog bilberry in 1 (Roberts 1934). In Iran, March–August, 8 stomachs analysed contained 55% (by number) ants Formicidae, 16% beetles, and very small numbers of other insects (some larval), spiders, molluscs, and seeds; of 287 food items measured, 59% less than 5 mm long, 33% 5–10 mm, 8% 10–15 mm (Cornwallis 1975). In Kazakhstan (south-central USSR), Acrididae, Formicidae, Carabidae, Tenebrionidae, and Curculionidae made up *c.* 80% (dry weight) of material from 98 stomachs (Ryabov 1968).

Diet in winter quarters mainly insects, including termites (Isoptera), ants, and beetles (Witherby *et al.* 1938*b*; Borrett and Jackson 1970). Prey of birds wintering exceptionally in Lincolnshire (England) included small snails *Helicella virgata* and *H. caperata*, worms, and pupae; also foraged intensively for insects among sheep dung (Posnett and Bates 1960).

In diet of young, larvae of Lepidoptera and Tipulidae often particularly important; to lesser extent, also spiders (Neufeldt 1961; Menzel 1964; Asbirk and Franzmann 1979; J Moreno). In central Sweden, caterpillars and spiders main food brought to nestlings 0–8 days old; for older young, proportion of spiders declined and Diptera increased. Broods less than 1 week old fed 10–15 times per hr; older broods fed twice as often. Number of items brought per visit increased with age of young, and proportion of small items (1–8 mm long) declined; prey 9–16 mm most numerous at all stages. (J Moreno.) On Skokholm, Tipulidae larvae fed preferentially to young measured *c.* 25 mm; other larvae and adult arthropods rarely exceeded 15 mm. Analysis of faeces also suggested young received larger items than adults themselves ate. (Brooke 1981, 1983.) In Breckland, analysis of 72 faecal samples of different age groups suggested beetles, grasshoppers, and spiders as major items in diet, with Diptera and larval Lepidoptera also important; compared with older young, faeces of nestlings 0–4 days old contained more items less than 6 mm long, caterpillars being the only large items; faeces of older nestlings included bees and large Diptera. 17 pellets of young more than 9 days old contained chiefly bees (mostly *Bombus*), though 4 consisted of single *Inachis io* caterpillar; mites *Erythraeus phalangoides*, large beetles, and (probably) grasshoppers also represented, and direct observation suggested many ants also eaten. Earthworms and other soil-dwelling invertebrates may be significant in wet summers. Nestlings of all ages in larger broods received fewer visits each than those in smaller broods, except for larger broods more than 10 days old which received proportionately more visits. Feeding rate increased as broods grew older. (Tye 1982, which see for feeding rates to broods of different sizes, and prey sizes delivered.) In Tien Shan (south-central USSR), in 95 collar-samples (presumably) obtained over 8 years, June–July, from 6 nests (age of chicks not given), adult and larval insects accounted for 88·1% (by number) of items (Orthoptera 37·2%, beetles 17·5%, Diptera 16·4%, Lepidoptera 10·4%, Hymenoptera 5·5%, bugs 1·1%), spiders 10·9%, woodlice 0·5%, molluscs 0·5%; only Lepidoptera occurred at all nests. Maximum feeding rate when young 1 week old. (Gubin and Kovshar' 1985.) In Kazakhstan, May–July, comparison of stomach contents of 28 adults and 18 independent fledged young showed young had far less varied diet (representatives of 20 arthropod families, compared with 47), including few aerial insects and few crepuscular forms (Ryabov 1968, which see for further details). In East Germany, from 16·00 to 17·30 hrs, 5 young fed 18 times at 2 days, 18 times at 4 days, 25 at 6 days, and 24 at 8 days (Menzel 1964). In north-east Greenland, feeding rate of *leucorhoa* increased with age of young: 258 visits per day at 1–2 days, 316 at 4–5 days, 317 at 7–8 days, 419 at 10–11 days (Asbirk and Franzmann 1979). On Røst island (Norway), feeding rate peaked sharply in 1st hour of day (see Wagner 1958).　　　　　　　　　　　DFV

Social pattern and behaviour. Widely studied. See especially Mildenberger (1943), Conder (1956), Berck (1961), Menzel (1964), Brooke (1979), Tye (1980, 1982), Moreno (1984*a*), and Carlson *et al.* (1985). For behaviour in winter quarters, see especially Leisler *et al.* (1983); for comparison with other species, see Panow (1974) and Panov (1978). The following based largely on material for Skokholm (Wales) and Alderney (Channel Islands). Account includes limited information on *leucorhoa*.

1. Mainly solitary, but birds of all ages often in parties on migration, sometimes forming larger assemblies particularly at island stopovers and in bad weather: e.g. on North Uist (Scotland), August, *c.* 150 together (Meinertzhagen 1954). In hard April weather, Cumbria (England), *c.* 250 in relatively small area for 2 days (Turnbull 1984). May also associate with Whinchat *Saxicola rubetra* and Black Redstart *Phoenicurus ochruros*. Birds of all ages and both sexes establish individual territories whenever localized for at least 2–3 days during dispersal from breeding territories, and later on migration stopovers; such territories often grouped. (P J Conder.) On Breckland (eastern England), July, juvenile territories of this sort *c.* 0·15–0·25 ha (Tye 1982). Similar temporary territories are held in late summer and autumn by adult migrants for up to 8 days (P J Conder), also on spring migration—see Simmons (1954) for Egypt and Vernon (1972) for Morocco—though Cornwallis (1975) found no evidence for passage territories in south-west Iran. On Skokholm and Alderney, nominate *oenanthe* and *leucorhoa* on spring passage defended territories of *c.* 100–300 m² for 2–7 days, in which they fed and rested (Conder 1949, 1956; P J Conder; see also Antagonistic Behaviour,

below). On Breckland, clusters of small territories, each *c.* 0·2–0·7 ha, held by birds of either sex on spring passage, usually for only a few days or even hours (Tye 1982). Short-term territories also held sometimes in winter quarters, e.g. after burning (Dittami 1981), but typical winter dispersion is of larger and more stable individual territories (Smith 1971, which see for review). In Kenya, sexes thus defend separate territories (see Antagonistic Behaviour, below), though 'pairs', perhaps neighbours, sometimes share territory (Leisler *et al.* 1983). Winter territory typically *c.* 2·5 ha (calculated from Leisler *et al.* 1983, which see for comparison with sympatric *Oenanthe*); in Zimbabwe, one territory *c.* 2 ha (Borrett and Jackson 1970). No information on fidelity to winter territory between years. BONDS. Mating system essentially monogamous, but occasionally polygynous (♂ plus 2 ♀♀). Paired ♂♂ sometimes promiscuous (see Heterosexual Behaviour, below). Monogamous pair-bond lasts for breeding season only, but renewed annually through strong bond to territory; bird of either sex holds territory and takes new partner if former mate fails to return at start of breeding season (Berck 1953; Conder 1956; see also Breeding Dispersion, below). On Skokholm, 2 records of ♀♀ changing mate for 2nd broods, in one case ♀ joining an already paired ♂ (Conder 1956). Polygyny more common for 2nd than for 1st broods (P J Conder). On Skokholm, 2nd-brood polygyny confined to neighbouring pairs (Conder 1956), and typically leads to overlapping nesting attempts. In various studies, the 2 nests of bigamous ♂♂ 75–350(–1400) m apart (Jenning 1954; Menzel 1964; Aro 1968; Brooke 1979; Kneis 1981). ♂ may divide care between his 2 broods in various ways, according partly to degree of synchrony: one ♂ helped to feed young of brood 1 until a few days before brood 2 hatched, then attended only brood 2; in another case, where laying more-nearly synchronous, ♂ fed both broods for a few days, then only later-hatched brood (Jenning 1954). Bigamous ♂ may help to raise 2 broods with one or both of his ♀♀. One case of extra ♂ helping to feed brood of 9 young in one nest (Axell 1954); extra ♂ feeding young also reported for *leucorhoa* (Sutton and Parmelee 1954). Only ♀ broods young, both sexes feeding them, though role of ♂ varies considerably, some giving no help (P J Conder). ♂ usually less attentive than ♀, but does most feeding if ♀ starts 2nd clutch, and all if ♀ dies (Tye 1982; see Relations within Family Group, below). One ♂ failed to rear young after ♀ killed (Ruthke 1954). Young fed for *c.* 2 weeks after leaving nest (Tye 1982; Moreno 1984a). Parents typically divide fledged brood for feeding (see Relations within Family Group). Age of first breeding 1 year (Conder 1956; Brooke 1979). BREEDING DISPERSION. Solitary and territorial. Size of territory related to population density. In Greenland, territories of *leucorhoa* at least *c.* 12–16 ha (Nicholson 1930); in Baffin Land (Canada), each of 4 nests at least *c.* 800 m apart (Sutton and Parmelee 1954). On Breckland, early-arriving ♂ nominate *oenanthe* take large territories, often up to 4 times the size they finally defend *c.* 3–4 weeks later; final size on average 2·73 ha (1·92–6·68), $n = 70$ (Tye 1982). Similar size in northern Pennines, England (Tye 1980). On Skokholm, with maximum population of 9 pairs, average 2·9 ha (Brooke 1979); with maximum 38 pairs, average of 99 1st-brood territories 1·5 ha (0·5–3·3); territories tend to be smaller where terrain broken by rocky outcrops (Conder 1956, which see for annual variation). Contiguous territories in Uppsala (Sweden) on average 1·2 ha (0·5–2·0), $n = 10$ (J Moreno and A Carlson). In Moscow region (USSR), territories apparently much smaller, 0·16–0·4 ha (Ptushenko and Inozemtsev 1968). On Skokholm, average distance between nests in 3 years when more than 30 pairs bred was 62–85 m (16–146); with only 14 pairs, 266 m (90–570), and pairs with no

immediate neighbours fed up to *c.* 250 m from nest (P J Conder). In Tuvinskaya ASSR (USSR), average 240 ± 17 m ($n = 16$) between nests, minimum 112 (Ivanitski 1980). Closest recorded nests 15 cm, the 2 broods (total 16 young, of 2 presumably monogamous pairs) grouping together (Saxby 1874). On Breckland, average density 35 pairs per km² (15–52) in suitable habitats varying from optimal short turf to ranker heath (Tye 1980). In upland regions in Britain, *c.* 1–10 pairs per km² (Williamson 1968; Robson and Williamson 1972; Massey 1978). In Uppsala, *c.* 26 pairs per km² (Carlson *et al.* 1985), 83 in optimal grazed habitat (J Moreno and A Carlson). In Luxembourg, allegedly 100 pairs per km² over 0·7 km² (Morbach 1934). In degenerate woodland, Morocco, 1·5 pairs per km² (Thévenot 1982). Territory all-purpose. On Breckland, serves for almost all feeding (Tye 1982); for strategy of territorial feeding, see Brooke (1981), also Relations within Family Group for partitioning associated with brood-division. Elsewhere, apparently some feeding outside territory: in Rheinland and Taunus mountains (West Germany), birds often feed in fields surrounding territory (Mildenberger 1943; Berck 1961). In cases of bigamy, ♂ ranges freely over whole territory, respective ♀♀ nesting apart and defending their 'own' half against each other (Tye 1982; P J Conder). A given pair does not usually lay 2nd clutch in same site as 1st, but occasionally does so if sites in short supply or if previous nest removed; site may be re-used after 1-year interval (P J Conder). At Dungeness (England) artificial sites re-used regularly in successive years (Axell 1954). In Sussex (England), ♀♀ which had been caught on nest and deserted bred up to 3·2 km away (Thomas 1925, 1926). Established breeders show marked fidelity to territory between years (e.g. Conder 1956, Hempel 1957): ♂♂ recorded returning for up to 4 successive years, ♀♀ for up to 5 years (Savinich 1983; P J Conder). However, may also move between years to (perhaps) better territory: on Skokholm, where low density favoured this, 6 of 7 ♂♂ moved between successive years, 5 to a better territory; of 10 ♀♀, 4 moved (Brooke 1979; see also Tye 1982). 1st-time breeders not markedly faithful to natal site (Savinich 1983). ROOSTING. During breeding season, ♀ roosts on nest from start of incubation until brooding ceases (see Relations within Family Group). At other times, both sexes roost singly in grass or heather tussock, burrow entrance, stone wall, etc. (P J Conder.) Each bird uses several roost-sites (Berck 1961). In winter quarters, frequently seeks shade of small bush, stone (etc.) in heat (Jackson and Sclater 1938), typically when temperature exceeds 30°C (Tye 1982). Active for much of day, notably early morning and late evening, feeding until almost dark (Berck 1961; see also Song-display, below), likewise in winter quarters (Bannerman 1936). For rhythm of feeding young in Arctic Norway, see Wagner (1958). In Greenland, both sexes of *leucorhoa* pair rested (on average) 02.39–07.25 hrs (Asbirk and Franzmann 1979, which see for details). Comfort behaviour includes sunning: adults and juveniles choose sheltered sunny bank to lie on their stomachs, head feathers ruffled, wings slightly spread to expose rump and flanks; dusting not recorded. Will bathe in shallow pools or puddles. (P J Conder.)

2. Typically a shy and wary bird throughout the year. In Zimbabwe, December, birds shunned close approach (Borrett and Jackson 1970), but wintering birds may be more approachable when associated with human settlements (e.g. Butler 1905). When approached to *c.* 30–100 m, often stands erect or flies off, landing on elevated perch or (often) hiding behind summit of rock outcrop and peering over it, only top of head showing. Statements that *leucorhoa* stands more erect and perches on trees more often than nominate *oenanthe* are wrong.

In anxiety, bobs body, wags tail, flicks wings exposing rump and tail pattern, and gives Tuc-calls—preceded by Weet-calls in breeding territory (see 4–5 in Voice); in greater anxiety (e.g. raptor nearby) bobbing particularly energetic and deep. (P J Conder.) Bobbing is a rapid (0·1 s or less) down-up movement of head and fore-body; tail-wagging begins with rapid flick down of slightly fanned tail, followed by slower up-down movement; similar bobbing in Isabelline Wheatear *O. isabellina* is slower, taking 0·2–0·3 s (Tye and Tye 1983). If attacked by raptor, will dive into nearest burrow (P J Conder). In Arabia, mostly ignores raptors, but if hard-pressed will seek refuge in bush or hole (Meinertzhagen 1954). For response to weasel *Mustela nivalis*, see Parental Anti-predator Strategies (below). FLOCK BEHAVIOUR. Regularly associates with conspecific birds on migration and at favourable feeding sites (typically short turf) outside breeding season. Following account of behaviour at stopover by P J Conder. On first arrival, individual-distance *c.* (1–)5–10 m. Such a group will feed facing mostly in the same direction, indicating some cohesion, but if disturbed scatter individually. When disturbance over, all return and re-group. Meinertzhagen (1954) described mass exodus of *c.* 150 birds from stopover in North Uist (Scotland), August: at *c.* 17.00 hrs, birds became restive, flying about up to *c.* 30 m above ground and being instantly joined by others, then landing again, 'twittering' the while; at dusk, all took off simultaneously. SONG-DISPLAY. Territorial-song (see 1a in Voice) given by ♂ mostly from favoured song-posts—usually some low eminence (e.g. stone, fence), occasionally telegraph wire or tree-top; also in Song-flight, less often in normal flight (Witherby *et al.* 1938*b*). Territorial-song rarely heard (except in intense territorial disputes) before ♂ is paired; thereafter given chiefly as self-advertisement, loudly and harshly in territorial defence, more softly in presence of mate (P J Conder). Following account of Song-flight of nominate *oenanthe* mostly from description by P J Conder. Performed during intervals in Flashing-display (see Antagonistic Behaviour, below) at any stage of encounter with conspecific or other intruder (see also Parental Anti-predator Strategies, below), and in pair-formation. Resembles that of pipit *Anthus*: bird flutters jerkily upwards at 60–80° (or 30–40°: Panow 1974), typically singing, wings beating rapidly in short bursts, tail fanned and flirted (i.e. momentarily spread wider). At peak of flight, wings do not beat continuously so that bird tends to dance up and down, but, *contra* some reports, stationary hovering apparently not a ritualized element of display (Conder 1954). In final phase, may dive obliquely back to ground, still singing, or precede dive by straight-line flight of up to *c.* 50 m or (typically in response to conspecific intruder) by circling for up to 10 s with deep slow wing-beats and fanned tail ('butterfly flight') before diving down; may also fly towards rival, turn back half-way and advance again, repeating sequence several times, such display often following sexual activity (Mildenberger 1943). Descent also described as slow angled glide (Panow 1974). Bird may return to former perch but often descends to within 1 m of ground, then flies at that height to another perch up to 300 m away (Menzel 1964). Height reached in Song-flight varies with wind strength, etc.: in hostile encounter, ascends usually to 2–3 m; in Song-duel with neighbouring resident, or when song directed at ♀, may reach 10(–30) m, exceptionally higher. Territorial-song and Song-flights rare at start of breeding season, except in fiercest territorial disputes between unpaired ♂♂; usually begin after pair-formation, peak just before hatching, then decline sharply, but increase again if ♀ starts 2nd clutch. Providing he has not begun moult, ♂ starts singing again within 2–3 days of 1st brood leaving nest. (P J Conder.) Frequency of Song-flights

increases significantly after mate's fertile period ends; ♂ then relaxes mate-guarding (see Heterosexual Behaviour, below) and more likely to intrude on and elicit Song-flights from neighbours (Carlson *et al.* 1985). In June, 2 ♂♂ performed 25 Song-flights in 2½ hrs (Berck 1961). No details of diurnal rhythm but, as with other activities (see Roosting), song not uncommonly heard in poor light of morning and evening and occasional singing at night reported (Menzel 1964). Various other kinds of song heard from one or both sexes: ♂ gives quieter Conversational-song (see 1b in Voice) in contact with mate. Subsong given by both sexes, but mostly by ♂, and in various forms. Loud Subsong (see 2b in Voice) given by ♂ demarcating territory before arrival of ♀, later by both sexes in territorial disputes, by ♂ and less often by ♀ in heterosexual encounters. Quiet Subsong (see 2a in Voice) given mostly in breeding season, e.g. by unpaired ♂♂ demarcating territories, by ♀ on nest to nearby mate, but also at other times of year, notably on migration by birds (including juveniles) resting or defending temporary territories. (Conder 1949; P J Conder.) Subsong, probably of this type, also reported in winter quarters (Borrett and Jackson 1970; Leisler *et al.* 1983), once in Zambia in July (Aspinwall 1973). ANTAGONISTIC BEHAVIOUR. (1) General. In breeding territory or temporary territory, both sexes highly aggressive towards conspecific birds, usually ♂ against ♂ and (less commonly) ♀ against ♀. Paired ♂ may join mate in expelling ♀ intruder from territory, but ♀ does not help mate when intruder is ♂. ♂ usually establishes territory immediately on return to breeding grounds, and keeps it until young independent, sometimes until autumn migration. (P J Conder.) ♂♂ especially aggressive during initial territorial demarcation, also during pair-formation and mate-guarding (J Moreno and A Carlson). Aggression wanes after start of incubation, and less for 2nd broods, though occasionally directed at trespassing 1st brood (Conder 1956). Juveniles remaining in breeding area constantly liable to attack, even from own parents (Tye 1982). ♀♀ most aggressive on first arrival and during pair-formation (J Moreno and A Carlson). Newly-arrived ♀ capable of defending previous year's territory against lone neighbour (though probably not against a pair) until former mate returns or new one is accepted. One ♀, arriving before mate, paired with new ♂ and both vigorously attacked her former mate on his return; however, latter was victorious after 2–3 days, re-claiming ♀ (Conder 1956). Status of nominate *oenanthe* in dominance hierarchy of sympatric *Oenanthe* well studied, both in and out of breeding season. In Tuvinskaya ASSR, dominant over Desert Wheatear *O. deserti* and *O. pleschanka*, but subordinate to (and rarely challenging) *O. isabellina* and Finsch's Wheatear *O. finschii* (Ivanitski 1980, which see for rates of confrontation). See also Eggebrecht (1943) and Panow (1974). Nominate *oenanthe* may also be dominant over *O. hispanica* (see Kinzelbach and Martens 1965, Mundy and Cook 1972). *O. o. leucorhoa* on passage capable of setting up temporary territory inside breeding territory of resident nominate *oenanthe* (P J Conder). In winter quarters, nominate *oenanthe* likewise dominant over *O. pleschanka* and subordinate to *O. isabellina* (Leisler *et al.* 1983), though aggression towards *O. isabellina* also reported (Dittami 1981, which see for relations with other *Oenanthe*; see also Sinclair 1978). In breeding season, both sexes may threaten and make flying attacks on wide variety of passerines trespassing near nest or dependent young (Sutton and Parmelee 1954; Berck 1961; Ivanitski 1980; Moreno 1984*a*). Sometimes also attacks larger birds (Mildenberger 1943; Berck 1961; P J Conder; see also Parental Anti-predator Strategies). In winter quarters, aggressive towards (e.g.) *S. rubetra*, Yellow Wagtail *Motacilla flava*, and Rock Thrush *Monticola saxatilis* (Dittami

1981; Leisler *et al.* 1983; G D Field). (2) Threat and fighting. Resident of either sex typically demarcates territory by Song-display (see above) and by aerial and ground threat-display in which aggressive intent signalled by varying exposure of plumage pattern and commonly by raising of head and tail; for full repertoire see Panow (1974) and Panov (1978). Following account mainly by P J Conder. Resident may fly fast, low, and direct to intercept intruder and, with Tuc-calls, chase it off; chases between neighbours may be long and tortuous with frequent changes of leader (Tye 1982). If intruder stays, owner usually switches to ground-threat, landing near rival and typically adopting Erect-posture: stands markedly upright, stretching neck upwards, and may bob body; usually gives Quiet Subsong if unpaired, Loud Subsong and Territorial-song if paired. In Erect-posture, resident hops a little way towards trespasser (Advancing-display) who usually retreats equivalent distance, followed by further advance of resident, and so on. If intruder stays, resident escalates threat. ♂ (not ♀) also commonly advertises to intruder (and vice versa) with Flashing-display (Fig A), often from some eminence: turns towards or

A

away from rival with head raised, exposing throat and breast, and giving Loud Subsong; tail is fanned and, every 1–2 s, flirted, while wings partially flicked out. Several variants occur, apparently indicating threat-appeasement conflict. Bird facing rival may expose fanned tail above back by bobbing or by crouching and lowering head so that bill almost touches ground. When facing away from rival, fanned tail is lowered and may be twisted towards rival not standing directly behind (Fig B).

B

These displays accompanied by pattern of advance (owner) and retreat (intruder) as described above, displaying bird apparently trying to overtake rival. Crouching ♂ sometimes rushes rival in 'rodent run' fashion, or, exceptionally, flies 10–15 m above ground with fluttering wings and dangling legs, lands, and

continues approaching intruder with crouched Flashing-display. Neighbours (♂–♂ or ♀–♀) may walk side by side along disputed territorial boundary with tails fanned and wings lowered to expose rump (Tye 1982). Birds displaying thus may be upright (and sing) or crouched; also reported apparently mock-feeding during this display (Panow 1974), but real feeding recorded by Tye (1982). Opponents may also pull and toss aside grass (Panow 1974). At higher intensity, resident threatens with Oblique-posture: crouches or bows, legs partly bent, wings drooped, tail upright and usually closed (Fig C); head may be

C

thrust forward and bill slightly open showing black mouth. (P J Conder.) Oblique-posture most often directed at larger avian intruders or against independent young pestering parents (P J Conder). Presence of intruder—usually (not always) one in shallow depression in ground—may elicit remarkable Dancing-display (see also Heterosexual Behaviour, below, for same display towards mate). Performed by ♂ only (though ♀ once 'danced' over Rock Pipit *Anthus spinoletta*: P J Conder) as follows: with fluttering wings (rendering underwing pattern conspicuous), and flirting tail, bird leaps rapidly and erratically from side to side in a 10–20(–200)-cm arc over intruder, scarcely touching ground on either side and giving Loud Subsong or Territorial-song (Turner 1950; P J Conder); may also apparently dangle and kick legs, and ruffle plumage, giving impression of whirling mass of feathers (Edwards *et al.* 1950; Smith and Hosking 1955). Display vigorous and rhythmic, e.g. bird 'rapidly jerking itself back as if it were caught on a piece of cotton' (Pettitt and Butt 1950). Intruder may respond with a threat-display. After Dancing-display, performer may (uncommonly) pick up grass or mock-feed. If intruder does not withdraw, performer often reverts to high-intensity variants of Flashing-display or Oblique-posture within 1–2 m of rival. After threat-display, resident (mostly ♂, sometimes ♀) may also perform Zigzag-flight: flies *c.* 1 m above ground on a fast and zigzagging but roughly circular path, sometimes figure-of-eight, apparently beating wings unevenly to produce erratic rolling flight, and displaying underwing pattern. Fight may ensue if intruder persistent: starts with resident flying at rival (for related attacks on mates, see Heterosexual Behaviour, below), both then usually flying up vertically for 2–3 m with tails fanned, pecking at each other and wing-cuffing. Occasionally, one bird is pinned on its back on ground. Intruder usually retreats after fight. Before fighting, ♀♀ typically adopt Erect-posture (accompanied by Quiet Subsong), or perform Advancing-display or some displacement-activity, e.g. mock-feeding, resting on stomach (as in sunbathing), or hopping rapidly in and out of burrows. On Skokholm, ♀-♀ fights more frequent and longer than ♂-♂ fights, possibly because ♀ lacks some of the threat-display associated with ♂'s plumage pattern (P J Conder); this difference in fighting frequency not evident on Breckland where fights rare and brief (Tye 1982), perhaps related to lower density. Temporary territories on migration defended by Quiet Subsong, Dominance-display, flying attack, chasing, Zigzag-flight, and Oblique-posture (Conder 1949; Tye 1982). During autumn migration, Iraq, frequent squabbles for song-posts (Moore and Boswell 1956). Appeasement/submissive postures not widely reported. ♂ trespassing on territory often crouches on stones making himself inconspicuous to owner (J Moreno and A Carlson). Residents reported flying high over or skirting neighbouring territories to reach feeding areas beyond, evidently to avoid eliciting attack (Mildenberger 1943; Berck 1961). For aggression in heterosexual context, see below. HETEROSEXUAL BEHAVIOUR. (1) General. On Skokholm, ♂♂ arrive on territories on average 4 days (1–11) before ♀♀ (Conder 1956; Brooke 1979). Older ♂♂ arrive first (Brooke 1979), older ♀♀ slightly before younger ♂♂ (Conder 1956). Members of previously established pair seem to accept each other again with little ceremony (P J Conder), though paired ♂♂ continue to show interest in other ♀♀ (see subsection 5, below). Gap between arrival and egg-laying *c.* 3 weeks (Brooke 1979). As in other *Oenanthe*, advertising-displays of ♂ in courtship similar to those in antagonistic encounters. (2) Pair-bonding behaviour. In early stages, ♂ vacillates between attracting and threatening ♀♀. Unpaired 1-year-old ♂, usually giving Quiet Subsong, flies in bouncy flight towards ♀ entering his territory, and lands before reaching her. ♀ usually takes off again with tail fanned, and lands a little further on, ♂ following (etc.). Uncommonly, ♂ adopts Enticement-posture in which (unlike Oblique-posture) legs are well bent, head and bill held low (throat partially obscured), and tail slightly fanned and lowered; aggressive plumage signals thus obscured. Posture accompanied by Sub-song or Territorial-song. (P J Conder.) In evidently similar posture (Fig D), adopted after Song-flight, ♂ runs brushing tail

D

along ground (Panow 1974). ♂ may then perform Song-flights or chase ♀ who keeps just ahead of him and within territorial limits; sexual chases (see also below) of this sort sometimes lead to Dancing-display (P J Conder), similar to antagonistic sort except that ♀, not intruder, is the close 'spectator'. As in aggressive context, spectator almost always stands in shallow depression (e.g. burrow-entrance, gap between tussocks, cleft between stones) whose sides ♂ can use as jumping platform (Selous 1901; Lloyd 1933; Conder 1950; Edwards *et al.* 1950; Monk 1950; Pettitt and Butt 1950). Once, in presence of ♀, 2 rival ♂♂ leaped over each other's heads (Selous 1901). After Dancing-display ♂ may adopt Ecstatic-posture, usually for 2–3 s: stands horizontally, head pointing at ♀, tail raised and closed,

E

and, as in Dancing-display, usually gives Loud Subsong (P J Conder). In another display (Fig E), apparently signalling some aggression, one ♂ stood erect, flicked carpal joints out and flirted tail prior to performing presumed variant of Dancing-display in which bird descended to roughly same spot each time (Panow 1974); same bird also shown bowing head (Panow 1974), perhaps in appeasement. After one Dancing-display, ♂ finally prostrated himself in front of ♀, head touching ground, wings and tail outspread; lay for a few seconds, his body quivering, then flew quietly away, accompanied by ♀ (Lloyd 1933). Dancing-display seems to stimulate ♀ who is often highly active afterwards, e.g. entering potential nest-sites or carrying nest-material (Smith and Hosking 1955). Dancing-display usually performed prior to egg-laying, rarely during incubation. Occasionally follows Greeting-display of ♀ (Fig F): bird giving

F

Quiet Subsong holds body and unfanned tail more or less horizontal, wings partly or fully outstretched and quivered sometimes in 2–3 bursts of 1 s each, but also continuously for up to 10 s; bird displaying thus sometimes moves forward a little way on tiptoe, and may occasionally take off briefly on quivering wings. Greeting-display may be performed by either sex, more often ♀, usually when pair meet at nest, or during pauses in chasing. (P J Conder.) (3) Nest-site selection. Apparently by ♀, though ♂ may show ♀ potential sites (see König 1964 for one sequence). Following account by P J Conder. Before ♀♀ arrive on breeding grounds, ♂♂ sometimes visit prospective sites, occasionally playing with pieces of vegetation (often unsuitable as nest-material) and even taking some into a site. Shortly after pair-formation, ♂ in some cases leads mate around territory for a day or so, flying *c.* 10 m ahead and waiting for her to hop up to him. In other pairs, ♂ follows ♀, occasionally displaying by 'pouncing', bobbing, and wing-flicking, giving impression of excitement. 1–2 days later,

♀ begins to inspect hole-sites, apparently listening first then entering hole for up to 20 s before visiting another; sometimes removes debris from burrows not eventually chosen for nesting. During these activities, ♂ plays sentinel role, often giving Quiet Subsong. Site selection behaviour by ♀ changes imperceptibly to nest-building (see subsection 7, below). (4) Courtship-feeding. Only one report and probably rare. Occasionally recorded by Brooke (1979) in observations of one pair during laying period and start of incubation, but thought to contribute less than 10% to ♀'s food intake. ♀ occasionally receives food from ♂ during nestling period, especially first 1-2 days (Mildenberger 1943), ♀ greeting ♂ at nest-hole entrance (P J Conder). No good evidence, however, that ♀ eats food herself. (5) Mate-guarding. ♂ guards mate carefully during her presumed fertile period (3-4 days before laying until 2nd egg laid), following her around within c. 2 m. At this time, ♂ also perches higher, presumably to enhance surveillance. ♀ in pre-fertile period sometimes approaches trespassing or neighbouring ♂♂, but is chased back to own territory by mate. (Conder 1956; Carlson et al. 1985.) (6) Mating. Occurs on ground, apparently usually in depression or nest-hole entrance as rarely seen in the open. Prelude may be Dancing-display, Zigzag-flight, or chasing, followed by mutual Greeting-display (Meeting-ceremony); ♀ then flies into (e.g.) burrow, and ♂ thought to copulate with her there. According to Panow (1974), preliminaries include Loud Subsong. One sequence (involving exceptionally prolonged chasing) as follows: pair chased each other for 35 min, comprising chases of 20-30 s alternating with pauses of 1-2 s when birds landed 1-2 m apart; chases became shorter and pauses longer, and during pauses ♀ performed Greeting-display and ♂ took off again, ♀ finally diving into burrow pursued by ♂; in such sequences, pair stay in burrow up to 2 min. (P J Conder.) In Greeting-display, used to solicit copulation, ♀'s whole body quivers (J Moreno and A Carlson). In a sequence on open ground, ♀ solicited, ♂ landed nearby and, after Meeting-ceremony, copulated with ♀. Afterwards, ♀ threw her head up vertically and quivered wings clear of body, while ♂ tail-wagged for c. 1 min. (P J Conder.) In each of 3 copulations between known mates, only 1 cloacal contact occurred, after which ♀ preened and ♂ flew to nearby perch; 2 of the matings occurred 1 day before, and the other 1 week before 1st egg laid (J Moreno and A Carlson). At Uppsala, paired ♂♂, whilst guarding their own mates, tried to copulate with other ♀♀ whenever possible; reacted to intruding ♀♀ by approach (39% of 36 cases), displaying to them (42%), attempting copulation (11%), or by attack (8%). ♀♀ occasionally 'visited' other ♂♂. (Carlson et al. 1985; see subsection 5.) Rarely, juveniles 3-4 weeks old attempt unsuccessfully to copulate with siblings (P J Conder). (7) Behaviour at nest. Nest-building by ♀. Unpaired ♂♂ occasionally take material into holes but no such holes known to be used subsequently for nesting. Paired ♂♂ occasionally bring material, bobbing and wing-flicking, but often drop it before entering nest-hole. ♀ sometimes builds in up to 4 holes for 2-3 days before concentrating on one. (P J Conder.) While ♀ builds, ♂ typically perches nearby and sings, also often performs Song-flights (Berck 1961). May show aggression (flying attack or pounce) to inactive ♀ at this time, also during nest-site selection: may fly just over her head (occasionally landing momentarily on her back), gliding the last 20-30 cm of approach, and giving Rattle-call (see 6 in Voice); ♀ may crouch or jump (P J Conder). ♂ shows little interest in nest during incubation, but begins visiting it more often near hatching (Tye 1982). While ♀ sitting, however, ♂ often perches nearby, giving Conversational-song, ♀ answering with soft Quiet Subsong (P

J Conder); for other calls of ♀ on nest, see 8 in Voice. When meeting outside nest-entrance, either or both of pair may perform Greeting-display (P J Conder). RELATIONS WITHIN FAMILY GROUP. Young brooded by ♀ for up to 5-6 days (P J Conder), ♂ never staying more than 2 min at nest at this stage (J Moreno). Eyes of young open at 6 days (Panow 1974), fully at 9 days (P J Conder). Both sexes feed young; on approaching nest, usually land on some eminence and scan surroundings before entering (J Moreno and A Carlson). Parents share nest-hygiene, eating faecal sacs for first 3 days, removing them thereafter, though sometimes allow faeces to accumulate near fledging. (P J Conder.) Faecal sacs usually dropped 30-50 m from nest (Brooke 1981; Gubin and Kovshar' 1985). Kneis (1983) reported parents carrying away addled eggs, and once perhaps dead chick. By 10th day, young can clamber on nest rim to defecate. Parent usually feeds chick nearest nest-entrance, but positions change so no chick monopolizes food. (P J Conder.) However, runts sometimes found cold and begging outside nest; probably jostled out by siblings (J Moreno). By 14 days, young quite mobile, moving in frog-like hops in burrow, begging noisily with bobbing action and wing-shivering, and receiving food either at nest-entrance or in nest (P J Conder, J Moreno). Account of behaviour outside nest based on Moreno (1984a). Brood leaves nest-hole over period of 2 days, starting at 14-16 days old. Young usually stay around nest-entrance until 18th day, then become increasingly dispersive, moving away 10-50 m and more, though continue to return to nest-area for roosting, sometimes even when independent (P J Conder). Usually wait on elevated places to be fed for 1 week after leaving nest, then become highly mobile, following parents around. In Uppsala, parents typically divide brood about equally from 3-4 days out of nest, and thereafter feed only the young in their care, each half of brood aggregating (in different parts of territory) to be fed. Young nearly always beg from 'their' parent; only 5% of 119 chases directed at 'wrong' parent, and never successful in securing food. (Moreno 1984a.) On Skokholm, contribution of ♂♂ varies, some feeding juveniles slightly less frequently than ♀, some feeding rarely or never (P J Conder). Young begin self-feeding 6-8 days out of nest (Gubin and Kovshar' 1985) but fed by parents for 14 days, and continue to beg up to 18 days (Moreno 1984a). According to Mildenberger (1943) and Panow (1974), 2nd clutch laid only when 1st brood independent, but Tye (1980, 1982) found 2nd clutch usually started before 1st brood independent, and similar strategy perhaps operates in leucorhoa (Nicholson 1930; Wynne-Edwards 1952). In nominate oenanthe, ♂ takes over feeding 1st brood, occasionally when still in nest, while ♀ starts incubating 2nd clutch (Tye 1980). ANTI-PREDATOR RESPONSES OF YOUNG. From c. 10-14 days, nestlings warned of danger by parents' calls (see 4-5 in Voice) shuffle into dark recesses of nest-hole and crouch facing away from light. When alarmed, fledged young continue to dive into holes, under tussocks, etc. (Berck 1961; P J Conder; J Moreno and A Carlson) and, up to c. 21 days, each brood-member has preferred refuge sites up to 70 m from nest. From c. 21 days, young fly from danger (P J Conder). In Taunus mountains (West Germany), fled from raptors, including Great Grey Shrike Lanius excubitor but not Buzzard Buteo buteo (Berck 1961). Juveniles express anxiety by bobbing and giving Tuc-calls (P J Conder). PARENTAL ANTI-PREDATOR STRATEGIES. (1) Passive measures. Sentinel ♂ uses alarm-calls to warn sitting ♀ (Mildenberger 1943; Berck 1961) who then usually hops quickly to rear of nest-hole and flattens herself against it. ♀ frequently deserts eggs early in incubation, rarely also young 1-2 days old. (P J Conder.) Sits more tightly later, and reported not leaving until almost touched or nest-hole knocked (Meinertzhagen 1938; Mildenberger 1943). When alarmed while feeding young, one ♂ *leucorhoa* crept away through crevices in rocks and emerged *c.* 2 m beyond nest (Nicholson 1930). (2) Active measures: against birds. One pair frequently attacked Magpies *Pica pica* (Ruthke 1954). On Skokholm, attacks Little Owl *Athene noctua* with Rattle-calls. When *A. noctua*, *B. buteo*, or Peregrine *Falco peregrinus* near nest of young, gives variant of Territorial-song (see 1a in Voice); may also perform Song-flight when any large bird threatens nest (P J Conder). (3) Active measures: against man. As intruder passes through territory, or threatens nest, owner may hover overhead, giving alarm-calls, or sometimes perform Song-flight (Conder 1954; P J Conder). May also try to distract attention by flying between different perches away from nest (J Moreno and A Carlson). In Greenland, pair of *leucorhoa* returned boldly to nest after intruder had withdrawn a few metres, making frequent visits to nest with peculiar slow gliding flight (Nicholson 1930), perhaps Song-flight. In Afghanistan, after intruder removed nest and eggs, pair returned and both feigned injury (no details) and fluttered off (Meinertzhagen 1938). (4) Active measures: against other animals. Most widespread response to ground predators appears to be mild mobbing. Fluttered over grass snake *Natrix natrix* and dropped down as if to strike. Often follows stoat *Mustela erminea* (P J Conder) and may harass it by hovering just overhead and calling (Schifferli and D'Alessandri 1971); on Breckland, both sexes mob stoats thus as soon as they enter territory (Tye 1982). Will also perform Flashing-display in presence of stoat (P J Conder). In Kent (England), August, sudden appearance of weasel *M. nivalis* caused nearest of 3 birds (presumed migrants) feeding on track to perform 'broken wing' distraction-lure display, fluttering along in front of predator for more than 1 min (Hindle 1979).

(Figs by K H E Franklin: A and C–F from illustrations in Panow 1974; B from drawing by J P Busby.)　　EKD, PJC

Voice. Used throughout the year, but especially in breeding season. Little information on geographical variation, but voice deeper further south (Chappuis 1969; see also song of *seebohmi*, below). According to Stadler (1952, which see also for musical notation), no dialects identified in Europe. Account based on extensive material provided by P J Conder for Skokholm (Wales). Unless otherwise indicated, account refers to nominate *oenanthe*. For additional sonagrams, see Bergmann and Helb (1982). Recording of *seebohmi* song includes a noise resembling Drumming of Snipe *Gallinago gallinago* (J Hall-Craggs) but significance not known.

CALLS OF ADULTS. (1) Song. (a) Territorial-song of ♂.

A short but vigorous, pleasantly modulated warble in which melodious, rather lark-like notes are mingled with harsh creaky and rattling sounds, including mimicry (Witherby *et al.* 1938b). Also described as a fast chatter of crunching creaking sounds preceded by a few 'weets' (Jonsson 1979). Mimicry common, versatile, and accurate. Saxby (1874) reported 10 species, including 5 waders (Charadrii), mimicked in Shetland. On Skokholm, Territorial-song and Subsong (see below) included mimicry of 38 species and races, of which 9 bred on the island, 10 were regular visitors, 16 occasional or irregular visitors, and 2 vagrants; also mimicked the squeak of a pulley wheel and distress-squeal of rabbit *Oryctolagus*. (P J Conder.) In Balkans, mimicry (perhaps by *libanotica*) includes Bill-clattering of White Stork *Ciconia ciconia* (Stadler 1952). On Skokholm, up to 4 species occasionally mimicked in one bout of song, and calls of migrants incorporated within hours of migrants' arrival. Song difficult to render, and highly variable in volume, composition, melodiousness, and rate of delivery (P J Conder), though every ♂ has several constant phrases in his repertoire (Bergmann and Helb 1982). Call-type units (notably calls 4–5, below) commonly incorporated singly or in groups (J Hall-Craggs), often starting phrase. Songs also vary markedly in length, this due partly to context. Panov (1978) distinguished short and long Territorial-songs, though continuum probably exists, varying with context (P J Conder). Available recordings and commentary by V C Lewis suggest that perched song (e.g. Fig I) may be short and given at long intervals, though longer phrases may intervene. One recording of perched 'song' effectively an assemblage of calls, including Meadow Pipit *Anthus pratensis*, and (Fig II) whistle simultaneous with rattle (thin vertical lines in sonagram); the 2 heavy vertical bars in Fig II are 'chack' sounds (see call 7) of another bird, presumably mate. (J Hall-Craggs.) Short songs also typical of low Song-flights, e.g. Fig III (apparent ascent) and Fig IV (apparent descent). In higher, undulating Song-flights, song apparently longer and more continuous: e.g. song shown in Fig V, described as musical, though somewhat repetitive, including notes like Skylark *Alauda arvensis* (V C Lewis). In Song-flights, usually 3–4(–10) phrases. Heightened aggression, including response to raptors, typified by peculiar twangy

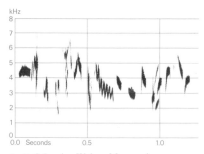

I　V C Lewis　Wales　May 1978

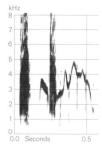

II　V C Lewis　Wales　April 1969

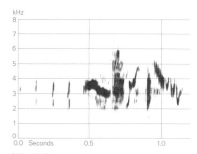

III　V C Lewis　Wales　May 1979

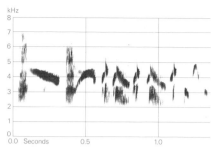

IV V C Lewis Wales May 1979

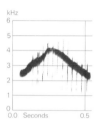

VI E D H Johnson Morocco April 1964

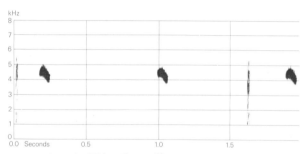

VIII V C Lewis Wales June 1968

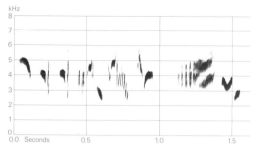

V V C Lewis Wales May 1978

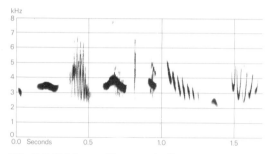

VII E D H Johnson Morocco April 1973

or vibrant sounds, some scratchy like shingle being rattled, some like Black Redstart *Phoenicurus ochruros*, preceded by rapidly repeated variants of call 5: e.g. when Peregrine *Falco peregrinus* threatening nestlings, renderings include 'zee-zee-widdly-ee', 'see-bree-yu', 'see-tiu-tiu', 'zwang-wuee', 'teng-whu', vibrant 'weng', twangy 'whuee-yu-yu', and 'eu-wirra' (P J Conder). Song of *seebohmi* audibly different from nominate *oenanthe*: according to J-C Roché, characteristically a few slow, low-pitched phrases, together with a 'crumpled paper' sound. Recordings show that song of *seebohmi* is more measured, melodious and sonorous; compared with nominate *oenanthe*, units longer and lower-pitched, mostly below 4 kHz, whereas much of nominate *oenanthe* song exceeds 4 kHz—also, pauses between units longer. One recording of perched song contains notably buzzy units; another (22 phrases in *c.* 46 s) contains chuckles (series of 'tuk' units) and snores (rattles) (E D H Johnson), also smoothly ascending and descending portamento whistling notes and terminal descending tinkling ripples. The rattle and portamento notes are given simultaneously, e.g. Fig

VI (compare Fig II). Similar diad structures also feature in display-flight song of *seebohmi*, e.g. 4th and 6th units in Fig VII show overlapping tones and hisses. (J Hall-Craggs.) (b) Conversational-song. Quieter and more musical than Territorial-song, usually lacking harsh scratchy and vibrant sounds. Typically given by ♂ in close contact with ♀, e.g. on ground outside nest when ♀ sitting. Renderings include a repeated 'eu twirra', a quiet 'zeewirrü', 'zeeü widdlü widdlü yü', and 'siu diddely diddly'. (P J Conder.) (2) Subsong. Given by both sexes for self-advertisement. Accounts in the literature are often insufficiently detailed to distinguish between 2 types, but distinction not always easy to make, and to some extent arbitrary. (a) Quiet Subsong. Most commonly used type. A varied, subdued scratchy warble, units delivered in rapid succession, including mimicry (see 1a, above). Given mostly by ♂ before arrival of ♀ in breeding season, but also by ♂ and ♀ in defence of territory on migration and in winter quarters. (P J Conder.) One song in Zambia, July, described as a quiet rattling (Aspinwall 1973). (b) Loud Subsong ('Battle Song': Panow 1974). Louder, harsher, and scratchier than Quiet Subsong, including mimicry (P J Conder). Described by Panov (1978) as incorporating a series of muffled clicks (presumably rattles), wailing, food-calls of young, and mimicry. Given by both sexes in threat-display, notably in territorial disputes, also in pair-formation (e.g. Dancing-display) and Greeting-display (P J Conder). (3) Bree-call. A vibrant, moderately high-pitched 'bree' used by both sexes for contact in variety of circumstances, but usually near nest, often when resting and occasionally in preliminaries to presumed copulation. Occasionally this, or similar sound used as prefix to song or Subsong when intruder present. (P J Conder.) (4) Tuc-call. Commonest call, given in various contexts by birds of both sexes throughout the year whenever excited or alarmed. Rendered 'tuc' (P J Conder) or a hard wooden 'tk' (Bergmann and Helb 1982). Paired birds in breeding season frequently combine it with call 5 (see below), e.g.'weet tuc tuc' (P J Conder); Fig VIII shows 'tk weet weet tk weet' (J Hall-Craggs). By mid-July, and during moult, call sounds more like 'tch', resembling food-calls of nestlings. In *leucorhoa*, a flatter 'tac' (P J Conder); in *seebohmi*, apparently 'tä' (Kleinschmidt 1905). (5) Weet-call. A 'weet' or 'yeet' of alarm and warning, given by both sexes once paired (P J Conder). Commonly combined with call 4 (see above). At highest intensity, when young directly threatened, a rapid succession (e.g. once 86 calls in 1 min) of Weet-calls may be given without Tuc-calls, accompanied by frequent deep bobbing and tail-wagging. Also used in Zigzag-flight and Dancing-display. By end of June, calls sounds more like 'chip'. (P J Conder.) (6) Rattle-call. A loud, rapid 'tetetetete' comprising 5-6 units, given by both ♂ and ♀ in flying attacks on conspecific or other birds, including raptors (P J Conder). Also used by either sex apparently frustrated by 'inappropriate' behaviour of mate: e.g. by ♂ when ♀ resting for lengthy spell during nest-building, or by ♀ when ♂ fails to attack intruder. (P J Conder.) (7) Tset-call. Protracted rhythmic series of 'tset-tset...', being a slightly accelerated more intense form of call 4, given when highly excited near nest or young, sometimes associated with distraction-lure display (Bergmann and Helb 1982, which see for sonagram). 2 call-units of this type, rendered 'chack', represented in Fig II by thick vertical lines spanning 1-8 kHz (J Hall-Craggs). (8) Nest-call. Very quiet 'prrt'. Sounds resembling short intake of air between pursed lips. Given by ♀ when incubating, and when nestlings a few days old, possibly also by ♂ to incubating ♀; no longer used once young able to see parents approaching down nest-hole, and last heard when young 10 days old (P J Conder).

CALLS OF YOUNG. Food-call a faint, thin 'ee ee ee' first heard at 3 days; a louder 'see see see' (P J Conder) or 'srri' (Bergmann and Helb 1982) from 4 days until fledging. From 11 days, a vibrant buzzing contact-call varying in intensity and pitch, rendered 'tchi' or 'tchie'; from 12th day variously 'tchee', 'tzee', 'ing', 'zweng', or 'zwang'. Also used out of nest when dependent young in separate refuge sites. After independence, 'bree', like adult call 3, given (e.g.) when 2-month-old bird solicited food from parents. From 13 days, young also begin to develop Tuc-call (adult call 4)—hard 'tchuk', suggesting plunger being drawn out of wet plug-hole; by 15 days, described as 'tac', flatter than adult Tuc-call. Juveniles start giving Quiet Subsong (call 2a) from *c.* 30 days. (P J Conder.)   EKD, PJC

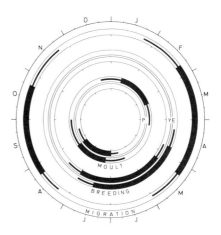

Breeding. Based mainly on account by P J Conder of studies on Skokholm (Wales). SEASON. Britain and north-west Europe: see diagram. South and central Europe: early May to June. Iceland: late May and June. Scandinavia: early to mid-May to early July. (Haartman 1969a; Makatsch 1976). SITE. In hole in wall, among stones or rocks, in burrow, or in ruined building; will use nest-box; also in holes in wide variety of man-made objects, e.g. pipes. Of 260 nests, Skokholm, 86% in burrows (mainly of rabbit *Oryctolagus*, also of Manx Shearwater *Puffinus puffinus* and Puffin *Fratercula arctica*) and 14% in walls, under rocks, or piles of slates. Favoured burrows 60-100 cm long, with entrance *c.* 10 cm diameter, usually partly grass-covered. Nest usually in small cavity *c.* $\frac{2}{3}$ of way down burrow; 61% of 119 nests 25-40 cm down burrow, leaving 10-15 cm at end, in which ♀ and mobile nestlings hide if burrow disturbed. Holes in walls *c.* 0-1 m above ground level. Open nests occur occasionally (Menzel 1964; Ludwig 1965). Most nests on Skokholm faced north-east or south-east quadrants, sheltered from prevailing winds. Nest: foundation (absent in nests in rock crevices) comprises large, untidy mass (up to 25 cm across) of dried stems of bracken *Pteridium*, heather *Calluna*, and other plants, plus grass and occasional large feathers; cup more tightly woven of finer grass stems and leaves, with some moss and lichen, *c.* 7 cm across and 5 cm deep, with walls 2·5 cm thick; lined with feathers and some hair. Sticks in foundation up to 20 cm long and 3·2 g in weight. Mean of 5 nests with foundation 106 g (93-121), and of 4 with no foundation 49 g (36-68). Building: by ♀. EGGS. See Plate 83. Sub-elliptical, smooth and not glossy; very pale blue, unmarked or with a few red-brown flecks at broad end. Nominate *oenanthe*, Skokholm: 21·0 × 15·8 mm (18·7-24·0 × 14·9-16·8), n=264; mean fresh weight 2·83 g (2·3-3·21), n=92. *O. o. leucorhoa*: 21·8 × 16·0 mm, no range given, n=34 (Witherby *et al.* 1938b). *O. o. seebohmi*: 21·6 × 15·9 mm (20·5-22·6 × 15·6-16·2), n=10; calculated weight 2·9 g (Schönwetter 1979). Clutch: 4-7 (2-9). Of 246 clutches, England: 2 eggs, less than 1%; 4, 8%; 5, 35%; 6, 45%;

7, 9%; 8, less than 1%; 9, less than 1%; mean 5·5. Declines through season on Skokholm: mean 6·2 in late April and early May, 5·7 mid-May, 5·5 late May, 5·4 early June, 4·8 mid-June, 4·3 late June. Of 48 clutches, Finland: 3 eggs, 2; 5, 8; 6, 22; 7, 15; 8, 1; mean 6·1 (Haartman 1969a). In Uppland (Sweden), mean 5·9 (n = 37); in Öland, mean 6·4 (n = 19) (J Moreno). 1–2 broods. Of 147 pairs on Skokholm, 47% laid 2 clutches. Normally only 1 brood in Scandinavia and Iceland (Haartman 1969a; Makatsch 1976). Replacements laid after clutch loss; minimum period to re-laying is 48 hrs. Eggs laid daily, though delays occur in cold weather. INCUBATION. 13·05 days (10–16), n = 31, on Skokholm. By ♀ only, though ♂ occasionally said to help (Witherby et al. 1938b); on Skokholm, ♂ entered burrow but only for less than 30 s. Begins with last or penultimate egg. Of 78 clutches, Skokholm, 35% hatched within 24 hrs, 47% within 48 hrs, and 18% within 72 hrs. In Swedish study, larger clutches tended to hatch asynchronously, smaller ones synchronously (J Moreno). Eggshells removed from nest; also damaged whole eggs, carried up to 50 m (Kneis 1983). YOUNG. Cared for and fed by both parents. Brooded by ♀ until 5–6 days old. Chiefly fed by ♀, but also variably by ♂, more for 1st brood than 2nd. FLEDGING TO MATURITY. On Skokholm, fledging period 15·0 days (10–21), n = 61, though most young leave actual nest in burrow and move around in it at c. 10 days. Become independent at c. 28–32 days. Age of first breeding 1 year. BREEDING SUCCESS. Of 165 clutches on Skokholm, all eggs hatched in 58%, all but 1 egg in 19%; of 153 nests, all nestlings fledged in 76%, all but 1 in 10%. On Skokholm and at Dungeness (southern England), 72% of 1071 eggs reared to fledging, with mean 4·2 young per pair (n = 187 pairs). Main causes of nest failure: desertion (including due to disturbance by observer), infertility, and predation by crows (Corvidae) and gulls (Larus). Of 98 nests in Breckland (eastern England), 28% predated, probably mostly by stoats *Mustela erminea*, though infertility (11·1% of 1st-clutch eggs, 7·5% of 2nd) contributed to losses; in one year, 53 pairs produced average 4·5 fledged young per pair (Tye 1980, which see for seasonal and habitat variations). One bigamous ♂ on Skokholm attended 4 broods (from 2 ♀♀) during the season, thus helping to raise 19 fledglings (Brooke 1979). On Hiddensee (East Germany), bigamous ♂ attended 3 broods from which 15 of 16 eggs hatched and 7–9 fledged (Kneis 1981).

Plumages (nominate *oenanthe*). ADULT MALE BREEDING. Forehead and broad supercilium white, often not sharply defined from crown. Crown, hindneck, mantle, scapulars, and back light grey or ash-grey, feathers with rather narrow and ill-defined olive-brown fringes along tips when plumage fresh, February–March, not concealing grey of feather-bases, but fringes soon wear off, last on lower mantle and scapulars (by about April) and on back (by May–June). Rump and upper tail-coverts white, faintly tinged cream February–March. Contrastingly black mask from feathering at nostril over lores and upper cheek to ear-coverts, not quite reaching upper rim of feathering round eye. Chin, throat, sides of neck, chest, and sides of breast cream-buff, yellow-buff, or warm buff, often almost white in narrow stripe below black mask, some dark grey of feather-bases showing on sides of breast; gradually paler cream-buff or cream-white on breast, flanks, and under tail-coverts, almost white on central belly and vent; when worn, June–July, underparts dirty white, only throat and sides of neck still distinctly buff. Tail-base white, tips black, most extensively so on t1, forming black ⊥-pattern: maximum extent of black on t1 36·3 (52) 33–40, exceptionally 45 or almost to base; minimum extent on t3–t5 and inner web of t6 17·5 (84) 15·5–20; maximum on outer web of t6 20·6 (84) 18–23, exceptionally (3 birds) up to 26; black tail-tips narrowly fringed grey or off-white when not too worn. Flight-feathers and tertials black, secondaries and innermost primaries with pale rufous-grey to off-white fringes c. 1 mm wide along tip, tips of other feathers and outer webs of secondaries sometimes with traces of narrow pale rufous or white fringes. All upper wing-coverts black, lesser and median coverts sometimes with traces of pale grey, off-white, or partly rufous fringes, tips of greater with traces of off-white or pale buff fringes less than 1 mm wide; tips of greater upper primary coverts with narrow pale grey-buff or off-white fringes generally less than 0·5 mm wide, even on inner coverts. Thus, all upperwing uniform black except for traces of pale fringes, most obvious along outer edges of secondaries and on tips of primary coverts and outer greater coverts. Under wing-coverts and axillaries white with large black feather centres, appearing heavily mottled when fresh and mainly dark when worn. *In heavily worn plumage*, June–July, often some dull dark grey of feather-bases visible, mainly on crown, mantle, and belly; white of supercilium sometimes partly worn off; upperwing black or brown-black, virtually without traces of pale fringes. ADULT FEMALE BREEDING. Upperparts down to back uniform pale olive-brown, usually with distinct grey tinge, caused by light grey feather-bases being partly visible or shining through; lower rump often slightly deeper olive-brown or buff-brown. Rump and upper tail-coverts white, tinged cream when plumage fresh February–April. Supercilium long and narrow; buff on front part, where indistinctly divided from forehead; sharper, narrower, and purer cream-white above eye; wider and rather sharply cream-white above ear-coverts. Lores dull black with buff mottling, forming indistinct narrow stripe; some buff specks just in front of and below eye, forming faint eye-ring. Upper cheek (back from below eye) and sides of neck buff-cinnamon or pale rufous-brown, merging into dark grey-brown, dark olive-brown, or black-brown ear-coverts, which are finely streaked pale buff. Chin buff, gradually deeper cinnamon-buff or pale cinnamon-brown on sides of throat, lower throat, chest, and sides of breast; flanks greyish-buff; breast, sides of belly, and under tail-coverts pink-buff or cream-buff; central belly and vent cream-white. Tail as adult ♂ breeding. Wing as adult ♂, but flight-feathers and tertials slightly browner black, upper wing-coverts dull black rather than deep black with slight gloss; outer webs of tertials and secondaries with broad (though rather abraded) rufous-cinnamon fringes (in adult ♂ breeding, paler and narrower), outer webs of primaries, greater upper wing-coverts, and greater upper primary coverts with narrow buff or off-white fringes (in ♂, virtually absent), lesser and median upper wing-coverts with slightly more extensive pale grey and pale brown fringes; under wing-coverts and axillaries with broader white feather-fringes, less extensively black. Thus, upperwing blackish like adult ♂ breeding, but slightly less deep and with more extensive brown variegation. *In worn plumage*,

May-July, upperparts duller and greyer, supercilium wider, off-white, but often less clear-cut; sides of neck, cheeks, throat, and chest paler buff, less rufous-cinnamon; belly, flanks, and under tail-coverts dirty white; upperwing black-brown, pale fringes largely worn off (least so on tips and outer webs of secondaries). ADULT MALE NON-BREEDING. Upperparts down to back grey as in adult ♂ breeding, but feather-tips more broadly olive-brown or slightly rusty-brown, largely or fully concealing grey unless heavily worn (from about November onwards); fresh feather-tips of lower mantle and scapulars sometimes washed off-white. Rump and upper tail-coverts white. Supercilium rather narrow, but clear-cut, white, extending into narrow line along base of upper mandible. Lores deep black, widening into patch in front of eye, often extending into narrow black line from just below and behind eye to shorter upper ear-coverts. Ear-coverts dark brown or dark olive-brown, paler buff-brown towards rear of cheeks, sometimes with some black admixed. Lower cheeks, throat, chest, and sides of breast either deep buff-cinnamon with slightly paler upper chin, or paler buff with more extensive white or cream-white on chin and upper cheek; buff deeper and less cream-yellow than in adult ♂ breeding. Remainder of underparts pale cinnamon-buff to pale cream-buff, palest on central belly and vent. Tail as in adult ♂ breeding, but tips with off-white fringes 1-2 mm wide. Wing similar to adult ♂ breeding, basically deep black, but plumage fresh and hence pale fringes more prominent: distinct pale rufous or off-white fringes along tips of primaries, broad rufous-cinnamon or cinnamon-buff fringes along tips and outer webs of tertials and greater upper wing-coverts, broad pale grey-buff or off-white fringes along tips of lesser and median coverts, narrow off-white edges along outer webs of primaries and greater upper primary coverts. ADULT FEMALE NON-BREEDING. Like adult ♀ breeding, but upperparts less olive-grey, more olive-brown or slightly rusty-brown, supercilium less pale and contrasting, dark stripe on lores less distinct; sides of neck and underparts often deeper and more extensively buff-cinnamon. Tail and wing as in adult ♀ breeding, but fresh and hence fringes broad: rufous-cinnamon to pale buff fringes slightly broader than those of adult ♂ non-breeding, slightly less contrasting with duller black or brown-black feather centres; broad but ill-defined fringes along lesser and median upper wing-coverts olive-brown (in adult ♂ non-breeding, narrow, pale grey-buff or off-white, and sharply defined from deep black centres). NESTLING. Down long and fairly dense, on upperparts only; dark grey (Witherby *et al.* 1938*b*). JUVENILE. Upperparts buff-brown, grey-brown, or hoary-grey, each feather with paler buff or dirty cream centre and dark brown or black fringe along tip, appearing spotted. Rump and upper tail-coverts white with some black on feather-tips. Sides of head mottled dark grey, grey-brown, and buff, sometimes with short pale buff supercilium and narrow eye-ring; ear-coverts darker brown with pale buff shaft-streak. Underparts rather variable (see e.g. Heinroth and Heinroth 1924-6 for variation of birds from same nest); ground-colour pale cream-buff to warm buff, usually deepest on throat and chest and palest on central belly and vent; cheeks, throat, chest, breast, and flanks with narrow dark arcs, appearing scalloped, sometimes extending up to chin and down to belly and under tail-coverts. Lesser and median upper wing-coverts dark olive-brown or dull black with triangular rufous or pale buff tips. Tail as adult, but rufous or grey-white fringes along feather-tips less sharply defined, 1-3 mm wide. Flight-feathers, tertials, greater upper wing-coverts, and greater upper primary coverts like adult ♀ non-breeding; outer webs and tips of tertials and greater coverts with rufous-cinnamon tips *c.* 2-3 mm wide; however, fringes along primary coverts

slightly wider than adult ♀ non-breeding, less clear-cut and more rufous-buff (less pale grey or grey-buff) than adult ♀. Under wing-coverts and axillaries buff with some grey at centres. FIRST ADULT NON-BREEDING. Like adult ♀ non-breeding, but juvenile tail, flight-feathers, greater upper wing-coverts, and greater upper primary coverts retained; these largely similar to adult feathers, but black on primary-tips less sharply divided from pale fringes; black on tail-tips often ends in point, and terminal fringe of uneven width (in adult ♀ non-breeding, black ends rounded, sometimes except for dark shaft, bordered terminally by even pale fringe; best seen on underside of t6); tertial coverts sometimes with trace of black fringe along tip; primary coverts with fringe along tip 1-1.5 mm wide, along outer web *c.* 1 mm wide (0.5-1 mm in adult). Sexes similar, but some ♂♂ have fairly distinct black stripe on lores (virtually absent in ♀); fringes along lesser and median upper wing-coverts of ♂ often slightly narrower, greyer, and more sharply defined than ♀ of any age; many indistinguishable. As in adult ♀ non-breeding, colour rather variable, upperparts sometimes greyish-brown or slightly olive, sometimes warmer brown or slightly rusty; depth and extent of rufous-cinnamon or buff on underparts variable. See also Bare Parts. FIRST ADULT MALE BREEDING. Like adult ♂ breeding, but olive-brown feather-tips of grey upperparts more extensive, never fully worn off, back in particular almost fully olive-brown; white on forehead often more restricted, less clear-cut; black ear-coverts mottled olive-brown, less pure black (unless heavily worn); sides of neck, throat, and chest deeper ochre-buff, less cream-yellow; juvenile tail, flight-feathers, tertials, and many or all upper wing-coverts retained, dark brown instead of black, fringes partly abraded and bleached to buff, but much more prominent than in adult ♂ breeding, where off-white and almost fully worn off; some wing-coverts occasionally new, contrastingly black (in adult, some also new, but no contrast as retained non-breeding coverts also black). FIRST ADULT FEMALE BREEDING. Like adult ♀ breeding; often indistinguishable. Upperparts often browner, less olive-grey, pale supercilium less distinct, black stripe on lore faint or absent. Part of juvenile feathers retained, as in 1st ♂ breeding, tail often browner and feathers less square-tipped than adult ♀ breeding, flight-feathers and (especially) upper wing-coverts browner, not dull black; tips of primaries and greater upper primary coverts more worn, more sharply pointed; fringes along tips of inner greater primary coverts often broader.

Bare parts. ADULT, FIRST ADULT. Iris dark brown. Bill, leg, and foot greyish-black or black; in 1st autumn, bill sometimes dark brown or black-brown with slightly paler base, leg and foot brown-black. In adult, inside of upper mandible grey-black or black; in 1st adult, partly yellow (Svensson 1984*a*). NESTLING. Mouth pale orange, no spots; gape-flanges very pale yellow. JUVENILE. Iris dark brown. Bill, leg, and foot dark brown, dark flesh-brown, or greyish-black. (Hartert 1903-10; Heinroth and Heinroth 1924-6; Witherby *et al.* 1938*b*; Dementiev and Gladkov 1954*b*; RMNH, ZMA.)

Moults. ADULT POST-BREEDING. Complete; primaries descendant. In Britain, starts with p1 between *c.* 20 June and late July, occasionally early August, mainly early July; completed after 40-50 days late July to early September, occasionally late September, mainly mid-August (Snow 1969*b*; Ginn and Melville 1983). Tail with coverts starts shortly after primaries, mainly from t1 outward but often rather irregular and many feathers grow simultaneously; completed within 3-4 weeks. Tertials and body start at about same time as tail; sides of breast and part of mantle, rump, and flanks first, head last, completed at primary moult score *c.* 40. Secondaries start when primary moult halfway through; last feathers (s5-s6) grow at same time as p9-p10 and bastard wing, sometimes completing after regrowth of primaries. Moults in or near breeding area, but some *leucorhoa* still have p9 and inner secondaries growing at start of autumn migration and some feathers of head sometimes old up to late September. (Williamson 1957; RMNH, ZMA.) Moult in *seebohmi* from North Africa and in *libanotica* from southern Yugoslavia, Iran, and Afghanistan from early or mid-June onwards, but in Mongolia not until early or mid-July (Stresemann 1920; Vaurie 1949; Paludan 1959; Piechocki and Bolod 1972; ZFMK). ADULT PRE-BREEDING. Partial: head, body, and occasionally some or all tertial coverts or 1-2 tertials. (December-)January-March, occasionally to early April (Stresemann 1920; Vaurie 1949). Occasionally, some parts of non-breeding plumage of head or body retained. POST-JUVENILE. Partial: head, body, median and often lesser upper wing-coverts, sometimes tertial coverts, occasionally 1-2 tertials. Starts at age of 5-8 weeks (Ginn and Melville 1983), completed in 3rd month (Heinroth and Heinroth 1924-6); timing depends strongly on hatching date. Starts early June to late August, completed from early July in south of range but mainly in first half of August elsewhere, including Greenland. 25% of 40 migrants of nominate *oenanthe*, Netherlands, September-October had suspended moult with scattered feathers on head, body, and lesser or median upper wing-coverts still juvenile (mainly part of ear-coverts, nape, tail-coverts, sides of neck and breast, and outer coverts), and this also frequent in migrant *leucorhoa*. (Stresemann 1920; Vaurie 1949; Dementiev and Gladkov 1954*b*; Williamson 1957; RMNH, ZFMK, ZMA.) FIRST PRE-BREEDING. Partial; moult sometimes more extensive than in adult pre-breeding, e.g. involving many median and some lesser and greater upper wing-coverts, often tertial coverts, and sometimes 1-2 tertials, but this then more probably a continuation of post-juvenile moult in winter quarters; moult on head and body often less extensive than in adult pre-breeding and moult occasionally fully suppressed (Stresemann 1920).

Measurements. ADULT, FIRST ADULT. Nominate *oenanthe*. Western Europe and north-west Africa, migrants; skins (ZMA). Bill (S) to skull, bill (N) to distal corner of nostril; exposed culmen on average 4·8 less than bill (S). Juvenile wing and tail refer to retained juvenile wing and tail of 1st adult.

WING AD	♂	98·1 (1·82; 20)	95-102	♀	94·7 (1·28; 9)	93-97
JUV		96·7 (2·18; 41)	93-101		94·0 (2·13; 26)	89-97
TAIL AD		54·5 (1·63; 20)	52-57		51·5 (2·09; 8)	49-54
JUV		52·8 (1·73; 41)	49-56		51·4 (1·42; 23)	48-53
BILL (S)		17·5 (0·75; 56)	16·4-18·8		17·1 (0·55; 32)	16·3-17·9
BILL (N)		9·6 (0·71; 41)	8·5-10·9		9·5 (0·44; 18)	8·9-10·3
TARSUS		27·2 (0·97; 60)	25·5-28·9		26·7 (0·94; 31)	25·4-28·2

Sex differences significant, except bill (N).

Wing and (for some populations) bill to skull; ages combined; all data refer to breeding birds and to skins, unless otherwise stated. Nominate *oenanthe*. (1) Faeroes, (2) Norway and Sweden, (3) Denmark, (4) West and East Germany (Salomonsen 1934). (5) Norway (Stresemann 1920). (6) Yamal peninsula, USSR, probably live birds (Danilov *et al.* 1984). (7) Tundra of European USSR, (8) Siberian tundra from Ob to Chukotsk (Portenko 1938). (9) Netherlands (RMNH, ZFMK, ZMA). *O. o. libanotica*. (11) migrants (RMNH, ZFMK, ZMA). (11) Spain and Balearic islands (ZFMK, ZMA). (12) Southern Yugoslavia (Stresemann 1920). (13) Rumania and Greece (RMNH, ZFMK, ZMA). (14) Crete (ZFMK). (15) Turkey (Kumerloeve 1969*a*,

1970*a*; ZFMK, ZMA). (16) Transcaspia east to Altai, USSR, (17) Iran and Transcaucasia (Portenko 1938). (18) Afghanistan (Paludan 1959). (19) Mongolia (Piechocki and Bolod 1972; Piechocki *et al.* 1982).

WING	(1)	♂ 99·7 (1·59; 23)	97-103	♀ 97·5 (1·91; 13)	95-101
	(2)	96·4 (1·47; 19)	94-99	91·8 (2·38; 6)	89-95
	(3)	96·2 (2·27; 19)	92-99	93·4 (1·74; 21)	91-97
	(4)	96·2 (1·60; 19)	92-99	93·7 (2·02; 9)	90-97
	(5)	96·3 (1·57; 10)	95-100	94·8 (3·10; 4)	92-99
	(6)	99·0 (— ; 8)	95-104	97·0 (— ; 8)	92-103
	(7)	95·4 (1·96; 19)	92-99	92·0 (1·72; 5)	90-95
	(8)	97·5 (2·21; 37)	93-105	94·7 (2·08; 8)	92-98
	(9)	95·2 (1·87; 19)	91-98	93·0 (2·69; 5)	89-96
	(10)	101·4 (0·96; 5)	100-103	96·1 (1·25; 4)	94-98
	(11)	96·3 (2·19; 15)	92-100	93·0 (1·92; 7)	91-96
	(12)	93·8 (2·36; 11)	90-97	91·2 (2·14; 6)	89-94
	(13)	96·2 (3·62; 12)	92-102	94·0 (5·39; 6)	88-98
	(14)	94·3 (1·88; 14)	91-97	88·8 (— ; 2)	88-89
	(15)	96·4 (0·99; 10)	95-98	92·5 (0·50; 5)	92-93
	(16)	97·0 (— ; 97)	93-102	93·2 (— ; 23)	89-96
	(17)	96·1 (— ; 35)	92-102	93·0 (— ; 16)	89-97
	(18)	97·4 (2·37; 7)	94-100	93·0 (— ; 3)	92-95
	(19)	99·3 (2·46; 23)	95-105	94·6 (1·14; 5)	93-96
BILL	(9)	17·9 (0·89; 15)	16·8-19·3	17·9 (0·72; 5)	17·5-18·6
	(10)	19·4 (0·76; 5)	18·7-20·6	18·6 (0·64; 4)	17·7-19·2
	(11)	18·5 (0·41; 15)	17·9-19·3	18·2 (0·62; 7)	17·5-18·9
	(13)	18·3 (0·82; 9)	17·2-19·4	18·0 (0·81; 6)	17·1-18·8
	(14)	18·6 (0·82; 14)	17·6-20·0	18·4 (— ; 2)	18·4-18·5
	(15)	18·7 (— ; 3)	18·5-18·9	18·6 (— ; 2)	18·2-18·9

Wing of live breeding birds, West Germany: ♂ 95·8 (94) 91-102, ♀ 92·1 (97) 87-96 (Bub 1975*a*, which see for details on influence of age on wing length). For East Germany, see Kneis and Benecke (1983). For Middle East, see Vaurie (1949).

O. o. leucorhoa. Wing: (1) Greenland (RMNH, ZMA); (2) Greenland, (3) Iceland, breeding birds (Salomonsen 1934); (4) migrants western Europe, identified by colour (RMNH, ZMA). Other measurements for (1) and (4) combined. Bill to skull; exposed culmen on average 5·0 less than bill to skull.

WING	(1)	♂ 105·2 (1·99; 18)	101-109	♀ 101·6 (1·63; 16)	99-105
	(2)	105·1 (1·93; 36)	102-110	103·4 (2·12; 37)	100-108
	(3)	102·6 (1·66; 49)	99-105	99·0 (1·80; 23)	96-103
	(4)	105·1 (1·79; 29)	102-110	101·7 (1·94; 17)	98-105
TAIL		56·9 (1·75; 23)	54-60	55·3 (1·48; 16)	53-58
BILL	(1)	18·1 (0·68; 17)	17·2-19·2	18·0 (0·62; 16)	17·2-19·0
	(4)	18·4 (0·84; 27)	17·3-19·7	18·1 (0·70; 16)	17·3-18·9
BILL	(N)	9·9 (0·56; 21)	9·0-10·8	9·7 (0·45; 15)	8·9-10·3
TARSUS		29·1 (0·73; 23)	28·3-30·7	28·5 (0·95; 16)	27·4-30·2

Sex differences significant for wing and tail. 1st adult with retained juvenile flight-feathers and tail combined with older birds above, though juvenile wing on average 0·4 shorter than adult and tail 1·1 shorter.

O. o. seebohmi. North-west Africa, summer; skins (ZFMK, ZMA).

WING		♂ 96·4 (2·25; 15)	93-100	♀ 93·8 (— ; 3)	91-96
BILL (S)		18·9 (0·83; 15)	17·6-20·2	18·6 (— ; 3)	18·0-18·9

Weights. Nominate *oenanthe*. Sexed birds. (1) Fair Isle, local breeders (Williamson 1958*a*). Netherlands: (2) April-May, (3) late August to early October (RMNH, ZMA). Napoli area (Italy): (4) April, (5) May (G L Lövei). (6) Yamal peninsula, USSR (Danilov *et al.* 1984). (7) Alaska, May-July (Irving 1960).

(1)	♂ 24·1 (2·38; 45)	—	♀ 23·8 (2·08; 35)	—
(2)	23·7 (3·00; 10)	21-30	23·9 (3·44; 5)	20-28

(3)	22·9 (2·75; 18)	18–28	23·3 (2·46; 12)	19–28
(4)	24·0 (1·46; 22)	19–27	22·3 (2·48; 45)	18–29
(5)	23·1 (1·45; 6)	21–25	20·7 (1·14; 8)	19–23
(6)	23·5 (— ; —)	22–26	23·3 (— ; —)	23–25
(7)	23·9 (— ; 15)	21–28	23·5 (— ; 6)	20–26

Unsexed birds. Skokholm (Wales): adult 25·2 (22) 21–30, juvenile and 1st adult 25·5 (172) 20–29 (Browne and Browne 1956). Malta: autumn 24·9 (3·4; 50) 19–34, spring 23·5 (2·9; 15) 18–28 (J Sultana and C Gauci). Ennedi (Chad), September: 20·5, 24·5 (Niethammer 1957). North-east Nigeria, March: 25·7 (4·81; 10) 20–33 (Ward 1963). Exhausted birds, Netherlands: 15·5 (0·82; 7) 14·8–16·8 (ZMA).

O. o. leucorhoa. Sexed birds. (1) Iceland and western Europe, April–October (Timmermann 1938–49; Herroelen 1970; ZMA). (2) Fair Isle (Scotland), on spring migration (Williamson 1958a).

(1)	♂	34·8 (4·86; 11)	26–41	♀ 26·8 (— ; 3)	25–28
(2)		31·0 (5·62; 30)	—	30·2 (4·41; 24)	—

Unsexed birds. Fair Isle. Spring: at arrival with favourable winds, 33·0 (38) 27–43; with unfavourable winds, 24·1 (16) 21–27 (Williamson 1958a); average at arrival 37·75 (Nisbet 1963). Autumn: at arrival with favourable winds, 25–27; with unfavourable winds 21–23 (Nisbet 1963) or (in early September) 22·7 (13) 19–27 (Williamson 1958a); on 21 September, 30·8 (2·99; 62) (Nisbet 1963). Exhausted birds: 17·1 (1·08; 6) 15·9–18·6 (Timmermann 1938–49; ZMA).

O. o. libanotica. Southern Spain, May: ♂ 22·4 (0·85; 4) 21–24 (ZFMK). Greece, Turkey, Armeniya (USSR), Iran, Afghanistan, Kazakhstan (USSR), Mongolia, and Manchuria (China), combined (Paludan 1938, 1940, 1959; Niethammer 1943; Makatsch 1950; Piechocki 1958; Schüz 1959; Nicht 1961; Kumerloeve 1968, 1969a, 1970a; Dolgushin *et al.* 1970; Rokitansky and Schifter 1971; Piechocki and Bolod 1972; Desfayes and Praz 1978; Piechocki *et al.* 1982); data from March–May and August–October include some migrant nominate *oenanthe.* (1) March–April, (2) May, (3) June, (4) July, (5) August, (6) September–October.

(1)	♂	24·7 (2·74; 13)	20–31	♀ 22·6 (— ; 3)	22–23
(2)		24·9 (2·35; 9)	22–28	23·8 (— ; 2)	23–25
(3)		24·1 (2·14; 41)	20–29	24·4 (2·80; 10)	19–27
(4)		24·4 (2·29; 13)	20–29	26·0 (— ; 3)	24–30
(5)		25·9 (1·45; 10)	24–28	22·8 (— ; 3)	20–25
(6)		27·3 (3·11; 7)	25–34	—	

Structure. Wing rather long, broad at base, tip pointed. 10 primaries: in nominate *oenanthe*, p8 longest, p9 1–5 shorter, p7 0·5–2, p6 5–10, p5 11–16, p4 15–20, p1 25–31; in *leucorhoa*, p9 1–5 shorter, p7 0·5–2·5, p6 6–10, p5 13–17, p4 18–22, p1 29–35. In both races, p10 reduced; 52–62 (nominate *oenanthe*) or 57–66 (*leucorhoa*) shorter than wing-tip, 2–7 shorter than longest upper primary covert. In nominate *oenanthe*, p9 shorter than p7 in 80% of birds, equal in 12%, longer in 8% (*n* = 203); in *leucorhoa*, shorter in 32%, equal in 22%, longer in 46% (*n* = 122) (Salomonsen 1934). Outer web of p7–p8 and inner of p8–p9 emarginated; frequently, outer web of p6 and inner of p7 also slightly emarginated. Tertials short. Tail relatively much shorter than other *Oenanthe* (but matched by Isabelline Wheatear *O. isabellina*), tip square; 12 feathers. Bill relatively short (shortest in *leucorhoa*); straight, but slightly decurved on tip of culmen, sometimes forming small hook; rather deep and wide at base, laterally compressed at tip. Nostril small, oval, partly covered by membrane above. Some rather long but soft bristles along base of upper mandible. Tarsus relatively long (relative length matched by *O. deserti-O. finschii-O. lugens* species-group, but exceeded by *O. isabellina* and Red-rumped

Wheatear *O. moesta*); slender. Toes rather short, slender. Middle toe with claw 18·3 (31) 17–20 in nominate *oenanthe* and *libanotica*, 19·4 (23) 18·5–20·5 in *leucorhoa*; outer toe with claw *c.* 67% of middle with claw, inner *c.* 63%, hind *c.* 78%. Claws rather short and sharp, slightly decurved.

Geographical variation. Slight and clinal on Eurasian continent, but difference of isolated races *seebohmi* of north-west Africa and *leucorhoa* from Iceland, Greenland, and adjacent Canada more marked. Birds from Fenno-Scandia and north European USSR south to Spain and Turkey all closely similar in wing length, only birds of Crete slightly smaller. Size very slightly greater towards east, both in *libanotica* towards Mongolia and in nominate *oenanthe* towards eastern Siberia and Alaska. Far more distinct cline of increasing size runs north-west from Britain through Faeroes and Iceland ending in large *leucorhoa* from Greenland and adjacent Canada; intermediate birds from Faeroes and southern Iceland sometimes separated as *schioeleri* Salomonsen, 1927, but Icelandic birds are close to those from Greenland in colour and size (Salomonsen 1934), and birds from Faeroes better included in nominate *oenanthe* (Haftorn 1971). Wing of *seebohmi* similar to populations of adjacent parts of Europe. Slight cline of increasing bill length towards south: bill shortest in Fenno-Scandia and northern Siberia, longest in Crete, Turkey, Iran, and Soviet central Asia, intermediate in Iberia and north-west Africa; in Iran and Soviet central Asia, bill of ♂ 19·5 (38) 17·5–21 (Vaurie 1949). Birds with very long bill occur on migration in Egypt and East Africa and these sometimes separated as *rostrata* (Hemprich and Ehrenberg, 1833); colour rather dark (in some approaching *leucorhoa*); sometimes thought to originate from Transcaucasia and Iran (Portenko 1938), but a few birds examined from there are rather pale, similar to *libanotica* from Turkey and Turkmeniya, and hence 'rostrata' perhaps a migrant from further north-east in Asia and not considered separable while breeding grounds uncertain. Upperparts of ♀ and 1st adult ♂ of *leucorhoa* in non-breeding (autumn) somewhat browner and more rusty (less grey) than in nominate *oenanthe* (but nominate *oenanthe* rather variable, darkest birds similar to paler *leucorhoa*); cheeks and underparts on average deeper and more extensively rusty-cinnamon; in breeding plumage (spring), ♀♀ of both races similar on upperparts, and ♀ *leucorhoa* often hardly darker on underparts. Adult ♂ (all year) and 1st adult ♂ breeding (in spring) of *leucorhoa* almost identical on upperparts to nominate *oenanthe* (some adult ♂ non-breeding *leucorhoa* tinged rusty, but others olive-brown like nominate *oenanthe*); sides of neck, throat, and chest distinctly deeper buff-cinnamon or rusty-buff, however, and this tinge often extending to breast and flanks when plumage not too worn; 1st adult generally darker below than adult, and thus some 1st adult nominate *oenanthe* similar to adult *leucorhoa*. Tail of *leucorhoa* on average rather more extensively black, maximum extent of black on t1 40·8 (37) 37–50, minimum extent on t3–t5 19·7 (39) 17·5–23, maximum on outer web of t6 22·7 (39) 19·5–27. *O. o. libanotica* from southern Europe through central Asia east to Transbaykalia and Mongolia rather poorly differentiated from nominate *oenanthe* and often not recognized, but maintained here for the following reasons. (1) Bill long, especially in east (see Measurements). (2) Plumage slightly paler, independent of abrasion: in breeding ♂, forehead more broadly white, upperparts slightly paler ash-grey, and chest paler cream than ♂ nominate *oenanthe* at same time of year, even in fresh plumage; in ♀ and non-breeding ♂, upperparts more sandy-grey, cheeks, throat, and chest slightly to distinctly paler buff or cream-buff. (3) Tail-tip has less black on average: in southern Europe, band on inner web of t3–t6 14·5 (24) 5–18 wide, maximum extent of black on outer web of t6 18·4 (23) 14–24; occasionally, white of feather-bases of t3–t5 reaches tip along shafts, interrupting black tail-band. (4) Breeding plumage of adult ♀ often similar to adult ♂ breeding: upperparts grey, forehead, and supercilium white, face-mask and upperwing deep black or brown-black with limited pale fringes; chin to chest often slightly deeper tawny-yellow than adult ♂, however; 1st adult breeding more or less intermediate between non-breeding and adult ♀ breeding. *O. o. libanotica* itself sometimes split into several races—*argentea* (Lönnberg, 1909) in Transbaykalia, *nivea* (Weigold, 1913) in southern Spain and Balearic islands, and *virago* Meinertzhagen, 1920, on Crete—but all these birds agree with typical *libanotica* from Levant in characters cited above. North-west African *seebohmi* similar to *libanotica*, but cheeks, chin, throat, and sides of breast of ♂ black (throat occasionally mixed buff); tail-tip has even less black, minimum extent on t3–t5 12·2 (18) 4–17, white occasionally reaching tail-tip along shaft; under wing-coverts virtually black; ♀ either

similar to ♂ (but ear-coverts often browner) or with sandy upperparts (like non-breeding *libanotica*), black stripe through lores, dark grey and buff mottled chin and throat, and black and white mixed under wing-coverts. For geographical variation, see also Hartert and Steinbacher (1932–8), Portenko (1938), Vaurie (1949), and Mackworth-Praed and Grant (1951).

May form superspecies with Somali Wheatear *O. phillipsi* from northern Somalia (Hall and Moreau 1970), but see *Bull. Br. Orn. Club* 1986, **106**, 104–11.

Recognition. ♂ with grey-backed upperparts easy to identify, but brown-backed ♀♀ often similar to Isabelline Wheatear *O. isabellina* and ♀♀ of Pied Wheatear *O. pleschanka* and Black-eared Wheatear *O. hispanica*: see Recognition of those species (pp. 770, 806, 819). In north-west Africa, ♀ *seebohmi* rather similar to ♀ Mourning Wheatear *O. lugens halophila*, but latter has shorter wing, longer tail, less black on t6, longer p10, and more rounded wing. CSR

Oenanthe pleschanka Pied Wheatear

PLATES 59, 64, and 65
[facing pages 760, 761, and 856]

DU. Bonte Tapuit FR. Traquet pie GE. Nonnensteinschmätzer
RU. Каменка-плешанка SP. Collalba pía SW. Nunnestenskvätta

Motacilla pleschanka Lepechin, 1770. Synonym: *Oenanthe leucomela*.

Polytypic. Nominate *pleschanka* (Lepechin, 1770), central Eurasia; *cypriaca* (Homeyer, 1884), Cyprus.

Field characters. 14·5 cm; wing-span 25·5–27·5 cm. Nominate *pleschanka* of distinctly lighter build than Wheatear *O. oenanthe* with slightly shorter bill and wings and 15% shorter legs, but with tail 5% longer and proportionately appearing more so; close in size to Finsch's Wheatear *O. finschii* and Mourning Wheatear *O. lugens*, though lighter-built and 10% shorter-billed than both and over 5% longer-winged than *O. finschii*; similar in size, weight, and structure to Black-eared Wheatear *O. hispanica* except for slightly longer wings. Cyprus race *cypriaca* 10% shorter-winged and 5% shorter-tailed than nominate *pleschanka*; smallest of all west Palearctic wheatears. Slight, round-headed and long-tailed wheatear; rather shy but with somewhat shrike-like perching and hunting behaviour. Rump and tail pattern basically as *O. oenanthe* but with longer black tips on outer tail-feathers. Adult ♂ breeding has silver-white crown and nape contrasting with black face and back (making bird appear broad-shouldered from behind). Under tail-coverts white. Black under wing-coverts of ♂ and ♀ obvious but not diagnostic. Sexes of nominate *pleschanka* markedly dissimilar in breeding plumage, of *cypriaca* less so. Juvenile separable. 2 races in west Palearctic, differing obviously in size and in plumage of ♀.

(1) Main race, nominate *pleschanka*. ADULT MALE. In spring (when plumage worn), crown and wide nape white, with faint silver or buff tone most obvious from behind

and becoming grey with wear. Face, chin, throat, upper breast, sides of neck, shoulder area, back, and wings black, though a few have white chin, throat, and upper breast. Lower breast, underbody, rump, and upper tail-coverts white, variably tinged buff. Axillaries and under wing-coverts black, contrasting with dusky under-surface to flight-feathers. Tail white, with black central feathers and black tips to outer ones, tips noticeably longer and extending further up tail towards outermost. From July, or not until September, plumage becomes far less contrasting as fresh feathers with brown, buff, and cream tips invade and sometimes cover basic pied pattern; bird becomes almost wholly buff and brown. Crown and nape buff-brown, with cream supercilium more or less obvious. Face and throat black, but hoary tips and scales obvious on rear cheeks and along border with breast; a few birds have buff-white throat. Back black-brown, copiously scaled, barred, or completely buff; wings black-brown, with all feathers tipped buff and cream creating freckled lesser wing-coverts, 2 wing-bars (on median and greater coverts), bold pale edges to tertials and inner secondaries, and pale tips to all other flight-feathers. Breast and underbody isabelline. Black under wing-coverts tipped white only at leading edge. Rump and vent cream, often looking duller than white-based tail. ADULT FEMALE. Except for rump and tail pattern, differs markedly from ♂. In spring, crown to back umber-brown, without grey or sandy tone and (at certain angles) showing dark mottling on mantle and back. Face pattern indistinct: short supercilium pale cream or buff, not contrasting sharply with crown or cheeks and tending to merge in front of eye with pale buff lore (making dark eye prominent); chin pale buff; throat and upper breast usually dusky, with black-brown feather-bases variably evident (rarely creating uniform dark bib, occasionally pale buff overall); chest rust- to grey-brown. Wings dark brown, retaining (except when very worn) paler buff-brown fringes and tips on larger coverts and inner flight-feathers. Rump and upper tail-coverts cream-white. Tail as ♂ but black marks toned brown. Underwing black-brown, contrasting with dusky undersurface to flight-feathers but less sharply than on ♂. In autumn becomes more uniform, as fresh grey to buff fringes and tips reduce or eliminate dark throat-patch, further reduce contrast of wings, and scallop upperparts. Throat cream-buff; chest-band tawny, flanks buff-brown, belly and vent cream-buff. New ear-coverts may form more obvious brown patch than in spring. JUVENILE. Resembles autumn ♀ but copiously spotted buff-white above and faintly scaled dark brown on chest. FIRST-YEAR MALE. In winter, intermediate in appearance between adult ♂ and ♀, but usually separable from ♀ by obvious 'shadow' of throat-patch and blacker back and wings. In spring, noticeably drabber than adult, with grey-brown centre to crown and nape (creating diffuse cream supercilium), brown-black back, browner wings (with pale tips retained), and buffier chest and rump. FIRST-YEAR FEMALE. Dubiously separable from adult, but birds with drab faded wings may be of this age. (2) Cyprus race *cypriaca*. ADULT MALE. Warmer-toned on underbody than most ♂ nominate *pleschanka*, with cinnamon shade from chest to under tail-coverts. In spring, deeper and more extensively black on chest-sides and back. ADULT FEMALE. In spring, quite unlike ♀ nominate *pleschanka*, resembling dull ♂ of own race, but separable by (when no too worn) much darker olive-brown crown-patch contrasting with pale buff frontal band, supercilium, and nape. JUVENILE. Darker than in nominate *pleschanka*, more obviously pale-spotted above, and much more heavily and extensively dark-scalloped below. All races and ages have bill and legs black.

Adult ♂ confusable in brief view with *O. hispanica* and *O. finschii* (though both these have pale backs), Red-rumped Wheatear *O. moesta* (which has different rump and tail pattern), Hooded Wheatear *O. monacha* and *capistrata* morph of Eastern Pied Wheatear *O. picata* with white crown and belly (which both have more black on chest and different tail pattern), and *O. lugens*. *O. lugens* is the only serious problem, though it often shows white in primaries in flight, and always has warm vent, less broadly black back with grey of nape extending downwards in centre, and much purer white breast, flanks, and belly; in fresh or little-worn plumage, also shows narrow white tip to tail (which does not have

black extending up outer feathers). Adult ♀ much more difficult to separate, but tail pattern eliminates *O. moesta* and *O. monacha* and almost uniform dark under wing-coverts eliminate *O. oenanthe*. Separation from *O. hispanica* best based on umber (not grey) tone to crown and back, duller (not rufous-brown) cheeks, usually more black on tips of outer tail-feathers, and slightly longer wings; from eastern *O. lugens* on umber (not light brown) tone to back, lack of rufous on crown and cheeks, and tail pattern. Throat patterns and underpart tones not reliable features, as all variable (due to moult and wear), but obvious speckles or patch on chest generally indicate *O. pleschanka* (and *O. finschii*) before *O. hispanica*. ♀♀ most safely identified through association with ♂♂. Immatures at times unidentifiable in the field, though most vagrant 1st-autumn birds identified in western Europe are ♂♂, already starting to show adult's plumage pattern. Occurrence of vagrant ♀♀ almost certainly hidden by difficulty of separation from *O. hispanica* and aberrant *O. oenanthe*. The whole subject is further bedevilled by hybridization of *O. pleschanka* with *O. hispanica* (see Geographical Variation). Flight freer than *O. oenanthe*, with length of tail contributing to looser, more flowing, less flitting action. Escape-flight usually long. Gait less free than *O. oenanthe*, essentially hopping (no proof that it uses loping run of *O. oenanthe*). Stands less upright on ground than *O. oenanthe*, due to shorter legs and longer tail, but on perch more erect, with longer tail often drooped giving attitude like shrike *Lanius*. Behaviour typical of genus but typically hunts from perch. Generally timid in breeding range but noticeably tame in winter quarters, adapting even to urban habitat in Africa.

Song of nominate *pleschanka* highly variable, consisting of fluting and whistling notes in rather lark-like series but often containing remarkable mimicry of other nearby breeding species; lacks frequent harsh notes of *O. oenanthe*. Song of *cypriaca* very different: comprises monotonous buzzing sounds. Commonest call a harsh 'zack-zack'.

Habitat. Approximately replaces Black-eared Wheatear *O. hispanica* in continental mid-latitudes of eastern sector of west Palearctic, through warmer temperate and steppe zones, largely in lowland but ranging up to 1800 m (and somewhat higher in Cyprus: Bannerman and Bannerman 1971). Extralimitally in Afghanistan often up to 2800 m and exceptionally to 3600 m on stony mountain slope above treeline. Typically occupies desolate stony stretches with scattered boulders and more seldom fallow fields at margin of small cultivated areas; shares this habitat with Eastern Pied Wheatear *O. picata* (Paludan 1959). In USSR, frequents declivities and stony mounds, stones of railway embankments, fissures in clay cliff-banks, and even human settlements (Dementiev and Gladkov 1954*b*). Occurs in more rugged and hilly areas than *O. hispanica*:

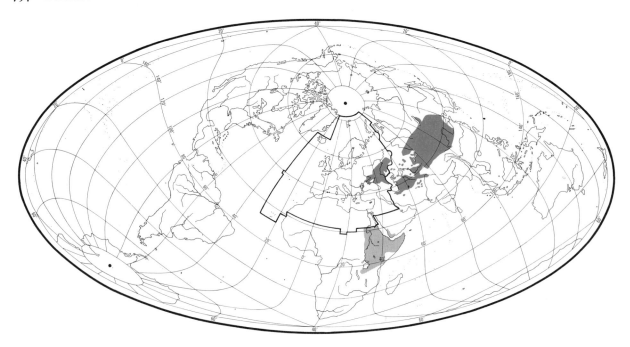

on broken terrain in stony and wooded steppes, steep river banks, earth cliffs, and rocky slopes and outcrops, usually where shrubs and small trees present (Harrison 1982). Will sing from top of a pine-tree (Bannerman and Bannerman 1971).

In Lebanon occurs on coastal cliffs (Vere Benson 1970). In Iran, breeds in variety of habitats: vast undulating steppe with short but fairly dense grass on pebbly ground amid occasional rock outcrops; poorer steppe with scattered bushes; fairly lush valleys at 1700–2000 m on south-facing slopes, along a stream bordered by crags and rocky juniper-clad slopes, where bottom land had up to 90% cover of grasses, thistles, and other herbaceous plants. One pair near village in area of mixed cultivation with gardens, poplar groves, springs, and stream in bottom of narrow valley with rocky sides. Frequent perches were walls, piles of stones, thistles, and low bushes; birds fed on ground, usually in stony areas with fairly dense vegetation. Shares preference of *O. hispanica* for prominent features such as rocks, boulders, walls, and well-grown vegetation, where annual precipitation exceeds 300 mm, but adapts to more broken, mountainous and desolate nature of much of range.

In winter, resorts to similar habitat at similar elevations in East Africa and south-west Arabia, but extends to dry cultivation with thorn hedges, animal enclosures, and human settlements, and even to gardens in towns (Cornwallis 1975).

Distribution. Few changes known.

YUGOSLAVIA. Bred 1966 (Siegner 1983). USSR. Armeniya: breeds occasionally (Kumerloeve 1969*d*). TURKEY. Bred 1964 and 1966 (Kumerloeve 1969*d*). LEBANON, SYRIA. Said to have bred near Beirut (Lebanon) and in Syria, but proof not known (Kumerloeve 1969*d*).

Accidental. Britain, Ireland, West Germany, Norway, Finland, Hungary, Italy, Greece, Libya, Malta.

Population. Nothing reported on population trends.

BULGARIA. Common on rocky coasts north of Varna (KM). CYPRUS. Abundant (Flint and Stewart 1983). USSR. Common, sometimes abundant, but distribution at times sporadic (Dementiev and Gladkov 1954*a*).

Movements. Migratory. Winters in eastern Africa and north-eastwards to south-west Arabia. Widespread and often common in Kenya (except coast), eastern Uganda, and north-east Tanzania (north of *c*. 4°S) (Britton 1980). Common in Sudan, chiefly in north and north-east, occasionally extreme south-east (Cave and Macdonald 1955). Winters commonly throughout much of Ethiopia (Urban and Brown 1971); described as abundant in Eritrea (Smith 1957), and very common and widespread in Somalia (Ash and Miskell 1983). Common in Aden (Browne 1950), and some probably winter in North Yemen (Cornwallis and Porter 1982).

Cyprus race, *cypriaca*, leaves Cyprus August–October, with peak in late August and September. There is 1 overwintering record. Passage appears to be on broad front with records from Egypt (Etchécopar and Hüe 1967) as far east as Syria (Kumerloeve 1969*b*) and Bethlehem, Israel (Hardy 1946); also recorded at Jiddah, Saudi Arabia, in March (Bates 1936–7). According to Vaurie (1959), winters in southern Sudan and Ethiopia; uncommon to frequent in open areas in north-east Ethiopia (Urban and Brown 1971). Returns to Cyprus

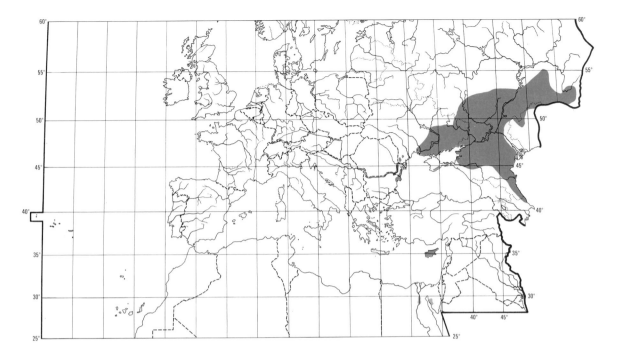

mainly late March and April, though recorded twice in February. Records of birds trapped by bird-limers showed that 57% caught 10 March–25 April were ♂♂, with ♂♂ arriving earlier (first on 10 March) than ♀♀ (first on 18 March) (Horner and Hubbard 1982). Some evidence of fidelity to natal area: 2 birds ringed as chicks were retrapped the following year at same site (Flint and Stewart 1983).

Main race, nominate *pleschanka*, to which remainder of account refers, also migrates from breeding grounds on broad front, eastern populations requiring south-west component. In Turkey, occurs on south coast having overflown central plateau where it is a scarce migrant, but mostly passes through eastern Turkey (*Orn. Soc. Turkey Bird Rep. 1966–75*). Common and widespread on passage in south-west Iran (Cornwallis 1975). Regularly migrates through Iraq (Allouse 1953), Afghanistan (Paludan 1959), and northern Baluchistan (Pakistan); occasional spring records in Kashmir and Punjab (Ali and Ripley 1973*b*). Common and widespread on passage (especially in spring) in Gulf states of Arabia (Bundy and Warr 1980), and indeed probably the commonest passage *Oenanthe* on Bahrain (Hammonds 1984), but less common in Oman (Walker 1981*a, b*). Passes across Saudi Arabia, with several inland records (Bates 1936–7).

Leaves breeding grounds from August but departures clearly protracted with birds still present in Samarkand (Uzbekistan, USSR) 25 October (Carruthers 1910) and at Su-fu (western China) 16 October (Ludlow and Kinnear 1933). In Kazakhstan, departs from north and high altitudes in August and September, but from south not until September or early October (Dolgushin *et al.* 1970). Passage through Turkey and Iraq mainly September–October. Recorded in Gulf states of Arabia from September to early November (Bundy and Warr 1980), with peak at Bahrain in mid- or late October (Hammonds 1984). Arrives Eritrea from early September (Smith 1957), and 1 record then from Dahlac archipelago in Red Sea (Mann 1971), but not generally recorded in Somalia until late September (M G Kelsey, from examination of skins at BMNH), Sudan until October (Hogg *et al.* 1984), and Kenya until mid-October (*Scopus* 1978–83, **2**-7). Arrives in southern Ethiopia later than Wheatear *O. oenanthe* (Benson 1946*b*). Passage recorded in November at Ngulia (south-east Kenya) may indicate 'step-migration' between early and later wintering areas or may represent late arrivals (Curry-Lindahl 1981).

Recorded until beginning of April in Kenya (1 record in late May) (*Scopus* 1978–83, **2**-7), and until 1st week of May in Eritrea (Smith 1957). Spring passage in Oman occurs mid-February to early April with peak in March (Gallagher and Woodcock 1980); in Persian Gulf, mid-February to mid-April (Bundy and Warr 1980), with peak in 3rd week of March on Bahrain (Hammonds 1984). Main passage in Turkey from end of March to mid-April (*Orn. Soc. Turkey Bird Rep. 1966–75*). Recorded chiefly in March–April in Iraq (Moore and Boswell 1956); still widespread in Iran in April (Erard and Etchécopar 1970). Passage through northern Baluchistan March–April (Ali and Ripley 1973*b*).

First arrivals on breeding grounds from mid- to late March, though once recorded 13 February in southern Caspian (Schüz 1959). Recorded in south-east Kazakhstan from late March, Syr-Dar'ya valley from early April, and

Aral Sea, northern Kazakhstan, and Orenburg (north of Caspian Sea) from mid- to late April or early May. Arrival at higher altitudes *c.* 1 month to 6 weeks later than in lowland regions (Grote 1939; Dolgushin *et al.* 1970). Early arrivals in Uzbekistan mid-March at low altitudes, April higher up (Carruthers 1910). Arrives in Tien Shan mid-March, Kansu (central China) late March; in Mongolia and northern China at end of March, though birds still arriving there in 2nd week of May. In Dobrogea (Rumania), occurs from early April. (Grote 1939.)

No long-distance ringing recoveries, but in Ethiopia of 55 birds ringed, 1 retrapped at same site the following year (Ash 1978). See also Social Pattern and Behaviour.

MGK, DFV

Food. Almost entirely insects. Taken mostly from bare ground but occasionally from low vegetation. Typically watches for prey from perch 1–1·15 m high and flies down to make capture, returning to same perch immediately, rarely spending time on ground (Ash 1955; Dolgushin *et al.* 1970; Cornwallis 1975; Knox and Ellis 1981). In Kenya, attracted to burnt grassland where prey easy to see (Dittami 1981). Low weight and long tail enable this species to perch on flimsy vegetation (Cornwallis 1975; Leisler *et al.* 1983). Of 66 perches used by 4 birds in south-west Iran, 45% on vegetation, 15% on boulders or rocky outcrops, 20% on stones or clods, and 20% on ground; 83% of 37 capture attempts involved short flight and 95% ended with prey being taken from bare ground (Cornwallis 1975). In Tsavo (Kenya), of 100 items captured, 64% taken from ground, 30% in flight, and 6% from bushes (Lack 1980). Recorded hovering over prey (Ash 1955). Large prey beaten against ground or stone before swallowing (Dolgushin *et al.* 1970).

The following recorded in diet. Invertebrates: damselflies (Lestidae), grasshoppers, etc. (Orthoptera: Acrididae, Gryllidae), earwigs (Dermaptera), termites (Isoptera), bugs (Cicadidae), ant-lion larvae (Myrmeleontidae), Lepidoptera (Geometridae), flies (Diptera: Muscidae, Bombyliidae), Hymenoptera (Ichneumonidae, ants Formicidae), beetles (Coleoptera: Carabidae, Tenebrionidae, Scarabaeidae, Chrysomelidae, Curculionidae), spiders (Araneae), mites (Acarina), woodlice (Isopoda), centipedes (Chilopoda), millipedes (Diplopoda), snails (Mollusca), and a small lizard (Lacertidae). Plant food includes fruits of wild cherry *Prunus*, mulberry *Morus*, *Pistacia*, seeds, and vegetative parts. (Morrison-Scott 1937; Witherby *et al.* 1938*b*; Dementiev and Gladkov 1954*b*; Bel'skaya 1961; Papadopol 1961; Kovshar' 1966; Survillo 1968; Dolgushin *et al.* 1970; Mambetzhumaev and Abdreimov 1972; Abdusalyamov 1973; Cornwallis 1975; *Cyprus orn. Soc.* (*1957*) *Rep.* 1981, **26**, 44; Dittami 1981; Kostin 1983.)

Ants and beetles form major part of diet while berries and grasshoppers may become seasonally important. In

Crimea (USSR), 24 stomachs, May–June, contained an unknown number of items comprising 98·7% (by number) insects. Of these, 70% Hymenoptera (62% ants), 20·1% beetles (5·9% Carabidae, 5·6% Curculionidae), 6·9% Lepidoptera (5·9% larvae), 1·0% Orthoptera, 1·0% Hemiptera, and a single cicada (Kostin 1983). For summer diet in Rumania (based on stomach analysis), see Papadopol (1961). In south-west Iran, 6 stomachs from March contained 100 items comprising 50% (by number) ants, 22% beetles, 5% Hemiptera, 3% insect larvae, 2% woodlice, 1% dragonflies, 1% Orthoptera, and 15% unidentified insects; 8 stomachs, September, contained 471 items comprising 59% ants, 5% berries, 4% beetles, 2% Hemiptera, 2% other Hymenoptera, 1% insect larvae, 1% seeds, and 25% unidentified insects; 3 stomachs, October, contained 45 items comprising 67% ants, 13% beetles, 7% Orthoptera, 7% Hemiptera, 2% seeds, and 4% unidentified insects; 11 stomachs, November, contained 213 items comprising 41% beetles, 16% ants, 12% Hemiptera, 8% termites, 4% other Hymenoptera, 4% spiders, 2% insect larvae, 1% mites, 1% seeds, and 11% unidentified insects (Cornwallis 1975). In south-east Turkmeniya (USSR), 20 stomachs contained mostly beetles (in 17), caterpillars (in 12), and ants (in 10); of 236 items, 54·8% (by number) ants, 21·6% caterpillars, 13·6% beetles (5% Tenebrionidae, 5% Curculionidae), 5·9% Hemiptera, 3·4% Myriapoda, 0.4% Orthoptera, and 0.4% spiders (Bel'skaya 1961). In Talasskiy Alatau mountains (Tien Shan, southern USSR), 12 stomachs from cultivated zone in spring comprised mainly Curculionidae and Carabidae by volume and frequency (each in 7 stomachs); also unidentified beetles (in 4 stomachs), Scarabaeidae (2), Hemiptera (3), ants (2), adult Lepidoptera (1), larval Lepidoptera (1), wasps (1), and seeds (1); 3 stomachs from 1st-years in alpine zone contained Acrididae (3), beetles (1), and single Lepidoptera and Hemiptera; 7 stomachs, August–October, contained mainly Acrididae (4) (Kovshar' 1966). Analysis of 98 stomachs from various sites in USSR showed main feeding methods, prey, and prey size to be same as Black-eared Wheatear *O. hispanica* (Loskot 1983), though no data given. In wintering birds and spring migrants, ants and beetles most important; all 16 stomachs from Jiddah (Saudi Arabia), January–April, contained ants and beetles; also Hymenoptera (in 8), Diptera (7), Hemiptera (4), mites (3), ant-lion larvae (2), and Lepidoptera adults (1) and larvae (1), spiders (1), and seeds (1) (Morrison-Scott 1937).

Young given adult and larval insects (Abdusalyamov 1973). In Cyprus, late May, adult also observed carrying small lizard (*Cyprus orn. Soc.* (*1957*) *Rep.* 1981, **26**, 44).

PJE

Social pattern and behaviour. Some aspects studied by Eggebrecht (1943) in Ukraine (southern USSR). Major studies, also in USSR, as follows: summary and comparison with other *Oenanthe* (Panow 1974); comparison with other *Oenanthe*, notably closely related Black-eared Wheatear *O. hispanica* (Panov and Ivanitski 1975*b*; Panov 1978), and Finsch's Wheatear *O. finschii* (Panov and Ivanitski 1975*a*). *O. p. cypriaca* differs vocally (see Voice) and perhaps in other respects, but no detailed studies.

1. Normally solitary in winter (e.g. Jackson 1938, Archer and Godman 1961*b*, Smith 1971), defending territory against potential food competitors. However, concentrations also occur in areas with abundant food, e.g. in recently burnt areas; for densities, see Dittami (1981). In Kenya, feeding territories average *c.* 0·8 ha—smaller than those of *O. oenanthe* (*c.* 2·5 ha) and Isabelline Wheatear *O. isabellina* (*c.* 2 ha). Both sexes normally hold separate territory though ♂ and ♀ (not definitely paired) recorded defending large territory together (see also Meinertzhagen 1920*a*, and Bonds, below). Where density low, *O. pleschanka*, *O. oenanthe*, and *O. isabellina* may use same area and overlap. Territories occupied before fires also held after them; other (marked) birds established territory in fire-affected area and remained there until migration departure. Birds highly territorial intraspecifically, ♀♀ tending to attack other ♀♀ more than ♂♂, while ♂♂ direct more attacks against ♀♀ than against other ♂♂. No attacks recorded against other migrant *Oenanthe* (*O. oenanthe*, *O. isabellina*), but often attacked by them; see also Breeding Dispersion and Antagonistic Behaviour (below). Local species attacked by *O. pleschanka* include Mourning Wheatear *O. lugens lugubris* and Fiscal Shrike *Lanius collaris*; also attacks (e.g.) Whinchat *Saxicola rubetra*, Rock Thrush *Monticola saxatilis*, Swallow *Hirundo rustica* (frequently), and Yellow Wagtail *Motacilla flava*—though often recorded showing no aggression near some of these nor near *O. isabellina* and *O. oenanthe*. (Dittami 1981; Leisler *et al.* 1983.) Fidelity to wintering site between years recorded (Curry-Lindahl 1981; see also Stoneham 1929). Loose territories apparently held also by migrants of both sexes in Iran (L Cornwallis) and in Egypt where Simmons (1951*c*) noted defence of territory against *O. lugens* and White-crowned Black Wheatear *O. leucopyga* for *c.* 12 days from mid-October. Shows marked tendency to flock on migration (Panov and Ivanitski 1975*a*); e.g. in Iran, frequently up to 15, mid-March (Sarudny and Härms 1926) and in Badkhyz (Turkmeniya, USSR), loose flocks of up to 30 during heavy passage (Loskot 1983); on Cyprus, up to 20 *cypriaca* noted in spring, 30–40 in mid-October (*Cyprus orn. Soc.* (1957) *Rep.* 1974, **19**, 62; 1977, **23**, 48). In Ukraine, recorded singly or in pairs from arrival, unlike compact parties of *O. oenanthe* (Frank 1952), while on Mangyshlak peninsula (east Caspian, USSR), not infrequently 2–4 ♂♂ *c.* 15–20 m apart (even briefly in same bush) and with little antagonism; similarly in autumn, only one case (over 6 days) of adult ♂ attacking juvenile (Panov and Ivanitski 1975*a*). May associate with other *Oenanthe* on migration: e.g. with *O. oenanthe* in Iran (Nielsen 1969). BONDS. Mating system evidently essentially monogamous, but in Altai mountains (USSR), ♂ apparently paired with 2 ♀♀ helped to feed both broods (nests *c.* 20 m apart); no antagonism between the ♀♀ who fed only their own young (Panow 1974). Also, ♀ extremely tolerant of strange ♂♂, moving freely beyond core of her own mate's territory and often close to centre of another whose owner may court her; after pair-formation, ♂ and ♀ tend to move about their own territory or beyond it independently (Panov and Ivanitski 1975*a*); see also part 2. No proof that pair-bond maintained outside breeding season, and records of winter song and association of ♂ and ♀ in single territory (Leisler *et al.* 1983; see above) are not firm evidence that pair-formation takes place before spring migration, although simultaneous arrival of both sexes on breeding grounds not uncommon (see Het-

erosexual Behaviour, below); one record (Kenya) of ♂ and ♀ apparently together in same area for over 4 months in winter (Meinertzhagen 1920*a*). Where ♂ experimentally removed, ♀ (with nest near completion) re-paired on same territory next day (Panov and Ivanitski 1975*b*); rapidity of re-pairing perhaps facilitated by presence of surplus ♂♂ as noted (e.g.) in small population on Kalmius river (Ukraine) by Eggebrecht (1943). Breeding ranges of *O. pleschanka* and *O. hispanica* are contiguous and the species hybridize to some extent where they meet. In Iran, hybrids constitute *c.* 65% of population near contact zone (Haffer 1977; Vuilleumier 1977). Hybridization also recorded in Bulgaria (Baumgart 1971; Panov 1978). On west Caspian coast (USSR), 61·8% of mixed population was *O. pleschanka*, 5·9% *O. hispanica*, 32·4% hybrids in Makhachkala, and 31·4%, 20·0%, 48·6% in Baku area (E N Panov). On Mangyshlak peninsula, 73·6% *O. pleschanka*, 2·9% *O. hispanica*, 23·5% hybrids; no apparent ecological isolation and virtually no behavioural barriers to hybridization (only small differences in calls and visual signals: see Voice and part 2). When ♀♀ experimentally removed, one ♂ *O. hispanica* re-paired 3 times successfully over 4 days, once within only 4 hrs; some ♀♀ involved probably *O. pleschanka* rather than *O. hispanica*. After ♂ *O. hispanica* shot, his mate re-paired with ♂ *O. pleschanka* within 6 hrs. For further details, including suggestion that ♂ plumage probably unimportant in determining choice of mate, see Panov and Ivanitski (1975*b*). In Hungary, where ♂ *O. hispanica* and *O. pleschanka* increasingly recorded in recent years, these birds find no conspecific ♀♀ and tend to associate with *O. oenanthe* pairs; ♂ *O. pleschanka* noted singing near nest of *O. oenanthe* (Keve 1967) and ♂♂ of *O. pleschanka* and *O. hispanica* recorded feeding young of *O. oenanthe* as nest-helpers (Panov 1974). In *O. pleschanka*, young normally fed by both sexes, including for certain period after fledging (Erard and Etchécopar 1970), though 1 fledged brood in Iran was fed entirely by ♀ over *c.* 30 min (Buxton 1921). According to Dolgushin *et al.* (1970), family breaks up when young independent—5–7 days after leaving nest (Abdusalyamov 1973); one *cypriaca* family stayed together *c.* 3 months after young left nest (*Cyprus orn. Soc.* (1957) *Rep.* 1974, **19**, 62). Age of first breeding not known. BREEDING DISPERSION. Solitary and territorial (e.g. Frank 1952), but often (more so than in *O. hispanica*) in dense neighbourhood groups (perhaps due to habitat constraints: Panov 1978), with small territories predominant (Loskot 1983). Similar dispersion with high density, particularly in hills and mountains, also typical of *cypriaca* (Christensen 1974; Flint and Stewart 1983). In Bulgaria, pairs *c.* 100 m apart along Black Sea cliffs (Mountfort and Ferguson-Lees 1961*b*). On Mangyshlak peninsula, nests average 126±18 m (*n* = 17), minimum 40 m, apart; average 104±10 m (*n* = 13) from *O. oenanthe* nests (Ivanitski 1980). In Tadzhikistan (USSR), pairs in 2 localities *c.* 800–1000 m apart, 1 pair per 5–6 ha (Abdusalyamov 1973). Territory defended by ♂; used for pair-formation, copulation, and nesting, while some feeding done outside (see below). In Ukraine, *c.* 40–60 m across. ♀♀ apparently following departing intruders (♂♂) usually flew not more than *c.* 100 m from future nest-site. Territory contained favourite song-posts where ♂ also spent much time (preening, etc.) and to which he returned after expelling intruder (Eggebrecht 1943). In one case, song-post of ♂ *cypriaca c.* 80 m from nest and this probably average distance (Ashton-Johnson 1961). Comparative studies in USSR have shown later-returning and later-breeding *O. pleschanka* to have looser territorial system than rigidly territorial *O. finschii* (migratory population) and *O. oenanthe*. Apart from marked tendency to form groups, evidence of reduced territoriality in *O. pleschanka* as follows.

Size of territory markedly variable but can be small (c. 0·3–0·5 ha) where density high. However, maximum considered by Loskot (1983) to be almost as in *O. hispanica* (see that species, p. 811) and average less than twice as small, rather than nearly 6 times (*contra* Panov 1978). Minimum significantly smaller (c. 3 times) at 0·3 ha (n = 20, in area of high *O. pleschanka* density) than in *O. finschii* and *O. isabellina* (Panov and Ivanitski 1975a; Ivanitski 1980). Unlike in (e.g.) *O. finschii* (migratory populations), no clearly marked territory limits: large overlap zones occur between territories, with only sporadic conflicts there; c. 50% of conflicts take place in small core area defended by ♂ against conspecific birds. (Panov and Ivanitski 1975a; Panov 1978; Ivanitski 1980.) Data on breeding densities as follows. In Cyprus, 5 ♂ *cypriaca* noted singing within c. 100 m of observer (Bannerman and Bannerman 1958). For indication of high density along cliffs in Bulgaria, see Baumgart (1971). In Ukraine, apparently at least 5 occupied and contiguous territories along c. 1 km of Kalmius river, and 2–3 surplus ♂♂ also present (Eggebrecht 1943; see Bonds, above). Area of c. 5 km² on Mangyshlak peninsula held 77 pairs, with average ratio 2 *O. pleschanka* : 1 *O. finschii*, locally even 3–4:1 (Panov and Ivanitski 1975a). Further east in Tuvinskaya ASSR (south of Krasnoyarsk, USSR), birds favour river cliffs and rocky outcrops (never more than c. 500 m from river), with 4–12 pairs per 10-km transect along belt 100 m wide (Ivanitski 1980). On Mangyshlak, *O. pleschanka* less strict in habitat choice than *O. finschii*, though also apparently prefers escarpment slopes, territories there being smaller than on loess cliffs. Early *O. pleschanka* ♂♂ forced to settle along escarpment ridge and on limits of large *O. finschii* territories where otherwise extreme aggression of that species somewhat reduced. *O. pleschanka* anyway more or less indifferent to attacks (see part 2) and also range widely in search of food, this allowing them to hold on to territories once chosen. As more *O. pleschanka* ♂♂ arrive, these able to penetrate into core of *O. finschii* territories whose owners now more tolerant, though their continued lower-level aggression probably regulates density of *O. pleschanka*. For details, see Panov and Ivanitski (1975a). Similarly in Tuvinskaya ASSR, low-ranking *O. pleschanka* (see Antagonistic Behaviour, below) can nest close to *O. oenanthe* by starting when peak of *O. oenanthe* social activities is past (Ivanitski 1980); see also (e.g.) Eggebrecht (1943). ROOSTING. Nocturnal and in most cases probably solitary, though little information. In Africa, wintering bird roosted in garage and on verandah ledge (Stoneham 1929). Family of *cypriaca* roosted together in cave for c. 3 weeks from fledging (*Cyprus orn. Soc.* (*1957*) *Rep.* 1974, **19**, 62). For diurnal activity rhythms on breeding grounds in Tadzhikistan, see Abdusalyamov (1973).

2. *O. pleschanka* and *O. hispanica* have homologous communication systems which show a number of significant differences from other well-studied *Oenanthe*. The following important postures, displays, etc. are characteristic of both: Courtship-display flights of ♂ and ♀ together (recorded only in these 2 species); raising of wings over back so that carpal joints almost touch, wing-tips lying along sides of tail; prolonged wing-shivering—typical of whole period of pair-consolidation, nest-building, etc., not just associated with copulation as in (e.g.) *O. finschii* and Eastern Pied Wheatear *O. picata*; Dancing-flight used more widely in heterosexual and antagonistic encounters than (e.g.) its restricted use in *O. finschii*. Differences between signal behaviour of *O. pleschanka* and *O. hispanica* mainly quantitative—i.e. in frequency and intensity of reactions. (Panov 1978.) For further details, see *O. hispanica* and below. *O. pleschanka* not especially shy in winter (Archer and Godman 1961b). *O. p. cypriaca* not at all shy on breeding

grounds (Bannerman and Bannerman 1958); nominate *pleschanka* rather shy, but equally conspicuous when breeding (e.g. Eggebrecht 1943, Hollom 1980), though singing ♂ may allow approach to c. 10–15 m (Kozlova 1930). One vagrant allowed man to within c. 4–5 m (Ash 1955); others gave alarm-calls (see 2a in Voice) when disturbed (Berthold 1955) or flew from perch to perch (Knox and Ellis 1981). Characteristic bobbing, wing-flicking, and tail-wagging often accompanied by calls 2, 3, and 5 (Eggebrecht 1943; Frank 1952; L Cornwallis: see Voice). FLOCK BEHAVIOUR. Only information relates to migrants resting 'close together' in Mongolia (Piechocki *et al.* 1982). SONG-DISPLAY. ♂ sings (see 1 in Voice) from perch (tree, rock, wire, etc.) and in flight. Although Song-flights mentioned for *cypriaca* (Bannerman and Bannerman 1958), that race apparently typically arboreal, perching readily on trees or other perches and singing (often persistently) from there (Christensen 1974; Bergmann and Helb 1982; Bergmann 1983). Bird has tail fanned when singing from rock (e.g. Kozlova 1930). For Song-flight, ascends steeply (almost vertically), hovers briefly at c. 25–40 m (see below), then tips to one side and glides down at angle (some undulations), with tail fanned, and singing (at first trill, then loud song); ends in steep plummet with closed wings, sometimes to take-off point where may sing after landing. ♂ (especially owner of small territory) usually flies high (averages much higher than *O. finschii*) and far beyond own territory, staying long over neighbours' (Panov and Ivanitski 1975a). Song-flight similar overall to *O. hispanica* (see Panow 1974 for illustrations), but performed less often. (Kozlova 1930; Eggebrecht 1943; Abdusalyamov 1973; Panov 1978.) In Ukraine, Song-flights much less frequent than in *O. oenanthe*, birds singing from high perch (Frank 1952); see also Raethel (1955) for Transcaucasia. In Mongolia, birds noted singing from tree or rock (Mauersberger *et al.* 1982), but Song-flights occur from May: in one case, vigorous wing-beats produced noise audible over some distance (Piechocki *et al.* 1982). Several *cypriaca* noted singing in 'chorus' in May (Bannerman and Bannerman 1958). Will sing until dusk well advanced (Grote 1937); in Tadzhikistan, at any time of day in peak breeding season (Abdusalyamov 1973). In USSR, ♂ normally starts to sing (see 1a, c in Voice) c. 3–4 days after occupying territory: initially quiet and sporadic, then loud and more frequent (Panov 1974). In Mongolia, song-period May–June, then resumption (not known for how long) in August (Kozlova 1930, 1933); in Tadzhikistan, continues to end of June (Abdusalyamov 1973). After main arrival in late March to April, *cypriaca* sings March–July, perhaps also later (Flint and Stewart 1983). Autumn song (perhaps Subsong) given normally from cover (Portenko 1954). Will sing also in winter. In Kenya, Subsong (see 1c in Voice) given from 17 February, from centre or edge of territory (see part 1) apparently not for territory demarcation; one singing ♂ in shared territory with ♀ (Leisler *et al.* 1983); also in Kenya, Jackson (1938) noted song only in March prior to departure while Smith (1971) heard none from wintering birds in Eritrea. For further use of song variants in antagonistic and heterosexual contexts, see below and 1 in Voice. ANTAGONISTIC BEHAVIOUR. (1) General. In USSR, many ♂♂ arrive and take up territories long before breeding starts; few disputes while ♂♂ alone in territories (Panow 1974). In Iran, 18 April, some birds paired and on territory though many still passing through (Erard and Etchécopar 1970); similar report from Mongolia, early May (Piechocki *et al.* 1982). In Ukraine, mid-May, ♂♂ highly territorial, almost uninterruptedly in conflict with neighbours or intruders (Frank 1952); *cypriaca* also extremely pugnacious, ♂♂ fighting even when no ♀ present (Bannerman and Bannerman 1958). Caged ♂ placed in centre of territory elicits extreme aggression from territory owner at any stage of breeding cycle (E N Panov). Overall however, *O. pleschanka* ♂♂ less aggressive than other *Oenanthe* on breeding grounds, during migration, and in winter, and thus differs especially from *O. finschii*. Arrives on Mangyshlak much later than *O. finschii*; main arrival from *c.* 8–10 April, but some ♂♂ take up territory as late as 5–6 May. Conflicts normally short and less frequent than in *O. finschii*, *O. isabellina*, *O. oenanthe*, and Desert Wheatear *O. deserti*. ♀ *O. pleschanka* even less aggressive than ♂. (Panow 1974; Panov and Ivanitski 1975a; Panov 1978; Ivanitski 1980.) (2) Threat and fighting. In Africa, winter territories demarcated mainly by various threat gestures (no details, but see below); birds sometimes also call (Leisler *et al.* 1983: see 2a in Voice). Breeding pair of *cypriaca* recorded using loud threat-calls (see 2a in Voice) to drive away intruding ♂ (Bannerman and Bannerman 1958). Patrolling of territory limits by neighbours normally restricted to high-density populations and to a few hours of first day of occupancy; interactions typically brief and usually involve only visual contact with some low-intensity plumage ruffling, sideways wing-flicking, bowing (perhaps same as bobbing), and slight raising of closed tail (Panov 1978, which see for illustrations). Intrusions (mostly by paired ♂♂) regular during pair-formation, but territory-owner normally shows no antagonism, continuing rather to court ♀ together with intruder. Stranger may stay 20(–60) min, but usually returns quickly to territory (Panov and Ivanitski 1975a). In territorial conflicts, rival ♂♂ will adopt Head-up posture (Fig A), with carpal joints raised and close together over back, and (unlike in *O. hispanica*) neck drawn in; also variants with tail more steeply raised, wings opened and held forward, back more hunched, etc. In Hunched-forward-threat posture (Fig B), neck again drawn in (compare *O. hispanica*, p. 813), and bird will flick wings sideways (Panow 1974; Panov 1978). Looking-down posture (Fig C) perhaps signals appeasement. In Tuvinskaya ASSR, analysis of 64 conflicts showed participation of ♀♀ in 13 (see below), ♂ antagonism directed against own mate (2), clear aggression between ♂♂ (21), raising or lowering of tail (18), tail-fanning (11), sleeking of head feathers (16), raising or lowering of bill (14), bobbing (10), wing-twitching and wing-spreading (17), and mock-feeding (3). Repertoire of postures and displays apparently less varied than in higher-ranking *O. finschii*, *O. isabellina*, and *O. oenanthe* (Ivanitski 1980). Dancing-flight (much as in *O. hispanica*) sometimes occurs in territorial conflicts: of 46 Dancing-flights observed by Panov (1978), only 5 in this context (see also Heterosexual Behaviour, below). Sometimes, rivals move sideways and, each time they stop, make rapid tripping steps on the spot and turn back on rival, looking over shoulder (Fig D); see *O. lugens* (p. 862) for similar behaviour. When unpaired ♂ attempted to win over already paired ♀, 2 ♂♂ entered rock crevices one after another, also adopted Hunched-forward-threat posture; in occasional tussles, became interlocked, rolling over ground (Panow 1974; Panov 1978). Similar threat reported for *cypriaca*, and rival ♂♂ also repeatedly leaped up breast-to-breast and fell down interlocked (Bannerman and Bannerman 1958). In small population with surplus of ♂♂ (see Bonds, above), ♂♂ constantly chased and fought (apparently over ♀♀); when chase developed after fight between 2 ♂♂, others would join in, singing in flight and after landing. After successful expulsion, ♂ tended to return to song-perch and sing loudly (Eggebrecht 1943). Even during long conflicts (3–7 hrs), ♀ will feed nearby unconcerned. During break from incubation, ♀ sometimes involved in brief conflict with ♀ neighbours, then adopting (e.g.) Threat-posture similar to that of ♂ in Fig B, though wings not drooped and back less hunched (Panow 1974; Panov and Ivanitski 1975a).

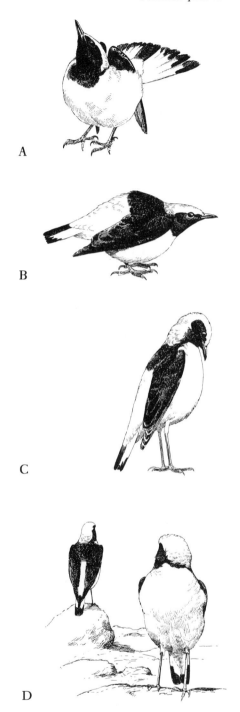

A

B

C

D

In Iran, ♀♀ (including territory-holding migrant) seen attacking individuals of both sexes (including adult and clearly subordinate 1st-winter ♂) (L Cornwallis). In Tuvinskaya ASSR, ♂ only once recorded attacking another *Oenanthe*—♂ *O. deserti*, this leading to fight; ♀♀ not seen to attack congeners. Strict dominance hierarchy (*O. pleschanka* lowest-ranking) exists (Ivanitski 1980). In Badkhyz, migrants attacked by Eastern Pied Wheatear *O. picata* immediately fly far off and do not return (Ivanitski 1982). Conflicts with *O. oenanthe* occur (e.g.

Panow 1974), but perhaps not initiated by *O. pleschanka*. *O. finschii* frequently intrudes into *O. pleschanka* territory and attacks owners (both sexes) which may react by wing-flicking, tail-raising, and calling (see 2a in Voice), i.e. as when disturbed by man at nest. Often, especially during incubation or nestling phase, simply flies off, though may retreat only 3-10 m, awaiting next attack. On the other hand, ♂ *O. pleschanka* launches frequent and vigorous attacks on other passerines, notably Black Redstart *Phoenicurus ochruros* (mainly ♀♀, while ♀ *O. pleschanka* tends to harass ♂♂ of *P. ochruros*)—also pipits *Anthus*, Lesser Whitethroat *Sylvia curruca*, Brown Accentor *Prunella fulvescens*, and others (Ivanitski 1980). HETEROSEXUAL BEHAVIOUR. (1) General. ♂♂ of *cypriaca* arrive on breeding grounds ahead of ♀♀ (Bannerman and Bannerman 1958). Nominate *pleschanka* ♂♂ arrive *c.* 7-10 days before ♀♀ (Dementiev and Gladkov 1954*b*) or, in south of USSR, 2-4 weeks before (Dolgushin *et al.* 1970; Panow 1974), though simultaneous arrival of both sexes not uncommon (Panov 1978). Pair-formation recorded in Ukraine (Eggebrecht 1943) and Mongolia (Kozlova 1930) in early May; also on Mangyshlak where *O. pleschanka* arrival starts when some *O. finschii* already incubating (Panov and Ivanitski 1975*a*); see Breeding Dispersion and Antagonistic Behaviour (above). Often quite long between pair-formation and start of building (Panow 1974). In dense population in Tuvinskaya ASSR, high degree of breeding synchrony (Ivanitski 1980). For synchronous breeding with *O. hispanica* on Mangyshlak, see that species (p. 813). 2-3 ♂♂ will display to 1 ♀ in succession or simultaneously (Panov and Ivanitski 1975*a*; Ivanitski 1981*a*). Other birds of both sexes normally present during pair-formation, such intrusions rarely effecting any change in behaviour of courting pair (see Antagonistic Behaviour, above). Pair-formation normally completed in single bout lasting *c.* 60-90 min. Courtship typically (more so than in *O. hispanica*) takes place at dusk. Behaviour of pair-members less co-ordinated than in *O. hispanica* (Panov 1978). For cases of antagonism by ♂ against mate, see Antagonistic Behaviour (above). (2) Pair-bonding behaviour and nest-site selection. As in *O. hispanica*, ♂ sings loudly from high perch before dawn, perhaps to attract migrating ♀♀ (Loskot 1983). Song-flights performed less than in *O. hispanica*, and first ♂-♀ meeting less likely to involve aerial display by ♂ (Panov 1978). ♂ may initiate proceedings by displaying at shallow holes (♂ *O. hispanica* tends to enter deep holes): adopts Head-up posture (also used later in ♂-♀ interactions, with ♂ singing according to Eggebrecht 1973) and sometimes waggles bill and moves tail up and down; or may hold head horizontal, back slightly hunched, tail depressed (variants and transitional forms also occur). ♀ does not approach ♂, but after 10-15 min ♂ makes constant attempts to approach her (unlike in *O. hispanica*, does not use Quiver-glide flight). ♂ may adopt Hunched-posture with head down (as in *O. hispanica*, Fig B, p. 813), or up, tail closed or fanned, breast lowered with wings drooped, etc. (see Panow 1974, Panov 1978). ♀ tolerates ♂'s close approach only minutes after first meeting (unlike in *O. hispanica*), though usually gives Threat-rattle (see 4 in Voice). ♀'s independent inspection of cavities usually begins *c.* 1 hr later and she takes little notice of ♂, though he may display near her at hole. Nest-site apparently selected by ♀; in one case, ♀ built 2 nests simultaneously though only one completed and used (Eggebrecht 1943); pair-members will hover briefly near chosen site, moving to left or right, ♂ above ♀ or vice versa (Panow 1974). Birds soon feed close together (2-3 m apart). Most later approaches by ♂ (usually in Low-horizontal posture: Fig E) lead to chases (rare earlier). Unlike ♀ *O. hispanica*, ♀ does not hide when chased (Panov 1978). Chases

E

typically fast and low, with wild curves and loops, ♂ singing or giving rattling calls (Eggebrecht 1943; L Cornwallis: see 3 in Voice). If unpaired ♀ appears near 2 neighbouring territories, both ♂♂ (usually without antagonism) will chase her over several hrs, ♀ then staying with ♂ whose territory she had favoured initially; 1-2 (even paired) ♂ neighbours may join in chase. For further evidence of ♂ and ♀ tolerating conspecific birds, see Bonds and Antagonistic Behaviour (above); see also Panov and Ivanitski (1975*a*) and Panov (1978). Dancing-flight during which ♂ typically sings (see 1c in Voice) is much as in *O. hispanica* (see that species, p. 813); may occur at first meeting, but also later during nest-building. One ♂ performed Dancing-flight to ♀ *O. picata* while his mate was incubating. For description of performance by ♂ *cypriaca*, see Witherby *et al.* (1938*b*). After Dancing-flight, ♂ assumes Head-up posture, sometimes moving head from side to side, or lowers breast (bowing forward) but keeps head up and fans tail, wing-tips laid on tail or held away from body. In Forward-tilt posture (Fig F), often quivers wings slightly and may move neck up

F

and down, sometimes changing (at last moment) to Cross-posture (same as *O. hispanica* Fig C, p. 814). Normally silent during such displays. ♂ and ♀ also perform Courtship-display flights at dusk as in *O. hispanica*. (Panow 1974; Panov 1978; E N Panov.) (3) Courtship-feeding. ♂ carrying insect once approached ♀ (near where nest later built): head held forward, wing-tips slightly drooped and quivered (probably Horizontal-posture, Forward-tilt posture, or similar); held insect close to ♀'s bill, but dropped it when she failed to react, then flew off to join chase of other ♂♂; ♀ did not eat food (Eggebrecht 1943). Neither this nor feeding of ♀ on nest by ♂ (see Erard and Etchécopar 1970) recorded by Panow (1974). (4) Mating. Usually preceded by ♂'s Song-flight according to Abdusalyamov (1973). ♀ recorded fleeing from ♂'s advances during nest-building; eventually flew forcefully at him so that he immediately desisted (Eggebrecht 1943). For song associated with copulation, see 1a in Voice. (5) Behaviour at nest. ♂ will sing from perch or in flight, but mostly keeps close to nest-building ♀ (who may also sing: see 1d in Voice), also when feeding; sometimes displays at her, e.g. in Head-up posture with markedly fanned tail after waiting for her at nest-entrance, then sometimes also performing Dancing-flight (Eggebrecht 1943; Panow 1974). However, unlike in *O. hispanica*, ♂ rarely accompanies ♀ during nest-building according to Panov (1978). ♂ normally stays as sentinel in territory during incubation (Abdusalyamov 1973). RELATIONS WITHIN FAMILY GROUP. No information on nestling phase. Young leave nest, probably still unable to fly, at 13-14 days—from 10-11 days if disturbed. Food-begging of fledglings as in *O. hispanica* (p. 814). Family stays near nest for up to *c.* 8 days (Survillo 1968; Ivanov 1969), though much longer period recorded in *cypriaca* (see Bonds, above). ANTI-PREDATOR RESPONSES OF YOUNG. No details, though brood of *cypriaca* out of nest and attended by parents were very inquisitive, even following observer at *c.* 4 m (Mason 1980). PARENTAL ANTI-PREDATOR STRATEGIES. ♀ *cypriaca* apparently a tight sitter (Bannerman and Bannerman 1958). Adults tending young out of nest very aggressive, attacking 'any other bird' which approached (Mason 1980). For further brief description of behaviour when disturbed, see subsection 2 in Antagonistic Behaviour (above); see also Voice.

(Figs by K H E Franklin: from drawings in Panow 1974 and Panov 1978.)   MGW

Voice. Much in common with Black-eared Wheatear *O. hispanica.* Song used in winter, and call 2 also given then (e.g. Leisler *et al.* 1983), though birds silent in Iraq on migration (Moore and Boswell 1956). *O. p. cypriaca* differs markedly from nominate *pleschanka* in song and apparently in at least one other call (see below). Classification of other calls mainly after Panov (1978) with additional data (notably for *cypriaca*) from other sources. For wing-noise in Song-flight, see Social Pattern and Behaviour. For additional sonagrams, see Panov (1978), Bergmann and Helb (1982), Sluys and Berg (1982), and Bergmann (1983).

CALLS OF ADULTS. (1) Song. In nominate *pleschanka*, short and longer variants of Territorial-song occur, also quiet and loud Subsong, but much overlap (Panov 1978); at least one variant given also by ♀ (see below). (a) Territorial- or Advertising-song. Short (normally of 6–12 units), varied, typical *Oenanthe* phrases comprising whistles and harder smacking or chattering, trilling and strangled sounds (Eggebrecht 1943; Panov 1978; Bergmann and Helb 1982). Frequency modulation (vibrato) producing buzzy quality as in *O. hispanica* is characteristic (Panov 1978), though phrase sometimes starts with 'bizzy' units while rest more or less tonal (Bergmann and Helb 1982). Uses wide frequency range, up to *c.* 8 kHz (Sluys and Berg 1982). Said by Jonsson (1982) to have livelier rhythm than *O. hispanica.* Song especially loud in flight according to Eggebrecht (1943), which see for many renderings. Mimicry of wide range of other bird species also typical (see, e.g., Kozlova 1930, Frank 1952, Panow 1974) and maximum number of units tends to occur when long song is copied (Panov 1978), e.g. as with mimicry of song of Short-toed Lark *Calandrella brachydactyla* (Fig I). Probable copy of *C. brachydactyla* (though louder) also evident in recording by C Chappuis (Rumania), but preceded by well-spaced 'bizz bizz chirrrrrr' (apparently another short song-phrase). In eastern Bulgaria, song mostly consists of simply structured phrases showing, unlike Wheatear *O. oenanthe*, little variety in length and sequence—'geretschiretschö' 'giretschiretsche(re)' (Schubert and Schubert 1982). Fig II shows one such phrase: harsh though less so than in *O. hispanica*, with more tonal sounds, but still reminiscent of Whitethroat *Sylvia communis*; sudden rise in pitch at end with 2 brief almost identical sounds perhaps not

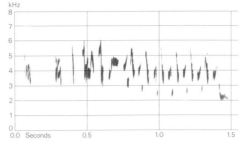

I   P A D Hollom   Iran   May 1977

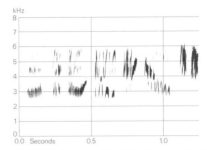

II   M Schubert/Swedish Radio Company (1981)   Bulgaria   May 1974

typical (J Hall-Craggs, M G Wilson). See also Schüz (1957*a*) who noted dull fluting 'uiduiduiduid' as well as quality like *S. communis.* Double calls (sometimes mimicry—e.g. of White Wagtail *Motacilla alba*) not infrequently interpolated between phrases or given at end of song (Schubert and Schubert 1982); see also Mauersberger *et al.* (1982). Series of short Territorial-songs (also with call 6) given in territory demarcation, also in pair-formation. Short but quiet songs changing to 1b precede copulation (Panov 1978). (b) With reduction of pauses, songs may merge into loud and long (continuous) variant. Melodious and long song, sometimes recalling Chaffinch *Fringilla coelebs* (Eggebrecht 1943) is perhaps this variant. In Ukraine (USSR), Frank (1952) noted attractive and varied song with recurring 'ta lie lie zwie' but lacking definite phrase structure; perhaps also belongs here. (c) Subsong. Of complex structure (perhaps involving simultaneous use of 2 internal tympanic membranes), comprising some tonal units and others with frequency modulation, also much mimicry. Most striking feature is inclusion of muffled clicks (see call 2a), sometimes in typewriter-like rattle (Panov 1978; E N Panov). In recording, probable Subsong initially a mixture of whistling chirrups like sparrow *Passer* ('tiuip' or 'tluip') and rasping 'zee(p)' or 'zee(k)' sounds (call 2a) which develop into irregular ratchet-like clicks in a final accelerando section containing also short rattles (see calls 2–3). Fig III shows 1 whistle-chirrup and 1 faint and 2 strong rasps (J Hall-Craggs, M G Wilson). Song in East Africa in March (Jackson 1938) and rapid sequence of high-pitched squeaky sounds given by West German vagrant (Berthold 1955) are perhaps Subsong. Quiet variant of Subsong given (e.g.) during pair-formation.

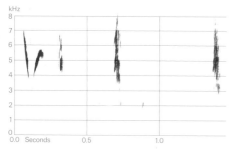

III   C Chappuis/Sveriges Radio (1972–80) Rumania May 1967

Louder, short or long bursts used in territory-demarcation ('aggressive song') or pair-formation ('courtship song') (Panov 1978). (d) In Ukraine, Eggebrecht (1943) noted quiet song of ♀ far more melodious than that of ♂ in Song-flight. (e) Song of ♂ *cypriaca*. Differs markedly from nominate *pleschanka*. Continuous or (sometimes) long phrases of rather monotonous, buzzing or harsh sawing 'bizz', 'dsid', or 'zee' units; noisy (pronounced frequency modulation) and reminiscent of grasshopper (Orthoptera) or cicada (Cicadidae) (Bannerman and Bannerman 1958; Christensen 1974; Bergmann and Helb 1982). Fig IV shows 5 'zee' units at *c.* 3 per s; more irregular units in Fig V delivered at fast rate of *c.* 5 per s (range

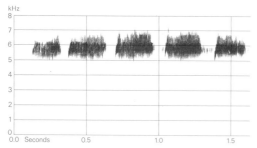

IV   J Gordon Cyprus May 1984

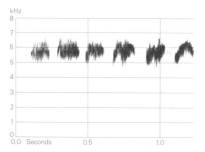

V   J Gordon Cyprus May 1984

3–6 per s: Bergmann 1983). In recording by H-H Bergmann, units of one song more like very rapid 'zee-zi-zee' (J Hall-Craggs, M G Wilson). Song lasts 2–10 s, average 8·4 s (Sluys and Berg 1982), or even up to 1 min (Bergmann 1983). Some variation also in pitch, though frequency range limited (*c.* 4–7 kHz); compare nominate *pleschanka*. Tonal elements and mimicry so characteristic of nominate *pleschanka* are virtually absent

in *cypriaca*, but tonal coda sometimes occurs: e.g. 3-unit figure (of more or less pure tones) at end of consistently short 'bizz' phrases by one bird (Bergmann 1983). No information on other song-variants (Subsong, etc.) in *cypriaca*. (2) Clicking calls. (a) Muffled click. Much overlap with calls 2b and 3. This or longer variants (see below) are the commonest everyday vocalization (see also Subsong), given in practically all social interactions, including as threat ('excitement call') both in winter (Leisler *et al.* 1983) and (by *cypriaca* pair) on breeding grounds (Bannerman and Bannerman 1958). As in many *Oenanthe*, bird with young (especially older nestlings or fledglings) signals alarm with series of muffled clicks, in mixed series with calls 2b or 5 (Panov 1978; E N Panov). Clicking calls (probably mostly muffled variant) rendered as follows: 'tschä' (Schüz 1957a); 'tschak' and (probably longer variant) harsh 'tschärr' (Berthold 1955); usually disyllabic 'schräpp 'schräpp' (Frank 1952); in recording by C Chappuis, a rasping 'zee(p)' (J Hall-Craggs); similarly in *cypriaca*, rasping 'dsed-dsed...' (Bergmann and Helb 1982) and recording of birds apparently alarmed by Magpie *Pica pica* (1st 2 units of Fig VI) also suggests rasping 'zzack' or 'chak' (J Hall-Craggs, M G Wilson). (b) Clear, loud click. 2 variants exist, one having energy concentrated in lowest part of frequency range; 2nd variant (presumably higher pitched) apparently like that of several other *Oenanthe*, including *O. hispanica* (Panov 1978), but no detailed information (nor for *O. hispanica*) on how this differs from call 2a. However, clicking 'tck tck' noted by Frank (1952) certainly suggests different timbre from same author's rendering of call presumed to be 2a. Series of clear clicks sometimes given in pair-formation (Panov 1978); see call 2a for alarm function. (3) Muffled clicks merging into series; a snoring sound. May also signal alarm and, at highest intensity, distress (Panov 1978). Calls probably of this type as follows: strangled 'trrrrr' and 'trrrrrrrrrrttl', the 2nd (with loud click at end) given during courtship chase (Eggebrecht 1943); dry 'trrlt' (Jonsson 1982). (4) Threat-rattle. Normally given by ♀ threatening approaching ♂, occasionally also in low-intensity ♂–♂ conflicts (Panov 1978). Sonagram from E N Panov indicates close-packed series of clicks, apparently lower pitched than in *O. hispanica* (M G Wilson). (5) Whistle-call. In sonagram (by E N Panov), Whistle-calls (combined with clicks in alarm) all descend in pitch, though one rises then falls (M G Wilson). Combined clicks and whistles rendered 'chrät chrät hji' or 'chrät chrät zjüe zjüe' (Frank 1952). 2 types of Whistle-call (apparently both from nominate *pleschanka*) distinguished by Jonsson (1982): 'psjiep' like Yellow Wagtail *M. flava*, and another like Little Ringed Plover *Charadrius dubius*; 'tzri' of Schüz (1957a) probably same as 1st of these. Whistle-call of *cypriaca* a short 'dj'ui' (Bergmann and Helb 1982: last unit of Fig VI, which shows markedly different pitch pattern from Whistle-calls of nominate *pleschanka* in sonagram by E N Panov—see

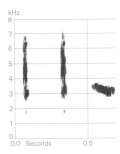

VI　J Gordon　Cyprus　May 1984

above). Recording from which Fig VI taken also includes loud, disyllabic whistle-chirrups like House Sparrow *Passer domesticus*: may be another alarm-call. (6) Rasp-call. Buzzy rasp like juvenile food-call (see below). Used mainly in high-intensity antagonistic and heterosexual encounters (Panov 1978). (7) Bugling-call. No description available, but Russian name suggests a (presumably subdued) crane *Grus* bugling. Characteristic only of *O. pleschanka* and *O. hispanica*. Occurs in wide range of contexts (sometimes as component of Subsong and its louder variant), including pair-formation and prior to copulation (Panov 1978). (8) Other calls. A 'hwitt hwitt' of unknown significance (Frank 1952); perhaps a variant of call 5, or mimicry.

CALLS OF YOUNG. Adult call 2a (or variant closer to that used by young *O. hispanica*) serves as contact-call. Food-call a raspy buzzing noise; in sonagram (E N Panov), apparently higher pitched than *O. hispanica*, and with distinct rise in pitch; call of *O. hispanica* long but apparently fairly steady in pitch. Differences may be more apparent than real, however (M G Wilson). For further details, see Panow (1974) and Panov (1978).

MGW

Breeding. SEASON. Cyprus: nests found from mid-April and into May (Bannerman and Bannerman 1958); *c.* 3-week difference between higher and lower ground (Flint and Stewart 1983). Ukraine (USSR): laying begins early May (Eggebrecht 1943), though fresh clutches recorded also in last third of month (Dementiev and Gladkov 1954*b*). In extralimital USSR, fledged young recorded in south and (despite later arrival) north of range from late May to early June; latest breeders are mountain populations—e.g. main fledging period in Altai last third of June, in Pamiro-Alay early to late July (Panow 1974). SITE. In hole in rock or bank, under stone, or sometimes in building. Nest: cup of dry grass and stems, lined with finer grasses, roots, and (perhaps not regularly: Panow 1974) wool or hair. Outer diameter 11–14 cm, height 6–7·5 cm, depth of cup 3·5–5·3 cm, and thickness of cup wall 2·5–3·5 cm (Panow 1974). Base and/or entrance platform (pathway) of small twigs typical (Panow 1974; Flint and Stewart 1983; Loskot 1983). Building: by ♀ (Eggebrecht 1943). EGGS. See Plate 83. Sub-elliptical, smooth and glossy; pale blue to green-blue, lightly speckled and spotted red-brown, sometimes

gathered at broad end. Nominate *pleschanka*: 19·6 × 15·2 mm (17·1–21·0 × 13·5–16·1), *n*=75; calculated weight 2·43 g. *O. p. cypriaca*: 19·3 × 14·6 mm (18·0–20·6 × 14·0–15·4), *n*=55; calculated weight 2·19 g (Schönwetter 1979). Clutch: (3–)4–6. Of 86 clutches, USSR: 3 eggs, 1·2%; 4, 11·6%; 5, 45·3%; 6, 41·9%; mean 5·28 (E N Panov). Normally one brood, but 2 recorded in *cypriaca* (Flint and Stewart 1983). INCUBATION. 13–14 days. By ♀ only (Panow 1974); in *cypriaca*, by both sexes according to Ashton-Johnson (1961). YOUNG. Cared for and fed by both parents. FLEDGING TO MATURITY. Fledging period 13–14 days (Survillo 1968). No further information.

Plumages (nominate *pleschanka*). ADULT MALE. In fresh plumage (autumn), forehead, crown, and nape light grey-brown or drab-brown, indistinctly bordered by narrow buff line across forehead and by narrow pale buff supercilium at sides; feathers of crown often slightly dusky grey at central tips, feather-bases white, but these hidden, usually except on nape. Mantle, scapulars, back, and sides of neck drab-brown, buff-brown, or dark olive-brown, slightly darker than crown; feather-bases black, but these largely or fully concealed. Rump and upper tail-coverts white with slight cream tinge. Sides of head (down from just above eye), chin, and throat black, feathers of chin, throat, rear cheek, and ear-coverts narrowly tipped cream or off-white (tip generally below 1 mm wide, on lower throat sometimes slightly wider). Rarely (so-called '*vittata*' morph), chin and throat buff like remainder of underparts, black restricted to mask from lores and upper cheeks across ear to sides of neck and scapulars. Sides of breast and upper flanks black, chest warm pink-cinnamon or buff, gradually paler cream-buff on breast, flanks, and under tail-coverts, cream-white on central belly and vent. Tail white, central pair (t1) black for distal $\frac{2}{3}$, tips of others with strongly varying amount of black; whatever the amount of black, all differ from other *Oenanthe* (except Black-eared Wheatear *O. hispanica*) in showing limited black on t3–t4, and then gradually more towards t1 and t6, with especially t1 and outer web of t6 showing much black. Of 22 birds examined, maximum extent of black on both webs of t1 37·5 (33–44) mm, on shafts often a few mm more; inner web of t3–t4 black for 3–13 mm, but in 7 birds fully white, bar or spot on outer web black for 10 (5–14) mm, and once fully white; outer web of t5 and inner of t6 black for 13 (10–18) mm, outer web of t6 for 26·5 (22–31) mm. Tail-tips faintly fringed white when very fresh. Flight-feathers, tertials, and all upper wing-coverts black, slightly more greyish on inner webs of flight-feathers; tips of secondaries and greater upper primary coverts narrowly fringed cream, outer webs very slightly so, soon wearing off; remainder of upper wing-coverts with narrow and even buff or cream-buff fringes along tips, widest (1–2 mm) on greater coverts. Axillaries and under wing-coverts black; smaller coverts along leading edge of wing narrowly fringed white. Abrasion has marked effect on plumage (least so on sides of head and neck, chin, and throat, which are moulted twice annually), but some individual variation in timing, some birds abraded February when others are still quite fresh. *In worn plumage* (spring), forehead, crown, and nape white (usually still spotted or washed grey-brown on crown in March); mantle, scapulars, and back black (often traces of brown feather-fringes up to March or early April); rump and upper tail-coverts white; sides of head and neck, chin, and throat fully deep black or (when freshly moulted mid-winter) with narrow cream or off-white feather-fringes on ear-coverts, rear cheeks, sides of neck, and throat up to March; underparts cream-buff to white, sometimes deeper cinnamon-buff on chest (when replaced in pre-breeding moult); fringes along flight-feathers and upper and under wing-coverts worn off, except for traces along tips of secondaries and greater upper primary coverts, hence all wing black. Thus, appearance buff-and-brown in autumn and winter, black-and-white in spring and summer. When heavily worn, late June or July, much grey of feather-bases sometimes visible among white of cap and underparts, occasionally crown fully grey. ADULT FEMALE. *In fresh plumage*, autumn and winter, upperparts drab brownish-grey or grey-brown; forehead slightly tinged buff, crown and nape slightly paler grey-brown; mantle, scapulars, and back with distinct pale cream-buff feather-fringes, appearing scalloped or mottled. Narrow and ill-defined supercilium cream-buff or pale buff, very faint above lores. Rump and upper tail-coverts white with slight cream-buff tinge. Lores and upper cheeks pale grey-brown, narrow and sometimes indistinct eye-ring pale buff. Ear-coverts dark brown, slightly darker than crown. Upper cheeks, chin, throat, and sides of neck pale pink-buff, cream-buff, or buff-white, sometimes with black feather-bases visible on ear-coverts, sides of neck, rear cheeks, and throat, giving mottled aspect. Band across chest light tawny-brown, merging into darker grey-brown on flanks but rather sharply demarcated from pink-buff to cream-white breast, belly, vent, and under tail-coverts. Tail as adult ♂, but black of feather-tips slightly browner. Flight-feathers, tertials, and upper wing-coverts black-brown, slightly paler grey-brown or drab-brown towards lesser coverts, slightly blacker on longest feather of bastard wing; fringes along tips and outer webs of flight-feathers, tertials, greater upper primary coverts, and greater upper wing-coverts buff, widest on tips and basal outer webs of secondaries and on tips of outer greater coverts; tips of median and lesser coverts with buff or pale cream-buff fringes. Axillaries and under wing-coverts dull black or dark grey with brown or buff feather-tips; paler than in adult ♂. *In worn plumage* (spring), upperparts uniform dark grey-brown without paler scalloping; crown sometimes with some paler buff mottling and often with darker shaft-streaks; rump and upper tail-coverts white. Supercilium cream-buff, poorly defined. Lores and front part of cheeks mottled grey-brown and buff. Sometimes a faint pale buff eye-ring. 2 morphs, more readily separable than in fresh plumage; pale-throated morph comparable with *vittata* morph of ♂, though (unlike ♂) intermediates occur also. In dark-throated morph, ear-coverts, sides of neck, rear of cheeks, and throat black or mottled black and pale buff or off-white, chin largely dirty white; this dark collar sometimes prominent, broad, and largely black, sometimes dark grey with heavy pale buff, pale grey, or off-white mottling and restricted to lower sides of neck and lower throat. In pale-throated morph, sides of neck, cheeks, chin, and throat uniform pale buff or off-white, ear-coverts dull brown. In all morphs, chest grey-brown, merging into buff-brown on centre and sometimes with rather distinct narrow dusky shaft-streaks at sides; rather sharply defined from dirty cream to off-white remainder of underparts. Flight-feathers, tertials, and all upper wing-coverts black-brown, buff or cream-buff edges and fringes partly worn off but usually just visible up to about June. See also Ash and Rooke (1956), Sluys and Berg (1982), and Recognition (below). NESTLING. No information. JUVENILE. Upperparts grey-brown, slightly paler and greyer than adult ♀; each feather with pale buff or grey-white centre and rather indistinct dark rim along tip. Rump and upper tail-coverts white or cream-white, some feather-tips tinged buff or narrowly fringed black. Indistinct paler grey-buff supercilium. Sides of head grey-brown, lores and cheeks closely mottled buff or dirty white, ear-coverts narrowly streaked off-white. Underparts cream-white; chest and sides of breast slightly darker, buff, with faint dark fringes along feather-tips. Tail as adult ♀, feather-tips narrowly fringed buff. Lesser and median upper wing-coverts dull grey-brown with rather triangular buff spot on tip, narrowly bordered dull black on tips. Flight-feathers, tertials, greater upper wing-coverts, and greater upper primary coverts as adult ♀, but buff fringes slightly broader, especially those along outer webs and tips of greater coverts and primary coverts, tending to form roughly triangular spot on tip of outer greater coverts and sometimes with narrow black fringe along tip of tertial coverts. FIRST ADULT MALE. Like adult ♂, but juvenile flight-feathers, tertials, greater upper primary coverts, and usually many greater upper wing-coverts retained, distinctly browner than adult and with buff fringes along outer webs and tips, contrasting in colour and wear with blackish new median and tertial coverts. In fresh plumage, rather similar to adult ♂, but white of feather-bases of crown and nape fully concealed and hardly any black feather-bases of mantle and scapulars visible; paler feather-fringes on sides of head and neck, chin, and throat slightly broader, less sharply defined (forming less regular arcs), often tinged sandy rather than white on chin and throat, where black almost concealed; buff fringes along outer webs and tips of flight-feathers, tertials, and upper wing-coverts distinct. *In worn plumage* (spring), crown and nape never as pure white as adult, crown often still largely grey-brown March, nape mottled brown; all mottled brown (especially nape) April–May, tinged brown with grey of feather-bases visible June. Mantle and scapulars fringed grey-brown or olive-grey until May (adult virtually black from April), fringes worn off but much grey of feather-bases visible June–July. Sides of head and neck, chin, and throat less pure black than adult, some off-white mottling present up to June, especially on ear-coverts, chin, and central throat. Buff fringes of flight-feathers and upperwing partly abraded but still prominent March–April (in adult, virtually worn off from April), traces still showing May, worn off June–July, when upperwing bleached and abraded to greyish-brown (adult black-brown). FIRST ADULT FEMALE. Like adult ♀, but part of juvenile feathering retained as in 1st adult ♂. Hardly any difference in colour of feathers and width of fringes between adult and retained juvenile feathers, and hence ageing difficult; tail-tips of 1st adult narrower, browner, and more worn than adult at same time of year, tips of flight-feathers and greater upper primary coverts more pointed and more abraded.

Bare parts. ADULT, FIRST ADULT. Iris brown or dark brown. Bill, leg, and foot black, occasionally dark brown in ♀. NESTLING. No information.

Moults. All data refer to nominate *pleschanka* except where stated. ADULT POST-BREEDING. Complete; primaries descendant. In USSR, some start early July and complete early August, but others start as late as early September (Dementiev and Gladkov 1954b). In Iran, Afghanistan, and Kashmir, starts about end of July, just about finished by end of August (Vaurie 1949). In Afghanistan, no moult in ♀ from 6 July, but 3 ♂♂ 10 July to 4 August were moulting (Paludan 1959). In Mongolia, moult not started by early July; primary moult score *c.* 11 and many body feathers growing in ♂ of 23 July; moult nearing completion in 5 birds 18–23 August (Piechocki and Bolod 1972). In Kansu (China), 2 in moult 1–11 August (Stresemann *et al.* 1937). ADULT PRE-BREEDING. Partial: lores, ear-coverts, chin, and (in *cypriaca*, but not in all nominate *pleschanka*) cheeks, throat, sides of neck, and chest; December–February

(-March) (Witherby *et al.* 1938*b*; Vaurie 1949; ZMA). Occasionally no moult, particularly in ♀ nominate *pleschanka*. POST-JUVENILE. Partial: head, body, lesser and median upper wing-coverts, and tertial coverts; occasionally a few tertials or inner greater upper wing-coverts; in *cypriaca*, sometimes many or all greater upper wing-coverts. Starts late June to early August, completed from mid-August onwards. (Dementiev and Gladkov 1954*b*; Vaurie 1949; RMNH, ZFMK, ZMA.) FIRST PRE-BREEDING. Like adult pre-breeding, but sometimes fewer feathers involved or (especially in ♀♀) no moult at all.

Measurements. ADULT, FIRST ADULT. Nominate *pleschanka*. All geographical range, all year; skins (RMNH, ZFMK, ZMA). Juvenile wing and tail refer to retained juvenile wing and tail of 1st adult. Bill (S) to skull, bill (N) to distal corner of nostril; exposed culmen on average 4·2 less than bill (S).

WING AD	♂ 94·9 (2·07; 20)	92–99	♀ 91·6 (1·38; 8)	90–95
JUV	93·3 (2·06; 22)	90–99	91·0 (2·40; 5)	88–94
TAIL AD	59·4 (2·30; 11)	56–64	58·8 (2·04; 8)	54–62
JUV	57·5 (1·97; 13)	54–61	58·5 (2·03; 4)	56–61
BILL (S)	16·8 (0·63; 41)	15·8–18·1	16·9 (0·62; 13)	15·9–17·9
BILL (N)	9·1 (0·48; 23)	8·2–9·6	9·4 (0·39; 12)	8·9–10·0
TARSUS	23·3 (0·58; 24)	22·3–24·3	23·2 (0·83; 12)	22·0–24·1

Sex differences significant for adult wing.

Wing. Ages combined. (1) Tien Shan to Altai, (2) Iran, (3) Afghanistan (Sluys and Berg 1982). (4) Afghanistan (Paludan 1959). (5) Kashmir and Tibet (Vaurie 1972). (6) Mongolia (Piechocki and Bolod 1972; Piechocki *et al.* 1982). (7) Kansu, China (Stresemann *et al.* 1937).

(1)	♂ 91·7 (1·5 ; 11)	88–94	♀ 89·7 (2·1 ; 6)	86–92
(2)	92·7 (3·3 ; 13)	88–100	88·6 (2·6 ; 6)	86–92
(3)	93·9 (2·3 ; 19)	89–97	90·6 (1·6 ; 12)	89–93
(4)	93·8 (— ; 22)	89–98	91·0 (— ; 14)	89–94
(5)	95·8 (— ; 13)	90–101	92·3 (— ; 7)	87–98
(6)	96·1 (2·30; 26)	91–101	93·7 (2·15; 12)	90–96
(7)	96·8 (— ; 8)	94–100	94·0 (— ; 5)	92–97

O. p. cypriaca. Cyprus, March–September; skins (ZFMK, ZMA). Exposed culmen on average 3·9 less than bill (S).

WING AD	♂ 86·0 (2·24; 6)	83–88	♀ 84·2 (1·44; 4)	82–86
JUV	84·4 (1·38; 9)	83–87	82·3 (1·86; 6)	80–85
TAIL AD	56·3 (1·15; 5)	54–58	56·0 (1·82; 4)	54–58
JUV	55·8 (1·81; 8)	53–59	52·8 (1·21; 6)	51–55
BILL (S)	16·3 (0·54; 13)	15·5–16·9	16·0 (0·73; 10)	15·1–17·1
BILL (N)	9·0 (0·40; 13)	8·4–9·6	8·9 (0·58; 10)	8·1–9·7
TARSUS	22·6 (1·09; 13)	20·8–23·9	22·5 (0·68; 10)	21·5–23·3

Sex differences significant for juvenile wing and tail.

Weights. ADULT, FIRST ADULT. Nominate *pleschanka*. (1) Afghanistan, late February to mid-April (Paludan 1959). Iran (Schüz 1959; Paludan 1940), Afghanistan (Paludan 1959), Kazakhstan, USSR (Dolgushin *et al.* 1970), Mongolia (Piechocki and Bolod 1972; Piechocki *et al.* 1982), and Kansu, China (Stresemann *et al.* 1937), combined: (2) March, (3) April, (4) May, (5) June, (6) July, (7) August–October.

(1)	♂ 18·9 (— ; 10)	17–22	♀ 18·0 (— ; 8)	16–20
(2)	21·3 (1·12; 7)	19–22	— (— ; -)	
(3)	18·3 (1·71; 9)	16–20	18·9 (— ; 3)	18–20
(4)	17·4 (1·09; 21)	15–20	19·9 (2·42; 9)	18–25
(5)	16·7 (1·38; 10)	15–19	18·8 (2·23; 6)	16–22
(6)	18·6 (1·59; 16)	16–21	18·3 (1·93; 5)	16–21
(7)	18·9 (1·70; 9)	16–22	20·6 (1·85; 8)	17–22

East Africa, winter: ♂♂ 17·5, 20; ♀♀ 20, 20 (RMNH, ZFMK). Kirgiziya (USSR): 22 ♂♂, 16–24; 15 ♀♀, 16–23 (Yanushevich *et al.* 1960).

Structure. Wing rather long, broad at base, tip bluntly pointed. 10 primaries: in nominate *pleschanka*, p8 longest, p9 4–7 shorter, p7 0–0·5, p6 1–4, p5 8–11, p4 14–17, p1 22–26; in *cypriaca*, p7 longest, p8 0–1 shorter, p9 4–10, p6 0·5–2, p5 6–9, p4 10–15, p1 16–21. In both races, p10 reduced; 2 shorter to 4 longer than longest upper primary covert, 47–59 (nominate *pleschanka*) or 42–51 (*cypriaca*) shorter than wing-tip. Outer web of p6–p8 and inner of p7–p9 emarginated. Tertials short. Tail rather short, tip square, 12 feathers. Bill relatively short (especially in nominate *pleschanka*), straight, slender; rather wide at base, laterally compressed at tip; tip of culmen slightly decurved. Nostrils rather small, oval, partly covered by membrane above. Some bristles and many fine hairs along base of mandibles. Tarsus relatively short, slender; toes rather short and slender. Middle toe with claw in both races 16·6 (21) 15–18; outer toe with claw *c.* 68% of middle with claw, inner *c.* 64%, hind *c.* 73%. Claws rather short, sharp, decurved.

Geographical variation. Slight within nominate *pleschanka*, wing length increasing towards east (see Measurements). Cyprus race *cypriaca* rather different in plumage: generally darker than nominate *pleschanka*, especially ♀ which is closely similar to ♂ (unlike nominate *pleschanka*); distinctly smaller (especially wing and tail), wing-tip more rounded, song different, and breeding range widely separated from nominate *pleschanka*, hence sometimes considered to be separate species (Sluys and Berg 1982). In spring and summer, adult ♂ *cypriaca* rather similar to adult ♂ nominate *pleschanka*, but black of *cypriaca* deeper, slightly glossy, more extensive on back (remaining white from rump to black of t1 2–2·5 cm in *cypriaca*, 3·5–4 cm in nominate *pleschanka*), and reaching slightly further down to upper chest; chest to under tail-coverts on average slightly deeper warm buff-cinnamon all over in March (paler pink-cinnamon with whitish belly in nominate *pleschanka*), paler buff-cinnamon or pink-cinnamon with whitish belly in April (in nominate *pleschanka*, chest and under tail-coverts pink-cream, remainder white); chest and under tail-coverts pink-buff and remainder whitish May–June (virtually all-white in nominate *pleschanka*); thus, relatively worn birds of *cypriaca* may resemble relatively fresh nominate *pleschanka* at same time of year; cap white in both, but traces of dull black feather-tips in *cypriaca* in March (grey-brown in nominate *pleschanka*). 1st adult ♂ *cypriaca* has mantle, scapulars, and back dull black in spring and summer, with some traces of dark brown in March only; white on rump (measured down to black of t1) 2–2·5 cm long (in nominate *pleschanka*, mantle and scapulars never fully black, but black-brown with broad grey-brown or olive-brown feather-fringes in March, dark brown or black-brown with traces of brown fringes May–June; white on rump 2·5–3·5 cm); chin and throat dull black (less deep than adult ♂, with some pale feather-fringes in March; in nominate *pleschanka*, more distinct pale fringes present up to June–July); colour of chest to under tail-coverts slightly deeper than adult ♂ in both races, but similarly affected by bleaching; cap white in both races, but crown (not forehead, supercilium, and nape) sometimes still largely black or brown-black in *cypriaca* in March–April (largely grey-brown in some nominate *pleschanka* at that time); retained juvenile feathers show same characters as in nominate *pleschanka* (see Plumages). ♀ *cypriaca* in spring and summer quite different from ♀ nominate *pleschanka*, being closely similar to ♂ and hence much darker than nominate *pleschanka* (however, palest 1st adult ♀ *cypriaca* about as dark as darkest adult ♀ nominate *pleschanka*). Adult ♀ *cypriaca* has mantle, scapulars, and back extensively black-brown, chin and throat dull black with some traces of pale feather-fringes, and underparts warm cinnamon-buff to pale buff, depending on wear; hence, closely similar to 1st adult ♂, but differs from both adult ♂ and 1st adult ♂ in dark brown, dark grey-brown, or dark olive-brown crown, bordered by pale grey-brown or buff forehead in front, by buff or off-white supercilium at sides, and by cream to off-white nape at rear; cap becomes paler through wear and some ♀♀ from May resemble 1st adult ♂♂ from March; by May–July (when cap of 1st adult ♂ virtually white), crown grey-brown or mottled grey and black, broadly bordered by white at sides, rear, and (sometimes) front. 1st adult ♀ in spring and summer similar to adult ♀, but cap more extensively dark grey-brown, supercilium and nape mottled brown, never as pure white as adult ♀ and less sharply defined; mantle, scapulars, and back dark olive-brown rather than black-brown; lores, cheeks, ear-coverts, chin, throat, and sides of neck show close olive-brown (at sides) or white (below) feather-fringes, appearing rather evenly barred black and pale in March, largely brown-black or black May–June (not as uniform dull black as adult ♀ or 1st adult ♂). Juvenile darker than juvenile nominate *pleschanka*: dark grey-brown to black-brown with pale spots on feather-centres on upperparts, buff with pronounced dark scalloping on underparts. In autumn and winter, both sexes more difficult to separate from ♂ and some ♀♀ of nominate *pleschanka*: upperparts of *cypriaca* sooty grey-brown or rusty-brown (♀ and 1st adult ♂) to sooty-black (adult ♂), with long and narrow supercilium (off-white or pale buff in adult, buff or rufous-buff in 1st adult) and white rump-patch 1·5–2·5 cm long; some white of feather-bases often visible on nape (most in ♂♂); sides of head and neck, chin, and throat black with brown (sides) or off-white (below) feather-fringes *c.* 3 mm wide (1st adult ♀, almost hiding black), 1–2 mm wide (adult ♀, 1st adult ♂), or 0·5 mm wide (adult ♂); underparts warm cinnamon-buff, chest with some olive-brown or sooty on feather-tips; pale fringes on upperwing buff (but virtually absent in adult ♂), soon bleaching to off-white and partly worn off. See also Christensen (1974). In all plumages, tail-tips have more black than in nominate *pleschanka*, but much variation. Maximum extent of black on both webs of t1 39 (20) 35–44 mm (on shaft often more or slightly less, 31–47 mm); minimum extent of black on t3–t5 mainly 16–18 mm, on average 16 (4–22) mm in 17 birds measured, but white reached tail-tip on inner webs of 2–3 tail-feathers in 2 adult ♂♂ and 1 1st adult ♀; maximum extent of black on outer web of t6 (excluding shaft) 32 (20) 26–38 mm. Smaller in size than nominate *pleschanka*, in particular wing (see Measurements).

Forms superspecies with Black-eared Wheatear *O. hispanica* (Stepanyan 1983); relatively narrow zone of overlap in northern Iran, central and eastern Transcaucasia, and on Mangyshlak peninsula in east Caspian Sea (overlap in south-west Iran, as suggested by Vaurie 1949, is unfounded: Cornwallis 1975; Haffer 1977). Extensive hybridization occurs in these overlap zones, and such populations show bewildering variety of plumages; see Panov and Ivanitski (1975b) for Mangyshlak and Haffer (1977) for northern Iran; for hybridization in small area of overlap in eastern Bulgaria, see Baumgart (1971). White-throated morph of ♂ *O. pleschanka* ('*vittata*') is probably a result of introgression of genes from *O. hispanica* into *O. pleschanka*: relatively common (up to 11%) in areas of pure *O. pleschanka* near hybridization zones, rare (*c.* 1%) further east (Panov and Ivanitski 1975b).

Recognition. ♂ rather similar to Mourning Wheatear *O. lugens* and Hooded Wheatear *O. monacha*; see Recognition of those species (pp. 867, 871). ♀ of nominate *pleschanka* is close to ♀ Wheatear *O. oenanthe*, but differs as follows. (1) Duller and greyer upperparts with more pronounced dull grey-brown chest-band. (2) Different tail pattern, with usually much more black on outer web of t6 than on tips of t3–t5: in ♀ *O. pleschanka*, minimum extent of black on t3–t5 on average 8 mm (0–13), maximum on t6 on average 28 mm (25–31); in *O. o. oenanthe*, average 18 mm (16–20) on t3–t5, on average 21 mm (18–23) on t6, but some *O. o. libanotica* show much less on t3–t5 and 14–24 mm on t6. (3) Size: tail longer but tarsus markedly shorter than *O. oenanthe*. ♂ in fresh autumn plumage distinctly darker and duller grey-brown above than more sandy-cinnamon ♂ *O. hispanica melanoleuca*. ♀ differs from ♀ *O. hispanica melanoleuca* only slightly and sometimes inseparable: measurements, wing formula, and tail-pattern similar; in fresh plumage, upperparts drab brown-grey, darker than *O. hispanica*, feathers of mantle, scapulars, and back with distinct buff fringes (upperparts virtually uniform in *O. hispanica*); chest-band tawny-brown, rather sharply divided from cream-white remainder of underparts (paler and less sharply divided in *O. hispanica*); in worn plumage, upperparts and chest-band even darker and browner than in *O. hispanica*, sides of chest and crown often with dark brown shaft-streaks (in *O. hispanica*, no streaks or only faint rufous ones); black of tail-tips often tends to end in sharp point at shaft, not as straight across feather as in *O. hispanica* (though *O. hispanica* has some difference between the webs in amount of black). Differences between ♀♀ of *O. pleschanka* and *O. hispanica* given by Sluys and Berg (1982) are by no means constant: especially in hybrid populations around Caspian Sea, identification on plumage colour or measurements is unreliable or even impossible (E N Panov).

CSR

Oenanthe hispanica Black-eared Wheatear

PLATES 59, 64, and 65
[facing pages 760, 761, and 856]

Du. Blonde Tapuit   Fr. Traquet oreillard   Ge. Mittelmeersteinschmätzer
Ru. Чернопегая каменка   Sp. Collalba rubia   Sw. Rödstenskvätta

Motacilla hispanica Linnaeus, 1758

Polytypic. Nominate *hispanica* (Linnaeus, 1758), north-west Africa, Iberia, southern France, northern and central Italy, and north-west Yugoslavia; *melanoleuca* (Güldenstädt, 1775) (synonym: *xanthomelaena*), southern Italy (Calabria, Puglia) and Yugoslavia east to Caspian region and Iran.

Field characters. 14·5cm; wing-span 25–27 cm. Distinctly lighter in build than Wheatear *O. oenanthe*, being similar in size and structure to Pied Wheatear *O. pleschanka pleschanka* except for slightly shorter wings.

Rather slim elegant wheatear, with long, conspicuous tail giving slimmer, lengthier outline than most others of similar plumage. Rump and tail pattern basically as *O. oenanthe* but black terminal band less uniformly broad, though more black along outer edge than in any other wheatear. Spring ♂ has wholly or partly black scapulars and wings more obviously divided by pale back and white rump and tail than any other wheatear (making bird appear narrow-shouldered from behind). ♀ has stronger pattern than many ♀ wheatears, having black wings contrasting boldly with sandy back and chest. Black under wing-coverts striking, particularly in western race which has paler undersurface to flight-feathers than eastern one (but this pattern also shown by other wheatears). Sexes markedly dissimilar in spring, less so in autumn. Juvenile separable. 2 races in west Palearctic; both races and both sexes have pale-throated and black-throated morphs, with pale-throated increasingly common towards east of range.

ADULT MALE. (1) West Mediterranean race, nominate *hispanica*. In spring (when plumage worn), forehead and fore-supercilium white; most of crown, nape, mantle, and inner scapulars sandy- or buff-cream, hardly contrasting with similarly toned underbody but dramatically offset by black face (and throat in black-throated morph) and black wings. Rump white; tail white with black ⊥-pattern, but terminal band much narrower and less regular in width than in *O. oenanthe*, with marked extension up outermost feathers. At close range, darker individuals show diffuse cream or white rear supercilium and similarly coloured line under cheek-patch. Under wing-coverts and axillaries black, contrasting with grey undersurface to flight-feathers. Almost white margins to inner webs of flight-feathers occasionally show through uppersurface of fully extended primaries and outer secondaries. From July, fresh feathers have foxy-buff to cream margins overlying basic plumage pattern: appearance changes less than in *O. pleschanka* but rear cheeks and border to throat-patch become hoary, while wings show 2 cream wing-bars (on median and greater coverts), broad warm buff fringes to secondaries and inner primaries, and pale buff tips to outer primaries; under wing-coverts flecked white; undersurface of flight-feathers slightly paler than before. (2) East Mediterranean and Middle East race, *melanoleuca*. Essentially as nominate *hispanica*, but upper- and underbody usually noticeably paler, cream or even cream-white (lacking sandy tone), scapulars wholly black (making pale division down back even narrower), and terminal band of tail often narrower or even broken next to central feathers. Most reliable features as follows: under wing-coverts black, hardly contrasting with dusky undersurface of flight-feathers; black frontal band over bill; larger black loral and cheek-patches (in pale-throated morph) or black throat-patch reaching further down on to upper breast (in black-throated morph). In autumn, plumage changes as in nominate *hispanica* but upperparts

pale grey-brown rather than warm buff, though tawny patches may show on mantle on darkest individuals; difference in face pattern remains visible; underwing pattern unchanged; palest individuals show almost white supercilium and throat and less buff vent. ADULT FEMALE. Both races. Lacks strong head markings of ♂, but upper- and underbody (sandy-buff to ochre-grey or ochre-brown) contrast strongly with scapulars and wings (dark brown-black, usually with some buff or cream fringes to larger feathers). Head pattern usually lacks any distinctive features, but grey-brown speckles may form dusky throat-mark in black-throated morph (though rarely as obvious as in *O. pleschanka*). Breast normally distinctly tawny contrasting with paler throat and rear underparts more distinctly than in *O. oenanthe*. Racial separation far more difficult than in ♂♂, with greyer cheeks and darker brown upperparts of *melanoleuca* also inviting confusion with *O. pleschanka*. In autumn, plumage changes as in ♂, with new fringes and tips to wing-feathers even more obvious and reducing overall plumage contrast; rump dull cream. JUVENILE. Both races. Resembles winter ♀ but upperparts spotted pale buff and chest and fore-flanks faintly scaled dark brown. FIRST-YEAR MALE. Both races. In autumn and winter, upperparts darker than adult ♂ and white face- and throat-marks more overlaid with buff fringes. In spring, closely resembles adult but upperparts remain sullied on crown and back, and duller black areas usually retain many pale fringes and tips. FIRST-YEAR FEMALE. Both races. Probably indistinguishable from adult though some with faded brown wings may be this age. Both races and all ages have bill and legs black.

Adult ♂ and spring adult ♀ easy to distinguish, but winter ♀ and immature subject to confusion, with some nominate *hispanica* approaching appearance of pale individuals of *O. oenanthe* and most *melanoleuca* closely resembling *O. pleschanka*. Dark under wing-coverts eliminate *O. oenanthe*, leaving separation from *O. pleschanka* as main problem and one not yet fully studied, though paler (sandier or greyer) tone of mantle, back, and scapulars, and absence of pale fringes on them, considered diagnostic of *O. hispanica* (see also Field Characters and Recognition of *O. pleschanka*, pp. 792 and 806). When perched or on ground, ♂ with sandy plumage can also suggest ♂ Desert Wheatear *O. deserti* but that species has all-black tail. Flight much as *O. oenanthe* but rather more agile, with lighter wing-beats and apparently looser tail. Gait, stance, and behaviour as *O. pleschanka*, with similarly frequent perching on vegetation. Quite tame.

Song comprise short, scratchy, warbling phrases, similar to *O. pleschanka pleschanka*. Commonest call a muffled click—'chek'.

Habitat. Breeds at lower middle latitudes in warm mainly continental Mediterranean and steppe regions, above 23°C July isotherm. Within these limits, largely replaces Wheatear *O. oenanthe* in habitats below *c*. 600 m (Voous

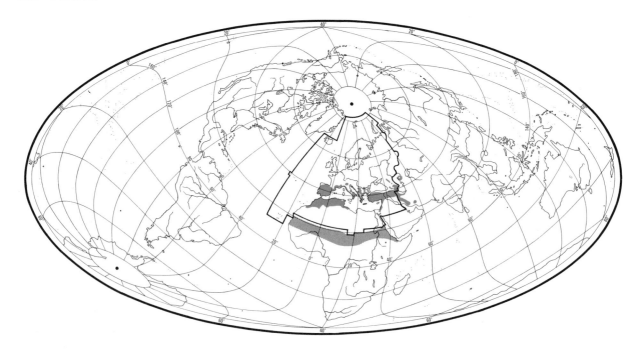

1960), but in USSR ascends in mountains to 2000-2300 m, also inhabiting steppes with rocky outcrops or stony hillocks and slopes, and cliff-like river banks (Dementiev and Gladkov 1954*b*). More generally in open or lightly wooded arid country (Peterson *et al.* 1983); also on warm rocky lowlands and stony ground, limestone hills, slopes with debris, dry river valleys, dry and stony fields, Mediterranean heaths with kermes oak *Quercus coccifera*, vineyards with stone banks, and dry cultivations (Voous 1960). In north-west Africa, resorts to broken, rocky, and denuded terrain up to 1000-1200 m, including Hauts Plateaux and Atlas Saharien as well as Atlantic coast (Heim de Balsac and Mayaud 1962). In south-west Iran, breeds in varied habitats at *c.* 1700-2500 m with annual precipitation of over 400 mm and usually richer vegetation than those occupied by other *Oenanthe*, including ample flush of spring herbage, shrublets, bushes and often trees, especially *Quercus* and *Juniperus*, sometimes forming woodland. Makes full use of elevated perches and all types of cover. (Cornwallis 1975.) Requires access to perches on shrubs, stones, banks, fences or other points of vantage. In winter resorts to *Acacia* and thorn-bush steppe and semi-desert country, rocky hills, and gardens (Moreau 1972). Normally has only limited contact with man. Like other *Oenanthe* flies as occasion demands, largely in lowest airspace.

Distribution. Some retreat of range in France.

FRANCE. Bred Burgundy 1885; possible retreat towards south (Yeatman 1976). JORDAN. Azraq: an apparently established pair present 1965, but breeding not proved (Wallace 1983*a*). MALTA. Bred 1982 (*Il-Merill* 1983, **22**, 17).

Accidental. Britain, Ireland, Channel Islands, Netherlands, West Germany, Norway, Sweden, Czechoslovakia, Austria, Hungary, Switzerland, Canary Islands.

Population. No reported population trends.

FRANCE. 1000-10 000 pairs (Yeatman 1976). SPAIN. Not scarce, locally numerous (AN). BULGARIA. Locally numerous (TM). USSR. Fairly common (Dementiev and Gladkov 1954*b*). LIBYA. Tripolitania: breeds sparingly (Bundy 1976).

Movements. Migratory. Winters in semi-desert and *Acacia* savanna belt across northern tropical Africa from Sénégal to Ethiopia.

Nominate *hispanica* (breeding south-west Europe and North Africa) winters south of *c.* 18°N, mainly in northern Sénégal (Morel and Roux 1966), south-west Mauritania (Browne 1982), and Mali where it is common and widespread as far south as 12½°N (Lamarche 1981); occurs also in south-west Niger (Koster and Grettenberger 1983), and in northern Nigeria where it is uncommon but regular (Elgood 1982).

Passage is on broad front: crosses south-west coast of Spain (Henty 1961), Moroccan coast (Thévenot *et al.* 1982), and north-west Mauritania (Browne 1982); recorded in small numbers in Malta, chiefly in spring (Sultana and Gauci 1982), and noted in desert of south-west Libya (Snow and Manning 1954); recorded throughout Algerian Sahara (Ledant *et al.* 1981) including south-east (Laferrère 1968).

Departure from breeding grounds starts in August. North African populations leave chiefly late August and September (Heim de Balsac and Mayaud 1962), although

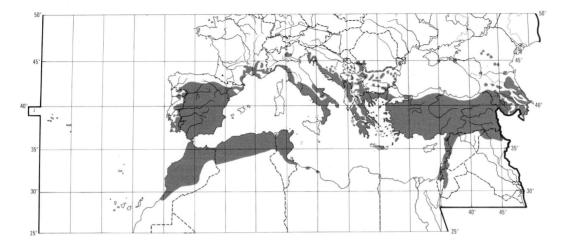

northern passage birds confuse the situation. Passage in Morocco starts in August but occurs mainly September–October with a few stragglers in November (Thévenot *et al.* 1982). On Malta, passage starts early August and usually over by early October (Sultana and Gauci 1982). Arrival on wintering grounds comparatively early, usually 2–3 weeks ahead of *O. oenanthe* in Sénégal with first arrivals in mid-September and full winter numbers present by mid-October (Morel and Roux 1966). Recorded in Mali from end of August (Lamarche 1981).

Spring passage in western Sahara and Morocco from late February, continuing through March and April (Heim de Balsac and Mayaud 1962). Last birds leave Sénégal by mid-April (Morel and Roux 1966). In Mali, loose parties often assemble in favourable areas during March–April prior to departure (Lamarche 1981). Passage through Tunisia from late March continuing into May with birds ringed at Cap Bon up to 2nd week of May (Bortoli *et al.* 1962). Spring passage recorded through Malta mid-March to mid-May with occasional records in June (Sultana and Gauci 1982). Arrivals north of Mediterranean mainly in April; recorded in small numbers on Balearic islands (Bannerman and Bannerman 1983), and in Camargue (southern France) in spring (Blondel and Isenmann 1981). Passage observed up to 2nd week of May in Coto Doñana (Ferguson-Lees 1956). One record of site-fidelity to winter quarters: ringed ♂ returned in successive years to a farm at Kano, Nigeria (Sharland 1967).

Eastern race, *melanoleuca* (breeding east from south-east Italy), tends to replace nominate *hispanica* on wintering grounds east of 0–5°E (Moreau 1961), although there is overlap in Mali where *melanoleuca* is uncommon and localized and not found south of 16°N (Lamarche 1981). 1 record from Nigeria (Elgood 1982). Common in northern and central Chad (Newby 1980) and Sudan where can occur as far south as 8°N (Cave and Macdonald 1955) but most common at 9–14°N west of Nile and at 10–15°N east of Nile (Hogg *et al.* 1984). Uncommon to frequent in west and north-east Ethiopia (Urban and Brown 1971), and winters regularly in some parts of Eritrea, chiefly in south (Smith 1957). Occasionally recorded wintering in Syria (Kumerloeve 1969*b*; Macfarlane 1978). Evidently only a straggler south to East Africa, with 2 records from Kenya (Backhurst *et al.* 1984; P B Taylor).

General direction of autumn passage lies slightly west of south, and overlaps with nominate *hispanica* on routes furthest west. Occurs in small numbers (chiefly in spring) on Malta (Sultana and Gauci 1982), and more commonly in both seasons in Tunisia (Moreau 1961); considerable passage in south-east of Algerian Sahara (Laferrère 1968). In Cyprus, fairly common in spring but scarce in autumn (Flint and Stewart 1983); noted as among most conspicuous migrants in spring in eastern Libya (Stanford 1953). Common in eastern Egypt in both spring and autumn (Moreau 1961). Occurs regularly at Elat (Israel) (Safriel 1968), and fairly common (particularly in spring) at Azraq (Jordan) (Clarke 1980). Described as abundant summer visitor and migrant in Lebanon (Vere Benson 1970), and regular in both seasons in Iraq (Allouse 1953). In Transcaucasia (USSR), birds head north-east on broad front, only a few migrating along western shore of Caspian Sea (Loskot 1983). In areas further south-east (e.g. Gulf states of Arabia, Oman, North Yemen), generally scarce or uncommon (Bundy and Warr 1980; Gallagher and Woodcock 1980; Cornwallis and Porter 1982). 2 recorded west of Socotra island (Gulf of Aden) flying south-west (Gush 1949).

Departure from breeding grounds August–September. Most pass through Aegean in September, a few in October (Watson 1964). In Turkey, passage from August; only a few September records from Black Sea, but present up to late September on central plateau and to mid-October on southern coast (*Orn. Soc. Turkey Bird Rep. 1966–75*). Only 16 autumn records from Cyprus 1956–70, mostly late August to October (Bannerman and Bannerman 1971; Flint and Stewart 1983). Passage in

Algeria chiefly mid-September to mid-October (Laferrère 1968). Passage occurs at Azraq from end of August (Clarke 1980) and at Elat from early September and throughout October (Safriel 1968). Passes through Iraq mid-September to mid-October (Ticehurst *et al.* 1925). Present within breeding-range in south-west Iran up to mid-September (Cornwallis 1975). Recorded in Gulf states of Arabia from August to early November, with exceptional record of 5 ♂♂ on Bahrain in June (Bundy and Warr 1980); occurs in Oman September-October (Gallagher and Woodcock 1980). Arrival on wintering grounds mainly October in Chad (Newby 1980), also in Mali, where somewhat later than nominate *hispanica* (Lamarche 1981). Arrives September-October in Sudan and Eritrea (Cave and Macdonald 1955; Smith 1957).

Stays on wintering grounds until March-April. Records in May in Mali (Lamarche 1981) and south-east Algeria (Laferrère 1968) indicate spring migration somewhat protracted in west of range. Recorded from late March at Cap Bon (Tunisia) with peak in mid-April (Deleuil 1956-66). Passage through Cyprus from early March to early May with heaviest passage late March to early April (Bannerman and Bannerman 1971; Flint and Stewart 1983). Arrives in Greece and Aegean from late March (Watson 1964), in Turkey from late March on south coast and early April on central plateau, Thrace, and south-east (*Orn. Soc. Turkey Bird Rep. 1966-75*). Passage through Middle East chiefly March-April (e.g. Safriel 1968, Clarke 1980, Bundy and Warr 1980). Arrives on breeding grounds in Iran early April (Cornwallis 1975). In USSR, arrives southern Armeniya and Azerbaydzhan late March or beginning of April, western Kopet-Dag (southern Turkmeniya) late March, and Mangyshlak (east Caspian) early April. Generally appears in southern USSR *c.* 35-40 days after the first Finsch's Wheatear *O. finschii* and Isabelline Wheatear *O. isabellina*, and 15-30 days after other *Oenanthe* including Pied Wheatear *O. pleschanka*. In Nakhichevanskaya ASSR (south of Caucasus), obvious passage was recorded one year in early May when most local birds already breeding. (Loskot 1983.)

♂♂ recorded earlier on spring passage and arrive on breeding grounds before ♀♀ (e.g. Moreau 1927). In Cyprus, examination of birds caught by bird-limers showed that 50% of ♂♂ had arrived by 7 April, though no ♀♀ recorded before 6 April and 50% not until 18 April (Horner and Hubbard 1982). ♀♀ generally arrive *c.* 10 days after ♂♂ in Nakhichevanskaya ASSR (Loskot 1983).   DFV, MGK

Food. Almost entirely insects. Taken mainly from bare ground or short vegetation up to 10 cm tall. Usually watches for prey from perch up to 3 m above ground; flies down to make capture, returning to same perch (Witherby 1903; Moore and Boswell 1956; Cornwallis 1975; Loskot 1983). Light weight allows it to perch on

flimsy vegetation unusable by other heavier *Oenanthe* (Cornwallis 1975; Loskot 1983). Of 271 perches used by 5 birds in south-west Iran, 40% on vegetation, 35% on boulders or rocky outcrops, 10% on stones or clods, and 15% on ground; 88% of 60 capture attempts involved a short flight and all ended with prey being taken from bare ground (Cornwallis 1975). Birds also make short flights from perches and catch prey in flight like flycatcher (Muscicapidae); when prey swarming, this becomes main feeding method (Moore and Boswell 1956; Loskot 1983). Recorded flycatching from river bank over water 1 hr after sunset (Rothschild and Wollaston 1902). Where vegetation sparse birds may forage over a large area by hopping on ground. Where dense vegetation inhibits such movement and perches are scarce, searches for prey from the air by hovering steadily at height of 1-1·5 m, most commonly in steady moderate wind. Prey hidden at base of rocks or vegetation may be flushed with sharp wing-beats (Loskot 1983). Small prey swallowed whole in one abrupt movement, always head first. Large prey beaten first against stone (Loskot 1983). For development of feeding skills in young, see Social Pattern and Behaviour.

The following recorded in diet. Invertebrates: mayflies (Ephemeroptera), dragonflies (Odonata: Calyopterygidae, Libellulidae), grasshoppers, etc. (Orthoptera: Gryllidae, Tettigoniidae, Acrididae, Tridactylidae, Tetrigidae, Oecanthidae), mantises (Mantidae), bugs (Hemiptera: Scutelleridae, Pentatomidae, Rhopalidae, Lygaeidae, Tingidae, Reduviidae, Nabiidae, Cydnidae, Cicadidae, Cicadellidae, Psyllidae), lacewings, etc. (Neuroptera: Ascalaphidae, Myrmeleontidae), Lepidoptera (Nymphalidae, Pieridae, Lasiocampidae, Pyralidae, Tortricidae, Arctiidae, Noctuidae, Lymantriidae, Geometridae, Yponomeutidae), flies (Diptera: Tipulidae, Asilidae, Syrphidae, Muscidae), Hymenoptera (Ichneumonoidea, Chalcidoidea, Proctotrupoidea, Chrysididae, ants Formicidae, bees Apoidea), beetles (Coleoptera: Carabidae, Hydrophilidae, Histeridae, Staphylinidae, Scarabaeidae, Byrrhidae, Melyridae, Bostrychidae, Dryopidae, Buprestidae, Elateridae, Meloidae, Tenebrionidae, Nitidulidae, Alleculidae, Coccinellidae, Cerambycidae, Chrysomelidae, Curculionidae), spiders (Araneae), harvestmen (Opiliones), woodlice (Isopoda), millipedes (Diplopoda), centipedes (Chilopoda), snails (Gastropoda). Plant material mainly fruits: of mulberry *Morus*, blackberry *Rubus*, cherry *Prunus*, sumach *Rhus*, and *Daphne*. (Witherby *et al.* 1938*b*; Blanchet 1951; Pek and Fedyanina 1961; Cornwallis 1975; Stepanyan 1978*b*; Loskot 1983.)

In south-west USSR, at 6 sites over 8 years, late April to mid-June, 47 stomachs contained 931 items of which 34·1% (by number) Hymenoptera (29·0% Formicidae), 22·9% beetles (8·9% Curculionidae, 5·5% Carabidae, 3·1% Scarabaeidae, 2·1% Chrysomelidae), 13·2% Lepidoptera (7·9% larvae, 2·0% Pieridae), 11·6% Hemiptera (4·0% Pentatomidae, 2·3% Lygaeidae), 6·7% Diptera,

6·6% Orthoptera, 1·5% spiders, 1·0% lacewings, 0·8% snails, 0·2% berries, and 4·5% others; after May, mulberries became more important for birds nesting near human settlements; size ranged from 3 mm (ants) to 40 mm (caterpillars and centipedes), but mostly 5-12 mm (Loskot 1983). In south-west Iran, fruit also becomes more important in late summer and diet narrows: 135 items from 11 stomachs comprised 42% (by number) ants, 14% beetles, 13% Hemiptera, 7% insect larvae, 7% bees and wasps, 5% other Hymenoptera, 3% Lepidoptera, 3% Orthoptera, 1% Diptera, 1% spiders, and 4% unidentified insects; in August, 77 items from 8 stomachs comprised 61% ants, 14% fruit and seeds, 10% Orthoptera, 7% Hemiptera, 3% beetles, 1% spiders, and 4% unidentified insects; in September, 65 items from 2 stomachs comprised 68% ants, 23% berries, 8% other Hymenoptera, and 1% unidentified insects; during March-June, modal length of prey was 5-10 mm, but during July-October mode fell to 2-5 mm (Cornwallis 1975). For further study emphasizing importance of beetles in diet, see Pek and Fedyanina (1961) for Kirgiziya (USSR). In winter quarters, said to feed on ants, beetles, a few Hemiptera and mites (Acarina), and some seeds (Witherby *et al.* 1938*b*).

Nestlings fed mainly caterpillars, Orthoptera, beetles, spiders, and a few Diptera and adult Lepidoptera. Near Megri (southern Armeniya), early June, older young (6-10 days) given significant proportion of small cherries (Loskot 1983). Feeding rate increased with age of nestlings from 5 to 30 feeds per hr (Loskot 1983) and from 7·5 per hr at 6 days old to 14·1 per hr at 11 days old (Aubrecht 1978). Feeding rates showed marked decrease during hottest part of the day, 13.00-15.00 hrs (Aubrecht 1978).   PJE

Social pattern and behaviour. Major study of *melanoleuca* in Transcaucasia and other south-western parts of USSR by Loskot (1983). For other important data, including comparisons with closely related (and in many respects similar) Pied Wheatear *O. pleschanka*, see sources cited under that species (p. 796).

1. Solitary and apparently territorial in winter (Salvan 1967-9; Smith 1971), but no detailed information. In Sénégal, shares habitat with Wheatear *O. oenanthe*, but density much lower: e.g. 1 *O. hispanica* to 10-15 *O. oenanthe* over *c.* 25 ha of savanna (Morel and Roux 1966). Winter site-fidelity between years recorded (Curry-Lindahl 1981). Occurs singly or in small, loose-knit parties prior to and during spring and autumn migration (Koenig 1924; Moreau and Moreau 1928; Chapman and McGeoch 1956; Lamarche 1981); for fuller details, see Loskot (1983) where loose association with other migrant *Oenanthe* also reported. In Tunisia small parties occasionally noted in early spring comprised only ♂♂ (Whitaker 1905) and in Iraq, in August of one year, adult ♂♂ found mainly above *c.* 1500 m, ♀♀ and immatures mainly on the plain (Moore and Boswell 1956). BONDS. Nothing to suggest mating system other than monogamous, but no details on this or on length of pair-bond. However, in south-west Iran, colour-ringed ♂ held same territory from May until mid-August by which time ♀ had left (Cornwallis 1975), suggesting pair-bond severed after

breeding. Both sexes feed young (e.g. Bannerman 1954, Aubrecht 1978); see also Relations within Family Group (below). For hybridization with *O. pleschanka*, see that species (p. 797). Age of first breeding 1 year (Loskot 1983). BREEDING DISPERSION. Solitary and territorial. Less tolerant of high density than *O. pleschanka* and territorial system stricter. Nevertheless, concentrations (neighbourhood groups: see below) with small territories do occur if (e.g.) suitable habitat restricted or relief impedes visual contact (Panov and Ivanitski 1975*b*; Panov 1978; Loskot 1983). In preferred USSR habitat of river valleys, irrigation ditches, and steep slopes of gorges, pairs frequently 100-120 m apart, locally 70-80 m. Densities tend to be lower where habitat overgrown; in sparse xerophytic (mainly juniper *Juniperus*) woods, pairs usually 300-500 m apart; 80-100 m on broken cliffs, river terraces, and steep slopes in such habitat (Panow 1974; Loskot 1983). Territory established and defended by ♂; used for pair-formation, copulation, nesting, and some feeding (see below). Size varies in different habitats. According to Panov (1978), not normally less than 1 ha, average in favoured habitat being 1·7 ha (1·0-2·4 ha, *n* = 8); compare *O. pleschanka* (p. 798). However, often difficult to determine size accurately as limits (sometimes marked by physical features) not clearly defined along whole course and not constant in time (Loskot 1983). Early-arriving ♂♂ (especially those older than 1 year) not infrequently have larger territory than that used later for nesting by pair: 2·12 ha (0·9-3·2, *n* = 16) initially, 0·95 ha (0·4-2·2, *n* = 28) for nesting. In early period after arrival of ♀, ♂ concentrates defence on small area around her rather than on territory limits. With start of building (perhaps earlier: see Panov 1978), nest-site becomes centre of pair's activity and limits of final territory established. New territory may be part of that initially occupied or even partly extend beyond its limits. (Loskot 1983.) In USSR, territories in favoured habitat may include not only slope (with crevices for nesting), but also 1-2 trees or bushes on adjoining damp meadow where food abundant. Density 2-3 times higher in such moist habitat than in adjacent arid terrain. On slopes, neighbouring territories may be tightly clumped, contiguous and compressed (see illustration in Panow 1974). Maximum 70-100 m wide on slope and strictly defended there; may extend *c.* 200-300 m in direction of meadow where boundaries of 'feeding territories' not so clearly delineated and less actively defended. Territories in dry areas more rounded and *c.* 150-200 m across. Later-arriving birds tend to establish territory between already existing ones whose owners forfeit part of theirs and clump together more. (Panow 1974.) Strip of neutral ground typically separates neighbouring territories (Loskot 1983). In Bulgaria, one territory occupied by another ♂ within a few hours of owner's death. 2 territories often have territory of *O. oenanthe* between them (Baumgart 1971). In Transcaucasia, late-arriving *O. hispanica* sometimes settle within large territories of Red-tailed Wheatear *O. xanthoprymna chrysopygia* and *O. finschii*, but mostly remain on periphery (Panow 1974); see also Antagonistic Behaviour (below) and *O. pleschanka* (p. 798). Some evidence of fidelity to nesting territory: 2 colour-ringed Iranian ♂♂ returned to occupy approximately same ground in successive years (Cornwallis 1975). In Guadalajara and Soria provinces (Spain), density 25-38 pairs per km² (Suárez Cardona 1977). 1·5 pairs per km² in degenerate woodland of Morocco (Thévenot 1982). In Azerbaydzhan (USSR), 4-26 birds per km² (Drozdov 1965, which see for habitat details). In Transcaucasia (Armeniya and Nakhichevanskaya ASSR), 1-8 pairs per km on transects; in one area, 5 ♂♂ in less than 3 ha. Numerous at foot of cliffs where large boulders (e.g. 18 pairs in *c.* 30 ha, 17 pairs along 3 km), also on edge of cultivated land (e.g. 2-6 pairs per km); 2·5 pairs per km in sparse (mainly *Juniperus*) woods. 2 areas in western Kopet-Dag held 5 pairs along 1·2 km and 2 areas along 11 km. In north-west Azerbaydzhan, *O. hispanica* recorded in small mixed neighbourhood groups with *O. pleschanka*. In one area by river, 21 pairs in 1·5 km², ♂♂ comprising 7 *O. hispanica*, 5 *O. pleschanka*, and 9 hybrids; in another year, 4·2 km² in same general area held 22 pairs (11, 7, 4). For proportions of hybrids in overlapping populations, see *O. pleschanka* (p. 797). (Loskot 1983.) ROOSTING. Nocturnal and probably solitary, but little information. July vagrant in Scotland apparently roosting in old bothy (Smith *et al.* 1970). Captive brood of 5 always roosted huddled together on perch (Loskot 1983). In Sudan, February–March, birds most active in evening, including up to 1 hr after sunset (Rothschild and Wollaston 1902). Rain-bathing recorded (Panow 1974); for characteristic posture, see Loskot (1983) who noted that birds will also hide in holes when it rains. Sun-bathing one of most frequently performed types of comfort behaviour in captive birds at 2–3 weeks old (Loskot 1983, which see for illustration).

2. Behaviour has much in common with *O. pleschanka*; see that species (p. 798). Can be shy and difficult to approach, though normally not flying far if disturbed (e.g. Witherby 1903, Whitaker 1905). Vagrants recorded as allowing approach to within *c.* 15–40 m (Baumann 1951; Andris 1974; Swelm 1974) or to within *c.* 3–5 m, then flying up to 20 m away (Björklund 1976; Duda 1978). When flushed near ground in southern Turkey, birds tended to fly to tree top (Hollom 1955). Swedish vagrant gave alarm-call (see 3 in Voice) when Great Grey Shrike *Lanius excubitor* nearby (Björklund 1978). Unlike (e.g.) *O. oenanthe*, does not bob when excited, tending rather to flick tail and slightly raise wings, then to wag tail slowly; captive birds showed hint of bobbing when wing-flicking marked (König 1964a); bobbing mentioned without further comment by Mountfort (1958). FLOCK BEHAVIOUR. In small flocks of spring migrants, birds tend to be 20–30(–50) m apart, but integrity of group still maintained: between short breaks for feeding, birds move on more or less as a flock, much as in *O. oenanthe* and Isabelline Wheatear *O. isabellina*. Where migrants gather, birds stay at least 3 m apart, usually over 5 m. Aggression confined to low-intensity distance threat: abrupt bows, tail-fanning, sideways wing-flicking, and sometimes also call 2 given (see Voice). No defence of individual-distance such as recorded in *O. oenanthe* (Loskot 1983); see Antagonistic Behaviour (below). SONG-DISPLAY. ♂ sings (see 1 in Voice) from perch (rock, overhead wire, etc.) spending much of day there and singing loudly; usually 1–3 favourite perches in open parts of territory (Panow 1974; Bergmann and Helb 1982; Loskot 1983). Also performs Song-flights (probably to attract mate: see below) over territory. With fluttering flight and tail widely fanned, ascends rather jerkily at *c.* 60° to *c.* 20–30(–50) m, singing loudly; makes slight deviations (1–2 m) to side. For up to 1½ (–2) min circles or flutters about over territory, making sharp turns (for illustration, see Panow 1974). Song-flight may appear slow, but actually performed quickly and often higher than other *Oenanthe* except *O. pleschanka* (Loskot 1983). In Turkey, Song-flight long, dancing, and as wide-flung as in Greenfinch *Carduelis chloris* (Hollom 1955). Descent usually rapid, with closed-wings plummet for last 5–10 m, bird typically landing on take-off point or another song-post. Stops singing during final dive, but may resume immediately after landing (Loskot 1983). For further description, see Schiebel (1908). Song-flight highly infectious so that 3–4 neighbours not uncommonly perform together. For role of Song-flight in territorial conflicts, see Antagonistic Behaviour (below). Most activity (4–6 Song-flights per hr) from *c.* 5–7 days after arrival, though ♂ pairing

up soon after arrival may perform no Song-flights at all (Loskot 1983). In Turkey, one ♂ repeatedly sang in flight while ♀ nest-building (Wadley 1951). Sings especially during *c.* 20–30 min before dawn (e.g. in Iraq, from 04·30 hrs in May: Moore and Boswell 1956), even performing Song-flight when still almost completely dark. As in other *Oenanthe*, such song perhaps serves to attract down migrating ♀♀ (Panow 1974; Loskot 1983). In Aegean, April, song from perch noted well into evening (Wettstein 1938); given also at night according to Reiser and Führer (1896). Pre-dawn song most melodious and intense before main arrival of ♀♀, but continues, gradually waning, to end of breeding cycle (Loskot 1983). In Montenegro (Yugoslavia), no song noted after eggs hatched (Reiser and Führer 1896). Song from wintering birds in Sudan perhaps associated with defence of territory (Smith 1971; see part 1). Migrants (*melanoleuca*) occasionally sang in Egypt in spring (Moreau and Moreau 1928) while apparent Subsong (see 1c in Voice) noted from feeding birds in Transcaucasia in October (Raethel 1955). For further use of song in antagonistic and heterosexual encounters, see below and 1 in Voice. ANTAGONISTIC BEHAVIOUR. (1) General. First ♂♂ to arrive on breeding grounds immediately occupy best territories. Usually 1–2 days elapse before active display and defence begin; most time spent on song-posts. Vigorous fights between territorial ♂♂ are a feature only of demarcation period; brief (border) disputes more likely once limits established. ♂♂ in conflict more excitable and flamboyant than *O. pleschanka*. Aggression declines markedly after pair-formation; ♂ spends less time on song-perch and only occasionally sings in flight; caged bird placed in centre of territory ignored. Few territorial disputes after building and none as a rule after hatching. ♀ generally less demonstrative than ♂ and shows little aggression. (Panow 1974; Panov 1978; Loskot 1983.) (2) Threat and fighting. Intruders usually fly rapidly across boundary straight for ♀ who tends to flee and hide (as she typically does during ♂–♂ conflicts). Territory-owner joins in chase and intruder is quickly expelled (owner staying within his territory) or retreats, each then not infrequently performing Song-flight over own territory. Straight after arrival, also later when new birds attempting to establish themselves and territory limits being redefined, prolonged conflicts (sometimes over hours) with actual fights may arise. One ♂ persistently entered neighbour's territory, eventually making substantial part a neutral zone where able to stay long without encountering active resistance. Territory limits normally respected such that birds tend not to cross line even when owner absent. However, frequent transgressions by ♂♂ occur when ♀♀ incubating and *O. hispanica* generally less strict about violations than (e.g.) the more aggressive *O. finschii*. Unpaired birds also spend much time patrolling and displaying in or close to neutral ground between territories: birds with territory *c.* 300–500 m from nearest neighbour would fly there every 2–3 hrs for *c.* 7–10 min, then return, singing loudly. Patrolling continued up to incubation where density high. Conflicts during territory demarcation involve fast, low chases, birds giving accelerated and muted song alternating with series of clicks merging into rattle (see 2–3 in Voice). Regular feature (also of ♂–♀ interactions: see Heterosexual Behaviour, below) are Quiver-glide flights (bouts of shallow wing-beats and gliding), sometimes including steep and stepped ascent with tail closed; not recorded in other *Oenanthe* except (rarely) in *O. pleschanka*. Also typical are long Song-flights performed by 2 ♂♂ simultaneously or one after another: fly at *c.* 10–20 m, singing incessantly. After landing, one (of 2 unpaired ♂♂) performed Dancing-flight (see Heterosexual Behaviour, below) near rival; in another case, birds fought. When threatening rival on ground

A   B

may adopt Head-up posture (Fig A, right), also moving tail up and down, Hunched-posture (Fig B), and many variants (for full details see Panov 1978). Occasional fights proper may start in the air and continue on ground, birds pecking furiously, claws interlocked, and singing loud Subsong without pause (see 1c in Voice). In later confrontation on established limit, opponents typically ruffle belly feathers, and will flick wings up, ground-peck, grass-pull, or enter one after another crevices in rocks (sometimes head-swaying while bowed forward: see Panow 1974), each then returning to territory. One conflict of 5–6 hrs between new ♂ and established territory-owner involved mostly Song-flights, also a damaging fight, and ended in victory for new arrival (see Panow 1974). When ♀ first arrives in an occupied territory, other unpaired ♂♂ will intrude in attempt to court her; ♀ usually hides while resident ♂ evicts intruders. Further disputes with ♀ as focus of attention may arise later when she makes longer flights in search of nest-site; highly agitated ♂ will attempt to drive his mate back if she inadvertently crosses limit, but with unpaired ♂♂ tending to stay near territory occupied by pair, chases develop (2–3 ♂♂ after ♀). Pair then normally return to territory, ♂ singing loudly; if ♀ hides, ♂♂ land within a few metres, then fairly quickly disperse. ♀ unlikely to intervene when ♂ in conflict with intruder, but just after pair-formation may attack and expel strange ♀, this sometimes involving chases, and song from both birds. Panov (1978) recorded conflict between 2 ♀♀ lasting more than 1 day, birds fighting fiercely on ground and in the air; perhaps attempt by one to re-occupy her territory of previous year (Loskot 1983). Like *O. pleschanka*, *O. hispanica* subordinate to other *Oenanthe*: e.g. usually leaves without resistance when confronted by *O. finschii* and *O. xanthoprymna chrsopygia* (Cornwallis 1975; Panov and Ivanitski 1975a); see also Raethel (1955). In northern Greece, once recorded chasing *O. oenanthe* (Cornwallis 1975). Unpaired and excited ♂♂ will perform Song-flight over territory if (e.g.) Swift *Apus apus* or swallow (Hirundinidae) appears; latter (Moore and Boswell 1956) and Redstart *Phoenicurus phoenicurus* sometimes attacked. (Panow 1974; Panov 1978; Loskot 1983.) HETEROSEXUAL BEHAVIOUR. (1) General. ♂♂ usually arrive on breeding grounds well before ♀♀; on Mangyshlak, first ♀♀ sometimes arrive with first ♂♂ (Panov and Ivanitski 1975b). A few days or even c. 1 month may elapse between territory occupation and pair-formation. In Nakhichevanskaya ASSR, pair-formation typically extends over c. 1 month (e.g. 1st ♂ 28 March, 1st ♀ 7 April; last arrivals and pair-formation 11 May). Synchrony in nest-building noted in a Spanish population (Suárez Cardona 1977). On Mangyshlak, marked degree of synchrony with *O. pleschanka* in pair-formation, nest-building, and laying (Panov and Ivanitski 1975b). Display (including copulation) often takes place in evening (tendency more marked in *O. pleschanka*), birds less active in early morning. Aerial-dance (see below) characteristic mainly of nest-building and laying periods, but noted up to hatching. (Panow 1974; Panov 1978; Loskot 1983.) (2) Pair-bonding behaviour and nest-site selection. Account based on Panow (1974), Panov (1978), and Loskot (1983). ♀ will fly through several territories, perhaps eventually selecting one with good supply of potential nest-sites. Often moves on: e.g. one occupied territory visited by 7 ♀♀, 17 April–3 May, the last staying. Searching ♀ conspicuous and occasionally sings. ♀ probably attracted by ♂'s Song-flights which are characteristic feature of pair-formation and period thereafter (see Song-display, above). According to Panov (1978), first meeting takes place in the air; however, this doubted by Loskot (1983) who also considered differences in pair-formation components between *O. hispanica* and *O. pleschanka* as detailed by Panov (1978) to be exaggerated. Data from Loskot (1983) suggest considerable variation in pair-formation: e.g. one sequence lacked first meeting in the air, chase, Dancing-flight of ♂ near ♀, and Nest-showing (see below). At least in some cases, ♂ changes from Song-flight to pursuit of ♀ which may encourage her to land in his territory. If ♀ flies on, ♂ will pursue her far beyond his own territory, owners of other territories crossed then also ascending into Song-flight. After landing, ♀ (sometimes singing, see 1d in Voice: Panow 1974) usually takes initiative, but her approach may cause ♂ to fly off. ♂ rarely approaches ♀ in fast flight as this always causes her to flee and (if chased) to hide (see Antagonistic Behaviour, above). Chases not all that frequent during pair-formation (Panov 1978), but occur during first few days after ♀'s arrival, typically before dawn and at dusk (Loskot 1983). ♂'s approach more likely to be in (usually silent) Quiver-glide flight. On ground, ♂ will then assume posture similar to Head-up posture, but head less steeply raised and breast not lowered, alternating with Dancing-flight near ♀ (for schematic depiction, see Panow 1974): typically performed in figure-of-eight c. 10 m across, bird flying fast and horizontally at 0·5–0·8 m if ♀ (as usually) on ground, vertically if ♀ on boulder, etc. ♂ sings constantly and loudly (or gives muted song and calls 2–3) and will quiver wings after landing in Head-up posture or similar. ♀ sometimes also sings, but threatens to peck ♂ if he tries to get close. ♂ may then ascend into Song-flight while ♀ sings briefly from perch. Nest-showing by ♂ often follows. ♀ moves about territory after ♂ or takes initiative, searching widely. (Panow 1974; Pineau and Giraud-Audine 1979; Loskot 1983.) Nest-showing ♂ will adopt Head-up and other postures in which wings spread (sometimes quivered) and held forward. Most striking is Cross-posture (not confined to Nest-showing, gaining rather in importance with increased intensity of ♂–♀ interactions): ♂, often perched (Fig C) will also swing body up and down through 15–60 (–70)°. Nest-showing may also involve (e.g.) song (from both sexes, ♂ also from within hole looking out), Dancing-flight of ♂, and picking up of material by ♂ (for further details see

C

D

Panov 1978, also Loskot 1983). ♂ tends to enter deep holes staying up to 1 min. ♀ may approach and stay motionless by entrance, ♂ then emerging to fly on to another hole for repeat performance, etc., or ♀ enters hole after ♂ has emerged and stays there long time. ♀ will also ruffle and shake feathers while perched on stone, look down (no details), bow deeply, abruptly fan tail and open wings, occasionally giving call 2, then suddenly fly 10–70 m away (Loskot 1983). Many ♂♂ stick to one hole, showing it repeatedly to ♀ even though she may meanwhile rove about in search of other sites. Final selection by ♀ (e.g. Suárez Cardona 1977). Compared with *O. finschii*, *O. hispanica* displays less by holes and sites less constant (Loskot 1983). When ♀ moving about and feeding in territory during first few days, ♂ stays within *c.* 5 m and occasionally sings briefly (Loskot 1983). Among *Oenanthe*, Courtship-flight by pair is known only in *O. hispanica* and *O. pleschanka*: performed in evening, birds flying low and slowly with shallow, winnowing wing-beats, ♂ *c.* 0·5–1 m behind ♀ (Panov 1978); usually ends with short chase and Dancing-flight of ♂ and ♀ (Loskot 1983). Pair-formation apparently completed in a number of bouts with overall duration (including pauses) of *c.* 2–3 hrs. Unlike in *O. pleschanka*, behaviour of ♂ and ♀ co-ordinated, birds mostly maintaining visual contact. Highly excited ♂ remaining long unpaired will display at (e.g.) ♀ Trumpeter Finch *Bucanetes githagineus*. (Panov 1974; Panov 1978; Loskot 1983.) (3) Courtship-feeding. Not certainly recorded, but ♂ once seen passing item (not definitely food) to ♀ during break from incubation (Panow 1974). (4) Mating. Long bouts of courtship, apparently typically at dusk, take place when nest completed. ♀ will indicate her readiness to copulate by adopting Spread-wings posture (Fig D), wings being also quivered. ♂ also quivers spread wings, then takes off (all with same small-amplitude wing-beats) and lands slowly by ♀ who allows him very close (almost touching) before running away a little, thereby initiating repeat performance, and so on. Usually followed by aerial pursuit, ♀ flying slowly with shallow wing-beats, and sometimes hovering. ♂ gives quiet and persistent Subsong throughout (Panow 1974); presumably leads to copulation eventually, but no details. In one (perhaps atypical daytime) case, ♂ performed Aerial-dance (probably typical prior to copulation) around perched ♀ who flew to ground and wing-shivered (presumably in Spread-wings posture), then crouched; ♂ copulated and flew off low in Quiver-glide for *c.* 40 m, ♀ joining him after *c.* 20 s inactivity (Loskot 1983). (5) Behaviour at nest. ♀ starts to build *c.* 3–10 days after site selected. In first days after pair-formation, ♂ will pick up and drop twigs, not take them to nest; accompanies ♀ while she is collecting material (more

so than in *O. pleschanka*) or follows her movements from perch, occasionally singing briefly. For further details of nest-building, see Loskot (1983). On seeing intruder, ♂ will change to loud song and immediately launch expelling attack. Eggs laid 1–6 days after completion of nest. In early part of incubation, ♀ leaves nest 2–3 times per hr for *c.* 7–10 min; from *c.* 6–7 days, usually once per hr. Typically uses low gliding flight when leaving nest; ♂ will give a rattle call (perhaps variant of calls 2–3: see Voice), approach ♀ in Quiver-glide and accompany her during brief bout of feeding. ♀ preens before returning, via 3–4 regular perches, to nest. ♂ on look-out sings less than before (especially around midday), but will display at other ♀♀ when his mate is incubating. (Panow 1974; Loskot 1983.) RELATIONS WITHIN FAMILY GROUP. Brood hatches within less than 24 hrs. Eyes of young open by day 5 (Loskot 1983, which see for further details of physical development). Wait for parents at nest-entrance when 10 days old, and usually leave at 11–12 days (Loskot 1983). Fledglings (also of *O. pleschanka*) beg for food with wings spread and slightly raised (less so than in Desert Wheatear *O. deserti*); also call with bill wide open, move head, and beat wings. Posture indicating refusal to take food (satiety) as in *O. deserti* (p. 826). In captive young, wing-flicking (see Food) develops by day 13. Can fly at 12 days, tackle quite large mobile prey at 14–16 days, and catch insects in the air at 18 days. Wild birds independent at 20–22 days. For comfort behaviour, see Roosting (above) and Loskot (1983). ANTI-PREDATOR RESPONSES OF YOUNG. Crouch in nest from 7 days and may leave if disturbed at 9 days. Captive young (19 days old) crouched motionless when Buzzard *Buteo buteo* flew over, not moving until it was out of sight (Loskot 1983). PARENTAL ANTI-PREDATOR STRATEGIES. (1) Passive measures. ♀ tends to leave nest as soon as man in view, and stays away until he leaves vicinity (Koenig 1895). (2) Active measures. At approach of man, sentinel ♂ will give alarm-calls (see 2–3 and 5 in Voice) from look-out near nest, also perform Song-flight (Reiser and Führer 1896), sometimes while moving towards intruder (Stadler 1952). Like some other *Oenanthe*, ♀ occasionally sings when disturbed at nest (see Panov 1978).

(Figs by K H E Franklin: from drawings in Panow 1974 and Panov 1978.)   MGW

Voice. Generally quiet outside breeding season; call 2 given then, but rarely (Moore and Boswell 1956; König 1964*a*; Björklund 1978); sometimes sings in winter (see Social Pattern and Behaviour). Repertoire and use of calls much as in Pied Wheatear *O. pleschanka* (Panov

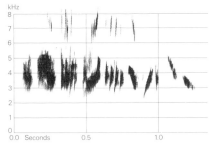

I   P A D Hollom   Israel–Lebanon border   April 1980

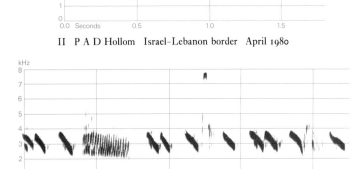

II   P A D Hollom   Israel–Lebanon border   April 1980

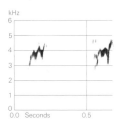

III   H-H Bergmann   Greece   May 1976

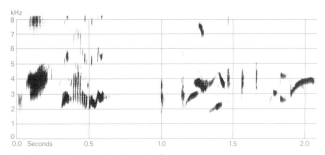

V   E D H Johnson   Turkey   April 1972

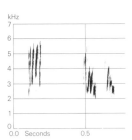

IV   H-H Bergmann   Greece   May 1976

VI   E D H Johnson   Turkey   April 1972

and Ivanitski 1975*b*; Panov 1978). For additional sona-grams, see Panov (1978) and Bergmann and Helb (1982).

CALLS OF ADULTS. (1) Song of ♂. Variants occur (as in *O. pleschanka*): short Territorial-song (at least this given also by ♀), longer version of same, and both quiet and louder Subsong, though many transitional forms (Panov 1978). (a) Territorial- (Advertising-)song. Close to main variant of *O. pleschanka*, harsh, scratchy quality of song in the 2 species making them readily dis-tinguishable from other *Oenanthe* (Loskot 1983; L Corn-wallis), but see below for variation. Typically comprises short, loud phrases of 7–11 units; lasts *c.* 1–1·5(–2) s. A cheerful, rapid warbling, rich and full-toned, rising and falling, and with characteristic dry, buzzy, or scratchy and raucous quality; complex units show much rapid modulation of frequency (vibrato) and amplitude. Unlike in Wheatear *O. oenanthe*, timbre dull, harsh, melancholy, and almost toneless, with gurgling and grating and occasional pure tones (Stadler 1952, which see for renderings). Also includes mimicry (see below) and therefore much variation within and between individuals,

though no evidence of constant differences between nominate *hispanica* and *melanoleuca*. Call-type units are interspersed at regular intervals between phrases, repeated frequently at start of phrase, or given separately; some are mimicry, e.g. of 'ch-si' of White Wagtail *Motacilla alba* in recording by E D H Johnson (J Hall-Craggs, P J Sellar), or of Skylark *Alauda arvensis*. (Panov 1978; Bergmann and Helb 1982; Jonsson 1982; Loskot 1983.) In Spain, Turkey, and Middle East, scratchy warbling like run-down of clockwork toy: high-pitched 'schwerr schwee schwee-oo' or 'cher-er-er chee-ou', also a dis-tinctive 'whit-whit' like Contact-call of Swallow *Hirundo rustica* (Hollom 1955, 1959; Mountfort 1958; P A D Hollom). Loud, spluttering harsh warbling phrases of this type evident in recording (Figs I–II). Both phrases shown end in 2 tones (lacking otherwise-typical frequency and/or amplitude modulation) over a descending pitch interval. 2nd phrase (Fig II) especially strongly suggests, after faint 'zit' or 'plit' sound (of which many in recording), buzzy descending phrase of Whitethroat *Sylvia communis* (J Hall-Craggs, M G Wilson). Similarly, in eastern Bulgaria, 6 rapid and high-pitched noisy units like loud but otherwise perfect copy of *S. communis*, with pitch descent after highest, loudest 2nd unit. Song-phrase preceded by 1–3 double units (often mimicry of Red-rumped Swallow *H. daurica*) or sometimes 2–3 groups of these *c.* 1 s apart, followed by pause of *c.* 2–3·5 s (Schubert and Schubert 1982). Fig III shows 2 mimicked Flight-calls of *H. daurica*, and at least the 1st call (perhaps all 3) in Fig IV is a copy of Contact-call of House Martin *Delichon urbica* (Bergmann and Helb 1982; M G Wilson). See also (e.g.) Moreau and Moreau (1928), Mackintosh (1944), Wadley (1951), Moore and Boswell (1956), Dandl (1957a), and Vere Benson (1970). Like *O. pleschanka*, an excellent mimic of wide range of bird species. In USSR, species most frequently copied are Swift *Apus apus* and hirundines (Hirundinidae) (Panow 1974; Loskot 1983). Series of short Territorial-songs with short pauses and including call 6 given during territory demarcation and (often with mimicry of *A. apus*) pair-formation. Quiet and rapid or loud and short Territorial-songs given also prior to copulation, but less commonly than variant 1b (Panov 1978). (b) Long (continuous) variant of Territorial-song given (e.g.) prior to copulation: loud, and typically incorporates call 7 (Panov 1978). Recording, perhaps of this variant, reveals succession of widely spaced harsh and singing 'chrrrrroooeee', then loud repeated units reminiscent of Song Thrush *Turdus philomelos*; Fig V shows rapid accelerating series of such units with a rattle and a few other harsh sounds (M G Wilson). (c) Subsong. In recording by P A D Hollom, quiet and rather rambling, garbled Subsong starts with slightly nasal 'bizz' sounds and continues through finch-like twitters and warbling to louder, purer units (M G Wilson); reminiscent of Twite *Carduelis flavirostris* (P J Sellar). Quiet and beautiful Subsong given by feeding birds in Transcaucasia (USSR), October, included mimicry of Bee-eater *Merops apiaster*, Crested Lark *Galerida cristata*, and *H. rustica* (Raethel 1955). Subsong directed at ♀ during pair-formation always contains clicks (Panow 1974: see call 2); in sonagram from E N Panov, mostly 2–3 clicks together. Apparently louder variant of Subsong (Fig VI) rendered 'chrreee hoorrrwooeee', followed by attractive tinkling mixed with wide-band, short-duration 'tk' clicks (J Hall-Craggs, M G Wilson). (d) ♀ also gives short Territorial-song; muffled and accelerated during pair-formation according to Loskot (1983). Classification of other calls mainly after Panov (1978). (2) Clicking calls. (a) Muffled click. General quality of call, contexts, and overlap with call 2b and 3 much as in *O. pleschanka* (p. 802). As in that species, this or longer variant (call 3) is commonest everyday vocalization. Rendered as follows: 'chek' (Loskot 1983); 'srèk' (Swelm 1974); dry 'dsed' (also call 5) given when disturbed (Bergmann and Helb 1982); in recording, short rasping 'dsched' (Fig VII); low grating 'gsch' or sometimes 'kscheup' (Jonsson 1982). See also Stadler (1952). (b) Clear, loud click.

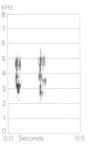

VII H-H Bergmann
Greece May 1976

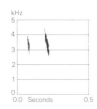

VIII H-H Bergmann
Greece May 1976

Apparently resembles one variant of *O. pleschanka* (see p. 802); separation from call 2a not clarified. (3) Muffled clicks (call 2a) merged into snore-like series (Panov 1978; Loskot 1983). The following presumably belong here: harsh, jarring 'schrrr' (Jonsson 1982); weak, buzzy 'trrrrrt' of alarm from vagrant (Björklund 1978). (4) Threat-rattle. Normally given by ♀ threatening approaching ♂. In sonagram from E N Panov, clicking component of rattle higher pitched and more bunched than in corresponding call of *O. pleschanka*, but not known if this a constant feature. (5) Whistle-call. Similar to *O. pleschanka*. Plaintive whistling 'füüid', typically descending in pitch (Witherby *et al.* 1983b; Bergmann and Helb 1982: Fig VIII), though sometimes rises then falls (E N Panov). Often combined with call 2 in alarm. See also Stadler (1952) and König (1964a). (6) Rasp-call. Like juvenile food-call (see below). A noise-type signal, but tonal variant occurs (Panov 1978). Contexts as in *O. pleschanka* (p. 803). (7) Bugling-call. As in *O. pleschanka*.

CALLS OF YOUNG. Loskot (1983) noted an initially quiet cheeping, louder by 5 days, developing into typical rasping food-call ('bzhee') by day 9, and still used after fledging. Panov (1978) and sonagrams from E N Panov indicate that (as in *O. pleschanka*) adult call 2 serves as contact-call of young, though given only by older nestlings of *O. hispanica*, sometimes in combination with a rasping contact-call typical of many *Oenanthe*; rasping food-call closely similar, but apparently longer (for details, see Panov 1978). 2 captive ♂♂ started to sing at 15 and 17 days, 2 ♀♀ (less regularly and persistently) from 25 and 28 days (Loskot 1983).   MGW

Breeding. SEASON. Algeria and Tunisia: eggs laid from end of April to early June (Heim de Balsac and Mayaud 1962). Spain: laying begins late April or early May (Bannerman 1954); in Guadalajara and Soria (north-central Spain), most clutches laid 16–31 May (Suárez Cardona 1977). Greece: from early May (Bannerman 1954). Armeniya (USSR): laying begins late April or early May (Dementiev and Gladkov 1954b; Loskot 1983). SITE. On ground in shallow hole, under stone, in thick vegetation, or (apparently more in nominate *hispanica* than in *melanoleuca*: Loskot 1983) at base of dense bush. Nest: cup of grass and moss, lined with finer material including hair. Outside diameter 12–12·5 cm, inner diameter 6 cm, depth 2 cm; weight 10 g (Makatsch 1976); see also Loskot (1983). Bits of twig often placed by nest (entrance), forming platform (Suárez Cardona 1977; Loskot 1983), perhaps also base, though this not recorded by Panow (1974). Building: by ♀, taking *c.* 1 week (Panow 1974); for details, see Suárez Cardona (1977) who noted that hollow (under bush) first cleared of plant debris; see also Loskot (1983). EGGS. See Plate 83. Sub-elliptical, smooth and glossy; pale blue, with fine markings of red-brown, sometimes forming cap at broad end. Markings less dense and clear in nominate *hispanica* than in *melanoleuca* (Panow 1974). Nominate *hispanica*: 19·9 × 15·2 mm (17·0–21·6 × 14·0–16·0), n = 156; calculated weight 2·46 g. *O. h. melanoleuca*: 19·7 × 15·0 mm (17·0–24·0 × 14·0–16·0), n = 150; calculated weight 2·4 g. (Schönwetter 1979). Clutch: 4–5(–6). Of 24 clutches, USSR: 4 eggs, 2; 5, 13; 6, 9 (Loskot 1983). Definitely or probably 2 broods (Harrison 1975; Makatsch 1976), though normally 1 throughout USSR range (Loskot 1983). Eggs laid daily. INCUBATION. 13–14 days, usually from last egg. By ♀ only (Bannerman 1954; Loskot 1983). YOUNG. Cared for and fed by both parents (Bannerman 1954). FLEDGING TO MATURITY. Fledging period 11–12 days. Become independent at 20–22 days (Loskot 1983). No further information.

Plumages (nominate *hispanica*). ADULT MALE. In fresh plumage (autumn), forehead, crown, mantle, and upper back deep cinnamon-buff or foxy-buff, forehead and line above lores usually paler cream-buff, crown, nape, and upper mantle with narrow grey feather-tips, appearing suffused sandy-grey. Scapulars variable; sometimes, all feathers deep black with rather narrow cinnamon-buff tips, but tips in some birds broad (especially on innermost), largely concealing black, in others inner scapulars largely cinnamon-buff and sometimes outermost also. Lower back, rump, and upper tail-coverts white; back suffused buff, remainder slightly cream. Narrow strip of feathers backwards from nostril black, not extending to base of culmen, continued into black lore, upper cheek, and ear-coverts; narrow eye-ring black, but black sometimes interrupted above eye. 2 distinct morphs: (1) *'aurita'* or pale-throated morph, with chin and throat cream-buff or cream, black restricted to contrasting mask from nostril to ear-coverts, and (2) *'stapazina'* or black-throated morph, in which chin and upper throat black, contiguous with black mask; black on throat usually does not reach as far back as on ear-coverts. A few birds are intermediate in throat pattern, mainly in north-west Africa; see Geographical Variation. Sides of neck, chest, breast, and flanks warm cinnamon-buff or pink-cinnamon, in black-throated birds often with cream-white gorget below black of throat; belly, vent, and under tail-coverts gradually paler cream-buff or pink-buff; sides of breast black. Tail strongly variable, white with variable amount of black on tip: in all birds, much black on central pair (t1) and outer web of t6, much less or none on t2–t5. In Spain, southern France, and central Italy, maximum extent of black on both webs of t1 (excluding shaft) 46·0 (12) 42–50 mm, on outer web of t6 30·0 (12) 28–33 mm; t3–t5 include 1–2 fully white feathers (1 bird), or are all white except for some black spots on one or both webs (5 birds), or have narrow black band across tip of all feathers, with minimum width of black

band 6·3 (6) 4–8 mm. In Algeria and Tunisia, 39·7 (15) 38–45 mm of t1 black, exceptionally 33–34 mm; 22·6 (15) 18–27 mm of outer web of t6 black; t3–t5 almost fully white (1 bird), with some black spots on tips (3 birds), or with full tail-band with narrowest width 12·5 (19) 7–18 mm. Hence, north-west African birds show less black on t1 and t6 but more on t3–t5 than birds from southern Europe. Flight-feathers and greater upper primary coverts black; outer webs and tips of secondaries narrowly fringed pink-buff or buff-white, tips of inner primaries and primary coverts narrowly edged off-white; inner webs of flight-feathers dark grey, sometimes with white border along basal inner edge. Tertials and upper wing-coverts deep black, often slightly glossy; longer lesser and median coverts usually narrowly tipped pale buff, greater coverts and especially tertials more broadly fringed pale buff or pink-cinnamon along tips and all or part of outer web. Under wing-coverts and axillaries black, tips of lesser coverts often pale buff or white, longer coverts dark grey. *In worn plumage* (spring), cinnamon colours bleached and fringes along wing-coverts and flight-feathers abraded. By about March, crown, nape, mantle, and inner scapulars rather orange-buff, less foxy-rufous than in fresh plumage; forehead and sometimes indistinct supercilium cream, lower throat, sides of neck, and (in pale-throated morph) chin and upper throat cream-white, chest and flanks buff, remainder of underparts cream-buff or cream, traces of pale buff fringes still visible on wing; in April–May, crown and mantle partly bleach to white, but some buff or orange-buff still present, especially on nape and upper mantle; underparts white, usually with cream tinge on chest, flanks, and under tail-coverts, wing fully black or with a few traces of off-white fringes. When heavily worn, June–July, some grey of feather-bases sometimes visible on crown, back, rump, or underparts. ADULT FEMALE. In fresh plumage (autumn), upperparts down to back warm sandy-cinnamon with slight grey tinge on crown and vinous-pink tinge on mantle and scapulars. Rump and upper tail-coverts pink-buff or cream. Supercilium long but faint, buff before eye, pink-cinnamon above and behind eye. Narrow and indistinct eye-ring buff. Lores grey-brown, contrasting slightly with paler front part of supercilium and chin. Rear of cheeks and lower ear-coverts pink-cinnamon, merging into grey-brown or tawny-brown on upper ear-coverts (which are often slightly darker than remaining sides of head). Chin and upper throat pink-buff or cream, some black of feather-bases sometimes visible in black-throated morph; sides of neck, chest, sides of breast, and upper flanks pink-cinnamon, merging into paler pink-buff or cream-buff of lower throat and of underparts down from breast and lower flanks. Tail as adult ♂; extent of black on t1 and t6 similar to ♂, but black on t3–t5 a few mm wider. Flight-feathers and greater upper primary coverts greyish-black; outer webs and tips of secondaries fringed rufous-cinnamon, outer webs and tips of primaries and primary coverts more narrowly pink-buff, pale buff, or pale grey-buff; fringes along primary coverts less than 1 mm wide, usually *c.* 0·5 mm. Tertials and upper wing-coverts dull black or brown-black with broad rufous-cinnamon or pale cinnamon fringes along outer webs and tips (narrower and more pale grey-buff along tips of lesser coverts). Under wing-coverts and axillaries dark grey, fringed pale buff along tips. *In worn plumage* (spring), forehead and crown grey-buff, mantle and scapulars sandy-buff or rufous-cinnamon, rump and upper tail-coverts cream or virtually white. Pink-buff supercilium and especially cream eye-ring more distinct than in worn plumage. Chin and throat either whiter, slightly contrasting with buff chest (pale-throated morph), or with variable amount of dark grey or black mottling, especially on upper throat, extending up to ear-coverts (dark-throated morph); many intermediate. Remainder of underparts cream-buff, cream, or off-white, not contrasting with chest. Tail and wing as in fresh plumage, but buff fringes on wing bleached and abraded, though generally well-visible in spring and not fully worn off until about July. NESTLING. Down long and fairly dense, on upperparts only; mouse-grey. For further development, see Loskot (1983). JUVENILE. Upperparts buff-brown, less vinous than fresh adult ♀, less sandy than worn adult ♀; faint paler buff shaft-streaks and spots on crown, nape, mantle, and scapulars, faint dark edges along feather-tips of crown, mantle, and scapulars. Rump and upper tail-coverts white, feathers loose. Sides of head and neck buff-brown, almost uniform, except for faintly paler eye-ring. Chin, throat, chest, sides of breast, and flanks buff, never black on throat; indistinct dark brown feather-tips on lower throat, chest, and upper flanks, these appearing faintly scalloped; feathering of remainder of underparts fluffy, dirty cream-white. Lesser and median upper wing-coverts grey-brown with large ill-defined buff spot on tip. Greater upper wing-coverts and tertials black-brown with broad (c. 2–3 mm) rufous-cinnamon fringes along outer webs and tips, similar to adult ♀ in fresh plumage, but sometimes broader, extending over whole outer web of inner tertials; flight-feathers as fresh adult ♀, but tips of flight-feathers more broadly fringed pink-buff; fringes along outer webs and tips of greater upper primary coverts often broader than in adult ♀, 1 mm wide or more. Under wing-coverts and axillaries dirty isabelline. FIRST ADULT MALE. Like adult ♂, but juvenile flight-feathers, some or all tertials, all tail, greater upper primary coverts, and variable number of outer greater upper wing-coverts retained, distinctly browner in adult and browner than new black neighbouring 1st adult median and tertial coverts; broadly fringed buff when plumage fresh, and fringes just visible as pale buff or off-white traces when plumage worn. Remainder of plumage as adult ♂, but crown, nape, and mantle often with grey-brown feather-tips, less uniform deep cinnamon-buff when fresh, appearing heavily mottled dusky when plumage worn; scapulars brown rather than black, fringes more broadly sandy-brown and not always fully worn off in spring; ear-coverts and (in black-throated morph) chin and upper throat with broad pale buff feather-tips, sometimes largely concealing black when plumage fresh and still showing as off-white mottling when worn; new lesser and median upper wing-coverts more broadly fringed buff, less uniform deep black when worn. FIRST ADULT FEMALE. Like adult ♀, but part of juvenile feathers retained, as in 1st adult ♂. Often difficult to distinguish, but note slightly greater wear of retained feathers, narrower and more frayed tips of primaries, tail-feathers, and greater upper primary coverts, and relatively wider fringes to primary coverts than adult at same time of year.

Bare parts. ADULT, FIRST ADULT. Iris brown or dark brown. Bill, leg, and foot black. NESTLING. Mouth deep yellow, no spots. Gape-flanges pale yellow. (Witherby et al. 1938b.) JUVENILE. Iris brown. Bill, leg, and foot flesh-grey, flesh-brown, or brown. (RMNH, ZFMK.)

Moults. ADULT POST-BREEDING. Complete; primaries descendant. In hispanica, starts with p1 late June or July, completed with p9–p10 and inner secondaries from mid-August in the small sample examined, and probably sometimes earlier. In northern Iran, single melanoleuca in final stage of moult on 25 July (Paludan 1940), one in Greece nearing completion 30 July (RMNH). In Iran, generally no moult May, one nearing completion 19 August (Vaurie 1949). ADULT PRE-BREEDING. Partial: sides of head, chin, and throat. December–February.

Occasionally, moult apparently partly or fully suppressed. POST-JUVENILE. Partial: head, body, lesser and median upper wing-coverts, tertial coverts, and 1–4(–6) inner greater upper wing-coverts (on average, 2·8 juvenile outer coverts retained, n = 21). Starts shortly after fledging and timing thus strongly variable; starts late June to early August, completed from July. FIRST PRE-BREEDING. As adult pre-breeding but more often partly or fully suppressed, apparently in melanoleuca in particular (of 14 examined, only 4 appeared to have moulted part or all of sides of head, chin, and throat).

Measurements. ADULT, FIRST ADULT. Nominate hispanica. Wing and bill to skull from (1) Iberia and southern France, (2) Tunisia and Italy; other data combined, March–September, skins (RMNH, ZMA). Bill (N) to distal corner of nostril; exposed culmen on average 4·5 less than bill to skull.

	♂		♀	
WING (1)	89·0 (1·68; 17)	86–92	88·3 (1·68; 5)	86–90
(2)	91·4 (2·07; 28)	89–96	87·1 (2·11; 11)	86–90
TAIL	59·3 (1·53; 29)	55–62	56·7 (1·83; 15)	54–60
BILL (1)	17·1 (0·95; 15)	15·9–18·3	17·1 (0·50; 5)	16·5–17·8
(2)	17·0 (0·67; 14)	16·1–17·9	17·2 (0·64; 11)	16·4–18·1
BILL (N)	9·7 (0·55; 29)	8·9–10·5	9·6 (0·45; 16)	9·0–10·3
TARSUS	23·3 (1·02; 29)	21·2–24·7	23·5 (0·70; 16)	22·4–24·6

Sex differences significant for wing (2) and tail. In both races, 1st adult with retained juvenile flight-feathers and tail combined with older birds, though juvenile wing on average 1·7 shorter than adult and tail 2·0 shorter.

O. h. melanoleuca. Wing and bill to skull of (1) Dalmatia (western Yugoslavia), (2) Greece and northern Turkey, (3) south-west Turkey and Levant; other data combined, March–August, skins (RMNH, ZMA).

	♂		♀	
WING (1)	92·0 (1·89; 21)	88–95	88·2 (2·47; 4)	86–91
(2)	89·9 (2·90; 9)	86–94	87·5 (—; 1)	—
(3)	88·8 (2·33; 19)	84–92	86·4 (0·75; 4)	85–87
TAIL	58·6 (2·51; 28)	55–64	56·0 (1·84; 9)	53–59
BILL (1)	16·9 (0·81; 20)	15·9–18·0	16·8 (0·75; 4)	16·2–17·6
(2)	17·2 (0·73; 9)	16·6–18·2	17·6 (—; 1)	—
(3)	16·0 (0·58; 19)	14·9–16·8	16·1 (0·19; 4)	15·9–16·4
BILL (N)	9·1 (0·47; 27)	8·2–9·9	9·3 (0·52; 8)	8·7–9·8
TARSUS	22·9 (0·78; 27)	21·7–24·2	23·0 (0·49; 9)	22·4–23·5

Sex differences significant for wing and tail. Wing: southern Yugoslavia, ♂ 91·0 (1·67; 21) 88–94, ♀ 87·4 (1·24; 12) 85–90 (Stresemann 1920); western Turkey, ♂ 90·8 (2·18; 36) 85–95, ♀ 85 (Weigold 1914); south-west Iran, ♂ 90·8 (2·8; 22), ♀ 89·0 (2·7; 22), northern Iran, adult ♂ 92·2 (1·98; 9) 90–95 (Haffer 1977).

Weights. Nominate hispanica. South-east Morocco, April: 14·6, 21·5 (BTO). Naples area (Italy): April–May, 17·8 (1·81; 14) 13·7–20·1; September, ♀ 26·0 (G Lövei).

O. h. melanoleuca. Turkey (Kumerloeve 1963, 1968, 1969a, 1970a; Rokitansky and Schifter 1971), Armeniya, USSR (Nicht 1961), Iran (Paludan 1938, 1940; Schüz 1959; Desfayes and Praz 1978), Mangyshlak peninsula, east Caspian Sea (Dolgushin et al. 1970), and Kuwait and Sjarjah (BTO), combined. (1) March–April, (2) May, (3) June–July.

	♂		♀	
(1)	16·4 (1·69; 15)	13–19	13·8 (—; 3)	12–15
(2)	17·5 (2·00; 5)	16–21	15·5 (—; 1)	—
(3)	19·5 (2·36; 8)	15–22	17·8 (—; 2)	17–18

Kenya, March: ♂ 18 (Backhurst et al. 1984). For development of nestling, see Loskot (1983); average on 1st day 2·8, on 10th 15·3, on 20th 18·5, on 30th 18·7, on 60th 19·3.

Structure. Closely similar to Pied Wheatear O. p. pleschanka, differing as follows. 10 primaries: in nominate *hispanica*, p7-p8 longest or either one occasionally 0·5 shorter than other, p9 4-7 shorter, p6 1-3, p5 7-10, p4 11-16, p1 18-26; *melanoleuca* similar, but p9 3-6 shorter than wing-tip, p6 1·5-3·5, p5 7-12, p4 12-18, p1 20-26. In both races, p10 reduced, 43-56 shorter than wing-tip, 0·5 shorter to 4·5 longer than longest upper primary covert. Outer web of p6-p8 and inner of p7-p9 emarginated. Middle toe with claw 16·7 (58) 15-18·5.

Geographical variation. Rather marked, with 2 distinct races differing mainly in general colour (both sexes) and in extent of black marks on face and throat (♂ only); also, some variation in tail pattern, in relative numbers of black- and white-throated morphs, and (very slightly) in size. Some individual variation, and difference between races not always marked, e.g. birds from humid coasts of Portugal darker than typical nominate *hispanica* from more arid Spain, approaching *melanoleuca* from eastern Europe in general colour and extent of black. Adult ♂ *melanoleuca* differs in particular from nominate *hispanica* in narrow strip of black extending along base of forehead: in nominate *hispanica*, black of lores extends narrowly to nostrils, but generally does not meet at base of culmen, in *melanoleuca*, black meets narrowly and forms black strip *c.* 1-3 mm wide. However, black strip occasionally absent in *melanoleuca* and one ♂ from Levant had head fully white except for black patch below and behind eye and black upper ear-coverts. Adult ♂ black-throated (*stapazina*) morph of *melanoleuca* more extensively black on throat than black-throated nominate *hispanica*: in nominate *hispanica*, black on chin and upper throat 17·5 (10) 14-21 mm long (measured along centre); in *melanoleuca*, chin and entire throat black, measuring 24 (10) 21-28 mm. See also Witherby *et al.* (1938*b*) and Svensson (1984*a*). In some nominate *hispanica*, throat pattern intermediate between typical black and pale forms: black of throat separated from black of lore by pale cheek, or chin and sides of throat black and cheek and central throat pale; no intermediates in *melanoleuca*. General colour of adult ♂ *melanoleuca* close to nominate *hispanica* in autumn, but *melanoleuca* with more distinct grey tinge on crown, nape, and mantle (nominate *hispanica* virtually uniform deep pinkish buff-cinnamon). In spring, difference more marked: by March-April, crown and hindneck of *melanoleuca* white with traces of brown-grey feather-tips, lower mantle tinged pale buff-cinnamon (in nominate *hispanica*, uniform darker buff-cinnamon); by May-June, crown, hindneck, and mantle of *melanoleuca* white except for slight cream tinge on outer mantle (in nominate *hispanica*, off-white with distinct cinnamon wash all over). Also, scapulars of *melanoleuca* black, of nominate *hispanica* sometimes partly or fully buff-cinnamon (though not as extensively pale as Desert Wheatear *O. deserti*); see also Ticehurst (1927). When heavily worn, in July, crown of ♂ *melanoleuca* occasionally fully grey, pale feather-tips worn off. No constant difference in colour of underparts. Extent of black on face and throat of 1st adult ♂ *melanoleuca* similar to adult ♂, but black strip along forehead sometimes narrow and less distinct, occasionally virtually absent; in fresh plumage, black largely concealed by buff feather-fringes (unlike adult). General colour of 1st adult ♂ more variable than adult ♂: in fresh plumage, some nominate *hispanica* tinged sandy-brown on crown and mantle (resembling fresh adult ♂ *melanoleuca*), and scapulars often dark brown rather than black, but in worn plumage usually similar to adult (except for retained juvenile feathers); in contrast to this, 1st

adult *melanoleuca* rather similar to adult in autumn (though underparts often deeper cinnamon-buff, less pink-buff, and upperparts often more buff-brown than greyish-buff), but quite different in spring, when forehead and crown remain grey-brown with variable white or grey mottling, nape and mantle become off-white with grey and buff mottling, and scapulars become black-brown with grey-brown or buff-brown suffusion; only rarely (e.g. in Levant) is 1st adult *melanoleuca* as black-and-white as adult, and birds from Asia Minor in particular remain greyish-brown on forehead, crown, nape, and mantle and rather deep buff on chest and flanks throughout spring and early summer, sometimes resembling adult ♀. Proportion of black-throated morph increases towards east, though with some local variation: in Algeria and Tunisia 45% black-throated (*n*=49), in Morocco 52% (*n*=35), in Iberia and France 67% (*n*=15), in Italy and western Yugoslavia 52% (*n*=31), in southern Yugoslavia 52% (*n*=23), in Asia Minor (mainly from Aydin along Menderes river) 86% (*n*=21), but in another sample from Priene along Menderes 31% (*n*=36), in Levant 54% (*n*=37), in Iran 64% (*n*=39), and on migration in Egypt 62% (*n*=68) (Weigold 1914; Stresemann 1920; Meinertzhagen 1930; Vaurie 1949; Mayr and Stresemann 1950; RMNH, ZMA; data combined). Adult and 1st adult ♀ *melanoleuca* distinctly darker on upperparts than ♀ nominate *hispanica*: in fresh plumage (autumn), upperparts buff-brown with grey tinge on crown (in nominate *hispanica*, sandy-cinnamon with vinous-pink tinge); in spring, forehead and crown grey-brown, nape, mantle, and scapulars brown, close to ♀ Wheatear *O. o. oenanthe* (in nominate *hispanica*, warm sandy-buff); difference on underparts less marked, but ♀ *melanoleuca* more often dark-throated than ♀ nominate *hispanica*. No constant difference in tail-pattern between *melanoleuca* and nominate *hispanica* (but see Plumages for variation of latter); maximum extent of black on t1 (excluding shaft) 41·6 (36) 38-48 mm in both sexes of *melanoleuca*, minimum on t3-t4 7·8 (44) 0-18 mm, maximum on outer web of t6 29·4 (37) 23-35 mm (exceptionally up to 43); in 23%, t3-t4 virtually fully white, in 29% white reaches tail-tip but with black spots on one or both webs, in 48% fully black band across tail-tip (*n*=48), and no apparent geographical variation. Birds from south-west Turkey and Levant rather small, especially bill (see Measurements).

Forms superspecies with Pied Wheatear *O. pleschanka*; for hybridization of *O. hispanica* and *O. pleschanka* on Mangyshlak peninsula, where pure *O. hispanica* relatively few, see Panov and Ivanitski (1975*b*); for plumages of hybrids and many other details in northern Iran, see Haffer (1977). See also *O. pleschanka* (p. 806).

Recognition. ♂ of pale-throated morph, juvenile, and ♀ *melanoleuca* are rather similar to Wheatear *O. oenanthe*, but upperparts of ♂ buff or white, not extensively grey, and tail pattern markedly different, with much black on outer web of t6 and limited amount or none on t3-t4 (in *O. oenanthe*, tail-band rather uniform in width). ♂ of black-throated morph and ♀ nominate *hispanica* resemble Desert Wheatear *O. deserti*: general colour close to *O. d. halophila*, but ♂ has black inner upper wing-coverts (buff or white in *O. deserti*), and both sexes have much less black on tail-tips. ♀ and some 1st adult ♂♂ of *melanoleuca* are closely similar to *O. pleschanka*; see Recognition of *O. pleschanka* (p. 806).   CSR

Oenanthe deserti Desert Wheatear

PLATES 60, 64, and 65
[between pages 760 and 761, and facing page 856]

Du. Woestijntapuit   Fr. Traquet du désert   Ge. Wüstensteinschmätzer
Ru. Пустынная каменка   Sp. Collalba desértica   Sw. Ökenstenskvätta

Saxicola deserti Temminck, 1825

Polytypic. *O. d. homochroa* (Tristram, 1859), North Africa; nominate *deserti* (Temminck, 1825), Levant; *atrogularis* (Blyth, 1847), Transcaucasia and Iran east through Kazakhstan and Afghanistan to Tien Shan, Altai, and Mongolia. Extralimital: *oreophila* (Oberholser, 1900), central Asia from Pamir and Himalayas through Sinkiang and Tibet east to Inner Mongolia.

Field characters. 14–15 cm; wing-span 24·5–29 cm. Averages smaller and appears dumpier than Wheatear *O. oenanthe oenanthe*, with less pointed wings but proportionately longer tail; noticeably smaller and slighter than Red-rumped Wheatear *O. moesta*, with shorter head, 5% shorter wings and 10% shorter legs. Rather small, round-headed, compact wheatear. All-black tail diagnostic. ♂ distinguished by black face and throat and white inner wing-coverts, ♀ by more uniform appearance than congeners (with only black-brown primaries and tail obvious). Sexes markedly dissimilar in spring, less so in autumn. Juvenile separable. 3 races occur in west Palearctic, westernmost North African birds most distinctive.

ADULT MALE. (1) North African and Levant races, *homochroa* and nominate *deserti*. In spring (when plumage worn), mostly sandy-grey to isabelline, with western birds usually tinged warm pink especially on head and neck, but strikingly marked by: (a) black face, throat, and upper border of breast; (b) black wings with black and white outer lesser coverts, increasingly white inner lesser, median, and greater coverts (forming bold pale patch at shoulder in flight), and noticeable cream or pale ochre fringes to all greater coverts, tertials, and innermost secondaries (only rarely wearing off completely); (c) predominantly buff-white rump and vent; (d) all-black tail (rarely showing white bases to outermost feathers). Axillaries black; under wing-coverts black with white fringes and tips but appearing solidly black only on leading edge; centre of underwing shows pale area formed by white coverts and pale grey bases of flight-feathers. From July (when plumage fresh), all black areas of plumage except tail obscured by white, cream, or isabelline fringes, making rear face and lower part of throat-patch hoary and reducing obviousness of shoulder-patch. (2) East Asian race (except Levant), *atrogularis*. Slightly larger than western races. Isabelline upperparts have overlying tone of brown rather than pink or sandy-grey; black throat may extend on to sides of neck toward shoulder. Separable from *homochroa* (but not all nominate *deserti*) by extensively white inner webs of flight-feathers which increase obviousness of pale under wing-coverts and show as pale lines on uppersurface of fully extended wing, particularly in easternmost birds. ADULT FEMALE. All races. Differs distinctly from ♂ in usually lacking dark face and throat (though dark brown feather-bases frequently visible on heavily worn birds) and in having wing-feathers always at least partly pale margined and lacking obvious pale shoulder-patch. Lower rump and upper tail-coverts strongly tinged buff, reducing contrast with duller, brown-black tail. All under wing-coverts and axillaries brown-black fringed and tipped white, thus almost uniformly pale, unlike ♂. Racial differences in wing pattern subdued (except in easternmost birds), but tone of upperparts matches that of ♂. Fresh autumn plumage even less contrasting: head, throat, and upperparts virtually uniform, only primaries and tail standing out. JUVENILE. All races. Resembles autumn adult ♀, but upperparts faintly spotted pale, throat and breast lightly scaled brown-buff, and rump whiter. FIRST-YEAR MALE. All races. In autumn, resembles autumn adult but more uniform, with black face and throat obscured by cream-white tips, and browner wings heavily overlaid with pale fringes and lacking more than pale buff shoulder-patch. In spring, as adult but less immaculate, with duller brown wings obvious at close range. FIRST-YEAR FEMALE. All races. Apparently as adult. All races at all ages have bill and legs black.

All-black tail quickly distinguishes it from Black-eared Wheatear *O. hispanica* and most other similarly coloured *Oenanthe*, but note that tail pattern and noticeably pale underwing closely matched by ♀ *O. moesta* and approached by Isabelline Wheatear *O. isabellina*; *O. moesta* also has pale upper wing-coverts. These 2 species both larger and stockier (especially the long-legged *O. isabellina*), and differ in head tone and pattern, back colour and pattern, and precise colours and pattern of rump and tail; see *O. isabellina* (p. 756) and *O. moesta* (p. 841). Flight essentially as *O. oenanthe* but more flitting and less fluent, with hint of *Saxicola* chat at times. Escape-flight usually long, often ending in bird hiding behind either plant or rock. Gait a hop. Stance usually half-upright, with round-topped head held well up even when neck retracted and this attitude contributing much to rather compact silhouette. Other behaviour much as *O. oenanthe* but ♂♂ on territory often perch prominently. Vagrants to western Europe not confined to coasts, even wintering by inland reservoirs.

Generally less vocal than congeners. Song a mournful,

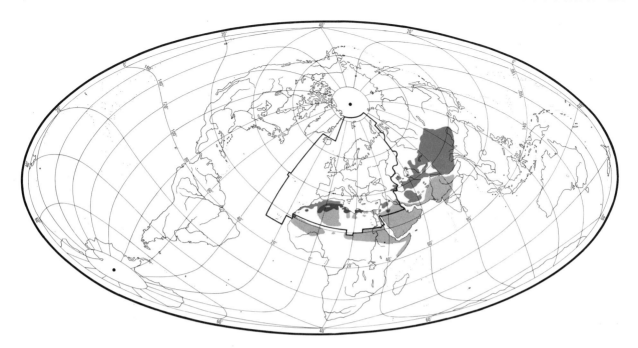

descending 'swee-you', repeated monotonously. Commonest call a piping 'huiie'.

Habitat. In lower middle latitudes, mainly continental, warm and arid, in steppe, Mediterranean, and desert zones; on wide variety of terrain from sea-level (especially in west of range) to high plateaux and even mountain summits extralimitally in Asia. In North Africa, occurs on Atlantic coast and on degraded steppe at edge of Sahara. In southern Morocco seeks flat ground avoiding both dunes and rocks (Heim de Balsac and Mayaud 1962), occupying coastal zone and preferring heath-type and shrubby habitat with (e.g.) *Tamarix*, *Salicornia*, and *Nitraria*; also river beds. Prefers stony or sandy soils, avoiding gravel tracts and pure desert, even where rich in insects. Ecologically replaces *O. hispanica* towards desert (Valverde 1957). In northern Sahara, often perches on low bush or shrub, and occasionally on telegraph wires, as well as rocks. Not a bird of bare sand-dunes, preferring stony plains covered with dwarf scrub, and scattered desert plants. Fond of old cultivation which has reverted to desert (Bannerman 1954). At Azraq (Jordan), breeds on *Nitraria* islands in deeply flooded marshes; also in silt-dune areas, patches of *Juncus* at edge of dried-out marshland, and in wadis with high or low cover and flat or rough ground (Hemsley and George 1966). Also at Azraq, classed as occupying more or less open stony desert, though with wadis, dry flats, and variable cover (Nelson 1973). In USSR, inhabits primarily low-lying places with elongated barchan sand-dunes or clay-covered steppeland, with stony soil of moraines carved by run-off gullies and ravines. In Altai, however, ascends to limits of alpine meadows. In Pamirs, resorts to rocky defiles and high mountain passes with scree and large boulders (Dementiev and Gladkov 1954*b*). In Afghanistan, rather numerous in high valleys above 3100 m (Paludan 1959). In south-west Iran, breeds on vast, open, sometimes gently undulating plains, typically silt or gravel with sparse or thin ground layer of steppe plants, sometimes in patches, and high proportion of scattered bushes (mainly *Zygophyllum altriplicoides*), or shrublets more than 30 cm high. These afford elevated perches valuable for defending territory, for detecting predators and prey, and for relief from excessively high temperatures at ground level, as well as providing shade and cover for nesting. Breeds almost entirely within cold steppe zone at *c*. 1400–2500 m. In winter, occurs on foothills below 800 m and coastal lowlands, sharing habitat with Isabelline Wheatear *O. isabellina*. Shares breeding grounds especially with Hume's Wheatear *O. alboniger*. (Cornwallis 1975.) Winters in India in wide arid open plains, preferring barren sandy wastes, but also occupying old cultivation, interspersed with barren patches and broken ground, either sandy or rocky. Perches on stones, mounds, and shrubs (Whistler 1941). Evidently adaptable to varying habitats over extensive range, but ecologically the name 'Desert' is misleading: penetrates little beyond desert fringe, and ready to ascend to high altitudes or to accommodate to well-vegetated habitats and beds of wadis or rivers, or even to relics of cultivation. Seems able to find similar conditions on migration and in winter, and to be unusually tolerant of human presence. Like other *Oenanthe*, flies freely in lower airspace as occasion arises.

Distribution. No information on range changes.

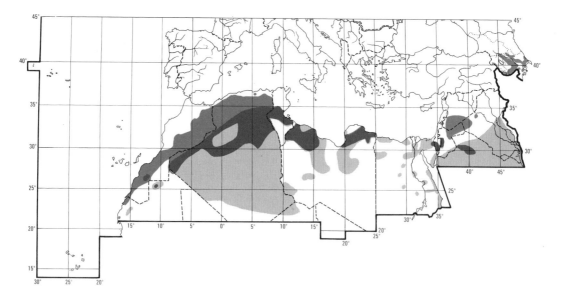

LIBYA. Southern limits uncertain (Bundy 1976). EGYPT. May breed in several areas, but proved only in Sinai (SMG, PLM, WCM).

Accidental. Britain, France, Netherlands, West Germany, Norway, Sweden, Finland, Italy (but annual Sicily), Greece, Turkey, Chad, Malta, Canary Islands, Madeira.

Population. No information on population trends.

MOROCCO. Common in south in several areas (Valverde 1957). LIBYA. Tripoli: common in most of semi-desert areas and locally on coast (Bundy 1976). IRAQ. Breeding Rutbah and apparently more numerous to west (Moore and Boswell 1956).

Movements. Most populations migratory, some only partially. Winters in Africa south to Sahel zone, in south-west Asia north to Syria and east to central India, and in eastern Himalayas. Frequent vagrant over large area north to Sweden, west to Canary Islands, and east to Kuril Islands (eastern USSR). Timing of movements and routes taken poorly known. Uncertainty over racial identification of many records makes lack of ringing data especially unfortunate. Like other *Oenanthe*, migration primarily nocturnal and as loose aggregations rather than compact flocks.

NORTH AFRICAN POPULATIONS (*homochroa*), breeding from Morocco east to Nile, range from migratory to sedentary, with migrants wintering south to Sahel zone. Vagrants assigned to this race have occurred in Canary Islands during February (3) and in Britain during October (1) and possibly November–January (1). Convenient to treat western populations (east to Tunisia) separately from eastern ones, though this division rather arbitrary. (1) Western populations. Mainly migratory, though southern parts of breeding ranges in Morocco, Algeria, and Tunisia

occupied all year. In Morocco, those in south-west more sedentary than those in south-east with few wintering east of about 7°W. Similar pattern in Wheatear *O. oenanthe* and Cream-coloured Courser *Cursorius cursor* (Smith 1965a). Northern parts of breeding range totally vacated with birds wintering south to Aïr and Ahaggar massifs in Saharan zone and beyond desert in Sahel. Main wintering concentrations appear to be in central Mauritania and Mali; Browne (1982) suggested *c.* 3·3 birds per km² in Mauritanian dry steppe. Frequently makes repeated local movements within wintering range, often in response to habitat changes caused by fires and agriculture (D G C Harper). Some records as far south as Bama ridge in Nigeria (11°31′N 13°14′E), but very scarce south of 15°N (Hall 1977; Lamarche 1983). On west coast, very scarce in Sénégal and only one record from Gambia (Jensen and Kirkeby 1980). South-east limits of wintering range unclear, although appears to reach Lake Chad (Vielliard 1971–2; Newby 1980). Most breeding sites in north vacated by end of September, by which time first arrivals occur in Mauritania and Mali (Lamarche 1981). Autumn records concentrated along coastal strip of western Sahara (Heim de Balsac and Heim de Balsac 1951; Curry-Lindahl 1981), although this probably reflects observer bias. Southernmost Nigerian records occurred January–February (Mundy and Cook 1972; Hall 1977). Birds appear back at breeding sites from mid-February, and spring passage probably completed by mid-April (Ledant *et al.* 1981). (2) Eastern populations. Libyan and Egyptian birds apparently more sedentary than western populations, with regular wintering throughout most of breeding range. However, passage observed in Libya (e.g. Snow and Manning 1954, Bundy 1976) more likely to involve local *homochroa* than immigrants of other races. Wintering range in east not clearly defined due to overlap with more numerous birds of other races (see below). Inspection of material at BMNH demonstrates that nominate *deserti* can only be separated from *homochroa* and some collections of *atrogularis* if large series of skins available (D G C Harper), and claims of wintering by *homochroa* as far south-east as Somalia (Portenko and Vietinghoff-Scheel 1971) require confirmation by ringing recoveries. Passage in Libya observed from early September to late October and from late February to early April (Bundy 1976).

LEVANT POPULATION (nominate *deserti*) partially migratory (Clarke 1980; Curry-Lindahl 1981). Proportion of individuals migrating and extent of wintering range totally obscured by potential for confusion between races (see above) and frequent treatment of *atrogularis* as synonym for nominate *deserti*. Most of birds seen on passage and wintering within breeding range of nominate *deserti* are probably residents of this race. In spring, stragglers reach further north into Lebanon than known breeding range (Macfarlane 1978).

CENTRAL ASIAN POPULATION (*atrogularis*), breeding from southern Caucasus and Iran north-east to Mongolia, migrates south and west to winter from western India through Arabia to north-east Africa. Regular straggler to west of migration route, especially in spring (implying reverse migration), reaching Libya (Moreau 1934) and even Britain. Apparently less regular as a vagrant to east of breeding range.

Common and widespread winter visitor to western India (east to *c.* 85°30′E and south to *c.* 18°30′N) and most of Arabian peninsula, though scarce in North Yemen (Cornwallis and Porter 1982). By contrast, wintering density very low on east of Persian Gulf and along Makran coast into Baluchistan. Indeed, extended passage periods revealed by recent studies (see below) suggest that many of these apparent wintering records require reassessment. Does not appear to winter in Palestine or Egypt and level of passage through east Mediterranean probably very low, although potential exists for confusion with nominate *deserti* (see above). North-west limit to passage through Middle East appears to be Iraq where heavy autumn passages recorded south of 32°N (e.g. Marchant 1963*a*), although weak passage observed along south coast of Caspian Sea (Feeny *et al.* 1968). South-east limit of passage appears to run through southern Saudi Arabia and Oman, with very few records from offshore islands, suggesting that, unlike *oreophila* (see below), it rarely makes long sea crossings. Winter range in Africa extends south-west to about 16°N 18°E in Chad, but birds much more common further east in arid plains of Sudan and on coastal plains of Ethiopia and Somalia. Most Sudanese records lie within 15–17°N (Hogg *et al.* 1984), but a few birds reach *c.* 14°N (Mackenzie 1955). Records from northern Uganda and Kenya do not seem to be substantiated (Britton 1980). Recent discovery of wintering around Mogadishu south to *c.* 1°53′N extends known range by 1000 km and suggests wintering through-

out coastal Somalia (Ash 1981). As large numbers reported in northern Somalia, seems likely that this poorly worked area contains important proportion of world population. Racial status of Somali birds controversial, but they average darker than Ethiopian birds (Ash 1981) and at least some are probably *oreophila* (see below).

In Semirech'e (Kazakhstan, USSR), post-breeding dispersal begins in August, developing into full scale emigration by end of month; birds have left by late September, stragglers remain until late October. Peak passage in Zaysan depression (Kazakhstan) not until late September, passage perhaps continuing even longer in southern USSR. By contrast to spectacular movements noted in east of breeding range, autumn passage in west not so obvious, extending from August to November. (Dementiev and Gladkov 1954*b*.) In detailed study in south-west Iran, Cornwallis (1975) found breeding grounds largely vacant from early October to mid-March, although a few birds wintered both at these and other nearby sites. First arrivals in Arabian peninsula and Africa occur in late August (Mann 1971; Bundy and Warr 1980), although peak arrivals in Somalia not until early October (Ash 1981). First birds reach wintering grounds in north-west Pakistan in early September but many Indian sites not occupied until November (Ali and Ripley 1973*b*).

Return passage starts relatively early, with first arrivals reaching south-west Iran by mid-January (Cornwallis 1975). Some Indian wintering sites deserted by mid-February, but African wintering grounds not completely vacated until late March. First arrivals in Turkmeniya and Bukhara during mid-March often involve large numbers of birds. Successive waves of arrivals move north and east, reaching north coast of Aral Sea by mid-April. (Carruthers 1949; Dementiev and Gladkov 1954*b*.) Arabian peninsula largely vacated by early May. In eastern Saudi Arabia, sharp decline in numbers of wintering birds during March, followed by small influx of migrants in late April and early May (G Bundy). Although suggestions of breeding in Arabia (e.g. Courtenay Thompson 1972) are not confirmed, there is a scattering of June–July records from well outside known breeding range including Bahrain, Saudi Arabia, Kuwait, and Iraq (Sage 1968; Bundy and Warr 1980).

CHINESE POPULATION (*oreophila*), breeding from Kashmir and Tibet north-east to Sinkiang and Inner Mongolia, winters in 2 disjunct areas. Largest runs from Makran coast south-west through Arabia, at least to Socotra; orthodox view that *oreophila* does not reach Africa (e.g. Portenko and Vietinghoff-Scheel 1971) challenged by descriptions of wintering birds in Somalia (Ash 1981; see above). Second wintering area lies south-east of Himalayas in Sikkim, Assam, Bhutan, and northern Cachar. Suggestion that some populations in north-west China are sedentary (Etchécopar and Hüe 1983) requires substantiation. Autumn passage poorly documented, but concentration of passage records at southern end of Arabian peninsula and regular wintering on Socotra (Moreau 1938; Bundy and Warr 1980; Walker 1981b) suggests frequent long-distance flights over Indian Ocean (unlike *atrogularis*). Details of timing poorly known, but probably resemble *atrogularis*. However, spring arrival in Soviet Pamirs relatively late, with first birds in 1st half of April and main passage not until early May (Dementiev and Gladkov 1954b). DGCH

Food. Most detailed information extralimital—from USSR (Bel'skaya 1961) and Iran (Cornwallis 1975). Diet predominantly insects, particularly ants, beetles, and larvae; occasionally spiders, worms, small lizards, and seeds. Takes food mainly from bare ground; sometimes from low vegetation or in flight like flycatcher (Muscicapidae). Of 39 items, 91% taken on bare ground and 9% picked off plants. Typically searches for food from low or elevated perch (up to *c.* 1·5 m high), launching flying attack usually against prey within 10 m. Light weight allows use of flimsier perches than most other *Oenanthe*. Sometimes, usually where few perches available, hunts on ground by making rapid dashes, mainly hopping rather than running. (Ticehurst 1922b; Schäfer 1938; Whistler 1941; Ali 1945; Edwards 1950; Valverde 1957; Dolgushin *et al.* 1970; Cornwallis 1975; Gallagher and Woodcock 1980). Occasionally hunts, in manner of Kestrel *Falco tinnunculus*, by hovering into wind a few metres above ground and diving on to prey (Valverde 1957; Abdusalyamov 1973). One ♂ ate smaller locust hoppers with ease, but 4th and 5th instars were beyond its capacity; often fanned wings before large insects (Smith 1971). However, one ♀ seen battering live lizard 12·5 cm long (Moore and Boswell 1956). ♂ once seen to extract worm from ground (Edwards 1950).

The following recorded in diet. Invertebrates: dragon-flies (Odonata), grasshoppers and locusts (Orthoptera: Acrididae), earwigs (Dermaptera), cockroaches and immature mantises (Dictyoptera: Blattidae, Mantidae), termites (Isoptera), bugs (Coreidae), larval ant-lions (Myrmeleontidae), adult and larval Lepidoptera, adult and larval Diptera, Hymenoptera (ants Formicidae, bees Apoidea), adult and larval beetles (Coleoptera: Carabidae, Scarabaeidae, Tenebrionidae, Chrysomelidae, Curculionidae), spiders (Araneae), Isopoda, Myriapoda, worms (probably Oligochaeta). Occasionally small lizards (Sauria, including Agamidae) and sometimes small seeds. (Ticehurst 1922b; Bates 1936-7; Morrison-Scott 1937; Witherby *et al.* 1938b; Trott 1947; Edwards 1950; Moore and Boswell 1956; Bel'skaya 1961; Pek and Fedyanina 1961; Kovshar' 1966; Survillo 1968; Dolgushin *et al.* 1970; Smith 1971; Abdusalyamov 1973; Cornwallis 1975; Mauersberger *et al.* 1982.)

In south-east Turkmeniya (USSR), April, 38 stomachs contained 589 items as follows: ants (71·4% by number, present in 82% of stomachs), adult Coleoptera (mostly Curculionidae, Scarabaeidae, and Tenebrionidae) (13·5%, 79%), larval Lepidoptera (10%, 53%), Myriapoda (1·9%, 18%), Hemiptera (1·8%, 18%), Dictyoptera (0·4%, 8%), spiders (0·4%, 5%), Orthoptera (0·3%, 5%), larval Coleoptera (0·3%, 3%), and bees and Solifugae (each 0·1%, 3%) (Bel'skaya 1961). In Tien Shan mountains (southern USSR), March–April and September–October, 6 stomachs contained similar insect remains (Kovshar' 1966). In south-west Iran, diet similar throughout the year but larvae important only during spring; 23 stomachs, January–May and November–December, contained 662 items as follows: ants (37%, 87%), insect larvae (March and May) (19%, 17%), Coleoptera (11%, 57%), Isoptera (9%, 13%), Hemiptera (4%, 39%), bees and wasps (2%, 30%), other Hymenoptera (10%, 39%), Orthoptera (2%, 26%), Neuroptera (1%, 9%), Diptera (1%, 22%), Isopoda (2%, 17%), spiders (1%, 13%), adult Lepidoptera (0·2%, 4%), seeds (0·3%, 9%); 61% of items 2–5 mm long, 34% 5–10 mm, 5% 10–15 mm, 0·5% 15–20 mm; items in diet similar in size and taxonomic composition to those available in environment, thus presumably an opportunistic feeder (Cornwallis 1975). Of 9 stomachs from near Jiddah (Saudi Arabia), mid-January to mid-April, ants in 6 (predominant in 4), Hemiptera in 5, larval Diptera in 2 (predominant in 1), adult Lepidoptera in 2, larval Lepidoptera in 2, larval ant-lion in 1, adult Diptera in 1, and seeds in 2 (Morrison-Scott 1937).

Little information on food of young. In Tadzhikistan (USSR) parents brought adult and larval insects (Abdusalyamov 1973); in Mongolia, larvae and a small lizard (Mauersberger *et al.* 1982). No information on feeding rates. LC

Social pattern and behaviour. Some aspects fairly well known but no comprehensive studies and much (e.g. about breeding behaviour) not fully documented. Studies in Soviet Central Asia by Mitropol'ski (1968a) and Ivanitski (1980) and in south-west Iran by Cornwallis (1975) and L Cornwallis; important reviews by Panow (1974) and Panov (1978) include detailed information from USSR.

1. Outside breeding season normally solitary and territorial (Ticehurst *et al.* 1921-2; Cornwallis 1975) but winter visitors to India sometimes in pairs (Ali 1955). Wintering birds in south-west Iran and north-east Libya often in loose aggregations with conspecific and other *Oenanthe*, these concentrations apparently not caused by habitat restrictions (L Cornwallis). Family parties reported from early April (Libya: Bundy and Morgan 1969) to mid-August (Siberia: Johansen 1954). Migrants in Soviet Central Asia sometimes in loose flocks of 5-15 or more birds (Sushkin 1938; Panow 1974). Except during short stops while on migration, both sexes usually hold individual territories which are defended against other members of same species (Smith 1971; Cornwallis 1975). During autumn and winter, territories sometimes occupied for short periods only—4-6 weeks or less (Cornwallis 1975). On coastal plain of Iran in winter, recorded having territories contiguous with those of Isabelline Wheatear *O. isabellina*, Finsch's Wheatear *O. finschii*, and Red-tailed Wheatear *O. xanthoprymna* (Cornwallis 1975);

in Libya, with Red-rumped Wheatear *O. moesta* (L Cornwallis). In these cases, however, *O. deserti* apparently occupying ground not wanted by other species as it is socially subordinate to them (see Antagonistic Behaviour, below). 5 territories on coastal plain of Iran *c.* 0·3–0·8 ha (L Cornwallis). Densities in winter generally low and distribution patchy: scattered examples the rule in Iraq (Ticehurst *et al.* 1921–2); 1–2 birds over 19 km around Hufuf in eastern Saudi Arabia (Ticehurst and Cheesman 1925); and during car-journeys through suitable country 0·03 per km over 2555 km on Gulf coast of Iran (L Cornwallis and D A Scott) and 0·03 ♂♂ per km over 130 km in north-east Libya (L Cornwallis). BONDS. No evidence to suggest that mating system other than monogamous. No information on duration of pair-bond or age of first breeding. Both parents care for nestlings and fledglings (e.g. Panow 1974, Mauersberger *et al.* 1982). Some evidence of brood-division by parents after fledging but no information on length of time before young independent (see Relations within Family Group, below). BREEDING DISPERSION. Solitary and territorial (e.g. Panow 1974, Cornwallis 1975, Ivanitski 1980), with tendency to form neighbourhood groups noted in northern Cyrenaica (Libya) (L Cornwallis). Territories normally large (e.g. one in Gobi Desert, Mongolia, *c.* 4 ha: Mauersberger *et al.* 1982) and widely dispersed, usually several hundred metres or more apart, e.g. in Soviet Central Asia (Panow 1974), Iran, Jordan, and northern Cyrenaica (L Cornwallis). In such conditions territorial boundaries of little significance as neighbouring pairs rarely come into contact. Denser population reported from central Pamirs where pairs only 100–200 m apart (Potapov 1966). In eastern Morocco, 15 censuses, each of *c.* 4 km² of suitable habitat, yielded 0·25–3·5 birds per km², average 1·6 per km² (Goriup 1983). In Emba river basin (USSR), 0·19 birds per km over 10 km (Neruchev and Makarov 1982). Vehicle transects through suitable country gave 1 ♂ per km over 9 km in Iranian Baluchistan, 0·3 pairs per km over 80 km in central Iran, and 0·7 pairs per km over 193 km in northern Cyrenaica (L Cornwallis). Use of territory all-purpose (L Cornwallis). ROOSTING. Often active in territory until dark (L Cornwallis) but roosting site not recorded. Vagrant to Britain reported sunning (King 1970).

2. Often extremely confiding, allowing approach to 3–4 m (e.g. Whitaker 1905, Heim de Balsac 1924, Archer and Godman 1961*b*, Panow 1974); in Eritrea (Ethiopia), ♂ on winter territory frequently entered tent and took locusts at Smith's (1971) feet. However, found to be shy by Meinertzhagen (1930) in Egypt and by Schäfer and Schauensee (1938) in Tibet. FLOCK BEHAVIOUR. No information. SONG-DISPLAY. Song (see 1 in Voice) given mainly by ♂ but also occasionally by ♀ (see 1*d* in Voice), e.g. in winter in Tunisia (Chappuis 1975), while feeding fledglings in Jordan (L Cornwallis) and once during territorial encounter between ♀♀ in Soviet Central Asia (Panow 1974). Territorial-song (see 1*a* in Voice) usually delivered from 2–3 favoured song-posts such as top of bush or rock (e.g. Whitaker 1905, Bannerman 1954, Panow 1974) and sometimes from ground, e.g. whilst feeding (Roché 1968; E D H Johnson). During breeding season, Song-flights often performed: ♂ usually ascends at moderate angle to height of 8–10 m and, singing most of the time, circles territory with measured, wide-amplitude wing-beats before closing wings and plunging back to earth. During Song-flight (Fig A), sometimes folds wings and, calling (see 3*b* in Voice), plummets precipitously with tail slightly spread before pulling out of dive and resuming normal Song-flight (Ali 1946; Mitropol'ski 1968*a*; Panow 1974; Mauersberger *et al.* 1982; L Cornwallis); such plummets during main part of Song-flight unique in *Oenanthe* (Panov 1978). In

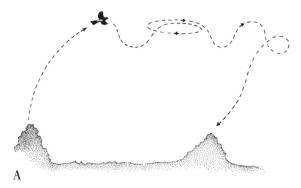

A

less typical Song-flight, ♂, singing continuously, rises straight up from song-post and hovers at no great height, sometimes with dangling legs, before plummeting back to earth, often to same perch (Koenig 1895; Bannerman 1954; Panow 1974; M G Wilson). Short bursts of song also uttered during courtship and aggressive encounters (Panow 1974; L Cornwallis: see 1*b* in Voice). Singing most intense during breeding season when ♂♂ often heard persistently throughout day (Ali 1946; Panow 1974) and sometimes also at night, e.g. in southern Altai (central Asia) throughout still, moonlit nights in April (Pleske 1889–94) and in Mongolia after dark in June (Piechocki *et al.* 1982). Weak song and Subsong (see 1*e* in Voice) heard widely outside breeding season (e.g. Whistler 1922, Browne 1950, Smith 1957, 1965*a*, Walker 1981*b*). ANTAGONISTIC BEHAVIOUR. (1) General. Territory-holders normally attack conspecific birds found trespassing, though territorial conflicts in breeding season rare over much of range as densities so low. In areas of greater density, however, conflicts fairly common early in breeding season. Intraspecific interactions most common between territory-holders and intruding migrants which are vigorously chased off (Panow 1974; Piechocki *et al.* 1982). Most territorial conflicts between ♂♂, but ♀ against ♀ noted on breeding grounds in USSR (Panow 1974). In aggressive encounters with other *Oenanthe*, subordinate to *O. isabellina*, Wheatear *O. oenanthe* (usually), *O. finschii*, *O. moesta*, and Mourning Wheatear *O. lugens*, but normally dominant over Pied Wheatear *O. pleschanka* (Cornwallis 1975; Ivanitski 1980; L Cornwallis). On breeding grounds in USSR, attacks reported by both sexes on *Oenanthe* and wide range of other small passerines (Ivanitski 1980). In least intense form of aggression, territory-owning ♂ makes little flights into air at approach of intruder which retreats. If intruder detected at distance, territory-holder makes long low flight towards intruder which retreats. Neighbours sometimes make short parallel flights along common boundary (L Cornwallis). More intense encounters described in detail by

B

Panow (1974), as follows. ♂♂ defending territories employ range of threatening postures (e.g. as in Fig B) which, except when tail raised vertically, are accompanied by constant or interrupted wagging up and down of fanned or closed tail and sometimes by Battle-song (see 1b in Voice). If more excited, opponents fly at one another and rising face to face in the air each attempts to claw the other. Long chases ensue during which abrupt calls given. When one lands, the other, with fanned tail, attacks it from above and bird on ground raises tail and briefly flicks wings. Sometimes one bird performs acrobatic display alongside the other involving repeated leaps up to several metres in the air (Fig C, left) or twists and turns from side to side near ground (Fig C, right). On Mangyshlak

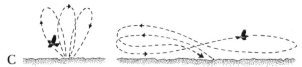

C

peninsula (east Caspian, USSR), in encounter with ♂ *O. pleschanka* which approached nest, ♂, calling continuously, lowered head, depressed tail, ruffled ochre feathers of back conspicuously, and advanced with mincing gait until intruder retreated (Mitropol'ski 1968a). In encounters between ♀♀, birds ruffle feathers, crouch slightly, and sway from side to side through 90°; birds usually silent but once song-phrase heard (Panow 1974: see 1d in Voice). HETEROSEXUAL BEHAVIOUR. (1) General. In USSR (and probably in other areas where migratory), ♂♂ normally return to breeding grounds shortly before ♀♀ and begin to sing and establish territories (Dementiev and Gladkov 1954b; Mitropol'ski 1968a; Panow 1974). (2) Pair-bonding behaviour. When ♀♀ arrive, ♂♂ follow them around and sometimes chase them (Koenig 1895; L Cornwallis); also use display and courtship-subsong to attract ♀♀ (Panow 1974; Panov 1978: see 1b in Voice). Displays often include (as in territorial disputes) wide fanning of tail which is wagged slowly up and down; sometimes also aerial acrobatics alongside ♀ (see above). Initially, ♀ often aggressive towards ♂ and threatens to lunge at him with bill if he approaches. (3) Courtship-feeding. None known. (4) Mating. No information. (5) Behaviour at nest. Nest-site chosen by ♀ with ♂ in attendance. ♀ inspects various potential sites while ♂, uttering quiet, strangled song-phrases similar to those heard during song-flights (see 1a in Voice), runs back and forth in front of each with head slightly raised, wagging fanned tail up and down—behaviour similar to that employed during territorial defence (Fig B, right). After selection of site, birds visit it several times per day but nest-building (by ♀ alone) does not begin for 10 days or more. In period preceding and during nest-building, characteristic Courtship-song (see 1c in Voice) and displays performed. When ♀ approaching nest-site sees ♂, she performs a number of impetuous flight manoeuvres before landing and disappearing into site. ♂, giving Courtship-calls (see 4 in Voice) and bowing forwards, stands by entrance and quivers partially opened wings, frequently wagging fanned tail up and down (Fig D); sometimes he half disappears into nest-hole leaving only raised tail visible. If ♀ stays long in

D

nest-site, ♂ flies to elevated perch, sings, sometimes picks up small stone, and after several aerial manoeuvres flies off. During incubation, ♂ keeps well away from nest. (Panow 1974.) RELATIONS WITHIN FAMILY GROUP. Little information about early life of young. On Mangyshlak peninsula, chicks *c.* 4 days old still had eyes closed (Mitropol'ski 1968a). Both parents feed young (Panow 1974) but at nest in Mongolia ♀ fed them twice as frequently as ♂ and was apparently also brooding as she spent up to 90 s in nest during each visit while ♂ spent only 5-18 s (Mauersberger *et al.* 1982). While still in nest, young give hunger-calls (after fledging, used also to help adults locate young). When adult arrives with food, young birds (if hungry) gape, wave heads about, flap wings (Fig E), and give loud food-calls. When satiated, fledglings bow heads and tremble wings slightly, holding them over back (Fig F). (For

E   F

details, including comparison with other *Oenanthe*, see Panow 1974.) At Azraq, some evidence of brood-division by parents: during 30-min watch, ♂ fed 2 fledglings while *c.* 20 m away the ♀ fed 2 others; young stayed more or less in same places and parents brought food to them (L Cornwallis). Not known when young become independent. ANTI-PREDATOR RESPONSES OF YOUNG. At Azraq, fledglings being fed by parents always remained close to dwarf shrubs, darting under them at approach of observer (L Cornwallis). PARENTAL ANTI-PREDATOR STRATEGIES. (1) Passive measures. During laying and incubation adults do not call at approach of man but simply slip away quietly (Survillo 1968). (2) Active measures. After hatching of young, adults become agitated and give alarm-calls (see 3 in Voice) at human intrusion (Survillo 1968). In Tibet, when nest with young approached by man, ♀ gave alarm-calls, flitted about anxiously wagging tail up and down, and repeatedly attacked and drove ♂ away from nest as if worried that he would give away its position (Ali 1946).

(Figs by D Rees: based on drawings in Panow 1974.)   LC

Voice. Generally less vocal than other *Oenanthe* (Witherby *et al.* 1938b; L Cornwallis). Geographical variation, at least in song, appears to be relatively slight.

CALLS OF ADULTS. (1) Song. (a) Territorial-song of ♂. A monotonous succession of brief (2-4 notes), rather stereotyped, mournful phrases delivered slowly (pauses between phrases *c.* 2-5 s) whilst perched or in Song-flight. Song alleged by Sarudny and Härms (1926) and Dolgushin *et al.* (1970) to include mimicry, but none found by Panow (1974), Panov (1978), or L Cornwallis, and simplicity of song argues against mimicry. Available recordings and descriptions (see below) indicate that nominate *deserti*, *homochroa*, and *atrogularis* may all vary around a similar phrase-type, roughly 'swee-you' (E K Dunn), likened to rusty hinge (L Cornwallis), while

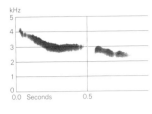

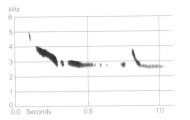

I   P A D Hollom
Morocco   March 1978

II   P A D Hollom
Morocco   March 1978

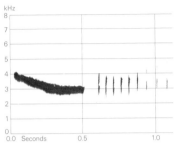

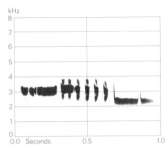

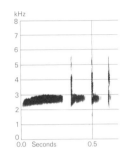

III   P A D Hollom   Morocco   March 1978   IV   P A D Hollom   Iran   April 1972   V   P A D Hollom   Iran   April 1972

song of extralimital *oreophila*—a monotonously repeated 'teee-ti-ti-ti' (Ali and Ripley 1973*b*)—is perhaps different from these. In Morocco, song of *homochroa* a mournful descending 'SWEE-you' or 'WHEE-you' (Fig I); sometimes tremulous in middle, thus roughly 'sweerer-you' (Fig II); sometimes incorporates alarm-type rattle (call 3b), thus 'swee-trrrrr' (Fig III). Frequency range typically 2–4 kHz, rattle up to 5 kHz (J Hall-Craggs). Other phrases by same bird are mild variations (e.g. with amplitude modulation) on the basic phrase exemplified by Fig I. For comparable rendering from Morocco, see Jonsson (1982), but for apparently quite different song see Witherby *et al.* (1938*b*). Song of *atrogularis* recorded in Iranian Baluchistan conforms to 'swee-you' type but has elaborate tremolo in middle of each phrase (J Hall-Craggs: Fig IV). On Mangyshlak peninsula (east Caspian, USSR), *atrogularis* gave a drawn-out 'eey-yu eey-yu' (Mitropol'ski 1968*a*). Recording of *atrogularis* in Mongolia (D Wallschläger) not markedly different. In Jordan, song of nominate *deserti* a very soft descending and mournful piping 'peu-h-h-hue' with tremulous quality, also (different bird) 'thrudeld-three'; in both cases, phrase followed by 'stone-rattling chirr' (Hollom 1959; P A D Hollom: see call 3b). Song in Lebanon and Jordan also described as a plaintive husky whistle 'peeoo' and a low warbling 'chittachit-cheeoo' (Vere Benson 1970). (b) Battle-song of ♂. During courtship and aggressive encounters, gives brief bursts of song in short stereotyped phrases, sometimes (in courtship) quiet and strangled like Subsong but otherwise similar to Territorial-song (Panow 1974). (c) Courtship-song of ♂. After nest-site selected, may give a distinctive quiet song: long varied phrase with notes differing greatly in pitch, and, most characteristically, a recurring clear rattle (Panow 1974). (d) Song of ♀. ♀ *atrogularis* once heard to give short song-phrase 'vaguely reminiscent' of ♂'s song when she attacked another ♀ intruding at nest; described as a melodious 'peiwi' followed by a threat-call (Panow 1974). In Tunisia, November, song of ♀ *homochroa* consisted of more uniform rolling phrases than that of ♂ (Chappuis 1975). (e) Subsong. In Dhofar (Oman), Subsong from *atrogularis* reported from mid-October; in spring, before migration, a repetitious rattling 'der lit lit lit derlit lit tac tac' and a warbling crackling 'tlui lui lui' interspersed with 'sweet' and 'tac' sounds (Walker 1981*b*). See also 1b–c for quiet songs used in courtship. (2) Contact-alarm call. A rather plaintive hoarse whistle according to Witherby *et al.* (1938*b*), though other renderings, apparently of this call, indicate fluting rather than hoarse quality: e.g. in Tunisia, 'huiie' (Lombard 1965), and a mournful piping a little like Bullfinch *Pyrrhula pyrrhula* (M G Wilson); a high-pitched 'pee pee pee' combined with call 3b (L Cornwallis). In Fig V, 1st unit 'wee' (J Hall-Craggs) is perhaps this call. (3) Threat- and alarm-calls. (a) Commonest sign of agitation a harsh 'tuk', much sharper than Wheatear *O. oenanthe* (Smith 1962–4). In Tunisia, rendered 'tsukk' (M G Wilson). During aerial chases, calls of ♂♂ described as quiet and abrupt, roughly 'zch zick kcha kcha' (Panow 1974). When intruder near nest, agitated ♀ evidently included call 3b to give repeated 'chuck-chrr' (Ali 1946). In Fig V, calls combined with call 2 (apparently), thus: 'weee tk tk t' (J Hall-Craggs). (b) Rattle-calls. Recording of alarmed bird (same bird from which Figs V and VIII derived) includes variety of rattling sounds, some a castanet-like clicking similar to series of 9 clicks at end of Fig III, others more ticking and with slight accelerando (Fig VI); finally, bird makes rattling chirping 'trrip' sounds (Fig VII). Rattles used to signal threat and warning when well-grown young endangered (Panow 1974). Other descriptions include

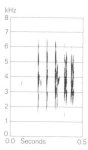

VI P A D Hollom
Iran April 1972

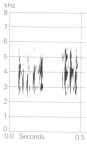

VII P A D Hollom
Iran April 1972

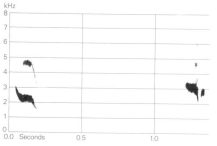

VIII P A D Hollom Iran April 1972

'cht-tt-tt' (Ali and Ripley 1973b), a scolding 'trrr trrr' preceded by call 2 (apparently), also a rapid ticking (L Cornwallis) and a chattering 'chek chek' (Smith 1962–4). Muffled chattering during dives in Song-flight (Panow 1974) are presumably related. (c) Other alarm-calls. Recording from which Figs V–VII derived also includes a number of sounds resembling sparrow *Passer*, notably a descending 'eeoo' (J Hall-Craggs: Fig VIII, 1st unit) and a chirp (Fig VIII, 2nd unit). (4) Courtship-calls. A short rattle or a quiet 'trüw trüw' given by ♂ displaying to ♀ as she approaches nest-site (Panow 1974). Not known how rattle differs from others listed above. (5) Other calls. In Iran, April, a twittering heard from ♂ (L Cornwallis), perhaps different from calls described here.

CALLS OF YOUNG. Panow (1974) distinguished (as in other *Oenanthe*) a food-call given whilst being fed, and a hunger-call given by nestlings soliciting food and by fledglings as a contact-call. Contact-call quite different from that of other young *Oenanthe* (Panow 1974); at Azraq (Jordan), rendered a quiet 'cluk cluk cluck' (L Cornwallis). EKD, LC

Breeding. SEASON. Algeria and Tunisia: eggs found 10 March–26 May (Heim de Balsac and Mayaud 1962). Middle East: eggs found April–May (Hüe and Etchécopar 1970). Kazakhstan (USSR): laying begins late April (Dementiev and Gladkov 1954b). SITE. In hole in ground, or among rocks, often in old rodent burrow. Nest: bulky cup of grass, dead leaves, and roots, lined with hair, feathers, and wool. 2 nests had outer diameter 12·5 and 13·0 cm, internal 9·5 and 6·5 cm, height 5·0 cm, and depth of cup 2·0 cm (Dementiev and Gladkov 1954b). Nest typically near hole entrance: of 12 nests, 3 were 5 cm from entrance, 4 up to 10 cm, 4 up to 25 cm, 1 35–40 cm (Survillo 1968). Building: by ♀ only (Panow 1974). EGGS. See Plate 83. Sub-elliptical, smooth and slightly glossy; pale blue, with variable red-brown specks and spots, often mainly at broad end forming ring. Nominate *deserti*: 19·9 × 14·9 mm (18·6–21·5 × 14·0–16·0), n=43; calculated weight 2·35 g (Schönwetter 1979). Clutch: 4–5 (3–6). Of 57 clutches, North Africa: 3 eggs, 7; 4, 36; 5, 14; mean 4·1 (Heim de Balsac and Mayaud 1962). In Africa, 2 broods probably common (Johns 1970). INCUBATION. 13–14 days. By ♀ only. (Panow 1974.)

YOUNG. Cared for by both parents (Panow 1974; Mauersberger *et al.* 1982). FLEDGING TO MATURITY. Young leave nest at 13–14 days (Panow 1974), 15–18 days (Johns 1970). No further information.

Plumages (*O. d. homochroa*). ADULT MALE. In fresh plumage, forehead, crown, and upper mantle pale grey-brown, sometimes with slight vinous tinge. Feathers at base of culmen and supercilium pink-buff; supercilium long, narrow in front of and above eye, wide above ear-coverts; sometimes hardly paler than forehead and crown. Lower mantle, scapulars, and back saturated pink-cinnamon, sometimes slightly tinged vinous. Rump and upper tail-coverts white with strong pink-buff tinge, most distinct on centre and on longer coverts. Sides of head (down from lores and ear-coverts and including narrow ring round eye), chin, and throat black, throat-feathers with fringes *c.* 1 mm wide along tips, not concealing black, cheeks and ear-coverts with faint pale fringes soon wearing off. Sides of neck and upper sides of breast black, feathers tipped pink-cinnamon, appearing mottled. Chest and upper flanks deep pink-buff with slight vinous tinge, rear flanks and under tail-coverts pale cream-pink, central belly and vent cream-white. Tail-base white (hidden underneath tail-coverts), terminal half black (except for narrow pale grey or white fringes along feather-tips); length of black on central pair (t1) 38·6 (16) 34–43 mm (sometimes a little black along shaft), on inner web of t3–t4 33·7 (16) 28–38 mm, on inner web of t6 37·0 (16) 33–40 mm (usually some mm more on outer webs). Flight-feathers and tertials greyish-black, deepest on outer webs and tips of flight-feathers; bases of inner webs of flight-feathers with long and narrow white wedge; tips of secondaries and inner primaries narrowly fringed grey-white, outer web of all flight-feathers narrowly pink-buff (except emarginated parts of outer primaries and on p8–p10); outer webs and tips of tertials with broad pink-cinnamon or rufous-cinnamon fringes 2–4 mm wide. Greater upper primary coverts black, tips with narrow and sharp white fringe *c.* 0·5 mm wide. Shorter and outer lesser coverts and most median coverts black, tips broadly fringed pink-cinnamon (soon bleaching to white and worn off), other lesser coverts pink-cinnamon or white, forming pale patch together with partly pale innermost median and greater coverts (largely hidden beneath scapulars at rest). Greater upper wing-coverts black, *c.* 2 mm wide tips and narrower outer edges pink-buff (soon bleaching to white, especially on outermost). Axillaries and under wing-coverts black, longer feathers broadly tipped white, shorter narrowly fringed white. *In worn plumage* (from about mid-winter), upperparts brighter pale sandy-cinnamon or yellowish pink-cinnamon, less grey and vinous; supercilium pure white, more distinct (mainly from just above eye backwards); rump and upper tail-coverts cream-white or pure white; sides of head, chin, throat, and shorter under wing-coverts pure black, without white fringes; chest and flanks paler pink-cream, central belly to under tail-coverts pink-white or white; pale edges, fringes, and tips of tail-feathers, flight-feathers, tertials, and upper wing-coverts worn off, upperwing thus deep black with distinct pale patch formed by pale cream or white longer lesser coverts and part of innermost median and greater coverts and with grey-brown tertials. When heavily worn, June–July, upperparts sometimes bleached to off-white, some dark grey or grey-brown of feather-bases of upperparts exposed (in particular on crown, nape, and back), belly dirty white, and tips of flight-feathers and tail browner. ADULT FEMALE. Upperparts similar to adult ♂, but mantle and scapulars usually less bright pink-cinnamon, more sandy-buff, and rump and upper tail-coverts sometimes more distinctly tinged cream-pink or pink-buff; supercilium and narrow eye-ring cream-buff or pale buff, rather indistinct. Sides of head, chin, and throat highly variable: either greyish-sandy, like crown, with indistinct dark spot in front of eye, darker brown or slightly rufous ear-coverts, and paler greyish sides of neck (pale-throated variant), or with black-mottled band over upper cheek to sides of neck (intermediate variant), but sometimes with black cheek, sides of neck, chin, and throat, usually with some pale buff feather-fringes (especially at gape and on chin), or even almost fully black (dark-throated variant) and then closely similar to adult ♂ (in particular because these dark-throated birds tend also to have blacker wings than pale-throated or intermediate birds). Of 15 adult ♀♀ examined (all races) 6 pale-throated, 4 intermediate, 5 dark-throated. Remainder of underparts like adult ♂, chest, sides of breast, and flanks often slightly darker pink-buff or pale buff-brown, sides of breast sometimes slightly tinged grey. Flight-feathers, greater upper primary coverts, and feathers of bastard wing greyish-black, outer webs and tips narrowly fringed pale sandy-buff or white (fringes even in width, 0·5–1 mm, not exceeding 1 mm on tip of inner greater upper wing-coverts). Tertials greyish-black or black-brown, outer webs strongly suffused pink-cinnamon. Lesser upper wing-coverts (except outermost) and inner median and greater coverts pinkish-sandy or sandy-buff, like mantle and scapulars, other upper wing-coverts greyish-black with broad sandy-buff fringes along outer webs and tips, tips of greater coverts with white fringe c. 1 mm wide. Under wing-coverts and axillaries like adult ♂, but black feather-bases more restricted, underwing appearing largely cream or white, except for mainly black longer inner coverts and black-spotted smaller coverts along leading edge of wing. *In worn plumage*, upperparts pale sandy-brown or greyish sandy-cinnamon, rump and upper tail-coverts white, belly dirty white; paler fringes of upperwing abraded, appearing largely black-brown or dark-brown with bleached pale sandy-cream patch on lesser and inner coverts. NESTLING. Down pale grey (Panow 1974). JUVENILE. Rather like adult ♀, but forehead, crown and nape grey-brown, mantle, scapulars, and back buff-brown, each feather with faintly paler centre and faint dusky edge along tip, appearing mottled. Rump and upper tail-coverts white with some buff suffusion. Supercilium only slightly paler than crown, indistinct; lores and upper cheek pale greyish-buff; eye-ring narrow, pale buff, often fairly distinct; ear-coverts buff-brown or brown with narrow and ill-defined pale streaks. Chest and sides of breast pink-buff or pale buff with pale cream-feather centres and ill-defined dusky tips, appearing mottled or scalloped; chin, throat, and underparts down from belly and flanks cream-white or off-white, fluffy. Tail as adult, but with ill-defined greyish-buff or pink-buff tips c. 1–2 mm wide; flight-feathers, greater upper primary coverts, and tertials with wider pink-buff, pale buff, or off-white fringes

along outer webs and tips: pale fringes on tips of outer primaries c. 1 mm wide (virtually absent in adult), on inner primary coverts 1–2 mm wide, pale buff (in adult, less than 1 mm and pure white). Lesser and median upper wing-coverts dull grey with broad and ill-defined pale buff tips (almost concealing grey); greater dull black or greyish-black with pink-buff or pale buff outer fringes and tips (in adult, fringes virtually absent along outer webs and pink-cream or white on tips). FIRST ADULT MALE. Like adult ♂, but juvenile flight-feathers, tertials, tail, greater upper primary coverts, and variable number of greater upper wing-coverts retained (see Moults), browner and relatively more worn than neighbouring new black upper wing-coverts; differences best visible in outer greater upper primary coverts and (if retained) outer greater secondary coverts which show pale fringes of c. 1–2 mm wide along outer webs (virtually no pale edge in adult) and 2–3 mm wide along tip (narrower in adult, especially on primary coverts). Head and body basically similar to adult, but pale fringes along feather-tips of black throat-patch wider, in autumn sometimes fully concealing black (least so on upper cheeks and ear-coverts), often still prominent by mid-winter, when throat-patch equally mixed black and off-white, and still showing as traces by March–April (in adult, usually fully black already by late autumn). FIRST ADULT FEMALE. Like adult ♀, but part of juvenile plumage retained, as in 1st adult ♂; best ageing character is greater upper primary coverts: in autumn, dark brown with pale buff fringes along tips over 1 mm wide (in adult, narrower, white); in spring, outer primary coverts brown, tips pointed and without pale fringes, but tips still broadly pale buff on innermost (in adult, greyish-black with narrow white fringes on all coverts still present); by summer, some indistinguishable. Head and body as in adult ♀, but sides of neck, cheeks, and throat black or mottled-black in only 4 of 25 examined.

Bare parts. ADULT, FIRST ADULT. Iris brown or dark brown. Bill, leg, and foot black. (ZFMK, ZMA.) NESTLING. Mouth yellow. Gape-flanges whitish. (Panow 1974.) JUVENILE. Iris brown. Bill dark horn-brown, base of lower mandible paler; gape-flanges pale yellow. Leg and foot dark brown or black. (ZFMK, ZMA.)

Moults. ADULT POST-BREEDING. Complete; primaries descendant. *O. d. homochroa* starts late June or July, completing from late July; *atrogularis* in Iran moults late June to late July, in Afghanistan early July to early September, in Mongolia about mid-July to mid- or late August; in Ladakh and Rupshu, *oreophila* moults July, but in Kashmir, Sinkiang (China), and Pamir most birds start in first half of August, with some from late July and a few others not until early September, completing late August to late September. (Vaurie 1949; Dementiev and Gladkov 1954*b*; Paludan 1959; Piechocki and Bolod 1972; RMNH, ZFMK, ZMA.) POST-JUVENILE. Partial: head, body, lesser and median upper wing-coverts, tertial coverts, and variable number of greater upper wing-coverts; on average, 2·9 juvenile outer greater coverts retained in *homochroa*, 4·9 in *oreophila*, $n = 10$ in each (ZMA); rarely, all coverts replaced (Svensson 1984*a*). Timing rather variable, depending on hatching. Starts mid-June (some *homochroa* and *atrogularis*) to late August (some *oreophila*), completed late July to mid-September (Vaurie 1949; Dementiev and Gladkov 1954*b*; RMNH, ZFMK, ZMA).

Measurements. ADULT, FIRST ADULT. *O. d. homochroa.* North-west Africa, all year; skins (RMNH, ZFMK, ZMA). Juvenile wing and tail refer to retained juvenile wing and tail of 1st adult. Bill (S) to skull, bill (N) to distal corner of nostril; exposed culmen on average 4·4 less than bill (S).

	♂		♀	
WING AD	91·2 (2·16; 14)	88-95	86·8 (1·81; 8)	84-90
JUV	89·4 (1·14; 9)	88-92	85·0 (2·27; 10)	81-88
TAIL AD	59·8 (2·39; 14)	56-63	58·1 (1·27; 8)	57-60
JUV	57·6 (2·79; 9)	54-62	56·8 (3·08; 10)	52-60
BILL (S)	17·9 (0·53; 22)	17·0-18·8	17·4 (0·74; 17)	16·2-18·4
BILL (N)	10·0 (0·53; 22)	9·0-11·0	9·9 (0·61; 17)	8·9-10·9
TARSUS	24·9 (0·86; 23)	23·5-26·5	24·8 (0·83; 18)	23·4-26·0

Sex differences significant for wing and bill (S).

Nominate *deserti*. Southern Egypt, northern Sudan, and Saudi Arabia, all year; skins (RMNH, ZFMK, ZMA). Ages combined.

	♂		♀	
WING	92·3 (1·07; 20)	88-96	89·5 (1·70; 6)	87-92
BILL (S)	17·4 (0·56; 20)	16·4-18·2	17·3 (0·48; 6)	16·7-17·9

Sex differences significant for wing. Tail 56·9 (2·10; 9) 54-61, tarsus 24·1 (0·67; 8) 23·8-25·5.

Breeders, Sinai: wing, ♂ 88·9 (1·96; 6) 86-92, ♀ 85; bill (S), ♂ 17·9 (0·18; 6) 17·6-18·1, ♀ 16·9 (ZFMK).

O. d. atrogularis. (1) Iran, Turkmeniya, and Uzbekistan (USSR), summer (RMNH, ZFMK, ZMA); (2) Mongolia, summer (Piechocki and Bolod 1972; Piechocki *et al.* 1982); (3) Afghanistan, migrants (mainly Seistan, March) (Paludan 1959). Bill measured to skull. All ages combined.

	♂		♀	
WING (1)	95·4 (2·12; 7)	91-98	88·8 (—; 3)	87-90
(2)	94·5 (2·15; 29)	91-98	91·5 (1·98; 13)	89-97
(3)	93·6 (2·67; 21)	89-99	91·2 (2·05; 8)	89-95
BILL (1)	17·8 (0·61; 7)	17·2-18·6	17·4 (—; 3)	17·2-17·5

Sex differences significant for wing. Wing of adult ♂, summer, eastern Iran: 94·0 (16) 91-96 (Vaurie 1949).

O. d. oreophila. Southern Sinkiang (China) and Kashmir, March-September; skins (RMNH, ZFMK, ZMA). Ages combined.

	♂		♀	
WING	100·1 (2·31; 16)	96-104	95·0 (0·94; 5)	94-96
TAIL	62·5 (2·87; 13)	58-68	58·5 (1·50; 5)	56-60
BILL (S)	18·0 (0·57; 15)	17·3-18·9	18·3 (0·75; 5)	17·3-19·0
BILL (N)	9·8 (0·43; 12)	9·3-10·6	10·0 (0·89; 5)	9·0-10·7
TARSUS	26·6 (0·55; 13)	25·9-27·6	26·1 (0·64; 6)	25·3-26·8

Sex differences significant for wing and tail.

Afghanistan, migrants, late September to November, wing: ♂ 101·3 (1·56; 11) 98-103, ♀♀ 90, 95 (Paludan 1959); adult ♂ 100·6 (11) 97-105 (Vaurie 1949). Tibet: ♂ 100·1 (81) 95-107, ♀ 95·6 (27) 92-101 (Vaurie 1972).

Weights. *O. d. homochroa.* Algeria, January: ♂ 26 (Niethammer 1963*b*).

O. d. atrogularis. Afghanistan (Paludan 1959), Kazakhstan (Dolgushin *et al.* 1970), Iran (Desfayes and Praz 1978), and Mongolia (Piechocki and Bolod 1972; Piechocki *et al.* 1982), combined: (1) March, (2) May, (3) June, (4) July-August, (5) September-October.

	♂		♀	
(1)	19·5 (1·74; 19)	17-22	17·1 (1·57; 7)	15-19
(2)	19·0 (1·58; 16)	16-23	20·6 (1·99; 7)	18-23
(3)	19·1 (1·31; 12)	18-22	20·2 (1·47; 6)	18-22
(4)	20·5 (1·34; 6)	19-22	20·4 (—; 2)	19-21
(5)	20·7 (1·20; 7)	19-23	18·8 (—; 2)	17-21

O. d. oreophila. Afghanistan, late September and October: ♂ 23·1 (1·93; 11) 18-34, ♀♀ 17, 17 (Paludan 1959).

Structure. Wing rather long, broad at base, tip bluntly pointed. 10 primaries: p8 longest, p9 2-6 shorter, p7 0-1 (rarely, p7 0·5 longer than p8), p6 2-4, p5 7-12, p4 12-18; in *homochroa*, nominate *deserti*, and *atrogularis*, p1 18-26 shorter, p10 44-56;

in *oreophila*, p1 23-30, p10 53-61. P10 reduced, 3 longer to 3 shorter than longest upper wing-covert. Outer web of p6-p8 and inner of p7-p9 emarginated. Tertials short. Tail rather short, tip square; 12 feathers. Bill rather short (relatively shortest in *oreophila*); straight and slender, but rather wide at base; tip of culmen slightly decurved. Nostrils small, oval; partly covered by membrane above. Some fairly long bristles at base of upper mandible, finer ones near nostril and on chin. Tarsus and toes rather long and slender. Middle toe with claw 16·0 (15) 15-17·5 in *homochroa*, *atrogularis*, and nominate *deserti*, 17·0 (10) 16-18 in *oreophila*; outer toe with claw *c.* 71% of middle with claw, inner *c.* 66%, hind *c.* 73%. Claws rather short, slender and sharp, slightly decurved.

Geographical variation. Marked, but largely clinal and only western race *homochroa* and eastern *oreophila* distinct, others variably intermediate. *O. d. homochroa* from North Africa is small, mantle and scapulars pink-cinnamon or slightly vinous, chest pink-buff, white wedge on inner webs of flight-feathers narrow, 2-4 mm wide at tips of coverts. In central part of distribution, from Levant east through Middle East and Afghanistan to Mongolia, 1-3 races recognized, all differing from *homochroa* in more saturated greyish-cinnamon or sandy-grey mantle and scapulars and deeper buff or cinnamon-buff chest; differ from each other only in size (slightly). They are: (1) nominate *deserti* from Levant, (2) *salina* (Eversmann, 1850) from plains of Iran and USSR from Urals and Turkmeniya east to foothills of Altai, and (3) *atrogularis* from eastern Tien Shan and Altai to Mongolia. Average wing length of ♂ *deserti* 92·3 (RMNH, ZFMK, ZMA), of *salina* 92·4 (Stepanyan 1978), 94·0 (Vaurie 1949), or 95·4 (RMNH, ZFMK, ZMA), and of *atrogularis* 94·6 (Stepanyan 1978) or 94·5 (Piechocki and Bolod 1972; Piechocki *et al.* 1982). Vaurie (1959) followed here in including *salina* in *atrogularis* and retaining nominate *deserti*, but inclusion of all in nominate *deserti* can equally well be advocated. Boundaries of central races with *homochroa* in the west and with *oreophila* in the east not sharply defined and still in dispute: e.g. small series of birds from former breeding population in Sinai inseparable in colour from *homochroa* (ZFMK; see also Meinertzhagen 1930) and even smaller in size (see Measurements), though usually considered to belong to nominate *deserti*, and breeders from Afghanistan either considered to be short-winged (like *atrogularis*) but with large amount of white on flight-feathers (like *oreophila*) (Paludan 1959) or long-winged without much white (Vaurie 1949). *O. d. oreophila* from mountains of Pamir, Kun Lun, and Nan Shan southwards similar in colour to nominate *deserti* and *atrogularis*, but distinctly larger and with much white on inner webs of flight-feathers, reaching shaft over basal $\frac{1}{3}-\frac{1}{2}$ of feather in ♂; tail often less extensively black than in other races, maximum extent of black on t1 38 (13) 35-42 mm (as in *homochroa*), minimum on t2-t5 30·2 (13) 23-38, maximum on t6 31·7 (13) 25-39 (ZMA).

Forms species-group with Finsch's Wheatear *O. finschii* and Mourning Wheatear *O. lugens*.

Recognition. Easily identified by large amount of black on tail-tips, forming square block rather than ⊥-mark. Largely black tail shared only by Red-rumped Wheatear *O. moesta*, which shows quite different plumage in both sexes and has more rounded wing and much longer bill and tarsus. Amount of black approached by Isabelline Wheatear *O. isabellina* (in fact, pattern of some *oreophila* similar to that species), which is larger, with relatively shorter tail and longer bill and tarsus, and which has paler and more uniform wing. CSR

Oenanthe finschii Finsch's Wheatear

Du. Finsch' Tapuit   Fr. Traquet de Finsch   Ge. Höhlensteinschmätzer
Ru. Черношейная каменка   Sp. Collalba oriental   Sw. Finschstenskvätta

Saxicola Finschii Heuglin, 1869

Polytypic. Nominate *finschii* (Heuglin, 1869), Turkey and Levant, south to Dead Sea region, east to Taurus mountains and south-east Turkey, grading into *barnesi* in Zagros mountains in Iran and probably in eastern Iraq; *barnesi* (Oates, 1890), eastern Turkey, Transcaucasia, eastern Caucasus, northern Iran, and from eastern shore of Caspian Sea and central Iran east to Syr Darya, Afghanistan, and western Pakistan.

Field characters. 14 cm; wing-span 25-27 cm. Western race slightly smaller than Wheatear *O. oenanthe oenanthe*, with more rounded wings; close in structure and size to Pied Wheatear *O. pleschanka* and Black-eared Wheatear *O. hispanica* but with more rounded wings; smaller (but heavier) than Mourning Wheatear *O. lugens*; thicker-legged than all. Small to medium-sized, often shy wheatear, with plumage pattern overlapping those of at least 3 congeners: ♂ most recalls black-throated *O. hispanica* but differs in black of throat being extended to meet black of wings; thus also suggests *O. pleschanka* and *O. lugens*. Underwing and tail pattern in ♂ can recall *O. lugens* but underwing duller. ♀ even less distinctive, but has black throat-band. Sexes markedly dissimilar in spring, less so in winter. Juvenile separable. 2 races in west Palearctic, difficult to separate in the field.

ADULT MALE. (1) Western race, nominate *finschii*. In spring (when plumage worn), top of head, hindneck, mantle, back, rump, and most of tail essentially white; birds which are not too worn are faintly buff to ochre-grey as far down as mantle. Face, throat, and upper breast (joining shoulder), all scapulars, and wings black, wings usually retaining obvious grey or dull white fringes to primary coverts, inner greater coverts, and inner secondaries and tertials. Upperparts thus have white central panel squeezed in by rather broad black lateral panels; white panel thus noticeably narrower than in most *O. hispanica* (due to completely black scapulars and union of black throat with shoulder). Black ⊥ on tail like that of *O. oenanthe* but at least 3 outermost pairs narrowly but distinctly tipped white, as in *O. lugens*. Underbody below upper breast dull white, washed buff on lower breast and more distinctly salmon-buff on vent and under tail-coverts, as in eastern races of *O. lugens*. Under wing-coverts and axillaries black, contrasting with dusky, basally white-streaked undersurface to outer flight-feathers. From July (when plumage fresh), loses full contrasts of spring plumage: upperparts from crown to back heavily tipped grey-isabelline and thus contrasting with white rump; throat becomes hoary and underparts generally less white with stronger vinous-buff tinge; wing-feathers heavily tipped and fringed cream and pale rufous-buff. (2) Eastern race, *barnesi*. Averages 10% larger and longer-winged than nominate *finschii*. Pale

plumage areas rather dirtier and buffier; under tail-coverts darker. ADULT FEMALE. (1) Nominate *finschii*. Except in tail pattern, quite unlike ♂ with confusing plumage colours recalling those of many other ♀ wheatears, especially *O. lugens halophila*. Head and whole back dull brown with grey tone, little marked except by indistinct buff-white supercilium and darker, more rufous cheeks, ending in black patch which usually extends downward to form mottled band across lower throat; occasionally, whole throat grey-brown. Wings darker grey-brown, with pale cream or buff fringes and tips to black-centred larger wing-coverts. Black on tail duller than in ♂. Rest of underparts buff-white, with buff under tail-coverts less isolated than on ♂. Under wing-coverts duller than in ♂, with partial buff suffusion sometimes creating illusion of paleness. Fresh autumn plumage much as at other times, but even darkest throats overlaid by grey and wings heavily pale-fringed. (2) *O. f. barnesi*. Closely resembles nominate *finschii* but head and back tinged sandy, not grey, and throat more often all-dark. JUVENILE. Both races. Resembles adult ♀ but has even broader pale fringes to wing-feathers. Lacks spotted and scaled appearance of most similar juvenile wheatears. Both races at all ages have bill and legs black.

Identification less studied than most other wheatears; appearance can be argued to fall within racial variation of *O. lugens*, most resembling palest (usually eastern) birds of that species, includes closely similar tail and underwing markings (see Geographical Variation). Flight often darting or banking, with sudden rise or falls and changes of direction. General behaviour similar to other *Oenanthe*. Often extremely wary, with long escape-flight; when breeding, hides persistently.

Song a scratchy warble but including rich fluty whistles. Calls include 'zik' and 'chek', combined in alarm with descending 'seep'.

Habitat. In continental lower-middle latitudes of south-east sector of west Palearctic, in dry warm temperate and steppe zones. Breeds in USSR on bare clay sands, rocky steppes, and ravines heaped with stones in low mountains, up to zone of pistachios at *c.* 1400-1600 m (Dementiev and Gladkov 1954*b*). On east coast of Caspian, occurs on rocky foothills and on plains, nesting in rodent holes

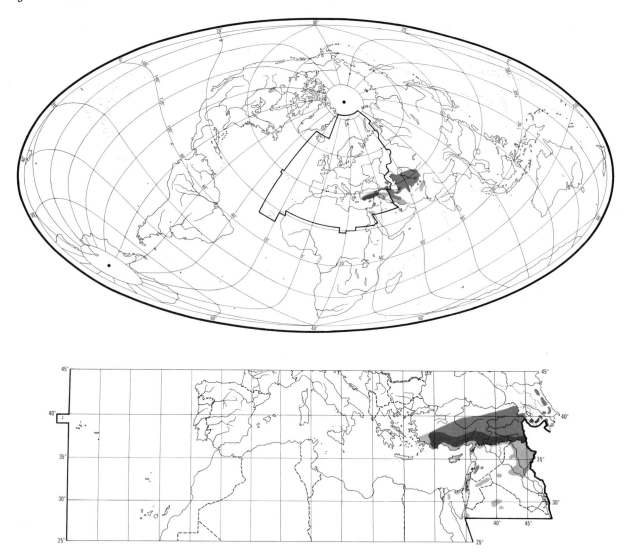

(Mitropol'ski 1968*b*). In Afghanistan, found only locally in north in foothills at 700-950 m, a typical habitat being within a canyon with outcrops of rock and lacking trees and bushes (Paludan 1959). In Baluchistan (Pakistan), frequent from mid-August along roadsides on high upland steppe-like terrain up to *c.* 2000 m, perching on *Artemisia* and standing on banks; habitat at this season shared with Isabelline Wheatear *O. isabellina* (E M Nicholson). In Lebanon, occurs on stony ground and outcrops of rock in mountain ravines or on hillsides, especially rock-rims in arid country (Vere Benson 1970). In Anatolia (Turkey), confined to rocky hillsides, preferring steep re-entrants; perches on prominent rocks (Wadley 1951). In mountains to east, found breeding up to 2400 m on bare rocky slopes or large boulder screes; less commonly in gullies and among scattered bushes (Gaston 1968).

In south-west Iran, all breeding territories studied included bare ground, boulder-strewn and rocky areas,

and gently sloping or flat land. In most, ground mainly stony, but birds often fed where stones were few. Plants were low-growing steppe species in patches. Breeds more often on gently sloping hillsides than high up on steep hillsides like Mourning Wheatear *O. lugens* with which, however, habitat overlaps. Winters widely from sea level to *c.* 1900 m in the steppe, sub-steppe and deforested parts of the xerophilous forest zones (Cornwallis 1975). Generally characteristic of degraded or rocky steppe with sparse shrub vegetation but no trees, and of narrow ravines and valleys with stony dry hillsides and rock outcrop; rarely occurs in cultivation and does not require a water source, but in winter may occur within cultivation as well as semi-desert. Not much in contact with man. Although darting undulating flight differs from congeners, it resembles them in confining itself to lower airspace.

Distribution. Little information on range changes.

TURKEY. May be expanding range slowly (Kumerloeve 1975). IRAQ. May breed near Kirkuk but no proof (Marchant and Macnab 1962).

Accidental. Kuwait.

Population. No information on trends.

USSR. Not abundant (Dementiev and Gladkov 1954*b*).

Movements. Partially migratory. Difficult to determine status in some parts of range, with wintering birds of unknown origin swelling local, probably resident, populations. Altitudinal differences in migratory status may occur: in south-west Iran, a summer visitor to Tang-i-Shul (2200 m) and other breeding grounds exceeding *c*. 1900 m but some may be resident lower down (Cornwallis 1975).

Wintering records from southern Turkmeniya, Transcaucasia, and Kazakhstan in USSR (Dementiev and Gladkov 1954*b*; Dolgushin *et al.* 1970; Portenko and Vietinghoff-Scheel 1971; see also below), but main west Palearctic wintering areas lie in southern Turkey and northern Iraq south to eastern Egypt and northern Saudi Arabia. Locally common on Cyprus (Flint and Stewart 1983). Winters regularly in Syria (Kumerloeve 1969*b*; Macfarlane 1978); fairly common at Azraq, Jordan (Clarke 1980), but very scarce in Gulf states of Arabia (Bundy and Warr 1980). Numerous and widespread in south-west Iran up to *c*. 1900 m, chiefly immigrants but probably including some residents; ♂♂ outnumber ♀♀ which probably tend to winter further south (Cornwallis 1975). In USSR, up to 70% of birds breeding in south-west Uzbekistan also winter there (Panov 1978); according to Mitropol'ski (1968*b*), if weather mild some ♂♂ and occasionally also ♀♀ overwinter near breeding grounds in (e.g.) Kyzylkum, southern Turkmeniya, and Uzbekistan, exceptionally to 45°N. For fidelity to winter territory, see Social Pattern and Behaviour.

Passage recorded on south coast of Turkey (*Orn. Soc. Turkey Bird Rep. 1966–75*) and a few (chiefly autumn) records at Elat, Israel (Safriel 1968). Spring passage recorded on Cyprus (Flint and Stewart 1983). Birds sometimes disperse irregularly for several weeks in late summer or early autumn (Portenko and Vietinghoff-Scheel 1971). Departure from breeding grounds probably from September (Portenko and Vietinghoff-Scheel 1971), though substantial influx into mid-Baluchistan (Pakistan) recorded from mid-August (E M Nicholson). Some populations, however, remain until late October (in central plateau of Turkey: *Orn. Soc. Turkey Bird Rep. 1966–75*) or early November (Mangyshlak peninsula in Kazakhstan: Dolgushin *et al.* 1970). Present on Cyprus chiefly from late September (Bannerman and Bannerman 1971); in Iraq from mid-October (Allouse 1953), Jordan from late October (Clarke 1980). Wintering birds arrive in south-west Iran from end of August to October (Cornwallis 1975).

Return passage starts late February and most wintering birds depart by mid-March. Arrival on breeding grounds recorded from early March on Mangyshlak, ♂♂ *c*. 2 weeks before ♀♀ (Dolgushin *et al.* 1970); arrives mid-March in Armeniya (Dementiev and Gladkov 1954*b*). MGK

Food. Mainly insects (especially ants and beetles); also some seeds and other plant material. In south-west Iran, 97% of all prey caught were on ground (*n* = 817). In breeding season, 52% of 25 feeding manoeuvres involved short flights, 24% involved run along ground, and 24% involved jabbing at ground, under stones (etc.) from stationary position. In winter (when more often on flat ground), 80% of 792 manoeuvres involved dashes along ground, 11% short flights, and 10% stationary. (Cornwallis 1975.) ♂ watched in eastern Saudi Arabia used different foraging techniques in different habitats: (1) in lightly vegetated desert, perched frequently on small shrubs and stones, rarely spending longer than a few seconds on each (unlike Mourning Wheatear *O. lugens* and White-crowned Black Wheatear *O. leucopyga*) before taking prey or hopping (sometimes flying) to new perch; (2) on bare stony desert, commonly alternated pauses (bobbing, and flicking tail each time) with hopping for a few metres (almost a rapid run). Bird also made 2 flycatching sallies, one to *c*. 1½ m above ground, other almost vertically to *c*. 6 m. (J Palfery.) Similar flycatching by ♂♂ also reported in Transcaucasia (Raethel 1955). In freezing temperatures in spring, most food probably taken from holes (etc.) in rocks (E N Panov). In south-west Iran, forages throughout the day in winter, with slight lull around midday, and maximum activity towards evening. At other times of year, fed mainly during morning and evening when prey most readily available. (Cornwallis 1975; L Cornwallis.)

In south-west Iran, 31 stomachs (5 from June–July, rest November–February) contained 2160 items: ants (Formicidae) in 90% (88% of total items), beetles (Coleoptera) in 61%; other items included Orthoptera, bugs (Hemiptera), termites (Isoptera), lacewings (Neuroptera), various insect larvae, spiders (Araneae), woodlice (Isopoda), seeds, and other plant material (Cornwallis 1975). In Tadzhikistan (USSR), 4 stomachs (2 February, 1 December, 1 July) contained beetles, flies (Diptera), Hymenoptera, etc. (Abdusalyamov 1973); in Kyzylkum, takes mainly spiders and larvae of moths (Lepidoptera); in Turkmeniya, takes ants, beetles, Orthoptera, Lepidoptera larvae, and occasionally also vertebrates and plant food (Dolgushin *et al.* 1970). In USSR, also recorded taking woodlice, rarely large beetles (Tenebrionidae); one ♂ took a large harvestman (Opiliones) and spent a long time dismembering it (E N Panov).

Food for young includes adult and larval beetles, caterpillars, and grasshoppers (Acrididae) (Panow 1974; L Cornwallis). BDSS, EKD

Social pattern and behaviour. Most important studies are Panow (1974), Panov and Ivanitski (1975a), and (especially) Panov (1978), all for (mostly extralimital) USSR. Account also includes detailed notes supplied by E N Panov.

1. Mostly solitary. Migratory birds and at least some sedentary ones defend individual territories outside breeding season. In Cyprus, apparently holds same territory all winter (Flint and Stewart 1983), and in southern Palestine one stayed within less than c. 0·5 ha from 20 October until at least January (Meinertzhagen 1920a). In south-west Iran, where wintering population comprises immigrants and perhaps some residents, both sexes hold individual territories throughout winter; no evidence that breeding territories retained. Average size of ♂♂'s territories varied from 3 ha (n=16) in good habitat to 5·8 ha (n=4) in poorer; ♀♀'s territories, always in poorer habitat, 5·3 ha (n=4). 3 out of 9 marked ♂♂ returned to same territory in 2 consecutive winters; one of the 3 also held the same territory the following (3rd) winter, as did 2 others not monitored in the intervening winter. (Cornwallis 1975; L Cornwallis.) In the partly sedentary population of south-west Uzbekistan (USSR), some pairs apparently stay together, occupying same home-ranges all year (boundaries not clear, and only core area defended), other pairs split up after breeding season and pair-members occupy individual home-ranges. A few juveniles (almost all ♂♂) also establish individual territories after breeding season. In October, 6 ♂♂, 4 ♀♀, and 2 juveniles in 2 km². 3 of the 6 ♂♂ shared their territories with ♀♀ (presumed mates from previous breeding season), rest solitary. Overall, of adults that had completed moult, 4 out of 16 ♂♂ and 4 out of 6 ♀♀ stayed paired in early October, rest solitary. (Panov 1978.) BONDS. Mating system essentially monogamous but occasionally polygynous; after ♀♀ start incubating, ♂♂ (mates) tend towards bigamy, sometimes pairing simultaneously with another ♀ in a 2nd territory (Panow 1974: see also Breeding Dispersion, below). Monogamous pair-bond evidently lasts only for breeding season in most populations (no information on renewal in successive years, but see below for site-fidelity), but in some sedentary populations, bond apparently maintained all year (see above). Only ♀ broods, while both sexes feed young, ♂ taking over feeding of 1st brood if ♀ starts 2nd clutch. Age of independence not known, but fledged brood remains together for a long time and, even after independence, 1st brood forms loose group with ♂ parent. (Panow 1974.) No information on age of first breeding. BREEDING DISPERSION. Solitary and territorial. In USSR, among sedentary southern populations pairs often isolated (e.g. in Tadzhikistan, pairs 1·5-2 km apart: Abdusalyamov 1973), but migratory populations further north form neighbourhood groups: e.g. dense concentrations on Mangyshlak peninsula, east Caspian (Panov and Ivanitski 1975a); in Nakhichevanskaya ASSR (south of Caucasus), fairly dense groups, each of 10-15 pairs, but often broken up by unoccupied ground. In sedentary populations, pairs occupy large home-ranges, often not contiguous. In migratory populations (high density), earliest-arriving ♂♂ acquire largest territories. (E N Panov.) On Mangyshlak, territory size inversely related to density, average 2·9 ha (0·7-10·0), n=20 (Panov and Ivanitski 1975a). In Transcaucasia, territories 106-230 m across (E N Panov). Territory of bigamous ♂, in which both ♀♀ bred, was 800 m long, one ♀ at each end (Panow 1974). Limited information on densities. In south-west Uzbekistan, 4·5 pairs per km² over 2 km² (Panov 1978). On Mangyshlak, c. 10 pairs per km² over 5 km² (Panov and Ivanitski 1975a). Territory of all-purpose type. Successive broods (within a season) never reared in same nest, though traditional sites may be re-used over the years, leading to accumulation of stones at entrance (see Breeding); ♀ seeks new site and starts building in the week after 1st brood fledges. (Panow 1974.) Little information on site-fidelity between years, but one ♂ returned to same territory and one ♀ returned to breed 150 m from previous year's territory (Cornwallis 1975). ROOSTING. In Iran, often uses rodent holes, entering at dusk and leaving soon after dawn (L Cornwallis). Adults sunbathe in prostrate posture with wings outspread (Panow 1974, which see for illustration). When captive young saw water for the first time they became excited but were very reluctant to enter it, though they dust-bathed in adjoining sand; bathed (presumably in water) very rarely during first 4 months of life. (E N Panov.)

2. In winter in Saudi Arabia, birds are typically wary and difficult to approach (J Palfery). In south-west Iran, approachable to c. 10 m (L Cornwallis). In hard winters, Transcaucasia, weakened birds were tame when visiting rubbish tips to feed (Raethel 1955). Often perches on prominent rocks, rarely bobbing (Wadley 1951). Bobbing also unusual in Saudi Arabia, though, on landing, bird occasionally makes shallow bob, flicking wings, and flicking and flirting (momentarily spreading wider) its tail; may also wag tail up and down slightly, or cock tail for a while before lowering it (J Palfery). These movements also confirmed by Panow (1974, which see for drawings) and E N Panov: intention-movements typically involve up-and-down movement of tail; in great excitement, tail raised vertically and even a little forward, in short jerks. Also typical, near a new and unfamiliar object, is a slow opening and forward movement of the wings. FLOCK BEHAVIOUR. None recorded. SONG-DISPLAY. Following account based on Panow (1974) and E N Panov. ♂ sings regularly during pair-formation, from perch and in Song-flight (see 1a-b in Voice). Young ♀♀ sing just as frequently as ♂♂ (see Heterosexual Behaviour, below, for song and Song-flights by ♀♀). Each ♂ has 2-3 preferred song-posts, usually on highest point of a hill. Singing ♂ runs back and forth in characteristic posture with tail widely spread and head raised (see Panow 1974, Fig 46a), then launches into Song-flight: ascends, singing, with shivering wing-beats and tail widely spread, then lands on another song-post. Sometimes hovers for a few seconds, and may describe 2-3 tight circles, fly in a flat curve, or drop a little way and then glide to new perch with wings fully extended. In course of Song-flight, Short Territorial-song (see 1a in Voice) often changes to longer variant (1b in Voice). On landing, again sings whilst running back and forth before ascending once more. Short display-flight, apparently of this kind, described by Wadley (1951) for Turkey: bird throws itself into the air up to (e.g.) c. 3 m and tumbles down to another rock. No details on duration of song-period, but see Heterosexual Behaviour (below) for period of pair-formation and thus likely song-period. In winter in south-west Iran, ♂♂ commonly give weak song in territorial defence (L Cornwallis). Subsong (see 1d in Voice) reported from Iraq in September (Moore and Boswell 1956), and in Transcaucasia in winter given from roofs of houses (Raethel 1955); see also Antagonistic Behaviour and subsection 4 of Heterosexual Behaviour (below) for further song variants. ANTAGONISTIC BEHAVIOUR. (1) General. From arrival on breeding grounds, both sexes defend territory fiercely against rivals, ♂ typically against ♂, ♀ against ♀; resident ♂ does not usually intervene to help mate in ♀-♀ disputes, nor vice versa. ♂♂ also dominant over all other Oenanthe in USSR range (Panov and Ivanitski 1975a). Of 252 attacks on wide variety of passerines, 92% were on 6 Oenanthe species, mostly Pied Wheatear O. pleschanka, Black-eared Wheatear O. hispanica, and Red-tailed Wheatear O. xan-

thoprymna. (E N Panov.) For dominance over, and territorial relationship with, *O. pleschanka*, see Panov and Ivanitski (1975a), also *O. pleschanka* (p. 798). In winter in south-west Iran, territory-holders residents fiercely attack passerine intruders including all *Oenanthe* except Hume's Wheatear *O. alboniger* (Cornwallis 1975). (2) Threat and fighting. Aggression signalled by elaborate repertoire of ground- and aerial-displays. On Mangyshlak, 41% of 51 disputes involved raising and lowering of tail, 35% tail-fanning, 45% raising or lowering of bill, 29% wing-flicking, 45% sleeking head plumage (Ivanitski 1980). Following account compiled from Panow (1974), Panov (1978, which see for illustrations of all threat-postures and -displays), and information from E N Panov. ♂ challenges any ♂ intruder in his territory, and fierce disputes common, especially during period of main arrival on breeding grounds. Typical sequence as follows. Resident flies directly towards rival, singing the while, and lands nearby. Before approaching rival, ♂ performs Forward Threat-display (Fig A): crouches forward with body

twisting and turning (possibly Aerial-dance: see Heterosexual Behaviour, below), singing incessantly (Long Territorial-song and Subsong). Aerial chasing comprised up to 70% of ♂-♂ confrontations which may continue, alternating between air and ground, for 6 hrs or more, at end of which opponent sometimes visibly exhausted. Fights occur occasionally (recorded in 2 out of 20 conflicts): rivals become interlocked, pecking, and whirling over the ground. Protracted ♂-♂ and ♀-♀ disputes characterized by Tchok- and Seep-calls (see 3-4 in Voice) and, between ♀♀, Zik-calls and Rattle-calls (see 2 and 5 in Voice). On Mangyshlak, of 51 disputes 16 involved ♀♀ (Ivanitski 1980). ♀-♀ disputes similar to those described (above) for ♂♂, and most often arise when ♀ tries to settle in territory already occupied by pair. Resident ♀ furiously confronts such intruders by combination of singing, short Song-flights, and threat-display (Fig C,

A

C

plumage ruffled, head tucked in, bill raised, and tail fanned and flicked rapidly upwards. At territory boundary, bowing indicates high-intensity threat. May also raise fore-body and head to expose black throat. If opponent undeterred, the 2 birds may approach to within 1-3 m of each other, staying parallel, and on slightly bent legs, make creeping runs in Horizontal Threat-display (Fig B): tail widely spread and

showing 2 threat-postures) in which, as in ♂'s displays, plumage ruffled and exposed to most intimidating effect. ♀♀ also commonly use Horizontal Threat-display (as in Fig B for ♂). Prior to nest-building, each of 2 mates of bigamous ♂ vigorously defended its section of the territory against other ♀♀. For aggressive elements in heterosexual encounters, see below. HETEROSEXUAL BEHAVIOUR. (1) General. On Mangyshlak, ♂♂ arrive first (6-13 March), occupy territories straight away, and start to sing and display. ♀♀ arrive some days later (Mitropol'ski 1968b). In Nakhichevanskaya ASSR, in 2 cases ♂♂ paired 1 and 2 days after arrival. In Uzbekistan (breeding and wintering area), some pairs form in October and not later than February. (E N Panov.) In study by Stepanyan (1970), ♂♂ older than 1 year paired in second half of January. (2) Pair-bonding behaviour and nest-site selection. Account based on information from E N Panov. When ♀ enters ♂'s territory, ♂ makes creeping run in Horizontal Threat-posture (Fig B), then Long Butterfly-flight (Fig D) in which bird flies straight and low

B

somewhat depressed, fore-body lowered but head may be raised and lowered. Rivals also run into holes, etc. Resident may also try to expel intruder in often-lengthy pursuit-flight, sometimes

D

E

from one raised perch to another, maintaining posture like Fig B in flight and giving intense Subsong. Similar display described from Turkey: level fluttering Song-flight across ravine, tail spread (Wadley 1951). Destination of flight is rocky recess previously selected by ♂ as potential nest-site. ♂ lands above recess, still in Horizontal Threat-posture, and enters it by creeping run, sometimes several times in succession (Nest-showing), and stays there up to 30 s. When ♀ arrives in territory she not infrequently gives short songs, sometimes Song-flight. Panow (1974) shows ♀ inviting courtship with posture similar to that of general excitement: jerks raised closed tail forwards and flicks partly drooped wings. When ♂ enters recess, ♀ usually flies up but stays 1–3 m away. Usually, ♂ then performs Long Butterfly-flight to another recess, and self-advertising and Nest-showing sequence is repeated. Typically, ♂ and ♀ do not approach each other closely at this stage, and close encounter may lead to aggression by ♀, then chasing by ♂, rarely culminating in renewed approach and Aerial-dance (Fig E, left): ♂ flies in swift circular flight near ♀, occasionally 'looping the loop' as he brushes his breast against rocky overhang under which ♀ perches. (3) Courtship-feeding. None reported. ♂ does not feed ♀ on nest (Panow 1974). (4) Mating. Takes place in concealment of rocky niche (not known if same one as that described above for Nest-showing); preceded by elaborate soliciting-display by ♂. Following account based on Panov (1978) and information from E N Panov. ♂ spends a long time flying around near mating site, either silently or giving variant of Short Territorial-song (see 1c in Voice). Sometimes performs Song-flight and also advertises with variety of displays, notably ruffling plumage and shivering drooped

wings (Fig F). ♂ (also ♀) may also mock-preen for lengthy periods. ♂ then flies to chosen site and runs about there, alternately entering and re-emerging, drawing attention by voice and display; gives a pre-copulatory song (see 1c in Voice) until ♀ joins him there, and switches to an erect posture, raises bill and lowers spread tail, similar to an aggressive upright posture. When ♀ arrives several minutes later, ♂ responds with perhaps ambivalent displays, notably apparent variant of Horizontal Threat-display (legs straight and tail somewhat raised) and Forward Threat-display. ♂ also performs Aerial-dance, then lands by ♀ and performs intense Soliciting-display (Fig E, right) with outspread shivering wings. Receptive ♀ adopts an invitatory posture: bows slightly, somewhat raises and flirts tail, and partly droops open wings (Fig E, right). ♂ then mounts, giving Zik-calls. After copulation, ♂ may give Seep-call as he flies away. ♀ may rebuff ♂ at end (perhaps also beginning) of her fertile period. Copulation usually occurs 17.00–20.30 hrs. (5) Behaviour at nest. In favourable weather, ♀ may start building within 1 week of pairing (Panow 1974). Only 1 record of nest being completed in site shown by ♂ to ♀ during pair-formation (E N Panov). Even after building (by ♀ alone) begun, change of site not uncommon. According to Panow (1974), construction of stone platform by *O. finschii* (see Breeding) is a vestigial ritualized behaviour pattern no longer serving any useful function. No help given by ♂ who usually sings some way off and only visits ♀ briefly and occasionally. In interactions with ♀ just before and during nest-building, ♂ commonly adopts Hunched-posture, not seen at other times: wings closed and apparently drawn up above line of back (Fig G). Towards end of nest-building, ♂ much more attentive to

F

G

♀, often flying to nest-entrance, waiting for her to emerge, and adopting Hunched-posture when she appears. Initially, ♀ resists close approach of mate, but he chases her frequently in the air, and contacts gradually become closer, leading eventually to copulation. Once ♀ incubating, ♂ commonly adopts variant of Hunched-posture with tail spread. When incubating ♀ leaves nest to feed, ♂ may accompany her and display, staying with her until she returns to nest. (Panow 1974.) RELATIONS WITHIN FAMILY GROUP. ♀ removes eggshells after hatching and drops them several tens of metres away. ♀ broods young almost constantly for first 3–4 days. ♂ does most feeding of young at this time, while still defending territory and performing Song-flights. On approaching nest with food, ♂ continues to sing loudly, lands by nest-entrance, looks about briefly, then enters. After first few days, ♀ ceases brooding and relinquishes almost all care of young to ♂ until they start becoming feathered, whereupon she takes equal share in feeding them. Shortly after leaving nest, young disperse, sometimes quite widely. (Panow 1974.) ♀ may alternate helping to feed fledged 1st brood with building nest for 2nd clutch and laying, though ♂ takes care of 1st brood once ♀ sitting (E N Panov). For development of vocal repertoire of young, see Voice; for comfort behaviour of young, see Roosting. ANTI-PREDATOR RESPONSES OF YOUNG. After young leave nest, each seeks and stays near individual refuge-hole into which it disappears on hearing parental alarm-calls (Panow 1974; E N Panov: see 3–4 in Voice). Hole sometimes has one or more 'emergency exits' (Panow 1974). PARENTAL ANTI-PREDATOR STRATEGIES. Pair usually very wary during incubation (Panow 1974). Both sexes give alarm-calls (see 3–4 in Voice) when nest or young threatened by man, also in presence of owl (Strigidae) (see Voice). ♀ (especially) gives prolonged Tchok-call (see 3 in Voice) when highly alarmed (E N Panov). No further information.

(Figs by K H E Franklin: A–B and F from Panov 1978; C and G from Panow 1974; D–E from E N Panov.) EKD

Voice. Used mostly in breeding season. No information on geographical variation. Account based on scheme provided by E N Panov (USSR) who also supplied additional sonagrams.

CALLS OF ADULTS. (1) Song. Given by both sexes but mainly ♂. Extremely varied, often melodious but interspersed with harsh sounds. Description of song as scratchy but including rich fluty whistles (Moore and Boswell 1956) is typical. Mimicry rare, but includes calls of wide variety of bird species and of tree-frog *Hyla* (E N Panov). 4 song-types provisionally distinguished by E N Panov, though marked intergradation occurs. (a) Short Territorial-song. Given from perch, mostly by ♂♂, also not uncommonly by ♀♀, especially young birds, in territorial disputes (Panow 1974). Clearly structured, each ♂ assembling his own repertoire of favourite 'notes' into brief songs (E N Panov). More chattering than Long Territorial-song (1b) (Jonsson 1982), and comparable with short song of Whitethroat *Sylvia communis* (Raethel 1955). Almost always preceded by 1–5 Tchok-calls (E N Panov: see call 3). Described as a quick thin warble with sweetness and great charm; one song rendered 'chik-chik-chik-cheeoowee', sometimes followed by harsh 'crr-tch' (Wadley 1951). Recordings from Iran (P A D Hollom) and Turkey (see below) likewise show that

phrases often start with 'zek' or rasping sounds (equivalent to 'tchok'), though some phrases start with tones and finish with rasping sounds; other well separated units in song may also be of call-type (J Hall-Craggs; see, e.g., call 6). Marked variation between successive phrases. Figs I–III show phrases from bout of bird whose song variously combines tones, clicks, rattles, rasps, hisses, and other noises. Fig I begins with 3 rasps and ends with tones; Fig II starts with a rattle and ends with rasps and finally 2 tones; Fig III shows 2 remarkable sneezing

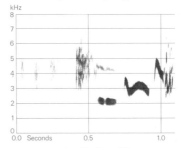

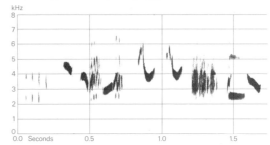

I C Chappuis Turkey May 1977

II C Chappuis Turkey May 1977

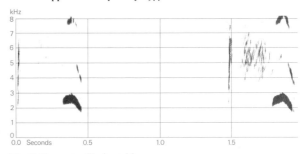

III C Chappuis Turkey May 1977

't-sss-ooo' phrases, each comprising respectively a click, a hiss, and a tone. Average length of 7 phrases by this bird 1·5 s, average interval between phrases *c.* 3·5 s, though phrases sometimes almost continuous. (b) Long Territorial-song. Given in Song-flight. Longer and less structured than Short Territorial-song, and typically lacks prominent 'tchok' units. (E N Panov.) Described as follows: a high-pitched and squeaky 'zee-widdy-widdy-widdy-widdy-tweee'; in one brief Song-flight, bird gave a repeated 'chiz-oo-wit' (not unlike White Wagtail *Motacilla alba*, but softer) on ascent, varied with 'ti-widdy-widdy' on descent (Wadley 1951). According to Jonsson (1982), short phrases resemble Black-eared Wheatear *O. hispanica*. See also call 6a. (c) Pre-copulatory songs. ♂ ready for copulation first gives rapid series of very quiet Short Territorial-songs. Once ♂ moves to site chosen for mating, and until ♀ appears, he switches to a long quiet monotonous song. (E N Panov.) (d) Subsong. Somewhat recalls Long Territorial-song but usually quiet and contains 'zik' calls (E N Panov). In recording (Fig IV), song, perhaps of this type, is

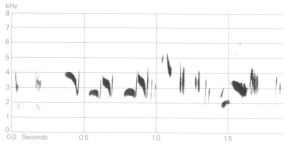

IV P A D Hollom Iran May 1977

pleasantly musical except for rather persistent interpolation of 'tk' sounds (appearing on sonagram as thin detached vertical lines), giving scratchy quality (J Hall-Craggs, P A D Hollom). Winter Subsong quiet, like winter song of Robin *Erithacus rubecula* (Raethel 1955). (2) Zik-call. A high-pitched (*c.* 8 kHz) 'zik zik zik' used by ♂ in various social interactions, notably during copulation, and introduced into Subsong (see above); also by ♀ during conflicts with other ♀♀ (E N Panov). Presumably a contact- or contact-alarm call. (3) Tchok-call. A 'tchok' (E N Panov), 'chek' or 'zek' (J Hall-Craggs), low-pitched 'trak-trak' (Raethel 1955). Equivalent call typical of *Oenanthe* repertoires. Given monotonously by both sexes (but especially by ♀) during disputes with rivals, also in alarm. Often included in Short Territorial-song (e.g. first 3 units of Fig I show low-intensity units of this type, also units preceding terminal tones in Fig II). In south-west Iran, high-pitched squeaking 'chep chup chup' from alarmed parents with 2 newly fledged young (L Cornwallis) is presumably high-intensity variant. In alarm, commonly combined with call 4, thus: 'seep chek seep chek chek' (J Hall-Craggs: Fig V, bird in presence of owl). Another recording by B N Veprintsev and V V Leonovich (Turkmeniya, USSR) shows same combination given by ♂ when recordist near nest. E N Panov described a long 'tchok' variant, given typically by ♀ in confrontations and when highly alarmed; sonagram supplied by E N Panov shows duration of this variant *c.* 0·3 s, compared with *c.* 0·1 s for normal Tchok-call. (4) Seep-call. A descending 'seep' of alarm, given by both sexes, commonly in combination with call 3 (see Fig V in which 1st and 3rd units are 'seep'). In recording by P A D Hollom (Iran) of pair with well-grown young, a quieter more plaintive 'cheep' or 'peep' (J Hall-Craggs). Seep-call also given by ♂ in pair-formation, when flying

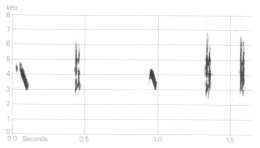

V C Chappuis Iran May 1977

away from ♀ after copulation, and in protracted disputes with other ♂♂ (E N Panov). (5) Rattle-call. A rattle of warning given by ♀ at approach of a strange ♂ (sometimes own mate) and in disputes with other ♀♀. No rendering given, but sonagram provided by E N Panov shows rattle not unlike 1st unit in Fig II. (6) Other calls. (a) Song of bird represented by Figs I–III includes strident ascending 'zeek zeek' sounds, followed by rising musical 'weet weet' (J Hall-Craggs: Fig VI). These are perhaps calls rather

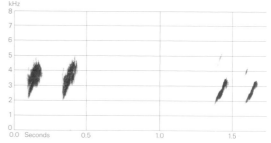

VI C Chappuis Turkey May 1977

than purely elements of song; 'zeek' is lower-pitched and longer than 'zik' (call 2) depicted in E N Panov's sonagrams. Squeaking 'weee' given by ♂ in Song-flight (Wadley 1951) is perhaps the same as 'weet' described from alarmed bird(s) at nest with well-grown young; perhaps a high-intensity warning- or alarm-call. (b) In pair-formation and other interactions with ♀, ♂ may give slightly modified juvenile food-call (E N Panov: see below). (c) Call described as 'too-tee too-tee' (Wadley 1951), not easily reconciled with any listed above.

Calls of Young. Sonagrams supplied by E N Panov indicate that food-calls resemble those of well-grown *O. oenanthe* nestlings (see p. 787), i.e. presumably a buzzing wheezing sound. Captive young first gave adult calls 3–4 at *c.* 1 month old; one started to sing at *c.* 7 weeks, and at 2 months both sexes sang frequently and loudly. (E N Panov.)   EKD

Breeding. Season. Turkey and Middle East: eggs laid April (Hüe and Etchécopar 1970). USSR: starts earlier than all other *Oenanthe*, in extreme south laying sometimes from mid-February, in north not before early April (Panow 1974). Site. In hole in rock outcrop, among stones, or in bank. Less commonly in rodent burrow, but in sandy desert, USSR, almost exclusively in burrows of

Rhombomys opimus (Panow 1974). Nest: flat saucer of grass and small twigs, usually lined with finer grass and hair, wool, feathers, etc. Usually an accumulation of stones (etc.) in nest-entrance and tunnel; 1 cavity contained 240 stones, total weight 720 g; measured up to 47 × 32 mm and weighed up to 10 g. (Panow 1974.) On average, nest 34 cm (15–75, $n=8$) from hole entrance (E N Panov). Building: by ♀ only (Panow 1974). EGGS. See Plate 83. Sub-elliptical, smooth and glossy; pale blue, with variable red-brown speckling or spotting. Nominate *finschii*: 22·1 × 16·2 mm (19·0–24·1 × 14·9–17·0), $n=30$; calculated weight 3·08 g (Schönwetter 1979). Clutch: 5(4–6). In USSR, sometimes 2 broods in north of range (Mitropol'ski 1968*b*), up to 3 in south (Bel'skaya 1965*b*). 2 broods recorded in south-west Iran (L Cornwallis). On Mangyshlak, 1st clutches of early spring averaged 4·9 eggs ($n=26$), 2nd clutches 5·6 ($n=6$) (Mitropol'ski 1968*b*). INCUBATION. 12–13 days. By ♀ only. (Panow 1974.) YOUNG. Cared for by both parents. Brooded by ♀ for 3–4 days. (Panow 1974.) FLEDGING TO MATURITY. Fledging period 15–16 days (Panov 1978). No further information.

Plumages (nominate *finschii*). ADULT MALE. In fresh plumage (autumn), forehead, crown, hindneck, mantle, and back cream-white, usually with slight pink-buff suffusion (in particular on crown and hindneck); central crown tinged pale grey and sometimes with faint dark shaft-streaks; white purest on forehead and sides of crown. Scapulars black. Back, rump, and upper tail-coverts pure white or white with pale cream-yellow tinge. Side of head (down from feathering at nostril and from upper rim of eye), side of neck, chin, throat, and side of breast black, broadly connected with black of scapulars, upperwing, and underwing. Thighs black with white mottling. Remainder of underparts white, often slightly tinged pale cream-yellow or pink-cream, especially on under tail-coverts. Tail-base white; tip of central pair of tail-feathers (t1) black for 31–39 mm; tip of t2–t6 black for 6–12 mm, but black more extensive on inner edge of t2 (15–25 mm) and sometimes on outer web of t6 (5–15 mm); black tips narrowly fringed white terminally (fringes 1–3 mm wide). Flight-feathers dull black with dark grey inner web; inner webs of inner primaries and outer secondaries often with narrow sharply defined submarginal white wedge or streak towards base; tips of secondaries and inner primaries narrowly fringed pale grey or white. Tertials and all upper wing-coverts (including primary coverts) deep black, often slightly glossy; tips of tertials narrowly fringed white when fresh. Axillaries and under wing-coverts black. *In worn plumage* (spring), white of head and body purer, less cream or yellowish, central crown sometimes with distinct grey patch; white fringes along tips of primaries, tertials, and tail (often except t6) worn off. When heavily worn (June–July), black of feather-bases sometimes visible on hindneck, mantle, back, or (less commonly) underparts, and then some birds confusingly similar to Mourning Wheatear *O. lugens* (see Recognition). ADULT FEMALE. Forehead, crown, hindneck, mantle, scapulars, and back pale brown-grey, usually with buff tinge on forehead, often purer ash-grey on sides of crown and hindneck, and slightly browner with faintly paler feather-tips on scapulars and back. Rump and upper tail-coverts white, often faintly suffused cream, particularly coverts. Supercilium and narrow eye-ring grey-white or buff-white, indistinctly defined from surrounding

feathering; extension of supercilium behind eye often hardly paler than sides of crown. Lores mottled grey and pale buff, upper cheeks and ear-coverts black with grey-brown or rufous-brown feather-tips. Chin and lower cheeks pale buff or cream-white; upper throat similar, but dark grey of feather-bases shining through; lower throat black with rather narrow brown-grey feather-tips (widest on central throat), forming mottled black half-collar extending up towards sides of neck behind ear-coverts, where gradually merging into brownish ash-grey upper sides of neck. Throat and sides of neck occasionally without black or dusky half-collar, uniform grey-brown (e.g. in 1 of 9 birds examined by Vaurie (1949), and these show paler upper cheeks and ear-coverts also; this morph common in *barnesi*). Remainder of underparts pale cream, almost white on central belly and vent, slightly pink-cream on under tail-coverts; sides of breast and thighs with variable amount of dark grey of feather-bases visible. Tail as in adult ♂, but black of tips often slightly browner, especially on outer feathers. Flight-feathers and tertials black-brown, inner webs usually with white submarginal streak or wedge, as in adult ♂; outer webs narrowly fringed sandy-grey or pale buff, tips more widely fringed sandy to off-white (except those of outer primaries). Lesser upper wing-coverts dark grey with pale brown-grey tips; median similar but centres black-brown or almost black; greater coverts, greater upper primary coverts, and feathers of bastard wing black-brown with rather narrow pale buff-grey or off-white fringe along outer web and tip. Under wing-coverts and axillaries dull black or brown-black; primary coverts fringed white, some other feathers variably suffused pale brown-grey. *In worn plumage* (spring), upperparts and sides of head and neck duller and darker sandy-brown; rump and upper tail-coverts fully white; pale eye-ring often more distinct (supercilium less so), dark upper cheeks, ear-coverts, and semi-collar down sides of neck to lower throat more distinct (if present); underparts down from chest and including under tail-coverts dirty white; pale edges and fringes of tail, flight-feathers, tertials, and upperwing bleached to off-white, partly or fully worn off. NESTLING. Down very pale silvery-grey (Panow 1974). JUVENILE. Closely similar to adult ♀, but wing-coverts and tertials markedly different. Grey of upperparts slightly more sandy, sometimes with faintly paler sandy or grey spots on feather-centres. Narrow but usually distinct pale eye-ring. Sides of neck, lower throat, chest, and upper flanks pink-buff, cream-buff, or pale buff, feathers with faint dusky tips (no dusky or black semi-collar); remainder of underparts including thighs and under wing-coverts pale cream-buff to off-white, feathering looser than in adult. Tail as adult, but black tail-tips often less sharply defined; flight-feathers as adult, but pale tips slightly broader. Lesser upper wing-coverts like mantle and scapulars; median and greater coverts, greater upper primary coverts, and tertials as adult, but pale tips markedly wider, up to c. 6 mm, rufous-buff or pink-buff when fresh, pale buff to off-white when bleached, tips of coverts forming distinct double wing-bar. Sexes similar, but ground-colour of tail, flight-feathers, tertials, and greater coverts black in ♂, dark brown in ♀; ♂ with blackish mask across lore, upper cheek, and shorter ear-coverts (♀ brown with pale mottling and streaking, not much darker than remaining side of head). Exceedingly similar to juvenile *O. lugens* and perhaps not always separable; differs in cream or virtually white under tail-coverts (in *O. lugens*, warmer pink-buff); black mask of ♂ apparently not as fully developed as in ♂ *O. lugens* (but only a few juvenile *O. finschii* examined). FIRST ADULT MALE. Like adult ♂, but juvenile flight-feathers, tertials, tail, greater upper primary coverts, and sometimes a few outer greater secondary coverts retained; flight-feathers, tertials, and tail only slightly browner and fringes at tips slightly wider and more buff than those of adult (affected by bleaching and wear at all ages, however), but difference in upper primary coverts and (if retained) outer greater coverts marked, these being black-brown with broad pale buff or white fringes, latter forming contrastingly pale panel on wing (in adult, uniform black or with very faint white edges only). FIRST ADULT FEMALE. Like adult ♀, but part of juvenile feathers retained, as in 1st adult ♂. Retained feathers similar in colour to adult ones (unlike 1st adult ♂), pale feather-fringes only slightly wider; ageing usually possible only by relatively greater wear of retained feathers, by more pointed greater upper primary coverts, and by narrower tail-feathers with less squarish tips.

Bare parts. ADULT, FIRST ADULT. Iris brown or dark brown. Bill brown-black or black in ♂, dark brown to horn-black in ♀. Mouth orange-yellow. Foot dark brown, brown-black, or black. (Dementiev and Gladkov 1954b; Ali and Ripley 1973b; BMNH, ZFMK.) NESTLING. Mouth yellow (E N Panov). JUVENILE. Like adult, but bill, leg, and foot greyish-black at fledging (darkest on culmen and toes); base of bill, rear of tarsus, and soles tinged flesh-grey (ZFMK, ZMA).

Moults. ADULT POST-BREEDING. Complete; primaries descendant. In barnesi, USSR, starts June, completed 20-31 July (Dementiev and Gladkov 1954b); others started mid-July to mid-August, completing late August or September (E N Panov). In Afghanistan, 3 ♂♂ in heavy moult 24-28 July (Paludan 1959). In Iran and Afghanistan, probably starts late June; usually over by mid- or late August, but ♂ from 31 August in eastern Iran in full moult, though 1 ♀ from that area had almost completed by July 29 (Vaurie 1949). No moult up to 28 June in Turkey; all moult completed in birds examined September-October, Turkey and Levant (BMNH, ZFMK, ZMA). POST-JUVENILE. Partial: head, body, lesser and median upper wing-coverts, and variable number of greater upper wing-coverts, tertials, and feathers of bastard wing. Of 20 examined, 10 replaced all greater coverts, 7 retained 1-2 outer coverts, 3 retained 5-6 coverts; 7 retained 1 tertial, 5 retained all tertials; 4 retained some longer feathers of bastard wing. Birds in full juvenile plumage examined late April to early July (BMNH, ZFMK, ZMA); moult just started in 2 ♂♂ from Afghanistan in late July and early August (Paludan 1959). In USSR, some already moulted on 22 June in Turkmeniya; moult started in Tadzhikistan on 28 June, completed up to late August in birds from 2nd broods; in Armeniya, completed about mid-August (Dementiev and Gladkov 1954b). In Iran and Afghanistan, probably late June to late August (Vaurie 1949).

Measurements. ADULT, FIRST ADULT. Nominate *finschii*. Central and southern Turkey and Dead Sea region, all year; skins (BMNH, ZFMK, ZMA). Bill (S) to skull, bill (N) to distal corner of nostril; exposed culmen on average 4·3 less than bill (S).

WING	♂ 88·7 (2·10; 17)	85-92	♀ 84·4 (1·27; 10)	82-87
TAIL	57·4 (2·68; 12)	52-61	53·9 (1·16; 6)	52-56
BILL (S)	18·4 (0·86; 11)	17·0-19·7	17·8 (0·61; 6)	17·0-18·6
BILL (N)	10·4 (0·80; 11)	9·1-11·6	10·2 (0·44; 6)	9·7-10·9
TARSUS	26·1 (0·69; 12)	24·7-27·3	25·1 (0·31; 6)	24·8-25·6

Sex differences significant, except bill. Cyprus and Levant, ♂: wing, 89·1 (14) 85-92; tail, 55·0 (14) 52-58 (Vaurie 1949).

O. f. barnesi. Eastern Turkey, eastern Caucasus, Turkmeniya (USSR), and north-east and eastern Iran, all year; skins (BMNH, ZFMK, ZMA).

WING	♂ 91·1 (1·68; 16)	89-96	♀ 87·6 (1·88; 5)	86-91
BILL (S)	18·0 (0·84; 11)	17·1-19·7	17·5 (0·62; 4)	17·1-18·4

Sex differences significant for wing. ♂♂: tail 60·0 (2·17; 11) 57-63, bill (N) 9·7 (0·64; 6) 8·8-10·6, tarsus 26·3 (0·48; 7) 25·9-27·3 (ZFMK, ZMA). Eastern Iran and Afghanistan, ♂: wing, 93·1 (23) 89-96; tail, 61·2 (23) 58-65 (Vaurie 1949).

In both races, 1st adult with retained juvenile wing and tail combined with older birds, though wing on average 1·7 shorter than in full adult and tail 0·5 shorter.

Weights. Nominate *finschii*. Central and southern Turkey: late April to June, ♂ 27·3 (2·75; 4) 24-30, ♀♀ 21, 32, sex unknown 24; October, ♂ 31 (Kumerloeve 1970a; ZFMK).

O. f. barnesi. Eastern Turkey and Armeniya (USSR), May-June: ♂ 27·8 (1·17; 6) 26-29 (Nicht 1961; ZFMK). Mangyshlak peninsula, east Caspian Sea: April, ♂ 25·6 (29) 21-28·5, ♀ 29·3 (10) 23·5-32·5; May, ♂ 23·7 (24) 20·5-26·5, ♀ 25·6 (14) 20·5-32·3 (Dolgushin et al. 1970). Afghanistan, ♂♂: May, 24; late July and early August, 26·0 (0·71; 5) 25-27 (Paludan 1959). Iran: winter, ♂ 19·5-29·5 (13), ♀ 20·5-24 (5); breeding ♀♀ 23-26 (5) (Ali and Ripley 1973b).

Structure. Wing rather short, broad at base, tip rather rounded. 10 primaries: p7 longest, p8 0-1·5 shorter, p9 4-8, p6 0·5-2, p5 3-7, p4 8-13, p3 12-17, p1 17-22. P10 reduced; 41-49 shorter than wing-tip, 2-7 longer than longest upper primary covert. Outer web of (p5-)p6-p8 and inner of (p6-) p7-p8 emarginated. Tail rather short, tip square; 12 feathers. Bill rather short, straight; rather heavy at base, laterally compressed at tip; tip of culmen slightly decurved, sometimes forming slight hook. Nostrils small, oval, partly covered by membrane above. Numerous fine hairs along base of bill; some longer bristles at base of upper mandible. Tarsus and toes rather long, strong. Middle toe with claw 17·9 (10) 16·5-19·5; outer toe with claw c. 73% of middle with claw, inner c. 67%, hind c. 80%. Claws short, rather strong, decurved.

Geographical variation. Rather slight; involves general colour and size. Nominate *finschii* from central Turkey and Taurus mountains south to Dead Sea region is smaller than *barnesi* from eastern Turkey and northern and central Iran eastwards (see Measurements, Ticehurst 1927, and Vaurie 1949); birds from Zagros mountains (south-west Iran) and perhaps those of neighbouring eastern Iraq are intermediate, with colour near nominate *finschii*, size near *barnesi* (Vaurie 1949). ♂ *barnesi* very similar to nominate *finschii*, but under tail-coverts slightly deeper cream-buff, and forehead, crown, mantle, and central back tinged pinkish cream-buff (in nominate *finschii*, pale buff-grey); races inseparable in colour when plumage worn. ♀ *barnesi* more sandy-buff on upperparts (less sandy-grey), sides of neck and throat usually buff, without grey or blackish half-collar, ear-coverts rufous rather than brown or greyish-black. Underparts tinged buff, especially under tail-coverts (♀ nominate *finschii* cream-white); in worn plumage, both races closely similar, and ♀ *barnesi* often has mottled grey throat-patch like ♀ nominate *finschii*. In both sexes, black on tail-tips slightly more extensive: 34-44 mm on t1, 18-28 mm on inner web of t2, 9-15 mm from outer web of t2 to inner web of t6, 12-20 mm on outer web of t6.

Closely related to Mourning Wheatear *O. lugens*, which overlaps with nominate *finschii* in Levant (*O. l. lugens*) and with *barnesi* in Zagros mountains of south-west Iran (*O. l. persica*). In *O. lugens*, sexes similar in races *O. l. lugens* and *O. l. persica*, and these resemble ♂ *O. finschii*, which differs in pale rather than black lower mantle and back, and in paler under tail-coverts; however, black of *O.lugens* sometimes mottled grey and buff of under tail-coverts affected by bleaching, and *O. finschii* has lower mantle and back often heavily mottled grey when plumage worn (especially in 1st adult); ♀ nominate *finschii* occasionally has largely black sides of head and throat resembling ♂ nominate *finschii* or ♀ *O. l. lugens*. North African race *O. l. halophila* of *O. lugens* (which does not overlap in range with *O. finschii*) is sexually dimorphic, ♀ closely resembling ♀ *O. finschii*. In view of close similarity between non-overlapping races and somewhat greater divergence in sympatric races, both species united in a species-group, together with Desert Wheatear *O. deserti*.

Recognition. Sometimes difficult to separate from *O. lugens* (see above), though races of *O. lugens* in overlap area with *O. finschii* average larger (see Measurements) and wing-tip of *O. lugens* slightly more pointed: p8 usually longest (p7 or p7-p8 in *O. finschii*), p5 6-10 shorter than wing-tip in *O. lugens*, 3-7 in *O. finschii* (see Structure). Best distinguishing character is

amount of white on inner webs of flight-feathers: in *O. lugens*, pronounced white wedge on most feathers, widest in nominate *lugens* where basal $\frac{2}{3}$ of secondaries fully white; in *O. finschii*, narrower submarginal wedge or stripe on basal inner webs, usually restricted to inner primaries and outer secondaries, but occasionally entirely absent, in particular in adult ♂. Both sexes rather similar to Black-eared Wheatear *O. hispanica*, but wing-tip of *O. finschii* shorter and more rounded, inner webs of flight-feathers often have sharply defined narrow white wedge, leg heavier, tarsus much longer, tail has less extensive black on outer web of t6, and tail has white fringe along tips of all feathers (if not too worn); ♂ differs also in having black continuous from sides of neck to scapulars (interrupted by white in *O. hispanica*); ♀ of nominate *finschii* (but not usually *barnesi*) differs in having black or grey half-collar from sides of neck down to throat (♀ *O. hispanica* sometimes black-throated, but black then more extensive towards chin, like patch of black-throated ♂ *O. hispanica*). CSR

Oenanthe moesta **Red-rumped Wheatear**

PLATES 61, 64, and 65
[between pages 760 and 761, and facing page 856]

Du. Roodstuittapuit   Fr. Traquet à tête grise   Ge. Fahlbürzelsteinschmätzer
Ru. Краснопоясничная каменка   Sp. Collalba de Tristram   Sw. Rödstjärtad stenskvätta

Saxicola moesta Lichtenstein, 1823

Polytypic. Nominate *moesta* (Lichtenstein, 1823), North Africa; *brooksbanki* Meinertzhagen, 1923, Middle East.

Field characters. 16 cm; wing-span 25-29 cm. Size close to Greenland race of Wheatear *O. oenanthe leucorhoa* though with slightly shorter, more rounded wings and 10% longer bill; noticeably larger than western races of Desert Wheatear *O. deserti*, with bigger bill and head and slightly longer, broader wings and tail. Rather bull-headed, compact wheatear with dull rufous rump and (like *O. deserti*) nearly all-dark tail. ♂ has grey crown and nape, black face, throat, shoulders, and back, pale-fringed wings, and dull underparts; lacks sharp contrasts of most ♂ wheatears. ♀ much more distinctive than most ♀ wheatears, with noticeably rufous crown and cheeks. Sexes dissimilar; marked seasonal variation in ♂. Juvenile ♂ separable. 2 races in west Palearctic, doubtfully distinguishable in the field.

ADULT MALE. Top of head, nape, and upper centre of mantle pale grey-brown merging into black back; narrow, diffuse white supercilium. Face, throat, upper breast, sides of neck and chest (extending under folded wing), lesser wing-coverts, scapulars, lower mantle, and back black. Foreparts thus recall many other wheatears (particularly Pied Wheatear *O. pleschanka* and Mourning Wheatear *O. lugens*), but wing pattern unique, with overall 'mealy' appearance comprising: (a) distinct white margins to median coverts forming obvious, quite sharp wing-bar; (b) wide whitish fringes to greater coverts creating pale panel (placed lower than that on shoulder of *O. deserti*); (c) cream to rufous-buff fringes to blackish-

centred tertials and secondaries producing 2nd pale panel; (d) obvious cream tips to primaries. In flight, pale fringes produce paler inner half to wing crossed by 2 even paler wing-bars. Rump, upper tail-coverts, and base of tail pale rufous-buff; remainder of tail dull black. Chest and underbody dull white. Under tail-coverts pale rufous-buff. Sides of chest, axillaries, and under wing-coverts black, contrasting with almost white inner webs of all flight-feathers except outer primaries, which form dusky wing-tip. From autumn, 'mealy' appearance increased by hoary throat-marks, pale cream fringes to back and scapulars, and even broader pale fringes to wing-feathers. ADULT FEMALE. Differs from all other similar ♀ wheatears in showing pale chestnut head, marked only by diffuse buff or cream supercilium and eye-ring and contrasting with cream throat, pale rufous-buff chest, and grey-buff back. Unlike ♂, under wing-coverts pale buff and not conspicuous. Rest of plumage as ♂ but paler and duller, with far less black on wings, usually paler rufous rump and under tail-coverts, and browner tail with more rufous at base. Wing-markings thus less obvious than on ♂, leading to possible confusion with ♀ *O. deserti* at distance, though bull head, large size, and narrow rufous base to tail remain clear differences. In fresh autumn plumage shows obvious broader pale fringes to wing-feathers. JUVENILE. Resembles adult ♀ but back faintly pale-spotted, head less chestnut (lacking obvious supercilium), and underbody indistinctly scaled brown. ♂ shows blackish

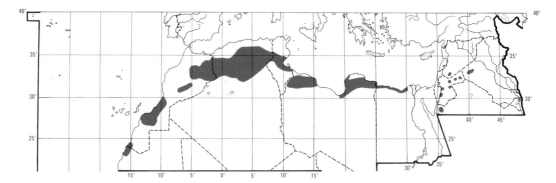

lore and ear-coverts. FIRST-YEAR MALE. Resembles adult ♂ but crown usually more rufous and mantle more grey-brown. Retains juvenile flight- and tail-feathers and some greater coverts. FIRST-YEAR FEMALE. As adult ♀ except for retained juvenile feathers. At all ages, bill and legs black.

Easy to identify given good view, with both sexes showing diagnostic wing, rump, and tail patterns and even ♀ possessing striking head colour. In brief glimpse or at distance, may be confused with *O. deserti* (which has similar rump and tail pattern but is smaller, without black back in ♂ or chestnut head in ♀ and immature). Flight distinctive, bird tending to drop from perch and skim over ground, with rather loose wing-beats producing floating progress recalling Skylark *Alauda arvensis*. Escape-flight not long, usually half-circular, with bird landing on ground rather than obvious perch. Gait a long hop. Rather quiet and undemonstrative but perches freely, holding body half-upright.

Song either a remarkable wavering but ascending warbling whistle (usually in display) or a thick throaty warble. Commonest call a hard, clicking 'prrt'.

Habitat. Across narrower belt of Afro-Asian lower-middle latitudes than Desert Wheatear *O. deserti*, also ranging from warm arid continental to oceanic climates, especially under Mediterranean influences. Frequents flat ground, often in vicinity of saline and barren areas, but not absolute desert (Etchécopar and Hüe 1967). Habitat almost identical with *O. deserti*. Will forage on wet sand (Heim de Balsac and Mayaud 1962). Intrudes into narrow belt of scrub between Mediterranean and western desert of Egypt (Moreau 1966). In southern Morocco, seems to use patches more densely vegetated than *O. deserti* would; occurs in rocky hills and between dense heath of *Atriplex* and *Nitraria* or scrub of *Tamarix* and salt shrubs (Valverde 1957). At Azraq (Jordan), appears confined to open flat areas with very sparse cover (Hemsley and George 1966). Distribution seems to owe its patchiness to availability of rodent holes (Smith 1965a). Likes small bushes to perch on (Smith 1962-4).

Distribution. No information on range changes.

EGYPT. Bred Sinai 1928; no records since (SMG, PLM, WCM). LIBYA. Limits uncertain (Bundy 1976), but widely distributed in north-east (L Cornwallis).

Accidental. Israel.

Population. No information on trends.

EGYPT. Fairly common (SMG, PLM, WCM). JORDAN. Azraq: 2-4 pairs (Wallace 1983a). LIBYA. Common in some areas (Bundy 1976); fairly common in north-east (L Cornwallis). ALGERIA. More common in west (EDHJ).

Movements. Mainly sedentary, though evidence for partial migration and local movements. Some southward movement by more northerly birds at least (Etchécopar and Hüe 1967). Almost none recorded October–April north of Hauts Plateaux in eastern Morocco (Brosset 1961); in Tunisia, rarer in winter than in March (Zedlitz 1909). ♂ recorded at Giza (Egypt) in January (Meininger and Mullié 1979).   DFV, MGK

Food. Mainly insects. In northern Cyrenaica (Libya), prey usually caught by hopping along ground and jabbing; also by perching higher, e.g. on top of bush, and pouncing at items on ground (L Cornwallis). Before eating large beetles (Coleoptera), typically hammers them on ground, and may reject wing-cases (Meinertzhagen 1940; L Cornwallis).

Grasshoppers (Acrididae), larval Lepidoptera, ants (Formicidae), and beetles (Coleoptera) recorded in diet, and, in Morocco at least, occasionally takes green plant material (Whitaker 1898; Meinertzhagen 1940, 1954; Valverde 1957; Jarry 1969). In Tunisia, scorpions *Scorpio maurus* formed bulk of prey brought by one pair to fledglings (Jarry 1969).   BDSS, EKD

Social pattern and behaviour. Not well known. Account includes notes supplied by L Cornwallis for northern Cyrenaica (Libya).

1. In winter, territorial dispersion maintained. In Morocco (Smith 1965a) and Libya (L Cornwallis), typically found in pairs which defend territory all year, but Bulman (1942), perhaps further into desert, found only single birds in Libya, September–February (see also Bonds, below). No information on size of winter territories. In southern Morocco, September,

3 road transect counts (total *c.* 300 km) yielded following densities: *c.* 1 per 10 km, *c.* 1 per 4 km, and *c.* 1 per 6-7 km (Knight 1975*b*). BONDS. No evidence for other than monogamous mating system. Pair-bond often, perhaps typically (see above), maintained outside breeding season. In winter, Cyrenaica, ♂ and ♀ consorted regularly, closely and amicably (L Cornwallis). Both parents feed young, and family bonds maintained for (unknown) time after young leave nest (L Cornwallis). Following observation perhaps indicates change in roles after fledging: at one site in Morocco, August, at least 2 pairs present, each ♀ accompanied by a single young bird while the ♂♂ were solitary (Sage and Meadows 1965). No information on age of first breeding. BREEDING DISPERSION. Solitary and territorial. Distribution apparently governed largely by availability of colonial rodent holes for nest-sites, producing scattered concentrations (Smith 1962-4, 1971; Bundy 1976). No information on size of territory. In Morocco, April-June, 9 counts along one 22-km transect gave on average 2 birds (Blondel 1962*a*). See above for higher densities outside breeding season. Territory evidently serves for courtship, nesting, and raising young. Nest-sites often re-used for subsequent broods (Heim de Balsac and Mayaud 1962; L Cornwallis). For defence of territory against other species, see Antagonistic Behaviour (below). ROOSTING. No information.

2. In Algeria, found to be shy by Koenig (1895), but not by Rothschild and Hartert (1911). In Tunisia, quite tame and confiding (Bannerman 1927), often allowing approach of man to within 'a few feet' before taking off (Whitaker 1898). In Algeria, when alarmed, ♂ and ♀ (pair) sometimes flew away, but more often disappeared into a burrow (at least sometimes a nest-hole) usually under a bush. Once, when man was digging for ♂ thus hidden, ♀ emerged from nearby hole and escaped; this strategy, facilitated by often-multiple exits to burrow system, said also to effect escape from lizard predators. (Tristram 1859.) FLOCK BEHAVIOUR. None known. SONG-DISPLAY. Both sexes sing. ♂ sings probably throughout year, ♀ perhaps mostly from winter until laying. ♂ gives Territorial-song (see 1a in Voice) perched on ground, bush, wires, etc. (L Cornwallis, J-C Roché); also in flight between perch-sites, during which tail typically fanned (Whitaker 1898). Apparent Territorial-song by ♀ also reported (Smith 1962-4). Subsong (see 1c in Voice) commonly given by both sexes outside breeding season: perhaps sometimes subdued Territorial-song since it may be given during aggressive pursuit-flights; one ♀ giving Subsong, November, was conspicuously drooping her wings. (L Cornwallis.) Courtship-song (see 1b in Voice) is given in the early stages and not normally once breeding begins—typically part of an elaborate Leap-frogging ceremony by pair-members thought to mark out territory (E D H Johnson), though perhaps also promotes pair-formation: pair move around territory on ground, leaping over each other by bounds of 5-10 m, usually singing in antiphonal duet (see 1b in Voice). Occasionally ♂ or ♀ alone may give Courtship-song. (E D H Johnson.) ANTAGONISTIC BEHAVIOUR. Little information for breeding season, but both sexes evidently help to defend territory throughout year. Following account after L Cornwallis. Outside breeding season, pair-members mostly share territory amicably, but ♂ once seen (November) to fly at ♀ (mate) who 'jinked' but did not move off. Disputes between ♂♂ sometimes involve pursuit-flights; once, both birds gave Subsong and tail-fanned (see also Song-display, above). Trespassers of other species may also be attacked. In October, ♂ chased off Crested Lark *Galerida cristata* 5 times; also two November records of Desert Wheatear *O. deserti* being attacked and chased, once by ♂, once by pair. (L Cornwallis.) In Morocco, November, ♂ similarly drove off

O. deserti and larks (Alaudidae) from vicinity of a nest-hole (Smith 1971). HETEROSEXUAL BEHAVIOUR. Little information. Pair-formation appears to occur in winter, or early spring, though maintenance of pair-bonds throughout year makes it difficult to establish timing with certainty, at least for new pairs. Leap-frogging ceremony (see above), accompanied by Courtship-song, may have pair-bonding function; account of this display by E D H Johnson (Morocco) and of probable Courtship-song in Morocco by Smith (1965*a*) both refer to February. Bulman (1942), who did not find ♂♂ singing until mid-February, stated that pair-formation begins at end of February. Other possible courtship-displays between established or prospective pair-members are as follows. In Morocco, mid-December, 'wing-fanning' attributed to courtship (Smith 1971). Record of ♂ opening wings to nearby ♀, mid-November (L Cornwallis), is presumably same display. Another ♂ seen 'wing-drooping' (L Cornwallis); for wing-drooping of ♀ giving Subsong, see Song-display (above). ♂ thought to take little, if any, part in incubation; not seen to enter or leave nest-hole, though he generally stays in immediate vicinity of nest (Whitaker 1898). No further information. RELATIONS WITHIN FAMILY GROUP. Little known. Nestlings fed by both parents (L Cornwallis). Observations by Tomlinson (1943) and Brosset (1961) and photographs by L Cornwallis indicate that young not yet able to fly stay near nest-hole entrance (see also below). Young recently out of nest continue to give food-calls (P A D Hollom). 2nd clutch often laid as soon as 1st brood fledged (Heim de Balsac and Mayaud 1962). ANTI-PREDATOR RESPONSES OF YOUNG. Rodent burrows serve as refuges for young; even independent young retreat to nest-burrows if alarmed (Brosset 1961); also seek refuge under piles of boulders (L Cornwallis). One young bird, not yet able to fly, kept darting out from burrow to watch human intruder, retreating hastily as soon as intruder moved (Tomlinson 1943). PARENTAL ANTI-PREDATOR STRATEGIES. (1) Passive measures. In central Sahara, incubating ♀♀ sit tight on approach of man (Heim de Balsac 1926). In Algeria, ♀ thus threatened gave Contact-alarm calls (see 2 in Voice), probably from nest-hole entrance (Lynes 1930). In Tunisia, one ♀ left nest when disturbed but stayed nearby (S Cramp and M G Wilson). (2) Active measures. When young threatened, parents demonstrate vocally, beginning as intruder approaches nest (Rothschild and Hartert 1911). When human intruder near well-grown young in one nest (see above), ♀ parent perched nearby giving variety of alarm-calls (Tomlinson 1943: see 3b in Voice). Various other alarm-calls (see 3a-b in Voice) heard when intruder near young recently out of nest (P A D Hollom). EKD

Voice. Used for much of year. Recordings and written sources suggest that repertoire less diverse than some other *Oenanthe*.

CALLS OF ADULTS. (1) Song. (a) Territorial-song. Given apparently by both sexes, probably less often by ♀. Evidently varies markedly in quality and timbre. In winter, Morocco, described by Smith (1962-4) as short and rather rattling in ♂, short and harsh in ♀. Perhaps similar quality in songs described by P A D Hollom (Jordan) as (e.g.) a thick throaty warble lacking much variety or strength. Other accounts report much more melodious songs: in Tunisia, song of ♂ described as short and rippling, singularly sweet and pathetic (Whitaker 1898); in Cyrenaica (Libya), a pleasant musical 'twee-

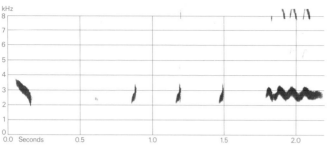

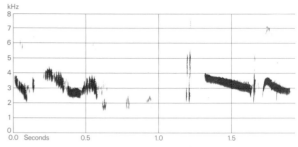

I C Chappuis Tunisia March 1971

II C Chappuis Tunisia March 1971

chirr-rur-rur-rur' from ♂, repeated slowly without much variation, the last phrase with rolled 'r' effect and 'cadence' (Bulman 1942). Our recordings suggest that such cadences are common, representing incorporation of whistling (as in Courtship-song: see 1b, below) into Territorial-song, though in Courtship-song whistling is always ascending, longer, and often louder: e.g. in recordings of presumed Territorial-song by P A D Hollom in March, Morocco, song typically a succession of short lazy warbling phrases (one phrase of 8-10 notes), either slightly descending, or alternately ascending and descending (E K Dunn, P A D Hollom). Another recording suggests considerable diversity in repertoire: Fig I shows 'siu wit wit wit', followed by a long (0.4 s) loud tone with a well-defined and relatively slow vibrato, this sequence and preceding series of 'siu' notes strongly resembling song of Song Thrush *Turdus philomelos*; elsewhere in same recording (Fig II), song has rough, squeaky but tonal beginning, followed by 3 quiet 'tuk' sounds, then some rather hard descending portamento tones; towards end of recording, song includes a 'churr' resembling sparrow *Passer* (J Hall-Craggs). (b) Courtship-song. Remarkable sound: a long wavering warbling whistle, rising progressively in pitch, aptly likened to whistling kettle coming to the boil (E D H Johnson). Described by Smith (1962-4, 1965) as an

extraordinary ascending 'whir whir whir whir whir'. Usually given in antiphonal duet by pair (each bird starting at bottom of scale as its partner reaches the top, but sometimes overlapping) during Leap-frogging ceremony; occasionally by ♂ or ♀ alone (E D H Johnson). Curtailed songs not uncommonly included in Territorial-song (see above). Analysis of sonagrams shows that ascent in pitch is by series of clearly audible, well defined steps, each essentially a tremolo; the very small steps within each tremolo also tend to rise in pitch. Recording (Fig III) begins with 3 'tik' sounds (not unlike those shown in middle of Fig II), followed by song 6·34 s long comprising *c.* 13 whistling steps, each *c.* 0·5 s, which show overall crescendo. In another recording (by C Chappuis, Tunisia), quiet rattling 'tik' sounds again precede song which comprises 9 whistling steps delivered at just over 3 per s, the odd-numbered steps lasting just less than 0·4 s, the even ones *c.* 0·3 s. In recording of duet, songs of presumed ♂ stronger, longer (up to 20 steps), and top of scale higher than (alternate) songs of presumed ♀. (J Hall-Craggs.) (c) Subsong. Curious sequence of creaking sounds, given by both sexes, sometimes including succession of units of contact-alarm type (see call 2), e.g. one ♂'s Subsong, began 'prrrt prrrt', followed by liquid falling creaking notes (L Cornwallis). (2) Contact-alarm call. Brief calls, typically

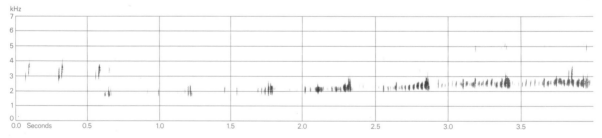

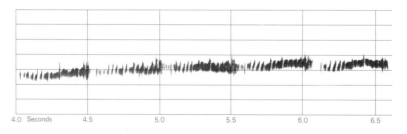

III E D H Johnson Morocco
February 1971

of hard or harsh clicking quality, given by both sexes as contact-call (L Cornwallis) and when nest and young threatened. Recording (Fig IV) shows repeated 'k-wik'

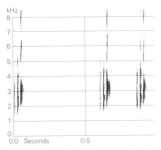

IV C Chappuis Morocco April 1976

calls, each 50–60 ms long (J Hall-Craggs), given by alarmed ♀. Lynes (1930) aptly described these calls as 'kreet kreet' like old mechanical lawnmower. In Morocco, calls of one ♂ in alarm a not very loud rasping 'krit' or 'krut', calls of ♀ quieter (P A D Hollom). In Libya, also variously rendered 'proup prrup prrup', 'prrt', 'thrup', or 'thrrrp' (L Cornwallis). (3) Other alarm-calls. (a) A 'choop' given by anxious parent with young out of nest (P A D Hollom, Jordan). In Morocco, 'choo' followed by a rattle, also 'tak' (Smith 1965), 'tuk', and 'chuk' (Smith 1962–4), are apparently the same or similar. In Figs II–III, 'tuk' and 'tik' units presumably also of this type. (b) In Morocco, alarm-call when recordist near fledged young a melodious whistled 'keeeyup', 'keeeyerp', or 'peeyou' like Ringed Plover *Charadrius hiaticula* (E K Dunn, J Hall-Craggs, P J Sellar: Fig V). In same context,

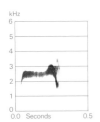

V P A D Hollom Morocco March 1978

♀ interspersed 'clicking notes' with 'whistles' (Tomlinson 1943), suggesting this whistling alarm-call may be given with calls 2 or 3.

CALLS OF YOUNG. Juvenile food-call a brief repeated grating note (P A D Hollom), very like adult call 2— 'trt' (J Hall-Craggs) or 'prp' (E K Dunn). By late May, Cyrenaica, independent young were giving 'fair imitations of their parents' song' (Tomlinson 1943). EKD

Breeding. SEASON. Algeria and Tunisia: prolonged, with eggs laid from 1st week of February and late nests in June (Heim de Balsac and Mayaud 1962). Jordan: newly-fledged young seen late April (Wallace 1984*b*).

SITE. Up to 2 m deep in hole in ground, usually of rodent or other small mammal, sometimes natural hole or hole in wall; entrance may be concealed under bush or root. Nest: cup of leaves, rootlets, and stems, lined with (e.g.) wool, hair, feathers, and not uncommonly snakeskin (Rothschild and Hartert 1911). Building: by ♀ (Heim de Balsac and Mayaud 1962). EGGS. See Plate 83. Sub-elliptical, smooth and slightly glossy. Nominate *moesta*: 23·4 × 16·3 m (23·0–24·0 × 15·5–17·0), $n = 7$; calculated weight 3·3 g (Schönwetter 1979). Clutch: 4–5. 2 broods, possibly 3; 2nd clutch laid immediately after 1st brood fledges (Heim de Balsac and Mayaud 1962). INCUBATION. Apparently by ♀ (Whitaker 1898). YOUNG. Cared for and fed by both parents (L Cornwallis). No further information.

Plumages (nominate *moesta*). ADULT MALE. Upperparts rather variable. Crown and hindneck usually dark ash-grey, sometimes pale cinnamon-brown, pale brown-grey, or dull rufous-brown, bordered at front and sides by pale cream-buff to cream-white forehead and supercilium; hindneck pale cinnamon-brown, pale ash-grey, or cream, gradually merging into deep black, sooty black, greyish-black, dark brown-grey, dark olive-grey, or olive-brown lower mantle, scapulars, and back. Rump rufous-cream, pale buff, or cream-white, upper tail-coverts pink-buff, pink-cinnamon or cinnamon-rufous. Side of head (down from lore, eye, and ear-coverts), chin, throat, and side of neck deep black, contiguous with black to olive-brown of scapulars and with black on sides of breast; feathers of chin and throat narrowly tipped white when plumage fresh. Underparts backwards from chest and flanks white with slight cream or pink-grey suffusion; upper flank with variable grey wash; under tail-coverts pink-buff or cinnamon-buff. Tips of tail-feathers extensively black (often slightly tinged olive or brown), tail-base pink-cinnamon or rufous-cinnamon; no white fringes to tail-tips. Black on central pair of tail-feathers (t1) 44–55 mm long; gradually less black towards t3–t4, 27–38 mm; increasing again towards outer web of t6, 35–45 mm. Flight-feathers and tertials greyish-black (darkest on tertials and tips, paler grey-white on bases of inner webs), outer webs and tips of tertials and outer webs of secondaries fringed pink-cinnamon, tips of secondaries and inner primaries fringed white, outer webs of primaries narrowly edged pale pink-buff or off-white. Greater upper primary coverts dark grey with narrow, even, and clear-cut pale grey to white fringe. Lesser upper wing-coverts and tertial coverts dark olive-brown to deep black (as variable as mantle and scapulars), tips usually fringed grey; median and greater coverts black-brown to sooty black, median with broadly white tip (sometimes extending into a point towards base at shaft), greater with white outer fringe, widening to broadly white tip, but only narrowly extending to inner web (partly tinged rufous on top when plumage fresh). Shorter feathers of bastard wing grey-black with broad white outer fringe, longest black with narrow white edge along tip. White tips of median and greater coverts either form double wing-bar (bars gradually narrower towards body) or, when white on outer webs of greater coverts very extensive, broad white wing-panel. Axillaries and under wing-coverts black, longer coverts with broad white tips, primary coverts largely white. *In worn plumage*, no marked changes; grey or rufous of crown more restricted, cream or off-white of forehead and supercilium broader, crown sometimes almost completely dirty white; hindneck paler; rufous of rump and upper tail-coverts bleached to pale pink-buff or buff-white, some rufous visible only on base of outer tail-feathers when tail fully spread; underparts backwards from chest cream-white or white, some dark grey of feather-bases sometimes visible, under tail-coverts cream-pink or off-white; pale fringes and tips of flight-feathers and tertials bleached to off-white and partly worn off; white tips of upper wing-coverts sometimes largely worn off. ADULT FEMALE. Forehead, crown, hindneck, and side of head down to upper cheek and ear-coverts rufous-cinnamon to deep cinnamon-brown, broken by paler cinnamon-buff lore and supercilium and by cream-buff eye-ring; some dark grey mottling just in front of and below eye. Cinnamon of hindneck gradually merging into brown-grey or pale cinnamon-brown on lower mantle, scapulars, and back, this in turn into pink-cinnamon or rufous-cinnamon on rump and tail-coverts. Lower cheek, side of neck, side of breast, chest, and flank pink-buff or pale greyish-buff, merging into pink-cream or cream-white on chin, throat, belly, and vent; under tail-coverts pink-cinnamon or pale rufous-cinnamon. Tail as adult ♂, but black of tail-tip browner and rufous of tail-base more extensive: tip of t1 dark for 40–48 mm, t3–t4 17–26 mm, t6 20–34 mm. Flight-feathers and tertials as adult ♂, but ground-colour dark grey-brown rather than blackish, especially tertials paler; greater upper primary coverts paler grey-brown, fringes broader, pale grey, less clear-cut; bastard wing grey-brown with white outer fringe, longest feather dark brown. Lesser upper wing-coverts grey or cinnamon-brown, like scapulars; median dark grey-brown with ill-defined pink-buff or pale buff tip; greater dark grey-brown with pink-buff outer web merging into broad pink-buff or pink-white tip. Axillaries and under wing-coverts pale grey-buff to off-white, some darker grey visible on bases of longer feathers. *In worn plumage*, top and sides of head deeper and more uniform rufous; mantle and scapulars slightly darker grey-brown; rump and upper tail-coverts sometimes slightly paler cinnamon or partly pale pink-buff; chin, throat, and sides of neck dirty grey-white; tail, flight-feathers, and upperwing duller brown, wing with fringes and tips bleached to pink-buff or off-white and partly worn off. NESTLING. No information. JUVENILE MALE. Upperparts rather like adult ♀, but top of head less rufous and upper tail-coverts paler: upperparts uniform buff-brown or greyish-cinnamon, sometimes with faint paler spots on feather centres of mantle and scapulars; lower rump and upper tail-coverts pale rufous-buff with much cream-pink of feather-bases visible; unlike adult ♂, forehead and sides of crown not paler than central crown, supercilium faint or absent. Lore, upper cheek, ear-coverts, feathering at gape, and side of chin greyish-black, forming dark mask (rather like juvenile ♂ of Mourning Wheatear *O. lugens*); rear of lower cheek and side of neck greyish-buff. Underparts backwards from central chin and throat paler buff or cream-white, feathers of chest and sides of breast slightly deeper pink-buff and with faint brown arc on tip. Tail as adult ♂, no pale tips. Flight-feathers as adult ♂, but outer primaries with *c.* 1 mm wide fringe along tip (virtually absent in adult). Tertials dark olive-brown or dark brown (blacker in adult ♂); fringes along outer edges and tips wider, less clear-cut, dull cinnamon-pink (when fresh) or off-white (when worn). Greater upper primary coverts paler than in adult ♂, grey-brown with rather broad and ill-defined off-white fringes (not narrow, clear-cut, and pale grey). Lesser upper wing-coverts sandy-grey; median grey with ill-defined buff tips 4–6 mm wide; greater dull brown-grey with broad cinnamon-pink (fresh) or off-white (worn) outer fringe and tip (slightly broader than adult). Axillaries and under wing-coverts grey, but most shorter coverts and primary coverts off-white. JUVENILE FEMALE. Like juvenile ♂, but no black on

sides of head; sides of head buff-brown, like crown, with rather faint paler supercilium and eye-ring; head like adult ♀, but less rufous. Tail as adult ♀, hence tail-tip not as extensively black as ♂; ground-colour of tail-tip, flight-feathers, tertials, and wing-coverts browner, less blackish. Only a few examined, not certain whether all juvenile ♂♂ show black mask or whether ♀♀ are always without it. FIRST ADULT MALE. Like adult ♂, but crown more often tinged rufous-brown rather than grey and mantle and scapulars more often grey-brown rather than blackish. Juvenile flight-feathers, many or all tail-feathers, 0–3 tertials, all greater upper primary coverts, and a variable number of outer greater secondary coverts retained, browner and more worn than neighbouring fresh feathers; tips of tail-feathers and primaries distinctly brown and worn by November–December (adult not until April); tertials and greater upper primary coverts pale brown with rather broad and ill-defined off-white fringe along outer web and tip (in adult, blacker with narrower and more clear-cut fringe, tip of primary coverts smooth until late spring). FIRST ADULT FEMALE. Like adult ♀, but part of juvenile feathers retained, as in 1st adult ♂. Differences in colour between old and new feathers often difficult to detect, and difference in abrasion sometimes slight, hence ageing often difficult. Tips of tail-feathers, primaries, and greater primary coverts worn and bleached by November–December, heavily worn March–April (in adult, still dark and smooth until about March).

Bare parts. ADULT, FIRST ADULT. Iris dark brown. Bill, leg, and foot black. (Hartert 1903–10; BMNH.) NESTLING. No information. JUVENILE. Iris brown. Bill, leg, and foot dull bluish-brown. (Hartert 1903–10.)

Moults. ADULT POST-BREEDING. Complete. None in moult examined; plumage completely new in birds from September, heavily worn but not moulting in May and early June (BMNH, RMNH, ZFMK, ZMA). POST-JUVENILE. Partial: head, body, lesser and median upper wing-coverts, many or all greater upper wing-coverts (occasionally 3–5 only), sometimes innermost tertial (rarely all), occasionally a few central tail-feathers. Timing probably highly variable, depending on hatching: birds examined in fully juvenile plumage from early April to late August (BMNH, RMNH, ZFMK, ZMA) and sometimes occur late March (Smith 1965a); none in moult examined, those seen from early September onwards all had moult completed.

Measurements. ADULT, FIRST ADULT. Nominate *moesta*. Algeria and Tunisia, September–June; skins (RMNH, ZFMK, ZMA). Bill (S) to skull, bill (N) to distal corner of nostril; exposed culmen on average 4·4 shorter than bill (S).

WING	♂ 92·1 (2·13; 22)	89–96	♀ 88·0 (1·47; 11)	86–90
TAIL	62·5 (2·40; 22)	57–67	60·0 (2·31; 11)	56–63
BILL (S)	20·1 (0·87; 22)	18·6–21·2	20·1 (0·70; 11)	19·2–21·3
BILL (N)	11·4 (0·81; 22)	10·1–12·6	11·5 (0·72; 11)	10·7–12·7
TARSUS	28·7 (0·86; 22)	27·2–29·8	27·7 (0·81; 11)	26·5–28·7

Sex differences significant, except bill. 1st adult with retained juvenile flight-feathers and often tail combined with older birds in table above, though wing on average 0·6 shorter and tail 2·0 shorter.

Weights. No information.

Structure. Wing rather short, broad at base, tip rounded. 10 primaries: p7 longest, p8 0–1 shorter, p9 3–7, p6 0·5–2, p5 2–5, p4 7–12, p3 11–16, p1 16–22. P10 reduced; 40–49 shorter than wing-tip, 2–11 longer than longest upper primary covert.

Outer web of p5–p8 and (slightly) inner of p6–p9 emarginated. Tail rather short, tip square; 12 feathers. Bill rather long, straight; rather wide at base, laterally compressed at tip; basal half of culmen straight, distal half gently decurved, tip sometimes forming slight hook. Many soft and fine bristles and hairs along base of both mandibles and on lores. Nostrils small, oval; partly covered by membrane above. Tarsus rather long, strong; toes rather short and strong. Middle toe with claw 18·4 (12) 17–19·5; outer toe with claw *c.* 75% of middle with claw, inner *c.* 68%, hind *c.* 84%. Claws short, strong, decurved.

Geographical variation. Very slight; birds from Middle East usually separated as *brooksbanki* on account of slightly larger bill, somewhat greyer (less black) mantle and scapulars, and whiter rump and upper tail-coverts of ♂, and greyer (less rufous) upperparts of ♀. However, colour of birds examined from Middle East fits within wide variation of nominate *moesta* from north-west Africa, and bill (though rather thick at base) on average only *c.* 1 mm longer; *brooksbanki* thus rather a poor race.   CSR

Oenanthe xanthoprymna Red-tailed Wheatear

PLATES 61, 64, and 65
[between pages 760 and 761, and facing page 856]

Du. Roodstaarttapuit   Fr. Traquet à queue rousse   Ge. Rostbürzelsteinschmätzer
Ru. Златогузая каменка   Sp. Collalba persa   Sw. Rödgumpad stenskvätta

Saxicola xanthoprymna Hemprich and Ehrenberg, 1833

Polytypic. Nominate *xanthoprymna* (Hemprich and Ehrenberg, 1833), south-east Turkey to Iranian Kordestan and (perhaps) western foothills of Zagros mountains (Iran) in Dezful and Shushtar areas; *chrysopygia* (De Filippi, 1863), Transcaucasia, and southern Transcaspia east to Afghanistan and Tadzhikistan, south to Zagros mountains and western Pakistan.

Field characters. 14·5 cm; wing-span 26–37 cm. Close in size and structure to Wheatear *O. oenanthe oenanthe*. Mountain-haunting wheatear, with general character much recalling *O. oenanthe* but showing diagnostic combination of rufous-chestnut rump and black ⊥-mark on tail; basal sides of tail chestnut or (in some nominate *xanthoprymna*) white. Crown and back dull brown, with face and throat fairly nondescript (*chrysopygia* and some ♀ nominate *xanthoprymna*) or solid black (♂ and most ♀ nominate *xanthoprymna*). Sexes usually similar; little seasonal variation. Juvenile similar to ♀. 2 races occur in west Palearctic; pale-throated birds may be difficult to allocate.

(1) Race of south-east Turkey to Iranian Kordestan, nominate *xanthoprymna*. ADULT MALE. Highly distinctive, with strongly rufous-chestnut rump (darkest of all wheatears, but becoming paler with wear) contrasting with white panels at either side of base of black tail. Upperparts and wings sandy-brown, little marked except by pale buff margins to larger wing-coverts and tertials. Head noticeably dark, with only narrow cream supercilium between dark crown and dull black face, throat, side of neck, and upper breast. Underbody buff-white, becoming rufous along flanks and strongly so on under tail-coverts so that white basal panels of tail sharply isolated. Axillaries and under wing-coverts fully black, contrasting sharply with long almost white bases to dusky-ended primaries and brown-ended secondaries. From July or August, fresh pale-tipped feathers produce paler brown back, wider pale margins on wings, and hoary lower and rear borders to black face and throat. ADULT FEMALE. Usually as ♂ (*contra* Heinzel *et al.* 1972), though black throat may be duller and some at least have chestnut (not

white) tail-base. Some birds have pale throat and so resemble *chrysopygia*, but darker with less obvious supercilium. JUVENILE. Not studied. FIRST WINTER AND SUMMER. As adult, but ♂ and black-throated ♀♀ have broader pale fringes on throat (until late winter). Either sex may have chestnut tail-base (see Geographical Variation). (2) Race of Caucasus to Afghanistan and Pakistan, *chrysopygia*. ADULT MALE. Far less distinctive than nominate *xanthoprymna*. Lateral basal panels on tail same colour as rump, and black on tail duller with narrower terminal band. Head lacks black face and throat, being marked only by broader dull white supercilium, pale brown lore, almost black rear eye-stripe, dark rufous-brown cheeks, and dull white chin and throat. Rest of plumage much as nominate *xanthoprymna* but upperparts sandier- or greyer-brown, especially on sides of neck and shoulder, and underparts buffier, with chestnut wash on breast and deeper rufous-chestnut tone on flanks and vent. Under wing-coverts buff-white, forming (with bases to flight-feathers) almost wholly pale underwing. Fresh autumn plumage differs as in nominate *xanthoprymna*. ADULT FEMALE. As ♂ but even duller. JUVENILE. Closely resembles adult ♀ (i.e. not spotted) and best distinguished by different degree of wear on wings and (if evident) incomplete growth. Both races at all ages have bill and legs black.

♂ and black-throated ♀ nominate *xanthroprymna* unmistakable. Pale-throated ♀ of that race and both sexes of *chrysopygia* less distinctive: confusion possible (but unlikely to persist) with Red-rumped Wheatear *O. moesta* (almost completely black tail, black back in ♂, chestnut head in ♀ and immature) or ♀ *O. oenanthe* (inhabiting similar breeding habitat in range of *O. xanthoprymna* but

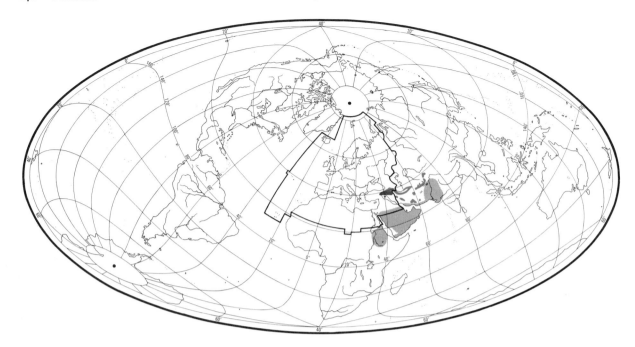

lacking rufous rump and vent). Flight, gait, stance, and behaviour much as *O. oenanthe* but more prone to perching on low vegetation. Hides in holes.

Song a loud warbling. Calls include (at least in *chrysopygia*) distinctive 'thrr thrr thrr', less harsh than other *Oenanthe*.

Habitat. Breeds in continental lower middle latitudes, in arid and steppe climates, mainly in upland and mountain regions, favouring steep and bare terrain with ready access for nesting and refuge to burrows, fissures, scree, and other sites offering convenient holes. In USSR, occurs on naked often windy rocky slopes and screes between *c.* 1200 and 3100 m (Dementiev and Gladkov 1954*b*). In Afghanistan during breeding season, common on stony mountain slopes and in dry fallow fields in desolate valleys between 2000 and 2700 m (Paludan 1959*a*). In Pakistan in summer, frequents arid rocky slopes, usually above 2100 m, especially near small perennial streams, perching upright on clods or rocks and less frequently on low bushes, and taking cover in gerbil holes (Ali and Ripley 1973*b*). In south-west Iran, breeds at high altitudes at *c.* 2500–4000 m; winters from sea-level to 2100 m in lower, deforested sub-steppe and warm steppe zones. Breeds on steep stony boulder-strewn hillsides, but many territories include more gently sloping areas, even of plain or valley bottom. Boulders and rock outcrops are important as song-posts, as look-outs for prey, and for cover from predators and adverse weather, as well as for nest-sites. Vegetation on territories studied was chiefly of widely-spaced montane dwarf shrubs, augmented in spring by flush of herbaceous plants which completely covered patches of ground; in some territories,

scattered bushes or even stunted trees. Flatter areas, where included, were favoured for foraging, and in winter most territories included flat or gently sloping ground, suggesting that these could afford better and more accessible food resources, and that need for rocky hillsides might be largely associated with breeding season requirements. Only congener ranging to equally high altitudes, Wheatear *O. oenanthe*, occupied much flatter sites, although equally arid and cool. (Cornwallis 1975.) In winter, occurs in flat stony and sandy semi-desert, with sparse bushes, especially near foot of rocky hillsides, from sea-level up to 3300 m; also on sand-dunes at edge of desert (Ali and Ripley 1973*b*).

In winter, in Eritrea (Ethiopia), locally common on rocky hills with light *Acacia* below 650 m (Moreau 1972). Occurs in arid bushy country, rarely true desert, perching frequently on bushes and telegraph wires (Etchécopar and Hüe 1967). ♂ in dry wadi at *c.* 400 m near Sudan-Egypt border in early March perched both inside and on top of *Acacia* tree crown, and also on rocks; ♀ not far away in hilly treeless terrain on rocks only (E M Nicholson). In Arabia in winter, seldom in absolute desert but always in arid situations, seldom entering cropland. Often among sparse bushes on which it perches freely, as on telegraph poles (Meinertzhagen 1954). Apparently avoids human contact, although not shy. Flies in lower airspace.

Distribution. TURKEY. No information on range changes. Map based on information supplied by M Kasparek. See also Kumerloeve *et al.* (1984).

Accidental. Syria, Israel, Libya.

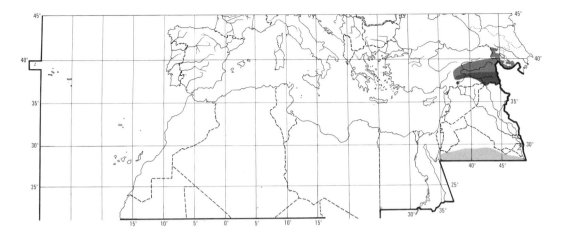

Population. No information on trends.

USSR. Numbers extremely low (Dementiev and Gladkov 1954*b*), though sometimes numerous in favoured habitat (Panow 1974).

Movements. Largely migratory, although some birds in south-west Iran perhaps make only altitudinal movements (Cornwallis 1975). Winters in eastern Africa and south-west Asia: rare in North Yemen (Cornwallis and Porter 1982), and only small numbers winter in Gulf states of Arabia, though fairly common in United Arab Emirates (Bundy and Warr 1980); common in south-west Iran (Cornwallis 1975).

Nominate *xanthoprymna* (breeding south-east Turkey east to Iranian Kordestan) winters in Sinai, eastern Egypt, and Sudan, and in south and centre of Arabian peninsula (Loskot and Vietinghoff-Scheel 1978). One early-December record from south-east Libya (Misonne 1974) is *c*. 1000 km west of known winter range (Loskot and Vietinghoff-Scheel 1978). Autumn passage September–October on broad front, probably SSW–SW; recorded then in Syria and Iraq; ♀ (probably nominate *xanthoprymna*) at Azraq (Jordan) on 9 November (Clarke 1980). Present in winter quarters October to mid-February or March (Loskot and Vietinghoff-Scheel 1978).

O. x. chrysopygia (breeding Transcaucasia and Zagros eastwards) is common winter visitor to Red Sea coast of Sudan and Eritrea, at least as far south as Zula; present mid-October to March (Cave and Macdonald 1955; Smith 1955*a*). Passage migrant and winter visitor to Arabian peninsula (including Oman), Iraq, and southern Iran north to *c*. 35°N (Loskot and Vietinghoff-Scheel 1978). Wintering population present in southern Iran from mid-September to end of March (Cornwallis 1975), though last leave north Iranian breeding grounds only in early November (Panow 1974). Occurs in Oman from late August to April (Gallagher and Woodcock 1980). Birds from eastern part of range winter mainly in southern Afghanistan, Pakistan, and north-west India (Ali and Ripley 1973*b*). MGK

Food. Almost all information extralimital and refers to eastern race *chrysopygia*; most detailed studies in Badakhshan, USSR (Loskot and Petrusenko 1974), and Iran (Cornwallis 1975). Diet largely insects, especially ants, beetles, and larval Lepidoptera. Typically finds food on bare ground using dash-and-jab technique or launches flying attacks from firm, elevated perch (e.g. stone or boulder—rarely vegetation). Food also picked off vegetation and dug out of ground with bill (e.g. beetle larvae). (Loskot and Petrusenko 1974; Cornwallis 1975.) In south-west Iran in winter (mainly in territories including steep and/or gently sloping ground with scattered stones, boulders, and dwarf shrubs), 55% of perches were on ground, 30% on stones or clods, 12% on boulders or rock outcrops, and 3% on vegetation ($n = 205$); 65% of movements (many in search of food) were dashes along ground, 30% short flights, and 5% long flights ($n = 201$); 45% of feeding actions employed a dash-and-jab technique (typically involving rapid hopping rather than running), 27% a short flight and pounce, and in 28% bird was stationary and pecking at ground, base of stone, or vegetation ($n = 58$); 84% of food items taken from bare ground, 9% from base of stone, and 7% from vegetation ($n = 58$) (Cornwallis 1975). Sometimes flushes insects from under stones by beating wings (wing-flashing) and calling (Loskot and Petrusenko 1974: see 3a in Voice), and, like many *Oenanthe*, will open wings and hold them forward on encountering large or unfamiliar prey (Panow 1974). Also takes insects in the air like flycatcher (Muscicapidae) (Panow 1974; L Cornwallis); in Badakhshan uses this technique especially during swarming of termites (Isoptera), ants (Formicidae), and chafers (Scarabaeidae), when successful pursuits of up to 25 m recorded (Loskot and Petrusenko 1974); bird in eastern Saudi Arabia made 3 sallies each to height of about 1 m (J Palfery); in Iran, one seen to rise *c*. 8 m, almost vertically, to take insect (P A D Hollom). In Tadzhikistan (USSR), feeds normally when sunny but when dull tends to stay near rivers frequently taking insects in flight over water (Abdusaiyamov 1973). Sometimes searches for prey by hovering like Kestrel *Falco tinnunculus* and then diving on to it (Loskot and Petrusenko 1974); in eastern Saudi Arabia this strategy used to feed in irrigated grassland over which 2 birds often hovered with spread tails, at height of 20–30 cm, dropping into long grass to take prey (J Palfery). Nominate *xanthoprymna* seems to hunt more from perch than *chrysopygia*, using flimsier vegetation; also frequently flycatches (L Cornwallis, M C Harrison). In south-west Iran in winter, took average 6·6 items per 5 min ($n = 12$ 5-min periods), rate increasing slightly towards dusk (L Cornwallis); in Badakhshan in breeding season, feeds most intensively 07.00–09.00 hrs and 17.00–19.00 hrs (Loskot and Petrusenko 1974).

The following recorded in diet. Invertebrates: grasshoppers, etc. (Orthoptera: Tettigoniidae, Acrididae), earwigs (Dermaptera), termites (Isoptera), bugs (Hemiptera: Scutelleridae, Pentatomidae, Coreidae, Lygaeidae, Tingidae, Reduviidae, nymphs of Cicadidae, Cicadellidae), larval ant-lions (Neuroptera: Myrmeleontidae), adult and larval Lepidoptera (Noctuidae, Notodontidae, Satyridae, Pieridae, Tortricidae, Coleophoridae, Geometridae), caddis flies (Trichoptera), adult and larval Diptera (Syrphidae, Stratiomyidae, Asilidae), Hymenoptera (sawflies Tenthredinidae, Ichneumonidae, Chrysididae, ants Formicidae, social wasps Vespidae, solitary wasps Sphecidae, bees Apoidea), adult and larval beetles (Coleoptera: Carabidae, Staphylinidae, Scarabaeidae, Byrrhidae, Buprestidae, Elateridae, Cantharidae, Meloidae, Tenebrionidae, larval Chrysomelidae, Curculionidae), spiders (Araneae), mites (Acarina), Myriapoda (including Scolopendromorpha), woodlice (Oniscidae), sandhoppers (Amphipoda). Vertebrates: once a small lizard (Lacertidae). Plant material: mainly seeds or fruits, including caltrops (Zygophyllaceae), madders (Rubiaceae), figworts (Scrophulariaceae), docks (Polygonaceae), daphnes (Thymelaeaceae), mulberry *Morus*; also fragments of leaves and stems mainly of grasses (Gramineae). (Meinertzhagen 1954; Abdusalyamov 1973; Loskot and Petrusenko 1974; Cornwallis 1975; Stepanyan 1978*b*; E N Panov.)

In southern Badakhshan in 2 successive breeding seasons, 18 May–10 July, 55 stomachs (including 7 fledglings) contained 1338 items including 48·4% (by number) ants, 8·0% other Hymenoptera, 24% adult (mostly) and larval Coleoptera (11·4% Curculionidae, 6·4% Carabidae), 9·4% adult (mostly) and immature Hemiptera (3·7% Lygaeidae, 2·6% Pentatomidae, 4·5% larval (mostly) and adult Lepidoptera, 1·7% spiders, and 1·4% seeds; fragments of leaves and stems (some dry) of herbaceous plants, mainly grasses (Gramineae), found in 9 stomachs, particularly those of fledglings—considered to have been ingested accidentally. (Loskot and Petrusenko 1974.) In south-west Iran, 5 stomachs taken on breeding grounds June–August and 17 taken on wintering grounds October–March, contained 901 items including 65·8% (by number) ants, 2·0% other Hymenoptera, 15·0% beetles, 8·9% seeds (December–January only), 3·3%

unidentified insects, 1·7% insect larvae, and 1·4% Hemiptera; 70·6% of items 0–5 mm long, 23·2% 5–10 mm, 5·2% 10–15 mm, 0·6% 15–20 mm, 0·5% 20–25 mm; one stomach, August, also contained berries of *Daphne acuminata* (Cornwallis 1975; L Cornwallis).

Information on food of young available only from Badakhshan; 61 collar-samples from nestlings 3–9 days old contained 108 items of which 33·3% (by number) Lepidoptera (31·5% larvae, largely Noctuidae), 20·4% beetles (14·8% Scarabaeidae), 14·8% ants, 8·3% Orthoptera, 7·4% Hemiptera, 5·6% termites, 5·6% spiders, 2·8% Diptera, and 0·9% larval ant-lions (Loskot and Petrusenko 1974).

LC

Social pattern and behaviour. Nominate *xanthoprymna* poorly known and account based almost entirely on studies of *chrysopygia* in southern USSR (Panow 1974; E N Panov); additional information from Cornwallis (1975) and L Cornwallis (Iran), and J Palfery (eastern Saudi Arabia).

1. Outside breeding season normally solitary and territorial (Smith 1971; Ali and Ripley 1973*b*; Cornwallis 1975; Gallagher and Woodcock 1980). In south-west Iran, 11 territories averaged 2·5 ha (*c.* 1·0–4·9). Densities in winter generally low, but distribution patchy and in some localities birds not uncommon (Cholmley 1897; Ticehurst 1922*b*; Ali 1955; J Palfery); in south-west Iran during car journeys through suitable country, 1 bird per 20 km over 214 km; in detailed surveys in 4 localities, 1·9–16 birds per km² over 2·5 km², average 7·9 per km² (L Cornwallis). Except during short stops on migration, both sexes normally hold individual territories which are defended against conspecific birds, socially subordinate *Oenanthe* and some other birds (see Antagonistic Behaviour, below). In south-west Iran, territories sometimes contiguous with those of Finsch's Wheatear *O. finschii* (10 records), Hume's Wheatear *O. alboniger* (8 records) and Desert Wheatear *O. deserti* (1 record), but apparently occupying ground not wanted by *O. finschii* and *O. alboniger* as socially subordinate to them (see Antagonistic Behaviour, below). (Cornwallis 1975; L Cornwallis.) No firm evidence of year-to-year fidelity to winter territory but in south-west Iran 2 cases of same territories being occupied in successive winters (Cornwallis 1975). BONDS. Mating system apparently monogamous. No information on duration of pair-bond or age of first breeding. Nestlings and fledglings fed by both parents (Erard and Etchécopar 1970; Abdusalyamov 1973; Helbig 1984; L Cornwallis). In Transcaucasia, ♀ seen re-building when 1st brood still dependent (Panow 1974); ♂ presumably cares for 1st brood once ♀ incubating 2nd clutch. Duration of family bonds not fully documented; in Tadzhikistan (USSR), young reported to become nomadic 5–8 days after fledging (Abdusalyamov 1973), but in eastern Turkey apparent family parties (of nominate *xanthoprymna*) seen as late as end of September (Kumerloeve *et al.* 1984; M Kasparek); in south-west Iran, at end of July, adult chased 3 juvenile members of his family party, apparently attempting to evict them from territory (L Cornwallis). BREEDING DISPERSION. Solitary and territorial (Loskot and Petrusenko 1974; Panow 1974; Cornwallis 1975; E N Panov). In Nakhichevanskaya ASSR (south of Caucasus), territories 200–300 m apart (Stepanyan 1978*b*); in eastern Turkey, 2 territories on opposite valley slopes separated by *c.* 250 m (Helbig 1984). Territory 200–400 m across (Panow 1974); in Badakhshan (southern Tadzhikistan), usually 1·2–1·8 ha (0·5–3·0) (Loskot and Petrusenko 1974). In USSR studies by E N Panov, territories typically overlap:

shared zone up to 30 m wide if territories small, up to 100 m if large; ♂♂ or breeding pairs commonly visit neighbouring territories (E N Panov). Breeding density varies considerably with quality of habitat: in Badakhshan, on rocky hillsides and screes, 10-20 pairs per km²; on terraces and side gorges of lower Shakhdarya river, average 40 pairs (20-70) per km²; adjacent to irrigated fields, where birds feed, up to 150-270 pairs per km² over 1·5-2 ha (Abdusalyamov 1973; Loskot and Petrusenko 1974); in south-west Iran, in rocky, mountainous terrain, 16 pairs per km² over 55 ha (L Cornwallis); in south-east Transcaucasia, up to 10 pairs along 1 km (Panow 1974). Both sexes defend territory (Panow 1974). Territory normally all-purpose but sometimes, especially when feeding young, birds collect food outside defended area, up to 700 m from nest (Loskot and Petrusenko 1974; L Cornwallis). ROOSTING. On breeding grounds, most active in morning and evening (until dusk well advanced). Sometimes active in very hot weather, but often seeks shade during heat of day in same sites (holes, crevices, under bushes) as used for nocturnal roosting (Abdusalyamov 1973; Loskot and Petrusenko 1974; E N Panov). Adults sun-bathe in prostrate posture with wings outspread (Panow 1974, which see for illustration). Sometimes sand-bathes (E N Panov).

2. Often quite tame and confiding (e.g. Taylor 1896, Christison and Ticehurst 1942), sitting motionless until approached (Moore and Boswell 1956). In eastern Saudi Arabia, birds generally wary, but more approachable in open country (J Palfery); in Jordanian desert, late April, bird allowed approach to within *c.* 20 m (Hollom 1959); for reports of wildness in open terrain, see (e.g.) Cholmley (1897) and Witherby (1903). When approached closely, tends to fly off low and discreetly, sometimes to perch *c.* 20 m away (Moore and Boswell 1956; Erard and Etchécopar 1970), or flies quickly out of sight (J Palfery). In India, winter, sometimes runs off to bolt down rodent hole when disturbed (Ali and Ripley 1973b). Perched bird (especially shortly after landing: J Palfery) typically wags tail slowly and constantly up and down, raising it smoothly to *c.* 10° (or 45° or more). More vigorous tail-raising (2-3 times) or -flicking, with simultaneous abrupt sideways flicking of wing-tips, may occur on landing after being disturbed (J Palfery, E N Panov). Constant wing-flicking likened to that of *Cercomela* chats (Meiklejohn 1948b). Apparently not found in other *Oenanthe*, but characteristic of *O. xanthoprymna* (in certain displays or when confronted by predator) is a gentle beating of raised or forward-moving wings, often accompanied by 'snoring' call (Panov 1978; E N Panov: see 3a in Voice). For illustration of main postures and transitional forms, also comparison with certain other *Oenanthe*, see Panov (1978). FLOCK BEHAVIOUR. No information. SONG-DISPLAY. ♂ sings (see 1 in Voice) from perch (e.g. bush, high rock) or in flight. Each ♂ usually has several (normally up to 3) favoured song-posts affording good view over territory, and will sing persistently from one such (e.g.) while ♀ feeding or nest-building. Subsong (see 1b in Voice) on breeding grounds also given while perched or moving about on rock, bush, etc. (Panow 1974; L Cornwallis, P A D Hollom, J Palfery, E N Panov). Song-flight closest to that of Eastern Pied Wheatear *O. picata* (of *picata* morph); sometimes sings while moving between perches. In Iran, early May, Song-flight performed at height of *c.* 10-15 m (P A D Hollom). Characteristically smooth and slow, with exaggerated wing-beats and smooth changes of height. Unlike other *Oenanthe*, sometimes hangs almost motionless in the air, with only slight movement of fully extended wings, legs often dangling; or makes smooth turns. Typically sings variants of Territorial-song during flight lasting up to *c.* 1 min.

On landing, usually adopts a horizontal posture and may then fan tail and lower it to ground. Song-flight rare during pair-formation, but otherwise common from first few days after arrival on breeding grounds up to nestling phase. (Panow 1974; Panov 1978; E N Panov.) No further information on song-period in any part of range. In eastern Turkey (nominate *xanthoprymna*), 19 June, song started at *c.* 04.00 hrs—earlier than in other bird species (M Kasparek). ANTAGONISTIC BEHAVIOUR. (1) General. On breeding grounds, generally highly aggressive towards strange conspecific birds. Territorial disputes frequent in late March but difficult to interpret as sexes not easy to distinguish. However, roles of ♂ and ♀ apparently as in (e.g.) *O. finschii*, with each attacking only other conspecific birds of its own sex (Panow 1974). In Iran, aggression noted between 2 pairs both feeding young (L Cornwallis). (2) Threat and fighting. In early period after arrival, ♂ neighbours commonly patrol border. In overlap zone (see Breeding Dispersion, above), may approach to within *c.* 5 m of each other and perform slow Parallel-walk (Fig A), sometimes also making short fluttering

A

flights. Birds usually silent and often mock-feed. May beat wings and aerial fight once recorded in this context. Where single bird intrudes into a territory, only 1 member of resident pair will attack it, while other keeps slightly apart. If 4 birds meet (presumably 2 pairs), 2 birds often fight, sometimes interlocking claws, tumbling over ground, and pecking, while other 2 only adopt various threat-postures (Figs B-C) or watch from some distance. Constant calling (chirping sounds: see 3a in Voice) during such conflicts is typical, or birds (both sexes) may sing loud variant of Subsong without interruption. Usually ends with chase, after which intruder(s) leave other's territory. (Panow 1974; E N Panov.) Furiously attacks both ♂ and ♀ Black-eared Wheatear *O. hispanica* which sometimes attempt to establish territory within that of *O. xanthoprymna* (Panow 1974). Winter territory-holders recorded defending territory against conspecific birds and certain other *Oenanthe*, including (isolated cases) Isabelline Wheatear *O. isabellina* (2 displaying birds half-heartedly attacked by *O. xanthoprymna*), ♀ *O. picata*, and Mourning Wheatear *O. lugens*, though these species normally dominate *O. xanthoprymna* (Ali and Ripley 1973b; Short and Horne 1981; L Cornwallis, M D Gallagher, J Palfery). In south-west Iran, winter, constantly in competition with *O. alboniger* and *O. finschii* to which subordinate. When attacked by *O. finschii*, usually retreats (sometimes only short distance, then flying off) and may give weak song; in one minor confrontation with *O. finschii*, both species gave apparent

B

C

threat-calls (see 3a in Voice) and both then retreated. Other bird species against which territory is defended include Shore Lark *Eremophila alpestris*, Tawny Pipit *Anthus campestris*, Bluethroat *Luscinia svecica*, Stonechat *Saxicola torquata*, Rock Thrush *Monticola saxatilis*, and Fieldfare *Turdus pilaris*. (Cornwallis 1975; L Cornwallis, J Palfery.) When feeding at favourable site outside territory in Badakhshan, showed no aggression towards conspecific birds or other species (Loskot and Petrusenko 1974). HETEROSEXUAL BEHAVIOUR. (1) General. In Tadzhikistan, pair-formation and start of breeding take place straight after arrival on breeding grounds (Abdusalyamov 1973). In Nakhichevanskaya ASSR, birds seen already paired on 26 March and nest-building noted there 29 March. In one late spring, ♂ occupied territory 7 March, started active song 20 March; pair-formation took place 23 March and nest-building on 2–3 and 7 April (Panow 1974; E N Panov). In northern Baluchistan, 'nuptial display' performed by ♂ for over 50 minutes, apparently not in presence of ♀, in mid-September (Christison and Ticehurst 1942); however, no details and may refer only to establishment of winter territory. Pair-formation generally as in other *Oenanthe*, though Bill-up posture (typical of, e.g., *O. hispanica*) does not occur in this context. Typical of ♂ *O. xanthoprymna* is Horizontal-posture with bill horizontal and tail closed. Significantly (at least for *chrysopygia*, in which sexes similar), courtship display of ♂ and ♀ identical, though ♀ (like rival ♂♂—see Antagonistic Behaviour, above) sometimes beats half-open wings (Panow 1974; E N Panov). (2) Pair-bonding behaviour and nest-site selection. Account based on Panow (1974) and additional data from E N Panov. When ♀ first appears in a territory, ♂ owner may perform various displays for up to 6 hrs. May approach her first in Song-flight, but flying in straight line and sometimes changing from Territorial-song to loud ('aggressive') variant of Subsong. On landing by ♀, ♂ typically assumes Horizontal-posture (Fig D, left). Occasionally, lowers fanned tail and creeps towards ♀ who may give threat-call (see 4 in Voice), though some ♀♀ silent and no song noted from ♀ during pair-formation. Sometimes, ♂'s approach elicits moderate aggression in ♀. Dancing-flight may occur at this stage: ♂ flies back and forth over ♀ in a series of low arcs (maximum height *c.* 1 m), almost touching ground to left and right of her (see also subsection 4 for variant of Dancing-flight). However, creeping approach of ♂ often causes ♀ to flee, ♂ then chasing her. Chases very common on first day of pair-formation, less so thereafter, but occur throughout whole period of nest-building. After a chase, ♀ usually flies outside territory and, during her absence, ♂ will sing variants of Territorial-song alternating with series of Whistle-calls (see 3b in Voice); often performs Song-flight (of varying duration), landing by crevices which are shown to ♀ (see below). Mock-preening and Looking-down are also not uncommon in this context. In Nest-showing ceremony, ♂ flies to crevice and there assumes Horizontal-posture, entering and leaving several times in succession. Normally gives Subsong and, while in cavity, may give Rasp-call (see 5 in Voice). When ♀ approaches, ♂ will fly to another crevice for similar performance, and so on. Bout of Nest-showing usually ends with ♂ chasing ♀. Birds generally more tolerant by end of ♀'s first day in territory, such that they tolerate close approach when inspecting crevices together. Fly about territory together, ♂ usually in lead. 2–3 days later may be in cavity together for nest-site selection. Not known which sex chooses site, but even after much effort to prepare one (see subsection 5, below), ♀ may reject it as unsuitable, and move up to *c.* 150 m. After pair-formation and during nest-building, sometimes in late dusk, ♀ not infrequently sings and approaches ♂, causing him to display at her: both birds call (see 6a in Voice) and stand with head bowed and back feathers ruffled; ♂ then moves in little leaps around ♀ (presumably different from Dancing-flight), now coming close to her, now moving further away (Fig E). Such a bout of courtship may lead to copulation, but usually ends in long chase, birds (presumably both sexes) calling incessantly (see 6b in Voice). (Panow 1974; E N Panov.) (3) Courtship-feeding. Not definitely recorded, but ♂ and ♀ sometimes come close together and will then Bill-touch (e.g. when ♀ takes break from incubation). On one occasion, food perhaps passed, though ♂ never recorded feeding ♀ on nest; once, ♂ passed nest-material to ♀ (Panow 1974; E N Panov). (4) Mating. During bout of courtship as described in subsection 2, ♂ usually attempts to approach ♀ from rear; if ♀ ready to copulate, she turns several times on the spot, until tail directed

D

E

F

at ♂. Frequently, preliminaries and actual copulation take place in small cave under rock overhang. ♂ sings there (variants of both Subsong and Territorial-song). Towards end of nest-building, ♂ ready to copulate will give Subsong in Wings-high posture (Fig F): raises wings and beats them slowly forwards and up, simultaneously raising and lowering widely fanned tail. When ♀ arrives, ♂ fans tail widely and runs into dark corner of cavity. If ♀ comes closer, ♂ will perform Dancing-flight over her: performs vertical leaps within cavity, *c*. 1 s apart. Singing all the while, ♂ then adopts Pre-copulatory posture (Fig G), also quivering widespread wings, and gently

G

crawls on to ♀; may give 'snore' call if copulation attempt frustrated. ♂ may attempt to copulate with lump of earth while ♀ incubating. (Panow 1974; E N Panov.) (5) Behaviour at nest. For up to 4 days prior to nest-building proper, ♀ carries small flat stones into nest-hole and deposits them there, sometimes to fill in cracks and make floor level, but others simply left by hole. For further details, see Panow (1974) and Breeding. ♂ often visits ♀ at nest and will display as described in subsection 2. May rapidly enter and leave nest-hole several times, until ♀ follows him in. Occasionally, ♂ carries stones into another favoured crevice. (Panow 1974.) RELATIONS WITHIN FAMILY GROUP. Young brooded for quite long time—if cold and wet, up to 5th day (when eyes open). Beg (not clear from what age) with slow or very fast wing-trembling, also by giving food-calls, each call accompanied by a bobbing movement. Refusal to take food (satiety) much as in other *Oenanthe* (for details, see Panow 1974; see also Voice). No information for wild birds on post-fledging phase, and the following apparently based on observations of captive birds: first attempt to self-feed at *c*. 2 weeks and no longer beg from *c*. 1½ months; first elements of courtship display behaviour (including copulation attempt) at *c*. 1 month, of territorial behaviour from *c*. 2 months (E N Panov). ANTI-PREDATOR RESPONSES OF YOUNG. No information. PARENTAL ANTI-PREDATOR STRATEGIES. Few details, all referring to man as potential predator. In Armeniya (USSR), adults flew

from rock to rock in apparent attempt to lure intruders away from young (Lyaister and Sosnin 1942). Can be bold when young threatened: e.g. may fly towards man and give alarm-calls (see 3 in Voice) from rock a few feet away, or call thus while hovering less than 1 m from intruder (Meiklejohn 1948*b*); one (probably ♀) of highly agitated pair with young in nest launched vigorous attack on observer trying to examine nest, while other bird sang continuously from perch *c*. 10 m away (Erard and Etchécopar 1970).

(Figs by K H E Franklin: from drawings in Panow 1974 and by E N Panov.)   LC, MGW

Voice. Relatively quiet outside breeding season in Iraq (Moore and Boswell 1956) and silent then in Eritrea (Smith 1971). In eastern Saudi Arabia in winter, birds gave Subsong and at least call 3a (J Palfery); call 3a also noted in Iran in winter (L Cornwallis). Data from most-detailed sources (Panow 1974; Panov 1978; E N Panov) refer to *chrysopygia*; not known whether nominate *xanthoprymna* significantly different.

CALLS OF ADULTS. (1) Song of ♂. Several variants but no sharp divisions between them. *Contra* Dementiev and Gladkov (1954*b*), mimicry virtually absent from all song variants, and in this respect closest to Finsch's Wheatear *O. finschii* (Panow 1974; Panov 1978; E N Panov). (a) Territorial- (Advertising-)song. Short phrases (7–16 units) separated by short pauses, or pauses reduced until phrases merge to form more continuous variant. Song of simple structure, with constant repetition of a melodious whistling motif and only slight variations. Attractive tonal quality of this (and Subsong—see below) recalls song of Blue Rock Thrush *Monticola solitarius* (Panow 1974; Panov 1978; E N Panov). In Iran, sweet, full, slow and serene warbling somewhat recalling Robin *Erithacus rubecula* (L Cornwallis; also Meiklejohn 1948*b*). Also in Iran, bird disturbed near nest gave rapidly delivered fluting phrase preceded by detached units in rhythm of Nuthatch *Sitta europaea* (Erard and Etchécopar 1970). Characteristic features of song are complex units ('trills'), notes with frequency modulation (vibrato) sometimes coupled with amplitude modulation, and overtones, and inclusion (at moment of maximum excitement) of peculiar 'snoring' sound (see call 3a); Rasp-calls (call 5) also quite often included. Long variant given during territory demarcation, also (typically with wailing sounds and 'snoring') during pair-formation. Prior to copulation, bird may change from short Territorial-songs (sometimes quiet and monotonous repetition of a single short phrase, or loud and varied) to Subsong. Territorial-song given in flight, also from perch or ground (Panow 1974; Panov 1978; E N Panov). In recording by C Chappuis, 2 short song-phrases connected by markedly contrasting Subsong, the latter so quiet as to be barely audible, and consequently not visible on sonagram (see discussion of Fig II, below): e.g. 1st part (0·84 s) comprises 'zizz' (perhaps a Rasp-call variant) followed by 3 main units of some tonal quality (rich, full-throated warbling), then very faint Subsong (c. 1 s) and 2nd loud section comprising 2 brief rasping or throaty clicks (call 3a) and 4 units not dissimilar to those in 1st part, i.e. portamento, frequency and amplitude modulated, sometimes buzzy in character, and sometimes with sub-units too subtle to be discernible to human ear. Some sounds rather strangled, and the more nasal units recall finch (Fringillidae), including a crooning like Twite *Carduelis flavirostris* (P J Sellar). Recording by P A D Hollom (Iran) comprises short phrases (apparently not connected by quiet Subsong) rendered as follows: 'weewee wee wee-urrr' or 'wee-urrr weee-eee' (M G Wilson); powerful and clear (with amphitheatre effect) and quite tuneful 'svilit boes-year' (in flight) and, from perch, 'see-wat-chew eeper', also 'watchew-era', and 'whee-chu chree' (P A D Hollom). One phrase (Fig I) from this recording is closely similar in form and content to 2nd loud part of song described above: starts with rasp of relatively low frequency (almost like Craking-call of Corncrake *Crex crex*: M G Wilson),

followed by portamento tone, unit with strong frequency modulation, complex unit (of 4 sub-units), and very rapid tremolo (or perhaps quite slow amplitude modulation) sounding hoarse and toneless. 1st part of another song (Fig II) rather distinct in starting with chattering (i.e. noisy quality, but not loud), followed by 2 quite loud rasps (call 3a), then descending scale which is entirely musical (no noise), until 'chip' end-flourish; 2nd part (following after c. 1 s of quiet Subsong) starts with fizzy buzz, then shows portamento tones and frequency and amplitude modulation (J Hall-Craggs). (b) Subsong. Given towards end of nest-building. More variable than 1a, though *M. solitarius* quality still evident. Includes quite high- and very low-pitched 'notes', melodious sounds alternating with 'snoring', wailing 'uuuu', fine and drawn-out whistles, grunting, tapping, ticking, and grating rasps (Panow 1974; Panov 1978; E N Panov). In eastern Saudi Arabia, winter Subsong of one bird a subdued warble like lark (Alaudidae), interspersed with low scratchy and buzzy sounds (like fly trapped under glass—perhaps a Rasp-call); another's like a series of calls (i.e. less sustained than 1st) and with quality of Blackbird *Turdus merula*—'sreeachup swee-ur', 1st unit rising, falling, then rising again, 2nd falling, also 'swee-ur swee-ur' and 'chreeup' (J Palfery). In Iraq, sweet Subsong with many notes like warbler (Sylviidae) lasted 10–45 s (Moore and Boswell 1956). (c) So-called 'aggressive song' or 'battle song'. Few details, but given (e.g.) in response to playback of Territorial-song, or ('courtship song') during pair-formation; may be incorporated with Territorial-song in Song-flight. Essentially a loud variant of Subsong (as in certain other *Oenanthe*) and sonagram (E N Panov) shows loud ascending whistles, and 'ripples' (vibrato) both steady and descending in pitch, also faint background rasps (J Hall-Craggs; M G Wilson). (2) Song of ♀. Usually shorter than ♂'s (but probably not otherwise significantly different) when given during antagonistic encounters. ♀ will also sing when taking initiative in courtship after pair-formation (Panow 1974). (3) Contact-alarm calls. (a) Clicking calls. Loud, with muffled and clear variants (Panov 1978), but no details on differences. In alarm at man or snake, a harsh rasping (croaking: Meiklejohn 1948b) or a clicking 'tchek', sometimes in series to form 'snore' (perhaps like a rapid, soft nasal clicking: L Cornwallis), or in alternating series with

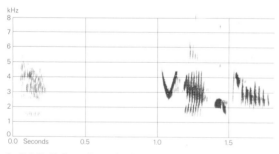

I P A D Hollom Iran April 1972

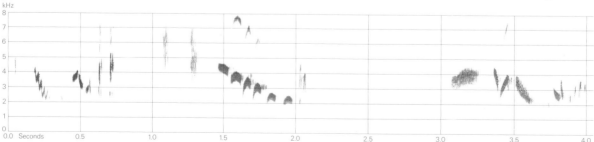

II C Chappuis Afghanistan May 1977

call 3b. Sonagram (E N Panov) indicates snore may also be series of longer rasps, perhaps thus indicating close relationship with call 5. Clicks or longer snore widely used: often incorporated into song (see above); snore sometimes given by ♂ when copulation attempt frustrated by ♀; 'tchek' may be combined with beating of wings to flush prey (see Food) (Loskot and Petrusenko 1974; Panow 1974; Panov 1978; E N Panov). Chirping sounds apparently serving as threat-call (Panow 1974) probably also belong here. Further renderings of presumed clicks and longer variants from study of *chrysopygia* in eastern Saudi Arabia: typical chat-like, low, grating 'grat grat', when hovering (perhaps to feed, or alarmed by snake); a dull, muffled, zizzing 'zrrp zrrp' (J Palfery). In Iran, alarmed birds gave constant chacking sounds (short dry 'zuk': P A D Hollom), also a 'thrrr', sometimes in high vertical flight (L Cornwallis); 'thrrr', presumably same as 'chirr chirr' noted in Jordan by Hollom (1959). (b) Whistle-calls. Sonagram (E N Panov) shows short whistles descending in pitch, sometimes steeply. Apart from alarm function, also used by ♂ in pair-formation (E N Panov, M G Wilson). In Iraq, a plaintive 'cheep' (Moore and Boswell 1956). 3a and 3b may be combined in alarm: 'thrrr thrrr si si' (L Cornwallis); 'zvee tuk' (P A D Hollom). (4) A 'tr-tr-tr-tr-tr' given, as in certain other *Oenanthe*, by ♀ (very rarely by ♂) as threat (E N Panov); not clear how (if at all) this differs from longer 3a variant. (5) Rasp-call. Resembles juvenile food-contact call. Given (e.g.) by ♂ during Nest-showing, also in song (E N Panov). Calls given when alarmed, also directed against intruding Bluethroat *Luscinia svecica* may belong here, or are perhaps variants of call 3a: harsh grating, scratchy 'chairz' or 'jairz', sometimes long, sometimes short and clipped (J Palfery) and similar (if not identical) 'tchazz' (Moore and Boswell 1956). (6) Other calls. (a) Quiet trill given during courtship after pair-formation (Panow 1974); perhaps refers to calls 3a or 4. (b) A 'zch wik...zch wik...' given during courtship chase (Panow 1974); perhaps a combination of calls 3a and 3b.

CALLS OF YOUNG. 2 rasping sounds of same type serve as contact- and food-calls. When begging, calls longer and pauses between calls reduced. Call harsher when refusing food in special posture (Panow 1974; Panov 1978; E N Panov). Development of calls in captive young as follows. Adult call 3a ('tchek') first given at end of 2nd week, when first attempts at singing also noted. Some ♂♂ will sing throughout day at 20 days, song then often starting with rasp. Adult call 3b first noted at 1½ months. Sing constantly and loudly at 3 months: sometimes like definitive long Territorial-song, and 'snoring' often interpolated. (E N Panov.) MGW

Breeding. SEASON. Eastern Turkey: nests with young recorded early to mid-June and 8 August; dependent fledged young seen 12 July and 8 August (Helbig 1984; Kumerloeve *et al.* 1984; M Kasparek). Transcaucasia

(USSR): April–June, sometimes from late March (Dementiev and Gladkov 1954*b*; Panow 1974). Northern Iran: well-incubated eggs and fledged young recorded 15 May, unincubated eggs 28 May (Zarudnyi 1896); timing dependent on altitude—in high Zagros of south-west Iran, fledged young noted second half of June and second half of July (Cornwallis 1975). Pamirs (southern USSR): breeds earlier than other *Oenanthe* of the region; many nests with young last 10 days of May, fledged young common mid-June, and nest-building noted 18 June (Panow 1974; E N Panov). Hindu Kush (Afghanistan): nest-building from first 10 days of April (Meinertzhagen 1938); breeding continuing late July (Hüe and Etchécopar 1970). SITE. In hole in rocks, among stones, or in wall of building, occasionally in burrow of Bee-eater *Merops apiaster*; nest 20–50(–90) cm from entrance. Nest: loosely constructed, fairly shallow cup of dry grass and coarser plant fibres, lined with finer fibres stripped from plant stems; sometimes on thick base of debris. Often with adjacent platform (less commonly also base) of small flat stones (from a few tens up to nearly 1000). Outer diameter of nest 11·0–15·0 cm, inner 6·0–9·5 cm, height 4·5–7·0 cm, depth of cup 3·2–5·0 cm. (Zarudnyi 1896; Panow 1974; Loskot and Vietinghoff-Scheel 1978; E N Panov.) Building: all work (including construction of platform of stones) by ♀ alone, taking at least 8 days (Panow 1974). EGGS. See Plate 83. Sub-elliptical, smooth and slightly glossy; very pale bluish-white, or pure white with barely discernible bluish tinge, unmarked or with sparse red-brown spotting, sometimes concentrated at broad end. Considerable variation even within a clutch. *O. x. chrysopygia*: 20·7 × 16·3 mm (19·0–22·4 × 15·6–16·9), calculated weight 2·95 g (Schönwetter 1979); 19·9–23·5 × 15·8–16·1 mm, $n=22$ (Panow 1974). Clutch: 4–6. 2 broods in *chrysopygia* (Panow 1974; Loskot and Vietinghoff-Scheel 1978). INCUBATION. 13 days. By ♀ alone (Panow 1974; Loskot and Vietinghoff-Scheel 1978). YOUNG. Cared for and fed by both parents (Panow 1974). No further information.

Plumages (*O. x. chrysopygia*). ADULT. Upperparts sandy-brown, merging into brownish-orange on rump and into rufous-cinnamon on upper tail-coverts; forehead, sides of crown, and often a band round hindneck tinged ash-grey. Broad supercilium white, poorly defined from sandy crown, but more sharply defined from dull black lores and brown-black upper ear-coverts, which form dark stripe through eye. Remainder of ear-coverts rufous-brown or dark brown, paler buff or dirty white towards frontal centre. Sides of neck pale ash-grey. Lower forecheeks, chin, and upper throat white; rear cheeks and lower throat white with pink-buff or grey-buff tinge; chest, sides of breast, and upper belly pink-buff or grey-buff with part of buff-brown feather-bases visible. Central belly and vent cream-white or white, flanks and under tail-coverts cinnamon-brown, deepest on under tail-coverts. Tail deep rufous-cinnamon; tip of central pair (t1) black for 26–36 mm, others for 12–18 mm, forming black ⊥-pattern; fringes along tips of outer feathers narrowly rufous. Flight-feathers grey-black with narrow buff fringes along outer webs (except emarginated parts of outer primaries) and broader pale buff fringes along tips. Greater upper primary coverts greyish-black with narrow pale buff fringe along outer web and tip. Lesser upper wing-coverts pale ash-grey, median dark brown-grey with ash-grey or sandy-grey fringes along tips. Greater upper wing-coverts and tertials greyish-black with broad cinnamon fringe along outer web and tip. Under wing-coverts and axillaries pale buff or white, some longer feathers with some black on bases visible. *In worn plumage*, rump and in particular upper tail-coverts bleached to pale cream-buff; lower throat tinged pale grey; chest more buff; flanks and under tail-coverts paler, tawny-buff; fringes of upper wing-coverts, tertials, flight-feathers, and tail bleached and partly or fully worn off; wing mainly brown-black with pale buff fringes along part of greater upper wing-coverts and tertials. Sexes similar. NESTLING. Down dense, up to 11 mm long; grey with warm flush; in 'kingi' (see Geographical Variation), pale grey or whitish (E N Panov). JUVENILE. Upperparts as adult, but slightly greyer and less saturated sandy-brown; rump and upper tail-coverts paler cinnamon-buff. Side of head pale sandy-buff, without pale supercilium and dark stripe through eye. Chin and throat dirty white; chest buff; flanks, belly, and vent pale cream; under tail-coverts pale cinnamon-buff. Lesser and median upper wing-coverts pale grey-brown with large pale buff triangle on tip. Flight-feathers, tertials, tail, greater upper wing-coverts, and greater upper primary coverts as adult, but innermost greater coverts sometimes with pale buff triangle on tip and fringes of greater primary coverts often slightly broader and less clear-cut. FIRST ADULT. Like adult, but juvenile flight-feathers, greater upper primary coverts, usually tertials, often tail, and sometimes a variable number of outer greater upper wing-coverts retained, these browner and more heavily worn than neighbouring fresh 1st adult feathers. Difference usually marked when some outer greater coverts retained: these sepia-brown with white fringe, contrasting with neighbouring inner coverts which are greyish-black with cinnamon fringe; when all greater coverts new, usually some difference in colour and wear between outer greater coverts and greater upper primary coverts. In spring, tips of primaries and greater upper primary coverts more worn than in adult; tail-feathers often slightly narrower and more worn than in adult, rufous of bases more strongly bleached.

Bare parts. ADULT, FIRST ADULT. Iris dark brown. Bill, leg, and foot black. Gape pink in one bird early in 2nd calendar year. Mouth black in *chrysopygia*. (Hartert 1910; Dementiev and Gladkov 1954b; E N Panov, BMNH.) NESTLING. Skin yellow-pink. Bill yellowish-pink. Mouth yellow. (E N Panov.) Gape-flanges of nominate *xanthoprymna* light orange at about fledging (Helbig 1984). JUVENILE. Iris dark brown. Bill, leg, and foot brown. (Dementiev and Gladkov 1954b; BMNH.)

Moults. Data for *chrysopygia* only. ADULT POST-BREEDING. Complete; primaries descendant. In USSR, starts in second half of June, completed late August or early September (Dementiev and Gladkov 1954b). In captivity, ♂ from Transcaucasia started 13 June, central tail-feathers first (E N Panov). 2 birds from northern Iran, late July and early August, in heavy moult, with all tail-feathers absent or growing and tertials shed (Paludan 1940). In Afghanistan, all birds examined from late August to late September in moult; moult almost completed from mid-September, though one from south-west Iran, 28 October, still with a few traces of moult (Vaurie 1949). POST-JUVENILE. Partial: head, body, lesser and median upper

wing-coverts, many or all greater upper wing-coverts (1–2 outer frequently retained), sometimes 1–3 tertials, and occasionally 1–8 tail-feathers. Timing highly variable, depending on fledging date (late May to mid-August: Dementiev and Gladkov 1954b; Helbig 1984; BMNH). In USSR, some still fully juvenile in late August when others have completed moult (Dementiev and Gladkov 1954b). Captive ♂ taken from nest in Transcaucasia, late April, started considerably later than Finsch's Wheatear *O. finschii* and Black-eared Wheatear *O. hispanica* taken in same period: tail moulted late August to late September, some wing-feathers mid-October to mid-November (E N Panov). In eastern Iran, late July, one had just started (outer scapulars, chest, and lesser upper wing-coverts first), 2 others had already finished; another from late August had completed moult (ZFMK). In south-west Iran, one in moult late August; in Afghanistan, moulting birds encountered on 21 June and in late July and August (Vaurie 1949).

Measurements. ADULT, FIRST ADULT. Both races combined as data basically similar; most data refer to nominate *xanthoprymna*; Iran, summer, and Iraq, Egypt, Sudan, Saudi Arabia, and Ethiopia, winter; skins (BMNH, ZFMK). Bill (S) to skull, bill (N) to distal corner of nostril; exposed culmen on average 4·3 less than bill (S).

	♂		♀	
WING	94·3 (1·85; 12)	92–98	89·9 (2·29; 7)	86–93
TAIL	58·8 (1·79; 12)	55–61	56·5 (2·31; 7)	54–59
BILL (S)	18·9 (0·41; 12)	18·5–19·6	18·4 (0·67; 7)	17·7–19·2
BILL (N)	10·8 (0·33; 12)	10·3–11·3	10·5 (0·44; 7)	10·1–11·1
TARSUS	25·1 (0·76; 12)	24·1–26·3	24·5 (0·86; 7)	23·6–25·6

Sex differences significant for wing and tail. Wing and tail of

PLATE 66. *Monticola saxatilis* Rock Thrush (p. 893): **1** ad ♂ breeding, **2** ad ♀ breeding, **3** 1st ad ♂ non-breeding fresh (1st autumn), **4** 1st ad ♂ non-breeding fresh (1st autumn), **5** juv.   *Monticola solitarius* Blue Rock Thrush (p. 903). Nominate *solitarius*: **6** ad ♂ worn (spring), **7** ad ♀ worn (spring), **8** 1st ad ♂ fresh (1st autumn), **9** juv. *M. s. longirostris*: **10** 1st ad ♂ fresh (1st autumn). (AH)

PLATE 67. *Zoothera sibirica* Siberian Thrush (p. 921): **1** ad ♂, **2** ad ♀, **3** 1st ad ♂ (1st year), **4** 1st ad ♀ (1st year).   *Turdus unicolor* Tickell's Thrush (p. 937): **5** ad ♂, **6** 1st ad ♂ (1st year).   *Turdus obscurus* Eye-browed Thrush (p. 964): **7** ad ♂ worn (spring), **8** 1st ad ♂ (1st winter). (AH)

PLATE 68. *Hylocichla mustelina* Wood Thrush (p. 924): **1** ad worn (spring), **2** 1st ad (1st year). *Catharus guttatus faxoni* Hermit Thrush (p. 926): **3** 1st ad (1st year). *Catharus ustulatus swainsonii* Swainson's Thrush (p. 929): **4** 1st ad (1st year). *Catharus minimus minimus* Gray-cheeked Thrush (p. 932): **5** 1st ad fresh (1st autumn). *Catharus fuscescens* Veery (p. 935). Nominate *fuscescens*: **6** 1st ad fresh (1st autumn). *C. f. fuliginosus*: **7** 1st ad fresh (1st autumn). (AH)

PLATE 69. *Turdus torquatus* Ring Ouzel (p. 939). Nominate *torquatus*: **1** ad ♂ worn (spring), **2** ad ♀ worn (spring), **3** 1st ad ♂ fresh (1st autumn), **4** 1st ad ♀ fresh (1st autumn), **5** juv. *T. t. amicorum*: **6** ad ♂ worn (spring), **7** 1st ad ♂ fresh (1st autumn). *T. t. alpestris*: **8** ad ♂ worn (spring), **9** ad ♀ worn (spring), **10** 1st ad ♂ fresh (1st autumn). (AH)

PLATE 74. *Turdus philomelos* Song Thrush (p. 989). Nominate *philomelos*: **1** ad worn (spring), **2** 1st ad fresh (1st autumn), **3** juv. *T. p. clarkei*: **4** ad worn (spring), **5** 1st ad fresh (1st autumn). *T. p. hebridensis*: **6** ad fresh (autumn), **7** juv. (AH)

PLATE 75. *Turdus viscivorus* Mistle Thrush (p. 1011). Nominate *viscivorus*: **1** ad worn (spring), **2** 1st ad (1st year), **3** juv. *T. v. deichleri*: **4** ad worn (spring). *Zoothera dauma aurea* White's Thrush (p. 914): **5** ad worn (spring), **6** 1st ad fresh (1st autumn), **7** juv. (AH)

PLATE 68. *Hylocichla mustelina* Wood Thrush (p. 924): **1** ad worn (spring), **2** 1st ad (1st year). *Catharus guttatus faxoni* Hermit Thrush (p. 926): **3** 1st ad (1st year). *Catharus ustulatus swainsonii* Swainson's Thrush (p. 929): **4** 1st ad (1st year). *Catharus minimus minimus* Gray-cheeked Thrush (p. 932): **5** 1st ad fresh (1st autumn). *Catharus fuscescens* Veery (p. 935). Nominate *fuscescens*: **6** 1st ad fresh (1st autumn). *C. f. fuliginosus*: **7** 1st ad fresh (1st autumn). (AH)

PLATE 69. *Turdus torquatus* Ring Ouzel (p. 939). Nominate *torquatus*: **1** ad ♂ worn (spring), **2** ad ♀ worn (spring), **3** 1st ad ♂ fresh (1st autumn), **4** 1st ad ♀ fresh (1st autumn), **5** juv. *T. t. amicorum*: **6** ad ♂ worn (spring), **7** 1st ad ♂ fresh (1st autumn). *T. t. alpestris*: **8** ad ♂ worn (spring), **9** ad ♀ worn (spring), **10** 1st ad ♂ fresh (1st autumn). (AH)

PLATE 70. *Turdus merula* Blackbird (p. 949). Nominate *merula*: **1** ad ♂, **2** ad ♀, **3** 1st ad ♂ (1st autumn), **4** 1st ad ♂ fresh (1st autumn) 'stockamsel' variety, **5** juv. *T. m. syriacus*: **6** ad ♀, **7** juv. *T. m. azorensis*: **8** ad ♂, **9** ad ♀. *T. m. mauritanicus*: **10** ad ♀. (AH)

PLATE 71. *Turdus naumanni* Dusky Thrush and Naumann's Thrush (p. 967). *T. n. eunomus* (Dusky Thrush): **1** ad ♂ worn (spring), **2** 1st ad ♂ (1st year). Nominate *naumanni* (Naumann's Thrush): **3** ad ♂ worn (spring), **4** 1st ad ♂ (1st year). *Turdus migratorius migratorius* American Robin (p. 1023): **5** ad ♂ worn (spring), **6** 1st ad ♂ fresh (1st autumn). (AH)

PLATE 72. *Turdus ruficollis* Black-throated Thrush and Red-throated Thrush (p. 970). *T. r. atrogularis* (Black-throated Thrush): **1** ad ♂ worn (spring), **2** ad ♀ worn (spring), **3** 1st ad ♂ fresh (1st autumn), **4** 1st ad ♀ fresh (1st autumn), **5** juv. Nominate *ruficollis* (Red-throated Thrush): **6** ad ♂ worn (spring), **7** ad ♀ worn (spring), **8** 1st ad ♂ fresh (1st autumn). (AH)

PLATE 73. *Turdus pilaris* Fieldfare (p. 977): **1** ad ♂ worn (spring), **2** ad ♀ worn (spring), **3** 1st ad ♂ fresh (1st autumn), **4** juv. *Turdus iliacus* Redwing (p. 1000). Nominate *iliacus*: **5** ad worn (spring), **6** 1st ad fresh (1st autumn), **7** juv. *T. i. coburni*: **8** ad worn (spring), **9** 1st ad fresh (1st autumn). (AH)

PLATE 74. *Turdus philomelos* Song Thrush (p. 989). Nominate *philomelos*: **1** ad worn (spring), **2** 1st ad fresh (1st autumn), **3** juv. *T. p. clarkei*: **4** ad worn (spring), **5** 1st ad fresh (1st autumn). *T. p. hebridensis*: **6** ad fresh (autumn), **7** juv. (AH)

PLATE 75. *Turdus viscivorus* Mistle Thrush (p. 1011). Nominate *viscivorus*: **1** ad worn (spring), **2** 1st ad (1st year), **3** juv. *T. v. deichleri*: **4** ad worn (spring).   *Zoothera dauma aurea* White's Thrush (p. 914): **5** ad worn (spring), **6** 1st ad fresh (1st autumn), **7** juv. (AH)

PLATE 76. *Monticola saxatilis* Rock Thrush (p. 893): **1–2** ad ♂, **3** ad ♀, **4** 1st ad ♂ non-breeding fresh (1st autumn). *Monticola solitarius solitarius* Blue Rock Thrush (p. 903): **5–6** ad ♂, **7** ad ♀. *Turdus torquatus torquatus* Ring Ouzel (p. 939): **8–9** ad ♂, **10** ad ♀. *Turdus merula merula* Blackbird (p. 949): **11–12** ad ♂, **13** ad ♀. (AH)

PLATE 77. *Turdus obscurus* Eye-browed Thrush (p. 964): **1–2** 1st ad (1st winter). *Turdus naumanni* Dusky Thrush and Naumann's Thrush (p. 967). *T. n. eunomus* (Dusky Thrush): **3–4** 1st ad (1st winter). Nominate *naumanni* (Naumann's Thrush): **5–6** 1st ad (1st winter). *Turdus ruficollis* Black-throated Thrush and Red-throated Thrush (p. 970). *T. r. atrogularis* (Black-throated Thrush): **7** ad ♂, **8–9** 1st ad (1st winter). Nominate *ruficollis* (Red-throated Thrush): **10** ad ♂, **11** 1st ad (1st winter). *Turdus migratorius migratorius* American Robin (p. 1023): **12** ad ♂, **13** 1st ad (1st winter). (AH)

adult and 1st adult combined, though retained juvenile wing of 1st adult on average 2·2 shorter than wing of older birds and tail 0·8 shorter.

O. x. chrysopygia. (1) Afghanistan (Paludan 1959); (2) Iran (Vaurie 1949).

WING (1)	♂ 93·8 (2·09; 12)	92–98	♀ 90·8 (1·89; 4)	88–92
(2)	93·2 (— ; 22)	90–98	90·3 (— ; 13)	89–93
TAIL (2)	58·4 (— ; 22)	56–62	57·8 (— ; 12)	55–60
BILL (S)(2)	19·5 (— ; 20)	19–21	19·4 (— ; 13)	18·5–20

Weights. *O. x. chrysopygia.* ADULT, FIRST ADULT, JUVENILE. Iran and Afghanistan. March: ♂ 25. May–June: ♂ 21·5 (0·58; 4) 21–22, ♀♀ 22, 22, 29 (laying). July and early August: ♂ 22·6 (1.04; 5) 20–25, ♀♀ 18·3, 19·7. September and early October: ♂ 22·0 (1·03; 4) 20–24, ♀ 21·7 (3) 20–24, unsexed 24, 27. (Paludan 1940, 1959; Desfayes and Praz 1978.)

Structure. Wing rather short, broad at base, tip rounded. 10 primaries: p7–p8 longest, p9 4–7 shorter, p6 0·5–2·5, p5 5–8, p4 10–14, p1 20–23. P10 reduced; 46–50 shorter than wing-tip, 1–7 longer than longest upper primary covert. Outer web of p6–p8 and inner of p7–p9 emarginated. Tertials short. Tail rather short, tip square; 12 feathers. Bill rather long, straight, slender; tip of culmen slightly decurved. Distinct bristles at base of upper mandible and near nostril. Leg and toes rather short, slender. Middle toe with claw 17·7 (8) 16·5–18·5; outer toe with claw *c.* 68% of middle with claw, inner *c.* 67%, hind 75%. Claws rather short, slender, decurved.

Geographical variation. Marked. Nominate *xanthoprymna* markedly different from *chrysopygia*: upperparts darker, sides of head, throat, and under wing-coverts black in ♂, rump and tail-coverts deeper rufous–cinnamon, and tail-base white (in *chrysopygia*, plumage almost uniform sandy-brown with rufous tail-base). The 2 races sometimes considered separate species, but some intergradation occurs and measurements and structure similar, hence treated here as single polytypic species, following Vaurie (1949), which see for discussion. Intermediates with

PLATE 78 (*facing*).
Zoothera dauma aurea White's Thrush (p. 914): **1–2** ad.
Zoothera sibirica Siberian Thrush (p. 921): **3–4** ad ♂, **5** ad ♀.
Hylocichla mustelina Wood Thrush (p. 924): **6–7** 1st ad (1st winter).
Catharus guttatus faxoni Hermit Thrush (p. 926): **8–9** 1st ad (1st winter).
Catharus ustulatus swainsonii Swainson's Thrush (p. 929): **10–11** 1st ad (1st winter).
Catharus minimus minimus Gray-cheeked Thrush (p. 932): **12–13** 1st ad (1st winter).
Catharus fuscescens fuscescens Veery (p. 935): **14–15** 1st ad (1st winter).
Turdus unicolor Tickell's Thrush (p. 937): **16–17** 1st ad (1st winter).
Turdus pilaris Fieldfare (p. 977): **18** ad ♂, **19–20** 1st ad ♂ (1st winter).
Turdus philomelos philomelos Song Thrush (p. 989): **21–22** 1st ad (1st winter).
Turdus iliacus iliacus Redwing (p. 1000): **23–24** 1st ad (1st winter).
Turdus viscivorus viscivorus Mistle Thrush (p. 1011): **25–26** 1st ad (1st winter). (AH)

black throat (like nominate *xanthoprymna*) and rufous tail (like *chrysopygia*) are sometimes separated as *cummingi* (Whitaker, 1899) and are stated to occur in area where nominate *xanthoprymna* and *chrysopygia* meet, but rufous tail perhaps normal for some pure nominate *xanthoprymna* (see below). Upperparts of adult ♂ nominate *xanthoprymna* dull drab-brown, sometimes with slight cinnamon tinge; rump and upper tail-coverts contrastingly deep rufous-cinnamon, long and narrow supercilium contrastingly white. Chin, throat, and side of head from lore downwards, area just above eye, and ear-coverts black, sides of breast greyish-black; remainder of underparts white (sometimes tinged pale vinous-grey or cream-buff when fresh), under tail-coverts deep cinnamon-rufous (sometimes pink-cinnamon). Lesser upper wing-coverts grey, outer largely black; median greyish-black with grey fringes, greater coverts and tertials black with broad vinous-buff fringes and tip. Flight-feathers and primary coverts darker than in *chrysopygia*, pale fringes more contrasting; under wing-coverts and axillaries black (buff or white in *chrysopygia*). ♀ apparently dimorphic: black-throated morph similar to ♂ but has some pale buff fringes to duller black feathers of throat (worn off by May-June); pale-throated morph similar to ♂ and ♀ *chrysopygia*, but ear-coverts and sides of throat darker grey-brown, supercilium less pronounced, some black of feather-bases on throat sometimes visible, rump and under tail-coverts sometimes deeper rufous-cinnamon, and upperparts darker brown (less sandy). 1st adult similar to adult, but part of juvenile plumage retained (as in *chrysopygia*) and (in ♂ and black-throated ♀♀) black feathers of chin and throat have broader pale buff fringes, in ♂ worn off by February-April. Tail-tip has slightly more black than in *chrysopygia*: 32–40 mm of tip of t1 black, 13–20 mm of t2-t6; remainder of tail stated to be white, but tail-base largely rufous in 2 (both 1st adults) of 8 ♂♂ examined, while of 5 black-throated ♀♀ tail fully or partly rufous in 4 (all 1st adult) and white only in 1 adult ♀; pale-throated ♀♀ apparently have tail-base rufous, feathers grading to white at point of implantation. Tail-colour thus perhaps linked with sex and age, as in Hooded Wheatear *O. monacha*. East from Khorasan (Iran) and Afghanistan, birds average slightly paler in fresh plumage than typical *chrysopygia* from further west; upperparts more sandy, underparts whiter, and these birds sometimes separated as *kingi* (Hume, 1871) (e.g. by Vaurie 1949), but difference very slight and recognition not warranted (Paludan 1959; BMNH).

Recognition. ♂ and black-throated morph of ♀ nominate *xanthoprymna* rather like ♂ Red-rumped Wheatear *O. moesta*, but latter has back and sides of neck black or dark grey (not drab brown) and tail more extensively black. Both sexes of *chrysopygia* and pale-throated morph of ♀ nominate *xanthoprymna* rather similar to ♀ *O. moesta* (but that bird more rufous on head, darker brown on back, and blacker on tail) and to ♀ *O. monacha* (but latter larger with hardly any dark on tail-tips except t1), and somewhat similar to Desert Lark *Ammomanes deserti*, which is structurally different. *O. x. chrysopygia* in summer, with bleached tail, closely similar to Isabelline Wheatear *O. isabellina*, but has plumage generally greyer, less sandy-buff. CSR

Oenanthe picata (Blyth, 1847) **Eastern Pied Wheatear**

FR. Traquet pie d'Orient   GE. Elster-Steinschmätzer

A polymorphic species (see Plates 62, 64, 65), breeding in uplands from south-west Caspian and central Iran east to Tadzhikistan and northern and western Pakistan, and wintering in Pakistan and northern India (Ali and Ripley 1973*b*; Flint *et al.* 1984); for eggs, see Plate 83. *Opistholeuca* morph (largely black) formerly regarded as occurring in west Palearctic as follows. Syria: several reported 1974–6 in lava outcrops south-east of Damascus (Macfarlane 1978). Jordan: said to be breeding and resident in black basaltic shield area in north (Clarke 1981; Wallace 1983*a*, *b*). In 1985, however, L Cornwallis found evidence (3 families) of alleged Jordanian *O. picata* paired and breeding with typical Mourning Wheatear *O. lugens*. Size,

plumage (see below and p. 867), and song consistent with *O. lugens* rather than with *O. picata*, and birds thus now considered to be a black morph of *O. lugens* analogous to cryptic black race *annae* of Desert Lark *Ammomanes deserti* which occurs in same region. Further, this black morph of *O. lugens* is, like typical conspecifics, apparently largely sedentary whereas *O. picata* is largely migratory. This interpretation that the Jordanian birds are all *O. lugens* is now supported by D I M Wallace and by A M Macfarlane, and the latter also considers Syrian birds part of the same population and thus subject to the same interpretation.

Oenanthe lugens **Mourning Wheatear**

PLATES 63, 64, and 65
[between pages 760 and 761, and facing page 856]

DU. Rouwtapuit   FR. Traquet deuil   GE. Schwarzrückensteinschmätzer
RU. Траурная каменка   SP. Collalba fúnebre   SW. Sorgstenskvätta

Saxicola lugens Lichtenstein, 1823

Polytypic. Nominate *lugens* (Lichtenstein, 1823), Egypt and Levant; *persica* (Seebohm, 1881), Iran, occasionally south to north-east Arabia in winter; *halophila* (Tristram, 1859), North Africa east to Libya. Extralimital: *lugentoides* (Seebohm, 1881), western Arabia from Taif to Yemen; *lugubris* (Rüppell, 1837), Ethiopia; 3 further races in Arabia and eastern Africa.

Field characters. 14·5 cm; wing-span 26–27·5 cm. Bulkier than Pied Wheatear *O. pleschanka*, with 10% longer bill and legs (also heavier but less obviously larger than extralimital Eastern Pied Wheatear *O. picata*, with 5% longer wings and 5–10% longer bill). Medium-sized, rock-haunting, strongly pied wheatear, with complex plumage variations and no obvious diagnostic characters except for combination of pale wing-panel (in flight) and (usually) pinkish or rusty under tail-coverts. ♂ patterned much as *O. pleschanka* except for wing-panel and longer white rump. ♀ like ♂ or much duller. Sexes similar in 2 races, dissimilar in 1; little seasonal variation. Juvenile separable. 3 races occur in west Palearctic, with North African *halophila* easily distinguished from other 2.

ADULT MALE. (1) Egyptian and Levant race, nominate *lugens*. Top of head, nape, and upper border of mantle white, usually with pale brown-grey tinge to crown, centre of nape, and lower edge except when very worn. Face, throat, upper breast, sides of neck, shoulder, lower mantle, upper back, scapulars, and folded wings black. Lower back, rump, and tail white, tail with black ⊥-mark but terminal band rather narrow and 4 outermost pairs of feathers narrowly tipped white. Underbody white except for contrasting pale buff under tail-coverts. Under wing-coverts and axillaries black, contrasting strongly with white inner webs of flight-feathers which are broad enough also to show on upper surface of extended wing

in pattern recalling Magpie *Pica pica*. From June, fresh feathers with white or buff fringes make crown and nape browner and add pale sandy margins to back, inner wing-coverts, tertials, and border of throat-patch, but change in appearance far less marked than in *O. pleschanka*. Basalt desert of southern Syria and northern Jordan is occupied largely by morph with head and underparts black except for vent and under tail-coverts (Wallace 1983*b*): formerly believed to be dark *opistholeuca* morph of *O. picata* (see above), but differs from it in having typical *O. lugens lugens* wing pattern (in *O. picata*, flight-feathers all-black above and, at palest, silvery-grey below). (2) Iranian race, *persica*. Crown and nape usually darker brown-grey, emphasizing diffuse white supercilium and border to nape; black terminal band to tail wider; vent and under tail-coverts usually fully buff, much darker than in nominate *lugens*. (3) West and central North African race, *halophila*. Crown usually almost completely white; inner webs of flight-feathers less broadly white (reducing contrast with black under wing-coverts) and not so visible from above; under tail-coverts pale rosy-buff, far less dark than in *persica* and most nominate *lugens*, and bleaching to white. ADULT FEMALE. (1) Nominate *lugens* and *persica*. Closely resembles ♂ (unlike ♀ *O. pleschanka*), but black plumage looks duller in direct comparison, though not approaching sooty-brown of *O. picata*. Fresh plumage differs as in ♂. (2) *O. l. halophila*.

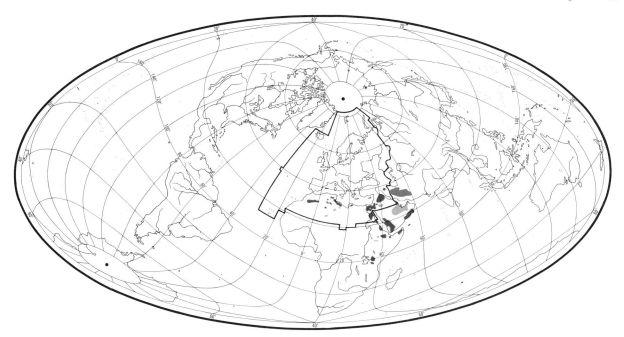

Differs markedly from ♂ and ♀ of nominate *lugens* and *persica*. Top of head, back, and lesser wing-coverts pale sandy- to grey-brown, contrasting with dull grey or black face and throat. Wings usually show at least some buff-white margins to coverts and tertials, even inner secondaries. Often shows pale cream supercilium, eye-ring, and patch on hindneck, and some pale individuals also have brown cheeks, with black rear surround, and grey-white throat (see Plumages). JUVENILE. All races. Essentially sandy-buff, streaked dull brown above and cream-buff below; faintly scaled grey on chest. ♂ has striking black mask on face. Juvenile of black morph nominate *lugens* is smoky in general coloration. FIRST WINTER AND SUMMER. All races. Resembles dull adult, with retained brown flight-feathers and tertials keeping pale fringes longer than in adult. All races at all ages have bill and legs black.

♂ and ♀ of Middle East races unmistakable, with extent of white on inner webs of flight-feathers much greater than any other similar wheatear and obvious in flight, and buff under tail-coverts usually discernible on ground in close view, so that confusion with *O. pleschanka* and extralimital *O. picata* (*capistrata* morph with white crown and belly) unlikely to persist. ♂ and ♀ of western race *halophila* far less distinctive, and ♀ particularly puzzling, requiring close study for separation from similar wheatears, though usually with more complete grey-black throat and rosy vent. Flight light and flitting, recalling *O. hispanica* more than Wheatear *O. oenanthe*. Escape-flight usually short. Gait and feeding behaviour typical of genus but less prone than *O. pleschanka* to perch on vegetation. Often shy but can be quite tame.

Song a pleasant warble, often interspersed with commonest call, 'chack chack'.

Habitat. Breeds along narrow band of lower middle latitudes in warm arid Mediterranean and subtropical desert climates, not generally extending either to sea-coast or beyond desert fringe, but being tied to rocky exposures, clay hill-slopes, and banks of wadis (Heim de Balsac and Mayaud 1962). In Red Sea Province of Egypt, inhabits most desolate wadis and rocky gorges, often in neighbourhood of Hooded Wheatear *O. monacha*. In Yemen, always occurs near rocky mountain slopes or steep hillsides, sometimes with considerable vegetation (Meinertzhagen 1954). Breeding territories studied in southwest Iran lay at *c.* 1700-2700 m, nearly all on steep, rocky, boulder-strewn hillsides, a minority also included more gentle slopes, but always predominantly stony and boulder-strewn; mostly also with piles of boulders or broken rock outcrops used as song-posts, look-outs for prey, cover from predators or adverse weather, nest-sites, and sources of shade. Habitat requirements overlap with those of Finsch's Wheatear *O. finschii*, and where both species co-exist latter usually occupies any gentle slopes, *O. lugens* becoming confined to steep hillsides. Apparently, however, cases of interspecific conflict generally averted by geographical separation. (Cornwallis 1975.) Found in desolate hills and ravines in Morocco where considered a winter visitor; usually seen on telegraph wires (Smith 1965a). Winter behaviour in Egypt studied in detail by Hartley (1949), as follows. Makes short migrations from dry wadis in mountains between Nile and Red Sea to winter quarters on spurs and shingle ridges where wadis

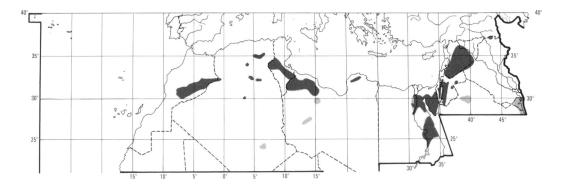

open out towards Nile; only exceptionally penetrates cultivation. Forages around edges of stones and bushes, pursuing flying insects and also entering clefts. Army huts served as song-posts and feeding places. Most songs were from perches *c.* 3·5-6 m high, on hut-tops, poles, stones, boulders, and sometimes bushes of camel-thorn *Zylla spinosa*. Winter temperature sometimes 30-36°C, and shady perches strongly preferred. Territories included a large abandoned quarry and, at a higher level, wide terraces of barren sand and shingle, with low precipices of calcareous sandstone between wadis. One strip of ground, more densely grown up with scrub, was contested with *O. monacha*. In the case studied, incursion of military camp into habitat perhaps traditionally used by *O. lugens* and congeners seems to have been accepted by them, although human settlements normally avoided. Flies normally in lower airspace, but in song-flight up to *c.* 70 m (Zachai 1984).

Distribution. No information on range changes.

IRAQ. Probably breeding in north (Moore and Boswell 1956). JORDAN. Azraq: breeding 35 km to south-east, 90 km to west, and on basalt shield to north (Wallace 1983*a*; L Cornwallis). (Largely black *Oenanthe* breeding on basalt shield north of Azraq and in adjoining southern Syria were formerly thought to be Eastern Pied Wheatear *O. picata* but are now known to be *O. lugens*: L Cornwallis; see p. 858). LIBYA. Tripolitania: southern limits uncertain (Bundy 1976). ALGERIA. Rather local (EDHJ).

Accidental. Lebanon.

Population. No information on trends.

EGYPT. Locally common (SMG, PLM, WCM). LIBYA. Tripolitania: breeds sparingly (Bundy 1976).

Movements. Palearctic populations partially migratory, with considerable variation between populations in proportion of birds migrating and in distances involved. Over-confident assignation of records to races has resulted in confusion over limits of each population's wintering range; potential also exists for confusion with Pied Wheatear *O. pleschanka* and Eastern Pied Wheatear *O. picata*. As with most *Oenanthe*, ringing recoveries would be invaluable. In most cases, wintering range represents slight extension of breeding range (often towards south-west), though breeding grounds of southern Iran almost entirely vacated, with these birds wintering in Arabia and possibly Egypt and Sudan.

NORTH AFRICAN populations (*halophila*), breeding from eastern Morocco to Libya, are partially migratory with migrants making short-distance movements. Some birds seen as far south as Aïr and Ahaggar massifs in January-February but probably most move less than 50 km (Curry-Lindahl 1981). A few wanderers reach Egypt between October and February, but a ♂ ascribed to *halophila* in BMNH shot in Palestine during March is not safely separable from nominate *lugens* which breeds there (D G C Harper). Although ♂♂ of *halophila* and nominate *lugens* potentially confusable, only the former race is sexually dimorphic. Therefore likely that eastern limit of wintering range of *halophila* reasonably well defined by present knowledge unless ♂♂ move further than ♀♀. Limited evidence suggests ♂♂ in fact more likely to remain near breeding sites (Smith 1971).

EAST MEDITERRANEAN populations (nominate *lugens*), breeding from eastern Egypt north through Palestine to Syria and northern Iraq, apparently less migratory than *halophila*. However, confusion possible with other races, especially *persica*, raising possibility that some individuals move further south than currently suspected. In Egypt, some post-breeding movement westward across Nile (as in Hooded Wheatear *O. monacha*) and more localized movements from dry wadis favoured for breeding to desert fringes (Hartley 1949; Short and Horne 1981; M G Kelsey). Palestinian birds especially similar in appearance to *persica*; see below. In detailed Israeli study (Zachai 1984), dispersal started late July, but most marked from September to late February. Some birds appeared to wander throughout non-breeding period, although net movement made by individuals over whole winter impossible to assess. Several wintering sites outside breeding range were in flat habitats not usually considered typical, and some wintering sites may be overlooked for this reason. Record of bird on Cyprus, July 1877, probably involved Finsch's Wheatear *O. finschii* rather than *O. lugens* (Flint and Stewart 1983).

SOUTH IRANIAN population (*persica*) primarily migratory, moving south-west into Arabia and (perhaps) Egypt and Sudan (see below). A few individuals may winter within breeding range but their number must be very small. Cornwallis (1975) found none in south-west Iran between mid-November and February, noting that return passage was under way by mid-February. Breeding sites vacated by early September, with some individuals holding temporary territories at sites where they had not bred before moving on. Small passage through Iraq reported, but no details or dates; possible that breeding occurs in Zagros mountains (Allouse 1953). Main wintering area appears to be Arabian peninsula and its offshore islands. Only one record from North Yemen (Cornwallis and Porter 1982) and appears to be similarly sparse at north-west end of Persian Gulf (Bundy and Warr 1980). Perhaps most frequent in eastern Saudi Arabia away from the coast (Bates 1936–7; G Bundy), but nowhere common (Jennings 1981a). Most records from Arabian peninsula between October and mid-March, although one Kuwaiti record in September and stragglers in Saudi Arabia during April (Bundy and Warr 1980). In eastern Saudi Arabia, small but regular February–March influx of grey and white ♀-type *Oenanthe*, apparently ascribable to *O. lugens* (G Bundy). Extent of wintering, if any, in rest of Middle East unclear due to risk of confusion with nominate *lugens*, especially those from Palestine. Certainly single bird from Palestine suggested to be *persica* best regarded as dubious (Hardy 1946). BMNH contains 3 '*persica*' skins collected November–February at 2 incompletely documented sites in Egypt and at Bin Shigub in Sudan (20°25′N 34°10′E); Sudan specimen can be equally well matched to nominate *lugens* skins in same collection (D G C Harper). Also disagreement about numerical status in Sudan, with reports ranging from 'rare' (Hogg *et al.* 1984) to 'not uncommon' (Cave and Macdonald 1955). All Sudanese records are early November to mid-February and north of 18°N, with suggestion of concentration around Red Sea hills.

Extralimital Arabian and Afrotropical races appear to be largely sedentary. However, a few records suggest movements into (or undetected breeding in) central Saudi Arabia by Yemen race *lugentoides* (King 1978b; Jennings 1981a). DGCH

Food. Mainly insects. Usually catches prey in sallies to ground from (preferably) firm perch, e.g. boulder or stone. Sometimes hops along ground in brief dash after prey, or jabs at ground or under stone from stationary position. (Cornwallis 1975.) Also creeps into crevices, and makes darting aerial-pursuits from low perch; seen to flutter steeply upward to capture small butterfly (Pieridae) (Hartley 1949). All food for young killed before

delivery—by breaking up on stone with bill, or by shaking from side to side (Zachai 1984). Will drink on occasion. Active from a little before sunrise to c. 20 min after sunset. (Hartley 1949.)

At Jiddah (Saudi Arabia), January–April, ants (Formicidae) and beetles (Coleoptera) occurred in all of 3 stomachs analysed, and ant-lion *Myrmeleon* larvae in 1 (Morrison-Scott 1937). In south-west Iran, March–November, 11 stomachs contained 422 items of which 85% (by number) ants, 3% other Hymenoptera, 5% beetles, and small numbers of grasshoppers, etc. (Orthoptera), termites (Isoptera), bugs (Hemiptera), flies (Diptera), spiders (Araneae), and plant material; of 316 food items, 76% less than 5 mm long, 19% 5–10 mm, 4% 10–15 mm, and 1% 15–20 mm (Cornwallis 1975). In central Morocco in autumn, takes mainly ants, also grasshoppers and beetles (Meinertzhagen 1940).

Young in Israel received chiefly insects, including termites, locusts, beetles, flies, larvae, and many caterpillars; also woodlice (Isopoda), scorpions (Scorpiones) and small dry red berries. Nestlings 5–6 days old received average 8·5 feeds per hr (♀ 4·4, ♂ 4·1); for fledglings 15–18 days old, average 11·4 feeds per hr (♀ 9·3, ♂ 2·1). (Zachai 1984.) BDSS, DFV

Social pattern and behaviour. Account based chiefly on study of wintering birds in Egypt by Hartley (1949), of breeding birds in Israel by Zachai (1984), and information supplied by L Cornwallis.

1. Not gregarious. Territorial during breeding season; at other times of year occurs in pairs in some areas, but more usually singly, ♂♂ and ♀♀ holding individual territories. In Oman, in pairs most of year (Gallagher and Rogers 1980); presumed pairs occur in Bahrain, October to early March (Gallagher and Rogers 1978). Juvenile with 2 adults seen in North Yemen in September (Phillips 1982). In Iran, families disperse after breeding season and only solitary birds occur (Cornwallis 1975); see Relations within Family Group (below). In Morocco, November, ♂♂ apparently solitary (Smith 1971). In Lower Egypt, individual winter territories c. 1·6 ha, and boundaries clearly demarcated, in apparently optimal habitat; territories larger in more desolate habitat and separated by tract of 'debated' land e.g. c. 45 × 180 m (Hartley 1949). Also record of 2 *O. lugens*, 1 White-crowned Black Wheatear *O. leucopyga*, and 1 Pied Wheatear *O. pleschanka* spaced out in individual territories in area c. 1500 × 400 m, i.e. c. 60 ha (Simmons 1951c). BONDS. No evidence for other than monogamous mating system. Occurrence of pairs most of year in Oman (Gallagher and Rogers 1980) suggests pair-bond may be maintained all year in some regions; no further information. Apparently only ♀ broods young; both ♂ and ♀ feed young in nest; young out of nest fed chiefly by ♀ (Zachai 1984); for further details, see Relations within Family Group. In East African *schalowi*, ♂♂ of 1st brood recorded helping at nest for later broods (Haas 1982b); adult helpers also recorded (Grimes 1976). No information on typical duration of family bonds (but see Relations within Family Group) or age of first breeding. BREEDING DISPERSION. Solitary and territorial. Territories serve for both nesting and feeding; boundaries well defined, birds rarely leaving own ground and doubling back if pursued by observer (Cornwallis 1975). In Sede Boqer (Israel), 10 territories

averaged 16·3 ha (9·4–23·4); adjacent territory of unpaired ♂ 4·7 ha; foraging mostly at some distance from nest, in one instance usually at c. 125 m, in another, up to c. 500 m (Zachai 1984). Record in Saudi Arabia of large lizard, several times body weight of O. lugens, sharing nest-hole; would sun itself at entrance, scurrying past nestlings to recess at back when parent bird returned with food (Jennings 1981a). ROOSTING. In Lower Egypt, late August and September, birds appeared to seek shaded perches at midday; 2 birds regularly observed deeply asleep c. 13.00–14.30 hrs in roof beams of ruined huts (Hartley 1949). No further information.

2. Often shy and wary (Whitaker 1895; Le Fur 1975), but at other times quite approachable, especially in breeding season (Witherby 1905; Heim de Balsac 1924; Hollom 1959), though keeping watchful eye on observer; in some areas frequents human habitations (Whitaker 1905). May take flight when alarmed, keeping c. 100 m ahead of human pursuer (Tristram 1859), or (e.g.) at appearance of raptor, dart behind rock, calling (L Cornwallis: see 4 in Voice), preferring largest available boulder (Meinertzhagen 1949c). FLOCK BEHAVIOUR. None reported. SONG-DISPLAY. In breeding season, ♂ sings (see 1 in Voice) either in Song-flight or from exposed perch (e.g. rock, bush). In Song-flight, ascends from perch in gentle, gliding curve, tail fanned, white underside of flight-feathers contrasting with black under wing-coverts and axillaries, and lands on same or different perch (L Cornwallis); may rise to c. 70 m, hovering with wings and tail outspread before gliding down (Zachai 1984). In short Song-flight, uses fast wing-beats and returns to same perch (L Cornwallis). Song in Israel rare during nestling period, but renewed in middle of fledging period in anticipation of 2nd brood (Zachai 1984). In Oman, where birds seen in pairs most of year, song begins February (Gallagher and Woodcock 1980). No winter song heard in Morocco by Smith (1971). However, in Lower Egypt, July–February, no month entirely without song; both ♂ and ♀ apparently sang, since no silent birds encountered. Song was most frequent when setting up territories August–September, decreasing thereafter; given in alert posture, chiefly from perch up to c. 6 m (higher perches available not used) and occasionally while flying between perches. Once, singing bird seen to make sudden bow, head craned forward and down and wings waved momentarily over back. Song began c. 15 min before sunrise and ended c. 10 min before sunset (at least ½ hr before foraging ended). From July to October, most song in early morning and in evening; from November to February, no marked early morning song, but noticeable increase in late afternoon. (Hartley 1949.) ANTAGONISTIC BEHAVIOUR. ♂ aggressive in defence of territory in breeding season (♀ does not participate); both sexes aggressive in defence of winter territories. For aggression of ♂ towards own family (and submissive posture of ♀), see Relations within Family Group. In Israel, hostility strongest before start of laying, declining towards time of hatching; ♂♂ regularly launch attacks from song-posts against conspecifics and other birds: Wheatear O. oenanthe, Black-eared Wheatear O. hispanica, Black Redstart Phoenicurus ochruros, Redstart P. phoenicurus, Blue Rock Thrush Monticola solitarius, and Blackstart Cercomela melanura (Zachai 1984). In territorial disputes in Iran, subordinate to Finsch's Wheatear O. finschii, Hume's Wheatear O. alboniger, and Red-tailed Wheatear O. xanthoprymna, and thus probably excluded from potential breeding sites (Cornwallis 1975, which see for other examples of confrontations with sympatric Oenanthe). In Egypt, very aggressive towards Hooded Wheatear O. monacha breeding nearby, and to migrating Oenanthe (Meinertzhagen 1954); recorded chasing O. xanthoprymna in low scrub, early February (Short and Horne

1981). In Saudi Arabia, March, tolerated intrusion by Isabelline Wheatear O. isabellina and Red-breasted Wheatear O. bottae (Meinertzhagen 1954). Outside breeding season, aggression most marked when setting up winter territories. In Lower Egypt, August–September, c. 7 recently-established O. monacha were gradually ousted from territories in limestone quarry by 5 incoming O. lugens, area being re-divided between them; one O. monacha retained small border territory. Not aggressive towards pair of Stonechats Saxicola torquata arriving November–December. (Hartley 1949.) Tolerant also of O. isabellina on adjacent territory in Iran, November (L Cornwallis). In Algeria, December, ♂ singing with tail spread and wings drooped was driven off by O. leucopyga (Niethammer 1954). In Cyrenaica (Libya), September, dispute with conspecific involved chase, Subsong (see 2 in Voice), and tail-fanning (L Cornwallis). In Lower Egypt, 85% of encounters with conspecifics took place within 3 months of arrival; hostility usually signalled by threat-postures from perches a few metres apart, with repeated 't'chut' calls (see 5 in Voice), at times interspersed with pursuit-flights (in 37% of 70 encounters). Song sometimes given during chases: in one encounter both birds sang loudly from perches between pursuits, and no 't'chut' calls heard; in others, territory-owner alone sang. Many encounters consisted simply of Flashing-display: small wing-flicks and bows in an alert, upright posture. Wings may also be spread widely at such times, bird facing opponent or turning its back; may also raise tail, displaying apricot-coloured under tail-coverts. Once, after brisk fluttering confrontation, birds faced each other c. 30 cm apart in Head-up posture: stretched heads upwards, bills pointing c. 30° from vertical, and ruffled feathers of lower breast and belly, exposing as much white as possible. Where stretch of debated land separates territories, bird may fly c. 120 m to engage rival. (Hartley 1949.) HETEROSEXUAL BEHAVIOUR. (1) Pair-bonding behaviour. No information. (2) Courtship-feeding. None observed. (3) Mating. In Israel, both ♂ and ♀ seen to solicit with a crouching posture (Fig A), perhaps Greeting-display (see subsection 4, below). ♂

A

circled towards back of crouching ♀ with mincing gait before mounting; ♀ sometimes solicited when ♂ some distance away; ♂ then flew across with hovering flight. Before and after copulating, ♂ gave Contact-calls (see 3a in Voice). (Zachai 1984.) In observation in Libya, ♂ adopted an upright posture and pranced around ♀ (who was crouching and wing-shivering), then attempted to mount her; after swift aerial-pursuit they landed and copulated (Bundy and Morgan 1969). In Israel, ♂ started singing and initiating copulation again when ♀ was feeding fledged 1st brood, but she rebuffed his advances (chasing him) and no 2nd brood recorded (Zachai 1984). (4) Behaviour at nest. In Israel, both ♂ and ♀ started exploring possible nest-sites from end of February or early March, looking into crevices with front of body lowered and tail raised; sometimes entered crevices for up to c. 1 min. Once site selected (not known by which sex) ♀ began bringing small flat stones in her bill to build paved platform (see Breeding). (Zachai 1984.) ♀ builds nest (Heim de Balsac and Mayaud 1962; Zachai 1984). Both ♂ and ♀ sometimes perform Greeting-display when meeting at nest-entrance: crouch, and vibrate wings in manner of begging fledgling (Zachai 1984). RELATIONS WITHIN FAMILY GROUP. Both adults feed young in nest (Sage and Meadows 1965; Jennings 1981*a*). Young give food-calls when parents approach (L Cornwallis). Nest-sanitation performed by both sexes (Zachai 1984); parent seen to fly away from nest with faecal sac in 'butterfly-flight' high over territory (L Cornwallis). In Israel, observations at 1 nest suggest only ♀ broods, visits of ♂ to nest being brief. Young left nest at 14-15 days, but unable to fly, and 'jumped around' nearby; fed chiefly by ♀, ♂ assisting mainly in middle of day. Parent with food landed where fledgling last seen; fledgling approached parent, lowered chest to ground and vibrated open wings, giving food-calls. On several occasions after being fed, fledged young turned 180° and deposited faecal sac in front of adult. If returning parents failed to find young, gave Contact-call (see 3a in Voice), then hopped about looking for young in crevices, etc.; sometimes froze for up to 1 min or longer, head bowed and tail raised (Fig B), at times opening and closing wings quickly or with

B

wings half open. ♂ ceased feeding young after 5 days out of nest, and began to sing and display. After 3-4 days out of nest, young able to fly 20 m. Self-feeding began on 4th day and gradually increased; young still with parents on 10th day, and precise age of independence not recorded. ♀ observed feeding one of 1st brood 49 days after fledging; in another case, ♀ fed one of 2nd brood 55 days after fledging; this presumably exceptional. On one occasion, young of adjacent territory approached boundary and begged for food, but elicited no response. (Zachai 1984.) In Iran, families disperse when young independent. 2 incidents suggest break-up of family may result from parental aggression: in one case, adult with 3 fledged young suddenly began to chase one as if to drive it from territory, chase continuing for 2 min; in the other, when 3 fledged young being fed regularly by ♀ and occasionally by ♂, ♂ chased ♀ round territory for ½ hr, ♀ uttering apparent Distress-calls (see 6 in Voice); once, ♂ uttered a few notes of song; when attacked, ♀ adopted a submissive posture with drooped wings. (Cornwallis 1975; L Cornwallis.) ANTI-PREDATOR RESPONSES OF YOUNG. In Israel, when snake took nestling, remaining 2 nestlings vacated nest (Zachai 1984). PARENTAL ANTI-PREDATOR STRATEGIES. (1) Passive measures. Adults described as extraordinarily bold in presence of observers, perching within 1-2 m and even entering nest-hole (Jourdain *et al.* 1915; Hollom 1959). (2) Active measures. Once, ♀ with eggs about to hatch flew out of nest-hole and hovered over observer's head (Lynes 1930). In Oman, ♂ defended fledgling from human intrusion with apparent Flashing-display (bowing, tail-raising, and wing-flicking), accompanied by alarm-calls (Gallagher and Woodcock 1980: see 4b in Voice). Will mob cats in low hovering flight with loud alarm-calls (see 3-4

in Voice) until cats driven off; other passerines (e.g. Desert Lark *Ammomanes deserti*) may join in mobbing (Zachai 1984).

(Figs by K H E Franklin: from drawings by G Zachai.) DFV

Voice. Varied repertoire, used for much of the year. Regional differences not known, but perhaps implied by variety of descriptions. For presumed bill-snapping, see below. No detailed study; fullest account by Hartley (1949) of winter song in Lower Egypt.

CALLS OF ADULTS. (1) Song. In winter territories, given by both sexes, and no differentiation noted by Hartley (1949). At end of February, Algeria, paired ♂ and ♀ both sang, and song of ♀ described by J-C Roché as deeper and slower than song of ♂. Accounts show song varied, but generally give impression of a pleasing warble, which may be loud or subdued; phrases, of varying length, often interspersed with pauses or with call-type units. Described in Saudi Arabia as loud bubbling musical whistled warble, with phrases of 6-20 notes (King 1978*b*). In Iran, on one occasion song was a pleasant warble interlaced with chacking (L Cornwallis); on another, a languid, easy but uneven delivery (P A D Hollom). According to J-C Roché, Algeria, song continuous, often imitative, during territorial disputes, but with long silences between phrases when perched. A thin subdued trill descending in pitch regarded as typical song in Libya (Bundy and Morgan 1969). Hartley (1949) found both continuous warbles and repeated phrases common in Lower Egypt during July-January, examples as follows: repetition of soft, evenly stressed 'tchru tchru wayt'; alternation of shrill clear notes, gratings, and squeals, audible at *c.* 80 m; shrill continuous song recalling Skylark *Alauda arvensis*, scarcely audible at *c.* 20 m; same bird also gave brief bill-clatter followed by trill, reminiscent of Swallow *Hirundo rustica*. Fig I shows the last of a

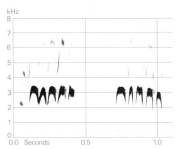

I E D H Johnson Algeria February 1969

sequence of 3 melodious rippling phrases with pauses of *c.* 8 s and 3 s; 2nd is very like 3rd, and 1st is briefer, rather similar to second half of others. Fig II shows excerpt from recording in which phrases reminiscent of Mistle Thrush *Turdus viscivorus* (basically tonal with good deal of amplitude modulation) are interspersed with very short clicks, like 't' sound produced by removing tip of tongue abruptly from roof of mouth (J Hall-Craggs); again there are pauses, and an occasional decisive 'fiu'. (2) Subsong. Often given October-February by birds

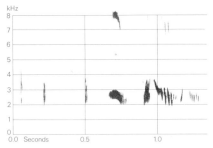

II   C Chappuis   Tunisia   February 1971

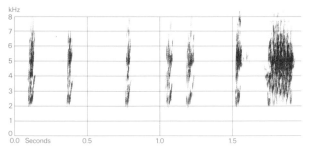

IV   C Chappuis   Tunisia   February 1971

wintering in Bahrain (Gallagher and Rogers 1978); recorded during territorial pursuits (L Cornwallis); described as a quiet warble in Libya (Bundy and Morgan 1969). Recording in Algeria shows sweet, unhurried, warbled phrases, some ending in a trill, among frequently repeated very brief 'tuk tek' calls of varying pitch; Fig III shows sequence 'tuk tak tek tik' (J Hall-Craggs), each

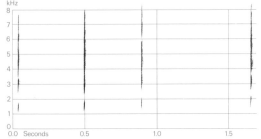

III   E D H Johnson   Algeria   February 1969

at slightly higher pitch. Recording by C Chappuis (Tunisia, February) is of much more jumbled Subsong: mixture of brief whistles, ascending and descending portamenti, frequently interspersed with 't' calls and grating calls. (3) Contact- and contact-alarm calls. Further study required to clarify whether various renderings refer to the same or different calls. (a) Metallic calls given before and after copulating, also while trying to locate fledglings; food-call of fledglings apparently similar or related (Zachai 1984). (b) A quiet 'chak chak' noted in Morocco by Smith (1965a); 'chek chek' (Hüe and Etchécopar 1970). Repeated 'chack' and high-pitched ventriloquial squeak given in agitation at nest (L Cornwallis). A repeated raspy 'kaak' (perhaps this call) is interspersed with high-pitched musical 'seek' or 'week' (King 1978b). (4) Alarm-calls. (a) Low, harsh 'zeeb' or 'dree' (Walker 1981b). (b) A double, staccato, grating 'krik krik' or 'chzak chzak' heard when ♂ defending fledgling (Gallagher and Woodcock 1980); perhaps an intensification of call 3b. Calls from alarmed ♀ in Tunisia are presumably related to this call: double and single grating 'krik' sounds, which gradually intensify (Fig IV) to more prolonged sound (reminiscent of tearing calico: J Hall-Craggs), before ebbing away. (c) High-pitched call given at appearance of Sparrowhawk *Accipiter nisus* (L Cornwallis). (5) Anger- and excitement-calls. A short,

incisive 't'chut t'chut' given frequently during perched confrontations and pursuit-flights; nervous 't'chut t'chut' also given when driven to territory boundary by human intruder (Hartley 1949). Dry little calls uttered in courtship-display and pursuit, Algeria, are perhaps similar (J-C Roché). (6) Distress-call. A shrill 'thrrr thrrr' uttered by ♀ during aggressive chase by mate (L Cornwallis); see Relations within Family Group.

CALLS OF YOUNG. Food-call of nestlings a loud incessant 'che che che' (L Cornwallis), of fledglings a continuous metallic buzzing 'bzrj', resembling call 3a of adult (Zachai 1984). Calls of young also described as nasal 'zjeep' (L Cornwallis) and slightly rasping 'buze' (P A D Hollom).

DFV

Breeding. SEASON. Oman: newly-fledged young found mid-April to mid-July (Walker 1981b). Jordan: eggs and week-old young found late April (Wallace 1984b). Israel: laying usually begins mid-March, continuing to early June (Zachai 1984). Egypt: eggs laid February–June (Meinertzhagen 1954). North-west Africa: eggs found early March to late April (Heim de Balsac and Mayaud 1962). SITE. Deep in crevice in rock, under boulder, in rodent burrow, or in hole in bank (Jourdain *et al.* 1915; Cornwallis 1975; Zachai 1984). Nest: shallow cup of vegetation, lined with (e.g.) hair, feathers, or wool (Zachai 1984; L Cornwallis). Outer diameter *c.* 11·0 cm, height *c.* 5·0 cm, depth of cup *c.* 2·5 cm (Zachai 1984). Building: by ♀, taking *c.* 2 weeks (Zachai 1984). Platform of small stones placed in front of, beneath, and around nest, in varying quantity—or none at all where crevice too narrow (Heim de Balsac 1926). If nest-site used for more than one season, platform may become dense. Stones *c.* 25 × 35 mm, weighing *c.* 5–7 g. (Zachai 1984.) EGGS. See Plate 83. Sub-elliptical, smooth and glossy; light green-blue, spotted red-brown, often with ring of larger blotches round broad end. Nominate *lugens*: 21·0–16·0 mm (19·6–22·2 × 15·2–16·8), *n*=26; calculated weight 2·87 g. *O. l. halophila*: 20·2 × 15·6 mm (19·0–20·8 × 14·5–16·4), *n*=21; calculated weight 2·63 g. (Schönwetter 1979.) Clutch: 4–5 (3–6). Of 26 clutches, north-west Africa: 3 eggs, 4; 4, 14; 5, 7; 6, 1; mean 4·2 (Heim de Balsac and Mayaud 1962). 2 broods reported in Egypt and Israel (Meinertzhagen 1954; Zachai 1984), and probably 2 broods in Oman (Gallagher and Woodcock 1980); no evidence of 2 broods in south-west Iran (Cornwallis 1975). INCUBATION. Probably by ♀ only (Zachai 1984); length of incubation period not recorded. YOUNG. Cared for and fed by both parents (Sage and Meadows 1965; Zachai 1984). FLEDGING TO MATURITY. Young leave nest at 14–15 days, before able to fly (Zachai 1984); duration of dependence not recorded, but see Social Pattern and Behaviour. No further information.

Plumages (*O. l. halophila*). ADULT MALE. In fresh plumage (autumn), forehead, crown, hindneck, and upper mantle pale brown-grey or pale ash-brown, feathers tipped cream-white; forehead, sides of crown, and upper mantle usually appearing paler, pale cream-grey; darker brown feather-centres on central crown showing through. Lower mantle, scapulars, and back black, feathers tipped sandy-grey (least so on outer scapulars). Rump and upper tail-coverts white with slight cream-pink suffusion. Side of head (down from nostril, lore, upper rim of eye, and ear-coverts), side of neck, chin, and throat black, feathers sometimes faintly edged grey-white; often some fine white filoplumes on chin and throat. Underparts down from chest white with pale cream-grey or pink-grey suffusion (deepest on chest and flanks); thighs mottled grey-brown, under tail-coverts deep cream-pink. Tail-base white, tail-tips black except for narrow white terminal rim (*c.* 1 mm wide on central pair, t1; *c.* 3 mm on t6); maximum extent of black on tip of t1 30–38 mm (shafts sometimes more extensively black), 18–27 mm on t2, 11–18 mm on outer webs of t3–t5, 9–13 on inner webs of t3–t6, 6–15 on outer web of t6 (*n*=21). Flight-feathers, tertials, and greater upper primary coverts sooty black, secondaries, inner primaries, and sometimes tertials and primary coverts narrowly fringed white on tip (fringe narrower than 0·5 mm on primary coverts). Inner web of flight-feathers with broad and contrasting white wedge along basal inner border, *c.* 3 mm wide on middle of primaries, *c.* 5 mm on middle of secondaries, almost reaching shafts at feather-base. Upper wing-coverts deep black, sometimes faintly edged sandy-grey on tips when plumage very fresh. Under wing-coverts and axillaries black, some smaller and outer coverts narrowly tipped white. *In worn plumage* (spring), forehead, crown, hindneck, and upper mantle bleached to cream-white or white; lower mantle, scapulars, sides of head and neck, chin, throat, and upper wing-coverts uniform deep black; underparts (including under tail-coverts) bleached to white. When heavily abraded, dark grey feather-bases often visible on central crown, hindneck, and (rarely) underparts; black of central lower mantle and back sometimes less uniform, variegated grey and dirty white; white fringes of tail-tips worn off. ADULT FEMALE. Forehead, crown, hindneck, mantle, scapulars, and back uniform pale sandy-brown, often slightly tinged rufous on crown, more greyish-sandy on scapulars and back; sides of crown and hindneck often slightly paler, cream-white or pale cream-grey, forming ill-defined pale supercilium (most distinct just above and behind eye) and pale neck-patch. Rump and upper tail-coverts cream-white or white, often not sharply demarcated from sandy-brown upper back. Often a narrow cream eye-ring. Lores dull grey with cream mottling and dark bristles. Remainder of sides of head and neck, chin, and throat strongly variable. In pale variants, ear-coverts rufous-brown, cheek and side of neck pale sandy-brown, chin and throat dirty cream-white or greyish-cream, often with mottled black-and-cream band down from behind ear-coverts over rear of cheek to central lower

throat; intermediate birds have ear-coverts and cheeks darker brown, with more uniform black band from rear of ear-coverts to lower throat and mottled dark grey chin and upper throat; dark variants have sides of head and neck down from just above eye dull black, usually mottled pale buff but sometimes as deep and uniform as ♂; chin and central throat usually with much cream-buff suffusion but sometimes largely black and then differing from ♂ by sandy-brown or dusky grey-brown upperparts and browner upper wing-coverts (forehead, sides of crown, and hindneck often paler than in 'normal' ♀, almost white). Remainder of underparts down from chest buff-white, suffused pink-grey on chest and flanks; under tail-coverts cream-pink or pale pink-buff. Tail as adult ♂, but black slightly browner and narrow fringes along tail-tips pale buff. Flight-feathers as adult ♂, but slightly browner and white wedges on inner webs of flight-feathers sometimes less clear-cut. Greater upper wing-coverts, greater upper primary coverts, and tertials black-brown; tips narrowly fringed white (usually less than 1 mm wide). Lesser and median upper wing-coverts dark grey-brown with broad sandy-grey fringes. In ♀♀ with largely black sides of head and throat, upper wing-coverts and tertials often almost black (but less deep and uniform than adult ♂). Under wing-coverts and axillaries grey-brown to greyish-black, tips pale buff or white to variable extent (tips widest in pale-throated variants; underwing almost black in dark-throated birds). *In worn plumage* (spring), upperparts paler and greyer, underparts (including under tail-coverts) dirty white; pale fringes along tips of tail, flight-feathers, tertials, and greater coverts worn off; usually traces of black throat-band and much grey mottling on chin and upper throat in pale-throated variants; forehead, supercilium, and hindneck almost white in dark-throated variants, lower mantle, scapulars, and wing-coverts black-brown, sides of head, chin, and throat almost uniform brown-black. NESTLING. No information. JUVENILE MALE. Forehead, crown, hindneck, upper mantle, and sides of neck sandy-buff with faint darker grey-brown streaks; sides of crown sometimes faintly paler sandy-cream, forming indistinct supercilium. Lower mantle and scapulars sandy-buff, often with much dark grey of feather-bases exposed (especially if plumage worn). Back to upper tail-coverts white. Lore, upper cheek, and ear-coverts black with faint buff mottling, forming contrasting black mask. Chin and throat pale buff or dirty white with faint grey feather-fringes or specks. Chest and sides of breast pale buff or cream-buff, sometimes with faint and narrow pale grey arcs. Remainder of underparts cream, grading to white on rear flanks and vent; under tail-coverts pink-cinnamon. Tail as in adult; feather-tips with 1–3 mm wide fringe. Flight-feathers and greater upper primary coverts like adult, but slightly browner and tips usually with whitish fringes *c.* 1 mm wide (except on outer primaries). Tertials black with pale buff to off-white tips *c.* 3–4 mm long. Lesser upper wing-coverts sandy-buff with grey-brown or dull black centres; median and greater coverts dull black with broad pale buff to cream-white tips (*c.* 3–4 mm long), forming distinct double wing-bar. Axillaries and under wing-coverts pale buff or cream-white. JUVENILE FEMALE. Like juvenile ♂, but without contrastingly black mask; lore and upper cheek mottled off-white and pale grey-buff, ear-coverts buff-brown with faint darker streaks; feather-centres of lower mantle and scapulars paler and less contrasting. FIRST ADULT MALE. Like adult ♂, but juvenile flight-feathers, greater upper primary coverts, tail (sometimes except some central feathers), 1–6 outer greater upper secondary coverts (mainly 2–3, rarely 0), and sometimes a few feathers of bastard wing retained. Juvenile flight-feathers and (especially) tertials and greater upper primary coverts browner than in adult and contrasting in colour with new black median coverts, broader rufous-buff to buff-white fringes along tips very conspicuous; juvenile outer greater secondary coverts and longer feathers of bastard wing (if retained) have similar wide fringes; outer webs of tertials and greater coverts narrowly fringed buff to off-white (unlike adult); tail-tips fringed buff (in adult, white). New 1st adult mantle and scapulars have sandy-brown suffusion on feather tips, less uniform and deep black than adult; central mantle and back sometimes largely sandy and grey, almost as pale as hindneck; chin and throat with slightly wider pale feather-fringes than adult, wearing off less quickly. FIRST ADULT FEMALE. Like adult ♀; colour of sides of head, chin, and throat equally variable. Retained juvenile feathers as in 1st adult ♂, but contrast in colour less marked; pale buff to off-white fringes along outer webs and tips of juvenile greater upper primary coverts, tertials, and (if retained) outer greater secondary coverts much wider than in adult, as in 1st adult ♂; fringes on tertials wear off but those on inner primary coverts and outer secondary coverts broad and distinct, even in worn plumage.

Bare parts. ADULT, FIRST ADULT. Iris brown or dark brown. Bill dark horn-brown or black. Mouth yellow. Leg and foot brown-black or black. (Hartert 1903-10; RMNH, ZFMK.) NESTLING. No information. JUVENILE. Iris brown. Bill horn-black with grey-flesh cutting edges and base of lower mandible. (ZFMK, ZMA.)

Moults. ADULT POST-BREEDING. Complete; primaries descendant. In Egypt, specimen of nominate *lugens* had just started (with some feathers of body) on 7 June, and one in full moult on 18 July (Steinbacher 1965); in Israel, 19 June, single bird with primary moult score 8 (p1 growing, p2-p5 shed, p6-p10 old), t1 growing, t2 shed, body heavily worn but not moulting; one from Sinai, 22 July, had moult score 31 (p1-p5 new, p6-p7 growing, p8-p10 old), t1-t4 new, t5-t6 growing, and body largely new (ZFMK). Moult not started in 2 *persica* from 23 June (ZFMK). In those *halophila* and nominate *lugens* examined, moult completed from early September onwards (RMNH, ZFMK, ZMA). POST-JUVENILE. Partial: head, body, lesser and median upper wing-coverts, tertial coverts, (0-)2-6 inner greater upper wing-coverts (on average, 1·5 old ones retained, *n*=27), 0-3 tertials (on average 0·8, *n*=23), usually all bastard wing, and occasionally some central tail-feathers. Starts shortly after fledging; birds examined in full juvenile plumage range from 16 May to 12 August (all races); one from 6 July had just started, one from 14 July in full moult; moult completed in birds examined from late August and early September (RMNH, ZFMK, ZMA).

Measurements. ADULT, FIRST ADULT. Nominate *lugens*. Eastern Egypt, Sinai, Israel, and Jordan, all year; skins (RMNH, ZFMK, ZMA). Juvenile wing and tail include retained juvenile wing and tail of 1st adult. Bill (S) to skull, bill (N) to distal corner of nostril; exposed culmen on average 4·5 shorter than bill (S).

WING AD	♂	96·1 (1·58; 10)	94-98	♀ 90·8 (1·60; 6)	89-94
	JUV	92·8 (2·20; 22)	89-96	89·9 (1·90; 17)	86-93
TAIL AD		64·6 (2·91; 7)	61-68	58·6 (0·63; 4)	58-62
	JUV	57·2 (3·45; 11)	53-62	61·0 (3·55; 5)	56-64
BILL (S)		18·8 (0·60; 16)	17·9-19·7	18·1 (0·55; 9)	17·6-19·1
BILL (N)		10·6 (0·82; 16)	9·6-11·8	10·3 (0·25; 9)	10·0-10·7
TARSUS		26·1 (1·12; 18)	24·9-28·3	25·6 (0·49; 9)	24·8-26·3

Sex differences significant, except bill (N) and tarsus.

O. l. persica. Iran, all year; skins (RMNH, ZFMK). Ages combined.

WING	♂	96·6 (1·72; 9)	94-99	♀ 90·6 (1·78; 5)	88-93
BILL (S)		19·3 (0·40; 5)	18·9-19·8	18·8 (— ; 3)	18·1-19·5

Sex differences significant for wing.

O. l. halophila. Algeria and Tunisia, all year; skins (RMNH, ZFMK, ZMA). Ages combined (but virtually all 1st adult).

WING	♂	91·8 (2·23; 17)	89-97	♀ 87·8 (1·50; 14)	85-90
TAIL		57·9 (2·74; 17)	54-63	56·5 (1·68; 14)	54-60
BILL (S)		18·4 (0·56; 17)	17·6-19·2	18·2 (0·81; 14)	17·1-19·3
BILL (N)		10·6 (0·64; 15)	9·7-11·5	10·6 (0·64; 14)	9·8-11·5
TARSUS		25·7 (0·76; 15)	24·4-26·9	24·9 (0·72; 14)	23·8-25·9

Sex differences significant for wing and tarsus.

Weights. *O. l. persica.* Zagros mountains (Iran), May: ♂ 23 (Desfayes and Praz 1978).
O. l. halophila. Central Algeria, November-December: ♂ 24·0 (3) 22-25, ♀ 21·0 (3) 19-22 (Niethammer 1963b; ZFMK).

Structure. Wing rather long, broad at base, tip bluntly pointed. 10 primaries: p8 longest, p9 4-10 shorter, p7 0-1, p6 1-3, p5 6-10, p4 11-16, p3 15-20, p1 20-27 (*halophila* and nominate *lugens*). P10 reduced; 43-52 shorter than wing-tip, 0-5 longer than longest upper primary covert. Outer web of p6-p8 and (faintly) inner of p7-p9 emarginated. Tail rather short, tip square; 12 feathers. Bill rather long, slender, straight; rather wide at base, laterally compressed at tip; tip of culmen decurved. Nostrils small, oval, partly covered by membrane above. Numerous fine bristles along base of both mandibles, some longer ones at base of upper mandible. Tarsus rather long, slender; toes rather short and slender. Middle toe with claw 17·5 (10) 16-19·5; outer toe with claw *c.* 71% of middle with claw, inner *c.* 67%, hind *c.* 77%. Claws rather short and blunt, slightly decurved.

Geographical variation. Marked, especially in ♀♀, as these are either similar to ♂ (nominate *lugens*, *persica*) or markedly different (*halophila*, described in Plumages, above). In extralimital races, underparts sometimes black or variable black-and-white (see, e.g., Mayr and Stresemann 1950); birds breeding in basalt deserts of southern Syria and northern Jordan, and formerly believed to be Eastern Pied Wheatear *O. picata* (see p. 858), are apparently also a largely black morph of *O. lugens lugens* (see Field Characters) which interbreeds with typical birds (L Cornwallis). Typical ♂ nominate *lugens* from eastern Egypt and southern part of Middle East closely similar to ♂ *halophila*, and inseparable in worn plumage (except for flight-feathers); in fresh plumage, nominate *lugens* on average slightly paler and greyer on central crown, underparts slightly paler cream-white; amount of black in tail about similar to *halophila*, maximum extent of black on t1 30-35 mm, on t2 16-23 mm, on outer web of t3-t5 8-15 mm, on inner web of t3-t6 7-12 mm, on outer web of t6 5-15 mm (*n*=15); main differences from *halophila* are deeper pink-cinnamon or rusty-buff under tail-coverts and more extensive white wedges on inner webs of flight-feathers, reaching shaft on middle portions of secondaries and on bases of primaries. Sexes similar, but black of mantle, scapulars, and throat of ♀ often slightly duller and browner, less deep black; central mantle and back of ♀ (and 1st adult ♂) often variegated pale grey-brown, grey, or off-white, especially if worn, sometimes approaching worn ♂ of Finsch's Wheatear *O. finschii*. *O. l. persica* from southern Iran similar to nominate *lugens*, but crown browner, under tail-coverts darker rufous-cinnamon, tail more extensively black (13-18 mm black on t3-

t6), less extensive white on inner webs of flight-feathers (about as much as in *halophila*), size larger (wing up to 100: Hartert and Steinbacher 1932–8); ♀ like ♂, but black slightly browner, as in nominate *lugens*. For extralimital races, see Vaurie (1949, 1950) and Meinertzhagen (1949*a*, 1954).

Recognition. Nominate *lugens*, *persica*, and ♂ of *halophila* rather like ♂ Pied Wheatear *O. pleschanka*, but tail-band more even in width, outer web of t6 less extensively black, under tail-coverts cinnamon-pink (bleached to off-white in worn plumage, however), and bill and tarsus usually longer. ♀ *halophila* often rather similar to ♀ Desert Wheatear *O. deserti*, but tail less black; some rather similar to black-throated ♀ Black-eared Wheatear *O. hispanica*, but tail-band more even in width and t6 usually less extensively black; see also Measurements. Black morph of nominate *lugens* differs from black *opistholeuca* morph of *O. picata* in having much white on inner webs of flight-feathers (none in *O. picata*); typically, nominate *lugens* also shows somewhat less extensive black on tip of t1 and t6, slightly longer and more sharply pointed wing-tip, and fewer primary emarginations—*O. picata* has outer web of p5–p8 and inner of (p5–)p6–p9 emarginated (compare Structure, above). For distinction from Finsch's Wheatear *O. finschii*, see that species (p. 841). CSR

Oenanthe monacha Hooded Wheatear

PLATES 63, 64, and 65
[between pages 760 and 761, and facing page 856]

Du. Monnikstapuit   Fr. Traquet à capuchon   Ge. Kappensteinschmätzer
Ru. Каменка-монашка   Sp. Collalba pechinegra   Sw. Munkstenskvätta

Saxicola monacha Temminck, 1825

Monotypic

Field characters. 17·5 cm; wing-span 29·5–30·5 cm. Noticeably larger than Mourning Wheatear *O. lugens*, with 15% longer bill, wings, and tail but 5–10% shorter legs, all contributing to more attenuated form, most obvious in length of rear body and tail. Long-billed, rather large-headed and lengthy wheatear. ♂ has distinctly long white crown, black foreparts, back, and wings, and white underbody, lower back, rump, and tail; tail has only black central line. ♀ has generally pale sandy-brown body and wings, and distinctive fawn-buff rump and tail-sides. Combination of pale underparts and tail pattern diagnostic in both sexes. Sexes dissimilar; no seasonal variation. Juvenile separable.

ADULT MALE. Most immaculate of all non-black wheatears. In worn plumage (from mid-winter), forehead, crown, and centre of nape white, sometimes becoming dull silver at rear. Face, throat, breast to below shoulder (and even along upper flanks in one Iranian bird), mantle, scapulars, and wings black without gloss. Underbody, lower back, rump, and upper tail-coverts white. Tail white with black central feathers and (though not usually visible) black corners to outermost feathers; rump and tail thus long-looking. Underwing black on coverts and axillaries, silver-grey on flight-feathers. In fresh plumage (from about July), throat and chest show hoary fringes, and mantle and scapulars show some dull grey tips. ADULT FEMALE. Palest of all ♀ wheatears, with unusual appearance recalling ♀ Eversmann's Redstart *Phoenicurus erythronotus*, Tawny Pipit *Anthus campestris*, and even Spotted Flycatcher *Muscicapa striata*. Head, back, and wings pale sandy- to grey-fawn, with grey tone most marked on worn bird; relieved by broad but indistinct cream supercilium and eye-ring, rufous cheeks (on some), and cream or pale buff margins and tips to wing-coverts, secondaries, and tertials. When fresh, tertials produce wing appearance of *A. campestris* and *M. striata*, but when worn leave wing more uniform drab brown though never as dark as in most ♀ wheatears. Rump fawn-buff, brighter and warmer than back but hardly contrasting with buff-white to pale rufous-buff tail, which has only dark brown central feathers and smudge on corners and next outermost feathers. Underparts off-white to yellow-cream, with pale buff suffusion strongest on throat, breast, flanks, and (becoming almost rufous) under tail-coverts. JUVENILE. Resembles adult ♀ but spotted cream-buff on upperparts and dully scaled or mottled black on sides of head and from throat to chest. Easily distinguished from other juvenile wheatears by tail colour and pattern. FIRST-WINTER MALE. Much duller than adult, with grey-brown suffusion on head, black plumage duller and more broadly fringed, rump and underbody distinctive pink-cream, and tail rufous-tinged and more smudged. Bill long and strong, black-horn; legs rather short, feet small, thin, and weak-looking, black.

♂ unmistakable in good view, with long, strong bill, immaculate pied plumage, long body and tail, and rather less erect carriage than other wheatears, producing distinctive appearance. Caution necessary only with suddenly retreating bird, in which length of rump and tail suggests Hume's Wheatear *O. alboniger* and tail pattern recalls White-crowned Black Wheatear *O. leucopyga*. ♀ almost as distinctive, with paler, more uniform appearance than other ♀ wheatears, suggesting several other passerines (see above) but actually larger than them; diagnostic tail pattern prevents confusion with dark-tailed Red-rumped Wheatear *O. moesta*. ♀ may also suggest

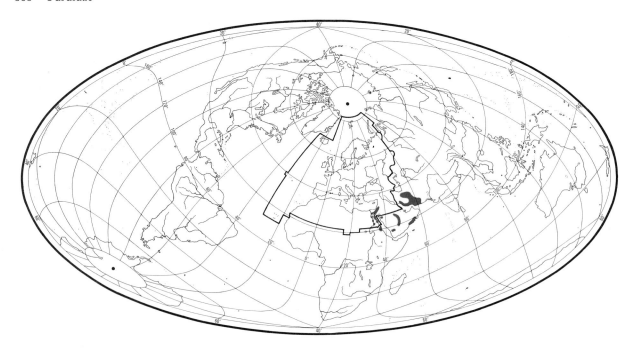

pale-throated race of Red-tailed Wheatear *O. xantho-prymna chrysopygia*, but that species' rump and tail much redder-orange with full ⊥-mark. Flight noticeably fluent and floating, with apparently loose wing-beats—due partly to plumage pattern (as in *O. alboniger*) and partly to wing-shape (see Structure); thus sometimes likened to butterfly, though better compared to large redstart *Phoenicurus*. Escape-flight long, with bird then hiding. Hops on boulders; relatively short legs give lower and more level stance than most wheatears. Behaviour apparently typical of genus but little known.

Song a sweet but rather quiet medley of whistles, warbles, and more chanted notes like those of *Turdus* thrushes. Usually silent otherwise, but calls include a harsh 'zack'.

Habitat. In arid hot lower middle and lower latitudes, demanding rocky hills of great barrenness; not found breeding west of the Nile (Moreau 1966). In Arabia, occurs in most desolate wadis and ravines, avoiding vegetation, and defends large territories against other insectivores owing to low insect populations. Needs rock faces with fissures or cavities as places of refuge if alarmed (Meinertzhagen 1954). In Pakistan, a bird of the most desolate desert ravines (Ali and Ripley 1973*b*). In southern Saudi Arabia, however, breeds in buildings by cultivation, and in Oman and Israel around small desert settlements (Bundy and Sharrock 1986).

In south-west Iran, relegated to hot arid wadis and badlands unacceptable to other *Oenanthe*, partly through social subordination and partly through capability for aerial pursuit of flying arthropod prey which, unlike ground-dwelling forms remain active during heat of day

in air where stress of high ground temperatures rising above 50°C can be avoided. Except for Hume's Wheatear *O. alboniger*, *O. monacha* is the only *Oenanthe* in south-west Iran occupying in summer lower areas of greatest heat, below *c.* 1400 m and down almost to sea-level. (Cornwallis 1975.)

Distribution. No information on range changes.
Accidental. Iraq, Kuwait, Cyprus.

Population. No information on trends.
EGYPT. Locally not uncommon (SMG, PLM, WCM).

Movements. Largely sedentary with a few individuals making short movements. Study hampered by low population density and patchy distribution.

In Egypt, regularly seen in low numbers west of Nile (e.g. at Thebes), November–February, though all Egyptian breeding records appear to be east of river (Meinertzhagen 1930; M G Kelsey). Less frequently recorded in northern Sudan (where it does not breed) during winter months (Cave and Macdonald 1955). Very scarce and irregular visitor to states on western side of Persian Gulf, late October to March, with one late-August record in United Arab Emirates (Gallagher and Rogers 1978; Bundy and Warr 1980; Loskot and Vietinghoff-Scheel 1981). In south-west Iran, probably resident but one mid-December record of 1st-year ♂ perhaps indicates dispersal from interior (Cornwallis 1975). Observed infrequently along Makran coast as far east as Kirthar mountains (*c.* 67°30′E); since these records are mostly October–March, some easterly dispersal from Iranian breeding grounds possible, but recent data from south-east

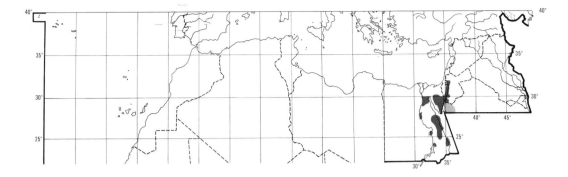

of range virtually non-existent (see Ali and Ripley 1973*b*, Loskot and Vietinghoff-Scheel 1981).

Only record of vagrancy during breeding season involves 2 which visited Cyprus in May 1875 (Flint and Stewart 1983). DGCH

Food. Poorly known; most available information extra-limital from Iran. Probably takes mainly insects and a few seeds. Searches for food from elevated perch (stone, post, etc.), launching flying attacks against prey on ground and (more so than other *Oenanthe*) in the air, sometimes pursuing prey to height of 50-100 m (Koenig 1924; Hartley 1949; Simmons 1952*a*; Cornwallis 1975; Jennings 1980; Bundy and Sharrock 1986). Of 16 attacks, 56% at prey on bare ground and 44% at flying insects (L Cornwallis). Taking both ground and aerial prey permits feeding throughout the day (see final paragraph), as ground prey relatively abundant during cooler periods of morning and evening while flying prey continue to be available during heat of day (Cornwallis 1975). In Arabia, sometimes takes large ticks from hides of camels and other livestock (Bundy and Sharrock 1986).

The following recorded in diet. Invertebrates: dragon-flies (Odonata), grasshoppers (Orthoptera), larval Neuro-ptera, adult Lepidoptera, Hymenoptera (including ants Formicidae), adult beetles (Coleoptera), large ticks (Acari: Ixodidae), and spiders (Araneae) (Hartley 1949; Corn-wallis 1975; Bundy and Sharrock 1986). Plant material: a few seeds found in one stomach (Desfayes and Praz 1978).

Combined contents of 2 stomachs collected in Iran, January and May, included 19 items of which 16% (by number) grasshoppers, 16% ants, 42% other Hymen-optera, 5% beetles, 5% larval Neuroptera, and 5% spiders (Cornwallis 1975); of these, 50% were 0-5 mm long, 22% 5-10 mm, 28% 10-20 mm (L Cornwallis).

No information on food of young. Nestlings in Iran were fed throughout the day at rate of 3-7 feeds per hr (Cornwallis 1975). LC

Social pattern and behaviour. Poorly known. Account based on scattered casual observations, systematic studies of winter territory in Egypt (Hartley 1949; Simmons 1951*c*), and an all-day watch at nest on Makran coast of Iran (L Cornwallis and R F Porter).

1. Usually solitary except during breeding season when pairs and family parties are seen (Jennings 1980; L Cornwallis and D A Scott). Territorial during breeding season and also at other times of year when individual ♂♂ and ♀♀ hold territories (Hartley 1949; Simmons 1951*c*). In winter territories described by Simmons (1951*c*), boundaries usually not sharply demarcated. Size of winter territory in Iran indicated by observation that probable residents were often wide-ranging, making flights of up to 1 km (L Cornwallis). BONDS. Mating system little studied but all evidence suggests monogamy. Duration of pair-bond unknown, but to south of Dead Sea, ♂♂ reported sometimes to breed before they have assumed perfect summer plumage, i.e. probably 1 year old (Dresser 1871-81). Details of parental care virtually unknown. Both parents fed young at nest in Iran (L Cornwallis and R F Porter). Family bonds maintained for unknown period after fledging (see Relations within Family Group, below). BREEDING DISPERSION. Solitary and territorial. No accurate measurements of territory size or dispersion but territories reported to be generally large and widely spaced (Dresser 1871-81; Meinertzhagen 1954; Erard and Etchécopar 1970). Pair feeding young in nest in Iran ranged (mostly while feeding) over *c.* 1 km², and 5 probably territorial ♂♂ seen in May along 8 km of track (L Cornwallis and R F Porter). Territorial behaviour not studied in detail but territories probably all-purpose. ROOSTING. No information.

2. Commonly shy and wary, quickly diving for cover among rocks at slightest alarm (e.g. Dresser 1871-81, Bulman 1942, Meinertzhagen 1954, Erard and Etchécopar 1970), or taking flight (L Cornwallis). Sometimes, however, extremely tame, allowing approach to within a few metres (Marchant 1941; L Cornwallis), and a ♀ once sat on Tragenza's (1955) shoulder. FLOCK BEHAVIOUR. None reported. SONG-DISPLAY. Song (see 1 in Voice) reported only from ♂ either while perched, often on favourite elevated song-post, or during Song-flights. Such flights observed in central Arabia by Jennings (1980) described as 'fluttering'. In 5 Song-flights in northern Iran, birds took off at an angle, circled over territory several times at *c.* 20-50 m, then descended at an angle. After one Song-flight with no ♀ in evidence, ♂ toured territory; made 2 further Song-flights (which involved tail-fanning and slow accentuated flapping of rigid wings) and *c.* ½ hr later chased ♀ (L Cornwallis). In central Arabia, song heard most often in April (Jennings 1980); in northern Iran from mid-March to end of April, mainly in the cooler morning hours (L Cornwallis). In southern Baluchistan (Iran), during dawn to dusk watch on nest where adults feeding young, ♂ did not perform Song-flight but on 6 out of 13 visits to nest with food he sang briefly near nest-hole before entering

(L Cornwallis and R F Porter). Song or Subsong reported from winter visitor to Bahrain (Gallagher and Rogers 1978). ANTAGONISTIC BEHAVIOUR. Reported by Meinertzhagen (1954) and Etchécopar and Hüe (1967) to guard territories jealously against all other wheatears *Oenanthe*. In Lower Egypt, however, ousted from recently established winter territories by Mourning Wheatear *O. lugens* which arrived later to winter in the area (Hartley 1949); in central Arabia, subordinate to White-crowned Black Wheatear *O. leucopyga* which displaces it from perches and often gives chase (Jennings 1980). No details of antagonistic behaviour on breeding grounds. HETEROSEXUAL BEHAVIOUR. See Song-display (above); no further information. RELATIONS WITHIN FAMILY GROUP. Poorly known. Young in nest in Iran gave food-calls; from dawn to dusk on one day, ♀ made 49 feeding visits, ♂ 13; ♀ removed 5 faecal sacs, ♂ 1 (L Cornwallis and R F Porter). After fledging, family parties formed: ♀ with 3 juveniles in northern Iran, late July (D A Scott); pair with 3 fledglings, Lower Egypt (Raw 1921); in central Arabia, adults with fledged young in September (Jennings 1980). ANTI-PREDATOR RESPONSES OF YOUNG. No information. PARENTAL ANTI-PREDATOR STRATEGIES. When observers approached nest with young in Iran, ♀ became agitated and gave repeated alarm-calls (L Cornwallis and R F Porter: see 2 in Voice). LC

Voice. Poorly known; no recordings available. Described as usually silent by Bundy and Sharrock (1986). Song heard mainly during breeding season but once reported in winter (see Social Pattern and Behaviour).

CALLS OF ADULTS. (1) Song. Given apparently only by ♂. Subdued but excessively sweet (Meinertzhagen 1930); a sweet medley of whistles and thrush-like notes (Jennings 1980); a melodious warbling recalling Blackbird *Turdus merula* (L Cornwallis and R F Porter); lasts *c.* 2 s (Bundy and Sharrock 1986). Subsong, heard once while ♂ sitting in shade, described as soft 'chuk chuk weez wez' (L Cornwallis). (2) Alarm-calls. When observers approached nest with young, ♀ gave a repeated anxious 'wit wit', and once a throaty 'prrupp prrupp' recalling Bee-eater *Merops apiaster* (L Cornwallis and R F Porter). (3) Other calls. A harsh 'zack' (Walker 1981b). Gives rattling calls during territorial disputes (Bundy and Sharrock 1986).

CALLS OF YOUNG. Chirruping food-calls from young in nest (L Cornwallis and R F Porter). LC

Breeding. Very little known. SEASON. Jordan: pair seen courting in late April; ♂ with food in bill seen late May (Etchécopar and Hüe 1967; Wallace 1984b). Egypt: young seen early June (Etchécopar and Hüe 1967). SITE. One nest in Iran deep in hole 2·5 m above base of north-facing bank of wadi (L Cornwallis and R F Porter). Nest: not described. EGGS. Not certainly described. Eggs claimed to be of this species (Plate 83) very pale blue with scattered tiny red specks at broad end, measuring 21·6 × 15·6 mm (Harrison 1975), but Schönwetter (1979) recorded 3 eggs measuring 19·3 × 15·9 mm (18·6–20·5 × 15·7–16·0); calculated weight 2·6 g. No further information.

Plumages. ADULT MALE. In fresh plumage, forehead and crown cream-white, nape white with vinous-grey suffusion. Mantle and scapulars black, feather-tips rather narrowly fringed pale vinous-grey. Back, rump, and upper tail-coverts white. Sides of head down from lores, line just above eye, and ear-coverts black; chin, throat, chest, and sides of neck and breast black; feather-tips of throat, chest, and sides of breast rather broadly tipped white (*c.* 1 mm on throat, *c.* 3 mm on lower chest). Remainder of underparts (backwards from breast and flanks) cream-white. Base of central pair of tail-feathers (t1) white, distal 40–45 mm black, except for pink-buff fringe 1–2 mm wide along tip, narrow white inner edge, and broader less clear-cut pale vinous-buff outer fringe. T2–t5 white except for narrow black shaft-streak on tip (sometimes entirely white); t6 white with dark streak *c.* 10–20 mm long on tip of outer web, fringed white along outer margin and tip (streak sometimes faint or mottled). Flight-feathers, tertials, and all upper wing-coverts deep black; longer tertials, secondaries, and inner primaries narrowly fringed white along tips; greater upper primary coverts faintly edged brown distally. Under wing-coverts and axillaries black; small coverts along leading edge of wing mottled black and white. *In worn plumage* (from about mid-winter), forehead, crown, and nape pure white, contrasting with uniform black mantle and scapulars; white fringes from chin to chest worn off, some traces of white remaining on sides of breast only; back to upper tail-coverts and breast to under tail-coverts white, but some black of feather-bases sometimes visible on back and belly. ADULT FEMALE. Forehead cream-buff; crown, nape, ear-coverts, mantle, scapulars, and back pale drab-brown with slight rufous tinge. Rump and upper tail-coverts buffish-pink. Lores, narrow ring round eye, and indistinct supercilium pale cream; cheeks rufous-buff. Entire underparts including under wing-coverts and axillaries warm pinkish-buff (paler on chin and throat, deeper rufous-cinnamon on under tail-coverts). Base of t1 pink-buff, remainder dark grey or greyish-black for 40–50 mm, narrowly fringed rufous-buff at sides and tip. T2–t6 pale rufous-cinnamon (in ♂, white), shafts often partly black near tips; (t3–)t4–t5 with dark olive-brown spot on tip 4–10 mm long (mainly on outer web); dark olive-grey streak on outer web of t6 *c.* 25 mm long, partly extending to inner web (hence, tail-tip more extensively dark than in adult ♂, but still less so than other brown-plumaged ♀ *Oenanthe*). Flight-feathers and tertials dark grey or blackish-grey; outer webs and tips of flight-feathers narrowly edged pale pink-buff (except emarginated parts), tertials broadly fringed pink-buff. Upper wing-coverts dark grey or blackish-grey, fringes of median and longer lesser coverts broadly pink-buff, those of greater coverts and greater primary coverts paler buff. *In worn plumage*, upperparts grey-brown, greyer and less sandy than in fresh plumage; rump and upper tail-coverts paler pink-buff; underparts dirty pale cream-yellow (more pink-buff on under tail-coverts), less saturated cream-pink than in fresh plumage; flight-feathers and upperwing browner, pale fringes bleached to pale buff and partly abraded; tail in spring as in fresh plumage, thus showing much rufous, but this bleached to cream by June–July. NESTLING. No information. JUVENILE. Like adult ♀, but feather-centres from crown to back and of lesser upper wing-coverts spotted cream-buff; sides of head, chin to chest, and sides of breast scaled dull black; for tail, see 1st adult below. Closely similar to other juvenile *Oenanthe*, but unique in showing limited amount of dark on tail-tip in combination with rufous suffusion to remainder of tail. FIRST ADULT MALE. In fresh plumage, white of forehead, crown, and nape strongly suffused pale grey-brown; mantle and scapulars duller black than in adult ♂, feathers more broadly fringed pale sandy-grey.

Back pale cream, rump and upper tail-coverts pink-cream (in adult ♂, white). Sides of head, chin to chest, and sides of breast dull black, feather-tips more broadly fringed with sandy-grey than in adult; remainder of underparts cream-pink, under tail-coverts pale rufous-cinnamon. Tail as adult ♂, but outer edges of t1-t5 strongly suffused rufous (gradually less so outwards); t1 largely brown-black (deep black in adult ♂); tip of outer web of t4 sometimes with small dusky spot, those of t2 and t5 often with dusky spot, outer web of t6 with dull black streak 15-35 mm long, extending for 5-10 mm on inner web and with partly rufous tip (in adult ♂, tail virtually white except t1 and small dark streak on outer web of t6). Lesser and median upper wing-coverts, tertial coverts, and new inner greater coverts black; longer lesser and median coverts have fringes 1-1·5 mm wide on tips, faint fringes on tertial and greater coverts also. Retained juvenile outer greater upper wing-coverts, tertials, and greater upper primary coverts dark grey-brown (black in adult ♂), strongly contrasting in colour and wear with neighbouring black new feathers; fringes along all coverts and tertials rather broadly pale buff or off-white and sometimes with white triangular spot on tip; old longest feather of bastard wing brown-black. Retained juvenile flight-feathers brown-black, tips and outer webs narrowly fringed pale buff or off-white (not as uniform deep black as adult ♂). *In worn plumage* (from about January), forehead, crown, nape, back to upper tail-coverts, and underparts backwards from breast not as pure white as adult ♂, retaining cream tinge all-over and showing grey tinge on nape and on longer upper tail-coverts; mantle, scapulars, chin to chest, and sides of breast with more and wider traces of pale fringes than in adult ♂; breast to under tail-coverts bleached to cream or dirty white; tail whiter than in fresh plumage, but still has traces of rufous tinge, and has more extensive dark suffusion on tip than adult ♂; retained juvenile flight-feathers, tertials, and outer upper wing-coverts contrastingly brown and with distinct pale buff or off-white fringes. FIRST ADULT FEMALE. Like adult ♀ and sometimes hardly distinguishable; part of juvenile plumage retained, as in 1st adult ♂, showing relatively greater wear than remainder of plumage, but colour similar to fresh feathering or only slightly paler brown (unlike 1st adult ♂); outer webs of t2-t5 often with 5-10 mm long dark olive-brown tips, and similar tips to inner webs of t2, t5, and t6; tip of outer web of t6 olive-brown for 20-40 mm.

Bare parts. ADULT, FIRST ADULT. Iris brown. Bill black or (at least in some ♀♀) dark brown. Leg and foot black. Mouth pale yellow. (Ali and Ripley 1973*b*; Desfayes and Praz 1978; BMNH, ZFMK). NESTLING, JUVENILE. No information.

Moults. ADULT POST-BREEDING. Complete; primaries descendant. Only 3 birds in moult examined (BMNH). Primary moult starts about late June or early July with p1, completed with p9-p10 mid- or late August. POST-JUVENILE. Partial: head, body,

lesser and median upper wing-coverts, tertial coverts, variable number of inner greater upper wing-coverts, and (rarely) a few tertials. Probably starts soon after fledging; completed in birds examined from mid-September and later (BMNH, ZFMK). Juveniles in full moult recorded early September in southern Iran (Vaurie 1949).

Measurements. ADULT, FIRST ADULT. Mainly Egypt, some Dead Sea valley, Saudi Arabia, Bahrain, Oman, and Iran, all year; skins (BMNH, ZFMK). Bill (S) to skull, bill (N) to distal corner of nostril; exposed culmen on average 6·1 less than bill (S).

	♂		♀	
WING	105·6 (2·92; 13)	102-111	101·1 (2·81; 10)	98-104
TAIL	68·8 (2·86; 13)	65-74	68·0 (2·25; 10)	65-72
BILL (S)	22·6 (0·72; 13)	21·4-23·6	22·4 (0·68; 10)	21·6-23·4
BILL (N)	12·6 (0·46; 12)	12·0-13·2	12·7 (0·76; 10)	11·7-14·0
TARSUS	24·0 (0·63; 13)	23·0-25·1	23·3 (0·58; 10)	22·5-23·9

Sex differences significant for wing and tarsus. 1st adult with retained juvenile wing and tail combined with older birds in table above, though wing of 1st adult on average 2·6 shorter than full adult and tail 1·6 shorter.

Weights. Iran: January, ♀ 22·5 (Ali and Ripley 1973*b*); May, ♂ 23 (L Cornwallis); June, ♀ 18 (Desfayes and Praz 1978).

Structure. Wing rather short, broad at base, tip bluntly pointed. 10 primaries: p8 longest, p9 4-8 shorter, p7 0-1, p6 3-5, p5 9-13, p4 16-21, p1 25-32. P10 reduced, 57-62 shorter than wing-tip, 0-4 longer than longest upper primary covert. Outer web of p6-p8 and inner of p7-p9 emarginated. Tertials short. Tail rather long, tip square; 12 feathers. Bill markedly long and slender, straight except for slightly decurved tip of culmen; wide and flattened at base, laterally compressed at tip. Tarsus and toes markedly short and slender, relatively more so than in any other Palearctic *Oenanthe*. Middle toe with claw 17·2 (7) 16-18; outer toe with claw *c.* 72% of middle with claw, inner *c.* 69%, hind *c.* 79%. Claws rather short, slender, decurved.

Recognition. White-hooded ♂ rather similar to ♂♂ of Mourning Wheatear *O. lugens*, Pied Wheatear *O. pleschanka*, and *capistrata* morph of Eastern Pied Wheatear *O. picata*, and to both sexes of White-crowned Black Wheatear *O. leucopyga*, but *O. lugens*, *O. pleschanka*, and *O. picata* are distinctly smaller and have more extensive black on tail-tip, while black of throat does not extend to chest; size and tail-pattern of *O. leucopyga* similar to *O. monacha*, but bill of *O. leucopyga* shorter and belly and vent black, not white. ♀ with rufous tail rather similar to ♀ Red-rumped Wheatear *O. moesta* and to both sexes of pale-throated race of Red-tailed Wheatear *O. xanthoprymna chrysopygia*, but tail-tip of *O. monacha* rufous with some olive-brown suffusion, not black; also, wing and bill of *O. monacha* longer and tarsus shorter. CSR

Oenanthe alboniger Hume's Wheatear

PLATES 62, 64, and 65
[between pages 760 and 761, and facing page 856]

DU. Hume's Tapuit   FR. Traquet de Hume   GE. Schwarzkopfsteinschmätzer
RU. Белочерная каменка   SP. Collalba de Hume   SW. Svartvit stenskvätta

Saxicola Alboniger Hume, 1872

Monotypic

Field characters. 17 cm; wing-span 29-30·5 cm. Distinctly larger than Eastern Pied Wheatear *O. picata*, with 25% longer and stouter bill, heavier and more rounded head, 10% longer wings, larger feet, and (due to different plumage pattern) apparently longer back and tail. Rather large, long-billed, bull-headed, and long-backed wheatear, with rather erect carriage and totally glossy black and pure white plumage. Roundness of head exaggerated by restriction of black on underparts to chin and throat. Tail pattern a classic ⊥, with white rump and tail-sides extended by white lower back into longer area of white than on any other wheatear. Combination of all-black head and long white rump and tail diagnostic. Sexes closely similar; no seasonal variation. Juvenile scarcely separable.

ADULT MALE. Head, chin, throat, sides of neck, mantle, scapulars, and wings intense glossy black with faint blue sheen showing at close range. Lower back, rump, and upper tail-coverts white, forming obvious divide between bases of tertials (often showing even when perched) as well as usual patch above tail; extent of white rump usually much larger and longer (up to 50 mm long) than in *O. picata* (up to 38 mm). Tail white with narrow but sharp black ⊥-pattern. Breast and rest of underparts white. Underwing shows black axillaries and coverts, contrasting with dusky-grey flight-feathers. ADULT FEMALE. Usually duller than ♂, but difference only visible in close comparison. JUVENILE. As adult, but black areas brownish and not glossy.

Lack of white crown sufficient to distinguish it from several other similar wheatears including *capistrata* morph of *O. picata* (with white crown and belly). However, easily confused with *picata* morph of *O. picata* (black crown and white belly); hence beware similarity of (1) worn *O. alboniger* to fresher ♂ *O. picata* and (2) juvenile or 1st-year *O. alboniger* to ♀ and juvenile *O. picata*. Note that *O. picata* (1) is smaller and more compact; (2) has more black on lower throat and upper breast (centre of black throat up to 38 mm from point of chin, only 25 mm in *O. alboniger*); (3) has black-white border on underbody rather straight (downcurving in *O. alboniger*); (4) has shorter white rump. Flight free and floating, with wing-beats appearing loose (though this due as much to plumage pattern as to wing-action). Escape-flight usually quite short, bird landing on rock or ridge. Rather less active than *O. picata*, sitting on perches for longer.

Song loud and cheerful, lacking discordant notes of most wheatears: 'chew-de-dew-twit'. Calls include short, sharp whistle given 3-4 times; in alarm, uses harsh, grating monosyllable and quiet 'chit-tit-tit'.

Habitat. Breeds commonly where hills outcrop from plains, and in valleys and ravines in warm steppe zone, from sea-level to *c.* 1900 m in steppe, sub-steppe, and deforested zones of lower middle to lower continental latitudes near fringe of west Palearctic. Requires mainly bare ground for feeding, usually gently sloping or flat, although high up on steep rocky hillsides less steep patches sometimes acceptable. In all cases, steep, rocky hillsides or broken cliffs essential, serving as song-posts, look-outs for prey, sources of shade, perches for loafing, backgrounds of cryptic value, refuges from predators, roosts, and nest-sites. Ground surface may vary from predominantly stone and gravel to stoneless silt or marl, both being included in most territories. Vegetation may include some oak woodland, but usually ranges from widely scattered steppe plants to medium density low cover. Thus at steepest end of spectrum of *Oenanthe* habitats, overlapping markedly with Hooded Wheatear *O. monacha*, which can, however, exist in badland areas affording relatively little shade compared with cliffs and steep slopes required by *O. alboniger*. Overlap also occurs with Mourning Wheatear *O. lugens* in cases where *O. lugens* includes in its territory gently sloping ground as well as steep hillsides or cliffs. However, overlap limited in south-west Iran by higher altitude of breeding of *O. lugens*—between 1500 and 2700 m (Cornwallis 1975). In Arabia, occurs at foot of most barren hills, nesting in rock clefts rather than rodent holes (Meinertzhagen 1954).

Non-breeding habitat in south-west Iran was in great majority of cases similar to breeding habitat, but on Gulf coastal plain, about 1 km from nearest hills, use was made in hot weather of shady lower branches of *Acacia* and of shady ground beneath, cooled by sea breeze (Cornwallis 1975).

Distribution. No information on range changes.

IRAQ. Breeding in east (Sage 1960*a*); 4 seen further south near Iranian border (Moore and Boswell 1956).

Accidental. Kuwait.

Population. No information on trends.

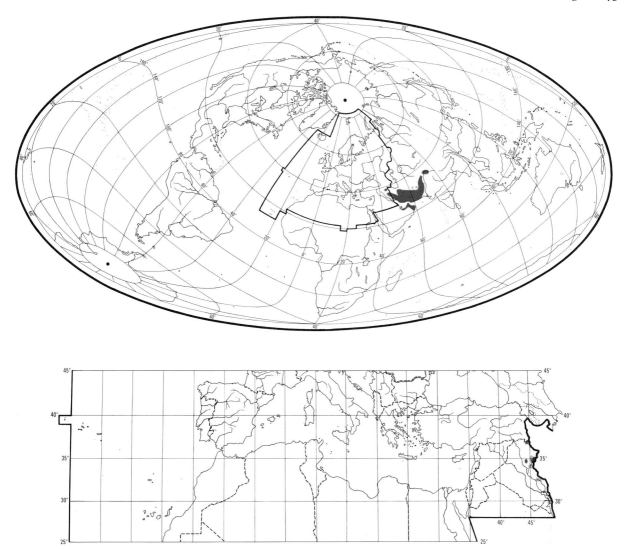

Movements. Information extremely limited; appears to be basically resident with (sometimes at least) some local movement outside breeding season. For review, see Loskot and Vietinghoff-Scheel (1980).

The only detailed study is that by Cornwallis (1975) in south-west Iran. Little seasonal change in either geographical or altitudinal distribution, but not known if breeding sites occupied all year. Dispersal started in late June while adults would be expected to be moulting (see Moults), suggesting that juveniles involved. A few sites occupied only during winter (mainly October–February); one marked ♀ vacated such an area in late February and returned there by end of September, indicating that at least some adults disperse.

Status in other parts of range confused by paucity of dated records (Johnson 1958a; Erard and Etchécopar 1970). Very scarce winter visitor to states on western side of Persian Gulf between late September and mid-

January, but no records for Saudi Arabia (Bundy and Warr 1980: Jennings 1981a). Resident in United Arab Emirates (Bundy and Warr 1980). Birds seen at Saiq in Oman during October were probably dispersing (Gallagher 1977). Vertical displacement suggested to occur in east of range (Ali and Ripley 1973b). DGCH

Food. Mainly insects; some seeds. Taken mainly from open ground, usually by dash-and-jab technique or by aerial-pursuit, launching attack on prey from low perch (as in Mourning Wheatear *O. lugens*). Of 116 feeding sorties, 45% were dash along ground, 39% short sally and pounce, and 16% stationary; 93% were to open ground, 3% into vegetation, 3% into the air, and 2% to base of stone. (Cornwallis 1975.)

In south-west Iran, 36 guts from throughout the year contained 1337 items: ants (Formicidae) in 81% of guts (59% of total items), beetles (Coleoptera) in 89%; other items included bees (Apoidea), bugs (Hemiptera), termites (Isoptera), lacewings (Neuroptera), grasshoppers, etc. (Orthoptera), flies (Diptera), lizards of up to 7 cm, scorpions (Scorpiones), mites (Acarina), spiders (Araneae), and seeds (Cornwallis 1975). In eastern Iran in summer, takes mainly insects, especially large grasshoppers (Acrididae); in autumn, guts also contained seeds, especially Leguminosae (sometimes exclusively); probably also takes the small desert lizard *Eremias* (Zarudnyi 1903).   BDSS

Social pattern and behaviour. Little known. Account compiled from scattered casual observations and systematic studies in south-west Iran (Cornwallis 1975; L Cornwallis).

1. Outside breeding season normally solitary and territorial. In south-west Iran, 6 territories averaged 5·8 ha (*c.* 2·6–12·7) in relatively well-watered highlands but considerably larger in arid foothills (Cornwallis 1975; L Cornwallis). Territories defended throughout the year against all intruding *Oenanthe* and other similar birds; interspecific territorial boundaries noted with Finseh's Wheatear *Oenanthe finschii* ($n = 11$) and Red-tailed Wheatear *Oenanthe xanthoprymna* ($n = 8$) (see Antagonistic Behaviour, below). Fidelity to territory in successive winters shown by colour-ringed ♀ and probable in unmarked bird. BONDS. Mating system apparently monogamous. Duration of pair-bond and age of first breeding unknown. A colour-ringed ♀ in south-west Iran successfully mated with different ♂♂ in successive seasons in neighbouring territories but marked mate of 1st season not found in 2nd (Cornwallis 1975). Both parents feed young; family parties seen April–June (see Relations within Family Group, below). BREEDING DISPERSION. Solitary and territorial. Territory all-purpose (L Cornwallis). In south-west Iran, breeding territories smaller in more productive highlands (one typical territory *c.* 1·1 ha) than in arid foothills (one area held *c.* 6 pairs per km² including one territory of at least 8·5 ha; nearby, one territory at least 21 ha) (L Cornwallis). In Khirthar range (Pakistan), 1 pair per *c.* 1·6 km (Ticehurst 1922*b*). ROOSTING. Often active until dark (L Cornwallis) but no details of roosting.

2. Generally fairly tame, frequently inquisitive (Baker 1924; L Cornwallis). In Oman often visits human habitations (Gallagher and Woodcock 1980). FLOCK BEHAVIOUR. None reported. SONG-DISPLAY. Song (see 1 in Voice) normally delivered from prominent rock, often high up on cliff; once from top of bush. Song-flight recorded twice: once in breeding season, after chasing Mourning Wheatear *O. lugens*, a probable ♂ launched Song-flight from cliff top out over adjacent valley; once in winter, after encounter with *O. xanthoprymna*, bird sang while flying over its teritory (L Cornwallis). In Iran, song mainly in cooler hours of morning and evening during breeding season from mid-March to end May but weak song heard occasionally outside breeding season in July–September and January–February (L Cornwallis, D A Scott). In Oman, song noted in April and brief song in July (Gallagher 1977). No evidence that ♀ sings. ANTAGONISTIC BEHAVIOUR. Breeding and non-breeding territories guarded jealously against all intruding *Oenanthe*, and other similar birds also attacked, though Desert Lark *Ammomanes deserti* and Crested Lark *Galerida cristata* usually ignored (Baker 1924; Meinertzhagen 1954; Cornwallis 1975; L Cornwallis). Commonest form of aggression (recorded 29 times throughout the year), is fast, low, direct, determined flight towards intruder which retreats: song delivered during 2 such attacks and 1 preceded by Threat-calls (see 2a in Voice). On

12 occasions, mainly in breeding season, attack followed by pursuit-flight. On 3 occasions, probably on territorial boundaries, attacked bird stood ground, attacker sat nearby for some minutes, then both drifted back into own territories. In one boundary dispute in August between conspecifics, attacked bird simply flinched, attacker then sat on nearby rock, uttered a few phrases of song, fanned tail, and parted folded wings, showing white of rump and back. In January, bird seen to hover over an *O. xanthoprymna* while calling (see 2b in Voice). When ♂ *O. finschii* in glass box introduced experimentally into winter territory of ♀, she immediately approached, calling loudly (call 2a) and opening wings, and persistently attacked glass for 20 min after which intruder was removed. Once, trespass by ♀ *O. finschii* on winter territory deterred by burst of song. (L Cornwallis.) HETEROSEXUAL BEHAVIOUR. Little known. ♂ displaying to ♀ crouched with head and tail held low exposing much white on arched back; then both went into hole under boulder. ♀ once observed playing with nest-material, then dropped it, collected more, and entered hole in rock-face accompanied by ♂ who had been singing nearby; shortly afterwards both came out of hole, with ♀ still carrying material. (L Cornwallis.) Will ruffle white feathers of back in display, but no details (Gallagher and Woodcock 1980). RELATIONS WITHIN FAMILY GROUP. Little information. Both parents feed nestlings; at nest in south-west Iran adults invariably flew directly into nest-hole from considerable distance (Erard and Etchécopar 1970; L Cornwallis). After fledging, both parents continue to feed young, which give food-calls. Family groups of up to 2 adults and 5 fledglings recorded 27 March–21 April in Oman (Walker 1981*b*) and 26 April–8 June in Iran. (Blanford 1876; Erard and Etchécopar 1970; L Cornwallis, D A Scott.) ANTI-PREDATOR RESPONSES OF YOUNG. No information. PARENTAL ANTI-PREDATOR STRATEGIES. When Walker (1981*b*) approached fledglings in Oman, an adult intervened with wings thrown forward, fluttering them rapidly; held tail open and depressed with rump fully exposed.   LC

Voice. Not studied in detail. Song heard mainly during breeding season; otherwise rather quiet.

CALLS OF ADULTS. (1) Song. Given apparently only by ♂. In Oman, a varied, far-carrying 'chiroochiri-chirrichiri' (Gallagher and Woodcock 1980); cheerful, rounded 'chew-de-dew-twit' on rising scale (Walker 1981*b*). In Iran and Afghanistan, regularly repeated, short musical phrases often almost stereotyped and preceded by quiet introductory note, e.g. 'dze', 'kooi', 'pepeepoo'. Singing bouts often last several minutes; average number of phrases per minute 10·7, $n = 5·8$; average length of phrase 1·7 s (0·8–4·2), $n = 44$; average interval between phrases 3·9 s (2·6–9·8), $n = 35$; frequency range of one song 2·5–4·25 kHz. In south-west Iran, song has sweet, rather plaintive thrush-like quality similar to Mourning Wheatear *O. lugens* which is also common in the area (Fig I). In south-east Iran, song phrases louder and less plaintive: 'koooi pi-ri-tlu-ee' (J Hall-Craggs), sometimes recalling bulbul *Pycnonotus* (Fig II) and sometimes busier with character more like Blackcap *Sylvia atricapilla* (Fig III). This trend apparently continues eastward into Afghanistan from where recordings comprise busier, harsher phrases, e.g. 'ki pui tee che-ri-oo chi chi' (Fig IV), resembling Whitethroat *Sylvia communis*

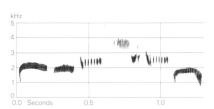

I   L Cornwallis   South-west Iran   May 1971

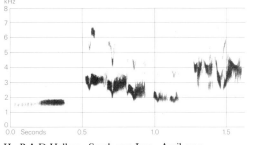

II   P A D Hollom   South-east Iran   April 1972

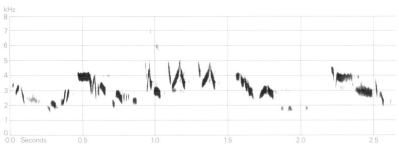

III   P A D Hollom   South-east Iran   April 1972

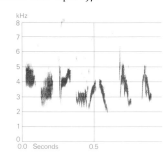

IV   C Chappuis   Afghanistan   May 1977

or Eastern Pied Wheatear *O. picata* (L Cornwallis, J Hall-Craggs, P J Sellar). (2) Threat-calls. (a) Regularly repeated harsh 'chak chak chak' sometimes given if territory infringed (Gallagher and Woodcock 1980; L Cornwallis). (b) Once during territorial encounter with *O. xanthoprymna*, bird hovered over it uttering a fine 'wioo wioo wioo' (L Cornwallis). (3) Alarm-calls. When nest or young threatened, harsh grating calls reported from India (Baker 1924), quiet 'chit-tit-tit' from Oman (Walker 1981*b*), and quiet, almost musical 'ti-ti-te' from Iran (P A D Hollom). (4) Other calls. Short, sharp, high-pitched whistle uttered 3-4 times (Baker 1924); 'teriki-treek trooti-trooti-tree', etc. (Gallagher and Woodcock 1980).

CALLS OF YOUNG. Food-call of fledglings a regular, rasping 'dzee dzee dzee', accentuated when adult arrives (L Cornwallis, P A D Hollom: Fig V).   LC

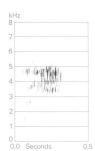

V   P A D Hollom   Iran   April 1971

Breeding. SEASON. Oman: fledged young seen late March and mid- to late April (Walker 1981*b*). Iran: eggs laid April-May (Hüe and Etchécopar 1970); in south-west, fledglings seen from second half of April to second half

of May (Cornwallis 1975). SITE. In hole or crevice in rock face, sometimes high up in cliff, or in wall of old building. Nest: shallow cup of twigs and plant stems, plastered with mud containing limestone chips; poorly lined with grass, hair, and feathers; platform of small stones in front of nest, shelving towards entrance of hole; one platform weighed 910 g (Ali and Ripley 1973*b*). No platform at 2 nests in Sind, Pakistan (Meinertzhagen 1954). Building: no information on role of sexes. EGGS. See Plate 83. Sub-elliptical, smooth and glossy; very pale blue-white, unmarked or with a few speckles of pale red. 23·0 × 16·3 mm (20·3-25·0 × 15·5-17·1), *n* = 24; calculated weight 3·25 g (Schönwetter 1979). Clutch: 4-5. YOUNG. Cared for and fed by both parents.

Plumages. ADULT. Head, upperparts down to upper back, and upper wing-coverts deep glossy bluish-black. Rump, lower back, and upper tail-coverts white; white extends 31-50 mm (*n* = 20) from black of back to black of central pair of tail-feathers (t1). Chin and throat glossy bluish-black, remainder of underparts backwards to under tail-coverts white (sometimes slightly suffused cream-pink). T1 black for 35-43 mm, base white; other tail-feathers white, tips black or brown-black for 11-16 mm (slightly more, 13-23 mm, on inner web of t2 and outer web of t6); in fresh plumage, tail-feathers narrowly tipped white. Flight-feathers and greater upper primary coverts deep brownish-black, coverts faintly tipped white in fresh plumage, secondaries and inner primaries faintly edged grey along tips. Axillaries and under wing-coverts deep black. Sexes similar; worn plumage similar to fresh plumage, except for some pale edges wearing off. NESTLING. No information. JUVENILE. Like adult, but much duller, more greyish brown-black, less glossy and bluish; feathering loose and fluffy, especially on vent and under tail-coverts; greater upper primary coverts faintly edged brown (not white) along tip when fresh; tips of tail-feathers narrowly bordered greyish, buff-brown, or pale buff (not white).

FIRST ADULT. Like adult; juvenile flight-feathers, tail, and greater upper primary coverts retained, but brown and buff fringes worn off and then similar to adult, except for often relatively greater wear and slightly paler brownish-black general coloration.

Bare parts. ADULT, FIRST ADULT. Iris brown or reddish-brown. Bill black, mouth orange-yellow. Leg and foot brown-black or black. (BMNH.) NESTLING. No information. JUVENILE. Iris and bill black, mouth yellow, leg and foot blackish (Desfayes and Praz 1978; P A D Hollom).

Moults. ADULT POST-BREEDING. Complete; primaries descendant. June–August (Ali and Ripley 1973b). No moult up to late May in India (BMNH). In Iran, no moult in early June, but birds in moult recorded 26 June, and in last stages of moult 18 July and 18 August (Vaurie 1949); one from late July had moult just completed (BMNH). In south-west Iran, moult recorded from end of April to end of June (Cornwallis 1975; L Cornwallis). POST-JUVENILE. Partial: head, body, and lesser and median upper wing-coverts; juvenile flight-feathers, tail, and greater upper primary coverts retained; no information on whether tertials or some or all greater upper wing-coverts replaced. In Iran, moult completed in one from mid-June (BMNH), another bird from mid-July still fully juvenile (ZFMK).

Measurements. ADULT, FIRST ADULT. Eastern Iraq, Oman, and Iran, October–July; skins (BMNH, ZFMK). Bill (S) to skull, bill (N) to distal corner of nostril; exposed culmen on average 5·1 less than bill (S).

	♂			♀		
WING	104·8	(2·47; 12)	102–111	99·7	(1·96; 10)	97–104
TAIL	66·2	(2·90; 12)	62–71	63·3	(1·93; 9)	60–66
BILL (S)	20·4	(0·86; 12)	18·8–21·5	20·0	(0·86; 9)	18·9–21·6
BILL (N)	11·5	(0·65; 12)	10·6–12·4	11·5	(0·67; 9)	10·7–12·4
TARSUS	27·0	(0·76; 11)	26·0–28·2	26·2	(0·73; 9)	25·4–27·4

Sex differences significant, except for bill. 1st adult and older birds combined, as retained juvenile wing of 1st adult on average only 0·4 shorter than full adult and tail 0·9 shorter.

Weights. Iran: April, ♂ 23; June, ♂ 26, ♀ 26 (Paludan 1938; Desfayes and Praz 1978); breeding, ♂ 23–27·5 (6), ♀ 23–27 (5); winter, ♂ 24–28·5 (8), ♀ 22–26·5 (7) (Ali and Ripley 1973b); ♂ 25·6 (1·5; 12) 23–28·5, ♀ 24·8 (1·9; 15) 22–28 (Cornwallis 1975). Oman, April: ♀ 23 (M D Gallagher, BMNH).

Structure. Wing rather short, broad at base, tip rounded. 10 primaries: p7–p8 longest, p9 5–8 shorter, p6 0·5–2, p5 5–7, p4 12–15, p1 22–28. P10 reduced, 48–56 shorter than wing-tip, 3–8 longer than longest upper primary covert. Outer web of p5–p8 and (faintly) inner of p6–p9 emarginated. Tertials short. Tail rather short, tip square; 12 feathers. Bill rather short, straight and slender, tip of culmen fine, slightly decurved. Long and stiff bristles at base of upper mandible and near nostrils. Leg and foot rather short, heavy. Middle toe with claw 18·8 (10) 18–20; outer toe with claw c. 72% of middle with claw, inner c. 70%, hind c. 84%. Claws rather short, strong, decurved.

Geographical variation. None.

Forms superspecies with White-crowned Black Wheatear O. leucopyga. CSR

Oenanthe leucopyga White-crowned Black Wheatear

PLATES 57, 64, and 65
[between pages 736 and 737, and facing pages 761 and 856]

DU. Witkruintapuit   FR. Traquet à tête blanche   GE. Saharasteinschmätzer
RU. Белогузая каменка   SP. Collalba negra de Brehm   SW. Vitkronad stenskvätta

Vitiflora leucopyga C L Brehm, 1855

Polytypic. Nominate *leucopyga* (C L Brehm, 1855), desert zone of northern Africa west of Nile valley, grading into *ernesti* in Nile valley and north-east Sudan; *ernesti* Meinertzhagen, 1930, Egypt east of Nile valley, Sinai, and Arabia north to Dead Sea area.

Field characters. 17 cm; wing-span 26·5–32 cm. Somewhat less plump than Black Wheatear O. leucura but with slightly longer, more pointed wings; distinctly larger than Eastern Pied Wheatear O. picata, with 15% longer wings. Rather large, oval-headed and oval-bodied, glossy black wheatear, with bold white rump and tail often showing only black central line. Adults of both sexes have white crown; juvenile and 1st-year have all-black head. Call distinctive. Sexes similar (but see Plumages); no seasonal variation. Juvenile separable. 2 races in west Palearctic, extremes separable in the field.

ADULT. (1) Race of North Africa east to Nile, nominate *leucopyga*. Largely black, glossy and faintly bluish. Crown white (with black reaching up to forehead and narrow area over eye), occasionally speckled black. White on rump and under tail-coverts, reaching forward to thighs and area between legs (thus more extensive than on O. leucura). Tail white with black central feathers and usually only small black marks on tips of outer feathers. Underwing black, with no obvious contrast. Bill long but more slender than in O. leucura, black; legs black. (2) Race of North Africa east of Nile and of Arabia, *ernesti*. Black plumage more heavily glossed blue or purple-blue than in nominate *leucopyga*; tail usually with obvious black spots on tips of outer feathers, noticeable in close view unlike those of nominate *leucopyga* (but not forming terminal band as in O. leucura or O. picata). Bill noticeably longer than that of nominate *leucopyga*. JUVENILE. Both races. Closely resembles adult but head all-black; plumage less glossy. Bill horn-brown, legs slate-grey. FIRST WINTER

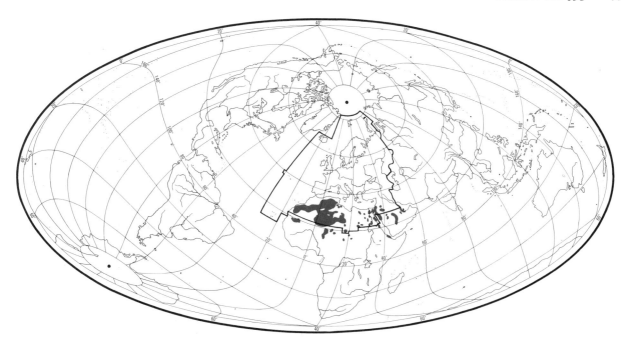

AND SUMMER. Both races. As adult except for all-black head and duller wings. Birds with mixed black and white crown are 1st-years—either undergoing normal moult into adult plumage or having grown some adult feathers after accidental feather loss prior to moult; all birds with all-black crown are juvenile or 1st-year (*contra* Brown 1986*b*).

Differs distinctly in underwing (all-black) and tail pattern (black restricted almost wholly to centre) from both *O. leucura* and *O. picata*; also much larger than *O. picata*. Flight free and floating, even recalling Swallow *Hirundo rustica* in its easy tumbles and sweeps along hillsides and over adjacent desert. Wings look less broad than in *O. leucura*. Escape-flight usually quite long; less often ascendant than *O. leucura*. Gait and behaviour much as *O. leucura*. Tame around human habitation but shy in desert.

Song loud, combining musical warbles with more discordant chanted notes like *Turdus* thrush. Calls include quiet 'trip-trip' (unlike piped tri-syllable of *O. leucura*); in alarm at nest, a rattled 'rrrrr'.

Habitat. Across Afro-Arabian lower middle latitudes, Mediterranean to subtropical and tropical. A true Saharan species characteristic of desert with less than 100 mm annual precipitation. Frequents the most impoverished localities, at all altitudes up to 3000 m in Ahaggar (Algeria) and Tibesti (Chad), especially rocky and sometimes earthen banks of wadis, but also oases. Comes within 100 km of Atlantic coast only at one point in southern Morocco (Heim de Balsac and Mayaud 1962), occurring in this region exclusively in rocky places and

at settlements; common on slopes but also reported on level ground 1 km from rocks; also on ruins (Valverde 1957). Sole passerine seen in 160 000 km² of low desert east of Aïr (Niger) and sole resident passerine at Egyptian desert oases; at home in desert wherever there is broken ground, and frequents houses in oases and cemeteries on their edges (Moreau 1966). In Chad, very common in semi-desert; in Eritrea (Ethiopia) on lava fields (Moreau 1972). In eastern Saudi Arabia, confined to cliffs and rocky hills with gullies, scree, or broken rocks (G Bundy). Found in the most inhospitable regions although locally becomes a village and house bird; normally requires rock crevices, banks, or spaces under boulders for nesting (Meinertzhagen 1954).

Distribution. No information on range changes.

ALGERIA. Found everywhere except the ergs proper (EDHJ). CHAD. Common south of Libyan border (Guichard 1955).

Accidental. Britain, Malta, Cyprus, Kuwait.

Population. No information on trends.

MOROCCO. Common in some areas in south (Valverde 1957). LIBYA. Common in more arid areas of Tripolitania and Fezzan; status in Libyan Desert uncertain (Bundy 1976). EGYPT. Locally common (SMG, PLM, WCM).

Movements. Largely sedentary throughout range, though some individuals or populations may make short-distance movements in winter: thus in some regions of Tunisia apparently recorded most frequently September–February which suggests breeding elsewhere (Thomsen

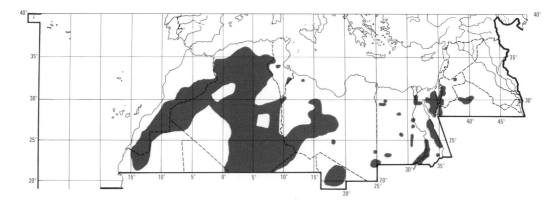

and Jacobsen 1979), and at Helwan (Egypt) present only in winter (Hartley 1949). Single records in Cyprus, March (Flint and Stewart 1983), and Malta, April (Sultana and Gauci 1982). Occasional records in Oman, Kuwait, and Qatar are in areas where breeding could occur and need not necessarily indicate movement within Arabian peninsula.   MGK

Food. Mainly insects, but diet notably diverse, including plant material and small reptiles. Catches prey in flight, on ground, or in bushes. Typically perches on low vantage points (shrub, stone, etc.) and drops down or sallies forth, up to 10(-50) m away, to take prey from ground in manner of a shrike *Lanius* (Valverde 1957; Gaston 1970; J Palfery); at moment of capture, frequently spreads wings, sometimes repeatedly, perhaps to confuse and entrap prey (J Palfery). Also sallies, sometimes steeply upwards, to catch aerial prey, or hops through foliage, gleaning insects or plucking berries (Geyr von Schweppenburg 1918; Fischman 1977; J Palfery); once hovered at branch (J Palfery). At other times, feeds on ground like Hoopoe Lark *Alaemon alaudipes*, tossing sand sideways and unearthing beetle larvae (Valverde 1957). Caterpillars and large beetles (Coleoptera) may be battered before eating, others (perhaps distasteful) dismembered (Brown 1986; J Palfery). In Niger, often attacks geckos (Gekkonidae), first tearing off and eating tail (Villiers 1950). In Israel, gleans ticks (Acarina) and other parasites from camels or their droppings (Fischman 1977). Around human habitations, takes flies and household scraps (Arnault 1926; Meinertzhagen 1954; Fischman 1977), e.g. seen pecking at rotten pomegranate *Punica* to extract seeds (Gurney 1871). In Saudi Arabia, feeding begins around sunrise and continues until *c.* 20 min after sunset; most active during first 3 hrs and last 2 hrs of day. Changes perch-site frequently: e.g. during 14 min, average 70 s (maximum 2 min 24 s) on any perch; at sunset, feeds hectically, flitting from bush to bush and scarcely pausing on perches. (J Palfery.)

The following recorded in diet in west Palearctic. Invertebrates: grasshoppers, etc. (Orthoptera: Acrididae, notably *Schistocerca gregaria*), aphids (Aphididae), adult and larval Lepidoptera, flies (Diptera), Hymenoptera (ants Formicidae, wasps Vespidae), adult and larval beetles (Tenebrionidae), ticks (Acarina) (and other parasites), worms (Oligochaeta). Also household scraps, e.g. small pieces of meat, crumbs, and flour. Plant material mostly fruits and soft seeds: of date *Phoenix*, pomegranate, sumach *Rhus oxycantha*, grasses (Gramineae). (Gurney 1871; Geyr von Schweppenburg 1918; Arnault 1926; Meinertzhagen 1940, 1954; Valverde 1957; Fischman 1977.) In Niger, also geckos (perhaps *Tarentola*) and viper *Cerastes* (Villiers 1950); in Sénégal, small lizards (Heim de Balsac and Heim de Balsac 1954).

In central Morocco, November, stomachs of 4 ♂♂ and 1 ♀ all contained ants, mixed with remains of beetles and grasshoppers (Meinertzhagen 1940). In Niger, July–August, stomachs contained flies, ants, and grass seeds (Villiers 1950). In Sinai, spring, diet largely insects, including caterpillars (Fischman 1977).

Little information on diet of young. In Saudi Arabia, nestling was fed on adult and larval Lepidoptera, mainly in early morning and late afternoon (J Palfery).   EKD

Social pattern and behaviour. Important studies by Pasteur (1956) in Morocco and, especially, Fischman (1977) in Sinai. Account includes detailed notes supplied by J Palfery for eastern Saudi Arabia.

1. Not markedly gregarious, but, like Black Wheatear *O. leucura*, young may remain with parents on parental territory during winter, members of family party feeding quite close together; in spring, young leave voluntarily or are evicted (Fischman 1977; see Antagonistic Behaviour, below). In other studies, winter territories apparently lack offspring, being occupied by pairs (e.g. Tristram 1859; see also Bonds, below) or single birds (Simmons 1951c; J Palfery). According to Smith (1971), 2(-3) birds together usually seen in winter, Morocco. Winter territory not sharply demarcated (Simmons 1951c). In Saudi Arabia, winter territories, held by single birds (apparently of either sex), evidently much smaller than, but partly including, those used in breeding season (see Breeding Dispersion, below); in October, combined extent of 2 adjoining territories 4·5 ha, a 3rd occupying *c.* 140 m of wadi of maximum width *c.* 40 m (J Palfery). In Egypt, when Blue Rock Thrush *Monticola solitarius* shifted territory, neighbouring *O. leucopyga* expanded its territory to include part of vacated area (Simmons 1951c). At Tamanrasset (Algeria) in winter, *c.* 60 birds in 2 km² oasis

(Gaston 1970). BONDS. No evidence for other than monogamous mating system. Pair-bond life-long (lasting until one partner dies or is evicted by rival); if mate dies, partner holds territory until it finds a new mate (Fischman 1977). Pair-bond maintained all year according to Fischman (1977), though this may vary regionally (see above). ♀ alone broods, both sexes feeding young. One record of 1st-brood young regularly visiting 2nd brood in nest, apparently helping with feeding and nest-sanitation. (J Palfery.) Young fed for *c.* 3 weeks after leaving nest. Age of first breeding 1 year, at least for ♀♀. (Fischman 1977.) BREEDING DISPERSION. Solitary and territorial. In A-Rabba plain (Sinai), spacing clearly determined by dispersion of human dwellings, every cluster of houses forming nucleus for a territory; topographical features (e.g. undulations, wadis) usually form natural territorial boundaries, and where no such features exist, boundaries are more contested (see Antagonistic Behaviour, below). In A-Raha plain (Sinai), *c.* 10–12 territories of about equal size in 2 km², average *c.* 0·4 ha each. (Fischman 1977.) In Saudi Arabia, territories much larger, 2 each at least *c.* 7 ha, another *c.* 9 ha (J Palfery). In Morocco, density 0·6 pairs per km², and, in transect counts, up to 1 bird per km (Blondel 1962a, which see for variations with habitat). Territory serves for nesting, roosting, and some feeding (J Palfery), but food also sought outside territory, e.g. pair in Jordan defended area up to 135 m from nest, but fed up to 150 m from it (Wallace 1984b). In Saudi Arabia, pair began to feed well outside territory after young hatched, at least doubling their former feeding area (J Palfery). In Pasteur's (1956) study, rather small, compressed territory thought to be used mainly for courtship, and defended by ♂, ♀ nesting *c.* 150–200 m away in buildings, and pair feeding elsewhere on common ground. In A-Raha also, neighbours sometimes fed together on neutral ground. New nest may be built for 2nd clutch while 1st brood still being fed. (Fischman 1977.) Nest-sites apparently re-used in successive years, but bird builds (or re-builds) new stone platform each year (George 1978: see Heterosexual Behaviour, subsection 3, below, also Breeding). ROOSTING. Nocturnal. Birds apparently roost alone, pair-members sometimes far apart. Typically under stones or in high narrow rock crevice on slopes within territory. (Fischman 1977.) Once eggs laid, incubating ♀ roosts on nest, until young a few days old (J Palfery). Birds approach rocky roost-sites at dusk, hopping from rock to rock (Fischman 1977) or flitting between bushes, entering site *c.* 20–30 min after sunset when almost dark (J Palfery). During hottest hours in middle of day, seek shade under bushes (Gaston 1970), boulders, ledges (etc.), where they loaf, preen, sleep, and sometimes feed; Subsong (see 2 in Voice) also heard. Pair-members sometimes loaf together but usually apart. (J Palfery.) In central Ahaggar (Algeria), winter, birds drank at pool most often between 07.00 hrs (1 hr after sunrise) and 11.00 hrs, less often after midday (Gaston 1970). In captivity, regularly drinks and bathes (Arnault 1926).

2. Often shy and wary in wilderness areas, shunning close approach, but birds frequenting human habitation quite tame, even entering buildings (Whitaker 1905; Laenen 1949–50; Meinertzhagen 1954; Arnould 1961). In Ennedi (Chad), curious by nature, approaching humans to watch them pass by (Gillet 1960). Commonly also attracted by unusual objects: e.g. pair in Morocco nearly flew inside car, and greatly interested in a slice of lemon thrown out by driver, 1 bird repeatedly 'wing-fanning' at it (Smith 1971). When man near roost-site at dusk, bird approaching was wary and gave alarm-calls (see 3a in Voice); landed in front of intruder, bobbing head downwards, cocking half-spread tail, and wing-flicking to expose rump (J

Palfery). In November, bird which had been giving Subsong flew towards and landed near approaching fox *Vulpes* and continued Subsong, adopting Crouch-posture (Fig A): bird

A

crouches, slightly lowers head, spreads and lowers tail, and ruffles rump plumage (G Bundy, J Palfery). On spotting raptor, bird instantly flees to rock, crouching in shadowed crevice (George 1978). FLOCK BEHAVIOUR. For family parties, see Introduction, part 1. No further information. SONG-DISPLAY. ♂ usually sings (see 1 in Voice) from top of a bush, cliff, etc. (e.g. Whitaker 1902, Geyr von Schweppenburg 1918), accompanied by spasmodic movements causing forebody to rock (Germain 1965). Singing ♂ also reported Bill-snapping (Fischman 1977). Each ♂ usually has 3–4 favourite song-perches in his territory. Also sings while flying normally between perches (J Palfery), and during Song-flight: bird launches itself from perch where it has been singing, and glides to another perch on stiff, sometimes quivering wings, with tail spread (Jennings 1980); also described as a slow, showy glide from one eminence to another, wing-tips curled upwards, tail spread, and legs dangling (Fig B). During one such flight of *c.* 40 m,

B

song-phrase given 3 times. (J Palfery.) In open desert, where perch-sites lacking, displaying bird first flies to suitable height, then parachutes down. On alighting after Song-flight, adopts Crouch-posture (Fig A) for a few seconds. (Jennings 1980.) Perched singing bird may also perform Crouch-display: shuffles or scurries forwards, singing the while, in Crouch-posture, dragging spread lowered tail across ground. Bird once then made short flight with rump feathers still raised, tail spread, and with wings almost horizontal and just flicked. (J Palfery.) In spring and summer, ♂♂ sing loudly from early morning, neighbours conducting Song-duels (Fischman 1977; J Palfery). In Algeria, February, began singing at 04.45 hrs (1¼ hrs before sunrise), strongly from 05.00–06.00 hrs; lull during day, with resurgence in the evening until 1 hr after sunset (Gaston 1970). Singing at night reported by Guichard (1955) and Panow (1974). In most places, birds begin singing in winter and continue until autumn; e.g. in Israel, begin at end of December and continue until September, ♀♀ and young also singing a little in winter (Fischman 1977). However, no winter song heard in Morocco by Smith (1971). In Saudi Arabia, song declined markedly from mid-April, apparently with progressive incubation, and virtually ceased after hatching; Subsong heard at various times of year (J Palfery). ANTAGONISTIC BEHAVIOUR.

Aggressive in defence of territory throughout the year but especially during breeding season. Outside breeding season, residents spend a lot of time standing guard on favoured vantage points in their territory (J Palfery). In winter, reported confronting and/or driving off Mourning Wheatear *O. lugens* (Niethammer 1954; J Palfery), Hooded Wheatear *O. monacha*, and *M. solitarius* (Simmons 1951*c*). In breeding season, aggressive towards conspecific birds, especially neighbours, also various other wheatears, larks (Alaudidae), and once each towards *M. solitarius* and House Sparrow *Passer domesticus* (J Palfery). During breeding season, defence mainly by ♂, sometimes apparently aided by ♀ (Pasteur 1956). ♂ advertised by singing (see above), also by 'beating the bounds' of his territory (George 1978). If trespasser approaches along ground, resident also drops to ground to await it; if rival comes to within ½–3 m, resident starts giving harsh alarm-calls and gaping widely. At close quarters, resident described as standing upright on tiptoe, beating his wings, and calling bill to bill. (Pasteur 1956.) Threatens rival with 2 displays (J Palfery): Crouch-display, sometimes accompanied by harsh alarm-calls, once apparently also some low clicking notes (see 5a in Voice); for calls associated with possibly aggressive Crouch-display outside breeding season, see 4a and 5b in Voice. Resident may also perform Advertising-display (Fig C): spreads and raises tail, bows

C

forwards to expose white crown, and shuffles around on the spot, presenting his tail in various directions. Advertising-display and other ♂–♂ confrontations not uncommonly lead to pursuit-flights, accompanied by excited calls (see 5a–b in Voice). When rivals face each other, one sometimes adopts Sleeked-upright posture (Fig D): stands very upright, bill pointing skywards,

D

and sleeks plumage, effecting slim appearance (J Palfery), perhaps signalling appeasement or submissiveness. Confrontation eliciting this display may be accompanied by various calls (see 5b, 5c, and 5e in Voice), though not known which bird gives them (J Palfery). Resident almost always dominant in disputes with rival who thus retreats, though neighbours occasionally fight at territorial boundaries (Fischman 1977), e.g. 2 ♂♂ rose squabbling *c.* 2 m into the air in fluttering flight (J Palfery). Fighting with non-conspecific intruders also occasionally reported (Hartley 1949; Simmons 1951*c*). Pair sharing winter territory with offspring forcibly evict them if reluctant to leave in spring, pecking and harassing them. Young may inherit a small part of natal territory, or occupy a vacant one, sometimes living on neutral ground until a vacancy arises nearby; may also join a single resident (of opposite sex) or sometimes evict a resident. (Fischman 1977.) HETEROSEXUAL BEHAVIOUR. (1) Pair-bonding behaviour. In Saudi Arabia, song, display, and pair-formation begin mid-January (G Bundy); in Morocco, 'sexual activity' begins in first 2 weeks of February (Blondel 1962*a*). In Israel, even earlier nest-prospecting (see subsection 2, below) suggests established pairs may start breeding preliminaries in winter. Precise sequence of pair-formation not known, but ♂'s display near ♀ may include Advertising-display (Wallace 1984*b*), Crouch-display, and perhaps a dipping-and-looping flight among the boulders where ♀ perched. Once paired, ♂ often gave low Subsong in presence of mate. Paired birds usually feed close together. (J Palfery.) (2) Nest-site selection. Following account after Fischman (1977). Both ♂ and ♀ began exploring likely nest-sites in their territory from the beginning of November. Not known which sex selects site, but ♂ seen entering a crevice several times and, on emerging, fluttering about opposite the opening, or standing near it and singing (perhaps Nest-showing). (3) Rock-carrying. Account based on Fischman (1977). Once nest-site chosen, ♀ begins bringing stones in her bill to build paved platform in approach to nest, sometimes extending outside nest-hole entrance. One ♀ brought 15–20 stones in 20–30 min, rested for 30–60 min, then continued construction. Carrying continues throughout daylight hours, with longer rests around midday. Older ♀♀ bring larger stones than younger ones (see Fischman 1977 for details, also Breeding), and building is perhaps a demonstration of ♀'s reproductive fitness to ♂ who typically stays nearby, singing. According to George (1978), which see for details, platform thought to cool nest by trapping air and dew. ♀ builds a more substantial platform for 1st than for subsequent nests of season (Fischman 1977). (4) Courtship-feeding. None reported. (5) Mating. ♂ apparently takes initiative, not uncommonly soliciting ♀ during pauses in her platform/nest-building activity (Fischman 1977). In one case, ♂, who had been singing, flew and landed near nest-building mate, adopted a low posture, and circled her excitedly 3–4 times. ♀ seemed to ignore ♂ who then tried to mount her; ♀ then flew off, ♂ following. (J Palfery.) Low posture also described by Pasteur (1956, which see for drawing): ♂ stands with body almost horizontal, his bill a few cm from ♀ who watches him passively. ♂ gives a harsh feeble call, barely opening his bill. After this brief encounter, either the 2 birds separate or ♂ attempts copulation, his wings beating for balance. After copulation (or attempt) ♀ flies off and ♂ follows playfully, the 2 birds describing tortuous flight-paths (perhaps dipping-and-looping described above). Display seems to be highly stimulating to neighbours who may join the chase, perhaps sometimes because displaying pair have infringed their territories. (Pasteur 1956.) (6) Behaviour at nest. Typically, ♀ builds nest and incubates, ♂ keeping away from nest until hatching

(Fischman 1977; J Palfery), but in one case ♂ recorded building (Thévenot *et al.* 1982). ♀ alternated incubation spells (up to 1 hr) with feeding breaks of 10–20 min. Once, when ♀ flew off nest, ♂ loafing nearby gave reedy calls (see 5c in Voice), followed by a short warble; as ♀ flew past him, he called again and 'shimmied' his shoulders and body from side to side. (J Palfery.) RELATIONS WITHIN FAMILY GROUP. At outset, ♀ broods day and night, less so as young develop, ceasing entirely by *c.* 7–9 days. Young begged to human intruder by gaping with noisy food-calls (see Voice). Both sexes feed young. (J Palfery.) Fledged young attended by one or both parents (Pasteur 1956). One case of adult tending 3 young, not well feathered, learning to fly (Pasteur 1956), suggests young may leave nest before fully fledged. ANTI-PREDATOR RESPONSES OF YOUNG. 6-week-old young sought refuge from human intruders in a rat-hole (Lynes 1925*a*). No further information. PARENTAL ANTI-PREDATOR STRATEGIES. (1) Passive measures. Secretive near nest, e.g. birds carrying nest-material dropped it when they spotted human intruder, then flew well beyond nest-site (Koenig 1924). (2) Active measures: against birds. ♀ once flew towards and swooped at Isabelline Shrike *Lanius isabellinus* in territory (J Palfery). (3) Active measures: against man and other animals. Pair studied in Saudi Arabia not very demonstrative towards man (J Palfery), but, in Israel, birds mob man (especially when handling young), snakes, and cats, indeed anything unfamiliar in territory: flutter closely and excitedly around source of disturbance, giving alarm-calls, frequently attracting other birds including other *Oenanthe* (Fischman 1977). In Morocco, ♀ flushed from nest flutters around it giving soft calls (Pasteur 1956). In Algeria, when fluttering young caught by human intruders at nest and released, food-bearing parents called incessantly as they flew overhead, apparently trying to lure young away from nest (Arnould 1961). In Egypt, March, bird hovered over, and made dive-attacks on snake *Psammophis sibilans*, attracting another *O. leucopyga* and several Trumpeter Finches *Bucanetes githagineus* (Koenig 1924).

(Figs by D Rees: from drawings by J Palfery.)   EKD

Voice. Complex repertoire, used for much of year. No complete study, though following account includes detailed notes supplied by J Palfery for eastern Saudi Arabia. Some indication of regional differences in song, but no precise information. Since mimicry common in song, and probably in calls also, some sounds evidently vary geographically. Calls, perhaps partly due to mimicry, remarkably diverse and difficult to classify (J Palfery), and some renderings (below) from different sources and attributed to same call may be genuinely different calls. Also, since calls readily incorporated into song, these not always easy to distinguish from song. Function of calls still harder to determine, many being given in apparently similar contexts, and division (especially 4–5, below) of alarm and threat contexts somewhat arbitrary, several calls applying to both. Bill-snapping sounds heard from ♂ during his Advertising-display (Fischman 1977), but some caution required as certain calls have clicking quality.

CALLS OF ADULTS. (1) Song of ♂. Loud, combining musical warbles with more discordant chanted notes like *Turdus* thrush (Wallace 1984*b*; D I M Wallace). Not unlike Black Wheatear *O. leucura* but higher pitched and less powerful (J-C Roché). Almost always called a warble,

but descriptions vary from pleasant or pleasing (Whitaker 1902; Serle 1943) in Algeria and Libya, to discordant in central Arabia (Jennings 1980). However, one individual might easily embrace these extremes, since song highly variable within as well as between individuals (Fischman 1977; J Palfery). According to J Palfery, recording of song from Morocco more fluid, melodious, complex, and continuous than any heard in Saudi Arabia. In Saudi Arabia, 1 ♂ had at least 4 songs, another had 11 (J Palfery). In Egypt, winter, at least 10 different song-types from one ♂ included a variety of whistles, pipings, chuckles, warbles, and combinations of these (Simmons 1951*c*). In Israel, song comprises many different phrases, frequently including mimicry of (e.g.) Scarlet Rosefinch *Carpodacus erythrinus*, larks (Alaudidae), partridge (presumably *Alectoris*), Tristram's Grackle *Onychognathus tristramii*, crows (Corvidae), goats, cats, dogs, braying donkey; will imitate human whistle, faithfully copying its intonations and trills (Fischman 1977). Mimicry, notably of larks, also reported in Saudi Arabia (J Palfery). Following details of song from J Palfery. Usually consists of fairly simple short phrases, each of 2–5 units. Phrases may be repeated but birds often switch from one to another, and also vary given phrases. One ♂ repeated apparently same phrase every 5–6 s during 2 min. Some songs, with slight variation, shared between individuals: (a) the most distinctive a rounded ringing almost bell-like 'tchulaleet tchulateeter'; (b) rapid lively 'twerchweeperdo twerchweeperdo', sometimes including a few squeaky notes and a rapid rattle similar to song of Lesser Whitethroat *Sylvia curruca*; (c) shrill, almost wavering squeal (given when disturbed and thus perhaps call 5b or 5c), subsequently expanded by adding some strong, rounded, rich fluting notes, entire sequence 'see-ee-ee-treeiu-pee-eeo'. 4 other songs selected here, from many described by J Palfery, to emphasize diversity: from one ♂, 'chip-sweear-chip', 'keeyar keeyar swee-whit' (first units having harsh quality of distant Herring Gull *Larus argentatus*), and a vibrant twanging 'trrue tructructruc'; from another ♂, a sweet high-pitched warble, 'te-weeti-te-weet te-weeti-teweeter'. Figs I–III, all from one bout of singing by one bird, show remarkable variation between phrases (which are often contiguous), e.g. 2nd phrase (Fig I) in recording shows short ascending scale near end; after pause of *c.* 5 s, 3rd phrase (Fig II) has buzzy rattling quality, upper limit of rattle units (2nd and 5th in sonagram) exceeding 7 kHz; 5th phrase (Fig III) essentially tonal (J Hall-Craggs). Sonagram of recording by P A D Hollom, Morocco, is not unlike Fig III. (2) Subsong of ♂. Weak, but not unmusical, perhaps more fluid and continuous than full song; one bird included rather thin, high-pitched whistles, some fluted warblings, and repeated notes, the whole reminiscent of Blackbird *T. merula* (J Palfery). ♂ which delivered song depicted in Figs I–III broke off into a soft low warbling Subsong; later, switched to perhaps loud Subsong which

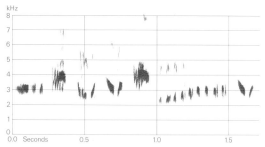

I E D H Johnson Algeria February 1968

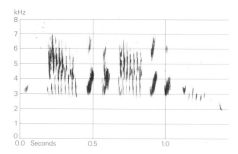

II E D H Johnson Algeria February 1968

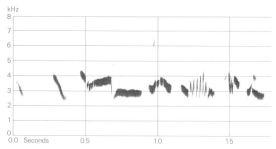

III E D H Johnson Algeria February 1968

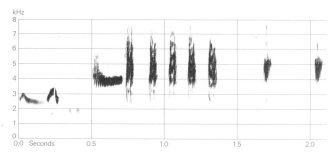

IV E D H Johnson Algeria February 1968

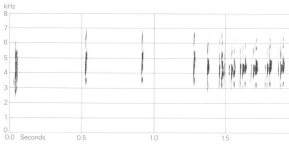

V E D H Johnson Algeria February 1968

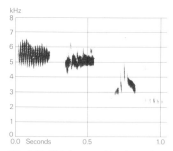

VI E D H Johnson Algeria February 1968

included various calls (see Figs IV–VI, below) and possible mimicry (J Hall-Craggs). Subsong once given by bird displaying at a fox *Vulpes* (G Bundy). (3) Contact-alarm calls. Commonest call a short, harsh, rather grating 'dzik', given by both sexes, typically just before going to roost, sometimes when alarming at human intruder. Also during territorial disputes, e.g. a soft 'zt zt'. Other renderings of probably same call a repeated scratchy 'zick' and a rather grating 'dzt'. (J Palfery.) In Fig IV, possible Subsong seems to include calls of this type, or similar: after fragment of song, 5 'tzet' units, then 2 'zrit' units (J Hall-Craggs). A quiet rather low 'trip-trip' (Wallace 1984*b*) perhaps belongs here. (4) Alarm-calls. (a) A repeated, strong, far-carrying 'hwee-weet', given by ♀ when human intruder approached nest; 'weet weet' and 'hwick hwick' in territorial disputes, latter by bird in Crouch-display and again when flying off (J Palfery). (b) Scolding ♀ followed call 4a with a grating 'jak jak' or 'drak drak'. On other occasions, a harsh 'jat' or 'jak-rat', a scratchy 'jaik' or 'jraik', a low screech 'jeek' or 'jreek' given by bird flitting ahead of human intruder.

Perhaps a drawn-out variant of call 3a. (J Palfery.) (c) A grating 'chairz', like a slightly longer, more slurred version of call 3a, given by bird on spotting human intruder. Low harsh 'chair' sometimes accompanies threatening Crouch-display. (J Palfery.) (d) A low rattle 'r-r-r-r-r' given in alarm when human intruder approached nest (Wallace 1984*b*). A rather squeaky 'trr-trr-trr...' (Fig V, rapidly repeated terminal units: J Hall-Craggs), incorporated in possible Subsong, perhaps this call. (5) Calls associated with territorial disputes (see also calls 3, 4a, and 4c). (a) A high-pitched jingling mixed with clicks when 2 birds faced each other, one in Sleeked-upright posture; at other times, notably in pursuit-flights, a series of low clicks (J Palfery). In Fig V, first 5 units are very short clicks, sounding rather like pebbles of different size being tapped together (J Hall-Craggs), and may be like the clicking part of this call. (b) High-pitched squeaks and squeals may accompany perched confrontations and pursuit-flights. A harsh, cat-like 'weeah' given by same bird (and in same encounter) that gave call 5a, also by a bird in apparent display-flight, perhaps in response to human presence. (J Palfery.) (c) A single 'srrrit' like rippling the teeth of a comb; thin high-pitched 'seet' when chasing Isabelline Wheatear *O. isabellina* possibly the same, also 'sree' (Hollom 1959), and a reedy 'zree-ree' from ♂ near mate when she flew off nest (J Palfery). In Fig VI, the 2 tonal units peaking at *c.* 6·5 kHz, and higher than any notes in song (J Hall-Craggs), may belong here; though pitch of the 2 units similar, quality quite different, the 1st showing rapid frequency modulation, the 2nd slow amplitude modulation (J Hall-Craggs). (d) A vibrant electronic buzz, associated with aerial chases (J Palfery). (e) In perched confrontations, creaking sounds like the conversational quacking of ducks (J Palfery).

CALLS OF YOUNG. Nestlings give cheeping food-calls (Fischman 1977). Well-feathered young in nest begged with noisy 'shreep shreep...' (J Palfery).　　EKD

Breeding. SEASON. North-west African Sahara: eggs found mid-February to late May, but main season probably mid-March to mid-April (Heim de Balsac and Mayaud 1962). Probably similar in Egypt and Sinai (Hüe and Etchécopar 1970). Jordan: young in nest and newly-fledged young seen in early May (Wallace 1984*b*). Eastern Saudi Arabia: eggs laid from mid-February (G Bundy). SITE. In hole (up to 25 cm deep) in rocks, under stones, in bank, or occasionally in wall of building. Nest: cup of dry grass, lined with wool and feathers, sometimes with base of twigs or bits of wood; approach often paved with platform of pebbles (see also Social Pattern and Behaviour); pebbles typically thin and flat, mean size 2 × 3 cm, weight 2–5(–10) g; 50–350 used, forming flat mosaic 3–4 layers deep and 10–40 cm wide (Fischman 1977; see also Geyr von Schweppenburg 1918, George 1978); platform not reported in all studies, and its inclusion apparently depends on availability of materials (Hüe and Etchécopar 1970). Building: both platform and nest mainly by ♀ (see Social Pattern and Behaviour). EGGS. See Plate 83. Sub-elliptical, smooth and glossy; very pale blue to whitish-blue, sparsely spotted red-brown at broad end. *O. l. ernesti*: 22·0 × 16·6 mm (20·2–23·4 × 13·5–17·2) *n* = 65; calculated weight 3·23 g. Nominate *leucopyga*: 21·6 × 15·7 mm (20·2–22·6 × 14·7–16·7), *n* = 9; calculated weight 2·85 g (Schönwetter 1979). Clutch: 3–5 (2–6). Of 31 clutches, north-west Africa: 3 eggs, 8; 4, 17; 4, 6; mean 3·9; varies with season and location (Heim de Balsac and Mayaud 1962). Sometimes double-brooded, pair sometimes building new nest or even incubating 2nd clutch while still feeding 1st brood (Fischman 1977; J Palfery). INCUBATION. By ♀, taking *c.* 2 weeks (Fischman 1977). Incubation spells of one ♀ 30–60 min (J Palfery). YOUNG. Cared for and fed by both parents. FLEDGING TO MATURITY. Fledging period *c.* 2 weeks, young becoming independent *c.* 3 weeks later (Fischman 1977). Age of first breeding 1 year, at least for ♀♀ (Fischman 1977).

Plumages (nominate *leucopyga*). ADULT MALE. Forehead and crown white, separated from base of upper mandible and from eye by narrow strip of black; occasionally, some black feathers mixed within white. Rump, upper and under tail-coverts, and broad tips on feathers of rear flanks, vent, and thighs white; remainder of head and body including lesser and median upper wing-coverts, axillaries, and under wing-coverts deep black, faintly glossy in some lights when plumage fresh. Tail white; distal half of central pair of feathers (t1) black for 40·9 (12) 27–48 mm (tip white for 3–6 mm when plumage fresh); t2–t5 fully white, except sometimes for small black spot or mottled grey patch on t2, rarely also on t3–t5; t6 either with black spot of *c.* 8 mm long near tip (usually on outer web), or with faint grey-mottled patch, or (in 6 of 12 examined) virtually white. Flight-feathers, tertials, greater upper wing-coverts, and greater upper primary coverts sooty greyish- or brownish-black, usually deepest and slightly glossy on outer webs and tips of feathers; greater coverts and primary coverts sometimes slightly fringed dark brown along tips. *In worn plumage*, black of body slightly duller and flight-feathers and primary coverts browner, in particular on tips. ADULT FEMALE. Virtually indistinguishable from adult ♂, but when series of both sexes compared (and perhaps also between partners of known pairs) ♀ on average slightly duller black, slightly browner and less glossy than adult ♂; t6 more often with black spot near tip (present in 7 of 9 examined), t3–t5 frequently with small black spots near tips; maximum extent of black on t1 49·0 (9) 44–54 mm. NESTLING. No information. JUVENILE. Rather like adult, but colour of forehead and crown dark, like mantle and sides of head, only rump, tail-coverts, vent, and tail-base white. Head, body backwards to back, flanks, and belly, and upper and under wing-coverts dull greyish-black (browner when plumage worn), distinctly less deep black than in adult and feathering softer and looser; belly and often some or all scapulars and lesser and median upper wing-coverts with narrow ill-defined white fringes or spots on feather-tips; tertials, secondaries, and inner primaries often narrowly fringed white along tips; greater upper wing-coverts and greater upper primary coverts narrowly tipped off-white or pale brown, tips sometimes broader towards tertial coverts and tending to be shaped as off-white triangular spots; black of flight-feathers and tail browner and less deep black than in adult. For tail pattern, see 1st adult below. FIRST ADULT MALE. Like adult ♂, but forehead and crown deep black with slight gloss (like mantle and sides of head); no white on head. Otherwise similar to adult ♂, but juvenile flight-feathers, tertials, tail, greater upper wing-coverts, and greater upper primary coverts retained, coverts distinctly browner and more worn than neighbouring 1st adult scapulars, median coverts, and (if new) tertial coverts; retained coverts and tertials often with traces of off-white fringes along tips and greater coverts sometimes with larger off-white triangles, but fringes sometimes brownish and narrow and then more difficult to age, apart from black crown. Maximum extent of black on t1 45·1 (24) 30–57 mm; t2 usually with paired black spots on tip, t3–t5 with black spot 3–6(–12) mm across on outer or both webs in about half of birds examined; t6 with black spot 4–12 mm long near tip of outer web, rarely with fully black subterminal band. FIRST ADULT FEMALE. Like 1st adult ♂: forehead and crown dark and juvenile tail, flight-feathers, and part of wing-coverts retained (unlike adult ♀). Black of head and body distinctly black-brown, less deep and glossy than 1st adult ♂ (difference visible already on growing feathers during post-juvenile moult); t1 black for 49·7 (12) 44–54 mm, t2–t6 with single or paired spots 5–12(–19) mm long near tips, as in 1st adult ♂, but t3–t5 fully white in 2 of 12 birds examined.

Bare parts. ADULT. Iris dark brown. Bill, leg, and foot black; soles blackish- or yellowish-horn. Inside of both mandibles greyish-black, mouth yellow. (BMNH, ZMA.) NESTLING. No information. JUVENILE. Iris brown. Bill horn-brown. Gape pink-yellow, including inside of mandibles; gape-flanges pale yellow. Leg and foot slate-grey. (Hartert 1910; RMNH, ZMA.) No information on when adult colours obtained.

Moults. ADULT POST-BREEDING. Based on *c.* 50 moulting birds (BMNH, ZFMK, ZMA). Complete; primaries descendant. In Ahaggar (southern Algeria), adult nominate *leucopyga* starts with shedding of p1 early April to early May, mainly *c.* 15 April; 1-year-olds start from late March, on average *c.* 20 days earlier than full adult (but sample small, only 4 1-year-olds in moult examined); moult completed with regrowth of p9-p10 in second half of July or first half of August (those completing early July probably 1-year-olds). In central Algeria and Morocco, moult starts about first half of May, completed early August. In *ernesti* from Sinai, moult half-way through (primary moult score 25) in July and first half of August, completed late August or September; probably starts June or early July, but none with primary score below 20 examined. Body and wing-coverts start at primary moult score 15-20: median upper wing-coverts first, followed by some feathers of forehead, mantle, and sides of breast, and a few scapulars and tertials; head, body, tertials, and wing-coverts largely completed at score *c.* 35. Start of tail rather variable; central pairs of feathers first, at primary score 0-15; completed with outermost at score 25-45. Secondaries moult during last stages of primary moult. POST-JUVENILE. Partial: head, body, lesser and median upper wing-coverts, usually tertial coverts, occasionally some inner greater upper wing-coverts, rarely a few tertials or tail-feathers. Starts soon after fledging and hence timing strongly variable. In Ahaggar, 35 birds examined from second half of April to late May comprised those in full juvenile plumage, birds in moult, and birds in full 1st adult at any date. In central Algeria and Sinai, birds in full juvenile encountered May-July, moulting birds June-August, and 1st adults from late May onwards; last birds completing moult in Sinai occurred late August. (BMNH, RMNH, ZFMK, ZMA.)

Measurements. Nominate *leucopyga*. Central and southern Algeria, all year; skins (RMNH, ZFMK, ZMA). Juvenile wing and tail include those of 1st adult. Bill (S) to skull, bill (N) to distal corner of nostril; exposed culmen on average 5·1 less than bill (S).

WING AD ♂	106·2 (1·88; 35)	102-110 ♀	99·6 (1·03; 8)	98-101
JUV	103·7 (2·06; 26)	99-107	97·7 (1·64; 16)	95-100
TAIL AD	68·6 (3·06; 14)	66-74	65·8 (—; 3)	64-67
JUV	68·4 (2·20; 24)	64-73	64·4 (1·86; 14)	61-67
BILL (S)	21·0 (0·67; 48)	19·6-22·4	19·9 (0·66; 18)	18·9-21·1
BILL (N)	11·7 (0·54; 21)	10·9-12·5	10·8 (0·18; 9)	10·6-11·2
TARSUS	27·3 (0·70; 35)	26·2-28·5	26·2 (0·55; 13)	25·5-27·1

Sex differences significant. Juvenile wing significantly shorter than adult. Nile valley of southern Egypt and northern Sudan: wing, ♂ 106·4 (7) 104-108, ♀ 100·3 (3) 94-105; bill (S), ♂ 21·5 (5) 20·9-21·8, ♀♀ 21·4, 22·2 (RMNH, ZFMK).

O. l. ernesti. Mainly Sinai, summer; some Dead Sea area, all year; skins (RMNH, ZFMK, ZMA).

WING AD ♂ 111·7 (2·53; 16) 107-115 ♀ 107·2 (1·96; 10) 104-111

BILL (S) 23·0 (0·57; 16) 21·9-23·9 22·2 (0·72; 11) 21·4-23·4

Sex differences significant.

Weights. Nominate *leucopyga*. Central and southern Algeria, late November-January: ♂ 29·6 (5·38; 5) 25-39, ♀ 25·0 (3) 24-26. Ennedi (northern Chad), May: ♂ 26. (ZFMK.)

Structure. Wing rather long, broad at base, tip rounded. 10 primaries: p7-8 longest or either one 0-1·5 shorter than other; p9 5-9 shorter than wing-tip, p6 1-2·5, p5 5-10, p4 10-18, p3 14-23, p1 21-28 (nominate *leucopyga*) or 28-34 (*ernesti*). P10 reduced, 52-62 (adult) or 48-57 (juvenile) shorter than wing-tip, 4·5 (12) 2-8 longer than longest upper primary coverts in adult, 5·7 (12) 4-10 in juvenile. Outer web of p6-p8 and (slightly) inner of (p6-)p7-p9 emarginated. Tertials short. Tail rather long, tip square or slightly rounded; 12 feathers, t6 2-4 shorter than t2-t3. Bill rather long, straight except for slightly decurved culmen-tip and distal cutting edges. Tarsus and toes relatively long and slender. Middle toe with claw 19·4 (21) 18-21; outer toe with claw *c.* 72% of middle with claw, inner *c.* 70%, hind *c.* 78%. Claws rather short and blunt, decurved.

Geographical variation. Slight; involves size, and depth of gloss on body. *O. l. ernesti* from Dead Sea region and Sinai largest, average wing of 16 adult ♂♂ 111·7, bill to skull 23·0; 2 from hills east of Cairo (Egypt) smaller (average wing 105·7) and thus similar to nominate *leucopyga*, though birds of eastern Egypt usually attributed to *ernesti*, being deeply glossy (see below). In Nile valley of southern Egypt and northern Sudan, from where nominate *leucopyga* described, average wing of 6 adult ♂♂ 106·9, bill 21·5; 26 adult ♂♂ from Ahaggar (southern Algeria) close to these (wing 106·4, bill 20·0), but 4 ♂♂ from Ennedi and Tibesti (northern Chad) distinctly smaller (wing 103·2, though bill on average 21·6), and 10 adult ♂♂ from central Algeria slightly smaller (average wing 105·5, bill 21·3). (BMNH, RMNH, ZFMK, ZMA.) See also Meinertzhagen (1930) and Niethammer (1955b). Gloss of *ernesti* stronger and more bluish than nominate *leucopyga* but some birds of either race intermediate. Tail of *ernesti* reported to show more black on tip than nominate *leucopyga* (e.g. Vaurie 1959), but birds examined (mainly from Sinai) had black more restricted: t3-t5 fully white in virtually all adults examined, t6 usually white in adult (except sometimes for faint grey spot) and in about half of 1st adults (in remainder, t6 has black spot on outer web). In nominate *leucopyga*, t3-t5 usually white in adult ♂ but partly marked with black spots up to *c.* 8 mm long in adult ♀ and 1st adult; t6 virtually white in about half of adult ♂♂ but almost always spotted in adult ♀ and 1st adult or sometimes with full black band. Some adults of nominate *leucopyga* show partial albinism on wing-coverts, outer upper primary coverts in particular sometimes fully white; apparently, this most frequently shown by birds of northern escarpment of Ahaggar plateau, where present in 5 of 7 examined (ZFMK). Occasional white feathers on throat or body and black feathers in white of cap occur in all populations.

Recognition. Adult is only Palearctic *Oenanthe* with white cap and black belly. Largely black juvenile and 1st adult similar to Black Wheatear *O. leucura* and *opistholeuca* morph of *O. picata*, but both show broader black tail-band and *O. picata* also distinctly smaller. CSR

Oenanthe leucura Black Wheatear

Du. Zwarte Tapuit   Fr. Traquet rieur   Ge. Trauersteinschmätzer
Ru. Белохвостая каменка   Sp. Collalba negra   Sw. Svart stenskvätta

Turdus leucurus Gmelin, 1789

Polytypic. Nominate *leucura* (Gmelin, 1789), southern Europe; *syenitica* (Heuglin, 1869), north-west Africa.

Field characters. 18 cm; wing-span 26–29 cm. Somewhat rounder-headed and bulkier than White-crowned Black Wheatear *O. leucopyga* but no larger, with wings slightly shorter and rounder. Rather large, big-headed, and deep-chested black wheatear with broad white rump and black ⊥ on white tail. Call distinctive. Sexes distinguishable at close range; no seasonal variation. Juvenile separable. 2 races in west Palearctic, distinguishable in the field.
    ADULT MALE. (1) South-west European race, nominate *leucura*. Largely black (slightly glossy when fresh but matt and tinged brown when worn), marked only by brown fringes to tertials and secondaries visible at close range. Rump and under tail-coverts white. Tail white with black ⊥-mark, but terminal band narrower than in Wheatear *O. oenanthe*. Underwing shows black coverts contrasting with grey-white inner fringes of flight-feathers. (2) North-west African race, *syenitica*. Distinctly browner than nominate *leucura* in both fresh and worn plumage, with more white tips on rear flanks and vent. Black ⊥ on tail bolder, with terminal band as wide as or wider than in *O. oenanthe*. ADULT FEMALE. (1) Nominate *leucura*. Duller than ♂, more sooty-brown than black; face and underparts show rich brown tips when fresh and more uniform brown colour when worn. Terminal tail-band may have ragged look due to division of black feather-tips with white. (2) *O. l. syenitica*. Duller than ♂ and ♀ nominate *leucura*, merely dark brown. On most, forehead and belly tinged grey and belly combines with white tips on vent to increase obviousness of pale area. Terminal tail-band always complete. JUVENILE. Both races. Plumage tone reflects black or brown tone of adult ♂ or ♀; upperparts faintly mottled with grey, but underparts less mottled with brown than in ♀. Racial difference in tail pattern as adult. FIRST WINTER. Both races. Distinguished from adult by retained dull brown wing-coverts and flight-feathers. At all ages, bill and legs black.
    Virtually free of confusion in Iberia, but easily confused in north-west Africa with wholly black-headed immature *O. leucopyga* at distance. Separation best based on tail pattern which in *O. leucopyga* lacks prominent terminal black band. Not known to overlap geographically with *opistholeuca* (black) morph of Eastern Pied Wheatear *O. picata*, but observer faced with vagrant black wheatear should not ignore that species, which has dangerously similar tail pattern. *O. picata* less bulky, with 20–35% shorter bill, smaller head, fractionally shorter wings, and

10% shorter legs; ♂ *opistholeuca* morph blacker than any *O. leucura* except ♂ nominate *leucura*. Flight thrush-like: exceptionally buoyant, with fluent beats of rather broad, round wings; action floating in level flight, but also makes bold tumbles down slopes and apparently effortless ascents up slopes and crags. Escape-flight usually short but markedly ascendant, ending with bird bouncing on to boulder or cliff-edge; longer and more level away from steep inclines. Gait a long or short hop. Carriage usually half-erect, but ♂ assumes upright stance on look-out perch. Behaviour includes exaggerated spreading of wings and tail.
    Song not loud, sounding distant even when delivered by close bird: begins with chortle, then brief mellow and melodious warble, then chortle, then further warble, and usually ending in chortling chatter, more sibilant than earlier phrases. Commonest call tri-syllabic, a quiet but penetrating 'pee-pee-pee', often used by paired birds in loose contact.

Habitat. In contrast to other west Palearctic *Oenanthe*, confined to west Mediterranean lower middle latitudes, largely under coastal and even oceanic influences rather than arid or continental, but in warm band of July isotherms 24–32°C (Voous 1960). In Spain, essentials for habitat are intense aridity, denuded soil, and presence of a rock-wall or equivalent (Juana 1980). Subject to these, occurs in variety of situations from Rock of Gibraltar and craggy sea-cliffs with boulders, to inland foothills and high sierras—in gorges and rocky or boulder-strewn places from sea-level to *c.* 2000 m, perching on trees and bushes where present (Witherby *et al.* 1938b; Bannerman 1954). Will inhabit stone quarries, screes, isolated crags, ruined buildings, ravines running down to coast, naked sea-cliffs, or well-vegetated steep valleys; sometimes forages on cultivation and frequents settlements (Valverde 1957). In northern Morocco, typical of Hauts Plateaux and ascends Atlas mountains from base to *c.* 3000 m, avoiding trees as much as true desert (Heim de Balsac and Mayaud 1962). Also avoids flat terrain, including wetlands, and infrequently in contact with man over most of range. Represents extreme manifestation of attachment of *Oenanthe* to rocky and perpendicular elements in habitat, and hints at ecological convergence with rock thrushes *Monticola* (Voous 1960).

Distribution. Has declined France.

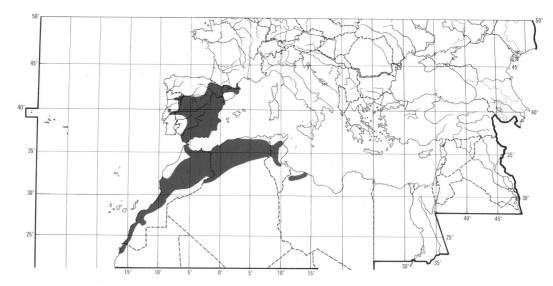

FRANCE. Probably bred Gard and Alpes Maritimes in 19th century, and until at least 1938 in Var (RC). EGYPT. Status not clear, but no proof of breeding (SMG, PLM, WCM). ITALY. No proof of breeding (PB, BM).

Accidental. Britain, Ireland, Norway, East Germany, Austria, Italy, Yugoslavia, Bulgaria, USSR, Malta.

Population. Has decreased France; little information from elsewhere.

FRANCE. Under 100 pairs (Yeatman 1976). About 16 pairs, perhaps further reduced by cold winter of 1985; decreasing (Beaufort 1983). SPAIN. Locally not scarce (AN). LIBYA. Common in Jebel Nafusa (Tripolitania), scarce elsewhere (Bundy 1976). MOROCCO. Common in some areas in south (Valverde 1957).

Movements. Generally sedentary, although some individuals disperse after breeding, and partial or total altitudinal migration occurs in some mountain regions, e.g. Atlas (Morocco) and Sierra Nevada (Spain) (Prodon 1985). Juvenile dispersal occurs in autumn (Prodon 1985), but no details available.

Most European birds (nominate *leucura*) appear to be extremely sedentary (Portenko and Vietinghoff-Scheel 1967; Prodon 1985). Detailed study of 7 pairs in south-east Spain found no evidence for dispersal (Richardson 1965). Only 2 reports suggest migration: majority of breeders from Malaga province appeared to vanish from November to February (Boxberger 1934), and 'passage in considerable numbers' recorded at Algeciras on 20 October (Stenhouse 1921).

North African populations (*syenitica*) arguably more dispersive, with sporadic records of wintering at sites not occupied during breeding season and observations of loose aggregations of up to 6 birds believed to be on passage (Smith 1971). Since racial identification unsatisfactory, origin of these birds remains unclear in absence of ringing data; possible that at least some records involve nominate *leucura*. Majority of North African breeding sites apparently occupied all year (e.g. Ledant *et al.* 1981).

In view of sedentary nature, has occurred as a vagrant over a surprisingly wide area, north to Shetland (Scotland) and Norway and east to Bulgaria—perhaps also to Egypt, though true status there unclear (see Distribution). Most such records are late August to January, although one reported at Emden (West Germany) in June (*Br. Birds* 1978, **71**, 254–8); bird recorded at Portnoo (Ireland), June 1964, may have been a White-crowned Black Wheatear *O. leucopyga* (D G C Harper). Origin of stragglers unknown; timing of records suggests that many may be dispersing juveniles.   DGCH

Food. Mainly insects. Prey usually caught on ground by 'hop and search' technique. May also fly from perch (e.g. rock, bush) to catch prey on ground. Will search around large rocks or probe cracks and holes for prey, and scratch for food under bushes or other vegetation. Observed to catch insect prey in flight showing short, rapid turning movements near to ground. May also fly steeply from ground or rock in pursuit of prey and once seen to make repeated swooping flights 30–70 m out from a cliff-top to take insects from the updraught. Wing-spreading observed in capture of large insects possibly to restrict escape. Large prey killed by repeated blows from bill. Long, heavy bill probably valuable in handling large prey and probing into dense bushes. Stiff, bristly feathers of chin, lores, and forehead probably give protection against thorns and stings of (e.g.) bees (Apoidea). Produces pellets 10–16 × 6–9 × 5–7 mm, $n=4$. (Valverde 1957; Richardson 1965; Prodon 1985.)

The following recorded in diet. Invertebrates: grasshoppers (Orthoptera: Acrididae), mantises (Dictyoptera: Mantidae), scale insects (Hemiptera: Coccoidea), adult and larval Lepidoptera, flies (Diptera), Hymenoptera (ants Formicidae, wasps Vespidae, honey bees *Apis mellifera*), beetles (Coleoptera: Cicindelidae, Carabidae, Scarabaeidae, Cerambycidae, Chrysomelidae), spiders (Araneae), millipedes (Diplopoda), scorpions (Scorpiones). Also takes lizards (Sauria) and, later in season, plant food including berries of *Asparagus acutifolius*, barberry *Berberis hispanica*, buckthorn *Rhamnus alpinus*, raspberry *Rubus idaeus*, olive *Olea*, *Smilax aspera*, and *Myrtus communis*. (Witherby *et al.* 1938*b*; Blanchet 1951; Ferguson-Lees 1960; Voous 1960; Géroudet 1963; Richardson 1965; Prodon 1985.)

Near Almeria (southern Spain), November–April, analysis of faeces and a few pellets showed food mostly beetles (apparently adults, taken through the period), especially *Chrysomela affinis c.* 7 mm long (present in most samples); ants and bees also important, and millipedes found in several samples. Stomach from Trevélez (southern Spain), February, contained remains of at least 7 beetles (one 22 mm long, another 25 mm) and 1 grasshopper. (Richardson 1965.) In Albères (eastern Pyrénées, France), takes beetles, especially *Timarcha* (Chrysomelidae), grasshoppers notably *Oedipoda* at end of summer, larval (during chick-rearing) and adult Lepidoptera, spiders, Hymenoptera, mantises, scorpions *Scolopendra cingulata*, and, later in year, berries (Prodon 1985). North African birds said by Witherby *et al.* (1938*b*) probably to take mostly ants, Orthoptera, and beetles, also Diptera. In central Morocco, takes ants and sometimes beetles (Meinertzhagen 1940).

Young near Almeria frequently given scorpions and lizards (Ferguson-Lees 1960); at another site in the area, food brought to nest included larvae of (apparently) Lepidoptera. Young fed throughout day but starting and finishing in partial darkness to correspond with availability of crepuscular beetles, especially *Chrysomela affinis* (Richardson 1965). Average 9·5 feeds per hr, 57% by ♂ (Prodon 1985). BDSS, EKD

Social pattern and behaviour. Important studies by Ferguson-Lees (1960), Richardson (1965), König (1966*a*), and Prodon (1985). For review of first 3 of these, see Panow (1974), and for useful early account, see Dresser (1871–81).

1. Not markedly gregarious, but more so than most other *Oenanthe*; in south-east Spain, loose groups of up to 6 birds commonly feed together at any time of year. In winter, feed singly, in pairs, or more often in groups of 3–5 (Richardson 1965; see also Flock Behaviour, below), these probably family parties (Prodon 1985; see Bonds, below), as also reported for White-crowned Black Wheatear *O. leucopyga*. Similar loose flock of 6 reported, Morocco, November (Smith 1971). In Albères (France), feeding pair or family party often accompanied by other Turdidae, especially Black Redstart *Phoenicurus ochruros*, also Blue Rock Thrush *Monticola solitarius* and Rock Thrush *M. saxatilis*; form loose flock in which *O. leucura* central and dominant (Prodon 1985). In Spain, feeding birds

sometimes followed around by Sardinian Warblers *Sylvia melanocephala*, Dartford Warblers *S. undata*, and Chiffchaffs *Phylloscopus collybita*, but no feeding commensalism apparent and significance of association not clear (Richardson 1965). BONDS. No evidence for other than monogamous mating system. Pair-bond probably life-long (Panow 1974) and (in common with some *O. leucopyga* and Red-rumped Wheatears *O. moesta*) maintained throughout the year, though loosely outside breeding season (Richardson 1965; König 1966*a*; Prodon 1985). ♀ alone broods, both ♂ and ♀ feeding young and performing nest-sanitation (Ferguson-Lees 1960; Richardson 1965). Young fed for up to 2 weeks after leaving nest, 'fully independent' 4 weeks after leaving (Richardson 1965; see also Relations within Family Group). In Albères, some young remain with parents up until December (Prodon 1985). Age of first breeding 1 year (Prodon 1985). BREEDING DISPERSION. Solitary and territorial. In some areas, however, minimal overt defence of boundaries, at least against conspecifics (but see Antagonistic Behaviour, below), makes size of territory almost impossible to determine (Richardson 1965). In Albères, territories large and partly overlapping (Prodon 1985). In south-east Spain, pairs fed and sang chiefly in nest-area, but also several hundred metres away, then overlapping with other pairs; 7 pairs in 2·25 km², nests on average 363 m (225–660) apart (Richardson 1965). In good habitat, Albères, 3 nests 400 m and 600 m apart (Prodon 1985). In Huesca (southern Pyrénées, Spain), 2 nests *c.* 150 m apart (Dorka *et al.* 1976). At Aium (southern Morocco) size of territory varied markedly with quality of habitat—in orchards 3 pairs in 1 km, on bare slopes 6 pairs in 7 km; at Etchera, 400 m between pairs (Valverde 1957). In eastern Morocco, 2·6 birds per km² in rocky habitat, 2 in 5 km of wadi (Blondel 1962*a*). In Tunisia, early May, *c.* 6 pairs in 14 km (M G Wilson). Birds (presumably often the same pair: see above) may re-use previous year's nest-site (König 1966*a*), and large accumulation of stones at nest (see Breeding, and Heterosexual Behaviour, below) suggests traditional use of favoured sites over many years (Ferguson-Lees 1960). ROOSTING. Nocturnal, typically in hole in rock. On 2 consecutive days in early January, Pyrénées, 3 birds (sexes not identified) shared roosting hole at foot of rock face; the 3 approached roost gradually up the rocky slopes, then flew one by one into hole within a few minutes of one another, *c.* 25 min after sunset (Dorka *et al.* 1976). In south-east Spain, ♂ roosted alone in a cave from autumn until end of January. At end of January, a pair roosted together in a cave where they later bred. During breeding season ♀ roosts on nest at night until young no longer need brooding (Richardson 1965.) While one bird (presumably ♀) roosted on nest, other bird (presumably ♂) roosted in hole *c.* 20 m away (Dorka *et al.* 1976). Active from morning until dusk (Richardson 1965). No further information.

2. Shy and wary, though more approachable in breeding season (Dresser 1871–81; Whitaker 1905). In alarm, bobs tail constantly (Ferguson-Lees 1960): after landing, or while hopping around, both sexes characteristically raise tail slowly and fan it, rendering white markings conspicuous. Occasionally performs Dancing-display (see Heterosexual Behaviour, below)—though less intensely than in courtship—as apparent displacement-activity. Dancing-display is equivalent to Acrobatic-display in other *Oenanthe*. Subsong (see 1d in Voice) also given in conflict situations. (König 1966*a*.) Sudden appearance of raptors stimulates alarm-calls (Dorka *et al.* 1976; Prodon 1985: see 4a–b in Voice). Pursued bird drops behind rocks and creeps into hole (Đixon 1882). FLOCK BEHAVIOUR. In feeding groups, individuals keep *c.* 5–50 m apart but move as a group to new areas (Richardson 1965). 3 birds which

roosted together in winter fed amicably together during the day (Dorka *et al.* 1976). SONG-DISPLAY. ♂ advertises with Territorial-song (see 1a in Voice), delivered in Song-flight resembling that of pipit *Anthus*. Almost always starts singing from exposed perch, usually rock, bush, or tree (Richardson 1965; König 1966a), then often ascends with fluttering wing-beats, spreads wings, fans tail, and sings while gliding to another perch (König 1966a). Also dangles legs during Song-flight (M G Wilson). After landing, ♂ usually performs Dancing-display (see Heterosexual Behaviour, below), accompanied by Courtship-song (König 1966a: see 1b in Voice). In south-east Spain, Territorial-song subsided in October, stopped late October until late November, started sporadically December, more common from January onwards. ♂♂ sang regularly and for long periods (up to 20 min per individual), from late January and even more intensely in February (peak pair-formation). In one ♂ at least, song declined late February once breeding under way. ♀ also sings briefly at times. (Richardson 1965: see 1c in Voice.) In Morocco, in year when breeding apparently late, singing started at beginning of February, peaked *c.* 10 March (Blondel 1962a). ♂♂ sing especially in morning and evening (Panow 1974); e.g. in early January, song heard shortly before going to roost, and again in the morning (Dorka *et al.* 1976). In recording by E D H Johnson (Algeria, February), evident Song-duels occur (see also below). ANTAGONISTIC BEHAVIOUR. Accounts differ. According to Richardson (1965), König (1966a), and Dorka *et al.* (1976), noticeably unaggressive towards conspecifics, and no obvious territorial defence at any time of year, but territorial disputes between pairs early in breeding season reported by Ferguson-Less (1960) and Prodon (1985); these mostly Song-duels, sometimes chases, at territorial boundaries and near nests. In France, young of the year may start defending territories thus in their 1st autumn. During breeding season, residents often very aggressive towards, and dominant over, other Turdidae in the neighbourhood, notably *M. solitarius*, *M. saxatilis*, and Black-eared Wheatear *O. hispanica*, which are driven off and pursued at length. (Prodon 1985.) Interspecific disputes occasionally observed in north-east Spain, especially with *M. solitarius* and *M. saxatilis*, which are vigorously attacked when they trespass (König 1966a). Also recorded chasing off (once each) Blackcap *S. atricapilla*, Blackbird *Turdus merula*, Hoopoe *Upupa epops* and *P. ochruros* (Richardson 1965), and in Tunisia, ♀ Redstart *P. phoenicurus* (M G Wilson). HETEROSEXUAL BEHAVIOUR. (1) General. Pair-formation intimately linked chronologically with Nest-showing (see below) and nest-site selection, same display apparently serving all these functions, so divisions (below) somewhat arbitrary. Collection of stones (Rock-carrying: see subsection 3, below, also Breeding) almost certainly important in nest-site selection and courtship (Ferguson-Lees 1960; Richardson 1965; König 1966a). Pairs usually maintain close and amicable contact throughout the year (König 1966a). (2) Pair-bonding behaviour

and nest-site selection. Apparent heterosexual chasing first seen late December and birds more obviously paired January–February (Richardson 1965). When 2 birds meet (Meeting-ceremony), they somewhat raise and lower slightly fanned tail (König 1966a). Before 2 birds, perhaps pair-members, entered winter roost (with a 3rd bird), they met briefly near roost, approached each other to within 30 cm, and performed Meeting-ceremony as follows: markedly ruffled body feathers and 'looked away' (i.e. averted looks) with slow raising and lowering of fanned tail (Dorka *et al.* 1976). Song-flight of ♂ often leads to ground-display; following account by König (1966a). After landing, ♂ usually performs Dancing-display, thought by Prodon (1985) to indicate attack-flee conflict: In Fig A (left), ♂ lowers breast, raises tail slowly, simultaneously fanning it (Fig B) and begins to ruffle belly feathers. All body plumage except crown is ruffled. Steeply raised, widely fanned tail is shivered up and down, while wing-tips, which lie over base of tail, are flicked (Fig A, middle). Display accompanied by Courtship-song, also by dancing (tripping) on the spot or in a small arc. Tail is gradually closed and slowly lowered, and plumage sleeked (Fig A, right). ♂ then starts Dancing-display anew, or flies off. Excited ♂ performs Dancing-display several times, and ever closer to prospective nest-site, finally outside nest-hole entrance (Nest-showing). ♀ observes ♂'s display usually for some time before approaching and finally perching near nest-entrance. ♂ repeats Dancing-display and finally slips into nest-hole. If site is in a vertical face, ♂ often flies to hole and clings in front of it, tail spread, before entering. Frequently enters and leaves hole, giving Subsong or Contact-calls (see 2 in Voice). Nest-showing ♂ also frequently picks up plant stems and carries them into hole or drops them in front of it. Usually ♀ inspects hole soon after and evidently makes final choice, rejecting some sites shown. ♂ usually shows mate several holes or else pair re-occupy their previous year's hole with little ceremony. (König 1966a.) (3) Rock-carrying. ♂ now increases his rate of carrying stones, also pieces of earth, wood, plant stems (etc.), usually placing them in or in front of nest-hole (Ferguson-Lees 1960; Richardson 1965, which see for other materials collected; König 1966a). Rock-carrying assumed to be part of ♂'s Nest-showing display (Ferguson-Lees 1960), equivalent to grass-carrying in Wheatear *O. oenanthe* (Richardson 1965). For discussion of this, and of other possible functions, see König (1966a). Stones are collected assiduously from nearby, e.g. during 25 min, ♂ ferried 42 stones to nest from 2–8(–10) m away, then 17 more after a pause of 25 min (Richardson 1965, which see for details of Rock-carrying rate). See Breeding for size of stones and numbers accumulated (over probably several years) at given nest-sites. According to Richardson (1965), ♀ attracted both by sight of stones and noise of ♂ dropping them, but Rock-carrying display considered primarily visual by König (1966a). ♀ takes no part early on but readily shows interest in these objects and starts carrying

A

B

·them into nest-hole (König 1966a). ♂ started Rock-carrying to an old nest-site on 19 February and his mate participated from 5 March (Richardson 1965). (4) Courtship-feeding. None reported. (5) Mating. Near nest-hole, both birds dance around each other (Dancing-ceremony), especially during nest-building (see below), ♀ giving a variety of sounds (see 1c in Voice), ♂ giving Courtship-song. Such display often, though not invariably, leads to copulation, ♂ flying suddenly on to ♀'s back. ♀ finally flies away and ♂ follows her. (König 1966a.) (6) Behaviour at nest. Both sexes share nest-building proper (i.e. as distinct from Rock-carrying) which, in south-east Spain, begins mid-February. Only ♀ incubates, though ♂ sings or feeds near nest. ♀ takes regular breaks after onset of incubation. (Richardson 1965.) Often, song of ♂ near nest stimulates ♀ to leave nest and pair then forage together nearby (Prodon 1985). ♂ may accompany ♀ back to nest (Ferguson-Lees 1960). RELATIONS WITHIN FAMILY GROUP. Hatching arouses great curiosity in ♂ (Prodon 1985). At one nest, young first fed by ♂, and he did most feeding for the next few days. When ♂ arrived to feed (but not brood) young, ♀ took a break from brooding. ♀ broods steadily less as young develop: on 1st day after hatching, ♀'s brooding spells on average $4\frac{1}{2}$ min, on 5th day less than 1 min (Prodon 1985). Once period of intense brooding over, ♀ helps ♂ in feeding young, and in nest-hygiene (Ferguson-Lees 1960). At one nest, ♂ on 3 occasions seen to carry faecal sacs at least 50 m away (Richardson 1965). Both parents sang on approach to nest and young kept up food-calls (see Voice) whenever parents in vicinity of nest (Ferguson-Lees 1960). According to Prodon (1985), approaching parents also give a special call (probably call 2 in Voice), this continuing after young leave nest. Young barely able to fly when they leave nest (Richardson 1965; Prodon 1985). When parent catches prey, all the young out of the nest run, beating their wings and calling, to solicit it (Dresser 1871–81). Young make short flights from 3rd day (Prodon 1985); partly self-feeding at 1 week after leaving nest, still occasionally fed by parents 2 weeks after leaving (Richardson 1965). ANTI-PREDATOR RESPONSES OF YOUNG. From 9 days old, young squirm into deep fissures in nest-hole when any attempt made to capture them. After leaving nest, continue to take refuge under large rocks. (Richardson 1965.) Alarm-call of parent stimulates young to hide instantly and, once disturbance passed, another call (not described) summons them out (Dresser 1871–81). PARENTAL ANTI-PREDATOR STRATEGIES. (1) Passive measures. ♀ a tight sitter, more so than *O. oenanthe* and *O. hispanica*, flushing only if nest-hole is tapped. In early incubation, ♂ calls ♀ off nest when anyone approaches (Ferguson-Lees 1960.) (2) Active measures: against man. When intruder at nest with young, parents (shy before hatching) made close approach and were highly demonstrative: ♂ tripped, danced, and sang on the spot, while ♀, closer than ♂, fluttered anxiously around intruder (Dresser 1871–81). Armitage (1935) confirmed that ♀ unusually bold, especially after young hatched, approaching nest closely in spite of presence there of human intruder. When intruder approaches nest containing young, ♀ may perform Dancing-display (Prodon 1985), presumably as described for ♂ (above) by Dresser (1871–81).

(Figs by K H E Franklin: based on drawings in König 1966a.)   EKD

Voice. Used for much of year. Repertoire complex. Fullest account by König (1966a) for north-east Spain. Regional differences, if any, not known.

CALLS OF ADULTS. (1) Song. Quite unlike songs of most other *Oenanthe* (Ferguson-Lees 1960), but sometimes not unlike White-crowned Black Wheatear *O. leucopyga* (see that species, p. 881). (a) Territorial-song. Given perched or in Song-flight. Frequently described as a brief, pleasing, mellow warble, rather like Blue Rock Thrush *Monticola solitarius* but quieter (e.g. Witherby *et al.* 1938b), shorter and sweeter (Ferguson-Lees 1960), slightly higher pitched and containing more harsh scratchy sounds (König 1966a). Quality also recalls Orphean Warbler *Sylvia hortensis* (Ferguson-Lees 1960). Not loud, sounding distant even when heard closely (D I M Wallace). According to Richardson (1965), songs 2–4 s long, exceptionally up to 10–12 s, given at intervals of 2–8 s; in recording by E D H Johnson (Algeria), 1 song continuous for more than 12 s. In analysis by Bergmann and Helb (1982, which see for additional sonagrams), phrases *c.* 1–2 s long, often beginning with 1–2 separate call-type units such as 'tschak', followed by more tonal, drawn-out, but mostly harsh sounding units. Introductory and terminal units often sibilant, churring, or chattering (Witherby *et al.* 1938b; Ferguson-Lees 1960). According to Stadler (1952, which see for details and comparison with other species), phrases typically comprise repeated descending 'triplets' (presumably 3 units) but pitch may rise and fall. Renderings include (roughly) 'tjockereu-keu-keke' (Jonsson 1982); 'kro zí tero tri rö' and 'krökrö zítero' (Meise 1959). One section (Fig I) of a rather

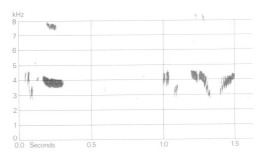

I E D H Johnson Algeria February 1969

disconnected series of calls, presumably low-intensity song, is rendered 'tlehwee tut fiu tiki-ti pleeooee' (J Hall-Craggs). Most of our recordings confirm that pattern is of short phrases given often at relatively long intervals, but quality and delivery vary markedly, suggesting mimicry at times. Some, perhaps typical, songs a brisk jangling husky warble, e.g. in Fig II, bird (that also gave sounds in Fig I) starts with small 'zizz' sounds and builds up to its favourite ascending note-type (like 'tlehwee' in Fig I), then ebbs away at the end; another phrase (Fig III) in same song begins with 3 quiet 'clicks' followed by a loud buzzing 'tzeeee' (J Hall-Craggs). Other songs much lazier and more rolling, almost melancholy: one recording (Morocco) rather like Robin *Erithacus rubecula*, with throaty, sometimes descending, purring trills 'treeioou' (P A D Hollom), these possibly twittering calls (J Hall-Craggs: see 6, below). (b) Courtship-song. Usually

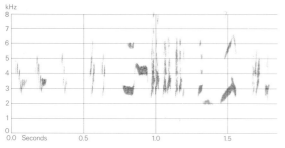

II E D H Johnson Algeria February 1969

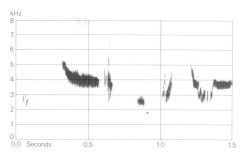

III E D H Johnson Algeria February 1969

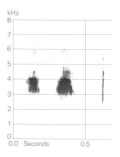

IV E D H Johnson Algeria
February 1969

begins with a number of 'jok' sounds and comprises purring, whinnying, and chuckling sounds, given typically during Dancing-display (König 1966a). Introductory 'tschak' units (see above) described by Bergmann and Helb (1982) perhaps appropriate to this song-type. (c) Song of ♀. ♀ may occasionally give a more scratchy, softer variant of (apparently) ♂'s Territorial-song, especially as a displacement-activity. Also occasionally gives harsher variant of ♂'s Courtship-song, e.g. during nest-building, when song described as creaking, hoarse, cawing, and trilling sounds. (König 1966a.) (d) Subsong. Distinctive, sometimes a low almost continuous warble, in or out of breeding season (Richardson 1965). A quiet chattering song given by both sexes in conflict situations and at other times ('functionless conversational song'), not infrequently in winter, notably also by Nest-showing ♂ (König 1966a). Also described as a scratchy warble with musical interludes, reminiscent of Whitethroat *Sylvia communis*; ♀'s Subsong briefer and includes more raucous alarm-calls (see below) than ♂'s (Ferguson-Lees 1960). (2) Contact-call. A quiet 'schrü', very similar to juvenile food-call, given (e.g.) by Nest-showing ♂ (König 1966a). Call described as brief, soft, and rolling, given by adults approaching nest or fledged young with food (Prodon 1985), presumably the same. See also call 4c for probable contact-alarm call. (3) Excitement-call. 'Krirr' sounds given when highly excited, and accompanied by bobbing (König 1966a). Sounds described by Bergmann and Helb (1982) as 'chrwä' and 'err' probably the same or related; first 2 units in Fig IV, rendered a whirring 'zweear zweear', thought to be this call (J Hall-Craggs). (4) Warning- and Alarm-calls. (a) Warning-call a plaintive nasal 'jöhb' or 'töt', similar to 'jihp' of *M. solitarius* but

lower pitched and harsher (König 1966a). Also rendered 'jüb', like Chaffinch *Fringilla coelebs*, given perched or on take-off (Stadler 1952; Bergmann and Helb 1982); scolding 'chääp' given by pair alarmed at presence of Little Owl *Athene noctua* (M G Wilson); a clear 'hiep' when Sparrowhawk *Accipiter nisus* near roost (Dorka *et al.* 1976). This presumably the call likened by J-C Roché to a bleating goat. (b) When disturbed, 'chack' sounds, typical of other *Oenanthe* (Ferguson-Lees 1960); also rendered 'tschek' (Bergmann and Helb 1982), a harsh 'tschäk-tschäk' (König 1966a). 'Tut' sound in Fig I, also 3rd unit in Fig IV, possibly this call. (c) An insistent oft-repeated high-pitched 'peeee', e.g. during territorial disputes (Ferguson-Lees 1960). In Fig V, strident pier-

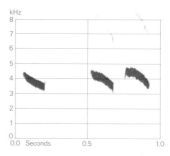

V P A D Hollom
Morocco March 1978

cing whistling 'pseeew' alarm-calls interrupt song (P A D Hollom). Also rendered a pure 'ii' or 'ie' (Bergmann and Helb 1982). In alarm, also rapidly repeated to give a piping 'pee-pee-pee-pee', sometimes interspersed with call 4b (Ferguson-Lees 1960); also described as 'hihihihi' and accompanied by bobbing (König 1966a). Quiet but penetrating trisyllabic 'pee-pee-pee' said to be commonest call of *O. leucura*, often used by paired birds in loose contact (D I M Wallace) and, on this evidence, as much a regular contact-call as one expressly indicating alarm. (5) Distress-call. A shrill screeching sound when handled (König 1966a). (6) Other calls. Recording by P A D Hollom of singing bird (see 1a, above), contains various finch (*Fringilla*)-like twitters, thought to be calls, some short and simple (Fig VI, 1st unit), others longer and more complex (Fig VI, 2nd unit). Same recording ends with 2 short strident 'tuwi tuwi' calls. (J Hall-Craggs.)

CALLS OF YOUNG. Food-calls of nestlings a loud, wheezy, querulous sound (Ferguson-Lees 1960); sounds described as a sort of loud, high-pitched chirring given

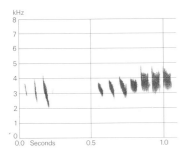

kHz
8
7
6
5
4
3
2
1
0
0.0 Seconds   0.5   1.0

VI P A D Hollom
Morocco March 1978

in and out of nest (Prodon 1985), presumably the same. Food-calls of fledged young similar to call 2 of adult. Fledglings also give call 4a of adult, also Subsong, shortly after leaving nest. (König 1966*a*.)   EKD

Breeding. SEASON. Eastern Pyrénées (France): eggs laid from mid-April (Prodon 1985). Southern Spain: eggs laid from mid-March (Richardson 1965). North-east Spain: latest eggs 1 July (König 1966*a*). Algeria and Tunisia: earliest eggs late February, most March or April to June (Heim de Balsac and Mayaud 1962). SITE. Hole in rock wall, cliff, cave, or man-made wall. Of 37 nests, Spain, 5 on ground (or even underground, e.g. in cave), 8 at up to 1 m above ground, 18 at 1–2 m, 5 at 2–3 m, and 1 over 4 m, mean 1·6 m (Richardson 1965). Nest: cup of dead grass and rootlets, incorporating feathers and wool; scant lining (Ferguson-Lees 1960), or lining integral with main structure; outer diameter *c.* 16 cm, inner diameter 7–8 cm, depth of cup 6–7 cm (Richardson 1965). Normally builds platform of small stones at sides of nest; can be 10–15 cm wide or even more, and incorporating several hundred stones, but in such cases undoubtedly built up over several years; varies from only 3–4 stones in sites without suitable building area to over 9000 covering *c.* 2 m² of cave floor (Richardson 1965); weight of 120 stones 6·83 g (1–18); one large stone 41 × 34 × 10 mm (Ferguson-Lees 1960). Building: both sexes build, ♀ perhaps doing most; ♂ brings stones for platform to start with but helped later by ♀ (Ferguson-Lees 1960; Richardson 1965; König 1966*a*, which see for discussion of functions of platform). EGGS. See Plate 83. Sub-elliptical, smooth and glossy; very pale blue to bluish-white, variably speckled red-brown, markings usually concentrated at broad end. Nominate *leucura*: 25·0 × 17·8 mm (22·0–26·6 × 16·5–19·4), *n* = 65; calculated weight 3·72 g (Schönwetter 1979). Clutch: 3–5(–6). Of 101 clutches, north-west Africa: 3 eggs, 15%; 4, 66%; 5, 19%; mean 4·0 (Heim de Balsac and Mayaud 1962). In Albères (eastern Pyrénées, France), mean of 21 clutches 4·57 (Prodon 1985). Larger clutches may also be more frequent in Spain (Heim de Balsac and Mayaud 1962). Regularly lays replacement clutches (Prodon 1985), in one case laid after 11–12 days (Richardson 1965). One brood, occasionally 2 (König 1966*a*), more often in early springs (Prodon 1985). INCUBATION. 14–18(–21) days

(Prodon 1985); *c.* 15–17 days (Richardson 1965). One late nest *c.* 12 days (König 1966*a*). By ♀ (Ferguson-Lees 1960). YOUNG. Cared for and fed by both parents; brooded by ♀ while small (Richardson 1965). FLEDGING TO MATURITY. Fledging period 14–15 days (Richardson 1965). In France, average for 5 broods 16·5 days (15–19). Age of first breeding 1 year. (Prodon 1985.) BREEDING SUCCESS. Only 4 of 10 nests, Spain, fledged young; losses of eggs and young attributed to snakes and lizards (Richardson 1965). In France, 10 of 17 nests fledged young, average brood 3·5; nest-failures thought due to snakes, small mammals, and Magpies *Pica pica* (Prodon 1985).

Plumages (*O. l. syenitica*). ADULT MALE. Upperparts backwards to upper rump and underparts backwards to vent and thighs entirely sooty black with faint brown tinge all over; black deepest on sides of head, throat, and chest, on sides of head sometimes slightly contrasting with browner forehead and crown. Lower rump and upper and under tail-coverts white; feathers at rear of flank tipped white. Tail white, distal part of central pair (t1) black for (30–)35–42 mm (maximum extent of black on average 37·4; *n* = 23); outer web of t2 to inner web of t6 with 8–13(–17) mm wide black band on tip, but often slightly more black on inner web of t2 and outer of t6; width of band on average 11·4 (*n* = 33), with maximum extent of black on outer web of t6 13·5 (25) 11–17 mm; in fresh plumage, black of tail-tips narrowly fringed white distally. Flight-feathers, tertials, and greater upper primary coverts brown-black, faintly and narrowly edged paler brown along outer webs and tips. Lesser and median upper wing-coverts sooty black, faintly tinged brown; greater coverts brown-black with faintly paler edges. Under wing-coverts and axillaries brownish-black. *In worn plumage*, black of tail, flight-feathers, and tertials slightly browner; white fringes along tips of tail-feathers worn off, browner edges of flight-feathers and greater upper primary coverts abraded. ADULT FEMALE. Like adult ♂, but all black replaced by dark brown and tips of brown feathers of head and body tinged sandy-brown or (in particular on back) faintly rufous; mantle, scapulars, and lesser upper wing-coverts usually darkest, forehead, crown and underparts more sandy-brown. Lower rump and tail-coverts white, as adult ♂; tail-pattern as ♂, flight-feathers and greater upper primary coverts browner than ♂, narrowly edged pale brown along outer edges and tips. *In worn plumage*, generally darker, as sandy-brown feather-tips wear off; upperparts often almost as dark as ♂, but ear-coverts and underparts distinctly browner (not almost black), many traces of sandy-brown or buff fringes remaining on chest, belly, vent, and flanks. NESTLING. Down long and plentiful, medium grey; on upperparts only (Witherby *et al.* 1938*b*). JUVENILE. Those of *syenitica* examined (5 ♂♂, 3 unsexed) sooty-black, like adult ♂, but 2 of nominate *leucura* (1 ♀, 1 unsexed) rufous-brown, like adult ♀; difference probably sexual rather than geographical. In both races, plumage shorter and looser than in adult, especially fluffy on rump, tail-coverts, and vent. In juvenile ♂, black parts of body slightly less deep, duller; ear-coverts, cheeks, and chin to breast with faintly paler grey-brown spots or fringes on feather-tips; in juvenile ♀, brown of upperparts slightly paler than in adult ♀ (more similar to brown of underparts of adult ♀), sandy- or rufous-brown fringes of underparts slightly narrower and less contrasting, lesser and median upper wing-coverts with rufous-buff spots on tips. Tail and flight-feathers as in adult, but black or brown slightly paler; for greater upper wing-coverts and greater upper primary coverts, see 1st adult below. FIRST ADULT MALE. Like adult ♂, but black of upperparts and upperwing with more pronounced brown tinge, somewhat contrasting with blacker lores and ear-coverts. Juvenile flight-feathers, greater upper primary coverts, nearly always tail and greater upper wing-coverts, and frequently tertial coverts retained; primaries and tail-tips paler brown than those of adult and more heavily worn at same time of year; greater upper wing-coverts and tertials with faint sandy-brown to off-white fringes along outer webs and tips (broader and paler than those of adult, but may wear off) and, occasionally, faint off-white triangular spots on tips; greater upper primary coverts distinctly browner than in adult ♂, narrowly but distinctly fringed off-white (unlike adult). FIRST ADULT FEMALE. Like adult ♀, but part of juvenile feathers retained, as in 1st adult ♂; retained secondaries and greater upper primary coverts with distinctly broader, paler, and often less clear-cut fringes along tips; greater upper wing-coverts with broader sandy-buff to off-white fringes, tending to form a triangular spot on tip of covert.

Bare parts. ADULT, FIRST ADULT. Iris dark brown. Bill, leg, and foot black. (Hartert 1903–10; BMNH.) NESTLING. Mouth yellow, no spots on tongue; gape-flanges very pale yellow. (Witherby et al. 1938b). JUVENILE. Iris dark brown. Bill dark greyish-horn, paler towards base of lower mandible; gape-flanges yellow-white. Tarsus dull grey with flesh tinge, toes slightly darker. (BMNH, RMNH.)

Moults. ADULT POST-BREEDING. Complete; primaries descendant. July–October (Witherby et al. 1938b). None in moult examined; 4 syenitica from early May heavily worn, but not moulting; 3 nominate leucura, late May and early June, not yet moulting; in both races, no moult from mid-October onwards. POST-JUVENILE. Partial: head, body, usually all lesser and median upper wing-coverts, shorter feathers of bastard wing, regularly (8 of 16 examined) 1–3 tertial coverts, occasionally a few inner greater upper wing-coverts or tertials, and rarely t1. Starts from early May (syenitica) or June (nominate leucura); completed early June to October. (Witherby et al. 1938b; BMNH, RMNH, ZFMK, ZMA.)

Measurements. ADULT, FIRST ADULT. O. l. syenitica. Tunisia and Algeria, October–May; skins (RMNH, ZFMK, ZMA.) Bill (S) to skull, bill (N) to distal corner of nostril; exposed culmen on average 4·6 less than bill (S).

WING	♂ 99·8 (2·81; 20)	96–105	♀ 94·1 (1·28; 11)	92–97
TAIL	66·3 (2·21; 19)	63–70	61·8 (2·00; 11)	59–66
BILL (S)	21·5 (0·77; 18)	20·4–23·0	20·9 (0·65; 11)	19·9–21·7
BILL (N)	12·5 (0·52; 18)	11·7–13·5	12·2 (0·50; 11)	11·4–12·9
TARSUS	27·6 (0·77; 23)	26·1–28·7	27·1 (0·56; 11)	26·4–28·2

Sex differences significant for wing and tail. Retained juvenile tail of 1st adult similar in length to tail of older birds, but retained juvenile wing on average 1·8 shorter than wing of full adult.

Nominate leucura. Only 3 of each sex examined, Spain and Sardinia; range of wing, ♂ 100–102, ♀ 94–98; tail, ♂ 67–74, ♀ 65–68, bill (S), ♂ 21·3–22·5, ♀ 20·8–22·2; tarsus, ♂ 27·9–28·6, ♀ 27·0–28·2 (RMNH, ZMA).

Weights. Nominate leucura Spain: ♂ c. 38 (Richardson 1965).
O. l. syenitica. Algeria, late October and early November, ♂♂: adult 44, 1st adult 37·5 (ZFMK).

Structure. Wing rather short, broad at base, tip rounded. 10 primaries: p7 longest, p8 and p6 equal to p7 or occasionally up to 1 shorter, p9 6–9 shorter than p7, p5 2–4, p4 8–11, p3 12–15, p1 17–21. P10 reduced, 43–52 shorter than wing-tip, 3–8 longer than longest upper primary covert. Outer web of p5–p8 and inner of (p6–)p7–p9 emarginated. Tertials short. Tail rather short, tip square; 12 feathers. Bill rather long, laterally compressed at tip, wide at base; straight, but tip of culmen and (slightly) cutting edges decurved. Nostrils rather small, oval, partly covered by frontal feathering; some fine bristles along base of upper mandible. Tarsus rather long and slender, toes rather short. Middle toe with claw 20·2 (17) 19–22; outer toe with claw c. 73% of inner with claw, inner c. 69%, hind c. 79%. Claws short and rather blunt, slightly decurved.

Geographical variation. Slight; mainly involves colour of body, perhaps slightly size (see Measurements), and amount of black on tail-tip. Adult ♂ nominate leucura from southern Europe distinctly blacker than adult ♂ syenitica from North Africa, without its brown tinge, sometimes slightly glossy above (but not as metallic as White-crowned Black Wheatear O. leucopyga); 1st adult ♂ nominate leucura slightly browner, however, close to adult ♂ syenitica. Both adult ♀ and 1st adult ♀ nominate leucura darker than ♀ syenitica, upperparts saturated deep brown or dark chocolate-brown (except for white lower rump and tail-coverts), feather-tips extensively tinged rusty-rufous (sandy-brown in syenitica), especially on back; sides of head and body and underparts from chin to vent brown-black with broad rusty fringes to feathers (again, less sandy-brown than ♀ syenitica). Black on tail-tips on average less extensive: t1 black for 35·1 (7) 30–42 mm, t2–t5 for 10·8 (6) 8–14 mm; in one 1st adult ♂ examined, white on t3–t5 extended to tip along shaft, splitting black band into 2 large spots; maximum extent of black on outer web of t6 11·6 (7) 9–15 mm. 4 birds from southern Morocco had tail-band 16–20 mm wide and therefore separated as riggenbachi (Hartert, 1909), but others inseparable from typical syenitica and hence riggenbachi not recognized (Hartert and Steinbacher 1932–8; Vaurie 1955a). White wing-patches occur exceptionally (K Koopman), perhaps due to partial albinism as in some populations of O. leucopyga.

Recognition. Largely black or brown-black body of O. leucura shared only with 1st adult O. leucopyga and with adult and 1st adult opistholeuca morph of Eastern Pied Wheatear O. picata. O. leucopyga differs in having white on tail-tips more extensive (except on t1, which is usually black for 38–57 mm in 1st adult), black band on t3–t5 absent (these feathers thus fully white) or present as black spot 3–8 mm across on one web only, t6 usually white with black spot 5–10 long on outer web (occasionally, however, t6 has full black band). Tail-pattern of O. picata similar to O. leucura, opistholeuca morph differing only in smaller size (wing 86–96, bill to skull 16–19), but breeding areas of O. leucura and O. picata well separated. CSR

Monticola saxatilis Rock Thrush

Du. Rode Rotslijster   Fr. Merle de roche   Ge. Steinrötel
Ru. Пёстрый каменный дрозд   Sp. Roquero rojo   Sw. Stentrast

Turdus saxatilis Linnaeus, 1766

Monotypic

Field characters. 18·5 cm; wing-span 33–37 cm. 10% smaller than Redwing *Turdus iliacus* and Blue Rock Thrush *M. solitarius*. Strong-billed, rather long-bodied, rock-haunting bird, with shape suggesting large short-tailed chat rather than thrush. ♂ essentially blue, black, and white above, with rufous body and tail; ♀ basically buff, mottled and scaled except on rufous tail. Flight chat-like, with floating action. Furtive when breeding. Sexes dissimilar; marked seasonal variation in ♂ only. Juvenile separable.

ADULT MALE BREEDING. From April, head and throat grey-blue, becoming blue-slate on mantle and scapulars and contrasting with barred or wholly white middle back; back further isolated by blue-slate rump, orange-chestnut upper tail-coverts and short, bright rufous-chestnut tail with brown centre. Wings essentially dark brown-black, with faint slate tone on lesser coverts and indistinct grey fringes and grey-buff tips to median and greater coverts and tertials. Underparts from upper breast to vent bright orange-chestnut, with buff-white tips creating faint scaling and barring on vent. Underwing also bright orange-chestnut, making most of bird in display-flight appear to 'flame' against bright sky. ADULT MALE NON-BREEDING. From late summer, loses boldly patterned appearance and becomes intensely mottled, scaled, and barred, nearly all feathers except tail having almost black subterminal bands and brown- to buff-white tips. Blue on head reduced to wash on crown, white on central back broken up (but still forming obvious dappled patch), white tips make underparts paler, and all wing feathers brightly fringed (with coverts scaled, like mantle and scapulars). ADULT FEMALE. Resembles adult ♂ non-breeding at all times. Upperparts generally paler, with no real blue and little slate tone on head and back, no white patch on central back, browner, paler-fringed wing feathers, and paler, more orange upper tail-coverts. Eye-ring and central chin and throat cream, giving head more chat-like appearance. Rest of underparts buff-chestnut in summer but duller, orange-buff in winter, with broader white tips and dark brown subterminal bars creating even scaling and mottling overall. Underwing and vent pale orange-buff, noticeably paler than even ♂ in winter. JUVENILE. Resembles ♀ but ground-colour of upperparts greyer, of underparts merely buff (lacking noticeable warmth), and of wings browner. Tips of many feathers also darker, brown or buff, making upperparts more mottled though underparts more strongly barred. At close range, median and innermost greater

coverts and tertials show buff spots. Compared with ♀, juvenile looks washed-out, with only tail showing bright tone. FIRST-WINTER MALE. From August, assumes head and body plumage with indications of adult ♂ colours, particularly on crown, mantle, central back, and sides of throat. FIRST-WINTER FEMALE. As adult. FIRST-SUMMER MALE. Duller than adult, with browner wings; buff tips to upper back feathers often retained, and usually some obvious black bars on underparts. In both sexes and at all ages, bill dark brown-horn and legs dark brown.

Unlike *M. solitarius*, ♀ and immature always pale-throated. Otherwise almost unmistakable among wild birds of west Palearctic but some risk of confusion with escaped congeners of similar appearance, so important to remember that ♀♀ and immatures of the following all suggest *M. saxatilis*: Blue-headed Rock Thrush *M. cinclorhynchus*, White-throated Rock Thrush *M. gularis*, Chestnut-bellied Rock Thrush *M. rufiventris* (all of Himalayan and south-east Asian mountains), Mottled Rock Thrush *M. angolensis*, and Little Rock Thrush *M. rufocinerea* (of south and east Africa). Flight usually low and floating, with full beats of wings as slow as *Turdus* thrush's but with short tail giving appearance of large chat; capable of considerable acceleration when escaping, shooting in among boulders, rounding bluffs suddenly (with plumage of both sexes cryptic in such flights) and plummeting down crags. Display-flight of ♂ has fluttering ascent with butterfly-like wing-beats and spread tail, and gentle 'parachute' descent with extended wings and tail. Stance strongly recalls large wheatear *Oenanthe*, with usually upright attitude; characteristically wags tail, with obvious upward jerk and loose flirt. Gait essentially hopping but long-paced, allowing even progress on flat ground and nimbleness over rocks. A ground bird but will perch on trees, buildings, and wires. Shy and solitary in breeding habitat, forming small groups on migration; relatively tame but again solitary and widely spaced in rocky savanna winter quarters. Most vagrants have occurred on coasts and isles.

Song loud and far-carrying: a mellow fluted and piped warble, softer and more flowing than *M. solitarius*. Commonest call 'chack-chack'. Rather silent in winter.

Habitat. Breeds in west Palearctic in lower middle latitudes in continental warm temperate, steppe and Mediterranean montane zones, on sunny, dry often stony hollows or terraces, preferably dotted with stunted trees

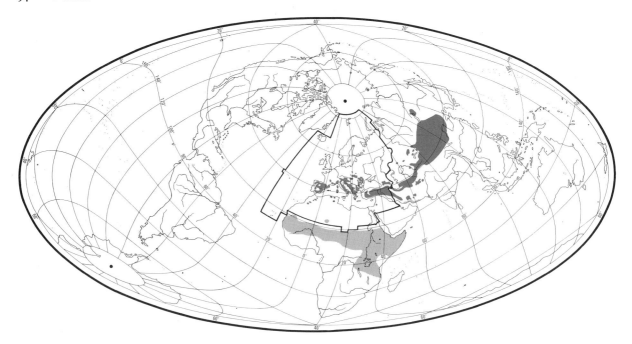

or shrubs serving as perches. In southern Switzerland, also on rocky heaths and in vineyards from 500 m, but mainly at 1500–2700 m. Forages over some distance from nest, down to hayfields and farmland, using rocks, walls, roofs of buildings, and bare branches or treetops as hunting look-outs (Glutz von Blotzheim 1962). In Spain, nests on barren hillsides with boulders and crags, chiefly at *c*. 1250–2300 m, above range of Blue Rock Thrush *M. solitarius* (Bannerman 1954), though they do overlap, e.g. in Sierra de Cazorla (M A Ogilvie). In Germany, as elsewhere towards north of European range, many sites occupied last century (e.g. ruined castles on the Rhine, heaps of debris) were deserted before or soon after its end, conceivably through some reduction in climatic suitability, but this seems questionable since other species were expanding northwards at that period (Niethammer 1937; Voous 1960). In North Africa, inhabits wild mountain ravines and valleys strewn with huge boulders and rocks of every size, with a few dwarf bushes and an occasional stunted tree. From *c*. 1900 or 2000 m to 3000 or above in Haut Atlas (Bannerman 1954; Heim de Balsac and Mayaud 1962). In USSR, breeds on rocky mountain slopes, scree, and stony hills, but sometimes among sparse juniper *Juniperus* or on grassland, from low hills with stony outcrops at 500–800 m, but most often at *c*. 1700–2000 m, and upwards to 3000 m (Dementiev and Gladkov 1954*b*). In Afghanistan, mostly found on rocky slopes at 2700–3000 m, but one seen in June along a riverside on outskirts of willow scrub (Paludan 1959). In winter in tropical west Africa, lives in savanna and in erosion areas with scattered low bushes and stony gullies, or better-wooded land, even gardens. At Mount Nimba (Guinea), on bare ground resulting from mining, and at

higher levels, and in Chad mainly at above 700 m. In Eritrea (Ethiopia), above 1300 m on open rocky moorland, on cliffs, and around buildings (Moreau 1972). Also in Nigeria on recently burnt areas in bush, and in desert on rocks or even small stones (Bannerman 1951). In Sierra Leone, perches readily in trees, flying into gallery-forest-edge trees when disturbed while feeding in the open (G D Field). Flies strongly and fast, but observed normally in lower airspace, using terrain for evasion. In Lebanon, scarcity attributed to shooting, even in mountains (Vere Benson 1970), but over most of range appears little affected by man.

Distribution. Has disappeared as breeder from West and East Germany, and range reduced in France, Czechoslovakia, and Austria.

FRANCE. Disappeared Autunois and Beaujolais since end of 19th century, some buildings in Burgundy *c*. 1917, and not seen regularly in Vosges after 1912 (Yeatman 1976). BELGIUM. Bred 19th century (Lippens and Wille 1972). WEST GERMANY. Formerly bred Niedersachsen 1885, Rheinland-Pfalz 19th century, and Hessen 19th century (AH). EAST GERMANY. Bred Zittauer Gebirge until 19th century (Makatsch 1981). CZECHOSLOVAKIA. Regular breeder in 19th century in Bohemia and more sites in Moravia (KH). AUSTRIA. Lowland populations in Wachau and eastern fringe of Alps extinct in 1950s and 1960s (HS, PP).

Accidental. Britain, Ireland, Netherlands, Belgium, West Germany, Denmark, Norway, East Germany, Canary Islands.

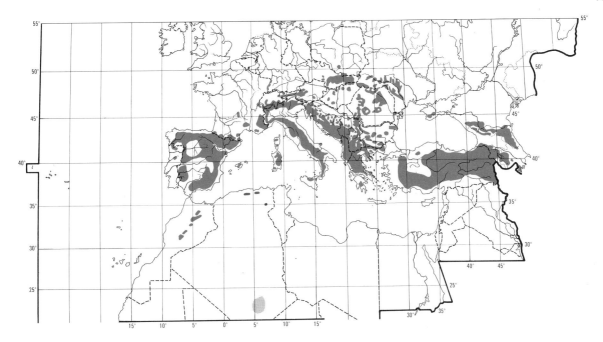

Population. Has declined in West and East Germany (see Distribution), Czechoslovakia, Italy, Rumania, Ukraine (USSR), and probably Poland. Possible increase in Switzerland. Volume of passage observed within Mediterranean basin seems to have fallen dramatically early this century (Sultana and Gauci 1982; Flint and Stewart 1983).

FRANCE. 1000–10 000 pairs (Yeatman 1976). POLAND. Probably more numerous in 19th century (Tomiałojć 1976). CZECHOSLOVAKIA. Slow decrease in 20th century, especially in west, more rapid after 1970 (KH). ITALY. Decreasing in south. Sardinia: 10–20 pairs (Schenk 1980). ALBANIA. Rare (EN). RUMANIA. Under 500 pairs; decreasing (WK). USSR. Ukraine: decreased (GGB).

Movements. Migratory. Most winter in Afrotropics, birds from eastern China travelling at least 7500 km from breeding to wintering grounds. A few birds appear to winter in Africa north of Sahara and in Arabian peninsula (Bundy and Warr 1980; Gallagher and Rogers 1980; Zink 1981). Claims of wintering in tropical Asia (e.g. Baker 1924) often repeated, but even recent claim by Etchécopar and Hüe (1983) is unsubstantiated (see below). Details of migration routes and timing obscured by infrequency with which passage detected. Nocturnal migrant, usually travelling singly or in loose aggregations, often with Blue Rock Thrush *M. solitarius*.

Main wintering area lies north and east of central African rain forests: from northern Nigeria and Cameroon (south to *c.* 8°30′N) east to Eritrea and from there south to at least 9°S in Tanzania (Zink 1981). In central Africa occurs as far south as 3°45′N 27°54′E in north-east Zaïre (Lippens and Wille 1976). Very few Ugandan records

west of 33°E in north or west of 34°E in east, although stragglers reach Akagera in Rwanda (1°30′S 30°45′E) (Vande Weghe 1979). 2 records from *Brachystegia* woodland in northern Zambia (*c.* 10°S), an area deserving closer scrutiny (Aspinwall 1977). Other southerly occurrences perhaps overlooked due to confusion (especially of ♀♀) with Mottled Rock Thrush *M. angolensis*, commonly found in *Brachystegia* woodland (even greater potential for confusion with Sentinel Rock Thrush *M. explorator*, which however not recorded north of 23°S). Strikingly few records, even on passage, west of Nigeria, though small numbers regular on passage in Sierra Leone, most leaving by early April (G D Field). Only 2 records from Ghana suggest wintering, and only 1 record at all from Gambia (Morel *et al.* 1983, *contra* Jensen and Kirkeby 1980). Origins of birds on different wintering grounds not known in absence of ringing recoveries or racial differentiation. Usually considered that those in eastern Africa mainly from Asia (Zink 1981), an arbitrary division followed in this account.

MEDITERRANEAN POPULATIONS of southern Europe and north-west Africa begin to disperse from breeding sites in August, most having left by late September. Claims of residency in Balearics (Bannerman and Bannerman 1983) appear to be incorrect. Breeders vacate Corsica in September, but further passage observed there in October (Thibault 1983). Appears to cross Sahara on broad front from Morocco to Sinai, but especially common in central section. However, autumn passage throughout Africa largely unobserved. Rarity of records in western Sahara suggests that breeding birds of west Mediterranean migrate south or east of south. Newby (1980) suggested that flight lines tend to be concentrated along rocky ridges (e.g. western rim of Chad basin). Reaches Chad mid-October (Newby 1980), Nigeria late November (Elgood *et al.* 1966). Occasional November–January records in Morocco, Ahaggar massif, Libya, and Egypt may indicate wintering north of Sahel zone by very small number (Bundy 1976; Ledant *et al.* 1981; Short and Horne 1981; Zink 1981). No quantitative data on wintering density in West Africa although widely regarded as less common than in eastern Africa. No data on timing of first departures from wintering grounds, although most sites south of Sahara vacated by mid-March with stragglers remaining until at least mid-April. Nearly all records west of Nigeria have occurred February–March (Morel *et al.* 1983) suggesting spring passage has more westerly bias, though this observation may simply reflect fact that passage is more visible in spring than autumn. Passage noted in Sahara and on North African coast March–May with peak in late March and early April (Bundy 1976; Thomsen and Jacobsen 1979; Ledant *et al.* 1981). However, first arrivals at southern breeding sites are usually in February, demonstrating that early passage in Africa overlooked. Northernmost European breeding sites (Switzerland) usually reached mid-April (Glutz von Blotzheim 1962). In some years, birds do not return to high-altitude sites (e.g. central Turkey) until early May (*Orn. Soc. Turkey Bird Rep. 1966–73*). Vagrants occur to north of breeding range as far as Shetland (Scotland) and Sweden; most such records occur May–June (e.g. 15 out of 18 British records) suggesting overshooting by late migrants; single records in April in Norway (Acklam 1970) and July in Netherlands (Zink 1981). Smaller peak of vagrancy records in late autumn perhaps due to reverse migration. Exceptional record of bird in south-east Britain from early February to early April may have been an escape.

ASIAN POPULATIONS breeding from east Mediterranean north-east to China, begin to leave breeding sites in August, with juveniles leaving ahead of adults (and usually claimed also to reach wintering grounds first). Passage continues in plains of Semirech'e (Kazakhstan, USSR) and Tadzhikistan until late September, but northern Baluchistan not vacated until at least late October (Dementiev and Gladkov 1954*b*). Fairly common on autumn passage to south of breeding range, especially in Pakistan. Most such records are August–September, but a few are later, especially in Sind where some in late November (Ali and Ripley 1973*b*). This prolonged passage casts considerable doubt on whether juvenile ♂ seen early November in Kiangsu (China) was wintering. This record appears to be basis for claim in Etchécopar and Hüe (1983) that wintering occurs in southern China; further confirmation required. Passage reaches Arabia by late August although most individuals appear to overfly. Distinctly uncommon in autumn throughout most of Middle East with concentration of records along south-east coast of Arabian peninsula and, to lesser extent,

eastern Mediterranean; peak passage appears to be early October but stragglers occur until late November. Scattered December–January records on Bahrain and Kuwait and at high altitude in Oman (Jabal al Akhdar) suggest sparse wintering population (Bundy and Warr 1980; Gallagher and Rogers 1980). First arrival in Ethiopia and Somalia in late August and early September, but peak passage not until early October (Benson 1946*b*; Smith 1957). Scarcity on passage in Sinai and Egypt combined with concentration of Sudanese records to west of Nile (Hogg *et al.* 1984) suggests that Asian birds tend to overfly this region or enter Africa further south and that breeding birds of east Mediterranean tend to migrate west of south. Equally possible that scale of passage across central Red Sea greatly underestimated. Birds reach northern Kenya by mid-October (exceptionally earlier) and southern Kenya and Tanzania by mid-November (Beesley 1972; Britton 1980). Evidence from Ngulia (3°00′S 38°13′E) and other sites on Kenyan-Tanzanian border for passage south continuing until at least mid-December. These movements interpreted as 'step' migrations between successive wintering areas (Curry-Lindahl 1981). Both Zambian records (see above) were in December. In Serengeti (Tanzania), sedentary after arrival, one marked bird returning to same spot for 3 successive years (Sinclair 1978). No quantitative data on wintering density; usually regarded as common in Kenya (see Zink 1981). Tanzanian wintering sites vacated by early March and most birds appear to leave Kenya later that month, although most sites not totally vacated until early April. Most leave Eritrea by late April (Smith 1957) although stragglers recorded until mid-May (Curry-Lindahl 1981). Passage across Sinai late March to late April (Dementiev and Gladkov 1954*b*; Strez 1984) and through Arabian peninsula late February to mid-May, with majority occurring in eastern Saudi Arabia in March (Gallagher and Woodcock 1980; G Bundy). Passage continues throughout April and early May over most of breeding range, with high-altitude sites (e.g. Uzbekistan, USSR) being occupied later than lower ground. Adult ♂♂ tend to arrive first (Meinertzhagen 1930; Ali and Ripley 1973*b*). In general, as for western Sahara, spring passage more obvious than autumn. However, reverse appears to be true for western Sudan, Oman, Pakistan, and north-west Indian peninsula (Ali and Ripley 1973*b*; Gallagher and Woodcock 1980; Hogg *et al.* 1984). In case of Sudan this may reflect greater aridity in spring; may also reflect more westerly bias suggested above to occur in Sahara during spring. Apparently reduced passage through Oman, Pakistan, and north-west India in spring, contrasting with situation to north-west of these areas (e.g. Bundy and Warr 1980), suggests different passage routes between seasons, with birds passing further to north in spring. However, some birds migrate so far to south in spring that long ocean crossings necessary: 2 ♂♂ shot well out to sea east of Kuria Muria islands at *c.* 17°N

60°E in late March (Moreau 1938). Record of vagrant in Seychelles (Penny 1974) may be extreme case of southerly tendency in autumn.　　　　DGCH

Food. Mostly large insects (especially beetles, Lepidoptera larvae, and Orthoptera). Feeds mainly by flying from perch (rock, tree, etc.) on to prey on ground; may eat several items while on ground, sometimes running or hopping a few metres between each before returning to perch (Farkas 1955; Beven 1969; Ali and Ripley 1973*b*; Schmidt and Farkas 1974; Lack 1980). In winter in Kenya, 96% of 116 items were from ground; during 13 of 27 pounces to ground, bird took several items before returning to perch (Lack 1980). Also takes aerial prey in brief flight from perch (Schmidt and Farkas 1974) and will take fruit direct from tree (Moreau 1943) and insects from cow dung (Meinertzhagen 1954). Flying prey usually brought back to perch but may be eaten in flight (Corti *et al.* 1949). Hammers beetles (Coleoptera) with bill before eating them. Lizards treated similarly, also shaken about, and one captive bird ate one 10 cm long, head first; often delivered to young in 2 pieces. (Koffan and Farkas 1956.)

Diet in west Palearctic includes the following. Invertebrates: damsel flies and dragonflies (Odonata), grasshoppers, etc. (Orthoptera: Gryllotalpidae, Gryllidae, Tettigoniidae, Acrididae), earwigs (Dermaptera), adult and (mostly) larval Lepidoptera including hairy ones (e.g. Saturniidae), flies (presumably Diptera), Hymenoptera (ants Formicidae, wasps Vespidae, bees Apoidea), adult beetles (Coleoptera: Carabidae, Scarabaeidae, Elateridae, Meloidae, Tenebrionidae, Chrysomelidae, Curculionidae), spiders (Araneae), centipedes and millipedes (Myriapoda), earthworms (Oligochaeta), small snails (Gastropoda). Also lizards *Lacerta*, etc., and small frogs (Anura). Plant material comprises fruits of *Spiraea*, cherry *Prunus*, rowan *Sorbus*, vine *Vitis*, currant *Ribes*, *Viburnum*, and elder *Sambucus*. (Csiki 1908; Dementiev and Gladkov 1954*b*; Farkas 1955; Glutz von Blotzheim 1962; Beven 1969; Schmidt and Farkas 1974; Kostin 1983.) In Crimea (USSR), June–September, 7 stomachs contained only insects: 44·7% (by number) beetles (Tenebrionidae most important), 40·2% Hymenoptera (22·3% wasps, 13·4% bees, 4·5% ants), 7·5% Tettigoniidae, 6·0% small caterpillars, and 1·5% earwigs (Kostin 1983). In Hungary, 3 stomachs, April and July, contained 16 beetles (mostly Scarabaeidae), 18 wasps and bees, 6 Orthoptera, and 1 millipede (Csiki 1908). See also Farkas (1955), Boehme (1958), Kovačević and Danon (1959), Pek and Fedyanina (1961), and Neufeldt (1966).

In winter, Zaïre, 4 stomachs all contained insects (1 contained only ants), also a large millipede and some fruit (Chapin 1953). In Kenya, stomachs contained small snails, a few Orthoptera, and occasionally berries (Meinertzhagen 1951*b*). Recorded taking mulberries *Morus* in Tanzania (Moreau 1943).

In Hungary, food brought to young comprised 30% Lepidoptera (especially caterpillars of *Saturnia*), 25% Orthoptera, 20% beetles, 5% Odonata, and 20% unidentified insects; also occasionally given small snails, lizards, and frogs (Farkas 1955). Food taken on Ala Dagh (Turkey), presumably intended largely for young, contained large proportion of caterpillars (Gaston 1968).　　　　DJB

Social pattern and behaviour. Major study in Hungary by Farkas (1955). For review, and comparison with other *Monticola*, see Schmidt and Farkas (1974).

1. Usually solitary. Small loose-knit flocks, notably of young birds, occur on spring migration (Farkas 1955). In winter, Tanzania, ♂ and ♀ probably defend separate territories (Sinclair 1978). BONDS. No evidence for other than monogamous mating system. Pair-bond presumably breaks down outside breeding season (see above), but may be renewed for several years on breeding grounds. 1 record of ♂ 'stealing' ♀ from another ♂. ♀ broods, while both members of pair feed young. One case of amicable and close association between 2 ♂♂ during successful breeding attempt (Cheylan 1973) suggests possible involvement of nest-helpers. Family bonds maintained for up to 3–4 weeks after leaving nest, ♂ playing especially prominent guardian role. (Farkas 1955.) In one case of 2 broods, ♂ ignored fledged young of 1st brood when 2nd brood in nest (Schmidt and Farkas 1974). Age of first breeding not known. BREEDING DISPERSION. Solitary and territorial. In northern Italy, 3 territories 8 ha, 10 ha, and 12·7 ha, density thus *c.* 10 pairs per km² (Saporetti 1981). In southern France, 1 territory *c.* 3·5 ha (Cheylan 1973). In Abruzzo (Italy), 1 singing ♂ per 600–1000 m (Carlo 1972). In Switzerland, pairs 200–500 m apart (Schifferli *et al.* 1982). In some secluded valleys, southern France, pairs sometimes only 500–800 m apart, usually much more (Yeatman 1976). Once, 2 nests 60 m apart (Bókai 1955). Territory may serve for courtship, nesting, and feeding (Schmidt and Farkas 1974), but areas used for courtship and feeding may be separate, sometimes kilometres apart (Farkas 1955). Birds usually move several hundred metres from nest to feed (Schmidt and Farkas 1974). If neighbouring territories fall vacant, residents may extend their own and make at least partial use of them; in successive years, pair thus bred in 3 formerly separate territories of which 2 contiguous, 1 *c.* 400 m away (Schmidt and Farkas 1974, which see for details). Marked site fidelity not uncommon, at least among ♂♂: 1 ♂ used same territory for 4 years (Bókai 1955; Farkas 1955). Nest-sites traditional over several years; in one case of 2 broods, 2nd nest 50 m from 1st (Schmidt and Farkas 1974). 1-year-olds tend to settle near natal area (Bókai 1955; Farkas 1955). ROOSTING. Solitary, high up in fissures in rocks (Farkas 1955; Bókai 1957). Captive young roosted nightly in same sites, never near one another nor their parents (Phillips 1910). Comfort behaviour includes bathing (Koffan and Farkas 1956).

2. Very shy, readily seeking cover of rocks if disturbed (Farkas 1955); ♂♂ bolder than ♀♀ (Bókai 1957). In winter, Sierra Leone, birds disturbed feeding in the open almost always fly for cover of trees (G D Field). On breeding grounds, tends to be first bird in vicinity to give warning (see 3a in Voice) of approaching raptor (Farkas 1955, which see for details). In distress, e.g. when unable to escape from predator, performs Wing-lifting display (Fig A), raising one wing (Farkas 1955; see also Antagonistic Behaviour, below). For other alarm responses, see Parental Anti-predator Strategies (below). FLOCK BEHAVIOUR. No information. SONG-DISPLAY. Unless otherwise

A

stated, following account after Farkas (1955), Koffan and Farkas (1956), and Schmidt and Farkas (1974), which together deal with same body of information. Song-display mainly by ♂, much less often ♀, given perched, in Song-flight, or normal flight. ♂ tends to have separate perches for singing and feeding. Subsong (see 1a in Voice) may be given from feeding perch, but full song (see 1a in Voice) only from song-perches. Singing bird usually sleeks plumage and may adopt a crouched posture. Song from perch often a response to intrusion by nearby ♂, leading to Song-duel (see also Antagonistic Behaviour, below). Song-duels said to be specially frequent in rainy weather (Bókai 1957). Bird often follows successful confrontation with a low variant of song (see 1a in Voice). Song also given when man or other intruder enters territory (see Parental Anti-predator Strategies, below). Song-flight begins from perch. ♂ takes off suddenly, initially staying low, then ascends steeply with slow powerful wing-beats. According to Jonsson (1982), ascent is fluttering like a lark (Alaudidae). Tail is fanned, back and head plumage ruffled. Bird begins singing during ascent, reaching maximum output at top of ascent where bird soars, flutters rapidly, and typically introduces mimicry (see 1b in Voice) into song; then suddenly plummets, not singing, for 15–20 m with wings and tail outspread (usually parachute descent: Saporetti 1981). Bird may make only a single ascent; flight path then usually undulating, but sometimes mostly horizontal, or with ascent markedly oblique (Saporetti 1981). One Song-flight, apparently of this kind, ended wih long glide of c. 40 m to a bush (Wadley 1951). More often, bird does not land after descent but, apparently using momentum of plummet, ascends for a 2nd Song-flight, less high, however, than 1st (Fig B).

B

Depending on intensity, bird repeats Song-flight 2–3 times or more. Song-flight always ends on perch, with low variant of song. Perched bird then reverts to full song and, after c. 10–15 min, performs another Song-flight. ♀ performs Song-flight rarely, and then only during nestling period; e.g. after feeding young, ♀ flies to a nearby perch in manner of Song-flight but less elaborate and with less fulsome song than ♂ (see 1c in Voice). ♀ may also display thus when released by handler, suggesting that Song-flight may signal general excitement. Song-flights begin upon arrival on breeding grounds, most

frequent during pair-formation, especially early morning and late afternoon. In winter, Kenya, Subsong reported, also quite loud song from captive ♂ (Beven 1969). Young ♂♂ start singing at end of summer (Turček 1967). Juveniles first heard singing (Subsong) at 8–10 weeks, full song at 12 weeks (Farkas 1955). Antagonistic Behaviour. Following account based largely on Farkas (1955) and Schmidt and Farkas (1974). Neighbouring residents seldom openly aggressive and both ♂♂ and ♀♀ often intrude on neighbouring territories with relative impunity. Rival neighbouring ♂♂ confront each other across territorial boundary, singing incessantly with ruffled plumage (Fig C). If

C

one tries to trespass, a fast erratic aerial chase results, accompanied by song, but no fight develops and intruder eventually retreats. However, resident ♂ confronts intruding 1-year-olds more vigorously: intruder usually the more aggressive, following resident around. Both birds perform Advertising-display (Fig D), similar to that of Blue Rock Thrush *M.*

D

solitarius (see that species—Fig C, p. 908): partly spread and droop wings, fan and shiver tail, sleek plumage on head but ruffle plumage on rest of body. In close encounter, rivals ruffle head plumage and Bill-snap or perform Song-duel in which they stretch their necks forward and weave their heads back and forth while giving Subsong. Rarely, one bird lunges with bill and rivals then run apart as if startled. Otherwise, confrontation interrupted by pauses of varying length in which rivals run around mock-feeding (e.g. peck furiously at stone) and mock-preening whilst watching each other intently. May also perch on bushes and begin another Song-duel, confrontation persisting sometimes for hours and intermittently for several days. Usually ends with retreat of younger bird. ♀ usually less aggressive, but in one territory ♀ markedly aggressive, singing and attacking intruders (Saporetti 1981). Threatening ♀ commonly Bill-snaps but does not shiver tail (Schmidt and Farkas 1974). Occasionally, signal threat with Wing-lifting display (Farkas 1955: Fig A). In breeding season, may defend territory against Wheatear *Oenanthe oenanthe* (Farkas 1955; Saporetti 1981). In winter, Sierra Leone, chases off *M. solitarius* in zone of habitat overlap (G D Field). Heterosexual Behaviour. (1) General. In spring, adult ♂♂ migrate first, followed soon after by ♀♀ (Meinertzhagen 1922, 1930). Most ♂♂ thus arrive on breeding grounds before ♀♀, sometimes up to 3 weeks earlier (Koffan and Farkas 1956), though some ♀♀ among first arrivals (Farkas 1955). ♂♂ take up territories immediately (Farkas 1955), though several weeks may then elapse before breeding (Schmidt and Farkas 1974). According to Cheylan (1973), in 2 cases ♂ and ♀ arrived on breeding grounds, southern France, already paired. (2) Pair-bonding behaviour. Courtship occurs throughout breeding season, continuing after breeding, and ending only with onset of moult (Farkas 1955). May occasionally begin in winter quarters (see subsection 3, below), and captive birds began in mid-February. Courtship comprises vocal and ground display (Farkas 1955; Schmidt and Farkas 1974, on which following account of ground display based). ♂ runs towards ♀, performing Advertising-display with neck extended markedly upwards, this presumably signalling appeasement as in *M. solitarius* (p. 908), and giving Subsong. When close to ♀, ♂ stops, and moves his neck and head from side to side (Snake-display), and moves sideways with tripping gait, singing incessantly. This display, prior to apparent copulation attempt, also reported by Magnenat (1962): ♂ perched with wings drooped and tail fanned, and stretched his neck with slow snake-like movements from left to right, and with contortions of the head, for 15-20 s. ♂ then performs Bowing-display: bends his legs, bows forwards, raising and shivering his fanned tail. Bird shown in drawing by E N Panov apparently performing this display (Fig E, right) does

E

not, however, have tail fanned. After bowing, ♂ again raises head and performs Snake-display, accompanied by rapid song. Ground display also includes Allopreening which may occur at any time during breeding season. ♂ display may lead directly to Mating (see below). During ♂-♂ confrontation, arrival of ♀ mate of one ♂ caused him to cease chasing and begin ground display, whereupon rival ♂ returned to his own territory. (3) Courtship-feeding. May occur at any time during breeding season (Farkas 1955), but no information on regularity. Also reported once outside breeding season: in Tanzania, March, ♂ offered mulberries *Morus* to ♀ who accepted them (Moreau 1943). ♂ occasionally feeds incubating ♀ (Farkas 1955; Schmidt and Farkas 1974: see also subsection 5, below). (4) Mating. Song-flight or ground display may lead directly to copulation. If ♀ crouches after ♂'s sequence of ground display, copulation follows immediately. If ♀ does not crouch, ♂ ceases ground display, runs away, takes off and performs Song-flight. ♀ may solicit by giving variant of song. (Farkas 1955; Schmidt and Farkas 1974: see 1c in Voice.) When ♂ pounced on ♀ (apparent copulation attempt) after ground display, ♀ escaped (Magnenat 1962). No further information. (5) Behaviour at nest. ♀ selects nest-site, accompanied by ♂ who sings from time to time. ♀ may continue nest-building after laying. Incubating ♀ leaving nest to feed, briefly early in morning and for longer around midday, usually accompanied by, and often chased by, ♂ mate. ♀ scarcely leaves nest for last 3 days before hatching, and food

she receives from sentinel ♂ then especially important. ♂ communicates with sitting ♀ by Subsong. 2 records of ♂ removing ♀'s faeces at this time. (Farkas 1955; Schmidt and Farkas 1974.) Relations within Family Group. Young fed by both parents. When ♀ brooding, ♂ passes food to her for transfer to young; ♂ often stays at nest for some time, singing softly. As young develop, ♀ broods them less (dependent on weather), and ♂ does increasingly more direct feeding. Arriving ♂ gives quiet song from surveillance point near nest, then flies silently to nest. Both parents perform nest-sanitation. Young leave nest unable to fly at 14-15 days, and disperse widely. Once fledged, brood re-groups and is fed and guarded by both parents, often mostly by ♂. Young quiver wings when begging and being fed. Especially when being fed from one side, frequently raise one wing, usually the one away from the parent (compare Fig A). Towards independence, wing-quivering gives way to a twitching movement which, along with tail-shivering and bowing, forms a complex of intention movements. Post-fledging care continues until onset of juvenile moult at 35-42 days. During juvenile moult brood does not completely disband, moving around independently but coming together at times during the day. (Farkas 1955.) Anti-predator Responses of Young. On hearing Warning-calls (see 3a in Voice), young crouch in nest (Schmidt and Farkas 1974). On leaving nest, hide under rocks, bushes (etc.) until able to fly (Farkas 1955). Parental Anti-predator Strategies. (1) Passive measures. ♀ sits tightly, especially in last 3 days before hatching when she can often be touched (Farkas 1955). (2) Active measures. Insufficient information to distinguish responses to different predators. On intrusion by man or other mammalian predator, ♂ sings softly and intermittently to warn mate and young. When young in nest (and not before), both sexes give Warning-calls as soon as intruder enters territory, frequency and urgency of calls increasing as young get older; once young leave nest, Warning-calls supplemented by, and, in greater danger, replaced by Alarm-calls (Phillips 1910; Bókai 1957: see 3b in Voice, also for variants), e.g. Warning-calls given when bird of prey distant, Alarm-calls when nearby (Farkas 1955; Schmidt and Farkas 1974). Alarm-calls accompanied by bobbing, tail-shivering, and occasionally wing-twitching (Schmidt and Farkas 1974).

(Figs by C Rose: A-C and E from drawings by E N Panov; D from drawing in Schmidt and Farkas 1974.) EKD

Voice. Freely used, especially in breeding season. For song outside breeding season, see Social Pattern and Behaviour. Unless otherwise stated, account based on Farkas (1955) and Schmidt and Farkas (1974); see also Bergmann and Helb (1982), especially for additional sonagrams. Bill-snapping used in threat against conspecifics (see Social Pattern and Behaviour).

Calls of Adults. (1) Song. Given by both sexes but mainly by ♂. Similar to Blue Rock Thrush *M. solitarius* (see that species, p. 909), but softer and more flowing. Similar timbre to Redwing *Turdus iliacus* (Jonsson 1982). Comprises melodious fluting phrases, often with obvious mimicry. 23 species reportedly mimicked; all ♂♂ include Chaffinch *Fringilla coelebs* song, and in one recording (Fig I) this rendered 'chip chip chip chip tell r-r-r-r-r tu tu wechoo tu wechoo'. Each ♂ has a repertoire of 3-6 phrases, each comprising 6-12 units. (a) Perched song of ♂. Relatively measured, with long pauses, thus slower and less continuous than type 1b (Bergmann and

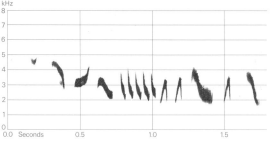

I P Szöke Hungary

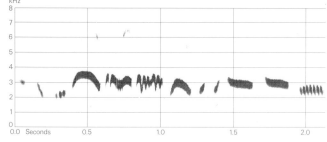

II J-C Roché/Cornell Laboratory of Ornithology (1977) France

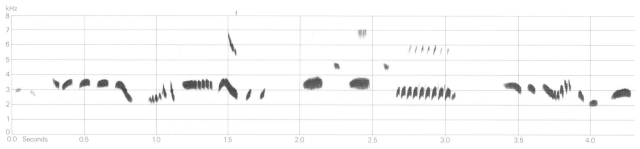

III J-C Roché/Cornell Laboratory of Ornithology (1977) France

Helb 1982: Fig II); seldom includes mimicry though Fig III, which depicts phrase of same song from which Fig II taken, shows, at c. 2 s from start, mimicry of rather 'buzzy' Great Tit *Parus major* (J Hall-Craggs). In confrontations, or when disturbed, bird repeats same phrase. After successfully expelling rival, ♂ typically gives 'organ' variant: a persistent, continuous, loud, low-pitched song. Subsong varied, continuous, and persistent; described as a quiet, twittering, rattling, gurgling sound with somewhat ventriloquial quality, including frequent mimicry. (b) Aerial song of ♂. A loud mixture of perched song and Subsong types, mostly given in Song-flight. Regularly includes mimicry, especially at summit of Song-flight; *F. coelebs* song often then given, switching to (e.g.) Nightingale *Luscinia megarhynchos* song on descent. (c) Song of ♀. ♀ gives organ variant (see 1a) of perched song, notably to solicit copulation and after feeding young; also rarely gives aerial song, and captive young ♀♀ gave Subsong. (2) Contact-alarm call. Single or short series of 'tak' sounds, often accompanied by tail-flicking, when mildly disturbed (Bergmann and Helb 1982); 'dack-dack'; e.g. when showing interest in potential food (Schmidt and Farkas 1974). (3) Warning- and Alarm-calls. (a) A plaintive mournful pipe, given by both sexes (Phillips 1910), not unlike Bullfinch *Pyrrhula pyrrhula* and rendered 'huep-huep-huep' (Bókai 1957). Also described as 'hüt-hüt-hüt-hüt' (Farkas 1955), a soft 'diu' (Jonsson 1982), a quiet pure 'jü' (Bergmann and Helb 1982). Soft 'uit-uit' by ♀ signalling mild anxiety (Farkas 1955) is presumably the same. (b) In greater alarm, an emphatic rapidly repeated variant of call 2, rendered 'schack-schack' (Bergmann and Helb 1982). May be combined with call 3a and other sounds, e.g.

'uit-schack-schack-schack' or 'frit-tschak-tschak-tschak'. (4) Discomfort- and Distress-calls. (a) A rather harsh 'rätsch', e.g. when hungry. (b) 'Scur-r-r' of captive ♀ apparently anxious to be fed (Phillips 1910). Drawn out soft rattling 'kschrrrr', like Woodchat Shrike *Lanius senator* (Jonsson 1982), perhaps the same or related.

CALLS OF YOUNG. Food-call a sharp 'ürr'. Contact-call of fledged young a quiet but penetrating 'zréii' or 'te-tre-träi', rising at the end (Fig IV); rather difficult

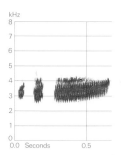

IV P Szöke and M Orszag
Hungary June 1967

for human to locate (Bergmann and Helb 1982). Young also give call 4a of adult. For development of song, see Social Pattern and Behaviour. EKD

Breeding. SEASON. Earliest eggs late April, main season May–June, apparently throughout range (Dementiev and Gladkov 1954b; Farkas 1955; Heim de Balsac and Mayaud 1962; Beven 1969; Makatsch 1976). SITE. Horizontal crevice in rock-face, wall, ruin, or crag, under boulder on steeply sloping ground, or occasionally in tree-hole. Nest: neat cup of grass, rootlets, and moss, lined with finer rootlets and moss. Building: by ♀ (Farkas 1955). EGGS. See Plate 83. Sub-elliptical, smooth and glossy;

pale blue, often unmarked, or with some faint speckles of red-brown at broad end. 26.0 × 18.2 mm (22.5–29.0 × 16.5–21.0), *n* = 100; calculated weight 4.95 g (Schönwetter 1979). Clutch: 4–5(–6). 1–2 broods. Replacements laid, sometimes twice, after clutch loss (Koffan and Farkas 1956). INCUBATION. 14–15 days. By ♀ only. YOUNG. Cared for and fed by both parents, though ♂ sometimes plays little part (Koffan and Farkas 1956). FLEDGING TO MATURITY. Fledging period 14–16 days. No further information.

Plumages. ADULT MALE BREEDING. Head and neck down to throat and upper mantle pale slate-blue or greyish cerulean-blue; eye-ring often slightly paler grey-blue, faint duller grey spot in front of eye. Lower mantle and scapulars rather variable, sooty-black, dark slate-blue, or grey-blue, usually with much white of feather-bases visible, especially on centre of mantle and inner scapulars. Back white, feathers often with traces of grey-blue fringes; lower sides of back and all rump slate-blue or grey-blue, longest feathers with pale tips. Upper tail-coverts bright rufous-cinnamon, sometimes with black spot on centre. Underparts backwards from chest entirely rufous-cinnamon, deepest, almost rufous-chestnut on chest and flanks, paler orange-rufous on central belly, vent, and under tail-coverts; traces of white feather-tips usually on central belly and vent, broader ones on under tail-coverts. Central pair of tail-feathers (t1) dark grey-brown or greyish black, basal half and sometimes part of shaft rufous-cinnamon; remainder of tail deep rufous-cinnamon, distal outer edge of t2 and t6 (and sometimes others, too) with dark streak. Flight-feathers, tertials, greater upper primary coverts, and greater upper wing-coverts black or brown-black, feathers with traces of even white fringes along outer edges and tips (in particular, along basal outer webs of primaries, along tips of greater coverts, and along inner primary coverts). Lesser and median upper wing-coverts black or slate-black with traces of off-white fringes along tips and slate-blue ones along outer webs. Under wing-coverts and axillaries deep rufous-orange, some small coverts along leading edge of wing tipped white. *In fresh plumage*, about February–March, all feather-tips of head, upperparts, and throat narrowly fringed sandy-grey or pale brown, each fringe narrowly bordered dull black subterminally; much grey-blue of head and mantle and white of back still visible; rufous of underparts partly hidden underneath white feather-tips. Pale fringes gradually wear off, traces remaining last on ear-coverts, back, rump and central belly to under tail-coverts. ADULT FEMALE BREEDING. Upperparts like adult ♂ non-breeding, grey-brown with pale buff or off-white spotting in fresh plumage (due to pale feather-tips), duller and more uniform grey-brown in worn plumage, when often showing black shaft-streak or small black spots on feather-tips; feathers of crown, hindneck, and mantle virtually without grey at base (unlike ♂ non-breeding); back and rump with buff or cream-white central spots or arrow-marks on bases (in ♂ non-breeding, broader pure white marks). Upper tail-coverts like ♂, but slightly paler rufous-cinnamon. Sides of head closely mottled cream or buff and grey-brown, sometimes with indistinct off-white supercilium; distinct narrow cream or off-white eye-ring; ear-coverts often slightly darker brown (in particular above and at rear), narrowly streaked warm buff or off-white. Ground-colour of chin and throat off-white, of remainder of underparts pale orange-buff or cream-buff with broad white feather-tips (in adult ♂, deeper rufous-cinnamon); fine dark arcs from chin to chest and bars on remainder of underparts as in adult ♂ non-breeding, but sometimes narrower,

less regular, and partly broken, underparts appearing paler and more uniform with limited amount of dark mottling. Tail as adult ♂. Flight-feathers and tertials dark brown or brown-black (darkest on tertials and feather-tips), tips and outer edges of tertials and secondaries and tips of primaries broadly fringed pale grey-buff to off-white; outer webs of primaries narrowly fringed sandy or off-white. Upper wing-coverts dark brown or black-brown, lesser with variable amount of grey tinge, tips and outer edges of all coverts (including primary coverts) fringed pale buff, sandy, or off-white. Axillaries and under wing-coverts rufous-orange. *In fresh plumage*, January–February, upperparts with pale spots on feather-tips, underparts sometimes with broad white feather-tips almost concealing subterminal black and rufous or buff bases; pale fringes of upperwing and flight-feathers slightly abraded. *In worn plumage*, all pale feather-fringes worn off, wing uniform dark brown except for some traces of off-white on tips of secondaries, outer greater coverts, and inner greater primary coverts. ADULT MALE NON-BREEDING. Differs from fresh adult ♂ breeding by largely brown head, upperparts, chin, and throat, almost without grey-blue at feather-bases, though often much white hidden on feather-bases of back and rump. Upperparts and sides of head grey-brown or olive-brown, feather-tips paler sandy-brown (paler and wider on mantle and scapulars, more pink-buff and appearing spotted on sides of head); hidden feather-bases grey (not bluish) on top of head, mantle, inner scapulars, and sometimes cheek, white on back and (less extensively so) rump. Upper tail-coverts rufous-cinnamon with paler tips, sometimes partly marked white. Indistinct stripe on lores and more distinct eye-ring cream-buff to off-white. Ground-colour of chin and central throat off-white, of sides and lower throat buff or cream-buff; each feather with black arc along tip, chin and throat appearing scaled. Remainder of underparts rufous-cinnamon, partly hidden underneath broad white feather-tips; chest-feathers with black subterminal arcs, appearing scaled (like throat), remainder with more regular narrow black bars (sometimes absent towards vent and under tail-coverts). Tail as in adult breeding. Flight-feathers, tertials, greater upper wing-coverts, and greater upper primary coverts brown-black or dull black, as in breeding, but broad and even fringes along outer webs sandy or pale grey-buff (paler, almost greyish-white on greater coverts; narrower on outer primaries), merging into broad off-white fringe along tip. Lesser and median upper wing-coverts as in breeding. *In worn plumage*, about November–December, pale fringes on upperparts and sides of head partly worn off, appearing more uniform grey-brown (some white sometimes visible on back); ground-colour of chin and throat paler, dark scales more contrasting, remainder of underparts more uniform rufous with more distinct black arcs and bars. ADULT FEMALE NON-BREEDING. Like adult ♀ breeding; head and body fresh and wing very fresh September–October; head and body worn but wing still fairly fresh about December, when pale fringes of wing-coverts and flight-feathers much broader than in worn adult ♀ breeding. NESTLING. Down long, dark bluish-grey (Witherby *et al.* 1938*b*); for development, see Schmidt and Farkas (1974). JUVENILE. Entire upperparts and sides of head and neck buff, each feather with narrow dark grey or brown-black fringe along tip, either appearing buff with dark scaling or dark grey-brown with large buff spots; upper rump gradually more rufous, lower rump and upper tail-coverts rufous-cinnamon, rump-feathers with blackish fringes. Ear-coverts dark brown with buff or rufous-brown streaks. Chin and throat pale buff, rather faintly scaled grey; chest backwards to belly and flanks warm buff with dull black scaling (buff paler and scaling narrower than on upperparts).

Under tail-coverts rufous-cinnamon, fluffy. Tail and flight-feathers as adult; greater upper primary coverts as adult, but fringes slightly broader, whiter, and less clear-cut. Lesser upper wing-coverts dull grey with narrow buff shaft-streak widening on feather-tip; median coverts dull black with narrow pale rufous or buff shaft-streak and a similarly coloured tip; greater coverts and tertials dull black with rufous or buff fringe of 2-3 mm wide along outer web and tip, sometimes forming triangular patch at shaft (apex to feather-centre). Axillaries and under wing-coverts rufous-buff. FIRST ADULT MALE NON-BREEDING AND BREEDING. Like breeding and non-breeding of adult ♂, but brown fringes on head and upperparts and white fringes on underparts slightly wider, less soon wearing off; in non-breeding, feather-bases on upperparts less extensively blue or white, ground-colour of chest, flanks, and belly paler rufous-cinnamon. Juvenile flight-feathers, tertials, greater upper primary coverts and many or all greater upper secondary coverts retained; usually browner and more worn than adult at same time of year; tail-feathers often narrower, primary coverts more sharply pointed; pale fringes along tips of retained feathers worn off at an earlier date than in adult, sometimes lost by December but occasionally still fresh April. Some birds show pale buff · or off-white triangular spots on tips of greater secondary coverts when slightly worn (or triangular indentations when heavily worn); these are easy to age, but others without such marks are hard to separate from adults. FIRST ADULT FEMALE NON-BREEDING AND BREEDING. Like breeding and non-breeding of adult ♀, but part of juvenile feathers retained, as in 1st adult ♂; characters as in 1st adult ♂; ageing often difficult.

Bare parts. ADULT. Iris brown or dark brown. Bill slate-black or black in ♂, dark brown or black-brown with slightly paler base of lower mandible in ♀. Leg and foot brown, dark brown, horn-black, or slate-black. NESTLING. Mouth dull ochre-yellow to bright orange-yellow; no spots. Gape-flanges pale yellow, creamy-white, or yellowish-white. JUVENILE. Iris brown. Bill yellowish flesh-grey. Leg and foot flesh-grey or slate-grey. FIRST ADULT. Like adult, but bill of ♂ dark brown or brown-black in 1st autumn. (Hartert 1910; Witherby *et al.* 1938b; BMNH, RMNH.)

Moults. ADULT POST-BREEDING. Complete; primaries descendant. Starts late June to early August with p1. Moult rapid, up to 4-5 primaries moulting at same time (e.g. p1-p3 new, p4-p7 growing, p8 just shed, p9-p10 old) and tail-feathers replaced simultaneously or nearly so (all feathers growing at same time, sometimes except for 1-2 old ones which are replaced later on). Moult completed with p9-p10 early August to early September. Tail shed from primary moult score 5-15; tertials, some wing-coverts, and sides of body at about same time as p1 or (occasionally) slightly earlier; tail, body, and wing-coverts largely completed at score 35-40, when secondaries moulting. (Stresemann 1920; Paludan 1940; Dementiev and Gladkov 1954b; Piechocki and Bolod 1972; Schmidt and Farkas 1974; BMNH, RMNH, ZFMK, ZMA.) ADULT PRE-BREEDING. Partial: head and body (sometimes except some feathers of central belly and vent or tail-coverts), sometimes a few inner lesser and median upper wing-coverts, occasionally a few tertial coverts. Starts mid-November or December, completed late January to early March. POST-JUVENILE. Partial: head, body, lesser and median upper wing-coverts, occasionally some tertial coverts, rarely some tertials or inner greater upper wing-coverts. Starts shortly after fledging, at age of 5-6 weeks (Schmidt and Farkas 1974); birds in fully juvenile plumage encountered late

June to early August, birds in moult late June to mid-August, birds with moult completed from late July (but mostly second half of August and early September). (Stresemann 1920; Stresemann *et al.* 1937; Dementiev and Gladkov 1954b; BMNH, RMNH, ZFMK, ZMA.) FIRST PRE-BREEDING. Extent as in adult pre-breeding. Starts late November–January, completed February–March. Subsequent moults as in adult.

Measurements. ADULT, FIRST ADULT. Wing (1) southern Europe (including data from Stresemann 1920), (2) Turkey, Levant, Armeniya (USSR), and northern Iran (including data from Paludan 1940 and Nicht 1961), (3) Afghanistan, Tien Shan, and Altai (including data from Paludan 1959), (4) Mongolia and Kansu, China (Stresemann *et al.* 1937; Piechocki and Bolod 1972; Piechocki *et al.* 1982), (5) whole geographical range, including Afrotropics, excluding data from literature; all year; skins (RMNH, ZFMK, ZMA). Bill (S) to skull, bill (N) to distal corner of nostril; exposed culmen on average 6·2 less than bill (S).

		♂		♀	
WING (1)		124·1 (3·27; 21)	119–129	120·0 (3·37; 10)	115–126
(2)		124·0 (4·41; 10)	118–131	121·4 (3·23; 5)	118–125
(3)		120·4 (2·05; 9)	117–123	121·1 (0·85; 4)	120–122
(4)		123·2 (1·95; 16)	120–126	117·8 (3·60; 9)	113–123
(5)		124·3 (3·49; 30)	118–131	120·7 (3·83; 15)	115–126
TAIL (5)		60·8 (3·00; 24)	56–65	62·0 (3·71; 12)	56–67
BILL (S) (5)		25·0 (1·10; 30)	23·2–27·2	25·2 (1·21; 13)	23·3–27·0
BILL (N) (5)		13·8 (0·77; 22)	12·5–15·0	14·2 (0·80; 13)	12·9–15·5
TARSUS (5)		28·3 (1·03; 26)	26·8–30·1	28·2 (0·84; 13)	27·1–29·6

Sex differences significant for wing (1), (4), and (5). 1st adult with retained juvenile flight-feathers and tail combined with older birds, though juvenile wing on average 1·8 shorter than adult wing, and tail on average 2·3 shorter.

Weights. Tanzania, January: ♂ 50·5 (Moreau 1944). Kenya, November–December: 48·0 (4·2; 78) 39–57 (D J Pearson). Nigeria: December, ♀ 47 (Fry 1970b); mid-March, 56; late March, 65, 72; injured, early April, 40·5 (Smith 1966). South-east Morocco, early April ♀ 41 (BTO). Southern Spain, May: ♂ 60·5, ♀ 55·5 (Niethammer 1957). Jordan, Kuwait, and Sharjah, on migration April and early May: ♂♂ 48, 48, 53; ♀♀ 34, 40 (BTO). Eastern Turkey, Armeniya (USSR), and Iran, June: ♂♂ 52·8, 53·5, 53·9, 59; juvenile ♀ 41 (Paludan 1940; Nicht 1961; Kumerloeve 1969a; Desfayes and Praz 1978).

Afghanistan (Paludan 1959) and Kazakhstan (Dolgushin *et al.* 1970): (1) May, (2) June, (3) July–August. Mongolia (Piechocki and Bolod 1972; Piechocki *et al.* 1982) and Kansu, China (Stresemann *et al.* 1937): (4) May, (5) June and early July.

	♂		♀	
(1)	51·5 (6·40; 10)	44–65	51·0 (— ; 3)	47–57
(2)	47·0 (3·15; 16)	40–54	52·4 (6·79; 6)	45–65
(3)	50·7 (3·54; 10)	46–56	48·6 (4·58; 8)	42–54
(4)	50·8 (1·50; 4)	49–52	58·5 (— ; 2)	57–60
(5)	49·9 (4·27; 13)	43–59	56·8 (4·27; 4)	54–63

Structure. Wing rather long, broad at base, tip bluntly pointed. 10 primaries: p8 longest, p9 0-2 shorter, p7 3-6, p6 9-13, p5 17-21, p1 36-44. P10 reduced, 74-86 shorter than wing-tip, 10-17 shorter than longest upper primary covert. Outer web of (p7-)p8 and inner of (p8-)p9 emarginated. Tertials short. Tail short, tip square; 12 feathers. Bill rather long, straight; wide at base, laterally compressed at tip; tapers gradually to sharp tip, but distal part of culmen decurved, often ending in fine hook. Nostril rather small, oval; partly covered by membrane above. Some short bristles at base of upper mandible; many fine hairs on lore, near nostril, and in front of eye. Leg and foot rather short and slender. Middle toe with claw 24·9 (10) 23·5–27; outer toe with claw *c.* 67% of middle with claw, inner *c.* 63%, hind *c.* 73%. Claws rather long and strong, decurved.

Geographical variation. Very slight. Some birds from central Asia on average paler grey on head than European ones, others slightly smaller, and therefore sometimes separated as *turkestanicus* Zarudny, 1918, but individual variation in colour everywhere large and difference in size slight (see Measurements). Birds from eastern Turkey separated as *coloratus* Stepanyan, 1964; stated to have more saturated colours than European birds and deep dark rusty-red underparts, but small sample examined from eastern Turkey inseparable from European birds (Kumerloeve 1969*a*; BMNH, ZFMK) and *coloratus* thus not recognized.

Forms superspecies with Little Rock Thrush *M. rufocinerea* from south-west Arabia and north-east Africa, Short-toed Rock Thrush *M. brevipes* from south-west Africa, Sentinel Rock Thrush *M. explorator* from South Africa (these 3 sometimes even considered races of *M. saxatilis*: Meinertzhagen 1951*b*, 1954), Mottled Rock Thrush *M. angolensis* from south-central Africa, Madagascar Rock Thrush *M. imerimus* and Sharpe's Rock Thrush *M. sharpei* from Madagascar, Blue-headed Rock Thrush *M. cinclorhynchus* from Himalayas, and White-throated Rock Thrush *M. gularis* from eastern Asia (Hall and Moreau 1970); for relationships, see also Schmidt and Farkas (1974).

CSR

Monticola solitarius Blue Rock Thrush

PLATES 66 and 76
[between pages 856 and 857]

Du. Blauwe Rotslijster   Fr. Merle bleu   Ge. Blaumerle
Ru. Синий каменный дрозд   Sp. Roquero solitario   Sw. Blåtrast

Turdus solitarius Linnaeus, 1758

Polytypic. Nominate *solitarius* (Linnaeus, 1758), southern Europe and North Africa east to Caucasus, Transcaucasia, Turkey, and Levant; *longirostris* (Blyth, 1847), north-east Iraq, Iran, Turkmeniya (USSR), and Afghanistan (except east). Extralimital: *pandoo* (Sykes, 1832), mountains of central Asia from eastern Afghanistan through Tien Shan to Dzhungarskiy Alatau and through Himalayas to mountains of central China; *philippensis* (Statius Müller, 1776), eastern Asia from Manchuria, southern Sakhalin, and Japan south to eastern China and Taiwan; *madoci* Chasen, 1940, Malay peninsula and Sumatra.

Field characters. 20 cm; wing-span 33–37 cm. 20% shorter than Blackbird *Turdus merula*; 10% longer than Rock Thrush *M. saxatilis*. Shape, stance, and habits much as *M. saxatilis*, but noticeably bulkier with longer tail. Plumage (in west Palearctic races) dark and relatively uniform at all ages and seasons: ♂ dusky blue, ♀ mottled dusky. Sexes dissimilar; little seasonal variation. Juvenile separable. 2 races in west Palearctic, not always separable in winter.

(1) Race of Europe and North Africa east to Caucasus, nominate *solitarius*. ADULT MALE. In spring, dull slate-blue, with black tone strongest on back, wings, and tail. Often looks dull black at distance or against light. Fresh autumn plumage less uniform, showing sandy fringes on upperparts, white tips to wing-coverts, brown tips to chest feathers, and buff-white tips on belly. ADULT FEMALE. Dark grey-brown, with little visible blue tone and more obviously marked than ♂. Sides of head and throat spotted brown-buff; rest of underparts and scapulars scaled with buff or buff-white. Wing-coverts and tertials noticeably fringed with pale grey-buff, with tips of greater coverts pale enough to create dull but quite obvious wing-bar. JUVENILE. Dark brown, with no blue tone and even more marked spotting and scaling than ♀. FIRST-YEAR MALE. In autumn, as adult but less blue, with broader and browner feather-fringes. In spring, as adult but wings less black. FIRST-YEAR FEMALE. As

adult. In both sexes and at all ages, bill dark brown-horn and legs black. (2) Iraqi, Iranian, and Afghan race, *longirostris*. Most ♂♂ paler and greyer than nominate *solitarius* and all ♀♀ and immatures noticeably so.

Unlike *M. saxatilis*, ♀ and immature always dark-throated. Otherwise almost unmistakable but could be confused with Blackbird *T. merula* (or even Starling *Sturnus vulgaris*) in brief glimpse. More serious risk of confusion is with escaped captives of own Far Eastern races, *pandoo* and *philippensis*; *pandoo* slightly smaller and rather darker than nominate *solitarius*, with ♀♀ less buff-toned below, while *philippensis* shows brighter blue tones and (on ♂) variable deep chestnut underbody and underwing. Further serious risk of confusion with escaped Himalayan whistling thrushes *Myiophoneus*, but these usually show bright blue sheeny patches on head and shoulders. Flight, gait, and behaviour much as *M. saxatilis* but appearance rather more thrush-like, with longer tail obvious both in flight and on ground.

Song a soft melodious, piping whistle, which (like *M. saxatilis*) recalls *T. merula* but has shorter phrases. Most characteristic call a liquid disyllabic 'uit-uit', recalling Nuthatch *Sitta europaea*.

Habitat. Breeds in west Palearctic in middle and lower middle latitudes in warm dry temperate, Mediterranean and steppe climatic zones, montane and coastal, rocky

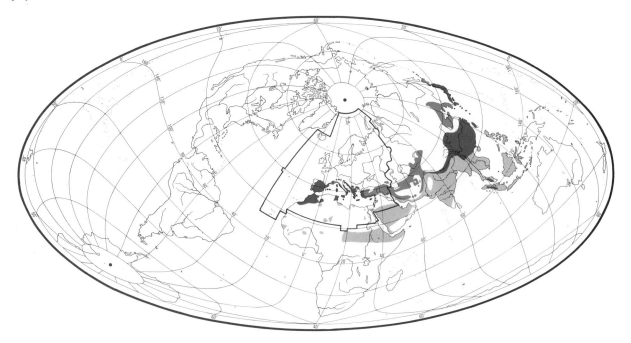

and nearly always in part precipitous. In western Europe, frequents 3 main habitat types—sea-cliffs or other rocky coastlines, mountain valleys and faces (usually below 800 m), and major structures (e.g. castles and ruins) in cities or other settlements. In Switzerland, after recent abandonment of many former breeding sites, now confined almost entirely to quarries of suitable aspect with high broken faces and adequate height and breadth, preferably with ample fringing vegetation and in neighbourhood of waterfalls; forages mainly on ground, but also takes prey in flight or on bushes (Glutz von Blotzheim 1962; Schifferli *et al.* 1982). On Mallorca, as in mainland Spain, frequents both mountain crags and coastal headlands (Parrack 1973). In parts of southern Europe where Blackbird *Turdus merula* absent, inhabits roofs of houses and churches, castles, and stone monuments, even in flat lowlands (Voous 1960). On Gibraltar, not restricted to upper part but breeds freely lower down, and sometimes seen in gardens (Lathbury 1970). In Morocco, breeds on Haut Atlas in rocky broken terrain at 2000-3000 m, but descends in winter even into towns (Heim de Balsac and Mayaud 1962). In Georgia (USSR), occurs at 1300-2500 m in steep rocky defiles with thorny shrubs; also in broad mountain valleys hemmed in by cliffs with heaps of scree or stones. Further east in USSR (extralimitally) occupies mountains from foothills to juniper zone, especially on precipitous rocky slopes above mountain streams, and on rocks strewn through forests, or stone fields with thick shrub growth and scattered trees, at low as well as high altitudes. In Far East even forages on beaches by sea, and in orchards and vegetable gardens. (Dementiev and Gladkov 1954b.) In Afghanistan, met with on a rocky mountain slope with scattered small bushes at 2100 m;

also in various valleys and breeding in rock walls at 3300 m (Paludan 1959).

In winter, habitat largely similar but descends to villages (even entering houses) in India (Ali and Ripley 1973b), and to desert edge in Egypt. In Africa, winters right down to coastal hills and cliffs (Moreau 1972) but in Sierra Leone supersedes Rock Thrush *M. saxatilis* on mountain summit slopes and ridges, latter occurring generally lower on open ground of plateaux (G D Field).

Distribution. Range decreased in France and Switzerland.

FRANCE. Range extended further to north earlier in 20th century (Yeatman 1976). POLAND. Bred 1967 (Tomiałojć 1976). CZECHOSLOVAKIA. Possibly bred 1875 (Brehm 1878), but not fully documented (KH). SWITZERLAND. Many former breeding sites abandoned (Glutz von Blotzheim 1962). EGYPT. Occasional breeding cannot be excluded (PLM, WCM).

Accidental. Belgium, West Germany, Sweden, East Germany.

Population. No information on trends.

FRANCE. 1000-10 000 pairs (Yeatman 1976). SWITZERLAND. About 20 pairs (RW).

Movements. Partially migratory, although extralimitally in eastern Asia primarily migratory. Vertical displacements common. Main wintering areas of migrants lie in North Africa. Poorly studied; proportion of birds migrating and distance travelled probably underestimated due to extreme shyness and low observer density in relevant areas. Migrates singly or in loose aggregations, primarily at

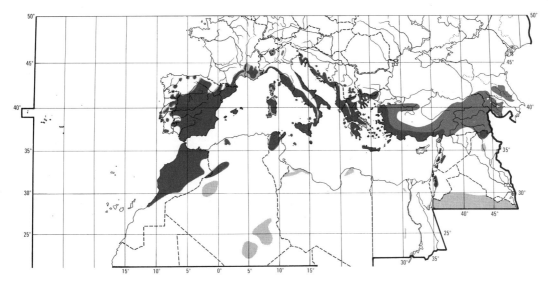

night, often with Rock Thrush *M. saxatilis*. Degree of overlap between wintering ranges and migration routes of nominate *solitarius* and those of *longirostris* confused.

WEST AND CENTRAL MEDITERRANEAN POPULATION of southern Europe and north-west Africa is partially migratory. No good data on proportions of birds wintering on breeding grounds, although likely that most individuals disperse at least short distances (J Hellmich). ♂♂ appear more likely to remain close to breeding sites than ♀♀ (e.g. Flint and Stewart 1983). Some high-altitude sites totally abandoned (e.g. Haut Atlas in Morocco, all Swiss breeding sites) and vertical displacements often noted (e.g. Munn 1931, Glutz von Blotzheim 1962, Walter 1965). In winter, appears widely in habitats rarely used for breeding, ranging from olive orchards in Spain (Muñoz-Cobo and Purroy 1980; F Holgado) to urban areas in Morocco (D G C Harper). 3 main wintering areas to south of breeding range; origins of birds in each area unknown. (1) Strip of up to 26 km south of North African breeding range and east along coast of Libya. Numbers in eastern section appear to be low (e.g. Bundy 1976). Heim de Balsac and Heim de Balsac (1951) suggested ♂♂ outnumbered ♀♀ in Tamrikat hills, Morocco (25°30′N 11°00′W). (2) Central Saharan massifs, notably Ahaggar and Tibesti. Occasional suggestions of breeding at these sites (e.g. Niethammer 1963*b*, Etchécopar and Hüe 1967) require substantiation. If (as likely) incorrect, these wintering sites are at least 1000 km from known breeding sites. (3) Coastal West Africa and east through Sahel zone to Lake Chad and Darfur massif (Sudan). True south-east limit unknown due to potential confusion with members of more easterly breeding populations; for discussion of wintering in Africa to east of Darfur massif see below. Winters further south in West Africa than once thought (e.g. by Fry 1970*b*, Greling 1972, Lamarche 1981), with scattered coastal records reaching south to

Mampong quarry, Accra plains (7°06′N 1°20′W); since this site occupied annually (Walsh and Grimes 1981), stragglers presumed to reach even further. Concentration of coastal records at Mampong and at Popenguine cliffs near Dakar (14°38′N 17°27′W) probably reflects observer bias, as may apparently increasing frequency at Popenguine (Morel *et al.* 1983). Inland records even more patchy, although suitable high terrain little known. No quantitative data on wintering density, although at several sites described as commoner than *M. saxatilis* (Field 1973; Morel *et al.* 1983).

Vagrants occur north of breeding range as far as Sweden. Most records in August or early September, suggesting reverse migration, although singles in April (Sweden), May (Hungary), and June (West Germany).

Timing of movements poorly known; passage rarely observed, even in Mediterranean basin, and difficult to distinguish between long-distance migration and local dispersal. Seasonal increases in numbers and distribution in breeding areas attributed to passage (e.g. Gibraltar: Cortés *et al.* 1980) or local dispersal (e.g. Malta: Sultana and Gauci 1982), often resulting in contradictory opinions about status in some areas (both above references cited other authors who had suggested opposite interpretation for that site). Ringing recoveries of Spanish (1) and Maltese birds (2) provide no evidence on movements since all 3 recovered close to ringing site within breeding range. Birds start vacating breeding site in August, but date of last departures obscured by residents. Wide scattering of passage records suggests migration on broad front, although distribution of records in Ouadi Rime Faunal Reserve (Chad) suggests flight-paths concentrated along rocky terrain (Newby 1980). Late date of arrivals in southern wintering areas may reflect slow passage: e.g. Popenguine records between late December and March, although some arrive mid-November (Morel *et al.* 1983).

However, main arrivals in North Africa and Saharan massifs occur in September, and main passage in Chad appears to occur in late September and early October (Newby 1980), suggesting that late arrivals further south result from delayed final stage in migration (Curry-Lindahl 1981). Spring passage perhaps more rapid than autumn, beginning mid-March and with most wintering areas vacated by late April or early May (J Hellmich). Passage within breeding areas during March (Munn 1921; J Hellmich), or even February (Flint and Stewart 1983), usually attributed to local dispersal, but may indicate that early departures from wintering areas overlooked (note early passage in Persian Gulf, below). Exceptional record of bird near Tripoli (Libya) in July (Bundy 1976).

MIDDLE EAST POPULATION breeding from Turkey to Afghanistan is partially migratory, although extensive areas at high altitude vacated almost entirely (e.g. central Turkey, much of Afghanistan). Likely that birds from this area, winter from Egypt south-east to Somalia and from there east along south Arabian and Makran coasts into Baluchistan; regular in winter in United Arab Emirates, scarce in eastern Saudi Arabia, and occasional in other states on western side of Persian Gulf (Bundy and Warr 1980; G Bundy). Most authors assume that east-west cline between *longirostris* (east) and nominate *solitarius* (west) is maintained on wintering grounds, though separation of single specimens usually impossible (D G C Harper) and exact wintering ranges will only be elucidated by ringing recoveries. In Egypt, Meinertzhagen (1930) claimed that, although very rare in autumn and winter, *longirostris* is the more abundant race on spring passage; requires confirmation. Both races appear to winter in Eritrea (Smith 1957), while those in Arabian peninsula appear closer to *longirostris* (D G C Harper).

Autumn passage poorly documented; in Turkey appears to start with limited post-breeding dispersal in August (Beaman *et al.* 1975). In USSR, breeding sites in Dagestan (eastern Caucasus) not vacated until at least mid-September; some October records in Armeniya. A single record at Nukhur (Turkmeniya, USSR) on 5 December might reflect wintering in an area which usually appears totally vacated (Dementiev and Gladkov 1954*b*). Scarce on autumn passage through Persian Gulf, mainly September–October (Bundy and Warr 1980). Reaches Egypt mid-September, although peak passage not until mid-October (P L Meininger). Arrives Sudan, Eritrea, and Somalia in early October; sparseness of records in southern Sudan (e.g. Cave and Macdonald 1955) might indicate that Ethiopian and Somalian wintering grounds mainly reached via Arabia rather than from north-west.

Return passage in Persian Gulf more noticeable than in autumn and starts in mid-February, with limited passage after March petering out in early May (Bundy and Warr 1980; Gallagher and Woodcock 1980). Over most of Middle East, peak passage rather later, late March to April (e.g. Allouse 1953, Macfarlane 1978, Strez 1984, P L Meininger). As with Mediterranean populations, dates of arrival on breeding grounds are obscured in partially migratory populations. Large-scale arrivals noted around Geok-Tepe (Turkmeniya, USSR) in mid-March, although most winter-vacated sites in Turkey and USSR not re-occupied until April–May (Dementiev and Gladkov 1954*b*).

Extralimital *pandoo* breeding from eastern Afghanistan to southern China appears to be primarily migratory, wintering in peninsular India and south-east Asia. South Chinese populations only partially migratory and disjunct population in Malaya resident. *M. s. philippensis* of eastern Asia ranges from totally migratory in north to resident in south. Passage periods of extralimital races similar to those of west Palearctic birds. Most birds wintering in urban areas of Hong Kong (*pandoo, philippensis*) are immatures (Webster 1975). DGCH, JH

Food. Account compiled from notes supplied by J Hellmich.

Mainly invertebrates, also lizards and plant material. Feeds on ground, by pouncing on prey from perch, and by making short chases after flying prey (Corti *et al.* 1949; Gallagher and Woodcock 1980). Predominance of these methods varies with time of year as relative abundance of different food types changes. In Spain, much aerial feeding observed in spring, more pouncing in summer, and more ground feeding in autumn and winter when flying prey scarce (J Hellmich). Will take fruits from ground or direct from plant (Meinertzhagen 1951*b*; J Hellmich). Beats large locusts (Orthoptera) before eating them (Meinertzhagen 1954) and, in captivity at least, also small mice (Amsler 1933). Lizards normally brought to young dead, and snakes and centipedes brought without head (Irby 1875; Himmer 1967; Hellmich 1984). Holds small items in bill-tip (Hellmich 1984); big worms and small lizards held in middle of body, bigger lizards behind head (Beven 1968; J Hellmich); folds snakes (Himmer 1967).

Diet in west Palearctic includes the following. Invertebrates: grasshoppers, etc., often large (Orthoptera: Acrididae, Gryllidae), adult and larval Lepidoptera (e.g. *Lymantria dispar*), occasionally flies (Diptera), Hymenoptera (especially ants Formicidae), beetles (Coleoptera), spiders (Araneae), small crustaceans, *Scolopendra* (Myriapoda), molluscs (especially snails Gastropoda), worms (Lumbricidae). Vertebrates: mainly lizards (Geckonidae, Lacertidae) up to 10 g in weight, less often snakes (Colubridae), and occasionally frogs or toads (Anura). Plant material (taken largely in autumn and winter) mainly seeds and fruits: of ivy *Hedera*, olive *Olea*, fig *Ficus*, vine *Vitis*, *Myrtus*, *Daphne*, hawthorn *Crataegus*, *Viburnum*, *Ephedra*, Umbelliferae. (Irby 1875; Brehm 1876; Koenig 1886; Lorenz and Hellmayr 1901; Amsler 1933; Ticehurst and Whistler 1938; Jahn 1942;

Tinner 1946; Heim de Balsac 1948; Moltoni 1948, 1949; Corti *et al.* 1949; Meinertzhagen 1951*b*; Moll Casasnovas 1957; Turček 1961; König 1966*b*; Beven 1968; Laferrère 1968; Porter *et al.* 1969; Toschi 1969; Tutman 1969; Sultana and Gauci 1970; Sutton and Gray 1972; König and König 1973; Arroyo and Herrera 1977; Ceballos and Purroy 1977; Sultana 1978; Roberts 1980*b*; Carlo 1983; Hellmich 1984; J Hellmich, J L Klein.) For data from outside west Palearctic, see also Meinertzhagen (1954), Pek and Fedyanina (1961), and Ali (1962).

No detailed studies of diet of adults.

Food of one brood of young at end of June comprised 62·7% (by weight) reptiles, 13·9% Orthoptera, 15·6% other insects, and 7·9% worms (Hellmich 1984). DJB

Social pattern and behaviour. Account based largely on material supplied by J Hellmich for western Spain.

1. Mostly solitary, rarely in small groups. In Punjab (India), April, 3-4 together, possibly for migration (Whistler 1926*a*). In winter, ♂ and ♀ defend separate territories (C König). One ♀ in Egypt defended territory regularly against various species of wheatear *Oenanthe*, also once against conspecific bird (Simmons 1951*c*; see also Antagonistic Behaviour, below). BONDS. Monogamous mating system. Pair-bond breaks down outside breeding season; not known if renewed in subsequent breeding seasons. 2 reports of hybridization with Rock Thrush *M. saxatilis* (Crespon 1844; Moltoni 1937). ♀ broods but both members of pair feed young. Family bonds maintained for some time after fledging. ♂ may look after fledged young of 1st brood while ♀ incubates 2nd clutch (König 1966*b*). Age of first breeding not known. BREEDING DISPERSION. Solitary and territorial with, however, considerable overlap between neighbours where density high; sometimes overlaps also with *M. saxatilis*. In southern Spain, one territory *c.* 2·5 ha (Witt 1973). Thought to need cliff area at least 100 m long by 40–60 m high (Schifferli *et al.* 1982). Topography makes size of territory difficult to determine: e.g. one territory in western Spain *c.* 1 ha in plan view, but contained vertical rock face 80 m long by 60 m high; thus actual surface area utilized substantially more than that simply delineated by boundaries. Territories may be extended, e.g. when raising young. In France, some pairs only 200–300 m apart, but usually more (Yeatman 1976). In western Spain, 5 ♂♂ in 8 ha, with minimum 75 m between nests. At Taormina (Sicily), at least 15 singing ♂♂ in *c.* 5 km² (Blaszyk 1972). In northern Sahara, 0·6 birds per km² (Blondel 1962*a*). Territory serves for courtship, nesting, and feeding, but not known if courtship also occurs outside territory. In Malta, same nest-site frequently used in consecutive years, though 'a pair' used 2 sites *c.* 200 m apart alternately in successive years (Sultana and Gauci 1970, 1982)—however, not stated whether pair ever comprised same 2 birds. ROOSTING. In sheltered sites on cliffs (Stiefel 1976). In southern France, December–February, ♀ roosted regularly at a site protected from wind and rain (Besson 1972). At Hildesheim (West Germany), August, vagrant bird roosted regularly in *Typha* used by various other passerines (Becker *et al.* 1964). In the Punjab, winter, frequently roosts under roofs of houses (Whistler 1926*a*). In breeding season, birds may loaf outside territory; off-duty birds sit at one of several preferred perches not far from nest. Comfort behaviour includes preening, bathing, also sun-bathing, sometimes in an upright posture with bill upwards facing sun, at other times prostrate with wings and tail widely

spread (J Hellmich, K Kräuter). Sometimes bird apparently seeks shade in rocks, trees, bushes, etc. In western Spain, activity pattern varies seasonally: at end of March–beginning of April, active 06.00–20.15 hrs, at end of June 05.30–21.15 hrs, at beginning September 06.15–19.45 hrs, at end of December–beginning January 08.00–18.30 hrs. In spring, loafing and comfort behaviour distributed evenly throughout the day, in summer mainly during hot hours of day, in autumn mostly 15.00–18.00 hrs, in winter apparently similar to autumn (see also Song-display, below).

2. Very shy throughout the year. On approach by man, bird on ground flushes immediately (Heim de Balsac 1948; Beven 1968) and may seek cover nearby (Corti *et al.* 1949) before flying off over 100 m away. If approached by aerial predator, either 'freezes' in cleft (etc.) in rock, or, with single beat of wings, swoops low over rock surface, wings close to body, to seek refuge, sometimes giving Contact-alarm call (Hellmich 1982; J Hellmich: see 2 in Voice). According to Gibb (1946), this call commonly accompanied by nervous bobbing, tail-spreading, and drooping of wings. For call of trapped bird when approached or handled, see 5 in Voice. See Voice also for Bill-snapping response to man, and other alarm-calls. FLOCK BEHAVIOUR. No information. SONG-DISPLAY. Mainly by ♂, but ♀ also sings (Simmons 1951*c*; J Hellmich: see 1 in Voice). ♀ sings from perch, ♂ either perched, in Song-flight, or normal flight. Perched ♂ sings from one of several exposed perches in territory (Beven 1968), sometimes apparently hidden from neighbours. Following account of Song-flight based on Hellmich (1982) and information from J Hellmich. During flight, song may be given continuously or with pauses. Mostly starts from a rock, dead tree (etc.), often beginning with upward glide, bird spreading wings and tail like bee-eater *Merops* for lift; less often uses beating ascent or downward glide. Following this, bird usually loses height in one of 3 ways: (a) angled downward glide with tail closed, wings slightly closed, speed increasing during descent; (b) almost vertical drop ('parachute phase'), wing- and tail-feathers widely spread and usually pressed downwards, effecting slow descent; (c) dropping at an angle ('pipit phase': Fig A), wings raised and slightly closed, tail

A

widely spread and lifted, legs dangling, achieving fairly rapid descent; (b) or (c) may change to (a), or all 3 types may change to horizontal glide (see Hellmich 1982 for possibilities). *Contra* Petersen *et al.* (1954) and Beven (1968), no 'vertical' display-flight known (not stated if this means ascent or descent); according to Blaszyk (1972) and J Hellmich, closest equivalent is near-vertical parachute descent. At end of Song-flight, bird glides upwards to a perch-site or (especially after parachute or pipit phases) brakes hard and lands softly. After landing, bird hops or runs a short way or stays still, then often makes wing movements similar to flicking of Pied Flycatcher *Ficedula hypoleuca* or singing Starling *Sturnus vulgaris* (Fig B). Seen from behind, wings appear to beat slowly and asymmetrically,

B

D

moving not only up and down but tips describing more-or-less circular path. Bird performs thus with varying intensity on landing after both Song-flights and feeding flights. Unlike Corti *et al.* (1949), Hellmich (1982) did not observe birds switching from Song-flight to feeding flight. Song-flights performed mostly in morning; in late evening, birds usually sing only from perches. In spring, western Spain, bird begins with perched song in morning, Song-flight activity peaking 08.00-10.00 hrs; resurgence of perched song 11.00-13.00 hrs and again (with some Song-flights) 17.00-20.00 hrs. In summer, song reduced and more evenly spread over the day. In autumn, Song-flights almost absent, and perched song given 07.00-10.00 hrs, and 18.00-19.30 hrs. One report (Steinbacher 1954) of singing in total darkness. Elsewhere, perched song (only occasionally Song-flights) used outside breeding season, especially in autumn, in defence of territory (Corti *et al.* 1949; Steinbacher 1956; Monk 1958). In Malta, song begins at end of January and continues until mid-May, declining thereafter; resurgence August-November, but less strongly than in spring (Gibb 1946; Sultana and Gauci 1982). In western Spain, song very soft and low mid-August, becoming markedly stronger towards end of September. Regularity of singing in winter requires further study. Only isolated reports of winter song (e.g. Stanford 1969, Grimes 1972, Dorka *et al.* 1976), and indications are that ♂ and ♀ sing little from late December to January or even February (J Hellmich, E K Dunn). In Egypt, song not heard until 10 March, and then feeble (Moreau and Moreau 1928, but see Simmons 1951*c*). ANTAGONISTIC BEHAVIOUR. ♂ defends territory, in or out of breeding season, by vocal and ground display. If neighbours meet at boundary of territories, perched resident ♂ performs Advertising-display (Fig C): crouches,

only 0·5 m apart, intruder looking away (Fig D) or retreating. If intruder does not respond submissively, resident may launch expelling-attack, pursuing intruder (J Hellmich). 2 ♀♀ reported conducting lengthy Song-duel, as between ♂♂ (E N Panov). In captivity, pair-members avoid close contact, and aggression between them increases towards end of breeding season (Vennekötter 1984; C König); once, captive ♀ killed mate after successful breeding (C König). Several reports of confrontations with *M. saxatilis*: in Lebanon, *M. saxatilis* trespassing on territory induced attack and pursuit (Meinertzhagen 1951*b*). In north-east Spain, where territories overlapped, ♂ hammered on head of trapped ♂ *M. saxatilis* (J Hellmich: see call 4b). In winter in Sierra Leone, however, is chased off by *M. saxatilis* in zone of habitat overlap (G D Field). Aggression also reported against *Sturnus vulgaris* (Wagner 1972), Rock Bunting *Emberiza cia*, Blackbird *Turdus merula*, Black Wheatear *Oenanthe leucura*, and Black Redstart *Phoenicurus ochruros* (Hellmich and Zwernemann 1983, which see for other interactions). In winter, Egypt, defence by territorial ♀ against (especially) White-crowned Black Wheatear *O. leucopyga* included singing near territorial boundary, aerial pursuit, and, once, a grappling encounter (Simmons 1951*c*). HETEROSEXUAL BEHAVIOUR. (1) Pair-bonding behaviour. Includes both ground and aerial display. Ground display complex; may be preceded by Song-flight, ♂ landing close to ♀, and adopting Sleeked-upright posture (Fig E): bill points upwards, otherwise similar to Appeasement-posture (Fig D). In captivity, ♂, on several occasions, directed Wing-lifting display (Fig F) at ♀, lifting

C

ruffles plumage all over body except head, then, with small steps, rapidly shuffles towards intruder, sometimes giving Subsong (see 1a in Voice). Drawings by E N Panov show that both ♂-♂ and ♀-♀ interact thus. Intruder, of either sex, may respond with Appeasement-posture (Fig D): stands upright and sleeks plumage. Both birds may then stand side by side, often

E

F

one wing only (C König). Sequence of ground courtship, described by J Hellmich, as follows: ♂ may creep forward in front of ♀ in a semi-circle, as in Advertising-display (see Antagonistic Behaviour, above), but with wings drooped and fanned tail raised above horizontal. ♂ then reverts to Sleeked-upright posture. ♂, behaving thus, sang excitedly (C König), and other calls, perhaps call 4, may be given. ♂ then performs Bowing-display: alternates frequently between holding head high, like Nuthatch *Sitta*, with bill open, and stooping until bill almost touches ground. Shortly afterwards, ♂, facing ♀, starts tripping towards her (not hopping, as in *M. saxatilis*: E N Panov), then back again, and performs Bowing-display, in the head-high position almost falling over backwards. (2) Courtship-feeding. None reported by J Hellmich but in Jerusalem (Israel), 2 birds stood facing each other, bobbing their tails up and down; ♂ then fed ♀ and they then turned back to back, tail-bobbing again (Anon 1946*b*). (3) Mating. Sequence of ground display described in subsection 1 (above) can lead to copulation, but not necessarily so; Song-flight may lead directly to copulation with no intervening ground display (Hellmich 1982; J Hellmich). In captivity, ♂ accompanied ground display with song, and copulation occurred immediately after (K Kraüter). Short loud burst of song heard immediately before copulation (J Hellmich). One sequence as follows: when ♀ singing on house, ♂ came in low, beating wings, in Song-flight, approached ♀ from behind; mounted with asymmetrical wing-beats (see Song-display) then flew away (Hellmich 1982). For probably same asymmetrical wing movement, see Blaszyk (1972). After copulating, ♀ preened then flew away (Hellmich 1982). Probably only one cloacal contact occurs. (4) Behaviour at nest. No information. RELATIONS WITHIN FAMILY GROUP. Young brooded for 3–4 days after hatching. Beg with food-calls (see Voice) and fed bill-to-bill by both parents. At one nest, ♂ more active than ♀, performing 63% of nest visits (Hellmich 1984, J Hellmich). ♂ may give a few short phrases of song on arrival at nest, and again on leaving (Beven 1968; Hellmich 1984). Both parents perform nest-sanitation, removing faeces and depositing them in exposed, elevated places, often at a regular site, e.g. electricity pylon (Hellmich 1984, J Hellmich). According to Sultana and Gauci (1970, 1982), young remain with and continue to be fed by parents for 'a few more days' after fledging; perhaps longer, however, and once at least 13 days

(J Hellmich). For food-calls of fledged young, see Voice. ANTI-PREDATOR RESPONSES OF YOUNG. Young handled on day before leaving nest gave a distress-call (see Voice). Just after leaving nest, captive young hid under boxes (K Kraüter). PARENTAL ANTI-PREDATOR STRATEGIES. (1) Passive measures: No information. (2) Active measures: against birds. ♂ and ♀ recorded attacking Spotless Starlings *Sturnus unicolor* within *c*. 40 m of nest (Hellmich and Zwernemann 1983). (3) Active measures: against man. When human intruder in vicinity of nest with young, ♂ and ♀ gave Warning-calls (see 4a in Voice) almost continuously, louder and more often when young near fledging. When intruder near nest, ♀ on ground regularly alternated Warning- and Contact-alarm calls, regularity breaking down when danger imminent.

(Figs by C Rose: after drawings by J Hellmich, C König, and E N Panov.)   JH, EKD

Voice. Used throughout the year, especially by ♂ in breeding season; regularity of vocalizations in winter requires further study (see Social Pattern and Behaviour). Voice deeper in south than in north of range (Chappuis 1969). ♀ once used Bill-snapping during attack on man (Heinroth and Heinroth 1924–6). Apart from synopsis by Bergmann and Helb (1982), no previous account of repertoire. Following provisional scheme devised mainly by J Hellmich in collaboration with H-H Bergmann and C König.

CALLS OF ADULTS. (1) Song. Given mainly by ♂. (a) Perched song of ♂. Deliberate, loud and melodious, recalling Blackbird *Turdus merula* or Mistle Thrush *T. viscivorus*, but phrases simple, short, and repetitive (Beven 1968). Also similar to Rock Thrush *M. saxatilis* (Harrison 1954); for comparison, see that species (p. 899). May include mimicry, e.g. of Willow Warbler *Phylloscopus trochilus* and Cirl Bunting *Emberiza cirlus* (Bergmann and Helb 1982). Far-carrying; once heard at more than 300 m (J Hellmich). Comprises short fluting phrases such as 'tju-sri tjurr-titi', often repeated at quite long intervals (Jonsson 1982). Phrases typically contain 3–5 units (Corti *et al.* 1949, which see for renderings), but length varies with context, and Fig I shows much longer phrase. Any phrase often given many times in a singing bout, but every ♂ has several different phrases in his repertoire (Bergmann and Helb 1982). In autumn, song similar but apparently lacks many of the fluting units; song heard in September simple and descending, similar to *P. trochilus* (Thiede 1983). Subsong usually low and warbling, sometimes containing many harsh notes, and relatively

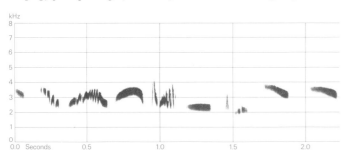

I J Hellmich Spain April 1982

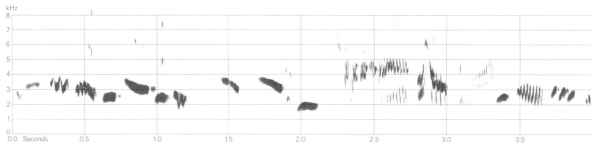

II J Hellmich Spain April 1982

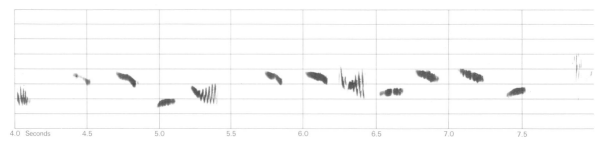

II cont.

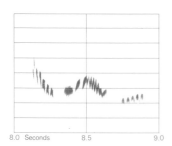

II cont.

continuous; may be given when threatening a conspecific, once by nesting bird when attacked by Crag Martin *Ptyonoprogne rupestris* (Hellmich and Zwernemann 1983). Soft Subsong also described as 'chattering' (Bergmann and Helb 1982), and a little like Starling *Sturnus vulgaris* (Jonsson 1982). 'Low and beautiful' song given just before copulation (K Kraüter) probably Subsong. (b) Aerial song of ♂. Compared with perched song, may be louder and phrases markedly longer, especially in Song-flight (Fig II). When singing in normal flight, however, phrases may be quite short, e.g. 'a musical chuckle' just before arriving at nest containing young (Beven 1968; see also Hellmich 1984). Short, very loud phrases given by ♂ flying in immediately before copulation (J Hellmich). (c) Song of ♀. Apart from being given less often, consistent differences from ♂ song not clearly established. Before attacking ♂, ♀ gave a harsh warbling phrase; perched ♀ gave similar song, but with a fluting 'tüdelit' (J Hellmich). Fig III depicts song of noisy harsh

warbling type. ♀ sang regularly in winter in defence of territory (Simmons 1951c, which see for details). (2) Contact-alarm call. Variously rendered a deep 'tac tac' (Harrison 1954; Beven 1968), 'tak-tak' (Bergmann and Helb 1982) and abrupt hard 'tchuk' (Gibb 1946), 'tchuc-tchuc' like *T. merula* (J Hellmich). Given by either sex on the ground or in the air, especially when disturbed, e.g. by presence of human intruder. In Fig IV, calls alternate with call 4 ('pee chuc chuc chuc pee pee': J Hall-Craggs) when man approached nest with young, more agitatedly as danger increased. (3) Courtship-calls. 'ssrrrt' given during ground display, not known if by ♂ or ♀ (see Social Pattern and Behaviour). Probably same sound described as a contact-call ending in a peculiar reeling 'ruisrsrsrsr' (Boxberger 1934). (4) Warning- and Alarm-calls. In addition to call 2 which often signals alarm, the following calls also used: (a) 'pee' or 'peet' (J Hellmich), or a very high-pitched plaintive 'peep' (Gibb 1946) given repeatedly in presence of intruder, e.g. man

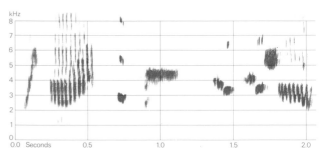

III C König France 1967

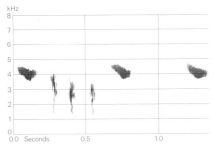

IV J Hellmich Spain April 1982

near nest. Same call also rendered a pure fluting 'düh' or 'jü' (Bergmann and Helb 1982), 'vivivi' or a disyllabic 'uit' rather like Nuthatch *Sitta europaea* (Géroudet 1963), a soft but penetrating whistling 'huid' (Jonsson 1982). In 2 of our recordings, ♂ calls at 2 different pitches (P J Sellar). Intensity and rate of delivery vary with degree of danger, e.g. louder and more persistent when nestlings ready to fledge. Typically, as in Fig IV, combined with call 2. (b) A harsh churring (Gibb 1946), a rattling 'trr', not uncommonly combined with call 2, e.g. 'trr zi tak-tak...'; at higher intensity, e.g. when fleeing danger, a hard rattling 'tschrr' (Bergmann and Helb 1982, which see for sonagram). When bird attacked a trapped ♂ *M. saxatilis*, an intense 'tchrr' heard; not known from which bird, however. A 'tschackerakack' (J Hellmich) and rasping anxious 'schrrackrr' (König 1966b) perhaps belong here. (c) Sibilant 'tsee' (Beven 1968) or high 'tsii' (Jonsson 1982); 'tzick' or 'tschik' like *T. merula* or anxiety-call of Corn Bunting *Miliaria calandra*, and thought likewise to signal fear and alarm (Becker *et al.* 1964). (5) Distress-call. A harsh screaming sound given by bird trapped and handled. (6) Other calls. Calls described by Lombard (1965) not reconcilable with those listed here: a deep 'hup' given several times; twice 'aipiié', similar to croaking alarm-call of Nightingale *Luscinia megarhynchos*.

CALLS OF YOUNG. Food-call of young in nest a buzzing sound, audible more than 70 m from nest once young well-grown. Fledged young beg with 'whirring' (or reeling) calls, sounding something like 'tsrrrr' (C König) or 'trrrrrrr' (J Hall-Craggs: Fig V); compare call 3 of

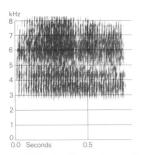

V C König Spain June 1964

adult. Distress-call of hand-held birds similar to adult call 5, but not as loud.   JH, EKD

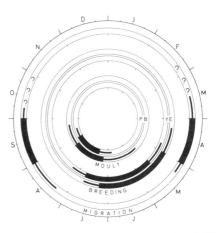

Breeding. Based largely on account by J Hellmich. SEASON. Iberia: see diagram. North-west Africa: eggs found late April to late May (Heim de Balsac and Mayaud 1962). Malta: laying begins late March (Sultana and Gauci 1982). Yugoslavia: late nest with 3 eggs, early July (Kollibay 1904). Palestine: nest with 4 eggs found 2 April (Dresser 1871-81). SITE. In hole or crevice in cliff, under overhanging rock, in cave or quarry, in wall of old building, occasionally in horizontal drainage pipe or hole in tree. 1-120 m above ground, mostly *c.* 3 m; 6 of 7 nests, Spain, at 2·3-4·8 m (J Hellmich); 1 at 30 cm above high water level of river (J Garzón). Nest: rather bulky but loosely built shallow cup of coarse dry grass, moss, and some roots, lined with softer and finer roots, and grasses, occasionally with feathers and plant down. Inner diameter 10-11 cm, cup up to 7 cm deep. Entrance sometimes hidden behind plants. Nest occasionally reconstructed and re-used in consecutive years (Sultana and Gauci 1982). Building: by ♀ only (Moll Casasnovas 1957; König 1970; Witt 1973; J Hellmich), but both reported building in Malta (Despott 1916; Sultana 1978; Sultana and Gauci 1982). EGGS. See Plate 83. Sub-elliptical, smooth and fairly glossy; very pale blue to blue-green, unmarked or with fine reddish, reddish-brown, or brown speckling and mottling, particularly at broad end; more speckled eggs in Spain, more unmarked ones in Greece (Rey 1912). 27·6 × 20·1 mm (25·4-30·4 × 18·3-21·2), *n* = 125 (Schönwetter 1979). Mean of 17 fresh eggs 5·4 g (4·2-6·2) (Mestre Raventós 1979). Clutch: 4-5 (3-6). Of 37 clutches: 3 eggs, 4; 4, 11; 5, 20; 6, 2; mean 4·54 (all available data, compiled by J Hellmich). 2 broods throughout main range, but perhaps only 1 brood in Switzerland at northern edge of range (Glutz von Blotzheim 1962). Replacement clutches laid after loss (König 1970). In captivity at least, eggs laid daily (Glas 1905; Vennekötter 1984). INCUBATION. 12-15 days. By ♀ (J Hellmich), or mainly ♀ (Sultana 1978). In captivity, begins with last egg (Glas 1905; Vennekötter 1984).

YOUNG. Cared for and fed by both parents. FLEDGING TO MATURITY. Fledging period *c.* 18 days (Makatsch 1976). Age of first breeding not known. BREEDING SUCCESS. No information.

Plumages (nominate *solitarius*). ADULT MALE. In fresh plumage (about September–December), upperparts slate-blue, feathers of forehead, crown, and nape tipped grey shading to sandy-brown terminally, feathers of mantle, scapulars, back, and (faintly) rump and upper tail-coverts with ill-defined dull black bar subterminally and pale sandy-brown tip; slate-blue largely concealed on top of head, but rump and upper tail-coverts often completely slate-blue. Lores and patch round and behind eye uniform slate-blue. Cheeks, ear-coverts, chin, and throat slate-blue with narrow brown feather-tips, remainder of underparts slate-blue with more distinct narrow dull black or dark grey subterminal bars (in particular on chest) and more contrasting narrow pale sandy or off-white feather-tips. Tail black; both sides of central pair (t1) and outer webs of others fringed slate-blue (widest near base). Flight-feathers black, tips faintly edged white, outer webs narrowly fringed slate-blue. Tertials, tertial coverts, greater upper wing-coverts, bastard wing, and greater upper primary coverts black with deep slate-blue tinge on outer webs and narrow sharply white edges along tips; edges generally less than 0·5 mm wide but up to *c.* 1 mm on tertials and sometimes virtually absent. Lesser and median upper wing-coverts slate-blue with darker centres, narrowly tipped white when newly-moulted. Under wing-coverts and axillaries dark grey with broad slate-blue tips. *In worn plumage* (about March–July), all brown, sandy, and black of feather-tips of head and body worn off; blue becomes brighter and more glossy azure-blue; white edges of flight-feathers and wing-coverts worn off. Flight-feathers and greater upper primary coverts black with distinct slate-blue on outer webs until about May, brownish-black with less distinct grey-brown tinge later on (slate-blue retained longest on secondaries and innermost primary coverts). When heavily worn, July–August, some grey of feather-bases visible on head and body. ADULT FEMALE. In fresh plumage, upperparts rather like adult ♂, but feather-tips more extensively grey-brown, largely concealing slate-blue of feather-bases (least so on scapulars, back, rump, and upper tail-coverts); feathers of mantle and scapulars with narrow black subterminal bars and off-white tips, as ♂. Sides of head and neck grey-brown, sometimes with strong slate-grey tinge; lores, upper cheeks and indistinct ring round eye finely mottled buff, ear-coverts narrowly streaked buff, lower cheeks and sides of neck more broadly streaked buff, each streak with round end and narrowly bordered black at tip; upper ear-coverts uniform dark grey-brown. Chin, throat, and chest buff; feathers of chin and upper throat narrowly edged black-brown at sides, feathers of lower throat and chest edged black-brown all round, buff feather centres showing as distinct pale drops. Remainder of underparts buff, narrowly and closely barred dull black or black-brown on breast, belly, flanks, and vent, more boldly barred black-brown on under tail-coverts. Rather much variation in colour of underparts (irrespective of age), some birds showing much slate-grey on feather-bases (in part suppressing buff and dark barring), in particular on breast and flanks. Tail black-brown, often with slight slate-blue tinge. Flight-feathers, tertials, tertial coverts, greater upper wing-coverts, and greater upper primary coverts black-brown, outer webs often slightly tinged slate-blue, tips of tertials and coverts with 0·5–1 mm wide fringes, tips of flight-feathers with faint white edges. Lesser upper wing-coverts grey-brown, sometimes with much

slate-blue on bases or largely slate-blue; median black-brown with off-white fringe and variable amount of slate-blue tinge. Under wing-coverts and axillaries grey-brown or black-brown, tips broadly fringed buff, longer feathers barred black-brown and buff. *In worn plumage*, brown feather-fringes on upperparts abraded, but still much brown remaining, blue duller than in adult ♂ and strongly tinged brown; sides of head darker, buff spots and streaks smaller; chin to chest less distinctly marked with drop-like spots, as dark feather-tips wear off; sides of breast and flanks almost uniform dark grey-brown or dull slate-grey, buff and black barring more restricted; upperwing almost entirely black-brown, except for slate-blue or grey-brown lesser coverts, traces of white feather-tips only remaining on some greater coverts and primary coverts. A few ♀♀ are closely similar to ♂, showing much glossy dark blue on head, body, and wing-coverts, but brown feather-fringes on body relatively broader than in adult ♂ (similar to 1st adult ♂, however), and chin, throat, and under tail-coverts mainly buff, as in normal ♀. NESTLING. Down black, on head and upperparts only (Harrison 1975). JUVENILE. Upperparts rather like adult ♀, but feathers paler and more greyish-brown without blue-grey subterminally, each with pale buff or off-white spot or short bar on centre, often narrowly bordered dusky black or brown on tip; spots most distinct on forehead, crown, and hindneck. Underparts buff or cream, feather-tips with narrow dark grey or brown arc on chin, throat, and chest (rather like adult ♀) and with dusky bar or mottling on remainder of underparts (bars much paler and less sharply defined than in adult ♀). Lesser and median upper wing-coverts like mantle and scapulars, tips with poorly defined pale buff or white fringe or bar. Feathering of rump, vent, and tail-coverts looser than in adult. For tertials, flight-feathers, tail, greater upper wing-coverts, and greater upper primary coverts, see 1st adult below. FIRST ADULT MALE. Like adult ♂, but dark feather-fringes on head and body on average slightly browner and broader, less soon wearing off, but much individual variation. Juvenile flight-feathers, tertials, greater upper primary coverts, greater upper secondary coverts, and usually tail and tertial coverts retained; coverts, tertials, and tail less extensively and paler greyish blue than adult, fully brown from about March (adult blue until May–June); tail-feathers narrower than in adult; white fringes along tips of greater primary coverts and outer greater secondary coverts broader, usually over 1 mm. FIRST ADULT FEMALE. Like adult ♀, and equally variable in extent of blue on subterminal parts of body-feathers. Part of juvenile feathers retained, as in 1st adult ♂; difference in colour between new 1st adult feathers and old juvenile ones not as marked as in 1st adult ♂, ageing sometimes difficult; retained greater upper primary coverts with more pointed and frayed tips than in adult, fringes along tip usually over 1 mm wide (but partly worn off in winter and spring); tail-feathers narrower; usually distinct contrast (in colour and abrasion) between new scapulars and old tertials.

Bare parts. ADULT. Iris dark grey-brown, hazel, dark brown, or black-brown. Bill black, sometimes with slight grey tinge; in ♀, sometimes tinged brown. Mouth pink-flesh or lemon-yellow, inside of both mandibles plumbeous-grey or black. Leg and foot very dark brown, greyish-black, or black; in ♀, sometimes dark brown; soles and edges of scutes of toes sometimes plumbeous-grey. NESTLING. No information. JUVENILE, FIRST ADULT. Like adult, but bill, leg, and foot of some ♀♀ brown in 1st autumn. Inside of both mandibles flesh-yellow. (Ali and Ripley 1973*b*; BMNH, RMNH, ZMA.)

Moults. ADULT POST-BREEDING. Complete; primaries descendant. In USSR, starts before brood disperses; single ♂ *longirostris* halfway through moult in late May, some *pandoo* and *philippensis* in moult early July but others not until late July and August (Dementiev and Gladkov 1954b). In Afghanistan, 2 of 3 *longirostris* moulting in second half of July and early August (Paludan 1959). Scanty data from west Palearctic (Italy, Greece, Turkey) point to start in second half of July and completion late August and early September (BMNH, RMNH, ZFMK); similarly, Himalayan populations of *pandoo* do not start as early as some USSR populations, instead moulting August-October (Ali and Ripley 1973b). POST-JUVENILE. Partial: head, body, lesser and median upper wing-coverts, occasionally tertial coverts (4 of 20 examined), and rarely some inner greater upper wing-coverts or central tail-feathers. Starts shortly after fledging, from late June in southern populations, but mainly July and early August elsewhere and some not until late August. Moult completed late July to mid-September. (Dementiev and Gladkov 1954b; BMNH, RMNH, ZFMK, ZMA.)

Measurements. ADULT, FIRST ADULT. Nominate *solitarius*. Iberian peninsula, Balearic islands, Alps, central Italy, and Balkans south to northern Greece, all year; skins (RMNH, ZFMK, ZMA). Bill (S) to skull, bill (N) to distal corner of nostril; exposed culmen on average 7·0 less than bill (S).

WING	♂ 128·0 (2·86; 17) 123–133	♀ 124·1 (2·99; 12) 120–128	
TAIL	85·2 (3·90; 13) 79–92	85·4 (3·01; 13) 79–88	
BILL (S)	30·4 (1·18; 17) 28·5–32·7	29·7 (1·43; 13) 28·2–31·8	
BILL (N)	17·2 (1·85; 11) 16·0–18·5	17·0 (0·97; 13) 15·8–18·5	
TARSUS	30·0 (0·77; 12) 28·8–31·2	29·7 (1·35; 13) 27·8–31·1	

Sex differences significant for wing. Birds from north-west Africa on average perhaps slightly larger, but only 2 of each sex examined (including a winter ♀ from Ahaggar): wing, ♂♂ 129, 131, ♀♀ 127, 129; bill (S), 32·0 (0·86; 4) 31–33 (RMNH, ZFMK).

Southern Greece, Cyclades, Crete, Rhodes, western and southern Turkey, Cyprus, and Levant, summer; skins (BMNH, RMNH, ZFMK).

WING	♂ 125·0 (2·43; 24) 124–129	♀ 120·7 (1·92; 19) 116–124	
TAIL	80·7 (3·08; 12) 76–85	78·8 (1·64; 9) 76–82	
BILL (S)	29·3 (1·53; 23) 26·7–31·3	29·3 (0·93; 18) 26·8–30·7	
BILL (N)	16·6 (1·01; 11) 15·4–17·8	16·8 (0·91; 8) 15·8–17·9	
TARSUS	29·7 (0·98; 13) 28·2–30·9	30·3 (0·89; 9) 28·4–31·2	

Sex differences significant for wing.

M. s. longirostris. Iran and Afghanistan, summer; skins (BMNH).

WING	♂ 123·0 (2·30; 14) 118–127	♀ 117·1 (1·60; 9) 114–119
BILL (S)	28·3 (1·16; 13) 26·7–30·1	27·7 (0·72; 9) 26·6–28·5

Sex differences significant for wing.

In all populations, 1st adult with retained juvenile wing and tail combined with older birds, though 1st adult wing on average 1·6 shorter than full adult and tail 2·7 shorter.

Weights. Nominate *solitarius*. Northern Nigeria, March: 70·5 (Fry 1970b). Southern Algeria, January: ♀ 69 (ZFMK). South-east Morocco, April: ♂ 57 (BTO). Southern Spain, May: ♀ 58 (Niethammer 1957; ZFMK). Malta: 57·0 (3·5; 26) 50·5–63·5 (J Sultana and C Gauci). Northern Greece, March: ♂ 64 (Makatsch 1950). Armeniya (USSR), mid-June: ♂ 55, laying ♀ 57 (Nicht 1961). *M. s. longirostris*. Afghanistan, July-August: ♂♂ 50, 53; ♀ 51 (Paludan 1959).

M. s. pandoo. South-east Afghanistan: late April, ♂ 43; June, ♂♂ 46, 48, 50; laying ♀ 53 (Paludan 1959). Kazakhstan (USSR): May-June, ♂ 45·4 (4·77; 5) 37–49; ♀ 49·2 (3) 45–54; July(-August), ♂ 45·1 (2·18; 6) 42–48, ♀♀ 45·7, 54·5 (Dolgushin et al. 1970).

M. s. philippensis. Manchuria, May-June: ♂ 51·3 (3) 49–54 (Piechocki 1958). Hopeh (China): ♂ 50 (39–56), ♀ 51 (46–60) (Cheng 1963).

Structure. Wing rather long, broad at base, tip bluntly pointed. 10 primaries: p8 longest, p9 5–8 shorter, p7 0–1·5, p6 4–7, p5 13–17, p4 17–23, p1 31–37. P10 reduced, 64–73 shorter than p8, 7 shorter to 1 longer than longest upper primary covert. Outer web of (p6-)p7–p8 and inner of (p7-)p8–p9 emarginated. Tertials short. Tail rather long, tip square or very slightly rounded; 12 feathers, t6 0–8 shorter than t2–t3. Bill rather long; heavy at base, laterally compressed at tip; distal part of culmen and cutting edges slightly decurved, tip of culmen ending in slight hook. Nostril rather small, rounded or oval, covered by rather broad membrane above. Some short bristles at base of upper mandible; finer hairs near nostril and in front of eye. Tarsus and toes short, strong. Middle toe with claw 25·4 (8) 24–26·5; outer toe with claw c. 69% of middle with claw, inner c. 64%, hind c. 71%. Claws short, strong, decurved.

Geographical variation. Rather slight in western and central part of range, where size gradually smaller towards east and colour gradually paler towards Afghanistan to become clinally darker again further east, but easternmost race (*philippensis*) conspicuously different. Typical nominate *solitarius* from Iberian peninsula, Balearic islands, Alps south to central Italy, and Balkans south to northern Greece and north-west Turkey large (see Measurements); ♀ dark, upperparts dark brown or dark olive-brown, underparts heavily marked, only chin uniform buff, throat and chest with small buff spots, dark bars on belly and flanks broad. *M. s. longirostris* from eastern Iraq, Iran, Turkmeniya, and Afghanistan distinctly smaller; ♂ slightly paler than ♂ nominate *solitarius* in fresh plumage, but similar when worn; ♀ and juvenile distinctly paler and greyer on upperparts, less brown; underparts of ♀ distinctly less heavily barred, throat largely uniform, pale spots on chest c. 4 mm wide rather than 2–3 mm, dark bars on belly less than 1 mm wide and browner (less black), ground-colour of underparts paler (in all races, ♀ also has variable amount of grey on feather-bases of underparts, also contributing to paler or dark appearance, but this variation individual, not geographical). Populations intermediate in size and colour between typical nominate *solitarius* and *longirostris* inhabit southern Greece, Cyprus, Levant, and southern and central Turkey east to Armeniya (USSR); sometimes separated as *behnkei* Niethammer, 1943, but difference from both nominate *solitarius* and *longirostris* too slight to warrant recognition, and here included in nominate *solitarius*, though inclusion in *longirostris* could equally well be advocated, as such intermediate ♀♀ are distinctly paler than Iberian birds (but not much paler than typical nominate *solitarius* from central Italy), and ♀♀ from Levant and northern Iraq are near *longirostris*. Populations inhabiting Sardinia, Sicily, Malta, and perhaps southern Italy (not examined) similar to 'behnkei' in size but near typical nominate *solitarius* in colour; North African birds are like nominate *solitarius* but apparently larger (see Measurements). *M. s. pandoo* from central Asia darker than nominate *solitarius*, not as pale as *longirostris*; ♂ in fresh plumage bright dark leaden-blue; ♀ colder brown than nominate *solitarius*, less buff; size slightly variable, but generally smaller than *longirostris*, wing of ♂ in USSR 118 (113–124), in

India 120·5 (116-126), in Tibet 122 (118-127) (Dementiev and Gladkov 1954*b*; Vaurie 1959, 1972; Dolgushin *et al.* 1970; BMNH). ♂ *philippensis* from eastern Asia differs from other races in rufous-chestnut breast, belly, and vent; ♀ darker and browner than ♀♀ of other races; differs also in relatively short tail and in being a long-distance migrant; wing of ♂ in northern populations 121-129, in southern ones 115-124 (Hartert and Steinbacher 1932-8); sometimes considered a different species, but intermediates between *philippensis* and *pandoo* (with mixed blue-and-chestnut bellies) occur in south-west and southern China. *M. s. madoci* from Malay peninsula and Sumatra rather small, wing-tip rounded; ♂ wholly blue. CSR

Zoothera dauma White's Thrush

PLATES 75 and 78
[between pages 856 and 857]

DU. Goudlijster FR. Grive dorée GE. Erddrossel
RU. Пестрый дрозд SP. Zorzal dorado SW. Guldtrast

Turdus Dauma Latham, 1790

Polytypic. *Z. d. aurea* (Holandre, 1825), USSR and Manchuria (China) east to lower Amur and Ussuri rivers. Extralimital: *toratugumi* (Momiyama, 1940), south-east Siberia south and east of *aurea*, southern Kuril Islands (Kunashir), Japan, and (perhaps this race) Korea; nominate *dauma* (Latham, 1790), Himalayas, southern Yunnan (China) and adjacent parts of Burma and south-east Asia, and (perhaps this race) east to Taiwan; 9-11 further races on Ryukyu Islands and from south-east Asia to Solomon Islands.

Field characters. 27 cm; wing-span 44-47·5 cm. Largest thrush of west Palearctic. As long as Mistle Thrush *Turdus viscivorus* but heavier about bill, head, and body and proportionately shorter-tailed. Structure unusual: bill long and heavy, head large, and wings relatively long (though bluntly pointed) in comparison with tail; combined with undulating flight, gives woodpecker-like appearance. Golden- or olive-buff above and yellow-white below, copiously scaled with black crescents. Underwing striped black and white. Sexes similar: no seasonal variation. Juvenile separable.

ADULT. Basically golden- or olive-buff on crown, back, rump, and chest, yellow-white on chin, throat, belly, vent, and under tail-coverts. Except on lores, front of ear-coverts, chin, rear of belly, vent, and under tail-coverts, feathers of head and body show bold black crescents on tips, these forming overlapping small and large scales, boldest on sides of chest and flanks. Wings basically golden-brown but with black-brown coverts (boldly buff-spotted on median, buff-tipped on greater and primary coverts) and blackish bases and black-brown tips to flight-feathers; these marks create irregular pattern, most noticeable features being buff spots on median coverts and contrast between buff-tipped primary coverts and black primary-bases. Underwing striking: white band along lesser coverts, black along other coverts, white along basal halves of all flight-feathers (except outermost primaries), and dusky-black on distal halves. Central tail-feathers olive-brown, others black-brown with increasingly white tips towards outermost. At distance, bird becomes more uniform in appearance, with olive tones in plumage seemingly enhanced, strengthening resemblance to woodpecker, particularly juvenile Green Woodpecker *Picus viridis*. With wear, bright buff tones of upperparts, wings, and tail-centre become greyer, underparts whiter.

JUVENILE. As adult, but ground-colours yellower and black feather-marks more spot- and bar-like than crescentic, with unmarked area on head more extensive, covering crown, hindneck, chin, and throat. At all ages, bill brown, with base of lower mandible obviously yellow-horn; legs pale yellow-brown.

Can be confused with white-spotted juvenile *T. viscivorus*, but that species never truly scaled (nor banded black and white under wing), and differs also in longer tail and narrower body. Main problems in identification are predilection for dense cover and difficulty in seeing diagnostic underwing pattern due to lack of prominent upstroke of wing. Flight recalls *T. viscivorus*, but bounding action produced by alternating bouts of deep wing-beats and closed wings is also strongly reminiscent of *Picus* woodpecker, particularly at long range. Gait apparently restricted to walk and loping run. Ground-loving, though at slightest disturbance flies up into dense foliage, swooping up to perch in characteristic glide.

Song a series of slow, melancholy, fluting whistles. On migration, exceptionally silent for a thrush, but may give churring call.

Habitat. Breeds in upper middle latitudes, mainly in boreal continental zone of taiga coniferous forest (Harrison 1982), largely within range of Siberian Thrush *Z. sibirica*, which apparently shows greater preference for neighbourhood of water. Habitat of *Z. dauma* includes dense spruce *Picea* along river valleys, adjoining mixed or broad-leaved stands on ridges or slopes, including open woods with larch *Larix*, birch *Betula*, and aspen *Populus*, often at headwaters of streams. See also Social Pattern and Behaviour. Habitat variable, including wooded steppes and closed forest with dense undergrowth; apparently differs in far-eastern maritime pro-

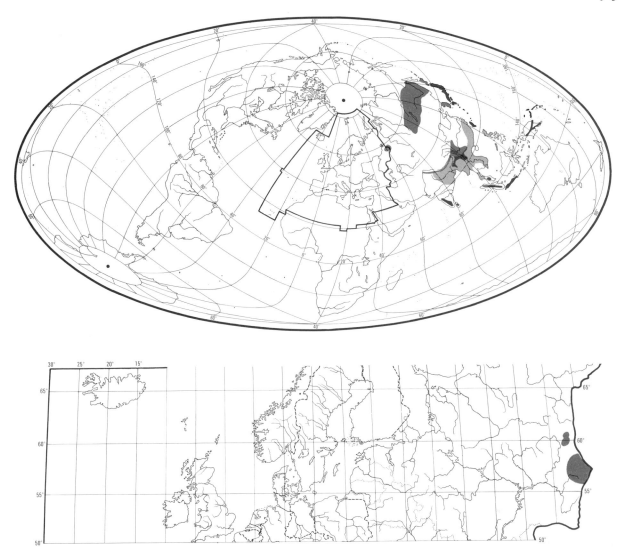

vinces (Dementiev and Gladkov 1954*b*), and again in Himalayas where it breeds to at least 3300 m, inhabiting densely forested hillsides on broken ground with outcrops of moss-covered rocks and scattered *Rubus*, *Berberis*, and similar bushes. Forages quietly among humus and dry leaves on floor, flying up to a tree when disturbed; nests in dense forest of deodar. In winter descends to foothills, also occurring in dense forest with grassy clearings, but sometimes on well-wooded banks of streams or edges of pastureland. Flies strongly within forest. (Ali 1949; Ali and Ripley 1973*b*.)

To even greater extent than most thrushes, combines dependence on ground feeding with need for overhead shelter, and protected nest-site within tree or bush cover. Consequent development of sustained and far-carrying territorial song, often sung from top of tall tree, contrasts with generally secretive and skulking behaviour on floor below.

Distribution. Range poorly known, especially in north (Dementiev and Gladkov 1954*b*). No information on changes.

Accidental. Iceland, Faeroes, Britain, Ireland, France, Belgium, Netherlands, West Germany, Denmark, Norway, Sweden, Finland, East Germany, Poland, Austria, Italy, Yugoslavia, Greece, Rumania, Spain.

Population. No information on trends.

USSR. Unclarified, due to skulking habits (Dementiev and Gladkov 1954*b*).

Movements. Varies from wholly migratory to sedentary in different parts of range.

Northern race *aurea* wholly migratory. Winters in Philippines, China south of Yangtze to Kwangtung and Yunnan, Hong Kong (where scarce with numbers varying considerably between years), Taiwan, Assam, and Indo-

China (Dementiev and Gladkov 1954b; Webster 1975; Schauensee 1984).

Migrates south-east from west of breeding range with passage across Sinkiang and north-west Mongolia (Dementiev and Gladkov 1954b). Populations breeding further east move south or south-west through north-east China (Wilder and Hubbard 1938) and Korea (Bocheński et al. 1981). Leaves breeding grounds from late August in Novosibirsk region (USSR) but rather later around Tomsk with some birds still present in October and latest record 4 November (Johansen 1954). Departs from Irkutsk (Lake Baykal) late September to early October (Reymers 1966). Recorded on passage in northern Tibet late September (Vaurie 1972) and in north-east China from mid-September (Hemmingsen and Guildal 1968), arriving in winter quarters from October (Viney and Phillipps 1983). Return passage from March in southern China (Caldwell and Caldwell 1931), though recorded in Hong Kong until late April (Webster 1975). Passage through north-east China mid-April to mid-May (Hemmingsen and Guildal 1968); recorded in the Gobi in mid-May (Piechocki et al. 1982). Arrives on breeding grounds in USSR mainly from mid-May at Irkutsk (Reymers 1966) and in Novosibirsk region, but migration still evident at Lake Chany in early June (Johansen 1954). Early records at Lake Khanka (north of Vladivostok) during second half of April (Dementiev and Gladkov 1954b) and at Tomsk from 22 April (Gyngazov and Milovidov 1977).

Japanese and Ussuriland race, toratugumi, migratory in north, partially migratory in Japan. Wintering range includes southern Japan, Ryukyu Islands, and Taiwan (Sonobe 1982; Schauensee 1984). Northern populations from Ussuriland (USSR) leave breeding grounds early September with stragglers remaining until early October (Panov 1973); recorded until 8 November on Kuril Islands (Nechaev 1969). Leaves northern and central Japan in October. In southern Japan, mainly a winter visitor (only a small breeding population). Returns to breeding grounds in central Honshu from second half of March (Jahn 1942), but does not arrive on Kuril Islands and Ussuriland until mid-April with passage until early May (Nechaev 1969; Panov 1973).

Himalayan and south Chinese race, nominate dauma, shows vertical displacement and partial migration (Inskipp and Inskipp 1985) with some populations probably sedentary (Schauensee 1984). Vagrant to Malay peninsula (Medway and Wells 1976).

Other races largely sedentary or show local movement or vertical displacement.

European records September–January and April, with April records perhaps of overwintering individuals. MGK

Food. Insects, worms, and berries. Feeds on ground, turning over leaves with bill. Flushes insects by suddenly opening wings and tail, and apparently brings worms to surface by raising itself up on toes and rapidly vibrating whole body for several seconds (Cooper 1959; Ali and Ripley 1973b).

In India, eats insects and their larvae and berries (Ali and Ripley 1973b), and bird in Nepal seen eating undescribed faeces (Rand and Fleming 1957). In Ussuriland (extreme eastern USSR), presumably in breeding season, stomachs contained earthworms (Oligochaeta), grasshoppers (Orthoptera), and small caterpillars (Vorobiev 1954). In eastern China, bird from October contained berries, and another from April contained beetles (Carabidae, Curculionidae) (Kolthoff 1932). See also Spasski and Sonin (1959), Nechaev (1965, 1969), Golovanova and Pukinski (1971), and Il'yashenko (1982).

In Korea, 7 items fed to young comprised 3 earthworms, 2 Cerambycidae larvae, and 2 adult insects (Won et al. 1968a). DJB

Social pattern and behaviour. Poorly known; information scattered and almost entirely extralimital. For closely related Australian species (see Voice), see Cooper (1959), Anon (1976), and Blakers et al. (1984).

1. Normally solitary outside breeding season, including during migration, though small flocks reported in both spring and autumn—e.g. 15 in Altai mountains (south-central USSR), early September (Ravkin 1973); in Amur region (eastern USSR), 'pairs' as well as small parties noted during spring migration (Il'yashenko 1982). See Jahn (1942) for Japan, Portenko (1954) for USSR, Hemmingsen and Guildal (1968) for China, Rand and Fleming (1957) and Fleming et al. (1976) for Nepal, and Ali and Ripley (1973b) for Indian subcontinent. Post-breeding densities in taiga of west and central Siberia detailed by Ravkin (1984). In Thailand where breeding and wintering populations occur, usually solitary or in pairs and shows little tendency to flock with other thrushes (Turdinae) (Lekagul et al. 1985); vagrant in West Germany similarly kept apart from Blackbird Turdus merula and Song Thrush T. philomelos (Vidal 1968). BONDS. Little information, but nothing to suggest mating system other than monogamous. ♂ does some brooding and young fed by both parents, including for indeterminate period when they leave nest (Nechaev 1965; Il'yashenko 1982). BREEDING DISPERSION. Solitary; not known if territorial. In Ussuriland (eastern USSR), density not high and pairs widely dispersed (Panov 1973). Apparent concentration of at least 6 singing ♂♂ within c. 0·5-km radius of inn in woods at foot of Mt Fuji, Japan (Jahn 1942). Secretive habits (see below) make censusing difficult, but following data on density available from various parts of range in USSR. In taiga of western and central Siberia, mainly in dense coniferous or mixed woods near water: 11–56 birds per km²; average in dense coniferous woods 10 birds per km² (Ravkin 1984); in Krasnoyarsk (central Siberia), 1·5 pairs per km² (Naumov 1960). For details of dispersion by habitat in southern taiga belt of central Siberia, see Reymers (1966); see also Panteleev (1972) for transect counts in marshy mixed woods along Ob' river. In Sikhote-Alin' mountains (eastern USSR), 2 birds per km² only in birch Betula ermani; less than 1 bird per km² in all other woodland habitats (Kuleshova 1976). On Kunashir (Kuril Islands), occurs in mixed coniferous and broad-leaved woods with lianas along mountain streams and on volcano slopes; generally scarce—e.g. 6 pairs along 5 km (Nechaev 1965). ROOSTING. No information. For nocturnal song, see below.

2. Usually rather shy and retiring (Rattray 1905; Jahn 1942;

Fleming *et al.* 1976), though not at all shy in winter in India according to Whistler (1926*a*). Sometimes allows close approach (3-10 m in West German vagrant: Vidal 1968), but not closer than 30-40 m when with young (Dementiev and Gladkov 1954*b*). Normally makes for cover when disturbed, flying up silently or sometimes making loud flutter like Hazel Grouse *Bonasa bonasia* (Seebohm and Sharpe 1902; Davis 1953; Dementiev and Gladkov 1954*b*; Vidal 1968; Ali and Ripley 1973*b*). In China, when feeding on ground, may skulk, almost creeping along with head lowered, then stopping for quite a while before moving on (Hemmingsen and Guildal 1968). This perhaps same as behaviour described in more detail from USSR and India: on sighting man, will run or fly low over ground, then suddenly stop and freeze (watching observer and apparently relying on crypsis), either on ground or low perch; will allow close approach (Whistler 1926*a*; Panov 1973). Sometimes flies through dense stand of trees if alarmed (Jahn 1942; Panov 1973). FLOCK BEHAVIOUR. 3-4 birds feeding together will fly to separate places of concealment if disturbed (Seebohm and Sharpe 1902). SONG-DISPLAY. ♂ sings (see 1 in Voice) from tree, less commonly from ground; in USSR, usually while perched motionless on top or high side-branch of tall tree, sometimes half-way up (Dementiev and Gladkov 1954*b*; Portenko 1954; Nechaev 1969); in Japan, often less than half-way up tree, usually half-hidden and never on tree-top according to Jahn (1942). Will turn head from side to side while singing, thus varying volume and making song difficult to locate (Dementiev and Gladkov 1954*b*; Golovanova and Pukinski 1971), and may fly intermittently from tree to tree (Panov 1973). Sings mostly in evening after sunset, at night, and in morning before dawn, though will do so at other times of day, especially in dull weather (Kobayashi 1956; Nechaev 1965; Fleming *et al.* 1976). According to Portenko (1954), song in USSR from mid-May mainly at night and dusk, later (June) extending into morning. In Nepal, will warble (perhaps Subsong) from tree-top in light rain (Fleming *et al.* 1976). In Japan, sings mainly late at night and in early morning; in April, chorus noted 03.00-04.00 hrs peaked before start of dawn; most birds silent by sunrise, only odd bird singing in morning. Song starts again in evening and may be given any time of night (Jahn 1942). In Amur region, towards end of incubation, ♂ sang for *c.* 10-15 min, usually within *c.* ½ hr of sunset. Song ceased with hatching but started again within 3 hrs of ♀'s being trapped (Il'yashenko 1982). Song-period in Ussuriland extends from arrival (mid- to late April) or (more regular) early May to early July (Panov 1973). In one year on Kunashir, birds sang from mid-April to mid-August (Nechaev 1965). In Hondo (Japan) starts occasionally from mid-March but main period only from mid-April (Jahn 1942). Song-period in Nepal early April to mid-June (Ali and Ripley 1973*b*). ANTAGONISTIC BEHAVIOUR. No information. HETEROSEXUAL BEHAVIOUR. For possible indication of pair-formation taking place before arrival on breeding grounds, see part 1. In Amur, during incubation, ♂ spent time on or near ground and not far from nest, birds maintaining contact vocally (Il'yashenko 1982: see 2a in Voice). For possibility of antiphonal duetting, see 1 in Voice. RELATIONS WITHIN FAMILY GROUP. In Ussuriland, ♀ almost constantly at nest while young small, ♂ bringing food for her and offspring (Golovanova and Pukinski 1971). Observations at 1 nest in Amur region showed ♀ to brood young assiduously during daylight hours for first few days. ♀ stood in nest every 1-1½ hrs for *c.* 5-7 min, giving contact-call (see 2a in Voice), with ♂ responding similarly; ♀ then rearranged feathers to uncover brood-patch and settled to brood. Summoned from nest by ♂ at *c.* 21.00 hrs, both parents then starting to feed young (1-2

per visit) and either remaining occasionally to brood; ♀ started renewed long stint of brooding from 03.40 hrs. When ♂ arrived with food at 04.00 hrs, ♀ rose, stood on nest-rim and helped ♂ to feed young (Il'yashenko 1982: Fig A). Young apparently

A

leave nest before able to fly (Panov 1973) and recorded still together with adult when able to fly up to *c.* 50 m between trees (Nechaev 1969). ANTI-PREDATOR RESPONSES OF YOUNG. No information. PARENTAL ANTI-PREDATOR STRATEGIES. (1) Passive measures. In Amur, ♀ sat tight when family of Nutcrackers *Nucifraga caryocatactes* near nest (Il'yashenko 1982). Often leaves nest quickly and quietly at approach of human intruder, flying off low, though may then watch from a certain distance (Portenko 1954; Kobayashi 1956; Vorobiev 1973). (2) Active measures. When with young, often much more agitated, flying about nearby, entering dense bushes near ground, and giving higher-intensity alarm-call (Portenko 1954; Panov 1973: see 2c in Voice).

(Fig by N McCanch: from photograph in Jahn 1942.) MGW

Voice. Generally rather quiet (Panov 1973; Vorobiev 1973; Anon 1976); true also of vagrants (Davis 1953; Vidal 1968); in China, no call given even when flushed (Hemmingsen and Guildal 1968). For wing-noise at take-off, see Social Pattern and Behaviour. Apart from song, only call 2b noted in Japan and apparently lacks alarm-rattles and ticking sounds of certain other thrushes (Turdinae) (Jahn 1942). Song shows considerable regional variation and recent study in Australia by Ford (1983) used sonagraphic analysis to support recognition of 2 species within *Z. dauma* complex: Bassian Thrush *Z. lunulata* (with race *cuneata*) and Russet-tailed Thrush *Z. heinei* (see also Cooper 1959, Holmes 1984, Simpson and Day 1984).

CALLS OF ADULTS. (1) Song of ♂. In *aurea* and *toratugumi*, completely unlike any other west Palearctic thrush: characteristically slow, haunting, melancholy, and rather monotonously repeated soft fluting whistles; loud (though this varies—see also Social Pattern and Behaviour) and carrying up to 0·5-1 km or more. May continue for hours, with pauses between whistles remaining roughly of equal length (Golovanova and Pu-kinski 1971; Vorobiev 1973). In Japan, drawn-out 'hjüi...' alternates with lower-pitched 'hjoü...' (Jahn 1942); also rendered 'pee-pyō', 1st syllable high, 2nd low (Yamashina 1982), and 'hé heyó' (Kobayashi 1956). As pitch varies individually, strange dissonances result if 2 or more birds overlap (Jahn 1942). In Ussuriland (eastern USSR), 1st of 2 whistles lower pitched and single whistle given less commonly according to Panov (1973); description from Kuril Islands (USSR) suggests one whistle long and clear, the other short and muffled (Nechaev 1969). Whistles likened to Advertising-call of Pygmy Owl *Glaucidium passerinum* (comparison appropriate only in case of lower-pitched tone: M G Wilson), and squeaky twittering or warbling, audible only at close quarters, is also interpolated (Dementiev and Gladkov 1954*b*). Such sounds not evident in recording from Japan (Boswall 1964), but in same country, Yamashina (1982) noted (probably same) very quiet and thin sound given after each double whistle and resembling (though higher pitched) terminal note in song of Siberian Thrush *Z. sibirica* and extralimital Red-bellied Thrush *Turdus chrysolaus*. In Japan, *c.* 10 or, during 'best song period', 12–15 double whistles given per min, otherwise only 1 per min (Jahn 1942; Yamashina 1982). Recording from Japan (Fig I) reveals repetition of 2 attractive whistles: 10 fuller sounding, low 'wheeoooooo' tones (sometimes slightly vibrato) at *c.* 2·0 kHz, followed in each case after average pause 3·75 s (2–6, *n* = 19) by a thinner, markedly higher-pitched (*c.* 4·5 kHz) somewhat strained 'weeeeeeeee' note. Low tone has quality of locomotive whistle, higher more reminiscent of whistling kettle (J Hall-Craggs, M G Wilson). In further description of same recording, Boswall (1964) noted drawn-out, pure

whistle with imperceptible start and crescendo followed by gradual diminuendo. In recording from USSR (Fig II), 10 high tones at *c.* 2·8 kHz counterbalanced by just 1 slightly lower-pitched tone at *c.* 2·5 kHz; pauses average 6·1 s (5–7, *n* = 10). Length of pauses between whistles suggests a rather different pattern from that indicated in renderings and use of 'syllables' by some authors cited above. Strong evidence indeed, particularly from Japanese recording, that this rather primitive 'song' may in fact be antiphonal duetting (see also call 2a), one bird giving the high tones, the other bird the lower tones: in recording by R Naumov, introduction of lower-pitched tone coincides with big drop in sound-pressure level and change in direction; in recording by R Nakatsubo, sound-pressure levels increasingly draw apart towards end and temporal patterning also less regular then, 2 pauses between tones being of only 2 s. Nevertheless, singing bird reported to turn head (see Social Pattern and Behaviour) and this may yet indicate that both tones are (sometimes) given by the same bird (J Hall-Craggs, M G Wilson). Song of *major* from Amami-Oshima island in Ryukyu archipelago (Japan) unlike *aurea* and *toratugumi*: typical thrush song (like *T. chrysolaus*) rendered 'piri piri kyo kyo' (Jahn 1942). This presumably only part of song and *major* may thus resemble other races of South Asia—notably nominate *dauma* (also apparently *neilgherriensis* and *imbricata*): rather slow song for a *Zoothera*, short phrases being usually given 2 or more times before moving on to next—'pur-loo-tree-lay; dur-lee-dur-lee; drr-drr-chew-you-we-eee' (Fleming *et al.* 1976). Also described as loud and of fine quality, like Mistle Thrush *T. viscivorus*, but more disconnected, with phrases separated by long pauses: 'chirrup...cheweee... chueu...wiow...we ep...chirrol...chup...chewee...wiop', or

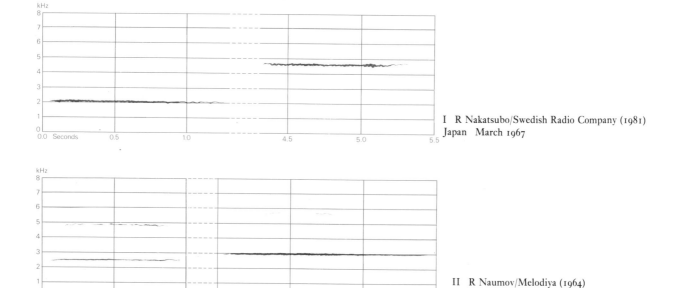

I   R Nakatsubo/Swedish Radio Company (1981)
Japan   March 1967

II   R Naumov/Melodiya (1964)
USSR   June 1961

a few fine notes connected by medley of squeaks and chuckles given rather monotonously for hours (Ali and Ripley 1973*b*); see also Seebohm and Sharpe (1902). (2) Contact-alarm calls. (a) Further study required to ascertain differences (if any) between this call and whistles of song (J Hall-Craggs). Quiet, plaintive whistle apparently resembling Bullfinch *Pyrrhula pyrrhula*. Serves as contact-call between pair-members: e.g. given every *c.* 5–10 s by ♀ from nest with young, ♂ responding similarly (Il'yashenko 1982); given also in Japan in winter and peculiar but melodious whistle (probably the same or closely related) by migrants (Dresser 1871–81; Naumann 1905). (b) Drawn-out 'zieh' like that of many *Turdus*, but higher pitched, longer and much sharper, even penetrating (Jahn 1942); 'tsi', short and quiet and serving as an alarm (Panov 1973) probably the same or closely related. (c) High-intensity alarm-call given at nest (perhaps more after hatching): muffled snoring or growling 'rra' or 'krrrua' (Panov 1973; also Portenko 1954); not like alarm-rattle of certain other thrushes (Dementiev and Gladkov 1954*b*), and churring mentioned with similar comment by Murton (1970*b*) probably also belongs here.

CALLS OF YOUNG. Contact-call 'tsrrrri' (Panov 1973).

MGW

Breeding. SEASON. Western Siberia (USSR): fresh eggs found 1 June, newly-fledged young in August (Johansen 1954). SITE. In fork of tree, 2–6 m up, occasionally on ground among stones and plants. Nest: base of dry ferns supporting cup of leaves, twigs, and moss, poorly plastered on inside with mud, thick lining of thin rootlets, grass, and leaves (Bocheński 1968). Average dimensions of nests, Japan: outer diameter 17–23 cm; inner diameter 12 cm; height 7–9 cm; depth of cup 4·5 cm (Kobayashi and Ishizawa 1932). EGGS. See Plate 83. Oval, smooth and moderately glossy; greenish-blue, almost covered with fine reddish speckling, though often 1 per clutch with stronger ground-colour and fewer and bolder markings (Witherby *et al.* 1938*b*). *Z. d. aurea*: 32·6 × 23·8 mm (30·0–36·0 × 21·8–25·3), *n* = 100; calculated weight 9·85 g (Schönwetter 1979). Clutch: 4–5 (Dementiev and Gladkov 1954*b*). One brood, possibly 2 (Makatsch 1976). INCUBATION. Apparently by ♀ alone (Il'yashenko 1982); period not known. YOUNG. Cared for and fed by both parents (Il'yashenko 1982). No further information.

Plumages (*Z. d. aurea*). ADULT. Feathers of forehead and crown dark olive with black crescentic tip and yellow-buff subterminal spot or bar, appearing closely barred black and buff or largely olive-black with buff spotting. Remainder of upperparts yellowish olive-brown or golden-olive, each feather with black crescent on tip (wider on lower mantle, scapulars, and upper tail-coverts, narrower on rump); feathers usually brighter yellow just proximal to black crescent and more olive towards base, in particular on central mantle, back, and rump. Lore and feathers at base of lower mandible pale greyish-buff; narrow eye-ring cream or off-white. Side of head and neck (from eye and gape backwards) yellow-buff or cream, each

feather with narrow black fringe along tip, fringes often indistinct on lower ear-coverts and wider just below and behind eye and on rear of ear-coverts; black on rear ear-coverts often coalescent, forming dark bar down side of head. Malar stripe rather narrow but usually distinct; black, sometimes broken into spots. Chin and upper throat cream or buff-white, marked with small black crescents on throat which become gradually larger towards chest. Lower throat and chest tawny-buff or yellow-buff, each feather with black crescent at tip; sides of breast olive with black crescents. Ground-colour of remainder of underparts white, breast and flanks with bold black crescents (each subterminally bordered by tawny-yellow), upper belly, sides of belly, and sometimes under tail-coverts with smaller black crescents. Central 2 pairs of tail-feathers (t1–t2) uniform brownish-olive (also on tip); t3–t5 black, partly fringed brownish-olive along outer web and with narrow clear-cut white or cream tip; t6–t7 brownish-olive with black suffusion on inner web and base, t6 with rather broad but ill-defined white tip, t7 with limited amount of white on tip or none at all. Flight-feathers and tertials dark brown-grey or greyish-black on inner webs, yellowish olive-brown on outer; dark grey of inner webs extends to outer webs on tips of all feathers and on middle portions of primaries. Tertials with yellow-buff fringe on outer tip, invading black on inner web. Greater upper primary coverts black with contrasting yellowish olive-brown fringes along outer web (not reaching tip). Lesser upper wing-coverts brownish-olive with narrow black terminal rim and variable amount of yellow-buff suffusion subterminally; median black with broad cream-buff tip extending into short and narrow shaft-streak, tip often narrowly edged black; greater black with broad brownish-olive fringe along outer web and narrower yellow-buff fringe on tip of outer web, just extending into black on tip of inner web and often forming rounded or triangular spot at shaft. Undersurface of flight-feathers, underwing, and axillaries with boldly contrasting pattern: basal halves of inner webs of secondaries and tertials as well as tips of greater under wing-coverts forming white band *c.* 3 cm wide on underwing, extending into cream band *c.* 2 cm wide on middle portions of inner webs of primaries (except on outer 2–3 primaries; band often just visible as a greyish-buff bar on upperwing also); bases of greater under wing-coverts, tips of median, and whole median under primary coverts contrastingly black, extending into dark grey band formed by greater under primary coverts, bases of inner primaries, and whole p8–p9; a 2nd white band formed by bases of median under wing-coverts; this band bordered in front by dark grey of bases of inner lesser coverts. *In worn plumage*, ground-colour of upperparts, t1, and fringes along outer web of flight-feathers somewhat less bright brownish-olive and golden-yellow, more pale greyish-olive; ground-colour of underparts and tips of median and greater upper wing-coverts purer white, less buff or cream. NESTLING. Down white at base with yellow tips (Il'yashenko 1982). JUVENILE. Rather like Mistle Thrush *Turdus viscivorus*, in particular on underparts, where spotted black instead of showing bold black dark crescents as adult. Upperparts pale brownish-olive, almost uniform on forehead, crown, and hindneck, marked with narrow cream shaft-streaks on mantle and inner scapulars and with wider streaks (*c.* 2 mm) on outer scapulars; centres of feather-tips of mantle and scapulars spotted black. Back and rump pale brownish-olive with faint pale shaft-streaks and dusky feather-tips. Upper tail-coverts with faint buff spots on tips. Side of head mottled cream-buff and black; narrow cream ring round eye, short and faint supercilium spotted cream and black. Ground-colour of chin, throat, chest, and flanks warm pink-buff or cream-buff, remainder of underparts paler, cream-white on central belly, vent, and under tail-coverts. Chest, sides of breast, and flanks marked with contrasting rounded black spots *c.* 4 mm across, smaller black spots on breast and sides of belly, faint olive-brown spots or streaks on under tail-coverts. Lesser upper wing-coverts like mantle and scapulars; median coverts brownish-olive with narrow buff shaft-streak ending in broadly triangular buff tip; greater coverts and tertials dark olive-grey with broad buff fringe along outer web, ending in triangular spot on tip, bordered black on tip of inner web; pattern of greater coverts, tertials, and greater upper primary coverts strongly variable, generally not distinguishable from those of adult. For tail, see First Adult below. FIRST ADULT. Like adult, but juvenile flight-feathers, tail, tertials, greater upper wing-coverts, greater upper primary coverts, and often longest feathers of bastard wing retained. No constant difference in pattern of retained coverts from those of adults, but often a slight contrast in shape, ground-colour, abrasion, and pattern between retained greater coverts and new tertial coverts, median coverts, or shorter feathers of bastard wing. Outer primaries and tail more worn than those of adult at same time of year, in particular tail distinctly worn at tip in spring (adult tail then still smoothly edged). Best ageing character is tail-tip; in adult, tip of t1–t2 fairly rounded except for small point at shaft, colour uniform olive; in 1st adult, tips taper more gradually to sharp point at shaft, tip marked with small buff spot (absent in 2 of 40 examined); buff tip sometimes worn off in spring (adult then still without wear); also, white tips of t3–t6 less clear-cut than those of adult.

Bare parts. ADULT, FIRST ADULT. Iris hazel-brown or brown. Eye-lids yellowish. Bill brown, darkest on culmen and tip, pale horn-brown, olive-brown, or yellow on base of lower mandible. Leg and foot brownish-flesh, pale yellow-brown, or pale horn-brown, soles paler flesh or yellowish. (Hartert 1903–10; Witherby *et al.* 1938*b*; BMNH, ZMA.) NESTLING. Skin of upperparts plumbeous-red, of belly yellowish-pink. Bill horn colour. Mouth and tongue yellow; gape-flanges yellowish-white. Claws white. (Il'yashenko 1982.) JUVENILE. No information.

Moults. ADULT POST-BREEDING. Complete; primaries descendant. Starts end of July (Il'yashenko 1982). One bird from USSR had some feathers of mantle and scapulars growing on 24 June, but others without moult until late July; one in full moult 24 August; one migrant bird had some growing feathers on body in early September (Dementiev and Gladkov 1954*b*). A few birds had some body feathers growing March–April (Witherby *et al.* 1938*b*), perhaps after accidental loss only, as many others did not show this (BMNH, RMNH, ZMA). No further information. POST-JUVENILE. Partial: head, body, lesser and median upper wing-coverts, usually tertial coverts, and occasionally a few inner greater upper wing-coverts. Starts with flanks, back, rump, and lesser coverts, as soon as flight- and tail-feathers full-grown; followed by head, scapulars, and upper tail-coverts; under tail-coverts shed just before departure on migration (Il'yashenko 1982). In USSR, in full moult in first half of July; body in 1st adult plumage but head largely juvenile in second half of July; moult finished by mid-August though one completed by 2 August and another by 26 July (Dementiev and Gladkov 1954*b*).

Measurements. ADULT, FIRST ADULT. Wing of (1) *aurea* from western and central Europe, autumn, and from Siberia west of Lake Baykal, summer (Meise and Seilkopf 1960; Ulfstrand 1960; BMNH, RMNH, ZMA); (2) *aurea* from eastern China,

autumn and winter (BMNH), (3) *aurea* from Taiwan and Japan, autumn and winter (BMNH, RMNH, ZMA), (4) *toratugumi* from Japan (Vaurie 1959; RMNH, ZMA); other measurements for *aurea* (1)–(3) only; bill (S) to skull, bill (N) to distal corner of nostril; exposed culmen on average 6·0 less than bill (S).

		♂		♀	
WING	(1)	172·2 (2·46; 7)	170–176	168·5 (3·70; 4)	164–172
	(2)	166·2 (4·51; 14)	159–174	161·8 (3·48; 8)	158–166
	(3)	165·0 (2·18; 15)	158–170	161·7 (1·86; 16)	158–165
	(4)	160·0 (— ; 8)	154–163	154·0 (— ; 2)	153–155
TAIL		109·1 (5·37; 18)	101–122	100·7 (3·86; 12)	95–106
BILL (S)		30·7 (1·15; 30)	28·7–32·9	30·3 (0·74; 24)	28·9–31·8
BILL (N)		16·5 (0·83; 15)	15·2–17·6	16·7 (0·42; 11)	16·1–17·5
TARSUS		36·6 (1·18; 18)	34·8–38·9	36·4 (1·39; 12)	33·9–39·4

Sex differences significant for tail. 1st adult with retained juvenile flight-feathers and tail combined with older birds, though wing of 1st adult on average 0·9 shorter than full adult and tail 3·5 shorter.

Weights. *Z. d. aurea*. Netherlands and Langeoog (West Germany), late September and October, 1st adult ♂♂: 172, 189, 192, 195 (all very fat), 98 (exhausted) (Kate 1933; Bosch 1947; Meise and Seilkopf 1960; RMNH, ZMA). Sweden, early October: ♂ 172 (Ulfstrand 1960). Finland, late September: 170 (Sulkava 1963). Kazakhstan (USSR), September: ♀ 181·6 (Dolgushin *et al.* 1970). Taiwan: November–December, ♂♂ 105, 123, ♀♀ 90, 100; January–February, ♂ 123·7 (3) 116–135, ♀ 122·3 (3) 105–150; March–April, ♂ 150·0 (20·0; 4) 120–160, ♀ 155·0 (3) 135–170 (RMNH). Mongolia, on migration, May: ♀♀ 120, 124 (Piechocki and Bolod 1972). USSR: ♂♂ 110, 146, 162; ♀ 128 (Dementiev and Gladkov 1954*b*).

Structure. Wing rather long, broad; tip bluntly pointed. 10 primaries: p8 longest, p9 2–6 shorter, p7 1–3, p6 8–15, p5 23–32, p4 32–41, p1 47–60. P10 reduced, 92–105 shorter than p8, 5–12 shorter than longest upper primary coverts. Outer web of (p6–)p7–p8 and inner of (p7–)p8–p9 emarginated. Tertials short. Tail rather short, tip rounded; 14 feathers (12 in some extralimital races), t7 10–15 shorter than t1. Bill long and heavy, straight at base, slightly decurved at tip (upper mandible particularly); nostril rather large, oval, partly covered by membrane above; some fine bristles along base of upper mandible. Tarsus and toes rather long, heavy. Middle toe with claw 31·1 (8) 28–34; outer toe with claw *c.* 71% of middle with claw, inner *c.* 69%, hind *c.* 83%. Claws rather long and heavy, slightly decurved.

Geographical variation. Marked; mainly involves width of black marks on head and body, ground-colour, wing-shape, number of tail-feathers, and (in particular) size. In general, marks heavier, ground-colour darker, wing rounder, and size smaller towards tropics, paler and larger again towards Australia. Wing of birds from Ryukyu Islands and northern Eurasia 153–176, of birds from Himalayas to Taiwan 127–150, of birds from southern India, Ceylon, and Indonesia 114–140. *Z. d. aurea* from east European USSR and western Siberia large (see Measurements, and Johansen 1954) and pale; towards east, size apparently gradually smaller, and thus probably merges into *toratugumi* from Ussuriland (USSR) and Japan, which is not only smaller but also has slightly deeper golden-olive upperparts and slightly broader black crescent marks. For survey of other races, see Deignan *et al.* (1964); for races of eastern Asia, see Mees (1977).

For related species in Australia, see Voice.

Recognition. Adult and 1st adult rather similar to Mistle

Thrush *Turdus viscivorus* but *Z. dauma* even larger (especially bill), upperparts and upperwing heavily marked with black (not uniform), and marks on underparts are short crescents or bars (not rounded or triangular spots). Juvenile rather similar to juvenile *T. viscivorus*, which is also rather golden-olive above and shows pale tail-tips; *Z. dauma* differs in darker and more distinctly marked head, upperparts, tail, and upperwing, and broader brownish-olive fringes along outer webs of flight-feathers; at all ages, contrasting pattern of dark grey/white/black/white/dark grey bands on underwing, unlike any *Turdus*.

CSR

Zoothera sibirica Siberian Thrush

PLATES 67 and 78
[between pages 856 and 857]

Du. Siberische Lijster   Fr. Grive sibérienne   Ge. Sibirische Drossel
Ru. Сибирский дрозд   Sp. Zorzal siberiano   Sw. Sibirisk trast

Turdus sibiricus Pallas, 1776

Polytypic. Nominate *sibirica* (Pallas, 1776), Siberia east to Sea of Okhotsk and Sea of Japan. Extralimital: *davisoni* (Hume, 1877), Sakhalin, southern Kuril Islands, and Japan.

Field characters. 22 cm; wing-span 34–36 cm. Close in size to Song Thrush *Turdus philomelos*, with all standard measurements overlapping. Rather flat-crowned thrush; adult ♂ mainly dark slate, ♀ and immature buff-brown, mottled and barred below, but in all plumages identifiable by prominent white or pale supercilium and tail-spots, and boldly barred underwing. Sexes dissimilar; no seasonal variation. Juvenile separable.

ADULT MALE. Head almost black, rest of upperparts slate-black and underparts slate-grey; larger feathers have paler margins when fresh which form dull mottling and scaling, most obvious on flanks. Conspicuous white supercilium, short in front of eye but long behind it. White centre of belly, white-barred vent, and white tail-corners may also show at close range. On take-off, tail shows white spots on all but central feathers, and in flight underwing shows 2 bold bars, formed by (a) white median and lesser primary coverts and (b) white tips to greater coverts, white secondary-bases, and large white spots across primaries. Bill black-horn; legs orange-yellow, duller in winter. ADULT FEMALE. Ground-colour of upperparts warm olive-brown, of underparts white, much suffused with buff on chest and flanks. Most obvious markings are: (a) face pattern, with long, narrow buff-white supercilium, dark brown eye-stripe, buff- and brown-spotted cheeks, dark black-brown malar stripe, and cream-white throat; (b) short, crescentic barring of chest and flanks, with dark brown marks appearing not as spots but as small smudges and deep bars (bars most obvious as individual marks on lower flanks); (c) almost white belly; (d) dense dark brown barring of vent. Tips of median and greater coverts ochre-buff, forming lines of spots rather than bars. Tail and underwing patterned as ♂ but dark areas brown, with front bar on underwing yellowish-buff and central one duller white. Bill black-brown with yellow base to lower mandible; legs yellow-brown. JUVENILE. Not studied in the field; resembles ♀ but more scaled. FIRST-WINTER MALE. From fledging,

quickly becomes dull blue-grey, particularly on back, rump, tail, and lower chest. Juvenile head and upper breast plumage less quickly lost, so that some birds look noticeably patchy with brown crown, ♂-like face pattern, and pale throat. Most juvenile wing-feathers retained, and wing thus mainly brown, with buff tips to greater coverts still obvious. Flanks initially cleaner than ♀, with short grey bars and crescents, later dull blue-grey with white spots rather than bars. Underwing pattern as ♀. Bill as adult ♀; legs dull brown.

Unmistakable in flight; only much larger and boldly scaled, buff-white White's Thrush *Z. dauma* shows similar underwing pattern (but note, however, that smaller *Catharus* thrushes show single pale band across bases of flight-feathers). On ground, adult and 1st-winter ♂ still unmistakable, but ♀ and juvenile ♂ confusing, recalling both *T. philomelos* and Redwing *T. iliacus* and best distinguished by close observation of underpart markings. White corners of folded tail not obvious from above in any plumage. Flight, gait, and stance little studied, but brief references suggest *Turdus*-like carriage and actions, unlike congeneric *Z. dauma*. Behaviour of one vagrant was close to that of migrant Blackbird *T. merula* (including tendency to skulk), but one wintering bird acted more like *Z. dauma* (using similar escape-flight into canopy). Gregarious on migration and in normal winter quarters, and vagrants will join flocks of other thrushes.

Apparently rather silent in winter quarters. Gruff squawk and a soft 'zit', recalling *T. philomelos*, recorded from vagrant.

Habitat. Breeds in east Palearctic, from upper to lower middle latitudes, overlapping to south with Dusky Thrush *Turdus naumanni* and marginally to north with White's Thrush *Z. dauma*, mainly in boreal coniferous taiga zone, largely lowland but partly montane. The sparse data suggest preference for dense stands of trees or shrubs, especially spruce *Picea* and broad-leaved species such as poplars *Populus* on moist ground in floodplains of rivers or in neighbourhood of water (Dementiev and Gladkov 1954*b*). In winter in India, frequents hill forest up to at least 1800 m; feeds on ground, flying up into trees when disturbed (Ali and Ripley 1973*b*).

Distribution. Breeds in Siberia from *c.* 85°E to Sea of Okhotsk and south to *c.* 55°N (boundary incompletely clarified); also in Sakhalin, northern Japan, and north-east China. Winters in India, south-east Asia, Borneo, and Greater Sunda and Andaman Islands.

Accidental. Britain, France, Belgium, Netherlands, West Germany, Norway, East Germany, Poland, Austria, Switzerland, Italy, and Malta.

Movements. Migratory.

Nominate *sibirica* migrates south from breeding range in eastern Siberia (or south-west from east of range) across Mongolia and China, mainly east of 100°E (Dementiev and Gladkov 1954*b*; Schauensee 1984). Uncommon on passage in Korea (Gore and Pyong-Oh 1971). Leaves breeding grounds from early September but some birds still present until mid-October (Dementiev and Gladkov 1954*b*; Panov 1973). Passage recorded in north-east China in early September (Hemmingsen and Guildal 1968) and arrives India from October (Ali and Ripley 1973*b*). Earliest record in Malay peninsula 18 October with passage until early December (Medway and Wells 1976); a ♀ netted on a hill at night on 21 January suggests possibility of mid-winter movements (Wells 1982).

Northward migration from late March with latest date in Malay peninsula 23 April, and 28 April at Trang, southern Thailand (Medway and Wells 1976). Recorded in flocks of up to 60 birds in April in Burma (Smythies 1953) and in small flocks with Eye-browed Thrush *Turdus obscurus* in March–April in north-west Thailand (Lekagul *et al.* 1985). Passage through north-east China mid- to late May (Hemmingsen and Guildal 1968), from late May in Korea (Gore and Pyong-Oh 1971), and in Mongolia up to early June with arrivals on breeding grounds from late May (Dementiev and Gladkov 1954*b*).

Eastern race *davisoni* (breeding Japan and Sakhalin) appears to follow similar migration route to nominate *sibirica*, both races occurring together on passage in eastern China (La Touche 1925–30) and occupying similar wintering range, e.g. both recorded in Burma and Malay peninsula (Smythies 1953; Medway and Wells 1976).

European records August–March (mainly October–February) include flocks of 17–18 birds in Poland in January and March (L Tomiałojć) and *c.* 25 birds in Hungary in mid-February (Warga 1955).   MGK

Voice. Call 2a given by migrating birds in China (Kolthoff 1932), but perhaps generally silent in winter as (e.g.) no reports of calls from India during that season (Ali and

Ripley 1973*b*). ♀ quiet at nest, e.g. not calling when leaving in panic at approach of Goshawk *Accipiter gentilis* (Panov 1973).

CALLS OF ADULTS. (1) Song of ♂. In Japan (*davisoni*), rather abruptly delivered short, loud phrases (or figures) separated by quite long pauses. Each phrase comprises fluted whistle followed by tremolo, or more twittering, squeezed, chirruping, or clicking sounds. Whistles typically double, though single, and perhaps triple units occur; variation also in pitch and timbre, with some whistles clear and high, others richer and deeper, occasionally mewing and plaintive. Renderings include 'kyorro-i-tsrr' (Yamashina 1982), 'hooweet-sirrr', 'heooweet-sirrr', 'wheeoo-sirrr', etc. (M G Wilson from recording by R Nakatsubo.) Twittering component alluded to above not obvious in another description from Japan (Jahn 1942), though short, rapidly delivered whistled phrases lacking variation much as already described: 'tjüelüt-tjüie-tjüelit-tjüö'. No indication that nominate *sibirica* of USSR differs significantly: rather monotonous 2-syllable fluting whistle and quiet twittering reported by (e.g.) Dementiev and Gladkov (1954*b*). In Ussuriland (eastern USSR), full song rendered 'tvee kyuvee keeyuvee...tveekakyuvee...' given with rather long pauses; sometimes more like 'choon...chveen' (Panov 1973). Sings from high in tree or other prominent perch (Panov 1973; Yamashina 1982); in Japan, typically for short period (*c.* 15 min) in early morning and late evening (Jahn 1942). (2) Contact-alarm calls. The following described, perhaps relating to no more than 3–4 distinct calls. Barely audible whistle given, when still far from nest, by adult approaching with food (Panov 1973). Delicate vibrating 'seep' or thin 'tseee' given when flushed and when migrating at night (Kolthoff 1932; Yamashina 1982). From vagrant: short 'zit', like Song Thrush *Turdus philomelos* but softer and perhaps purer; gruff squawk when flushed (Andrew *et al.* 1955). Chattering or chuckling sounds mentioned without details of context by Yamashina (1982). Dry rattle similar to Rattle-call of Mistle Thrush *Turdus viscivorus* given as alarm-call at nest (Panov 1973). Series of 'veetststs' sounds given by ♂ brooding young when ♀ long absent. Not dissimilar 'yueetsssss-veetsss', resembling song fragment, given at nest, apparently when bird becoming calmer after excitement (Panov 1973).   MGW

Plumages (nominate *sibirica*). ADULT MALE. Entire upperparts dark plumbeous-grey or bluish-grey, each feather with greyish-black centre and often with faint dusky border at tip. Distinct white supercilium; rather narrow in front of eye, reaching to about half-way along lore, not reaching bill-base; broad and reaching to side of nape behind eye. Side of head and neck, chin, throat, and chest plumbeous-grey, like upperparts, but sometimes with slight olive-brown tinge on feather-tips when plumage fresh; lore, feathering round lower half of eye, and ear-coverts almost black. Breast, flanks, and sides of belly slightly paler and less bluish grey than upperparts and chest, feather-centres paler, flanks and sides of belly often with whitish shaft-streaks, feathers of rear flank broadly tipped white. Central belly, vent, and under tail-coverts white, but often some grey of feather-centres visible on belly and vent and usually much dark grey on under tail-coverts, latter appearing boldly spotted dark grey and white. Tail black; central pair of feathers (t1) and fringes along outer webs of t2–t5 plumbeous-grey (most extensively so near tail-base); tip of (t3–)t5 with white triangular spot, larger and more rounded spot on t6 (mainly on inner web). Flight-feathers, tertials, and greater upper primary coverts black; primaries and primary coverts rather narrowly and sharply fringed blue-grey on outer web (except for emarginated parts of primaries); secondaries and tertials with broader and less clear-cut dark blue-grey borders along outer webs reaching shaft on tertials. Flight-feathers (except p9–p10 and inner tertials) with large and contrasting oblong white spot at about middle of inner web, showing as obvious white band on undersurface. Upper wing-coverts uniform dark plumbeous-grey or bluish-grey, except for slightly paler blue-grey fringe on median and greater coverts and for black inner webs of greater coverts. Axillaries dark grey with broad white tips; lesser under primary coverts white, median plumbeous, greater grey; remaining under wing-coverts white with narrow plumbeous band along leading edge of wing and broad plumbeous-black band across tips of median and bases of greater under wing-coverts; thus, underwing shows 2 broad white bands, one formed by tips of greater under wing-coverts in combination with white spots on inner webs of flight-feathers, another by white basal halves of median under wing-coverts and lesser under primary coverts. *In worn plumage*, upperparts slightly duller and less bluish grey, underparts slightly paler grey, white of central belly, vent, and tail-tips often more restricted, as partly worn off. ADULT FEMALE. Upperparts entirely olive-brown, sometimes with distinct grey tinge, in particular from lower mantle to upper tail-coverts. Supercilium long but narrow, buff or cream-white with olive-brown mottling, less distinct than in adult ♂. Narrow eye-ring cream or white, most distinct above eye, less so below, broken in front and behind. Lores and shorter and upper ear-coverts olive-brown. Cheeks buff or cream-white, faintly spotted olive-brown in front, more heavily towards rear, ending in brown-and-buff mottled lower ear-coverts; pale cheeks bordered by distinct brown-black or olive-brown malar stripe below and usually by similar bar behind (extending to rear of ear-coverts); often a buff bar behind ear-coverts, connecting rear supercilium with sides of throat. Chin and throat buff with white feather-bases; lower throat often spotted olive-brown. Chest, lower sides of neck, and sides of breast rufous-buff to cream, each feather with short olive-brown bar at tip and (usually hidden) olive-brown basal sides. Flanks and thighs olive-brown with whitish feather-centres, remainder of underparts cream-white or white; breast and sides of belly with short olive-brown bars or crescents, under tail-coverts with some olive-brown on basal sides (mainly hidden). Tail, flight-feathers, greater upper primary coverts, and underwing as adult ♂, but all plumbeous-grey replaced by olive-brown, black of inner webs of flight-feathers and primary coverts browner, less contrasting; pale spots on tips of outer tail-feathers and pale band across undersurface of flight-feathers tinged cream-buff. Upper wing-coverts olive-brown, like upperparts (lesser sometimes tinged grey); median and greater with rufous-buff (if fresh) to off-white (if worn) triangular or rounded spots on tips (usually largest on outer coverts, gradually smaller inwards), but spots sometimes absent on coverts with faint buff or grey-buff fringe along tips only. JUVENILE. Upperparts, sides of head, and lesser and median upper wing-coverts dark olive-brown with narrow ochre shaft-streaks. Chin and throat uniform buff, remainder of underparts rufous-ochre or buff, feathers of chest, breast, flanks, and sides of belly with dark fringe along tip, appearing scaled (not spotted as *Turdus* thrushes). (Dementiev and Gladkov 1954b; BMNH, RMNH.) For greater upper wing-coverts, tertials, tail, and flight-feathers, see 1st adult below. FIRST ADULT MALE. Usually markedly different from adult ♂, in particular on side of head and underparts. Upperparts dark plumbeous-grey, as adult ♂, but usually with olive-brown tinge on forehead and crown and occasionally elsewhere. Supercilium rather variable—sometimes rather short and narrow, buff-white with fine olive-brown mottling, sometimes long and distinct, with broad buff stripe on lore almost reaching nostril. Distinct buff ring round eye (absent in adult ♂). Lore, upper cheek, and ear-coverts brown-black, faintly mottled buff below eye and streaked pale buff on lower ear-coverts. Lower cheek buff with olive-brown spots, forming pale stripe backwards from lower mandible, bordered below by distinct dull black malar stripe. Side of neck and side of breast sometimes dark plumbeous-grey with olive-brown tinge, sometimes dark brown with limited grey suffusion and some buff mottling. Chin and upper throat buff, spotted dark olive-brown at sides, gradually merging into buff lower throat with more distinct spots, and this in turn merging into dark olive-brown chest, which is closely spotted or barred buff; chest often with slight plumbeous suffusion or black spotting. Breast, flanks, and sides of belly dark grey (often contrasting in colour with olive-brown chest); much white or pale grey of feather-bases visible on breast and sides of belly (appearing spotted or barred white) and with pale grey feather-bases with whitish shaft-streaks on flanks. Central belly and vent white; under tail-coverts white with some partly hidden dark grey spots. New lesser and median upper wing-coverts and (if any new) tertial coverts plumbeous, as adult; contrasting with olive-brown retained juvenile greater upper wing-coverts, but occasionally new coverts suffused olive-brown and then less contrasting; new median coverts occasionally with white spots on central tips. Retained juvenile flight-feathers, tertials, and greater upper primary coverts browner than in adult, fringes along outer webs and tips olive-brown (as in adult ♀); retained greater coverts often with buff or off-white spot on tip, in particular on innermost. Retained juvenile tail as adult, but greyer, less plumbeous-blue, and pale spot on tip of outer feathers less sharply defined. Underwing as adult ♂. Occasionally (2 of 19 examined), plumage of 1st adult ♂ similar to adult ♂, head and body largely plumbeous-grey, and retained juvenile wing-coverts and flight-feathers often then also tinged plumbeous-grey rather than olive-brown (in particular on flight-feathers); usually distinguishable by more extensively olive-brown crown, partly buff supercilium or chin, browner tertials, outer greater upper wing-coverts, or greater upper primary coverts, and less clear-cut white tips on outer tail-feathers; 3 other birds intermediate in colour, upperparts and sides of head as adult ♂ but chest and breast spotted white. FIRST ADULT FEMALE. Virtually indistinguishable from adult ♀; part of juvenile feathers retained, as in 1st adult ♂, but these not contrasting in colour with new feathers and hardly in abrasion; upperparts of some adult ♀♀ extensively tinged plumbeous, but a few 1st adults show this too, and some adult ♀♀ are as olive-brown as most 1st adult ♀♀. Both adult ♀ and 1st adult ♀ show shallow pale triangular spots on tips of greater upper wing-coverts, but spots sometimes absent (perhaps more often in adult ♀); certain ageing only possible for those 1st adults which show large and contrasting pale spots on tips of retained juvenile inner greater coverts, which then are contrasting with new uniform olive-brown tertial coverts; some 1st adults have narrow black rim along tip of retained juvenile coverts (absent in adult).

Bare parts. ADULT, FIRST ADULT. Iris brown or dark brown. Bill black with dark horn base of lower mandible in adult ♂, dark brown or black-brown with yellowish base to lower mandible in ♀ and in 1st adult ♂ in autumn and winter. Gape orange-yellow. Leg and foot of adult ♂ yellow or orange-yellow in summer, brownish-yellow or purplish-horn with dirty yellow rear tarsus and soles in winter; in 1st adult ♂, grey-brown or glossy brown in 1st autumn and winter, turning yellow in spring; in ♀, pale brown, yellow-brown, or yellow. (Andrew *et al.* 1955; Ali and Ripley 1973*b*; BMNH, RMNH, ZMA.) JUVENILE. No information.

Moults. ADULT POST-BREEDING. Complete; primaries descendant. No moult in birds examined October to early June. Single bird from USSR in full moult of wing and tail on 20 August (Dementiev and Gladkov 1954*b*). No further information. POST-JUVENILE. Partial: head, body, lesser and median upper wing-coverts, usually tertial coverts (in 24 of 27 examined), and exceptionally a few inner greater wing-coverts or tertials. Starts shortly after fledging when still near breeding grounds, but no further data on timing; no moult after arrival in winter quarters early October and later.

Measurements. ADULT, FIRST ADULT. Mainly Indonesia, winter; a few from breeding area and from western Europe; skins (RMNH, ZMA). Bill (S) to skull, bill (N) to distal corner of nostril; exposed culmen on average 4·9 shorter than bill (S).

	♂		♀		
WING	120·9 (2·41; 27)	117–127	118·6 (2·95; 22)	115–124
TAIL	80·0 (1·81; 21)	78–84	78·7 (3·10; 11)	73–82
BILL (S)	23·0 (1·16; 21)	21·2–24·6	23·6 (1·26; 11)	21·9–25·6
BILL (N)	13·2 (0·75; 20)	12·0–14·7	13·3 (0·40; 11)	12·6–13·9
TARSUS	29·2 (1·12; 20)	27·5–30·7	28·9 (0·86; 11)	27·8–30·3

Sex differences significant for wing and tail. 1st adult with retained juvenile wing and tail rather similar to older birds; e.g. wing, adult ♂ 120·8 (17) 117–126, 1st adult ♂ 121·1 (10) 117–127; tail, adult ♂ 81·3 (13) 78–84, 1st adult ♂ 80·0 (8) 78–82.

Weights. Nominate *sibirica*. USSR: ♂ 60, ♀ 70 (Dementiev and Gladkov 1954*b*). Mongolia, migrants from early June: ♀♀

52 (lean), 72 (Piechocki and Bolod 1972). Average of 8, Malay Peninsula, October: 71·2 (Nisbet 1968). Britain, October: adult ♂ 59·7 (Andrew *et al.* 1955).

Z. s. davisoni. Kuril Islands: ♂ 71, ♀ 77·2 (Nechaev 1969).

Structure. Wing rather long, broad at base, tip bluntly pointed. 10 primaries: p8 longest, p9 1–4 shorter, p7 2–5, p6 10–13, p5 17–20, p4 23–26, p1 34–40. P10 reduced, 69–77 shorter than p8, 5–13 shorter than longest upper primary covert. Outer web of p7–p8 and inner of (p7–)p8–p9 emarginated. Tertials short. Tail rather long, tip square or slightly rounded; 12 feathers, t6 2–10 shorter than t2–t3. Bill rather short, strong, straight; tip of upper mandible slightly decurved. Nostril rather large, oval, partly covered by membrane above. Many rather long and stiff bristles along base of upper mandible. Tarsus and toe rather long and slender. Middle toe with claw 26·6 (5) 25–29; outer toe with claw *c.* 71% of middle with claw, inner 66%, hind *c.* 76%. Claws rather short and blunt, slightly decurved.

Geographical variation. Slight, involving size and colour. *Z. s. davisoni* from Japan, Sakhalin, and southern Kuril Islands slightly larger than nominate *sibirica*, wing of ♂ 127·2 (4) 126–128 (RMNH, ZMA) or on average a few mm larger (Hartert 1903–10); general colour darker in both sexes, ♂ plumbeous-black on upperparts, sooty-black on crown, ear-coverts, cheeks, and chin (less uniform dark plumbeous-grey); white on lower belly and vent absent or restricted; white on tips of under tail-coverts less extensive; white spots on tips of outer tail-feathers smaller or virtually absent.

Recognition. ♂ unmistakable, but ♀ shows more resemblance to *Catharus* thrushes (which also show pale band at base of flight-feathers), in particular to Olive-backed Thrush *C. ustulatus*, differing in much larger size, more rufous-brown (less olive-grey) general tinge, more pronounced supercilium, and different pattern of spots on underparts. Size and colour of upperparts of ♀ also similar to Song Thrush *Turdus philomelos*, but sides of head of ♀ *Z. sibirica* more contrastingly marked, showing long (though mottled) pale supercilium and buff bar down rear of ear-coverts; ground-colour of underparts similar, but marks of *Z. sibirica* are short olive-brown bars or crescents, not bold black rounded or triangular spots; white or cream spots on tips of outer tail-feathers and white or cream band on undersurface of flight-feathers absent in *T. philomelos*.   CSR

Hylocichla mustelina Wood Thrush

PLATES 68 and 78
[between pages 856 and 857]

Du. Amerikaanse Boslijster   Fr. Grivette des bois   Ge. Walddrossel
Ru. Американский древесный дрозд   Sp. Zorzal charlo americana   Sw. Fläckskogstrast

Turdus mustelinus Gmelin, 1789

Monotypic

Field characters. 19 cm; wing-span 30–34cm. Only 15% smaller than Song Thrush *Turdus philomelos*, being noticeably larger than *Catharus* thrushes with more *Turdus*-like form obvious in larger bill and head and plumper body. Small Nearctic thrush with much more

robust appearance than *Catharus* thrushes, immediately suggesting small, bright, and clean *Turdus*. Red cap contrasts with otherwise mainly tawny-brown upperparts; underparts pure white covered with round black spots from lower throat to fore-belly and rear flanks. Obvious white eye-ring emphasizes large, dark eye. Flight action faster than *Turdus* thrushes. Sexes similar; no seasonal variation. Juvenile separable.

ADULT. Crown and nape bright russet, forming obvious cap; mantle russet-brown, and back and wings warm tawny-brown, but rump, upper tail-coverts, and tail colder brown, feathers fringed olive-grey. Face basically buff-white with short and narrow white fore-supercilium and bold eye-ring (forming 'spectacle' around large dark eye) which contrast with dark brown line from eye to bill, cheeks flecked and lined with black-brown, and quite broad, black-spotted malar stripe. Underparts virtually white (especially when worn), overlaid on sides of throat and from breast to fore-belly and over flanks with large and round black spots. Underwing mainly white, boldly barred dark grey across median coverts. JUVENILE. Plumage patterned as adult but 'red cap' less pronounced, with pale buff streaks on crown, mantle, and wing-coverts. Broad buff tips on greater coverts form obvious wing-bar. Underparts less clearly spotted. FIRST WINTER. Retains juvenile wing-marks. At all ages, bill dark brown-horn with yellow-buff base to lower mandible; legs pale flesh.

Unmistakable, being (1) larger and much more boldly and fully spotted below than any *Catharus* thrush, even rufous-crowned Veery *C. fuscescens*, and (2) smaller, much redder above (especially on head), and much whiter and more boldly spotted below than *T. philomelos*. For detailed comparison with other Nearctic thrushes, see p. 929. Flight like *Turdus* thrush but lighter and faster, with rapid turns and ducks into cover. Typically hops but also uses fast loping run. Skulks but less than *Catharus* thrushes. Like all Nearctic thrushes, fond of feeding in shade of woods, skulking in understorey of ground plants but moving out to feed.

Calls include a low, medium-pitched 'quirt' in alarm and a sharp 'pit pit', which may be extended into rapid 'pip-pip-pip-pip'.

Habitat. Breeds in temperate middle latitudes of Nearctic, mainly in lowland moist shady broad-leaved woodlands with ample undergrowth beneath tall trees, especially along streams and lake borders in swamps, keeping usually to ground and lower branches of trees (Pough 1949). Inhabits low cool damp forests, normally avoiding conifers, and preferring plenty of undergrowth and saplings beneath shady canopy, although since about 1890 sites in parks, gardens, and near dwellings have also been chosen, and it has also expanded into New England mountain valleys and slopes up to *c*. 600 m (Bent 1949). Sometimes found also on dry wooded hillsides (Forbush and May 1939). Winters in similar habitats.

Distribution. Breeds in USA from Dakota and Wisconsin, and in Canada from southern Ontario and south-west Quebec, south to Gulf coast and northern Florida.

Winters mainly from Mexico to Panama.

Accidental. Iceland: ♂, 23 October 1967 (AP).

Movements. Migratory. Moves south-west in autumn, and recorded as a migrant and wintering from south Texas south to Panama and north-west Colombia; winters mainly on Caribbean slopes. Galindo *et al.* (1963), mist-netting on Caribbean coast of Panama 1962–3, captured 84 in autumn compared with only 5 in spring. Outside main migration route, occurs casually west to Saskatchewan and California, and east to Cuba and Bahamas. Accidental in Nova Scotia, Bermuda, Puerto Rico, Curaçao, and Guyana (Godfrey 1966; American Ornithologists' Union 1983; Rappole *et al.* 1983).

Dispersal and autumn migration occur July–August, with few records on breeding grounds after September. Spring migrants reach southern USA in late March and southern Canada by early May (Robbins *et al.* 1983). Vagrants on Bahamas and Cuba recorded early October and early April (Bond 1961). TL-E

Voice. See Field Characters.

Plumages. ADULT. Forehead, crown, and nape tawny-rufous, slightly greyish or olive near base of upper mandible, brightest on crown, duller on nape, where merging into cinnamon-brown of mantle, scapulars, and sides of neck. Back olive-brown, rump and upper tail-coverts greyish-olive or light olive. Lores and stripe below eye pale buff or off-white, finely mottled dull grey (most extensive in front of eye). Narrow but distinct eye-ring white or pale cream, often narrowly interrupted just in front and behind eye. Short and rather indistinct supercilium backwards from eye mottled pale cinnamon-brown and off-white. Ear-coverts narrowly streaked dull grey or dark brown and off-white. Lower cheeks cream-white or white, partly spotted black (particularly near base of lower mandible), separated from white chin and upper throat by distinct black malar stripe (latter sometimes mottled white at rear). Lower sides of neck, lower throat, chest, and sides of breast pale buff or cream-buff, marked with bold and rounded dull black spots 3–4 mm across. Remainder of underparts white, boldly spotted dull black on flanks and sometimes upper belly and sides of belly; spots paler olive-brown and less clear-cut towards rear of flanks; some shorter under tail-coverts partly tipped olive. Tail olive-brown or greyish-olive. Tertials and outer webs of flight-feathers cinnamon-brown (on primaries slightly less bright than mantle and scapulars); inner webs of flight-feathers dark grey or greyish-black, slightly paler grey towards bases. Upper wing-coverts cinnamon-brown, like mantle and scapulars; fringes along outer webs of greater coverts and greater primary coverts slightly brighter cinnamon, inner webs of these duller brown-grey, tips uniform cinnamon-brown. Axillaries and greater under wing-coverts white with ill-defined dark grey spots on tips; lesser coverts white, median coverts contrastingly dark grey, forming dark bar; small coverts along leading edge of wing dark grey or olive-grey. *In worn plumage* (spring), similar to fresh autumn plumage, but crown tends to be paler rufous-cinnamon; mantle and scapulars paler cinnamon-brown; ground-colour of underparts whiter, bold spots even more contrasting; upper wing-coverts, tertials, and flight-feathers slightly more olive-brown. JUVENILE. Like adult, but crown, nape, mantle, and scapulars indistinctly streaked pale tawny or buff. Ear-coverts buff with distinct dull black streaks; cheeks cream-white except for fine dusky spots below eye; distinct black malar stripe. Spots on underparts less clear-cut, smaller, and duller black or olive-black, spots of flanks greyish and not as numerous as in adult. Lesser and median upper wing-coverts have narrow tawny-cinnamon shaft-streak, ending in large triangular tawny spot or fringe on feather-tip; greater upper wing-coverts and tertials have broad tawny-rufous tip or triangular patch. FIRST ADULT. Like adult, but juvenile flight-feathers, tail, tertials, greater upper wing-coverts, greater upper primary coverts, and often all or part of median upper wing-coverts retained; flight-feathers and tail as in adult, except for slightly greater wear, and tail-tips often more sharply pointed; tips of retained upper wing-coverts and tertials with rather contrasting tawny-cinnamon or tawny-buff fringe or triangular spot (in adult, uniform cinnamon brown or olive-brown); both webs of greater upper primary coverts often suffused black-brown on tips.

Bare parts. ADULT, FIRST ADULT, JUVENILE. Iris brown or dark brown. Bill dark horn, basal half of lower mandible flesh, pale flesh, yellow-buff, or pale flesh-brown. Legs and foot flesh or pale flesh. (Ridgway 1901–11; RMNH, ZMA.)

Moults. ADULT POST-BREEDING. Complete; primaries descendant. July–August; tail of some birds very short late July (Bent 1949), indicating simultaneous regrowth of feathers. POST-JUVENILE. Partial: head, body, lesser upper wing-coverts, and variable number of tertial coverts and median upper wing-coverts. Starts at c. 6 weeks old; completed late July to mid-September. (Bent 1949; BMNH, RMNH.)

Measurements. ADULT, FIRST ADULT. South-east Canada and north-east USA, summer; skins (BMNH, RMNH, ZMA). Bill (S) to skull, bill (N) to distal corner of nostril; exposed culmen on average 4·6 less than bill (S).

	♂		♀	
WING	110·8 (3·65; 16)	106–116	106·8 (2·41; 8)	103–110
TAIL	69·2 (2·14; 16)	67–74	67·1 (2·26; 8)	63–70
BILL (S)	20·6 (0·86; 16)	19·3–21·9	19·5 (0·69; 8)	18·5–20·6
BILL (N)	11·8 (0·66; 16)	11·1–12·7	11·2 (0·52; 8)	10·4–12·0
TARSUS	31·6 (0·83; 17)	30·5–33·2	30·9 (0·89; 8)	29·3–32·3

Sex differences significant for wing and bill (S). 1st adult with retained juvenile flight-feathers and tail combined with older birds, as juvenile wing on average only 0·5 shorter than adult and tail 0·9 shorter.

Weights. Eastern USA, April–September: ♂ 48·8 (4·56; 7) 44–54, ♀ 48·3 (3·40; 12) 43–54 (Baldwin and Kendeigh 1938; Norris and Johnston 1958; Graber and Graber 1962; Mengel 1965; Stewart and Skinner 1967; ZMA). Louisiana (USA), spring: 44·8 (4·76; 52) (Rogers and Odum 1966). Georgia (USA), early October: ♂ 56·0 (17) 48–71, ♀ 59·6 (10) 53–72 (Johnston and Haines 1957). New Jersey (USA), autumn: 48·1 (8) 43–52 (Murray and Jehl 1964). Panama, autumn: 42·6 (3·18; 24) (Rogers and Odum 1966). Curaçao (Antilles), October: exhausted ♂ 36 (ZMA). Belize: ♂ 51·1 (April), ♀ 46·2 (October) (Russell 1964). Mexico, January : ♂ 57·4 (RMNH).

Structure. Wing rather short, broad at base, tip bluntly pointed. 10 primaries: p8 longest, p9 and p6 3–5 shorter, p7 0–1, p5 9–14, p4 14–19, p1 23–29. P10 reduced, 60–69 shorter than wing-tip, 0–7 shorter than longest upper primary covert. Tertials short. Outer web of p6–p8 and (slightly) inner of p7–p9 emarginated. Tail rather short, tip straight; 12 feathers. Bill rather long and stout, about half length of head; straight, but tip of culmen gently decurved; bill close in shape to *Turdus* thrushes, not as short as in *Catharus*. Nostril small, oval, bordered by broad membrane above and behind. Some fairly long and stiff bristles along base of upper mandible. Leg and toes long, slender. Middle toe with claw 23·9 (6) 22–25; outer toe with claw c. 70% of middle with claw, inner c. 62%, hind c. 71%. Claws short, strong, decurved.

Recognition. Differs from *Catharus* thrushes in larger size, longer bill, and bold rounded black spots on chest and flanks (see Table A, p. 929). Hermit Thrush *C. guttatus* also rather heavily spotted black below, but spots more triangular and not extending backwards from breast. Unique in having upperpart colour grading from tawny-cinnamon on crown to olive on rump and tail; Veery *C. fuscescens* often rather rufous on crown, but duller and browner than *H. mustelina* and remainder of upperparts (including tail) uniform tawny-brown; *C. guttatus* shows opposite colour gradient, being gradually more rufous from olive-brown crown to cinnamon-brown tail-coverts and tail.   CSR

Catharus guttatus Hermit Thrush

PLATES 68 and 78
[between pages 856 and 857]

DU. Heremietlijster   FR. Grivette solitaire   GE. Einsiedlerdrossel
RU. Дрозд-отшельник   SP. Zorzal comun americano   SW. Eremitskogstrast

Muscicapa guttata Pallas, 1814. Synonym: *Hylocichla guttata*.

Polytypic. *C. g. faxoni* (Bangs and Penard, 1921), southern Canada and north-east USA from James Bay and central Nova Scotia west to central Yukon (Canada), north-east British Columbia, and southern Alberta. Extralimital: *crymophilus* (Burleigh and Peters, 1948), Newfoundland and southern Labrador west to James Bay and northern Nova Scotia; nominate *guttatus* (Pallas, 1814), southern Alaska and south-west Yukon, south to south-central British Columbia; 5–7 further races in western North America.

Field characters. 17 cm; wing-span 25–28·5 cm. 5% larger than Thrush Nightingale *Luscinia luscinia*; slightly smaller than other *Catharus* thrushes; 20% smaller than any Palearctic *Turdus* thrush. Smallest North American thrush with form and upperparts recalling *L. luscinia* but underparts spotted on breast and dappled on flanks; contrast of olive-brown back with chestnut tail, rump, and upper tail-coverts, and habit of cocking and slowly lowering tail are distinctive. Flight like nightingale *Luscinia*. Sexes similar; little seasonal variation. Juvenile separable. At least 8 races in Nearctic, of which 2 may reach west Palearctic but are not separable in the field.

ADULT. Race of central and south-east Canada and north-east USA, *faxoni*. Crown, nape, back, and wings umber-brown, with faint tawny tone in some lights and with rufous margins to primaries forming faint panel when fresh; crown and back more olive when worn. Rump and (especially) upper tail-coverts and tail rufous-chestnut, contrasting with tertials and back. Face basically pale olive-brown, with paler buff-mottled lore, dull white to buff eye-ring, brown mottling on cheeks, and narrow but distinct malar stripe of black spots. Breast (to shoulder) and flanks washed buff to pale brown-grey, breast quite heavily spotted black-brown and flanks dully spotted and dappled olive-brown; vent and under tail-coverts off-white. Underwing grey, with cream-buff bases to flight-feathers forming central panel. JUVENILE. Duller than adult. Upperparts (including wing-coverts) duskier and more olive, distinctly spotted buff-white. Underparts more extensively and more intensely dark-spotted. FIRST WINTER. Resembles adult but retains juvenile wing and tail-feathers. At all ages, bill dark brown-horn with pale flesh base to lower mandible; legs pale flesh.

From above or behind, may suggest *L. luscinia* or Nightingale *L. megarhynchos* to Palearctic observer, with form only slightly bulkier and flight action and behaviour similar, but clear view of face and spotted underparts allows instant identification to genus, while within *Catharus* contrast between back and rump/tail and persistent cocking and lowering of tail are diagnostic. For detailed comparison with other Nearctic thrushes, see Recognition. Flight less flapping than *Turdus*, with rather flitting wing-beats, reminiscent of *Luscinia*. Carriage fairly upright, with raised tail also recalling *Luscinia*—but appears neckless at times. Gait a hop. Not shy but secretive, escaping into low canopy or thick ground-cover. Adapted to less damp habitat than Swainson's Thrush *Catharus ustulatus* and Veery *C. fuscescens*; more likely to be seen on open grass near cover.

Commonest call a low 'chuck'.

Habitat. Breeds in upper and middle latitudes of North America from lowlands up to mountain forest zone of Engelmann spruce *Picea engelmanni* at *c.* 3000 m in Colorado (Niedrach and Rockwell 1939). In Canada, inhabits pure or mixed coniferous woodlands, varying from wooded bogs and swamps to dry sandy and sparse jack pine *Pinus banksiana*; also clearings after logging, and second growth after fire, with standing dead trees used as song-posts; frequents forest floor and lowest branches of trees. In Great Plains of USA, favours shady, moist coniferous or mixed forests (Johnsgard 1979). Although hardy, seeks to avoid snow cover. Breeds in wet *Sphagnum* bogs surrounded by larches and spruce, or on hot dry barren ground, rarely far from forest edge. Various mountain races breed at up to *c.* 3500 m, in coniferous forests up to treeline. Neighbourhood of streams, especially in ravines, often preferred. In winter in California, found in chaparral and woods, in stream-side thickets, and in shrubbery of city gardens, but avoids open meadows, fields, and hillsides. In Mexico, winters in small valleys and by streams in forests, and by city park lawns which are watered each day. Sometimes, however, winters within breeding range by wooded swamps and elsewhere (Bent 1949). Also occurs in mainly broad-leaved woodlands, dry hillsides and uplands, and rocky brush-grown pastures (Pough 1949).

Distribution. Breeds from central Alaska east across much of forested Canada to Newfoundland and south to southern California, northern New Mexico, central Minnesota, central Pennsylvania, western Maryland, and Long Island. Winters from Massachusetts, Pennsylvania, Kentucky, southern Missouri, Oklahoma, and Texas south to Florida and Gulf coast, Baja California, Guatemala, and El Salvador.

Accidental. Iceland, Britain, West Germany, East Germany.

Movements. Almost wholly migratory (including almost all *faxoni* and *crymophilus*), some south-western populations perhaps partly resident (American Ornithologists' Union 1957). In Chicago, autumn migrants noted 20 September to end of October (Annan 1962). Data from 3941 birds ringed over 10 years at 5 sites (northern California coast, southern Ontario, western Pennsylvania, Rhode Island coast, south-east Massachusetts coast) showed migration started in last week of September, peaked in mid-October, and declined through November. Age-ratios at each site varied little between years, yearly fluctuations in numbers were generally independent from site to site, and yearly variations of timing (median and termination dates) were of the order of 2 days. Proportion of adults highest at the inland Canadian site, lowest on coast (6% at Rhode Island site), suggesting that adults undertook longer flights and immatures shorter with greater tendency to be drifted to coasts (Smith and Schneider 1978; see also Ralph 1975). On Bermuda, recorded late October to November in autumn (Wingate 1973).

Main wintering areas are central and southern USA and west coast from southern British Columbia south to Mexico, Guatemala, El Salvador, and northern Bahamas. Some winter as far north as southern Canada. Spring migration is from early April in southern part of breeding range to May in north (Robbins *et al.* 1983), with birds leaving Bermuda late March to early April (Wingate 1973).

Within Nearctic, accidental records include Southampton Island (northern Hudson Bay), Cuba, and Jamaica (Godfrey 1976; American Ornithologists' Union 1957; Rappole *et al.* 1983). TL–E

Voice. See Field Characters.

Plumages (*C. g. faxoni*). ADULT. Upperparts, sides of neck, tertials, and upper wing-coverts cinnamon-brown or umber-brown, slightly mottled buff at base of upper mandible. Rump brighter cinnamon, upper tail-coverts and tail rufous-cinnamon or rufous-brown, contrastingly brighter rufous than remainder of upperparts. Lores mottled pale buff and dull black; spot in front of eye dull black; stripe over eye greyish olive-brown, slightly paler and greyer than crown. Ear-coverts dark olive-brown, narrowly streaked pale buff; upper cheeks mottled pale buff and olive-brown. Narrow pale buff eye-ring. Lower cheeks, chin, and throat pale buff or cream; distinct dull black malar stripe down sides of throat; front part of lower cheeks and sides of throat sometimes finely spotted dull black or olive-brown. Malar stripe broken into dull black spots on lower sides of neck, merging into boldly spotted chest; spots of chest dull black, rather triangular in shape, 3–4 mm wide, not extending to sides of breast and flanks, but continued into less contrasting dull olive-brown spots on breast and upper belly. Ground-colour of chest pale cream-buff. Sides of breast, flanks, and sides of vent pale olive-brown or buff-brown, merging into white on central belly and vent; under tail-coverts cream-buff, pale buff, or buff-white. Outer webs of flight-feathers, greater upper primary coverts and longest feather of bastard wing brown, tinged tawny-rufous along edges (paler and more tawny than tertials, mantle, and scapulars); inner webs dark grey or greyish-black, but basal $\frac{1}{3}$–$\frac{1}{2}$ of secondaries and inner primaries contrastingly pink-buff or rufous-cream, well-visible on underside of spread wing. *In worn plumage* (spring), upperparts slightly more olive-brown, less dark umber-brown or cinnamon (sometimes rather like upperparts of Gray-cheeked Thrush *C. minimus*, but rump to tail rufous, not similar to remainder of upperparts as in *C. minimus*); ground-colour of underparts white or dirty white, except for faint cream or buff tinge from chin to chest and on under tail-coverts and for pronounced olive wash on flanks, sides of vent, and thighs; olive-brown spots on upper belly (if any) sometimes partly worn off; marks on sides of head indistinct, except for black malar stripe; narrow pale buff eye-ring sometimes virtually absent, in particular when plumage heavily worn. JUVENILE. Upperparts and sides of head sepia or olive-brown (except for rufous rump, tail-coverts, and tail), each feather with large buff-white spot on centre bordered by black along tip. Underparts white with faint buff tinge, spotted deep black on sides of neck, across chest, and on flanks and vent; throat, chest, upper belly, and flanks with faint dark arcs on feather-tips. Lesser upper wing-coverts like upperparts; median coverts and tertials darker olive-brown, greater coverts black-brown, each with paler olive or rufous-tawny fringe along tip and often with small buff or pale buff spot or larger triangle on centre of tip; this buffish spot usually most pronounced on inner greater coverts and frequently absent on other coverts and tertials; occasionally, entirely absent. (Dwight 1900; ZMA.) Tail-feathers as 1st adult. FIRST ADULT. Like adult, but juvenile flight-feathers, tail, greater upper primary coverts, and some or all tertials and greater upper wing-coverts retained, contrasting in wear with neighbouring fresh feathers and often more distinctly worn than adult at same time of year. Tail-feathers often narrower and more sharply pointed than adult. 2–6 outer greater upper wing-coverts juvenile; often a pale spot or triangle

on centre of tip and pale fringe along distal part of outer web (colour buff to off-white, depending on bleaching), but spot or triangle sometimes completely worn off in spring, leaving indentation at feather-tip (sometimes small and hard to see); some birds (7 of 28 examined) without spots, though greater coverts still juvenile, and these hard to distinguish from adult, apart from greater wear of retained juvenile feathers.

Bare parts. ADULT, FIRST ADULT. Iris brown or dark brown. Bill dark horn-brown or black, basal cutting edges of upper mandible and basal half of lower yellowish-flesh, lilac-flesh, pale grey-flesh, or flesh grading to yellowish at extreme base. Leg flesh, lilac-flesh, pale grey-flesh, or brownish-flesh with paler rear side; upper surface of toes often slightly darker, soles pale flesh. (Ridgway 1901–11; RMNH.) JUVENILE. Bill, leg, and foot dull pink-buff (Bent 1949).

Moults. ADULT POST-BREEDING. Complete; primaries descendant. August–September. POST-JUVENILE. Partial: head, body, lesser and median upper wing-coverts, usually tertial coverts, 0–4 inner greater upper wing-coverts, and occasionally tertials. Starts early July to early September, completed from early August onwards. (Dwight 1900; BMNH, RMNH.)

Measurements. *C. g. faxoni*. ADULT, FIRST ADULT. Eastern USA and southern Ontario (Canada), April–October; skins (BMNH, RMNH, ZMA). Bill (S) to skull, bill (N) to distal corner of nostril; exposed culmen on average *c.* 4·0 less than bill (S).

	♂		♀	
WING	96·7 (1·84; 20)	93–100	92·4 (1·25; 21)	89–94
TAIL	67·0 (2·72; 20)	63–71	63·6 (1·52; 21)	60–66
BILL (S)	17·9 (0·47; 20)	17·4–18·7	17·5 (0·59; 21)	16·7–18·4
BILL (N)	9·7 (0·50; 21)	9·1–10·6	9·6 (1·61; 21)	8·7–10·4
TARSUS	30·9 (0·92; 20)	29·4–32·9	30·3 (0·95; 19)	29·1–31·9

Sex differences significant for wing and tail. 1st adult with retained juvenile flight-feathers and tail combined with older birds, as juvenile wing on average only 0·8 shorter than full adult, juvenile tail 1·0 shorter.

Weights. *C. g. faxoni*. Eastern USA. January–February: ♀♀ 32, 32. April–May: ♂ 32·1 (3) 31–33, ♀ 28·6 (3) 27–30. October–November: ♂ 29·8 (1·94; 4) 28–32, ♀ 31·4. (Baldwin and Kendeigh 1938; Mengel 1965; Stewart and Skinner 1967.)

Structure. Wing rather short, broad at base, tip fairly rounded. 10 primaries: p8 longest, p9 0·5–1 shorter, p7 1–2, p6 5–6, p5 11–13, p4 15–17, p1 20–22. P10 reduced, 49–56 shorter than wing-tip, 3 shorter to 1 longer than longest upper primary covert. Outer web of p5–p8 and inner of p7–p9 emarginated. Tertials short. Tail rather short, tip square; 12 feathers. Bill rather short, strong; straight, but tip of culmen decurved. Nostril oval, partly covered by frontal feathering at base and by narrow membrane above. Some stiff and long bristles at base of upper mandible; smaller bristles near nostril and on chin. Tarsus and toes long and slender. Middle toe with claw 21·4 (8) 20–23; outer toe with claw *c.* 67% of middle with claw, inner *c.* 63%, hind *c.* 73%. Claws rather long, strong, decurved.

Geographical variation. Marked and complex; involves size, relative length of bill and tarsus, colour of upperparts, sides of body, and chest, and amount of spotting on underparts. Most variation clinal, boundaries between races often hard to draw; outline developed here mainly follows Aldrich (1968),

Table A Some identification characters of small North American *Hylocichla* and *Catharus* thrushes.

		Wood Thrush *Hylocichla mustelina*	Hermit Thrush *Catharus guttatus faxoni*	Swainson's Thrush *Catharus ustulatus swainsonii*	Gray-cheeked Thrush *Catharus minimus minimus*	Veery *Catharus fuscescens fuscescens*
Upperparts and tail		Contrast between rufous of crown to back and olive-brown of rump to tail	Contrast between olive-brown of crown to back and rufous of rump to tail	Uniform olive-green or olive-brown	Uniform dull dark olive-grey	Uniform cinnamon-brown to olive-brown
Eye-ring		Narrow, but distinct, white or pale cream	Narrow, pale buff, sometimes faint	Broad, distinct, buff	Narrow, indistinct, whitish, often only present behind eye or absent	Narrow, indistinct, pale buff, often only present behind eye or absent
Chest-spots		Large, black	Rather large, (brown-)black	Rather small, brown-black	Rather small, black	Small or rather small, brown
Flanks		White, boldly spotted black	Uniform buff-brown	Uniform olive-brown	Extensively grey, sometimes with darker spots	Uniform light (olive-)grey
Emarginated outer webs		p6–p8	p5–p8	p7–p8	p7–p8	p6–p8
Distance from wing-tip(p8) to other primary-tips (mm)	p9	3–5	0·5–1	1·5–4	1·5–4	2–4
	p6	3–5	5–6	7–12	7–9	5–7
	p5	9–14	11–13	13–18	14–18	10–14
	p1	23–29	20–22	27–30	28–33	23–29
Wing length (mm)		108·8 (103–116)	94·5 (89–100)	99·7 (93–107)	104·5 (98–111)	99·6 (93–105)
Tarsus length (mm)		31·3 (29–33)	30·6 (29–33)	28·0 (26–31)	30·4 (29–33)	29·8 (28–32)

who recognized 10 races, but number much debated. Eastern races of medium size (average wing 93–95), but most western races smaller (average wing 86–90), except for races of Rocky Mountains and mountains of eastern California (average wing 97–103). *C. g. faxoni* the most widespread race of Canada and north-east USA: upperparts cinnamon-brown or umber-brown, but 2 colour morphs recognizable, differing slightly in general hue, one more rufous, other more greyish (Aldrich 1968). Easternmost race *crymophilus* (potential vagrant to Europe) generally darker than *faxoni*, upperparts duller brown or olive-brown. In west, races of humid coasts of Alaska, British Columbia, and north-west USA generally dark, and races of arid western interior USA rather paler. For details, see Aldrich (1968).

Recognition. See Table A. *C. guttatus* and Veery *C. fuscescens* are the only *Catharus* with rufous rump, upper tail-coverts, and tail; however, upperparts of *C. fuscescens* similar to colour of rump and tail (in *C. guttatus*, upperparts dull umber-brown or cinnamon-brown, contrasting with rufous rump and tail); *C. fuscescens* lacks distinct pale eye-ring; cheeks, lower throat, and chest of *C. fuscescens* deeper pink-buff or tawny-buff than in *C. guttatus*, chest-spots smaller and less contrasting, malar stripe and chest-spots olive or brown rather than dull black.　CSR

Catharus ustulatus Swainson's Thrush (Olive-backed Thrush)

PLATES 68 and 78
[between pages 856 and 857]

Du. Dwerglijster　Fr. Grivette à dos olive　Ge. Zwergdrossel
Ru. Свэнсонов дрозд　Sp. Zorzal chico　Sw. Gråbrun dvärgtrast

Turdus ustulatus Nuttall, 1840. Synonym: *Hylocichla ustulata*.

Polytypic. *C. u. swainsonii* (Tschudi, 1845), Newfoundland and north-east USA, west to eastern Alberta. Extralimital: *almae* Oberholser, 1898, Alaska, Yukon, western Mackenzie, and central Alberta, south to mountains of Colorado and north-west Utah; nominate *ustulatus* (Nuttall, 1840), coastal ranges of western North America from Juneau (Alaska) to north-west Oregon; *oedica* Oberholser, 1899, from inland northern Washington and coastal south-west Oregon to California.

Field characters. 18 cm; wing-span 27–30 cm. About 10% larger than Thrush Nightingale *L. luscinia* averages slightly smaller than Gray-cheeked Thrush *C. minimus* but all measurements overlap. Small thrush with form recalling *Luscinia*, olive upperparts and buff to dusky-white underparts spotted on throat and breast; closely similar to *C. minimus* in general appearance but with usually less grey, even faintly russet tinge to upperparts, bold buff eye-ring and buff ground-colour to cheeks, and buffier ground-colour to throat, breast, and flanks. Only occasionally raises tail like Hermit Thrush *C. guttatus*. Sexes similar; little seasonal variation. Juvenile separable. 4 races in Nearctic, 1 reaching west Palearctic.

ADULT. Eastern north American race, *swainsonii*. Upperparts, wings, and tail uniform olive, faintly russet when fresh and never as grey as northern population of *C. minimus*. Face basically buff, with conspicuous pale buff lore and broad eye-ring, brown-mottled cheek, and narrow, black-spotted malar stripe. Throat to shoulder basically pale buff, becoming more tawny or yellow on breast and more olive- to grey-brown on flanks; breast quite heavily spotted black-brown but flanks only faintly dappled and mottled dusky; vent and under tail-coverts white. Underwing dull olive-brown, with cream tips of greater coverts and cream bases to inner webs of flight-feathers forming pale band along centre. JUVENILE. Browner than adult, buff-spotted above and with short bars (as well as spots) below. FIRST WINTER. Resembles adult except for retention of juvenile feathers in wing and tail, and thus may show pale spots on wing-coverts. At all ages, bill dark brown-horn with pale buff-flesh base to lower mandible; legs grey-flesh.

From above or behind, might suggest dull *L. luscinia* to Palearctic observer but larger and tail not obviously redder than rest of upperparts. Not difficult to distinguish from *C. guttatus* (which is slightly smaller, lacks buff tone to face, and shows markedly rufous rump and tail) or Veery *C. fuscescens* (same size, with upperparts of western race *C. fuscescens salicicolus* close in tone to more russet-backed *C. ustulatus*, but not strongly spotted on breast, lacks obvious eye-ring, and has different call); dullest *C. ustulatus* do, however, invite confusion with *C. minimus*, with close attention to face pattern and call essential to distinction—see *C. minimus* (p. 932). For detailed comparison with this and other Nearctic thrushes, see p. 929. Flight, gait and behaviour much as *C. guttatus* but cocks and lowers tail more slowly and far less often. Vagrant most likely to be found in bottom of thicket.

Commonest call an emphatic 'whit' or 'wick'; also a short, high-pitched 'heep' or 'queep' in flight.

Habitat. Breeds in higher and middle latitudes of Nearctic within forest zone, predominantly coniferous and lowland, but in south of range up to *c*. 3000 m in Rocky Mountains. In northern New England and eastern Canada, inhabits mixed growth on borders of spruce and fir forests, with some birches, preferring lower and damper parts of forest, especially near streams, but often breeding also in dry upland coniferous woods, to *c*. 800 m or more. In Michigan, favours conifer stands where lowest branches have died and dropped off leaving relatively open floor zone up to nearly 2 m; also occupies dense alder thickets. In British Columbia, however, favours lowland poplar woods and willow thickets. In the west, nests more often in broad-leaved trees and bushes, while in New England normally in spruce or balsam fir. On spring migration also occurs in brushy fields, edges of brushy swamps and cattail marshes, orchards, gardens, parks, and shrubbery near farmhouses; at this season often feeds in treetops (Bent 1949). In plains states, breeds in heavily shaded coniferous forests with streams or springs and relatively open undergrowth allowing foraging on forest floor. In hills (up to *c*. 2200 m) prefers higher cooler spruce forests to pines (Johnsgard 1979). In Alaska, differs from Gray-cheeked Thrush *C. minimus* in generally keeping to coniferous woods, rather than willow or alder brush outside them (Murie 1963). In Canada, however, occupies second growth after fire or logging (Godfrey 1966) and this also noted in West Virginia after felling of mountain spruces (Bent 1949). In winter in Venezuela, occurs in wooded habitats up to *c*. 2300 m, in forest, second growth, open woodland, plantations, and brushy areas, foraging from dark undergrowth to middle heights (Schauensee and Phelps 1978).

Distribution. Breeds from central Alaska east across forested Canada to Newfoundland, and south to California, Colorado, Minnesota, Wisconsin, northern Michigan, southern Ontario and New Hampshire, and Maine; also in Appalachians. Winters from Mexico to Argentina.

Accidental. Iceland, Britain, Ireland, Belgium, West Germany, Norway, Finland, Austria, Italy, USSR.

Movements. Migratory. Moves through southern Canada and USA (scarce in south-west), less commonly through Bahamas and west Caribbean islands, to winter from Mexico to north-west Argentina (American Ornithologists' Union 1983; Rappole *et al.* 1983). Analysis of 5079 birds ringed over 10 years showed proportion of adults among migrants was higher inland than on coasts, with only 2% adults at a Rhode Island site; no common year-to-year fluctuation in numbers at the various ringing sites; immatures arrived earlier and left later than adults (Smith and Schneider 1978). On Caribbean coast of Panama, Galindo *et al.* (1963) caught more *C. ustulatus* than any other thrush, more in autumn (1279) than spring (365). Spring migrants apparently retrace autumn routes.

First captures of autumn migrants were noted from Ontario to California and Massachusetts in late August; median (mid-September) and last dates (late October) also very similar (Smith and Schneider 1978); corroborated by Annan (1962) in Chicago. Migrants reach western West Indies (where rare) 12 October to 28 November (Bond 1961). Regular on Bermuda in early October (Wingate 1973), and present in Venezuela November–March (Schauensee and Phelps 1978). In spring, recorded (rarely) in western West Indies 19 March–10 May. Most birds reach breeding grounds by early May (Bond 1961; Robbins *et al.* 1983).   TL–E

Voice. See Field Characters.

Plumages (*C. u. swainsonii*). ADULT. Upperparts dark greenish-olive or olive-brown, sometimes slightly greyish on back and rump, often with slightly darker feather-centres visible on crown and outer scapulars. Distinct stripe from nostril to upper front corner of eye and bold eye-ring warm buff; stripe bordered below by black and pale buff mottling on lores and lower front corner of eye. Upper cheeks and sides of head from ear-coverts backwards dull olive-brown, mottled buff on cheeks and finely streaked buff on shorter and lower ear-coverts. Lower cheeks and lower sides of neck buff, partly marked with indistinct olive-brown spots, bordered below by dull black or brown-black malar stripe (broken by buff at rear). Chin and upper throat pale buff; lower throat and chest deeper buff or tawny-yellow, marked with small triangular black spots on lower throat and with gradually broader, more rounded, and paler dull black or dark olive-brown spots on chest. Sides of breast, flanks, and thighs olive-brown or greyish-olive (paler than upperparts), marked with indistinct darker olive-brown spots on sides of breast. Remainder of underparts white, marked with olive spots at border of chest and upper flanks. Tail dark greyish-olive or olive-black, like upperparts or slightly darker. Flight-feathers, tertials, and upper wing-coverts dark greenish-olive or olive-brown, like upperparts, often slightly brighter along outer edges of primaries, greater upper primary coverts, and greater upper wing-coverts; darker greyish-black or olive-black on centres of wing-coverts and on inner webs and tips of flight-feathers and primary coverts; bases of inner webs of secondaries and inner primaries cream, forming pale band at undersurface. Under wing-coverts and axillaries greyish olive-brown, tips of greater under wing-coverts, base of median coverts and axillaries, and part of under primary coverts cream or white. *In worn plumage*, spring and summer, upperparts, tail, and wing duller and greyer, less greenish; buff of sides of head (including eye-ring) and in particular chin to chest paler, more cream-buff. JUVENILE. Upperparts olive-brown, forehead, crown, hindneck, mantle, scapulars, and lesser upper wing-coverts with buff shaft-streak, often ending in wider buff spot on each feather-tip; rump and upper tail-coverts less distinctly marked buff. Underparts as adult but marks bar-shaped rather than spot-like, except sometimes on sides of throat and lower throat; sides of breast, flanks, and sometimes central breast and sides of belly marked with olive-brown or grey-brown bars or spots. Under tail-coverts pale grey-buff. Flight-feathers, tertials, greater upper primary coverts, and median and greater upper wing-coverts like adult, but see 1st adult below. FIRST ADULT. Like adult, but juvenile flight-feathers, tail, greater upper primary coverts, usually tertials and greater upper wing-coverts, and occasionally some median and tertial coverts retained, these sometimes more worn than adult at same time of year, tail-feathers usually narrower and more sharply pointed at tip than in adult. 10 of 27 1st adults examined had distinct buff, tawny, or off-white triangular spots on tips of some or all greater coverts and (if retained) tertial coverts, median coverts, or tertials (adults wholly uniform olive-brown); 10 others had spots small or showed rufous-olive, pale olive, or buff fringe on greater coverts only, sometimes also with narrow buff shaft-streak on inner greater or (if retained) median or tertial coverts; occasionally, some small buff tips on tail-feathers.

Remaining 7 had retained feathers similar in colour to adult, differing slightly in shape and abrasion only. In another sample, 9 of 42 examined did not show pale spots on wing-coverts (Payne 1961).

Bare parts. ADULT, FIRST ADULT. Iris bright brown, deep brown, or dark brown. Bill dark horn-brown, horn-black or plumbeous-black, basal half of lower mandible and basal cutting edges of upper pale yellowish-horn, pale flesh-horn, yellow-pink, or lilac-pink. Mouth yellowish. Leg and foot pink-flesh, purple-flesh, pink-grey, or pale brown-grey, toes often slightly darker, dull grey-flesh or pale flesh-brown; rear of tarsus and soles lighter, pale grey, flesh-grey, or yellowish-grey. (Ridgway 1901–11; Dol 1977; RMNH, ZMA.) JUVENILE. No information.

Moults. ADULT POST-BREEDING. Complete; primaries descendant. August–September; completed in migrants examined from 10 September onwards (Bent 1949; BMNH, RMNH). POST-JUVENILE. Partial: head, body, lesser upper wing-coverts, often some or all median upper wing-coverts, sometimes tertial coverts, and occasionally some greater upper wing-coverts or tertials (RMNH, ZMA). Starts about mid-August (Bent 1949), but probably sometimes earlier as already completed in migrants from early September (RMNH, ZMA).

Measurements. *C. u. swainsonii.* ADULT, FIRST ADULT. Eastern North America, summer, and migrants Central America, autumn; skins (BMNH, RMNH, ZMA). Bill (S) to skull, bill (N) to distal corner of nostril; exposed culmen on average 4·0 less than bill (S).

	♂			♀		
WING	101·9	(3·13; 17)	94–107	97·5	(2·41; 21)	93–102
TAIL	66·5	(3·03; 14)	62–72	62·4	(2·27; 18)	58–66
BILL (S)	17·3	(0·90; 16)	16·4–18·6	16·3	(0·53; 21)	15·2–16·9
BILL (N)	9·2	(0·60; 13)	8·4–9·9	8·7	(0·44; 17)	7·9–9·4
TARSUS	28·7	(1·32; 14)	27·3–31·1	27·4	(0·91; 18)	25·7–28·6

Sex differences significant. 1st adult with retained juvenile flight-feathers and tail included with older birds in table above, though juvenile wing on average 1·9 shorter than fully adult (or 1·4: Child 1969) and tail on average 2·5 shorter.

Weights. ADULT, FIRST ADULT. Mainly *swainsonii*, but some samples include *almae*. New Jersey, autumn: 28·5 (2·54; 138) 23–37 (Murray and Jehl 1964). Michigan, autumn: 31·4 (23) (Child 1969). Illinois, mainly September: ♂ 32·2 (74), ♀ 30·8 (51) (Graber and Graber 1962). South Carolina, 30 September: 41·1 (21) (Child 1969). Georgia, early October: ♂ 39·4 (19) 32–50, ♀ 37·7 (17) 33–42 (Johnson and Haines 1957). Florida, autumn: 38·9 (110) (Child 1969). Belize, October: ♂ 32·7 (2·89; 5) 30–37, ♀ 30·0 (2·19; 7) 27–33 (Roselaar 1976). Panama, 13 October: 33·5 (3·34; 12) 28–39 (Rogers 1965). Panama, April: 29·6 (218) (Leck 1975). Panama and Belize, early May: 35·0 (6·19; 4) 31–44 (Russell 1964; Leck 1975). Texas and Louisiana, mid-April to early May: 26·7 (92) (Child 1969). Kentucky, late April to mid-May: ♂ 31·8 (4·11; 6) 28–37, ♀ 33·1 (3) 27–38 (Mengel 1965). See also Irving (1960) and Drury and Keith (1962).

Structure. Wing rather long, broad at base, tip bluntly pointed. 10 primaries: p8 longest, p9 1·5–4 shorter, p7 0–2, p6 7–12, p5 13–18, p4 17–21, p1 27–30. P10 reduced, 54–63 shorter than wing-tip, 0–5 shorter than longest upper primary covert. Outer web of p7–p8 and (slightly) inner of p8–p9 emarginated. Bill short. Middle toe with claw 20·6 (12) 19·5–21·6; outer toe with claw *c.* 68% of middle with claw, inner *c.* 62%, hind *c.*

73%. Remainder of structure as in Hermit Thrush *C. guttatus* (p. 928).

Geographical variation. Slight, mainly involving colour, of upperparts in particular. Following account based on Bond (1963). Eastern race *swainsonii* rather rufous-olive on upperparts, ground-colour of sides of head and all chest tinged buff; recognition of *clarescens* (Burleigh and Peters, 1948) for birds from Newfoundland and Nova Scotia (as in, e.g., American Ornithologists' Union 1957) not warranted. *C. u. almae* from Alaska to central Alberta and south through Rocky Mountains greyish-olive or brownish-olive in autumn, not as rufous as *swainsonii*; upperparts in spring generally greyer and less olive than *swainsonii*; ground-colour of chest and sides of head paler, less buff, chest-spots darker than in *swainsonii*; northern birds from range of *almae* sometimes stated to be greyer and hence separated as *incana* (Godfrey, 1952), but recognition not warranted. Nominate *ustulatus* from northern part of west coast deep rufous on upperparts in autumn, brown-umber in spring, not as olive as *swainsonii* nor as grey as *almae*; *oedica* from

southern part of west coast like nominate *ustulatus*, but upperparts paler and greyer, though more buffy and less olive than in *swainsonii*.

Recognition. Uniform olivaceous upperparts, tail, and upperwing in combination with warm buff ground-colour of sides of head (including broad eye-ring), throat, and chest are good characters for this species (see Table A, p. 929). Upperparts of Gray-cheeked Thrush *C. minimus* also olive, but darker and less greenish, ground-colour of sides of head, throat, and chest cream or off-white, not warm buff, and eye-ring absent or showing only as a narrow grey-and-white ring. Hermit Thrush *C. guttatus*, which often also shows a narrow buff eye-ring, differs from present species in more olive-brown upperparts with contrasting rufous rump and tail and in different wing formula. Some individuals of Veery *C. fuscescens* are also rather olive above and rather deep buff on throat and chest, but upperparts generally browner, sides of head grey without pale eye-ring, and spots on chest less contrastingly brown instead of black. CSR

Catharus minimus Gray-cheeked Thrush

PLATES 68 and 78
[between pages 856 and 857]

Du. Grijswangdwerglijster   Fr. Grivette à joues grises   Ge. Grauwangendrossel
Ru. Малый дрозд   Sp. Zorzal carigris   Sw. Gråkindad trast

Turdus minimus Lafresnaye, 1848. Synonyms: *Hylocichla minima*, *H. aliciae*.

Polytypic. Nominate *minimus* (Lafresnaye, 1848), north-east Siberia and Alaska through Canada to Labrador and Newfoundland; *bicknelli* (Ridgway, 1882), north-east USA and eastern Canada north to Nova Scotia, north shore of Gulf of St Lawrence, and southern Labrador.

Field characters. 18 cm; wing-span 28·5–32 cm. Slightly larger than Swainson's Thrush *C. ustulatus*, with up to 10% longer legs; size difference most obvious with northern race (nominate *minimus*); largest of genus. Small thrush with form recalling nightingale *Luscinia*, grey-olive upperparts, and grey-white underparts spotted on throat and breast; closely similar to *C. ustulatus* in general appearance but with grey- rather than green-toned upperparts (never looking warm) in post-juvenile plumages, usually less obvious eye-ring, and only faintly buff-grey ground-colour to breast and flanks. Does not cock tail like *C. ustulatus*. Voice includes one call unlike any of *C. ustulatus*. Sexes similar; no seasonal variation. Juvenile separable. 2 races in Nearctic, of which either may reach west Palearctic; northernmost, nominate *minimus* (most grey above) described here (see also Geographical Variation).

ADULT. Upperparts dull, faintly grey-olive (less olive- and brown-toned than south-eastern race, *bicknelli*), distinctly colder-toned than *C. ustulatus*, never having that species' russet or tawny tinge. Face basically grey, lores and cheeks paler than crown and finely mottled off-white, with almost concolorous, narrow, and indistinct

eye-ring (unlike *C. ustulatus*) and narrow, black-spotted malar stripe, more sharply defined against white throat than *C. ustulatus*. Sides of breast and flanks basically buff- to ochre-grey, suffusion colder and paler in tone than *C. ustulatus*; breast quite heavily spotted black-brown but flanks merely dappled or mottled dusky brown; throat, belly, and under tail-coverts white. Underwing olive-grey, with broad white tips on greater secondary coverts forming pale central band. When worn, whole bird even greyer above and whiter below. JUVENILE. Upperparts olive-brown, lacking cold tone of adult and showing extensive buff streaks and some buff spots; face pale buff spotted olive and black, usually without distinct eye-ring. Underparts more heavily marked than adult, with black spots, bars, and scales from chest to belly and flanks. FIRST WINTER. Resembles adult but retained juvenile greater wing-coverts and tertials may show terminal buff spots and fringes, appearing as faint wing-bar on coverts. At all ages, bill dark brown-horn with grey-flesh base to lower mandible; legs dusky flesh.

Likely to suggest grey-backed chat to Palearctic observer, with form and flight like nightingale, though bulkier and with rather larger wings. Clear sight of malar stripe and spotted breast removes this confusion, allowing rapid identification to genus, though specific diagnosis must be careful, since (apart from face pattern) browner individuals come close in appearance to greyer individuals of *C. ustulatus*. Field guides have stressed difference in eye-ring as best basis for identification (see above) but some immatures show more prominent and whiter eye-rings than adults, leaving colour of lores and cheeks (always buff in *C. ustulatus*, never with more than grey-brown or off-white ground-colour in post-juvenile *C. minimus*) as most visible diagnostic character. In northernmost birds of nominate *minimus*, cheeks often dusky, distinctly flecked or lined white and thus greyer than those of southern nominate *minimus* and *bicknelli*. Also important to recognize that throat and centre of breast whiter in *C. minimus* than *C.ustulatus*; *C. minimus* also has breast-spots somewhat smaller, forming sparser bib on most birds. For detailed comparison with this and other *Catharus* and *Hylocichla* thrushes, see p. 929. Flight, gait, and carriage as *C. ustulatus* but apparently never cocks tail. Exceptionally shy and skulking.

Commonest call a long mono- or disyllable, 'wheu', 'quee-a', or 'vee-a', down-slurred and sounding higher pitched and more nasal than similar call of Veery *C. fuscescens* and quite different from short monosyllable of *C. ustulatus*.

Habitat. Breeds from high latitudes of east Palearctic and Nearctic to middle latitudes of Nearctic, in boreal forests of tamarack *Larix laricina* and black spruce *Picea mariana*, giving way to open tundra with stunted trees such as dwarf willow *Salix nana*, reaching northern limits of arboreal vegetation (Pough 1949; Dementiev and Gladkov 1954*b*). In Alaska, a characteristic bird of willow and alder fringes along small streams of tundra and thickets of large valleys, as well as among stunted spruces; sings from top twig of willow or small spruce, feeding almost entirely on ground (Gabrielson and Lincoln 1959). In Canada, also occupies dense stunted spruce on coastal islands and near treeline on mountains (Godfrey 1966). Throughout northern fringe of stunted spruces, willows, and alders of arctic Canada and Alaska, finds suitable habitat beyond recognized treeline. Has been observed, when alarmed, retreating to a boulder-strewn beach and hiding under large rocks. Ascends well above 1000 m in New England. On migration, occurs wherever there is sufficient cover, in or on edge of woodland, and in thickets along streams, roadside shrubbery, village gardens, and city parks (Bent 1949). In winter in Venezuela, occurs at up to 3000 m in damp thickets, clearings, open woodland, forest edge, and in undergrowth to mid-tree levels (Schauensee and Phelps 1979).

Distribution. Breeds from north-east Siberia, northern Alaska, and from near treeline across Canada south to British Columbia, Saskatchewan, Quebec, and New-foundland; also disjunct populations further south in north-west Canada and northern USA. Winters in northern South America.

Accidental. Iceland, Britain, Ireland, West Germany, Norway.

Movements. Migratory. Autumn migration primarily through eastern North America (from Bermuda in the east to Great Plains and eastern Texas), Bahamas, Greater Antilles, and southern Central America, mainly eastern slopes (American Ornithologists' Union 1983), though apparently scarce in Mexico, Belize, Guatemala, Honduras, Costa Rica, and Panama (Rappole *et al.* 1983). Winters in northern South America (Colombia, Venezuela, Guyana, Ecuador), with a few birds of south-eastern race, *bicknelli*, on Hispaniola in Caribbean (Bond 1961). Analysis of 1393 birds ringed in autumn over 10 years showed proportion of adults among migrants in North America was higher inland than on coasts, with only 2% adults on Rhode Island coast (Smith and Schneider 1978). On Caribbean coast of Panama, Galindo *et al.* (1963) caught 264 migrants in autumn over 2 years, but none in spring. Route apparently more easterly in spring than autumn (American Ornithologists' Union 1983); also more concentrated, e.g. no spring records from Bermuda (Wingate 1973).

In northern USA and southern Ontario (Canada), autumn migrants recorded early September to mid-October, peaking late September and early October (Annan 1962; Smith and Schneider 1978). Occurs throughout October on Bermuda (Wingate 1973). Present in Venezuela September–May (Schauensee and Phelps 1978). Reaches central USA by 1 May and northern Alaska by 1 June (Robbins *et al.* 1983).   TL-E

Voice. See Field Characters.

Plumages (nominate *minimus*). ADULT. Upperparts entirely dark olive, greyish-olive, or sepia-olive; feather centres of forehead and crown slightly darker, rump often slightly paler and with slight green or brown tinge. Lores and narrow eye-ring closely mottled dull olive-grey and off-white, rear part of eye-ring often uniform off-white. Cheeks and ear-coverts dull greyish-olive, cheeks streaked and mottled cream-white or pale buff, ear-coverts with narrow off-white shaft-streak; upper part and rear of ear-coverts virtually uniform dark olive. Upper sides of neck olive, like upperparts; lower sides cream-buff, spotted olive. Chin and throat uniform cream-white or white, bordered at sides by black malar stripe (often broken into spots), and marked with increasing number of black triangular spots from throat downwards. Chest pale buff or cream, marked with rather triangular dull black or olive-black spots 2–3 mm wide (sometimes restricted, forming a spotted gorget); sides of breast olive or greyish-olive with indistinct darker marks. Flanks and thighs virtually uniform grey or olive-grey, remainder of underparts white, marked with variable number of grey or olive-grey spots or short round-ended bars on breast and upper sides of belly. Tail like upperparts or slightly more olive-brown. Flight-feathers and greater upper primary coverts dull black or greyish-black on inner webs and tips, olive on outer web, grading towards brighter olive-brown on fringe; basal inner webs of secondaries and inner primaries with large cream-white patch, forming pale band on undersurface of flight-feathers. Tertials and remainder of upperwing olive like upperparts, centres of lesser and median coverts often slightly darker, inner webs of greater coverts and of longer feathers of bastard wing virtually black. Under wing-coverts and axillaries dull olive-grey, greater secondary coverts broadly tipped white, part of shorter primary coverts fringed white. *In worn plumage* (spring and summer), olive of upperparts (including tail and upperwing) colder and more greyish; sides of head virtually uniform olive-grey, except for fine off-white mottling on lores and cheeks, faint white partial eye-ring at rear of eye, and faint white shaft-streaks on shorter ear-coverts; ground-colour of lower cheeks, lower sides of neck, chin, throat, and chest paler, cream or almost white; grey spots on breast and sides of belly sometimes partly worn off. JUVENILE. Upperparts greenish olive-brown; crown, mantle, scapulars, back, rump, and lesser and median upper wing-coverts with narrow pale buff shaft-streaks, sometimes widening to large triangular or rounded spot on feather-tips. Sides of head pale buff with olive and black spots and streaks; no distinct eye-ring. Black malar stripe. Underparts white with faint buff tinge on chest and sides of breast; throat and chest spotted black, tending to barring or scaling on upper belly and flanks. Generally greyer and less buff than juvenile Olive-backed Thrush *C. ustulatus swainsonii*. (Dwight 1900; Ridgway 1901–11.) Flight-feathers, tail, tertials, greater upper wing-coverts (including tertial coverts), and greater upper primary coverts as in adult, but greater coverts and sometimes tertials often have small buff spots at centre of tip and usually buff-brown fringe along distal part of outer web; tail-feathers sometimes narrower and with more sharply pointed tip than adult. FIRST ADULT. Like adult, but juvenile flight-feathers, tail, tertials, greater upper wing-coverts (often except some or all tertial coverts), greater upper primary coverts, and occasionally some median upper wing-coverts retained. Greater coverts have narrow pale buff to off-white spot on tips and sometimes traces of buff fringe along distal part of outer web (most pronounced on innermost, where contrasting with uniform brighter olive tertial coverts, if these new), but spot or fringe occasionally absent even in fresh autumn plumage (e.g. in 5 of 43 examined by Payne 1961 and 4 of 23 in BMNH, RMNH, and ZMA) and sometimes largely or fully worn off in spring plumage. Retained juvenile flight-feathers, tail, and tertials sometimes more heavily worn than those of adult in same time of year, but not always so; juvenile tail usually narrower and more sharply pointed than in adult, but character useful in direct comparison only and some birds intermediate.

Bare parts. ADULT, FIRST ADULT. Iris dark brown or dark chocolate-brown. Bill dark horn-brown or horn-black, basal ½–⅔ of lower mandible and basal cutting edges of upper mandible lilac, pale flesh-horn, pale flesh-grey, whitish-horn, or yellow-horn. Mouth and tongue chrome-yellow; gape-flanges and inside of upper mandible yellow in 1st adult in autumn. Leg pale pink-horn, light horn, pale brown, pale brown-grey, or dark grey-brown, paler yellow or flesh at rear; foot grey-brown, soles dull grey-yellow or greyish-flesh. (Ridgway 1901–11; Clafton 1963; Grubb 1966; Ree 1974a; RMNH, ZMA.) JUVENILE. No information.

Moults. ADULT POST-BREEDING. Complete; primaries descendant. August; completed before migration started (Bent 1949).

POST-JUVENILE. Partial: head, body, lesser and most or all median upper wing-coverts, often some or all tertial coverts, and occasionally a few tertials. August; birds examined from early and mid-September had finished moult. (Bent 1949; BMNH, RMNH, ZMA.)

Measurements. ADULT, FIRST ADULT. Nominate *minimus*. Mainly migrants from north-east USA, September–October and May, a few from breeding and winter areas; skins (BMNH, RMNH, ZMA). Bill (S) to skull, bill (N) to distal corner of nostril; exposed culmen on average 4·2 less than bill (S).

WING	♂ 106·7 (2·64; 18) 101–111	♀ 102·3 (2·25; 13) 98–107	
TAIL	71·7 (3·22; 17) 67–76	67·4 (1·92; 13) 64–70	
BILL (S)	17·8 (0·76; 16) 16·9–18·7	17·8 (0·49; 12) 17·2–18·7	
BILL (N)	9·5 (0·43; 15) 8·9–10·2	9·6 (0·45; 12) 8·9–10·3	
TARSUS	30·8 (1·04; 16) 29·7–33·1	30·0 (0·76; 12) 28·9–31·3	

Sex differences significant for wing and bill. 1st adult with retained juvenile flight-feathers and tail combined with older birds in table above, though juvenile wing on average 2·2 shorter than in full adult and tail 2·7 shorter. Wing, ♂ 103–112, ♀ 100–107 (Williamson 1954).

C. m. bicknelli. Appalachian mountains, USA, summer; skins (Ridgway 1901–11).

WING	♂ 92·1 (10) 88–98	♀ 88·6 (7) 85–93
TAIL	66·5 (10) 62–70	64·2 (7) 60–68
TARSUS	28·5 (10) 26–30	29·0 (7) 27–30

Wing, ♂ 89–101, ♀ 90–96 (Williamson 1954).

Weights. ADULT, FIRST ADULT. Mainly nominate *minimus*, but migrants may include some *bicknelli*. Northern Alaska and Yukon (Canada), May–June: ♂ 30·6 (14) 26–34, ♀ 30·6 (3·45; 7) 27–36 (Irving 1960). New Jersey, autumn: 28·6 (32) 24–35 (Murray and Jehl 1964). Ohio, September: 31·7 (Baldwin and Kendeigh 1938). Illinois, mainly September: ♂ 33·4 (33), ♀ 31·1 (24) (Graber and Graber 1962). Kentucky, September: ♀ 29 (Mengel 1965). Georgia, early October: ♂ 42·8 (14) 30–50, ♀ 38·3 (18) 30–45 (Johnston and Haines 1957). Belize: October, ♂♂ 33, 35, ♀ 31·4 (Roselaar 1976); May, ♂ 31·1 (Russell 1964). Venezuela, November: 30, 30·5 (Thomas 1982). Kentucky, May: ♂ 35·1 (4·05; 4) 31–39, ♀ 40 (Mengel 1965). Britain, October: 24·9 (Williamson 1954). Norway, October: 26·9 (Ree 1974a).

C. m. bicknelli. Britain, October: ♂ 22·4 (Clafton 1963).

Structure. Wing rather long, broad at base, tip bluntly pointed. 10 primaries: p8 longest, p9 1·5–4 shorter, p7 1–2, p6 7–9, p5 14–18, p4 18–21, p1 28–33. P10 reduced, 59–66 shorter than wing-tip, 2–9 shorter than longest upper primary covert. Outer web of p7–p8 and inner of p8–p9 emarginated. Middle toe with claw 21·0 (8) 20–22; outer toe with claw *c.* 66% of middle with claw, inner *c.* 60%, hind *c.* 74%. Claws rather short and slender, decurved. Remainder of structure as in Hermit Thrush *C. guttatus* (p. 928).

Geographical variation. Marked in size, slight in colour. *C. m. bicknelli* from north-east USA and neighbouring Canada distinctly smaller than nominate *minimus* from remainder of Canada, Alaska, and north-east Siberia, though some overlap in size, especially if sex not known (see Measurements). Upperparts of *bicknelli* slightly browner on average, olive-sepia or sepia-brown (Ridgway 1901–11). Following American Ornithologists' Union (1957), *aliciae* (Baird, 1858), Siberia to Labrador, considered inseparable from nominate *minimus*.

Recognition. Uniform olive upperparts and tail matched only by Swainson's Thrush *C. ustulatus*, which is usually slightly paler and more greenish (see Table A, p. 929). In *C. ustulatus*, lore and broad eye-ring deep buff (autumn) or pale buff (spring) (in *C. minimus*, mottled grey and off-white, eye-ring often uniform off-white at rear of eye only); ground-colour of chest and streaks on ear-coverts buff or pale buff (in *C. minimus*, cream to off-white), ground-colour of chin, throat, and chest warm buff (autumn) or cream-buff (spring) (in *C. minimus*, chin and throat off-white and chest pale buff or cream in autumn, off-white in spring); flanks and spots on breast olive-brown (in *C. minimus*, grey or olive-grey). Other *Catharus* decidedly browner above; also, Veery *C. fuscescens* has spots on chest much less distinct, and Hermit Thrush *C. guttatus* has contrastingly rufous rump and tail and is decidedly smaller than nominate *minimus* (but exactly the same size as *bicknelli* apart from slightly larger bill and longer leg). CSR

Catharus fuscescens Veery

PLATES 68 and 78
[between pages 856 and 857]

Du. Veery   FR. Grivette fauve   GE. Wilsondrossel
RU. Вертлявый дрозд   SP. Zorzal solitario   SW. Rostskogstrast

Turdus Fuscescens Stephens, 1817. Synonym: *Hylocichla fuscescens*.

Polytypic. Nominate *fuscescens* (Stephens, 1817), south-east Canada and north-east USA from Nova Scotia, southern Quebec, and south-east Ontario south to Appalachian mountains. Extralimital: *fuliginosus* (Howe, 1900), south-central Quebec and south-west Newfoundland; *salicicolus* (Ridgway, 1882), southern Canada and USA from south-east Ontario and Michigan west to British Columbia and Colorado.

Field characters. 17 cm; wing-span 28–31·5 cm. 5% larger than Nightingale *Luscinia megarhynchos*, with 10% longer wings; close in size to Swainson's Thrush *C. ustulatus*. Small thrush, with slightly longer tail than congeners; upperparts uniform tawny-brown recalling *L. megarhynchos*, underparts buff and indistinctly spotted on chest recalling Thrush Nightingale *L. luscinia*. Within genus, warm uniform colour of upperparts and lack of well-marked dark-spotted chest diagnostic. Sexes similar; some seasonal variation. Juvenile separable. 3 races in Nearctic, of which 2 may reach west Palearctic but these hardly separable in the field.

ADULT. Race of south-east Canada and north-east USA, nominate *fuscescens*. Upperparts warm tawny-brown (but duller when worn, with olive tone suggesting *C. ustulatus*); tail slightly brighter, almost chestnut. Face basically pale grey-buff, with prominent dark eye but otherwise lacking obvious features except for brown surround to rear cheek and narrow, dark brown-spotted malar stripe; off-white lore and narrow eye-ring visible at close range but not particularly noticeable. Rear cheeks, sides of throat and breast suffused yellow-buff, with dull, diffuse olive- to tawny-brown spots, forming obvious patch on chest; flanks suffused grey, lightly dappled below breast but appearing uniform at any distance; centre of throat, belly, vent, and under tail-coverts white. Wings brown, with smaller coverts uniform with back but larger ones with paler fringes and tips (forming faint wing-bars); flight-feathers brown-black, fringed warm brown on tertials. Underwing grey, with broad white tips to greater coverts forming central panel. JUVENILE. Darker than adult, almost umber-brown above; dully spotted with tawny to cream on crown and back and much more heavily marked below, with obvious spots from throat to flanks and even belly; pale buff spots on larger wing-coverts. FIRST WINTER. Resembles adult, but retained worn juvenile flight-feathers, tertials, and greater coverts visible at close range. At all ages, bill dark brown-horn with pale flesh base to lower mandible; legs pale pink-brown.

As likely to be mistaken for Nightingale *Luscinia megarhynchos* as another *Catharus* thrush, with duller, least spotted birds also recalling *L. luscinia* in brief glimpse; distinguished, however, by bulkier, rather neckless form, faintly indicated wing-bars (shared with other *Catharus* but most obvious on this species), and well-indicated malar stripe. For detailed comparison with other *Catharus* and *Hylocichla* thrushes, see p. 929. Gait, carriage, and behaviour as Hermit Thrush *C. guttatus* but lacks that species' pronounced tail movement.

Commonest call a low 'phew' or 'view', down-slurred and often extended to 'whee-u', then recalling higher-pitched disyllable of Gray-cheeked Thrush *C. minimus*.

Habitat. Breeds in Nearctic middle latitudes, mainly in temperate lowlands but ascends suitably wooded mountain slopes, in central Appalachians preferring stands of hemlock *Tsuga mertensiana*, and moist bottoms to dry ridges. Having the most southerly range of North American thrushes except Wood Thrush *Hylocichla mustelina*, is correspondingly less attached to coniferous forest (Bent 1949). In north of range in southern Canada, however, occupies mixed as well as wholly broad-leaved woodland, especially where it is more open with broad-leaved undergrowth; also frequents second growth, willows *Salix*, or alders *Alnus* along lakes and streams

(Godfrey 1966). On Great Plains, also occupies some coniferous forests, including swamps of tamarack *Larix laricina*, but prefers moist wooded ravines and river-bottom forests with wet ground, nearby running water, and well-developed understoreys (Johnsgard 1979). In Rocky Mountains, ascends above 2500 m, usually in thickets and underbrush near water (Niedrach and Rockwell 1939). Prefers wooded areas open enough to encourage fairly dense undergrowth of shrubs or ferns, and thickets of birches and other second growth, as well as willow and alder swamps, borders of streams and lakes, and sometimes even dry brushy hillsides; forages mostly on forest floor (Pough 1949). Often lives during summer in woods of mixed oak *Quercus* or pine *Pinus*, on dry hillsides or even on summits of low hills (Forbush and May 1939). Winters in Venezuela on edge of rain forest up to 950 m and in second growth (Schauensee and Phelps 1978).

Distribution. Breeds from interior British Columbia east to southern Newfoundland and south to north-east Arizona, South Dakota, Minnesota, Alleghenies, Georgia, New York, and Long Island. Winters in northern South America.

Accidental. Britain, Sweden.

Movements. Migratory. From most parts of breeding range, autumn migration is essentially south-east, migrants passing through eastern USA west to Rocky Mountains and central Texas and occurring on Caribbean slopes of Central America and on Bahamas and Cuba. Winters in South America in Colombia, Ecuador, Venezuela, Guyana, and north-west Brazil (American Ornithologists' Union 1983; Rappole *et al.* 1983). Analysis of 879 birds ringed in North America in autumn over 10 years showed proportion of adults highest in Ontario and Pennsylvania, lowest on east coast (5% on Rhode Island coast); numbers at the various study sites varied independently from year to year (Smith and Schneider 1978). Spring migration more direct, and concentrated largely across Caribbean, with fewer migrants in Central America than in autumn (e.g. Galindo *et al.* 1963). Only 1 Bermuda record, in autumn (Wingate 1973).

Autumn migration across northern and eastern USA occurs mid-August to mid-October, peaking early September (Smith and Schneider 1978); finishes late September in Chicago (Annan 1962). Present in Venezuela October–April (Schauensee and Phelps 1978). Occurs in spring in central USA from 1 May, and arrives Chicago 1–25 May (Bond 1961; Annan 1962; Robbins *et al.* 1983). TL-E

Voice. See Field Characters.

Plumages (nominate *fuscescens*). ADULT. In fresh plumage, upperparts entirely cinnamon-brown. Sides of head pale grey-brown or olive-grey, finely streaked white or pale buff on upper

cheeks and ear-coverts; lores and indistinct ring round eye off-white with dusky grey or olive-grey mottling. Rear part of cheeks, lower sides of neck, and sides of breast yellow-buff or tawny-yellow, streaked dark olive-brown on lower cheeks (forming ill-defined malar stripe), spotted olive-brown or tawny-brown elsewhere. Chin and upper throat uniform pale buff or cream-white, lower throat and chest pink-buff, tawny-yellow, or tawny-buff; sides of throat with olive-brown or tawny-brown streaks, grading to ill-defined olive-brown or tawny-brown triangular spots on chest. Flanks and thighs pale olive-grey (darkest on feather-tips), remainder of underparts white, often with faint grey rounded-triangular spots on breast and sometimes on upper sides of belly. Tail cinnamon-brown or tawny-brown, similar to upperparts or slightly darker. Flight-feathers, tertials, and greater upper primary coverts greyish-black or brownish-black, outer webs broadly fringed cinnamon-brown or olive-brown, shorter tertials largely cinnamon-brown or tawny-brown. Bases of inner webs of secondaries and inner primaries off-white. Upper wing-coverts olive-brown, tips and fringes along outer webs slightly paler, cinnamon-brown or tawny-brown. Under wing-coverts and axillaries grey; greater under wing-coverts broadly tipped white, smaller primary coverts partly white, axillaries partly suffused white. *In worn plumage*, underparts as in fresh plumage, but buff of chest sometimes bleached to paler yellow-buff, spots appearing slightly more clear-cut; upperparts and tail either as in fresh plumage or decidedly duller and more olive, without cinnamon or tawny tinge, not unlike Swainson's Thrush *C. ustulatus*; upperparts rather like Hermit Thrush *C. guttatus*, but tail similar to upperparts, not contrastingly rufous. JUVENILE. Upperparts and sides of head deep umber-brown or olive-brown with slight rufous tinge, each feather with contrasting tawny-olive to cream spot on centre and dusky edge along tip; spots large and almost rounded on crown and mantle, more elongate on forehead and outer scapulars. Rump and upper tail-coverts tinged rufous. Sides of head mottled and streaked buff and dark olive-brown; mottled dull black malar stripe at sides of throat. Underparts white, strongly tinged tawny-olive or cream-buff on cheeks and throat, less so on chin, chest, sides of belly, and vent; throat rather heavily spotted dusky; chest and sides of breast with broad black spots, darker and more contrasting than spots of adult; spots gradually narrower towards belly, flanks, and vent, where reduced to narrow bars; under tail-coverts uniform cream-white. Lesser and median upper wing-coverts black-brown, each with narrow buff shaft-streak ending in rounded spot at tip. Flight-feathers, tail, tertials, and remaining upper wing coverts tawny olive-brown, as adult, but greater upper wing-coverts and tertials edged tawny-olive and faintly tipped dusky, greater sometimes with pale buff triangular spot on tip. (Dwight 1900; RMNH.) FIRST ADULT. Like adult, but juvenile flight-feathers, tertials, and greater upper wing-coverts (often except tertial coverts), and greater upper primary coverts retained; these feathers contrast somewhat in colour and wear with neighbouring fresh feathers, but difference often hard to see; juvenile greater upper wing-coverts and tertials with tawny-olive or buffish fringe along tip, usually fairly distinct in fresh autumn plumage but sometimes absent (Brewer 1972; BMNH) and often difficult to trace when worn; occasionally, large pale triangular spot on inner greater coverts. Juvenile tertials and tail-feathers on average narrower than in adult, tips of tail-feathers more sharply pointed, but these characters not fully reliable.

Bare parts. ADULT, FIRST ADULT. Iris brown or dark brown. Upper mandible and tip of lower dark horn-brown to brown-black, remainder of lower mandible pale horn-yellow, flesh-yellow, or pale horn with violet tinge. In 1st adult, October, flanges at base of lower mandible yellow, mouth bright yellow. Leg and foot pale flesh-brown, light brown, or pale purplish-brown, rear of tarsus and soles paler, horn-white or pale flesh-pink. (Alsopp 1972; BMNH, RMNH.) JUVENILE. Iris dull grey-brown. Bill dusky brown, base of lower mandible dull flesh. Leg and foot greyish-flesh. (RMNH.)

Moults. ADULT POST-BREEDING. Complete; primaries descendant. July–August (Bent 1949). Moult not started in birds examined from 1st week of July; outer primaries still growing in one from 26 August; moult completed in some from early September (BMNH). POST-JUVENILE. Partial: head, body, lesser and some or all median upper wing-coverts, often some or all tertial coverts, occasionally some tertials. Mid-July to August; completed in some examined from mid-August (BMNH, RMNH).

Measurements. ADULT, FIRST ADULT. Nominate *fuscescens*. South-east Canada and north-east USA, May–September; skins (BMNH, RMNH, ZMA). Bill (S) to skull, bill (N) to distal corner of nostril; exposed culmen on average 4·3 shorter than bill (S).

WING	♂ 102·4 (1·77; 19)	99–105	♀ 96·9 (2·78; 13)	93–101
TAIL	69·4 (2·35; 19)	66–73	65·2 (2·18; 13)	62–69
BILL (S)	17·7 (0·55; 18)	17·1–18·6	17·1 (0·70; 12)	16·3–17·9
BILL (N)	9·7 (0·45; 18)	9·3–10·4	9·3 (0·41; 12)	8·6–9·8
TARSUS	30·3 (1·12; 19)	28·4–32·4	29·2 (0·93; 13)	27·5–30·5

Sex differences significant. 1st adult with retained juvenile flight-feathers and tail combined with adult, though juvenile wing on average 1·4 shorter than in older birds and tail 1·9 shorter.

Weights. Races combined. North Carolina, early July: ♂ 29·0 (3) 26–31 (Stewart and Skinner 1967). New Jersey, autumn: 30·2 (57) 24–39 (Murray and Jehl 1964). Illinois and Kentucky, September: ♂♂ 32, 34·2, 33·5(2); ♀ 30·5 (Graber and Graber 1962; Mengel 1965). Lake Erie (Ontario), early May: 31·5 (80) 26–37 (Hussell 1969, which see for fluctuations during nocturnal migration). Kentucky, late April and early May: ♂ 38·3, ♀ 30·4 (Mengel 1965). Minimum weight recorded in Panama: 16·3 (Rogers and Odum 1966). Sweden, probably ♀ (wing 94), September: *c.* 30 (Edholm *et al.* 1980).

Structure. Wing rather long, broad at base, tip bluntly pointed. 10 primaries: p8 longest, p9 2–4 shorter, p7 0·5–2, p6 5–7, p5 10–14, p4 15–19, p1 23–29. P10 reduced, 57–64 shorter than wing-tip; 2–7 shorter than longest upper primary covert. Outer web of p6–p8 and inner of (p7–)p8–p9 emarginated. Middle toe with claw 21·0 (5) 20–22; outer toe with claw *c.* 70% of middle with claw, inner *c.* 63%, hind *c.* 76%. Claws rather short, sharp, decurved. Remainder of structure as in Hermit Thrush *C. guttatus* (p. 928).

Geographical variation. Rather slight, clinal; involves colour only, and bridged in part by marked individual variation. Nominate *fuscescens* from north-east USA and south-east Canada tawny-rufous on upperparts; *salicicolus* (synonym: *subpallidus* Burleigh, 1959) from further west duller, darker, and more olive-grey on upperparts, less tawny; malar stripe and spots on chest slightly broader and darker grey, not as tawny-brown and triangular as in nominate *fuscescens*; rather similar to Swainson's Thrush *C. ustulatus*, but eye-ring faint or absent. *C. f. fuliginosus* from south-central Quebec and Newfoundland like nominate *fuscescens*, but upperparts darker and browner, with less tawny or cinnamon tinge; chest and usually lower throat heavily spotted with broad arrow-shaped brown marks, rather similar to *C. ustulatus*; upperparts without olive-grey, unlike *salicicolus*; many inseparable from nominate *fuscescens*. (Ridgway 1901–11; Bent 1949; Godfrey 1966; Alsopp 1972; BMNH.)

Recognition. See Table A (p. 929). Upperparts rather variable, cinnamon-brown or tawny-brown, not contrasting with colour of tail; colour of back to tail as in Wood Thrush *H. mustelina*, but crown, mantle, and scapulars not bright rufous-cinnamon like *H. mustelina*; crown to back as *C. guttatus*, but rump and tail not contrastingly rufous; thus, upperparts uniformly coloured, as in Gray-cheeked Thrush *C. minimus* and Swainson's Thrush *C. ustulatus*, but not as dark and dull olive as in these species, though some *C. fuscescens* are near *C. ustulatus*. Spots on chest paler and less clear-cut than in other *Catharus*; flanks greyer and less olive-brown or buff-brown than *C. guttatus* and *C. ustulatus*, but not as dark and extensively grey as *C. minimus*; sides of head greyish, not as warm buff as *C. guttatus* and *C. ustulatus* and without their buff eye-ring, but not as grey and heavily streaked as *C. minimus*.   CSR

Turdus unicolor Tickell's Thrush

PLATES 67 and 78
[between pages 856 and 857]

DU. Tickells Lijster   FR. Merle unicolore   GE. Einfarbdrossel
RU. Одноцветный дрозд   SP. Zorzal unicolor   SW. Gråtrast

Turdus Unicolor Tickell, 1833

Monotypic

Field characters. 21 cm; wing-span 33–38 cm. About 30% smaller than Blackbird *T. merula*, with proportionately shorter wings; 10% smaller than Eye-browed Thrush *T. obscurus* but with similar proportions. A small and slight thrush, with paler plumage than any other occurring in west Palearctic; ♂ mainly blue-grey, with white underbody, ♀ olive-brown above and on breast, with pale face, brown-streaked white throat, tawny flanks, and white underbody. Underwing orange-buff. Sexes dissimilar; no seasonal variation. Juvenile separable.

ADULT MALE. Head, neck, back, wings, and tail plain ash- to blue-grey, with faint brown tone strongest on tail and creating slight contrast with rump; cheeks below eye and throat slightly paler. Breast and flanks pale grey, contrasting with white underbody and vent. Wings grey-brown; under wing-coverts orange-buff. Bill orange-yellow; narrow yellow eye-ring; legs brown-yellow. ADULT FEMALE. Unusually for a thrush, more strongly patterned than ♂, with head, back, wings, and tail less grey, more olive-brown; fore-face pale, with buff lore and indistinct supercilium and eye-ring. Throat white, with brown speckles and streaks on sides below cheeks; broad breast-band of pale grey-brown to olive-brown, streaked dark brown; flanks tawny, showing even at rest. Belly and under tail-coverts white; underwing paler orange-buff than ♂. Bare parts duller than ♂. JUVENILE. Both sexes. Darker brown above than ♀, with pale spots and shaft-streaks except on rump; wing-coverts streaked and tipped olive-buff, with obvious bar across greater coverts. More strongly marked below than ♀, with ochre tone to white throat, belly, and under tail-coverts and olive-buff underwing; sides of throat, breast, and flanks ochre-brown, heavily barred with black-brown; obvious malar stripe running into chest-marks. FIRST-YEAR MALE. As adult ♂, but mantle, crown, and sides of head brownish, chin more extensively white and bordered by malar stripe, chest washed olive with dark streaks in centre, and outer greater coverts tipped pale. FIRST-YEAR FEMALE. As adult ♀, but outer greater coverts tipped pale.

♂ virtually unmistakable but might be confused at distance with ♀♀ of paler races of *T. merula* (but that species much larger, with streaked throat and different behaviour and calls). ♀ far trickier, inviting confusion with dull ♀ and immature *T. obscurus* (see that species, p. 964). Flight light and fast, escaping as *T. obscurus* to nearby tree canopy. Gait alternates short fast hops and mincing steps; often cocks head to one side during pauses. Not markedly gregarious, forming only small flocks on passage and in winter.

Contact- and alarm-call 'juk-juk'.

Habitat. Breeds in Indian subcontinent in alpine low middle latitudes at 1500–1800 m and up to 2700 m, in broad-leaved open forest with grassy carpet or thin undergrowth, and in willow groves, orchards, and gardens. Although found in Kashmir in any wooded area, and even in heavy mixed forest or open scrub, has conspicuously adapted to settlements, foraging on lawns and flowerbeds, singing in tree-tops, and nesting lower in trees. Shifts in winter to similar habitats at lower level. (Bates and Lowther 1952; Ali and Ripley 1973*b*.)

Distribution. Breeds in Himalayas from Chitral to eastern Nepal and possibly Sikkim. Winters south to 18–22°N and in Assam and Manipur.

Accidental. West Germany: ad ♂, specimen, Helgoland, 15 October 1932 (Drost 1933).

Movements. Short-distance migrant. Arrives on breeding grounds at end of March and in April, leaving in September–October to move east along Himalayas to winter quarters lying along foothills from Kangra to Arunachal Pradesh, and in north-east peninsular India, Bangladesh, and northern Baluchistan (Pakistan). Recorded at Bharatpur November–February and at Khandala in November, and stragglers seen at 3000 m at Ladakh in April and at Sambhar Lake and Mount Abu in September. (Ali and Ripley 1973*b*.) An unlikely species to reach western Europe. LAB

Voice. See Field Characters.

Plumages. ADULT MALE. Crown, sides of head, and rest of upperparts blue-grey. Chin whitish. Throat pale grey. Chest and flanks grey (paler and less blue than upperparts). Central belly, vent, and under tail-coverts white. Tail grey-brown with blue-grey wash (as upperparts). Wings grey-brown, outer webs of feathers with blue-grey wash. Under wing-coverts orange-buff. ADULT FEMALE. Crown and mantle olive-brown, lower back and rump greyer. Ear-coverts brown with white shaft-streaks. Chin and throat whitish with lateral dark brown streaks forming well-marked malar stripe. Chest pale grey-brown, feathers with dark brown spots on centres. Flanks pale brown, washed orange. Belly and under tail-coverts white. Tail dark grey-brown. Wings dark brown; under wing-coverts pale orange-buff. JUVENILE. Crown, sides of head, and rest of upperparts dark brown, feathers with buff shaft-streaks. Throat buff with conspicuous blackish malar stripe. Chest and flanks orange-buff, spotted sooty black; spots heaviest on chest where black is confluent with malar stripe; spotting reduced on rear flanks where orange wash is strongest. Belly, vent, and under tail-coverts white. Tail dark brown. Wings dark brown, lesser coverts with buff shaft-streak widening to triangular spot near tip; greater coverts tipped buff. FIRST ADULT MALE. Crown and mantle grey-brown with strong olive wash; lower back and rump blue-grey. Ear-coverts dull brown with white shaft-streaks. Chin and throat whitish with dark brown streaks at sides, forming well-marked malar stripe. Chest grey with olive wash, feathers on centre with dark shaft-streaks. Flanks smoky-grey. Belly, vent, and under tail-coverts white. Tail grey-brown with variable blue-grey wash. Wings dull brown, outer greater coverts with pale tips. Under wing-coverts as adult. FIRST ADULT FEMALE. Like adult ♀, but outer greater wing-coverts with pale tips.

Bare parts. ADULT MALE. Iris dark brown or red-brown; eye-ring yellow. Bill in breeding season yellow or orange-yellow, in non-breeding season yellow suffused with dusky, especially at base and tip. Leg and foot dull yellow. ADULT FEMALE. Like adult ♂, but bill duller, in breeding season with dusky clouding especially at base, and eye-ring dull yellow. JUVENILE. Iris dark brown. Bill brown, yellow towards gape. Leg and foot yellowish. FIRST ADULT MALE AND FEMALE. Like adult, but bill darker and eye-ring duller in 1st autumn and winter. (Drost 1933; Ali and Ripley 1973*b*; BMNH.)

Moults. ADULT POST-BREEDING. Complete; primaries descendant. August–September, before leaving breeding area. POST-JUVENILE. Partial: head, body, tail (perhaps not always), lesser, median, and some inner greater upper wing-coverts, and sometimes one or more tertials. Before migration, mostly August–September. (Ali and Ripley 1973*b*; BMNH.)

Measurements. Mainly Himalayas (all year), a few from peninsular India (winter); skins (BMNH). Juvenile wing is retained juvenile wing of 1st adult. Bill from basal corner of nostril; bill to skull on average 6·5 longer, exposed culmen 3·5.

WING AD ♂	124·4 (1·65; 10)	122–127	♀ 118·0 (2·35; 9)	115–121
JUV	118·9 (2·21; 11)	116–123	115·1 (2·81; 10)	110–119
TAIL AD	88·2 (2·15; 10)	85–91	81·3 (2·12; 9)	78–85
BILL AD	15·6 (1·01; 9)	14–17	14·9 (0·35; 8)	14–15
TARSUS	31·0 (0·82; 10)	30–32	29·8 (1·09; 9)	28–31

Sex differences significant, except for bill.

Weights. Helgoland (West Germany), October: ♂ 72 (Drost 1933). Nepal, mainly October: ♂ 68, ♀ 68·2 (3·95; 6) 64–75

(Diesselhorst 1968*a*). October–March: 63·0 (12) 57–75 (Ali and Ripley 1973*b*). ♂♂ 58·0, 58·4, 66·7 (BMNH).

Structure. Wing rather long, broad at base, tip moderately pointed. 10 primaries: p7–p8 longest or either one 0–1 shorter than other; p9 6–10 shorter than longest, p6 2–5, p5 7–10, p4 10–15, p1 26–31. P10 reduced, 63–71 shorter than wing-tip, 1 shorter to 4 longer than longest upper primary covert. Outer web of p6–p8 and inner of p7–p9 emarginated. Tail rather long, tip square or slightly rounded; 12 feathers. Bill rather slender. Middle toe 28·8 (4) 28–30; outer and inner toe with claw both *c.* 64% of middle with claw, hind *c.* 67%. Remainder of structure as in Blackbird *T. merula* (p. 963).   DWS

Turdus torquatus Ring Ouzel

PLATES 69 and 76
[between pages 856 and 857]

DU. Beflijster   FR. Merle à plastron   GE. Ringdrossel
RU. Белозобый дрозд   SP. Mirlo de collar   SW. Ringtrast

Turdus torquatus Linnaeus, 1758

Polytypic. Nominate *torquatus* Linnaeus, 1758, Scandinavia, Britain, Ireland, and Brittany; *alpestris* (Brehm, 1831), mountains of central and southern Europe; *amicorum* Hartert, 1923, Caucasus and eastern Turkey east to Kopet-Dag (Turkmeniya, USSR).

Field characters. 23–24 cm; wing-span 38–42 cm. Slightly smaller and rather less stocky than Blackbird *T. merula*, but with 10% longer wings. Medium-sized, restless thrush; round-headed but otherwise rather attenuated, having noticeably long, sharp-cornered tail. Differs from all other west Palearctic thrushes in combination of dark plumage, pale chest-band (except in juvenile) and wing-panel, and more or less prominent scaling of underparts. Voice distinctive. Sexes dissimilar; distinct seasonal variation. Juvenile separable. 3 races in west Palearctic, western race separable from other 2.

ADULT MALE. (1) British and Scandinavian race, nominate *torquatus*. From May, sooty- to brown-black, blackest on lower underparts and tail and strikingly marked by large white crescent across chest, and long, pale wing-panel formed by silver-grey and grey-brown fringes of both coverts and flight-feathers. At close range, some brown or dull grey mottled fringes may show on rump, upper tail-coverts, and lower flanks. Underwing grey, with dark-mottled coverts. Fresh autumn plumage characterized by scaling caused by brown-grey, grey, or grey-white feather-fringes and -tips; pattern least obvious on head and nape, most obvious on lower throat, body below chest-band, and (especially) vent; chest-band duller, with smudgy brown tips producing cream or even buff tone. Bill orange-yellow in spring, with brown tip; at other times, duller and with brown also at base. Legs brown. (2) Central and south European race, *alpestris*. Differs distinctly from nominate *torquatus*. Scaling of fresh plumage more prominent, and even in spring retains

many white or grey fringes on lower breast, belly, vent, and underwing. Fringes of greater coverts and both webs of flight-feathers broader and whiter than in nominate *torquatus*, making wing-panel even more striking. (3) South-west Asian race, *amicorum*. Wing-panel even more striking than on *alpestris*, with very broad, virtually white fringes on larger coverts and flight-feathers. Body plumage lacks obvious scales, however, as in nominate *torquatus*, though white chest-band deeper than in that race. ADULT FEMALE. All races. Browner than ♂, with duller scaling always more fully retained, especially in nominate *torquatus*; chest-band narrower and duller. Underparts of *alpestris* may look mainly white. In fresh autumn plumage, browner than autumn ♂, with scaling duller but broader above and whiter and broader below; chest-band narrower and darker, with brown tips sufficiently wide to overlap and make band rather inconspicuous except from head on; throat noticeably streaked with white, especially in *alpestris*. Bare parts as ♂ but bill duller and with less yellow. JUVENILE. All races. Paler brown than ♀ and juvenile *T. merula*. Less obviously scaled than any adult plumage due to rustier feather-fringes. Throat noticeably buff-white, spotted brown. No chest-band, that area being closely barred and mottled black-brown, buff, and white, and merging with similarly patterned underbody which lacks obvious scaling except for diffuse pattern on longest flank and vent feathers which also show white shaft-streaks. Underpart markings whitest on *alpestris*. Bill horn-brown with some yellowish on lower mandible. Legs pale brown. FIRST-YEAR MALE. All races. By

September, some birds become indistinguishable from winter adult but most show whiter marks on chin, duller chest-band, whiter or more obvious scaling on vent, and duller wing-panel (due to retention of juvenile feathers). In spring, as adult but wing-panel duller. FIRST-WINTER FEMALE. All races. Some indistinguishable from winter adult ♀, but chest-band usually apparent only at close range, being closely barred brown (not showing white feather-bases) and thus dark enough to separate white streaking on throat from pale scaling on underbody.

Adult ♂ and ♀ can only be confused with partial albino *T. merula* with white chest-band. Immatures more open to confusion with *T. merula*, so important to recognize that *T. torquatus* has: (1) distinctive, rather rakish form differing from *T. merula* in longer and more pointed wings, long, sharp-cornered tail, and slim elliptical body (thus looks far less compact and even tail-heavy in flight); (2) wings always paler than *T. merula*, with pale fringes individually invisible in flight but still casting grey-brown appearance over whole area above and below; (3) distinctive voice (see below). Flight easy, rapid, and direct, with little undulation and noticeable acceleration; lacks hesitant action of *T. merula*, being noticeably bold in plummeting down inclines. Wing-beats appear to flicker, with length of wing-point more obvious in flight silhouette than in any other west Palearctic thrush. Stance and gait much as *T. merula*, with similar tail-cocking on landing. Noticeably wild, and escape-flight when flushed at short range often long and panicky. Migrants stay closer to cover and leave it earlier than other thrushes. Often perches conspicuously, however. Gregarious on migration but occurs in smaller numbers than other thrushes, rarely lingering at stopovers. Often heard before seen.

Song has timbre of *T. merula* or Mistle Thrush *T. viscivorus*, but less elaborate: a series of phrases of 2–4 notes separated by pauses. Calls include diagnostic loud rattling 'tac-tac-tac', notably hard in tone and often extended into long chattering 'tac-tac-tac-tac-a-tac-a-tac-ker-tac-ker-' by escaping bird; also piping sounds, chuckles (in short phrase recalling Fieldfare *T. pilaris* but more abrupt and higher pitched), and trills (including low-pitched, slightly quavering 'rheep' frequently given by migrants).

Habitat. Breeds in upper and middle latitudes, largely oceanic upland in former and continental montane in latter, tolerating exposure to high winds and rainfall, but generally avoiding ice and persistent snow. Habitat of nominate *torquatus* in north-west Europe typically open moorland or fell with rarely more than sparse and stunted trees, occasionally at sea-level but normally at 250 m or higher, in Scotland up to 1200 m. Most nesting territories in Britain include small crags, gullies, screes, boulders, or broken ground, as well as sloping or flat areas, often of heather, with a few trees or bushes, although in one study 2% of 297 nests were in trees (Flegg and Glue

1975). Coastal cliffs and inshore islands also sometimes used, as well as artefacts such as walls, quarries, and mine-shafts. Increase in population and range of Blackbird *T. merula*, which overlaps lower habitats, is suspected as cause of recent decrease (Sharrock 1976; Fuller 1982). In Scandinavia, favours similar open areas on fells above pine forests, but often nests in birch or spruce trees, even in plantations (Bannerman 1954). In Switzerland, however, *alpestris* breeds normally above 1100–1300 m in conifer woodlands on shady and moist slopes, but avoids deeper forest, preferring margins near moist open grass or moors, as well as park-like well-lit stands or groves, avalanche tracks, and accessible patches of short grass or moist ground in or near woods, among which spruce *Picea* and silver fir *Abies* most favoured (Glutz von Blotzheim 1962). In Carpathians (USSR), breeds chiefly in conifer forest, somewhat less in beech *Fagus*, from *c.* 250 m to treeline and krummholz zone. In Caucasus and Armeniya, in upper forest belts, rhododendron thickets, and juniper *Juniperus* zone, and in shrubs on cliffs and sides of ravines, but nests more on ground or rocks than in vegetation (Dementiev and Gladkov 1954b). On migration in England, halts regularly on chalk downs (White 1789) and wherever berries or soft fruits available, chiefly in hedgerows or tall bushes, but not excluding woods, and in open country near sea (Witherby *et al.* 1938a). Winters mainly in north-west Africa, especially on Atlas Saharien on dry and bare slopes or crests with *Juniperus oxycedrus* or *J. phoenicea*, only exceptionally reaching edge of Sahara (Heim de Balsac and Mayaud 1962).

Distribution. Marked decline in Ireland and smaller one in Sweden; small increases in France and Czechoslovakia.

FAEROES. Bred 1981–1982 (Bloch and Sørensen 1984). BRITAIN. In 19th century scattered breeding in lowland England (Parslow 1973; Sharrock 1976). IRELAND. Marked decrease in range; in 1900, bred in 27 of 32 counties (Ussher and Warren 1900). FRANCE. Now breeding in Brittany (Yeatman 1976). BELGIUM. Bred 1940, 1973 and 1975–1976 (PD). NETHERLANDS. May have bred mid-19th century but no proof (CSR). DENMARK. Bred 1935 (TD). SWEDEN. Formerly bred northern part of west coast, but declined in 20th century and last bred there 1966 (LR). EAST GERMANY. Bred Erzgebirge 1975, after long absence, and Fichtelberg 1976 (SS). CZECHOSLOVAKIA. Increased and breeding at lower altitudes 1950–75 (KH). ALBANIA. Probably breeds, but no proof (EN). GREECE. Only a few tens of pairs (GIH). USSR. Has bred rarely Estonia and Latvia and perhaps Lithuania (HV). TURKEY. Breeds sparingly (Vittery *et al.* 1971). ALGERIA. Breeds Djurdjura (EDHJ).

Accidental. Iceland, Portugal, Hungary, Syria, Israel, Kuwait, Egypt, Canary Islands.

Population. Declined Britain and Ireland; some increases in alpine area of Bayern (West Germany).

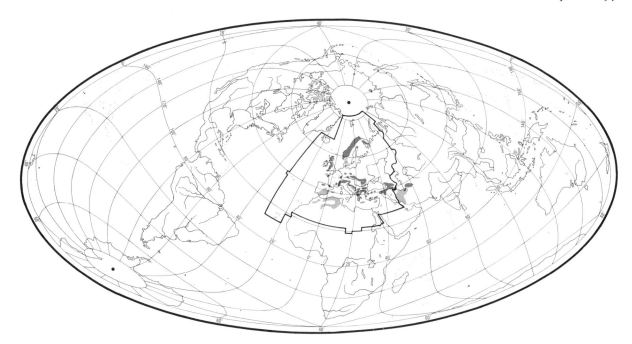

BRITAIN, IRELAND. 8000–16 000 pairs. Has declined Scotland and perhaps south Wales; marked decrease in Ireland. Changes perhaps due to spread of Blackbird *T. merula* or to climate; in southern Scotland, competition with Mistle Thrush *T. viscivorus* may be significant. (Parslow 1973; Sharrock 1976; Durman 1978.) FRANCE. 1000–10 000 pairs (Yeatman 1976). WEST GERMANY. About 9000–25 000 pairs, mainly in Bayern, where increasing in alpine areas (Bauer and Thielcke 1982). SWEDEN. About 5000 pairs (Ulfstrand and Högstedt 1976). FINLAND. About 20 pairs (Merikallio 1958).

Oldest ringed bird 8 years 2 months (Rydzewski 1978).

Movements. Migratory to resident. In winter, birds centred around Mediterranean and Middle East, though some also (of unknown origin) in northern France (R Cruon), and there are records of birds in winter at lower altitudes than breeding areas almost throughout range of both European races (even north of Arctic Circle) and occasional winter records outside breeding range, e.g. Denmark, Belgium. Degree of overlap of winter ranges of the races not known (Zink 1981).

NORTH EUROPEAN AND BRITISH RACE, nominate *torquatus*, known to winter in southern Spain and north-west Africa, and also recorded from Greece (Vaurie 1959) and Malta (Sultana and Gauci 1982). In north-west Africa, both European races winter mainly in Atlas mountains, particularly on mountain tops of Atlas Saharien from Tunisia to Morocco (Blondel 1962*b*; Isenmann 1972).

Southward migration starts in September, birds leaving Scandinavia on south to SSW heading towards southern France, some moving along North Sea coast of West Germany and Netherlands; pass Helgoland (northern West Germany) mainly mid-September to October. Regular but sparse passage through Shetland and Fair Isle, Scotland. (Durman 1976; Zink 1981.) Young birds from Britain and some birds ringed in Britain on autumn passage have been recovered in France, and a few autumn birds in Italy (Zink 1981). Birds of both European races arrive in north-west Africa from mid-October but main arrival is from mid-November (Niethammer 1955; Blondel 1962*b*).

In spring, nominate *torquatus* starts leaving North Africa in March–April (Blondel 1962). Scandinavian birds appear to migrate further to west in spring than in autumn: west of a line from Helgoland to Garonne, 23 spring and only 7 autumn recoveries of Scandinavian birds; east of that line, 7 spring recoveries and 45 autumn (Zink 1981). Thus spring passage on Fair Isle also apparently more marked than autumn (Durman 1976), and there are more trapped in spring on Helgoland than in autumn (Vauk 1965). No such loop migration detectable for birds from Britain and Alps.

Arrival in England begins from 2nd week of March and passage through Fair Isle extends from early April with bulk in May, ♂♂ *c*. 10 days earlier than ♀♀ (Durman 1976). Arrives on Norwegian breeding grounds April–May (Zink 1981).

SOUTH EUROPEAN AND WEST TURKISH RACE *alpestris* winters in south of breeding range (some birds thus apparently only short-distance migrants or perhaps resident or making only altitudinal movements) as well as in north-west Africa (see above), Malta (scarce), and Cyprus (Vaurie 1959; Sultana and Gauci 1982; Flint and Stewart 1983). Winter recoveries (including also some British-ringed birds) concentrated in western Alps, in-

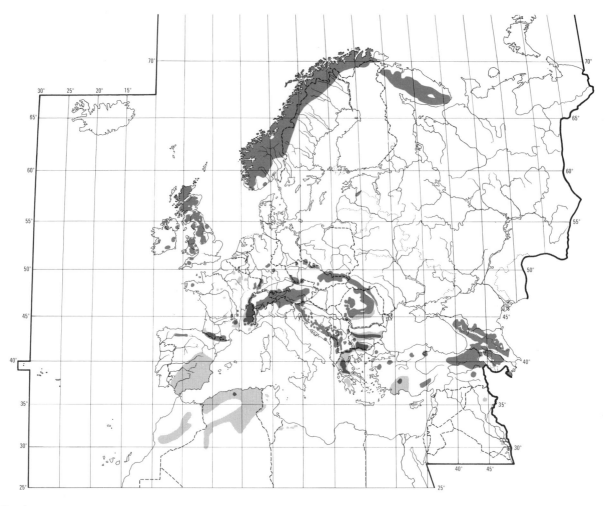

dicating south to south-west movement of juveniles, but this reflects heavy hunting pressure there rather than use of region as main wintering area. Single recovery of bird ringed as nestling in southern Poland showed SSE movement to Bulgaria, and another from Austria moved south-east to Dalmatia (Yugoslavia). Autumn migrants ringed in Alpine passes Col de Bretolet and Col de la Golèze moved between west and south-east, majority south-west to south. Whereas these perhaps include some nominate *torquatus*, majority (at least) are *alpestris*. The only 2 recoveries of birds ringed in eastern Alps and Balkans indicated wintering to south-east rather than to south-west in north-west Africa. (Zink 1981.)

In Carpathians, some adult ♂♂ move to higher altitudes immediately after breeding (Dementiev and Gladkov 1954*b*). Occurs on passage at Bosphorus (western Turkey) late August to October, mostly late September and early October (*Orn. Soc. Turkey Bird Rep.* 1966–73), and on Malta from mid-September together with smaller numbers of nominate *torquatus* (Sultana and Gauci 1982). Present on Cyprus October–March, but numbers there peak in November (Flint and Stewart 1983), suggesting some

onward passage. Returns to breeding grounds in Switzerland mid-March to mid-April (Zink 1981), and Ukraine (south-west USSR) mid-March or April (Dementiev and Gladkov 1954*b*).

CAUCASUS, EAST TURKISH, AND TURKMENIYAN RACE *amicorum* presumably winters largely in Iran and southern Turkmeniya; also in Asia Minor according to Dementiev and Gladkov (1954*b*). Birds leave breeding grounds in late September and October (Dementiev and Gladkov 1954*b*), but no detailed information on movements. Stragglers recorded in Libya 30 September–1 March (mainly on coast, also in desert), on Mediterranean coast of Egypt between 25 October and 9 December, and in Nile valley to northern Sudan in December (Butler 1905; Goodman and Watson 1984). Birds presumably of this race recorded in Gulf states of Arabia November to early March (Bundy and Warr 1980). Arrives back in northern Caucasus from early March or early April, occupying foothills at first but extending higher as snow melts. In Armeniya, leaves plains in March. (Dementiev and Gladkov 1954*b*.)

LAB, DJB

Food. In spring and early summer, adult and larval insects and earthworms; at other times, mainly fruit. Feeds on ground and in trees and bushes (Witherby *et al.* 1938*b*). When collecting food for young will accumulate pile of prey on ground before carrying it all off (Géroudet 1963).

Diet in west Palearctic includes the following. Invertebrates: earwigs (Dermaptera), bugs (Hemiptera), adult and larval Lepidoptera (e.g. Noctuidae), adult and larval flies (Diptera: Tipulidae, Bibionidae, Muscidae), Hymenoptera (sawflies Tenthredinidae, ants Formicidae), adult and larval beetles (Coleoptera: Carabidae, Silphidae, Staphylinidae, Geotrupidae, Scarabaeidae, Byrrhidae, Elateridae, Cantharidae, Cryptophagidae, Lathridiidae, Cerambycidae, Chrysomelidae, Curculionidae), spiders (Araneae), millipedes (Diplopoda), small snails (Gastropoda), slugs (Pulmonata), earthworms (Oligochaeta). Also lizards *Lacerta* (once, *c.* 5–6 cm long: Isenmann 1985) and salamander *Salamandra*. Plant material mainly fruits: of juniper *Juniperus*, bramble *Rubus*, strawberry *Fragaria*, hawthorn *Crataegus*, cherry *Prunus*, rowan *Sorbus*, elder *Sambucus*, currant *Ribes*, buckthorn *Rhamnus*, alder buckthorn *Frangula*, *Viburnum*, olive *Olea*, vine *Vitis*, *Lantana*, mistletoe *Viscum*, ivy *Hedera*, bilberry *Vaccinium*, crowberry *Empetrum*. (Csiki 1908; Meinertzhagen 1930; Schuster 1930; Witherby *et al.* 1938*b*; Chapman and McGeoch 1956; Marisova and Vladyshevski 1961; Korodi Gál 1970.)

In spring in Ukraine (USSR), eats mainly larval Diptera, larval and adult Scarabaeidae, and larval Lepidoptera; in summer, takes adult beetles (70% of diet), earthworms, and millipedes, etc., sometimes also lizards; from mid-summer until departure, mainly berries and other plant material (Marisova and Vladyshevski 1961). In Scotland in spring and early summer, takes mainly adult and larval insects and worms; later on, also fruit (Watson 1972). See also Krayser (1971) and Knolle *et al.* (1973). In Algeria in winter, apparently eats mainly juniper berries (Heim de Balsac 1931).

Food of young in western Carpathians (Rumania), studied from 924 items in 179 collar-samples, comprised 41·7% (by weight) Lepidoptera (largely larvae), 28·8% earthworms, 24·6% beetles (largely adults), 4·0% Hymenoptera (largely ants), and 0·9% others; average daily food consumption 2·5 g at 1 day, 10 g at 5 days, 19 g at 10 days, 25 g at 15 days (Korodi Gál 1970). See also Ware (1961). DJB

Social pattern and behaviour. Important study of *alpestris* in western Carpathians (Rumania) by Korodi Gál (1970). No comprehensive published accounts of nominate *torquatus* but entry includes material supplied by R F Durman from studies in Pentland Hills (Midlothian, Scotland).

1. Usually gregarious outside breeding season. At end of breeding season, gathers into flocks for feeding, often associating loosely with other *Turdus* (Westphal 1975; Elst 1984). Such association continues in winter quarters, notably with Song

Thrush *T. philomelos*, Redwing *T. iliacus*, and Mistle Thrush *T. viscivorus* (Heim de Balsac 1931; Snow 1952; Smith 1965*a*): e.g. in December, Turkey, several flocks of 30–60 in company with *T. viscivorus* (Vittery *et al.* 1972). Migrates singly or in small flocks apparently of both sexes, and usually of not more than 5(–10) birds (Labitte 1947; Marisova and Vladyshevski 1961; Korodi Gál 1970). In Belgium, of 25 records of autumn passage, 10 were single birds, 14 were flocks of up to 5 birds, 1 of 50 birds; ♂:♀ ratio 1·8:1 (Elst 1984). In Berlin area, 66% singles, 12% of 2 birds, 17% 3, 5% more than 3; once 15 together—this exceptional inland, but flocks of 20–30 not uncommon on North Sea coast (Westphal 1975). In spring, of 300 records, majority single birds, 100 flocks of 2–5, 10 of 6–10, 2 of 20, 1 of 30, 1 of 80; flocking most prevalent during peak migration (mid-April), proportion of 'pairs' increasing from end of April (Elst 1984; see also Flock Behaviour, below). BONDS. No evidence for other than monogamous mating system. No information on duration of pair-bond, though presumably breaks down outside breeding season. 2 reports of hybridization with Blackbird *T. merula*; in both cases latter was ♂ (Kirkpatrick 1907; Bonhote 1910). See also Heterosexual behaviour, subsection 4 (below). Both sexes recorded brooding and caring for young (Flegg and Glue 1975) but ♀ usually performs most, if not all incubation and brooding (Marisova and Vladyshevski 1961; Korodi Gál 1970; Durman 1978). Both sexes care for fledglings when no 2nd clutch started, but, if ♀ starts 2nd clutch, ♂ tends 1st brood while she builds new nest, lays, and incubates. Parents typically feed young for up to 12 days after fledging, once up to 36 days. Age of first breeding 1 year. (R F Durman.) BREEDING DISPERSION. Solitary and territorial, though tends to form neighbourhood groups (Géroudet 1963). For apparent establishment of temporary territories on spring migration, see Wolsey (1982) and Antagonistic Behaviour (below). No hard information on size of territory though nests in Pentland Hills regularly 160–200 m apart (Durman 1978). Territory serves for courtship, nesting, and feeding, though birds commonly feed to varying extent on neutral ground outside it, e.g. 6 ♂♂ once fed amicably together (R F Durman). In Erzgebirge (East Germany), feed typically up to 150 m from nest, one pair ranging 300–400 m (Holupirek 1977). In Harz mountains (West Germany), 500–700 m between nest and feeding grounds; seen occasionally feeding with *T. merula* in grassy area *c.* 100 m from nest (Knolle *et al.* 1973). Owing to formation of neighbourhood groups (often along sides and floor of steep valleys), density estimates vary greatly with boundaries chosen. In Pentland Hills, 14–16 1st-brood nests in *c.* 2·8 km of valley (R F Durman). In Derbyshire (England), density relatively high and (e.g.) 1 nest had 3 others within radius of *c.* 100–300 m (Hems 1966). On Dartmoor (south-west England), up to 5 singing ♂♂ along *c.* 400 m (Ware 1961). In Yorkshire (England), at least 10 nests in *c.* 1·3 km² (Hosking and Newberry 1946). In Meirionnydd (Wales), 36 pairs in 112 km², locally up to 1·3–1·5 pairs per km² (Hope Jones 1979). In France, 20–90 pairs per km² (Dejonghe 1984). In Switzerland, 90 pairs per km² in pine habitat; in Jura mountains, average 22 pairs per km²; in Chablais de Haute-Savoie, 37 pairs per km² (Schifferli *et al.* 1982). In Ukraine, in best habitat (summer grazing near forest edge), pairs up to 100 m apart, average 20–30 pairs per km² in spruce, 10–20 in beech (Marisova and Vladyshevski 1961). In Cumbria (England), nests commonly re-used for 2nd clutches and in successive years (up to 4 recorded) (Robson 1975). In Pentland Hills, ♀ typically builds new nest for 2nd clutch, in 4 cases *c.* 40–100 m from 1st. If nest predated, pair usually take a new territory nearby and re-nest there. (Durman 1978.) In Scotland often breeds close to nest of Golden Eagle *Aquila chrysaetos* (Gordon 1912) or Peregrine *Falco peregrinus* (Watson 1972), suggesting possible positive attraction. In Pentland Hills, at least some ♂♂ faithful to former breeding territory, and young also tend to return to natal area to breed (Durman 1978; R F Durman). ROOSTING. Nocturnal and presumably alone in breeding season, communally outside breeding season. On breeding grounds, *alpestris* roosts habitually in conifers (Witherby *et al.* 1938*b*). In late August, Lauderdale (Scotland), flock of *c.* 30 roosted in heather *Calluna* (Baxter and Rintoul 1953). In winter quarters, *amicorum* and *alpestris* roost in rocks but not in trees (Meinertzhagen 1930, 1954). Activity pattern in Finnish Lapland apparently similar to *T. iliacus* there: main 'roosting' spell before midnight, but birds feed actively during the hours immediately after midnight; one ♀ watched on nest often slept with head under wing (Pulliainen *et al.* 1981, which see for preening rates of sitting ♀). In winter, Algeria, goes singly (sometimes in twos and threes) to water-holes to drink; drinks rapidly then retires immediately to perch-sites (Heim de Balsac 1931). In Oberbayern (West Germany), 3 ♂♂ in flock of 14 performed bathing movements on snow-covered ground (Pfeifer 1967*a*).

2. Often much shyer and less approachable than other *Turdus* (Witherby *et al.* 1938*b*), but this varies. In autumn flocks, Berlin, usually rather shy, not allowing approach to less than 50–100 m (Westphal 1975). On spring passage in France, sometimes very shy, at other times approachable to within 5–6 m (Labitte 1947). In breeding season, may be very confiding when feeding (Holupirek 1977). Like *T. merula*, often flushes at very last moment when disturbed by human intruder; commonly dashes low downhill, dodging out of sight wherever possible (Witherby *et al.* 1938*b*), often taking refuge behind crest of rocky outcrop (Jonsson 1979). FLOCK BEHAVIOUR. After 2-day stopover by hundreds of birds in late September on Isle of May (Scotland), 1 ♂ rose high into the air giving a chuckling call (probably 2a in Voice) and spiralled ever higher; rest of flock followed suit and flew off, evidently continuing migration (Baxter and Rintoul 1953). While loosely accompanying other Turdidae in autumn, usually fly up separately (Westphal 1975). Winter flocks more noisy than most *Turdus*, giving a loud chuckle when feeding and alarmed (Meinertzhagen 1954: see 2 in Voice). SONG-DISPLAY. ♂ sings from exposed elevated perch in territory, e.g. tree-top, rock, wall, heather clump, also often in flight (Witherby *et al.* 1938*b*; R F Durman: see Antagonistic Behaviour, below). On Dovre Fjeld (Norway), a number seen climbing into the air, and singing, like Meadow Pipit *Anthus pratensis* (Collett and Olsen 1921). For Song-duel, see Antagonistic Behaviour (below). Sings mostly early morning and evening (Holupirek 1977), sometimes late into dusk (Watson 1972), occasionally at night (Witherby *et al.* 1983*b*). In Finnish Lapland, silent around 23.00–02.00 hrs (Pulliainen *et al.* 1981). Full song said to be given only at start of breeding season, thereafter simpler song (see 1 in Voice for details). Song-period late March until early July (Witherby *et al.* 1938*b*). Song largely ceases end of June/early July (Géroudet 1963; Watson 1972). Flock-members sometimes sing on spring migration (Jánossy and Török 1977; Elst 1984). No reports of singing in winter quarters. ANTAGONISTIC BEHAVIOUR. Little information. ♂♂ markedly aggressive to conspecifics in breeding season, regularly chasing one another in pursuit-flights. Commonly 2 ♂♂ together on the ground, follow each other in the manner of *T. merula*, pursuer with tail raised high and head raised as if to display white throat crescent; often display thus in presence of ♀. (R F Durman.) Resident ♂ threatens rival by crouching low and flying up at rival with much fluttering; said occasionally to lock feet (Simms 1978), but this never seen by R F Durman.

After chasing off intruder or predator, ♂ often sings while flying back to favourite song-post. When recording of rival song broadcast in occupied territory, resident always flew swiftly towards source, giving 'tac' Alarm-calls. (R F Durman: see 4a in Voice.) In dispute over temporary territories on spring migration, 2 ♂♂ chased each other to and fro along wall, frequently coming into contact, once fluttering up together to *c.* 1 m; quarrelled noisily for 15 min before 1 flew *c.* 200 m downstream. Next evening, rivals conducted Song-duel for 1 hr (of observation), and appeared to be holding discrete territories flanking a common boundary. (Wolsey 1982.) In Pentland Hills, markedly aggressive towards *T. viscivorus*, but not *T. merula* (Durman 1978). For apparent aggression towards ♀ mate, see Heterosexual Behaviour (below). HETEROSEXUAL BEHAVIOUR. (1) General. ♂♂ tend to arrive on breeding grounds before ♀♀ (Korodi Gál 1970; Durman 1976; Elst 1984), in Ukraine 3–5 days before; majority paired by *c.* 1 month later (Marisova and Vladyshevski 1961). (2) Pair-bonding behaviour. Little detailed information. ♂ *torquatus* follows ♀ along ground, both with heads erect, ♀ advancing a little, ♂ following until almost abreast of her, whereupon she advances again, and so on; ♂ gives subdued twittering (see 1 in Voice). ♂ sometimes hopped ahead of ♀, turning suddenly to face her and then, crouching, would suddenly fly at her; ♀ either flew aside to avoid ♂, or fluttered into the air facing ♂ (Fig A), then both

A

dropped to the ground again to resume 'parading'. ♂ may also expose white throat crescent to ♀ and raise his tail. (Witherby *et al.* 1938*b*; see also Antagonistic Behaviour, above.) (3) Courtship-feeding. Not reported. (4) Mating. Only one detailed sequence reported: at 09.00 hrs, 3 May, pair close together; ♂ became excited and 'appeared to go through the motions of feeding on worms but at an apparently speeded up rate' (presumably mock-feeding), then chased ♀ to a rock scree and continued to chase her from rock to rock. ♂ then mounted ♀ and chased off an intruding ♂ before visiting vicinity of nest. (R F Durman.) Copulation occurs especially during nest-building (Korodi Gál 1970). ♂ twice tried to copulate with ♀ *T. merula* when she was off her nest, ♀ successfully evading him (George 1977). (5) Behaviour at nest. Both members of pair seen nest-prospecting along edges of heather; site for 2nd nest of season often remarkably similar to 1st (R F Durman). In western Carpathians, incubating ♀ left nest only once per day, usually in afternoon, to feed and drink. ♂ sang intensively near nest, and fed ♀ on nest, but never incubated (Korodi Gál 1970). Incubating ♀ left nest unattended 11% of the time, even when it was snowing. When ♀ off nest, ♂ often sat on rim but did not incubate. Noisy near and on nest, ♀ even calling while incubating, notably when disturbed by human intruder. Sitting ♀ also lunged at and caught insects. (Pulliainen *et al.* 1981, which see for diurnal frequencies of shifting, preening, and settling on nest.) Brooding ♀ greeted ♂ vocally (Ware 1961: see 2 in Voice, also below). RELATIONS WITHIN FAMILY GROUP. Brooding regime evidently varies with region, climate (etc.). In Pentland Hills, young brooded by day until 3 days old (R F Durman). In western Carpathians, brooded (only by ♀) regularly for first 5 days; at this time fed by ♂, thereafter by both sexes. Still some brooding by ♀ after 6 days, especially in morning and evening. From 10-14 days, brooded only at night or during rain. Young beg with food-calls (see Voice), eyes opening at 7 days. (Korodi Gál 1970.) Food-bearing ♂ announced approach to nest by singing; on arrival, brooding ♀ greeted him vocally and stood aside to let him deliver bulk of meal to young, then she distributed any 'left-overs' (presumably food not immediately accepted) (Ware 1961: Fig B). In one nest, well-grown young

B

brooded after every feed (Knolle *et al.* 1973). Young leave nest at 14-15 days, barely able to fly, and hide in surrounding vegetation (Géroudet 1963; Knolle *et al.* 1973). In Pentland Hills, usually hide singly under heather, often near nest (R F Durman). Give food-calls (Fox 1900; see Voice). When young newly fledged, parents return repeatedly to nest with food, and young therefore return there to be fed (R F Durman). In July, families quit breeding territories to join up with others for feeding elsewhere (Géroudet 1963). In Ukraine, commonly move up to subalpine zone for berry food (Marisova and Vladyshevski 1961). ANTI-PREDATOR RESPONSES OF YOUNG. When adder *Vipera berus* slid over nest, all but one of brood scattered and hid in heather; remaining chick 'froze' and was killed (Wenner 1933). Newly fledged young scared by sheep flew weakly over water, landed and, beating wings, swam to the bank and sought refuge (R F Durman). If approached, recently fledged young usually fly *c.* 50 m and plunge into undergrowth (Fox 1900). Bittern-posture (see p. 995) also recorded (Zucchi 1975). PARENTAL ANTI-PREDATOR STRATEGIES. (1) Passive measures. ♂ often gives Subsong (see 1 in Voice) to warn incubating ♀ of approaching danger (R F Durman). ♀ a tight sitter: if disturbed early in incubation she left with a lot of noise, but later had to be forced off eggs; ♂ is attracted by ♀'s 'screeches', leading to combined attack on intruder (Korodi Gál 1970; see below). ♀ brooding well-grown young did not leave nest until human intruder climbed tree to within 2 m of nest; returned immediately after intruder left (Knolle *et al.* 1973). ♂ gives 'tac' Alarm-calls when Kestrel *Falco tinnunculus*, Merlin *F. columbarius*, or Short-eared Owl *Asio flammeus* in vicinity of nest (R F Durman). (2) Active measures: against birds. Both sexes give alarm-calls (see 4 in Voice) while boldly attacking and driving off Corvidae, Buzzard *Buteo buteo* (Witherby *et al.* 1938*b*; Knolle *et al.* 1973; Simms 1978), and *F. tinnunculus* (etc.) (Fox 1900; Géroudet 1963; R F Durman); for lack of aggression towards *F. tinnunculus* see Holupirek

(1977). Sparrowhawk *Accipiter nisus* which took nestling elicited alarm-calls from parents perched on tree-top; when raptor flew off with nestling, both parents pursued it, giving alarm-calls (Lunn 1984). (3) Active measures: against man and other animals. Typically bold and aggressive in defence of nest and young against man and other predators (Witherby *et al.* 1938*b*) but individual variation occurs: some highly demonstrative, others relatively passive; in passive pair, ♂ the least active (Holupirek 1977), and Korodi Gál (1970) found ♀ far more aggressive than ♂. On one occasion, ♀ flew to nearby tree and gave repeated Warning-calls (Holupirek 1977: see 3 in Voice). On approach of human intruder to nest, ♂ may fly off, giving 'tac' Alarm-calls, then often makes gradual approach; as he lands, flicks tail steeply upwards, more emphatically than *T. merula*, almost like Wheatear *Oenanthe oenanthe* (Watson 1972). When intruder searching for nest, alarm-calls (4a and 4b in Voice) of ♂ waxed and waned with proximity of intruder to nest; when intruder climbed tree to nest containing week-old brood, ♂ mounted sustained attack, diving to brush intruder while ♀ perched nearby (Fontaine 1975). ♀ will peck at hand touching eggs (Korodi Gál 1970). Dive-attacks and buffeting or brushing intruder on head, back, and arms not uncommon (Witherby *et al.* 1938*b*; Korodi Gál 1970), once when intruder ringing young (Knolle *et al.* 1973). When fledged young nearby, parents usually land on low tree (etc.) well away from them and give 'tac' Alarm-calls (Jonsson 1979). Mobile distraction-lure display of disablement type reported: bird reeled and tumbled on the ground; this apparently extreme, however, and more often, with hard-set eggs or young, bird 'fluttered a few yards in a lazy sort of fashion' (Witherby *et al.* 1938*b*). Distraction-displays recorded rarely, and considered exceptional (Simms 1978). Will also vigorously attack squirrels (Sciuridae) (Marisova and Vladyshevski 1961). Makes flying attacks to within 1 m on stoat *Mustela erminea*, accompanied by 'tac' alarm-calls (R F Durman).

(Figs by N McCanch: A from drawing in Simms 1978; B from photograph in Ware 1961.) EKD

Voice. Wide variety of calls, freely used, especially in breeding season. No information on possible racial differences. For musical notation, see Stadler (1926); for additional sonagrams, Bergmann and Helb (1982).

CALLS OF ADULTS. (1) Song of ♂. A repetition of melancholy, plaintive phrases of 2-4 simple, monotonous, fluting piping notes; distinct pause between phrases (Witherby *et al.* 1938*b*; Simms 1978), as if hesitating to proceed (Watson 1972). In recording (Fig I), phrase

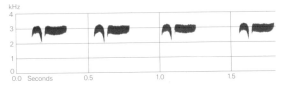

I V C Lewis Wales May 1970

rendered 'pi-ree pi-ree pi-ree . . .', another in same song consists of 3 monosyllabic whistled tones: 'pee pee pee' (J Hall-Craggs). Descriptions of other songs include 'tchouee-tchouee-tchouee', also 'ter-WEE ter-WEE ter-WEE' (Simms 1978); 'tru tru tru' (Dejonghe 1984); 'trruu-trruu-trruu-tjii-tjii-tjii' and 'tu-li tu-li tviv tviv'

(Haftorn 1971, which see for other examples, also Christiani 1942). Structure of phrases reminiscent of Song Thrush *T. philomelos*, but in timbre more like Mistle Thrush *T. viscivorus* or Blackbird *T. merula* (Bergmann and Helb 1982). Rather like *T. philomelos* but steadier, slower, and more interrupted, lacking the vigour, *élan*, and richness of that species. Each ♂ has relatively small repertoire; often repeats a phrase 4–5 times, then another, until—when repertoire exhausted—he returns to 1st phrase. (Géroudet 1963). Repertoires of 3 birds analyzed by Ince and Slater (1985) contained 2–4 song types. Full song, said to be given only at start of breeding season (V C Lewis), as above but interspersed with chuckling or warbling sounds (Fig II) apparently like

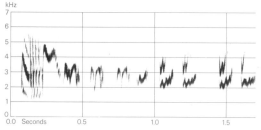

II   V C Lewis   Wales   May 1970

call 2a, below (Simms 1978). Full song also incorporates short melodious phrases (Watson 1972), and simple phrases described as ending with a tit (*Parus*)-like note or a twittering like Redwing *T. iliacus* (Jonsson 1979). In one recording, song rich, melodious, and elaborate, like *T. philomelos* at outset (P J Sellar); Fig III shows phrase ending in tremolo/trill. Subsong a warbling sound interspersed with chuckling calls (P J Sellar). Gentle

subdued twittering given by ♂ in ground courtship (Simms 1978) perhaps Subsong. (2) Contact-calls. (a) On ground or when taking off, a dry 'tjuck' or 'tjuck-uck-uck' (Jonsson 1982); a chuckling rattle 'tchook-tchook-tchuc' (Simms 1978); in Fig IV, 'ti-ti-tjuck' (J Hall-Craggs). This presumably the chuckling sound reported from migrating birds (Baxter and Rintoul 1953) and winter feeding flocks (Meinertzhagen 1954). (b) In flight, including migration, 'dchèrrr' or 'dcharrr' (Géroudet 1963); also rendered a rolling 'tjuirr' (Jonsson 1982), 'tschwierr' (Gatter 1976). Trilling 'tchurree' given on take-off and containing an element of alarm probably this call, and presumably that said by Witherby *et al.* (1938*b*) to resemble 'ordinary note' of Dunlin *Calidris alpina* (Simms 1978). (c) A 'ssierk' (Gatter 1976), 'tschirk' (Holupirek 1977), or 'zrrp' (Westphal 1975); given in flight. (d) Other calls: occasionally a soft 'irig' (Gatter 1976). Calls ('skirls of pleasure') of ♀ greeting ♂ at nest (Ware 1961) presumably contact-calls. (3) Warning-call. A thin, high-pitched 'ssii', as in *T. merula* call for aerial predator (Géroudet 1963; J Hall-Craggs), or 'SEEeee' (P J Sellar: Fig V, half speed). Quiet, long-drawn, penetrating 'zieh' from perched ♀ when intruder at nest (Holupirek 1977). (4) Alarm-calls. (a) A rapid 'tac-tac-tac', not unlike call 2a but louder and harder (Witherby *et al.* 1938*b*), given by both sexes, perched or in flight; 'clack clack', like rattling 2 pebbles together, at times with a wheezy twanging prefix (Watson 1972). Rather metallic and somewhat like *T. merula* but deeper and mellower, more like Wheatear *Oenanthe oenanthe* (Simms 1978). (b) In greater alarm, call 4a runs into a loud rattling chatter (Witherby *et al.* 1938*b*; Bergmann and Helb 1982). In recording by V C Lewis, timbre varies from hard dry

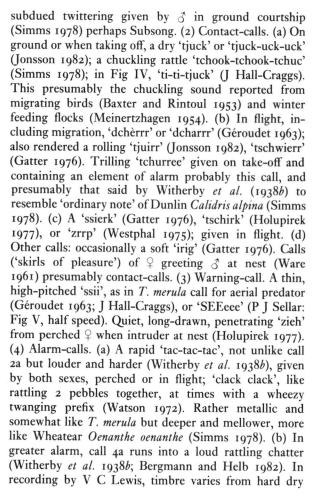

III   Roché (1970)   France   April 1962

IV   V C Lewis   Wales   June 1973

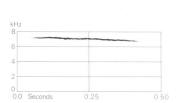

V   P A D Hollom   Wales   May 1978

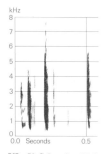

VI   V C Lewis   Wales   June 1973

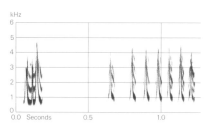

VII   V C Lewis   Wales   June 1973

rattle (Fig VI) to a more fruity whinnying sound; in Fig VII a tittering preceded by 'tschr' (J Hall-Craggs). Hawk-like tittering, softer than equivalent rattle of *T. merula* (Watson 1972), evidently this variant. (c) In extreme alarm, a harsher 'chrèchrèchrèchrè' (Géroudet 1963). In recording by V C Lewis of bird alarming at Magpie *Pica pica*, calls of this type have a distinct nasal quality (P J Sellar). Call described as a 'fast screech' given when chasing Corvidae (Simms 1978) probably of this type. (d) Anxiety may be signalled by 'vrèvrèvrè' like Whitethroat *Sylvia communis* (Géroudet 1963); possibly not markedly different from calls (e.g. 4b) listed above.

CALLS OF YOUNG. Food-calls of fledged young similar to those of young *T. merula* (Watson 1972; V C Lewis). When young first leave nest they have a peculiar twittering call, not very unlike song of *O. oenanthe* (Fox 1900).

EKD

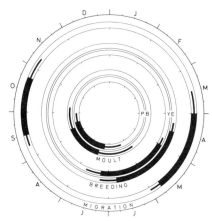

Breeding. SEASON. Britain and Ireland: see diagram (Flegg and Glue 1975). Scandinavia: from early May to end of June in south, and from late May to early August in north (Haartman 1969*a*; Makatsch 1976). Alps: similar to Britain and Ireland (Makatsch 1976). SITE. In Britain (nominate *torquatus*), on or close to ground in low vegetation, or on rock-ledge or in crevice, rarely in tree. Of 297 sites: 9% on ground, 36% off ground in vegetation under 45 cm tall (mainly heather *Calluna*, bilberry *Vaccinium*, and bracken *Pteridium*), 4% in vegetation 45-90 cm tall, 2% in trees up to 3 m above ground, and 49% on ledge or in crevice, mainly close to ground, but including 7% in potholes or mine shafts down to 5 m below ground. Nest almost always has backing or flanking wall or cliff or overhang of earth bank (Flegg and Glue 1975). In Poland (*alpestris*) all of 26 nests in trees (mainly coniferous), placed close to trunk and supported by twigs or 1-2 thicker branches. Mean height 3·5 m (1-16), with 21 in range 1-5 m (Bocheński 1968). In Caucasus, USSR (*amicorum*), nests normally in crevices of rocks (Dementiev and Gladkov 1954*b*). Nest: comprised of 3 parts—thick and compact external layer of twigs; thin, and sometimes incomplete, plastering of mud mixed with broken grass

leaves and moss, covering bottom and lower part of walls; thick lining (normally concealing mud) of delicate grass blades or, occasionally, rootlets. Rim of wall thickened with strongly woven grass leaves, stalks, and twigs, to 3 cm in width. Humus sometimes used instead of earth for plastering. Mean dimensions of 20 nests (*alpestris*); outer diameter 17·6 cm (15·5-20·5); inner diameter 10·2 cm (9·3-11·3); height 12·2 cm (9·5-20·0); depth of cup 6·2 cm (5·0-7·5). Substantially larger than 2 nests of nominate *torquatus* from Wales, with outer diameter 14 cm, inner 9-9·3, height 6-6·6, and depth 4-4·5. (Bocheński 1968.) Building: by ♀ (R F Durman). EGGS. See Plate 84. Sub-elliptical, smooth and glossy; pale blue, with evenly distributed small red-brown, reddish-purple, and purplish-grey blotches; sometimes with reddish wash overall. Nominate *torquatus* 30·2 × 21·5 mm (26·2-33·5 × 19·6-23·4), $n = 100$; calculated weight 7·4 g. *T. t. alpestris*: 30·5 × 22·3 mm (28·9-34·0 × 20·3-24·0), $n = 65$; calculated weight 8·1 g (Schönwetter 1979). Clutch: 4 (3-6). Of 79 clutches, Britain: 3 eggs, 10; 4, 56; 5, 12; 6, 1; mean 4·1. Possible variation with altitude, though samples small, with mean 4·3 for 7 clutches below 230 m, 4·2 for 24 clutches from 230-300 m, and 4·0 for 20 clutches above 300 m (Flegg and Glue 1975). Clutch size apparently similar throughout range, though data limited. Eggs laid at intervals of 18-20 hrs (Korodi Gál 1970). 1-2 broods, 2 probably more regular in south of range. INCUBATION. 12-14 days. Mostly by ♀ (Flegg and Glue 1975); in *alpestris*, apparently by ♀ only (Marisova and Vladyshevski 1961; Korodi Gál 1970). Begins with last egg (Marisova and Vladyshevski 1961) or penultimate egg (Korodi Gál 1970). Hatching takes place over 24 hrs (Korodi Gál 1970). YOUNG. Cared for and fed by both parents. Brooded intensively for first 6-8 days. (Flegg and Glue 1975.) For variations in brooding regime, see Social Pattern and Behaviour. FLEDGING TO MATURITY. Fledging period 14-16 days. Young independent after 12 days, sometimes later. Age of first breeding 1 year. (R F Durman.) BREEDING SUCCESS. Of 79 clutches in Britain, 24 lost completely before hatching, mainly through desertion and predation (Flegg and Glue 1975). Of 183 eggs in Rumania, 5% addled, 3% predated by squirrels *Sciurus vulgaris* and 2% by Nutcracker *Nucifraga caryocatactes*; of 164 which hatched, 79% fledged and remainder destroyed by squirrels and *N. caryocatactes* (Korodi Gál 1970).

Plumages (nominate *torquatus*). ADULT MALE. In fresh plumage (autumn), head and most of body sooty-black; feathers of upperparts with dull brown fringes, those of underparts with whitish fringes; under tail-coverts with tendency to white shaft-streaks. Broad band across upper chest white, feathers fringed brown to variable extent. Tail sooty-black. Wings sooty-black, like body; flight-feathers with fine whitish outer edges, and upper wing-coverts finely edged whitish. Under wing-coverts mainly whitish with sooty subterminal chevron-shaped marks; undersurface of flight-feathers grey, becoming silvery basally. *In worn plumage* (spring and summer), pale feather-fringes of head and body mostly lost by abrasion, producing a clearer contrast between sooty-black plumage and white band across chest. ADULT FEMALE. In fresh plumage (autumn), head and body sooty-brown; feathers of upperparts with pale olive-brown fringes, those of underparts with whitish fringes; variable number of throat-feathers sometimes with broad sandy-white fringes, giving noticeably pale-throated appearance. Broad band across upper chest dirty white, feathers with pale sooty-brown fringes. Under tail-coverts broadly fringed whitish and with variable whitish central streaks. Tail sooty-brown. Wing sooty-brown, like body; flight-feathers and upper wing-coverts with pale grey fringes; underwing as ♂. *In worn plumage* (spring and summer), feather-fringes on head and body largely worn off, resulting in clearer contrast between mainly sooty-brown plumage and dirty white chest-band. NESTLING. Down buff, fairly long and plentiful. JUVENILE. Crown, sides of head, and rest of upperparts sooty-brown, feathers with variable paler brown fringes; feathers of mantle and scapulars with variable whitish shaft-streak. Chin and throat pale buff with sooty spots, especially at sides; rest of underparts spotted sooty-black and pale buff, each feather mainly dark with a pale subterminal transverse bar. FIRST ADULT MALE. Like adult ♂, but pale feather-fringes wider than those of adult at same time of year and white chest-band less distinct. Retained juvenile outer greater upper wing-coverts with broad white edges; new 1st adult inner coverts more narrowly edged greyish-white (but contrast not always apparent, in particular in spring and summer when heavily worn). FIRST ADULT FEMALE. Like adult ♀, but pale feather-fringes wider and pale chest-band less distinct, grey with sooty suffusion, sometimes almost absent. Contrast between retained juvenile outer and new 1st adult inner greater coverts as in 1st adult ♂.

Bare parts. ADULT. Iris dark brown. Bill mostly green-yellow or yellow, tip and base of upper mandible brown-black; yellow of adult more extensive than in 1st adult, ♂♂ yellower than ♀♀, and yellow brighter and more extensive in spring and summer than in winter; bill of adult ♂ in spring orange-yellow with some brown on tip. Leg and foot dark brown. NESTLING. Mouth deep yellow, no spots. Gape-flanges yellowish-white. JUVENILE. Iris dark brown. Bill horn-brown, tending to yellowish in middle of lower mandible. Leg and foot pale brown.

Moults. ADULT POST-BREEDING. Complete; primaries descendant. Late June to September, before leaving breeding area. POST-JUVENILE. Partial: head, body, lesser and median upper wing-coverts, and variable number of inner greater upper wing-coverts (3–5 old ones usually retained). July–September, before migration. (Witherby *et al.* 1938b; Ginn and Melville 1983.) In *alpestris*, about ⅓ of migrants still partly in juvenile plumage in second half of August, all virtually completed by first half of September (Jenni 1984).

Measurements. Nominate *torquatus*. Mainly Britain and France, on autumn passage; skins (mainly BMNH). Juvenile wing is retained juvenile wing of 1st adult. Bill to basal corner

of nostril; bill to skull on average 9·7 longer, exposed culmen 3·3 longer.

WING AD	♂ 141·7 (2·36; 10)	138–145	♀ 139·0 (2·79; 10)	136–143
JUV	140·9 (1·60; 10)	138–143	136·3 (1·83; 10)	134–139
TAIL AD	112·3 (3·33; 10)	108–116	108·7 (3·30; 10)	106–114
BILL AD	16·4 (0·52; 10)	16–17	16·6 (0·69; 10)	16–18
TARSUS	33·7 (0·67; 10)	33–35	33·7 (0·82; 10)	32–35

Sex differences significant for juvenile wing.

T. t. alpestris. Alps and Pyrénées, breeding season; ages combined; skins (mainly BMNH).

WING	♂ 140·0 (2·49; 10)	136–144	♀ 136·2 (1·14; 10)	134–138
TAIL	108·8 (2·35; 10)	105–113	105·2 (2·10; 10)	101–108

Sex differences significant.

Southern Yugoslavia, March–April: ♂ 140·0 (3·50; 10) 136–146, ♀♀ 134, 140 (Stresemann 1920).

T. t. amicorum. Iran and Sinai; ages combined. Wing ♂ 139·5 (4) 137–144; tail ♂ 113·5 (4) 111–116 (BMNH). USSR: wing, ♂ 140·5 (23) 134–151, ♀ 135·5 (8) 132–137 (Dementiev and Gladkov 1954b).

Weights. Nominate *torquatus*. Norway, on migration: 95–120 (5) (Haftorn 1971). Helgoland, on migration: 110·8 (34) 92–138 (Weigold 1926). Britain and Netherlands, on migration: April–May, ♂♂ 90, 129, ♀ 106; September–November, ♂ 119·6 (5·22; 5) 113–126, ♀ 114·4 (10·83; 7) 97–126 (BTO, ZMA). Exhausted, Netherlands: ♂ 51·5, ♀ 70 (ZMA). *T. t. alpestris.* Single ♂♂: 94 (BMNH), 115·4 (Dementiev and Gladkov 1954b). *T. t. amicorum.* Eastern Turkey, May: ♂ 103 (Kumerloeve 1968).

Structure. Wing rather long, broad at base, tip bluntly pointed. 10 primaries: p8 longest, p9 5–8 shorter, p7 0–2, p6 5–7, p5 17–21, p4 23–29, p1 37–45. P10 reduced, narrow and pointed; 79–89 shorter than p8, 6–13 shorter than longest upper primary covert. Tail long, 12 feathers; tail-tip square, but t6 4–7 shorter than others. Outer web of p6–p8 and inner of p7–p9 emarginated. Middle toe with claw 29·7 (7) 26–33; outer toe with claw *c.* 71% of middle with claw, inner *c.* 67%, hind *c.* 72%. Remainder of structure as in Blackbird *T. merula* (p. 963).

Geographical variation. Mainly involves width of pale feather-fringes and chest-band, the 3 races differing distinctly. *T. t. alpestris*, breeding in central and southern Europe, differs from nominate *torquatus* in much broader pale edges to body feathers (not wearing off as in nominate *torquatus*) and more extensive whitish edges to wing-feathers; underparts of ♀ sometimes white with limited number of dark marks. *T. t. amicorum*, breeding from Anatolia and Caucasus to Transcaspia, has body similar to nominate *torquatus* except for broader white chest-band, but white fringes of upper wing-coverts and flight-feathers much broader than in both nominate *torquatus* and *alpestris*, especially on secondaries; rest of outer webs of tertials tend to pale grey. DWS

Turdus merula **Blackbird**

Du. Merel   Fr. Merle noir   Ge. Amsel
Ru. Чёрный дрозд   Sp. Mirlo común   Sw. Koltrast

Turdus Merula Linnaeus, 1758

Polytypic. Nominate *merula* Linnaeus, 1758, Europe, except southern Balkans, Greece, and southern USSR; *aterrimus* (Madarász, 1903), Balkans south from southern and eastern Yugoslavia and Rumania, southern Ukraine, Crimea, and Caucasus, south to Crete, Taurus mountains (Turkey), and northern slopes of Elburz in northern Iran; *syriacus* Hemprich and Ehrenberg, 1833, south-east Turkey, Levant, Iraq, and Iran, south of *aterrimus*, east to Turkmeniya (USSR); *mauritanicus* Hartert, 1902, north-west Africa; *cabrerae* Hartert, 1901, Madeira and western Canary Islands; *azorensis* Hartert, 1905, Azores; *intermedius* (Richmond, 1896), central Asia from Afghanistan east through Sinkiang to Zaidam area (China) and north through Tien Shan to Dzhungarskiy Alatau, reaching Iraq in winter. Extralimital: *maximus* (Seebohm, 1881), Himalayas; *simillimus* Jerdon, 1839, south-west India; *c.* 6 further races in southern Asia.

Field characters. 24–25 cm; wing-span 34–38·5 cm. At least 50% bulkier and longer-tailed than Starling *Sturnus vulgaris*; 5% longer and 10% stockier than Ring Ouzel *T. torquatus*. Medium-sized, round-headed, rather long-tailed, noisy thrush, differing from all other west Palearctic *Turdus* in uniformly or mainly dark plumage—black in ♂, umber or rufous-brown in ♀, and rufous-brown in juvenile. Only obvious features are yellow bill in ♂, pale throat in ♀, and streaking and mottling in juvenile. Flight wavering, with fast bursts of wing-beats; rather long, rounded tail obvious. Song rich and fluting. Sexes dissimilar; little seasonal variation. Juvenile separable. 7 races occur in west Palearctic, differences varying from slight to marked and largely restricted to ground-colour of ♀ and immature; size and structure differ slightly in island and peripheral mainland forms.

(1) Race of Europe except Balkans, Greece, and southern USSR, nominate *merula*. ADULT MALE. Glossy to matt jet-black, sheen declining with wear; worn wings noticeably brown-toned. By November, most have completely fresh glossy black plumage, relieved only by silver-grey tone of inner webs to flight-feathers (noticeable on spread wing). Subject to frequent loss of plumage pigment; white-splashed, piebald, and wholly white birds are not rare, though pale grey-brown and other forms are. Eyelids (forming conspicuous ring at close range) and bill orange-yellow. ADULT FEMALE. Head and body dark umber-brown (sometimes with marked rufous tone to underparts), wings, rump, and tail brown-black. Upperparts unmarked but face and underparts show variable, subtle pattern of (a) faint, dull grey fore-supercilium, (b) brown shaft-streaks on cheeks, (c) pale grey or buff-grey throat with umber or rufous brown streaks, (d) quite marked mottling on upper breast with many feathers showing dark brown notch or triangle on tip, and (e) variable dull white edges and dark brown notches elsewhere. A few birds have breast markings unusually distinct on noticeably pale pink-buff ground-colour; such a pattern may suggest other thrushes. Underwing dark brown, with rufous margins to coverts.

Plumage becomes duller and less rufous below with wear. By November, most have completely fresh plumage, being at their darkest and most rufous, with underpart pattern most pronounced. Bill dark brown-horn, with dull yellow usually visible by gape and sometimes invading whole bill except tip, even becoming orange-yellow like ♂. JUVENILE. Resembles adult ♀ but much more rufous in general appearance and with markings on both upper- and underparts. Head, mantle, scapulars, and feathers of forewing have obvious rufous streaks, particularly striking on innermost greater coverts (producing discrete set of pale bright lines, more distinct than buff tips). Back, rump, and upper tail-coverts more uniform, with only scattered rufous streaks and tips. Underparts basically white- to rufous-buff, with pale throat, rough malar stripe, quite densely spotted chest, openly barred or notched flanks, and densely barred vent. Separation of ♂ from ♀ untrustworthy, but ♂ usually darker, especially on wings and tail, and with more strongly rufous streaks. FIRST-YEAR MALE. By October (in migrant populations) or early December (in residents), plumage distinctively dull and dusky, lacking gloss and total uniformity of adult ♂. Difference most obvious on wings and underbody which appear paler than head, back, and tail, having brown-grey fringes to feathers. Ageing in the field much assisted by dark brown-horn bill and retention of juvenile outer greater coverts and flight-feathers. Some show pale grey throat like ♀. FIRST-YEAR FEMALE. Far less easy to age than 1st-year ♂: usually paler than adult ♀ but only safely distinguished by retained pale-tipped outer greater coverts. At all ages, legs medium or dark brown. (2) Race of Balkans, Greece, south European USSR, and Turkey, *aterrimus*. Marginally smaller and weaker-billed than nominate *merula*. Both sexes duller, with ♀ noticeably less brown above and greyer below. (3) Middle East race, *syriacus*. Bill rather longer and more slender than in nominate *merula*. ♂ slatier; ♀ distinctly greyer, less rufous below, with scarcely streaked throat. (4) North-west African race, *mauritanicus*. Heavier-billed and up to 10% longer-tailed than nominate *merula*. ♀ greyer and sootier, appearing dun below, and usually having yellow bill. (5) Atlantic Islands races, *cabrerae* and *azorensis*. Marginally smaller and rounder-winged than nominate *merula*. ♂ glossier and deeper black; ♀ also blacker, especially below on *cabrerae*. (6) Central Asian race *intermedius*. Largest race reaching west Palearctic, with heavy bill. Both sexes dull, like *aterrimus*.

Adult ♂ in normal plumage unmistakable but high incidence of plumage aberrations and less uniform dress of ♀ and immature bring in whole range of pitfalls. Of these, most serious are confusion of (1) adult with Blue Rock Thrush *Monticola solitarius* (smaller, shorter-tailed, blue-toned in ♂, mottled and barred on dusky in ♀, with bill and legs always black), (2) dilute ♂ or greyer ♀ with Tickell's Thrush *T. unicolor* (see that species, p. 937), (3) white-throated partial albino ♂ and flying adult and immature (with paler flight-feathers) with *T. torquatus* (see that species, p. 939), and (4) strongest-marked ♀♀ and immature with immature American Robin *T. migratorius* (see that species, p. 1023). Best answer to these pitfalls is full knowledge of *T. merula*'s plumage variation. Normal flight rapid and agile, with easy movement through dense cover, 'shooting' acceleration, and quite long glides over open spaces in breeding habitat; wing-beats occur in obvious wavering bursts. Migratory flight also distinctive, with characteristic bursts of fluttered wing-beats producing 'shooting' surges of speed along fairly level track or dramatic plummeting dive; lacks slow rise and fall of larger thrushes (e.g. Fieldfare *T. pilaris*) and always conveys hint of 'panic'. All flight actions involve much angling of body plane, and movement and spreading of tail which is frequently cocked on landing, even by newly arriving migrant. Flight silhouette distinctive, with elliptical outline of round head and long body and tail suddenly sprouting whirring 'blobs' of wings and 'fan' of tail; at all times, wings and rather long tail show noticeably rounded tips, unlike other thrushes, and wings held forward more. Stance highly variable but usually less upright than most thrushes, with tail often held horizontally or slightly up and frequently flicked and cocked, and head carried up but usually without much extension of neck; wings often drooped. Singing ♂ may adopt noticeably erect stance with tail lowered. Runs and hops quickly, with typical start-stop-start progress when feeding; movements alert and sprightly. Although often tame, spending long periods in the open around human habitation, essentially skulking in behaviour and given to sudden panic and noisy retreat to dense cover. Gregarious on migration but not in winter.

Song a rich, mellow, fairly low-pitched warble, with characteristically languid delivery and fluted quality; lacks 'shouted' phrases of other thrushes (e.g. Song Thrush *T. philomelos*) and often tails off into creaky, chuckling notes. Calls loud and varied: most distinctive are low 'pook' and 'chook' sounds often given in warning and half-alarm situations (and accompanied by simultaneous tail-cocking and wing-flicking), and hysterical chatter and screaming rattle in full alarm; calls of ♂ usually more insistent and querulous than those of ♀. Groups join in distinctive 'chink'-ing chorus at dawn and dusk.

Habitat. Exceptionally diverse, including dense woodland, varied types of farmland, heaths, moors, some wetlands, and settled sites including inner cities. Found in middle and overlapping to lower middle and upper latitudes of west Palearctic, including oceanic islands and coasts as well as milder boreal and temperate continental regions (Voous 1960). Given shelter, will tolerate wet, windy, and cool situations better than very warm and dry ones; prefers moisture and shade, with ample access to bare ground, layers of dead leaves or short grass and herbage, even where overshadowed by low bushes and shrubs or tree canopy; avoids distances from cover exceeding c. 100-200 m. In woodland, at home in all layers, but least frequently in canopy, and forages mainly on ground except when berry crops ripe. Favours richer soils and hardly discriminates between tree species, whether broad-leaved or coniferous, provided ground suitably moist with sufficient clear patches and undergrowth. Rare, however, in Scottish pinewoods and in some plantations of alien conifers, and not suited by certain stages of woodland management. Although outnumbering Song Thrush *T. philomelos* in various types of common British woodlands, preponderance is now much greater still on farmland and in gardens. Nevertheless, in both summer and winter, a characteristic woodland species (Yapp 1962). Since start of 19th century, however, has come to rate equally high over much of western Europe as a garden and parkland species, inhabiting even the largest inner cities, wherever trees and bushes comprise even a modest part of land cover. Outside human settlements, modified habitats such as farms, roadsides, railway embankments, and orchards offer hardly less attractive habitats. In Britain, enough farm hedgerows survive to maintain the species on practically every lowland farm. While extensive moorlands, mountains, and crop fields may be avoided, the species nevertheless maintains a presence on such marginal areas as sand-dunes, patches of moorland with some cover, sea-cliffs with suitable vegetation, and various types of wetlands (Sharrock 1976), although apparently indifferent to neighbourhood of standing or running water (except in drier parts of USSR) and to altitude, occurring at least up to 1200-1500 m in Switzerland and Czechoslovakia, and to 1800 m in Armeniya (USSR). In Canary Islands, breeds from gardens at sea-level through orchards, ravines, and high laurel, chestnut, or pine forest, or tree-heath *Erica*, to altitudes of c. 1250 m where only a few bushes and ferns grow; there lives like a rock thrush *Monticola*, perching on stones or huge boulders (Bannerman 1963).

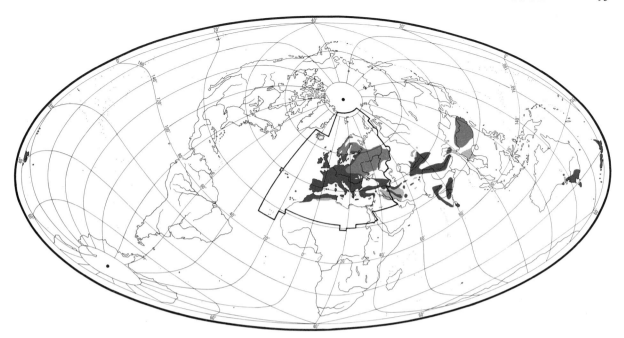

Distribution. Has colonized Faeroes and is spreading in Sweden, Finland, Czechoslovakia, Poland, Baltic republics, Israel, and (to lesser extent) Scotland.

ICELAND. Has bred (AP). FAEROES. First bred 1947 (Williamson 1970). BRITAIN. Spread to Scotland, notably Shetland, in 20th century, and into city-centres (Parslow 1973). IRELAND. Has spread to islands and coastal regions in west (Parslow 1973; Sharrock 1976). SWEDEN. Much of spread north of 60½°N in last 50 years (LR). FINLAND. Marked spread north since 1913 (Merikallio 1958; see also Spencer 1975*b* and, for map, Haartman 1973). Bred Korppoo 1981 (H Mikkola). CZECHOSLOVAKIA. Spread in east in Slovakia (Turček 1960). POLAND. Spreading east; increased tendency to breed in towns (Turček 1960; Tomiałojć 1976). USSR. Baltic republics: spread, with increased tendency to breed in towns (GGB). Leningrad region: spread, with increased tendency to breed in towns (Nankinov 1981). ISRAEL. Spread greatly in last 20 years; now common in towns and settlements throughout (HM, UP). CANARY ISLANDS. Spread to dry south on all islands (KWE).

Accidental. Spitsbergen, Bear Island, Jan Mayen, Kuwait.

Population. Increasing Faeroes, Netherlands, Finland, USSR, and Italy. Increased tendency to breed in towns, e.g. London (Cramp and Tomlins 1966), Warsaw (Luniak 1970), Rumania, and Baltic towns of USSR.

FAEROES. 50–70 pairs, increasing (Bloch and Sørensen 1984). BRITAIN, IRELAND. Possibly over 7 million pairs; increasing (Sharrock 1976). FRANCE. Over 1 million pairs (Yeatman 1976). BELGIUM. About 500 000 pairs (Lippens and Wille 1972). NETHERLANDS. 575 000–850 000 pairs

(Teixeira 1979). Has increased (CSR). WEST GERMANY. 8–18 million pairs (Rheinwald 1982). DENMARK. Relatively stable 1978–82 (Vickholm and Väisänen 1984). SWEDEN. 1·5 million pairs (Ulfstrand and Högstedt 1976). FINLAND. Great increase (Merikallio 1958); increased 1978–83 (Vickholm and Väisänen 1984). CZECHOSLOVAKIA. No changes reported. Tendency to breed in towns spreading from west to east (KH). SWITZERLAND. Increased (Géroudet 1983). GREECE. Declined due to hunting (GIH). RUMANIA. Becoming urbanized (VC). USSR. Baltic republics: increased (GGB). Leningrad region: considerable increase since 1980s (Mal'chevski and Pukinski 1983).

Survival. Britain: annual mortality 58% in 1st year (from 1 August), 38% in 2nd, 50% in 3rd, 40% in 4th and 5th (Lack 1943); annual mortality 54% in 1st year of life (from 1 August), 40% in 2nd (Lack 1946*b*); annual mortality after end of 1st calendar year 44 ± 1·5%, irrespective of age, with no significant sex difference; possibly lower in north; varied annually from 34% in 1933–4 to 69% in 1928–9 (Coulson 1961); at Oxford, annual juvenile mortality 59% (Snow 1958*b*); annual mortality decreased from *c.* 50% in 1951–2 to *c.* 32% in 1960–1, apparently not due to weather (Snow 1966*b*); annual adult mortality in London 41·8 ± 1·0%, in rural southern England 34·9 ± 0·5% (Batten 1973); mortality highest March–June; traffic and predation by cats have played progressively bigger role in mortality (Batten 1978); annual adult mortality 41% in ♂♂, 60% in ♀♀ (Naylor 1978). Belgium: annual adult mortality 69% (Verheyen 1958); annual adult mortality 52·2 ± 2·3%, or 45·8 ± 2·5% excluding those shot or otherwise killed by man; annual juvenile mortality 12·4 ± 1·9% (Van

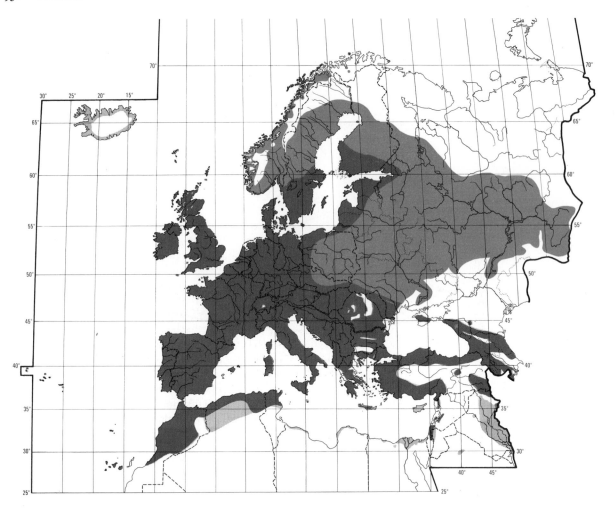

Steenbergen 1971). France: annual urban mortality 25%
in ♂♂, 33% in ♀♀ (Ribaut 1964). West Germany: annual
mortality 49%, for urban birds 28% (Erz 1964). Finland:
annual overall mortality 58-59% (Haukioja 1969).
Czechoslovakia: mortality in 1st year of life 68·4%, in
2nd year 56·3%, in 3rd year 30% (Beklová 1972);
mortality in 1st year 72% (Havlín 1961). Oldest ringed
bird 20 years 3 months (Rydzewski 1978).

Movements. Resident and migratory, with northern
populations moving south or west to winter in southern
or western Europe chiefly within boundaries of breeding
range.
　A proportion of birds from Scandinavia, Britain,
Ireland, Denmark, and Germany winter in western
Europe whilst those from further south winter in southern
Europe. Some birds from central European breeding
populations also move south and south-west in winter.
In Belgium, parts of central Europe, and in Britain and
Ireland, almost ¾ of breeding population may be resident,
but in Netherlands proportion much less. Some nominate
merula from southern Sweden and Finland move west or

even north-west into Norway whilst others including
Norwegian birds move south-west to Britain, Ireland,
Denmark, Netherlands, Belgium, and north-west France.
(Ashmole 1962.) Small numbers wintering in Iceland also
thought to be of Scandinavian origin (Gudmundsson
1951). Those departing from Low Countries and West
Germany move west or north-west into Britain, Ireland,
western France, and northern Spain (Ashmole 1962).
Birds from southern Scandinavia cross either directly to
Britain (Norwegian birds predominating in Scotland) or
may arrive indirectly having first moved south to Denmark,
northern Germany, or Helgoland, as recoveries of birds
from Sweden and Finland are more southerly in England
(Ashmole 1962). Considerable numbers of British breeders
from Scotland and northern England move to Ireland
whilst some from southern England go to north-west
France (Ashmole 1962; Snow 1978). Birds from Germany,
Poland, Czechoslovakia, Switzerland, and Hungary winter
mainly to south or south-west in central and western
France, Spain, Portugal, Balearics, Corsica, northern
Italy, and Malta (Rendahl 1960b; Ashmole 1962; Sultana
and Gauci 1982). Regular winter influxes also into

Gibraltar, Morocco, Algeria, and Tunisia, birds keeping predominantly to Mediterranean coast (Bannerman 1954; Smith 1965a; Cortés *et al.* 1980). Passage migrants in spring and autumn in southern Portugal and Gibraltar are evidence in some winters of more distant movements than so far detected by ringing recoveries (Cary 1973; Cortés *et al.* 1980). Those breeding in western USSR also move south or south-west but most reportedly winter within the country (Dementiev and Gladkov 1954b), though recoveries in Cyprus of 3 birds from Ryazan and Sea of Azov (Flint and Stewart 1983) and regular winter increase in population there and in Turkey (*Orn. Soc. Turkey Bird Rep. 1970–5*) suggest some emigration.

Some *aterrimus* from south-east Europe, Caucasus, and Turkey winter in Egypt (see below) and (according to Vaurie 1959) in Cyprus, but no other indication of movement. Movements of north-west African *mauritanicus* little known; many clearly resident (Smith 1965a), but records from southern Morocco (Valverde 1957) and southern Tunisia (Thomsen and Jacobsen 1979) indicate some southward wandering. Regular (occasionally common) in Libya, October–March, mostly in western half but one record from desert at Serir (Bundy 1976); *mauritanicus* perhaps involved but nominate *merula* recorded all across North Africa including Egypt where it has occurred late October to early March (Goodman and Ames 1984). *T. m. aterrimus* present in Egypt October to mid-April, and *syriacus* (breeding Levant and south-east Turkey east to Iran) occurs there November–March, mostly along Mediterranean (Goodman and Watson 1984); *syriacus* also recorded in Sinai (Etchécopar and Hüe 1967), and 2 records (neither subspecifically identified) February–March in western desert (Goodman and Ames 1983). One record from Sudan at Port Sudan on Red Sea coast (Goodman and Watson 1984). In Jordan, winter passage migrant in Azraq area (Clarke 1980; Conder 1981; Wallace 1982). Rare and irregular winter visitor to Gulf states, mid-November–March (Bundy and Warr 1980). Populations from further east are resident or altitudinal or medium-distance migrants.

Chiefly a nocturnal migrant with many killed on North Sea crossings at lighthouses. Autumn arrivals of up to 2000 birds occur on east coast of England, but usually only small flocks seen arriving at dawn; inland movements not often observed but some overland passage takes place along lines of hills (Simms 1978). In contrast to other *Turdus*, little affected by severe winters and hard-weather movements are unusual (Snow 1966a), although some movement from birds on high ground to valleys takes place (Bannerman 1954). In northern Europe, more ♀♀ than ♂♂ migrate since ♂♂ are more aggressive and territorial (Simms 1978). In Britain and Ireland, has marked tendency to return to breed in locality where reared: this applies to 72% of summer recoveries (Werth 1947). Also evidence from ringing that some adult birds migrate in one winter but not in another (Ashmole 1962).

Autumn movements begin late September with main passage in October and early November, but juveniles disperse from breeding areas during July and early August; directions usually random, but a proportion of 1st-year birds in Britain move north-west (Ashmole 1962; Snow 1966a). Autumn departures from Bashkiria (south-west Urals, USSR) begin in second half of September, and last birds said to leave Tula region (south of Moscow) in mid-October, though passage in nearby Ryazan region continues into early November; further south, in Kharkov region, movements occur into mid-November and around Zhdanov on Sea of Azov to end of November. Occurs as passage migrant around Belovezh'e (western Belorussiya) August–October, but Semirech'e area (south-east Kazakhstan) not deserted until December. (Shnitnikov 1949; Dementiev and Gladkov 1954b.) Some movement, especially across North Sea, throughout winter with regular influxes into Helgoland (northern West Germany), Fair Isle (Shetland), east coast of England, and Irish coast. Considerable increase in number of Swedish recoveries in Britain and Ireland in January–February. Recoveries of birds on autumn passage have shown some instances of reversed migration: thus, birds ringed on Helgoland have been recovered within 24 days in Denmark, Norway, and Fair Isle, while others ringed on Fair Isle recovered within 10 days back in Norway, and another at Spurn Head (eastern England) back in north-west Germany within 4 days. (Ashmole 1962.)

Return movement to north or north-east begins late February, with main passage in March and early April and some birds still on passage in early May by which time residents in central and southern Europe (including southern Britain) are breeding. Arrives in central Sweden late April. (Ashmole 1962.) In European USSR, arrives from early March in south to early May at north of range. In Tadzhikistan (south-central USSR), returns to higher altitudes from mid-February onwards. (Dementiev and Gladkov 1954b.) Overshooting birds regularly found north of breeding range in Finland (Merikallio 1958). Vagrants recorded from Greenland (Bent 1949) and, in November, at Montreal, Canada (McNeil and Cyr 1971).

PC

Food. Mainly insects and earthworms; also fruit from late summer to early winter. Feeds largely on ground throughout year, though also in trees and bushes (e.g. Beven 1959). Foraging behaviour (also of Song Thrush *T. philomelos*) on grass meadow described as follows: typically makes series of straight-line movements (hopping or running) separated by short pauses after which bird may change direction; *T. philomelos* and ♀ *T. merula* both made longer movements between pauses than did ♂ *T. merula* (respectively, 44·9 ± SE2·7, 42·1 ± SE2·6, 33·8 ± SE1·7 cm), and ♀ *T. merula* moved across meadow at higher overall rate than *T. philomelos* or ♂ *T. merula* (respectively, 12·9 ± SE0·8, 9·1 ± SE0·6, 8·6 ± SE0·7 cm per s). Only *c.* 10–17% of foraging time on meadows spent actually moving, remainder probably spent scanning for or handling prey. Following capture of earthworm, apparently searches immediate area more thoroughly than if no prey found, this presumably being adaptation to worms' clumped distribution. (Smith 1974.) For searching strategies see Greenwood and Harvey (1978), and for development of search image see Lawrence (1985b). Often searches for food among leaf litter, etc., flicking loose material aside with bill, or seizing it momentarily to throw it aside, and at same time may bring foot forward to level of head and scratch backward; can dig thus through 5–7 cm of snow (Snow 1958b). Once seen holding twig *c.* 8 cm long in bill and using it to sweep clear 900 cm² of snow 4–5 cm thick in order to feed on ground below (Priddey 1977). Pulls earthworms out of ground, apparently usually locating them by seeing tip protruding from burrow, but can also detect and locate by ear invertebrates moving under soil (e.g. will suddenly dig to extract larva from well below surface). In December–January, recorded catching earthworms at rate of 1–2 per min. Often pulls out more than it can eat, and leaves some on surface. Eats small snails (Gastropoda) and will rob larger ones from Song Thrush *T. philomelos* when it has extracted them from shells, but apparently unable to open them itself. Sometimes washes food before eating. Swallows worms whole; for young, usually chops them into pieces or mangles them, often dragging them over soil first (perhaps to remove slime or to add grit to diet of young). (Snow 1958b.) Recorded snatching berry from tree as bird flew past it (Wilkinson 1985), also impaling moth on barbed wire in order to remove wings and feed body to young (Wilson 1972), and feeding on faeces (or items within them) of fox *Vulpes* and dog (Barnes 1985; Radford 1985b). Will feed on ground being disturbed by mole *Talpa* (Taapken 1977; Sharrock 1982). Often steals food from birds of own and other species, rushing up and driving other bird away; if it can hear but not see *T. philomelos* hammering snail, it will approach to steal it as soon as hammering sound stops (i.e. when *T. philomelos* has opened it). The only smaller species not normally robbed is Starling *Sturnus vulgaris*, which is dominant (except in flight) over *T. merula* in such disputes (though juvenile *T. merula* often attempt to rob them). (Snow 1958b; Naylor 1974.) Average length of feeding visit to *Prunus mahaleb* trees in south-east Spain, July, was 79 ± SD38 s (*n* = 42), taking 5·8 ± SD4·4 fruit per visit (*n* = 24), and ingesting 5·5 ± SD4·0 fruits per min (*n* = 44) (Herrera and Jordano 1981).

Diet in west Palearctic includes the following. Invertebrates: springtails (Collembola), mayflies (Ephemeroptera), damsel flies (Odonata: Lestidae), grasshoppers, etc. (Orthoptera: Gryllotalpidae, Tettigoniidae, Acrididae), earwigs (Dermaptera), cockroaches (Dictyoptera), bugs (Hemiptera: Scutelleridae, Pentatomidae, Nabiidae, Miridae, Cicadidae, Coccoidea), lacewings, etc. (Neuroptera: Hemerobiidae, Chrysopidae, Rhaphidiidae), scorpion flies (Mecoptera), adult and larval Lepidoptera (Nymphalidae, Lycaenidae, Hepialidae, Tortricidae, Coleophoridae, Tineidae, Notodontidae, Arctiidae, Noctuidae, Geometridae), adult and larval flies (Diptera: Tipulidae, Bibionidae, Stratiomyidae, Tabanidae, Asilidae, Empididae, Syrphidae, Tachinidae, Calliphoridae), adult, larval, and pupal Hymenoptera (sawflies Tenthredinidae, Ichneumonoidea, Cynipoidea, ants Formicidae, wasps Vespidae, bees Apoidea), adult and larval beetles (Coleoptera: Carabidae, Hydrophilidae, Silphidae, Staphylinidae, Scarabaeidae, Byrrhidae, Elateridae, Cantharidae, Tenebrionidae, Coccinellidae, Cerambycidae, Chrysomelidae, Curculionidae), spiders (Araneae), harvestmen (Opiliones), woodlice (Isopoda), millipedes (Diplopoda), centipedes (Chilopoda), pseudoscorpions (Pseudoscorpioidea), small snails (Gastropoda), molluscs including slugs (Pulmonata), earthworms (Oligochaeta). Also kitchen scraps, small fish, newts *Triturus*, frogs and tadpoles (Anura), lizards *Lacerta*, snake *Natrix*, nestling birds, dead adult of own species, and (apparently) mouse (Muridae). Plant material mainly fruits, also seeds: juniper *Juniperus*, yew *Taxus*, mistletoe *Viscum*, barberry *Berberis*, bramble *Rubus*, strawberry *Fragaria*, rose *Rosa*, rowan *Sorbus*, cherries *Prunus*, apple *Malus*, pear *Pyrus*, hawthorn *Crataegus*, currant *Ribes*, bilberry *Vaccinium*, strawberry tree *Arbutus*, honeysuckle *Lonicera*, nightshades *Solanum*, privet *Ligustrum*, elder *Sambucus*, buckthorn *Rhamnus*, alder buckthorn *Frangula*, olive *Olea*, *Pistachia*, snowberry *Symphoricarpos*, vine *Vitis*, ivy *Hedera*, dogwood *Cornus*, *Viburnum*, *Smilax*, maize *Zea*. (Csiki 1908; Raines 1955; May 1962; Tutman 1962; Pickess 1965; Jenks 1970; Vauk and Wittig 1971; Meineck 1973; Havlín 1977; King 1979; Pavan 1980; Taylor 1980a; Herrera 1984a; Coleman 1985; Stephan 1985; Török 1985.)

Diet in suburban Oxford (England) has 5 main components: earthworms, eaten all year, mostly late summer and autumn; other concealed items, mostly immobile or inactive invertebrates from leaf litter, taken all year; fruits, taken July–December (especially October–November), when they form chief food, at least in volume; caterpillars and adult insects from above ground, taken in late spring and summer; food from man, especially in hard winter weather and in summer droughts (Snow 1958b; see also Collinge 1924–7). Havlín (1977) found diet similar; see that study also for comparison with diet of *S. vulgaris*. On Helgoland (West Germany), diet of 200 spring migrants comprised 60% invertebrates (mostly snails, including marine species, and beetles) and 40% seeds provided by man (Vauk and Wittig 1971). In experimental study with captive birds, found not to prefer plant food at any season; birds fed all-fruit diet lost weight and died, but addition of 2–3 g of beetle larvae per day was sufficient to maintain weight (Berthold 1976). Most seeds eaten are not digested (Havlín 1961, which see for experimental study). On Scilly (England), often takes nectar from puja flowers (Hunt 1985). See also Florence (1914), Tutman (1952, 1962), Ruppert and Langer (1955), Havlín (1961), Pek and Fedyanina (1961), Ptushenko and Inozemtsev (1968), Dyrcz (1969), Kiss *et al.* (1978), Meier and Furrer (1979), Jordano (1982), Kostin (1983), and Herrera (1984*a*); for review, see Stephan (1985).

Young in deciduous woodland, Oxford (England), given chiefly caterpillars, but in suburban gardens mostly earthworms with adult insects increasing in importance through breeding season (Snow 1958*b*). In Hungary, young given mainly earthworms, adult beetles, Lepidoptera larvae, and Diptera (Török 1981, 1985). See also Collinge (1924-7) and Pfeifer and Keil (1959).   DJB

Social pattern and behaviour. Well studied, especially in north and west of range. Major studies in Scandinavia (Lind 1955, and by J Karlsson), Germany (Steinbacher 1953; Heyder 1953; Messmer and Messmer 1956), Britain (Venables and Venables 1952; Snow 1958*b*; Edwards 1983), and of introduced population in New Zealand (Gurr 1954). For recent review, see Stephan (1985).

1. Tendency to gregarious behaviour outside breeding season strongly dependent on migratory/resident status of individuals concerned. In areas providing winter food supply (e.g. gardens) territory-holders generally remain in territories all year, ♂ and ♀ outside breeding season tending to occupy different areas within former breeding territory. Non-migratory 1st-years in autumn and early winter may occupy small, temporary territories within territory of established pair, usually related to temporarily abundant food supply. Migratory birds moderately gregarious outside breeding season, typically migrating in loose flocks and feeding in groups of up to 10-20. Resident individuals may leave territories temporarily to feed gregariously on concentrated food sources such as fruit, and in very cold weather territories may break down completely. Both migratory and resident individuals often roost gregariously, especially in winter in colder parts of range. BONDS. Mating system monogamous, though a few exceptional cases of bigamy recorded (Snow 1958*b*). In areas with resident populations, established pairs usually remain together in successive breeding seasons, as long as both partners survive. Most new pairs formed in late winter and early spring, but also at any time throughout breeding season if one member of an existing pair dies. Young birds may form temporary pairs in 1st autumn and early winter, but these break up and breeding pairs involving young birds usually formed in early spring (Snow 1958*b*). Both ♂ and ♀ breed first at 1 year old, usually beginning somewhat later in season than older birds (Snow 1958*a*). ♀ alone broods young. Both ♂ and ♀ feed young in nest and for *c.* 3 weeks after fledging. Brood division habitual, starting soon after young leave nest (see Relations within Family Group, below). BREEDING DISPERSION. Solitary and territorial. In densest (usually suburban) populations, up to *c.* 7 pairs per ha recorded and nests may be as little as 10 m apart, but usually not less than 20-30 m (Campbell 1953; Snow 1958*b*). Suburban densities over wider areas *c.* 200-300 pairs per km[2] (Batten 1973; Karlsson and Källander 1977). Woodland populations *c.* 100 pairs per km[2] (Snow 1958*b*; Batten 1973). Very sparse breeding populations in open country as low as 1 pair per km[2] (Shetland: Venables and Venables 1952). For extensive data on densities in East Germany, see

Stephan (1985). Territory essential for pair-formation and breeding, but usually provides only part of food for young, especially for later broods when earthworms less easily accessible and much foraging done on (e.g.) damper areas outside territory or at neighbouring fruit sources. ROOSTING. Nocturnal. In thick cover, usually bushes or small trees, often against walls, either in or outside territory; rarely (records from Germany) in reeds (Berndt and Tautenhahn 1951). Roosting in territory usually solitary, or ♂ and ♀ close together. In areas where thick cover restricted (e.g. conifer clumps or thickets in otherwise more open areas), especially in winter half of year, communal roosting regular and may draw birds from wide area; may involve a few birds or up to several hundred; one roost in south-east England held up to perhaps 2000 birds, along with smaller numbers of other *Turdus* (Simms 1978). For roosting times, as measured by pre- and post-roosting calls (5a in Voice), in relation to dawn and dusk throughout year, see Haarhaus (1973) and Stephan (1985). From autumn to spring, gatherings of up to *c.* 20 birds near thick cover occur in early morning and late afternoon, with ritualized aggressive behaviour (Morley 1937; Lack 1941; Stephan 1985; see Antagonistic Behaviour, below); apparently related to roosting, such assemblies probably consisting of local territory-holders and intruding birds from surrounding areas. Comfort behaviour (preening, bathing, sunning) typical of passerines; when sunning, becomes unwary, allowing unusually close approach; for details and photographs, see Teager (1967). Anting behaviour typical of passerines; for descriptions, see Williams (1947*b*), Logan Home (1954), and Tenison (1954).

2. Generally shy in woodland and other habitats unfrequented by man, but often confiding in towns and suburbs. Reacts to mild disturbance with 'chook' calls (see 4a in Voice), to sudden disturbance with Alarm-rattle (see 4b in Voice), to ground predators with 'pook' calls (see 3a in Voice), and to flying predators with warning 'seee' calls (see 3b in Voice), but considerable individual variation in such behaviour. Mobbing behaviour, especially that directed at owls (Strigiformes), is accompanied by persistent Chink-calls, usually initiated by ♀ and drawing birds from neighbouring territories (Lack and Light 1941). FLOCK BEHAVIOUR. For communal display, see Roosting (above). For flock Contact-call, see 2 in Voice. SONG-DISPLAY. Song (see 1 in Voice) delivered from perch which, except very early in season, is usually high (J Hall-Craggs); occasionally sings in flight between perches. Exclusively by ♂ (rarely, apparently abnormal ♀♀: Snow 1958*b*), during establishment and maintenance of territory and as advertisement for ♀ by unpaired ♂, especially after loss of mate. Song-period several months, from late winter to end of breeding season; time of onset much dependent on weather, stimulated by mild and damp conditions. In northern Europe, may start late December if weather mild, rarely even earlier; more usually from early or mid-February. Regularly begins earlier in towns than in country (Lack 1944; Bedford 1945). Early in season, given mainly in late afternoon, gradually extending back into earlier part of day; dawn song begins a little later in season than afternoon song. During period of full song, dawn song typically begins *c.* 1 hr before sunrise, and evening song ends around sunset (J Karlsson). For diurnal variation in delivery and quality of song, see 1a in Voice. Output related to stage of individual's nesting cycle: at maximum when ♀ incubating, decreasing at hatching and remaining at low level during feeding of young (Snow 1958*b*). During nestling period, song may stop completely for up to 1 week (J Hall-Craggs). ANTAGONISTIC BEHAVIOUR. Territorial individuals of both sexes generally aggressive towards conspecific birds: in breeding season mainly

towards birds of same sex, outside breeding season more or less equally to birds of either sex. Intruders usually driven out without marked aggressive display from territory-owner, retreating in characteristic Sleeked-upright posture (Fig A)

A

with head held high, crown feathers raised to give peaked profile, and body plumage sleeked; this submissive posture often associated with flight-intention Contact-call. Antagonistic behaviour between more evenly matched birds (usually of same sex) generally occurs at or near territory boundaries, especially when new territories being established, and involves special postures and behaviour. Aggressive displays include: stretching of neck upwards, with bill pointing a little above horizontal and slightly opened (Fig B, right); ruffling of neck and body

B

feathers; lowering of wings, and depressing and fanning of tail (Fig B, left). Aggressive 'seee' call (see 5b in Voice) may be uttered. Displaying bird may make short runs to and fro beside opponent, head lowered at beginning of each run (Fig C, right) and raised at end (Fig C, left). ♂ may pick up and carry a leaf about during aggressive display, sometimes brandishing it 'like a talisman' (Howard 1952; also Shanks 1953); this behaviour apparently an individual trait, of erratic occurrence, as not reported in some intensively studied populations. Settled territory-holders meeting at territory borders often run c. 1 m back and forth, beside one another, without other display or in only mildly aggressive posture. Fighting not uncommon, usually in early spring when competition for territories most intense. Birds face each other, then clash with bill and claws, often rising fluttering 1 m in the air before falling back to ground where they may separate and squat facing each other

with tail depressed and fanned and bill open; or may remain interlocked, pecking and clawing. During attacks, attacked bird typically has tail more widely fanned than attacker (Snow 1958b). Such fights usually soon broken off, but may continue many minutes and occasionally lead to death. Aggression may continue after death, victor persistently pecking and mutilating the dead victim (Hall-Craggs 1977; Erlemann 1983). Submissive postures in antagonistic encounters include sleeking of body plumage with raising of crown feathers (see above), also Hunched-posture (Fig D) with legs bent and head retracted,

D

E

and distinct Tail-up posture (Fig E) with legs stretched and whole rear half of body and tail pointing almost vertically up; last 2 postures characteristic of birds not very alarmed and not intending to take flight, and sometimes accompanied by Chink-call. Pursuit-flights involving up to 6 birds occur occasionally during breeding season (probably only when food shortage compels birds to feed close together): fly fairly close, with much calling (see 5c in Voice), but then separate (J Hall-Craggs). HETEROSEXUAL BEHAVIOUR. (1) General. Heterosexual displays often not well-marked in settled pairs, especially those persisting from previous year, except for pre-copulatory display. ♂ of settled pair usually dominant over ♀ outside breeding season, ♀ dominant during breeding season. (2) Pair-bonding behaviour. Advertising-display of ♂, associated with pair-formation (but very variable in intensity and frequency of performance), involves the following postures and movements. ♂ stretches head forward with crown feathers partially erected and bill open; ruffles body feathers, especially

C

F

rump, forming conspicuous hump; fans and lowers tail (Fig F, left). If on ground, ♂ moves round ♀ or runs back and forth in front of her, lowering head at beginning of each run; may jump up and twirl round between runs. If in tree, ♂ remains stationary or occasionally shifts perch; bowing more prominent and may develop into rhythmic up and down movement of whole forepart of body, sometimes with marked side-to-side swaying component, bill touching or wiping perch at bottom of bow. Display usually accompanied by low-volume Courtship-song (see 1e in Voice). Sometimes adopts Advertising-posture in flight (with Courtship-song) flying with slow wing-beats from perch to perch near ♀. Unpaired ♂♂ occasionally perform courtship displays, not fully developed, in absence of ♀. Mutual billing ceremony between ♂ and ♀ reported (Macdonald 1984) but evidently rare: each bird touches other's bill for up to 2 s with partly opened mandibles. (3) Courtship-feeding. Reported by Sauerbrei (1926) but not recorded in any detailed study and needs confirmation; certainly not normal. Report by Manning (1946) almost certainly erroneous (A W G Manning). For discussion, see Tucker (1946). (4) Mating. Initiated by ♀'s adoption of Soliciting-posture (Fig F, right): points half opened bill almost vertically up, holds tail up at about same angle, sleeks body plumage, and stretches legs; may give Courtship-song (see 1f in Voice) (Steinbacher 1953). ♂ usually adopts intense Advertising-posture before mounting. Copulation rarely seen considering the bird's abundance in suburban areas; brief, and mainly takes place in early morning, 1–5 days before laying of 1st egg. (5) Behaviour at nest. Both sexes prospect for nest-sites, often in company, ♀ usually the more assiduously. ♂ 'shows' possible sites to ♀, who often inspects a site immediately after he has visited it; final choice apparently normally made by ♀ (Snow 1958b; Berndt 1962). In Sweden, ♂ often holds green leaf in bill while prospecting (J Karlsson); this not seen in Britain. Nest-building: by ♀ alone, but ♂ often accompanies building ♀ and may extend territory if chosen nest-site is near or beyond territory boundary. RELATIONS WITHIN FAMILY GROUP. ♀ cleans nestlings soon after hatching; also helps weak nestlings to get free from shell. During first few hours ♀ feeds young on saliva; ♂ chased away from nest (Messmer and Messmer 1956). Later, both parents feed young. Nestlings brooded by ♀ alone—almost continuously for first 3–4 days (except when ♀ foraging), then decreasingly until by 8–9 days only at night or in bad weather. Nestlings beg upwards for 1st week, in response to tactile stimuli (usually shaking of nest by arriving parent), then after eyes open (at *c.* 8 days) orientate open bills towards parent's head (Tinbergen and Kuenen 1939). Food-bearing ♀ gives Feeding-call (see 8a in Voice) to encourage young to gape. From 14th day, gaping accompanied by simultaneous wing-movements. (Messmer and Messmer 1956.) Nest-sanitation by both parents; faecal pellets swallowed for first few days, then carried away and dropped well away from nest. Dead young removed, if not too large to carry, and dropped near territory border. Both parents may shade young in exposed nests in very hot weather, and once reliably reported

soaking breast and belly feathers in water before going to nest, young then drinking from feathers (Handmann 1973). Young may leave nest at 9 days old if disturbed, though normally at *c.* 13 days; departure of undisturbed young usually spontaneous. Fledglings cared for by both parents, who divide brood soon after they have left nest (Edwards 1985). When ♀ starts next clutch, ♂ usually takes over care of whole family, but ♀ may occasionally care for one fledgling while building new nest and incubating new clutch (Snow 1958b). After last brood of season, ♂ and ♀ divide family and both continue to feed young until independence. Young normally beg only from parent that is feeding them; remain in thick cover for *c.* 7 days after leaving nest, then begin to move out into open, often hopping to meet parents approaching with food; after *c.* 2 weeks, range widely over parents' territory. Fed by parent(s) for *c.* 21 days, then usually leave territory (not driven away but apparently in search of suitable feeding areas) (Snow 1958b). ANTI-PREDATOR RESPONSES OF YOUNG. Young in nest show alarm from 7 days, crouching silently if disturbed; remain still and silent in response to parents' warning-calls (see 3 in Voice). At *c.* 9 days, scream when handled by human (and presumably if attacked by predator) and sudden disturbance may cause them to 'explode' prematurely from nest. Main method of defence of recently fledged young is to remain in thick cover, and become still and silent in response to parents' warning-calls. Gaping at attacker with bill wide open effective against adult conspecifics (Thielcke 1964) and perhaps against some potential predators. PARENTAL ANTI-PREDATOR STRATEGIES. (1) Passive measures. Reactions to humans as potential predators well developed especially in woodland populations; suburban birds usually too used to presence of humans to treat them as predators. ♀ typically slips silently off nest when intruder many metres away, sits tightly only when eggs near hatching or nestlings small. Warning 'seee' call uttered by parent away from nest, apparently directed at mate; 'pook' calls uttered by parents with nestlings, causing them to keep still and silent in nest. ♂ sometimes warns incubating ♀ of danger by giving low-volume song near nest, changing to normal alarm-calls if danger increases (J Karlsson). (2) Active measures. Seen especially in suburban birds with little fear of man. ♀ often defends nest aggressively: if incubating or brooding, remains on nest and pecks at intruder's hand; if not on nest, may attack in flight, usually from behind, striking intruder's head with foot and giving loud Chink-calls (Snow 1958b; Böhr 1960; Tebbutt 1968; Beven 1980). Similar attacks also made on Jay *Garrulus glandarius* approaching nest, by both ♂ and ♀; birds may grapple and fall to ground interlocked (Goodwin 1953a). Distraction-display less common. ♀ with young in nest may perform mobile distraction-lure display: lands on ground and runs to and fro in crouched posture with plumage ruffled; recorded distracting cat from fledged young by performing display of disablement type, fluttering just above ground, as if injured (Beven 1946). ♂ not reported attacking humans at nest, but may chase smaller mammals from vicinity (Christie 1985).

(Figs by N McCanch: A and C–F from drawings in Snow 1958b; B from drawing in Simms 1978.) DWS

Voice. Used throughout the year. Full song restricted to breeding season; Subsong usual and loud song occasional in autumn; Subsong also occurs in winter when weather mild and food not scarce. Loud song from ♀♀ recorded in captive birds but occurs only abnormally in the wild (Snow 1958b). Aggressive Bill-snapping occurs (e.g. Bergmann and Helb 1982); probably more common from ♀♀ and sometimes directed also at birds of other species (J Hall-Craggs).

CALLS OF ADULTS. (1) Song. Well documented, with studies by Messmer and Messmer (1956), Snow (1958b, with detailed bibliography), Thielcke-Poltz and Thielcke (1960), Hall-Craggs (1962, 1969, 1976, 1984), Tretzel (1965, 1967), Todt (1967a, b, 1968, 1970, 1975a, b), Morgan et al. (1974), Thorpe and Hall-Craggs (1976). However, extreme complexity, due to individual inventiveness and pronounced (probably life-long) capacity for learning, give scope for many more detailed, long-term studies. (a) Territorial-song of ♂. Each adult ♂ has large repertoire of song-phrases exhibiting great individual variation—despite reciprocal learning between neighbouring territory-holders resulting in long-term local dialects. No known innate patterns of time or pitch patterning, yet song of T. merula easily recognizable through: (i) overall variety, with general lack of contiguous repetition at all levels (note or unit, figure, phrase) except during practising stages and sometimes in evening when same phrase may occur twice or even several times in succession; (ii) timbre—often called 'flute-like' which is admissible for initial and medial figures of phrases that comprise tones of relatively long duration (0·1–0·4 s) and fall within frequency range c. 1·5–4·0 kHz (Fig I, 0–1·6 s), but the many portamenti (gliding tones), high rates of frequency and amplitude modulation, and sounds involving the independent but simultaneous use of the two internal tympaniform membranes (Fig I, 1·7–2·2 s) are not characteristic of the classical flute. Final figures (codas) sometimes omitted but always included in brilliant dawn song: upper frequency limit extended to over 8 kHz and contain many noisy units, latter often copied from vocalizations of other species. Mimicry of many bird species and cat recorded (see Hall-Craggs 1984, also for sonagrams). Witchell (1896) named 11 species whose calls were copied by T. merula in a 'faint voice'. Mechanical sounds also copied (Howard 1956; Snow 1958b; Thielcke-Poltz and Thielcke 1960). Much mimicry is delivered sotto voce and passes unnoticed, but human whistles often sung loudly in initial or medial figures, or constitute entire phrase; Tretzel (1967) described such a phrase being transposed upwards by about a 5th and

then being passed from bird to bird, suffering gradual alteration with increasing distance from model. In 1st breeding season, young birds learn song figures from neighbours; in early stages, same figure may be sung many times in succession (16 consecutive repetitions heard by R E Jellis and J Hall-Craggs, with different succeeding figures attached). Whole phrases comprising selected combinations soon established but reorganization and further learning continue throughout season. Sentences (chains of phrases uttered in preferred order but lacking usual pauses between) are formed by some—probably experienced—birds. Series of phrases (as sentences but with pauses interpolated) also occur (Hall-Craggs 1962; Todt 1967a, b, 1968a, b, 1970, 1974, 1975a, b), but capacity for such integration varies much between individuals and may also depend on age of bird, stage of breeding season, and time of day. Song learning probably continues throughout life. One wild bird recorded through 5 successive years added to repertoire each year, the new material often acquired from songs of new neighbours including 1st-years; little original material discarded or forgotten so repertoire size increases throughout (J Hall-Craggs). Amount of time spent singing and mode of delivery is related to activity of ♀; thus during nest-building and nestling phase, song is sporadic, fragmented, and localized according to activity, but during incubation ♂ sings loud and long near nest (Messmer and Messmer 1956). Diurnal pattern imposed on this, as follows. (i) Pre-dawn song, from roost, is quiet and of rather slow delivery (Hillstead 1945) but is not Subsong. (ii) Dawn song is brilliant, prolonged, and has shortened pauses between phrases due partly to the complete and sometimes much extended codas. Except very early in season, uttered usually from high song-posts and declines gradually after 10–30 min (occasionally as much as 50 min). Amount much reduced, or bird even silent, when feeding young. (iii) Daytime song variable and dependent on stage of breeding cycle, weather, food supply, and such activities as preening and foraging, when pauses between phrases are necessarily long. (iv) In pre-roost singing, bird often 'beats the bounds' of territory, singing rather briefly at each song-post. Song is loud, phrases often curtailed and repeated in uncharacteristic manner. It seems to have an aggressive

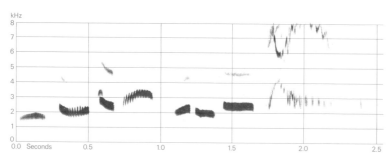

I C Weismann Denmark May 1970

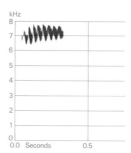

II J Hall-Craggs
Channel Islands May 1979

component lacking in the structured dawn and 'incubating' songs (J Hall-Craggs); this supports the view of Messmer and Messmer (1956) that the pre-roost song is purely territorial (their 1-year-old captive birds in auditory isolation never sang at twilight presumably because they had no audible territorial rivals). It follows that most loud, structured Territorial-song has an advertising function and is not purely aggressive (e.g. Snow 1958*b*). (b) Subsong of ♂. A quiet and continuous warbling, in which figures occur out of context but connected by improvisatory material comprising calls and fragments of songs of other species and formalized versions of own calls, especially gentle 'bubbling' versions of the Alarm-rattle (call 4b). Apparently not used in communication but serves as practice for full song without wasting energy. (c) Subdued-song. Intermediate between full autumn song and Subsong, thus louder than Subsong; rather hurried and jumbled, interspersed with soft Alarm-rattles (Snow 1958*b*). (d) Rambling-song of ♂. Uncommon and apparently restricted to warm, still afternoons during breeding season. Occurs when ♂ in full song gradually eliminates inter-phrase pauses and sings continuously but loudly, mixing normal and curtailed phrases with mimicry and other strange sounds. To listener familiar with the individual's repertoire it resembles a review of stored phrases and rarely-used copies of conspecific and other vocalizations. Usually reverts gradually to normal Territorial-song. (e) Courtship-song of ♂. A quiet 'strangled' song, made up of Alarm-rattles, rough warbles, and subdued snatches of what sounds like Territorial-song (Snow 1958*b*; Howard 1952). (f) Courtship-song of ♀. ♀ soliciting copulation utters strained, half-suppressed song (Messmer and Messmer 1956). ♀'s response to courting ♂, a high, rapid tinkling titter (Witherby *et al.* 1938*b*), may be the same. (2) Contact-call. A 'ssrie' (Messmer and Messmer 1956), 'sri' (Bergmann and Helb 1982), or 'seep' (Snow 1958*b*). Duration rarely more than 0·3 s. High-pitched (6–8 kHz), but sounds lower than calls 3b and 5b and has slightly rough or rolled 'r' quality due to pronounced frequency modulation at rate of 24–30 per s and ranging over 1·0–1·5 kHz (Fig II). Used to maintain contact during flight and, when perched, to stimulate flight. Also given by subordinate or trespassing birds about to take flight and consequently serves to inhibit

attack by dominant or territory-holding birds (Snow 1958*b*). (3) Warning-calls. (a) A 'pook' (Snow 1958*b*), 'djück' (Messmer and Messmer 1956; Bergmann and Helb 1982), or 'kop' (P J Sellar); see Fig III and compare Fig Va with which call is often confused. Used mainly to alert young to presence of ground predator; repeated until predator moves away. (b) A 'seee' (Snow 1958*b*) or 'ziep' (Thielcke 1970; Messmer and Messmer 1956); a thin, high descending tone (6–10 kHz), duration *c.* 0·5 s with very gradual onset and cessation (Fig IV), giving minimal indication of caller's position. Used to warn of aerial predator and repeated while danger remains; much like similar calls by other small passerines and recognized by them (Marler 1959); compare Fig V in Ring Ouzel *T. torquatus* (p. 946). Used in slightly modified and less repetitious form when nest approached by potential predator. See also call 5b. (4) Alarm-calls. (a) Low-pitched 'chook' (Snow 1958*b*), 'tuc' (P J Sellar), 'tuk' (J Hall-Craggs), or 'duck' (Messmer and Messmer 1956); see Fig Va. Often confused with call 3a (compare Fig III) from which it is distinct. Expresses mild alarm or protest; given by birds in unfamiliar or insecure situations. Uttered by ♀ if watched when nest-building or prospecting for nest-site (Snow 1958*b*). These calls often initiate Alarm-rattle when reiterated with some acceleration and ascending pitch, or may change to 'pink' or 'chink' (Fig Vb) if bird becomes aggressive or more excited. (b) Alarm-rattle. (i) Typical form ('Zetern'). Sudden shrill chatter of well-defined pattern but varying greatly in detail; usually prefaced and sometimes ended with 'tuk' calls (4a) producing rather elaborate form—'tuk tuk tuk tchook-a tchWEE tcheWEE cheWEE cheWEE-cheWEEcheWEEcheWEEcheWEE tchook chook' (Fig VI). Given when suddenly but mildly alarmed or surprised, also in excitement, e.g. during not very serious flight chase. Function in communication not clear. (ii) Neighing form ('Wiehern'). A 'gi-gi-gi-gi' (Fig VII), associated by Messmer and Messmer (1956) with playful chasing during courtship. One ♂ used Zetern commonly on taking flight (especially when leaving song-post) and in flight at almost any time of day; same bird used Wiehern almost always in evening during flights between song-posts and when chasing another ♂ (J Hall-Craggs). Formalized version of rattle used by some birds as regular coda to song-phrase

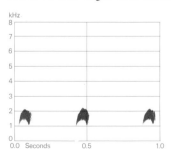

III J Hall-Craggs
Channel Islands May 1980

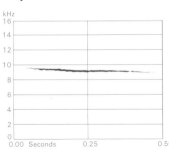

IV J Hall-Craggs
Channel Islands May 1980

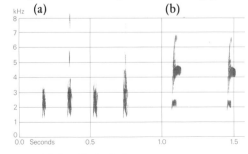

V P J Sellar England June 1971

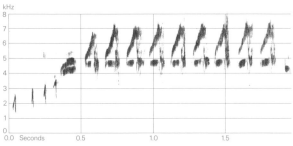

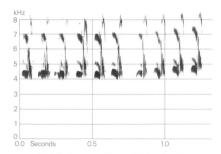

VI J Hall-Craggs Channel Islands May 1980

VII J Hall-Craggs Channel Islands May 1980

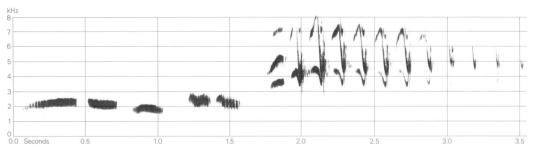

VIII J Hall-Craggs Channel Islands May 1980

(Fig VIII); this more nearly matches Zetern than Wiehern, but all forms exhibit reiterated units produced by 2 internal tympaniform membranes. (5) Aggressive-calls. (a) A repeated metallic 'chink', 'mik' (Snow 1958*b*), 'tix' (Messmer and Messmer 1956), 'pink', or 'tink'; analyses reveal a number of subtle variants of this call (Figs IX–X) which, along with rate of calling, may be linked to degree of aggression which is generally only moderate. Especially associated with mobbing (Fig XI) and going to and leaving roost. In territorial situations usually given by dominant bird but subordinate bird may call thus when trying to establish itself. Most intense at good winter roosts where many strange birds intrude (Snow 1958*b*). As spring approaches, dusk and dawn 'chinking' gradually replaced by song (Baan 1953). 'Tix' used in territorial quarrels and as assembly call (Messmer and Messmer 1956). (b) Aggressive 'seee'. Thin, piercing, high-pitched 'seee' (Snow 1958*b*); a drawn-out, anxious 'eeeee' of low volume (Nicholson 1951); see Fig XII. Most clearly aggressive of all calls, given by combative (usually dominant) birds in taking up and maintaining territory—between bouts of fighting or posturing, or by bird which has just acquired territory. Never heard from a submissive bird (Snow 1958*b*). Duration of recording examples *c*. 0·3–0·5 s; frequency range *c*. 6–8 kHz. In contrast to alarm 'seee' (call 3b), aggressive 'seee' ascends sharply at start and levels off to steady tone at *c*. 8 kHz; thus sounds higher pitched than former which often falls well below 8 kHz at end. (c) Aggressive screaming. During aggressive group pursuit-flights (see Social Pattern and Behaviour), birds utter noisy shrieking and screamed, congested scraps of song-phrases (J Hall-Craggs). (6) Distress-calls. (a) Screaming of handled nestling once caused parent to scream too. Screaming attracts other *T. merula* which may then mob predator and give captive a chance to escape (Snow 1958*b*). (b) Adult in pain gives similar sound to 'weeping' of nestling (Messmer and Messmer 1956; see below). (7) Courtship-calls of ♂. Messmer and Messmer (1956) described a 'ziep' from displaying bird during pair-formation, also a trill when

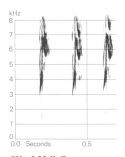

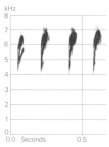

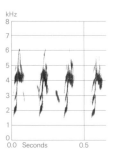

IX J Hall-Craggs
Channel Islands May 1980

X V C Lewis
England April 1961

XI P J Sellar
England May 1972

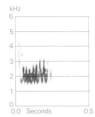

XII P J Sellar England January 1973

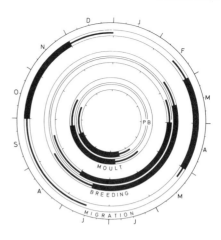

XIII P J Sellar
England June 1971

pursuing ♀; latter comprises repeated steeply descending portamenti (4·5–3·5 kHz) at rate of *c.* 10 per s. (8) Feeding-calls (apparently given by ♀ only). (a) Quiet, short duck-like sound elicits gaping from young in nest (Messmer and Messmer 1956). (b) P J Sellar has recorded 2 types of trill from ♀ approaching nest with food for young: (i) 'trrierp', *c.* 1·5 to 5 kHz or more, with *c.* 20 syllables per s; (ii) 'tyerrp', *c.* 1·5–3·0 kHz, *c.* 40 syllables per s (Fig XIII).

CALLS OF YOUNG. The following based on Messmer and Messmer (1956). From hatching until day 4, birds call 'hip'; gradually replaced by a disyllabic 'hiëp' during days 4–10. After 10th day, 'hiëp' merges into food-call 'trii', and 'hip' and 'hiëp' heard only after or between feeds. Fledglings' locatory call 'dschö döt' appears before they leave nest and is used as food-call for first 3 days after fledging. A 2nd locatory call, 'zit', used from 15th day. From day 16, 'dschö döt' and 'zit' combine to form the later food-call 'dschö zi zi döt' which, from 18th day, replaces 'trii'. Subsequently, components of 'dschö zi zi döt' develop: 'zi' becomes 'zri', and finally contact-call 'ssrie' (see adult call 2); 'dschö' and 'döt' pass through intermediate form ('döck') and become 'duck' ('chook' or 'tuk', as in adult call 4a) expressing insecurity, mild protest, or some measure of excitement. Fig XIV shows

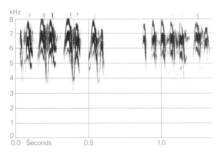

XIV J Hall-Craggs Channel Islands May 1980

juvenile calls at *c.* 26 days: 'tsi-tsi-tsi-tsi tsrri-tsi'. From 6th day, nestlings in discomfort utter a weeping 'ö ö ö ö ö'. Scream in distress (Snow 1958*b*). Messmer and Messmer (1956) described innate juvenile song in wild and aviary birds starting by age of 3 weeks; imitative song begins not later than 7 weeks and innate song disappears by 1st spring.   JH–C

Breeding. SEASON. Rather little variation across range. Western Europe and Britain: see diagram; in Britain, timing significantly correlated with latitude, up to 2 weeks later in Scotland than in southern England (Myres 1955). Finland: laying begins mid-April in advanced springs, but main period usually from last week of April; north of 62°N, first eggs early to mid-May (Haartman 1969*a*). Czechoslovakia: laying begins last 10 days of April, annual variation significantly correlated with climatic factors (Pikula and Beklová 1983). Canary Islands: eggs found March–June, mainly May–June (Bannerman 1963). SITE. Typically against trunk of small tree or bush supported by small branches and twigs, or among branches; frequently in or on wall, outside or inside building, among pile of brushwood or other debris, or occasionally on ground, usually sloping bank but sometimes flat, when normally well concealed; recorded using old nest of Magpie *Pica pica* and tree-cavity. In Czechoslovakia, 78% of 1739 sites in trees and shrubs, remainder in human artefacts. Of 1589 nests in Czechoslovakia, mean height above ground 1·9 m (0–15), with 94% at 0–4 m and only 1·1% on ground (Pikula and Beklová 1983). Of 146 nests in Poland, mean height above ground 2·3 m (0–12·5) with 90% at 0–4 m (Bocheński 1968). Nest: substantial cup of grass, straw, small twigs and other plant material, usually on foundation of moss, occasionally incorporating decoration of paper, foil, etc.; plastered inside with mud (rarely not) and lined with fine grass. In Czechoslovakia, mean outer diameter 15·2 cm (12·0–26·0), *n* = 533, inner diameter 9·8 cm (7·5–15·0), *n* = 554, height 12·4 cm (5·0–30·0), *n* = 540, and depth of cup 6·7 cm (4·0–14·0), *n* = 545 (Pikula and Beklová 1983); lining *c.* 1 cm thick (Bocheński 1968). Building: mainly by ♀, occasionally assisted by ♂; of 62 nests, most took 10–14 days (1–16) (Pikula and Beklová 1983). EGGS. See Plate 84. Sub-elliptical, smooth and glossy; usually pale greenish-blue, mottled and speckled light red-brown, though varying from brownish tint all over to unmarked. Nominate *merula*: 29·3 × 21·4 mm (25·0–34·5 × 18·5–23·5), *n* = 250; calculated weight 7·2 g. *T. m. azorensis*:

30·3 × 21·7 mm (27·3-33·3 × 19·9-22·6), $n = 100$: calculated weight 7·6 g. No significant differences found between sizes of other races, though Spanish birds and *cabrerae* similar to central European nominate *merula*, and *mauritanicus* similar to *azorensis*; samples of other races too small to judge (Makatsch 1976; Schönwetter 1979). Clutch: 3-5 (2-6); clutches of up to 9 recorded but perhaps always by 2 ♀♀. Of 1904 clutches from southern England: 2 eggs, 3%; 3, 27%; 4, 52%; 5, 18%; 6, 1%; mean 3·87; mean size varied through season from 3·4 ($n = 397$) in March to peak 4·56 ($n = 466$) in mid-May, falling to 3·61 ($n = 100$) in June-July; also varied with temperature and rainfall, especially early in season; clutches of 1st-year ♀♀ smaller (3·38, $n = 40$) than older birds (3·81, $n = 59$) (Snow 1955a, 1958a). Of 1601 clutches from Czechoslovakia: 2 eggs, less than 1%; 3, 16%; 4, 53%; 5, 29%; 6, 2%; 7, less than 1%; 8, less than 1%; mean 4·2; varied within season (smallest at beginning and end) and between years (probably because of weather), but no statistically significant variation found in relation to geographical or physical position of nest (Pikula and Beklová 1983). 2-3 broods regular (except in north of range, where 3 rare), with 4(-5) recorded in (e.g.) Britain (Mayer-Gross and Perrins 1962); mean attempts per year 2·87 ± SE0·16 in London suburb, 2·76 ± SE0·08 in rural area (Batten 1973). Eggs laid daily, mean 2·4 days (0-14), $n = 193$ after nest completed; period longer at start of season (Snow 1958a; Pikula and Beklová 1983). INCUBATION. In Czechoslovakia, 12·6 days (10-19), $n = 192$ (Pikula and Beklová 1983). In Britain, mean period 13·7 days in March nests, 13·1 days in May, 12·7 days in June ($n = 123$) (Snow 1958a). By ♀ only, though ♂ recorded on nest (Witherby *et al.* 1938b). Begins gradually, depending on final clutch size; hatching occurs over 1-2 days (Snow 1958a). YOUNG. Cared for and fed by both parents; brooded while small. FLEDGING TO MATURITY. Fledging period 13·6 days (10-19), $n = 302$ (Pikula and Beklová 1983). Young fed by both parents for up to 3 weeks, mean 19·6 days (15-25, $n = 22$), mainly by ♂ for earlier broods when ♀ re-nesting; ♀ feeds last brood of season for mean 20·5 days (15-24), ♂ for 12·4 days (0-24), $n = 11$ for both (Snow 1958a). Age of first breeding 1 year (Snow 1958a). BREEDING SUCCESS. Of 1428 nests in Britain, at least 1 young hatched in 56% and at least 1 young fledged in 41% (Snow 1955b). Older adults more successful than 1-year-olds (Snow 1958a). Of 6664 eggs laid in 1601 nests in Czechoslovakia, 35·7% lost before fledging (affecting 47·8% of nests), with predation, addled eggs, and desertion the main causes (Pikula and Beklová 1983). Mean overall breeding success in London suburb 39·46 ± SE2·17%, and mean 3·46 ± SE0·26 nestlings reared per year per pair; in rural areas, mean overall success 30·40 ± SE1·11%, and mean 2·61 ± 0·13% nestlings reared per year per pair (Batten 1973). Of 146 eggs laid, southern Sweden, 54% hatched; 39% of original number survived to fledging, 32% for a

further 5 days, and 27% to independence at about day 16 (Ebenman and Karlsson 1984).

Plumages (nominate *merula*). ADULT MALE. Whole plumage sooty-black or pure black with silky, non-metallic sheen; flight-feathers generally less black, more sooty, than rest of plumage. ADULT FEMALE. Crown and all upperparts dark earth-brown to brown-black. Sides of head dark earth-brown to brown-black, lores and upper cheeks mottled grey-buff or brown, ear-coverts finely streaked off-white. Chin and throat evenly streaked pale grey or off-white and brown or black-brown, pale ground-colour shading to deep rufous-brown on chest and breast, where feather centres show triangular dark brown or black spots. Belly and flanks duller brown-grey, feathers often with paler grey or dull rufous edges; vent and under tail-coverts dark brown-grey to dull black, much as upperparts. Tail and wing dark earth-brown to dull black, flight-feathers often slightly paler, more olive. Some individual variation: some birds almost as black as adult ♂, but chin and throat streaked grey and chest and breast clouded brown, others paler olive-brown with warmer and more extensively rufous-brown chest and breast. NESTLING. Down fairly long but rather scanty, dark grey or pale buffish-grey. JUVENILE. Upperparts and sides of head warm brown or rich earth-brown, feathers with pale shaft-streaks. Chin and throat pale buff with dark feather-tips. Chest rufous buff, feathers with dark brown to blackish tips, giving heavily spotted appearance; belly similar but dark spots smaller, appearing paler; under tail-coverts dark brown with pale shaft-streaks and rufous tips. Tail earth-brown to blackish-brown. Flight-feathers dark earth-brown to blackish-brown. Lesser and median upper wing-coverts dark brown with buff central streaks widening to broad buff tips; greater coverts similar but pale central streaks gradually reduced towards outermost. Sexes similar, but ♂♂ usually distinguishable by generally darker, less rufous colour, and blacker flight-feathers and tail. FIRST ADULT MALE. Like adult ♂ but body usually duller, with more or less well-marked dull brown fringes to feathers, especially of underparts; chin sometimes pale, approaching ♀ colour; juvenile tail, flight-feathers, tertials, greater upper primary coverts, and often variable number of outer greater upper wing-coverts retained, duller and browner than in adult, contrasting with black 1st adult coverts; greater coverts sometimes with traces of pale spots on tips. FIRST ADULT FEMALE. Like adult ♀, but upperparts and upper wing-coverts slightly paler earth-brown (not as black as some adult ♀♀); flanks, vent, and under tail-coverts paler grey; throat and chest more extensively rufous; occasionally, throat to belly pale vinous-chestnut with rather limited number of dark spots and birds with this often show vinous-chestnut supercilium. Part of juvenile feathers retained, as in 1st adult ♂, these paler than in adult ♀, olive-brown rather than grey; retained outer greater coverts sometimes with traces of pale spots on tips and usually differing in colour, shape, and abrasion from adjacent median coverts and (if new) inner greater and tertial coverts.

Bare parts. ADULT MALE. Iris dark brown; eyelids yellow. Bill yellow or orange-yellow. Leg and foot dark brown. ADULT FEMALE. Iris dark brown; eyelids yellow. Bill variable; dark brown, with little or no yellow, to almost completely yellow or orange-yellow, at least partly dependent on age and season. Leg and foot dark brown. NESTLING. Mouth deep yellow, no spots. Gape-flanges yellowish-white. JUVENILE. Iris brown. Bill flesh-grey, darkest on culmen and tip. Leg dull flesh-grey. Eyelids and gape-flanges dull orange-yellow. FIRST ADULT MALE. Like adult ♂, but bill wholly dark in late summer and autumn, remaining dark into winter; becoming marked with yellow usually in the course of late winter and early spring (individually variable), usually fully yellow by beginning of breeding season. FIRST ADULT FEMALE. Like adult ♀, but bill dark, yellow not developing until first breeding season. The above refers to northern populations of nominate *merula*; in southern races, bill becomes partly or fully yellow considerably earlier in young birds of both sexes.

Moults. ADULT POST-BREEDING. Complete; primaries descendant. In Britain, starts with p1 late May to late August, *c.* 15 days after end of breeding; completed with regrowth of p9–p10 after *c.* 66–87 days, early August to late October (Ginn and Melville 1983). Moult probably on average earlier in southern part of range. POST-JUVENILE. Partial: head, body, lesser and median upper wing-coverts, tertial coverts, variable number of greater upper wing-coverts, and occasionally some tertials, tail-feathers, or shorter feathers of bastard wing. Starts 4–6 weeks after fledging, mainly July–August; generally completed by October. For extent, sequence, and timing in various parts of species' range see Sommerfeld (1930), Gurr (1954), Snow (1958b, 1969b), Richter (1972), Baillie and Swann (1980), Jenni (1984), and Svensson (1984a).

Measurements. Nominate *merula*. Southern England, summer; skins (BMNH). Juvenile wing is retained juvenile wing of 1st adult. Bill to basal corner of nostril; to skull, on average 8·5 longer, exposed culmen 3·8 longer.

WING AD ♂ 129·0 (1·56; 10) 127–132 ♀ 123·3 (1·89; 10) 121–127
JUV 125·3 (1·42; 10) 124–128 120·7 (3·16; 9) 116–126
TAIL AD 108·8 (2·53; 10) 105–112 106·0 (3·13; 10) 101–112
BILL AD 17·3 (0·67; 10) 16–18 17·1 (0·74; 10) 16–18
TARSUS 34·2 (1·30; 9) 32–36 33·2 (0·92; 10) 32–34
Sex differences significant for wing.

Wing: Sweden, (1) adult, (2) 1st adult; West and East Germany, (3) adult, (4) 1st adult; Mallorca, (5) ages combined (Stresemann 1920; Jordans 1950; Eck and Geidel 1971; ZMA).

(1) WING ♂ 133·9 (9) 126–140 ♀ 128·2 (2) 125–131
(2) 130·2 (13) 125–136 127·7 (10) 124–131
(3) 131·9 (65) 124–136 126·4 (33) 119–131
(4) 128·0 (107) 120–134 125·6 (43) 116–128
(5) 119·8 (12) 116–126 — (5) 113–116

T. m. aterrimus. Wing: southern Yugoslavia, (1) adult, (2) 1st adult (Stresemann 1920); (3) Turkey, northern Iran, and Armeniya (USSR), ages combined (Paludan 1940; Schüz 1959; Kumerloeve 1961, 1969a, 1970a; Nicht 1961; Vauk 1973a; ZMA).

(1) WING ♂ 128·0 (3·16; 6) 124–133 ♀ 124·3 (2·07; 6) 122–128
(2) 124·7 (2·11; 18) 121–128 120·5 (1·98; 13) 117–123
(3) 127·5 (2·81; 11) 122–131 123·0 (6·04; 7) 115–134

T. m. azorensis. Azores, all year; skins (AMNH, BMNH).

WING AD ♂ 122·3 (2·14; 28) 117–127 ♀ 119·1 (2·33; 16) 113–123
JUV 120·6 (2·83; 16) 115–125 115·6 (2·92; 9) 112–122
TAIL AD 98·7 (2·26; 19) 94–102 95·9 (2·75; 15) 90–101
BILL AD 17·7 (0·76; 25) 16–19 17·7 (0·75; 15) 16–18·5
TARSUS 34·6 (0·78; 25) 33–36 33·8 (1·11; 15) 32–36
Sex differences significant for wing and tail.

Wing and tail. (1) Nominate *merula*, central Spain; (2) *aterrimus*; (3) *syriacus*; (4) *mauritanicus*, Morocco; (5) *mauritanicus*, northern Algeria and Tunisia; (6) *cabrerae*; (7) *intermedius*; (8) *maximus* (Hartert 1910; Jordans 1950; Dementiev and Gladkov 1954b; Paludan 1959; Vaurie 1959, 1972; Ali and Ripley 1973b; ZMA).

	WING ♂	WING ♀	TAIL
(1)	122–130	112–124	100–115
(2)	120–138	118–125	95–115
(3)	125–136	120–125	110–120
(4)	(121–)126–132	118–127	95–130
(5)	119–124	—	105–110
(6)	121–130	115–124	94–110
(7)	130–143	127–139	118–133
(8)	144–167	144–161	107–120

Weights. Nominate *merula*. Southern England: (1) January–February, (2) March–April, (3) May–June, (4) September–October, (5) November–December (BTO).

		♂		♀	
(1) AD		105·9 (198) 81–148		108·2 (100) 87–135
(2)		96·4 (256) 82–125		99·8 (100) 83–129
(3)		94·6 (193) 80–122		97·1 (50) 81–128
(4)		102·6 (229) 80–124		97·7 (100) 85–129
(5)		113·8 (422) 86–149		106·8 (100) 82–129
(1) 1ST AD		109·5 (216) 90–135		104·1 (100) 80–140
(2)		94·7 (194) 80–122		98·7 (100) 70–122
(3)		94·2 (141) 78–110		97·3 (50) 84–119
(4)		100·4 (598) 83–122		97·5 (100) 80–116
(5)		108·9 (615) 83–148		108·1 (100) 89–140

Norway: March–April, 93·2 (18) 80–108; July–August 99·2 (18) 88–113; September–November, 105·3 (53) 71–130 (Haftorn 1971). Portugal, July–September: 86·1 (5·25; 23) 76–97 (C J Mead). Malta, winter: 83·0 (10·0; 53) 61–110 (J Sultana and C Gauci). Victims of frost, England and Wales: ♂ 61·2 (6·84; 18) 51–73, ♀ 64·3 (7·6; 12) 52–79 (Harris 1962; Ash 1964). For weight curves throughout winter and influence of weather, see Swann (1980); for weight of ♀ during laying, see Pikula (1976); for migrants Denmark, see Krüger (1940). For growth of nestling, see Snow (1958a).

T. m. aterrimus. Greece, Turkey, Armeniya (USSR), and northern Iran, combined; February–July (Paludan 1940; Makatsch 1950; Schüz 1959; Nicht 1961; Kumerloeve 1968, 1969a, 1970a; Rokitansky and Schifter 1971; Vauk 1973).

♂ 88·9 (9·65; 9) 69–98 ♀ 96·6 (5·46; 5) 88–102

T. m. intermedius. Kazakhstan (Dolgushin *et al.* 1970) and Afghanistan (Paludan 1959), combined: (1) July–August, (2) October–March.

(1) ♂ 95·6 (5·59; 9) 84–102 ♀ 88·5 (2·81; 4) 85–92
(2) 99·2 (15·6; 4) 86–121 96·5 (9·06; 5) 85–106

Structure. Wing rather short, broad at base; tip bluntly pointed in northern populations, more rounded in southern ones. 10 primaries: in nominate *merula*, p7 longest, p8 (0–)1–4 shorter, p9 10–15, p6 0(–2), p5 2–7, p4 10–16, p3 15–23, p1 21–31; in sedentary southern and island populations, p6 often longest. In all races, p10 reduced, 63–80 shorter than wing-tip, 8 shorter to 1 longer than longest upper primary covert. Outer web of p5–p8 and inner of (p6–)p7–p9 emarginated. Tertials short. Tail rather long, 12 feathers; tip square, but t6 5–11 shorter than t1–t4. Bill rather short, straight; rather wide and deep at base, laterally compressed at tip; culmen slightly decurved at base, more strongly at tip, cutting edges straight or slightly decurved. Nostril rather large, oval; partly covered by frontal feathering at basal corner and by narrow membrane above. Rather stiff bristles along base of upper mandible, longest towards gape. Tarsus and toes rather short, strong. Middle toe with claw 28·9 (10) 27–31; outer toe with claw *c.* 81% of middle with claw, inner *c.* 80%, hind *c.* 74%. Claws long, rather slender, moderately decurved.

Geographical variation. Clinal, except for some island populations; mainly involves colour of ♀ and size. Boundaries between continental races rather arbitrary. In general, size smaller towards south, but mountain birds have longer wings than lowland ones: e.g. *mauritanicus* from Moroccan lowlands and northern Algeria and Tunisia distinctly smaller than birds from Middle Atlas, *aterrimus* from Greece and western Turkey smaller than birds from eastern Turkey, Caucasus area, and northern Iran, and montane *intermedius* and *maximus* from central Asia are largest races. *T. m. azorensis* from Azores is about as small as *mauritanicus*, but wing more rounded than both *mauritanicus* and rather small *cabrerae* from Madeira and western Canary Islands. In nominate *merula*, some variation in colour of ♀ (see, e.g., Eck and Geidel 1971), but individual variation marked and too much overlap in colour to separate *pinetorum* Brehm, 1831, in central Europe, *ticehursti* Clancey, 1938, from eastern Scotland, or *mallorcae* Jordans, 1950, from Mallorca. Nominate *merula* gradually merges into *mauritanicus* in central and southern Spain and into *aterrimus* in south-east Europe, Ukraine, and Crimea. In *mauritanicus*, ♂ slightly deeper black than nominate *merula*, but ♀ distinctly darker and greyer all over; in *aterrimus*, ♂ slightly duller black, ♀ duller and greyer than ♀ nominate *merula*. *T. m. syriacus* from south-east Turkey, Iraq, and Iran (south of Elburz) merges into *aterrimus* in colour; generally paler than *aterrimus*, ♂ more slaty, ♀ distinctly greyer all over and less rufous below. *T. m. intermedius* from central Asia occurs occasionally in southern Iraq in winter (Vaurie 1959); colour like *aterrimus*, but larger, especially bill; replaced in Himalayas by even larger *maximus*, in which both sexes dark brown or black-brown. ♂♂ of *cabrerae* and *azorensis* from Atlantic islands are dark and glossy, deeper black than in nominate *merula*; ♀♀ darker black-brown, in particular in *cabrerae*, which shows less white on throat than *azorensis*; most birds from Hierro and Palma darker than *cabrerae* from other Canary Islands and Madeira, and perhaps separable as *agnetae* Volsøe, 1949 (Niethammer 1958). *T. m. simillimus* and allied races of southern Asia are very different in plumage, and although customarily considered races of *T. merula* are doubtfully conspecific.   DWS

Turdus obscurus Eye-browed Thrush

PLATES 67 and 77
[between pages 856 and 857]

Du. Vale Lijster   Fr. Merle obscur   Ge. Weissbrauendrossel
Ru. Оливковый дрозд   Sp. Zorzal rojigris   Sw. Gråhalsad trast

Turdus obscurus Gmelin, 1789

Monotypic

Field characters. 23 cm; wing-span 36–38 cm. Slightly larger than Redwing *T. iliacus* (though often appears smaller), with bill proportionately shorter (though longer relative to head). Rather small and rakish thrush, with general character most like *T. iliacus*. Grey- to olive-brown above, with obvious white supercilium, eye-ring, and throat; upper breast grey but chest and flanks orange or buff, contrasting with white belly and vent. Sexes closely similar; no seasonal variation. Juvenile separable. Monotypic, but western birds much less saturated than eastern ones which approach Red-bellied Thrush *T. chrysolaus* of eastern Asia in appearance; description here based on observations of western birds in Siberia and on British vagrants.

ADULT MALE. Head, face, and upper breast basically smoky grey-brown, with darker edges to crown, black lore, and black-brown rear eye-stripe emphasizing long white supercilium (cleaner but less deep than in *T. iliacus*) and obvious white crescent below eye; lower cheeks mottled grey with dark grey-brown surround usually creating moustache; chin and short throat white, contrasting with upper breast and only rarely flecked dark grey-brown. Head often appears flat-crowned, with sloping forehead extended by seemingly long bill. Head pattern and head and bill shape create rather frowning, yet surprised expression. Upper breast and side and back of neck slightly paler than head but usually still with marked slate-grey tone; rest of upperparts less grey, more olive-brown, appearing paler than head in some lights. Below breast, band of orange-rufous or apricot of variable depth across chest, usually complete and extending down flanks; rest of underparts from centre of lower breast basically white, creating similar pattern and contrast with flanks as adult *T. iliacus*. Under wing-coverts pale grey, contrasting with pale dusky undersurface of flight-feathers. Uppersurface of flight-feathers and tail dark olive-brown, with no obvious pattern on flight-feathers and only indistinct white patches on outer tail-feathers, usually not visible. Bill blackish-brown with yellow base (more extensive in summer); legs yellowish-brown. ADULT FEMALE. Resembles ♂ but usually shows white-streaked cheeks, less obvious moustache (due to dark brown streaks on lower throat), and less grey, more olive or brown tone on head, making it almost uniform with rest of upperparts. Colour of chest and flanks duller, less apricot, even brown-buff on some. Underwing pale grey-buff. JUVENILE. Basic plumage pattern as adult ♀ but upperparts and wing-coverts with ochraceous streaks or tips (appearing pale-spotted), cheeks dark-spotted, and underparts paler, with fulvous or buff suffusion on breast and flanks and overlying dark brown spots, most obvious on breast. Bill mostly dark, but paler than adult initially. FIRST YEAR. Resembles adult ♀ but probably with pale tips of greater coverts forming indistinct wing-bar and other retained juvenile wing-coverts showing pale streaks at close range; some indication that ♂ shows greyer head and warmer chest-band and flanks (see Plumages).

Unmistakable when seen well at close range, with lack of obvious spots (except in juvenile), grey underwing, and uniformity of basic plumage colours ruling out confusion with Naumann's Thrush *T. naumanni naumanni* (distinctly mottled black on cheeks and sides of neck and rufous on flanks, underwing, and tail). Less distinctive at distance, when size, appearance, flight, and behaviour strongly recall *T. iliacus*. Flight fast, with rakish silhouette; over long distance, direct and not undulating, recalling Starling *Sturnus vulgaris*. Has habit of escaping by ascent to tree canopy or highest branches like *T. iliacus*, not ducking into lower cover like Song Thrush *T. philomelos*. Hops and runs with frequent pauses, like other small thrushes; carriage quite upright, with tail frequently held higher than *T. iliacus* and legs then appearing rather long. Most active in early morning and evening. Gregarious in normal winter range, occasionally mixing with Black-throated and Red-throated Thrushes *T. ruficollis* but usually keeping to forest.

Song clear and rich, mostly a repetition of 2–3 ringing notes. Calls include: thin, quiet 'zip-zip', 'plip', or 'tlip', recalling pipit *Anthus*; soft 'tchuck' or 'tchick'; in full alarm, loud 'ke(w)k' recalling *T. iliacus* but less hoarse; also other calls resembling Black-throated Thrush *T. ruficollis atrogularis*.

Habitat. Breeds in upper middle and higher latitudes of east Palearctic, in continental boreal lowland and montane habitats, from fringe of taiga in larch *Larix* to dense forests of spruce *Picea* and fir *Abies*, especially in sheltered valleys or near water (Harrison 1982). In Far East, more in mixed forests and broad-leaved groves in river floodplains, on lakeshores, and by marine inlets, ranging from sea-level to *c.* 800–1500 m and occasionally up to 2400 m, sometimes occurring in arable land, orchards, and cities. In winter, occurs in wooded montaneregions, in damp gullies above *c.* 1000 m in Malaysia, but also in woods, open country, and gardens at lower elevations. In Indian subcontinent, winters in open forest, not venturing into open country beyond; feeds mostly on ground, but also on berry-laden shrubs, resting in shady trees during day, and taking refuge in them when disturbed (Ali and Ripley 1973*b*).

Distribution. Breeds Siberia east from Yenisey to Sea of Okhotsk, south to Lake Baykal and possibly to Mongolia. Winters in southern Japan, Taiwan, Philippines, and Palau and from south-east Asia to north-east India and Bangladesh.

Accidental. Britain, Belgium, Netherlands, West Germany, Norway, Finland, East Germany, Poland, Czechoslovakia, Italy, Malta.

Movements. Migratory. Passes through Mongolia, Manchuria, Korea, and China to winter in eastern India, Burma, Vietnam, southern China, and Japan south to Malay peninsula, Philippines, and Indonesia. Stragglers recorded in Arabia, West Bengal, Nepal, Alaska, and western Europe.

Departure from Yenisey occurs from beginning of September, and birds move to some extent along Yenisey past Krasnoyarsk down to Minusinsk. Departures also occur in September from eastern Mongolia, and in last third of September migrants common around Lake Khalkhin. Spring passage occurs mainly 21–31 May in Ussuriland and 20 May–5 June on Kamchatka. Recorded at end of May around Krasnoyarsk. Arrives on breeding grounds usually from second half of May and in northwest in early June, although arrival on Yenisey below Arctic Circle recorded 6 May. (Seebohm 1879; Dementiev and Gladkov 1954*b*.) LAB

Voice. At least 2 types of contact-alarm call given outside breeding season (see below).

CALLS OF ADULTS. (1) Song of ♂. Recording by M Schubert from Mongolia reveals pure-toned and attractive whistles and a more strangled, twittering or warbling component reminiscent of Blackbird *T. merula*. Short phrases (or figures) separated by pauses delivered in slow and deliberate fashion and song lacks variety. One whistle has rich yodelling quality recalling *Tringa* wader: 'h(e) elooeet', rather mournful at times (presumably the melodious, full-sounding, arpeggiated triad of Mauersberger *et al.* 1982); another more like 'dweep'. Pattern sometimes 1st whistle-2nd whistle-twittering (Schubert 1982; M G Wilson). In recording from Tuva (USSR) by B N Veprintsev, song more varied with loud, forceful, often vibrato-type 2-syllable whistles recalling Song Thrush *T. philomelos*, though delivery slower, figures more disjunct, and no obvious immediate repetition; at times suggests harsh Greenshank *Tringa nebularia*. Warbling component (see above) thinner and more hissing and fizzing, but also with some rich, more chattering sounds (M G Wilson). For further brief descriptions, see (e.g.) Grote (1925), Löw (1976), and (with recording) Svensson (1984*b*). (2) Contact-alarm calls. (a) A 'dack-dack' or 'tacktack' when alarmed (Jahn 1942; Löw 1976). Perhaps some variation in timbre as chuckling sounds reported by Yamashina (1982), these being soft and pleasant, like Black-throated Thrush *T. ruficollis atrogularis*, when feeding (Ali and Ripley 1973*b*); vagrant also gave 'tchup' or 'tchuck' from cover (Clugston 1981). (b) Contact-call apparently given mainly in flight, e.g. constantly by members of small migrant flocks flying between trees in Krasnoyarsk (USSR), late May (Naumov and Burkovskaya 1959): 'zieh' (Jahn 1942; Yamashina 1982), resembling *T. merula* (Mauersberger *et al.* 1982) or quite harsh 'seee' like Redwing *T. iliacus* (Clugston 1981); thin 'zip-zip' like pipit *Anthus* given by wintering birds in

India when flushed and making for tree (Ali and Ripley 1973*b*) probably the same. MGW

Plumages. ADULT MALE. In fresh plumage (autumn), head, neck, and throat grey, except for variable olive-brown wash on forecrown, narrow white supercilium, black lores bordered white at gape and below eye, black malar stripe, white chin, and faint black streaks on upper throat adjacent to chin. Rest of upperparts olive-brown. Chest and flanks tawny-buff to orange-brown, belly and centre of breast white; under tail-coverts white, a few feathers with indistinct brown streaks. Tail dark olive-brown; t6 with variable whitish bar at end of inner web, t5 sometimes narrowly edged whitish at tip. Upper wing-coverts and outer webs of flight-feathers olive-brown, much as upperparts; inner webs of flight-feathers darker. Under wing-coverts pale grey-brown with white edges. *In worn plumage* (spring and summer), olive-brown feather-fringes of crown tend to be lost through abrasion and whole plumage usually paler; in particular, grey of head region and tawny-buff of underparts faded. ADULT FEMALE. Rather variable. Most individuals differ from adult ♂ in having crown and sides of head olive-brown with little or no grey tinge; much of throat as well as chin whitish with dark streaks, separated from buff of chest by variable olive-grey band; breast and flanks paler and duller orange-brown, less tawny. Some have head, neck and throat predominantly grey, approaching adult ♂, but distinguished by more extensive white on centre of throat; all intermediates occur. *In worn plumage*, plumage usually paler, especially buff of breast and flanks faded. JUVENILE. Crown, mantle, and scapulars olive-brown, each feather with pale central streak and darker tip, giving strongly spotted appearance; back unspotted; rump and upper tail-coverts with pale shaft-streak or pale spots on feather-tips. Sides of head, chin, and throat whitish or pale buff with dark feather-tips and well-marked sooty malar stripes. Chest and flanks pale buff, breast and belly white; feathers with dark spots on tips, heaviest on chest, gradually smaller on lower breast and flanks, much reduced or absent on central belly and vent. Tail as adult. Wings olive-brown, as in adult; lesser and median upper wing-coverts with pale buff tips, and tertials with buff tips on outer webs. FIRST ADULT MALE. Like adult ♂, but crown more strongly washed rich olive-brown, whole throat region white with more or less well-marked dark malar stripe and dark streaks on throat. Juvenile flight-feathers, tail, tertials, greater upper primary coverts, and variable number of greater upper secondary coverts retained; greater coverts (especially inner) and sometimes tertials with large pale triangular spots on tips (sometimes largely worn off in spring). See also Svensson (1984*a*). FIRST ADULT FEMALE. Like adult ♀, but head olive-brown without trace of grey, and white of chin and throat washed pale yellow-brown; buff of breast and flanks sometimes fading to drab grey-brown in spring and summer. Part of juvenile feathers retained, as in 1st adult ♂, greater coverts and sometimes tertials showing pale spots on tips.

Bare parts. ADULT, FIRST ADULT. Iris brown. In breeding season, upper mandible blackish brown, cutting edge more or less yellow, lower mandible yellow with dark tip; bill darker in rest of year, yellow reduced in extent and duller. Leg and foot yellowish brown, light brown, or dark fleshy-brown. JUVENILE. Iris brown. Bill paler than adult in early autumn, becoming darker, mostly dark horn in 1st winter, but much variation. Leg and foot yellowish-brown. (Dementiev and Gladkov 1954*b*; Ali and Ripley 1973*b*; BMNH.)

Moults. ADULT POST-BREEDING. Complete; primaries descendant. Starts late July, completed before leaving breeding area. POST-JUVENILE. Partial: head, body, lesser, median, and some inner greater upper wing-coverts. Completed before autumn migration. (Dementiev and Gladkov 1954*b*; BMNH.)

Measurements. Whole geographical range, all year; skins (BMNH). Juvenile wing is retained juvenile wing of 1st adult. Bill from basal corner of nostril; to skull, on average 7·8 longer, exposed culmen on average 3·1 longer.

WING AD	♂ 126·2 (2·97; 10)	123–129	♀ 122·5 (2·68; 10)	119–127
JUV	121·3 (2·11; 10)	119–125	121·0 (3·77; 10)	115–126
TAIL AD	90·6 (3·69; 10)	86–98	84·5 (3·57; 10)	80–92
BILL AD	14·6 (0·52; 10)	14–15	14·6 (0·70; 10)	13–15
TARSUS	31·6 (0·73; 9)	31–33	30·5 (1·08; 10)	29–32

Sex differences significant for wing, tail, and tarsus.

Weights. USSR: ♂ 69 (Dementiev and Gladkov 1954*b*). Kuril Islands, USSR: ♀♀ 66, 75 (Nechaev 1969). On migration, north-east China: ♂ 74 (13) 61–117, ♀ 70 (9) 50–110 (Shaw 1936). Western China: ♂ 58 (Rensch 1924). On migration, Mongolia: ♂ 67 (May), ♀ 70 (August) (Piechocki and Bolod 1972). India, winter: ♂♂ 65–80 (9); ♀♀ 59–75 (8) (Ali and Ripley 1973*b*). Norway, November: ♂ 86 (Holgersen 1962).

Structure. Wing rather long, broad at base, tip fairly pointed. 10 primaries: p8 longest, p9 4–7 shorter, p7 0–2, p6 6–9, p5 17–22, p4 24–28, p1 36–41. P10 reduced; 76–81 shorter than p8, 7–12 shorter than longest upper primary covert. Outer web of p6–p8 and inner of p7–p9 emarginated. Tail rather long, tip square; 12 feathers, t6 2–6 shorter than t2–t3. Middle toe with claw 26·4 (5) 24–28; outer toe with claw *c.* 71% of middle with claw, inner *c.* 66%, hind *c.* 70%. Remainder of structure as in Blackbird *T. merula* (p. 963), but bill and tarsus relatively shorter.

Geographical variation. None described in detail, but see Field Characters.

Forms superspecies with Pale Thrush *T. pallidus* from south-east Siberia and Manchuria and with Red-bellied Thrush *T. chrysolaus* from Kuril Islands, Sakhalin, and Japan (Stepanyan 1983). DWS

Turdus naumanni Dusky Thrush and Naumann's Thrush

Du. Bruine Lijster   Fr. Grive à ailes rousses   Ge. Rostflügeldrossel
Ru. Дрозд Науманна   Sp. Zorzal de Naumann   Sw. Bruntrast

Turdus naumanni Temminck, 1820

Polytypic. *T. n. eunomus* Temminck, 1831 (Dusky Thrush), central Siberia, from Pur, Taz, and lower Yenisey rivers east to Kamchatka; nominate *naumanni* Temminck, 1820 (Naumann's Thrush), south of *eunomus* (but with overlap in west of range), from middle and lower Nizhnyaya Tunguska east to Maya and Aldan rivers, south to Sayan mountains.

Field characters. 23 cm; wing-span 36–39 cm. Larger than Redwing *T. iliacus* (wings and tail nearly 15% longer). Medium-sized, robust thrush, with rather stout bill and body. Diagnostic combination of broad pale supercilium and much rufous or chestnut under and on wings and on rump. Northern race (Dusky Thrush) spotted and barred black below, southern race (Naumann's Thrush) spotted rufous. Sexes closely similar; some seasonal variation. 2 races occur in west Palearctic, easily distinguishable (though intermediates occur).

ADULT MALE. (1) South-east Siberian race, nominate *naumanni* (Naumann's Thrush). Upperparts olive-grey to grey-brown, dusky on crown and variably mottled chestnut on back, becoming warmer and then bright rufous on rump and tail; central tail-feathers dark brown. Wings dark brown, with rufous-grey fringes to coverts and inner secondaries and tertials. Face quite strongly marked, with buff-white supercilium, dark brown eye-stripe, dull brown buff-streaked cheeks, malar stripe of black speckles, and pale rufous-white throat. Breast pink-red, with faint mottling and conspicuous line of black spots on side of lower neck; flanks also pink-red, with stronger white mottling; belly and under tail-coverts rufous-white. Underwing red-chestnut, with duller rufous tone spreading over most of flight-feathers. From September, fresh tips and margins to feathers produce less contrasted plumage, with more white on underparts breaking up dull red breast and flanks. Bill horn-brown, with cutting edges and base of lower mandible yellow, giving bill stout appearance; yellow more extensive in summer; legs brown. (2) North Siberian race, *eunomus* (Dusky Thrush). Basic plumage pattern much as nominate *naumanni* but differing distinctly in: (a) darker, brown head and back, back roughly scaled and strongly mottled black and chestnut; (b) more chestnut rump; (c) mainly brown-black tail without full rufous tone; (d) more red-chestnut fringes to wing-feathers producing a wide warm-coloured panel when folded; (e) much darker, almost black rear cheeks, emphasizing cream supercilium which is often broken by smudge above eye; (f) cream-buff throat, extreme upper breast, and broad collar from under rear cheeks to nape; (g) lines of black spots on sides of throat; (h) noticeable gorget of rusty- or brown-black spots, looking uniform at distance or when white tips worn; (i) much more strongly marked flanks, with brown-

black spots and chevrons always dappled or splashed in pattern and extending under tail, sometimes creating 2nd incomplete chest-band. ADULT FEMALE. Both races. Resembles ♂ but duller, less grey or chestnut. Differs most obviously in (a) more spotted chin and centre of throat, (b) more suffused, less spotted breast and flanks, and (c) paler chestnut edges to scapulars and wing-coverts, reducing richness of wing-panel. JUVENILE. Both races. Duller and browner even than ♀, with all main marks less contrasting, particularly noticeable in (a) restricted buff-brown rump, (b) dull chestnut-brown wing-panel, (c) less distinct face pattern, with spotted supercilium, (d) paler rufous underwing, and (e) pale streaking and spotting of upperparts and wing-coverts. Bill pale initially, then all-dark. FIRST-WINTER MALE. Both races. Resembles adult ♂ but upperparts less scaly. Best aged by pale tips to retained juvenile wing-coverts and less cream-buff lower throat and half-collar.

Both races unmistakable at close range. Most telling characters of nominate *naumanni* are dull pink-red suffusion and mottling of underparts and red-chestnut tail, brighter than rump. Best features of *eunomus* are bold face pattern (broad supercilium and dark rear cheek-patch), pale half-collar, bright chestnut wing-panel and rump (rump warmer than tail), and black-splashed underparts. In brief glimpse or at longer range, however, both races liable to be confused with other species. Nominate *naumanni* may suggest Red-throated Thrush *T. ruficollis ruficollis* (but lacks its markedly uniform plumage colours, instead always showing mottled flanks and pale supercilium), Eye-browed Thrush *T. obscurus* (but lacks its grey-brown upper breast and uniform upperparts), and American Robin *T. migratorius* (but is smaller and lacks its dusky upperparts, black-streaked white throat, and fully brick-red underparts). *T. n. eunomus* can recall juvenile Fieldfare *T. pilaris* (but lacks its grey rump, weak face pattern, and more evenly spotted underparts) and *T. iliacus*. *T. n. eunomus* liable to be passed over as being *T. iliacus* since face patterns very similar, breast markings of *T. iliacus* form gorget on adults, and dark rufous flanks of *T. iliacus* can simulate wing-panel of *eunomus*, while dark rufous underwing common to both in all plumages. Note, however, that *T. iliacus* lacks scaly black marks on flanks and rear underbody, dark mottling on back, and rufous rump, all these being present on *eunomus*. Appearance in flight most recalls Song Thrush *T. philomelos*, being plump and not slight and rakish like *T. iliacus*; flight occasionally looks as leisurely as *T. pilaris*. Escape-flight often long, with bird taking its time before returning to previous position. Gait and behaviour on ground recall both *T. philomelos* and *T. pilaris*, with noticeably uptilted bill at times. Tamer than *T. pilaris*, associating with and tending to dominate smaller thrushes. Gregarious in normal range.

Song very similar in phrasing and tone to *T. philomelos* but less chanted and strong. Calls varied, recalling Blackbird *T. merula* and *T. iliacus*. Nominate *naumanni* (as spring migrant) heard to give: insistent 'swer(k)-swer(k)-swer(k)' in alarm, with tone suggesting Kestrel *Falco tinnunculus*; 'zeep(r)' when taking off, with thin rasping tone like *T. iliacus*; conversational 'que-que-que(k)'. *T. n. eunomus* (as spring migrants and in winter) heard to give: 'kvereg' in alarm, with squawky tone like *T. merula*; 'spirr' in flight, with piercing quality like Starling *Sturnus vulgaris*; 'tack-tack-tack-tack' when going to roost.

Habitat. Breeds in higher latitudes of east Palearctic, mainly in lowlands from fringe of tundra through taiga and wooded steppe, in riverain woods of willow *Salix* and poplar *Populus*, with dense tall bushy undergrowth; also in birch *Betula* and alder *Alnus* and more rarely in larch *Larix* along terraces and low scrub. Occurs in thinly wooded regions and on outskirts of forest. In south of range (nominate *naumanni*), breeds infrequently in river valleys with mature mixed woods, and more typically on slopes of mountains with sparse larch cover, or in tall pine and larch woods. On migration in Ussuriland (USSR), occurs in scrub oak and on old stubble fields; elsewhere in floodplain vegetation. In winter in Japan, lives in fields, taking refuge when disturbed in trees. (Dementiev and Gladkov 1954*b*; Yakhontov 1976.) In India, feeds mostly on ground in winter, in open fields, grasslands, and thinly wooded country at 900–3000 m (Ali and Ripley 1973*b*).

Distribution. Breeds in northern and central Siberia from treelimit south to Lake Baykal and east to Pacific (Anadyr', Kamchatka, Sea of Okhotsk, and perhaps Sakhalin). Winters in Japan, Korea, China, Ryu Kyu Islands, and Taiwan in small numbers to northern Burma and Assam.

Accidental. Faeroes, Britain, France, Belgium, Netherlands, West Germany, Denmark, Norway, Finland, East Germany, Poland, Czechoslovakia, Austria, Switzerland, Yugoslavia, Italy, Israel, Kuwait, Cyprus.

Movements. Migratory. North Siberian race, *eunomus*, migrates through Kuril Islands, Sakhalin, Hokkaido, Korea, Manchuria, and Mongolia to winter in southern Japan, Taiwan, southern China, north-west Thailand, northern Burma, Assam (India), and occasionally Nepal and Pakistan; some birds apparently winter as far north as middle Yenisey (USSR) in some years (Johansen 1954; Lekagul *et al.* 1985). South Siberian race, nominate *naumanni*, migrates through Mongolia and Korea to winter in China north to southern Manchuria and south to Yangtse river and in Korea, with a few in Japan and Taiwan.

Departure of *eunomus* from Kamchatka takes place in first half of September. Noted around Krasnoyarsk (south-central Siberia) in early October and in Ussuriland from second half of September to mid-October. In central Japan, large numbers arrive in middle zone of mountains from end of October or early November, remaining there until snowfall. Common in winter quarters until April. Autumn departures of nominate *naumanni* from Yakutsk occur throughout September. In Lesser Khingan mountains, passage migrants observed mid-September to mid-October. In south-east USSR, passes Khor river region in first half of October and noted on Askold Island in first half of November. (Dementiev and Gladkov 1954*b*.)

Spring passage of nominate *naumanni* starts mid-March in far-eastern USSR, tending to be 2–3 days earlier than *eunomus* (Yakhontov 1976), though *eunomus* said not to appear at Lake Khanka (south-east USSR) until last third of April, being recorded there up to mid-May. Passage of nominate *naumanni* noted as continuing through April on Lefu river (south-east USSR), with birds arriving on breeding grounds from early to late May. More-northerly *eunomus* arrives on breeding grounds at Krasnoyarsk near end of May and at Anadyr' (north-east USSR) from 17 May (Dementiev and Gladkov 1954*b*).

Birds occurring in northern Europe are chiefly *eunomus*, those in south nominate *naumanni*. About 50 records in western Europe from 1800 to 1965; 29 out of the 37 dated occurrences were October–January (Machalska *et al.* 1967). LAB

Voice. Contact-alarm calls used outside breeding season (see below, and Field Characters). In mixed flocks, *eunomus* reported as noisier than nominate *naumanni*, frequently giving call 2a (Portenko 1981).

CALLS OF ADULTS. (1) Song of ♂. Perhaps some variation between races, but material insufficient for detailed comparison. Only available recording (by S Palmér and N Noréhn) of nominate *naumanni* is of July vagrant in Norway and may be of subdued variant or Subsong: brief phrases separated by pauses of several seconds comprise whistle starting thin and developing through marked crescendo (and apparently also fall in pitch) to harsher rattling warbles; strongly reminiscent of Redwing *T. iliacus* (M G Wilson). Although song of *eunomus* sometimes likened to *T. iliacus* (more melodious but less varied according to Yakhontov 1976), detailed descriptions suggest a more powerful and more varied performance than in nominate *naumanni*. *T. n. eunomus* said by Yakhontov (1976) to have loud continuous song (carrying *c.* 300 m) separated by pauses of 1–3 min. Beautiful lilting whistles and fluting notes apparently predominate and produce rich quality recalling nightingale *Luscinia* or other *Turdus* such as Blackbird *T. merula* or Song Thrush *T. philomelos* (even rudimentary song given in Japan, late April, like *T. philomelos*: Jahn 1942). More warbling, chirping or 'trilling' (like *T. iliacus*) sounds sometimes interpolated, as are alarm- or excitement-calls (see call 2b). Oft-repeated units include 'ki-chyur', 'chi-kyur', and (at end of song) 'chi-yo' (Portenko 1981). In Lena valley, USSR (*eunomus* according to Portenko 1981), typical variant (given at night or in dull weather during day) as follows: 'veet tyulir-tyulir fru-fru fir-fee veet-veet tyulir-tyulir che-che-che-che veet-veet-veet fru-fru pryupee-pryupee' (Kapitonov and Chernyavski 1960). Sings from tree (Flint *et al.* 1984). (2) Contact-alarm calls. (a) In Japan, chattering 'quäwäg' or 'kwet kwet' given by *eunomus* at take-off and when flying long distance, including while migrating at night (Jahn 1942). Recording of winter flock in Japan by R Nakatsubo reveals curiously strained, melancholy, at times slightly harsh 'wäwä' or similar, rather like Huit-call of Fieldfare *T. pilaris* (see p. 986); volume and timbre vary—some calls more musical, some very squeaky; apparently disyllabic (M G Wilson). See also (e.g.) Kolthoff (1932), Fiebig (1983), and Flint *et al.* (1984). (b) Harder alarm- or warning-call of both races a 'chok-chok' (Yakhontov 1976). Further descriptions: 'chak-chak' (*eunomus*), similar to *T. pilaris* (Dementiev and Gladkov 1954*b*), with laughing quality (Kolthoff 1932); 'tsepit-chak-chak' (Flint *et al.* 1984); hard 'täck-täcktäck' given by *eunomus* when going to roost (Jahn 1942). (c) A 'spirr' like Starling *Sturnus vulgaris* also given by *eunomus* at take-off (Jahn 1942). In recording by R Nakatsubo (see above), 3 thin, high-pitched hissing sounds apparently the same, and 'kwee' mentioned without detail by Yamashina (1982) perhaps also (M G Wilson). (d) Apparently higher-intensity alarm or fright-call of *eunomus* said by Dementiev and Gladkov (1954*b*) also to resemble *T. pilaris*.   MGW

Plumages (*T. n. eunomus*). ADULT MALE. In fresh plumage, feathers of crown, mantle, and scapulars black with olive-brown fringes, giving scaly appearance; back and rump similar, but olive-brown fringes less sharply demarcated and feathers with variable rufous-brown wash. Supercilium creamy-white. Lores and ear-coverts blackish. Chin and throat creamy or buffy-white; side of throat and upper throat with fine dark spots and streaks, tending to form thin malar stripe on side. Creamy throat distinctly demarcated from black chest and flanks; feathers of chest with white fringes, which become broader towards rear of flanks, giving markedly scaly appearance; dark parts of flank-feathers with variable rufous-brown tinge. Centre of breast and belly whitish, unspotted. Under tail-coverts white with mainly hidden rufous-brown feather-bases. Tail dark sooty brown, feathers fringed chestnut-brown at base of outer webs. Lesser upper wing-coverts sooty with olive-brown fringes and rufous suffusion; median and greater coverts rufous-brown with narrow and ill-defined whitish fringes along tips. Secondaries sooty brown, outer edges rufous-brown becoming whitish towards tip; outer webs of tertials wholly rufous. Primaries and primary coverts

sooty-brown, with bases and outer edges rufous except at tip; basal $\frac{2}{3}$ of inner webs of primaries rufous-brown. Axillaries, under wing-coverts, and undersurface of flight-feathers rufous; coverts along leading edge of wing with white fringes. *In worn plumage* (spring and summer), brown feather-fringes of upper-parts, white fringes of underparts, and most of dark throat-spots lost by abrasion, giving a more contrasting pattern of black, white, and rufous-brown; upper chest almost uniform black, bordered by white band on lower chest; breast and flanks boldly spotted black. ADULT FEMALE. Like adult ♂, but upperparts usually browner, throat more spotted, dark feather-centres of underparts browner, and upper wing-coverts with less extensive rufous fringes; in worn plumage, abrasion of feather-fringes gives more contrasting pattern on underparts. JUVENILE. Crown dark brown with fine pale shaft-streaks. Rest of upperparts brown, spotted; feathers with pale buff centres and dark tips, pale centres of scapulars especially large and well-defined. Chin and throat pale buff with sooty malar stripes; chest and flanks slightly deeper buff, heavily spotted sooty black; belly whitish with dark spots becoming smaller towards vent. Tail dark brown. Wings dark brown; coverts with broad rufous edges and pale tips, flight-feathers with rufous outer edges. FIRST ADULT MALE. Like adult ♂, but feathers of upperparts generally with less well defined scaling, giving more clouded appearance. Juvenile flight-feathers, tail, tertials, and greater upper primary coverts, and variable number of greater upper secondary coverts retained; rufous brown on wings paler, and greater coverts with more prominent pale tips. See also Svensson (1984*a*). FIRST ADULT FEMALE. Like adult ♀, but scaling of upperparts less marked, darker parts of feathers generally paler and duller; chest and flanks with broader pale feather-fringes and reduced dark feather-centres, giving noticeably paler appearance overall. Retained juvenile feathers as in 1st adult ♂.

Bare parts. ADULT, FIRST ADULT. Iris dark brown. Bill dark horn-brown, black on tip, yellow on basal $\frac{1}{2}$–$\frac{2}{3}$ of lower mandible; yellow more extensive in breeding season, extending to cutting edge and central section of upper mandible. Mouth yellow. Leg and foot dark brown. JUVENILE. Iris dark brown. Bill pale at fledging, becoming dark by 1st autumn; base of lower mandible paling to light brown or dull yellow by end of 1st winter. Leg and foot brown. (Coates 1960; BMNH.)

Moults. ADULT POST-BREEDING. Complete; primaries descendant. July–September, before leaving breeding area. POST-JUVENILE. Partial: head, body, lesser, median, and some inner greater upper wing-coverts. August–October, before autumn migration.

Measurements. *T. n. eunomus*. Whole geographical range, all year; skins (BMNH). Juvenile wing is retained juvenile wing of 1st adult. Bill to basal corner of nostril; to skull, on average 7·8 longer, exposed culmen on average 2·4 longer.

WING AD ♂	132·2 (2·86; 10)	128–136	♀ 125·5 (2·22; 10)	122–129
JUV	128·5 (1·84; 10)	126–132	126·2 (2·78; 10)	122–130
TAIL AD	94·0 (2·05; 10)	91–97	90·0 (3·30; 10)	84–93
BILL AD	15·6 (0·70; 10)	14–16	15·1 (0·74; 10)	14–16
TARSUS	33·0 (0·67; 10)	32–34	32·1 (0·74; 10)	31–33

Sex differences significant for adult wing, tail, and tarsus.

Nominate *naumanni*. Similar to *eunomus* or wing up to *c.* 5 longer (Hartert 1903–10; Dementiev and Gladkov 1954*b*; BMNH).

Weights. *T. n. eunomus*. USSR: ♂♂ 78, 83·5, 106; ♀♀ 87, 88 (Dementiev and Gladkov 1954*b*). Kuril Islands (USSR): ♂♂ 78, 79, 87; ♀ 78 (Nechaev 1969). Western China, December–May:

71·6 (7) 62–82 (Rensch 1924). On migration, north-east China: ♂ 74 (15) 63–88; ♀ 68 (11) 50–80 (Shaw 1936). Mongolia, on migration May–June: ♂♂ 57, 67; ♀ 60; sex unknown 57 (Piechocki and Bolod 1972). Exhausted ♂, Netherlands, November: 44 (ZMA). Intermediate ♂, France, November: 70 (Cheylan 1980).

Nominate *naumanni*. Mongolia, May: ♂ 63 (Piechocki and Bolod 1972). USSR: ♀♀ 74, 75, 81 (Dementiev and Gladkov 1954*b*).

Structure. Wing rather long, broad at base, tip fairly pointed. 10 primaries: p8 longest, p9 3–7 shorter, p7 0·5–1·5, p6 4–8, p5 16–21, p4 25–29, p1 36–43. P10 reduced; 78–85 shorter than p8, 9–16 shorter than longest upper primary covert. Outer web of p6–p8 and inner of (p7–)p8–p9 emarginated. Tail rather long, tip square or slightly rounded; t6 3–9 shorter than t2–t3. Middle toe with claw 26·1 (10) 23–29; outer toe with claw *c.* 72% of middle with claw, inner *c.* 68%, hind *c.* 73%. Remainder of structure as in Blackbird *T. merula* (p. 963), but bill relatively shorter.

Geographical variation. 2 distinct races (sometimes treated as separate species), relations between which are not fully understood. Compared with *eunomus*, nominate *naumanni* (breeding south of *eunomus*, south and east from south-central Siberia) has black parts of plumage replaced by olive-grey on upperparts and by rufous-brown on underparts, and tail mainly rufous. In west of range, breeding areas overlap, and individuals intermediate in colour are not uncommon there and in winter quarters; these have been attributed to hybridization but such intermediates also found in areas where no chance of hybridization, e.g. on Anadyr' river far from breeding range of nominate *naumanni* (Dementiev and Gladkov 1954*b*). Westernmost population of *eunomus*, breeding west from Anabar and Vilyuy basins, averages slightly larger than typical *eunomus* from eastern Siberia, and these birds sometimes separated as *turuchanensis* Johansen, 1954: wing of ♂♂ in west 130–138, in east 125–130 (Johansen 1954).

Forms species-group with Black-throated Thrush *T. ruficollis*, and sometimes considered conspecific with it (Portenko 1981, which see for details). DWS

Turdus ruficollis Black-Throated Thrush and Red-throated Thrush

PLATES 72 and 77
[between pages 856 and 857]

Du. Zwartkeellijster/Roodkeellijster Fr. Grive à gorge noire/Grive à gorge rousse
Ge. Schwarzkehldrossel/Rotkehldrossel Ru. Темнозобый дрозд
Sp. Zorzal papinegro/Zorzal papirrojo Sw. Taigatrast

Turdus ruficollis Pallas, 1776

Polytypic. *T. r. atrogularis* Jarocki, 1819 (Black-throated Thrush), east European USSR east through central Siberia to Nizhnyaya Tunguska river, south to Tarbagatai, Altai, and Tannu Ola; nominate *ruficollis* Pallas, 1776 (Red-throated Thrush), south-east of *atrogularis*, from Altai east to Transbaykalia, north to about 60°N on upper Nizhnyaya Tunguska and Lena.

Field characters. 25 cm; wing-span 37–40 cm. Close in size to Blackbird *T. merula*, but tail 20% shorter. Medium-sized, bulky thrush, with well-balanced form most recalling *T. merula* but behaviour and plumage pattern somewhat reminiscent of Fieldfare *T. pilaris*. Head and upperparts pale grey- to umber-brown; throat and chest dark, underbody dull white, and underwing rufous-buff. ♂ of western race (Black-throated Thrush) has black chest and black-brown tail, ♂ of eastern race (Red-throated Thrush) has dull red chest and tail; ♀♀ and immatures less contrasting. Sexes dissimilar; some seasonal variation. Juvenile separable. 2 races occur in west Palearctic, easily distinguishable (though intermediates occur).

ADULT MALE. (1) Central Asian race, nominate *ruficollis*. Crown, rear cheeks, back, and wings smoky-brown, with greyer or more olive tone according to angle of light, contrasting slightly with paler, less brown rump; rump contrasts quite markedly with dull red, dark-brown-centred tail. Supercilium dull red, eye-stripe brown, fore-cheeks, throat, and chest (down to shoulder and fore-flanks) dull red; rest of underparts grey-white. Upperwing uniform with back but all coverts and inner secondaries show paler margins, these being most obvious against dark centres of tertials; underwing rufous-buff. Bill black-horn, with all but tip of lower mandible dark yellow; legs brown-yellow. ADULT FEMALE. Duller than ♂, with more olive tone to upperparts and dull red foreparts paler. Chest either mottled with white and spotted with black-brown or wholly cream with rusty tinge and no spots; throat bordered with moustache of dark brown spots; crown lacks dark streaks. Rest of plumage as ♂ but usually tinged cream or rusty below chest. (2) Central USSR race, *atrogularis*. ADULT MALE. Differs distinctly from ♂ nominate *ruficollis* in having fore-face, throat, and chest black, tail wholly dark brown, and flanks grey, with smoky-brown spots on flanks and rear of underbody. Contrast of paler, greyer rump with back and tail more marked than in nominate *ruficollis*; crown and nape streaked black-brown. Underwing orange-buff. From August, fresh dull white feather tips produce more scaled or spotted appearance on throat and breast but this lost by mid-winter. Bill rather paler than in nominate *ruficollis*; legs dusky-grey, less yellow. ADULT FEMALE. More olive-toned above than ♂, with virtually unstreaked crown, indistinct grey-buff supercilium, chin, throat, and

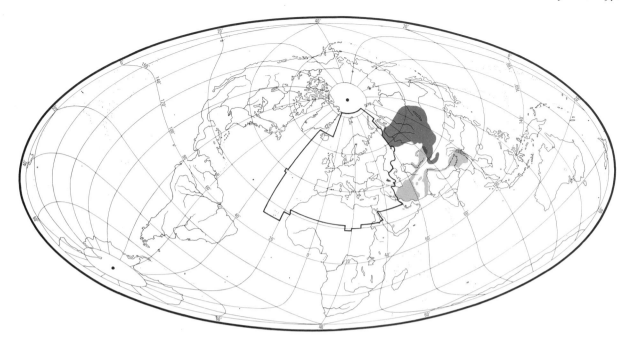

cheeks white with dark brown streaks, and chest buff with black-brown streaks and spots. Flanks darker than ♂, with stronger smoky-brown suffusion. Fringes of wing-feathers buffier than ♂, and rusty inner webs of tail-feathers may show when spread. JUVENILE. Both races. Resembles adult ♀ but quite markedly streaked grey-white and close-barred dark brown above, and spotted and closely barred dark brown on throat, chest, and flanks. Wing-coverts streaked and tipped grey-white on lesser and median, tipped buff-white on greater. Bill dark horn. FIRST WINTER. Both races. ♂ intermediate in appearance between adult ♂ and ♀; ♀ like adult ♀. Both difficult to age except for retention of juvenile wing-coverts and incomplete face and chest pattern, with larger grey-white tips to feathers. Note that individual birds of any age may be intermediate between the 2 races, some *atrogularis* having pronounced degree of dull red in ground-colour of breast and on outer tail-feathers.

Adult of both races distinctive, with dark chest, rather uniform upperparts, and pale rear underbody producing pattern only reminiscent of Eye-browed Thrush *T. obscurus* (much smaller, with obvious white supercilium and orange-buff lower breast and flanks). However, less typically marked ♀♀ of nominate *ruficollis* and immatures of both races may invite confusion with American Robin *T. migratorius* (see p. 1023) and (particularly) Naumann's Thrush *T. naumanni naumanni* (see p. 967). Flight action over distance more like *T. pilaris* than *T. merula*, with bursts of strong wing-beats and momentary closures of wings producing direct, little-undulating progress; within cover and when feeding, more like *T. merula*, with similar habit of cocking tail on landing. Escapes into tree canopy when alarmed. Gait as *T. merula* but with markedly long

hops and more upright carriage, sometimes suggesting wheatear *Oenanthe*. Highly gregarious, mixing freely with other thrushes.

Song resembles Song Thrush *T. philomelos*. In alarm, gives throaty 'which-which-which', recalling chuckles of *T. merula* but softer in tone. Flight-call a thin 'see', similar to Redwing *T. iliacus* and continental race of *T. philomelos*.

Habitat. Breeds in central and marginally in west Palearctic in upper to middle latitudes from lowlands and boreal continental to montane temperate zones, with marked ecological differences between northern race *atrogularis* and south-eastern nominate *ruficollis*. Nominate *ruficollis* inhabits sparse mountain forests, mossy scrub tundra above them, taiga on plateaux, and bottomland forests by mountain rivers. *T. r. atrogularis* frequents borders and open parts of various types of forest including low swampy taiga with sunny glades or clearings due to fire or other influences; also riversides, sparse dry woods or clusters of (e.g.) larches *Larix* scattered over subalpine steppe, and in south of range up mountains to c. 2000–2200 m. On Yenisey river, found to favour pine forest, but locally elsewhere breeds in groves of birch or poplar, and also in thickets of buckthorn *Rhamnus*. Forages on ground in the open, but flies up to trees when disturbed; likes to be by water. In mountain regions, after young fledge, parties ascend to zone of alpine shrubs, or of subalpine sparse woodland. A few remain to winter in breeding range near springs and unfrozen streams, especially by towns or villages. (Dementiev and Gladkov 1954*b*.) Most winter in Himalayas and adjoining mountains up to 3000–4200 m, descending to plains under

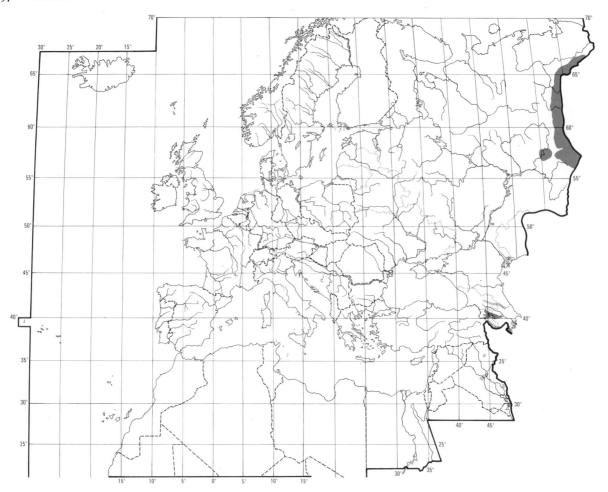

stress of weather, and resorting to cultivation, stubble fields, pastures, grassy slopes, fallow land with sparse scrub, and edges of forest (Ali and Ripley 1973*b*).

Distribution and population. No information on changes.

USSR. Common (Dementiev and Gladkov 1954*b*).

Accidental. Britain, France, Belgium, Netherlands, West Germany, Denmark, Norway, Sweden, Finland, East Germany, Poland, Czechoslovakia, Austria, Italy, Greece, Spain, Bulgaria, Rumania, Iraq, Israel, Egypt.

Movements. Migratory, but some stay to winter in Siberia in years with good berry crop (Johansen 1954). West Siberian race *atrogularis* migrates through northern Iran, Afghanistan, and Mongolia to winter abundantly all across Indian subcontinent as well as from Arabia and Iraq to Burma and south-west China north to Turkmeniya and Himalayas, sometimes north to southernmost Kazakhstan; also winters in Altai and Sayan mountains and occasionally as far north as Tomsk (Johansen 1954). East Siberian race, nominate *ruficollis*, migrates through Mongolia and China west to Sinkiang and Tibet to winter

in Afghanistan, northern Pakistan, northern Kashmir to Assam, northern Burma, and China mainly in north and west (Ticehurst 1926-7; Dementiev and Gladkov 1954*b*).

Southward migration of *atrogularis* begins at end of August or early September, continuing to early October at Tomsk. Autumn departures around Krasnoyarsk occur mid-October and passage observed at Semipalatinsk up to end of October. In eastern part of western Siberia, migration proceeds more or less due south along Ob' and Irtysh valleys. In Urals and in western part of western Siberia, birds migrate towards south-east. (Dementiev and Gladkov 1954*b*; Johansen 1954.) Arrives on high ground in northern parts of Indian subcontinent in October, descending to lower valleys and plains as winter progresses (Ali and Ripley 1973*b*). Present in eastern Saudi Arabia (usually scarce, but numbers variable) November to early March (G Bundy). Spring passage of *atrogularis* is both diurnal and nocturnal with record of bird-strike at night at 3300 m on 7 April near Alma-Ata (Sema 1974). In India return passage is March–April, over shorter period than autumn migration (Ali and Ripley 1973*b*). Migrating birds observed at Semipalatinsk and Zaysansk as early as 21 March and large numbers seen in northern Altai 5–10 April. At Tomsk, where wintering sometimes occurs, migrants arrive on 15–20 April. (Johansen 1954.) Appears around Omsk in early May and around Krasnoyarsk in mid-April (Dementiev and Gladkov 1954b).

Little detail for nominate *ruficollis*. In the Altai at *c.* 2000 m, birds that finish nesting early begin to move to lower levels in mid-July, with bulk of population following in August (Johansen 1954). Leaves breeding grounds from late September and early October. In Mongolia in Kentei mountains, spring passage begins 16–19 April and continues until 10 May. Occurs central Gobi in late April and arrives then in south-west Transbaykalia. Appears around Irkutsk in mid-April (Dementiev and Gladkov 1954b).

Vagrants in west Palearctic are mostly *atrogularis*. In Britain and Ireland (all *atrogularis*), most occur late autumn or winter; may be birds which have not moved south of breeding grounds until onset of severe weather (Simms 1978).   LAB

Food. Invertebrates and berries. Feeds on ground by hopping and pausing, cocking head to one side like (e.g.) Song Thrush *T. philomelos* but less exaggeratedly than Blackbird *T. merula*; often perches briefly on logs, molehills, etc. (D J Brooks). Perches in trees and bushes to take berries (Ali and Ripley 1973b; Gallagher and Woodcock 1980). Will feed in water of streams, taking aquatic invertebrates (Dementiev and Gladkov 1954b).

Diet includes the following. Invertebrates: grasshoppers, etc. (Orthoptera: Tettigoniidae, Acrididae), earwigs (Dermaptera), termites (Isoptera), bugs (Hemiptera), adult and larval Lepidoptera (Noctuidae, Geometridae), larvae and eggs of flies (Diptera: Tipulidae, Tabanidae, Muscidae), Hymenoptera (Tenthredinidae, ants Formicidae), adult and larval beetles (Coleoptera: Carabidae, Geotrupidae, Histeridae, Staphylinidae, Scarabaeidae, Elateridae, Tenebrionidae, Cerambycidae, Chrysomelidae, Curculionidae), spiders (Araneae), woodlice (Isopoda), small molluscs, earthworms (Oligochaeta). Plant material mainly fruits and some seeds: of spruce *Picea*, juniper *Juniperus*, docks and knotweeds (Polygonaceae), cherries *Prunus*, Leguminosae, sea buckthorn *Hippophae*, *Eleagnus*, *Berberis*, *Nitraria*, *Zizyphus*, *Magnolia*, and grasses (Gramineae); also nectar. (Isakov and Vorobiev 1940; Pek and Fedyanina 1961; Dolgushin *et al.* 1970; Ali and Ripley 1973b; Ogorodnikova 1980.)

In Kirgiziya (USSR), 75 birds contained mainly invertebrates, especially adult and larval beetles and larval Lepidoptera (Pek and Fedyanina 1961). Apparently eats largely or wholly berries and seeds in winter (e.g. Ludlow and Kinnear 1933, Ivanov 1945, Fleming 1968, Morioka and Sakane 1981), though one stomach from north-east Burma, March, contained only invertebrates (Kinnear 1934). For further data (mostly scanty), see Altum (1866), Grote (1925, 1935a), Jones (1947–8), Yanushevich *et al.* (1961), and Kekilova (1970).

Food of young in Tomsk area (USSR) mainly earthworms, also caterpillars and beetles (Moskvitin 1969). In central Siberia, mainly earthworms (Reymers 1966). At one nest in Chita region (USSR), food comprised 93·4% insects (mainly Diptera), also spiders and earthworms (Ogorodnikova 1980).   DJB

Social pattern and behaviour. Most aspects poorly known; for summary, including comparisons with Dusky/Naumann's Thrush *T. naumanni*, see Portenko (1981).

1. Often in flocks outside breeding season, birds also concentrating where food abundant (e.g. Ivanov 1945). In valleys of Nepal, *atrogularis* sometimes recorded singly and flocks (5–50 birds) normally loose-knit, never so compact as (e.g.) migrating Fieldfare *T. pilaris* or Redwing *T. iliacus* (Diesselhorst 1968a; Fleming *et al.* 1976). Small loose flocks noted also in Oman (Gallagher and Woodcock 1980) and frequently of 3–30 in Indian subcontinent, building up to *c.* 50 prior to spring departure (Ali and Ripley 1973b); flocks of up to several hundred or thousand recorded in Kazakhstan, Tadzhikistan, and Altai mountains (USSR) in winter and spring (Sushkin 1938; Ivanov 1969; Dolgushin *et al.* 1970; Stakheev 1979), though 'pairs' (as well as small flocks) noted in Semirech'e (Kazakhstan), December–February (Shnitnikov 1949). Migrates singly or in small flocks (Dresser 1871–81; Kozlova 1930; Grote 1935; Reymers 1966), and report in Bamberg (1911) suggests that small parties occur on breeding grounds when breeding quite far advanced. Flocking occurs from July, increasingly from mid-month (Moskvitin 1969). In Baluchistan, small flocks noted high up in juniper *Juniperus* forest late September, with 'swarms', nearly all adult ♂♂, by early October (Ticehurst 1926–7); see also Meinertzhagen (1920a). Intermingling of nominate *ruficollis* and *atrogularis* frequently recorded: in Altai, most contact between the races in period of post-breeding dispersal, least during spring and autumn migration, and only sporadic in winter (Stakheev 1979). Frequently associates outside breeding season also with other *Turdus*: in Kazakhstan, mixed flocks formed with Mistle Thrush *T. viscivorus* and *T. pilaris* during post-breeding dispersal (Dolgushin *et al.* 1970), in Indian subcontinent with Tickell's Thrush *T. unicolor* and Eye-browed Thrush *T. obscurus* (Ali and Ripley 1973b); Naumann (1905) noted association of nominate *ruficollis* with *T. naumanni*. BONDS. Nothing to indicate mating system other than monogamous. For report of 'pairs' in winter, see above. According to Portenko (1981), not infrequently in pairs on migration, suggesting that pair-formation sometimes takes place before arrival on breeding grounds—though Moskvitin (1969) noted that ♂♂ of *atrogularis* arrive ahead of ♀♀ in Tomsk area. Interbreeding occurs within whole *T. ruficollis*–*T. naumanni* complex; see, especially, Portenko (1981), also Seebohm and Sharpe (1902) and Meeth (1967). No information on role of sexes in care of young. In southern USSR/Mongolia, family parties noted 16–20 July when young already full-grown (Dementiev and Gladkov 1954b). BREEDING DISPERSION. Solitary and territorial, though nominate *ruficollis* is said to nest sometimes in well-dispersed small colonies like *T. pilaris* (Bamberg 1911; Grote 1925), while another report claimed this does not occur, pairs rarely nesting close together (see Portenko 1981). *T. r. atrogularis* normally solitary, even where several pairs in same locality (Popham 1898; Dementiev and Gladkov 1954b); however, in Tomsk area, nests not infrequently 20–30 m apart where feeding conditions favourable and, in 'mixed thrush colonies' (no details), average 12% of all nests (Moskvitin 1969). In Tuva (south of Krasnoyarsk, USSR), nests of nominate *ruficollis* and *atrogularis* c. 30–40 m apart (Berman

and Zabelin 1963). Densities in various parts of USSR as follows. Only west Palearctic study—in Pechora basin and adjoining Urals—revealed 4·8 birds per km² of montane birch *Betula*, 17-25·4 birds per km² of spruce *Picea* with green moss or herbaceous ground cover, 35·6 birds per km² in montane willow *Salix*, and 48 per km² in montane larch *Larix* (Estafiev 1981). On eastern slopes of central Urals, 4 pairs per km² in mature spruce and fir *Abies* (Danilov 1958). For densities in Narym (Ob' river valley), see Panteleev (1972). In Salairskiy kryazh (mountains, east of Novosibirsk), willow by water favoured—136·2 birds per km², up to 40 or more per km² in depths of mixed woods (Chunikhin 1965). In southern taiga around Tomsk and Anzhero-Sudzhensk, 22-27 pairs per km², further north on Tym and Yeloguy rivers 5-30 pairs per km²; mainly in dense fir, spruce, and pine *Pinus*, with some poplar *Populus*; low density otherwise (Moskvitin 1969); for further counts along Yeloguy, see Rogacheva (1962); for Krasnoyarsk, see Naumov (1960). In central Siberia, 10-15 pairs per km² with regenerated burnt-over areas favoured (see Reymers 1966); in northern taiga of western Siberia, highest density (25 birds per km²) in dense riverine taiga, fewer (2 birds per km²) in burnt-over *Pinus*; further details in Vartapetov (1984). In Altai, 40-80 birds (*atrogularis*) per km² in lower forest zone on shore of lake; at another lake, 32 nominate *ruficollis* per km² (Stakheev 1979); see also Sushkin (1938) and Ravkin (1973). Further examples for nominate *ruficollis*: in Tuva, 13 birds per km² in sparse *Pinus* (Berman and Kolonin 1967); in *Larix* east of Lake Baykal, 0·4-1 birds per km² (Izmaylov and Borovitskaya 1967). ROOSTING. Nocturnal and, at least outside breeding season, communal; in thickets and densely foliaged trees (Grote 1925; Ali and Ripley 1973b; Piechocki *et al.* 1982). In northern Urals, active during light night in late June (Portenko 1937). Feeds mainly in morning and, in hot weather of late summer, seeks out denser and darker parts of wood. Bathes readily (Grote 1925; Portenko 1981).

2. Usually rather shy in and outside breeding season. Nominate *ruficollis* in Mongolia shy after arrival, less so later in season (Kozlova 1930). Will fly up into tree if disturbed, sometimes giving alarm-call (see 2a in Voice), and may fly off if flushed (Dresser 1871-81; Seebohm and Sharpe 1902; Grote 1925; Portenko 1937; Meeth 1967; Ali and Ripley 1973b; Gallagher and Woodcock 1980). Sometimes allows a close approach—e.g. after flying up to perch (Moore and Boswell 1956)—and tame in very cold weather (Dresser 1871-81). Vagrant in Shetland would stop and raise head when approached, staring intently often for several minutes if observer stayed at c. 15 m; usually retreated by running, flying only when pressed (Davis 1958a). Usually raises tail and may also flick wings after landing (Meinertzhagen 1930; Fleming and Traylor 1964). FLOCK BEHAVIOUR. No details. SONG-DISPLAY. ♂ sings (see 1 in Voice) from tree (e.g. Reymers 1966). Will sing on spring migration, also (low-intensity) in early September if weather fine. During incubation, ♂ of nominate *ruficollis* will sing far into the night. Vigorous song noted in western Urals 20-24 June (Portenko 1981). In Tomsk area, song-period not long, extending from late April up to hatching (23 May-21 June) (Moskvitin 1969). In Mongolia, song noted from early June after heavy passage mid-April to c. 10 May (Kozlova 1930). No song apparently in winter quarters (Fleming and Traylor 1964). ANTAGONISTIC BEHAVIOUR. (1) General. In Tomsk area, territories occupied from c. 21 April, though peak passage continues for c. 1 week more (Moskvitin 1969). On Chadobets river (tributary of Angara, north-east of Krasnoyarsk), arrival from late April, passage continuing to c. 6-8 May and territories occupied by mid-May (Reymers 1966). (2)

Threat and fighting. Nominate *ruficollis* quarrelsome on breeding grounds, any conspecific bird coming close being driven off immediately (Bamberg 1911). In Iraq, winter, *atrogularis* also typically pugnacious, driving away other *Turdus* on same lawn or flower bed (Chapman and McGeoch 1956). HETEROSEXUAL BEHAVIOUR. ♂♂ first to arrive on breeding grounds in Tomsk area (from c. 10-14 April); main laying period 9 June-30 June (Moskvitin 1969). In Mongolia, pair-formation in nominate *ruficollis* takes place in early June (Kozlova 1930); perhaps this late because refers to montane population (Portenko 1981). RELATIONS WITHIN FAMILY GROUP. No information on nestling phase. Young leave nest at 11-12 days (Moskvitin 1969), hiding initially in dense trees or in vegetation on ground (Grote 1935). ANTI-PREDATOR RESPONSES OF YOUNG. No information. PARENTAL ANTI-PREDATOR STRATEGIES. In Mongolia, nominate *ruficollis* shy, not coming near to feed nestlings while observer present (Kozlova 1930). Such behaviour not necessarily typical: both nominate *ruficollis* and *atrogularis* can be quite bold, flying from tree to tree near nest or fluttering about close to intruder and giving loud alarm calls (Popham 1898; Grote 1925, 1935; Portenko 1937, 1981: see 2c in Voice).   MGW

Voice. Rather quiet on migration and in winter (Naumann 1905; Moore and Boswell 1956; Davis 1958a). Differences appear to exist between nominate *ruficollis* and *atrogularis* in song (based on analysis of only 1 recording for each race) and transcriptions suggest this perhaps true also of other calls, though more information required (see below).

CALLS OF ADULTS. (1) Song of ♂. In *atrogularis*, a few whistled notes given at a time and thus like Song Thrush *T. philomelos*; considerable variation (Popham 1898). Quiet song (still developing or perhaps Subsong) of captive *atrogularis* likened to that of Blackbird *T. merula* (Naumann 1905). Recording of vagrant *atrogularis* reveals song overall similar to *T. philomelos*, though in part recalling *T. merula*. Prominent are loud whistled sounds of varying timbre—e.g. quite deep and rich 'hweet-a' (Fig I), often given 3 times in succession, and powerful, drawn-out and slightly vibrato 'h-weeeeeee' notes not unlike telephone bell similarly repeated (Fig II). Some thinner and more squeezed, almost strangled, warbling and chirruping sounds also evident as in Fig III which shows phrase more like *T. merula*, melodic at beginning, and with decorative coda added. *Contra* Popham (1898), repetition in manner of *T. philomelos* apparently typical; apart from frequent contiguous repetition of single notes, similar to *T. philomelos* also in interpolation of decorative but quiet phrases between bouts of repetition (J Hall-Craggs, M G Wilson). Simple song of nominate *ruficollis* alleged to be less melodious than *T. philomelos*, whistled sounds being (largely) absent and hissing and squeaking sounds frequently interspersed (Dementiev and Gladkov 1954b; Portenko 1981). Schubert (1982) noted loud 2-part song in almost melodic intervals. Analysis of same recording (27 phrases) as follows. In form quite unlike *atrogularis*, though notes similar in structure and frequency range; small, well-organized repertoire (larger, more varied in *atrogularis*) of 3 figures (very few minor exceptions), each of 2-4 units. Short figures combined in definite patterns and constituent figures used as either

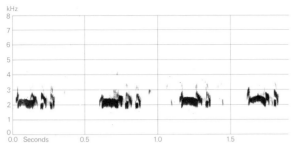

I S Palmér/Sveriges Radio (1972–80) Sweden May 1977

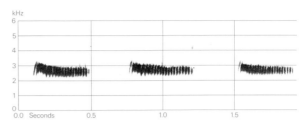

II S Palmér/Sveriges Radio (1972–80) Sweden May 1977

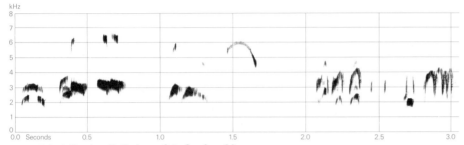

III S Palmér/Sveriges Radio (1972–80) Sweden May 1977

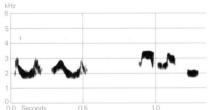

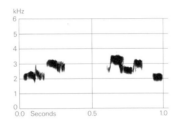

IV Schubert (1982)
Mongolia June 1979

V Schubert (1982)
Mongolia June 1979

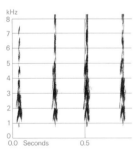

VI B N Veprintsev and V V Leonovich/Sveriges Radio (1972–80)
USSR June 1975

introductory or final figures in resultant phrases. Of 2 initial figures, type (i) used only as such: see Fig IV in which phrase rendered 'chooee chooee whee-oo-ee oo'. Initial figure type (ii) twice follows type (i), but Fig V shows the commoner pattern with type (ii) preceding same (apparently the only) final figure as in Fig IV: 'hoo-eee whee-oo-ee oo'. Of 27 phrases only 14 full, others comprising a single figure—usually initial figure type (i) (J Hall-Craggs, M G Wilson). (2) Contact-alarm calls. Some calls difficult to ascribe owing to multiplicity of transcriptions which in many cases not supported by contextual or other detail; overlap probably exists. Some duplication likely in following necessarily arbitrary scheme. (a) A 'chuck', 'tjuk', 'chik', or 'schäck' given singly, 2 close together (e.g. 'schnäckschäck' in nominate *ruficollis*) or in series of 3 or more; Fig VI shows 4 quite soft, not harsh units, suggesting 'tchk', 'tcht' 'tj(u)k' or 'quit' (J Hall-Craggs; M G Wilson). Often likened to chuckling chatter (rattle) of Ring Ouzel *T. torquatus* (e.g. Naumann 1905, Meinertzhagen 1930, Moore and Boswell 1956) or sometimes to Fieldfare *T. pilaris* (Mauersberger *et al.* 1982), though Naumann (1905) considered it higher pitched than *T. pilaris* and Dolgushin *et al.* (1970) much coarser. In vagrant nominate *ruficollis*, like rapid alarm-call of *T. merula*, but higher pitched (Meeth 1967) and another report referred to ♂'s contact-call being like Mistle Thrush *T. viscivorus* (Bamberg 1911). In *atrogularis* (though nominate *ruficollis* apparently same) throaty, chuckling 'which-which-which' of alarm like that of *T. merula*, but much softer (Ali and Ripley 1973*b*); rather harsh 'squeetch', otherwise similarly described (Smythies 1953), is probably the same. In Nepal, contact-call rendered 'puk peer-up' (Fleming *et al.* 1976); at least part of this call probably same as others here described. Call 2a probably serves as contact-call, also expressing general excitement and, at least in some cases, alarm. (b) Probably closely related to call 2a (perhaps only a variant) and apparently expressing alarm. Characteristic 'tack-tack' (sometimes in longer series) given commonly; loud in alarm (Meinertzhagen 1930; Smythies 1953; Mauersberger *et al.* 1982). Further descriptions as follows: 'tsak-tsak-tsak' (Dolgushin *et al.* 1970); 'jak' like Redwing *T. iliacus* from captive *atrogularis* (Naumann 1905); abrupt 'chap', 'kak' and 'cha' sounds given by alarmed nominate *ruficollis* when man approaching nest (Portenko 1981). (c) *T. r. atrogularis* alarmed at nest gave 'chit chit chéet' (Popham 1898); similar 'pee-wit chip-chi-chip' given by alarmed birds of same race in Nepal (Fleming *et al.* 1976). In Mongolia, birds (*atrogularis*) in small feeding flock called once like *T. merula* (presumably call 2a) and otherwise gave rapid 'hetetetet hetetetet' and very hoarse 'retet riepriep' (Piechocki *et al.* 1982); probably the same or closely related. Rapid 'wiwiwi' given (also in Mongolia) by nominate *ruficollis* at take-off (Piechocki *et al.* 1982) perhaps also belongs here. (d) Thin 'seee' or 'ziep' like *T. iliacus* or *T. philomelos* apparently also a contact-call (Naumann 1905; Ali and Ripley 1973b). Grote (1935) likened call given by migrating birds (also at night) to that of *T. pilaris* in similar situation; this presumably Huit-call of *T. pilaris* (see that species, p. 986) and thus also best classified here. (e) A 'schrieh schrieh' like *T. pilaris* given (as well as calls 2a–b) by nominate *ruficollis* in Mongolia (Mauersberger *et al.* 1982). (3) Distress-call of captive *atrogularis* like *T. merula* (Naumann 1905).

CALLS OF YOUNG. No information.   MGW

Breeding. SEASON. North-west USSR: main laying period May (Dementiev and Gladkov 1954b; Johansen 1954). SITE. On low stump, or in tree, close to ground. Nest: external layer of grass stems and leaves, a thick rim of stalks, thickly plastered with mud, and inner lining of finer grass. Dimensions of 1 nest of *atrogularis*: outer diameter 13·0 × 13·5 cm, inner diameter 10·2 × 10·5 cm, height 12·0 cm, depth 7·5 cm (Bocheński 1968). Building: no information on role of sexes. EGGS. See Plate 83. Sub-elliptical, smooth and glossy; light blue with fine reddish-brown speckling and mottling, sometimes thicker at broad end. *T. r. atrogularis*: 29·4 × 21·4 mm (27·4–32·0 × 19·6–23·0), *n* = 47; calculated weight 7·2 g (Schönwetter 1979). Clutch: 5–6(4–7). One brood (Dementiev and Gladkov 1954b). INCUBATION. 11–12 days in *atrogularis* (Moskvitin 1969). In nominate *ruficollis*, by ♀ only (Dementiev and Gladkov 1954b). No further information.

Plumages (*T. r. atrogularis*). ADULT MALE. In fresh plumage (autumn), forehead, crown, rear part of side of head, and rest of upperparts grey-brown, feathers of forehead and crown with blackish centres. Lores, area immediately round eye, cheeks, chin, throat, sides of neck, and chest black, feathers with conspicuous pale fringes. Rest of underparts dull white, flanks washed buffy grey and sometimes indistinctly streaked dark grey; under tail-coverts grey-white with some dark-brown at feather-bases. Tail dark grey-brown above, paler below. Upperwing grey-brown; outer edges of feathers paler, tending to olive-grey on upper wing-coverts and secondaries. Axillaries and under wing-coverts contrastingly warm orange-buff, whitish

along margin of wing. Exceptionally, all head and neck black (*relicta* morph). *In worn plumage* (spring and summer), pale fringes of face and chest worn off, appearing uniform black. ADULT FEMALE. Crown and upperparts like adult ♂ but slightly browner, less grey. Sides of head grey-brown, with narrow whitish supercilium variable in extent; lores and area round eye darker and more sooty. Chin and throat variable; in some birds, like adult ♂ but with broader whitish feather-fringes, appearing streaked; in others, whitish to pale rufous-buff, with broad dark streaks at sides and narrower streaks at centre, leaving uniform white or pink-buff area on chin and central throat. Feathers of chest sooty black with broad pale buff fringes, giving spotted or mottled appearance; rest of underparts dull white, clouded with grey-brown laterally, faintly streaked with dark brown except for central belly and vent. Under tail-coverts white, feathers with dull brown marks near bases. Tail and wing like adult ♂. *In worn plumage*, pale feather-fringes on face and chest partly abraded, dark marks on throat and chest more conspicuous, on chest forming heavily black-mottled band. NESTLING. No information. JUVENILE. Upperparts olive-brown or grey-brown; crown-feathers with dark tips, feathers of mantle, scapulars, and back with dark tips and pale buff shaft-streaks, and feathers of rump and upper tail-coverts with pale fringes. Sides of head olive-brown with narrow buff supercilium. Chin, throat, and chest pale buff, feathers at sides of throat with dark streaks and those of chest with sooty spot on tip. Breast and belly whitish, feathers with small dark tips gradually narrowing towards vent. Tail dull olive-brown. Wing dull olive-brown, outer edges of secondaries buff or rufous buff, tertials with variable amount of white on tip. Greater upper wing-coverts with buff edges and whitish tips; lesser and median coverts with buff tips narrowing to short shaft-streaks. Axillaries and under wing-coverts warm orange-buff. FIRST ADULT MALE. Rather variable. Either like adult ♂, but black of head and chest with broader pale fringes (especially on chin and throat), or very similar to adult ♀, chin and throat with conspicuous pale streaks. Juvenile flight-feathers, tail, tertials, greater upper primary coverts, and variable number of outer greater secondary coverts retained, more worn than neighbouring fresh feathers; outer coverts and tertials with rufous or sandy outer border, bleaching to white. *In worn plumage*, pale fringes of face and chest partly worn off, appearing blacker, though not as uniform as adult ♂. See also Svensson (1984a). FIRST ADULT FEMALE. Variable. Generally like adult ♀, but head and underparts paler, pale supercilium more marked, and feathers of chest mainly pale buff-grey with restricted dark central area. Retained juvenile feathers as in 1st adult ♂; outer greater coverts with rufous to off-white outer borders.

Bare parts. ADULT, FIRST ADULT. Iris dark brown. Bill mainly dark horn or blackish, basal half of lower mandible and adjacent cutting edge of upper mandible yellow or orange-yellow. Leg and foot variable, pale brown, yellow-grey, brown-grey, or dark horn-brown; soles pale. NESTLING. No information. JUVENILE. Iris dark brown. Bill pale at fledging, soon darkening to dark horn; yellow at base of lower mandible developing by 1st autumn. Leg and foot pale at fledging, usually darkening in 1st autumn and winter.

Moults. ADULT POST-BREEDING. Complete; primaries descendant. Late June to late August or early September, before leaving breeding area. POST-JUVENILE. Partial: head, body, lesser, median, and variable number of inner greater upper wing-coverts. Starts late June to early August (depending on hatching date) and completed by start of autumn migration.

(Dementiev and Gladkov 1954*b*; Piechocki and Bolod 1972; BMNH.)

Measurements. *T. r. atrogularis*. Mainly Himalayas, winter to early spring; skins, mainly BMNH. Juvenile wing is retained juvenile wing of 1st adult. Bill to basal corner of nostril; to skull, on average 8·8 longer, exposed culmen on average 3·2 longer.

	JUV			
WING AD ♂	136·8 (2·53; 10)	134–142	♀ 133·3 (2·26; 10)	130–136
	133·5 (1·58; 10)	132–136	130·0 (2·21; 10)	125–133
TAIL AD	101·1 (3·25; 10)	97–107	99·1 (2·02; 1 96–103	
BILL AD	15·1 (0·74; 10)	14–16	15·1 (0·60; 9)	14–16
TARSUS	33·4 (1·17; 10)	32–35	32·9 (0·99; 10)	31–34

Sex differences significant for wing.

Nominate *ruficollis*. Similar to *atrogularis* (Dementiev and Gladkov 1954*b*; Vaurie 1972; BMNH). Wing, Mongolia and China: ♂ 139·7 (2·90; 11) 135–143, ♀ 135·5 (3·78; 6) 131–142 (Stresemann *et al.* 1937; Piechocki and Bolod 1972).

Weights. *T. r. atrogularis*. USSR: ♂ 90·1 (6) 84·2–102; ♀ 85·7 (5) 80–97 (Dementiev and Gladkov 1954*b*). Kazakhstan: November–February, ♂ 88·7 (3) 81–96, ♀♀ 88, 92; March–April, ♂ 91·5 (6·54; 11) 83–103, ♀ 87·1 (6·84; 10) 75–96; May–June, ♂ 84·5 (6·81; 4) 75–91, ♀ 81; September–October, ♂ 95·0 (6·57; 12) 82–105, ♀ 93·0 (3) 87–101 (Dolgushin *et al.* 1970). Kuwait, November–February: ♂ 90·6 (19·4; 5) 66–110 (V A D Sales, BTO). India, November–March: 77·1 (21) 59–94 (Ali and Ripley 1973*b*). Afghanistan: April, ♂ 78, ♀♀ 83, 96; September, ♂ 83, ♀ 72 (Paludan 1959). South-west Siberia, September: ♀ 68 (Havlín and Jurlov 1977). Mongolia: May, ♂ 71; August, ♂♂ 82, 87, ♀♀ 82, 94 (Piechocki and Bolod 1972). Ladakh, October–November: ♂ 88, ♀ 75·7 (7·16; 5) 65–82

(Delany *et al.* 1982). Norway, December: ♀ 100 (Haftorn 1971).

Nominate *ruficollis*. USSR: ♂♂ 76, 79; ♀ 63 (Dementiev and Gladkov 1954*b*). Mongolia: May–June, ♂ 82·5 (6·50; 6) 76–90, ♀ 93·0 (6·60; 5) 85–103; early August, ♂ 87·2 (5·93; 5) 78–94 (Piechocki and Bolod 1972).

Structure. Wing rather long, broad at base, tip fairly pointed. 10 primaries: p8 longest, p9 3–11 shorter, p7 0·5–2·5, p6 4–10, p5 16–22, p4 24–31, p1 37–47. P10 reduced, 78–90 shorter than p8, 8–15 shorter than longest upper primary covert. Outer web of p6–p8 and inner of (p7–)p8–p9 emarginated. Tail rather long, tip square; 12 feathers. Middle toe with claw 26·9 (10) 25–29; outer and hind toe with claw both *c.* 72% of middle with claw, inner *c.* 64%. Remainder of structure as in Blackbird *T. merula* (p. 963), but bill and tarsus relatively short.

Geographical variation. 2 quite distinct races (sometimes considered different species: Stepanyan 1978*a*), which overlap and regularly hybridize in area of overlap. Compared with *atrogularis*, nominate *ruficollis* (breeding south-east of *atrogularis*, from Altai east to Transbaykalia) has black of head, throat, and chest replaced by russet-brown; upperparts rather paler, and tail extensively rufous. Intermediates locally more numerous than the typical forms; reported from Altai, western Sayan, and Nizhnyaya Tunguska (Dementiev and Gladkov 1954*b*); for appearance, see Stepanyan (1982). Populations of *atrogularis* from northern Urals are sometimes separated as *vogulorum* Portenko, 1981; said to have upperparts more olive-brown than in rest of range (Portenko 1981), but difference very slight. Sometimes considered conspecific with Dusky Thrush *T. naumanni* (see Portenko 1981).   DWS

Turdus pilaris Fieldfare

PLATES 73 and 78
[between pages 856 and 857]

Du. Kramsvogel   Fr. Grive litorne   Ge. Wacholderdrossel
Ru. Рябинник   Sp. Zorzal real   Sw. Björktrast

Turdus pilaris Linnaeus, 1758

Monotypic

Field characters. 25·5 cm; wing-span 39–42 cm. 5–10% shorter than Mistle Thrush *T. viscivorus* but proportionately longer-tailed; 20–25% longer than Redwing *T. iliacus*. Large, bold, long-tailed, often noisy thrush, with rather rakish form both on ground and in the air. Plumage more boldly variegated and richly coloured than any other west Palearctic thrush, with blue-grey head, vinous-chestnut back, grey rump, and almost black tail obvious on ground, and heavily speckled breast and flanks, white vent, and black undertail obvious from below. Combination of grey rump, black tail, and white underwing diagnostic. Flight characteristically leisurely. Commonest call diagnostic. Sexes closely similar; little seasonal variation. Juvenile separable.

ADULT MALE. Forehead and crown blue-grey with black-brown streaks; long but rather indistinct grey-white to cream supercilium; lore and patch under eye black; cheeks grey. Nape blue-grey, and thus, together with crown, contrasting with dark rufous-chestnut, black-mottled back, scapulars, and fore-wing-coverts and black-

brown, rufous-edged and grey-tipped greater coverts and secondaries. Rump grey, less blue than crown and nape, and so contrasting even more strongly with inner wings and with brown-black primaries and tail. Chin cream-white but throat, chest, and flanks warm, almost orange-buff, boldly marked with brown-black speckles and spots on throat and chest and fat chevrons on flanks; rest of underparts and underwing white. From August, fresh plumage shows even bluer tone on back of head and rump, darker back (looking almost purple-chestnut in some lights), and buffer, less contrastingly spotted underparts. ADULT FEMALE. As ♂ but duller, with brown tinge to crown and nape, fewer streaks on crown, less rich tone to back, and paler, less heavily marked underparts. JUVENILE. Much duller than adult, with only grey rump and black-brown tail providing obvious relief to dull-spotted upperparts and less dark-chested, more evenly spotted underparts. Face pattern more pronounced than adult, with cream-buff supercilium and cheeks contrasting with grey-brown crown and dark eye-stripe.

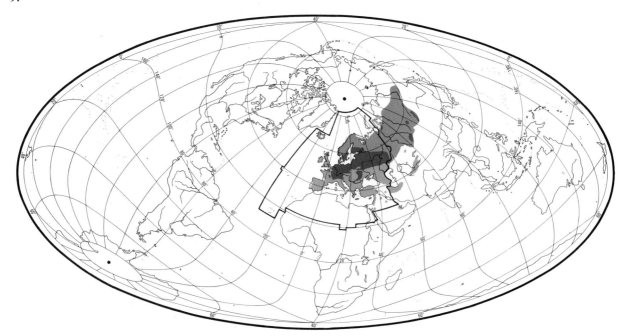

Wing-coverts show pale buff-white streaks and tips. FIRST WINTER. Resembles adult but duller, with less pure blue-grey tone to crown, nape, and rump and less chestnut on wing. White tips and shaft-streaks show on retained juvenile greater coverts. Bill yellow on breeding ♂, orange-yellow with dusky tip and culmen in winter ♂, ♀, and immature; legs brown.

Unmistakable; no other west Palearctic thrush shows as much contrast between rump and tail, and white underwing shared only by *T. viscivorus*. Flight noticeably loose and leisurely, with bursts of wing-beats alternating with brief glides on extended wings and short 'shooting' glides with wings closed. Action recalls *T. viscivorus* but wing-closures shorter and makes more changes in angle of tail; outline also rather more angular. Escape-flight less panicky than other gregarious thrushes, ending not by ducking into cover but in birds festooning bare branches of 'safe tree', often in company with *T. iliacus*. Gait and carriage much as *T. viscivorus*, but with plumage pattern exaggerating both upright stance and tail length. Wary, liking wide horizons in winter but bold around nest, mobbing any intruder. Gregarious at all seasons, with passage movements often involving hundreds or thousands of birds.

Noisy at all seasons. Song little developed: at best a feeble string of chuckles, whistles, squeaks, and normal calls; subsong more guttural and warbling. Commonest (diagnostic) call a harsh, aggressive-sounding cackle 'chacker chack chack' or 'chac-chac-chac-chack'; also gives a soft, drawn-out 'seeh' and a squawk.

Habitat. Breeds in middle and higher latitudes of west Palearctic, in subarctic, boreal, and temperate zones, in woods of birch *Betula*, pine *Pinus*, spruce *Picea*, alder *Alnus*, and mixed species, usually in open growth or on fringes of moist areas with grass cover; often along rivers or in groups of trees in fens or bogs, in sheltered but cool and humid situations (Voous 1960). In Scandinavia, nests above 1000 m where juniper *Juniperus* and dwarf birch afford sufficient shelter; also on rocky outcrops in folds of exposed fells, or on stony slopes with a few straggling bushes, resembling habitat of Ring Ouzel *Turdus torquatus*, and even on open tundra beyond treeline. In such respects resembles American Robin *T. migratorius* (Bent 1949) and similarly has, in Norway especially, taken to nesting in towns, parks, orchards, and gardens as well as tree-lined streets (Bannerman 1954). In Switzerland prefers medium altitudes, where open corridors along watercourses and shelterbelts or groves of trees afford a suitable blend, commonly to *c.* 1300-1500 m, but apparently not often above 1700 m except on migration (Glutz von Blotzheim 1962). Uses trees as nest-sites and for seasonal berry foods but is rather a ground-feeder and to some extent a bird of open country and cultivated land, tending to move away from trees after breeding season and to occupy different habitats according to current conditions rather than on a regular schedule. In Britain in winter, prefers fringes between open and wooded country, needing big rough fields (including cropfields) and avoiding forests, moorlands, wetlands, and (except in hard weather) human settlements. Attraction of opportunities to feast on berry crops is overriding at the right season. In contrast to usual habit, however, will often perch and at times sing on tree-tops, and habitually flies at 100 m or more above ground.

Distribution. Has colonized and spread in Britain,

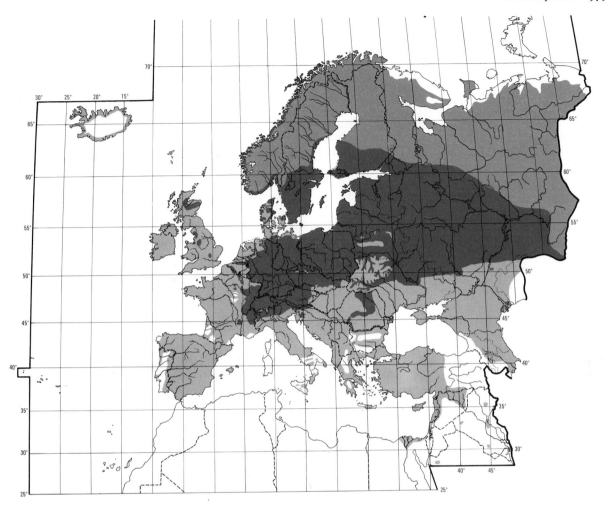

France, Belgium, Netherlands, West Germany, Denmark, East Germany, Austria, Switzerland, and Rumania; spreading in Sweden, Poland, Czechoslovakia, and Italy; breeds occasionally in Iceland. Expansion began after 1750 (Rommel 1953).

ICELAND. Bred several years in 1950s and a few pairs nested recently (AP). BRITAIN. First bred Orkney 1967, annually on Shetland 1968–70, then on mainland south to Staffordshire (Parslow 1973; Sharrock 1976). FRANCE. First bred 1953; spread later (Arnhem 1969; Valet 1973; Riols 1982). BELGIUM. First bred 1967 (Arnhem 1967, 1969); still spreading (Liedekerke 1976). NETHERLANDS. Bred 1903, 1905, 1925, 1936, and regularly from 1972 (CSR). WEST GERMANY. First bred in second half of 19th century (Sharrock 1976) and still spreading west (Rommel 1953; Peitzmeier 1964; Pfeifer 1967b; Kooiker 1982). DENMARK. First bred 1960 (Pedersen 1966); still spreading slowly (Skov 1970; TD). SWEDEN. Slow expansion into Skåne since *c*. 1972 (LR). EAST GERMANY. See Rommel (1953). POLAND. Until early 19th century, only in north-east; spread in 1820s until early 20th century, when some decline. Expanding again in some

regions, e.g. south-east (Tomiałojć 1976; AD). CZECHO-SLOVAKIA. Spread in south after 1950 (Stollmann 1964; KH). AUSTRIA. Spreading south-east (HS, PP). SWITZ-ERLAND. First bred 1923 (Géroudet 1963). Spreading especially now in alpine regions, occasionally to 2000 m (Arnhem 1969; Furrer 1977; RW). ITALY. Range increasing in Alps (Brichetti 1982). YUGOSLAVIA. First bred 1975 (Gregori 1977). RUMANIA. Bred Moldavia 1966–71 (Munteanau 1971). Spreading in Carpathians and Transylvania (VC). USSR. In 20th century, spread south and west in west of range (GGB).

Accidental. Spitsbergen, Bear Island, Jan Mayen, Kuwait, Tunisia, Libya, Algeria, Morocco, Madeira.

Population. Has increased in countries colonized (see Distribution), also in Poland and Czechoslovakia. Has decreased in Estonia and Latvia (USSR).

BRITAIN. No more than 6 pairs annually (Sharrock 1976); 1–4 pairs breeding 1973–7 and 1979–82 (Spencer *et al.* 1985), 3–11 pairs 1983 (Spencer *et al.* 1986). FRANCE. Under 10 000 pairs (Yeatman 1976). BELGIUM. About 100 nests (Lippens and Wille 1972). Walloon: *c*.

10 000 pairs (Leprince 1985). NETHERLANDS. 54–75 pairs 1977 (CSR). WEST GERMANY. 110 000 to 3 million pairs (Rheinwald 1982); latter figure much too high (AH). DENMARK. Several hundred pairs (TD). SWEDEN. About 15 million pairs (Ulfstrand and Högstedt 1976). FINLAND. About 560 000 pairs (Merikallio 1958). Decreased 1978–83 (Vickholm and Väisänen 1984). POLAND. Marked increase in west (AD). CZECHOSLOVAKIA. Rapid increase after 1950 (KH). SWITZERLAND. Increasing (RW). USSR. Estonia: decreased recently (HV). Latvia: decreased since 1950 (HV). Declined in last 35 years in some other areas, e.g. south of Moscow and in Gor'kiy region (GGB). Probably increased in Voronezh, though fluctuating (Wilson 1976).

Mortality. Switzerland: annual mortality 60–70%, independent of age (Furrer 1977). Finland: annual overall mortality 61–65% (Haukioja 1969). Oldest ringed bird 18 years (Rydzewski 1978).

Movements. Migratory, though in some years of winter abundance of food some resident or move only short distances (Ashmole 1962). Winters mainly in western, central, and southern Europe, Turkey, and Iran, also south to Canary Islands and Persian Gulf states. Usually reaches southernmost parts of Europe only in bad winters and rarely occurs on North African coast in good numbers (e.g. Bundy 1976)—though regular in Nile delta and also recorded in Wadi Natrun and elsewhere in lower Egypt (Goodman and Watson 1984). Some birds remain to winter in Siberia, southern Finland, southern Sweden and along south and west coasts of Norway below 65°N (Johansen 1954; Ashmole 1962) and in smaller numbers in Iceland and Faeroes (Salomonsen 1950–1).

Birds from Finland and southern Sweden move west into Norway then south (Ashmole 1962; Alerstam 1976), wintering predominantly in Belgium, France, Germany and northern Italy. Those from northern Sweden and Norway move south-west into Britain, Ireland, central Europe, south-west France, and Spain, whilst those from Germany, Poland, and Switzerland winter in southern France and northern Italy although usually avoiding Mediterranean coast (Ashmole 1962). Fenno-Scandian populations mix freely in winter with birds from central Europe (Ashmole 1962). East Palearctic birds winter mainly in Black Sea–Caspian region (Johansen 1954).

Birds flock prior to departure, becoming increasingly restless and making local movements. Spring passage generally more visible than in autumn with some impressive continual movements of birds often totalling several thousand passing along lines of hills or valleys in a matter of hours (Simms 1978). Sudden movements of large numbers as a consequence of severe weather are commonplace across the entire wintering range and even early nests can be abandoned at the onset of a late cold spell (Ashmole 1962). Large scale mid-winter irruptions into Sweden and Finland (Svärdson 1957; Tyrväinen 1970, 1975) and Estonia (Jõgi 1966) have been noted.

Individuals do not necessarily return to same area in successive winters with some subsequently recovered in winter up to 1600 km distant (Ashmole 1962).

Southward migration begins late September or early October and continues into November, although large flocks moving south or south-west have arrived in Britain in December and January apparently unrelated to severe weather (Simms 1978). Leaves northern Urals suddenly about mid-August (Johansen 1954). Scandinavian birds cross North Sea on broad front with dawn arrivals noted along entire east coast of Britain (Myres 1964). This broad front approach, usually below 1600 m, results in some birds travelling far out into the Atlantic and flocks have been seen from weather ships 890 km due west of Orkney (Lack 1962). Dawn ascent and subsequent south-easterly re-oriented movement of migrating *Turdus* over North Sea may be an adaptation towards eventually making safe landfall (Lack 1962); this dawn ascent has not been detected in spring return movements (Lack 1963; Myres 1964).

Return often begins early, birds wintering in south-central Europe making partial return movements in February. By February, recoveries of Norwegian birds are concentrated in northern France, Belgium, and Britain, and these birds evidently remain in that area longer than those from Finland, which by March are recovered further east in Europe (and many have probably returned to Finland by then) (Ashmole 1962). Flow of returning birds steadily increases during March but some have penetrated well beyond Arctic Circle in Norway by early March (Bannerman 1954). Main arrivals in Norway from mid-April and in Sweden and Finland from late April. In Leningrad region of west-central USSR large flocks arrive mid-March and continue until late April. In Siberia, it is among earliest migrants to arrive with large numbers in Tomsk and Arkhangel'sk areas by mid-April (Dementiev and Gladkov 1954b); others, presumably from the most northerly populations, pass through in first half of May when Tomsk population already breeding (Johansen 1954). First recorded on Yenisey, south of Arctic Circle on 10 May whilst those from the Pechora region do not arrive until mid-May (Johansen 1954). Most wintering areas are deserted by May although some (possibly 1st-year) birds linger in places as far south as Italy, and cold and wet weather can induce some to stay well south into June (Ashmole 1962).   PC

Food. Wide range of invertebrates; also fruits from late summer to early winter. Feeds on ground and in trees and bushes. Feeding methods described by Tye (1981) for Redwing *T. iliacus* (p. 1004) are also used by *T. pilaris*, though searches litter much less often and never seen at cow pats. Will turn over clods of earth, etc. (even stones up to 10 cm across), and scratch through snow to take food beneath (Lübcke and Furrer 1985). To take flying insects will fly up high in the air like Starling *Sturnus vulgaris* (Simms 1978). Recorded entering shallow water to take fish, also pecking at dead fish (Lübcke and Furrer 1985). When collecting food for young, adult usually eats small items itself; will collect items in a pile before carrying them off (Lübcke and Furrer 1985).

Diet in west Palearctic includes the following. Invertebrates: dragonflies (Odonata), crickets (Orthoptera: Gryllidae), bugs (Hemiptera: Pentatomidae, Miridae), scorpion flies (Mecoptera), larval Lepidoptera (Psychidae, Noctuidae, Geometridae), adult and larval flies (Diptera: Tipulidae, Tabanidae, Asilidae, Muscidae), adult and larval Hymenoptera (sawflies Tenthredinidae, Ichneumonoidea, Proctotrupoidea, ants Formicidae), adult and larval beetles (Coleoptera: Carabidae, Dytiscidae, Hydrophilidae, Staphylinidae, Scarabaeidae, Byrrhidae, Elateridae, Cantharidae, Dermestidae, Tenebrionidae, Cerambycidae, Chrysomelidae, Curculionidae), spiders (Araneae), harvestmen (Opiliones), millipedes (Diplopoda), snails (Gastropoda), slugs (Pulmonata), earthworms (Oligochaeta), leeches (Hirudinea). Also fish up to 7 cm long and apparently nestling birds. Plant material mainly fruits, also seeds: juniper *Juniperus*, yew *Taxus*, pinks (Caryophyllaceae), mistletoe *Viscum*, barberry *Berberis*, bramble *Rubus*, strawberry *Fragaria*, rose *Rosa*, rowan *Sorbus*, cherries *Prunus*, apple *Malus*, pear *Pyrus*, *Cotoneaster*, hawthorn *Crataegus*, currant *Ribes*, bilberry *Vaccinium*, crowberry *Empetrum*, elder *Sambucus*, buckthorn *Rhamnus*, sea buckthorn *Hippophae*, snowberry *Symphoricarpos*, vine *Vitis*, holly *Ilex*, *Viburnum*, sedges (Cyperaceae), and grasses (Gramineae); also shoots of grasses, buds of alder *Alnus*, and buds and catkins of birch *Betula*. (Csiki 1908; Collinge 1924–7; Meidell 1936; Béres and Molnár 1964; Tyrväinen 1970; Pénzes 1974; Lübcke and Furrer 1985.)

In southern Norway (mostly upland areas), 62 stomachs contained largely invertebrates in both spring and autumn: in spring, invertebrates mostly earthworms and adult beetles, plant food largely berries of heaths and juniper; in autumn, invertebrates largely beetles (mostly adults), harvestmen, and ants, plant material almost wholly berries (especially crowberry, heaths, and rowan) (Meidell 1936). In June–July near Moscow (USSR), 15 stomachs contained 97 items comprising 36% (by number) beetles (largely Curculionidae), 15% Hymenoptera, 14% larval Lepidoptera, 6% Diptera (mainly larvae), 5·5% earthworms, 4·5% centipedes, 5% other invertebrates, and 14% plant material (Ptushenko and Inozemtsev 1968). From September to May in Britain, diet found to comprise 37·5% insects, 14·5% earthworms, 4·5% slugs, 2·5% other invertebrates, 36% fruit and seeds, and 5% other plant material (Collinge 1924–7). In Cambridgeshire, October to mid-February, unlike *T. iliacus*, took more subsurface items (56·3% of total 608 items) than surface items (43·7%); in late February and March, when availability of larger insects on ground surface and vegetation increased, took relatively more surface items (75·4%) (Tye 1981). In Finland in one winter when unusually high numbers of birds present, food almost

wholly berries of rowan; also other fruits, buds, and catkins (Tyrväinen 1970). See also Florence (1914), Kovačević and Danon (1952), Mal'chevski (1959), Tutman (1962), Korenberg *et al.* (1972), Lübcke (1975b), Kiss *et al.* (1978), Kostin (1983), and review by Lübcke and Furrer (1985).

Food of young in southern Norway largely beetles (mostly adults), with fewer Lepidoptera larvae and earthworms (Meidell 1936). DJB

Social pattern and behaviour. Well studied. See especially Steinfatt (1941a), Paccaud (1947, 1952a, b), Hohlt (1957a), Lübcke (1975a), Haas (1980, 1982a), and, for recent review, Lübcke and Furrer (1985). Account includes detailed notes by T Slagsvold and C G Wiklund.

1. Most gregarious of west Palearctic *Turdus* outside breeding season (see also Roosting, below), with flocks of up to several thousand on migration (Lübcke and Furrer 1985). Birds resident on breeding grounds outside breeding season often solitary (C G Wiklund), except where food localized (T Slagsvold). In winter of exceptionally good food supply, Finland, flock size increased from typically less than 50 birds in autumn to often 500(−1000) in mid-January, and decreased thereafter, though small flocks occurred until late February (Tyrväinen 1970). For solitary feeding territories (one of radius 4–5 m) outside breeding season, see Lübcke and Furrer (1985). After breeding season, independent young form flocks initially with few adults, but proportion increases with time (Lübcke and Furrer 1985). In Switzerland, young form substantial flocks from mid-June, mixed with adults August–October (Paccaud 1947). In West Germany, post-breeding flocks include residents and immigrants (Lübcke 1975a). Migrant flocks often associated with Redwing *T. iliacus* (e.g. Lübcke and Furrer 1985). Also mixes regularly with *T. iliacus* for feeding, and variously with Song Thrush *T. philomelos* (Paccaud 1947, 1952b), Starling *Sturnus vulgaris* (Wiklund 1983), and Brambling *Fringilla montifringilla* (T Slagsvold). BONDS. Monogamous mating system (C G Wiklund); no polygyny recorded (Hohlt 1957a). Pair-bond maintained only for breeding season (Haas 1982a; C G Wiklund), and breeding by same pair in 2 successive years apparently exceptional (Lübcke and Furrer 1985). During breeding season, birds (sex not specified) re-pair within 4–5 days of losing mate (Lübcke 1975a). For mixed pairs with Blackbird *T. merula* and *T. iliacus*, see Fontaine (1956). ♀ alone broods, but both sexes feed young (Lübcke and Furrer 1985). When ♀ of one pair was killed, ♂ raised 8-day-old young on his own (Hohlt 1957a). If ♀ starts 2nd clutch, ♂ takes over care of 1st brood (Paccaud 1947; Lübcke and Furrer 1985); ♂ apparently helps to feed 2nd brood while also tending fledged 1st brood (Paccaud 1947). Parents continue feeding young for 1–2 weeks after fledging (Lübcke and Furrer 1985; C G Wiklund); young of late broods remain longer with parents (Lübcke and Furrer 1985). Lübcke and Furrer (1985) reported a strange juvenile, which had previously begged successfully from a pair, brooding their young while they were absent; at times, adults had to push the 'helper' aside to feed their own young. Age of first breeding not known. BREEDING DISPERSION. Colonial or solitary, but often no clear distinction (A Håland, T Slagsvold) and, though most birds in colonies of up to 40–50 pairs, dispersion varies regionally. In south of range, more often solitary, and colonies smaller, than in north: e.g. in southern Sweden, *c.* 25% or less of birds solitary, and colonies rarely exceed 20 pairs; in northern Sweden, *c.* 97% in colonies, commonly of 20–30 pairs (Wiklund 1983; C G Wiklund; see also Wiklund 1982, Wiklund and

Andersson 1980). In Zuid-Limburg (Belgium), average 4 pairs per colony ($n = 51$) (Ganzevles and Tillo 1982). In Westfalen (West Germany), 1960-3, average 4·5 ($n = 137$) (Peitzmeier 1964). In East Germany, 3-30 (Sauer 1978). Colony size also smaller in poorer habitat (T Slagsvold): e.g. in rich alder *Alnus* woodland near Trondheim (Norway), over 6 years, average size of 9 colonies 33·3 pairs (10-55) (Slagsvold 1980*a*); in subalpine birch *Betula* forest, Budal (central Norway), 5·2 pairs ($n = 16$ colonies, 10 years) (Hogstad 1983). Given colony may vary markedly in size between years, e.g. in Halle (East Germany), 10-45 pairs over 9 years (Gülland *et al.* 1972). Size of territory highly variable, depending on density: in centre of colony, may be confined to nest-tree and look-out perch, whereas solitary breeders may defend *c.* 1 ha (Paccaud 1952*a*; Hohlt 1957*a*). In one French colony, parents vigorously defended area within *c.* 30 m of nest (Fournier 1983). Territory serves for nesting and some courtship though most pair-formation takes place elsewhere. All feeding done on neutral ground outside territory: e.g. *c.* 300 m away (Fournier 1983), up to 4 km (Valet 1973). In East Germany, no records of 2 nests in same tree (Sauer 1978). In Norway, 2 nests seldom in same tree; not often closer than 10 m, usually 20-30 m apart (T Slagsvold). Distance between nests commonly 5-30 m (Wiklund and Andersson 1980). Large colonies may thus occupy several ha. In Eder Valley (West Germany), where colonies average less than 5 pairs, greatest density 16-23 pairs in 400-500 m² (Lübcke 1975*a*). In a park in West Germany, 75 pairs per km² (Lucan *et al.* 1974). In Switzerland, where main colonies in small isolated woods, 10-15 pairs may occupy 0·5 ha; average density in Zurich Canton (370 km²) 1-2 pairs per km² (Schifferli *et al.* 1982); in Wauwiler Moos, 50-70 pairs per km² (Lübcke and Furrer 1985). In Uppland (Sweden), 2·1 pairs per km² (Olsson 1947). In Leningrad (USSR), 6-8 pairs per km² (Nankinov 1970). In Tornio (Finland), 20-121 pairs per km² in different urban habitats (Huhtalo and Järvinen 1977); in a park in south-west Finland, 590 pairs per km² (Haartman 1971). In Budal, 1966-83, average 18 pairs per km² in subalpine birch forest; from 1966-72, 16 pairs per km² in subalpine forest of pine *Pinus* and birch (O Hogstad). According to Glutz von Blotzheim (1962), often uses 1st nest for 2nd clutch; few confirmed records, however, and as a rule builds new nest for 2nd clutch (Lübcke 1975*a*; Lübcke and Furrer 1985). In Hessen (West Germany), nests reported moving up to 3·5 km after losing 1st clutch. In East Germany, nests tend to be more often solitary for 2nd or replacement clutches (Sauer 1978). Not uncommonly uses a nest or nest-site from a previous season (Lübcke and Furrer 1985): of 203 nests, 13·8% used more than 1 season, but no evidence that same birds involved (Lübcke 1975*a*). Colony sites often change between years (Wiklund 1977; Furrer 1979*a*), and nest-site fidelity low (Lübcke 1975*a*); in Czechoslovakia, 6 marked adults recovered 2-30 km from former breeding site (Bürger 1977). Fidelity of young to natal area even lower (Lübcke and Furrer 1985) but may vary regionally, e.g. for southern Norway, Haftorn (1971) believes young commonly return to natal areas to breed: of 51 ringed and controlled, 58·8% found up to 5 km from natal site, 17·6% each up to 10 km and up to 50 km, rest much further away. Commonly associates with other species for breeding, notably passerines evidently attracted by defensive advantages of proximity to *T. pilaris* (see also Parental Anti-predator Strategies, below); also associates amicably with shrikes *Lanius* and falcons *Falco* (e.g. Hogstad 1981). In particular, *F. montifringilla* and *T. iliacus* nest close to *T. pilaris* (Slagsvold 1980*b*, which see for list of other associates). *T. pilaris* shown to afford protection to *F. montifringilla* and Chaffinch *F. coelebs* (Slags-

vold 1979). *Lanius* may nest in heart of *T. pilaris* colonies, often showing preference for nesting thus. In Ajoie (Switzerland), of 36 nests of Great Grey Shrike *L. excubitor*, 75% near *T. pilaris* colonies (Bassin 1983). Association with *L. excubitor* also reported by Paccaud (1952*b*), Tratz (1953), Diesselhorst (1956), Peitzmeier (1956), Hohlt (1957*b*), and Thielcke (1957). For association with Red-backed Shrike *L. collurio* see Hohlt (1957*b*); for Woodchat Shrike *L. senator* see Blanc (1956), Thielcke (1957), Glayre (1960), and Hirschfeld (1969); for Lesser Grey Shrike *L. minor* see Blanc (1956) and Hirschfeld (1969). Often develops close nesting association with Merlin *Falco columbarius* (Wiklund 1977, 1979, 1982; Hogstad 1981) and Kestrel *F. tinnunculus* (Hohlt 1957*b*; Bezzel 1961; Lübcke 1976). In northern Sweden, 70·8% of colonies near breeding *F. columbarius* (Wiklund 1982); breeding success of latter benefited thereby (Wiklund 1977, 1979), though association thought to be mutually advantageous (Hohlt 1957*a*; Wiklund 1979). ROOSTING. Outside breeding season, from June or July onwards, roosts communally in young conifers and other evergreens, tall hedgerows, and thorns (Witherby *et al.* 1938*b*; Simms 1978); also osier beds (Labitte 1937) and once in heather (Simms 1978). On Danube (Austria), in reeds, along with *T. merula* and Yellowhammers *Emberiza citrinella* (Aschenbrenner 1959). According to Lübcke and Furrer (1985) usually occupies shrub layer, lower than *T. iliacus* (*contra* Witherby *et al.* 1938*b*), sometimes less than 1 m from ground (Labitte 1955), but no evidence for statement in Witherby *et al.* (1938*b*) that ground roosting is widespread. In Britain, roosts commonly *c.* 300, but reports of 1000, once *c.* 15 000-20 000 comprising *c.* 60% *T. pilaris*, 40% *T. iliacus* (Simms 1978). At Oset (south central Sweden), studied July-December, *T. pilaris* increased in numbers from July (100 birds) to peak, of *c.* 200, August-September, fluctuating thereafter; most birds apparently of local origin (Gyllin and Källander 1980). Arrival at one winter roost as follows (Labitte 1955): birds gathered in flocks of variable size to return to roost. A few arrived shortly before dusk and assembled in highest tree-top, then suddenly dived, along with bulk of flock, into the copse, sometimes in silence, sometimes with a short sharp call (perhaps Chack-call: see 2 in Voice). When the weather was bad, November-December, all the birds arrived together. (Labitte 1955.) See Aschenbrenner (1959) for similar sequence. At roost in Switzerland, awakening at 04.00 hrs (28 June) signalled by Rattle-calls (call 6) of one bird, soon followed by general clamour (Paccaud 1947). For resting behaviour of mixed feeding flocks, see Berg-Schlosser (1979). During breeding season, roosts alone within territory or nearby (C G Wiklund). During breeding season, ♀ starts sitting on nest, staying there at night, before incubation begins (Steinfatt 1941*a*). In 6 cases, ♀ roosted on nest from 2nd egg. When young *c.* 11-12 days old, ♀ starts roosting on edge of nest. After fledging, adults and young roost separately, e.g. young recorded sitting in a row together in tree-top, parents resorting to communal roost which young eventually join. (Hohlt 1957*a*.) In northern Sweden, birds active throughout the day (C G Wiklund); rest before midnight, feed most actively shortly after midnight (Brown 1963). Comfort behaviour includes bathing, often by flocks of up to 40 in shallow water at edge of river, in pools (etc.), usually in the afternoon, sometimes quite late in the year. Sunning reported (C G Wiklund), also sandbathing by captive fledgling, but no anting (Lübcke and Furrer 1985).

2. Not especially shy. Birds flushed from communal roost remain silent until all are out of roost and heading in same direction (Labitte 1955). On 9 September, 7 birds from loose flock attacked Red Kite *Milvus milvus* and migrants seen to make silent dive-attacks on Sparrowhawk *Accipiter nisus*. ♀ A.

nisus approaching mixed *Turdus* flock caused *T. pilaris* to flee *en masse* into hedge with Chack-calls (Lübcke and Furrer 1985, which see for comparison of alarm reaction with other *Turdus*). For Distress-call when handled, see 9 in Voice. FLOCK BEHAVIOUR. In mixed *Turdus* flocks, *T. pilaris* and Mistle Thrush *T. viscivorus* defend largest individual distance (Berg-Schlosser 1979). For dominance relations with other *Turdus* see Antagonistic Behaviour (below). Resting ♂♂ in spring flocks commonly give Subsong (see 1c in Voice), stimulating same in others, leading to chorus, often joined by *T. iliacus*. Chack-calls also common among flock-members, especially migrants in flight, and when manoeuvring to land or when flushed. In loose flying flock, migrants give Huit-call. (Haas 1980: see 3 in Voice.) In Turkey, large flock, apparently continuing migration after stopover, rose, bunched, circled up to *c.* 150 m, then flew off (Wadley 1951). For heterosexual behaviour in spring passage flocks, see below. SONG-DISPLAY. ♂ frequently sings (see 1a in Voice) in display-flight, especially at start of breeding season (Paccaud 1952a), mostly in early morning and evening (Haas 1980). Song-flight distinctive, not undulating like normal flight but horizontal, with slower more deliberate wing-beats (Lübcke and Furrer 1985). Resembles display of Greenfinch *Carduelis chloris*, with wings held stiffly out, but not fully extended, between beats, giving impression of impeded progress (Witherby *et al.* 1938b). Flight-song considered unlikely to serve function of territorial song in other passerines, this function replaced in *T. pilaris* by Territorial-call (4 in Voice) which serves to attract ♀♀ and rebuff other ♂♂ (Lübcke and Furrer 1985). Song-flights performed in variety of circumstances which suggest general excitement: flying to and from roost, entering or leaving territory, accompanying ♀ during nest-building, after confrontations with rivals or predators (Paccaud 1952a; Lübcke and Furrer 1985; see also Parental Anti-predator Strategies, below). ♂ also gives perched Courtship-song (1b in Voice) but only during brief pairing period, and thus rarely heard (Lübcke and Furrer 1985; see Heterosexual Behaviour, below). Subsong heard from ♂♂ singly or several together in nesting area; birds singing thus described jerking head in various directions, forehead sometimes ruffled, wings and tail vibrating and twitching in rhythm with song (Hohlt 1957a). ANTAGONISTIC BEHAVIOUR. At start of breeding season, ♂♂ markedly aggressive near nest-sites and surrounding area (Paccaud 1952a; Håland 1984). Following account based mainly on Hohlt (1957a). Territorial disputes usually develop in intensity during pre-laying period. Initially, resident gives sporadic Chack-calls as he flies at rivals; later, gives almost uninterrupted Chack-calls while performing Threat-display (Fig A): adopts a horizontal

A

posture, ruffles plumage, and variously flicks and shakes fanned tail, lowers and twitches slightly open wings, also takes to high perches and, in silence, chases rivals well beyond territory boundaries, eliciting chase from other residents whose territories

are infringed. Once ♀♀ arrive, Chack-calls give way to Territorial-calls, and resident curtails expelling-flights at territorial limits. Bird reported approaching rival in Song-flight and driving it off with Territorial-call (Haas 1980). Residents attack rivals from rear, but if rival turns, both flutter upwards in tussle, sometimes dropping to ground interlocked. Bill-snapping occasionally heard. Once boundaries settled, fighting less common but ♂♂ remain aggressive until young fledge. ♀♀ participate in territorial defence after pair-formation, apparently challenging other ♀♀, as in ♂–♂ disputes. ♀ aggression usually wanes when they start feeding young. (Hohlt 1957a.) Birds may also defend temporary feeding-territories, both in and out of breeding season (Lübcke and Furrer 1985). Commonly aggressive towards *T. viscivorus* and *T. merula*, less often towards (smaller) *T. philomelos* and *T. iliacus* (Hohlt 1957a). Dominant over all of these except *T. viscivorus* (Simms 1978). In mixed feeding flocks, Devon (England) during hard winter, *T. pilaris* repeatedly confronted conspecific birds, also *T. iliacus* and *T. merula*; bird faced rival in Threat-display (see above) and after a few seconds, raised bill almost vertically (Fig B),

B

then suddenly lowered head and tail, rushed at and displaced rival (Goodfellow 1965). Similar food-fighting, sometimes with Starlings *Sturnus vulgaris*, included nasal 'wu' calls (see 2 in Voice), squealing (perhaps call 9), and Bill-snapping (P J Sellar). HETEROSEXUAL BEHAVIOUR. (1) General. ♂♂ usually arrive on breeding grounds *c.* 1 week before ♀♀ (Paccaud 1952a; Hohlt 1957a). In Switzerland, young birds arrive later than older birds (Paccaud 1952a). In southern Norway, pre-laying period *c.* 2–3 weeks (Håland 1984). (2) Pair-bonding behaviour. Some birds establish pair-bond during spring migration (Hohlt 1957a; Freitag 1972). Territorial disputes and nest-site searching recorded in passage flocks (Hohlt 1957a). Pair-formation seldom reported (Steinfatt 1941a), taking place very quickly, e.g. Courtship-song heard only on 1 day (Haas 1980). Preliminaries to pair-formation take place mostly on open ground, sometimes far from nesting area. Before pair-formation, ♀♀ in small groups are attracted to Territorial-calls, followed by Gnüg-calls (5 in Voice) of ♂♂. ♂ that fails to attract ♀ gives only Gnüg-calls. If ♀ lands near ♂, he falls silent and performs apparent Threat-display (as above); if ♀ reacts similarly, ♂ drives her off. If ♀ remains passive, tail propped on ground and breast high, ♂ with tail raised may approach ♀ and perform Advertising-display (Hohlt 1957a). Advertising-display performed on ground or, more often, as described here (Paccaud 1952a: Fig C) in tree: ♂ near ♀ hops through branches in rather slow laboured fashion, crouching, tail widely fanned and lowered, sometimes with tip pressed against branches, rump feathers erect, head drawn in, bill down, crown feathers may be ruffled, and wings stiffly drooped and held away from body at carpal joints, exposing white coverts. ♂ thus hops about, twisting and turning near ♀, moving around her in ever-

C

decreasing circles, usually giving Courtship-song throughout. In southern West Germany, pair-formation mostly in 2nd half of March (Hohlt 1957a). (3) Courtship-feeding. Not reported, but see Behaviour at nest (below). (4) Mating. Usually near nest, in nest-tree or look-out tree (Steinfatt 1941a); typically less than 1 m from prospective nest-site (Paccaud 1952a). No marked preliminaries. Either sex may solicit. ♂ solicits by landing near ♀ after Song-flight or by normal approach. ♀ solicits by approaching ♂ and crouching rigidly. (Hohlt 1957a.) Squatting ♀ once gave a 'special' call, a little reminiscent of song, to reluctant ♂ (Paccaud 1952a). One sequence as follows: ♀ flew to fork in tree, hovered briefly and landed; ♂ followed quickly and immediately mounted her with frenzied beating of wings. Mating repeated again shortly after when ♀ moved to nearby branch, then both flew and landed a little further off in the tree with typical twitching of wings and tail. Pair suspected to have mated during several subsequent returns to same fork in tree. (Paccaud 1952a.) Several records of ♂ following ♀ to nest when she was carrying nest-material, and attempting copulation; ♂ mounted several times before copulating which took several seconds (Haas 1982a). If attempt unsuccessful, ♂ gives a quiet whistling sound (Hohlt 1957a). Copulation most frequent from 3 days before 1st egg up to 3 days before last egg (Lübcke and Furrer 1985), and, for 2nd clutch, may begin before 1st brood leaves nest (Hohlt 1957a), once when 1st brood 6–7 days old (Freitag 1972). (5) Nest-site selection. May take a few hours or several weeks. ♀ seeks site for 2nd clutch after 1st brood fledges. ♂ takes initiative but ♀ makes final choice. ♂ first leads ♀ to various sites ('showing'), giving Territorial-call to attract her. ♀ then assumes active role, and ♂ follows her. ♂ especially aggressive to other birds at this time. ♀ creeps over prospective site a couple of times and, with tail steeply raised, makes rotating movements, also plucks off and tosses aside any obstructing foliage. (Hohlt 1957a.) Rotating movements by ♂ seen twice (Lübcke 1975a). Captive pair crept around branches in a hunched posture with heads withdrawn and tails lowered, both birds squatting on branch when apparently suitable site found (Haas 1982a). Material may be brought to several sites before one finally chosen (Hohlt 1957a). (6) Behaviour at nest. ♂ accompanies ♀ while she collects nest-material, performing sentinel role (Paccaud 1952a), perhaps mate-guarding. Sauer (1978) once saw ♂ relieving incubating ♀, but incubation typically by ♀ alone. Incubating ♀ periodically flies 50–100 m from nest to feed. ♂ regularly visits incubating ♀ and exceptionally feeds her on nest. (Lübcke and Furrer 1985.) RELATIONS WITHIN FAMILY GROUP. Brood may take up to 3 days to hatch (Hohlt 1957a). ♀ alone broods, typically stopping at night before young 1 week old, later if weather bad (Paccaud 1947; Hohlt 1957a). ♀ shields young from sun, and food-bearing ♂ once spread wing to shelter ♀ and young from rain (Hohlt 1957a, which see for illustration). Both parents feed young, ♀ taking lesser share until she stops brooding (Paccaud 1947; Hohlt 1957a). ♂ giving Flight-song on approach to nest indicates prospect of 2nd clutch (Hohlt 1957a). ♂ occasionally passes food to ♀ for transfer to young;

♀ then said to adopt a begging posture with wing-shivering and gives 'sieh' sounds (Haas 1982a: see 10a in Voice). For distribution of food in relation to hatching order, see Rydén and Bengtsson (1980). ♀ seen regurgitating water for nestlings, her body jerking with the effort, tail low, wings trembling (Fournier 1983). Faecal sacs initially swallowed by parents, by brooding ♀ until 7th day, later increasingly carried away (Steinfatt 1941a; Hohlt 1957a). At one nest ♂ collected faecal sacs c. 3 times per hr, ♀ c. 6 times (Fournier 1983). Adults also remove dead young from nest and nest-area (Wiklund 1982a). Eyes of young open at 5 days (Haas 1982a), c. 8 days according to Sauer (1978). Young increasingly active from 10th day, exercising wings and giving food-calls (Favarger 1948; see Voice). Near fledging, young become restless and clamber on to branches around nest, returning, however, to roost (Pfeiffer 1965). May leave nest over 1–3 days, and start begging loudly from edge of nest (Haas 1982a). Able to fly, but poorly, on first leaving nest (Paccaud 1947). Fledged young will beg from strange adults, which respond by threat; ♀ incubating 2nd clutch may threaten and peck 1st brood. Young stay near nest, but well dispersed, for c. 4 days until flying well, then follow parents to communal feeding grounds where young form flocks ('crèches'). Adults signal intention to feed young by approaching them, then hopping away, sometimes repeatedly until young follow. After independence (see Bonds, above), young may beg unsuccessfully, once at 50 days. (Hohlt 1957a.) ANTI-PREDATOR RESPONSES OF YOUNG. Nestlings crouch if threatened; from c. 10 days threat-gape and may bill-snap audibly (Hohlt 1957a). After c. 9 days, often leave nest prematurely if disturbed by man (Hogstad 1983); leap out and, with distress-calls (see Voice) attract parents. Once out of nest, young 'freeze' when warned vocally by parents of approaching human intruder, adopting a horizontal posture if on the ground (Lübcke and Furrer 1985). Recently fledged young on tree-perch may adopt Bittern-posture (Goethe 1973; Zucchi 1975), as in T. philomelos (see that species, p. 995), once maintaining posture for 10 min (Hunziker-Lüthy 1970). When handled, young give distress-calls (Gebhardt 1951; Pedersen 1966) or threat-gape. In alarm, young c. 1 month old hide in loose groups in dense foliage, droop wings, and give variant of Chack-call (Arnhem 1969; see Voice). PARENTAL ANTI-PREDATOR STRATEGIES. (1) Passive measures. ♀ a tight sitter, sometimes allowing herself to be touched (Lübcke and Furrer 1985). Distance to which intruder can approach nest without flushing ♀ declines from 5–10 m during incubation to 2 m while brooding (Paccaud 1947). Adults have different calls (7 and 8 in Voice) to warn of approach of respectively aerial and ground predators (Lübcke and Furrer 1985). For presumed protection gained by nesting near shrikes Lanius and falcons (Falconidae), see Breeding Dispersion (above). (2) Active measures: general. Renowned for launching bold dive-attacks on predators, accompanied by Rattle-calls and well-aimed defecation. For review, see Mester (1976). Defecation long known in Scandinavia, but only in recent decades have reports increased from central Europe; assumed by Furrer (1975) to be a spreading behaviour pattern, though incidence affected by range expansion, observer awareness, stage of breeding, and individual variation. Typically, defecation most frequent when pair have well-grown nestlings or newly fledged young. Colonial nesters defecate more often than solitary birds, due partly to higher individual attack rate (Wiklund 1983), though Haas (1980) found no difference in attack rate. This defence capability probably a major factor favouring colonial breeding (Hohlt 1957b; Lübcke 1975a; Andersson and Wiklund 1978; Wiklund and Andersson 1980). For further details, see below. (3) Active measures: against birds. Widely reported attacking all Corvidae, especially Carrion Crow *Corvus corone*, and variety of raptors (Lübcke and Furrer 1985, which see for list of species attacked). Occasionally mobs *Lanius* (Paccaud 1952*b*; Haas 1982*a*) and gulls *Larus* (Tulloch 1968, 1969; Ritzel 1978). Reported making defecating attacks on raptor silhouette on window (Reichholf 1977). Aggression varies: sometimes alarmed bird flies off and hides, or frequently stays relatively close and scolds (Haas 1980). Response also varies with size of predator: e.g. in Belgium, simply chases Magpie *Pica pica* and Jay *Garrulus glandarius*, but mobs and defecates on *C. corone* and Buzzard *Buteo buteo* (Pfeiffer 1980; see also below). Cuckoo *Cuculus canorus* was confronted with a threat-posture, then chased (Gülland *et al.* 1972). In Switzerland, highly agitated groups chased Sparrowhawk *Accipiter nisus* and Hobby *Falco subbuteo* over long distance; less dangerous predators occasionally chased by 1 or 2 birds (Paccaud 1952*b*). Sequence of response, especially for Corvidae, as follows. One or more birds fly towards approaching predator and escort it along edge of colony, only attacking if it poses greater threat (Lübcke and Furrer 1985). In colony, only pairs whose territories are infringed by passing *C. corone* give chase (Paccaud 1947). Single *C. corone* elicited attacks from only the 'guarding' birds but groups of *C. corone* mobbed by whole colony (Baier 1978). Attacks on *C. corone* concentrated from behind and on the head (Baier 1974*b*), accompanied by much scolding and violent pecking (Paccaud 1952*b*). *P. pica* and *G. glandarius* subjected to ramming attacks from side (Pfeiffer 1965). Response fiercest near nest; when predator spotted, birds fly scolding to look-outs and launch attacks (Lübcke and Furrer 1985). Attack-flight direct, with shallow rapid wing-beats; often swoops low over predator, braking and veering off at last moment, defecating as it does so (Löhrl 1983: Fig D, which shows defecating attack on stuffed and mounted Tawny Owl *Strix aluco*). At moment of defecation, Rattle-calls change to an intense screech (Lübcke and Furrer 1985).

Defecation usually accompanies first few dive-attacks, less as assault continues (Wiklund 1983). Approaching from behind, will also defecate at flying predators (Löhrl 1983). Attacks persist until predator retreats, but by then its plumage may be so soiled and matted that it is grounded, and several reported dying as a result (Bezzel 1975; Furrer 1979*b*; Pfeiffer 1980; Löhrl 1983). (4) Active measures: against man. Less often defecates on human intruders (and other ground predators) than on raptors (Herren 1969). Apparently more often attacks thus in northern parts of range (Löhrl 1983, which see for possible explanation). In Switzerland, adults with well-grown young in nest perched close to intruder, raising and lowering their tails and slightly drooping their wings, giving Rattle-calls on approach of intruder, switching to Ground-predator calls (8 in Voice) when danger imminent. Only pair whose nest is threatened mobs intruder, neighbours staying alert on site. (Paccaud 1947.) Similar response shown by both sexes to human intruder inspecting nest. Nest defence increased from laying to fledging. (Andersson *et al.* 1980, which see for individual variation.) For evident Song-flight by ♂ after intruder ringed young, Inverness-shire (Scotland), see Weir (1970). (5) Active measures: against other animals. Reported attacking cats, dogs, and squirrels *Sciurus*. Distinct Bill-snapping heard as birds swooped low over back of cat or squirrel (Hohlt 1957*a*). For violent attack on squirrel on ground, see Klaas (1974). Dog thought less likely to elicit defecating attack than raptor (Herren 1969).

(Figs by N McCanch: A from drawing in Simms 1978; B from drawing in Goodfellow 1965; C from drawing in Paccaud 1952*a*; D from photograph in Löhrl 1983.)   EKD

Voice. Freely used. Major study by Haas (1980). See also Hohlt (1957*a*). Unless otherwise acknowledged, following account based on that given by H-H Bergmann and H-W Helb in Lübcke and Furrer (1985). For additional sonagrams, see Lübcke and Furrer (1985) and Haas (1980). Bill-snapping sounds given to threaten conspecific and other birds in disputes over (e.g.) territories and food.

CALLS OF ADULTS. (1) Song of ♂. A feeble warble interrupted by wheezes and chuckles (Simms 1978). A number of variants (a–c, below), distinguished according to contact and mode of delivery but all basically a weak twittering, warbling, babbling, or chattering (etc.), and inferior to (e.g.) Song Thrush *T. philomelos* or Blackbird *T. merula*, notably in lacking clear fluting sounds. (a) Flight-song. Most striking song type, typically given in display-flight: a rapid, more or less shrill twittering, comprising phrases of varying length but may sound continuous. Likened to song of Swallow *Hirundo rustica* but more powerful and abridged (Arnhem 1969). Similar to types b and c (below) but faster. Often begins with call-notes (especially calls 2 and 6), slightly separated from twitter that follows: variants of calls may be incorporated subsequently. In Fig I, 2 loud Rattle-calls (call 6) precede a twittering phrase (J Hall-Craggs). Descriptions of song (not distinguished as type a or b) include a high tuneless 'took-took-tcheree' and 'took-took-cherri-weeoo' (Simms 1978); 'tschäk tschäk tscherr tscherr' and 'tschäk tschäck tschirr tschirr' (Steinfatt 1941*a*). Another song described as 'srui (or 'svrui') srui

D

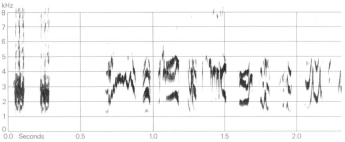

I Roché (1970) Finland May 1968

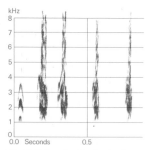

II V C Lewis England January 1969

III P J Sellar England
January 1985

IV S Carlsson Sweden
September 1983

srui srui' (Paccaud 1947). Marked frequency modulation (as in Fig I) occurs (Lübcke and Furrer 1985, which see for details). Units often relatively long, but rarely exceeding 0·3 s, and divisible into sub-units. Intervals between units average 70·5 ms (n = 14). For function of Flight-song and contexts in which it occurs, see Social Pattern and Behaviour. (b) Courtship-song. Quiet, muffled or strangled chattering sounds, given with bill closed by perched ♂ performing Advertising-display near ♀. Similar to Flight-song, but quieter and less differentiated, lacking units of simple structure and the rapid rhythmical frequency modulation of other song types. Rarely heard. (c) Subsong. Low, rather guttural warbling (Witherby et al. 1938b). A more or less continuous twittering with quite long pauses (e.g. c. 1 s or more) between units and groups of units. Compared with Flight-song, tempo therefore slower, also quieter, and frequency modulation less. Given mainly by resting birds in spring flocks (see Social Pattern and Behaviour for details), also sometimes in autumn and winter (Witherby et al. 1938b; Simms 1978): e.g. at roost, March, a trilling 'tchirrick-tchirrick' running into the 'twitter-warble'—a series of subdued coarse twitterings, like H. rustica, but lacking the wide frequency range of that species; followed by several clear units of rising pitch, rendered 'trit-trit-trit-trit' (Simms 1978). (2) Chack-call. Commonest call, used for low excitement contact-alarm. Given typically in a series of varying length. In flight, a constant 'tchack-tchack' (Simms 1978). 1st unit often different from rest, and sounds weaker, e.g. 'dja-dsch-dschak' (Lübcke and Furrer 1985, which see for other sequences); also 'cha-cha-cha-chack' (Witherby et al. 1938b), a staccato sharp 'ter-tchack-tchack-tchack', occasionally interspersed with

a softer 'tchuck-tchuck', from perched, perhaps slightly uneasy, birds (Simms 1978). In Fig II, 'chack' sounds rather weak. Call often heard from flocks when manoeuvring to land or, in rapid series, when flushed (Haas 1980). In disputes over winter food, birds in Threat-display gave deep, resonant, quiet 'wu wu' (Fig III: P J Sellar), apparently strangled low-pitched variant. In alarm commonly mixed with call 6; then a more agitated 'tjetjetjetje' or, in aerial demonstration, 'tjouic tjouic' (Paccaud 1947). (3) Huit-call. A high-pitched whistle, rendered 'huit' (Haas 1980) or often soft 'hüid'. At close quarters may sound like a slightly hoarse nasal 'wäid'. In Fig IV, 'huee huit' (J Hall-Craggs). According to Haas (1980), serves as distant contact-call between members of loose flying flock. (4) Territorial-call. Usually disyllabic 'dschrät-schrätt' or 'trätt-trätt'; each unit comprises c. 6 very short noisy sub-units. Often given in long series, at intervals of several seconds, by both sexes in defence of territory, also by ♂♂ to attract ♀, thus serving function of song in most other passerines. May be followed by call 5. (5) Gnüg-call. A tonal, or at least melodious 'gnüg' sound, also described as hiccoughing sound, given by ♂ after call 4 after being forced to edge of colony when trying to attract a ♀. ♂♂ which fail to attract ♀ eventually give only Gnüg-calls. Song (presumably Courtship-song) may be interspersed with these calls. (Hohlt 1957a.) Calls heard near fledged young around colony, rendered 'djö', thought to be related. (6) Rattle-call. A series of 2–3 'trt' or 'trat' (Fig V, also first

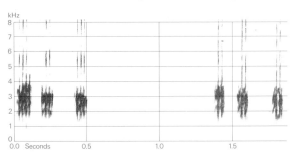

V Roché (1970) Finland May 1968

2 units in Fig I) sounds, signalling moderate disturbance, e.g. man approaching colony; then serves to warn mate and young. Commonly mixed with call 2. At higher intensity, when colony-members mobbing and defecating

on intruder, marked variation occurs in number of syllables, e.g. 'trt-trt-trrrt-trrt'. At moment of defecation, call gives way to an intense screech (see call 9). (7) Aerial-predator call. A high pitched pure whistle 'zieh', difficult for human to locate. Soft prolonged 'seeh', not so high-pitched as *T. philomelos* (Witherby *et al.* 1938*b*). See also call 10a for evidently similar sound. (8) Ground-predator call. A single rather muffled sound, similar to that given by *T. merula*, rendered a soft 'quok' (Haas 1980), or harder 'tock' (Hohlt 1957*a*); may also sound like 'dack', 'dag' or 'dog', though last thought by Hohlt (1957*a*) to encourage young to gape. (9) Distress-call. A drawn-out rising shrill screech which also takes the form of shorter more abrupt units, given when acutely threatened, e.g. when handled. Squeals heard during fighting (P J Sellar) presumably related. (10) Other calls. (a) 'Sieh' calls given by ♀ adopting a begging-posture when ♂ passes food to her for transfer to young (Haas 1982*a*). 'Special call', said to be a little reminiscent of song, given by ♀ soliciting copulation (Paccaud 1952*a*), perhaps related. (b) Quiet whistling sound given by ♂ if copulation unsuccessful (Hohlt 1957*a*). (c) Sharp powerful 'tsie' or 'tsrie' of unknown function, given in flight by adults and young (Paccaud 1947) not obviously reconcilable with above list, but possibly call 3. (d) Rising 'gezeek', apparently similar to call of fledged young (Simms 1978).

CALLS OF YOUNG. For development of vocalizations in nestlings, see Korbut (1982). Food-call of nestlings a thin high-pitched chirp (Lübcke and Furrer 1985), rendered an almost plaintive 'tchêêê' (Arnhem 1969), like House Sparrow *Passer domesticus* (Weir 1970). When closely threatened, young may bill-snap (Hohlt 1957*a*) and give a distress-call like that of adult. After fledging, food- and contact-calls variously 'zrii-zri' or 'pssie' or 'qui-quick'. Pitch evidently varies (see sonagrams in Haas 1980 and in Lübcke and Furrer 1985), often combined with noisy sounds, e.g. 'trat-zri-zri', in faster longer series when being fed. Haas (1980) also reports a 'satiety' call. Call 2 of adult given by fledged young as a somewhat muffled 'tjoc' or 'tjoic' (Paccaud 1947); a hard 'tack', later a softer 'döck' (Hohlt 1957*a*), in alarm 'tchèèèc', softer than adult (Arnhem 1969). EKD

Breeding. SEASON. Scandinavia: see diagram. Central Europe: laying begins early April. Lapland: laying begins late May and early June (Haartman 1969*a*; Makatsch 1976). SITE. In tree, placed in crotch of branch against trunk, or on side branch; exceptionally on ground or in depression among rock. Of 121 nests, Poland, mean height above ground 9·4 m (1–25) (Bocheński 1968). Of 119 nests, West Germany, mean height above ground 7·4 m (2–19·5) (Hohlt 1957*a*). Of 750 nests, West Germany, mean height above ground 7·5 m (0–21), declining slightly through season; no preferred compass orientation but usually facing away from surrounding vegetation (Haas

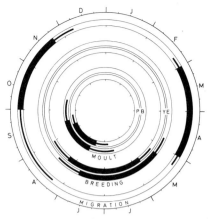

1982*a*). Nest: bulky though compact structure with outer parts of grass reinforced with twigs, roots, etc., lined with thick layer of mud, and inner lining of fine grasses and a few roots; woven ring of grass glued with mud, rarely with a little horse hair and fur. Mean dimensions of 37 nests: external diameter 14·5 mm (12·5–17·5), internal diameter 9·9 mm (8·0–12·0), height 12·9 mm (9·0–20·0), depth of cup 6·8 mm (4·5–8·5); cup usually elliptical or oval, less often circular (Bocheński 1968). Building: by ♀ with ♂ in attendance (Haas 1982*a*). EGGS. See Plate 84. Sub-elliptical, smooth and glossy; pale blue, very variably marked with red-brown speckles, often very heavily, less often blotched, sometimes forming cap at broad end. 28·8 × 21·0 mm (26·0–34·0 × 18·5–23·4), $n = 250$ (Schönwetter 1979). Weight varies with clutch size: in 4-egg clutches, 5·63 ± SE0·11 g ($n = 8$); in 5-egg clutches, 6·53 ± 0·10 ($n = 55$); in 6-egg clutches, 6·37 ± 0·07 ($n = 90$); in 7-egg clutches, 5·96 ± 0·04 ($n = 14$) (Otto 1979). Clutch: 5–6 (3–7). Of 116 clutches, West Germany: 4 eggs, 13%; 5, 45%; 6, 40%; 7, 2%; mean 5·2 (Lübke 1975*a*). Of 372 clutches, Finland: 3 eggs, 3%; 4, 11%; 5, 31%; 6, 51%; 7, 4%; mean 5·43, increasing slightly from south to north (Haartman 1969*a*). Mean of 858 clutches from literature search, 4·93 (Haas 1982*a*). 1–2 broods; *c.* 20% of 664 pairs laid 2 clutches (Haas 1982*a*). Eggs laid daily. INCUBATION. 10–13 days. Mean 12·0 ± SE0·6 days, $n = 15$ (Haas 1982*a*). By ♀ only. Start of incubation variable, often with 3rd egg, sometimes earlier, or not until clutch complete; hatching often asynchronous (Haas 1982*a*). YOUNG. Cared for and fed by both parents, in approximately equal shares (Hohlt 1957*a*). FLEDGING TO MATURITY. Fledging period 12–15 days; mean 12·9 (11–15), $n = 12$ (Haartman 1969*a*). Become independent at *c.* 30 days, but seen begging for food at 50 days (Hohlt 1957*a*). BREEDING SUCCESS. Of 346 eggs laid, West Germany, 61·8% hatched and 48·0% fledged, equalling 0·9 young reared per pair laying (Hohlt 1957*a*). Of 159 clutches, West Germany, 52·1% produced fledged young (Lübke 1975*a*). Of 758 nests with eggs, West Germany, 40·1% produced fledged young, with mean 1·8 young per nest, and 4·57 young per successful nest (Haas 1982*a*).

Plumages. ADULT MALE. In fresh plumage, forehead, crown, nape, and most of sides of head grey, slightly washed brownish to variable extent; crown-feathers with bold black blotch on centre; indistinct supercilium from nostrils over eye greyish-white to creamy buff; lores and small patch in front of eye and below eye black. Mantle and scapulars deep cinnamon-brown, feathers with obscure blackish centres; back, rump, and upper tail-coverts grey, faintly washed brownish to variable extent, a little paler than head and without dark feather centres. Chin buff. Throat and sides of neck buff with black streaks, small and fine on centre and heaviest at sides; feathers of sides of neck and chest heavily marked with black. Ground-colour of chest warm rufous-buff, shading to whitish on belly and rear of flanks; feathers of sides of breast and flanks with broad black arrow-marks or spots and pale fringes, appearing scaled; belly unmarked. Under tail-coverts whitish, feathers with dark markings as on flanks but reduced in extent. Tail brownish-black, outer feathers with variable amount of grey on tips, most noticeable on underside. Primaries blackish with pale grey outer edges; secondaries mainly cinnamon-brown (as mantle and scapulars) on outer webs, tending to grey toward tip and with narrow pale edge, remainder of feathers blackish, like primaries. Upper wing-coverts cinnamon-brown, as mantle and scapulars; outer greater-coverts tend to grey and all feathers with indistinct pale grey outer fringes and tips. Axillaries and under wing-coverts white. *In worn plumage*, pale feather-fringes partly lost by abrasion, head and back appearing purer grey, black feather centres on crown, chest, sides of breast, and flanks more conspicuous, and buff of throat, chest, and flanks paler. ADULT FEMALE. Like adult ♂, but black of feather centres of crown usually narrower and more attenuated at tip; mantle and scapulars sometimes duller earth-brown (less rufous), feather-centres dark grey-brown or black-brown (not black); tail-feathers dull dark brownish-olive (not largely blackish); buff of chest on average paler (Svensson 1984a; BMNH, ZMA). NESTLING. Down buff, fairly long and plentiful; for development, see Haas (1982a). JUVENILE. Like adult, but grey areas of head and upperparts tinged brown and mantle paler brown, each feather with sharply defined pale buff central streak; chin and throat whitish, unspotted; chest and flanks heavily spotted with rounded sooty-black spots (in adult, scaly); upper wing-coverts and tertials with well-defined whitish fringes; greater coverts as 1st adult. FIRST ADULT. Like adult, but head and back more extensively washed with olive-brown or olive (almost wholly obscuring grey in some ♀♀), chestnut of mantle washed olive, and dark centres of crown-feathers reduced (least so in ♂). Juvenile flight-feathers, tail, tertials, greater upper primary coverts, and usually a variable number of outer greater upper secondary coverts retained; outer greater coverts usually duller brown than fresh neighbouring inner ones, often shorter and less smoothly tipped, and sometimes with trace of white fringe along tip or white triangular spot on tip of outer web. See also Svensson (1984a).

Bare parts. ADULT. Iris dark brown. In breeding season, bill yellow or orange-yellow, sometimes with small dark tip; at other times, upper mandible and tip of lower mandible extensively suffused black-brown; bill of ♀ duller than ♂. Leg and foot dark brown. NESTLING. Mouth yellow, no spots. Gape-flanges yellow-white. JUVENILE. Iris dark brown. Bill yellowish with dark brown tip. Leg and foot yellow-brown, becoming dark brown in 1st autumn.

Moults. ADULT POST-BREEDING. Complete; primaries descendant. July–September, before leaving breeding area. In Finland,

average duration of primary moult 51 days (Ginn and Melville 1983). POST-JUVENILE. Partial: head, body, lesser, median, and some (usually 4–6) inner greater coverts. Mainly August–September, completed before start of autumn migration, but early individuals begin July and late ones continue into October (Lidauer 1983).

Measurements. Western Europe, all year; skins (BMNH). Juvenile wing is retained juvenile wing of 1st adult. Bill to basal corner of nostril; to skull, on average 8·8 longer, exposed culmen on average 2·6 longer.

WING AD	♂ 146·8 (3·68; 10)	139–152	♀ 142·6 (3·27; 10)	136–147
JUV	142·6 (3·98; 10)	136–151	140·8 (2·30; 10)	137–144
TAIL AD	108·4 (1·78; 10)	105–112	104·4 (2·76; 10)	101–108
BILL AD	15·7 (0·95; 10)	14–17	15·5 (0·85; 10)	14–17
TARSUS	33·6 (1·43; 10)	31–36	33·6 (0·97; 10)	32–35

Sex differences significant for wing and tail.

Wing of freshly dead birds, Netherlands, October–April (ZMA).

AD	♂ 152·0 (3·63; 27)	145–159	♀ 146·1 (3·09; 23)	142–153
JUV	147·8 (3·62; 44)	141–156	145·2 (3·11; 51)	137–152

Weights. England, November–December: ♂ 114·9 (15) 105–132, ♀ 116·6 (17) 101–141 (BTO). Netherlands, killed at lighthouses (ZMA):

OCT	♂ 95·4 (13·0; 10)	80–115	♀ 88·8 (9·7; 12)	76–106
NOV	107·7 (9·3; 31)	81–120	103·7 (10·2; 38)	83–128
DEC–MAR	102·6 (10·9; 12)	83–120	102·8 (9·6; 11)	85–118

Helgoland (West Germany), migrants: 91 (24) 79–106 (Weigold 1926). Norway, all year: ♂ 106·7 (30), ♀ 103·4 (28) (Haftorn 1971). South-west Siberia, autumn: 108·5 (8·52; 8) 95–122 (Havlín and Jurlov 1977).

Kazakhstan, USSR (Dolgushin et al. 1970).

OCT–MAR	♂ 105·4 (6·0; 4)	94–114	♀ 110·0 (11·0; 7)	91–125
APR–SEP	102·0 (5·4; 4)	96–109	105·2 (9·3; 4)	96–117

Frost-killed birds, Skomer (Wales), January: ♂ 57·6 (8) 53–60, ♀ 58·5 (17) 53–66 (Harris 1962). Frost-killed, Netherlands, winter: ♂ 58·5 (4·99; 18) 47–64, ♀ 58·2 (5·30; 15) 50–66 (ZMA). See also Hope Jones (1962), Lübcke (1980), and Swann (1980).

Structure. Wing rather long, broad at base, tip fairly pointed. 10 primaries: p8 longest, p9 3–8 shorter, p7 0–2, p6 5–10, p5 17–25, p4 27–36, p1 42–52. P10 reduced; 84–98 shorter than p8, 7–15 shorter than longest upper primary covert. Outer web of p6–p8 and inner of (p7–)p8–p9 emarginated. Tail rather long, tip square or very slightly rounded; 12 feathers, t6 4–10 shorter than t1–t2. Middle toe with claw 28·6 (10) 26–31; outer and hind toe with claw both c. 70% of middle with claw, inner c. 68%. Remainder of structure as in Blackbird *T. merula* (p. 963), but bill and tarsus relatively shorter and bill relatively thicker at base.

Geographical variation. Slight, involving colour and size, but bridged by marked individual variation and hence no races recognized. Southern populations from central Europe east to central European USSR tend to be slightly smaller than typical nominate *pilaris* from northern Europe (wing of ♂ 138–147 in central Europe, 142–148 in central European USSR) and to have spots on underparts more rusty-brown (less black), and mantle and scapulars more red-brown; thus sometimes separated as *subpilaris* Brehm, 1831. Populations east from Yenisey on average larger (wing of ♂ 146–153), paler grey on head and rump, paler rufous on mantle and scapulars, and less deep buff on underparts; sometimes separated as *tertius* Johansen, 1936. (Johansen 1954; Vaurie 1959.)

DWS

Turdus philomelos **Song Thrush**

Du. Zanglijster   Fr. Grive musicienne   Ge. Singdrossel
Ru. Певчий дрозд   Sp. Zorzal común   Sw. Taltrast

Turdus philomelos C L Brehm, 1831. Synonym: *Turdus ericetorum*.

Polytypic. Nominate *philomelos* C L Brehm, 1831, most of continental range of the species; *clarkei* Hartert, 1909, Britain (except for parts occupied by *hebridensis*), Ireland, western and central Netherlands and Belgium, and north-west and western France; *hebridensis* Clarke, 1913, Outer Hebrides and Skye (Scotland). Extralimital: *nataliae* Buturlin, 1929, perhaps central Siberia or (perhaps this race) south-west Iran (see p. 1000).

Field characters. 23 cm; wing-span 33–36 cm. Slightly larger than Redwing *T. iliacus*, with stockier build, noticeably blunter wing-point, and rump and tail slightly longer in proportion; 15% smaller than Mistle Thrush *T. viscivorus*, with proportionately shorter tail. Medium-sized thrush, with well-balanced form, upright carriage, brown-toned upperparts, and boldly spotted underparts. Within west Palearctic *Turdus*, has diagnostic combination of faint face pattern and golden-buff underwing. Sexes similar; slight seasonal variation. Juvenile separable. 3 races in west Palearctic, paler towards east.

ADULT. (1) Main continental European race, nominate *philomelos*. Upperparts rather pale, grey-washed brown, lacking any warm tone, tinged olive on rump and upper tail-coverts and umber on tail and crown; at close range, shows black-brown centres to tertials, buff-spotted tips to median coverts, and buff tips to greater coverts. Face buff- to grey-brown, indistinctly marked: cream fore-supercilium, eye-ring, and flecks on lores and central and lower cheeks, with darker brown surround on edges of cheeks (particularly at rear), pale cream moustachial stripe, and black-brown malar streak which contrasts with almost white throat. Rest of underparts essentially cream-white (with pale buff suffusion most marked on sides of chest and along flanks), liberally marked with (usually) evenly spaced black-brown fan-shaped spots and blobbed streaks which spread out from base of malar stripe across chest and fore-belly and over flanks. Spotting does not end sharply on rear flanks, so no conspicuous pale patch there (unlike *T. iliacus*), but pale cream, virtually unmarked vent and under tail-coverts are noticeable. Axillaries and under wing-coverts golden-buff, glowing when caught in light and contrasting with dusky-buff undersurface of flight-feathers and dark flanks. Wear makes upperparts greyer (particularly from nape over back), and ground-colour of underparts whiter (with increased contrast of spots). Intergrades with *clarkei*. For unusually dark form, see below. (2) Race of Britain (except Outer Hebrides and Skye), Ireland, and adjacent European mainland, *clarkei*. Upperparts brown, with rufous tone on crown and tail and faintly tawny hue elsewhere, having overall warm tone. Underparts suffused yellow- to olive-buff, this tone spreading over whole chest and flanks and combining with denser and more diffuse

spotting to make bird darker-chested and -flanked. Underwing tawny-buff. Wear removes full rufous tone of upperparts, dulls both ground-colour and spots on underparts, and makes tips of wing-coverts paler. (3) Race of Outer Hebrides and Skye, *hebridensis*. Upperparts umber-brown, more earthy and much less rufous than in typical *clarkei*, with grey wash on rump and upper tail-coverts but no olive tone as in nominate *philomelos*. Underparts suffused buff, on flanks smoky-brown; boldly marked with almost black spots. Thus appears as dark on chest and flanks as on upperparts and (head on, at distance) almost wholly black-brown across chest. Underwing rufous-buff, distinctly darker than in other races. Wear has no noticeable effect on colours. JUVENILE. All races. Resembles adult in ground-colour and plumage pattern but easily distinguished by small dark feather-tips on crown and face, buff spots on mantle, tawny- or yellow-buff streaks and spots on forewing (these forming conspicuous bars on tips of median and greater coverts), and smaller, less diffuse spots and streaks on underparts. Rump washed tawny-buff; flanks less olive or buff; vent duller, with buff suffusion. Pale marks and dark spots most accentuated on *hebridensis*. FIRST-YEAR. All races. From September, closely resembles adult and only distinguished at close range by retained juvenile greater coverts, with buff shaft-streaks. In all races, bill black-brown with flesh-yellow base to lower mandible; legs pale flesh.

Commonest and most widespread spotted thrush in Fenno-Scandia and temperate Europe, with little risk of confusion with *T. viscivorus* (much larger and longer-tailed), *T. iliacus* (smaller and more patterned), Siberian Thrush *Zoothera sibirica* (similarly sized but with barred underwing), Nearctic thrushes *Hylocichla* and *Catharus* (much smaller and less spotted). However, one problematic racial identification: namely, regular autumn occurrence on North Sea coasts of a small form dark enough above to recall *hebridensis* but differing from that race in cleaner ground-colour to underparts which have heavy but neat marks (*Fair Isle Bird Obs. Bull.* 1950, 8, 7; Flamborough Ornithological Group). Origin of these birds not known, though occurrences are associated with other species from north or central Europe. Flight typical of smaller *Turdus*, with clear affinity to that of wheatears *Oenanthe* and nightingales *Luscinia* but more powerful and faster; wing-beats fast, with obvious bursts producing acceleration but not marked undulations of larger thrushes, with wings completely closed only momentarily; agile in confined spaces, making rapid turns and then ducking into cover. Alternates short runs or series of hops with pauses and alert posture. Flicks wings and tail in excitement. Not markedly gregarious except on passage, and widely scattered, rather than in obvious flocks, in winter. Feeds closer to cover than *T. iliacus* and larger *Turdus* thrushes; often secretive in winter.

Song ringing, clear, musical, and vigorous; comprises phrases of 1–4 syllables in rhythmic series or broken up with more-warbling passages (but these never having richness of Blackbird *T. merula* nor wild skirling quality of *T. viscivorus*). Full song frequently sustained and often delivered from conspicuous treetop perch, particularly in early spring; far-carrying, with loudest parts audible at 1 km or more. Subsong, much used by interacting ♂♂, a low twittering warble recalling *T. iliacus* (but without occasional fluting notes or frequent performances in chorus). Commonest call 'sipp', more punctuated but far less penetrating and drawn-out than similar call of *T. iliacus*, used both on flushing and in flight; also gives quieter 'tic' in flight. Alarm-call sharper and more explosive: 'check' or, more usually, 'tchuk-tchuk'; often repeated and slightly higher pitched than similar call of *T. merula*. Continental nominate *philomelos* also gives thin 'seep' in flight, close to commonest call of *T. iliacus* but still with terminal punctuation; not certain if this call also used by other races.

Habitat. In upper and middle latitudes of west and central Palearctic, both continental and oceanic, largely temperate but also boreal and marginally subarctic. Tolerates cool, humid, and windy but not arid, very warm, nor persistently frosty and snowy climate. Birds can exist almost anywhere where trees or bushes accompany open grassland, patches of dead leaves under trees, or moist ground supporting ample invertebrate food organisms, and accordingly of neutral or higher pH value. Predominantly in lowlands and valleys, but in Switzerland ascends to treeline (1600–2200 m) in conifer forest and in USSR to 1200 m, reaching birch-scrub zone on Kola peninsula. Basically a bird of primitive forests, both coniferous and broad-leaved, where beech *Fagus*, birch *Betula*, and oak *Quercus* give some shade and humidity. Requires ample undergrowth, either of young spruce *Picea* or fir *Abies* or (increasingly in recent times) of equivalent deciduous cover. Modern conversion of lowlands to agricultural, industrial, and urban uses has stimulated, especially in western Europe, switch to small woodlands, parklands, hedgerows, railway embankments, roadsides, cemeteries, gardens, and even interiors of cities. Densities in these habitats often higher than in semi-natural woodlands, some of which (e.g. ash *Fraxinus*,

beech, birch, yew *Taxus*) are relatively little-occupied. Presence of snails *Helix* and earthworms *Lumbricus* appears to be of critical significance, and this is often associated with lime-rich soil. Remains, however, one of the commonest species in British woodlands of all types and sizes, and maintains itself in such marginal areas as heath-covered inshore islands and sparse hillside birchwoods, although recently overtaken nationally in numbers and in extent of distribution by Blackbird *T. merula*. Forages more under trees and bushes and less far out into open fields than other thrushes, such as Mistle Thrush *T. viscivorus*, and tends to make shorter flights, mainly in lower airspace. Migrates only when and as far as is essential, at cost of occasionally heavy mortality when caught by severe weather. Habitat in winter differs from that in breeding season only insofar as high or outlying areas and woodland are vacated. On the whole, human developments have proved beneficial to the species. (Nicholson 1951; Dementiev and Gladkov 1954*b*; Voous 1960; Glutz von Blotzheim 1962; Yapp 1962; Parslow 1973; Sharrock 1976; Simms 1978; Fuller 1982.)

Distribution. BRITAIN, IRELAND. Formerly bred Shetland regularly; some islands off Irish west coast have been colonized (Parslow 1973). FRANCE. Corsica: no proof of breeding (Thibault 1983). ITALY. Expanding in hills at 100–200 m (PB). ALBANIA. No proof of breeding (EN). USSR. Baltic: beginning to breed in towns (HV).

Accidental. Iceland, Faeroes (possibly annual), Azores, Madeira.

Population. Declined Britain and Ireland.

BRITAIN, IRELAND. Over 1 million pairs; apparently declined 1920–30 (Parslow 1973). Fluctuates with hard winters; *c.* 3·5 million pairs, but *c.* 1 million pairs after 2 hard winters; still declining in Ireland (Sharrock 1976). FRANCE. Under 1 million pairs (Yeatman 1976). BELGIUM. About 150 000 pairs (Lippens and Wille 1972). NETHERLANDS. 100 000–160 000 pairs 1976–7 (Teixeira 1979). Increased (CSR). WEST GERMANY. 1·1–2·2 million pairs (Rheinwald 1982). DENMARK. Slight increase 1978–82 (Vickholm and Väisänen 1984). SWEDEN. About 35 million pairs (Ulfstrand and Högstedt 1976). FINLAND. About 640 000 pairs (Merikallio 1958). Increased in north after 1930 (Haartman 1973); slight increase 1978–83 (Vickholm and Väisänen 1984). POLAND. Marked decline in urban habitats (Tomiałojć 1976; AD). CZECHOSLOVAKIA. No changes reported (KH).

Survival. Britain: mortality in 1st year of life (from 1 August) 53%, in 2nd year 40% (Lack 1946*b*). Finland: annual overall mortality 54% (Haukioja 1969). Oldest ringed bird 13 years 9 months (Rydzewski 1958).

Movements. Mostly resident but northern populations partially or entirely migratory; more birds move if weather severe. In contrast to (e.g.) Redwing *T. iliacus* and

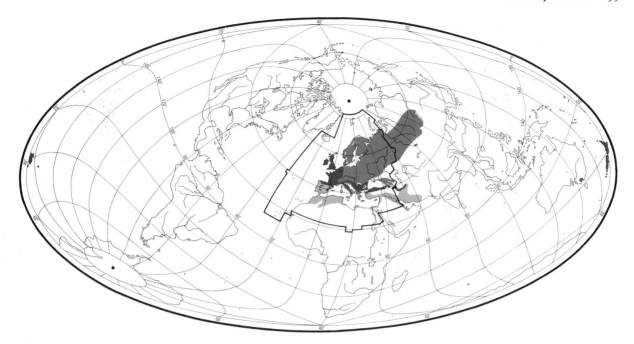

Fieldfare *T. pilaris*, populations show strong affinity to regular wintering areas (Ashmole 1962). Most nominate *philomelos* from Fenno-Scandia, Germany, Switzerland, Poland, and USSR are migratory, moving south-west or south-east through Europe to winter in southern England, France (mainly towards south-west), Spain, and Portugal (Witherby *et al.* 1938*b*; Dementiev and Gladkov 1954*b*; Ashmole 1962). Those from furthest north, especially 1st-year birds, winter furthest south to Canary Islands, Morocco, Algeria, Tunisia, Libya, and Cyprus; also 1 record from Sénégal (Roux 1959). Birds from Denmark, Netherlands, Belgium, and north-east France are partially resident with most others moving only short distances south or south-west, though considerable numbers from Netherlands winter in Britain and Ireland (Ashmole 1962). Birds from Switzerland move south, with recoveries in southern France along Mediterranean coast into northern Italy, and in Corsica and Libya (Ashmole 1962). Birds from east-central Europe winter correspondingly east of birds from Fenno-Scandia and western Europe: mainly in Italy, Yugoslavia, Greece, Balkans, and Cyprus; also a very few recoveries in Balearics, Spain, and Portugal (Ashmole 1962; Pikula 1972). Populations from further east presumably winter largely from eastern Mediterranean to Iran, with birds of unknown origin also apparently common in some years November–March in Kuwait and eastern Saudi Arabia (Bundy and Warr 1980), Oman (Gallagher and Woodcock 1980), and Eritrea (Smith 1957). Small numbers may now winter in Egypt on Mediterranean coast, Suez canal, and northern Red Sea coast and south along the Nile into Sudan (Cave and Macdonald 1955; Short and Horne 1981).

Many birds breeding Britain and Ireland (*clarkei*) winter north-west France, northern Spain, and Portugal to Balearics—also collected in Algeria according to Meinertzhagen (1932), but this considered by Etchécopar and Hüe (1967) to require confirmation; many from northern Britain winter in Ireland (Ashmole 1962). Of British population that winters in France, Spain, and Portugal, birds from northern Britain are recovered further south than those from southern Britain; majority of these are 1st-years, as only half of the adult population but $\frac{2}{3}$ of the 1st-years are migratory (Lack 1943). In early part of winter, migrant populations from Britain, Belgium, Netherlands, and Denmark are chiefly in north-west France, north of Scandinavian birds (in southern Spain and Portugal), but by January birds from Low Countries have also moved south-west into Iberia (Ashmole 1962). Birds from Outer Hebrides and Skye (*hebridensis*) are largely sedentary but some move to Ireland (Witherby *et al.* 1938*b*; Ruttledge 1975).

Night migrants over North Sea regularly attracted to lighthouses and killed in considerable numbers (Simms 1978). Coastal arrivals of flocks not uncommonly seen, but visible migration (especially inland) otherwise unusual. Southward departures in autumn begin in August but main passage September to early November; movement of birds into Ireland continues even into February (Ashmole 1962). Birds from Fenno-Scandia move south-west on broad front, fringe of movement (or birds drifted west) passing through eastern Britain (Myres 1964). Siberian birds depart mid-September while populations further south remain until November, in southern central Asia, Kazakhstan, and Turkmenistan; rarely recorded on passage outside breeding range (Johansen 1954). Birds wintering around Mediterranean arrive mid-October with

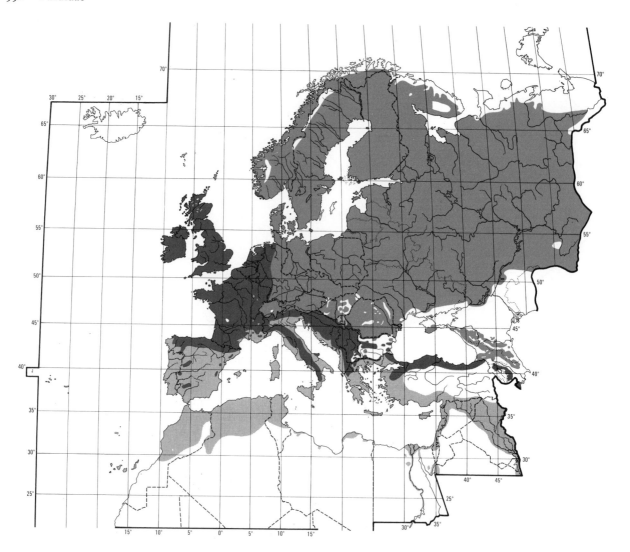

frequent influxes until mid-April. During severe weather over Europe, large-scale mid-winter arrivals occur regularly in North Africa (Smith 1965*a*).

Returning birds leave North Africa late March to mid-April (Ashmole 1962) and passage also occurs then in Caucasus and Ukraine (Dementiev and Gladkov 1954*b*; Johansen 1954). Northward movements from Portugal, Spain, western France, and through Britain and Ireland also at this time. Movement through Netherlands, Helgoland, and Denmark begins March and continues to mid-May. Finnish birds back on breeding grounds by mid-April and those in northern Sweden by early May (Ashmole 1962), whilst arrivals in north of range in USSR east to Yenisey occur late April to early May (Johansen 1954). In early May, northern breeding birds are passing through areas where local birds are already breeding.   PC

Food. Wide range of invertebrates; also fruit from late summer to winter. For foraging behaviour on grass meadow, see Blackbird *T. merula* (p. 953). Searches for food in ground litter, described as follows in captive birds: makes rapid sideways sweeping movements with bill, usually 3–8 movements in succession, occasionally almost continuous for up to 1½ min; sometimes scratches with foot simultaneously, though such movement only slight; also pulls material obliquely backward with bill (Henty 1976). Deals with snails (Gastropoda) by beating them against any hard surface, often a stone ('anvil'): holds snail in bill, either by lip of shell or by snail itself, through shell's opening; holds head to left or right, brings head and neck rapidly down, turning head at same time; usually only a part of shell removed thereby, and bird then flicks out snail's body, picks it up, and wipes it on ground before eating it (Morris 1954; Simms 1978); behaviour apparently largely innate (Herring 1984; Henty 1986, which see for review of snail-smashing in birds generally). Juvenile seen beating cherry against concrete before pecking at it (Radford 1967*a*), and one bird opened stone of cherry by same means (Thomas 1970). Recorded associating with mole *Talpa* to feed in leaf litter and earth disturbed by it (McCanch and McCanch 1982; Sharrock 1982). Once seen to steal earthworm from conspecific bird (Naylor 1974). Recorded chipping small pieces of ice and snow from top of wall and swallowing them (Harding 1986).

Diet in west Palearctic includes the following. Invertebrates: grasshoppers, etc. (Orthoptera: Gryllidae, Tettigoniidae, Acrididae, Tetrigidae), earwigs (Dermaptera), bugs (Hemiptera: Pentatomidae, Miridae, Pyrrhocoridae, Coreidae, Membracidae, Cicadidae), adult and larval lacewings, etc. (Neuroptera: Hemerobiidae, Chrysopidae, Raphidiidae), scorpion flies (Mecoptera), adult and larval Lepidoptera (Nymphalidae, Pieridae, Tortricidae, Notodontidae, Noctuidae, Geometridae), adult and larval flies (Diptera: Tipulidae, Bibionidae, Syrphidae, Tachinidae, Calliphoridae, Muscidae), Hymenoptera (sawflies Tenthredinidae, Ichneumonoidea, ants Formicidae), adult and larval beetles (Coleoptera: Carabidae, Silphidae, Staphylinidae, Lucanidae, Geotrupidae, Scarabaeidae, Byrrhidae, Elateridae, Cantharidae, Tenebrionidae, Dermestidae, Coccinellidae, Cerambycidae, Chrysomelidae, Curculionidae), spiders (Araneae), harvestmen (Opiliones), mites (Acari), woodlice (Isopoda), sandhoppers (Amphipoda), millipedes (Diplopoda), centipedes (Chilopoda), snails (Gastropoda), slugs (Pulmonata), earthworms (Oligochaeta). Also recorded taking lizard (Chater 1965), attacking slow-worm *Anguis* (Husband 1967), trying to eat shrew (Soricidae) (Deeble 1972), and eating bird faeces (Boyle 1970) and an egg of own species (Harper *et al.* 1972). Plant material mainly fruits, also seeds: juniper *Juniperus*, yew *Taxus*, mistletoe *Viscum*, barberry *Berberis*, bramble *Rubus*, strawberry *Fragaria*, rowan *Sorbus*, cherries *Prunus*, *Cotoneaster*, currant *Ribes*, bilberry *Vaccinium*, crowberry *Empetrum*, elder *Sambucus*, spindle *Euonymus*, olive *Olea*, *Lonicera*, sea buckthorn *Hippophae*, vine *Vitis*, ivy *Hedera*, dogwood *Cornus*, *Viburnum*; also spruce *Picea* needles, clover *Trifolium*, and turnip. (Csiki 1908; Siivonen 1939; King 1960*b*; Baker 1970; Raiss 1976; Török 1985.)

In suburban Oxford, earthworms taken largely December–March (up to 94% of *c.* 900 feeding records); snails taken mostly July–September (up to 62%) and in February when snow made earthworms more difficult to catch; caterpillars important in June (72%), and fruit September–November (up to 93%) (Davies and Snow 1965; see also for relative availabilities of different food items). In Britain, 84 stomachs (apparently from throughout year) contained 35·5% insects (largely adult and larval beetles), 15% earthworms, 5% slugs and snails, 1·5% other invertebrates, 41·5% fruits and seeds, and 1·5% grass, bread, etc. (Collinge 1924–7). On Helgoland (West Germany), 244 stomachs of spring migrants contained 890 invertebrate items: 34·6% (by number)

snails, 28·3% adult insects (mostly beetles), 16·4% larval insects (mostly beetles and Diptera), 9·7% slugs, 9·3% earthworms, and 1·8% others; 12·7% of stomachs also contained plant material (Raiss 1976). Autumn migrants on Holy Island (England) appeared to eat largely snails (Richards 1977). In winter in Córdoba (Spain), diet largely fruit: 130 stomachs contained 69–82% (monthly averages, by volume) plant material, 41–60% olives; animal material largely beetles and snails (Manzanares 1983). In autumn and winter around Montpellier (southern France) also mainly fruit, and this mainly grapes (in 64% of stomachs containing fruit) and juniper berries (in 45%) (Debussche and Isenmann 1985*b*). For important compilation of early studies, see Siivonen (1939); see also Florence (1914), Gil (1927, 1928), Lebeurier and Rapine (1941), Kovačević and Danon (1950–1), Tutman (1950, 1962), Boehme (1958), Mal'chevski (1959), Ptushenko and Inozemtsev (1968), Korodi Gál (1969), Kiss *et al.* (1978), and Kostin (1983).

In Negev desert (Israel), winter 1972 (when numbers of birds not unusually high), snails subject to unusually heavy predation by *T. philomelos* (also some Mistle Thrushes *T. viscivorus* involved): over *c.* 55 days, ate *c.* 65% of population of one species and *c.* 10% of population of another (Shachak *et al.* 1981). Will concentrate on one morph of a snail species out of proportion to its abundance: thus 99% of *Ariata arbustorum* taken in one study were of banded-and-spotted morph, though that morph comprised only 77% of local population (Reichholf 1980). For some other studies of predation on snails, see Cameron (1969), Schlegel and Schlegel (1974), Richardson (1975), Roos (1976), Wardhaugh (1984), and Sueur (1985). For predation on *Pieris* caterpillars, see Baker (1970).

In Britain, 38 stomachs of nestlings contained 38 Lepidoptera larvae, 9 Diptera larvae, 5 Elateridae larvae, 4 Noctuidae larvae, 4 spiders, and remains of earthworms and slugs (Collinge 1924–7). In Hungary, young given more earthworms and fewer beetles than young of *T. merula* (Török 1985). Adult recorded feeding leaves of thrift *Armeria* to fledged young (Radford 1973).   DJB

Social pattern and behaviour. Moderately well but not intensively studied. Major general ecological study (Finland) by Siivonen (1939); study of territory (Oxford, southern England) by Davies and Snow (1965), and of parental behaviour by Hartley (1967).

1. Outside breeding season more or less solitary or in small feeding or roosting aggregations, except during migration when regularly in large but loosely coordinated flocks. In areas where populations partially resident, many ♂♂ and smaller numbers of ♀♀ occupy individual winter territories, territories of ♂♂ being more or less the same as previous or subsequent breeding territories; 31 ♂ and 3 ♀ winter territories recorded in Oxford study area (Davies and Snow 1965). Occupation of winter territories dependent on availability of food; tend to be abandoned during spells of severe weather (Davies and Snow 1965). BONDS. Monogamous mating system. In Oxford, new

pairs formed each year, apparently as result of ♀'s tendency to migrate and not return to same area in following year (Davies and Snow 1965). Pair-formation from early spring (late February in southern England); process gradual and sometimes interrupted by cold weather, often not completed until shortly before nesting begins (Davies and Snow 1965). Age of first breeding apparently 1 year: specimens of 1-year-olds of both sexes recorded with gonads in breeding condition (BMNH). ♀ alone broods young; both sexes feed young in nest and for *c.* 3 weeks after fledging. BREEDING DISPERSION. Solitary and territorial. Territory boundaries not as well-defined as in (e.g.) Blackbird *T. merula*, and even when territorial behaviour at height birds spend periods away from territory, especially for feeding. Territory apparently essential for pair-formation and breeding, but only part of food for young obtained within territory. Territory size variable, down to *c.* 0·2 ha in optimal garden habitat, mostly 0·4-0·6 ha in Oxford Botanic Garden (Davies and Snow 1965). Woodland territories highly variable, e.g. 1·5-6 ha in southern Finland (Siivonen 1939), in France usually 2-3 ha (1-6) (Géroudet 1954). Densities highest in rich garden habitats with abundant nest-sites, e.g. 170-280 pairs per km² in 3 ha of Oxford Botanic Garden. Much lower in woodland and other habitats: 27 per km² in Oxfordshire oak *Quercus* wood (Davies and Snow 1965); up to 43 per km² in Surrey (England) oak wood (Beven 1963); in southern Finland, up to 40 pairs per km² in spruce *Picea* or mixed woodland with spruce (apparently the main natural breeding habitat in northern Europe), up to 10 pairs per km² in broad-leaved woodland with rich bush layer, up to 5 pairs per km² in pine *Pinus* woodland (Siivonen 1939). In Oslo region (Norway), 68 ♂♂ per km² (Slagsvold 1973*b*). ROOSTING. In thick cover, especially bushes; in Finland especially in low, thick spruce; in areas where higher cover is lacking, in heather and other low scrub (e.g. Hebrides, Scotland); locally in coastal sand-dunes. Solitary or in pairs, or in small groups if site suitable. Paired ♂ and ♀ regularly roost together from early spring (Labitte 1937). Suitable cover within territory may attract neighbouring territory-holders which may then be persistently but unsuccessfully chased by residents (Davies and Snow 1965). Birds noisy and excited before entering communal roost, calling (see 3c in Voice) or giving harsh, hurried song, flying from one high perch to another (Spencer 1953). Occasionally roost in company with other *Turdus*, especially in large mixed roosts. Captive roosting birds always perch on one leg (Siivonen 1939). Comfort behaviour not studied in detail but similar to that of *T. merula* and other *Turdus*. For anting, see Gough (1947) and Wells (1951).

2. Generally shy in woodland and other habitats unfrequented by man; more confiding in towns and suburbs, some individuals becoming very tame. Shy woodland birds, when on ground, react to sudden disturbance (by man) with characteristic behaviour: fly up with weak Tsip-call (see 2a in Voice) to low perch, and there watch intruder from semi-concealed position; then hop up to higher position in tree and from there fly down to vanish in low undergrowth; do not, however, move far but stay still and are difficult to see (Siivonen 1939). Rapid flight with Tsip-call characteristic of disturbed birds in other situations. Mobbing behaviour (not studied in detail) similar to *T. merula*, accompanied by loud, sustained Alarm-chatter (see 3c in Voice). FLOCK BEHAVIOUR. Migrants give variety of contact-calls (see 2a-c in Voice), perhaps varying to some extent with race and region. Unspecified type of subsong (see 1 in Voice) sometimes given chiefly in autumn and early spring by chorus of a dozen or more birds together, much like Redwing *T. iliacus*, but generally low down in bushes, etc. (Witherby *et*

al. 1938*b*). SONG-DISPLAY. By ♂ only. Song (see 1 in Voice) usually delivered from perch, often tree-top or other high vantage point, occasionally from ground. Exceptionally (perhaps displacement behaviour) singing birds may repeatedly pull at twigs near perch between phrases (Soper 1972). Usual loud Territorial-song (see 1a in Voice), given from tree perch, has dual function: to proclaim and maintain ownership of territory, and to advertise unpaired ♂ to potential mates. Also 3 other main types of song (see 1b-d in Voice), and subdued song occasionally given by ♂ remaining in territory in autumn (September and early October), also by unestablished ♂♂ (Davies and Snow 1965). In areas where populations totally migratory, song begins immediately after arrival of ♂♂ on breeding grounds, depending mainly on latitude (e.g. February-March at 45-50°N, March-April at 57½-60°N, May at 67½-70°N: Siivonen 1939); ends about mid-July in all areas. Where some ♂♂ resident (e.g. southern England), song begins when territories re-established in late autumn, usually late October or November; becomes intermittent in winter, restricted to periods when weather mild; intensifies in early spring, then declines sharply from May to end of song-period (Davies and Snow 1965; Simms 1978). During breeding season, most sustained song given by ♂♂ that have remained unpaired or have lost mates; song output of breeding ♂♂ highest at beginning of each nesting cycle (Slagsvold 1973*a*; Ince 1982). Strong tendency for song to be concentrated in early morning and evening. In southern Finland, song given in early morning only when birds first arrive after migration, evening song beginning later; from early June, song continues throughout night, daytime period without song (Siivonen 1939). In northern Ireland, song at first (late December or early January) also in early morning only, extending to later in day; continues throughout day during period of maximum song (late January and early February); by mid-March, again confined to early morning and evening, evening song continuing to end of song-period (Burkitt 1935). Song regular from migratory birds in winter quarters (from early October to late March near Rome: Alexander 1917); migrants arriving Kent (southern England) in late November began singing almost at once and showing apparent territorial behaviour (Harrison 1944). ANTAGONISTIC BEHAVIOUR. Territorial individuals of both sexes generally aggressive towards intruders. Aggressive displays, notably Forward-threat posture (Fig A), include the following

A

elements: crouching with legs flexed, facing opponent, bill open, tail depressed and fanned; body plumage ruffled, exhibiting spots on breast conspicuously; sometimes accompanied by 'subsong' (Simms 1978)—perhaps Alarmed-song (see 1c in Voice). Equally matched antagonists may fight, fluttering up in the air, then squatting in horizontal posture on ground. One report of a bird twice plucking red geranium petal during aggressive encounter and jumping up and down holding it (Cornish 1950). Appeasement/submissive posture apparently undescribed. HETEROSEXUAL BEHAVIOUR. (1) Pair-bonding behaviour. Pair-formation takes place in territory after establishment of territory by ♂; not accompanied by special display. During early stage, ♀ may visit many different territory-holding ♂♂ for short periods, associating and feeding with each; later, ♀ tends to return to same territory and remain for longer. Process may be interrupted by cold weather (Davies and Snow 1965). (2) Courtship-feeding. Not recorded, but ♂ may feed sitting ♀ (see below). (3) Mating. ♂'s precopulatory display is striking, and includes the following elements (e.g. Harvey 1946, Blume 1967, Radford 1975): head held up vertically with bill wide open (or bill may be opened and closed), breast feathers ruffled, wings partly extended, tail fanned and depressed; in this posture, ♂ runs beside or in front of ♀, or circles her, or may rotate slowly with short steps, silent or uttering subdued song. Mounting follows immediately. No special soliciting posture by ♀ recorded. After mating, ♀ once seen to peck into open bill of ♂ (Ellison 1931). (4) Nest-site selection and behaviour at nest. Nest-site selection, apparently by ♀ only, involves testing possible sites, squatting in site, rotating with tail raised, and hopping around site; this behaviour may continue for 3–9 days (Pokrovskaya 1968). Nest-building by ♀ only. Incubation by ♀ only, but ♂ attentive, staying near nest and even perching on nest-rim when ♀ sitting; ♂ occasionally takes food to ♀ on nest (see also Relations within Family Group, below). When ♀ off nest, ♂ may stand in nest-cup but does not settle on eggs. (Hartley 1967.) RELATIONS WITHIN FAMILY GROUP. Nestlings brooded by ♀ only, intensively (at least 75% of daytime) for first 2–3 days, then decreasingly, depending on temperature, finally only at night. ♂ also shelters nestlings against heavy rain. (Hartley 1967.) Nestlings at first beg upwards in response to tactile stimuli (shaking of nest); when eyes open (at *c.* 8 days) direct open bill towards head of parent (Tinbergen and Kuenen 1939). Both parents feed young, mainly ♂ for first few days; part of food usually passed to ♀, who may eat some of it (occasionally all). For call of parent arriving at nest with food, see 5 in Voice. Nestlings' faeces mostly swallowed by parents, in last 3–4 days some carried away; in final day before fledging, some left on edge of nest. Dead young thrown out, falling below nest, or carried and dropped several metres from nest (Davis 1967). Fledging period usually 13–14 days, but young may leave nest earlier (from 9 days) if disturbed. Undisturbed young creep and flutter from nest, usually more or less at same time, and remain in surrounding cover for several days; occasionally return to spend night in nest (Siivonen 1939; Hartley 1967). Remain in thick cover for 12–14 days, then begin to emerge and follow parent, at first in morning only; continue to be fed for *c.* 22 days after fledging. Division of brood between parents not reported. After independence, young remain in territory for about a week, then leave (not driven away by parents). (Hartley 1967.) ANTI-PREDATOR RESPONSES OF YOUNG. Not well studied, but generally similar to *T. merula* (see p. 957). If disturbed and scattered, fledglings utter croaking sound, otherwise usually silent. In addition, threatened juvenile may adopt Bittern-posture with bill pointing upwards (Clarkson 1981: Fig B). PARENTAL ANTI-PREDATOR STRATEGIES.

B

(1) Passive measures. Best developed in woodland birds, suburban birds tending to react more aggressively to disturbance of nest. On approach of intruder, incubating ♀ first settles deeper into nest, then (if intruder approaches closer) hops silently off nest, dropping from perch to perch, finally to ground, before flying away (Siivonen 1939). ♂ keeps guard near nest, but not reported to give warning-calls to ♀. (2) Active measures: against birds. Jay *Garrulus glandarius* approaching nest may be attacked, parents pressing home attack with great boldness (Goodwin 1953a), and other large birds near nest, even if not a threat (e.g. Collared Dove *Streptopelia decaocto*), may be similarly attacked (Hollick 1980; Zentzis 1983). Mobbing-attacks accompanied by Alarm-chatter calls. (3) Active measures: against man. ♀ with young commonly reacts to intruder with intense threat behaviour: perches near nest in Forward-threat posture, Bill-snapping and/or giving Alarm-chatter calls; also bill-wipes on perch (e.g. Took 1944, Evans 1950, Field 1950). ♀ once recorded attacking intruder with discharge of faeces (Siivonen 1939). Only one detailed report of distraction-lure display of disablement type: bird landed on ground near nest in which young being ringed, then stumbled away, dragging half-opened wings (Cawkell 1947b). ♂ generally much more wary than ♀, remaining at distance (Siivonen 1939).

(Figs by N McCanch: A from drawing in Simms 1978; B from photograph in Clarkson 1981.)   DWS

Voice. Freely used throughout the year. No major study of whole repertoire, but song analysed in detail by Ince (1982), which see for additional sonagrams, also Bergmann and Helb (1982). Bill-snapping sounds given when threatening human intruder at nest (D W Snow). Differences between races require further study: according to Witherby *et al.* (1938b), all notes of *hebridensis* appear softer than *clarkei*; see also calls 1–2 (below).

CALLS OF ADULTS. (1) Song of ♂. A number of variants, according to context and purpose, described by Siivonen (1939), though the following names are not all his. Relationship between kinds of subdued song (types 1c–d) not clear, these often referred to collectively as 'subsong', and further research needed. (a) Territorial-song. Loud, clear, vigorous succession of simple but musical phrases distinguished by their repetitive character, great variety, and clear enunciation; compared with *clarkei*, *hebridensis* less shrill and more varied and rambling, with less repetition of well-marked phrases (Witherby *et al.* 1938b). More penetrating and less rich than Blackbird *T. merula*, lacking wild skirling quality of Mistle Thrush *T. viscivorus* (D I M Wallace). Also described as a loose sequence of very loud, mostly pure-toned and polysyllabic units (Bergmann and Helb 1982). Fluting notes are mixed with less pleasant, harsh impure sounds and noises, also mimicry (Géroudet 1963). Mimics songs and calls of a wide variety of passerines and non-passerines, notably waders (Charadriiformes) (Simms 1978, which see for full list for, presumably, *clarkei*). In Britain, will also mimic 'trimphone' trill (Slater 1983). For examples of mimicry by nominate *philomelos*, see Ferry and Frochot (1964), Lhoest (1981), and Bergmann and Helb (1982). Each unit is repeated *c.* 2–4 or more times, thus (e.g.) 'filip filip filip codidio codidio quitquitquit tittitt tittitt tèrèrèt tèrèrèt tèrèrèt kvièt kvièt kvièt kvièt...'. In Fig I, a typical repetition

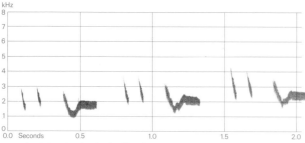

I   A G Field   England   April 1963–5

segment 'ti-ti-huwee ti-ti-huwee ti-ti-huwee'; another repetition (not shown) by this bird rendered 'cheri-too cheri-too cheri-too'. But song of same bird includes much more complex sequences, e.g. phrase in Fig II consists of: 1st segment, roughly 'kru krree', repeated 4 times, then, after pause of 0·8 s, 'k-t-k-te', followed by 3 low 'g-d' castanet-like units, and 4 much higher-pitched units rendered 'pi-chi-pi-chi'; there follows a very quiet rattle, and finally 3 high thin 'peeep' units. In song of bird just before going to roost, pauses scarcely exist and division into discrete phrases arbitrary: one sonagram (Fig III) of this song shows 2nd of 2 'si tiu peeoo' figures, followed by 3 short musical figures difficult to render, finally 2 purring units. (J Hall-Craggs.) Following analysis of song of 3 birds after Ince (1982); see also Ince and Slater (1985). Average frequency 3·77 kHz (2·48–5·07), sharing with Fieldfare *T. pilaris* highest average frequency and greatest structural variability of European *Turdus*. Songs are poorly defined temporally, and may follow one another without pause, or at intervals of several seconds; overall

effect is of almost continuous song. Average length of song 1·38 s, average interval 1·24 s. According to Ince (1982), possesses largest repertoire of any European *Turdus*, 104–219 'songs', average *c.* 130. Some song types recur frequently, at intervals of *c.* 30 s, others rarely, and some transitions occur much more often than expected by chance, leading to natural grouping of song types. *Contra* some descriptions, however, it is exceptional for a given song type to be repeated consecutively, and repetition is mainly of units and sub-units within the song. (Ince 1982; Ince and Slater 1985.) One bird repeated only 2 songs in sequence of 85 songs, another repeated 60 in 203 (Marler 1959). (b) Battle-song. A special form of Territorial-song, and intergrading with it, given from the ground as well as from perch during periods of intense rivalry with other ♂♂; described as very loud, 'nervous' sounding, varied, and containing warning-calls (Siivonen 1939). (c) Alarmed-song. A subdued 'confused' sounding song, similar to Courtship-song (see 1d, below), given by ♂ close to intruder at nest with young (Siivonen 1939). 'Subsong' described as a confused low warbling like Robin *Erithacus rubecula* with a curious twittering quality, which sometimes accompanies Forward-threat posture (Simms 1978), is presumably of this type. So-called low amplitude 'Whisper Song', induced by playback of Territorial-song and thus interpreted as response of apparently both paired and unpaired birds to rivals in territory (Ince 1982), probably the same: compared with Territorial-song, pitch more variable but typically higher; most sub-units different from those in Territorial-song; given in apparently great agitation, often in twilight (Ince 1982). (d) Courtship-song. Subdued, highly variable song, carrying only *c.* 10 m, given by ♂ near ♀ during courtship (Siivonen 1939; Ince 1982). (2) Contact-alarm calls. (a) Tsip-call. Commonest call a short

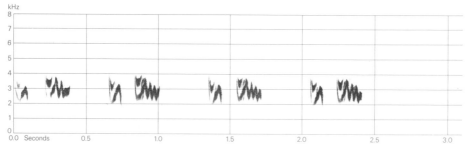

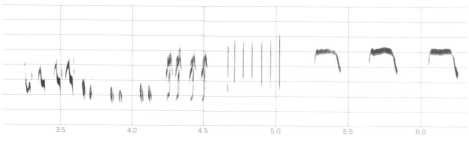

II   A G Field   England   April 1963–5

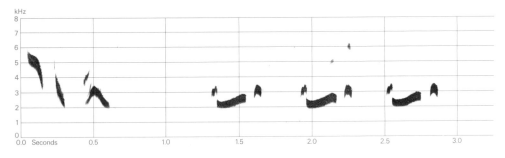

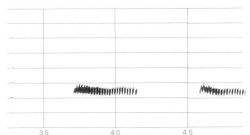

III P J Sellar England May 1982

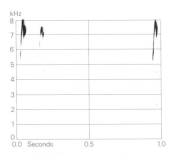

IV P J Sellar England January 1979

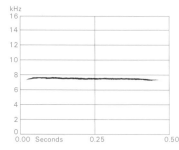

V P J Sellar England April 1979

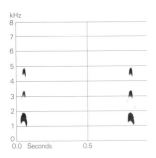

VI V C Lewis England April 1965

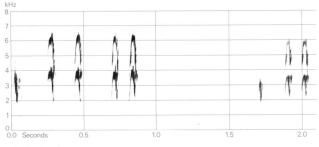

VII P J Sellar England April 1984

'tsip' (e.g. Géroudet 1963, Simms 1978). Also rendered a thin 'sipp' or 'tick' not unlike Seeip-call of Redwing *T. iliacus* (see call 2a of that species—also call 2c, below, which is much more confusable with *T. iliacus*), but not so penetrating or prolonged, given chiefly in flight and when flushed (Witherby *et al.* 1938*b*), often during nocturnal migration (Géroudet 1963); short, high-pitched 'zip' or 'zick', sometimes also disyllabic (Bergmann and Helb 1982, which see for sonagram). Apparently same call (Fig IV: 'zipzip zip') given to warn off conspecifics and other *Turdus* at feeding station in severe weather (P A D Hollom, P J Sellar); thus may signal, in conflict or in flight, 'not a good idea to approach me any closer' (J Hall-Craggs). (b) Zili-call. A disyllabic 'zili' given regularly by spring migrants in south Caspian region (Iran); not simply a disyllabic 'zip' call (see 2a, above), since the unusually metallic 1st syllable of 'zilip' was sometimes given separately and then sounded different from the normal 'zip'; 'zilip' rarely heard in Europe. (Schüz 1959.) (c) A thin, far-carrying 'seep' (D I M Wallace) or 'seeh', very like call 2a of *T. iliacus*, heard from flock-members of nominate *philomelos* migrating in autumn, and readily distinguishable from the drier shorter call 2a (above) which a few individuals also gave (Hollyer 1972). Perhaps confined to nominate *philomelos* and, according to Simms (1978), distinguishable from *T. iliacus* to the practised ear. (3) Warning- and alarm-calls. (a) A thin 'siih' given when raptor passes by (Géroudet 1963), also 'SEEep' in presence of stuffed owl (P J Sellar); in Fig V, from latter sequence of calls, sonagram is narrow band at half-speed playback and begins with slight initial ascent to a little over 7 kHz, descending slightly and gradually to the end, duration just over 0·45 s; not unlike equivalent call of *T. merula* (J Hall-Craggs: see that species, call 3b). (b) A subdued 'djük' or 'dukduk' in low intensity alarm (Bergmann and Helb 1982). Also rendered 'kuk kuk' (Fig VI), closely similar to 'pook'-call of *T. merula* (J Hall-Craggs: see call 3a and Fig III of that species). For other renderings, see Witherby *et al.* (1938*b*), Géroudet (1963), and Simms (1978). (c) Alarm-chatter. In greater excitement, a rapid succession of units rendered 'tschi' and much higher pitched than call 3b, producing a high intensity chatter or rattle, not as shrill as *T. merula* (Bergmann and Helb 1982). Rendered 'tictictictix', often heard before roosting (Géroudet 1963). Recordings by P J Sellar show call to be a muffled 'tuk tuk tuk' variant when going to roost, not unlike alarm 'chook'-call of *T. merula* (J Hall-Craggs: see call 4a and Fig Va of that species); at roost, more strident 'chik' series (Fig VII shows 'tuk-chik-chik-chik-chik tuk-chik-chik': J Hall-Craggs). Still more rapid and agitated 'chik' series given when mobbing Magpie *Pica pica* (P J Sellar, J Hall-Craggs). Often interspersed with Bill-snapping when human intruder threatening nest (D W Snow). (4) Distress-call. When handled, a shrill screeching 'kschri kschri' (Bergmann and Helb 1982), a high-pitched wavering scream, once also heard when seized by Short-eared Owl *Asio flammeus* (Simms 1978). (5) Other calls. In our recordings, parent arriving at nest with young gave quiet 'jup jup' sounds, not unlike pre-roosting 'tuk' calls (see 4c, above). Also similar to feeding-call of ♀ *T. merula* (see call 8a of that species, p. 961).

CALLS OF YOUNG. Food-call of small young a 'chip' sound; call of well-grown nestlings a rapidly repeated, hoarser, sibilant 'tschiptschiptschip'. Contact-call of juvenile a querulous 'treeep' (Witherby *et al.* 1938*b*); in recording by P A D Hollom, apparently the same food-call of fledgling rendered 'zrrit' (J Hall-Craggs), given regularly at intervals of *c.* 4·5-6·5 s. If disturbed and scattered, fledged young may give a deep frog-like croak (D W Snow).   EKD

Breeding. SEASON. Britain and western Europe: see diagrams; season prolonged and nests recorded in virtually all months of year. Central and eastern Europe: laying begins mid- to late April (Makatsch 1976). Finland: laying from late April in south and mid-May in north (Haartman 1969*a*). SITE. In trees and shrubs, often against trunk supported by twigs or branch, or among dense twigs; also in creepers on wall, on ledge, in bank, and on ground among thick vegetation. Mean height of 196 nests, Poland, 2·5 m (0-8) (Bocheński 1968). Nest: neat structure of twigs, grass, and some moss, loose towards outside, compacted towards inside; thickly lined with hard plaster material made from mud, dung, and

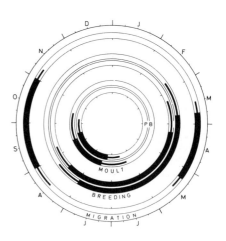

(especially) rotten wood, often mixed with leaves. Mean dimensions of 68 nests, Poland: external diameter 14·8 mm (9·3-18·0); internal diameter 9·1 mm (8·0-10·3); height 11·7 mm (7·5-23·0); depth of cup 6·7 mm (5·5-9·5) (Bocheński 1968). Building: by ♀, producing smoothed lining by pressing with breast. EGGS. See Plate 84. Sub-elliptical, smooth and slightly glossy; bright pale blue, lightly spotted and speckled dark purple-brown or black; rarely unmarked. Nominate *philomelos*: 27·1 × 20·4 (23·5-30·8 × 18·7-22·0), n=250; calculated weight 6·0 g. *T. p. clarkei*: 27·5 × 20·8 mm (25·0-31·1 × 19·2-23·0), n=150; calculated weight 6·2 g. *T. p. hebridensis*: 28·5 × 20·6 mm (26·2-30·8 × 19·5-21·6), n=18; calculated weight 6·45 g (Schönwetter 1979). Clutch: 3-5 (2-6). Of 1155 clutches, Britain: 2 eggs, 3%; 3, 13%; 4, 57%; 5, 26%; 6, 1%; mean 4·7; mean varies through season—3·9 in March, 4·3 April-May, 3·8 in June (Snow 1955*a*). Mean of 264 clutches, Finland, 4·8 (Haartman 1969*a*). 2-3(-4) broods, not more than 2 in north of range. Eggs laid daily. INCUBATION. 13·4 days (10-17), n=79 (Silva 1949). By ♀ only, or very rarely assisted by ♂. Begins with last egg; hatching synchronous. YOUNG. Cared for and fed by both parents. FLEDGING TO MATURITY. Fledging period 13·2 days (11-17), n=172 (Silva 1949). Become independent shortly after. Age of first breeding 1 year. BREEDING SUCCESS. Of 739 eggs laid, Britain, 71% hatched. Of 1034 young hatched, 78% fledged; overall fledging success 55% (Silva 1949). Of 816 nests, Britain, 50% hatched at least some young, and 36% fledged at least some young (Snow 1955*b*).

Plumages (*T. p. clarkei*). ADULT. In fresh plumage, forehead, crown, and all upperparts warm brown, slightly more olive-brown on rump and upper tail-coverts. Line from base of bill to above eye and eye-ring pale buff. Ear-coverts olive-brown with pale buff streaks and sooty brown rear border. Lores, thin moustachial stripe, and broader malar stripe blackish, lower cheeks and sides of neck pale buff with dusky mottling (rear of cheeks) and spots (neck). Throat and chin pale buff, with dark streaks at sides, grading into deeper golden-buff on chest and flanks with larger tear-shaped sooty spots. Belly whitish in centre, pale buff at sides, with sooty spots, as on chest, becoming smaller in centre. Vent and under tail-coverts white or very pale buff; thighs and lateral under tail-coverts washed or spotted olive-brown or grey-brown. Tail warm brown. Flight-feathers, tertials, and upper wing-coverts warm brown, median coverts with rounded-triangular pale buff spot on tip and greater coverts with buff fringe along tip of outer web. Axillaries and under wing-coverts orange-buff. *In worn plumage*, spots on wing-coverts reduced through wear and buff of underparts paler. NESTLING. Down golden-buff, fairly long. JUVENILE. Forehead, crown, mantle, and scapulars warm brown, feathers with indistinct pale buff shaft-streaks. Feathers of back warm brown with buff central streak and sooty brown tip. Rump and upper tail-coverts olive-brown, unspotted. Stripe from base of bill to above eye pale buff. Ear-coverts buff, feathers at rear with sooty black tips forming dark rear border. Cheeks pale buff with indistinct dark malar stripe. Throat and chin pale buff with small dark spots at sides, merging into deeper buff with larger dark spots on chest and flanks (but spots considerably smaller than in adult). Belly buffish white with small dark spots. Vent and under tail-coverts dull buffish white. Tail warm brown; for shape, see First Adult (below). Wings warm brown, inner secondaries with pale tips; greater, median, and lesser upper wing-coverts with buff shaft-streak, narrowing to a point towards base and widening towards tip to form a broad buff spot. FIRST ADULT. Like adult, but juvenile flight-feathers, tail, tertials, greater upper primary coverts, and variable number of outer greater secondary coverts retained, often slightly paler brown on outer webs than neighbouring fresh feathers; retained greater secondary coverts with larger and more triangular pale spot on tip than neighbouring 1st adult feathers; tail-feathers more sharply pointed. *In worn plumage*, contrast in size and shape of spots on greater coverts less apparent, as spots wear off; notch at tip of worn retained juvenile coverts deeper, but ageing often impossible when whole feather-tip abraded in late spring and summer. See also Svensson (1984a).

Bare parts. ADULT. Iris dark umber, dark hazel, or dark brown. Bill mainly black-brown, base of lower mandible yellow or flesh-yellow. Leg and foot pale flesh-colour. NESTLING. Mouth golden-yellow; no spots. Gape-flanges pale yellow. JUVENILE. Like adult, but bill paler at fledging, soon darkening to adult colour.

Moults. ADULT POST-BREEDING. Complete; primaries descendant. In Britain, starts with p1 mainly late June to mid-August, occasionally from mid-June or up to early September. Primary moult lasts *c.* 50 days. Completed with p9–p10 mainly mid-August to late September, occasionally early August or up to mid-October. (Snow 1969b; Ginn and Melville 1983). In USSR, moult August–September(–October) (Dementiev and Gladkov 1954b). POST-JUVENILE. Partial: head, body, lesser and median upper wing-coverts, tertial coverts, a variable number of inner greater upper wing-coverts (sometimes all), and rarely a few tertials or tail-feathers. Starts shortly after fledging, completed July–October. See also Jenni (1984).

Measurements. *T. p. clarkei.* Southern England, all year; skins (BMNH). Juvenile wing is retained juvenile wing of 1st adult. Bill to basal corner of nostril; to skull, on average 7·5 longer, exposed culmen on average 2·5 longer.

WING AD	♂	115·7 (1·34; 10)	114–118	♀	114·1 (2·73; 10)	111–119
JUV		114·0 (2·26; 10)	111–118		112·6 (2·37; 10)	109–116
TAIL AD		85·2 (3·26; 10)	78–89		83·5 (2·55; 10)	81–87
BILL AD		14·0 (0·67; 10)	13–15		14·5 (0·53; 10)	14–15
TARSUS		32·1 (0·86; 10)	31–33		32·3 (1·16; 10)	31–34

Sex differences not significant.

Nominate *philomelos*. Wing, Sweden and Norway, ♂: 116 (10) 113–120 (Vaurie 1959). On migration and in winter, Turkey, Iran, and south-west Siberia; skins and live birds: 118·7 (2·19; 15) 114–122 (Schüz 1959; Kumerloeve 1961, 1970a; Havlín and Jurlov 1977).

Wing of freshly dead birds, Netherlands: nominate *philomelos*, migrants, May, 1st adult ♂ 119·4 (2·60; 72) 113–125, 1st adult ♀ 117·7 (2·28; 115) 113–123; mixed sample of *clarkei* and nominate *philomelos*, all year, ♂ 118·5 (3·08; 192) 109–126, adult ♀ 118·2 (1·84; 41) 114–123, 1st adult ♀ 116·7 (3·05; 180) 108–125 (ZMA).

Weights. *T. p. clarkei.* England (BTO).

JAN–FEB	84·8 (26)	69–101	SEP–OCT	74·4 (38)	61–94
MAR–APR	77·8 (52)	68–96	NOV–DEC	88·2 (29)	75–107
MAY–JUN	74·0 (46)	65–83			

Nominate *philomelos*. Norway: July–August, 63·9 (5); September–October, 75·4 (18) 61–100 (Haftorn 1971). On migration, Helgoland: 67·3 (100) 50–83 (Weigold 1926). West Germany: ♂ 72·2 (15) 63–78, ♀ 65 (3) 60–70 (Niethammer 1971). Netherlands, migrants, May: ♂ 73·2 (5·10; 72) 62–84, ♀ 71·0 (4·22; 115) 60–82 (ZMA). Turkey and northern Iran, March: ♂ 71·6 (6·84; 4) 62–77 (Schüz 1959; Kumerloeve 1970a). Kazakhstan (USSR): April–May, ♂ 69·8 (5·50; 4) 65–77; October, ♀ 90·1 (Dolgushin *et al.* 1970). South-west Siberia, September: 71·8 (4·38; 5) 66–78 (Havlín and Jurlov 1977). For variation of ♀ during laying, see Pikula (1976); for development of nestling, see Pikula (1971).

Race unknown. Migrants southern England: 73·9 (10·2; 50) 62–112 (Ash 1964). Gibraltar: autumn, 70·5 (5·97; 14); spring, 74·0 (1·32; 18) (Finlayson 1981). Malta, autumn and winter: 66·0 (6·0; 52) 50–71 (J Sultana and C Gauci). Netherlands, lighthouse victims (ZMA):

SEP	♂	68·4 (4·98; 12)	60–76	♀	66·2 (4·84; 11)	57–73
OCT		68·9 (6·00; 129)	56–89		66·6 (6·63; 140)	52–89
NOV		71·3 (8·05; 9)	62–84		70·6 (7·85; 26)	56–88
DEC–FEB		74·4 (11·6; 7)	62–98		76·2 (7·50; 5)	64–83
MAR–APR		69·5 (5·81; 13)	59–79		71·5 (5·85; 23)	61–82

Frost-killed, Wales and southern England: 45·9 (18) 40–59 (Harris 1962; Ash 1964). Netherlands, exhausted and freshly dead, mainly winter: ♂ 47·1 (4·31; 13) 42–54, ♀ 42·3 (3·58; 7) 38–45 (ZMA).

Structure. Wing rather long, broad at base, tip fairly pointed. 10 primaries: p7–p8 longest or either one 0–1 shorter than other, p9 and p6 both 3–6 shorter than longest, p5 12–18 shorter, p4 17–27, p1 26–35. P10 reduced; 65–78 shorter than longest primary, 5–13 shorter than longest upper primary covert. Outer web of p6–p8 and (rather faintly) inner web of (p7–)p8–p9 emarginated. Tail rather long, tip square or very slightly rounded; 12 feathers, tip bluntly (adult) or sharply pointed (juvenile); t6 2–6 shorter than t2–t3. Middle toe with claw 26·3 (10) 25–28; outer toe with claw *c.* 71% of middle with claw, inner *c.* 66%, hind *c.* 72%. Remainder of structure as in Blackbird *T. merula* (p. 963), but bill relatively short.

Geographical variation. Clinal: main trend towards increasing paleness of plumage running from Hebrides through Scotland, England, and continental Europe to USSR, and minor trend towards increasing darkness of plumage from north to south in western continental Europe (Vaurie 1959; see, however, Voous 1959 and Niethammer and Wolters 1970). 3 races recognized in Europe, each a more or less well-marked stage on east-west cline. *T. p. hebridensis* from Outer Hebrides and Isle of Skye is darker earth-brown above than *clarkei*, except for greyish rump and upper tail-coverts; underparts have more numerous, larger, and blacker spots contrasting more with paler buff ground-colour; flanks darker smoke-brown. Populations intermediate between *hebridensis* and *clarkei* inhabit western Scotland and south-west Ireland (Witherby *et al.* 1938*b*), sometimes separated as *catherinae* Clancey, 1938. *T. p. clarkei* from remainder of Britain and Ireland and adjacent parts of European continent characterized by warm brown or rufous-brown upperparts; rump and upper tail-coverts olive-brown or slightly rufous, not as grey as *hebridensis* and nominate *philomelos*; chest and flanks rather deep yellow-buff; underparts profusely spotted. Populations from narrow zone through westernmost West Germany, south-east Netherlands, eastern Belgium, and probably north-east and central France intermediate between *clarkei* and nominate *philomelos*. Typical nominate *philomelos* from Scandinavia, Poland, and eastern Rumania east to western Siberia and Caucasus area greyer olive-brown on upperparts, rump and upper tail-coverts olive-grey without rufous or brown tinge; buff ground-colour of underparts paler, more cream, less extensive. For intermediate populations, see also Meinertzhagen (1948). No appreciable variation in size in west Palearctic. Birds at extreme eastern end of cline in central Siberia separated as *nataliae*; not considered valid by recent authors (e.g. Johansen 1954, Stepanyan 1978*a*), but birds collected March-April in Zagros mountains of south-west Iran appear similar to *nataliae* and are distinctly larger than other races, wing ♂ 122 (11) 118-128; not known whether these are migrant *nataliae* from central Siberia or local breeders (Vaurie 1959).

Forms superspecies with Chinese Song Thrush *T. mupinensis* from central China. DWS

Turdus iliacus Redwing

PLATES 73 and 78
[between pages 856 and 857]

Du. Koperwiek   Fr. Grive mauvis   Ge. Rotdrossel
Ru. Белобровик   Sp. Zorzal alirrojo   Sw. Rödvingetrast

Turdus iliacus Linnaeus, 1766. Synonym: *Turdus musicus*.

Polytypic. Nominate *iliacus* Linnaeus, 1766, northern Eurasia; *coburni* Sharpe, 1901, Iceland and Faeroes.

Field characters. 21 cm; wing-span 33–34·5 cm. Noticeably slighter and more rakish than Song Thrush *T. philomelos*, with slightly longer and more pointed wings but slightly shorter, sharp-cornered tail; slightly smaller than Dusky/Naumann's Thrush *T. naumanni* and Eyebrowed Thrush *T. obscurus*. Rather small, slight, restless thrush, with striped head and spots on underbody; red-chestnut underwing and flanks combine with rather dark upperparts to provide rather dark, distinctive but not diagnostic appearance. Adult has chest, most of flanks, and sides of belly well marked, but rear flanks and vent noticeably white; immature has less obvious rufous flanks and fuller pattern of spots and streaks. Silhouette and action in flight recall Starling *Sturnus vulgaris*. Commonest call distinctive. Sexes similar; some seasonal variation. Juvenile separable. 2 races in west Palearctic, distinguishable with care.

Adult. (1) Main race, nominate *iliacus*. Upperparts rather dark, slightly warm umber-brown, lacking strong olive tone except on rump and upper tail-coverts and relieved only by dark-flecked crown and distinctly paler fringes to tertials and wing-coverts. Head and face boldly patterned, with long, deep supercilium cream-white to buff, contrasting noticeably with dark edges to crown and black-brown lores and dark brown cheeks (with paler flecks below lores); cheeks contrast with white to buff-white cheek-surround; below this, well-marked malar stripe of dark spots. Important to recognize that *T. iliacus* has boldest supercilium and cheek-surround of all common west Palearctic thrushes, with clarity of supercilium only fully matched by Siberian Thrush *Zoothera sibirica* and Dusky Thrush *T. naumanni eunomus*. Chest suffused yellowish-buff on sides and marked with dark brown streaks coalescing to form loose gorget above much paler, almost white belly, which, like red-chestnut flanks, is sparsely spotted dark brown, marks tending to form straight lines and not evenly splashed areas. Deep rear flanks, vent, and under tail-coverts cream-white, unmarked and forming conspicuous pale area, particularly from behind (and overhead in flight, when it contrasts with dusky undertail). Underwing noticeably dark, with red-chestnut coverts and axillaries merging with flanks and contrasting little with dusky flight-feathers. (2) Race of Iceland and Faeroes, *coburni*. Up to 5% longer than nominate *iliacus* and noticeably bulkier. Darker above, with smoky-olive tone on crown, back, and rump and strongly buff tone to pale head markings, but best distinguished by more heavily marked underparts, with suffusion on chest less buff, more olive-brown, and dark brown spots and streaks merging more on chest and flanks; under tail-coverts darker. Face pattern still well-marked but no increase in contrast since supercilium buffier. Underwing and flanks dark red-chestnut. Thus looks noticeably swarthier than nominate *iliacus* (but beware poor light making that race look darker than usual). Juvenile. Both races. Resembles dark adult, differing most in whiter supercilium, patchier cheeks, buffier suffusion on throat and chest, much less obvious chestnut flanks and underwing (these usually appearing no darker than rufous-buff), and much more heavily and more evenly spread spots and streaks on throat, sides of neck, chest, and flanks (thus no obvious gorget, and area of markings on rear flanks has much less distinct rear edge—hence also no marked white patch on rear body). Upperparts flecked dark and spotted pale, as in juvenile *T. philomelos*. FIRST YEAR. Both races. Resembles adult but more easily aged than *T. philomelos*, with pale tips to tertials and greater coverts obvious. At all ages, bill black-brown, with base of lower mandible yellow-flesh; legs yellow-to flesh-brown in nominate *iliacus*, darker, horn-brown in *coburni*.

Adult easily distinguished from adult *T. philomelos*, with face pattern, chestnut underwing and flanks, and white rear body obvious in comparison, and not to be confused with any other common thrush. *T. iliacus* nevertheless a tricky species, with (1) adult suggesting *T. obscurus* (close in size, with not dissimilar basic plumage pattern, differing at distance only in lack of spotted gorget and pale underwing), *T. naumanni eunomus* (larger, with less marked underparts and greater rufous suffusion on underparts, wings, rump, and tail), and ♀ Siberian Thrush *Zoothera sibirica* (similar face pattern, but larger and with barred underwing obvious in flight), and (2) immature recalling *T. naumanni eunomus*, *Z. sibirica*, and *T. philomelos*. Flight recalls *T. philomelos* but faster, with sharp wing-point and tail-corners contributing to rakish silhouette which suggests *S. vulgaris*, and rather more undulating track, with periods of closed wings between bouts of wing-beats, also recalling *S. vulgaris*. Migrating flocks perform spectacular tumbling into cover. Gait and general behaviour as *T. philomelos* but stance rather less upright. Normally much less in cover than *T. philomelos*, frequently sharing open feeding grounds with Fieldfare *T. pilaris*. Markedly gregarious, forming large flocks on passage and in winter. Shy, even wild, on passage, flocks readily taking fright and flying up to canopy until danger past.

Song less developed than *T. philomelos* and variable in content; commonest phrase contains 3–4 fluted notes (sometimes faintly disyllabic or with others interspersed) and characteristic throaty chuckle at end recalling Blackbird *T. merula* but without richness of that species; also has richer, sweeter song, with sections recalling *T. philomelos* but lower-pitched and less 'shouted'. Subsong noticeably babbling, with twittering warble interspersed with fluted notes; often given by wintering birds in chorus. All forms of song less loud and far-carrying than other common west Palearctic thrushes, sounding muted even at close range. Commonest call a thin, quite low-pitched but nevertheless slightly rasped, penetrating, and prolonged 'seez', 'seeih', or 'seeip', ending less sharply than similar call of *T. philomelos*; remarkably

far-carrying, and often heard from night sky in late autumn and winter. In excitement or alarm, gives a harsh, slightly rattled 'chittuck' or 'chi-tic-tick'. Other calls generally more varied than *T. philomelos*: e.g. loud, sharp 'kewk' when surprised and a quieter but still abrupt 'tchop' when feeding.

Habitat. Breeds in upper and upper middle latitudes of west Palearctic, mainly in subarctic and arctic lowlands and uplands, but avoiding snow and ice, and exposed chilly or stormy situations. Likes cover of birch or mixed woodland, often with many pines and spruces, especially along rivers and on floodlands, but also in low thickets of scrub birch, dwarf willow, and juniper, preferably on swampy ground. In Iceland breeds in broken, often rock-strewn country, commonly in birch scrub but also on ground sites among rocks even where scrub almost absent (Bent 1949). In northern Europe beyond treeline, will also nest on ground or in banks; in Norway, extends down to margins of fjords and up to birch woodlands on fells. In USSR, breeds on burnt-over patches or clearings in forests; on tundra in thickets of willow, and in taiga among spruce, but avoids denser stands of forest (Dementiev and Gladkov 1954b). In Scandinavia and Iceland, also nests locally in town parks and gardens. Recent colonists in Scotland have nested in grounds of large private houses, in hedgerows, on edges of oakwoods, in hillside birchwoods, and in variety of other sites combining open ground with some suitable cover. Sings from high and prominent perch, e.g. rock, tree, overhead wire. In winter in Britain, shows preference for good grassland and for smaller and less rough fields than Fieldfare *T. pilaris*. Often mingles with *T. pilaris* but usually keeps somewhat closer to cover of trees and bushes, of any height, and is more ready to occupy playing fields or open spaces in parks, provided there is not too much disturbance. Moist ground, even floodland, often preferred. Stubble and root fields, thorn hedges, and open woods also favoured as winter habitats, but in hard weather will even use inner city gardens with berried shrubs. In Bern (Switzerland), winters very sparsely and normally below 800 m, in contrast to Blackbird *T. merula* which winters regularly up to *c.* 1000 m and not uncommonly to 1350 m or higher (Lüps *et al.* 1978). In Belgium, snow or hard weather lead to large concentrations on the coast, though this does not avert high mortality unless a thaw follows within at most 1 week (Lippens and Wille 1972). In Morocco, seen in hundreds in fruiting trees in Haut Atlas valleys in December and in fruiting olives on scarp of Hauts Plateaux in January; also winters in cedars (Heim de Balsac and Mayaud 1962; Smith 1965a). Although among least robust of Palearctic Turdidae, and vulnerable to mass mortality, remarkably successful in maintaining high population in cool or cold northern and middle latitudes by ability to find vacant ecological niches and to shift promptly, as circumstances

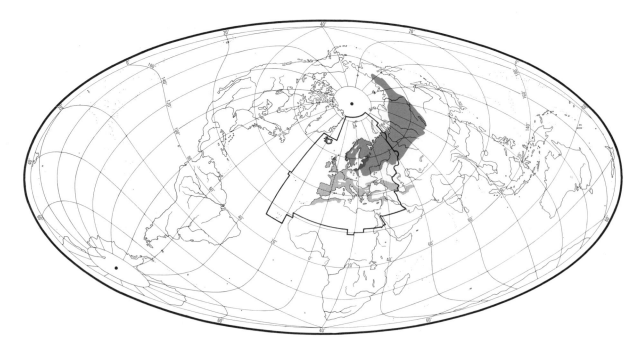

dictate, to find fresh food resources and shelter. Readiness to fly long distances in middle and upper airspace plays significant role in this capability.

Distribution. Has colonized Britain and bred occasionally in Belgium, Denmark, East Germany, Czechoslovakia, and Italy. Spreading in Sweden and USSR.

BRITAIN. First bred 1925, then in 17 years to 1966 and annually since 1967 (Sharrock 1976). IRELAND. Attempted to breed 1951 (Parslow 1973). WEST GERMANY. Breeding suspected (AH). DENMARK. First bred 1967, *c.* 10–50 breeding records since (TD). SWEDEN. Expanded rapidly to south in 1950s and 1960s; continuing (LR). EAST GERMANY. Bred 1904 and 1966–7 (Maschke 1969; Makatsch 1981). CZECHOSLOVAKIA. First bred 1960; more occasional breeding after 1970 (KH). AUSTRIA. Bred 1939, 1974, 1977 (LS). ITALY. Breeds occasionally in north, perhaps escaped birds (PB). USSR. Since *c.* 1930, has spread south of 53°N in Lithuania, Belorussiya, and Ukraine (GGB). Leningrad region: considerable increase in 20th century (Mal'chevski and Pukinski 1983). Voronezh region: spreading south (Wilson 1976).

Accidental. Spitsbergen, Bear Island, Jan Mayen, Kuwait, Tunisia, Libya, Azores, Madeira.

Population. Has increased Iceland, Finland, Poland, and southern USSR.

ICELAND. Increased over last 50 years though little change since 1960; starting to breed in towns (Timmermann 1934; AP). FAEROES. 10–20 pairs; increasing (Bloch and Sørensen 1984). BRITAIN. Possibly *c.* 300 pairs 1972 (Sharrock 1976). Apparently declined since, with 33–62 pairs 1982 (Spencer *et al.* 1985) and 17–68

pairs 1983 (Spencer *et al.* 1986). SWEDEN. About 1 million pairs (Ulfstrand and Högstedt 1976). FINLAND. About 570 000 pairs; fluctuating (Merikallio 1958). Has increased considerably recently, especially in south (OH), but little change 1978–83 (Vickholm and Väisänen 1984). POLAND. Scarce in north-east, extremely scarce elsewhere (Tomiałojć 1976). Increasing but probably stable in recent years (AD). USSR. Increased in south (see Distribution).

Survival. Finland: annual overall mortality 57–58% (Haukioja 1969). Oldest ringed bird 18 years 9 months (Rydzewski 1978).

Movements. Migratory or partially migratory. Winter range of whole population only just extends outside west Palearctic, so east Siberian birds must travel at least 6500 km WSW to reach winter quarters.

Iceland and Faeroes race, *coburni*, winters in Scotland, Ireland, western France, and Iberia. Ringing recoveries suggest birds from eastern Iceland winter mainly north of Loire river (France), those from western Iceland mainly south of this. Since early 1930s, has also wintered increasingly in Iceland, particularly in larger towns (Zink 1981).

British and mainland Eurasian race, nominate *iliacus*, winters in western Europe south from Scotland, coastal Norway, and south-east Baltic area, and around Mediterranean, Black, and south Caspian Seas. Ringed birds wintering in east Mediterranean and between Black and Caspian Seas are mainly of Finnish origin. Numbers wintering in North Africa fluctuate markedly between years, while birds recorded on migration occur as far south as 29°N at Goulimine in Morocco, so the species probably winters even further south. (Zink 1981.) A

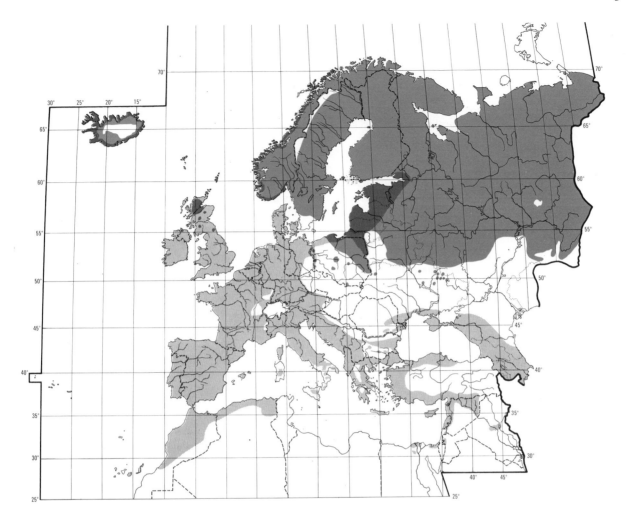

sporadic winter visitor to Egypt along Mediterranean coast and in Nile delta and Wadi Natroun, recorded 30 October to 18 March. Also occasional along coastal belt of Libya in winter. (Goodman and Watson 1984.) Small numbers occur in Levant, Asia Minor, and Cyprus. Very scarce in Gulf states of Arabia (Bundy and Warr 1980).

Autumn movement out of Sweden and Norway occurs late September to mid-November, sporadically to December. Recoveries suggest most birds leave Fenno-Scandia on broad front, heading between south and west. The spread possible is graphically illustrated by 2 brood-siblings from Finland, one recovered in Wales, other in south-east France, both in February of their 1st winter (Jones 1961). Smaller proportion of birds from Finland have been recovered to south-east in Aegean area, Asia Minor, Levant, and near Caspian Sea. Juveniles from Latvia, Estonia, and Leningrad region move between WSW and SSW, September–October, to winter from England to Portugal and southern Italy. (Zink 1981.) Departure from Tomsk (south-central Siberia) starts at end of August and a few remain to late October (Johansen

1954). In north of USSR range, leaves Lapland 1–20 October (average 9 October over 6 years), Arkhangel'sk 10 September–10 October (but in drought year of 1950 departure started 17 August), and Turukhansk (central Siberia) mid-September. Around Kharkov (50°N, European USSR) appears at end of September or early October with main passage at end of month and a few to mid-November, and in Caucasus region from mid-September to mid-November. (Dementiev and Gladkov 1954*b*.) Arrives in France from late September, weather and feeding conditions there determining when birds move on to Iberia: may reach Spain and Portugal in November, but in some years not until January (Zink 1981).

There are a number of recoveries in western Europe (e.g. Britain, France, West Germany) of birds ringed in breeding season or autumn at 30–80°E in USSR, as well as one from Tomsk in Siberia (85°E) recovered in southern Norway. A few birds ringed in autumn in western Europe have been recovered to the east the following winter, e.g. Fair Isle and Berlin to Greece,

France to Sardinia, Switzerland to Levant, and Netherlands to Turkey. These may have originated east of their ringing sites and changed direction. Only a few recoveries demonstrate year-to-year winter site-fidelity, and there are many to show that birds may winter in widely different localities in different winters, e.g. many birds ringed in Britain in winter have been recovered in subsequent winters in Italy, Greece, and localities at 30–45°E 30–45°N. (Zink 1981.)

Autumn migration of *coburni* is mainly south-east or SSE to Britain, Ireland, western France, Spain, and Portugal. Passage through Fair Isle slightly later than for nominate *iliacus* (Prato *et al.* 1980). Some birds perhaps fly to Iberia directly, across the sea, or by way of Ireland, as indicated by recoveries in north-west Spain and one at sea to south-west of Ireland (Zink 1977).

According to Zink (1981), no detectable differences between spring and autumn migration routes for either race. However, Prato *et al.* (1980) concluded that North Sea crossing favoured in autumn but return passage may occur to south-east of Britain in spring. Spring migration through Fair Isle begins late February or early March and continues to mid-May (Williamson 1965*b*). Arrives in southern Norway at end of March and in first half of April, but not usually until first half of May in eastern Finnmark (Zink 1981). In USSR, leaves Azerbaydzhan lowlands from mid-March to mid-April, northern Caucasus March–April, in Kharkov region (50°N) appears late March or early April, and present until late April. In Minsk region (54°N) present from early April; in Pskov region (58°N) arrives from 10 April to mid-May, average 19 April over 14 years. Around Arkhangel'sk (65°N) arrives 15 April–7 May, and in Russian Lapland average arrival date 8 May over 9 years, earliest 22 April, latest 17 May. Further east in European USSR, recorded in large numbers at Buguruslan (53°N) in second half of April, and present in central Urals in late April. (Dementiev and Gladkov 1954*b*.) At Tomsk, average arrival 30 April over many years (25 April–6 May); *c.* 4 days earlier at Tobol'sk (58°N 68°E), with passage at peak in first days of May (Johansen 1954).   LAB

Food. Wide variety of invertebrates, in autumn and winter also berries. Feeds on ground and in trees and bushes. In foraging on open ground in winter, runs or hops in short bursts, usually 1–5 paces, halting between each run to scan ground in immediate vicinity; if potential food item seen, may take a few steps towards it. For surface items, usually pecks at it immediately, though may first pause briefly to cock head on one side; occasionally makes more than one peck. For subsurface prey in hard ground, stops near potential prey, hesitates, often cocking head on one side and sometimes taking short step backwards or sideways, then stabs downwards with bill; often makes several pecks, sometimes followed by sideways flick to remove soil. Also sweeps bill sideways

to remove loose material; will walk through ground litter searching continually thus, and uses same technique while standing by cattle dung to search for larvae in it. (Tye 1981.) Said to open snails by striking them with bill on 'thinner or spiral end', and will wipe slugs on bare earth or grass to remove slime (Simms 1978).

Diet in west Palearctic includes the following. Invertebrates: mayflies (Ephemeroptera), dragonflies (Odonata), crickets (Orthoptera: Gryllotalpidae, Gryllidae), bugs (Hemiptera: e.g. Psyllidae), larval and pupal Lepidoptera (e.g. Noctuidae), adult and larval flies (Diptera: Tipulidae, Bibionidae), adult and larval Hymenoptera (sawflies Symphyta, Ichneumonoidea, ants Formicidae), adult and larval beetles (Coleoptera: Carabidae, Geotrupidae, Scarabaeidae, Byrrhidae, Elateridae, Cerambycidae, Chrysomelidae, Curculionidae), spiders (Araneae), sandhoppers (Amphipoda), small crabs (Decapoda), millipedes (Diplopoda), small molluscs (including snails, slugs, and bivalves), earthworms (Oligochaeta), marine worms (Nereidae). Plant material mainly fruits, also seeds: of pine *Pinus*, yew *Taxus*, juniper *Juniperus*, ivy *Hedera*, holly *Ilex*, bramble *Rubus*, strawberry *Fragaria*, rose *Rosa*, apple *Malus*, pear *Pyrus*, *Cotoneaster*, cherries *Prunus*, hawthorn *Crataegus*, rowan *Sorbus*, currants *Ribes*, buckthorn *Rhamnus*, alder buckthorn *Frangula*, heaths (Ericaceae), *Viburnum*, elder *Sambucus*, madder *Rubia*, vine *Vitis*, and root crops. Will also, though infrequently, eat kitchen scraps. (Csiki 1908; Schuster 1930; Witherby *et al.* 1938*b*; Kovačević and Danon 1952; Dementiev and Gladkov 1954*b*; Hartley 1954; Korenberg *et al.* 1972; Arheimer 1978*b*; Simms 1978.)

In summer, takes wide variety of invertebrates, as available; little plant material (e.g. Novikov 1952, Korenberg *et al.* 1972). In 3 areas of southern Britain, 86% of 166 records of fruit-eating involved hawthorn (Hartley 1954). In January in Mediterranean France, 87% of 23 birds contained fleshy fruits (mostly grapes and juniper), 35% contained arthropods (millipedes, beetles, caterpillars), and 30% gastropod molluscs (Debussche and Isenmann 1985*a*). In Cambridgeshire, October to mid-February, unlike Fieldfare *T. pilaris*, took more surface items (66.9% of total 263 items) than subsurface items (33.1%) (Tye 1981). Autumn migrants on Holy Island (England) appeared to eat largely snails (Richards 1977). See also Florence (1914), Mal'chevski (1959), Tutman (1962), Uspenski *et al.* (1962), Ptushenko and Inozemtsev (1968), and Kostin (1983).

Young at 11 nests in northern Sweden given largely earthworms (77–96% of 3199 feedings over 5 years); other food mostly Diptera and Lepidoptera larvae (Arheimer 1978*b*). In southern Finland, 67% of 102 feedings to young comprised small earthworms, 27% adult insects (largely mayflies, also beetles, dragonflies, Diptera), and 5% insect larvae; earthworms comprised 86% of feedings up to June, then declined, though still most important item (Tyrväinen 1969). See also Davies and Rowell (1956). DJB

Social pattern and behaviour. No comprehensive studies, but important information from Joroinen (southern Finland) by Tyrväinen (1969). For useful early account of *coburni*, see Hantzsch (1905). Account includes notes supplied by Y Espmark and T Bjerke (Norway).

1. Gregarious outside breeding season. From July and early August, family parties gradually join into flocks (Ptushenko and Inozemtsev 1968). Typically migrates in loose flocks, also singly, aggregating more at stopovers (e.g. Mal'chevski and Pukinski 1983). During spring migration, Scandinavia, resting flocks of up to several hundred occur (Y Espmark; see also Flock Behaviour, below). In autumn, Fair Isle (Scotland), migrant flocks of nominate *iliacus* sometimes of several hundreds, often in company with Fieldfare *T. pilaris*, Blackbird *T. merula*, Song Thrush *T. philomelos*, and Brambling *Fringilla montifringilla* (Williamson 1958c). For age composition of migrant flocks, see Prato *et al.* (1980). In winter, continues to mix freely with other thrushes *Turdus*, notably with *T. pilaris* (e.g. Witherby *et al.* 1938b), in Algeria with Mistle Thrush *T. viscivorus* and Ring Ouzel *T. torquatus* (Snow 1952). See also Roosting (below). BONDS. Monogamous mating system. Pair-bond maintained for 2nd brood or re-nesting, but breaks down after breeding season (Tyrväinen 1969; Y Espmark and T Bjerke). Though no proof of previous breeding together, 'pair' overwintered in town in Iceland then bred the following spring (Timmermann 1934). For 2 alleged hybrids with *T. pilaris*, see Collett (1898). Both sexes brood and feed young (Tyrväinen 1969). Young fed for 6–10 days after leaving nest (Ptushenko and Inozemtsev 1968). Age of first breeding not known. BREEDING DISPERSION. Essentially solitary, though in optimum habitat may form loose colonies (Y Espmark): e.g. in south-east Finland, 8 nests in *c.* 750 m² of young spruce *Picea* plantation (Tiainen 1977); minimum distance between nests not given but less than previous record of 25 m for Finland. For other distances between nests of conspecifics, and from other species, see below. Nest-territory relatively small, usually *c.* 0·5 ha, never more than 1 ha (Tyrväinen 1969). According to Tyrväinen (1969), feeds mostly within this area, but other sources (Ptushenko and Inozemtsev 1968; Y Espmark) indicate most feeding outside nest-territory. Thus, in Moscow region (USSR), nest-territory up to 490–630 m², 'feeding territory' up to 1·2–1·5 ha (Ptushenko and Inozemtsev 1968). Role of territory in courtship not known. In Trondheim area (Norway), density in 2 years 6·3 and 11·4 pairs per km² (Bjerke *et al.* 1985). In Uppland (Sweden), up to 1·4 pairs per km² (Olsson 1947). In southern Finland, 17·9 pairs per km² overall, varying from 3·3 in '*Vaccinium* type mixed (broadleaf conifer) stand' to 94 in 'grass-herb broadleaf stand' (Tyrväinen 1969, which see for other densities over large areas of Finland). In forest of ash *Fraxinus* and alder *Alnus*, Białowieża (Poland), 10 pairs per km² (Tomiałojć *et al.* 1984). In Kareliya (USSR), 1 pair per km² on dry mineral soils, 4·3 pairs per km² on moist mineral soils (Lehtonen 1943). In Moscow region, 50–55 singing ♂♂ along 10 km, pairs 80–300 m apart, density 15–20 pairs per km² (Ptushenko and Inozemtsev 1968). In suburban parks, Leningrad (USSR), 50–250 pairs per km² (Mal'chevski and Pukinski 1983). For densities in various parts of Yamal peninsula (USSR), see Danilov *et al.* (1984). Builds new nest for 2nd clutch, sometimes close to 1st (e.g. Maschke 1969), but in Joroinen birds typically moved outside 1st territory, soon after 1st brood left, to start 2nd clutch, usually *c.* 100 m (10–700,

$n = 12$) away from 1st nest; unpaired ♂♂ also shifted territory at this time (Tyrväinen 1969). Fidelity of young to natal area apparently rather low; higher for adults. Thus, in Oslo area (Norway), 21% of birds marked as adults and 4% as nestlings were controlled the following year, on average 320 m (10–800) from previous breeding or natal site, i.e. within former song dialect area (Bjerke and Bjerke 1981: see 1 in Voice). Not uncommonly nests in colonies of *T. pilaris* (see that species, p. 982); north of Urals, mixed colonies occur, containing almost equal numbers of both species, although *T. iliacus* just as often nests outside *T. pilaris* colonies (Danilov and Tarchevskaya 1962). In park, south-west Finland, minimum distances from nests of other *Turdus* as follows: *T. pilaris* 10 m, *T. merula* 5 m, *T. philomelos* 3 m (Haartman 1971). ROOSTING. Nocturnal and communal outside breeding season, hundreds often gathering at traditional sites (Labitte 1937): thick shrubberies (especially evergreens), plantations, thickets, dense hedgerows, etc.; in hedges, usually perches lower than *T. pilaris* according to Witherby *et al.* (1938b), higher according to Lübcke and Furrer (1985). Said often to share roost also with Starlings *Sturnus vulgaris* (Stubbs 1909). Once, *c.* 80 in *Phragmites* reedbed (Rolls 1972). In Dorset (England), *c.* 1200 in low willows *Salix* and brambles *Rubus* along disused railway track (Hughes 1977). Maximum catchment radius of large winter roost in Northumberland (England) *c.* 20 km. Flocks often assembled at favourable feeding sites, especially in afternoon, and from *c.* 15.00 hrs, started progressing towards roost in twos and threes, halting here and there *en route* and joining up with others before continuing. On arrival at roost, settle immediately in trees, with much calling (see 3a in Voice) and do not usually leave again unless disturbed (see part 2, below). In morning, birds leave discreetly, alone or in small groups, halting frequently on way to feeding grounds. (Phillipson 1937.) Song reported at night from heart of roost (Simms 1978). In France, December–January, birds reached roost at *c.* 16.00 hrs in groups of 15–20, perching high and descending gradually into thicker middle and lower layers (Labitte 1937). Little information for breeding season but one brooding ♀ slept with her head tucked into scapulars for 44 out of 151 min, at other times dozed with head forward (Davies and Rowell 1956). During breeding season in Lapland, birds rest before midnight, and most active (feeding) in the hours immediately after midnight (Swanberg 1951; Davies and Rowell 1956; Brown 1963; Peiponen 1970). In Swedish Lapland, apparently most active 22.00–04.00 hrs (Fredriksson *et al.* 1973). Anting reported (Simms 1978), but no further information on comfort behaviour.

2. Loose feeding flocks in winter typically disperse to trees if disturbed (Witherby *et al.* 1938b). If flushed from winter roost, fly out *en masse* if not too dark, in ones and twos in darkness (Labitte 1937). Flock thus disturbed sometimes performs massed aerial evolutions, not unlike *S. vulgaris* (Seebohm and Sharpe 1902). No further information. FLOCK BEHAVIOUR. For behaviour associated with roosting, see above. Prior to nocturnal migration, flocks commonly give apparent Rattling alarm-call (Mal'chevski and Pukinski 1983: call 3b). Seeip-call (see 2a in Voice) typically used in flight by migrating flock members to maintain contact, also by feeding flocks in autumn (Y Espmark) as flight-intention call (Bergmann and Helb 1982). During spring migration, members of loafing flocks often give Subsong (see 1b in Voice), occasionally full song (see 1a in Voice) which elicits a further chorus of Subsong from others (Rosenberg 1953; Tyrväinen 1969). Subsong also heard from winter flocks (Witherby *et al.* 1938b). SONG-DISPLAY. ♂ sings from high exposed perch, e.g. rock, tree, overhead wire (Sharrock 1976), occasionally from the ground (Daukes 1932).

During breeding season, individual ♂♂ differ greatly in singing activity, and some relatively silent (Bjerke and Bjerke 1981). In Scotland, ♂ sang with increased vigour after nest robbed (Daukes 1932). In Britain, Subsong may be heard in autumn, winter, and spring (Stubbs 1909; see also above). On arrival on breeding grounds, ♂♂ move over quite a large area, giving full song from outset (Ptushenko and Inozemtsev 1968). Singing activity peaks shortly before laying of 1st egg, then declines to minimum at fledging (Lampe 1983; Mal'chevski and Pukinski 1983). A 2nd lower peak may occur 3–4 weeks after 1st peak, coinciding with start of 2nd clutch (Slagsvold 1977; Mal'chevski and Pukinski 1983; T Bjerke). In Scandinavia, song may be given throughout summer and autumn, though declines gradually in rate of output, completeness, and loudness (Y Espmark). In Norwegian Lapland, June–July, birds sang intermittently throughout the night (Brown 1963). In Swedish Lapland, June–July, ♂♂ began singing around midnight, stopped at c. 07.00 hrs (Davies and Rowell 1956). Swanberg (1951) found similar commencement of singing. In Joroinen (southern Finland), May, ♂♂ do not sing at night but begin suddenly and vigorously sometimes before sunrise; in June, ♂♂ sing very little but are occasionally heard throughout the night (Tyrväinen 1969). For development of song in young, see Voice. ANTAGONISTIC BEHAVIOUR. Neighbouring ♂♂ commonly perform song-duels in which the Twittering segment of the song (see 1a in Voice) tends to be prolonged (Y Espmark). Birds seldom trespass on another's territory. If rival intrudes, resident ♂ sings louder and sometimes approaches. If rival continues to advance, resident comes to within a few metres and sings quietly. Challenged thus, intruders always withdraw. Direct confrontations rare. (Tyrväinen 1969.) Occasional chases and fights, accompanied by Rattling alarm-calls and Bill-snapping, occur in vicinity of nest (Bjerke et al. 1985; Y Espmark). For defence of 2 nests, see Relations within Family Group (below). Bill-snapping also heard at winter roost (P A D Hollom). In winter feeding flocks, bird confronts rivals with Threat-posture (Fig A): raises forebody, droops wings, sometimes until tips

A

touch ground, fans tail, and gives alarm-calls (E K Dunn, P J Sellar: see 3 in Voice). HETEROSEXUAL BEHAVIOUR. Little known. (1) General. In Moscow region, pair-formation takes place within 3–5 days and nest-building within 7–8 days of arrival of ♂♂ (Ptushenko and Inozemtsev 1968). In Novgorod region (USSR), similarly c. 1 week between arrival and start of nesting (Dementiev and Gladkov 1954b). (2) Pair-formation. Other than song-display of ♂, no information. (3) Courtship-feeding. Not reported away from nest; Tyrväinen (1969) never saw ♂ feeding ♀. ♂ coburni said to feed ♀ on nest (Hantzsch 1905). Once, ♂ nominate iliacus who apparently regularly fed ♀ (but not said how often or over what span of nesting), arrived when brooding ♀ perched at edge of nest. Both bowed and wiped their bills together, than ♂ crouched beside ♀ and fed her. Both birds then fed the young. Response of ♂ interpreted as possibly courtship-feeding ♀ at early stage in 2nd nesting attempt

(Davies and Rowell 1956), but ♂ perhaps giving ♀ food for transfer to young. (4) Mating. No information. (5) Behaviour at nest. Nest-site selection similar to that of T. philomelos (Pokrovskaya 1968: see that species, p. 995). Contra some published reports, only ♀ incubates (Arheimer 1978c). When ♀ off nest, ♂ often stands in nest but does not incubate (Tyrväinen 1969; Pulliainen 1982); ♂ thought thus to retard cooling of eggs, to help prevent detection of nest from above (Pulliainen 1982), and to discover when eggs have hatched (Tyrväinen 1969). When ♀ incubating, ♂ often stays near nest, singing quietly. Shortly before hatching, ♂ begins to visit nest more often; ♀ almost always leaves nest when ♂ approaches within 0·5–3 m. (Tyrväinen 1969.) On becoming aware of approaching mate, brooding ♀ typically raised head, looked quickly about, and flew straight off nest, ♂ relieving her directly (Davies and Rowell 1956; see also Relations within Family Group, below). RELATIONS WITHIN FAMILY GROUP. In first few days after hatching, parents alternate brooding. As soon as one parent arrives at nest with food, other leaves, so that young scarcely uncovered for 3(–5) days (Danilov and Tarchevskaya 1962; Tyrväinen 1969). Thereafter, parents, especially ♂, brood less though may continue at night even when young well grown (Tyrväinen 1969). If arriving ♂ had no food for brood, he settled immediately, otherwise fed young, 'clucking' the while, before brooding (Davies and Rowell 1956). Eyes of young open at 4–5 days. Both parents feed nestlings about equally (Tyrväinen 1969). Bird sometimes brings a meal of worms, drops them in nest, then distributes them to young (Mal'chevski and Pukinski 1983). Adult settling to brood always eats faecal sacs, otherwise carries them some distance from nest (Davies and Rowell 1956). Not seen to carry faeces from nest until young 8 days old. Age at which young leave nest may vary with latitude (Tyrväinen 1969). After leaving nest, brood-members stay together, maintaining contact by calls (Davies and Rowell 1956; Mal'chevski and Pukinski 1983: see Voice). Young unable to fly on first leaving nest, staying mostly on ground, but 1–2 days later can fly 10–50 m and land quite high up. Disperse further 2–3 days after leaving nest, frequently settling in small area, usually a favourable feeding site. Several broods may use such an area simultaneously, but stay apart from one another. (Tyrväinen 1969.) Parents entice young to follow by approaching with food, then flying off (Mal'chevski and Pukinski 1983). Commonly feed young for up to 1 week after leaving nest, and both parents may accompany them for a further week. ♂ alone may accompany some broods, but ♀ never seen alone with brood. ♀ may lay again only a few days after 1st brood fledged, and ♂ seems likely to feed 1st brood while ♀ incubates 2nd clutch. (Tyrväinen 1969.) ♀ builds 2nd nest while still attending young in 1st, and both nests may be defended simultaneously (Mal'chevski and Pukinski 1983). ANTI-PREDATOR RESPONSES OF YOUNG. Alarmed older nestlings crouch in nest, recently fledged young crouch silently in vegetation (Y Espmark and T Bjerke). PARENTAL ANTI-PREDATOR STRATEGIES. (1) Passive measures. Incubating ♀ said to sit more closely than Fieldfare T. pilaris (Simms 1978), but this variable: before hatching, some ♀♀ sit very tight and immobile, others slip away quietly (Bjerke et al. 1985). (2) Active measures: against birds. No information. (3) Active measures: against man. When human intruder enters nest-area, ♂ and ♀ give alarm-calls (see 3a in Voice), rate of which increases during breeding cycle; also Bill-snap and occasionally fly at intruder (Fig B), but rarely make physical contact (Bjerke et al. 1985). e.g. at nest where eggs just hatching, bird flew directly at intruder, veering off abruptly at 2–3 m from target (Maschke 1967). At another nest, ♂ flew straight at and low over head of intruder, and,

B

when 2nd nest started, began attacking when intruder 30 m away (Maschke 1969). When several fledged broods assembled in small area, parents (especially ♂) mobbed intruder incessantly, flying to within a few m and giving Rattling alarm-calls (Tyrväinen 1969). (4) Active measures: against other animals. Twice, breeding birds seen to attack vole *Clethrionomys rufocanus* near nest: flew down into undergrowth and pecked vigorously at it (Davies and Rowell 1956).

(Fig A by N McCanch from photograph in Burton 1985; Fig B by J P Busby from photograph in Minton 1970.)   EKD

Voice. Freely used throughout the year. Song dialects (see below) better studied than in any other *Turdus*. Bill-snapping serves as threat in confrontations with conspecifics and other intruders (see Social Pattern and Behaviour). Following scheme based on detailed notes from Y Espmark and T Bjerke. Unless otherwise stated, descriptions refer to nominate *iliacus*. For additional sonagrams, see Bjerke and Bjerke (1981), and Bergmann and Helb (1982).

CALLS OF ADULTS. (1) Song of ♂. (a) Full song. Comprises often quite long phrases separated by intervals of average 3·69 s (0–13·3, *n* = 500: Bjerke and Bjerke 1981). Phrases usually divided into 2 structurally different segments, though series of each segment may also be given on its own (see below). Introductory segment shows relatively little variation within, but much between, individuals; consists of a varying number (1–11 in southern Norway: Bjerke and Bjerke 1981; 3–10 in northern Sweden: Espmark 1981; up to 15 in 'another area' in Sweden: Y Espmark) of similar or different units,

usually ascending or descending in pitch through the segment (Y Espmark). Introductory units described as short, monotonously melancholy fluting (Bergmann and Helb 1982), but may vary geographically (see below). Renderings include 'trUi trUi trUi' and 'tee (or 'tee-TU') tiruppi-tiruppi-tiruppi' (Witherby *et al.* 1938*b*); a reedy, conversational 'teetra-teetra-teetra' (Simms 1978). Average duration of Introductory segment in southern Norway 875 ms (390–1680), average range 2·79–4·45 kHz (Bjerke and Bjerke 1981). Some ♂♂ alternate between 2–3 distinct Introductory segments (Y Espmark). In full song (Fig I of apparent nominate *iliacus*), Introduction (5 units) followed by a lengthy terminal twittering segment described as fast, more varied, more complex, and of lower intensity, than Introduction (Bjerke and Bjerke 1981); a poor, low, chuckling warble, in quality not unlike shorter chuckling ending of Blackbird *T. merula* song, sometimes with squeaky notes (Witherby *et al.* 1938*b*); a quiet chattering sequence of variable length, including rattling scratchy units (Bergmann and Helb 1982). Twittering segment varies markedly, both within and between individuals, in general composition, pitch, and duration (0–34·3 s, southern Norway, these extremes occurring in song of a given bird: Bjerke and Bjerke 1981). According to Hantzsch (1905), Twittering of *coburni* similar to *T. pilaris* but louder and clearer, sometimes lasts for several minutes without interruption. In central Norway, average duration of Twittering segment was 3·5 s before and 1·1 s after 1st egg laid (Lampe 1983). This decline typical after laying (T Bjerke) and accounts for distinction by some earlier sources (Witherby *et al.* 1938*b*; Fisher *et al.* 1944) of a 'pre-mating song'. Considerable geographical variation occurs in song structure. According to P J Sellar, Introductory segment of *coburni* more chirpy or twittery (e.g. Fig II), that of nominate *iliacus* clearer and more flute-like (e.g. Fig III, of bird which sang only Introductory segments; Fig IV, another phrase by same bird, illustrates rolled 'r' effect of modulated tones: J Hall-Craggs). However, according to Hantzsch (1905), *coburni* may also start song with clear fluting units. Within given race, song type pattern often homogeneous within relatively small (dialect) areas, separated by sharp boundaries which are stable over time and in space (Bjerke 1974, 1980; Bjerke and Bjerke 1981, which see for geographical variation in, and sonagrams

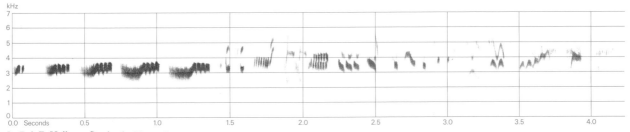

I   P A D Hollom   Scotland   May 1981

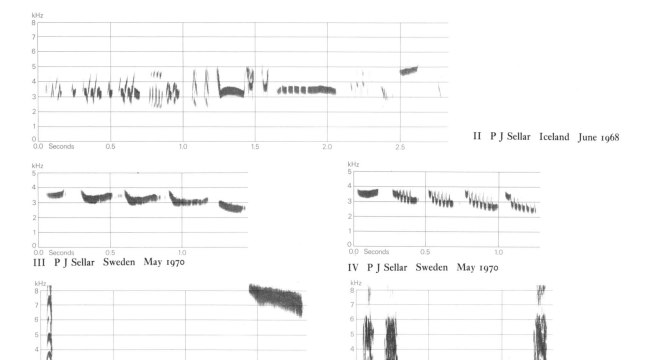

II   P J Sellar   Iceland   June 1968

III   P J Sellar   Sweden   May 1970

IV   P J Sellar   Sweden   May 1970

V   V C Lewis   England   October 1977

VI   P J Sellar   England   January 1985

of, Introductory segments). In southern Norway, average size of 27 dialect areas 41·5 km² (2-158·8). Within given dialect area, dominant song type shared by almost all ♂♂; some may sing secondary song type which, on dialect boundary, may belong to adjacent area. (Bjerke and Bjerke 1981.) Elsewhere, song pattern may be less uniform locally, and dialect boundaries more diffuse, e.g. northern Sweden and central Norway (Espmark 1981, 1982). (b) Subsong. Chorus heard from flocks resting on migration, and in winter quarters, dominated by twittering Subsong (Witherby *et al.* 1938b). (2) Contact-calls. (a) Seeip-call. The well-known, far-carrying, soft thin 'seeih' or 'seeip' (Fig V, 2nd unit), habitually given by nocturnal migrants; more drawn-out and penetrating than call 2a of Song Thrush *T. philomelos* (Witherby *et al.* 1938b), but confusable with call 2c of *T. p. philomelos* (see that species, p. 997). Also used as flight-intention call in autumn feeding flocks (Bergmann and Helb 1982). Other renderings include 'srieh' (J Hall-Craggs), 'tsiiiiih' (Y Espmark), 'seeng' (Simms 1978), and a slightly harsh, slightly descending 'ziih' (Bergmann and Helb 1982). (b) An abrupt 'tchup' or 'tchep', given in winter by birds when feeding and going to roost (Simms 1978); also rendered 'gjüg' (Bergmann and Helb 1982), 'chic', 'chittuck', or 'chittick' (Witherby *et al.* 1938b). In Fig V (1st unit), call rendered 'juk' (J Hall-Craggs) or 'kuk'

given in mixed *Turdus* feeding flock in hard winter weather (P J Sellar). For other calls in same flock, see below. (c) A growling or snarling low-pitched 'grrr' sound, audible at a few metres, given by parents landing on nest-rim with food for young (Y Espmark and T Bjerke). (3) Alarm- and Threat-calls. (a) A series of short sharp 'jypp jypp' sounds, resembling woodpecker (Picinae), given when moderately disturbed and before source of disturbance is located (Y Espmark). (b) Rattling alarm-call. A series of snarling 'trrrt trrt trrrrt' sounds (Y Espmark) with a sharp metallic quality, reminiscent of, but stronger than, Wren *Troglodytes troglodytes* (Daukes 1932). Signals greater danger and alarm than call 3a; given when predator near or at nest, and being mobbed, also in disputes with conspecifics (Y Espmark). (c) Rasping sounds (Fig VI) during disputes in winter feeding flock (P J Sellar). (d) Squealing-call. *In extremis* during feeding confrontations (as above), squealing sounds given in combination with calls 3a-c, sometimes breaking into Subsong (P J Sellar). Presumably indicates intense anger, threat, or distress, according to circumstances.

CALLS OF YOUNG. Food-call of young out of nest when approached by food-bearing parents a high-pitched twitter. Contact-call of fledged young a series of twittering sounds, occasionally interspersed with clucks, chirps, and melodious notes. (Y Espmark and T Bjerke.) This description probably accounts for report by Mal'chevski and Pukinski (1983) of young first attempting to sing at 17–18 days, this gradually developing into Subsong during autumn and winter. Juvenile call said to be a husky 'chucc' (Witherby *et al.* 1938*b*). EKD

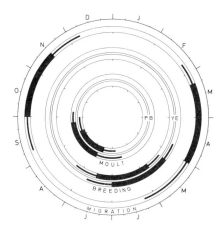

Breeding. SEASON. Scandinavia: see diagram; only slightly later in far north (Haartman 1969). Iceland: laying begins mid-May (Timmermann 1938–9). SITE. On ground under bushes, or in thick vegetation, tree, bush, or stump. Of 451 nests, Swedish Lapland, 34% in tree, 28% on ground in vegetation, 17% in stump, 13% in juniper *Juniperus* bush, 8% on ground under juniper bush. Of 360 nests, same area, 45% 0–10 cm above ground, 14% at 10–50 cm, 23% 0·5–1 m, 13% 1–2 m, 3% 2–3 m, 3% over 3 m (Arheimer 1973). Nest: bulky cup with outer layer of twigs, grass, and moss, plastered inside with mud, plus some fragments of vegetation, with inner lining of fine grass stems and leaves; often very thin, but with thicker rim, up to 2 cm. Mean measurements of up to 16 nests, Poland: external diameter 13·5 cm (11·5–16·0); internal diameter 8·4 cm (6·8–9·5); height 9·3 cm (7·0–11·0); depth of cup 5·4 cm (4·5–6·8) (Bocheński 1968). Building: by ♀ only (Tyrväinen 1969). EGGS. See Plate 84. Sub-elliptical, smooth and glossy; pale blue to greenish-blue, profusely marked with fine red-brown speckling and mottling, though marks often small and indistinct. Nominate *iliacus*: 25·9 × 19·0 mm (22·0–29·0 × 17·5–20·6), *n*=240; calculated weight 14·9 g. *T. i. coburni*: 26·0 × 19·5 mm (23·5–29·0 × 18·0–21·0), *n*=38; calculated weight 15·5 g (Schönwetter 1979). Clutch: 4–6 (3–7). Of 261 clutches, Swedish Lapland: 4 eggs, 4%; 5, 40%; 6, 53%; 7, 3%; mean 5·5 (Arheimer 1973). Of 110 clutches, Finland: 3 eggs, 1%; 4, 17%; 5, 52%; 6, 29%, 7, 1%; mean 5·1 (Tyrväinen 1969). 2 broods normal, and replacements also laid (Tyrväinen 1969), but, despite some reports, no good evidence for 2 broods in northern and alpine Scandinavia (Arheimer 1978*a*). Eggs

laid less than 24 hrs apart; clutches of 5–6 eggs laid in 4–5 days and therefore at all times of day and night (Arheimer 1978*c*). INCUBATION. 12·7 ± SE0·9 days (12–13) for last egg in 28 nests (Arheimer 1978*c*). 11·9 ± SE0·26 days (10–14) for 14 nests, egg not specified (Tyrväinen 1969). By ♀ only, but ♂ may enter nest when ♀ leaves and sit or stand over eggs without incubating (Tyrväinen 1969; Pulliainen 1982). Usually begins with penultimate egg; in 6-egg clutches, first 4 hatch synchronously, within a day, followed closely by 5th, then 6th one day later (Arheimer 1978*c*). YOUNG. Cared for and fed by both parents. Brooded more or less continuously for first 3 days after hatching, declining thereafter (Tyrväinen 1969). FLEDGING TO MATURITY. Fledging period 10·0 ± SE0·05 days (8–12), *n*=282 (Arheimer 1979). Mean 10·6 days (9–13), *n*=26, from 1st hatching to last young leaving nest, mean 9·5 days from last hatching to last fledging, and mean 10·2 days from 1st hatching to 1st fledging (Tyrväinen 1969). Become independent from 14 days after fledging. ♂ will continue to feed 1st brood young while ♀ incubating 2nd brood (Tyrväinen 1969). Age of first breeding not known. BREEDING SUCCESS. Of 259 nests, Swedish Lapland, 32% completely destroyed, mainly by crows (Corvidae); of 286 eggs, 81% hatched; of 181 hatched young, 69% fledged; overall success 62·1% (Arheimer 1973). Of 285 eggs laid, Finland, 69% hatched; of 340 young, 73% fledged; overall success 50·4% (Tyrväinen 1969).

Plumages (nominate *iliacus*). ADULT. Forehead, crown, and rest of upperparts olive-brown, crown with obscure dark feather centres, rump and upper tail-coverts more strongly olive than rest. Broad supercilium from base of bill to behind eye pale buff or off-white; sometimes slightly interrupted above eye. Lores and ear-coverts dark sooty brown, latter with fine pale shaft-streak. Chin and throat whitish; sides of neck whitish shading to buff at rear, with sooty brown malar stripe expanding at rear to heavy streaking on side of neck, and variable fine streaks in centre of throat. Chest pale buff to cream-white with obscure dark brown streaks; flanks rufous, obscurely streaked or spotted with dark brown. Belly white, unspotted in centre, streaked olive-brown at sides. Vent and under tail-coverts white, latter with some olive-brown spots, mainly at sides. Tail dark olive-brown. Upperwing dark olive-brown, feathers with paler olive-brown or rufous-olive outer edges; tertials with narrow and obsolete whitish tips mainly confined to outer web; greater upper wing-coverts with faint and ill-defined buff edge along tip. Axillaries and under wing-coverts rufous, similar in colour to flanks. *In worn plumage*, upperparts slightly greyer brown; dark marks on underparts more contrasting. NESTLING. Down fawn-coloured, long. JUVENILE. Upperparts like adult, but mantle, scapulars, and back spotted, each feather with pale buff shaft-streak or wedge and sooty tip. Sides of head and throat like adult. Rest of underparts like adult but with clearly defined sooty spots (not streaks), largest on chest and flanks and smaller on belly; flanks paler rufous. Tail as adult (but see First Adult, below). Upperwing as adult, but tertials with larger, angular pale buff or whitish spot on tip, greater upper wing-coverts with buff tips, and lesser and median coverts with buff shaft-streak, narrowing to a point towards base and widening towards tip to form a broad buff triangular spot. Axillaries and under wing-coverts pale rufous, like flanks. Closely similar to juvenile Song Thrush *T. philomelos*, but with distinct cream-white supercilium. FIRST ADULT. Like adult, but juvenile flight-feathers, tail, tertials, greater upper primary coverts, and usually outer greater secondary coverts retained; tail-feathers more sharply pointed than in adult, tertials and outer coverts with larger more triangular pale buff or off-white spots on tip. In worn plumage, pale spots wear off, but size and shape of remaining notch often still indicative of age unless whole feather-tip heavily worn. See also Svensson (1984*a*).

Bare parts. ADULT, FIRST ADULT. Iris dark brown. Bill black-brown to black, base of lower mandible and basal cutting-edge of upper mandible yellow or orange-yellow. Leg and foot flesh-pink or pale flesh-grey; in *coburni*, horn-brown with yellow-pink rear and soles. NESTLING. Mouth yellow, no spots. Gape-flanges yellow-white or ivory-white. JUVENILE. Like adult, but base of bill sometimes duller, greyish-flesh or brownish-flesh. (Witherby *et al.* 1938*b*; Williamson 1958*c*; BMNH, ZMA.)

Moults. ADULT POST-BREEDING. Complete; primaries descendant. Late June to late September. In Finland, some start whilst still feeding young; duration of primary moult *c.* 52 days at 61°N, *c.* 40 days at 70°N (Ginn and Melville 1983). Completed before leaving breeding area. POST-JUVENILE. Partial: head, body, lesser, median, and some inner or (rarely) all greater upper wing-coverts, and rarely some tertials or tail-feathers. July-September, completed before migration.

Measurements. Nominate *iliacus*. British Isles, September-May; skins (BMNH). Juvenile wing is retained juvenile wing of 1st adult. Bill to basal corner of nostril; to skull, on average 8·0 longer, exposed culmen on average 3·3 longer.

	♂		♀	
WING AD	116·1 (1·79; 10)	114-119	114·3 (3·80; 10)	110-121
JUV	114·9 (2·56; 10)	112-119	114·2 (1·75; 10)	111-118
TAIL AD	81·8 (1·99; 10)	79-85	79·9 (2·71; 9)	75-83
BILL AD	14·6 (0·70; 10)	14-16	14·3 (0·67; 10)	13-15
TARSUS	29·3 (0·82; 10)	28-30	28·7 (0·82; 10)	28-30

Sex differences not significant.

Fair Isle (Scotland), live birds, autumn: wing 116·7 (2·88; 537) 109-126, tail 79·9 (3·51; 323) 69-89, bill to skull 20·9 (1·21; 437) 16-25, tarsus 30·3 (1·25; 437) 26-34 (Williamson 1958*c*). Wing of freshly dead birds, Netherlands: ♂ 119·5 (3·01; 103) 113-129, ♀ 116·9 (2·67; 119) 111-127 (ZMA); even largest bird with wing 129 (fresh) or 128 (skin) still nominate *iliacus* (K Williamson).

T. i. coburni. Iceland, spring and summer; ages combined; skins (BMNH, ZMA).

	♂		♀	
WING	123·0 (3·12; 15)	116-127	120·4 (3·30; 10)	116-127
TAIL	85·0 (3·01; 15)	81-90	84·7 (3·90; 10)	80-93
BILL	14·1 (0·55; 15)	12·5-14·7	14·1 (0·47; 10)	13·5-15·0
TARSUS	31·1 (0·76; 15)	30·3-32·9	30·9 (0·64; 10)	30·0-32·0

Sex differences not significant.

Fair Isle (Scotland), live birds, autumn: wing 121·3 (2·79; 246) 113-133, tail 85·2 (3·11; 216) 78-93, bill to skull 21·4 (1·15; 230) 17-24; tarsus 32·4 (1·11; 231) 30-36 (Williamson 1958*c*).

Weights. Nominate *iliacus*. Norway, autumn (mainly October): 68·4 (143) 50-88, once 43 (Haftorn 1971). Yamal peninsula (USSR), summer: ♂♂ 62, 74; ♀♀ 71, 73 (Danilov *et al.* 1984). On migration, Helgoland: 60·3 (53) 47-77 (Weigold 1926).

Netherlands. Killed at lighthouses: (1) October, (2) November, (3) December, (4) January-February, (5) March, (6) April and early May. (7) Exhausted birds, found dead or dying, mainly during frost. (ZMA.)

	♂		♀	
(1)	61·6 (6·07; 160)	47-80	60·8 (5·16; 122)	46-76
(2)	65·4 (4·75; 122)	55-78	62·8 (4·80; 163)	50-76
(3)	63·1 (6·15; 16)	52-76	60·6 (6·48; 17)	50-74
(4)	63·8 (3·16; 8)	57-68	59·4 (6·21; 17)	49-74
(5)	59·5 (1·72; 12)	56-62	59·9 (5·47; 20)	48-68
(6)	63·3 (4·45; 5)	58-68	62·4 (4·71; 34)	52-77
(7)	36·0 (2·58; 17)	31-41	38·0 (2·76; 27)	32-44

Britain (perhaps including some *coburni*) (BTO).

OCT	67·4 (174) 50-82		JAN-FEB	67·9 (128) 49-85
NOV-DEC	69·6 (166) 47-88		MAR-APR	66·6 (20) 57-76

Malta, winter: 60·0 (5·5; 18) 47-70 (J Sultana and C Gauci). Frost-killed birds, Skomer (Wales), January: ♂ 37·8 (55) 33-44, ♀ 37·1 (73) 31-46 (Harris 1962). Fair Isle, autumn: 65·5 (4·45; 68) (Williamson 1958*c*).

T. i. coburni. South-west Iceland, spring (mainly April): ♂ 78 (14) 66-88, ♀♀ 70, 74 (Timmermann 1938-49). Fair Isle, late September and October: 71·0 (5·90; 145) (Williamson 1958*c*). Exhausted, at sea, October and April: ♂♂ 35, 36, 38, 40; ♀♀ 37, 43 (ZMA). Helgoland, ♂♂: October, 61·5; April, 90 (Bub 1975*b*).

Structure. Wing rather long, broad at base, tip fairly pointed. 10 primaries: p8 longest, p7 0-2 shorter (rarely, p8 1 shorter than p7), p9 3-8 shorter, p6 5-8, p5 15-21, p4 22-28, p1 33-42; no difference between races. P10 reduced; 71-85 shorter than wing-tip, 8-16 shorter than longest upper primary covert. Outer web of p6-p8 and inner of p7-p9 emarginated. Tail rather long, tip square; 12 feathers. Middle toe with claw of nominate *iliacus* 25·9 (8) 25-27, of *coburni* 28·5 (6) 28-29; outer toe with claw *c.* 72% of middle with claw, inner *c.* 73%, hind *c.* 74%. Remainder of structure as in Blackbird *T. merula* (p. 963), but bill and tarsus relatively shorter.

Geographical variation. *T. i. coburni* from Iceland and Faeroes differs from nominate *iliacus* in slightly larger size (see Measurements), slightly darker upperparts, and in particular in darker underparts: black streaks on throat heavier; breast, sides of belly, flanks, and under tail-coverts washed more extensively with olive-brown; brown marks on chest tend to merge together and extend further down to breast and sides of belly; sides of head, chest, and under tail-coverts strongly tinged buff; also, leg and foot darker (see Bare Parts). DWS

Turdus viscivorus **Mistle Thrush**

PLATES 75 and 78
[between pages 856 and 857]

Du. Grote Lijster   Fr. Grive draine   Ge. Misteldrossel
Ru. Деряба   Sp. Zorzal charlo   Sw. Dubbeltrast

Turdus viscivorus Linnaeus, 1758

Polytypic. Nominate *viscivorus* Linnaeus, 1758, west Palearctic east to western Siberia, except Corsica, Sardinia, and north-west Africa; *deichleri* Erlanger, 1897, north-west Africa, Corsica, and Sardinia. Extralimital: *bonapartei* Cabanis, 1860, central Siberia and mountains of central Asia west to Kopet-Dag (Turkmeniya) and south to western Himalayas, intergrading with nominate *viscivorus* along Ob and Irtysh rivers.

Field characters. 27 cm; wing-span 42–47·5 cm. 25–30% larger-bodied than Song Thrush *T. philomelos*, with proportionately longer wing and tail contributing to marked attenuation of rear body and markedly heavy-chested and long silhouette; slightly longer and much bulkier than Fieldfare *T. pilaris*. Large thrush, with bold, upright carriage emphasized by length of tail, powerful bill, and sturdy legs; combination of olive-grey-brown upperparts, large discrete black spots on underparts, and white underwing diagnostic. Noticeable pale area on lore and around eye; corners of tail almost white, obvious from behind. Flight not undulating despite frequent complete closure of wings. Sexes similar; little seasonal variation. Juvenile separable. 2 races in west Palearctic, but only 1 described here (see also Geographical Variation).

ADULT. Main European race, nominate *viscivorus*. Upperparts brown with marked grey and yellow-buff tones producing pale olive appearance overall; conspicuous cream or grey-white fringes to olive larger wing-coverts and tertials, browner flight-feathers (particularly secondaries), and olive tail-coverts; fringes on wing form noticeably paler area on greater coverts and secondaries when folded. Tail olive grey-brown, with 2 outermost pairs of feathers noticeably tipped white and so forming pale corners. Face lacks any supercilium but has dark eye emphasized by white-buff on lores and front of ear-coverts and pale grey-brown on rear cheeks which have dark brown surround; moustachial and malar stripes not obvious. Ground-colour of underparts essentially white, with pale, clean, yellow-buff suffusion obvious only on chest, and large black spots (wedge-shaped on chest, more rounded elsewhere) scattered overall as far as vent but in noticeably irregular pattern (producing close overlay of black marks on some areas and wide spacing on others). Under wing-coverts white, contrasting with grey flight-feathers; undertail dull grey, with pale, almost white corners, not contrasting markedly with vent. JUVENILE. Differs distinctly from adult, being liberally marked with white spots and short black-brown chevrons on buffer ground of crown, sides of neck, back, and rump; conspicuously streaked white on small wing-coverts, and less heavily black-spotted below; face paler, with lores, cheeks, and throat almost white, and black malar stripe more obvious. Thus looks noticeably pale and variegated, lacking uniform crown, back, rump, and tail. FIRST-YEAR. Resembles adult but retained juvenile feathers on rump, scapulars, and forewing often show pale shaft-streaks, while retained larger wing-coverts and flight-feathers have pale buff fringes; suffusion of breast and flanks usually deeper buff. At all ages, bill dark horn-brown, with buff-yellow base to lower mandible; legs pale yellow-horn.

Large size, greyer upperparts, and white underwing immediately rule out smaller spotted *Turdus*. Juvenile traditionally mistaken for White's Thrush *Zoothera dauma*, but plumage pattern actually very different, lacking bold, heavy barring on underparts and black-and-white-banded underwing. Flight powerful and direct, with bursts of loose wing-beats giving noticeable momentum to heavily built and long-tailed bird; closes wings after each burst (producing long elliptical silhouette) but flight not undulating. Often flies rather high, even in breeding season reaching 30 m or more. Has more bounding action when taking off or flying low, and may momentarily suggest other non-related species, e.g. Green Woodpecker *Picus viridis* or cuckoo *Cuculus* or *Coccyzus*. Gait as *T. philomelos*, but more aggressive at shared food source or near nest. Flicks tail and wings in excitement. Wary in winter. Migrates in much smaller parties than other thrushes; forms loose associations of families in autumn but usually solitary in winter.

Song loud and challenging, phrases typically of 3–6 notes with notable skirling quality; overall rather monotonous compared with *T. philomelos* and (especially) Blackbird *T. merula* but volume greater than either, carrying up to 2 km. Sings persistently early in spring, and usually from higher perch than even *T. philomelos*. Subsong less frequent than in *T. philomelos* and Redwing *T. iliacus*: merely a faint version of full song except for additional rapid warbled passage. Commonest calls essentially chattering: ticking 'churr' in flight, a dry, staccato 'tuck-tuck-tuck' on perch (accompanied by synchronized wing-flicks), a louder, more excited rattle in alarm, and an even more intense, scolding rattle when mobbing.

Habitat. Breeds in west Palearctic, historically in more

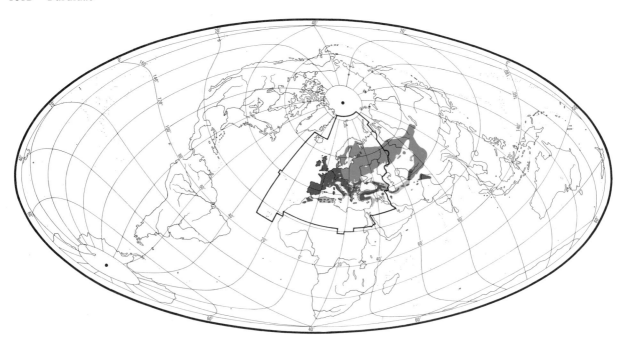

continental upper and middle latitudes than other *Turdus*, but has during last 200 years extended into oceanic zone, especially in Britain and Ireland. Despite bold and robust appearance and conduct in wind and rain, is vulnerable to cold, especially snow and ice, and avoids arid and very warm areas. In mountain regions prefers middle altitudes, *c*. 800–1800 m in Europe. In boreal, temperate, and Mediterranean zones, widely but thinly distributed where there is conjunction of stands of tall trees with open grassland or short herbage, not too close to human settlements and with ready access to seasonal berry fruits. Avoids dense forests, but also treeless or sparsely wooded areas, broken or bare terrain, and wetlands. Has recently, in parts of range, overcome reluctance to inhabit urban parks and gardens, and has simultaneously expanded range as well as diversity of habitats, assisted in some areas by afforestation and planting for amenity. Appears unattracted by water or by shrubs or bushes of 1–3 m high, and as a ground feeder prefers areas such as pastures, meadows, and large clearings not overhung by taller ·vegetation. Holding extensive territory, requires commanding song-posts, usually on tree-tops; often favours avenues, spinneys, plantations, and park trees. As an early breeder, free to shift locally by midsummer to alternative habitats such as upland grasslands or woods of yew *Taxus* and other favoured berry crops, including in some parts of range (e.g. Switzerland) mistletoe *Viscum album*, availability of which is locally a determinant of winter habitat (Glutz von Blotzheim 1962).

In Morocco, nests in woods of cedar *Cedrus atlantica*, Aleppo pine *Pinus halepensis*, oak and holm oak *Quercus ilex* at 1500–1700 m (Heim de Balsac and Mayaud 1962). At fringes of range, in moorlands or on islands, will occupy treeless sites, where nesting must be in walls or on ground.

A strong flier, rising freely above lower airspace, and unusually capable of surveying extensively alternative habitats. (Nicholson 1951; Bannerman 1954; Dementiev and Gladkov 1954*b*; Voous 1960; Yapp 1962; Parslow 1973; Sharrock 1976; Fuller 1982; Harrison 1982.)

Distribution. Marked spread in Britain and Ireland and, more recently, in Netherlands and Austria; slight spread in Norway. Man-made habitat colonized in north-west Europe (Snow 1969*a*).

BRITAIN. Marked spread in first half of 19th century; rare in northern England and Scotland at end of 18th century, then spread in 19th and 20th centuries (Alexander and Lack 1944). Has bred irregularly Orkney and Lewis (Parslow 1973). IRELAND. First bred 1807 and colonized most of country by 1850 (Alexander and Lack 1944). NETHERLANDS. Rare before *c*. 1870, restricted to north-east; since 1900–20 has spread to all provinces, but spreading halted 1970–5 and now stable or declining (CSR). NORWAY. Some spread (VR). AUSTRIA. Spreading and colonizing lowlands (HB, PP). IRAQ. Probably breeds in north (Moore and Boswell 1956). MADEIRA. Bred 1983 (W R P Bourne).

Accidental. Iceland, Faeroes, Kuwait, Israel, Egypt, Azores.

Population. Marked increase in Britain, Ireland, and Netherlands, but decrease in Finland.

BRITAIN, IRELAND. Has probably increased recently, including spread into gardens and parks, after major increase in 19th century (Cramp and Tomlins 1966;

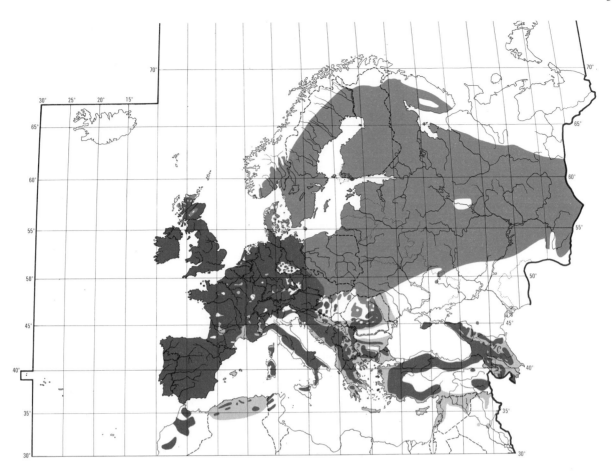

Parslow 1973). About 300 000-600 000 pairs (Sharrock 1976). FRANCE. Under 1 million pairs (Yeatman 1976). BELGIUM. About 42 000 pairs (Lippens and Wille 1972). NETHERLANDS. 20 000-30 000 pairs 1976-7 (Teixeira 1979). WEST GERMANY. 130 000-280 000 pairs (Rheinwald 1982). SWEDEN. About 300 000 pairs (Ulfstrand and Högstedt 1976). FINLAND. About 125 000 pairs (Merikallio 1958). Marked recent decrease (OH). For north, see Väisänen (1983). POLAND. No increase (AD). CZECHOSLOVAKIA. No changes reported (KH). GREECE. A few tens of pairs (GIH).

Survival. Britain: annual adult mortality 48%; mortality of fledged young to end of 1st calendar year 62% (Snow 1969*a*). Oldest ringed bird 9 years 6 months (Rydzewski 1978).

Movements. Varies from migratory in north and east of range to sedentary or dispersive in west and south, though even in least migratory populations some birds make substantial movements. Some central and south European birds winter within breeding range of the relatively sedentary *deichleri* of north-west Africa, Corsica, and Sardinia: apart from specimens obtained in Tunisia (as far south as Gabes) and in Morocco, proof exists

with single ringing recoveries in Morocco and Sardinia of nominate *viscivorus* ringed Austria and Yugoslavia respectively. Main winter range of central European and Scandinavian birds extends from Belgium through western and southern France to north-east Spain south to $39\frac{1}{2}°N$ and west to $2\frac{1}{2}°W$. West and south of this, recoveries of central European and Scandinavian birds are fewer: 5 in Spain, 4 in Portugal, and 1 in Morocco. (Zink 1981.) Occasional winter records to east and north of normal winter range in Scandinavia and central Europe: to *c.* 61°N in Norway and 59°N in Sweden (Rendahl 1960*b*), and in central Europe north and east to Minsk, western USSR (Pax 1925; Tischler 1941; Dementiev and Gladkov 1954*b*). No evidence that long-distance migration linked primarily with especially hard winters: of 20 recoveries south and west of Ebro river (north-east Spain), 12 were October–November, i.e. when onset of a hard winter normally not yet detectable (Zink 1981). Rarely, invades coastal belts of Libya and Egypt in good numbers (Bundy 1976; Goodman and Watson 1984).

British birds tend to be sedentary or make movements of less than 50 km, though proportion travel considerably further. Thus of 35 ringed as nestlings or juveniles and recovered in 1st autumn or winter, 22 had remained within 50 km, 3 had moved 100–300 km within Britain, and 10 had moved further to France (Snow 1969a).

Autumn migration in Britain mainly August–November although juveniles and adults may form flocks in July and start to move south. Record of 280 flying west on 12 September at Limpsfield Common (Surrey, England) was a far larger number than usually reported for movement in Britain (Simms 1978). In Ireland, large immigration occurs on all coasts September–November (Ruttledge 1966). Long-distance recoveries of British birds ringed as nestlings indicate movements between SSW and SSE to France or south-west to Ireland, though some birds from southern England move east. Recoveries of birds from continental Europe and southern Scandinavia suggest autumn migration takes place mostly in south-west to SSW direction, though several recoveries of birds ringed in Finland and the only one of a bird ringed in northern Sweden show movement to south-east. Thus a suggestion of migratory divide in Gulf of Bothnia area. Birds trapped and recovered within Italian peninsula also show south-easterly route but this may be due to leading line effect of Italian west coast. (Zink 1981.)

Few birds cross North Sea and thus rare on Fair Isle (Williamson 1965b), perhaps because breeding within Norway confined largely to south-east. Even on Helgoland (West Germany) only small numbers occur (Vauk 1972a). However, bird ringed as nestling in Estonia (western USSR) was recovered in south-east England (Zink 1981).

In spring, first birds appear on central European breeding grounds in February. In central Sweden, average arrival date 27 March over 22 years (Rendahl 1960b).

East Palearctic race *bonapartei* resident in Himalayas, though moving to lower altitudes in winter. Partially migratory in USSR, with some wintering in southern Kazakhstan and Tadzhikistan. (Dementiev and Gladkov 1954b; Dolgushin *et al.* 1970; Kovshar' 1979.)   LAB

Food. Wide variety of invertebrates, in autumn and winter also berries. Feeds on ground (though not usually in undergrowth) and in trees and bushes. Never seen by Simms (1978) to strike snails against hard objects in order to break them open. To take flying insects will fly up high in the air like Starling *Sturnus vulgaris* (Simms 1978). Will fly up to *c.* 1·5 m from ground, without hovering, to take berries (Peitzmeier 1949). For defence of feeding trees in winter, see Social Pattern and Behaviour.

Diet in west Palearctic includes the following. Invertebrates: grasshoppers, etc. (Orthoptera: Gryllidae, Acrididae), earwigs (Dermaptera), bugs (Hemiptera: Pentatomidae, Cercopidae), adult and (mostly) larval Lepidoptera (Pieridae, Coleophoridae), adult and larval flies (Diptera: e.g. Tipulidae), ants (Formicidae), adult and larval beetles (Coleoptera: Cicindelidae, Carabidae, Histeridae, Staphylinidae, Geotrupidae, Scarabaeidae, Elateridae, Cantharidae, Lampyridae, Curculionidae),

spiders (Araneae), millipedes (Diplopoda), snails (Gastropoda), slugs (Pulmonata), earthworms (Oligochaeta). Also young birds (see final paragraph). Plant material mainly fruits, also seeds: juniper *Juniperus*, yew *Taxus*, knawel *Scleranthus*, mistletoe *Viscum*, *Celtis*, *Loranthus*, barberry *Berberis*, bramble *Rubus*, strawberry *Fragaria*, rose *Rosa*, rowan *Sorbus*, cherries *Prunus*, apple *Malus*, pear *Pyrus*, hawthorn *Crataegus*, *Parthenocissus*, currant *Ribes*, bilberry *Vaccinium*, elder *Sambucus*, buckthorn *Rhamnus*, alder buckthorn *Frangula*, snowberry *Symphoricarpos*, vine *Vitis*, holly *Ilex*, ivy *Hedera*, dogwood *Cornus*, *Viburnum*; also shoots of grasses, flowers of sycamore *Acer*, moss, and fungus. (Csiki 1908; Florence 1914; Schuster 1930; Witherby *et al.* 1938b; Kovačević and Danon 1952; Huber 1953; Niethammer 1955c; Béres and Molnár 1964.)

In Britain, 52 stomachs from throughout the year contained 55% invertebrates (36% insects, mostly adult and larval beetles, 14% earthworms, 3·5% slugs and snails, 1·5% others) and 45% plant material (24% wild fruits and seeds, 16·5% cultivated fruit, 4·5% other plant material) (Collinge 1924–7). In 3 areas of southern Britain, of 250 records of fruit-eating, 44% yew, 25% holly, 18% hawthorn, 10% ivy, and 3% bilberry (Hartley 1954). In central Morocco, December, 4 birds contained caterpillars, beetles, grasshoppers, and berries, including 1 bird full of holly berries (Meinertzhagen 1940). In Yugoslavia, bird from July contained grasshoppers, beetles, and Hemiptera, and 2 others from February and August contained fruits of mistletoe and dogwood (Kovačević and Danon 1952). See also Ptushenko and Inozemtsev (1968), and Kostin (1983). For importance of mistletoe in diet, see Hardy (1969). For depletion through the winter of fruits on defended and undefended trees, see Snow and Snow (1984). For predation on snails in Negev desert (Israel), see Song Thrush (p. 993).

Stomachs of 16 nestlings from Britain contained only invertebrates, including 19% adult and larval Diptera, 16·5% beetles (mostly larvae), 15·5% earthworms, 10·5% slugs, and 8·4% Lepidoptera larvae (Collinge 1924–7). See also Prokofieva (1983). Recorded killing young Dunnock *Prunella modularis*, *T. philomelos*, and Blackbird *T. merula* and feeding them to young (Witherby *et al.* 1938b).   DJB

Social pattern and behaviour. Moderately well known, though no comprehensive studies. Account includes material on extralimital *bonapartei* in Tien Shan, USSR (Kovshar' 1966, 1979; Kovshar' and Gavrilov 1973).

1. Moderately gregarious at times outside breeding season, but also occurs singly and in pairs then. In Britain, typically nomadic and far-ranging in winter, perhaps within large home-range (Hartley 1955), though pairs in Sussex markedly sedentary on 'territories' from mid-December (Wheatley 1984). Some birds in Eure-et-Loir (France) also thought to hold winter territories (Labitte 1952a). Recent study in Britain has shown that some single birds and pairs defend fruit-sources, especially holly *Ilex aquifolium*, hawthorn *Crataegus monogyna*, and

buckthorn *Rhamnus catharticus*, from October–November through winter to early spring, also (e.g.) ivy *Hedera* or several clumps of mistletoe *Viscum* for short periods. Each tree (usually a free-standing holly of moderate size with other suitable look-out perches nearby) defended by single bird or pair, with pair-members probably in neighbouring trees. Defended tree generally avoided by other birds in vicinity and zone round it (territory may extend up to *c.* 100 m from tree) kept clear of other *Turdus* thrushes for most part, though success of defence varies according to weather: in mild winter, birds spend time on grassland within *c.* 100 m of tree, though feed more nearby if possible, and will visit tree, but take little or no food from it. Spend most time near tree in mid-winter and defend it successfully through to start of breeding season if conditions mild, though severe weather brings invading flocks of other *Turdus* which overwhelm defenders. Much of defence takes place when *T. viscivorus* mainly taking other food (from nearby trees or ground); tree-fruit thus conserved can, if slowly depleted, be food-source in spring and early summer. (Snow and Snow 1984.) See also Antagonistic Behaviour in part 2 (below). Flocks typically form in Britain from late summer; at least some in late July are of independent young (in *bonapartei*, such birds disperse far following break-up of family: Kovshar' 1979). From late summer to early winter, frequently in family parties and small, rather loose nomadic flocks or (particularly for feeding on migration) larger aggregations of 100 or more. Flocks tend to break up by mid-winter (Witherby *et al.* 1938*b*; Bannerman 1954; Simms 1978). In northern West Germany, spring, recorded migrating singly or in pairs (Westerfrölke 1970), but at Randecker Maar in south of country, fairly dense flocks and long lines of up to 50 birds occur (Gatter 1976). In USSR, migrates mainly singly or in small parties in spring, more often in small flocks in autumn; flock of 1500 birds reported south of Gor'kiy in September (Dementiev and Gladkov 1954*b*). In Mecklenburg, often associates with other *Turdus* at time of departure (Klafs and Stübs 1977); in Italy, with Fieldfare *T. pilaris* during exceptional influx of both species (see Casati 1937); in USSR, family parties of *T. viscivorus* and *T. pilaris* said to merge in first half of July, though no association later when flocks nomadic (Dementiev and Gladkov 1954*b*); see also Antagonistic Behaviour (below). In Tien Shan, winter, frequently seen with Black-throated Thrush *T. ruficollis atrogularis* (Kovshar' 1966), but not clear how close such associations are, and in Eure-et-Loir Labitte (1952*a*) found *T. viscivorus* not gregarious with other *Turdus*. BONDS. Nothing to suggest mating system other than monogamous. Little substantial information on duration of pair-bond: birds reported as 'mostly in twos' during one winter in eastern England (Hartley 1955) and study of winter territoriality in England (see above) strongly suggests some pair-bonds persist outside breeding season, though territorial and courtship behaviour anyway starts with break-up of flocks by mid-winter (Simms 1978), and in Eure-et-Loir pair-formation (or at least association in twos) often takes place from early January, sometimes from December (Labitte 1952*a*); see also Heterosexual behaviour (below). In Denmark, ♂♂ apparently migrate *c.* 8 days ahead of ♀♀ in spring (Krüger 1938) and in *bonapartei*, pair-formation takes place on breeding grounds soon after ♂♂ start singing (Kovshar' and Gavrilov 1973); see Song-display (below). Young brooded by ♀ (sometimes also ♂: P A D Hollom) and fed by both parents (more by ♂ in early stages), including for *c.* 2 weeks (or more) after they leave nest, though ♂ takes over sole responsibility for 1st brood if ♀ re-lays. 1st brood apparently maintain some contact with parents while 2nd brood being reared (Ryves 1928; Witherby *et al.*

1938*b*; Hartley 1955); according to Géroudet (1963), 1st brood eventually join up with parents and 2nd brood to form enlarged family unit, though this not confirmed by Raevel (1981). In *bonapartei*, family breaks up not later than 3 weeks after young leave nest (Kovshar' 1979). See also Relations within Family Group (below). Age of first breeding not known. BREEDING DISPERSION. Normally solitary and strictly territorial, though some reports of apparent colonies. In northern Jutland (Denmark), 63·7% of birds solitary, 36·3% colonial (*n* = 133 pairs); colonies of 2–7, average 3·17 pairs, and nests sometimes only a few metres apart (Møller 1978*c*). In Netherlands, sometimes 2–3 pairs in orchard and nests then *c.* 40 m apart (Brink 1936). Near Bielefeld (West Germany), nests initially at least 300 m apart, but with range expansion and increasing density, reduced to *c.* 100 m (Peitzmeier 1942; Stein 1952). Nests of *bonapartei* usually not less than 100 m apart; where density high, sometimes only *c.* 15–50 m (Kovshar' 1966; Kovshar' and Gavrilov 1973). Territories typically do not overlap (Wheatley 1984). Defended mainly by ♂, but ♀ also involved, especially against predators (Raevel 1981). Size apparently varies with density. In Britain, where density generally low (only *c.* 10% of that of Blackbird *T. merula* or Song Thrush *T. philomelos*, even where conditions favourable, and usually much less), individual pairs range over wide area (Snow 1969*a*) and actual territory generally large (Sharrock 1976)—'enormous' compared with *T. merula* or *T. philomelos* (Hartley 1955): from *c.* 0·6 ha in wooded parkland to 'several acres' in woods and open country (Simms 1978). ♂ of pair nesting in garden, southern England, usually sang outside garden, moving about within *c.* 300-m radius (Marples 1937). From data of Schiermann (1934) for Brandenburg (East Germany), area available per pair in pines *Pinus* calculated at 5·1 ha, but this unlikely to correspond to size of territory (Stein 1952). Area used (including for feeding) by pair at Bailleul (northern France) *c.* 15–17 ha; bounded on 2 sides by busy roads, on 3rd by building; 4th limit, defined only by presence of neighbouring pair, fluctuating and more or less defended and respected; most constant and aggressive defence reserved for area within *c.* 50 m of nest (Raevel 1981). Peitzmeier (1949) noted that nesting territory and feeding area normally contiguous in parkland and suggested feeding area perhaps also defended; however, where nests close together (see above), adults (like *T. pilaris*) tend to feed together on common ground (Verheyen 1947). This pattern apparently exists in part of Tien Shan, where nesting territory small, but birds feed almost exclusively outside it, flying up to *c.* 1 km (Kovshar' 1966, 1979*a*). Densities from various parts of range as follows. In Britain, average 1·9 pairs per km[2] on farmland, 4·9 pairs per km[2] in woodland; few census plots have density exceeding 10 pairs per km[2] (Sharrock 1976); one suburban area of 202 ha held 1 pair of *T. viscivorus* compared with 12 of *T. philomelos* and 180 of *T. merula* (Simms 1978, which see for further details and comparisons; see also Stein 1952); sessile oak *Quercus petraea* wood in North Wales held 13 pairs per km[2] (Gibbs and Wiggington 1973); see also (e.g.) Benson and Williamson (1972) and Glue (1973). In Switzerland at higher altitudes in larch *Larix*, up to 30 pairs per km[2]; below *c.* 700 m altitude, 5 pairs or less per km[2] (Schifferli *et al.* 1982); see also Glutz von Blotzheim (1962). Near Bielefeld, 1–21 pairs in 2·5 km[2] over 9 years i.e. (excluding 21 pairs of 1 year) average 8·4 pairs per km[2] (Peitzmeier 1942; Stein 1952). In Mecklenburg (East Germany), 3–7 pairs per km[2] of coniferous woodland, 4–13 pairs per km[2] of broad-leaved woods (Klafs and Stübs 1977). Pure pine up to 100 years old in Brandenburg (East Germany) holds average 4(1–8) pairs per km[2], 14–16 pairs per km[2] where also some broad-leaved trees (Rutschke 1983). For further examples from East Germany, see (e.g.) Saemann (1974) and Möckel and Möckel (1975). Sparsely distributed also in Sweden where (e.g.) 0·7 pairs per km² of forest in Uppland (Olsson 1947) and in southern Finland where 1 pair per km² of spruce *Picea* and other conifers, 2 pairs per km² of deciduous groves and fens (Palmgren 1930; Stein 1952). Further east in Moscow region (USSR), 8–12 singing ♂♂ noted along 10 km and 0·2–0·3 pairs per km² in pinewood (Ptushenko and Inozemtsev 1968). In Tien Shan, 10 occupied nests of *bonapartei* recorded in 3 ha (Kovshar' 1966); see comments on territory size (above). Further data from extra-limital parts of USSR in (e.g.) Ravkin (1984) and Vartapetov (1954). In woods of Morocco, 5·2 pairs per km² (Thévenot 1982). Builds new nest for 2nd brood but remains within same territory (Ryves 1928) or occasionally uses same nest for 1st and 2nd broods (Brink 1936; Witherby *et al.* 1938b). ♀ *bonapartei* recovered *c.* 100–200 m from ringing site over 2 following years; of 86 nestlings ringed, 3 returned to natal area to breed, *c.* 400–600 m away (Kovshar' 1979). Sometimes nests within a few metres of other *Turdus* (Kovshar' and Gavrilov 1973; Raevel 1981). Often nests in same tree as Chaffinch *Fringilla coelebs* (also Goldfinch *Carduelis carduelis*: Raevel 1981), and this thought by Mayaud (1952b, which see for references) to be of mutual benefit, due to vigilance of *F. coelebs* and aggression of *T. viscivorus* (see also *T. pilaris*, p. 982, for similar associations). ROOSTING. Nocturnal. In Britain, autumn parties roost communally in trees (especially evergreens), also tall hedges, bushes, and thick brambles *Rubus*. Usually solitary or in pairs from December (Witherby *et al.* 1938b). In northern West Germany, over 2 years, up to 40 birds roosted in late summer in small, mainly broad-leaved wood (not same each night), using oak or beech *Fagus*; arrived in apparent family parties from various directions and before dusk (Peitzmeier 1942). In Eure-et-Loir, winter (all weathers), roosts mainly in ivy on old trees at 3–6(–12) m, also in clumps of mistletoe at 6–15 m, but sometimes uses leafy oaks. Always solitary in mistletoe, but 2 birds sometimes roost in same ivy, though at different levels; maximum 5–6 in area *c.* 100 m square. Go to roost *c.* 16·00–16·30 hrs, November–January, chosen sites being occupied regularly each night at about same time of year. 1 bird used same roost from end of September to February. Reluctant to leave roost once installed. (Labitte 1937, 1952a.) Active anting recorded (Simms 1978).

2. Normally rather wary and suspicious of man, shyer than *T. philomelos* in winter and shunning close approach. Usually flies up to tree and gives Rattle-call (see 3a in Voice) when disturbed or, in open country, will fly off (Brink 1936; Witherby *et al.* 1938b; Labitte 1952a; Dementiev and Gladkov 1954b). Sometimes less shy: e.g. allowed approach to 'a few feet' in Oulmes (Morocco), autumn (Meinertzhagen 1940), and Horváth (1977) noted greater tameness with increased use of urban habitat in Hungary. Contrasting boldness in defence of nest and young well known (see Parental Anti-predator Strategies, below). When excited, will down-flick tail, sometimes also wings, then (as also when tail-fanning) giving call 5. When mobbing owl (Strigiformes), gives calls 3a–b, working up to peak of intensity (Witherby *et al.* 1938b; Simms 1978). FLOCK BEHAVIOUR. Roving birds tend to scatter in small parties rather than forming compact flock. Flocks usually at most cohesive in morning, then split up to reassemble for roosting. If numbers congregate, birds generally well dispersed for feeding, individuals leaving randomly; also take off singly when disturbed (Bannerman 1954; Géroudet 1963; Simms 1978). SONG-DISPLAY. ♂ sings (see 1 in Voice) normally from high perch, typically tree-top; exceptionally from ground (Witherby *et al.* 1938b; Simms 1978) and not infrequently in flight (Peitzmeier

1970; Radermacher 1970b). Typically stays long on one song-post before changing position (Ince 1982); see, however, Marples (1937). In Thüringer Wald (East Germany), late June, bird moving about on ground and feeding stopped and adopted an erect posture to sing part of song-phrase (Münch 1956). Several descriptions of Song-flight, especially from West Germany, and considered normal phenomenon, at least in open (park) landscape, March–April (February–May), taking place, like song generally, in almost any weather. Excited song from perch before take-off into Song-flight apparently typical, though perhaps sometimes starts singing only when airborne. Recorded flying and singing over several hundred metres and continuing after landing in tree, also repeatedly traversing field in irregular loops at height of *c.* 25–40 m. At least in some cases, Song-flight apparently has territory-marking function, and is particularly associated with period of territory establishment and 1st brood: singing ♂ seen flying behind intruding pairs and following single bird (at *c.* 50 m) for some distance. (Mester 1957a, b; Stichmann 1958; Greve 1969; Peitzmeier 1970; Radermacher 1970b, 1973.) Song-flight also occurs in *bonapartei* (Kovshar' 1979). Actively migrating birds will sing in mild weather, February–March (Westerfrölke 1970). In southern England, start averages 24·7 min before sunrise (*n* = 10), while ending less regular at 24·6 min before to 11·3 min after sunset; 11 hrs of almost continuous song per day (Marples 1937, which see for diagram). High output early in day from paired and unpaired birds (no significant differences), normally declining through morning (Ince 1982). In Britain, main song-period is late December or January to early June (peak February–March and much reduced during incubation and nestling periods); exceptional (or only Subsong) late August to mid-September, also mid-October; irregular, but fairly frequent, late November to mid-December. No song noted in Italy in winter (Witherby *et al.* 1938b; Ince 1982; also Ryves 1928, Simms 1978). Song is given mainly by unpaired ♂♂ and probably serves mainly for mate-attraction (Wheatley 1984; see also Ince 1982). In West Germany sings mostly from late January or February and in September–October (Müller-Using 1967; Gatter 1968; Kroymann 1968; Peitzmeier 1968). Resident *bonapartei* start to sing in early March, migrants soon after arrival (from late March); during April–May peak, will sing virtually all day, but most in morning (earliest 04.22 hrs) and (especially in May) in evening (latest 21.00 hrs). Up to 1117 songs per hr in April. Sings mainly up to start of building, May–June song probably from failed breeders or after 1st brood fledged (Kovshar' and Gavrilov 1973; Kovshar' 1979). ANTAGONISTIC BEHAVIOUR. (1) General. Rather unsocial and quarrelsome and intolerant of competitors when feeding on berries (Brink 1936; Witherby *et al.* 1938b). Dominates other *Turdus* (e.g. when defending fruit-sources) probably due particularly to large size, but perhaps also to other morphological and plumage features including certain resemblance to Sparrowhawk *Accipiter nisus* in swooping attack (Snow and Snow 1984). However, according to Simms (1978), does not attack, nor is attacked by *T. pilaris* (but see below). For timing of territory occupation, see Bonds and Song-display (above). (2) Threat and fighting. Most detailed information relates to defence of food-sources. Territorial bird spends long periods perched in fruit-tree and longer on nearby look-out. Persistent chasing common October–November. Main competitors (other *Turdus*) driven away immediately, territory-owner sometimes coming from edge of territory to expel them, often (not always: Lloyd 1984) chasing them even when only feeding on ground. Territorial bird usually silent in expelling attack, but will give constant Rattle-calls in conflict, also when attacks unsuccessful (e.g. when overwhelmed by sheer numbers).

Attack rates vary markedly depending on food-supply and pressure from intruders: successful defence of holly involved 1·4 attacks per hr October–November, 3·1 per hr in December, 1·3–0·7 per hr, January–March; up to 30 attacks per hr recorded against Starling *Sturnus vulgaris* on ivy. (Snow and Snow 1984.) As in other *Turdus*, territorial ♂ commonly gives Whisper-song (for discussion of Whisper-song, Subsong, etc., see 2a–c in Voice) in response to conspecific intruder; perhaps both sing thus in close encounter. In response to playback of song tends to change perch immediately (some birds fall silent), then (after playback) starts Whisper-song, normally giving 4–8 bouts (each of 5–20 s) before changing back to normal song (Ince 1982). Bird recorded running about on tree branch and giving Subsong had plumage (especially of head and neck) sleeked, and head lowered with neck extended. After several runs, stayed silent and crouched, head and neck as before. Arrival of another bird from below led to fierce fight, involving pecking and short upward leaps, each assault lasting *c.* 2–3 s; when one flew off, other chased closely (Suffern 1965). Protracted and noisy ♂–♂ chases through tree branches are typical territorial feature (Witherby *et al.* 1938*b*). In dispute with Hawfinch *Coccothraustes coccothraustes* at drinking place, *T. viscivorus* adopted Threat-posture and 'screamed' (perhaps higher-intensity variant of call 3c: see Voice) at opponent (Fig A). When *C. coccothraustes* attacked with loud Bill-snapping,

A

B

T. viscivorus leaped into the air (Fig B) and counter-attacked using feet (Smith and Hosking 1955). Marked aggression shown towards Kestrel *Falco tinnunculus* coming to roost on rock-face where perhaps nest of *T. viscivorus* (Praz 1979). HETEROSEXUAL BEHAVIOUR. (1) General. In Moscow area, pair-formation reported to take place *c.* 5–7 days after first birds arrive (Ptushenko and Inozemtsev 1968). In *bonapartei*, follows soon after start of song (see Song-display, above); birds already paired (and ♂♂ singing) on arrival at breeding grounds in subalpine Tien Shan (Kovshar' and Gavrilov 1973; Kovshar' 1979). A number of observations suggest that a certain element of threat persists between ♂ and ♀ partners even after

pair-formation (see subsection 3, and Relations within Family Group, below). For incipient reproductive behaviour, notably nest-building, of captive juveniles, see Goodwin (1954). (2) Pair-bonding behaviour. In presumed heterosexual encounter, 1 bird gave Subsong, other (in same tree) sang loudly, took off and continued to sing in flight; then returned, 1st bird (still in tree) giving Subsong, and both eventually departed (Almond 1931). Presumed ♂ also recorded making rapid hopping ascent of tree in which ♀ perched, then flying round her in wide circles, revealing white underwings. Landing of ♂ causes ♀ to fly to another tree, pursued by ♂, etc. Alternatively, following noisy (Rattle-calls) 'bouncing about' in tree by 2 pairs (perhaps some antagonistic element) and departure of 1, presumed ♂ recorded flying to another tree and singing there, chase developing when this bird's presumed mate joined it, birds heading for tree where nest located in previous year (Hartley 1955). Chases can involve noisy and excited calling from both birds (Labitte 1952*a*) or (perhaps more often) are quiet and rather leisurely in contrast to antagonistic pursuits, being typical low-intensity sexual display (not leading to copulation). Flight reported to differ markedly from normal: 'bobbing and bucketing', recalling Jay *Garrulus glandarius*, with curious breaks in level or little half-loops in otherwise more or less straight path; birds have head and (sometimes) tail raised. 1 bird may fly close behind and slightly above other (Hartley 1955), or both maintain same speed and stay apart while chasing ceaselessly and for long time through trees (Labitte 1952*a*). If pursuer closes, pursued bird takes evasive action, but without haste; birds will also change roles. During pauses between pursuits, presumed ♀ will open and rapidly shiver wings. ♂ will also fan tail, revealing white tips, and spread wings (Fig C), though unlike in other *Turdus*, rump feathers not ruffled.

C

(Witherby *et al.* 1938*b*; Hartley 1955.) In encounter between 2 birds on lawn, late January, 1 hopped towards other and adopted a horizontal posture, with ruffled breast and flank feathers (regular part of sexual display: Witherby *et al.* 1938*b*), lowered head, raised and slightly fanned tail and ruffled vent feathers; repeatedly assumed this posture (each time after 7–10 hops), directing cloaca at stationary companion. Apparently higher-intensity display involved more marked ruffling of breast feathers (spots prominent) and conspicuous rapid trembling of feathers around cloaca. Bird made final arcing approach to within *c.* 2 m of hitherto stationary companion, then turned away to be followed by other bird. Both flew off when disturbed (White 1971). (3) Courtship-feeding and Mating. ♂ feeds ♀ on nest during incubation (e.g. Brooks-King 1946, Ryves 1946, Brown 1947), but feeding also occurs earlier and sometimes associated with copulation: e.g. ♂ fed ♀ in tree *c.* 150 m from completed (but empty) nest and copulated 3 times in quick succession (Williams 1947*a*); ♀ of pair on ground accepted and ate proffered food, copulation then taking place 4 times in fairly quick succession, and birds facing each other with bill wide open (see below) between each mounting (Boyd 1946). In another case where birds on ground, several of ♂'s attempts to mount frustrated by ♀'s moving forward and (once) by apparent intervention of *T. philomelos*. ♂, first adopting an upright posture, finally mounted crouched ♀ at least 6 times in a row; end abrupt (Hartley 1955). ♀ *bonapartei* recorded adopting soliciting-posture during building, but no copulation took place (Kovshar' 1979). Captive ♀, apparently stimulated to nest-build by this activity in another presumed ♀, then flew to perch and solicited by giving very soft, inward variant of Rattle-call and moving wings ('intermediate between typical wing-flirting and shivering'). Presumed ♂ approached but before he mounted, both exchanged gaping (perhaps threat); copulation attempt frustrated by ♀'s falling or flying from perch (Goodwin 1954). (4) Nest-site selection. By ♀ (Ptushenko and Inozemtsev 1968), 1–2 days before building (Ryves 1928). (5) Behaviour at nest. Building by ♀ (Ryves 1928; Kovshar' 1979) or both sexes (Brooks-King 1946; Labitte 1952*a*); see Raevel (1981) for details. ♂ usually nearby on high perch and will warn of any danger, also accompany ♀, including later during incubation (Ryves 1928; Ptushenko and Inozemtsev 1968; Raevel 1981). ♀ said by Ryves (1928) to employ special flight with wings widely spread when flying down for fresh material (confirmation required). Incubating ♀ once seen to be called off nest by ♂ (no details, but see 4 in Voice) who then flew to nest and incubated for *c.* 6 min (Ryves 1928, 1943). RELATIONS WITHIN FAMILY GROUP. In *bonapartei*, 3 of brood usually hatch on 1 day, 4th (normally runt) on next, but hatching also recorded over 3 days (Kovshar' and Gavrilov 1973). Studies in Britain (nominate *viscivorus*) indicate young brooded (apparently mainly by ♀; some done by ♂, however: P A D Hollom) for first 4–5(–6) days, especially closely in poor weather—e.g. stint of 106 min in heavy rain. Brooding only occasional thereafter and discontinued, even at night, at 9–10 days, except in bad weather. ♀ will make short excursions for food, but most early feeding by ♂, ♀ eating some food brought by him (Ryves 1928; Hartley 1955; Kovshar' and Gavrilov 1973; Radford 1980). Young fed bill-to-bill (Witherby *et al.* 1938*b*). Coot-call (see 4 in Voice) given by adult in contact with young, perhaps also arousing them for feeding (P A D Hollom). Adults normally fly straight to nest (Kovshar' 1966). Faecal sacs swallowed by adults (including by brooding ♀: Kovshar' 1966) or carried away far (Hartley 1955). Eyes of young fully open by 7th day (Kovshar' 1966, which see for details of physical development). ♀ recorded flying from nest (where she had brooded 6-day-old young) to tree at *c.* 20 m where joined by ♂; both adults gave apparently aggressive Rattle-calls and, between bursts, turned to face each other 4 times, each time bill-touching (bill closed or partly open) 4–5 times; continued 1 min or more, birds then flying off with loud Rattle-calls (Radford 1980). Young leave nest at 14–16(–17) days, with variation probably due in part to weather (Ryves 1928, Niethammer 1937; Witherby *et al.* 1938*b*; Hartley 1955). Feeding rate drops noticeably just prior to fledging (Raevel 1981). Clumsy, often falling off branches, when first leave (Géroudet 1963). Move initially into crown of nest-tree (Ptushenko and Inozemtsev 1968), but fairly rapidly disperse, seeking separate hiding places in foliage. Give food-calls from there and wait for parent to come with food or may emerge, fly to parent to take food and then retreat; will also follow parent to feeding grounds, begging with food-calls and wing-shivering (Raevel 1981). Brood of 3 leaving nest at 16–17 days old able to fly 30–40 m or more and used high perches; left area, presumably to embark on nomadic life typical of *T. viscivorus* (and unlike *T. merula* and *T. philomelos* which perch lower or on ground and remain near nest *c.* 3 weeks)

(Hartley 1955). Other reports refer to young being able to fly 'reasonably well' at 20 days (Niethammer 1937) or *c.* 1 week after leaving nest (Kovshar' 1979). In Cornwall (south-west England), brood stayed close to nest (in garden) and fed regularly by parents 12 days after leaving (Ryves 1928); in *bonapartei*, fed for up to 16–18 days, gradually moving up to *c.* 100 m from nest (Kovshar' 1979). Young thus independent at *c.* 1 month old (Géroudet 1963). ♀ will start 2nd-brood nest before 1st brood leave nest (Brink 1936) or build it rapidly *c.* 1 week after 1st brood fledge (Ryves 1928), so that *c.* 6–12 days between fledging of 1 brood and start of re-laying (Snow 1969*a*). ANTI-PREDATOR RESPONSES OF YOUNG. No information on nestlings. Like other *Turdus* thrushes (see, e.g., *T. philomelos*, p. 995), fledged young will adopt Bittern-posture when threatened (Goethe 1973); see also Parental Anti-predator Strategies (below). Bird disturbed while feeding gave alarm-calls (not described) and adopted Bittern-posture immediately on landing in tree. Stayed put while observer walked round it, then (when man at *c.* 15 m), slowly 'unfroze', looked about, and flew off low, calling loudly. Camouflage-effect negligible as bird clearly visible against green of leaves (Wagner 1982). PARENTAL ANTI-PREDATOR STRATEGIES (1) Passive measures. Can be a tight-sitter, sometimes allowing approach to within a few metres (Brink 1936; Labitte 1952*a*), and staying even when man starts to climb tree. If disturbed when approaching nest, will fly on past, then approach furtively later (Kovshar' 1966). Posture adopted by bird alarmed at nest (Fig D) perhaps same as

D

Bittern-posture of fledged young. Pair having detected man hiding near nest, circled overhead, refusing to visit nest; eventually abandoned young which perished overnight (Labitte 1952*a*). (2) Active measures: general. Notoriously bold in defence of nest and young, launching often prolonged attacks similar to *T. pilaris*; defecation used, but apparently rarely (e.g. Niethammer 1937, Peitzmeier 1962, Simms 1978). (3) Active measures: against birds. Attacks various Corvidae, especially Magpie *Pica pica* (which can be a serious predator), also Carrion Crow *Corvus corone*, though not when these predators actually at *T. viscivorus* nest (Brink 1936; Peitzmeier 1962). *P. pica* and *C. corone* attacked by ♂ *bonapartei* at *c.* 100 m from nest, ♀ assisting when predators *c.* 20–30 or 50–70 m away; at site where *P. pica* numerous and nest of *T. viscivorus* rather exposed, attacks probably ensured successful fledging (Kovshar' 1979). Recorded striking and killing Jackdaw *C. monedula* near nest, knocking Barn Owl *Tyto alba* off perch (Gurney 1903), and will attack other quite large birds (perched or flying past)—

e.g. forced Buzzard *Buteo buteo* to leave (Peitzmeier 1962) and chased *A. nisus* furiously, giving loud Rattle-calls (Brink 1936). (4) Active measures: against man. Includes attacks of type described above, and recorded pecking man's hand and drawing blood, but will also dash about calling loudly. When young bird fallen from nest picked up, adults gave only quiet Rattle-calls from a distance (Brink 1936; Peitzmeier 1962; Simms 1978); in another similar situation, call 3c, then higher-intensity variant in attack (Hesse 1909; Voigt 1933). Man working *c.* 50 m from nest mobbed initially by ♂, then by this bird and a probable neighbour, these being joined by 1st ♂'s mate; when strange bird left, pair continued (Raevel 1981). In *bonapartei*, some variation depending on age of young: when nest containing small young examined, parents gave Rattle-calls from tree *c.* 5–10 m away; when young older, attacks (often both adults) persistent and effective, birds veering away at *c.* 1 m or less (Kovshar' 1966). (5) Active measures: against other animals. Cats, dogs, foxes *Vulpes* and squirrels *Sciurus* subjected to attacks as described (Géroudet 1963; Simms 1978).

(Figs by N McCanch: A–B from photographs in Smith and Hosking 1955; C from drawing in Simms 1978; D from photograph in Hosking and Lowes 1947.)　　MGW

Voice. Considered by Chappuis (1969) to be lower pitched in south of range than in north. Loud song given from high perch is a prominent feature early in year; call 3a also well known, others much less so. For extra sonagrams, see Isaac and Marler (1963), Simms (1978), Bergmann and Helb (1982), Wheatley (1984), and, also for comparison with other *Turdus*, Ince (1982).

CALLS OF ADULTS. (1) Song of ♂. Normal song typically loud (further-carrying than Song Thrush *T. philomelos*), wild, shrill, and skirling, comprising short phrases, separated by pauses, of 3–6 (up to 12 in available recordings: J Hall-Craggs) characteristically rich, pure, and challenging fluted whistles (most units tone-like) with quality of Blackbird *T. merula* (Marples 1937; Witherby *et al.* 1938*b*; Dementiev and Gladkov 1954*b*; Simms 1978; Bergmann and Helb 1982; Ince 1982). Short, quieter coda (Subsong) sometimes given at end of phrase (Bergmann and Helb 1982); see also call 2. Song given in southern England, January, comprised repetition of 3 phrases, always in same order: 'mairidoit', 'teeawti', 'tawteeteetaw'; later variations included 'quick-quick-quick' after 1st phrase, change of 2nd to 'tee-awtiawti' or 'teeawtitooee', and introduction (by April) of 'pweeepweeepwee' (Marples 1937). For further renderings of song-phrases, see French study by Hertzog (1954) who noted a 'diüiüijü' sometimes like (in timbre) Crested Lark *Galerida cristata* and, at times, pitch descent towards end of song. Mimicry rare in study of 4 birds in Sussex (England) (Wheatley 1984). Recording (Fig I) reveals song-phrase similar to *T. merula*; similarity more pronounced than usual in another more tuneful phrase (Fig II) which also shows decided frequency and amplitude modulation not apparent in Fig I. Song (Fig I) typical of *Turdus* with loud introductory section of rich, expansive whistles followed by quiet warbling coda. Similarity to *T. merula* more obvious in sonagram than

to ear; audibly different perhaps because introductory section relatively short compared with whole (though coda often missing or inaudible, e.g. in Fig II) and frequency range throughout more limited, that of whistled notes apparently not exceeding 1 octave (J Hall-Craggs, M G Wilson). Concentration of energy in relatively narrow band (for details, see Ince 1982) and consequent slight pitch fluctuations make song comparatively monotonous (Bergmann and Helb 1982). Each phrase sometimes given 2–3 times in succession (Marples 1937) and repetition of same or closely similar phrases from rather small repertoire (see below) further contributes to impression of monotony (Witherby *et al.* 1938*b*). As in *T. merula*, unit repetition rare (Ince 1982), though can be a conspicuous feature (Bergmann and Helb 1982), and exceptionally units always repeated, sometimes in triplets (Isaac and Marler 1963). Low mean frequency values are characteristic of song, which propagates well in forested habitat; as bird also sings from high perch, possible ground attenuation of low-frequency sounds in song is avoided. Unlike *T. philomelos*, sings with marked temporal definition, pauses between songs being almost always longer than any within a given song; occasionally, 2 songs juxtaposed such that gap between them considerably shorter than normal. Songs short (average 1·36 s, *n* = 3 birds), with pauses between them averaging 2·25 s (Ince 1982); all pauses more than 0·6 s, most more than 1·1 s (Isaac and Marler 1963; also Witherby *et al.* 1938*b*). Songs thus delivered at rapid rate (Bergmann and Helb 1982): 14–21 per min (Simms 1978; also Hertzog 1954). Sequential organization of song investigated by Isaac and Marler (1963): songs begun by only a few of possible syllable types (5 types account for 87·5%); 'syllables' (defined here as groups of notes separated by longer than 45 ms) within a given song tend to follow in a definite order. Nor is order of songs random, and given song normally followed by a few out of the bird's total repertoire (of 10–20 song-types: Ince 1982). General temporal patterning and method of sequence formation apparently the same for 5 birds studied. Marked sharing of syllable types also noted in songs of 4 birds from one locality (Isaac and Marler 1963); similarly in sonagrams of 2 birds in another study, but songs not entirely alike (Ince 1982). With practice, easy to recognize individuals by their song (Ince 1982). (2) Subsong and other quiet song variants. (a) Subsong. As noted above, sometimes occurs as coda to, or may precede, more tonal part of song-phrase. Combination of Rattle-calls (see below) and Subsong noted in August (Peitzmeier 1968). Given in heterosexual, antagonistic, and other contexts (Suffern 1965*a*). Quiet but hurried and excited warbling suggestive of Garden Warbler *Sylvia borin* and lasting 2–3 s (Almond 1931); unmusical chattering, in some ways similar to song of Fieldfare *T. pilaris* (Voigt 1933). Units of Subsong become harsh and impure, the rambling utterance suggesting imperfectly formed full song (Witherby *et al.*

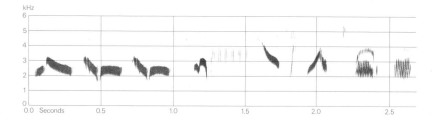

I P J Sellar Scotland May 1973

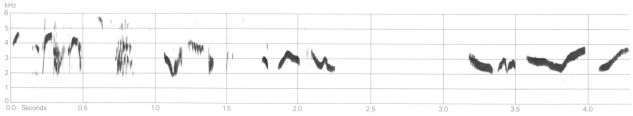

II P J Sellar England April 1975

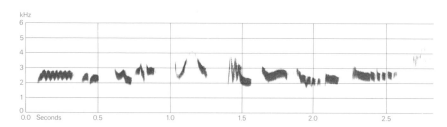

III P J Sellar England February 1973

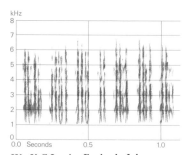

IV V C Lewis England July 1972

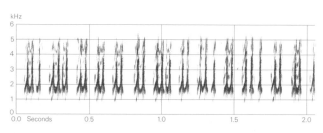

V V C Lewis England April 1969

1938*b*; Simms 1978). Fig III shows presumed Subsong with typical wide frequency range (lowest *c*. 1·5 kHz, while some quieter sounds almost 6 kHz); tonal section following a pause of over 0·9 s ends (apparently uncharacteristically) in sudden pitch ascent on the last 2 notes—'hui huee' (J Hall-Craggs). (b) Whisper-song. Significantly different from normal song, more so than in *T. merula* or *T. philomelos*. In Whisper-song elicited by playback, mean frequency increased and becomes more variable (*c*. 1·5–4 kHz, or up to 8 kHz) while mean length of units decreases to half that of normal song (0·07 s versus 0·14 s) (Ince 1982). Whisper-song clearly has much in common with Subsong and further study required to determine if (constant) differences exist. In recording by P J Sellar of probable Whisper-song in apparent heterosexual interaction (perhaps as prelude to copu-

lation), quiet, intimate warbling much less well-defined (structured) than Subsong in Fig III (J Hall-Craggs); warbling, trills, fizzy sounds and twittering, also deeper sounds like *T. merula*, curiously attenuated whistle like Starling *Sturnus vulgaris*, and harder 'dip' sounds (M G Wilson). (c) Low-volume and high-frequency song given by ♂ when close to mate (compare 'strangled song' of *T. merula*) perhaps identical to Whisper-song (Ince 1982). (3) Contact-alarm calls. (a) Rattle-call. Variations in loudness, timbre, and duration occur. Familiar harsh churring rattle or chatter, like drawing piece of wood over coarse teeth of comb (Witherby *et al.* 1938*b*). Loud, dry, slightly muffled 'khrr' (Dementiev and Gladkov 1954*b*), 'rrr' (Jonsson 1978*a*), or phrase-type sequences of hard rattling 'trrr' or 'tzrr' sounds, sometimes more clicking-rattling, but always in series (Bergmann and

Helb 1982). Recording (Fig IV) reveals irregular pattern of 'karrrrr' and 'kerrrrr' sounds, variation also evident in timbre and speed of delivery (M G Wilson). Rattle-call given in flight, but also when perched and when flushed. Calls tend to be louder and more rattling when alarmed or excited (Witherby *et al.* 1938*b*): e.g. short calls of warning developing into high-intensity 'terterter' of alarm (man near fledged young); same timbre and rate of delivery as in *T. pilaris* (Voigt 1933); rhythmic 'tschrt-schr...', probably equivalent of alarm-rattle in other *Turdus* (Bergmann and Helb 1982). In recording (Fig V), bird extremely alarmed at presence of Jay *Garrulus glandarius* near nest gives rapidly repeated 'churr' sounds in unbroken pattern, producing rather mechanical effect (M G Wilson). (b) Harsh guttural croaking sounds sometimes combined with call 3a in high-intensity excitement (Witherby *et al.* 1938*b*; Simms 1978). Series of 'gra' sounds, lower pitched than typical call 3a, given by adult with fledged young (Voigt 1933); apparently expressing at least mild alarm and probably same or related. Recording of adult in contact with young reveals buzzy and twangy 'zerrrrr', like raucous corvine croak (J Hall-Craggs, P J Sellar: Fig VI); perhaps also belongs

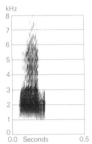

V C Lewis England June 1972

here, but could be rough variant of call 4. (c) Low-pitched, muffled 'arrr' or 'arrre' (perhaps only variant of call 3b) given by adults when young bird on ground approached by observers; developing at higher intensity into 'arrrihst', with marked pitch ascent during attacks (Hesse 1909; Voigt 1933). Loud, rather nasal 'kkhee' sounds becoming a squeal (Dementiev and Gladkov 1954*b*); perhaps same as higher-intensity variant. (d) A 'quiz' presumably of alarm on leaving bush and seeing observer (Mester 1957*a*); difficult to reconcile with other descriptions. (4) Coot-call. In recording by P A D Hollom, 'qwuk', or often like 'kow(k)', 'kut', or 'kewk' of Coot *Fulica atra*. Sometimes given in quite rapid series, some calls rather explosive and squeaky, others slightly wavering or disyllabic. Used as contact-call between pair-members and between parent and young (including while brooding); also once while nest-building (P A D Hollom, M G Wilson). (5) Dry staccato 'tuck-tuck-tuck' used in breeding season (Witherby *et al.* 1938*b*; Simms 1978); perhaps same as or related to call 4, but no details of context or function. (6) Soft 'seeih' like *T. iliacus* (Witherby *et al.* 1938*b*); no further details and not noted by Simms (1978). (7) Distress-call. Sharp scream (Simms 1978).

CALLS OF YOUNG. In recording by P A D Hollom, faint 'pseeoo' sounds given by small nestlings, developing into stronger, rapid cheeping at *c.* 1 week, and sounding like flock of Greenfinches *Carduelis chloris* at 9 days (M G Wilson). Hesse (1909) noted a frequently uttered 'pitepitepit' and, on sighting observer, a rattling 'pirrrr' (perhaps expressing alarm). Rattle given by recently fledged juvenile in recording by P A D Hollom thinner and squeakier than typical adult call 3a, rather liquid and 'notes' more clearly separated (M G Wilson). Plaintive and sharp food-calls given by fledglings (Raevel 1981). Disyllabic 'piiie' given by young bird fallen from nest (Hesse 1909), and juveniles will screech when handled (Peitzmeier 1962). Rather quiet and brief bursts of rudimentary song (2-4 in a row) given by juvenile *bonapartei* in Tien Shan (USSR), late August and late September (Kovshar' 1979).   MGW

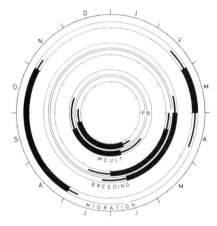

Breeding. SEASON. Western and central Europe: see diagram. Britain: earliest nests late February (Campbell and Ferguson-Lees 1972). North Africa: laying from late March (Heim de Balsac and Mayaud 1962). Finland: earliest nests late April, most begun in first half of May (Haartman 1969*a*). SITE. On stout branch against trunk of tree, or on fork of horizontal branch; also on ledge of building (including inside ruin), cliff-face, or bank. Earliest nests are more often in coniferous trees than deciduous (Campbell and Ferguson-Lees 1972). Mean height of 1137 nests, Finland, *c.* 5·0 m (0-20), but actual mean probably higher as lower nests easier to find (Haartman 1969*a*). Mean of 22 nests, Poland 9·5 m (1-21) (Bocheński 1968). Nest: large cup comprising 3 layers—outermost of sticks, grass, moss, and roots, loosely woven then compacted by middle layer of mud, often containing grass and leaves and some rotten wood which sometimes penetrates to outside of nest; thicker inner lining of fine grasses, sometimes with pine needles (Bocheński 1968). Mean outer diameter 18·5 cm (15·0-24·0), *n* = 9; inner diameter 10·2 cm (8·6-14·0), *n* = 12; height 9·3 cm (8·4-11·0), *n* = 9; depth of cup 6·1 cm (5·5-7·2), *n* = 9 (Bocheński 1968). Building: by ♀, often accompanied by ♂ according to Haartman (1969a); by both sexes according to Labitte (1952a); takes 8–13 days (Labitte 1952a), 4–5 days (*bonapartei*: Kovshar' 1966), or 7–8 days, though 2nd nest in 3 days (Ryves 1928). EGGS. See Plate 84. Sub-elliptical, smooth and glossy; pale blue, green-blue, or buff-tinted, spotted and blotched red-brown to purple, markings sometimes gathered at broad end. 30·2 × 22·3 mm (26·5–34·0 × 20·2–24·0), $n = 250$; calculated weight 7·8 g (Schönwetter 1979). Clutch: 3–5(–6). Of 137 clutches, Britain: 2 eggs, 2%; 3, 23%; 4, 61%; 5, 15%; mean 3·92 (Snow 1955b). Of 144 clutches, Belgium, 2 eggs, 3%; 3, 11%; 4, 62%; 5, 21%; 6, 3%; mean 4·08 (Verheyen 1953). 2(–3) broods, occasionally in same nest; replacements occur (Witherby *et al.* 1938b; Labitte 1952a; P A D Hollom). Eggs laid daily, in morning, starting fairly soon after completion of nest (Kovshar' and Gavrilov 1973) or 1 week or more after (Ryves 1928). INCUBATION. 12–15 days. By ♀; rarely assisted by ♂. Begins with last egg. During incubation, ♀ leaves nest 8–10 times daily for breaks of up to 20 min. (Niethammer 1937; Ryves 1928, 1943.) Hatching nearly synchronous. YOUNG. Cared for and fed by both parents. FLEDGING TO MATURITY. Fledging period 12–15 days. Become independent *c.* 2 weeks later (Simms 1978). ♂ continues to feed 1st brood while ♀ lays 2nd clutch (Makatsch 1976). Age of first breeding not known. BREEDING SUCCESS. In Britain, 40% of 435 nests produced at least 1 fledged young (Snow 1969a).

Plumages (nominate *viscivorus*). ADULT. Forehead, crown, mantle, and scapulars greyish olive-brown; back and rump similar but with a golden-brown wash; upper tail-coverts similar, but feathers edged whitish. Lores and feathers in front of and below eye greyish-white, distinct eye-ring buff-white; cheeks and ear-coverts pale buff with dark brown and black spots, forming a dark patch below rear corner of eye and blackish rear border of ear-coverts. Malar stripe sooty black. Chin and throat whitish or pale buff, with small sooty spots increasing in size towards cheeks and lower throat. Rest of underparts whitish with variable golden-buff wash (strongest on chest, flanks, and vent); chest with large and triangular sooty spots, similar but more rounded spots on remainder of underparts, almost coalescent on sides of breast; spots smaller and sparser on centre of belly. Under tail-coverts whitish, washed golden-buff. Flight-feathers and tertials grey-brown with pale olive-brown outer edge and narrow whitish edge to tip; base of inner web white. Upper wing-coverts grey-brown, median and greater edged and tipped pale buff or off-white, outer web of greater fringed buff. Greater upper primary coverts black-brown with pale olive-brown outer fringe. Axillaries, under wing-coverts, and marginal coverts at bend of wing white. Tail grey-brown, three outer feathers (t4–t6) with variable amount of white on tip of inner web and less extensive white on outer web; white most extensive on t6, usually forming a long wedge on inner web, less extensive on t5, and much reduced on t4. *In worn plumage*, upperparts paler and greyer, ground-colour of underparts whiter, less buff; fringes of flight-feathers and upper wing-coverts paler. NESTLING. Down buffish-white; plentiful and fairly long. JUVENILE. Like adult, but head noticeably pale, finely spotted and barred brown and with conspicuous blackish rear border to ear-coverts; feathers of upperparts with pale buff

centres and blackish tips, giving three-toned spotted pattern, most marked on mantle, inner scapulars, and back; chin and throat uniform white, spots on remainder of underparts small; secondaries, tertials, and all upper wing-coverts with broad buff outer edges; lesser, median, and inner greater upper wing-coverts with broad and contrasting buff shaft-streaks. FIRST ADULT. Like adult, but underparts usually deeper buff and juvenile flight-feathers, tail, tertials, greater upper primary coverts, 3–6(–9) outer greater upper secondary coverts, and occasionally a few feathers on body retained; retained outer greater coverts with off-white tip to outer web, tips wider and coverts shorter and more worn than new inner coverts (if any) which are edged light olive-buff; juvenile tail-feathers more narrowly pointed than adult. See also Svensson (1984a).

Bare parts. ADULT, FIRST ADULT. Iris dark brown. Bill dark horn-brown, yellowish at base of lower mandible. Leg and foot yellowish brown or light olive-brown. NESTLING. Mouth bright yellow, no spots; gape-flanges pale yellow. JUVENILE. Similar to adult.

Moults. ADULT POST-BREEDING. Complete; primaries descendant. Starts late May to late June, sometimes before breeding finished (Ginn and Melville 1983). Average duration of primary moult *c.* 90 days. Moult completed with p9–p10 and some secondaries mid-August to early October, mainly early September. (Snow 1969b; Ginn and Melville 1983.) POST-JUVENILE. Partial: head, body, lesser, median, and sometimes a number (usually 3–6) of inner greater upper wing-coverts. Start variable, depending on date of hatching; from May onwards, but late birds start August and continue into October.

Measurements. Nominate *viscivorus*. England, all year; skins (mainly BMNH). Juvenile wing is retained juvenile wing of 1st adult. Bill from basal corner of nostril; to skull, on average 9·7 longer, exposed culmen on average 4·4 longer.

	♂		♀	
WING AD	153·9 (4·43; 10)	150–162	150·6 (2·01; 10)	148–154
JUV	148·5 (1·84; 10)	147–151	149·5 (2·37; 10)	146–154
TAIL AD	113·8 (3·77; 10)	109–120	111·1 (3·84; 10)	107–120
BILL AD	15·9 (0·57; 10)	15–17	15·5 (0·85; 10)	14–17
TARSUS	33·4 (0·70; 10)	32–34	32·7 (0·82; 10)	32–34

Sex differences not significant.

Wing. Sweden: ♂ 149–164 (38), ♀ 142–162 (25) (Svensson 1984a). East and West Germany, Hungary, and France: 154·2 (3·69; 21) 149–162 (Stresemann 1920; ZMA). Netherlands: adult ♂, 158·2 (2·06; 14) 156–163; 1st adult ♂, 155·6 (3·08; 14) 150–164; adult ♀, 156·1 (2·91; 7) 151–160; 1st adult ♀, 154·4 (4·21; 8) 148–161 (ZMA). Southern Yugoslavia and northern Greece: ♂ 152·7 (3·59; 7) 150–160, ♀ 152·2 (2·50; 4) 149–155 (Stresemann 1920; Makatsch 1950). Turkey, Caucasus, and northern and south-western Iran: ♂ 153·5 (2·67; 8) 149–156, ♀♀ 150, 155·5, unsexed 155·3 (3·33; 6) 151–161 (Stresemann 1920; Paludan 1938, 1940; Kumerloeve 1961, 1968). Central Spain and Portugal: ♂ 149·8 (1·97; 12) 147–152, ♀ 149·6 (3·76; 9) 146–156 (Jordans 1950; ZMA).

T. v. deichleri. North-west Africa, ages combined; skins (BMNH).

	♂		♀	
WING	149·4 (2·13; 9)	147–154	149·4 (2·45; 8)	145–153
BILL	16·4 (0·58; 9)	15·5–17·0	16·3 (0·71; 8)	15·0–17·0

Sex differences not significant.

North-west Africa, wing: ♂ 154·8 (14) 150–160 (Vaurie 1959). Corsica, wing 152·6 (4·30; 8) 148–159 (Stresemann 1920; ZMA).

T. v. bonapartei. Wing. ♂ 168·5 (20) 162–176 (Vaurie 1959). Afghanistan: 165·0 (5·20; 5) 158–170 (Paludan 1959). USSR: 163·4 (35) 156–171 (Dementiev and Gladkov 1954*b*).

Weights. Nominate *viscivorus*. Great Britain, unsexed (BTO).

APR–AUG	119·2 (7·44; 16) 100–128	DEC–FEB	139·1 (10·9; 14) 128–167
SEP–NOV	127·4 (5·59; 15) 117–139	MAR	131·3 (10·3; 5) 117–148

Sexed birds, all geographical range: ♂ 112·4 (10·4; 14) 96–134, ♀ 123·2 (12·8; 5) 106–140 (Makatsch 1950; Willgohs 1955; Schüz 1959; Dolgushin *et al.* 1970; Eck 1970; ZMA). Turkey and Iran, April–August: 102·1 (6·97; 4) 94–108 (Paludan 1938, 1940; Kumerloeve 1961, 1968). Frost-killed birds, England and Wales: ♂ 73·9 (2·17; 6) 72–77, ♀ 68·0 (4·64; 5) 62–75 (Harris 1962; Ash 1964).

T. v. bonapartei. Kazakhstan (USSR), Afghanistan, and Mongolia, combined (Paludan 1959; Dolgushin *et al.* 1970; Piechocki and Bolod 1972).

NOV–FEB	♂ 134·2 (15·6; 5) 110–146	♀ 133·1 (18·7; 7) 93–149
MAR–APR	116·8 (15·6; 4) 100–130	123·3 (11·1; 6) 110–139
MAY–JUN	120·7 (7·5; 9) 105–130	127·8 (14·3; 10) 107–140
JUL–SEP	129·2 (14·2; 8) 110–150	128·0 (5·5; 5) 122–136

Structure. Wing rather long, broad at base, tip bluntly pointed. 10 primaries; in northern and central Europe, p8 longest, p9 3–8 shorter, p7 0–1, p6 3–7, p5 15–23, p4 27–37, p1 46–54; in southern Europe, p7 often longest or p7–p8 about equally long, p9 3–11 shorter, p6 1–6 shorter, p5 14–19, p4 26–31, p1 43–51. P10 reduced; 82–103 shorter than wing-tip, 7–18 shorter than longest upper primary covert. In northern and central Europe, outer web of p6–p8 (sometimes faintly p5) and inner of (p6–)p8–p9 emarginated; in southern Europe, outer web of p5 and inner of p6–p7 more often emarginated. Tail rather long, tip square or very slightly rounded; 12 feathers, t6 3–8 shorter than t2–t3. Middle toe with claw 27·3 (8) 24–31; outer toe with claw *c.* 73% of middle with claw, inner *c.* 70%, hind *c.* 76%. Remainder of structure as in Blackbird *T. merula* (p. 963), but bill relatively shorter and thicker and tarsus and toes relatively much shorter.

Geographical variation. Clinal throughout much of west Palearctic range; also considerable individual variation in plumage colour. Demarcation of races thus difficult and largely arbitrary, resulting in disagreement between authorities. Following Vaurie (1959) and Stepanyan (1978*a*), only nominate *viscivorus*, *deichleri*, and *bonapartei* recognized here. A main cline of decreasing saturation of plumage colours, including reduction of spotting on underparts, runs from north-west (Scotland) to east (western Siberia) and south-east (south-east Europe and Asia Minor). Races described along this line, most of which now considered invalid, include (in order of decreasing colour saturation): '*precentor*' Clancey, 1950, Scotland, Ireland, Isle of Man, and perhaps north-west England; nominate *viscivorus* (*sensu stricto*), south-east England, Denmark, and western Poland, south to northern Yugoslavia, mainland Italy, Sicily, and northern Spain; '*jordansi*' Eck, 1970, central and southern Spain and Portugal; '*jubilaeus*' Lucanus and Zedlitz, 1917, Fenno-Scandia, European USSR, and through Rumania south to Dalmatia and Greece; '*uralensis*' Zarudny, 1918, Urals and western Siberia. Size varies little in north of range (see Measurements), but south-eastern populations tend to be small and some races have been described partly on allegedly smaller size: '*bithynicus*' Keve, 1943, Aegean area, and '*loudoni*' Zarudny, 1912, south-east Transcaucasia. '*T. v. tauricus*' Portenko, 1954, from Crimea is also rather small (Vaurie 1959), but not enough to warrant separation from nominate *viscivorus* (Stepanyan 1978*a*). According to Vaurie (1959), *deichleri* of north-west Africa and '*reiseri*' Schiebel, 1911, from Corsica and Sardinia are not part of the continental cline. Both are paler and greyer than nominate *viscivorus* and hence both united under the name *deichleri*, though north-west African birds have longer and more attenuated bills, while those of Sardinia and Corsica have bill similar to nominate *viscivorus*. *T. v. bonapartei* from central Siberia south to western Himalayas, Kopet-Dag (Turkmeniya), and north-east Iran is markedly larger, paler, and greyer than nominate *viscivorus*; they probably intergrade in extreme east of west Palearctic and *bonapartei* perhaps occurs in Middle East in winter. DWS

Turdus migratorius **American Robin**

PLATES 71 and 77 [between pages 856 and 857]

DU. Roodborstlijster   FR. Merle migrateur   GE. Wanderdrossel
RU. Странствующий дрозд   SP. Robín americano   SW. Vandringstrast

Turdus migratorius Linnaeus, 1766

Polytypic. Nominate *migratorius* Linnaeus, 1766, Alaska (except south-east), Canada (except coastal British Columbia and area occupied by *nigrideus*), and north-east USA. Extralimital: *nigrideus* Aldrich and Nutt, 1939, northern Quebec, Labrador, and Newfoundland; *achrusterus* (Batchelder, 1900), south-east USA; *caurinus* (Grinnell, 1909), west coast of North America from south-east Alaska to north-west Oregon; 3 further races in south-west USA, Great Plains area, and Mexico.

Field characters. 25 cm; wing-span 35–39·5 cm. Length similar to that of Blackbird *T. merula* but head and body heavier. Quite large, robust thrush; mainly dark grey above and brick-red below, with bold but broken white eye-ring and white vent. Sexes somewhat dissimilar; no seasonal variation. Juvenile separable. 1 race vagrant to west Palearctic.

ADULT MALE. Largely dark grey above (olive-toned when fresh), but with almost black head and broken white eye-ring; larger wing-coverts show almost black centres and grey fringes, tertials blackish with narrow white fringes, and grey rump somewhat paler than almost black, narrowly white-tipped and -fringed tail. Chin and throat white, streaked black; breast to fore-belly and flanks brick-red, with white scaly feather-margins on flanks which reduce through wear; belly-centre to under tail-coverts white. Underwing pale brick-red. Bill yellow-horn, with dark tip; legs black-brown. ADULT FEMALE. As ♂, but browner above, with markings less obvious; throat less streaked and breast duller; flanks more broadly-scaled white. JUVENILE. Resembles ♀ but breast spotted black, while back mottled and streaked. Bill and legs dark horn. FIRST WINTER. Like dull adult, with head and back more olive and lower underparts more scaly.

Unmistakable, though in brief view, ♀ and immature might be confused with Eye-browed Thrush *T. obscurus*, or Naumann's Thrush *T. naumanni naumanni* (both noticeably smaller, with obvious supercilium and far less orange or red on underparts) or with Red-throated Thrush *T. ruficollis ruficollis* (which has red on face to chest and on tail, not from chest to vent). As befits a highly migratory species, flight powerful with action recalling both *T. merula* (at ground level) and Fieldfare *T. pilaris* (on migration). Gait typical of genus but nervous on ground and movements noticeably restless, often accompanied by tail- and wing-flicking. Behaviour like that of (migratory) *T. merula*; thus fairly confiding and aggressive where established but furtive and wild on migration. Vagrants occur mostly on islands and coasts but occasionally inland (in winter).

Calls similar in tone and structure to those of *T. merula*. Include challenging trisyllabic 'kwik kwik kwik' or 'tut tut tut' (given with jerk of tail); low-pitched 'tchook-tchook-tchook' and 'seech each-each-each' given in anxiety; quiet, disyllabic 'pit pit'; quiet, monosyllabic 'tseep' and 'sssp'.

Habitat. Probably the most fully known example of modern transformation from a natural habitat to one associated closely with man, recorded over *c.* 300 years in middle and lower-middle Nearctic temperate latitudes. Before European settlement, preponderant vegetation cover of primitive forests and long-grass prairies allowed only limited suitable habitat, principally in open northern woods, mountains at *c.* 1600–3700 m, small meadows at foot of conifer forests, burnt-over patches, and saline flats and other barren ground up to at least 1 km from nearest trees. In natural woodlands, wild fruits from undergrowth of open sections contribute at least half of diet. Forest clearance and supersession by farmland and human settlements has not only vastly expanded areas of mixed open land and bush or tree cover but has also expanded moist grassland on soft soils rich in invertebrate prey (above all earthworms) which flourish best on irrigated cultivation, watered lawns, and low-lying grassland near watercourses or standing water, conditions during spring and early summer being critical. Western races were adapted to heavy coniferous forest in high-rainfall regions, and to situations beyond treeline at heads of rivers and on high passes, as in Mount McKinley

National Park, Alaska (Murie 1963). Along Yukon river, frequents willow and alder thickets. In Canada, favours open and broken parts of forested areas, second growth after felling and burning, edges, and planted woodlands on farms with adjoining open fields (Godfrey 1966). Settlements with gardens containing grassy lawns and shrubberies, including suburbs and even inner cities, have not only been colonized but have become concentrations of higher density than was attainable in earlier natural habitat. In areas of high density, communal roosts occur in broad-leaved or sometimes mixed woodlands, swamps, and reedbeds, most often in winter but sometimes in summer, even in lilac thicket in a city garden. Recorded foraging in summer by retreating waves on an ocean beach, as well as on sand-dunes and salt-marshes and at fishing villages along shore. In winter in Mexico frequents open bushy forests of oak, pine, and cypress. (Bent 1949; Pough 1949.)

Distribution. Breeds over much of North America to treeline or just beyond and south to southern Mexico. Winters from southern Canada south to southern Florida and Guatemala.

Accidental. Iceland, Ireland, Britain, France, Belgium, West Germany, Norway, East Germany, Czechoslovakia, Austria.

Movements. Migratory over most of its range, scale and distance of migration varying with berry crops and severity of weather. Winters casually in much of breeding range, but primarily in USA (numbers increasing towards south) and in Bahamas, Cuba, Mexico, and Guatemala (American Ornithologists' Union 1983; Rappole *et al.* 1983). Alaskan birds leave breeding grounds in August (late records in early September), central Canada in September, and northern USA in October. Some extreme northern populations may not arrive in northern USA wintering areas until late December. (Bent 1949.) Present on Bermuda late October or early November to March (Wingate 1973), and on Bahamas and western Cuba late October to April (Bond 1961).

Late spring migrants leave Florida early to mid-April, but breeding birds arrive earlier in much of range: e.g. late February in central USA, early March in Vancouver Island and New England, mid- to end of March in North Dakota and Utah, mid-May in Newfoundland and Alaska. Ringing recoveries show general north-south movements channelled by southern USA coastline, some north-eastern breeders wintering as far west as Texas (Bent 1949; T Lloyd-Evans).   TL-E

Voice. See Field Characters.

Plumages (nominate *migratorius*). ADULT MALE. In fresh plumage, forehead, crown, sides of head, and nape deep black, feathers of hindcrown and nape with olive-grey fringes. Two white spots or lines above eye (formed by white posterior margin of lores and by rear part of upper eye-lid) and white line below eye (formed by white lower eye-lid). Rest of upperparts olive-grey, somewhat purer grey on rump and upper tail-coverts. Chin white. Throat white, feathers heavily streaked black especially at sides of throat, becoming almost all-black at sides of lower throat. Chest, flanks, and belly brick-red, feathers with narrow white fringes; vent white. Under tail-coverts white with grey lateral wedge-shaped markings. Tail sooty-black narrowly edged grey; outermost pair of feathers (t6) with white tip, most extensive on inner web, t5 with reduced white on inner web, t4 sometimes with narrow white tip. Upperwing dark brownish black, feathers edged olive-grey, as upperparts; lesser and median upper wing-coverts with narrow whitish or sometimes rufous tips, greater coverts with whitish tips, and tertials with narrow pale fringes. Axillaries and under wing-coverts brick-red. *In worn plumage*, head blacker and rest of upperparts greyer, as olive-grey feather-fringes wear off; underparts purer brick-red, white feather-fringes reduced or worn off. ADULT FEMALE. Similar to ♂, but generally much duller, grey of upperparts paler and browner, head browner with more extensively greyish fringes, and underparts paler, with broader white feather-fringes which do not fully wear off. JUVENILE. Crown and sides of head sooty-brown (nearly black in ♂, more olive in ♀); white spots round eye less well defined than in adult, sometimes pale buff or forming full buff eye-ring. Rest of upperparts dull brown, feathers with pale shaft-streaks and blackish tips. Chin and throat white or pale buff, with blackish stripes on cheek. Chest and sides of breast buff, variable in depth of colouring, usually deepening to rich tawny buff on flanks; feather-bases of chest and all central belly, and vent white. Chest, breast, and upper flanks marked with rounded sooty-black spots; smaller spots on vent and lower flanks. Tail as adult. Upperwing as adult, but lesser and median coverts with pale buff shaft-streak widening to pale triangular spot on tip, greater coverts with triangular whitish spot on tip, and tertials with more prominent white tip. FIRST ADULT. Sex for sex, similar to adult, but generally paler and duller, white feather-fringes more prominent below, and lesser wing-coverts with buff or olive tips. Juvenile flight-feathers, tail, tertials, greater upper primary coverts, and variable number of outer greater secondary coverts retained, browner than neighbouring new wing-coverts; outer greater coverts with more sharply defined white spot on tip of outer web, worn off in spring.

Bare parts. ADULT, FIRST ADULT. Iris dark brown. In breeding season, bill orange-yellow, except for dusky tip of culmen; in non-breeding season, mostly horn-colour, ridge of culmen blackish with yellow or orange-yellow confined to basal part of lower mandible. Mouth bright yellow. Leg and foot dark horn or blackish brown. JUVENILE. Iris dark brown. Bill, leg, and foot yellowish-brown at fledging, soon darkening to dark horn.

Moults. ADULT POST-BREEDING. Complete; primaries descendant. Mainly August-September. POST-JUVENILE. Partial: head, body, lesser and median upper wing-coverts, and variable number of inner greater upper wing-coverts. August-October, timing depending largely on date of hatching.

Measurements. Nominate *migratorius*. North-east USA and south-east Canada, mainly summer; skins (AMNH, BMNH). Juvenile wing includes retained juvenile wing of 1st adult. Bill to basal corner of nostril; to skull, on average 7·8 longer, exposed culmen on average 3·6 longer.

WING AD	♂ 131·2 (3·26; 11)	127–136	♀ 126·8 (3·88; 10)	122–132
JUV	128·4 (3·06; 10)	123–133	125·3 (3·65; 10)	120–131
TAIL AD	101·2 (3·74; 10)	96–108	96·9 (3·02; 9)	93–102
BILL AD	15·4 (0·53; 9)	15–16	15·2 (0·79; 10)	14–16
TARSUS	33·0 (1·05; 10)	31–34	32·0 (1·25; 10)	30–34

Sex differences significant for adult wing and tail.

Weights. Nominate *migratorius*, but migrants may include some *nigrideus*. Northern Alaska and north-west Canada, May-September: ♂ 77·0 (22) 59–91, ♀ 80·8 (8) 72–94 (Irving 1960). Coastal New Jersey (USA), autumn migrants: 72·2 (5·76; 49) 65–84 (Murray and Jehl 1964). Averages Ohio, April-October: ♂ 79·0 (15), ♀ 78·5 (7), juvenile 69·3 (37) (Baldwin and Kendeigh 1938). Alabama, February: ♂ 87 (Stewart and Skinner 1967). Kentucky (in part perhaps including some *achrusterus*): June-July, ♂ 71·2 (5·76; 7) 64–81, ♀ 77·2 (3) 66–34; September-October, ♂ 84·6 (2·71; 4) 82–87, ♀ 84·2 (4·73; 4) 80–89 (Mengel 1965). See also Manning (1982).

Structure. Wing rather long, broad at base, tip bluntly pointed. 10 primaries: p7 longest, p8 0–4 shorter, p9 4–11, p6 1–6, p5 7–13, p4 16–22, p1 30–37. P10 reduced; 72–83 shorter than p7, 2–10 shorter than longest upper primary covert. Outer web of (p5-)p6-p8 and (slightly) inner of (p6-)p7-p9 emarginated. Tail rather long, tip square or very slightly rounded; 12 feathers, t6 2–8 shorter than t2–t3. Middle toe with claw 26·8 (6) 25–28; outer toe with claw *c.* 70% of middle with claw, inner *c.* 66%, hind *c.* 69%. Structure closely similar to Blackbird *T. merula* (p. 963), including rather rounded wing-tip (in west Palearctic, shared only with Tickell's Thrush *T. unicolor* and Mistle Thrush *T. viscivorus*, wings of other *Turdus* more sharply pointed); bill relatively slightly shorter than *T. merula* (but relatively longer than other west Palearctic thrushes, except *T. unicolor*, Ring Ouzel *T. torquatus*, and Dusky Thrush *T. naumanni*), tarsus more distinctly shorter.

Geographical variation. Considerable. Mainly involves depth of pigmentation (including extent of white on outer tail-feathers) and size. Nominate *migratorius* from north and north-east of species' range is one of the darker races, though exceeded in depth of pigmentation by *nigrideus* of eastern Canada (in which black of head of ♂ extends to mantle) and by *caurinus* of west coast. Inland populations generally paler; western and southern ones virtually without white tips to outer tail-feathers. Size gradually increases towards west, but both in west and east gradually smaller towards south, races from south-east USA and southern Mexico smallest. An extremely pale form in Baja California, *confinis* Baird, 1864, has sometimes been given specific status. DWS

REFERENCES

ABDUSALYAMOV, I A (1973) *Fauna Tadzhikskoy SSR 19 Ptitsy* 2. Dushanbe. ABS, M (1963) *Bonn. zool. Beitr.* 14, 1-128; (1964) *Auspicium* 2, 87-8; (1970) In Grzimek, B (ed) *Grzimek's animal life encyclopaedia* 9, 171-81. New York. ACKLAM, G (1970) *Sterna* 9, 97-100. ADAMS, L E G (1957) *Bird Study* 9, 28-33. ADAMS, R G (1952) *Br. Birds* 45, 138; (1966) *Devon Birds* 19, 30. ADAMYAN, M S (1963*a*) *Ornitologiya* 6, 238-45; (1963*b*) *Izv. Akad. Nauk Armyan. SSR Biol. nauki* 16 (7), 69-83. ADOLFSSON, K (1984) *Anser* 23, 163-8. ADRIAENSEN, F and DHONDT, A A (1984) *Bird Study* 31, 69-75. AGATHO, B (1960-1) *Publ. natuurh. Genootsch. Limburg* 12, 97-175. AHARONI, J (1928) *Beitr. Fortpfl. Vögel* 4, 172-4; (1931) *Beitr. Fortpfl. Vögel* 7, 161-6. AHLÉN, I (1972) *Vår Fågelvärld* 31, 9-15. AICHHORN, A (1969) *Verh. dt. zool. Ges.* 32, 690-706. AIREY, A F (1951) *Br. Birds* 44, 287. ALABASTER, B H (1946) *Br. Birds* 39, 344. ALDER, J (1957) *Br. Birds* 50, 267-9; (1963) *Br. Birds* 56, 73-6;·(1966) *Birds* 1, 6-9. ALDER, L P and JAMES, C M (1951) *Br. Birds* 44, 280. ALDRICH, J W (1968) *Proc. US natn. Mus.* 124, 1-33. ALERSTAM, T (1974) *Vår Fågelvärld* 33, 44-8; (1976) *Oikos* 17, 457-75. ALEXANDER, B (1898*a*) *Ibis* (7) 4, 74-118; (1898*b*) *Ibis* (7) 4, 277-85. ALEXANDER, C J (1917) *Br. Birds* 11, 98-102. ALEXANDER, C J and ALEXANDER, H G (1908) *Br. Birds* 1, 367-72. ALEXANDER, H G (1924) *Br. Birds* 17, 304-5; (1927) *Ibis* (12) 3, 245-83. ALEXANDER, W B (1953) *J. Min. Agric.* 40, 8-12. ALEXANDER, W B and LACK, D (1944) *Br. Birds* 38, 42-5, 62-9, 82-8. ALI, S (1945) *The birds of Kutch.* Bombay; (1946) *J. Bombay nat. Hist. Soc.* 46, 286-308; (1949) *Indian hill birds.* Bombay; (1955) *J. Bombay nat. Hist. Soc.* 52, 735-802; (1962) *The birds of Sikkim.* Madras; (1968) *The book of Indian birds.* Bombay; (1969) *Birds of Kerala.* London; (1977) *Field guide to the birds of the eastern Himalayas.* Delhi. ALI, S and RIPLEY, S D (1971) *Handbook of the birds of India and Pakistan* 6; (1972) 5; (1973*a*) 8; (1973*b*) 9. Bombay. AL'KEMEIER, F (1985) *Ökol. Vögel* 7, 155. ALLEN, R C R (1947) *Br. Birds* 40, 213. ALLEYN, W F, BERGH, L M J VAN DEN, BRAAKSMA, S, HAAR, T J F A TER, JONKERS, D A, LEYS, H N, and STRAATEN, J VAN DER (1971) *Avifauna van Midden-Nederland.* Assen. ALLINSON, M D (1979) *Birds Cornwall* 48, 76. ALLOUSE, B E (1953) *Iraq nat. Hist. Mus. Publ.* 3. ALMÁSY, G von (1896) *Aquila* 3, 209-16. ALMOND, W E (1931) *Br. Birds* 25, 103; (1956) *Br. Birds* 49, 183. ALPERS, H (1942) *Beitr. Fortpfl. Vögel* 18, 34. AL-RAWY, M A and KAINADY, P V G (1976) *Bull. Basrah nat. Hist. Mus.* 3, 77-87. ALSOPP, E M P (1972) *Br. Birds* 65, 45-9. ALSTRÖM, P (1985) *Br. Birds* 78, 304-5. ALTHEN, J (1950-1) *Jber. Vogelk. Beob. Untermain* 24, 33-4. ALTUM, B (1866) *J. Orn.* 14, 423-6. AMADON, D and JEWETT, S G (1946) *Auk* 63, 541-59. AMANN, F (1949) *Orn. Beob.* 46, 187-90. AMERICAN ORNITHOLOGISTS' UNION (1957) *Check-list of North American birds*, 5th edn. Baltimore; (1983) *Checklist of North American birds*, 6th edn. Lawrence. AMES, P L (1971) *Bull. Peabody Mus. nat. Hist.* 37; (1975) *Bonn. zool. Beitr.* 26, 107-34. AMMERMANN, D (1975) *Anz. orn. Ges. Bayern* 14, 296-9. AMSLER, M (1933) *Avic. Mag.* (4) 11, 90. ANDERSON, E M (1915) *Condor* 17, 145-8. ANDERSON, R M (1909) *Auk* 26, 10-12. ANDERSSON, G (1963) *Vår Fågelvärld* 22, 290-2. ANDERSSON, I (1982) *Fåglar Närke* 5 (1), 4-19. ANDERSSON, I S and WESTER, S A L (1971) *Ornis scand.* 2, 75-9; (1972) *Ornis scand.* 3, 39-43; (1973) *Ornis scand.* 4, 153-6. (1975) *Fauna och Flora* 70, 253-65; (1976) *Vår Fågelvärld* 35, 279-86. ANDERSSON,

M (1981) *Ecology* 62, 538-44. ANDERSSON, M and WIKLUND, C G (1978) *Anim. Behav.* 26, 1207-12. ANDERSSON, M, WIKLUND, C G, and RUNDGREN, H (1980) *Anim. Behav.* 28, 536-42. ANDREW, D G, NELDER, J A, and HAWKES, M (1955) *Br. Birds* 48, 21-5. ANDREWS, P (1981) *Br. Birds* 74, 266-7. ANDRIS, K (1974) *Anz. orn. Ges. Bayern* 13, 95. ANGUS, D D (1980) *J. Gloucs. Nat. Soc.* 31, 393. ANNAN, O (1962) *Bird-Banding* 33, 130-7. ANON (1939) *Bull. Br. Orn. Club* 59, 145-9; (1946*a*) *Bull. Jerusalem Nat. Club* 2 (27), 3-4; (1946*b*) *Bull. Jerusalem Nat. Club* 2 (27), 5-6; (1952) *Br. Birds* 45, 373; (1959) *Oiseaux de France* 9 (1), 4-14; (1960) *Br. Birds* 53, 241-3; (1970*a*) In Gooders, J (ed) *Birds of the world*, 1784-5; (1970*b*) 1804-6. London; (1976) *Reader's Digest complete book of Australian birds.* Sydney; (1977) *Ardeola* 23, 231-5. APLIN, O V (1896) *Zoologist* (3) 20, 121-33; (1916) *Zoologist* (4) 20, 22-4. APPERT, O (1951) *Orn. Beob.* 48, 170-1; (1970) *Orn. Beob.* 67, 37-40. ARAGÜÉS, A and HERRANZ, A (1983) *Br. Birds* 76, 57-62. ARAÚJO, A and RUFINO, R (1981) *Migradores paleárticos nas Ilhas Selvagens.* Centro de Estudos de Migrações e Protecção de Aves, Lisbon. ARCHBOLD, M E (1971) *Bull. E. Afr. nat. Hist. Soc.*, 198-9. ARCHER, G and GODMAN, E M (1961*a*) *Birds of British Somaliland and the Gulf of Aden* 3; (1961*b*) 4. Edinburgh. ÅRESTRUP, W C and MØLLER, A P (1980) *Dansk orn. Foren. Tidsskr.* 74, 149-52. ARHEIMER, O (1973) *Vår Fågelvärld* 32, 1-10; (1978*a*) *Anser Suppl.* 3, 15-30; (1978*b*) *Anser Suppl.* 3, 31-46; (1978*c*) *Vår Fågelvärld* 37, 297-312; (1979) *Vår Fågelvärld* 38, 23-38; (1982) *Vår Fågelvärld* 41, 249-60; ARKHIPENKO, E V, PANOV, E N, and RASNITSYN, A P (1968) *Problemy evolyutsii* 1, 208-11. ARMITAGE, J (1935) *Br. Birds* 29, 94. ARMSTRONG, E A (1944) *Br. Birds* 38, 70-2; (1950) *Ibis* 92, 384-401; (1952) *Ibis* 94, 220-42; (1953*a*) *Br. Birds* 46, 37-50; (1953*b*) *Auk* 70, 127-50; (1954) *Ibis* 96, 1-30; (1955) *The Wren.* London; (1956) *Ibis* 98, 430-7. ARMSTRONG, E A and THORPE, W H (1952) *Br. Birds* 45, 98-101. ARMSTRONG, E A and WESTALL, P R (1953) *Ibis* 95, 143-5. ARMSTRONG, E A and WHITEHOUSE, H L K (1977) *Biol. Rev.* 52, 235-94. ARMSTRONG, T E and ROBERTS, B B (1956) *Polar Rec.* 8 (52), 4-12; (1958) *Polar Rec.* 9 (59), 90-6. ARNAULT, C (1926) *Oiseau* 7, 156-60; (1929) *Oiseau* 10, 477-9. ARNDT, H (1979) *Gef. Welt* 103, 48-50; (1981) *Falke* 28, 274-7. ARNHEM, R (1967) *Aves* 4, 117-22; (1969) *Gerfaut* 59, 77-109. ARNOLD, P and FERGUSON, W (1962) *The birds of Israel.* Haifa. ARNOULD, M, BARDIN, P, CANTONI, J, CASTAN, R, DELEUIL, R, and VIRÉ, F (1959) *Mém. Soc. Sci. nat. Tunisie* 4; (1961) *Oiseau* 31, 140-52. ARO, M (1968) *Ornis fenn.* 45, 16-18. ARRIGONI DEGLI ODDI, E (1902) *Manuale di Ornitologia italiana.* Milan. ARROYO, B and HERRERA, C M (1977) *Fauna Ibérica* 7, 106-9. ARVEY, M D (1951) *Univ. Kansas Mus. nat. Hist. Publ.* 3, 475-530. ASBIRK, S (1976) *Vidensk. Medd. dansk nat. Foren.* 139, 147-77. ASBIRK, S and FRANZMANN, N-E (1979) *Dansk orn. Foren. Tidsskr.* 73, 95-102. ASCHENBRENNER, L (1959) *Egretta* 2, 46-8. ASH, J S (1955) *Br. Birds* 48, 130-2; (1962) *Br. Birds* 55, 44-6; (1964) *Br. Birds* 57, 221-41; (1969) *Ibis* 111, 1-10; (1973) *Ibis* 115, 267-9; (1978) *Bird ringing in Ethiopia* 7. US natn. Mus. Washington; (1980) In Johnson, D N (ed) *Proc. Pan-Afr. orn. Congr.* 4, 199-208; (1981) *Scopus* 5, 35-6. ASH, J S, FERGUSON-LEES, I J, and FRY, C H (1967) *Ibis* 109, 478-86. ASH, J S and ROOKE, K B (1956) *Br. Birds* 49, 317-22. ASHBY, C B (1942) *Br. Birds* 35, 201-5. ASHFORD, R W (1970) *Bull. Niger. orn. Soc.* 77, 24-6. ASHFORD,

W J (1915) *Br. Birds* 9, 154; (1922) *Br. Birds* 15, 264-8. ASHMOLE, M J (1962) *Ibis* 104, 314-46, 522-59. ASHTON-JOHNSON, J F R (1961) *Ool. Rec.* 35, 33-9. ASPINWALL, D (1975) *Bull. Zambian orn. Soc.* 7 (2), 39-67. ASPINWALL, D R (1973) *Bull. Br. Orn. Club* 95, 46-8; (1977) *Bull. Zambian orn. Soc.* 9, 26-8. AUBRECHT, G (1978) *Egretta* 21, 61-8. AUSTIN, O L and KURODA, N (1953) *Bull. Mus. comp. Zool.* 109 (4). AVERIN, Y V and GANYA, I M (1970) *Ptitsy Moldavii* 1. Kishinev. AXELL, H E (1954) *Bird Notes* 26, 38-41. AXELL, H E, MAKEPEACE, P J, FFENNELL, H, and FFENNELL, J (1965) *Br. Birds* 58, 344-6. AZZOPARDI, J and BONETT, G (1980) *Il-Merill* 20, 26.

BAAKE, W (1982) *Falke* 29, 373-6. BAAN, G VAN DER (1953) *Levende Nat.* 10, 193-9. BABAEV, K (1967) *Ornitologiya* 8, 333. BABENKO, V G (1984) *Ornitologiya* 19, 171-2. BACHMANN, H (1983) *Vogel Umwelt* 2, 180-1. BACKHURST, D, PEARSON, D J, and RICHARDS, D K (1984) *Scopus* 8, 50-1. BAER, W (1909) *Orn. Monatsschr.* 34, 33-44. BAGENAL, T B (1958) *Bird Study* 5, 83-7. BAGGOTT, G K (1969) *Bristol Orn.* 1, 59-70; (1970) *Bird Study* 17, 45-6; (1973) *Chew Valley Ring. Stn. Rep.* 4, 18-21. BAGNALL-OAKELEY, R P (1955) *Br. Birds* 48, 36-8; (1961) *Br. Birds* 54, 39-41; (1968) *Br. Birds* 61, 313-14. BÄHRMANN, U (1950) *Vogelwelt* 71, 95. BAIER, E (1974a) *Gef. Welt* 98, 204-6; (1974b) *Orn. Mitt.* 26, 177-8; (1977) *Orn. Mitt.* 29, 44-5; (1978) *Orn. Mitt.* 30, 119-20. BAILLIE, S R and SWANN, R L (1980) *Ring. Migr.* 3, 21-6. BAILLY, J-B (1853) *Ornithologie de la Savoie* 2. Paris. BAIRLEIN, F, BECK, P, FEILER, W, and QUERNER, U (1983) *Ibis* 125, 404-7. BAKER, E C S (1924) *The fauna of British India: birds* 2; (1926) 3. London; (1934) *The nidification of birds of the Indian Empire* 3. London. BAKER, K (1977) *Bird Study* 24, 233-42. BAKER, R R (1970) *J. Zool. Lond.* 162, 43-59. (1982) *Migration: paths through time and space.* London; (1984) *Bird navigation: the solution of a mystery?* London. BALÁT, F (1960) *Zool. Listy* 9, 257-64; (1961) *Zool. Listy* 10, 135-46; (1962) *Zool. Listy* 11, 131-44; (1964a) *Zool. Listy* 13, 281-2; (1964b) *Zool. Listy* 13, 305-20; (1974) *Zool. Listy* 23, 343-56. BALCELLS, E and FERRER, F (1968) *Publ. Cent. Pirenaico Biol. exp.* 2, 153-7. BALDWIN, S P and KENDEIGH, S C (1938) *Auk* 55, 416-67. BALL, G F (1983) *Auk* 100, 998-1000. BALLANCE, D K and LEE, S L B (1961) *Ibis* 103a, 195-204. BAMBERG, O (1911) *Verh. int. orn. Kongr.* 5, 332-4. Berlin. BANNERMAN, D A (1914) *Ibis* (10) 2, 38-90, 228-93; (1919) *Ibis* (11) 1, 291-321; (1927) *Ibis* (12) 3 suppl.; (1936) *The birds of tropical West Africa* 4; (1939) 5; (1951) 8. London; (1953) *The birds of the British Isles* 2; (1954) 3. Edinburgh; (1963) *Birds of the Atlantic islands* 1. Edinburgh. BANNERMAN, D A and BANNERMAN, J W M (1953) *Trav. Inst. sci. chérifien Sér. Zool.* 10; (1958) *Birds of Cyprus.* London; (1965) *Birds of the Atlantic islands* 2; (1966) 3; (1968) 4. Edinburgh; (1971) *Handbook of the birds of Cyprus and migrants of the Middle East.* Edinburgh; (1983) *The birds of the Balearics.* London. BANNERMAN, D A and VELLA-GAFFIERO, J A (1976) *Birds of the Maltese archipelago.* Valletta. BARABASH-NIKIFOROV, I and SEMAGO, L L (1963) *Psitsy yugo-vostoka Chernozemnogo tsentra.* Voronezh. BARASH, D P (1975) *Wilson Bull.* 87, 367-73. BARBER, M (1975) *Avic. Mag.* 81, 191-5. BARBER-STARKEY, F (1909) *Br. Birds.* 3, 7-11. BARK JONES, R and HARTLEY, P H T (1945) *A list of the birds of Aden and the Aden Protectorate.* Unpubl. MS, Edward Grey Inst., Oxford Univ. BARLOW, J C, KLAAS, E E, and LENZ, J L (1963) *Condor* 65, 438-40. BARNARD, A (1963) *Br. Birds* 56, 295-6. BARNES, J (1985) *Br. Birds* 78, 513. BARRENTINE, C D (1980) *J. Fld. Orn.* 51, 368-71. BARRETT, J H (1947) *Br. Birds* 40, 150. BARRETT, J H, CONDER, P J, and THOMPSON, A J B (1948) *Br. Birds* 41, 162-6. BARTLETT, E (1970) *Br. Birds* 63, 179-80. BARTONEK, J C (1968) Ph D Thesis. Wisconsin Univ.

BARTONEK, J C and HICKEY, J J (1969) *Condor* 71, 280-90. BÄSECKE, K (1955) *Vogelwelt* 76, 61-4. BASSETT, B B (1953) *Br. Birds* 46, 379. BASSIN, P (1983) *Nos Oiseaux* 36, 333, 352. BATES, G L (1924) *Ibis* (11) 6, 648-719; (1933-4) *Ibis* (13) 3, 752-80, (13) 4, 61-79, 213-39, 439-66, 685-717; (1936-7) *Ibis* (13) 6, 531-56, 674-712, (14) 1, 47-65, 301-21; (1937) *Ibis* (14) 1, 786-830. BATES, R S P and LOWTHER, E H N (1952) *Breeding birds of Kashmir.* London. BATESON, P P G and NISBET, I C T (1961) *Ibis* 103a, 503-16. BATTEN, L A (1973) *Bird Study* 20, 251-8; (1976) *Proc. roy. Irish Acad.* 76B, 285-313; (1978) *Bird Study* 25, 23-32. BATTEN, L A, PRESTT, I, DENNIS, R H, and THE RARE BREEDING BIRDS PANEL (1979) *Br. Birds* 72, 363-81. BATTEN, T A (1943) *Ostrich* 13, 238-9. BAUDOIN, G (1980) *Alauda* 48, 153-4. BAUER, S and THIELCKE, G (1982) *Vogelwarte* 31, 183-391. BAUMANN, R (1951) *Nos Oiseaux* 21, 68-9. BAUMGART, W (1971) *Beitr. Vogelkde.* 17, 449-56. BAXTER, E V and RINTOUL, L J (1918) *Ibis* (10) 6, 247-87; (1937) *Scott. Nat.* 225, 93-101; (1947) *The migration of Waxwings into Scotland 1946.* Scott. Orn. Club Publ., Edinburgh; (1953) *The birds of Scotland* 1. Edinburgh. BAZIEV, Z K and CHUNIKHIN, S P (1963) *Ornitologiya* 6, 235-7. BEAL, F E L (1918) *US Dept. Agric. Bull.* 619. BEAMAN, M (1978) *Orn. Soc. Turkey Bird Rep. 1974-5.* Sandy. BEAMAN, M and KNOX, A G (1981) *Br. Birds* 74, 182-5. BEAMAN, M, PORTER, R F, and VITTERY, A (1975) *Orn. Soc. Turkey Bird Rep. 1970-3.* Sandy. BEASON, R C (1984) *J. Fld. Orn.* 55, 489-90. BEASON, R C and FRANKS, E C (1974) *Auk* 91, 65-74. BEAUDOIN, C (1976) *Alauda* 44, 77-90. BEAUFORT, DE (1950) *Livre rouge des espèces menacées en France* 1 *Vertébrés.* Secrétariat de la Faune et de Flore, Paris. BECKER, P (1974) *Beobachtungen an palaärktischen Zugvögeln in ihrem Winterquartier Südwestafrika.* Windhoek; (1984) *Mitt. orn. Ver. Hildesheim* 8, 158-61. BECKER, P, FEINDT, P, and ROSEMEYER, P (1964) *J. Orn.* 105, 352-4. BÉDÉ, P (1918) *Rev. fr. Orn.* 10, 224-5. BEDFORD, DUKE OF (1945) *Br. Birds* 38, 256. BEECHER, M D and BEECHER, I M (1979) *Science* 205, 1282-5. BEECHER, W J (1953) *Auk* 70, 270-333. BEENEN, H (1970) *Angew. Orn.* 3, 118-22. BEER, M, and GÖSSLING-BEDNAREK, A (1973) *Vogelwelt* 94, 21-6. BEER-HEINZELMANN, E (1956) *Orn. Beob.* 53, 113. BEESLEY, J S (1972) *J. E. Afr. nat. Hist. Soc.* 132, 1-30. BEHLE, W H (1942) *Univ. Calif. Publ. Zool.* 46, 205-316; (1943) *Auk* 60, 216-21. BEKLOVÁ, M (1972) *Zool. Listy* 21, 337-46; (1976) *Zool. Listy* 25, 147-55. BELCHER, C F (1930) *The birds of Nyasaland.* London. BELIK, V P (1977) *Ornitologiya* 13, 187-8. BELL, A P (1965) *Br. Birds* 58, 21. BEL'SKAYA, G S (1961) *Izv. Akad. Nauk Turkmen. SSR Ser. biol.* (3), 76-9; (1965a) *Izv. Akad. Nauk Turkmen. SSR Ser. biol.* (2), 64-73; (1965b) *Sbornik Novosti Orn.,* 23-4; (1973) *Izv. Akad. Nauk Turkmen. SSR Ser. biol.* (3), 56-61; (1974) *Fauna Ekol. ptits Turkmen.* 1, 18-33; (1979) *Izv. Akad. Nauk Turkmen. SSR Ser. biol.* (5), 7-19. BENEDEN, A VAN (1934) *Gerfaut* 24, 186-90; (1938) *Gerfaut* 28, 1-23; (1946) *Gerfaut* 36, 34; (1950) *Gerfaut* 40, 107-19. BENEDEN, A VAN and HUXLEY, J S (1951) *Br. Birds* 44, 127-8. BENGTSON, S-A (1970) *Bird Study* 17, 260-8; (1971) *Ornis fenn.* 48, 77-92; (1975) *Ornis fenn.* 52, 1-4; BENNETT, C J L (ed) (1977) *Cyprus orn. Soc.* (1957) *Rep.* 23; (1982) *Cyprus orn. Soc.* (1957) *Rep.* 26. BENSON, C W (1946a) *Ibis* 88, 25-48; (1946b) *Ibis* 88, 180-205. BENSON, C W and BENSON, F M (1977) *Birds of Malawi.* Limbe. BENSON, C W, BROOKE, R K, DOWSETT, R J, and IRWIN, M P S (1971) *The birds of Zambia.* London. BENSON, G B G and WILLIAMSON, K (1972) *Bird Study* 19, 34-50. BENT, A C (1942) *Bull. US natn. Mus.* 179; (1948) *Bull. US natn. Mus.* 195; (1949) *Bull. US natn. Mus.* 196; (1950) *Bull. US natn. Mus.* 197. BENTZIEN, D (1983) *Hamburger avifaun. Beitr.* 19, 135-8. BERCK, K H (1953) *Vogelring* 22, 17-18; (1961) *Vogelwelt* 82, 109-12.

BEREGOVOY, V E (1970) *Zool. Zh.* **49**, 898-902. BÉRES, J and MOLNÁR, P (1964) *Aquila* **69-70**, 57-70. BERETZK, P (1967) *Aquila* **73-4**, 197-8. BERG, A B VAN DEN (1980) *Dutch Birding* **1**, 22-3; (1984) *Dutch Birding* **6**, 102-5. BERG, M VAN DEN (1975) *Limosa* **48**, 158-62. BERG, M VAN DEN and OREEL, G J (1985) *Br. Birds* **78**, 176-83. BERGER, A J (1953) *Bird-Banding* **24**, 19-20. BERGER, M and KIPP, M (1966) *Natur Heimat* **26**, 52-61. BERGER, W (1960) *Falke Suppl.* **4**, 20-7; (1967) *J. Orn.* **108**, 320-7. BERGMAN, G (1948) *Vår Fågelvärld* **7**, 57-67. BERGMAN, S (1935) *Zur Kenntnis nordostasiatischer Vögel.* Stockholm. BERGMANN, H-H (1977a) *Publ. Wiss. Film Sekt. Biol.* (10) 8/E2255. Göttingen; (1977b) *Publ. Wiss. Film Sekt. Biol.* (10) 10/E2257. Göttingen; (1977c) *Publ. Wiss. Film Sekt. Biol.* (10) 11/E2258. Göttingen; (1977d) *Publ. Wiss. Film Sekt. Biol.* (10) 14/E2261. Göttingen; (1983) *Cyprus orn. Soc.* (1969) *Rep.* **8**, 41-54. BERGMANN, H-H and HELB, H-W (1982) *Stimmen der Vögel Europas.* Munich. BERGMANN, H-H and WEISS, J (1974) *Z. Tierpsychol.* **35**, 403-17. BERG-SCHLOSSER, G (1975) *Anz. orn. Ges. Bayern* **14**, 273-95; (1979) *Anz. orn. Ges. Bayern* **18**, 23-35. BERLIOZ, J (1950) *Bull. Mus. Hist. nat. Paris* (2) **22**, 209-11. BERMAN, D I and KOLONIN, G V (1967) *Ornitologiya* **8**, 267-73. BERMAN, D I and ZABELIN, V I (1963) *Ornitologiya* **6**, 153-60. BERNARD, P (1904) *Ornis* **12**, 565-7. BERNDT, R (1933) *Beitr. Fortpfl. Vögel* **9**, 138; (1938) *Beitr. Fortpfl. Vögel* **14**, 30-1; (1944) *Ber. Ver. schles. Orn.* **29** (1/4), 28-34; (1958) *Vogelwarte* **19**, 211-12; (1962) *Vogelwelt* **83**, 70-4; (1982) *Vogelwelt* **103**, 189-90. BERNDT, R and BERNDT, A (1942) *Beitr. Fortpfl. Vögel* **18**, 130-4. BERNDT, R and TAUTENHAHN, W (1951) *J. Orn.* **93**, 64-5. BERNDT, R and WINKEL, W (1979) *Vogelwelt* **100**, 55-69. BERNIS, F (1965) *Ardeola* **11**, 158-9; (1971) *Aves Migradoras Ibericas* **7-8**. Madrid. BERNIS, F, GARCÍA RODRÍGUEZ, L, and JUANA, E DE (1973) *Ardeola* **19**, 29-30. BERTHET, G (1947) *Alauda* **15**, 257-8. BERTHOLD, P (1955) *J. Orn.* **96**, 421; (1976) *J. Orn.* **117**, 145-209; (1977) *J. Orn.* **118**, 204-5; (1983) *J. Orn.* **124**, 117-31. BESER, H J (1974) *Charadrius* **10**, 81-90. BESHIR, E S A (1978) *Proc. vert. Pest Conf.* **8**, 220-3. BESSON, J (1966) *Alauda* **34**, 70-2; (1972) *Alauda* **40**, 101. BEST, J R (1977) *Bull. E. Afr. nat. Hist. Soc.*, 39-40. BEVEN, G (1959) *London Nat.* **38**, 64-73; (1963) *Br. Birds* **56**, 307-23; (1964) *London Nat.* **43**, 86-109; (1968) *Br. Birds* **61**, 303-7; (1969) *Br. Birds* **62**, 23-5; (1970) *Br. Birds* **63**, 294-9; (1976) *London Nat.* **55**, 64-73; (1980) *Br. Birds* **73**, 35-6. BEYER, L K (1938) *Wilson Bull.* **50**, 122-37. BEZZEL, E (1961) *Vogelwelt* **82**, 184-5; (1966) *Anz. orn. Ges. Bayern* **7**, 847-54; (1975) *J. Orn.* **116**, 488-9. BEZZEL, E, LECHNER, F, and RANFTL, H (1980) *Arbeitsatlas der Brutvögel Bayerns.* Greven. BEZZEL, E, and LÖHRL, H (1972) *Anz. orn. Ges. Bayern* **11**, 282-7. BEZZEL, E and STIEL, K (1977) *Anz. orn. Ges. Bayern* **16**, 1-9. BIBBY, C J (1978) *Bird Study* **25**, 87-96; (1979) *J. Zool. Lond.* **188**, 557-76. BIBER, O (1973) *Orn. Beob.* **70**, 147-56; (1978) *Jb. naturhist. Mus. Stadt Bern* **6**, 133-42; (1984) *Orn. Beob.* **81**, 1-28. BIBER, J-P and LINK, R (1975) *Nos Oiseaux* **33**, 149-53. BIBIKOV, D I and BIBIKOVA, V A (1955) *Zool. Zh.* **34**, 399-407. BIJLSMA, R G (1977) *Limosa* **50**, 127-36; (1978a) *Limosa* **51**, 107-21; (1978b) *Limosa* **51**, 122-31; (1978c) *Veldorn. Tijdschr.* **1**, 126-35; (1982) *Vogelwarte* **31**, 423-7. BILBY, H A (1957) *Bull. Br. Orn. Club* **77**, 5-7. BILLEN, G and TRICOT, J (1977) *Aves* **14**, 101-13. BIRCHER, R K (1980) *J. Gloucs. Nat. Soc.* **31**, 405. BIRD, C G and BIRD, E G (1935) *Ibis* (13) **5**, 837-55. BIRKHEAD, M E (1981) *Ibis* **123**, 75-84. BISHTON, G (1984) *Br. Birds* **77**, 486-7; (1985) *Bird Study* **32**, 113-15. BJÄRVALL, A (1967) *Vår Fågelvärld* **24**, 294-300. BJERKE, T (1971) *Sterna* **10**, 97-116; (1974) *Sterna* **13**, 65-76. BJERKE, T, ESPMARK, Y, and FONSTAD, T (1985) *Ornis scand.* **16**, 14-19. BJERKE, T K (1980) *Fauna Norveg.* (C) *Cinclus* **3**, 73-9. BJERKE, T K and BJERKE, T H

(1981) *Ornis scand.* **12**, 40-50. BJÖRKLUND, S (1978) *Vår Fågelvärld* **37**, 59-62. BJÖRNFORS, G and GÖTMARK, F (1981) *Fåglar Västkosten* **15**, 16-25. BLAIR, H M S (1936) *Ibis* (13) **6**, 280-308, 429-59, 651-74. BLAKE, K F (1980) *Hobby* **6**, 47. BLAKERS, M, DAVIES, S J J F, and REILLY, P N (1984) *The atlas of Australian birds.* Melbourne. BLANC, T (1956) *Nos Oiseaux* **23**, 165-8. BLANCHARD, D H (1969) *Ethiopia: its culture and its birds.* San Antonio. BLANCHET, A (1923) *Rev. fr. Orn.* **15**, 83-7; (1951) *Mém. Soc. Sci. nat. Tunisie* **1**; (1955) *Mém. Soc. Sci. nat. Tunisie* **3**. BLANFORD, W T (1876) *Eastern Persia* **2**. London. BLANKENAGEL, H-J and SEITZ, B-J (1983) *Ökol. Vögel* **5**, 217-30. BLASZYK, P (1963) *J. Orn.* **104**, 168-81; (1972) *Vogelkdl. Ber. Niedersachs.* **4**, 51-2. BLASZYK, P and STEINBACHER, G (1954) *Bonn. zool. Beitr.* **5**, 49-67. BLEM, C R (1979) *Wilson Bull.* **91**, 135-7. BLOCH, D and SØRENSEN, S (1984) *Yvirlit yvir Føroya fuglar.* Tórshavn. BLONDEL, J (1962a) *Terre Vie* **3**, 209-51; (1962b) *Alauda* **30**, 1-29; (1963) *Alauda* **31**, 22-6; (1967) *Alauda* **35**, 83-105, 163-93. BLUME, D (1966) *Luscinia* **39**, 65-8; (1967) *Ausdrucksformen unserer Vögel.* Wittenberg Lutherstadt. BLÜMEL, H (1980) *Falke* **27**, 104. BOASE, H (1926) *Br. Birds* **20**, 20-2; (1952) *Br. Birds* **45**, 317-20. BOCHENSKI, Z (1968) *Acta Zool. Cracov.* **13**, 349-440. BOCHENSKI, Z, OLES, T, and TOMEK, T (1981) *Acta Zool Cracov.* **25**, 13-32. BOCK, A, MESTER, H, and PRÜNTE, W (1961) *J. Orn.* **102**, 228-30. BOCK, W J (1962) *Auk* **79**, 425-43. BOCK, W J and FARRAND, J (1980) *Amer. Mus. Novit.* **2703**. BODENHAM, K L (1944) *Bull. zool. Soc. Egypt* **6**, 26-32; (1945) *Bull. zool. Soc. Egypt* **7**, 21-47. BODENSTEIN, G (1953) *Orn. Mitt.* **5**, 191. BOEHME, I R (1982) In Merauskas, P (ed) *Ekologicheskie issledovaniya i okhrana ptits Pribaltiyskikh respublik*, 102-4. Kaunas. BOEHME, R L (1958) *Uchen. zap. Severo-Osetinsk. gos. ped. Inst.* **23** (1), 111-83; (1959) *Ornitologiya* **2**, 208. BÖHM, W and STROHKORB, O (1964) *Beitr. Vogelkde.* **10**, 235-6. BÖHME, I R and BÖHME, R L (1986) *Zool. Zh.* **65**, 378-86. BÖHR, H-J (1960) *Vogelwelt* **81**, 151-2; (1962) *Bonn. zool. Beitr.* **13**, 50-114. BÓKAI, B (1955) *Aquila* **59-62**, 459-60; (1957) *Aquila* **63-4**, 354. BOLEY, A (1932) *Vogelzug* **3**, 17-21. BOLLE, C (1857) *J. Orn.* **5**, 258-92. BOLSTER, R C (1922) *J. Bombay nat. Hist. Soc.* **28**, 1132. BOND, G M (1963) *Proc. US natn. Mus.* **114**, 373-87. BOND, J (1961) *Birds of the West Indies.* Boston. BONHOTE, J L (1910) *Br. Birds* **4**, 116-17. BOOTH, C G (1967) *Br. Birds* **60**, 420. BORMAN, F W (1928) *Bird Notes News* **13**, 1-4, 27-30. BOROVITSKAYA, G K (1972) *Ornitologiya* **10**, 328-9. BOROWSKI, S (1966) *Przeglad Zool.* **10**, 62-4. BORRETT, R P (1973) *Ostrich* **44**, 145-8. BORRETT, R P and JACKSON, H D (1970) *Bull. Br. Orn. Club* **90**, 124-9. BORRETT, R P and WILSON, K J (1970) *Ostrich Suppl.* **8**, 333-41. BORRMANN, K (1974) *Falke* **21**, 67. BORTOLI, L, CANTONI, J, DELABRE, J, and SCHOENENBERGER, A (1962) *Cent. Baguage Tunis* **4**. BOSCH, G (1947) *Limosa* **20**, 234. BÖSENBERG, K (1960) *Prob. Angew. Orn.* **30**, 53-62. BOSMANS, R and MOREAUX, F (1977) *Gerfaut* **67**, 395-412. BOSTANZHOGLO, B N (1911) *Mat. pozn. fauny flory Ross. Imp. otd. zool.* **11**, 1-410. BOSWALL, J H R (1952) *Skokholm Bird Obs. Rep. 1951*, 28-31; (1964) *Br. Birds* **57**, 183-4; (1966a) *Br. Birds* **59**, 100-6; (1966b) *Urban roosting by the Pied Wagtail.* Unpubl. MS, Edward Grey Inst, Oxford Univ.; (1966c) *Bull. Br. Orn. Club* **86**, 131-40. BOSWELL, C and NAYLOR, P (1957) *Iraq nat. Hist. Mus. Publ.* **13**, 16. BOTTOMLEY, S (1978) *Br. Birds* **71**, 84. BOUBIER, M (1925) *Monographie biologique du Rossignol du muraille.* Basel. BOULDIN, L E (1959) *Br. Birds* **52**, 141-9; (1968) *Bird Study* **15**, 135-46; (1971) *Bird Life* **7**, 80-4. BOURNE, W R P (1955) *Ibis* **97**, 508-56; (1959) *Ibis* **101**, 170-6. BOXBERGER, L VON (1934) *J. Orn.* **82**, 185-209. BOYD, A W (1933) *Br. Birds* **26**, 255-6; (1935) *Br. Birds* **29**, 3-21; (1936) *Br. Birds* **30**, 98-116; (1945) *Br. Birds* **38**, 353-4; (1946) *Br.*

Birds **39**, 88; (1949) *Br. Birds* **42**, 244. BOYD, A W and THOMSON, A L (1937) *Br. Birds* **30**, 278-87. BOYLE, G (1970) *Br. Birds* **63**, 178. BRACKENBURY, J (1978) *Ibis* **120**, 526-8. BRANDT, H (1963) *Anz. orn. Ges. Bayern* **6**, 546-50. BREHM, A E (1876) *Gefangene Vögel*. Leipzig. BREMOND, J-C (1966) Doct. Sci. Nat. Thesis. Faculté des Sciences de Paris; (1967) *Proc. int. orn. Congr.* **14**, 217-29. Oxford; (1968) *Terre Vie* **22**, 109-220. BREWER, A D (1972) *Br. Birds* **65**, 359. BRICHE, L (1962) *Alauda* **30**, 68-9. BRICHETTI, P (ed) (1982) *Riv. ital. Orn.* **52**, 3-50. BRIGGS, K B (1984) *Br. Birds* **77**, 569. BRINK, J N VAN DEN (1936) *Org. Club nederl. Vogelk.* **9**, 122-30. BRITISH ORNITHOLOGISTS' UNION (1971) *The status of birds in Britain and Ireland*. Oxford. BRITTON, P L (ed) (1980) *Birds of East Africa*. Nairobi. BRITTON, P L and BRITTON, H A (1977) *Scopus* **1**, 109-11. BRITTON, P L and DOWSETT, R J (1969) *Ostrich* **40**, 55-60. BROAD, R A (1977) *Scott. Birds* **9**, 301-2. BROAD, R A, HOLMES, P, and ROBERTSON, I S (1973) *Scott. Birds* **7**, 263-4. BROADBENT, J (1969) *Bull. Niger. orn. Soc.* **6**, 33. BROCK, J (1970) In Gooders, J (ed) *Birds of the world* **6**, 1793-4. London. BROEKHUYSEN, G J (1952) *Ostrich* **23**, 134-5; (1953) *Ostrich* **24**, 148-152; (1961) *Bokmakierie* **13**, 5-6; (1964) *Ardea* **52**, 140-65. BROEKHUYSEN, G J and BROWN, A R (1963) *Ardea* **51**, 25-43. BROEKHUYSEN, G J and STANFORD, W P (1954) *Ostrich* **25**, 99. BROMBACH, H (1982) *Vogelwelt* **103**, 153; (1984) *Vogelwelt* **105**, 105-9. BRONSART, G von (1954) *Vogelwelt* **75**, 239. BROOK, A (1912) *Br. Birds* **5**, 294-6. BROOKE, M de L (1979) *J. Anim. Ecol.* **48**, 21-32; (1981) *J. Anim. Ecol.* **50**, 683-96; (1983) *Anim. Behav.* **31**, 304-5; BROOKE, R K (1956) *Ostrich* **27**, 88; (1971) *Honeyguide* **66**, 19-26; (1972) *Bull. Br. Orn. Club* **92**, 53-7; (1973) *Ibis* **115**, 606; (1974) M. Sc. Thesis. Natal Univ; (1975) *Honeyguide* **81**, 19-21. BROOKE, R K and VERNON, J C (1961a) *Ostrich* **32**, 51-2; (1961b) *Ostrich* **32**, 128-33. BROOKS, E A (1908) *Auk* **25**, 235-6. BROOKS-KING, M (1946) *Br. Birds* **39**, 179. BROOM, D M, DICK, W J A, JOHNSON, C E, SALES, D I, and ZAHAVI, A (1976) *Bird Study* **23**, 267-79. BROSSET, A (1956) *Alauda* **24**, 161-205; (1957) *Alauda* **25**, 196-208; (1959) *Alauda* **27**, 36-60; (1961) *Trav. Inst. sci. chérifien Sér. Zool.* **22**, 7-155; (1971a) *Biol. Gabonica* **7**, 423-60; (1971b) *Alauda* **39**, 127-31. BROUWER, H and DAALDER, R (1982) *Vogeljaar* **30**, 57-62. BROWN, A J (1986a) *Br. Birds* **79**, 136. BROWN, B J (1986b) *Br. Birds* **79**, 221-7. BROWN, J L (1978) *Ann. Rev. Ecol. Syst.* **9**, 123-55. BROWN, K (1948) *Br. Birds* **41**, 214. BROWN, L H and NEWMAN, K B (1974) *Ostrich* **45**, 194-5. BROWN, P E and GOODYEAR, R T (1948) *Br. Birds* **41**, 387. BROWNE, P W P and HARLEY, B H (1953) *Br. Birds* **46**, 265. BROWN, R G B (1963) *Ibis* **105**, 63-75. BROWN, R H (1924) *Br. Birds* **17**, 183-4; (1947) *Br. Birds* **40**, 52. BROWNE, K, and BROWNE, E (1956) *Br. Birds* **49**, 241-57. BROWNE, P W P (1950) *Ibis* **92**, 52-65; (1982) *Malimbus* **4**, 69-92. BRUCE, S (1948) *Scott. Nat.* **60**, 6-7. BRUCH, A, ELVERS, H, POHL, C, WESTPHAL, D, and WITT, K (1978) *Orn. Ber. Berlin (West)* 3 suppl. BRUCKER, J W (1976) *Oxford orn. Soc. Rep.* 1975, 43-8. BRUDERER, B (1979) *Orn. Beob.* **76**, 293-304. BRUDERER, B and MUFF, J (1979) *Orn. Beob.* **76**, 229-34. BRUDERER, B and HIRSCHI, W (1984) *Orn. Beob.* **81**, 285-302. BRUHN, J F W and JEFFREY, B (1958) *Ardeola* **4**, 109-17. BRUNS, H (1957) *Orn. Mitt.* **9**, 241-53; (1959) *Orn. Mitt.* **11**, 57. BRUNS, H and HEINRICH, M (1968) *Orn. Mitt.* **20**, 117-33. BRUSEWITZ, G (1980) *Wings and seasons*. Stockholm. BRYANT, D M (1973) *J. Anim. Ecol.* **42**, 539-64; (1975a) *Ring. Migr.* **1**, 33-6; (1975b) *Ibis* **117**, 180-216; (1978a) *Ibis* **120**, 16-26; (1978b) *Ibis* **120**, 271-83; (1979) *J. Anim. Ecol.* **48**, 655-75. BRYANT, D M and TURNER, A K (1982) *Anim. Behav.* **30**, 845-56. BRYANT, D M and WESTERTERP, K R (1980a) *Ardea* **68**, 91-102; (1980b) *Proc. int. orn. Congr.* **17**, 292-9. Berlin. BRZOZOWKSI, A (1984) *Notatki orn.* **25**, 83-4. BUB, H (1955)

Ibis **97**, 25-37; (1963) *Vogelwarte* **22**, 85-93; (1975a) *Vogelkdl. Ber. Niedersachs.* **7**, 85-8; (1975b) *Orn. Mitt.* **27**, 58-61; (1977) *Beitr. Naturkde. Niedersachs.* **30**, 46. BUB, H, ECK, S, and HERROELEN, P (1981) *Stelzen, Pieper und Würger*. Wittenberg Lutherstadt. BUB, H, ECK, S, HERROELEN, P, LIEDEL, K, NOLL, W, STORSBERG, K, and WINKLER, R (1984) *Seidenschwanz, Wasseramsel, Zaunkönig, Braunellen, Spötter, Laubsänger, Goldhähnchen*. Wittenberg Lutherstadt. BUB, H and HERROELEN, P (1981) *Lerchen und Schwalben*. Wittenberg Lutherstadt. BUB, H and KLINGS, M (1968) *Auspicium* **3**, 69-95. BUCHET, R and JOUGLEUX, C (1979) *Héron* **1**, 66. BULMAN, J F H (1942) *Bull. zool. Soc. Egypt* **4**, 5-12. BUNDY, G (1976) *The birds of Libya*. Br. Orn. Union, London; (1985) *Br. Birds* **78**, 93-5; (1986) *Sandgrouse* **7**, 43-6. BUNDY, G and MORGAN, J H (1969) *Bull. Br. Orn. Club* **89**, 139-44, 151-9; BUNDY, G and SHARROCK, J T R (1986) *Br. Birds* **79**, 120-3. BUNDY, G and WARR, E (1980) *Sandgrouse* **1**, 4-49. BUNN, D S (1963) *Br. Birds* **56**, 152-3. BUNNI, M K and SIMAN, H Y (1978) *Bull. nat. Hist. Res. Cent. Baghdad* **7** (2), 21-6; (1979) *Bull. nat. Hist. Res. Cent. Baghdad* **7** (3), 73-82. BUNYARD, P F (1936) *Bull. Br. Orn. Club* **56**, 108-9. BÜRGER, P (1977) *Acta Sci. Nat. Mus. Bohem. merid. České Budějovice* **17**, 67-88. BURKHARD, S (1985) *Die Amsel*. Wittenberg Lutherstadt. BURKITT, J P (1924a) *Br. Birds* **17**, 294-303; (1924b) *Br. Birds* **18**, 97-103; (1925a) *Br. Birds* **18**, 250-7; (1925b) *Br. Birds* **19**, 120-4; (1926) *Br. Birds* **20**, 91-101; (1935) *Br. Birds* **28**, 364-7. BÜRKLI, W (1974) *Orn. Beob.* **71**, 172; (1977) *Orn. Beob.* **74**, 75-7; (1983) *Orn. Beob.* **80**, 295. BURLEIGH, T D (1942) *Occ. Pap. Mus. Zool. Louisiana State Univ.* **11**, 179-83. BURRI, H (1960) *Orn. Beob.* **57**, 158-9. BURTON, J F (1970) *Br. Birds* **63**, 85. BURTON, P J K (1971) *Bull. Br. Orn. Club* **91**, 108-9. BURTON, R (1985) *Bird behaviour*. London. BURTT, E H, JR (1977) *Anim. Behav.* **25**, 231-9. BUSCHE, G (1982) *Orn. Mitt.* **34**, 185-96. BUSCHE, G and MEYER, D (1978) *Vogelwarte* **29**, 254-61. BUSSE, H (1980) *Beitr. Vogelkde.* **26**, 362-3. BUSSE, P (1972) *Acta Orn.* **13**, 193-241; BUSSMANN, J (1940) *Orn. Beob.* **37**, 51-9. BUTLER, A L (1905) *Ibis* (8) **5**, 301-401. BUTLIN, S M (1940) *Br. Birds* **34**, 108-9. BUTTERFIELD, A (1953) *Br. Birds* **46**, 65-6. BUTURLIN, S A (1929) *Sistematicheskie zametki o ptitsakh Severnogo Kavkaza*. Makhachkala. BUXTON, E J M (1946) *Br. Birds* **39**, 73-6; (1947) *Br. Birds* **40**, 50; (1961) *Br. Birds* **54**, 432-3; (1975) *Br. Birds* **68**, 299-300. BUXTON, J (1945) *Quart. J. For.* **39**, 103-9; (1950) *The Redstart*. London. BUXTON, P A (1921) *J. Bombay nat. Hist. Soc.* **27**, 844-82.

CALDWELL, H R and CALDWELL, J C (1931) *South China birds*. Shanghai. CAMERON, E S (1908) *Auk* (2) **25**, 39-56. CAMERON, R A D (1969) *J. Anim. Ecol.* **38**, 547-53. CAMERON, R A D and CORNWALLIS, L (1966) *Ibis* **108**, 284-7. CAMPBELL, B (1953) *Rep. Trans. Cardiff Nat. Soc.* **81**, 4-65; (1980) *Br. Birds* **73**, 366. CAMPBELL, B and FERGUSON-LEES, J (1972) *A field guide to birds' nests*. London. CAMPBELL, K (1971) *Bird Life* **7**, 36. CAMPBELL, L W (1932) *Wilson Bull.* **44**, 118-19. CANO, A (1960) *Ardeola* **6**, 320-3. CARLO, E A DI (1972) *Riv. ital. Orn.* **42**, 1-160; (1983) *Gli Uccelli Ital.* **8**, 137-9. CARLSON, A, HILLSTRÖM, L, and MORENO, J (1985) *Ornis scand.* **16**, 113-20. CARMAN, T (1973) *Bird Life* Jan-Mar, 35. CARR, D (1968) *Br. Birds* **61**, 416-17. CARRUTHERS, D (1910) *Ibis* (9) **4**, 436-75; (1949) *Beyond the Caspian: a naturalist in central Asia*. Edinburgh. CARTER, F E (1961) *Br. Birds* **54**, 245. CARY, P (1973) *A guide to birds of southern Portugal*. Lisbon. CASATI, A (1937) *Riv. ital. Orn.* (2) **7**, 225-36, 285-91. CASEMENT, M B (1966) *Ibis* **108**, 461-91. CASSIDY, M (1971) *Bird Life* **7**, 35-6. CASTAN, R (1958) *Alauda* **26**, 56-62. CATLEY, G P (1981) *Br. Birds* **74**, 443; (1982) *Br. Birds* **75**, 33. CATZEFLIS, F (1978) *Nos Oiseaux* **34**,

287-302. CAVE, F O and MACDONALD, J D (1955) *Birds of the Sudan*. Edinburgh. CAWKELL, E M (1947*a*) *Br. Birds* **40**, 213; (1947*b*) *Br. Birds* **40**, 249; (1950) *Br. Birds* **43**, 374. CAWKELL, E M and MOREAU, R E (1963) *Ibis* **105**, 156-78. CEBALLOS, P and PURROY, F J (1977) *Pájaros de nuestros campos y bosques*. Madrid. CEDERWALL, G and SVENAEUS, S (1973) *Vår Fågelvärld* **32**, 128-30. ČERNY, W (1944) *Orn. Monatsber.* **52**, 46-7. ČERNY, W, PIČMAN, J, PITHART, K, and PIVONKA, P (1970) *Sylvia* **18**, 123-33. CHABOT, F (1932) *Oiseau* (2) **2**, 499-519. CHAMPERNOWNE, A W (1908) *Br. Birds* **2**, 202. CHANDLER, R J (1979) *Br. Birds* **72**, 299-313; (1980) *London Bird Rep. 1979*, 85-90. CHANTREY, D F and WORKMAN, L (1984) *Ibis* **126**, 366-71. CHAPIN, J P (1953) *Bull. Amer. Mus. nat. Hist.* **75**A (3). CHAPMAN, E A and McGEOCH, J A (1956) *Ibis* **98**, 577-94. CHAPPELL, B M A (1949) *Br. Birds* **42**, 87. CHAPPUIS, C (1969) *Alauda* **37**, 59-71; (1975) *Alauda* **43**, 427-74; (1976) *Alauda* **44**, 475-95. CHARLEMAGNE, E V (1912) *Orn. Vestnik* **3**, 306-7. CHARLWOOD, R H (1973) *Br. Birds* **66**, 398-9. CHASEN, F N (1921) *Ibis* (11) **3**, 185-227. CHATER, A O (1965) *Br. Birds* **58**, 513. CHAVIGNY, J DE and LE DÛ, R (1938) *Alauda* **10**, 91-115. CHAWORTH-MUSTERS, J L (1939) *Ibis* (14) **3**, 269-81. CHEESMAN, R D and SCLATER, W L (1935) *Ibis* (5) **5**, 594-622. CHEKE, A S (1967) *Ibis* **109**, 442-4; CHEKE, R A (1985) In Morgan, B J T and North, P M (eds) *Statistics in ornithology*, 13-24. Berlin. CHENG, TSO-HSIN (1963) *China's economic fauna: birds*. Peiping. CHERNEL, S VON (1919) *Aquila* **26**, 132. CHERNIKIN, E M (1976) *Byull. Mosk. obshch. ispyt. prir. otd. biol.* **81** (6), 135-6. CHERNYSHOV, V M (1981) In Yurlov, K T (ed) *Ekologiya i biotsenoticheskie svyazi pereletnykh ptits Zapadnoy Sibiri*, 138-60. Novosibirsk; (1982) In *Razmeshchenie i chislennost' pozvonochnykh Sibiri*, 84-110. Novosibirsk. CHEYLAN, G (1973) *Alauda* **41**, 85-9; (1980) *Alauda* **48**, 258. CHILD, G I (1969) *Auk* **86**, 327-38. CHOLMLEY, A J (1897) *Ibis* (7) **3**, 196-209. CHRISTEN, W (1983) *Orn. Beob.* **80**, 210; (1984) *Orn. Beob.* **81**, 73-4. CHRISTEN, W and JENNY, M (1983) *Orn. Beob.* **80**, 299-300. CHRISTENSEN, P V (1981) *Dansk orn. Foren. Tidsskr.* **75**, 47-50. CHRISTENSEN, S (1974) *Ornis scand.* **5**, 47-52. CHRISTIAENS, R and NELISSEN, R (1970) *Gerfaut* **60**, 227. CHRISTIAN, R J B (1965) *Ool. Rec.* **39** (1), 11-17. CHRISTIANI, A (1942) *Dansk orn. Foren. Tidsskr.* **36**, 57-8. CHRISTIE, D A (1985) *Br. Birds* **78**, 354. CHRISTISON, A F P (1941) *The birds of northern Baluchistan*. Quetta. CHRISTISON, A F P and TICEHURST, C B (1942) *J. Bombay nat. Hist. Soc.* **43**, 478-87. CHRISTOLEIT, E (1928) *Beitr. Fortpfl. Vögel* **4**, 41-7; (1929) *Ber. Ver. schles. Orn.* **15**, 8-12. CHUNIKHIN, S P (1965) *Ornitologiya* **7**, 76-82. CHURCH, H F (1958) *Bird Study* **5**, 87-9. CILIA, V (1978) *Il-Merill* **19**, 11. CLAFTON, F R (1963) *Br. Birds* **56**, 192-3; (1971) *Br. Birds* **64**, 320. CLANCEY, P A (1943) *Bull. Br. Orn. Club* **63**, 6-7; (1943) *Ibis* **85**, 95-7; (1944) *Br. Birds* **38**, 134; (1948) *Bull. Br. Orn. Club* **68**, 54-6; (1949) *Limosa* **22**, 369-70; (1950*a*) *Bonn. zool. Beitr.* **1**, 39-42; (1950*b*) *Dansk orn. Foren. Tidsskr.* **44**, 41-4; (1961) *Bull. Br. Orn. Club* **81**, 147-8; (1964) *The birds of Natal and Zululand*. Edinburgh; (1970) *Durban Mus. Novit.* **8**, 325-51; (1978) *Bonn. zool. Beitr.* **29**, 148-64; (1980) *South African orn. soc. checklist of southern African birds*. SAOS. CLANCEY, P A and JORDANS, A VON (1950) *Auk* **67**, 361-3. CLAPHAM, C S (1964) *Ibis* **106**, 376-88. CLAPP, R B, KLIMKIEWICZ, M K, and FUTCHER, A G (1983) *J. Fld. Orn.* **54**, 123-37. CLARK, F and MCNEIL, D A C (1980) *Ibis* **122**, 27-42. CLARK, R B (1947) *Br. Birds* **40**, 34-43; (1948) *Br. Birds* **41**, 244-6; (1949) *Br. Birds* **42**, 337-46. CLARK, R J (1975) *Wildl. Monogr.* **47**. CLARK, W D and KARR, J R (1979) *Wilson Bull.* **91**, 143-5. CLARKE, J E (1980) *Sandgrouse* **1**, 50-70; (1981) *Sandgrouse* **2**, 98-9. CLARKE, W E (1902) *Ann. Scott. nat. Hist.* **42**, 118; (1915) *Scott. Nat.* **46**, 291-6. CLARKSON, J R (1981) *Br. Birds* **74**, 267. CLEGG, M (1984) *Br. Birds* **77**, 361. CLODIUS, G (1894) *Orn. Monatsschr.* **19**, 136-7. CLUGSTON, D L (1981) *Br. Birds* **74**, 527-8. COATES, B J (1960) *Br. Birds* **53**, 275-6. COCKBAIN, R P (1958) *Br. Birds* **51**, 310. CODY, M L and CODY, C B J (1972) *Condor* **74**, 473-7. COHEN, E (1963) *Birds of Hampshire and the Isle of Wight*. Edinburgh. COLEMAN, J R (1985) *Br. Birds* **78**, 511-12. COLLAR, N J and STUART, S N (1985) *Threatened birds of Africa and related islands*. Cambridge. COLLETT, R (1898) *Ibis* (7) **4**, 317-19. COLLETT, R and OLSEN, Ø (1921) *Norges fugle*. Kristiania. COLLINGE, W E (1924-7) *The food of some British wild birds*. York. COLLINS, D R (1984) *Br. Birds* **77**, 467-74. COLLMAN, J R and CROXALL, J P (1967) *Ibis* **109**, 359-72. COLQUHOUN, M K (1940) *Br. Birds* **33**, 274-5. COLSTON, P R (1982) *Bull. Br. Orn. Club* **102**, 106-14. COMMISSIE VOOR DE NEDERLANDSE AVIFAUNA (1970) *Avifauna van Nederland*. Leiden. COMTE, A (1928) *Bull. Soc. Zool. Genève* **3** (7), 20-31. CONDER, P J (1948) *Br. Birds* **41**, 181-2; (1949) *Ibis* **91**, 649-55; (1950) *Br. Birds* **43**, 299; (1954) *Br. Birds* **47**, 76-9; (1956) *Ibis* **98**, 453-9; (1969) *Birds* **2**, 291-3; (1979) *Br. Birds* **72**, 2-4; (1981) *Sandgrouse* **2**, 22-32. CONGREVE, W M (1936) *Ool. Rec.* **16**, 73-8; (1945) *Ibis* **87**, 107-8; (1950) *Br. Birds* **43**, 17. CONSTANT, P and EYBERT, M-C (1980) *Nos Oiseaux* **35**, 349-60. COOKE, C H (1951) *Br. Birds* **44**, 62. COOPER, R P (1959) *Austral. Bird Watcher* **1**, 1-5. COOPMAN, L (1919) *Gerfaut* **5-9**, 48-58, 79-87. COPE, D A (1985) *Br. Birds* **78**, 111. CORLEY SMITH, G T and BERNIS, F (1956) *Ardeola* **3**, 115-25. CORMACK, R S (1954) *Br. Birds* **47**, 445. CORNISH, A V (1950) *Br. Birds* **43**, 155-6. CORNWALLIS, L (1975) Ph D Thesis. Oxford Univ. CORNWALLIS, L and PORTER, R F (1982) *Sandgrouse* **4**, 1-36. CORNWALLIS, R K (1961) *Br. Birds* **54**, 1-30. CORNWALLIS, R K and TOWNSEND, A D (1968) *Br. Birds* **61**, 97-118. CORTÉS, J E (1977) *Bull. Gibraltar orn. Soc.* **2** (1), 6-7; (1982) *Alectoris* **4**, 26-9. CORTÉS, J E, FINLAYSON, J C, MOSQUERA, M A J, and GARCIA, E F J (1980) *The birds of Gibraltar*. Gibraltar. CORTI, U A (1959) *Die Vogelwelt der Alpen* **5**; (1961) **6**; (1965) **7**. Chur. CORTI, U A, MELCHER, R, and TINNER, T (1949) *Schweiz. Arch. Orn.* **2**, 193-212. COTRON, G and PRODON, R (1977) *Nos Oiseaux* **34**, 129-30. COTTAM, C and HANSON, H C (1938) *Field Mus. nat. Hist. zool. Ser.* **20** (31), 405-26. COTTRILLE, B D (1950) *Wilson Bull.* **62**, 134-5. COULSON, J C (1956*a*) *Bird Study* **3**, 119-32; (1956*b*) Ph D Thesis. Durham Univ; (1961) *Bird Study* **8**, 89-97. COURTENAY THOMPSON, F C W (1972) *J. Saudi Arab. nat. Hist. Soc.* **1** (3), 15-17. COWLEY, E (1977) *Avic. Mag.* **83**, 185-8; (1979) *Bird Study* **26**, 113-16; (1983) *Bird Study* **30**, 1-7. COWPER, C N L (1973) *Scott. Birds* **7**, 302-6. COX, A H M (1922) *Br. Birds* **15**, 293-4. COX, S and INSKIPP, T (1978) *Br. Birds* **71**, 209-13. CRACRAFT, J (1981) *Auk* **98**, 681-714. CRAMP, S (1970*a*) *Br. Birds* **63**, 239-43. CRAMP, S (1970*b*) In Gooders, J (ed) *Birds of the world*, 1933-4. London. CRAMP, S and CONDER, P (1970) *Ibis* **112**, 261-3. CRAMP, S and GOODERS, J (1967) *London Bird Rep.* **31**, 93-8. CRAMP, S and TOMLINS, A D (1966) *Br. Birds* **59**, 209-32. CRAMP, S and WARD, J H (1934) *J. Anim. Ecol.* **3**, 1-7. CRESPON, J (1844) *Faune méridionale* **1**. Nîmes. CRETTÉ DE PALLUEL, A (1903-4) *Ornis* **12**, 141-2. CREUTZ, G (1941) *Vogelzug* **12**, 144-51; (1952) *J. Orn.* **93**, 174; (1959*a*) *Orn. Mitt.* **11**, 29-31; (1959*b*) *Falke* **6**, 88-93; (1961) *Falke* **8**, 304-13; (1962) *Orn. Mitt.* **14**, 64-6; (1966) *Die Wasseramsel*. Wittenberg Lutherstadt; (1974) *Falke* **21**, 402-9; (1980) *Abh. Ber. Nat. Mus. Forsch. Görlitz* **53** (7), 1-14. CRIDDLE, N (1920) *Can. Fld.-Nat.* **34**, 14-16. CROCKETT, D B and NICKELL, W P (1955) *Jack-Pine Warbler* **33**, 86. CROOK, J H (1965) *Symp. zool. Soc. Lond.* **14**, 181-218. CROUSAZ, G DE (1961) *Nos Oiseaux* **26**, 78-104; (1966) *Nos Oiseaux* **28**, 161-8. CROWE, T M, REBELO, A G, LAWSON, W J, and MANSON, A J (1981) *Ibis* **123**,

336-45. Csicsáky, M (1978) *J. Orn.* **119**, 249-64. Csiki, E (1904) *Aquila* **11**, 270-317; (1908) *Aquila* **15**, 183-206; (1909) *Aquila* **16**, 139-44. Ctyroky, P (1958) *Sylvia* **15**, 232-4; (1972) *Bull. Iraq nat. Hist. Mus.* **5** (3), 1-8. Cube, A (1950) *Vår Fågelvärld* **9**, 177-80. Cullen, J M, Guiton, P E, Horridge, G A, and Peirson, J (1952) *Ibis* **94**, 68-84. Cumming, W D (1899) *J. Bombay nat. Hist. Soc.* **12**, 760-5; (1902) *J. Bombay nat. Hist. Soc.* **14**, 611-12. Curmi, E (1977) *Il-Merill* **18**, 27. Currie, P W E (1949) *London Bird Rep.* **13**, 33-4. (1965) *Ibis* **107**, 253. Currier, N and Howorth, M (1957) *Br. Birds* **50**, 76-7. Curry, P J and Sayer, J A (1979) *Ibis* **121**, 20-40. Curry-Lindahl, K (1958) *Ark. Zool.* **11** (33), 541-57; (1963a) *Ostrich* **34**, 99-101; (1963b) *Proc. int. orn. Congr.* **13**, 960-73; (1964) *Ibis* **106**, 255-6; (1981) *Bird migration in Africa.* London. Cvitanić, A (1962) *Larus* **14**, 147-53. Czajkowski, M (1973) *Nos Oiseaux* **32**, 99-102. Czikeli, H (1975) *Egretta* **18**, 23. Czikeli, H and Knötzsch, G (1979) *Ökol. Vögel* **1**, 159-63.

D'Abreu, E A (1918) *Rec. Nagpur Mus.* **2**. Dallas, J E S (1928) *London Nat. 1927*, 19-20. Dallmann, M (1977) *Anz. orn. Ges. Bayern* **16**, 153-70. Damm, K (1976) *Vogelkdl Hefte Edertal.* **2**, 136-45. Dandl, J (1957a) *Aquila* **63-4**, 355; (1957b) *Aquila* **63-4**, 369. Danilov, N N (1958) *Zool. Zh.* **37**, 1898-903. Danilov, N N, Ryzhanovski, V N, and Ryabitsev, V K (1984) *Ptitsy Yamala.* Moscow. Danilov, N N and Tarchevskaya, V A (1962) *Ornitologiya* **4**, 142-53. Dare, P J (1953) *Devon Birds* **6**, 25-8. Dathe, H (1933) *Orn. Monatsber.* **41**, 145-7; (1952) *Beitr. Vogelkde.* **2**, 15-32; (1975) *Beitr. Vogelkde.* **21**, 493-4; (1983) *Beitr. Vogelkde.* **29**, 244. Dathe, H and Neufeldt, I A (eds) (1983) *Atlas der Verbreitung palaearktischer Vögel* **11**; (1984) **12**. Berlin. Daukes, A H (1932) *Br. Birds* **26**, 132-4. Davenport, H S (1922) *Br. Birds* **16**, 126-8. David, A and Oustalet, M E (1877) *Les oiseaux de la Chine.* Paris. Davidson, J (1898) *Ibis* (7) **4**, 1-42. Davies, N B (1976) *J. Anim. Ecol.* **45**, 235-53; (1977) *J. Anim. Ecol.* **46**, 37-57; (1980) *Ardea* **68**, 63-74; (1981a) *Anim. Behav.* **29**, 529-34; (1982) *Br. Birds* **75**, 261-7; (1983) *Nature* **302**, 334-6; (1985) *Anim. Behav.* **33**, 628-48. Davies, N B and Houston, A I (1981) *J. Anim. Ecol.* **50**, 157-80. Davies, N B and Lundberg, A (1984) *J. Anim. Ecol.* **53**, 895-912. Davies, P W and Snow, D W (1965) *Br. Birds* **58**, 161-75. Davies, S (1981b) *Ring. Migr.* **3**, 173-80. Davies, S J J F (1958) *Bird Study* **5**, 184-91. Davies, S J F and Rowell, C H F (1956) *Bird Study* **3**, 242-8. Davis, J G (1932) *Br. Birds* **25**, 333. Davis, P (1953) *Br. Birds* **56**, 455; (1958a) *Br. Birds* **51**, 195-7; (1958b) *Br. Birds* **51**, 198; (1964) *Br. Birds* **57**, 214-16; (1965) *Bird Study* **12**, 151-69; (1966) *Br. Birds* **59**, 353-76.Davis, P and Hope Jones, P (1958) *Br. Birds* **51**, 356-7. Davis, P G (1975) *Br. Birds* **68**, 77-8; (1982) *Bird Study* **29**, 73-9. Davis, T A W (1967) *Br. Birds* **60**, 91. Davola, J (1972) *Ochrana Fauny* **6**, 151-8. Dawson, J P (1976) *Br. Birds* **69**, 273. Day, D H (1975) *Ostrich* **46**, 192-4. De Bont, A F (1957) *Gerfaut* **47**, 127-34; (1962) *Gerfaut* **52**, 298-343. Debout, G (1981) *Cormoran* **4** (22), 123-41. De Braey, L (1946) *Gerfaut* **36**, 133-93. Debussche, M and Isenmann, P (1985a) *Bird Study* **32**, 152-3; (1985b) *Rev. Ecol.* **40**, 379-88. Deeble, M (1972) *Bird Life* Apr-Jun, 31. Deignan, H G (1945) *Bull. US natn. Mus.* **186**. Deignan, H G, Paynter, R A, Jr, and Ripley, S D (eds) (1964) *Peters' check-list of birds of the world* **10**. Cambridge, Mass. Dejonghe, J-F (1984) *Les oiseaux de montagne.* Maisons-Alfort. Dejonghe, J F and Czajkowski, M A (1983) *Alauda* **51**, 27-47. Dekeyser, P L (1956) *Mém. Inst. Fr. Afr. noire* **48**, 79-141. Delacour, J (1943) *Zoologica* **28**, 17-28; (1946) *Oiseau* **16**, 7-36. Delacour, J and Amadon, D (1949) *Ibis* **91**, 427-9. Delacour, J and Mayr, E (1946) *Birds of the Philippines.* New York. Delany, S, Chadwell, C,

and Norton, J (1982) In *Univ. Southampton Ladakh Exped. Rep.* (1980), 1-153. Southampton Univ. Deleuil, R (1913) *Rev. fr. Orn.* **5**, 2-5; (1954) *Oiseau* **24**, 189-96; (1956-66) *Bull. Soc. Sci. nat. Tunisie* **9-10**, 7-17. Delius, J D (1963) *Z. Tierpsychol.* **20**, 297-348; (1965) *Ibis* **107**, 466-92; (1969) *Behaviour* **33**, 137-78. Del-Nevo, A and Ewins, P J (1981) *Bird watching in Nepal.* Unpubl. MS, Edward Grey Inst., Oxford Univ. Dementiev, G P (1934) *Oiseau* (2) **4**, 591-625. Dementiev, G P and Gladkov, N A (1954a) *Ptitsy Sovetskogo Soyuza* **5**; (1954b) **6**. Moscow. Denby, C A and Phillips, A P (1976) In *Univ. Southampton Ladakh Exped. Rep.* (1976), 31-63. Southampton Univ. Dennis, R H (1967) *Br. Birds* **60**, 161-6. Dennis, R H and Wallace, D I M (1975) *Br. Birds* **68**, 238-41. Denny, J (1952) *Br. Birds* **45**, 373. Dent, G (1907) *Br. Birds* **1**, 89-90. Desfayes, M (1975) *Rev. zool. afr.* **89**, 505-35. Desfayes, M and Praz, J C (1978) *Bonn. zool. Beitr.* **29**, 18-37. Despott, G (1916) *Zoologist* (4) **20**, 161-81. Devillers, P (1964) *Gerfaut* **54**, 376-88. Devillers, P and Esbroeck, J van (1974) *Oiseau* **44**, 185. Dewar, D (1902) *The birds in the wood.* London; (1923) *Himalayan and Kashmiri birds.* London. Dharmakumarsinhji, R S (1976) *J. Bombay nat. Hist. Soc.* **72**, 557. Dick, W and Holupirek, H (1978) *Falke* **25**, 308-12. Dick, G and Sackl, P (1985) *Ökol. Vögel* **7**, 197-208. Diesselhorst, G (1938) *Beitr. Fortpfl. Vögel* **14**, 224-5; (1939) *Beitr. Fortpfl. Vögel* **15**, 216; (1956) *Vogelwelt* **77**, 80-4; (1957) *Vogelwelt* **78**, 195-6; (1968a) *Khumbu Himal* **2**. Innsbruck; (1968b) *J. Orn.* **109**, 396-401. Dijk, A J van and Os, B L J van (1982) *Vogels van Drenthe.* Assen. Dijk, J van (1975) *Limosa* **48**, 86-99. Dillery, D G (1965) *Auk* **82**, 281. Dillon, O W (1959) *N. Amer. Wildl. Conf. Trans.* **24**, 374-82. Dittami, J (1981) *Vogelwarte* **31**, 177-8. Dittberner, H and Dittberner, W (1969) *Milu* **2**, 495-618; (1979) *Orn. Jber. Mus. Hein.* **4**, 3-18; (1984) *Die Schafstelze.* Wittenberg Lutherstadt. Dittberner, W and Dittberner, H (1959) *Falke* **6**, 178. (1973) *Orn. Mitt.* **25**, 216-18. Dittmann, E (1925) *Mitt. Ver. sächs. Orn.* **1** suppl., 21-6; (1927) *Mitt. Ver. sächs. Orn.* **2**, 27-30. Dixon, C (1882) *Ibis* (4) **6**, 550-79; (1885) *Ibis* (5) **3**, 69-97. Dobler, E and Stadelmann, F (1975) *Egretta* **18**, 23-4. Dodsworth, P T L (1914) *J. Bombay nat. Hist. Soc.* **22**, 798-800. Doerbeck, F (1966) *Vogelwelt* **87**, 120-2. Dohle, W, Jüde, H-D, Sturhan, D, and Goethe, F (1957) *J. Orn.* **98**, 119-21. Dol, J H van der (1977) *Br. Birds* **70**, 550-2. Dolgushin, I A, Korelov, M N, Kuz'mina, M A, Gavrilov, E I, Gavrin, V F, Kovshar', A F, Borodikhin, I F, and Rodionov, E F (1970) *Ptitsy Kazakhstana* **3**. Alma-Ata; Dolgushin, I A, Korelov, M N, Kuz'mina, M A, Gavrilov, E I, Kov'shar, A F, and Borodikhin, I F (1972) *Ptitsy Kazakhstana* **4**. Alma-Ata. Domaniewski, J (1925) *Prace Zool. Pol. Panstw. Muz. Przyrodn.* **4**, 85-125. Dorka, V (1966) *Vogelwelt* **87**, 23-4. Dorka, V, Dorka, U, and Haas, V (1976) *Verh. orn. Ges. Bayern* **22**, 467-71. Dornbusch, M (1968) *Vogelwelt* **89**, 43-5; (1981) *Beitr. Vogelkde.* **27**, 73-99. Dorst, J and Pasteur, G (1954) *Oiseau* **24**, 248-66. Dorzhiev, T Z (1983) In Kuchin, A P (ed) *Ptitsy Sibiri*, 170-2. Gorno-Altaisk. Dosseter, L J (1944) *London Bird Rep. 1943*, 28. Douaud, J (1956) *Alauda* **24**, 146-7; (1957) *Alauda* **25**, 241-66. Doughty, J (1970) *Avic. Mag.* **76**, 227-30. Dove, R S and Goodhart, H J (1955) *Ibis* **97**, 311-40. Dowsett, R J (1965a) *Bull. Br. Orn. Club* **85**, 150-2; (1965b) *Ostrich* **36**, 32-3; (1971) *Bird Study* **18**, 53-4. Dowsett, R J and Fry, C H (1971) *Ibis* **113**, 531-3. Draulans, D and Vessem, J van (1982) *Ibis* **124**, 347-51. Dresser, H E (1871-81) *A history of the birds of Europe.* London. Drost, I (1949) *Beitr. Naturkde. Niedersachs.* **1**, 12-14. Drost, R (1933) *Orn. Monatsber.* **41**, 22-3; (1935) *Vogelzug* **6**, 67-72; (1948) *Vogelwarte* **15**, 18-28. Drost, R and Desselberger, H (1932)

Vogelzug 3, 105-15. DROST, R and SCHÜZ, E (1940) *Vogelzug* 11, 145-61. (1952) *Vogelwarte* 16, 95-8. DROZDOV, N N (1965) *Ornitologiya* 7, 166-99. DROZDOV, N N and ZLOTIN, R I (1962) *Ornitologiya* 5, 193-207. DRURY, W H, JR (1961) *Bird-Banding* 32, 1-46. DRURY, W H and KEITH, J A (1962) *Ibis* 104, 449-89. DUBOIS, A D (1935) *Condor* 37, 56-72; (1936) *Condor* 38, 49-56. DUDA, E (1978) *Egretta* 21, 74-5. DUHART, F and DESCAMPS, M (1963) *Oiseau* 33 no. spéc. DUNAEVA, T N and KUCHERUK, V V (1941) *Byull. Mosk. obshch. ispyt. prir. otd. zool.* NS 4 (19), 5-80. DUNBAR, M J (1955) In Kimble, G H T and Good, D (eds) *Amer. geog. Soc. spec. Publ.* 32. DUNCAN, N (1981) *Bird Study* 28, 186. DUNSHEATH, M H and DONCASTER, C C (1941) *Br. Birds* 35, 138-48. DuPONT, J E and RABOR, D S (1973a) *Nemouria* 9, 1-63; (1973b) *Nemouria* 10, 1-111. DUPUY, A (1966) *Oiseau* 36, 256-68. DURANGO, S (1936) *Fauna och Flora* 31, 274-8; (1949) *Ibis* 91, 140-3. DURMAN, R F (1976) *Bird Study* 23, 197-205; (1978) *Edinb. Ring. Group. Rep.* 5, 24-7. DWIGHT, J (1900) *Ann. New York Acad. Sci.* 13, 73-360. DYBBRO, T (1976) *De danske ynglefugles udbredelse.* Copenhagen. DYMOND, J N (1980) *The birds of Lundy.* Plymouth. DYMOND, J N and THE RARITIES COMMITTEE (1976) *Br. Birds* 69, 321-68. DYRCZ, A (1969) *Ekol. Polska* 17A, 735-93; (1976) *Notatki orn.* 17, 79-92; (1977) *Ibis* 119, 215.

EAST, M (1980) *Ibis* 122, 517-20; (1981a) *Ibis* 123, 223-30; (1981b) *Ornis scand.* 12, 230-9; (1982) *Ornis scand.* 13, 85-93. EBENMAN, B and KARLSSON, J (1984) *Ann. zool. fenn.* 21, 249-51. EBBUTT, T (1947) *Br. Birds* 40, 148-9. EBER, G (1960) In Stresemann, E and Portenko, L A (eds) *Atlas der Verbreitung palaärktischer Vögel* 1. Berlin. EBER, G and SZIJJ, J (1960) In Stresemann, E and Portenko, L A (eds) *Atlas der Verbreitung palaärktischer Vögel* 1. Berlin. ECCLES, L (1955) *Br. Birds* 48, 421-2. ECK, S (1970) *Zool. Abh. Staatl. Mus. Tierkde. Dresden* 30, 135-6; (1975a) *Zool. Abh. Staatl. Mus. Tierkde. Dresden* 33, 223-4; (1975b) *Beitr. Vogelkde.* 21, 21-30. ECK, S and GEIDEL, B (1971) *Zool. Abh. Staatl. Mus. Tierkde. Dresden* 30, 161-71. EDGAR, W H and ISAACSON, A J (1974) *Ann. appl. Biol.* 76, 335-7. EDHOLM, M, GRANDBERG, B, and LUNDBERG, A (1980) *Vår Fågelvärld* 39, 137-8. EDQVIST, J (1979) *Vår Fågelvärld* 38, 47. EDWARDS, G, HOSKING, E, and SMITH, S (1949) *Br. Birds* 42, 13-19; (1950) *Br. Birds* 43, 9-10. EDWARDS, G R (1950) *Br. Birds* 43, 179-83. EDWARDS, P J (1985) *Ibis* 127, 42-59. EDWARDS, S B (1980) *Br. Birds* 73, 416. EFTELAND, S (1975) *Sterna* 15, 11-16; (1983) *Vår Fuglefauna* 6, 252-4. EFTELAND, S and KYLLINGSTAD, K (1984) *Fauna. Norveg. (C) Cinclus* 7, 7-11. EGGEBRECHT, E (1937) *J. Orn.* 85, 636-76; (1939) *Orn. Monatsber.* 47, 109-17; (1943) *Orn. Monatsber.* 51, 127-35. ELFSTRÖM, T (1979) Ph D Thesis. Göteborg Univ. ELGOOD, J H (1959) *Ostrich Suppl.* 3, 306-16; (1982) *The birds of Nigeria.* London. ELGOOD, J H, FRY, C H, and DOWSETT, R J (1973) *Ibis* 115, 1-45. ELGOOD, J H, SHARLAND, R E, and WARD, P (1966) *Ibis* 108, 84-116. ELKINS, N (1983) *Weather and bird behaviour.* Calton. ELKINS, N and ETHERIDGE, B (1974) *Br. Birds* 67, 376-87; (1977) *Ring. Migr.* 1, 158-65. ELLISON, C S S (1931) *Br. Birds* 25, 55-6. ELMS, N E G (1972) *Br. Birds* 65, 126 ELOFSON, O (1968) *Vår Fågelvärld* 27, 346-8. ELSNER, C (1951) *J. Orn.* 93, 65. ELST, D VAN DER (1984) *Aves* 21, 65-77. ELTON, C (1928) *Br. Birds* 21, 266-7. ELVERS, H (1972) *Orn. Mitt.* 24, 175. EMLEN, S T and DEMONG, N J (1975) *Science* 188, 1029-31. EMLEN, S T and ORING, L W (1977) *Science* 197, 215-23. EMLEY, D W (1985) *Br. Birds* 78, 110. EMMRICH, R (1971) *Zool. Abh. Staatl. Mus. Tierkde. Dresden* 32, 57-67; (1975) *Beitr. Vogelkde.* 21, 102-10. ENDES, M (1969-70) *Acta biol. Debrecina* 7-8, 161-8. ENDES, M (1970) *Die Kurzzehenlerche.* Wittenberg Lutherstadt; (1972) *Bull. Br.*

Orn. Club 92, 149-51. ENDES, M, HORVÁTH, L, and HÜTTLER, B (1967) *Acta Zool. Cracov.* 12, 379-91. ENEMAR, F (1980) *Vår Fågelvärld* 39, 231-6. ENGELEN, G D (1979) *Vogeljaar* 27, 91. ENGELMOER, M, ROSELAAR, K, BOERE, G C, and NIEBOER, E (1983) *Ring. Migr.* 4, 245-8. ENGELS, W L (1940) *Univ. Calif. Publ. Zool.* 42, 341-400. ENGLAND, M D (1978a) *Br. Birds* 71, 88; (1978b) *Br. Birds* 71, 258-66. ENNION, E A R and ENNION, D (1962) *Ibis* 104, 158-68. ERARD, C (1959) *Nos Oiseaux* 25, 13-16; (1966) *Oiseau* 36, 4-51. ERARD, C and ETCHÉCOPAR, R-D (1970) *Mém. Mus. natn. Hist. nat. (A)* 66. ERARD, C and JARRY, G (1973) *Bull. Br. Orn. Club* 93, 139-40. ERARD, C and LARIGAUDERIE, F (1972) *Oiseau* 42, 81-169. ERARD, C and NAUROIS, R DE (1973) *Bull. Br. Orn. Club* 93, 141-2. ERARD, C and YEATMAN, L (1967) *Oiseau* 37, 20-47. ERIKSSON, A, HEINO, K, and HOLMA, O (1976) *Lintumies* 11, 120. ERLANGER, C VON (1899) *J. Orn.* 47, 309-74, 449-532. ERLEMANN, P (1983) *Orn. Mitt.* 35, 214. ERN, H (1966) *J. Orn.* 107, 310-14. ERNST, S and THOSS, M (1975) *Falke* 22, 305-11. ERZ, W (1964) *Z. wiss. Zool.* 170, 1-111. ESILEVSKAYA, M A (1967) *Ornitologiya* 8, 347-50; (1968) *Vestnik Zool.* (2), 68-72. ESPMARK, Y (1981) *Vår Fågelvärld* 40, 81-90; (1982) *Fauna Norveg. (C) Cinclus* 5, 73-83. ESTAFIEV, A A (1981) *Sovremennoe sostoyanie, raspredelenie i okhrana avifauny taezhnoy zony basseyna r. Pechory.* Syktyvkar; (1979) In Labutin, Y V (ed) *Migratsii i ekologiya ptits Sibiri,* 142-3. Yakutsk. ETCHÉCOPAR, R-D and HÜE, F (1957) *Oiseau* 27, 309-34; (1967) *The birds of North Africa.* Edinburgh; (1983) *Les oiseaux de Chine de Mongolie et de Corée: passereaux.* Paris. EVANS, P R (1966a) *Skokholm Bird Obs. Rep.,* 22-7; (1966b) *J. Zool. Lond.* 150, 319-69; (1968) *Br. Birds* 61, 281-303. EVANS, W (1901) *Ann. Scott. nat. Hist.* 37, 12-15. EVANS, W I (1950) *Br. Birds* 43, 337-8. EVERETT, M J (1967) *Scott. Birds* 4, 534-48. EVERETT, M J and HAMMOND, N (1975) *Br. Birds* 68, 118-19.

FAIRBANK, R J (1980a) *Br. Birds* 73, 314; (1980b) *Br. Birds* 73, 415-16. FAIRON, J (1971) *Gerfaut* 61, 146-61. FALLY, J (1984) *Ökol. Vögel* 6, 169-74. FARINA, A (1978) *Avocetta* 2, 35-46; (1979) *Monitore Zool. ital.* NS 13, 203. FARIS, R C (1937) *Irish Nat. J.* 6, 199-200. FARKAS, T (1954) *Aquila* 55-8, 303-4; (1955) *Vogelwelt* 76, 164-80. FARMER, R (1979) *Malimbus* 1, 56-64. FARRAND, J, JR (ed) (1983) *The Audubon Society master guide to birding.* New York. FATIO, V and STUDER, T (1907) *Catalogue des oiseaux de la Suisse* 4. Geneva. FAVARGER, J (1948) *Nos Oiseaux* 19, 278-84. FEARE, C J (1970) *Oecologia* 5, 1-18. FEDUCCIA, A (1974) *Auk* 91, 427-9; (1975) *Misc. Publ. Univ. Kansas Mus. nat. Hist.* 63, 1-34; (1977) *Syst. Zool.* 26, 19-31. FEDYUSHIN, A V and DOLBIK, M S (1967) *Ptitsy Belorussii.* Minsk. FEENY, P P (1959) *Bird Migr.* 1, 153-8. FEENY, P P, ARNOLD, R W, and BAILEY, R S (1968) *Ibis* 110, 35-86. FEINDT, P, GÖTTGENS, F, and GÖTTGENS, H (1956) *Beitr. Naturk. Niedersachs.* 9, 83-9. FELDMANN-LUTERNAUER, H and FELDMANN-LUTERNAUER, A (1978) *Vögel Heimat* 48, 130. FERGUSON-LEES, I J (1956) *The birds of the Coto Doñana, south Spain April-May 1956.* Duplicated; (1960) *Br. Birds* 53, 553-8; (1962) *Br. Birds* 55, 37-42; (1968a) *Br. Birds* 61, 312; (1968b) *Br. Birds* 61, 525-6; (1969) *Br. Birds* 62, 110-15; (1970a) In Gooders, J (ed) *Birds of the world,* 1783-4; (1970b) 1789-91; (1970c) 1792; (1970d) 1794-6; (1970e) 1796-7. London. FERRERO CANTISÁN, J J, NEGRO BALMASEDA, J J, and ROMÁN ALVAREZ, J A (1983) *Alytes* 1, 363-8. FERRY, C (1947) *Alauda* 15, 209-20. FERRY, C, DESCHAINTRE, A, and FERRY, F (1969) *Jean le Blanc* 8, 52-62. FERRY, C and FROCHOT, B (1964) *Jean le Blanc* 3, 41-2. FIEBIG, J (1983) *Mitt. zool. Mus. Berlin* 59, Suppl. Ann. Orn. 7, 163-87. FIELD, G D (1973) *Bull. Br. Orn. Club* 93, 81-2. FIELD, J (1950) *Br. Birds* 43, 18. FILIPASCU, A (1964) *Aquila*

69-70, 159-67. FINCHER, F (1963) *Br. Birds* 56, 222. FINLAYSON, J C (1978) *Alectoris* 1, 23-9; (1980) *Ring. Migr.* 3, 32-4; (1981) *Ibis* 123, 88-95. FIRSOVA, L V and LEVADA, A V (1982) *Ornitologiya* 17, 112-18. FISCHER, W (1974) *Falke* 21, 66. FISCHMAN, L (1977) *Israel Land Nat.* 2, 101-6. FISHER, C (1979) *Br. Birds* 72, 38. FISHER, D J (1978) *Br. Birds* 71, 223. FISHER, J (1949) *Bird Notes* 23, 253-60. FISHER, J, MORLEY, A, and VENABLES, L S V (1944) *Br. Birds* 37, 177-8. FITTER, R S R (1944) *London Bird Rep. 1943*, 17-19; (1946) *Br. Birds* 39, 207-11; (1947) *Br. Birds* 40, 267; (1948) *Bird Notes* 23, 185-8; (1965) *Br. Birds* 58, 481-92; (1971) *Br. Birds* 64, 117-24; (1976) *Br. Birds* 69, 9-15. FJELDSÅ, J (1973a) *Ornis scand.* 4, 55-86; (1973b) *Sterna* 12, 161-217. FLADE, M (1979) *Vogelkdl. Ber. Niedersachs.* 11, 75-6. FLEGG, J J M and GLUE, D E (1975) *Bird Study* 22, 1-8. FLEISCHMANN, B (1977) *Gef. Welt* 101, 82-4. FLEMING, R L (1968) *Pavo* 6, 1-11. FLEMING, R L, SR, FLEMING, R L, JR, and BANGDEL, L S (1976) *Birds of Nepal.* Kathmandu. FLEMING, R L and TRAYLOR, M A (1964) *Fieldiana Zool.* 35, 495-558. FLEMING, T H (1981) *Ibis* 123, 463-76. FLETCHER, T B and INGLIS, C M (1924) *Birds of an Indian garden.* Calcutta. FLINKS, H and PFEIFER, F (1984) *Vogelwelt* 105, 41-51. FLINT, J H (1976) *Naturalist* 937, 76. FLINT, P R and STEWART, P F (1983) *The birds of Cyprus.* Br. Orn. Union, London. FLINT, V E (1962) *Ornitologiya* 4, 186-9. FLINT, V E, BOEHME, R L, KOSTIN, Y V, and KUZNETSOV, A A (1984) *A field guide to birds of the U.S.S.R.* Princeton. FLORENCE, L (1912) *Trans. Highland agric. Soc. Scot.* (5) 24, 180-219; (1914) *Trans. Highland agric. Soc. Scot.* (5) 26, 1-74. FLOWER, U (1969) *Br. Birds* 62, 157-8. FLÜCK, D and FLÜCK, H (1984) *Orn. Beob.* 81, 72-3. FLUMM, D S (1977) *Br. Birds* 70, 298-300. FONTAINE, S (1975) *Aves* 12, 160-1. FONTAINE, V (1955) *Fauna och Flora* 50, 225-33. FORBES-WATSON, A D (1983) *Br. Birds* 76, 535. FORBUSH, E H (1929) *Rep. Massachusetts Dep. Agr.* 3, 350-2. FORBUSH, E H and MAY, J B (1939) *A natural history of American birds of eastern and central North America.* New York. FORD, J (1983) *Emu* 83, 141-51. FORGES, G DES (1959) *Br. Birds* 52, 390. FOSCHI, F (1978) *Gli Uccelli Ital.* 3, 192-3. FOUARGE, J (1971) *Aves* 8, 192. FOURNIER, A (1983) *Héron* 4, 1-19. FOWLER, J A, BLAKESLEY, D, and MILLER, C J (1984) *Br. Birds* 77, 361-2. FOX, W S (1900) *Zoologist* (4) 4, 1-10. FRAINE, R DE (1982) *Wielewaal* 48, 112-13. FRANCIS, D M (1980) *Ring. Migr.* 3, 4-8. FRANÇOIS, J (1975) *Alauda* 43, 279-93. FRANDSEN, J (1982) *Birds of the south western Cape.* Cape Town. FRANK, F (1952) *J. Orn.* 93, 138-41. FRANKEVOORT, W and HUBATSCH, H (1966) *Unsere Wiesenschmätzer.* Wittenberg Lutherstadt. FRANKUM, R G (1955) *Br. Birds* 48, 235. FREDRIKSSON, S, JACOBSSON, S, and SILVERIN, B (1973) *Vår Fågelvärld* 32, 245-51. FREDRIKSSON, S and SVENSSON, S (1984) *Vår Fågelvärld* 43, 42. FREITAG, F (1942) *Beitr. Fortpfl. Vögel* 18, 10-12; (1943) *Beitr. Fortpfl. Vögel* 19, 133-7; (1972) *Luscinia* 41, 306-8. FRELIN, C (1983) *Alauda* 51, 11-26. FREMMING, O R (1984) *Vår Fuglefauna* 7, 197-204. FRIEDMANN, H (1962) *Los Angeles Co. Mus. Contrib. Sci.* 59, 1-27. FROST, R A (1972) *Br. Birds* 65, 483-4. FROST, R A, HERRINGSHAW, D and MCKAY, C R (1982) *Br. Birds* 75, 89-90. FRY, C H (1961) *Ibis* 103a, 291-3; (1966a) *Bull. Niger. orn. Soc.* 3 (10), 47; (1966b) *Bull. Niger. orn. Soc.* 3 (12), 98-9; (1970a) In Gooders, J (ed) *Birds of the world*, 1830-2. London; (1970b) *Ostrich Suppl.* 8, 239-63; FRY, C H, ASH, J S, and FERGUSON-LEES, I J (1970) *Ibis* 112, 58-82. FRY, C H, BRITTON, P L, and HORNE, J F M (1974) *Ibis* 116, 44-51. FRY, C H, FERGUSON-LEES, I J, and DOWSETT, R J (1972) *J. Zool. Lond.* 167, 293-306. FRYCKLUND, I (1980) *Fåglar Uppland* 7, 15-27. FUCHS, E (1970) *Orn. Beob.* 67, 3-14; (1972) *Orn. Beob.* 69, 302-3; (1979) *Orn. Beob.* 76, 235-46. FULK, G W (1967) *Wilson Bull.* 79, 344-5. FULLER, R J (1982) *Bird habitats in Britain.* Calton. FULLER, R J and GLUE, D E (1977) *Bird Study* 24, 215-28. FURRER, R K (1975) *Orn. Beob.* 72, 1-8; (1977) *Orn. Beob.* 74, 37-53; (1979a) *J. Orn.* 120, 86-93; (1979b) *Orn. Mitt.* 31, 141-5.

GABRIELSON, I N and LINCOLN, F C (1959) *The birds of Alaska.* Harrisburg. GADGIL, M and ALI, S (1976) *J. Bombay nat. Hist. Soc.* 72, 716-27. GALBRAITH, H (1977) *Ring. Migr.* 1, 184-6; (1979) *Scott. Birds* 10, 180-1. GALBRAITH, H and BROADLEY, B (1980) *Ring. Migr.* 3, 62-4. GALBRAITH, H, MITCHELL, A B, and SHAW, G (1981) *Bird Study* 28, 53-9. GALBRAITH, H and TYLER, S J (1982) *Ring. Migr.* 4, 9-14. GALINDO, P, MENDEZ, E, and ADAMES, A J (1963) *Bird-Banding* 34, 202-9. GALLAGHER, M and WOODCOCK, M W (1980) *The birds of Oman.* London. GALLAGHER, M D (1977) *J. Oman Stud. spec. Rep.* 1, 27-58. GALLAGHER, M D and ROGERS, T D (1978) *Bonn. zool. Beitr.* 29, 5-17; (1980) *J. Oman Stud. spec. Rep.* 2, 347-85. GALLOWAY, B, HOWEY, D, and PARKIN, D T (1961) *Br. Birds* 54, 73. GAMBLE, P H (1952) *Br. Birds* 45, 373. GANGULI, U (1975) *A guide to the birds of the Delhi area.* New Delhi. GANYA, I M, LITVAK, M D, and KUKURUZYANU, L S (1969) *Voprosy ekol. prakt. znach. ptits mleko.* Moldavii 4, 26-54. GANZEVLES, W and TILLO, P VAN (1982) *Vogeljaar* 30, 288-90. GARDNER-MEDWIN, D and MURRAY, J (1958) *Ibis* 100, 313-18. GARLING, M (1926) *Beitr. Fortpfl. Vögel* 2, 176-7; (1944) *Beitr. Fortpfl. Vögel* 20, 120-3. GARNETT, R M (1930) *Br. Birds* 23, 339-40; (1932) *Br. Birds* 25, 223. GARRICK, A S (1981) *New Zealand J. Ecol.* 4, 106-14. GARSON, P J (1978) D Phil Thesis. Oxford Univ.; (1980a) *Bird Study* 27, 63-72; (1980b) *Anim. Behav.* 28, 491-502. GARSON, P J and HUNTER, M L JR (1979) *Ibis* 121, 481-7. GASS, F (1975) *Orn. Beob.* 72, 84. 503g GASTON, A J (1968) *Ibis* 110, 17-26; (1970) *Bull. Br. Orn. Club* 90, 53-60, 61-6. GÄTKE, H (1891) *Die Vogelwarte Helgoland.* Braunschweig. GATTER, W (1968) *Orn. Mitt.* 20, 252; (1970) *Vogelwelt* 91, 1-11; (1976) *Vogelwelt* 97, 201-17. GAUGRIS, Y, PRIGOGINE, A, and VANDE WEGHE, J-P (1981) *Gerfaut* 71, 3-39. GAVRILOV, E I (1971) *Zool. Zh.* 50, 599-602; (1973) *Trudy zapoved. Kazakh.* 3, 59-70. GAVRILOV, E I and KOVSHAR', A F (1970) *J. Bombay nat. Hist. Soc.* 67, 14-25. GEBHARDT, L (1951) *J. Orn.* 93, 64. GEBHARDT, L and SUNKEL, W (1954) *Die Vögel Hessens.* Frankfurt am Main. GEE, J P (1984) *Malimbus* 6, 31-66. GEISSBÜHLER, W (1949) *Orn. Beob.* 46, 158. GENGLER, J (1903) *Verh. orn. Ges. Bayern* 4, 96-101. GEORG, P V and AL-RAWY, M (1970) *Bull. Iraq nat. Hist. Mus.* 4, 3-20. GEORG, P V and VIELLIARD, J (1970) *Bull. Iraq nat. Hist. Mus.* 4, 61-85. GEORGE, D E P (1977) *Scott. Birds* 9, 349-50; GEORGE, P V (1965) *J. Bombay nat. Hist. Soc.* 62, 160. GEORGE, U (1978) *In the deserts of this earth.* London. GEPTNER, V G (1958) *Ornitologiya* 1, 131-43. GERBER, R (1949) *Urania* 12, 441; (1953) *Gefiederte Sänger.* Leipzig. GERMAIN, M (1965) *Oiseau* 35, 117-34. GERMAIN, M, DRAGESCO, J, ROUX, F, and GARCIN, H (1973) *Oiseau* 43, 212-59. GERMOGENOV, N I (1982) In Labutin, Y V (ed) *Migratsii i ekologiya ptits Sibiri*, 74-87. Novosibirsk. GÉROUDET, P (1942) *Nos Oiseaux* 16, 185-92; (1956) *Nos Oiseaux* 23, 225-33; (1957a) *La vie des oiseaux: les passereaux* 3. Neuchâtel; (1957b) *Nos Oiseaux* 24, 109-17; (1961) *La vie des oiseaux: les passereaux* 1. Neuchâtel; (1962) *Nos Oiseaux* 26, 165-79; (1963) *La vie des oiseaux: les passereaux* 2. Neuchâtel; (1983) *Nos Oiseaux* 37, 53-64. GEYR VON SCHWEPPENBURG, H (1910) *Orn. Jahrb.* 21, 52-4; (1918) *J. Orn.* 66, 121-76; (1940) *Beitr. Fortpfl. Vögel* 16, 190-1; (1942) *Beitr. Fortpfl. Vögel* 18, 97-101. GIBB, J (1946) *Br. Birds* 39, 354-7; (1948) *Br. Birds* 41, 2-9, 34-40; (1956) *Ibis* 98, 506-30. GIBB, J and GIBB, C (1951) *Br. Birds* 44, 158-63. GIBBS, R G and WIGGINGTON, M J (1973) *Nat. Wales* 13, 158-62. GIL, A (1927) *Bol. real Soc.*

española *Hist. nat.* **27**, 81-96; (1928) *Bol. real Soc. española Hist. nat.* **28**, 171-94; (1944) *Bol. real Soc. española Hist. nat.* **42**, 177-97, 459-69, 553-64. GILL, E L (1936) *A first guide to South African birds.* Cape Town. GILLER, F (1955) *Vogelwelt* **76**, 180-4. GILLET, H (1960) *Oiseau* **30**, 99-134. GILLHAM, E H (1955) *Br. Birds* **48**, 549-50. GINN, H B (1975) *J. Orn.* **116**, 263-80. GINN, H B and MELVILLE, D S (1983) *Moult in birds.* Tring. GIZENKO, A I (1955) *Ptitsy Sakhalinskoy oblasti.* Moscow. GLADKOV, N A (1957) *Ibis* **99**, 269-74; (1962) *Ornitologiya* **4**, 15-28. GLADKOW, N A (1941) *J. Orn.* **89**, 124-56. GLADSTONE, H S (1910) *The birds of Dumfriesshire.* London. GLAS, J (1905) *Gef. Welt* **34**, 210-11. GLAYRE, D (1960) *Nos Oiseaux* **25**, 250-3; (1980) *Nos Oiseaux* **35**, 246. GLEGG, W E (1941) *Ibis* (14) **5**, 556-610. GLIEMANN, L (1976) *Falke* **23**, 134-5. GLOE, P (1982) *Ökol. Vögel* **4**, 209-11. GLOWACKI, J (1977) *Przeglad Zool.* **21**, 60-2. GLUE, D E (1973) *Br. Birds* **66**, 461-72; (ed) (1982) *The garden bird book.* London. GLUSHCHENKO, Y N (1981) In Litvinenko, N M (ed) *Redkie ptitsy Dal'nego Vostoka,* 25-33. Vladivostok. GLUTZ VON BLOTZHEIM, U N (1955) *Orn. Beob.* **52**, 152-7; (1962) *Die Brutvögel der Schweiz.* Aarau; (1966) *Orn. Beob.* **63**, 93-146; (1981) *Orn. Beob.* **78**, 212-14. GNIELKA, R (1969) *Orn. Mitt.* **21**, 179-88, 205-7. GODFREY, W E (1966) *The birds of Canada.* Ottawa. GODIN, J, GODIN, J, and LOISON, M (1977) *Aves* **14**, 88-9. GODIN, J and LOISON, M (1978) *Héron* **4**, 55-73. GODMAN, F DU C (1872) *Ibis* (3) **2**, 158-77. GOERTZ, A (1960) *Orn. Mitt.* **12**, 222. GOETHE, F (1933) *Mitt. Vogelwelt* **32**, 103-9. (1934) *Vogelzug* **5**, 183-8; (1973) *Vogelwelt* **94**, 27-8; (1977) *Orn. Mitt.* **29**, 195-6. GOLOVANOVA, E N (1967) *Ornitologiya* **8**, 342-4. GOLOVANOVA, E N and PUKINSKI, Y B (1971) *Puteshestvie v mir ptits.* Leningrad. GOODBODY, I M (1950) *Br. Birds* **43**, 265-71. GOODFELLOW, P F (1965) *Devon Birds* **18**, 3-5. GOODLIFFE, V H (1969) *Br. Birds* **62**, 284. GOODMAN, S M and AMES, P L (1983) *Sandgrouse* **5**, 82-96. GOODMAN, S M and STORER, R W (1985) *Bull. Br. Orn. Club* **105**, 84-5. GOODMAN, S M and WATSON, G E (1983) *Bull. Br. Orn. Club* **103**, 101-6; (1984) *Gerfaut* **74**, 145-61. GOODPASTURE, K A (1950) *Migrant* **21**, 37-41. GOODWIN, D (1950) *Br. Birds* **43**, 372; (1953a) *Br. Birds* **46**, 193-200; (1953b) *Br. Birds* **46**, 348-9; (1954) *Br. Birds* **47**, 81-3. GOODWIN, S H (1905) *Condor* **7**, 98-100. GÖRANSSON, G, HÖGSTEDT, G, KARLSSON, J, KÄLLANDER, H, and ULFSTRAND, S (1974) *Vår Fågelvärld* **33**, 201-9. GÖRANSSON, G and KARLSSON, J (1978) *Anser Suppl.* **3**, 90-5. GORDEEV, Y I (1963) *Ornitologiya* **6**, 469-70. GORDON, N J (1962) *Bird Migr.* **2**, 116-18. GORDON, S (1912) *The charm of the hills.* London; (1942) *Br. Birds* **36**, 73-4. GORE, M E J (1968) *Ibis* **110**, 165-96; (1981) *Birds of the Gambia.* Br. Orn. Union, London. GORE, M E J and PYONG-OH, W (1971) *The birds of Korea.* Seoul. GORIUP, P D (ed) (1983) *The Houbara Bustard in Morocco.* Rep. Al-Areen Wildlife Park, Bahrain/ICBP. GÖRNER, M (1971) *Falke* **18**, 225-7; (1978) *Falke* **25**, 282; (1981) *Mitt. zool. Mus. Berlin* **57**, Suppl. *Ann. Orn.* **5**, 63-70. GÓRSKI, W (1982) *Notatki orn.* **33**, 3-13. GOSZCZYNSKI, J (1981) *Ekol. Polska* **29**, 431-9. GOUGH, K (1947) *Br. Birds* **40**, 117. GOULLIART, A, ELVY, R J, OLIVER, P J, WHEELER, C E, and WILKINS, A C (1968) *Alauda* **36**, 123-4. GOULLIART, A, LEGRAND, B, and RICHARD, A (1965) *Alauda* **23**, 327-9. GOUTTENOIRE, G (1955) *Alauda* **23**, 1-64. GRABER, R G and GRABER, J W (1962) *Wilson Bull.* **74**, 74-88. GRÄFE, F, REQUATE, H, and VAUK, G (1962) *J. Orn.* **103**, 399-400. GRÄFF, H (1975) *Gef. Welt* **99**, 98-9. GRANT, C H B and MACKWORTH-PRAED, C W (1958) *Bull. Br. Orn. Club* **78**, 18; (1982) *Bull. Br. Mus. (nat. Hist.) Zool.* **1** (9). GRANT, P J (1968) *Bird Study* **15**, 106-7; (1972) *Br. Birds* **65**, 287-90; (1980) In Sharrock, J T R (ed) *The frontiers of bird identification,* 114-19. London. GRAY, D B (1973) *Bird Study* **20**, 80-2; (1974) *Bird Study* **21**, 280-2.

GREAVES, R H (1941) *Ibis* (14) **5**, 459-62. GRECH, J (1981-3) *Il-Merill* **22**, 16. GREEN, A A (1983) *Malimbus* **5**, 17-30. GREEN, D (1978a) *Br. Birds* **71**, 83-4. GREEN, R E (1978b) *J. Anim. Ecol.* **47**, 913-28; (1980) *J. appl. Ecol.* **17**, 613-30. GREEN, R H and MOLLISON, B C (1961) *Emu* **61**, 223-36. GREENWAY, J C and VAURIE, C (1958) *Breviora* **89**. GREENWOOD, J J D (1968) *Br. Birds* **61**, 524-5. GREENWOOD, P J and HARVEY, P H (1978) *Anim. Behav.* **26**, 1222-36. GREGORI, J (1977) *Larus* **29-30**, 83-8. GREIG-SMITH, P W (1979a) D Phil Thesis. Sussex Univ.; (1979b) *Ibis* **121**, 501-4; (1980) *Anim. Behav.* **28**, 604-19; (1981) *Behav. Ecol. Sociobiol.* **8**, 7-10; (1982a) *Ornis scand.* **13**, 225-31; (1982b) *Ornis scand.* **13**, 232-8; (1982c) *Anim. Behav.* **30**, 245-52; (1982d) *Anim. Behav.* **30**, 299-301; (1982e) *Ibis* **124**, 72-6; (1982f) *Bird Study* **29**, 162-4; (1983) *Behaviour* **86**, 215-36; (1984) *Ornis scand.* **15**, 11-15; (1985) *J. Zool. Lond.* **205**, 453-65. GREIG-SMITH, P W and QUICKE, D L J (1983) *Bird Study* **30**, 47-50. GREINER, R (1953) *Orn. Mitt.* **5**, 215. GRELING, C DE (1972) *Bull. Br. Orn. Club* **92**, 24-7. GRENQUIST, P (1947) *Ornis fenn.* **24**, 47-52. GRENQVIST, P (1935) *Ornis fenn.* **12**, 100-4; (1934) *Kócsag* **6**, 89-93. GRESSEL, J and PETERSEN, B (1981) *Vogelwelt* **102**, 106-7. GREVE, K (1969) *Orn. Mitt.* **21**, 196. GRIFFIN, D R (1965) *Bird migration.* London. GRIFFITHS, C I and ROGERS, T D (1975) *An interim list of the birds of Masirah Island, Oman.* RAF Masirah (duplicated). GRIMES, L (1972) *The passerine list of birds of the Accra Plains, Ghana.* Unpubl. MS, Edward Grey Inst, Oxford Univ.; (1976) *Ostrich* **47**, 1-15. GRISCOM, L and HARPER, F (1915) *Auk* **32**, 369. GRÖBBELS, F (1909) *Orn. Monatsber.* **17**, 114-17; GROEBBELS, F (1940) *Beitr. Fortpfl. Vögel* **16**, 64-5; (1950) *Orn. Abh.* **5**, 3-16. GROSSKOPF, G (1968) *Die Vögel der Insel Wangerooge.* Jever. GRÖSSLER, K (1959) *Orn. Mitt.* **11**, 208. GROTE, H (1919) *J. Orn.* **67**, 337-83; (1925) *Falco Suppl.*; (1932) *Beitr. Fortpfl. Vögel* **8**, 31; (1934) *Beitr. Fortpfl. Vögel* **10**, 99-105; (1935a) *Beitr. Fortpfl. Vögel* **11**, 215-17; (1935b) *Zool. Garten* NS **8**, 52-9; (1936) *Beitr. Fortpfl. Vögel* **12**, 133-9, 195-206; (1937) *Orn. Monatsber.* **45**, 114-34; (1939) *Orn. Monatsber.* **47** (2), 54-7. GRUBB, P (1966) *Scott. Birds* **4**, 310-12. GRÜLL, A (1981) *J. Orn.* **122**, 259-84. GRUNER, D (1977) *Bonn. zool. Beitr.* **28**, 77-81. GUBIN, B M and KOVSHAR', A F (1985) *Ornitologiya* **20**, 53-9. GUBITZ, C (1982) *Anz. orn. Ges. Bayern* **21**, 87-95; (1983) *Anz. orn. Ges. Bayern* **22**, 177-96. GUDMUNDSSON, F (1951) *Proc. int. orn. Congr.* **10**, 502-14. Uppsala; (1970) *Surtsey Res. Prog. Rep.* **5**, 1-9. GUERMEUR, Y (1977) *Ar Vran* **7**, 1-13. GUERMEUR, Y and MONNAT, J-Y (1980) *Ar Vran* **8**, 1-240. GUICHARD, G (1937) *Alauda* **9**, 368; (1960) *Oiseau* **30**, 239-45; (1963) *Oiseau* **33**, 183-8. GUICHARD, K M (1955) *Ibis* **97**, 393-424; (1957) *Ibis* **99**, 106-14. GUITIÁN RIVERA, J, SÁNCHEZ CANALS, J L, CASTRO LORENZO, A DE, and BAS LÓPEZ, S (1980) *Ardeola* **25**, 181-91. GULDI, R (1965) *Orn. Mitt.* **17**, 146. GÜLLAND, H, HIRSCHFELD, H, and HIRSCHFELD, K (1972) *Beitr. Vogelkde.* **18**, 174-206. GUNDELWEIN, E (1955) *Falke* **2**, 215. GUNTEN, K VON (1957) *Schweizer Nat.* **23**, 6-8; (1961) *Orn. Beob.* **58**, 13-34; (1963) *Orn. Beob.* **60**, 1-11. GUNTEN, K VON and SCHWARZENBACH, F H (1962) *Orn. Beob.* **59**, 1-22. GÜNTHER, R (1972) *Falke* **19**, 339-43. GUNZINGER, E (1983) *Orn. Beob.* **80**, 211-12. GURNEY, J H (1871) *Ibis* (3) **1**, 68-86; (1903) *Zoologist* (4) **7**, 121-38. GURR, L (1954) *Ibis* **96**, 225-61. GUSH, G H (1949) *Ibis* **91**, 526. GUTSCHER, H (1958) *Vogelwarte* **19**, 256. GÜTH, K (1956) *Orn. Mitt.* **8**, 77. GWINNER, E (1958) *Vogelwelt* **79**, 114. GYLLIN, R and KÄLLANDER, H (1980) *Vår Fågelvärld* **39**, 75-84. GYNGAZOV, A M and MILOVIDOV, S P (1977) *Ornitofauna Zapadno-Sibirskoy ravniny.* Tomsk.

HAAR, H (1975) *Egretta* **18**, 22. HAARHAUS, D (1973) *J. Orn.*

114, 71-8. HAARTMAN, L VON (1969a) *Comm. Biol. Soc. Sci. Fenn.* **32**; (1969b) *Ornis fenn.* **46**, 1-12; (1971) *Ornis fenn.* **48**, 93-100; (1973) In Farner, D S (ed) *Breeding biology of birds*, 448-81. Washington. HAAS, V (1980) Ph D Thesis. Tübingen Univ.; (1982a) *Ökol. Vögel* **4**, 17-58; (1982b) *Abstr. Congr. int. orn.* **18**, 204-5. HAAS, W (1969) *Alauda* **37**, 28-36. HACHISUKA, M U (1924) *Ibis* (11) **6**, 771-3; (1927) *A handbook of the birds of Iceland*. London. HAENSEL, J (1977) *Beitr. Vogelkde.* **23**, 9-30. HAFFER, J (1977) *Bonn. zool. Monogr.* **10**, 1-64. HAFTORN, S (1957) *Årbok Kgl. Norske Vidsk. Selsk. Mus.* 1956-7, 5-14; (1959) *Sterna* **3**, 229-37; (1971) *Norges Fugler*. Oslo. HÄGGLÖF, M (1954) *Vår Fågelvärld* **13**, 114-15, 120. HÅGVAR, S and ØSTBYE, E (1976) *Norw. J. Zool.* **24**, 53-64. HAKALA, J and TENOVUO, J (1968) *Ornis fenn.* **45**, 131-7. HÅLAND, A (1984) *Ann. zool. fenn.* **21**, 405-10. HALL, B P (1961) *Bull. Br. Mus. (nat. Hist.) Zool.* **7**, 245-89; (1963) *Bull. Br. Orn. Club* **83**, 133-4. HALL, B P and MOREAU, R E (1970) *An atlas of speciation in African passerine birds*. London. HALL, P (1977) *Bull. Niger. orn. Soc.* **13** (43), 15-36. HALLCHURCH, T T (1981) *Army Birdwatching Soc. Exped. Cyprus (Oct. 1980)* Rep. Dorset. HALL-CRAGGS, J (1962) *Ibis* **104**, 277-300; (1969) In Hinde, R A (ed) *Bird vocalizations*, 367-81. Cambridge; (1976) *Biophon* **4** (2), 6-7; (1977) *Br. Birds* **70**, 300; (1978) In Merson, J (ed) *Investigating music*, 8-20 (with tape). Australian Broadcasting Commission; (1979) *Condor* **81**, 185-92; (1984) *Wildl. Sound* **4** (7), 20-9. HALLER, W (1934) *Orn. Beob.* **31**, 169-71. HALLER, W and HUBER, J (1937) *Orn. Monatsber.* **45**, 81-2 HAMILTON, E (1978) *J. Edinb. nat. Hist. Soc.*, 39-40. HAMMER, U (1977) *Orn. Mitt.* **29**, 62-3. HAMMERSCHMIDT, R (1966) *Orn. Mitt.* **18**, 235. HAMMLING, J (1909) *Orn. Monatsber.* **17**, 129-38; (1914) *Orn. Monatsber.* **22**, 73-5. HAMMOND, J (1912) *J. agric. Sci.* **4**, 380-409. HAMMONDS, E (1984) *Wildl. Bahrain* **3**, 9-73. HANCOCK, M (1965) *Br. Birds* **58**, 155-6; (1969) *Br. Birds* **62**, 285. HANDMANN, W (1973) *Falke* **20**, 174. HANEDA, K and OBUCHI, J (1967) *Misc. Rep. Yamashina Inst. Orn. Zool.* **5**, 72-84. HANMER, D B (1977) *Nyala* **3** (2), 44; (1979) *Scopus* **3**, 81-92. HANNOVER, R (1977) *Vogelk. Hefte Edertal* **3**, 51-5. HANSSEN, O J (1984) *Vår Fuglefauna* **7**, 188-96. HANTGE, E and SCHMIDT-KOENIG, K (1958) *J. Orn.* **99**, 142-59. HANTZSCH, B (1905) *Beitrag zur Kenntnis der Vogelwelt Islands*. Berlin. HARBER, D D (1948) *Br. Birds* **41**, 348. HARDING, B D (1986) *Br. Birds* **79**, 405. HARDMAN, J A (1974) *Ann. appl. Biol.* **76**, 337-41. HARDY, A R (1977) Ph D Thesis. Aberdeen Univ. HARDY, E (1946) *A handlist of the birds of Palestine*. Unpubl. MS, Edward Grey Inst., Oxford Univ.; (1964) *Birds Illust.* **10**, 56-7; (1969) *Bird Study* **16**, 191-2. HARLEY, B (1955) *Br. Birds* **48**, 188. HARPER, D G C (1984a) Ph D Thesis. Cambridge Univ.; (1984b) *Ring. Migr.* **5**, 101-4; (1985a) *Ibis* **127**, 262-6; (1985b) *Anim. Behav.* **33**, 466-80; (1985c) *Anim. Behav.* **33**, 862-75; (1985d) *Anim. Behav.* **33**, 876-84. HARPER, D, HARPER, W, and HARPER, E (1973) *Bird Life* Oct-Dec, 32. HARPER, J and HARPER, L (1974) *Bull. E. Afr. nat. Hist. Soc.*, 113-15. HARREVELD, A P VAN (1985) *Dutch Birding* **7**, 106-7. HARRIS, M P (1962) *Br. Birds* **55**, 97-103. HARRISON, C J O (1966) *Ibis* **108**, 573-83; (1970a) In Gooders, J (ed) *Birds of the world*, 1798-9. London; (1973) *Br. Birds* **66**, 225-7; (1975) *A field guide to the nests, eggs and nestlings of European birds*. London; (1982) *An atlas of the birds of the western Palearctic*. London. HARRISON, C J O and FORSTER, J (1959) *Bird Study* **6**, 60-8. HARRISON, J G (1943) *Br. Birds* **36**, 36-7; (1944) *Br. Birds* **37**, 234-5; (1970b) In Gooders, J (ed) *Birds of the world*, 1782-3. London. HARRISON, J M (1954) *Bull. Br. Orn. Club* **74**, 96; (1962) *Bull. Br. Orn. Club* **82**, 75. HARRISON, J M D (1952) *Br. Birds* **45**, 368. HARRISSON, T H and BUCHAN, J N S (1934) *J. Anim. Ecol.* **3**, 133-45; (1936) *Scott. Nat.* **217**, 9-21. HART,

K E (1984) *Scott. Birds* **13**, 116. HARTERT, E (1902) *Novit. Zool.* **9**, 322-39; (1910) *Die Vögel der palaärktischen Fauna* 1. Berlin; (1913) *Novit. Zool.* **20**, 1-76; (1915) *Novit. Zool.* **22**, 61-79; (1921) *Novit. Zool.* **28**, 78-141; (1921-2) *Die Vögel der palaärktischen Fauna* 3. Berlin; (1923) *Novit. Zool.* **30**, 1-32; (1926) *Bull. Soc. Sci. nat. Maroc* **5**, 271-304. HARTERT, E and STEINBACHER, F (1932-8) *Die Vögel der palaärktischen Fauna, Ergänzungsband*. Berlin. HARTHAN, A J (1934) *Br. Birds* **28**, 50-2. HARTLEY, P H T (1941) *Br. Birds* **34**, 256-8; (1946a) *Br. Birds* **39**, 44-7; (1946b) *Br. Birds* **39**, 142-4; (1948) *Ibis* **90**, 361-81; (1949) *Ibis* **91**, 393-413; (1954) *Br. Birds* **47**, 97-107; (1955) *Trans. Suffolk Nat. Soc.* **9** (2), 158-62; (1967) *Birds* **1**, 271-3. HARVEY, B F (1946) *Br. Birds* **39**, 247-8. HARVEY, G H (1923) *Br. Birds* **17**, 84. HARVEY, H J (1973) *Br. Birds* **66**, 448. HARWOOD, J and HARRISON, J (1977) *Bird Study* **24**, 47-53. HASLAM, S H (1844) *Zoologist* (1) **2**, 564. HASSE, H (1965) *Orn. Mitt.* **17**, 192. HASSON, O (1978) MA Thesis. Tel-Aviv Univ.; (1983) *Israel Land Nat.* **8**, 104-9. HAUKIOJA, E (1969) *Ornis fenn.* **46**, 171-8; (1971) *Ornis fenn.* **48**, 101-16. HAUKIOJA, E and KALINAINEN, P (1969) *Ann. Rep. orn. Soc. Pori* **2**, 75-8; (1972) *Ann. Rep. orn. Soc. Pori* **3**, 5-16. HAUN, M (1930) *Beitr. Fortpfl. Vögel* **6**, 79-81; (1931) *Beitr. Fortpfl. Vögel* **7**, 135-8. HAURI, R (1966) *Orn. Beob.* **63**, 223-6; (1968) *Orn. Beob.* **65**, 192-4. HAUSMANN, S (1983) *Verh. orn. Ges. Bayern* **23**, 515-18; HAVLÍN, J (1961) *Zool. Listy* **10**, 243-8; (1977) *Folia Zool.* **26**, 45-56. HAVLÍN, J and JURLOV, K T (1977) *Acta Sci. Nat. Brno* **11** (2), 1-50. HAWTHORN, I (1971) *Ringers' Bull.* **3** (9), 9-11; (1974) *Bird Study* **21**, 88-91; (1975a) *Bird Study* **22**, 19-23; (1975b) *Bird Study* **22**, 84; (1980) *Shetland Bird Rep. 1979*, 44-7. HAWTHORN, I, CROCKFORD, R, SMITH, R G, and WESTON, I (1976) *Bird Study* **23**, 301-3. HAWTHORN, I and MEAD, C J (1975) *Br. Birds* **68**, 349-58. HAWTHORN, I, WESTON, I, CROCKFORD, R, and SMITH, R G (1971) *Bird Study* **18**, 27-9. HAYNES, V M (1980) *Br. Birds* **73**, 104-5. HEER, E (1978) *Orn. Mitt.* **30**, 233. HEERDE, H (1982) *Beitr. Naturkde. Wetterau* **2**, 11-24. HEGAZI, E M (1981) *J. agric. Sci.* **96**, 497-501. HEIM DE BALSAC, H (1924) *Rev. fr. Orn.* **8**, 338-57, 372-92, 411-22; (1925) *Rev. fr. Orn.* **9**, 170; (1926) *Mém. Soc. Hist. nat. Afr. Nord* 1; (1931) *Alauda* **3**, 250-6; (1948) *Alauda* **16**, 75-96; (1949-50) *Alauda* 17-18, 183-5; HEIM DE BALSAC, H and HEIM DE BALSAC, T (1949-50) *Alauda* 17-18, 129-43; (1951) *Alauda* **19**, 19-39, 97-112; (1954) *Alauda* **22**, 145-205. HEIM DE BALSAC, H and MAYAUD, N (1951) *Alauda* **19**, 137-51; (1962) *Les oiseaux du nord-ouest de l'Afrique*. Paris. HEINROTH, O and HEINROTH, M (1924-6) *Die Vögel Mitteleuropas* 1; (1931) 4. Berlin-Lichterfelde. HEINZEL, H, FITTER, R, and PARSLOW, J (1972) *The birds of Britain and Europe with North Africa and the Middle East*. London. HELBIG, A (1984) *Bonn. zool. Beitr.* **35**, 57-69. HELLMICH, J (1982) *Orn. Mitt.* **34**, 211-16; (1983) *Orn. Mitt.* **35**, 301-3; (1984) *Orn. Mitt.* **36**, 9-16. HELLMICH, J and ZWERNEMANN, A (1983) *Vogelwelt* **104**, 161-8. HELMINEN, M (1958) *Ornis fenn.* **35**, 51-64. HÉMERY, G, NICOLAU-GUILLAUMET, P, and THIBAULT, J-C (1979) *Oiseau* **49**, 213-30. HEMMINGSEN, A M and GUILDAL, J A (1968) *Spolia zool. Mus. haun.* **28**. HEMPEL, C (1957) *Vogelwarte* **19**, 25-36. HEMPEL, C and REETZ, W (1957) *Vogelwarte* **19**, 97-119. HEMS, H A (1966) *Br. Birds* **59**, 107-8. HEMSLEY, J H and GEORGE, M (1966) *Azraq Desert National Park, Jordan, Draft Management Plan*. Duplicated (IBP/CT, London). HENLE, K (1983) *Vogelwarte* **32**, 57-76. HENNEMANN, W (1921) *Orn. Monatsber.* **29**, 103-5. HENNY, C J (1972) *US Bur. Sport Fish. Wildl. Wildl. Res. Rep.* 1. HENTY, C J (1961) *Ibis* **103a**, 28-36; (1975) *Doñana Acta Vert.* **2**, 215-20; (1976) *Wilson Bull.* **88**, 497-9; (1980) *Forth Nat. Hist.* **5**, 67-71; (1986) *Br. Birds* **79**, 277-81. HENZE, O (1943) *Vogelschutz gegen Insektenschaden in der Forstwirtschaft*. Munich; (1958)

Vogelwelt **79**, 17-19. HEREWARD, A C (1979) *Ring. Migr.* **2**, 113-17. HERKLOTS, G A C (1967) *Hong Kong birds*. Hong Kong. HERREN, H (1969) *Orn. Beob.* **66**, 22. HERRERA, C M (1977) *Doñana Acta Vert.* **4**, 35-59; (1978a) *Doñana Acta Vert.* **5**, 61-71; (1978b) *Ibis* **120**, 542-5; (1981) *Oikos* **36**, 51-8; (1984a) *Ecol. Monogr.* **54**, 1-23; (1984b) *Ardeola* **30**, 77-81. HERRERA, C M and JORDANO, P (1981) *Ecol. Monogr.* **51**, 203-18. HERRERA, C M and RODRIGUEZ, M (1979) *Ring. Migr.* **2**, 160. HERRING, J (1984) *Br. Birds* **77**, 365. HERRMANN, J (1985-6) *Gef. Welt* **109**, 337-9, **110**, 15-17. HERROELEN, P (1959) *Gerfaut* **49**, 11-30; (1960) *Gerfaut* **50**, 87-99; (1970) *Gerfaut* **60**, 278-86; (1979) *Orn. Brabant* **82**, 2-7. HESS, A (1920) *Orn. Monatsber.* **28**, 35-6; (1927) *Orn. Beob.* **24**, 249. HESSE, E (1909) *J. Orn.* **57**, 322-65; (1917) *Orn. Monatsber.* **25**, 143-4. HEUGLIN, M T VON (1869) *Ornithologie Nordost-Afrika's* I. Kassel. HEWSON, R (1967) *Br. Birds* **60**, 244-52; (1969) *Bird Study* **16**, 89-100. HEYDER, D (1950) *Zool. Garten* NS **17**, 242-9; (1980) *Beitr. Vogelkde.* **26**, 122-4. HEYDER, R (1934) *Mitt. Ver. sächs. Orn.* **4**, 199-209; (1953) *Die Amsel*. Wittenberg Lutherstadt. HICKLING, R A O (1959) *Ibis* **101**, 497-502. HIGUCHI, H and HIRANO, T (1983) *Tori* **32**, 1-11. HILDÉN, O (1965) *Ann. zool. fenn.* **2**, 53-75; (1974) *Ornis fenn.* **51**, 10-35. HILDÉN, O and KOSKIMIES, P (1984) *Lintumies* **19**, 15-25. HILLMAN, A K K and YOUNG, C M A (1977) *Ibis* **119**, 206-7. HILLSTEAD, A F C (1945) *The Blackbird*. London. HILPRECHT, A (1954) *Nachtigall und Sprosser*. Wittenberg Lutherstadt. HIMMER, K H (1967) *Gef. Welt* **91**, 188-92. HINDE, R A (1956) *Ibis* **98**, 340-69. HINDEMITH, J (1972) *Vogelwelt* **93**, 71-2. HINDLE, C (1979) *Br. Birds* **72**, 38. HINSCHE, A (1958) *Beitr. Vogelkde.* **6**, 159-71; (1960) *Beitr. Vogelkde.* **7**, 129-32. HIRSCHFELD, K (1969) *Apus* **1**, 270-6. HOBBS, J T (1950) *Br. Birds* **43**, 256-7. HODGSON, C J (1978) *Br. Birds* **71**, 313-14. HODSON, N L (1962) *Br. Birds* **55**, 166. HOEHL, O (1941) *Beitr. Fortpfl. Vögel* **17**, 30-1. HOELZEL, A R (1986) *Ibis* **128**, 115-27. HOESCH, W and NIETHAMMER, G (1940) *J. Orn.* **88** suppl. HOFFMANN, H J (1951) *Br. Birds* **44**, 387. HOFFMAN, K (1959) *J. Orn.* **100**, 84-9. HOFFMANN, B (1917) *Verh. orn. Ges. Bayern* **13**, 61-73; (1936) *Verh. orn. Ges. Bayern* **21**, 65-70. HOFFMANN, L (1958) *Br. Birds* **51**, 321-49. HOFFMANN, M (1976) *Regulus* **12**, 25. HOGG, P (1974) *Ibis* **116**, 466-76. HOGG, P, DARE, P J, and RINTOUL, J V (1984) *Ibis* **126**, 307-31. HOGSTAD, O (1977) *Sterna* **16**, 57-60; (1981) *Fauna norv.* (C) *Cinclus* **5**, 1-4; (1983) *Ibis* **125**, 366-9. HÖGSTEDT, G (1969) *Medd. SKOK* **8**, 28-31; (1978) *Ornis scand.* **9**, 193-6. HOHL, J (1981) *Vögel Heimat* **52**, 17. HOHLT, H (1957a) *J. Orn.* **98**, 71-118; (1957b) *Vogelwelt* **78**, 48-53. HÖHN, E O (1951) *Can. Fld.-Nat.* **65**, 168-9. HOLDEN, P (1985) *Br. Birds* **78**, 403-4. HOLGERSEN, H (1962) *Sterna* **5**, 27-30; (1982) *Sterna* **17**, 85-123. HOLLICK, K (1980) *Br. Birds* **73**, 417. HOLLOM, P A D (1930) *Br. Birds* **23**, 248-9; (1955) *Ibis* **97**, 1-17; (1959) *Ibis* **101**, 183-200; (1966) *Br. Birds* **59**, 502; (1980) *The popular handbook of rarer British birds*. London; (1985) *Br. Birds* **78**, 240. HOLLYER, J N (1972) *Br. Birds* **65**, 170; (1975) *Kent Bird Rep.* **22**, 84-95. HOLMBERG, L (1952) *Vår Fågelvärld* **11**, 130. HOLMBRING, J-Å (1972) *Vår Fågelvärld* **31**, 16-19. HOLMBRING, J-Å and KJEDEMAR, H (1968) *Vår Fågelvärld* **27**, 97-121. HOLMES, D A (1974) *Bull. Niger. orn. Soc.* **10** (37), 28-36. HOLMES, G (1984) *Austral. Bird Watcher* **10**, 164-6. HOLMES, P F (1939) *Br. Birds* **32**, 350-1. HOLT, E G (1960) *Br. Birds* **53**, 313-14. HOLUPIREK, H (1977) *Beitr. Vogelkde.* **23**, 161-76. HÖLZINGER, J (1969) *Anz. Ges. Bayern* **8**, 610-24; (1972) *Mitt Bad. Landver. Naturkde. Naturschutz* NF **10** (3), 583-92. HÖLZINGER, J, KNÖTZSCH, G, KROYMANN, B, and WESTERMANN, K (1970) *Anz. orn. Ges. Bayern* **9**, suppl. HOMANN, P (1960) *J. Orn.* **101**, 195-224. HOOGLAND, J L and SHERMAN, P W (1976) *Ecol. Monogr.* **46**,

33-58. HOOKER, T (1958) *Ibis* **100**, 446-9. HOPE JONES, P (1961) *Br. Birds* **55**, 178-81; (1979) *Nat. Wales* **16**, 267-9. HORNBUCKLE, J (1978) *Br. Birds* **71**, 590-1. HORNBY, H E (1973) *Honeyguide* **74**, 30-1, 37. HORNE, E (1924) *Br. Birds* **17**, 306-7. HORNER, K O and HUBBARD, J P (1982) *Cyprus orn. Soc.* (1969) *Rep.* **7**, 54-104. HORSFIELD, H K (1915) *Scott. Nat.* **44**, 263-4. HORST, F (1941) *Beitr. Fortpfl. Vögel* **17**, 63-71. HORSTKOTTE, E (1962) *Ber. naturw. Ver. Bielefeld* **16**, 107-65; (1965) *Ber. naturw. Ver. Bielefeld* **17**, 67-145; (1969) *J. Orn.* **110**, 62-70; (1971) *Orn. Mitt.* **23**, 125-9. HORVÁTH, L (1956) *Bull. Br. Orn. Club* **76**, 132; (1958) *Bull. Br. Orn. Club* **78**, 124-5; (1959a) *Acta Zool. Acad. Sci. Hung.* **5**, 353-67; (1959b) *Ann. Hist.-Nat. Mus. Nat. Hungarici* **51**, 451-81; (1977) *Aquila* **83**, 167-71. HORVÁTH, L, KEVE, A, and MARIAN, M (1964) *Ann. Hist.-Nat. Mus. Nat. Hungarici* **56**, 529-28. HOSKING, E and LOWES, H (1947) *Masterpieces of bird photography*. London. HOSKING, E J and NEWBERRY, C W (1946) *More birds of the day*. London; (1949) *Birds in action*. London. HOSTIE, P (1937) *Gerfaut* **27**, 204-5. HÖTKER, H (1982) *Vogelwelt* **103**, 1-16. HÖTKER, H and SUDFELDT, C (1978) *Vogelwelt* **99**, 189-90; (1979a) *Vogelwelt* **100**, 112-13; (1979b) *J. Orn.* **120**, 324-5; (1982a) *J. Orn.* **123**, 183-201; (1982b) *Vogelwelt* **103**, 178-87. HOUSTON, A I, McCLEERY, R H, and DAVIES, N B (1985) *J. Anim. Ecol.* **54**, 227-39. HOWARD, L (1952) *Birds as individuals*. London; (1956) *Living with birds*. London. HUBATSCH, H (1983) *Charadrius* **19**, 23-6. HUBBARD, J (1969) *Cyprus orn. Soc.* (1957) *Rep.* **15**, 23-5. HUBER, J (1953) *Orn. Beob.* **50**, 29; (1954a) *Larus* **8**, 76-95; (1954b) *Orn. Beob.* **51**, 136-7. HUDEC, K (1957) *Zool. Listy* **20**, 197-214. HUDEC, K and ŠTASTNY, K (1979) *Egretta* **22**, 18-26. HUDSON, R (1979) *Bird Study* **26**, 204-12. HÜE, F (1952) *Alauda* **20**, 261-4. HÜE, F and ETCHÉCOPAR, R-D (1958) *Terre Vie* **105**, 186-219; (1970) *Les oiseaux du proche et du moyen orient*. Paris. HUGHES, A (1977) *Bird Study* **24**, 195. HUGHES, S W M (1972) *Sussex Bird Rep.* **24**, 68-79; (1975) *Sussex Bird Rep.* **27**, 64-7; (1980) *Br. Birds* **73**, 414-15. HUHTALO, H and JÄRVINEN, O (1977) *Bird Study* **24**, 179-85. HUI-FRÜH, M (1975) *Vögel Heimat* **45**, 173-4. HULTSCH, H (1981) *Verh. dtsch. zool. Ges.* **241**, 240. HULTSCH, H and TODT, D (1981) *Behav. Ecol. Sociobiol.* **8**, 183-8. HUMPHREY, P S and PARKES, K C (1959) *Auk* **76**, 1-31. HUND, K (1974) *Orn. Mitt.* **26**, 151; (1976) *Orn. Mitt.* **28**, 169-78. HUND, K and PRINZINGER, R (1979a) *Ökol. Vögel* **1**, 133-58; (1979b) *Vogelwarte* **30**, 107-17; (1985) *J. Orn.* **126**, 15-28. HUNT, D (1985) *Confessions of a Scilly birdman*. London. HUNZIKER-LÜTHY, G (1970) *Orn. Beob.* **67**, 299; (1973) *Orn. Beob.* **68**, 223. HUSBAND, C I (1967) *Br. Birds* **60**, 168-9. HUSSELL, D J T (1969) *Auk* **86**, 75-83. HUTCHINSON, C (1932) *Vasculum* **18**, 19-20. HUTSON, H P W (1954) *The birds about Delhi*. Kirkee. HUTTON, A E (1978) *Wilson Bull.* **90**, 396-403. HUXLEY, J S (1949) *Br. Birds* **42**, 185-6. HYDE-PARKER, T (1938) *Bird Notes News* **18**, 3-4.

IJZENDOORN, A L J VAN (1950) *The breeding-birds of the Netherlands*. Leiden. IL'YASHENKO, V Y (1982) *Ornitologiya* **17**, 183-4. ILYICHEV, V D and FLINT, V E (1982) *Ptitsy SSSR*. Moscow. IMPE, J VAN (1971) *Ardeola* **15**, 82-5. INBAR, R (1977) *Birds of Israel*. Tel-Aviv; (1982) *Israel Land Nat.* **7**, 111-15. INCE, S A (1982) *D Phil Thesis*. Sussex Univ. INCE, S A and SLATER, P J B (1985) *Ibis* **127**, 355-64. INCLEDON, C S L (1968) *Br. Birds* **61**, 550-3. INGLIS, I R, FLETCHER, M R, FEARE, C J, GREIG-SMITH, P W, and LAND, S (1982) *Ibis* **124**, 351-5. INGRAM, C (1955) *Oiseau* **25**, 147-50. INGRAM, G C S, SALMON, H M, and TUCKER, B W (1938) *Br. Birds* **32**, 58-63. INNES, J L C (1981) *Wicken Fen Group Rep.* **10**, 16-19. INOZEMTSEV, A A (1962) *Nauch. dokl. vyssh. Shk. Biol. nauki* **2**, 55-7; (1963) *Ornitologiya* **6**, 101-3. INSKIPP, C and INSKIPP, T (1985) *A guide* to the birds of Nepal. London. INSLEY, H and WOOD, J B (1972) Nat. Wales 13, 165-73. IOALE', P and BENVENUTI, S (1982) Avocetta 6, 63-74. IRBY, L H L (1875) The Ornithology of the Straits of Gibraltar. London. IRISOV, E I (1967) Ornitologiya 8, 355-6. IRVINE, G and IRVINE, D (1974) Bull. E. Afr. nat. Hist. Soc., 158-9. IRVING, L (1960) Bull. US natn. Mus. 217. IRWIN, M P S (1958) Occ. Pap. natn. Mus. S. Rhodesia 22B, 198-201; (1960) Bull. Br. Orn. Club 80, 61-4; (1978) Honeyguide 96, 21-2; (1981) The birds of Zimbabwe. Salisbury. ISAAC, D and MARLER, P (1963) Anim. Behav. 11, 179-88. ISAKOV, Y A and VOROBIEV, K A (1940) Trudy Vsesoyuz. orn. zapoved. Gassan-Kuli 1, 5-159. ISENMANN, P (1965) Alauda 33, 248-9; (1972) Sterna 11, 256-7; (1985) Alauda 53, 231-2. IVANITSKI, V V (1978) Zool. Zh. 57, 1555-65; (1980) Zool. Zh. 59, 587-97, 739-49; (1981a) Zh. obshch. Biol. 42, 708-20; (1981b) Zool. Zh. 60, 1212-21; (1981c) Nauch. dokl. vyssh. Shk. Biol. nauki 7, 53-8; (1981d) Priroda 3, 54-63; (1982) Zool. Zh. 61, 71-81. IVANOV, A I (1945) Izv. Tadzhik. Fil. Akad. Nauk SSSR 6, 36-59; (1969) Ptitsy Pamiro-Alaya. Leningrad. IVASHCHENKO, A A (1982) In Merauskas, P (ed) Ekologicheskie issledovaniya i okhrana ptits Pribaltiyskikh respublik, 35-8. Kaunas. IZMAYLOV, I V and BOROVITSKAYA, G K (1967) Ornitologiya 8, 192-7.

JACKSON, F J (1938) Birds of Kenya Colony and the Uganda Protectorate 2. London. JACKSON, R and LONG, J (1973) Hampshire Bird Rep. 1972, 59-68. JACKSON, R D (1958) Irish Nat. J. 12, 229-36. JAHN, H (1942) J. Orn. 90, 7-302. JAMES, B and JAMES, C M (1955) Br. Birds 48, 373-4. JÁNOSSY, L and TÖRÖK, J (1977) Aquila 83, 307. JARRY, G (1969) Oiseau 39, 112-20; (1980) Oiseau 50, 277-94. JARRY, G and LARIGAUDERIE, F (1971) Bull. Br. Orn. Club 91, 32; (1974) Oiseau 44, 62-71. JÄRVINEN, A (1978) Ornis fenn. 55, 69-76; (1981) Ornis fenn. 58, 129-31. JÄRVINEN, A and PRYL, M (1980) Kilpisjärvi Notes 4, 1-7. JÄRVINEN, A, JÄRVINEN, L, PIETIÄINEN, H, and PRYL, M (1980) Lintumies 15, 129-34. JÄRVINEN, O (1980) Birds 8 (4), 30-1; JARVIS, D (1973) Bird Life Oct-Dec, 29. JEFFREY, B and WALLACE, D I M (1956) Br. Birds 49, 454-6. JEHL, H (1974) Alauda 42, 397-405. JENKS, S (1970) Bird Life 6, 128. JENNER, E W C (1945) Br. Birds 38, 237-8. JENNI, L (1981) Orn. Beob. 78, 52-3; (1984) Orn. Beob. 81, 183-213. JENNI, L and WINKLER, R (1983) Orn. Beob. 80, 203-7. JENNING, W (1954) Vår Fågelvärld 13, 167-71. JENNINGS, M C (1980) Sandgrouse 1, 71-81; (1981a) The birds of Saudi Arabia: a check-list. Whittlesford, Cambridgeshire; (1981b) Birds of the Arabian Gulf. London; (1981c) J. Saudi Arab. nat. Hist. Soc. 2 (1), 8-14. JENNY, O (1946) Orn. Beob. 43, 194. JENSEN, J V and KIRKEBY, J (1980) The birds of the Gambia. Aarhus. JOBSON, G J (1978) Br. Birds 71, 312-13. JÖGI, A (1966) Soobshch. Pribalt. Kom. Izuch. Migr. Ptits 4, 128-35. JOHANSEN, H (1946) Dansk orn. Foren. Tidsskr. 40, 121-42; (1952) J. Orn 92, 1-105, 145-204; (1954) J. Orn. 95, 319-42; (1955) J. Orn. 96, 58-91. JOHANSEN, O (1984) Fauna 37, 53-5. JOHANSSON, S (1980) Vår Fågelvärld 39, 103. JOHNS, R (1970) In Gooders, J (ed) Birds of the world, 2005-7. London. JOHNSGARD, P A (1979) Birds of the Great Plains. Lincoln, Neb. JOHNSON, E D H (1961) Br. Birds 54, 213-25; (1971a) Br. Birds 64, 201-13, 267-79; (1971b) Bull. Br. Orn. Club 91, 103-7. JOHNSON, H S (1933) Wilson Bull. 45, 114-17. JOHNSON, I G (1970) Bird Study 17, 297-319. JOHNSON, L R (1958a) Iraq nat. Hist. Mus. Publ. 16, 1-32. JOHNSON, O W (1958b) Jack-Pine Warbler 36, 173-7. JOHNSSON, T (1981) Anser 20, 203-4. JOHNSTON, A F (1985) Br. Birds 78, 242. JOHNSTON, D W and HAINES, T P (1957) Auk 74, 447-58. JOHNSTON, T L, BLEZARD, E, and ELLISON, N F (1943) North-west. Nat. 18, 206-7. JON, Z (1970) Sylvia 18, 242-4. JONES, A E (1947-8) J. Bombay nat. Hist. Soc. 47, 117-

25, 219-49, 409-32. JONES, J W and KING, G M (1952) Br. Birds 45, 400-1. JONES, M (1965) Br. Birds 58, 309-12. JONES, P H (1961) Oiseau 31, 193-213. JONES, R E (1955) Br. Birds 48, 458. JONSSON, L (1978a) Birds of lake, river, marsh and field. Harmondsworth; (1978b) Birds of sea and coast. Harmondsworth; (1979) Birds of mountain regions. Harmondsworth; (1982) Birds of the Mediterranean and Alps. London. JORDANIA, R (1970) Falke 17, 184-5. JORDANIA, R G (1974) Ornitologiya 11, 371-3. JORDANO, P (1982) Oikos 38, 183-93. JORDANS, A VON (1924) J. Orn. 72, 145-70; (1928) Novit. Zool. 34, 262-336; (1933) Anz. orn. Ges. Bayern 2, 250-66; (1950) Syllegom. biol., 165-81. JORDANS, A VON and STEINBACHER, J (1941) Ann. naturhist. Mus. Wien 52, 200-44; (1943) Senckenbergiana 26, 72-86; (1948) Senckenbergiana 28, 159-86. JØRGENSEN, O H (1970) Dansk orn. Foren. Tidsskr. 64, 70-7; (1976) Ornis scand. 7, 13-20; (1977) Dansk orn. Foren. Tidsskr. 71, 121-38. JOST, O (1969) J. Orn. 110, 71-8; (1972) Luscinia 41, 298-301; (1975) Bonn. zool. Monogr. 6. JOUARD, H (1926) Gerfaut 16, 57-66. JOURDAIN, F C R, WALLIS, H M, and RATCLIFF, F R (1915) Ibis (1O) 3, 133-69. JÓZEFIK, M (1962) Acta Orn. 7, 69-87. JUANA, E DE and SANTOS, T (1981) Alauda 49, 1-12. JUANA ARANZANA, E DE (1980) Atlas ornitologico de la Rioja. Logroño. JUKEMA, J and RIJPMA, U (1984) Limosa 57, 91-6. JUNG, K (1967) Vogelwelt 88, 181-4. JUON, M (1968) Orn. Beob. 65, 194.

KAINADY, P V G (1976) Bull. Basrah nat. Hist. Mus. 3, 95-100. KAISER, W (1961) Naturschutzarb. Mecklenb. 4, 19-35. KALABÉR, L (1978) Héron 4, 89-104. KALBE, L (1961) Vogelwelt 82, 174-9. KANUŠČÁK, P and KUBÁN, V (1969) Sb. slov. národ. Múz. Přír. Ved. 15, 153-8. KAPITONOV, V I and CHERNYAVSKI, F B (1960) Ornitologiya 3, 80-97. KARANJA, W K (1982) D Phil Thesis. Oxford Univ. KAREILA, R (1961) Ornis fenn. 38, 65-72. KARLSSON, J and KÄLLANDER, H (1977) Ornis scand. 8, 139-44. KARLSSON, L, PERSSON, K, and WALINDER, G (1986) Anser 25, 15-28. KARVIK, N-G (1952) Vår Fågelvärld 11, 76-80. KASPAREK, M (1976) Vogelwelt 97, 121-32. KASTEPOLD, T and KABAL, R (1982) Väljaspool Eesti 2. Tallinn. KATE, C G B TEN (1933) Org. Club nederl. Vogelk. 5, 118-21. KEICHER, K (1983) Ökol. Vögel 5, 203-16. KEKILOVA, A F (1969) Izv. Akad. Nauk Turkmen. SSR Ser. biol. 4, 43-50; (1970) Izv. Akad. Nauk Turkmen. SSR Ser. biol. 4, 72-5. KELSO, L (1931) Condor 33, 60-5. KEMPF, C (1976) Oiseaux d'Alsace. Strasbourg; (1982) Alauda 50, 278-85. KENDRICK, J S (1985) Br. Birds 78, 353. KENNEDY, C H (1913) Condor 15, 135-6. KENNEDY, P G, RUTTLEDGE, R F, and SCROOPE, C F (1954) The birds of Ireland. Edinburgh. KENTISH, B J (1976) M Sc Thesis. Durham Univ. KERAUTRET, L (1979) Héron 4, 37-47. KERSTEN, M, PIERSMA, T, SMIT, C, and ZEGERS, P (1983) Rep. Neth. Morocco Exped. 1981. Texel. KEVE, A (1950) Larus 3, 55-62; (1952) Larus 4-5, 74-83; (1958) Bull. Br. Orn. Club 78, 48-51; (1967) Beitr. Vogelkde. 13, 135; (1978) Anz. orn. Ges. Bayern 17, 225-37. KEYSERLINGK, A VON (1937) Orn. Monatsber. 45, 185-8. KHOKHLOVA, N A (1960) Ornitologiya 3, 259-69. KIERSKI, W (1934) Beitr. Fortpfl. Vögel 10, 188. KIIS, A (1981) Dansk orn. Foren. Tidsskr. 75, 144-6. KING, B (1953) Br. Birds 46, 415-16; (1954) Br. Birds 47, 444; (1955) Br. Birds 48, 548-9; (1956) Br. Birds 49, 502-3; (1958) Br. Birds 51, 121-2; (1960a) Br. Birds 53, 200; (1960b) Br. Birds 53, 224-5; (1966) Br. Birds 59, 501; (1967) Br. Birds 60, 342; (1968) Br. Birds 61, 315; (1970) Br. Birds 63, 37-8; (1978a) Br. Birds 71, 463; (1978b) J. Saudi Arab. nat. Hist. Soc. 1 (21), 3-24; (1979) Bristol Orn. 12, 70; (1980) Amer. Birds 34, 317-18; (1983) Bristol Orn. 16, 40-1; (1986) Br. Birds 79, 340. KING, B and LADHAMS, D E (1970) Bristol Orn. 1 (3), 127-8. KING, B and PENHALLURICK,

R D (1977) *Br. Birds* **70**, 341. KING, B, WOODCOCK, M, and DICKINSON, E C (1975) *A field guide to the birds of south-east Asia.* London. KINNEAR, N B (1934) *J. Bombay nat. Hist. Soc.* **37**, 347-68. KINZELBACH, R (1967) *J. Orn.* **108**, 65-70; (1969) *Bonn. zool. Beitr.* **20**, 175-81. KINZELBACH, R and MARTENS, J (1965) *Bonn. zool. Beitr.* **16**, 50-91. KIRIKOV, S V (1974) *Byull. Mosk. obshch. ispyt. prir. otd. biol.* **79** (3), 133-7. KIRKMAN, F B (1913) *Zoologist* (4) **17**, 229. KIRKPATRICK, T (1907) *Br. Birds* **1**, 153. KISHCHINSKI, A A (1980) *Ptitsy Koryakskogo nagor'ya.* Moscow. KISHCHINSKI, A A, TOMKOVICH, P S, and FLINT, V E (1983) In Flint, V E and Tomkovich, P S (eds) *Rasprostranenie i sistematika ptits*, 3-76. Moscow. KISS, J B, REKASI, J, and STERBETZ, I (1978) *Avocetta* **1** (2), 3-18. KIST, J and WALDECK, K (1961) *Limosa* **34**, 6-11. KISTYAKIVSKI, O (1926) *Trav. Mus. Zool. Kiev* **1**, 69-77. KITSON, A R (1979) *Br. Birds* **72**, 94-100. KITTENBERGER, K (1960) *Aquila* **66**, 53-87. KIVRIKKO, K E (1947) *Suom. Linnut.* **1**, 398-402. KLAAS, C (1952) *Natur Volk* **82**, 9-14; (1974) *Gef. Welt* **98**, 200. KLAFS, G and STÜBS, J (1977) *Die Vogelwelt Mecklenburgs.* Jena. KLAWITTER, J and LENZ, M (1967) *Orn. Mitt.* **19**, 36-8. KLEIN, E (1933) *Orn. Monatsber.* **41**, 60. KLEIN, W and SCHAACK, K (1972) *Luscinia* **41**, 277-97. KLEINER, A (1936) *Mitt. Königl. Naturwiss. Inst. Sofia* **9**, 69-80; (1937) *XII Congr. int. Zool. Lisbonne 1935*, 1805-24; (1939) *Festschr. Prof. E. Strand* **5**, 365-84. KLEINSCHMIDT, O (1905) *Falco* **1**, 30-5; (1908) *Berajah, Zoographia infinita. Erithacus.* Halle. (1931) *Berajah, Zoographia infinita. Motacilla Sulphurea.* Halle. KLIEBE, K (1970) *Orn. Mitt.* **22**, 172. KLÍMA, M and URBÁNEK, B (1958) *Zool. Listy* **7**, 24-37. KLIMMEK, F (1950) *Vogelwelt* **71**, 145-8, 191-5. KLOSE, A (1978) *Anz. orn. Ges. Bayern* **17**, 332-3. KLUG-ANDERSEN, B (1984) *Dansk orn. Foren. Tidsskr.* **78**, 41-4. KLUIJVER, H N, LIGTVOET, J, OUWELANT, C VAN DEN, and ZEGWAARD, F (1940) *Limosa* **13**, 1-51. KNECHT, S (1960) *Anz. orn. Ges. Bayern* **5**, 525-56. KNECHT, S and SCHEER, U (1971) *Bonn. zool. Beitr.* **22**, 275-96. KNEIS, P (1981) *Orn. Jber. Mus. Hein.* **5-6**, 81-7; (1982) *Ber. Vogelwarte Hiddensee* **3**, 55-81; (1983) *Beitr. Vogelkde.* **29**, 119-20. KNEIS, P and BENECKE, H-G (1983) *Falke* **30**, 50-3 KNEIS, P and LAUCH, M (1983) *Zool. Jb.* **87**, 381-90. KNIGHT, P J (1975a) In Dick, W J A (ed) *Oxford and Cambridge Mauritanian Exped. 1973 Rep.*, 52-61. Cambridge. KNIGHT, P J (1975b) In Pienkowski, M W (ed) *Studies on coastal birds and wetlands in Morocco 1972*, 51-4. Univ. East Anglia, Norwich. KNOLLE, F, KUNZE, P, and ZANG, H (1973) *Vogelkdl. Ber. Niedersachs.* **5**, 65-76. KNORR, E (1927) *Orn. Monatsber.* **35**, 109-10. KNOWLTON, G F (1944) *Auk* **61**, 137-8. KNOX, A G and ELLIS, P (1981) *Br. Birds* **74**, 185-7. KNYSTAUTAS, A and LIUTKUS, A (1982) *In the world of birds.* Vilnius. KOBAYASHI, K and ISHIZAWA, T (1932) *The eggs of Japanese birds* **2**. Rokko, Kobe. KOBAYASHI, K (1956) *Ool. Rec.* **30**, 48-50. KOCH, C (1977) *Orn. Beob.* **74**, 204-5; (1983) *Orn. Beob.* **80**, 293-5. KOCH, E L (1948-9) *Jber. Vogelk. Beob. Untermain* **22**, 40. KOCH, J C (1930) *Org. Club nederl. Vogelk.* **3**, 101-4; (1936) *Org. Club nederl. Vogelk.* **9**, 76-9. KOEFOED, A (1935) *Dansk orn. Foren. Tidsskr.* **29**, 107-8. KOENIG, A (1886) *J. Orn.* **34**, 487-524; (1888) *J. Orn.* **36**, 121-298; (1890) *J. Orn.* **38**, 257-488; (1893) *J. Orn.* **41**, 13-105; (1895) *J. Orn.* **43**, 113-238, 257-321, 361-457; (1924) *J. Orn.* **72** suppl. KOERSVELD, E VAN (1951) *Proc. int. orn. Congr.* **10**, 592-4. KOFFÁN, K (1960) *Acta Zool. Acad. Sci. Hung.* **6**, 371-412. KOFFAN, K and FARKAS, T (1956) *Br. Birds* **49**, 268-71. KOHL, S, SZOMBATH, Z, KONYA, I, and GOMBOS, A (1975) *Nymphaea* **3**, 191-200. KOLBE, H (1963) *Vogelwelt* **84**, 84-90. KOLLIBAY, P (1904) *J. Orn.* **52**, 80-121. KOLTHOFF, K (1932) *Medd. Göteborgs Mus. Zool. Avd.* **59**. KOLUNEN, H and VIKBERG, P (1978) *Ornis fenn.* **55**, 126-31. KONDELKA, D (1970) *Sylvia* **18**, 229-30; (1978) *Folia Zool.* **27**, 37-45; (1980) *Acta Mus.*

Silesia Opavà **29**, 189-90. KÖNIG, C (1964a) *J. Orn.* **105**, 200; (1964b) *J. Orn.* **105**, 201; (1966a) *Vogelwelt* **87**, 182-8; (1966b) *Europäische Vögel* **1**; (1970) **3**. Stuttgart. KÖNIG, I and KÖNIG, C (1973) *Ardeola* **19**, 49-55. KÖNIGSTEDT, D (1980) *Beitr. Vogelkde.* **26**, 303. KÖNIGSTEDT, D and MÜLLER, H E J (1983) *Beitr. Vogelkde.* **29**, 77-88. KÖNIGSTEDT, D and ROBEL, D (1973) *Larus* **25**, 121-2; (1983) *Mitt. zool. Mus. Berlin* **59**, Suppl. Ann. Orn. **7**, 127-49. KONING, F J (1982) *Limosa* **55**, 115-20. KOOIKER, G (1978) *Orn. Mitt.* **30**, 277; (1982) *Vogelkdl. Ber. Niedersachs.* **14**, 38-44. KORBUT, V V (1982) *Zool. Zh.* **61**, 265-77. KORENBERG, E I, RUDENSKAYA, L V, and CHERNOV, Y I (1972) *Ornitologiya* **10**, 151-60. KORODI GÁL, I (1969) *Communicări Zool.*, 33-50; (1970) *Trav. Mus. Hist. nat. Grigore Antipa* **10**, 307-29. KORODI GÁL, J (1965) *Zool. Abh. Staatl. Mus. Tierkde. Dresden* **28**, 95-102. KORODI GÁL, J and GYÖRFI, A (1958) *Studii Cercet. Biol.* **9**, 59-68. KORODI GÁL, J and NAGY, Z (1965) *Zool. Abh. Mus. Tierk. Dresden* **28**, 113-25. KORTNER, W (1981) *Anz. orn. Ges. Bayern* **20**, 177-80. KOSHELEV, A I (1980) *Trudy Biol. Inst. Sib. otd. Akad. Nauk SSSR* **44**, 234-40. KOSTER, S H and GRETTENBERGER, J F (1983) *Malimbus* **5**, 62-72. KOSTIN, Y V (1983) *Ptitsy Kryma.* Moscow. KOVAČEVIĆ, J and DANON, M (1952) *Larus* **4-5**, 185-217; (1959) *Larus* **11**, 111-30. KOVSHAR', A F (1962) In *Voprosy ekologii* **4**, 119-21. Kiev; (1966) *Ptitsy Talasskogo Alatau.* Alma-Ata; (1979) *Pevchie ptitsy v subvysokogor'e Tyan'-Shanya.* Alma-Ata; (1981) *Osobennosti razmnozheniya ptits v subvysokogor'e.* Alma-Ata. KOVSHAR', A F and GAVRILOV, E I (1973) *Trudy zapoved. Kazakh.* **3**, 41-58. KOZENÁ, I (1975) *Zool. Listy* **24**, 149-62; (1979) *Folia Zool.* **28**, 337-46; (1980) *Folia Zool.* **29**, 143-56. KOZHEVNIKOVA, R K (1962) *Ornitologiya* **5**, 320-1. KOZLOVA, E V (1930) *Ptitsy yugo-zapadnogo Zabaykal'ya severnoy Mongolii i tsentral'noy Gobi.* Leningrad; (1933) *Ibis* (13) **3**, 301-32; (1945) *Zool. Zh.* **24**, 299-308; (1952) *Trudy Zool. Inst. Akad. Nauk SSSR* **9**, 966-1028; (1975) *Ptitsy zonal'nykh stepey i pustyn' Tsentral'noy Azii.* Leningrad; (1981) *Trudy Zool. Inst. Akad. Nauk SSSR* **102**, 56-61. KRAMER, H (1968) *Orn. Mitt.* **20**, 168-9. KRAMPITZ, H E (1952) *Vogelwelt* **73**, 81-92. KRAUSS, W (1972) *Anz. orn. Ges. Bayern* **11**, 54-7. KRAYSER, E (1971) *Orn. Mitt.* **23**, 220. KREBS, J R and DAVIES, N B (1981) *An introduction to behavioural ecology.* Oxford. KRECHMAR, A V (1966) In Ivanov, A I (ed) *Biologiya Ptits*, 185-312. Moscow. KRECHMAR, A V, ANDREEV, A V, and KONDRATIEV, A Y (1978) *Ekologiya i rasprostranenie ptits na severo-vostoke SSSR.* Moscow. KREUTZER, M (1973) Doct. Thesis. Univ. René Descartes, Paris; (1974a) *Rev. Comp. Anim.* **8**, 270-86; (1974b) *Rev. Comp. Anim.* **8**, 287-95. KRIVITSKI, I A (1962) *Ornitologiya* **4**, 208-17. KRIWANEK, H (1965) *Falke* **12**, 162-8. KROHN, H (1915) *Orn. Monatsber.* **23**, 147-51. KRONEISL-RUCNER, R (1960) *Larus* **12-13**, 41-9. KROODSMA, D E (1980) *Condor* **82**, 357-65. KRÖSCHE, O (1976) *Egretta* **19**, 64; (1979) *Orn. Mitt.* **31**, 247-9. KROYMANN, B (1967) *Vogelwelt* **88**, 170-3; (1968) *Orn. Mitt.* **20**, 253. KRÜGER, C (1938) *Dansk orn. Foren. Tidsskr.* **32**, 53-84; (1940) *Dansk orn. Foren. Tidsskr.* **34**, 114-53. KRÜGER, S (1967) *Beitr. Vogelkde.* **12**, 412-14; (1970) *Falke* **17**, 158-63; (1980) *Falke* **27**, 348-51. KUHN, H (1955) *Vogelwarte* **18**, 32. KUHNEN, K (1978) *Vogelwelt* **99**, 161-76; (1985) *J. Orn.* **126**, 1-13. KULESHOVA, L V (1976) *Ornitologiya* **12**, 26-54. KUMERLOEVE, H (1954) *Orn. Mitt.* **6**, 35; (1961) *Bonn. zool. Beitr.* **12** spec. vol.; (1963) *Alauda* **31**, 110-36, 161-211; (1964) *Alauda* **32**, 97-104; (1966) *Falke* **13**, 282; (1968) *Istanbul Univ. Fen. Fakült. Mecmuasi B* **32**, 79-213; (1969a) *Istanbul Univ. Fen. Fakült. Mecmuasi B* **34**, 245-312; (1969b) *Alauda* **37**, 43-58, 114-34; (1969c) *J. Orn.* **110**, 324-5; (1969d) *Ibis* **111**, 238-9; (1970a) *Istanbul Univ. Fen. Fakült. Mecmuasi B* **35** (3-4), 85-160; (1970b) *Beitr. Vogelkde.*

16, 239-49; (1970c) *Orn. Mitt.* **22**, 83; (1974) *J. Orn.* **115**, 371-2; (1975) *Bonn. zool. Beitr.* **26**, 183-98. KUMERLOEVE, H, KASPAREK, M, and NAGEL, K-O (1984) *Bonn. zool. Beitr.* **35**, 97-101. KUMMER, J (1983) *Beitr. Vogelkde.* **29**, 118. KUMMERLÖWE, H and NIETHAMMER, G (1934) *Alauda* **6**, 298-307. KUNKEL, P (1974) *Z. Tierpsychol.* **34**, 265-307. KURODA, N (1925) *Avifauna of the Ryukyu Islands.* Tokyo. KUX, Z and WEISZ, T (1977) *Acta Mus. Moraviae Brno* **62**, 153-68. KUZ'MENKO, V Y (1977) *Vestnik Zool.* (4), 32-7. KUZNETSOV, A A (1962) *Ornitologiya* **5**, 215-42. KUZNIAK, S (1967) *Acta Orn.* **10**, 177-211. KYDYRALIEV, A (1959) *Ornitologiya* **2**, 209-13.

LABHARDT, A (1984) *Orn. Beob.* **81**, 233-47. LABITTE, A (1934) *Oiseau* **4**, 740-2; (1937) *Oiseau* **7**, 85-104; (1944) *Oiseau* **14**, 165-76; (1947) *Alauda* **15**, 248-51; (1952a) *Alauda* **20**, 21-30; (1952b) *Oiseau* **22**, 261-82; (1955) *Oiseau* **25**, 168-71; (1957a) *Oiseau* **27**, 59-71; (1957b) *Oiseau* **27**, 143-9; (1958) *Oiseau* **28**, 39-52. LABITTE, A and LANGUETIF, A (1962) *Oiseau* **32**, 57-73. LACK, D (1932) *Br. Birds* **25**, 301-2; (1938) *Br. Birds* **32**, 23-4; (1939) *Proc. zool. Soc. Lond.* **109a**, 169-219; (1940a) *Br. Birds* **33**, 262-70; (1940b) *Ibis* (14) **4**, 299-324; (1940c) *Auk* **57**, 169-78; (1943) *Br. Birds* **36**, 166-75; (1944) *Br. Birds* **38**, 116; (1946a) *Br. Birds* **39**, 98-109, 130-5; (1946b) *Br. Birds* **39**, 258-64; (1946-7) *Bull. Br. Orn. Club* **67**, 51-4; (1948a) *Ibis* **90**, 252-79; (1948b) *Br. Birds* **41**, 98-104, 130-7; (1949) *Br. Birds* **42**, 147-50; (1951) *Ibis* **93**, 629-30; (1954a) *The natural regulation of animal numbers.* Oxford; (1954b) *Ibis* **96**, 312-14; (1962) *Br. Birds* **55**, 139-58; (1963) *Ibis* **105**, 1-54; (1965) *The life of the Robin.* London; (1966) *Ibis* **108**, 141-3; (1968) *Ecological adaptations for breeding in birds.* London. LACK, D and LIGHT, W (1941) *Br. Birds* **35**, 47-53. LACK, D and SILVA, E T (1949) *Ibis* **91**, 64-78. LACK, D and SOUTHERN, H N (1949) *Ibis* **91**, 607-26. LACK, H L (1941) *Br. Birds* **35**, 54-7. LACK, P C (1980) D Phil Thesis. Oxford Univ.; (1983) *J. Anim. Ecol.* **52**, 513-24. LACK, P C and QUICKE, D L J (1978) *Scopus* **2**, 86-91. LAENEN, J (1949-50) *Alauda* **17-18**, 169-79. LAFERRÈRE, M (1953) *Alauda* **21**, 64; (1968) *Alauda* **36**, 260-73. LAKHANOV, Z L (1966) *Nauch. Trudy Samarkand. Univ.* **156**, 63-4. LAMARCHE, B (1981) *Malimbus* **3**, 73-102. LAMBERT, A (1957) *Ibis* **99**, 43-68. LAMBERT, F J (1965) *Br. Birds* **58**, 221-2. LAMBERTINI, M (1981) *Avocetta* **5**, 65-86. LAMPE, H (1983) MS Thesis. Trondheim Univ. LANDMANN, A and LANDMANN, C (1978) *Anz. orn. Ges. Bayern* **17**, 247-65. LANGE, H (1951) *Dansk orn. Foren. Tidsskr.* **45**, 34-43. LANGSLOW, D R (1977) *Bird Study* **24**, 169-78. LANZ, H and WIGGER, F (1976) *Orn. Beob.* **73**, 32-3. LATHBURY, G (1970) *Ibis* **112**, 25-43. LA TOUCHE, J D D (1925-30) *A handbook of the birds of eastern China.* London. LAVAUDEN, L (1924) *Oiseaux. Voyage de M. Guy Babault en Tunisie: résultats scientifiques.* Paris. LAWRENCE, E S (1985a) *J. Anim. Ecol.* **54**, 965-75; (1985b) *Anim. Behav.* **33**, 929-37. LEA, D and BOURNE, W R P (1975) *Br. Birds* **68**, 261-83. LEA, A M and GRAY, J T (1935-6) *Emu* **34**, 275-92, **35**, 63-98, 145-78, 251-80, 335-47. LEBEURIER, E and RAPINE, J (1935a) *Oiseau* **5**, 258-83; (1935b) *Oiseau* **5**, 462-80; (1936a) *Oiseau* **6**, 86-103; (1936b) *Oiseau* **6**, 466-79; (1941) *Oiseau* **11**, 104-18. LEBRETON, P (1968) *Alauda* **36**, 36-51. LECK, C F (1975) *Bird-Banding* **46**, 201-3. LEDANT, J-P and JACOB, J-P (1980) *Gerfaut* **70**, 95-103. LEDANT, J-P, JACOB, J-P, JACOBS, P, MALHER, F, OCHANDO, B, and ROCHÉ, J (1981) *Gerfaut* **71**, 295-398. LEE, J A and BROWN, W, JR (1979) *Chat* **43**, 19-20. LEE, S L B (1963) *Ibis* **105**, 493-515. LEEGE-JUIST, O (1906) *Falco* **2**, 35-7. LEES, J (1949) *Ibis* **91**, 79-88, 287-99. LE FUR, R (1975) *Alauda* **43**, 317-19. LEHMANN, E VON (1951) *Bonn. zool. Beitr.* **2**, 225-7. LEHNER, A (1975) *Orn. Beob* **72**, 83. LEHTONEN, L (1943) *Ornis fenn.* **20**, 33-58. LEINONEN, M (1973a) *Ornis*

fenn. **50**, 53-82; (1973b) *Ornis fenn.* **50**, 126-33; (1974a) *Ann. zool. fenn.* **11**, 276-82; (1974b) *Ornis fenn.* **51**, 110-16. LEISLER, B (1968) *Egretta* **11**, 6-15. LEISLER, B, HEINE, G, and SIEBENROCK, K-H (1983) *J. Orn.* **124**, 393-413. LEKAGUL, B, ROUND, P D, and KOMOLPHALIN, K (1985) *Br. Birds* **78**, 2-39. LELEK, A and HAVLÍN, J (1957) *Zool. Listy* **20**, 177-83. LENNERSTEDT, I (1973) *Ornis scand.* **4**, 17-23. LENZ, M (1969) *Falke* **16**, 17-22; (1970) *Orn. Mitt* **22**, 189-90. LENZ, M, HINDEMITH, J, and KRÜGER, B (1972) *Vogelwelt* **93**, 161-80. LEONOVICH, V V (1962) *Byull. Mosk. obshch. ispyt. prir. otd. biol.* NS **67** (2), 121-4; (1977) *Ornitologiya* **13**, 91-4. LEPRINCE, P (1985) *Aves* **22**, 153-68. LESER, H and SCHNORR, R (1960) *J. Orn.* **101**, 500-1. LEUSCHNER, C (1974) *Hamburger avifaun. Beitr.* **12**, 1-16. LEUTHOLD, W (1973) *Bull. E. Afr. nat. Hist. Soc.*, 111-12. LEUZINGER, H (1955) *Orn. Beob.* **52**, 77-82. LÉVÊQUE, R (1964) *Alauda* **32**, 227-8. LEYS, H N and WILDE, J J F E DE (1970) *Levende Nat.* **73**, 66-72. LHOEST, S (1981) *Aves* **18**, 83. LIDAUER, R M (1983) *Vogelwarte* **32**, 117-22. LIEDEKERKE, R DE (1970) *Aves* **7**, 122; (1976) *Aves* **13**, 243-56. LINCOLN, F C (1917) *Auk* **34**, 341. LIND, E A (1960) *Ann. zool. Soc. Vanamo* **21**; (1962) *Ann. zool. Soc. Vanamo* **23**; (1963) *Ann. zool. Soc. Vanamo* **25**; (1964) *Ann. zool. fenn.* **1**, 7-43. LIND, H (1955) *Dansk orn. Foren. Tidsskr.* **49**, 76-113. LINDELL, L, WALINDER, G, BENGTSSON, D, and PETTERSSON, J (1978) *Vår Fågelvärld* **37**, 69-72. LINDNER, E (1919) *Verh. orn. Ges. Bayern* **14**, 148-9. LINDSAY-BLEE, M (1939) *Countryside*, **510**. LINDSTRÖM, Å, BENSCH, S, and HASSELQUIST, D (1985) *Vår Fågelvärld* **44**, 197-206. LINK, R and RITTER, M (1973) *Orn. Beob.* **70**, 184. LINSENMAIR, K E (1960) *Kosmos* **56**, 190-4. LIPPENS, L and WILLE, H (1972) *Atlas des oiseaux de Belgique et d'Europe occidentale.* Tielt; (1976) *Les oiseaux du Zaïre.* Tielt. LIVERSIDGE, R (1968) *Ostrich* **39**, 223-7. LLETGET, A G (1945) *Trab. Inst. Cienc. Nat. José de Acosta* **1**, 133-346. LLOYD, B (1933) *Trans. Herts. nat. Hist. Soc.* **19**, 135-9. LLOYD, M D (1984) *Br. Birds* **77**, 616. LOBACHEV, Y S and KAPITONOV, V I (1968) *Byull. Mosk. Obshch. Ispyt. prir. otd. Biol.* **73** (3), 17-25. LOBKOV, E G (1983) *Ornitologiya* **18**, 13-22. LOCKLEY, R M (1952) *Ibis* **94**, 144-57. LOGAN HOME, W M (1954) *Br. Birds* **47**, 312. LÖHRL, H (1944) *Beitr. Fortpfl. Vögel* **20**, 97; (1957a) *Vogelwelt* **78**, 102; (1957b) *Vogelwelt* **78**, 155-7; (1962a) *Vogelwelt* **83**, 116-22; (1962b) *J. Orn.* **103**, 492; (1969) *J. Orn.* **110**, 327; (1971) *Vogelwelt* **92**, 58-66; (1976) *Vogelwelt* **97**, 132-9; (1983) *J. Orn.* **124**, 271-9. LÖHRL, H and DORKA, V (1981) *Ökol. Vögel* **3**, 1-6. LÖHRL, H and GUTSCHER, H (1973) *J. Orn.* **114**, 399-416. LOMBARD, A L (1965) *Alauda* **33**, 206-35. LONG, J L (1981a) *Introduced birds of the world.* Newton Abbot. LONG, R (1981b) *Br. Birds* **74**, 327-44. LONG, R C (1959) *Ostrich* **30**, 136-7; (1961) *Ibis* **103a**, 131-3. LONGSTAFF, T G (1932) *J. Anim. Ecol.* **1**, 119-42. LOPE REBOLLO, F DE (1981) *Alauda* **48**, 99-112; (1982) *Doñana Acta Vert.* **8**, 313-18; (1983) *Alauda* **51**, 81-91. LÓPEZ IBORRA, G (1983) *Alytes* **1**, 373-92. LORENZ, K (1932) *Vogelzug* **3**, 4-10. LORENZ, L VON and HELLMAYR, C E (1901) *J. Orn.* **49**, 230-45. LORENZ, W (1954) *Orn. Mitt* **6**, 32. LOSKE, K-H (1982) *J. Orn.* **123**, 106; (1983a) *Beih. Veröff. Nat. Land. Bad.-Württ.* **37**, 43-52; (1983b) *Beih. Veröff. Nat. Land. Bad.-Württ.* **37**, 79-87; (1984) *Vogelwelt* **105**, 51-60; (1985) *Ökol. Vögel* **7**, 135-54. LOSKOT, V M (1981) *Trudy Zool. Inst. Akad. Nauk SSSR* **102**, 62-71; (1983) *Trudy Zool. Inst. Akad. Nauk SSSR* **116**, 79-107. LOSKOT, V M and PETRUSENKO, A A (1974) *Vestnik Zool.* **8** (5), 59-65. LOSKOT, W M and VIETINGHOFF-SCHEEL, E VON (1978) In Dathe, H and Neufeldt, I A (eds) *Atlas der Verbreitung palaärktischer Vögel* **7**; (1981) In Dathe, H and Neufeldt, I A (eds) *Atlas der Verbreitung palaärktischer Vögel* **9**. Berlin. LOUETTE, M (1981) *The birds of Cameroon: an annotated check-list.* Brussels. LOVATY, F (1985)

Nos Oiseaux 38, 27-31. LOVELL, H B (1944) *Auk* 61, 648-50. LØVENSKIOLD, H L (1964) *Avifauna svalbardensis*. Oslo. LOW, D J (1961) *Br. Birds* 54, 362. LÖW, W (1976) *Gef. Welt* 100, 68-9. LOWE, A R (1979) *Br. Birds* 72, 89-94. LOWE, P R (1947) *Proc. zool. Soc. Lond.* 117, 109-14. LOWE, V P W (1980) *J. Zool. Lond.* 192, 283-93. LÜBCKE, W (1975a) *J. Orn.* 116, 281-96; (1975b) *Vogelk. Hefte Edertal* 1, 82-7; (1976) *Vogelk. Hefte Edertal* 2, 91-8; (1980) *Beitr. Naturkde. Niedersachs.* 33, 147-52; (1981) *Vogelk. Hefte Edertal* 7, 90-105. LÜBCKE, W and FURRER, R (1985) *Die Wacholderdrossel*. Wittenberg Lutherstadt. LUBIÁN, J M and MORENO, J (1974) *Ardeola* 20, 378-9. LUCAN, V, NITSCHE, L, and SCHUMANN, G (1974) *Vogelwelt des Land- und Stadtkreises Kassel*. Kassel. LUCCA, C DE (1969) *Ibis* 111, 322-37. LUDLOW, A R (1966) *Ibis* 108, 129-32. LUDLOW, F and KINNEAR, N B (1933) *Ibis* (13) 3, 440-73; (1944) *Ibis* 86, 176-208. LÜDTKE, A (1966) *Anz. orn. Ges. Bayern* 7, 855-7. LUDWIG, E (1965) *Orn. Mitt* 17, 85. LUDWIG, H (1973) *Gef. Welt* 97, 39. LUNAIS, B (1983) *Oiseau* 53, 182-3. LUNDBERG, A and EDHOLM, M (1982) *Br. Birds* 75, 583-5. LUNDBERG, P, BERGMAN, A, and OLSSON, H (1981) *J. Orn.* 122, 163-72. LUNAU, C (1952) *Vogelwelt* 73, 138. LUNIAK, M (1969) *Acta Orn.* 11, 445-60; (1970) *Acta Orn.* 12, 177-208; (1971) *Acta Orn.* 13, 17-113. LUNN, J (1984) *Magpie* (3), 55. LÜPS, P, HAURI, R, HERREN, H, MÄRKI, H, and RYSER, R (1978) *Orn. Beob.* 75 suppl. LÜTKEN, E (1964) *Dansk orn. Foren. Tidsskr.* 58, 166-92. LUTTIK, R and WATTEL, J (1979) *Limosa* 52, 191-208. LYAISTER, A F and SOSNIN, G V (1942) *Materialy po ornitofaune Armyanskoy SSR (Ornis Armeniaca)*. Erevan. LYNES, H (1920) *Ibis* (11) 2, 260-301; (1924) *Ibis* (11) 6, 648-719; (1925a) *Ibis* (12) 1, 71-131, 344-416, 541-90, 757-97; (1925b) *Mém. Soc. Sci. nat. Maroc* 13; (1930) Diary of Algerian expedition. Unpubl. MS, Edward Grey Inst., Oxford Univ.; (1933) *Mém. Soc. Sci. nat. Maroc* 36. LYULEEVA, D S (1974) *Trudy Zool. Inst. Akad. Nauk SSSR* 55, 101-41.

MCATEE, W L (1905) *US Dept. Agric. Biol. Survey Bull.* 23, 1-37. MCCANCH, N V and MCCANCH, M (1982) *Br. Birds* 75, 90. MCCLURE, H E (1974) *Migration and survival of the birds of Asia*. Bangkok. MCCLUSKEY, R (1972) *Avic. Mag.* 78, 20-2. MACDONALD, D (1968) *Scott. Birds* 5, 176; (1984) *Br. Birds* 77, 159-60. MACDONALD, J D (1973) *Birds of Australia*. Sydney. MCDONALD, K (1974) *Bokmakierie* 26, 14. MACDONALD, M A (1978) *Bull. Niger. orn. Soc.* 14, 66-70. MACFARLANE, A M (1963) *Ibis* 105, 319-26; (1978) *Army Bird-watching Soc. per. Publ.* 3. MCGEOCH, J A (1963) *Ardea* 51, 244-50. MCGINN, D B (1979) *Scott. Birds* 10, 221-9; MCGINN, D B and CLARK, H (1978) *Bird Study* 25, 109-18. MACHALSKA, J, KANIA, W, and HOŁYNSKI, R (1967) *Notatki orn.* 8, 25-32. MÄCHLER, G (1947) *Vögel Heimat* 17, 85-7; (1975) *Vögel Heimat* 45, 187-90. MACKENZIE, J M D (1954) *Scott. Nat.* 66, 146-54. MACKENZIE, P Z (1955) *Sudan Notes Rec.* 36, 1-4. MACKINTOSH, D R (1944) *Zool. Soc. Egypt Bull.* 6 suppl., 10-14. MACKOWICZ, R (1970) *Acta Zool. Cracov.* 15, 61-160. MACKWORTH-PRAED, C W and GRANT, C H B (1951) *Ibis* 93, 234-6; (1960) *Birds of eastern and north eastern Africa* 2. London; (1963) *Birds of the southern third of Africa* 2. London; (1973) *Birds of west central and western Africa* 2. London. MCLACHLAN, G R and LIVERSIDGE, R (1970) *Roberts' birds of South Africa*. Cape Town. MCLAUGHLIN, R L and MONTGOMERIE, R D (1985) *Auk* 102, 687-95. MCLEAN, D D (1936) *Condor* 38, 16-17. MACLEAN, G L (1985) In Campbell, B and Lack, E (eds) *A dictionary of birds*, 155-6. Calton. MACLEAY, K N G (1960) *Univ. Khartoum nat. Hist. Mus. Bull.* 1. MACLEOD, J G R, MURRAY, C D'C, and MURRAY, E M (1953) *Ostrich* 24, 118-20. MACLEOD, J G R, MURRAY, E M, and MURRAY, C D'C (1952) *Ostrich* 23, 16-25. MCNEIL, D

and CLARK, F (1977) *Bird Study* 24, 130-1. MACNEIL, R and CYR, A (1971) *Auk* 88, 919-20. MADGE, S G (1967) *Devon Birds* 20, 3-5. MAGEE, J D (1965) *Bird Study* 12, 83-9. MAGNENAT, D (1962) *Nos Oiseaux* 26, 216. MAHER, W J (1979) *Ibis* 121, 437-52. MAINWOOD, A R (1972) *Scott. Birds* 7, 57. MAKATSCH, W (1950) *Die Vogelwelt Macedoniens*. Leipzig; (1958) *Vogelwelt* 79, 1-9; (1965) *Die Vögel in Wald und Heide*. Berlin; (1976) *Die Eier der Vögel Europas* 2. Radebeul; (1981) *Verzeichnis der Vögel der Deutschen Demokratischen Republik*. Leipzig. MALBRANT, R and RECEVEUR, P (1955) *Oiseau* 25, 87-101. MAL'CHEVSKI, A S (1959) *Gnezdovaya zhizn' pevchikh ptits*. Leningrad. MAL'CHEVSKI, A S and PUKINSKI, Y B (1983) *Ptitsy Leningradskoy oblasti i sopredel'nykh territoriy* 2. Leningrad. MALE, A E (1975) *Br. Birds* 68, 77. MAMBETZHUMAEV, A M and ABDREIMOV, T (1972) In Reymov, R (ed) *Ekologiya vazhneyshikh mlekopitayushchikh i ptits Karakalpakii*, 200-12. Tashkent. MANN, C F (1971) *Bull. Br. Orn. Club* 91, 41-6. MANNING, A (1946) *Br. Birds* 39, 26. MANNING, T H (1982) *Can. J. Zool.* 60, 3143-9. MANSER, G E and OWEN, D F (1949) *Br. Birds* 42, 244. MANUEL, F (1949) *Nos Oiseaux* 20, 109-13. MANUEL, F and BEAUD, P (1982) *Nos Oiseaux* 36, 277-81. MANZANARES, M (1983) *Alytes* 1, 369-72. MARCHANT, J H (1982) *Bird Study* 29, 143-8. (1984) *BTO News* 134, 7-10. MARCHANT, J H and HYDE, P A (1980) *Bird Study* 27, 183-202. MARCHANT, S (1938) *Br. Birds* 31, 338-40; (1942) *Ibis* (14) 6, 137-96; (1953) *Ibis* 95, 38-69; (1961) *Bull. Iraq nat. Hist. Mus.* 1 (4), 1-37; (1962) *Bull. Iraq nat. Hist. Mus.* 2 (1), 1-40; (1963a) *Ibis* 105, 369-98; (1963b) *Ibis* 105, 516-57. MARCHANT, S and MACNAB, J W (1962) *Bull. Iraq nat. Hist. Inst.* 2 (3), 1-48. MARIÁN, M (1968) *Tiscia* 4, 127-39. MARIEN, D (1951) *Amer. Mus. Novit.* 1482. MARION, L (1977) *Alauda* 45, 257-63. MARISOVA, I V and VLADYSHEVSKI, D V (1961) *Zool. Zh.* 40, 1240-5. MARKUS, M B (1963) *Ostrich* 34, 110-11; (1967) *Bull. Br. Orn. Club* 87, 17-23; (1974) *Ibis* 116, 232. MARLER, P (1959) In Bell, P R (ed) *Darwin's biological work: some aspects reconsidered*, 150-206. Cambridge. MARLER, P and HAMILTON, W J (1966) *Mechanisms of animal behavior*. New York. MARPLES, G (1935) *Br. Birds* 29, 22-5; (1937) *Br. Birds* 30, 305-6; (1940) *Br. Birds* 33, 294-303. MARSHALL, A J and WILLIAMS, M C (1959) *Proc. zool. Soc. Lond.* 132, 313-20. MARSHMAN, P (1977) *Br. Birds* 70, 503-4. MARTELLI, C (1976) *Riv. ital. Orn.* 46, 172-4. MARTENS, J (1971) *Vogelwarte* 26, 113-28. MARTIN, A C, ZIM, H S, and NELSON, A L (1951) *American wildlife and plants*. New York. MARTIN, T M (1980) *Br. Birds* 73, 354-5. MARUYAMA, N, KAWANO, M, ATSUMI, H, UEKI, K, and NEZU, W (1972) *Tori* 91, 37-50. MASCHKE, H-J (1967) *Falke* 14, 160-1; (1969) *Falke* 16, 310-13. MASON, C F and LYCZYNSKI, F (1980) *Bird Study* 27, 1-10. MASON, C W and MAXWELL-LEFROY, H (1912) *Mem. Dept. Agric. India Ent. Ser.* 3. MASON, E A (1953) *Bird-Banding* 24, 91-100. MASON, R L (1980) *Royal Air Force orn. Soc. J.* MASSEY, M E (1972) *Brecon. Birds* 3, 39-42; (1978) *Bird Study* 25, 167-74. MASTERSON, A (1971) *Honeyguide* 82, 43. MATHER, J (1971) *Br. Birds* 66, 447-8. MATHEWS, G F (1864) *Naturalist* 1, 88-90. MATHEY-DUPRAZ, A (1926) *A travers le Maroc*. Colombier. MATOUŠEK, F (1960) *Acta nat. Mus. Slov.* 6, 69-71. MATTHEWS, G V T (1968) *Bird navigation*. Cambridge. MATTHIESSEN, J N (1973) *Emu* 73, 191-3. MATVEJEV, S D and VASIĆ, V F (1973) *Catalogus Faunae Jugoslaviae* 4 (3). Aves. Ljubljana. MATYUSHKIN, E N and KULESHOVA, L V (1972) *Ornitologiya* 10, 182-93. MAU, K G (1972) *Gef. Welt* 96, 198-9. MAUERSBERGER, G (1964) *J. Orn.* 105, 345-6; (1971) *J. Orn.* 112, 438-50; (1979) *Falke* 26, 126-7; (1980) *Mitt. zool. Mus. Berlin* 56, *Suppl. Ann. Orn.* 4, 77-164; (1982) *Mitt. zool. Mus. Berlin* 58, 129-40; (1983) *Mitt. zool. Mus. Berlin* 59, *Suppl. Ann. Orn.* 7, 47-83. MAUERSBERGER,

G, WAGNER, S, WALLSCHLÄGER, D, and WARTHOLD, R (1982) *Mitt. zool. Mus. Berlin* **58**, 11–74. MAWBY, P J (1961) *Naturalist* **877**, 45–9. MAY, A (1962) *Beitr. Naturk. Niedersachs.* **15**, 23. MAY, D J (1947) *Br. Birds* **40**, 49–50; (1948) *Br. Birds* **41**, 310–11. MAYAUD, N (1931) *Alauda* (2) **3**, 511–52; (1932) *Gerfaut* **22**, 1–7; (1938) *Alauda* **10**, 116–36, 305–23; (1941–5) *Alauda* **13**, 72–89; (1952a) *Alauda* **20**, 1–20; (1952b) *Alauda* **20**, 31–8; (1952c) *Alauda* **20**, 65–79. MAYER, F (1983) *Seevögel* **4**, 45–6 MAYER-GROSS, H and PERRINS, C M (1962) *Br. Birds* **55**, 189–90. MAYES, W E (1935) *Br. Birds* **29**, 56. MAYO, L M (1931) *Emu* **31**, 71–6. MAYR, E (1956) *Br. Birds* **49**, 115–19; (1969) *Principles of systematic zoology.* New York. MAYR, E and AMADON, D (1951) *Amer. Mus. Novit.* **1496**. MAYR, E and BOND, J (1943) *Ibis* **85**, 334–41. MAYR, E and GREENWAY, J C (1956) *Breviora* **58**, 1–11. MAYR, E and STRESEMANN, E (1950) *Evolution* **4**, 291–300. MEAD, C J (1962) *Sussex Bird Rep.* **15**, 27–32; (1970a) *Bird Study* **17**, 229–40; (1970b) In Gooders, J (ed) *Birds of the world*, 1814–18. London; (1975) *Ring. Migr.* **1**, 57; (1979a) *Bird Study* **26**, 99–106; (1979b) *Bird Study* **26**, 107–12; (1980) *Bird Study* **27**, 51–3; (1984a) *Robins*. London; (1984b) *BTO News* **133**, 1. MEAD, C J and HARRISON, J D (1979a) *Bird Study* **26**, 73–86; (1979b) *Bird Study* **26**, 87–98. MEAD, C J and PEPLER, G R M (1975) *Br. Birds* **68**, 89–99. MEAD, C J and WATMOUGH, B R (1976) *Bird Study* **23**, 187–96. MEADEN, F (1964) *Avic. Mag.* **70**, 191–5; (1970) *Avic. Mag.* **76**, 9–15. MEADEN, F M and HARRISON, C J O (1965) *Br. Birds* **58**, 206–8. MEADE-WALDO, E G (1889) *Ibis* (6) **1**, 1–13. MEADOWS, B S (1969) *London Bird Rep.* **34**, 72–9. MEDVEDEV, S I and ESILEVSKAYA, W A (1973) *Izv. Akad. Nauk Turkmen. SSR Ser. biol.* **5**, 85–8. MEDWAY, LORD (1973) *Ibis* **115**, 60–86. MEDWAY, LORD and WELLS, D R (1976) *The birds of the Malay Peninsula* **5**. London. MEES, G F (1977) *Zool. Meded. Rijksmus. Nat. Hist. Leiden* **51**, 243–64. MEETH, P (1967) *Limosa* **40**, 1–5. MEEUS, H (1982) *Wielewaal* **48**, 43–7. MEIDELL, O (1936) *Nytt Mag. Naturvid.* **76**, 163–236; (1961) *Nytt Mag. Zool.* **10**, 5–47. MEIER, H (1954) *Orn. Beob.* **51**, 12–19. MEIER, M M and FURRER, R K (1979) *Angew. Orn.* **5**, 109–27. MEIKLEJOHN, M F M (1937) *Br. Birds* **31**, 85; (1948a) *Br. Birds* **41**, 117; (1948b) *Ibis* **90**, 76–86. MEIKLEJOHN, R F (1930) *Ibis* (12) **6**, 560–4. MEINECK, S (1973) *Bird Life* Oct–Dec, 30. MEINERTZHAGEN, R (1920a) *Ibis* (11) **2**, 132–95; (1920b) *Ibis* (11) **2**, 195–259; (1921) *Ibis* (11) **3**, 621–71; (1922) *Ibis* (11) **4**, 1–74; (1924a) *Ibis* (11) **6**, 87–101; (1924b) *Ibis* (11) **6**, 601–25; (1925a) *Ibis* (12) **1**, 305–24; (1925b) *Ibis* (12) **1**, 600–21; (1930) *Nicoll's birds of Egypt.* London; (1932) *Ibis* (13) **2**, 349; (1934) *Ibis* (13) **4**, 528–71; (1935) *Ibis* (13) **5**, 110–51; (1938) *Ibis* (14) **2**, 480–520, 671–717; (1940) *Ibis* (14) **4**, 106–36, 187–234; (1948) *Bull. Br. Orn. Club* **68**, 18–33; (1949a) *Bull. Br. Orn. Club* **69**, 104–8; (1949b) *Bull. Br. Orn. Club* **69**, 110; (1949c) *Ibis* **91**, 465–82; (1951a) *Proc. zool. Soc. Lond.* **121**, 81–132; (1951b) *Ibis* **93**, 443–59; (1953) *Bull. Br. Orn. Club* **73**, 41–4; (1954) *Birds of Arabia.* Edinburgh. MEININGER, P L and MULLIÉ, W C (1979) *Bull. orn. Soc. Middle East* **3**, 12–14. MEISE, W (1959) *Abh. Verh. Naturwiss. Verh. Hamburg* NF **3**, 86–104. MEISE, W and SEILKOPF, H (1960) *Abh. Verh. Naturwiss. Verh. Hamburg* NF **4**, 71–5. MEITZ, P (1972) *Gef. Welt* **96**, 220. MELCHER, R (1951) *Orn. Beob.* **48**, 168–70. MELCHIOR, E (1973) *Regulus* **11**, 54–5; (1975) *Regulus* **11**, 249–53. MELTOFTE, H (1975) *Medd. Grønland* **191** (9), 1–72. MENDELSOHN, J M (1973) *Ann. Transvaal Mus.* **28**, 79–89. MENGEL, R M (1952) *Auk* **69**, 273–83; (1965) *Orn. Monogr. AOU* **3**. MENZEL, H (1964) *Der Steinschmätzer.* Wittenberg Lutherstadt; (1967) *Regulus* **47**, 61–2; (1968) *Vogelwelt* **89**, 48–9; (1971) *Der Gartenrotschwanz.* Wittenberg Lutherstadt; (1983) *Der Hausrotschwanz.* Wittenberg Lutherstadt; (1984) *Die Mehlschwalbe.* Wittenberg Lutherstadt. MERI-

KALLIO, E (1946) *Ann. zool. Soc. Zool. Bot. Fenn. Vanamo* **12** (1–2); (1958) *Fauna Fenn.* **5**. MERIWANI, Y N (1973) *Ibis* **115**, 285. MERRITT, W, GREENHALF, R R, and BONHAM, P F (1970) *Sussex Bird Rep.* **22**, 68–80. MESSMER, E and MESSMER, I (1956) *Z. Tierpsychol.* **13**, 341–441. MESTER, H (1957a) *Vogelwelt* **78**, 103; (1957b) *Vogelwelt* **78**, 166; (1957c) *Vogelwelt* **78**, 185–9; (1959) *Orn. Mitt.* **11**, 153–6; (1969) *J. Orn.* **110**, 487–92; (1971) *Bonn. zool. Beitr.* **22**, 28–89; (1976) *Orn. Beob.* **73**, 99–108. MESTER, H and PRÜNTE, W (1965a) *Beitr. Vogelkde.* **10**, 441–7; (1965b) *J. Orn.* **106**, 460–1. MESTRE RAVENTÓS, P (1979) *Ocells del Penedès* **2**. Vilafranca. MEY, F (1974) *Abh. Ber. Nat. Mus. Mauritianum* **8** (3), 319–24. MICHAELIS, H J (1977) *Gef. Welt* **101**, 205. MICHEL, J (1917) *Orn. Jahrb.* **28**, 1–18. MIKKOLA, H (1973) *Br. Birds* **66**, 3–12. MIKKOLA, K (1984) *Lintumies* **19**, 154–67. MIKKOLA, K and MIKKOLA, H (1976) *Savon Luonto* **4**, 49–56. MILDENBERGER, H (1943) *Orn. Monatsber.* **51**, 6–12; (1950) *Bonn. zool. Beitr.* **1**, 11–20. MILLER, J (1972) *Bird Life* Jan–Mar, 36. MILLS, D H (1956) *Ibis* **98**, 137. MILNE, B S (1959) *Br. Birds* **52**, 281–95. MINTON, C D T (1960) *Br. Birds* **53**, 132–3. (1970) In Gooders, J (ed) *Birds of the world*, 2037–40. London. MISONNE, X (1974) *Gerfaut* **64**, 41–73. MITROPOL'SKI, O V (1968a) *Trudy Inst. Zool. Alma-Ata* **29**, 64–6; (1968b) *Trudy Inst. Zool. Alma-Ata* **29**, 67–70. MOCCI DEMARTIS, A (1973) *Alauda* **41**, 35–62. MÖCKEL, R and MÖCKEL, W (1975) *Mitt. IG Avifauna DDR* **8**, 85–90. MOEED, A (1975) *Notornis* **22**, 135–42. MOFFAT, C B (1931) *Br. Birds* **24**, 364–6. MOHR, H (1958) *Orn. Mitt.* **10**, 7–9. MÖHRING, G (1958) *Falke* **5**, 119; (1964) *Beitr. Vogelkde.* **10**, 230–2. MOISEEV, A P (1980) In Kovshar', A F (ed) *Biologiya ptits Naurzumskogo zapovednika*, 120–6. Alma-Ata. MOLL CASASNOVAS, J (1957) *Las aves de Menorca.* Palma. MØLLER, A P (1974a) *Flora og Fauna* **80**, 74–80; (1974b) *Dansk orn. Foren. Tidsskr.* **68**, 81–6; (1978a) *Dansk orn. Foren. Tidsskr.* **72**, 189–96; (1978b) *Nordjyllands Fugle.* Klampenborg; (1978c) *Dansk orn. Foren. Tidsskr.* **72**, 60–1; (1979) *Dansk orn. Foren. Tidsskr.* **75**, 248–9; (1982) *Ibis* **124**, 339–43; (1983a) *Bird Study* **30**, 134–42; (1983b) *Holarctic Ecol.* **6**, 95–100; (1984a) *Bird Behav.* **5**, 110–17; (1984b) *Ornis scand.* **15**, 43–54. MOLTONI, E (1935) *Riv. ital. Orn.* **5**, 127–76; (1937) *Riv. ital. Orn.* **7**, 206–12; (1945) *Riv. ital. Orn.* **15**, 33–78; (1948) *Riv. ital. Orn.* **18**, 74–86; (1949) *Riv. ital. Orn.* **19**, 86–8; (1952) *Riv. ital. Orn.* **22**, 59–61; (1969) *Riv. ital. Orn.* **39**, 1–25; (1971) *Riv. ital. Orn.* **41**, 150–60. MOLTONI, E and BRICHETTI, P (1976) *Riv. ital. Orn.* **46**, 24–32; (1978) *Riv. ital. Orn.* **48**, 65–142. MONK, J F (1950) *Br. Birds* **43**, 10; (1958) *Publ. Inst. Zool. Dr Augusto Nobre* **63**, 7–24. MONTAGNA, W (1943) *Auk* **60**, 210–15. MONTAGU, G (1831) *Ornithological dictionary.* London. MOODY, C (1955) *Br. Birds* **48**, 184. MOON, H J (1923) *Br. Birds* **17**, 59. MOORE, A (1960) *Field* August. MOORE, H J and BOSWELL, C (1956) *Iraq nat. Hist. Mus. Publ.* **10**; (1957) *Iraq nat. Hist. Mus. Publ.* **12**. MORATH, S (1979) *Gef. Welt* **103**, 129–31. MORBACH, J (1934) *Vögel der Heimat* **6**. Esch-Alzette. MOREAU, R E (1927) *Ibis* (12) **3**, 210–45; (1928) *Ibis* (12) **4**, 453–75; (1934) *Ibis* (13) **4**, 595–632; (1937) *Tanganyika Notes Rec.* **4**, 1–34; (1938) *Proc. zool. Soc. Lond.* (A) **108**, 1–26; (1939) *Bull. Br. Orn. Club* **59**, 145–9; (1940) *Auk* **57**, 313–25; (1941) *Bull. Inst. Egypt.* **23**, 247–61; (1944) *Ibis* **86**, 16–29; (1953) *Ibis* **95**, 329–64; (1961) *Ibis* **103a**, 373–427, 580–623; (1962) *Ostrich* **33** (4), 42; (1965) *Br. Birds* **58**, 222–3; (1966) *The bird faunas of Africa and its islands.* London; (1969) *Ibis* **111**, 621–4; (1972) *The Palaearctic-African bird migration systems.* London. MOREAU, R E and DOLP, R M (1970) *Ibis* **112**, 209–28. MOREAU, R E and MONK, J F (1957) *Ibis* **99**, 500–8. MOREAU, R E and MOREAU, W M (1928) *Ibis* (12) **4**, 233–52; (1939a) *Br. Birds* **33**, 95–7; (1939b) *Br. Birds* **33**, 146–51; (1949–50) *Alauda* **17–18**, 66–9. MOREAU, W M

(1943) *Ibis* 85, 103. MOREL, G and ROUX, F (1966) *Terre Vie* 113, 19-72, 143-76; (1973) *Terre Vie* 27, 523-50. MOREL, G J, MONNET, C, and ROUCHOUSE, C (1983) *Malimbus* 5, 1-4. MOREL, G J and NDAO, B (1978) *Oiseau* 48, 281-2. MORENO, J (1983) *Acta Univ. Upsaliensis* 704; (1984a) *Auk* 101, 741-52; (1984b) *J. Anim. Ecol.* 53, 147-59. MORGAN, B J T, SIMPSON, M J A, HANBY, J P, and HALL-CRAGGS, J (1976) *Behaviour* 56, 1-43. MORGAN, R A (1979) *Bird Study* 26, 129-32. (1982) *Bird Study* 29, 67-72. MORGAN, R A and DAVIS, P G (1977) *Bird Study* 24, 229-32. MORGAN, R A and GLUE, D E (1981) *Bird Study* 28, 163-8. MORIOKA, H and SAKANE, T (1981) *Tori* 29, 129-46. MORK, K (1975) *Sterna* 14, 131-4. MORLEY, A (1937) *Br. Birds* 31, 34-41; (1940) *Br. Birds* 34, 65. MORONY, J J, JR, BOCK, W J, and FARRAND, J (1975) *Reference list of the birds of the world.* New York. MOROZOV, V V (1984) *Ornitologiya* 19, 30-40. MORRIS, D (1954) *Br. Birds* 47, 33-49. MORRISON-SCOTT, T C S (1937) *Proc. zool. Soc. Lond.* (A) 107, 51-70. MOSKVITIN, S S (1969) *Mat. 5 Vsesoyuz. orn. Konf.* 2, 424-7. Ashkhabad; (1972) *Ornitologiya* 10, 173-81. MOSS, D (1975) *Edinb. Ring. Group Rep.* 3, 8-15. MOUNTFORT, G R (1935) *Br. Birds* 29, 145-8; (1954) *Ibis* 96, 111-15; (1958) *Portrait of a wilderness.* London; (1965) *Portrait of a desert.* London. MOUNTFORT, G and FERGUSON-LEES, I J (1961a) *Ibis* 103a, 86-109; (1961b) *Ibis* 103a, 443-71. MOUSLEY, H (1916) *Auk* 33, 281-6. MOUTON, J (1984) *Héron* 1, 81-93. MUFF, J (1977) *Orn. Beob.* 74, 205. MÜHL, K (1955) *Vogelwelt* 76, 103-4; (1958) *Vogelwarte* 19, 254-5. MULLARNEY, K (1980) *Irish Birds* 1, 541-5. MÜLLER, A (1879) *J. Orn.* 27, 304-8; (1929) *Anz. orn. Ges. Bayern* 2, 24-5. MÜLLER, H (1983) *Beitr. Vogelkde.* 29, 245-6. MÜLLER, H E J (1982) *Falke* 29, 78-85. MÜLLER, T (1974) *Beitr. Vogelkde.* 20, 487. MÜLLER-USING (1967) *Orn. Mitt.* 19, 63. MUMFORD, R E (1964) *Misc. Publ. Mus. Zool. Univ. Michigan.* 125. MÜNCH, H (1956) *Orn. Mitt.* 8, 30. MUNDY, P J and COOK, A W (1972) *Bull. Niger. orn. Soc.* 9, 60-76. MUNKEJORD, A (1981) *Fauna norv.* (C) *Cinclus* 4, 69-75. MUNN, P W (1894) *Ibis* (6) 6, 39-77; (1921) *Ibis* (11) 3, 672-719; (1925) *Ibis* (12) 1, 39-47; (1931) *Novit. Zool.* 37, 53-132. MUÑOZ-COBO, J and PURROY, F I (1980) *Proc. VI int. Congr. Bird Census Work*, 185-9. MUNSTERHJELM, L (1922) *Medd. Göteborgs Mus. Zool. Avd.* 13. MUNTEANU, D (1966) *Bull. Br. Orn. Club* 86, 97; (1971) *Studii Commun. Muz. Ştiinţ. Nat. Bacău* (Zool.), 255-61. MURDOCH, J (1885) *Report in Polar Exped. Point Barrow Alaska* 4, 91-176. MURIE, A J (1963) *Birds of Mount McKinley Nat. Park, Alaska.* MURPHY, R C (1924) *Bull. Amer. Mus. nat. Hist.* 50, 211-78. MURR, F (1923) *Verh. orn. Ges. Bayern* 15, 329-46. MURRAY, B G and JEHL, J R (1964) *Bird-Banding* 35, 253-63. MURILLO, F and SANCHO, F (1969) *Ardeola* 13, 129-37. MURTON, R K (1970a) In Gooders, J (ed) *Birds of the world*, 1849-51; (1970b) 2020-1. London. MYASOEDOVA, O M (1965) *Ornitologiya* 7, 481-2. MYERS, G R and WALLER, D W (1977) *Auk* 94, 596. MYRBACH, H (1975) *Monticola* 4, 10-11. MYRBURGH, N and STEYN, P (1979) *Bokmakierie* 31, 64. MYRES, M T (1955) *Bird Study* 2, 2-24; (1964) *Ibis* 106, 7-51.

NADLER, T (1974) *Falke* 21, 12-17. NAGY, E (1934) *Aquila* 38-41, 153-63. NANKINOV, D N (1970) *Vestnik Leningr. gos. Univ.*, 90-5; (1981) *Ornitologiya* 16, 176-7. NATORP, O (1920) *Orn. Monatsber.* 28, 15-17; (1925a) *Orn. Monatsber.* 33, 65-8; (1925b) *Orn. Monatsber.* 33, 143-5; (1928) *Beitr. Fortpfl. Vögel* 4, 136-9; (1929) *Orn. Monatsber.* 37, 65-70. NAUMANN, J A (1823) *Naturgeschichte der Vögel Deutschlands* 3. Leipzig; (1900) (ed Hennicke, C R) *Naturgeschichte der Vögel Mitteleuropas* 3; (1901) (ed Hennicke, C R) 4; (1905) (ed Hennicke, C R) 1. Gera. NAUMOV, R L (1960) *Ornitologiya* 3, 200-11; (1962) *Ornitologiya* 5, 135-43. NAUMOV, R L and BURKOVSKAYA, T E

(1959) *Ornitologiya* 2, 180-3. NAUROIS, R DE (1969a) *Mém. Mus. natn. Hist. nat.* (A) 56, 1-293; (1969b) *Bull. Inst. fond. Afr. noire* (A) 31, 143-218; (1974) *Alauda* 42, 111-16. NAYLOR, A K (1974) *Wicken Fen Group Rep.* 6, 13-21; (1978) *Wicken Fen Group Rep.* 10, 19-22. NAZARENKO, A A (1968) In Vorontsov, N N (ed) *Problemy evolyutsii* 1, 195-201. Novosibirsk; (1978) *Zool. Zh.* 57, 1743-5. NAZAROV, Y N (1981) In Litvinenko, N M (ed) *Redkie ptitsy Dal'nego Vostoka*, 67-73. Vladivostok. NECHAEV, V A (1965) *Ornitologiya* 7, 122-9; (1969) *Ptitsy yuzhnykh Kuril'skikh ostrovov.* Leningrad. NELDER, J A (1948) *Br. Birds* 41, 85. NELSON, B (1973) *Azraq: desert oasis.* London. NELSON, T H (1907) *The birds of Yorkshire.* London. NEOPHYTOU, P (ed) (1974) *Cyprus orn. Soc.* (1969) *Rep.* 4. NERUCHEV, V V and MAKAROV, V I (1982) *Ornitologiya* 17, 125-9. NESENHÖNER, H (1956) *Ber. naturw. Ver. Bielefeld* 14, 128-67. NETHERSOLE-THOMPSON, C and NETHERSOLE-THOMPSON, D (1940a) *Br. Birds* 34, 109; (1940b) *Br. Birds* 34, 137-8; (1943) *Br. Birds* 37, 88-94. NETHERSOLE-THOMPSON, D (1932) *Ool. Rec.* 12, 3-8. NEUB, M (1967) *Orn. Mitt.* 19, 25-31. NEUFELDT, I A (1956) *Zool. Zh.* 35, 434-40; (1961) *Zool. Zh.* 40, 416-26; (1966) *Trudy Zool. Inst. Akad. Nauk SSSR* 39, 120-84; (1970) *Trudy Zool. Inst. Akad. Nauk SSSR* 47, 111-81. NEUFELDT, I A and SOKOLOV, B V (1960) *Ornitologiya* 3, 236-50. NEUMANN, E (1943) *Beitr. Fortpfl. Vögel* 19, 93-6. NEUMANN, J (1978) *Orn. Mitt.* 30, 233-4. NEWBY, J E (1980) *Malimbus* 2, 29-50. NEWMAN, K (1983) *Newman's birds of southern Africa.* Johannesburg. NEWSTEAD, R (1908) *J. Board Agric. Suppl.* 15 (9). NEWTON, A (1861) *Ibis* 3, 92-106. NICE, M M (1943) *Trans. Linn. Soc. New York* 6, 1-328. NICHOLSON, C F (1981) *Br. Birds* 74, 441-2. NICHOLSON, E M (1930) *Ibis* (12) 6, 280-313; (1951) *Birds and men.* London; (1973) In Lovejoy, D (ed) *Land use and landscape planning*, 287-97. London. NICHOLSON, E M and KOCH, L (1936) *Songs of wild birds.* London. NICHT, M (1961) *Zool. Abh. Staatl. Mus. Tierkde.* Dresden 26, 79-99. NICOLAI, J (1976) *Vogelwarte* 28, 274-8. NICOLAU-GUILLAUMET, P (1965) *Vie Milieu* 16 (2), 1159-74; (1971) *Oiseau* 41, 182-3; (1972) *Oiseau* 42, 74-5. NICOLL, M (1980) *Rep. Tay Ring. Group 1978-9*, 40-4. NICOLL, M J (1906) *Zoologist* (4) 10, 463-7. NIEDERFRINIGER, O (1973) *Br. Birds* 66, 121-3. NIEDRACH, R J and ROCKWELL, R B (1939) *The Birds of Denver and mountain parks.* Denver. NIELSEN, B P (1969) *Dansk orn. Foren. Tidsskr.* 63, 50-73. NIETHAMMER, G (1937) *Handbuch der deutschen Vogelkunde* 1. Leipzig; (1942) *Ann. naturhist. Mus. Wien* 53, 5-59; (1943) *J. Orn.* 91, 167-238; (1950) *Syllegomena Biologica*, 267-86. Leipzig; (1954) *Vogelwarte* 17, 194-6; (1955a) *J. Orn.* 96, 411-17; (1955b) *Bonn. zool. Beitr.* 6, 29-80; (1955c) *Vogelwarte* 18, 22-4; (1957) *Bonn. zool. Beitr.* 8, 230-47; (1958a) *Beitr. Vogelkde.* 6, 79-87; (1958b) *Bull. Br. Orn. Club* 78, 87; (1963a) *J. Orn.* 104, 258-60; (1963b) *Bonn. zool. Beitr.* 14, 129-50; (1970) *Vogelwarte* 25, 356-7; (1971) *J. Orn.* 112, 202-26. NIETHAMMER, G and LAENEN, J (1954) *Alauda* 22, 25-31. NIETHAMMER, G and WOLTERS, H E (1970) *Zool. Abh. Staatl. Mus. Tierkde.* Dresden 31, 263-8. NIETHAMMER, J (1967) *J. Orn.* 108, 119-64. NIKOLAUS, G (1981) *Scopus* 5, 121-4; (1983) *Scopus* 7, 15-18; (1984) *Scopus* 8, 38-42. NIKOLAUS, G and PEARSON, D J (1982) *Scopus* 6, 17-19. NILSSON, N (1944) *Vår Fågelvärld* 3, 131-7. NISBET, I and ELTIS, W A (1951) *Br. Birds* 44, 62-3. NISBET, I C T (1957) *Ibis* 99, 228-68; (1963) *Bird-Banding* 34, 139-59; (1968) *Ibis* 110, 352-4. NISBET, I C T and SMOUT, T C (1957) *Ibis* 99, 483-99. NITECKI, C (1969) *Notatki orn.* 10, 1-8. NITSCHE, G (1967) *Egretta* 10, 32-3. NOAKES, D (1984) *Br. Birds* 77, 323. NØHR, H, BRAAE, L, and KLUG-ANDERSEN, B (1983) *Dansk orn. Foren. Tidsskr.* 77, 95-106. NÖHRING, R (1943) *Orn. Monatsber.* 51, 1-4. NORRIS, A S (1964) *Ibis* 106, 531-2. NORRIS,

C A (1960) *Bird Study* 7, 129-84. NORRIS, R A and JOHNSTON, D W (1958) *Wilson Bull.* 70, 114-29. NORTON, W J E (1958) *Ibis* 100, 179-89. NORUP, S (1963) *Dansk orn. Foren. Tidsskr.* 57, 110-17. NOSKOV, G A and GAGINSKAYA, A R (1977) *Ornitologiya* 13, 190-1. NOVAL, A (1974) *Asturnatura* 2, 33-42. NOVIKOV, G A (1952) *Trudy Zool. Inst. Akad. Nauk SSSR* 9, 1133-54; (1975) *Biologiya lesnykh ptits i zverey.* Moscow. NUMME, G (1976) *Sterna* 15, 177-8.

OATLEY, T B and SKEAD, D M (1982) *Lammergeyer* 15, 65-74. OBERHOLSER, H C (1902) *Proc. US natn. Mus.* 245 (1271), 801-83. O'CONNOR, R J (1961) *Br. Birds* 54, 362-3. ODDIE, W E (1982) *Br. Birds* 75, 96. OELKE, H (1960) *Orn. Mitt.* 12, 105-10; (1968a) *J. Orn.* 109, 25-9; (1968b) *Vogelwelt Beih.* 2, 39-46; (1975) *Vogelk. Ber. Niedersachs.* 7, 19-31. OESER, R (1966) *Beitr. Vogelkde.* 11, 342-3. OESER, R E and MARTIN, A (1983) *Beitr. Vogelkde.* 29, 17-18. OGGIER, P-A (1979) *Nos Oiseaux* 35, 85. OGILVIE-GRANT, W R (1902) *Ibis* (8) 2, 393-470. OGORODNIKOVA, L I (1979) *Nauch. Doklad. vyssh. Shk. Biol. Nauk* 1, 106; (1980) In Shurakov, A I (ed) *Gnezdovaya zhizn' ptits*, 115-18. Perm'. OKIA, N O (1976) *Ibis* 118, 1-13. OLDFIELD, A G (1952) *Br. Birds* 45, 69-70. OLIOSO, G (1974) *Alauda* 42, 226-30. OLIVER, P J (1979) *Br. Birds* 72, 36. OLIVER, W R B (1955) *New Zealand birds.* Wellington. OLIVIER, G (1959) *Br. Birds* 52, 61. OLNEY, P J S (1965) *Int. Union Game Biol. Trans.* 6, 309-22. ÖLSCHLEGEL, H (1985) *Die Bachstelze.* Wittenberg Lutherstadt. OLSON, S L, JAMES, H F, and MEISTER, C A (1981) *Bull. Br. Orn. Club* 101, 339-46. OLSSON, V (1947) *Vår Fågelvärld* 6, 93-125. OMEL'KO, M A (1979) *Ornitologiya* 14, 219-21. OORDT, G J VAN (1950) *Ardea* 37, 179-81. OO-U-KIJO (1936) *Bot. Zool. Tokyo* 4, 2065-75. ORIANS, G H (1969) *Amer. Nat.* 103, 589-603; (1971) In Farner, D S and King, R J (eds) *Avian Biology* 1, 513-46. New York. ORMEROD, S J (1985a) *Ibis* 127, 316-31; (1985b) *Naturalist* 110, 99-103. ORMEROD, S J and PERRY, K W (1985) *Irish Birds* 3, 90-5. ORMEROD, S J and TYLER, S J (1986) *Bird Study* 33, 36-45. ORMEROD, S J, TYLER, S J, and LEWIS, J M S (1985) *Bird Study* 32, 32-9. ORR, N W (1976) *Br. Birds* 69, 265-71. ORR, Y (1970) *Condor* 72, 476-8. OSMASTON, B B (1925) *Ibis* (12) 1, 663-719; (1927) *J. Bombay nat. Hist. Soc.* 32, 134-53. OSTAPENKO, V A (1981) *Ornitologiya* 16, 179-80. ÖSTERLÖF, S and STOLT, B-O (1982) *Ornis scand.* 13, 135-40. OTTO, C (1979) *Ornis scand.* 10, 111-16. OWEN, D F (1969) *Ardea* 57, 77-85. OWEN, J (1958) *Ibis* 100, 515-34. OWEN, J H (1919) *Br. Birds* 13, 23.

PACCAUD, O (1947) *Nos Oiseaux* 19, 1-22; (1952a) *Nos Oiseaux* 21, 188-95; (1952b) *Nos Oiseaux* 21, 195-6. PAEVSKI, V A (1971) *Trudy Zool. Inst. Akad. Nauk SSSR* 50, 1-110; (1977) *Zool. Zh.* 56, 753-61. PAIGE, J P (1960) *Ibis* 102, 520-5. PAKENHAM, R H W (1979) *The birds of Zanzibar and Pemba.* Br. Orn. Union, London. PALM, B (1951) *Dansk orn. Foren. Tidsskr.* 45, 29-33. PALMER, A H (1895) *The life of Joseph Wolf: animal painter.* London. PALMER, E M and BALLANCE, D K (1968) *The birds of Somerset.* London. PALMER, K H (1983) *London Bird Rep.* 47, 106-21. PALMGREN, P (1930) *Acta zool. fenn.* 7; (1935) *Ornis fenn.* 12, 107-21. PALUDAN, K (1938) *J. Orn.* 86, 562-638; (1940) *Danish Sci. Invest. Iran* 2, 11-54; (1959) *Vidensk. Medd. dansk nat. Foren.* 122. PANCHENKO, S G (1976) *Vestnik Zool.* (6), 24-7. PANOV, E N (1973) *Ptitsy yuzhnogo Primor'ya.* Novosibirsk; (1978) *Mekhanizmy kommunikatsii u ptits.* Moscow. PANOV, E N and IVANITSKI, V V (1975a) *Zool. Zh.* 54, 1357-70; (1975b) *Zool. Zh.* 54, 1860-73. PANOV, E N, KOSTINA, G N, and GALICHENKO, M V (1978) *Zool. Zh.* 57, 569-81. PANOW, E N (1974) *Die Steinschmätzer der Nördlichen Palaärktis.* Wittenberg Lutherstadt. PANTELEEV,

P A (1972) *Ornitologiya* 10, 161-72. PAPADOPOL, A (1961) *Commun. Acad. Bucuresti* 11, 1213-22. PARKER, A C (1960) *London Bird Rep.* 23, 74-5. PARNELL, B W (1976) *Honeyguide* 87, 38. PARRACK, J D (1973) *The naturalist in Majorca.* Newton Abbot. PARRINDER, E R and PARRINDER, E D (1944-5) *Br. Birds* 38, 362-9. PARSLOW, J L F (1967) *Br. Birds* 60, 177-202; (1969) *Ibis* 111, 48-79; (1973) *Breeding birds of Britain and Ireland.* Berkhamsted. PARSONS, A J (1976) *Bird Study* 23, 287-93. PARSONS, T and REID, D (1975) *Ring. Migr.* 1, 56. PASSBURG, R E (1959) *Ibis* 101, 153-69; (1966) *Some additional bird notes from Iran.* Unpubl. MS, Edward Grey Inst., Oxford Univ. PASSOW, A and PASSOW, H (1913) *Mitt. Vogelwelt* 13, 3-6. PASTEUR, G (1956) *Bull. Soc. Sci. Nat. Phys. Maroc* 36, 165-84; (1958) *Oiseau* 28, 73-6. PASTUKHOV, V D (1961) *Zool. Zh.* 40, 1536-42. PASZKOWSKI, W (1969) *Orn. Mitt.* 21, 60. PĀTKAI, I (1966) *Aquila* 71-2, 244. PATTEN, C J (1912) *Irish Nat.* 21, 125-30. PĀTZOLD, R (1971) *Heidelerche und Haubenlerche.* Wittenberg Lutherstadt; (1979) *Falke* 26, 118-25; (1981) *Falke* 28, 114-23; (1983) *Die Feldlerche.* Wittenberg Lutherstadt; (1984) *Der Wasserpieper.* Wittenberg Lutherstadt. PAVAN, P (1980) *Riv. ital. Orn.* 50, 227-8. PAULER, K (1972) *Egretta* 15, 55-60. PAULL, D E (1968) *Br. Birds* 61, 312. PAULSEN, B E (1978) *Vår Fuglefauna* 1, 81-7. PAVLOVA, N R (1962) *Ornitologiya* 4, 122-31. PAYNE, R B (1961) *Wilson Bull.* 73, 384-6. PAYNE, R B and PAYNE, K (1977) *Z. Tierpsychol.* 45, 113-73. PAYNTER, R A, JR (1957) *Breviora* 71, 1-15. PAX, F (1925) *Wirbeltierfauna von Schlesien.* Berlin. PEAKALL, D B (1956) *Br. Birds* 49, 135-9. PEARSON, D J (1971) *Ibis* 113, 173-84; (1972) *Ibis* 114, 43-60; (1984) *Scopus* 8, 18-23. PEARSON, D J and BACKHURST, G C (1973) *Ibis* 115, 589-91; (1976) *Ibis* 118, 78-105. PEDERSEN, E T (1966) *Dansk orn. Foren. Tidsskr.* 60, 95-100. PEDROLI, J-C (1975) *Rev. suisse Zool.* 82, 712-16; (1976) *Bull. Soc. Neuchât. Sci. nat.* 99, 33-44; (1978) *Ornis scand.* 9, 168-71. PEDROLI, J-C and GRAF-JACCOTTET, M (1978) *Alauda* 46, 171-6. PEETERS, J (1979) *Veldorn. Tijdschr.* 2, 122-8. PEIPONEN, V A (1959) *Arch. Soc. Zool. Bot. Fenn. Vanamo* 12, 146-55; (1960) *Ornis fenn.* 37, 69-83; (1970) *Proc. Helsinki Symp. UNESCO 1970*, 281-7. PEITZMEIER, J (1942) *J. Orn.* 90, 311-22; (1949) *Beitr. Naturk. Niedersachs.* 6, 4-8; (1956) *Vogelwelt* 77, 54-6; (1962) *Vogelwelt* 83, 81; (1964) *J. Orn.* 105, 149-52; (1968) *Orn. Mitt.* 20, 253; (1970) *Orn. Mitt.* 22, 84. PEITZMEIER, J and WESTERFRÖLKE, P (1960) *J. Orn.* 101, 365. PEK, L V and FEDYANINA, T F (1961) In Yanushevich, A I (ed) *Ptitsy Kirgizii* 3, 59-118. Frunze. PEKLO, A M and SOPYEV, O S (1980) *Vestnik Zool.* (3), 47-52. PELTZER, R (1967a) *Regulus* 47, 9-11; (1967b) *Regulus* 47, 118-20. PENCE, D B and CASTO, S D (1975) *Wilson Bull.* 87, 75-82. PENNY, M (1974) *The birds of Seychelles.* London. PENOT, J (1948) *Oiseau* 18, 141-51. PENRY, E H (1979) *Bull. Zambian orn. Soc.* 11 (1), 33-4. PÉNZES, A (1974) *Aquila* 78-9, 197-8. PEPLER, G (1966) *Ringers' Bull.* 2 (10), 9-10. PÉREZ PADRÓN, F (1983) *Las aves de Canarias.* Tenerife. PERRY, K W (1983) *Irish Birds* 2, 272-7. PERSSON, C (1973) *Dansk orn. Foren. Tidsskr.* 67, 25-34; (1977) *Ornis scand.* 8, 97-9; (1978) *Anser Suppl.* 3, 199-212; (1979) *Vår Fågelvärld* 38, 48-9. PERTTULA, U (1945) *Ornis fenn.* 22, 122-9. PETERS, H (1959) *Egretta* 2, 42-4. PETERSEN, A J (1955) *Wilson Bull.* 67, 235-86. PETERSEN, F D (1972) *Dansk orn. Foren. Tidsskr.* 66, 97-107. PETERSEN, P C and MUELLER, A J (1979) *Bird-Banding* 50, 69-70. PETERSON, R, MOUNTFORT, G, and HOLLOM, P A D (1983) *A field guide to the birds of Britain and Europe.* London. PETERSON, R T (1947) *A field guide to the birds.* Cambridge, Mass.; (1960) *A field guide to the birds of Texas.* Cambridge, Mass. PETERSON, R T and CHALIF, E L (1973) *A field guide to Mexican birds.* Boston. PETHON, P (1968) *Nytt Mag. Zool.* 15, 44-9. PETTERSSON, J (1983) *Vår Fågelvärld* 42, 333-42. PETTETT,

A (1975) *Br. Birds* 68, 45-7. PETTITT, R G and BUTT, D V (1950) *Br. Birds* 43, 298. PFEIFER, S (1965) *Orn. Mitt.* 17, 199-202; (1967a) *Orn. Mitt.* 19, 240; (1967b) *Orn. Mitt.* 19, 258; (1974) *Orn. Mitt.* 26, 185. PFEIFER, S and KEIL, W (1959) *Luscinia* 32, 13-18. PFEIFFER, W (1980) *Aves* 17, 45-7. PFLUGBEIL, A (1951) *Orn. Mitt.* 3, 137-8. PFROMM, G (1931) *Vogelzug* 2, 139. PHILLIPS, A R, HOWE, M A, and LANYON, W E (1966) *Bird-Banding* 37, 153-71. PHILLIPS, A R and LANYON, W E (1970) *Bird-Banding* 41, 190-7. PHILLIPS, B N (ed) (1986) *ICBP Study Rep.* 8. PHILLIPS, B T (1949) *J. Bombay nat. Hist. Soc.* 46, 487-500; (1950) *J. Bombay nat. Hist. Soc.* 47, 84-102. PHILLIPS, J S (1968) *Bird Study* 15, 104-5; (1970) *Bird Study* 17, 320-4; (1976) *Bird Study* 23, 57-78. PHILLIPS, J S and GREIG-SMITH, P (1980) *Bird Study* 27, 255-6. PHILLIPS, N R (1982) *Sandgrouse* 4, 37-59. PHILLIPS, R (1910) *Avic. Mag.* (3) 1, 324-8. PHILLIPS, W W A (1953) *Ibis* 95, 142. PHILLIPSON, N R (1937) *Br. Birds* 30, 343-5. PICKESS, B P (1965) *Br. Birds* 58, 386. PICKWELL, G (1947) *Auk* 64, 1-14. PICKWELL, G B (1931) *Trans. Acad. Sci. St Louis* 27. PIECHOCKI, R (1958) *Zool. Abh. Staatl. Mus. Tierkde. Dresden* 24, 105-203. PIECHOCKI, R and BOLOD, A (1972) *Mitt. zool. Mus. Berlin* 48, 41-175. PIECHOCKI, R, STUBBE, M, UHLENHAUT, K, and SUMJAA, D (1982) *Mitt. zool. Mus. Berlin* 58, Suppl. *Ann. Orn.* 6, 3-53. PIIPARINEN, T and TOIVARI, L (1958) *Ornis fenn.* 35, 65-70. PIKULA, J (1971) *Zool. Listy* 20, 281-91; (1972) *Zool. Listy* 21, 359-81; (1976) *Zool. Listy* 25, 65-72. PIKULA, J and BEKLOVÁ, M (1983) *Acta Sci. Nat. Brno* 17 (7). PIMM, S L (1970) *Bird Study* 17, 49-51; (1972) *Bird Study* 19, 116. PINEAU, J and GIRAUD-AUDINE, M (1977) *Alauda* 45, 75-104; (1979) *Trav. Inst. Sci. Rabat Sér. Zool.* 38. PITMAN, C R S (1921) *Ool. Rec.* 1, 15-38. PIZZEY, G (1980) *A field guide to the birds of Australia.* Sydney. PLATH, L (1977) *Falke* 24, 280-1. PLESKE, T (1889-94) *Wissenschaftliche Resultate der von N. M. Przewalski nach Central-Asien unternommenen Reisen.* St Petersburg. (1928) *Mem. Boston. Soc. nat. Hist.* 6, 111-485. PLUCINSKI, A (1956) *Orn. Mitt.* 8, 41-3; (1973) *Orn. Mitt.* 25, 135-41. PLUMMER, M V (1977) *Southwest. Nat.* 22, 147-8. PODARUEVA, V I (1979) In Labutin, Y V (ed) *Migratsii i ekologiya ptits Sibiri*, 170-2. Yakutsk. POHL, R (1922) *Orn. Monatsber.* 30, 129. POKROVSKAYA, I V (1968) *Ibis* 110, 571-3. POLATZEK, J (1908) *Orn. Jahrb.* 19, 161-97. PONOMAREVA, T S (1974) *Ornitologiya* 11, 404-7. POPHAM, H L (1898) *Ibis* (7) 4, 489-520. POPOV, V A (1978) *Ptitsy Volzhsko-Kamskogo kraya.* Moscow. PORTENKO, L A (1937) *Fauna ptits vnepolyarnoy chasti severnogo Urala.* Moscow; (1938) *Izv. Akad. Nauk SSSR. otd. mat. estestvenn. nauk,* 1057-62; (1939) *Trudy Nauch.-issled. inst. polyarn. zemledeliya zhivotnovod. i promysl. khoz.* 5, 5-211; (1954) *Ptitsy SSSR* 3; (1960) 4. Moscow; (1973) *Ptitsy Chukotskogo poluostrova i ostrova Vrangelya* 2. Leningrad; (1981) *Trudy Zool. Inst. Akad. Nauk SSSR* 102, 72-109. PORTENKO, L A and VIETINGHOFF-SCHEEL, E VON (1967) In Stresemann, E, Portenko, L A, and Mauersberger, G (eds) *Atlas der Verbreitung palaärktischer Vögel* 2. Berlin; (1971) In Stresemann, E, Portenko, L A, and Mauersberger, G (eds) *Atlas der Verbreitung palaärktischer Vögel* 3. Berlin; (1974) In Stresemann, E, Portenko, L A, Dathe, H, and Mauersberger, G (eds) *Atlas der Verbreitung palaärktischer Vögel* 4. Berlin; (1976) In Dathe, H (ed) *Atlas der Verbreitung palaärktischer Vögel* 5. Berlin. PORTENKO, L A and WUNDERLICH, K (1977) In Dathe, H (ed) *Atlas der Verbreitung palaärktischer Vögel* 6. Berlin. PORTER, R F (1983) *Sandgrouse* 5, 45-74. PORTER, R F, SQUIRE, J E, and VITTERY, A (eds) (1969) *Orn. Soc. Turkey Bird Rep. 1966-7.* Sandy. PORTIG, F (1942) *Beitr. Fortpfl. Vögel* 18, 176. POSLAVSKI, A N (1963) *Ornitologiya* 6, 195-203; (1974) *Ornitologiya* 11, 238-52. POSNETT, G L A and BATES, A D (1960) *Countryside* 18,

536-8. POTAPOV, R L (1966) *Trudy Zool. Inst. Akad. Nauk SSSR* 39, 3-119. POTAPOVA, E G and PANOV, E N (1977) *Zool. Zh.* 56, 743-52. POTOROCHA, V I (1972) *Ornitologiya* 10, 378-9. POUGH, R H (1949) *Audubon bird guide: small land birds.* New York. POULDING, R H (1969) *Bristol Orn.* 1, 75-6. POULSEN, H (1956) *Dansk orn. Foren. Tidsskr.* 50, 267-98. POUNDS, H E (1942) *Br. Birds* 36, 94; (1944) *Br. Birds* 37, 234. PRATO, S DA (1981) *Bird Study* 28, 60-2. PRATO, S R D DA, PRATO, E S DA, and CHITTENDEN, D J (1980) *Ring. Migr.* 3, 9-20. PRAZ, J-C (1976) *Nos Oiseaux* 33, 257-64; (1979) *Nos Oiseaux* 35, 141. PRENN, F (1929) *Orn. Monatsber.* 37, 33-5; (1937) *J. Orn.* 85, 577-86. PRESTON, G (1975) *Bull. E. Afr. nat. Hist. Soc.,* 112-13. PREUSS, N O (1959) *Dansk orn. Foren. Tidsskr.* 53, 1-19; (1978) *Feltornithologen* 20, 201-2. PRICE, K (1973) *Br. Birds* 66, 281. PRICE, M P (1961) *Br. Birds* 54, 100-6. PRIDDEY, M W (1977) *Br. Birds* 70, 262-3. PRIEST, C D (1935) *The birds of Southern Rhodesia* 3. London. PRITCHARD, A (1950) *Br. Birds* 43, 156. PRODON, R (1982) *Alauda* 50, 176-92; (1985) *Alauda* 53, 295-305. PRÖGER, H (1979) *Falke* 26, 80-5. PROKOFIEVA, I V (1962) *Ornitologiya* 4, 99-102; (1972a) *Uchen. zap. Leningr. gos. ped. Inst.* 392, 129-48; (1972b) *Uchen. zap. Leningr. gos. ped. Inst.* 392, 149-51; (1980) *Ornitologiya* 15, 89-93; (1981) In Kumari, E (ed) *Tez. dokl. XI Pribalt. orn. Konf.,* 173-5. Tallinn. PRÖSCHL, R (1981) *Falke* 28, 273. PROZESKY, O P M (1970) *A field guide to the birds of southern Africa.* London. PRUSKA, M (1980) *Acta Orn.* 17, 321-31. PRYCE-JONES, N (1972) *Bird Life* Jul-Sep, 29. PRYL, M (1980) *Ornis fenn.* 57, 33-9, 82-7. PRYTHERCH, R (1981) *Bristol Orn.* 14, 135-6. PTUSHENKO, E S and INOZEMTSEV, A A (1968) *Biologiya i khozyaystvennoe znachenie ptits Moskovskoy oblasti i sopredel'nykh territoriy.* Moscow. PUKINSKI, Y B (1975) *Po taezhnoy reke Bikin.* Moscow. PULLIAINEN, E (1977) *Aquilo Ser. Zool.* 17, 1-6; (1982) *Aquilo Ser. Zool.* 21, 6-8. PULLIAINEN, E, BALÁT, F, OJANEN, M, and ORELL, M (1982) *Ekológia* 1, 345-52. PULLIAINEN, E, ESKONEN, H, and HIETAJÄRVI, T (1981) *Ornis fenn.* 58, 175-6. PULLIAINEN, E, HELLE, P, and TUNKKARI, P (1981) *Ornis fenn.* 58, 21-8. PULLINGER, T (1971) *Bird Life* 7 (2), 68-9. PULYAKH, V (1977) In Voinstvenski, M A (ed) *Tez. dokl. VII Vsesoyuz. orn. Konf.* 1, 306-7. Kiev. PURCHON, R D (1948) *Proc. zool. Soc. Lond.* 118, 146-70. PURVEY, M (1985) *Scott. Birds* 13, 230. PUTNAM, L S (1949) *Wilson Bull.* 61, 141-82. PÜTTMANN, R (1973) *Anthus* 10, 39-44. PUZACHENKO, Y G (1968) *Ornitologiya* 9, 370-1. PYCRAFT, W P (1909) *Br. Birds* 3, 121-2.

RABØL, J (1974) *Dansk orn. Foren. Tidsskr.* 68, 5-14; (1981) *Ornis scand.* 12, 89-98. RABØL, J and NOER, H (1973) *Vogelwarte* 27, 50-65. RÁCZ, B (1907) *Aquila* 14, 327. RACZYNSKI, J and RUPRECHT, A L (1974) *Acta Orn.* 14, 25-38. RADDE, G and WALTER, A (1889) *Ornis* 5, 1-128. RADERMACHER, W (1967) *Orn. Mitt.* 19, 114-15; (1970a) *Charadrius* 6, 7-23; (1970b) *Orn. Mitt.* 22, 21; (1973) *Orn. Mitt.* 25, 94-5; (1977) *Orn. Mitt.* 29, 51; (1982) *Charadrius* 18, 134-5. RADETZKY, J (1970) *Falke* 17, 112-15. RADFORD, A P (1967a) *Br. Birds* 60, 372; (1967b) *Br. Birds* 60, 374; (1970) *Br. Birds* 63, 342-3; (1973) *Br. Birds* 66, 231; (1975) *Br. Birds* 68, 210; (1978) *Br. Birds* 71, 133; (1980) *Br. Birds* 73, 417; (1981) *Br. Birds* 74, 266; (1984) *Br. Birds* 77, 568; (1985a) *Br. Birds* 78, 353-4; (1985b) *Br. Birds* 78, 512. RADIG, K (1914) *Orn. Monatsber.* 22, 122-5. RADU, D (1975) *Studii Commun. Muz. Ştiinţ. Nat. Bacău (Biol. anim.)* 8, 259-68. RAETHEL, S (1955) *J. Orn.* 96, 419-21. RAEVEL, P (1981) *Héron* 4, 1-10. RAINES, R J (1945) *Br. Birds* 38, 297; (1955) *Br. Birds* 48, 185; (1962) *Ibis* 104, 490-502. RAISS, R (1976) *Zool. Anz.* 196, 201-11. RAKHILIN, V K (1979) In Labutin, Y V (ed) *Migratsii i ekologiya ptits Sibiri,* 40-2.

Yakutsk. RALPH, C J (1975) Ph D Thesis. Johns Hopkins Univ., Baltimore. RAMSAR CONVENTION ON WETLANDS OF INTERNATIONAL IMPORTANCE (1971) Ramsar, Iran. RAMSAY, L N G (1914) *Ibis* (10) **2**, 365-87. RAND, A L (1958) *Fieldiana Zool.* **35**, 145-220. RAND, A L and FLEMING, R L (1957) *Fieldiana Zool.* **41**, 1-218. RANDLA, T (1963) *Orn. Kogumik* **3**, 69-76. RANKIN, M N and RANKIN, D H (1940) *Irish Nat. J.* **7**, 273-82. RAPPE, A (1960) *Gerfaut* **50**, 209-22; (1978) *Gerfaut* **68**, 217-27. RAPPOLE, J H, MORTON, E S, LOVEJOY, T E, and RUOS, J L (1983) *Nearctic avian migrants in the Neotropics.* US Dept. Interior, Washington DC. RASHEK, V L (1965) *Ornitologiya* **7**, 486-7. RASHKEVICH, N A (1965) *Zool. Zh.* **44**, 1532-7. RASMUSSEN, C A (1921) *Dansk orn. Foren. Tidsskr.* **15**, 25-36. RATHBUN, S F (1920) *Auk* **37**, 458-60; RATTRAY, R H (1905) *J. Bombay nat. Hist. Soc.* **16**, 657-63. RAUSTE, V and SALONEN, V (1978) *Ornis fenn.* **55**, 84-5. RAVKIN, Y S (1973) *Ptitsy severo-vostochnogo Altaya.* Novosibirsk; (1984) *Prostranstvennaya organizatsiya naseleniya ptits lesnoy zony.* Novosibirsk. RAW, W (1921) *Ibis* (11) **3**, 238-64. RAYFIELD, P A (1941) *Br. Birds* **34**, 186-8. RAYNER, M (1962) *Ibis* **104**, 415-16. RAYNOR, G S (1979) *Bird-Banding* **50**, 124-44. REBOUSSIN, R (1928) *Rev. fr. Orn.* **20**, 343-5. REE, V (1974a) *Sterna* **13**, 191-7; (1974b) *Sterna* **13**, 257-68; (1977) *Fauna* **30**, 41-7. REEVES, E (1954) *Br. Birds* **47**, 443. REHBERG, H P (1970) *Falke* **17**, 137. REICHENOW, A (1904-5) *Die Vögel Afrikas* **3**. Neudamm. REICHHOLF, J (1968) *Anz. orn. Ges. Bayern* **8**, 294-5; (1977) *Orn. Beob.* **74**, 77-8; (1980) *Bonn. zool. Beitr.* **30**, 404-9; (1984) *Anz. orn. Ges. Bayern* **23**, 89-98. REICHHOLF-RIEHM, H (1972) *Anz. orn. Ges. Bayern* **11**, 190-3. REICHHOLF-RIEHM, H and REICHHOLF, J (1973) *Vogelwelt* **94**, 191. REID, S G (1887-8) *Ibis* (5) **5**, 424-35. REISER, O (1905) *Materialien zu einer Ornis Balcanica* **3**. Vienna. REISER, O and FÜHRER, L von (1896) *Materialien zu einer Ornis Balcanica* **4**. Vienna. RENDAHL, H (1960a) *Vogelwarte* **20**, 282-7; (1960b) *Ark. Zool.* (2) **13**, 1-71; (1964) *Ark. Zool.* **16**, 279-313; (1967) *Ark. Zool.* (2) **20**, 381-408. RENDAHL, H and VESTERGREN, G (1959) Vogelwarte 20, 162-3. RENSCH, B (1924) *Abh. Ber. Mus. Tierkde. Völkerkde. Dresden* **16**, 46-57. RENTSCH, L (1975) *Falke* **22**, 139. RETTIG, K (1970) *Orn. Mitt.* **22**, 103; (1974) *Orn. Mitt.* **26**, 25-6; (1983) *Beitr. Vogel. Insektenwelt NW Ostfriesland* **14**, 2-8. REY, E (1908) *Orn. Monatsschr.* **33**, 221-31; (1910a) *Orn. Monatsschr.* **35**, 225-34; (1910b) *Orn. Monatsschr.* **35**, 248-54; (1912) *Die Eier der Vögel Mitteleuropas.* Gera. REY, E and REICHERT, A (1908) *Orn. Monatsschr.* **33**, 189-97. REYMERS, N F (1966) *Ptitsy i mlekopitayushchie yuzhnoy taygi sredney Sibiri.* Moscow. REYNOLDS, J F (1971) *Bull. E. Afr. nat. Hist. Soc.*, 70-1; (1974) *Br. Birds* **67**, 70-6. RHEINWALD, G (1971) *Charadrius* **7**, 114-20; (1973) *Bonn. zool. Beitr.* **24**, 374-86; (1975) *Vogelwelt* **96**, 221-4; (1977) *Bonn. zool. Beitr.* **28**, 299-303; (1979) *Vogelwelt* **100**, 85-107; (1982) *Brutvogelatlas der Bundesrepublik Deutschland.* Bonn. RHEINWALD, G and GUTSCHER, H (1969) *Vogelwarte* **25**, 141-7. RHEINWALD, G, GUTSCHER, H, and HÖRMEYER, K (1976) *Vogelwarte* **28**, 190-206. RHEINWALD, G and SCHULZE-HAGEN, K (1972) *Charadrius* **8**, 74-81. RIBAUT, J-P (1964) *Rev. suisse Zool.* **71**, 815-902. RIBBECK, K (1904) *Z. Oologie (Berlin)* **14**, 8-10. RICHARD, A (1938) *Nos Oiseaux* **14**, 97-105. RICHARD, F (1968) *Aves* **5**, 190-1. RICHARDS, A J (1977) *Bird Study* **24**, 53-4. RICHARDS, B A (1945) *Br. Birds* **38**, 353. RICHARDS, B A and GOODWIN, D (1950) *Br. Birds* **43**, 300-1. RICHARDSON, A M M (1975) *Proc. malac. Soc. Lond.* **41**, 481-8. RICHARDSON, F (1965) *Ibis* **107**, 1-16. RICHARDSON, R A (1948) *Br. Birds* **41**, 306-7; (1956) *Br. Birds* **49**, 503. RICHTER, A (1972) *Orn. Beob.* **69**, 31-2. RICHTER, H (1953) *J. Orn.* **94**, 68-82; (1954) *Beitr. Vogelkde.* **3**, 251-8; (1955a) *Beitr. Vogelkde.* **4**, 139-42. (1956) *Beitr. Vogelkde.* **5**, 163-8. RICHTER, W (1955b) *Vogelwarte* 18,
31-2. RIDDIFORD, N and FINDLEY, P (1981) *Seasonal movements of summer migrants.* Tring. RIDGWAY, R (1901-11) *Bull. US natn. Mus.* **50** (1-5). RIES, A (1908) *Verh. orn. Ges. Bayern* **8**, 47-96. RIOLS, C (1978) *Alauda* **46**, 183; (1982) *Oiseau* **52**, 290-3. RINGLEBEN, H (1935) *Orn. Monatsber.* **43**, 73-7; (1948) *Vogelwarte* **15**, 40-1; (1949) *Beitr. Naturkde. Niedersachs.* **1** (5), 15-20; (1953) *Beitr. Naturkde. Niedersachs.* **6**, 82-6; (1957) *Ber. naturhist. Ges. Hannover* **103**, 91-100. RINNHOFER, G (1965) *Beitr. Vogelkde.* **11**, 115-16; (1970) *Beitr. Vogelkde.* **15**, 185-93; (1972) *Falke* **19**, 80-1; (1974) *Falke* **21**, 60-1. RIPLEY, S D (1948) *J. Bombay nat. Hist. Soc.* **47**, 622-7; (1951) *Postilla* **9**, 1-11; (1952) *Postilla* **13**, 1-48; (1958) *Postilla* **35**, 1-12. RIPLEY, S D and BOND, G M (1966) *Smithson. misc. Coll.* **151** (7), 1-37. RIPLEY, S D and RABOR, D S (1958) *Bull. Peabody Mus. nat. Hist.* **13**. RITTINGHAUS, H (1948) *Vogelwarte* **15**, 37-9. RITZEL, L (1978) *Beitr. Naturk. Niedersachs.* **31**, 21-2. RJABOW, W F (1968) *Falke* **15**, 112-18. ROBBINS, C S, BRUUN, B, and ZIM, H S (1983) *Birds of North America.* New York. ROBEL, D (1974) *Beitr. Vogelkde.* **20**, 157-8; (1981) *Beitr. Vogelkde.* **27**, 222-4. ROBERSON, D (1980) *Rare birds of the west coast of North America.* Pacific Grove. ROBERT, J-C (1979) *Alauda* **47**, 213-14. ROBERT, J-C and TOULON, D (1984) *Aves* **21**, 105-8. ROBERTS, A H N (1980a) *Birds Durham 1979*, 79-87. ROBERTS, B (1934) *Ibis* (13) **4**, 239-64. ROBERTS, J L (1980b) *Bonn. zool. Beitr.* **31**, 20-37. ROBERTS, T S (1955) *A manual for the identification of the birds of Minnesota and neighbouring states.* Minneapolis. ROBERTSON, K D (1983) *Br. Birds* **76**, 538. ROBERTSON, I S (1977) *Br. Birds* **70**, 237-45. ROBSON, B (1975) *BTO News* **75**, 6-7. ROBSON, J E (1967) *Br. Birds* **60**, 254-5. ROBSON, R W (1956) *Bird Study* **3**, 170-80; (1972) *Br. Birds* **65**, 303. ROBSON, R W and WILLIAMSON, K (1972) *Bird Study* **19**, 202-14. ROCHE, J (1958) *Inst. Rech. Sahar. Univ. Alger.*, 151-65. ROCHÉ, J-C (1968) *A sound guide to the birds of north-west Africa.* Institut Echo, Aubenas-les-Alpes. ROEVER, J W DE (1980) *Dutch Birding* **1** (4), 118. ROGACHEVA, E V (1962) *Ornitologiya* **5**, 118-34. ROGERS, A E F and GAULT, L N (1968) *Nat. Wales* **11**, 15-19. ROGERS, D T (1965) *Bird-Banding* **36**, 115-16. ROGERS, D T, JR and ODUM, E P (1966) *Wilson Bull.* **78**, 415-33. ROGERS, M J (1981a) *Br. Birds* **74**, 181-2; (1981b) *Br. Birds* **74**, 228-9; (1984) *Br. Birds* **77**, 361. ROGERS, M J and THE RARITIES COMMITTEE (1983) *Br. Birds* **76**, 476-529. ROGERS, T D and GALLAGHER, M D (1973) *Birds of Bahrain.* Unpubl. MS, Edward Grey Inst., Oxford Univ. ROGGE, D (1966) *Beitr. Vogelkde.* **12**, 162-88. ROKITANSKY, G (1959) *Egretta* **2**, 41-2. ROKITANSKY, G and SCHIFTER, H (1971) *Ann. naturhist. Mus. Wien* **75**, 495-538. ROLLIN, N (1931) *Scott. Nat.* **188**, 47-54; (1943a) *Br. Birds* **36**, 146-50; (1943b) *Br. Birds* **37**, 85-7; (1956) *Br. Birds* **49**, 218-21. ROLLS, J C (1972) *Br. Birds* **65**, 126-7. ROLLS, J C and ROLLS, M J (1977) *Br. Birds* **70**, 393. ROMMEL, K (1953) *Vogelring* **22**, 90-135. ROOKE, K B (1947) *Ibis* **89**, 204-10. ROOS, G (1978) *Anser* **17**, 133-8; (1983) *Anser* **22**, 1-26; (1984a) *Anser* **23**, 1-26; (1984b) *Anser* **23**, 149-62. ROOS, G T DE (1976) *Levende Nat.* **79**, 44-6. ROOTSMÄE, I (1981) *Eesti Loodus* **24**, 401-2. ROPE, G T (1889) *Zoologist* **13**, 184. ROSAIR, D B (1975) *Br. Birds* **68**, 248. ROSE, L N (1982) *Br. Birds* **75**, 382. ROSELAAR, C S (1976) *Bull. zool. Mus. Univ. Amsterdam* **5**, 13-18. ROSENBERG, E (1953) *Fåglar i Sverige.* Stockholm. ROSENSON, J (1946) *Vår Fågelvärld* **5**, 91-2. ROSHARDT, P A (1927) *Orn. Beob.* **24**, 129-34. ROSKELL, J (1982) *Devon Birds* **35**, 90-1. ROTENBERRY, J T (1980) *Ecol. Monogr.* **50**, 93-110. ROTHSCHILD, N C and WOLLASTON, A F R (1902) *Ibis* (8) **2**, 1-33. ROTHSCHILD, LORD and HARTERT, E (1923) *Novit. Zool.* **30**, 79-88. ROTHSCHILD, W and HARTERT, E (1911) *Novit. Zool.* **18**, 456-550; (1914) *Novit. Zool.* **21**, 180-204. ROUND, P D and Moss, M (1984) *Bird Study* **31**, 61-8.

ROUND, P D and WALSH, T A (1981) *Sandgrouse* **3**, 78-83. ROUNTREE, F R G (1977) *Br. Birds* **70**, 361-5. ROUX, F (1959) *Bull. Mus. Hist. nat. Paris* **31**, 334-40; (1960) *Alauda* **28**, 181-7. ROWAN, M K (1968) *Ostrich* **39**, 76-84. ROWAN, W (1919) *Trans. Herts. nat. Hist. Soc.* **17**, 81-2. ROWNTREE, M H (1943) *Bull. zool. Soc. Egypt* **5**, 18-32. RUCNER, D (1960) *J. Orn.* **101**, 310-15. RUDEBECK, G (1955) *Ibis* **97**, 572-80. RUITER, C J S (1941) *Ardea* **30**, 175-214. RUPPERT, K and LANGER, R (1955) *Vogelwelt* **76**, 93-102. RUSSELL, S M (1964) *Orn. Monogr. AOU* **1**. RUSTAMOV, A K (1954) *Ptitsy pustyni Kara-Kum*. Ashkhabad; (1958) *Ptitsy Turkmenistana* **2**. Ashkhabad. RUSTAMOV, E A (1982) *Ornitologiya* **17**, 91-7. RUTHKE, P (1938) *Beitr. Fortpfl. Vögel* **14**, 67; (1941) *Beitr. Fortpfl. Vögel* **17**, 106; (1954) *Vogelwelt* **75**, 188-91; (1971) *Vogelwelt* **92**, 191-2. RUTSCHKE, E (ed) (1983) *Die Vogelwelt Brandenburgs*. Jena. RUTTLEDGE, R F (1961) *Br. Birds* **54**, 327-8; (1966) *Ireland's birds*. London; (1973) *Br. Birds* **66**, 449; (1975) *A list of the birds of Ireland*. Dublin. RUWET, J-C (1965) *Les oiseaux des plaines et du lac-barrage de la Lufira supérieure (Katanga méridional)*. Liège Univ. RYABOV, V F (1949) *Trudy Naurzum. gos. zapoved.* **2**, 153-232; (1967) *Nauch. dokl. vyssh. Shk. Biol. nauki* (2), 18-22; (1968) *Vestnik Mosk. gos. Univ.* (6), 22-6. RYABOV, V F and MOSALOVA, N I (1967a) *Vestnik Mosk. Univ. (Biol. Pochvoved.)* **1**, 12-18; (1967b) *Nauch. dokl. vyssh. Shk. Biol. nauki* (6), 48-52. RYABOV, V F and SAMORODOV, Y A (1969) *Byull. Mosk. obshch. ispyt. prir. otd. biol.* **74** (5), 42-9. RYDÉN, O and BENGTSSON, H (1980) *Z. Tierpsychol.* **53**, 209-24. RYDZEWSKI, W (1978) *Ring* **96-7**, 218-62. RYVES, B H (1928) *Br. Birds* **22**, 31-3; (1943) *Br. Birds* **37**, 10-16; (1946) *Br. Birds* **39**, 179.

SACHER, G (1980a) *Thür. Orn. Mitt.* **25**, 17-20; (1980b) *Thür. Orn. Mitt.* **26**, 13-15. SAEMANN, D (1974) *Beitr. Vogelkde.* **20**, 12-41. SAEZ-ROYUELA, R (1952) *Bol. real Soc. española Hist. nat.* **50**,153-4. SAFRIEL, U (1968) *Ibis* **110**, 283-320. SAGE, B L (1954) *Br. Birds* **47**, 404-5; (1956a) *Bull. Br. Orn. Club* **76**, 32-3; (1956b) *Br. Birds* **49**, 182; (1960a) *Ardea* **48**, 160-78; (1960b) *Br. Birds* **53**, 265-71; (1964a) *Br. Birds* **57**, 327-8; (1964b) *Br. Birds* **57**, 435-6. SAGE, B L and JENKINS, A R (1956) *Br. Birds* **49**, 41-2. SAGE, B L and MEADOWS, B S (1965) *Bull. Soc. Sci. Nat. Phys. Maroc* **45**, 191-233. SAGITOV, A K and BAKAEV, S B (1980) *Ekologiya gnezdovaniya massovykh vidov ptits yugo-zapadnogo Uzbekistana*. Tashkent. ST PAUL, R (1975) *Notornis* **22**, 273-82. SALES, D (1972) *BTO News* **55**, 5-6. SALIMOV, K V (1977) In Sagitov, A K (ed) *Voprosy ekologii i morfologii zhivotnykh*, 84-90. Samarkand. SALMEN, H (1930) *Aquila* **36-7**, 127-8. SALOMONSEN, F (1933) *J. Orn.* **81**, 100-7; (1934) *Oiseau* **4**, 223-37; (1950-1) *Grønlands fugle*. Copenhagen; (1967) *Dansk orn. Foren. Tidsskr.* **61**, 151-64; (1971) *Dansk orn. Foren. Tidsskr.* **65**, 11-19; (1979) *Dansk orn. Foren. Tidsskr.* **73**, 191-206. SALT, G and HOLLICK, F S J (1946) *J. exp. Biol.* **23**, 1-46. SALVAN, J (1961) *Oiseau* **31**, 163; (1967-9) *Oiseau* **37**, 255-84, **38**, 53-85, 127-50, 249-73, **39**, 38-69. SAMORODOV, Y A (1967) *Vestnik Mosk. gos. Univ.* **3**, 28-9; (1968) *Ornitologiya* **9**, 371-3. SAMMALISTO, L (1956) *Ornis fenn.* **33**, 1-19; (1958) *Ann. Acad. Sci. Fenn.* (A) IV *Biol.* **41**; (1961) *Br. Birds* **54**, 54-69; (1968a) *Ann. zool. fenn.* **5**, 196-206; (1968b) *Trav. Mus. Hist. nat. Grigore Antipa* **9**, 529-47; (1971) *Ann. Acad. Sci. Fenn.* (A) IV *Biol.* **182**. SAMUEL, D E (1971a) *Wilson Bull.* **83**, 284-301; (1971b) *Auk* **88**, 839-55. SANDEMAN, G L (1978) *Scott. Birds* **10**, 23. SANTOS JÚNIOR, J R DOS (1959) *An. Fac. Ciências Porto* **41**, 126-8; (1960) *An. Fac. Ciências Porto* **42**, 5-16. SAPORETTI, F (1981) *Avocetta* **5**, 147-50. SARUDNY, N and HÄRMS, M (1926) *J. Orn.* **74**, 1-52. SASSI, M and ZIMMER, F (1941) *Ann. naturhist. Mus. Wien* **51**, 236-346. SASVÁRI-SCHÄFER,

L (1966) *Alauda* **34**, 200-9. SÁTORI, J (1942) *Aquila* **46-9**, 446-8. SAUER, F and SAUER, E (1960) *Bonn. zool. Beitr.* **11**, 41-86. SAUER, J (1978) *Falke* **25**, 422-3. SAUERBREI, F (1926) *Orn. Monatsschr.* **51**, 65-6. SAVINICH, I B (1983) In Kumari, E (ed) *Tez. dokl. XI Pribalt. orn. Konf.*, 188-90. Tallinn. SAXBY, H L (1874) *The birds of Shetland*. Edinburgh. SAXENBERGER, O (1905) *Orn. Monatsschr.* **30**, 479. SCEBBA, S and VITOLO, A (1983) *Gli Uccelli Ital.* **8**, 249-51. SCHÄFER, E (1938) *J. Orn.* **86** suppl. SCHÄFER, E and SCHAUENSEE, R M DE (1938) *Proc. Acad. Nat. Sci. Philadelphia* **90**, 185-260. SCHAUENSEE, R M DE (1984) *The birds of China*. Oxford. SCHAUENSEE, R M DE and PHELPS, W H (1978) *A guide to the birds of Venezuela*. Princeton. SCHELBERT, B (1984) *Orn. Beob.* **81**, 76-7. SCHENK, H (1980) *Lista rossa degli uccelli della Sardegna*. Parma. SCHERNER, E R (1982) *Vogelkdl. Ber. Niedersachs.* **14**, 14-15. SCHERNER, E R and WILDE, O (1972) *Vogelkdl. Ber. Niedersachs.* **4**, 15-18. SCHERRER, B and DESCHAINTRE, A (1970) *Jean le Blanc* **9**, 77-84. SCHIEBEL, G (1908) *Orn. Jahrb.* **19**, 1-30. SCHIERMANN, G (1934) *J. Orn.* **82**, 455-86; (1943) *Beitr. Fortpfl. Vögel* **19**, 13-18. SCHIFFERLI, A (1961a) *Orn. Beob.* **58**, 125-33; (1961b) *Rev. suisse Zool.* **68**, 143-5; (1965) *Rap. 1965 Stat. Orn. Suisse Sempach*; (1968) *Orn. Beob.* **65**, 38-42; (1969) *Orn. Beob.* **66**, 24; (1975) *Orn. Beob.* **72**, 207-34. SCHIFFERLI, A and D'ALESSANDRI, P (1971) *Orn. Beob.* **68**, 231. SCHIFFERLI, A, GÉROUDET, P, and WINKLER, R (1982) *Verbreitungsatlas der Brutvögel der Schweiz*. Sempach. SCHIFFERLI, L (1972) *Orn. Beob.* **69**, 257-74. SCHILDMACHER, H (1966) *Wir beobachten Vögel*. Jena. SCHLEGEL, S and SCHLEGEL, J (1974) *Beitr. Vogelkde.* **20**, 485-7. SCHLEICH, H-H and WUTTKE, M (1983) *Natur Mus.* **113**, 33-44. SCHLOSS, W (1982) *Auspicium* **7** (3), 169-83. SCHMIDT, E (1967a) *Aquila* **73-4**, 147-59; (1967b) *Aquila* **73-4**, 203; (1967c) *Beitr. Vogelkde.* **12**, 377-86; (1970) *Das Blaukehlchen*. Wittenberg Lutherstadt; (1973) *Aquila* **76-7**, 193; (1975) *Gef. Welt* **99**, 15-16. SCHMIDT, E and FARKAS, T (1974) *Der Steinrötel*. Wittenberg Lutherstadt. SCHMIDT, G (1960) *Mitt. faun. Arbeitsgem. Schlesw.-Holst.* (2) **13** (4), 51-9; (1967d) *Sterna* **7**, 400-1. SCHMIDT, K (1966) *Orn. Mitt.* **18**, 187. SCHMIDT, K and HANTGE, E (1954) *J. Orn.* **95**, 130-73. SCHMIDT, R K (1964) *Ostrich* **35**, 122. SCHMIDT, W (1908) *Orn. Monatsschr.* **33**, 243-6. SCHMIDT-KOENIG, K (1956) *Vogelwarte* **18**, 185-97. SCHMIED, A (1969) *Publ. Wiss. Film* AII (6) E939/1966. Göttingen. SCHMITT, C and STADLER, H (1915) *Verh. orn. Ges. Bayern* **12**, 165-73; (1918) *J. Orn.* **66**, 220-34. SCHMITZ, J P (1980) *Regulus* **60**, 40. SCHMUTTERER, H (1969) *Pests of crops in northeast and central Africa*. Stuttgart. SCHOENNAGEL, E (1972) *Orn. Mitt.* **24**, 106.SCHÖNFELD, M (1962) *Beitr. Vogelkde.* **7**, 451; (1972) *Beitr. Vogelkde.* **18**, 435-6; (1975a) *Beitr. Vogelkde.* **21**, 357-8; (1975b) *Beitr. Vogelkde.* **21**, 358-9. SCHÖNWETTER, M (1967) *Handbuch der Oologie* **1**; (1979) **2**. Berlin. SCHUBERT, G and SCHUBERT, M (1982) *Falke* **29**, 366-72. SCHUBERT, M (1982) *Stimmen der Vögel Zentralasiens* (2 LP discs). Eterna. SCHUBERT, W (1979) *Vogelwelt* **100**, 151-5. SCHÜCKING, A (1963) *Orn. Mitt.* **15**, 183. SCHULZE, H-U and SCHULZE, V (1977) *Orn. Mitt.* **29**, 225. SCHULZE, K-H (1971) *Beitr. Vogelkde.* **17**, 459. SCHUMANN, H (1959) *Orn. Mitt.* **11**, 204-5. SCHÜMPERLIN, W (1984) *Orn. Beob.* **81**, 164-6. SCHUSTER, L (1930) *J. Orn.* **78**, 273-301; (1944) *Beitr. Fortpfl. Vögel* **20**, 60-1; (1953) *Vogelwelt* **74**, 128-33. SCHÜZ, E (1933) *Vogelzug* **4**, 1-21; (1934) *Vogelzug* **5**, 9-18; (1956) *Vogelwarte* **18**, 169-77; (1957a) *Orn. Beob.* **54**, 9-33; (1957b) *Vogelwelt* **78**, 73-82; (1959) *Die Vogelwelt des südkaspischen Tieflandes*. Stuttgart. SCHWAGER, G and GÜTTINGER, H R (1984) *J. Orn.* **125**, 261-78. SCHWARZ, M (1949) *Orn. Beob.* **46**, 29-39; (1956) *Orn. Beob.* **53**, 61-72. SCHWARZE, E (1978) *Orn. Mitt.* **30**, 233. SCHWARZE, H (1975) *Corax* **5**, 143-6. SCHWEINSTEIGER, H (1938) *Beitr.*

Fortpfl. Vögel **14**, 210-11. SCLATER, W L and MACKWORTH-PRAED, C (1918) *Ibis* (10) **6**, 416-76, 602-720. SCLATER, W L and MOREAU, R E (1933) *Ibis* (13) **3**, 187-219. SCOTT, R E (1962a) *Bird Migr.* **2**, 118-20; (1962b) *Br. Birds* **55**, 87-8; (1965) *Vår Fågelvärld* **24**, 156-71. SECKER, H L (1955) *Emu* **55**, 104-7. SEEBOHM, H (1879) *Ibis* (4) **3**, 1-18; (1901) *The birds of Siberia*. London. SEEBOHM, H and HARVIE BROWN, J A (1876) *Ibis* (3) **6**, 105-26. SEEBOHM, H and SHARPE, R B (1902) *A monograph of the Turdidae or family of thrushes*. London. SEEL., D C and WALTON, K C (1974) *Bird Study* **21**, 282; (1979) *Ibis* **121**, 147-64. SEGUIN-JARD, E (1922) *Rev. fr. Orn.* **14**, 295-6. SELANDER, R K (1965) *Amer. Nat.* **99**, 129-41. SELIGMAN, O R and WILLCOX, J M (1940) *Ibis* (14) **4**, 464-79. SELLAR, P (1974) *Fair Isle Bird Obs. Rep.* **27**, 4. SELOUS, E (1901) *Bird watching*. London. SEMA, A M (1974) *Ornitologiya* **11**, 409-10. SEMADAM, G (1967) *Aquila* **73-4**, 198. SEMAGO, L L (1974) In Böhme, R L and Flint, V E (eds) *Mat. VI Vsesoyuz. orn. Konf.* **2**, 128-30. Moscow. SEMAGO, L L, SARYCHEV, V S, and IVANCHEV, V P (1984) *Ornitologiya* **19**, 187-8. SENK, R (1962) *Vogelwelt* **83**, 122-7. SENK, R, SENK, A, and WÖRNER, H (1972) *Vogelwarte* **26**, 314. SERLE, W (1939-40) *Ibis* (14) **3**, 654-99, (14) **4**, 1-47; (1943) *Ibis* **85**, 101-2; (1949) *Ostrich* **20**, 70-85; (1955) *Ostrich* **26**, 115-27; (1956) *Ibis* **98**, 307-11; (1957) *Ibis* **99**, 628-85. SERLE, W and MOREL, G J (1977) *A field guide to the birds of West Africa*. London. SEVERINGHAUS, S R and BLACKSHAW, K T (1976) *A new guide to the birds of Taiwan*. Taipei. SHACHAK, M, SAFRIEL, U N, and HUNUM, R (1981) *Ecology* **62**, 1441-9. SHANKS, R (1953) *Notornis* **5**, 202-3. SHANNON, G R (1974) *Br. Birds* **67**, 502-11. SHAPOVAL, A P (1981) *Zool. Zh.* **60**, 1866-8; (1982) *Vestnik Zool.* (4), 76-7. SHARLAND, M (1967) *Bull. Niger. orn. Soc.* **4** (16), 23; (1979) *Malimbus* **1**, 43-6. SHARLAND, R E and HARRIS, B J (1961) *Niger. Field* **26**, 65-8. SHARPE, R B (1886) *Ibis* (5) **4**, 475-93; (1890) *Catalogue of the Passeriformes or perching birds in the collection of the British Museum* **13**. London. SHARROCK, J T R (1964) *Br. Birds* **57**, 10-24; (1969) *Bird Study* **16**, 17-34; (1970) *Br. Birds* **63**, 313-24; (1972) *Br. Birds* **65**, 381-92; (1976) *The atlas of breeding birds in Britain and Ireland*. Tring; (1980a) *Br. Birds* **73**, 106-7; (1980b) *Br. Birds* **73**, 233; (1982) *Br. Birds* **75**, 90. SHARROCK, J T R and BROMFIELD, G N (1968) *Cape Clear Bird Obs. Rep.* **9**, 49-50. SHARROCK, J T R and DALE, M B (1964) *Br. Birds* **57**, 37-40. SHARROCK, J T R and THE RARE BREEDING BIRDS PANEL (1983) *Br. Birds* **76**, 1-25. SHARROCK, J T R and SHARROCK, E M (1976) *Rare birds in Britain and Ireland*. Berkhamsted. SHAW, G (1974) *Tay Ring. Group Rep. 1973*, 20-1; (1978) *Bird Study* **25**, 149-60; (1979a) *Bird Study* **26**, 66-7; (1979b) *Bird Study* **26**, 171-8. SHAW, TSEN-HWANG (1936) *The birds of Hopei province*. Peking. SHEPPERD, J (1953) *Br. Birds* **46**, 68-9. SHEVCHENKO, V L (1969) *Mat. 5 Vsesoyuz. orn. Konf.* **2**, 715-17. Ashkhabad; (1979) In Sagitov, A K and Kashkarov, D Y (eds) *Ekologiya gnezdovaniya ptits i metody ee izucheniya*, 236-7. Samarkand. SHEVCHENKO, V L, KAYMASHNIKOV, V I, and ANDREEVA, T A (1969) *Zool. Zh.* **48**, 270-83. SHIELDS, W M (1984) *Anim. Behav.* **32**, 132-48. SHIRIHAI, H (1986) *Br. Birds* **79**, 186-97. SHIRT, D B (1983) *Bustard Stud.* **1**, 57-68. SHISHKIN, V S (1976) *Zool. Zh.* **55**, 402-7; (1980) *Zool. Zh.* **59**, 1204-16; (1982) *Ornitologiya* **17**, 83-90. SHNITNIKOV, V N (1949) *Ptitsy Semirech'ya*. Moscow. SHOOTER, P (1970) *Br. Birds* **63**, 158-63. SHORT, H L and DREW, L C (1962) *Amer. Midl. Nat.* **67**, 424-33. SHORT, L L, JR (1964) *Pavo* **2**, 26-36. SHORT, L L and HORNE, J F M (1981) *Sandgrouse* **3**, 43-61. SIBLET, J-P and TOSTAIN, O (1984) *Nos Oiseaux* **37**, 284-8. SIBLEY, C G (1970) *Bull. Peabody Mus. nat. Hist.* **32**; (1974) *Emu* **74**, 65-79; (1985) In Campbell, B and Lack, E (eds) *A dictionary of birds*, 150-1. Calton. SIBLEY, C G and AHLQUIST,

J E (1980) *Proc. int. orn. Congr.* **17**, 1215-20. Berlin; (1981a) *Oiseau* **51**, 189-99; (1981b) *J. Orn.* **122**, 369-78; (1982) *J. Yamashina Inst. Orn.* **14**, 122-30; (1984) *Auk* **101**, 230-43; (1985a) *Emu* **85**, 1-14; (1985b) *Proc. int. orn. Congr.* **18**, 83-121. Moscow. SIBLEY, C G, CORBIN, K W, AHLQUIST, J E, and FERGUSON, A (1974) In Wright, C A (ed) *Biochemichal and immunological taxonomy of animals*, 89-176. London. SIBLEY, C G and SHORT, L L, JR (1959) *Ibis* **101**, 177-82. SIEBER, O (1980) *Z. Tierpsychol.* **52**, 19-56; (1982) *Orn. Beob.* **79**, 25-38. SIEGFRIED, W R (1968) *Ostrich* **39**, 105-29. SIEGNER, J (1983) *Verh. orn. Ges. Bayern* **23**, 529. SIIVONEN, L (1935) *Ornis fenn.* **12**, 89-99; (1939) *Ann. zool. Soc. Zool. Bot. Fenn. Vanamo* **7** (1); (1941) *Ann. zool. Soc. Zool. Bot. Fenn. Vanamo* **8** (6). SIKORA, S (1980) *Roczniki Akad. Roln. Poznán* **122**, 55-9. SILSBY, J (1980) *Inland birds of Saudi Arabia*. London. SILVA, E T (1949) *Br. Birds* **42**, 97-111. SIMAN, H Y and BUNNI, M K (1976a) *Bull. nat. Hist. Res. Cent. Baghdad* **7** (1), 142-66; (1976b) *Bull. nat. Hist. Res. Cent. Baghhad* **7** (1), 187-90; (1978) *Bull. nat. Hist. Res. Cent. Baghhad* **7** (2), 69-83. SIMEONOV, S D and MICHEV, T M (1980) *Ecology Bulg. Acad. Sci.* **7**, 84-93. SIMEONOW, S D (1968) *J. Orn.* **109**, 57-61; (1969) *J. Orn.* **110**, 499-500. SIMKIN, G N (1981) *Ornitologiya* **16**, 73-83. SIMKIN, G N and STEINBACH, M V (1984) *Ornitologiya* **19**, 135-45. SIMMONS, K E L (1949) *Br. Birds* **42**, 24; (1951a) *Br. Birds* **44**, 92-3; (1951b) *Br. Birds* **44**, 369-72; (1951c) *Ibis* **93**, 407-13; (1952a) *Ardea* **40**, 67-72; (1952b) *Ibis* **94**, 358-9; (1954) *Ardea* **42**, 140-51; (1960) *Br. Birds* **53**, 314-15; (1965) *Bull. Br. Orn. Club* **85**, 161-8; (1966) *J. Zool. Lond.* **149**, 145-62; (1970) In Gooders, J (ed) *Birds of the world*, 1779-81. London; (1974) *Br. Birds* **67**, 413-37; (1985a) In Campbell, B and Lack, E (eds) *A dictionary of birds*, 19; (1985b) 101-5; (1985c) 144-5; (1985d) 569-70. Calton; (1985e) *Br. Birds* **78**, 508; (1986) *The sunning behaviour of birds*. Bristol. SIMMS, E (1978) *British thrushes*. London. SIMON, P (1965) *Gerfaut* **55**, 26-71. SIMPSON, D M (1984) *Sea Swallow* **33**, 53-8. SIMPSON, J H (1971) *Scott. Birds* **6**, 338. SIMPSON, K and DAY, N (1984) *The birds of Australia*. South Yarra. SIMPSON, T (1970) *Br. Birds* **63**, 177. SINCLAIR, A R E (1978) *Ibis* **120**, 480-97. SITTERS, H P (1986) *Br. Birds* **79**, 105-16. SKAR, H-J, HAGEN, A, and ØSTBYE, E (1972) *Norw. J. Zool.* **20**, 51-9. SKEAD, D M (1966) *Ostrich* **37**, 10-16. SKEAD, D M and SKEAD, C J (1970) *Ostrich* **41**, 247-51. SKOOG, I (1973) *Vår Fågelvärld* **32**, 131-2; (1981) *Vår Fågelvärld* **40**, 193-7. SKOV, H (1970) *Dansk orn. Foren. Tidsskr.* **64**, 271-2. SKUTCH, A F (1976) *Parent birds and their young*. Austin. SLAGSVOLD, T (1973a) *Norw. J. Zool.* **21**, 139-58; (1973b) *Norw. J. Zool.* **21**, 159-72; (1977) *Ornis scand.* **8**, 197-222; (1979) *Fauna norv.* (C) *Cinclus* **2**, 65-9; (1980a) *Ornis scand.* **11**, 92-8; (1980b) *J. Anim. Ecol.* **49**, 523-36. SLATER, P (1974) *A field guide to Australian birds*. Sydney. SLATER, P J B (1983) *Anim. Behav.* **31**, 308-9. SLINGS, Q L (1982) *Limosa* **55**, 59-60. SLUYS, R and BERG, M VAN DEN (1982) *Ornis scand.* **13**, 123-8. SMEENK, C (1969) *Limosa* **42**, 27-31. SMETANA, N M (1980) In Kovshar', A F (ed) *Biologiya ptits Naurzumskogo gosudarstvennogo zapovednika*, 105-14. Alma-Ata. SMETANA, N M and GUSEVA, V S (1981) *Ornitologiya* **16**, 88-92. SMIRENSKI, S M (1979) *Ornitologiya* **14**, 196-7. SMITH, E (1978) *Br. Birds* **71**, 313. SMITH, E J (1979) *Sorby Rec.* **17**, 29-30. SMITH, E M, SMITH, R W J, and STARK, D M (1970) *Scott. Birds* **6**, 214-15. SMITH, H C (1942) *Notes on birds of Burma*. Simla. SMITH, J N M (1974) *Behaviour* **48**, 276-302. SMITH, K D (1951) *Br. Birds* **44**, 113-17; (1955a) *Ibis* **97**, 65-80; (1955b) *Ibis* **97**, 480-507; (1957) *Ibis* **99**, 1-26, 307-37; (1960a) *Ibis* **102**, 536-44; (1962-4) Unpubl. diaries of Moroccan expeditions. Edward Grey Inst., Oxford Univ.; (1965a) *Ibis* **107**, 493-526; (1968a) *Ibis* **110**, 90-1; (1968b) *Ibis* **110**, 452-92; (1971) *Bird Study* **18**,

71-9. SMITH, K G and SCHNEIDER, D C (1978) *US Fish Wildl. Serv. Rep.* 14-16-0009-029. SMITH, M J and GRAVES, H B (1978) *Behav. Biol.* **23**, 355-72. SMITH, M Q (1960*b*) *Ibis* **102**, 576-83. SMITH, R P and ORMEROD, S J (1986) *Bird Study* **33**, 138-9. SMITH, S (1950) *The Yellow Wagtail.* London; (1960*c*) *Br. Birds* **53**, 301-3. SMITH, S and HOSKING, E (1955) *Birds fighting.* London. SMITH, V W (1965*b*) *Bull. Niger. orn. Soc.* **2**, 26-34; (1966) *Ibis* **108**, 492-512. SMITH, V W and EBBUTT, D (1965) *Ibis* **107**, 390-3. SMOGORZHEVSKI, L A and KOTKOVA, L I (1973) *Vestnik Zool.* **7** (3), 34-9. SMYTHIES, B E (1953) *The birds of Burma.* Edinburgh; (1981) *The birds of Borneo.* Kuala Lumpur. SMYTHIES, B E and HARRISSON, T (1956) *Ibis* **98**, 535-7. SNELL, M L (1963) *Ostrich* **34**, 36-9. SNOW, B K and SNOW, D W (1982) *J. Yamashina Inst. Orn.* **14**, 281-92; (1984) *Ibis* **126**, 39-49. SNOW, D W (1952) *Ibis* **94**, 473-98; (1953) *Ibis* **95**, 376-8; (1955*a*) *Bird Study* **2**, 72-84; (1955*b*) *Bird Study* **2**, 169-78; (1958*a*) *Ibis* **100**, 1-30; (1958*b*) *A study of Blackbirds.* London; (1965) *Bird Study* **12**, 135-42; (1966*a*) *Bird Study* **13**, 237-55; (1966*b*) *Nature* **211**, 1231-3; (1969*a*) *Bird Study* **16**, 34-44; 544cegi (1969*b*) *Bird Study* **16**, 115-29; (1978) *Ring. Migr.* **2**, 52-4. SNOW, D W and MANNING, A W G (1954) *Alauda* **22**, 1-24. SNOW, D W, OWEN, D F, and MOREAU, R E (1955) *Ibis* **97**, 557-71. SNOW, D W and SNOW, B K (1983) *Bird Study* **30**, 51-6. SOKAL, R R and ROHLF, F J (1969) *Biometry.* San Francisco. SOKOŁOWSKI, J (1964) *Przeglad Zool.* **8**, 349-59. SOMEREN, V D VAN (1958) *A bird watcher in Kenya.* Edinburgh. SOMEREN, V G L VAN (1916) *Ibis* (10) **4**, 373-472; (1934) *J. E. Afr. Uganda nat. Hist. Soc. Suppl.* **4**, 27-8; (1956) *Fieldiana Zool.* **38**. SOMEREN, V G L VAN and SOMEREN, G R C VAN (1949) *Uganda J.* **13** suppl. SOMMERFELD, E (1930) *Anz. orn. Ges. Bayern* **2**, 60-9. SONIN, V D and ANUCHINA, N F (1979) In Labutin, Y V (ed) *Migratsii i ekologiya ptits Sibiri*, 183-4. Yakutsk. SONNABEND, H (1958) *Vogelwarte* **19**, 212. SONOBE, K (ed) (1982) *A field guide to the birds of Japan.* Tokyo. SOPER, E A (1972) *Br. Birds* **65**, 259. SOPYEV, O S (1981) *Priroda* **9** (793), 58. SORJONEN, J (1974) *Ornis Karelica* **1**, 16-27; (1977) *Ornis fenn.* **54**, 101-7; (1983*a*) *Ornis fenn. Suppl.* **3**, 81-3; (1983*b*) *Ornis scand.* **14**, 278-88. SOTAVALTA, O (1956) *Ann. zool. Soc. Zool. Bot. Fenn. Vanamo* **17** (4). SOUTHERN, H N (1937) *Br. Birds* **31**, 2-4. SOVERI, J (1940) *Acta zool. fenn.* **27**. SOVINEN, M (1952) *Ornis fenn.* **29**, 27-35. SPAEPEN, J (1953) *Gerfaut* **43**, 178-230; (1957) *Gerfaut* **47**, 17-43. SPAEPEN, J and CAUTEREN, F VAN (1961) *Gerfaut* **51**, 148-55; (1962) *Gerfaut* **52**, 275-97; (1968) *Gerfaut* **58**, 24-77. SPARKS, J H (1961) *Br. Birds* **54**, 431. SPASSKI, A A and SONIN, M D (1959) *Ornitologiya* **2**, 184-7. SPEEK, B J (1972) *Limosa* **45**, 145-73. SPENCER, K G (1953) *Naturalist*, 153-4; (1975*a*) *Countryside* summer, 482-5; (1978) *Br. Birds* **71**, 89. SPENCER, R (1975*b*) *Bird Study* **22**, 177-90. SPENCER, R and HUDSON, R (1978) *Ring. Migr.* **1**, 189-252; (1981) *Ring. Migr.* **3**, 213-56; (1982) *Ring. Migr.* **4**, 65-128. SPENCER, R and THE RARE BREEDING BIRDS PANEL (1985) *Br. Birds* **78**, 69-92; (1986) *Br. Birds* **79**, 53-81. SPJØTVOLL, Ø (1970) *Sterna* **9**, 163-74. STAAV, R (1975) *Vår Fågelvärld* **34**, 212-20; (1978) *Fauna och Flora* **73**, 89-93. STADLER, H (1926) *Ber. Ver. schles. Orn.* **12**, 22-38; (1928*a*) *Orn. Beob.* **25**, 53-6; (1928*b*) *Verh. orn. Ges. Bayern* **18**, 107-31; (1931) *Verh. orn. Ges. Bayern* **19**, 331-59; (1952) *Larus* **4-5**, 149-84; (1957) *Larus* **9-10**, 193-207; (1959) *Larus* **11**, 131-9. STADLER, H and SCHMITT, L (1913) *Ardea* **2**, 109-15; (1915) *Ardea* **4**, 104-8. STAHLBAUM, G (1978) *Falke* **25**, 91. STAINTON, J M (1978) *Br. Birds* **71**, 130-1. STAKHEEV, V A (1979) In Labutin, Y V (ed) *Migratsii i ekologiya ptits Sibiri*, 184-6. Yakutsk. STALLCUP, W B (1961) *J. grad. Res. Cent. S. Methodist Univ.* **29**, 43-65. STANFIELD, J P (1973) *Ibis* **115**, 606. STANFORD, J K (1941) *Ibis* (14) **5**, 353-78; (1953) *Ibis* **95**, 316-28; (1954) *Ibis* **96**, 449-73, 606-24. STANFORD, W (1969) *Notes on the birds of Oman and Trucial States.* Unpubl. MS, Edward Grey Inst., Oxford Univ. STATON, J (1950) *Br. Birds* **43**, 300. STEGMANN, B (1928) *J. Orn.* **76**, 496-503; (1932) *Orn. Monatsber.* **40**, 54; (1936) *J. Orn.* **84**, 58-139. STEGMANN, B K (1935) *Compt. Rend. Acad. Sci. URSS* **3**, 45-8. STEIN, G H W (1952) *J. Orn.* **93**, 158-71. STEIN, H (1968) *Mitt. IG Avifauna DDR* **1**, 29-39. STEINBACHER, G (1948) *Orn. Mitt.* **1**, 47-8; (1953) *Biol. Abh.* **5**. STEINBACHER, J (1952) *Vogelwelt* **73**, 197-208; (1954) *Vogelwelt* **75**, 18-20; (1956) *Vogelwelt* **77**, 1-12; (1965) *Senckenbergiana Biol.* **46**, 429-59; (1981) *Gef. Welt* **105**, 121. STEINER, H M (1971) *Egretta* **14**, 55-6. STEINFATT, O (1937) *Verh. orn. Ges. Bayern* **21**, 139-54; (1938) *Orn. Monatsber.* **46**, 65-76; (1939*a*) *Dtsch. Vogelwelt* **64**, 34-9; (1939*b*) *Orn. Monatsber.* **47**, 38-46; (1941*a*) *J. Orn.* **89**, 204-12; (1941*b*) *J. Orn.* **89**, 393-403; (1954) *J. Orn.* **95**, 22-37, 245-62. STEINIGER, F (1955) *Vogelwelt* **76**, 134-7. STEJNEGER, L (1905) *Smithson. misc. Coll.* **47**, 421-30. STENHOUSE, J H (1921) *Ibis* (11) **3**, 573-94. STEPANYAN, L S (1967) *Ornitologiya* **10**, 97-107; (1970) *Uchen. zap. Mosk. gos. ped. Inst.* **394**, 102-50; (1977) *Zool. Zh.* **56**, 1834-8; (1978*a*) *Sostav i raspredelenie ptits fauny SSR, Passeriformes.* Moscow; (1978*b*) In Sudilovskaya, A M and Flint, V E (eds) *Ptitsy i presmykayushchiesya*, 25-51. Moscow; (1982) *Zh. obshch. Biol.* **43**, 589-95; (1983) *Nadvidy i vidy-dvoniki v avifaune SSSR.* Moscow. STEPHAN, B (1970) *Mitt. zool. Mus. Berlin* **46**, 339-437; (1978) *Falke* **25**, 175; (1985) *Die Amsel.* Wittenberg Lutherstadt. STEPHENSON, G C and DORAN, T M J (1982) *Br. Birds* **75**, 290. STERBETZ, I (1971) *Allattani közl.* **58**, 171-2. STEURI, T, FELLER, V, and SCHMID, H (1979) *Orn. Beob.* **76**, 136. STEWART, A G (1971) *Scott. Birds* **6**, 448. STEWART, P A (1931) *Auk* **54**, 324-32. STEWART, P A and SKINNER, R W (1967) *Wilson Bull.* **79**, 37-42. STEYN, P (1971) *The birds of Zambia.* London. STICHMANN, W (1958) *Vogelwelt* **79**, 113-14. STIEFEL, A (1976) *Ruhe und Schlaf bei Vögeln.* Wittenberg Lutherstadt. STOKES, M E (1947) *Irish Nat. J.* **9**, 49-50. STOLLMANN, A (1964) *Aquila* **69-70**, 195-8. STOLT, B-O (1971) *Vår Fågelvärld* **30**, 201-2. STOLT, B-O and MASCHER, J W (1962) *Vogelwarte* **21**, 319-26. STONEHAM, H F (1929) *Bateleur* **1**, 56-7; (1931) *Ibis* (13) **1**, 701-12. STONEHOUSE, B (1971) *Animals of the Arctic: the ecology of the far north.* London. STONER, D (1926) *Auk* **43**, 198-213; (1935) *Auk* **52**, 400-7; (1936) *Roosevelt Wildl. Ann.* **4** (2), 126-233; (1942) *Bird-Banding* **13**, 107-10. STONER, D and STONER, L C (1941) *Auk* **58**, 52-5. STORER, R W (1971) In Farner, D S, King, J R, and Parkes, K C (eds) *Avian Biology* **1**, 1-18. New York. STRAHM, J (1953) *Orn. Beob.* **50**, 41-8; (1954) *Nos Oiseaux* **22**, 187-96; (1956) *Nos Oiseaux* **23**, 257-65; (1963) *Nos Oiseaux* **27**, 61-6. STRANGEMAN, P J (1975) *London Bird Rep.* **40**, 78-84. STREICHERT, J (1984) *Beitr. Naturk. Niedersachs.* **37**, 24-47. STREMKE, A and STREMKE, D (1980) *Orn. Rdbr. Meckl.* **22**, 69-77. STRESEMANN, E (1910) *Orn. Monatsschr.* **35**, 119-20; (1920) *Avifauna Macedonica.* Munich; (1926) *Orn. Monatsber.* **34**, 131-9; (1927-34) *Handbuch der Zoologie* **7** (2). Berlin; (1928) *J. Orn.* **76**, 313-411; (1931) *J. Orn.* **79**, 128-32; (1943) *J. Orn.* **91**, 448-514; (1948) *Orn. Ber.* **1**, 193-227; (1956*a*) *J. Orn.* **97**, 44-72; (1956*b*) *J. Orn.* **97**, 441; (1957) *J. Orn.* **98**, 123; (1963) *Auk* **80**, 1-8. STRESEMANN, E, MEISE, W, and SCHÖNWETTER, M (1937) *J. Orn.* **85**, 375-576. STRESEMANN, E and STRESEMANN, V (1966) *J. Orn.* **107** spec. vol.; (1968*a*) *J. Orn.* **109**, 17-21; (1968*b*) *J. Orn.* **109**, 475-84. (1969) *J. Orn.* **110**, 39-52. STREZ, S (1984) *Tzufit* **2**, 68-98. STRICKLAND, M J and GALLAGHER, M D (1969) *A guide to the birds of Bahrain.* Muharraq. STUBBS, F J (1909) *Zoologist* (4) **15**, 361-6. STUTTARD, P and WILLIAMSON, K (1971) *Bird Study* **18**, 9-14. SUÁREZ, F, SANTOS, T, and TELLERÍA, J L (1982) *Gerfaut* **72**, 231-5. SUÁREZ

CARDONA, F (1977) *Ardeola* **23**, 63-79. SUDHAUS, W (1964) *Vogelwelt* **85**, 44-6; (1965) *Vogelwelt* **86**, 69-77; (1966a) *Corax* **1**, 129-44; (1966b) *Orn. Mitt.* **18**, 131-4; (1970) *Beitr. Vogelkde.* **15**, 340-3; (1972) *Orn. Mitt.* **24**, 231-6; (1974) *Beitr. Vogelkde.* **20**, 461-6. SUEUR, F (1982) *Alauda* **50**, 148; (1985) *Nos Oiseaux* **38**, 77-9. SUFFERN, C (1951) *Br. Birds* **44**, 387-8; (1965a) *Br. Birds* **58**, 23; (1965b) *Br. Birds* **58**, 300. SUGDEN, L G (1973) *Can. Wildl. Serv. Rep.* **24**. SULKAVA, S (1963) *Aquilo Ser. Zool.* **1**, 24-5. SULTANA, J (1978) *Heritage* **20**, 382-3. SULTANA, J and GAUCI, C (1970) *Malta Year Book 1970*, 339-46. Valletta; (1975) *Il-Merill* **15**, 4; (1982) *A new guide to the birds of Malta*. Valletta. SUMMERS, D D B (1953) *Br. Birds* **46**, 69-70. SUMMERS-SMITH, D and LEWIS, L R (1952) *Bird Notes* **25**, 44-8. SUNDKVIST, H (1979) *Vår Fågelvärld* **38**, 106. SUNKEL, W (1952) *Orn. Mitt.* **4**, 217-22. SURVILLO, A V (1968) *Trudy Inst. Zool. Alma-Ata* **29**, 71-5. SUSCHKIN, P P (1914a) *J. Orn.* **62**, 557-607; (1914b) *Orn. Vestnik* **5**, 3-43. SUSHKIN, P P (1925) *Proc. Boston Soc. nat. Hist.* **38**, 1-55; (1938) *Ptitsy Sovetskogo Altaya* **2**. Moscow. SÜSS, K-H (1970) *Beitr. Vogelkde.* **15**, 346-7. SUTHERLAND, W J and BROOKS, D J (1981) *Sandgrouse* **3**, 87-90. SUTTER, E (1960) *Orn. Beob.* **57**, 159. SUTTON, G M (1927) *Wilson Bull.* **39**, 131-41; (1932) *Mem. Carnegie Mus. Pittsburgh* **12** (2) sect. 2. SUTTON, G M and PARMELEE, D F (1954) *Condor* **56**, 295-306; (1955) *Bird-Banding* **26**, 1-18. SUTTON, R W W and GRAY, J R (1972) In Vittery, A and Squire, J E (eds) *Orn. Soc. Turkey Bird Rep. 1968-9*, 186-205. Sandy. SUZUKI, S, TANIOKA, K, UCHIMURA, S, and MARUMOTO, T (1952) *J. Agric. Meteorol.* **7**, 149-51. SVÄRDSON, G (1957) *Br. Birds* **50**, 314-434. SVENSSON, L (1974) *Vår Fågelvärld* **33**, 48; (1977) *Vår Fågelvärld* **36**, 48-52; (1982) *Vår Fågelvärld* **41**, 267-8; (1984a) *Identification guide to European passerines*. Stockholm; (1984b) *Soviet birds*. Stockholm (cassette and booklet). SVENSSON, S (1963) *Vår Fågelvärld* **22**, 161-81; (1969) *Vår Fågelvärld* **28**, 236-40. SWAINE, C M (1932) *Br. Birds* **25**, 333. SWANBERG, P O (1951) *Fauna och Flora* **46**, 111-36. SWANN, R L (1975) *Bird Study* **22**, 93-8; (1980) *Ring. Migr.* **3**, 37-40. SWANSON, G A and BARTONEK, J C (1970) *J. Wildl. Mgmt.* **34**, 739-46. SWELM, N D VAN (1980) *Dutch Birding* **2**, 28-31. SWEREW, M D (1927) *Uragus* **5** (4), 7-9. SWYNNERTON, C F M (1907) *Ibis* (9) **1**, 30-74. SYMONDS, J H (1914) *Avic. Mag.* (3) **5**, 259-61. SZABÓ, L (1976) *Aquila* **82**, 247-8. SZIJJ, L (1954) *Aquila* **55-8**, 305. SZULC-OLECH, B (1964) *Bird Study* **12**, 1-7.

TAAPKEN, J (1977) *Vogeljaar* **25**, 19. TADIĆ, Z (1975) *Larus* **26-8**, 111-15. TALBOT, J N (1974) *Honeyguide* **79**, 43. TALLOWIN, J and YOUNGMAN, R E (1978) *Br. Birds* **71**, 182-3. TAMMELIN, H (1975) *Lintumies* **10**, 107. TARASHCHUK, V I (1953) *Ptitsy polezashchitnykh nasazhdeniy*. Kiev. TAUCHNITZ, H (1977) *Apus* **4** (1), 9-14. TAVERNER, J H (ed) (1976) *Hampshire Bird Rep. 1975*, 8-35; (ed) (1979) *Hampshire Bird Rep. 1978*, 11-45. TAYLOR, D W, DAVENPORT, D L, and FLEGG, J J M (1981) *The birds of Kent*. Meopham. TAYLOR, E C (1896) *Ibis* (7) **2**, 477-82. TAYLOR, I R and MACDONALD, M A (1979) *Bull. Br. Orn. Club* **99**, 29-30. TAYLOR, J (1980a) *Br. Birds* **73**, 356. TAYLOR, J S (1942) *Ostrich* **13**, 148-56; (1964) *Ostrich* **35**, 66. TAYLOR, L S (1954) *Br. Birds* **47**, 85. TAYLOR, M (1984) *Ring. Migr.* **5**, 65-78. TAYLOR, P B (1979) *Scopus* **3**, 80; (1980b) *Scopus* **4**, 72. TAYLOR, R J F (1953) *Sterna* **10**, 3-36. TAYLOR, W P (1925) *Auk* **42**, 349-53. TEAGER, C W (1939) *Countryside*, 495-6; (1967) *Br. Birds* **60**, 361-3. TEBBUTT, C F (1968) *Br. Birds* **61**, 33. TEIDEMAN, S J (1946) *Br. Birds* **39**, 54. TEIXEIRA, R M (ed) (1979) *Atlas van de Nederlandse broedvogels*. Deventer. TELLERÍA, J L (1981) *La migracion de las aves en el Estrecho de Gibraltar* **2**. Madrid. TEMME, M (1974) *Bonn. zool. Beitr.* **25**, 292-6.

TENISON, W P C (1954) *Br. Birds* **47**, 312. TERRILL, L M (1917) *Wilson Bull.* **29**, 130-41. TESCHEMAKER, W E (1912a) *Avic. Mag.* (3) **3**, 273-80; (1912b) *Avic. Mag.* (3) **3**, 293-7, 330-5; (1913) *Avic. Mag.* (3) **4**, 323-7. THALMANN, E (1981) *Vögel Heimat* **52**, 18-19. THANNER, R VON (1905) *Orn. Jahrb.* **16**, 50-66. THEISS, N (1972) *Orn. Mitt.* **24**, 27-31; (1973) *Orn. Mitt.* **25**, 231-40. THÉVENOT, M (1982) *Oiseau* **52**, 21-86, 97-152. THÉVENOT, M, BEAUBRUN, P, BAOUAB, R E, and BERGIER, P (1982) *Docum. Inst. Sci. Rabat* **7**. THIBAULT, J-C (1983) *Les oiseaux de la Corse*. Ajaccio. THIEDE, W (1975) *Misc. Rep. Yamashina Inst. Orn.* **7**, 330-2; (1983) *Vogelwelt* **104**, 217-22. THIEDE, W and THIEDE, U (1974) *Vogelwelt* **95**, 88-95. THIELCKE, G (1957) *Vogelwelt* **78**, 69; (1965) *Vogelwelt* **86**, 155; (1970) *Vogelstimmen*. Berlin. THIELCKE, G and THIELCKE, H (1964) *Vogelwelt* **85**, 46-53. THIELCKE-POLTZ, H and THIELCKE, G (1960) *Z. Tierpsychol.* **17**, 211-44. THIMM, F (1973) *J. comp. Physiol.* **84**, 311-34. THIMM, F, CLAUSEN, A, TODT, D, and WOLFFGRAMM, J (1974) *J. comp. Physiol.* **93**, 55-84. THIOLLAY, J-M (1977) *Alauda* **45**, 343. THOM, A S (1947) *Br. Birds* **40**, 20-1. THOMAS, B T (1982) *Bull. Br. Orn. Club* **102**, 48-52. THOMAS, J F (1925) *Br. Birds* **19**, 98; (1926) *Br. Birds* **20**, 24; (1934) *Br. Birds* **27**, 231-2; (1934) *Br. Birds* **28**, 171-2; (1936) *Br. Birds* **29**, 244-5; (1937a) *Br. Birds* **30**, 293-4; (1937b) *Br. Birds* **31**, 234-5; (1938) *Br. Birds* **32**, 233-6; (1940) *Br. Birds* **33**, 335-6. THOMAS, M (1970) *Bird Life* **6**, 129. THOME, J A (1926) *Norsk orn. Tidsskr.* (2) **7**, 266-71. THOMPSON, W (1849) *The natural history of Ireland* **1**. London. THOMSEN, P and JACOBSEN, P (1979) *The birds of Tunisia*. Copenhagen. THÖNEN, W (1948) *Orn. Beob.* **45**, 33-9. THONNERIEUX, Y (1980) *Bièvre* **1** (2), 73-4. THORPE, W H (1961) *Bird-song*. Cambridge. THORPE, W H and HALL-CRAGGS, J (1976) In Bateson, P P G and Hinde, R A (eds) *Growing points in ethology*, 171-89. Cambridge. THORPE, W H and PILCHER, P M (1958) *Br. Birds* **51**, 509-14. TIAINEN, J (1977) *Ornis fenn.* **54**, 95; (1981) *Vogelwarte* **31**, 178-80. TIAINEN, J and YLIMAUNU, J (1984) *Lintumies* **19**, 26-9. TICEHURST, C B (1922a) *Ibis* (11) **4**, 151-8; (1922b) *Ibis* (11) **4**, 605-62; (1923) *Ibis* (11) **5**, 1-43; (1926) *J. Bombay nat. Hist. Soc.* **31**, 91-119; (1926-7) *J. Bombay nat. Hist. Soc.* **31**, 687-711, 862-81, 32, 64-97; (1927) *Ibis* (12) **3**, 65-74. (1938) *Ibis* (14) **2**, 338-41. TICEHURST, C B, BUXTON, P A, and CHEESMAN, R E (1921-2) *J. Bombay nat. Hist. Soc.* **28**, 210-50, 381-427, 650-74, 937-56. TICEHURST, C B and CHEESMAN, R E (1925) *Ibis* (12) **1**, 1-31. TICEHURST, C B, COX, P Z, and CHEESMAN, R E (1925) *J. Bombay nat. Hist. Soc.* **30**, 725-33. TICEHURST, C B and WHISTLER, H (1938) *Ibis* (14) **2**, 717-46. TICEHURST, N F (1912) *Br. Birds* **6**, 170-6. TICEHURST, N F and TICEHURST, C B (1902) *Zoologist* (4) **6**, 261-77. TIMMERMANN, G (1934) *J. Orn.* **82**, 319-24; (1935) *Greinar Vísindafél. Íslendinga* **1**, 14-47; (1938-49) *Die Vögel Islands*. Reykjavik. TINBERGEN, N (1961) *Br. Birds* **54**, 122. TINBERGEN, N and KUENEN, D J (1939) *Z. Tierpsychol.* **3**, 37-60. TINNER, T (1946) *Vögel Heimat* **17**, 4-10. TISCHLER, F. (1917) *Orn. Monatsschr.* **42**, 185-9; (1941) *Die Vögel Ostpreussens* **1**. Königsberg. TISCHLER, W (1955) *Synökologie der Landtiere*. Stuttgart. TODD, W E C (1957) *Condor* **59**, 211. TODT, D (1967a) *Z. Naturforsch.* **22b**, 997; (1967b) *Z. Naturforsch.* **22b**, 997-8; (1968a) In Mittelstaedt, H (ed) *Kybernetik 1968*, 465-85. Munich; (1968b) *Naturwiss.* **55**, 450; (1970a) *Verh. dt. zool. Ges.* **64**, 249-52; (1970b) *Z. vergl. Physiol.* **66**, 294-317; (1971) *Z. vergl. Physiol.* **71**, 262-85; (1974) *Z. Naturforsch.* **29c**, 157-60; (1975) *J. comp. Physiol.* **98**, 289-306. TODT, D, HULTSCH, H, and HEIKE, D (1979) *Z. Tierpsychol.* **51**, 23-35. TODT, D and WOLFFGRAMM, J (1975) *Biol. Cybernetics* **17**, 109-27. TOFT, J and CHRISTENSEN, P V (1977) *Feltornithologen* **19**, 84-5. TOMEK, T (1984) *Acta Zool. Cracov.* **27** (2), 19-46. TOMIAŁOJĆ, L (1976) *Birds of Poland*.

Warsaw. TOMIAŁOJĆ, L, WESOŁOWSKI, T, and WALANKIEWICZ, W (1984) *Acta Orn.* **20**, 241-310. TOMKOVICH, P S and SOROKIN, A G (1983) In Flint, V E and Tomkovich, P S (eds) *Rasprostranenie i sistematika ptits*, 77-159. Moscow. TOMLINSON, W (1943) *Bull. zool. Soc. Egypt* **5**, 8-17. TOOBY, J (1947) *Br. Birds* **40**, 290-7. TOOK, G E (1944) *Br. Birds* **38**, 36; (1947) *Br. Birds* **40**, 83-4. TOOK, J M E (1972) *Cyprus orn. Soc.* (*1957*) *Rep.* **18**, 40-9. TÖRÖK, J (1981) *Opusc. Zool. Budapest* **17-18**, 145-56; (1985) *Opusc. Zool. Budapest* **21**, 105-35. TOSCHI, A (1949) *Avifauna italiana.* Firenze; (1961) *Riv. ital. Orn.* **31**, 179-80. TOURNIER, H (1973) *Nos Oiseaux* **32**, 93-9. TRACY, N (1925) *Br. Birds* **18**, 217. TRAGENZA, L A (1955) *The Red Sea mountains of Egypt.* London. TRATZ, E P (1953) *Vogelwelt* **74**, 58. TRAYLOR, M A (1963) *Publ. cult. Co. Diament. Angola* **61**; (1977) *Bull. Mus. comp. Zool.* **148**, 129-84. TREE, A J (1966) *Ostrich* **37**, 184-90. TRETZEL, E (1965a) *Z. Tierpsychol.* **22**, 784-809; (1965b) *Zool. Anz. Suppl.* **29**, 367-80; (1966) *Beitr. Vogelkde.* **11**, 276-95; (1967) *Z. Tierpsychol.* **24**, 137-61. TREVOR, E M (1965) *Br. Birds* **58**, 511. TRICOT, J (1965) *Aves* **2**, 97-125. TRILLMICH, F (1968) *Orn. Beob.* **65**, 195. TRIMINGHAM, J S (1956) *Naturalist* **857**, 45-6. TRIPLET, P (1981) *Oiseau* **51**, 323-8. TRISTRAM, H B (1859) *Ibis* (1) **1**, 277-301; (1859-60) *Ibis* (1) **1**, 153-62, 415-35, (1) **2**, 68-83, 149-65; (1865) *Ibis* (2) **1**, 67-83; (1867) *Ibis* (2) **3**, 73-97. TROST, C H (1972) *Auk* **89**, 506-27. TROTT, A C (1947) *Ibis* **89**, 77-98. TROTTER, W D C (1970) *Oiseau* **40**, 160-72. TRUCKLE, W H (1968) *Hampshire Bird Rep. 1967*, 22-36. TSCHUSI ZU SCHMIDHOFFEN, V VON (1905-10) *Ornis* **13**, 1-56; (1922) *J. Orn.* **70**, 49-56. TUBBS, C R (1954) *Br. Birds* **47**, 208-9. TUCKER, B W (1946) *Br. Birds* **39**, 88-9; (1947) *Br. Birds* **40**, 335-7. TUCKER, B W, TICEHURST, N F, and BOYD, A W (1948) *Br. Birds* **41**, 387. TUCKER, J J (1972) *Bull. Zambian orn. Soc.* **4**, 29-30; (1981) *Br. Birds* **74**, 525-6. TULLOCH, R J (1968) *Scott. Birds* **5**, 218-19; (1969) *Br. Birds* **62**, 36-7. TURČEK, F J (1960) *Orn. Mitt.* **12**, 172-3; (1961) *Ökologische Beziehungen der Vögel und Gehölze.* Bratislava; (1967) *Orn. Mitt.* **19**, 39. TURCEK, J F (1965) *Orn. Mitt.* **17**, 192. TURNBULL, S C (1984) *Br. Birds* **77**, 159. TURNER, A K (1982) *Anim. Behav.* **30**, 862-72. TURNER, A K and BRYANT, D M (1979) *Bird Study* **26**, 117-22. TURNER, B C (1950) *Br. Birds* **43**, 59. TURNER, E L (1914) *Br. Birds* **7**, 320-1; (1915) *Br. Birds* **9**, 102-8. TURRIAN, F (1980) *Nos Oiseaux* **35**, 246-7. TUSCHL, H (1985) *Gef. Welt* **109**, 102-4. TUTMAN, I (1950) *Larus* **3**, 353-60; (1952) *Larus* **4-5**, 218-19; (1962) *Larus* **14**, 169-79; (1969) *Vogelwelt* **90**, 1-8. TYE, A (1980) *Bull. Ecol.* **11**, 559-69; (1981) *Wilson Bull.* **93**, 112-14; (1982) Ph D Thesis. Cambridge Univ. TYE, A and TYE, H (1983) *Br. Birds* **76**, 427-37. TYLER, L and TYLER, S J (1972) *Wilts. arch. nat. Hist. Soc. J.* **67** (A), 16-20. TYLER, S J (1970) *Hampshire Bird Rep. 1969*, 37-40; (1972) *Bird Study* **19**, 69-80; (1979) *Ring. Migr.* **2**, 122-31. TYRER, M J and MORAN, F T (1977) *Scott. Birds* **9**, 350. TYRVÄINEN, H (1969) *Ann. zool. fenn.* **6**, 1-46; (1970) *Ann. zool. fenn.* **7**, 349-57; (1975) *Ornis fenn.* **52**, 23-31.

UDDLING, Å (1955) *Vår Fågelvärld* **14**, 112-15. UHL, F (1929) *Anz. orn. Ges. Bayern* **2**, 34-6; (1930) *Anz. orn. Ges. Bayern* **2**, 91-4. ULFSTRAND, S (1960) *Vår Fågelvärld* **19**, 339-40. ULFSTRAND, S and HÖGSTEDT, G (1976) *Anser* **15**, 1-32. ULLRICH, B (1972) *Vogelwarte* **26**, 289-98. UMRIKHINA, G S (1970) *Ptitsy Chuyskoy doliny.* Frunze. UNFRICHT, T and UNFRICHT, W (1983) *Gef. Welt* **107**, 305-6. UNIVERSITY OF SOUTHAMPTON (1982) *Ornithological Project (Upper Indus) 1981-2 prelim. Rep.* Unpubl. MS, Dept. Biology, Southampton Univ. URBAN, E K and BROWN, L H (1971) *A checklist of the birds of Ethiopia.* Addis Ababa. URBANIAK, W and ZATWARNICKI, T

(1979) *Notatki orn.* **20**, 65-7. USHAKOVA, M M and USHAKOV, V A (1976) *Ornitologiya* **12**, 248. USSHER, R J and WARREN, R (1900) *The birds of Ireland.* London. USPENSKI, S M, BOEHME, R L, PRIKLONSKI, S G, and VEKHOV, V N (1962) *Ornitologiya* **5**, 49-67.

VADER, W (1971) *Br. Birds* **64**, 456-8. VÄISÄNEN, R A (1983) *Aureola* **8**, 58-65. VALET, G (1973) *Alauda* **41**, 253-64. VALLÉE, J-L (1983) *Bièvre* **5**, 119. VALVERDE, J A (1957) *Aves del Sahara Español.* Madrid; (1958) *Br. Birds* **1**, 1-23. VANDE WEGHE, J-P (1979) *Gerfaut* **69**, 29-43. VAN HECKE, P (1965a) *Gerfaut* **55**, 124-41; (1965b) *Gerfaut* **55**, 146-94; (1979a) *J. Orn.* **120**, 12-29; (1979b) *J. Orn.* **120**, 265-79; (1980) *Vogelwelt* **101**, 99-114, 140-53; (1981) *J. Orn.* **122**, 23-35. VAN STEENBERGEN, G (1971) *Gerfaut* **61**, 3-13. VAN VEGTEN, J A and SCHIPPER, W J A (1968) *Ardea* **56**, 194-5. VARTAPETOV, L G (1984) *Ptitsy taezhnykh mezhdurechiy Zapadnoy Sibiri.* Novosibirsk. VASIĆ, V F (1985) *Proc. Fauna SR Serbia* **3**, 193-205. VAUCHER, C (1955) *Alauda* **23**, 182-211. VAUGHAN, R (1979) *Arctic summer.* Shrewsbury. VAUK, G (1972a) *Die Vögel Helgolands.* Hamburg; (1972b) *J. Orn.* **113**, 105-6; (1973a) *Beitr. Vogelkde.* **19**, 225-60; (1973b) *Vogelwelt* **94**, 146-54. VAUK, G and SCHRÖDER, H (1972) *Zool. Anz.* **188**, 291-300. VAUK, G and WITTIG, E (1971) *Vogelwarte* **26**, 238-45. VAURIE, C (1949) *Amer. Mus. Novit.* **1425**; (1950) *Ibis* **92**, 540-4; (1951a) *Bull. Amer. Mus. nat. Hist.* **97** (5), 431-526; (1951b) *Amer. Mus. Novit.* **1485**; (1951c) *Amer. Mus. Novit.* **1529**; (1953) *Amer. Mus. Novit.* **1640**; (1954) *Amer. Mus. Novit.* **1672**; (1955a) *Amer. Mus. Novit.* **1731**; (1955a) *Amer. Mus. Novit.* **1733**; (1955b) *Amer. Mus. Novit.* **1751**; (1955c) *Amer. Mus. Novit.* **1753**; (1957) *Amer. Mus. Novit.* **1832**; (1958) *Amer. Mus. Novit.* **1869**; (1959) *Birds of the Palearctic fauna: passeriformes.* London; (1972) *Tibet and its birds.* London. VENABLES, L S V and VENABLES, U M (1952) *Ibis* **94**, 636-53. VENABLES, W A (1981) *Br. Birds* **74**, 98-9. VENNEKÖTTER, J (1984) *Voliere* **7**, 49-51. VEPSÄLÄINEN, K (1968) *Ann. zool. fenn.* **5**, 389-95. VERBEEK, N A M (1967) *Wilson Bull.* **79**, 208-18; (1970) *Auk* **87**, 425-51; (1973) *Condor* **75**, 287-92; (1981) *Ornis scand.* **12**, 37-9; (1984) *J. Orn.* **125**, 333-4. VERE BENSON, S (1970) *Birds of Lebanon and the Jordan area.* London. VERHEYEN, R (1947) *Les passereaux de Belgique* **2**. Brussels; (1952) *Gerfaut* **42**, 92-124; (1953) *Gerfaut* **43**, 231-61; (1956a) *Gerfaut* **46**, 95-102; (1956b) *Gerfaut* **46**, 143-50; (1957) *Les passereaux de Belgique* **1**. Brussels; (1958) *Gerfaut* **48**, 5-14. VERHEYEN, R and GRELLE, G LE (1950) *Gerfaut* **40**, 124-31. VERKERK, W (1961) *Limosa* **34**, 225-7. VERNON, C (1975) *Honeyguide* **86**, 15-16. VERNON, J D R (1972) *Alauda* **40**, 307-20. VEYSEY, C M (1961) *Br. Birds* **54**, 122. VICKHOLM, M and VÄISÄNEN, R A (1984) *Lintumies* **19**, 2-12. VIDAL, H (1968) *Orn. Mitt.* **20**, 194. VIE, A (1975) *Passer* **10**, 63-84. VIELLIARD, J (1971-2) *Alauda* **39**, 227-48, **40**, 63-92. VIETINGHOFF-RIESCH, A VON (1955) *Die Rauchschwalbe.* Berlin; (1961) *Orn. Mitt.* **13**, 96. VIETINGHOFF-SCHEEL, E VON (1974) In Stresemann, E, Portenko, L A, Dathe, H, and Mauersberger, G (eds) *Atlas der Verbreitung palaärktischer Vögel* **4**. Berlin; VILLIERS, A (1950) *Mém. Inst. Fr. Afr. noire* **10**, 345-85. VINCENT, A W (1946-9) *Ibis* **88**, 462-77, **89**, 163-204, **90**, 284-312, **91**, 111-39, 313-44, 483-507, 660-88. VINCENT, J (1935) *Ibis* (13) **5**, 485-529. VINEY, C and PHILLIPPS, K (1983) *New colour guide to Hong Kong birds.* Hong Kong. VIT, R (1979) *Gef. Welt* **103**, 76. VITALE, D and BRÉMOND, J-C (1979) *Biol. Behav.* **4** (4), 303-10. VITTERY, A, PORTER, R F, and SQUIRE, J E (eds) (1971) *Check list of the birds of Turkey.* Orn. Soc. Turkey. Sandy. VITTERY, A, SQUIRE, J E, and PORTER, R F (1972) *Orn. Soc. Turkey Bird Rep. 1968-9.* Sandy. VOGT, W (1944) *Orn. Beob.* **41**, 36-43. VOIGT, A (1933) *Exkursionsbuch* zum Studium der Vogelstimmen. Leipzig. VOIPIO, P (1970) Ornis fenn. **47**, 15-19. VOLLBRECHT, K (1939) Beitr. Fortpfl. Vögel **15**, 217-18. VOLLNHOFER, P (1906) Erdészeti Kisérletek **8**, 1-81. VOLSØE, H (1951) Vidensk. Medd. dansk nat. Foren. **113**, 1-153. VON DER DECKEN, H H (1972) Abh. Landesmus. Naturkde. Münster **34** (4), 103-9. VOOUS, K H (1950) Treubia **20**, 647-56; (1959) Ardea **47**, 28-41; (1960) Atlas of European birds. London; (1969) Anz. orn. Ges. Bayern **8**, 630-1; (1977) List of recent Holarctic bird species. London; (1985) In Campbell, B and Lack, E (eds) (1985) A dictionary of birds, x-xvii. Calton. VOROBIEV, K A (1954) Ptitsy Ussuriyskogo kraya. Moscow; (1959) Ornitologiya **2**, 115-21; (1963) Ptitsy Yakutii. Moscow; (1968) Ornitologiya **9**, 164-8; (1973) Zapiski ornitologa. Moscow; (1980) Ornitologiya **15**, 194-6. VRONSKI, N V (1977) Ornitologiya **13**, 12-21. VRYDAGH, J-M (1952) Gerfaut **42**, 153-64. VTOROV, P P (1967) Ornitologiya **8**, 254-61. VUILLEUMIER, F (1977) Terre Vie **31**, 459-88; (1979) Ornithological observations during the fall 1979 Nile cruise of the American Museum of Natural History. Unpubl. MS, Edward Grey Inst., Oxford Univ.

WADEWITZ, O (1957a) Falke **4**, 41-3; (1957b) Falke **4**, 151-3. WADLEY, N J P (1951) Ibis **93**, 63-89. WAGNER, G (1958) Orn. Beob. **55**, 37-54; (1972) Falke **19**, 65. WAGNER, U (1982) Ökol. Vögel **4**, 208-9. WAHLSTEDT, J (1970) In Gooders, J (ed) Birds of the world, 1922-6. London. WAIT, W E (1931) Manual of the birds of Ceylon. London. WALKER, C F and TRAUTMAN, M B (1936) Wilson Bull. **48**, 151-5. WALKER, F J (1981a) Sandgrouse **2**, 33-55; (1981b) Sandgrouse **2**, 56-85. WALKINSHAW, L H (1966) Bird-Banding **37**, 227-57. WALLACE, D I M (1955) Br. Birds **48**, 337-40; (1964) Ibis **106**, 389-90; (1965) Br. Birds **58**, 337-41; (1969) Bull. Niger. orn. Soc. **6** (22), 45-9; (1982) Sandgrouse **4**, 77-99; (1983a) Sandgrouse **5**, 1-18; (1983b) Sandgrouse **5**, 102-4; (1984a) Br. Birds **77**, 363-5; (1984b) Sandgrouse **6**, 24-47. WALLIS, E A (1919) Br. Birds **12**, 180-1. WALLIS, H M and PEARSON, C E (1912a) Bull. Br. Orn. Club **29**, 104-5; (1912b) Bull. Br. Orn. Club **29**, 105-6. WALLSCHLÄGER, D (1978) Mitt. zool. Mus. Berlin **54**, Suppl. Ann. Orn. **2**, 173-81; (1983) Mitt. zool. Mus. Berlin **59**, Suppl. Ann. Orn. **7**, 117-25; (1984) Mitt. zool. Mus. Berlin **60**, Suppl. Ann. Orn. **8**, 37-56. WALPOLE-BOND, J (1938) A history of Sussex birds. London WALRAVENS, M and LANGHENDRIES, R (1985) Aves **22**, 3-34. WALSH, J F and GRIMES, L G (1981) Bull. Br. Orn. Club **101**, 327-34. WALTER, H (1965) J. Orn. **106**, 81-105. WALTERS, J (1971) Bird Study **18**, 31-3. WALTERS, M (1981) Sandgrouse **2**, 86-90. WALTHER, H J (1972) Beitr. Vogelkde. **18**, 455-6. WALTON, K C (1979) Ibis **121**, 325-9; (1984) Bird Study **31**, 39-42. WAMMES, D F, BOERE, G C, and BRAAKSMA, S (1983) Limosa **56**, 231-42. WARD, M D (1956) Starfish **9**, 13-17. WARD, P (1963) Ibis **105**, 109-11; (1964) Ibis **106**, 370-5. WARDHAUGH, A A (1984) Br. Birds **77**, 365-6. WARE, E H (1961) Devon Birds **14**, 18-21. WARGA, K (1926) Aquila **32**-3, 289-90; (1929) Aquila **34**-5, 155-83; (1939a) Aquila **42**-5, 410-528; (1939b) Aquila **42**-5, 529-42; (1939c) J. Orn. **87**, 54-60; (1955) Aquila **59**-62, 446-7. WARHAM, J (1964) Ibis **106**, 260-1. WARTMANN, B (1980) Orn. Beob. **77**, 241-4. WATERS, E (1964) Br. Birds **57**, 49-64. WATKINSON, E (undated) A guide to bird-watching in Mallorca. Stockholm. WATSON, A (1946) Br. Birds **39**, 282; (1973a) Br. Birds **66**, 505-8. WATSON, D (1972) Birds of moor and mountain. Edinburgh. WATSON, G E (1961a) J. Orn. **102**, 301-7; (1961b) Postilla **52**; (1962) Bull. Br. Orn. Club **82**, 9-18; (1964) Ph D Thesis. Yale Univ; (1973b) A serological and ectoparasite survey of migrating birds in north-east Africa. Smithsonian Inst., Washington. WATZINGER, A (1914) Orn. Jahrb. **25**, 45-7. WAUGH, D R (1978) Ph D Thesis. Stirling Univ.; (1979) Bird Study **26**, 123-8.

WEBER, E (1970a) Falke **17**, 7; (1970b) Falke **17**, 137. WEBER, H (1973) Falke **20**, 136. WEBSTER, M A (1975) An annotated checklist of the birds of Hong Kong. Hong Kong. WEIBULL, P (1979) Anser **18**, 51. WEIGOLD, H (1912) J. Orn. **60**, 365-410; (1914) J. Orn. **62**, 57-93; (1926) Wiss. Meeresunters. NF **15** (3) article 17. WEIMANN, R (1938) Ber. Ver. schles. Orn. **23**, 1-14. WEINZIERL, H (1955) Vogelwarte **18**, 31. WEIR, D N (1970) Scott. Birds **6**, 212-13. WEISE, W (1971) Beitr. Vogelkde. **17**, 167-8. WELLER, L G (1947) Br. Birds **40**, 248-9. WELLS, D R (1982) Malay Nat. J. **36**, 61-85. WELLS, T P (1951) Br. Birds **44**, 21. WENNER, M V (1933) Br. Birds **27**, 176. WENNRICH, G (1979) Orn. Mitt. **31**, 88-9. WERNER, K-W (1975) Gef. Welt **99**, 189-91 WERTH, I (1947) Br. Birds **40**, 328-30. WESOŁOWSKI, T (1981) J. Anim. Ecol. **50**, 809-14; (1983) Ibis **125**, 499-515. WESTERFRÖLKE, P (1970) Orn. Mitt. **22**, 220; (1971) Orn. Mitt. **23**, 156. WESTPHAL, D (1975) Orn. Mitt. **27**, 84-8. WETMORE, A (1960) Smithson. misc. Coll. **139** (11). WETTSTEIN, O VON (1938) J. Orn. **86**, 9-53. WHEATLEY, J J (1973) Surrey Bird Rep. 1972, 66-75. WHEATLEY, K (1984) Ph D Thesis. Sussex Univ. WHISTLER, H (1916) Ibis (10) **4**, 35-118; (1919) J. Bombay nat. Hist. Soc. **26**, 770-5; (1920) J. Bombay nat. Hist. Soc. **27**, 403-5; (1922) Ibis (11) **4**, 259-309; (1925a) Ibis (12) **1**, 152-208; (1925b) J. Bombay nat. Hist. Soc. **30**, 701-2; (1926a) Ibis (12) **2**, 521-81; (1926b) Ibis (12) **2**, 724-83; (1941) Popular handbook of Indian birds. London. WHITAKER, J I S (1895) Ibis (7) **1**, 85-106; (1896) Ibis (7) **2**, 87-99; (1898) Ibis (7) **4**, 125-32; (1902) Ibis (8) **2**, 643-56; (1905) The birds of Tunisia. London. WHITE, C A (1967a) Br. Birds **60**, 372-3. WHITE, C M N (1961) A revised check list of African broadbills, pittas, larks, swallows, wagtails and pipits. Lusaka; (1962) A revised check list of African shrikes, orioles, drongos, starlings, crows, waxwings, cuckoo-shrikes, bulbuls, accentors, thrushes and babblers. Lusaka. WHITE, D J B (1967b) Br. Birds **60**, 523. WHITE, G (1789) The natural history of Selborne. London. WHITE, M F (1971) Br. Birds **64**, 505. WHITE, W W (1941) Br. Birds **34**, 179. WHITEHEAD, C H T (1909) Ibis (9) **3**, 214-84. WHITEHOUSE, H L K and ARMSTRONG, E A (1953) Behaviour **5**, 261-88. WHITEHOUSE, S M (1978) Br. Birds **71**, 462-3. WHITELEGG, J R (1961) Br. Birds **54**, 430-1. WICHT, U VON (1978) Anz. orn. Ges. Bayern **17**, 79-98. WIDMER, P N (1979) Orn. Beob. **76**, 291. WIESEMES, H (1982) Natur Umwelt **13**, 8-9. WIKLUND, C G (1977) Vår Fågelvärld **36**, 260-5; (1979) Ibis **121**, 109-11; (1982) Behav. Ecol. Sociobiol. **11**, 165-72; (1983) Ph D Thesis. Göteborg Univ. WIKLUND, C G and ANDERSSON, M (1980) Ibis **122**, 363-6. WILCOCK, J (1964) Ibis **106**, 101-9; (1965) Ibis **107**, 316-25. WILDER, G D and HUBBARD, H W (1938) Birds of north-eastern China. Peking. WILKINSON, D M (1985) Br. Birds **78**, 242. WILLCOX, D R C and WILLCOX, B (1978) Ibis **120**, 329-33. WILLFORD, A H (1925) Br. Birds **18**, 293-5. WILLGOHS, J F (1955) Univ. Bergen Årbok 1954 Naturvitensk. Rekke **7**, 1-16. WILLIAMS, J (1984) Honeyguide **30**, 77-8. WILLIAMS, J G (1941) Ibis (14) **5**, 245-64. WILLIAMS, J G and ARLOTT, N (1980) A field guide to the birds of East Africa. London. WILLIAMS, R E (1947a) Br. Birds **40**, 52; (1947b) Br. Birds **40**, 84-5. WILLIAMSON, K (1940) Br. Birds **33**, 252-4; (1941) Br. Birds **34**, 221-2; (1947a) Ibis **89**, 105-17; (1947b) Ibis **89**, 513; (1947c) Ibis **89**, 514-15; (1951a) Br. Birds **44**, 197-9; (1951b) Ibis **93**, 599-601; (1953) Br. Birds **46**, 210-12; (1954) Br. Birds **47**, 266-7; (1955a) Br. Birds **48**, 26-9; (1955b) Br. Birds **48**, 382-403; (1957) Bird-Banding **28**, 129-35; (1958a) Br. Birds **51**, 209-32; (1958b) Br. Birds **51**, 369-93; (1958c) Ibis **100**, 582-604; (1959a) Br. Birds **52**, 138-40; (1959b) Bird Migr. **1**, 88-91; (1961) Bird Migr. **1**, 235-40; (1962) Bird Migr. **2**, 131-59; (1965a) Fair Isle and its birds. Edinburgh; (1965b) Br. Birds **58**, 493-504; (1968) Field Stud. **2**, 651-68; (1970) The Atlantic

Islands 1. London; (1971) *Bird Study* 18, 222-5; (1977) *Bull. Br. Orn. Club* 97, 60-1. WILLIAMSON, K and FERGUSON-LEES, I J (1955) *Br. Birds* 48, 358-62. WILSON, E O (1975a) *Sociobiology: the new synthesis.* Harvard. WILSON, M G (1975b) *Bristol Orn.* 8, 110; (1976) *Bristol Orn.* 9, 127-52; (1979) *Br. Birds* 72, 42-3; (1984) *Dutch Birding* 6, 143. WILSON, P J (1972) *Bird Life* Jan-Mar, 36; WILSON, S B (1887) *Ibis* (5) 5, 130-50. WINDSOR, D and EMLEN, S T (1975) *Condor* 77, 359-61. WINGATE, D B (1973) *A checklist and guide to the birds of Bermuda.* Bermuda. WINKLER, R (1975) *Orn. Beob.* 72, 119-20; (1979) *Orn. Beob.* 76, 291-2. WINSTANLEY, D, SPENCER, R, and WILLIAMSON, K (1974) *Bird Study* 21, 1-14. WINTERBOTTOM, J M (1967) *Ostrich* 38, 116-22. WIPRÄCHTIGER, P (1971) *Orn. Beob.* 68, 87-8; (1978) *Orn. Beob.* 75, 276-7. WIRDHEIM, A (1981) *Halmstads Orn. Klubb Årsskrift* 20, 44-6. WITCHELL, C A (1896) *The evolution of bird song.* London. WITHERBY, H F (1901) *Ibis* (8) 1, 237-78; (1903) *Ibis* (8) 3, 501-71; (1905) *Ibis* (8) 5, 179-99; (1907) *Ibis* (9) 1, 74-111; (1920) *Br. Birds* 14, 92; (1928) *Ibis* (12) 4, 385-436, 587-663. WITHERBY, H F, JOURDAIN, F C R, TICEHURST, N F, and TUCKER, B W (1938a) *The handbook of British birds* 1; (1938b) 2; (1941) 5. London. WITSACK, W (1968) *Naturkdl. Jber. Mus. Hein.* 3, 47-66; (1969) *Naturkdl. Jber. Mus. Hein.* 4, 61-75: WITT, H (1973) *Orn. Mitt.* 25, 19-21. WITT, K (1976) *Vogelkdl. Ber. Niedersachs.* 8, 41-4; (1982) *Vogelwelt* 103, 90-111; (1983) *Orn. Ber. Berlin (West)* 8 (1), 29-46; (1984a) *Orn. Ber. Berlin (West)* 9 suppl.; (1984b) *J. Orn.* 125, 465-71. WOLFF, G (1941) *Beitr. Fortpfl. Vögel* 17, 28-9. WOŁK, E (1964) *Acta Orn.* 8, 125-38. WOLSEY, R P S (1982) *Br. Birds* 75, 382. WOLTERS, H E (1975-82) *Die Vogelarten der Erde.* Hamburg. WON, P-O, WOO, H-C, HAM, K-W, and CHUN, M-Z (1968a) *Misc. Rep. Yamashina Inst. Orn. Zool.* 5, 363-9. WON, P-O, WOO, H-C, HAM, K-W, CHUN, M-Z, and PARK, Y-S (1968b) *Misc. Rep. Yamashina Inst. Orn. Zool.* 5, 259-77. WOOD, B (1974) *Bull. Niger. orn. Soc.* 11 (39), 19-26; (1978) *Ring. Migr.* 2 (1), 20-6; (1979) *Ibis* 121, 228-31; (1982) *J. Zool. Lond.* 197, 267-83. WOOD, J B (1976) Ph D Thesis. Aberdeen

Univ. WOODELL, S R J (1979) *Dansk orn. Foren. Tidsskr.* 73, 177-9. WORTELAERS, F (1942) *Gerfaut* 32, 60-1. WRIGHT, H W (1921) *Auk* 38, 59-78. WUNDERLICH, K and VIETINGHOFF-SCHEEL, E VON (1975) *Beitr. Vogelkde.* 21, 487. WÜRTELE, K (1969) *J. Orn.* 110, 327. WYNNE-EDWARDS, V C (1952) *Auk* 69, 353-91. WYSS, H (1947) *Orn. Beob.* 44, 219-24.

YAKHONTOV, V D (1976) *Zool. Zh.* 55, 1263-5. YAMASHINA, Y (1982) *Birds in Japan.* Tokyo. YANUSHEVICH, A I, TYURIN, P S, YAKOVLEVA, I D, KYDYRALIEV, A, and SEMENOVA, N I (1960) *Ptitsy Kirgizii* 2. Frunze. YAPP, W B (1962) *Birds and woods.* London. YEATMAN, L J (1967) *Oiseau* 37, 145-6; (1971) *Histoire des oiseaux d'Europe.* Paris; (1976) *Atlas des oiseaux nicheurs de France.* Paris.

ZABELIN, V I (1976) *Ornitologiya* 12, 68-76. ZACHAI, G (1984) *Tzufit* 2, 44-56. ZAHAVI, A (1971) *Ibis* 113, 203-11. ZANG, H (1972) *Vogelkdl. Ber. Niedersachs.* 4, 87-8; (1975) *Auspicium* 5, 369-76; (1979) *Auspicium* 6, 411-15; (1981) *J. Orn.* 122, 153-62. ZARUDNYI, N A (1888) *Zap. Ross. Akad. Nauk* 57 suppl. 1, 1-338; (1896) *Mat. pozn. fauny flory Ross. Imp. otd. zool.* 2; (1903) *Ptitsy vostochnoy Persii.* St Petersburg. ZARUDNYI, N A and BIL'KEVICH, S I (1918) *Izv. Zakasp. Muz.* 1, 1-48. ZASYPKIN, M Y (1981) *Ornitologiya* 16, 100-14. ZEDLER, W (1963) *Anz. orn. Ges. Bayern* 6, 571-2. ZEDLITZ, O VON (1909) *J. Orn.* 57, 121-211, 241-322; (1911) *J. Orn.* 59, 1-92. ZENTZIS, K (1983) *Orn. Mitt.* 35, 215. ZHORDANIA, R G and GOGILASHVILI, G S (1976) *Acta Orn.* 15, 323-38. ZIEGLER, G (1966) *J. Orn.* 107, 187-200. ZIMIN, V B (1972) *Zool. Zh.* 51, 770-2. ZINK, G (1969) *Bonn. zool. Beitr.* 20, 191-9; (1973) *Der Zug europäischer Singvögel* 1; (1975) 2. Möggingen; (1977) *Vogelwarte* 29 suppl., 44-54; (1980) In Johnson, D N (ed) *Proc. Pan-Afr. orn. Congr.* 4, 209-13; (1981) *Der Zug europäischer Singvögel* 3; (1985) 4. Möggingen. ZLOTIN, R I (1968) *Ornitologiya* 9, 158-63. ZUCCHI, H (1973) *Orn. Mitt.* 25, 249; (1974) *Orn. Mitt.* 26, 120-1; (1975) *Vogelwelt* 96, 226-9.

CORRECTIONS

CORRECTIONS TO VOLUME I

Contents. *Pterodroma* petrels are in the wrong order. Correct sequence (Voous 1977) is: *Pterodroma neglecta* **Kermadec Petrel**, *Pterodroma mollis* **Soft-plumaged Petrel**, *Pterodroma hasitata* **Black-capped Petrel**, *Pterodroma leucoptera* **Collared Petrel**.

Page 13. Introduction: Food. Column 2, line 1. Amend to read '...(Olney 1965*b*).'

Page 26. Introduction: Voice. Column 2, paragraph 2, last line. Amend to read '...(Fig 12).'

Page 62. *Gavia adamsii* White-billed Diver. **Distribution.** Accidental. Add 'Italy'.

Page 129. Accounts for *Pterodroma* petrels are in the wrong order. Correct sequence (Voous 1977) is: *Pterodroma neglecta* **Kermadec Petrel**, *Pterodroma mollis* **Soft-plumaged Petrel**, *Pterodroma hasitata* **Black-capped Petrel**, *Pterodroma leucoptera* **Collared Petrel**.

Page 215. *Phalacrocorax nigrogularis* Socotra Cormorant. **Breeding.** Lines 7–8. Delete 'dark umber-brown spots and blotches at large end'.

Page 374. *Cygnus olor* Mute Swan. **Population.** Paragraph 3, line 3. Amend '2·8 years' to read '4·8 years'.

Page 432. *Branta leucopsis* Barnacle Goose. **Distribution.** Accidental. Add 'Italy'.

Page 538. *Anas discors* Blue-winged Teal. **Distribution.** Accidental. Add 'Italy: 1948, 1952, 1975 (P Brichetti).'

CORRECTIONS TO VOLUME II

Page 67. *Neophron percnopterus* Egyptian Vulture. **Food.** Second column, lines 2–4 from end. Amend to read: '... this behaviour may or may not be confined to certain local populations (Alcock 1970; Veselovský 1970; Boswall 1977). In Aragon...'. New reference is BOSWALL, J (1977) *Bull. Br. Orn. Club* 97, 77–8.

Page 366. *Falco peregrinus* Peregrine. **Food.** Second column, line 8 from end. Amend to read: '... thrushes and *S. vulgaris* 16·3%. If prey assessed by weight ...'.

Page 379. *Falco pelegrinoides* Barbary Falcon. **Distribution.** Accidental. Add 'Italy'.

Page 418. *Tetrao tetrix* Black Grouse **Movements.** Column 2, lines 6–13. Delete 'Long-distance...Höglund 1952).'

Page 436. *Tetrao urogallus* Capercaillie. **Movements.** Column 1, paragraph 2, lines 5–10. Delete '8 long-distance...Höglund

1952). Long-distance movements noted in Sweden 1943–4 by Notini (1947), later quoted in good faith by others (e.g. Höglund 1952), were based on false data and original paper considered fraudulent (K Borg and C-F Lundevall); see also *Zeitschr. Jagdwissensch.* 1955, 1, 59–64.

Page 588. *Porphyrula alleni* Allen's Gallinule. **Field characters.** Column 2, paragraph 2, line 8. Amend to read '... duller bare parts...'

Page 619. *Grus grus* Crane. **Habitat.** Column 1, line 3. Amend to read '... holm oaks *Quercus ilex*...'

Page 621. *Grus grus* Crane. **Food.** Column 1, lines 7–8. Amend to read '... holm oak *Quercus ilex*...'

CORRECTIONS TO VOLUME III

Page 59. *Recurvirostra avosetta* Avocet. Voice. Fig I caption to read 'I S Palmér/Sveriges Radio (1972)...'

Page 91. *Cursorius cursor* Cream-Coloured Courser. **Field characters.** Lines 1–3. Amend to read '**Field characters.** 22–24 cm (legs 7–8 cm); wing-span 51–57 cm. About size of Greater Sand Plover *Charadrius leschenaultii*; shorter-tailed...'

Page 243. *Chettusia gregaria* Sociable Plover. **Food.** Column 1, line 3 from below to read '...and Gladkov...'

Page 333. *Calidris melanotos* Pectoral Sandpiper. **Distribution.** Accidental. Add 'Italy'.

Page 352. *Calidris maritima* Purple Sandpiper. **Voice.** Fig I caption to read '...N Norehn...'. Figs II–III captions to read '...J P Lomholt...'

Page 379. *Micropalama himantopus* Stilt Sandpiper. **Habitat.** Paragraph 1, lines 2–3 from below. Amend to read '...near Least Sandpiper *Calidris minutilla* on dry...'

Page 541. *Tringa stagnatilis* Marsh Sandpiper. **Movements.** Column 2, line 7 from below. Amend 'Kufa' to read 'Kufra'.

Page 817. *Larus argentatus* Herring Gull. **Distribution.** Column 2, line 17. Amend '1908' to read '1968'.

References:
Page 893. Column 2, lines 6–7. Amend to read '...Lietuvos paukščiai...'

Page 901. Column 1, line 26. Amend 'RUTTLEDGE, R P' to read 'RUTTLEDGE, R F', and insert '(1961) *Bird Study* 8, 2–5;'.

Plate 14. Caption. Amend to read '...178): **5-6** ad breeding, **7** downy...'

Plate 16. Caption. Amend to read '...**6** ad breeding,...'

CORRECTIONS TO VOLUME IV

Contents. Re-name 'Colaptes auratus Yellow-shafted Flicker' as 'Northern Flicker'.

Page 2. Acknowledgements. Column 2. Add 'FRANCE P Nicolau-Gillaumet' and 'MALTA J Sultana, C Gauci'.

Page 3. Acknowledgements. Column 2. Amend 'A P Müller' to read 'A P Møller'; amend 'G Rinnhoffer' to read 'G Rinnhofer'.

Page 33. Sterna maxima Royal Tern. **Breeding.** Last line. Amend to read 'young gathered on strandline (Dupuy 1975).'

Page 43. Sterna bengalensis Lesser Crested Tern. **Distribution.** Accidental. Add 'Italy'.

Page 44. Sterna bengalensis Lesser Crested Tern. **Movements.** Column 1, line 4. Delete 'Hallam' and insert 'Holmden and Hammonds'.

Page 50. Sterna sandvicensis Sandwich Tern. **Population.** Column 2, line 14. Amend '7 pairs' to read '53 pairs'.

Page 56. Sterna sandvicensis Sandwich Tern. **Social pattern and behaviour.** Column 2, line 29. Amend 'A' to read 'B'.

Page 105. Sterna repressa White-cheeked Tern. **Movements.** Paragraph 1, line 3 from below. Delete 'Hallam' and insert 'Holmden and Hammonds'.

Page 147. Chlidonias niger Black Tern. **Movements.** Column 1, line 2 from below, and column 2, paragraph 2, line 4 from below. Amend 'Marigoni' to read 'Martignoni'.

Page 155. Chlidonias leucopterus White-winged Black Tern. Heading. Add North American name 'N. AM. White-winged Tern'.

Page 158. Chlidonias leucopterus White-winged Black Tern. **Food.** Column 2, paragraph 2, line 21. Amend 'Gibson' to read 'Sibson'.

Page 224. Alle alle Little Auk. **Social pattern and behaviour.** Column 2, lines 16-17. Amend 'King 1972b' to read 'Brown 1972' (and remainder of sentence should not be italicized).

Page 226. Alle alle Little Auk. **Voice.** Paragraph 2, lines 10-11. Amend 'King 1972b' to read 'Brown 1972'.

Page 231. Fratercula arctica Puffin. Heading. Amend North American name to read 'N. AM. Atlantic Puffin'.

Page 254. Pterocles senegallus Spotted Sandgrouse. World map. Delete breeding and wintering for North and South Yemen.

Page 259. Pterocles exustus Chestnut-bellied Sandgrouse. Heading. Line 4. Amend to read 'Pterocles...'

Page 278. Syrrhaptes paradoxus Pallas's Sandgrouse. **Distribution.** Accidental. Add 'Italy'.

Page 299. Columba oenas Stock Dove. **Distribution.** Paragraph 3. Delete 'Syria, Iraq,' (regular in winter in both these countries, as shown on map).

Page 314. Columba palumbus Woodpigeon. **Distribution.** Last line. Delete ', Jordan' (regular there in winter, as shown on map).

Page 393. Clamator glandarius Great Spotted Cuckoo. **Distribution.** Column 1, line 6. Amend 'PF' to read 'PB'.

Page 394. Clamator glandarius Great Spotted Cuckoo. **Food.** Paragraph 3, lines 7-9. Delete '5-6....(Labitte 1940).' (refers to Jay Garrulus glandarius).

Page 402. Cuculus canorus Cuckoo. Heading. Add North American name 'N. AM. Common Cuckoo'.

Page 432. Tyto alba Barn Owl. Heading. Add North American name 'N. AM. Common Barn-Owl'.

Page 496. Surnia ulula Hawk Owl. Heading. Add North American name 'N. AM. Northern Hawk-Owl'.

Page 548. Strix butleri Hume's Tawny Owl. **Distribution.** Line 7. Amend '1976' to read '1936'.

Page 550. Strix uralensis Ural Owl. Heading. Amend Russian name to read 'RU. Длиннохвостая неясыть'.

Page 554. Strix uralensis Ural Owl. **Social pattern and behaviour.** Column 2, line 9 from below. Amend 'Fig A' to read 'Fig B'.

Page 555. Strix uralensis Ural Owl. **Social pattern and behaviour.** Column 1, line 4. Amend 'Fig B' to read 'Fig A'.

Page 556. Strix uralensis Ural Owl. **Social pattern and**

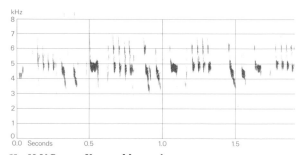

II   H-U Reyer   Kenya   May 1976

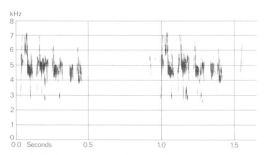

III   H-U Reyer   Kenya   May 1976

behaviour. Column 1, line 18 from below. Amend 'Fig A' to read 'Fig B'.

Page 622. *Caprimulgus europaeus* Nightjar. **Population.** Penultimate line. Amend to read '...Czajkowski...'

Page 623. *Caprimulgus europaeus* Nightjar. Map. Add red (breeding) to Menorca.

Page 658. *Apus apus* Swift. **Distribution.** Accidental. Delete 'Malta,' (common there: Sultana and Gauci 1982).

Page 713. *Alcedo atthis* Kingfisher. **Population.** Last line. Amend to read '...Czajkowski...'

Page 765. *Coracius garrulus* Roller. **Distribution.** Column 2, line 3. Amend 'PD' to read 'PB'.

Page 728. *Ceryle rudis* Pied Kingfisher. **Voice.** Figs II-IV (shown here) were omitted.

Page 803. *Jynx torquilla* Wryneck. Map. Add red (breeding) to Mallorca; delete from Menorca.

Page 813. *Colaptes auratus* Yellow-shafted Flicker. Amend **'Yellow-shafted'** to read **'Northern Flicker'**.

Page 826. *Picus viridis* Green Woodpecker. **Population.** Last line. Amend to read '...Czajkowski...'

Page 853. *Sphyrapicus varius* Yellow-bellied Sapsucker. Heading. Plate page-reference to read '[facing pages 831 and 879]'.

Page 856. *Dendrocopos major* Great Spotted Woodpecker.

Heading. Plate page-reference to read '[facing pages 854 and 879]'.

Page 874. *Dendrocopos syriacus* Syrian Woodpecker. Heading. Plate page-reference to read '[facing pages 854 and 879]'.

Page 882. *Dendrocopos medius* Middle Spotted Woodpecker. Heading. Plate page-reference to read '[facing pages 855 and 879]'.

Page 891. *Dendrocopos leucotos* White-backed Woodpecker. Heading. Plate page-reference to read '[facing pages 855 and 879]'.

Page 901. *Dendrocopos minor* Lesser Spotted Woodpecker. Heading. Plate page-reference to read '[between pages 878 and 879]'.

Page 913. *Picoides tridactylus* Three-toed Woodpecker. Heading. Plate page-reference to read '[between pages 878 and 879]'.

References:
Page 925. Insert 'AUSTIN, O L, JR (1961) *Birds of the world.* New York.'

Page 927. Insert 'BROWN, B J (1972) *Br. Birds* **65**, 397-8.'

Page 927. Column 1, line 16 from below. 'MARIGONI' to read 'MARTIGNONI'.

Page 930. Column 1, line 8. Amend to read '...CZAJKOWSKI...'

Page 932. Delete 'GIBSON, R B (1954) *Notornis* **6**, 43-7'.

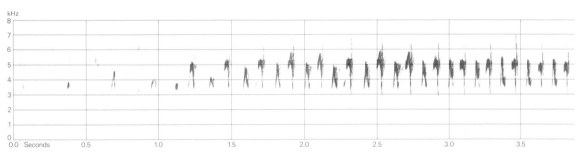

IV H-U Reyer Kenya May 1976

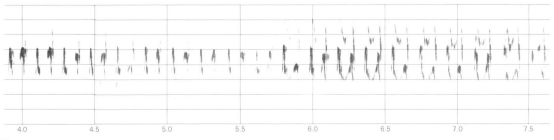

IV *cont.*

Page 935. Add 'HOLMDEN, M AND HAMMONDS, E (1980) *Wildlife in Bahrain* 1, 17–52.'

Page 937. Line 12 from below. KING, B. Delete '*a*' in '(1972*a*)'. Delete '(1972*b*)...397–8;'

Indexes: English Names. Pages 958–9. Add as follows.
Barn-Owl, Common, 432, **43**, **53**; egg, **97**

Cuckoo, Common, 402, **38**, **42**; eggs, **96**

Flicker, Northern, 813 (and delete 'Flicker, Yellow-shafted, 813')

Hawk-Owl, Northern, 496, **47**, **53**; egg, **97**

Puffin, Atlantic, 231, **21**, **22**, **23**; egg, **92** (and delete 'Puffin, Common, 231, **21**, **22**, **23**; egg, **92**')

Plate 65. Caption. Amend to read: '...**3** juv, **4** nestling. Nominate *atthis*: **5** ad ♂. (NA)'.

INDEXES

Figures in **bold type** refer to plates

SCIENTIFIC NAMES

ENGLISH NAMES

NOMS FRANÇAIS

DEUTSCHE NAMEN

EGG PLATES

PLATE 79

Eremopterix nigriceps Black-crowned Finch Lark, 4 eggs
(left)
Ammomanes cincturus Bar-tailed Desert Lark, 4 eggs
(right)

Ammomanes deserti Desert Lark, 4 eggs
(left)
Chersophilus duponti Dupont's Lark, 4 eggs
(right)

Alaemon alaudipes Hoopoe Lark, 3 eggs
(left)
Rhamphocoris clotbey Thick-billed Lark, 3 eggs
(right)

Melanocorypha calandra Calandra Lark, 4 eggs
(left)
Melanocorypha bimaculata Bimaculated Lark, 4 eggs
(right)

Melanocorypha leucoptera White-winged Lark, 4 eggs
(left)
Melanocorypha yeltoniensis Black Lark, 4 eggs
(right)

Calandrella brachydactyla Short-toed Lark, 4 eggs
(left)
Calandrella rufescens Lesser Short-toed Lark, 4 eggs
(right)

Galerida cristata Crested Lark, 4 eggs
(left)
Galerida theklae Thekla Lark, 4 eggs
(right)

Lullula arborea Woodlark, 4 eggs
(left)
Eremophila alpestris Shore Lark, 4 eggs
(right)

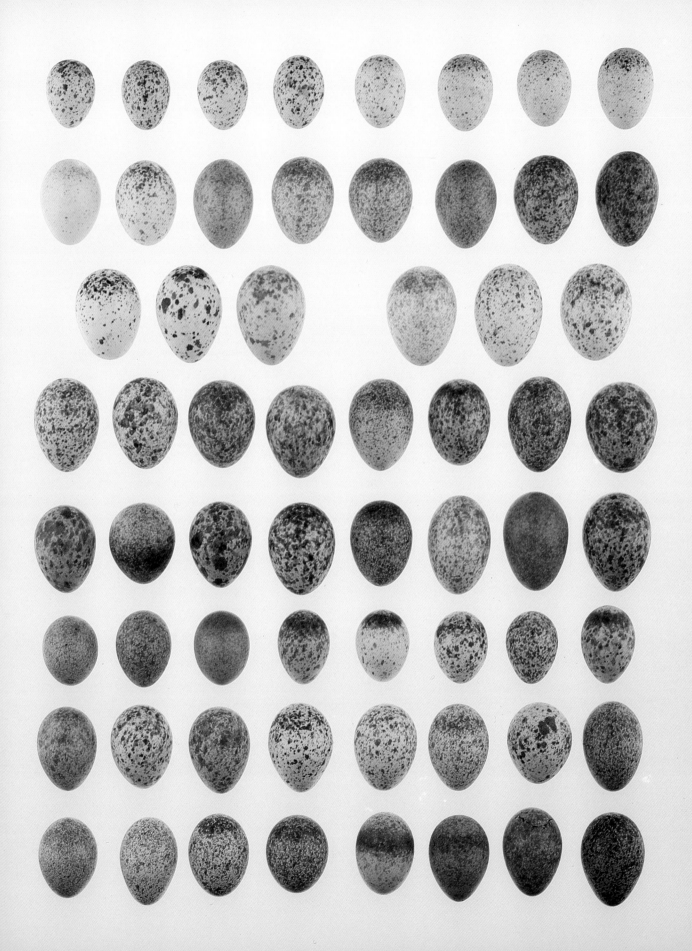

PLATE 80

Alauda arvensis Skylark, 8 eggs

Eremophila bilopha Temminck's Horned Lark, 4 eggs

Riparia paludicola Brown-throated Sand Martin, 2 eggs
(far left)
Riparia riparia Sand Martin, 2 eggs (centre left)
Delichon urbica House Martin, 2 eggs (centre right)
Hirundo daurica Red-rumped Swallow, 2 eggs (far right)

Hirundo rustica Swallow, 3 eggs
(left)
Ptyonoprogne fuligula African Rock Martin, 3 eggs
(centre)
Ptyonoprogne rupestris Crag Martin, 3 eggs
(right)

Anthus campestris Tawny Pipit, 4 eggs
(left)
Anthus berthelotii Berthelot's Pipit, 4 eggs
(right)

Anthus similis Long-billed Pipit, 4 eggs
(left)
Anthus hodgsoni Olive-backed Pipit, 4 eggs
(right)

Anthus gustavi Pechora Pipit, 4 eggs
(left)
Anthus cervinus Red-throated Pipit, 4 eggs
(right)

Anthus pratensis Meadow Pipit, 8 eggs

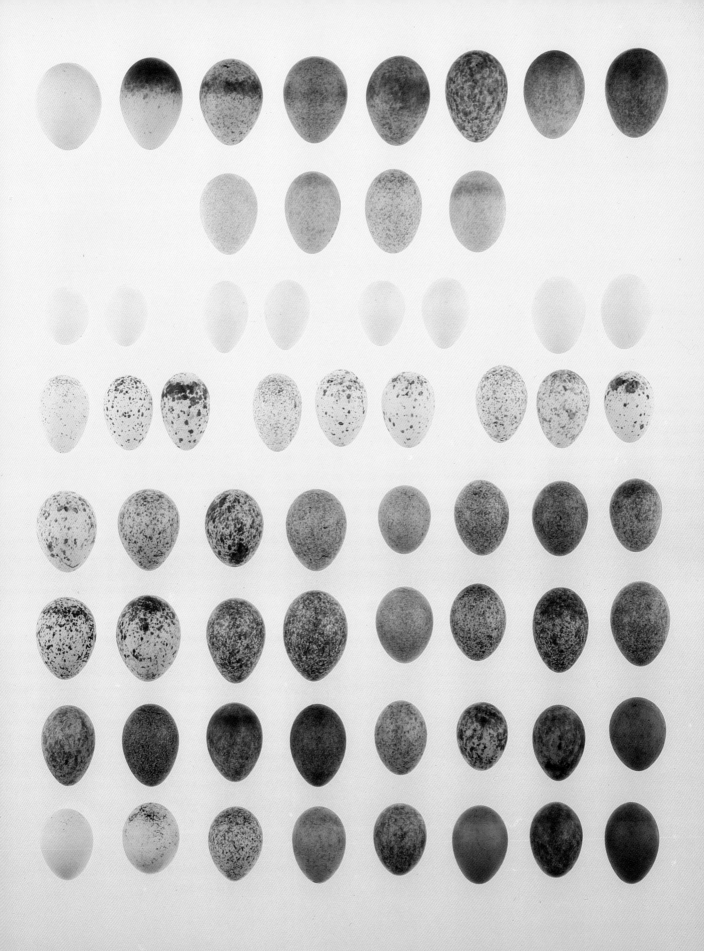

PLATE 81

Anthus trivialis Tree Pipit, 16 eggs

Anthus spinoletta Rock Pipit and Water Pipit, 4 eggs

Motacilla flava Yellow Wagtail, 8 eggs

Motacilla cinerea Grey Wagtail, 8 eggs

Motacilla citreola Citrine Wagtail, 4 eggs
(left)
Motacilla alba Pied Wagtail and White Wagtail, 4 eggs
(right)

Pycnonotus leucogenys White-cheeked Bulbul, 3 eggs
(left)
Pycnonotus xanthopygos Yellow-vented Bulbul, 2 eggs
(centre)
Pycnonotus barbatus Common Bulbul, 3 eggs
(right)

Hypocolius ampelinus Grey Hypocolius, 2 eggs
(left)
Bombycilla garrulus Waxwing, 3 eggs
(centre)
Cinclus cinclus Dipper, 2 eggs
(right)

Troglodytes troglodytes Wren, 4 eggs

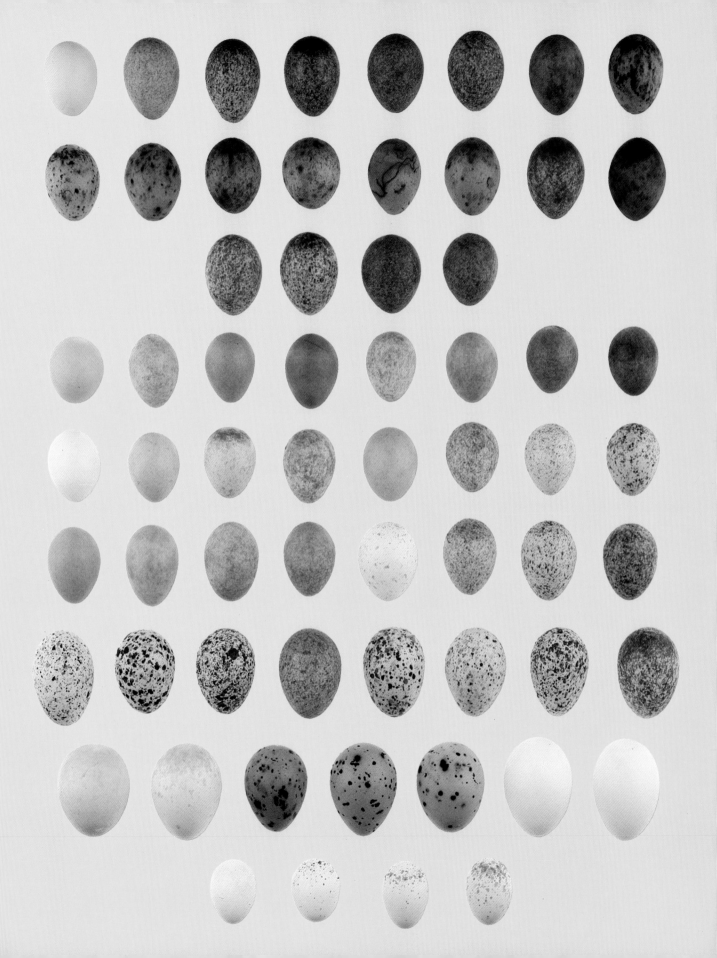

PLATE 82
Prunella modularis Dunnock, 2 eggs (far left)
Prunella montanella Siberian Accentor, 2 eggs (centre left)
Prunella collaris Alpine Accentor, 2 eggs (centre right)
Prunella atrogularis Black-throated Accentor, 2 eggs (far right)

Cercotrichas galactotes Rufous Bush Robin, 4 eggs
(left)
Erithacus rubecula Robin, 4 eggs
(right)

Luscinia luscinia Thrush Nightingale, 4 eggs
(left)
Luscinia megarhynchos Nightingale, 4 eggs
(right)

Luscinia svecica Bluethroat, 4 eggs
(left)
Luscinia calliope Siberian Rubythroat, 4 eggs
(right)

Tarsiger cyanurus Red-flanked Bluetail, 4 eggs
(left)
Irania gutturalis White-throated Robin, 4 eggs
(right)

Phoenicurus ochruros Black Redstart, 4 eggs
(left)
Phoenicurus phoenicurus Redstart, 4 eggs
(right)

Phoenicurus moussieri Moussier's Redstart, 4 eggs
(left)
Cercomela melanura Blackstart, 4 eggs
(right)

Saxicola rubetra Whinchat, 4 eggs
(left)
Saxicola torquata Stonechat, 4 eggs
(right)

Saxicola dacotiae Canary Islands Stonechat, 4 eggs

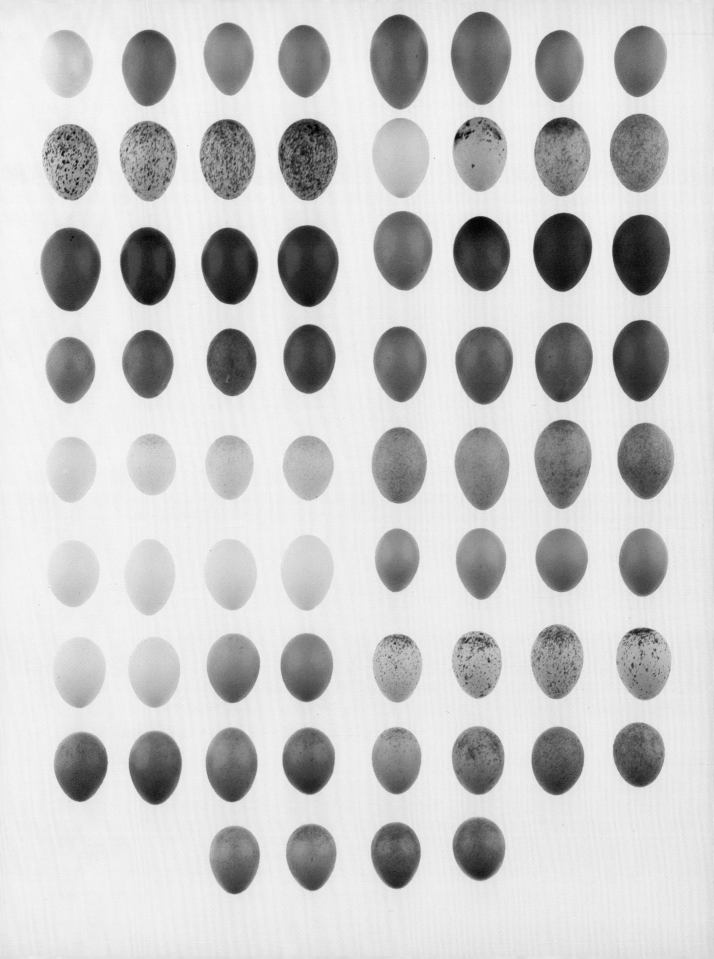

PLATE 83
Oenanthe isabellina Isabelline Wheatear, 3 eggs
(left)
Oenanthe oenanthe Wheatear, 3 eggs
(right)

Oenanthe pleschanka Pied Wheatear, 3 eggs
(left)
Oenanthe hispanica Black-eared Wheatear, 3 eggs
(right)

Oenanthe deserti Desert Wheatear, 3 eggs
(left)
Oenanthe finschii Finsch's Wheatear, 3 eggs
(right)

Oenanthe moesta Red-rumped Wheatear, 3 eggs
(left)
Oenanthe xanthoprymna Red-tailed Wheater, 3 eggs
(right)

Oenanthe picata Eastern Pied Wheatear, 3 eggs
(left)
Oenanthe lugens Mourning Wheatear, 3 eggs
(right)

Oenanthe monacha Hooded Wheatear, 2 eggs (far left)
Oenanthe alboniger Hume's Wheatear, 1 egg (centre left)
Oenanthe leucopyga White-crowned Black Wheatear, 2 eggs
(centre right)
Oenanthe leucura Black Wheatear, 2 eggs (far right)

Monticola saxatilis Rock Thrush, 3 eggs
(left)
Monticola solitarius Blue Rock Thrush, 3 eggs
(right)

Turdus ruficollis Black-throated Thrush and Red-throated
Thrush, 3 eggs
(left)
Zoothera dauma White's Thrush, 3 eggs
(right)

PLATE 84
Turdus torquatus Ring Ouzel, 6 eggs

Turdus merula Blackbird, 6 eggs

Turdus pilaris Fieldfare, 6 eggs

Turdus iliacus Redwing, 6 eggs

Turdus philomelos Song Thrush, 6 eggs

Turdus viscivorus Mistle Thrush, 6 eggs

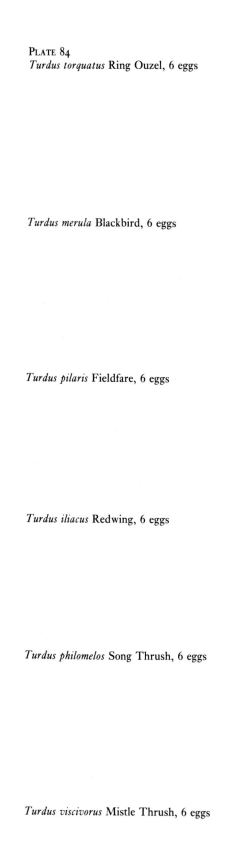

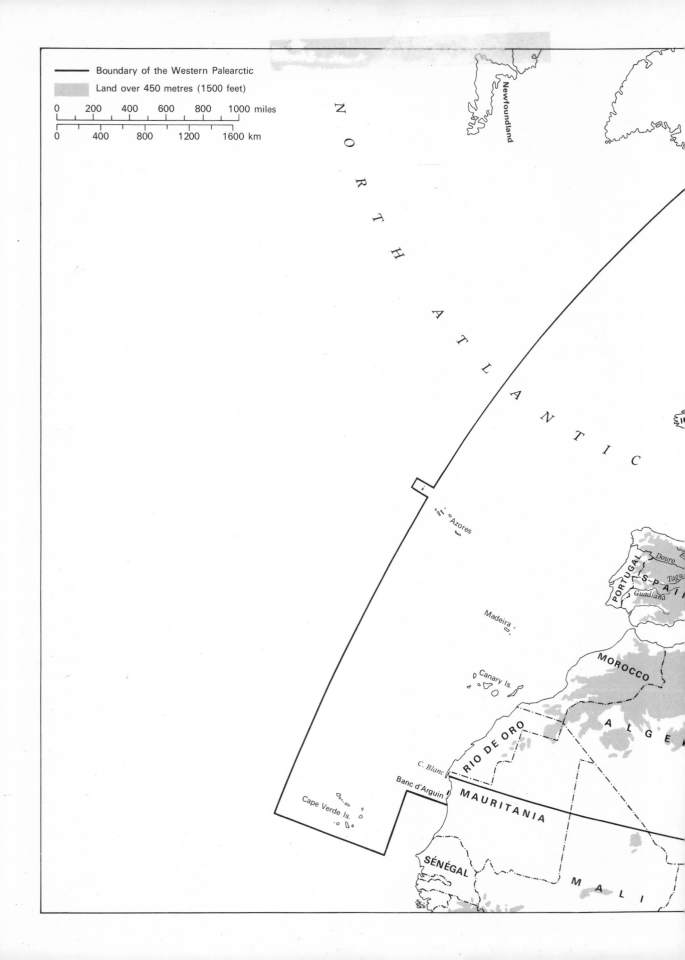

0   200   400   600   800   1000 miles

0   400   800   1200   1600 km

NORTH ATLANTIC

Newfoundland

Azores

Madeira

Canary Is.

PORTUGAL

SPAIN

Douro

Tagu

Guadiana

MOROCCO

ALGE

RIO DE ORO

C. Blanc

Banc d'Arguin

MAURITANIA

Cape Verde Is.

SÉNÉGAL

MALI